CRC/

Handbook of Data on Organic Compounds

VI: ED

RC WEAST | 2E | 1989

2nd Edition

Add (1) WEAST, RC

Volume I

Editors

Robert C. Weast
Jeanette G. Grasselli

CRC Press, Inc.
Boca Raton, Florida

Library of Congress Cataloging-in-Publication Data
Handbook of data on organic compounds.

Rev. ed. of: CRC handbook of data on organic compounds. c1985.
Includes index.
1. Organic compounds—Handbooks, manuals, etc.
I. Weast, Robert C., 1916- . I. Grasselli, Jeanette G. III. CRC handbook of data on organic compounds.
QD257.7.H36 1988 547 88-7485
ISBN 0-8493-0420-2 (set)
ISBN 0-8493-0421-0 (v. 1)
ISBN 0-8493-0422-9 (v. 2)
ISBN 0-8493-0423-7 (v. 3)
ISBN 0-8493-0424-5 (v. 4)
ISBN 0-8493-0425-3 (v. 5)
ISBN 0-8493-0426-1 (v. 6)
ISBN 0-8493-0427-X (v. 7)
ISBN 0-8493-0428-8 (v. 8)
ISBN 0-8493-0429-6 (v. 9)

This book represents information obtained from authentic and highly regarded sources. Reprinted material is quoted with permission, and sources are indicated. A wide variety of references are listed. Every reasonable effort has been made to give reliable data and information, but the author and the publisher cannot assume responsibility for the validity of all materials or for the consequences of their use.

Direct all inquiries to CRC Press, Inc., 2000 Corporate Blvd., N.W., Boca Raton, Florida, 33431.

International Standard Book Number 0-8493-0420-2 (set)
International Standard Book Number 0-8493-0421-0 (Volume I)
International Standard Book Number 0-8493-0422-9 (Volume II)
International Standard Book Number 0-8493-0423-7 (Volume III)
International Standard Book Number 0-8493-0424-5 (Volume IV)
International Standard Book Number 0-8493-0425-3 (Volume V)
International Standard Book Number 0-8493-0426-1 (Volume VI)
International Standard Book Number 0-8493-0427-X (Volume VII)
International Standard Book Number 0-8493-0428-8 (Volume VIII)
International Standard Book Number 0-8493-0429-6 (Volume IX)

Library of Congress Card Number 88-7485

Printed in the United States

PREFACE

Chemists and engineers today have a greater need than ever before for easily accessible reference data. It is well recognized that the total number of organic compounds is growing rapidly, and physical data or information about them is often widely scattered throughout the literature. The second edition of the *CRC Handbook of Data on Organic Compounds* (HODOC II) is a reference source for physical and spectral data on the 26,000 most common organic compounds. This book incorporates not only a substantial revision of the first edition, but also now includes spectral data and/or references for many compounds. It therefore provides a comprehensive and unique database for identifying unknown compounds. It was our intent to provide a convenient, easy-to-use format on a reliable and important body of data for the industrial, academic, or government scientist.

All of the data for each compound in HODOC II appear in one location. The compounds are presented in alphabetical order by Chemical Abstracts Service (CAS) Chemical Index name. The IUPAC name, all common synonyms, the CAS number and the Beilstein number are given along with a HODOC II number for working with the various indexes. The Chemical Abstracts Service guide to naming chemical substances is reproduced in Volume I. This guide also includes molecular skeletons, stereochemistry, and chemical structural diagrams from CA index names. The molecular formula, molecular weight (calculated from the 1985 International Atomic Weights of the elements), line formula, and, in some cases, structure are given. The most recent physical property data available are listed and include the boiling point, melting point, density, refractive index, specific rotation, solubility, color, and crystalline form.

We have tried to present spectral information of uniformly high quality in a useful format. Our goal was to be concise and display the spectral information along with the other physical constants and information on each compound in the database. Careful coding criteria were established, the most reliable spectral reference catalogs were utilized, and a very sincere effort was made to evaluate the data as they were coded. It is worth noting, however, that the spectral data were coded with emphasis on selecting values characteristic of compound structure, not necessarily striving for a complete representation of the original spectrum.

A critical and truly unique feature of HODOC II is the extensive set of indexes which have been generated from the database. For example, there are 100,000 synonyms in the Synonym Index or an average of 4 synonyms per entry. There are also indexes for molecular formula, molecular weight, melting and boiling points, and 15 spectral indexes. The spectral indexes are for infrared, Raman, ultraviolet, H NMR, C NMR, and mass spectra, and these contain numerical indexes of source curves and special spectral data indexes subdivided by spectral region, chemical class, or band intensities (UV). Therefore, HODOC II can be used to obtain information, including spectral data, on approximately 26,000 compounds, or conversely, to identify an unknown from some characteristic physical constant or from spectral data.

Successful utilization of HODOC II depends upon recognition of the basic philosophy inherent in its organization and content. It is widely accepted that spectral data are of immense value in solving analytical problems, or in determining identity or structure of unknown materials. The enhanced status of spectroscopic methods should not contribute to a disregard for the importance of other physical constants of a substance. Certainly the most rapid and unambiguous identification of an unknown is often achieved by using several different physical properties. Toward this aim, this work provides a basic reference collection of physical constants and spectral data on organic compounds to facilitate such identifications. We well recognize that with the massive amount of data handled to assemble this book, errors or omissions are likely to be found. Corrections, criticism, and suggestions for improvement from users will be very welcome. It should be noted that HODOC II is a computer generated compendium. Therefore, some anomalies in generally accepted conventions for hyphenation of words, alphabetical sorting, etc. exist.

We intend to keep this extraordinary database current by publishing annual supplements. These will add new compounds, new data, and/or updated information on compounds listed in HODOC II. The supplements and this second edition of HODOC II will be available in computer-readable format as well as in bound volumes.

Many people have been involved in the preparation of HODOC II. We gratefully acknowledge the significant contribution of Chemical Abstract Services in providing Chemical Abstract numbers and names, and for verifying compound structures. Special recognition should be given to Jim Brody, Tom York, David Westbrook, Barbara Caras, Sandy Pearlman, Glenna Orsbin, and Robert Pitman of CRC Press, Inc., who had the overall responsibility for organizing the volumes of data and publishing the book. They did a superb job.

Jeanette G. Grasselli
Robert C. Weast

THE EDITORS

Robert C. Weast is currently the Editor-in-Chief of the *CRC Handbook of Chemistry and Physics* and has edited the handbook continuously since 1952. He received his doctorate degree from the University of Illinois, and in 1945 became a member of the faculty at Case Western Reserve University, Department of Chemistry and Chemical Engineering, Cleveland, Ohio. Within 2 years he was acting head of the department. In 1962, Dr. Weast resigned his professorship to accept a Vice Presidency in research with the Consolidated Natural Gas Service Company, and from 1972 to 1980 was the American Representative on the International Gas Union's Committee on Industrial Gas Utilization. Dr. Weast has since retired from the Consolidated Natural Gas Service Company and is currently Vice Chairman of the Board and the Director of Corporate Planning for Perfection Corporation. He is a Fellow in the American Association for the Advancement of Science, a member of Alpha Chi Sigma (the Chemical professional society), Sigma Xi (the honorary research society), the American Chemical Society, and CODATA.

Jeanette G. Grasselli is a Distinguished Visiting Professor at Ohio University. She was formerly Director of Research and Analytical Sciences at the Research and Development Department of BP America, Cleveland, Ohio. Dr. Grasselli obtained her B. S. degree in chemistry at Ohio University, her M. S. at Case Western Reserve University, and has received D. Sc. (Hon.) degrees from Ohio University (1978) and Clarkson University (1986). She has 1 patent, 80 publications, and is co-editor of 8 books in the field of molecular spectroscopy. In 1970 she served as the National President of the Society for Applied Spectroscopy. She was active in organizing FACSS (The Federation of Analytical Chemistry and Spectroscopy Societies) and has been Chairman and Secretary of its Governing Board. She is on the Visiting Committee for the National Institute of Science and Technology (NBS), and serves on the Energy Research Advisory Board of the Department of Energy and the Chemical Sciences and Technology Board of the National Research Council. Dr. Grasselli is a Member of the U.S. National Committee for the International Union of Pure and Applied Chemistry (IUPAC), and is Secretary of the Spectroscopy Commission. She has received numerous awards and has been recognized for her many achievements in science, including the Anachem Award from the Detroit Association of Analytical Chemists, the Williams-Wright Award from the Coblentz Society, and the Garvan Medal of the American Chemical Society

Handbook of Data on Organic Compounds, 2nd edition

Table of Contents — Volumes I to IX

PHYSICAL PROPERTIES

HODOC Number

Abbreviation: HODOC No

HODOC is an acronym for *Handbook of Data on Organic Compounds*. The HODOC number is a numeric designation starting with 00001. The 00001 indicates the compound is the first compound in HODOC II, etc.

CAS Number

Abbreviation: CAS No

Data from the Chemical Abstracts Service (CAS) Registry System have been loaded into one file containing more than 7 million substances and more than 10 million names. The size of the file increases every week with routine updates. The original file contained CAS Registry Numbers only for compounds in the literature after 1965. Newly cited chemical compounds and substances from the current Pre-1965 Registration Project result in the addition of 10,000 to 15,000 substances to the CAS Registry File per week. Each unique chemical substance in the CAS Registry File has a unique Registry Number. CAS Registry Numbers of all compounds in HODOC II have been verified against CAS Registry Numbers in the CAS Registry File. Users will note that the CAS Registry Number always has from 2 to 6 digits to the left of the first dash, 2 numbers between the first and second dash, and only one number to the right of the second dash. Thus, a number might be 42-26-1 or 93673-25-0, but never 45-321-1 or 45-32-11.

Beilstein Number

Abbreviation: Beilstein No

All references are from the fourth edition of Beilstein's *Handbuch der Organische Chemie*. Nearly all of the references were checked by Beilstein personnel. Use of Beilstein is not difficult, but sometimes requires reading the instructions on its use. For example: 3^4,250 indicates the compound is listed on page 250 of the fourth supplement to Volume 3. Beilstein continues to be expanded. The fifth edition is in preparation and is being published in English.

Additional informational material on Beilstein is available upon request free of charge from:

Springer-Verlag KG	Springer-Verlag New York, Inc.
Heidelberger Platz 3	175 Fifth Avenue
D-1000 Berlin 33	New York, NY 10010

Name

The compound name listed for the compound is the name approved by the Chemical Abstracts Service as of late 1988. It is that name one would obtain first if one were to enter the Chemical Abstracts Service computerized databank via the CAS Registry Number. The Chemical Abstracts Service guide to naming chemical substances is reproduced in Volume I. This guide also includes molecular skeletons, stereochemistry, and chemical structural diagrams from CA index names.

Synonyms

The most common synonym for each compound is listed in the database. In addition, up to 100 synonyms known to the compilers and editors are included in the Synonym Index. All synonyms are listed in alphabetical order. For example, there are nine synonyms listed in the Synonym Index for the compound benzenamine, *N*-methyl (CAS Number 100-61-8). These are as follows: aniline, *N*-methyl; anilomethane; (methylamino)benzene; *N*-methylaniline; *N*-methylbenzenamine; methylphenylamine; *N*-methylamine; *N*-monomethylaniline; *N*-phenylmethylamine. Each synonym is followed by a HODOC II number which refers the searcher to the proper compound in the database where spectral information and physical properties for the compound can also be found.

Line Formula

Abbreviation: Line Form

The line formula presents the atoms in the way they are arranged in the molecule. Frequently, this is done by arranging the atoms in groups. For example, for ethanoic acid (CAS Number 64-19-7): CH_3COOH. Another example, *n*-octane (CAS Number 111-65-9), is written as $CH_3CH_2CH_2CH_2CH_2CH_2CH_2CH_3$, whereas *iso*-octane or pentane, 2,2,4-trimethyl (CAS Number 540-84-1), is written as $CH_3C(CH_3)_2CH_2CH(CH_3)CH_3$. Each of these last two named compounds has the molecular formula C_8H_{18}.

Structure

Structures, or as called by some persons, structural formulae, are presented when the systems are complicated or there is need to prevent ambiguity. Many structures are presented in the main database as part of the information on the compound. The Chemical Abstracts Service guide to naming chemical substances is reproduced in Volume I. This guide also includes molecular skeletons, stereochemistry, and chemical structural diagrams from CA index names.

Molecular Formula

Abbreviation: Mol Form

The molecular formula gives the types and numbers of atoms present in a molecule. Thus, the molecular formula for ethanoic acid (frequently called acetic acid) is

$C_2H_4O_2$. This is not the same as the *empirical formula* which lists the atoms in their simplest ratio. For ethanoic acid this would be CH_2O.

Molecular Weight

Abbreviation: M_r

Molecular weights for compounds were calculated using the 1985 International Atomic Weights of the elements. In HODOC II the *molecular weight* is treated the same as the *formula weight*. There are neither molecular weights nor formula weights listed for certain naturally occurring polymers, i.e., starch, whose degree of polymerization is unknown.

Boiling Point

Abbreviation: bp

The boiling point is given in °C at normal atmospheric pressure (760 mmHg, which equals 760 torr, or 101,325 pascals) unless indicated by a superscript to the numerical value for the boiling point. Thus, 58.4^{60} indicates the boiling point of 58.4°C was determined at a pressure of 60 mmHg which is equal to 60 torr. Pascals do not appear as a superscript for any data in this book. For some compounds there are two or more boiling points listed, with each boiliing point having been determined at a different pressure.

Melting Point

Abbreviation: mp

The melting point of a compound is listed in °C. Presence of the letters "d" or "dec" with the numerical value in the melting point column indicates the compound decomposes at the temperature listed. No distinction is made whether the decomposition occurs before the melting point is reached or if the compound decomposes during melting. Many compounds are salts or hydrates. Some of these lose an acid or water when heated. Such a reaction is indicated by -HCl (if the salt is a hydrochloride) or $-H_2O$ (if the compound is a hydrate) for the temperature at which the loss begins to occur.

Density

Abbreviation: *d*

Density is mass per unit volume. In SI units it is expressed as kg/m^3. However, most of the data in this book were obtained from publications wherein the density was expressed as g/cm^3. Unless indicated otherwise, density values are assumed to have been determined at laboratory room temperature. If the temperature was reported in the literature at which the determination was made, the density is followed by a superscript. Thus, 2.6372^{20} indicates the substance has a density of 2.6372 g/cm^3 at 20°C. If the density of the material is compared to the density of a reference material, frequently water, the number resulting from this ratio is called the *relative density*, formerly called *specific gravity*. *Relative density* is defined as the density of a substance relative to the density of some reference substance. For liquids or solids it is frequently the ratio of the density of the substance (usually at 20°C) to the density of water at its maximum density. The maximum density of water is at 3.98°C. However, most measurements are reported as being carried out at 4°C. The difference in density of water at 4°C and that at 3.98°C is less than usually necessary for relative density determinations. A *relative density* would be printed as $2.3480^{20/4}$. This number indicates the density of the substance was determined at 20°C and that this density was compared to that of water at 4°C. The ratio of the two densities is then reported as $2.3480^{20/4}$. It is understood that when the number 4 appears in the denominator of the superscript the reference material is water. If a substance other than water were used as the reference substance, that substance would be listed in the superscript.

Refractive Index

Abbreviation: *n*

The refractive index, which is also known as the refractive constant, is a function of the wavelength of light used to measure the refractive index. Thus the wavelength of light should be specified. The refractive index is usually determined using yellow light, the sodium D line, wavelength 589.3 nm. The numerical value for the refractive index should have a subscript indicating the wavelength of light used for the measurement. However, since the sodium D-line is most generally used, the subscript is either missing in some literature or appears as D. Thus 1.3446, $1.3446_{589.3}$, and 1.3446_D are identical. The temperature at which the refractive index is measured is frequently listed in the literature as a superscript to the numerical value. Thus, $1.5678^{23/D}$ is the numerical value of the refractive index of a compound when the measurement was made at 23°C and the sodium D-line was the light used. Refractive index is a ratio of the speed of electromagnetic radiation, nearly always light, in free space to the speed of the same radiation in another medium. Therefore, the number is dimensionless.

Specific Rotation

Abbreviation: α

Specific rotation, sometimes called optical rotation, is an optical property of a compound which is influenced by the wavelength of light used during the measurement, the length of the path through which the light passes, the particular solvent used to prepare a solution of the compound being studied, the concentration of the solution expressed in terms of the solute, and the temperature of the experiment. The path length of light through the experimental apparatus is usually 10 decimeters. This length is not always reported in the literature. Therefore, one needs to assume that if the length of the tube used in the experiment is not specified, the tube was 10 decimeters in length if the results are listed as specific rotation. Other controlled variables are nearly always listed. For example, $[\alpha]^{20/D} = +58°$ (c = 10% in C_2H_5OH) indicates that when 10 g of compound were dissolved in 100 g of C_2H_5OH and a beam of monochromatic light of wavelength 589.3 nm (the wavelength of the sodium D-line) passed through a tube of 10 decimeters in length containing the aforementioned solution at 20°C, the beam of light was rotated 58° to the right. If the light beam had been rotated to the left, a minus sign would have been in front of the numerical value of the rotation. Absence of square brackets around the α indicates the measurement was made without disolving the compound in a solvent, that is, the experiment was performed neat.

Color

The color and crystalline form of many compounds are listed in the database. The solvent from which the crystalline form is obtained is given in parentheses. Abbreviations for names of the colors and crystalline form are listed near the beginning of the database in the section *Abbreviations and Symbols*.

Solubility

Abbreviation: *s*

Quantitative data on solubilities rarely appear in the literature so a relative scale was established: 1 = insoluble, 2 = slightly soluble, 3 = soluble, 4 = very soluble, 5 = miscible, 6 = decomposes.

SPECTRAL DATA

Infrared, Raman, Ultraviolet, Nuclear Magnetic Resonance (^1H and ^{13}C), and Mass Spectral Data are given in the database. Either a reference to a spectral source curve or a coded representation of the spectrum is given for every compound. A complete listing of the reference collections used is given before the Numerical Index of Source Curves for every technique

Each curve was evaluated during the encoding procedure to guard against inclusion of impurity peaks or of anomalous absorptions. When more than one spectrum of the same compound was available, the one judged to be of highest quality was used. The coding of the spectral data was according to the following criteria.

Infrared

Abbreviation: IR

All absorption bands characteristic of the specific structure (i.e., the common group frequencies), regardless of intensity, were coded. In addition, at least one strong band in each micrometer or 100 cm^{-1} interval, from 3800 to 250 cm^{-1}, was coded. Spectra were coded in either micrometers or wavenumbers, depending on the format of the reference curve. Data in micrometers were converted by computer to wavenumbers for listing in the database and the Indexes (see Wavelength-Wavenumber Conversion Table). A micron symbol, μ, is shown before the spectrum reference number for every converted spectrum. Peak positions were read to ±0.1 μm or ±10 cm^{-1}. When a coder was unfamiliar with the structure and could not select the "most characteristic" bands, all medium to strong bands were coded.

Raman

The Raman spectra from the early API collection (#1-500) and those from the *Analytical Chemistry (AC)* reference article were obtained with a mercury arc source. Only the ten strongest bands in the spectra were coded. Spectra were only recorded from 1700 to 150 cm^{-1} in *AC* and this is indicated by an asterisk (*) before the first coded band. All other Raman spectra were generated using a laser source. These were coded over the range 4000 to 50 cm^{-1}. All recognized group frequencies and most significant bands were listed. In the Sadtler spectra only the parallel polarized curve was coded.

Ultraviolet

Abbreviation: UV

All bands, their molar absorption coefficients (in parentheses, when given), and the solvent used were coded. The wavelength range was from 170 to 600 nm and peaks were read to ±1 nm. When the spectrum showed vibrational fine structure, only the peak centers, characteristic of the electronic transitions, were read. If only a cut-off of absorption appeared, the notation b, for the peak below this wavelength, was used.

Nuclear Magnetic Resonance

Abbreviation: NMR

^1H — The proton chemical shifts, in ppm (δ), for specific protons or recognizable groups were coded to ±0.1 ppm over the range 0 to 15 ppm referenced to tetramethylsilane (TMS). When complex spectra due to second order effects or overlapping resonances were encountered, the range (indicated by connecting chemical shifts with a dash) was recorded. The solvent in which the spectrum was obtained was also coded, if known.

^{13}C — The carbon chemical shifts, in ppm (δ), for specific carbons or recognizable groups were coded to ±0.1 ppm over the range 0 to 200 ppm referenced to TMS. The solvent in which the spectrum was obtained was also coded, if known.

Mass Spectroscopy

Abbreviation: MS

The eight most abundant peaks, the parent peak (if not among the eight most abundant), and the relative intensities (in parentheses) were coded and are listed in order of decreasing intensity. If the parent peak was not observed it is listed with an intensity of (0). All data are from magnetic sector spectrometers and were taken at normal (~70 ev) ionizing voltages.

m/e

The mass number or the m/e value of an ion is the quotient of the mass of that ion (given in atomic mass units) and its positive charge (number of electrons lost during ionization). One observes the value of m/e in HODOC II approximates closely to the molecular weight (M_r) listed for the compound. The molecular weight of a compound was calculated from the 1985 International Atomic Weights whereas the m/e was calculated from the atomic mass of the most abundant naturally occurring isotope of the element. Thus, when calculating the m/e ratio, one uses for hydrogen, 1.007825; for carbon, 12.0000; for nitrogen, 14.003074; for oxygen, 15.994915; for sulfur, 31.972070, for chlorine, 34.968852; etc. These atomic masses of the isotopes are presented in most books which contain data on the elements and their isotopes. Among these are the *CRC Handbook of Chemistry and Physics*, 69th edition, or *Table of Isotopes*, C. M. Lederer and V. S. Shirley, 7th edition, Wiley-Interscience. The m/e values listed in the database of HODOC II are for ions which have lost one electron.

Wavelength–Wavenumber Conversion Table

Wavelength, μm	Wavenumber, cm^{-1} 0	1	2	3	4	5	6	7	8	9
2.0	5000	4975	4950	4926	4902	4878	4854	4831	4808	4785
2.1	4762	4739	4717	4695	4673	4651	4630	4608	4587	4566
2.2	4545	4525	4505	4484	4464	4444	4425	4405	4386	4367
2.3	4348	4329	4310	4292	4274	4255	4237	4219	4202	4184
2.4	4167	4149	4132	4115	4098	4082	4065	4049	4032	4016
2.5	4000	3984	3968	3953	3937	3922	3906	3891	3876	3861
2.6	3846	3831	3817	3802	3788	3774	3759	3745	3731	3717
2.7	3704	3690	3676	3663	3650	3636	3623	3610	3597	3584
2.8	3571	3559	3546	3534	3521	3509	3497	3484	3472	3460
2.9	3448	3436	3425	3413	3401	3390	3378	3367	3356	3344
3.0	3333	3322	3311	3300	3289	3279	3268	3257	3247	3236
3.1	3226	3215	3205	3195	3185	3175	3165	3155	3145	3135
3.2	3125	3115	3106	3096	3086	3077	3067	3058	3049	3040
3.3	3030	3021	3012	3003	2994	2985	2976	2967	2959	2950
3.4	2941	2933	2924	2915	2907	2899	2890	2882	2874	2865
3.5	2857	2849	2841	2833	2825	2817	2809	2801	2793	2786
3.6	2778	2770	2762	2755	2747	2740	2732	2725	2717	2710
3.7	2703	2695	2688	2681	2674	2667	2660	2653	2646	2639
3.8	2632	2625	2618	2611	2604	2597	2591	2584	2577	2571
3.9	2564	2558	2551	2545	2538	2532	2525	2519	2513	2506
4.0	2500	2494	2488	2481	2475	2469	2463	2457	2451	2445
4.1	2439	2433	2427	2421	2415	2410	2404	2398	2392	2387
4.2	2381	2375	2370	2364	2358	2353	2347	2342	2336	2331
4.3	2326	2320	2315	2309	2304	2299	2294	2288	2283	2278
4.4	2273	2268	2262	2257	2252	2247	2242	2237	2232	2227
4.5	2222	2217	2212	2208	2203	2198	2193	2188	2183	2179
4.6	2174	2169	2165	2160	2155	2151	2146	2141	2137	2132
4.7	2128	2123	2119	2114	2110	2105	2101	2096	2092	2088
4.8	2083	2079	2075	2070	2066	2062	2058	2053	2049	2045
4.9	2041	2037	2033	2028	2024	2020	2016	2012	2008	2004
5.0	2000	1996	1992	1988	1984	1980	1976	1972	1969	1965
5.1	1961	1957	1953	1949	1946	1942	1938	1934	1931	1927
5.2	1923	1919	1916	1912	1908	1905	1901	1898	1894	1890
5.3	1887	1883	1880	1876	1873	1869	1866	1862	1859	1855
5.4	1852	1848	1845	1842	1838	1835	1832	1828	1825	1821
5.5	1818	1815	1812	1808	1805	1802	1799	1795	1792	1789
5.6	1786	1783	1779	1776	1773	1770	1767	1764	1761	1757
5.7	1754	1751	1748	1745	1742	1739	1736	1733	1730	1727
5.8	1724	1721	1718	1715	1712	1709	1706	1704	1701	1698
5.9	1695	1692	1689	1686	1684	1681	1678	1675	1672	1669
6.0	1667	1664	1661	1658	1656	1653	1650	1647	1645	1642
6.1	1639	1637	1634	1631	1629	1626	1623	1621	1618	1616
6.2	1613	1610	1608	1605	1603	1600	1597	1595	1592	1590
6.3	1587	1585	1582	1580	1577	1575	1572	1570	1567	1565
6.4	1563	1560	1558	1555	1553	1550	1548	1546	1543	1541
6.5	1538	1536	1534	1531	1529	1527	1524	1522	1520	1517
6.6	1515	1513	1511	1508	1506	1504	1502	1499	1497	1495
6.7	1493	1490	1488	1486	1484	1481	1479	1477	1475	1473
6.8	1471	1468	1466	1464	1462	1460	1458	1456	1453	1451
6.9	1449	1447	1445	1443	1441	1439	1437	1435	1433	1431
7.0	1429	1427	1425	1422	1420	1418	1416	1414	1412	1410
7.1	1408	1406	1404	1403	1401	1399	1397	1395	1393	1391
7.2	1389	1387	1385	1383	1381	1379	1377	1376	1374	1372
7.3	1370	1368	1366	1364	1362	1361	1359	1357	1355	1353
7.4	1351	1350	1348	1346	1344	1342	1340	1339	1337	1335

Wavelength–Wavenumber Conversion Table (continued)

Wavelength, μm	Wavenumber, cm^{-1} 0	1	2	3	4	5	6	7	8	9
7.5	1333	1332	1330	1328	1326	1325	1323	1321	1319	1318
7.6	1316	1314	1312	1311	1309	1307	1305	1304	1302	1300
7.7	1299	1297	1295	1294	1292	1290	1289	1287	1285	1284
7.8	1282	1280	1279	1277	1276	1274	1272	1271	1269	1267
7.9	1266	1264	1263	1261	1259	1258	1256	1255	1253	1252
8.0	1250	1248	1247	1245	1244	1242	1241	1239	1238	1236
8.1	1235	1233	1232	1230	1229	1227	1225	1224	1222	1221
8.2	1220	1218	1217	1215	1214	1212	1211	1209	1208	1206
8.3	1205	1203	1202	1200	1199	1198	1196	1195	1193	1192
8.4	1190	1189	1188	1186	1185	1183	1182	1181	1179	1178
8.5	1176	1175	1174	1172	1171	1170	1168	1167	1166	1164
8.6	1163	1161	1160	1159	1157	1156	1155	1153	1152	1151
8.7	1149	1148	1147	1145	1144	1143	1142	1140	1139	1138
8.8	1136	1135	1134	1133	1131	1130	1129	1127	1126	1125
8.9	1124	1122	1121	1120	1119	1117	1116	1115	1114	1112
9.0	1111	1110	1109	1107	1106	1105	1104	1103	1101	1100
9.1	1099	1098	1096	1095	1094	1093	1092	1091	1089	1088
9.2	1087	1086	1085	1083	1082	1081	1080	1079	1078	1076
9.3	1075	1074	1073	1072	1071	1070	1068	1067	1066	1065
9.4	1064	1063	1062	1060	1059	1058	1057	1056	1055	1054
9.5	1053	1052	1050	1049	1048	1047	1046	1045	1044	1043
9.6	1042	1041	1040	1038	1037	1036	1035	1034	1033	1032
9.7	1031	1030	1029	1028	1027	1026	1025	1024	1022	1021
9.8	1020	1019	1018	1017	1016	1015	1014	1013	1012	1011
9.9	1010	1009	1008	1007	1006	1005	1004	1003	1002	1001
10.0	1000.0	999.0	998.0	997.0	996.0	995.0	994.0	993.0	992.1	991.1
10.1	990.1	989.1	988.1	987.2	986.2	985.2	984.3	983.3	982.3	981.4
10.2	980.4	979.4	978.5	977.5	976.6	975.6	974.7	973.7	972.8	971.8
10.3	970.9	969.9	969.0	968.1	967.1	966.2	965.3	964.3	963.4	962.5
10.4	961.5	960.6	959.7	958.8	957.9	956.9	956.0	955.1	954.2	953.3
10.5	952.4	951.5	950.6	949.7	948.8	947.9	947.0	946.1	945.2	944.3
10.6	943.4	942.5	941.6	940.7	939.8	939.0	938.1	937.2	936.3	935.5
10.7	934.6	933.7	932.8	932.0	931.1	930.2	929.4	928.5	927.6	926.8
10.8	925.9	925.1	924.2	923.4	922.5	921.7	920.8	920.0	919.1	918.3
10.9	917.4	916.6	915.8	914.9	914.1	913.2	912.4	911.6	910.7	909.9
11.0	909.1	908.3	907.4	906.6	905.8	905.0	904.2	903.3	902.5	901.7
11.1	900.9	900.1	899.3	898.5	897.7	896.9	896.1	895.3	894.5	893.7
11.2	892.9	892.1	891.3	890.5	889.7	888.9	888.1	887.3	886.5	885.7
11.3	885.0	884.2	883.4	882.6	881.8	881.1	880.3	879.5	878.7	878.0
11.4	877.2	876.4	875.7	874.9	874.1	873.4	872.6	871.8	871.1	870.3
11.5	869.6	868.8	868.1	867.3	866.6	865.8	865.1	864.3	863.6	862.8
11.6	862.1	861.3	860.6	859.8	859.1	858.4	857.6	856.9	856.2	855.4
11.7	854.7	854.0	853.2	852.5	851.8	851.1	850.3	849.6	848.9	848.2
11.8	847.5	846.7	846.0	845.3	844.6	843.9	843.2	842.5	841.8	841.0
11.9	840.3	839.6	838.9	838.2	837.5	836.8	836.1	835.4	834.7	834.0
12.0	833.3	832.6	831.9	831.3	830.6	829.9	829.2	828.5	827.8	827.1
12.1	826.4	825.8	825.1	824.4	823.7	823.0	822.4	821.7	821.0	820.3
12.2	819.7	819.0	818.3	817.7	817.0	816.3	815.7	815.0	814.3	813.7
12.3	813.0	812.3	811.7	811.0	810.4	809.7	809.1	808.4	807.8	807.1
12.4	806.5	805.8	805.2	804.5	803.9	803.2	802.6	801.9	801.3	800.6
12.5	800.0	799.4	798.7	798.1	797.4	796.8	796.2	795.5	794.9	794.3
12.6	793.7	793.0	792.4	791.8	791.1	790.5	789.9	789.3	788.6	788.0
12.7	787.4	786.8	786.2	785.5	784.9	784.3	783.7	783.1	782.5	781.9
12.8	781.3	780.6	780.0	779.4	778.8	778.2	777.6	777.0	776.4	775.8
12.9	775.2	774.6	774.0	773.4	772.8	772.2	771.6	771.0	770.4	769.8
13.0	769.2	768.6	768.0	767.5	766.9	766.3	765.7	765.1	764.5	763.9
13.1	763.4	762.8	762.2	761.6	761.0	760.5	759.9	759.3	758.7	758.2

Wavelength–Wavenumber Conversion Table (continued)

Wavelength, μm	Wavenumber, cm⁻¹ 0	1	2	3	4	5	6	7	8	9
13.2	757.6	757.0	756.4	755.9	755.3	754.7	754.1	753.6	753.0	752.4
13.3	751.9	751.3	750.8	750.2	749.6	749.1	748.5	747.9	747.4	746.8
13.4	746.3	745.7	745.2	744.6	744.0	743.5	742.9	742.4	741.8	741.3
13.5	740.7	740.2	739.6	739.1	738.6	738.0	737.5	736.9	736.4	735.8
13.6	735.3	734.8	734.2	733.7	733.1	732.6	732.1	731.5	731.0	730.5
13.7	729.9	729.4	728.9	728.3	727.8	727.3	726.7	726.2	725.7	725.2
13.8	724.6	724.1	723.6	723.1	722.5	722.0	721.5	721.0	720.5	719.9
13.9	719.4	718.9	718.4	717.9	717.4	716.8	716.3	715.8	715.3	714.8
14.0	714.3	713.8	713.3	712.8	712.3	711.7	711.2	710.7	710.2	709.7
14.1	709.2	708.7	708.2	707.7	707.2	706.7	706.2	705.7	705.2	704.7
14.2	704.2	703.7	703.2	702.7	702.2	701.8	701.3	700.8	700.3	699.8
14.3	699.3	698.8	698.3	697.8	697.4	696.9	696.4	695.9	695.4	694.9
14.4	694.4	694.0	693.5	693.0	692.5	692.0	691.6	691.1	690.6	690.1
14.5	689.7	689.2	688.7	688.2	687.8	687.3	686.8	686.3	685.9	685.4
14.6	684.9	684.5	684.0	683.5	683.1	682.6	682.1	681.7	681.2	680.7
14.7	680.3	679.8	679.3	678.9	678.4	678.0	677.5	677.0	676.6	676.1
14.8	675.7	675.2	674.8	674.3	673.9	673.4	672.9	672.5	672.0	671.6
14.9	671.1	670.7	670.2	669.8	669.3	668.9	668.4	668.0	667.6	667.1
15.0	666.7	666.2	665.8	665.3	664.9	664.5	664.0	663.6	663.1	662.7
15.1	662.3	661.8	661.4	660.9	660.5	660.1	659.6	659.2	658.8	658.3
15.2	657.9	657.5	657.0	656.6	656.2	655.7	655.3	654.9	654.5	654.0
15.3	653.6	653.2	652.7	652.3	651.9	651.5	651.0	650.6	650.2	649.8
15.4	649.4	648.9	648.5	648.1	647.7	647.2	646.8	646.4	646.0	645.6
15.5	645.2	644.7	644.3	643.9	643.5	643.1	642.7	642.3	641.8	641.4
15.6	641.0	640.6	640.2	639.8	639.4	639.0	638.6	638.2	637.8	637.3
15.7	636.9	636.5	636.1	635.7	635.3	634.9	634.5	634.1	633.7	633.3
15.8	632.9	632.5	632.1	631.7	631.3	630.9	630.5	630.1	629.7	629.3
15.9	628.9	628.5	628.1	627.7	627.4	627.0	626.6	626.2	625.8	625.4
16.0	625.0	624.6	624.2	623.8	623.4	623.1	622.7	622.3	621.9	621.5
16.1	621.1	620.7	620.3	620.0	619.6	619.2	618.8	618.4	618.0	617.7
16.2	617.3	616.9	616.5	616.1	615.8	615.4	615.0	614.6	614.3	613.9
16.3	613.5	613.1	612.7	612.4	612.0	611.6	611.2	610.9	610.5	610.1
16.4	609.8	609.4	609.0	608.6	608.3	607.9	607.5	607.2	606.8	606.4
16.5	606.1	605.7	605.3	605.0	604.6	604.2	603.9	603.5	603.1	602.8
16.6	602.4	602.0	601.7	601.3	601.0	600.6	600.2	599.9	599.5	599.2
16.7	598.8	598.4	598.1	597.7	597.4	597.0	596.7	596.3	595.9	595.6
16.8	595.2	594.9	594.5	594.2	593.8	593.5	593.1	592.8	592.4	592.1
16.9	591.7	591.4	591.0	590.7	590.3	590.0	589.6	589.3	588.9	588.6
17.0	588.2	587.9	587.5	587.2	586.9	586.5	586.2	585.8	585.5	585.1
17.1	584.8	584.5	584.1	583.8	583.4	583.1	582.8	582.4	582.1	581.7
17.2	581.4	581.1	580.7	580.4	580.0	579.7	579.4	579.0	578.7	578.4
17.3	578.0	577.7	577.4	577.0	576.7	576.4	576.0	575.7	575.4	575.0
17.4	574.7	574.4	574.1	573.7	573.4	573.1	572.7	572.4	572.1	571.8
17.5	571.4	571.1	570.8	570.5	570.1	569.8	569.5	569.2	568.8	568.5
17.6	568.2	567.9	567.5	567.2	566.9	566.6	566.3	565.9	565.6	565.3
17.7	565.0	564.7	564.3	564.0	563.7	563.4	563.1	562.7	562.4	562.1
17.8	561.8	561.5	561.2	560.9	560.5	560.2	559.9	559.6	559.3	559.0
17.9	558.7	558.3	558.0	557.7	557.4	557.1	556.8	556.5	556.2	555.9
18.0	555.6	555.2	554.9	554.6	554.3	554.0	553.7	553.4	553.1	552.8
18.1	552.5	552.2	551.9	551.6	551.3	551.0	550.7	550.4	550.1	549.8
18.2	549.5	549.1	548.8	548.5	548.2	547.9	547.6	547.3	547.0	546.7
18.3	546.4	546.1	545.9	545.6	545.3	545.0	544.7	544.4	544.1	543.8
18.4	543.5	543.2	542.9	542.6	542.3	542.0	541.7	541.4	541.1	540.8
18.5	540.5	540.2	540.0	539.7	539.4	539.1	538.8	538.5	538.2	537.9
18.6	537.6	537.3	537.1	536.8	536.5	536.2	535.9	535.6	535.3	535.0
18.7	534.8	534.5	534.2	533.9	533.6	533.3	533.0	532.8	532.5	532.2
18.8	531.9	531.6	531.3	531.1	530.8	530.5	530.2	529.9	529.7	529.4
18.9	529.1	528.8	528.5	528.3	528.0	527.7	527.4	527.1	526.9	526.6

Wavelength–Wavenumber Conversion Table (continued)

Wavelength, μm	Wavenumber, cm^{-1} 0	1	2	3	4	5	6	7	8	9
19.0	526.3	526.0	525.8	525.5	525.2	524.9	524.7	524.4	524.1	523.8
19.1	523.6	523.3	523.0	522.7	522.5	522.2	521.9	521.6	521.4	521.1
19.2	520.8	520.6	520.3	520.0	519.8	519.5	519.2	518.9	518.7	518.4
19.3	518.1	517.9	517.6	517.3	517.1	516.8	516.5	516.3	516.0	515.7
19.4	515.5	515.2	514.9	514.7	514.4	514.1	513.9	513.6	513.3	513.1
19.5	512.8	512.6	512.3	512.0	511.8	511.5	511.2	511.0	510.7	510.5
19.6	510.2	509.9	509.7	509.4	509.2	508.9	508.6	508.4	508.1	507.9
19.7	507.6	507.4	507.1	506.8	506.6	506.3	506.1	505.8	505.6	505.3
19.8	505.1	504.8	504.5	504.3	504.0	503.8	503.5	503.3	503.0	502.8
19.9	502.5	502.3	502.0	501.8	501.5	501.3	501.0	500.8	500.5	500.3

ABBREVIATIONS AND SYMBOLS

[α], α	specific rotation and column head
>	above, more than
<	below, less than
∞	soluble in all proportions
ABBIA	*Archives of Biochemistry and Biophysics*
abs	absolute
ac	acid
Ac	acetyl
AC	Analytical Chemistry reference to spectra
ace, Ace	acetone
$aceD_6$	acetone deuterated
$aceF_6$	hexafluoro acetone
AcOEt	ethyl acetate
Ac_2O	acetic anhydride act
ACSAA	*Acta Chemica Scandivanica*
al	alcohol[1]
ALD	*The Aldrich Handbook of Organic Chemicals and Biochemicals,* Aldrich Chemical Co., Inc., New York; Aldrich Library of Infrared Spectra; Aldrich Library of Mass Spectra; Aldrich Chemical Co., Milwaukee, WI
ALDN	Aldermaston, *8 peak Index of Mass Spectra,* U.K.
alk	alkali
Am	amyl (pentyl)
AmOH	amyl alcohol
amor	amorphous
AMS	*Archives of Mass Spectral Data,* John Wiley & Sons, New York
ANCPA	*Annales de Chimie (Paris)*
anh	anhydrous
API	American Petroleum Institute spectral collection
aq	aqueous
ARKEA	*Arkiv foer Kemi*
as	asymmetric
atm	atmospheres
b	below
BCSJA	*Bulletin of the Chemical Society of Japan*
BIJOA	*Biochemical Journal*
bipym	bipyramidal
bk	black
bl	blue
bp	boiling point
BPSMA	*Bulletin de l'Academie Polonaise des Sciences*
br	brown
BSCFA	*Bulletin de la Societe Chimique de France*
BSCJA	*Bristol Chamber of Commerce Journal*
bt	bright
Bu	butyl
BuOH	butyl alcohol
bz, Bz	benzene
c	percentage concentration
ca	about
CA	*Chemical Abstracts*
CAS	Chemical Abstracts Service
CCCCA	*Collection of Czechoslovak Chemical Communications*
cello	cellosolve
CHBEA	*Chemische Berichte*
CHINA	*Chemistry and Industry*
chl, Chl	chloroform
CJCHA	*Canadian Journal of Chemistry*
co	columns
COB	Coblentz Society spectral collection
col	colorless
con, conc	concentrated
cor	corrected
COREA	*Comptes Rendus Hebdomadaires des Seances de l'Academie des Sciences*
ctc	carbon tetrachloride
cw	continuous wave
cy, cyhex	cyclohexane
cyD_{12}	cyclohexane deuterated
D	line in the emission spectrum of sodium (subscript), i.e., $[n]_D$
d, D	dextro
d	as column head, density
DANKA	*Doklady Akademii Nauk S.S.S.R.*
dec	decomposes
DFB	Dollish et al., *Characteristic Raman Frequencies of Organic Compounds,* John Wiley & Sons, New York
dil	diluted
diox	dioxane
dk	dark
dl, DL	racemic[2]
dlq	deliquescent
DMF	dimethyl formamide
$DMFd_6$	dimethyl formamide deuterated
DMS	Documentation of Molecular Spectroscopy spectral collection, Butterworth Scientific Publication, London WC2, England
DMSO	dimethyl sulfoxide
$DMSOd_6$	dimethyl sulfoxide deuterated
efl	efflorescent
Et	ethyl
eth	ether[3]
EtOH	ethyl alcohol
Et_2O	ethyl ether
exp	explodes
extrap	extrapolated
fl	flakes

flr	fluorescent
FO	Friedel and Orchin, *Ultraviolet Spectra of Aromatic Compounds*, John Wiley & Sons, New York
fp	freezing point
FT	Fourier transform
fum	fuming
GCITA	*Gazzetta Chimica Italiana*
gel	gelatinous
gl	glacial
gr	green
gran	granular
gy	gray
HBCP	*Handbook of Chemistry and Physics*, CRC Press, Inc., Boca Raton, Florida
HCACA	*Helvetica Chimica Acta*
hcni	hydrochloride not indicated
hex	hexagonal
hni	hydrate not indicated
HOAc	acetic acid
HODOC II	acronym for *Handbook of Data on Organic Compounds*, 2nd edition
hp	heptane
hx	hexane
hyd	hydrate
hyg	hygroscopic
i	insoluble
i-	iso-
ign	ignites
IJOCA	*Indian Journal of Chemistry*
in	inactive
infus	infusible
ini	isomer not indicated
IR	infrared spectral reference as source of physical constants
irid	iridescent
iso	isooctane
JACSA	*Journal of the American Chemical Society*
JCS, B	*Journal of the Chemical Society, Section B*
JCSOA	*Journal of the Chemical Society (London)*
JJ	Johnson and Jankowski, *Carbon-13 NMR Spectra*, John Wiley & Sons, New York
JMOSA	*Journal of Molecular Spectroscopy*
JOCEA	*Journal of Organic Chemistry*
JPRAA	*Journal de Physique, et le Radium*
l, *L*	levo
LA	Levy and Nelson, *Carbon-13 NMR for Organic Chemists*, John Wiley & Sons, New York
LANG	*Absorption Spectra in the UV and Visible Region*, Academic Press, New York
lf	leaf
lig	ligroin
liq	liquid
lo	long
lpet	light petroleum ether
lt	light
m-	meta-
M	molar concentration
M_r	molecular weight, column head
MCA	Manufacturing Chemists Association Research Project, Thermodynamics Research Center, Texas A&M University, College Station, TX 77843
mcl	monoclinic
Me	methyl
MeCN	acetonitrile
MeOH	methyl alcohol
met	metallic
micr	microscopic
min	mineral
MOCHA	*Monatshefte für Chemie und Verwandte Teile Anderer Wissenschaften*
mod	modification
mp	melting point
MS	Mass spectral reference as source of physical constant
mut	mutarotatory
n-	normal chain
n	refractive index (n_D), column head
N	normal (concentration of solute in solvent)
N-	nitrogen[4]
naph	naphthalene
NBS	National Bureau of Standards EPA-NIH Mass Spectra Data Base, NSRDS-NBS-63
nd	needles
NRC	National Bureau of Standards spectral collection
o-	ortho-
oct	octahedral
OES	Phillip et al., *Organic Electronic Spectral Data*, John Wiley & Sons, New York
og	orange
OPSUA	*Optics and Spectroscopy (USSR — English)*
ord	ordinary
org	organic
orh	orthorhombic
os	organic solvent
p-	para-
pa	pale
par	partial
peth	petroleum ether
Ph	phenyl
pH	hydrogen ion concentration
PhCl	chlorobenzene
$PhNH_2$	aniline
$PhNO_2$	nitrobenzene
pk	pink
pl	plates
poly D	polysol D

pr	prisms
Pr	propyl
PrOH	propyl alcohol
purp	purple
pwd	powder
py, Py	pyridine
pym	pyramids
pyr, pyrim	pyrimidine
rac-	racemic
rect	rectangle
res	resinous
rh	rhombic
rhd	rhombohedral
RTCPA	*Recueil des Travaux Chimiques des Pays-Bas*
s	solubility, column head[5]
s	soluble
s-	secondary[6]
SAD	Sadtler Research Laboratories spectral collection
SADG	Sadtler Research Laboratories IR grating collection
SADP	Sadtler Research Laboratories IR prism collection
sat	saturated
sec-	secondary[6]
sl, sli	slight
sni	stability not indicated
sol	solid
solu	solution
solv	solvent
SPACA	*Spectrochimica Acta*
st	stable
STOT	Stothers, *Carbon-13 NMR Spectroscopy,* Academic Press, New York, 1972
sub	sublimes
sulf	sulfuric acid
sym-	symmetrical
syr	syrup
t-	tertiary[6]
tcl	triclinic
tert-	tertiary[6]
Tet	*Tetrahedron*
tetr	tetragonal
TETRA	*Tetrahedron*
tfa	trifluoroacetic acid
thf, THF	tetrahydrofuran
to, tol	toluene
tr	transparent
trg	trigonal
μ	micron
undil	undiluted
uns-	unsymmetrical
unst	unstable
vac	vacuum
vap	vapor
var	variable
VAR	Varian Associates NMR spectra collection
vic-	vicinal
visc	viscous
vt	violet
w	water
wh	white
WILEY	*Atlas of Mass Spectral Data,* John Wiley & Sons, New York
wr	warm
wx	waxy
ye	yellow
xyl	xylene
ZEELA	*Zeitschrift für Elektrochemie*
ZPCBA	*Zeitschrift für Physikalische Chemie, Abteilung B*
ZPCHA	*Zeitschrift für Physiologische Chemie, Hoppe-Seylers*

1 Generally means ethyl alcohol.

2 *D,L* generally means configuration and *d,l* mean optical rotation. However, there are many examples in the chemical literature for which the meaning of these symbols is ambiguous and/or interchangable.

3 Generally means diethyl ether.

4 *N* indicates a position in the molecule.

5 Solubility: 1, insoluble; 2, slightly soluble; 3, soluble; 4, very soluble; 5, miscible; 6, decomposes.

6 *s-* and *sec-*, or *t-* and *tert-*, are used as convenient.

CHEMICAL ABSTRACTS

NAME GUIDE

101. Foreword. Although the account which follows describes in considerable detail the selection of substance names for *Chemical Abstracts (CA)* Indexes, it is not a nomenclature manual. It has the more restricted aim of enabling a user of CA indexes to proceed from the structure of an individual chemical compound to the place in the current *Chemical Substance Index* where the particular index name and any associated index entries will be found. This is the identical operation performed by a *CA* indexer when he assigns an index name to a new or previously unnamed substance. What follows, in fact, is a comprehensive summary of *CA* substance indexing policies, which cover not only conventional organic and inorganic compounds but other completely defined substances entered in the *Chemical Substance Index* and given CAS Registry Numbers. These substances include specific chemical elements, alloys, minerals, mixtures, polymers, enzymes, polysaccharides, and elementary particles.

The chemical nomenclature used by CAS has developed in parallel and generally in accordance with the rules published by the International Union of Pure and Applied Chemistry (IUPAC). Although these rules provide unambiguous text equivalents for the great majority of substances, equally acceptable alternative rules within the present IUPAC system often lead to two or more unambiguous names. This causes no difficulty in normal scientific communication, but is totally unacceptable in a formal, rigidly controlled, alphabetic listing such as the *CA Chemical Substance Index*. Here the names must be not only unambiguous, unique, and totally reproducible, but selected so as to bring the names of structurally related substances into juxtaposition in the alphabetical listing. They must be equally derivable by index users searching for information about individual substances and by those who prepare the index. It is also desirable that both should be able to use mechanical aids in name generation and retrieval. Lengthy and expensive training courses for humans, and elaborate programming for machines, must be avoided.

A major revision of *CA* index names was carried out in 1972 as the Ninth Collective Index period began. Most trivial names were dropped; exceptional treatment for various classes of substances was discontinued. Where, because of the stereochemical complexity of a natural product name, this name was retained as a "stereoparent" (see ¶ 202), diagrams were furnished in the *Chemical Substance Index* to aid interpretation of index entries. The 1972 nomenclature revision and the reasons for its adoption are set forth in greater detail in the Ninth Collective *Index Guide* and in a journal article (*J. Chem. Doc.* **14**(1), 3-15 (1974)).

The preferred *CA* index names for chemical substances have been, with very few exceptions, continued unchanged since that date. The minor changes in name-selection policies for the Twelfth Collective Index period (1987-1991) affect inorganic compounds, peptides, polymers and ring systems. These changes, as well as the major policy changes made in 1972 and the few revisions of the Tenth (1977-1981) and Eleventh (1982-1986) Collective Index periods are reviewed in Section G (¶¶ 225-293) of this Appendix.

The main part of the *Index Guide* must be consulted before any search is conducted in the *CA Chemical Substance* and *General Subject Indexes*.

The arrangement of sections in "Selection of Index Names for Chemical Substances" is as follows:

The arrangement within each of these sections is indicated by a key at the beginning of the section.

In the development of CAS policies for index names of chemical substances, no new nomenclature systems have been devised. Adaptation of current IUPAC rules to the specific needs of a highly ordered alphabetical index, not arbitrary coinage of new terms, has been the approach taken. It continues to be recognized by CAS that, while a unique name is needed for an index, and that this name, and the CAS Registry Number, are invaluable aids for substance identification, the use of this invariant index name for citation throughout every context in the scientific community is neither practicable nor desirable. But international agreement in chemical nomenclature, as embodied in the rules of IUPAC, IUB, and other organizations, continues to be of the greatest importance in restricting the *arbitrary* proliferation of substance names. References to individual rules which have formed the basis of CAS policies recorded in the sections that follow have not been cited, but the selective bibliography of the nomenclature of chemical substances which constitutes Section J contains a comprehensive list of current accepted rules.

102. Acknowledgement. Chemical Abstracts Service acknowledges the large contribution made by Cecil C. Langham in helping to develop and record *CA* name-selection policies for the Eighth Collective period (1967-1971) during the years immediately preceding his retirement in 1969. Dr. Langham's work constituted an invaluable starting point for the revised name-selection policies introduced in 1972 and continued in the current 1982-1986 indexing period.

A. NOMENCLATURE SYSTEMS AND GENERAL PRINCIPLES

103. Introduction. Many names may be employed in scientific publications for a single compound. Even so simple a compound as $H_2NCH_2CH_2OH$ can be named 2-Aminoethanol, 2-Aminoethyl alcohol, 2-Hydroxyethylamine, β-Hydroxyethylamine, 2-Hydroxyethanamine, 1-Amino-2-hydroxyethane, etc., all of which describe it unambiguously; often, the less systematic ("trivial") name Ethanolamine may be encountered. For more complex compounds, the number of possible names increases exponentially.

In these circumstances, selection rules are essential if a single preferred name for citation in an index of chemical names is to be determined for each identifiable substance; lacking such a single name, information regarding the substance becomes scattered in the index. Beyond this, it is desirable that the selection rules bring chemically related substances close together in the index, and that they should be as consistent and as free from exceptions as possible. The main part of the *Index Guide* provides cross-references from trivially named index headings employed before *CA* Volume 76 (1972), as well as for most of the trivial or author names of chemical substances encountered in the scientific literature.

Because of the special interest attached to ring systems, all but the most common rings receive additional index entries when the preferred index name appears elsewhere in the *Chemical Substance Index* (see ¶ 163, below).

104. Inversion of names. Ordering in the *Chemical Substance Index* is based on the *index heading parent* (1), which is often made up of a basic skeleton name, e.g., "**Butane**," plus a suffix denoting the principal function, e.g., "**-sulfonic acid**." A locant, e.g., "**1-**," fixing its position is also often necessary. Following a comma (the *comma of inversion*) the substituents (2) are expressed in alphabetical order, e.g., "**2,4-diamino-3-chloro-**," and the *modification* (3), now printed in boldface, completes the name by citing any derivatives of the principal function, e.g., "**ethyl ester**", and stereochemical information (see ¶ 203) if appropriate. The uninverted name of the *acid* is 2,4-diamino-3-chloro-1-butanesulfonic acid, and of the *ester*, ethyl 2,4-diamino-3-chloro-1-butanesulfonate. The latter appears in the *Chemical Substance Index* as follows:

(1) (2)

1-Butanesulfonic acid, 2,4-diamino-3-chloro-
ethyl ester

(3)

Also appearing in the boldface headings of certain very well-known substances are subdivision terms describing properties, etc., of the compound itself, or classifying certain derivatives of it, e.g., **reactions, esters** (see Appendix II, ¶ 10B).

105. Name selection principles (see also ¶ 138). In choosing **1-Butanesulfonic acid** as the heading parent in the example above, rather than, for example, 1,3-Butanediamine, an order of precedence of chemical functions and compound classes (¶ 106) was followed. In this hierarchy, sulfonic acids are ranked higher than amines. In the example cited, so-called substitutive nomenclature (¶ 130) was the type of nomenclature used. Generally, a preferred index name is determined by proceeding as follows until a decision is reached:

(*a*) determine the most senior compound class;
(*b*) determine the type of nomenclature that is appropriate;
(*c*) determine the preferred index heading parent;
(*d*) name the remainder of the structure as substituents, and/or as functional derivatives by modification phrases;
(*e*) choose between alternatives where more than one unambiguous name is still possible.

The remainder of Section IVA is devoted to detailing the application of these rules.

106. Order of precedence of compound classes, in descending order:

(*a*) Free radicals and compounds for which substituent prefixes are unavailable, e.g., **Sydnone** (¶ 201), **Sulfur diimide** (¶ 200).

(*b*) Cationic compounds: coordination cations, onium, aminium, ylium cations.

(*c*) Neutral coordination compounds, including metallocenes.

(*d*) Anionic compounds, e.g., **Borate(1-).**

(*e*) Acids: peroxy acids, expressed as principal groups, in the order of the parent acids; acids, expressed as principal groups, in the order carbon,[1] sulfur, selenium, tellurium; acids, expressed as functional parent compounds (¶ 130), in the order carbon (including **Carbonic acid** and **Formic acid**; see ¶ 183), chalcogen, nitrogen, phosphorus, arsenic, antimony, silicon, and boron.

(*f*) Acid halides and related species, first in the order of the parent acid (see (*e*), above); then, for each acid, in the order fluoride, chloride, bromide, iodide, azide, isocyanate, isothiocyanate, isocyanide, cyanide (for non-carbon acid residues only).

(*g*) Amides, in the same order as the parent acids (see (*e*), above).

(*h*) Nitriles, in the same order as the parent acids (see (*e*), above).

(*i*) Aldehydes, Thials, Selenals, Tellurals.

(*j*) Ketones, Thiones, Selones, Tellones.

(*k*) Alcohols and Phenols (of equal rank), Thiols, Selenols, Tellurols.

(*l*) Hydroperoxides.

(*m*) Amines.

(*n*) Imines. (This is the lowest compound class expressed by a functional suffix; all the following classes are considered to be nonfunctional. For the ranking of nonfunctional cyclic and acyclic skeletons, see ¶ 138.)

(*o*) Nitrogen compounds: **Formazan**; heterocyclic; acyclic (other than "a"-named chains; see ¶ 127), e.g., **Triazane, Diazene, Hydrazine, Hydroxylamine, Thiohydroxylamine.**

(*p*) Phosphorus compounds: heterocyclic, acyclic (other than "a"-named chains; see ¶ 127), e.g., **Diphosphine, Phosphine oxide, Phosphine sulfide, Phosphine imide, Phosphorane, Phosphine.**

(*q*) Arsenic compounds (in similar order).

(*r*) Antimony compounds (in similar order).

(*s*) Bismuth compounds (in similar order).

(*t*) Boron compounds: carbapolyboranes, hetero polyboranes, polyboranes, heterocyclic, **Borane.**

(*u*) Silicon compounds: heterocyclic, acyclic (other than "a"-named chains; see ¶ 127), e.g., **Disiloxane, Disilathiane, Trisilane, Disilane, Silane**. Note that the order is determined first by the total number of skeletal atoms, then by the presence of oxygen, sulfur, etc; see ¶ 128.

(*v*) Germanium compounds (in similar order).

(*w*) Tin compounds (in similar order).

(*x*) Lead compounds (in similar order).

(*y*) Oxygen compounds other than "a"-named chains (see ¶ 127): heterocyclic; acyclic polyoxides, e.g., **Trioxide, Peroxide.**

(*z*) Sulfur compounds: heterocyclic; acyclic polysulfides and their oxides, e.g., **Trisulfone, Trisulfide, Disulfone, Disulfoxide, Disulfide.**

(*aa*) Selenium and tellurium compounds (in similar order).

(*bb*) Carbon compounds: carbocyclic, acyclic hydrocarbons.

107. Spelling. Webster's Third New International Dictionary[2] has been accepted as the authority for most words; e.g., sulfur (not sulphur); aluminum (not aluminium). Elision of vowels is often practiced in combining the segments of names: e.g., in **Butanone** and disiloxanyl the final "e" of the basic skeleton name has been dropped; in **Oxazepine** an "a" has been omitted twice, after "oxa" and before "ep"; and "a" is often omitted before a multiplied "amine" or "one" suffix, as in **Benzenetetramine** and **Cyclohexanehexone**; the terminal "o" of acenaphtho, benzo, naphtho, and perylo, and the terminal "a" of cyclobuta, etc., are elided before vowels, e.g., **Benz[*cd*]indole, 5*H*-Cyclobut[*f*]indene**. Other examples of elision are -imidic (not -imidoic) acids; -imidamides (not imidoamides); -thiones (not thioones); and -hydrazonamides (not -hydrazonoamides). Examples will be found in later paragraphs. In a few cases, the vowel "o" is added for euphony, e.g., **Carbonothioic acid** (not Carbonthioic acid).

Elision of entire syllables is now uncommon. Remaining examples include methoxy, ethoxy, propoxy, butoxy, phenoxy (not (methyloxy), etc.) radicals, and the thienyl (not thiophene-yl) radical. **Carbamic acid** is an elided form of Carbonamidic acid (¶ 183); **Sulfamic acid** is used in place of Sulfuramidic acid, and **Sulfamide** instead of Sulfuramide or Sulfuric diamide. The suffix "-carboxylic acid" undergoes various forms of elision in formation of replacement names, e.g., "-carbothioamide."

108. Punctuation in chemical names is frequently of great importance in removing ambiguities and in differentiating one substance from another. Lower case italic Roman letters are used in fusion prefixes (¶ 151) in ring system names, and in *as*- and *s*-**Indacene**; capital italics such as *N*-, *O*-, *P*-, *S*-, are locants indicating substitution on these hetero atoms; *H*- denotes indicated or added hydrogen (¶¶ 135, 136); italic Arabic numerals are locants for atoms in abnormal valency states (¶ 158) and for "labeled" atoms (¶ 220); italic words and syllables are used in modifications to express isomeric oxides, e.g., *thiono*-oxide at a thio acid heading parent, and in stereochemical descriptors (¶ 203), e.g., *erythro*-, *tetrahedro*-. The small capitals D-, L-, and DL-, are configurational descriptors (¶ 203); like italic letters, they are disregarded in placing chemical names in order until Roman letters have been alphabetized.

The "comma of inversion" has been mentioned above (¶ 104). Other commas are used between individual locants in index heading parents, substituents, and modifications. Different *types* of functional derivatives are

---

[1] **Acetic acid** and **Benzoic acid** are ranked here as though they were named Ethanoic acid and Benzenecarboxylic acid, respectively.

[2] G. and C. Merriam Co., Springfield, Massachusetts, 1961 and 1966.

separated by commas in the text of the modification.
Examples:

(*a*) **hydrazone phenylhydrazone**

(*b*) **diethyl ester, sodium salt**

(*c*) **ethyl methyl ester, oxime**

Hyphens at the end of the set of substituents in the inverted part of a boldface heading signify that no space is intended when the name is uninverted for use in textual matter. Conversely, absence of a hyphen after substituents at headings such as **Disulfide**, **Hydroperoxide**, **Peroxide**, indicates that a space appears at that point in the uninverted name.
Examples:

Acetic acid, chloro-	(index name)
Chloroacetic acid	(uninverted name)
Disulfide, bis(2-chloroethyl)	(index name)
Bis(2-chloroethyl) disulfide	(uninverted name)

Hyphens separate locants from the words and syllables of a name; when used between locants, the intention is to indicate that such locants refer to different parts of the name; e.g., in **Acetamide, *N*-2-naphthalenyl-**, the "*N*-" places the 2-naphthalenyl substituent on the nitrogen of the heading parent, **Acet= amide**.
Periods separate ring size descriptors in Von Baeyer and spiro names, e.g., **Bicyclo[3.2.0]heptane**. Colons separate sets of locants already related to one another; if a further step is called for, semicolons are employed.
Examples:

1,4:5,6-Dimethanonaphthalene
Benzo[1'',2'':3,4;5'',4'':3',4']dicyclobuta[1,2-*a*:1',2'-*a*']diindene

109. Enclosing marks are placed around compound substituent radicals and around and within complex radicals (¶ 162). Their presence or absence frequently removes ambiguity, especially when locants are omitted through lack of precise structural information.
Examples:

$MeSiH_2Cl$ — **Silane, chloromethyl-**

$ClCH_2SiH_3$ — **Silane, (chloromethyl)-**

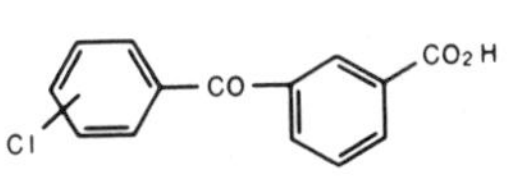

Benzoic acid, 3-(chlorobenzoyl)-

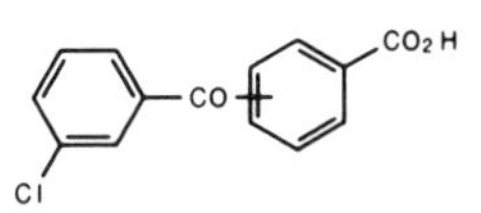

Benzoic acid, (3-chlorobenzoyl)-

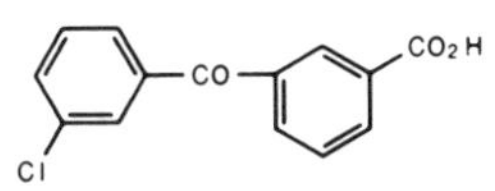

Benzoic acid, 3-(3-chloro= benzoyl)-

Parentheses are placed around compound substituents like "(chloro= methyl)", above; in a case like (chloromethylamino), it is to be understood that both the chlorine atom and the unsubstituted methyl group are substituents of the amino group, i.e., $Cl(CH_3)N$-. The alternative structure, $ClCH_2NH$- is named [(chloromethyl)amino], which is a *complex* substituent prefix. Parentheses are used around simple radicals when they are preceded by "bis," "tris," etc. (¶ 110), e.g., bis(methylene), tris(decyl). They are used also to separate locants of the same kind which would otherwise be separated only by hyphens, to indicate the second atom involved in double-bond formation when it is not the next in the numbered pathway, to enclose parts of a heading parent, to set off added hydrogen, and to enclose multiplied terms in modi= fications, ion terms that would otherwise be ambiguous, Ewens-Bassett num= bers (¶ 215), descriptive terms and ratios, and parts of synonym line formulas.
Examples:

Benzoic acid, 4-(2-naphthalenyl)-
Bicyclo[4.2.0]oct-1(6)-ene
Butane(dithioic) acid (see ¶ 165)
1(2*H*)-Naphthalenone
bis(inner salt)
(disulfate) (from **Disulfuric acid**)
iron(3+) salt
compd. with benzenamine (1:1)
acetate (salt)
Thioperoxydiphosphoric acid ($[(HO)_2P(O)]_2S_2$)

Brackets enclose complex substituent prefixes and derivative terms, as well as Von Baeyer and spiro ring size designations (already described above). They are also employed around a ring-assembly name when it is followed by a principal-group suffix or forms part of a radical name.
Examples:

[1,2'-Binaphthalene]-2-carboxylic acid
[1,1'-biphenyl]-4-yl

Brackets enclose structural features of bridges or component rings when the enclosed locants are not applicable to the total system.
Examples:

4a,9a-[2]Butenoanthracene	(the "2" locates the double bond in the buteno bridge)
4*H*-[1,3]Oxathiolo[5,4-*g*]benzo= xazole	(formed by fusion of 1,3-oxathiole with benzoxazole; in the total ring system, the oxygen and sulfur atoms of the oxathiole ring are in the 6- and 8-positions respectively)

When a multiplicative index name is uninverted, brackets are placed around the heading parent.
Examples:

Acetic acid, 2,2'-oxybis-	(index name)
2,2'-Oxybis[acetic acid]	(uninverted name)
Benzoic acid, 4,4'-methylene= bis[2-chloro-	(index name)
4,4'-Methylenebis[2-chlorobenzoic acid]	(uninverted name)

Brackets are sometimes needed for functional terms in modifications, es= pecially following locants or multiplicative prefixes.
Examples:

***S*-[(dodecylthio)methyl] ester**
bis[(2,4-dinitrophenyl)hydrazone]

110. Multiplicative prefixes. Generally, prefixes derived from the Greek (di, tri-, etc.) are used, rather than the Latin (bi-, ter-, etc.); exceptions are nona- (not ennea-) for nine, and undeca- (not hendeca-) for eleven. (For lists of Latin and Greek prefixes, see ¶ 309.) The Latin prefixes bi-, ter-, etc., are used for ring assemblies, and bi- is employed in the term "bimol. mono= anhydride."
The prefixes bis-, tris-, tetrakis-, etc., are used for compound and complex radicals and functional derivatives, and to avoid misunderstanding in other cases, especially with names beginning with replacement terms like "aza" or "oxa", fusion prefixes like "benzo" or "naphtho," or compound fusion prefixes like "cyclopentapyrido." They are used always in multiplying a heading parent.
Examples:

bis(methylene)
bis(2-aminoethyl)
[1,2-ethanediylbis(oxymethylene)]
bis(*O*-methyloxime)
bis(cyclohexaneacetate)
tris(dihydrogen phosphate)
bis(aziridinyl)
bis(diazo)
bis([1,1'-biphenyl]-4-yl)
bis(bicyclo[2.2.1]hept-2-yl)
tris(decyl)
tetrakis(1-aziridinyl)
bis(anhydrosulfide)
bis(benz[*a*]anthracen-1-yl)
Benzo[1,2-*c*:3,4-*c*']bis[1,2,5]= oxadiazole
Biscyclopenta[5,6]pyrido[4,3-= *b*:3',4'-*c*]pyridine
Benzoic acid, 2,2'-silylenebis-
Phosphonic acid, 1,4-phenyl= enebis-

111. "Mono" is only rarely employed in index heading parents (an exam= ple is **Peroxymonosulfuric acid**) but is needed to express functional deriv= atives of polyfunctional heading parents. It is not used if a locant is necessary, or when all functions are modified by the same class of derivative, or when only one functional group is present. The term "hydrogen" in an uninverted ester name precludes the use of "mono."
Examples:

MeC(=NOH)Me — **2-Propanone oxime**

$MeO_2C(CH_2)_2CO_2H$ — **Butanedioic acid monomethyl ester**

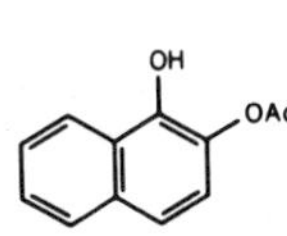

1,2-Naphthalenediol 2-acetate

$EtOP(O)(OMe)_2$ — **Phosphoric acid ethyl dimethyl ester**

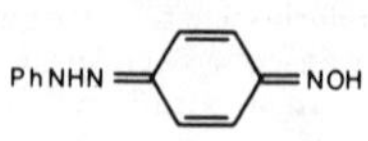

2,5-Cyclohexadiene-1,4-dione mono(phenylhydrazone), oxime (note that the term "mono" is not repeated unless needed)

EtOP(O)(OH)$_2$

Phosphoric acid monoethyl ester (index name)

Ethyl dihydrogen phosphate (uninverted name)

112. **Functional derivatives** of the principal reactive chemical groups of systematically-named index heading parents are cited in the modification; these derivatives, as defined for indexing purposes, are restricted to acyclic anhydrides, esters, hydrazides, hydrazones, and oximes. Other derivatives, such as semicarbazones, azines, acetals, and cyclic esters, are named in other ways, e.g., substitutively at the highest functional heading parent, as substituted hydrazones, etc., or as heterocycles, as detailed in Section D, below.

Functional derivatives of subsidiary functions (those not expressed by the suffix of the heading parent) are cited in the main boldface heading as com= pound or complex substituents.

Examples:

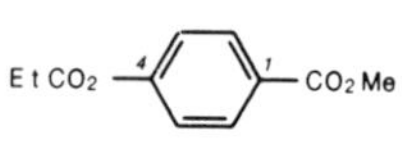

Benzoic acid, 4-(1-oxopropoxy)- methyl ester (not Benzoic acid, 4-hydroxy-, methyl ester, propanoate)

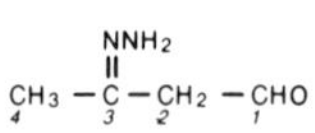

Butanal, 3-hydrazono- (not Butanal, 3-oxo-, 3-= hydrazone)

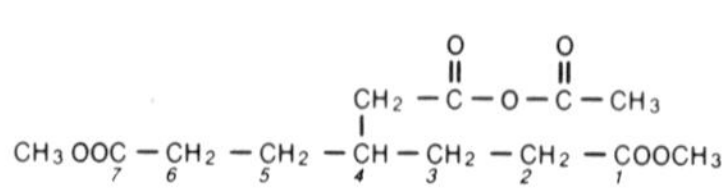

Heptanedioic acid, 4-[2-(acet= yloxy)-2-oxoethyl]- dimethyl ester (not Heptanedioic acid, 4-(carboxymethyl)-, 4-anhydride with acetic acid, dimethyl ester)

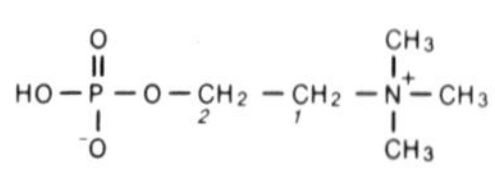

Ethanaminium, *N,N,N*-tri= methyl-2-(phosphonooxy)- hydroxide, inner salt (not Ethanaminium, 2-hydroxy-*N,N,N*-trimethyl-, dihydrogen phosphate (ester), inner salt)

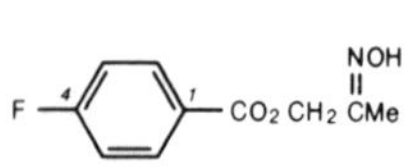

Benzoic acid, 4-fluoro- 2-(hydroxyimino)propyl ester (not Benzoic acid, 4-fluoro-, 2-oxopropyl ester, oxime)

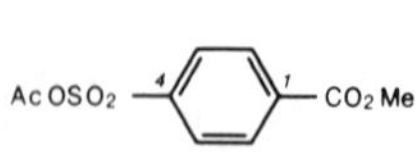

Benzoic acid, 4-[(acetyloxy)= sulfonyl]- methyl ester (not Benzoic acid, 4-sulfo-, *S*-anhydride with acetic acid, methyl ester)

113. **Order of citation of derivative terms** in modifications. The nor= mal order is:

(*a*) ionic terms relating to the heading parent, e.g., "chloride" at an "-aminium" heading;

(*b*) functional derivatives in the order: anhydrides, esters, hydrazides, hydrazones, oximes; multiplicative terms are cited before simple terms, e.g., 1,3-propanediyl dimethyl ester, otherwise alphabetic order is followed;

(*c*) additive terms describing fragments covalently attached to the index heading compound, e.g., *N*-oxide;

(*d*) ionic terms, e.g., ion(1-), radical ion(1-), then metal salts, followed by other salts alphabetically, e.g., acetate, hydrochloride;

(*e*) other additive terms describing portions of the molecular structure not covalently attached, e.g., compd. with..., hydrate, mixt. with..., polymer with....

114. **Locants.** When a choice is necessary, italic Roman letters are placed before Greek letters, and **Arabic numerals** are placed last, e.g., *As, N, P, S*, α, β, γ, 1, 2, 3. (For the Greek **alphabet see** ¶ 310.)

Unprimed locants are followed by primed locants, then by doubly primed locants, etc., e.g., *N, N', S*, α, 1', 2, 2', 2'', 3. Low numbering of indices (superscript Arabic numbers) and application of primes are not considered until regular numerical locants have been chosen.

Examples:

MeP(=NMe)(NHMe)$_2$

Phosphonimidic diamide, *N,N',N'',P*- tetramethyl-

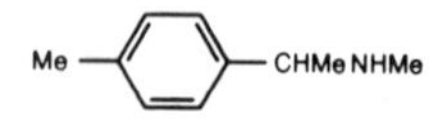

Benzenemethanamine, *N*,α,4-tri= methyl-

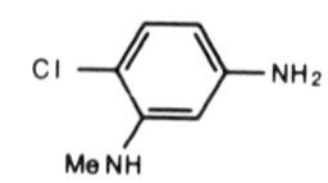

1,3-Benzenediamine, 4-chloro-N^3- methyl- (not 1,3- Benz= enediamine, 6-= chloro-N^1-methyl-)

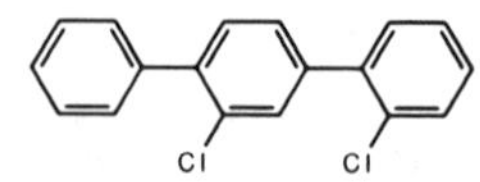

1,1':4',1''-Terphenyl, 2',2''-dichloro- (not 1,1':4',1''-Terphenyl, 2,3'-dichloro-)

Locants for unsaturation in compounds named as index heading parents are always cited when the compound contains three or more skeletal atoms, except for monocyclic hydrocarbons with one multiple bond and no suffix.

Examples:

HN=NNH$_2$ (1, 3)	**1-Triazene**
H$_2$C=C=CH$_2$	**1,2-Propadiene**
H$_2$C=CHC≡CH (1, 4)	**1-Buten-3-yne**
H$_2$C=CH$_2$	**Ethene**

Cyclohexene

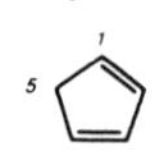

1,3-Cyclopentadiene

Locants denoting ring junctions of ring assemblies are always cited except for two-component assemblies of cycloalkenes, cycloalkadienes, etc.

Examples:

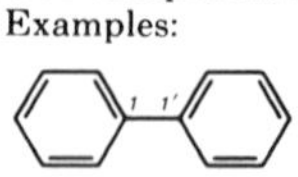

1,1'-Biphenyl

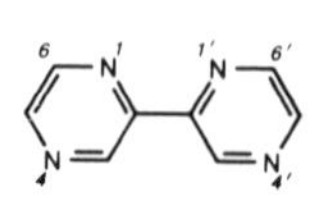

2,2'-Bipyrazine

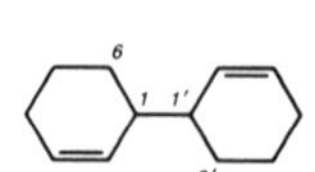

Bi-2-cyclohexen-1-yl

115. **Locants for substituent suffixes** of index heading parents are always cited if it is necessary to cite locants in the molecular skeleton for unsaturation, hetero atoms, indicated hydrogen, spiro or ring assembly junc= tions, bridges (in fused systems), or isotopic labeling. Locants for fusion sites, e.g., "[3,4-*d*]" (¶ 151), are disregarded in this connection. Locants are not cited for Geneva suffixes which terminate a chain, e.g., "-oic acid," "-dial," except when a monofunctional suffix is not at the 1-position of an "a"-named chain. Locants are always cited with cycloalkene names to which suffixes are attached.

Examples:

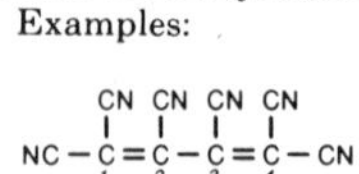

1,3-Butadiene-1,1,2,3,4,4-hexacar= bonitrile

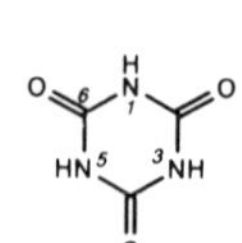

1,3,5-Triazine-2,4,6(1*H*,3*H*,5*H*)-= trione

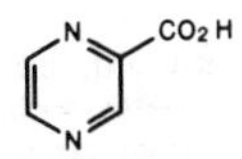

Pyrazinecarboxylic acid

$HO_2CN=NCO_2H$ — **Diazenedicarboxylic acid**

$H_2C=CHCO_2H$ — **2-Propenoic acid**

$CH_3CH_2OCH_2CH_2OCH_2CH_2OCH_2CH_2OCH_2CH_2COOH$ (1 2 3 4 5 6 7 8 9 10 11 12 13 14 15)

3,6,9,12-Tetraoxapentadecan-15-oic acid

2-Cyclopropen-1-one

116. **Locants in substituent prefixes (radicals).** Locants are assigned for multiple bonds in all unsaturated radicals containing three or more skeletal atoms. Locants are always cited for free valencies involving more than one position of a skeleton, except for the trivial names hydrazo, azino, azo, azoxy (¶ 193).

Examples:

$HC{\equiv}CCH{=}$ (3, 1) — 2-propynylidene

$-N{=}NNH-$ (1, 3) — 1-triazene-1,3-diyl

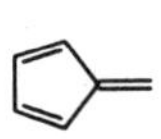
2,4-cyclopentadien-1-ylidene

$-C{\equiv}CC{\equiv}C-$ — 1,3-butadiyne-1,4-diyl

$-CH_2CH_2-$ — 1,2-ethanediyl

$=C=C=$ — 1,2-ethenediylidene

1,2-cyclopropanediyl

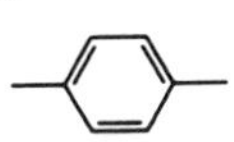
1,4-phenylene

Locants are not cited for free valencies of radicals which have lost hydrogen from only one skeletal atom of an acyclic, or saturated monocyclic, homogeneous molecular skeleton.

Examples:

$MeCH_2-$ — ethyl

$HN=N-$ — diazenyl

HP–P– / HP–PH (ring) — tetraphosphetanyl

$H_2C=CHCH_2CH=$ (4, 1) — 3-butenylidene

$HN=NNH-$ (3, 1) — 2-triazenyl

Locants are cited for free valencies of radicals from unsaturated homogeneous monocyclic systems, and from saturated heterogeneous monocyclic systems with cyclo...ane, cyclo...ene, etc., names.

Examples:

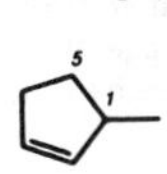
2-cyclopenten-1-yl

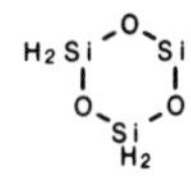
cyclotrisiloxan-2-ylidene

Locants for free valencies are cited for monocyclic radicals which possess locants for indicated hydrogen, hetero atoms, unsaturation, spiro or ring assembly junctions or bridges (in fused systems).

Examples:

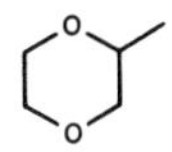
1,4-dioxan-2-yl

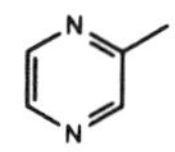
pyrazinyl (not 2-pyrazinyl)

117. **Locants for substituents** on index heading parents and parent radicals are cited when the parent names possess locants for substituent suffixes, unsaturation, hetero atoms, indicated hydrogen, spiro or ring assembly junctions, or bridges (in fused systems). Locants are dispensed with for "hydro" prefixes in fully saturated ring systems unless ambiguity could result, e.g., because of a remaining etheno or other unsaturated bridge.

Examples:

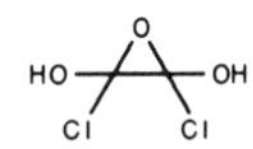
2,3-Oxiranediol, 2,3-dichloro-

Oxirenol, chloro-

$MeCOCOCH_2Cl$ — **2,3-Butanedione, 1-chloro-**

$H_2NCONHMe$ — **Urea, methyl-**

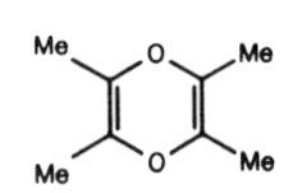
1,4-Dioxin, 2,3,5,6-tetramethyl-

$PhC{\equiv}CC{\equiv}C-$ (4, 1) — (4-phenyl-1,3-butadiynyl)

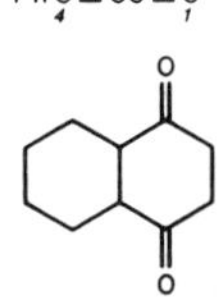
1,4-Naphthalenedione, octahydro-

When one or more locants are needed for substituents on a heading parent or parent radical, all are cited.

Examples:

$F_3CCF_2CFICF_3$ — **Butane, 1,1,1,2,2,3,4,4,4-nonafluoro-3-iodo-**

$CF_3-CF(CF_3)-$ — [1,2,2,2-tetrafluoro-1-(trifluoromethyl)ethyl]

$HO_2CCHClSCHClCO_2H$ — **Acetic acid, 2,2'-thiobis[2-chloro-**

118. **Locants in multiplicative nomenclature** are cited for the positions of attachment on the heading parent if it consists of more than one skeletal atom, e.g., **Benzene, Acetic acid**, or contains an additional position for substitution, e.g., **Silanamine, Phosphonamidic acid**. Locants are otherwise not cited for heading parents which contain only one skeletal atom, e.g., **Methane, Silane, Methanone**, or for heading parents which are functional parent compounds (¶ 130), e.g., **Phosphonic acid**.

Examples:

PhOPh — **Benzene, 1,1'-oxybis-**

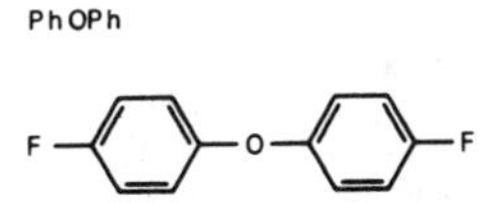
Benzene, 1,1'-oxybis[4-fluoro-

$HO_2CCH_2NHNHCH_2CO_2H$ — **Acetic acid, 2,2'-hydrazobis-**

$HO_2CNHSiH_2NHCO_2H$ — **Carbamic acid, silylenebis-**

MeSMe — **Methane, thiobis-**

$HOSiH_2CH_2SiH_2OH$ — **Silanol, methylenebis-**

$H_2NSiH_2CH_2SiH_2NH_2$ — **Silanamine, 1,1'-methylenebis-**

119. **Locants for functional derivatives** are used when needed to define the structure unambiguously. The terms "mono," "di," etc., are preferred to locants, but if a locant is necessary, "mono" is omitted.
Examples:

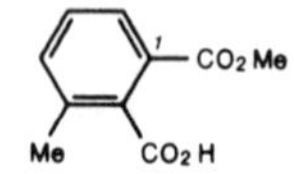

1,2-Benzenedicarboxylic acid, 3-methyl-
1-methyl ester

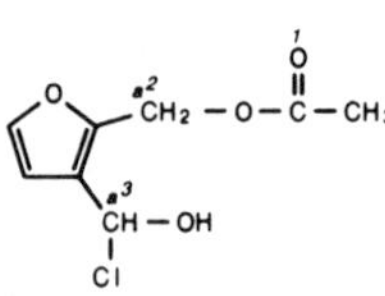

2,3-Furandimethanol, α^3-chloro-
α^2-acetate

MeOP(O)(OH)OP(O)(OH)OMe — **Diphosphoric acid**
***P,P'*-dimethyl ester**

$HO_2C(CH_2)_2CONHNH_2$ — **Butanedioic acid**
monohydrazide

MeC(=NOH)CMe=NNHPh — **2,3-Butanedione**
mono(phenylhydrazone), oxime

Locants for all derivatives are cited if one or more are needed.
Example:

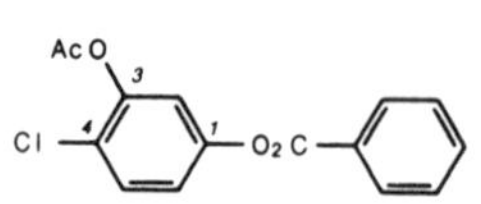

1,3-Benzenediol, 4-chloro-
3-acetate 1-benzoate

Locants are used in modifications for index headings that express the same (or similar) functions in both the index parent and the substituents when the latter are not derivatized. This avoids confusion with former *CA* index names.
Examples:

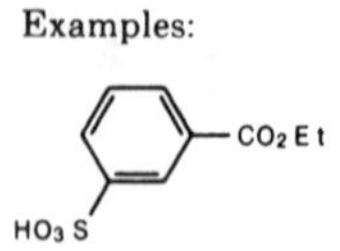

Benzoic acid, 3-sulfo-
1-ethyl ester

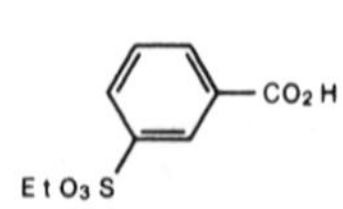

Benzoic acid, 3-(ethoxysulfonyl)-
(formerly Benzoic acid, 3-sulfo-, 3-ethyl ester)

The locant for an additive term such as "oxide" is an Arabic number when a nitrogen, phosphorus, sulfur, etc., atom of a ring in the index heading parent is involved; otherwise a letter locant (*N*-, *S*-, etc.) is employed.
Locants are not employed for ionic modification terms, e.g., salts such as "monosodium salt," "hydrochloride."

120. **Locants for indefinite compounds** must often be omitted. In addition, such compounds can sometimes be named only by departing from the regular name selection policies, e.g., by citation of the principal group in the modification instead of as a suffix of the heading parent, citation of a functional derivative in the modification rather than as a substituent, use of "mono" instead of the (unknown) locant, replacement of a numerical locant by the indefinite aromatic locant "*ar*-," or inclusion of question marks in a set of otherwise known locants.
Examples:

Benzenediamine
Benzoic acid, dichloro-
Naphthalene, 2,2'-(1,4-phenylene)bis-
disulfo deriv.
Benzeneacetic acid, 2-carboxy-
monomethyl ester
1,2-Ethanediol, 1-phenyl-
monocarbamate
Naphthalene, *ar*-chloro-1,2,3,4-tetrahydro-
Benzenemethanol, *ar*-amino-
[1,1'-Biphenyl]-*ar,ar'*-dicarboxylic acid
Benzene, 1,2,?-trimethyl-

The italic word "*or*" is used with substituent prefixes (but never with index heading parents) when the number of alternative structures cannot be misinterpreted.
Examples:

Quinoline, 2-chloro-3(*or* 4)-methyl-
Naphthalene, 1(*or* 2)-ethyl-2(*or* 1)-methyl-

When one or more substituent prefixes are in known positions and the remainder in unknown positions, lowest locants are used for the former.
Example:

Naphthalene, chloro-2-methyl-

An indefinite name like **Piperidine, 2(*or* 4)-bromo-4(*or* 2)-chloro-** cannot be used because this name could be held to include the 2-bromo-2-chloro- and 4-bromo-4-chloro-isomers; in such cases locants are usually omitted.

121. **Alphabetization** of substituent prefixes affects the position in the index where an inverted chemical substance name will be found. Simple prefixes are placed in alphabetic order according to their names; only then are multiplicative prefixes (di-, tri-, etc.) placed in front of each as required, and locants inserted; e.g., an index compound in which two nitro groups, three bromine atoms, and a chlorine atom are present receives the substituent name "tribromochlorodinitro," and the substituents so arranged will be found together with an appropriate index parent, such as **Naphthalene**, alphabetized in accordance with *all* the Roman letters in the complete name. The total name with locants, e.g., **Naphthalene, 2,5,8-tribromo-3-chloro-1,6-dinitro-**, will be preceded in the list of index entries by both **Naphthalene, nitro-** and **Naphthalene, tetrachloro-**.
Compound and complex substituent prefixes (radicals) are constructed on similar alphabetic principles and then arranged by their first letters (which may have been derived from multiplicative prefixes within the radicals) in the total name. This name is then placed in the index as described above, *all* letters being alphabetized. When the letters are all identical, arrangement depends on locants.
Examples:

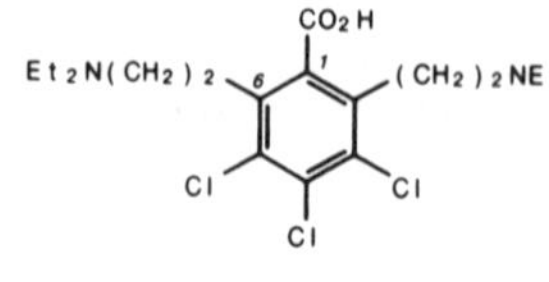

Benzoic acid, 3,4,5-trichloro-2,6-bis[2-(diethylamino)ethyl]- (letter "c" is placed before "d", the initial letter of the complex radical)

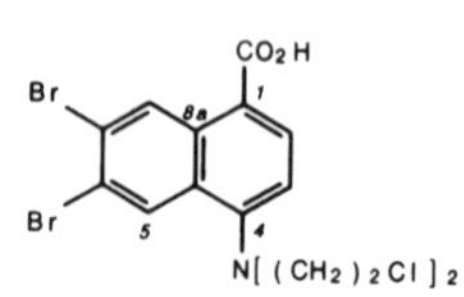

1-Naphthalenecarboxylic acid, 4-[bis(2-chloroethyl)amino]-6,7-dibromo- (the multiplicative prefix "bis-" is part of the name of the complex substituent prefix and this is therefore placed before "bromo")

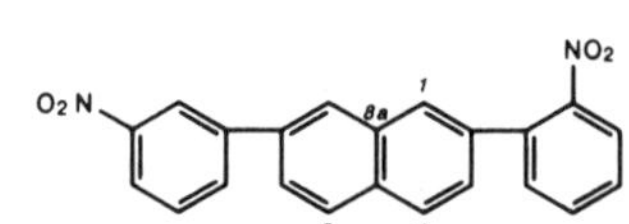

Naphthalene, 2-(2-nitrophenyl)-7-(3-nitrophenyl)- (the radicals are placed in order, according to the locants they contain, before locants relating to the heading parent are inserted)

122. **Tautomeric compounds.** Index names are normally based on the precise structures shown or described in the author's original document. Tautomerism (ready interconvertibility of isomers) in certain types of compounds causes a serious problem in index-name selection, the issue here being not which name to select for a given molecular structure but which diagram to name for a given tautomeric system if scattering at different names of information about what is essentially a single substance is to be avoided.

Common (trivial) names for most nitrogenous tautomeric systems are cross-referred to preferred systematic names.

Examples:

Adenine.	See *1H-Purin-6-amine*
Barbituric acid.	See *2,4,6(1H,3H,5H)-Pyrimidinetrione*
Caffeine.	See *1H-Purine-2,6-dione, 3,7-dihydro-1,3,7-tri= methyl-*
Carbostyril.	See *2(1H)-Quinolinone*
Cytosine.	See *2(1H)-Pyrimidinone, 4-amino-*
Guanine.	See *6H-Purin-6-one, 2-amino-1,7-dihydro-*
Hydantoin.	See *2,4-Imidazolidinedione*
Melamine.	See *1,3,5-Triazine-2,4,6-triamine*
Theophylline.	See *1H-Purine-2,6-dione, 3,7-dihydro-1,3-dimethyl-*
Uracil.	See *2,4(1H,3H)-Pyrimidinedione*
Uric acid.	See *1H-Purine-2,6,8(3H)-trione, 7,9-dihydro-*
Xanthine.	See *1H-Purine-2,6-dione, 3,7-dihydro-*

Unless a structure at variance with the index name selected for a particular tautomer, e.g., **2,4-Pyrimidinediol** as the structure of Uracil, is emphasized by an author, information on tautomeric systems is collected in *CA Chemical Substance Indexes* at the preferred names to which the cross-references lead. Index names for related tautomers not possessing common names are selected by similar principles.

The necessity for many tautomeric structures to be redrawn to accord with selected *CA* index names is obviated by computer "normalization" algorithms in the CAS Registry System. In the normalization process, the different structural diagrams for a single tautomeric system (of one of the types ex= pressed in the cross-references above) are recognized as equivalent and stored in identical machine-language representations. They share a single unique CAS Registry Number (see Appendix II, ¶ 13) and *CA Chemical Substance Index* name.

When the author of an original document emphasizes a tautomeric structure different from that represented by the usual preferred index name, the nor= malization algorithm is bypassed and a different Registry Number and index name are assigned.

The structural requirements for the normalization process and the rules for selecting unique *CA Chemical Substance Index* names are as follows:

I. *Requirements for normalization of structures.* Tautomeric structures represented by the following equilibrium:

$$M=Q-ZH \rightleftharpoons HM-Q=Z$$

are normalized, i.e., recognized as equivalent in the CAS Registry System, when the following requirements are met:

(a) Q = C, N, S, P, Sb, As, Se, Te, Br, Cl or I with any acceptable valency for the individual elements.

(b) M and Z = any combination of trivalent N and/or bivalent O, S, Se or Te atoms.

(c) The bonds involved in tautomerization may be in an acyclic chain or in a ring system or partly in both.

(d) The end-points, M and Z, may be in adjacent rings of a fused ring system, but a nitrogen atom which occupies a fusion point in such a system cannot take part in tautomerization.

(e) The hydrogen atom of the tautomeric system may be replaced by deuterium or tritium.

(f) Two or more systems of the form shown above may be linked through a common atom, whereby a proton can be considered to migrate along the chain.

Example:

—NH—C=N—C=N—C=N— ⇌ —N=C—NH—C=N—C=N— ⇌

—N=C—N=C—NH—C=N— ⇌ —N=C—N=C—N=C—NH—

Replacement in the generalized formula above by specific elements affords normalized tautomeric systems such as the following:

—NH—C=N— ⇌ —N=C—NH

(amidine tautomerism)

—NH—N=N— ⇌ —N=N—NH—

(diazoamino tautomerism)

—NH—C=O(S, Se, Te) ⇌ —N=C—O(S, Se, Te)H

(lactam-lactim tautomerism)

—NH—P=N— ⇌ —N=P—NH—

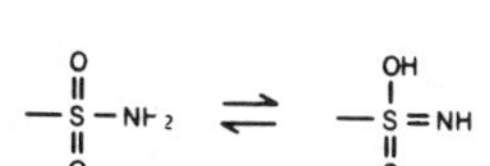

A unique *CA Chemical Substance Index* name is selected for each com= pound containing a normalized tautomeric structure by application *first* of the structural rules which follow, and then of the nomenclature rules (see III, below). For tautomeric pyrazole derivatives and for tropolones, *CA* selects a single preferred structure and index name, and assigns a single CAS Registry Number, even though these systems do not conform to the general equilibrium illustrated above and are not currently normalized by the CAS Registry Sys= tem. Lowest locants are employed successively in index parents and sub= stituents in these cases.

Examples:

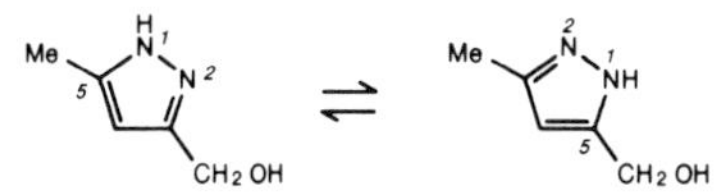

1*H*-Pyrazole-3-methanol, 5-methyl- (principle: lowest locant for prin= cipal group) NOT **1*H*-Pyrazole-5-methanol, 3-methyl-**

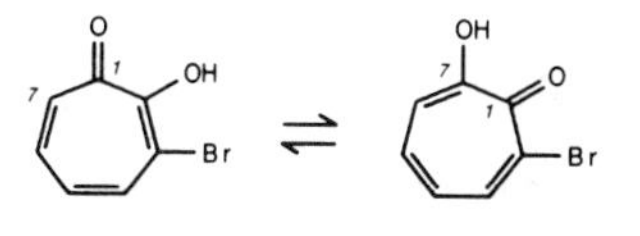

2,4,6-Cycloheptatrien-= 1-one, 3-bromo-2-= hydroxy- (principle; lowest locants for sub- stituents) NOT **2,4,6-Cycloheptatrien-= 1-one, 2-bromo-7-= hydroxy-**

For phosphonic-phosphorous and phosphinic-phosphonous acid tau= tomers, see ¶ 197.

II. *Structural rules.* These rules are used to select the particular structure of a tautomer from which the unique *CA Chemical Substance Index* name is then derived. They are applied before the nomenclature rules (see III, below) are considered.

1. The *maximum number of oxygen (or other chalcogen) atoms are doubly bonded to the central atom Q.* If it is necessary to make a choice, the descending order of precedence for double-bond formation is O, S, Se, Te.

Examples:

$CH_3-C(=O)-NH_2$ NOT $CH_3-C(OH)=NH$

$HS-P(=S)(SH)-NH_2$ NOT $HS-P(SH)(SH)=NH$

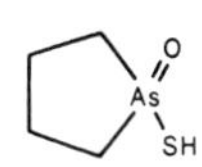 NOT 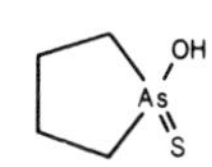

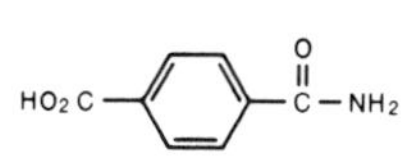 NOT 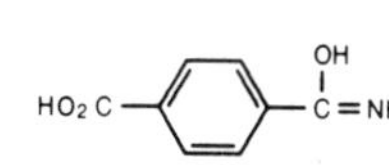

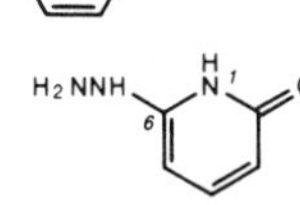 NOT 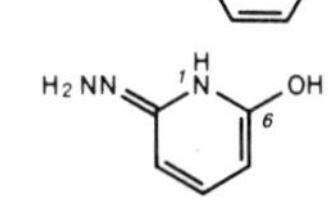

2. For a compound which contains a nitrogenous heterocycle bonded to one or more acyclic nitrogen atoms in such a way that a normalized tautomer results, the preferred structure is that which has the maximum number of single bonds between the heterocycle and the nitrogen atoms. *This rule is not applied* when the nitrogen atom is part of a hydrazone or oxime which is a derivative of the principal function (*see nomenclature rule III, (a) below*).

Examples:

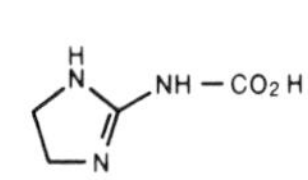 NOT 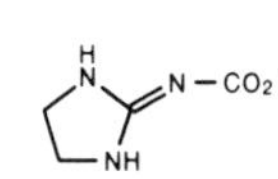

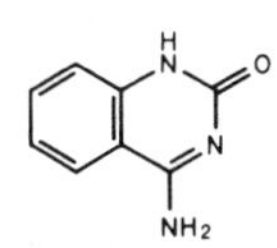 NOT 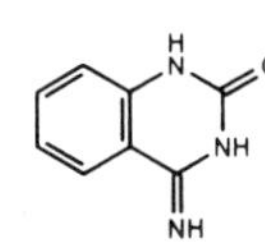

When it is necessary to choose between amino and hydrazino, amino is preferred.

Example:

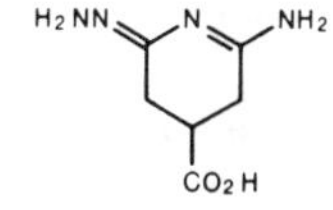 NOT 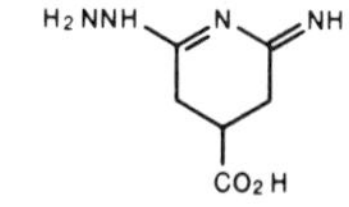

3. **For a compound which contains a completely *acyclic***

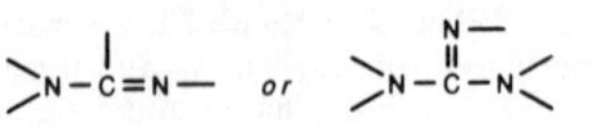

fragment which is capable of being structured with an unsubstituted imino (=NH) group, that tautomeric form is preferred over any other, even when the fragment is bonded to an amino, hydrazino, or other nitrogen-containing radical.

Examples:

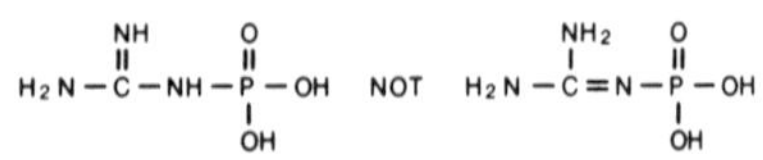

III. *Nomenclature rules.* When the structural rules, above, are insufficient to enable a choice of preferred tautomeric form to be made, the regular rules of index name selection (see ¶ 138) are then applied. Nomenclature rules applicable to tautomeric systems are as follows:

(a) The preferred index heading parent expresses the maximum number of the principal chemical functional group.

Examples:

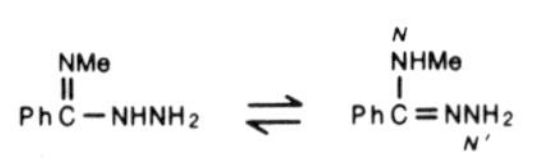

Benzenecarboximidic acid, *N*-methyl-hydrazide NOT **Benzenecarbohydrazonamide, *N*-methyl-**

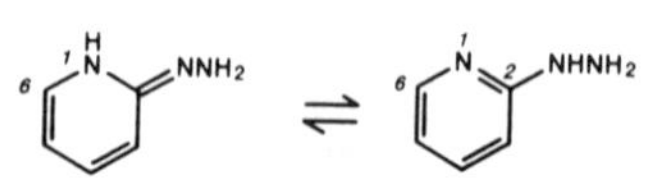

2(1*H*)-Pyridinone hydrazone NOT **Pyridine, 2-hydrazino-**

(NOTE: Hydrazones and oximes of principal functional groups are exceptions to structural rule II. 2, above.)

(b) The lowest locant for indicated hydrogen (see ¶ 135) is expressed in the preferred index name for nitrogenous heterocycles which are normalized, whether such heterocycles are expressed in the index heading parent, substituent prefix, or index modification.

Examples:

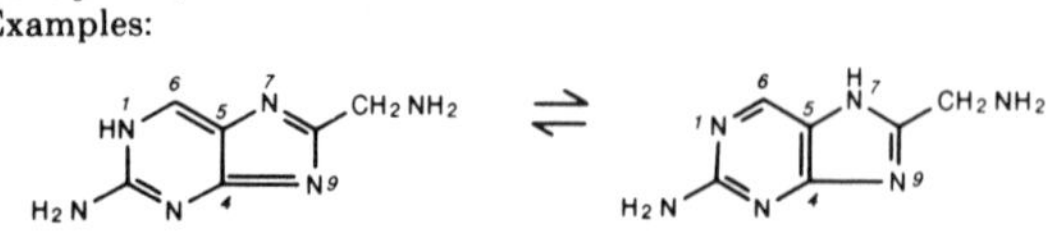

1*H*-Purine-8-methanamine, 2-amino- NOT **7*H*-Purine-8-methanamine, 2-amino-**

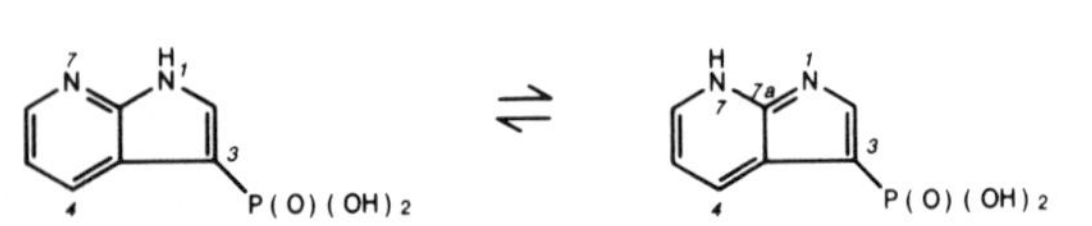

Phosphonic acid, 1*H*-pyrrolo[2,3-*b*]pyridin-3-yl- NOT **Phosphonic acid, 7*H*-pyrrolo[2,3-*b*]pyridin-3-yl-**

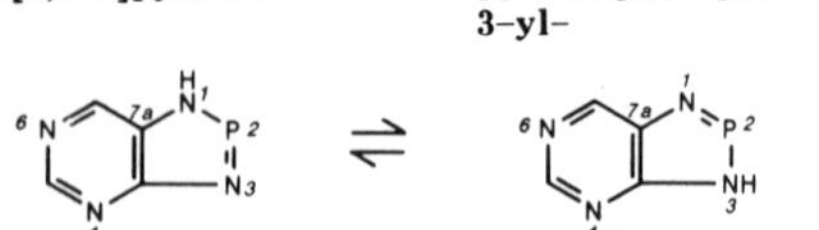

1*H*-1,3,2-Diazaphospholo[4,5-*d*]pyrimidine NOT **3*H*-1,3,2-Diazaphospholo[4,5-*d*]pyrimidine**

(c) Lowest locants are expressed for (i) suffixes, i.e., principal groups of index heading parents and free valencies of substituent radicals, and (ii) multiple bonds in parents and radicals.

Examples:

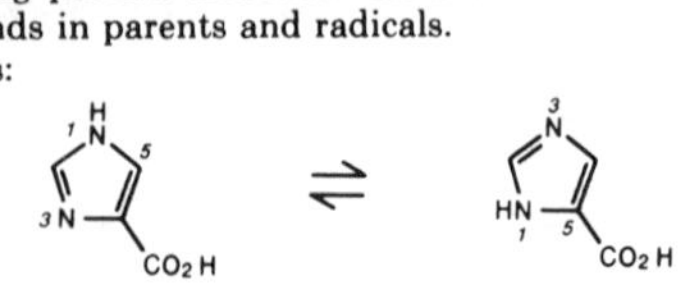

1*H*-Imidazole-4-carboxylic acid NOT **1*H*-Imidazole-5-carboxylic acid**

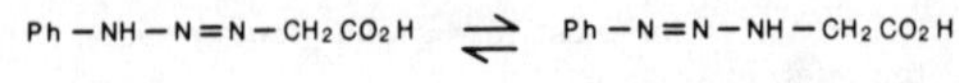

Acetic acid, (3-phenyl-1-triazenyl)- NOT **Acetic acid, (3-phenyl-2-triazenyl)-**

(d) The maximum number of substituents are cited as prefixes in the index heading.

Example:

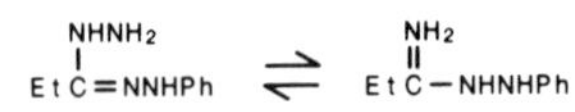

Propanehydrazonic acid, *N*-phenyl-hydrazide NOT **Propanehydrazonic acid 2-phenylhydrazide**

(e) Substituent prefixes and added hydrogen (see ¶ 136) are assigned lowest locants (see ¶ 137) to denote positions of attachment to an index heading parent or a substituent radical.

Examples:

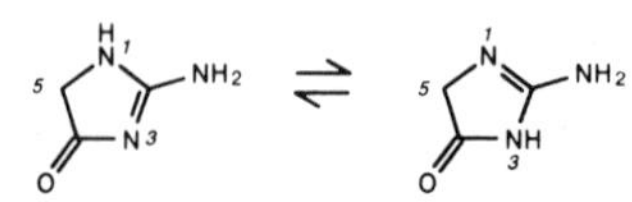

4*H*-Imidazol-4-one, 2-amino-1,5-dihydro- NOT **4*H*-Imidazol-4-one, 2-amino-3,5-dihydro-**

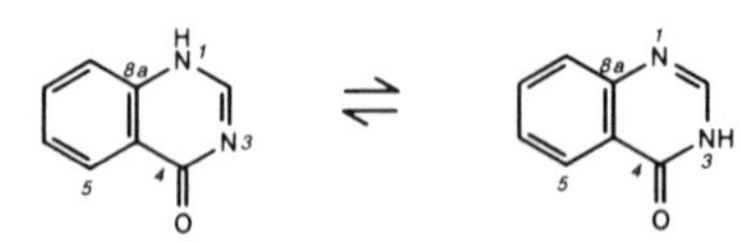

4(1*H*)-Quinazolinone NOT **4(3*H*)-Quinazolinone**

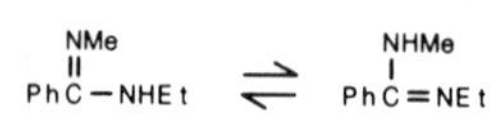

Benzenecarboximidamide, *N*-ethyl-*N'*-methyl- NOT **Benzenecarboximidamide, *N'*-ethyl-*N*-methyl-**

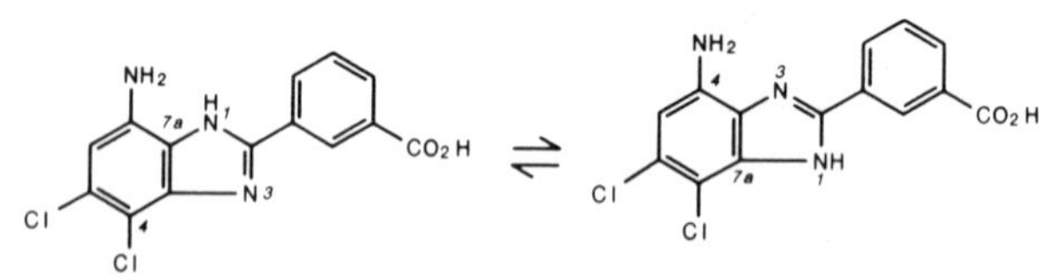

Benzoic acid, 3-(7-amino-4,5-dichloro-1*H*-benzimidazol-2-yl)- NOT **Benzoic acid, 3-(4-amino-6,7-dichloro-1*H*-benzimidazol-2-yl)-**

(f) The index heading parent expresses the maximum number of occurrences by use of multiplicative nomenclature (see ¶ 125).

Example:

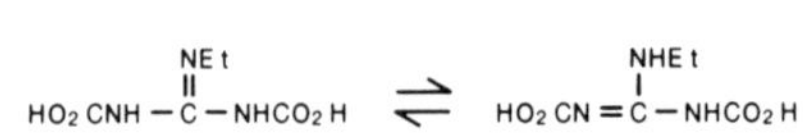

Carbamic acid, (ethylcarbonimidoyl)bis- NOT **Carbamic acid, [(carboxyamino)(ethylamino)methylene]-**

(g) The preferred *CA* index name is that which occurs earliest in the index arrangement as determined by the rules for ordering of *Chemical Substance Index* entries (see Appendix II, ¶ 10C).

Examples:

Acetic acid, (ethylamino)(methylimino)- NOT **Acetic acid, (ethylimino)(methylamino)-**

123. Additive nomenclature embraces molecular structures whose several component parts are considered to be added together without replacement (substitution) of atoms (usually hydrogen). It includes coordination names (¶ 215), conjunctive nomenclature (¶ 124) and binary names of inorganic compounds (¶ 219).

Examples:

Copper, dichlorobis(methanamine)–
Benzeneethanol
Sodium chloride (NaCl)

The construction of additive names often involves indicated or added hydrogen (¶¶ 135, 136) in that part of the molecule known as the heading parent, the addition of hydro "substituents", or the use of additive terms such as "oxide" or "sulfide" in the modification. In a few cases the additive term becomes a part of the heading parent.

Examples:

9(10*H*)–Anthracenone
Naphthalene, 1,2,3,4–tetrahydro–
Pyridine
1–oxide
Phosphine imide

For salts and molecular addition compounds, see ¶¶ 192, 198.

124. Conjunctive nomenclature allows a cyclic molecular skeleton to be included as a part of the heading parent name even though the principal chemical group is separated from the ring by an acyclic chain. Larger molecules may be named thereby as heading parents and more compounds of similar structure can be collected at a given ring system name. Moreover, the major requirement of substitutive nomenclature, that the principal group be expressed in the heading parent as a suffix, is fulfilled.

A conjunctive name is employed when any ring system (including a polyhedral borane) is attached by single bonds to one or more saturated acyclic hydrocarbon chains, each of which bears only one functional substituent corresponding to the principal chemical group of the compound. When a second or third such substituent is present on the chain, a conjunctive name may still sometimes be employed so long as the resulting index heading parent does not express more than a single function in each chain and other principles are not violated (see the final example below). It is always implied that the chemical functional group is at one end of the acyclic chain and the ring system is at the other.

Examples:

$PhCH_2OH$ — **Benzenemethanol**

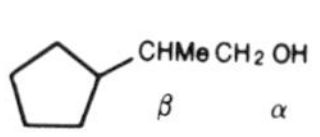

Cyclopentaneethanol, β–methyl–

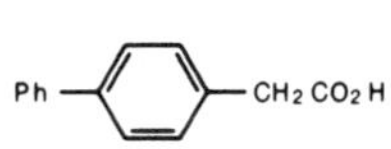

[1,1'–Biphenyl]–4–acetic acid

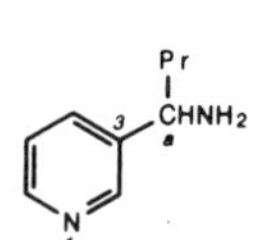

3–Pyridinemethanamine, α–propyl–

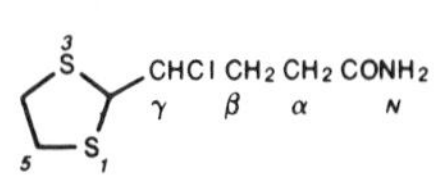

1,3–Dithiolane–2–butanamide, γ–chloro–

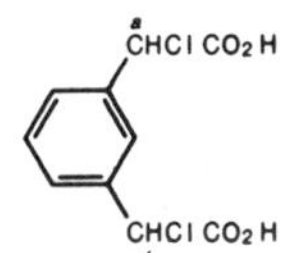

1,3–Benzenediacetic acid, α,α'–dichloro–

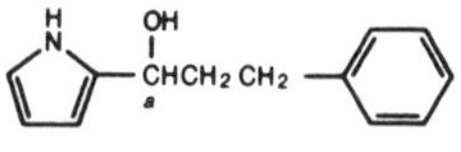

1*H*–Pyrrole–2–methanol, α–(2–phenylethyl)– (the heading parent which expresses the preferred ring system, not the preferred acyclic chain, is chosen)

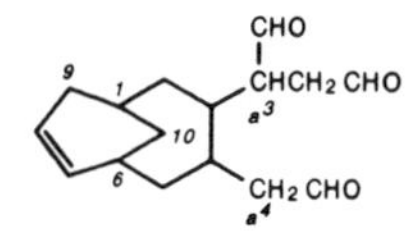

Bicyclo[4.3.1]dec–7–ene–3,4–diacetaldehyde, α³–(2–oxoethyl)– (not Bicyclo[4.3.1]dec–7–ene–3–propanol, β–formyl–4–(2–oxoethyl)–; the preferred heading parent expresses the maximum number of principal groups and, because it also expresses a ring system, is preferred to **Butanedial)**

A conjunctive name is not permissible under the following conditions and the regular rules of substitutive nomenclature apply: 1) when a *double* bond joins the ring to the functional acyclic chain; 2) when a conjunctive index parent would express two or more functional groups in a single acyclic chain; 3) when the acyclic chain is unsaturated, or contains hetero atoms; and 4) when a conjunctive name would fail to express the maximum number of principal chemical groups.

Examples:

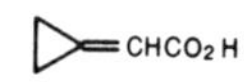

Acetic acid, cyclopropylidene–

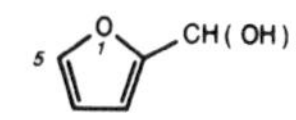

Methanediol, 2–furanyl–

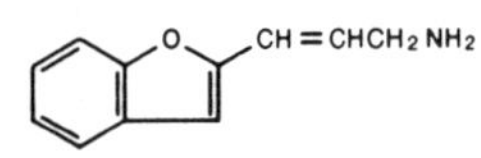

2–Propen–1–amine, 3–(2–benzofuranyl)–

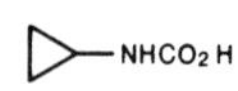

Carbamic acid, cyclopropyl–

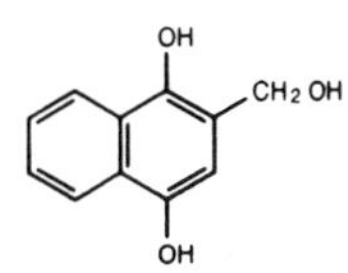

1,4–Naphthalenediol, 2–(hydroxymethyl)–

125. Multiplicative nomenclature employs polyvalent radicals by which a multiplicity of occurrences of an index heading parent in a compound may be expressed.

Example:

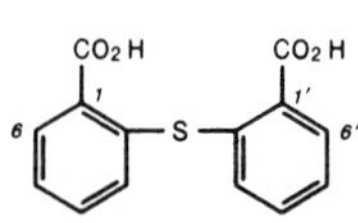

Benzoic acid, 2,2'–thiobis–

In the example above, a simple one–part polyvalent radical, thio, was employed as a multiplier; other such radicals are oxy, –O–; methylene, $-CH_2-$; 1,4–phenylene, $1,4-C_6H_4=$; imino, –NH–; nitrilo, –N=; and 1,3–disiloxanediyl, $-SiH_2OSiH_2-$. In general, any simple multivalent radical may be used as a multiplicative radical, and may itself be substituted; e.g., (methylimino), $-N(CH_3)-$; (1–methyl–1,3–propanediyl), $-CH(CH_3)CH_2CH_2-$.

Multicomponent radicals may be used as multipliers if they contain a central one–part multivalent radical (simple, compound, or complex) around which all other multivalent radicals are so arranged that the sequence of atoms and bonds in each path is identical as one proceeds outwards. There is no restriction in the number of components that may comprise the total multiplying radical, so long as their use results in an unambiguous total name. (Use of "ylidene" radicals in certain situations, e.g., in [iminobis(cyclohexylidenemethylene)], leads to ambiguity and is therefore avoided.)

Examples of permissible multiplying radicals:

$-CH_2OCH_2-$ — [oxybis(methylene)]

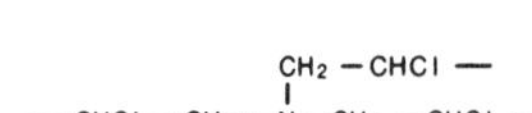

[nitrilotris(1–chloro–2,1–ethanediyl)]

$-OCH_2CHClO-$ — [(1–chloro–1,2–ethanediyl)bis(oxy)]

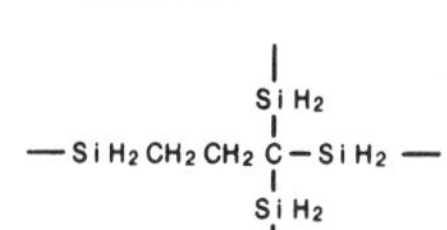

(1–propanyl–3–ylidynetetrasilylene)

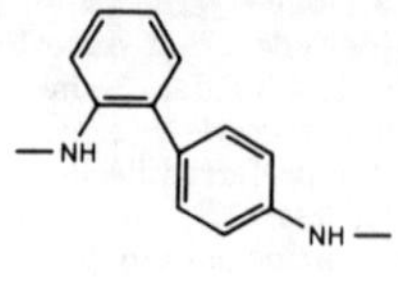

([1,1'-biphenyl]-2,4'-diyldiimino)

Examples of combinations of multivalent radicals which are not used as multipliers:

—CH_2NH— —$CH_2CH{=}NCH_2CH{=}$

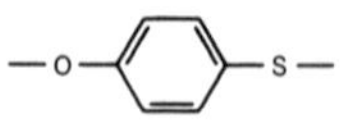

—$CHClCH_2PHCHClCH_2$—

—CH_2OO—

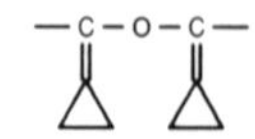

The naming of multiplying radicals is accomplished by citing first the central unit. This is followed by a prefix, e.g., "di" or "bis", denoting the number of "radial" series generated by the central unit. The remaining terms of the name, in the form of radicals, are cited in appropriate enclosing marks as necessary. The entire multiplying radical is set off by further enclosing marks, which are preceded by locants "placing" the radical at the proper point of attachment on each index heading parent. Multivalent radicals other than central units are numbered (if there is a choice) from the heading parents towards the center of the complete radicals, and the locant relating to the position closest to the heading parent is cited last.
Examples:

—$CHClCH_2OCH_2CHCl$— [oxybis(1-chloro-2,1-ethanediyl)]

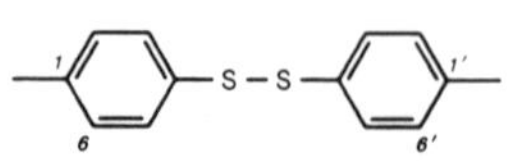

(dithiodi-4,1-phenylene)

—$CH_2CH{=}CHSeCH{=}CHCH_2$— (selenodi-2-propene-3,1-diyl)

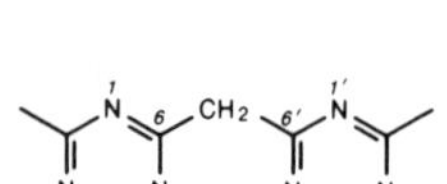

[methylenebis(1,3,5-triazine-6,2,4-triyl)] (the 2- and 4-positions are equally close to the heading parent and are therefore cited in normal sequence)

Carbonyl groups which are part of carbon chains are expressed as oxo substituents; chalcogen, imino, and hydrazono analogs are expressed as thioxo, selenoxo, telluroxo, imino, and hydrazono substituents.
Examples:

—$CH_2NHCOCH_2CONHCH_2$— [(1,3-dioxo-1,3-propanediyl)bis(iminomethylene)]

—COCO— (1,2-dioxo-1,2-ethanediyl)

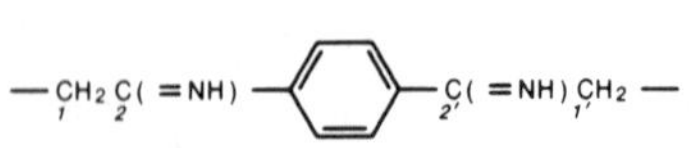

[1,4-phenylenebis(2-imino-2,1-ethanediyl)]

When carbonyl groups and their analogs are not part of an acyclic carbon chain, the names carbonyl, carbonothioyl, carbonimidoyl, etc., are employed.
Examples:

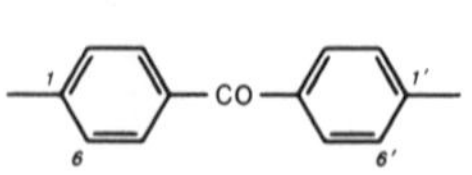

(carbonyldi-4,1-phenylene)

—NHC(=NH)NH— (carbonimidoyldiimino)

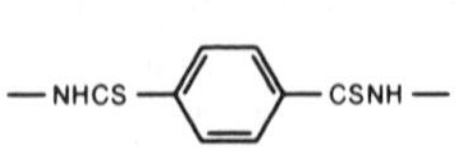

[1,4-phenylenebis(carbonothioylimino)]

Valid multiplying radicals are used only when the entire compound is symmetrical around the central unit of such a radical; i.e., the radical must be attached to the heading parent by bonds of the same type (single, double, or triple) and at equivalent positions, and this parent must be identical with regard to positions of principal groups (and their functional derivatives) and other substituents. Whether or not such other substituents are present, the terms "bis," "tris," etc., are employed after the multiplying radical, not "di," "tri," etc. If other substituents are present, they are cited as regular substitutive radicals after an opening bracket (which, perforce, is left unclosed).
Examples:

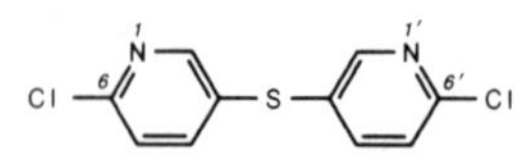

Pyridine, 3,3'-thiobis[6-chloro- (not Pyridine, 5,5'-thiobis[2-chloro-; multiplicative radicals are given lowest locants) (The uninverted form of this name is 3,3'-Thiobis[6-chloropyridine].)

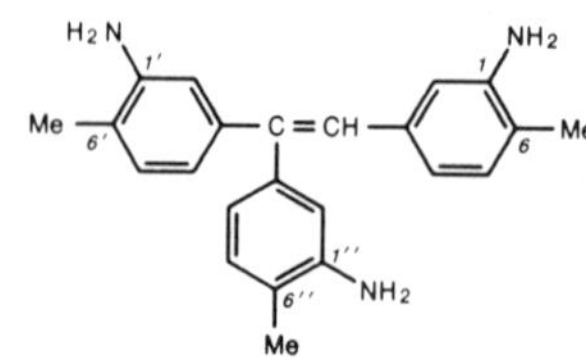

Benzenamine, 3,3',3''-(1-ethenyl-2-ylidene)tris[6-methyl-

The principles of multiplicative nomenclature are applied only after the index heading parent has been chosen, and after other principles, e.g., centrality (¶ 138), have been applied. When more than one multiplicative name is possible, that one is used which multiplies the greatest number of index heading parents, and then, if a choice is still necessary, that one which appears earliest in the alphabetic sequence of index entries. The number of occurrences of the parent is not increased by arbitrary breaking of the skeleton from which the multiplying radical is derived.
Example:

$HOCH_2CH_2OCH_2SiH_2OSiHOSiH_2CH_2OCH_2CH_2OH$ (with $OSiH_2CH_2OCH_2CH_2OH$ on the central Si)

Ethanol, 2,2'-[[3-[[[(2-hydroxyethoxy)methyl]silyl]oxy]-1,5-trisiloxanediyl]bis(methyleneoxy)]bis- (not Ethanol, 2,2',2''-[silylidynetris(oxysilylenemethyleneoxy)]tris-)

126. Radicofunctional nomenclature is used by *CA* in only a few cases, for disulfides, hydroperoxides, and peroxides (¶¶ 200, 196). Radicofunctional names express the compound type, e.g., "peroxide," usually as a separate word. When inverted, the substituents are not followed by a hyphen unless multiplicative nomenclature (¶ 125) is used with **Hydroperoxide**; e.g., **Disulfide, ethyl methyl**; **Hydroperoxide, 1-methylethyl**; **Hydroperoxide, cyclohexylidenebis-**.

127. Replacement nomenclature ("a" nomenclature) is used for certain heterocyclic ring systems (¶ 146) and also, sometimes, for heteroorganic acyclic compounds. This nomenclature is limited to cases in which carbon atoms have been replaced in chains and rings by nonmetals and/or elements of which the hydrides are *CA* index heading parents, i.e., P, As, Sb, Bi, Si, Ge, Sn, Pb, B.

Requirements for its use in expressing acyclic chains are as follows:

(*a*) A minimum of four hetero units must be present, none of which may be all or a part of a functional chemical group to be expressed in the index heading parent as the preferred functional compound class (i.e., as a functional suffix or as a functional index compound such as **Carbonic** or **Phosphonic acid**). A hetero unit is defined as an isolated hetero atom or a series of consecutive hetero atoms, alike or different, that may be expressed as a unit, such as by a bivalent radical name. Examples: -S- (thio); -S-S- (dithio); -N=N- (azo); $-SiH_2-O-SiH_2-$ (1,3-disiloxanediyl); $-SiH_2-NH-SiH_2-$ (1,3-disilazanediyl). The above are all single hetero units, but -HP-NH-, -S-O-, and $-O-SiH_2-O-$, are not.

(*b*) The "a" name must not be lower in order of precedence than the name obtained by regular substitutive nomenclature, i.e., it must express at least as many principal functions of at least equal rank.

(*c*) All hetero atoms must be in their standard valency state, or else the abnormal valency must be expressible unambiguously by use of "oxide," etc., terms.

(*d*) The chain may be terminated only by C, P, As, Sb, Bi, Si, Ge, Sn, Pb, or B.

Acyclic "a" nomenclature is employed for organic chains containing silicon or metal atoms, polyesters, anhydrides, amides, polyamides, polyalkylene glycols, and condensed carbonic acid derivatives. It is not used for peptides or polymers, or (if it can be avoided) for chains containing no carbon atoms. Otherwise, if the above criteria are met, an "a" name is always selected.

128. Replacement prefixes for the elements most frequently found in carbon chains are set out in descending order of precedence in Table I. (The order is the reverse of that shown in ¶ 215 for coordinated elements.)

TABLE I
REPLACEMENT PREFIXES IN DESCENDING ORDER OF PRECEDENCE

Element	*Substitutive Valence*	*Prefix*
Oxygen	II	Oxa
Sulfur	II	Thia
Selenium	II	Selena
Tellurium	II	Tellura
Nitrogen	III	Aza
Phosphorus	III	Phospha
Arsenic	III	Arsa
Antimony	III	Stiba
Bismuth	III	Bisma
Silicon	IV	Sila
Germanium	IV	Germa
Tin	IV	Stanna
Lead	IV	Plumba
Boron	III	Bora

The replacement prefixes are placed in descending order of precedence ahead of the name of the carbon skeleton with locants to indicate the positions of the atoms replaced. Lowest locants are assigned to the most preferred hetero atoms, not to functional groups or unsaturation.

Replacement nomenclature is also employed for acyclic substituent prefixes (radicals) when the above requirements are fulfilled. In this case, lowest locants are assigned to free valency positions; i.e., the "a" names are based on the carbon chain radical names, but the free valency locant ("1") is always cited.

Examples:

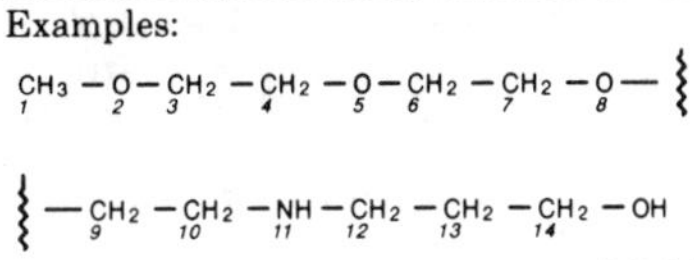

2,5,8-Trioxa-11-azatetradecan-=14-ol

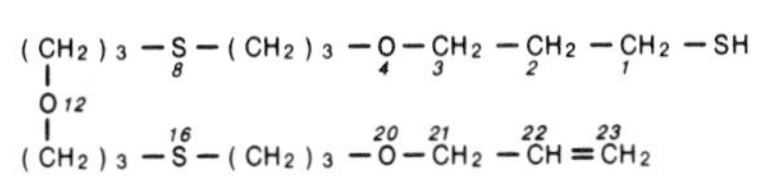

4,12,20-Trioxa-8,16-dithiatricos-22-ene-1-thiol

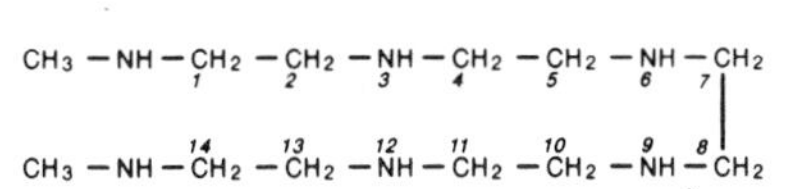

3,6,9,12-Tetraazatetradecane-=1,14-diamine, N,N'-dimethyl-

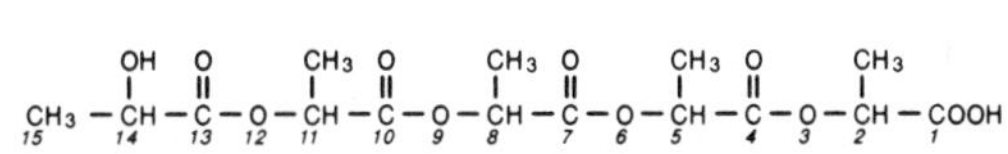

3,6,9,12-Tetraoxapentadecanoic acid, 14-hydroxy-2,5,8,11-=tetramethyl-4,7,10,13-tetraoxo-

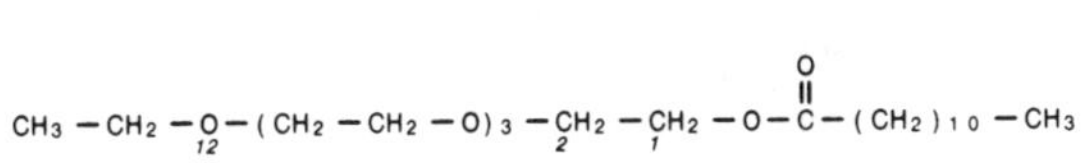

Dodecanoic acid 3,6,9,12-tetraoxatetradec-1-yl ester

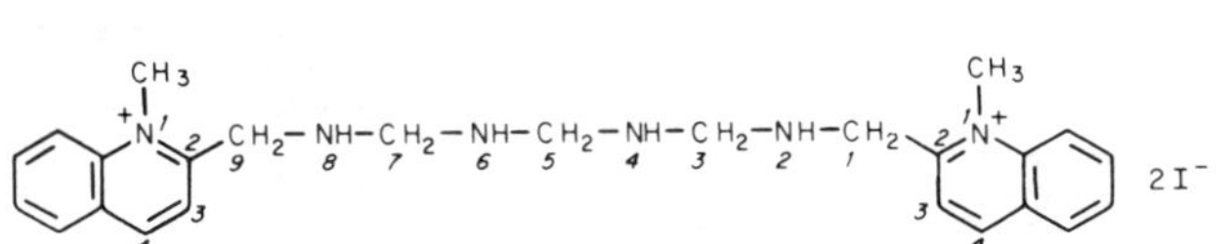

Quinolinium, 2,2'-(2,4,6,8-tetra=azanonane-1,9-diyl)bis[1-=methyl- diiodide

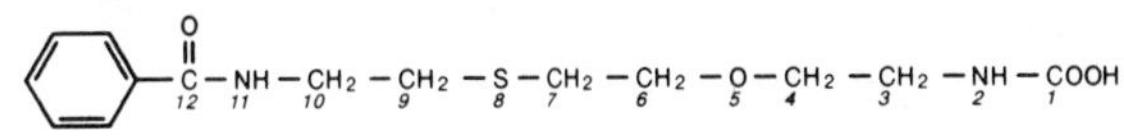

5-Oxa-8-thia-2,11-diazadodecanoic acid, 12-oxo-12-phenyl-

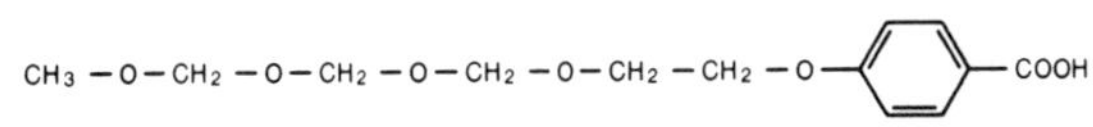

Benzoic acid, 4-(3,5,7,9-tetraoxa=dec-1-yloxy)-

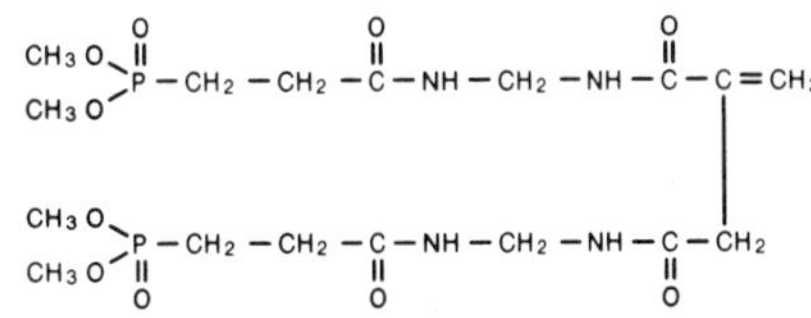

Phosphonic acid, (8-methylene-3,7,10,14-tetraoxo-4,6,11,13-tet=raazahexadecane-1,16-diyl)bis-tetramethyl ester

129. Replacement nomenclature for functions is a method of de=scribing the replacement of hydroxyl and oxo functional groups by nitrogen, chalcogens, halogens, or halogenoids such as isocyanato. The replacement may be carried out in substituent suffixes, e.g., -thioic acid from -oic acid; in substituent prefixes, e.g., carbonimidoyl (¶ 134) from carbonyl, and phos=phinothioyl from phosphinyl (¶ 197); and in functional parent compounds (¶ 130), e.g., **Phosphonimidodithioic acid** from **Phosphonic acid**.

Replacement of hydroxyl in compounds and radicals is denoted by the following affixes (the final "o" is often elided): amido (for $-NH_2$), azido (for $-N_3$), chlorido (for -Cl) (and similarly for other halo atoms), cyanatido (for -OCN), hydrazido (for $-NHNH_2$), isocyanatido (for -NCO), (isothiocyanatido) (for -NCS), and (thiocyanatido) (for -SCN). Seleno and telluro analogs are named analogously.

Replacement of oxo is denoted by the affixes hydrazono (for $=NNH_2$), imido (for =NH), thio (for =S), etc. Replacement of a hydroxyl and an oxo together by ≡N is denoted by nitrido. Peroxy acids are named by use of the affixes peroxo, (thioperoxo), and (dithioperoxo).

These affixes are combined, in alphabetical order, with the functional suffix of compound names based on molecular skeletons. The systematic names **Benzenecarboxylic**, **Ethanoic**, and **Methanoic acids**, not the trivial names **Benzoic**, **Acetic**, and **Formic acids**, are used as the parents for functional replacement nomenclature.

Examples:

$MeCS_2H$	**Ethane(dithioic) acid** (not Acetic acid, dithio-)
PhC(=NH)OH	**Benzenecarboximidic acid** (the tautomeric **Benzamide** is preferred, except for esters and anhydrides; see ¶ 122)
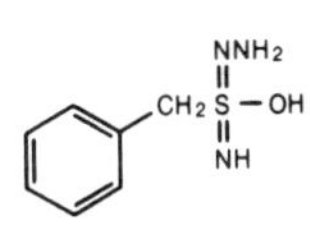	**Benzenemethanesulfonohydrazo=nimidic acid** (derived from the conjunctive name **Benzene=methanesulfonic acid**)
	**Ethane(dithioperoxoic) acid**

In replacement names from phosphorus and arsenic functional parent compounds (¶ 197), all the affixes except hydrazono may be employed as part of the suffix. For **Carbonic acid** and its relatives (¶ 183) all but hydrazido and nitrido may be used.

Examples:

$ClP(O)(OH)_2$	**Phosphorochloridic acid**
$MeAs(NH_2)OH$	**Arsonamidous acid**, *As*-**methyl-**
$P(S)(NH_2)_3$	**Phosphorothioic triamide**
$(HO)_2C{=}NNH_2$	**Carbonohydrazonic acid**
$(HOO)_2CO$	**Carbonodiperoxoic acid**

In replacement names from other mononuclear acids and from condensed nuclear acids (anhydrides), the affixes are used as nondetachable prefixes cited at the beginning of the heading parents. Because multiplicative prefixes are rarely used for thio and other chalcogen prefixes, ambiguity is resolved by synonym line formulas in the boldface index headings.

Examples:

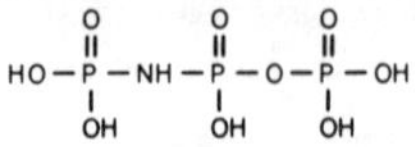
Imidotriphosphoric acid

HO_2CNHCO_2H **Imidodicarbonic acid**

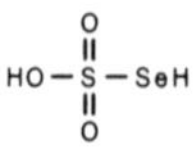
Selenosulfuric acid (H_2SO_3Se)

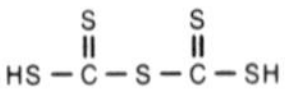
Thiodicarbonic acid [(HS)C(S)]$_2$S

Nondetachable prefixes are used in a few other cases.
Examples:

$ClSO_3H$	**Chlorosulfuric acid**
$S(O)(=NH)(NH_2)_2$	**Imidosulfamide**
$(H_2N)_2CS$	**Thiourea**
H_2NSH	**Thiohydroxylamine**
AcOSAc	**Thioperoxide, diacetyl**
EtOSH	**Thiohydroperoxide, *O*–ethyl**

130. **Substitutive nomenclature**, in which hydrogen atoms are replaced by other atoms or chemical groups, is of paramount importance among no= menclature systems because of its versatility. *Substitutive parent compounds*, which are real or hypothetical compounds whose names imply the presence of replaceable hydrogen atoms, are of two kinds:

(*a*) *Functional parent compounds* have names which express a chemical function but are not based on a molecular skeleton. Substitutive examples include **Arsonic acid, Imidodicarbonic acid, Carbamic acid** and **Phos= phonamidic chloride**. Their substituents are always expressed as prefixes, never as suffixes.

(*b*) *Molecular–skeleton parent compounds* are nonfunctional. They are chains or rings of atoms with only hydrogen atoms attached and possessing names which express or imply the substitutive valency and bonding of the skeletal atoms. (**Methane** and monoatomic hydrides of the Group IVA ele= ments and the Group VA elements (except nitrogen) are treated as molecular skeletons.) Examples include **Ethane, Distannane, Diazene, 3,6,9,12,15– Pentaoxaheptadecane, Cyclohexane, Morpholine**, **Phosphine, Stannane.** They are transformed into index heading parents by appending, as a suffix, the substituent which represents the principal functional group of the com= pound; other substituents are expressed as prefixes.

131. **Substituent suffixes** of molecular–skeleton parent compounds are chosen to represent the principal chemical functional group (or groups) in accordance with the order of precedence of compound classes (¶ 106). When no suffix is available to represent the preferred compound class, either a substitutive functional parent compound (¶ 130) is used as a heading parent, or another system of nomenclature, e.g., coordination or radicofunctional, is adopted.

Examples:

$EtP(O)(OH)_2$	**Phosphonic acid, ethyl**– (Phosphonic acid is a functional parent com= pound.)
$PhSF_3$	**Sulfur, pentafluorophenyl**– (a coordination name)
F_3COOCF_3	**Peroxide, bis(trifluoromethyl)** (a radicofunctional name)

The particular suffixes used for various classes of compounds in descending order of precedence from acids through imines (¶ 106) are described in the sections of this Guide dealing with these classes. Only one class may be expressed as a suffix in a single index heading parent; less preferred classes are denoted by prefixes. Multiplicative prefixes are employed to indicate the number of principal groups present.

Examples (in descending order of compound classes):

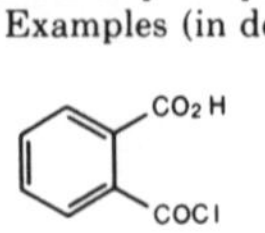
Benzoic acid, 2–(chloro= carbonyl)–

$EtCO_2H$ **Propanoic acid**

$MeCHMeSO_3H$ (1 2) **2–Propanesulfonic acid**

$MeCOCH_2CSMe$ (1 5) **2–Pentanone, 4–thioxo–**

$HOSiMe_2OSiMe_2OH$ **1,3–Disiloxanediol, 1,1,3,3–tetra= methyl–**

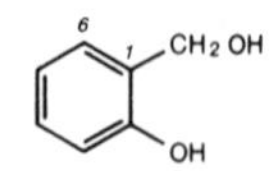
Benzenemethanol, 2–hydroxy–

$CH_3CH_2NH_2$ **Ethanamine**

132. **Substituent prefixes** (commonly called "radicals") are employed to denote atoms and chemical groups attached to an index heading parent. The following substituents are never expressed as suffixes; they may be termed "compulsory" or "mandatory" prefixes: astato (At–), astatyl (AtO_2–), azido (N_3–), bromo (Br–), chloro (Cl–), chlorosyl (OCl–), chloryl (O_2Cl–), diazo (N_2–), fluoro (F–), iodo (I–), iodosyl (OI–), iodyl (O_2I–), isocyanato (OCN–), isocyano (CN–), nitro (O_2N–), *aci*–nitro ((HO)(O)N=), nitroso (ON–), perchloryl (O_3Cl–). In addition, all thio, sulfinyl, and sulfonyl radicals, (RS–), (RS(O)–), and ($RS(O)_2$–), and their seleno and telluro analogs, are mandatory substituent prefixes; so are hydrocarbon radicals and other radicals derived from molecular skeletons, e.g., ethyl, furanyl, disiloxanediyl, when attached to a more preferred heading parent.

Radicals may be simple, compound, or complex. A compound radical is made up of two or more simple radicals, e.g., (chlorothio), (diaminomethyl). A complex radical is composed of a simple radical to which at least one compound radical is attached; e.g., [(chloromethyl)amino], [1–(trichloro= methyl)–2–butenyl]. In these examples, amino and 2–butenyl are parent radicals, and methyl (in both cases) is a subsidiary radical. This procedure may be repeated indefinitely. (Chlorothio) is obtained by *addition* of the two components, (aminomethyl) by *substitution* of methyl by amino. Substitution is the preferred method when a substitutive simple radical is available; e.g., (aminomethylene) is (NH_2CH=), not (NH_2CH_2–). Substitution in certain radicals, including the following, is not permitted: hydroxy, mercapto, selenyl, telluryl, hydroperoxy, sulfeno, diazenyl, formyl, carboxy, sulfo, phosphono, and carbonothioyl.

Examples:

$Me(CH_2)_4O$—	(pentyloxy)	NOT	(pentylhydroxy)
PhN=N—	(phenylazo)	NOT	(phenyldiazenyl)
ClCO—	(chlorocarbonyl)	NOT	(chloroformyl)
PhO_2C—	(phenoxycarbonyl)	NOT	(phenylcarboxy)
$(MeO)_2P(O)$ —	(dimethoxyphosphinyl)	NOT	(dimethylphosphono)
MeOCS —	(methoxythioxomethyl)	NOT	(methoxycarbonothioyl)

All compound radicals are enclosed in parentheses. Simple radicals are so enclosed when two locants of like type fall together; e.g., **Benzoic acid, 3– (4–pyridinyl)**–; for clarity, when one locant has been omitted in the name of an indefinite compound; e.g., **1,2–Propanediol, 3–(thienyl)**–; and when "bis" or "tris" has been employed to remove ambiguity (see ¶ 110); e.g., tris(decyl), bis(benzanthracenyl), bis(azepinyl). Brackets are used in complex radicals, e.g., [2–(dimethylamino)ethoxy]. Spaces in a name often permit the dropping of one set of enclosing marks around radicals in substituents and modifications; e.g., **Disulfide, 2–chloroethyl ethyl; Propanoic acid, 2–ethylbutyl ester**.

133. **Compound radicals.** Selection of names for compound radicals is usually simple, but, when chain branching is present, can sometimes become perplexing. The following rules are successively applied; ((*a*) through (*e*) lead to selection of the preferred parent radical, (*f*) through (*h*) to a particular occurrence of this radical).

(*a*) Greatest number of acyclic hetero atoms.

(*b*) Greatest number of skeletal atoms.

(*c*) Greatest number of most preferred acyclic hetero atoms (see Table I, ¶ 128).

(*d*) Greatest number of multiple bonds.

(*e*) Lowest locants in the simple radical for replacement atoms in "a" names, then for multiple bonds of any kind, and finally for double bonds.

(*f*) Greatest number of substituents attached to the simple radical.

(*g*) Lowest locants for such substituents.

(*h*) Earliest index position of the total radical as it appears within the index name.

Examples (the italic letters on the left indicate the particular rule (above) that is exemplified):

(*a*)

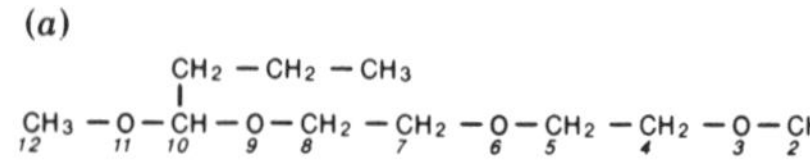
(10–propyl–3,6,9,11–tetraoxadodec–1–yl)

(*a*) 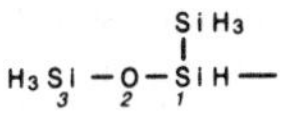 (1–silyldisiloxanyl)

(*b*) 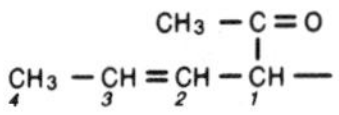 (1–acetyl–2–butenyl)

(*b*) 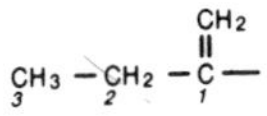 (1–methylenepropyl)

(*b*) 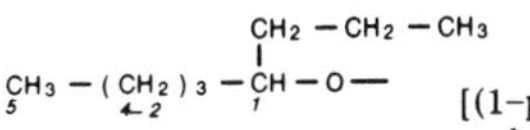 [(1–propylpentyl)oxy](not (1–butylbutoxy))(the selection principles are applied before elision is performed; another example follows)

(*b*) 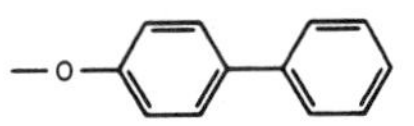 ([1,1'–biphenyl]–4–yloxy)(not (4–phenylphenoxy))

(*b*) 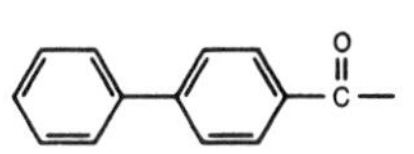 ([1,1'–biphenyl]–4–ylcarbonyl)(not (4–phenylbenzoyl))(the simple radical benzoyl is retained for (phenylcarbonyl) only when the rules lead to this name)

(*b*) 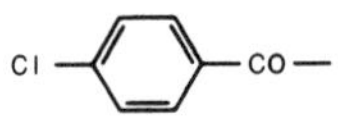 (4–chlorobenzoyl)(not [(4–chlorophenyl)carbonyl])

(*b*) 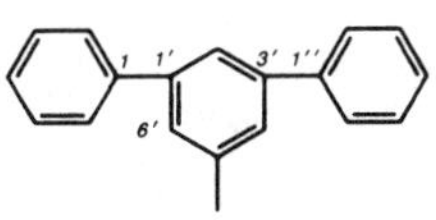 [1,1':3',1''–terphenyl]–5'–yl

(*c*) 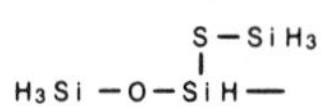 [1–(silylthio)disiloxanyl]

(*c*) 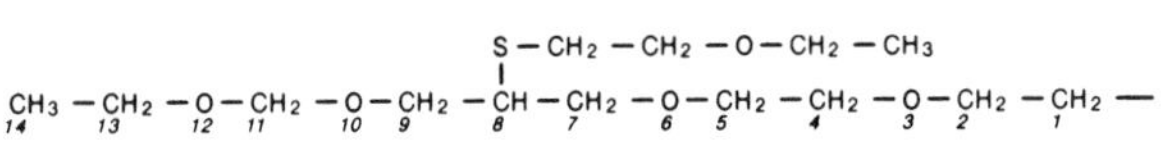

[8–[(2–ethoxyethyl)thio]–3,6,10,12–tetraoxatetradec–1–yl]

(*d*) 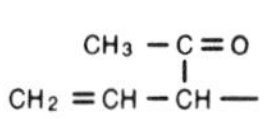 (1–acetyl–2–propenyl)

(*d*) 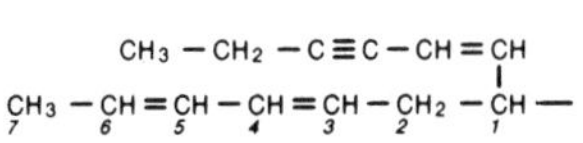

[1–(1–hexen–3–ynyl)–3,5–heptadienyl] (double bonds are preferred over an equal number of triple bonds)

(*e*) 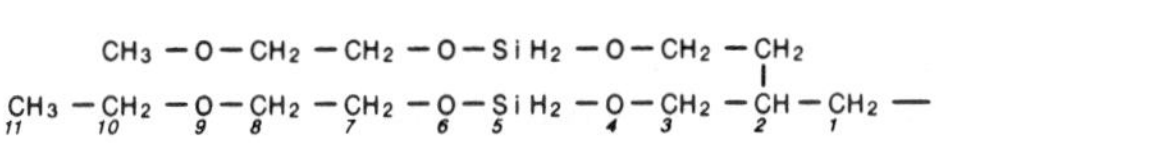

[2–(3,5,8–trioxa–4–silanon–1–yl)–4,6,9–trioxa–5–silaundec–1–yl]

(*e*) 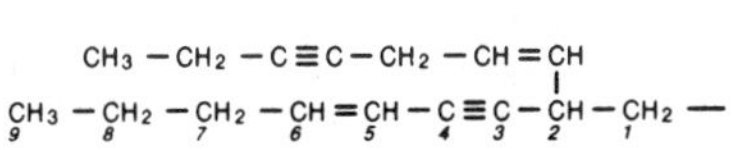

[2–(1–hepten–4–ynyl)–5–nonen–3–ynyl]

(*e*) 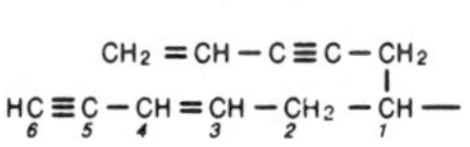

[1–(4–penten–2–ynyl)–3–hexen–5–ynyl]

(*f*) 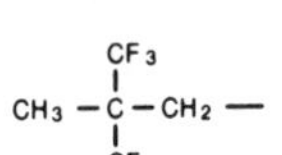 [3,3,3–trifluoro–2–methyl–2–(trifluoromethyl)propyl]

(*f*) 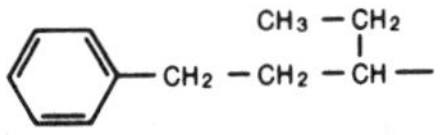 (1–ethyl–3–phenylpropyl)(not [1–(2–phenylethyl)propyl])

(*g*) 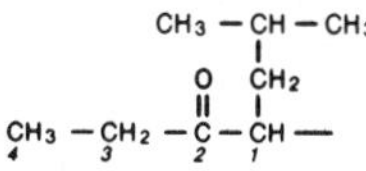 [1–(2–methylpropyl)–2–oxobutyl](not [3–methyl–1–(1–oxopropyl)butyl])

(*g*) 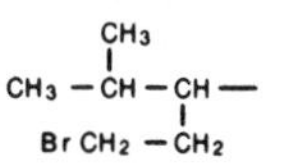 [1–(2–bromoethyl)–2–methylpropyl] (not [3–bromo–1–(1–methylethyl)propyl])

(*h*) 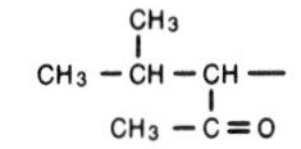 (1–acetyl–2–methylpropyl)(not [1–(1–methylethyl)–2–oxopropyl])

134. Carbonyl radicals which form part of a carbon chain are expressed by oxo substituents on the chain; the only exceptions are carboxy (–C(O)OH) and acetyl (–C(O)CH_3) radicals. The latter is used whenever (1–oxoethyl) would otherwise be called for. All chalcogen, imido and hydrazono analogs of carbonyl in a chain are treated similarly by use of thioxo, selenoxo, telluroxo, imido, and hydrazono radicals, except for chalcogen analogs (but not imido, etc., analogs) of carboxy; e.g., (HS(S)C–) is named (dithiocarboxy). Replacement analogs of acetyl are named (1–iminoethyl), (1–thioxoethyl),etc. Acyl radical names other than acetyl and benzoyl, e.g., propanoyl, are not used for substituents; neither are amido radicals, e.g., acetamido.

Examples:

HO_2CCH_2 —	(carboxymethyl)
$HCOCH_2$ —	(2–oxoethyl)(not (formylmethyl))
EtCO—	(1–oxopropyl)
$ClCOCH_2$ —	(2–chloro–2–oxoethyl)
H_2NCH_2CO —	(aminoacetyl)(glycyl is permitted in peptide nomenclature)
ClCOCO—	(chlorooxoacetyl)
AcNH—	(acetylamino)
EtCONH—	[(1–oxopropyl)amino]
PhC(=NH)$(CH_2)_2$CS— (4 ... 1)	(4–imino–4–phenyl–1–thioxobutyl)
MeC(=NH)NH—	[(1–iminoethyl)amino]

Isolated carbonyl radicals, other than carboxy, are expressed as carbonyl (as a doubling radical, or when both free valencies are attached to a single atom), formyl (if unsubstituted), benzoyl (if attached to a phenyl group which is not itself attached to another phenyl), or as a compound radical in which carbonyl is the parent.

Examples:

HO_2CCO—	(carboxycarbonyl)
H_2NCO—	(aminocarbonyl)(not carbamoyl)
ClCO—	(chlorocarbonyl)(not chloroformyl)
—NHCONH—	(carbonyldiimino)

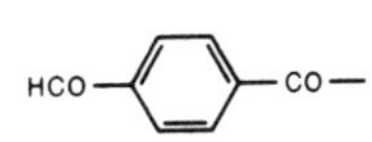 (4–formylbenzoyl)

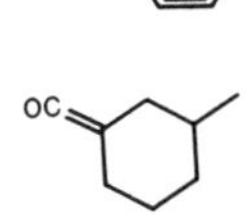 (3–carbonylcyclohexyl)

Replacement analogs of isolated carbonyl groups (other than chalcogen analogs of carboxy) are named as thioxo, imino, etc., derivatives of methyl, unless both free valencies are attached to a single atom, or the radical is being used multiplicatively (¶ 125), in which case carbonimidoyl, carbonohydrazonoyl, carbonothioyl, etc., are employed.

Examples:

HS_2CNH—	[(dithiocarboxy)amino]

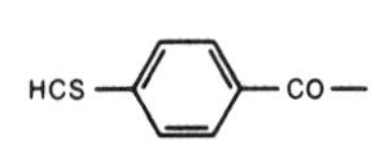 [4–(thioxomethyl)benzoyl]

HOC(=NH)—	(hydroxyiminomethyl)
H_2NC(=NH)NH—	[(aminoiminomethyl)amino]
HN=C=N—	(carbonimidoylamino)

135. Indicated hydrogen is a designation comprising a locant followed immediately by an italic capital *H* placed before a ring system name to express the position of each of the saturated atoms necessary for formation of a definable, stable ring system. Thus, Pyrrole always has one saturated atom (an atom not connected to either of its neighbors by a double bond) and, according to the position of this atom, the compound is named as follows:

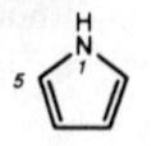

1*H*-Pyrrole

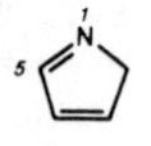

2*H*-Pyrrole

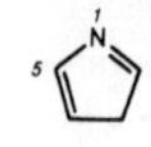

3*H*-Pyrrole

In the *Chemical Substance Index* only a single illustrative structural diagram is provided for each ring system, viz., the diagram which shows the saturated center(s) in the lowest-numbered nonangular position(s).

Tetrahydropyrrole has the trivial name **Pyrrolidine**: dihydropyrroles are named as derivatives of that pyrrole which has indicated hydrogen at the lowest numbered position consistent with the structure. Other monocyclic hetero systems are named in the same way. Hydrogen on a single ring atom between two bivalent hetero atoms is not indicated in the name.

Examples:

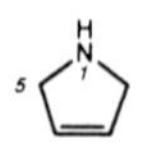

1*H*-Pyrrole, 2,5-dihydro- (not 2*H*-Pyrrole, 1,5-dihydro-)

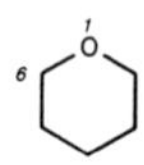

2*H*-Pyran, tetrahydro- (not 4*H*-Pyran, tetrahydro-)

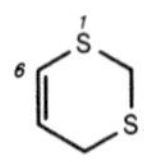

4*H*-1,3-Dithiin (not 2*H*,4*H*-1,3-Dithiin)

The lowest locants for nonangular positions of fused ring systems are normally cited for indicated hydrogen.

Example:

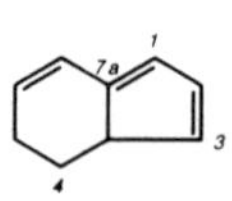

4*H*-Indene, 3a,5-dihydro- (not 3a*H*-Indene, 4,5-dihydro- or 5*H*-Indene, 3a,4-dihydro-)

Indicated hydrogen is assigned to angular or nonangular positions when needed to accommodate structural features, e.g., a bridge, spiro junction or ring-assembly junction, if that form of the ring system can exist.

Examples:

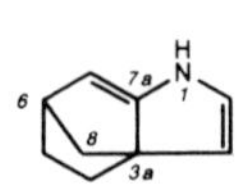

3a,6-Methano-3a*H*-indole, 1,4,5,6-tetrahydro-

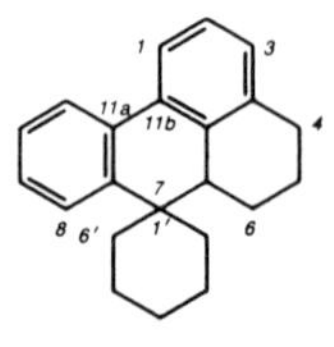

Spiro[7*H*-benz[*de*]anthracene-7,1'-cyclohexane], 4,5,6,6a-tetrahydro-

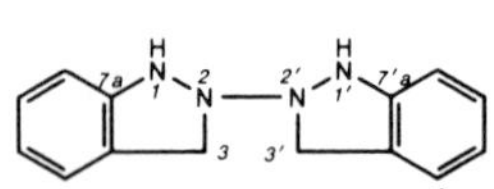

2,2'-Bi-2*H*-indazole, 1,1',3,3'-tetrahydro-

When a bridge requires hydrogen to be added, but indicated hydrogen of the parent system cannot be used for that purpose, the lowest locant, or a locant to accommodate a principal function, is chosen for the parent ring, and *additional indicated hydrogen* is cited in the name ahead of the bridge designation.

Example:

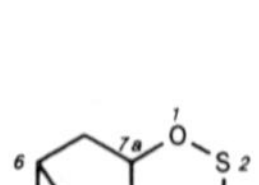

4H-3a,6-Methano-3H-1,2-benzoxathiole, tetrahydro- (3a*H*-1,2-Benzoxathiole cannot exist; the lowest available locant is therefore cited and the "extra" hydrogen for the bridge cited as additional indicated hydrogen, not in the "added" hydrogen form (see below), 3a(4*H*),6-Methano....)

After structural requirements have been met, indicated hydrogen is chosen to accommodate principal functions or (in a cyclic radical) free valencies, so long as the number of indicated hydrogens cited equals or exceeds the number of principal groups or free valencies that must be accommodated. For the usual case of a ring which requires a single indicated hydrogen for its existence, a single principal function or free valency is accommodated, but a polyfunctional compound is named at the ring system with lowest nonangular indicated hydrogen. (Functions on bridges are disregarded in applying this rule.)

Examples:

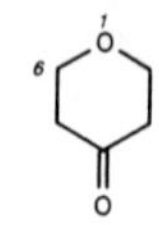

4*H*-Pyran-4-one, tetrahydro-

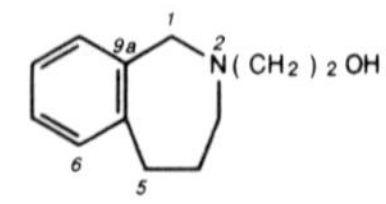

2*H*-2-Benzazepine-2-ethanol, 1,3,4,5-tetrahydro-

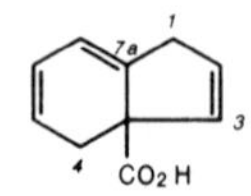

3a*H*-Indene-3a-carboxylic acid, 1,4-dihydro-

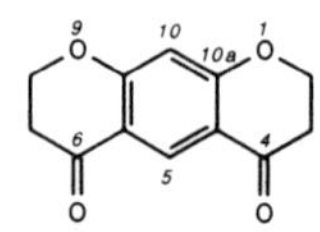

4*H*,6*H*-Benzo[1,2-*b*:5,4-*b*']dipyran-4,6- dione, 2,3,7,8-tetrahydro-

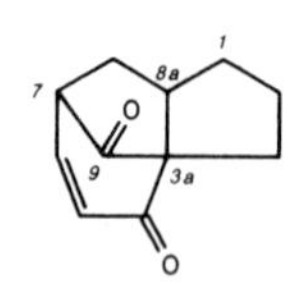

4*H*-3a,7-Methanoazulene-4,9-dione, 1,2,3,7,8,8a-hexahydro- (not 1*H*-3a,7a-Methanoazulene-4,9(7*H*)-dione, 2,3,8,8a-tetrahydro-) (the ketone function on the bridge is ignored in choosing indicated hydrogen to accommodate the single remaining function)

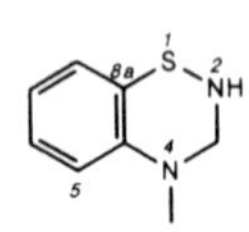

(2,3-dihydro-4*H*-1,2,4-benzothiadiazin-4-yl (not 2*H*-1,2,4-benzothiadiazin-4(3*H*-yl)

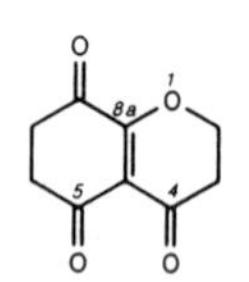

2*H*-1-Benzopyran-4,5,8(3*H*)-trione, 6,7-dihydro- (not 4*H*-1-Benzopyran-4,5,8-trione, 2,3,6,7-tetrahydro-)

136. **Added hydrogen** is hydrogen which is added to a ring system in the same operation as, but in a position different from, hydrogen added to accommodate structural features of a ring system, e.g., bridges, or spiro or ring-assembly junctions, or principal groups of a heading parent, or free valencies of a parent radical, when indicated hydrogen (¶ 135) is either not needed for the ring system itself or cannot be chosen to accommodate them. It differs from indicated hydrogen in being expressed as a locant and capital italic *H* in parentheses immediately following the locant for the principal function or other accommodated structural feature, e.g., "2(1*H*)-." Use of added hydrogen permits expression of a principal function, etc., in a heading parent instead of as a substituent. Thus, **1-Naphthalenone** cannot exist without partial hydrogenation of the naphthalene ring system; a name such as **Naphthalene, 1,2-dihydro-1-oxo-** violates the rule that the principal function be expressed as a suffix. Therefore, two hydrogen atoms are added in one operation to provide the name **1(2*H*)-Naphthalenone,** in which the "added" (or "extra") hydrogen is at the 2-position.

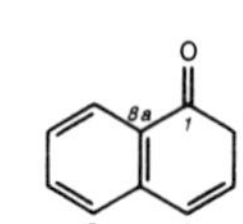

When principal functions or free valencies require added hydrogen, it is assigned to the lowest-numbered available angular or nonangular position; e.g., **1(2*H*)-Naphthalenone, 3,4-dihydro-** (not 1(4*H*)-Naphthalenone, 2,3-dihydro-); **2(4a*H*)-Naphthalenone, 5,6,7,8-tetrahydro-**. When the ring system requires indicated hydrogen and it cannot be assigned to accommodate a principal group or free radical, it has preference over added hydrogen for lowest locants. When a *pair* of principal groups, e.g., "-dione," are expressed by a heading parent, added hydrogen is not cited unless necessary, it being understood that only sufficient hydrogen has been added to accommodate the functions.

Examples:

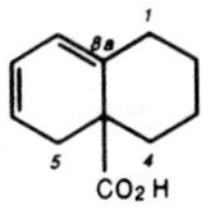

4a(2*H*)-Naphthalenecarboxylic acid, 1,3,4,5-tetrahydro- (a 4a(1*H*)-isomer cannot exist)

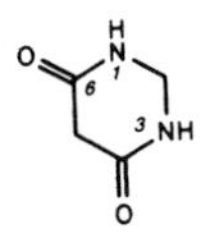

4,6(1*H*,5*H*)-Pyrimidinedione, dihydro-

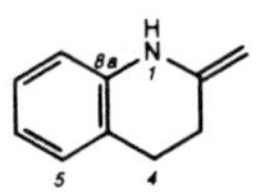

(3,4-dihydro-2(1*H*)-quinolinylidene) (not (1,4-dihydro-2(3*H*)-quinolinylidene))

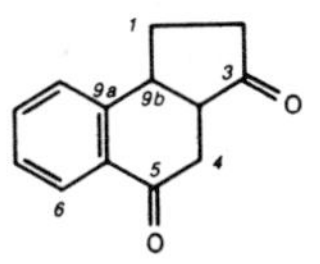

1*H*-Benz[*e*]indene-3,5(2*H*,4*H*)-dione, 3a,9b-dihydro-

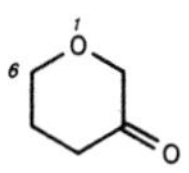

2*H*-Pyran-3(4*H*)-one, dihydro- (3*H*-Pyran cannot exist)

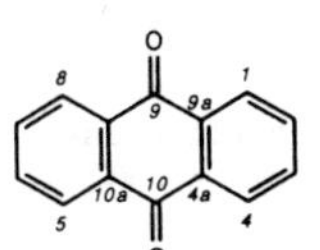

9,10-Anthracenedione (not 9,10=(9*H*,10*H*)-Anthracenedione)

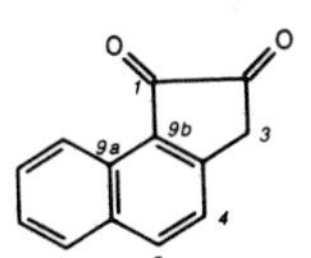

1*H*-Benz[*e*]indene-1,2(3*H*)-dione (not 3*H*-Benz[*e*]indene-1,2-dione) (low numbering of indicated hydrogen is observed, even if added hydrogen must then be cited)

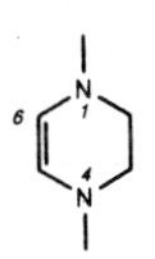

(2,3-dihydro-1,4-pyrazinediyl) (not (1,4(2*H*,3*H*)-pyrazinediyl))

Added hydrogen cited when hydrogen is required elsewhere for spiro and ring-assembly junctions is assigned (in descending order of preference) (*a*) to accommodate another spiro or ring-assembly junction, (*b*) to accommodate principal groups or free valencies, or (*c*) to lowest-numbered available posi=tions.

137. Numbering of molecular skeletons. Lowest locants for a set of principal groups, substituents, etc., are always preferred. The set, e.g., 7,6,1,2,4, is compared with another (alternative) set, e.g., 2,3,7,8,5, by rearranging them both in ascending numerical sequence: 1,2,4,6,7 and 2,3,5,7,8. The set with the maximum number of lowest locants is preferred (in this case, 1,2,4,6,7).

Example:

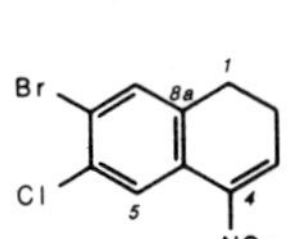

Naphthalene, 7-bromo-6-chloro-1,2-di=hydro-4-nitro- (not Naphthalene, 2-bromo-3-chloro-7,8-dihydro-5-nitro-)

Lowest locants for various kinds of structural features in cyclic and acyclic molecular skeletons are assigned, in order, to:

(*a*) hetero atoms (except for "a"-named radicals, see ¶¶ 127, 161);
(*b*) indicated hydrogen;
(*c*) principal groups or (for radicals) free valencies;
(*d*) multiple bonds;
(*e*) substituent prefixes;
(*f*) the substituent prefix cited earliest in the name.

Examples:

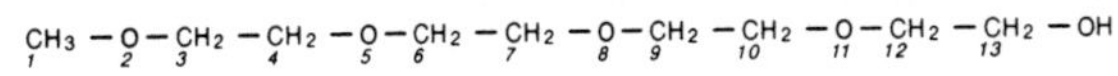

2,5,8,11-Tetraoxatridecan-13-ol (not 3,6,9,12-Tetraoxatridecan-1-ol)

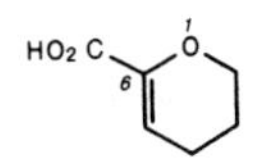

2*H*-Pyran-6-carboxylic acid, 3,4-=dihydro- (not 4*H*-Pyran-2-carboxylic acid, 5,6-dihydro-)

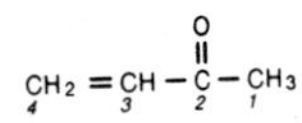

3-Buten-2-one (not 1-Buten-3-one)

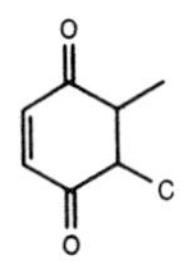

(6-chloro-2,5-dioxo-3-cyclohexen-1-yl)

2-Hexen-4-yne, 6,6,6-trifluoro- (double bond given preference over triple bond)

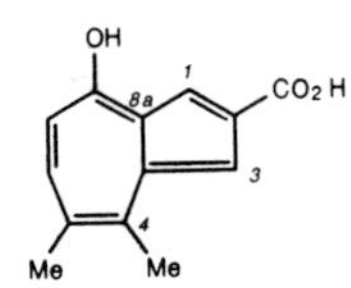

2-Azulenecarboxylic acid, 8-hydroxy-4,5-dimethyl-

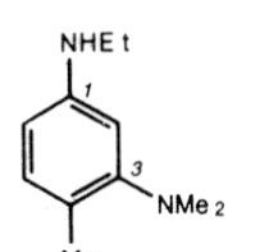

1,3-Benzenediamine, N^1-ethyl-N^3,=N^3,4-trimethyl- (not 1,3-Benzene=diamine, N^3-ethyl-N^1,N^1,6-trimethyl-; see ¶ 114)

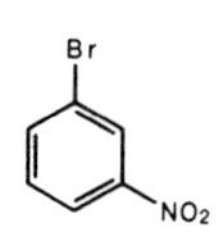

Benzene, 1-bromo-3-nitro- (not Benzene, 3-bromo-1-nitro-)

138. Index name selection. Most organic compounds have names based on molecular skeletons, e.g., **Propanoic acid** (from propane); **1,3-Dioxan-2-amine** (from 1,3-dioxane). Procedures for selecting the preferred name of this kind for index use are described in this section (see also ¶ 105).

Selection of a heading parent name based on a molecular skeleton is made by successive application of the following principles until a decision is reached.

(*a*) Greatest number of the principal chemical functional group.

(*b*) Preferred atomic content of the molecular skeleton in accordance with the order of precedence of compound classes (¶ 106). The heading parent should express at least one occurrence of an atom appearing earliest in the following list: N, P, As, Sb, Bi, B, Si, Ge, Sn, Pb, O, S, Se, Te. (This principle is used to decide between a cyclic and an acyclic parent, but is not applied to choices between "a" names or between ring systems. When acyclic and cyclic skeletons of the same compound class are present, a cyclic parent is preferred.)

(*c*) Preferred ring system. The choice between ring systems for use as heading parents is based on the following criteria, applied successively until a decision is reached. The senior ring system should:

(1) be a nitrogenous heterocycle;
(2) be a heterocycle;
(3) contain the largest number of rings;
(4) be a cyclic system occurring earliest in the following list of systems; spiro, bridged fused, bridged nonfused (Von Baeyer), fused;
(5) contain the largest individual ring (applies to fused carbocyclic sys=tems);
(6) contain the greatest number of ring atoms;

(7) contain the greater number of ring atoms common to two or more rings (applies to Von Baeyer ring systems); thus

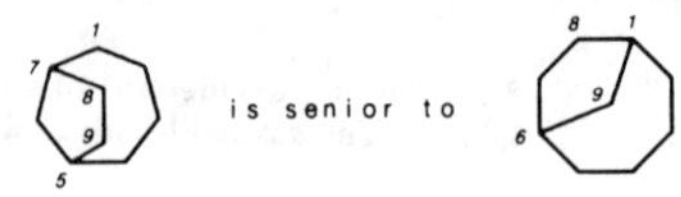

(8) contain lowest locants for bridges;
(9) contain the largest number of hetero atoms;
(10) contain the most preferred hetero atom other than nitrogen, according to the order in Table I, ¶ 128, i.e., O, S, Se, Te, P, As, Sb, Bi, Si, Ge, Sn, Pb, B.
(11) possess the most linear arrangement of rings (thus, **Anthracene** is senior to **Phenanthrene**);
(12) possess the lowest locants for hetero atoms assigned according to the rules (¶¶ 146, 152);
(13) express the lowest state of hydrogenation; thus, **Benzene** is preferred over **Cyclohexane**, **Pyridine** over **Piperidine**;
(14) express the lowest locant for indicated hydrogen.
Note: These criteria differ from those employed in selecting base components for fused systems (¶ 150).
(*d*) Greatest number of acyclic hetero atoms.
(*e*) Largest index heading parent.
(*f*) Greatest number of most preferred acyclic hetero atoms, according to the order of precedence in ¶ 128, above; i.e., O, S, Se, Te, N, P, As, Sb, Bi, Si, Ge, Sn, Pb, B.
(*g*) Greatest number of multiple bonds.
(*h*) Lowest locants in the heading parent successively for hetero atoms, principal groups (suffixes), all multiple bonds, double bonds.
If the preferred heading parent occurs more than once in the total compound, further principles must be applied, as follows:
(*i*) Centrality. For three or more occurrences of the heading parent, at least one of which must be nonterminal, the basis of the name is the central occurrence (or one of the central pair—both, when multiplicative nomenclature is permitted—if the total number is even) in the linear arrangement which comprises all or part of the maximum number of occurrences.
(*j*) Maximum number of substituent prefixes.
(*k*) Lowest locants for substituents on the heading parent.
(*l*) Multiplication of heading parents; when there is a choice of multiplicative names, that one is chosen which multiplies the largest number of occurrences of the index heading parent.
(*m*) Earliest index position of the total name.
When the preferred name is based on a molecular skeleton other than a ring system, an additional index entry is made at the name of each ring system present in the molecular structure except for certain commonly-occurring systems (see ¶ 163).
Examples (the italic letters on the left indicate the particular rules (above) that are exemplified):

(*a*) 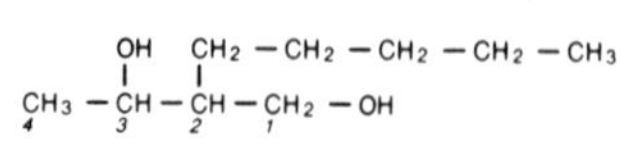

1,3-Butanediol, 2-pentyl-

(*a*) 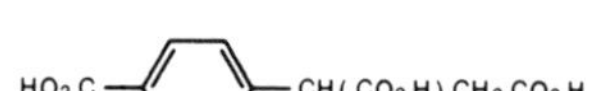

Butanedioic acid, (4-carboxyphenyl)-

(*b*) 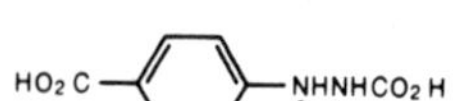

Hydrazinecarboxylic acid, 2-(4-carboxyphenyl)-

(*b*) 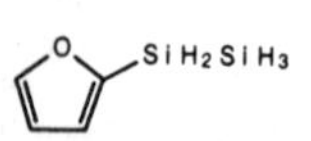

Disilane, 2-furanyl-

(*b*) PhN=NH

Diazene, phenyl- (not Benzenamine, *N*-imino-) (homogeneous hetero chains are never broken to obtain a higher function or more preferred parent)

(*b*) 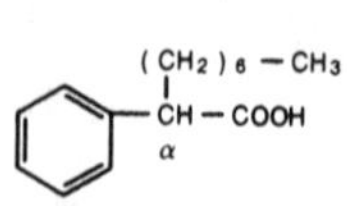

Benzeneacetic acid, α-heptyl- (not Nonanoic acid, 2-phenyl-)

(*b*) 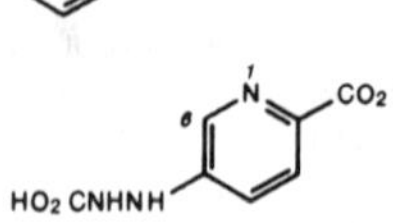

2-Pyridinecarboxylic acid, 5-(2-carboxyhydrazino)- (the *number* of preferred hetero atoms in the cyclic and acyclic chains is disregarded)

(*c*) 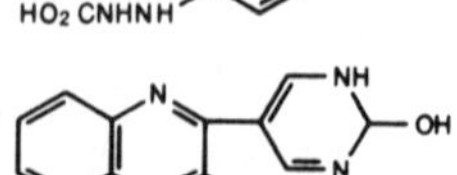

3-Quinolinol, 2-(1,2-dihydro-2-hydroxy-5-pyrimidinyl)-

(*d*) 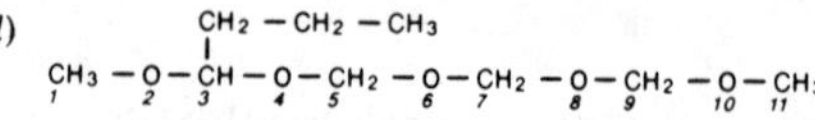

2,4,6,8,10-Pentaoxaundecane, 3-propyl-

(*e*) 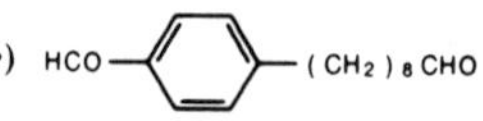

Benzenenonanal, 4-formyl- (not Benzaldehyde, 4-(9-oxononyl)-)

(*e*) 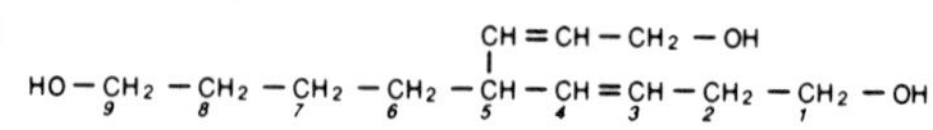

3-Nonene-1,9-diol, 5-(3-hydroxy-1-propenyl)- (not 2,5-Octadiene-1,8-diol, 5-(4-hydroxybutyl-)

(*e*) 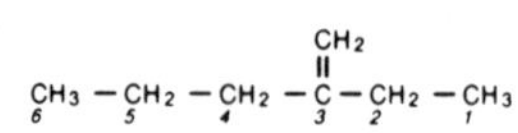

Hexane, 3-methylene- (not 1-Pentene, 2-ethyl-)

(*e*) 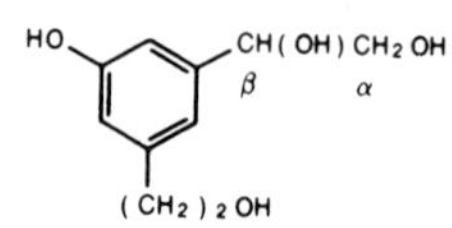

1,3-Benzenediethanol, β,5-dihydroxy- (see ¶ 124)

(*f*) $H_3\underset{3}{Si}\underset{2}{O}\underset{1}{Si}H_2SiH_2SSiH_3$

Disiloxane, disilathianyl-

(*f*) 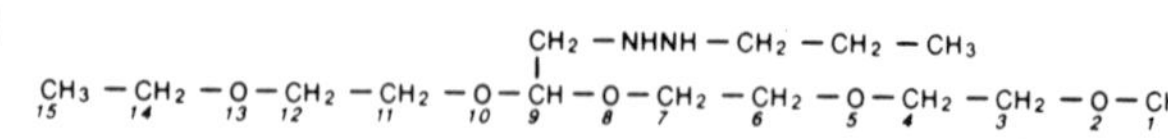

2,5,8,10,13-Pentaoxapentadecane, 9-[(2-propylhydrazino)methyl]-

(*g*) 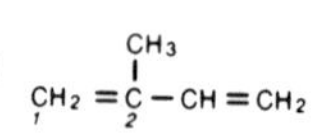

1,3-Butadiene, 2-methyl-

(*g*) 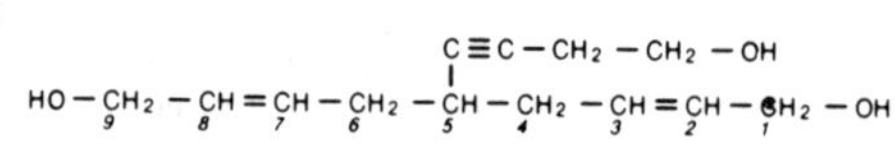

2,7-Nonadiene-1,9-diol, 5-(4-hydroxy-1-butynyl)- (double bonds are preferred over an equal number of triple bonds)

(*h*) 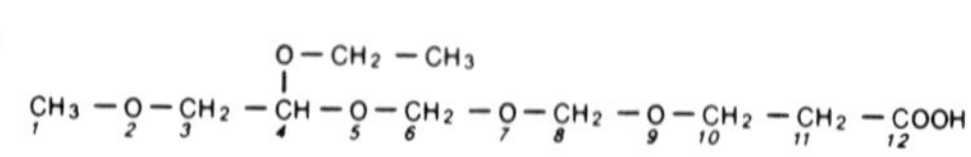

2,5,7,9-Tetraoxadodecan-12-oic acid, 4-ethoxy- (not 3,5,7,9-Tetraoxadodecan-12-oic acid, 4-(methoxymethyl)-)

(*h*) 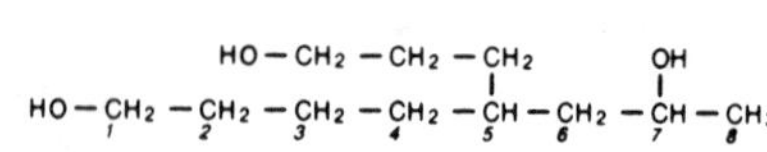

1,7-Octanediol, 5-(3-hydroxypropyl)- (not 1,8-Octanediol, 4-(2-hydroxypropyl)-)

(*h*) 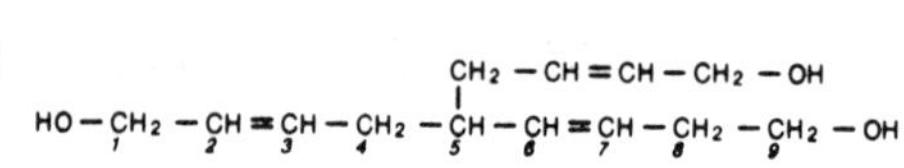

2,6-Nonadiene-1,9-diol, 5-(4-hydroxy-2-butenyl)- (not 2,7-Nonadiene-1,9-diol, 5-(4-hydroxy-1-butenyl)-)

(*i*) 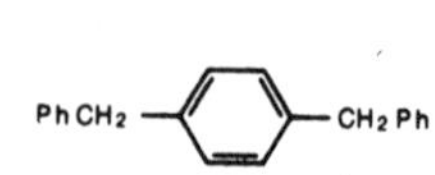

Benzene, 1,4-bis(phenylmethyl)-

(*i*) $Cl_2CHCH_2O(CH_2)_2O(CH_2)_2Cl$

Ethane, 1-(2-chloroethoxy)-2-(2,2-dichloroethoxy)-

(*i*)

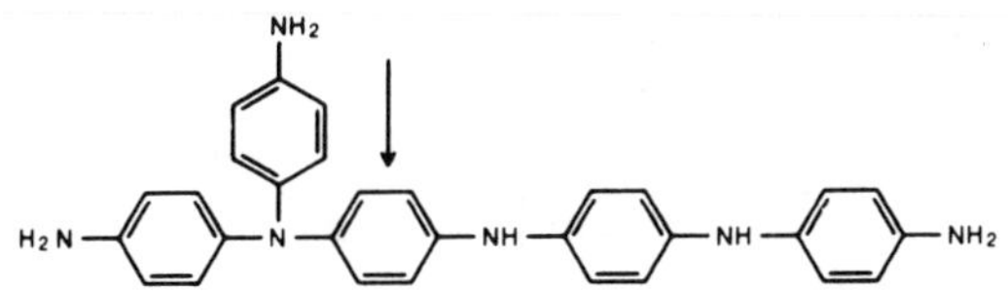

1,4-Benzenediamine, *N,N*-bis(4-= aminophenyl)-*N'*-[4-[(4-amino= phenyl)amino]phenyl]- (there are five (not four) overlapping occurrences of the heading parent centered on the one marked by the arrow)

(*i,j*)

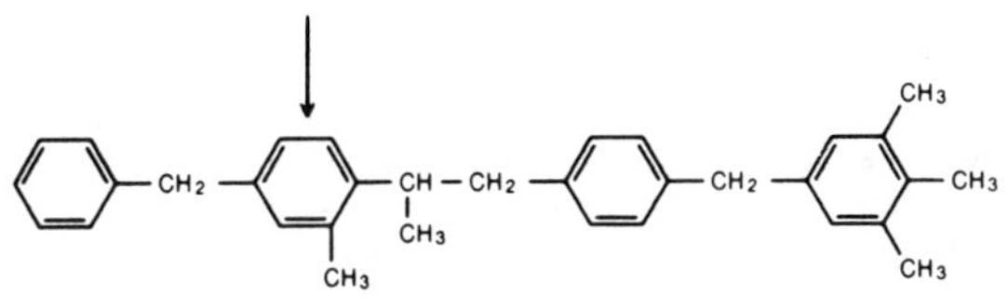

Benzene, 2-methyl-1-[1-methyl-2-[4-[(3,4,5-trimethylphenyl)= methyl]phenyl]ethyl]-4-(phe= nylmethyl)- (of the two central occurrences of the preferred heading parent, the one with the maximum number (three) of substituents is chosen; note that the centrality principle prevents choice of the tet= rasubstituted parent on the right)

(*i,l*)

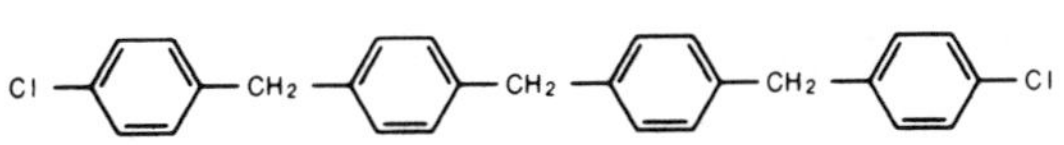

Benzene, 1,1'-methylenebis[4-= [(4-chlorophenyl)methyl]-

(*i,l*)

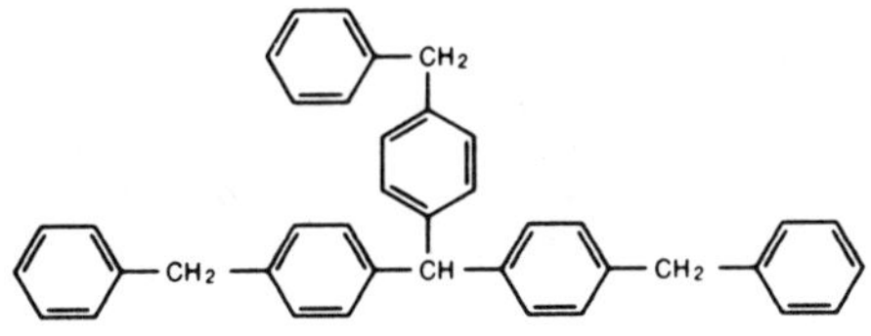

Benzene, 1,1',1''-methylidynetris= [4-(phenylmethyl)- (when multi= plication of the heading parent is permitted, occurrences directly at= tached to the central connecting radical are included in the same operation if possible)

(*j*)

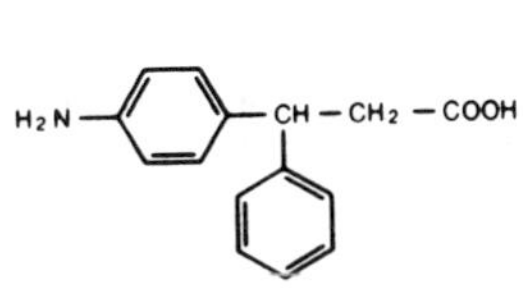

Benzenepropanoic acid, 4-amino-β-phenyl-

(*j*)

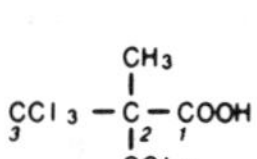

Propanoic acid, 3,3,3-trichloro-2-= methyl-2-(trichloromethyl)- (not Propanoic acid, 2,2-bis(trichloro= methyl)-) (like treatment of like groups is discontinued (¶ 255))

(*k*)

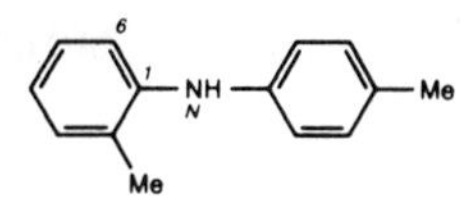

Benzenamine, 2-methyl-*N*-(4-methylphenyl)-

(*k*)

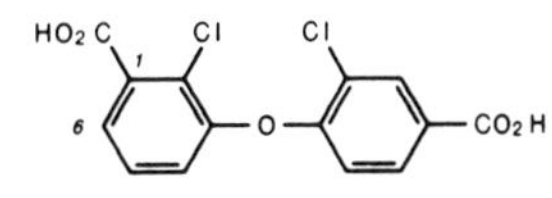

Benzoic acid, 3-(4-carboxy-2-chlorophenoxy)-2-chloro-

(*l*)

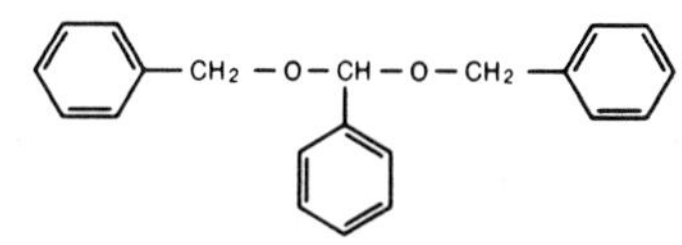

Benzene, 1,1'-[(phenylmethylene)= bis(oxymethylene)]bis- (note that all occurrences of benzene are terminal and that the principle of centrality is therefore not applicable)

(*l,m*)

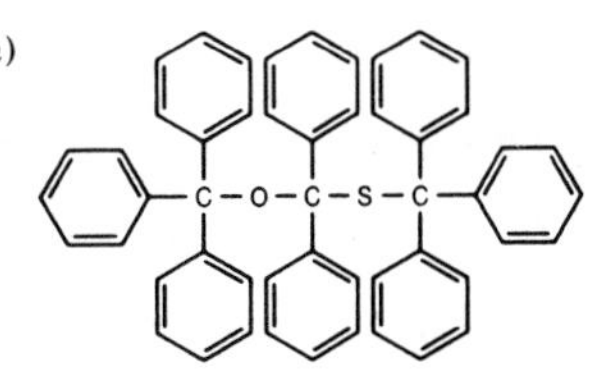

Benzene, 1,1',1''-[[[diphenyl= (triphenylmethoxy)methyl]thio]= methylidyne]tris- (not Benzene, 1,1',1''-[[diphenyl[(triphenylmethyl)= thio]methoxy]methylidyne]tris-)

(*m*) $F_3CCF(CCl_3)CO_2H$ (locants 3 2 1)

Propanoic acid, 2,3,3,3-tetrafluoro-2-(trichloromethyl)- (not Propa= noic acid, 3,3,3-trichloro-2-fluoro-2-(trifluoromethyl)-)

(*m*)

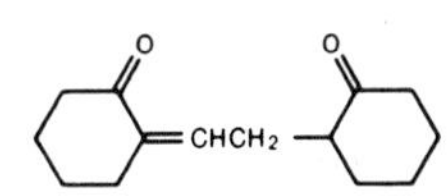

Cyclohexanone, 2-[2-(2-oxocyclo= hexyl)ethylidene]- (not Cyclo= hexanone, 2-[2-(2-oxocyclo= hexylidene)ethyl]-)

139. Subtractive nomenclature employs the suffixes "-ene" and "-yne" to signify removal of hydrogen pairs from saturated molecular skeletons; removal of hydrogen atoms from an aromatic system is denoted by "dehydro." Examples:

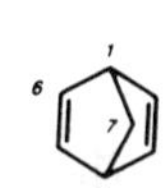

Bicyclo[2.2.1]hepta-2,5-diene

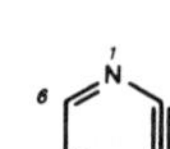

Pyridine, 2,3-didehydro- (not Pyri= dyne)

Functional class names such as "anhydride," "ester," "oxime," imply loss of water and may strictly be regarded as subtractive terms. Prefixes such as "deoxy," "nor," and "anhydro," are sometimes employed in names for classes of stereoparents (see Section E), but never in general index nomenclature.

140. Introduction. A molecular skeleton is defined for purposes of name selection as a chain or ring of atoms in which the number of hydrogen atoms attached to each skeletal atom is (usually) implied or (occasionally) explicitly stated by citation of the substitutive valency and bonding of the skeletal atoms. Monoatomic hydrides of Group IVA and VA elements (except nitrogen) are also treated as molecular skeletons; so are boron hydrides, but because of their unusual nature these are discussed separately (¶ 159).

Examples:

CH_4 — **Methane**

AsH_5 — **Arsorane**

$H_3SiOSiH_2OSiH_2$ — **Trisiloxane**

H_3SnSnH_3 — **Distannane**

$CH_3 \cdot CH_2 \cdot O \cdot CH_2 \cdot CH_2 \cdot O \cdot CH_2 \cdot CH_2 \cdot O \cdot CH_2 \cdot CH_2 \cdot O \cdot CH_2 \cdot CH_3$ (numbered 14 to 1)

3,6,9,12–Tetraoxatetradecane

Benzene

Pyrrolidine

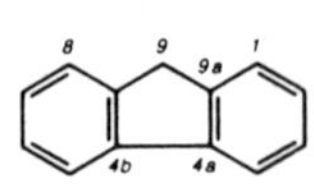

9*H*–Fluorene

Bicyclo[2.2.1]hept–2–ene

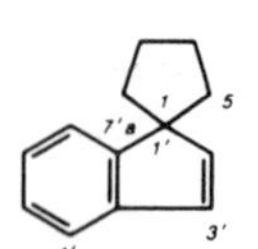

Spiro[cyclopentane–1,1'–[1*H*]=indene]

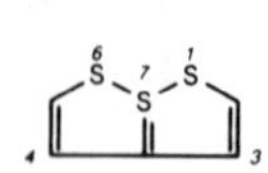

7λ⁴–[1,2]Dithiolo[1,5–*b*][1,2]di=thiole

Molecular skeletons are nonfunctional substitutive parent compounds. Substituents denoting a preferred compound class are generally cited as suf=fixes and other substituents as prefixes. Molecular skeletons, alone or in combination with a substituent suffix, are *index heading parents.*

This part of the manual deals with the formation of index names for structures that consist solely of one or more molecular skeletons.

141. Acyclic hydrocarbons. Saturated unbranched alkanes containing one through four carbon atoms are named **Methane, Ethane, Propane,** and **Butane.** Higher members of the class are named by adding the termination "–ane" to the appropriate multiplicative term, as, **Nonane** for C_9H_{20}, **Hexa=decane** for $C_{16}H_{34}$, **Eicosane** for $C_{20}H_{42}$, **Heneicosane** for $C_{21}H_{44}$, and **Tri=triacontane** for $C_{33}H_{68}$.

Unsaturated unbranched acyclic hydrocarbons (unbranched alkenes, alka=dienes, alkynes, etc.) are named by replacing the ending "–ane" by "–ene" (for a single double bond), "–adiene" for two double bonds, "–yne" for a single triple bond, etc. Combinations, e.g., "–enyne," "–trienediyne," are employed when both bond types are present. Low numbering (¶ 137) is employed for the *set* of multiple bonds; when there is a choice, double bonds are preferred over triple bonds. When principal groups (functional suffixes) are present, these and not the multiple bonds are preferred for low numbering.

Examples:

MeCH=CHCH₂CH=CH₂ — **1,4–Hexadiene** (not 2,5–Hexadiene)

$H_2C{=}C{=}CH_2$ — **1,2–Propadiene**

$MeCH_2C{\equiv}CCH_2Me$ — **3–Hexyne**

$HC{\equiv}CC{\equiv}CH$ — **1,3–Butadiyne**

$HC{\equiv}CCH_2CH{=}CH_2$ — **1–Penten–4–yne**

$MeCH{=}CH(CH_2)_3C{\equiv}CH$ — **6–Octen–1–yne** (not 2–Octen–7–yne)

$H_2C{=}CHCH_2OH$ — **2–Propen–1–ol** (not 1–Propen–3–ol)

142. Organic hetero chains containing at least four hetero units (¶ 127) are given "a" names, i.e., replacement names based on the hydrocarbon skeleton by use of "oxa," "thia," etc., prefixes. Lowest locants (¶ 137) are assigned to all hetero atoms regardless of type and then to preferred hetero atoms (Table I, ¶ 128). These locants are *not* affected by the presence of a principal chemical group. Unsaturation is indicated as for the hydrocarbon chain, with lowest locants compatible with the low numbering of the hetero atoms.

Examples:

$CH_3 \cdot CH_2 \cdot O \cdot CH_2 \cdot CH_2 \cdot O \cdot CH_2 \cdot CH_2 \cdot O \cdot CH_2 \cdot CH_2 \cdot O \cdot CH_3$ (numbered 13 to 1)

2,5,8,11–Tetraoxatridecane

$CH_3{-}O{-}SiH_2{-}CH_2{-}CH_2{-}SiH_2{-}S{-}CH_3$ (numbered 1 to 8)

2–Oxa–7–thia–3,6–disilaoctane (not 7–Oxa–2–thia–3,6–disilaoctane)

$CH_3{-}CH_2{-}SiH_2{-}CH_2{-}CH_2{-}SiH_2{-}CH_2{-}CH_2{-}SiH_2{-}CH_2{-}CH_2{-}SiH_2{-}CH_2{-}CH_3$ (numbered 1 to 14)

3,6,9,12–Tetrasilatetradecane

$H_2C{=}CHCH_2[O(CH_2)_2]_3OMe$ — **2,5,8,11–Tetraoxatetradec–13–ene**

$H_2C{=}CH(OCH{=}CH)_3OCH{=}CH_2$ — **3,6,9,12–Tetraoxatetradeca–1,4,7,10,13–pentaene**

143. Homogeneous hetero chains are considered to include the mono=nuclear hydrides **Phosphine** (PH_3), **Phosphorane** (PH_5), **Arsine** (PH_3), **Arsorane** (AsH_5), **Stibine** (SbH_3), **Bismuthine** (BiH_3), **Silane** (SiH_4), **Ger=mane** (GeH_4), **Stannane** (SnH_4), and **Plumbane** (PbH_4). Chains composed of two or more of these hydride residues are named by prefixing the hydride name with "Di," "Tri," etc.

Examples:

H_2BiBiH_2 — **Dibismuthine**

H_4PPH_4 — **Diphosphorane**

$H_3SnSnH_2SnH_3$ — **Tristannane**

$H_3Si(SiH_2)_3SiH_3$	**Pentasilane**

Saturated nitrogen chains are named **Hydrazine** (not Diazane) for H_2N=NH_2, **Triazane** for H_2NNHNH_2, **Tetrazane** for $H_2NNHNHNH_2$, etc. Unsaturation is denoted by use of the subtractive suffixes "-ene" and "-yne." Locants are employed as for hydrocarbon chains (¶¶ 114, 141).

Examples:

$HAs=AsH$	**Diarsene**
$H_2Si=SiH_2$	**Disilene**
$H_2NN=NNH_2$	**2-Tetrazene**
$HN=NN=NH$	**1,3-Tetrazadiene**

144. Heterogeneous hetero chains in which any one of the Group IVA elements (silicon, germanium, tin, or lead) alternates with chalcogen are given "oxane," "thiane," etc. names.

Examples:

$H_3SiOSiH_2OSiH_3$ (locants 1 2 3 4 5)	**Trisiloxane**
$H_3GeSeGeH_3$ (locants 1 2 3)	**Digermaselenane**
$H_3SnSSnH_2SSnH_3$ (locants 1 2 3 4 5)	**Tristannathiane**

Chains of alternating atoms of a Group IVA element and nitrogen, e.g., SiH_3NHSiH_3, SnH_3NHSnH_3, are not named Disilazane, Distannazane, etc. Instead, the amine function is recognized; thus, Disilazane is indexed at **Silanamine, *N*-silyl-**. However, substituent radicals derived from such heterogeneous chains containing nitrogen, e.g., disilazanyl for SiH_3NHSiH_2-, are employed in the presence of higher functions (see ¶ 161).

"A" names for hetero chains are avoided unless carbon substituents can be included in the chain. Otherwise the preferred parent is a homogeneous hetero chain, hydride, or element name.

Examples:

$H_3GeOSiH_2SPbH_3$	**Silane, (germyloxy)(plumbylthio)-**
$H_3GeOPbH_2SeMe$ (locants 1 2 3 4)	**2-Oxa-4-selena-1-germa-3-plumbapentane**

145. Monocyclic hydrocarbons (cycloalkanes, cycloalkenes, etc.) are named by attaching the prefix "cyclo" to the name of the acyclic hydrocarbon with the same number of carbon atoms. Unsaturation is expressed by use of "-ene" and "-yne" in place of "-ane" as for the acyclic analogs. No locant is employed for a single multiple bond; lowest locants (¶ 137) are cited when two or more multiple bonds are present. The trivial name **Benzene** is used for 1,3,5-Cyclohexatriene.

Examples:

Cyclohexane

1,3-Cyclopentadiene

Cyclohexene

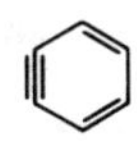

1,3-Cyclohexadien-5-yne (not Benzyne)

146. Monocyclic hetero systems.

(*a*) *Rings of three through ten members containing nonmetallic hetero atoms (except silicon)* are named systematically by the (extended) Hantzsch-Widman system or at trivial names. Table II supplies the stems for the systematic names, which are completed by adding replacement prefixes for the hetero atoms in the order set out in Table I, ¶ 128, e.g., oxa, thia, aza, together with locants and multiplicative prefixes denoting the position and number of each. A locant for a single hetero atom is not cited. The preferred hetero atom is numbered "1." This means the *set* of locants may not be the lowest possible, as defined in ¶ 137. The letter "a" of replacement prefixes is elided before another vowel in Hantzsch-Widman names; e.g., **Dioxazole**, not Dioxaazole.

Examples:

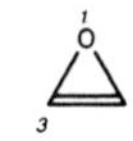

Oxirene

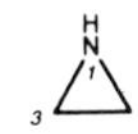

Aziridine

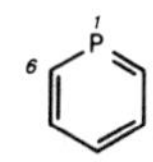

Phosphorin

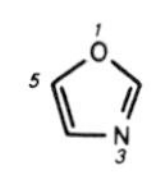

Oxazole (not 1,3-Oxazole, because the 1,2-isomer has the trivial name Isoxazole)

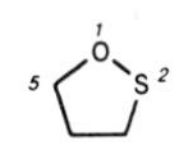

1,2-Oxathiolane (all locants are placed ahead of the name; cf. ¶ 146(*c*), below)

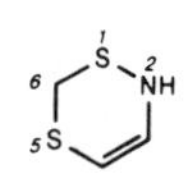

2*H*-1,5,2-Dithiazine (not 1*H*-2,4,1-Dithiazine; not 6*H*-1,3,6-Dithiazine; not 4*H*-1,5,4-Dithiazine) (The numbering must begin with a sulfur atom and proceed in the direction that gives lowest numbers to the *remaining hetero atoms*.) (The "2*H*" signifies indicated hydrogen (see ¶ 135)).

TABLE II
HANTZSCH-WIDMAN STEMS FOR MONOCYCLIC HETERO SYSTEMS OF THREE THROUGH TEN MEMBERS[1]

No. of members in the ring	*Rings containing nitrogen*		*Rings containing no nitrogen*	
	Unsaturated[2]	*Saturated*	*Unsaturated*[2]	*Saturated*
3	-irine	-iridine	-irene	-irane
4	-ete	-etidine	-ete	-etane
5	-ole	-olidine	-ole	-olane
6	-ine[3]	[4]	-in[3]	-ane[5,6]
7	-epine	[4]	-epin	-epane
8	-ocine	[4]	-ocin	-ocane
9	-onine	[4]	-onin	-onane
10	-ecine	[4]	-ecin	-ecane

[1]The symbols denoting the ring sizes for 3, 4, 7, 8, 9, 10 members are derived from numerical prefixes as follows: "ri" from *tri*; "et" from *tet*ra; "ep" from *hep*ta; "oc" from "*oc*ta; "on" from *no*na; and "ec" from *dec*a.

[2]Corresponds to the maximum number of noncumulative double bonds when the hetero atoms have the substituent valencies given in Table I, ¶ 128.

[3]When the Hantzsch-Widman prefixes "phospha," "arsa," "stiba" or "germa" are immediately followed by the Hantzsch-Widman stems "-in" or "-ine," they are replaced by the prefixes "phosphor," "arsen," "antimon," or "german," respectively.

[4]Saturation is expressed by detachable prefixes such as "tetrahydro-," "hexahydro-," etc. The prefix "perhydro-" is not used.

[5]This stem is not used for saturated hetero systems based on the elements silicon, germanium, tin, or lead. Saturation of these rings is indicated by detachable prefixes such as "tetrahydro-," "-hexahydro-," etc., when Hantzsch-Widman names are used.

[6]Saturation of six-membered hetero systems based on the elements boron or phosphorus is denoted by the stem "-inane."

Table II indicates those cases in which special endings are employed for fully saturated monocyclic hetero systems: e.g., **Azetidine,** not Azete, tetrahydro–. Special names for *partially* saturated ring systems are discontinued.

Example:

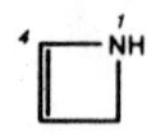

Azete, 1,2-dihydro– (not 2-Azetine)

Certain five– and six–membered monocyclic hetero systems, both saturated and unsaturated, are indexed at trivial names. These names are set out in Table III in the order of the corresponding Hantzsch–Widman names, which are *not* used for indexing. Fully hydrogenated five–membered rings are given "–olidine" names, as for systematically named systems. Special names for partially hydrogenated five–membered rings e.g., **2–Pyrroline,** were discontinued in 1972; they are now named as dihydro derivatives of the fully unsaturated rings. Indicated hydrogen (see ¶ 135) is necessary in some rings to describe the location of the saturated skeletal atom, e.g., **1*H*–Pyrrole, 2*H*–Pyrrole, 3*H*–Pyrrole.** Presence of a triple bond in addition to the maximum number of noncumulative double bonds is expressed by the subtractive prefix "didehydro."

Example:

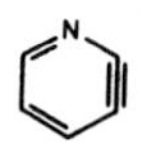

Pyridine, 2,3–didehydro–

TABLE III
MONOCYCLIC HETERO SYSTEMS WITH TRIVIAL NAMES

Hantzsch–Widman name	*Index name*
Azine	**Pyridine**
Azine, hexahydro–	**Piperidine**
Azole	**Pyrrole** (1*H*–, 2*H*–, or 3*H*–)
Azolidine	**Pyrrolidine**
1,2–Diazine	**Pyridazine**
1,3–Diazine	**Pyrimidine**
1,4–Diazine	**Pyrazine**
1,4–Diazine, hexahydro–	**Piperazine**
1,2–Diazole	**Pyrazole** (1*H*–, 3*H*–, or 4*H*–)
1,3–Diazole	**Imidazole** (1*H*–, 2*H*–, or 4*H*–)
1,2–Diazolidine	**Pyrazolidine**
1,3–Diazolidine	**Imidazolidine**
1,2,5–Oxadiazole	**Furazan**
2*H*–1,4–Oxazine, tetrahydro–	**Morpholine**
1,2–Oxazole	**Isoxazole**
Oxin	**Pyran** (2*H*– or 4*H*–)
Oxole	**Furan**
1,2–Selenazole	**Isoselenazole**
Selenole	**Selenophene**
Tellurole	**Tellurophene**
2*H*–1,4–Thiazine, tetrahydro–	**Thiomorpholine**
1,2–Thiazole	**Isothiazole**
Thiole	**Thiophene**
1,3,5,2,4,6–Triazatriborine, hexahydro–	**Borazine**
1,3,5,2,4,6–Trioxatriborinane	**Boroxin**
1,3,5,2,4,6–Trithiatriborinane	**Borthiin**

(*b*) *Rings of three through ten members containing antimony, tin, lead, germanium or bismuth atoms in addition to carbon atoms* are indexed at Hantzsch–Widman names. Partially saturated and fully saturated six–membered ring systems of this type containing germanium, lead or tin are named on the basis of the unsaturated rings. When a ring is composed entirely of the same metallic atoms, a Hantzsch–Widman name is used if unsaturated, otherwise "Cyclo" is placed ahead of the name of the acyclic compound. Heterocycles containing metallic atoms other than the above five are indexed by coordination nomenclature (¶ 215).

Examples:

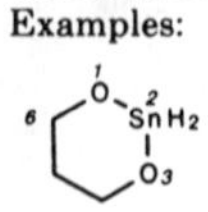

4*H*–1,3,2–Dioxastannin, dihydro–

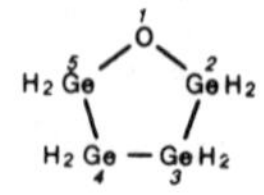

Oxatetragermolane

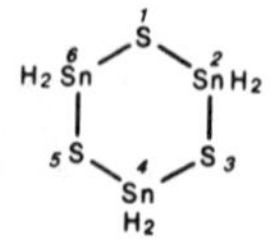

1,3,5–Trithia–2,4,6–tristannacyclohexane

(*c*) *Rings of more than ten members not containing silicon atoms* are indexed at organic replacement ("a") names. Unsaturation is indicated by "–ene" and "–yne" suffixes.

Examples:

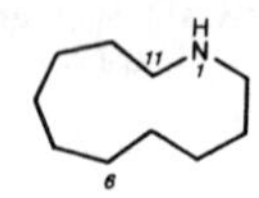

Azacycloundecane

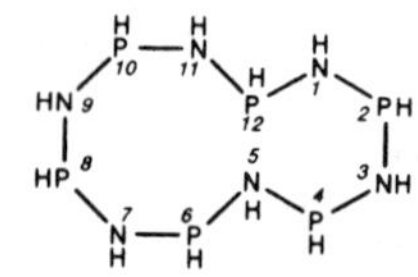

1,3,5,7,9,11–Hexaaza–2,4,6,8,10,12–hexaphosphacyclododecane

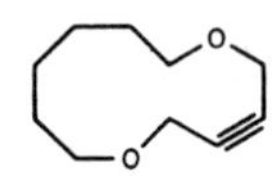

1,6–Dioxacyclododec–3–yne

(*d*) *Rings containing silicon* in general are indexed at replacement ("a") names based on the cyclic hydrocarbons. Systems comprising only silicon atoms or silicon atoms alternating with nitrogen or one of the chalcogens are given "Cyclosila–" names.

Examples:

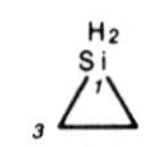

Silacyclopropane

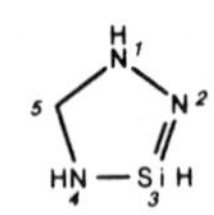

1,2,4–Triaza–3–silacyclopent–2–ene

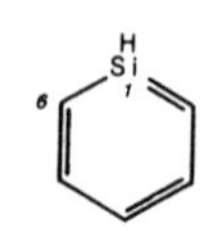

Silabenzene

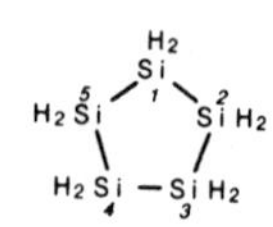

Cyclopentasilane

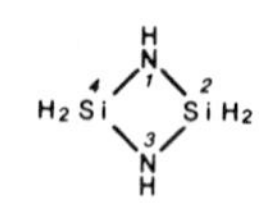

Cyclodisilazane

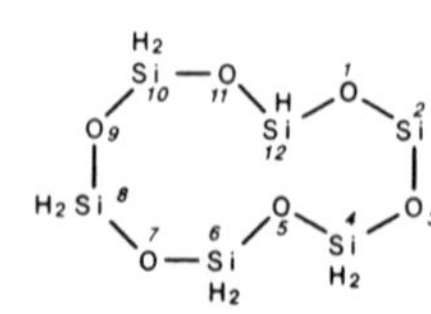

Cyclohexasiloxane

147. Polycyclic systems may be subdivided into four classes as follows:

(*a*) *Fused systems* contain at least two rings of five or more members and only "ortho–" or "ortho– and peri–" fusions (see below).

(*b*) *Bridged systems* are monocyclic or fused systems with valence bonds, atoms, or chains connecting different parts of the structure.

(*c*) *Spiro systems* have pairs of rings (or ring systems) with only one common atom.

(*d*) *Ring assemblies* have pairs of rings (or ring systems) connected by single bonds.

In the following sections, methods of naming ring systems of all these types will be described. The names of more complicated cases are built up from base components which may be described as "fundamental" systems.

148. Fundamental fused carbocycles with ortho–fusions only, e.g., **Naphthalene,** have adjoining rings with only two atoms in common; they thus have *n* common faces and 2*n* common atoms. An ortho– and peri– fused system contains a ring which has two, and only two, atoms in common with each of two or more rings, the total system containing *n* common faces and *fewer than* 2*n* common atoms.

Examples:

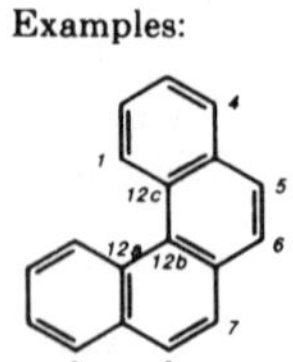

An ortho–fused system
(3 common faces; 6 common atoms)

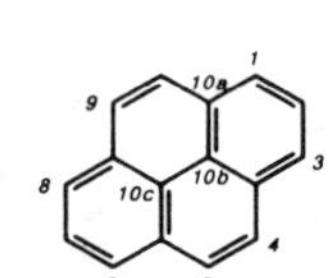

An ortho– and peri–fused system
(5 common faces; 6 common atoms)

Systems of five or more ortho–fused benzene rings are named by the "acene" system if the arrangement is linear, by the "phene" system if one central angular site is present.
Examples:

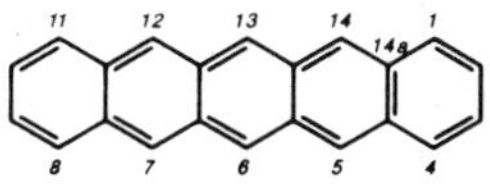

Pentacene

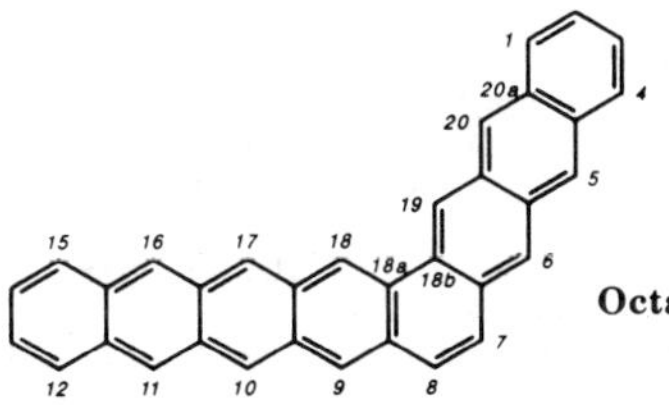

Octaphene

Names ending in "–alene" are employed for bicyclic fused systems, and in "–phenylene" for systems built up from benzene rings fused to alternate sides of a monocyclic hydrocarbon. (Analogous "–naphthylene" names have been used for corresponding 2,3–fusion systems of naphthalene.)
Examples:

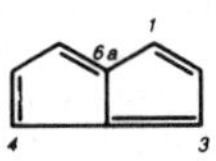

Pentalene

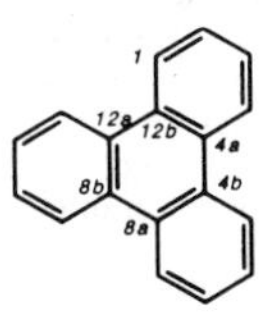

Triphenylene

Table IV lists the names of trivially–named fundamental fused carbocycles in *ascending* order of preference for adoption as base components in the naming of more complex fused hydrocarbon systems. Also included are some of the names discussed immediately above. (The order is based on the rules described in ¶¶ 138, 150.) The ring analyses describe the number of component rings and the number of atoms each ring contains. Diagrams of these rings, which show the preferred orientation and numbering, are displayed in the *Ring Systems Handbook*. Diagrams justified by current entries are also provided in the semiannual and collective *Chemical Substance Indexes*.

The following ring systems require citation of indicated hydrogen (¶ 135) to complete the name: **Indene, Fluorene, Phenalene, Trindene.**

TABLE IV
FUNDAMENTAL FUSED CARBOCYCLES IN ASCENDING ORDER OF PRECEDENCE FOR USE AS BASE COMPONENTS IN FUSED SYSTEMS

Name	*Ring analysis*
Pentalene	C_5–C_5
Indene	C_5–C_6
Naphthalene	C_6–C_6
Azulene	C_5–C_7
Heptalene	C_7–C_7
Biphenylene	C_4–C_6–C_6
***as*–Indacene**	C_5–C_5–C_6
***s*–Indacene**	C_5–C_5–C_6
Acenaphthylene	C_5–C_6–C_6
Fluorene	C_5–C_6–C_6
Phenalene	C_6–C_6–C_6
Phenanthrene	C_6–C_6–C_6
Anthracene	C_6–C_6–C_6
Trindene	C_5–C_5–C_5–C_6
Fluoranthene	C_5–C_6–C_6–C_6
Acephenanthrylene	C_5–C_6–C_6–C_6
Aceanthrylene	C_5–C_6–C_6–C_6
Triphenylene	C_6–C_6–C_6–C_6
Pyrene	C_6–C_6–C_6–C_6
Chrysene	C_6–C_6–C_6–C_6
Naphthacene	C_6–C_6–C_6–C_6
Pleiadene	C_6–C_6–C_6–C_7
Picene	C_6–C_6–C_6–C_6–C_6
Perylene	C_6–C_6–C_6–C_6–C_6
Pentaphene	C_6–C_6–C_6–C_6–C_6
Pentacene	C_6–C_6–C_6–C_6–C_6
Tetraphenylene	C_6–C_6–C_6–C_6–C_8
Hexaphene	C_6–C_6–C_6–C_6–C_6–C_6
Hexacene	C_6–C_6–C_6–C_6–C_6–C_6
Rubicene	C_5–C_5–C_6–C_6–C_6–C_6–C_6
Coronene	C_6–C_6–C_6–C_6–C_6–C_6–C_6
Trinaphthylene	C_6–C_6–C_6–C_6–C_6–C_6–C_6
Heptaphene	C_6–C_6–C_6–C_6–C_6–C_6–C_6
Heptacene	C_6–C_6–C_6–C_6–C_6–C_6–C_6
Pyranthrene	C_6–C_6–C_6–C_6–C_6–C_6–C_6–C_6
Ovalene	C_6–C_6–C_6–C_6–C_6–C_6–C_6–C_6–C_6–C_6

149. Fundamental fused heterocycles often have trivial names, e.g., **Cinnoline, Xanthene.** Others belong to one or another semisystematic sys= tem. A linear set of three fused six–membered rings with the same hetero element in both unfused positions of the central ring are given "–anthrene" names.
Example:

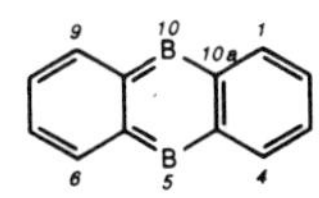

Boranthrene

A similar set with different hetero elements in these positions is given a "Pheno–" name containing the organic replacement terms in the usual order (Table I, ¶ 128) and the ending "–in" (or "–ine" if nitrogen, phosphorus or arsenic is included).
Example:

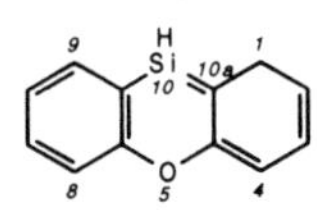

1*H*–Phenoxasilin

An exception is **Phenazine** for the analog that contains nitrogen in *both* central positions.

Arsenic and phosphorus analogs of fused nitrogen heterocycles (**Indole, Quinoline,** etc.) are named as follows: **Arsindole, Isoarsindole, Arsinoline, Isoarsinoline, Phosphindole, Isophosphindole, Phosphinoline, Iso= phosphinoline, Arsanthridine, Acridarsine, Acridophosphine, Phen= arsazine, Phenophosphazine.**

The replacement of the oxygen in **Xanthene** by sulfur or selenium has been denoted by the appropriate chalcogen functional replacement prefix: **Thio= xanthene, Selenoxanthene.**

In the "benzo" system, bicyclic fused heterocyclic systems containing a benzene ring and a ring named by the Hantzsch–Widman system are indexed by prefixing the latter name by "Benz–" or "Benzo–." Indicated hydrogen, if necessary, and locants are placed in front of the complete name. Similar names are used when benzene is fused to a monocycle with a trivial name (unless the bicyclic system itself has a trivial name).
Examples:

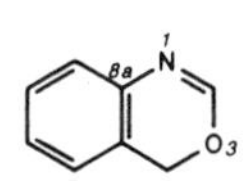

4*H*–3,1–Benzoxazine (not 4*H*–Benz= [*d*]oxazine)

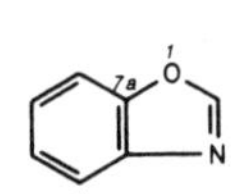

Benzoxazole (locants are not cited; the isomers are named 1,2– and 2,1–Benzisoxazole)

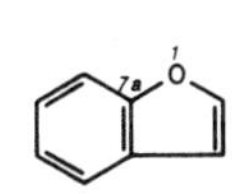

Benzofuran (the isomer is named Isobenzofuran)

Such "benzo" names are not usually adopted as base components of fused systems when only hydrocarbon rings are fused to the benzene portion.

When benzene is fused to a heterocyclic ring containing more than ten skeletal atoms, "Benzo–" or "Benz–" is placed ahead of the replacement ("a") name of the saturated ring and the ending changed to "–in" (or "–ine" if nitrogenous) to indicate the maximum number of noncumulative double bonds. (Saturated positions other than those occupied by indicated hydrogen are denoted by hydro substituents.)
Example:

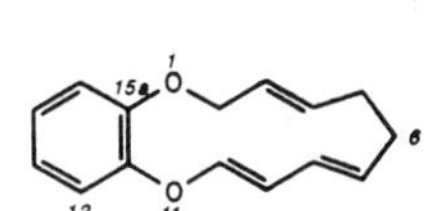

2*H*–1,11–Benzodioxacyclotridecin, 5,6–dihydro–

Trivially named fundamental fused hetero systems are listed in Table V. Also included are some of the systems discussed immediately above, as well as a selection of monocyclic hetero systems to help illustrate the ascending order of priority for adoption as base components in complex fused systems.

The following trivially named ring system is employed as a heading parent, but has not been adopted as a base component.

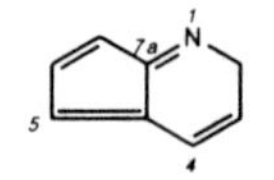

Pyrindine (2*H*-1-isomer shown)

Fused systems containing only silicon and carbon skeletal atoms are indexed at "Sila-" replacement names if the corresponding hydrocarbon has a funda= mental name (Table IV, ¶ 148).

TABLE V
FUNDAMENTAL HETEROCYCLES IN ASCENDING ORDER OF PRECEDENCE FOR USE AS BASE COMPONENTS IN FUSED SYSTEMS[1]

Index Name	*Ring Analysis*
Isoarsindole[2]	C_4As-C_6
Arsindole[2]	C_4As-C_6
Isoarsinoline	C_5As-C_6
Arsinoline	C_5As-C_6
Arsanthridin	$C_5As-C_6-C_6$
Acridarsine	$C_5As-C_6-C_6$
Arsanthrene	$C_4As_2-C_6-C_6$
Isophosphindole[2]	C_4P-C_6
Phosphindole[2]	C_4P-C_6
Isophosphinoline	C_5P-C_6
Phosphinoline	C_5P-C_6
Tellurophene	C_4Te
Selenophene	C_5Se
Selenanthrene	$C_4Se_2-C_6-C_6$
Thiophene	C_4S
Thianthrene	$C_4S_2-C_6-C_6$
Furan	C_4O
Pyran[2]	C_5O
Isobenzofuran	C_4O-C_6
Xanthene[2]	$C_5O-C_6-C_6$
Phenoxantimonin	$C_4OSb-C_6-C_6$
Phenoxarsine[2]	$C_4AsO-C_6-C_6$
Phenoxaphosphine[2]	$C_4OP-C_6-C_6$
Phenoxatellurin	$C_4OTe-C_6-C_6$
Phenoxaselenin	$C_4OSe-C_6-C_6$
Phenoxathiin	$C_4OS-C_6-C_6$
Pyrrole[2]	C_4N
Imidazole[2]	C_3N_2
Pyrazole[2]	C_3N_2
Isothiazole	C_3NS
Isoxazole	C_3NO
Pyridine	C_5N
Pyrazine	C_4N_2
Pyrimidine	C_4N_2
Pyridazine	C_4N_2
Pyrrolizine[2]	C_4N-C_4N
Indolizine	C_4N-C_5N
Isoindole[2]	C_4N-C_6
Indole[2]	C_4N-C_6
Indazole[2]	$C_3N_2-C_6$
Purine[2]	$C_3N_2-C_4N_2$
Isoquinoline	C_5N-C_6
Quinoline	C_5N-C_6
Quinolizine[3]	C_5N-C_5N
Phthalazine	$C_4N_2-C_6$
Naphthyridine[4]	C_5N-C_5N
Quinoxaline	$C_4N_2-C_6$
Quinazoline	$C_4N_2-C_6$
Cinnoline	$C_4N_2-C_6$
Pteridine	$C_4N_2-C_4N_2$
Carbazole[2]	$C_4N-C_6-C_6$
Phenanthridine	$C_5N-C_6-C_6$
Acridine	$C_5N-C_6-C_6$
Perimidine[2]	$C_4N_2-C_6-C_6$
Phenanthroline[5]	$C_5N-C_5N-C_6$
Phenazine	$C_4N_2-C_6-C_6$
Anthyridine	$C_5N-C_5N-C_5N$
Phenarsazine	$C_4AsN-C_6-C_6$
Phenophosphazine	$C_4NP-C_6-C_6$
Phenotellurazine[2]	$C_4NTe-C_6-C_6$
Phenoselenazine[2]	$C_4NSe-C_6-C_6$
Phenothiazine[2]	$C_4NS-C_6-C_6$
Phenoxazine[2]	$C_4NO-C_6-C_6$
Thebenidine	$C_5N-C_6-C_6-C_6$
Quindoline[2]	$C_4N-C_5N-C_6-C_6$
Quinindoline[2]	$C_4N-C_5N-C_6-C_6$
Phthaloperine[2]	$C_5-C_4N_2-C_6-C_6-C_6$
Acrindoline[2]	$C_4N-C_5N-C_6-C_6-C_6$
Triphenodithiazine	$C_4NS-C_4NS-C_6-C_6-C_6$
Triphenodioxazine	$C_4NO-C_4NO-C_6-C_6-C_6$
Phenanthrazine	$C_4N_2-C_6-C_6-C_6-C_6-C_6$
Anthrazine	$C_4N_2-C_6-C_6-C_6-C_6-C_6$

[1]The order of precedence is based first on the presence or absence of nitrogen, then upon the nature of the (other) hetero atoms (see Table I, ¶ 128). For fused heterocycles, this order (for base-component selection) is distinct from that used to determine seniority of a total ring system in an index name (¶ 138).

[2]Citation of indicated hydrogen (¶ 135), e.g., 1*H*-**Pyrrole,** 2*H*-**Pyrrole,** is necessary when these component names are used alone.

[3]Because of established usage, **Quinolizine** is favored over **Isoquinoline** and **Quinoline** as a base component.

[4]**Naphthyridine** requires the locants 1,5-, 1,6-, 1,7-, 1,8-, 2,6- or 2,7- to define the position of the nitrogen atoms.

[5]**Phenanthroline** requires the locants 1,7-, 1,8-, 1,9-, 1,10-, 2,7-, 2,8- or 2,9- to define the position of the nitrogen atoms.

150. Selection of a base component is facilitated by use of Tables IV, ¶ 148 and V, ¶ 149; the appropriate component listed *latest* in these Tables is used. The system must contain at least two rings of five or more atoms, although such rings need not be directly fused to one another; i.e., they may be joined by a smaller ring, as in **Cyclobutadicyclopentene.** The criteria for base components differ markedly (in the case of heterocyclic systems) from those described (¶ 138) for a preferred ring system in a compound containing more than one. A base component of lower preference is used if the fusion procedure is not possible on the preferred component; as a last resource, an organic replacement ("a") name based on the fused hydrocarbon is employed.

The preferred base component should:

(a) be a heterocycle;
(b) be a nitrogenous heterocycle;
(c) be a nonnitrogenous heterocycle containing a hetero atom of highest precedence (see Table I, ¶ 128);
(d) contain the greatest number of rings;
(e) contain the largest individual ring; **Benzindene** (not Cyclopenta= naphthalene) is an exception based on established usage;
(f) contain the greatest total number of hetero atoms;
(g) contain the greatest variety of hetero atoms, e.g., one nitrogen and one oxygen rather than two nitrogens;
(h) contain the greatest number of hetero atoms of highest precedence;
(i) possess the most linear structure;
(j) have the lowest locants for hetero atoms (before fusion).

151. Index names for fused systems, other than fundamental systems which possess their own names (Tables IV, ¶ 148 and V, ¶ 149), are formulated from the names of the components. Cycloalkanes may be adopted as base components by invariant use of the "-ene" suffix. This denotes a maximum number of noncumulative double bonds; e.g., **Cyclooctene** as part of a fused system is not meant to imply the presence of a *single* double bond; instead, *saturated* carbon atoms are indicated by "hydro" prefixes. *Fusion locants* for the base component comprise lower case italic letters assigned sequentially to all sides, beginning with the side "1,2" as denoted by the usual peripheral locants. (See the *Ring Systems Handbook* for a complete set of ring system diagrams, including base components, complete with such locants.) If more than twenty-six letters are required, subsequent alphabets of the form a_1, b_1, c_1, . . . etc., are adopted. Locants for the fusion prefixes (derived from the less preferred fundamental ring systems) comprise the normal peripheral numerical locants. When a choice is possible, lowest alphabetic and numerical locants are cited. When one or both types of locants are unnecessary they are usually omitted. Numerical and letter locants are separated by a hyphen, and the locant set is bracketed.

Example:

Benzene

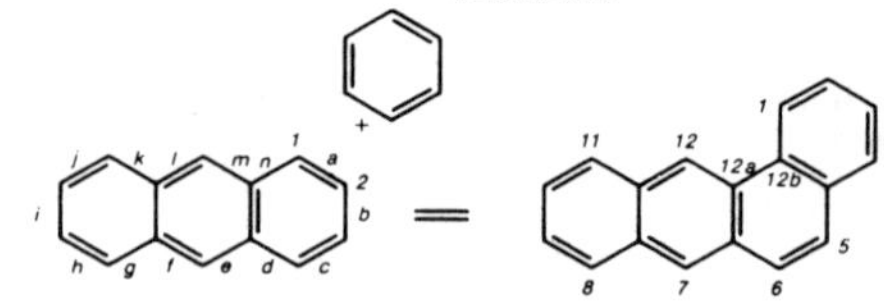

Anthracene (base component) **Benz[*a*]anthracene**

In this example, a locant defining the fusion site on **Benzene** is unnecessary. The "1,2" side of **Anthracene** is lettered "a" and the lettering proceeds around every side back to the 1-position. The fused system is then oriented (¶ 152) and renumbered.

Angular positions of base components involved in fusion are not cited.
Example:

Naphthalene

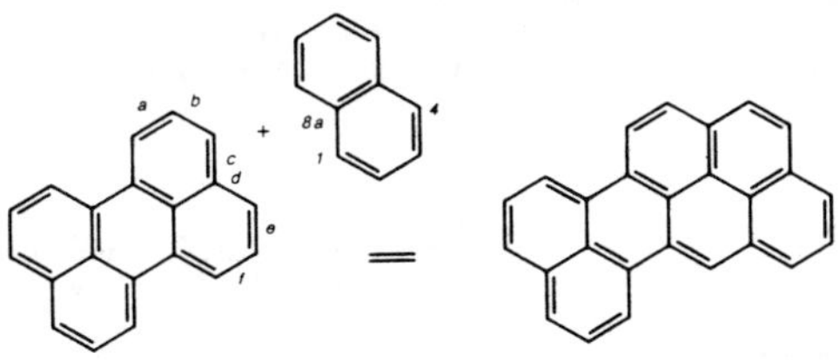

Perylene **Naphtho[8,1,2-*bcd*]perylene** (not "[8,8a,1,2-*bcd*]")

When a hetero atom is shared by two or more rings, it is expressed in all the components. When the order of lettering of the base component proceeds in the direction opposite to numbering in the fusion-prefix component, nu= merical locants for the latter are reversed.

Example:

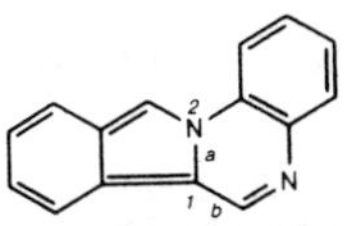

Isoindolo[2,1-*a*]quinoxaline (not "[2,3-*a*]"; not "[1,2-*c*]")

Fusion prefixes are placed in alphabetical order and the earliest cited prefix is given preference for lowest letter locant. When two or more fusion prefixes are identical, as in "Dibenzo-" systems, the letter locants are separated by commas, e.g., "[*a,j*]."

Example:

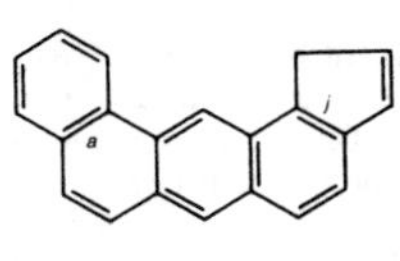

Benzo[*a*]cyclopent[*j*]anthracene

A form of multiplicative name is employed for fused systems different from that described for general substitutive nomenclature. Multiplication proceeds in steps, with "di," "tri" repeated as necessary (not "bis," "tris," etc., except to avoid ambiguity). Serially primed letters are used for fusion sites on the second, third, etc., base components and the locant sets are separated by colons. When a base component is fused to a central component and to another component, lowest letters (when a choice must be made) relate to the central fusion site.

Examples:

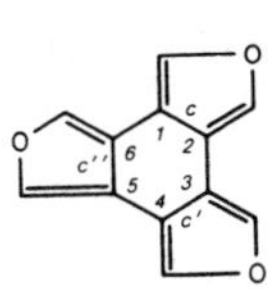

Benzo[1,2-*c*:3,4-*c*':5,6-*c*'']trifuran

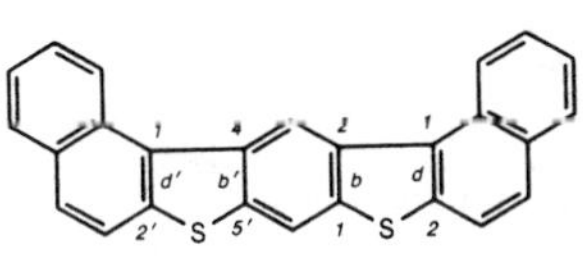

Dinaphtho[1,2-*d*:1',2'-*d*']benzo=[1,2-*b*:5,4-*b*']dithiophene (not Dinaphtho[2,1-*b*:2',1'-*b*']benzo=[1,2-*d*:5,4-*d*']dithiophene)

Ring systems fused to base components are designated primary components; a ring system (other than a base component) fused only to a primary compo= nent is a secondary component, and primed numerical locants are used to denote its fusion sites. Primed and unprimed locant sets are separated by colons. Lowest locants are used for the site closest to (or fused directly to) the base component. Doubly primed locants are needed (*a*) when the sec= ondary component is centrally located with identical primary and base com= ponents on both sides, and (*b*) when tertiary components are present.

Examples:

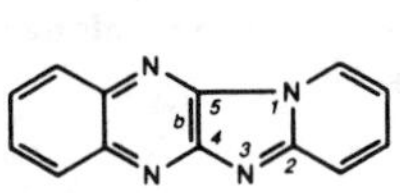

Pyrido[1',2':1,2]imidazo[4,5-*b*]=quinoxaline

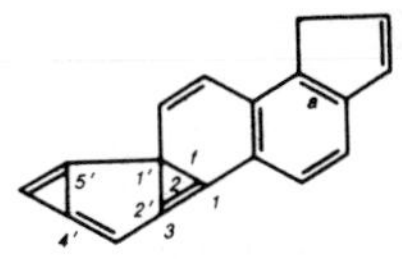

NOT

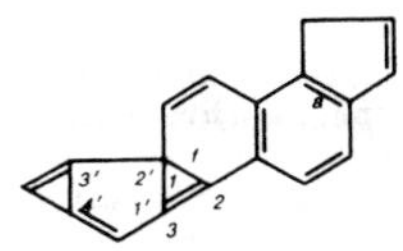

Cyclopenta[*a*]cyclopropa[4',5']=cyclopenta[1',2':2,3]cyclo=propa[1,2-*f*]naphthalene (not Cyclopenta[*a*]cyclopropa[3',4']=cyclopenta[2',1':1,3]cyclopropa=[2,1-*f*]naphthalene.) (The lo= cants 1,2 in the preferred name are lower than 2,1 in the incor= rect name. It should be noted that the use of 1',2' requires that the cyclopentane ring be num= bered the long way around.)

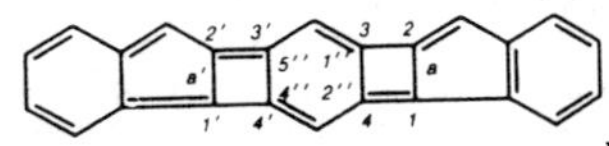

Benzo[1'',2'':3,4;5'',4'':3',4']di=cyclobuta[1,2-*a*:1',2'-*a*]di=indene (Note that a semicolon separates locant sets which al= ready contain colons.)

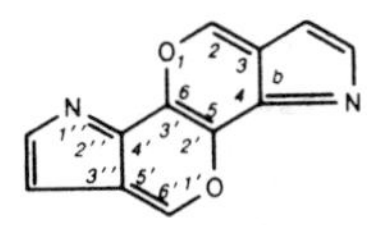

Pyrrolo[2'',3'':4',5']pyrano=[2',3':5,6]pyrano[4,3-*b*]pyrrole

152. Orientation and numbering of fused systems.

(*a*) *Hydrocarbons.* The component rings are normally drawn as regular polygons. The cyclopropane ring may point left or right, and cyclopentane and cycloheptane rings may point up or down. The total system is oriented so that (*a*) a maximum number of rings are in a horizontal row, and (*b*) a maximum number are above and to the right. If further choice is necessary, then (*c*) a minimum number of rings should be in the lower left quadrant. Numbering begins at an atom not engaged in fusion in the most counterclockwise position of the uppermost ring furthest to the right. Angular positions are not counted; their locants, when needed, are derived from those of the preceding nonangular positions by addition of the lower-case Roman letters, "a," "b," etc. Interior atoms are numbered last by addition of letters to the highest available nu= merical locant in a continuous pathway, a clockwise route being followed whenever a choice presents itself; any remaining interior atoms are then numbered similarly from the next highest available numerical locant.

Examples:

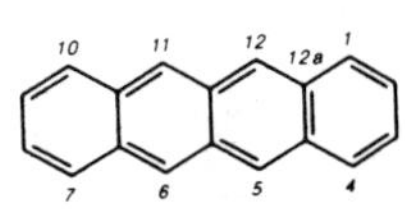

Naphthacene

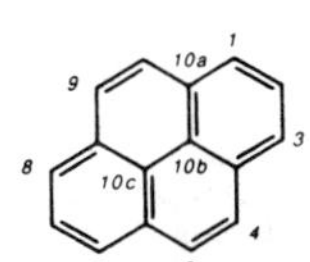

NOT

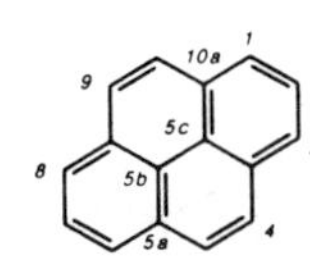

Pyrene

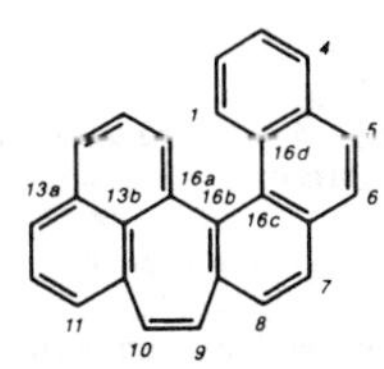

NOT

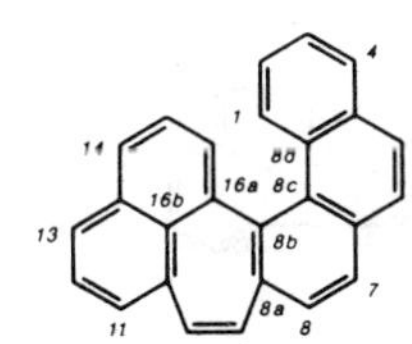

Naphtho[1',8':3,4,5]cyclohepta[1,2-*c*]phenanthrene

When a further choice is needed for orientation and numbering, carbon atoms at angular positions are assigned lowest numbers.

Examples:

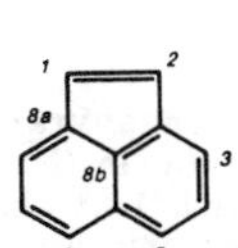

NOT

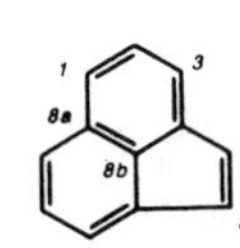

Acenaphthylene (Note: 2a,5a,8a,8b is lower than 3a,5a,8a,8b)

Indicated hydrogen is assigned the lowest locant.
Example:

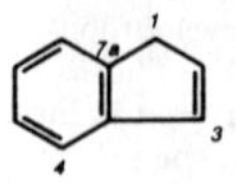

1*H*-**Indene** (not 3*H*-Indene)

The following fused carbocyclic compounds have special numbering systems: **Anthracene, Phenanthrene, Cyclopenta[*a*]phenanthrene** (steroid numbering) and the Cyclopropacyclopenta[*a*]phenanthrenes.

(*b*) *Heterocycles.* The ring systems are oriented as for hydrocarbons. When a choice is necessary, lowest locants are assigned to (*a*) all hetero atoms; (*b*) most preferred hetero atom (Table I, ¶ 128); (*c*) carbon atoms common to two or more rings; (*d*) positions bearing indicated hydrogen; (*e*) an angular rather than a nonangular atom of the same hetero element. The ring is then numbered as for hydrocarbons, except that hetero atoms common to two or more rings are counted. Interior atoms are numbered last, following the shortest path from the highest previous number.
Examples:

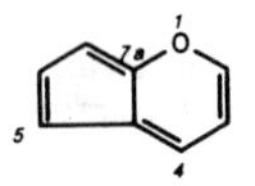

Cyclopenta[*b*]pyran

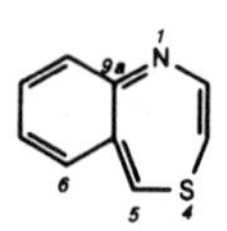

4,1-Benzothiazepine

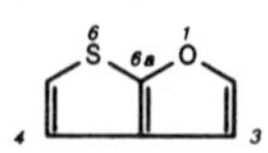

Thieno[2,3-*b*]furan

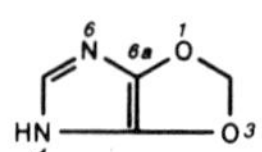

4*H*-**1,3-Dioxolo[4,5-*d*]imidazole**

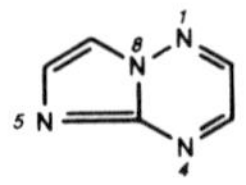

NOT

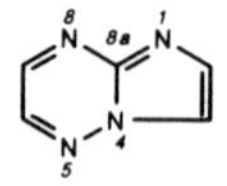

NOT

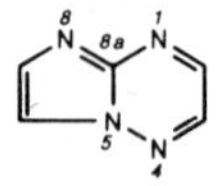

Imidazo[1,2-*b*][1,2,4]triazine

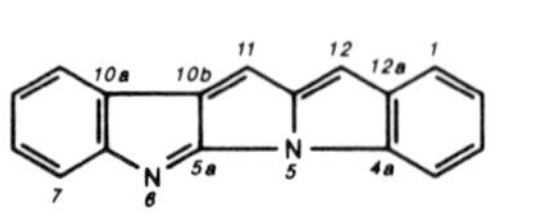

NOT

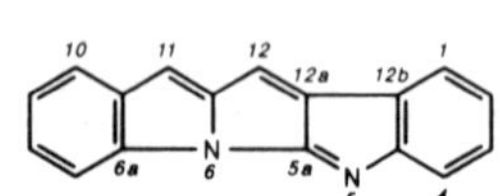

Pyrrolo[1,2-*a*:5,4-*b*']diindole

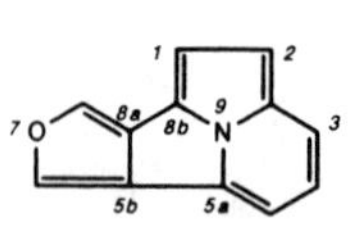

Furo[3,4-*a*]pyrrolo[2,1,5-*cd*]indolizine

The following heterocyclic systems have special numbering systems: **Acridine, Carbazole, Purine, Xanthene,** and Epoxy- and Epithiocyclopenta[*a*]phenanthrenes (steroid numbering).

Indicated hydrogen for fused carbocyclic and heterocyclic systems is normally cited, if there is a choice, at the lowest nonangular position, unless a saturated angular atom is required to accommodate a principal function or free valency (see ¶ 135). Indicated hydrogen of component systems is ignored in constructing a fused ring name, and is reassigned if it is still needed in the final system.
Example:

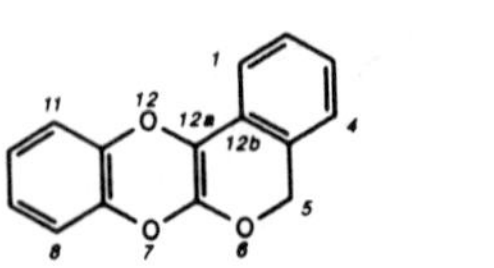

5*H*-**[2]Benzopyrano[3,4-*b*][1,4]benzodioxin** (Note: The locants "2" and "1,4," which relate to the components, are bracketed to indicate that they do not conform to the peripheral numbering of the total system)

153. Replacement ("a") nomenclature for fused systems is employed when fusion names fail to express all interfaces (fusion sites) between component systems. This occurs when two or more components that are expressed as prefixes are fused to one another as well as to the base component. "A" names are also used for silicon replacement in a carbocycle that has a trivial name. Indicated hydrogen of the parent carbocycle is ignored, but is cited, if needed for the "a" name, ahead of the replacement prefixes.
Examples:

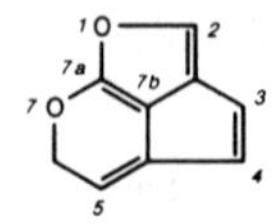

6*H*-**1,7-Dioxacyclopent[*cd*]indene** (not 1,7-Dioxa-6*H*-cyclopent[*cd*]indene)

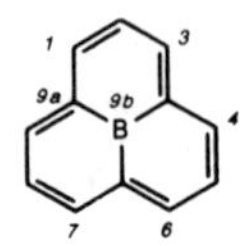

9b-Boraphenalene

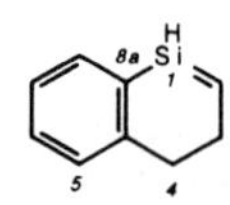

1-Silanaphthalene, 3,4-dihydro-

Saturation of double bonds in fused systems is denoted by hydro prefixes which are given lowest locants; e.g., **Naphthalene, 1,2,3,4**(*not* 5,6,7,8)-**tetrahydro-**. Triple bonds are indicated by "didehydro."

154. Bridged fused systems are fused ring systems that possess atomic bridges or valence bonds which connect two or more parts of the system without creating or extending a fused system. They are named by adding bridge prefixes (in alphabetical order if different types are present) to the fused system names.

Simple bivalent bridges include methano ($-CH_2-$), ethano ($-CH_2-CH_2-$), etheno ($-CH=CH-$), propano ($-CH_2CH_2CH_2-$), 2-buteno ($-CH_2CH=CHCH_2-$), and benzeno ($-C_6H_4-$). Trivalent bridges, e.g., metheno ($-CH=$), 1-propanyl-3-ylidene ($-CH_2CH_2CH=$), and tetravalent bridges are also employed; locants for positions of attachment on the fused system are cited in the same order as free-valency locants of the radicals. Bridges from monocyclic hydrocarbons other than benzene are named as for the fusion prefixes, except that "*endo*-" is used with them to avoid ambiguity, e.g., "*endo*-cyclopenta." Simple hetero bridges include epoxy (-O-), epithio (-S-), imino (-NH-), epidioxy (-O-O-), and -silano- ($-SiH_2-$). Heterocyclic rings may also be used as bridges.
Example:

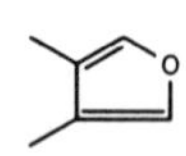

3,4-furano (cf. "furo" for the fusion prefix)

When locants are used for the bridge itself, e.g., 2-buteno, 3,4-furano-, they are placed in brackets within the bridged system name.

Compound bridges are named by combination of simple bridges beginning at the terminal position which gives the preferred (*a*) cyclic bridge (¶ 138), (*b*) hetero atom (Table I, ¶ 128), (*c*) chain, (*d*) alphabetic order.
Examples:

$-NH(CH_2)_2-$ (iminoethano)

-OSO- (epoxythioxy)

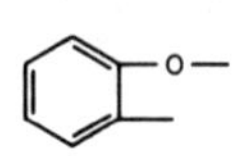

(epoxy[1,2]benzeno)

Indicated hydrogen of a fused system is cited, if possible, to accommodate a bridge. When this is unnecessary or impossible, the lowest-numbered nonangular indicated hydrogen is cited for the fused system, and additional indicated hydrogen, when needed, is cited ahead of the bridge locants.
Examples:

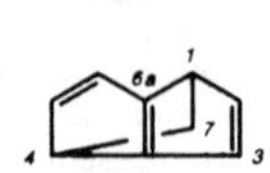

1,4-Methanopentalene, 1,4-dihydro-

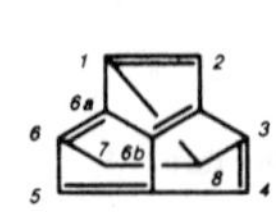

6,1,3-Ethanylylidenecyclopenta[*cd*]pentalene

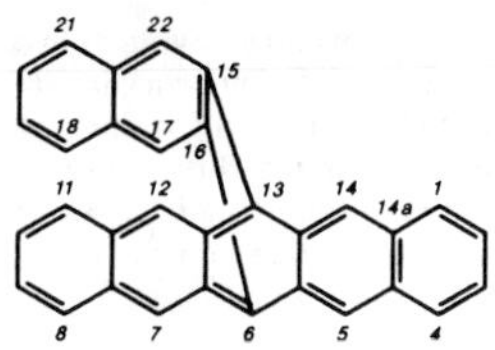

6,13[2',3']-Naphthalenopentacene

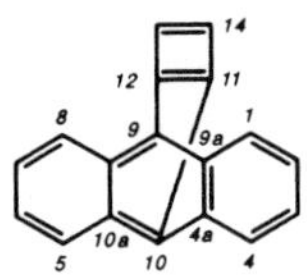

9,10[1',2']-*endo*-Cyclobutanthracene

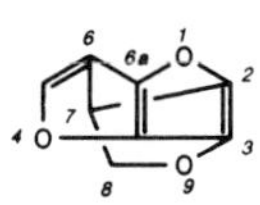

3,2,6-(Epoxyethanylylidene)furo=[3,2-*b*]furan

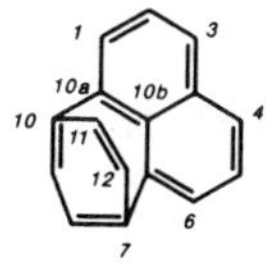

1*H*-3a,7-Ethanoazulene

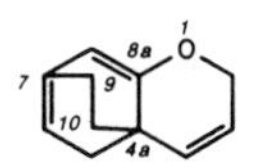

5*H*-4a,7-Ethano-2*H*-1-benzopyran (4a*H*-1-Benzopyran cannot exist)

Numbering of bridged fused systems is based on the regular numbering of the parent fused system. Lowest locants are assigned to bridgehead positions and the bridge atoms are numbered from the end nearest the highest numbered position of the parent fused system. In cyclic bridges, e.g., benzeno, *endo*-cyclobuta, the shorter bridge is first numbered, and then the rest of the ring in the same direction. If possible, hetero atoms in bridges are numbered low.

Examples:

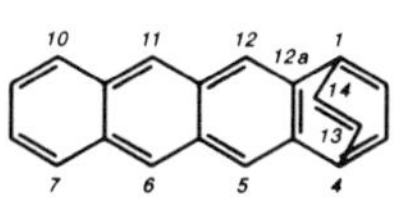

1,4-Ethenonaphthacene

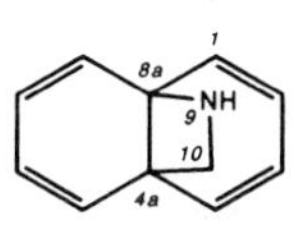

8a,4a-(Iminomethano)naphthalene (not 4a,8a-(Iminomethano)=naphthalene)

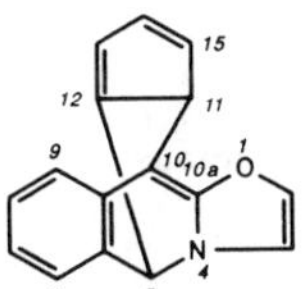

11*H*-5,10[1',2']-*endo*-Cyclopent-5*H*-oxazolo[3,2-*b*]isoquinoline

The exceptional numbering employed for bridged cyclopenta[*a*]=phenanthrenes is shown in the following example. Steroid numbering is used for positions 1 through 17. When the methyl groups normally numbered 18 and 19 are transformed into methano bridges, their locants are retained. Other bridges are numbered 20 and upward.

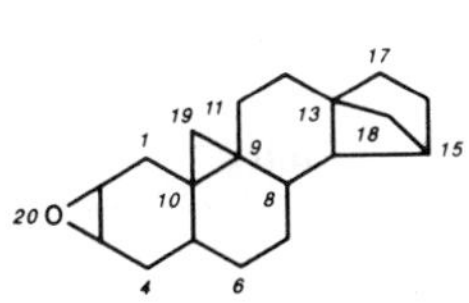

Criteria for the naming of bridged fused systems are applied successively as follows:

(*a*) The unbridged system contains the maximum number of (i) rings, (ii) skeletal atoms.

(*b*) The bridges are as simple as possible; e.g., two simple bridges are pre=ferred to one compound bridge, and saturated bridges are preferred to un=saturated ones.

(*c*) The unbridged system has the highest precedence according to ¶ 138(*c*).

Examples:

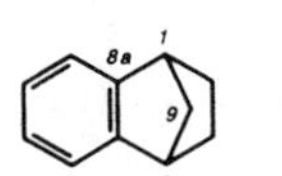

1,4-Methanonaphthalene, 1,2,3,4-tetrahydro- (not 1,3-Ethano-1*H*-indene, 2,3-dihydro-)

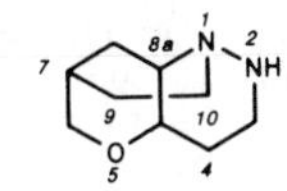

NOT

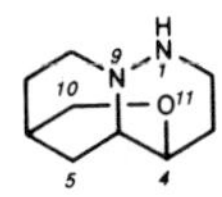

1,7-Ethano-1*H*-pyrano[3,2-*c*]=pyridazine, octahydro- (not 4,6=(Epoxymethano)-1*H*-pyrido[1,2-*b*]pyridazine, octahydro-)(The fused ring system **Pyranopyridazine** is preferred to **Pyridopyridazine**; see ¶ 138.)

Fused carbocyclic and nonnitrogenous heterocyclic systems with simple imino bridges are named by use of the termination "-imine" with appropriate locants. Such a name requires addition of a regular functional suffix when appropriate. (Nitrogen heterocycles with imine bridges are named as bridged fused heterocycles in the usual manner.)

Example:

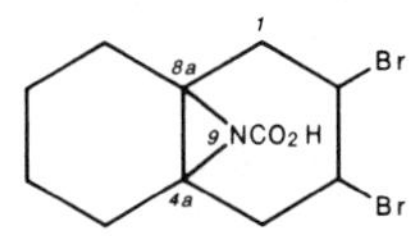

Naphthalen-4a,8a-imine-9-carboxylic acid, 2,3-dibromo=octahydro-

155. Von Baeyer nomenclature. This was first developed to name alicyclic hydrocarbons containing two rings. It has been extended to all bridged systems which cannot be treated as fused or bridged fused systems. Von Baeyer names for hydrocarbons are formed by prefixing to the name of the acyclic hydrocarbon with the same number of carbon atoms "Bicyclo-," "Tricyclo-," etc., terms, followed by a set of numerals, separated by periods and bracketed, which describes in descending sequence the number of atoms in each bridge. The system is numbered from one bridgehead *via* the other bridgehead(s) and back, always choosing the longest route. The system is numbered along the same route, ending with the smallest bridge, numbered from the bridgehead with the highest locant.

Example:

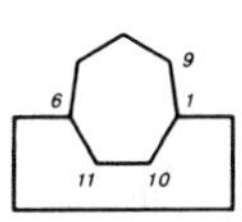

Bicyclo[4.3.2]undecane

For tricyclo- and higher hydrocarbon systems, superscripts are employed to indicate the positions of secondary bridges.

Example:

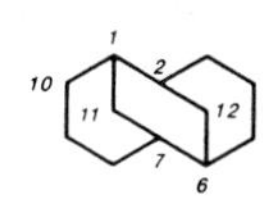

Tricyclo[5.3.1.$1^{2,6}$]dodecane

When more than one Von Baeyer name is possible for a hydrocarbon, the choice is determined by the following principles, applied successively until a decision is reached.

(*a*) The main ring contains the maximum number of atoms, two of which must serve as bridgeheads for the main bridge.

(*b*) The main bridge is as large as possible.

(*c*) The main ring is divided as symmetrically as possible by the main bridge.

(*d*) Lowest superscripts (regardless of order of citation) are cited.

Examples:

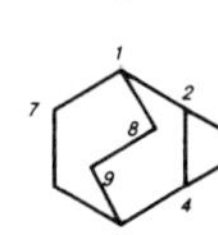

Tricyclo[3.2.2.$0^{2,4}$]nonane (not Tri=cyclo[2.2.2.$1^{2,3}$]nonane)

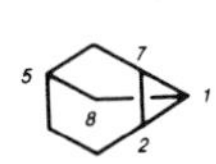

Tricyclo[3.2.1.$0^{2,7}$]octane (not Tri=cyclo[2.2.2.$0^{2,6}$]octane)

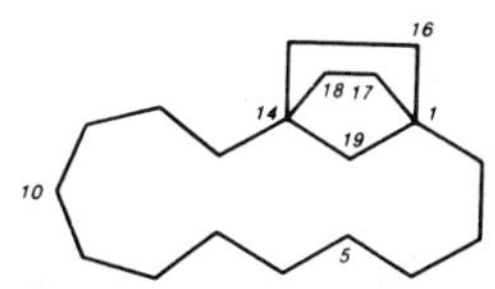

Tricyclo[12.2.2.$1^{1,14}$]nonadecane

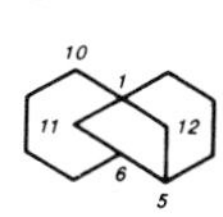

Tricyclo[4.4.1.$1^{1,5}$]dodecane (not Tricyclo[5.3.1.$1^{1,6}$]dodecane)

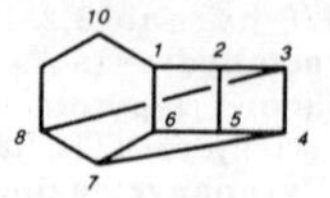

Pentacyclo[4.4.0.$0^{2,5}$.$0^{3,8}$.$0^{4,7}$]decane (not Pentacyclo[4.4.0.$0^{4,9}$.$0^{5,8}$.$0^{7,10}$=]decane; not Pentacyclo=[4.4.0.$0^{2,5}$.$0^{3,10}$.$0^{4,9}$]decane; not Pentacyclo[4.4.0.$0^{2,9}$.$0^{3,8}$.$0^{7,10}$]decane)

Unsaturation is denoted by "-ene" and "-yne" suffixes. A second locant in parentheses is cited for a double bond at a bridgehead when it does not proceed to the next atom in the numbered path. Multiple-bond locants are determined by the following criteria applied successively:

(*a*) The numbering proceeds in a clockwise sequence.

(*b*) The cases in which both locants for a double bond are cited are mini=mized.

(*c*) Lowest locants are employed.

Examples:

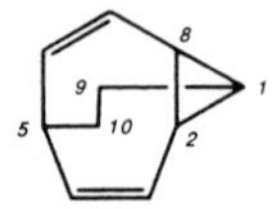

Tricyclo[3.3.2.$0^{2,8}$]deca-3,6-diene

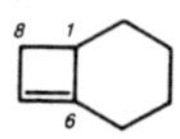

Bicyclo[4.2.0]oct-6-ene (not Bicy=clo[4.2.0]oct-1(8)-ene)

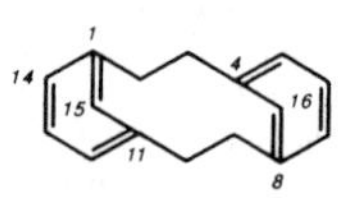

Tricyclo[9.3.1.$1^{4,8}$]hexadeca-1=(15),4,6,8(16),11,13-hexaene (not Tricyclo[9.3.1.$1^{4,8}$]hexadeca-4,6,8=(16),11(15),12,14-hexaene)

Von Baeyer names for *heterocyclic systems* are formed from the hydro=carbon names by use of replacement (oxa, thia, aza, etc.) prefixes and lowest locants for hetero atoms in the order of Table I, ¶ 128. Unsaturation is denoted as for hydrocarbons.

Examples:

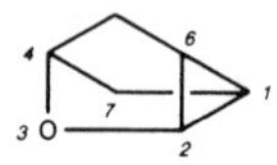

3-Oxatricyclo[2.2.1.$0^{2,6}$]heptane (not 5-Oxatricyclo[2.2.1.$0^{2,6}$]heptane; not 2-Oxatricyclo[2.2.1.$0^{3,5}$]heptane)= (low numbering for bridge takes precedence over hetero atom)

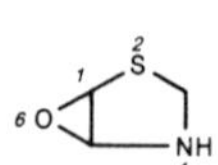

6-Oxa-2-thia-4-azabicyclo[3.1.0]=hexane (not 6-Oxa-4-thia-2-azabicyclo[3.1.0]hexane)

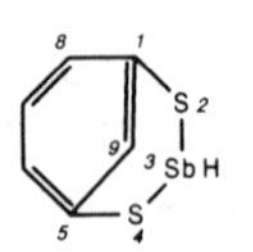

2,4-Dithia-3-stibabicyclo[3.3.1]=nona-1(9),5,7-triene (not 2,4-Dithia-3-stibabicyclo[3.3.1]nona-1(8),5(9),6-triene)

Saturated bridged systems containing only silicon atoms, or silicon atoms alternating with nitrogen or one of the chalcogens, are given Bicyclosilazane, Tricyclosiloxane, etc., names. Regular Von Baeyer numbering is employed.

Examples:

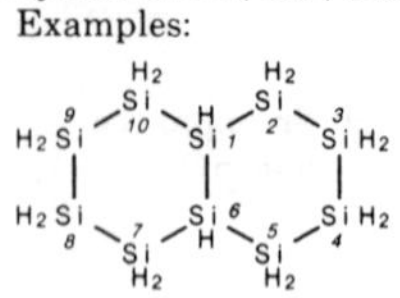

Bicyclo[4.4.0]decasilane

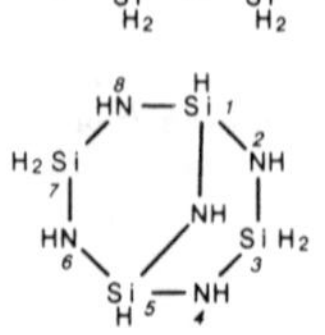

Bicyclo[3.3.1]tetrasilazane

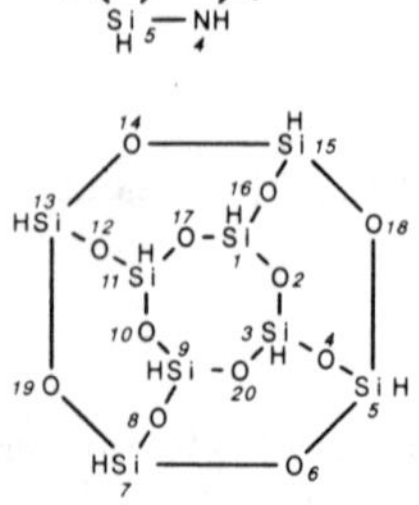

Pentacyclo[9.5.1.$1^{3,9}$.$1^{5,15}$.$1^{7,13}$]octa=siloxane

156. Spiro systems contain pairs of rings or ring systems having only one atom (a "spiro atom") in common. In the simplest monospiro systems, in which two alicyclic rings share an atom, the name is formed by prefixing to the acyclic hydrocarbon name the term "Spiro" and numerals (separated by periods) in *ascending* sequence to define the number of atoms in each ring linked to the spiro atom. The numbering begins at the atom *next to* the spiro atom in the smaller ring and proceeds around that ring, through the spiro atom and around the larger ring.

Example:

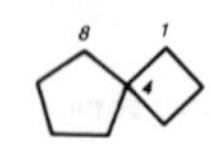

Spiro[3.4]octane

This system is extended to dispiro and higher systems. Numbering begins next to a terminal spiro atom and proceeds in such a way as to give the spiro atoms lowest locants.

Example:

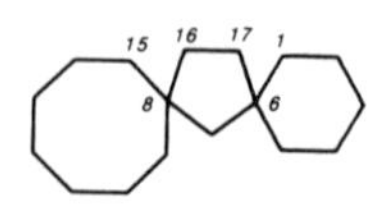

Dispiro[5.1.7.2]heptadecane (note that the numbering path corre=sponds to the bracketed sequence)

Heterocyclic analogs are named by "a" nomenclature. The hetero atoms are given locants as low as are compatible with the ring numbering.

Example:

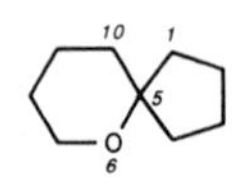

6-Oxaspiro[4.5]decane

Unsaturation is expressed by "-ene" and "-yne" suffixes.

Example:

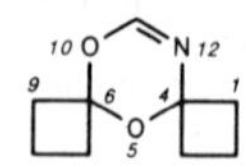

5,10-Dioxa-12-azadispiro[3.1.3.3]=dodec-11-ene

Saturated spiro systems containing only silicon atoms or silicon atoms alternating with nitrogen or one of the chalcogens are given Spirosilazane, Spirosiloxane, etc., names.

Example:

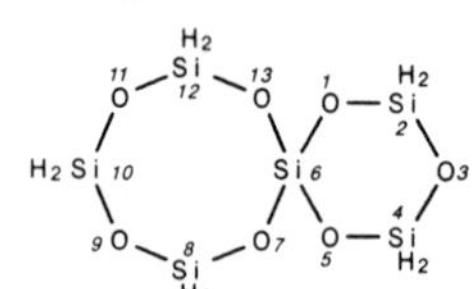

Spiro[5.7]hexasiloxane

Monospiro systems containing at least one fused or bridged component are named by placing the component names in brackets in alphabetical order and prefacing them with "Spiro." The position of the spiro atom is denoted by two locants, separated by a comma, related to the two components. Primes are used for the component cited second. Indicated hydrogen (¶ 135) is assigned, where possible, to accommodate the spiro unions. Locants related to the components but not to the total spiro system are bracketed to avoid ambiguity.

Example:

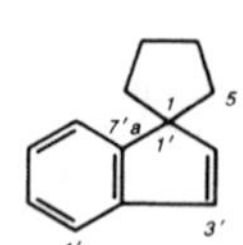

Spiro[cyclopentane-1,1'-[1*H*]=indene] (not Spiro[cyclopentane-1,1'-1'*H*-indene])

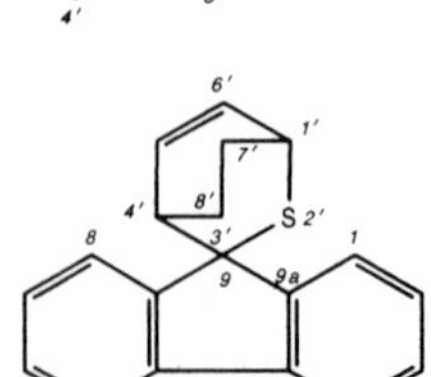

Spiro[9*H*-fluorene-9,3'-[2]thia=bicyclo[2.2.2]oct[5]ene]

Added hydrogen (¶ 136) is cited in parentheses in the usual way, but with a primed locant if it does not relate to the component cited first. It is assigned the lowest available locant unless a different one can be used to accommodate a principal group.

Example:

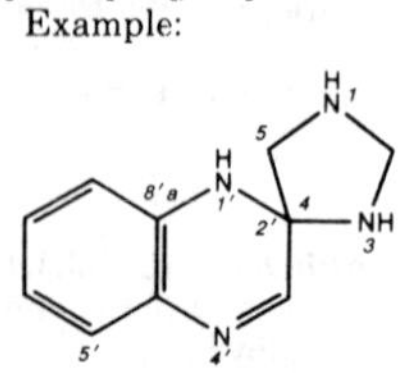

Spiro[imidazolidine-4,2'(1'*H*)-quinoxaline]

Monospiro systems containing identical fused components are given "Spirobi–" names. The component name is bracketed if it is preceded by locants or is itself made up of fusion components. Added hydrogen is cited in parentheses following the spiro locants.
Examples:

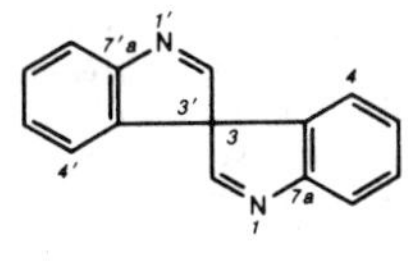

3,3'-Spirobi[3*H*-indole]

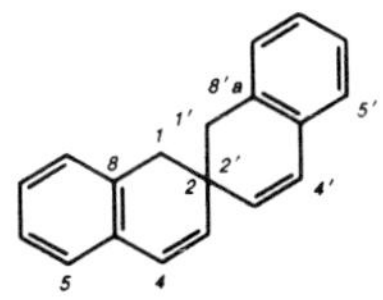

2,2'(1*H*,1'*H*)-Spirobinaphthalene

Di–, tri–, etc., spiro systems containing at least one fused or bridged com= ponent are named by extension of these policies. If terminal components are identical, citation is determined by earliest index position of the complete name. Serially primed locants are used for successive components.

"Branched" polyspiro systems in which a single component is surrounded by three or more identical components are named by citing the central com= ponent (which is assigned plain locants) first and multiplying the identical (terminal) components. When two terminal components of a "branched" spiro system are identical, and one different, they are cited in alphabetical order (as usual) and the term "bis" is applied as appropriate.
Examples:

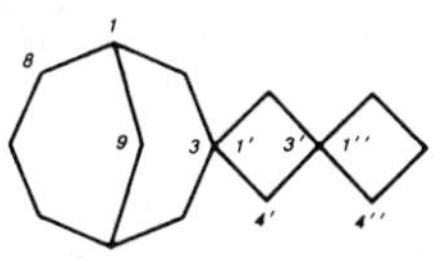

Dispiro[bicyclo[3.3.1]nonane–3,1'-cyclobutane-3',1''-cyclo=butane]

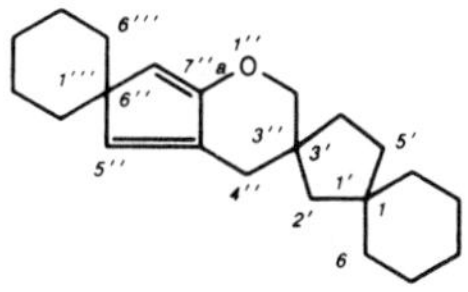

Trispiro[cyclohexane-1,1'-cyclo=pentane-3',3''(2''*H*)-cyclopenta=[*b*]pyran-6''(4''*H*),1'''-cyclo=hexane]

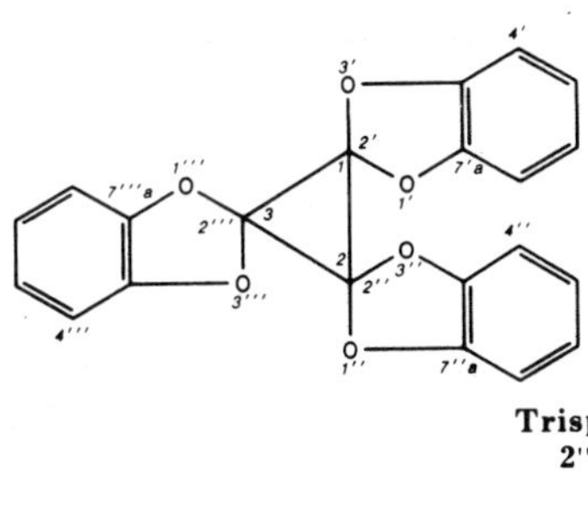

Trispiro[cyclopropane-1,2':2,2'':3,=2'''-tris[1,3]benzodioxole]

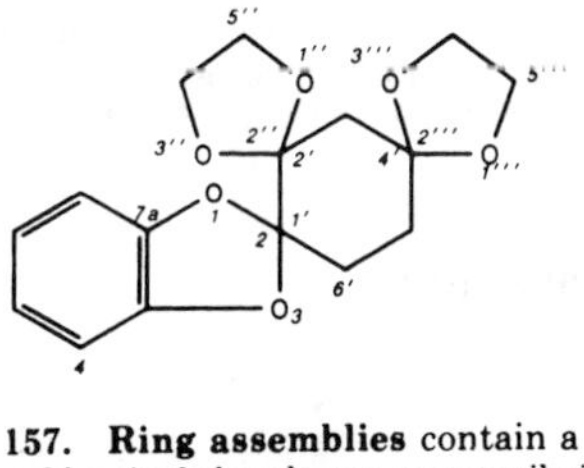

Trispiro[1,3-benzodioxole-2,1'-cyclohexane-2',2'':4',2'''-bis=[1,3]dioxolane]

157. **Ring assemblies** contain a multiplicity of the same cyclic system joined by single bonds, not necessarily in equivalent positions. They are treated as molecular skeletons in substitutive nomenclature and rank just above the component ring. Except for assemblies of benzene, and two-component as= semblies of cycloalkanes, cycloalkenes, and hetero systems with "cyclo" names (see below), they are named by prefixing the component names with the terms Bi–, Ter–, Quater–, Quinque–, Sexi–, Septi–, Octi–, Novi–, Deci–, Undeci–, etc.

Locants are placed ahead of the name to define the points of attachment. These locants are as low as possible, compatible with fixed numbering (ex= pressed or implied) of the components, *including "–ene" and "–yne" suffixes.*
Examples:

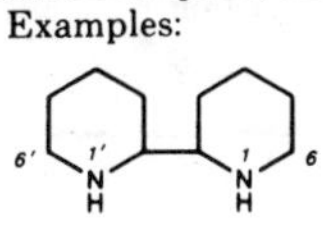

2,2'-Bipiperidine

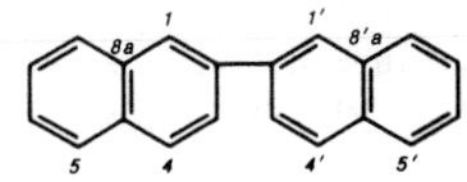

2,2'-Binaphthalene

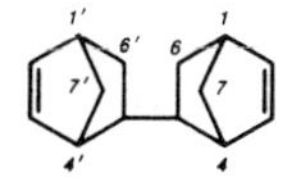

5,5'-Bibicyclo[2.2.1]hept-2-ene (not 2,2'-Bibicyclo[2.2.1]hept-5-ene)

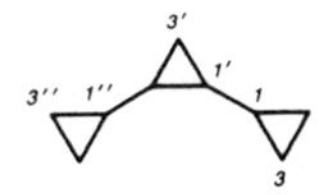

1,1':2',1''-Tercyclopropane

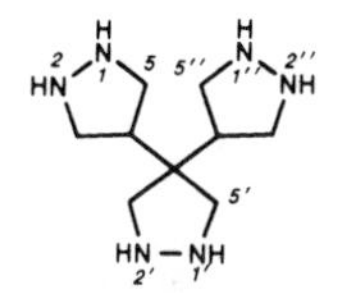

4,4':4',4''-Terpyrazolidine

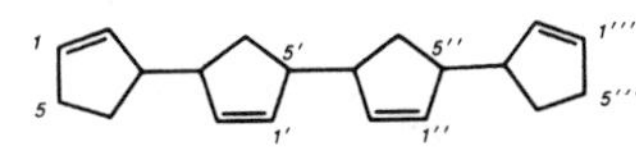

3,3':5',3'':5'',3'''-Quatercyclopentene

Indicated hydrogen (¶ 135) is assigned, where possible, to points of at= tachment. When indicated hydrogen is cited in different positions for different components, a ring-assembly name is not used. Added hydrogen (¶ 135) is cited immediately after the locant to which it relates.
Example:

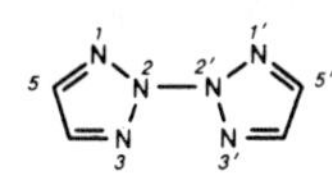

2,2'-Bi-2*H*-1,2,3-triazole

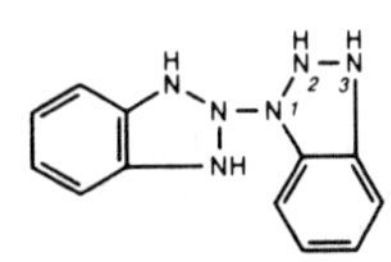

1*H*-Benzotriazole, 1-(1,3-dihydro-2*H*-benzotriazol-2-yl)-2,3-di=hydro– (not 2*H*-Benzotriazole, 2-(1*H*-benzotriazol-1-yl)-2,3-di=hydro–; not 1(3*H*),2'-Bi-2*H*-benzotriazole, 1',3'-dihydro–)

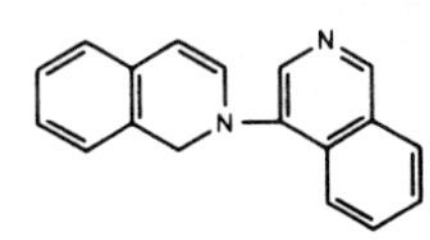

2(1*H*),4'-Biisoquinoline

Linear benzene assemblies (*polyphenyls*) are named by prefacing "phenyl" with the appropriate term (Bi–, Ter–, etc.). Arabic numeral locants are cited in all cases for points of attachment. Two-component assemblies of monocyclic hydrocarbons and of hetero systems with "cyclo" names, e.g., **Cyclopenta=stannane, Cyclotrisiloxane,** are named from the *radicals*, and locants for the points of attachment are cited only when the radical has no locant for the free valency. Two-component ring-assembly names from unsaturated ("–enyl") radicals are formed only when the unsaturation is symmetrical with respect to the points of attachment.
Examples:

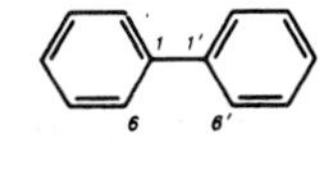

1,1'-Biphenyl

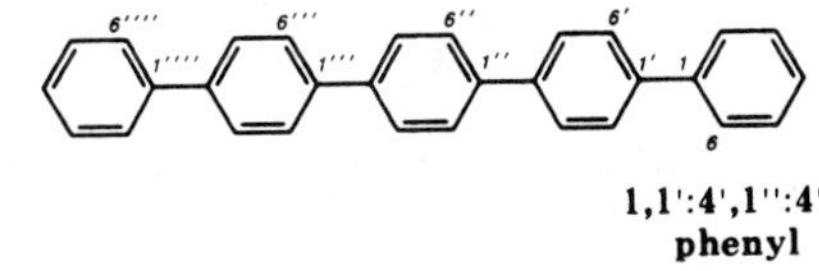

1,1':4',1'':4'',1''':4''',1''''-Quinque=phenyl

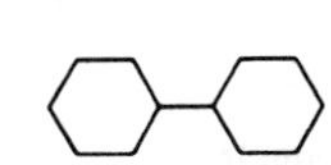

1,1'-Bicyclohexyl (not Bicyclohex-1-yl).

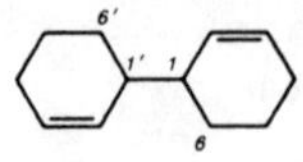

Bi-2-cyclohexen-1-yl (not 1,1'-Bi-2-cyclohexen-1-yl)

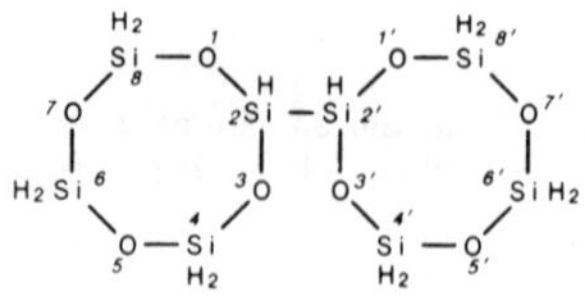

Bicyclotetrasiloxan-2-yl

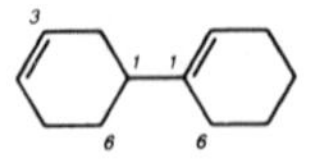

Cyclohexene, 1-(3-cyclohexen-1-yl)- (not Cyclohexene, 4-(1-cyclohexen-1-yl)- (lowest locant for the substituent prefix is preferred (¶ 138))

158. **Neutral hetero atoms with abnormal valencies** in ring systems are expressed (*a*) by additive terms, e.g., "oxide," "sulfide," in the modification; (*b*) by the greek letter "λ" followed by a superscript numeric (*c*) by the prefix "hydro" (in molecular skeletons not treated normally in substitutive nomen= clature); (*d*) by superscript Roman numerals attached to the italicized element symbol, e.g. S^{IV}; (*e*) by a combination of methods *a*, *b*, *c* or *d* above.

Examples:

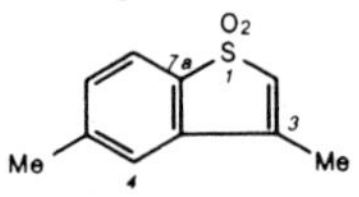

Benzo[*b*]thiophene, 3,5-dimethyl-1,1-dioxide

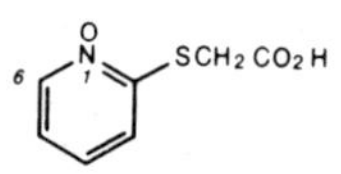

Acetic acid, (2-pyridinylthio)- *N*-**oxide** (not 1-oxide; not 1'-oxide)

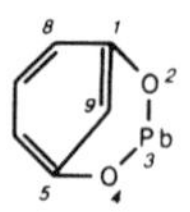

2,4-Dioxa-3λ^2-plumbabicyclo= [3.3.1]nona-1(9),5,7-triene

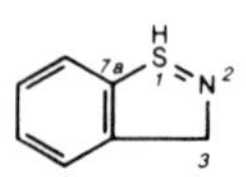

1λ^4-3*H*-1,2-Benzisothiazole

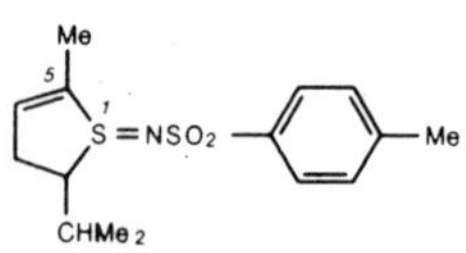

Thiophene, 1,1,2,3-tetrahydro-5-methyl-2-(1-methylethyl)-1-[[(4-methylphenyl)sulfonyl]im= ino]- (because the abnormal va= lency cannot be expressed in a substituent by an approved method, the higher (sulfonamide) function is "overstepped")

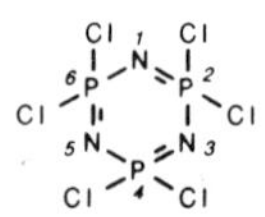

1,3,5,2,4,6-Triazatriphosphorine, 2,2,4,4,6,6-hexachloro-2,2,4,4,6,= 6-hexahydro-

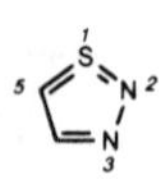

1,2,3-Thia(S^{IV})diazole

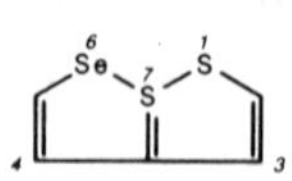

7λ^4-[1,2]Dithiolo[5,1-*e*][1,2]thiase= lenole

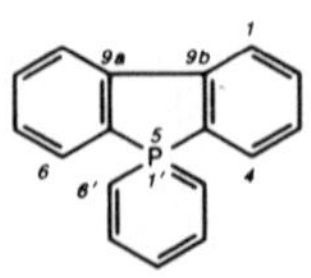

Spiro[5*H*-dibenzophosphole-5,1'λ^5-phosphorin]

In a few cases, abnormal valency is implied rather than expressed, e.g. cyclic iodine systems, in which iodine is usually trivalent.

Example:

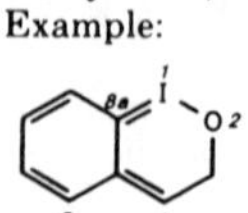

3*H*-1,2-Benziodoxin

159. **Boron molecular skeletons.** Because the number of hydrogen atoms in neutral and anionic boron hydrides often bears no simple relationship to the number of boron atoms, borane names must express the number of both. (The single exception is **Borane** itself, which represents BH_3.) The *Ring Systems Handbook* should be consulted for structural diagrams of the neutral polyboranes of established structure. Diagrams justified by current index entries are displayed in the *Chemical Substance Indexes.* (In these diagrams, the lines do not represent electron-pair bonds but indicate the geometry of the structures.) Neutral boron hydrides, real or hypothetical, are treated as molecular skeletons in substitutive nomenclature, **Borane** and the diboranes as heteroacyclic compounds, the higher hydrides as heterocyclic compounds. **Borane(1)** is BH, **Borane(2)** is BH_2. In higher boranes, the number of boron atoms is expressed by multiplicative prefixes.

Examples:

B_2H_4	**Diborane(4)**
B_2H_6	**Diborane(6)**
B_3H_7	**Triborane(7)**
B_4H_{10}	**Tetraborane(10)**
B_5H_9	**Pentaborane(9)**
B_6H_{10}	**Hexaborane(10)**
$B_{10}H_{14}$	**Decaborane(14)**

Until structures have been determined, author terms such as "*iso*" and "*neo*" are cited in index modifications to differentiate isomers.

Numbering of polyboranes in *CA* indexing is based on the "Rules for the Nomenclature of Boron Compounds" adopted by the American Chemical Society,[1] which may be consulted for further details. For "closed" polyboranes (those whose boron skeletons are polyhedra with triangular faces throughout) the numbering begins with the boron atom at the head of the largest axis of highest order, then proceeds sequentially to the planes which intersect this axis. Boron atoms in each plane are numbered clockwise unless lowest locants for substituents demand anticlockwise numbering. On succeeding planes, num= bering begins at the boron atom immediately "below" the lowest-numbered one on the previous plane, or the one nearest to it in the direction of num= bering.

Example:

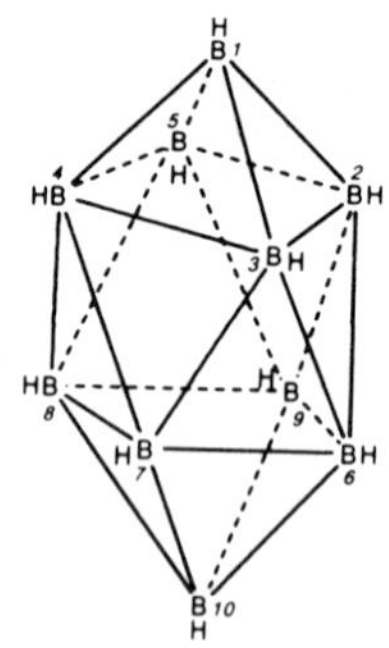

Decaborane(10)

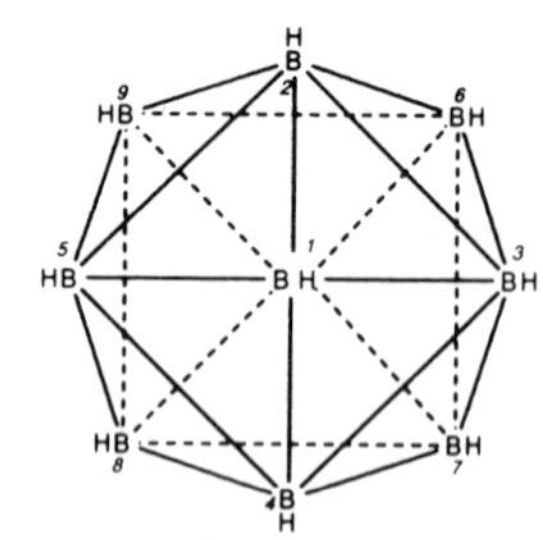

planar projection

(Note that atom "6" (in the lower plane) is nearest to atom "2" in the direction of numbering.)

For "open" polyboranes (those with incomplete polyhedral boron skeletons) the rules are more complex. A planar projection, as viewed from the open side, is numbered so that interior boron atoms have lowest locants, beginning at the "center" or "apex." Each atom set is numbered in the same direction.

Examples:

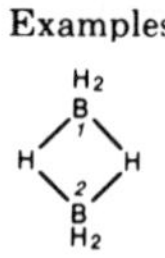

Diborane(6)

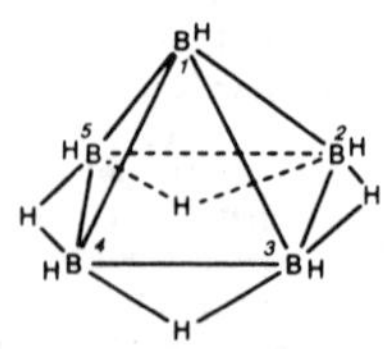

Pentaborane(9)

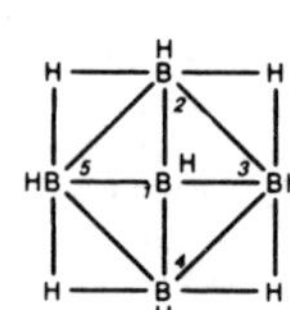

planar projection

[1]*Inorg. Chem.* **1968**, 7(10), 1945–65.

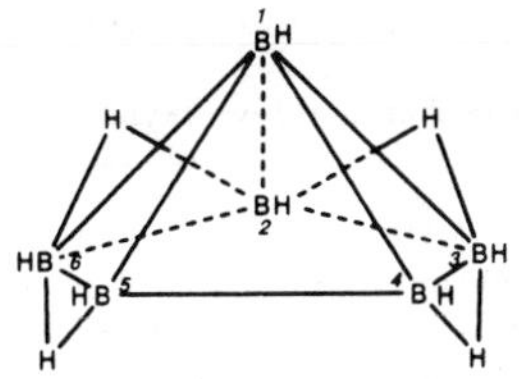

Hexaborane(10)

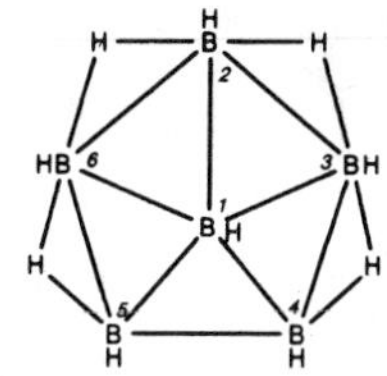

planar projection

Some polyboranes can be named as derivatives of simpler polyboranes. Thus, a bimolecular polyborane, i.e., a two-component "ring" assembly in which both skeletons are identical, can be named as follows:

1,1'-Bipentaborane(9)

When the various parts of the structure are not identical, the general principles of substitutive nomenclature are applied, and a polyborane radical is used for the less preferred skeleton, e.g., **Decaborane(10), 2-octaboran(8)-1-yl-**.

Polyboranes joined along an edge, or with a triangular face in common, are named like fused ring systems; e.g., **Decaborano(14)[5',6':5,6]decaborane(14), Undecaborano[2',7',11':1,2,3]dodecaborane(17)**.

160. "Hetero" polyboranes are boron hydride skeletons in which boron atoms have been replaced by those of other elements, notably carbon. Re= placement prefixes, e.g., "carba," "phospha," are employed with the polyborane name, and the number of hydrogen atoms attached to the skeleton expressed in parentheses after the name. Structural diagrams for compounds of estab= lished structure can be found in the *Ring Systems Handbook* and, when justified by current entries, in the *Chemical Substance Index*. Numbering is as for the parent polyboranes, with lowest compatible locants assigned to the replacement atoms.

Examples:

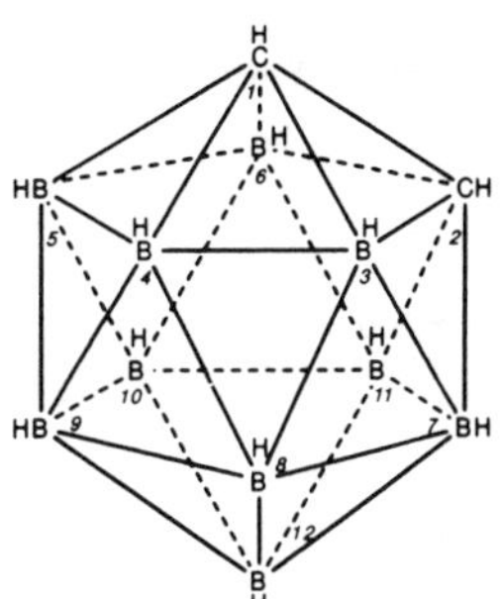

1,2-Dicarbadodecaborane(12)

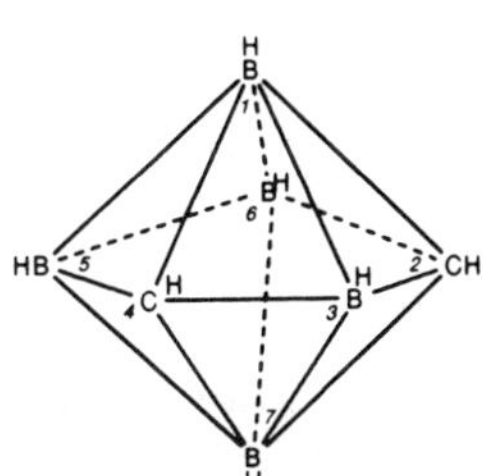

2,4-Dicarbaheptaborane(7)

161. Substituent prefixes (radicals) derived from molecular skele= tons are used very frequently in substitutive nomenclature. Their names are based on the skeleton names and may be classified accordingly as radicals from (*a*) monoatomic skeletons, (*b*) hydrocarbon chains, (*c*) organic hetero ("a"-named) chains, (*d*) homogeneous hetero chains, (*e*) heterogeneous hetero chains, (*f*) carbocycles, (*g*) heterocycles, (*h*) ring assemblies, (*i*) polyboranes. Combination of simple radicals to form compound and complex radicals is performed by application of principles described earlier (¶ 133). (See also "Illustrative List of Substituent Prefixes," which constitutes Section H (¶ 294).

(*a*) *Monoatomic radicals* from borane, methane, silane, germane, stan= nane, and plumbane are named by replacing "-ane" by "-yl," "-ylene," and "-ylidyne" to denote the loss of one, two, or three hydrogen atoms. The final "e" of the hydrides arsorane and phosphorane may be replaced by "-yl," "-ylidene," and "-ylidyne." Stibine and bismuthine may have the "-ine" ending replaced by "-ino," "-ylene," and "-ylidyne." Phosphine and arsine may have the "-ine" replaced by "-ino," "-inidene," and "-inidyne." The -tetrayl suffixes indicate loss of all hydrogen from Group IVA monoatomic hydrides.

Examples:

H_2B-	boryl	HC≡	methylidyne
H_2C=	methylene	HSi≡	silylidyne
=Sn=	stannanetetrayl	HAs=	arsinidene
H_2P-	phosphino	As≡	arsinidyne
HP=	phosphinidene	H_2Sb-	stibino
P≡	phosphinidyne	HSb=	stibylene
H_4P-	phosphoranyl	Sb≡	stibylidyne
H_3P=	phosphoranylidene	H_2Bi-	bismuthino
H_2P≡	phosphoranylidyne	HBi=	bismuthylene
H_2As-	arsino	Bi≡	bismuthylidyne

(*b*) *Acyclic hydrocarbon radicals* are named from the skeletons by re= placing "-ane," "-ene," and "-yne" suffixes by "-yl," "-enyl," and "-ynyl," (for monovalent radicals); by "-diyl," "-triyl," "-enediyl," "-ynediyl," etc., for divalent radicals with hydrogen removed at more than one position; and by "-ylidene" and "-ylidyne" to indicate two or three atoms lost at one position. (Methylene is an exception.) Locants are not cited for monovalent radicals (the free valency position is always "1"), but unsaturated positions are always indicated for chains of three or more atoms. Free valencies (single or multiple) in one or two positions of an acyclic chain are always terminal, otherwise a compound radical name is employed. When three or more positions have free valencies, two of them must be terminal.

Examples:

$MeCH_2$-	ethyl
H_2C=$CHCH_2$-	2-propenyl (not allyl)
EtCH=	propylidene
PrC≡	butylidyne
H_2C=C=	ethenylidene
H_2C=CHCH=	2-propenylidene
MeC≡CCH_2CH=CHCH=	2-hepten-5-ynylidene
-$(CH_2)_2$-	1,2-ethanediyl
-CH_2CH=	1-ethanyl-2-ylidene
-CH=CH-	1,2-ethenediyl
-$(CH_2)_3$-	1,3-propanediyl
=C=C=C=	1,2-propadiene-1,3-diylidene
—CH_2CH(—)CH_2—	1,2,3-propanetriyl
—CH_2C(=)CH_2—	1,3-propanediyl-2-ylidene

(*c*) *Organic heteroacyclic ("a") radicals* are used when the requirements (¶ 127) are met. The numbering of the parent radical (not necessarily that of the molecular skeleton) is retained. A single free valency is hence always in the 1-position, and *this locant is always cited*. When there is still a choice, lowest locants are assigned to hetero atoms, then to most preferred hetero atoms (Table I, ¶ 128), then to unsaturation (with double bonds preferred).

Examples:

Me[O$(CH_2)_2$]$_3$OCH_2CH_2—	3,6,9,12-tetraoxatridec-1-yl (not [2-[2-[2-(2-methoxyeth= oxy)ethoxy]ethoxy]ethyl])
MeSiH_2CH_2SiH_2CH=CHSiH_2CH_2SiH_2CH=	2,4,7,9-tetrasiladec-5-en-1-yl= idene
—CH_2CH_2[S$(CH_2)_2$]$_2$SCH=CHNHCH_2CH_2	3,6,9-trithia-12-azatetradec-10-ene-1,14-diyl

(*d*) *Homogeneous heteroacyclic radicals* are named analogously to acyclic hydrocarbon radicals, except that only the "e" of "ane" suffixes of heteroacyclic skeleton names is replaced by "-yl."

Examples:

HN=N-	diazenyl (the substituted radical is named azo)
H_2NNHNH-	triazanyl
HN=NNH-	2-triazenyl

$H_2NN{=}NN{=}N-$	1,3-pentazadienyl
H_2PPH-	diphosphinyl
$-PHPH-$	1,2-diphosphinediyl
$=AsAs=$	1,2-diarsinediylidene
$-Sb{=}Sb-$	1,2-distibenediyl
H_2NNH-	hydrazino
$H_2NN=$	hydrazono
$-NHN=$	1-hydrazinyl-2-ylidene
$-NHNH-$	hydrazo (to different atoms) hydrazi (to the same atom)
$-N{=}N-$	azo (to different atoms) azi (to the same atom)
$=NN=$	azino

(*e*) *Heterogeneous heteroacyclic radicals* are named analogously to hy= drocarbon radicals.
Examples:

$H_3SiOSiH_2-$	disiloxanyl
$H_3SiOSiH=$	disiloxanylidene
$H_3SnOSn\equiv$	distannoxanylidyne
$-SiH_2NHSiH_2-$	1,3-disilazanediyl
$-SiH_2OSi(-)HOSiH_2-$	1,3,5-trisiloxanetriyl

(*f*) *Cyclic hydrocarbon radicals.* Loss of one or two hydrogen atoms from a single cycloalkane carbon atom is denoted by replacement of "ane" of the ring name by "-yl" and "-ylidene," respectively. The implied locant ("1" in all cases) is not expressed. In radicals from cycloalkenes, cycloalkadienes, etc., only the final "e" is replaced by the radical suffix, and locants for unsaturation and the free valency (always "1") are all cited. Loss of hydrogen at more than one position is expressed by the suffixes "-diyl" (not "-ylene"), "-diylidene," "-enylylidene," etc. All locants are cited, and locants for free valencies are assigned lowest locants.
Examples:

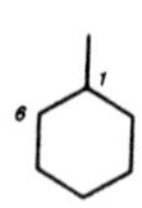
cyclohexyl

2-cyclopropen-1-yl

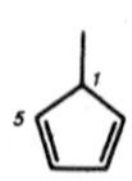
2,4-cyclopentadien-1-yl

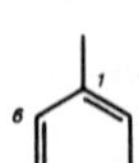
1,5-cyclohexadien-3-yn-1-yl

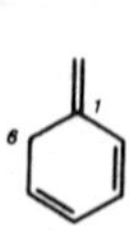
2,4-cyclohexadien-1-ylidene

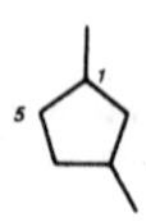
1,3-cyclopentanediyl (not 1,3-cyclo= pentylene)

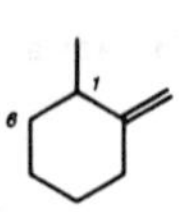
1-cyclohexanyl-2-ylidene (not 1-cyclohexyl-2-ylidene)

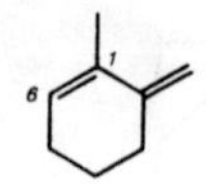
6-cyclohexen-1-yl-2-ylidene

Radicals from *benzene* are named phenyl (C_6H_5-), 1,2-, 1,3-, and 1,4-phenylene ($-C_6H_4-$), 1,2,3-benzenetriyl, 1,2,3,4-benzenetetrayl, etc.

Fused hydrocarbon radicals are assigned the lowest locants compatible with the fixed numbering of the ring system. Indicated hydrogen (¶ 135) necessary for the existence of the ring is assigned to the lowest nonangular position unless it can be located to accommodate a monovalent radical in an angular position or an "ylidene" radical in a nonangular position. Added hydrogen (¶ 136) is cited immediately after the radical locant and is assigned the lowest available angular or nonangular position.
Examples:

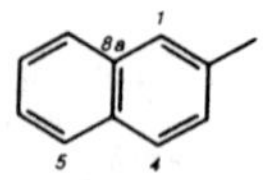
2-naphthalenyl (not 2-naphthyl)

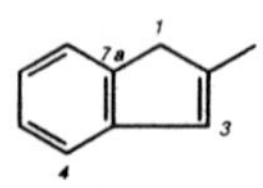
1*H*-inden-2-yl

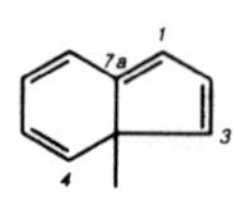
3a*H*-inden-3a-yl (a hydro derivative would be named, e.g., 2,3-di= hydro-3a*H*-inden-3a-yl, not 2*H*-inden-3a(3*H*)-yl)

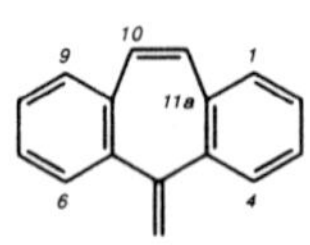
5*H*-dibenzo[*a*,*d*]cyclohepten-5-ylidene

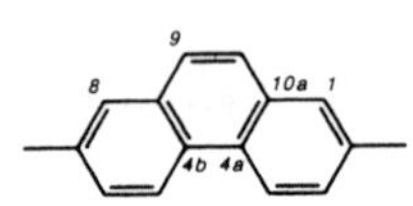
2,7-phenanthrenediyl

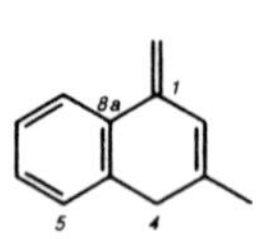
3-naphthalenyl-1(4*H*)-ylidene (not 2-naphthalenyl-4(1*H*)-ylidene)

Von Baeyer and spiro radicals follow similar principles. Free valency locants are assigned lowest locants compatible with ring-system numbering and are preferred over locants for unsaturation.
Examples:

bicyclo[2.2.2]oct-5-en-2-yl (not bi= cyclo[2.2.2]oct-2-en-5-yl)

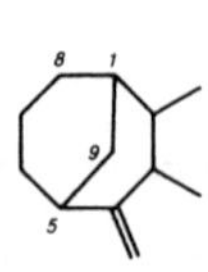
bicyclo[3.3.1]nonane-2,3-diyl-4-ylid= ene

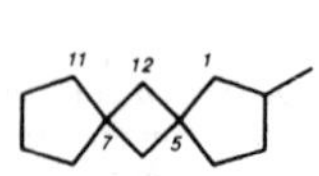
dispiro[4.1.4.1]dodec-2-yl

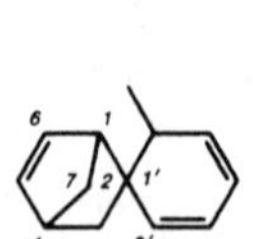
spiro[bicyclo[2.2.1]hept-5-ene-2,1'-[3,5]cyclohexadien]-2'-yl (note: lowest locants are assigned, in order, to spiro atoms, free valencies, all multiple bonds, double bonds)

(*g*) *Heterocyclic radicals* from ring systems not named by organic re= placement ("a") nomenclature are named analogously to fused-hydrocarbon radicals (above). Radicals involving free valencies at hetero atoms in abnormal valency states are not normally employed, unless the valency is expressed by indicated hydrogen (¶ 158). Instead, the compound is named at the abnor=

mal-valency parent (see last example below). The shortened form thienyl is used for thiophene radicals and all their fused derivatives; selenophene-yl (not selenophenyl) is the name of the seleno analog.

Examples:

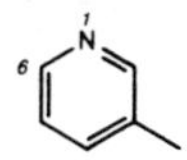
3-pyridinyl (not 3-pyridyl)

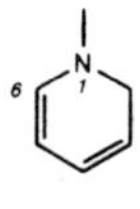
1(2*H*)-pyridinyl

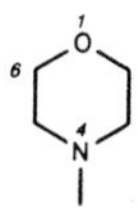
4-morpholinyl (not morpholino)

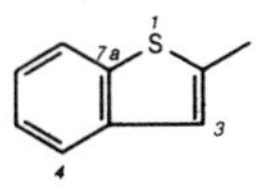
benzo[*b*]thien-2-yl (not benzo[*b*]=thiophen-2-yl)

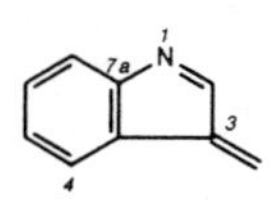
3*H*-indol-3-ylidene (a hydro derivative would be named, e.g., 1,2-di=hydro-3*H*-indol-3-ylidene, not 1*H*-indol-3(2*H*)-ylidene)

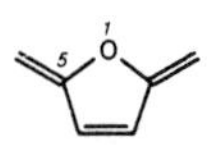
2,5-furandiylidene

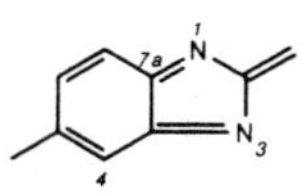
2*H*-benzimidazol-5-yl-2-ylidene

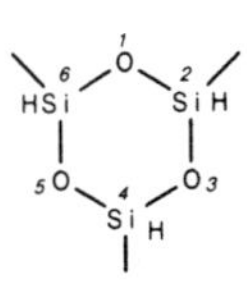
cyclotrisiloxane-2,4,6-triyl

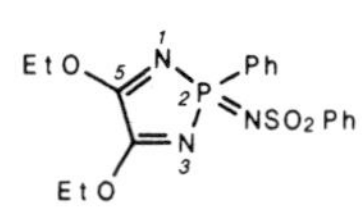
2*H*-1,3,2-Diazaphosphole, 4,5-di=ethoxy-2,2-dihydro-2-phenyl-2-[(phenylsulfonyl)imino]- (not Benzenesulfonamide, *N*-(4,5-di=ethoxy-2-phenyl-2*H*-1,3,2-diaza=phosphol-2(2*H*)-ylidene)-)

"A"-named cyclic radicals are named like the hydrocarbon parents from which they have been derived by atom replacement. The hetero atoms receive lowest locants (cited or implied), then the free valencies (cited just ahead of the radical suffix).

Examples:

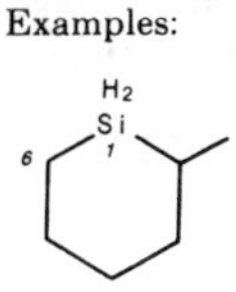
silacyclohex-2-yl (not 2-silacyclohexyl)

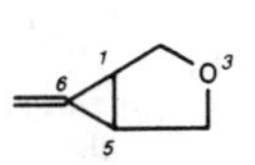
3-oxabicyclo[3.1.0]hex-6-ylidene

(*h*) *Ring-assembly radicals* are derived by bracketing the assembly name, eliding a final "e" if "y" is to follow, and appending the radical endings "-yl," "-ylidene," "-diyl," etc., as appropriate. The free valencies need not be on the terminal rings of the assembly.

Examples:

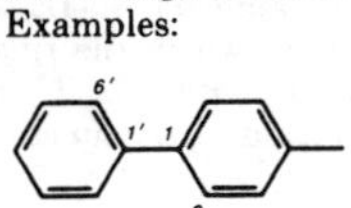
[1,1'-biphenyl]-4-yl (not 4-biphenylyl)

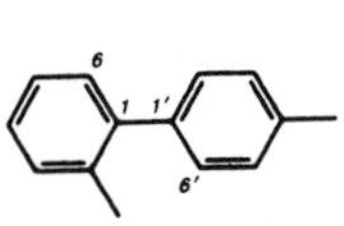
[1,1'-biphenyl]-2,4'-diyl (not 2,4'-biphenylylene)

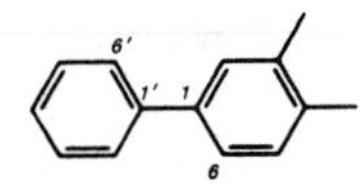
[1,1'-biphenyl]-3,4-diyl (not (4-phenyl-1,2-phenylene))

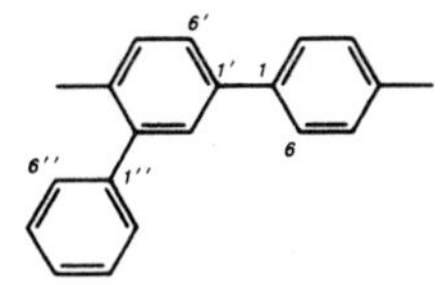
[1,1':3',1''-terphenyl]-4,4'-diyl (not (2-phenyl[1,1-biphenyl]-4,4'-diyl))

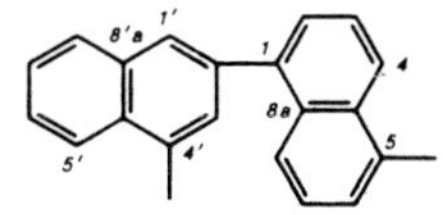
[1,2'-binaphthalene]-4',5-diyl

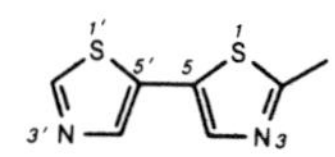
[5,5'-bithiazol]-2-yl (not [5-(5-thia=zolyl)-2-thiazolyl])

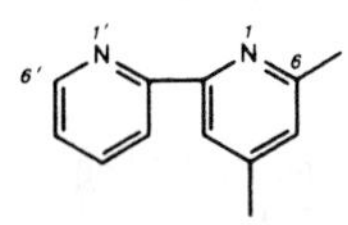
[2,2'-bipyridine]-4,6-diyl (not [6-(2-pyridinyl)-2,4-pyridinediyl])

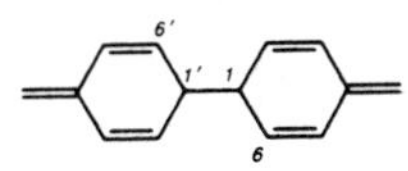
[bi-2,5-cyclohexadien-1-yl]-4,4'-diylidene

(*i*) *Polyborane radicals* are formed from polyborane and hetero poly=borane names by citing the usual radical suffixes after the parenthetical designation of the number of hydrogen atoms, with elision of the final "e" before "y." Requirements for locant citation with diborane radicals are the same as for the analogous acyclic hydrocarbon radicals.

Examples:

diboran(4)yl	1-pentaboran(9)yl
diboran(6)yl	1,2-dicarbadodecaboran(12)-1-yl
diboran(4)ylidene	6,9-decaborane(10)diyl
1,2-diboran(4)diyl	

162. Compound and complex radicals from molecular skeletons are named by principles described earlier (¶ 133).

Examples:

Me_2CH-	(1-methylethyl)
$H_2C{=}CMe-$	(1-methylethenyl) (the unsaturated parent radical is preferred)

$$\overset{11}{CH_3}-\overset{10}{CH}=\overset{9}{CH}-\overset{8}{CH}=\overset{7}{CH}-\underset{6}{\overset{CH=CH-C\equiv C-CH_3}{CH}}-\overset{5}{CH}=\overset{4}{CH}-\overset{3}{CH}=\overset{2}{CH}-\overset{1}{CH_2}-$$

[6-(1-penten-3-ynyl)-2,4,7,9-unde=catetraenyl]

PhCH=CH-	(2-phenylethenyl) (not styryl)
PhC≡	(phenylmethylidyne) (not benzylidyne)

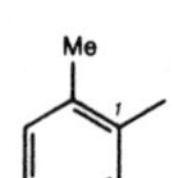
(2-methylphenyl) (not *o*-tolyl)

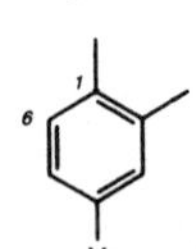
(4-methyl-1,2-phenylene) (not 4-methyl-*o*-phenylene)

Radicals from branched polyphenyls are chosen by application of the fol=lowing principles successively until a decision is reached:

(*a*) longest chain of rings containing all of the free valencies, which need not be on terminal rings;

(*b*) lowest locants in the radical name for (1) ring junctions, and (2) free valencies;

(*c*) maximum number of substituent prefixes;

(*d*) lowest locants for substituent prefixes;

(*e*) earliest index position of the radical name.

Examples (the letters on the left refer to the principles above):

(*a*)

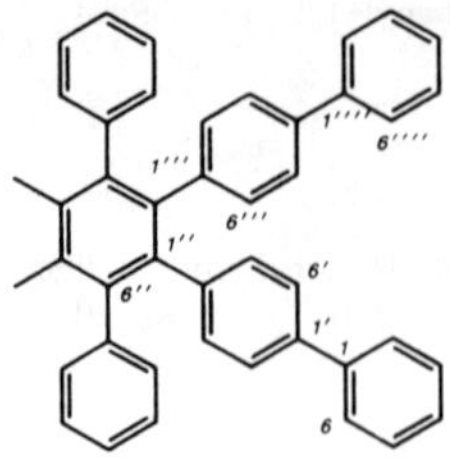

(3'',6''-diphenyl[1,1':4',1'':2'',1''':4''',= 1''''-quinquephenyl]-4'',5''-diyl)

(*b*) (1)

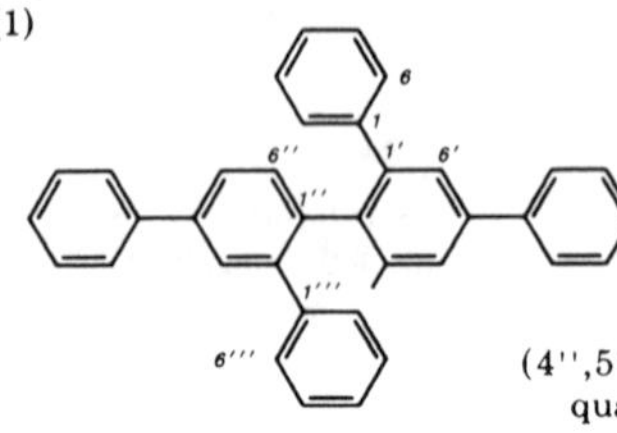

(4'',5'-diphenyl[1,1':2',1'':2'',1'''-quaterphenyl]-3'-yl)

(*b*) (2)

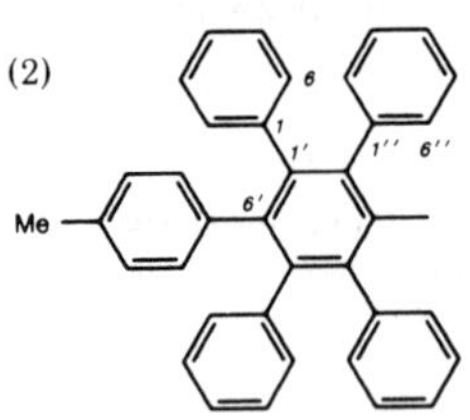

[6'-(4-methylphenyl)-4',5'-diphenyl= [1,1':2',1''-terphenyl]-3'-yl)

(*c*)

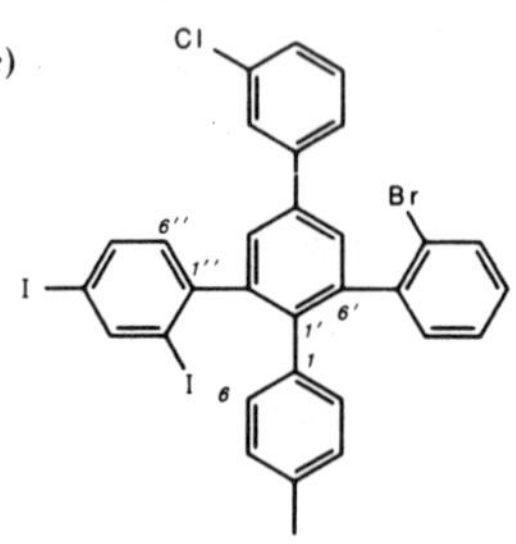

[6'-(2-bromophenyl)-4'-(3-chloro= phenyl)-2'',4''-diiodo[1,1':2',1''-terphenyl]-4-yl]

(*d*)

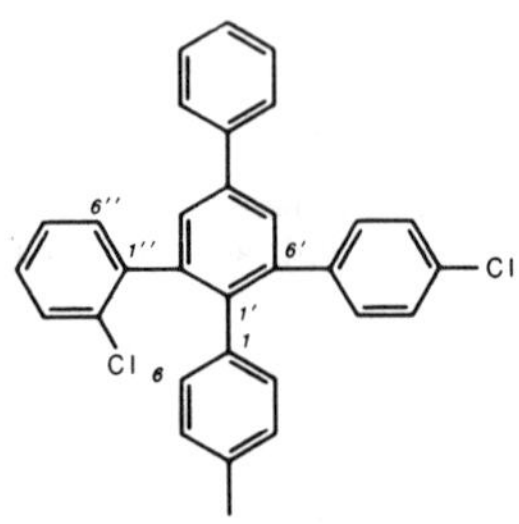

[2''-chloro-6'-(4-chlorophenyl)-4'-phenyl[1,1':2',1''-terphenyl]-4-yl]

(*e*)

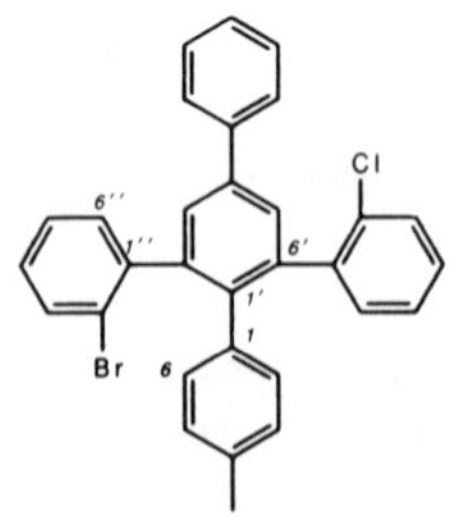

[2''-bromo-6'-(2-chlorophenyl)-4'-phenyl[1,1':2',1''-terphenyl]-4-yl](not [6'-(2-bromophenyl)-2''-chloro-4'-phenyl[1,1':2',1''-ter= phenyl]-4-yl])

163. Molecular skeletons as index heading parents. These two en= tities coincide when no suffix, expressing a principal group, is added to the skeleton name, either because such a group is absent, or because it is attached to a hetero atom of the skeleton which changes it from a functional into a nonfunctional group.

Examples:

MeMe	**Ethane**
$PhNO_2$	**Benzene, nitro-** ("nitro" is a man= datory prefix (¶ 132))
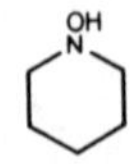	**Piperidine, 1-hydroxy-** ("hydroxy" is considered nonfunctional when attached to a hetero atom other than silicon)
$Me_3SiCOMe$	**Silane, acetyltrimethyl-** (not Eth= anone, 1-(trimethylsilyl)-: an acyclic ketone function is not rec= ognized unless it has two carbon linkages)

Choice of a preferred index heading parent has already been described (¶ 138). In the case of a branched polyphenyl, the criteria are the same as for the derived radicals (¶ 162), except that principal groups replace free valencies.

Example:

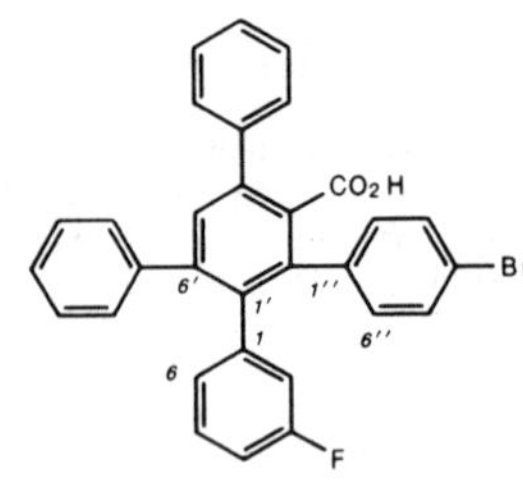

[1,1':2',1''-Terphenyl]-3'-carboxylic acid, 4''-bromo-3-fluoro-4',6'-diphenyl- (a decision is reached by choosing lowest locants for the **Terphenyl** skeleton and then for the principal group)

Because of the importance attached to uncommon ring systems in chemical compounds, index entries are made by *CA* at the names of all but the most frequently occurring rings (see Table VA) for each compound in which they occur whenever the preferred index name does not appear at the ring name.

Example:

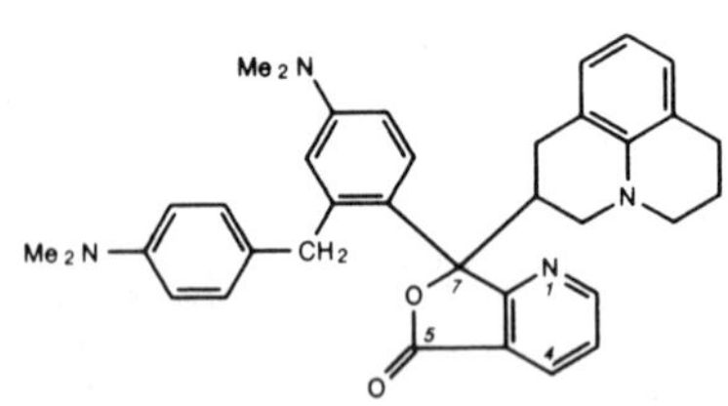

Furo[3,4-*b*]pyridin-5(7*H*)-one, 7-[4-(dimethylamino)-2-[[4-(dimethylamino)phenyl]= methyl]phenyl]-7-(2,3,6,7-tetrahydro-1*H*,5*H*-benzo[*ij*]= quinolizin-2-yl)- [*55912-74-8*]

1*H*,5*H*-**Benzo[*ij*]quinolizine furo[3,4-*b*]pyridin-5(7*H*)-one deriv.** [*55912-74-8*]

The preferred index entry is placed as usual at the name which expresses the highest function (a cyclic ketone in this case), and this name falls in the index close to the illustrative structural diagram for **Furo[3,4-*b*]pyridine**; no additional index entry for this ring is considered necessary. The compound also contains two benzene rings, but no index entry appears at **Benzene**, which is listed in Table VA as a commonly occurring ring not requiring additional entries. However, the compound also contains an uncommon ring, represented in the preferred index name by the substituent prefix "**1*H*,5*H*-benzo[*ij*]= quinolizin-2-yl-**". An additional entry therefore appears at **1*H*,5*H*-Benzo[*ij*]quinolizine** for this and other compounds in which the ring is present. The index modification, in this case "furo[3,4-*b*]pyridin-5(7*H*)-one deriv.," indicates the complete index heading parent (including the functional suffix, if any) at which the full name can be found.

TABLE VA
RING SYSTEMS FOR WHICH ADDITIONAL INDEX ENTRIES ARE NOT MADE

Acenaphthylene
Acridine
Anthracene
1–Azabicyclo[2.2.2]octane
8–Azabicyclo[3.2.1]octane
Azepine
Azetidine
Aziridine
Azulene
Benzene
Benzimidazole
Benz[*e*]indene
Benzocycloheptene
1,3–Benzodioxole
Benzofuran
1–Benzopyran
2–Benzopyran
Benzothiazole
Benzo[*b*]thiophene
Benzotriazole
Benzoxazole
Bicyclo[2.2.1]heptane
Bicyclo[4.1.0]heptane
Bicyclo[3.1.1]heptane
Bicyclo[3.1.0]hexane
Bicyclo[2.2.2]octane
Carbazole
Chrysene
Cyclobutane
Cycloheptane
Cyclohexane
Cyclooctane
Cyclopentane
Cyclopenta[*a*]phenanthrene
Cyclopropane
Dibenzo[*a,d*]cycloheptene
Dibenzofuran
1,3–Dioxane
1,4–Dioxane
1,3–Dioxolane
1,3–Dithiolane
3,9a–Ethano–9,9b–(iminoethano)=
  phenanthro[4,5–*bcd*]furan
Fluoranthene
Fluorene
Furan
Imidazole
Imidazolidine
10,4a–(Iminoethano)phenan=
  threne
8,9c–(Iminoethano)phenanthro=
  [4,5–*bcd*]furan
Indene
Indole
Indolizine
Isobenzofuran
Isoindole
Isoquinoline
Isothiazole
Isoxazole
Isozazolidine
4,7–Methanoindene
Morpholine
Naphthacene
Naphthalene
1,3,4–Oxadiazole
Oxazole
Oxazolidine
Oxirane
Pentalene
Phenalene
Phenanthrene
1,10–Phenanthroline
Phenazine
Phenothiazine
Phthalazine
Picene
Piperazine
Piperidine
Pteridine
Purine
Pyran
Pyrazine
Pyrazole
Pyrazolidine
Pyrene
Pyridazine
Pyridine
Pyrimidine
Pyrrole
Pyrrolidine
Pyrrolizine
Quinazoline
Quinoline
Quinoxaline
Tetrazole
4–Thia–1–azabicyclo[3.2.0]=
  heptane
5–Thia–1–azabicyclo[4.2.0]=
  octane
1,3,4–Thiadiazole
Thiazole
Thiazolidine
Thiophene
1,2,4–Triazine
1,3,5–Triazine
1,2,4–Triazole
Tricyclo[3.3.1.1^{3,7}]decane
Xanthene

Prior to January 1984 (*CA* Volume 100) only certain forms of those ring systems containing indicated hydrogen (e.g., 1*H*–Indole but not 2*H*–Indole) were included in the list above. Now the list is unspecific; no additional index entires are made for Azepine, Indole, etc., in any indicated hydrogen form.

Derivatives of cyclic stereoparents other than cyclic carbohydrates, nu= cleosides, and nucleotides are also given additional index entries when the main entry appears at a different name (see ¶ 202). Additional entries at systematic ring system names also appear for ring–modified stereoparents; thus, the substance indexed at **Pregnane–18,20–diol**, cyclic ethylidene acetal, also receives an entry at 1*H*,5*H*–**Naphth[2′,1′:4,5]indeno[1,7a–*e*][1,3]dioxepin.**

164. Introduction. In substitutive nomenclature, a principal chemical group is that substituent of a molecular skeleton which is selected for ex= pression as a suffix. Only one kind of substituent may be cited as a suffix, viz., the most senior one as determined by the Order of Precedence of Compound Classes (¶ 106); all other substituents are expressed as prefixes (radicals) which may be simple, compound, or complex (¶¶ 132, 133). For functional com= pounds, the molecular skeleton name together with its suffix constitutes an index *heading parent*. The locants for suffixes are placed in front of the heading parent name unless locants for indicated hydrogen, hetero atoms (in "a" names), unsaturation, fusion sites, etc., are present, in which case they are placed just before the suffix.

In the following paragraphs, compound classes expressed as substituent suffixes are discussed in descending order of precedence.

165. Acids expressed as substituent suffixes on molecular skeletons include carboxylic, sulfonic, sulfinic, selenonic, and telluronic acids and their functional replacement analogs, such as peroxy, imidic and thio acids. For **Carbonic acid** and its relatives (including **Carbamic acid** and **Formic acid**) see ¶ 183.

(*a*) *Carboxylic acids* are named by the Geneva ("–oic") or "–carboxylic" system. The "–oic acid" suffix is employed for acyclic mono– and dicarboxylic acids of carbon chains, including "a"–named acids; the "–carboxylic acid" suffix is used for acyclic polycarboxylic acids and compounds in which the carboxyl group is attached to a ring, a monoatomic hydride, or a heteroacyclic chain. The trivial names **Acetic acid** and **Benzoic acid** are retained for these two acids and their substituted derivatives. (The amides, acid chlorides, etc., are named similarly, but organic replacement analogs are named systematically, e.g., **Benzenecarboximidic acid** (not Benzimidic acid.)

Examples:

$MeCH_2CO_2H$ — **Propanoic acid** (not Propionic acid)

$Me(CH_2)_3CO_2H$ (5 … 1) — **Pentanoic acid** (not Valeric acid)

$H_2C{=}CHCO_2H$ (3 … 1) — **2–Propenoic acid** (not Acrylic acid)

$Me(CH_2)_4CH{=}CHCH_2CH{=}CH(CH_2)_7CO_2H$ (18, 13, 9, 1) — **9,12–Octadecadienoic acid**

HO_2CCO_2H (1 2) — **Ethanedioic acid** (not Oxalic acid)

$HO_2C(CH_2)_3CO_2H$ (1 … 5) — **Pentanedioic acid** (not Glutaric acid)

$H_3C{-}CH_2{-}O{-}CH_2{-}CH_2{-}O{-}CH_2{-}CH_2{-}O{-}CH_2{-}CH_2{-}O{-}CH_2{-}CH{=}CH{-}(CH_2)_7{-}COOH$ (1 2 3 4 5 6 7 8 9 10 11 12 13 14 15 … 23) — **3,6,9,12–Tetraoxatricos–14–en–23–oic acid** (The hetero atoms are given preference for lowest locants; the locant for the "–oic acid" suffix is cited if it is not "1.")

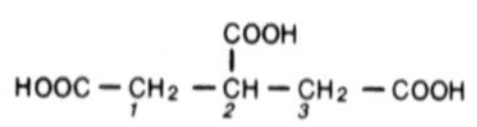

1,2,3–Propanetricarboxylic acid (numbering excludes the carboxyl groups)

$(HO_2C)_2C{=}C(CO_2H)_2$ — **Ethenetetracarboxylic acid**

$HO_2CNHNHCO_2H$ (1 2) — **1,2–Hydrazinedicarboxylic acid** (not Bicarbamic acid)

H_2PCO_2H — **Phosphinecarboxylic acid** (not Formic acid, phosphino–)

$HO_2CSnH_2OSnH_2CO_2H$ (1 2 3) — **1,3–Distannoxanedicarboxylic acid**

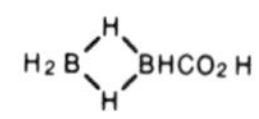

Diborane(6)carboxylic acid

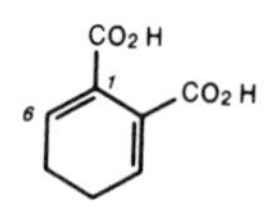

2,6–Cyclohexadiene–1,2–dicarboxylic acid (lowest locants for principal groups are preferred)

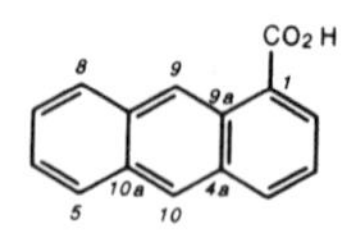

1–Anthracenecarboxylic acid

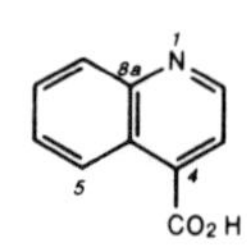

4–Quinolinecarboxylic acid

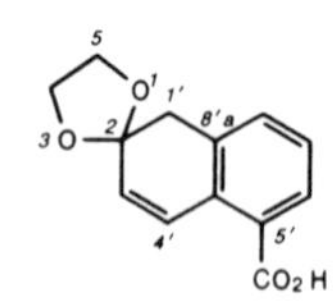

Spiro[1,3–dioxolane–2,2'(1'*H*)–= naphthalene]–5'–carboxylic acid

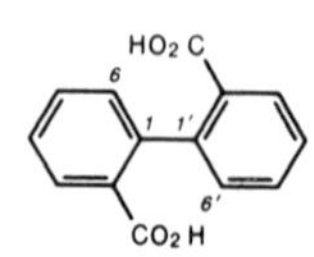

[1,1'–Biphenyl]–2,2'–dicarboxylic acid

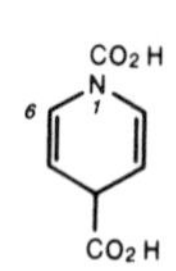

1,4(4*H*)–Pyridinedicarboxylic acid (the added hydrogen is expressed after the final locant but relates to the 1–carboxyl group; see ¶ 136)

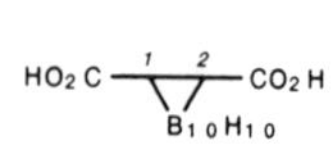

1,2–Dicarbadodecaborane(12)–1,2–= dicarboxylic acid (the 1,2–bond does not possess single–bond character)

(*b*) *Sulfonic, sulfinic and sulfenic acids* and their selenium and tellurium analogs are expressed by appending the appropriate suffix to the name of the molecular skeleton. Mono– and diacids of these series, unlike "–oic acids," above, do not need to occupy terminal positions on a chain.

Examples:

$EtSO_3H$ — **Ethanesulfonic acid**

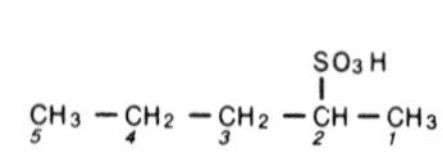

2–Pentanesulfonic acid

$PhSO_3H$ **Benzenesulfonic acid**

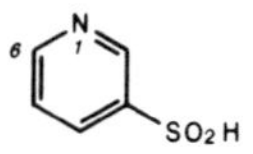
3-Pyridinesulfinic acid

$MeSOH$ **Methanesulfenic acid**

$HN{=}NNHSO_3H$ (locants 3, 1) **2-Triazene-1-sulfonic acid**

H_2NNHSO_2H **Hydrazinesulfinic acid**

$CH_3-CH_2-O-CH_2-CH_2-O-CH_2-CH_2-O-CH_2-CH_2-O-CH_2-CH(SO_3H)-CH_3$ (locants 1–15)

3,6,9,12-Tetraoxapentadecane-14-sulfonic acid (hetero atoms are preferred over principal groups for lowest locants)

H_2PSO_3H **Phosphinesulfonic acid**

$PhSeO_3H$ **Benzeneselenonic acid**

$CH_3-CH_2-CH_2-CH_2-CH(SeO_2H)-CH_3$ (locants 6–1)

2-Hexaneseleninic acid

$CH_3-CH_2-CH(TeO_2H)-CH_3$ (locants 4–1) **2-Butanetellurinic acid**

$PhTeOH$ **Benzenetellurenic acid**

(*c*) *Imidic and hydrazonic acids.* Names for these are formed from the parent carboxylic, sulfonic, sulfinic, selenonic, etc., acid names by functional replacement nomenclature (¶ 129). Some modification of the formal endings is made; thus, an "-oic acid" becomes an "imidic acid," not an "-imidoic acid." **Acetic acid** affords systematically named Ethanoic acid replacement analogs, while **Benzoic acid** is treated as Benzenecarboxylic acid in a similar manner. The suffixes appended to the molecular skeleton name in each case are as follows:

Parent Acid	*Imidic acid*	*Hydrazonic acid*
-oic	-imidic	-hydrazonic
-carboxylic	-carboximidic	-carbohydrazonic
-sulfonic	-sulfonimidic	-sulfonohydrazonic
-sulfinic	-sulfinimidic	-sulfinohydrazonic

The group $-S(:NH)_2OH$ is named by the suffix "-sulfonodiimidic acid," and $-S(:NH)(:NNH_2)OH$ by "-sulfonohydrazonimidic acid." Selenium and tellurium acids are named analogously.

Note: Imidic acids are tautomeric with amides; except for derivatives in which an acid proton has been replaced, e.g., esters and anhydrides, amides are preferred for index entries; see ¶ 122. Hydrazonic acids are tautomeric with hydrazides, which are preferred as index entries.

Examples:

$MeC(=NH)OH$ **Ethanimidic acid** (not Acetimidic acid)

$PhC(=NH)OH$ **Benzenecarboximidic acid** (not Benzimidic acid)

$Me(CH_2)_4C(=NH)OH$ (locants 6, 1) **Hexanimidic acid**

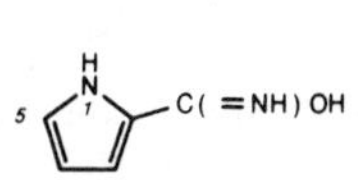
1*H*-**Pyrrole-2-carboximidic acid**

$HOC(=NH)(CH_2)_4C(=NH)OH$ **Hexanediimidic acid**

$H_2NNHC(=NH)OH$ **Hydrazinecarboximidic acid**

$MeC(=NNH_2)OH$ **Ethanehydrazonic acid**

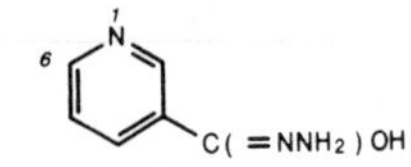
3-Pyridinecarbohydrazonic acid

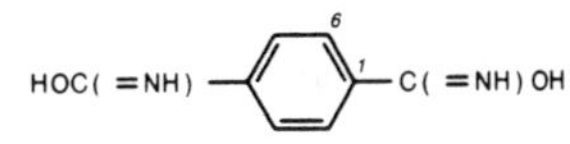
1,4-Benzenedicarboximidic acid

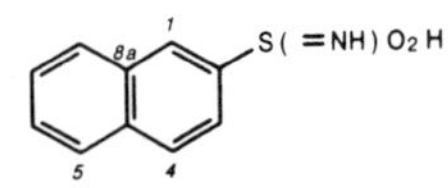
2-Naphthalenesulfonimidic acid

$EtS(=NH)_2OH$ **Ethanesulfonodiimidic acid**

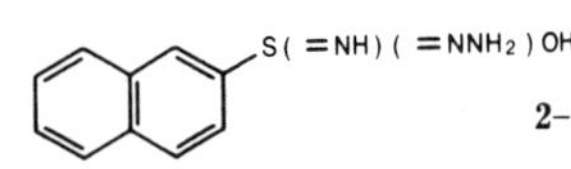
2-Naphthalenesulfonohydrazonimidic acid

$EtS(=NH)OH$ **Ethanesulfinimidic acid**

$PhSe(=NH)O_2H$ **Benzeneselenonimidic acid**

(*d*) *Peroxy acids* are named by use of the suffixes "-peroxoic acid" (from "-oic acid") and "-carboperoxoic acid" (from "-carboxylic acid"). **Acetic acid** and **Benzoic acid** afford systematically named peroxy acids. Sulfonic, sulfinic, and sulfenic acids and their selenium and tellurium analogs are named by use of the suffixes "-sulfonoperoxoic acid," etc. Combinations with imidic and hydrazonic suffixes are made, in alphabetic order, by the normal rules.

Examples:

$CH_3-C(=O)-OOH$ **Ethaneperoxoic acid** (not Peroxyacetic acid)

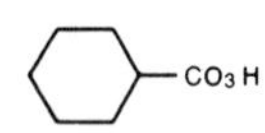
Cyclohexanecarboperoxoic acid

$HO_3C(CH_2)_2CO_3H$ **Butanediperoxoic acid**

$MeC(=NH)OOH$ **Ethanimidoperoxoic acid** (not Peroxyacetimidic acid)

$PhC(=NH)OOH$ **Benzenecarboximidoperoxoic acid** (not Peroxybenzimidic acid)

Peroxy analogs of acids expressed as heading parents, e.g., **Carbonoperoxoic acid** (¶ 183), **Phosphoroperoxoic acid** (¶ 197), are ranked with the acids, e.g., **Carbonic acid**, **Phosphoric acid**, from which they are derived.

(*e*) *Thio acids* derived from carboxylic, sulfonic, sulfinic, and sulfenic acids and their imidic, hydrazonic, and peroxy replacement analogs are named by incorporating "thio" (or "dithio") into the suffixes of the oxygenated acid names. The terms "seleno" and "telluro" are used similarly when appropriate. Selenonic, telluronic, etc., acids are handled like sulfonic acids. The names do not distinguish between replacement of oxygen in =O and -OH groups in the unesterified acids. This information is usually given in the ester name, or by a substituent prefix.

Examples:

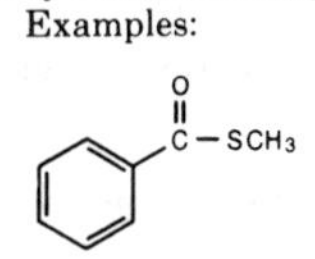
Benzenecarbothioic acid *S*-methyl ester

$CH_3CH_2O-C(=S)CH_2CH_2C(=O)-OH$ (locants 4, 3, 2, 1) **Butanoic acid, 4-ethoxy-4-thioxo-** (for order of precedence of acid groups, see ¶ 167)

When a specific name for a single form is imperative, an italicized element symbol is used in the heading parent, e.g., **Ethanethioic *O*-acid** for $CH_3C(S)OH$.

Replacement by two sulfur atoms in a monocarboxylic acid named by the suffix "-oic acid" is denoted by the suffix "-(dithioic) acid" and by one sulfur atom in each of the two groups in a "-dioic acid" by "-bis(thioic) acid." Ambiguity is absent from "-carboxylic" names, and parentheses are therefore not employed for the analogous "-carbodithioic acid" and "-dicarbothioic acid" suffixes.

Examples:

$Me(CH_2)_2COSH$ — **Butanethioic acid**

$H_2C{=}CHCS_2H$ (3, 1) — **2-Propene(dithioic) acid**

$PhCOSH$ — **Benzenecarbothioic acid** (not Benzoic acid, thio-)

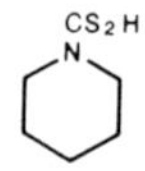

1-Piperidinecarbodithioic acid

$HOSCCH_2COSH$ — **Propanebis(thioic) acid**

$HS_2CCH_2CH{=}CHCH_2CS_2H$ (1, 6) — **3-Hexenebis(dithioic) acid**

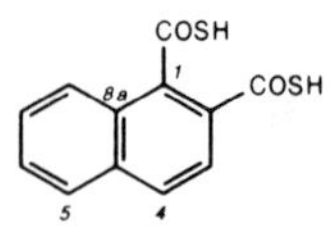

1,2-Naphthalenedicarbothioic acid

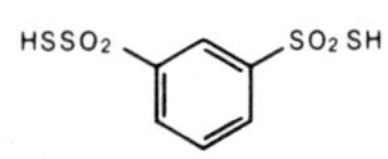

1,3-Benzenedisulfonothioic acid

$MeCH[S(S)SH]Me$ (1, 2, 3) — **2-Propanesulfinodithioic acid**

$MeSSH$ — **Methanesulfenothioic acid**

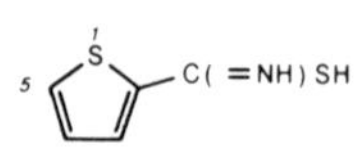

2-Thiophenecarboximidothioic acid

$EtS(=NH)OSH$ — **Ethanesulfonimidothioic acid**

When different numbers of sulfur atoms replace oxygen in the functional groups of polyacids, the groups of higher sulfur content are expressed as substituent prefixes.

Examples:

$HOSCCH_2CH_2CO_2H$ — **Propanoic acid, 3-(thiocarboxy)-**

$EtOSCCO_2H$ — **Acetic acid, ethoxythioxo-**

Thio peroxy acids are named by similar principles; the replacement affixes "(thioperoxo)" and "(dithioperoxo)" are placed, in alphabetic order with other terms such as "imido" and "hydrazono," in the "-oic," "carboxylic," "sulfonic," etc., suffixes of the appropriate parent acids.

Examples:

$MeCOSSH$ — **Ethane(dithioperoxoic) acid**

$PhC(=NH)OSH$ — **Benzenecarboximido(thioperoxoic) acid**

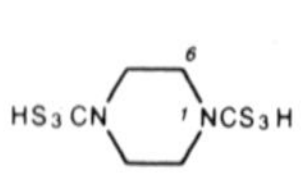

1,4-Piperazinedicarbo(dithioperoxo)=thioic acid

$PhSO_2SSH$ — **Benzenesulfono(dithioperoxoic) acid**

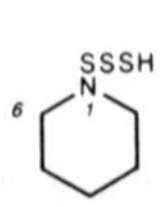

1-Piperidinesulfeno(dithioperoxic) acid

$MeCHMeSSOH$ (1, 2) — **2-Propanesulfeno(thioperoxoic) acid**

$MeCSeOH$ — **Ethaneselenoic acid**

Benzenecarboselenothioic acid

166. Acyclic acids with cyclic substituents may usually be named by conjunctive nomenclature (¶ 124).

Examples:

$PhCH_2CH_2CO_2H$ (β, α) — **Benzenepropanoic acid** (not Hydrocinnamic acid)

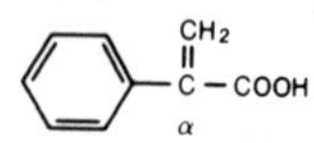

Benzeneacetic acid, α-methylene- (not Atropic acid)

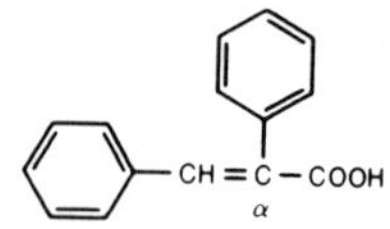

Benzeneacetic acid, α-(phenylmeth=ylene)- (not Benzene-2-propenoic acid, α-phenyl-; conjunctive nomenclature is not used with unsaturated acyclic acids (¶ 124))

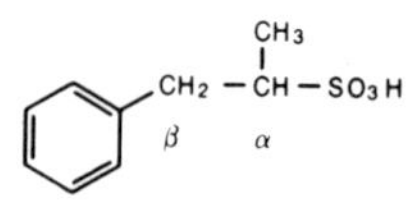

Benzeneethanesulfonic acid, α-methyl-

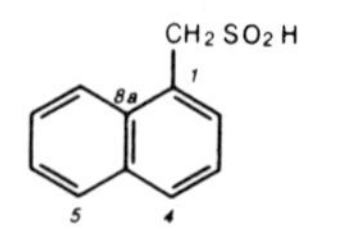

1-Naphthalenemethanesulfinic acid

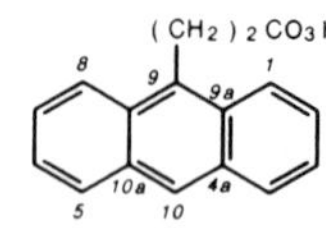

9-Anthracenepropaneperoxoic acid

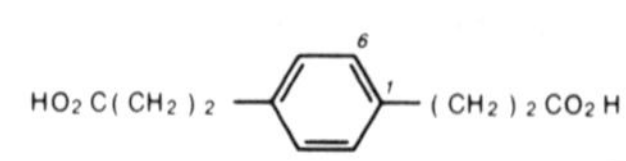

1,4-Benzenedipropanoic acid

When conjunctive names are impermissible, e.g., for unsaturated and poly=functional acyclic acids, acids attached to rings by a double bond, and acids of noncarbon chains, the cyclic group is expressed as a substituent.

Examples:

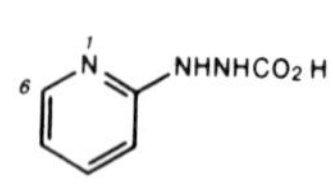

Hydrazinecarboxylic acid, 2-(2-=pyridinyl)-

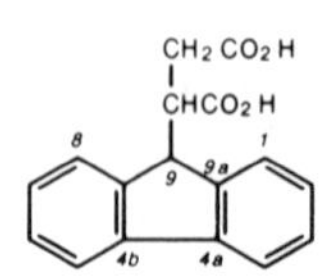

Butanedioic acid, 9*H*-fluoren-9-yl-

Epoxy derivatives of acids are named as oxirane and oxetane derivatives.

Example:

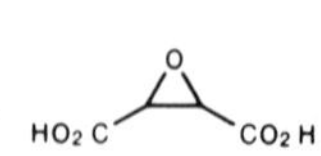

2,3-Oxiranedicarboxylic acid (not Butanedioic acid, epoxy-)

167. Order of precedence of acids. Acid suffixes are the most preferred of all non-cationic substituent suffixes as principal groups (¶ 106), but only one type of acid suffix may be expressed in a heading parent. Less preferred acid functions are cited as substituent prefixes. The choice is made in ac=cordance with the following hierarchy, listed in order of descending precedence:

(*a*) *Peroxy acids.* (Among peroxy acids, the choice depends on the nature of the parent acid as described in (*b*) through (*i*), below.) (See also the separate ranking of peroxy carbonic and peroxy phosphorus acids at ¶¶ 183, 197.)

(*b*) *Carboxylic acids*, followed by thio, seleno, and telluro analogs, in that order. The preferred acid group contains the maximum number of preferred chalcogens, oxygen being the most preferred. (For **Carbonic acid, Formic acid,** etc., see ¶ 183.)

(*c*) *Carbohydrazonic acids*, followed by chalcogen analogs (see (*b*)).

(*d*) *Carboximidic acids*, likewise.

(*e*) *Sulfonic acids*, followed by chalcogen and nitrogen analogs in the order of (*b*), (*c*), and (*d*).

(*f*) *Sulfinic acids*, likewise.

(*g*) *Sulfenic acids*, likewise.

(*h*) *Selenonic, seleninic,* and *selenenic acids*, as for sulfonic acids.

(*i*) *Telluronic, tellurinic,* and *tellurenic acids*, likewise.

Examples:

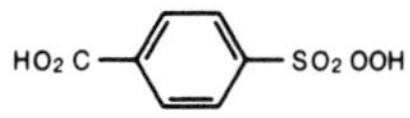

Benzenesulfonoperoxoic acid, 4-carboxy-

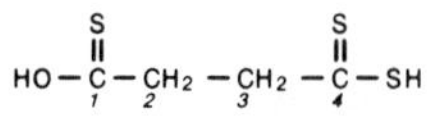

Propanethioic acid, 3-(dithiocarboxy)-

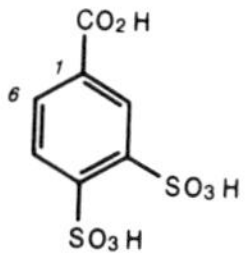

Benzoic acid, 3,4-disulfo-

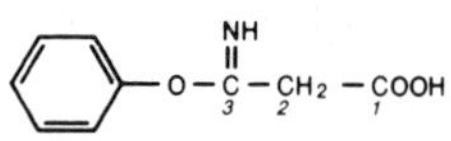

Propanoic acid, 3-imino-3-phenoxy-

When two or more like acid groups are attached to different molecular skeletons, or when one or more are attached to a branched skeleton, the preferred index name is selected according to the usual rules (¶ 138).

Examples:

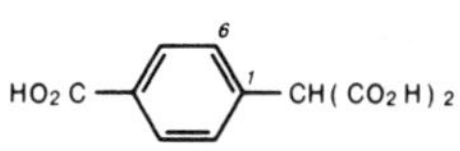

Propanedioic acid, (4-carboxyphenyl)- (principle: maximum number of the principal group)

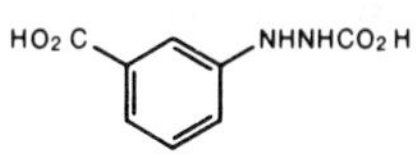

Hydrazinecarboxylic acid, 2-(3-carboxyphenyl)- (principle: hetero acyclic parent preferred)

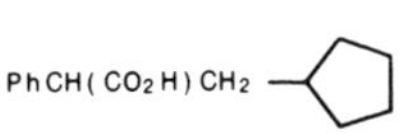

Benzeneacetic acid, α-(cyclopentylmethyl)- (principle: preferred ring system)

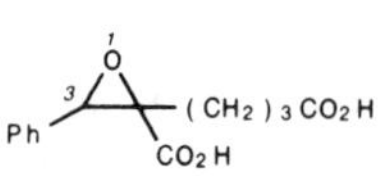

Oxiranebutanoic acid, 2-carboxy-3-phenyl- (principle: largest heading parent)

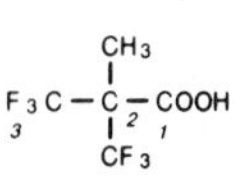

Propanoic acid, 3,3,3-trifluoro-2-methyl-2-(trifluoromethyl)- principle: maximum number of substituent prefixes)

Aldehydic, amic, anilic, hydroxamic, hydroximic, nitrolic, and nitrosolic acids are indexed as compounds of mixed function (see ¶ 228). So are trivially named hydroxy and oxo acids, e.g., Glycolic acid, Acetoacetic acid, and amino acids, e.g., Sulfanilic acid, other than those which are of biological significance (¶ 205).

168. Acid radicals derived, by removal of hydroxyl groups, from acids expressed as suffixes are named as compound and complex radicals. Acyl radicals, e.g., propionyl, naphthoyl, acetimidoyl, are no longer used as substituent prefixes; the only exceptions in general index nomenclature are acetyl (CH_3CO–) and benzoyl (C_6H_5CO–). Amino acid radicals, e.g., glycyl, L-alanyl, are restricted to use in peptide and depsipeptide names (¶ 206).

Radicals derived from monocarboxylic acids are in general named as (1-oxoalkyl) or (arylcarbonyl); carboximidic acids afford (1-iminoalkyl) and (aryliminomethyl) radicals; carbothioic acids give (1-thioxoalkyl) and (arylthioxomethyl) radicals (carbonimidoyl and carbonothioyl are used only as multiplicative radicals (¶ 125) and in cases where both bonds are attached to the same atom); sulfonic, sulfinic, and sulfenic acid radical names are based on the parent radicals "sulfonyl," "sulfinyl," and "thio."

Examples:

EtCO—	(1-oxopropyl)
MeCS—	(1-thioxoethyl)(not thioacetyl)
HN=CH—	(iminomethyl)(not formimidoyl)
—COCOCO—	(1,2,3-trioxo-1,3-propanediyl)
$HO_2C(CH_2)_2CO$—	(3-carboxy-1-oxopropyl)
HOC(=NH)C(=NH)—	(2-hydroxy-1,2-diiminoethyl)

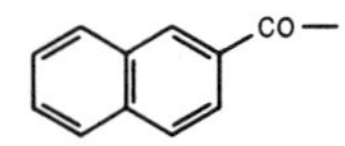

(2-naphthalenylcarbonyl)

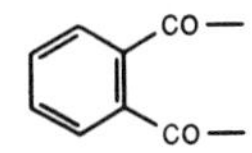

(1,2-phenylenedicarbonyl)

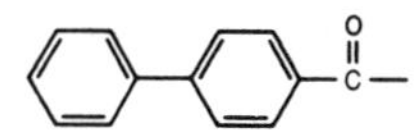

([1,1'-biphenyl]-4-ylcarbonyl)(not (4-phenylbenzoyl)(see ¶ 133))

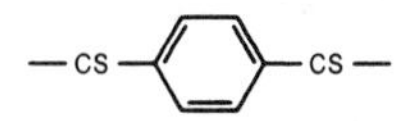

(1,4-phenylenedicarbonothioyl)

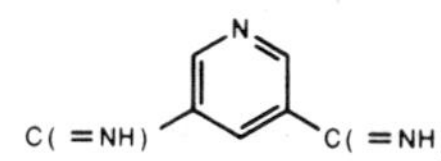

(3,5-pyridinediyldicarbonimidoyl)

$PhSO_2$—	(phenylsulfonyl)
PhSO—	(phenylsulfinyl)
PhS(=NH)—	(*S*-phenylsulfinimidoyl)
PhS—	(phenylthio)

169. Functional derivatives of acids. In the absence of higher functions or more preferred compound classes (¶ 106), *esters* are indexed in the modification, usually at the acid name, sometimes at the alcohol (see ¶ 185). *Hydrazides* are likewise indexed at the acid name (¶ 189). *Hydrazones, azines,* and *semicarbazones* are named at hydrazonic acid parents ($RC(OH){:}NNH_2$) (¶ 165). *Oximes* of carboxylic acids are given *N*-hydroxy imidic acid names; *hydrates* and *acetals* (ortho carboxylic acids and their diesters) are indexed as alcohols (or thiols).

Examples:

$MeC(OH)_3$	**1,1,1-Ethanetriol** (not Orthoacetic acid)
$MeC(OMe)_2OH$	**Ethanol, 1,1-dimethoxy-**

170. Acid halides. In this category are now included the halogenides, in which the hydroxyl groups of acids are replaced by -NC, -NCO, -NCS, -NCTe, $-N_3$, and (in acids other than carbon acids) -CN groups. They are named by placing the halide (etc.) term in the heading as a separate word following an acid term which ends as follows for various acid classes:

Acid suffix	*Acid halide suffix*
-carboxylic	-carbonyl
-carbohydrazonic	-carbohydrazonoyl
-carbothioic	-carbothioyl
-carboximidic	-carboximidoyl
-oic	-oyl
-hydrazonic	-hydrazonoyl
-thioic	-thioyl
-imidic	-imidoyl
-sulfonic	-sulfonyl
-sulfonimidic	-sulfonimidoyl

Examples:

$Me(CH_2)_3CONCS$	**Pentanoyl isothiocyanate**

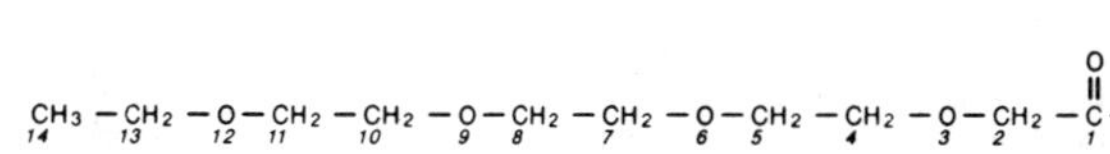

3,6,9,12-Tetraoxatetradecanoyl fluoride

$H_2NNHCOCl$	**Hydrazinecarbonyl chloride**

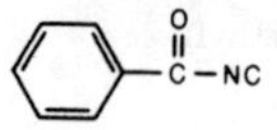

Benzoyl isocyanide

PhC(=NH)Br — **Benzenecarboximidoyl bromide**

PhCSCl — **Benzenecarbothioyl chloride**

$CH_3-CH_2-CH_2-CH_2-SO_2-NCO$ — **1-Butanesulfonyl isocyanate**

PhS(O)(=NH)Cl — **Benzenesulfonimidoyl chloride**

$ClCO(CH_2)_2COCl$ — **Butanedioyl dichloride** (not Butanedioyl chloride)

$OCNSO_2CH_2(CH_2)_4CH_2SO_2NCO$ — **1,6-Hexanedisulfonyl diisocyanate**

$PhCH_2CH_2CON_3$ — **Benzenepropanoyl azide**

Halides, etc., of peroxy acids are generally indexed as anhydrides, anhydrosulfides, etc., with halogen or halogenoid "oxo" acids such as **Hypochlorous** or **Thiocyanic acid**. Halides, etc., of (thioperoxy)sulfenic acids are indexed at **Disulfide**.

Examples:

ClCOSCl — **Carbonochloridothioic acid anhydrosulfide with thiohypochlorous acid**

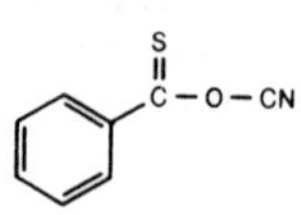

Benzenecarbothioic acid anhydride with cyanic acid

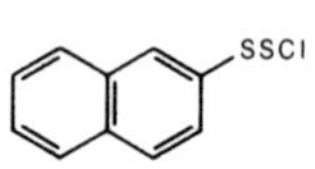

Disulfide, chloro 2-naphthalenyl

When more than one acid halide residue is present in a compound, only one type is named in the heading parent. This is chosen by consideration first of the hierarchy of the parent acids (¶ 167) and then, if a further choice is necessary, of the following list of halides and halogenides (in descending order of precedence): -F, -Cl, -Br, -I, $-N_3$, -NCO, -NCS, -NCSe, -NCTe, -NC, -CN.

Examples:

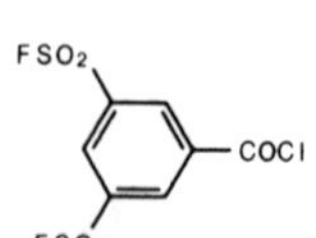

Benzoyl chloride, 3,5-bis(fluorosulfonyl)-

$FCO(CH_2)_2COI$ — **Butanoyl fluoride, 4-iodo-4-oxo-**

Functional derivatives of acid chlorides are indexed similarly to those of the parent acids (¶ 169).

171. **Amides** are named by modification of the parent acid suffixes, thus:

-oic acid	becomes	-amide
-carboxylic acid	becomes	-carboxamide
-carbohydrazonic acid	becomes	-carbohydrazonamide
-carbothioic acid	becomes	-carbothioamide
-carboximidic acid	becomes	-carboximidamide
-sulfonic acid	becomes	-sulfonamide

Secondary and tertiary amides are named as primary amides with *N*-substituents. Anilides, toluidides, etc., are indexed as *N*-aryl amides.

Examples:

$MeCONH_2$ — **Acetamide**

$HN=NHCONH_2$ — **Diazenecarboxamide**

$MeCSNH_2$ — **Ethanethioamide**

$H_2NCS(CH_2)_4CSNH_2$ — **Hexanedithioamide**

$H_2NC(=NH)CH_2C(=NH)NH_2$ — **Propanediimidamide**

$PhC(=NNH_2)NH_2$ — **Benzenecarbohydrazonamide**

$MeS(=NH)_2NH_2$ — **Methanesulfonodiimidamide**

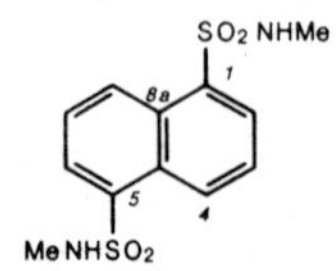

1,5-Naphthalenedisulfonamide, *N*,*N*'-dimethyl-

PhCSNHPh — **Benzenecarbothioamide, *N*-phenyl-**

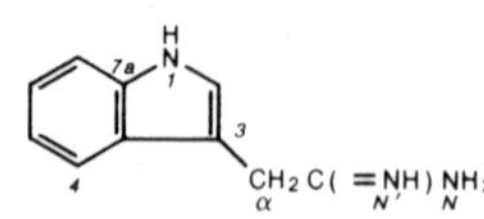

1*H*-Indole-3-ethanimidamide

MeCONHCOMe — **Acetamide, *N*-acetyl-** (not Diacetamide)

$PhSO_2NHSO_2Ph$ — **Benzenesulfonamide, *N*-(phenylsulfonyl)-** (not Dibenzenesulfonamide)

Amides of peroxy acids and thio peroxy acids are indexed as *O*-acyl derivatives of **Hydroxylamine** and **Thiohydroxylamine** (¶ 193).

Amide radicals are named as compound or complex radicals based on "amino" or "imino," with acid radicals (¶ 168) as substituents.

Examples:

AcNH—	(acetylamino)(not acetamido)
BzN=	(benzoylimino)
$Me(CH_2)_4CONH-$	[(1-oxohexyl)amino](not hexanamido)
$PhSO_2NH-$	[(phenylsulfonyl)amino](not benzenesulfonamido)
PhC(=NH)NH—	[(iminophenylmethyl)amino]

The radicals above are employed as substituents when a more preferred amide or a higher function, e.g., an acid or acid chloride, is present in part of the molecule attached to the amide by way of the nitrogen atom. Other attachments call for use of amino and oxo, or (aminocarbonyl), (aminosulfonyl), etc., radicals.

Examples:

$H_2NCO(CH_2)_2CO_2H$ — **Butanoic acid, 4-amino-4-oxo-**

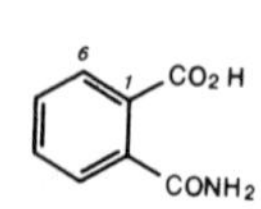

Benzoic acid, 2-(aminocarbonyl)-

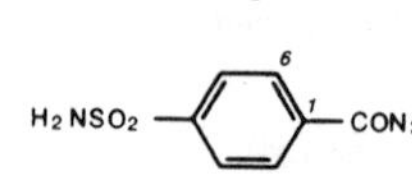

Benzoyl azide, 4-(aminosulfonyl)-

$H_2NCSCH_2CONHMe$	**Propanamide, 3-amino-*N*-methyl-3-thioxo-**
$AcNHSO_2Ph$	**Acetamide, *N*-(phenylsulfonyl)-**

Replacement ("a") nomenclature is employed where applicable (¶ 127) for naming polyamides, but not peptides (¶ 206).
Example:

$H_2NCOCH_2CONH(CH_2)_2NHCOCH_2CONH(CH_2)_2NHCOCH_2CONH_2$

4,7,11,14-Tetraazaheptadecanediamide, 3,8,10,15-tetraoxo-

Amides incorporated in a ring system are indexed at the ring name and ranked according to the nature of the suffix, if any (¶ 106), not as amides.
Examples:

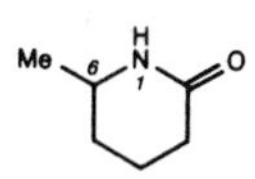

2-Piperidinone, 6-methyl- (ranked as a ketone)

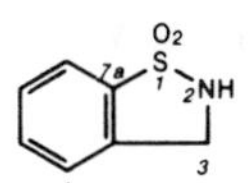

1,2-Benzisothiazole, 2,3-dihydro-1,1-dioxide (ranked as a non-functional cyclic nitrogen skeleton)

Amides of which only the nitrogen atom forms part of a ring system are indexed as *N*-acyl derivatives at the ring name. They are regarded as "unexpressed amides" and rank just below an "expressed amide" *of the same type of acid.* Lower functions are disregarded unless they can be expressed as suffixes at the name of the ring of which the amide nitrogen is part. Res= idues of formic acid, carbonic acid and related compounds are expressed at the nitrogen heterocycle by such suffixes as "-carboxaldehyde," "-carboxylic acid," "carboxamide," etc., and the compounds are ranked accordingly.
Examples:

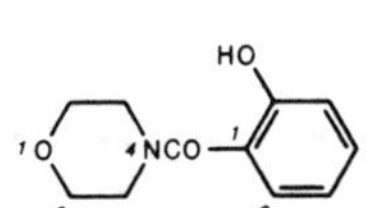

Morpholine, 4-(2-hydroxybenzoyl)-

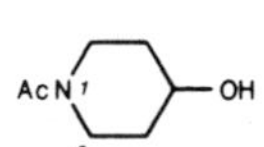

4-Piperidinol, 1-acetyl-

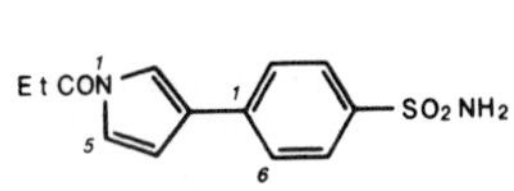

1*H*-Pyrrole, 3-[4-(amino= sulfonyl)phenyl]-1-(1-oxo= propyl)- (unexpressed carboxamide is favored over expressed sulfon= amide)

Amides of amino acids with trivial names are named systematically except in peptide nomenclature (¶ 206).
Example:

$H_2NCH_2CONH_2$	**Acetamide, 2-amino-** (not Glycin= amide)

Oximes of amides (amidoximes) (R-C(:NOH)NH_2) are tautomeric with *N*-hydroxy carboximidamides (R-C(:NH)NHOH) and are indexed at imid= amide names:
Example:

$Me(CH_2)_4C(=NH)NHOH$	**Hexanimidamide, *N*-hydroxy-**

172. Nitriles (RC≡N) are indexed at names derived from "-carboxylic" and "-oic" acid names by use of "-carbonitrile" and "-nitrile" suffixes, re= spectively.
Examples:

MeCN	**Acetonitrile**
HC≡CCH=CHCN	**2-Penten-4-ynenitrile**
NCCN	**Ethanedinitrile** (not Cyanogen)
H_3GeCN	**Germanecarbonitrile**
H_2NNHCN	**Hydrazinecarbonitrile**
$(NC)_2C=C(CN)_2$	**Ethenetetracarbonitrile**
PhCN	**Benzonitrile**

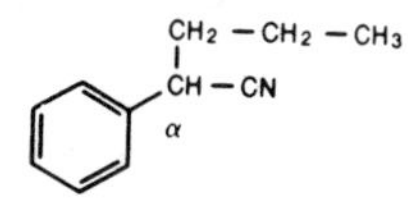

Benzeneacetonitrile, α-propyl-

Compounds of the type RC≡N⁺-N̄R' are indexed at **Nitrilimine.**
Example:

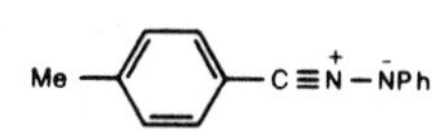

Nitrilimine, *C*-(4-methyl= phenyl)-*N*-phenyl-

In the presence of higher functions, nitriles are always expressed as cyano radicals.
Examples:

$NCCH_2CONH_2$	**Acetamide, 2-cyano-** (not Propanamide, 3-nitrilo-)
MeCH(CN)COCl	**Propanoyl chloride, 2-cyano-**

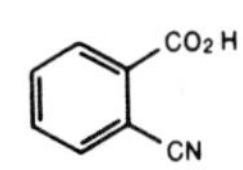

Benzoic acid, 2-cyano-

173. Aldehydes, RCHO, are named from "-carboxylic" and "-oic" acids by use of "-carboxaldehyde" and "-al" suffixes, respectively.
Examples:

AcH(MeCHO)	**Acetaldehyde**
MeCH=CHCHO	**2-Butenal**
$HCOCH_2CH=CHC≡CCH_2CHO$	**3-Octen-5-ynedial**
$H_2P(O)CHO$	**Phosphinecarboxaldehyde oxide**

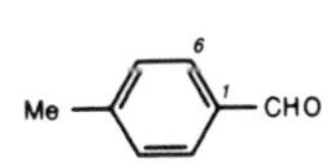

Benzaldehyde, 4-methyl-

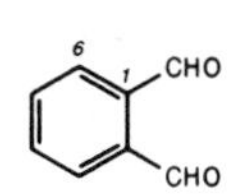

1,2-Benzenedicarboxaldehyde

$PhCH_2CHO$	**Benzeneacetaldehyde**

Chalcogen analogs of aldehydes are given "-thial," "-selenal," "-carbo= thioaldehyde," etc., names.
In the presence of more highly ranked compound classes (¶ 106) or more preferred aldehydes, the -CHO group is expressed by formyl (if it does not form part of an acyclic carbon chain) or by a terminal oxo radical. For thio aldehydes, the equivalent radicals are (thioxomethyl) and a terminal thioxo radical.
Examples:

$HCOCH_2CO_2H$	**Propanoic acid, 3-oxo-**

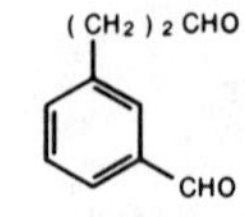

Benzenepropanal, 3-formyl-
(principle: largest heading parent)

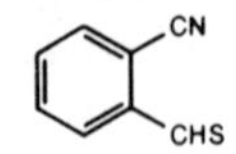

Benzonitrile, 2-(thioxomethyl)-

$TeCH(CH_2)_2CHS$ — **Butanethial, 4-telluroxo-**

Substitution of the aldehydic hydrogen atom by radicals derived from molecular skeletons is avoided; but radicals such as nitro and nitroso (or, for Formaldehyde and its chalcogen analogs only, a single sulfonyl or sulfinyl) are permitted.

Examples:

MeC(=NOH)NO — **Acetaldehyde, 1-nitroso-**
oxime (not Acetonitrosolic acid)

$BzNO_2$ — **Benzaldehyde, α-nitro-**

174. Ketones, RC(:O)R', and their chalcogen analogs are named by use of the characteristic suffixes -one, -thione, -selone, and -tellone. (The last two classes must be differentiated from selenones and tellurones, which contain the noncarbon groups $-SeO_2-$ and $-TeO_2-$, respectively.) In acyclic ketones the group must generally be in a nonterminal position of a carbon or "a"-named chain and have two carbon attachments. Ketenes, in which a terminal unsaturated carbon bears a chalcogen atom, are exceptions; they are named by use of the same suffixes as ketones. Conjunctive names are not employed for acyclic ketones attached to ring systems; instead, the cyclic portion is expressed as a substituent of the acyclic ketone parent (in which the oxo group may occupy the 1-position). When an acyclic ketone with two cyclic substituents (attached at carbon atoms) consists only of a single carbon atom with a chalcogen attached, the heading parents **Methanone, Methanethione, Methaneselone** and **Methanetellone** are employed.

Examples:

MeCOMe (1 2 3) — **2-Propanone** (not Acetone)

$MeCH_2OSiH_2(CH_2)_2CO(CH_2)_2SiH_2OCH_2Me$ (1 3 4 7 10 11 13) — **3,11-Dioxa-4,10-disilatridecan-7-one**

$MeCOCH_2COCH=CH_2$ (1 6) — **5-Hexene-2,4-dione** (not 1-Hexene-3,5-dione)

$H_2C=C=O$ — **Ethenone** (not Ketene)

$Me(CH_2)_3CH=C=O$ (6 1) — **1-Hexen-1-one**

$MeCSeCH_2Me$ (1 4) — **2-Butaneselone**

$MeCOCH_2CSCH_2Me$ (1 6) — **2-Hexanone, 4-thioxo-** (not 2,4-Hexanedione, 4-thio-)

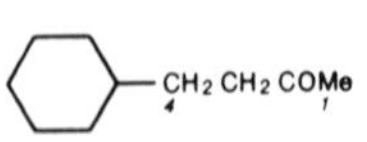

2-Butanone, 4-cyclohexyl-

$PhCO(CH_2)_4Me$ — **1-Hexanone, 1-phenyl-** (not Hexanal, 1-phenyl-)

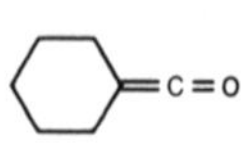

Methanone, cyclohexylidene-

Ph_2CO — **Methanone, diphenyl-** (not Benzophenone)

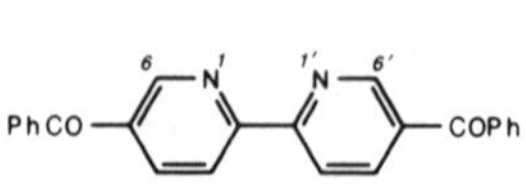

Methanone, [2,2'-bipyridine]-5,5'-diylbis[phenyl- (a multiplicative name (¶ 125))

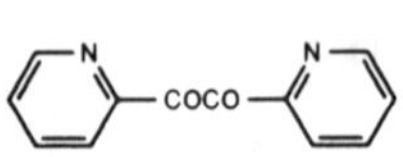

Ethanedione, di-2-pyridinyl-

Cyclic ketones (including those with neighboring hetero atoms) are named by appending "-one," "-thione," etc., suffixes to the ring names. Indicated hydrogen (required to form the ring system itself) is often chosen (where possible) to accommodate a single ketone group (¶ 135). In other cases, added hydrogen (¶ 136) is introduced at some other position of the ring system when the ketone suffix is attached.

Examples:

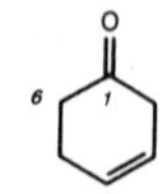

3-Cyclohexen-1-one

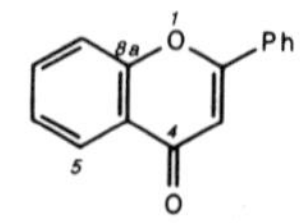

4*H*-1-Benzopyran-4-one, 2-phenyl-
(not Flavone)

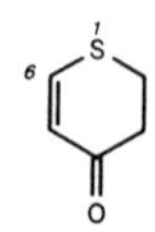

4*H*-Thiopyran-4-one, 2,3-dihydro-
(not 2*H*-Thiopyran-4(3*H*)-one)

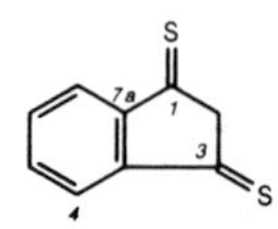

1*H*-Indene-1,3(2*H*)-dithione

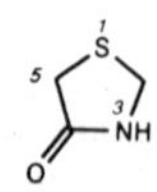

4-Thiazolidinone

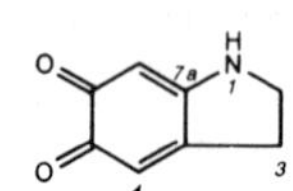

1*H*-Indole-5,6-dione, 2,3-dihydro-
(the two oxo substituents can be added as a pair without hydrogen being added elsewhere to the ring system)

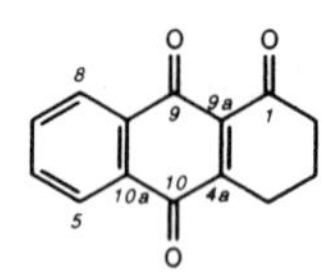

1,9,10(2*H*)-Anthracenetrione, 3,4-dihydro- (the added hydrogen is cited in the lowest-numbered available position)

In the presence of higher functions or more preferred ketones, oxo, thioxo, selenoxo, and telluroxo radicals, are used. The =CO group, when not part of an acyclic chain or a ring, is expressed as carbonyl; the chalcogen analogs (carbonothioyl, carbonoselenoyl, and carbonotelluroyl) are used when they are not part of a chain, and when, in addition, they are either bonded to a single atom or used as multiplicative radicals. The trivially named radicals benzoyl, for (oxophenylmethyl), and acetyl, for (1-oxoethyl), are used when appropriate.

Examples:

$MeCOCH_2COBr$ — **Butanoyl bromide, 3-oxo-**
(not Acetyl bromide, acetyl-)

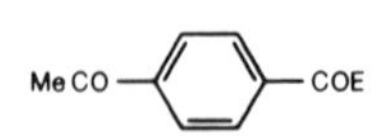

1-Propanone, 1-(4-acetylphenyl)-

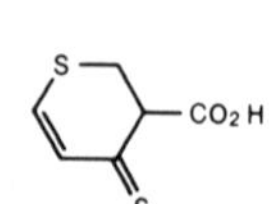

2*H*-Thiopyran-3-carboxylic acid, 3,4-dihydro-4-thioxo-

$HCOCH_2COCO_2H$ (4 1) — **Butanoic acid, 2,4-dioxo-** (not Propanoic acid, 3-formyl-2-oxo-)

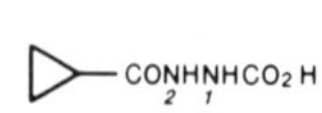

Hydrazinecarboxylic acid, 2-(cyclopropylcarbonyl)-

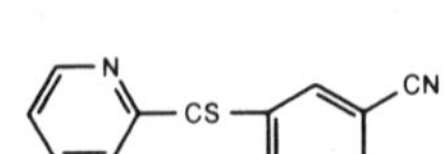

Benzonitrile, 3-(2-pyridinylthioxomethyl)- (not Benzonitrile, 3-(2-pyridinylcarbonothioyl)-)

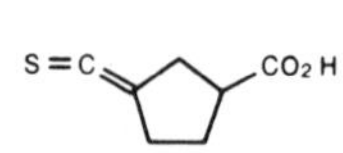

Cyclopentanecarboxylic acid, 3-carbonothioyl-

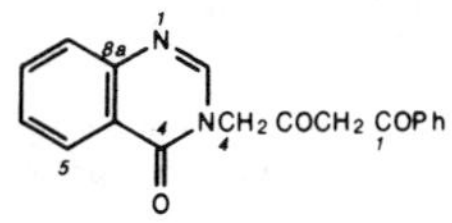

1,3-Butanedione, 4-(4-oxo-3(4*H*)-quinazolinyl)-1-phenyl- (principle: maximum number of the principal chemical functional group (¶ 138))

Carbonyl groups and their chalcogen analogs in heterocycles are named and ranked as ketones, thiones, etc., even when a neighboring ring atom is not carbon (see above), but acyclic carbonyl (etc.) groups attached to cyclic or acyclic hetero atoms (including silicon) are not recognized as ketones, and the functionality of the group is disregarded. It is expressed, not as a suffix, but as an appropriate substituent prefix (acetyl, oxo, carbonyl, (thioxomethyl), etc.) of a heading parent.

Examples:

AcNHN=NH — **1-Triazene, 3-acetyl-** (not Ethanone, 1-(2-triazenyl)-)

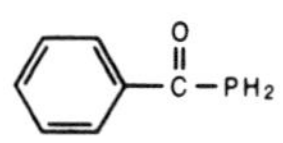

Phosphine, benzoyl-

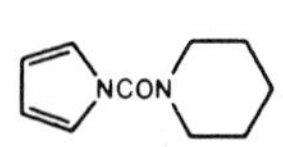

Piperidine, 1-(1*H*-pyrrol-1-ylcarbonyl)- (an unexpressed amide (¶ 171))

$EtCOSiH_2OH$ — **Silanol, (1-oxopropyl)-**

$H_2P(S)CSPh$ — **Phosphine sulfide, (phenylthioxomethyl)-**

175. **Alcohols (and phenols)** and their chalcogen analogs (thiols, selenols, and tellurols) are expressed by the suffixes -ol, -thiol, -selenol, and -tellurol, attached to a carbon or silicon atom of a molecular skeleton name. The only trivial name employed in *CA* indexes for a compound of this class is **Phenol** (for Benzenol). Phenols as a class are treated precisely like alcohols, the choice of index name for a compond containing alcoholic and phenolic groups depending on the usual rules (¶ 138). Alcoholic groups and their analogs are expressed as hydroxy, mercapto, selenyl, and telluryl prefixes on more preferred heading parents.

Examples:

EtOH — **Ethanol** (not Ethyl alcohol)

$H_2C=CHCH_2OH$ — **2-Propen-1-ol** (not Allyl alcohol)

MeCH(OH)Me — **2-Propanol** (not Isopropyl alcohol)

$H_2C=CHCH_2CH(SH)CH_2Me$ — **5-Hexene-3-thiol**

$MeSiH_2OH$ — **Silanol, methyl-**

$Me(CH_2)_2CH_2TeH$ — **1-Butanetellurol**

$CH_2-(CH_2)_{11}-S-CH_2-CH_2-O-CH_2-CH_2-O-CH_2-CH_2-O-CH_2-CH_2-OH$

24 12 11 10 9 8 7 6 5 4 3 2 1

3,6,9-Trioxa-12-thiatetracosan-1-ol

PhSeH — **Benzeneselenol**

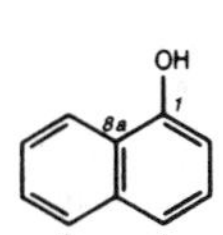

1-Naphthalenol (not 1-Naphthol)

$CH_3-CH(SH)-CH(SH)-CH_3$ — **2,3-Butanedithiol**

$HO-SiH_2-SiH_2-OH$ — **1,2-Disilanediol**

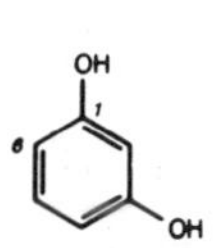

1,3-Benzenediol

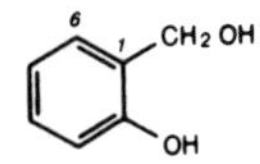

Benzenemethanol, 2-hydroxy- (not Phenol, 2-(hydroxymethyl)-) (principle: largest parent)

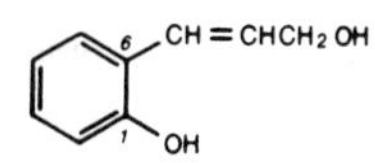

Phenol, 2-(3-hydroxy-1-propenyl)- (not 2-Propen-1-ol, 3-(2-hydroxyphenyl)-)

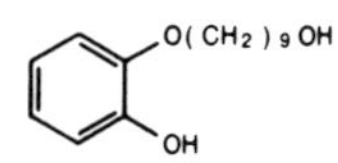

Phenol, 2-[(9-hydroxynonyl)oxy]- (principle: a cyclic parent is preferred (¶ 138))

$HSCH_2CH_2OH$ — **Ethanol, 2-mercapto-**

Hydroxy derivatives of thiophene, selenophene, and tellurophene are indexed at **Thiophene-ol**, etc., (the final "e" of the parents being left unelided to avoid confusion with chalcogen analogs of **Phenol**). The locant is placed immediately before the suffix.

Example:

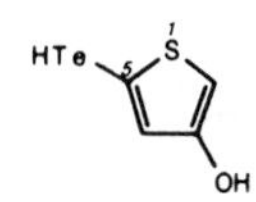

Thiophene-3-ol, 5-telluryl-

Hydroxy, mercapto, selenyl, and telluryl groups attached to hetero atoms other than silicon are always expressed as prefixes (unless they form part of an acid functional parent compound (¶ 130)). "Esters" of such groups are also expressed as prefixes.

Examples:

H_2NNHSH — **Hydrazine, mercapto-**

Me_2BOH — **Borinic acid, dimethyl-**

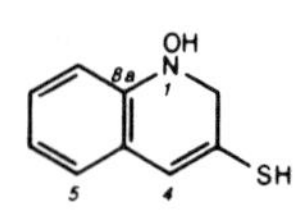

3-Quinolinethiol, 1,2-dihydro-1-hydroxy-

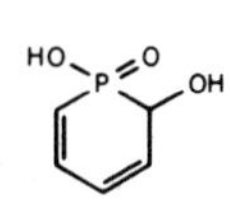

2-Phosphorinol, 1,2-dihydro-1-hydroxy-
1-oxide

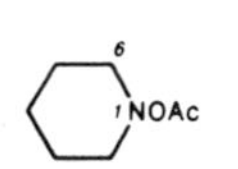

Piperidine, 1-(acetyloxy)-

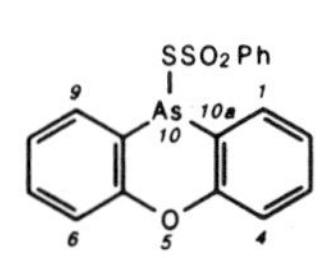

10*H*-Phenoxarsine, 10-[(phenylsulfonyl)thio]-

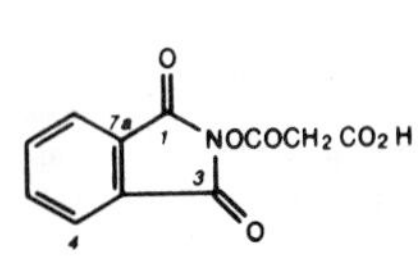

Propanoic acid, 3-[(1,3-dihydro-1,3-dioxo-2*H*-isoindol-2-yl)oxy]-3-oxo-

176. **Amines** are always named as primary amines, RNH_2, or their *N*-derivatives, by attaching the suffix "-amine" to the name of a molecular skeleton, cyclic or acyclic. Attachment may be at a carbon or hetero atom. Trivial names, e.g., Aniline, and radicofunctional names, e.g., Methylamine, are not used. Conjunctive names (¶ 124) are employed for monoamines where appropriate.

Examples:

$MeNH_2$ — **Methanamine** (not Methylamine)

$MeCH(NH_2)Me$ — **2-Propanamine** (not Isopropylamine)

$H_2C=CHCH(NH_2)Me$ — **3-Buten-2-amine**

H_3PbNH_2 — **Plumbanamine**

H_2PPHNH_2 — **Diphosphinamine**

$PhNH_2$ — **Benzenamine** (not Aniline)

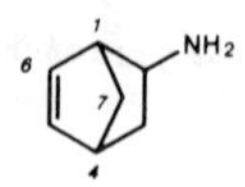

Bicyclo[2.2.1]hept-5-en-2-amine

$H_2NBiHNH_2$ — **Bismuthinediamine**

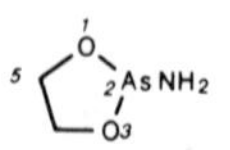

1,3,2-Dioxarsolan-2-amine

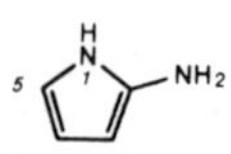

1*H*-Pyrrol-2-amine

$H_2NCH_2CH_2NH_2$ — **1,2-Ethanediamine** (not Ethylene= diamine)

$H_2NCH_2CH(NH_2)Me$ — **1,2-Propanediamine**

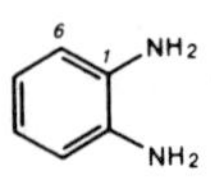

1,2-Benzenediamine (not 1,2-Phenylenediamine)

$PhCH_2CHMeNH_2$ — **Benzeneethanamine, α-methyl-**

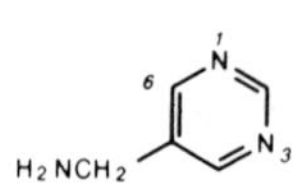

5-Pyrimidinemethanamine

When higher functions (¶ 106) or more preferred amine parents are present, the prefix "amino" is employed.
Examples:

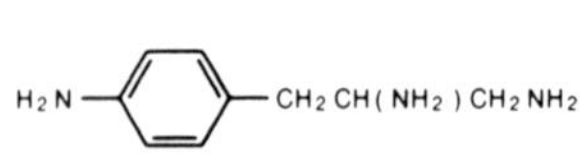

1,2-Propanediamine, 3-(4-amino= phenyl)- (principle: maximum number of the principal group)

$H_2N(CH_2)_5SiH_2NH_2$ — **Silanamine, 1-(5-aminopentyl)-** (principle: hetero-atom parent preferred)

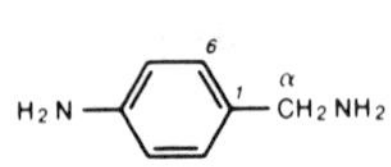

Benzenemethanamine, 4-amino- (principle: largest heading parent)

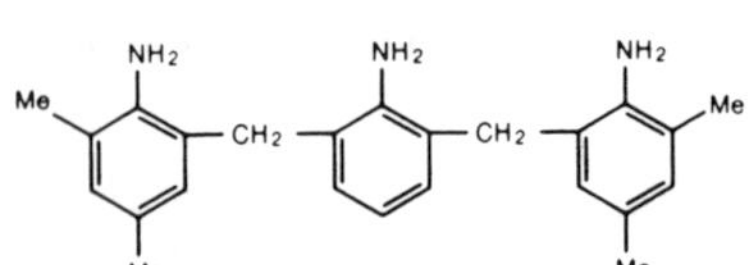

Benzenamine, 2,6-bis[(2-amino-3,5-dimethylphenyl)methyl]- (principle: centrality)

Secondary and tertiary amines, RR'NH and RR'R''N, are named as de= rivatives of primary amines by application of the usual criteria (¶ 138).
Examples:

$MeCH_2NHCH_2Me$ — **Ethanamine, *N*-ethyl-** (not Di= ethylamine)

$(Me_2CH)_3N$ — **2-Propanamine, *N*,*N*-bis(1-methylethyl)-** (not Triisopropyl= amine)

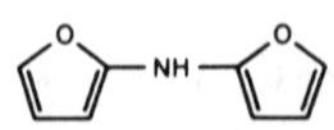

2-Furanamine, *N*-2-furanyl-

$H_2N-CH_2-CH_2-CH_2-NH-CH_2-CH_2-CH_2-NH_2$

1,3-Propanediamine, *N*-(3-amino= propyl)-

$H_3SiNHSnH_3$ — **Silanamine, *N*-stannyl-** (principle: preferred hetero atom)

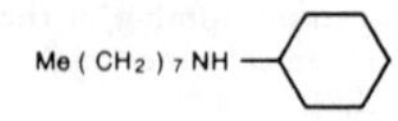

Cyclohexanamine, *N*-octyl- (prin= ciple: a cyclic parent is preferred)

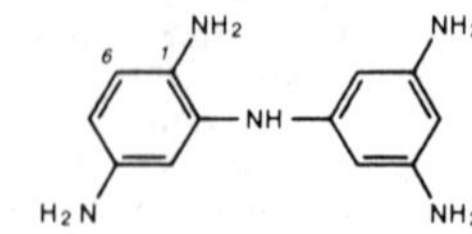

1,2,4-Benzenetriamine, N^2-(3,5-diaminophenyl)- (principle: lowest locants for principal groups)

Me_2NONMe_2 — **Methanamine, *N*,*N*'-oxybis[*N*-methyl-** (principle: multiplication)

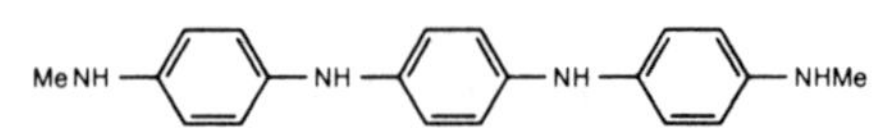

1,4-Benzenediamine, *N*,*N*'-bis= [4-(methylamino)phenyl]- (principle: centrality)

Replacement names are employed for acyclic secondary and tertiary amines provided that the "-amine" suffix of the "aza" name expresses at least the same number of such groups as the conventional name and that other requirements (¶ 127), e.g., the presence of a minimum of four hetero units in the molecular skeleton, are satisfied.
Examples:

$EtNHCH_2CH_2NH[(CH_2)_2NH]_3CH_2CH_2NHEt$

3,6,9,12-Tetraazatetradecane-1,14-diamine, *N*,*N*'-diethyl-

$CH_3-NH-CH_2-CH_2-NH-CH_2-CH_2-NH-CH_2-CH_2-NH-CH_2-CH_2-NH-CH_3$

1,2-Ethanediamine, *N*-[2-(methy= lamino)ethyl]-*N*'-[2-[[2-(me= thylamino)ethyl]amino]ethyl]- (not 3,6,9-Triazaundecan-1,11-diamine, *N*,*N*'-dimethyl-; this or= ganic replacement name expresses only three hetero units) (not 2,5,8,11,14-Pentaazapentadecane; this name does not express the amino groups as a substituent suf= fix)

Schiff bases (anils, azomethine compounds) contain the -N=C-grouping and are therefore both amines and imines. They are indexed as amines in the absence of higher functions.
Example:

$PhN=CHPh$ — **Benzenamine, *N*-(phenyl= methylene)-**

N-Hydroxy amines are named as such, not as hydroxylamine derivatives. Amine oxides are named by citation of the additive term "*N*-oxide" in the modification.
Examples:

$PhNHOH$ — **Benzenamine, *N*-hydroxy-**

$EtN(O)Et_2$ — **Ethanamine, *N*,*N*-diethyl-*N*-oxide**

177. **Imines** are ranked as the lowest compound class named by use of a functional suffix. The "-imine" suffix is attached to a cyclic or acyclic mo= lecular skeleton (at a carbon or hetero atom). Indicated and added hydrogen (¶¶ 135, 136) for cyclic imines are assigned as for the analogous ketones (¶ 174). *N*-Alkyl, *N*-aryl, etc., imines are indexed as amines (¶ 176). Conjunctive nomenclature is used for imines when the molecular skeleton to which a single function is attached is itself connected to a ring system by a single bond.
Examples:

$MeCH=NH$ — **Ethanimine** (not Ethylidenimine)

$MeC(=NH)Me$ — **2-Propanimine**

$HP=NH$ — **Phosphinimine**

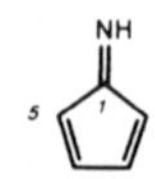

2,4-Cyclopentadien-1-imine

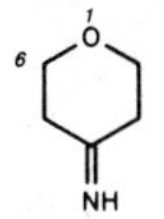

4*H*-**Pyran-4-imine, tetrahydro-** (not 2*H*-Pyran-4(3*H*)-imine, dihydro-)

HN=C=NH

Methanediimine (not Carbodiimide)

$Me_2CHN{=}C{=}NCHMe_2$

2-Propanamine, *N,N'*-**methanetetraylbis-** (not Methanediimine, bis(1-methylethyl)-)

MeCH=N(O)Cl

Ethanimine, *N*-**chloro-** *N*-**oxide**

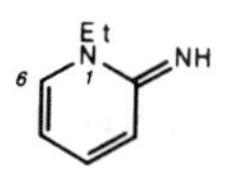

2(1*H*)-Pyridinimine, 1-ethyl-

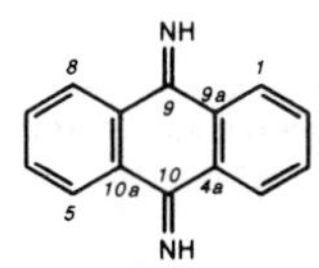

9,10-Anthracenediimine

$Ph_2C{=}NSSN{=}CPh_2$

Benzenemethanimine, *N,N'*-**dithiobis[**α-**phenyl-** (principle: multiplication of a conjunctive name)

In the presence of any other chemical function expressible as a suffix, imines are described by substituent prefixes. The =NH group is named imino; the =C=NH group is expressed as carbonimidoyl in a multiplying radical or when attached to a single atom; the -CH=NH group is named (iminomethyl) (not formimidoyl) unless the methyl group is part of an acyclic carbon chain. Examples:

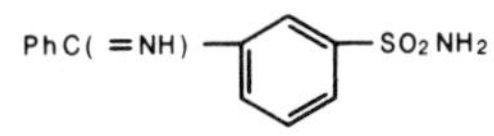

Benzenesulfonamide, 3-(iminophenylmethyl)-

$HN{=}CHCH_2CN$

Propanenitrile, 3-imino-

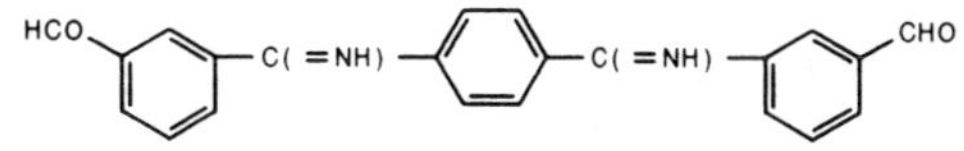

Benzaldehyde, 3,3'-(1,4-phenylenedicarbonimidoyl)bis-

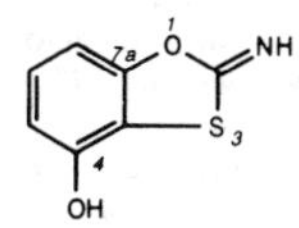

1,3-Benzoxathiol-4-ol, 2-imino-

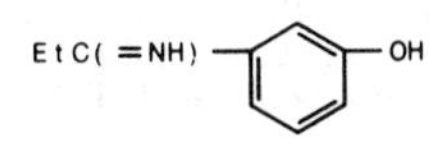

Phenol, 3-(1-iminopropyl)-

178. Introduction. Previous sections have dealt with general principles and with the naming of specific compounds by combination of a molecular skeleton name with a suffix which describes the principal chemical function. The present section discusses nonfunctional compound classes (those that fall below imines in order of precedence (¶ 106)), and classes of compounds such as esters, free radicals, ions, addition compounds, oxo acids, and carbonic acid relatives, which are named by application of principles already discussed. The order is alphabetic by class.

179. Anhydrides of acid groups, at least one of which is expressed as a functional suffix ("-oic acid," "-carboxylic acid," "-sulfonic acid," etc.), are indexed, if cyclic, at heterocycle names and, if acyclic, either at "a" names (¶ 127) or at acid heading parents with the term "anhydride" in the modification. (Acyclic anhydrides of certain mononuclear "oxo" acids, e.g., **Carbonic acid, Phosphonic acid,** are indexed at such headings as **Dicarbonic acid, Tri=phosphonic acid.**)

Replacement ("a") names are used for acyclic anhydrides when the suffix expresses no lower functionality than the heading parent of the regular sub=stitutive name and the other requirements (¶ 127), e.g., that the molecular skeleton contain at least four hetero units, are satisfied.

Example:

$HO_2CCH_2COOSO_2CH_2SO_2OCOCH_2CO_2H$ (locants 1, 3, 4, 5, 7, 8, 11)

4,8-Dioxa-5,7-dithiaundecanedioic acid, 3,9-dioxo-5,5,7,7-tetraoxide

Symmetrical anhydrides of monobasic organic acids are indexed at the acid heading parent with the simple term "anhydride" in the modification. Un=symmetrical anhydrides of monobasic acids, at least one of which is organic, are indexed at the name of the preferred acid (¶¶ 167, 138) with an "anhydride with" phrase in the modification. Anhydrides of **Hydrazinecarboxylic acid** and related compounds are given special treatment (see the final example below).

Examples:

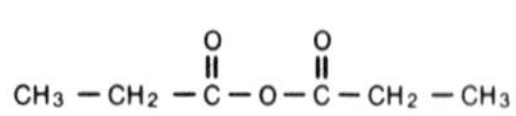

Propanoic acid anhydride

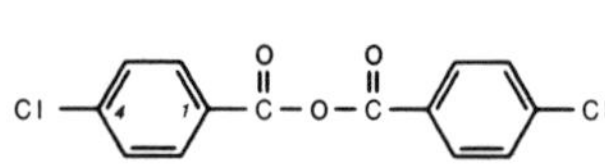

Benzoic acid, 4-chloro- anhydride

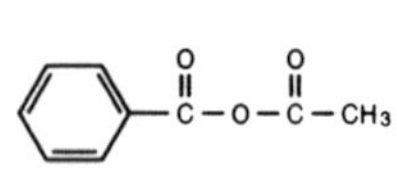

Benzoic acid anhydride with acetic acid

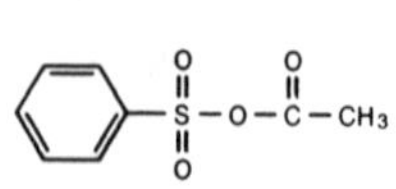

Acetic acid anhydride with benzenesulfonic acid

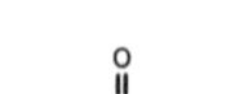

Acetic acid anhydride with nitrous acid

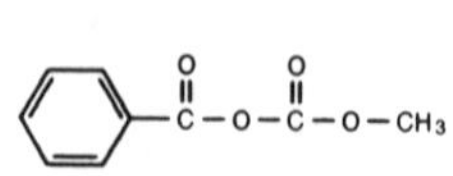

Benzoic acid anhydride with methyl hydrogen carbonate (the ester name is given in its uninverted form (¶ 185); not Benzoic acid, anhydride with methoxyformic acid)

Dicarbonic dihydrazide, 2-methyl- (see also ¶ 183)

Anhydrides of monobasic organic acids with polybasic acids are indexed at the name of the preferred acid. The terms "mono," "di," etc., are used to indicate the number of molecules of water lost in anhydride formation, and locants must sometimes be cited. When three or more different acids are involved, the "anhydride with" term is repeated.

Examples:

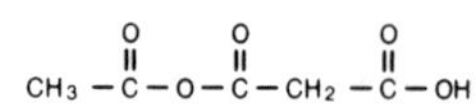

Propanedioic acid monoanhydride with acetic acid

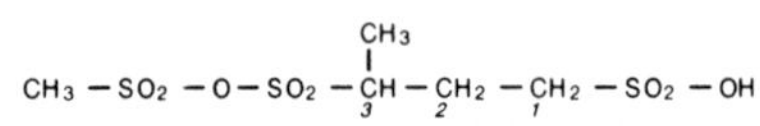

1,3-Butanedisulfonic acid 3-anhydride with methane=sulfonic acid (not "3-mono=anhydride with...")

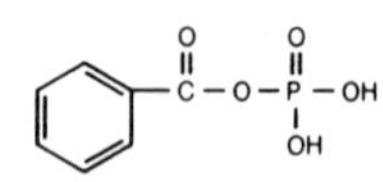

Benzoic acid monoanhydride with phosphoric acid

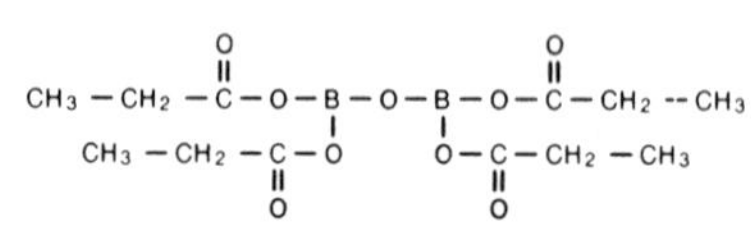

Propanoic acid tetraanhydride with boric acid ($H_4B_2O_5$)

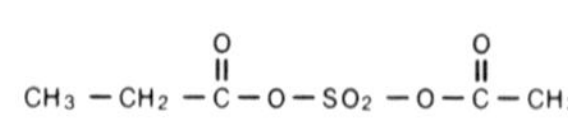

Propanoic acid monoanhydride with sulfuric acid, anhydride with acetic acid

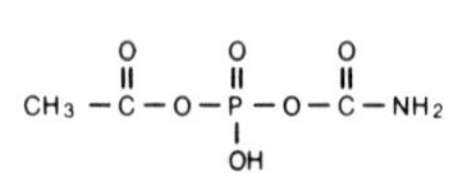

Acetic acid monoanhydride with phosphoric acid, monoanhydride with car=bamic acid (repetition of "mono" is necessary here)

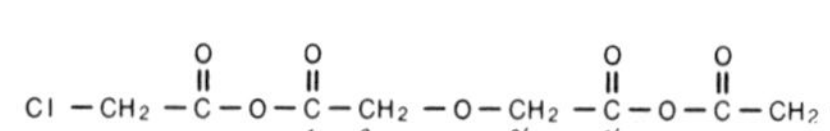

Acetic acid, 2,2'-oxybis- dianhydride with chloroacetic acid (principle: centrality (¶ 138))

Butanedioic acid monoanhydride with cyanic acid

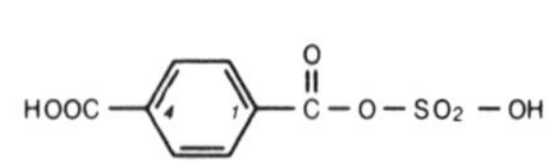

1,4-Benzenedicarboxylic acid monoanhydride with sulfuric acid

Acyclic anhydrides of different polybasic acids sometimes require ratios.

Examples:

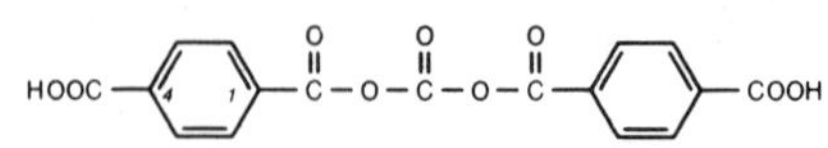

1,4-Benzenedicarboxylic acid anhydride with carbonic acid (2:1)

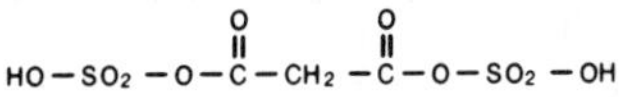

**Propanedioic acid
anhydride with sulfuric acid
(1:2)**

Anhydrides of a single polybasic organic acid have such terms as "bimol. monoanhydride" and "trimol. dianhydride" in the modification, with locants when necessary.

Examples:

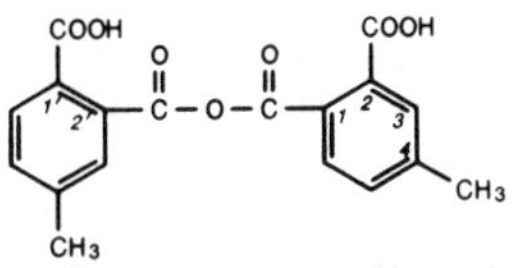

**1,2-Benzenedicarboxylic acid, 4-methyl-
bimol. 1,2'-monoanhydride**

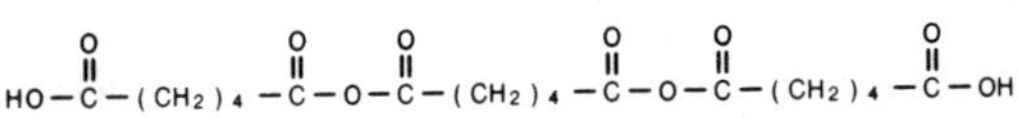

**Hexanedioic acid
trimol. dianhydride**

When the acid group which has undergone anhydride formation would have been expressed (in its unmodified form) as a substituent of the preferred heading parent, the anhydride is likewise expressed as a (more complex) substituent of the same parent. Anhydride formation of an acid group ex= pressed as a substituent in an "anhydride with" phrase is treated similarly.

Examples:

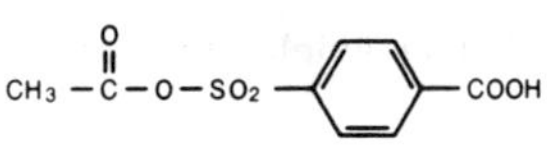

Benzoic acid, 4-[(acetyloxy)= sulfonyl]-

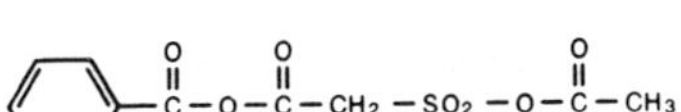

**Benzoic acid
anhydride with [(acetyloxy)= sulfonyl]acetic acid**

Chalcogen analogs of acyclic anhydrides are indexed like anhydrides. When the oxygen atom connecting the acid residues has been replaced, the terms "anhydrosulfide," "anhydroselenide," and "anhydrotelluride" are used, and the sulfur, selenium, or tellurium is indicated in the names of *both* acid components.

Examples:

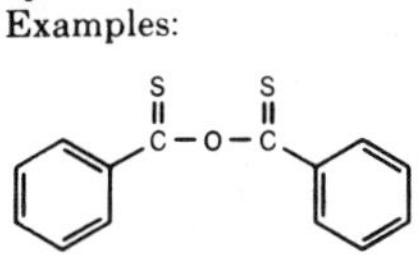

**Benzenecarbothioic acid
anhydride** (not "anhydrosulfide")

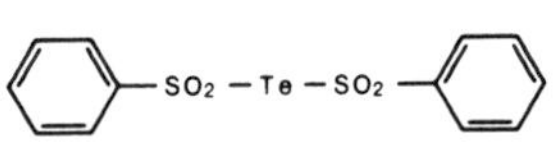

**Benzenesulfonotelluroic acid
anhydrotelluride**

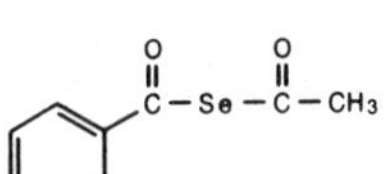

**Benzenecarboselenoic acid
anhydroselenide with ethane= selenoic acid**

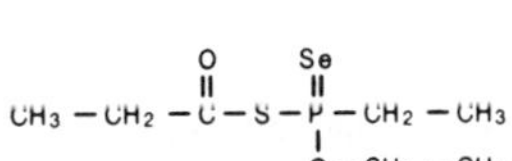

**Propanethioic acid
anhydrosulfide with *O*-ethyl hydrogen ethylphosphonosel= enothioate**

MeC(=NOH) SCN

**Ethanimidothioic acid, *N*-hydroxy-
anhydrosulfide with thiocyanic acid**

Symmetrical anhydrides of the monobasic "oxo" acids **Formic, Phos= phinic, Arsenic, Phosphinous,** and **Arsinous acids** (and their substituted derivatives) are indexed by citation of the simple term "anhydride" in the modification. Symmetrical anhydrides of the dibasic "oxo" acids **Phosphonic, Arsonic, Phosphonous,** and **Arsonous acids** and their substituted deriva= tives are indexed at Di-, Tri-, etc., acid headings.

Example:

MeAs (OH) OAs (OH) Me

Diarsonous acid, dimethyl-

Anhydrides of **Borinic acid** are indexed at **Borane.** Symmetrical anhy= drides of **Boronic acid** are indexed at the acid names with "bimol. mono= anhydride," etc., in the modification.

Examples:

PhBMeOBMePh

Borane, oxybis[methylphenyl-

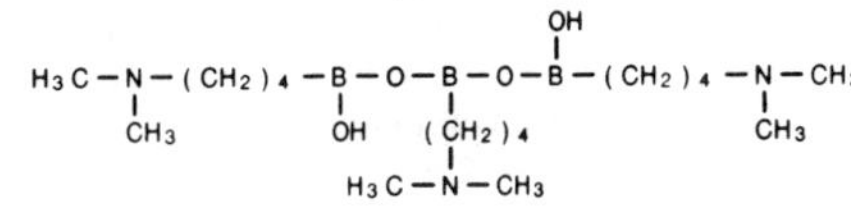

**Boronic acid, [4-(dimethyl= amino)butyl]-
trimol. dianhydride**

Unsymmetrical anhydrides of "oxo" acids in general are indexed by use of "anhydride with" phrases at the preferred acid component heading parents, but cyanic acid anhydrides with mononuclear arsenic and phosphorus acids are expressed by means of "cyanatido" replacement affixes or by the class term "cyanate."

Examples:

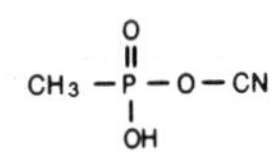

Phosphonocyanatidic acid, methyl-

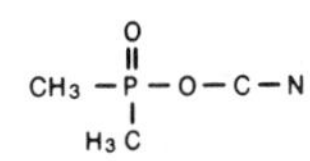

Phosphinic cyanate, dimethyl-

Anhydrosulfides of "oxo" acids are generally named analogously by use of "anhydrosulfide" terms. Other chalcogens are treated similarly. When sulfur, etc., replaces oxygen in **Diphosphonic acid** and similar compounds, the nondetachable prefixes "Thio," etc., are employed. The number of sulfur atoms is not indicated in the name; instead, a synonym line formula is always cited.

Example:

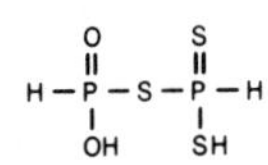

**Thiodiphosphonic acid
((HO)HP(O)SHP(S)SH)**

The peroxy analogs of this kind of "oxo" acid are indexed (with synonym line formulas) at such headings as **Thioperoxydiarsonic acid ([(HO)H= As(S)]$_2$S$_2$).** Anhydrides of mononuclear peroxy "oxo" acids are generally named at **Peroxide, Disulfide,** etc.

Examples:

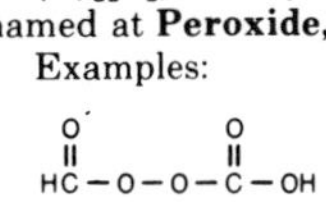

Peroxide, carboxy formyl

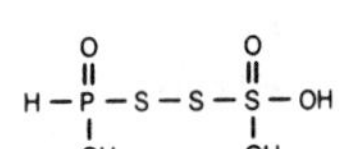

Disulfide, hydroxyphosphinyl sulfo

Cyclic anhydrides, anhydrosulfides, etc., are indexed like other heterocyclic compounds.

Examples:

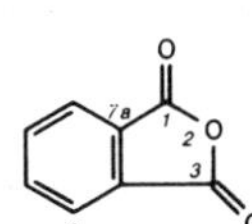

1,3-Isobenzofurandione (not 1,2-= Benzenedicarboxylic acid, cyclic anhydride)

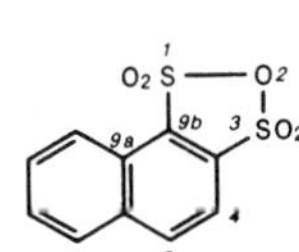

Naphth[1,2-*c*][1,2,5]oxadithiole 1,1,3,3-tetraoxide (not 1,2- Naphthalenedisulfonic acid, cyclic anhydride)

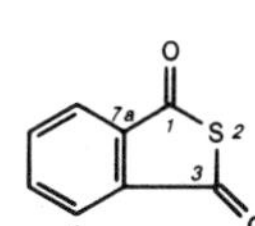

Benzo[*c*]thiophene-1,3-dione

180. Anions. Index names for anions are required as sole entries when anions themselves are being studied, and as additional entries in the indexing of salts (¶ 198). Anions are often expressed differently as modification terms at cationic heading parents.

Anions from unsubstituted **Ethyne, Arsine, Phosphine, Stibine, Silane** (Si^{4-} only) and **Hydrazine** are named **Acetylide, Arsenide, Phosphide, Antimonide, Silicide,** and **Hydrazide.** Synonym line formulas are used, e.g., **Acetylide** (C$_2^{2-}$), except for **Hydrazide** (which is H$_2$NNH$^-$) and **Silicide,** and for **Arsenide, Phosphide,** and **Antimonide** when all hydrogens have been lost.

Anions derived from compounds with names based on substitutive parent compounds (¶ 130) other than those just described are named at the heading parent for the neutral compound with a modification term such as "ion(1-)," "ion(2-)," or (if indefinite) "ion (neg)." Anions from esters of "oxo" acids are named similarly.

Examples:

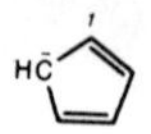

1,3-Cyclopentadiene ion(1-)

$MeCH_2CH_2O^-$

1-Propanol ion(1-) (not Propoxide)

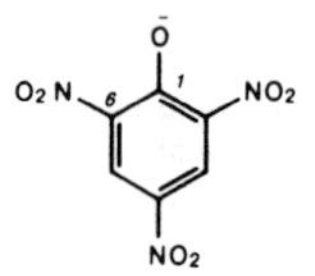

Phenol, 2,4,6-trinitro- ion(1-)

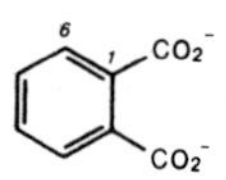

1,2-Benzenedicarboxylic acid ion(2-)

Me_2As^-

Arsine, dimethyl- ion(1-)

$EtP(O)(OH)O^-$

Phosphonic acid, ethyl- ion(1-)

$PhOSO_3^-$

Sulfuric acid monophenyl ester, ion(1-)

$MeO_3SCH_2CH_2SO_3^-$

1,2-Ethanedisulfonic acid monomethyl ester, ion(1-)

Certain resonance-stabilized anions and cations containing hetero atoms are indexed by *CA* at names corresponding to preferred canonical structures. In the same manner as the analogous tautomeric compounds (¶ 122), anions are normalized, i.e., recognized as equivalent, by machine programs, regardless of how the structures are shown in the original documents. Each ion is assigned a single CAS Registry Number (see Appendix II, ¶ 13) and a unique *CA* index name.

Resonance-stabilized anions of the general formula

$$M{=}Q{-}Z^- \longleftrightarrow {}^-M{-}Q{=}Z$$

in which Q = C, N, S, P, Sb, As, Se, Te, Br, Cl or I, and M and Z represent any combination of trivalent N and/or bivalent O, S, Se or Te atoms are normalized in this way. The formula is analogous to that for normalized tautomeric compounds (¶ 122), with a negative charge replacing the hydrogen atom, and the requirements described for them apply equally to normalized anions. The names are derived by the same structural and nomenclature rules, and are identical except for addition of the index modification term "ion(1-)". Negative ions from tautomeric pyrazole and tropolone systems, though not normalized by the CAS Registry System, are handled like the uncharged tautomers.

Examples:

Propanamide ion(1-) NOT **Propanimidic acid ion(1-)**

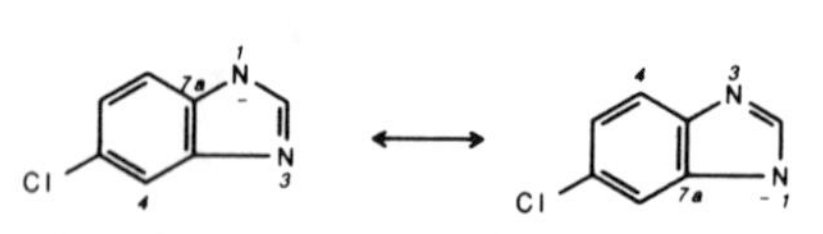

1*H*-Benzimidazole, 5-chloro- ion(1-) NOT **1*H*-Benzimidazole, 6-chloro- ion(1-)**

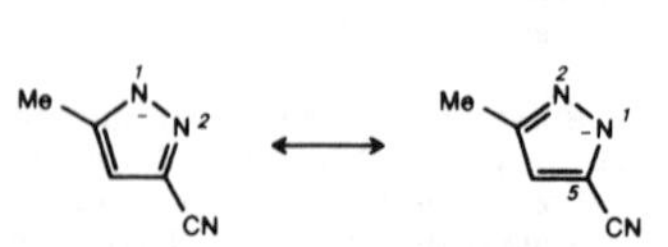

1*H*-Pyrazole-3-carbonitrile, 5-methyl- ion(1-) NOT **1*H*-Pyrazole-5-carbonitrile, 3-methyl- ion(1-)**

Radical ions are named at the neutral compound heading with "radical ion" terms in the modification.

Examples:

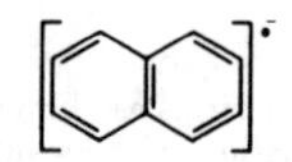

Naphthalene radical ion(1-)

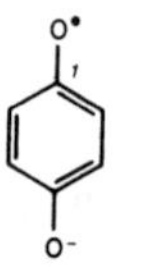

2,5-Cyclohexadiene-1,4-dione radical ion(1-)

At cationic index headings, e.g., **Ethanaminium** (see ¶¶ 184, 198), anions are expressed by "-ide" or "-ate" terms, as described below, or by "salt with" phrases.

Modification terms for *unsubstituted* carbanions from acyclic and monocyclic hydrocarbons are derived by adding "-ide" to the hydrocarbon name after elision of the final "e," e.g., "benzenide," "cyclopentadienide." Unsubstituted acids expressed as principal groups (e.g., carboxylic and sulfonic acids) afford anions which are named by "-ate" terms in the modification, e.g., "acetate," "1,2-benzenedicarboxylate." **Phosphonic acid, Carbamic acid,** and other substitutive functional parent acids, whether substituted or not, also provide anions which are named in modifications by means of "-ate" terms, e.g., "phosphonate," "dimethylcarbamate." Similar terms are used for anions from partial esters of polybasic "oxo" acids.

Examples:

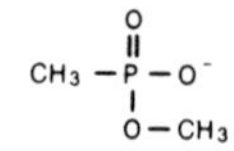

methyl methylphosphonate

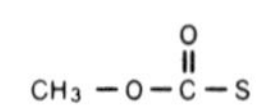

***O*-methyl carbonothioate**

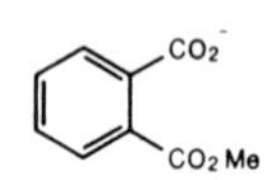

methyl 1,2-benzenedicarboxylate

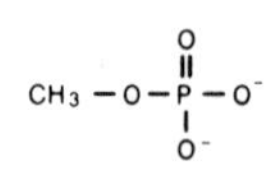

methyl phosphate

Anions from the *unsubstituted* alcohols and phenols **Methanol, Ethanol, 1-Propanol, 1-Butanol,** and **Phenol** are named by "-oxide" terms, e.g., "propoxide," "phenoxide." Loss of hydrogen from the mercapto group of unsubstituted **Benzenethiol** is expressed as "benzenethiolate."

In all other cases, anion names at cation headings are replaced by "salt with" phrases; it is to be understood that, in a complete salt name, a ratio would always be added when known.

Examples:

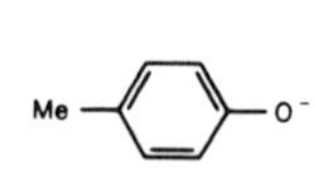

salt with 4-methoxyphenol (not "4-methoxyphenoxide")

PhNH—

salt with benzenamine

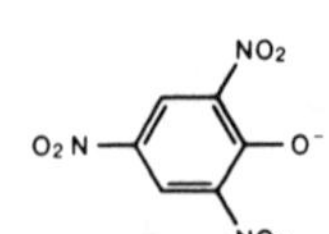

salt with 2,4,6-trinitrophenol (not "picrate")

181. **Antimony and Bismuth compounds,** are conveniently discussed together because of the close similarity in the indexing treatment of their derivatives. Antimony and bismuth are metals (¶ 215), and their salts are named as such, not as cyclic or acyclic molecular skeletons. (Prior to *CA* Volume 95 (see ¶ 101), antimony was classed as a nonmetal for indexing purposes; now, it and bismuth are treated alike.)

Hydrides of trivalent antimony and bismuth are named **Stibine** and **Bismuthine,** respectively; polymolecular saturated and unsaturated hydride chains have names such as **Distibine, Distibene, Tribismuthine.** The mononuclear oxide heading parents **Stibine oxide** and **Bismuthine oxide** are employed, as are also the analogous names for the sulfides, etc., and imides. In heterocyclic compounds the valency is understood to be three unless an abnormal valency can be expressed in the name (¶ 158).

Examples:

Et_3Bi

Bismuthine, triethyl-

$EtSbCl_2$	**Stibine, dichloroethyl–**
Ph_2BiI	**Bismuthine, iododiphenyl–**
$(F_3C)_2\underset{2}{Sb}\underset{1}{Sb}Me_2$	**Distibine, 1,1–dimethyl–2,2–bis=(trifluoromethyl)–**
$PhSb=SbPh$	**Distibene, diphenyl–**
$Ph_2Sb(O)OH$	**Stibine oxide, hydroxydiphenyl–**
$H_2\underset{1}{Sb}(O)\underset{N}{NH_2}$	**Stibinamine 1–oxide**
$Et\underset{1}{Bi}H_2=\underset{N}{N}Me$	**Bismuthine imide, 1–ethyl–*N*–methyl–**
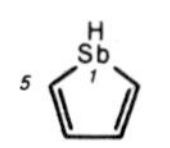	1*H*–**Stibole**
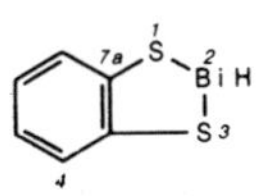	**1,3,2–Benzodithiabismole**

Heterocyclic antimony and bismuth compounds without functional suffixes are ranked in accordance with the seniority of ring systems (¶ 138). Non=functional acyclic antimony substitutive parent compounds follow arsenic compounds in order of precedence (¶ 106) and are followed in turn by bismuth and then boron parents. Within each class the order is determined by the number of hetero atoms, then unsaturation, size and additive hetero atoms, as illustrated by the following descending order of antimony compounds: **Tristibine, Distibene, Distibine, Stibine oxide, Stibine sulfide, Stibine imide, Stibine.** In the presence of more preferred compound classes, the following substituent prefixes are employed. (The substituent prefixes stiboso (–SbO), stibo (–SbO_2), stibinico (=Sb(O)OH), and stibono (–Sb(O)$(OH)_2$) were used prior to *CA* Volume 95 (see ¶ 101).)

	Substituent Prefix		*Substituent Prefix*
–SbH_2	stibino	–BiH_2	bismuthino
=SbH	stibylene	=BiH	bismuthylene
≡Sb	stibylidyne	≡Bi	bismuthylidyne
–Sb=Sb–	1,2–distibenediyl		

Examples:

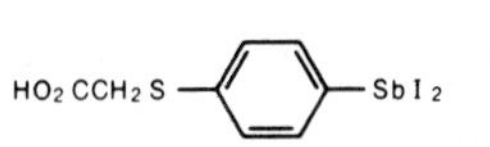	**Acetic acid, [[4–(diiodostibino)=phenyl]thio]–**
$Ph\underset{N}{N}=BiI$	**Benzenamine, *N*–(iodo=bismuthylene)–**

Trihydroxy and hydroxy oxo derivatives of **Stibine, Bismuthine, Stibine oxide, Bismuthine oxide** and their chalcogen analogs are given binary oxide, hydroxide, etc., names (with synonym line formulas) such as **Antimony hy=droxide ($Sb(OH)_3$), Bismuth hydroxide ($Bi(OH)_3$), Antimony hydroxide oxide (Sb(OH)O)** and **Bismuth hydroxide oxide (Bi(OH)O).** (Antimonic acid headings were used prior to *CA* Volume 95 (see ¶ 101).)

Halo, alkoxy, and aryloxy derivatives of **Stibine, Bismuthine** and their oxides are so named; amino derivatives are named at **Stibinamine, Stibine=diamine, Bismuthinamine,** etc.

Examples:

$MeOSbCl_2$	**Stibine, dichloromethoxy–**
$(HO)_2Sb(O)NH_2$	**Stibinamine, 1,1–dihydroxy– 1–oxide**

Arsenic compounds. See *Phosphorus and Arsenic compounds* (¶ 197).

182. Boron compounds. For the naming of neutral boron hydrides and replacement ("a") analogs (hetero polyboranes) see ¶¶ 159, 160. Except for hydroxyl groups attached to boron (¶ 175), principal groups on such hydrides are expressed as suffixes in the regular way, and conjunctive names are adopted with those known to have closed polyhedral structures; e.g., **Diborane(4)=tetramine** and **1,2–Dicarbadodecaborane(12)–1,2–diethanol.**

Acyclic carbon chains containing boron atoms are given "a" names if the requirements (¶ 127) are met.

Example:

$\underset{1}{Me}\underset{2}{B}(Me)CH_2\underset{4}{O}(CH_2)_2\underset{7}{O}CH_2\underset{9}{B}(Me)\underset{10}{Me}$	**4,7–Dioxa–2,9–diboradecane, 2,9–dimethyl–**

Heterocyclic boron compounds and their derivatives are named by the usual procedures (¶ 146). Intramolecular coordination bonds between boron atoms and other hetero atoms are ignored in naming; thus, a zwitterionic ring bond between boron and phosphorus in the last example below is disregarded, and the monocyclic system is named. (A monocycle entirely dependent on such a bond is named as an acyclic compound.)

Examples:

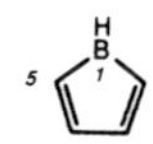	1*H*–**Borole**
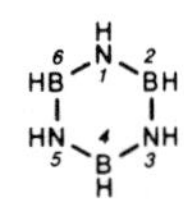	**Borazine**
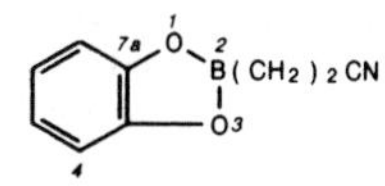	**1,3,2–Benzodioxaborole–2–propan=enitrile**
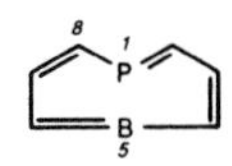	**1,5–Phosphaborocine**

Boron molecular skeletons fall between nonfunctional bismuth and silicon compounds in the order of precedence of compound classes (¶ 106). Within the boron class, the descending order is carbapolyboranes, hetero polyboranes, polyboranes, heterocyclic boron compounds, and finally **Borane.** When more preferred groups or molecular skeletons are present, boron substituent prefixes (¶ 161) are used.

Examples:

Me_3B	**Borane, trimethyl–**
H_2SbBH_2	**Stibine, boryl–**
$Me_2\underset{1}{B}\underset{N}{N}HMe$	**Boranamine, *N*,1,1–trimethyl–** (principle: heteroatom molecular skeleton preferred)
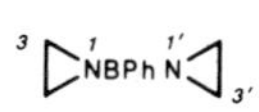	**Aziridine, 1,1'–(phenylborylene)bis–**
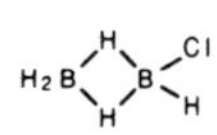	**Diborane(6), chloro–**

When the position of substituent suffixes or prefixes cannot be related to the accepted numbering of polyboranes and hetero polyboranes (as illustrated in the *Ring Systems Handbook* and current *Chemical Substance Index*), no numerical locants are used, but capital italic letters may be cited to denote substitution on a "hetero" atom in a hetero polyborane.

Examples:

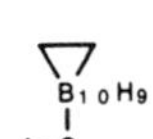	**1,2–Dicarbadodecaborane(12), (acetyloxy)–**
$B_{10}C_2Me_2H_{10}$	**Dicarbadodecaborane(12), *C,C'*–=dimethyl–**

Substitution of *bridging hydrogen* of a polyborane or hetero polyborane is indicated by the prefix "*μ*" (mu); when necessary, the locants of the boron-atom bridgeheads are cited.

Examples:

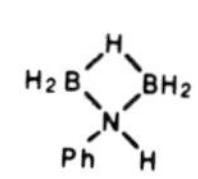	**Diborane(6), [*μ*–(phenylamino)]–**
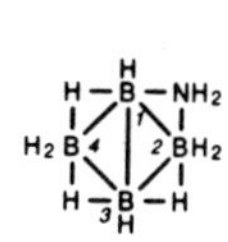	**Tetraborane(10), 1,2–*μ*–amino–**

Cyclic derivatives of polyboranes (other than **Diborane(4)**) and hetero polyboranes formed by replacement of non-bridging hydrogen atoms by bi= valent radicals are named by citing such radicals as substituents.
Examples:

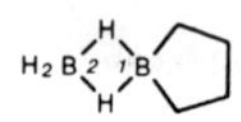

Diborane(6), 1,1-(1,4-butanediyl)-

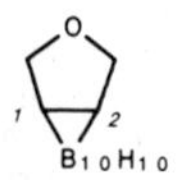

1,2-Dicarbadodecaborane(12), 1,2-[oxybis(methylene)]-

Hydroxy derivatives of **Borane** have acid names as follows:

$B(OH)_3$	**Boric acid (H_3BO_3)**
$HB(OH)_2$	**Boronic acid**
$H_2B(OH)$	**Borinic acid**

Boronic and borinic acids have replaceable hydrogen atoms attached to boron and are used as substitutive parent compounds. Their esters, anhydrides, salts, and hydrazides are named in the usual way, but their acid halides and amides are named as **Borane** and **Boranamine** derivatives, respectively. Chalcogen analogs are named by use of the detachable prefixes thio, seleno, and telluro.
Hydroxy derivatives of **Diborane(4)**, are now indexed at that index heading parent.
Examples:

MeB(OMe)OEt	**Boronic acid, methyl-** **ethyl methyl ester**
PhBMeOH	**Borinic acid, methylphenyl-**
$PhB(SH)_2$	**Boronic acid, phenyldithio-**
$AcOBMe_2$	**Acetic acid** **anhydride with dimethylborinic acid**
$PhB(NHNH_2)_2$	**Boronic acid, phenyl-** **dihydrazide**
$(HO)_2BB(OH)_2$	**Diborane(4), tetrahydroxy-** (prior to *CA* Volume 95 the name Hypoboric acid was used)
$B(NH_2)_3$	**Boranetriamine**

Boric acid (H_3BO_3) is not a substitutive parent. Esters and anhydrides are indexed as functional derivatives; hydrazides are indexed at **Hydrazine**.
Examples:

$(HO)_2BOPr$	**Boric acid (H_3BO_3)** **monopropyl ester**
$(H_2NNH)_3B$	**Hydrazine, 1,1',1''-borylidynetris-**

Addition compounds of neutral boranes are named as molecular coordi= nation compounds (see ¶ 215). (Prior to *CA* Volume 95 (see ¶ 101), they were indexed at the component names (¶ 192).)
Example:

$[B(NMe_3)H_3]$ — **Boron, (*N,N*-dimethyl-= methanamine)trihydro-** (*T*-4)- (preferred index name) (formerly indexed at Methanamine, *N,N*-dimethyl-, compd. with borane (1:1), and at Borane, compd. with *N,N*-di= methylmethanamine (1:1))

Oligomeric boranamines which are linear or unspecified are indexed at the monomer name with "dimer," "trimer," etc., in the modification. Cyclic dimers are named as μ-derivatives of **Diborane(6).** Monocyclic trimers, tetramers, etc., are given ring names in which the abnormal valencies of hetero atoms are expressed (see ¶ 158).
Examples:

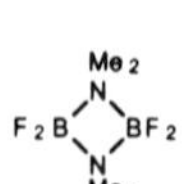

Diborane(6), bis[μ-(dimethyl= amino)]tetrafluoro-

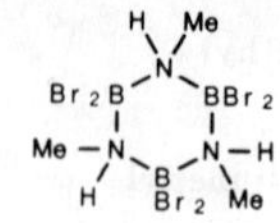

Borazine, 2,2,4,4,6,6-hexabromo-= 1,2,3,4,5,6-hexahydro-1,3,5-= trimethyl-

Ionic boron compounds are indexed by coordination nomenclature (see ¶ 215) at such names as **Borate(2-), decahydrodeca-** (for $[B_{10}H_{10}]^{2-}$), and **Boron(1+), diamminedihydro-** (for $[BH_2(NH_3)_2]^{1+}$). Acidic polyboranes and hetero polyboranes are named as complex acids (¶ 215); e.g., **Borate(2-), decahydrodeca-, dihydrogen** (not Decaborane(12)). Prior to the Eleventh Collective Index period, the special term "borata" denoted a tetrahedral borate anion attached *to carbon atoms* in a heterocyclic ring system. Now, compounds of this type are named by coordination nomenclature at such index headings as **Borate(1-).**

183. Carbonic acid and relatives, with a few trivially named exceptions, are indexed by the principles of replacement nomenclature for functions (¶ 129), based on the names **Carbonic acid** (for $(HO)_2C{=}O$) and **Formic acid** (for HCO_2H). Trivial names employed in *CA* indexes are: Formyl halides and halogenides (except the cyanide, which is indexed at **Acetonitrile, oxo-**), **Formamide, Formaldehyde, Hydrocyanic acid, Urea, Guanidine, Cyanic acid, Thiocyanic acid, Selenocyanic acid, Tellurocyanic acid.** These trivially named compounds are ranked with the appropriate class (acid, amide, etc.) as **Carbonic acid** derivatives, which fall below derivatives of acids named as principal groups (carboxylic, sulfonic, etc.) and above inorganic "oxo" acids (**Hypochlorous acid, Phosphonic acid,** etc.) (see ¶ 106).
Examples:

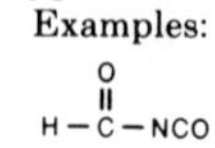

Formyl isocyanate

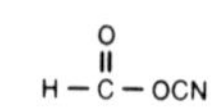

Formic acid **anhydride with cyanic acid**

HCONHMe — **Formamide, *N*-methyl-**

Analogs (imidic, hydrazonic, peroxy, chalcogen) of **Formic acid, Form= amide,** etc., are named systematically as methanoic acid analogs, but are ranked as compounds related to **Formic acid.**
Examples:

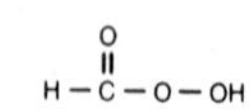

Methaneperoxoic acid

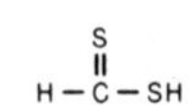

Methane(dithioic) acid (not Formic acid, dithio-)

$HCSNH_2$	**Methanethioamide** (not Formamide, thio-)
HC(=NH)NCO	**Methanimidoyl isocyanate**
HCS_2CN	**Methane(dithioic) acid** **anhydrosulfide with thiocyanic acid**

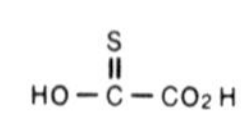

Formic acid, (thiocarboxy)- (not Methanethioic acid, carboxy-)

Replacement of the nuclear hydrogen atom in formic acid compounds by radicals derived from molecular skeletons leads to carboxylic acids, carbo= thioamides, etc., expressed as suffixes on the skeleton names.
Formaldehyde analogs have systematic names based on **Methane,** e.g., **Methanimine** (for $CH_2{=}NH$) and **Methanethial** (for $CH_2{=}S$). (**Methane= thione** (¶ 174) is employed only as the index heading parent for thio ketones with two cyclic thio substituents and for cyclic thio ketenes.) Replacement of hydrogen in formaldehyde by carbon skeletons leads to larger aldehydes and to ketones; they and their analogs are named by the usual principles of sub= stitutive nomenclature (¶¶ 173, 174). Replacement by nitrogen, halogen, etc., leads to compounds which are often named as **Formic** or **Carbonic acid** derivatives.
Examples:

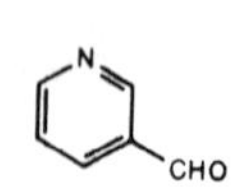

3-Pyridinecarboxaldehyde

Ph_2CO	**Methanone, diphenyl-**
PhC(=NH)Me	**Benzenemethanimine, α-methyl-**
$HCONH_2$	**Formamide**
H_2NCOCl	**Carbamic chloride**
$Cl_2C{=}NH$	**Carbonimidic dichloride**

Carbonic acid analogs in which oxygen is replaced by halogen, halogenoid, chalcogen, or nitrogen atoms or groups (¶ 129) (except cyano or a single hydrazino) are given functional replacement names. The replacement of one hydroxyl by amino leads formally to Carbonamidic acid, $H_2NC(O)OH$, but it and its analogs are named at the approved abbreviated forms **Carbamic acid, Carbamothioic acid,** etc.
Examples:

$(HO)_2C{=}NH$ — **Carbonimidic acid** (the tautomeric **Carbamic acid** is preferred in indexing unless both acid hydrogen atoms have been replaced.)

$H_2NC({=}NH)OH$ — **Carbamimidic acid** (not Carbon= amidimidic acid) (**Urea** is preferred in indexing in the absence of a covalent acid derivative.)

$ClCO_2H$ — **Carbonochloridic acid**

$NCCO_2H$ — **Carbonocyanidic acid**

$(HOO)_2CO$ — **Carbonodiperoxoic acid**

$H_2NC({=}NH)SSH$ — **Carbamo(dithioperox)imidic acid**

Cl_2CS — **Carbonothioic dichloride**

$ClCOBr$ — **Carbonic bromide chloride**

Carbonic diamide is named **Urea**, and its chalcogen analogs are indexed at **Thiourea** (not Urea, thio–), etc. The monohydrazide of carbonic acid is **Hydrazinecarboxylic acid,** but the dihydrazide is named **Carbonic di= hydrazide.** The monocyanide of **Carbonic acid** is **Acetonitrile, oxo–**; the dicyanide is **Propanedinitrile, oxo–**; the trivial name **Guanidine** is employed for Carbonimidic diamide.
Example:

$MeNHC({=}NH)NHEt$ (locants N′, N″, N) — **Guanidine,** ***N*–ethyl–*N′*–methyl–**

Carbonic acid linear polyanhydrides are named at **Dicarbonic acid, Tricarbonic acid,** etc. Carbonimidic, carbonoperoxoic, and carbonimido= peroxoic acid anhydrides are named in the same manner. When all the anhydride oxygens have been replaced by –OO– or –NH– groups, nondetach= able "peroxy–" and "imino–" prefixes are cited ahead of the name along with multiplicative prefixes. Chalcogen analogs are treated similarly, except that synonym line formulas form part of the name, and multiplicative prefixes are not cited. When both acid groups have been replaced by amide or acid halide functions, appropriate names are derived, but when only one hydroxyl group has been replaced, or different functions are present, choice of a simpler parent is made. Longer chains can often be indexed by replacement ("a") nomen= clature (¶ 127).
Examples:

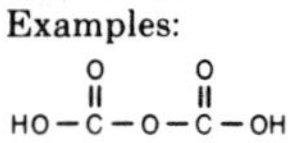
Dicarbonic acid

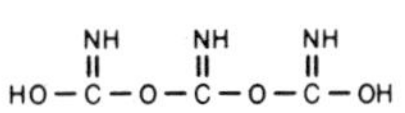
Tricarbonimidic acid

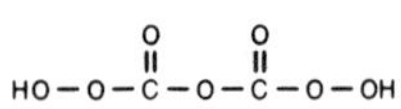
Dicarbonodiperoxoic acid

HO_2COOCO_2H — **Peroxydicarbonic acid**

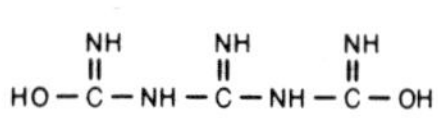
Diimidotricarbonimidic acid

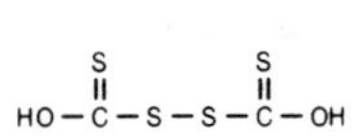
Thioperoxydicarbonic acid **($[(HO)C(S)]_2S_2$)**

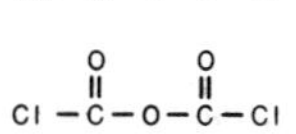
Dicarbonic dichloride

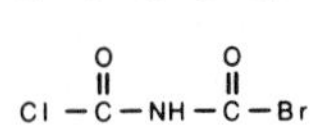
Imidodicarbonic bromide chloride

$H_2NNHCONHCONHNH_2$ — **Imidodicarbonic dihydrazide**

$ClCONHCO_2H$ — **Carbamic acid, (chlorocarbonyl)–**

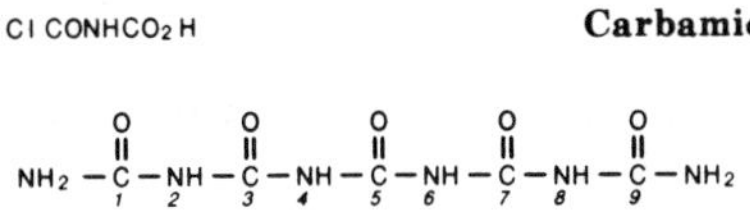
2,4,6,8–Tetraazanonanediamide, 3,5,7–trioxo–

Carbonic acid and its relatives are placed in the order of precedence of compound classes just below acids expressed as suffixes attached to molecular skeleton names, e.g., sulfonic acids (¶ 106). Within this subclass of acids, they are ranked by the following criteria, applied successively until a decision is reached: (*a*) number of acid groups; (*b*) number of nuclear carbon atoms; (*c*) precedence of atoms directly attached to nuclear carbon atoms (see Table I, ¶ 128); (*d*) number of most preferred hetero atoms directly attached to nuclear carbon atoms; (*e*) order of priority of other atoms or groups attached to nuclear carbon atoms. A partial list in descending order is: **Peroxydicarbonic acid, Dicarbonic acid, Imidodicarbonic acid, Carbonoperoxoic acid, Carbonic acid, Carbonimidic acid, Carbonochloridic acid, Carbamic acid, Formic acid, Cyanic acid, Thiocyanic acid.** (Chalcogen analogs of each acid im= mediately follow it in descending order of increasing replacement of oxygen by sulfur, selenium, and tellurium.) Acid chlorides, amides, etc., are ranked within their own classes in a similar order.

Carbamic acid derivatives with cyclic substituents are not assigned con= junctive names. In the presence of higher functions, including any acid ex= pressed as a suffix, the carbamic acid residue is indicated by a (carboxyamino) radical. Its replacement analogs are named in the usual manner. Its hydrazides are indexed at **Hydrazinecarboxamide.**
Examples:

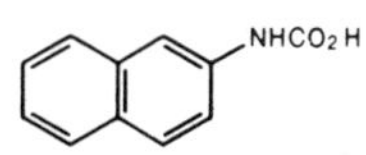
Carbamic acid, 2–naphthalenyl– (not 2–Naphthalenecarbamic acid)

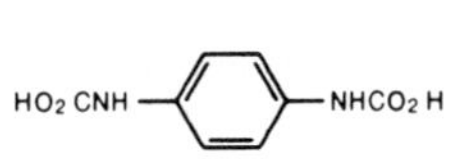
Carbamic acid, 1,4–phenylenebis– (not Carbamic acid, *N,N′*–1,4–phenylenebis– (see ¶ 118))

Me_2NCS_2H — **Carbamodithioic acid, dimethyl–**

$PrNHC({=}NH)OEt$ — **Carbamimidic acid, propyl–** **ethyl ester**

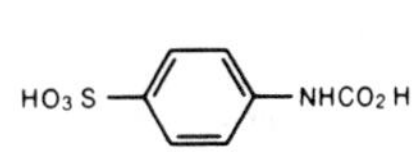
Benzenesulfonic acid, 4–(carbox= yamino)–

Cyanic acid and its chalcogen analogs are treated as mononuclear "oxo" acids; their esters and anhydrides are named in the usual way. Isocyanic acid and its analogs are not used in general index nomenclature; their esters and anhydrides are named like halogen compounds. The acids themselves, and their salts, are indexed at **Cyanic acid, Thiocyanic acid,** etc. The amide, H_2NCN, is named **Cyanamide.**
Examples:

$MeOCN$ — **Cyanic acid** **methyl ester**

$H_3CC(O)SCN$ — **Ethanethioic acid** **anhydrosulfide with thiocyanic acid**

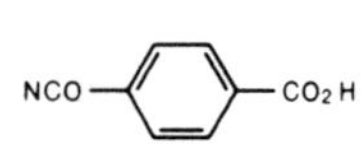
Benzoic acid, 4–cyanato–

$PhNCO$ — **Benzene, isocyanato–** (not Isocyanic acid, phenyl ester)

$H_2C{=}CHCONCO$ — **2–Propenoyl isocyanate**

$PhNHCN$ — **Cyanamide, phenyl–** (not Benzene= carbamonitrile)

Numerical and italic letter locants are used with carbonic acid relatives to place substituents. Unless a locant is expressed in the heading parent, none is cited for a substituent prefix unless needed to specify an isomer (¶ 117). Locants are not usually employed with monosubstituted **Guanidine** and **Carbamimidic acid.**
Examples:

$MeNHCONH_2$ (locants N, N′) — **Urea, methyl–**

$MeNHCSNHMe$ (locants N, N′) — **Thiourea,** ***N,N′*–dimethyl–**

$H_2NNHC({=}NH)SEt$ — **Hydrazinecarboximidothioic acid, ethyl ester**

$H_2NNHCONHNH_2$ (locants 2, 1, 1', 2') — **Carbonic dihydrazide**

$H_2NC(=NH)NH_2$ (locants N, N'', N') — **Guanidine**

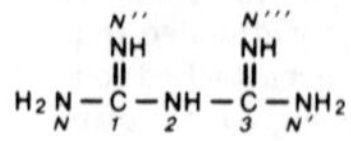

Imidodicarbonimidic diamide

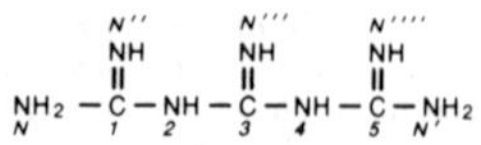

Diimidotricarbonimidic diamide

184. **Cations**, including carbonium ions (which possess an electron-deficient, tricoordinate carbon atom), "ium" ions (defined for index nomenclature purposes as resulting from addition of a proton to a saturated carbon atom or of one covalent substituent other than hydrogen to a fully substituted hetero atom), and radical cations related to these two classes are described here. (For salts of which they are components, see ¶ 198.) When a proton is added to a hetero atom, the resulting compound is named as a salt (if the anion is known) or by such modification phrases as "conjugate acid" or "conjugate monoacid" (if two or more heteroatoms are present) at the neutral component.

Examples:

$MeOH \cdot HF$ (not $[MeOH_2]^+ + F^-$) — **Hydrofluoric acid**
compd. with methanol (1:1)
Methanol
compd. with hydrofluoric acid (1:1) (additional *Chemical Substance* and *Formula Index* entry)

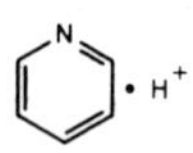

Pyridine
conjugate acid

Carbon cations formally derived by *addition* of a proton to a saturated carbon atom are rarely described. They are named at the molecular skeleton name by use of a modification term such as "monoprotonated."

Example:

$MeCH_4+$ — **Ethane**
monoprotonated

Orthodox carbonium compounds are named from the hydrocarbon (or other parent) *radical* by addition of "-ium."

Examples:

$MeCH_2+$ — **Ethylium**

Me_2CH^+ — **Ethylium, 1-methyl-** (not 2-Propylium

$MeCO^+$ (locants 2, 1) — **Ethylium, 1-oxo-** (not Acetylium)

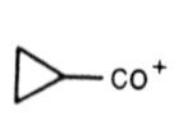

Methylium, cyclopropyloxo- (not Cyclopropylcarbonylium)

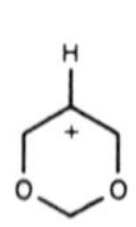

1,3-Dioxan-5-ylium

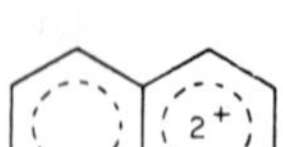

Naphthalenediylium, dihydro-

Acyclic *nitrogen cations* derived from amines attached to parent molecular skeletons are named by converting the preferred "-amine" name (¶ 176) to "-aminium" and expressing the remaining quaternizing groups as substituent prefixes.

Examples:

Me_4N^+ — **Methanaminium, *N,N,N*-trimethyl-**

Me_2N^+O — **Methanaminium, *N*-methyl-*N*-oxo-**

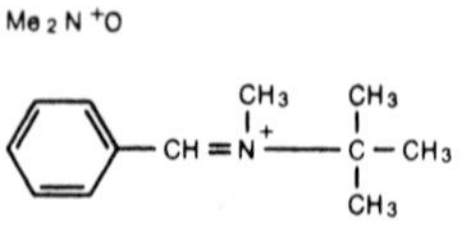

2-Propanaminium, *N*,2-dimethyl-*N*-(phenylmethylene)- (principle: largest *amine* parent)

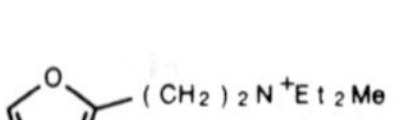

2-Furanethanaminium, *N,N*-diethyl-*N*-methyl- (from a conjunctively-named amine)

$Pr_3N^+(CH_2)_2N^+Pr_3$ — **1,2-Ethanediaminium, *N,N,N,N',N',N'*-hexapropyl-** (principle: maximum number of the preferred substituent suffix)

$MeC\equiv N^+Me$ — **Methanaminium, *N*-ethylidyne-**

Quaternary ammonium cations not included in the heading parent are expressed as substituted ammonio, $^+NH_3$-, iminio, $^+NH_2$=, and nitrilio, $^+NH\equiv$, radicals.

Example:

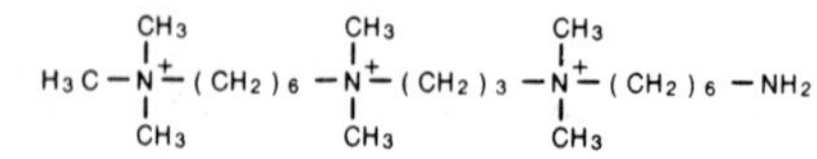

1,6-Hexanediaminium, *N*-[3-[(6-aminohexyl)dimethylammonio]propyl]-*N,N,N',N',N'*-pentamethyl- (principle: largest heading parent)

Aminium and diaminium names may be derived from amines named by replacement ("a") nomenclature. In addition, cationic centers in the "a" name of a molecular skeleton may be expressed, if such a name is permitted (see ¶ 127), by use of the "azonia" replacement prefix.

Example:

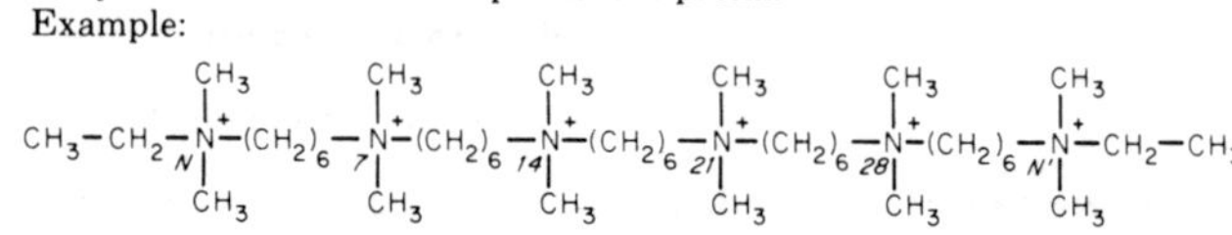

7,14,21,28-Tetraazoniatetratriacontane-1,34-diaminium, *N,N'*-diethyl-*N,N,N',N'*,7,7,14,14,21,21,28,28-dodecamethyl-

Iminium names are employed when the quaternary nitrogen atom is attached to one or more bivalent radicals derived from a molecular skeleton, and analogous monovalent radicals are absent. The names are derived from those of the preferred imines (¶ 177).

Example:

$Ph_2C=N^+=CPh_2$ — **Benzenemethaniminium, *N*-(diphenylmethylene)-α-phenyl-**

Diazonium compounds contain the $-N_2^+$ group attached to a substitutive parent compound. When such a parent is a molecular skeleton, the suffix "-diazonium" is employed; otherwise **Diazonium** is the heading parent. The corresponding substituent prefix is "diazonio."

Examples:

$MeCOCH(N_2^+)COMe$ (locants 1, 3, 4, 5) — **3-Pentanediazonium, 2,4-dioxo-**

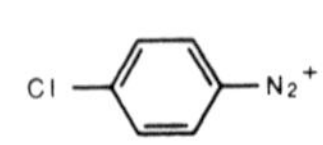

Benzenediazonium, 4-chloro-

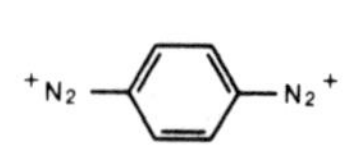

1,4-Benzenebis(diazonium)

$MeNHN_2^+$ — **Diazonium, (methylamino)-**

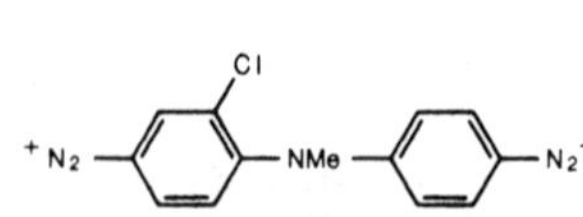

Benzenediazonium, 3-chloro-4-[(4-diazoniophenyl)methylamino]-

Cations from acyclic nitrogen molecular skeletons are named in a similar manner by use of the "-ium" suffix. Lowest locants are given to cationic centers when a choice is necessary.

Examples:

$PhN^+Me_2NH_2$ — **Hydrazinium, 1,1-dimethyl-1-phenyl-**

$H_2NN^+Me_2NH_2$ (locants 1, 2, 3) — **Triazanium, 2,2-dimethyl-**

$MeN=NN^+Me_3$ (locants 3, 2, 1) — **2-Triazenium, 1,1,1,3-tetramethyl-**

$Me_3N^+N^+Me_3$ — **Hydrazinium, hexamethyl-** (not Hydrazinium, 1,1,1,2,2,2-hexamethyl-)

Acyclic cations from nonnitrogenous hetero atoms are indexed at **Phos=phonium, Arsonium, Stibonium, Bismuthonium, Oxonium, Sulfonium, Iodonium,** etc. The corresponding substituent prefixes are phosphonio, ar=sonio, etc. **Diphosphine,** H_2P–PH_2, affords **Diphosphinium.**

Examples:

$Me_3P^+(CH_2)_2P^+Me_3$ — **Phosphonium, 1,2–ethanediylbis=[trimethyl–** (analogous compounds with one or more noncarbon attach–ments to phosphorus are indexed at **Phosphorus(1+)** (see ¶ 215))

$Cl_2PP^+Me_3$ — **Diphosphonium, 2,2–dichloro–1,1,1–trimethyl–**

Me_4As^+ — **Arsonium, tetramethyl–**

Me_3S^+ — **Sulfonium, trimethyl–** (the corre=sponding *S*–oxide is named **Sulf=oxonium, trimethyl–**)

Ph_2I^+ — **Iodonium, diphenyl–** (the corre=sponding *I*–oxide is named **Iodo=nium, diphenyl–, *I*–oxide**)

Cyclic cations from rings with "a" names are named by replacing the aza, oxa, thia, etc., replacement terms by the cationic prefixes azonia, oxonia, thionia, etc. They are cited in the same order as the corresponding neutral terms (Table **I** ¶ 128); when the same element is present in neutral and cationic forms, alphabetic order of the terms is followed, but the cationic center is preferred for lowest locants if a choice is available.

Examples:

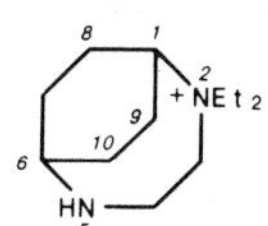

5–Aza–2–azoniabicyclo[4.2.2]=decane, 2,2–diethyl–

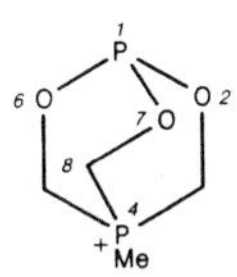

2,6,7–Trioxa–1–phospha–4–phos=phoniabicyclo[2.2.2]octane, 4–=methyl–

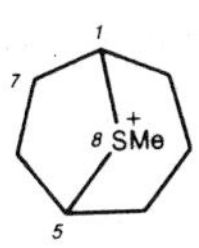

8–Thioniabicyclo[3.2.1]octane, 8–methyl–

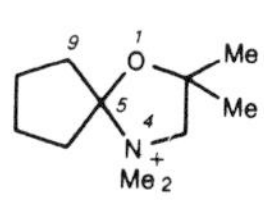

1–Oxa–4–azoniaspiro[4.4]nonane, 2,2,4,4–tetramethyl–

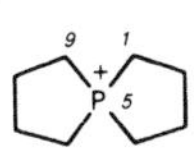

5–Phosphoniaspiro[4.4]nonane

Cationic monocyclic and fused ring systems, other than those with "a" names, are indexed at heading parents derived from the heterocycle name by addition of an "–ium" suffix. Lowest locants are preferred for indicated hydrogen, e.g., 1*H* (never 3*H*)–**Benzimidazolium**; a cationic center is num=bered lower than a neutral hetero atom of the same element if a choice is available.

Examples:

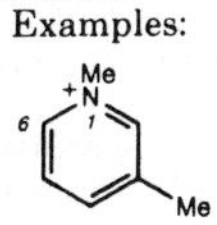

Pyridinium, 1,3–dimethyl–

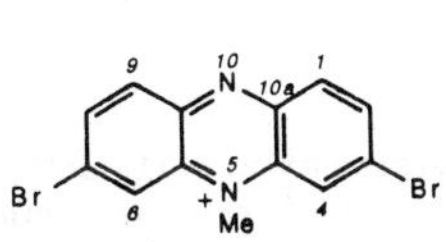

Phenazinium, 3,7–dibromo–5–methyl– (not Phenazinium, 2,8–dibromo–10–methyl–)

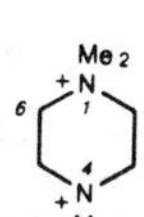

Piperazinium, 1,1,4,4–tetramethyl–

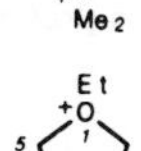

Furanium, 1–ethyltetrahydro–

In all the examples above, the charge of the cationic atom was derived from an additional covalent substituent at a ring hetero atom already saturated to the covalency limit. Similar names are used (for rings other than those with "a" names) when the covalency limit is exceeded within the ring system itself, no "external" substitution being involved. The "–ium" name is derived from the uncharged heterocycle, indicated hydrogen, if present, being removed. If indicated hydrogen is needed for the cationic system, it is added at the lowest available nonangular position. The trivial names **Furylium, Pyrylium, Thiopyrylium** (and their benzo analogs), and **Xanthylium** (from **Xanthene**) are used. When other hetero atoms are present in the ring system, a locant before the "–ium" suffix defines the position of the cationic center.

Examples:

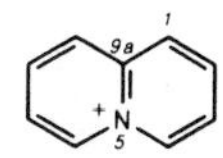

Quinolizinium

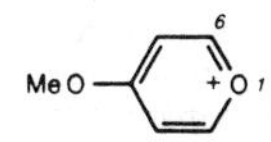

Pyrylium, 4–methoxy– (from 2*H*– or 4*H*–**Pyran**)

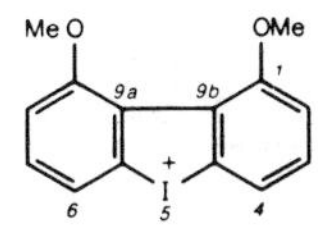

Dibenziodolium, 1,9–dimethoxy–

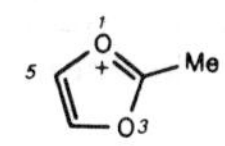

1,3–Dioxol–1–ium, 2–methyl–

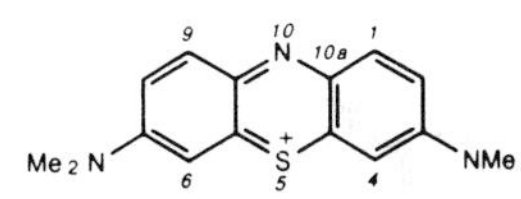

Phenothiazin–5–ium, 3,7–bis(di=methylamino)–

Cationic heterocyclic monospiro compounds, other than those from "a"–named rings, have the "–ium" term placed in the name in accordance with the nature of the component rings. When the cationic hetero atoms are in nonspiro positions, the "–ium" suffix is appended to the appropriate component. If the spiro and cationic centers coincide, or if both component rings contain such a center, the "–ium" is attached only to the second component (if the com=ponents are different). For "a"–named components, "azonia," "oxonia," etc., terms are used as appropriate; in addition, an "–ium" suffix is attached to the second component if the spiro atom is cationic and the second component does not have an "a" name.

Examples:

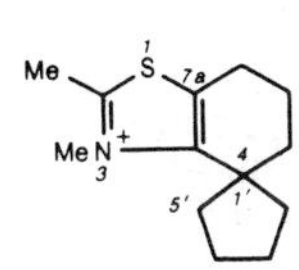

Spiro[benzothiazolium–4(5*H*),1'–=cyclopentane], 6,7–dihydro–2,3–=dimethyl–

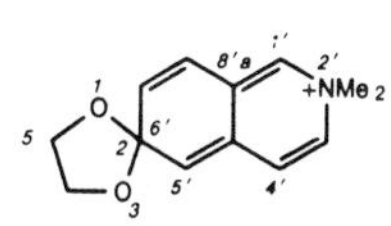

Spiro[1,3–dioxolane–2,6'(2'*H*)–=isoquinolinium], 2',2'–dimeth=yl–

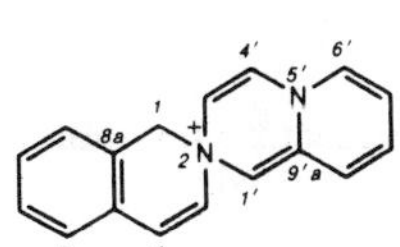

Spiro[isoquinoline–2(1*H*),2'–[2*H*]=pyrido[1,2–*a*]pyrazinium]

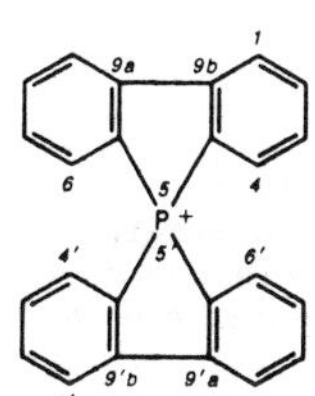

Spirobi[5*H*–dibenzophospholium]

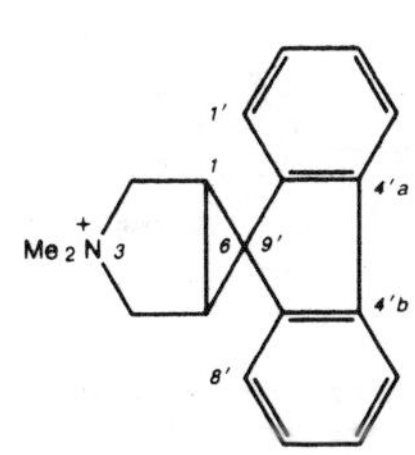

Spiro[3–azoniabicyclo[3.1.0]hex=ane–6,9'–[9*H*]fluorene], 3,3–di=methyl–

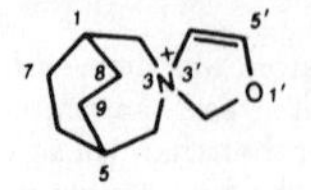

Spiro[3-azoniabicyclo[3.2.2]no= nane-3,3'(2*H*)-oxazolium]

Dispiro compounds with other than "a"-named components and containing two cationic spiro atoms have the "ium" suffix appended to the terminal, i.e., the first and third, component names.

Several classes of *resonance-stabilized cations* containing nitrogen atoms are either normalized by the CAS Registry System or treated as though they were; i.e., to avoid scattering of information in *CA* indexes, each such cation, regardless of how its structure is drawn, is assigned a unique CAS Registry Number and unique *CA Chemical Substance Index* name (cf. ¶¶ 122, 180). Normalized cations have the general structure:

in which R, R', R'', and R''' represent any atom or group except hydrogen. The structure may be acyclic or wholly or partly within a ring system, and two or more structures may be linked through a common atom to permit further "migration" of the cationic center. The unique *CA Chemical Substance Index* name is derived in general by regular nomenclature rules (see ¶ 138).

Examples:

Ethanaminium, *N*-[(di= methylamino)methyl= ene]-*N*-ethyl- (principle: prefer= red index parent) NOT **Methanaminium, *N*-[(di= ethylamino)methylene]-= *N*-methyl-**

1*H*-Benzimidazolium, 1-= methyl-3-phenyl- (principle: lowest locants for sub- stituents) NOT **1*H*-Benzimidazolium, 3-= methyl-1-phenyl-**

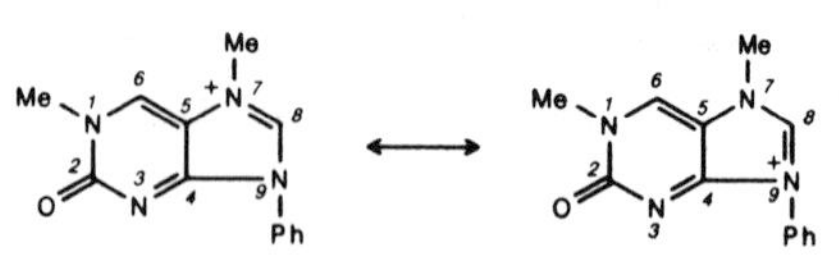

1*H*-Purinium, 2,9-di= hydro-1,7-dimethyl-= 2-oxo-9-phenyl- (principle: lowest locant for indicated hydrogen, then for cationic center) NOT **1*H*-Purinium, 2,7-di= hydro-1,7-dimethyl-= 2-oxo-9-phenyl-**

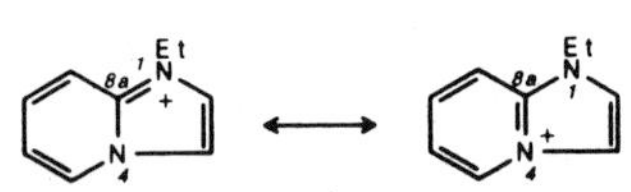

Imidazo[1,2-*a*]pyrid= inium, 1-ethyl- (principle: the neutral ring, quaternized by substitution, is preferred) NOT **1*H*-Imidazo[1,2-*a*]pyridin-= 4-ium, 1-ethyl-**

When at least one canonical form of a resonance-stabilized cation represents *a hydrogen atom* as being attached to a positively charged nitrogen atom, it is named as a neutral compound with an index modification term such as "conjugate monoacid" or (if the anion is known) as a salt such as "hydro= chloride".

Example:

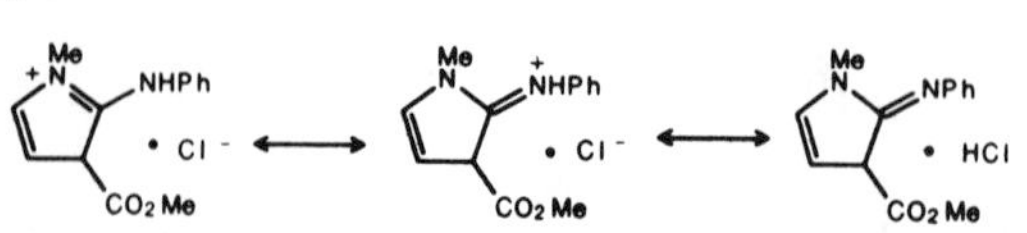

(The hydrochloride salt is the preferred structure.)

Cationic prefixes are used when more preferred compound classes (¶ 106) are present, or when different cations, or additional occurrences of the same cation, cannot be included in the heading parent. Heteroacyclic cationic prefixes are based on "-onio" radicals such as ammonio and iminio (for singly and doubly bound nitrogen, respectively), diazonio, sulfonio, phosphonio, and iodonio. Cyclic cationic radicals from rings, not named by "a" nomenclature, containing a single hetero atom at which the free valency is located are named by changing the "-ium" ending of the cation name to "-io"; e.g., pyridinio, 2*H*-pyranio, phenanthridinio. In all other cases, heteroacyclic and heterocyclic cationic prefixes are derived from the cation name by adding "-yl," "-diyl," etc., to the "-ium" suffix with locants (low-numbered if there is a choice) to indicate the points of attachment.

Examples:

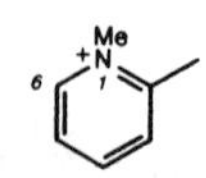

(1-methylpyridinium-2-yl)

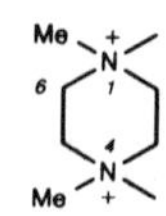

(1,4-dimethylpiperazinium-1,4-diyl)

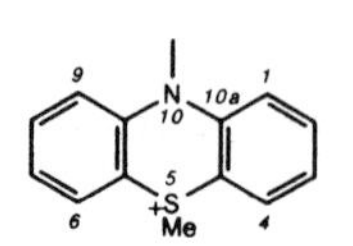

(5-methyl-10*H*-phenothiazinium- 10-yl)

$Me_2NN{=}NN^{+}Me_2$ (locants 4 3 2 1) — (1,1,4,4-tetramethyl-2-tetrazenium- 1-yl)

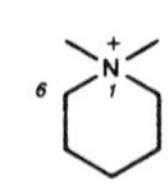

piperidinium-1-ylidene

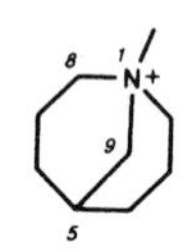

1-azoniabicyclo[3.3.1]non-1-yl

Carbon cationic prefixes are named by addition of "-yl" to the cation name. Locants are cited (except for methane prefixes) for both the cationic center and the point of attachment, with lowest locants for the latter.

Examples:

H_2C^{+}— — methyliumyl

$MeC^{+}H$ — — 1-ethylium-1-yl

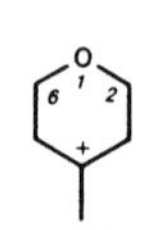

(tetrahydro-4*H*-pyran-= 4-ylium-4-yl)

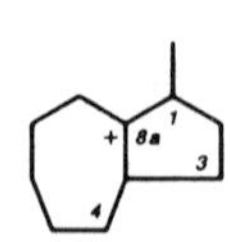

(octahydro-8a(1*H*)-= azulenylium-1-yl)

Cationic compounds are ranked in the following descending order of pre= cedence of elements: C, N, P, As, Sb, Bi, O, S, Se, Te, F, Cl, Br, I. When more than one cationic center is present, the general rules of substitutive nomen= clature (¶ 138) are applied, the cationic centers being considered as principal chemical groups for this purpose.

Examples:

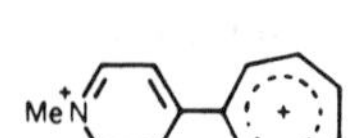

Cycloheptatrienylium, (1-methyl= pyridinium-4-yl)- (principle: carbon cation preferred)

$Me_3N^{+}(CH_2)_3S^{+}Me_2$

1-Propanaminium, 3-(dimethyl= sulfonio)-*N*,*N*,*N*-trimethyl- (principle: nitrogen cation prefer= red)

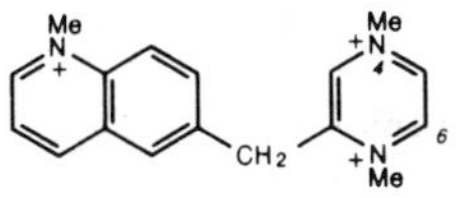

Pyrazinium, 1,4-dimethyl-2-[(1-methylquinolinium-6-yl)=methyl]- (principle: maximum number of cationic centers)

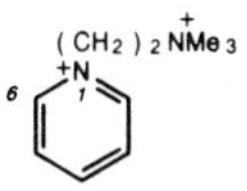

Pyridinium, 1-[2-(trimethyl=ammonio)ethyl]- (principle: ring parent preferred)

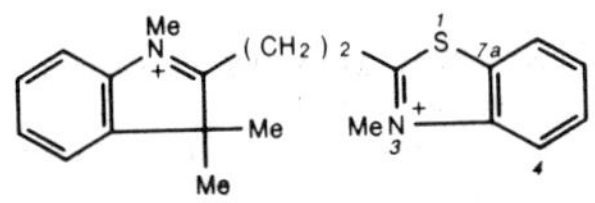

Benzothiazolium, 3-methyl-2-[2-(1,3,3-trimethyl-3*H*-indoli=um-2-yl)ethyl]- (principle: pre=ferred ring)

Cationic compounds of *indefinite structure* are named by use of alternative locants if possible, e.g., **Quinazolinium, 1(*or* 3)-methyl-, iodide;** otherwise as molecular addition compounds of the neutral compounds with a term such as "compd. with iodomethane (1:1)" (not "monomethiodide") in the modifi=cation and additional *Chemical Substance* and *Formula Index* entries at the less preferred component, e.g., **Methane, iodo-.**

Cationic free radicals from carbonium ions are named like the carbon cationic prefixes (above); e.g., **Methyliumyl; 4-Cyclohexylium-1-yl.** Isolated nitrogen atoms afford the **Ammoniumyl** cationic free radical. Loss of an electron from a hetero atom of a molecular skeleton or isolated chalcogen atom leads to cationic free radicals such as **Hydraziniumyl** ($H_2NNH_2{\bullet}^+$), **Thio=pheniumyl,** and **Sulfoniumyl.** When the free radical and cationic centers are separated, the "-yl" suffix is preceded by a locant, if known. When the free radical and cationic center are in different parent molecular skeletons, the free radical supplies the heading parent.

Examples:

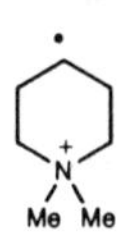

Piperidinium-4-yl,
1,1-dimethyl-

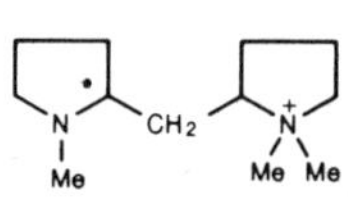

2-Pyrrolidinyl,
2-[(1,1-dimethyl=pyrrolidinium-2-yl)=methyl]-1-methyl-

Delocalized radical cations are described at the index name of the corre=sponding neutral compound by the modification term "radical ion" followed by a cationic Ewens-Bassett number in parentheses.

Example:

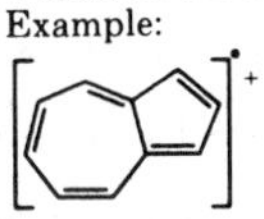

Azulene
radical ion(1+)

185. Esters, other than cyclic esters, of principal functions are named as such for indexing purposes, usually at the acid component name. Cyclic esters are named as heterocyclic ring system derivatives. Esters of hydroxy, mercapto, carboxy, sulfo, etc., groups expressed as substituents of heading parents are expressed as compound or complex substituents, not as modification terms at the same heading parent. To be recognized as an ester in *CA* indexing, a C-O, C-S, C-Se, or C-Te bond must be present. Thus, ethyl cyanate and ethyl perchlorate are considered to be esters, but ethyl azide, ethyl isocyanate, and ethyl chloride are not.

Examples:

EtOCN	**Cyanic acid** **ethyl ester**
$EtOClO_3$	**Perchloric acid** **ethyl ester**
EtN_3	**Ethane, azido-**
EtNCO	**Ethane, isocyanato-**
EtCl	**Ethane, chloro-**

To permit information on esterified alcohols and thiols to be more readily found in the *Chemical Substance Index*, the usual rules of index name se=lection are modified for esters. The chemical functionality (¶ 106) of certain very common acids ("Class I" acids) is disregarded for the purpose of naming their esters, unless the "alcoholic" component is also very common, and the entry is made instead at the uncommon alcohol, despite its lower functionality. The "Class I" acids comprise: **Acetic acid; Benzenesulfonic acid; Ben=zenesulfonic acid, 4-methyl-; Benzoic acid** and its monoamino, mononitro, and dinitro derivatives; **Boric acid (H_3BO_3); Carbamic acid; Carbamic acid, methyl-; Carbamic acid, phenyl-; Carbonic acid; Formic acid; Meth=anesulfonic acid; Nitric acid; Phosphoric acid; Phosphorodithioic acid; Phosphorothioic acid; Phosphorous acid; Propanoic acid; Sulfuric acid; Sulfurous acid.** All other acids, including isotopically labeled forms of "Class I" acids, belong to "Class II."

Esters of "Class I" monobasic acids with "Class II" alcohols and thiols are indexed at the latter. Esters of "Class I" acids with "Class I" alcohols and thiols are indexed at the acids. "Class I" alcohols are: **Benzeneethanol; Benzen=emethanol; 1-Butanol; 1-Butanol, 2-ethyl-; 2-Butanol; Cyclohexanol; 1-Decanol; 1-Dodecanol; Ethanol; Ethanol, 2-(diethylamino)-; Ethanol, 2-(dimethylamino)-; Ethenol; 1-Heptanol; 1-Hexanol; 1-Hexanol, 2-ethyl-; Methanol; 1-Nonanol; 1-Octadecanol; 1-Octanol; 1-Pentanol; Phenol** and its monochloro, monomethyl, and mononitro derivatives; **1-Propanol; 1-Propanol, 2-methyl-; 2-Propanol; 2-Propanol, 2-methyl-; 2-Propen-1-ol.** The list of "Class I" thiols is completely analogous to the "Class I" alcohol list; the individual index names are also analogous, e.g., **Benzeneethanethiol; 2-Propanethiol, 2-methyl-;** except that the **Phenol** analog is **Benzenethiol.** All selenols and tellurols belong to "Class II."

Examples:

$AcO(CH_2)_2Ph$	**Acetic acid** **2-phenylethyl ester**
$MeOSO_3H$	**Sulfuric acid** **monomethyl ester** (the uninverted name is methyl hydrogen sulfate)
PhSP(S)(OMe)OH	**Phosphorodithioic acid** ***O*-methyl *S*-phenyl ester**

In each of the examples above, a "Class I" acid is esterified by a "Class I" alcohol or thiol. When the acid belongs to "Class I" and the alcohol or thiol to "Class II," the entry is found at the alcoholic component name, except when the acid is polybasic and the alcohols differ from one another, in which case the acid heading is chosen.

Examples:

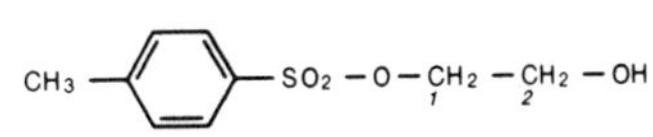

1,2-Ethanediol
mono(4-methylbenzene=sulfonate)

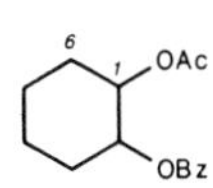

1,2-Cyclohexanediol
acetate benzoate

$HCO_2CH_2CH_2O_2CH$	**1,2-Ethanediol** **diformate** (the uninverted name is 1,2-ethanediyl diformate)
$[Cl(CH_2)_2O]_3P$	**Ethanol, 2-chloro-** **phosphate (3:1)** (all the "Class II" alcoholic components are alike here; the ratio is necessary to indicate a neutral ester of known composition; the mono- and diesters are named with **"dihydrogen phosphite"** and **"hydrogen phosphite,"** respec=tively in the modification)
$HO(CH_2)_2OP(OMe)_2$	**Phosphorous acid** **2-hydroxyethyl dimethyl ester** (the entry is made at a "Class I" acid, not at the "Class II" alcohol **1,2-Ethanediol,** because the alcoholic components are unlike)

When an ester of the type illustrated immediately above has a more preferred acid residue present in the molecule, the "Class I" polybasic acid becomes a substituent.

Example:

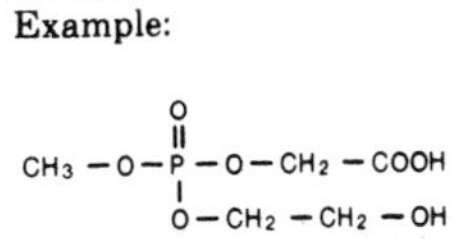

Acetic acid, [[(2-hydroxyethoxy)=methoxyphosphinyl]oxy]- (not Acetic acid, hydroxy-, 2-hydroxy=ethyl methyl phosphate)

If an "-ate" term in a modification could be interpreted to denote either an ester or a salt, the parenthetical expression "(ester)" is added. This is done whenever the index heading expresses any nitrogen atom or any trivalent

phosphorus or arsenic atom. When a locant is needed in the modification, the "(ester)" term is dropped.
Examples:

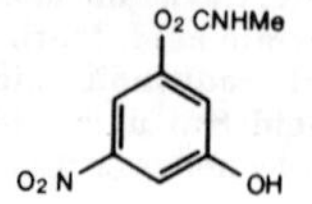

1,3-Benzenediol, 5-nitro- mono(methylcarbamate) (ester)

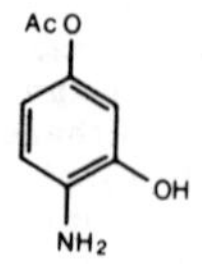

1,3-Benzenediol, 4-amino- 1-acetate

Esterified substituents of the index heading parent are expressed as sub= stituents, not as modification phrases.
Examples:

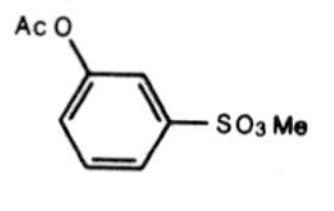

Benzenesulfonic acid, 3-(acetyl= oxy)-methyl ester (not Benzene= sulfonic acid, 3-hydroxy-, methyl ester acetate)

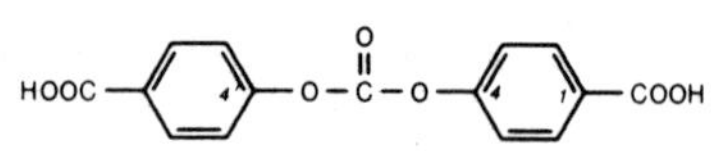

Benzoic acid, 4,4'-[carbonylbis= (oxy)]bis- (not Benzoic acid, 4-hydroxy-, carbonate (2:1))

The phrase "ester with" is usually avoided in general index nomenclature for compounds of known structure, even when the alcoholic component con= tains a function higher than alcohol, but it is employed when an acid which requires a line formula is cited in a modification; a ratio is usually needed also unless "monoester with," "diester with," etc., phrases can be employed un= ambiguously.
Examples:

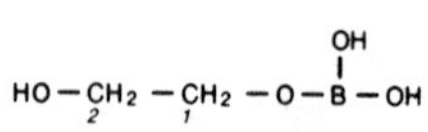

1,2-Ethanediol monoester with boric acid (H_3BO_3)

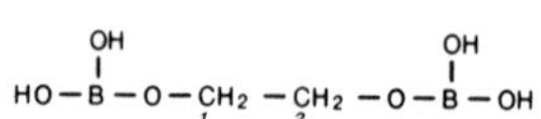

1,2-Ethanediol ester with boric acid (H_3BO_3) (1:2)

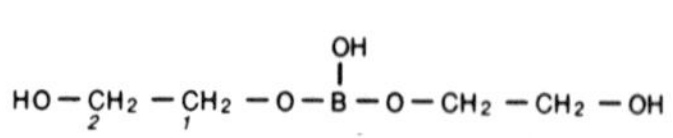

1,2-Ethanediol ester with boric acid (H_3BO_3) (2:1)

The word "hydrogen," with multiplicative prefixes if necessary, is used with an "-ate" (or "-ite") term derived from a polybasic acid cited in a modification to denote unesterified acid groups. One or more "hydrogen" terms may be replaced by radicals to denote further esterification not expressed in the heading parent. When all acid groups of a polybasic acid have been esterified, "hydrogen" cannot be cited; therefore, if the precise structure is known, a ratio is placed after the "-ate" term. If more than one such term is necessary, the complete "-ate" phrases are cited (without commas) in alphabetical order.
Examples:

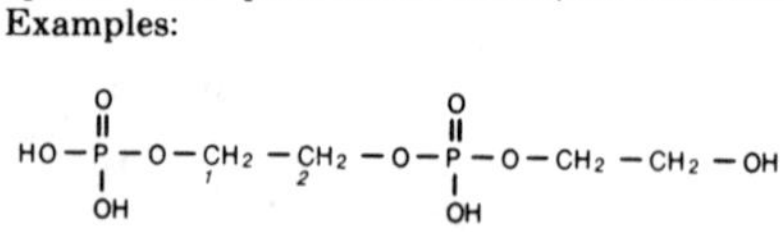

1,2-Ethanediol dihydrogen phosphate 2-hy= droxyethyl hydrogen phosphate (principle: maximum number of functional derivatives)

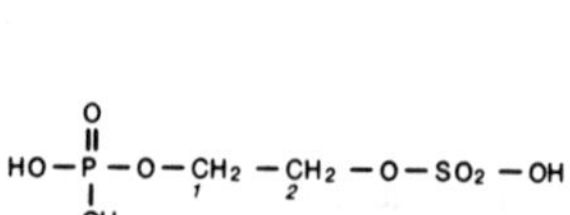

1,2-Ethanediol dihydrogen phosphate hydrogen sulfate

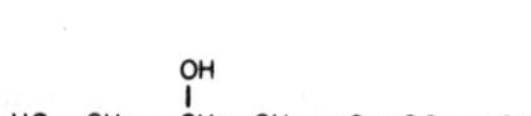

1,2,3-Propanetriol 1-(hydrogen sulfate)

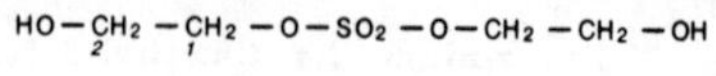

1,2-Ethanediol sulfate (2:1)

Esters are named by replacement ("a") nomenclature if the requirements (¶ 127) are met.
Example:

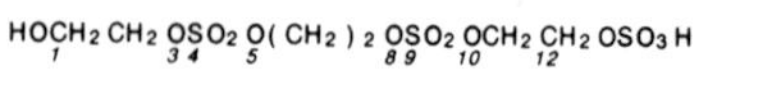

3,5,8,10-Tetraoxa-4,9-dithiado= decane-1,12-diol mono(hydrogen sulfate), 4,4,9,9- tetraoxide

Multiplicative radicals in modifications are cited ahead of nonmultiplicative.
Example:

$(MeO)_2BO(CH_2)_3OB(OMe)_2$

Boric acid (H_3BO_3) 1,3-propanediyl tetramethyl ester

Although locants to differentiate between principal and subsidiary groups are strictly not necessary in ester modifications (because esterification of the latter is not expressed in the modification), they are nevertheless cited to preclude misinterpretation.
Examples:

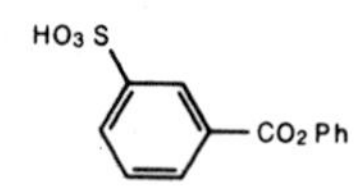

Benzoic acid, 3-sulfo- 1-phenyl ester

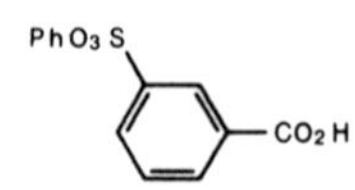

Benzoic acid, 3-(phenoxysulfonyl)-

Esterification of an alcoholic component cited in the index modification is expressed as a substituent of the esterifying radical.
Example:

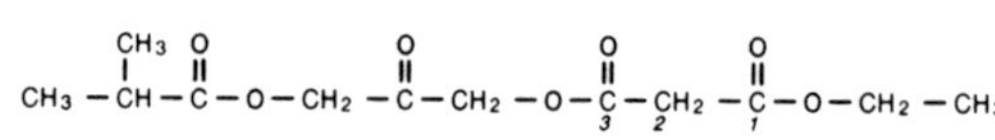

Propanedioic acid ethyl 3-(2-methyl-1-oxoprop= oxy)-2-oxopropyl ester (not Propanedioic acid, ethyl 3-hydroxy-2-oxopropyl ester, 2-methylpropanoate)

The choice of a preferred index name for a complex ester depends on the normal criteria (¶ 138), first on the preferred acid class (peroxoic, carboxylic, carboximidic, sulfonic, carbonic, etc.) then on the preferred heading parent and the particular occurrence of such a parent if it occurs more than once. If there are three or more occurrences, the principle of centrality is invoked; if there are only two, often the preferred name is the one appearing earliest in index sequence.
Example:

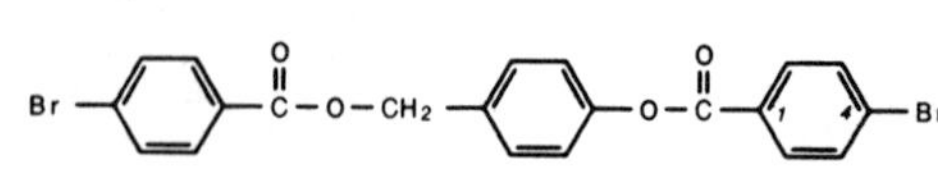

Benzoic acid, 4-bromo- 4-[[(4-bromobenzoyl)oxy]= methyl]phenyl ester (not Benzoic acid, 4-bromo-, [4-[(4-bromo= benzoyl)oxy]phenyl]methyl ester)

When functions higher in precedence than acids are present, all esters are expressed as substituents.
Example:

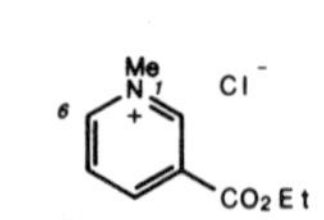

Pyridinium, 3-(ethoxycarbonyl)- 1-methyl- chloride

Peroxy acid esters and their chalcogen analogs are named in the usual way, with "*OO*," "*OS*," "*SeO*," "*SS*," etc., locants used when necessary along with

the usual "*O*," "*S*," etc., locants.
Examples:

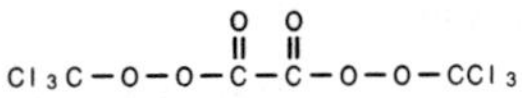

Ethanediperoxoic acid bis(trichloromethyl) ester

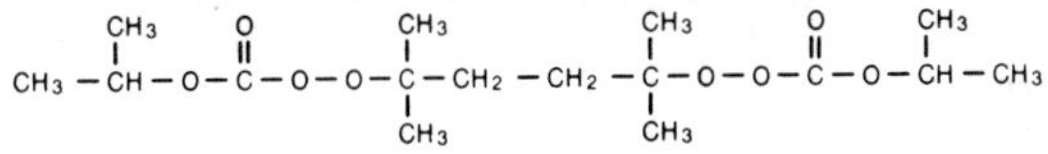

Carbonoperoxoic acid ***OO,OO'*-(1,1,4,4-tetramethyl-1,4-butanediyl) *O,O'*-bis(1-methyl= ethyl) ester**

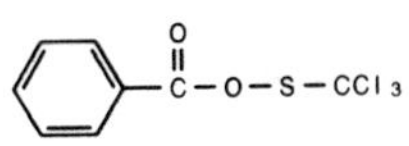

Benzenecarbo(thioperoxoic) acid ***OS*-(trichloromethyl) ester**

EtSSOMe

Ethanesulfeno(thioperoxoic) acid ***SO*-methyl ester**

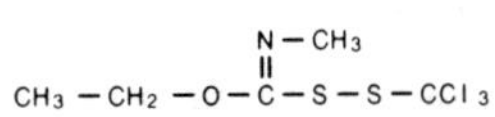

Carbono(dithioperox)imidic acid, methyl- ***O*-ethyl *SS*-(trichloromethyl) ester**

Esters of *boric acids* and their chalcogen analogs are indexed regularly, except for those of Hypoboric acid and cyclic metaboric acids, $(HBO_2)_n$, which are named as derivatives of the molecular skeletons.
Examples:

(EtO)$_2$BB(OMe)$_2$

Diborane(4), 1,1-diethoxy-2,2-= dimethoxy-

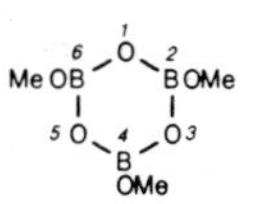

Boroxin, trimethoxy-

(BuS)$_2$BOMe

Thioboric acid ((HO)(HS)$_2$B) ***S,S*-dibutyl *O*-methyl ester**

Esters of *ortho acids* and their peroxy and chalcogen analogs are indexed like ethers, sulfides, peroxides, hydroperoxides, alcohols, etc. Ortho acid names are not used as heading parents.
Examples:

MeC(OMe)$_3$ — **Ethane, 1,1,1-trimethoxy-** (not Orthoacetic acid, trimethyl ester)

PhC(SMe)$_3$ — **Benzene, [tris(methylthio)methyl]-**

MeC(OEt)$_2$OH — **Ethanol, 1,1-diethoxy-**

MeC(OEt)$_2$OOCMe$_3$ — **Peroxide, 1,1-diethoxyethyl 1,1-= dimethylethyl**

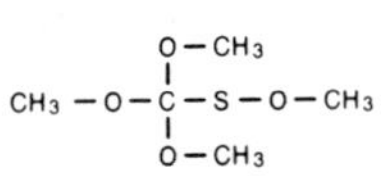

Methanesulfenic acid, trimethoxy-methyl ester

Oxides of thio esters are named not as ester derivatives but at the preferred molecular skeleton or other heading parent with sulfinyl or sulfonyl prefixes.
Examples:

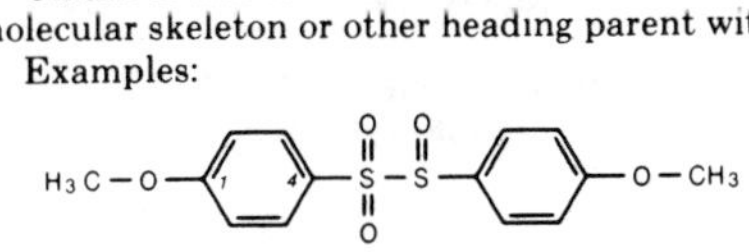

Benzene, 1-methoxy-4-[[(4-meth= oxyphenyl)sulfinyl]sulfonyl]-

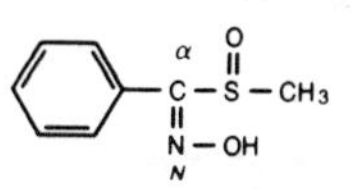

Benzenemethanimine, *N*-hydroxy-*α*-(methylsulfinyl)-

Glycerides, esters of **1,2,3-Propanetriol** (glycerol), are indexed at the name of the preferred acid unless only "Class I" acids are present.
Examples:

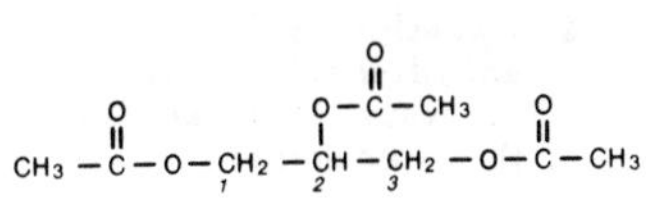

1,2,3-Propanetriol triacetate

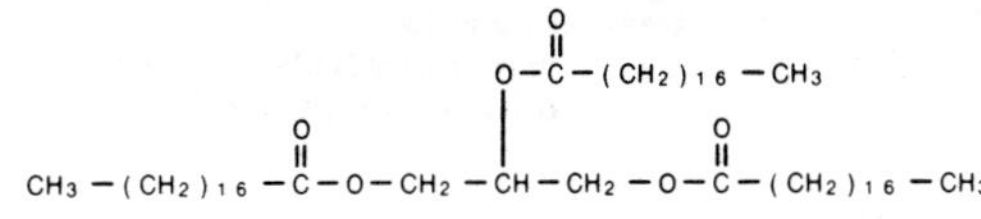

Octadecanoic acid 1,2,3-propanetriyl ester

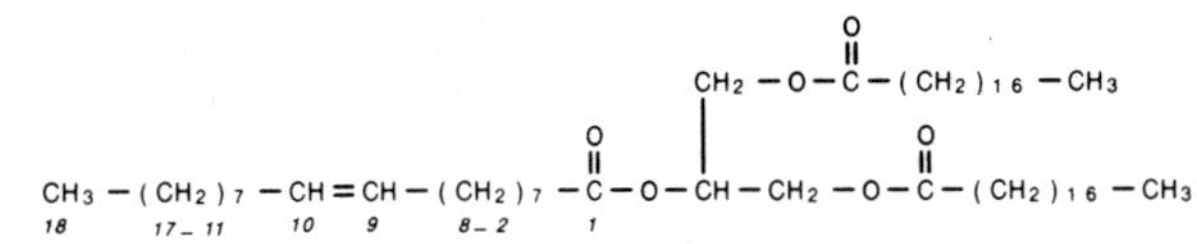

9-Octadecenoic acid 2-[(1-oxooctadecyl)oxy]-1-[[(1-oxooctadecyl)oxy]methyl]= ethyl ester (principle: unsaturated acid preferred)

Urethanes are esters of **Carbamic acid** and its derivatives. *Xanthic acids* are *O*-esters of **Carbonodithioic acid.**
Examples:

PhNHCO$_2$Et — **Carbamic acid, phenyl-ethyl ester**

PhOCS$_2$H — **Carbonodithioic acid *O*-phenyl ester**

Cyclic esters and lactones are indexed as heterocycles.
Examples:

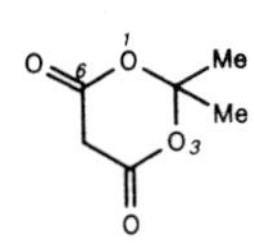

1,3-Dioxane-4,6-dione, 2,2-di= methyl- (not Propanedioic acid, cyclic 1-methylethylidene ester)

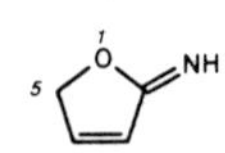

2(5*H*)-Furanimine (not 2-Buten= imidic acid, 4-hydroxy-, γ-lactone)

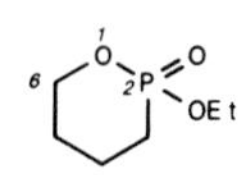

1,2-Oxaphosphorinane, 2-ethoxy-2-oxide (not Phosphonic acid, (4-hydroxybutyl)-, monoethyl ester, δ-lactone)

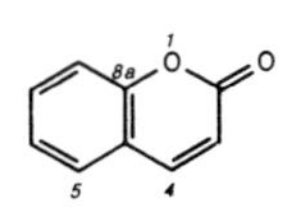

2*H*-1-Benzopyran-2-one (not Coumarin; not 2-Propenoic acid, 3-(2-hydroxyphenyl)-, δ-lactone)

186. Ester-anhydrides are named by combining the policies for anhy= drides and esters (¶¶ 179, 185). Cyclic anhydride and cyclic ester components are named in accordance with heterocyclic nomenclature. Anhydride terms precede ester terms where this can be done unambiguously (but see the last example, below). Esters cited in modifications are named in the uninverted form unless the acid is one for which a synonym line formula is required.
Examples:

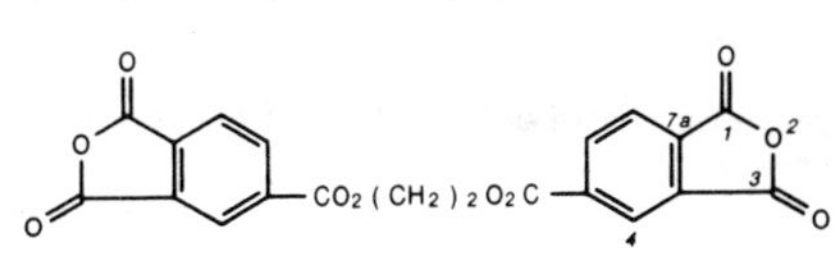

5-Isobenzofurancarboxylic acid, 1,3-dihydro-1,3-dioxo-1,2-ethanediyl ester

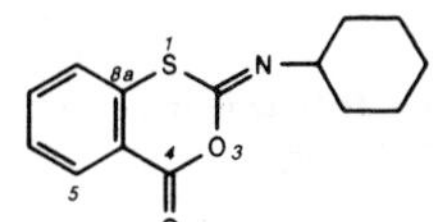

4*H*-3,1-Benzoxathiin-4-one, 2-(cyclohexylimino)- (not Carbonimidothioic acid, cyclohexyl-, anhydride with 2-mercaptobenzoic acid, cyclic ester)

EtOP(OH)OAc

Acetic acid
monoanhydride with ethyl dihydrogen phosphite

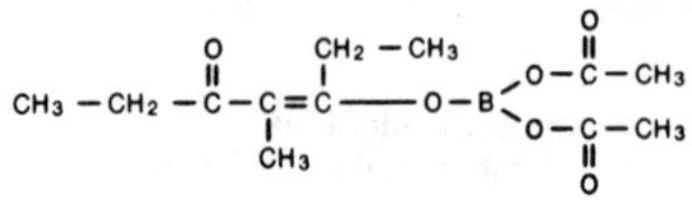

Acetic acid
dianhydride with boric acid (H_3BO_3) mono(1-ethyl-2-methyl-3-oxo-1-pentenyl) ester

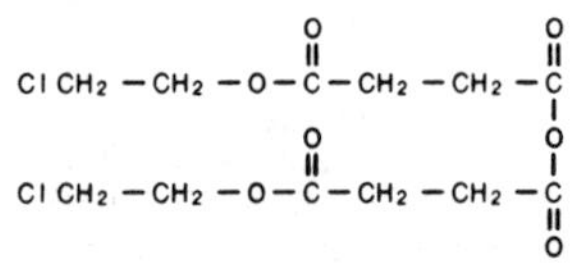

Butanedioic acid
bimol. monoanhydride, bis(2-chloroethyl) ester

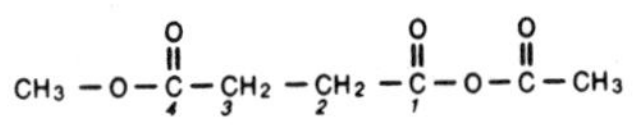

Butanedioic acid
monoanhydride with acetic acid, methyl ester

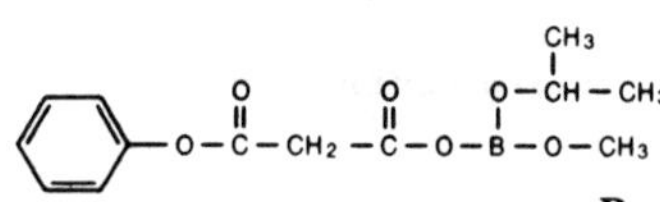

Propanedioic acid
monophenyl ester, anhydride with boric acid (H_3BO_3) mono= methyl mono(1-methylethyl) ester (for the purposes of subdi= vision (see Appendix II, ¶ 10B), this compound is considered an anhy= dride, even though the ester term is cited first)

187. **Free radicals** are highest in the order of precedence of compound classes (¶ 106) and functional groups, if present, are named as substituents. Most, but not all, free radical names coincide with the names of substituent prefixes (¶¶ 132, 133).
Examples:

Et• — **Ethyl**

PhCH2• — **Methyl, phenyl-**

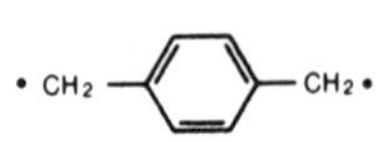
Methyl, 1,4-phenylenebis-

BrC⁝ — **Methylidyne, bromo-**

EtO2CCO• (2 1) — **Ethyl, 2-ethoxy-1,2-dioxo-** (not Acetyl, ethoxyoxo-)

H2P• — **Phosphino**

F2Si: — **Silylene, difluoro-**

Ph2NN(CHO)• (2 1) — **Hydrazyl, 1-formyl-2,2-diphenyl-**

Free radicals that might be considered as derived from **Borane** are indexed at **Borane(1)** and **Borane(2).**
Examples:

FHB• — **Borane(2), fluoro-** (not Boryl, fluo= ro-)

FB: — **Borane(1), fluoro-** (not Boron fluoride (BF))

Free radicals from ammonia, amides, or amines are named as **Amidogen,** H_2N·, **Imidogen,** HN:, and their derivatives.

F2N• — **Amidogen, difluoro-** (not Nitrogen fluoride (NF_2))

PhCH(CO2H)N: — **Imidogen, (carboxyphenylmethyl)-**

The heading parent **Nitroxide** is employed for the free radical H_2N-O·.
Example:

Me2NO• — **Nitroxide, dimethyl**

Free radicals from hydroxyl and hydroperoxy groups that are attached to a molecular skeleton, including (acyloxy) radicals (see below), are indexed at "-oxy" and "-dioxy" heading parents with systematically named nondetach= able prefixes. Chalcogen analogs are named similarly at "-thio" and "-dithio" heading parents.
Examples:

MeO• — **Methoxy** (not Oxy, methyl-)

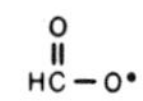
Methoxy, oxo- (not Formyloxy)

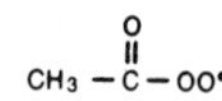
Ethyldioxy, 1-oxo-

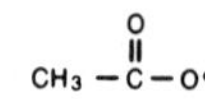
Ethoxy, 1-oxo- (not Acetyloxy)

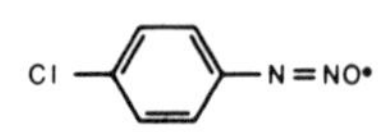
Diazenyloxy, (4-chlorophenyl)- (not (Phenylazo)oxy, 4-chloro-)

MeS• — **Methylthio**

Acyl free radicals (derived from aldehydes or acids) are named as α-oxo derivatives of the alkyl parents.
Examples:

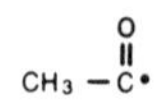
Ethyl, 1-oxo- (not Acetyl)

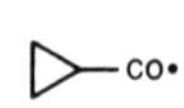
Methyl, cyclopropyloxo- (not Cyclopropylcarbonyl)

H2NCO• — **Methyl, aminooxo-** (not Amino= carbonyl)

Free radicals from sulfonic and sulfinic acids are named as derivatives of alkyl and aryl sulfonyl and sulfinyl radicals and the analogous oxy radicals.
Examples:

CH3-SO2-O• — **(Methylsulfonyl)oxy**

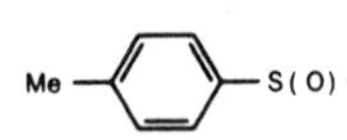
Phenylsulfinyl, 4-methyl-

The following chalcogen hydride free radicals are employed: **Hydroxyl,** HO•; **Hydroperoxo,** HOO•; **Mercapto,** HS•; **Selenyl,** HSe•.

188. **Halogen and halogenoid compounds.** (See also *Acid halides,* ¶ 170). The halogen and halogenoid "oxo" acids are hypohalous acids, HOX (e.g., **Hypochlorous acid,** HOCl); halous acids, HOXO; halic acids, $HOXO_2$; per= halic acids, $HOXO_3$; **Cyanic acid,** HOCN; and **Fulminic acid,** HONC. Their esters and anhydrides are named in the regular way. They rank below acids, e.g., carboxylic, sulfonic, expressed as suffixes on molecular skeleton names (¶ 106), and in their presence are expressed as cyanato, (iodyloxy), (chloryloxy), etc., radicals.
Examples:

PhOF — **Hypofluorous acid**
phenyl ester

MeCO(CH2)2OBrO2 — **Bromic acid**
3-oxobutyl ester

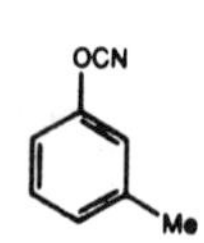
Cyanic acid
3-methylphenyl ester

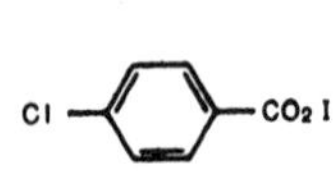
Benzoic acid, 4-chloro-
anhydride with hypoiodous acid (not Benzoyl hypoiodite, 4-chloro-)

Propanethioic acid
anhydrosulfide with thiohypo= chlorous acid (not Propanoyl thiohypochlorite)

NCOCH2CO2H — **Acetic acid, cyanato-**

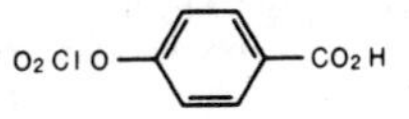

Benzoic acid, 4–(**chloryloxy**)–

The groups –X, –XO, –XO_2, –XO_3, –NC, –NCO, –N_3, and their chalcogen analogs, e.g., –XSe; –NCS, are normally expressed as substituent prefixes when attached to a molecular skeleton; the radicals chloro, chlorosyl, chloryl, per= chloryl, isocyano, isocyanato, isothiocyanato, azido, etc., are used (see the Illustrative List of Substituent Prefixes in Section H (¶ 294)).
Examples:

$CHCl_3$ — **Methane, trichloro**– (not Chloroform)

$PhIO_2$ — **Benzene, iodyl**–

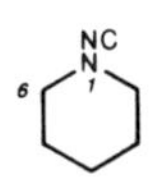

Piperidine, 1–isocyano–

$Me_2Si(NCO)_2$ — **Silane, diisocyanatodimethyl**– (not Isocyanic acid, dimethylsilylene ester)

Groups such as –$I(OH)_2$, –$ClBr_2$, are not expressed by (dihydroxyiodo) and (dibromochloro) radicals; instead, coordination nomenclature (¶ 215) based on the central halogen atom is employed.
Example:

$PhI(OH)_2$ — **Iodine, dihydroxyphenyl**–

189. Hydrazides of principal acid groups are expressed by hydrazide terms in the index modification except for (*a*) sulfenic acid hydrazides, R–S–$NHNH_2$, which are indexed at **Hydrazine** with a "thio" substituent prefix; and (*b*) hydrazides of **Hydrazinecarboxylic acid,** for which the heading parent **Carbonic dihydrazide,** $(H_2NNH)_2CO$, (¶ 183) is employed. Hydra= zides of subsidiary groups, e.g., of carboxyl substituents on aminium com= pounds, are indexed as substituents by use of hydrazino radicals. Cyclic hydrazides are indexed as heterocycles. Replacement ("a") names are used when the requirements (¶ 127) are satisfied.
Examples:

$PhCSNHNH_2$ — **Benzenecarbothioic acid hydrazide**

AcNHNHAc — **Acetic acid 2–acetylhydrazide** (not Hydrazine, 1,2–diacetyl–)

$HO_2CCH_2CONHNHPh$ — **Propanedioic acid mono(2–phenylhydrazide)** (not Propanoic acid, 3–oxo–3–(2–= phenylhydrazino)–)

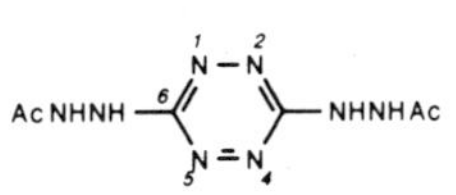

Acetic acid 2,2′–(1,2,4,5–tetrazine–3,6–diyl)= dihydrazide

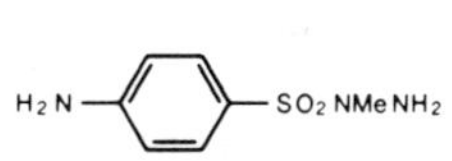

Benzenesulfonic acid, 4–amino– 1–methylhydrazide

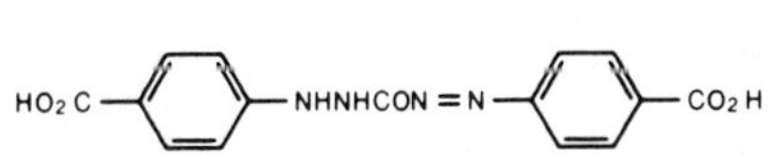

Diazenecarboxylic acid, (4–carboxyphenyl)– 1–[2–(4–carboxyphenyl)= hydrazide] (principle: hetero acyclic parent preferred)

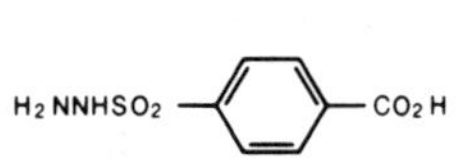

Benzoic acid, 4–(hydrazino= sulfonyl)–

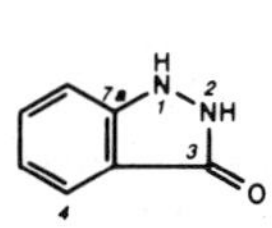

3*H*–Indazol–3–one, 1,2–dihydro– (not Benzoic acid, 2–hydrazino–, cyclic hydrazide)

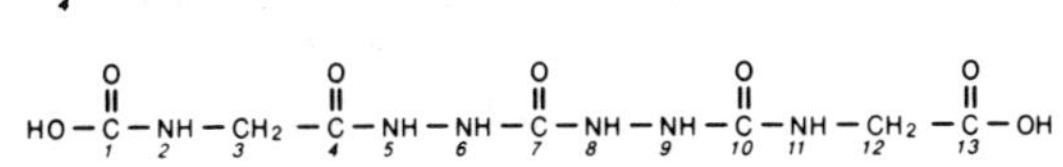

2,5,6,8,9,11–Hexaazatridecanedioic acid, 4,7,10–trioxo–

190. Hydrazones of ketone and aldehyde principal groups are expressed by hydrazone terms in the index modification except for hydrazones of car= bonyl groups of acids, acid halides, and amides; these are usually indexed as hydrazonic acids (¶ 169) and their halides and amides. Hydrazones of **Di= azenecarboxaldehyde, Diazenecarboxylic acid,** and related compounds are indexed as derivatives of **Formazan** (¶ 193). Azines are indexed as "–ylidene" hydrazones; osazones as dihydrazones (of adjacent carbonyl groups); cyclic hydrazones as heterocycles. In the presence of compound classes more pre= ferred than the aldehyde or ketone bearing the hydrazone, a hydrazono sub= stituent is cited.
Examples:

$Me_2C{=}NNH_2$ — **2–Propanone hydrazone**

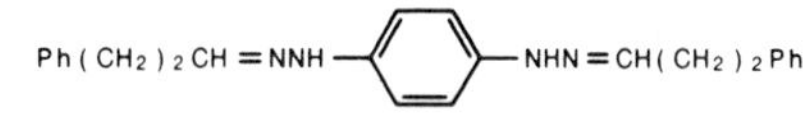

Benzenepropanal 1,4–phenylenedihydrazone

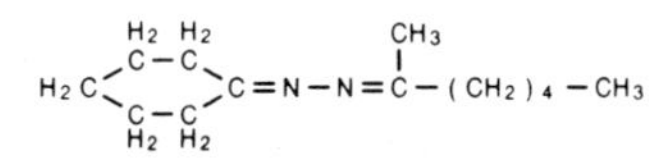

Cyclohexanone (1–methylhexylidene)hydrazone (not 2–Heptanone, azine with cyclohexanone) (principle: ring preferred to chain)

$PhNHN{=}CHSCH_2CO_2H$ — **Acetic acid, [[(phenylhydrazono)= methyl]thio]**– (not Acetic acid, mercapto–, formate, phenyl= hydrazone)

$MeNHN{=}CPhCPh{=}NNHMe$ — **Ethanedione, diphenyl– bis(methylhydrazone)**

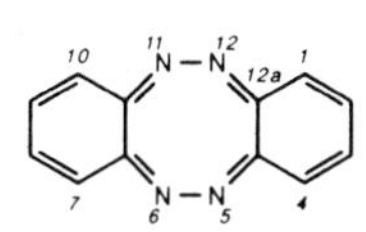

Dibenzo[*c*,*g*][1,2,5,6]tetrazocine (not 3,5–Cyclohexadiene–1,2–dione, bimol. cyclic azine)

Phosphazines, which contain the fragment =C=N–N=P, are indexed as phosphoranylidene hydrazones.
Example:

$MeCH{=}NN{=}PPh_3$ — **Acetaldehyde (triphenylphosphoranylidene)= hydrazone**

Semicarbazones, isosemicarbazones, carbohydrazones, and *semi= oxamazones* of systematically named compounds are expressed, not by these modification terms or as substituted hydrazones, but by the general principles of substitutive nomenclature on the basis of the highest function present.
Examples:

$Me_2C{=}NNHCSNH_2$ (2 1) — **Hydrazinecarbothioamide, 2–(1– methylethylidene)**– (not 2– Propanone, thiosemicarbazone)

$PhCH{=}NNHCNHOEt$ — **Hydrazinecarboximidic acid, (phenylmethylene)– ethyl ester** (not Benzaldehyde, 3– ethylisosemicarbazone)

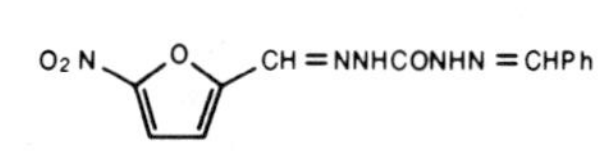

Carbonic dihydrazide, [(5–nitro– 2–furanyl)methylene](phenyl= methylene)– (not 2–Furan= carboxaldehyde, 5–nitro, carbo= hydrazone with benzaldehyde)

$PhCH{=}NNHCOCONH_2$ — **Acetic acid, aminooxo– (phenylmethylene)hydrazide** (not Benzaldehyde, semioxamazone)

191. Imides are indexed as heterocycles with "–one" suffixes and are accordingly ranked as ketones; in the presence of higher ranking compounds, the imide is expressed as a heterocyclic radical. The "diimide" of orthocarbonic acid, HN=C=NH ("carbodiimide") is indexed at **Methanediimine,** and its derivatives are often to be found at amine or amide headings.

Examples:

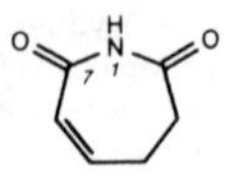

1*H*-**Azepine-2,7-dione, 3,4-dihydro-** (not 2-Hexenimide)

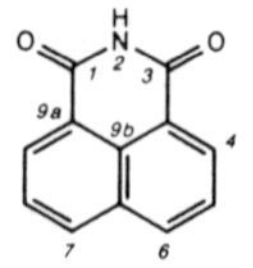

1*H*-**Benz[*de*]isoquinoline-1,3=**
2*H*)-dione (not 1,8-Naphthalenedicarboximide)

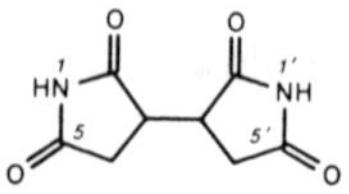

[3,3'-Bipyrrolidine]-2,2',5,5'-tetrone (not 1,2,3,4-Butanetetracarboxylic 1,2:3,4-diimide

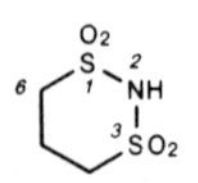

4*H*-**1,3,2-Dithiazine, dihydro-1,1,3,3-tetraoxide** (not 1,3-Propanedisulfonimide)

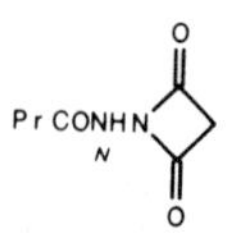

Butanamide, *N*-**(2,4-dioxo-1-azetidinyl)-**

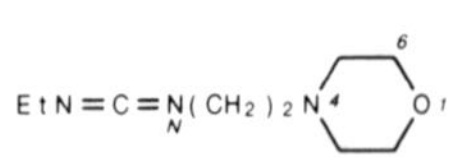

4-Morpholineethanamine, *N*-**(ethylcarbonimidoyl)-**

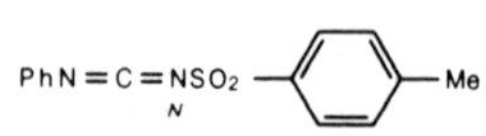

Benzenesulfonamide, 4-methyl- *N*-**(phenylcarbonimidoyl)-**

For sulfur imides, see ¶ 200; for **Phosphine imide,** see ¶ 197.

192. Molecular addition compounds of neutral components are generally indexed in the *Chemical Substance* and *Formula Indexes* at the name and formula of each component. (The formula headings used are those of the components.) Some common components are not indexed unless all other components are also "common" or cannot be related to compound classes described in the "Order of Precedence of Compound Classes" (¶ 106); however, unesterified acids in the following list are indexed when not components of salts with bases. These common components are:

Acetic acid
Acetic acid, trifluoro-
Benzene, 1,3,5-trinitro-
1,3-Benzenediol, 2,4,6-trinitro-
Benzenesulfonic acid, 4-methyl-
Butanedioic acid, 2,3-dihydroxy- (all stereoisomers)
2-Butenedioic acid (*E*)-
2-Butenedioic acid (*Z*)-
Carbamimidothioic acid, phenylmethyl ester
Cyclohexanamine
Cyclohexanamine, *N*-**cyclohexyl-**
Ethanamine, *N,N*-**diethyl-**
Ethanedioic acid
Methanesulfonic acid
Phenol, 2,4,6-trinitro-
1,2,3-Propanetricarboxylic acid, 2-hydroxy-
3*H*-**Pyrazol-3-one, 2,4-dihydro-5-methyl-4-nitro-2-(4-nitrophenyl)-**

In addition, compounds with *water* and *ammonia* are not indexed at these names; they are expressed as "hydrate" and "ammoniate," with the prefixes hemi (for $^1/_2$), mono, sesqui (for $^3/_2$), di, tri, etc., up to pentadeca. Other solvates are indexed as molecular addition compounds. Often the solvate component receives the preferred entry; e.g., **Ethanol, compd. with pyridine (1:1)** (an additional entry appears at **Pyridine, compd. with ethanol (1:1)**). Crystal forms of organic compounds containing solvents of crystallization are indexed only as the unsolvated species except when properties of the crystals themselves are being studied. "Hydrates" of carbonyl compounds are indexed as *gem*-diols; e.g., "acetaldehyde hydrate" is indexed as **1,1-Ethanediol.**

Ozonides of known structure are indexed by regular nomenclature; ozonides of unsaturated compounds are indexed at the compound headings with "ozonide" modification terms when the structures are unknown; "ozonides" of other compounds, e.g., phosphorus acid esters, are expressed as "compd. with ozone" and a ratio.

Bisulfite addition compounds are named as salts of specific hydroxy sulfonic acids when the structures are known or can reasonably be assumed, otherwise as a molecular addition compound of the carbonyl compound with a phrase such as "compd. with sodium hydrogen sulfite" and a ratio (if known) in the modification. An additional entry appears at **Sulfurous acid, compounds, monosodium salt, compd. with....** (This is an exception; an "oxo" acid salt ranks higher than an aldehyde or ketone (¶ 106) and would normally receive the preferred index entry.)

Diels-Alder adducts (diene adducts) of unknown constitution are indexed like molecular addition compounds, except that the "compd. with" phrase is replaced by "adduct with."

Catena compounds (cyclic compounds with interlocking rings) are indexed at the components with a "catena compd. with" phrase and a ratio.

The "preferred" index name (for molecular addition compounds that receive more than one) is that name to which cross-references in the *Index Guide* direct the reader from trivial names, the name given precedence in the CAS Registry System, and therefore the name which, after uninversion of the index heading if necessary, may be used among *CA* users in general discussions and reports. (To the index user in search of information, all the index names for an individual compound are of equal value.) For molecular addition compounds of stereoparents and their derivatives (¶ 203 I) with nonstereoparents, the preferred index entry appears at the former heading. In the absence of a stereoparent, the preferred index name is that which describes, in order of decreasing preference:

(*a*) a component other than a common component (see list above) selected according to the Order of Precedence of Compound Classes (¶ 106);

(*b*) a common component highest in the same order;

(*c*) a component which does not belong to any compound class described in the "Order of Precedence of Compound Classes," according to the earliest alphabetic position of the index name.

Examples ((*a*) is the preferred index name in each case):

1. (*a*) **2,5-Cyclohexadiene-1,4-dione, 2,3,5,6-tetrachloro-**
compd. with coronene (no ratio cited because unknown; functional compound preferred to cyclic hydrocarbon)

(*b*) **Coronene**
compd. with 2,3,5,6-tetrachloro-2,5-cyclohexadiene-1,4-dione

2. (*a*) **Anthracene**
compd. with 2,4,6-nitrophenol (2:1) (only entry; the cyclic hydrocarbon is preferred to a common compound of higher rank)

3. (*a*) **Methane, sulfinylbis-**
compd. with iodine (1:1) (by "iodine," the molecular form I_2 is implied; atomic forms are indicated, when necessary, by phrases such as "**compd. with at. chlorine (1:2)**" and an entry, in such a case, at **Chlorine, atomic**)

(*b*) **Iodine**
compd. with sulfinylbis[methane] (1:1) (note that multiplied heading parents are bracketed in the uninverted names)

193. Nitrogen compounds. Cyclic nitrogen compounds, including lactams, sultams, and cyclic hydrazones, oximes, etc., are indexed at heterocyclic molecular skeletons (see *Section B*). Nitrogen-containing functional derivatives include imidic acids, amides, amines, imines, etc., for which the appropriate paragraph should be consulted. **Hydroxylamine** (see below) is a substitutive functional parent compound (¶ 130). Acyclic nitrogen skeletons, alone or with principal groups expressed as suffixes, are employed as heading parents for indexing purposes. Some groups, including azido, nitro, nitroso, and isocyano, are always expressed as substituent prefixes (¶ 132). *aci*-Nitro, HON(O)=, is also a mandatory prefix, and it may be substituted; e.g., (propyl-*aci*-nitro) is $CH_3CH_2CH_2ON(O)=$; (benzoyl-*aci*-nitro) is $C_6H_5C(O)-ON=(O)=$;

Diazene, HN=NH; **1-Triazene,** NH=N-NH_2; **Triazane,** NH_2-NH-NH_2, etc., are molecular skeleton heading parents to which the principal groups except hydroxy (and its chalcogen analogs), amino, and imino can be suffixed. **Hydrazine** is used as a trivial name for Diazane and its derivatives. Except for hydrazides (¶ 189) and hydrazones (¶ 190), alkyl, aryl and acyl derivatives are indexed at these nitrogen parents, which rank just below **Formazan** and nitrogen heterocycles as the highest class of nonfunctional compounds (¶ 106). Within the class, seniority depends first on the number of hetero atoms, then on maximum unsaturation. When more preferred compound classes (including all those expressed by means of functional suffixes) are present, the nitrogen skeletons are named as substituent prefixes, some of which have trivial names.

Examples (See also ¶ 161):

HN=N—	diazenyl (only when unsubstituted; otherwise azo is used)
—N=N—	azo (when attached to different atoms)
—N=N(O)—	azoxy (When used as a substituted monovalent radical, the position of the oxide (if known) is indicated by an *NNO*- or *ONN*-locant (see example below).)
H_2NNH—	hydrazino

$H_2NN=$	hydrazono
$-NHNH-$	hydrazo (when attached to different atoms)
$-NHN=$	1–hydrazinyl–2–ylidene
$=NN=$	azino
$H_2NNHNH-$	triazanyl
$H_2NN=N-$	1–triazenyl
$HN=NNH-$	2–triazenyl
$-N=NNH-$	1–triazene–1,3–diyl
$H_2NNHNHNH-$	tetrazanyl
$=NNHNHN=$	1,4–tetrazanediylidene

Hydrazides (¶ 189) and hydrazones (including azines) (¶ 190) are excluded from the following examples.

Examples:

H_2NNHCO_2H	**Hydrazinecarboxylic acid**
$HO_2CNHNHCO_2H$	**1,2–Hydrazinedicarboxylic acid**
$H_2NNHCONHMe$	**Hydrazinecarboxamide,** ***N*–methyl–**
Et_2NNHEt	**Hydrazine, triethyl–**
H_2NNHOH	**Hydrazine, hydroxy–**

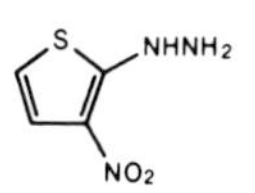

Hydrazine, (3–nitro–2–thienyl)– (principle: preferred hetero atom in heading parent (¶ 138))

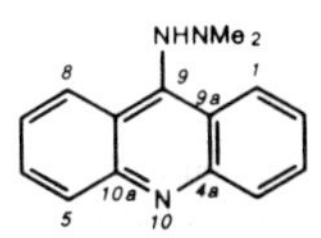

Acridine, 9–(2,2–dimethyl=hydrazino)– (principle: cyclic parent preferred (¶ 138))

$PhN=NSO_3H$	**Diazenesulfonic acid, phenyl–**
$PhN=NCONH_2$	**Diazenecarboxamide, 2–phenyl–**

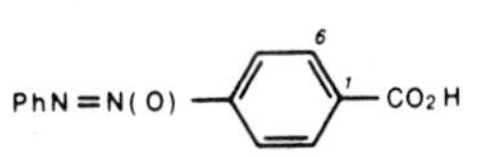

Benzoic acid, 4–(phenyl–*NNO*–az=oxy)–

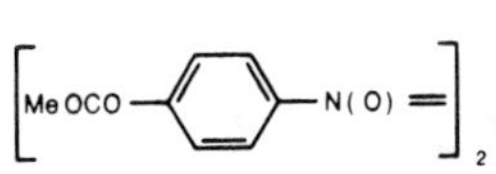

Benzoic acid, 4,4'–azobis– dimethyl ester, *N,N'*–dioxide

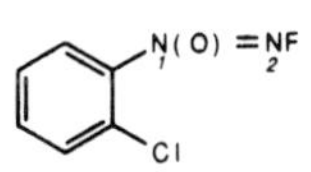

Diazene, (2–chlorophenyl)fluoro– 1–oxide

$H_2NN=NAc$	**1–Triazene, 1–acetyl–**
$MeSO_2NHNHNHC(=NH)Me$	**Triazane, 1–(1–iminoethyl)–3–=(methylsulfonyl)–**

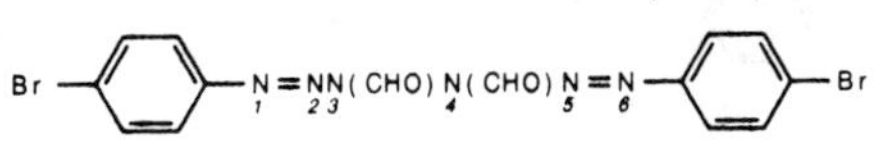

1,5–Hexazadiene–3,4–dicarbox=aldehyde, 1,6–bis(4–bromo=phenyl)–

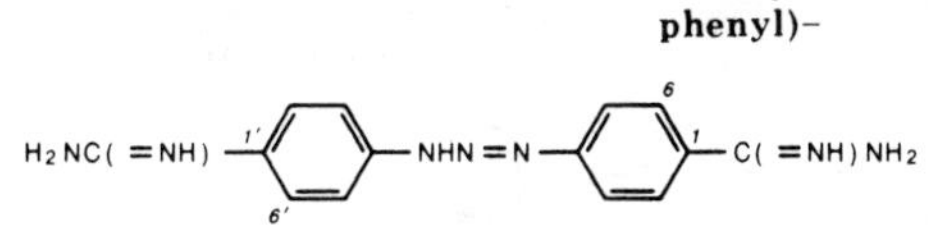

Benzenecarboximidamide, 4,4'–=(1–triazene–1,3–diyl)bis–

Formazan is a trivial name employed to index the tautomeric structure $NH{=}NCH{=}NNH_2$ (the hydrazone of **Diazenecarboxaldehyde**) and its de=rivatives. It is ranked highest of all nonfunctional nitrogen compounds, below imines but above nitrogen heterocycles (¶ 106). The chain is never broken to include skeletal atoms in a heading parent that expresses a more preferred compound class. The double bonds in **Formazan** are assigned lowest locants. Substituent radicals derived from it are as follows:

$-N=NCH=NNH_2$	1–formazano
$HN=NCH=NNH-$	5–formazano
$HN=N\overset{\vert}{C}=NNH_2$	formazanyl
$HN=N\overset{\vert}{C}=NHNH-$	3,5–formazandiyl
$-N=NCH=NN=$	1–formazanyl–5–ylidene
$HN=N\overset{\vert}{C}=NN=$	3–formazanyl–5–ylidene
$-N=N\overset{\vert}{C}=NNH-$	1,3,5–formazantriyl

Principal chemical groups are expressed as suffixes, except for hydroxy (and chalcogen analogs), amino, and imino, on nitrogen atoms. Other substituents are expressed as prefixes.

Examples:

$PhN=NC(CO_2H)=NNHPh$	**3–Formazancarboxylic acid, 1,5–=diphenyl–**
$PhN=NC(CH_2CO_2H)=NNHPh$	**Acetic acid, (1,5–diphenylforma=zanyl)–**

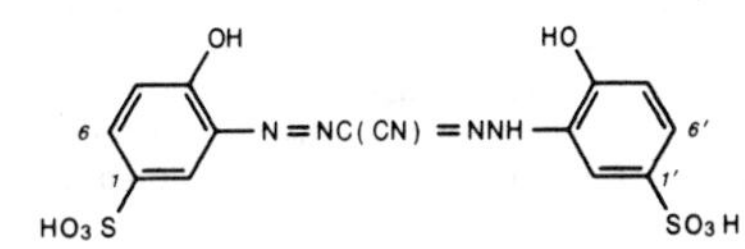

Benzenesulfonic acid, 3,3'–(3–cyano–1,5–formazandiyl)bis[4–hydroxy–

1–Acyl derivatives of **Formazan** are cited as substituent prefixes; 5–acyl derivatives are indexed as hydrazides, 5–alkylidene derivatives as hydrazones, 3–acyl derivatives at 1–alkanone or **Methanone** heading parents.

Examples:

$AcN=NCH=NNHPh$	**Formazan, 1–acetyl–5–phenyl–**

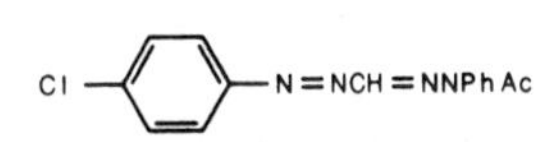

Acetic acid [[(4–chlorophenyl)azo]=methylene]phenylhydrazide

$PhN=NC(COMe)=NNHPh$	**Ethanone, 1–(1,5–diphenylforma=zanyl)–**

Hydroxylamine, H_2NOH; **Thiohydroxylamine,** H_2NSH, etc., are sub=stitutive parent compounds; i.e., substituent prefixes but (usually) not sub=stituent suffixes may be attached to them. *N*–Ylidene derivatives are oximes (¶ 195), imidic acid derivatives (¶ 165), etc.

In general, *N*–alkyl and *N*–aryl derivatives of **Hydroxylamine** are indexed as amines; *N*–acyl derivatives as amides; *S*–amino derivatives of **Thio=hydroxylamine** are indexed as sulfenamides. When the nitrogen atom is unsubstituted, *O*–derivatives are usually indexed at **Hydroxylamine,** etc. The only suffixes employed are "sulfonic acid," "sulfonyl chloride," etc., when attached to the chalcogen atoms; the locant ("*O*" "*S*," etc.) is placed just ahead of the suffix.

Examples:

H_2NOSO_3H	**Hydroxylamine–*O*–sulfonic acid**
H_2NOAc	**Hydroxylamine, *O*–acetyl–**
H_2NOCO_2H	**Hydroxylamine, *O*–carboxy–**
$H_2NOCONH_2$	**Hydroxylamine, *O*–(amino=carbonyl)–**

BuONH2 — **Hydroxylamine, *O*-butyl-**

$H_2NO(CH_2)_2ONH_2$ — **Hydroxylamine, *O,O'*-1,2-ethane=diylbis-**

EtSNH2 — **Ethanesulfenamide**

PhNHOH — **Benzenamine, *N*-hydroxy-**

AcNHSH — **Acetamide, *N*-mercapto-**

$HONHCO_2H$ — **Carbamic acid, hydroxy-**

$HONHCONH_2$ — **Urea, hydroxy-**

$PhSO_2NHOH$ — **Benzenesulfonamide, *N*-hydroxy-**

Hydroxylamine and its chalcogen analogs rank lowest among the non=functional nitrogen parents. In the presence of more preferred compound classes, the radicals (aminooxy), (aminothio), (hydroxyamino), etc., are em=ployed.

Examples:

$HONHCOCO_2H$ — **Acetic acid, (hydroxyamino)oxo-**

$H_2NOCH_2NHNHNH_2$ (1 2 3) — **Triazane, 1-[(aminooxy)methyl]-**

194. Organometallic compounds. Organic derivatives of germanium, tin, lead, antimony, and bismuth are indexed at hydride heading parents (see ¶¶ 181, 199).

Cyclic derivatives of these metals with standard substitutive valencies are indexed as heterocycles.

Example:

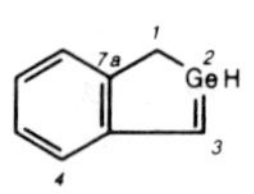

1*H*-2-Benzogermole

Acyclic metal derivatives, other than acetylides (¶ 219) and those indexed at metal hydride names, are indexed at the metal element name. Such names are ranked as neutral coordination compounds (¶ 215) and all functional compounds are expressed as substituent prefixes. When all metal atoms cannot be expressed by the heading parent, elements other than those of Groups IVA and VA are expressed as "-io" radicals, e.g., sodio, magnesio, aurio, lithio, when the metal replaces a single hydrogen. No hydrogen or other "substituent" is implied by these radical names, and atoms or groups attached to them must be expressed.

Examples:

Li-Ph — **Lithium, phenyl-**

$Me_3P^+CH=CHCH_2HgBr \cdot Br^-$ — **Phosphonium, [3-(bromo=mercurio)-1-propenyl]=trimethyl-bromide**

195. Oximes contain a bivalent hydroxylamine residue, =N-OH. When this residue has replaced oxygen in carboxylic acids, acid halides, and amides, the compounds are named as *N*-hydroxy derivatives of imidic acids (¶ 165), imidoyl halides (¶ 170), and imidamides (¶ 171). Cyclic oximes are indexed at heterocyclic parents, e.g., **Isoxazole.**

Acyclic oximes derived from aldehydes and ketones expressed as princi=pal-group suffixes are expressed by "oxime" terms in the index modifications at the carbonyl-containing heading parents. When the carbonyl group is expressed as a substituent, it is replaced by the (hydroxyimino) radical, HO-N=. *O*-Alkyl, *O*-aryl and *O*-acyl oximes are named as such in the modification or as an "-oxyimino" radical. Oximes derived from **Hydroxylamine-*O*-sulfonic acid** (¶ 193) are named as substituents of that heading parent.

Examples:

$H_2C=NOH$ — **Formaldehyde oxime**

$Ph_2C=NOH$ — **Methanone, diphenyl- oxime**

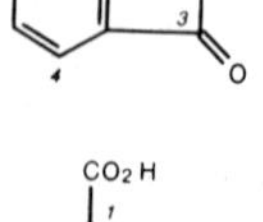

1,3-Isobenzofurandione monooxime

Benzoic acid, 3-[(hydroxyimino)=methyl]-

MeCH=NOMe — **Acetaldehyde *O*-methyloxime**

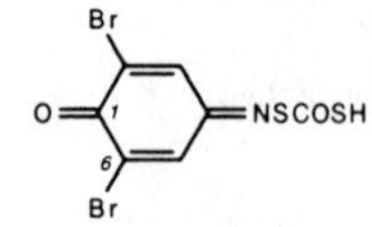

2,5-Cyclohexadiene-1,4-dione, 2,6-dibromo-4-[*S*-(thiocarboxy)thiooxime]

$MeCH=NOSO_3Me$ — **Hydroxylamine-*O*-sulfonic acid, ethylidene- methyl ester**

196. Oxygen compounds include a wide range of compound classes: acids, anhydrides, esters, alcohols, salts, metal oxides, etc. These are discussed in other sections and paragraphs. Cyclic oxygen compounds are indexed as heterocycles; these include cyclic esters, anhydrides, ethers, and oximes, as well as lactones, sulfones, etc.

Examples:

Oxirane, 2,2-diethyl- (not Butane, 1,2-epoxy-2-ethyl-)

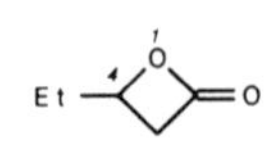

2-Oxetanone, 4-ethyl- (not Penta=noic acid, 3-hydroxy-, β-lactone; see ¶ 185)

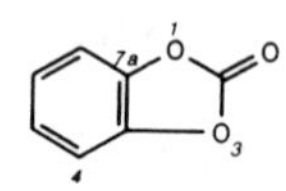

1,3-Benzodioxol-2-one (not Carbonic acid, cyclic 1,2-phenylene ester; see ¶ 185)

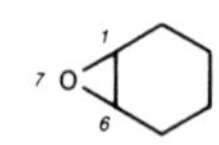

7-Oxabicyclo[4.1.0]heptane (a Von Baeyer ring system; see ¶ 155)

Acyclic ethers (including acetals and ortho esters) are indexed by replacement ("a") nomenclature if the requirements (¶ 127) are satisfied.

Example:

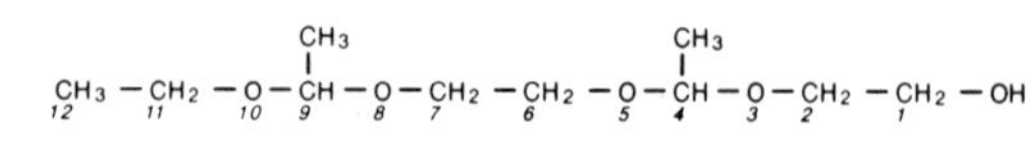

3,5,8,10-Tetraoxadodecan-1-ol, 4,9-dimethyl- (not Acetaldehyde, 1,2-ethanediyl ethyl 2-hydroxyethyl diacetal)

When "a" names are not permitted, acyclic ethers are indexed at heading parents (functional parent compounds, hydrocarbons, etc.) by use of "oxy" radicals, including the elided radicals (¶ 107) methoxy, ethoxy, propoxy, bu=toxy, and phenoxy.

Examples:

EtOMe — **Ethane, methoxy-** (not Ether, ethyl methyl)

Et_2O — **Ethane, 1,1'-oxybis-** (not Ethyl ether)

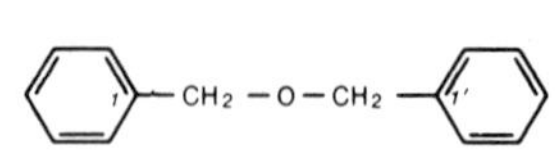

Benzene, 1,1'-[oxybis(methy=lene)]bis-

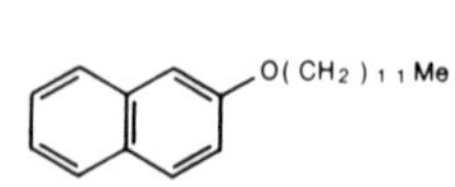

Naphthalene, 2-(dodecyloxy)- (principle: ring preferred to chain (¶ 138))

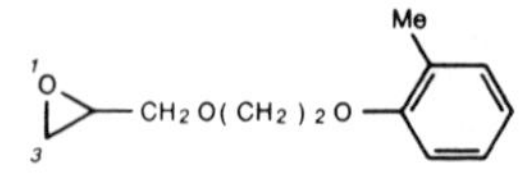

Oxirane, [[2-(2-methylphenoxy)=ethoxy]methyl]- (principle: pre=ferred ring)

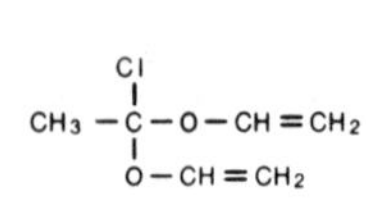

Ethene, 1,1'-[(1-chloro=ethylidene)bis(oxy)]bis- (prin=ciple: unsaturation)

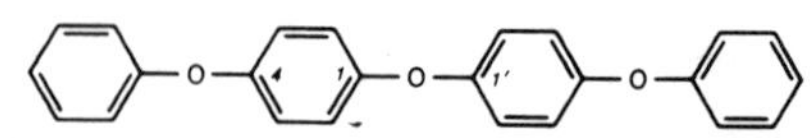

Benzene, 1,1'-oxybis[4-phenoxy- (principles: centrality and multi=plication)

$CH_3-CH_2-O-CH_2-CH_2-S-S-CH_2-CH_2-O-CH_2-CH_3$

Disulfide, bis(2-ethoxyethyl) (an "a" name cannot be used; only three hetero units (¶ 127) are present)

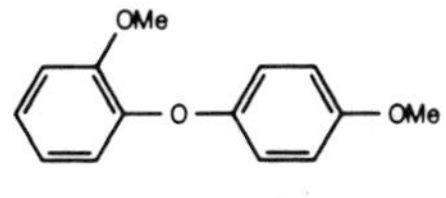

Benzene, 1-methoxy-2-(4-meth=oxyphenoxy)-principle: lowest locants for substituent prefixes)

Hydroperoxides, ROOH, follow alcohols (including phenols) in order of precedence (¶ 106). They are indexed at the radicofunctional heading parent **Hydroperoxide**. No hyphen follows the substituent radical (¶ 108) unless the heading parent is being multiplied by a prefix such as "bis." In the presence of more preferred compound classes, or when the HOO- group is attached to a hetero atom other than silicon, a hydroperoxy radical is employed. The chalcogen analogs R-S-OH and R-S-SH are indexed as sulfenic acids and sulfenothioic acids, respectively (¶ 165). **Thiohydroperoxide** is the heading parent for R-O-SH compounds.

Hydrotrioxide, Hydrotetraoxide, etc., heading parents are treated in a similar manner, and hydrotrioxy and hydrotetraoxy radicals can be used as substituent prefixes.

Examples:

EtOOH

Hydroperoxide, ethyl (the uninverted name is Ethyl hydroperoxide)

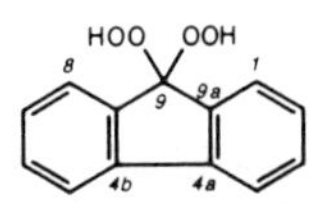

Hydroperoxide, 9*H*-fluoren-9-yl=idenebis-

$HOOCH_2COPh$

Ethanone, 2-hydroperoxy-1-phenyl-

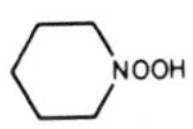

Piperidine, 1-hydroperoxy-

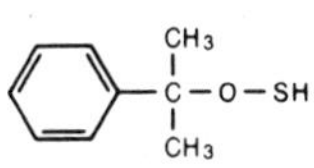

Thiohydroperoxide, *O*-**(1-methyl-1-phenylethyl)**

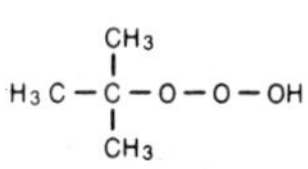

Hydrotrioxide, 1,1-dimethylethyl

Acyl hydroperoxides are peroxy acids (¶ 165).

Peroxides, R-O-O-R' (R and R' = alkyl or acyl), are named at the radi=cofunctional heading parent **Peroxide** unless a replacement ("a") name (¶ 127) is appropriate. Cyclic peroxides are indexed as heterocycles and are ranked in accordance with the seniority of ring systems (¶ 138). Acyclic peroxides rank with the nonfunctional oxygen parents (¶ 106) and in the presence of a more preferred compound class are expressed by "dioxy" radicals. Acyl alkyl (and acyl silyl) peroxides are esters of peroxy acids. Symmetrical diacyl peroxides derived from "oxo" acids are named at acid, acid halide, etc., heading parents such as **Peroxydiphosphonic acid.** Dithio peroxides are indexed at **Di=sulfide** (¶ 200); R-S-O-R' compounds are named as esters of sulfenic acids (¶ 165). The corresponding diacyl compounds are indexed at **Thioperoxide.**

Examples:

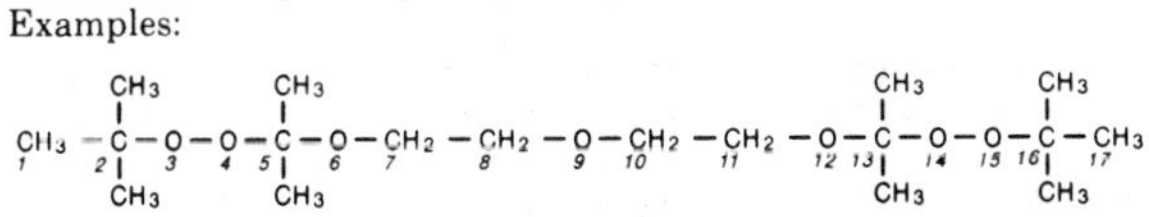

3,4,6,9,12,14,15-Heptaoxaheptade=cane, 2,2,5,5,13,13,16,16-octa=methyl-

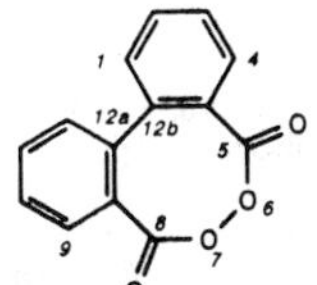

Dibenzo[*d,f*][1,2]dioxocin-5,8-dione

$Me_2CHOOCHMe_2$

Peroxide, bis(1-methylethyl)

EtOOMe

Peroxide, ethyl methyl (the unin=verted name is Ethyl methyl per=oxide)

Peroxide, diacetyl

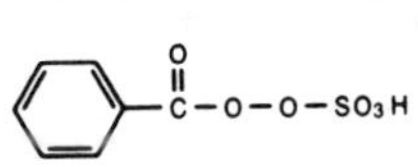

Peroxide, benzoyl sulfo

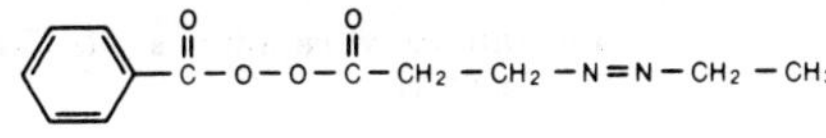

Diazene, [3-(benzoyldioxy)-3-=oxopropyl]ethyl-

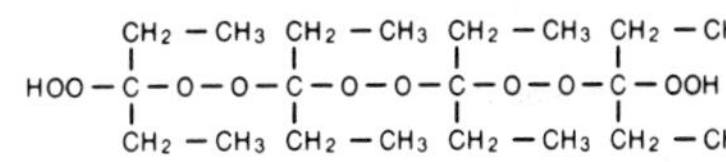

Hydroperoxide, [dioxybis[(1-ethyl=propylidene)dioxy(1-ethyl=propylidene)]]bis-

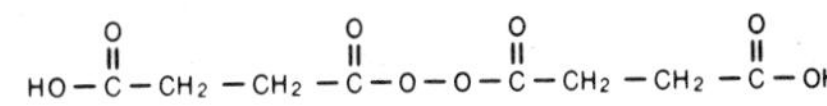

Butanoic acid, 4,4'-dioxybis[4-oxo-

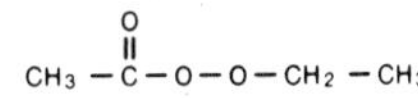

Ethaneperoxoic acid ethyl ester

MeOOF

Peroxyhypofluorous acid methyl ester

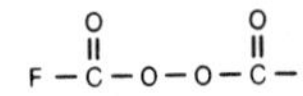

Peroxydicarbonic difluoride

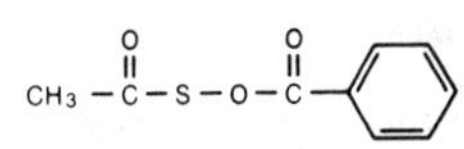

Thioperoxide, *S*-**acetyl** *O*-**benzoyl**

The heading parents **Trioxide, Tetraoxide,** etc., are treated analogously and, in the presence of more preferred compound classes, radicals such as (methyltrioxy), CH_3-O-O-O-, are used.

Example:

MeOOOMe

Trioxide, dimethyl

197. Phosphorus and arsenic compounds. The fundamental phos=phorus and arsenic "oxo" acids are shown in Table VI. By use of replacement nomenclature (¶ 129) in which oxo and hydroxyl groups are replaced by other chalcogen atoms or nitrogen groups, a large number of acids, acid halides, amides, etc., can be named. **Phosphorous acid** is the heading parent when all three hydroxyl groups are involved in anhydride, ester, or salt formation; otherwise the tautomeric **Phosphonic acid** is the preferred heading parent. Similarly, **Phosphinic acid** is preferred over the tautomeric **Phosphonous acid. Metaphosphorous acid** is the name used for the unsubstituted acid and its ester and salts, but **Phosphenous acid** is the fundamental name from which replacement analogs are derived. **Metaphosphoric** and **Phosphenic acids** stand in the same relationship, except that the special names **Meta=phosphimic acid, Thiometaphosphimic acid,** etc., are used for $(HO)P=(O)(=NH)$ (or $(HO)_2P \equiv N$) and its chalcogen analogs.

TABLE VI
FUNDAMENTAL MONONUCLEAR PHOSPHORUS AND ARSENIC OXO ACIDS

"Trivalent" acids

$(HO)_3P$	**Phosphorous acid**	$(HO)_3As$	**Arsenous acid**
$(HO)_2HP$	**Phosphonous acid**	$(HO)_2HAs$	**Arsonous acid**
$(HO)H_2P$	**Phosphinous acid**	$(HO)H_2As$	**Arsinous acid**
HOPO	**Phosphenous acid** **Metaphosphorous acid** **(HPO_2)**	HOAsO	**Arsenenous acid**

"Pentavalent" acids

$(HO)_3PO$	**Phosphoric acid**	$(HO)_3AsO$	**Arsenic acid** **(H_3AsO_4)**
$(HO)_2HPO$	**Phosphonic acid**	$(HO)_2HAsO$	**Arsonic acid**
$(HO)H_2PO$	**Phosphinic acid**	$(HO)H_2AsO$	**Arsinic acid**
$HOPO_2$	**Phosphenic acid** **Metaphosphoric acid** **(HPO_3)**	$HOAsO_2$	**Arsenenic acid**

Examples:

$(PhNH)_2P(O)OH$

Phosphorodiamidic acid, *N,N'*-**di=phenyl-**

$Me_2NNHP(O)(OH)_2$	**Phosphorohydrazidic acid, 2,2-dimethyl–**
$MeNHP(OH)(=NH)NHMe$	**Phosphorodiamidimidic acid,** ***N,N'*–dimethyl–**
Ph_2AsOH	**Arsinous acid, diphenyl–**
$EtNHAs(O)(OH)_2$	**Arsenamidic acid, ethyl–**

Esters and anhydrides are indexed by the usual rules. **Phosphorous, Phosph -ic, Phosphorothioic** and **Phosphorodithioic acids** are "Class I" acids; their esters with a single alcohol or thiol of "Class II" (¶ 185) are indexed at the alcohol name, but "mixed" esters are indexed at the phosphorus acid unless a more preferred compound class (e.g., carboxylic acid, aminium compound) is present. All cyclic esters are indexed as heterocycles.
Examples:

$(EtO)_2PO(CH_2)_2NH_2$	**Phosphorous acid 2–aminoethyl diethyl ester**
$[Cl(CH_2)_2O]_3P$	**Ethanol, 2–chloro– phosphite (3:1)**
$HP(O)(OH)OMe$	**Phosphonic acid monomethyl ester**
$EtP(O)(OEt)SEt$	**Phosphonothioic acid, ethyl–** ***O,S*–diethyl ester**
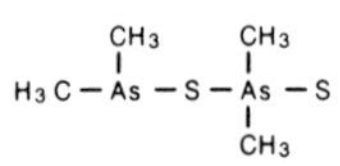	**Arsinodithioic acid, dimethyl– anhydrosulfide with dimethylarsinothious acid**

Nonacidic mononuclear arsenic and phosphorus names are derived from the acid names in the usual way by expressing the highest function as a class term, e.g., "amide," "chloride," and the other replacement terms in alphabetical order within the main part of the name. The descending order of precedence for class terms is: hydrazide, halide (all considered equivalent, and cited in alphabetical order), azide, amide, cyanide, isocyanide, cyanate, thiocyanate (etc.), isocyanate, isothiocyanate (etc.), nitride, imide. Nonacidic analogs of phosphoric acid in which an oxo and hydroxyl group have been replaced by ≡N, are named at **Phosphonitrile** headings. **Phosphoryl** heading parents are employed for halides and halogenoids of the general structure X_3PO.
Examples:

$(H_2N)_3P{=}NH$	**Phosphorimidic triamide**
$Cl_3P{=}NH$	**Phosphorimidic trichloride**
$(H_2N)Cl_2PS$	**Phosphoramidothioic dichloride**
$(H_2N)BrClPO$	**Phosphoramidic bromide chloride**
$(OCN)ClFPO$	**Phosphorisocyanatidic chloride fluoride**
$(H_2N)_2(N_3)PO$	**Phosphorodiamidic azide**
$(H_2N)_2(NC)PO$	**Phosphorocyanidic diamide**
$(H_2NNH)F_2PO$	**Phosphorodifluoridic hydrazide**
$(H_2N)_3P$	**Phosphorous triamide**
Cl_3P	**Phosphorous trichloride**
ClF_2P	**Phosphorous chloride difluoride**
$(NC)H_2P$	**Phosphinous cyanide**
$(SCN)H_2P$	**Phosphinous isothiocyanate**
H_2NPS	**Phosphenothious amide**
$ClP{=}NH$	**Phosphenimidous chloride**
$(H_2N)ClHP{=}NH$	**Phosphonamidimidic chloride**
$Cl_2P{\equiv}N$	**Phosphonitrile chloride** (not Phosphoronitridic dichloride)
$(H_2N)_3AsO$	**Arsenic triamide**
$(H_2N)ClHAs{=}NH$	**Arsonamidimidic chloride**
$(NCS)_2HAsO$	**Arsonic dithiocyanate**
$(H_2N)H_2As{=}NH$	**Arsinimidic amide**
Cl_3As	**Arsenous trichloride**
Cl_2HAs	**Arsonous dichloride**
$(NC)H_2As$	**Arsinous cyanide**
$ClAs{=}NH$	**Arsenenimidous chloride**
$BrAsO$	**Arsenenous bromide**
Cl_3PO	**Phosphoryl chloride**
ClF_2PO	**Phosphoryl chloride fluoride ($POClF_2$)**
$(OCN)_3PO$	**Phosphoryl isocyanate**
Cl_3PS	**Thiophosphoryl chloride**
$OP{\equiv}N$	**Phosphoryl nitride**

Esters and anhydrides of cyclic oligomers of metaphosphorus acids are indexed as heterocycles.
Examples:

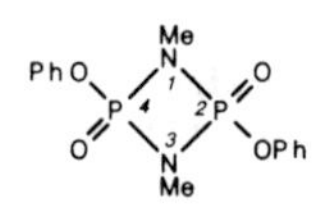	**1,3,2,4–Diazadiphosphetidine, 1,3–dimethyl–2,4–diphenoxy– 2,4–dioxide**
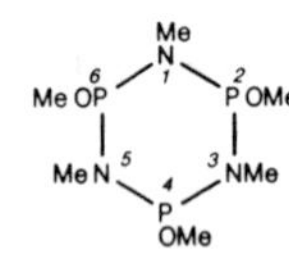	**1,3,5,2,4,6–Triazatriphosphorine, 1,2,3,4,5,6–hexahydro–2,4,6–trimethoxy–1,3,5–trimethyl–**
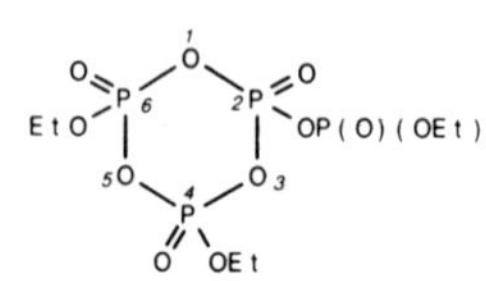	**1,3,5,2,4,6–Trioxatriphosphorinane, 2–[(diethoxyphosphinyl)oxy]– 4,6–diethoxy– 2,4,6–trioxide**

Phosphorus and arsenic molecular skeletons other than those named as acids and acid analogs are indexed as heterocycles or at acyclic hydride names: **Phosphine**, PH_3; **Arsine**, AsH_3; **Phosphorane**, PH_5; **Arsorane** AsH_5; **Diphosphine**, $H_2P\text{–}PH_2$; **Triarsine**, $H_2As\text{–}AsH\text{–}AsH_2$; etc. Oxo, thio (etc.), and imino derivatives of **Phosphorane** and **Arsorane** are indexed at additive heading parents.
Examples:

H_3PO	**Phosphine oxide**
H_3AsSe	**Arsine selenide**
$H_3P{=}NH$	**Phosphine imide**
HPO_2	**Phosphine oxide, oxo–** (not Phosphorane, dioxo–)

Oxides, etc., of **Diphosphine, Diarsine, Triphosphine,** etc., are indexed by use of "oxide" and "sulfide" terms, with locants if necessary, in the modification. Acid derivatives of one of the phosphorus atoms are indexed at acid parents when names based on the phosphorus chains are impracticable.
Example:

$PhPHP(O)H_2$	**Diphosphine, phenyl– 2–oxide**
$Et_3P(Cl)P(O)Cl_2$	**Phosphonic dichloride, (chlorotriethylphosphoranyl)–**

Derivatives of **Phosphine** and other phosphorus and arsenic molecular skeleton parents are named substitutively by use of suffixes and prefixes.
Examples:

$(HO)_2P(O)CO_2H$	**Phosphinecarboxylic acid, dihydroxy– oxide**
F_4PNHMe	**Phosphoranamine, 1,1,1,1–tetrafluoro–*N*–methyl–**
Me_2AsCO_2H	**Arsinecarboxylic acid, dimethyl–**
$EtAsMe_2{=}NH$	**Arsine imide, *As*–ethyl–*As,As*–dimethyl–**

$(EtO)_5P$	**Phosphorane, pentaethoxy-**
$Ph_3As(OAc)_2$	**Arsorane, bis(acetyloxy)triphenyl-**
HPO	**Phosphine, oxo-**
MeAsS	**Arsine, methylthioxo-**
MePHPHMe	**Diphosphine, 1,2-dimethyl-**
$[PhAsMe]_2$	**Diarsine, 1,2-dimethyl-1,2-di=phenyl-**
MeAs=AsMe	**Diarsene, dimethyl-**

Polyphosphorus acids (anhydrides of the mononuclear acids) and their replacement analogs do not follow the rules described for the mononuclear acids. They are given traditional inorganic names, and many cross-references will be found at mononuclear names (see also ¶ 219). (Polyarsenic acids are named analogously.)
Examples:

$(HO)_2P(O)OP(O)(OH)_2$	**Diphosphoric acid**
$(HO)_2P(O)P(O)(OH)_2$	**Hypophosphoric acid**
$(HO)HP(O)OP(O)(OH)_2$	**Isohypophosphoric acid**
$(HO)_2POP(O)(OH)_2$	**Diphosphoric(III,V) acid**
$(HO)_2POP(OH)_2$	**Diphosphorous acid**
$(HO)_2P(O)NHP(O)(OH)_2$	**Imidodiphosphoric acid**
$(HO)_2P(O)OP(O)(OH)NHP(O)(OH)OP(O)(OH)_2$	P'-**Imidotetraphosphoric acid**
$(H_2N)_2P(O)NHP(O)(OH)_2$	*P,P*-**Diamidoimidodiphosphoric acid**

Line formulas are used with thio, seleno, and telluro analogs of polyphosphorus acids to indicate the number and positions of sulfur (etc.) atoms. Thio, etc., prefixes are placed at the beginning of the name without multiplicative terms; e.g., **Thiodiphosphorous acid ($[(HO)_2P]_2S$); Thiodiphosphoric(III,V) acid ($(HO)_2POP(S)(OH)_2$); Thioperoxydiphosphoric acid ($[(HO)_2P(S)]_2S_2$); Thioimidodiphosphoric acid ($[(HO)_2P(S)]_2NH$).**

When all acid groups have been replaced, the compounds are named as halides, amides, etc. The phosphoric halides and halogenoids are named by phosphoryl terms, the phosphoric amides by single-word heading parents. The phosphorous analogs of both classes are indexed at Phosphorous binary heading parents with a multiplicative prefix for the class term.
Examples:

$Cl_2P(O)OP(O)Cl_2$	**Diphosphoryl chloride** (the phos=phorous analog is named **Di=phosphorous tetrachloride**)
$(H_2N)_2POP(NH_2)_2$	**Diphosphorous tetraamide** (the phosphoric analog is named **Di=phosphoramide**)
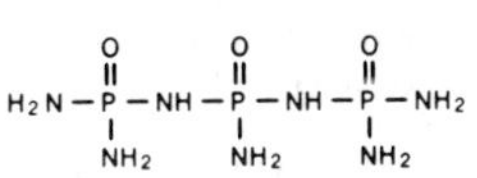	**Diimidotriphosphoramide**

In the order of precedence of compound classes (¶ 106), phosphorus and arsenic acids are functional heading parents, which fall below the acids ex=pressed as principal functions (carboxylic, sulfonic, etc.) but above the highest nonacid group (acid halides). Phosphorus acids as a group rank just above arsenic acids. Within each group, seniority is dependent on (*a*) the greatest number of acid groups; (*b*) the greatest number of nuclear phosphorus or arsenic atoms; (*c*) the highest oxidation state (5+) of these atoms; (*d*) the most preferred atoms (Table I, ¶ 128) attached to the nuclear atoms; (*e*) the greatest number of such preferred atoms; (*f*) the nature of less preferred atoms. A partial list of acids in descending order of precedence is: **Triphosphoric, Diphosphoric, Imidodiphosphoric, Diphosphorous, Phosphoroperoxoic, Phosphoric, Phosphorothioic, Phosphorodithioic, Phosphorous, Phos=phorochloridic, Phosphorohydrazidic, Phosphoramidic, Phosphonic, Phosphinic, Phosphinous acid.** (Chalcogen analogs of each acid immediately follow it in descending order of increasing replacement of oxygen by sulfur, selenium, and tellurium.) For the choice between tautomeric pairs such as **Phosphorous** and **Phosphonic acids**, see also above.

Nonacid analogs are ranked according to their class, e.g., halide, amide; then on the multiplicity of their class, e.g., a diamide is preferred over a monoamide; then on the remaining criteria set out above for the acids.

In the presence of more preferred compound classes, phosphorus and ar=senic acids and hydrides are expressed by the following substituent radicals (* denotes a radical which requires bis, tris, etc., (and parentheses) instead of di, tri, etc., when more than one is present):

H_2P-	*phosphino	H_2As-	*arsino
$H_2P(O)-$	*phosphinyl	$H_2As(O)-$	*arsinyl
$H_2P(S)-$	phosphinothioyl	$H_2As(S)-$	arsinothioyl
$H_2P(=NH)-$	phosphinimyl	$H_2As(=NH)-$	arsinimyl
$HP(\equiv N)-$	phosphononitridyl	$HAs(\equiv N)-$	arsononitridyl
H_4P-	*phosphoranyl	H_4As-	*arsoranyl
$HP=$	phosphinidene	$HAs=$	arsinidene
$HP(O)=$	*phosphinylidene	$HAs(S)=$	arsinothioylidene
$P(O)\equiv$	*phosphinylidyne	$As\equiv$	arsinidyne

The above are substitutive; the following are not, and can be used only as simple radicals (¶ 132):

$P(O)-$	phosphoroso	$As(O)-$	arsenoso
$P(O)_2-$	phospho	$As(O)_2-$	arso
$(HO)_2P(O)-$	phosphono	$(HO)_2As(O)-$	arsono
$-P(O)(OH)-$	phosphinico (multiplying radical only)	$-As(O)(OH)-$	arsinico (multiplying radical only)

Radicals from polynuclear arsenic and phosphorus hydrides are formed like the carbon analogs.
Examples:

H_2PPH-	diphosphinyl
-PHPH-	1,2-diphosphinediyl
=PP=	1,2-diphosphinediylidene
-P=P-	1,2-diphosphenediyl
-As=As-	1,2-diarsenediyl
$-PH(PH)_2PH-$	1,4-tetraphosphinediyl

198. Salts. *Metal salts* of acids and of other compounds having re=placeable hydrogen on chalcogen atoms (whether expressed in the heading parent, in the substituents, or in a previous modification phrase) are indexed by use of salt terms in the modification. The terms mono, di, and tri, Ew=ens-Bassett numbers (¶ 215), and ratios, are used as required (unless infor=mation is lacking), but locants are never cited. In the presence of hetero atom groups capable of forming chelate rings, salts of aluminum, beryllium, gallium, indium, magnesium, thallium, and the transition metals are named by coor=dination nomenclature (¶ 215).
Examples:

AcOH • Na	**Acetic acid sodium salt**
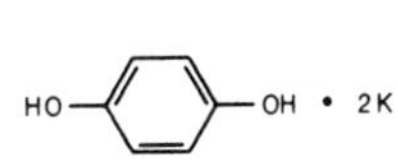	**1,4-Benzenediol dipotassium salt**
PhP(O)(OH)OEt • K	**Phosphonic acid, phenyl- monomethyl ester, potassium salt**
3 MeOH • Bi^{3+}	**Methanol bismuth(3+) salt**
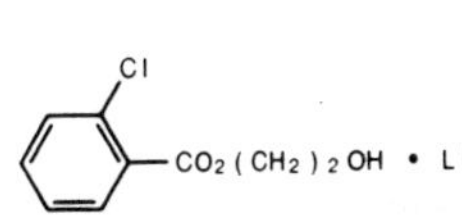	**Benzoic acid, 2-chloro- 2-hydroxyethyl ester, lithium salt**

When a metal (except Sb^{3+}, Bi^{3+}, Ge^{4+}, Sn^{4+}, or Pb^{4+}) has replaced hydrogen atoms attached to nitrogen, phosphorus, arsenic or antimony in molecular skeletons, they also are named as salts, e.g., **Piperidine, potassium salt; Acetamide, monosodium salt; Phosphine, cyclohexyl-, monolithium salt;** but unsubstituted **Hydrazine, Phosphine,** and **Arsine** afford the binary heading parents **Sodium hydrazide (NaN_2H_3); Lithium arsenide (LiH_2As); Aluminum phosphide**; etc.

Except for those cyclic derivatives of germanium, tin, lead, antimony, and bismuth which are treated as heterocycles (¶ 194), cyclic metal salts of two or more compounds, as well as mixed acyclic salts, are named by coordination nomenclature (¶ 215). Metal salts of radical anions and of certain delocalized structures, e.g., β-dicarbonyl compounds and nitro alkanes, are indexed by citing the ion term in the modification followed by the metal name.

Examples:

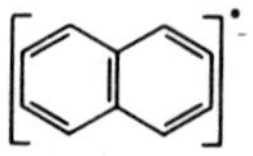

Naphthalene
  radical ion(1−),
  sodium

$O_2N-\bar{C}H-NO_2 \cdot Na^+$

Methane, dinitro–
  ion(1−), sodium

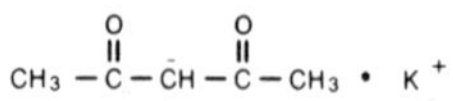

2,4–Pentanedione
  ion(1−), potassium

Uninverted names for metal salts of acids and esters are formed by citing first the metal, then the ester radical, then "hydrogen" (if any acidic group remains unaccounted for), and finally an "–ate" or "–ite" term. When the acid is one that requires a line formula, or has acidic sites, e.g., hydroxyl or amino substituents, elsewhere than in the principal chemical group, the acid name is placed first.

Examples:

sodium acetate
potassium hydrogen carbonate
calcium ethyl phosphate
zinc bis(methyl sulfate)
copper(2+) diacetate
boric acid (H_3BO_3) monopotassium salt
2–hydroxybenzoic acid monosodium salt
4–aminobenzoic acid calcium salt (2:1)

Amine, phosphine, etc., salts of acids, alcohols, thiols, etc., in general are indexed in the *Chemical Substance* and *Formula Indexes* at the names and formulas of both components. Exceptions are salts of ammonia, indexed only as ammonium salts at the acid headings, and the "common components" of molecular addition compounds (¶ 192) together with the following common acids: hydriodic, hydrobromic, hydrochloric, hydrofluoric, nitric, nitrous, perchloric, phosphoric, phosphorous, sulfuric, sulfurous. Entries are not made for these components except for salts that contain *only* "common components," in which case a single entry based on the rules for molecular addition com= pounds (¶ 192) is made. At the acid heading, a "compd. with" phrase appears in the modification.

Example:

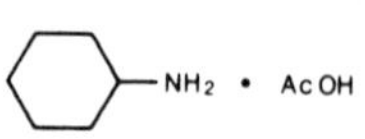

Acetic acid
  compd. with cyclohexanamine (1:1)

At index headings for nitrogenous substances, and those containing trivalent phosphorus or arsenic, an "–ate", "–ite", or "–ide" term is normally cited in the modification, but when the acid has a synonym line formula, or is an acid ester indexed at a "Class II" alcohol name, a "compd. with..." phrase is used instead. Sometimes the designation "(salt)" must be added to an "–ate" term to differentiate it from an ester.

Examples:

$Me_2AsH \cdot HCl$

Arsine, dimethyl–
  hydrochloride

$PhNH_2 \cdot H_3PO_4$

Benzenamine
  phosphate (1:1)

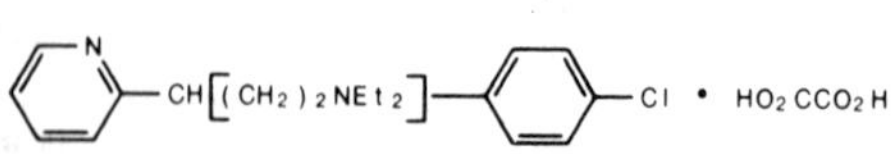

2–Pyridinepropanamine, γ–(4–chlorophenyl)–*N*,*N*–diethyl–
  ethanedioate (1:1)

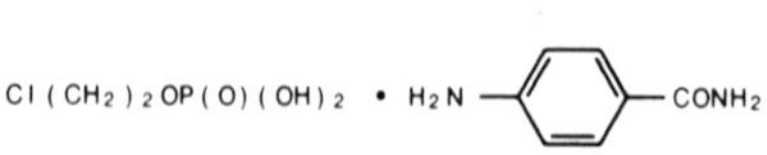

Benzamide, 4–amino–
  compd. with 2–chloroethyl di= hydrogen phosphate (1:1) (pre= ferred index name)
Ethanol, 2–chloro–
  dihydrogen phosphate, compd. with 4–aminobenzamide (1:1) (additional index name)

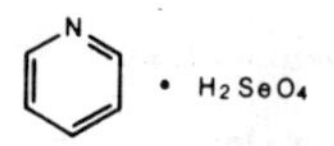

Selenic acid
  compd. with pyridine (1:1) (pre= ferred index name)
Pyridine
  selenate (1:1) (additional index name)

$\underset{4}{Me}CH_2\underset{2}{C}H(NH_2)\underset{1}{C}H_2OH \cdot HO_2CCO_2H$

1–Butanol, 2–amino–
  ethanedioate (1:1) (salt)

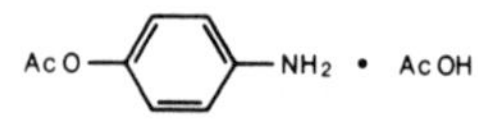

Phenol, 4–amino–
  acetate (ester), acetate (salt)

For salts of amines, etc., with alcohols and thiols, a "compd. with" phrase and a ratio are used at both component headings.

Examples:

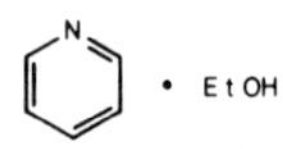

Ethanol
  compd. with pyridine (1:1) (preferred index name)
Pyridine
  compd. with ethanol (1:1) (additional index name)

Salts of aminium compounds and other substitutive cations are indexed by combining the "–ium" heading and the anion term, named according to the rules already described, (¶¶ 180, 184). Both components are indexed in the *Chemical Substance* and *Formula Indexes* except for the following anions:

(*a*) halide and halogenide simple anions, e.g., bromide, cyanide, thio= cyanate, azide;

(*b*) chalcogenide anions, e.g., hydroxide, sulfide;

(*c*) anions derived from "Class I" acids, alcohols, and thiols (¶ 185), nitrous acid, and perchloric acid;

(*d*) methyl sulfate ($MeOSO_3^-$).

(*e*) tetrafluoroborate(1−) (BF_4^-), tetraphenylborate(1−) (BPh_4^-), and hexafluorophosphate(1−) (PF_6^-)

Ternary salts of this kind are indexed by citing the terms in alphabetical order in the modification, cations first, then anions. In molecular addition compounds, the "–ium" salt is cited in uninverted form. In each example, below, the preferred index name is cited first; less preferred *Chemical Sub= stance* and *Formula Index* names (if any) are then described.

Examples:

$Me_4N^+ \cdot Br^-$

Methanaminium, *N*,*N*,*N*–trimethyl–
  bromide

$Me_4N^+ \cdot MeCOS^-$

Methanaminium, *N*,*N*,*N*–trimethyl–
  ethanethioate
Ethanethioic acid
  ion(1−), *N*,*N*,*N*–trimethyl= methanaminium

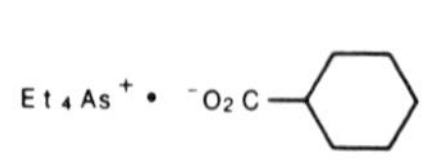

Arsonium, tetraethyl–
  cyclohexanecarboxylate
Cyclohexanecarboxylic acid
  ion(1−), tetraethylarsonium

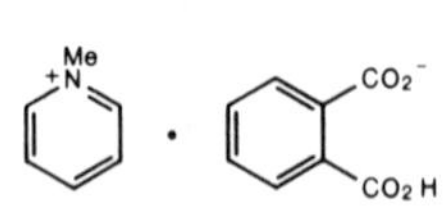

Pyridinium, 1–methyl–
  1,2–benzenedicarboxylate (1:1)
1,2–Benzenedicarboxylic acid, compounds
  ion(1−), 1–methylpyridinium

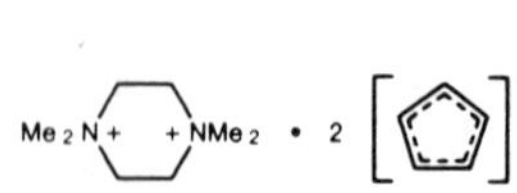

Piperazinium, 1,1,4,4–tetramethyl–
  dicyclopentadienide
1,3–Cyclopentadiene
  ion(1−), 1,1,4,4–tetramethyl= piperazinium (2:1)

$Me_4N^+ \cdot {}^-NHPh$

Methanaminium, *N*,*N*,*N*–trimethyl–
  salt with benzenamine (1:1)
Benzenamine
  ion(1−), *N*,*N*,*N*–trimethyl= methanaminium

$Me_3N^+CH_2OH \cdot ClO_4^-$

Methanaminium, 1–hydroxy–*N*,= *N*,*N*–trimethyl–
  perchlorate (salt) (note the use of the "(salt)" designation to resolve any ambiguity, even though the ester would have been named as a substituent)

Piperazinium, 1,1-dimethyl- salt with trifluoroacetic acid (1:1), hydrochloride
Acetic acid, trifluoro-, compounds ion(1-), 1,1-dimethyl= piperazinium, hydrochloride

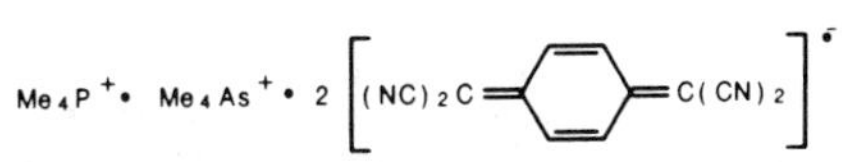

Phosphonium, tetramethyl- tetramethylarsonium, salt with 2,2'-(2,5-cyclohexadiene-1,4-diylidene)bis[propanedinitrile] (1:1:2)
Arsonium, tetramethyl- tetramethylphosphonium, salt with 2,2'-(2,5-cyclohexadiene-1,4-diylidene)bis[propanedi= nitrile] (1:1:2)
Propanedinitrile, 2,2'-(2,5-cyclo= hexadiene-1,4-diylidene)bis- radical ion(1-), tetramethyl= arsonium tetramethyl= phosphonium (2:1:1)

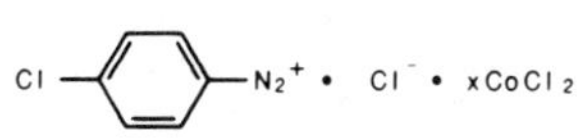

Benzenediazonium, 4-chloro- chloride, compd. with cobalt chloride ($CoCl_2$)
Cobalt chloride ($CoCl_2$) compd. with 4-chlorobenzene= diazonium chloride
(Diazonium "double salts" of known ratio are indexed as "trichloro= cobaltate(1-)," "tetrachloro= cobaltate(2-)," etc.)

199. **Silicon, germanium, tin, and lead compounds** are indexed either as heterocycles; as derivatives of acyclic hydride parents such as **Germane** or **Disilane**, or special acyclic parents such as **Tristannathiane** or **Tetrasi= loxane**; or as silicic acids.

Heterocyclic compounds are indexed by the usual methods (¶¶ 146, 149).
Examples:

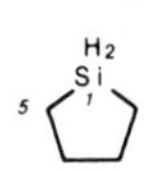

Silacyclopentane

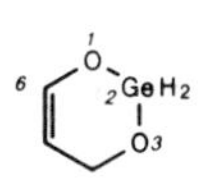

4*H*-1,3,2-Dioxagermanin

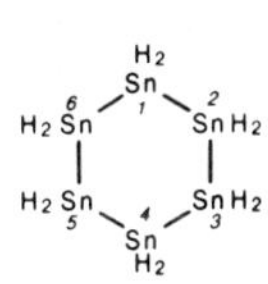

Cyclohexastannane

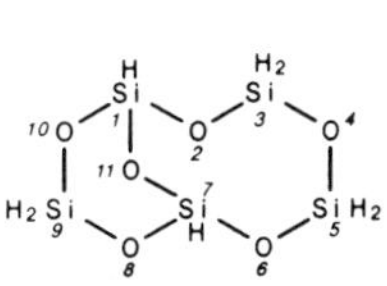

Bicyclo[5.3.1]pentasiloxane

Acyclic linear chains of tetravalent hydrides of silicon, germanium, tin, and lead are molecular skeletons with names derived from the mononuclear hy= drides: **Silane**, SiH_4; **Germane**, GeH_4; **Stannane**, SnH_4; **Plumbane**, PbH_4.
Examples:

H_3SiSiH_3	**Disilane**
$H_3GeGeH_2GeH_3$	**Trigermane**
$H_3Sn(SnH_2)_{11}SnH_3$	**Tridecastannane**
$H_3SiSiH_2SiH(SiH_3)SiH_2SiH_3$	**Pentasilane, 3-silyl-**

When a single element of this series alternates with a single chalcogen in a linear acyclic chain, "-ane" names which express the nature of the alternating elements and number of silicon (etc.) atoms are employed.
Examples:

$(H_3Si)_2O$	**Disiloxane**
$(H_3Si)_2S$	**Disilathiane**
$(H_3Ge)_2Se$	**Digermaselenane**
$(H_3Sn)_2Te$	**Distannatellurane**
$(H_3SnO)_2SnH_2$	**Tristannoxane**
$H_3Si(OSiH_2)_2OSiH_3$	**Tetrasiloxane**

An acyclic chain of alternating silicon and nitrogen atoms is not named as a silazane; instead the functionality of the nitrogen is recognized. Thus, disi= lazane, $H_3SiNHSiH_3$, is indexed at **Silanamine, *N*-silyl-**. However, silazanyl radicals (see below) are employed in the presence of higher functions.

Replacement ("a") names are used as parents when the chain contains carbon atoms and the other requirements (¶ 127) are met.
Example:

$MeO(CH_2)_2OCH_2SiMe_2OSiMe_2CH_2O(CH_2)_2OMe$ (locants 1 2 5 7 8 9 11 14 15)

2,5,8,11,14-Pentaoxa-7,9-disilapen= tadecane, 7,7,9,9-tetramethyl-

Derivatives of all these parents are named by the usual substitutive methods except for silicic acids and germanium, tin, and lead hydroxides and salts (see below). A hydroxyl group on a germanium, tin, or lead atom is expressed as a hydroxy prefix, but by the suffix "-ol" on a silicon atom when it is a principal group. *O*-Acyl silanols are named as esters.
Examples:

$SiCl_4$	**Silane, tetrachloro-**
Me_3SiCN	**Silanecarbonitrile, trimethyl-**
Me_3SnCN	**Stannanecarbonitrile, trimethyl-**
Me_3SiOH	**Silanol, trimethyl-**
$Et_3SiOSiEt_2OH$	**Disiloxanol, pentaethyl-**
H_3GeNH_2	**Germanamine**
$(HOSiMe_2)_2CH_2$	**Silanol, methylenebis[dimethyl-**
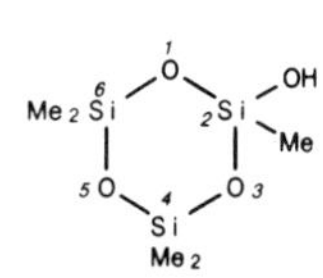	**Cyclotrisiloxanol, pentamethyl-**
$ClSnMe_2OSnMe_2Cl$ (1 2 3)	**Distannoxane, 1,3-dichloro-1,1,3,3-tetramethyl-**
$PhPb(O)OPb(O)Ph$ (1 2 3)	**Diplumboxane, 1,3-dioxo-1,3-di= phenyl-**
$(Me_3Si)_2Se$	**Disilaselenane, hexamethyl-**
Me_2PbO	**Plumbane, dimethyloxo-**
F_3SiSiH_3 (1 2)	**Disilane, 1,1,1-trifluoro-**
$Et_3PbPbEt_3$ (1 2)	**Diplumbane, hexaethyl-**
$PhPb(O)OH$	**Plumbane, hydroxyoxophenyl-**
$HOGeH_2OGeH_2OH$ (1 2 3)	**Digermoxane, 1,3-dihydroxy-**

(MeO)$_3$SiMe **Silane, trimethoxymethyl-**

Me$_3$SiOAc **Silanol, trimethyl- acetate**

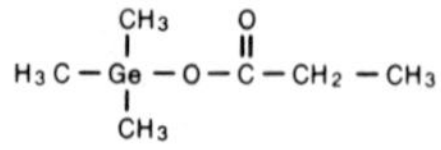

Germane, trimethyl(1-oxoprop= oxy)-

Substituent prefixes (radicals) from silicon, germanium, tin, and lead mo= nonuclear hydrides are formed by adding the suffixes "-yl," "-ylene," and "-ylidyne" to the appropriate stem, or "-tetrayl" to the hydride name: silyl, H_3Si–; germylene, H_2Ge=; stannylidyne, HSn≡; plumbanetetrayl, =Pb=. Radicals from polynuclear hydrides and "oxanes," etc., are derived as for hydrocarbons.

Examples:

H$_3$SiSiH$_2$ —	disilanyl
—GeH$_2$GeH$_2$— (1, 2)	1,2-digermanediyl (formerly diger= manylene)
H$_3$SnSnH$_2$SnH =	tristannanylidene
≡SnSn≡ (1, 2)	1,2-distannanediylidyne
H$_3$SiOSiH$_2$ —	disiloxanyl
H$_3$SiNHSiH$_2$ —	disilazanyl
—PbH$_2$OPbH$_2$— (1, 2, 3)	1,3-diplumboxanediyl
—SiH$_2$OSiHOSiH$_2$— (1, 3, 5)	1,3,5-trisiloxanetriyl

The nonfunctional silicon, germanium, tin, and lead hydrides fall (in that order) between boron and oxygen heading parents in order of precedence of compound classes (¶ 106). When a principal group is expressed as a suffix, the compound is classed accordingly. Thus **Silanol** is ranked with the alcohols, and the silicon "chain" places it above all carbon-chain monohydric alcohols, cyclic or acyclic (¶ 138).

Examples:

Me$_3$SiOSiMe$_2$CH$_2$OH **Methanol, (pentamethyl= disiloxanyl)-** (not 3-Oxa-2,4-di= silapentan-1-ol, 2,2,4,4-tetra= methyl-)

Benzenamine, 4-disilanyl-

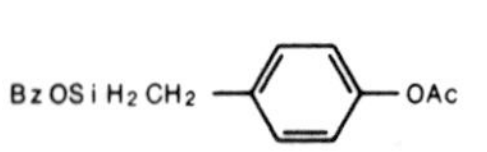

Silanol, [[4-(acetyloxy)phenyl]= methyl]- benzoate (not Phenol, 4-[[(ben= zoyloxy)silyl]methyl]-, acetate)

Et$_3$GeCH$_2$(CH$_2$)$_4$CH$_2$NH$_2$ (6, 1) **1-Hexanamine, 6-(triethylgermyl)-**

H$_3$SiOSiH$_2$SSiH$_3$ (3 2 1) **Disiloxane, (silylthio)-**

Silicic acids are silanes and siloxanes (and chalcogen analogs) in which all hydrogen atoms have been replaced by hydroxyl and oxo groups (or their chalcogen analogs). Their esters and anhydrides are named as usual. Cyclic derivatives are named as heterocycles.

Examples:

(EtO)$_4$Si **Silicic acid (H_4SiO_4) tetraethyl ester**

(Me$_2$CHO)$_3$SiSH **Thiosilicic acid (H_4SiO_3S) *O,O,O*-tris(1-methylethyl) ester**

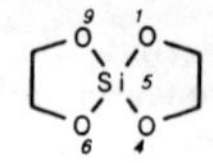

1,4,6,9-Tetraoxa-5-silaspiro[4.4]= nonane (not 1,2-Ethanediol, cyclic diester with silicic acid (H_4SiO_4))

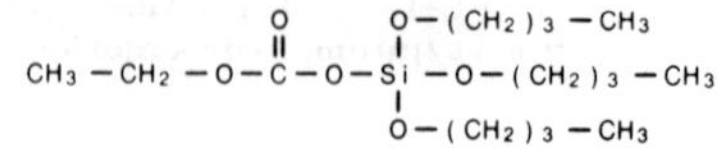

Carbonic acid monoethyl ester, anhydride with silicic acid (H_4SiO_4) tributyl ester (see also ¶ 186)

Germanium, tin, and lead analogs of silicic acids are named at binary hydroxide and hydroxide oxide heading parents. Symmetrical derivatives of **Germane, Stannane,** and **Plumbane,** in which all hydrogen atoms have been replaced by groups derived from alcohols, thiols, etc., or acids, are named as metal salts.

Examples:

Ge(OH)$_4$ **Germanium hydroxide (Ge(OH)$_4$)**

SnO(OH)$_2$ **Tin hydroxide oxide (Sn(OH)$_2$O)**

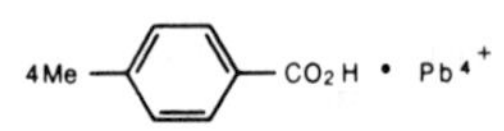

Benzoic acid, 4-methyl- lead(4+) salt

200. Sulfur, selenium, and tellurium compounds are often named similarly to oxygen compounds, but sometimes, e.g., in sulfonic and sulfenic acids (¶ 165), the sulfur, etc., acts as a nuclear atom in a functional group of which no oxygen analog is known. Because sulfur, selenium, and tellurium are treated identically, "thio," "sulfur," "sulfide," etc., in the following discussion may invariably be replaced by the corresponding selenium or tellurium term.

Cyclic sulfur compounds are named as heterocycles; sulfur alternating with silicon, germanium, tin, or lead atoms forms the silathianes, etc. (¶ 199). Sulfur analogs of anhydrides are usually named as anhydrosulfides, but the anhy= drosulfides of acids named as functional parent compounds are often indexed at polynuclear acid headings, e.g., **Thiodicarbonic acid.**

Examples:

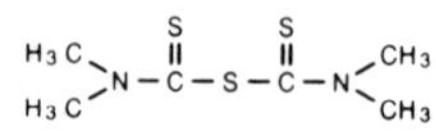

Thiodicarbonic diamide ([H$_2$NC(S)]$_2$S), tetramethyl-

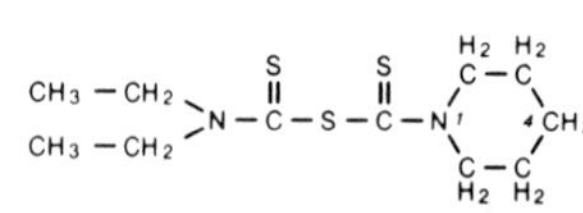

1-Piperidinecarbodithioic acid anhydrosulfide with diethyl= carbamodithioic acid

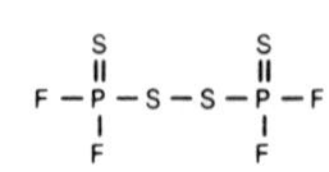

Thioperoxydiphosphoric tetra= fluoride ([F$_2$P(S)]$_2$S$_2$)

The additive term "sulfide" is used in modifications when sulfur is attached to saturated Group VA elements in molecular skeletons. The locant relating to the parent is used when the parent contains the heteroelement sulfide; otherwise, an element symbol provides the locant.

Examples:

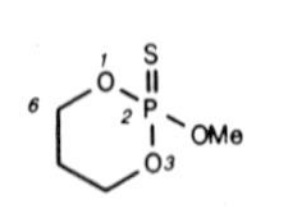

1,3,2-Dioxaphosphorinane, 2-methoxy- 2-sulfide

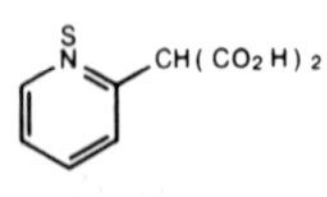

Propanedioic acid, 2-pyridinyl- *N*-sulfide

Oxides of doubly-bound sulfur atoms in principal groups are named in the modification with an "*S*" locant.

Examples:

MeC(=S=O)NH$_2$ **Ethanethioamide *S*-oxide**

MeCH$_2$C(=S=O)CH$_2$(CH$_2$)$_2$Me (1, 3, 4, 7) **3-Heptanethione *S*-oxide**

Phosphine sulfide, Arsine sulfide, and **Stibine sulfide** (¶¶ 181, 197) are substitutive parent compounds, but the sulfides of **Diphosphine, Triarsine,** etc., are named by modification terms.

Acyclic sulfides, sulfoxides, and sulfones, containing one or more isolated sulfur atoms, including thio acetals and thio ortho esters, are named (like ethers (¶ 196)) as substituents of hydrocarbons and other heading parents by use of thio radicals, unless a replacement ("a") name (¶ 127) is permissible. In the latter, oxide terms in the modification are employed to express sulfinyl (–S(O)–) and sulfonyl (–S(O)$_2$–) groups. The analogs of "thio" are "seleno" and "telluro"; those of "sulfinyl" are "seleninyl" and "tellurinyl"; those of "sulfonyl" are "selenonyl" and "telluronyl."

Examples:

$CH_3-S-CH_2-S-CH_2-S-CH_2-S-CH_3$ (numbered 1–9)
2,4,6,8-Tetrathianonane

$CH_3-CH_2-S(O)-CH_2-CH_2-S(O)-CH_2-CH_2-S(O)-CH_2-CH_2-S(O)-CH_2-CH_3$ (numbered 1–14)
3,6,9,12-Tetrathiatetradecane 3,6,9,12-tetraoxide

$MeSeCH_2(CH_2)_2CO_2H$
Butanoic acid, 4-(methylseleno)-

$PhCOCH_2CH_2TePh$
1-Propanone, 1-phenyl-3-(phenyltelluro)-

$H_2NCH_2CH_2SeCH_2CH_2NH_2$
Ethanamine, 2,2′-selenobis-

$CH_3-CH_2-S(O_2)-C(CH_3)_2-S(O_2)-CH_2-CH_3$
Propane, 2,2-bis(ethylsulfonyl)- (principle: largest heading parent (¶ 138))

$CH_3-CH_2-Te(O)-CH_3$
Ethane, (methyltellurinyl)- (principle: largest heading parent)

$Me(CH_2)_6SO_2Ph$
Benzene, (heptylsulfonyl)- (principle: ring preferred over chain)

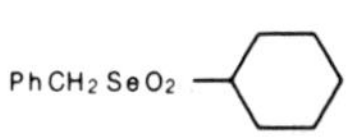

Benzene, [(cyclohexylselenonyl)methyl]- (principle: preferred ring system)

$H_2C=CHSCH_2SCH=CH_2$
Ethene, 1,1′-[methylenebis(thio)]bis- (principles: largest heading parent and multiplication)

$H_2C=CHSCHMeOEt$
Ethene, [(1-ethoxyethyl)thio]- (principle: maximum number of unsaturated bonds)

$Cl_2CHCH_2SCH_2CH_2SCH_2CH_2SCH_2CHCl_2$
Ethane, 1,1′-thiobis[2-[(2,2-dichloroethyl)thio]- (principles: centrality and multiplication)

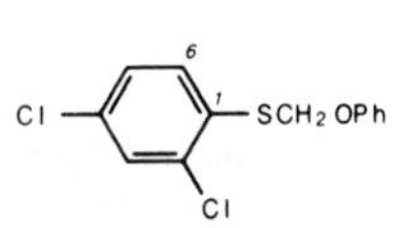

Benzene, 2,4-dichloro-1-[(phenoxymethyl)thio]- (principle: maximum number of substituent prefixes)

$Ph(CH_2)_2SPh$
Benzene, [(2-phenylethyl)thio]- (not Benzene, [2-(phenylthio)ethyl]-) (principle: earliest index position)

Hydrodisulfides are named as dithioperoxoic acids, as sulfenothioic acids, or as thio analogs of mononuclear peroxy "oxo" acids. The oxides R–S(O)–SH and R–S(O)$_2$–SH are named as sulfinothioic and sulfonothioic acids.

Examples:

BzSSH — **Benzenecarbo(dithioperoxoic) acid**

EtSSH — **Ethanesulfenothioic acid**

$(HO)_2P(O)SSH$ — **Phosphoro(dithioperoxoic) acid**

$HOSO_2SSH$ — **Thioperoxymonosulfuric acid ((HO)(HSS)SO$_2$)**

The compounds R–S–S–SH, R–S(O)–S–SH, R–S(O)$_2$–S–SH, are named as sulfeno(dithioperoxoic) acids and their sulfino and sulfono analogs.

Example:

PhSSSH — **Benzenesulfeno(dithioperoxoic) acid**

Hydrotrisulfide is a radicofunctional heading parent for acyl derivatives, and **Hydrotetrasulfide,** etc., for alkyl, aryl, and acyl derivatives.

Example:

AcSSSH — **Hydrotrisulfide, acetyl**

Cyclic polysulfides are named as heterocycles; acyclic heterocycles fulfilling the requirements (¶ 127) of replacement nomenclature are indexed at "thia" names. Other acyclic disulfides, trisulfides, etc., are indexed at **Disulfide** and similar heading parents unless they are esters of dithioperoxoic acids or chalcogen analogs of peroxy condensed mononuclear acids. (Examples of these exceptions are illustrated below.) Hydrogen polysulfides in which both of the terminal hydrogen atoms have been replaced by sulfo groups constitute the polythionic acids; thus, **Tetrathionic acid** is HO_3S–S–S–SO_3H.

Examples:

$MeCH_2S(CH_2)_2OSSO(CH_2)_2SCH_2Me$ — **6,9-Dioxa-3,7,8,12-tetrathiatetradecane**

$Me_2CHSSCHMe_2$ — **Disulfide, bis(1-methylethyl)**

$PhCH_2SeSeCH_2Ph$ — **Diselenide, bis(phenylmethyl)**

$(EtSSCH_2)_2$ — **Disulfide, 1,2-ethanediylbis[ethyl**

PhSSSPh — **Trisulfide, diphenyl**

AcSSAc — **Disulfide, diacetyl**

$[Et_2P(Se)Se]_2Se$ — **Triselenide, bis(diethylphosphinoselenoyl)**

EtOSSOEt — **Disulfide, diethoxy**

AcSSCl — **Disulfide, acetyl chloro**

$Cl(CH_2)_2SSSCl$ — **Trisulfide, chloro 2-chloroethyl**

FC(O)SSCN — **Disulfide, cyano fluorocarbonyl** (not Formyl fluoride, (cyanodithio)-)

The heading parents **Disulfoxide, Trisulfone,** etc., are employed for chalcogen compounds in which the same number of oxygen atoms is attached to each skeletal atom. When all oxide atoms are on one sulfur, the compounds are named whenever possible as esters or anhydrides of thio analogs of sulfur acids. When different numbers of oxide atoms are attached to skeletal sulfur atoms, substitutive nomenclature is employed in which thio, sulfinyl, and sulfonyl radicals are cited separately.

Examples:

$[MeS(O)]_2$ — **Disulfoxide, dimethyl**

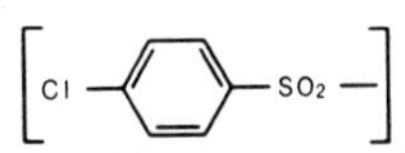

Disulfone, bis(4-chlorophenyl)

$PhSO_2SPh$ — **Benzenesulfonothioic acid *S*-phenyl ester**

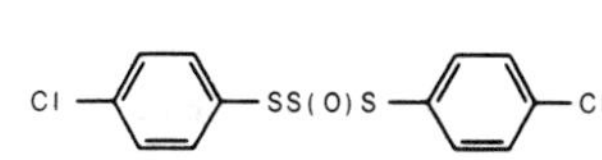

Thiosulfurous acid (H_2S_3O) *S,S*-bis(4-chlorophenyl) ester

$HO-C(S)-S-S(O)_2-S-C(S)-OH$
Carbonodithioic acid bis(anhydrosulfide) with thiosulfuric acid ($H_2S_3O_2$)

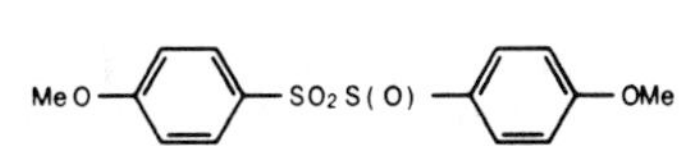

Benzene, 1-methoxy-4-[[(4-methoxyphenyl)sulfinyl]sulfonyl]- (not Benzene, 1-methoxy-4-[[(4-methoxyphenyl)sulfonyl]sulfinyl]-) (principle: earliest index position)

Cyclic sulfides, etc., are ranked according to the rules for ring systems; acyclic sulfur, selenium, and tellurium heading parents follow, in that order, the oxygen heading parents (**Peroxide,** etc.). Within each element group, the descending order is illustrated as follows: **Trisulfide, Disulfone, Disulf=oxide, Disulfide.** In the presence of more preferred compound classes, dithio, trithio, diseleno, disulfonyl, etc., radicals are used as substituents.
Examples:

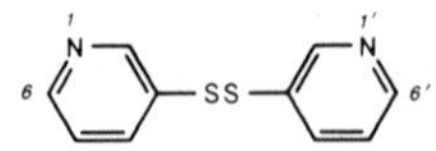

Pyridine, 3,3′-dithiobis-

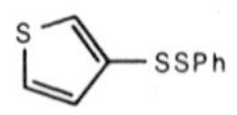

Thiophene, 3-(phenyldithio)- (principle: preferred ring system)

$HO_3SCH_2CH_2SS(O)(CH_2)_2SO_3H$

Ethanesulfonic acid, 2-[[(2-sulfo=ethyl)sulfinyl]thio]- (not Eth=anesulfonic acid, 2-[[(2-sulfo=ethyl]thio]sulfinyl]- (principle: earliest index position)

Acyl alkyl and acyl aryl disulfides are named as esters of either dithio=peroxoic acids or peroxy mononuclear "oxo" acids.
Examples:

$CH_3-C(=O)-S-S-CH_3$

Ethane(dithioperoxoic) acid
***SS*-methyl ester**

$PhSO_2SSPh$

Benzenesulfono(dithioperoxoic) acid
***SS*-phenyl ester**

$HOSO_2SSMe$

Thioperoxymonosulfuric acid ((HO)(HSS)SO_2)
***SS*-methyl ester**

Symmetrical diacyl disulfides derived from polybasic mononuclear "oxo" acids (except **Sulfuric acid** and **Sulfurous acid**; see above) are named at Thioperoxy headings.
Example:

$HO-CH_2-CH_2-NH-C(=S)-S-S-C(=S)-NH-CH_2-CH_2-OH$

Thioperoxydicarbonic diamide ([(H_2N)C(S)]$_2S_2$), *N,N′*-bis(2-hydroxyethyl)-

Sulfilimine, H_2S=NH, and **Sulfoximine,** $H_2S(O)$=NH, are substitutive heading parents. The selenium and tellurium analogs are named **Selenil=imine, Tellurilimine, Selenoximine** and **Telluroximine**. They are ranked just below imine suffix compounds (¶ 106) but *N*-derivatives containing higher chemical functions must often be indexed at the chalcogen parents for lack of an acceptable name for the H_2S=, etc., radicals. However, higher functions in chalcogen-linked derivatives can be recognized by use of other chalcogen-nitrogen parents or by sulfonimidoyl, sulfinimidoyl, etc., radicals.
Examples:

$Et_2S=NSO_2Ph$

Sulfilimine, *S,S*-diethyl-*N*-(phe=nylsulfonyl)-

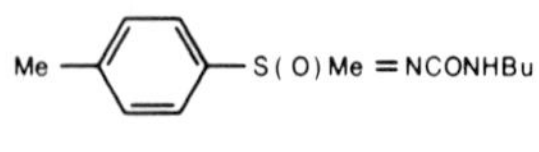

Sulfoximine, *N*-[(butylamino)=carbonyl]-*S*-methyl-*S*-(4-methylphenyl)-

PhS(OH)=NPh

Benzenesulfinimidic acid, *N*-phenyl-

$(MeO)_2S=NPh$

Imidosulfurous acid, phenyl-dimethyl ester

$MeS(O)(=NH)CH_2CH_2CH(NH_2)CO_2H$

Butanoic acid, 2-amino-4-(*S*-methylsulfonimidoyl)-

Sulfur diimide, HN=S=NH, and **Sulfur triimide,** HN=S(=NH)=NH, are substitutive heading parents. They are never expressed as substituents.
Example:

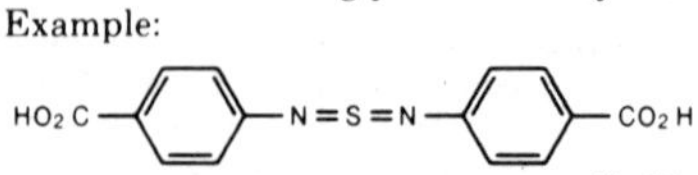

Sulfur diimide, bis(4-carboxy=phenyl)-

Sulfimide, O_2S=NH, and **Thionyl imide,** OS=NH, are substitutive heading parents. They follow the imines in order of precedence (¶ 106) and, when necessary, are expressed by sulfonyl (O_2S=) (sulfinylamino) (OS=N-), etc., radicals.
Examples:

$O_2S=NMe$

Methanamine, *N*-sulfonyl-

$PhSO_2N=SO$

Benzenesulfonamide, *N*-sulfinyl-

Amidosulfenyl chloride (H_2N-S-Cl) is ranked with mononuclear "oxo" acid halides and can be expressed by (chlorothio) and [(chlorothio)amino] radicals when necessary.
Example:

$PhSO_2NMeSCl$

Amidosulfenyl chloride, methyl=(phenylsulfonyl)- (not Benzene=sulfonamide, *N*-(chlorothio)-*N*-methyl-)

Mixed sulfide-selenides, etc., with three contiguous chalcogen atoms are usually named by compound and complex radicals; e.g., [selenobis(thio)] for -S-Se-S-. When only two different contiguous chalcogen atoms are present, the compound is an ester of a sulfenic (etc.) acid analog.
Example:

PhSSePh

Benzenesulfenoselenoic acid phenyl ester

Acyclic tetravalent and hexavalent sulfur compounds containing at least one substituent derived from a molecular skeleton (¶ 130) that cannot be named by general index nomenclature as described above, or as "oxo" acid derivatives, are given coordination names (¶ 215).
Examples:

$PhSF_5$

Sulfur, pentafluorophenyl-
(*OC*-6-21)-

Ph_2TeCl_2

Tellurium, dichlorodiphenyl-
(*T*-4)-

201. Zwitterionic compounds have internally compensating ionic cen=ters. When the cationic center is known, the compound is named either as a "hydroxide, inner salt" or an an "ylide." Meso-ionic compounds are indexed as **Sydnone** derivatives or by means of the expression "meso-ionic didehydro deriv."

Ylides have a compensating carbanion *adjacent to* the cationic center. They are named at the "-ium" heading (except **Phosphonium**) with the suffix "-ide" added to the hydrocarbon radical to express the anion. Phosphonium ylides are indexed as ylidene derivatives of **Phosphorane** (at this heading or by use of a phosphoranylidene radical).
Examples:

$Me_3N^+CH_2^-$

Methanaminium, *N,N*-dimethyl-methylide

$PhCOC^-HS^+Me_2$

Sulfonium, dimethyl-2-oxo-2-phenylethylide

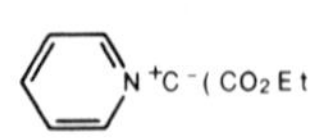

Pyridinium 2-ethoxy-1-(ethoxycarbonyl)-2-oxoethylide

$Ph_3P=CHPh$

Phosphorane, triphenyl(phenyl=methylene)- (preferred index name)

Zwitterionic compounds other than ylides are usually named at the "-ium" heading with "hydroxide" as the salt phrase in the modification. A molecule of water, comprising the hydroxide anion and a hydrogen atom attached to a hetero atom, is "removed" by use of the phrase "inner salt." (For removal of two molecules of water, the appropriate expression is "bis(inner salt)".) A phosphonium compound of this type is indexed as a phosphonium hydroxide inner salt only when such a zwitterionic structure is emphasized or discussed by the author. Otherwise only a neutral **Phosphorane** entry is made.
Examples:

$Me_3N^+CH_2CO_2^-$

Methanaminium, 1-carboxy-*N*,=*N,N*-trimethyl-hydroxide, inner salt

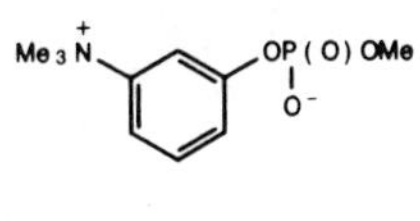

Benzenaminium, 3-[(hydroxy=methoxyphosphinyl)oxy]-*N*,=*N*,*N*-trimethyl- hydroxide, inner salt

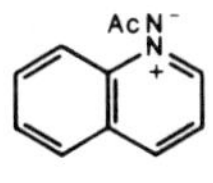

Quinolinium, 1-(acetylamino)- hydroxide, inner salt

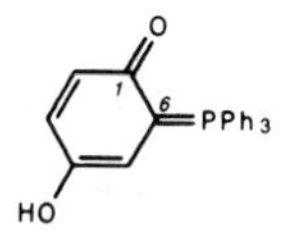

2,4-Cyclohexadien-1-one, 4-hydroxy-6-(triphenylphospho=ranylidene)- (the alternative ionic structure is named **Phosphonium, (2,5-dihydroxyphenyl)tri=phenyl-**, hydroxide, inner salt)

Cationic "ium" compounds with internally compensating borate anions are named where possible by coordination nomenclature (¶ 215); the term "borata" is avoided.

Example:

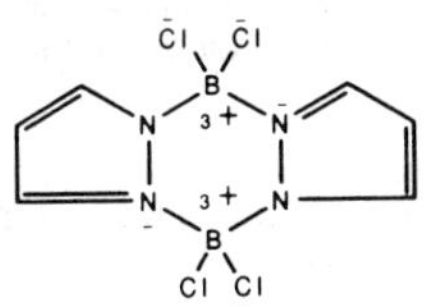

Boron, tetrachlorobis[μ-(1*H*-pyrazolato-*N*¹:*N*²)]di- (not 4*H*,4*H*,8*H*,8*H*-7a,8a-Diaza-3a,4a-diazonia-4,8-diborata-*s*-indacene, 4,4,8,8-tetrachloro-)

Meso-ionic compounds derived from 1,2,3-oxadiazolidin-5-one and the corresponding 5-imine by loss of two hydrogen atoms are indexed at the trivial names **Sydnone** and **Sydnone imine**. All substituents are expressed as prefixes, and conjunctive nomenclature (¶ 124) is not employed. Other meso-ionic compounds are named, if possible, at an "-ium" parent by use of the term "hydroxide, inner salt" in the modification.

Examples (the structures shown are possible resonance forms):

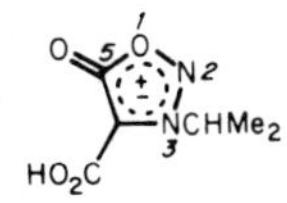

Sydnone, 4-carboxy-3-(1-methyl=ethyl)-

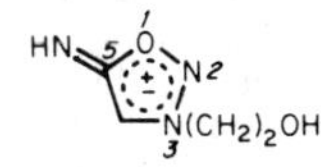

Sydnone imine, 3-(2-hydroxy=ethyl)-

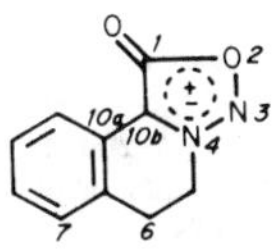

[1,2,3]Oxadiazolo[4,3-*a*]iso=quinolin-4-ium, 5,6-dihydro-1-hydroxy-, hydroxide, inner salt

202. Introduction. A discussion of stereochemical descriptors for systematically named compounds, modified stereoparents, and coordination compounds follows this introduction. The remainder of Section E is directed to various classes of stereoparents. A stereoparent is an index heading parent the name of which implies specific stereochemical information. Illustrative structural diagrams for stereoparents that have received entries in current indexing are provided in the *Chemical Substance Index*; these diagrams indicate both the stereochemistry and the numbering systems from which locants are derived.

It is convenient to divide natural products into four classes according to the methods by which they are indexed. The classification has been most completely worked out for alkaloids, discussed more fully in ¶ 204, below.

Class A contains those substances, of little or no stereochemical complexity, which are indexed systematically rather than at stereoparents.

Class B stereoparents represent groups of natural products which have cyclic molecular skeletons in common. The base structure is derived by removal of all chemical functions but retention of hydrocarbon substituents and the pattern of hydrogenation; the name is derived from the trivial names of the related natural products, and author numbering is adopted if possible. General substitutive nomenclature, as illustrated in previous sections of this introduction, is used to convert the stereoparents into complete index names by addition of prefixes, suffixes and conjunctive terms. Examples of Class B stereoparents are **Prostane** and **Thromboxane**, where prostaglandins and their "homo-oxa" analogs are indexed, and **Picrasane**:

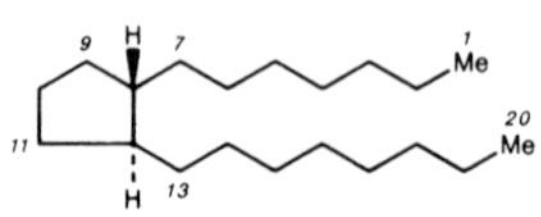

Prostane ((*1S-trans*)-*1-heptyl-2-octylcyclopentane*)

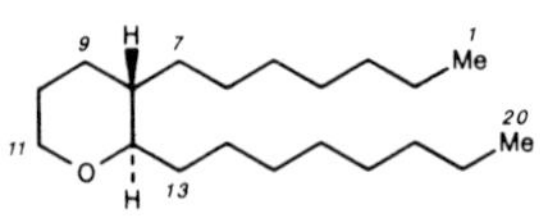

Thromboxane (*homo-11a-oxaprostane, (2R-trans)-3-heptyltetrahydro-2-octyl-2H-pyran*)

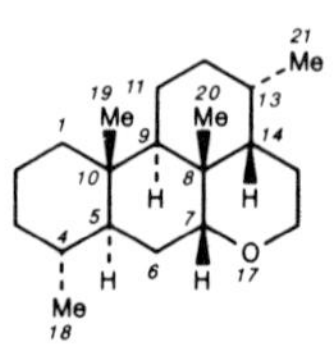

Picrasane ([*3S-(3α,3aβ,6aβ,7aα,8α,11aβ,11bα,11cβ)*]-*hexadecahydro-3,8,11a,11c-tetramethylphenanthro[10,1-bc]pyran*)

Class C stereoparents imply both stereochemistry and chemical functionality. Other functions (including higher functions) are expressed as substituents. Examples will be found in the discussion of alkaloids (¶ 204). Miscellaneous Class C stereoparents include **[13]Cytochalasan, Kromycin,** and **Leucomycin V.** Their derivatives are indexed like alkaloids.

Class D stereoparents possess incompletely elucidated structures. If sufficient information is available, they are named systematically and no stereochemical descriptor is assigned. When no systematic name is possible, the trivial name in the original document is used.

Because treatment of stereoparents by the general rules of substitutive nomenclature can sometimes lead to loss of stereochemical information, special rules for their derivatives are employed. Thus, esters and semicarbazones (and their chalcogen analogs) are named as derivatives of the stereoparents, not at "Class II" acid names or as **Hydrazinecarboxamide** derivatives. Esters formed by an acidic and an alcoholic stereoparent are indexed at the stereoparent that contains the highest chemical function.

Examples:

Androstan-17-ol
  butanoate, (5α,17β)- (not Butanoic acid, (5α,17β)-androstan-17-yl ester)

L-**Valine**
  2′-ester with adenosine

Molecular addition compounds and salts of stereoparents with nonstereoparents are indexed at the name and formula of each component. The preferred *CA* name is the one that employs the stereoparent in the index heading parent. Addition compounds and salts containing more than one stereoparent derivative are assigned preferred index names based on chemical functionality of the stereoparents.

Example:

L-**Proline**
  compd. with (8α,9R)-**cinchonan-9-ol (1:1)** (preferred index name)
Cinchonan-9-ol
  (8α,9R)-, **compd. with** L-**proline (1:1)** (additional index name)

Acyclic *acetals* of functions expressed by "-one" suffixes or "oxo" substituents of stereoparents are indexed as dialkoxy derivatives. Cyclic acetals of stereoparent diols with formaldehyde are named as [methylenebis(oxy)] derivatives of the steroparent; other cyclic acetals are named similarly if subsidiary functions of the stereoparent are involved, but by modification terms if a principal chemical function has been acetalized. Lactones of principal functions are also named in the modification. In these and other cases of ring modification, cross-references (in the *Index Guide*) or additional entries (in the *Chemical Substance Index*) will be found for the ring systems formed by the derivatives.

Examples:

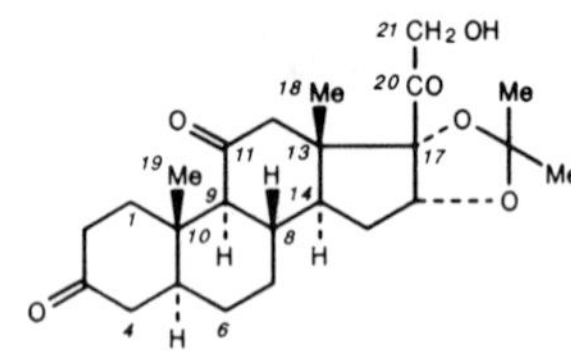

Pregnane-3,11,20-trione, 21-hydroxy-16,17-[(1-methylethylidene)bis(oxy)]- (5α,16α)-
1*H*-**Naphth[2′,1′:4,5]indeno[1,2-*d*][1,3]dioxole**
  pregnane-3,11,20-trione deriv. (additional *Chemical Substance Index* entry)

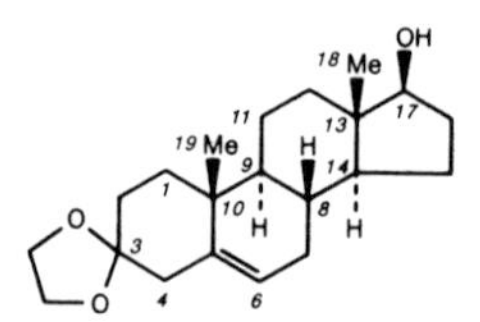

Androst-5-en-3-one, 17-hydroxy-cyclic 1,2-ethanediyl acetal, (17β)-
Spiro[3*H*-cyclopenta[*a*]phenanthrene-3,2′-[1,3]dioxolane]
  androst-5-en-3-one deriv. (additional *Chemical Substance Index* entry)

Restrictions in the use of conjunctive names (¶ 124), lack of suitable radicals, etc., sometimes prevent use of a cyclic stereoparent (alkaloid, steroid, terpene, or cyclic peptide) as a heading parent, and it must instead be expressed as a substituent of a parent which includes the principal function. In such cases, an additional entry is made at the fundamental stereoparent (without unsaturation or functional suffixes) with a modification which directs the index user to the preferred entry.

Example:

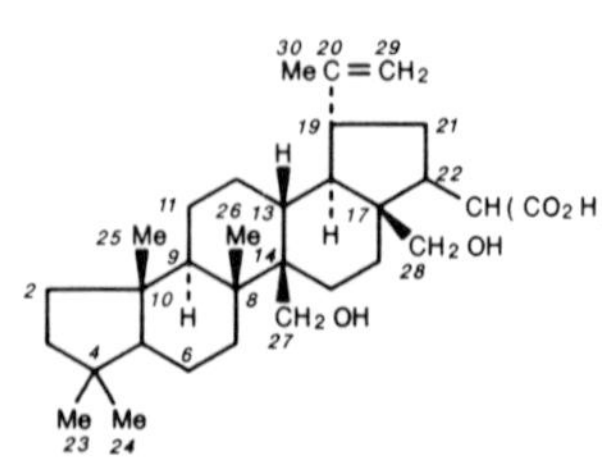

Propanedioic acid, [(14β)-27,28-dihydroxy-*A*(1)-norlup-20(29)-en-22-yl]- (preferred index name; conjunctive names derived from dibasic acyclic acids are not permitted (¶ 124))
A(1)-**Norlupane**
  propanedioic acid deriv. (additional *Chemical Substance Index* entry for stereoparent)
Dicyclopenta[*a*,*i*]phenanthrene
  propanedioic acid deriv. (additional *Chemical Substance Index* entry for ring system)

When two or more stereoparents are covalently attached to one another the compound is normally indexed at the stereoparent that expresses (*a*) the largest number of highest chemical functions, (*b*) the most preferred skeleton. When this is not possible, an additional entry is made at the preferred stereoparent.

Example:

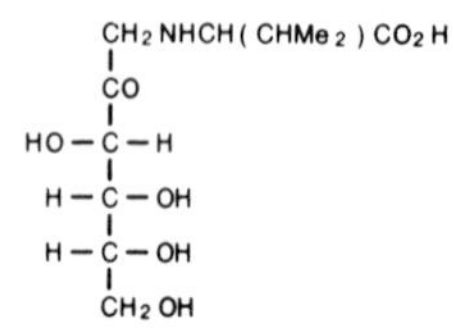

D-**Fructose, 1-[(1-carboxy-2-methylpropyl)amino]-1-deoxy-** (S)- (an acceptable fructose radical is unavailable) (preferred index name)
L-**Valine**
  D-**fructose deriv.** (additional *Chemical Substance Index* entry)

Access to cyclic stereoparents is available by way of the systematic ring system names. If these are not known, the ring analysis may be consulted in the *Index of Ring Systems*; here the systematic and stereoparent names will be found. (A complete set of cyclic and acyclic stereoparents is provided in the *Index Guide* beginning with the 1985 edition.) Modified stereoparent ring systems, e.g., cyclosteroids and steroid fusion derivatives such as **Androst-16-eno[16,17-*b*]azete,** are indexed also at systematic ring system names, and both types of name will be found in the *Index of Ring Systems* in sufficient detail for the *Chemical Substance Index* entries to be found.

Citation of additional and modified stereochemistry at stereoparent and stereoparent derivative names is discussed in the following section.

203. Stereochemistry is expressed in *CA* index names[1] by three methods:

I. The stereochemistry of *systematically named* compounds, including some natural products with only one or two chiral centers, e.g., most carbohydrates with four or fewer carbon atoms, simple alkaloids and terpenes, and some steroid degradation products, is expressed in the modification, following all other structural information such as **"ethyl ester,"** and ahead of descriptive terms related to a specific abstract, such as "spectrum of." The terms used in this "systematic stereochemistry" are the main subject of the present section.

II. *Natural products*, especially those with a multiplicity of chiral centers, are conveniently indexed at fundamental trivial names (*stereoparents*) where they and their derivatives can be found at headings familiar to users in the alkaloid, carbohydrate, steroid, and terpene fields. The concept of stereoparents has been discussed above; the separate natural product classes to which it is applied are the subjects of ¶¶ 204-212.

III. *Coordination compounds* require specialized descriptors to describe the arrangement of ligands around the central atom. These are discussed in the final part of this paragraph.

I. *Systematically named compounds.* In the long-established system for carbohydrate derivatives, two terms are used, e.g., D-gluco-, in which one center is defined absolutely by D and three other centers are related to it by "*gluco*" to describe the set of four centers absolutely. Analogously, among systematic stereochemical descriptors the absolute terms *R* and *S* are employed for a single "reference" center, and the following relative terms are cited under various conditions as defined below: *cis*, *trans*, *exo*, *endo*, *syn*, *anti*, α, β, *R**, *S**, *E*, and *Z*.

Assignment of the absolute terms *R* and *S* depends on the priority ranking of atoms or groups attached to the stereochemical element whose chirality ("handedness") is to be determined by the Cahn-Ingold-Prelog Sequence Rule.[2,3] This ranking depends first on the descending order of atomic number of the atoms directly attached to the chiral center; thus, for bromochlorofluoroiodomethane the order is I, Br, Cl, F. In the following diagrams these atoms are represented by *a*, *b*, *c*, and *d*, respectively. The least preferred atom or group, *d*, is represented by a dotted line to indicate it is to be considered to be below the plane of the paper, while *a*, *b*, and *c* are to be imagined to project toward the viewer at an angle. (The analogy of an automobile steering wheel with three radial bars is a useful one to visualize.) *R* is assigned to a clockwise (right-handed) sequence of *a*, *b*, *c*, while *S* denotes a counterclockwise (anticlockwise) sequence.

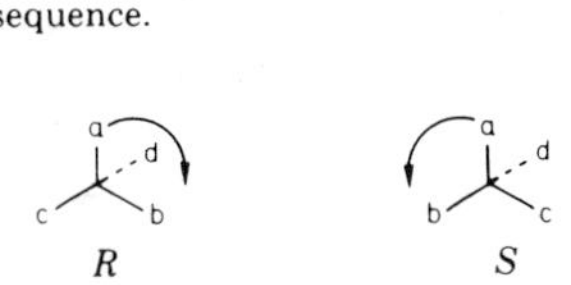

Alternative ways of drawing chiral diagrams are often more convenient: either (a) at most two bonds (as indicated by ordinary lines) are shown to be in the plane while one projects forward and one backward (as indicated by a wedge and a dotted line, respectively), or (b) two bonds are shown projecting forward and two backward, as in the diagrams below. In both instances it must be remembered that the atom or group of lowest priority, *d*, must be oriented away from the viewer so that the clockwise or counterclockwise arrangement of *a*, *b*, and *c* is correctly observed.

In organic compounds it is generally necessary to compare the ranking of two or more carbon bonds. This is done by proceeding outward *one step at a time* until a decision is reached.

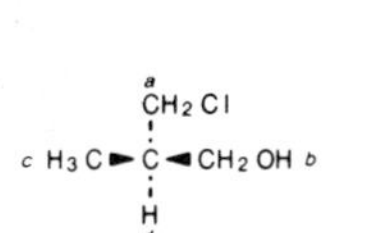

In the structure above, H clearly has the lowest priority, but it is necessary to establish the ranking of the three carbon groups. This is done by arranging the atomic numbers of *the most senior substituent* on each carbon in descending order: Cl, O, H. This elicits the absolute stereochemical descriptor *R*.

[1]For a more extensive discussion of the stereochemical descriptors found in *CA* index names, see J. E. Blackwood and P. M. Giles, Jr., *J. Chem. Inf. Comput. Sci.*, **1975**, 15, 67-72; M. F. Brown, B. R. Cook, and T. E. Sloan, *Inorg. Chem.*, **1975**, 14, 1273-8; *ibid.* **1978**, 17, 1563-8.

[2]International Union of Pure and Applied Chemistry. "Nomenclature of Organic Chemistry, Sections, A, B, C, D, E, F and H." Pergamon Press, Oxford, 1979.

[3]R. S. Cahn, C. K. Ingold, and V. Prelog, *Angew. Chem., Int. Ed. Engl.* **1966**, 5, 385-415. For a recent modification in the treatment of cyclic pathways, see V. Prelog and G. Helmchen, *ibid.*, **1982**, 21, 567-83.

Branching, especially branching close to the chiral center, raises the priority ranking of alkyl groups. Multiple bonds of all kinds are handled by "duplicating" or "triplicating" both of the atoms connected by the bonds; thus

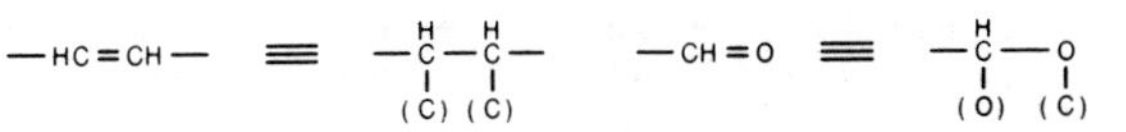

Only the immediate atoms are replicated; the procedure is not carried further. Aryl, e.g., phenyl, naphthalenyl, radicals are handled in a similar manner.

The following groups are arranged in descending order of Sequence Rule priority: $-SCH_3$, $-SH$, $-OCOCH_3$, $-OCH_2CH_3$, $-OCH_3$, $-OH$, $-NO_2$, $-N(CH_3)_2$, $-NH_2$, $-COOH$, $-COCH_3$, $-CHO$, $-CH_2OH$, $-C{\equiv}N$, $-CH_2NH_2$, $-C_6H_5$, $-C{=}(CH_3)_2CH_2CH_3$, $-C(CH_3){=}CH_2$, $-CH(CH_3)CH_2CH_3$, $-CH_2CH_2CH_3$, $-CH_2CH_3$, $-CH_3$. It is essential to rank branched chains by a rigorous procedure if assignment errors are to be avoided. In

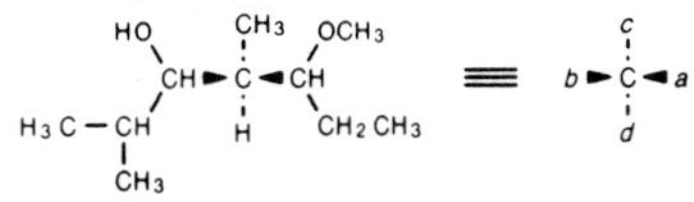

it can be immediately perceived that H has the lowest rank (*d*) and that of the three carbon bonds CH_3 has rank *c*. Proceeding outward to the left and right from the chiral enter we observe that each carbon is attached to one oxygen, one carbon, and one hydrogen. The rule to be observed here is that we should now proceed *just one step along the senior* (*highest-priority*) *branches* (in this instance through the oxygens) to find H on the left, C (preferred) on the right. The fact that, in the *junior* branch, the priorities are reversed, with C attached to two C's and an H on the left, to only one C and two H's on the right, is disregarded, because this stage in the process is never reached. Only when a one-step search of the senior branches results in a tie is the junior branch inspected *to the same extent*. (This would have been necessary, with a consequent assignment of *S* instead of *R*, if $-OCH_3$ replaced $-OH$ in the diagram, even if, at the same time, $-OCH_2CH_3$ replaced $-OCH_3$ on the right.)

In complicated cases it is sometimes necessary to construct an exploration table or tree diagram in assigning *R* and *S* descriptors. A useful summary of Sequence Rule procedures is included in Section E of the IUPAC rules,[2] but consultation of the papers of Cahn, Ingold and Prelog, especially their 1966 paper,[3] may be necessary for resolution of the most difficult cases.

Relative descriptors of various kinds are used as follows:

(*A*) *cis* and *trans* are restricted to cyclic structures with two stereogenic atoms (an atom is stereogenic when interchange of two of the atoms or groups attached to it produces a nonidentical compound). An example is **1,2-Cyclopentanediol.** The *cis* isomer has the senior groups (as defined by the Sequence Rule) on the same side of the reference plane.

(*B*) *endo* and *exo* are used only for ring positions on the X and Y bridges of bicyclo[X.Y.Z]anes (¶ 155) in which X ≥ Y > Z > 0. The *exo* isomer has the configuration in which the senior substituent is on the same side of the reference plane as the Z bridge (see diagram below).

(*C*) *syn* and *anti* are restricted to ring positions on the Z bridge of a bicyclo[X.Y.Z]ane. The *syn* isomer has the senior substituent oriented towards the X bridge.

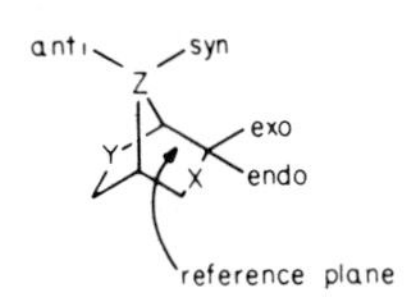

(*D*) α and β are employed for ring positions of cyclic compounds. In the diagram below, the senior groups at the three stereogenic centers (by the Sequence Rule) are a, c, e; the junior groups are b, d, f. The α-side of the reference plane is that side on which the preferred substituent lies at the lowest-numbered stereogenic position; c lies on the same side as a, so both are assigned α descriptors; e lies on the opposite side of the reference plane and is assigned a β descriptor; hence: 1α, 2α, 4β. It should be noted that this usage differs from that for cyclic stereoparents (¶ 203II, below), in which "α-" means "below the plane" and denotes absolute configuration.

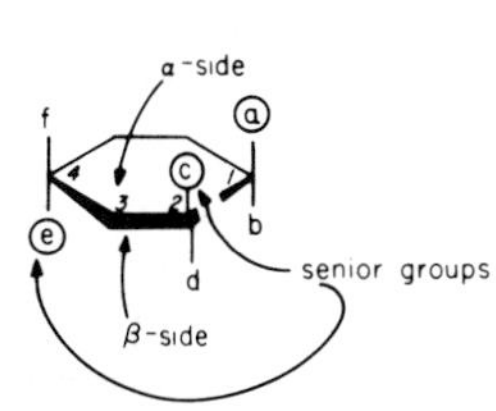

(*E*) *R** and *S** are similar to *R* and *S* but are relative in that the lowest numbered (or first cited) stereogenic atom is arbitrarily assigned an *R** descriptor.

(*F*) *E* and *Z* (from German: "entgegen" (opposite) and "zusammen" (together)) are geometrical stereodescriptors for substances having achiral elements resulting from double bonds. In a compound

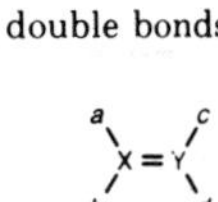

the double bond can be considered to be in a vertical plane, and a, b, c, and d in a horizontal plane. When the senior atom or group at X (a) and the senior atom or group at Y (c) are on the same side of the vertical reference plane, the descriptor Z is cited; its isomer is assigned an E descriptor.[4]

(*G*) The optical rotation descriptors (+) and (−) indicate the sign of rotation of plane-polarized light at the sodium-D line. Racemates are assigned a (±) descriptor.

These descriptors, alone or in combination, are employed to express the total stereochemical information for a chemical substance as follows:

(*1*) *Absolute configuration.* When only one chiral element is present, the absolute descriptor *R* or *S* is cited in parentheses and followed by a hyphen thus: (*R*)−. When two or more chiral elements are present, the reference center is assigned an *R* or *S* descriptor. This reference center, for cases described in 2(*a, b, e, f, g* and *h*) below, is the lowest-numbered chiral center; for 2(*c*) it is the first cited chiral center to which the α/β descriptor applies; for 2(*d*) it is the center which has the highest-ranking substituent according to the Sequence Rule.

(*2*) *Relative configuration* (for substances containing two or more stereogenic atoms).

(*a*) When only two stereogenic atoms are present in an eight-membered or smaller ring, and the configurations of both are known, *cis* or *trans* is cited as the relative descriptor. (*R** or *S** is employed for internal spiro centers named as ortho fusions and for rings of more than eight members.)

(*b*) *endo, exo, syn,* and *anti* are used for bicyclo[X.Y.Z]ane compounds, in which X ≥ Y > Z > 0, with no other known centers. (However, when these terms could be ambiguous, i.e., when (*a*) a ring of more than 8 atoms is present, (*b*) the bridgehead atoms are not in the normal configuration, or (*c*) there is a double bond at a bridgehead, the relative descriptors *R** and *S** are cited instead.)

(*c*) For two or more stereogenic centers in a ring system to which neither 2(*a*) nor 2(*b*) applies, α and β are cited.

(*d*) When two chiral centers are not in the same ring system, *R** and *S** are employed without locants. Two centers of like chirality are indicated by the relative descriptor *R*,R**, while two centers of unlike chirality are indicated by *R*,S**.

(*e*) For substances with three or more chiral elements to which neither 2(*b*) nor 2(*c*) applies, *R** and *S** are used with appropriate locants.

(*f*) The descriptors *E* and *Z*, for configurations about double bonds, are cited in descending order of seniority of the double bonds when used alone or with *R** and *S** as in 2(*d*).[4] When other descriptors are present, locants are cited with *E* and *Z*.

(*g*) When both cyclic and acyclic chiral elements are present, rules 2(*c–f*) are invoked. When 2(*c*) is applicable the ring is the parent from which the descriptor is derived; the first cited α-center is the reference center and is implicitly *R**.

(*h*) When chiral elements are present in two ring systems (or two components of a spiro system), the parent is the system (or component) containing the greater number of stereogenic centers, or is the preferred system (¶ 138), or affords the lower descriptor set, or contains the preferred group according to the Sequence Rule.

(*i*) When achiral elements present in a ring can be described by rules 2(*a–c*), any chiral element outside this ring is described by rule 1.

(3) *Optical rotation* (descriptors (+), (−), (±)).

(*a*) When the absolute stereochemistry is described by rules 1 and 2, the sign of rotation is omitted.

(*b*) When the substance has only one chiral element, and this is not defined by the author, the sign of rotation is cited to distinguish enantiomers.

(*c*) When the substance has more than one chiral element, the sign of rotation, or (±)−, is cited together with a relative descriptor.

The following examples show only the stereochemical descriptors (as cited in index modifications at substance names). The absolute descriptor is cited first, followed by the relative descriptors, if any, in parentheses. The entire descriptor set is then bracketed and cited in the modification.

Examples:

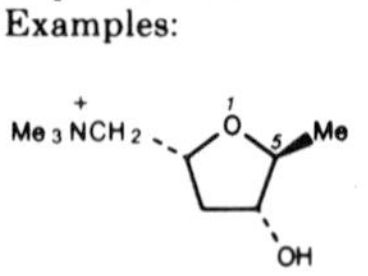

[2R-(2α,4α,5β)]-

[The first cited chiral center is assigned the absolute descriptor *R*; the relationship of the three stereogenic centers is then expressed by the parenthetical expression.]

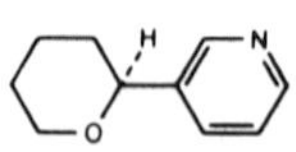

(S)-

[The chirality of the single element is expressed according to the Sequence Rule.]

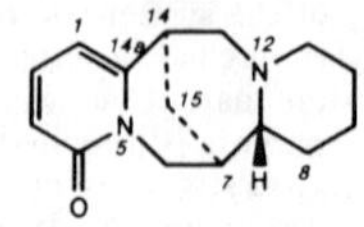

[7R-(7α,7aβ,14α)]-

[The first cited chiral center is assigned the absolute descriptor *R*; the relationship of the three stereogenic centers is then expressed. At positions 7 and 14, the methano bridge is treated as if it were a substituent; at position 7a, the hydrogen atom is considered the highest ranking substituent.]

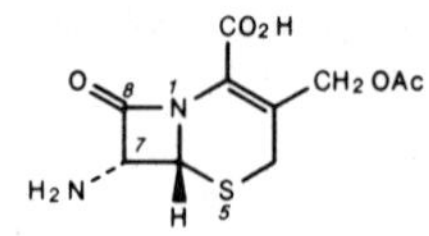

(6R-trans)-

[The reference center is assigned as usual; the two stereogenic centers are related by *trans*.]

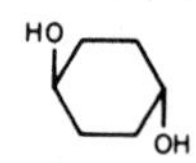

trans-

[No absolute stereochemistry is possible; the two achiral stereogenic centers are related by *trans*.]

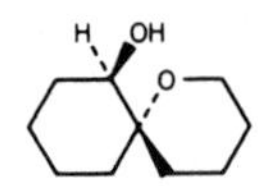

[The two stereogenic centers are related by *trans*.]

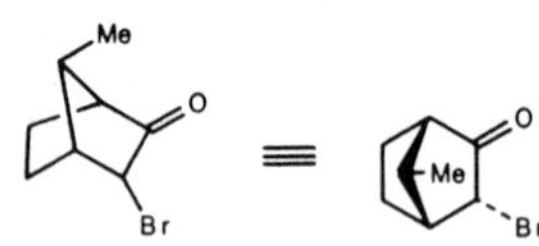

[1S-(endo,syn)]-

[No locants are needed in the relative portion of the descriptor; *endo* and *exo* precede *syn* and *anti* in the relative descriptor.]

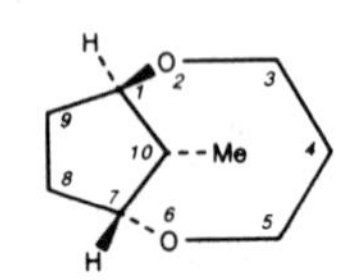

[1R-(1R*,7R*,10R*)]-

[Because the bridgehead atoms have an abnormal configuration, a *syn* or *anti* descriptor would be ambiguous.]

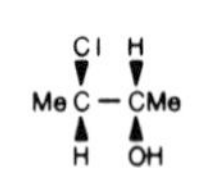

[S-(R*,R*)-]

[The *R*,R** relative descriptor indicates that the two centers have like chirality; the *S* absolute descriptor signifies that the reference center (the carbon atom to which the highest ranking atom (Cl) is attached) has *S* chirality.]

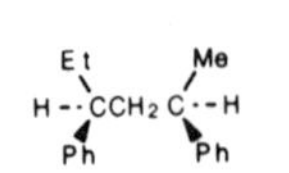

[R-(R*,S*)]-

[The *R*,S** relative descriptor indicates that the two centers have unlike chirality; the *R* absolute descriptor signifies that the reference center (the carbon atom to which the

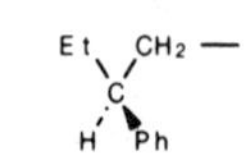

radical is attached) has *R* chirality.]

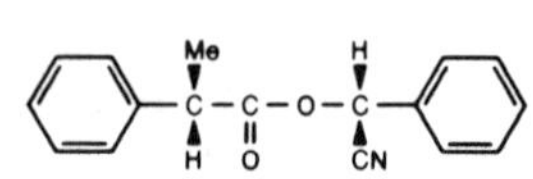

[R-(R*,S*)]-

[The *R* absolute descriptor refers to the center with the O attached, even though that center is not in the parent of the *CA* index name.]

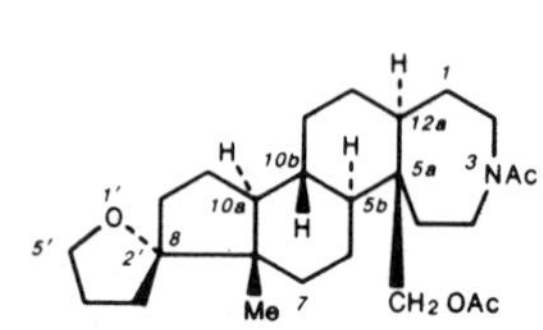

[5aR-(5aα,5bβ,7aα,8β,10aβ,10bα,12aβ)]-

[4]Blackwood, J. E., Gladys, C. L., Petrarca, A. E., Powell, W. H., and Rush, J. E., "Unique and Unambiguous Specification of Stereoisomerism about a Double Bond in Nomenclature and Other Notation Systems", *J. Chem. Doc.*, 8, 30–32 (1968).

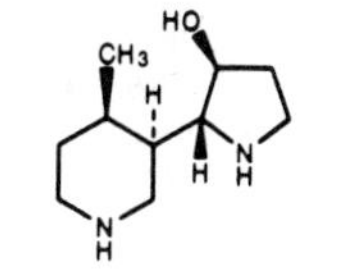

[3α(2R*,3S*),4α]-

[The preferred ring (see ¶ 138) is used as the parent ring.]

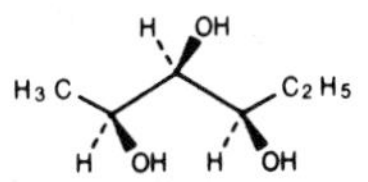

Abs.

[2S-(2R*,3S*,4S*)]-

[Locants are cited for the three stereogenic centers.]

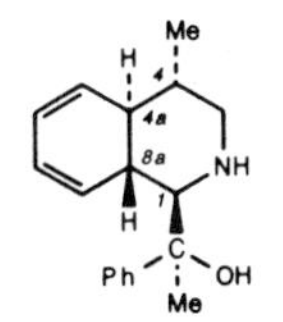

[1R-[1α(S*),4β,4aβ,8aα]]-

[The chiral acyclic atom possesses S* chirality when the reference center is R*. The S* is parenthesized following the locant and stereochemistry of the point of attachment.]

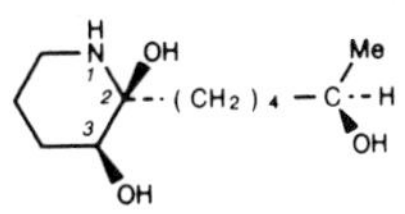

[2S-[2α,2(S*),3α]]-

[Note the repetition of the locant "2" because a chiral center is present on the *less* preferred substituent.]

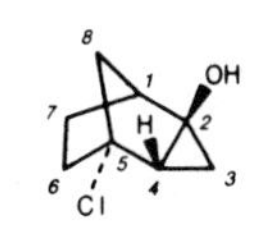

(1α,2α,4α,5β)-

[The bridge (atom 8) is treated as a substituent and is the preferred attachment at position 1. The hydroxyl group at 2 and the hydrogen at 4 are treated as if they were at ring fusion positions; at 5, the chlorine is preferred to the carbon bridge.]

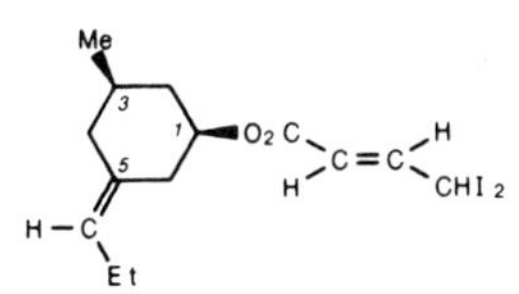

[1α(E),3α,5Z]-

(a relative descriptor)

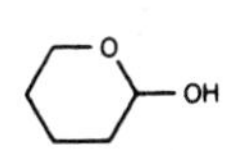

(-)-

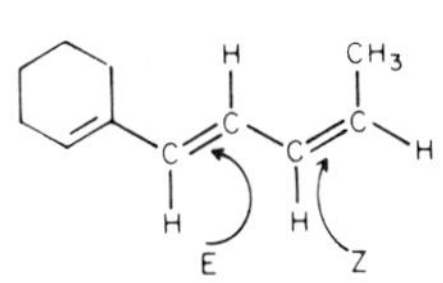

Rel

(E,Z)-

[The *E* double bond has the higher ranking radical attached and is therefore cited first.]

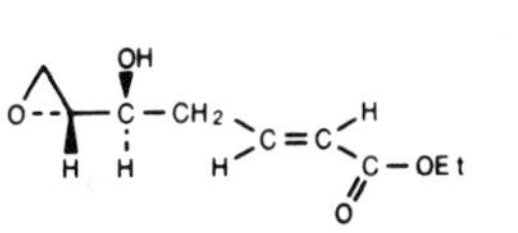

[R-[R*,S*-(E)]]-

[The relative chiral descriptors are separated from geometrical descriptors as shown.]

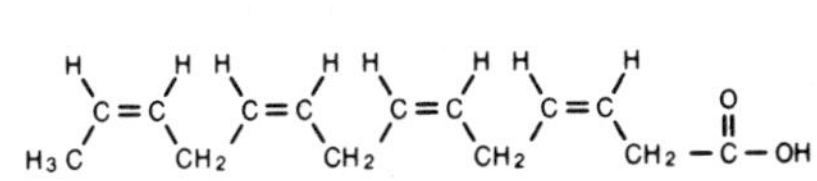

(all-Z)-

[The descriptor "(*all-E*)" or "(*all-Z*)" is cited when four or more centers of like stereochemistry are present.]

The stereochemistry of some compounds cannot be described completely by the rules above. The presence of stereochemical information in an original document is indicated in the index entries for such compounds by the mod= ification term "stereoisomer."

Examples:

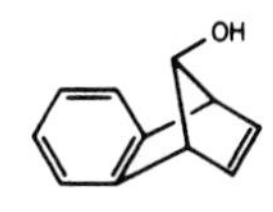

stereoisomer

[This achiral stereogenic system can only be described by *syn* or *anti*, but these descriptors are not allowed for this ring system.]

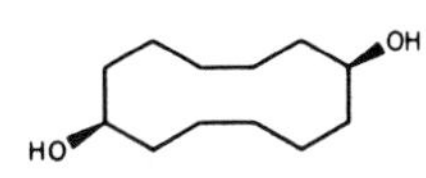

stereoisomer

[This achiral stereogenic system can only be described by the α/β system, which is not allowed in rings of this size. Had the system been chiral, R* and S* could have been used.]

Molecular addition compounds, *mixtures* and *polymers* of components with stereogenic elements have the stereochemical descriptors cited with the re= spective components; a descriptor for the entire addition compound, etc., is cited last, if known.

Examples (only the preferred index entry is shown here for each compound):

Cyclohexanol, 4-amino-
trans-, **acetate (salt)**

Propanoic acid, 2-fluoro-
(*R*)-, **compd. with** (1*R-trans*)-**2-aminocyclohexanol (1:1)**

Piperidine, 2-methyl-
(*R*)-, **compd. with trifluoroborane (1:1)**, (-)-

1,2-Cyclobutanedicarbonyl dichloride
trans-, **polymer with** *cis-***2,5-dimethylpiperazine**

Although the polymer in the final example above probably has *cis-trans* stereochemistry, these terms are not repeated at the monomer headings; the structural repeating unit entry (¶ 222), however, would cite the descriptors in the order dictated by that heading, e.g., [*cis*,(1α,2β,4α)]-.

Normally, no stereochemical descriptor is cited for compounds with only partially known relative stereochemistry, but descriptors for known centers are used in the following cases:

(*i*) Keto-enol systems with stable configurations elsewhere in the mol= ecule.

(*ii*) Penicillins (4-thia-1-azabicyclo[3.2.0]heptane derivatives) and cephalosporins (5-thia-1-azabicyclo[4.2.0]octane derivatives) in which acyclic centers are often of unknown configuration.

(*iii*) Compounds containing stereogenic *hetero* atoms of unknown con= figuration, e.g., sulfones, trivalent phosphorus, or quaternary nitrogen.

No descriptor is cited for a mixture of diastereoisomers whether or not the stereochemistry of the components is known; thus, "*dl-α-*tocopherol" when used to indicate a diastereoisomeric mixture is indexed at its systematic name without a stereochemical designation.

II. *Stereoparents* have been discussed in ¶ 202. Their use enables lengthy stereochemical descriptors to be dispensed with. Cross-references and synonyms relating stereoparents to the corresponding systematic names appear (as stereospecifically as possible) in the *Index Guide*, and author names for stereoparents and derivatives are also cross-referred.

A stereochemical descriptor is cited in an index modification at a stereo= parent heading to express differences between the stereoparent illustrative diagram and the structure of the reported substance. In the diagrams the configurations on ring-system substituents are indicated by dotted lines for α and "wedges" for β. Variations for specific derivatives are indicated by citing appropriate α and β terms with the locants of the modified centers. Variant stereochemistry for acyclic, bridge and spiro centers is indicated by *E* and *Z* or by *R* and *S* as described for systematic stereochemistry, above. If a variation is not reported, the normal stereochemistry is considered to be retained. When the modified stereochemistry is unknown, e.g., indicated by a wavy line in a structural diagram, the descriptor ξ (xi) is assigned to that position. De= scriptors are also cited for added configurations (when the stereoparent dia= gram shows no stereochemistry at a center). When the added stereochemistry is in a stereoparent substituent or modification, the descriptor is cited as a prefix, e.g., [(5α,16*E*)-androstan-16-ylidene]-. When the center is in a sys= tematically named substituent or modification term, the descriptor is placed in the modification in the usual position (after all other structural information), where it is identified by a locant relating to the stereoparent, e.g., **Preg= nane-3,20-dione, 5-hydroxy-6-(3-hydroxy-1-propenyl)-**, [5α,6β(*Z*)]-.

Examples (the stereoparent diagram is compared with an author's diagram for a derivative in each case):

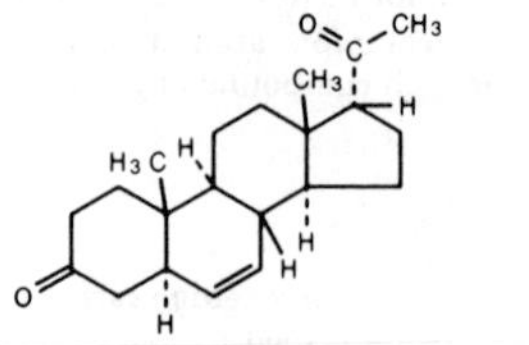

A pregnane derivative

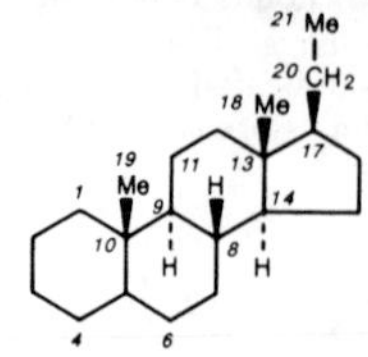

Pregnane (stereoparent)

[The configuration at positions 8, 9, 10, 13 and 14 correspond; position 5 has added stereochemistry; position 17 has modified stereochemistry. The stereochemical descriptor at **Pregn-6-ene-3,20-dione** is (5α,17α)-.]

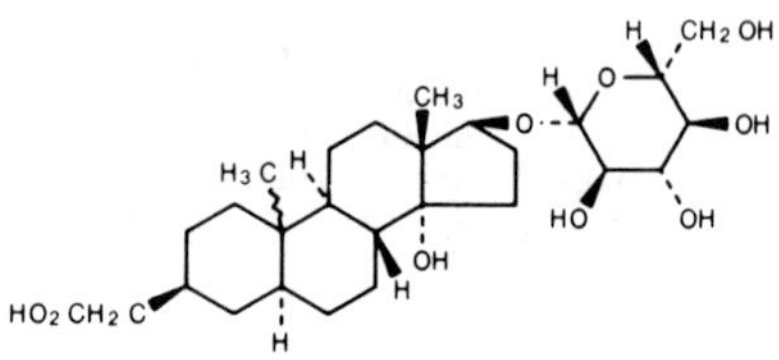

An androstane derivative

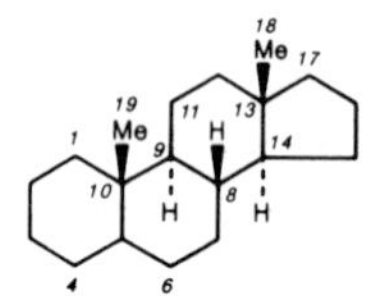

Androstane (stereoparent)

Positions 8, 9, and 13 are normal; position 14 is substituted, but has normal configuration. Position 10 has unknown modified stereochemistry; positions 3, 5 and 17 have added stereochemistry and the substituent at 17 is a stereoparent (β-D-glucopyranosyl radical).

The stereochemical descriptor is (3β,5α,10ξ,17β)-. [The stereochemistry within the stereoparent radical is expressed by the substituent radical name.]

III. *Coordination compounds.* The stereochemistry of mononuclear complexes is expressed by special descriptors; for polynuclear coordination compounds only the term "stereoisomer" is employed in CA indexes. The discussion that follows is necessarily brief; for a more detailed explanation the review by T. E. Sloan[5] should be consulted.

For coordination numbers 1, 2, and 3, no nuclear stereochemistry is possible. Ligands are assigned systematic descriptors according to the rules in I above. If the chirality of a 3-coordinate tetrahedral complex is reported, it is described as *R* or *S* by application of the Sequence Rule as described above. The special coordination descriptors comprise:

(a) a *symmetry site term* to describe the molecular geometry about the nuclear atom;

(b) a *configuration number* to identify the atoms on each axis and plane of the system;

(c) a *chirality symbol* to differentiate members of enantiomeric pairs (when the structure has no reflection symmetry); and

(d) a *ligand segment* (when ligands have stereochemistry of their own) cited as described in ¶ 203I, above.

Symmetry site terms are comprised of one- to three-letter abbreviations to describe the geometry around the central atom combined with the coordination number. They are generally based on information reported by an author, but some assumptions are made in indexing: square planar for 4-coordinate Pd(II), Pt(II), Rh(I), Ir(I), and Au(III); octahedral for all 6-coordinate compounds in the absence of contrary information; tetrahedral for 4-coordinate complexes of (*a*) the representative (main-group) elements; (*b*) Cd, Co(I), Cu(I), Fe(II), Hg(II), Ni(0), Pd(0), Pt(0), and Zn; (*c*) complexes of Fe(II), Fe(III), Mn(II), and Ni(II) with four attached halogens; (*d*) complexes of Co(II), Ti(II), Ti(III), and Ti(IV) with four monodentate ligands of any kind; and (*e*) substances indexed at **Antimony(1+)**, **Arsenic(1+)**, **Bismuth(1+)**, and **Phosphorus(1+)** (not **Phosphonium**, see ¶ 184). Americyl, neptunyl, plutonyl and uranyl (MO_2^{n+}) groups in mononuclear 6, 7, 8, and 9 coordinate complexes are assumed to be *trans*.

Planar ring systems as ligands impose their geometry on the central atom; thus a zinc-porphine complex is square planar, not tetrahedral. Symmetry site terms are not required for 4- and 6-coordinate anionic coordination complexes containing nonmetallic central atoms and identical monodentate ligands: they can be assumed to be tetrahedral and octahedral, respectively. The terms are omitted also when information is lacking.

T-4	tetrahedral
SP-4	square planar
TB-5	trigonal bipyramidal
SP-5	square pyramidal
OC-6	octahedral
TP-6	trigonal prismatic
PB-7	pentagonal bipyramidal
OCF-7	octahedral faced monocapped
TPS-7	trigonal prismatic square faced monocapped
CU-8	cubic
SA-8	square antiprismatic
DD-8	dodecahedral
HB-8	hexagonal bipyramidal
OCT-8	octahedral trans-bicapped
TPT-8	trigonal prismatic triangular faced bicapped
TPS-8	trigonal prismatic square faced bicapped
TPS-9	trigonal prismatic square faced tricapped
HB-9	heptagonal bipyramidal

[5] T. E. Sloan in *Topics in Inorganic and Organometallic Stereochemistry*, ed. G. Geoffroy, John Wiley & Sons, Inc., New York, **1981**, 1-36.

Configuration numbers depend first upon application of the Sequence Rule to determine order of seniority (priority) of atoms coordinated to the central atom. When constitutionally equivalent atoms are present, the same priority number is assigned to each; thus in $Ma_2b_2c_2$, the order is 1,1,2,2,3,3. When a choice of configuration number is possible, preference is given to the atom of lower priority (higher numerical value). Also, some arbitrary tie-breaking is necessary with equivalent sets of polydentate ligand atoms. Chiral ligands, otherwise identical, are ranked with the *R*-form above the *S*-form. The chirality symbols *C*— for clockwise — and *A*— for anticlockwise (counterclockwise)—are used to denote absolute stereochemistry in coordination compounds, except that *R* and *S* are used for tetrahedral complexes, and Δ (delta) and Λ (lambda) for octahedral complexes containing cis-bis(monodentate)-bis(bidentate) ligands and tris(bidentate) ligands, respectively. Chirality symbols are placed after configuration numbers, which are determined and cited as follows:

(*a*) *Tetrahedral complexes* (*T*-4) are assigned no configuration number. The chirality symbol *R* or *S* is assigned as for organic compounds with a single chiral center.

(*b*) *Square planar complexes* (*SP*-4) have the ligating atoms at the corners of a square. The rank number (2, 3, or 4) of the atom diagonally opposite the senior atom (1) is cited as the configuration number. Three isomers are possible when all four ligands are different.

Example:

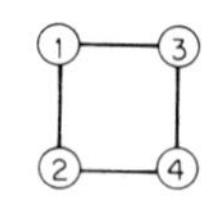

Configuration number = 4

Stereochemical descriptor = (*SP*-4-4)-

(No chirality symbol is required)

(*c*) *Trigonal bipyramidal complexes* (*TB*-5). The configuration number comprises the rank numbers for the atoms at the ends of the major axis cited without punctuation in descending order of priority. If the complex has no reflection symmetry, i.e., if the axial atoms are dissimilar, the order of atoms in the plane perpendicular to the axis is expressed by *C* (clockwise ascending sequence of numerals, equivalent to descending priority) or *A* (anticlockwise) as viewed from the highest priority ligand.

Example:

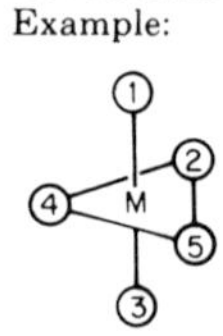

Configuration number = 13

Chirality symbol = A

Stereochemical descriptor = (*TB*-5-13-*A*)-

(When all five coordinating atoms are different, 10 configurations are possible. Each is chiral, affording 20 possible stereoisomers.)

(*d*) *Square pyramidal complexes* (*SP*-5) have a lone coordinating atom on the principal axis and four atoms in a square planar configuration at right angles to this axis. The first digit of the configuration number is the priority number of the lone atom; the second digit is the priority number of the atom situated *diagonally opposite* to the most senior coordinating atom in the plane. The chirality symbol is derived by viewing the plane from the position of the lone axial atom and tracing a path around the square plane from the most senior atom to the next most senior atom present. If this path is clockwise, *C* is the assigned chirality symbol; if anticlockwise, the symbol is *A*.

Example:

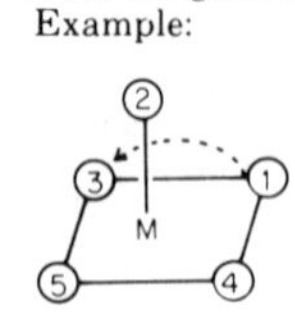

Configuration number = 25

Chirality symbol = A

Stereochemical descriptor = (*SP*-5-25-*A*)-

(When all coordinating sites are unlike, 15 configurations are possible. Each is chiral, affording 30 possible stereoisomers.)

(*e*) *Octahedral systems* (*OC*-6) for which *C* and *A* can be used as chirality symbols (see above) have configuration numbers of which the first digit is derived from the priority number of the atom opposite to atom 1 and the second digit from the priority number opposite to the senior atom in the plane perpendicular to the axis containing atom 1. When no reflection symmetry is present, the chirality symbol is derived by viewing the plane from atom 1 and tracing a path clockwise (*C*) or anticlockwise (*A*) from the most senior atom to the next most senior atom.

Examples:

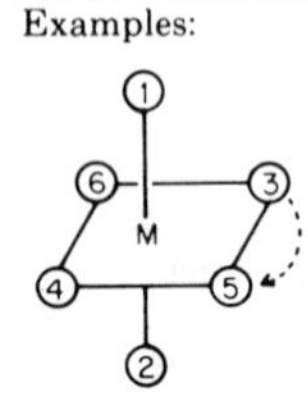

Configuration number = 24

Chirality symbol = C

Stereochemical descriptor = (*OC*-6-24-*C*)-

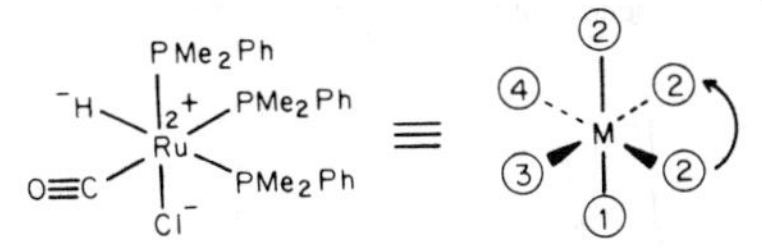

(The priorities of the ligand atoms are determined by arranging their atomic numbers (17,15,6 and 1) in descending order thus: Cl,P,C and H. There are three identical ligands of order 2; one is on the main axis (opposite ligand 1), the others are in the transverse plane opposite 3 and 4; the higher number is cited. The transverse pathway from the preferred 2 to the neighboring (less preferred) 2 is clockwise when viewed from atom 1.)

Configuration number = 24

Stereochemical descriptor = (*OC*-6-24-*C*)-

For octahedral complexes with two or three bidentate ligands oriented in a skew configuration, the helicity symbols Δ and Λ are employed instead of chirality symbols, they are related to right-handed and left-handed spirals, respectively, as follows:

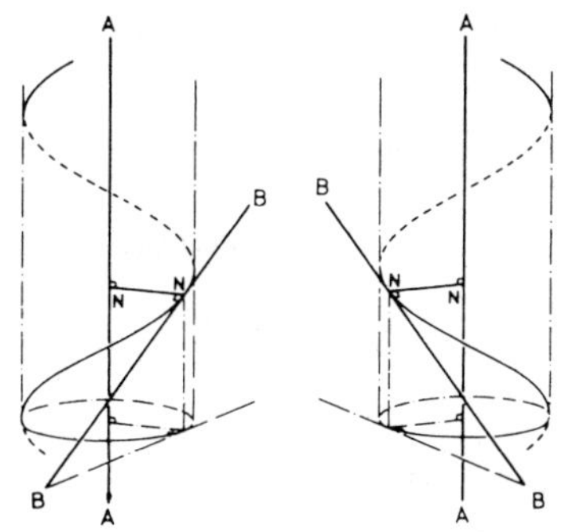

(*a*) Δ (*b*) Λ

In diagram (*a*) above, BB is a tangent to the right-handed helix that has axis AA and radius NN; (*b*) shows its mirror image, a left-handed helix. Turning each diagram through 90 degrees *in either direction* results in the simplified diagrams (*c*) and (*d*), respectively.

(*c*) (*d*)

AA and BB have now lost their respective identities as axis and tangent and can be considered either as two tangents or as two segments of a double helix. BB is in front of AA in both diagrams; in (*c*), when AA is horizontal, BB descends to the right, in (*d*) to the left.

In the octahedral complex, pictured as a regular octahedron in (*e*) below, the lines AA and BB represent two bidentate ligands oriented as in (*c*) above and represented by the helicity symbol Δ. (A third bidentate ligand, if it were present at CC, would not affect the helicity.)

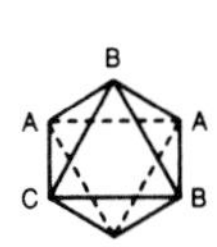

(*e*)

The helicity symbol replaces the chirality symbol in octahedral complexes of this type. In the complex named in abbreviated fashion as $[Co(1,2\text{-}pn)_3]^{3+}$ the ligands themselves have chirality (denoted by *R* or *S*), and their geometry around the cobalt atom is described by use of "1" (for bonding to a secondary carbon) and "2" (for bonding to CH_2). The helicity can be seen to correspond to the diagram above and the symbol Δ is therefore appropriate.

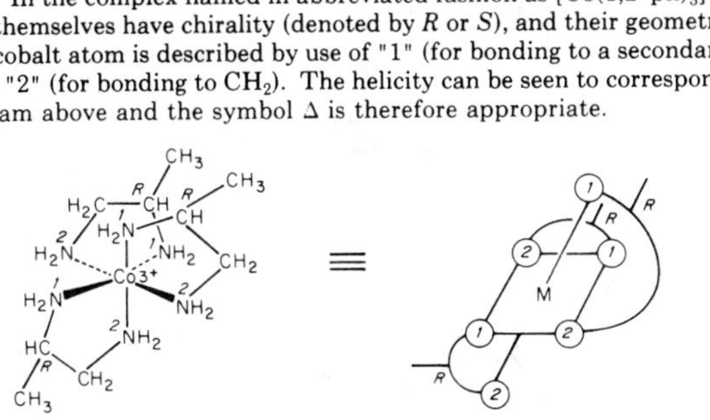

Configuration number = 21

Chirality symbol = Δ

Stereochemical descriptor = (*OC*-6-21-Δ-(*R*),(*R*),(*R*)]-

Octahedral complexes containing identical bis(tridentate) ligands are assigned geometric and chirality symbols in the same manner as normal octahedral complexes. However, in order to distinguish between enantiomers it is necessary to add primes to the donor atoms of one ligand. The primes are retained in the configuration number and this distinguishes between enantiomers.

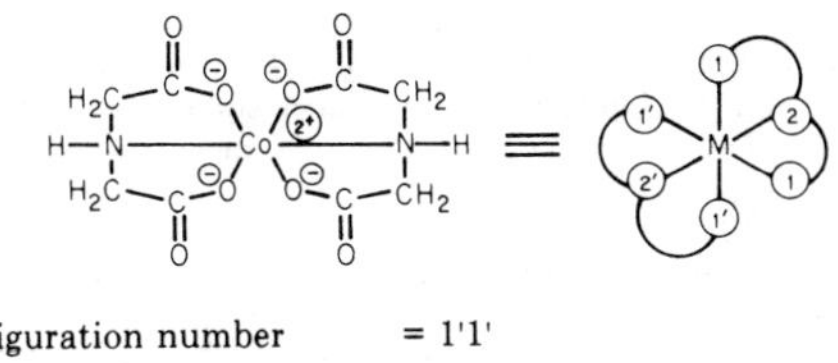

Configuration number = 1'1'

Stereochemical descriptors = (*OC*-6-1'1')-

(*f*) *Trigonal prismatic complexes* (*TP*-6). The configuration number is obtained by citing the priority numbers of the three atoms opposite to (eclipsed by) the preferred triangular face, i.e., the face containing the maximum number of ligating atoms of highest priority (lowest numerals). These numbers are cited to correspond to the ascending numerical order of the respective eclipsing atoms. A chirality symbol, *C* or *A*, denotes the direction of numerical progression *of the eclipsed atoms*.

Example:

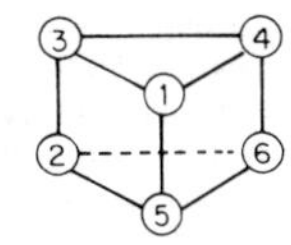

Configuration number = 526

Chirality symbol = A

Stereochemical descriptor = (*TP*-6-526-*A*)

(*g*) Stereochemical descriptors were extended to 7-, 8-, and 9-coordinate complexes in *CA* indexes in 1977.[6]

In the following table, all 12 model polyhedra for 7-, 8-, and 9-coordinate complexes are shown, with examples of priority numbers for ligands from which configuration numbers are derived.

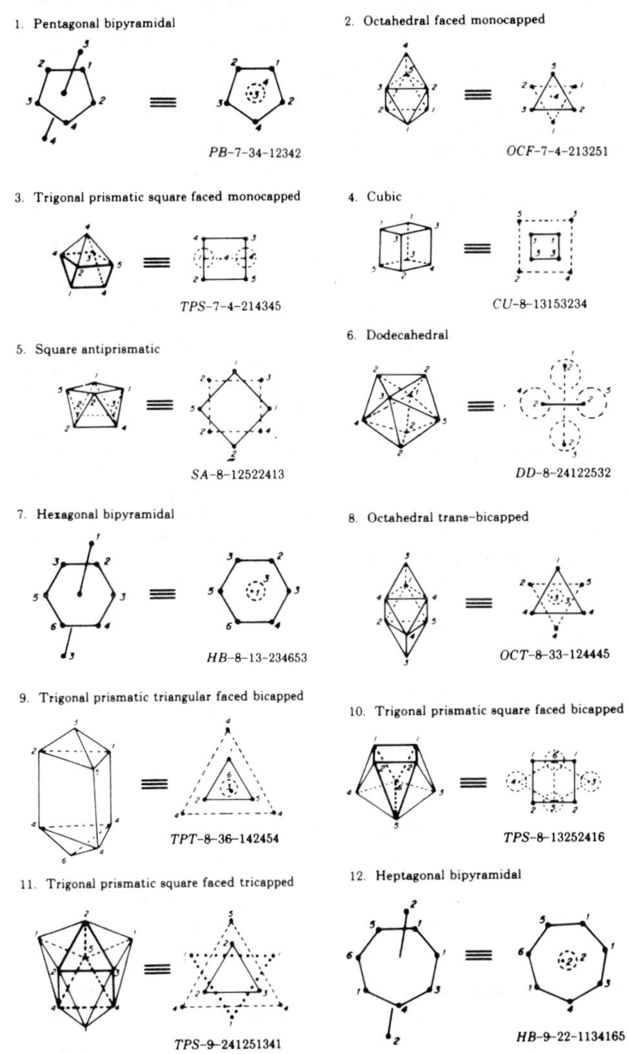

[6] M. F. Brown, B. R. Cook, and T. E. Sloan, *Inorg. Chem.*, **1978**, *17*, 1563-8.

The configuration number is assigned by orienting the model structure with the highest-order axis in the vertical plane. The model structure is then viewed from the highest priority ligating atom on the highest-order axis or from a point on the highest-order axis above the most preferred end or terminal plane perpendicular to the axis. The most preferred end or terminal plane is that end plane which either contains the greatest number of atoms, contains the greatest number of highest priority ligating atoms, or is adjacent to a plane containing the greatest number of highest priority ligating atoms. (Note that the *OCF*-7 (octahedral faced monocapped), *TPS*-7 (trigonal prismatic square faced monocapped), and *TPS*-8 (trigonal prismatic square faced bicapped) model structures are of low symmetry and have only one correct orientation.)

The configuration numbers for the model structures coded *PB*-7, *OCF*-7, *TPS*-7, *HB*-8, *OCT*-8, *TPT*-8, and *HB*-9 begin with the priority numbers of the ligating atom(s) on the highest-order axis and are given in lowest numerical order sequence. These priority numbers are separated from the remainder of the configuration number by a hyphen. The remaining portion of the configuration number is derived by viewing the structure from the highest-priority ligating atom on the highest-order axis, or from the axial ligating atom located above the preferred plane, and by citing the priority number of the ligating atom with the highest priority in the plane adjacent to that atom. The configuration number is then completed by continuing to cite the priority numbers of the ligating atoms in sequence as they are encountered, either clockwise or anticlockwise around the projection of the model structure, alternating between planes where necessary, when viewing from the highest priority atom. For those model structures with eclipsed pairs of ligating atoms, the priority numbers are given in pairs with the priority number for the preferred atom in the first plane followed by that for the eclipsed atom. The remaining priority numbers are given as they are encountered, either clockwise or anticlockwise around the projection of the model structure as viewed from the preferred end of the highest-order axis. For compounds in which clockwise and anticlockwise citations of the configuration number give two different configuration numbers (generally chiral compounds), the correct one is the lower numerical sequence as determined at the first point of difference. The remaining five model structures, *CU*-8, *SA*-8, *DD*-8, *TPS*-8, and *TPS*-9, do not have a ligating atom on the highest-order axis and thus do not have a distinct portion of the configuration number set off by a hyphen. These model structures are viewed from a point on the highest-order symmetry axis above the preferred terminal plane as defined previously. The configuration number is derived by first citing the priority number of the preferred ligating atom in the preferred end plane and then citing the priority number of the ligating atom it eclipses, if one exists. In the next step, one proceeds clockwise or anticlockwise around the projection of the model structure, giving the priority numbers of the ligating atoms as they are encountered, alternating between planes when necessary. Again, the clockwise or anticlockwise direction is chosen to give the lowest-order numerical sequence for the configuration number as determined at the first point of difference.

When there are two or more equivalent bidentate or tridentate ligands and the same priority numbers thus occur in equivalent ligands, the ties are broken by identically priming all the CIP priority numbers of the ligating atoms within a ligand to determine both the configuration number and the chirality symbol. In those complexes with symmetrical polydentate ligands, tetradentate, hexadentate (including symmetrical macrocyclic) ligands, etc., ties between equivalent ligating atoms are broken by priming the ligating atom priority numbers in chelating groups or pairs, thereby reducing the polydentate ligand to groupings of equivalent bidentate or tridentate ligands. When two or more nonequivalent tie-breaking choices exist for the coordination polyhedra of coordination numbers 7, 8, and 9, the tie is resolved by (a) assigning the lowest priming to the preferred symmetry axis or plane, and (b) assigning the lowest priming to give the lowest numerical value to the configuration number at the point of difference. Primes are restricted to the configuration number and are not included except when absolutely necessary, e.g., for octahedral complexes containing two identical tridentate ligands and trigonal-prismatic complexes containing two or more identical polydentate ligands and for all 7, 8 and 9 coordinate complexes. The use of primes is exemplified by the first example below, which is explained in some detail. The remaining examples show only the structures and total stereochemical descriptors.

Examples:

Pentagonal bipyrimidal (PB-7)

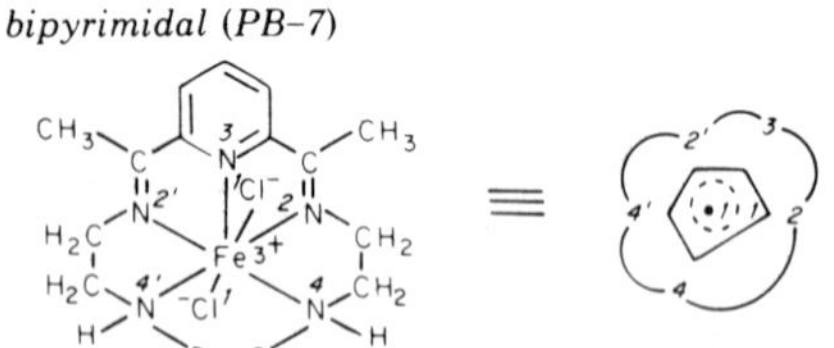

Configuration number = 11-232'4'4

CA Index name (in uninverted form) =

(*PB*-7-11-232'4'4)-dichloro(2,13-dimethyl-3,6,9,12,18-pentaazabicyclo[12.3.1]octadeca-1(18),2,12,14,16-pentaene-N^3,N^6,N^9,N^{12},N^{18})iron(1+)

The ligands of highest atomic number (chlorine) are numbered 1. An exploration table (not shown) may be needed to establish the order of priorities (2,3, and 4) of the nitrogen ligands.[7] No chirality symbol (*C* or *A*) is needed. The name appears in the *CA Chemical Substance Index* at **Iron(1+).**

[7]See Ref. 3, pp. 391-5.

Octahedral faced monocapped (OCF-7)

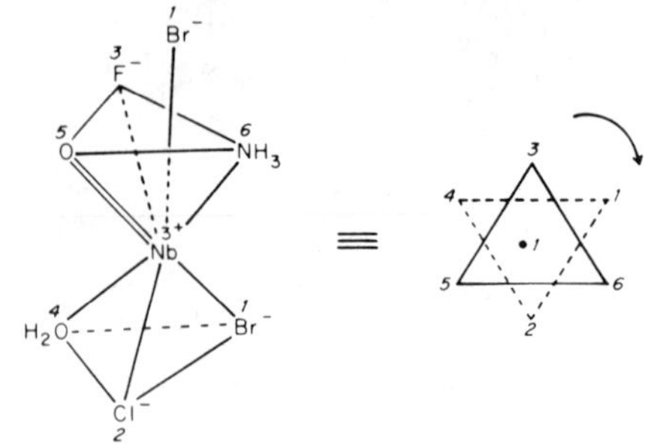

(*OCF*-7-1-316254-*C*)-

Trigonal prismatic square faced monocapped (TPS-7)

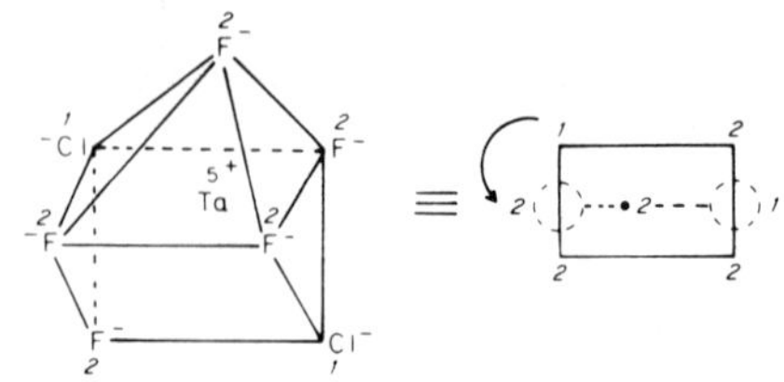

(*TPS*-7-2-122212-*A*)-

(The direction must be selected so as to cite the priority number of the atom in the next lower plane.)

Cubic (CU-8)

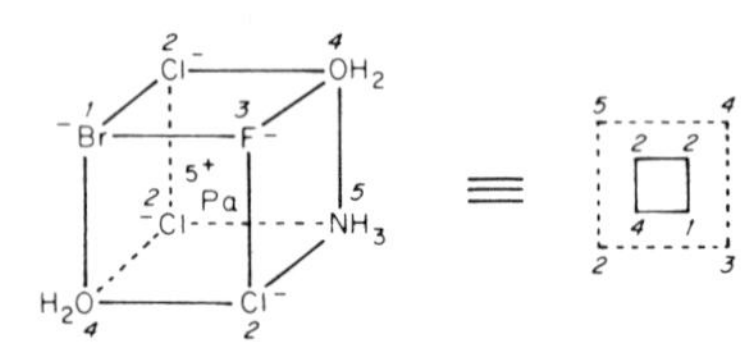

(*CU*-8-13242542-*A*)-

(Each face must be tested to determine the preferred terminal plane. As drawn in this example, the preferred face (terminal plane) is on the left side of the cube. Consequently the cube must be re-oriented before assigning the descriptor.)

Square antiprismatic (SA-8)

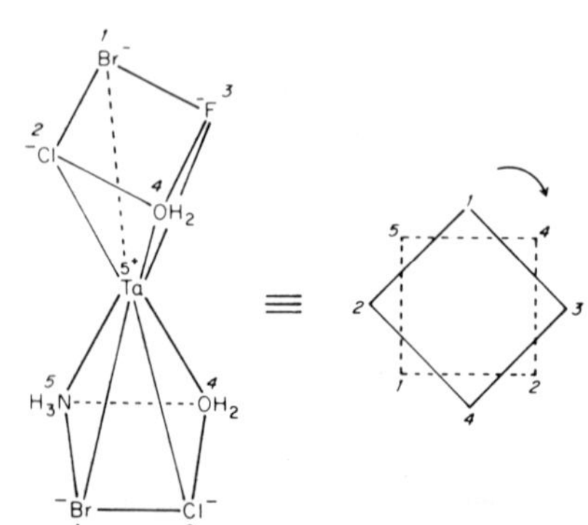

(*SA*-8-14324125-*C*)-

(The preferred orientation must be determined before the descriptor is assigned.)

Dodecahedral (DD-8)

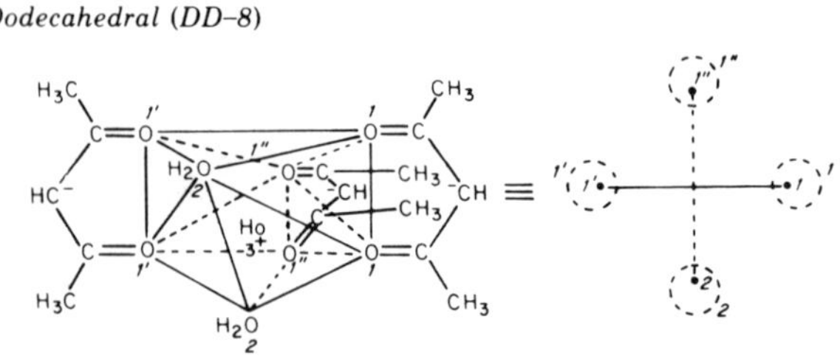

(*DD*-8-111"1"1'1'22)-

(The preferred orientation must be determined before the de= scriptor is assigned; also, the pathway is selected to pass through the atom of highest priority (lowest numerical value) in the next lower plane.)

Hexagonal bipyramidal (HB-8)

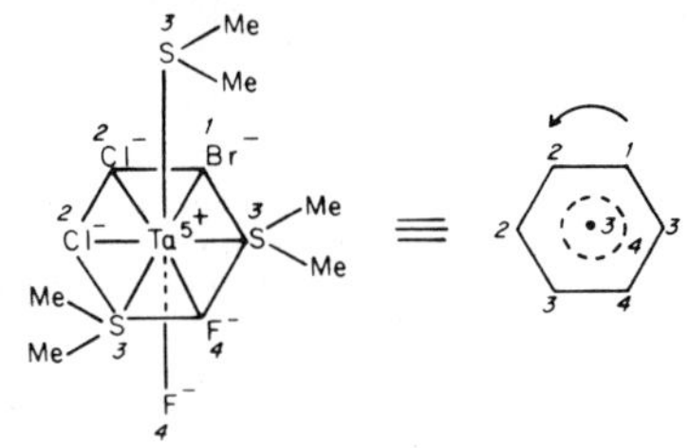

(*HB*–8–34–122343–*A*)–

Octahedral trans-bicapped (OCT-8)

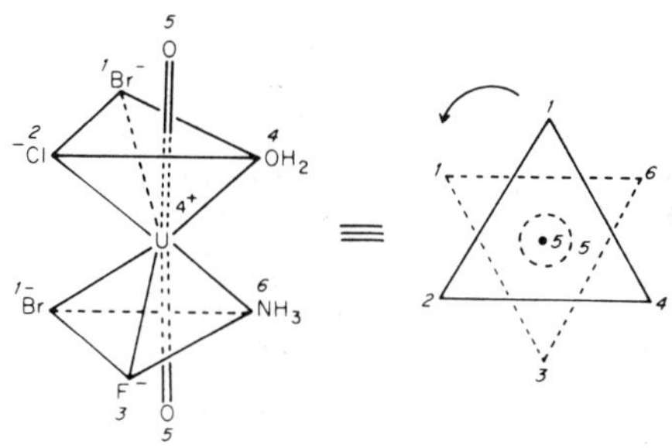

(*OCT*–8–55–112346–*A*)–

(The preferred orientation must be determined before the de= scriptor is assigned.)

Trigonal prismatic triangular faced bicapped (TPT-8)

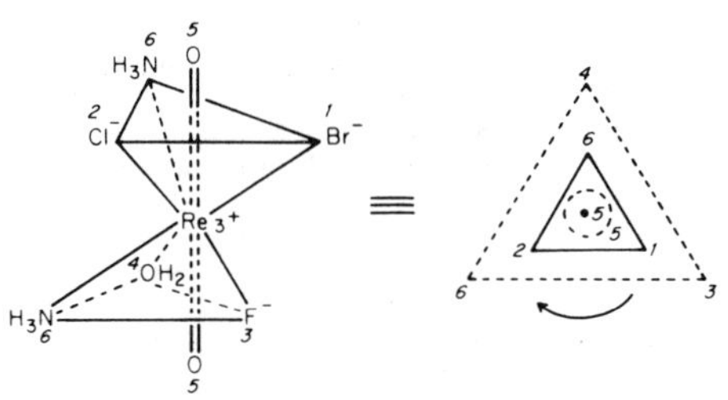

(*TPT*–8–55–132664–*C*)–

(The preferred orientation must be determined before the de= scriptor is assigned.)

Trigonal prismatic square faced bicapped (TPS-8)

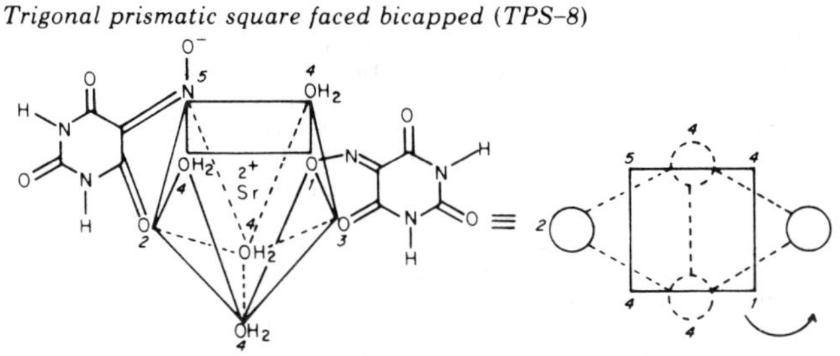

(*TPS*–8–13445244–*A*)–

(The pathway is selected to pass through the atom of highest priority (lowest numerical value) in the next lower plane.)

Heptagonal bipyramidal (HB-9)

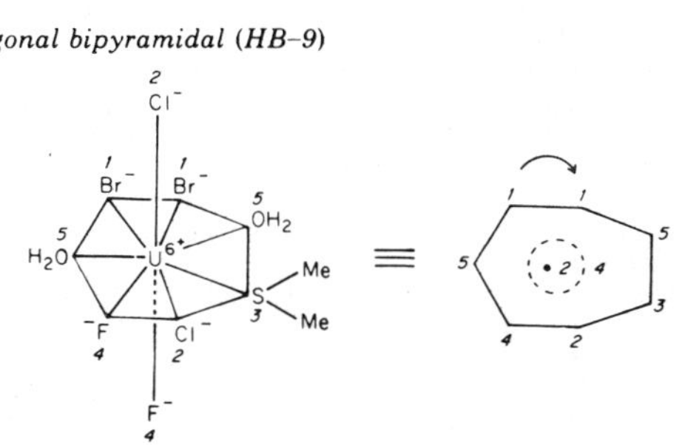

(*HB*–9–24–1153245–*C*)–

Trigonal prismatic square faced tricapped (TPS-9)

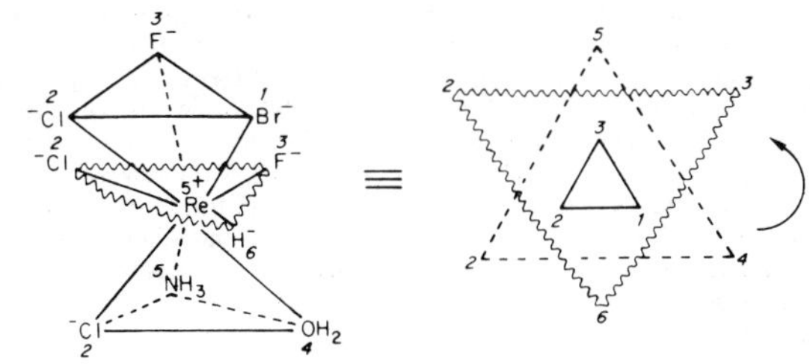

(*TPS*–9–143352226–*A*)–

Mononuclear coordination complexes containing ligands which require the use of η do not follow the stereochemical rules for other mononuclear coor= dination complexes. When the author indicates stereochemistry for the total complex, the term "stereoisomer" is used as the STE and an * used as the DES. An exception is made for metallocenes.

Metallocenes, when asymmetrically disubstituted, are not superimposable on their mirror images; they can therefore exist as enantiomers. The absolute configuration of the chiral center of highest priority is determined by the Sequence Rule and cited as *R* or *S* when specific information is stated in the original document. (The configuration of the other four centers is then fixed.)

Metallocenes which have stereochemistry only in the substituents have that indicated as described by the rules in I above.

Partial stereochemistry is not cited for coordination compounds.

204. Alkaloids are divided into classes largely in accordance with their stereochemical complexity. Class A alkaloids include substances containing only one chiral center, or none at all, or whose stereochemistry, typically restricted to a single ring system, is easily defined. Alkaloids of this class are indexed systematically; cross-references are found at the alkaloid names.

Example:

Tropine
See *8-Azabicyclo[3.2.1]octan-3-ol, 8-methyl-, endo-*

Class B alkaloids possess more complex stereochemistry. They have been organized as derivatives of a single stereoparent of known absolute configu= ration common to several alkaloids. (Class C alkaloids are those which have not yet been organized in this way, but whose absolute configuration is known.) Class B stereoparents are illustrated in the *Chemical Substance Index* when justified by current entries. New Class B stereoparents are created when justified at any time within the collective indexing period. The following is a partial list of those currently employed for alkaloids:

Aconitane	**Hasubanan**
Ajmalan	**Ibogamine**
Aspidofractinine	**Lycopodane**
Aspidospermidine	**Lycorenan**
Berbaman	**Lythran**
Cevane	**Morphinan**
Cinchonan	**Oxayohimban**
Condyfolan	**Oxyacanthan**
Corynan	**Rheadan**
Corynoxan	**Sarpagan**
Crinan	**Securinan**
Curan	**Senecionan**
Eburnamenine	**Strychnidine**
Emetan	**Thalicberan**
Ergoline	**Thalidasan**
Ergotaman	**Thalman**
Erythrinan	**Tubocuraran**
Ethenomorphinan	**Tubulosan**
Formosanan	**Vobasan**
Galanthan	**Yohimban**

Deviations in stereochemistry for specific derivatives of Class B alkaloids are indicated in the modification (¶ 203 II). The prefixes, suffixes, conjunctive names, etc., of regular substitutive nomenclature are employed in naming derivatives. Radicals, e.g., aconitanyl, are formed from the stereoparent names in the usual way. Ring modifications are expressed by prefixes such as cyclo, homo, nor, and seco, on the heading parent, and illustrative diagrams are provided for stereoparents e.g., **11,20–Cycloaspidospermidine,** formed by their use. Carboxylic acids are formed from Class B stereoparents by oxidation of an existing carbon to form an "-oic acid" or by addition of a carboxyl group to form a "-carboxylic acid"; the latter type of name is preferred if a choice is presented. Removal of hydrogen is indicated by "dehydro" terms in the inverted part of the heading.

Class C alkaloids are indexed at stereoparent names supplied in the liter= ature, the numbering system is that most commonly adopted in original documents, and the numbering is extended, if necessary, to include all positions except acyclic hetero atoms. The latter are denoted by italic element symbols, with superscripts, if necessary, derived from the lowest numbered atom to which the hetero atom is attached, e.g., N^6, O^2. Illustrative diagrams in the *Chemical Substance Index* at Class C alkaloid names, e.g., **Alstonisidine, Meloscandonine, Pleiocorine, Vincaleukoblastine,** indicate the stereo= chemistry and numbering systems. When a structural relationship, including stereochemistry, becomes apparent for a group of Class C alkaloids, they may be reassigned to Class B by creation of a stereoparent system stripped of functional groups and common to all. Derivatives of Class C alkaloids are named, when sufficient information is presented, by modifying the stereoparent name. Suffixes are not used except for "-oic acid," to denote oxidation of a methyl group or removal of an ester group. "Hydro" terms are employed to indicate saturation of positions; "de" terms, e.g., "dehydro," "deoxy," "deepoxy," "deoxo," "demethyl," "de(methoxycarbonyl)," denote removal of atoms or groups, which may be replaced by other groups; e.g., **Pleiocorine, 16'-de(methoxycarbonyl)-16'-(hydroxymethyl)-**. Higher functions attached to Class C stereoparents are indexed as substituents. When an alkaloid derivative can be named at two or more Class C stereoparents, the decision depends in descending order of preference on (*a*) removal of the fewest different substituents; (*b*) removal of substituents of lowest molecular weight; (*c*) retention of the stereoparent of highest molecular weight; (*d*) addition of prefixes related to the lowest number of smallest substituents; (*e*) the least stereochemical change; (*f*) the lowest locants for substituents; (*g*) the earliest index positions of the index name.

Derivatives of Class C alkaloids that meet the criteria for Class A alkaloids are named systematically by substitutive nomenclature. When a Class C stereoparent structure contains an oxo group, derivatives of the group, e.g., oximes or hydrazones, are expressed in the modification; otherwise they are expressed as substituents, e.g., as (hydroxyimino) or hydrazono radicals. Esters are named analogously in the modification or by (acyloxy) or [(alkyloxy)carbonyl] substituents. Stereochemical deviations of Class C alkaloid derivatives are expressed as the final term in the modification in the usual way (¶ 203 II).

Class D alkaloids are those reported without structures or without configurations. Those of unknown structure are indexed only at the author name. Alkaloids of known structure but unknown stereochemistry are named: (*a*) systematically if they conform to Class A criteria or are of completely novel structure, or (*b*) at a Class B stereoparent with the stereo modification ξ for the unknown asymmetric centers. A cross reference and synonym will be found at each such trivial alkaloid heading in the *Index Guide*. Each Class D alkaloid is reassigned to Class B or Class C when sufficient information becomes available, and the index name and cross reference are then revised accordingly.

If two or more structures are suggested for a Class D alkaloid, cross-references from the trivially named index heading to all the systematic names will be found in the *Index Guide*, and all the synonyms appear at the latter headings. When two or more trivial names are suggested for the same Class D alkaloid, index entries are made at all of them, and "see also" cross-references leading from each to all the others will be found in the *Index Guide*. Later, the cross-references run to the single preferred name.

Quaternary alkaloids are either (*a*) alkaloids that contain a cationic nitrogen atom, or (*b*) alkaloids that have been quaternized, e.g., by methylation of a tertiary nitrogen group. Names of (*a*) are completed by citing the anion (if known) in the modification, e.g., "chloride"; (*b*) are usually named by adding "ium" to the neutral alkaloid name, and indicating the quaternizing substituents as in regular substitutive nomenclature; e.g., **Morphinanium, 17,17-dimethyl-**, iodide by quaternization of **Morphinan**). When the location of the quaternary center is unknown, the "ium" ending is omitted and the quaternary compound is named as an addition compound; e.g., **Emetan,** compd. with iodomethane (1:1), with an additional index entry at **Methane, iodo-,** compd. with emetan (1:1). Nonquaternary derivatives of the alkaloids described in (*a*), above, are named by subtractive prefixes, e.g., demethyl, or by "dihydro" terms.

Removal of a methyl group from the carbon skeleton of an alkaloid is indicated by "nor" with the locant of the carbon atom lost, but removal of a methyl group from a hetero atom is expressed by "demethyl."

Ring contraction and enlargement in alkaloids is denoted by "nor" and "homo" respectively with a capital italic letter to indicate the ring affected, e.g., *C*-**Noribogamine.** The locant for the carbon atom lost is cited, when necessary, in parentheses immediately after the ring designation. (No more than two atoms are removed or added in this way.) The locant for the added carbon atom is derived from the highest existing locant in that ring by addition of a lowercase "a." When a carbon atom is added between two bridgehead positions, its "a" locant is cited in parentheses immediately after the letter designation of the ring involved (see also Steroids, ¶ 211).

Ring closure may be effected by insertion of heteroatoms (other than nitrogen) as bridges while maintaining stereoparent nomenclature, e.g., **Aspidospermidine, 19,21-epoxy-**. Such a bridge may also include one carbon atom, e.g., **6,14-Ethenomorphinan, 7,5-(methyleneoxy)-**. Replacement of one carbon atom by a heteroatom other than nitrogen is permitted, with creation of a new stereoparent; *e.g.*, **8-Oxaergoline,** but insertion of additional nitrogen or replacement of existing nitrogen is not permitted.

Seco alkaloids are formed by ring cleavage and addition of hydrogen at the resulting terminal groups. Such compounds can often be named in more than one way. The earliest cited available name is chosen from the following alternatives: (*a*) a systematic name (when the Class A specifications are met); (*b*) a Class B name; (*c*) a seco-Class B name; (*d*) a Class C name; (*e*) a seco-Class C name. Cross references will be found at "seco" names and author's trivial names not employed as heading parents.

Examples:

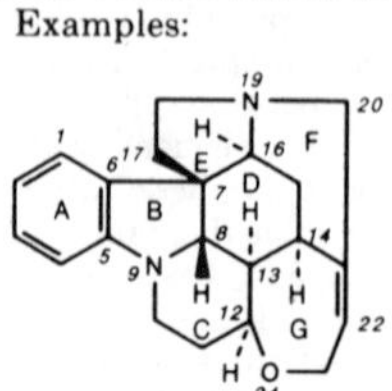

Strychnidine (stereoparent)

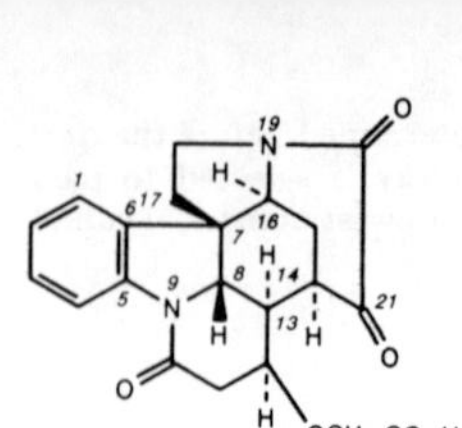

21,22(21*H*,22*H*-)-Secostrychnidin-22-oic acid, 10,20,21-trioxo- (cross-reference at Strychninonic acid; cross-reference from **12*H*-1,12-Ethano-9*H*-pyrido[1,2,3-*lm*]pyrrolo[2,3-*d*]carbazole** to **21,22(21*H*,22*H*)-Secostrychnidine)**

Ring-fused derivatives of alkaloid stereoparents are named systematically.

Degradation products of alkaloids in which ring cleavage and removal of large portions of the molecule still leave intact some rings and the original stereochemical relationships are named, when possible, as derivatives of a smaller alkaloid stereoparent. When this is impossible and the structure is known, degradation products of Class C alkaloids are named systematically, but, for those of Class D alkaloids, only the author's names are employed.

For steroidal alkaloids, see ¶ 211.

205. Amino acids. The following biologically significant amino acids are stereoparents:

Alanine	**Isoleucine**
β-Alanine	**Isovaline**
Alloisoleucine	**Leucine**
Allothreonine	**Lysine**
Arginine	**Methionine**
Asparagine	**Norleucine**
Aspartic acid	**Norvaline**
Cysteine	**Ornithine**
Cystine	**Phenylalanine**
Glutamic acid	**Proline**
Glutamine	**Serine**
Glycine	**Threonine**
Histidine	**Tyrosine**
Homocysteine	**Tryptophan**
Homoserine	**Valine**

The following amino carboxylic, amino sulfonic, and amino arsonic acids, indexed prior to 1972 at their trivial names, are now named either systematically or as derivatives of an amino acid listed above:

Trivial name	*CA Index Name*
Allocystathionine	**Homocysteine, *S*-(2-amino-2-carboxyethyl)-**
Anthranilic acid	**Benzoic acid, 2-amino-**
Arsanilic acid (3 isomers)	**Arsonic acid, (aminophenyl)-**
Carnosine	**Histidine, *N*-β-alanyl-**
Creatine	**Glycine, *N*-(aminoiminomethyl)-*N*-methyl-**
Cystathionine	**Homocysteine, *S*-(2-amino-2-carboxyethyl)-**
Ethionine	**Homocysteine, *S*-ethyl-**
Hippuric acid	**Glycine, *N*-benzoyl-**
Lanthionine	**Cysteine, *S*-(2-amino-2-carboxyethyl)-**
Metanilic acid	**Benzenesulfonic acid, 3-amino-**
Panthothenic acid	**β-Alanine, *N*-(2,4-dihydroxy-3,3-dimethyl-1-oxobutyl)-** (see ¶ 224)
Sarcosine	**Glycine, *N*-methyl-**
Sulfanilic acid	**Benzenesulfonic acid, 4-amino-**
Taurine	**Ethanesulfonic acid, 2-amino-**
Thyronine	**Tyrosine, *O*-(4-hydroxyphenyl)-**
Thyroxine	**Tyrosine, *O*-(4-hydroxy-3,5-diiodophenyl)-3,5-diiodo-**

The configurational descriptors D- and L- are placed before the stereoparent names; DL- is cited for the optically inactive mixture or racemic form; thus, L-Leucine; D-Valine; DL-Phenylalanine. In the absence of contrary evidence, the L-isomer is assumed for amino acid stereoparents, except **Glycine** and **β-Alanine,** which possess no asymmetric center, and **Alloisoleucine** and **Allothreonine** (see below). When the original document clearly indicates that an amino acid is synthetic, it is indexed as the DL-form. The same assumptions are made for their radicals, which are employed in peptide nomenclature, and for derivatives such as esters, salts, and *N*-, *O*-, and *S*-derivatives indexed at the stereoparent. No assumptions are made for carbon-substituted derivatives, or for derivatives named systematically or isolated from bacterial sources. Systematically named amino acids of known stereochemistry are assigned the absolute descriptors *R* and *S* in the modification. Of the diastereoisomeric pairs, assumptions in favor of **Isoleucine** over **Alloisoleucine,** and of **Threonine** over **Allothreonine** are made. The relative descriptors *erythro* and *threo* are cited in the modifications of diastereoisomeric derivatives of amino acid stereoparents to define the second asymmetric center, e.g., L-**Aspartic acid, 3-hydroxy-**, *erythro*-.

Phenylalanine is a stereoparent employed for Alanine, 3-phenyl-. It is treated in substitutive nomenclature as though it were the conjunctive name Benzenealanine. The naturally occurring isomer is L-**Phenylalanine,** which affords the radical L-phenylalanyl. Derivatives containing higher functions or more preferred ring systems are named systematically according to the regular rules.

Example:

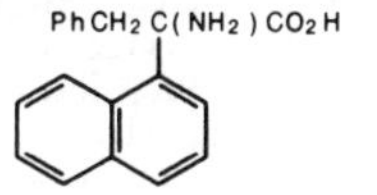

1-Naphthaleneacetic acid, ***α*-amino-*α*-(phenylmethyl)-**

Radicals derived from the names of amino acid stereoparents, e.g., glycyl, L-alanyl, L-phenylalanyl, are employed only in naming peptides (¶ 206). In other situations, systematically named radicals are used; e.g., (aminoacetyl) instead of glycyl. The phenylalanyl radical is considered to be a specially-named one-part radical; no parentheses are placed around it when it is unsubstituted. Stereoparent radical names are derived by replacing the final "-ine" by "-yl" (an exception is **Cysteine,** which affords cysteinyl); **Tryptophan** gives tryptophyl; **Aspartic acid** and **Glutamic acid** radicals are as follows:

$-COCH_2CH(NH_2)CO-$	aspartoyl
$HO_2CCH_2CH(NH_2)CO-$	*α*-aspartyl
$-COCH_2CH(NH_2)CO_2H$	*β*-aspartyl
$-CO(CH_2)_2CH(NH_2)CO-$	glutamoyl
$HO_2C(CH_2)_2CH(NH_2)CO-$	*α*-glutamyl
$-CO(CH_2)_2CH(NH_2)CO_2H$	*γ*-glutamyl

Amino acids are given less preference for special treatment than other stereoparents. They are ranked just above the unsubstituted parent acids; thus, **Alanine** (2-aminopropanoic acid) is ranked above **Propanoic acid** but below **Butanoic acid.** All carbon-substituted derivatives of **Glycine** and ***β*-Alanine** are indexed at systematic names, but the carbon-substituted *radicals* are permitted in peptide nomenclature.

Examples:

$BzNHCH_2CO_2H$ (N 2 1) — **Glycine, *N*-benzoyl-**

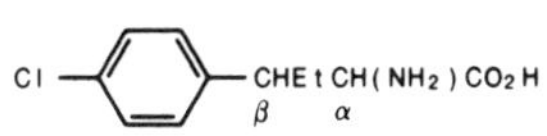

L-Phenylalanine, 4-chloro-*β*-ethyl- *threo-* (not L-Norvaline, 3-(4-chlorophenyl)-, *threo-*)

$Me(CH_2)_5CMe(NH_2)CO_2H$ — **Octanoic acid, 2-amino-2-methyl-** (not Alanine, *α*-hexyl-)

$PhCH(NH_2)CH_2CO_2H$ — **Benzenepropanoic acid, *β*-amino-** (*R*)- (not *β*-Alanine, *β*-phenyl-, (*R*)-)

O-Substituted derivatives of the hydroxy amino acids (**Allothreonine, Homoserine, Serine, Threonine,** and **Tyrosine**) and *S*-derivatives of the mercapto amino acids (**Cysteine** and **Homocysteine**) are named at those stereoparents, the alkyl derivatives as substituents, e.g., **Serine, *O*-methyl-,** and the acyl derivatives as esters in the modification, e.g., **L-Cysteine, acetate (ester).** The stereoparent **Methionine,** $CH_3SCH_2CH_2CH(NH_2)COOH$, is the *S*-methyl derivative of **Homocysteine.** Derivatives in which this methyl group is substituted are indexed as **Homocysteine** derivatives. *S*-Oxide derivatives of **Cysteine, Homocysteine,** and **Methionine** are indexed as sulfinyl and sulfonyl derivatives of simpler parents.

Derivatives of the branched-chain acids **Isoleucine, Isovaline, Leucine,** and **Valine,** in which the methyl side chain is substituted are indexed as derivatives of the linear acid (**Butanoic acid** or **Norvaline**) if locants for the side chain would otherwise be needed. Conjunctive names are not formed from amino acid stereoparents attached to ring systems; instead, such a combination is usually indexed at a systematic conjunctive name with an "*α*-amino" substituent. Derivatives of **Arginine,** $H_2NC(:NH)NH(CH_2)_3CH(NH_2)COOH$, substituted in the guanidino group are indexed as derivatives of **Ornithine,** $H_2N(CH_2)_3CH(NH_2)COOH$.

Histidine (R = H) is capable of existing in the following tautomeric forms:

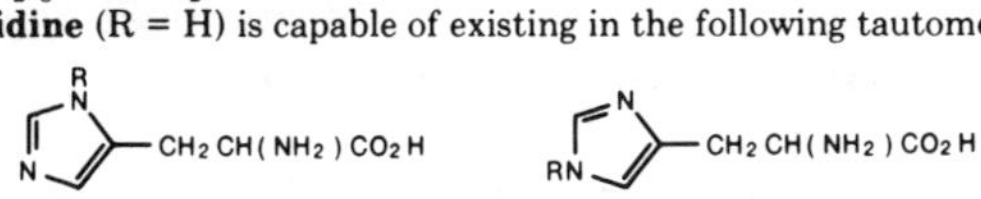

(*a*) (*b*)

When the tautomers are stabilized by substitution on a nitrogen atom of the ring, two isomers result. The side chain has been arbitrarily assigned to the 4-position of the ring and, if R = ethyl, the structures shown above are indexed:

(*a*) **L-Histidine, 3-ethyl-**
(*b*) **L-Histidine, 1-ethyl-**

It should be noted that while (*b*) conforms to the correct treatment for **1*H*-Imidazole,** the numbering of (*a*) violates the rules (**3*H*-Imidazole** is not used as a heading parent) and is adopted only for these special cases. A·1-substituent is assumed in indexing indefinite 1- or 3-derivatives of **Histidine.**

Esters of amino acid stereoparents with nonstereoparents are indexed at the amino acid names; e.g., **L-Alanine,** 4-carboxyphenyl ester. Esters with other stereoparents are indexed at the parent which represents the greatest number of highest functions. Usually only one index entry is made for an ester. Locants are used with ester terms when necessary.

Examples:

$EtO_2C(CH_2)_2CH(NH_2)CO_2Me$ — **L-Glutamic acid 5-ethyl 1-methyl ester**

$HO_2CNHCHMeCO_2Et$ — **L-Alanine, *N*-carboxy- 1-ethyl ester** (the locant "1" is cited for clarity, even though the *N*-ethyl ester is indexed as **L-Alanine, *N*-(ethoxycarbonyl)-** (¶ 185))

Amides of amino acids are indexed at systematic names, e.g., **Acetamide, 2-amino-** (not Glycinamide), but peptide amides (¶ 206) are indexed at *C*-terminal amino acid names, e.g., **Glycinamide, L-alanyl-.**

Anhydrides of amino acids are indexed at the preferred acid heading parent with "anhydride" or "anhydride with" terms in the modification. Cyclic anhydrides and lactones are indexed as heterocycles. Hydrazides are indexed at acid headings with "hydrazide" terms in the modifications. Molecular addition compounds of nonstereoparents with amino acids named as stereoparents receive preferred index names at the latter, but *Chemical Substance* and *Formula Index* entries will be found also at the other component(s). The preferred name of an addition compound of two or more stereoparents is chosen on the basis of the maximum number of highest functions in the stereoparent.

Beryllium, magnesium, aluminum, gallium, indium, thallium, and transition-metal salts of amino acids are indexed as coordination compounds (¶ 215) if sufficient information is presented, and the special stereochemical descriptors (¶ 203 III) for such compounds are then adopted; additional entries appear in the *Chemical Substance Index* for amino acid ligands other than **Alanine, Glycine,** and a few common substituted glycine derivatives. Other salts are indexed at such names as **Glycine,** sodium salt; **L-Phenylalanine,** methyl ester, hydrochloride.

Example:

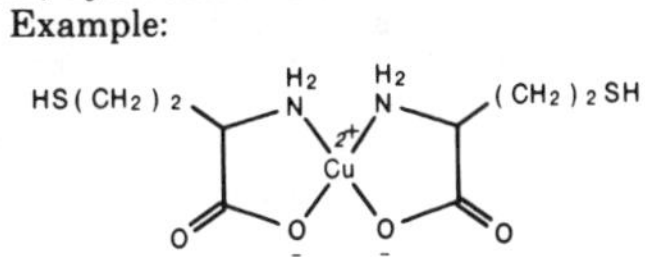

Copper, bis(L-homocysteinato-*N*,*O*)- (*T*-4)-
L-Homocysteine copper complex (additional *Chemical Substance Index* entry)

206. Peptides are generally named by use of amino acid stereoparents (¶ 205) and stereoparent radicals; the L-isomers are assumed for indexing purposes in the absence of contrary information. Trivial names of some peptides and all proteins are employed, and special "Cyclo" names are used for tetra- and higher cyclic peptides. The systematic ring names are covered by cross-references in the *Index Guide* or by additional entries in the *Chemical Substance Index* and *Index of Ring Systems.*

Linear peptides of two through twelve units are indexed at "peptide systematic names." The heading parent is the *C*-terminal amino acid, and all enclosing marks and locants for points of attachment of the amino acid radicals are cited. *N*-Derivatives of the radicals are expressed as substituents. Esters of carboxy groups of aspartic and glutamic acid residues are cited in the modification; *O*- and *S*-acyl derivatives of seryl, cysteinyl, etc., radicals are expressed by use of (*O*-acetyl-L-seryl), etc., radicals; when a hydroxyl- or mercapto-group-containing amino acid is the heading parent, the ester, e.g., "acetate," is cited in the modification.

Examples:

H-L-Met-L-Asp(NH_2)-L-Pro-D-Phe-L-Phe-OH

L-Phenylalanine, *N*-[*N*-[1-(N^2-L-methionyl-L-asparaginyl)-L-prolyl]-D-phenylalanyl]-

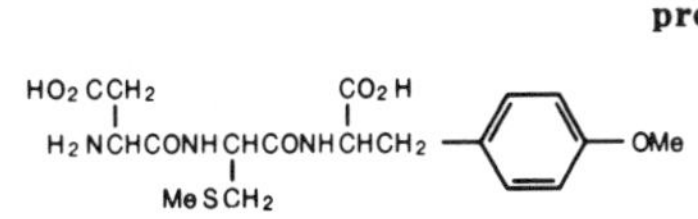

L-Tyrosine, *N*-(*N*-L-*α*-aspartyl-*S*-methyl-L-cysteinyl)-*O*-methyl-

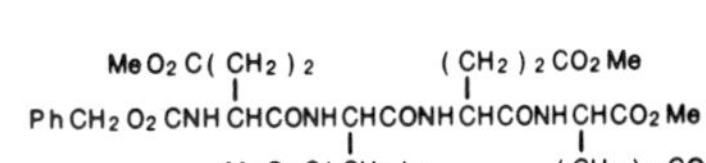

L-Glutamic acid, *N*-[*N*-[*N*-[*N*-[(phenylmethoxy)carbonyl]-L-*α*-glutamyl]-L-*α*-glutamyl]-L-*α*-glutamyl]- pentamethyl ester

Linear tridecapeptides and larger compounds are indexed at *amino-acid-sequence names*, similar to peptide systematic names (above) but with om=

ission of enclosing marks and locants for points of attachment of amino acid radicals; however, Greek-letter locants are cited for aspartyl and glutamyl radicals, and "N^6–" and "N^5–" are used with **Lysine** and **Ornithine**, respectively, and their radicals, when the attachment is not on the α-amino group (N^2–position). Substituents and functional derivatives are indexed as for peptide systematic names, and enclosing marks are used for compound and complex radicals by the usual rules of substitutive nomenclature.

Example:

L-Valine, *O*–**(1,1-dimethylethyl)**–*N*–**[(phenylmethoxy)carbonyl]–L-threonyl-*O*-(1,1-dimethylethyl)–L-seryl–L-α-aspartyl-D-α-glutamyl-*O*-methyl-L-tyrosyl-*S*-(phenylmethyl)-L-cysteinyl-4-fluoro-L-phenylalanyl-*N*-methyl-L-valylglycyl-*O*-acetyl-L-seryl-3-iodo-L-tyrosyl-N^6-formyl-L-lysyl-N^5-formyl-L-ornithylglycyl–**

Peptide amides receive amino-acid-sequence names with the *C*-terminal amide as heading parent.

Example:

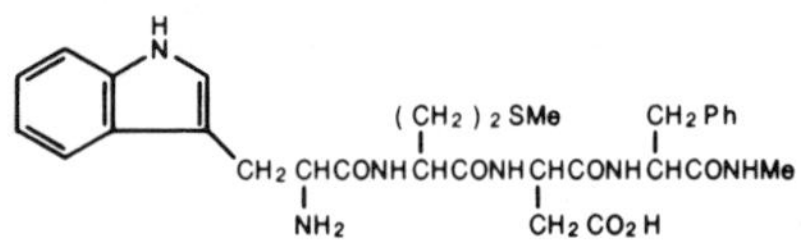

L-Phenylalaninamide, L-tryptophyl-L-methionyl-L-α-aspartyl-*N*-methyl–

The trivially named radicals "α-aminobutyryl" and "α,γ-diaminobutyryl" are converted to "2-aminobutanoyl" and "2,4-diaminobutanoyl" respectively in peptide (and peptide amide) names; amino-acid-sequence names are employed, with substituents placed ahead of the complete radical name, e.g., **L-Alanine, glycyl-L-valyl-L-threonyl-N^4-formyl-D-2,4-diaminobutanoyl–**.

Branched-chain peptides may comprise N^5– or N^6–derivatives of **Ornithine** or **Lysine,** respectively, or *O*– or *S*–derivatives of hydroxy or mercapto amino acids (heterodetic homomeric peptides). Both types are named as substituted derivatives of the parent linear peptides.

Examples:

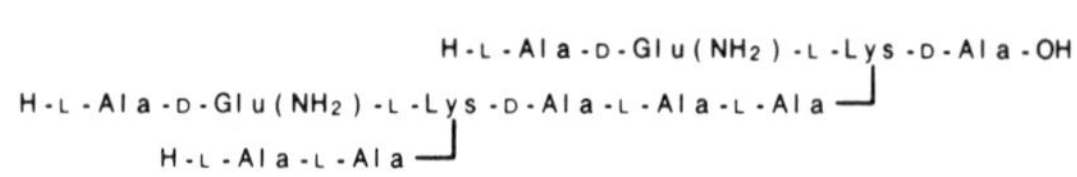

D-Alanine, *N*-[N^6-[*N*-[*N*-[N^6-(*N*-L-alanyl-L-alanyl)-N^2-(N^2-L-alanyl-L-glutaminyl)-L-lysyl]-D-alanyl]-L-alanyl]-L-alanyl]-N^2-(N^2-L-alanyl-D-glutaminyl)-L-lysyl]–

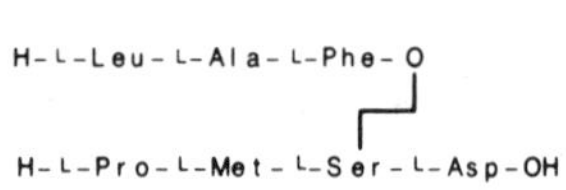

L-Aspartic acid, *N*-[*O*-[*N*-(*N*-L-leucyl-L-alanyl)-L-phenylalanyl]-*N*-(*N*-L-prolyl-L-methionyl)-L-seryl]–

Higher functions in peptides and peptide amides are expressed as substituents. When the *C*-terminal residue is not an amino acid or amino acid amide, but an acid-related compound such as an alcohol, aldehyde, or nitrile, the adjacent amino acid amide residue is adopted as heading parent, and the terminal group expressed as an *N*-(cyanomethyl), *N*-(2-oxoethyl), etc., substituent.

Cyclic di- and tripeptides are indexed at systematic heterocycle names, with cross-references at "Cyclo" names (see below).

Example:

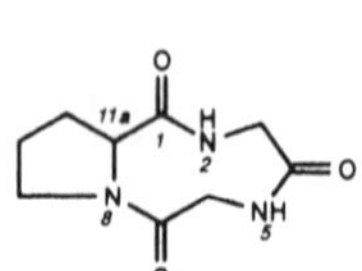

1*H*-Pyrrolo[1,2-*a*][1,4,7]triazonine-1,4,7-trione, octahydro- (cross-reference from Cyclo(glycylglycylprolyl))

Cyclic tetra- and higher peptides are assigned "Cyclo" names formed by citing the amino acid radicals (with substituents) as for an amino-acid-sequence name, placing it in enclosing marks, and prefixing it with "Cyclo". The order of citation is by lowest alphabetical order of *parent* amino acid residues (substituents are generally disregarded, but 2-aminobutanoyl and 2,4-diaminobutanoyl are ordered at "a" and "d," respectively). An additional systematic ring-system entry is made.

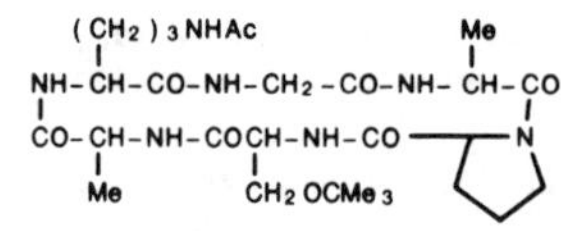

Cyclo[L-alanyl-N^5-acetyl-D-ornithylglycyl-L-alanyl-D-prolyl-*O*-(1,1-dimethylethyl)-L-seryl] (note that the substituent "acetyl" is disregarded in determining preferred citation of the peptide sequence; the lowest alphabetical sequence of parent-radical initial letters is "a, o, g, a, p, s")

Pyrrolo[1,2-*a*][1,4,7,10,13,16]hexaazacyclooctadecine cyclic peptide deriv. (additional *Chemical Substance Index* entry)

When ring formation has taken place by way of a side-chain peptide linkage, the acyclic peptide is named and the linkage expressed in the modification by a phrase of the type "cyclic (10→4)-peptide" in which the lactam linkage is denoted by amino-acid-residue locants (numbered from the *N*-terminus) with the acid end of the linkage cited first.

Peptide lactones are expressed by modification terms such as "ψ-lactone," in which the Greek letter indicates the size of the lactone ring (secondary alphabets, e.g., α_1, β_1, . . ., are employed for large rings). An additional cut appears, as usual, at the systematic ring system name. Peptides containing disulfide linkages are indexed at the reduced (cysteinyl-group-containing) form with terms such as "(2→2')-disulfide," "cyclic (1→6)-disulfide" and "(5→3')-disulfide with ..." in the modification.

N-(Peptidyloxy) derivatives of nitrogenous heterocycles are indexed at the heterocycle names, with additional *Chemical Substance Index* entries at the peptide names.

Depsipeptides contain ester linkages with hydroxy acids as well as amide (peptide) linkages with amino acids. Those with four or more acid residues, of which at least two are amino acid residues, and of which at least one hydroxy acid residue is nonterminal, are indexed by "depsipeptide nomenclature"; otherwise a systematic name is employed. A depsipeptide name is based on the *C*-terminal amino acid (a *C*-terminal hydroxy acid is expressed in the modification as a carboxyalkyl ester) and indexed by methods analogous to those for amino-acid-sequence names. Trivially named hydroxyacyl radicals (lactoyl, glycoloyl, etc.) are given systematic acyl radical names in this area.

Example:

HOCHMeCONHCHMeCO$_2$CHMeCONHCHMeCO$_2$H

L-Alanine, D-lactoyl-L-alanyl-D-2-hydroxypropanoyl–

Cyclic di- and tridepsipeptides are indexed as heterocycles. Cyclic tetra- and higher depsipeptides are indexed at "Cyclo..." names analogous to those for cyclic peptides.

Naturally occurring biologically active peptides with five or fewer amino acid units are indexed as described above. Those with six to fifty units are assigned the trivial names commonly used in the literature. Cross-references at peptide sequence names appear in the *Index Guide*, and illustrative diagrams appear after the structure and stereochemistry have been fully established. For cysteine-containing natural peptides, only the reduced form is illustrated (**Oxytocin** and **Vasopressin**, below, are exceptions).

Examples:

H-L-Arg-L-Pro-L-Pro-Gly-L-Phe-L-Ser-L-Pro-L-Phe-L-Arg-OH
1 2 3 4 5 6 7 8 9

Bradykinin (cross-reference from L-Arginine, N^2-[*N*-[1-[*N*-[*N*-[*N*-[1-(1-L-arginyl-L-prolyl)-L-prolyl]glycyl]-L-phenylalanyl]-L-seryl]-L-prolyl]-L-phenylalanyl]–)

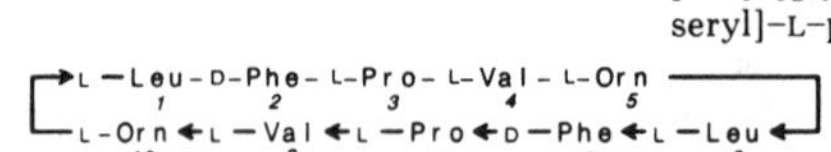

Gramicidin S (cross-reference from Cyclo(L-leucyl-D-phenylalanyl-L-prolyl-L-valyl-L-ornithyl-L-leucyl-D-phenylalanyl-L-prolyl-L-valyl-L-ornithyl))

H-L-Cys-L-Asp(NH$_2$)-L-Cys-L-Lys-L-Ala-L-Pro-
1 2 3 4 5 6

L-Glu-L-Thr-L-Ala-L-Leu-L-Cys-L-Ala-L-Arg-
7 8 9 10 11 12 13

L-Arg-L-Cys-L-Glu(NH$_2$)-L-Glu(NH$_2$)-L-His-NH$_2$
14 15 16 17 18

Apamin (reduced) (cross-reference from **L-Histidinamide,** L-cysteinyl-L-asparaginyl-L-cysteinyl-L-lysyl-L-alanyl-L-prolyl-L-α-glutamyl-L-threonyl-L-alanyl-L-leucyl-L-cysteinyl-L-alanyl-L-arginyl-L-arginyl-L-cysteinyl-L-glutaminyl-L-glutaminyl–)

Because the location of disulfide bonds in natural apamin has been determined the following cross-reference and synonym appear in the *Index Guide*:

Apamin (reduced)
cyclic(1→11),(3→15)-bis(disulfide)-see *Apamin*

Apamin (*apamin (reduced) cyclic(1→11),(3→15)-bis(disulfide)*)

Species variations are dealt with by citing the name of the species as a homograph definition after the heading; furthermore, when variations between species are minor, one substance is adopted as the standard, and the others named by reference to it. An illustrative diagram for the amino acid sequence of **Fibrinopeptide B (human)** is shown in the *Chemical Substance Index*; it has L-phenylalanine at position 10 and L-serine at position 12. The *Index Guide* has the following cross-reference to indicate where a species variant is indexed.

Fibrinopeptide B (orangutan)
See *Fibrinopeptide B (human), 10-L-leucine-12-glycine-*

Another set of cross-references can be found at **β-Melanocyte-stimulating hormone (pig).** Sometimes, natural peptides with *different* trivial names can be related in this way; thus, a cross-reference leads from Kallidin to **Brady=kinin,** N^2-**L-lysyl-.**

When species variations affect one-half or more of the amino-acid residues, each variant is an independent stereoparent with its own illustrative diagram; e.g., **Calcitonin (pig reduced); Calcitonin (human reduced).**

Examples have been cited above of how replacement of one amino acid residue by another is indicated at a trivially named peptide heading. The names of the new amino acids are cited in numerical (not alphabetical) order as substituents; e.g., **1-34-Gastrin I (pig), 2-L-serine-10-L-alanine-**. In addition, the chain of the reference compound can be *extended* at either end, either by substitution of the *N*-terminal α-amino group or by citing additional terms with locants derived from the highest present by addition of "a," "b," etc. In a similar manner, *insertion* of amino acid units between existing ones is indicated by "*endo*" terms with "a" locants derived from the lower of the two neighboring units (e.g., "3a" indicates insertion between "3" and "4"). Removal of a unit is indicated by a "de" term such as "4-desulfo" to indicate removal of only a modifying group in a unit, or "6-de-L-glutamic acid" to express removal of an entire unit.

Examples:

Bradykinin, N^2-**L-lysyl-** (*N*-terminal extension)

Glucagon (pig), *N*-**acetyl-** (*N*-terminal extension)

Bradykinin, 9a-L-valine-9b-L-lysine- (*C*-terminal extension)

Fibrinopeptide B (human), 14a-D-serinamide- (*C*-terminal extension)

Bradykinin, 6a-*endo***-L-alanine-** (insertion)

Caerulein, 4-desulfo- (removal of modifying group)

Caerulein, 4-de(*O***-sulfo-L-tyrosine)-** (removal of unit)

Combinations of various types of structure modifiers are indicated by citing them in numerical order. When the modifying operations total at least one-half the number of units in the original parent, a regular peptide name is employed.

The prefix "*enantio*" denotes reversal of configuration of all amino acid residues; thus, *enantio*-**Bradykinin** contains only the D-forms of the amino acid residues, but in the same sequence as in **Bradykinin.** The new parent is used if more than half of the units have been replaced by their enantiomers; unchanged units are expressed by replacement nomenclature. The prefix "*retro*" denotes a reversal of the amino acid sequence; thus *retro*-**Gramicidin S** has the structure of **Gramicidin S** (see diagram above) but with the acid-amine linkages running from right to left. The combined prefix "*enantio-retro-*" indicates reversed configuration and reversed sequence.

In **Angiotensin I** and **Angiotensin II,** the amino acid in the 5-position, if known, is cited in the inverted part of the heading, e.g., **Angiotensin I, 5-L-isoleucine-.**

Oxytocin and **Vasopressin** derivatives in the disulfide form are named at these headings with replacement, etc., terms as usual, but reduced analogs are indexed at amino acid sequence names, not at **Oxytocin (reduced)** and **Vasopressin (reduced).** The amino acid at position 8 of Vasopressin is always specified if known.

α^{1-39}-**Corticotropin (pig)** is a stereoparent reference compound (see the diagram in the *Chemical Substance Index*) on the basis of which minor var=iants, e.g., α^{1-39}-**Corticotropin (human),** are named ("α" indicates that this was the first principle isolated; the superscripts give the range of amino acid units). Fragments containing more than twelve sequential units receive names such as α^{1-24}-**Corticotropin;** α^{11-39}-**Corticotropin.** An amino acid sequence name is employed for fragments of fewer than twelve units, but cross-refer=ences at the alternative names, e.g., α^{1-10}-**Corticotropin,** appear in the *Index Guide.*

Insulin contains two peptide chains, an A-chain of 21 amino acid units and a B-chain of 30 units, connected at two points by disulfide bridges. The species reference compound is **Insulin (ox).** Cross-references appear in the *Index Guide* at the amino-acid-sequence names of the two chains. Replacement names for insulins that retain the disulfide bridges and have fewer than 11 and 15 residue changes in the A- and B-chains, respectively, are formed in the usual way. Locants have superscript letters to indicate to which chain they belong; e.g., Insulin (human) is cross-referred to **Insulin (ox),** 30^B-**L-threo=nine-.** The reduced forms of the individual insulin chains are also stereo=parents; internal disulfide bonds are expressed in the modification, and cross-references to the bovine reference compounds are made as usual.

Example:

Insulin (sheep-A reduced)
See *Insulin (ox-A reduced), 9-glycine-*

Actinomycins are a class of natural products that contain peptide chains attached to non-peptide moieties. Actinomycin D (sometimes called C_1) is the reference compound, on the basis of which others are named.

Example:

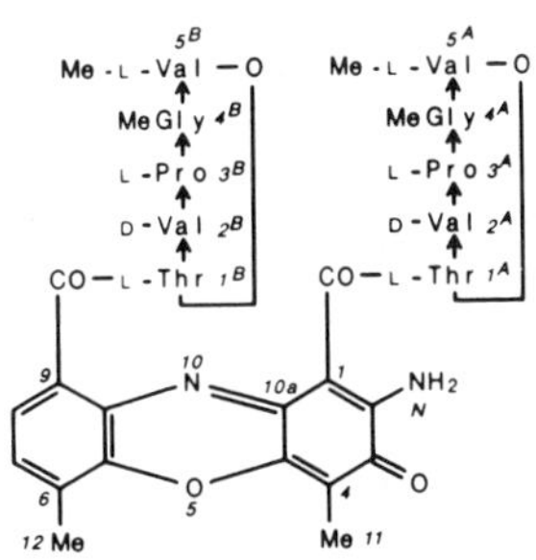

Actinomycin D (cross-references from 3*H*-Phenoxazine-1,9-dicarbox=amide, 2-amino-*N,N'*-bis[hexade=cahydro-2,5,9-trimethyl-6,13-bis=(1-methylethyl)-1,4,7,11,14-pentaoxo-1*H*-pyrrolo[2,1-*i*]=[1,4,7,10,13]oxatetraazacyclo=hexadecin-10-yl]-4,6-dimethyl-3-oxo-; and 1*H*-Pyrrolo[2,1-*i*]=[1,4,7,10,13]oxatetraazacyclo=hexadecine)

The "Me-L-Val" symbol denotes *N*-methyl-L-valine and "Me-Gly" rep=resents *N*-methylglycine (sarcosine). Substitution in the phenoxazine ring is expressed (alphabetically) before amino-acid replacement, etc., (numerically) in the peptide moiety.

Examples:

Actinomycin D, 8-bromo-7-chloro-

Actinomycin D, 2-deamino-2-hydroxy-3^A**-D-proline-**4^A**-D-nor=leucine-**

Removal of a methyl group is indicated by "nor-" or "dinor" prefixes, e.g., **11,12-Dinoractinomycin D.** When one or other or both of the peptide lactone rings are opened, the stereoparents **Actinomycin-**5^A**-oic D acid; Actinomycin-**5^B**-oic D acid;** and **Actinomycindioic D acid** are formed. Derivatives are named as for Actinomycin D.

Homopolymers of amino acids are indexed by polymer procedures (¶ 122); e.g., L-**Alanine,** homopolymer. The structural repeating unit is indexed at **Poly[imino(1-methyl-2-oxo-1,2-ethanediyl)],** (*S*)-. The stereochemical assumptions are the same as for the monomeric amino acids. Peptides of indefinite molecular weight with repeating peptide sequences are named as homopolymers at the peptide monomer names and at structural repeating unit names.

207. Proteins are arbitrarily defined for indexing purposes as natural peptides containing more than fifty amino acid units, but some exceptions are made. A protein for which a complete amino acid sequence has been reported is indexed as a chemical substance at the trivial name with auxiliary infor=mation, e.g., the species name, cited in parentheses in the heading after the name.

Newly identified components of a specifically named protein for which the literature provides no adequate method of differentiation are denoted by the numerals 2, 3, etc., appended after the auxiliary information in parentheses.

Protein fragments are named by prefixing the trivial name with two num=bers relating to the range of units represented; e.g., **1-124-Somatotropin (ox reduced).** Replacement names, as for peptides, are used for analogs.

208. Carbohydrates. Carbohydrate stereoparents are defined for in=dexing purposes as polyhydroxy acids, aldehydes, ketones, alkanes, and their derivatives, with a skeleton of five or more carbon atoms, more than half of which must be attached to oxygen (or another chalcogen) or nitrogen, and at least one of these attachments must be by a single bond to a nonterminal carbon atom. At least half of the nonterminal carbon atoms must be asym=metric. Carbohydrate stereoparents existing in the cyclic hemiacetal form must contain at least three asymmetric carbon atoms, including the anomeric carbon atom. Except for uronic and ulosuronic acids, the open-chain is numbered to give the highest function the lowest possible locant. Illustrative structural diagrams related to current index entries will be found in the *Chemical Sub=stance Index* for open-chain forms and common cyclic forms of carbohydrate stereoparents. Stereoparents representing the following carbohydrates and their derivatives are employed as heading parents:

Pentoses:	**Arabinose, Lyxose, Ribose, Xylose.**
Hexoses:	**Allose, Altrose, Galactose, Glucose, Gulose, Idose, Mannose, Talose.**
2-Hexuloses:	**Fructose, Psicose, Sorbose, Tagatose.**
Acids:	**Ascorbic acid (¶ 224), Muramic acid, Neuraminic acid.**

Monosaccharides that exist in the open form are indexed at the stereo=parents. Higher functions are expressed as substituents or in the modification; radicals from open-chain sugars are not used. Derivatives of the cyclic forms are indexed at the highest function if a glycosyl radical can be used to express the sugar, which is ranked as a polyhydric alcohol unless a higher function, e.g., a nonglycosidic ketone or aldehyde, is expressed as a suffix.

When a second stereoparent is represented in a structure, the parent of higher functionality is chosen in indexing; if this cannot be done, because of lack of a carbohydrate radical or other reason, an additional entry of a general nature appears in the *Chemical Substance Index* (see ¶ 202).

Configurational descriptors, D and L, are placed before stereoparent names, including monosaccharide semisystematic names (see below); DL is employed for racemic mixtures. Examples: D-**Glucose**; L-*arabino*-**Hexose, 2-deoxy-**; DL-**Ribose.** In the Fischer projections of the open forms, the carbon skeleton is displayed vertically, with locant "1" at the top. The D-series can then be recognized by the hydroxyl group to the *right* of the bottom nonterminal asymmetric carbon atom.

Examples:

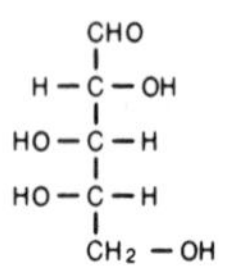

L-Arabinose

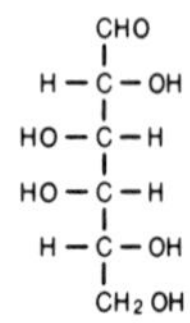

D-Galactose

Configurational prefixes, derived from trivial aldose names, are used in semisystematic carbohydrate nomenclature; for one asymmetric carbon atom, e.g., >CHOH, >$CHOCH_3$, >$CHNH_2$, *glycero* is the prefix used (with D or L as described above); with two, *erythro* and *threo*; with three, *arabino*, *lyxo*, *ribo*, and *xylo*; with four, *allo*, *altro*, *galacto*, *gluco*, *gulo*, *ido*, *manno*, and *talo*. (The centers need not be on neighboring carbon atoms.) For more than four consecutive centers, combinations of these terms are used.

Examples ("X" is the group with the lowest-numbered carbon atom):

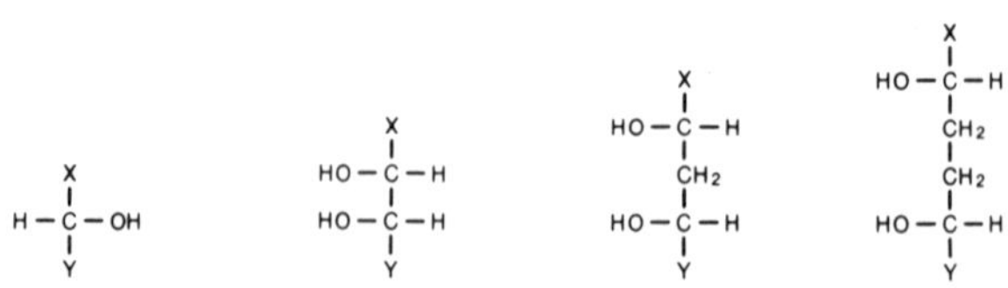

D-*glycero*- L-*erythro*-

In choosing a monosaccharide stereoparent name (when such a choice is necessary), (*a*) the oxo (or higher) function is numbered low (except the carboxyl group in uronic and ulosuronic acids); (*b*) the name appearing earliest alphabetically is chosen; (*c*) "D" is preferred to "L"; (*d*) the anomeric prefix "α" is preferred to "β"; (*e*) lowest locants for substituents are employed; (*f*) the lowest locant is used for the first-cited substituent.

Systematic carbohydrate names, employed for monosaccharides other than those with trivial names, are based on stem names which express the size and function; the stem names for open-chain aldoses of five or more carbon atoms are **Pentose, Hexose, Heptose, Octose, Nonose,** etc.; corresponding ketose stems are **Pentulose, Hexulose,** etc. (Additional terms are used to indicate ring size if necessary; see below.) The stem names are preceded by stereochemical descriptors. 2-Hexuloses have trivial names (see above); other ketoses are named systematically, e.g., L-*erythro*-**2-Pentulose.** For meso-forms, the D- and L- configurational symbols are not needed; e.g., *erythro*-**3-Pentulose.** Diketoses have names of the type D-*threo*-**2,4-Hexodiulose**; ketoaldoses have "-osulose" names such as D-*ribo*-**Hexos-3-ulose**; dialdoses have names such as D-*gluco*-**Hexodialdose**; chalcogen analogs are indexed by use of thio, etc., prefixes, e.g., D-**Glucose, 1-thio-**; replacement of hydroxyl by hydrogen is denoted by "deoxy" terms.

Substitution on carbon with prior removal of a hydroxyl group requires citation of "deoxy" and the substituent prefix in alphabetic order, e.g., D-**Glucose, 2-(acetylamino)-2-deoxy-.** When the existing hydrogen is substituted on a carbon atom already carrying an oxygen (or other chalcogen) or nitrogen group in the open or cyclic forms, the italic capital letter *C* is employed, e.g., D-**Ribose, 3-*C*-(nitromethyl)-.** When both the hydrogen and hydroxyl are replaced by nonchalcogen, nonnitrogen substituents, the stereochemistry is expressed by a Sequence Rule descriptor at the systematic saccharide name, e.g., D-*ribo*-**Hexose, 3-bromo-3-chloro-3-deoxy-,** (3*S*)-.

Substitution of hydrogen on a hydroxyl group of a carbohydrate is denoted by "*O*" locants (or by "*S*", etc., if chalcogen replacement has also taken place).

Examples:

D-**Mannose, 2,3,4,5,6-penta-*O*-methyl-**

D-**Glucose, 2-*S*-ethyl-2-thio-**

D-**Glucitol, 1-*O*-ethyl-1-*C*-(phenylamino)-**

Cyclic acetals employ bivalent radicals with "*O*" terms, e.g., D-**Ribose, 2,3:4,5-bis-*O*-(1-methylethylidene)-.**

Oximes and hydrazones are expressed in the modification, osazones similarly as dihydrazones of osuloses. Hydrates, acyclic acetals, hemiacetals and esters are also named in the modification.

Examples:

D-**Xylose**
diethyl mercaptal

D-**Xylose**
2,3-diacetate 4,5-dibenzoate

Cyclic hemiacetal forms of monosaccharides are indexed by inclusion in the name of a term indicating the ring size, if known; five, six, and seven-membered rings, are denoted by "-furanose," "-pyranose," and "-septanose" suffixes. Two anomers, named "α" and "β," result on ring formation, and these Greek letters are employed as anomeric prefixes, which are placed ahead of the configurational descriptors (D or L) in the name.

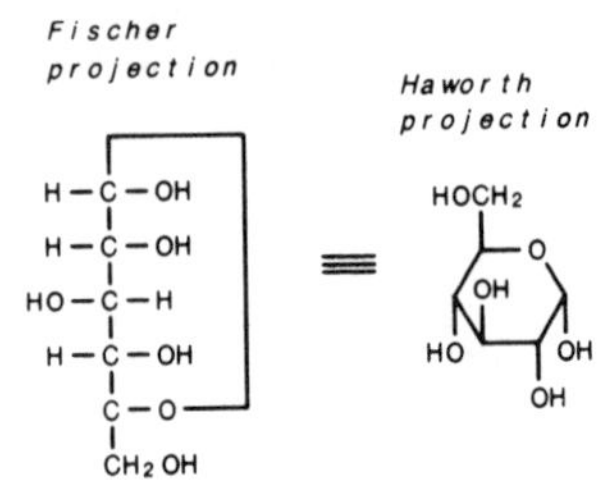

α-D-**Glucopyranose**

The following examples show the derivation of names for ketoaldoses when the ring closure involves (*a*) the aldehyde, and (*b*) the keto function.

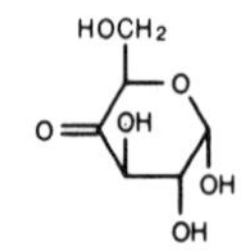

(*a*) α-D-*xylo*-**Hexopyranos-4-ulose**

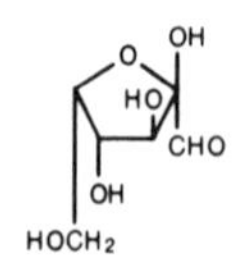

(*b*) α-L-*xylo*-**Hexos-2-ulo-2,5-furanose**

More complicated cases, e.g., cyclic forms of dialdoses and higher ketoses, are named in accordance with rules (lowest locants for the keto group, alphabetically preferred configurational prefixes, etc.) already described. Locants, e.g., "-6,10-" are inserted before the ring-size descriptor in the name, and "α" or "β" is also cited if the anomeric carbon is numbered lower than the reference carbon atom in the ring being formed, otherwise *R* or *S* is used.

Nitrogen analogs of cyclic sugars are named similarly to the acyclic forms, e.g., β-D-**Ribofuranose, 4-amino-4-deoxy-** (the locant of the highest-numbered carbon atom in the ring is cited).

Intramolecular *anhydrides* are denoted by terms such as "1,5-anhydro" in which the locants define the pair of hydroxyl groups involved. The sulfur analogs are named with "dideoxy-epithio" terms. Intermolecular anhydrides are polysaccharides (below); intermolecular dianhydrides are named by citing the word "dianhydride" after the name(s) of the parent monosaccharides (an aldose precedes a ketose); two pairs of locants define the positions of the anhydride linkages.

Example:

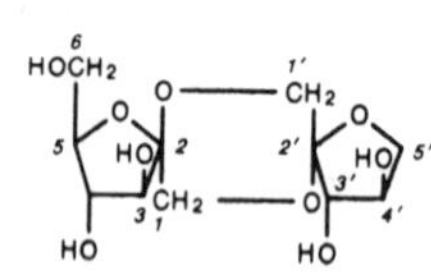

β-D-**Fructofuranose-**β-D-*threo*-**2-pentulofuranose 1,2':2,1'-dianhydride**

1,6,9,13-Tetraoxadispiro[4.2.4.2]-tetradecane
β-D-fructofuranose β-D-*threo*-2-pentulofuranose 1,2':2,1'-dianhydride deriv. (additional *Chemical Substance Index* entry for ring system)

Carbohydrate *acids* are named by characteristic suffixes, as indicated below:

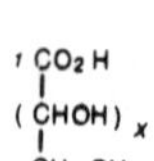

-aldonic acid

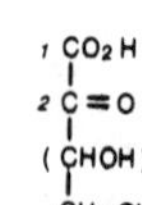

-ulosonic acid

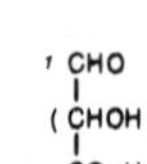

-uronic acid

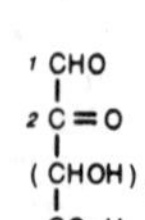

-ulosuronic acid

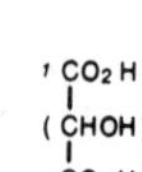

-aldaric acid

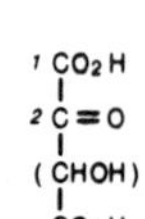

-ulosaric acid

The stems of aldaric, aldonic and uronic acid names are derived from trivial (common) aldose names whenever possible. Acids containing a keto function are named from the systematic stem names. Acid halide, amide, and nitrile names are derived as usual. Esters, hydrazides, salts, lactones, and lactams are expressed in the modification.
Examples:

D-**Gluconoyl chloride**

L-**Ribonic acid**
γ-lactone

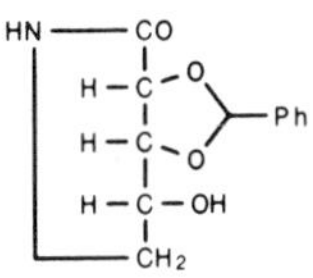

D-**Ribonic acid, 5-amino-5-deoxy-2,3-*O*-(phenylmethylene)-**
δ-lactam

1,3-Dioxolo[4,5-*c*]pyridine
D-ribonic acid deriv.
(additional *Chemical Substance Index* entry for ring system)

The trivial names **Muramic acid** and **Neuraminic acid** are used for the structures shown below; the acyl radicals muramoyl and neuraminoyl, and the glycosyl radicals muramosyl and neuraminosyl, are also employed in indexing. (**Isomuramic acid** is treated similarly; it is the muramic acid (*S*)-epimer about the carbon atom bearing the carboxyl group.)
Examples:

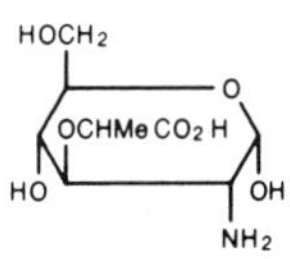

α-**Muramic acid**
(the (*R*)-epimer)

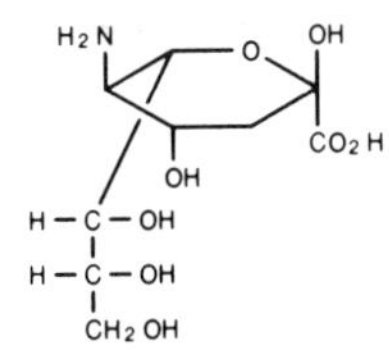

β-**Neuraminic acid**

Glycosyl radicals derived from trivially and systematically named sac= charides by removal of the anomeric (hemiacetal) hydroxyl group have the suffix "-osyl," e.g., β-D-mannopyranosyl (if the ring size is known), D-mannosyl, α-D-*arabino*-hexopyranos-2-ulos-1-yl. Uronic acid radicals are derived by replacing the "-ic acid" by "-osyl," e.g., β-D-glucopyranuronosyl; acid derivatives, e.g., uronamides, yield "-uronamidosyl" radicals, etc.

Glycosyl halides, isocyanates, etc., without functions higher than a polyol in the unsubstituted saccharide have names containing the "-osyl" radical; otherwise the heading parent expresses the higher function.
Examples:

α-D-**Mannopyranosyl bromide**

β-D-**Glucopyranuronic acid, 1-chloro-1-deoxy-**

Glycosylamines are named by similar principles.
Examples:

α-D-**Glucopyranosylamine**

α-D-**Glucopyranuronamide, 1,1′-iminobis[1-deoxy-**

Glycosides are mixed acetals derived by replacement of the hydrogen atom on the anomeric (glycosidic) hydroxyl group of a saccharide by a group derived from an alcohol or phenol. Glycoside heading parents, e.g., α-D-**Ribo= furanoside,** are employed. In the inverted part of the heading, the radical derived from the alcoholic or phenolic aglycone is cited, followed by a space. Then any substituents of the saccharide are expressed in the usual way. When functions higher than the saccharide (which is normally a polyol) are present in the aglycone, a heading parent which expresses the greatest number of highest functions is chosen, provided that an appropriate glycosyl radical name can be employed to express the saccharide as a substituent. When the fun= tionality is equal in the two portions of the molecule, the stereoparent is preferred. Glycoside parents are also derived from uronic acids and ulosonic acids, e.g., α-D-**Mannopyranosiduronic acid**; α-D-*arabino*-2-**Hexulo= furanosidonic acid.**
Examples:

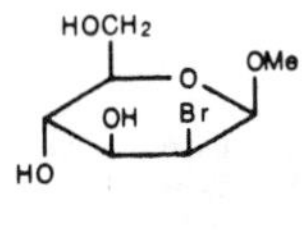

β-D-**Mannopyranoside, methyl 2-bromo-2-deoxy-**

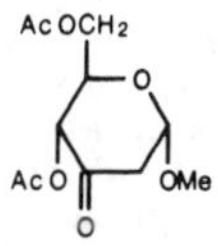

α-D-*erythro*-**Hexopyranosid-3-ulose, methyl 2-deoxy-**
diacetate

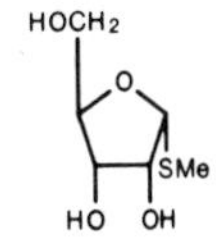

α-D-**Ribofuranoside, methyl 1-thio-**

Glycosides of heterocycles are indexed at the saccharide headings unless the heterocyclic aglycone contains a higher function or the glycosyl attachment is at a nitrogen atom. *C*-Glycosides (glycosyl derivatives with carbon-carbon bonds) of ring systems are indexed at the higher function.
Examples:

D-**Glucitol, 1,5-anhydro-1-*C*-phenyl-**
(1*R*)-

4*H*-**1-Benzopyran-4-one, 8-β-D-glucopyranosyl-2-phenyl-**

A naturally occurring glycoside is indexed at the systematic (or stereo= parent) name if the structure, including the ring-size of the saccharide, is known, whether or not all the stereochemistry has been elucidated. (A glu= coside is assumed to be the β-D-glucopyranoside, a "rhamnoside" the 6-deoxy-α-L-mannopyranoside.)

N-Glycosyl derivatives of heterocycles are usually indexed at the hetero= cycle names (see also Nucleosides and Nucleotides, ¶ 210), but a saccharide containing an expressed aldehyde or higher functional suffix is preferred.
Examples:

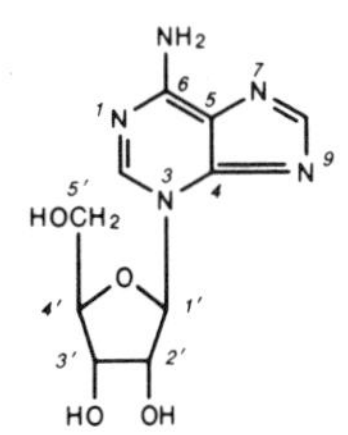

3*H*-**Purin-6-amine, 3-β-D-ribo= furanosyl-**

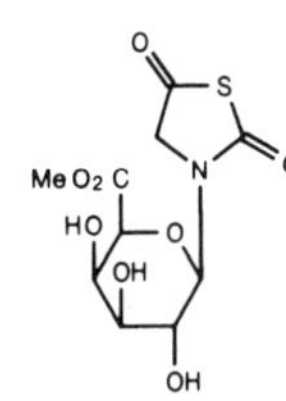

β-D-**Galactopyranuronic acid, 1-deoxy-1-(2,5-dioxo-3-thiazoli= dinyl)-**
methyl ester

Alditols are polyhydric alcohols derived mainly by reduction of aldehydes. They are named by changing the suffix "-ose" to "-itol." Since the end groups are alike, they are more symmetrical than saccharides; the preferred heading parent is the one in which the trivial name (or the configurational prefix (*erythro*, etc.) of a semisystematic name) and then the configurational de= scriptors (D and L) expresses the lowest alphabetic order. Hence Gulitol is cross-referred to **Glucitol,** and D-*glycero*-L-*gulo* is preferred over D-*glycero*-D-*ido*. Lowest locants for substituents are considered only after the heading parent has been determined.
Example:

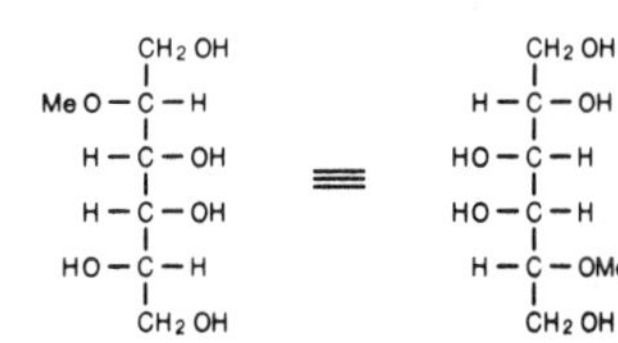

D-**Galactitol, 5-*O*-methyl-** (not L-Galactitol, 2-*O*-methyl-)

The *meso* alditols (**Ribitol, Xylitol, Allitol,** and **Galactitol**) do not require D or L unless they are substituted unsymmetrically, in which case the D-form is chosen. Radical names are not formed from alditols (or from open-chain saccharides). Higher chemical functions than the alditol are expressed as substituents of the alditol parent.

Branched-chain monosaccharides are generally named as derivatives of the linear saccharides, selecting first the highest function and then the longest chain, but the trivial name **Apiose** is used.
Example:

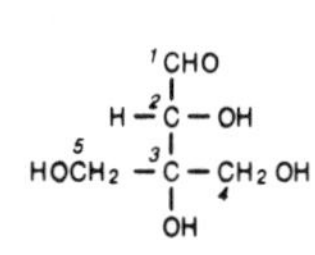

D-**Apiose**

Olefinic monosaccharides are expressed by "en" in the stereoparent. De= struction of the anomeric center leads to an anhydro unsaturated alditol with the multiple bond assigned the lowest possible locant after the alphabetic choice (D before L, α before β) has been made. Geometrical isomerism about

a double bond, when known, is indicated by *E* and *Z* descriptors.
Examples:

D-*threo*-**Hex-2-enose, 2-(acetyl=amino)-2,3-dideoxy-**
(*Z*)-

9*H*-**Purin-6-amine, 9-(2,3-dideoxy-** *β*-D-*erythro*-**hex-2-enopyranosyl)-**

D-*xylo*-**Hex-1-enitol, 1-*C*-amino-1,4-anhydro-2-deoxy-2-(phenylamino)-**

Oligosaccharides are defined for indexing purposes as compounds which, when hydrolyzed, afford not more than ten molecules of monosaccharides. Above this limit, with a few exceptions, polysaccharide nomenclature (below) is employed. In nonreducing oligosaccharides, all the units are linked by glycosidic bonds; reducing oligosaccharides have at least one unsubstituted anomeric hydroxyl group. Oligosaccharides of known structure are cross-re=ferred to carbohydrate names which express the monosaccharide content. The open-chain form is used in naming reducing oligosaccharides in the absence of anomeric prefixes or contrary derivative information.

Trisaccharides and higher members are named by use of "arrow" nomen=clature, in which arrows and locants indicate the direction and position of linkages from glycosylic carbon atoms to hydroxylic carbon atoms. The re=ducing monosaccharide unit is named as the stereoparent. Other substituents of the parent are named last. Cross-references at oligosaccharide trivial names are found in the *Index Guide*.
Examples:

Lactose.	See D-*Glucose, 4-O-β-D-galactopyranosyl-*
Sucrose.	See *α-D-Glucopyranoside, β-D-fructofuranosyl*
Cellotriose.	See D-*Glucose, O-β-D-glucopyranosyl-(1→=4)-O-β-D-glucopyranosyl-(1→4)-*
Stachyose.	See *α-D-Glucopyranoside, β-D-fructofuranosyl O-α-D-galactopyranosyl-(1→6)-O-α-D-galactopyranosyl-(1→6)-*

Polysaccharides are often indexed at common names, e.g., **Agar, Amylo=pectin, Cellulose, Starch.** Homoglycans (polysaccharides derived from one type of sugar residue) are named by converting the "-ose" suffix of an aldose or ketose, or the "-ic acid" of a glycuronic acid, to "-an," e.g., *β*-D-**Mannan,** (1→4)-; L-**Arabinan**; D-**Glucuronan.** Heteroglycans contain more than one kind of saccharide; the name, e.g., **Glucuronoarabinoxylan,** expresses the saccharide with the higher chemical function first; if the functions are alike, longer chain and then alphabetic order decide. Esters and most other deriv=atives are cited in the modification. Synthetic homopolymers and copolymers are indexed by polymer nomenclature, e.g., L-**Arabinose,** homopolymer.

209. Cyclitols are cycloalkanes in which a hydroxyl group is attached to each of three or more ring atoms. Cyclitols of the cyclohexane series constitute the *inositols*; those that contain at least five asymmetric centers in the ring, of which at least three are directly bonded to oxygen (or another chalcogen) or nitrogen, are given **Inositol** stereoparent names; the others are named systematically as cyclohexanepolyols. Relative stereochemistry is expressed by special italicized configurational prefixes. The eight **Inositol** stereoparents, arranged according to the number and position of hydroxyl groups on the same side of the cyclohexane ring (as indicated by the numerical locants, which are *not* used in indexing) are:

cis-**Inositol**	(1,2,3,4,5,6)
epi-**Inositol**	(1,2,3,4,5)
allo-**Inositol**	(1,2,3,4)
myo-**Inositol**	(1,2,3,5)
muco-**Inositol**	(1,2,4,5)
neo-**Inositol**	(1,2,3)
chiro-**Inositol**	(1,2,4)
scyllo-**Inositol**	(1,3,5)

Choice of a preferred index name for an inositol derivative depends on (*a*) the alphabetically preferred configurational prefix in the order: *allo, chiro, cis, epi, muco, myo, neo, scyllo*; (*b*) the preferred configurational descriptor (D over L); (*c*) the lowest set of locants for substituent prefixes (¶ 137). The configurational descriptor D or L may be determined by numbering the hy=droxyl groups (or replacement groups) which lie on *one side* of the ring be=ginning at each possible "position 1." If the groups, e.g., the 1,2,4-groups in *chiro*-**Inositol,** are up when numbering clockwise, or down when numbering counterclockwise, the descriptor is L; if the groups are up in numbering counterclockwise or down in numbering clockwise, the descriptor is D. The D-stereoparent is chosen regardless of high locants for substituent prefixes. The D and L descriptors are used only when optical activity is present and can be determined from the original document. DL is used when the author indicates a racemic mixture.
Example:

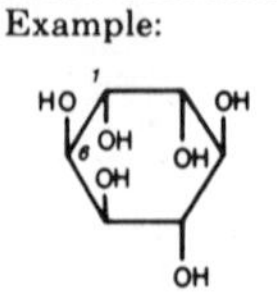

D-*chiro*-**Inositol**

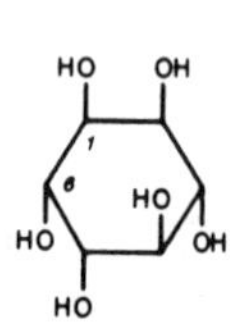

L-*chiro*-**Inositol**

Common names for inositols, e.g., **Scyllitol,** are cross-referred in the *Index Guide* to preferred index names. Ethers are named as *O*-derivatives; chalcogen analogs are expressed by "thio," etc., in the inverted part of the heading. Esters are expressed by modification terms, e.g., "hexaacetate." Glycosides are in=dexed at the **Inositol** stereoparent with *O*-glycosyl substituent prefixes. 1,2,3,4,5-Cyclohexanepentols (quercitols) are indexed at the alphabetically preferred **Inositol** parent by use of a "deoxy" prefix, and D or L determined as described above, e.g., D-*chiro*-**Inositol, 2-deoxy-** (not D-*chiro*-Inositol, 5-deoxy; not D-*muco*-Inositol, 2-deoxy-; not L-*muco*-Inositol, 1-deoxy-). Replacement of one, two, or three hydroxyl groups, with retention of config=uration, is expressed by "deoxy" terms and appropriate radicals. The capital italic letter *C* is used to denote replacement of hydrogen on a carbon atom to which a hydroxyl group is also attached, e.g., *scyllo*-**Inositol, 1-*C*-methyl-.**

Inosamines and their *N*-acyl (including carboxy) derivatives are usually indexed at such names as *neo*-**Inositol, 2-amino-2-deoxy-**, but **Streptamine** is the stereoparent for 1,3-diamino-1,3-dideoxy-*scyllo*-inositol and derivatives.
Example:

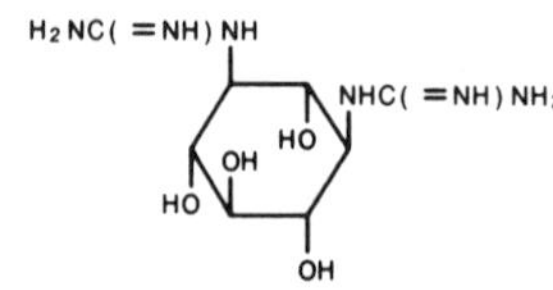

Streptamine, *N,N'*-bis(amino=iminomethyl)-

Inositols containing functions higher than hydroxy (except *N*-acyl and *N*-carboxy inosamines) are named systematically.
Example:

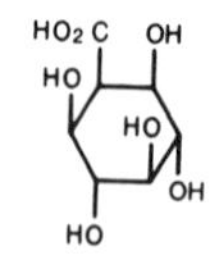

Cyclohexanecarboxylic acid, 2,3,4,5,6-pentahydroxy-
(1*α*,2*α*,3*β*,4*α*,5*β*,6*α*)-

Inosose is the stereoparent for 2,3,4,5,6-pentahydroxycyclohexanone ste=reoisomers.
Example:

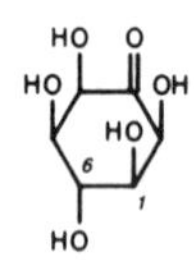

epi-**3-Inosose** (not *muco*-3-Inosose)

210. Nucleosides and **Nucleotides** are hydrolytic products of nucleic acids; chemically, nucleosides are *N*-glycosyl derivatives of heterocyclic bases, principally purine and pyrimidine, and nucleotides are esters of nucleosides with phosphoric acid and polyphosphoric acids.

The bases are named systematically, with the tautomerism resolved for indexing purposes in favor of highest expressed function, etc. (¶ 122).

Trivial Name	*Index Name*
Adenine	**1*H*-Purin-6-amine**
Cytosine	**2(1*H*)-Pyrimidinone, 4-amino-**
Guanine	**6*H*-Purin-6-one, 2-amino-1,7-dihydro-**
Hypoxanthine	**6*H*-Purin-6-one, 1,7-dihydro-**
Thymine	**2,4(1*H*,3*H*)-Pyrimidinedione, 5-methyl-**
Uracil	**2,4(1*H*,3*H*)-Pyrimidinedione**
Xanthine	**1*H*-Purine-2,6-dione, 3,7-dihydro-**

Trivial names for seven common nucleosides and the related nucleotides are employed as stereoparents; these nucleosides are:

Purine derivatives	*Pyrimidine derivatives*
Adenosine	**Cytidine**
Guanosine	**Thymidine**
Inosine	**Uridine**
Xanthosine	

The seven stereoparents (above) are heading parents for derivatives, in=cluding *N*-acyl derivatives, e.g., **Adenosine, *N*-acetyl-.** When a chemical function higher than carboxamide (¶ 106) is present, an index heading parent which expresses it is chosen, unless lack of a suitable radical precludes ex=pression of the nucleoside residue as a substituent prefix (this occurs when the higher function is attached to the saccharide at any position other than "1'"). Replacement of oxygen in the saccharide moiety or on the heterocycle by sulfur is expressed by a "thio" prefix, and the regular methods of carbohydrate nomenclature (¶ 208) are employed for other saccharide modifications, in=

cluding ester formation other than phosphates.

Examples:

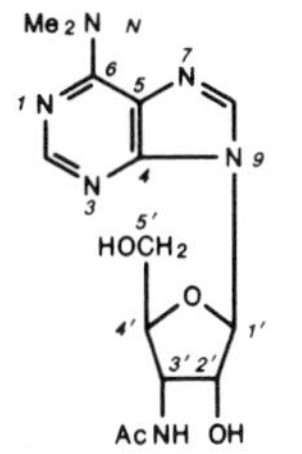

Adenosine, 3'-(acetylamino)-3'-deoxy-*N*,*N*-dimethyl-

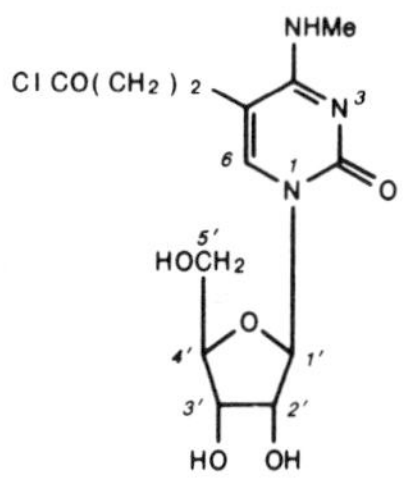

5-Pyrimidinepropanoyl chloride, 1,2-dihydro-4-(methylamino)-2-oxo-1-β-D-ribofuranosyl- (not Cytidine, 5-(3-chloro-3-oxopropyl)-*N*-methyl-)

Nucleotides are indexed at stereoparents descriptive of the type and position of the phosphate residue on the sugar moiety. Esters with phosphoric acid, H_3PO_4, are "-ylic acid" stereoparents. Each isomer has its own heading, e.g., **5'-Adenylic acid** for the 5'-(dihydrogen phosphate) of **Adenosine**. When more than one such acid residue is present, the lowest locant is chosen for the stereoparent, e.g., **3'-Adenylic acid,** 5'-(dihydrogen phosphate). Cyclic phosphates are expressed in the modification at the nucleoside name by such phrases as "cyclic 3',5'-(hydrogen phosphate)."

Nucleoside esters with unsubstituted linear polyphosphoric acids are indexed at stereoparents which express the complete ester name, e.g., **Guanosine 3'-(trihydrogen diphosphate).** An unspecified isomer is assumed to be the 5'-isomer. Di- and triesters with the same acid are indexed at the lowest numbered isomer heading. Substituted derivatives and esters with nonphosphorus acids are named like those of the nucleotides.

Example:

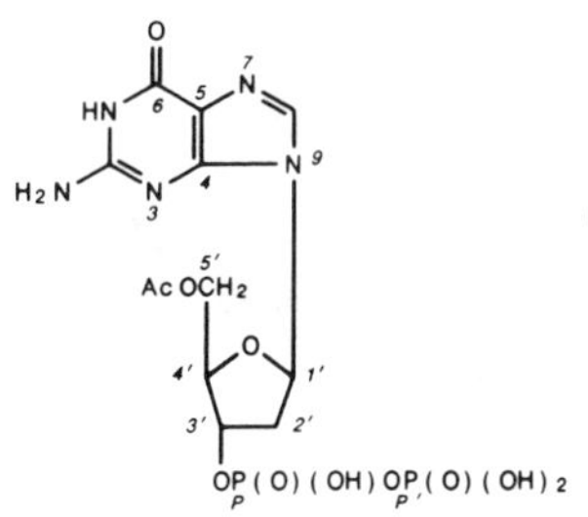

Guanosine 3'-(trihydrogen diphosphate), 2'-deoxy-5'-acetate

Coenzyme A is employed as a stereoparent for adenosine 5'-(trihydrogen diphosphate) 3'-(dihydrogen phosphate) 5'-[3-hydroxy-4-[[3-[(2-mercaptoethyl)amino]-3-oxopropyl]amino]-2,2-dimethyl-4-oxobutyl] ester; functional derivatives, notably esters such as "*S*-acetate," are also indexed at this stereoparent.

Esters and molecular addition compounds of nucleoside and nucleotide stereoparents, including "-ylic acids," with nonstereoparents are indexed at the stereoparents regardless of functionality.

Example:

Inosine 5'-(trihydrogen diphosphate) 2'-(dihydrogen phosphate), 5' → 5'-ester with 3-(aminocarbonyl)-1-β-D-ribofuranosylpyridinium hydroxide, inner salt

Oligonucleotides contain not more than twenty nucleotide units connected by phosphoric acid residues. Higher-molecular-weight compounds are indexed as polynucleotides (see Polymers ¶ 222) or at the headings **Deoxyribonucleic acid** or **Ribonucleic acid.** Phosphate links attached to the preferred stereoparent of an oligonucleotide form parts of "-ylyl" radical substituents, but additional (free) phosphate ester groups allow choice of a nucleotide, rather than a nucleoside, as heading parent.

The choice of stereoparent is decided successively by (*a*) nucleotide preferred to nucleoside; (*b*) the nucleotide appearing first in the order: **Xanthylic, Guanylic, Inosinic, Adenylic, Thymidylic, Uridylic, Cytidylic, Isocytidylic acid;** (*c*) the lowest locant in the stereoparent, e.g., "3'-" rather than "5'-"; (*d*) the corresponding nucleosides in descending order similar to (*b*); (*e*) maximum number of substituents; (*f*) lowest locants of substituents at the first point of difference. Arrows are employed with locants, as in oligosaccharide nomenclature (¶ 208), to indicate progression and points of attachment on saccharide residues beginning at the end of the chain furthest from the preferred stereoparent. Esters and substituents of "-ylyl" radicals are named as substituents without regard to higher functions. Nonnucleotidylyl esters of the stereoparent are expressed in the modification. When a cyclic ester is part of the polynucleotide linkage, the locants for the ester are separated by a hyphen, and the arrow to the point of attachment follows in the usual way, e.g., **Adenylic acid, adenylyl-(2'-3'→5')-.**

Example:

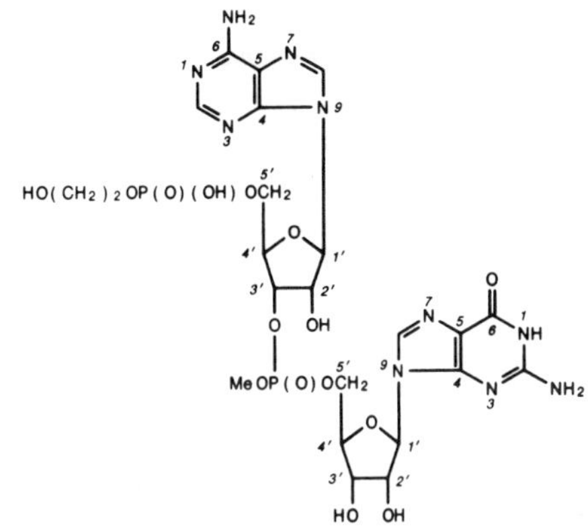

5'-Adenylic acid, *P*-methylguanylyl-(5'→3')-mono(2-hydroxyethyl) ester

Intermolecular cyclic nucleotides are named like the linear compounds, but with the term "cyclic nucleotide" in the modification.

211. Steroids are cyclopenta[*a*]phenanthrene derivatives that are indexed as stereoparents (¶ 202) at trivial names that imply stereochemistry. The following steroid numbering system is employed:

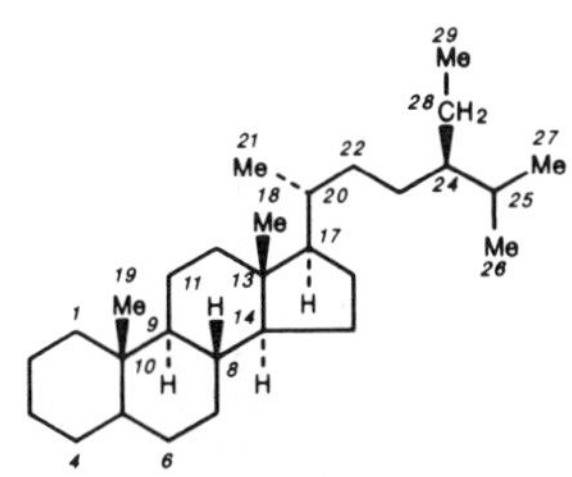

When two methyl groups are present at the 4-position, as well as one at the 14-position, along with a 1,5-dimethylhexyl substituent at position 17, the compound is indexed as a terpene stereoparent (¶ 212); otherwise the steroid or terpene stereoparent is employed that requires least modification.

The hydrogenated ring system without substituents is **Gonane; Estrane** has a methyl group only at the 13-position; the compound with only a 10-methyl group is named as **18-Norandrostane.** The following stereoparents are derivatives of **Androstane** (hexadecahydro-10,13-dimethyl-1*H*-cyclopenta[*a*]phenanthrene) with various side chains at the 17-position:

$-CH_2Me$	**Pregnane**
$-CHMePr$	**Cholane**
$-CHMe(CH_2)_3CHMe_2$	**Cholestane**
$-CHMe(CH_2)_2CHMeCHMe_2$	**Ergostane**
$-CHMe(CH_2)_2CHEtCHMe_2$	**Stigmastane**

The implied configurations in all cases are **8β,9α,10β,13β,14α.** The configuration at the 5-position, when known, is cited as added stereochemistry in the modification. **Pregnane** is **17β; Cholane** and **Cholestane** are **17β,20*R*; Ergostane** is **17β,20*R*,24*S*; Stigmastane** is **17β,20*R*,24*R*.** Illustrative diagrams appear for all these fundamental stereoparents to accompany the entries in the *Chemical Substance Index.*

Unsaturation in steroids is expressed by "en" and "yn" infixes, e.g., **Estra-1,3,5(10)-triene; Androst-1-en-16-ol, (5α)-.** The first example requires two locants for the last cited double bond; this situation is avoided where possible, with steroids containing benzenoid rings, by rearrangement of bonds; e.g., **Gona-5,7,9-triene** (not Gona-5(10),6,8-triene).

Cyclosteroids contain an additional ring formed by a valence bond between two existing ring positions or between a ring position and an angular methyl group or side-chain atom. They are indexed at such stereoparents as **3,5-Cyclopregnane** and **9,19-Cycloandrostane,** for which diagrams showing the implied configurations are displayed in the *Chemical Substance Index* when required by current entries.

Bufanolide and **Cardanolide** are stereoparents with a δ-lactone and a γ-lactone ring, respectively, attached to the 17-position of **Androstane.** The configurations are as for **Cholane** except for **14β** instead of **14α.** When a carboxylic acid or higher function is present the lactone ring is opened and named in the modification.

Example:

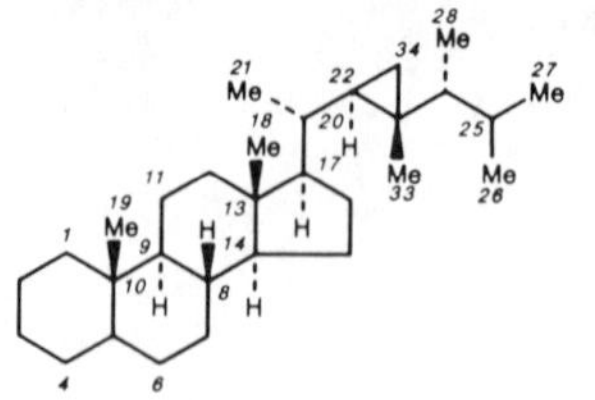

24–Norchol–20(22)–ene–19,23–dioic acid, 3,5,14,21–tetrahydroxy– γ–lactone, (3β,5β,14β)–
(cross–reference from Card–20=(22)–enolid–19–oic acid)

Gorgostane is a newly introduced stereoparent with a cyclopropane ring in the side chain. It has the numbering and configuration shown below; derivatives are named in the usual way; e.g., Gorgosterol is cross–referred to **Gorgost–5–en–3–ol,** (3β)–
Example:

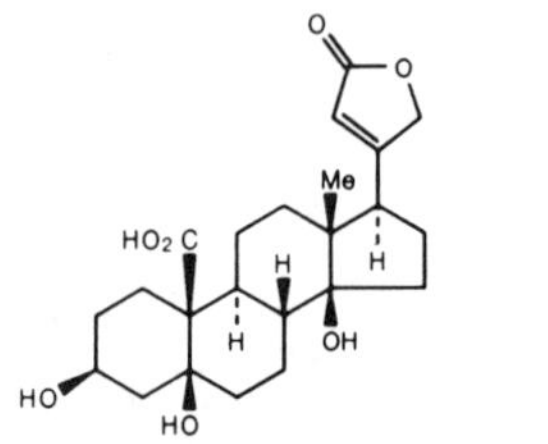

Gorgostane
(stereoparent diagram)

Furostan is a stereoparent which represents 16β,22–epoxycholestane; this has a furan ring fused to the steroid system in the 16,17–position. **Spirostan** is a 16,22:22,26–diepoxycholestane derivative. Configurations at the 5–, 22–, and 25–positions of furostan and at the 5– and 25–positions of spirostans are expressed in the modifications if known.
Examples:

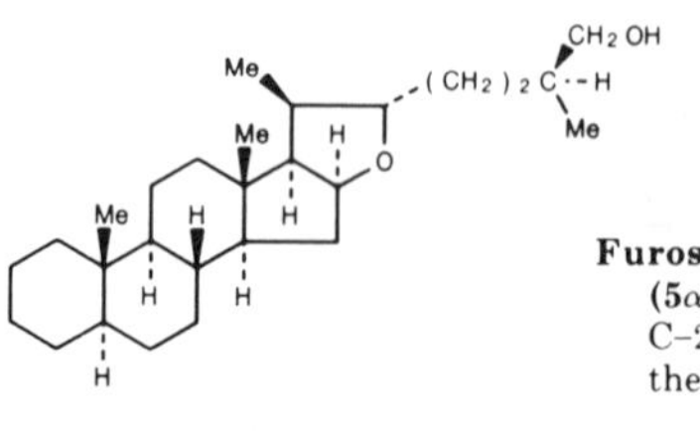

Furostan–26–ol
(5α,20β,22α,25S)– (configuration at C–20 is reversed from that shown in the stereoparent diagram)

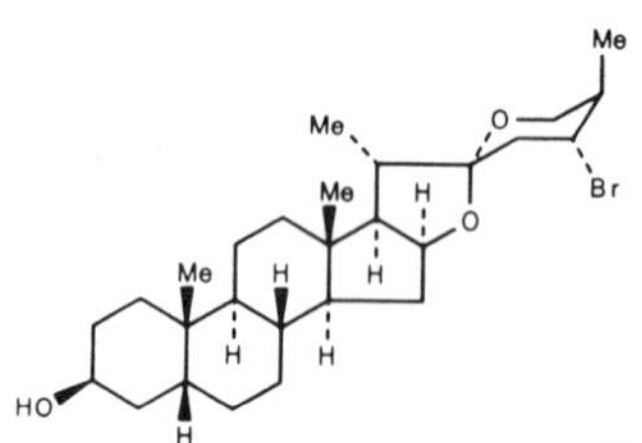

Spirostan–3–ol, 24–bromo–
(3β,5β,24R,25R)– (configurations for added stereochemistry at C–3 and C–24 are indicated: there is no modified stereochemistry)

Elimination of an acyclic carbon atom (or methyl group) from a steroid is expressed by the prefix "nor" with the locant for the carbon removed; (but 21–Norpregnane is indexed at **Androstane, 17–methyl–**). Elimination of one or two carbon atoms from the C–17 side chain of C_{27} and larger steroids may be expressed by "nor" and "dinor," unless a smaller stereoparent is thereby obtained; e.g., **21,27–Dinorcholestan–26–oic acid; 26,27–Dinorcholestane;** but **Cholane–24–carboxylic acid** (not 26,27–Dinorcholestan–25–oic acid). Removal of C–18 and C–19 by "nor" terms is independent of treatment of the C–17 side chain; hence **18,19–Dinorpregnane** (not Gonane, 17–ethyl–).

Ring contraction and expansion are expressed by "nor" and "homo" prefixes with ring locants "A" through "D"; the modified ring systems are new ster=eoparents, illustrated by their own diagrams. On ring contraction, the original locants are retained, with elimination of the highest numbered position of the particular ring. On ring expansion, "a," "b," etc., are added to the highest locant of the enlarged ring to provide for the new atoms. Addition of a carbon atom between two bridgeheads is indicated by citation of the new locant after identification of the ring or rings affected; thus, *AB*(*10a*)–Homo–; *BC*(*8a*)–Homo–; *C*(*14a*)–Homo–; *B*(*9a*)–Homo–; *CD*(*13a*)–Homo–. Insertion of a methylene group between C–13 and C–17 gives *D*(*17a*)–**Homopregnane**. *D*–**Homopregnane** is the compound with the side chain at C–17*a*.

The following limits are placed on the use of "homo" and "nor": (*a*) the stereoparent must contain at least one angular methyl group or a steroidal group on ring *D*; (*b*) not more than two ring methylene groups can be affected; (*c*) not more than two rings can be altered.

Ring scission is expressed by the "*seco*" prefix and two locants to indicate the positions between which the ring has been opened.

Example:

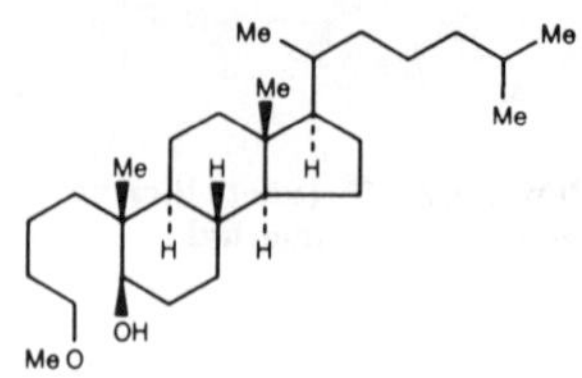

4,5–Secocholestan–5–ol, 4–methoxy–
(5β)–

A combination of "seco" and "cyclo," to express ring scission and ring formation, can be employed to alter orientation of rings in the steroid nucleus. When the same result can be obtained by use of "homo–nor," this is employed.
Examples:

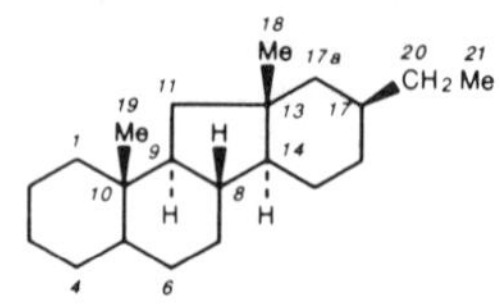

D(*17a*)–**Homo**–*C*–**norpregnane**
(stereoparent)

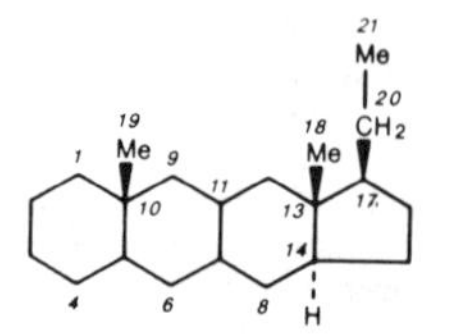

7,11–Cyclo–8,9–secopregnane
(stereoparent)

Replacement ("a") nomenclature (¶ 127) may be used for heterosteroids, e.g., **2–Thiaandrostane,** and may be combined with "homo" and "nor" treatment, e.g., **3–Aza–*A*–homoandrostane,** but not more than one atom may be replaced; otherwise the preferred index entry is made at the systematic name, with a cross–reference at the replacement ("a") name.

Steroidal alkaloids containing exocyclic nitrogen are indexed at steroid stereoparents. Thus Irehine is cross–referred to **Pregn–5–en–3–ol, 20–(di=methylamino)–**, (3β,20S)–. Kurchi alkaloids are named as derivatives of the stereoparent **Conanine**; implied configurations are the same as for **Pregnane,** with the addition of 20β.
Example:

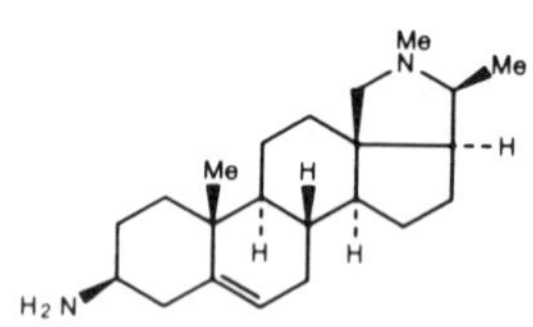

Con–5–enin–3–amine
(3β)– (cross–reference from Conamine)

The solasodine–tomatidine and solanidine groups of *Solanum* alkaloids are indexed at **Spirosolane** and **Solanidane** stereoparents, respectively.
Example:

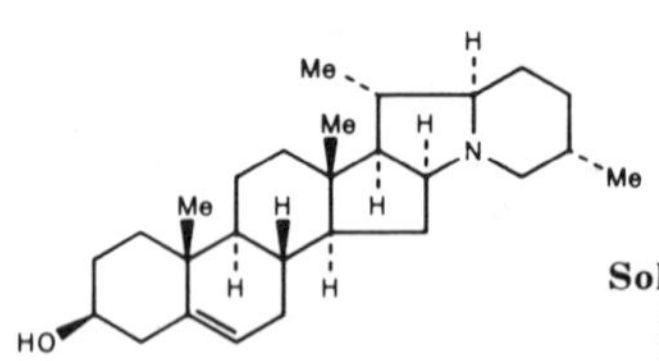

Solanid–5–en–3–ol
(3β)– (When C–5 is asymmetric, its configuration, when known, is cited.)

Veratrum alkaloids are indexed, according to type, at the stereoparents **Veratraman** and **Cevane**. The Salamandra alkaloids are indexed as deriv=atives of **3–Aza–*A*–homoandrostane**. Buxus alkaloids are indexed at such headings as **9,19–Cyclopregnane** and *B*(*9a*)–**Homo–19–norpregnane.** As usual, all these stereoparents are illustrated in the *Chemical Substance Index* when related entries appear there.

Radicals derived from steroid stereoparents are employed when the highest function is present in a nonsteroid part of the molecule (except the esters, semicarbazones and additional compounds are always indexed at the stereo=parents (¶ 202)). Radical names are formed in the usual way, and the free valency can be at an acyclic atom; e.g., estra–1,3,5(10)–triene–17–yl, cholest–4–en–26–yl. No radical names are formed from cardanolides or bufanolides. Configurations are cited ahead of the entire substituted radical; e.g., [(3β,5α)–3–hydroxyandrostan–5–yl]–. When the preferred index name does not appear at the stereoparent, an additional *Chemical Substance Index* entry is placed there. For acetals of steroids, see ¶ 202.

Steroidal lactones are indexed at the open forms, with an additional entry in the *Chemical Substance Index* and *Index of Ring Systems* at the systematic ring system name of the lactone. When a "Homooxa..." name is possible for a lactone, it is preferred over the "Seco–acid, hydroxy–, lactone" name.

Steroids with -O- and -OO- bridges are indexed by use of "epoxy" and "epidioxy" substituents at the stereoparents, with aditional entries for the systematic names. When the -NHNH- group is attached to a single skeletal atom, the "hydrazi" radical is employed. Steroid stereoparents may be adopted as components of spiro systems (¶ 156), e.g., **Spiro[androst-4-ene-2,1'-cyclopropan]-3-one**.

Fused steroid names are used on a restricted basis. The steroid component is named first when fusion is to a heterocycle; the reverse order occurs with less preferred carbocycles. Each system retains its own numbering and the nonsteroid is given primed locants. When saturation is expressed at positions common to both components, steroid locants are cited if possible.

Examples:

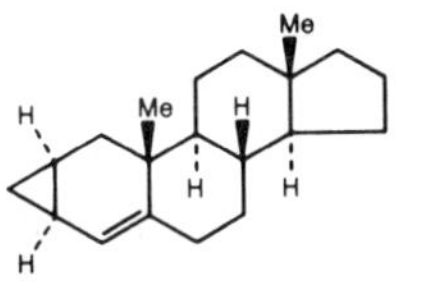

3'*H*-Cyclopropa[2,3]androsta-2,4-diene, 2,3-dihydro- (2α,3α)-

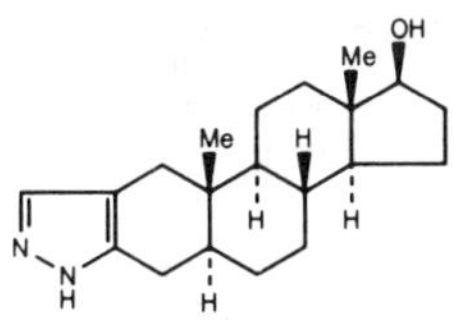

2'*H*-Androst-2-eno[3,2-*c*]pyrazol-17-ol (5α,17β)-

212. Terpenes contain repeating isopentane units:

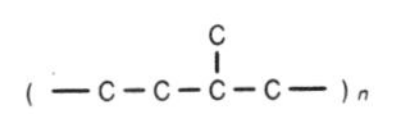

Mono-, sesqui-, di-, sester-, tri-, and tetraterpenes contain two, three, four, five, six, and eight such units; tetraterpenes are usually called carotenes (see below). Most mono-, sesqui-, and diterpenes are indexed by the principles of systematic substitutive nomenclature; stereoparents are employed for terpenes that contain four or more rings, or possess three or more elements of stereochemistry, at least one of which is associated with a bridged or side-chain center, or a center in a ring of twelve or more members. Illustrative diagrams for terpene stereoparents, exhibiting the numbering systems and partial or complete configurations, appear as needed in the *Chemical Substance Index*. Cross-references from trivially named terpenes, including all heading parents appearing in *CA* indexes prior to Volume 76, are supplied in the *Index Guide*, e.g., **Geraniol**. See *2,6-Octadien-1-ol, 3,7-dimethyl-, (E)-*.

Examples:

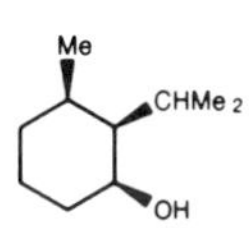

Cyclohexanol, 3-methyl-2-(1-methylethyl)- [1*S*-(1α,2α,3α)]- (cross-reference from *o*-Neoisomenthol)

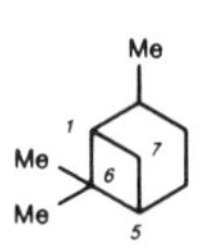

Bicyclo[3.1.1]heptane, 2,6,6-trimethyl- (cross-reference from Pinane)

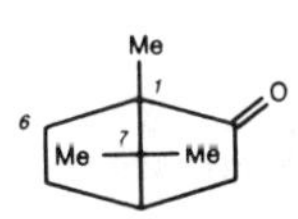

Bicyclo[2.2.1]heptan-2-one, 1,7,7-trimethyl- (cross-reference from Camphor)

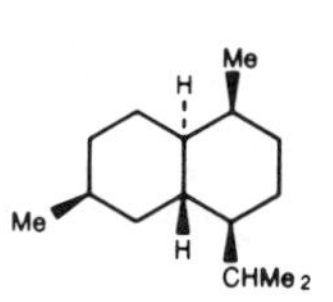

Naphthalene, decahydro-1,6-dimethyl-4-(1-methylethyl)- [1*S*-(1α,4α,4aα,6α,8aβ)]- (cross-reference from Cadinane)

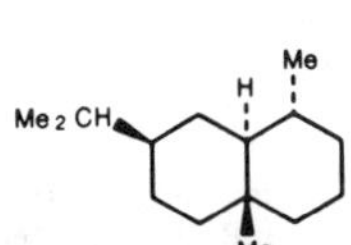

Naphthalene, decahydro-1,4a-dimethyl-7-(1-methylethyl)- [1*R*-(1α,4aβ,7β)]- (cross-reference from Eudesmane)

The stereoparent **Trichothecane**, a sesquiterpene "epoxide," is used in indexing, and other derivatives of sesquiterpenes and diterpenes may be used as stereoparents when they meet the criteria. **Gibbane** is a partially stereospecific stereoparent employed in naming gibberellic acid derivatives. Two tetracyclic diterpene stereoparents are **Kaurane** and **Atisane**. The class of sesterterpenes (C_{25}) is represented by the stereoparent **Ophiobolane**. Cyclic triterpenes include **Gammacerane, Lanostane, Lupane, Oleanane,** and **Ursane**.

Abnormal configurations are expressed by "α" or "β" at angular carbon atoms in ring systems, and by "*R*" and "*S*" when there is inversion on a bridge or side-chain. Additional stereochemistry produced by substitution is expressed similarly. All terms cited with a stereoparent are considered to be absolute, not relative. The geminate methyl groups of triterpenes have no implied configuration associated with the locants; "α" or "β" is cited when substitution in the methyl groups makes this necessary.

Because lowest locants are assigned to principal groups, in accordance with general systematic nomenclature, the author's numbering may be changed in indexing, especially with a symmetrical stereoparent such as **Gammacerane.**

Example:

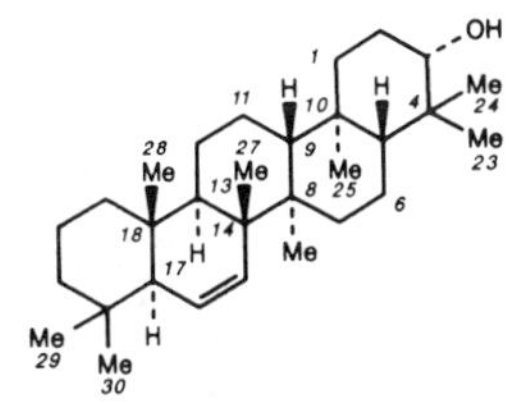

Gammacer-15-en-3-ol (3β)- (not Gammacer-6-en-21-ol, (21α)-)

Ring contraction is expressed by "nor" and citation both of the ring affected and of the locant for the expelled carbon atom. When the methyl groups (C-23 and C-24) at the 4-position of Lupane, etc., are lost, a "trinor" term is used, e.g., ***A*(4),23,24-Trinorlupan-3-one**. The highest available locant is always eliminated when one of a set of equivalent carbon atoms is lost.

"Friedo," preceded by *D*, *D:A*, *D:B*, or *D:C* as prefixed to pentacyclic triterpene stereoparents indicates that angular methyl groups have shifted from their normal positions as follows:

D	14 to 13
D:C	14 to 13 8 to 14
D:B	14 to 13 8 to 14 10 to 9
D:A	14 to 13 8 to 14 10 to 9 4 to 5

The initial numbering of the rings and methyl groups is retained. Friedo structures are used as stereoparents even though they still possess the normal ring system.

Examples:

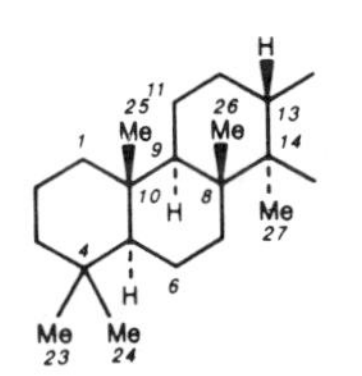

Normal arrangement

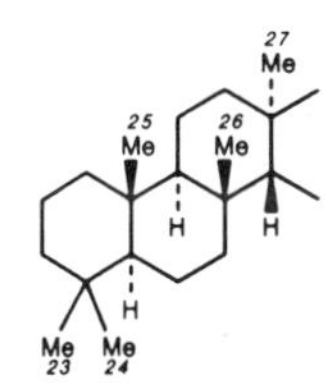

[*D*-Friedo-]

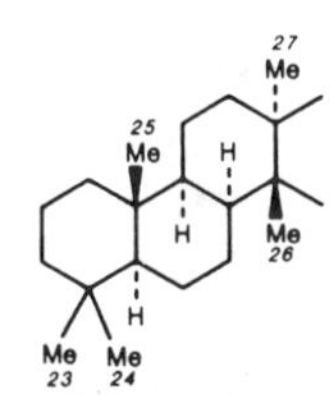

[*D*:*C*-Friedo-]

The prefix "neo-" preceded by an italicized capital *A* indicates that ring *A* has undergone the following rearrangement:

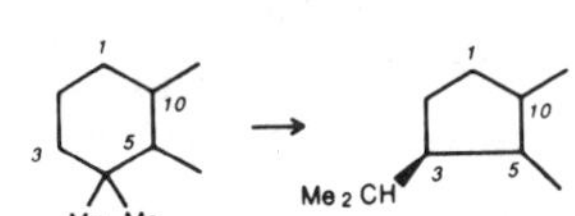

The prefixes "*A*:*B*-Neo-," "*A*:*C*-Neo-" and "*A*:*D*-Neo-" indicate the "*A*-Neo-" ring alteration plus migration of angular methyl groups located in positions 10, 8, and 14, respectively. In these transitions the methyl groups have been shifted from their normal positions as follows:

A:*B*-Neo-	*A*-Neo-alteration 10 to 5
A:*C*-Neo-	*A*-Neo-alteration 10 to 5 8 to 9
A:*D*-Neo-	*A*-Neo-alteration 10 to 5 8 to 9 14 to 8

As with "friedo" modifications, the changes are cumulative in proceeding from the normal arrangement of the fundamental structure to the *A:D*–Neo–modification. The initial numbering of the fundamental structure and the methyl groups which have migrated is retained. Neoterpenes and friedoneoterpenes (e.g., *D:C*–**Friedo–*A*–neooleanane**) are stereoparents.

Examples:

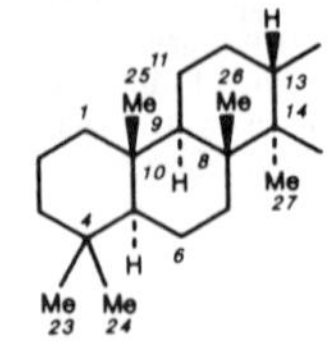

Normal arrangement

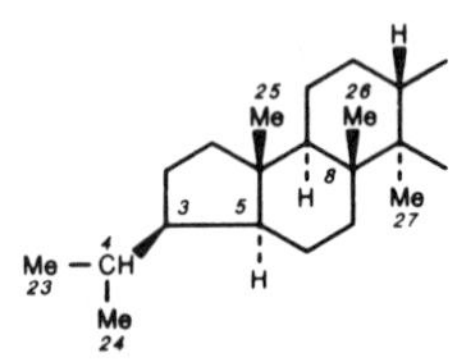

A-Neo-

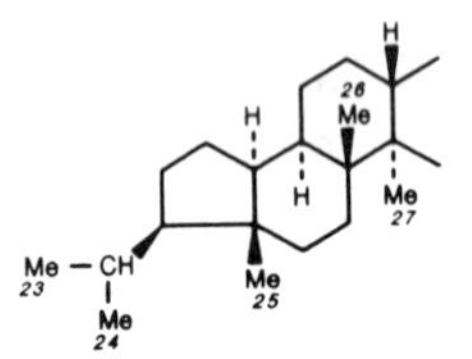

A: *B*-Neo-

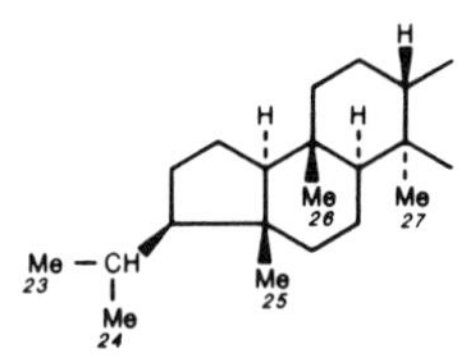

A: *C*-Neo-

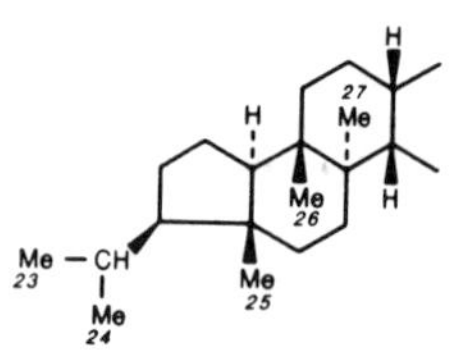

A: *D*-Neo-

In the gammacerane skeleton, symmetry of the structure leads to the additional modifications *A'*–Neo– and *B'*:*A'*–Neo–, as follows:

Normal arrangement

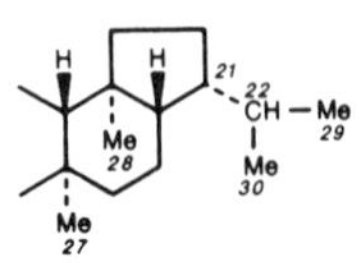

A'-Neo-

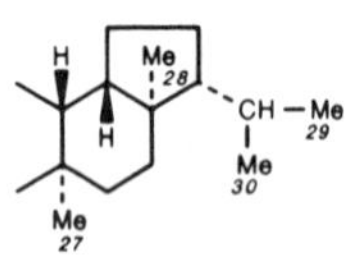

B' : *A'*-Neo-

The prefix "cyclo" indicates a valence bond forming an extra ring, e.g., in **9,19–Cyclolanostane** a "fused" cyclopropane ring is formed. Ring enlargement is expressed by "homo" to which is prefixed a capital italic locant to indicate the ring enlarged; when the added carbon is introduced between two bridgehead atoms, its locant (derived from the higher neighboring existing locant) is cited; e.g., ***E*–Homo–30–norurs–21–ene**; ***C*–Homoursane**; ***C*(*14a*)–Homoursane.**

Ring scission is expressed by "seco" terms; the terpene numbering is unchanged (cf. Steroids, ¶ 211); e.g., **3,4–Secooleana–4(23),12–diene–3,29–dioic acid, 11–oxo–**, (18α,20β)–. Ring scission may cause a stereoparent "seco" derivative to be named systematically in accordance with the requirements above.

"Abeo" terpenes are those in which a bond has migrated. Thus, **5(4→3)=Abeolupane** is lupane in which the 5,4–bond has been replaced by a 5,3–bond, contracting ring A from six to five members. The carbon atom C–4 and its attached methyl groups are eliminated. "Abeo" names are not used by *CA* in indexing; instead, they are cross–referred in the *Index Guide* to the preferred names (in the case cited, to ***A*(4),23,24–Trinorlupane.**)

When two or more modifiers are employed, they are cited in alphabetical order, e.g., ***D*:*A*–Friedo–*A*(1)–nor–3,4–secooleanan–3–oic acid.**

Principal groups are expressed as "–oic acid," "–al," "–one," "–ol," etc., by the procedures of general substitutive nomenclature. Terpene radicals are formed in the usual way, e.g., **(27,28–dihydroxy–*A*(1)–norlup–20(29)–en–22–yl).** Configurational terms are placed ahead of the complete radical (cf. Steroids, ¶ 211). When a terpene radical must be used, an additional entry appears at the stereoparent, e.g., ***A*(1)–Norlupane.**

Lactones of triterpene acids are indexed at the acid names; cyclic acetals are likewise expressed in the modification at stereoparent headings, with an additional *Chemical Substance Index* and *Index of Ring Systems* entry at the systematic parent.

Carotenoids comprise carotenes (hydrocarbons) and xanthophylls (their oxygenated derivatives) in which eight isoprene units are joined in such a manner that the arrangement is reversed at the center of the molecule, whereby the two central substituent methyl groups are in 1,6–relationship and the remaining methyl groups in 1,5–relationship. The class also includes some rearranged and degraded compounds, but excludes Retinol and related C_{20}–compounds (see below).

General entries are found at **Carotenoids** in the *General Subject Index*. Specific substances are indexed at **Carotene**:

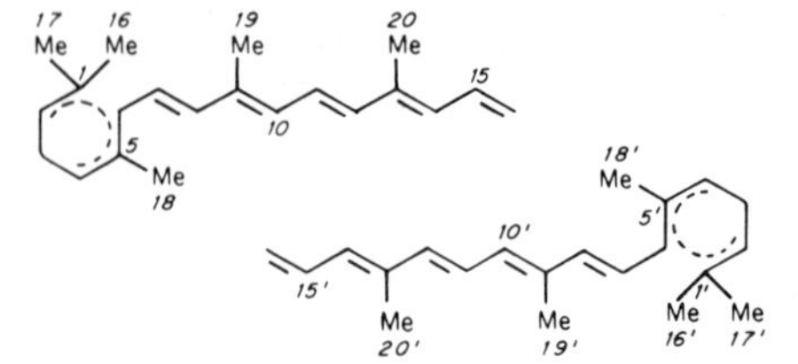

Each dotted curved line represents two double bonds or their equivalent; thus, individual compounds may have acyclic end groups with two double bonds, or terminal cyclohexenyl groups, or one of each. By extension, **Carotene** is used as the stereoparent representing dihydro and tetradehydro derivatives which possess cyclopentyl and phenyl end–groups. Two Greek letter prefixes on the **Carotene** stereoparent, cited in (Greek) alphabetical order, are employed to express the terminal structures of specific compounds:

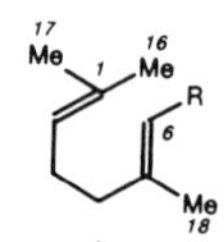

ψ

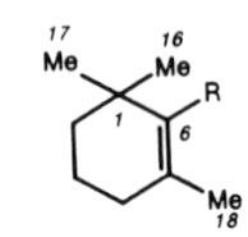

β

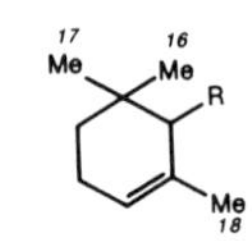

ε

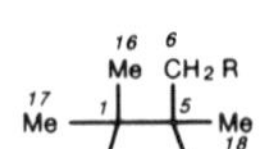

κ

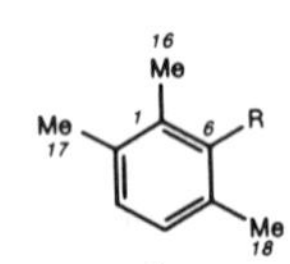

φ

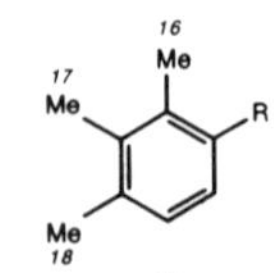

χ

α–Carotene is now indexed at **β,ε–Carotene,** β–carotene is **β,β–Carotene**; γ–carotene is **β,ψ–Carotene**; and lycopene is **ψ,ψ–Carotene.** The plain (unprimed) locants are assigned to the end of the molecule related to the first–cited Greek letter.

"Hydro" and "dehydro" express addition and subtraction of hydrogen; oxygen functions are expressed by "–oic acid," "–al," "–one," "–ol," etc., in the usual way. Oxygen bridges are indicated by "epoxy," it being understood that the oxygen replaces one hydrogen atom at each bridgehead; thus, an "epoxide" is an epoxy–dihydro derivative.

Example:

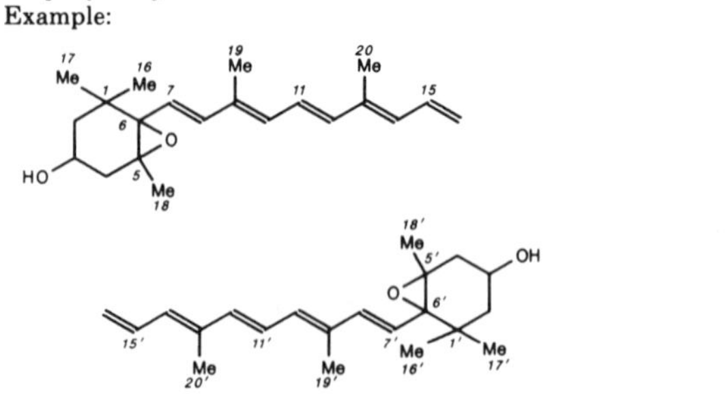

β,β–Carotene–3,3'–diol, 5,6:5',6'–diepoxy–5,5',6,6'–tetrahydro–

Addition of water (H, OH) or methanol (H, OCH_3) to a double bond is expressed by "dihydro–hydroxy" and "dihydro–methoxy." When the "–ol" suffix can be used only for *some* of the hydroxy groups, the "dihydro–hydroxy" method is employed for all.

Carotenoids with diosphenol end–groups are indexed as the keto tautomers; e.g., **β,β–Carotene–3,3',4,4'–tetrone.** For carotenoids with identical end groups, lowest locants are assigned to the principal groups, if present, and then to prefixes; e.g., **ε,ε–Carotene–2,3'–dione, 2'–methoxy–** (not ε,ε–Carotene–2',3–dione, 2–methoxy–).

"Retro" carotenes have undergone a shift of the alternating single–double bond system by one position. A prefix such as "4,7'–*retro*–" is applied to the stereoparent to express this situation; the first cited locant shows where a proton has been gained, the second, where a proton has been lost.

"Apo" carotenes are oxidative degradation products retaining more than one–half of the carotene structure. "Apo" and "diapo" (unitalicized) prefixes are employed to indicate loss of skeletal atoms at one or both ends of the molecule. The end that is shortened is considered to have had a ψ (acyclic) end–group. For example, **2'–Apo–β,ψ–carotenal** has an intact β (cyclohexene) end group and one that has lost its 1' (and attached 16' and 17') skeletal atoms, and the carbon atom at the point of scission has been oxidized to –CHO.

Oxidative degradation products containing less than one–half of the original carotene skeleton are indexed systematically; e.g., **2,4–Pentadienal, 3–**

methyl-5-(2,6,6-trimethyl-1-cyclohexen-1-yl)-.

Stereochemistry of carotenoids is of two types. The geometrical configuration around double bonds is implied by the stereoparent stem **Carotene** to be *all-trans*. Abnormal configuration at one or more positions is indicated in the modification by terms such as "15-*cis*-." Absolute configuration at chiral centers is expressed by "*R*" or "*S*."

Example:

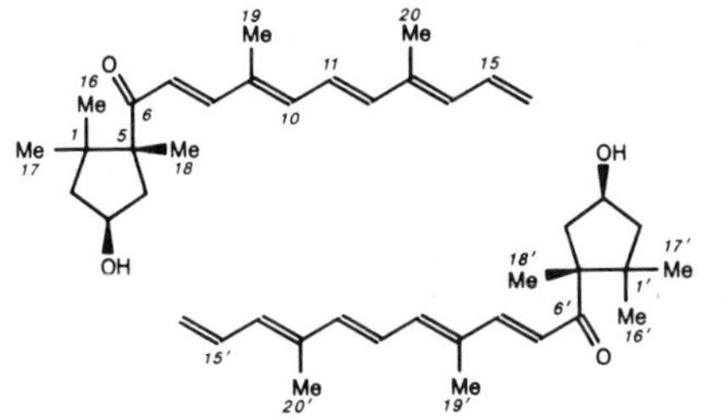

κ,κ-**Carotene-6,6'-dione, 3,3'-dihydroxy-**
(3*S*,3'*S*,5*R*,5'*R*)-

Retinol and its relatives are indexed at the stereoparent:

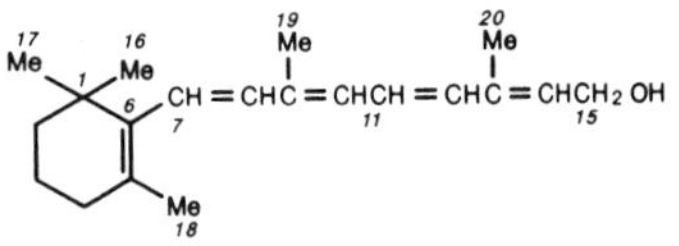

Retinol (stereoparent)

Illustrative diagrams for **Retinol** and **Retinoic acid** appear in the *Chemical Substance Index*. Cross-references between retin names and Vitamin A terms are found in the *Index Guide*. Vitamin A_1 is indexed at **Retinol**; Vitamin A_2 at **Retinol, 3,4-didehydro-**. Hydrocarbon derivatives are indexed at systematic names. At **Retinol,** functions higher than the alcohol are expressed as substituents, e.g., **Retinol, 4-oxo-**. The *all-trans* configuration is assumed; if abnormal stereochemistry is reported, it is expressed in the modification by terms such as "13-*cis*-".

213. Introduction. The special classes of chemical substances discussed in this section exclude those natural products and their derivatives, largely indexed at stereoparents, which were the subject of the previous section. The topics covered here fall into three groups: inorganic (alloys, coordination compounds, elementary particles, and general inorganic compounds); biological (enzymes, porphyrins and bile pigments, and vitamins); and special subjects not chemically related to any particular class (dyes, isotopes, mixtures, and polymers). The arrangement is alphabetic.

214. Alloys, including those with nonmetallic and gaseous components, are indexed and registered by CAS as specific chemical substances. The percentage composition is expressed in the modification in descending order of abundance of metals, either as individual round numbers or as ranges. Components present in amounts below 0.1% are ignored. The metal in greatest amount is indexed at an "element alloy base" heading parent, all other com= ponents of 1% or greater amount at "element alloy nonbase" headings; index entries for minor (less than 1%) components are made at the element (not the element nonbase) heading if justified by the emphasis of the original document.

Example (an alloy of 69.93% copper, 17.45% nickel, 10.29% aluminum, 1.50% cobalt, 0.83% iron):

Copper alloy, base
Cu 70,Ni 17,Al 10,Co 1.5,Fe 0.8

Nickel alloy, nonbase
Cu 70,Ni 17,Al 10,Co 1.5,Fe 0.8 (additional index entry)

(Entries are also to be found at **Aluminum alloy, nonbase,** and **Cobalt alloy, nonbase** headings with the same modification.)

Binary alloys of 50–50% composition are indexed at the base heading for the metal with the earliest alphabetical element *symbol*, and at the nonbase heading for the other, e.g., **Iron alloy, base** and **Antimony alloy, nonbase.**

When ranges, rather than specific compositions, are supplied in the original document for all or some components, they are cited thus: Cu 68–73,Ni 16–18,Al 9–10,Co 1–2,Fe 0.7–1. Alloys of unknown percentage composition are entered only at nonbase headings, unless the base component is known; the symbols are then cited in alphabetical order, except that the base element is placed first if known; e.g., Al,Co,Fe,Ni or (for a cobalt base alloy) Co,Al,Fe,Ni.

Alloys with trade names or code designations (other than standards) sometimes vary in reported composition; therefore, the numerical values are omitted from the entries and from cross-references at the trade names and code designations in the *Index Guide* except in those cases for which the composition is known to be invariant (see below). The base metal symbol is placed first and the others in alphabetical order; finally, the trade name or code designation is cited in parentheses. If only the base metal is known, the parenthetical expression alone is cited in the modification.

Examples:

Copper alloy, base
Cu,Fe,Ni (Cunife I) (cross-reference at Cunife I)
(also entries at **Iron alloy, nonbase,** and **Nickel alloy, nonbase,** with the same modification)

Nickel alloy, base
(Permalloy) (cross-reference at Permalloy)

Numerical values are included in entries for trade-named alloys of fixed composition and for those identified by the following U.S. standards: AA, CDA, AISI, ASTM, SAE, AMS, and AWS. The international standards of ISO and the standards of most other countries are also recognized for the purpose of *CA* indexing of alloys.

Cermets are alloys containing nonmetals. They are indexed at metal alloy headings and at nonmetal headings with "alloy" in the modification.

Example (a cermet containing tungsten carbide (WC) 94%, cobalt 6%):

Tungsten carbide (WC)
alloy, WC 94,Co 6
Cobalt alloy, nonbase
WC 94,Co 6 (additional index entry)

Alloys containing gases (0.1% or more) are indexed similarly.

Example:

Nickel alloy, base
Ni 99,H 1
Hydrogen, compounds
alloy, Ni 99,H 1

Steel is a heading parent for iron–carbon alloys so described and either containing a minimum of 97% iron or having no disclosed compositions. The steel composition or a designation (such as AISI C1017) which implies the composition is cited in the modification.

Examples:

Steel
(AISI C1017)

Steel
Fe 98,Mn 1.5,Si 0.5,C 0.1

Isotopically labeled alloys are distinguished by the mass number on the element symbol.

Example:

Aluminum alloy, base
Al 95,^{235}U 5

Uranium alloy, nonbase
Al 95,^{235}U 5 (additional index entry)

215. Coordination compounds are molecules or ions in which a central atom (in polynuclear compounds, more than one) has atoms or groups of atoms, called ligands, attached to it to the extent of its coordination valency. The central atom may be of any element, but it is usually a metal atom. The metals comprise all elements except the following: Ar, As, At, B, Br, C, Cl, F, He, H, I, Kr, Ne, N, O, P, Rn, Se, Si, S, Te, Xe. (Prior to *CA* Volume 95 (see ¶ 101), antimony was also indexed as a nonmetal.) Compounds of nonmetals are generally not indexed as coordination compounds, but there are several ex= ceptions, e.g., tetra– and hexavalent sulfur compounds (¶ 200), and borates containing ligands other than oxo and hydroxy. Hydride derivatives of an= timony, bismuth, germanium, tin, and lead are named not as coordination compounds but as covalent derivatives of the hydride heading parents, e.g., **Stibine, Germane** (¶¶ 181, 199).

A *ligand* is any atom or group of atoms, charged or neutral, that is attached to the central atom of a coordination complex. The atoms of a ligand that are attached to the central atom are called coordinating atoms regardless of the type of bonding involved. A ligand with more than one coordinating site is described as multidentate; when more than one such site is engaged with a single central atom, it is a chelate ligand; when it is coordinated with two or more central atoms it is a bridging ligand.

Neutral compounds with monoatomic ligands (except hydride) and mono= nuclear carbonyls and nitrosyls are given binary salt names. Common oli= gomeric salts are named in the monomeric form, e.g., **Aluminum chloride ($AlCl_3$)**, but studies of the dimers, etc., are indexed additionally at the oli= gomeric names, e.g., **Aluminum, di–μ–chlorotetrachlorodi–**.

Coordination compounds may be anionic, neutral, or cationic, in accordance with the oxidation state of the central atom and the number and nature of the ligands; when the latter are derived from the molecular skeletons of substitutive nomenclature, including nonmetal hydrides, by loss of hydrogen, they are considered negative in computing the total charge of the complexes containing them. This charge is expressed by a Ewens–Bassett number, e.g., (3+) or (2–), placed after the heading parent derived from the name of the central element; thus, **Copper(1+), Borate(2–)**. Absence of a Ewens–Bassett number indicates that the complex expressed by the *complete boldface heading* is neutral. Ligands are cited in alphabetical order and their ligating atoms are identified, if possible, by italic capital–letter locants, e.g., *N,N',N'',O*^{1},*O*^{2}, cited in al= phabetical order. These *letter locants* are used only when all the ligating atoms can be adequately identified by primes or indices (superscript numerals or letters) related directly to the regular locants belonging to the index parent from which the ligand is derived, e.g., [2,3–dihydroxybutanedioato(2–)–*O*^{2},*O*^{3}]. Letter locants are not cited with (*a*) regular substituent prefixes, e.g., methoxy, phenyl, 2–pyridinyl; (*b*) other sigma–bonded ligands in which hetero atoms are either absent or uninvolved in the coordination; (*c*) monoatomic ligands, e.g., hydro, iodo; (*d*) ligands with unspecified bonding; (*e*) neutral and monoanionic organic ligands derived from cyclic or acyclic compounds containing only one hetero atom capable of coordination, e.g., phosphorous trichloride, pyridine, *N,N*–dimethylmethanamine, *N*–methylmethanaminato, provided that the hetero atom is coordinated but no carbon atom is also bonded; (*f*) the following simple polyatomic inorganic ligands: amido, ammine, aqua, selenyl, telluryl, hydroxy, imido, mercapto, phosphino, and (except sometimes when present as bridges) azido, carbonyl, dinitrogen, dioxygen, nitrosyl, peroxy, and carbo= nothioyl.

In addition to the *Chemical Substance* and *Formula Index* entries for coordination compounds at the central element names (or the derived "–ate" terms), entries will be found in the *Chemical Substance Index* at the names of the ligands or of the compounds from which such ligands are derived by loss of hydrogen, except for the common ligands listed below. The modifications at these headings are general in form, but they supply the names of the coordinating elements and also cite the CAS Registry Number which, by way of the *CAS Registry Handbook—Number Section* or CAS ONLINE, leads directly to the preferred name and other information.

Example:

Benzenamine, 4-ethyl-
chromium complex [*31826-83-2*] (the preferred entry is at **Chro=mate(1-), bis(4-ethylbenzenamine)tetrakis(thiocyanato-*N*)-**)

The precise method of construction of the preferred entry will be detailed in what follows. However, if it is understood that "-ato" in a ligand name denotes loss of a proton, it can be deduced already that a central chromium atom coordinates two neutral molecules of 4-ethylbenzenamine (¶ 176) and four anionic thiocyanate groups, each of which has lost a proton. Because the complex carries overall only a unit negative charge, as expressed by the index parent **Chromate(1-)**, the metal is evidently in an oxidation state of +3 and the structure of the compound is as follows:

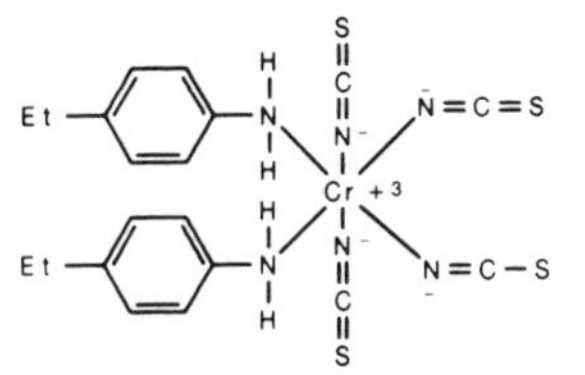

Common ligands for which additional entries are not regularly made at specific substance names (except for studies of unusual bonding modes, etc.) are listed below. Only the ligands named as in the list are excluded from index coverage; thus, neither a phenyl nor a neutral benzene ligand receives an additional index entry at **Benzene**, but phenylene and benzenetriyl ligands are entered there. However, letter locants are not specified in the list, so acetato-*O* and acetato-*O,O'* ligands, for example, are equally excluded from additional index coverage at **Acetic acid**. Indexing notes at **Benzene**, **Acetic acid**, etc. in the current *Index Guide* explain these policies. The common halo ligand bromo (Br^-) is not covered at the index parent **Bromide**, but the tetrabromo(2-) ligand receives an entry at **Bromide** (Br^{2-}) and phosphido is covered at **Phosphide**. All isotopically labeled ligands are given index entries; thus, the (chloro-Cl^{37}) ligand is indexed at **Chloride**-Cl^{37}.

acetato
acetonitrile
acetyl
L-alaninato
amido
N-(2-aminoethyl)-1,2-ethane=diamine
ammine
aqua
benzenamine
benzene
1,2-benzenediamine
benzenethiolato
benzoato
2,2'-bipyridine
N,N-bis(carboxymethyl)glycina=to(3-)
bromo
1,3-butadiene
butoxy
butyl
carbonothioyl
carbonyl
N-(carboxymethyl)glycinato(2-)
chloro
cyanato
cyano
2,4-cyclopentadien-1-yl
diethylcarbamodithioato
1-1-dimethylethyl
N,N-dimethylformamide
N,N-dimethylmethanamine
1,4-dioxane
1,3-diphenyl-1,3-propanedionato
1,2-ethanediamine
ethanedioato(2-)
[*N,N'*-1,2-ethanediylbis[*N*-=(carboxymethyl)glycinato]](4-)
[2,2'-[1,2-ethanediylbis(nitrilo=methylidyne)]bis[phenolato]]=(2-)
ethanethiolato
ethene
ethenyl
ethoxy
ethyl
fluoro
glycinato
1,1,1,5,5,5-hexafluoro-2,4-pen=tanedionato
hydro
hydroxy
2-hydroxybenzoato
imido
1*H*-inden-1-yl
iodo
methanamine
methanethiolato
methoxy
methyl
1-methylethyl
N-methylmethanamine
1-methylpropyl
2-methylpropyl
2-methylpyridine
3-methylpyridine
4-methylpyridine
nitrato
nitrito
nitrosyl
oxo
1,1'-oxybis[ethane]
2,4-pentanedionato
1,10-phenanthroline
phenoxy
phenyl
1-phenyl-1,3-butanedionato
piperidine
2-propenyl
propoxy
propyl
pyridine
quinoline
8-quinolinolato
sulfato(2-)
sulfinylbis[methane]
tetrahydrofuran
thiocyanato
thiourea
thioxo
tributyl phosphate
triethylphosphine
triethyl phosphite
4,4,4-trifluoro-1-(2-thienyl)-=1,3-butanedionato
trimethylphosphine
trimethyl phosphite
triphenylphosphine
triphenyl phosphite
urea

(Note: "hydro" and "sulfato(2-)" were added to the list in 1982.)

Cationic complexes are highest in order of precedence (¶ 106) among co=ordination compounds. (A positive element heading, e.g., **Phosphorus(1+)**, is ranked just above the corresponding "-onium" heading, e.g., **Phosphonium**.) Neutral complexes are lower in the order of precedence, followed by anionic coordination compounds. For mixed polynuclear complexes, the preferred element (or derived "-ate") heading is that appearing earliest in the following list: Rn, Xe, Kr, Ar, Ne, He, Fr, Cs, Rb, K, Na, Li, H, Ra, Ba, Sr, Ca, Mg, Be, (Lr to Ac), (Lu to La), Y, Sc, Hf, Zr, Ti, Ta, Nb, V, W, Mo, Cr, Re, Tc, Mn, Os, Ru, Fe, Ir, Rh, Co, Pt, Pd, Ni, Au, Ag, Cu, Hg, Cd, Zn, Tl, In, Ga, Al, B, Pb, Sn, Ge, Si, C, Bi, Sb, As, P, N, Po, Te, Se, S, O, At, I, Br, Cl, F.

Many ligand names correspond to those of the substituent prefixes of general nomenclature. They include hydro, chloro (etc.), oxo, thioxo, hydroxy, mercapto, peroxy, diazenyl, sulfo. Among ligand names which differ are the following: superoxido (O_2-), amido (H_2N-), imido (HN=), chloramido (ClH=N-), nitrido (N≡), phosphido (P≡), arsenido (As≡).

"Ato" ligand names are employed to express loss of hydrogen from inorganic "oxo" acids named as "-ic acids" and their chalcogen analogs. Those named as "-ous acids" correspondingly afford "-ito" ligands. Loss of one, two, etc., protons from polybasic acids is expressed by Ewens-Bassett numbers (loss of one hydrogen atom from most other classes of compounds is denoted simply by omission of a Ewens-Bassett number). Letter locants (see above) are sometimes employed. "Ato" and "ito" ligand names are always placed in enclosing marks.

Examples:

$^-$O-PF(O)-O$^-$	[phosphorofluoridato(2-)]
O_3ClO^-	(perchlorato)
$^-OSO_2O^-$	[sulfato(2-)]
$^-$OS(O)O$^-$	[sulfito(2-)]
^-O_2N-	(nitrito-*N*)(not nitro)
$^-$O=NO-	(nitrito-*O*)(not nitrito)
$^-$O-N=N-O$^-$	[hyponitrito(2-)]
$OP(O^-)_3$	[phosphato(3-)]

The following cyanato radicals are employed:

$^-$OCN-	(cyanato-*N*)(not isocyanato)
$^-$NCO-	(cyanato-*O*)(not cyanato)

and similarly for thiocyanato, etc. Oxo acids which require synonym line formulas at their own heading parents are given more specific names when converted into ligands, e.g., [monothiosulfato(2-)], [trimetaborato(3-)].

Anionic carbon-attached ligands containing no anionic hetero-atom at=tachments to the metal, and acyl ligands containing no anionic hetero-atom attachments except for the acyl portion, are given names identical with those of normal substituent prefixes, e.g., ethyl, 2-pyridinyl, 1*H*-imidazol-2-yl, (phenylsulfonyl). Phosphino, phosphinyl and (phenylazo) radicals are also used. Anionic ligands derived from heterocycles by loss of a proton only from a hetero atom in the ring, and containing no additional hetero atoms in the ring or substituents, are also named by substituent prefixes, e.g., 1-piperidinyl. All other anionic heterocyclic ligands are given "ato" names; they include porphines, phthalocyanines, corrins, other heterocycles containing rings of more than ten members, and heterocyclic ligands containing more than one hetero atom (including substituents). Unsubstituted phenoxy and also methoxy through (dodecyloxy) radicals are used to express these ligands, but the chalcogen analogs and the substituted oxy radicals are named (chloro=methanolato), (2-propanolato) (not (1-methylethoxy)), (ethanethiolato), etc. Organic "-ic acids" and organophosphorus acids afford ligand names in which "-ic acid" or "-ous acid" is replaced by "-ato" or "-ito", respectively. Ligands from esters are named in uninverted form. The "ato" suffix denotes loss of one proton in all cases unless a Ewens-Bassett number, e.g., "(2-)," is cited.

Examples (letter locants are added to many of these names in particular coordination names; e.g., (acetato-*O,O'*)):

$^-O_2C^-$	(carboxylato) (this is an exception: not [carboxylato(2-)])
AcO^-	(acetato)
$^-CH_2CO_2^-$	[acetato(2-)]
Ph_2(O)P-	(diphenylphosphinyl)

$Ph_2P(O)O^-$ (diphenylphosphinito-*O*)

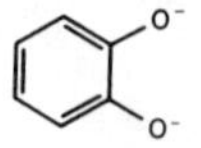
[1,2-benzenediolato(2−)]

$^-O_2C(CH_2)_3CO_2^-$ [pentanedioato(2−)]

$AcO(CH_2)_2O^-$ (2-hydroxyethyl acetato)(an ester is named in its uninverted form)

$MeO\text{-}S(O)_2\text{-}O^-$ (monomethyl sulfato-*O'*) (derived from methyl hydrogen sulfate)

$H_2N\text{-}CH_2CO_2Me$ (methyl glycinate) (derived from glycine methyl ester)

Metal radical names (¶ 194) such as aluminio, aurio, and sodio, are used in the presence of higher compound classes or more preferred metals when the metal replaces a single hydrogen, but not to "place" a metal directly on another metal (see polynuclear complexes, below); e.g., **Ethyl, 2-sodio-** (a free-radical name); **Potassium, (5-lithio-1-naphthalenyl)-**. Because metal radicals carry no implication of valency, hydrogen and other attached atoms and groups must be expressed, e.g., dihydroaluminio for H_2Al-.

π-bonded ligands, and ligands attached to metal atoms by both π and σ bonds, are named by prefixing the Greek letter η (eta). (Delocalized bonds are "fixed" for the purpose of naming the ligand so that lowest locants are given first to the point of attachment, then for unsaturation, etc.) Locants (if available) for atoms contributing to the ligand-metal bonding are cited ahead of "η" and the entire descriptor so formed is enclosed in parentheses; alter= natively, if all the skeletal atoms of an unsubstituted conjugated acyclic or cyclic molecule or radical are involved, a superscript denoting the number of such atoms is cited immediately after "η" and the locants are omitted. Ligands with negative charges on adjacent hetero atoms are named instead as the neutral unsaturated compounds. σ-bonded ligands are named as regular radicals, e.g., 2,4-cyclopentadien-1-yl. (The name cyclopentadienyl is used without locants in the absence of further information.)

Examples:

[(1,2,3-η)-2,4-cyclopentadien-1-yl]

($η^5$-2,4-cyclopentadien-1-yl)

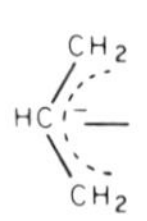
($η^3$-2-propenyl)

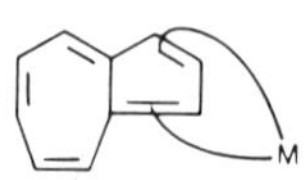
[(1,2,3,3a-η)-azulene]

[(1,2,3-η)-4-oxo-2-cyclohexen-1-yl]-

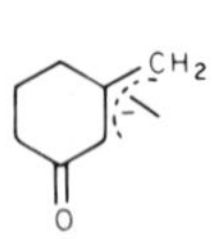
($η^3$-2-methylene-6-oxocyclohexyl) (The choice is based on the pre= ferred parent radical; the locants "α,1,2-" cannot be used because "α-" does not relate to the parent radical.)

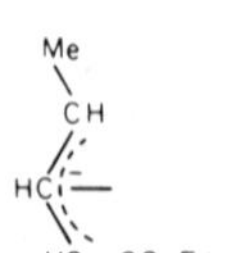
[(1,2,3-η)-4-ethoxy-1-methyl-4-oxo-2-butenyl]

In polynuclear compounds (see below) a colon partitions the locants into sets to indicate bonding to different metal atoms.

Examples:

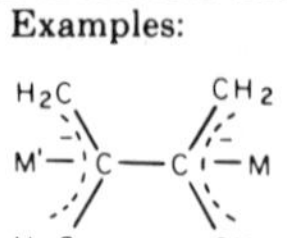
[μ-($η^3$:$η^3$)-2,3-bis(methylene)-1,4-bu= tanediyl)]

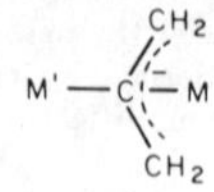
[μ-[(2-η:$η^3$)-1,2-propadiene]]

Cationic molecules (¶ 184) are named in the unchanged (but uninverted) form when present as ligands, except that a Ewens-Bassett number denotes the charge when necessary; e.g., oxonium, [hydrazinium(1+)], [hydrazinium= (2+)], (1-methyl-4-aza-1-azoniabicyclo[2.2.2]octane). When protons have been lost from hetero atoms of such ligands, "ato" names are used.

Example:

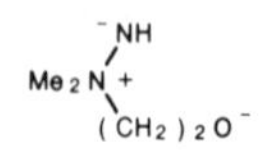
[1-(2-hydroxyethyl)-1,1-dimethyl= hydraziniumato(2−)]

Enclosing marks are used with all "ato" ligand names, and with ligands, like (cyano-C), that have letter locants; "bis," etc., (rather than "di," etc.) are used to indicate two or more, e.g., bis(benzoato). When a Ewens-Bassett number is also present, the "ato" name is bracketed, e.g., [tetrahydroborato(1−)]. When such a ligand name includes a multiplicative radical (¶ 125), it is closed off and followed by the Ewens-Bassett number, which in this case may be (1−), and the total name bracketed.

Example:

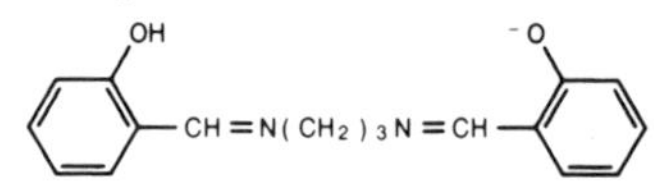
[[2,2'-[1,3-propanediylbis(nitrilo= methylidyne)]bis[phenolato]](1−)]

Ligands from dioximes, trihydrazones, etc., are expressed by enclosing the name and then citing the Ewens-Bassett number (which may be (1−) in this case also) to indicate the number of protons lost; e.g., [(2,3-butanedione dioximato)(1−)].

Neutral ligands are enclosed by parentheses or brackets (as required), except for ammine, aqua, carbonyl, and nitrosyl, e.g., (dinitrogen), (phosphine), (benzene); they are multiplied by "bis," "tris," etc.

Anionic mononuclear complexes are indexed at "-ate" headings derived from the name of the central element, e.g., **Borate, Cuprate.** The *total* charge of the complex anion is cited by a Ewens-Bassett number, and in the inverted part of the name the ligands are cited in alphabetical order. All cations are named in the modification in alphabetical order, except that hydrogen is placed last. Multiplicative terms are used with univalent cations if necessary, but when multivalent cations are present, ratios are employed for all. For metals of variable valency, Ewens-Bassett numbers are cited after their names. Ligands, with the exceptions outlined above, receive additional *Chemical Substance Index* entries.

Examples:

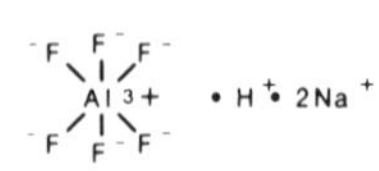
Aluminate(3−), hexafluoro-
disodium hydrogen, (*OC*-6-11)- (For the significance of the final term, see ¶ 203III.)

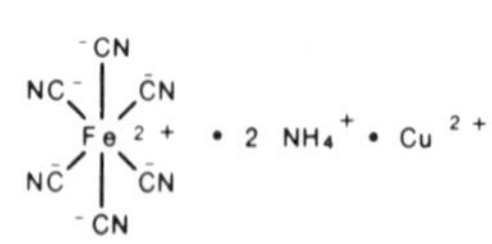
Ferrate(4−), hexakis(cyano-*C*)-
ammonium copper(2+) (1:2:1), (*OC*-6-11)- (Note that the ratio is expressed in the same order as the index name, i.e., first the heading, then the modification terms.)

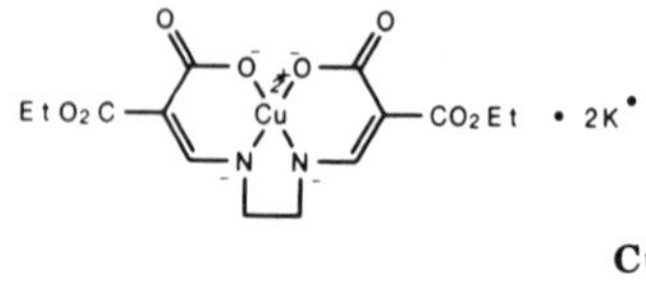
Cuprate(2−), [[1,1'-diethyl 2,2'-[1,2-ethanediylbis(imino= methylidyne)]bis[propane= dioato]](4−)]-
dipotassium (Note that ring systems formed entirely by coordination are not recognized as such in index names.)

Propanedioic acid, 2,2'-[1,2-ethanediylbis(iminomethylidy= ne)]bis-
1,1'-diethyl ester, copper complex

Anionic mononuclear complexes which contain simple anions as well as simple cations are named at the complex anion heading as shown above. The cations are named in the modification, followed by the simple anions. Thus **Borate(1−), tetrahydro−,** europium(2+) bromide (1:1:1). When a complex or organic cation is present, this receives the preferred entry.

Example:

$Li_4[Co(NH_3)_6]_8[Fe(CN)_6]_7$

Cobalt(3+), hexaammine−
(*OC*−6−11)−, lithium (*OC*−6−11)−hexakis(cyano−*C*)ferrate(4−) (8:4:7)

Ferrate(4−), hexakis(cyano−*C*)−
(*OC*−6−11)−, (*OC*−6−11)−hexaamminecobalt(3+) lithium (7:8:4)

When more than one complex anion is present, the choice is determined first by the preferred central element (see above); then by the highest numerical value, e.g., **Antimonate(3−)** preferred over **Antimonate(1−)**; finally, by the earliest index position of the entire entry.

Example:

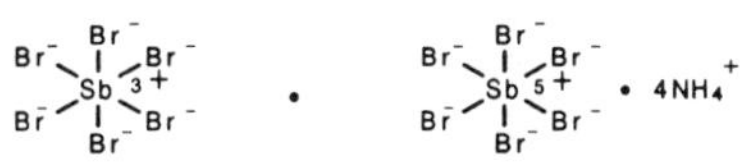

Antimonate(3−), hexabromo−
(*OC*−6−11)−, ammonium (*OC*−6−11)−hexabromoantimonate(1−) (1:4:1)

Antimonate(1−), hexabromo−
(*OC*−6−11)−, ammonium (*OC*−6−11)−hexabromoantimonate(3−) (1:4:1)

Complexes that contain ligands with *uncoordinated acid functions*, e.g., carboxylic acids or inorganic oxo acids, are named as though all acidic protons were dissociated from the uncoordinated groups; the cations and acid hydrogens are then cited in the modification.

Examples:

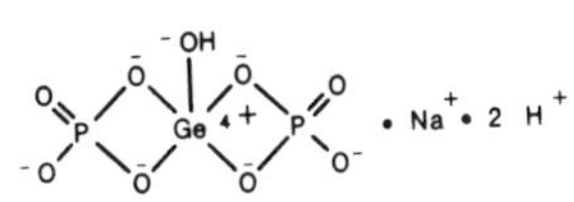

Germanate(3−), hydroxybis[phosphato(3−)−*O*,*O*']−
sodium dihydrogen

Phosphoric acid
germanium complex

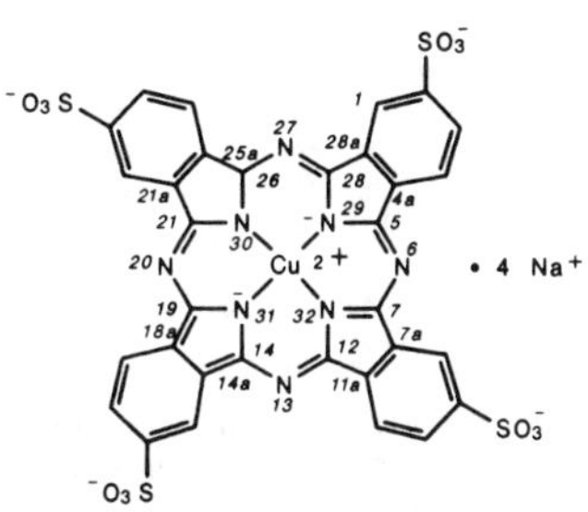

Cuprate(4−), [29*H*,31*H*−phthalocyanine−2,9,16,23−tetrasulfonato(6−)−N^{29},N^{30},N^{31},N^{32}]−
tetrasodium, (*SP*−4−1)− (not Copper, [29*H*,31*H*−phthalocyanine−2,9,16,23−tetrasulfonato(2−)−N^{29},N^{30},N^{31},N^{32}]−, tetrasodium salt)

29*H*,31*H*−Phthalocyanine−2,9,16,23−tetrasulfonic acid
copper complex

Cationic mononuclear complexes are indexed at the central element names with a Ewens−Bassett number to indicate the total charge on the cation. The ligands are cited in the inverted part of the heading in alphabetical order, the associated anions (if any) in the modification; ion terms, e.g., "ion(1+)," are not cited. Univalent anions are prefixed by "di," "tri," etc., if necessary; ratios are used with multivalent anions and with mixtures of anions and cations, the first numeral relating to the heading cation, the others to the modification terms in sequence. The *Formula Index* headings exclude the atoms expressed by modification terms. Additional *Chemical Substance* and *Formula Index* entries will be found for the anions with the following exceptions: "Class I" acids (¶ 185); **Ethanol; Methanol; Carbonic, Ethanedioic, Hydrocyanic,** and **Cyanic acids** and their chalcogen analogs and isomers; **Hydrazoic acid;** monoatomic anions; "methyl sulfate"; tetrafluoroborate(1−), tetraphenylborate(1−), and hexafluorophosphate(1−).

Example:

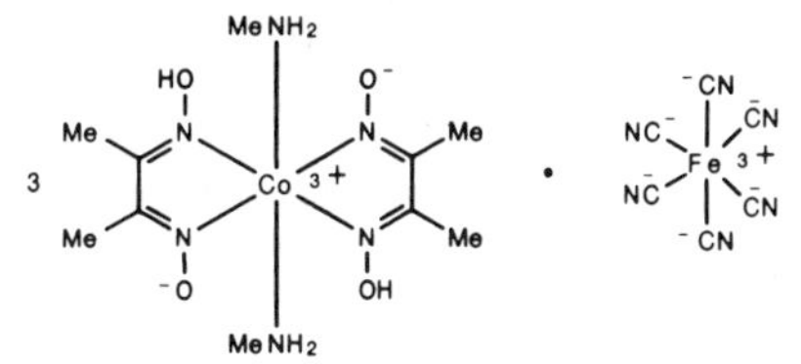

Cobalt(1+), bis[(2,3−butanedione dioximato)(1−)−*N*,*N*']bis(methanamine)−
(*OC*−6−11)−, (*OC*−6−11)−hexakis(cyano−*C*)ferrate(3−) (3:1)

Ferrate(3−), hexakis(cyano−*C*)−
(*OC*−6−11)−, (*OC*−6−11)−tris[bis(2,3−butanedione dioximato)(1−)−*N*,*N*']bis(methanamine)cobalt(1+)]

2,3−Butanedione
dioxime, cobalt complex

Cationic complexes containing both complex mononuclear cations and simple cations are indexed at the names of the former; associated complex anions are also indexed as usual.

Example:

$Li_2H_2[Co(NH_3)_6]_8[Fe(CN)_6]_7$

Cobalt(3+), hexaammine−
(*OC*−6−11)−, lithium hydrogen (*OC*−6−11)−hexakis(cyano−*C*)ferrate(4−) (8:2:2:7)

Ferrate(4−), hexakis(cyano−*C*)−
(*OC*−6−11)−, (*OC*−6−11)−hexaamminecobalt(3+) lithium hydrogen (7:8:2:2)(additional *Chemical Substance* and *Formula Index* entry)

When more than one complex cation is present in a compound, the choice of preferred index entry is based on the nature of the central atom (see above for precedence list of elements), e.g., the heading parent **Platinum(2+)** is preferred over **Palladium(3+).** If the elements are alike, the complex with the higher positive charge is preferred. Additional entries for other complex cations, and for anions and ligands, appear as usual.

Example:

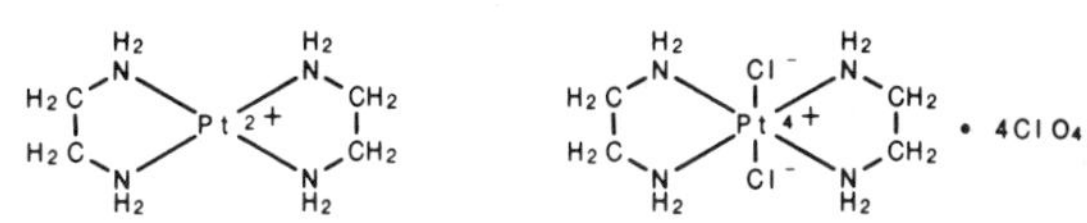

Platinum(2+), bis(1,2−ethanediamine−*N*,*N*')−, (*SP*−4−1)−, dichlorobis(1,2−ethanediamine−*N*,*N*')platinum(2+)
perchlorate (1:1:4)

Platinum(2+), dichlorobis(1,2−ethanediamine−*N*,*N*')−, (*SP*−4−1)−bis(1,2−ethanediamine−*N*,*N*')platinum(2+)
perchlorate (1:1:4)

Neutral mononuclear coordination complexes are indexed at the central element names without a Ewens−Bassett number.

Example:

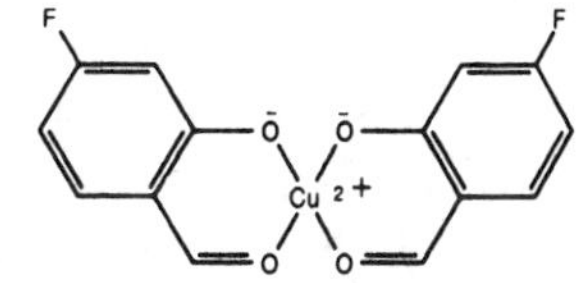

Copper, bis(4−fluoro−2−hydroxybenzaldehydato−*O*,*O*')−

Benzaldehyde, 4−fluoro−2−hydroxy−
copper complex

Neutral complexes with internal charge compensation, e.g., with ligands which are ylides, inner salts, etc., are named according to structures provided in the original documents.

Example:

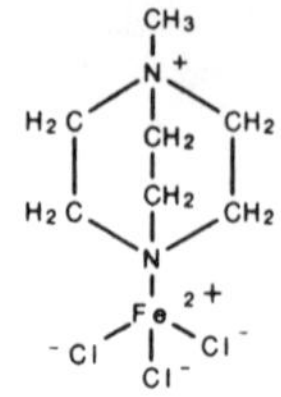

Iron, trichloro(1-methyl-4-aza-1-azoniabicyclo[2.2.2]octane-*N*[4])-
4-Aza-1-azoniabicyclo[2.2.2]octane, 1-methyl-
iron complex
1,4-Diazabicyclo[2.2.2]octane, 4-aza-1-azoniabicyclo[2.2.2]octane deriv.
iron complex

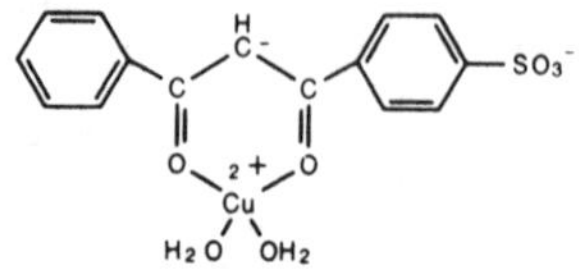

Copper, diaqua[4-(1,3-dioxo-3-phenylpropyl)benzenesulfonato(2–)-*O*[4],*O*[4']]-
Benzenesulfonic acid, 4-(1,3-dioxo-3-phenylpropyl)-
copper complex

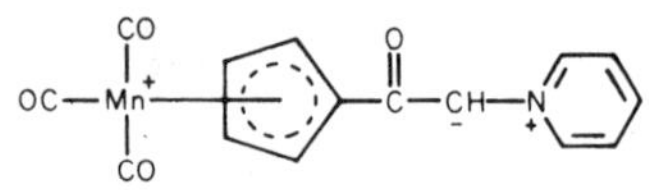

Pyridinium, 2-oxo-2-[η^5-(tricarbonylmanganio)-2,4-cyclopentadien-1-yl]ethylide
Manganese, tricarbonyl[(1,2,3,4,5-η)-1-(pyridinoacetyl)-2,4-cyclopentadien-1-yl]-
ylide

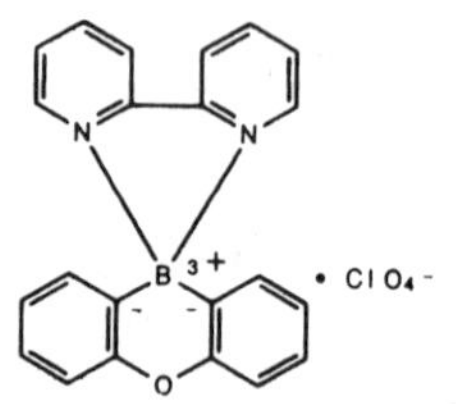

Boron(1+), (2,2'-bipyridine-*N*,*N'*)(oxydi-2,1-phenylene)-(*T*-4)-
perchlorate
Benzene, 1,1'-oxybis-
boron complex

Metallocenes (dicyclopentadienyl metal complexes in which all carbon atoms of the cyclopentadiene rings contribute to the metal–ligand bondings) are indexed at the parents **Ferrocene, Nickelocene, Cobaltocene, Osmocene, Ruthenocene,** etc.

Example:

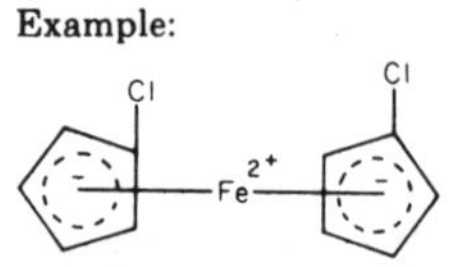

Ferrocene, 1,1'-dichloro-

When additional ligands or ring systems more complex than cyclopentadiene are present, the compounds are indexed at the names of the central metals. All positions of each cyclopentadiene ring are considered equivalent, and substituents are assigned lowest locants without regard to attachment of the metal. In order of precedence, metallocenes are ranked with the neutral coordination complexes (¶ 106), **Ferrocene,** for example, being placed just below **Iron** as an index parent. Radical names, e.g., ferrocenyl, 1,1'-ferrocenediyl, can be formed.

Example:

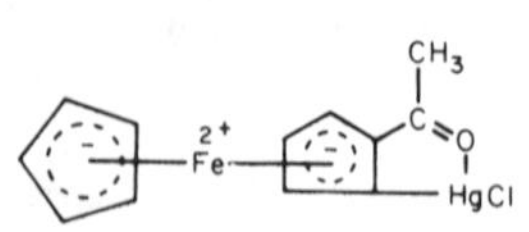

Mercury, (2-acetylferrocenyl-*C*,*O*)chloro-
Ferrocene, acetyl-
mercury complex

Suffixes are not attached to metallocene heading parents, and conjunctive names are not derived from them; thus, **Ferrocene, carboxy-** (not Ferrocenecarboyxlic acid); **Ferrocene, 1,1'-bis(carboxymethyl)-** (not 1,1'-Ferrocenediacetic acid). Bridged derivatives are named by use of bivalent substituent prefixes.

Example:

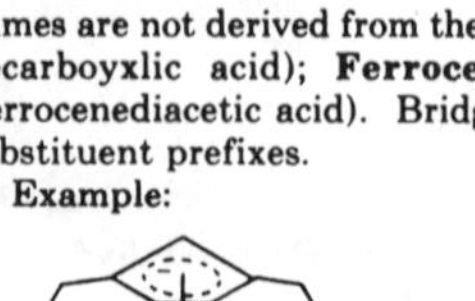

Ferrocene, 1,1':3,3'-bis(1,3-propanediyl)-

The cationic analog of **Ferrocene** is named **Ferrocenium.**

Example:

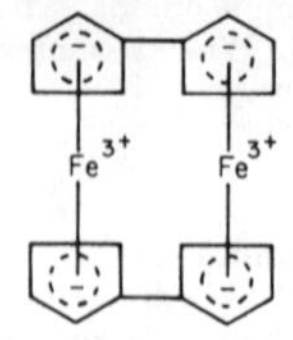

1,1'':1',1'''-Biferrocenium(2+)

Polynuclear coordination complexes may have (*a*) direct linkages between the central metal atoms; (*b*) bridging ligands connecting them; (*c*) both kinds of linkages. A bridging ligand is designated by the Greek letter μ (mu); when it binds more than two central atoms, a subscript is cited with it, e.g., μ_3. When the ligand name itself requires enclosing marks, an extra set is employed for the μ-ligand, otherwise none are used; e.g., μ-chloro, [μ-(acetato-*O*:*O'*)]. A μ-ligand is cited just ahead of the same ligand without the prefix, disregarding in the ordering process any accompanying letter locants, e.g., [μ-(acetato-*O*:*O'*)](acetato-*O*)-. Colons within letter-locant sets indicate the distribution of bonds from a bridging ligand to individual metal atoms.

Bridging ligands are not treated as multiplying radicals; i.e., the total of nonbridging ligands on all nuclear atoms is indicated directly. The number of nuclear atoms is indicated, after citation of all ligands, by "di," "tri," etc. Letter-locant sets are partitioned by colons. The cyclic structure of an oligomeric complex is indicated by "*cyclo*" as the first term in the modification. Direct linkages between like or unlike metals are always indicated, if known, by such terms as "(*Fe–Fe*)" or "(*Co–Re*)," whether or not bridging ligands are present. (In more complicated cases the total number of various bondings is expressed in terms such as "(3*Co–Co*)(3*Co–Fe*).") For "cluster" compounds, in which three or more like central atoms form a core, the geometrical descriptors *triangulo*, *tetrahedro*, or *octahedro* are assigned if the information is available.

The entry is made at the name (or derived "-ate" term) of the preferred metal, e.g., **Cobalt, Cobalt(2+), Cobaltate(1–)**; other metals with their associated ligands are named as complex ligands of the preferred atom(s). These complex ligands are placed in a single alphabetical sequence with bridging and nonbridging ligands bonded to the preferred metal. (Metal radicals, e.g., cobaltio, mercurio, are not used in naming polynuclear complexes.)

Examples:

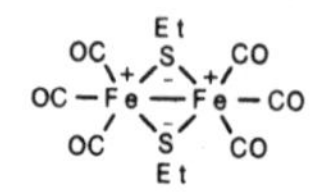

Iron, hexacarbonylbis[μ-(ethanethiolato)]di-
(*Fe–Fe*)

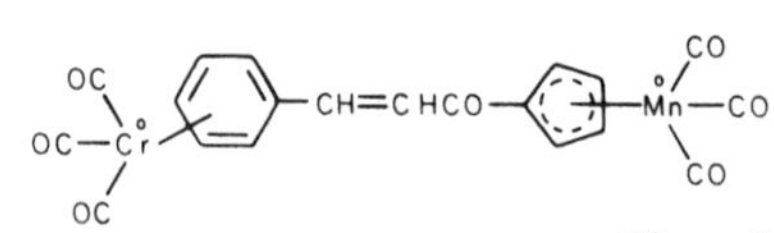

Chromium, tricarbonyl[μ-[η^5:η^6-1-(3-phenyl-1-oxo-2-propenyl)-2,4-cyclopentadien-1-yl]](tricarbonylmanganese)-
2-Propen-1-one, 1-(1,3-cyclopentadien-1-yl)-3-phenyl-
chromium-manganese complex
(additional *Chemical Substance Index* entry)

$(MeHg)_4S^{2+}$

Mercury(2+), tetramethyl-μ_4-thioxotetra-

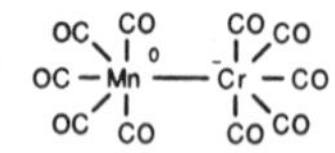

Chromate(1–), pentacarbonyl(pentacarbonylmanganate)-
(*Cr–Mn*)
(Note: The "-ate" ending is used in all central atoms names in *anionic* complexes)

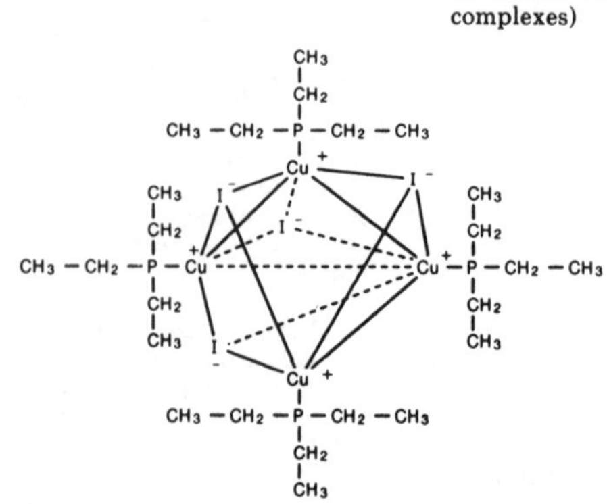

Copper, tetra-μ_3-iodotetrakis(triethylphosphine)tetra-
tetrahedro

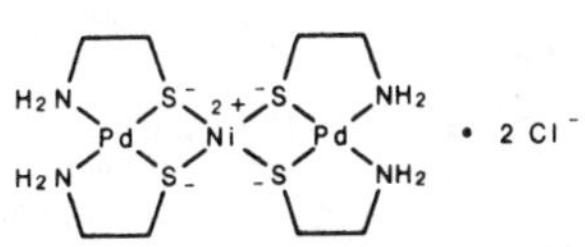

Palladium(2+), tetrakis[μ-(2-aminoethanethiolato-*N*,*S*:*S*)]=(nickel)di- dichloride
Ethanethiol, 2-amino- nickel-palladium complex

Polynuclear complexes with *extended structures* caused by bridging ligands are indexed at monomer headings, with a modification term such as "homo=polymer." Bridging ligands within the monomeric unit are named as usual, but those which bind the units to one another are not named by use of the "μ" prefix.

Examples:

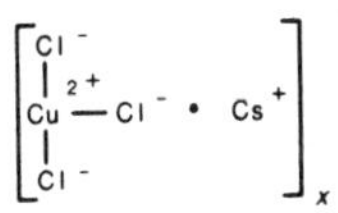

Cuprate(1-), trichloro-, cesium homopolymer

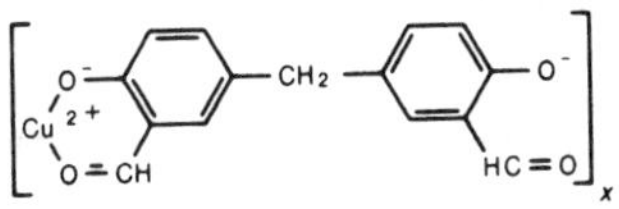

Copper, [[3,3'-methylenebis[6-hydroxybenzaldehydato]]=(2-)-O^1,O^6]- homopolymer
Benzaldehyde, 3,3'-methylenebis=[6-hydroxy- copper complex, homopolymer

Coordination compounds of indefinite structure are indexed (if the formula is unknown) at the organic ligand (if one is present) and at each metal "compounds" heading. These entries in the *Chemical Substance Index* are of a general nature and describe the entire series of complexes studied.

Example:

Pyridine, 2,6-dimethyl-
platinum complexes [if complexes with no other metal were studied]

Platinum, compounds
2,6-dimethylpyridine complexes [or "alkylpyridine complexes" or "nitrogen heterocycle complexes," etc., as appropriate]

A polynuclear coordination complex of known formula but unknown structure is indexed at the heading parent for the most preferred central element (or its "-ate" term). The ligands are cited in the inverted part of the heading in alphabetical order, followed by the less preferred metals (or their "-ate" terms) in descending order of precedence.

Examples:

$AlTiCl_2Et_2(OH)_2$	**Titanium, dichlorodiethyldi=hydroxy(aluminum)-**
$[AgAuCl_6]^{2-}$	**Aurate(2-), hexachloroargentate-**
$[Nb_6O_3(SO_4)_{12}]^{8-}$	**Niobate(8-), trioxododecakis[sul=fato(2-)]hexa-**
$[Cu_4(OH)_5[N[(CH_2)_2OH]_3]_4]^{3+}$	**Copper(3+), pentahydroxy=tetrakis[2,2',2''-nitrilotris[eth=anol]]tetra-** **Ethanol, 2,2',2''-nitrilotris- copper complex**

216. Dyes of established structure are indexed at their systematic names by the regular rules of substitutive nomenclature. For most classes of dyes a cross-reference will be found in the *Index Guide* at the *Colour Index* (C.I.) name, and the C.I. name is also used as a synonym (in the *Index Guide* and *Chemical Substance Index*) at the systematic name. The *Colour Index*[1] is employed as the chief reference source for the chemical constitution of trade-named dyes.

A dye of unknown structure is indexed at the C.I. name if this is available, otherwise only at a trade name or other designation used in the original document. When the *Colour Index* shows an indefinite structure, the sys=tematic name is still used if this can be done by omission of one or more locants or by use of alternative locants, e.g., "5(*or* 6)-chloro-."

Example:

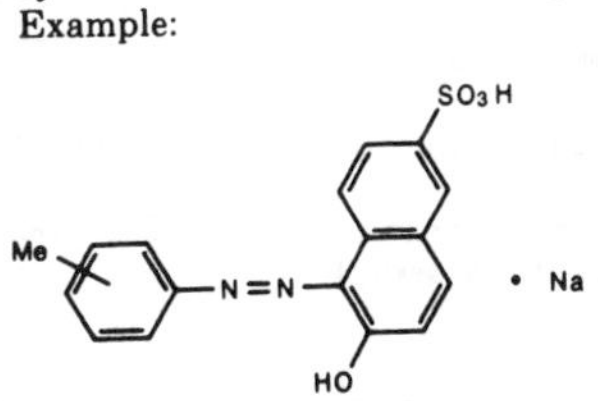

2-Naphthalenesulfonic acid, 6-hydroxy-5-[(methylphenyl)azo]- monosodium salt (cross-reference at C.I. Acid Orange 16)

[1]*Colour Index*, Third Edition, Second Revision, Society of Dyers and Co=lourists, Bradford, England, and American Society of Textile Chemists and Colorists, Research Triangle Park, N.C., 1982, 7 vols.

Deliberate mixtures (¶ 221) of dyes with known structures are indexed at each component with a "mixt. with" modification. Reaction products are not considered to be deliberate mixtures.

When an author supplies a structure for a trade-named dye which has been assigned a C.I. name but no constitution in the *Colour Index*, an entry will be found at the systematic name for the author's structures, and a separate entry, in the *Chemical Substance Index* only, at the C.I. name. The cross-reference continues to be based on *Colour Index* information; i.e., it runs from the trade name to the C.I. name as usual. If a trade name (or common name) is not listed in the *Colour Index*, the author's information (a C.I. number or name, or a structure) is used to make the index entry, cross-references and synonym.

Cyanine dyes (methines) are indexed systematically. Because the location of the charge may be shown differently in different documents, the preferred ring system (¶ 138) is selected as the site of the quaternized center in every case and, if further criteria are needed, the structure named is chosen to conform to the principles of maximum number of substituents, lowest locants, and earliest index position.

Example (name in original document: 3'-ethyl-3-methyl-8-phenoxythia=carbocyanine iodide):

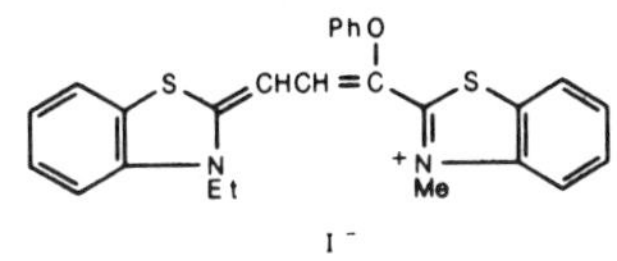

Benzothiazolium, 2-[3-(3-ethyl-2=(3*H*)-benzothiazolylidene)-1-phenoxy-1-propenyl]-3-methyl- iodide

Fluoresceins, phthaleins, and sulfonephthaleins have tautomeric structures in which a heterocyclic ring may be considered closed or opened, with gener=ation of a carboxyl or sulfo group.

Example (fluorescein):

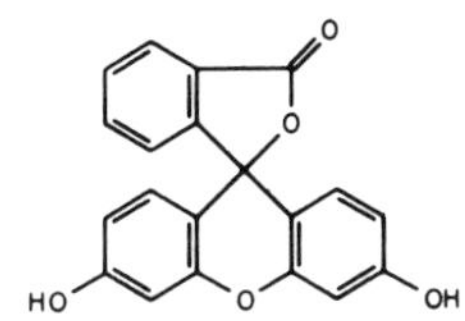

Spiro[isobenzofuran-1(3*H*),9'-[9*H*]xanthen]-3-one, 3',6'-di=hydroxy-

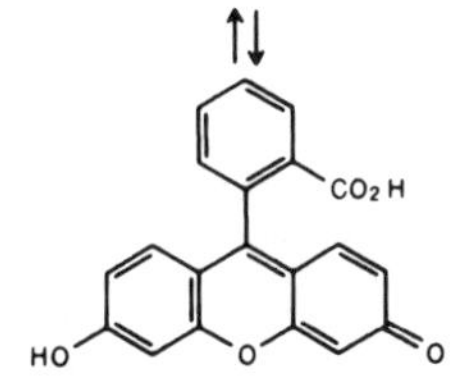

Benzoic acid, 2-(6-hydroxy-3-oxo-3*H*-xanthen-9-yl)-

These compounds and their derivatives, including metal salts, are named in the closed forms unless a covalently bound derivative of the acid group, e.g., an ester or amide, is present. The preferred (closed) forms are indexed as shown by the following cross-references:

Fluorescein.	See *Spiro[isobenzofuran-1(3H),9'-[9H]=xanthen]-3-one, 3',6'-dihydroxy-*
Phenolphthalein.	See *1(3H)-Isobenzofuranone, 3,3-bis(4-hydroxyphenyl)-*
Sulfonefluorescein.	See *Spiro[3H-2,1-benzoxathiole-3,9'-[9H]=xanthene]-3',6'-diol, 1,1-dioxide*
Phenolsulfonephtha=lein.	See *Phenol, 4,4'-(3H-2,1-benzoxathiol-3-ylidene)bis-, S,S-dioxide*

Azo dyes from pyrazolines and from acetoacetanilide derivatives are indexed as the keto rather than the enol forms.

Cationic resonance-stabilized diamino derivatives of phenazine, phen=oxazine, phenothiazine, xanthene, etc., are generally indexed as possessing the cation within the ring.

Lakes are pigments which comprise insoluble salts of soluble dyes. They are indexed at the systematic names, e.g., as calcium or barium salts at the acid heading parent, if the constitution is known. The structure of a transition-metal lake is usually not precisely defined; when it is, a coordination name (¶ 215) is used; otherwise it is indexed at a C.I. name (if such a name has been assigned to the lake itself) or at trade names.

New dyes, as reported in patent specifications, are indexed at systematic names, including coordination names (for metal complexes of precisely stated structure).

Zinc chloride double salts cause an indexing problem, both because the same dye can be used in the zinc-containing and zinc-free forms, and because the toxicity of soluble zinc salts renders misindexing a serious matter. The presence of zinc is never assumed; it is excluded from cross-references at C.I.

names and included in index entries for a specific trade–named dye only when it is listed in the *Colour Index* as a zinc salt, or when its presence is indicated by an original document. When a ratio is known, the appropriate entry can be made both at the cationic heading, e.g., **Benzenediazonium,** and at the anion, e.g., **Zincate(1–), trichloro–**, or **Zincate(2–), tetrachloro–**. Otherwise it is indexed as a molecular addition compound (¶ 192) of the "–ium" chloride and **Zinc chloride ($ZnCl_2$)** with no ratio cited at either heading.

Sulfur dyes seldom have a known constitution. The preferred entry is a C.I. name, if available; otherwise the trade name is indexed, or, as a last resort, the starting material, e.g., **Phenol, 4–amino–**, sulfur dyes from.

217. Elementary particles and their atomic and molecular states are indexed as chemical substances. Elementary particle names follow a systematic scheme. In this scheme, a given letter (roman or greek, upper or lower case) is used in the context of a given set of quantum numbers. This scheme replaces the haphazard assignment of letters used previously.[2] These particles may be subdivided as follows:

(*a*) Particle classes and particles without symbols or masses; e.g., **Boson, Fermion, Lepton, Hadron, Baryon, Meson, Hyperon, Nucleon, Tachyon, Quark.** These terms are employed as heading parents.

(*b*) Particles with symbols but no reported mass; e.g., Σ–Hyperon, π–Meson, Muon, Alpha particle, Beta particle, Neutron, Proton, Neutrino, Deuteron, Tau particle. The index headings have the form **Meson,** *π*, in which the particle symbol appears in the heading after the name.

(*c*) Particles with unique symbols and masses; e.g., Hyperon Λ(2100), Positive Muon, π^+–Meson(140). The index entries are of the form **Muon,** μ^+(106) and **Meson,** π^+(140).

(*d*) Particle resonances without reported masses; e.g., Meson resonance $K^+\ \pi$, Nucleon resonance Δ, Nucleon resonance N^*. The index entry corre= sponds to the information available, e.g., **Nucleon,** resonance; **Nucleon,** resonance Δ. For multiple resonances, more than one *Chemical Substance Index* entry appears; e.g., **Meson,** *K*, resonance $K\pi\pi$; **Meson,** *π*, resonance $K\pi\pi$.

(*e*) Particle resonances with reported masses.

Examples of index entries:

Nucleon
  resonance N^***(2040)**

Meson, *K*
  resonance K^+K^+**(1280)**

Meson, *K*
  resonance *KN***(1780)** (preferred entry)

Nucleon
  resonance *KN***(1780)** (additional entry)

(*f*) Particles composed of nuclei of elements of less than 98% natural abundance; e.g., lithium–6, boron–10, are indexed in the modification at the element headings. Specific particles from isotopes of hydrogen and helium–3 and –4 are indexed at **Proton, Deuteron, Triton, Tau particle,** and **Alpha particle,** respectively. The zero–charge nucleon is indexed at **Neutron.**

Examples of index entries. (Notation in the original document: ^{115}In= (^{7}Li,3n)^{119m}*Te*):

Lithium, reactions
  isotope of mass 7, indium–115 bombardment by

Indium, reactions
  isotope of mass 115, bombardment of, by lithium–7

Tellurium, properties
  isotope of mass 119, nuclear energy levels of, metastable, from lithium–7 bombardment of indium–115

Neutron
  from lithium–7 bombardment, of indium–115

In particle–particle and particle–nuclei interactions, all new particles and nuclei are indexed. otherwise, interactions of known particles and nuclei are indexed according to the author's emphasis.

(*g*) Particles of nuclei of elements of at least 98% natural abundance. Such particles are not named at the beginning of the modification at the element name.

Examples of index entries. (Notation in the original document: ^{63}Cu= (^{12}C,*αp*)):

Carbon, reactions
  ions of carbon–12, copper–63 bombardment by

Copper, reactions
  isotope of mass 63, bombardment of, by carbon–12

Proton
  from carbon–12 bombardment, of copper 63

Alpha particle
  from carbon–12 bombardment, of copper 63

(*h*) Antielements and antiparticles are indexed at the element and par= ticle names with "Anti–" as a prefix; the symbol (the usual one with a vinculum added) is cited in the heading after a comma, and the isotope number of the antielement is given in the modification; e.g., **Antihelium,** $\bar{\mathrm{He}}$, isotope of mass 4; **Antiproton,** $\bar{p}$. The antielectron is indexed at **Positron,** and the negative muon and positive pion antiparticles are named as the particles of opposite charge, e.g., **Meson,** π^+; **Muon,** μ^- (106).

(*i*) Atomic and molecular states of elementary particles. General studies are indexed at such headings as **Mesonic atom,** *K*–**Mesonic atom,** and **Muonic molecule.** Specific atomic and molecular states are indexed at the element name ("compounds" category) or particle name.

Examples of index entries:

1. The original document describes the *K*Be mesonic atom:

Meson, *K*
  mesonic atom with beryllium (*K*Be) (preferred index entry)

Beryllium, compounds
  K–**mesonic atom (*K*Be)** (additional entry)

2. The original document describes the exotic combination of muonium with chlorine:

Muon, *μ*
  leptonic mol. with chlorine ($\mu^+e^-Cl^0$) (preferred index entry)

Chlorine
  leptonic mol. ($\mu^+e^-Cl^0$) (additional entry)

The atomic and molecular states of elementary particles are assigned the preferred entry at the particle name, with an additional entry for the atom or molecule.

(*j*) **Muonium** and **positronium** are used for the bound states (μ^+e^-) and (e^+e^-).

(*k*) Hypernuclei from hyperon interaction with atomic nuclei are indexed in the *Chemical Substance Index* at the element name.

Example:

Helium
  hypernucleus of helium–5

218. Enzymes are indexed by *CA* at the names recommended by the Enzyme Commission of the International Union of Biochemistry (IUB)[3] and Supplements[4] as far as is compatible with CA nomenclature and indexing practices. Each specific enzyme is assigned a CAS Registry Number which is cited after the preferred index name in the *Chemical Substance Index*, where entries are to be found, and also in the *Index Guide* at **Enzyme Commission,** where E.C. designations are listed as part of cross–references leading to the preferred names. The E.C. designations are four–figure sets, e.g., E.C. 2.6.1.2 for Alanine aminotransferase, in which the first figure denotes one of the six main classes based on the type of reaction catalyzed by the enzyme, the second and third figures indicate further classification, and the final figure is a unique enzyme identification number. Whereas the E.C. cross–references lead directly from designation to enzyme name, the reverse route can also be taken by the user. At **Dehydrogenase** in the *Index Guide* appears the cross–reference "For related subclasses, see E.C. 1." Such subclasses include **Reductase, Oxidase, Hydrogenase,** and **Hydroxylase,** all with E.C. 1 designations.

Names for enzymes may comprise (*a*) a trivial substrate name combined with an action term, e.g., Lactate dehydrogenase, which is indexed at **De= hydrogenase, lactate** (this is E.C. 1.1.1.27, at which designation a cross–reference can be found); (*b*) a substrate name and the suffix "–ase," e.g., Adenosine triphosphatase, which is indexed at **Phosphatase, adenosine tri–**; (*c*) a name, similar to those in (*b*), but from which a heading parent expressing the enzyme action cannot be readily separated, e.g., **Asparaginase,** which is indexed at this name; and (*d*) other single terms which may describe the source, e.g., **Papain,** or the action, e.g., **Lysozyme** or **Chymotrypsin**; such enzymes (mainly proteinases) are indexed at these names.

Qualifying phrases, which form part of the total enzyme name, appear in parentheses in the boldface heading after the main part of the name and express coenzyme specificity, secondary enzyme activity, secondary product formed, etc. Sometimes more than one type of information is expressed in this way.

Examples:

Dehydrogenase, malate (decarboxylating)

Synthetase, acyl coenzyme A (guanosine diphosphate–forming)

Dehydrogenase, glyceraldehyde phosphate (nicotinamide adenine dinucleotide phosphate) (phosphorylating)

---

[2]*The Review of Particle Properties* (Particle Data Group) is recognized as an authority in this field. It is published in alternate years in *Review of Modern Physics* and *Physics Letters B.*

[3]"Enzyme Nomenclature: Recommendations (1978) of the International Union of Biochemistry" (Academic Press, New York, 1979).

[4]See ¶ 297, item 13, below.

General studies of enzymes, and new enzymes, will be found at **Enzymes** in the *General Subject Index*. When complete primary structural information (amino acid sequence) is reported for an enzyme, an entry with as much source specificity as possible is made, e.g., **Pepsin A (human pancreas)**, and a separate CAS registration linked to the structural information is prepared.

219. Inorganic compounds are indexed at names based on United States usage and IUPAC rules[5]. For isotopically labeled inorganic compounds, see ¶ 220.

Elements of atomic number higher than 103, prior to 1982, were indexed at such headings as **Element 104**. They are now named by a combination of three syllables derived from the atomic number, the numerals 0–9 corresponding to the syllables nil, un, bi, tri, quad, pent, hex, sept, oct, and enn. The final "n" of enn is elided before nil, and the "i" of bi and tri before ium, which is the invariant suffix.

Example:

Unnilpentium (formerly Element 105; cross-references from Bohrium, Hahnium, and Nielsbohrium)

The initial letters of the first three syllables of the names provide the symbols from which synonym line formulas are derived for binary compounds of these superheavy elements.

Example:

Ununnilium fluoride ($UunF_4$) (formerly Element 110 fluoride (1:4))

These index headings are not subdivided into categories (see Appendix II, ¶ 10B) as is done for elements 1 through 103.

Elements reported or discussed without specific reference to atomic number are indexed at **Elements** in the *General Subject Index*. Class headings in the same index, e.g., **Group IVA elements, Alkaline earth metals, Halogens, Helium–group gases, Platinum metals, Transition metals,** are employed for general studies of varying specificity.

Molecular forms of the elements are indicated in the modification, except that hydrogen and its isotopes, and nitrogen, oxygen, and the halogens are assumed to be diatomic (unless the term "atomic" is cited). Examples: **Sulfur,** mol. (S_8); **Oxygen; Hydrogen,** atomic.

Uninverted salt–type heading parents are used to index binary compounds, double salts, mixed salts, and certain other ionic and simple covalent compounds. The electropositive (cationic) constituents are named first in alphabetical order, followed by electronegative (anionic) constituents in the following sequence: Rn, Xe, Kr, Ar, Ne, He, B, Si, C, As, P, N, H, Te, Se, S, At, I, CN, SCN, Br, Cl, O, O_2^{2-}, F, N_3^{1-}, OCN. (Elements not listed here are treated as metals; prior to the Tenth Collective Index (1977–1981) antimony was included among the nonmetals.) For monoatomic anions, the ending "–ide" is cited. When a binary compound contains only nonmetals, the same sequence determines the order of the total name, e.g., Silicon carbide (SiC).

A synonym line formula follows the name of each uninverted salt–type heading parent. (Prior to the Tenth Collective Index (1977–1981), line formulas did not appear with unambiguous names, e.g., **Sodium chloride**, **Zinc sulfide**.) Beginning with the Twelfth Collective Period (1987–1991), line formulas may be expressed with decimals or numerical ranges as well as integers.

Example:

Sodium tungsten oxide ($Na_{0.37}WO_3$)

Hydrides of the metals antimony, bismuth, germanium, tin, and lead are indexed at hydride names such as **Germane** (¶ 199). Halides are indexed as derivatives at these headings, e.g., **Stannane, tetrachloro–** (not Tin chloride ($SnCl_4$)); **Plumbane, diiodo–**. Hydrides, halides, etc., of most other elements are indexed at headings such as **Sodium hydride (NaH); Thorium hydride iodide ($ThHI_3$).** The hydrides of nitrogen have trivial names: **Ammonia,** for NH_3; **Amidogen,** for NH_2; **Hydrazine,** for N_2H_4; and **Hydrazoic acid,** for HN_3. Some derivatives of ammonia also have trivial names: **Chloramine,** for NH_2Cl; **Fluorimide,** for NHF_2; **Hydroxylamine,** for $HONH_2$; **Sulfamide,** for $SO_2(NH_2)_2$; **Sulfimide,** for SO_2NH. The cyclic trimer of sulfimide is indexed at the heterocyclic name: **1,3,5,2,4,6–Trithiatriazine,** 1,1,3,3,5,5–hexaoxide.

Chalcogen hydride names include **Water,** for H_2O; **Hydrogen peroxide (H_2O_2); Hydrogen trioxide,** for H_2O_3; **Hydroxyl,** for HO; **Hydroperoxo,** for HO_2; **Hydrotrioxo,** for HO_3; **Selenyl,** for HSe; and **Hydrogen sulfide (H_2S).**

Salts of hydric cations such as **Ammonium, Phosphonium, Nitric acidium,** are indexed at uninverted salt names with the "–ium" term cited first, or at "oxo" acid headings with "–ium salt" terms in the modification.

Examples:

Ammonium chloride ($(NH_4)Cl$)

Phosphonium cyanide ($(PH_4)CN$)

Oxonium chloride ($(OH_3)Cl$)

Perchloric acid, compounds
  ammonium salt (named like a metal salt of the same acid, see below)

Salts of **Hydrazine** and **Hydroxylamine** are named as molecular addition compounds (¶ 192), e.g., **Hydrazine, sulfate (2:1).**

---

[5]International Union of Pure and Applied Chemistry, "Nomenclature of Inorganic Chemistry," 2nd ed., Buterworths, London, 1971.

Binary oxides are indexed at headings such as **Sodium oxide (Na_2O), Iron oxide** (with a synonym line formula, if known), **Chlorine oxide (ClO_2).** Uninverted peroxide, superoxide, and ozonide headings are also employed, e.g., **Sodium peroxide (Na_2O_2).** The oxides CO, CO_2, SO_2, SO_3, and SiO_2 are indexed at **Carbon monoxide, Carbon dioxide, Sulfur dioxide, Sulfur trioxide,** and **Silica,** respectively.

Mixed–metal oxides containing only the oxide (O^{2-}) anion are indexed at oxide headings, with the metals cited in alphabetic order, e.g., **Iron zinc oxide (Fe_2ZnO_4).** Hydroxides of one or more metals are indexed at metal hydroxide headings, e.g., **Chromium hydroxide ($Cr(OH)_3$).** Mixed–metal oxides, hydroxides, and oxide–hydroxides that also contain a Group IA and/or Group IIA metal are named at oxy–anion headings with cation terms in the modification, e.g., **Stannate ($Sn(OH)_6^{2-}$), dipotassium,** (*OC*–6–11)–; **Tantalate ($Ta_6(OH)O_{18}^{7-}$), heptapotassium; Molybdate(2–), tetra–μ–oxotetraoxoferratedi–, disodium.** Hydroxide–oxides that contain no Group IA or IIA metals are indexed at such headings as **Manganese hydroxide oxide (= Mn(OH)O).** The index heading parents **Chromic acid (H_2CrO_4), Chromic acid ($H_2Cr_2O_7$), Manganic acid (H_2MnO_4)** and **Permanganic acid ($HMnO_4$)** are retained. Hydroperoxides are indexed at peroxide headings with synonym line formulas, e.g., **Sodium peroxide ($Na(HO_2)$),** for Na(OOH). Mixed peroxide, hydroperoxide, ozonide, and superoxide headings are also employed. Sulfides, selenides, and tellurides are named like the oxides. The compounds CS_2 and COS are indexed at **Carbon disulfide** and **Carbon oxide sulfide (COS),** respectively.

Uninverted salt–type headings are used for halides, amides, arsenides, azides, borides, carbides, cyanides, hydrazides, hydroxylamides, imides, nitrides, phosphides, and silicides. Synonym line formulas are used to differentiate the hydrazides ($H_3N_2^{1-}$) and ($H_2N_2^{2-}$); for special cases such as iodide (I_3^{1-}), chloride (HCl_2^{1-}), and fluoride (HF_2^{1-}); and for arsine, phosphine, and stibine which have lost one or two protons, e.g., **Arsenide (HAs^{2-}).** Otherwise it is understood the anions are monoatomic and have lost all protons, e.g., **As^{3-}, P^{3-}, Si^{4-}.** Examples: **Sodium chloride (NaCl), Potassium iodide (KI), Potassium iodide ($K(I_3)$), Calcium niobium fluoride** [formula unknown]. **Sodium phosphide ($Na(H_2P)$), Cyanogen chloride ($(CN)Cl$).**

In these compounds, and the oxides, sulfides, etc., discussed earlier, all metals are considered to be cationic. Nonmetals are placed in cationic or anionic sequence to balance the charges. If this violates the electronegativity sequence described earlier, all metals are named as cations and all nonmetals as "–ide" anion terms.

Examples:

Aluminum gallium arsenide ($(Al,Ga)As$)

Cadmium silicon phosphide ($CdSiP_2$) (not Cadmium phosphide silicide)

Thallium phosphide selenide (TlPSe) (not Thallium selenium phosphide)

Mercury iodide phosphide ($Hg_3I_4P_2$)

Salts of interhalogen anions are indexed as coordination compounds (¶ 215), e.g., **Iodate(1–), dichloro–,** sodium, for $Na[ICl_2]$. Acetylides are indexed at names such as **Sodium acetylide ($Na(C_2H)$).** The common name **Calcium carbide ($Ca(C_2)$)** is the single exception. Metal derivatives of *substituted* **Ethyne** and of other alkynes are indexed by organometallic nomenclature (¶ 194).

Examples:

BrC≡CAg	**Silver, (bromoethynyl)–** **Ethyne, bromo–** **silver complex** (additional *Chemical Substance Index* entry)
EtC≡CNa	**Sodium, 1–butynyl–** **1–Butyne** **sodium complex** (additional *Chemical Substance Index* entry)

Graphite compounds are indexed at the **Graphite** index heading parent and at the name of the other component(s) as molecular addition compounds. Prior to *CA* Volume 95 (see ¶ 101), binary headings such as **Graphitic acid, Graphite nitrate**, etc., were employed.

Example:

C · 1/10 HNO_3	**Nitric acid, compd. with graphite (1:10)** **Graphite, compd. with nitric acid (10:1)**

The following metal–oxide radical names are sometimes used: americyl (O_2Am– and O_2Am=), chromyl (O_2Cr=), neptunyl (O_2Np– and O_2Np=), permanganyl (O_3Mn–), perrhenyl (O_3Re–), pertechnetyl (O_3Tc–), plutonyl (O_2Pu– and O_2Pu=), titanyl (O_2Ti=), uranyl (O_2U– and O_2U=), vanadyl, (OV–, OV=, and OV≡), and zirconyl OZr=). These names, and those for the thio, seleno, and telluro analogs, are used for the ions, e.g., **Vanadyl ion(2+),** and to complete the salt names at coordination anion headings (¶ 215). They are not used for their compounds with simple inorganic salts, nonmetal oxo acids or organic acids.

Example:

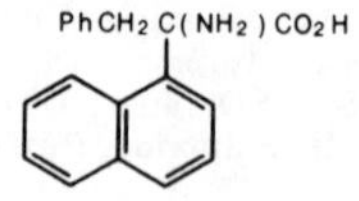

1-Naphthaleneacetic acid, α-amino-α-(phenylmethyl)-

Radicals derived from the names of amino acid stereoparents, e.g., glycyl, L-alanyl, L-phenylalanyl, are employed only in naming peptides (¶ 206). In other situations, systematically named radicals are used; e.g., (aminoacetyl) instead of glycyl. The phenylalanyl radical is considered to be a specially-named one-part radical; no parentheses are placed around it when it is unsubstituted. Stereoparent radical names are derived by replacing the final "-ine" by "-yl" (an exception is **Cysteine,** which affords cysteinyl); **Tryptophan** gives tryptophyl; **Aspartic acid** and **Glutamic acid** radicals are as follows:

$-COCH_2CH(NH_2)CO-$	aspartoyl
$HO_2CCH_2CH(NH_2)CO-$	α-aspartyl
$-COCH_2CH(NH_2)CO_2H$	β-aspartyl
$-CO(CH_2)_2CH(NH_2)CO-$	glutamoyl
$HO_2C(CH_2)_2CH(NH_2)CO-$	α-glutamyl
$-CO(CH_2)_2CH(NH_2)CO_2H$	γ-glutamyl

Amino acids are given less preference for special treatment than other stereoparents. They are ranked just above the unsubstituted parent acids; thus, **Alanine** (2-aminopropanoic acid) is ranked above **Propanoic acid** but below **Butanoic acid.** All carbon-substituted derivatives of **Glycine** and **β-Alanine** are indexed at systematic names, but the carbon-substituted *radicals* are permitted in peptide nomenclature.

Examples:

Bz NHCH2 CO2H (N 2 1)

Glycine, *N*-benzoyl-

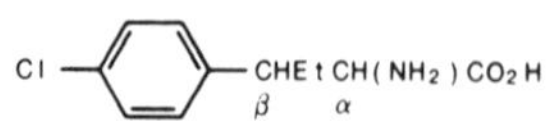

L-**Phenylalanine, 4-chloro-β-ethyl-**
threo- (not L-Norvaline, 3-(4-chlorophenyl)-, *threo*-)

Me(CH2)5 CMe(NH2) CO2H

Octanoic acid, 2-amino-2-methyl- (not Alanine, α-hexyl-)

PhCH(NH2) CH2 CO2H

Benzenepropanoic acid, β-amino- (*R*)- (not β-Alanine, β-phenyl-, (*R*)-)

O-Substituted derivatives of the hydroxy amino acids (**Allothreonine, Homoserine, Serine, Threonine,** and **Tyrosine**) and *S*-derivatives of the mercapto amino acids (**Cysteine** and **Homocysteine**) are named at those stereoparents, the alkyl derivatives as substituents, e.g., **Serine, *O*-methyl-,** and the acyl derivatives as esters in the modification, e.g., L-**Cysteine, acetate (ester).** The stereoparent **Methionine,** $CH_3SCH_2CH_2CH(NH_2)COOH$, is the *S*-methyl derivative of **Homocysteine.** Derivatives in which this methyl group is substituted are indexed as **Homocysteine** derivatives. *S*-Oxide derivatives of **Cysteine, Homocysteine,** and **Methionine** are indexed as sulfinyl and sulfonyl derivatives of simpler parents.

Derivatives of the branched-chain acids **Isoleucine, Isovaline, Leucine,** and **Valine,** in which the methyl side chain is substituted are indexed as derivatives of the linear acid (**Butanoic acid** or **Norvaline**) if locants for the side chain would otherwise be needed. Conjunctive names are not formed from amino acid stereoparents attached to ring systems; instead, such a combination is usually indexed at a systematic conjunctive name with an "α-amino" substituent. Derivatives of **Arginine,** $H_2NC(:NH)NH(CH_2)_3CH(NH_2)COOH$, substituted in the guanidino group are indexed as derivatives of **Ornithine,** $H_2N(CH_2)_3CH(NH_2)COOH$.

Histidine (R = H) is capable of existing in the following tautomeric forms:

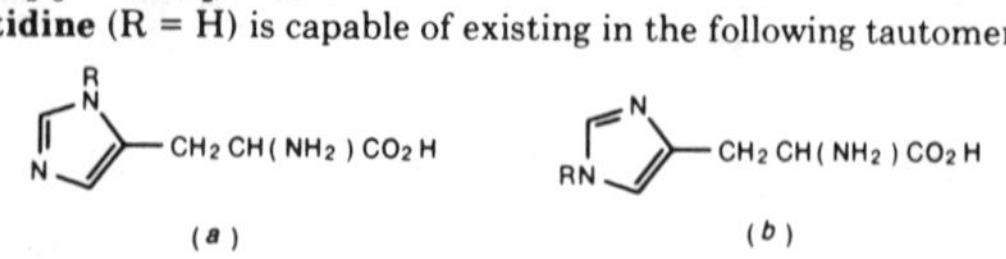

(a) (b)

When the tautomers are stabilized by substitution on a nitrogen atom of the ring, two isomers result. The side chain has been arbitrarily assigned to the 4-position of the ring and, if R = ethyl, the structures shown above are indexed:

(*a*) L-**Histidine, 3-ethyl-**
(*b*) L-**Histidine, 1-ethyl-**

It should be noted that while (*b*) conforms to the correct treatment for 1*H*-**Imidazole,** the numbering of (*a*) violates the rules (3*H*-**Imidazole** is not used as a heading parent) and is adopted only for these special cases. A 1-substituent is assumed in indexing indefinite 1- or 3-derivatives of **Histidine.**

Esters of amino acid stereoparents with nonstereoparents are indexed at the amino acid names; e.g., L-**Alanine,** 4-carboxyphenyl ester. Esters with other stereoparents are indexed at the parent which represents the greatest number of highest functions. Usually only one index entry is made for an ester. Locants are used with ester terms when necessary.

Examples:

$EtO_2C(CH_2)_2CH(NH_2)CO_2Me$

L-**Glutamic acid**
5-ethyl 1-methyl ester

$HO_2CNHCHMeCO_2Et$

L-**Alanine, *N*-carboxy-**
1-ethyl ester (the locant "1" is cited for clarity, even though the *N*-ethyl ester is indexed as L-**Alanine, *N*-(ethoxycarbonyl)-** (¶ 185))

Amides of amino acids are indexed at systematic names, e.g., **Acetamide, 2-amino-** (not Glycinamide), but peptide amides (¶ 206) are indexed at *C*-terminal amino acid names, e.g., **Glycinamide,** L-**alanyl-.**

Anhydrides of amino acids are indexed at the preferred acid heading parent with "anhydride" or "anhydride with" terms in the modification. Cyclic anhydrides and lactones are indexed as heterocycles. Hydrazides are indexed at acid headings with "hydrazide" terms in the modifications. Molecular addition compounds of nonstereoparents with amino acids named as stereoparents receive preferred index names at the latter, but *Chemical Substance* and *Formula Index* entries will be found also at the other component(s). The preferred name of an addition compound of two or more stereoparents is chosen on the basis of the maximum number of highest functions in the stereoparent.

Beryllium, magnesium, aluminum, gallium, indium, thallium, and transition-metal salts of amino acids are indexed as coordination compounds (¶ 215) if sufficient information is presented, and the special stereochemical descriptors (¶ 203 III) for such compounds are then adopted; additional entries appear in the *Chemical Substance Index* for amino acid ligands other than **Alanine, Glycine,** and a few common substituted glycine derivatives. Other salts are indexed at such names as **Glycine,** sodium salt; L-**Phenylalanine,** methyl ester, hydrochloride.

Example:

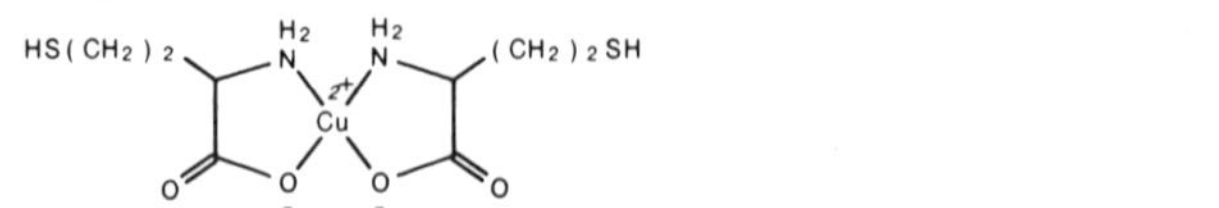

Copper, bis(L-**homocysteinato-*N*,*O*)-**
(*T*-4)-
L-**Homocysteine**
copper complex (additional *Chemical Substance Index* entry)

206. Peptides are generally named by use of amino acid stereoparents (¶ 205) and stereoparent radicals; the L-isomers are assumed for indexing purposes in the absence of contrary information. Trivial names of some peptides and all proteins are employed, and special "Cyclo" names are used for tetra- and higher cyclic peptides. The systematic ring names are covered by cross-references in the *Index Guide* or by additional entries in the *Chemical Substance Index* and *Index of Ring Systems.*

Linear peptides of two through twelve units are indexed at "peptide systematic names." The heading parent is the *C*-terminal amino acid, and all enclosing marks and locants for points of attachment of the amino acid radicals are cited. *N*-Derivatives of the radicals are expressed as substituents. Esters of carboxy groups of aspartic and glutamic acid residues are cited in the modification; *O*- and *S*-acyl derivatives of seryl, cysteinyl, etc., radicals are expressed by use of (*O*-acetyl-L-seryl), etc., radicals; when a hydroxyl- or mercapto-group-containing amino acid is the heading parent, the ester, e.g., "acetate," is cited in the modification.

Examples:

H-L-Met-L-Asp(NH2)-L-Pro-D-Phe-L-Phe-OH

L-**Phenylalanine, *N*-[*N*-[1-(*N*²-L-methionyl-L-asparaginyl)-L-prolyl]-D-phenylalanyl]-**

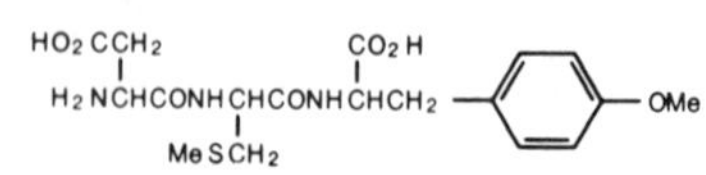

L-**Tyrosine, *N*-(*N*-L-α-aspartyl-*S*-methyl-L-cysteinyl)-*O*-methyl-**

MeO2C(CH2)2 (CH2)2CO2Me
PhCH2O2CNHCHCONHCHCONHCHCONHCHCO2Me
MeO2C(CH2)2 (CH2)2CO2Me

L-**Glutamic acid, *N*-[*N*-[*N*-[*N*-[(phenylmethoxy)carbonyl]-L-α-glutamyl]-L-α-glutamyl]-L-α-glutamyl]-**
pentamethyl ester

Linear tridecapeptides and larger compounds are indexed at *amino-acid-sequence names*, similar to peptide systematic names (above) but with om=

Examples:

$H_2N(HO)S$	**Amidosulfoxylic acid**
H_2NSCl	**Amidosulfenyl chloride**
HO(NC)SO	**Cyanosulfurous acid**
Cl(NC)SO	**Thionyl chloride cyanide**
$(HO)_2S(S)$ (*or* HO(HS)S(O))	**Thiosulfurous acid ($H_2S_2O_2$)**
$Cl(H_2N)S{=}NH$	**Amidoimidosulfurous chloride**
$HO(N_3)SO_2$	**Azidosulfuric acid**
$Cl(HO)SO_2$	**Chlorosulfuric acid**
$HO(H_2N)SO_2$	**Sulfamic acid**
Cl_2SO_2	**Sulfuryl chloride**
$F(ON)SO_2$	**Nitrososulfonyl fluoride**
$HO(H_2N)S(O){=}NH$	**Imidosulfamic acid**
$F_2N(H_2N)SO_2$	**Sulfamide, *N,N*–difluoro–**
$Cl(HO)S(O){=}NH$	**Chloroimidosulfuric acid**
$Cl_2S(O)S$	**Thiosulfuryl chloride ($(S_2O)Cl_2$)**
$(H_2N)_2S(O){=}NH$	**Imidosulfamide**
$HOO(HO)SO_2$	**Peroxymonosulfuric acid**
$HO(HSS)SO_2$	**Thioperoxymonosulfuric acid ($(HO)(HSS)SO_2$)**

Polynuclear sulfur and selenium acids have such names as **Disulfuric acid** (for $(HO)S(O_2)O(O_2)S(OH)$), **Diselenious acid** (for (HO)Se(O)O(O)Se(OH), **Trisulfuric acid**, etc. Analogs are named by use of nondetachable prefixes and replacement of the acid term (if all hydroxyl groups are absent).
Examples:

$(N_3)SO_2OSO_3H$	**Azidodisulfuric acid**
$HO(SO_2NH)_2SO_3H$	**Diimidotrisulfuric acid**
$H_2NSO_2NHSO_3H$	**Amidoimidodisulfuric acid**
HO_3SOOSO_3H	**Peroxydisulfuric acid ($[(HO)S(O)_2]_2(O)_2$)**
$ClSO_2OSO_2OSO_2Cl$	**Trisulfuryl chloride**
$H_2NSO_2OSO_2NH_2$	**Disulfamide**
$(ClSO_2)_2NNH_2$	**1,1–Hydrazinedisulfonyl dichloride**

Phosphorus and arsenic acids. For mononuclear phosphorus and arsenic acids, both nonsubstitutive, e.g., **Phosphoric acid, Arsenous acid,** and substitutive, e.g., **Phosphonic acid,** see ¶ 197. There are no official rules for naming polynuclear phosphorus and arsenic acids; they are indexed by *CA* at names traditionally used in the literature. Pyrophosphoric acid and its analogs and derivatives are indexed at **Diphosphoric acid** names, and similarly for Pyrophosphorous acid. Specific esters of **Metaphosphoric acid** ($H_3P_3O_9$) are indexed at the heterocyclic parent **1,3,5,2,4,6–Trioxatriphosphorinane** (¶ 197), and other meta acids are treated similarly. Mixed polynuclear phos= phoric–phosphorous acids are indexed at headings which include Stock num= bers (parenthetical Roman numerals) to indicate the sequence of trivalent and pentavalent phosphorus atoms. When all hydroxyl groups have been replaced, the acid term gives way to acid halide, amide, etc., suffixes. Amides of poly= phosphorous acids are indexed by replacing the word "acid" by "amide" preceded by a multiplicative prefix; amides of polyphosphoric acids have names in which "–ic acid" is simply replaced by "–amide," e.g., **Diphosphorous tetraamide,** for $[(H_2N)_2P]_2O$, and **Peroxydiphosphoramide,** for $[(H_2N)_2P{=}(O)]_2O_2$. P,P', etc., locants are used to designate the position of replacement amido and imino, bridging imido, etc., groups.

Arsenic analogs are named analogously to the phosphorus examples below, except that "Hypo–" and "Isohypo–" names and headings in which mixed valencies are denoted by "III" and "V" terms are not employed.
Examples:

$(HO)_2P(O)OP(O)(OH)_2$	**Diphosphoric acid**
$(HO)_2P(O)OP(O)(OH)OP{=}(O)(OH)_2$	**Triphosphoric acid**
$(HO)_2P(O)P(O)(OH)_2$	**Hypophosphoric acid**
$HO(H)P(O)OP(O)(OH)_2$	**Isohypophosphoric acid**
$(HO)_2POP(OH)_2$	**Diphosphorous acid**
$(HO)_2POP(O)(OH)_2$	**Diphosphoric(III,V) acid**
$(HO)_2POP(O)(OH)OP(OH){=}OP(O)(OH)_2$	**Tetraphosphoric(III,V,III,V) acid**
$(HO)_2P(O)NHP(O)(OH)_2$	**Imidodiphosphoric acid**
$[(HO)_2P(O)OP(O)(OH)]_2NH$	***P'*–Imidotetraphosphoric acid**
$Cl_2P(O)OP(O)(OH)Cl$	**Trichlorodiphosphoric acid**
$H_2N(HO)P(O)OP(O)(OH)_2$	**Amidodiphosphoric acid**
$(HO)_2POP(O)(OH)NH_2$	***P'*–Amidodiphosphoric(III,V) acid**
HO(HS)P(O)SP(O)(OH)SH	**Thiodiphophoric acid ($[(HO)(HS)P(O)]_2S$)**
$(HO)_2P(S)SSP(S)(OH)_2$	**Thioperoxydiphosphoric acid ($[(HO)_2P(S)]_2S_2$)**
$H_2N(HO)POP(S)(OH)_2$	***P*–Amido–*P'*–thiodiphosphoric= (III,V) acid**
$Cl(H_2N)P(O)OP(O)Cl_2$	**Amidodiphosphoryl chloride**
$Cl_2P(O)OP(O)(Cl)F$	**Diphosphoryl chloride fluoride ($Cl_2P(O)OP(O)ClF$)**
$[(HO)_2P(O)]_3N$	**Nitridotriphosphoric acid**
$[(H_2N)_2P(O)NH]_2P({=}NH)NH_2$	***P'*–Iminodiimidotriphosphoramide**

For further examples, see ¶ 197.
Nitrogen oxo acids are indexed at the following headings:

HONO	**Nitrous acid**
$HONO_2$	**Nitric acid**
HON=NOH	**Hyponitrous acid**
HON=N(O)OH	**Hyponitric acid**

Halide and halogenoid derivatives are indexed at nitryl and nitrosyl names, e.g., **Nitrosyl chloride ((NO)Cl).** The amides of nitric and nitrous acid are **Nitramide** and **Nitrosamide,** respectively; organic derivatives are named as *N*–nitro and *N*–nitroso amines.
Halogen oxo acids include:

HOCl	**Hypochlorous acid**
HOBrO	**Bromous acid**
$HOClO_2$	**Chloric acid**
$HOIO_2$	**Iodic acid (HIO_3)**
$(HO)_2IO_2$	**Iodic acid (H_2IO_4)**
$HOClO_3$	**Perchloric acid**
$HOIO_3$	**Periodic acid (HIO_4)**
$(HO)_5IO$	**Periodic acid (H_5IO_6)**

Mixed halides are indexed at such names as **Chloryl fluoride ($(ClO_2)F$).**
Silicon oxo acids in *CA* indexing are named as **Silicic acid** with one of the following synonym line formulas: **H_2SiO_3, H_4SiO_4, $H_2Si_2O_5$, $H_6Si_2O_7$**. The same index heading is employed for cyclic acids of the general formulas $H_{2n}Si_nO_{3n}$ and $H_{2n}Si_{2n}O_{5n}$ in which *n* is equal to or greater than 3. The compound $(HO)_3SiSi(OH)_3$ is named **Disilanehexol**. (See also ¶ 199.)

Anhydrides of inorganic oxo acids with carboxylic, sulfonic, etc., acids are indexed at the organic acid heading (¶ 179); mixed inorganic anhydrides which contain neither free nor esterified hydroxyl groups are given uninverted salt names.
Examples:

$ClONO_2$	**Chlorine nitrate ($Cl(NO_3)$)**
$B(OPO_2)_3$	**Boron metaphosphate ($B(PO_3)_3$)**

When free (or esterified) hydroxyl groups are present, the index entry is made at the preferred acid heading with an "anhydride with" phrase is the index modification.

Examples:

$EtOSO_2OClO_3$	**Perchloric acid** anhydride with ethyl hydrogen sulfate
$HOSO_2OSi(OH)_2OSO_2OH$	**Sulfuric acid** anhydride with silicic acid (H_4SiO_4) (2:1)

Mixed anhydrides of inorganic peroxy and thio (etc.) acids are indexed similarly, with use of "anhydrosulfide with" terms when appropriate, but anhydrides with a peroxy, dithio, etc., linkage are indexed at **Peroxide, Disulfide,** etc.

Example:

$HO_3SSSP(O)(OH)_2$	**Disulfide, phosphono sulfo**

Metal (and helium-group element) *salts* of inorganic oxo acids and their analogs are indexed at the oxo heading if only one anion is present (otherwise the salt is indexed at its own heading, e.g., **Potassium phosphate sulfate ($K_2H_3(PO_4)(SO_4)$)**. A Ewens–Bassett number is cited with a metal of variable valence, and the stoichiometry of the salt is indicated by "mono," multiplicative prefixes, or ratios, when necessary.

Examples:

$3\ HClO_3 \cdot Al$	**Chloric acid** **aluminum salt**
$3\ HNO_3 \cdot Ag$	**Nitric acid** **silver(1+) salt (3:1)**
$H_2SO_4 \cdot Ca$	**Sulfuric acid** **calcium salt (1:1)**
$HOSO_2OSO_2OH \cdot Na$	**Disulfuric acid** **monosodium salt**
$2\ H_2NSO_2OH \cdot Sn$	**Sulfamic acid** **tin(2+) salt (2:1)**
$HN(SO_2F)_2 \cdot Ag$	**Imidodisulfuryl fluoride** **silver(1+) salt**
$H_3PO_4 \cdot NH_3 \cdot Mg$	**Phosphoric acid** **ammonium magnesium salt (1:1:1)**
$2\ HClO_4 \cdot Xe$	**Xenon perchlorate ($Xe(ClO_4)_2$)**

Alums, chrome alums, etc., are indexed (*a*) at oxo acid names, e.g., Sulfuric acid, compounds, aluminum potassium salt (2:1:1), dodecahydrate, (*b*) at "ium" headings, e.g., **Methanaminium,** *N,N,N*–**trimethyl–**, aluminum sulfate (1:1:2) or (*c*) (if they contain conjugate acids of nitrogen bases) at names of acids and bases, e.g., **Sulfuric acid,** aluminum salt (2:1), compd. with guanidine (1:1), with an additional entry at **Guanidine,** compd. with aluminum sulfate (1:1:2).

Anions are indexed at names usually identical with those employed in binary cation–anion headings (see above), e.g., **Hydride**, for H^{1-}; **Carbide**, for C^{4-}; **Carbide (C_2^{2-})**; **Hydrazide ($H_3N_2^{1-}$)**. Anions from inorganic oxo acids are indexed at "ate" and "ite" headings with the retained hydrogen atoms or derivatives expressed in the modification; an ion term is expressed for esters and salts. Oxo anions of metals have synonym line formulas; other polyatomic anions are assigned coordination anion names.

Examples:

$H_2PO_4^-$	**Phosphate** **dihydrogen**
$EtSO_4^-$	**Sulfuric acid,** **monoethyl ester, ion(1–)**
AlO_2^-	**Aluminate (AlO_2^{1-})**
NO_3^{2-}	**Nitrate(2–), trioxo–**

Anions of elements without "–ide" names are indexed at the element names, with "ion" modification terms followed by a symbol to show the charge and composition, e.g., ion (Re^{1-}), ion (Hg_2^{2-}), ion (S_8^{1-}).

Cations. The cation of hydrogen in the gas phase is indexed at **Proton** (¶ 217). Examples of other cation names are **Sulfonium,** for SH_3^{1+}; **Chloronium,** for ClH_2^{1+}; **Ammonium,** for NH_4^{1+}; **Ammoniumyl,** for NH_3^{1+}; **Hydrazinium(1+),** for $N_2H_5^{1+}$; **Diphosphinium(2+),** for $P_2H_6^{2+}$; **Thiosulfuric acidium(1+) ($H_3S_2O_3^{1+}$); Plutonyl ion(1+),** for PuO_2^{1+}; **Nitryl ion,** for NO_2^{1+}; **Magnesium(1+), diiodo–,** for MgI_2^{1+}; **Helium(1+), hydro–,** for HeH^{1+}. Cations of elements are indexed at element headings with the ion specified in the modification, e.g., **Carbon,** ion (C_2^{1+}).

Minerals are indexed at traditional and new mineral names as correlated by the Commission on New Minerals and Mineral Names of the International Mineralogical Association and reported in *The American Mineralogist.* Other references used to standardize the spelling and transliteration of mineral and rock names, as well as to standardize mineral formulas in the *CA* indexes are:

(*a*) American Geological Institute, *Glossary of Geology.* Washington, D.C., 1972, 857 pp.

(*b*) E. S. Dana, *A Textbook of Mineralogy*, 4th ed., revised and enlarged by W. E. Ford. John Wiley and Sons, New York, 1958, 851 pp.

(*c*) J. D. Dana and E. S. Dana, *System of Mineralogy*, 7th ed., entirely rewritten and enlarged by C. Palache, H. Berman, and C. Frondel, John Wiley and Sons, New York, Vol. I, 1944; Vol. II, 1951; Vol. III, 1962.

(*d*) R. W. Fairbridge, *The Encyclopedia of Geochemistry and Environmental Sciences.* (Volume IVA of *The Encyclopedia of Earth Sciences* series). Van Nostrand Reinhold Co., New York, 1972, 1321 pp.

(*e*) M. Fleischer, *Glossary of Mineral Species*, Mineralogical Record, Tucson, Arizona, 1980, 192 pp.

(*f*) M. H. Hey, *An Index of Mineral Species and Varieties*, 2nd. rev. ed., Trustees of the British Museum (Natural History), London, 1955, 728 pp. Appendixes, 1963, 135 pp; 1974, 168 pp.

(*g*) C. Hintze, *Handbuch der Mineralogie*, Supplement 2, *Neue Mineralien und Neue Mineralnamen.* Walter de Gruyter, Berlin, 1960, 958 pp.

(*h*) C. M. Rice, *Dictionary of Geological Terms*, Edwards Brothers, Ann Arbor, Mich., 1961, 465 pp.

(*i*) H. Strunz, *Mineralogische Tabellen*, 4th. rev. ed., Akademische Verlagsgesellschaft Geest und Portig K.–G., Leipzig, 1966, 560 pp.

(*j*) P. W. Thrush, *A Dictionary of Mining, Mineral, and Related Terms.* U.S. Department of the Interior, 1968, 1269 pp.

Minerals of definite composition have a synonym line formula cited after the name, e.g., **Chromatite ($Ca(CrO_4)$)**. When no mineral formula is given in the original document, the assumed formula is that given in reference (*e*), above. Cross-references from less preferred to more preferred index names appear in the *Index Guide.* New minerals are indexed at names and formulas supplied in original documents and collected in each issue of the *General Subject Index* at **Minerals, new.**

When one or more elements have replaced the whole or part of the original element in a mineral, and no new mineral name has been coined, the replacing element is named in the modification at the original heading.

Examples:

$BaSi_2O_5$	**Sanbornite ($Ba(Si_2O_5)$)**
$Li_2Si_2O_5$	**Sanbornite** **lithium ($Li_2(Si_2O_5)$)**
$LiNaSi_2O_5$	**Sanbornite** **lithium sodium ($LiNa(Si_2O_5)$)**
Zn_2SiO_4	**Willemite ($Zn_2(SiO_4)$)**
$ZnMgSiO_4$	**Willemite** **magnesium ($ZnMg(SiO_4)$)**

When partial replacement has occurred but stoichiometric information is lacking, or when unusual elements are present, adjectival terms for these elements are cited in the modification in the *Chemical Substance Index.* The adjectival terms all end in "–an," and are derived from the English or Latin names of the elements, e.g., aluminian, beryllian, hydrogenian, aurian (from gold), sodian, zincian. The forms "–oan" and "–ian" are used for the lower and higher oxidation states of arsenic, copper (cuproan and cuprian), iron (ferroan and ferrian), lead (plumboan and plumbian), manganese, mercury, and uranium.

Example:

$(Ba,Li)Si_2O_5$	**Sanbornite** **lithian**

Intermetallic compounds are indexed like molecular addition compounds (¶ 192).

Example:

$AlGe_2$	**Aluminum** **compd. with germanium (1:2)** **Germanium** **compd. with aluminum (2:1)** (additional *Chemical Substance Index* entry)

220. Isotopes. The isotopes of hydrogen of atomic masses two and three are indexed at **Deuterium** and **Tritium,** respectively. Isotopes of other elements with a natural abundance less than 98% are indexed at the usual element names with a phrase in the modification: "isotope of mass..." When the isotope is metastable, this word appears in the descriptive part of the modification, e.g., "formation of metastable." Anions of nonmetals are named in the index headings, metal anions and all cations in the modification.

Examples:

Chloride ($^{38}Cl^{1-}$)
Hydride–*d* (not Deuteride)
Tin
  isotope of mass 120 ($^{120}Sn^{1-}$)
Chlorine
  isotope of mass 37 ($^{37}Cl^{1+}$)

Molecular forms of isotopic elements are named in the index heading or indexed with "mol." or "ion" terms in the modification; "mol. with" phrases are employed for mixed hydrogen isotopes.

Examples:

Iodide ($^{131}I_3{}^{1-}$)

Nitrogen
mol. ($N^{15}N$) (a labeled atom is cited *after* an unlabeled atom of the same element)

Helium
mol. ($^3He_2$)

Deuterium
ion ($D_2{}^{1+}$)

Tritium
mol. with hydrogen (HT)

Isotopically labeled organic compounds are indexed by the Boughton system by placing the symbol for the isotope (with a subscript numeral to indicate the number of isotopic atoms) after the name or after the relevant portion of the name; in either case, locants are cited if necessary. The locants (except Greek letters) and symbols are in italics, and hyphens are used to separate them from one another and from the remainder of the names. (In the following examples "*d*" has been used, but the tritium-labeled compounds are named quite analogously by citation of "*t*." When both "*d*" and "*t*" are cited, they are separated by a hyphen, e.g., **Ethane**-*1-d-2-t*.)

In labeled heading parents with single-word names and no hydrogen-containing principal group, the *d* is placed after the parent. No locant is needed if the positions are equivalent or are fully deuterated.

Examples:

MeD	**Methane**-*d*
CD_4	**Methane**-d_4
F_3CCH_2D	**Ethane**-*d*, **2,2,2-trifluoro**- (not Ethane-*2-d*, 1,1,1-trifluoro-)
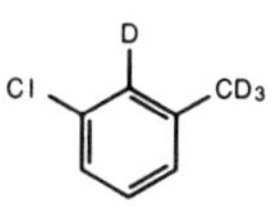	**Benzene**-*d*, **2-chloro-6-(methyl**-d_3)- (not Benzene-*2-d*, 1-chloro-3-(methyl-d_3)-)
$MePD_2$	**Phosphine**-d_2, **methyl**-
$(D_2N)_2CO$	**Urea**-d_4

Fully deuterated alcohols, amines, or imines are indexed similarly, but partially deuterated compounds of these classes have the "*d*" symbol placed after the appropriate part(s) of the name, with locants if necessary.

Examples:

$DOND_2$	**Hydroxylamine**-d_3
D_3CCD_2OD	**Ethanol**-d_6
$(D_3C)_2NH$	**Methan**-d_3-**amine**, *N*-(**methyl**-d_3)-
H_3SiND_2	**Silanamine**-d_2
DNHOD	**Hydroxyl**-*d*-**amine**-*d*
$D(CH_2)_2NH_2$	**Ethan**-*2-d*-**amine**
MeOD	**Methanol**-*d*
$MeSO_3CH_2D$	**Methan**-*d*-**ol methanesulfonate** (isotopically labeled "Class I" alcohols (¶ 185) belong to "Class II")
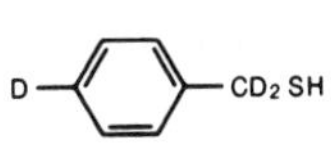	**Benzene**-*4-d*-**methane**-*α,α*-d_2-**thiol**
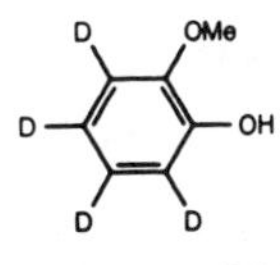	**Phen**-*2,3,4,5*-d_4-**ol, 6-methoxy**- (not Phen-*3,4,5,6*-d_4-ol, 2-methoxy-)
$MeCH(ND_2)CO_2D$	DL-**Alanine**-*N,N,1*-d_3
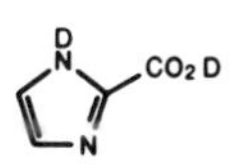	1*H*-**Imidazole**-*1-d*-**2-carboxylic acid**-*d*

When a heading parent contains locants the labeling of each part is expressed separately and locants for the labeling are cited.

Example:

$DOCD_2CD_2OD$	**1,2-Ethane**-*1,1,2,2*-d_4-**diol**-d_2

The "*d*" symbol is placed after the appropriate word in multiword headings for classes not yet discussed, e.g., **Acetic**-d_3 **acid**-*d*; **Acetyl**-d_3 **chloride.** In most other cases the "*d*" is cited after the complete heading parent, and conventional locants or italicized words are often necessary to denote the labeled position.

Examples:

DCH_2CDO	**Acetaldehyde**-*1,2*-d_2
$D(CH_2)_2CONHD$	**Propanamide**-*N,3*-d_2
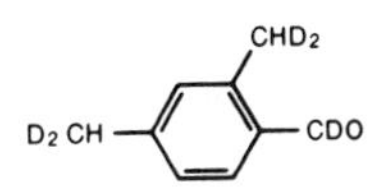	**Benzaldehyde**-*formyl-d*, **2,4-di(methyl**-d_2)- (not "bis(methyl-d_2)"; the labeled and unlabeled compound names are kept as similar as possible)

Enclosing marks are used with a labeled radical if it is preceded by a locant which expresses its attachment to a heading parent or parent radical, but not if such a locant belongs to the radical itself.

Examples:

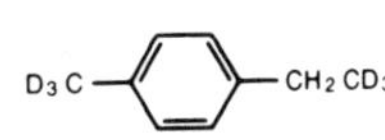	**Benzene, 1-(ethyl**-*2,2,2*-d_3)-**4-(methyl**-d_3)-
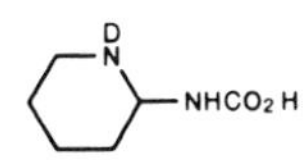	**Carbamic acid, 2-piperidinyl**-*1-d*-

Addition of deuterium, alone or with hydrogen, to a ring system is indicated by hydro "substituents" and the "*d*" symbol. Where there is a choice, deuterium is expressed in the heading parent.

Example:

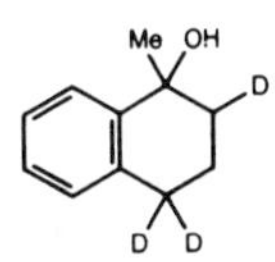	**1-Naphthalen**-*2,4*-d_2-**ol, 1,2,3,4-tetrahydro**-*4-d*-**1-methyl**-

Labeled protonated species (¶ 184) are expressed in index modifications by terms such as "conjugate monoacid-*d*" and "monoprotonated-*d*".

When locants are expressed in a heading parent or parent radical for unsaturation, hetero atoms, indicated hydrogen, spiro or ring-assembly junctions, bridges (in fused ring systems), suffixes, or points of attachment (in radicals), locants are cited for the labeled positions, whether or not their use would otherwise be necessary.

Example:

$(D_3C)_2CO$	**2-Propanone**-*1,1,1,3,3,3*-d_6

Exceptions are "*d*" and "*t*" terms placed after the word "acid" and after the suffixes of amines, imines, alcohols, etc. (see above), for which locants are seldom employed.

The examples above have dealt with organic compounds labeled with deuterium (tritium-labeling is handled similarly). Isotopes of elements other than hydrogen in organic compounds are expressed by the appropriate symbols. Nomenclature is similar, except that the instances in which the isotopic symbol appears within the name are restricted to (*a*) names comprising more than one word, (*b*) acids and acid derivatives with "carboxylic", "sulfonic," etc., names, and (*c*) conjunctive names.

Examples (labeled atoms are indicated by asterisks):

$MeC(\dot{O})\dot{O}H$	**Acetic**-$^{17}O_2$ **acid**
$PhC(\dot{O})OMe$	**Benzoic**-^{18}O **acid** ^{16}O-**methyl ester**
$Ph\dot{C}H_2\dot{C}O_2H$	**Benzeneacetic**-*carboxy,α*-$^{14}C_2$ **acid**
$MeC\dot{O}F$	**Acetyl**-^{17}O **fluoride**
$\dot{O}=C=\dot{O}$	**Carbon dioxide**-$^{18}O_2$

H3GeCN | **Germane-**^{74}*Ge*-**carbonitrile-**^{15}*N*

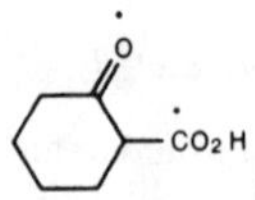

Cyclohexanecarboxylic-^{14}*C* **acid, 2-(oxo-**^{17}*O*)-

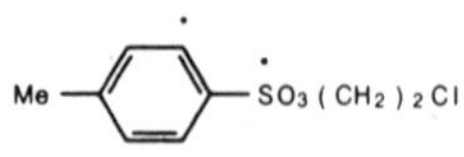

Benzene-2-^{14}*C*-**sulfonic-**^{35}*S* **acid 2-chloroethyl ester** (labeled "Class I" acids (¶ 185) belong to "Class II")

In general, for all other compounds isotopic labeling of a heading parent is expressed after the name, although, as in the first example below, a combination of these policies with those for "*d*" (or "*t*") is sometimes necessary. Symbols for different isotopic elements which fall together are cited in alphabetical order and separated by hyphens. Labeling of radicals is expressed after the individual simple radical names.
Examples:

OCH2CH2OH	**Ethan-**2-*d*-**ol-**1-^{14}*C*
MeCONH2	**Acetamide-**1-^{13}*C*-^{15}*N*
MeCONDCH2CHDMe	**Acetamide-**N-*d*-^{15}*N*, **N-(propyl-**2-*d*)-

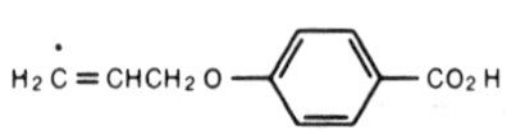

Benzoic acid, 4-(2-propenyl-3-^{14}*C*-**oxy)-**

Apart from isotopes of "hydro" (see above), a multiplicity of substituents which are identical except for labeling are named separately. Multiplicative nomenclature is not employed with unsymmetrically labeled parents; instead, that heading parent is chosen which contains (*a*) the maximum number of isotopic atoms, (*b*) the alphabetically earliest isotope symbol.
Examples:

(H3C)2SiMe2	**Silane, dimethyldi(methyl-**^{13}*C*)-
Cl2C:	**Methylene, chloro-**^{35}*Cl*-**chloro-**^{37}*Cl*-
Me2CHNHCDMe2	**2-Propanamine-**2-^{14}*C*, **N-(1-methylethyl-**1-*d*)- (not **2-Propan-**2-*d*-**amine, N-(1-methylethyl-**1-^{14}C)-)

For derivatives named in modifications, labeling is expressed by such terms as "oxime-^{15}*N*," "calcium-^{44}*Ca* salt," "di(hydrate-d_2)." For labeled hydrates and ammoniates, additional index entries will be found at **Water-**d_2, **Ammonia-**^{15}*N*, etc. Ammonium-d_4 salts are treated like the analogous metal salts.

A name containing isotopic symbols is always employed for a labeled compound, regardless of how little of the labeled species is present so long as its nature is known. When the number of labeled atoms is unknown, the heading for the unlabeled compound is indexed and the modification contains a phrase such as "labeled with deuterium" or "labeled with chlorine-37."
Examples:

Acetamide, labeled with carbon-14

1-Propene, 1,1,2,2,3,3,3-hexachloro-, labeled with chlorine-36

When only the positions of the isotopic atoms are unknown, the labeling is indicated, if possible, in the appropriate part of the name without citation of locants. Otherwise a "labeled with..." modification is used.
Examples:

1-Propene-^{14}*C*$_2$, **2-methoxy-**

Benzoic acid, 4-(ethyl-^{14}*C*)-

Benzoic acid, 4-ethyl-, labeled with carbon-14

Isotopically labeled *inorganic compounds* are named by procedures similar to those above when the unlabeled compounds have unambiguous names.
Examples:

DNHT	**Ammonia-***d*-*t*
$P(SiD_3)_3$	**Phosphine, tri(silyl-**d_3)-
$DP(O)(O^-)_2$	**Phosphonic-***d* **acid ion(2-)**
$H_3{}^{32}PO_4$	**Phosphoric-**^{32}*P* **acid**
$H_3PO_4 \cdot 3\ {}^{22}Na$	**Phosphoric acid tri(sodium-**^{22}*Na*) **salt**
$HDSO_4$	**Sulfuric acid-***d*
^{35}ClT	**Hydrochloric-**^{35}*Cl* **acid-***t*
^{18}OSO	**Sulfur dioxide-**^{18}*O*
$D_2{}^{18}O$	**Water-**d_2-^{18}*O*

In other cases labeled inorganic compounds are differentiated by synonym line formulas which contain isotope symbols. Isotopic atoms are cited after unlabeled atoms of the same element.
Examples:

Hydrogen sulfide (D_2S)

Cobalt iron oxide ($CoFe_2O_3{}^{18}O$)

Molybdenum carbonyl ($Mo(CO)_4({}^{13}CO)_2$)

Nitrogen fluoride ($N^{15}NF_2$)

Sodium chloride ($^{24}NaCl$)

Sodium sulfide ($Na_2{}^{35}S$)

Uranium chloride ($^{213}UCl_3$)

Line formulas containing isotopic symbols are cited also for compounds which do not require them in the unlabeled form, e.g., **Sodium chloride ($^{24}NaCl$).**

Salts of inorganic oxo acids with labeled *cations* are named in the modification by such terms as "strontium-^{90}*Sr* salt," "ammonium-d_4 salt," if no ratio is employed for the unlabeled salt; otherwise the ratio is replaced by a labeled synonym line formula. Such formulas are also needed to indicate unsymmetrical labeling of anions.
Examples:

Phosphoric acid
  calcium salt ($Ca_2{}^{44}Ca(PO_4)_2$)

Phosphoric-^{32}*P* **acid**
  calcium salt ($Ca_3(PO_4)({}^{32}PO_4)$) (the unlabeled acid is not indexed)

Compounds with indefinite labeled structures are indexed at the unlabeled headings with "labeled with" modification phrases.

Labeled *alloys* are indexed like the unlabeled counterparts but with isotope symbols included in the elemental composition at each heading, e.g., **Aluminum alloy, base, Al 95, ^{233}U 5.** For intermetallic compounds, synonym line formulas are cited in the modification instead of a ratio, e.g., **Iron, compound with uranium ($Fe_2{}^{235}U$)**, with an additional entry at **Uranium.**

Labeled *minerals* are indexed at the systematic and mineral names, as well as at the name of the unlabeled mineral if it is studied. Labeling is indicated at the mineral name by synonym line formulas which contain isotope symbols or by modification phrases such as "labeled with boron-10" or "deuterated hydrate."

Coordination compounds (¶ 215) containing isotopic nuclear atoms are indexed at the labeled element or "ate" term, e.g., **Copper-**^{64}*Cu*, **Borate(1-)-**^{10}*B*. Ligand entries have modifications such as "copper-64 complex." Labeled ligands have the isotopic symbols appended; the ligand names are enclosed in parentheses if multiplicative prefixes are needed.
Examples:

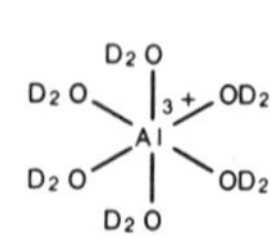

Aluminum(3+), hexa(aqua-d_2)-
  (*OC*-**6-11**)- (see also ¶ 203 III)
Water-d_2
  aluminum complex (additional *Chemical Substance Index* entry)

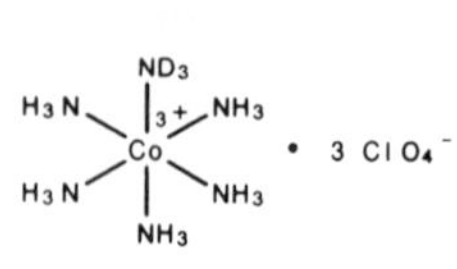

Cobalt(3+), pentaammineammine-d_3-
  (*OC*-**6-22**)- **triperchlorate**
Ammines-d_3
  cobalt (class entry in *General Subject Index*; no entry appears at the unlabeled **Ammines** heading)

Isotopes of hydro ligands, like those of hydro "substituent" prefixes for labeled organic compounds, are expressed by "*d*" or "*t*" terms placed after citation of the hydro set.

Example:

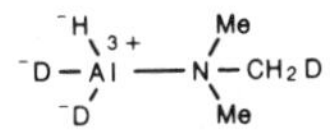

Aluminum, (*N,N*-dimethyl= methan-*d*-amine)trihydro-d_2- (*T*-4)
Methan-*d*-amine, *N,N*-dimethyl- aluminum complex (additional *Chemical Substance Index* entry)
Hydride-*d*
aluminum complex (additional *Chemical Substance Index* entry)

221. **Mixtures** are indexed like individual chemical substances, and are assigned CAS Registry Numbers. They are indexed at the name of each significant component; all other such components are expressed in the mod= ification of the main entry; solvents, fillers, and inactive trace materials are disregarded. In the Tenth Collective Index, the modification term at index entries other than the main one reads "mixt. contg." The complete name can be found by way of the CAS Registry Number. In addition to commercial products with trade or trivial names, mixtures reported by authors to have particular properties or uses are also indexed. Mixed reactants and impure reaction products are not indexed in this way; nor are unresolved natural products, mixed stereoisomers other than pure racemates, solid solutions, commercially available mixtures of structural isomers, column fractions from distillations or chromatography, catalyst systems, or welding fluxes. The percentage composition is not expressed in the index entry; thus, in the modification term "mixt. with copper(2+) sulfate (1:1)," the ratio refers to the copper sulfate, not to the mixture. Cross-references appear at trade and trivial names encountered in original documents.

The preferred index entry is made at a stereoparent (¶ 202) if possible, e.g., D-**Glucose** or **Pregn-4-ene-3,20-dione.** Choice among stereoparents and nonstereoparents depends on functionality, as for molecular addition com= pounds (¶ 192) and polymers (¶ 222). Components are cited in each modifi= cation in alphabetical order, esters and salts being expressed in their uninverted format. Mixtures containing ill-defined, nonregistrable substances such as egg white are indexed as usual, with the **Egg white** entry appearing in the *General Subject Index*. The "compounds" functional subdivision (Appendix II, ¶ 10B) is used for components that have subdivided headings.

Example:

Glycine, compounds
mixt. with egg white and sodium chloride (NaCl) (preferred *Chemical Substance Index* entry; cross-reference at Yumol)
Sodium chloride (NaCl), compounds
mixt. contg. (additional *Chemical Substance Index* entry)
Egg white
mixt. contg. (*General Subject Index* entry)

Mixtures of ten or more indexable components are registered as usual, but, because of the excessive length of index entries which list the components, they are indexed only at the mixture name. "See also" cross-references at the component names in the *Index Guide* lead the index user to the entries at the mixture heading.

A mixture of salts of a single parent is indexed only once, although all salts are registered.

Example:

Carrageenan, hydrogen sulfate, calcium salt, mixt. with potassium and sodium salts

222. **Polymers.** Classes of polymers, natural and synthetic, are indexed in the *General Subject Index* (see Appendix III, ¶ 30).

Specific polymers are named on the basis of the monomers from which they are formed and/or on the basis of their structure, as represented by a structural repeating unit (SRU). Since original documents do not always provide suf= ficient structural information to allow generation of the SRU name, the method most frequently used for describing polymeric substances is by citation of the component monomers. A few commercial polymers, each of which accounts for a large number of index entries, are indexed only at the SRU-based systematic polymer name. (Cross-references at the monomer names appear in the *Index Guide*). Systematic nomenclature is discussed first in the fol= lowing paragraphs, followed by monomer-based polymer nomenclature.

Systematic (SRU) nomenclature for polymers has been adapted from the system developed by the Committee on Nomenclature of the Division of Polymer Chemistry of the American Chemical Society[6]. Names derived by this system, in addition to monomer-based entries, are cited for polymers whose structural repeating units are well-documented or can confidently be assumed. "Expected," "idealized," or "drawn-for-convenience" SRUs are not given systematic polymer names. Occasionally, when there is no information on the component monomers in the original document, an entry derived from the SRU name is the only index entry available.

The SRU is named by citation of one or more multivalent radicals of regular substitutive nomenclature. Many of these radical names will be found in the Illustrative List of Substituent Prefixes (Section H, ¶ 294); others are supplied in the various sections dealing with classes of compounds from which the radicals are derived. The SRU name is enclosed in parentheses or brackets, and prefixed by the term "Poly". Each multivalent radical retains its own numbering and is oriented, if possible, so that the point of attachment written at the left end of the repeating unit is assigned the lowest possible number. This permits the naming of the SRU in a directional manner, reading from left to right. The largest possible multivalent radicals are chosen as all or part of the name, and naming proceeds from left to right, starting with the most preferred multivalent radical. (See the following sections for the choice of most preferred radical.) Unsaturation and substituents are indicated by appropriate locants. Functional derivatives, such as esters and hydrazones, which are an integral part of the repeating unit are expressed by prefixes rather than by modification terms, and are numbered as low as possible while preserving the preferred names of the parent radicals. Salts of acids, anions of quaternary "-onium" compounds, and oxides of hetero atoms which are an integral part of the repeating unit are cited following the name of the SRU. The number of free valencies between units is minimized; i.e., unsaturated radicals are preferred.

Polymers of unspecified length and chains of reported "average" length, are named by the methods described above. The prefix "oligo-" is not used to differentiate polymers of relatively low molecular weight from high polymers. When, however, the number of structural repeating units is exactly specified, the oligomer is usually named according to the principles of substitutive nomenclature.

Examples:

$(-CH_2-)_n$	**Poly(methylene)**
$(-CHMeCH_2-)_n$	**Poly(1-methyl-1,2-ethanediyl)**
$(-CH=CH-)_n$ (not $(=CHCH=)_n$)	**Poly(1,2-ethenediyl)** (not Poly(1,2-ethanediylidene))
$[-(CO)_2(CH_2)_2-]_n$	**Poly(1,2-dioxo-1,4-butanediyl)** (not Poly(1,4-dioxo-1,4-butanediyl))
$(-CH=CHCHMeCH_2-)_n$	**Poly(3-methyl-1-butene-1,4-diyl)**

For more complex examples, further criteria for arranging the components of an SRU are required. The descending order of priority of citation (and of structuring of the SRU) is (*a*) heterocyclic rings, (*b*) acyclic hetero atoms in the order: O, S, Se, Te, N, P, As, Sb, Bi, Si, Ge, Sn, Pb, B, Hg, (*c*) carbocyclic rings, (*d*) acyclic carbon chains. If substituents are present, otherwise identical parent radicals in the SRU are chosen by the principles, in turn, of maximum number, lowest locants, and earliest alphabetic order of substituents. The shortest path (smallest number of atoms) is taken from the most preferred multivalent radical to another occurrence of the same radical (if present) within the SRU, then to the next most preferred radical, and so on.

Examples:

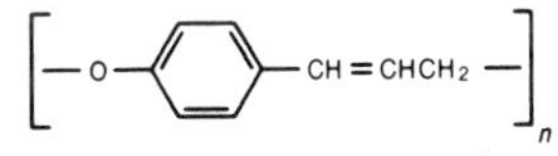

Poly(oxy-1,4-phenylene-1-propene-1,3-diyl)

$[-NHCO(CH_2)_2-]_n$ — **Poly[imino(1-oxo-1,3-propanediyl)]**

$[-(SO_2)_2(CH_2)_2-]_n$ — **Poly(disulfonyl-1,2-ethanediyl)**

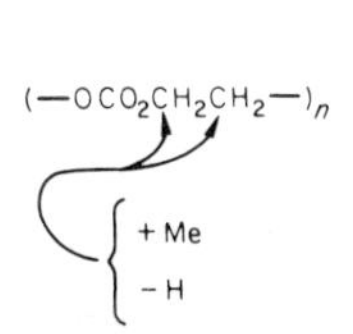

Poly[oxycarbonyloxy(methyl-1,2-ethanediyl)] (the position of the methyl group is not assumed in this polyester from carbonic acid and 1,2-propanediol)

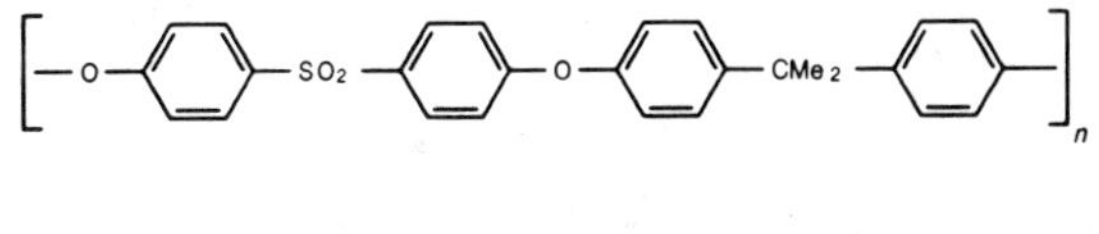

Poly[oxy-1,4-phenylenesulfonyl-1,4-phenyleneoxy-1,4-phenyle= ne(1-methylethylidene)-1,4-phenylene] (With equal numbers of atoms between the oxygen atoms in two possible arrangements, the preferred path includes the other hetero atom, sulfur, as early as possible.)

[6] "A Structure-based Nomenclature for Linear Polymers", *Macromolecules* **1968**, 1(3), 193-8. The recently published IUPAC rules are in full agreement with CAS practice; see *Pure Appl. Chem.* **1976**, 48, 373-85. The IUPAC term "constitutional repeating unit" (CRU) corresponds to *CA*'s SRU.

$[-OCH_2SNH(CH_2)_2OCH_2SCH_2NHCH_2-]_n$

Poly(oxymethylenethioimino-1,2-ethanediyloxymethylenethio= methyleneiminomethylene) (With equal distances between the two oxygen atoms and between the oxygen and sulfur atoms, the di= rection is determined by the shor= test distance between the oxygen atom and the hetero atom of third preference (nitrogen).)

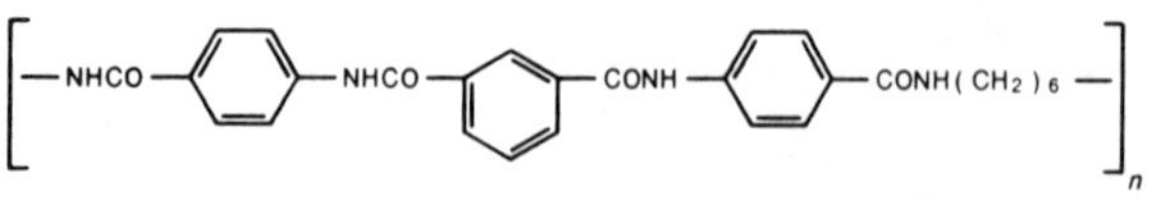

Poly(iminocarbonyl-1,4-phenyle= neiminocarbonyl-1,3-phenyle= necarbonylimino-1,4-phenyle= necarbonylimino-1,6-hexane= diyl) (The citation proceeds in the increasing order of distances be= tween the nitrogen atoms: 5,5,5, and 6 intervening carbon atoms.)

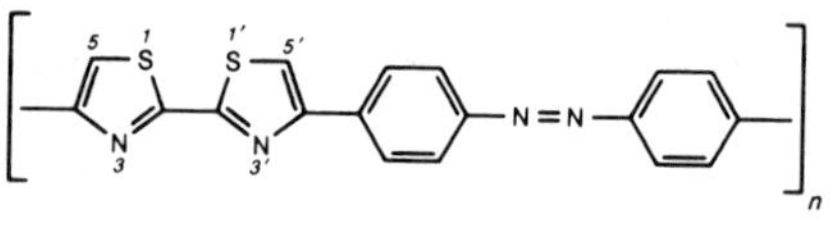

Poly([2,2'-bithiazole]-4,4'-diyl-1,4-phenyleneazo-1,4-phenyle= ne)

$[-NHCOCH(CO_2Pr)-]_n$ **Poly[imino[1-oxo-2-(propoxy= carbonyl)-1,2-ethanediyl]]** (not Poly[imino(1-carboxy-2-oxo-1,2-ethanediyl)], propyl ester)

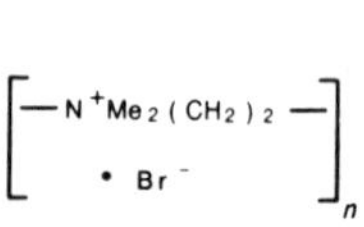

Poly[(dimethyliminio)-1,2-ethanediyl bromide] (the ionic derivative term is included in the SRU)

When SRUs are bridged only by metals, systematic polymer nomenclature is not used; instead, the substance is indexed either at the monomeric salt name or by coordination nomenclature (¶ 215), with a modification phrase, in either case, such as "homopolymer" or "polymer with" (see below).

End groups, when known, are specified by means of appropriate radical names, together with Greek letters "α-" and "ω-" expressed in the index modification. The α-end group is the group attached to the left end of the SRU when the structure is ordered by the specified rules; it is cited first, regardless of alphabetic order.

Examples:

$Cl-(CH_2)_n-CCl_3$ **Poly(methylene)** *α*-**chloro-***ω*-**(trichloromethyl)-**

$Cl_3C-(CF_2-CH_2-)_nCl$ **Poly(1,1-difluoro-1,2-ethanediyl)** *α*-**(trichloromethyl)-***ω*-**chloro-**

Linear double-stranded ("ladder") polymers may sometimes be named as a chain of quadrivalent radicals. Two pairs of locants, separated by a colon, indicate the distribution of bonds.

Example:

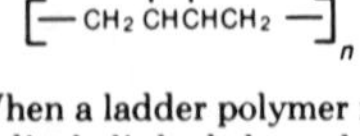

Poly(1,4:2,3-butanetetrayl)

When a ladder polymer must be named as an SRU of one or more quadrivalent radicals linked through one or more bivalent radicals (here, these terms are extended to mean radicals attached to four or two different atoms, not only to radicals with four or two free valence bonds) the direction of citation is from the most favored quadrivalent radical by the shortest path to the next most favored quadrivalent radical, and so on; then toward the most favored bivalent radical. Rings are broken (*a*) to minimize the number of free valencies of the total "mer," (*b*) to maximize the number of most preferred hetero atoms in the ring system, (*c*) to maintain intact the most preferred ring system (¶ 138). End groups, when known, are identified by *α* and *α'* (at the left terminus as the structure is drawn) and by *ω* and *ω'* (at the right terminus) as locants for substituent prefixes in the modification, e.g., *α*,*α'*-dihydroxy-*ω*,*ω'*-dihydro-.

Examples:

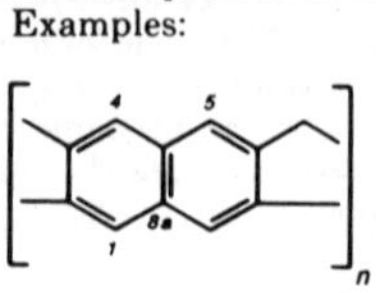

Poly(2,3:6,7-naphthalenetetrayl-6-methylene)

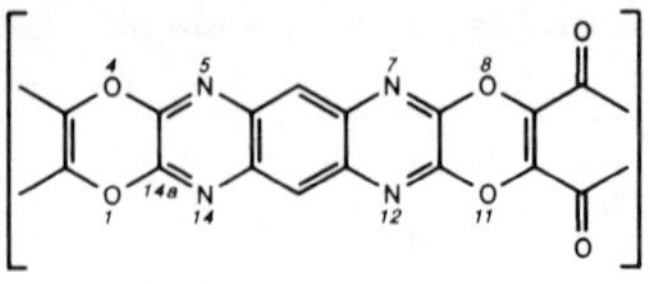

Poly([1,4]dioxino[2,3-*b*]-1,4-dioxi= no[2',3':5,6]pyrazino[2,3-*g*]= quinoxaline-2,3:9,10-tetrayl-9,10-dicarbonyl) (The same number of free valences can be expressed by breaking the oxygen ring or the partially saturated hy= drocarbon ring; the latter course keeps intact the maximum number of heterocyclic rings.)

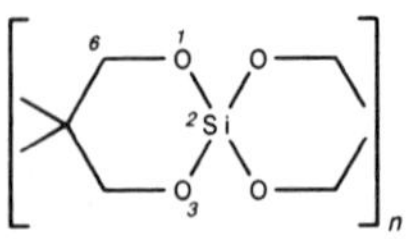

Poly[1,3-dioxa-2-silacyclohexane-5,2-diylidene-2,2-bis(oxy= methylene)] (not Poly[1,3-dioxa-2-silacyclohexane-2,5-diylidene-5,5-bis(methyleneoxy)] (The di= rection is determined by the shor= test path from the hetero atom in the ring to the acyclic hetero atom.)

Linear polymers composed of SRUs within SRUs, e.g.,

$$[-[O(CH_2)_2]_m-O_2C(CH_2)_4CO_2-]_n$$

are not assigned systematic names; they are indexed at the monomer names only.

The previous paragraphs described a systematic nomenclature for polymers of well characterized structure, and for polymers whose structural repeating unit (SRU) can be confidently assumed. The latter are restricted to (*a*) polyamides from a dibasic acid (or derivative) and a diamine, or from an amino acid or lactam; (*b*) polyesters from a dibasic acid (or derivative) and a dihydric alcohol, or from a hydroxy acid or lactone; (*c*) polyurethanes from a diiso= cyanate and a dihydric alcohol; and (*d*) polycarbonates from carbonic acid (or an ester or halide) and a dihydric alcohol.

Polymers manufactured from known monomers are generally indexed at the monomer names whether or not systematic (SRU) entries are also made. An exception is the treatment of a few very common industrial polymers, e.g., nylon 6, nylon 66, terephthalic acid polymer with ethylene glycol, which, to preclude inordinate repetition of a large number of index entries at various names, are cross-referred in the *Index Guide* from monomer names to SRU names.

Example:

Terephthalic acid
See 1,4-Benzenedicarboxylic acid

1,4-Benzenedicarboxylic acid, polymers
polymer with 1,2-ethanediol—see *Poly(oxy-1,2-ethanediyloxy= carbonyl-1,4-phenylenecarbonyl)*

A cross-reference appears also at **1,4-Benzenedicarboxylic acid, esters, dimethyl ester, polymer with 1,2-ethanediol;** and two corresponding cross-references at **1,2-Ethanediol, polymers.**

Polymers from a single monomer are indexed at the monomer name with the term "homopolymer" cited in the modification. (The terms "peptides," "polyamides," and "polyesters" are not used for specific homopolymers at monomer headings.)

Examples:

$(BuCH=CH_2)_n$ **1-Hexene** **homopolymer** (only index entry)

$[H_2C=CMeCO_2(CH_2)_{12}Me]_n$ **2-Propenoic acid, 2-methyl-tridecyl ester, homopolymer** (only index entry)

$[H_2N(CH_2)_{10}CO_2H]_n$ **Undecanoic acid, 11-amino-homopolymer** (the *Formula Index* entry appears at $C_{11}H_{23}NO_2$; a systematic entry is also made at the assumed SRU name: **Poly[imi= no(1-oxo-1,11-undecanediyl)],** $(C_{11}H_{21}NO)_n$)

Polymers formed from two or more monomers are indexed at each monomer name with the modification term "polymer with" followed by the other mo= nomer names in alphabetical order. The preferred index name is determined by the usual rules for selection of a heading parent (¶ 138) but a stereoparent is preferred over a nonstereoparent, e.g., D-**Glucose,** polymer with butanedioic acid. For identical heading parents, the choice is determined by (*a*) maximum number of substituents, (*b*) lowest locants of substituents, (*c*) maximum number of occurrences of the index heading parent (in a multiplicative name), (*d*) earliest index position of the index heading. When the choice is dependent on modification terms, it is determined as follows: (*a*) underivatized heading preferred over derivatives cited in the modification; thus, a free acid is preferred over an ester; (changes in format caused by elevation of modification terms into the heading for purposes of subdivision (Appendix II, ¶ 10B) are ignored in applying this rule; thus **Acetic acid** is preferred over **Acetic acid ethenyl ester**); (*b*) class of derivative in the descending order: anhydride, ester, hydrazide, hydrazone, oxime; (*c*) largest number of (most preferred) derivative; thus, monoester preferred over dioxime preferred over monooxime; (*d*) lowest expressed locants of derivative terms; thus, for **1,2,4-Benzenetricarboxylic acid,** a 1,2-diester is preferred over a 1,4-diester; (*e*) the earliest alphabetical order; thus, "ethyl ester" preferred over "propyl ester."

Examples:

1(*a*) **1-Heptene**
polymer with 1-hexene (preferred index entry)

1(*b*) **1-Hexene**
polymer with 1-heptene (additional *Chemical Substance* and *Formula Index* entry)

2(*a*) **2-Propenoic acid**
butyl ester, polymer with 1-ethenyl-4-methylbenzene and 2,5-furandione (preferred index entry)

2(*b*) **2,5-Furandione**
polymer with butyl 2-propenoate and 1-ethenyl-4-methylbenzene (additional *Chemical Substance* and *Formula Index* entry)

2(*c*) **Benzene, 1-ethenyl-4-methyl-**
polymer with butyl 2-propenoate and 2,5-furandione (additional *Chemical Substance* and *Formula Index* entry)

Peptides of established structure are indexed at systematic peptide names such as **Glycine,** *N*-(*N*-**glycylglycyl**)- (see ¶ 206). Peptides of unknown structure are indexed as polymers.

Example:

$[HO_2CCH(NH_2)(CH_2)_2CO_2CH_2Ph]_n$

L-**Glutamic acid**
5-(phenylmethyl) ester, homopolymer

and the assumed structural repeating unit:

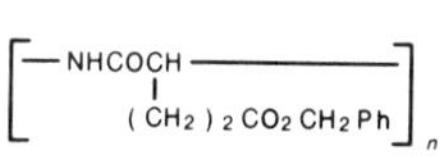

Poly[imino[1-oxo-2-[3-oxo-3-(phenylmethoxy)propyl]-1,2-ethanediyl]]
(*S*)-

When more than one amino acid is present and the sequence is unknown, each amino acid is indexed, and a "polymer with . . ." phrase is cited in the modification.

Alternating, block and graft polymers are distinguished from random polymers by indexing as copolymers at the monomer names. The term "alternating", "block", or "graft" (or a combination thereof) is cited in a special modification after all other structural information but before descriptive phrases relating to properties, uses, etc. Differentiation between the polymeric substrate and the applied monomer is not made; e.g., whether 1-hexene is grafted on 1-heptene homopolymer or vice versa, the preferred index entry is at **1-Heptene,** polymer with 1-hexene, graft, with an additional entry at **1-Hexene,** polymer with 1-heptene, graft. The term "random" is not employed by *CA* in indexing specific polymers.

Polymers of specific compounds with classes of compounds are indexed at the specific monomers in the *Chemical Substance Index* and at the class name, e.g., **Aldehydes** or **Nitriles,** in the *General Subject Index*.

Example:

1-Hexene, 5-methyl-
polymer with unsatd. nitriles
Nitriles
polymers of unsatd., with 5-methyl-1-hexene

Formaldehyde homopolymers containing only oxymethylene repeating units are indexed only at the heading **Poly(oxymethylene)** in the Chemical Substance Index. Commercial and impure polyformaldehydes are indexed at **Polyoxymethylenes** (a plural class heading in the *General Subject Index*) unless author emphasis is centered on **Paraformaldehyde.** Formaldehyde copolymers are indexed as polymers formed from two or more monomers (except **Formaldehyde** in its polymers with **Phenol**, **Urea**, or **1,3,5-Triazine-2,4,6-triamine** which are cross-referred in the *Index Guide*.

Oligomers of definite structure are indexed as specific compounds by the regular principles of index nomenclature. When the precise structure is not known but the number of units is specific, "dimer," "trimer," etc., is expressed in the modification at the name of the monomer. The term "oligomeric" may be cited after "homopolymer" or "polymer with . . ." if this aspect is stressed in the original document.

Telomers with a definite structure are named systematically.

Examples:

$Cl-(CH_2)_6-CCl_3$ — **Heptane, 1,1,1,7-tetrachloro-**

$Cl-(CF_2-CH_2)_n-CHCl_2$ — **Poly(1,1-difluoro-1,2-ethanediyl)** *α*-**chloro**-*ω*-**(dichloromethyl)-**

Telomers of unknown structure are indexed as copolymers with the term "telomer with . . ." cited in the modification.

Example:

1-Hexene
telomer with tetrachloromethane
Methane, tetrachloro-
telomer with 1-hexene

Post-treated polymers are described by modification terms after the polymer has been described.

Examples:

2-Propenoic acid
homopolymer, sodium salt

Benzenamine, 4-ethenyl-
polymer with ethenylbenzene, hydrochloride

2-Propenoic acid, 2-methyl-
decyl ester, homopolymer, hydrolyzed

Poly[1-(4-sulfophenyl)-1,2-ethanediyl]
propyl ester

Ethenol (vinyl alcohol) polymers that are indexed at that heading are exceptions (they are prepared by hydrolysis of ethenol ester polymers). Also, acetals of ethenol polymers are entered in the *General Subject Index* at the trivially named class term **Vinyl acetal polymers** with a modification, e.g., "with butanal," or "with 4-chlorobenzaldehyde," to indicate the type of acetal. Cross-references or additional entries (in the *Chemical Substance Index*) are found at the aldehyde names.

Molecular addition compounds of polymers are indexed in the usual way (¶ 192), e.g., **Ethenesulfonic acid,** homopolymer, compd. with 4-ethenylpyridine homopolymer, with an additional entry for the other component.

Polyethylene and polypropylene *glycols* are indexed as specific compounds when the precise structure is reported and not more than ten repeating units are present, e.g., **3,6,9,12,15-Pentaoxaheptadecane-1,17-diol, hexamethyl-** (a cross-reference will be found in the *Index Guide* at **Hexapropylene glycol**). The position of methyl substituents is not assumed; if it is reported, locants are cited. Glycol polymers of eleven units or more are indexed at **Poly(oxy-1,2-ethanediyl),** *α*-**hydro**-*ω*-**hydroxy-**, and **Poly[oxy(methyl-1,2-ethanediyl)],** *α*-**hydro**-*ω*-**hydroxy-**. The class term **Polyoxyalkylenes** in the *General Subject Index* is used when the nature of the alkanediyl group is unknown, and for general studies. When it is known, the SRU can be named specifically; thus, **Poly[oxy(2-phenyl-1,3-propanediyl)],** *α*-**hydro**-*ω*-**hydroxy-**. Polyethylene-polypropylene glycols are named as copolymers: **Oxirane, methyl-**, polymer with oxirane, with an additional *Chemical Substance* and *Formula Index* entry at **Oxirane.**

Esters and ethers of polyalkylene glycols containing not more than ten repeating units are indexed by the regular principles of substitutive nomenclature. For larger polymers, the ester or ether is expressed in the modification at the SRU-based heading. Mono derivatives are cited in the *α*-position, with *ω*-hydroxy at the other terminus. Dissimilar diesters or diethers are expressed in alphabetical order. Ester-ethers have the acyl group assigned to the *α*-position, and the ether at the *ω*-end. When appropriate information is lacking, an ester or ether term is cited without locant. When multiplying radicals must be employed, they are given *α*-locants.

Examples:

$Me(CH_2)_{16}CO[O(CH_2)_2]_nOH$ — **Poly(oxy-1,2-ethanediyl)** *α*-**(1-oxooctadecyl)**-*ω*-**hydroxy-**

$Ac[O(CH_2)_2]_nO_2CCMe=CH_2$ — **Poly[oxy-1,2-ethanediyl]** *α*-**acetyl**-*ω*-**[2-methyl-(1-oxo-2-propenyl)oxy]-**

$H_2C=CHCO[O(CH_2)_2]_nO(CH_2)_{11}Me$ — **Poly(oxy-1,2-ethanediyl)** *α*-**(1-oxo-2-propenyl)**-*ω*-**(dodecyloxy)-**

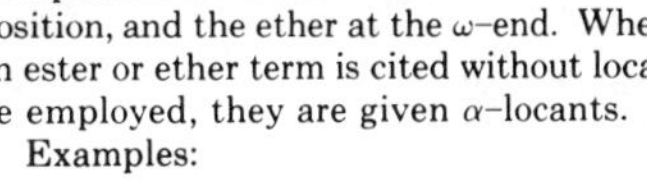

Poly(oxy-1,2-ethanediyl)
α,α'-**1,4-phenylenebis[***ω***-[(1-oxooctadecyl)oxy]-**

Polynucleotides (cf. Nucleotides, ¶ 210) are indexed as "homopolymer" or "polymer with . . ." at the "-ylic acid" monomers (5'-isomers).

Examples:

5′-Adenylic acid
homopolymer

5′-Adenylic acid, cytidylyl-(5′→3′)-
homopolymer

5′-Guanylic acid
polymer with 5′-adenylic acid and 5′-cytidylic acid (additional *Chemical Substance* and *Formula Index* entries at **5′-Adenylic acid** and **5′-Cytidylic acid**)

Polynucleotides in which different primary chains are associated intermolecularly by hydrogen bonds are indexed as molecular addition compounds at each component (the modification term "complex" is used in this special case); e.g., **5′-Adenylic acid, homopolymer, complex with 5′-uridylic acid homopolymer** (1:1).

In more complicated cases, a combination of policies is applied.
Examples:

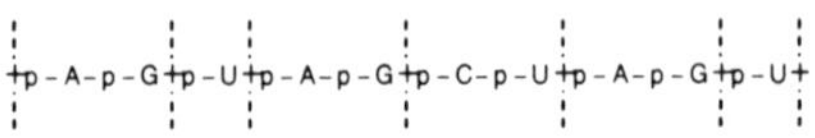

5′-Adenylic acid, guanylyl-(5′→3′)-
polymer with 5′-cytidylic acid and 5′-uridylic acid (additional *Chemical Substance* and *Formula Index* entries at **5′-Cytidylic acid** and **5′-Uridylic acid)**

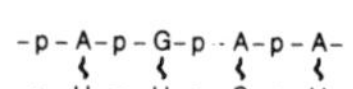

5′-Guanylic acid
polymer with 5′-adenylic acid, complex with 5′-uridylic acid, polymer with 5′-cytidylic acid (1:1) (three additional *Chemical Substance* and *Formula Index* entries at the names of the other components)

Trade names are frequently used for polymers and are cross-referred to specific polymers if the components can be structurally defined. If the polymer components are unknown, or only partially known, or cannot structurally be defined, an index citation is made at the trade name and at the appropriate polymer class name.

Natural *rubber* is indexed at **Rubber, natural**; **Gutta-percha**; **Balata**; and **Chicle**; treated-rubber headings include **Rubber, chlorinated** and **Rubber hydrochloride.** When a synthetic rubber is indexed at **Rubber, butadiene**; **Rubber, isoprene**; etc., no additional entry will be found, but for studies indexed at **Rubber, synthetic**, **Rubber**, **Urethane**, or at **Rubber, polysulfide** (see Appendix III, ¶ 30) information will also be found, when available, at the appropriate chemical compound or polymer heading, with "rubber" cited in the index modification.
Examples:

Rubber, synthetic
bromoethylene-styrene
(*General Subject Index* entry)
Benzene, ethenyl-
polymer with bromoethene, rubber
(preferred *Chemical Substance Index* entry)
Ethene, bromo-
polymer with ethenylbenzene, rubber
(additional *Chemical Substance Index* entry)

General subject headings for fibers are discussed in Appendix III, ¶ 30. For studies indexed at **Rayon, Acetate fibers, Polyamide fibers,** etc., no additional index entries will be found, but for studies of fibers of specific chemical composition entered at **Synthetic fibers,** additional information is entered at the appropriate chemical substance name.
Example:

Synthetic fibers
adipic acid-butanediol (*General Subject Index* entry)
Hexanedioic acid
polymer with 1,4-butanediol, fiber (additional *Chemical Substance Index* entries at **1,4-Butanediol** and at the SRU name)

Stereochemistry of polymers is expressed by special modification terms when the necessary information is reported; such terms include *isotactic, syndiotactic, threo-diisotactic, erythro-diisotactic,* and *disyndiotactic.* The term "atactic" (for a random configuration) is not employed by *CA* in indexing specific polymers.

In addition to the special terms above, polymer stereochemistry is defined, when appropriate, by the regular descriptors *E, Z, R, S, R**, and *S** (¶ 203 I). For commercial elastomeric polymers, which cannot generally be assumed to be stereochemically homogeneous, terms such as "1,2-configuration," and "cis-1,4-configuration" are included in the descriptive portions of index modifications, but are not made part of the unique preferred *CA* index names.

223. Porphyrins and **Bile pigments.** The porphyrins embrace all cyclic tetrapyrroles in which single methene groups link pairs of pyrrole rings. The parent **21*H*,23*H*-Porphine** is used, with a suffix to express principal groups if present, for all derivatives, including hydrogenated derivatives, unless an author emphasizes the absence of hydrogen at the 21- and 23-positions.
Example:

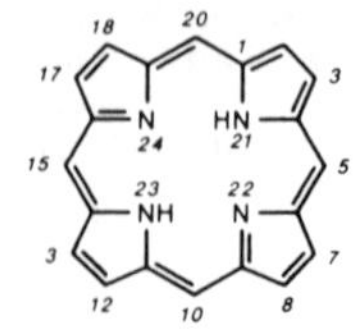

21*H*,23*H*-**Porphine**

Porphyrin analogs containing additional hetero atoms are indexed by organic replacement nomenclature (¶ 127) at such names as **21*H*-5-Thiaporphine** and **21*H*,23*H*-5,15-Diazaporphine**. The trivial name **21*H*,23*H*-Porphyrazine** is employed instead of 21*H*,23*H*-5,10,15,20-Tetraazaporphine, and its fused tetrabenzo derivative is **29*H*,31*H*-Phthalocyanine**. Radicals are formed in the regular way; e.g., 21*H*,23*H*-porphin-2-yl, 29*H*,31*H*-phthalocyanine-2,9,17,24-tetrayl.

The metal complexes, including the biologically important porphyrin iron and magnesium complexes (hemes and chlorophylls) are indexed as coordination compounds (¶ 215) with italic letter locants to indicate coordinating atoms.
Example:

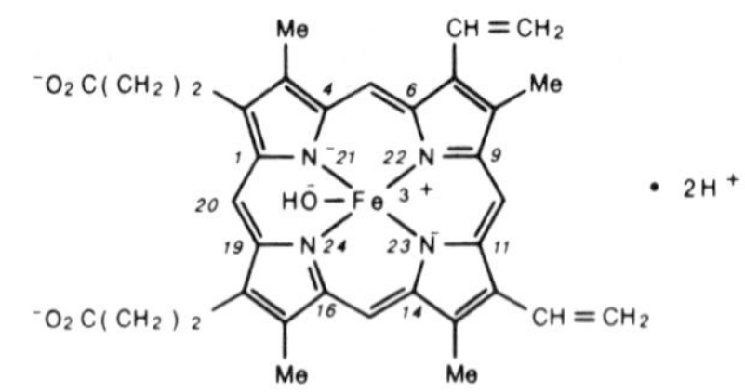

Ferrate(2−), [7,12-diethenyl-3,8,13,17-tetramethyl-21*H*,23*H*-porphine-2,18-dipropanoato(4−)-N^{21},N^{22},N^{23},N^{24}]hydroxy-, dihydrogen, (*SP*-5-13)-
(cross-reference from Hematin)

Phorbine is a cyclopentaporphine derivative:

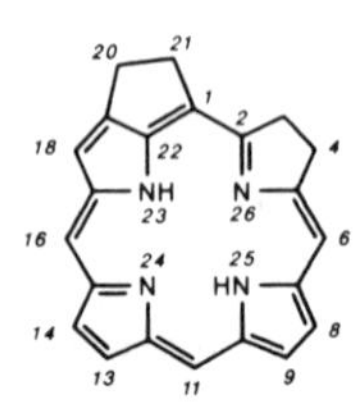

Phorbine

Indicated hydrogen is not expressed with Phorbine, which contains six saturated centers. Removal of hydrogen is indicated by dehydro terms, e.g., **Phorbine, 3,4-didehydro-**. Metal complexes, radicals, etc., are named as for the analogous porphine derivatives.
Example:

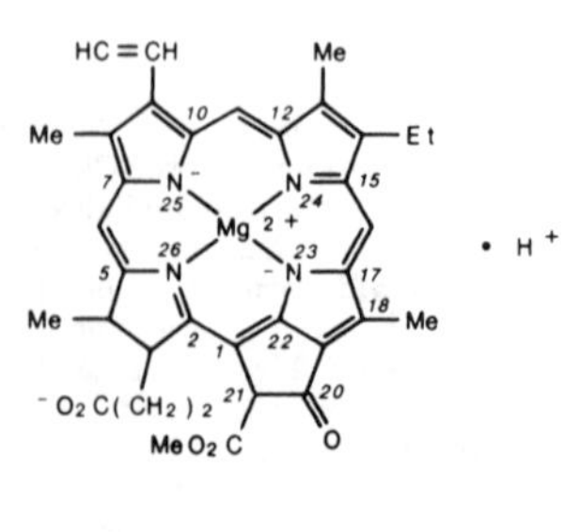

Magnesate(1−), [9-ethenyl-14-ethyl-21-(methoxycarbonyl)-4,8,13,18-tetramethyl-20-oxo-3-phorbinepropanoato(3−)-N^{23},N^{24},N^{25},N^{26}]-hydrogen, [*SP*-4-2-[3*S*-(3α,4β,21β)]]-
(cross-reference from Chlorophyllide a)

Bile pigments are indexed at **21*H*-Biline,** the numbering system of which derives from that of porphyrin (the carbon atom involved in ring closure, C-20, is omitted from the numbering system).

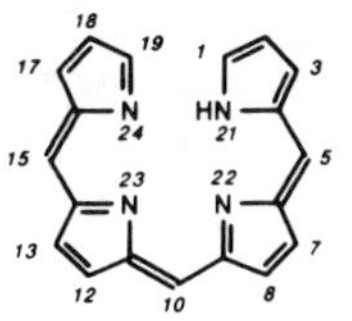

21*H*-Biline

All derivatives, including hydrogenated derivatives, are indexed at **21*H*-Biline** unless hydrogen is absent at the 21-position, in which case the 22*H*-parent is preferred over lower positions. The (*all-Z*) stereochemistry shown in the diagram above is assumed. The 1,19-dihydroxy derivatives are tautomeric with the dioxo compounds, which are preferred in indexing.

Example:

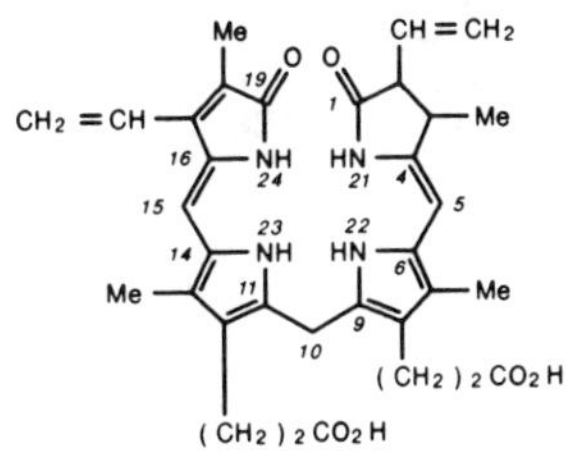

21*H*-Biline-8,12-dipropanoic acid, 2,17-diethenyl-1,10,19,22,23,24-hexahydro-3,7,13,18-tetramethyl-1,19-dioxo-
(cross-reference from Bilirubin)

For derivatives of **Corrin,** see **Vitamin B$_{12}$** (¶ 224).

224. Vitamins, being structurally diverse, are indexed by more than one method. The headings **Vitamin B, Vitamin K,** etc., are used to index groups of compounds having similar physiological activities, especially for biological studies. Such headings as **Vitamin B$_4$, Vitamin B$_6$, Vitamin F,** may each refer to one or more specific compounds or to nothing more than an ill-defined vitamin activity. They are used in indexing when employed in original documents, and CAS Registry Numbers are assigned to them, but "see also" cross-references in the *Index Guide* lead from these names to more specific related headings whenever possible. Vitamin names occasionally still encountered, e.g., **Vitamin P$_4$, Vitamin U,** but not discussed in the present account will usually be found cross-referred in the *Index Guide* to the preferred index names.

Vitamin A is a heading parent employed for the group of A vitamins and for vitamin A activity in general. Vitamin A$_1$ is indexed at the carotenoid heading **Retinol** (¶ 212); Vitamin A$_2$ is indexed at **Retinol, 3,4-didehydro-**. Related trivially named heading parents are **Retinoic acid** and **Retinal.** In all these cases, an *all-trans*-configuration about the double bonds is implied by the names. Abnormal stereochemistry is specifically cited in the modification.

Vitamin B is a heading parent employed for the B-complex and vitamin B activity in general. **Vitamin B$_1$** and related compounds are indexed at the systematic names. (The trivial name Thiamine, not used as a *CA* index heading, implies presence of a chloride anion.)

Example:

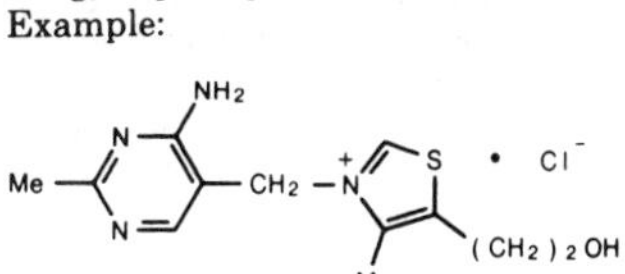

Thiazolium, 3-[(4-amino-2-methyl-5-pyrimidinyl)methyl]-5-(2-hydroxethyl)-4-methyl-chloride
(cross-references from Aneurine; Thiamine; and Vitamin B$_1$)

The diphosphoric (pyrophosphoric) acid ester of Vitamin B$_1$ (cocarboxylase) is indexed by expressing the ester as an "a"-named substituent; i.e., the 5-(2-hydroxyethyl) group becomes 5-(4,6,6-trihydroxy-3,5-dioxa-4,6-diphosphahex-1-yl)-, with "*P,P'*-dioxide" cited in the index modification.

Vitamin B$_2$ contains a ribitol residue and is indexed at the stereoparent **Riboflavin**:

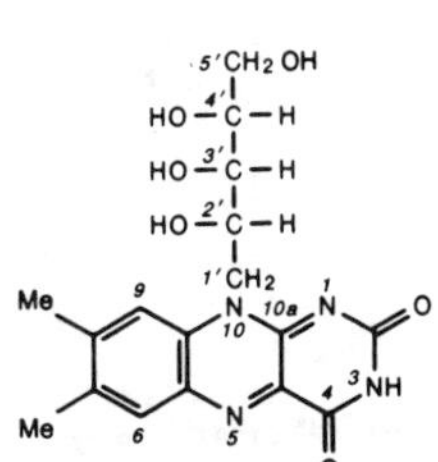

Riboflavin
(cross-references from Benzo[*g*]pteridine-2,4(3*H*,10*H*)-dione, 7,8-dimethyl-10-(D-*ribo*-2,3,4,5-tetrahydroxypentyl)-; Lactoflavin; D-Ribitol, 1-deoxy-1-(3,4-dihydro-7,8-dimethyl-2,4-dioxobenzo[*g*]pteridin-10(2*H*)-yl)-; Vitamin B$_2$; Vitamin G)

Plain locants are employed for ring substituents, primed locants for the ribitol moiety. Binary headings analogous to those for nucleotides (¶ 210) are used for phosphates, diphosphates (pyrophosphates), etc., e.g., **Riboflavin 5'-(dihydrogen phosphate).** Functional derivatives of **Riboflavin** are expressed in the modification, e.g., 2',3',4'-triacetate, other derivatives as substituents. Analogs in which ribitol is replaced by other alditols are indexed at the alditol stereoparent names, e.g., D-**Galactitol, 1-deoxy-1-(3,4-dihydro-7,8-dimethyl-2,4-dioxobenzo[*g*]pteridin-10(2*H*)-yl)-** (cross-reference in the *Index Guide* at Galactoflavin).

Vitamin B$_3$ is indexed at **3-Pyridinecarboxamide**; Vitamin B$_5$ is a little-used term for Pantothenic acid, which is indexed at *β*-**Alanine:**

$HOCH_2CMe_2CH(OH)CONH(CH_2)_2CO_2H$

β-**Alanine,** *N*-**(2,4-dihydroxy-3,3-dimethyl-1-oxobutyl)-**, (*R*)-
(cross-reference from Pantothenic acid) (The (*R*)-isomer is assumed unless otherwise stated in the original document.)

Vitamin B$_6$ is a heading parent employed for studies in which the precise compound is not further specified. The individual compounds and their derivatives are indexed systematically at pyridine headings.

Examples:

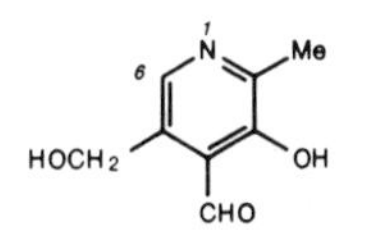

4-Pyridinecarboxaldehyde, 3-hydroxy-5-(hydroxymethyl)-2-methyl-
(cross-reference from Pyridoxal)

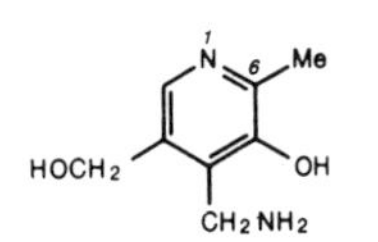

3-Pyridinemethanol, 4-(aminomethyl)-5-hydroxy-6-methyl-
(cross-reference from Pyridoxamine)

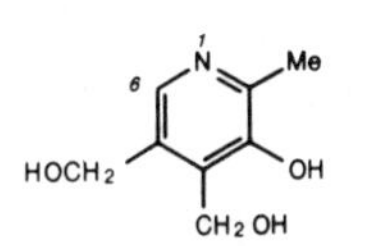

3,4-Pyridinedimethanol, 5-hydroxy-6-methyl-
(cross-references from Pyridoxine; Pyridoxol)

Vitamin B$_{12}$ is an index heading used for the vitamin itself:

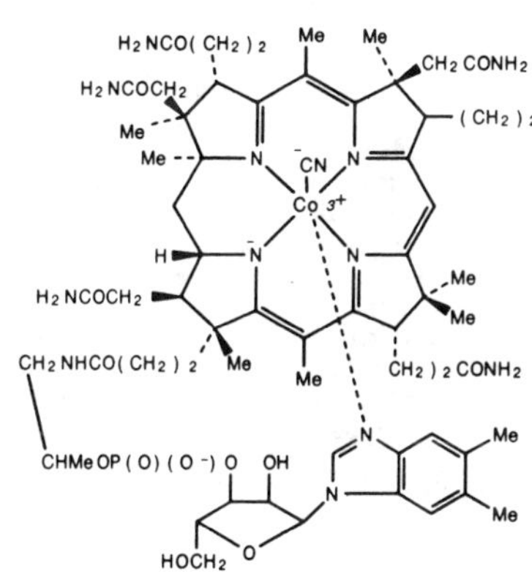

Vitamin B$_{12}$
(cross-references from Cobalamin, cyano-; Cobamin; Cobinamide, cyanide hydroxide, dihydrogen phosphate (ester), inner salt, 3'-ester with 5,6-dimethyl-1-*α*-D-ribofuranosyl-1*H*-benzimidazole; and Cyanocobalamin)

The terms Cyanocobalamin and Cobalamin are not used in index names. Derivatives, other than functional derivatives, of Vitamin B_{12} are indexed largely in accordance with rules of the International Union of Pure and Applied Chemistry,[7] which should be consulted for details.

The fundamental ring system is Corrin, which is numbered like Porphyrin, the locant "20" being omitted to facilitate analogous treatment of derivatives; the methene bridges which remain are numbered "5," "10," and "15." The index heading (in the *General Subject Index*) for this class of compounds is Corrinoids, and specific compounds are generally indexed at the largest ap= propriate heading parent chosen from the following: **Cobyrinic acid, Cobinic acid, Cobamic acid,** or the corresponding amides.

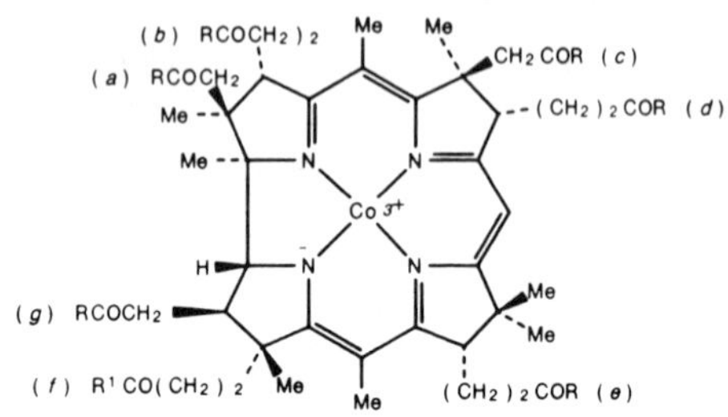

R=R¹=OH	Cobyrinic acid
R=R¹=NH2	Cobyrinamide acid
R=OH; R¹=NHCH2CHMeOH	Cobinic acid (in Cabinamide, R=NH2)
R=OH; R¹=NHCH2CHMe OP(O)(O⁻)O	Cobamic acid (in Cobamide, R=NH2)

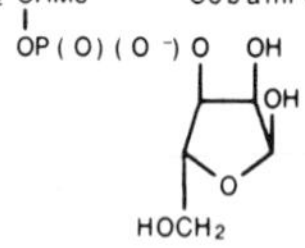

In **Cobamic acid** and **Cobamide,** the structures (above) have been shown as hydroxide inner salts; the cobalt(3+) nucleus is also coordinated with the nitrogen atoms with loss of one proton overall. The compound therefore still possesses a 1+ charge. Similarly, **Cobyrinic** and **Cobinic acids** (and their amides), which lack the phosphate group, have a 2+ charge, and can form dihydroxides, cyanide hydroxides, etc. Partial amides of **Cobyrinic acid** are designated by use of the small italic locants shown in the diagram above; e.g., **Cobyrinic acid–***abcdeg***–hexamide,** dicyanide. Substitution on amide groups is indicated by *N* with a superscript, e.g., N^f–(carboxymethyl)–. In modifi= cations at corrinoid headings, the various types of derivatives are cited in the following order: (*a*) groups linked to the cobalt atom, e.g., *Co*–ammine, *Co*–methyl deriv., *Co*–(pyridine); followed by anions, e.g., cyanide, hydroxide, acetate (salt) hydroxide; and then "hydrate" if present; (*b*) lactones and lactams formed between acetic or propanoic residues and existing or added hydroxyl and amino groups; (*c*) esters of the corrin moiety, e.g., dihydrogen phosphate (ester); (*d*) "inner salt" (for zwitterionic structures); (*e*) further ester terms for a polybasic acid residue such as phosphoric acid, e.g., "3–ester with. . . . "

When the cobalt is in a 2+ rather than a 3+ oxidation state, Ewens-Bassett numbers (¶ 215) or Stock numbers are cited in the heading; e.g., **Cobin= amide–***Co(1+)*, *Co*–**ethyl deriv., monohydrate** (the "1+" designation belongs to the entire cobinamide parent, and results from coordination of the Co(2+) atom with loss of a proton from corrin; the modification term renders the total compound neutral); **Cobamide–***Co***(II)** (the Stock number is employed because no charge remains; a proton has been lost from corrin and another from the phosphoric acid group by inner salt formation).

Analogs of **Cobyrinic acid,** etc., in which cobalt is replaced by another metal are named by substituting a suitable term for "Co–" in the original name. Thus, the iron(2+) and nickel(3+) analogs of cobyrinic acid are named **Fer= robyrinic acid** and **Nickelibyrinic acid,** respectively. The hydrogen analog is **Hydrogenobyrinic acid.**

Vitamin C is indexed at L–**Ascorbic acid**:

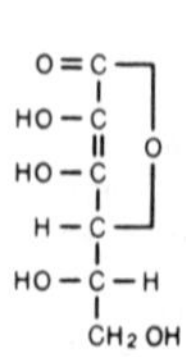

L–**Ascorbic acid**
(cross-references from L–*threo*–Hex–2–enonic acid, γ–lactone; Vitamin C)

Derivatives are named in accordance with the rules for carbohydrates (¶ 208), e.g., L–**Ascorbic acid, 5,6–***O***–(1–methylethylidene)–**; L–*threo*–**2,3–Hexodiulosonic acid,** γ–lactone (cross-reference from L–Dehydroascorbic acid).

[7] *Biochemistry* **1974**, 13, 1555–60; *Pure Appl. Chem.* **1976**, 48, 495–502.

Vitamin D is an index heading used for vitamin D activity in general. Vitamin D_1 is a molecular addition compound of Vitamin D_2 and (3β,9β,10α)–ergosta–5,7,22–trien–3–ol (lumisterol). Vitamins D_2 and D_3 are indexed as stereoparents in accordance with the rules for secosteroids (¶ 211).

Examples:

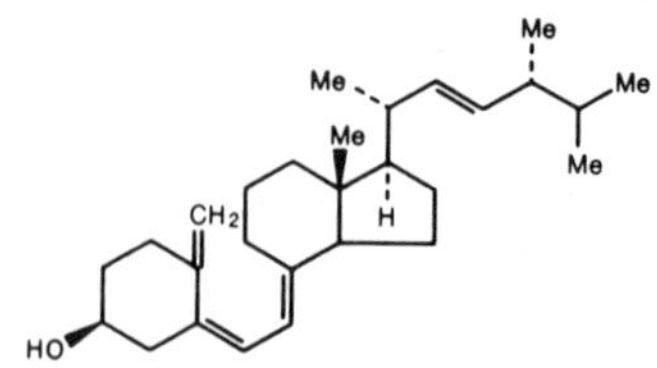

9,10–Secoergosta–5,7,10(19),22–tet= raen–3–ol
(3β,5*Z*,7*E*,22*E*)– (cross-references from Calciferol, Ergocalciferol, and Vitamin D_2)

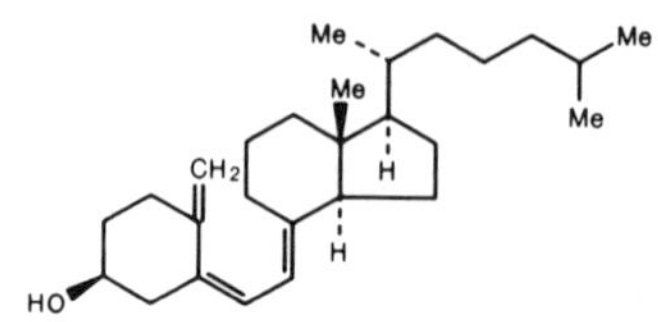

9,10–Secocholesta–5,7,10(19)–trien–3–ol
(3β,5*Z*,7*E*)– (cross-references from Cholecalciferol and Vitamin D_3)

Vitamin E is used as a heading for Vitamin E activity in general. Indi= vidual compounds, notably α–tocopherol, exhibiting this activity are indexed systematically.

Example:

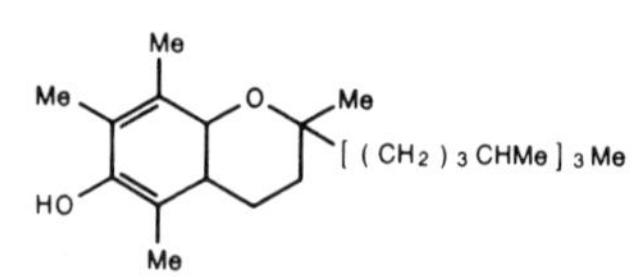

2*H*–**1–Benzopyran–6–ol, 3,4–di= hydro–2,5,7,8–tetramethyl–2–(4,8,12–trimethyltridecyl)–, [2*R*–[2*R**(2*R**,8*R**)]]–**
(cross-reference from α–Tocopherol; "see also" cross-reference at Vi= tamin E)

Vitamin K is used as a group heading; the specific compounds are named systematically.

Examples:

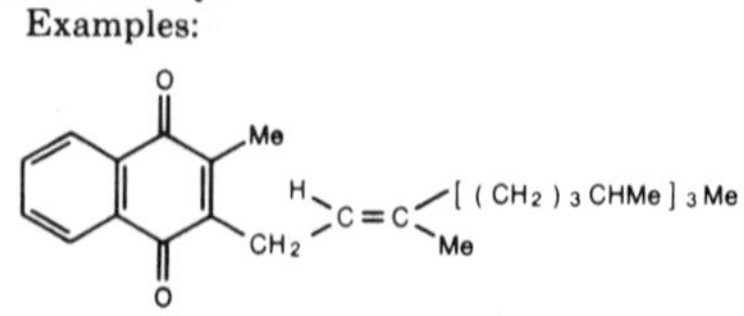

1,4–Naphthalenedione, 2–methyl–3–(3,7,11,15–tetramethyl–2–hex= adecenyl)–
[*R*–[*R,*R**–(*E*)]]–**
(cross-references from Phyllo= quinone and Vitamin $K_{1(20)}$)

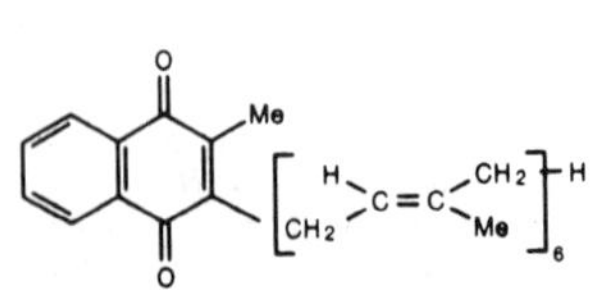

1,4–Naphthalenedione, 2–(3,7,11,15,19,23–hexamethyl–2,6,10,14,18,22–tetracosahexae= nyl)–3–methyl–
(*all*–*E*)– (cross-references from Farnoquinone, and Vitamin $K_{2(30)}$)

225. Introduction. The current name-selection policies for chemical substances described in Sections A through F above were introduced in 1972 and have been continued virtually unchanged since that date. These policies, which are now well-established, enable the index user to derive preferred *CA* names for substances of specific molecular structure and thereby to find information about them in current and recent indexes. However, in carrying searches back beyond 1972, difficulties can arise. The current *CA* name of our compound has been determined; but where will we find this same substance in older *CA* indexes? Cross-references in these indexes are helpful, but of course none exist that lead from current names to those in Eighth (and earlier) Collective Indexes. As a last resource, the Formula Indexes can be consulted. But a simpler method, especially in searching for *groups* of compounds, is to read the appropriate paragraphs of this section. Although published originally (in the Volume 76 Index Guide under the title "Ninth Collective Index Changes") to alert index users to prospective changes, it will now be found most serviceable to remind retrospective searchers of the main nomenclature changes occurring in 1972. For example, ¶ 228 (below) indicates that peroxoic acids, in Volume 75 and earlier indexes, will be found at Peroxy headings, that aldehydic, anilic, and hydroxamic acid headings were also employed, and that many acids were indexed at trivial names; ¶ 270 explains that cationic com= pounds currently indexed at aminium headings were formerly entered at **Ammonium**; and ¶ 281 gives examples of current and former names for cyclic epoxides.

The changes in name selection rules described in this section are 1972 changes unless otherwise specified. The very few revisions found necessary in 1977 and 1982 mainly affect inorganic compounds and are described in ¶¶ 239, 257, and 273. The changes to coordination compounds, inorganic com= pounds, molecular addition compounds, peptides, polymers, ring systems and salts effective in 1987 are described in their respective sections.

The arrangement of subjects is alphabetic, and references are supplied in each case to the paragraphs in Sections A through F where a complete account of current policies will be found. The index (Section M) may also be consulted.

Headings found in current indexes are shown in boldface within the fol= lowing paragraphs.

226. Acetals are indexed like ethers (¶ 196); thus, Acetone, diethyl acetal became **Propane, 2,2-diethoxy-**. Cyclic acetals, except those of stereoparents (¶ 202), are indexed at the ring names. Epoxides are likewise indexed at the names of the appropriate ring systems, e.g., **Oxirane.**

227. Acid halide names (¶ 170) were affected by the discontinuance of most trivial acid names. Isocyanates, isothiocyanates, and isocyanides are now named like the halides. In the following examples, names prior to 1972 are shown in parentheses.

Benzenepropanoyl chloride	(Hydrocinnamoyl chloride)
Butanethioyl fluoride	(Butyryl fluoride, thio-)
Benzenesulfonyl isocyanate	(Benzenesulfonic acid, anhydride with isocyanic acid)

228. Acids. Carbonic acid and related compounds including **Formic acid** (¶ 183) are now relegated to a position below "organic" acids and above inorganic "oxo" acids (¶ 106).

Among carboxylic acids, the only trivial names retained were **Acetic acid** and **Benzoic acid.** (When oxygen in the functional group of these two acids is replaced, systematic functional replacement names, e.g., **Ethanimidic acid, Benzenecarbothioic acid,** are employed.) All other organic acids (carboxylic, sulfonic, sulfinic, etc.) are named systematically (¶ 165); e.g., **Propanoic acid** (formerly Propionic acid); **Benzoic acid, 2-hydroxy-** (Salicylic acid); **2-Bu= tenedioic acid** (*E*)- (Fumaric acid); **Benzenesulfonic acid, 4-amino-** (Sulfanilic acid).

Use of replacement affixes (¶ 129) for modified carboxylic, sulfonic, etc., groups was extended for Volume 76. In the following examples, the names formerly used are shown in parentheses:

Ethanimidothioic acid	(Acetimidic acid, thio-)
3-Pyridinecarbohydrazonic acid methyl ester	(Nicotinic acid, methyl ester, hydrazone)
Butanediperoxoic acid	(Peroxysuccinic acid)
2-Propanesulfeno(thioper= oxoic) acid	(Peroxy-2-propanesulfenic acid, thio-)

Aldehydic, amic, anilic, hydroxamic, hydroximic, nitrolic, and nitrosolic acids are now named as derivatives of carboxylic acids, amides, etc.

Examples:

1-Naphthalenecarboxylic acid, 8-formyl-	(Naphthalaldehydic acid)
Propanoic acid, 3-amino-3-oxo-	(Malonamic acid)
Butanoic acid, 4-oxo-4-(phenylamino)-	(Succinanilic acid)
Acetamide, *N***-hydroxy-**	(Acetohydroxamic acid)
Acetaldehyde, 1-nitro- oxime	(Acetonitrolic acid)

See also **Oxo acids** (¶ 273).

229. Alcohols and Phenols (¶ 175) are treated as of equal seniority in the order of chemical functions; previously, alcohols (in which the hydroxyl group is formally attached to a saturated carbon atom) were placed just ahead of phenols. Now, the preference of cyclic over acyclic hydrocarbons (¶ 138) often causes a change of nomenclature unless a conjunctive name can be used.

Examples:

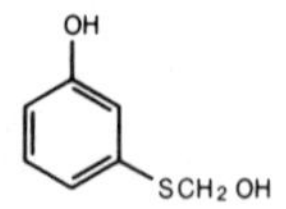

Phenol, 3-[(hydroxymethyl)thio]- (formerly Methanol, [(*m*-hydroxyphenyl)thio]-

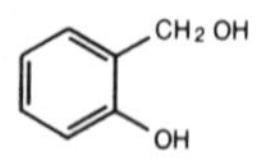

Benzenemethanol, 2-hydroxy- (formerly Benzyl alcohol, *o*-hydroxy-) (the preferred heading parent is larger than **Phenol**)

Radicofunctional names previously used for unsubstituted Ethyl through Dodecyl alcohol were replaced by **Ethanol, 1-Propanol,** etc. Iso-, *sec-* and *tert-*names are no longer used, except to index "isooctanol," etc., when no further information is provided. Isopropyl alcohol is now indexed at **2-Propanol**; *sec*-Butyl alcohol at **2-Butanol**; and *tert*-Butyl alcohol at **2-Propanol, 2-methyl-**. **Phenol** was retained as a heading parent, but all other trivially named phenols, including polyhydric benzene derivatives, are named systematically at such names as **Phenol, 2-methyl-** (formerly *o*-Cresol) and **1,3,5-Benzenetriol. Naphthalenol** (formerly Naphthol) and **Anthracenol** (formerly Anthrol) are now spelled out.

230. Aldehydes (¶ 173) are named systematically by use of the suffixes "-al" and "-carboxaldehyde," except for the three trivial names **Acetaldehyde, Formaldehyde,** and **Benzaldehyde.** Chalcogen analogs are all named systematically. Examples (previous index names are on the right):

Propanal	(Propionaldehyde)
Benzaldehyde, 2-hydroxy-3-methyl-	(2,3-Cresotaldehyde)
Benzenecarbothioaldehyde	(Benzaldehyde, thio-)
Butanedithial	(Succinaldehyde, dithio-)

231. Alkaloids (¶ 204). Many alkaloids indexed prior to 1972 at trivial names are now indexed systematically. The remaining trivial headings (stereoparents) are limited to alkaloids exhibiting stereochemical complexity. Class A alkaloids (named systematically) therefore comprise a much larger group than previously, when it was restricted to compounds containing no asymmetric center. Classes B and C are largely unchanged, except that the number of Class B alkaloid stereoparents ("systematic" alkaloid names) has been greatly increased with a consequent reduction in the number of Class C alkaloid stereoparents (trivial names for alkaloids of known constitution, including all stereochemistry). Class D alkaloids (those not fully elucidated) are indexed at systematic names without stereodescriptors if sufficient information is available, otherwise at author names or as Class B stereoparents with partial stereochemical information.

232. Alloys (¶ 214) are now indexed and registered as specific chemical substances. The components, if known, are cited as symbols in the modification with their percentage compositions (when present to the extent of 0.1% or more). Cross-references from trade names and some code designations have been provided (without percentage compositions); e.g., **Alnico V.** See *Iron alloy, base, Fe,Al,Co,Cu,Ni,Si,Ti (Alnico V)*. This alloy is also indexed at "nonbase" headings, e.g., **Cobalt alloy, nonbase,** for the other component elements (down to a limit of 1% composition), with the order of terms in the modification unchanged. For alloys of unknown percentage composition, only the "nonbase" headings for the components are employed.

233. Amides (¶ 171) are named systematically by use of the suffixes "-amide" and "-carboxamide." The three remaining exceptions are **Formamide, Acetamide,** and **Benzamide,** but their chalcogen and imidic analogs are now named systematically; for example, Acetamide, thio- is named as **Ethanethioamide.** Anilides are now indexed as *N*-phenyl amides. Secondary and tertiary amides are named at the preferred primary amide name. Thio analogs of sulfonamides are named as sulfonothioamides and sulfonodithioamides. See also **Urea** (¶ 292).

234. Amidines (¶ 171) are indexed systematically as imidamides, i.e., as amides of imidic acids; thus, Formamidine in Volume 76 became **Methanimidamide.** Amidoximes are named as *N*-hydroxy imidamides.

235. Amines (¶ 176) are all named systematically; Aniline is indexed at **Benzenamine**; Methylamine became **Methanamine.** Derivation of amine names for amino derivatives of nitrogen heterocycles is now permitted; Pyridine, 2-amino- (pre-1972 name), is now **2-Pyridinamine**; and Piperidine, 4-(2-aminopropyl)-, became **4-Piperidineethanamine, α-methyl-**. Secondary and tertiary amines are named as derivatives of the preferred primary amines, e.g., **Ethanamine, *N*-ethyl-** (formerly Diethylamine). Amino derivatives of the hydrides **Borane, Phosphorane,** and **Stannane** are now indexed at **Boranamine, Stannanediamine,** etc.

236. Amino acids (¶ 205) which are biologically significant are usually indexed at the trivial (or "common") names which are now classed as stereoparents. A few trivial names of α-amino carboxylic acids (Allocystathionine, Carnosine, Creatine, Cystathionine, Ethionine, Hippuric acid, Lanthionine, Pantothenic acid, Sarcosine, Thyronine, Thyroxine) and all trivial names of amino sulfonic acids were discontinued. The configurational descriptor is now placed in the heading as a prefix, thus: L-**Leucine.** The new stereoparent **Phenylalanine** is now employed, instead of Alanine, phenyl-, for $C_6H_5CH_2CH(NH_2)COOH$. Amino acid radicals are now employed only in naming peptides. *O*- and *S*-Derivatives of hydroxyl- and mercapto-group-containing amino acids, e.g., **Serine** and **Cysteine,** are now indexed at those heading parents. Esters of stereoparent amino acids with systematically named hydroxy compounds are indexed at the stereoparents. An ester with another stereoparent is indexed at the component possessing the higher function. Normally, only one entry is now made for each ester (¶ 247).

237. Anhydrides (¶ 179) of unsubstituted monobasic organic acids prior to Volume 76 (1972) were named in the heading, e.g., Acetic anhydride; now they are cited in the modification at the acid heading. Anhydrides of organic acids with inorganic monobasic oxo acids are named at the organic acid heading with an "anhydride with..." phrase in the modification, not at such former headings as Acetyl nitrate, but anhydrides of organic acids with isocyanic and isothiocyanic acids are treated like the corresponding acid halides. The term "bimol. monoanhydride" is employed when appropriate, with locants before the second word if necessary, at polybasic organic acid headings.

Cyclic anhydrides are now indexed as diones at the oxygen heterocycle names. Heading parents such as **Dicarbonic, Dicarbonimidic,** and **Tricarbonic acid** are used for carbonic acid anhydrides and their analogs (¶ 183). Anhydrides of phosphonic, phosphorous, arsonic, and arsonous acids are now treated like phosphoric acid anhydrides, e.g., **Diphosphonic acid.** "A" names are used for anhydrides when the requirements (¶ 127) are met.

238. Azo and **azoxy compounds** (¶ 193) in the absence of higher functions (e.g., formazans, nitrogen heterocycles, or functions expressible by a suffix) are indexed at **Diazene,** HN=NH (formerly Diimide). Azobenzene thus became **Diazene, diphenyl-**, and Naphthalene-2-*NNO*-azoxymethane became **Diazene, methyl-2-naphthalenyl-, 1-oxide.**

239. Boron compounds (¶¶ 159, 182). Cyclic compounds in which ligands bridge two boron atoms are named as coordination compounds; for example, Diazoniadiboratacyclobutane

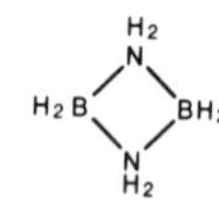

is now indexed at **Diborane(6), di-μ-amino-**. Derivatives of polyboranes are given low numbers for substituents regardless of the direction of numbering. Boronic acid is treated as an independent index heading parent; thus, Benzeneboronic acid was renamed **Boronic acid, phenyl-** in 1972.

Since the Tenth Collective Index period (1977–81), most molecular addition compounds of boron have been indexed as coordination compounds. From the same date, Hypoboric acid, $((HO)_2BB(OH)_2)$, has been indexed at **Diborane(4), tetrahydroxy-**.

240. Carbohydrates (¶ 208). Glycosides of known constitution are now indexed at the systematic names; they are ranked as polyhydric alcohols, not as aldehydes or ketones, and when attached to a carbon atom of an aglycone containing a higher chemical function are indexed at a name which expresses that function. *N*-Glycosyl derivatives of heterocycles have continued to be indexed at the heterocyclic parents.

Radical (substituent prefix) names are no longer formed from alditols.

Oligosaccharides indexed prior to Volume 76 at common names, e.g., Sucrose, are now named by systematic carbohydrate nomenclature.

The names Erythrose and Threose were discontinued except in the naming of polysaccharides. "Arrow" nomenclature was extended to all tri- and higher oligosaccharides, including those indexed at "-oside" heading parents.

Configurational descriptors (D-, L-, and DL-) are now always placed ahead of the heading parent; e.g., D-**Glucopyranoside.**

The index parent Cellulose acetate was discontinued; entries since 1972 have been made instead at **Cellulose, esters,** with modifications such as **"monoacetate," "diacetate,"** and (for the indefinite ester) **"acetate."**

241. Carbonic acid and relatives (¶ 183) underwent name changes in several respects. **Carbonic acid** itself was retained but is now classed below all "organic" acids, including carboxylic and sulfonic acids, and above inorganic oxo acids (¶ 106). It is the parent of numerous acids, amides, acid halides, etc., derived from it by the use of affixes and suffixes. Thus, Carbonic acid, thio-, became **Carbonothioic acid**; Imidocarbonic acid, $HN{=}C(OH)_2$, became **Carbonimidic acid**; Formic acid, chlorothio-, hydrazone, $H_2NN{:}C(Cl)SH$, became **Carbonochloridohydrazonothioic acid. Cyanic acid, Thiocyanic acid,** etc., were retained. **Carbamic acid** was kept as an acceptable abbreviated version of Carbonamidic acid; the abbreviation is also employed when other affixes are present, as indicated in the following list of revised names.

Carbamic acid	H_2NCO_2H
Carbamohydrazonic acid	$H_2NC({=}NNH_2)OH$
Carbamimidic acid	$H_2NC({=}NH)OH$
Carbamothioic acid *S*-ethyl ester	$H_2NC(O)SEt$
Carbamo(dithioperox)imidic acid	$H_2NC({=}NH)SSH$

Carbonic acids with a single hydrazide residue replacing an acid group are indexed, not as Carbonohydrazidic acid derivatives, but at **Hydrazine** index heading parents such as **Hydrazinecarboxylic acid.** The dihydrazide is named **Carbonic dihydrazide.** Phosgene, $C({:}O)Cl_2$, is indexed at **Carbonic dichloride.**

Carbonic acid is now a "Class I" acid in the naming of esters (¶¶ 185, 247) as are **Formic acid, Carbamic acid, Carbamic acid, methyl–**, and **Carbamic acid, phenyl–.**

242. Coordination compounds (¶ 215). Beginning in Volume 76 (1972), the stereochemistry (when known) of mononuclear coordination complexes has been described in the index name modification by capital italic letter symbols (¶ 203 III).

The ligating atoms of ligands, with certain limitations, are denoted by capital italic element-symbol locants for mononuclear complexes and for nonbridging ligands of polynuclear complexes. Use of these locants has resulted in a change in some simple inorganic ligand names; thus nitrito became (nitrito-*O*) and nitro became (nitrito-*N*).

Delocalized ligands are now denoted by the "hapto" convention, in which the Greek letter eta (η) is employed, along with locants to indicate the particular ligand atoms involved, or else a superscript indicating the total number of such atoms (when all are involved).

Compounds containing more than one complex anion are indexed at the preferred anion while other anions are cited with (simple) cations in the modification.

Ligands containing ester groups now normally include the esters in the ligand names. The rules for choice of a radical name for a ligand, rather than a name ending in "-ato," underwent some modification. Enclosing marks are employed around all "-ato" ligand names, and additional enclosing marks are added for bridging ligands when the ligand name itself requires enclosing marks; e.g., [μ-(acetato-*O:O'*)].

Beginning in Volume 106 (1987) salts of dithio(seleno, telluro) organic acids and polythio(seleno, telluro) arsenic and phosphorus acids with coordination metals will be structured and named at the chelated forms.

Metal "radical" names, e.g., "aluminio," are no longer employed for attachment of one metal to another in a binuclear complex, or as multiplicative radicals (¶ 194).

Numerical prefixes gave way to numerical ratios when mixed cations and anions (or at least one multivalent ion) are expressed in the modification. The terms "ion(1+)," "ion(2+)," etc., are not cited in modifications.

243. Diazo compounds, including diazohydroxides, etc., are now often named at **Diazene** headings (¶ 193).

Example:

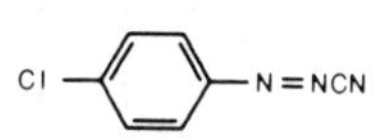

Diazenecarbonitrile, (4-chlorophenyl)– (formerly Benzenediazocyanide, *p*-chloro–)

The unsubstituted radical HN=N– is now named diazenyl (formerly diazeno).

244. Dyes (¶ 216) of known constitution are now always indexed at their systematic names. Previously, azo dyes and a few other classes were given C.I. (*Colour Index*) names for indexing purposes. Mixed dyes are indexed (like other mixtures, ¶ 221) at the names of their components when these are known. C.I. names are employed, when available, for dyes of unknown composition.

245. Elementary particles (¶ 217) are now indexed and registered as chemical substances.

Examples:

Lepton
Meson, π^- (140)
Muon, μ^+ (106)
Nucleon
  resonance N*(2040)
Positron
  leptonic mol. with chlorine ($e^+e^-Cl^0$)
Proton

246. Enzymes (¶ 218) are now indexed as chemical substances to which CAS Registry Numbers are assigned. At the index heading **Enzyme Commission** in the *Index Guide*, a list of E.C. numbers will be found from which cross-references lead to the *CA* index names. These are inverted if they consist of more than one word.

Examples:

Dehydrogenase, lactate
Phosphatase, adenosine tri–
Asparaginase
Papain

The heading may contain further information regarding secondary activities, etc.; thus: **Dehydrogenase, malate (decarboxylating).**

247. Esters (¶ 185). Some changes were made in the list of "Class I" (i.e., common) acids. The following acids were added:

Benzoic acid, amino– (all isomers)
Carbamic acid, methyl–
Carbonic acid
Phosphorodithioic acid
Phosphorothioic acid

Inorganic oxo acids other than **Boric acid (H_3BO_3), Nitric, Phosphoric, Phosphorous, Sulfuric,** and **Sulfurous acids** were removed from the Class I list and therefore are now preferred as the heading parents for esters with Class I and Class II alcohols. (Additional entries for esters were discontinued.)

Cyclic esters are named as heterocycles. Esters of stereoparents with systematically named acids are indexed at the stereoparents (¶ 202).

The use of a radical term ending in "-yl ester" is now preferred to "ester with...," even when the alcoholic component contains a chemical function higher than alcohol (see ¶ 250), unless the latter phrase is followed by a stereoparent, as in "5'-ester with adenosine," or by an acid requiring a synonym line formula, as in "triester with boric acid (H_3BO_3)." When the "-ate" form of a polybasic acid name is cited in the modification at an alcohol heading, all free acid groups are denoted by the word "hydrogen," e.g., **Cyclohexanol, 4-bromo–, dihydrogen phosphate.**

Esters of substituents of index heading parents are expressed as substituents. Thus, the acetate of **Benzenesulfonic acid, 4-hydroxy–** is named **Benzenesulfonic acid, 4-(acetyloxy)–,** and the methyl ester of **Pyridinium, 3-carboxy-1-methyl–, chloride,** is indexed at **Pyridinium, 3-(methoxycarbonyl)-1-methyl–, chloride.**

Ortho esters are named like ethers, i.e., as alkyloxy and aryloxy substituents. "A" names are used for esters where appropriate (¶ 127).

248. Ethers (¶ 196) are named as substituents of hydrocarbons and other index heading parents; Ether as a heading parent was discontinued. Thus, Ethyl ether became **Ethane, 1,1'-oxybis–.** Polyethers may often be named by "a" nomenclature (¶ 127). Methylenedioxy derivatives of benzene and its hydrogenated derivatives are named at **1,3-Benzodioxole.**

249. Formazan (¶ 193) was retained but is now ranked just below imines as the highest of the nitrogen compound classes lacking functional suffixes. Formazan radicals are unchanged.

250. Functional derivatives (¶ 112) (now restricted to esters, hydrazides, hydrazones, and oximes) of principal chemical functions have continued to be expressed in the index modifications at boldface index headings, but derivatives of subsidiary functions are usually included in the substituent prefixes which follow the comma of inversion in an inverted heading. Hence Benzoic acid, *p*-hydroxy–, ethyl ester, acetate, is now named **Benzoic acid, 4-(acetyloxy)–, ethyl ester** (¶ 185). Esters are always expressed by use of an "-yl ester" phrase when possible, not by an "ester with" phrase, unless a stereoparent or an acid which requires a synonym line formula is being cited in a modification. Prior to Volume 76, "ester with" was used when the alcoholic component contained a higher functon; now, "ester with glycolic acid" becomes "carboxymethyl ester."

251. Guanidine and relatives. Guanidine (¶ 183) was retained as an index heading parent for Carbonimidic diamide with the locants shown (numerical locants were formerly employed):

$$\underset{N}{H_2N}C(=\underset{N''}{N}H)\underset{N'}{N}H_2$$

It is ranked below **Urea** (carbonic diamide).

Biguanidine is now named **1,2-Hydrazinedicarboximidamide** with the locants shown:

$$\underset{N}{H_2N}C(=\underset{N'}{N}H)\underset{1}{N}H\underset{2}{N}HC(=\underset{N'''}{N}H)\underset{N''}{N}H_2$$

It ranks as the amide of a diimidic acid of a nitrogen acyclic parent, e.g., above **Benzenedicarboximidamide** but below **Benzamide.**

Biguanide and Triguanide are now named systematically as **Imidodicarbonimidic diamide** and **Diimidotricarbonimidic diamide,** respectively.

252. Hydrazine (¶ 193) was retained as a trivial name for Diazane, and functional suffixes may be appended to it; thus, **Hydrazinecarboxylic acid,** H_2NNHCO_2H (formerly Carbazic acid), ranks above monocarboxylic acids derived from carbon skeletons.

Hydrazides (¶ 189) and hydrazones (¶ 190) of principal chemical functions expressed in the heading parent are cited in the modification as before, but these derivatives of functions expressed as substituents are now "named through" as in the case of esters (¶¶ 247, 250), by use of hydrazino and hydrazono radicals.

Diacyl hydrazines are now indexed not at **Hydrazine** but as acyl hydrazides of the preferred acid (¶ 189). Semicarbazones of nonstereoparents are no longer indexed as such but as derivatives of **Hydrazinecarboxamide,** either as an index heading parent or as an appropriate substituent. Semioxamazones are treated as alkylidene hydrazides of **Acetic acid, aminooxo–,** RCH:NNHC(:O)C(:O)NH_2. The heading parent **Carbonic dihydrazide** (¶ 183) was introduced for $H_2NNHC(O)NHNH_2$ (formerly Carbohydrazide) and its derivatives, including compounds previously indexed as carbohydrazones of oxo parents.

253. Hydroxylamine (¶ 193) was retained as a heading parent; it ranks low among nonfunctional nitrogen parents between **Hydrazine,** H_2NNH_2, and **Thiohydroxylamine,** H_2NSH (formerly Hydrosulfamine). *N*-Acyl derivatives of **Hydroxylamine** are now indexed at amide headings; *N*-alkyl derivatives at amine names. In the absence of *N*-substituents, *O*-derivatives are usually indexed at **Hydroxylamine.** Alkylidene derivatives are indexed as oximes of the corresponding carbonyl compounds unless higher functions are present, in which case a (hydroxyimino) substituent prefix is employed (see ¶ 195).

254. Imines (¶ 177) are indexed by means of the suffix "-imine" appended to the molecular skeleton, which may now be a nitrogen heterocycle; e.g., **4(1*H*)-Pyridinimine** (formerly Pyridine, 1,4-dihydro-4-imino–). Cyclic imines are indexed at the ring names; e.g., **Aziridine** (formerly Ethylenimine). Imines rank below amines and all other compounds expressed by functional suffixes. 1-Iminoalkyl radicals are so named, not as imidoyl radicals; hence, formimidoyl has become (iminomethyl), and propionimidoyl is indexed as (1-iminopropyl).

255. Index name selection policies for chemical substances (¶ 138) were revised for Volume 76 and subsequent volumes as follows:

(*a*) "Like treatment of like things" was abandoned.

(*b*) The principle of "complexity" (a measure of the number of parentheses and brackets in a name) is no longer employed.

(*c*) The principle of "lowest locants" of substituents on an index heading parent (¶ 137, 138) was introduced on a regular basis.

(*d*) The principle of "centrality" is now applied when a compound contains a sequence of three or more occurrences of the same heading parent.

(*e*) Conjunctive nomenclature (¶ 124) was extended for the first time to benzene with a *single* acyclic functional substituent, as in **Benzenemethanol** (names such as **Benzenediacetic acid** were used previously). It was discontinued for compounds in which ring attachment is by a double bond, and for **Carbamic** and **Sulfamic acids,** unsaturated acyclic functional compounds, and acyclic difunctional compounds.

(*f*) Multiplicative nomenclature (¶ 125) based on hydrocarbons is now permitted, as in **Ethane, 1,1'-oxybis-**, and **Benzene, 1,1'-[1,2-ethanediylbis(thio)]bis-**. The terms "bis," "tris," etc., not "di-," "tri-," etc., are always employed in multiplying a parent. Greater symmetry than previously is now demanded of compounds for which multiplicative names may be used.

256. Indicated hydrogen (¶ 135) is always cited, never implied; thus, Indene became 1*H*-**Indene,** and Fluorene is now named **9*H*-Fluorene.** Hydrogen not required for formation of a ring system, but "added" at the same time as a functional (or radical) suffix is now described as "added hydrogen" (¶ 136). In tautomeric systems (¶ 122), lowest locants for indicated hydrogen are normally preferred; thus **1*H*-Purine,** not 7*H*-Purine, is indexed in the absence of information to the contrary.

257. Inorganic compounds (¶ 219) are largely indexed at names previously used. Binary names for derivatives of elements possessing hydride names were changed to derivatives of the hydrides; e.g., **Silane, tetrachloro-** (formerly Silicon chloride ($SiCl_4$)); **Borane, trifluoro-** (formerly Boron trifluoride). When the compound is the halide of a recognized acid, it is so named; e.g., **Phosphorous trichloride** (formerly Phosphorus chloride (PCl_3)). Pyrophosphoric acid, $(HO)_2P(O)OP(O)(OH)_2$, was renamed **Diphosphoric acid,** and analogous compounds are named similarly; for example, **Diphosphoryl chloride,** $Cl_2P(O)OP(O)Cl_2$; ***P,P'*-Diamidodiphosphoryl fluoride,** $(H_2N)F=P(O)OP(O)F(NH_2)$; ***P'*-Amidodiphosphoric(III,V) acid,** $(HO)_2POP(O)(O=H)(NH_2)$.

Since the Tenth Collective Index period (1977–81), antimony has been classed as a metal, and synonym line formulas have been cited in index headings with *all* binary salt-type names, e.g., **Sodium chloride (NaCl)**. Graphite derivatives have been indexed as molecular addition compounds instead of at the pre-1977 headings **Graphitic acid, Graphite nitrate**, etc.

Beginning in 1982, elements of atomic number 104 and above have been indexed at names derived from these numbers, e.g., **Unnilquadium** (formerly Element 104).

Starting in 1987 inorganic line formulas may be expressed with decimals or numerical ranges as well as integers, e.g. **Aluminum gallium arsenide ($Al_{0.15}Ga_{0.85}As$).**

See also ¶¶ 239, 273.

258. Inositols (¶ 209) are now indexed at the separate stereoparents derived by combining individual configurational prefixes with the name Inositol, e.g., *myo*-**Inositol;** *scyllo*-**Inositol**. These names in turn are preceded by the configurational descriptors D- and L-. When a choice must be made for derivatives, the earliest alphabetic prefix is preferred; then the prefix D- is selected, rather than L-; finally, lowest locants are assigned to substituent prefixes. **Inosose** and **Streptamine** are retained as stereoparents.

259. Iodine compounds (¶ 188) with abnormal valencies are indexed by coordination nomenclature (¶ 215); e.g., **Iodine, dichlorophenyl-** (formerly Benzene, (dichloroiodo)-.

260. Isocyanides (¶ 188). Alkyl isocyanides, RNC, are now indexed substitutively; acid isocyanides are treated like acid halides. For example, Isopropyl isocyanide became **Propane, 2-isocyano-; 2-Naphthalenecarbonyl isocyanide** was formerly indexed at 2-Naphthoic acid, anhydride with hydroisocyanic acid.

261. Ketene (¶ 174) is now indexed systematically at **Ethenone**; Ketene, thio- at **Ethenethione.** Cyclic ketenes are named as cycloalkylidene (etc.) derivatives of the new heading parent **Methanone,** H_2CO. Unsubstituted H_2CO is still indexed at **Formaldehyde**, but unsubstituted CH_2S is named **Methanethial**. In the presence of chemical functions higher than ketone, the =CO radical is named carbonyl (if the carbon atom does not form part of an acyclic chain) or as (oxoalkyl). The isolated =CS radical is named carbonothioyl if both free valencies are attached to a single atom, or if it is used in a multiplying prefix; otherwise as (thioxomethyl).

262. Ketones (¶ 174). All trivial names (including Acetone, Chalcone, Benzil, Benzoin, and all flavone, quinone, phenone and naphthone headings) were abandoned. The heading parent Ketone was discontinued. **Methanone,** $H_2C{:}O$, is used to index ketones having two ring systems directly attached. The analogous headings **Methanethione,** $H_2C=S$, etc., were introduced. Acetone was changed to **2-Propanone,** Acetophenone to **Ethanone, 1-phenyl-**, Acetophenone, thio- to **Ethanethione, 1-phenyl-**, Anthraquinone to **9,10-Anthracenedione,** Benzil to **Ethanedione, diphenyl-**, Flavone to **4*H*-1-Benzopyran-4-one, 2-phenyl-**, and Uracil to **2,4(1*H*,3*H*)-Pyrimidinedione.** For a discussion of this last name, see ¶ 289A.

The ketone function is no longer overstepped; hence **Ethanone, 1,1'-(1,5-naphthalenediyl)bis-** (formerly Naphthalene, 1,5-diacetyl-).

263. Locants (¶¶ 114, 137). The locants *o*-, *m*-, and *p*- were replaced by Arabic numerals in all cases; *as*-, *s*-, and *v*- were likewise replaced, except in *as*- and *s*-**Indacenes** and their fused derivatives. Locants for unsaturation are now always cited for molecular skeletons of three or more atoms except for monocyclic hydrocarbons containing one multiple bond. Polyvalent radicals with free valencies located at two or more positions now have all locants cited; thus, ethylene became 1,2-ethanediyl. In ring assemblies, locants for positions of attachment are always cited except for two-component assemblies of cycloalkenes, cycloalkadienes, etc. Thus, 4-Biphenylamine became **[1,1'-Biphenyl]-4-amine.**

In multiplicative names, locants are now cited to indicate positions of attachment of the multiplying radical to the heading parent, unless such a parent (*a*) contains only one skeletal atom, e.g., **Methanone, Silane;** (*b*) is a functional parent compound (¶ 130) possessing replaceable hydrogen at only one position, e.g., **Carbamic acid, Formic acid, Phosphonic acid;** or (*c*) is a radicofunctional heading parent, e.g., **Disulfide.**

Special rules whereby locants for *fully halogenated compounds* and radicals were sometimes omitted are now discontinued.

Urea is now assigned the locants *N*- and *N'*- in place of 1- and 3- (¶ 183).

When locants are required for derivatives, e.g., esters, cited in modifications, *all* such locants are used (¶ 119).

Letter locants are now employed with ligand names in coordination nomenclature to define the ligating atoms (¶ 215).

264. Metallocenes (¶ 215), including **Ferrocene** and its derivatives, are now considered to be neutral coordination complexes. Suffix-type names and conjuctive names are no longer used; hence Ferrocenecarboxylic acid is now indexed at **Ferrocene, carboxy-**; and 1,1'-Ferrocenediacetic acid at **Ferrocene, 1,1'-bis(carboxymethyl)-.**

265. Mixtures (¶ 221) are now indexed and registered as individual chemical substances if they are considered significant in their own right, e.g., pharmaceutical and pesticide mixtures which possess trade names, or mixtures which are emphasized in an original document as possessing special properties. Solvents, fillers, and inactive trace components are disregarded. Ratios of components are not cited. Cross-references from trade names, etc., are provided; e.g., **Terracoat.** See *1,2,4-Thiadiazole, 5-ethoxy-3-(trichloromethyl)-*, mixt. with pentachloronitrobenzene; an index entry appears also at **Benzene, pentachloronitro-.**

265A. Molecular addition compounds (¶ 192). Two additions were made to the list of common components beginning in 1987. The new additions are Acetic acid, trifluoro- and Ethanamine, *N,N*-diethyl-.

266. Multiplicative prefixes (¶ 110). Bis-, tris-, etc., are used in multiplicative names instead of di-, tri-, etc., even when the heading parent is not otherwise substituted. Bis-, tris-, etc., are used with methylene in every circumstance in general index nomenclature (but "di-*O*-methylene" is used with sugar names), and with oxy and thio, etc., to denote a multiplicity of single chalcogen atoms; thus, [methylenebis(oxy)] denotes $-O-CH_2-O-$, while (methyldioxy)- denotes CH_3-O-O-. Bi-, etc., names for ring assemblies were discontinued for assemblies of rings joined by double bonds. All ring-assembly names are enclosed in brackets when followed by a suffix; derived radicals are treated similarly. Binaphthyl and Bianthryl become **Binaphthalene** and **Bianthracene** (see ¶ 281). Bi-, etc., names derived from acyclic compounds, e.g., Bicarbamic acid, Triguanide, are discontinued.

267. Nitriles (¶ 172). Only the trivial names **Acetonitrile** and **Benzonitrile** are retained; otherwise, changes in nitrile names are almost entirely parallel to those for the carboxylic acids (¶ 165) from which they and their names are derived. Thus, Hydrocinnamonitrile became **Benzenepropanenitrile. Cyanamide** was retained for H_2NCN, but conjunctive "-carbamonitrile" names were discontinued.

268. Nitrone, $H_2C=NH=O$, was renamed **Methanimine,** *N*-oxide (¶ 177); *N*-alkyl nitrones are indexed as oxides at amine index heading parents; *N*-acyl nitrones at amide headings.

269. Nucleosides and Nucleotides (¶ 210) have continued to be indexed at trivial names, e.g., **Adenosine, Cytidine,** although the purine and pyrimidine bases from which they are derived are now named systematically; e.g., Cytosine is indexed at **2(1*H*)-Pyrimidinone, 4-amino-** (¶ 122). *N*-Acyl derivatives of nucleosides are indexed at the nucleoside parents instead of at amide names, but higher functions have continued to be indexed at the parents which express them.

Substituents are now expressed at nucleotide headings, e.g., **5'-Uridylic acid, 2'-amino-2'-deoxy-3'-thio-.** Mixed phosphate esters are indexed at the preferred phosphate heading rather than at the plain nucleoside name; e.g., **Inosine 5'-(trihydrogen diphosphate),** 2'-(dihydrogen phosphate). Nucleosides and nucleotides are considered to be stereoparents (¶ 203), and their esters with systematically named substances are therefore indexed at nucleoside or nucleotide index heading parents.

270. Onium compounds (¶ 184). Ammonium compounds are named from the corresponding preferred amines by use of "-aminium" as a suffix and expression of the remaining groups as *N*-substituents; e.g., **Ethanaminium, *N*-ethyl-*N,N*-dimethyl-,** iodide (formerly Ammonium, diethyldimethyl-, iodide). Naming of cyclic quaternary nitrogen compounds and other cationic species, e.g., sulfonium compounds, is largely unchanged.

271. Order of precedence of compound classes (¶ 106) has continued to play an important role in the selection of a preferred index name. Changes include the following:

(*a*) more emphasis is placed on the presence of a principal chemical functional group (expressed as a suffix attached to a molecular skeleton). Thus **Ethanamine** (a functional compound) is ranked above **Quinoline** (a nonfunctional molecular skeleton).

(*b*) **Ferrocene** and other metallocenes are ranked with neutral coordination complexes.

(*c*) Peroxy acids expressed as principal groups are ranked, *as a class,* higher than all other acids. Previously, each such acid was placed just above the corresponding parent acid. Carbonic acid and related compounds are placed below those acids, e.g., carboxylic and sulfonic, expressed as functional suffixes, and above inorganic "oxo" acids. Derivatives of **Arsonic** and **Boronic acids** are named like those of other inorganic "oxo" acids with replaceable nuclear hydrogen atoms; thus, Benzenearsonic acid has become **Arsonic acid, phenyl-**. Derivatives of isocyanic acid and its chalcogen analogs are named substitutively by use of the prefixes isocyanato, isothiocyanato, etc.

(*d*) Alcohols and phenols now rank as equal in precedence.

(*e*) Ether, Sulfide, Selenide, and Telluride are no longer employed as heading parents and are therefore not listed in the Order of Precedence. Single chalcogen atoms are named substitutively (by use of oxy, thio, etc., radicals) or at organic replacement ("a") names.

(*f*) Phosphorus compounds without functional suffixes are ranked together (following nitrogen compounds). Formerly, phosphine oxide was placed above amines with phosphine and phosphorane following them. The same policy is adopted in descending order for nonfunctional heading parents derived from arsenic, antimony, bismuth, boron, silicon, germanium, tin, and lead. Then follow nonfunctional oxygen parents in the order: oxygen heterocycles, acyclic polyoxides **(Trioxide, Peroxide)**, acyclic "oxa" names. Other chalcogens are ranked below oxygen in the order: sulfur, selenium, tellurium. Lowest in order of precedence are nonfunctional carbon skeletons.

(*g*) In each class of nonfunctional compounds, ranked in the order of the most preferred heteroatom it contains, cyclic compounds are preferred over acyclic, and acyclic parents named by "a" nomenclature are less preferred than other acyclic parents of that hetero-atom class. Among carbon compounds, similarly, carbocycles are preferred to acyclic carbon chains, regardless of length. Unsaturated skeletons are preferred to saturated skeletons with the same number and type of skeletal atoms; thus, **Benzene** is preferred to **Cyclohexane** (no policy change), and **Pyridine** is preferred to **Piperidine** (reversal of policy).

272. Organometallic compounds (¶ 194). Binary compounds of element hydrides, e.g., **Arsorane,** AsH_5, **Plumbane,** PbH_4, **Germane,** GeH_4, are indexed at those headings. Thus, Lead chloride ($PbCl_4$) became **Plumbane, tetrachloro-**. (In the carbon series, Carbon tetrachloride became **Methane, tetrachloro-**.) Amino derivatives of hydrides are now assigned "-amine" names, e.g., **Germanediamine,** and other functional suffixes are also employed, e.g., **Bismuthinecarboxylic acid,** but "-ol" to express a hetero-atom-attached hydroxyl is used only with carbon and silicon. "A" names must (as usual) be employed when appropriate.

273. Oxo acids (¶ 219). Some changes (¶ 185) were made in 1972 to the content of the "Class I" list for esters of inorganic "oxo" acids. Pyrophosphoric acid, $(HO)_2P(O)OP(O)(OH)_2$, was renamed **Diphosphoric acid.** The new heading parent **Diphosphonic acid** was previously indexed at Phosphonic acid, anhydride. Since 1972, arsenic mononuclear acids have been treated analogously to phosphorus acids.

Examples:

Arsenic acid (H_3AsO_4)	$(HO)_3AsO$
Arsonic acid	$(HO)_2HAsO$
Arsinic acid	$(HO)H_2AsO$
Arsenenic acid	$HOAsO_2$

Cyclic esters of arsenic and phosphorus acids are indexed at the ring names.

Since the Tenth Collective Index period (1977-81), metal oxo acids (including oxo acids of antimony) and their salts have been indexed as coordination compounds or as mixed salts. The only exceptions are **Chromic acid (H_2CrO_4), Chromic acid ($H_2Cr_2O_7$), Manganic acid (H_2MnO_4)** and **Permanganic acid ($HMnO_4$)**, which were retained as index heading parents. A few little-used nonmetallic oxo acid names, e.g., **Mesosulfuric acid**, were also discontinued.

274. Peptides (¶ 206), other than biologically active peptides with trivial names, containing two through twelve amino-acid residues are now indexed at the name of the *C*-terminal unit in every case; peptide amides are treated analogously. No change was made in the treatment of larger peptides. Some trivial acyl names were changed starting in the twelfth collective period (1987-1991). Thus α-aminobutyryl became 2-aminobutanoyl and lactoyl became 2-hydroxypropanoyl. Cyclic peptides of two or three units are indexed at systematic ring-system names, larger compounds at "Cyclo" names such as **Cyclo[L-alanyl-N^5-acetyl-D-ornithylglycyl-L-alanyl-D-propyl-*O*-(1,1-dimethylethyl)-L-seryl]** in which the units are cited in the order of occurrence, with preference given to the lowest alphabetic arrangement of *parent* amino-acid radicals. Prior to 1987 the term "Cyclic" was used instead of "Cyclo". Additional entries are made at ring-system names. Peptides with fewer than four units and which contain disulfide linkages are now indexed, like larger peptides, at cysteine rather than cystine names.

Linear depsipeptides are now usually indexed at the *C*-terminal *amino*-acid names; terminal hydroxy acids are expressed as carboxyalkyl esters. Cyclic depsipeptides are indexed at ring names or "Cyclo" names in analogy with peptides.

Naturally-occurring biologically active peptides and depsipeptides of six to fifty units are indexed at the trivial names; those of five or fewer units are named like other peptides at the *C*-terminal amino acid or "Cyclo" name; trivially named stereoparents such as **Bradykinin** and **Gramicidin S** have continued to be employed for substances containing six through fifty units. Small species variations in trivially-named peptides are now interrelated by using one form as the reference compound; e.g., **Fibrinopeptide B (orangutan).** See *Fibrinopeptide B (human), 10-L-leucine-12-glycine-*; similar cross-references are now made between trivial names of closely-related peptides and corticotropin sequences; e.g., **Kallidin.** See *Bradykinin, N^2-L-lysyl-*; **$\alpha^{1,39}$-Corticotropin (human).** See *$\alpha^{1,39}$-Corticotropin (pig), 31-L-serine-*.

Esters of hydroxy and mercapto groups in amino-acid radicals of peptides are now expressed as substituents; e.g., *O*-acetyl-L-seryl. Functional derivatives of carboxy substituents are expressed by radicals which include the carboxyl group; e.g., *N*-[(phenylmethoxy)carbonyl].

Such former headings as Angiotensins were made singular in Volume 76. The indexing of *Oxytocins* and *Vasopressins* was further refined. Bovine insulin is now the reference compound for insulins, with the name **Insufin (ox);** cross-references in the *Index Guide* show the species relationships; thus, **Insulin (human-B reduced).** See *Insulin (ox-B reduced), 30-L-threonine-*. **Actinomycin D** is the reference stereoparent to which other actinomycins are related for indexing purposes.

Proteins (¶ 207) are indexed at trivial names, with accompanying species information.

Synthetic peptides are indexed like polymers; the term "polymer with" is used instead of "peptide with" at amino-acid monomer names.

275. Peroxides (¶ 196). Binary headings were discontinued; Isopropyl peroxide became **Peroxide, bis(1-methylethyl).** The heading **Peroxide** is a nonfunctional oxygen compound; in the presence of more preferred compound classes, such radicals as (ethyldioxy) are employed. **Hydroperoxide** is a functional oxygen compound ranking just above amines; such names as Peroxyacetyl hypochlorite are replaced by **Peroxide, acetyl chloro,** etc.

276. Phosphorus compounds (¶ ¶ 197, 219). Functional suffixes are now employed with phosphorus hydride names; e.g., **Phosphinecarboxylic acid** (formerly Formic acid, phosphino-) and **Phosphoranamine** (formerly Phosphorane, amino-). The phosphoro radical, -P=P-, was renamed 1,2-diphosphenediyl; phosphino was retained, but diphosphino, H_2P-PH-, was renamed diphosphinyl, and diphosphinetetrayl became 1,2-diphosphinediylidene. **Phosphorothioic** and **Phosphorodithioic acids** were added to **Phosphoric** and **Phosphorous acids** in "Class I" (¶ 185) for the indexing of esters. Pyrophosphoric acid was renamed **Diphosphoric acid. Diphosphonic acid** became the new name for **Phosphonic acid,** anhydride.

277. Polymers (¶ 222). Polymers whose component monomers are known are now indexed at the names of each of the component monomers as actually reported in the original document. Names based on structural repeating units (SRUs) have continued to be used as additional index names for polymers of well characterized or assumed structure. In the cases of very common trade-named industrial polymers of known composition for which cross-references are used, the SRU name is preferred. Thus, **1,4-Benzenedicarboxylic acid, polymers,** polymer with 1,2-ethanediol. See *Poly(oxy-1,2-ethanediyloxycarbonyl-1,4-phenylenecarbonyl)*; **Nylon 6.** See *Poly[imino(1-oxo-1,6-hexanediyl)]*. At monomer names, the terms "polymers," "polyamides," "polyesters," etc., were replaced by "homopolymer" except for general studies; "polymer with" was retained for copolymers named at monomer headings and replaced such terms as "polyamide with", "polyester with", etc.; "copolymers" is an acceptable general term. Block, graft, and alternating copolymers are registered and indexed with their own specific Registry numbers starting in the twelfth collective period (1987-1991). The terms "block", "graft" and "alternating" are part of the index name for these kinds of polymers.

Phenol condensation products are now indexed at **Phenol polymers,** with "polymer with formaldehyde" or other appropriate term cited in the modification. Urea condensation products are named analogously at **Urea polymers.**

278. Porphyrins and Bile pigments (¶ 223) were affected by the new rule that indicated hydrogen of ring systems is always cited. **21*H*,23*H*-Porphine** and **21*H*,23*H*-Porphyrazine** are the forms preferred for these and related substances unless an original document emphasizes the absence of hydrogen at these positions. The **29*H*,31*H*-**form of **Phthalocyanine** is preferred; the heading **21*H*-Biline** is employed unless hydrogen is absent from that position, in which case **22*H*-Biline** is indexed. (An (*all-Z*) form is now assumed for Biline.)

Radicals. See Substituent prefixes (¶ 287).

279. Replacement ("a") nomenclature (¶ 127), in which replacement of carbon by heteroatoms is indicated by terms such as "aza," "oxa," and "thia," was unchanged for cyclic skeletons, but was amended for acyclic compounds, primarily to eliminate exceptions for certain classes, including silicon compounds. Now, "a" names are used whenever four or more "hetero units" (isolated hetero atoms, homogeneous hetero chains, or groups for which simple radical names are employed, such as disiloxanediyl) are present, so long as the resulting name, including functional suffixes, does not express a lower functionality than the alternative conventional name. The hetero atoms must not be in an abnormal valency state (unless this valency can be readily expressed), and the "a"-named chain must not be terminated by a nitrogen or chalcogen atom. ("A" names are not used for peptides, polymers or purely inorganic chains.)

280. Replacement nomenclature for functions (¶ 129) was extended to various classes of acids and acid derivatives, as outlined by the following examples (former names are in parentheses):

Ethanethioic acid	(Acetic acid, thio-)
Ethane(dithioic) acid	(Acetic acid, dithio-)
Ethaneperoxoic acid	(Peroxyacetic acid)
Butanebis(thioic) acid	(Succinic acid, 1,4-dithio-)
Carbonothioic acid	(Carbonic acid, thio-)
Carbamodithioic acid	(Carbamic acid, dithio-)
Methanimidamide	(Formamidine)
Imidodicarbonic acid	(Imidodicarboxylic acid)
Diimidotricarbonimidic diamide	(Triguanide)
1,3-Benzenedisulfonothioic acid	(*m*-Benzenedisulfonic acid, 1,3-dithio-)

281. Ring systems (¶¶ 145-157). Fused oxireno and thiireno derivatives of hydrocarbon rings are now indexed according to the general rule that the base component should be a heterocycle; thus, Naphthalene, 2,3-epoxy- has become **Naphth[2,3-*b*]oxirene.** Adamantane and its replacement derivatives are indexed at **Tricyclo[3.3.1.1^{3,7}]decane, 2-Thiatricyclo[3.3.1.1^{3,7}]decane,** etc. (¶ 155). The locants *o*-, *m*-, *p*-, and *v*- in ring names were replaced by numerals; *as*- and *s*- are used only with **Indacene** and its fused derivatives; thus, *m*-Dioxane became **1,3-Dioxane;** *p*-Terphenyl became **1,1':4',1''-Terphenyl;** *s*-Triazine became **1,3,5-Triazine;** and **Benz[*e*]-*as*-indacene** is unchanged. Some trivial names were replaced; thus, Quinuclidine was renamed **1-Azabicyclo[2.2.2]octane.** Radical names from these ring systems were changed correspondingly.

Special names for partially hydrogenated ring systems, e.g., Acridan, Indan, Indoline, Phthalan, Pyrroline, were abandoned; thus Indan became **1*H*-Indene, 2,3-dihydro-.** Norbornane, Norcarane, and Norpinane are now indexed at the systematic "Bicyclo-" names.

Ring systems containing metals other than antimony, tin, lead, germanium, or bismuth are now indexed by coordination nomenclature.

In ring-assembly names, points of attachment are now cited, except for two-component assemblies of cycloalkenes. Thus, **1,1'-Biphenyl, 1,2'-Binaphthalene** (formerly 1,2'-Binaphthyl), **Bi-2-cyclohexen-1-yl.** Ring assemblies in which component rings are joined by double bonds are now indexed at a component name, e.g., **Cyclopentane, cyclopentylidene-** (formerly Bicyclopentylidene).

The Hantzsch-Widman system (¶ 146) was extended, beginning in 1987, to partially or fully saturated rings, formerly named by organic replacement nomenclature, e.g. **Stannolane** (formerly Stannacyclopentane).

Non-standard valence states (¶ 158) of certain ring heteroatoms were denoted by the greek letter lambda (λ) starting in 1987, e.g. **1λ^4,3λ^4,5λ^4,7λ^4-1,3,5,7,2,4,6,8-Tetrathiatetrazocine** (formerly 1*H*,3*H*,5*H*,7*H*-1,3,5,7,2,4,6,8-Tetrathiatetrazocine).

Names for cyclic systems containing ring heteroatoms in non-standard valence states described by the "hydro" prefix (¶ 158) were changed beginning in 1987 to allow the expression of principal suffixes and radicals, e.g. **2*H*-1,4-Selenazin-2-one, 1,1-dihydro-1,1-dimethyl-** (formerly 2*H*-1,4-Selenazine, 1,1-dihydro-1,1-dimethyl-2-oxo-).

281A. Salts (¶ 198). Beginning with the twelfth collective period (1987-1991) salts of substitutive cations no longer have additional index entries at the following common anions: **Borate(1-), tetrafluoro-; Borate(1-), tetraphenyl-; Phosphate(1-), hexafluoro-.**

282. Silicon compounds (¶ 199). Former binary names for silicon halides are now indexed at **Silane;** for example, Silicon chloride ($SiCl_4$), is now indexed at **Silane, tetrachloro-.** Functional suffixes are now appended to **Silane, Disilane, Disiloxane,** etc., but conjunctive names are not formed from them. **Silanol** and **Silanediol** now rank above all carbon-skeleton monohydric and dihydric alcohols, respectively, by reason of their hetero atom content (¶ 106). Similarly, **Silanamine** ranks above **Benzenamine,** etc.

Silthianes are now spelled out as silathianes; 1,3-disiloxanediyl, 1,5-trisilathianediyl, 1,7-tetrasilazanediyl, etc., all rank as single hetero units (¶ 127) in the formation of "a" names. A total of four such units (of any kind) are now necessary for a compound to be named by replacement nomenclature.

Silicic acids now all belong to "Class II" in the naming of esters (¶ 185). Since the Tenth Collective Index period, the **Silicic acid** heading has been generally restricted to mono- and dinuclear oxo acids of silicon.

283. Spelling (¶ 107). Elided forms of radicals were largely discontinued. Nitramino became (nitroamino); acetoxy became (acetyloxy); naphthyl became naphthalenyl. Ethoxy, methoxy, propoxy, butoxy, phenoxy, and thienyl were retained, but branched-chain radicals are now named as derivatives of the straight-chain alkoxy parents; thus, isopropoxy became (1-methylethoxy). Elided forms of heading parents (Pyridol; Pyridone) are now spelled out **(Pyridinol; Pyridinone).** Punctuation in compound names is unchanged.

284. Stereochemistry (¶ 203). Major revisions were made in this area for the Ninth Collective Index Period (1972-76). The concept of stereoparents (¶ 285) was introduced for those substances, mainly natural products, containing complex stereochemistry and not readily indexed at systematic names. For other substances, the stereochemistry is expressed in the name by use of relative and absolute descriptors, including *cis*- and *trans*-; *endo*- and *exo*-; α- and β-; *E*- and *Z*-; *R*- and *S*-; and *R**- and *S**-. The Sequence Rule of Cahn, Ingold, and Prelog is applied in the assignment of these descriptors. A special system was developed for coordination compounds (¶ 203 III).

285. Stereoparents (¶ 202) are heading parents whose names imply stereochemistry, as indicated by structural diagrams shown at these names in the *Chemical Substance Index*. Synonyms and cross-references for the systematic names are always provided. The stereochemistry implied by the stereoparent, and shown in the diagram, may be augmented or selectively reversed in the name. Stereoparents include biologically significant amino acids, carbohydrates, and cyclic natural products (alkaloids, steroids, terpenes) within certain limitations. Stereoparents are preferred to systematically named substances of higher functionality in the naming of esters, mixtures, molecular addition compounds, and polymers. Derivatives of stereoparents are kept at the stereoparent names as much as practicable, but substituents (¶ 287) are named systematically in accordance with the changes in general index nomenclature.

286. Steroids (¶ 211). Stereochemical descriptors, e.g., "5α-," which formerly appeared as a prefix at steroid names, were combined in Volume 76 with descriptors required for suffixes and substituent prefixes and are now cited in parentheses in their own field.

Cyclogonanes were discontinued *as the names of ring systems;* they are now cross-referred to systematic ring names. Cyclogonane *stereoparents* are retained, with stereochemistry defined by illustrative diagrams in the *Chemical Substance Index* and the *Index Guide*.

Trivial names for steroids, including Cholesterol, Ergocalciferol, and Testosterone, were discontinued as heading parents.

Use of "homo" (to denote ring enlargement), "nor" (to denote ring contraction), and "seco" (to denote ring fission), are now further restricted.

Functional derivatives of steroids are indexed at the steroid names, either in the substituents (for derivatives of subsidiary functions) or in the modifications (for derivatives of the principal functions). Acyclic acetals are expressed by alkyloxy or aryloxy substituents (¶ 196); cyclic acetals with formaldehyde by methylenebis(oxy) radicals; other cyclic acetals by a modification phrase such as "cyclic 1,2-ethanediyl acetal" (if a principal group has been acetalized) or by an alkanediylbis(oxy) or alkylidenebis(oxy) substituent. Steroidal lactones are "opened" to permit the steroid stereoparent to be used; the lactone is then cited in the modification.

287. Substituent prefixes (radicals) (¶¶ 132, 161). Groups always expressed as substituents now include isocyano, isocyanato, isothiocyanato, etc., and groups terminating in oxy, thio, sulfinyl, sulfonyl, and their analogs, such as seleno and telluronyl.

Changes in radical names include iodosyl (instead of iodoso) for -IO; and iodyl (instead of iodoxy) for $-IO_2$. Morpholino was changed in Volume 76 to 4-morpholinyl; piperidino to 1-piperidinyl; the unsubstituted diazeno radical, HN=N-, to diazenyl; *p*-phenylene, etc., to 1,4-phenylene, etc.; *v*-phenenyl to 1,2,3-benzenetriyl; benzyl to (phenylmethyl); styryl to (2-phenylethenyl); *p*-tolyl to (4-methylphenyl); 2,4-xylyl to (2,4-dimethylphenyl); and similarly for other radicals containing a benzene ring and one or more acyclic chains. Thenyl became (thienylmethyl) and furfuryl became (2-furanylmethyl).

Ring-assembly radicals are now based on the ring assembly names, as, [1,2'-binaphthalen]-8'-yl and [1,1'-biphenyl]-4,4'-diyl (¶ 161).

Acyl radicals in substitutive names were replaced by substituted radicals (exceptions are acetyl, benzoyl, carbonyl, and, when unsubstituted, formyl). Hence propionyl became (1-oxopropyl); acetimidoyl became (1-iminoethyl); succinoyl became (1,4-dioxo-1,4-butanediyl). The three radicals carbonimidoyl, -C(:NH)-; carbonohydrazonoyl, $-C(:NNH_2)-$, and carbonothioyl, -C(:S)-, were introduced for use in multiplicative nomenclature and for cases in which both free valencies are attached to the same atom. In other cases, (iminomethyl) (formerly formimidoyl), (thioxomethyl), etc., are employed.

Replacement ("a") names for acyclic radicals (¶ 128) were introduced under the same restrictions as for heading parents. The free valencies, not the hetero atoms, are preferred for lowest locants, which are always cited; e.g., 4,12-dioxa-7,9-dithiatetradec-1-yl.

Compound and complex radicals are now constructed in accordance with revised rules which reflect the new policies for compounds. The most important changes were emphasis on hetero-atom content of the parent radical, abandonment of "like treatment of like things" and the "complexity" principle, elimination of preference for unsaturated acyclic radicals regardless of size, consistent application of the principle of "lowest locants" for substituents on the parent radical, and adoption of more systematic radical names.

A list of substituent prefixes (radicals) as revised for the Ninth Collective Index Period (1972-1976) constitutes Section H (¶ 294).

288. Sulfur compounds (¶ 200). **Sulfide** was discontinued as a heading parent in Volume 76, as were such binary headings as Phenyl sulfide. Compounds containing one or more single sulfur atoms are now indexed substitutively by use of thio radicals or at "thia"-named index parents (¶ 127). Thus, Sulfide, ethyl phenyl, became **Benzene, (ethylthio)-,** and Propyl sulfide is now indexed at **Propane, 1,1'-thiobis-.** The retained headings **Disulfide, Trisulfide,** etc., rank as nonfunctional acyclic sulfur parents, below oxygen compounds and above selenium compounds. Within the class, a partial descending order is: heterocyclic sulfur compounds, e.g., **Thiophene;** "a"-named acyclic hetero systems, e.g., **2,6,9,12,13-Pentathiapentadecane;** acyclic trisulfide, disulfone, disulfoxide, disulfide.

1,2-Episulfides are named as **Thiirane** or **Thiirene** derivatives; 1,3-episulfides as **Thietane** or **Thiete** derivatives. Acyclic mercaptals and mercaptoles are now named like sulfides by use of thio radicals.

Selenium and tellurium compounds are treated in strict analogy with sulfur compounds.

289. Sulfones and **Sulfoxides** (¶ 200). **Sulfone** was discontinued as a heading parent, as were binary sulfone headings like Isopropyl sulfone, now indexed at **Propane, 2,2'-sulfonylbis-.** Sulfonyl radicals are used for all compounds containing single SO_2 groups, e.g., **Butane, 1-chloro-3-(methylsulfonyl)-** (formerly Sulfone, 3-chloro-1-methylpropyl methyl). **Disulfone, Trisulfone,** etc., are retained. Sulfoxides are indexed in an analogous manner by use of sulfinyl radicals. Acyclic skeletons containing four or more sulfur units may be indexed at "thia"- names with oxide terms in the index modification. Cyclic sulfones and sulfoxides are indexed at ring names; e.g., **Thiirane,** 1,1-dioxide (formerly Ethylene sulfone).

289A. Tautomers (¶ 122). To avoid scattering of information in the index, *CA*, aided by new machine programs, now indexes certain common tautomeric systems at single preferred index names, regardless of the particular structures presented in original documents. The most common tautomers handled in this way include compounds containing the nitrogenous skeletons N-N-N, N-C-N, N-C-O, and N-C-S, as well a certain phosphorus and sulfur acids and amides. In general, the preferred index name expresses the tautomeric form in which a double bond extends to an oxygen atom, rather than to a sulfur or nitrogen; after this condition has been satisfied, the preferred name usually expresses the maximum number of highest functions. Cross-references in the *Index Guide* reflect these policies; thus, Uracil is cross-referred to **2,4(1*H*,3*H*)-Pyrimidinedione,** not to 2,4-Pyrimidinediol.

290. Terpenes (¶ 212) are now named at stereoparent names (¶ 285) only when they contain four or more rings or possess complex stereochemistry. Hence, many trivial names formerly employed as index headings are now cross-referred to systematic names; e.g., **Eudesmane.** See *Naphthalene,* *decahydro-1,4a-dimethyl-7-(1-methylethyl)-*, [1*R*-(1α,4aβ,7β,8aα)]-.

Illustrative diagrams for terpene stereoparents, e.g., **Gammacerane, La=nostane,** showing the stereochemistry for each, are provided in the *Chemical Substance Index* and the *Index Guide*. New stereoparents, formed by ring enlargement, contraction, scission, or addition, are adopted as they are needed. When ring scission results in a structure which does not qualify for a terpene stereoparent name, systematic nomenclature is used.

Cyclic acetals of terpenes are expressed at stereoparent names, either in the modification or by use of appropriate substituents, in analogy with steroids (¶ 286). Lactones are expressed in the modification at terpene acid headings.

291. Thiols (¶ 175) have been affected only by the general changes in indexing policy (¶ 255), including those for conjunctive names, abandonment of trivial names, and discontinuance of like treatment of like things. Thus, α-Toluenethiol became **Benzenemethanethiol;** *p*-Cymene-2-thiol became **Benzenethiol, 2-methyl-5-(1-methylethyl)-;** and Ethanethiol, 1,2-di=phenyl-, became **Benzeneethanethiol, α-phenyl-.**

292. Urea (¶ 183) was retained as a heading parent; Urea, thio-, and its derivatives are now indexed at **Thiourea.** In both cases, the locants 1- and 3- were replaced by *N*- and *N'*-. Urea ranks as an amide of **Carbonic acid,** which is now placed below the "organic" acids, including carboxylic and sul=fonic acids. Acyclic acyl ureas are therefore indexed at "organic" amide names. Pseudourea was renamed **Carbamimidic acid**. In the following list, former names are shown in parentheses:

Urea, *N'*-ethyl-*N,N*-di=methyl-	(Urea, 3-ethyl-1,1-dimethyl-)
Acetamide, *N*-(amino=carbonyl)-	(Urea, acetyl-)
Ethanethioamide, *N*-(ami=nocarbonyl)-	(Urea, (thioacetyl)-)
Urea, (aminoiminomethyl)-	(Urea, amidino-)
1,2-Hydrazinedicarboxamide	(Biurea)
Imidodicarbonic diamide	(Biuret)
Urea, *N,N'*-diphenyl-	(Carbanilide)
Diimidotricarbonic diamide	(Triuret)
Carbamimidothioic acid, methyl-, ethyl ester	(Pseudourea, 2-ethyl-3-methyl-2-thio-)

293. Vitamins (¶ 224). Index headings such as Vitamin B, Vitamin F, are used for general discussions of vitamin activity and for ill-defined substances named by authors. Otherwise, many specific vitamin names have been cross-referred to more systematic names. The stereoparent **Retinol** is used for Vitamin A_1 (Vitamin A_2 is indexed at **Retinol, 3,4-didehydro-); Ribo=flavine (Riboflavin** beginning in Volume 86) for Vitamin B_2; and L-**Ascorbic acid** for Vitamin C. **Vitamin B_{12}** is the only specific vitamin name retained as a heading parent; related compounds have continued to be indexed at names such as **Cobinamide** and **Cobyrinic acid.**

(The equals (=) directs the user from the boldface name to the current *CA* name; absence of the sign indicates that the boldface name is correct.)

abietamido = [[[1,2,3,4,4a,4b,5,6,10,10a-decahydro-1,4a-dimethyl-7-(1-methylethyl)-1-phenanthrenyl]carbonyl]amino] $C_{19}H_{29}CONH-$

acenaphthenyl = (1,2-dihydroacenaphthylenyl) $(C_{12}H_9)-$

1,2-acenaphthenylene = (1,2-dihydro-1,2-acenaphthylenediyl) $-(C_{12}H_8)-$

1-acenaphthenylidene = 1(2*H*)-acenaphthylenylidene $(C_{12}H_8)=$

acetamido = (acetylamino) AcNH-

acetenyl = ethynyl HC≡C-

acetimido = (acetylimino) or (1-iminoethyl) AcN= or MeC(=NH)-

acetimidoyl = (1-iminoethyl) MeC(=NH)-

acetoacetamido = [(1,3-dioxobutyl)amino] $MeCOCH_2CONH-$

1,2-ethenediyl -CH=CH-

acetoacetyl = (1,3-dioxobutyl) $MeCOCH_2CO-$

acetohydroximoyl = [1-(hydroxyimino)ethyl] MeC(=NOH)-

acetonyl = (2-oxopropyl) $MeCOCH_2-$

acetonylidene = (2-oxopropylidene) MeCOCH=

acetoxy = (acetyloxy) AcO-

acetyl Ac (MeCO-)

acetylene = 1,2-ethanediylidene =CHCH=

acridanyl = (9,10-dihydroacridinyl) $(C_{13}H_{10}N)-$

acryloyl = (1-oxo-2-propenyl) $H_2C=CHCO-$

acrylyl = (1-oxo-2-propenyl) $H_2C=CHCO-$

adamantyl = tricyclo[3.3.1.1^{3,7}]decyl $(C_{10}H_{15})-$

adamantylene = tricyclo[3.3.1.1^{3,7}]decanediyl $-(C_{10}H_{14})-$

adipaldehydoyl = (1,6-dioxohexyl) $HCO(CH_2)_4CO-$

adipamoyl = (6-amino-1,6-dioxohexyl) $H_2NCO(CH_2)_4CO-$

adipaniloyl = [1,6-dioxo-6-(phenylamino)hexyl] $PhNHCO(CH_2)_4CO-$

adipoyl = (1,6-dioxo-1,6-hexanediyl) $-CO(CH_2)_4CO-$

adipyl = (1,6-dioxo-1,6-hexanediyl) $-CO(CH_2)_4CO-$

alaninamido = [(2-amino-1-oxopropyl)amino] $MeCH(NH_2)CONH-$

alanyl[1] = (2-amino-1-oxopropyl) $MeCH(NH_2)CO-$

β-**alanyl**[1] = (3-amino-1-oxopropyl) $H_2N(CH_2)_2CO-$

aldo[2] = oxo O=

alloisoleucyl[1] = (2-amino-3-methyl-1-oxopentyl) $EtCHMeCH(NH_2)CO-$

allophanamido = [[[(aminocarbonyl)amino]carbonyl]amino] $H_2N(CONH)_2-$

allophanoyl = [[(aminocarbonyl)amino]carbonyl] $H_2NCONHCO-$

allothreonyl[1] = (2-amino-3-hydroxy-1-oxobutyl) $MeCH(OH)CH(NH_2)CO-$

allyl = 2-propenyl $H_2C=CHCH_2-$

β-**allyl** = (1-methylethenyl) $H_2C=CMe-$

π-**allyl** = (η^3-2-propenyl)

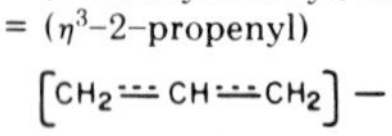

allylidene = 2-propenylidene $H_2C=CHCH=$

ambrosan-6-yl = [decahydro-3a,8-dimethyl-5-(1-methylethyl)-4-azulenyl] $(C_{15}H_{27})-$

amidino = (aminoiminomethyl) $H_2NC(=NH)-$

amidoxalyl = (aminooxoacetyl) $H_2NCOCO-$

amino H_2N-

(aminoamidino) = (aminohydrazonomethyl) or (hydrazinoiminomethyl) $H_2NC(=NNH_2)-$ or $H_2NNHC(=NH)-$

(aminoiminophosphoranyl) = (*P*-aminophosphinimyl) $H_2NPH(=NH)-$

ammonio $H_3{}^+N-$

amoxy = (pentyloxy) $Me(CH_2)_4O-$

amyl = pentyl $Me(CH_2)_4-$

tert-**amyl** = (1,1-dimethylpropyl) $EtCMe_2-$

amylidene = pentylidene BuCH=

anilino = (phenylamino) PhNH-

anisal = [(methoxyphenyl)methylene] $MeOC_6H_4CH=$

anisidino = [(methoxyphenyl)amino] $MeOC_6H_4NH-$

anisoyl = (methoxybenzoyl) $MeOC_6H_4CO-$

anisyl = (methoxyphenyl) or [(methoxyphenyl)methyl] $MeOC_6H_4-$ or $MeOC_6H_4CH_2-$

anisylidene = [(methoxyphenyl)methylene] $MeOC_6H_4CH=$

anthranilamido = [(2-aminobenzoyl)amino] $2\text{-}H_2NC_6H_4CONH-$

anthraniloyl = (2-aminobenzoyl) $2\text{-}H_2NC_6H_4CO-$

anthranoyl = (2-aminobenzoyl) $2\text{-}H_2NC_6H_4CO-$

anthraquinonyl = (9,10-dihydro-9,10-dioxoanthracenyl) $(C_{14}H_7O_2)-$

anthraquinonylene = (9,10-dihydro-9,10-dioxoanthracenediyl) $-(C_{14}H_6O_2)-$

anthroyl = (anthracenylcarbonyl) $(C_{14}H_9)CO-$

anthryl = anthracenyl $(C_{14}H_9)-$

anthrylene = anthracenediyl $-(C_{14}H_8)-$

antimono = 1,2-distibenediyl -Sb=Sb-

antipyrinyl (antipyryl) = (2,3-dihydro-1,5-dimethyl-3-oxo-2-phenyl-1*H*-pyrazol-4-yl)

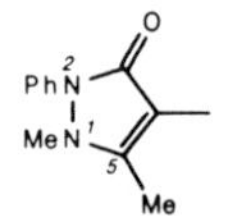

antipyroyl = [(2,3-dihydro-1,5-dimethyl-3-oxo-2-phenyl-1*H*-pyrazol-4-yl)carbonyl]

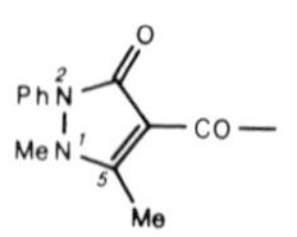

apocamphanyl = (7,7-dimethylbicyclo[2.2.1]heptyl) $(C_9H_{15})-$

apotrichothecanyl = (decahydro-3a,6,8a,8b-tetramethyl-1*H*-cyclopenta[*b*]benzofuranyl) $(C_{15}H_{25}O)-$

arginyl[1] = [2-amino-5-[(aminoiminomethyl)amino]-1-oxopentyl] $H_2NC(=NH)NH(CH_2)_3CH(NH_2)CO-$

arseno = 1,2-diarsenediyl -As=As-

arsenoso OAs-

arsinico[3,4] HOAs(O)=

arsinimyl $AsH_2(=NH)-$

arsino AsH_2-

arsinothioyl $AsH_2(S)-$

arsinyl $AsH_2(O)-$

arsinylidene AsH(O)=

arso O_2As-

arsono[4] $(HO)_2As(O)-$

arsononitridyl AsH(≡N)-

arsoranyl AsH_4-

arsoranylidyne $AsH_2≡$

arsylene = arsinidene AsH=

arsylidyne = arsinidyne As≡

asaryl = (2,4,5-trimethoxyphenyl) $2,4,5\text{-}(MeO)_3C_6H_2-$

asparaginyl[1] = (2,4-diamino-1,4-dioxobutyl) $H_2NCOCH_2CH(NH_2)CO-$

asparagyl = asparaginyl

aspartoyl[1] = (2-amino-1,4-dioxo-1,4-butanediyl) $-COCH_2CH(NH_2)CO-$

aspartyl = aspartoyl or unspecified aspartyl (see below)

α-**aspartyl**[1] = (2-amino-3-carboxy-1-oxopropyl) $HO_2CCH_2CH(NH_2)CO-$

β-**aspartyl**[1] = (3-amino-3-carboxy-1-oxopropyl) $HO_2CCH(NH_2)CH_2CO-$

astato At-

astatoxy = astatyl O_2At-

atisanyl (from atisane) $(C_{20}H_{33})-$

atropoyl = (1-oxo-2-phenyl-2-propenyl) $H_2C=CPhCO-$

azelaoyl = (1,9-dioxo-1,9-nonanediyl) $-CO(CH_2)_7CO-$

azelaaldehydoyl = (1,9-dioxononyl) $HCO(CH_2)_7CO-$

azi[5] (see also azo) -N=N-

azido N_3-

(azidoformyl) = (azidocarbonyl) N_3CO-

azino =NN=

azo[6] (see also azi) -N=N-

azoxy -N(O)=N-

benzal = (phenylmethylene) PhCH=

benzamido = (benzoylamino) BzNH-

benzenesulfenamido = [(phenylthio)amino] PhSNH-

benzenesulfonamido = [(phenylsulfonyl)amino] $PhSO_2NH-$

benzenyl = (phenylmethylidyne) PhC≡

benzhydryl = (diphenylmethyl) Ph_2CH-

benzhydrylidene = (diphenylmethylene) $Ph_2C=$

benzidino = [(4'-amino[1,1'-biphenyl]-4-yl)amino] $4\text{-}(4\text{-}H_2NC_6H_4)C_6H_4NH-$

benziloyl = (hydroxydiphenylacetyl) $Ph_2C(OH)CO-$

(3-benziloylpropyl) = (5-hydroxy-4-oxo-5,5-diphenylpentyl) $Ph_2C(OH)CO(CH_2)_3-$

benzimidazolinyl = (2,3-dihydro-1*H*-benzimidazolyl) $(C_7H_7N_2)-$

2-**benzimidazolyl** = 1*H*-benzimidazol-2-yl $(C_7H_5N_2)-$

benzimido = (benzoylimino) or (iminophenylmethyl) BzN= or PhC(=NH)-

benzimidoyl = (iminophenylmethyl) PhC(=NH)-

benzofuryl = benzofuranyl $(C_8H_5O)-$

benzohydroximoyl = [(hydroxyimino)phenylmethyl] PhC(=NOH)-

o-**benzoquinon-3-yl** = (5,6-dioxo-1,3-cyclohexadien-1-yl) $(C_6H_3O_2)-$

p-**benzoquinon-2,5-ylene** = (3,6-dioxo-1,4-cyclohexadiene-1,4-diyl) $-(C_6H_2O_2)-$

benzoselenophene-yl C_8H_5Se-

benzoxy = (benzoyloxy) BzO-

benzoyl Bz (PhCO-)

(benzoylacetyl) = (1,3-dioxo-3-phenylpropyl) $PhCOCH_2CO-$

(benzoylformyl) = (oxophenylacetyl) PhCOCO-

benzyl = (phenylmethyl) $PhCH_2-$

benzylidene = (phenylmethylene) PhCH=

benzylidyne = (phenylmethylidyne) PhC≡

(benzyloxy) = (phenylmethoxy) $PhCH_2O-$

(benzylselenyl) = [(phenylmethyl)seleno] $PhCH_2Se-$

bicarbamoyl = (hydrazodicarbonyl) -CONHNHCO-

bicyclo[1.1.0]butylene = bicyclo[1.1.0]butanediyl $-(C_4H_4)-$

biphenylyl = [1,1'-biphenyl]yl PhC_6H_4-

biphenylene = [1,1'-biphenyl]diyl $-(C_{12}H_8)-$

biphenylylene = [1,1'-biphenyl]diyl $-(C_{12}H_8)-$

bismuthino BiH_2-

bismuthylene BiH=

bismuthylidyne Bi≡

[2,2'-bithiophen]-5-yl $C_8H_5S_2-$

2-**bornyl** = (1,7,7-trimethylbicyclo[2.2.1]hept-2-yl) $(C_{10}H_{17})-$

3-**bornylidene** = (4,7,7-trimethylbicyclo[2.2.1]hept-2-ylidene) $(C_{10}H_{16})=$

borono[4] $(HO)_2B-$

boryl BH_2-

borylene BH=

borylidyne B≡

bromo Br-

1,3-butadienediylidene = 1,3-butadiene-1,4-diylidene =C=CHCH=C=

butadiynylene = 1,3-butadiyne-1,4-diyl -C≡CC≡C-

2-**butenylene** = 2-butene-1,4-diyl $-CH_2CH=CHCH_2-$

butoxy BuO-

sec-**butoxy** = (1-methylpropoxy) EtCHMeO-

tert-**butoxy** = (1,1-dimethylethoxy) Me_3CO

butyl Bu $(Me(CH_2)_3-)$

butyl$^\beta$ = (2-methylpropyl) Me_2CHCH_2-

butyl$^\gamma$ = (1,1-dimethylethyl) Me_3C-

sec-**butyl** = (1-methylpropyl) EtCHMe-

tert-**butyl** = (1,1-dimethylethyl) Me_3C-

1,4-butylene = 1,4-butanediyl $-(CH_2)_4-$

sec-**butylidene** = (1-methylpropylidene) EtCMe=
(butyloxy) = butoxy BuO–
butynedioyl = (1,4-dioxo-2-butyne-1,4-diyl) –COC≡CCO–
butyryl = (1-oxobutyl) PrCO–

cacodyl = (dimethylarsino) Me_2As–
cadinan-1-yl = [octahydro-4,7-dimethyl-1-(1-methylethyl)-4a(2*H*)-naphthalenyl] $(C_{15}H_{29})$–
2-**camphanyl** = (4,7,7-trimethylbicyclo[2.2.1]hept-2-yl) $(C_{10}H_{17})$–
camphoroyl (from camphoric acid) = [(1,2,2-trimethyl-1,3-cyclopentanediyl)dicarbonyl] $-CO(C_8H_{14})CO-$
5-**camphoryl** (from camphor) = (4,7,7-trimethyl-5-oxobicyclo[2.2.1]hept-2-yl) $(C_{10}H_{15}O)$–
canavanyl = [2-amino-4-[[(aminoiminomethyl)amino]oxy]-1-oxobutyl] $H_2NC(=NH)NHO(CH_2)_2CH(NH_2)CO$–
caprinoyl = (1-oxodecyl) $Me(CH_2)_8CO$–
caproyl (from caproic acid) = (1-oxohexyl) $Me(CH_2)_4CO$–
capryl (from capric acid) = (1-oxodecyl) $Me(CH_2)_8CO$–
capryloyl (from caprylic acid) = (1-oxooctyl) $Me(CH_2)_6CO$–
caprylyl (from caprylic acid) = (1-oxooctyl) $Me(CH_2)_6CO$–
carbamido = [(aminocarbonyl)amino] H_2NCONH–
carbamoyl = (aminocarbonyl) H_2NCO–
carbamyl = (aminocarbonyl) H_2NCO–
carbanilino = [(phenylamino)carbonyl] PhNHCO–
carbaniloyl = [(phenylamino)carbonyl] PhNHCO–
carbazimidoyl = (hydrazinoiminomethyl) $H_2NNHC(=NH)$–
carbazol-9-yl = 9*H*-carbazol-9-yl $(C_{12}H_8N)$–
carbazoyl = (hydrazinocarbonyl) H_2NNHCO–
carbethoxy = (ethoxycarbonyl) EtO_2C–
carbobenzoxy = [(phenylmethoxy)carbonyl] $PhCH_2O_2C$–
carbonimidoyl[7] (see also (iminomethyl)) –C(=NH)–
carbonothioyl[7] (see also (thioxomethyl)) –CS–
carbonyl –CO–
(carbonyldioxy) = [carbonylbis(oxy)] $-OCO_2-$
(1-carbonylethyl) = (methyloxoethenyl) O=C=CMe–
(carbonylmethyl) = (oxoethenyl) O=C=CH–
(carbonylmethylene) = (1-oxo-1,2-ethanediyl) or (oxoethenylidene) $-COCH_2-$ or O=C=C=
carboxy[4] HO_2C–
(carboxyformyl) = (carboxycarbonyl) HO_2CCO–
(5-carboxyvaleryl) = (5-carboxy-1-oxopentyl) $HO_2C(CH_2)_4CO$–
carnosyl = (*N*-β-alanylhistidyl)[1] or [2-[(3-amino-1-oxopropyl)amino]-3-(1*H*-imidazol-4-yl)-1-oxopropyl]

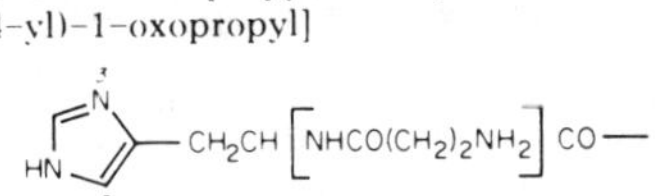

caronaldehydoyl = [(3-formyl-2,2-dimethylcyclopropyl)carbonyl]

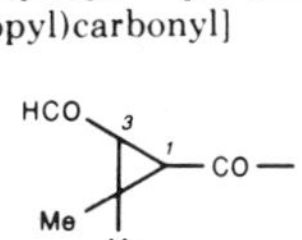

carvacryl = [2-methyl-5-(1-methylethyl)phenyl]

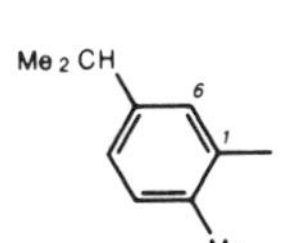

carvomenthyl = [2-methyl-5-(1-methylethyl)cyclohexyl]

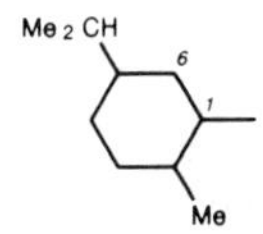

10-**caryl** = [(7,7-dimethylbicyclo[4.1.0]hept-3-yl)methyl] $(C_9H_{15})CH_2$–
cathyl = [(ethoxycarbonyl)oxy] $EtOCO_2$–
cedranyl = (octahydro-3a,6,8,8-tetramethyl-1*H*-3a,7-methanoazulenyl) $(C_{15}H_{25})$–
cetyl = hexadecyl $Me(CH_2)_{15}$–
chaulmoogroyl (from chaulmoogric acid) = [13-(2-cyclopenten-1-yl)-1-oxotridecyl] $C_5H_7(CH_2)_{12}CO$–
chaulmoogryl (from chaulmoogryl alcohol) = [13-(2-cyclopenten-1-yl)tridecyl] $C_5H_7(CH_2)_{13}$–
chloro Cl–
(chloroformyl) = (chlorocarbonyl) ClCO–
(chloroglyoxyloyl) = (chlorooxoacetyl) ClCOCO–
(chlorooxalyl) = (chlorooxoacetyl) ClCOCO–
chlorosyl OCl–
chloryl O_2Cl–
cholesteryl (from cholesterol) = cholest-5-en-3-yl (from cholestene) $(C_{27}H_{45})$–
choloyl (from cholic acid) = (3,7,12-trihydroxy-24-oxocholan-24-yl) $(HO)_3(C_{23}H_{36})CO$–
chromanyl = (3,4-dihydro-2*H*-1-benzopyranyl) (C_9H_9O)–
cinchoninoyl (from cinchoninic acid) = (4-quinolinylcarbonyl) $(4-C_9H_6N)CO$–
cinnamal = (3-phenyl-2-propenylidene) PhCH=CHCH=
cinnamenyl = (2-phenylethenyl) PhCH=CH–
cinnamoyl = (1-oxo-3-phenyl-2-propenyl) PhCH=CHCO–
cinnamyl = (3-phenyl-2-propenyl) $PhCH=CHCH_2$–
cinnamylidene = (3-phenyl-2-propenylidene) PhCH=CHCH=
citraconimido = (2,5-dihydro-3-methyl-2,5-dioxo-1*H*-pyrrol-1-yl)

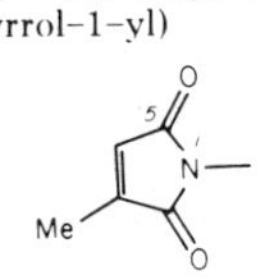

citraconoyl = (2-methyl-1,4-dioxo-2-butene-1,4-diyl) –COCMe=CHCO–
conaninyl (from conanine) $(C_{22}H_{36}N)$–
cresotoyl (from cresotic acid) = (hydroxymethylbenzoyl) $HO(Me)C_6H_3CO$–
cresoxy = (methylphenoxy) MeC_6H_4O–
cresyl = (hydroxymethylphenyl) or (methylphenyl) $HO(Me)C_6H_3$– or MeC_6H_4–
cresylene = (methylphenylene) $-(MeC_6H_3)-$
crotonoyl = (1-oxo-2-butenyl) MeCH=CHCO–
crotonyl = (1-oxo-2-butenyl) MeCH=CHCO–
crotyl = 2-butenyl $MeCH=CHCH_2$–
cumal = [[4-(1-methylethyl)phenyl]methylene] $4-(Me_2CH)C_6H_4CH=$
cumenyl = [(1-methylethyl)phenyl] $Me_2CHC_6H_4$–
cumidino = [[4-(1-methylethyl)phenyl]amino]- $4-(Me_2CH)C_6H_4NH$–
cuminal = [[4-(1-methylethyl)phenyl]methylene] $4-(Me_2CH)C_6H_4CH=$
cuminyl = [[4-(1-methylethyl)phenyl]methyl] $4-(Me_2CH)C_6H_4CH_2$–
cuminylidene = [[4-(1-methylethyl)phenyl]methylene] $4-(Me_2CH)C_6H_4CH=$
cumoyl = [4-(1-methylethyl)benzoyl] $4-(Me_2CH)C_6H_4CO$–
cumyl[8] = [(1-methylethyl)phenyl] $(Me_2CH)C_6H_4$–
α-**cumyl** = (1-methyl-1-phenylethyl) $-PhCMe_2-$
cyanamido = (cyanoamino) NCNH–
cyanato NCO–
cyano NC–
cyclodisiloxan-2-yl

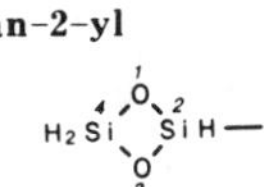

cyclohexadienylene = cyclohexadienediyl $-C_6H_6-$
cyclohexanecarboxamido = [(cyclohexylcarbonyl)amino] $C_6H_{11}CONH$–
1,2-**cyclohexanedicarboximido** = (octahydro-1,3-dioxo-2*H*-isoindol-2-yl)

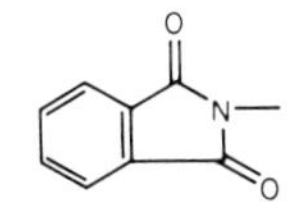

cymyl = [methyl(1-methylethyl)phenyl] $Me(Me_2CH)C_6H_3$–
cysteinyl[1] = (2-amino-3-mercapto-1-oxopropyl) $HSCH_2CH(NH_2)CO$–
cysteyl = (3-sulfoalanyl)[1] or (2-amino-1-oxo-3-sulfopropyl) $HO_3SCH_2CH(NH_2)CO$–

decanedioyl = (1,10-dioxo-1,10-decanediyl) $-CO(CH_2)_8CO-$
decanoyl = (1-oxodecyl) $CH_3(CH_2)_8CO$–
decasiloxanylene = 1,19-decasiloxanediyl $-SiH_2(OSiH_2)_8OSiH_2-$
desyl = (2-oxo-1,2-diphenylethyl) PhCOCHPh–
diarsenyl HAs=As–
diarsinetetrayl = 1,2-diarsinediylidene =AsAs=
diarsinyl H_2AsAsH–
1,2-**diazenediyl** = azi or azo –N=N–
diazeno = diazenyl[4] (see also azo) HN=N–
diazo[5] N_2=
diazoamino = 1-triazene-1,3-diyl –NHN=N–
diazonio N_2^+–
dibenzothiophene-yl = dibenzothienyl $(C_{12}H_7S)$–
diborane(4)tetrayl = 1,2-diborane(4)diylidene =BB=
1,2-**dicarbadodecaboran(12)-1-yl** $(C_2H_{11}B_{10})$–
digermanylene = 1,2-digermanediyl $-GeH_2GeH_2-$
digermthianyl = digermathianyl $H_3GeSGeH_2$–
diglycoloyl = [oxybis(1-oxo-2,1-ethanediyl)] $-COCH_2OCH_2CO-$
dimethyliminio Me_2N^+=
dioxy[9] (see also epidioxy) –OO–
1,2-**diphosphinediyl** –PHPH–
diphosphinetetrayl = 1,2-diphosphinediylidene =PP=
diphosphino = diphosphinyl H_2PPH–
diseleno[9] (see also epidiseleno and seleninoselenoyl) –SeSe–
disilanoxy = (disilanyloxy) H_3SiSiH_2O–
disilanyl H_3SiSiH_2–
disilanylene = 1,2-disilanediyl $-SiH_2SiH_2-$
disilazanoxy = (disilazanyloxy) $H_3SiNHSiH_2O$–
disilazanyl $H_3SiNHSiH_2$–
2-**disilazanyl** = (disilylamino) $(H_3Si)_2N$–
disiloxanediylidene = 1,3-disiloxanediylidene =SiHOSiH=
disiloxanoxy = (disiloxanyloxy) $H_3SiOSiH_2O$–
disiloxanylene = 1,3-disiloxanediyl $-SiH_2OSiH_2-$
disilthianoxy = (disilathianyloxy) $H_3SiSSiH_2O$–
distannanylene = 1,2-distannanediyl $-SnH_2SnH_2-$
distannthianediylidene = 1,3-distannathianediylidene =SnHSSnH=
disulfinyl[6] –S(O)S(O)–
dithio[9] (see also epidithio and sulfinothioyl) –SS–
(dithiobicarbamoyl) = (hydrazodicarbonothioyl) –CSNHNHCS–
(dithiohydroperoxy) = (thiosulfeno)[4] HSS–
dodecanoyl = (1-oxododecyl) $Me(CH_2)_{10}CO$–
duryl = (2,3,5,6-tetramethylphenyl) $2,3,5,6-Me_4C_6H$–
durylene = (2,3,5,6-tetramethyl-1,4-phenylene) $-(2,3,5,6-Me_4C_6)-$

enanthoyl = (1-oxoheptyl) $Me(CH_2)_5CO$–
enanthyl = (1-oxoheptyl) $Me(CH_2)_5CO$–
epidioxy[10] (see also dioxy) –OO–
epidiseleno[10] (see also diseleno and seleninoselenoyl) –SeSe–
epidithio[10] (see also dithio and sulfinothioyl) –SS–
epioxy = epoxy (see also oxy and oxo) –O–
episeleno[10] (see also seleno and selenoxo) –Se–
epithio[10] (see also thio and thioxo) –S–

epoxy[10] (see also oxy and oxo) –O–
(epoxyethyl) = oxiranyl

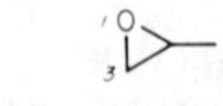

(2,3-epoxypropyl) = (oxiranylmethyl)

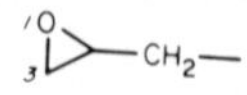

eremophilan-1-yl = [decahydro-4,4a-dimethyl-5-(1-methylethyl)-1-naphthalenyl] ($C_{15}H_{27}$)–
ethanediylidene = 1,2-ethanediylidene =CHCH=
ethene = 1,2-ethanediyl –CH_2CH_2–
ethinyl = ethynyl $HC\equiv C$–
ethoxalyl = (ethoxyoxoacetyl) EtO_2CCO–
ethoxy EtO–
(ethoxycarbonyl) EtO_2C–
(1-ethoxyformimidoyl) = (ethoxyiminomethyl) EtOCH(=NH)–
(ethoxyphosphinyl) EtOPH(O)–
ethyl Et ($MeCH_2$–)
ethylene = 1,2-ethanediyl –CH_2CH_2–
[ethylenebis(nitrilodimethylene)] = [1,2-ethanediylbis[nitrilobis(methylene)] $(-CH_2)_2N(CH_2)_2N(CH_2-)_2$
(ethylenedioxy) = [1,2-ethanediylbis(oxy)] –$O(CH_2)_2O$–
ethylidene CH_3C=
ethylidyne $CH_3C\equiv$
1-ethylium-1-ylidene CH_3C^+=
(ethyloxy) = ethoxy EtO–
(ethylselenyl) = (ethylseleno) EtSe–
(ethylthio) EtS–
eudesman-8-yl = [decahydro-5,8a-dimethyl-2-(1-methylethyl)-2-naphthalenyl] ($C_{15}H_{27}$)–

farnesyl (from farnesol) = (3,7,11-trimethyl-2,6,10-dodecatrienyl) $Me_2C=CH(CH_2)_2CMe=CH(CH_2)_2CMe=CHCH_2$–
fenchyl = (1,3,3-trimethylbicyclo[2.2.1]hept-2-yl) ($C_{10}H_{17}$)–
1,1'-ferrocenediyl –$(C_5H_4)Fe(C_5H_4)$–
fluoranyl = (3-oxospiro[isobenzofuran-1(3*H*),9'-[9*H*]xanthen]yl) ($C_{20}H_{11}O_3$)–
fluoren-9-ylidene = 9*H*-fluoren-9-ylidene ($C_{13}H_8$)=
fluoro F–
formamido = (formylamino) HCONH–
1,5-formazandiyl –N=NCH=NNH–
1-formazano H_2NN=CHN=N–
5-formazano HN=NCH=NNH–
1,3,5-formazantriyl

$-N{=}N\overset{|}{C}{=}NNH-$

formazanyl HN=NC(=NNH_2)–
formazyl = (1,5-diphenylformazanyl) PhN=NC(=NNHPh)–
formimidoyl = (iminomethyl) CH(=NH)–
(1-formimidoylformimidoyl) = (1,2-diiminoethyl) HC(=NH)C(=NH)–
(formimidoylformyl) = (iminoacetyl) CH(=NH)CO–
(formimidoylmethyl) = (2-iminoethyl) HC(=NH)CH_2–
formyl[4] HCO–
fucosyl = (6-deoxygalactosyl) ($C_6H_{11}O_4$)–
fumaraniloyl = [1,4-dioxo-4-(phenylamino)-2-butenyl] PhNHCOCH=CHCO–
fumaroyl = (1,4-dioxo-2-butene-1,4-diyl) –COCH=CHCO–
furfural = (2-furanylmethylene)

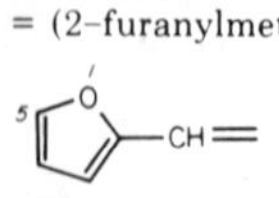

furfuryl = (2-furanylmethyl)

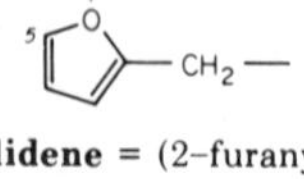

furfurylidene = (2-furanylmethylene)

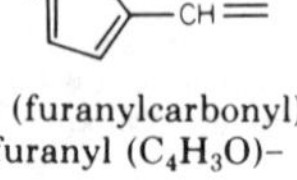

furoyl = (furanylcarbonyl) (C_4H_3O)CO–
furyl = furanyl (C_4H_3O)–

galloyl = (3,4,5-trihydroxybenzoyl) 3,4,5-$(HO)_3C_6H_2CO$–
gentisoyl = (2,5-dihydroxybenzoyl) 2,5-$(HO)_2C_6H_3CO$–
geranyl (from geraniol) = (3,7-dimethyl-2,6-octadienyl) $Me_2C=CH(CH_2)_2CMe=CHCH_2$–
germacran-6-yl = [5,9-dimethyl-2-(1-methylethyl)cyclodecyl] ($C_{15}H_{25}$)–
germanetetrayl =Ge=
germyl H_3Ge–
germylene H_2Ge=
germylidyne $HGe\equiv$
gibbanyl (from gibbane) ($C_{15}H_{23}$)–
(glucosyloxy) ($C_6H_{11}O_5$)O–
glutaminyl[1] = (2,5-diamino-1,5-dioxopentyl) $H_2NCO(CH_2)_2CH(NH_2)CO$–
glutamoyl[1] = (2-amino-1,5-dioxo-1,5-pentanediyl) –$CO(CH_2)_2CH(NH_2)CO$–
glutamyl = glutamoyl or unspecified glutamyl (see below)
α-**glutamyl**[1] = (2-amino-4-carboxy-1-oxobutyl) $HO_2C(CH_2)_2CH(NH_2)CO$–
γ-**glutamyl**[1] = (4-amino-4-carboxy-1-oxobutyl) $HO_2CCH(NH_2)(CH_2)_2CO$–
glutaryl = (1,5-dioxo-1,5-pentanediyl) –$CO(CH_2)_3CO$–
glyceroyl = (2,3-dihydroxy-1-oxopropyl) $HOCH_2CH(OH)CO$–
glyceryl = 1,2,3-propanetriyl –$CH(CH_2-)_2$
glycidyl = (oxiranylmethyl)

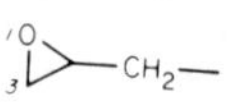

glycinamido = [(aminoacetyl)amino] H_2NCH_2CONH–
glycinimidoyl = (2-amino-1-iminoethyl) H_2NCH_2C(=NH)–
glycoloyl = (hydroxyacetyl) $HOCH_2CO$–
glycolyl = (hydroxyacetyl) $HOCH_2CO$–
glycyl[1] = (aminoacetyl) H_2NCH_2CO–
glyoxalinyl = imidazolyl ($C_3H_3N_2$)–
glyoxalyl = imidazolyl ($C_3H_3N_2$)–
glyoxylimidoyl = (1-imino-2-oxoethyl) HCOC(=NH)–
glyoxyloyl = (oxoacetyl) HCOCO–
(glyoxyloylmethyl) = (2,3-dioxopropyl) $HCOCOCH_2$–
glyoxylyl = (oxoacetyl) HCOCO–
guaiacyl = (4-hydroxy-3-methoxyphenyl) 4-OH-3-$(CH_3O)C_6H_3$– or (2-methoxyphenyl) 2-$(MeO)C_6H_4$–
guaian-8-yl = [decahydro-1,4-dimethyl-7-(1-methylethyl)-6-azulenyl] ($C_{15}H_{27}$)–
guanidino = [(aminoiminomethyl)amino] H_2NC(=NH)NH–
(guanidinoazo) = [3-(aminoiminomethyl)-1-triazenyl] H_2NC(=NH)NHN=N–
guanyl = (aminoiminomethyl) H_2NC(=NH)–

heptadecanoyl = (1-oxoheptadecyl) $Me(CH_2)_{15}CO$–
heptanamido = [(1-oxoheptyl)amino] $Me(CH_2)_5CONH$–
heptanedioyl = (1,7-dioxo-1,7-heptanediyl) –$CO(CH_2)_5CO$–
heptanoyl = (1-oxoheptyl) $Me(CH_2)_5CO$–
hexadecanoyl = (1-oxohexadecyl) $Me(CH_2)_{14}CO$–
2,4-hexadiynylene = 2,4-hexadiyne-1,6-diyl –$CH_2C\equiv CC\equiv CCH_2$–
hexamethylene = 1,6-hexanediyl –$(CH_2)_6$–
hexanedioyl = (1,6-dioxo-1,6-hexanediyl) –$CO(CH_2)_4CO$–
hexanethioyl = (1-thioxohexyl) $Me(CH_2)_4CS$–
hippuroyl = (*N*-benzoylglycyl)[1] or [(benzoylamino)acetyl] $BzNHCH_2CO$–
hippuryl = (*N*-benzoylglycyl)[1] or [(benzoylamino)acetyl] $BzNHCH_2CO$–
histidyl[1] = [2-amino-3-(1*H*-imidazol-4-yl)-1-oxopropyl]

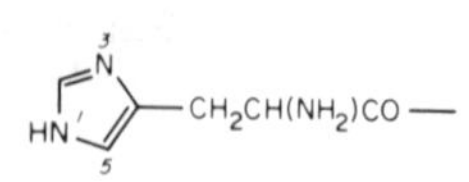

homocysteinyl[1] = (2-amino-4-mercapto-1-oxobutyl) $HS(CH_2)_2CH(NH_2)CO$–
homomyrtenyl = [2-(6,6-dimethylbicyclo[3.1.1]hept-2-en-2-yl)ethyl] (C_9H_{13})CH_2CH_2–

homopiperonyl = [2-(1,3-benzodioxol-5-yl)ethyl] ($C_7H_5O_2$)CH_2CH_2–
homoseryl[1] = (2-amino-4-hydroxy-1-oxobutyl) $HO(CH_2)_2CH(NH_2)CO$–
homoveratroyl = [(3,4-dimethoxyphenyl)acetyl] 3,4-$(MeO)_2C_6H_3CH_2CO$–
homoveratryl = [2-(3,4-dimethoxyphenyl)ethyl] 3,4-$(MeO)_2C_6H_3CH_2CH_2$–
hydantoyl = [*N*-(aminocarbonyl)glycyl][1] or [[(aminocarbonyl)amino]acetyl] $H_2NCONHCH_2CO$–
hydnocarpoyl (from hydnocarpic acid) = [11-(2-cyclopenten-1-yl)-1-oxoundecyl] $(C_5H_7)(CH_2)_{10}CO$–
hydnocarpyl (from hydnocarpyl alcohol) = [11-(2-cyclopenten-1-yl)undecyl] $(C_5H_7)(CH_2)_{10}CH_2$–
hydracryloyl = (3-hydroxy-1-oxopropyl) $HO(CH_2)_2CO$–
hydratropoyl = (1-oxo-2-phenylpropyl) PhCHMeCO–
hydrazi[5] (see also hydrazo) –NHNH–
1,2-hydrazinediylidene = azino =NN=
hydrazino H_2NNH–
1-hydrazinyl-2-ylidene –NHN=
hydrazo[6] (see also hydrazi) –NHNH–
hydrazono H_2NN=
hydrocinnamoyl = (1-oxo-3-phenylpropyl) $Ph(CH_2)_2CO$–
hydrocinnamyl = (3-phenylpropyl) $Ph(CH_2)_3$–
hydroperoxy[4] HOO–
(hydroperoxyformyl) = (hydroperoxycarbonyl) HOOCO–
hydroxamino = (hydroxyamino) HONH–
hydroximino = (hydroxyimino) HON=
hydroxy[4] HO–
hydroxyl = hydroxy[4] HO–
(hydroxyphosphinyl) HOPH(O)–
hygroyl = (1-methylprolyl)[1] or [(1-methyl-2-pyrrolidinyl)carbonyl]

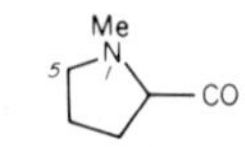

imidazolidyl = imidazolidinyl ($C_3H_7N_2$)–
imidazolinyl = (dihydro-1*H*-imidazolyl) ($C_3H_5N_2$)–
imidocarbonyl = carbonimidoyl[7] –C(=NH)–
(imidocarbonylamino) = (carbonimidoylamino) HN=C=N–
imino HN=
(3-iminoacetonyl) = (3-imino-2-oxopropyl) $HN=CHCOCH_2$–
(iminodisulfonyl) = [iminobis(sulfonyl)] –SO_2NHSO_2–
(iminomethyl) (see also carbonimidoyl) HN=CH–
(iminonitrilo) = 1-hydrazinyl-2-ylidene –NHN=
(iminophosphoranyl) = phosphinimyl H_2P(=NH)–
indanyl = (2,3-dihydro-1*H*-indenyl) (C_9H_9)–
indenyl = 1*H*-indenyl (C_9H_7)–
1-indolinyl = (2,3-dihydro-1*H*-indol-1-yl) (C_8H_8N)–
2-indolinylidene = (1,3-dihydro-2*H*-indol-2-ylidene) (C_8H_7N) =
indyl = 1*H*-indolyl (C_8H_6N)–
iodo I–
iodoso = iodosyl OI–
iodoxy = iodyl O_2I–
isoallyl = 1-propenyl MeCH=CH–
isoamoxy = (3-methylbutoxy) $Me_2CH(CH_2)_2O$–
isoamyl = (3-methylbutyl) $Me_2CH(CH_2)_2$–
sec-**isoamyl** = (1,2-dimethylpropyl) $Me_2CHCHMe$–
isoamylidene = (3-methylbutylidene) Me_2CHCH_2CH=
isobornyl (from isoborneol) = (1,7,7-trimethylbicyclo[2.2.1]hept-2-yl) ($C_{10}H_{17}$)–
isobutenyl = (2-methyl-1-propenyl) $Me_2C=CH$–
isobutoxy = (2-methylpropoxy) Me_2CHCH_2O–
isobutyl = (2-methylpropyl) Me_2CHCH_2–
isobutylidene = (2-methylpropylidene) Me_2CHCH=

isobutyryl = (2-methyl-1-oxopropyl) Me_2CHCO–
isocrotyl = (2-methyl-1-propenyl) $Me_2C{=}CH$–
isocyanato OCN–
isocyano CN– ⊖ ⊕
isodiazenyl N≡NH– or N=NH–
isodiazenylidene[6] (see also diazo)
isohexyl = (4-methylpentyl) $Me_2CH(CH_2)_3$–
isohexylidene = (4-methylpentylidene) $Me_2CH(CH_2)_2CH$=
2-isoindolinyl = (1,3-dihydro-2*H*-isoindol-2-yl) (C_8H_8N)–
isoleucyl[1] = (2-amino-3-methyl-1-oxopentyl) $EtCHMeCH(NH_2)CO$–
isonicotinoyl = (4-pyridinylcarbonyl)

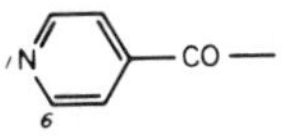

isonipecotoyl = (4-piperidinylcarbonyl)

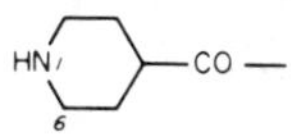

isonitro = *aci*-nitro HON(O)=
isonitroso = (hydroxyimino) HON=
1-isopentenyl = (3-methyl-1-butenyl) $Me_2CHCH{=}CH$–
isopentyl = (3-methylbutyl) $Me_2CH(CH_2)_2$–
isopentylidene = (3-methylbutylidene) Me_2CHCH_2CH=
isophthalal = (1,3-phenylenedimethylidyne) $1,3\text{-}C_6H_4(CH{=})_2$
isophthalaldehydoyl = (3-formylbenzoyl) $3\text{-}(HCO)C_6H_4CO$–
isophthaloyl = (1,3-phenylenedicarbonyl) $1,3\text{-}C_6H_4(CO\text{–})_2$
isophthalylidene = (1,3-phenylenedimethylidyne) $1,3\text{-}C_6H_4(CH{=})_2$
isopropenyl = (1-methylethenyl) $H_2C{=}CMe$–
isopropoxy = (1-methylethoxy) Me_2CHO–
isopropyl = (1-methylethyl) Me_2CH–
isopropylidene = (1-methylethylidene) Me_2C=
(isopropylidenedioxy) = [(1-methylethylidene)bis(oxy)] $\text{-}OCMe_2O$–
isosemicarbazido = [(aminohydroxymethylene)hydrazino] $H_2NC(OH){=}NNH$–
isothiocyanato SCN–
isothiocyano = isothiocyanato SCN–
isovaleryl = (3-methyl-1-oxobutyl) Me_2CHCH_2CO–
isovalyl[1] = (2-amino-2-methyl-1-oxobutyl) $EtCMe(NH_2)CO$–
isoviolanthrenylene = (9,18-dihydrodinaphtho[1,2,3-*cd*:1',2',3'-*lm*]perylenediyl)-$(C_{34}H_{18})$–

kauranyl (from kaurane) $(C_{20}H_{33})$–
kauranylene = kauranediyl (from kaurane) $\text{-}(C_{20}H_{32})$–
kaurenyl (from kaurene) $(C_{20}H_{31})$–
keto[11] = oxo O=

labdan-15-yl = [5-(decahydro-2,5,5,8a-tetramethyl-1-naphthalenyl)-3-methylpentyl] $(C_{20}H_{37})$–
lactosyl = (4-*O*-β-galactopyranosyl-β-D-glucopyranosyl) $(C_6H_{11}O_5)O(C_6H_{10}O_5)$–
lactoyl = (2-hydroxy-1-oxopropyl) $MeCH(OH)CO$–
lanostenylene = lanostenediyl (from lanostane) $\text{-}(C_{30}H_{50})$–
lauroyl = (1-oxododecyl) $Me(CH_2)_{10}CO$–
leucyl[1] = (2-amino-4-methyl-1-oxopentyl) $Me_2CHCH_2CH(NH_2)CO$–
levulinoyl = (1,4-dioxopentyl) $MeCO(CH_2)_2CO$–
linalyl (from linalool) = (1-ethenyl-1,5-dimethyl-4-hexenyl) $Me_2C{=}CH(CH_2)_2CMe(CH{=}CH_2)$–
linolelaidoyl = (1-oxo-9,12-octadecadienyl) $Me(CH_2)_4CH{=}CHCH_2CH{=}CH(CH_2)_7CO$–
linolenoyl = (1-oxo-9,12,15-octadecatrienyl) $EtCH{=}CHCH_2CH{=}CHCH_2CH{=}CH(CH_2)_7CO$–
γ-**linolenoyl** = (1-oxo-6,9,12-octadecatrienyl) $Me(CH_2)_4CH{=}CHCH_2CH{=}CHCH_2CH{=}CH(CH_2)_4CO$–
linoleoyl = (1-oxo-9,12-octadecadienyl) $Me(CH_2)_4CH{=}CHCH_2CH{=}CH(CH_2)_7CO$–

lupanyl (from lupane) $(C_{30}H_{51})$–
lysyl[1] = (2,6-diamino-1-oxohexyl) $H_2N(CH_2)_4CH(NH_2)CO$–

maleoyl = (1,4-dioxo-2-butene-1,4-diyl) -COCH=CHCO–
malonaldehydoyl = (1,3-dioxopropyl) $HCOCH_2CO$–
malonamoyl = (3-amino-1,3-dioxopropyl) H_2NCOCH_2CO–
malonaniloyl = [1,3-dioxo-3-(phenylamino)propyl] $PhNHCOCH_2CO$–
malonimido = (2,4-dioxo-1-azetidinyl)

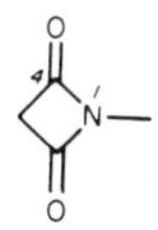

malonyl = (1,3-dioxo-1,3-propanediyl) $\text{-}COCH_2CO$–
maloyl = (2-hydroxy-1,4-dioxo-1,4-butanediyl) $\text{-}COCH(OH)CH_2CO$–
maltosyl = (4-*O*-α-D-glucopyranosyl-β-D-glucopyranosyl) $(C_6H_{11}O_5)O(C_6H_{10}O_5)$–
mandeloyl = (hydroxyphenylacetyl) PhCH(OH)CO–
p-**menth-2-yl** = [2-methyl-5-(1-methylethyl)cyclohexyl] $2,5\text{-}Me(Me_2CH)C_6H_9$–
p-**menth-3,5-ylene** = [5-methyl-2-(1-methylethyl)-1,3-cyclohexanediyl]

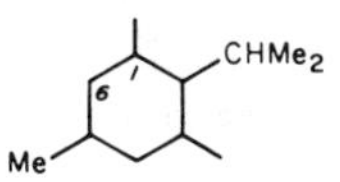

mercapto[4] HS–
mesaconoyl = (2-methyl-1,4-dioxo-2-butene-1,4-diyl) -COCMe=CHCO–
mesidino = [(2,4,6-trimethylphenyl)amino] $2,4,6\text{-}Me_3C_6H_2NH$–
mesityl = (2,4,6-trimethylphenyl) $2,4,6\text{-}Me_3C_6H_2$–
α-**mesityl** = [(3,5-dimethylphenyl)methyl] $3,5\text{-}Me_2C_6H_3CH_2$–
mesoxalyl = (1,2,3-trioxo-1,3-propanediyl) -COCOCO–
mesyl = (methylsulfonyl) $MeSO_2$–
metanilamido = [[(3-aminophenyl)sulfonyl]amino] $3\text{-}H_2NC_6H_4SO_2NH$–
metanilyl = [(3-aminophenyl)sulfonyl] $3\text{-}H_2NC_6H_4SO_2$–
methacryloyl = (2-methyl-1-oxo-2-propenyl) $H_2C{=}CMeCO$–
methallyl = (2-methyl-2-propenyl) $H_2C{=}CMeCH_2$–
methanetetrayl =C=
methene = methylene H_2C=
methenyl = methylidyne HC⋮
methionyl[1] = [methylenebis(sulfonyl)] or [2-amino-4-(methylthio)-1-oxobutyl] $\text{-}SO_2CH_2SO_2$– or $MeS(CH_2)_2CH(NH_2)CO$–
methoxalyl = (methoxyoxoacetyl) MeO_2CCO–
methoxy MeO–
(methoxycarbonyl) MeO_2C–
methyl Me (H_3C)–
[(methyldithio)sulfonyl] $MeSSSO_2$–
methylene H_2C=
(methylenedioxy) = [methylenebis(oxy)] $\text{-}OCH_2O$–
[(methylenedioxy)phenyl] = 1,3-benzodioxol-*ar*-yl $(CH_2O_2)C_6H_3$–
(methylenedisulfonyl) = [methylenebis(sulfonyl)] $\text{-}SO_2CH_2SO_2$–
methylidyne HC≡
methyliumylidene C^+H=
methylol = (hydroxymethyl) $HOCH_2$–
(methyloxy) = methoxy MeO–
(1-methyl-2*H*-pyranium-2-yl)

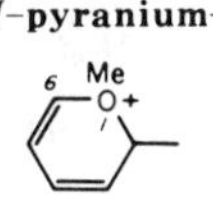

(1-methylpyridinium-2-yl)

(methylselenyl) = (methylseleno) MeSe–
(methylthio) MeS–
(methyltelluro) MeTe–
(methyltrioxy) MeOOO–
morpholino = 4-morpholinyl

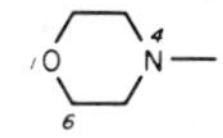

myristoyl = (1-oxotetradecyl) $Me(CH_2)_{12}CO$–

naphthal = (naphthalenylmethylene) $(C_{10}H_7)CH$=
naphthalimido = (1,3-dioxo-1*H*-benz[*de*]isoquinolin-2(3*H*)-yl) $(C_{12}H_6NO_2)$–
naphthenyl = (naphthalenylmethylidyne) $(C_{10}H_7)C$≡
naphthionyl = [(4-amino-1-naphthalenyl)sulfonyl] $4,1\text{-}H_2NC_{10}H_6SO_2$–
naphthobenzyl = (naphthalenylmethyl) $(C_{10}H_7)CH_2$–
naphthothiophene-yl = naphthothienyl $(C_{12}H_7S)$–
naphthoxy = (naphthalenyloxy) $(C_{10}H_7)O$–
naphthoyl = (naphthalenylcarbonyl) $(C_{10}H_7)CO$–
naphthyl = naphthalenyl $(C_{10}H_7)$–
naphthylene = naphthalenediyl $\text{-}(C_{10}H_6)$–
naphthylidene = naphthalenylidene $(C_{10}H_8)$=
(naphthylnaphthyl) = [binaphthalen]yl $C_{10}H_7C_{10}H_6$–
nazyl = (naphthalenylmethyl) $(C_{10}H_7)CH_2$–
neopentyl = (2,2-dimethylpropyl) Me_3CCH_2–
neophyl = (2-methyl-2-phenylpropyl) $PhCMe_2CH_2$–
neryl (from nerol) = (3,7-dimethyl-2,6-octadienyl)- $Me_2C{=}CH(CH_2)_2CMe{=}CHCH_2$–
nicotinimidoyl = (imino-3-pyridinylmethyl) $(C_5H_4N)C({=}NH)$–
nicotinoyl = (3-pyridinylcarbonyl)

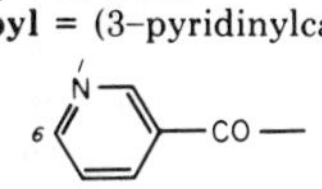

nipecotoyl = (3-piperidinylcarbonyl)

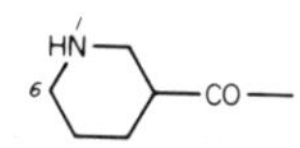

nitramino = (nitroamino) O_2NNH–
aci-**nitramino** = (*aci*-nitroamino) HON(O)=N–
nitrilio HN^+≡
nitrilo N≡
(nitrilophosphoranyl) = phosphononitridyl HP(≡N)–
nitro O_2N–
aci-**nitro** HON(O)=
nitrosamino = (nitrosoamino) ONNH–
nitrosimino = (nitrosoimino) ONN =
nitroso ON–
(nitrothio) O_2NS–
nonanedioyl = (1,9-dioxo-1,9-nonanediyl) $\text{-}CO(CH_2)_7CO$–
nonanoyl = (1-oxononyl) $Me(CH_2)_7CO$–
norbornyl = bicyclo[2.2.1]heptyl (C_7H_{11})–
norbornylene = bicyclo[2.2.1]heptanediyl $\text{-}(C_7H_{10})$–
norcamphanyl = bicyclo[2.2.1]heptyl (C_7H_{11})–
norcaryl (from norcarane) = bicyclo[4.1.0]heptyl (C_7H_{11})–
norpinyl (from norpinane) = bicyclo[3.1.1]heptyl (C_7H_{11})–
norleucyl[1] = (2-amino-1-oxohexyl) $Me(CH_2)_3CH(NH_2)CO$–
norvalyl[1] = (2-amino-1-oxopentyl) $Me(CH_2)_2CH(NH_2)CO$–

octadecanoyl = (1-oxooctadecyl) $Me(CH_2)_{16}CO$–
octanedioyl = (1,8-dioxo-1,8-octanediyl) $\text{-}CO(CH_2)_6CO$–
octanoyl = (1-oxooctyl) $Me(CH_2)_6CO$–
tert-**octyl** = (1,1,3,3-tetramethylbutyl) $Me_3CCH_2CMe_2$–
oenanthyl = (1-oxoheptyl) $Me(CH_2)_5CO$–
oleananyl (from oleanane) $(C_{30}H_{51})$–
oleoyl = (1-oxo-9-octadecenyl) $Me(CH_2)_7CH{=}CH(CH_2)_7CO$–
ornithyl[1] = (2,5-diamino-1-oxopentyl) $H_2N(CH_2)_3CH(NH_2)CO$–

oxalaldehydoyl = (oxoacetyl) HCOCO–
oxalyl = (1,2-dioxo-1,2-ethanediyl) –COCO–
oxamido = [(aminooxoacetyl)amino] $H_2NCOCONH-$
oxamoyl = (aminooxoacetyl) $H_2NCOCO-$
oxamyl = (aminooxoacetyl) $H_2NCOCO-$
oxaniloyl = [oxo(phenylamino)acetyl] PhNHCOCO–
oxazolinyl = (dihydrooxazolyl) $(C_3H_4NO)-$
oximido = (hydroxyimino) HON=
oxo[5] (see also epoxy and oxy) O=
(oxobornyl) = (trimethyloxobicyclo[2.2.1]heptyl) $(C_{10}H_{15}O)-$
(oxoboryl) OB–
(1-oxoethyl) = acetyl Ac (MeCO–)
(oxoethylene) = (1-oxo-1,2-ethanediyl) $-COCH_2-$
(oxophenylhydrazino) = (nitrosophenylamino) PhN(NO)–
(oxophenylmethyl) = benzoyl Bz (PhCO–)
(oxopyridinylmethyl) = (pyridinylcarbonyl) $(C_5H_4N)CO-$
(2-oxotrimethylene) = (2-oxo-1,3-propanediyl) $-CH_2COCH_2-$
(2-oxovinyl) = (oxoethenyl) OC=CH–
oxy[9] (see also epoxy and oxo) –O–
[oxybis(methylenecarbonylimino)] = [oxybis=[(1-oxo-2,1-ethanediyl)imino]] $-NHCOCH_2OCH_2CONH-$

palmitoyl = (1-oxohexadecyl) $Me(CH_2)_{14}CO-$
pantothenoyl = *N*-(2,4-dihydroxy-3,3-di=methyl-1-oxobutyl)-*β*-alanyl[1] or [3-[(2,4-di=hydroxy-3,3-dimethyl-1-oxobutyl)amino]-1-oxopropyl] $HOCH_2CMe_2CH(OH)CONH(CH_2)_2CO-$
perlargonoyl = (1-oxononyl) $Me(CH_2)_7CO-$
pelargonyl = (1-oxononyl) $Me(CH_2)_7CO-$
pentadecanoyl = (1-oxopentadecyl) $Me(CH_2)_{13}CO-$
pentamethylene = 1,5-pentanediyl $-(CH_2)_5-$
3-pentanesulfonamido = [[(1-ethylpropyl)=sulfonyl]amino] Et_2CHSO_2NH-
1,3-pentazadieno = 1,3-pentazadienyl $H_2NN{=}NN{=}N-$
2-pentenediylidyne = 2-pentene-1,5-diylidyne $\equiv CCH{=}CHCH_2C\equiv$
tert-**pentyl** = (1,1-dimethylpropyl) $EtCMe_2-$
pentyl $Me(CH_2)_4-$
pentylidyne BuC⁝
perchloryl O_3Cl-
perseleno = seleninoselenoyl Se=Se=
perthio = sulfinothioyl S=S=
phenacyl = (2-oxo-2-phenylethyl) $PhCOCH_2-$
phenacylidene = (2-oxo-2-phenylethylidene) PhCOCH=
phenanthrothiophene-yl = phenanthrothienyl $(C_{16}H_9S)-$
phenanthryl = phenanthrenyl $(C_{14}H_9)-$
phenanthrylene = phenanthrenediyl $-(C_{14}H_8)-$
phenenyl = benzenetriyl $C_6H_3\equiv$
phenethyl = (2-phenylethyl) $PhCH_2CH_2-$
phenethylidene = (2-phenylethylidene) $PhCH_2CH=$
phenetidino = [(ethoxyphenyl)amino] $(EtO)C_6H_4NH-$
phenetyl = (ethoxyphenyl) $(EtO)C_6H_4-$
phenoxy PhO–
phenyl Ph $(C_6H_5)-$
phenylalanyl[1] (from phenylalanine) $PhCH_2CH{=}(NH_2)CO-$
(phenylarsinico) = (hydroxyphenylarsinyl) PhAs(O)(OH)–
[(phenylazo)imino] = (3-phenyl-2-tri=azenylidene) PhN=NN=
(phenylbenzoyl) = ([1,1'-biphenyl]ylcarbonyl) PhC_6H_4CO
(phenyldiazenyl) = (phenylazo) PhN=N–
phenylene $-(C_6H_4)-$
[phenylenebis(azo)] $-N{=}NC_6H_4N{=}N-$
[phenylenebis[azo(methylimino)]] = [phenyl=enebis(1-methyl-2-triazene-3,1-diyl)] $-NMeN{=}NC_6H_4N{=}NNMe-$
[phenylenebis(1-oxo-1-ethanyl-2-ylidene)] = [phenylenebis(2-oxo-2-ethanyl-1-ylidene)] $=CHCOC_6H_4COCH=$
(phenylenedimethylene) = [phenylenebis=(methylene)] $-CH_2C_6H_4CH_2-$
(phenylenedimethylidyne) $=CHC_6H_4CH=$
(phenylenedioxy) = [phenylenebis(oxy)] $-OC_6H_4O-$
(phenylglyoxyloyl) = (oxophenylacetyl) PhCOCO–
phenylidene = cyclohexadienylidene $(C_6H_6)=$
(phenylimidocarbonyl) = (phenylcarbon=imidoyl)[7] PhN=C=
(phenyloxalyl) = (oxophenylacetyl) PhCOCO–
(phenyloxy) = phenoxy PhO–
(phenylphenoxy) = ([1,1'-biphenyl]yloxy) PhC_6H_4O-
(phenylsulfenyl) = (phenylthio) PhS–
(phenylsulfinyl) PhS(O)–
(*S*-phenylsulfonimidoyl) PhS(O)(=NH)–
phorbinyl (from phorbine) $(C_{22}H_{17}N_4)-$
phosphinico[3,4] HOP(O)=
phosphinidene HP=
phosphinidyne P≡
phosphinimyl $H_2P({=}NH)-$
phosphino H_2P-
phosphinothioyl $H_2P(S)-$
phosphinothioylidene HP(S)=
phosphinyl $H_2P(O)-$
phosphinylidene HP(O)=
phosphinylidyne P(O)≡
phospho O_2P-
phosphono[4] $(HO)_2P(O)-$
(phosphonoformyl) = (phosphonocarbonyl) $(HO)_2P(O)CO-$
phosphononitridyl HP(≡N)–
phosphoranyl H_4P-
phosphoranylidene $H_3P=$
phosphoranylidyne $H_2P\equiv$
phosphoro = 1,2-diphosphenediyl –P=P–
phosphoroso OP–
phthalal = (1,2-phenylenedimethylidyne) $1,2-C_6H_4(CH=)_2$
phthalaldehydoyl = (2-formylbenzoyl) $2-(HCO)C_6H_4CO-$
phthalamoyl = [2-(aminocarbonyl)benzoyl] $2-(H_2NCO)C_6H_4CO-$
phthalanyl = (1,3-dihydroisobenzofuranyl) $(C_8H_7O)-$
phthalidyl = (1,3-dihydro-3-oxo-1-isobenzo=furanyl) $(C_8H_5O_2)-$
phthalidylidene = (3-oxo-1(3*H*)-isobenzo=furanylidene) $(C_8H_4O_2)-$
phthalimido = (1,3-dihydro-1,3-dioxo-2*H*-iso=indol-2-yl) $(C_8H_4NO_2)-$
phthalocyaninyl[1] (from phthalocyanine) $(C_{32}H_{17}N_8)-$
phthaloyl = (1,2-phenylenedicarbonyl) $1,2-C_6H_4(CO-)_2$
phthalylidene = (1,2-phenylenedimethylidyne) $1,2-C_6H_4(CH=)_2$
phyllocladanyl = kauranyl (from kaurane) $(C_{29}H_{33})-$
phytyl = (3,7,11,15-tetramethyl-2-hexadecenyl) $Me[CHMe(CH_2)_3]_3CMe{=}CHCH_2-$
picolinoyl = (2-pyridinylcarbonyl)

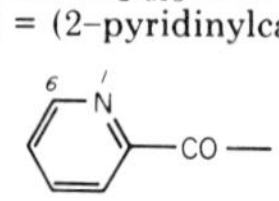

picryl = (2,4,6-trinitrophenyl) $2,4,6-(O_2N)_3C_6H_2-$
pimeloyl = (1,7-dioxo-1,7-heptanediyl) $-CO(CH_2)_5CO-$
4-pinanyl (from pinane) = (4,6,6-trimethyl=bicyclo[3.1.1]hept-2-yl) $(C_{10}H_{17})-$
pinanylene = (trimethylbicyclo[3.1.1]heptane=diyl) $-(C_{10}H_{16})-$
pipecoloyl = (2-piperidinylcarbonyl)

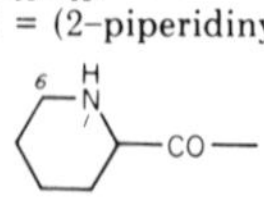

piperidino = 1-piperidinyl

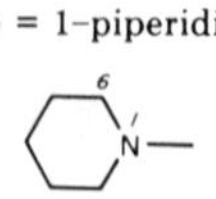

piperidyl = piperidinyl $(C_5H_{10}N)-$
piperidylidene = piperidinylidene $(C_5H_9N)=$
piperonyl = (1,3-benzodioxol-5-ylmethyl) $3,4-(CH_2O_2)C_6H_3CH_2-$
piperonylidene = (1,3-benzodioxol-5-yl=methylene) $3,4-(CH_2O_2)C_6H_3CH=$
piperonyloyl = (1,3-benzodioxol-5-ylcarbonyl) $3,4-(CH_2O_2)C_6H_3CO-$
pivaloyl = (2,2-dimethyl-1-oxopropyl) Me_3CCO-
pivalyl = (2,2-dimethyl-1-oxopropyl) Me_3CCO-
plumbanetetrayl =Pb=
plumbyl H_3Pb-
plumbylene $H_2Pb=$
plumbylidyne HPb≡
podocarpan-13-yl = (tetradecahydro-4b,8,8-trimethyl-2-phenanthrenyl) $(C_{17}H_{29})-$
porphinyl (from porphine) $(C_{20}H_{13}N_4)-$
pregna-5,16-dien-21-yl (from pregnadiene) $(C_{21}H_{31})-$
prenyl = (3-methyl-2-butenyl) $Me_2C{=}CHCH_2-$
prolyl[1] = (2-pyrrolidinylcarbonyl)

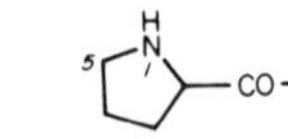

2-propanesulfonamido = [[(1-methylethyl)=sulfonyl]amino] Me_2CHSO_2NH-
propargyl = 2-propynyl $HC\equiv CCH_2-$
propenyl = 1-propenyl MeCH=CH–
2-propenyl $CH_2{=}CH{-}CH_2-$
propenylene = 1-propene-1,3-diyl $-CH{=}CHCH_2-$
propenylidene = 1-propenylidene MeCH=C=
propioloyl = (1-oxo-2-propynyl) HC≡CCO–
propiolyl = (1-oxo-2-propynyl) HC≡CCO–
propionamido = [(1-oxopropyl)amino] EtCONH–
propionyl = (1-oxopropyl) EtCO–
(propionyldioxy) = [(1-oxopropyl)dioxy] EtC(O)OO–
propionyloxy = (1-oxopropoxy) $EtCO_2-$
propoxy PrO–
propyl Pr $(CH_3CH_2CH_2-)$
sec-**propyl** = (1-methylethyl) Me_2CH-
propylene = (1-methyl-1,2-ethanediyl) $-CHMeCH_2-$
propylidene EtCH=
propylidyne EtC≡
(propyloxy) = propoxy PrO–
protocatechuoyl = (3,4-dihydroxybenzoyl) $3,4-(HO)_2C_6H_3CO-$
pseudoallyl = (1-methylethenyl) $H_2C{=}CMe-$
pseudocumidino = [(2,4,5-trimethylphenyl)=amino] $2,4,5-Me_3C_6H_2NH-$
as-**pseudocumyl** = (2,3,5-trimethylphenyl) $2,3,5-Me_3C_6H_2-$
s-**pseudocumyl** = (2,4,5-trimethylphenyl) $2,4,5-Me_3C_6H_2-$
v-**pseudocumyl** = (2,3,6-trimethylphenyl) $2,3,6-Me_3C_6H_2-$
pseudoindolyl = 1*H*-indolyl $(C_8H_6N)-$
pteroyl = [4-[[(2-amino-1,4-dihydro-4-oxo-6-pteridinyl)methyl]amino]benzoyl] $(C_{14}H_{11}N_6O_2)-$
2*H*-pyranio

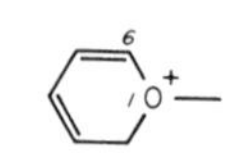

2*H*-pyran-2-ylium-2-yl

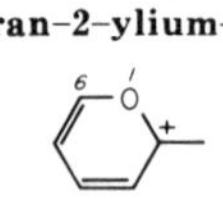

pyrazolidyl = pyrazolidinyl $(C_3H_7N_2)-$
pyrazolinyl = (dihydropyrazolyl) $(C_3H_5N_2)-$
pyridinio

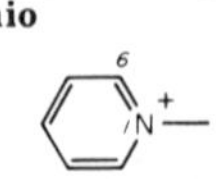

pyridyl = pyridinyl $(C_5H_4N)-$
pyroglutamoyl = (5-oxoprolyl)[1] or [(5-oxo-2-pyrrolidinyl)carbonyl]

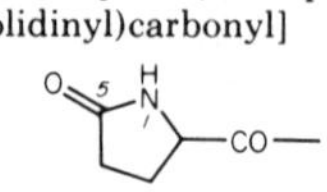

pyromucyl = (2-furanylcarbonyl)

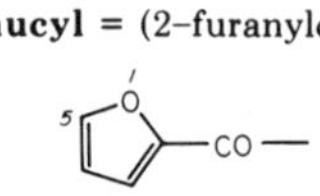

pyrrolidyl = pyrrolidinyl $(C_4H_8N)-$
pyrrolinyl = (dihydropyrrolyl) $(C_4H_6N)-$
pyrrol-1-yl = 1*H*-pyrrol-1-yl $(C_4H_4N)-$
pyrroyl = (py·rolylcarbonyl) $(C_4H_3N)CO-$

pyrryl = pyrrolyl (C_4H_4N)–
pyruvoyl = (1,2-dioxopropyl) MeCOCO–

p-**quaterphenylyl** = [1,1':4',1'':4'',1'''-quater=phenyl]yl ($C_{24}H_{17}$)–
quinaldoyl = (2-quinolinylcarbonyl) (2-C_9H_6N)CO–
quinolyl = quinolinyl (C_9H_6N)–
quinonyl = (dioxocyclohexadienyl) ($C_6H_3O_2$)–
quinuclidinyl = 1-azabicyclo[2.2.2]octyl ($C_7H_{12}N$)–

α-**resorcyloyl** = (3,5-dihydroxybenzoyl) 3,5-$(HO)_2C_6H_3CO$–
β-**resorcyloyl** = (2,4-dihydroxybenzoyl) 2,4-$(HO)_2C_6H_3CO$–
γ-**resorcyloyl** = (2,6-dihydroxybenzoyl) 2,6-$(HO)_2C_6H_3CO$–
rhamnosyl = (6-deoxymannopyranosyl)

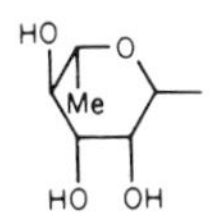

ricinelaidoyl = (12-hydroxy-1-oxo-9-octadece=nyl) $Me(CH_2)_5CH(OH)CH_2CH=CH(CH_2)_7CO$–
ricinoleoyl = (12-hydroxy-1-oxo-9-octadecenyl) $Me(CH_2)_5CH(OH)CH_2CH=CH(CH_2)_7CO$–
rosan-6-yl = (2-ethyltetradecahydro-2,4a,8,8-tetramethyl-9-phenanthrenyl) ($C_{20}H_{35}$)–

salicyl = [(2-hydroxyphenyl)methyl] 2-$HOC_6H_4CH_2$–
salicylidene = [(2-hydroxyphenyl)methylene] 2-HOC_6H_4CH=
salicyloyl = (2-hydroxybenzoyl) 2-HOC_6H_4CO–
sarcosyl = (*N*-methylglycyl)[1] or [(methyl=amino)acetyl] $MeNHCH_2CO$–
sebacoyl = (1,10-dioxo-1,10-decanediyl) –$CO(CH_2)_8CO$–
seleneno[4] HOSe–
selenino[4] HOSe(O)–
seleninoselenoyl Se=Se=
seleninyl OSe=
seleno[9] (see also episeleno and selenoxo) –Se–
selenocyanato NCSe–
selenono[4] $(HO)SeO_2$–
selenonyl O_2Se=
selenophenyl = selenophene-yl (C_4H_3Se)–
selenoxo[5] (see also episeleno and seleno) Se=
selenyl[4] HSe–
semicarbazido = [2-(aminocarbonyl)hydrazino] $H_2NCONHNH$–
semicarbazono = [(aminocarbonyl)hydrazono] $H_2NCONHN$=
senecioyl = (3-methyl-1-oxo-2-butenyl) $Me_2C=CHCO$–
seryl[1] = (2-amino-3-hydroxy-1-oxopropyl) $HOCH_2CH(NH_2)CO$–
siamyl = (1,2-dimethylpropyl) $Me_2CHCHMe$–
silanetetrayl =Si≡
siloxy = (silyloxy) H_3SiO–
silyl H_3Si–
silylene H_2Si=
silylidyne HSi≡
sorboyl = (1-oxo-2,4-hexadienyl) MeCH=CHCH=CHCO–
spirohex-1-yl (C_6H_9)–
spirostanyl (from spirostane) ($C_{27}H_{43}O_2$)–
stannanetetrayl =Sn=
stannyl H_3Sn–
stannylene H_2Sn=
stannylidyne HSn≡
stearoyl = (1-oxooctadecyl) $Me(CH_2)_{16}CO$–
stibino H_2Sb–
stibyl = stibino H_2Sb–
stibylene HSb=
stibylidyne Sb≡
styrene = (1-phenyl-1,2-ethanediyl) –$CHPhCH_2$–
styrolene = (1-phenyl-1,2-ethanediyl)–$CHPhCH_2$–
styryl = (2-phenylethenyl) PhCH=CH–
suberoyl = (1,8-dioxo-1,8-octanediyl) –$CO(CH_2)_6CO$–

succinaldehydoyl = (1,4-dioxobutyl) $HCO(CH_2)_2CO$–
succinamoyl = (4-amino-1,4-dioxobutyl) $H_2NCO(CH_2)_2CO$–
succinamyl = (4-amino-1,4-dioxobutyl) $H_2NCO(CH_2)_2CO$–
succinaniloyl = [1,4-dioxo-4-(phenylamino)=butyl] $PhNHCO(CH_2)_2CO$–
succinimido = (2,5-dioxo-1-pyrrolidinyl)

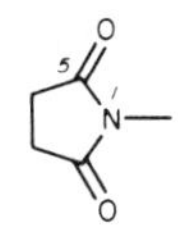

succinyl = (1,4-dioxo-1,4-butanediyl) –$CO(CH_2)_2CO$–
sulfamino = (sulfoamino) $HOSO_2NH$–
sulfamoyl = (aminosulfonyl) H_2NSO_2–
sulfamyl = (aminosulfonyl) H_2NSO_2–
sulfanilamido = [[(4-aminophenyl)sulfonyl]=amino] 4-$H_2NC_6H_4SO_2NH$–
sulfanilyl = [(4-aminophenyl)sulfonyl] 4-$H_2NC_6H_4SO_2$–
sulfeno[4] HOS–
sulfhydryl = mercapto[4] HS–
sulfinimidoyl HN=S=
sulfino[4] HOS(O)–
sulfinothioyl S=S=
sulfinyl OS=
sulfo[4] HO_3S–
sulfonimidoyl HN=S(O)=
sulfonodiimidoyl $(HN=)_2S$=
sulfonyl –SO_2–
sulfuryl = sulfonyl –SO_2–

tartaroyl = (2,3-dihydroxy-1,4-dioxo-1,4-bu=tanediyl) –COCH(OH)CH(OH)CO–
tartronoyl = (2-hydroxy-1,3-dioxo-1,3-propanediyl) –COCH(OH)CO–
tauryl = [(2-aminoethyl)sulfonyl] $H_2N(CH_2)_2SO_2$–
telluro[6] (see also telluroxo) –Te–
telluroxo[5] (see also telluro) Te=
telluryl[4] HTe–
terephthalal = (1,4-phenylenedimethylidyne) 1,4-$C_6H_4(CH=)_2$
terephthalaldehydoyl = (4-formylbenzoyl) 4-$HCOC_6H_4CO$–
terephthalamoyl = [4-(aminocarbonyl)benzoyl] 4-$(H_2NCO)C_6H_4CO$–
terephthalaniloyl = [4-[(phenylamino)=carbonyl]benzoyl] 4-$(PhNHCO)C_6H_4CO$–
terephthaloyl = (1,4-phenylenedicarbonyl) 1,4-$C_6H_4(CO–)_2$
terephthalylidene = (1,4-phenylenedi=methylidyne) 1,4-$C_6H_4(CH=)_2$
m-**terphenylyl** = [1,1':3',1''-terphenyl]yl ($C_{18}H_{13}$)–
terphenylylene = [terphenyl]diyl –($C_{18}H_{12}$)–
tetradecanoyl = (1-oxotetradecyl) $Me(CH_2)_{12}CO$–
tetramethylene = 1,4-butanediyl –$(CH_2)_4$–
1,4-tetraphosphinediyl –$(PH)_4$–
tetrasiloxanylene = 1,7-tetrasiloxanediyl –$SiH_2(OSiH_2)_2OSiH_2$–
tetrathio[12] –SSSS–
tetrazanediylidene = 1,4-tetrazanediylidene =$N(NH)_2N$=
tetrazanylene = 1,4-tetrazanediyl –$(NH)_4$–
1-tetrazeno = 1-tetrazenyl $H_2NNHN=N$–
thenoyl = (thienylcarbonyl) (C_4H_3S)CO–
thenyl = (thienylmethyl) (C_4H_3S)CH_2–
thenylidene = (thienylmethylene) (C_4H_3S)CH=
(thenyloxy) = (thienylmethoxy) (C_4H_3S)CH_2O–
thexyl = 1,1,2-trimethylpropyl Me_2CHCMe_2–
thianaphthenyl = benzo[*b*]thienyl (C_8H_5S)–
thiazolidyl = thiazolidinyl (C_3H_6NS)–
thiazolinyl = (dihydrothiazolyl) (C_3H_4NS)–
[(5-thiazolylcarbonyl)methyl] = [2-oxo-2-(5-thiazolyl)ethyl]

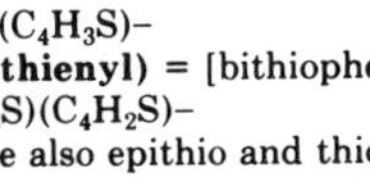

thienyl (C_4H_3S)–
(thienylthienyl) = [bithiophen]yl (C_4H_3S)(C_4H_2S)–
thio[9] (see also epithio and thioxo) –S–

(thioacetonylidene) = (2-thioxopropylidene) MeCSCH=
thioacetyl = (1-thioxoethyl) MeCS–
(thioarsenoso) = (thioxoarsino) S=As–
(thiobenzoyl) = (phenylthioxomethyl) PhCS–
(thiocarbamoyl) = (aminothioxomethyl) H_2NCS–
thiocarbamyl = (aminothioxomethyl) H_2NCS–
(thiocarbonyl) = carbonothioyl[7]–CS–
(thiocarboxy)[13] HOSC–
thiocyanato NCS–
thiocyano = thiocyanato NCS–
(thioformyl) = (thioxomethyl) HCS–
(thiohexanoyl) = (1-thioxohexyl) $Me(CH_2)_4CS$–
thiohydroperoxy = sulfeno[4] HOS– or (mercap=tooxy)[4] HSO–
(thiohydroxy) = mercapto[4] HS–
thiomorpholino = 4-thiomorpholinyl

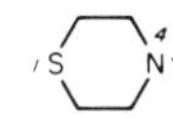

(thionitroso) SN–
thionyl = sulfinyl –SO–
(thiophenacyl) = (2-phenyl-2-thioxoethyl) $PhCSCH_2$–
thiophene-yl = thienyl (C_4H_3S)–
(thiophosphono) = (hydroxymercapto=phosphinyl (HO)(HS)P(O)–
(thioseleneno)[4] HSSe–
(thiosulfeno)[4] HSS–
(thiosulfo)[13] (HO_2S_2)–
thioxo[5] (see also epithio) S=
(thioxoarsino) SAs–
(thioxomethyl) (see also carbonothioyl) S=CH–
thiuram = (aminothioxomethyl) H_2NCS–
threonyl[1] (2-amino-3-hydroxy-1-oxobutyl) $MeCH(OH)CH(NH_2)CO$–
4-thujyl = [2-methyl-5-(1-methylethyl)bicyclo=[3.1.0]hex-2-yl] ($C_{10}H_{17}$)–
thymyl (from thymol) = [5-methyl-2-(1-methylethyl)phenyl]

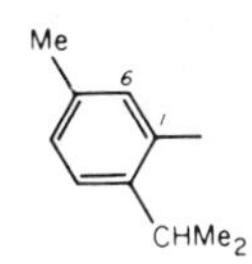

thyronyl = [*O*-(4-hydroxyphenyl)tyrosyl][1] or [2-amino-3-[4-(4-hydroxyphenoxy)phenyl]-1-oxopropyl] [4-$(4-HOC_6H_4O)C_6H_4]CH_2CH(NH_2)CO$–
toloxy = (methylphenoxy) MeC_6H_4O–
p-**toluenesulfonamido** = [[(4-methylphenyl)=sulfonyl]amino] 4-$MeC_6H_4SO_2NH$–
toluidino = [(methylphenyl)amino] MeC_6H_4NH–
toluoyl = (methylbenzoyl) MeC_6H_4CO–
toluyl = (methylbenzoyl) MeC_6H_4CO–
tolyl = (methylphenyl) MeC_6H_4–
α-**tolyl** = (phenylmethyl) $PhCH_2$–
tolylene = (methylphenylene)–(MeC_6H_3)–
α-**tolylene** = (phenylmethylene) PhCH=
tosyl = [(4-methylphenyl)sulfonyl] 4-$MeC_6H_4SO_2$–
triazano = triazanyl H_2NNHNH–
1-triazeno = 1-triazenyl $H_2NN=N$–
s-**triazin-2-yl** = 1,3,5-triazin-2-yl

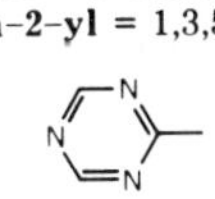

trichothecanyl (from trichothecane) ($C_{15}H_{25}O$)–
tridecanoyl = (1-oxotridecyl) $Me(CH_2)_{11}CO$–
(trimethylammonio) Me_3N^+–
(trimethylarsonio) Me_3As^+–
trimethylene = 1,3-propanediyl –$(CH_2)_3$–
(1,3,3-trimethyl-2-norbornyl) = (1,3,3-tri=methylbicyclo[2.2.1]hept-2-yl) ($C_{10}H_{17}$)–
(trimethylphosphonio) Me_3P^+–
triseleno[12] –SeSeSe–
trisilanylene = 1,3-trisilanediyl –$(SiH_2)_3$–
trisiloxane-1,3,5-triyl = 1,3,5-trisiloxanetriyl –$SiH(OSiH_2–)_2$
trithio[12] –SSS–
trityl = (triphenylmethyl) Ph_3C–
tropanyl = (8-methyl-8-azabicyclo[3.2.1]octyl) ($C_8H_{14}N$)–
tropoyl = (3-hydroxy-1-oxo-2-phenylpropyl) $HOCH_2CHPhCO$–
tryptophyl[1] = [2-amino-3-(1*H*-indol-3-yl)-1-oxopropyl] (C_8H_6N)$CH_2CH(NH_2)CO$–

tyrosyl[1] = [2-amino-3-(4-hydroxyphenyl)-1-oxopropyl] 4-$HOC_6H_4CH_2CH(NH_2)CO$-

undecanoyl = (1-oxoundecyl) $Me(CH_2)_9CO$-
uramino = [(aminocarbonyl)amino] H_2NCONH-
ureido = [(aminocarbonyl)amino] H_2NCONH-
ureylene = (carbonyldiimino) -NHCONH-
(ureylenediureylene) = [carbonylbis(hydrazo=carbonylimino)] -NHCONHNHCONHNHCONH-
ursanyl (from ursane) ($C_{30}H_{51}$)-

valeryl = (1-oxopentyl) BuCO-
valyl[1] = (2-amino-3-methyl-1-oxobutyl) $Me_2CHCH(NH_2)CO$-
vanillal = [(4-hydroxy-3-methoxyphenyl)=methylene] 4,3-$HO(MeO)C_6H_3CH$=
vanilloyl = (4-hydroxy-3-methoxybenzoyl) 4,3-$HO(MeO)C_6H_3CO$-
vanillyl = [(4-hydroxy-3-methoxyphenyl)=methyl] 4,3-$HO(MeO)C_6H_3CH_2$-
vanillylidene = [(4-hydroxy-3-methoxy=phenyl)methylene] 4,3-$HO(MeO)C_6H_3CH$=
vanilmandeloyl = [hydroxy(4-hydroxy-3-=methoxyphenyl)acetyl] 4,3-$HO(MeO)C_6H_3CH(OH)CO$-
veratral = [(3,4-dimethoxyphenyl)methylene] 3,4-$(MeO)_2C_6H_3CH$=
veratroyl = (3,4-dimethoxybenzoyl) 3,4-$(MeO)_2C_6H_3CO$-
o-**veratroyl** = (2,3-dimethoxybenzoyl) 2,3-$(MeO)_2C_6H_3CO$-
veratryl = [(3,4-dimethoxyphenyl)methyl] 3,4-$(MeO)_2C_6H_3CH_2$-
o-**veratryl** = [(2,3-dimethoxyphenyl)methyl] 2,3-$(MeO)_2C_6H_3CH_2$-
veratrylidene = [(3,4-dimethoxyphenyl)=methylene] 3,4-$(MeO)_2C_6H_3CH$=
vinyl = ethenyl H_2C=CH-
vinylene = 1,2-ethenediyl -CH=CH-
vinylidene = ethenylidene H_2C=C=

xanthen-9-yl = 9*H*-xanthen-9-yl ($C_{13}H_9O$)-
xanth-9-yl = 9*H*-xanthen-9-yl ($C_{13}H_9O$)-
xenyl = [1,1'-biphenyl]-4-yl 4-PhC_6H_4-
xylidino = [(dimethylphenyl)amino] $Me_2C_6H_3NH$-
xyloyl = (dimethylbenzoyl) $Me_2C_6H_3CO$-
xylyl = (dimethylphenyl) $Me_2C_6H_3$-
xylylene = [phenylenebis(methylene)] -$CH_2C_6H_4CH_2$-

NOTE: In addition to the following comments it must be understood that stereochemical information is not provided for systematically named natu=ral-product radicals in the list above. For example, the phytyl radical is named as 3,7,11,15-tetramethyl-2-hexadecenyl. Because the total stereochemistry of a phytyl compound is expressed separately in the index modification, and may be influenced by the presence of other chiral centers, it would be mis=leading to supply it with the radical name. However, the cross-reference at **Phytol** in the *Index Guide* includes the stereodescriptor [*R*-[*R**,*R**-(*E*)]]-.

---

[1] This prefix is used in peptide nomenclature.
[2] This prefix may be used in a generic sense, e.g., aldoxime.
[3] This prefix is used only as a multiplying radical.
[4] This prefix is used only when unsubstituted.
[5] This prefix is used when both free valencies are attached to the same atom.
[6] This prefix is used when the free valencies are attached to different atoms which are usually not otherwise connected.
[7] This prefix is used as a multiplying radical or when both free valencies are attached to the same atom.
[8] The prefix "cumyl" has been used in the recent literature to mean α-cumyl.
[9] This prefix is used when the free valencies are attached to different atoms which are not otherwise connected.
[10] This prefix is used when the free valencies are attached to different atoms which are otherwise connected.
[11] This prefix may be used in a generic sense, e.g., ketoxime.
[12] This prefix is used to denote a series of chalcogen atoms in a chain or an indefinite structure.
[13] This prefix is not used when the hydrogen atom has been substituted by another atom or group if a definite structure can be determined.

J. SELECTED BIBLIOGRAPHY OF NOMENCLATURE OF CHEMICAL SUBSTANCES

295. Introduction. The development of systematic chemical nomenclature is shared by a number of organizations. In the United States the American Chemical Society (ACS) established a Committee on Nomenclature and Notation as early as 1886, followed in 1911 by the ACS Committee on Nomenclature, Spelling, and Pronunciation, now known as the ACS Committee on Nomenclature. In addition, special subject subcommittees exist under several ACS divisions.

Internationally, nomenclature commissions of the International Union of Pure and Applied Chemistry (IUPAC) and the International Union of Biochemistry (IUB) approve and publish detailed recommendations.[1]

Before becoming officially recommended nomenclature policy, a typical path for a proposal originating in the United States has been as follows: an idea (suggested by an author, editor, committee member, etc.) is submitted to a subcommittee of subject specialists, then to an ACS divisional committee, the ACS Committee on Nomenclature, and finally to the appropriate nomenclature commission of the International Union of Pure and Applied Chemistry.

References to current nomenclature rules of IUPAC, IUB, and ACS are listed below. A selection of references to older, superseded rules and to significant proposals of individual authors is also included because of the occasional use of such nomenclature in current chemical literature, and the need in retrospective searching. These older references also provide a historical perspective illustrating the precedents on which modern chemical nomenclature is based.

Additional nomenclature information may be obtained from the Director of Nomenclature, Chemical Abstracts Service.

IUPAC and IUB Nomenclature Rules and Recommendations

296. Organic chemistry

1. International Union of Pure and Applied Chemistry. Organic Chemistry Division. Commission on the Nomenclature of Organic Chemistry, *Nomenclature of Organic Chemistry, Sections A, B, C, D, E, F, and H,* 1979 edition, Pergamon Press, Oxford, 1979, 559 pp. (Section A, Hydrocarbons; Section B, Fundamental Heterocyclic Systems; Section C, Characteristic Groups Containing C, H, O, N, Halogen, S, Se, and/or Te; Section D, Organic Compounds Containing Elements Which Are Not Exclusively C, H, O, N, Halogen, S, Se, and Te (Provisional Recommendations 1978); Section E, Stereochemistry (Recommendations 1974); Section F, General Principles for the Naming of Natural Products and Related Compounds (Provisional Recommendations 1975); Section H, Isotopically Modified Compounds (Approved Recommendations 1978). This edition contains changes from previous editions of various sections. The first edition of Sections A and B (1957) was published with comments in *J. Am. Chem. Soc.* **1960**, 82, 5545–74.[2] The first edition of Section C (1965) was published in *Pure Appl. Chem.* **1965**, 11, 1–260. A summary of the changes that were made in Sections A, B, and C for the combined 1971 edition of these sections appears in *J. Chem. Doc.* **1972**, 12, 132–8. An earlier version of Section D was published in *IUPAC Inf. Bull. Append.* No. 31, August, 1973. The contents of the first edition of Section E appeared in *Pure Appl. Chem.* **1976**, 45, 11–30, and Section E was previously published as tentative rules in *IUPAC Inf. Bull.* No. 35, June, 1969, pp. 36–80, and in *J. Org. Chem.* **1970**, 35, 2849–67.[2] Section F appeared earlier in *IUPAC Inf. Bull. Append.* No. 53, December, 1976, and in *Eur. J. Biochem.* **1978**, 86, 1–8. The contents of the first edition of Section H appeared in *Pure Appl. Chem.* **1979**, 51, 353–80 and Section H was published previously as provisional recommendations in *IUPAC Inf. Bull. Append.* No. 62, July, 1977, and in *Eur. J. Biochem.* **1978**, 86, 9–25.

2. ——, "Revision of the Extended Hantzsch–Widman System of Nomenclature for Heteromonocycles (Recommendations 1982)", *Pure Appl. Chem.* **1983**, 55, 409–16. Published previously as provisional recommendations in *Pure Appl. Chem.* **1979**, 51, 1995–2003.

3. ——, "Treatment of Variable Valence in Organic Nomenclature (Lambda Convention) (Recommendations 1983)", *Pure Appl. Chem.* **1984**, 56, 769–78. Published previously as provisional recommendations entitled "The Designation of Non–Standard Classical Valence Bonding in Organic Nomenclature" in *Pure Appl. Chem.* **1982**, 54, 217–27.

4. ——, "Extension of Rules A–1.1 and A–2.5 Concerning Numerical Terms Used in Organic Chemical Nomenclature (Recommendations 1986)", *Pure Appl. Chem.* **1986**, 58, 1693–6. Published previously as provisional recommendations in *Pure Appl. Chem.* **1983**, 55, 1463–6.

5. International Union of Pure and Applied Chemistry. Organic Chemistry Division. Commission on the Nomenclature of Organic Chemistry. International Union of Pure and Applied Chemistry and International Union of Biochemistry. Commission on Biochemical Nomenclature, "Definitive Rules for Nomenclature of Steroids", *Pure Appl. Chem.* **1972**, 31, 283–322. Also available in book form from Butterworths, London, 1972. Published previously as tentative rules in *Biochemistry* **1969**, 8, 2227–42; (amendments *ibid.* **1971**, 10, 4994).[2] (These publications supersede rules in *J. Am. Chem. Soc.* **1960**, 82, 5577–81.[2])

6. ——, "Nomenclature of Carotenoids (Rules Approved 1974)", *Pure Appl. Chem.* **1975**, 41, 405–31 (amendments *Biochemistry* **1975**, 14, 1803–4). Published previously as tentative rules in *Biochemistry* **1971**, 10, 4827–37. (These publications supersede rules in *J. Am. Chem. Soc.* **1960**, 82, 5583–4.[2])

7. ——, "Rules for Cyclitol Nomenclature (Recommendations 1973)". *Biochem. J.* **1976**, 153, 25–31. Published previously as tentative rules in *J. Biol. Chem.* **1968**, 243, 5809–19.

8. ——, "Tentative Rules for Carbohydrate Nomenclature, Pt. 1", *Biochemistry* **1971**, 10, 3983–4004 (errata *ibid.* **1971**, 10, 4995) (see also ¶ 302, item 14 below).

9. International Union of Pure and Applied Chemistry. Organic Chemistry Division. Commission on Physical Organic Chemistry, "Glossary of Terms Used in Physical Organic Chemistry (Recommendations 1982)", *Pure Appl. Chem.* **1983**, 55, 1281–371. Published previously as provisional recommendations in *Pure Appl. Chem.* **1979**, 51, 1725–801.

10. ——, "Nomenclature for Straightforward Transformations (Provisional)", *Pure Appl. Chem.* **1981**, 53, 305–21.

297. Biochemistry

1. International Union of Biochemistry, *Biochemical Nomenclature and Related Documents*, The Biochemical Society, London, 1978, 223 pp. A compendium of recommendations on biochemical nomenclature prepared for the International Union of Biochemistry by its Committee of Editors of Biochemical Journals with the full support of the International Union of Pure and Applied Chemistry. The compendium includes Sections E, F, and H of the IUPAC Organic Rules, ¶ 296, item 1; nomenclature recommendations given as items 5–8, ¶ 296; and many of the items listed below, in addition to the following: "Isotopically Labeled Compounds. Common Biochemical Practice" and "Recommendations for the Nomenclature of Human Immunoglobulins" (the latter also published in *Eur. J. Biochem.* **1974**, 45, 4–6).

2. International Union of Pure and Applied Chemistry and International Union of Biochemistry. Commission on Biochemical Nomenclature, "Nomenclature of Phosphorus–Containing Compounds of Biochemical Importance (Recommendations 1976)", *Eur. J. Biochem.* **1977**, 79, 1–9.

3. International Union of Pure and Applied Chemistry and International Union of Biochemistry. Joint Commission on Biochemical Nomenclature, "Nomenclature of Tetrapyrroles (Recommendations 1978)", *Pure Appl. Chem.* **1979**, 51, 2251–304. An amended version appears in *Eur. J. Biochem.* **1980**, 108, 1–30. An abbreviated version by P. Karlson was published in *Hoppe–Seyler's Z. Physiol. Chem.* **1981**, 362, vii–xii.

4. ——, "Conformational Nomenclature for Five and Six–Membered Ring Forms of Monosaccharides and their Derivatives (Provisional Recommendations)", *Pure Appl. Chem.* **1981**, 53, 1901–6. Published as "Recommendations 1980" in *Eur. J. Biochem.* **1980**, 111, 295–8. (a preliminary version was published: *J. Chem. Soc., Chem. Commun.* **1973**, 505–8. See ¶ 302, item 18)

5. ——, "Nomenclature of Branched–Chain Monosaccharides (Provisional Recommendations)", *Pure Appl. Chem.* **1982**, 54, 211–15. Published as "Recommendations 1980" in *Eur. J. Biochem.* **1981**, 119, 5–8 (corrections: *ibid.* **1982**, 125, 1).

6. ——, "Nomenclature of Unsaturated Monosaccharides (Provisional Recommendations)", *Pure Appl. Chem.* **1982**, 54, 207–10. Published as "Recommendations 1980" in *Eur. J. Biochem.* **1981**, 119, 1–3 (corrections: *ibid.* **1982**, 125, 1).

7. ——, "Abbreviated Terminology of Oligosaccharide Chains, Provisional", *Pure Appl. Chem.* **1982**, 54, 1517–22. Published as "Recommendations 1980" in *J. Biol. Chem.* **1982**, 257, 3347–51.

8. ——, "Polysaccharide Nomenclature (Provisional)", *Pure Appl. Chem.* **1982**, 54, 1523–6. Published as "Recommendations 1980" in *J. Biol. Chem.* **1982**, 257, 3352–4.

9. ——, "Symbols for Specifying the Conformation of Polysaccharide Chains (Provisional)", *Pure Appl. Chem.* **1983**, 55, 1269–72 and as

"Recommendations 1981," in *Eur. J. Biochem.* **1983**, 131, 5–7.

Amino Acids, Peptides, Proteins, and Nucleic Acids

10. International Union of Pure and Applied Chemistry and International Union of Biochemistry. Commission on Biochemical Nomenclature, "Abbreviated Nomenclature of Synthetic Polypeptides (Polymerized Amino Acids) (Revised Rules)", *Pure Appl. Chem.* **1973**, 33, 437–44. Published previously as revised recommendations in *Biochemistry* **1972**, 11, 942–4.

11. ———, "Abbreviations and Symbols for Nucleic Acids, Polynucleotides and their Constituents (Rules Approved 1974)", *Pure Appl. Chem.* **1974**, 40, 277–90. Published previously as revised rules in *Biochemistry* **1970**, 9, 4022–7.

12. ———, "Nomenclature of Peptide Hormones (Recommendations 1974)", *Biochemistry* **1975**, 14, 2559–60.

13. International Union of Pure and Applied Chemistry and International Union of Biochemistry. Joint Commission on Biochemical Nomenclature. "Nomenclature and Symbolism for Amino Acids and Peptides (Recommendations 1983)," *Pure Appl. Chem.* **1984**, 56, 595–624. A combination and revision of four previously published recommendations: (a) "Nomenclature of α–Amino Acids (Recommendations 1974)", *Biochemistry* **1975**, 14, 449–62; published previously as provisional recommendations in *IUPAC Inf. Bull. Append.* No. 46, September 1975 (these publications supersede rules published in *J. Am. Chem. Soc.* **1960**, 82, 5575–7[2]); (b) "Symbols for Amino Acid Derivatives and Peptides (Rules Approved 1974)", *Pure Appl. Chem.* **1974**, 40, 315–31; published previously as recommendations in *Biochemistry* **1972**, 11, 1726–32; (c) "Naming Synthetic Modifications of Natural Peptides (Definitive Rules)", *Pure Appl. Chem.* **1972**, 31, 647–53; published previously as tentative rules in *Biochemistry* **1967**, 6, 362–4; (d) "A One–Letter Notation for Amino Acid Sequences (Definitive Rules)", *Pure Appl. Chem.* **1972**, 31, 639–45; published previously as tentative rules in *Biochemistry* **1968**, 7, 2703–15.

14. ———, "Abbreviations and Symbols for the Description of the Conformation of Polypeptide Chains (Provisional)", *Pure Appl. Chem.* **1983**, 55, 1273–80 and as "Recommendations 1982" in *Eur. J. Biochem.* **1983**, 131, 9–15. Previously published as "Rules Approved 1974" in *Pure Appl. Chem.* **1974**, 40, 291–308, and as tentative rules in *Biochemistry* **1970**, 9, 3471–9.

15. ———, "Nomenclature of Glycoproteins, Glycopeptides, and Peptidoglycans (Recommendations 1985)", *J. Biol. Chem.* **1987**, 262, 13–8.

16. International Union of Biochemistry. Nomenclature Committee, "Nomenclature of Iron–Sulfur Proteins (Recommendations 1978)", *Eur. J. Biochem.* **1979**, 93, 427–30 (corrections: *ibid.* **1979**, 95, 369 and **1979**, 102, 315). Published previously as recommendations in *Biochemistry* **1973**, 12, 3582–3.

17. ———, "Nomenclature for Incompletely Specified Bases in Nucleic Acid Sequences (Recommendations 1984)", *Eur. J. Biochem.* **1985**, 150, 1–5 (corrections: *ibid.* **1986**, 157, 1).

Enzymes

18. International Union of Biochemistry, *Enzyme Nomenclature, 1984*, Academic Press, New York, 1984, 650 pp.; Supplement 1: Corrections and Additions, *Eur. J. Biochem.* **1986**, 157, 1–26. This edition includes corrections and additions to "Enzyme Nomenclature, Recommendations 1978" published as four supplements in *Eur. J. Biochem.* **1980**, 104, 1–4; **1981**, 116, 423–35; **1982**, 125, 1–13; **1983**, 131, 461–72.

19. ———, "Multienzyme Proteins", *Trends. Biochem. Sci.* **1979**, 4, N275.

20. International Union of Pure and Applied Chemistry and International Union of Biochemistry. Commission on Biochemical Nomenclature, "The Nomenclature of Multiple Forms of Enzymes (Recommendations 1976)", *Eur. J. Biochem.* **1978**, 82, 1–3. Published previously as "Rules Approved 1974" in *Pure Appl. Chem.* **1974**, 40, 309–14.

21. International Union of Biochemistry. Nomenclature Committee, "Symbolism and Terminology in Enzyme Kinetics (Recommendations 1981)", *Eur. J. Biochem.* **1982**, 128, 281–91.

Lipid, Vitamins, Coenzymes, and Related Compounds

22. International Union of Pure and Applied Chemistry and International Union of Biochemistry. Commission on Biochemical Nomenclature, "The Nomenclature of Lipids (Recommendations 1976)", *Eur. J. Biochem.* **1977**, 79, 11–21. Previously published as recommendations in *J. Biol. Chem.* **1967**, 242, 4845–9 (amendments: *ibid.* **1970**, 245, 1511).

23. ———, "Trivial Names of Miscellaneous Compounds of Importance in Biochemistry (Tentative Rules)", *J. Biol. Chem.* **1966**, 241, 2987–8. (Supersedes "Definitive Rules for the Nomenclature of the Vitamins", *J. Am. Chem. Soc.* **1960**, 82, 5581–3.[2])

24. ———, "Nomenclature of Quinones with Isoprenoid Side Chains (Recommendations 1973)", *Eur. J. Biochem.* **1975**, 53, 15–18. Published previously as tentative rules in *J. Biol. Chem.* **1966**, 241, 2989–91. (Supersedes Rule V–4 in *J. Am. Chem. Soc.* **1960**, 82, 5581–3.[2])

25. ———, "Nomenclature and Symbols for Folic Acid and Related Compounds (Tentative Rules)", *J. Biol. Chem.* **1966**, 241, 2991–2. (Supersedes Rule V–13 in *J. Am. Chem. Soc.* **1960**, 82, 5581–3.[2])

26. ———, "Nomenclature of Corrinoids (Rules Approved 1975)", *Pure Appl. Chem.* **1976**, 48, 495–502. Published previously as recommendations in *Biochemistry* **1974**, 13, 1555–60 and as tentative rules in *J. Biol. Chem.* **1966**, 241, 2992–4. (Supersedes Rule V–15 in *J. Am. Chem. Soc.* **1960**, 82, 5581–3.[2])

27. ———, "Nomenclature for Vitamins B_6 and Related Compounds (Recommendations 1973)", *Biochemistry* **1974**, 13, 1056–8. Published previously as tentative rules in *Biochemistry* **1970**, 9, 4019–21. (Supersedes Section M–7 in *J. Biol. Chem.* **1966**, 241, 2987–8; see item 23 above.)

28. International Union of Pure and Applied Chemistry and International Union of Biochemistry. Joint Commission on Biochemical Nomenclature, "Nomenclature of Tocopherols and Related Compounds (Recommendations 1981)", *Pure Appl. Chem.* **1982**, 54, 1507–10 and in *Eur. J. Biochem.* **1982**, 123, 473–5. Published previously as "Recommendations 1973" in *Eur. J. Biochem.* **1974**, 46, 217–19 and as provisional recommendations in *IUPAC Inf. Bull. Append.* No. 47, September, 1975. (Supersedes Section M–3 in *J. Biol. Chem.* **1966**, 241, 2987–8; see item 23 above.)

29. ———, "Nomenclature of Vitamin D (Provisional)", *Pure Appl. Chem.* **1982**, 54, 1511–6. Published as "Recommendations 1981" in *Eur. J. Biochem.* **1982**, 124, 223–7.

30. ———, "Nomenclature of Retinoids (Provisional)", *Pure Appl. Chem.* **1983**, 55, 721–6. Published as "Recommendations 1981" in *Eur. J. Biochem.* **1982**, 129, 1–5.

Miscellaneous

31. International Union of Pure and Applied Chemistry and International Union of Biochemistry. Commission on Biochemical Nomenclature, "Abbreviations and Symbols for Chemical Names of Special Interest in Biological Chemistry (Revised Tentative Rules 1965)", *Biochemistry* **1966**, 5, 1445–53.

32. ———, "Abbreviations and Symbols", *Eur. J. Biochem.* **1977**, 74, 1–5.

33. International Union of Nutritional Sciences. Committee 1/I. Nomenclature, "Generic Descriptors and Trivial Names for Vitamins and Related Compounds (Recommendations 1976)", *Nutrition Absts. and Revs.*, Series A: *Human and Experimental* **1978**, 48, 831–5.

298. Inorganic chemistry

1. International Union of Pure and Applied Chemistry. Inorganic Chemistry Division. Commission on the Nomenclature of Inorganic Chemistry, *Nomenclature of Inorganic Chemistry*, 2nd ed., Butterworths, London, 1971, 110 pp. This edition revises and extends the first edition (1957 Rules), published in 1959. (The 1957 Inorganic Rules were published with comments in *J. Am. Chem. Soc.* **1960**, 82, 5523–44.[2]) The 1971 edition incorporates "Tentative Proposals for Nomenclature of Absolute Configurations Concerned with Six–Coordinated Complexes Based on the Octahedron", *Inorg. Chem.* **1970**, 9, 1–5;[2] π–Complexes (Tentative), *IUPAC Inf. Bull.* No. 35, June, 1969, pp. 31–35; and Iso– and Heteropolyanions (Rules 5.1 and 5.2, Tentative), *IUPAC Inf. Bull.* No. 35, June, 1969, pp. 22–27.

2. ———, "Nomenclature of Inorganic Boron Compounds", *Pure Appl. Chem.* **1972**, 30, 681–710 (see also ¶ 302, item 16 below).

3. ———, "How to Name an Inorganic Substance. A Guide to the Use of Nomenclature of Inorganic Chemistry 2nd ed., 1971", Pergamon Press, New York, 1977.

4. ———, "Nomenclature of Hydrides of Nitrogen and Derived Cations, Anions, and Ligands (Recommendations 1981)", *Pure Appl. Chem.* **1982**, 54, 2545–52. Published previously as provisional recommendations in *IUPAC Inf. Bull.* **1978**, (2), 151–60.

5. ———, "Recommendations for the Naming of Elements of Atomic Number Greater than 100 (Definitive)", *Pure Appl. Chem.* **1979**, 51, 381–4. Previously published as "Recommendations for the Naming of Elements of Atomic Numbers Greater Than 105 (Provisional)" in *IUPAC Inf. Bull. Append.* No. 55, December, 1976.

6. ———, "Nomenclature of Inorganic Chemistry: II.1. Isotopically Modified Compounds (Recommendations 1981)", *Pure Appl. Chem.* **1981**, 53, 1887–900. Previously published as provisional recommendations in *Pure Appl. Chem.* **1979**, 51, 1981–94.

299. Macromolecular chemistry

1. International Union of Pure and Applied Chemistry. Physical Chemistry Division. Commission on Macromolecules. Subcommission on Nomenclature, "Report on Nomenclature in the Field of Macromolecules (1951)", *J. Polym. Sci.* **1952**, 8, 257–77. (modifications and additions (tentative), *J. Polym. Sci., Part B: Polymer Letters*, **1968**, 6, 257–60).

2. ———, "Report on Nomenclature Dealing with Steric Regularity in High Polymers", *Pure Appl. Chem.* **1966**, 12, 645–56 [Authors often cited as M. L. Huggins, G. Natta, V. Desreux, and H. Mark]; published previously as a preliminary version in *J. Polym. Sci.* **1962**, 56, 153–61.

3. International Union of Pure and Applied Chemistry. Division of Applied Chemistry. Plastics and High Polymers Section, "Recommendations for Abbreviations of Terms Relating to Plastics and Elastomers", *Pure Appl. Chem.* **1969**, 18, 583–9.

4. International Union of Pure and Applied Chemistry. Macromolecular Division. Commission on Macromolecular Nomenclature, "Nomenclature of Regular Single–Strand Organic Polymers (Rules Approved 1975)", *Pure Appl. Chem.* **1976**, 48, 373–85. Published previously as provisional recommendations in *Macromolecules* **1973**, 6, 149–58[2] and in *IUPAC Inf. Bull. Append.* No. 29, November, 1972 (see also ¶ 302, item 15 below).

5. ———, "List of Standard Abbreviations (Symbols) for Synthetic Polymers and Polymer Materials 1974", *Pure Appl. Chem.* **1974**, 40, 473–6.

6. ———, "Basic Definitions of Terms Relating to Polymers (1974)", *Pure Appl. Chem.* **1974**, 40, 477–91.

7. ———, "Stereochemical Definitions and Notations Relating to Polymers (Recommendations 1980)", *Pure Appl. Chem.* **1981**, 53, 733–52. Published previously as provisional recommendations in *Pure Appl. Chem.* **1979**, 51, 1101–21.

8. ———, "Source–based Nomenclature for Copolymers (Recommendations 1985)", *Pure Appl. Chem.* **1985**, 57, 1427–40.

9. International Union of Pure and Applied Chemistry. Macromolecular Division. Commission on Macromolecular Nomenclature and Inorganic Chemistry Division. Commission on Nomenclature of Inorganic Chemistry, "Nomenclature for Regular Single–Strand and Quasi–Single–Strand Inorganic and Coordination Polymers (Recommendations 1984)", *Pure Appl. Chem.*

1985, 57, 149–68. Published previously as provisional recommendations in *Pure Appl. Chem.* **1981**, 53, 2283–302.

300. Miscellaneous

1. International Union of Pure and Applied Chemistry. Interdivisional Committee on Nomenclature and Symbols, "Use of Abbreviations in the Chemical Literature (Recommendations 1979)", *Pure Appl. Chem.* **1980**, 52, 2229–32. Previously published as provisional recommendations jointly with the Physical Chemistry Division Commission on Molecular Structure and Spectroscopy in *IUPAC Inf. Bull. Append.* No. 58, July, 1977

2. International Union of Pure and Applied Chemistry. Analytical Chemistry Division, *Compendium of Analytical Nomenclature (Definitive Rules 1977)*, Pergamon Press, Oxford, 1978, 223 pp.

3. International Union of Pure and Applied Chemistry. Analytical Chemistry Division. Commission on Analytical Nomenclature, "Guide to Trivial Names, Trade Names, and Synonyms for Substances Used in Analytical Nomenclature", *Pure Appl. Chem.* **1978**, 50, 339–70. Previously published as provisional recommendations under the title "List of Trivial Names and Sy= nonyms (for Substances Used in Analytical Chemistry)" in *IUPAC Inf. Bull. Append.* No. 45, September, 1975.

4. International Union of Pure and Applied Chemistry. Physical Chemistry Division. Commission on Symbols, Terminology and Units, *Manual of Symbols and Terminology for Physicochemical Quantities and Units (1979 edition)*, *Pure Appl. Chem.* **1979**, 51, 1–41. Replaces 1973 edition published by Butterworths, London, 1975. The 1969 edition appeared in *Pure Appl. Chem.* **1970**, 21, 1–44. (Supersedes rules in *J. Am. Chem. Soc.* **1960**, 82, 5517–22.[2])

5. International Union of Pure and Applied Chemistry. Physical Chemistry Division. Commission on Colloid and Surface Chemistry, "Chemical Nomenclature and Formulation of Compositions of Synthetic and Natural Zeolites (Definitive)", *Pure Appl. Chem.* **1979**, 51, 1091–100. Previously published as provisional recommendations jointly with the Inorganic Chemistry Division Commission on Nomenclature of Inorganic Chemistry in *IUPAC Inf. Bull. Append.* No. 41, January, 1975.

6. International Union of Pure and Applied Chemistry; "Compendium of Chemical Terminology (IUPAC Recommendations), V. Gold, K. L. Loening, A. D. McNaught, P. Sehmi, Compilers, Blackwell Scientific Publications, Oxford, 1987, 456 pp.

American Chemical Society Nomenclature

301. Chemical Abstracts Index Nomenclature

1. American Chemical Society. Chemical Abstracts Service, "The Naming and Indexing of Chemical Compounds from Chemical Abstracts" [a reprint of the Introduction to the *Subject Index* to Volume 56 (January–June, 1962)], Chemical Abstracts Service, Columbus, Ohio, 1962. A previous edition was reprinted from the Introduction to the 1945 *Subject Index* (*Chemical Abstracts* **1945**, 39, 5867–975) which was combined with revisions introduced in 1957 (*Subject Index* **1957**, 51, 1R–28R).

2. ——, "Combined Introductions: Subject, Formula, Ring System, and HAIC Indexes to Volume 66, January–June, 1967", Chemical Abstracts Service, Columbus, Ohio, 1968.

3. ——, "Naming and Indexing of Chemical Substances for *Chemical Abstracts* during the Ninth Collective Period (1972–1976)" [a reprint of Section IV (Selection of Index Names for Chemical Substances) from the Volume 76 *Index Guide*], Chemical Abstracts Service, Columbus, Ohio, 1973.

4. ——, "Selection of Index Names for Chemical Substances" in the Volume 76–85 Cumulative *Index Guide* (1972–1976), Chemical Abstracts Service, Columbus, Ohio, 1977, Appendix IV.

5. ——, "Naming and Indexing of Chemical Substances for *Chemical Abstracts*" [a reprint of Appendix IV (Selection of Index Names for Chemical Substances) from the 1977 *Index Guide*], Chemical Abstracts Service, Co= lumbus, Ohio, 1978.

6. ——, "Selection of Index Names for Chemical Substances" in the 1977–1981 *Index Guide*, Chemical Abstracts Service, Columbus, Ohio, 1982, Appendix IV.

7. ——, "Naming and Indexing of Chemical Substances for *Chemical Abstracts*" [a reprint of Appendix IV (Chemical Substance Index Names) from the 1982 *Index Guide*], Chemical Abstracts Service, Columbus, Ohio, 1982.

8. ——, "Chemical Substance Index Names" in the 1984 *Index Guide*, Chemical Abstracts Service, Columbus, Ohio, 1984, Appendix IV.

9. ——, "Naming and Indexing of Chemical Substances for *Chemical Abstracts*" [a reprint of Appendix IV (Chemical Substance Index Names) from the 1985 *Index Guide*], Chemical Abstracts Service, Columbus, Ohio, 1985.

10. A. M. Patterson, L. T. Capell, and D. F. Walker, *The Ring Index*, 2nd ed., American Chemical Society, Washington, D.C., 1960. Introduction. pp. ix–xxvi; Suppl. 1, 1963, Introduction, pp. ix–xii.

11. J. E. Blackwood, C. L. Gladys, K. L. Loening, A. E. Petrarca, and J. E. Rush, "Unambiguous Specification of Stereoisomerism about a Double Bond", *J. Am. Chem. Soc.* **1968**, 90, 509–10.[2]

12. J. E. Blackwood, C. L. Gladys, A. E. Petrarca, W. H. Powell, and J. E. Rush, "Unique and Unambiguous Specification of Stereoisomerism about a Double Bond in Nomenclature and Other Notation Systems", *J. Chem. Doc.* **1968**, 8, 30–2.[2]

13. K. L. Loening, W. Metanomski, and W. H. Powell, "Indexing of Polymers in *Chemical Abstracts*", *J. Chem. Doc.* **1969**, 9, 248–51.

14. N. Donaldson, W. H. Powell, R. J. Rowlett, Jr., R. W. White, and K. Yorka, "*Chemical Abstracts* Index Names for Chemical Substances in the Ninth Collective Period (1972–76)", *J. Chem. Doc.* **1974**, 14, 3–15.[2]

15. R. J. Rowlett, Jr. and F. A. Tate, "A Computer-Based System for Handling Chemical Nomenclature and Structural Representations", *J. Chem. Doc.* **1972**, 12, 125–8.[2]

16. R. J. Rowlett, Jr. and D. W. Weisgerber, "Handling Commercial Product Names at Chemical Abstracts Service", *J. Chem. Doc.* **1974**, 14, 92–5.

17. J. E. Blackwood and P. M. Giles, Jr., "*Chemical Abstracts* Stereo= chemical Nomenclature of Organic Substances in the Ninth Collective Period (1972–1976)", *J. Chem. Inf. Comput. Sci.* **1975**, 15, 67–72.[2]

18. M. F. Brown, B. R. Cook, and T. E. Sloan, "Stereochemical Notation in Coordination Chemistry. Mononuclear Complexes", *Inorg. Chem.* **1975**, 14, 1273–8.[2]

19. M. F. Brown, B. R. Cook, and T. E. Sloan, "Stereochemical Notation in Coordination Chemistry. Mononuclear Complexes of Coordination Numbers Seven, Eight, and Nine", *Inorg. Chem.* **1978**, 17, 1563–8.[2]

302. American Chemical Society Committee Reports

1. American Chemical Society, "Recommendations of the Nomenclature Committee of the Organic Division of the ACS to the Nomenclature Committee of the National Society", *J. Am. Chem. Soc. Proc.* **1931**, 40–1.

2. ——, "The Pronunciation of Chemical Words", *Ind. Eng. Chem., News Ed.* **1934**, 12, 202–5.[2]

3. ——, "Nomenclature of the Hydrogen Isotopes and Their Compounds", *Ind. Eng. Chem., News Ed.* **1935**, 13, 200–1.[2]

4. ——, "Report of the Committee for the Revision of the Nomencla= ture of Pectic Substances", *Chem. Eng. News* **1944**, 22, 105–6.

5. ——, "Nomenclature of Carotenoid Pigments", *Chem. Eng. News* **1946**, 24, 1235–6.

6. ——, "The Naming of *cis* and *trans* Isomers of Hydrocarbons Containing Olefin Double Bonds", *Chem. Eng. News* **1949**, 27, 1303.[2]

7. ——, "The Designation of 'Extra' Hydrogen in Naming Cyclic Compounds", *Chem. Eng. News* **1949**, 27, 1303.[2]

8. ——, "The Naming of Geometric Isomers of Polyalkyl Monocyclo= alkanes", *Chem. Eng. News* **1950**, 28, 1842–3.[2]

9. ——, "The Report of the ACS Nomenclature, Spelling, and Pron= unciation Committee for the First Half of 1952", *Chem. Eng. News* **1952**, 30, 4513–26.

A. "Arene and Arylene", p. 4513.[2]
B. "Halogenated Derivatives of Hydrocarbons", pp. 4513–4.[2]
C. "The Use of 'Per' in Naming Halogenated Organic Compounds", pp. 4514–5.[2]
D. "The Use of 'H' to Designate the Positions of Hydrogens in Almost Completely Fluorinated Organic Compounds", p. 4515.[2]
E. "Organic Compounds Containing Phosphorus", pp. 4515–22.[2]
F. "Organosilicon Compounds", pp. 4517–22.[2] (These rules su= persede the ACS 1946 rules which appeared in *Chem. Eng. News* **1946**, 24, 1233–4.)
G. "Nomenclature of Natural Amino Acids and Related Substances", pp. 4522–6.[2] (These rules were reprinted in *J. Am. Chem. Soc.* **1960**, 82, 5575–7. They replaced the 1947 ACS rules which appeared in *Chem. Eng. News* **1947**, 25, 1364–7. See also IUPAC–IUB rules, ¶ 297, item 13.)

10. ——, "A New General System for the Naming of Stereoisomers", G. E. McCasland, included with the "Report of the Advisory Committee on Configurational Nomenclature", July 14, 1953, H. B. Vickery, Chairman, to the Nomenclature Committee, Division of Organic Chemistry, American Chemical Society.

11. ——, "Addendum to Definitive Rules for the Nomenclature of Natural Amino Acids and Related Substances", *J. Org. Chem.* **1963**, 28, 291–3.[2]

12. ——, "A Proposed System of Nomenclature for Terpene Hydro= carbons", *Chem. Eng. News* **1954**, 32, 1795–7.[2]

13. ——, "A System of Nomenclature for Terpene Hydrocarbons, Acyclics, Monocyclics, Bicyclics", *Advances in Chemistry Series* No. 14, 1955, 1–98.

14. ——, "Rules of Carbohydrate Nomenclature", *J. Org. Chem.* **1963**, 28, 281–91.[2] (This report was the result of a cooperative effort by the No= menclature Committee of the Division of Carbohydrate Chemistry of the American Chemical Society and a British Committee on Carbohydrate No= menclature [a Subcommittee of the Publications Committee of the Chemical Society, London]. They are sometimes referred to as the "Anglo-American Rules" and were accepted by the Council of the American Chemical Society in 1962. They replaced the ACS 1953 rules which appeared in *Chem. Eng. News* **1953**, 31, 1776–82, which in turn replaced the ACS 1948 Carbohydrate Nomenclature Rules published in *Chem. Eng. News* **1948**, 26, 1623–9. These rules precede the tentative IUPAC carbohydrate rules. See ¶ 296, item 8.)

15. ——, "A Structure-Based Nomenclature for Linear Polymers", *Macromolecules* **1968**, 1, 193–8. These rules precede the IUPAC rules; see ¶ 299, item 4.

16. ——, "The Nomenclature of Boron Compounds", *Inorg. Chem.* **1968**, 7, 1945–64. These rules precede the IUPAC rules; see ¶ 298, item 2.

17. ——, "ACS Committee on Nomenclature Recommendations", *J. Chem. Doc.* **1973**, 13, 45 (News and Notes) (On the use of the diagonal, the hyphen, and the dash in chemical names).

18. American Chemical Society. Division of Carbohydrate Chemistry. Nomenclature Committee and the British Carbohydrate Nomenclature Com= mittee, "Rules for Conformational Nomenclature for Five- and Six-Membered Rings in Monosaccharides and Their Derivatives", *J. Chem. Soc., Chem. Commun.* **1973**, 505–8. These rules precede the IUPAC recommendations; see ¶ 297, item 4.

19. J. A. Young, "Revised Nomenclature for Highly Fluorinated Organic Compounds", *J. Chem. Doc.* **1974**, 14, 98–100.

20. W. C. Fernelius and W. H. Powell, "Confusion in the Periodic Table

of the Elements", *J. Chem. Educ.* **1982**, 59, 504–8.

21. K. L. Loening "Recommended Format for the Periodic Table of the Elements", *J. Chem. Educ.* **1984**, 61, 136.

History and Development of Chemical Nomenclature

303. General

1. E. J. Crane, "The Standardization of Chemical Nomenclature", *J. Chem. Educ.* **1931**, 8, 1335–40.[2] (Contains references.)
2. A. D. Mitchell, *British Chemical Nomenclature*, Edward Arnold & Co., London, 1948, 156 pp.
3. E. J. Crane, "Chemical Nomenclature in the United States" in *Chemical Nomenclature*, American Chemical Society, Washington, D.C., 1953, 55–64.
4. A. M. Patterson, *Words About Words*, American Chemical Society, Washington, D.C., 1957, 86 pp. (A compilation of the column by A. M. Patterson which appeared in *Chem. Eng. News* 1951–1956.)
5. L. T. Capell, "Effect of Changes in Nomenclature on the Use of Indexes" in *Searching the Chemical Literature*, American Chemical Society, Washington, D.C., 1961, 58–66. (*Advances in Chemistry Series* No. 30.)
6. J. D. Scott and K. L. Loening, "Inorganic–Organic Nomenclature", *J. Chem. Doc.* **1964**, 4, 66–9.
7. O. C. Dermer, G. Gorin, and K. L. Loening, "The Standardization of Chemical Language", *Int. J. Social. Lang.* **1976**, 11, 61–83.
8. R. S. Cahn and O. C. Dermer, *Introduction to Chemical Nomenclature*, 5th, ed., Butterworths, London, 1979, 200 pp.
9. K. L. Loening, "Nomenclature" in *Kirk–Othmer Encyclopedia of Chemical Technology*, 3rd ed., John Wiley and Sons, New York, N.Y., 1981, Vol. 16, pp. 28–46; 2nd ed., Vol. 14, pp. 1–15.
10. R. Lees and A. F. Smith eds., *Chemical Nomenclature Usage*, Ellis Horwood, Ltd., Chichester, England, 1983, 172 pp.

304. Organic Chemistry

Geneva Rules

1. A. Pictet, "Le Congres International de Geneve pour la Reforme de la Nomenclature Chimique", *Arch. Sci. Phys. Nat.* **1892**, [3] 27, 485–520.
2. M. Hanriot, "Congres de Nomenclature Chimique, Geneve 1982", *Bull. Soc. Chim. Paris Suppl.* **1892**, s. 3, 7, XIV–XXIV.
3. F. Tiemann, "Ueber die Beschluesse des Internationalen, in Genf von 19 bis 22 April 1892, versammelten Congresses zur Regelung der chemischen Nomenklatur", *Ber. Dtsch. Chem. Ges.* **1893**, 26, 1595–631.

Liege Rules

1. International Union of Chemistry. Commission on the Reform of the Nomenclature of Organic Chemistry, "Definitive Report, Liege Meeting, 1930". (Translation with comments by A. M. Patterson, *J. Am. Chem. Soc.* **1933**, 55, 3905–25. Appeared also in English in *J. Chem. Soc.* **1931**, 1607–16.)
2. International Union of Chemistry. Commission on the Reform of the Nomenclature of Organic Chemistry, "Full Text of the New Organic Nomenclature Rules". Reported by A. M. Patterson, *Science* **1938**, 87, 215–6. [Rules 34 and 49 of the Liege Report are given in full.]

Amsterdam Rules

1. International Union of Pure and Applied Chemistry. Organic Chemistry Division. Commission on the Nomenclature of Organic Chemistry, "Nomenclature of Organosilicon Compounds", *Compt. Rend. de la Quinzieme Conference*, Amsterdam, 1949, 127–32.[2]
2. ———, "Changes and Additions to the Definitive Report [of the Liege Meeting]", *Compt. Rend. de la Quinzieme Conference*, Amsterdam, 1949, 132–85.[2]

Proposals and Reviews

1. A. Hantsch and J. H. Weber, *Ber. Dtsch. Chem. Ges.* **1887**, 20, 3118–32.
2. O. Widman, *J. Prakt. Chem.* **1888**, 38, 185–201.
3. A. Baeyer, "Systematik und Nomenklatur Bicyclischer Kohlenwasserstoffe", *Ber. Dtsch. Chem. Ges.* **1900**, 33, 3771–5.
4. E. Buchner und W. Weigand, "Bornylen und Diazoessigester [Nebst einer Nomenklatur Tricyclischer Kohlenstoff–Ringsysteme nach Adolf von Baeyer]", *Ber. Dtsch. Chem. Ges.* **1913**, 46, 2108–17.
5. A. M. Patterson and C. E. Curran, "A System of Organic Nomenclature", *J. Am. Chem. Soc.* **1917**, 39, 1623–38.
6. A. M. Patterson, "Proposed International Rules for Numbering Organic Ring Systems", *J. Am. Chem. Soc.* **1925**, 47, 543–61.
7. ———, "The Nomenclature of Parent Ring Systems", *J. Am. Chem. Soc.* **1928**, 50, 3074–87.
8. ———, "The Nomenclature of Organic Compounds of Complex Function", *Rec. Trav. Chim.* **1929**, 48, 1012–17.
9. P. E. Verkade, "La revision recente de la nomenclature des combinaisons organiques", *Rec. Trav. Chim.* **1932**, 51, 185–217.
10. A. M. Patterson and L. T. Capell, "The Ring Index", Reinhold, New York, 1940, pp. 20–7.
11. L. T. Capell and K. L. Loening, "Principles of Organic Nomenclature" in *Treat. Anal. Chem.*, ed. I. M. Kolthoff and P. J. Elving, Interscience, New York, 1965, Vol. 11, Part II, Section B, 1–44.
12. D. R. Eckroth, "Method for Manual Generation of Correct von Baeyer Names of Polycyclic Hydrocarbons", *J. Org. Chem.* **1967**, 32, 3362–5.
13. A. D. McNaught, "The Nomenclature of Heterocycles" in *Advances in Heterocyclic Chemistry*, ed. A. R. Katritzky and A. J. Boulton, Academic Press, New York, N.Y., 1976, Vol. 20, pp. 175–247.
14. N. Lozac'h, A. L. Goodson, and W. H. Powell, "Nodal Nomenclature – General Principles", *Angew. Chem.* **1979**, 91, 951–64; *Angew. Chem., Int. Ed. Engl.* **1979**, 18, 887–99.
15. A. L. Goodson, "Graph Based Chemical Nomenclature. 1. Historical Background and Discussion", *J. Chem. Inf. Comput. Sci.* **1980**, 20, 167–72; "2. Incorporation of Graph–Theoretical Principles into Taylor's Nomenclature Proposal", *ibid.* **1980**, 20, 172–6.
16. K. L. Loening, "Hydrocarbons: Nomenclature" in *Kirk–Othmer Encyclopedia of Chemical Technology*, 3rd ed., John Wiley and Sons, Inc., New York, N.Y., 1980, Vol. 12, pp. 892–900; 2nd ed., 1966, Vol. 11, pp. 288–293.
17. D. Eckroth, "Nitrogen's Hydrido Oxo Acids", *J. Chem. Inf. Comput. Sci.* **1983**, 23, 157–9.
18. A. L. Goodson, "Application of Graph–based Chemical Nomenclature to Theoretical and Preparative Chemistry," *Croat. Chem. Acta* **1983**, 56, 315–24.
19. N. Lozac'h and A. L. Goodson, "Nodal Nomenclature II – Specific Nomenclature for Parent Hydrides, Free Radicals, Ions, and Substituents", *Angew. Chem.* **1984**, 96, 1–15; *Angew. Chem., Int. Ed. Engl.* **1984**, 23, 33–46.
20. A. D. McNaught and P. A. S. Smith, "Nomenclature of Heterocyclic Compounds" in *Comprehensive Heterocyclic Chemistry*, ed. A. R. Katritzky and C. W. Rees, Pergamon Press, Oxford, England, 1984, pp. 7–45.
21. A. L. Goodson and N. Lozac'h, "Advantages of Nodal Nomenclature for Uniquely Identifying Atoms in Chemical Nomenclature", *Croat. Chem. Acta* **1986**, 59, 547–63.

Books

1. R. Stelzner and H. Kuh, *Nomenklatur–Fragen in der Organischen Chemie*, G.m.b.H., Leipzig–Berlin, 1921.
2. A. P. Terent'ev, A. N. Kost, A. M. Tsukerman, and V. M. Potapov, *Nomenklatura Organicheskikh Soedinenii, Obzor, Kritika, Predlozheniya* (Nomenclature of Organic Compounds, Review, Comments, Proposals), Izdatel'stvo Akademii Nauk SSSR, Moscow, 1955, 302 pp.
3. N. Lozac'h, *La Nomenclature en Chimie Organique*, Masson et Cie Editeurs, Paris, 1967 (*Monographs de Chimie Organique*, Vol. VI; Complements au *Traite de Chimie Organique*, 294 pp.)
4. J. H. Fletcher, O. C. Dermer, and R. B. Fox, *Nomenclature of Organic Compounds, Principles and Practice*, American Chemical Society, Washington, D.C., 1974, 337 pp. (*Advances in Chemistry Series* No. 126.)
5. D. Hellwinkel, *Die Systematische Nomenklatur der Organischen Chemie*, 3rd edition, Springer–Verlag, Berlin, 1982, 170 pp.
6. P. E. Verkade, *A History of the Nomenclature of Organic Chemistry*, ed. F. C. Alderweireldt, et. al., D. Reidel Publishing Co., Dordrecht, The Netherlands, 1985, 528 pp. A collection of translated articles which appeared as follows in *Bull. Soc. Chim. Fr.*: I. "La naissance de la nomenclature de Geneve (1892) et des propositions de nomenclature de Saint–Etienne (1897) et la litterature afferente", *Bull. Soc. Chim. Fr.* **1966**, 1807–12; II. "Le developpement de la nomenclature de chimie organique pendant la periode de 1889–1897", *ibid.*, **1967**, 4009–20; III. "Faits et evenements relatifs a la nomenclature de la chimie organique pendant la periode 1897–1911", *ibid.*, **1968**, 1358–67; IV. "L'activite de l'Association Internationale de Societes Chimiques dans le domaine de la nomenclature de la chimie organique, 1911–1914", *ibid.*, **1969**, 3877–81; V. "Activites individuelles dans le domaine de la nomenclature de la chimie organique pendant la periode de 1914 a 1930", *ibid.*, **1969**, 4297–308; VI. "La creation de la nomenclature de Liege", *ibid.*, **1970**, 2739–46; VII. "Le contenu de la nomenclature de Liege compare a celui de la nomenclature de Geneve", *ibid.*, **1971**, 1634–47; VIII. "Faits et evenements relatifs a la nomenclature de la chimie organique pendant la periode de 1930 a 1946", *ibid.*, **1971**, 4299–307; IX. "Sur l'activite de la Commission Internationale pendant la periode de 1946 a 1957", *ibid.*, **1973**, 1961–71; X. "Sur la contenu des Sections A et B de l'*IUPAC Nomenclature of Organic Chemistry*", *ibid.*, **1975**, 555–70; XI. "Sur le procede de Combes (1892) pour la nomenclature des composes acycliques a fonction complexe", *ibid.*, **1975**, 1119–26; XII. "Sur l'histoire de la creation de la Section C de l'*IUPAC Nomenclature of Organic Chemistry*", *ibid.*, **1975**, 2029–36; XIII. "Un Livre Roumain tres remarquable sur la nomenclature de la chimie organique, publie en 1913 par C. I. Istrati", *ibid.* **1976**, 1445–64; XIV. "Sur le contenu de la Section C de l'*IUPAC Nomenclature of Organic Chemistry*", *ibid.*, **1977**, 457–78; XV. "L'oeuvre de Siboni, sur la nomenclature de la chimie", *ibid.*, **1978**, II, 13–31; XVI. "Historique de la Section D de l'*IUPAC Nomenclature of Organic Chemistry*", *ibid.*, **1979**, 215–29.
7. *Handbuch zur Anwendung der Nomenclatur Organisch–chemischer Verbindungen*, ed. Wolfgang Liebscher, Akademie–Verlag, Berlin, 1979.
8. "Guiding Principles for Coining U.S. Adopted Names" in *USAN and the USP Dictionary of Drug Names*", ed. M. C. Griffiths, United States Pharmacopeial Convention, Inc., Rockville, Maryland, 1984, Appendix I, pp. 621–4.
9. K. Hirayama, *The HIRN System: Nomenclature of Organic Chemistry*, Springer–Verlag, Berlin, 1984, 149 pp.
10. J. Heger, *Ako Tvorit Nazvy Organickych Zlucenin* (How to Name Organic Compounds), Slovenske Pedagogicke Nakladatel'stvo, Bratislava, 1985, 163 pp.

305. Biochemistry

1. Reports of the International Union of Chemistry. Commission on the Reform of the Nomenclature of Biological Chemistry:
   - C. Belgium, "La nomenclature de chimie biologique: une reforme heureuse", *Schweiz. Apoth. Z.* **1928**, 66, 193–7.
   - G. Bertrand, "Project de reforme de la nomenclature de chimie biolo=

gique (rapport presente a la Commission Internationale en 1922)", *Bull. Soc. Chim. Biol.* **1923**, 5, 96–109.

G. Bertrand, "Propositions relatives a la nomenclature des glucides", *Bull. Soc. Chim. Biol.* **1927**, 9, 854–6.

M. Bridel, "La reforme de la nomenclature de chimie biologique. Etat de la question a la fin de 1926", *Bull. Soc. Chim. Biol.* **1926**, 8, 1211–16.

2. J. E. Courtois, "Work of Commission on Nomenclature of Biological Chemistry" in *Chemical Nomenclature*. American Chemical Society, Washington, D.C., 1953, 83–94. (*Advances in Chemistry Series* No. 8.)

3. W. E. Cohn "Representation of Macromolecules and Polymers of Biological Importance", *J. Chem. Doc.* **1969**, 9, 235–41.

4. N. A. Nelson, "Prostaglandin Nomenclature," *J. Med. Chem.* **1974**, 17, 911–8.

5. B. M. Benjamin, "Nomenclature in Genetic Engineering", *CBE Views* **1986**, 9, 48–50.

306. Inorganic Chemistry

1. A. Werner, *New Ideas on Inorganic Chemistry*, translated by E. P. Headley from 2nd German Ed., Longmans, Green, and Co., London, 1911. (See also A. Werner, *Z. Anorg. Chem.* **1897**, 14, 23, and A. Werner and A. Klein, *Z. Anorg. Chem.* **1897**, 14, 28.)

2. A. Stock, "Einige Nomenklaturfragen der Anorganischen Chemie", *Z. Angew. Chem.* **1919**, 32(I), 373–6.

3. R. J. Meyer and A. Rosenheim, "Die Vorschlaege der Deutschen Nomenklaturkommission fuer Anorganische Chemie", *Z. Angew. Chem.* **1925**, 38, 713–5. (See also *Chem. Weekbl.* **1926**, 23, 93–6.)

4. M. Delepine, "Revision of Inorganic Chemical Nomenclature, Report of the IUPAC Committee for the Reform of the Nomenclature of Inorganic Chemistry", *Chem. Weekblad.* **1926**, 23, 86–93. (Minutes of the meeting of the Commission for the Reform of Inorganic Chemical Nomenclature, October 6–7, 1925, appear in *Chem. Weekbl.* **1926**, 23, 96–9.)

5. E. J. Crane, "Report of the Committee on Nomenclature in Inorganic Chemistry, Washington, D.C., September, 1926", *Chem. Weekbl.* **1926**, 23, 486–7.

6. W. P. Jorissen, H. Bassett, A. Damiens, F. Fichter, and H. Remy, "Rules for Naming Inorganic Compounds. Report of the Committee of the International Union of Chemistry for the Reform of Inorganic Chemical Nomenclature, 1940", *J. Am. Chem. Soc.* **1941**, 63, 889–97. Also published in *J. Chem. Soc.* **1940**, 1404–15. (See also R. J. Meyer, "Report on the Nomenclature of Inorganic Compounds", *Chem. Weekbl.* **1936**, 33, 722–9.)

7. J. D. Scott, "The Need for Reform in Inorganic Chemical Nomenclature", *Chem. Rev.* **1943**, 32, 73–97.

8. R. V. G. Ewens and H. Bassett, "Inorganic Chemical Nomenclature", *Chem. Ind.* **1949**, 27, 131–9.

9. H. Bassett, "Some General Principles of Inorganic Chemical Nomenclature" in *Chemical Nomenclature*, American Chemical Society, Washington, D.C., 1953, 5–8. (*Advances in Chemistry Series* No. 8.)

10. W. C. Fernelius, "Nomenclature of Coordination Compounds and Its Relation to General Inorganic Nomenclature" in *Chemical Nomenclature*, American Chemical Society, Washington, D.C., 1953, 9–37. (*Advances in Chemistry Series* No. 8.)

11. W. C. Fernelius, "Principles of Inorganic Nomenclature", in *Treatise on Analytical Nomenclature*, ed. I. M. Kolthoff and P. J. Elving, Interscience, New York, 1961, Vol. 1, Pt. 2, Section A.

12. W. C. Fernelius, "Nomenclature of Coordination Compounds: Present Status", *J. Chem. Doc.* **1964**, 4, 70–83; **1965**, 5, 200–5.

13. P. M. McDonnell and R. F. Pasternak, "Line-Formula Notation System for Coordination Compounds", *J. Chem. Doc.* **1965**, 5, 56–60.

14. W. C. Fernelius, "The Nomenclature of Coordination Compounds from Pre-Werner Times to the 1966 IUPAC Report" in *Werner Centennial*, American Chemical Society, Washington, D.C., 1967, 147–60. (*Advances in Chemistry Series* No. 62.)

15. R. M. Adams, "Inorganic Nomenclature in 1966, Problems and Progress", *J. Chem. Doc.* **1967**, 7, 67–71.

16. F. A. Cotton, "Proposed Nomenclature for Olefin-Metal and Other Organometallic Complexes", *J. Am. Chem. Soc.* **1968**, 90, 6230–2.

17. W. C. Fernelius, "Present Status of Inorganic Chemical Nomenclature", *J. Chem. Inf. Comput. Sci.* **1981**, 21, 213–18.

18. J. B. Casey, W. J. Evans, and W. H. Powell, "A Descriptor System and Principles for Numbering Closed Boron Polyhedra with at Least One Rotational Symmetry Axis and One Symmetry Plane", *Inorg. Chem.* **1981**, 20, 1333–41; "A Descriptor System and Suggested Numbering Procedures for Closed Boron Polyhedra Belonging to D_n, T, and C_s Symmetry Point Groups", *ibid.* 3556–61.

19. J. B. Casey, W. J. Evans, and W. H. Powell, "Structural Nomenclature for Polyboron Hydrides and Related Compounds. 1. Closed and Capped Polyhedral Structures", *Inorg. Chem.* **1983**, 22, 2228–35; "2. Nonclosed Structures", *ibid.* 2236–45; "3. Linear *conjuncto*-Structures", *ibid.* **1984**, 23, 4132–43.

20. W. C. Fernelius, "Some Reflections on the Periodic Table and Its Use", *J. Chem. Educ.* **1986**, 63, 263–6.

307. Macromolecular chemistry

1. M. L. Huggins, "Macromolecular Nomenclature: General Background and Perspective", *J. Chem. Doc.* **1969**, 9, 230–1.

2. H. K. Livingston and R. B. Fox, "Nomenclature of Organic Polymers", *J. Chem. Doc.* **1969**, 9, 232–4.

3. N. M. Bikales, "Polymer Nomenclature in Industry", *J. Chem. Doc.* **1969**, 9, 245–7.

4. R. B. Fox, "Nomenclature of Macrocyclic Compounds by Sequential Citation", *J. Chem. Inf. Comput. Sci.* **1984**, 24, 266–71.

308. Stereochemistry

Organic Compounds

1. R. S. Cahn, C. K. Ingold, and V. Prelog, "Specification of Molecular Chirality", *Angew. Chem., Int. Ed. Engl.* **1966**, 5, 385–414 (errata, **1966**, 5, 511); *Angew. Chem.* **1966**, 78, 413–47.

2. R. S. Cahn, "An Introduction to the Sequence Rule", *J. Chem. Educ.* **1964**, 41, 116–25 (errata, **1964**, 41, 508).[2]

3. R. S. Cahn, C. K. Ingold, and V. Prelog, "The Specification of Asymmetric Configuration in Organic Chemistry", *Experientia* **1956**, 12, 81–94.

4. R. S. Cahn and C. K. Ingold, "Specification of Configuration about Quadricovalent Asymmetric Atoms", *J. Chem. Soc.* **1951**, 612–22.

5. E. L. Eliel, "Recent Advances in Stereochemical Nomenclature", *J. Chem. Educ.* **1971**, 48, 163–7.

6. Beilstein Institut, "Stereochemical Conventions in the Beilstein Handbook of Organic Chemistry", *Beilsteins Handbuch der Organischen Chemie*, Vierte Auflage, Drittes and Viertes Ergänzungswerk, Einundzwanzigster Band, Dritter Teil, Springer-Verlag, Berlin, 1978, pp. xx–xxviii.

7. V. Prelog and G. Helmchen, "Basic Principles of the CIP-System and Proposals for a Revision", *Angew. Chem.* **1982**, 94, 614–31; *Angew. Chem., Int. Ed. Engl.* **1982**, 21, 567–83.

8. K. Mislow and J. Siegel, "Stereoisomerism and Local Chirality", *J. Am. Chem. Soc.* **1984**, 106, 3319–28.

9. J. H. Brewster, "On the Distinction of Diastereoisomers in the Cahn-Ingold-Prelog (RS) Notation", *J. Org. Chem.* **1986**, 51, 4751–3.

Inorganic Compounds

1. W. Thewalt, K. A. Jensen, and C. E. Schaeffer, "A Nomenclature Symbolism for Chiral and Achiral Isomers of Bridged Inorganic Complexes", *Inorg. Chem.* **1972**, 11, 2129–36.

2. T. E. Sloan and D. H. Busch, "Stereochemical Description and Notation for Coordination Systems" in *Stereochemistry of Optically Active Transition Metal Compounds*, ed. B. E. Douglas and Y. Saito, American Chemical Society, Washington, D. C., 1980, 397–419. (*ACS Symposium Series* No. 119.)

3. T. E. Sloan, "Stereochemical Nomenclature and Notation in Inorganic Chemistry" in *Topics in Inorganic and Organometallic Stereochemistry*", ed. G. Geoffroy, John Wiley & Sons, Inc., New York, 1981, 1–36.

4. T. Damhus and C. E. Schaeffer, "Three Reference Systems for Chirality Specification, Application, Geometric Properties, and Mutual Relationships", *Inorg. Chem.* **1983**, 22, 2406–12.

5. M. Brorson, T. Damhus, and C. E. Schaeffer, "Exhaustive Examination of Chiral Configurations of Edges on a Regular Octahedron: Analysis of the Possibilities of Assigning Chirality Descriptors within a Generalized Δ/Λ System", *Inorg. Chem.* **1983**, 22, 1569–73.

---

[1](a) K. L. Loening, "International Cooperation on Scientific Nomenclature", *J. Chem. Doc.* **1970**, 10, 231–6. (b) K. L. Loening, "Terminology Guidelines and Their Application in the International Union of Pure and Applied Chemistry" in: *International Co-operation in Terminology Work* ed. M. Krommer-Benz, K. G. Saur, Munich, 1986, pp. 92–116 (*Infoterm Series* 8).

[2]Also available as a reprint from the Marketing Communications Department, Chemical Abstracts Service, P.O. Box 3012, Columbus, Ohio 43210.

309. Miscellaneous chemical prefixes. The following list of prefixes most often encountered in the chemical literature (though not necessarily employed in *CA* index names) is intended to supplement Section H (Illustrative List of Substituent Prefixes); items in Section H are therefore not repeated here. Also excluded are prefixes derived from individual element or compound names, e.g., phospho–, ferri–, aceto–, oxa–, thiazolo–. Prefixes usually italicized are so shown.

The lower–case Greek alphabet is supplied in a separate list below (¶ 310). Greek and Latin multiplicative prefixes are also listed separately (¶ 311).

ac– abbreviation of *alicyclic*; as, *ac*–amino derivatives of Tetralin. Cf. *ar*–.

ace– from *acetylene*; as, *ace*naphthene.

aci– the *acid* form: as *aci*–acetoacetic ester, $CH_3C(OH):CHCOOC_2H_5$; *aci*–nitro group HON(O):.

aldo–, ald– from *aldehyde*; as, *aldo*hexose, *ald*oxime.

allo– (Greek *allos*, other) indicating *close relation*; as, *allo*–telluric acid, *al=loocimene* (a structural isomer of ocimene). Specifically, designating the more stable of two geometrical isomers; as, *allo*maleic acid (fumaric acid).

amphi– (Greek, both, around) relating to both sides or both kinds; as, *am=phi*–naphthoquinone (2,6–naphthalenedione), *amphi*phile.

andro–, andr– (Greek) relating to *man, male*; as *andro*gen, *andro*sterone.

ang– having an *angular* alignment of rings; as, *ang*–2',3'–naphth–1,2–anthracene (pentaphene). *Cf. lin*–.

anhydro– (Greek *anhydros*, without water) denoting *abstraction of water, anhydride of*; as, *anhydro*glucose. Cf. *dehydro*–.

antho–, anth– (Greek) of *flowers*; as, *antho*cyanin, *anth*oxanthin.

anthra–, anthr– (Greek, *anthrax*, coal) of *coal* or *anthracene*; as, *anthr*acite, *anthra*quinone, *anthra*pyrrole.

anti– (Greek, against) *opposite, opposed to*; as, *anti*oxidant; specifically, equivalent to *trans*– (which see) as, *anti*–benzaldoxime.

apo– (Greek, from) denoting formation from, or relationship to, another com=pound; as, *apo*morphine, $C_{17}H_{17}NO_2$ (morphine is $C_{17}H_{19}NO_3$).

ar– abbreviation of *aromatic*; as, *ar*–derivatives of Tetralin.

as– abbreviation of *asymmetric*; as, *as*–trichlorobenzene (1,2,4–$Cl_3C_6H_3$).

benzo–, benz– of *benzene*; as, *benz*oic; specifically, denoting fusion of a benzene ring; as, *benzo*quinoline.

bi– (Latin) *twice, two, double*; specifically: (a) in double proportion; as, *bi*carbonate (no longer considered good usage); (b) denoting the doubling of an organic radical or molecule; as, *bi*phenyl, $C_6H_5C_6H_5$; *bi*pyridine, $C_5H_4N.C_5H_4N$.

bicyclo– of *two rings*; specifically, designating certain *bicyclic* bridged com=pounds, as; *bicyclo*[2.2.1]heptane.

bili– (Latin) of *bili*rubin, as, *bili*verdin.

bisnor–, dinor– indicating removal of two CH_2 groups; as, *bisnor*cholanic acid or *dinor*cholanic acid $C_{22}H_{36}O_2$ (cholanic acid is $C_{24}H_{40}O_2$).

bufo–, buf– (Latin *bufo*, toad) derived from the *toad*; as, *bufo*talin.

chole–, cholo, chol– (Greek) of *bile*; as, *chole*sterol, *chol*ine.

chromo– (Greek *chroma*, color) *color, colored*; as, *chromo*phore, *chromo*protein, *chromo*isomer (a colored isomer of a colorless compound).

chryso–, chrys– (Greek) *gold, golden yellow, yellow*; as *chryso*phanic acid, *chrys*azin.

cincho–, cinch– of *cinchoma* or *cinchonine*; as *cincho*meronic acid, *cinch*onan.

cis– (Latin, on this side) an isomer in which certain atoms or groups are on the *same side* of a plane; as, *cis*–1,4–cyclohexanediol.

copro– (Greek) of *dung* or *excrement*; as, *copro*sterol.

cyclo– (Greek *kyklos*, circle) of ring structure, *cyclic*; as, *cyclo*hexane.

D– denoting configurational relationship to D–glyceraldehyde; as, D–fructose.

d– (a) abbreviation of *dextro* or *dextrotatory*; as, *d*–strychnine; (b) less properly = D.

de–, des– (Latin) indicating *removal* of something from the molecule; as, *de*oxybenzoin, $C_6H_5CH_2COC_6H_5$ (benzoin with one oxygen atom removed).

dehydro– denoting (a) *removal of hydrogen*; as, *dehydro*cholic acid; (b) sometimes, *removal of water*; as, *dehydro*mucic acid.

dextro– (Latin *dexter*, right) rotating the plane of polarization to the right; as, *dextro*pinene. Abbreviation, *d*; as, *d*–valine.

dl–; *d,l*– denoting a *racemic form*. Cf. *dextro*–, *levo*–.

dvi– (Sanskrit, two, twice) designating provisionally an element of the same family, in the *second place beyond*; as *dvi*–manganese (rhenium).

eka– (Sanskrit, one) designating provisionally an element of the same family, in the *first place beyond*; as, *eka*–manganese (technetium).

endo– (Greek, within) indicating an *inner position*, specifically: (a) in the ring and not in a side chain; (b) attached as a bridge within a ring; as, 1,4–*endo*–methylenecyclohexane (bicyclo[2.2.1]heptane).

epi– (Greek, upon, on, to) denoting (a) the 1,6–positions in naphthalene; as, *epi*–dichloronaphthalene; (b) in aldoses and related compounds identity of structure except arrangement about the α–carbon atom; as, *epi*rhamnose (epimer of rhamnose); (c) a bridge connection; as, 9,10–*epi*dioxyanthracene (anthracene 9,10–peroxide).

ergo–, ergot– relating to *ergot*; as, *ergo*sterol, *ergot*amine.

erythro–, erythr– (Greek) *red*; as *erythro*mycin, *erythr*osine.

eso– (Greek, within) denoting immediate attachment to a ring atom. Cf. *exo*–.

etio–, aetio– (Greek *aitia*, cause) denoting a *degradation product*; as, *etio=*cholanic acid, *etio*cobalamin.

exo– (Greek) *outside, out of*; as, *exo*toxin (an excreted toxin); specifically, denoting attachment in a side chain. Cf. *endo, eso*–.

flavo–, flav– (Latin *flavus*, yellow) *yellow*; as, *flavo*protein, *flav*one; specifically, designating certain series of coordination compounds.

fuco–, fuc– of *fucus* (a seaweed); as, *fuco*xanthin, *fuc*ose.

gala–, galacto–, galact– (Greek, milk, milky) relating to: (a) *milk*; as, *galac*tase, *galact*ose; (b) *galactose*; as, *galacto*cerebroside, *galacto*lipin.

gallo–, gall– relating to *gallnuts* or *gallic acid*; as, *gallo*tannic acid, *gallo=*catechin.

gem– abbreviation of *geminate* (said of two groups attached to the same atom); as, a *gem*–diol (e.g., 1,1–ethanediol), the *gem*–dimethyl grouping in camphor.

gluco–, gluc– (a) of *glucose*; as, *gluco*pyranose, *gluc*uronic acid; (b) less properly = *glyco*–, *glyc*–.

glyco–, glyc– (Greek) *sweet*, or relating to *sugars* or *glycine*; as, *glyco*gen, *glyco*side, *glyco*cholic acid.

hemato–, hemat–, hemo–, hem– (also haemato–, etc.) (Greek) of *blood* or its color; as, *hemato*porphyrin, *hemat*ein, *hemo*globin, *hem*in.

hetero–, heter– (Greek *heteros*, other) *other, different*; as, *hetero*poly acids, *hetero*cyclic.

holo– (Greek) *whole, complete*; as, *holo*cellulose; *holo*phosphoric acid, H_5PO_5.

homo– (Greek) *same, similar*; as, *homo*cyclic, *homo*logous (differing by an increase of CH_2; as, *homo*phthalic acid).

hydro–, hydr– (Greek) (a) denoting presence or addition of *hydrogen*; as, *hydro*chloric, *hydr*acrylic; (b) sometimes, relating to *water*; as *hydr*ate.

hyo– (Greek) of *swine*; as, *hyo*deoxycholic acid (from hog bile), *hyo*scyamine (from *Hyoscyamus* (hog bean)).

hypo– (Greek, under, beneath) indicating a lower (or the *lowest*) state of oxidation; as, *hypo*chlorous acid, *hyp*oxanthine.

i– abbreviation of (a) *inactive*; as, *i*–tartaric acid; (b) *iso*–; as, *i*–pentane.

iso– (Greek) *equal, alike*; as, *iso*mer; usually, denoting an *isomer* of a com=pound; as, *iso*cyanic acid; specifically, denoting an isomer having a single, simple branching at the end of a straight chain; as, *iso*pentane, $(CH_3)_2CH=CH_2CH_3$.

keto–, ket– from *ketone*; as, *keto*hexose, *ket*oxime.

L– configurationally related to L–glyceraldehyde; as, L–fructose.

l– (a) abbreviation of *levo* or *levorotatory*; as, *l*–strychnine; (b) less properly = L–.

lano– (Latin) of *wool*; as, *lano*sterol (from wool fat).

leuco–, leuc– (Greek) *colorless, white*; as, *leuc*ine; specifically, a colorless reduced derivative of a triphenylmethane dye; as, *leuco*methylene blue.

levo–, laevo– (Latin *laevus*, left) rotating the plane of polarization to the left; as, *levo*valine. Abbreviation, *l*–; as, *l*–valine.

lin– denoting a straight, *linear* alignment of rings; as *lin*–naphthanthracene (pentacene). Cf. *ang*–.

litho–, lith– (Greek *lithos*, stone) related to *stone* or *calculus*; as, *lith*arge, *litho*cholic acid (from gallstones).

luteo– (Latin *luteus*) *orange–yellow, brownish yellow*; as *luteo*lin; specifically, the coordination compounds $[M(NH_3)_6]Cl_3$.

m– abbreviation of *meta* (sense c).

meso–, mes– (Greek) *middle, intermediate*; specifically; (a) an intermediate hydrated form of an inorganic acid; as, *meso*periodic acid, H_3IO_6; (b) op=tically inactive owing to internal compensation; as, *meso*tartaric acid; (c) (abbreviated μ–) centrally substituted; as, *meso*–chloroanthracene (9–chloroanthracene), *meso*–phenylimidazole (2–phenylimidazole); (d) (ab=breviated *ms*–) centrally fused; as, *meso*naphthodianthrene (phenanthro=[1,10,9,8–*opqra*]perylene).

meta–, met– (Greek) indicating *changed relation*; specifically, designating (a) a low hydrated form of an acid (usually that derived from the "ortho" form by loss of one molecule of water); as, *meta*phosphoric acid, HPO_3; (b) a closely related compound (sometimes, a polymer); as, *met*aldehyde (trimer of ordinary (acet)aldehyde); (c) (abbreviated *m*–) the 1,3–positions in benzene; as, *m*–xylene.

n– abbreviation of *normal* (unbranched); as, *n*–butane, *n*–butyl.

naphtho–, naphth– (Greek *naphtha*) relating to *naphthalene*; as, *naphtho=quinone*.

neo– (Greek) *new*; designating new or recent; as, *neo*arsphenamine; (of a hydrocarbon) having one carbon atom connected directly to four others; as, *neo*pentane, $(CH_3)_4C$.

nor– from *normal*; (a) a lower homologue; as, *nor*camphane (of which camphane is a trimethyl derivative); (b) a normal (straight–chain) isomer; as, *nor=*leucine.

o– abbreviation of *ortho* (sense d).

oligo–, olig– (Greek *oligos*, small) meaning few; as, *oligo*saccharide.

ortho– (Greek, straight, right, true) (a) the fully hydrated form of an acid; as, *ortho*nitric acid, H_5NO_5; (b) the highest–hydrated stable form; as, *ortho=*phosphoric acid, H_3PO_4; (c) the common or symmetrical molecular form of an element; as, *ortho*hydrogen; (d) (abbreviated *o*–) the 1,2–positions in benzene; as, *o*–xylene.

p– abbreviation of *para* (sense c).

para–, par– (Greek, beside, alongside of, beyond) indicating a relationship; as *para*xanthine (1,7–dimethylxanthine). Specifically, (a) a higher hydrated form of an acid; as, *para*periodic acid, H_3IO_5 (preferably called orthoper=iodic); (b) a polymer; as, *par*aldehyde; (c) (abbreviated *p*–) the 1,4–positions in benzene; as, *p*–xylene.

per– (Latin) *complete, thorough, extreme*; (a) the highest (or a high) state of oxidation; as, *per*chloric acid, manganese *per*oxide (better, dioxide); (b) presence of the peroxide group (O_2); as, barium *per*oxide, *per*benzoic acid, *per* acid (better, peroxy acid); (c) exhaustive substitution or addition; as, *per*chloroethylene, C_2Cl_4; *per*hydronaphthalene, $C_{10}H_{18}$.

peri– (Greek, around, about) (a) the 1,8–positions in naphthalene; as *peri*–dinitronaphthalene (b) in polycyclic ring systems, fusion of a ring to two or more adjoining rings: as, *peri*naphthindene (phenalene).

peroxy– containing the *peroxide* group (O_2); as, *peroxy*monosulfuric acid, $HOSO_2OOH$.

pheno–, phen– (from *phene*, benzene) related to *phenyl* or *benzene*; as, *phe*=*nacyl*; specifically, an anthracene analogue having two hetero atoms in the central positions; as, *phen*azine, *pheno*thiazine.

phloro–, phlor– relating to *phlorizin*; as, *phloro*glucinol, *phlor*etin.

phthalo–, phthal– relating to *phthalic acid*; as, *phthalo*cyanine, *phthal*ide.

phyllo–, phyll– (Greek *phyllos*) of *leaves*; as, *phyllo*porphyrin.

phyto–, phyt– (Greek) relating to *plants*; as, *phyto*sterol, *phyto*hormones.

picro–, picr– (Greek) *bitter*; as, *picro*toxin, *picr*ic acid.

pino–, pin– (Latin *pinus*, pine) relating to *pine* or *pinene*; as, *pin*ic acid, *pino*carvone.

poly– (Greek) many; as, *poly*mer, *poly*sulfide, *poly*saccharide.

pro– (Greek, before) a precursor; as *pro*enzyme, *pro*vitamin.

proto–, prot– (Greek) *first*; specifically, designating: (a) first in an inorganic series; as, *prot*oxide (lowest in oxygen content); (b) parent or immediate antecedent; as, *prot*actinium, *proto*chlorophyll.

pseudo–, pseud– abbreviated ψ or *ps* (Greek, false) indicating *resemblance* to, or *relation* (especially isomerism) with; as, ψ–cumene, *pseud*aconitine, *pseudo*base. Abbreviation: as, ψ–cumene.

purpuro– (Latin *purpura*, purple) indicating *purple* or *red* color; as, *purpu*=*ro*gallin.

pyo– py– (Greek) relating to *pus*; as, *pyo*cyanine.

pyro–, pyr– (Greek *pyr*, fire) indicating *formation by heat*; as, *pyro*cinchonic acid, *pyr*ene; specifically, designating an acid derived from two molecules of an "ortho" acid by loss of $1H_2O$; as, *pyro*phosphoric acid, $H_4P_2O_7$ ($2H_3PO_4$–H_2O).

pyrrolo–, pyrro–, pyrr– containing the *pyrrole* ring; as, *pyrrolo*pyridine, *pyr*=*ro*coline.

reso–, res– relating to *resorcinol*; as, *res*orufin, *res*azurin.

rhodo–, rhod– (Greek *rhodon*, rose) *rose–red*; as, *rhodo*porphyrin, *rhod*amine.

s– abbreviation of (a) *symmetric(al)*; (b) *secondary*.

sapo– (Latin, soap) relating to *soap* or *soap bark*; as *sapo*nin, *sapo*genin.

sec– abbreviation of *secondary*; as, *sec*–butyl.

seco– (Latin *secare*, to cut) denoting ring cleavage; as, 16,17–*seco*androstane.

sub– (Latin, under, below) denoting (a) a *low proportion* (or deficiency, as in a basic salt); as, *sub*iodide, *sub*oxide, aluminum *sub*acetate.

super– (Latin, above, over) denoting a *high proportion* (or superfluity, as in an acid salt); as, *super*oxide (peroxide), *super*phosphate.

sym– abbreviation of *symmetric(al)*; as *sym*–dichloroethylene (1,2–dichloro=ethene).

syn– (Greek, with, together) equivalent to *cis* (which see); as, *syn*–benzald=oxime.

t– abbreviation of *tertiary*; as, *t*–butyl.

tauro– (Latin *taurus*, bull) relating to *bulls* or to *taurine*; as, *tauro*cholic acid, *tauro*cyamine.

tere– (Latin *terebinthus*, terebinth) relating to *terebene* or *terpenes*; as, *ter*=ephthalic, *tere*santalic.

tert– abbreviation of *tertiary*; as, *tert*–butyl.

thymo–, thym– relating to (a) *thyme*; as, *thymo*quinone, *thym*ol; (b) the *thymus*; as, *thymo*nucleic.

trans– (Latin, across) an isomer in which certain atoms or groups are on *opposite sides* of a plane; as, *trans*–cinnamic acid.

uns–, *unsym*– abbreviations of *unsymmetrical*; as, *uns*–dichloroethane, CH_3CHCl_2.

uro–, ur– (Greek *ouro*–, *our*–) relating to *urine* or *urea*; as, *uro*bilin, *ur*ethane, *ur*ic acid.

urso–, urs– (Latin *ursus*, bear) relating to *bears* or the *bearberry*; as, *urs*olic acid.

v– abbreviation of *vicinal*; as, *v*–triazine.

verdo– (French, *verd*, *vert*, green) indicating *green* color; as, *verdo*hemin.

vic– abbreviation of *vicinal*; as, *vic*–triazole.

xantho–, xanth– (Greek) *yellow*; as, *xantho*toxin, *xanth*ic acids, *xanth*ine.

xylo–, xyl– (Greek *xylon*, wood) relating to *wood*, *xylene*, or *xylose*; as, *xyl*an, *xyl*idine, *xylo*quinone, *xylo*caine.

zymo–, zym– (Greek *zyme*, leaven) relating to a *ferment* or *fermentation*; as, *zymo*sterol, *zym*ase.

310. Greek alphabet. Lower-case Greek letters are employed in the chemical literature to number carbon chains and to indicate the size of lactone rings (a γ–lactone generally contains a furan ring, a δ–lactone a pyran ring, and so on). In *CA* names, Greek letters are reserved for the *acyclic* portion of conjunctive index parents (see ¶ 124), while cyclic portions are numbered with arabic numbers. Of Greek capital letters, Δ (delta) is sometimes encountered in the literature to denote a double bond); T (tau) indicates a triple bond. Some lower-case letters have additional meanings; φ (phi) is a shorthand version of "phenyl" or "Ph"; see also "meso" and "pseudo" in ¶ 309.

α–	alpha	ν–	nu
β–	beta	ξ	xi
γ–	gamma	o	omicron
δ–	delta	π–	pi
ϵ–	epsilon	ρ–	rho
ζ	zeta	σ–	sigma
η–	eta	τ–	tau
θ–	theta	υ–	upsilon
ι–	iota	φ	phi
κ–	kappa	χ–	chi
λ–	lambda	ψ–	psi
μ–	mu	ω–	omega

311. Multiplicative prefixes. In *CA* index names, Greek prefixes are preferred, except for sesqui– (for one and one–half), nona– (for nine), and undeca– (for eleven). The terms hemi– (Greek) and sesqui– (Latin) are em=ployed by *CA* only in hydrate and ammoniate names (see ¶ 192). For use of the special terms bis–, tris–, tetrakis–, etc., with complex terms and to avoid ambiguity, see ¶ 110.

	Greek	*Latin*
1/2	hemi–	semi–
1	mono–, mon–	uni–
1 1/2		sesqui–
2	di–	bi–
3	tri–	tri–, ter–
4	tetra–, tetr–	quadri–, quadr–, quater–
5	penta–, pent–	quinque–, quinqu–
6	hexa–, hex–	sexi–, sex–
7	hepta–, hept–	septi–, sept–
8	octa–, oct–, octo–, octi–	
9	ennea–, enne–	nona–, non–, novi–
10	deca–, dec–, deci–	
11	hendeca–, hendec–	undeca–, undec–
12	dodeca–, dodec–	
13	trideca–, tridec–	
14	tetradeca–, tetradec–	
15	pentadeca–, pentadec–	
16	hexadeca– hexadec–	
17	heptadeca–, heptadec–	
18	octadeca–, octadec–	
19	nonadeca–, nonadec–	
20	eicosa–, eicos–	
21	heneicosa–, heneicos–	
22	docosa–, docos–	
23	tricosa–, tricos–	
24	tetracosa–, tetracos–	
25	pentacosa–, pentacos–	
26	hexacosa–, hexacos–	
27	heptacosa–, heptacos–	
28	octacosa–, octacos–	
29	nonacosa–, nonacos–	
30	triaconta– triacont–	
31	hentriaconta–, hentriacont–	
32	dotriaconta–, dotriacont–	
33	tritriaconta–, tritriacont–	
40	tetraconta–, tetracont–	
50	pentaconta–, pentacont–	
60	hexaconta–, hexacont–	
70	heptaconta–, heptacont–	
80	octaconta–, octacont–	
90	nonaconta–, nonacont–	
100	hecta–, hect–	
101	henhecta–, henhect–	
102	dohecta–, dohect–	
110	decahecta–, decahect–	
120	eicosahecta–, eicosahect–	
132	dotriacontahecta–, dotriacontahecta–	
200	dicta–, dict–	
300	tricta[1]	
400	tetracta[1]	
1000	kilia[1]	

[1] International Union of Pure and Applied Chemistry, Organic Chemistry Division, Commission on Nomenclature of Organic Chemistry, "Extension of Rules A–1.1 and A–2.5 Concerning Numerical Terms Used in Organic Chemical Nomenclature (Recommendations 1986)", *Pure Appl. Chem.* **1986**, 58, 1693–6.

312. Introduction. The foregoing sections of Appendix IV are concerned with selection of index names for chemical substances, but the user of *CA* printed indexes and computer-readable files often needs to proceed from a *CA* index name to the structure diagram of a chemical substance. The aim of this section is therefore to describe succinctly the form of *CA* index names, to illustrate the procedure for deriving a chemical structural diagram from a *CA* index name, and to show where in CAS publications structural data can be found.

Sources of structural data include the *Ring Systems Handbook*, the *Index Guide*, and (when information concerning a substance has been published) the Volume and Collective issues of the *Chemical Substance Index*. The reader is also referred to other paragraphs of Appendix IV, when appropriate, for more detailed discussion of complex subjects. It is not the purpose of this section to cover every subject. Discussion of subjects not covered in this section can be found by reference to the Index at the end of Appendix IV.

313. Form of CA index names. The complexity of chemical substances generally dictates the complexity of *CA* index names. The *CA* index name consists of up to five fragments, namely, the *index heading parent* alone or followed successively by the *substituent*, *modification*, and *stereochemistry* fragments (compare ¶104), as necessary. Some index heading parents are followed by *synonym line formulas* (compare ¶¶219, 315.II). Every name contains an index heading parent and it is by citing the *parent* first (i.e., in an inverted format) that the names of related chemical substances are listed together in the printed indexes.

314. Deriving a chemical structural diagram from a *CA* index name proceeds by taking each name fragment sequentially, converting it into a structural fragment, and then placing each structural fragment in its proper position in the diagram, using the appropriate positional information. The complete derivation of a chemical structural diagram may require up to four steps, depending upon the complexity of the chemical substance, each based on one of the four possible *CA* index name fragments. This process is illus= trated in the remaining paragraphs of this section.

315. Index heading parents include the largest or most important mo= lecular skeleton and (when present) the highest function in a chemical sub= stance (see ¶¶ 130, 164).

I. Index heading parents that stand alone.

The simplest *index heading parents* may be illustrated by examples such as the following:

Butane

Chlorine

2–Hexene

Nickel

The names of acyclic hydrocarbons are discussed in ¶ 141 and the multipli= cative terms that indicate the number of carbon atoms in a hydrocarbon are listed in ¶ 311. Element names are discussed in ¶ 219.

Some index heading parents are made up of two parts: a *basic skeleton name* and a *principal functional suffix*. Principal functional suffixes are discussed in detail in Section C (¶¶ 164–177).

Example:

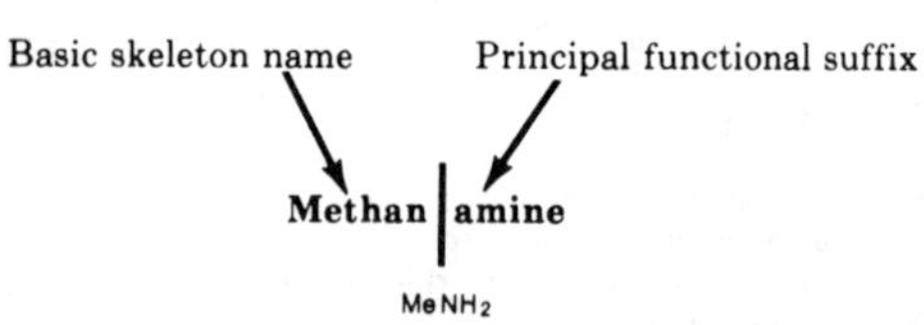

The basic skeleton is the one–carbon unit "Methane" (¶ 141), the final "e" being elided before a vowel. The principal (here, the only) functional group is the amine (¶ 176).

Example:

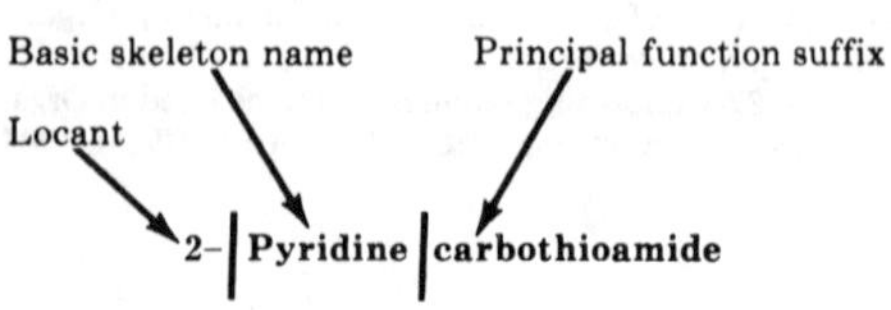

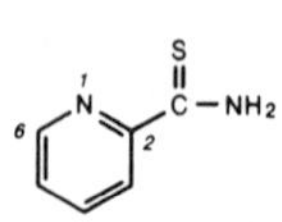

The *Chemical Substance Index* contains the following entry for pyridine.

Pyridine

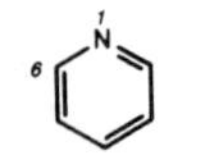

The carbothioamide principal function suffix is discussed in detail with other amide groups in ¶¶ 171 and 233. The locant, 2, indicates where the carbo= thioamide group is attached to the pyridine ring (¶ 115).

Another type of two–part name, known as a "conjunctive name" (see ¶ 124), is a combination of the name of a cyclic molecular skeleton and the name(s) of one or more identical acyclic hydrocarbon chains, each terminated by the same functional group.

Example:

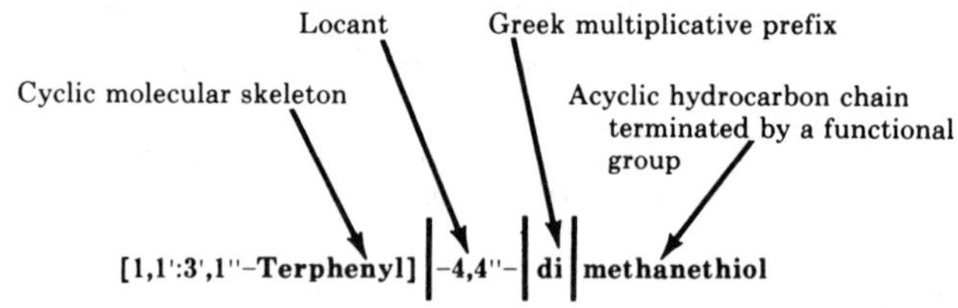

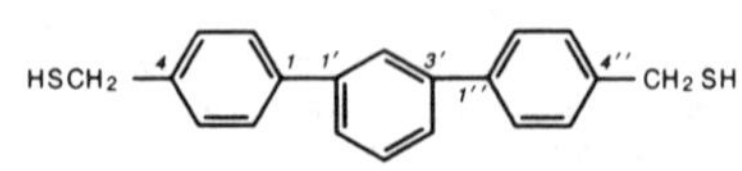

The structure of the phenyl group is found in the Illustrative List of Sub= stituent Prefixes (¶ 294). That "ter" a Latin prefix meaning "three", is found in ¶ 311. The first set of locants (1,1':3',1'') indicates that the three benzene rings are connected as shown. The second set of locants (4,4'') shows that the two methanethiol groups (see ¶¶ 141 and 175) are attached as shown to the first and third benzene rings. Such "ring assembly" names are described in ¶ 157.

Polymers are named either on the basis of the monomers from which they are formed (see ¶ 317) or on the basis of their structure, as represented by a structural repeating unit (compare ¶ 222). In the latter, the multivalent radicals are cited in sequence, following the term "Poly", and can be drawn from left to right. Each radical retains its own numbering, and is oriented if possible so that the point of attachment at the left of the radical is assigned the lowest possible locant.

Example:

Poly[(4,4–dimethyl–2,5–dioxo–1,3–imidazolidinediyl)–1,4–phenyl= eneoxy–1,4–phenylene]

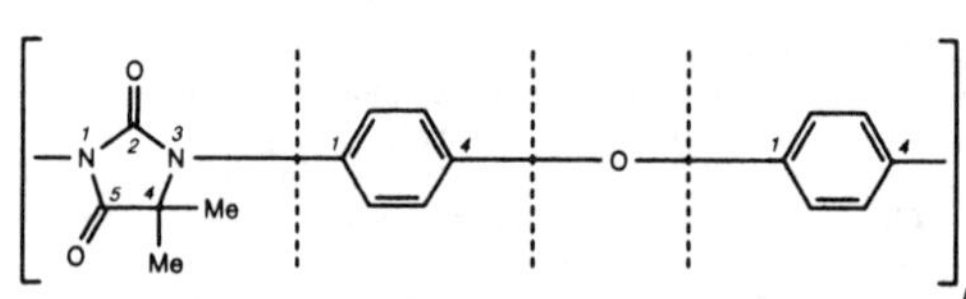

In this chemical structural diagram, the multivalent radicals are separated by dashed vertical lines and the numbering of each radical is shown, where appropriate. The structures of the radicals (e.g., oxy, phenylene) are found in ¶ 294 or can be inferred from the chemical structural diagrams associated with parent names (e.g., imidazolidine) in the *Ring Systems Handbook* or the *Chemical Substance Index*. The meanings of the methyl and oxo terms can also be determined from ¶ 294.

II. Index heading parents with synonym line formulas.

Synonym line formulas are molecular formulas which follow the index heading parent. They are printed in boldface and are enclosed in parentheses. They are often useful for resolving ambiguity, where two or more substances, usually inorganic (see ¶ 219), have the same name.

Examples:

Aluminum calcium titanium oxide (Al_2CaTiO_6)

Aluminum calcium titanium oxide ($Al_2Ca_4Ti_2O_{11}$)

Lack of a synonym line formula in such a case means that the author did not provide the necessary specificity in the original document.

Example:

Aluminum calcium titanium oxide

III. Index heading parents with chemical structural diagrams. A *chemical structural diagram* is provided in the *Index Guide* and *Chemical Substance Index* where the structure of an inorganic substance (¶ 219) may not be readily apparent from the name.
Example:

Phosphorodihydrazidothioic acid

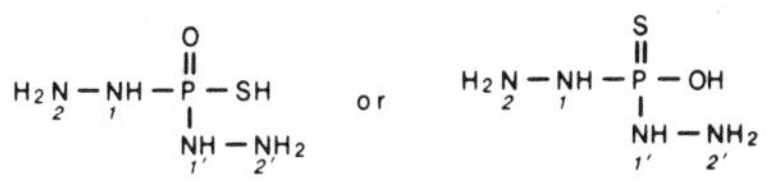

A similar procedure is followed for other types of index heading parents such as *cage parents, ring parents,* and *stereoparents.* Here, however, the structural diagrams are published in the *Ring Systems Handbook* as well as the *Chemical Substance Index.*
Examples:

Cage Parent:

1,12-Dicarbadodecaborane(12)

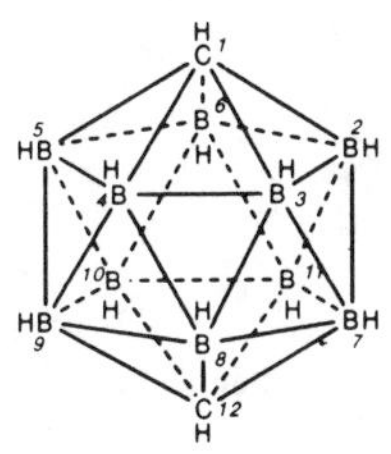

Ring Parent:

Phenanthrene

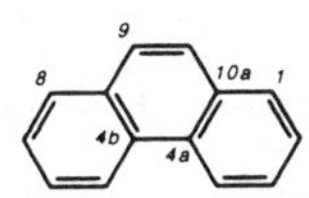

As stated in ¶ 313, the complexity of chemical substances generally dictates the complexity of *CA* index names. This is particularly true for natural products whose names must imply standard orientation and stereochemical representation as well as structure and numbering (compare ¶ 318). Such complexity makes it desirable to use a simple, nonsystematic ("trivial") name, or *stereoparent name,* as the index heading parent and to define the *stereoparent name* by means of a chemical structural diagram.
Examples:

Acyclic stereoparent:

ψ,ψ-Carotene (*(all-E)-2,6,10,14,19,23,27,31-octamethyl-2,6,8,10,12,14,16,18,20,22,24,26,30-dotriacontatridecaene)*

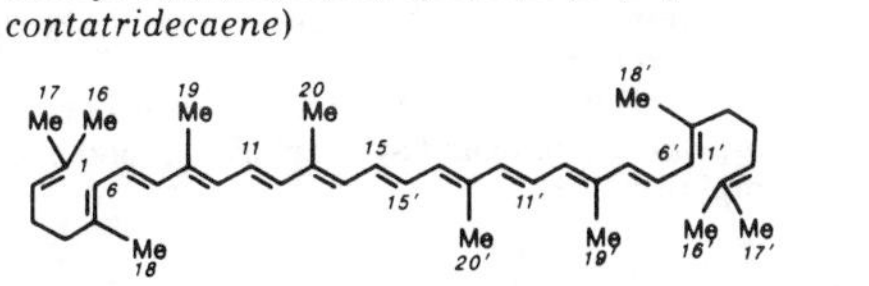

Cyclic stereoparent:

Rifamycin IV (*[9S-(9R*,12E,14R*,15S*,16R*,17S*,18S*,19S*,20R*,21R*,22E,24Z)]-16-(acetyloxy)-5,6,18,20-tetrahydroxy-14-methoxy-7,9,15,17,19,21,25-heptamethyl-9,4-(epoxypentadeca[1,11,13]trienimino)-1H-benzofuro[5,4-g]indazole-10,26-(9H)-dione)*

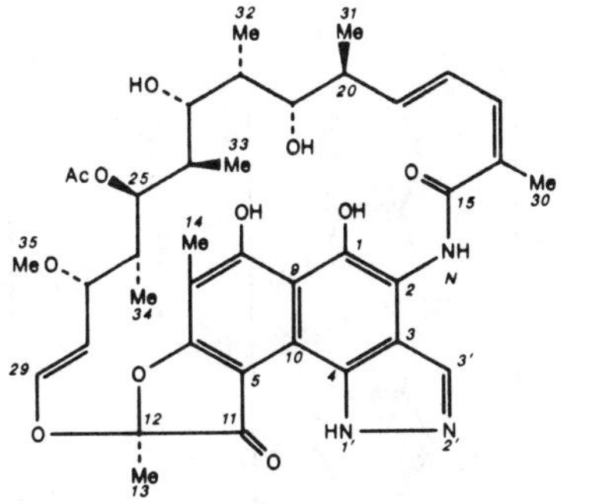

316. Substituent prefixes (see ¶ 10A) follow the "comma of inversion" (see ¶ 104) in *CA* index names and their structures can be determined from ¶ 294, the Illustrative List of Substituent Prefixes.
Example:

2*H*-Pyran-2,4(3*H*)-dione
—, 3-[[(4-aminophenyl)amino]phenylmethylene]-6-phenyl-

The structure of the chemical substance represented by this name is derived by first obtaining the structure of **2*H*-Pyran** from the *Chemical Substance Index* or, better, from the *Ring Systems Handbook.* The Ring Name Index of the *Ring Systems Handbook* reveals that the Ring File (RF) number for **2*H*-Pyran** is RF 1520. Entry RF 1520 in the *Ring Systems Handbook* is as follows:

RF 1520 289-66-7
2*H*-Pyran
C_5H_6O
T6O BHJ

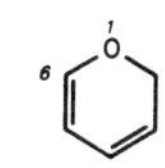

The *index heading parent* contains two functional groups which ¶ 174 identifies as ketone groups. However, putting an oxygen on position 4 requires "adding" a hydrogen at position 3. The indicated and added hydrogen terms (see ¶¶ 135 and 136) in the index heading parent thus define the bonding in the ring (the bonding could also be defined by "3,4-dihydro" but that would be part of the *substituent* fragment of the name and not of the index heading parent, which must be able to stand alone). The complete index heading parent (**2*H*-Pyran-2,4(3*H*)-dione**) is therefore:

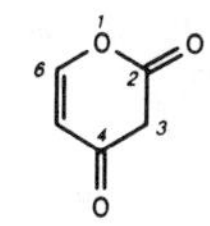

The *substituent* fragment of the name indicates that a simple substituent (i.e., "phenyl", see ¶ 294) is attached to the ring at position 6:

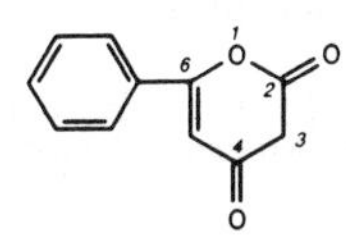

A complex substituent is attached to the ring at position 3. The second phenyl group is attached through the methylene group (see ¶ 294) to position 3 to yield:

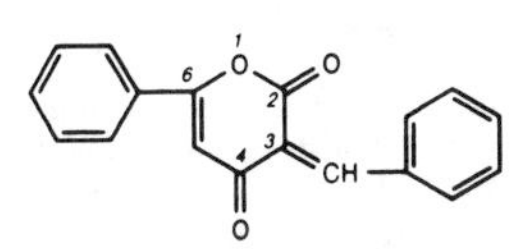

The third phenyl group has an amino group (¶ 294) attached to the 4-position and is itself attached through an amino group to the methylene group of the partial structure. The complete structure is therefore:

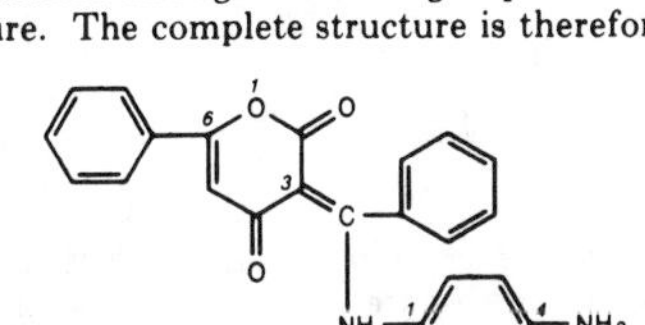

317. Modifications of the principal functions or other groups follow the substituent(s) (¶ 104). Where derivatives of more than one functional group must be named, the derivative terms are cited in the order described in ¶ 113.

Example:

2*H*-Benzo[*g*]indazole-2-propanamine
—, 3,3a,4,5-tetrahydro-*N,N*-dimethyl-3-phenyl-trihydrochloride

Only the form **1*H*-Benzo[*g*]indazole** is illustrated in the *Chemical Substance Index* and the *Ring Systems Handbook*:

1*H*-Benz[*g*]indazole

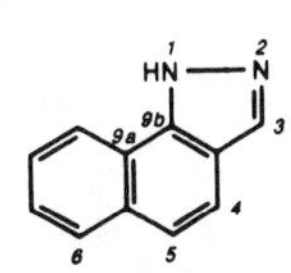

The structure is converted into the 2*H*–isomer and the propanamine group (¶¶ 141 and 176) is attached at the 2–position. (It should be noted that, in conjunctive nomenclature, the function is located at the end of the acyclic chain furthest from the ring system of the index heading parent (see ¶ 124).)

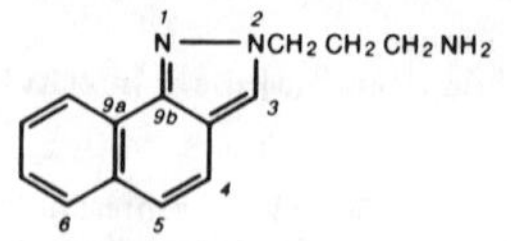

The four hydrogens are added to the 3, 3a, 4, and 5 positions to saturate the two double bonds; then, when the phenyl and two methyl substituents (see ¶ 294) are attached, the structure becomes:

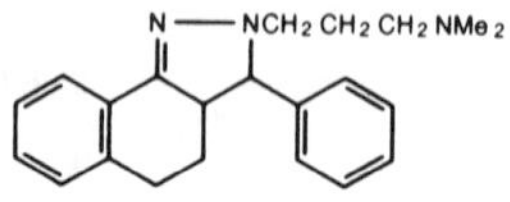

The "trihydrochloride" modification completes the structure:

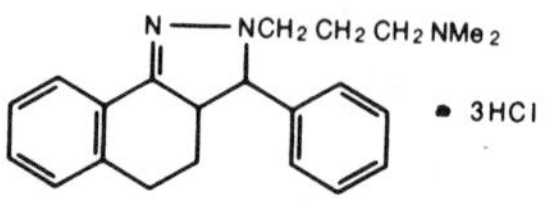

The modification is also used in naming *polymers* (¶ 222) on the basis of the monomers from which they are formed.
Example:

1–Butene
—, 2,3–dimethyl–
homopolymer

The structure of 1–Butene is deduced from ¶ 141. When the two methyl groups (¶ 294) are attached at positions 2 and 3, the structure becomes:

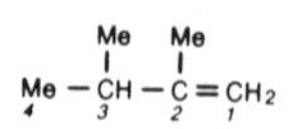

The homopolymer is represented as follows:

$(Me_2CHCMe=CH_2)_x$

318. Stereochemistry is the last structural information described in a *CA* name. The various symbols used to describe the spatial arrangement of atoms are discussed in ¶ 203.
Example:

8–Azabicyclo[3.2.1]octan–3–ol
—, 8–methyl–
exo–

The structure corresponding to the basic skeleton name, principal functional suffix, and substituent is determined as described above. The stereochemistry of the hydroxy group (the methyl group rocks back and forth about the nitrogen atom, eliminating the effects of asymmetry there) is denoted by "*exo*–", the meaning of which is determined from ¶ 203 I. The structure is therefore:

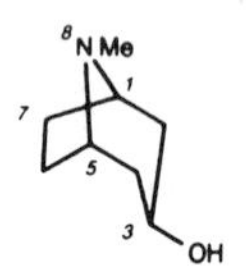

As stated in ¶ 315, *stereoparent names* (see ¶ 203 II) in the *Chemical Substance Index* imply (in the absence of cited stereochemical descriptors) a standard structure, including specific stereochemistry, as illustrated by an accompanying structural diagram complete with the numbering system from which locants for substituents and derivatives are derived. With the exception of monosaccharide and some peptide stereoparents, italicized systematic names (and in some cases one or more trivial names) appear as synonyms in par= entheses immediately following the (preferred) boldface stereoparent name. In the following examples these systematically named synonyms (1) illustrate the various types of stereochemical descriptors found in *CA* indexes and (2) show how the use of stereoparents as the preferred *CA* names for many natural products avoids citation of (often necessarily complicated) stereochemical descriptors.
Examples:

ψ,ψ–Carotene (*(all-E)–2,6,10,14,19,23,27,31–octa= methyl–2,6,8,10,12,14,16,18,20,22,24,26,30–dotria= contatridecaene)*

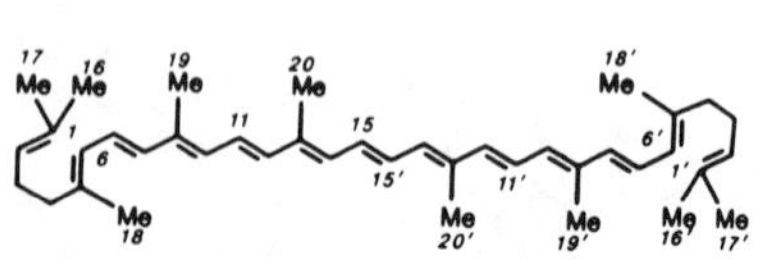

Olivomycin D (*[2S–[2α,3β(1R*,3R*,4S*)]]–6–[[3–O–= acetyl–2,6–dideoxy–4–O–(2,6–dideoxy–4–O–methyl–= α–D–lyxo–hexopyranosyl)–β–D–lyxo–hexopyrano= syl]oxy]–2–[[2,6–dideoxy–3–O–(2,6–dideoxy–β–D–= arabino–hexopyranosyl)–β–D–arabino–hexopyrano= syl]oxy]–3–(3,4–dihydroxy–1–methoxy–2–oxo= pentyl)–3,4–dihydro–8,9–dihydroxy–1(2H)–anthracenone)*

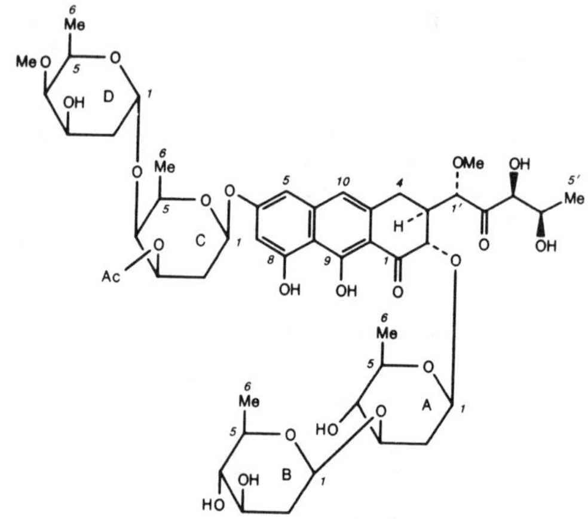

In the first example, the stereochemical descriptor (*all–E*) indicates that the "senior" groups (as defined by the Sequence Rule) attached to the double bonds are all in a "*trans*" relationship (see ¶ 203 I).

The stereochemical prefix of the second example, [2*S*–[2α,3β= (1*R**,3*R**,4*S**)]], illustrates the use of absolute and relative stereochemical terms (¶ 203 I) to describe the stereochemistry of a complex substance. The first term (2*S*) establishes the location of the "reference" center as position 2 of the anthracene ring system and the absolute stereochemistry at that point as determined by application of the Sequence Rule. The next two terms (2α,3β) define the stereochemistry at position 3 of the anthracene ring system in relation to the "reference" center which, for this purpose, is arbitrarily assigned the relative descriptor "α". (In other words, the substituents at positions 2 and 3 lie on opposite sides of the plane of the ring system.) The remaining three terms (1*R**,3*R**,4*S**) define the stereochemistry within the acyclic group at position 3, also in relation to the reference center. (The α,β convention is not used for the acyclic group because the chain is free to rotate and thus α and β are meaningless.) The *R** descriptors show that positions 1 and 3 of the acyclic group have the same chirality (in this case, *S*) as the "reference" center, while the *S** descriptor shows that the chirality of position 4 is opposite to that of the "reference" center. **Olivomycin D** contains further chiral centers within the disaccharide substituent at position 2, but these are described not by the stereochemical description set that prefaces the univerted systematic name but by α– and β–D–*arabino* and *lyxo* terms in the substituent radical names, as explained by the carbohydrate rules (¶ 208). It is quite evident from this example that use of the stereoparent name **Olivomycin D** in *CA* indexes, accompanied by a stereospecific structural diagram, obviates citation of all these descriptors. It also makes unnecessary the derivation of a structural diagram from the index name.

Stereochemical descriptors in the CA index names of *coordination com= pounds* reflect the geometry of ligand attachments around one or more central metal atoms (compare ¶ 215).
Example:

Nickel(1+)
—, bis[[2–(diphenylstibino)phenyl]diphenylarsine–= *As,Sb*]iodo–
(*T*–4)–tetraiodonickelate(2–) (2:1)

In this example, the central atom is nickel. The structure of the two large ligands can be determined by reference to ¶ 294. The italicized element symbols (*As,Sb*) identify the ligating atoms. The stereochemistry symbol (*T*–4) indicates that the anion is tetrahedral (see ¶ 203 III). The structure of this coordination compound is therefore:

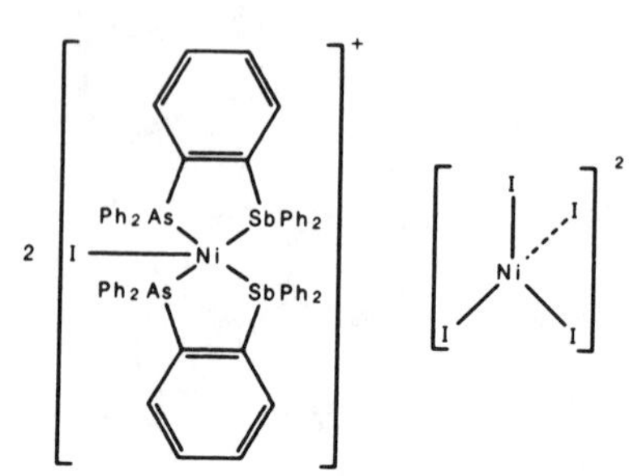

M. INDEX

The references are to paragraphs, not to pages. Trivial and former *CA* names, general terms and name fragments are listed along with current *CA* index names. Locants and other numerals, etc., have generally been omitted.

HODOC No: 00001
CAS No: 6373-11-1
Beilstein No: 7^2, 757
Name: 1,2-Aceanthrylenedione
Synonym: 1,2-Aceanthrenedione
Structure:

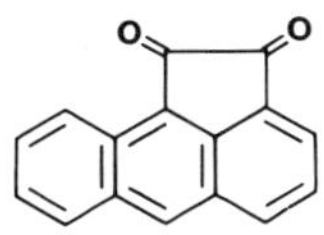

Mol Form: $C_{16}H_8O_2$
M_r: 232.22
bp: sub
mp: 270
Color: red pr (Bz)
m/e: 232.05

HODOC No: 00002
CAS No: 92378-92-2
Beilstein No: 7^3, 2096
Name: 3-Acenaphthenecarboxaldehyde
Synonym: Acenaphthene-3-aldehyde
Mol Form: $C_{13}H_{10}O$
M_r: 182.21
mp: 100
Color: nd (eth-peth)
m/e: 182.07

HODOC No: 00003
CAS No: 5428-95-5
Name: 1-Acenaphthenone, 2-dimethylaminomethyl-8-methoxy-2*a*,3,4,5-tetrahyd ro
Synonym: Carbamic acid, thio-, *S*-ester with 2-mer- captoacetanilide
Mol Form: $C_{16}H_{21}NO_2$
M_r: 259.33
IR: **μCOB 658** 3030 2940 2780 1700 1610 1590 1490 1450 1330 1270 1180 1100 1040 980 850 790
m/e: 259.16

HODOC No: 00004
CAS No: 892-78-4
Name: 1-Acenaphthenone, 2-piperidinomethyl-8-methoxy-2*a*,3,4,5-tetrahydro
Synonym: Benzamide, *N*-[α-(thiocarbamoyl)benzyl]-
Mol Form: $C_{19}H_{25}NO_2$
M_r: 299.39
mp: 77-8
IR: **μCOB 660** 3030 2940 1700 1590 1490 1450 1370 1270 1220 1110 1040 810
m/e: 299.19

HODOC No: 00005
CAS No: 40745-44-6
Beilstein No: 12^2, 764
Name: 1-Acenaphthylenamine, 1,2-dihydro-
Synonym: 1-Acenaphthenamine
Mol Form: $C_{12}H_{11}N$
M_r: 169.21
bp: sub
mp: 135
Color: cr (peth)
***s*:** Bz 4; EtOH 4
m/e: 169.09

HODOC No: 00006
CAS No: 55939-13-4
Beilstein No: 12^1, 3210
Name: 3-Acenaphthylenamine, 1,2-dihydro-
Synonym: 3-Acenaphthenamine
Mol Form: $C_{12}H_{11}N$
M_r: 169.21
mp: 81.5
Color: pl (al), nd (peth)
***s*:** EtOH 4; Chl 4
m/e: 169.09

HODOC No: 00007
CAS No: 4657-97-0
Beilstein No: 12^3, 3212
Name: 4-Acenaphthylenamine, 1,2-dihydro-
Synonym: 4-Acenaphthenamine
Mol Form: $C_{12}H_{11}N$
M_r: 169.21
mp: 87
Color: nd (al, w)
***s*:** Bz 4; lig 4
m/e: 169.09

HODOC No: 00008
CAS No: 208-96-8
Beilstein No: 5^4, 2138
Name: Acenaphthylene
Synonym: Cyclopenta[de]naphthalene
Structure:

Mol Form: $C_{12}H_8$
Mr: 152.18
bp: $156\text{-}50^{28}$
mp: 92-3
***d*:** $0.8988^{16/2}$
***s*:** H_20 1; al 4; eth 4; bz 4; chl 2
IR: COB 7227 3065 3045 1930 1810 1730 1620 1495 1480 1460 1415 1390 1225 1205 1180 1150 1080 1030 970 910 865 830 775 720 550
UV: SAD 3631 339(4510) 333(4830) 322(10900) 311(8660) 275(2470) 265(2800) 229 MeOH
H NMR: SAD 16066 7.0 7.3 7.4-7.8 $CDCl_3$
C NMR: STOT 92 124.5 127.5 128.0 128.6 128.9 129.9 140.2
MS: ALDN 108 152(100) 76(17) 151(17) 153(16) 150(11) 63(7) 75(5) 154(3)
m/e: 152.06

HODOC No: 00009
CAS No: 2435-82-7
Beilstein No: 5^3, 2595
Name: Acenaphthylene, 1-(1(2H)-acenaphthylenylidene)-1,2-dihydro-
Synonym: Δ1,1′-Biacenaphthene
Structure:

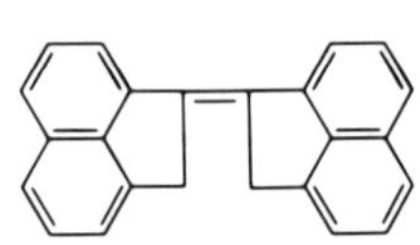

Mol Form: $C_{24}H_{16}$
Mr: 304.37
mp: 277
Color: red-ye pr or nd (Bz)
***s*:** bz 3; CS_2 3
UV: OES 5-807 286(3162) 254(3467) 216(7079) CY
m/e: 304.13

HODOC No: 00010
CAS No: 480-72-8
Beilstein No: 5^2, 1385
Name: Acenaphthylene, 1,2,2a,3,4,5-hexahydro-
Synonym: Acenaphthene, 2a,3,4,5-tetrahydro-
Mol Form: $C_{12}H_{14}$
Mr: 158.23
bp: 246
mp: 12
***s*:** chl 3
IR: µSADP 15887 3030 2940 2860 1610 1450 1430 1330 1250 1050 820 760 720
UV: SAD 4846 267 cyhex
H NMR: SAD 12759 0.8-2.6 2.5-3.0 6.6-7.1 CCl_4
MS: WILEY 985 130(100) 158(34) 129(30) 115(23) 128(15) 131(11) 141(7) 77(6)
m/e: 158.11

HODOC No: 00011
CAS No: 5209-31-4
Beilstein No: 5^4, 1839
Name: Acenaphthylene, 3-bromo-1,2-dihydro-
Synonym: Acenaphthene, 3-bromo-
Mol Form: $C_{12}H_9Br$
Mr: 232.10
mp: 78
Color: nd (MeOH)
m/e: 231.99

HODOC No: 00012
CAS No: 2051-98-1
Beilstein No: 5^1, 276
Name: Acenaphthylene, 5-bromo-1,2-dihydro-
Synonym: Acenaphthene, 5-bromo-
Mol Form: $C_{12}H_9Br$
Mr: 232.10
bp: 335
mp: 52
***d*:** $1.4392^{52/4}$
***n*:** 1.6565^{54}
Color: pl (al)
***s*:** H_2O 1; al 2; chl 2
IR: SADG 18389 3040 2940 1620 1600 1490 1440 1420 1390 1360 1070 830 810 760
UV: SAD 20922 326(2330) 312(5260) 299(8470) 289(7130) 230(68800) MeOH
H NMR: SAD 8587 3.3 7.1 7.3 7.4-7.7 7.8 $CDCl_3$
MS: WILEY 1644 153(100) 152(80) 232(58) 234(56) 76(48) 151(25) 76.5(23) 63(20)
m/e: 231.99

HODOC No: 00013
CAS No: 5209-33-6
Beilstein No: 5^4, 1837
Name: Acenaphthylene, 5-chloro-1,2-dihydro-
Synonym: Acenaphthene, 5-chloro-
Mol Form: $C_{12}H_9Cl$
M_r: 188.64
bp: 319^{770}, 163^{13}
mp: 70.5
***d*:** $1.1954^{70/4}$
***n*:** 1.6169^{100}
Color: pl, nd (al)
m/e: 188.04

HODOC No: 00014
CAS No: 83-32-9
Beilstein No: 5^4, 1834
Name: Acenaphthylene, 1,2-dihydro-
Synonym: Acenaphthene
Structure:

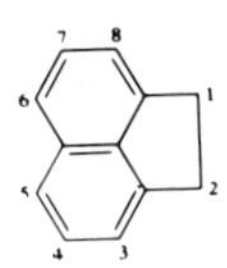

Mol Form: $C_{12}H_{10}$
M_r: 154.20
bp: 279
mp: 96.2
***d*:** $1.0242^{99/4}$
***n*:** 1.6048^{95}
***s*:** H_2O 1; al 2; bz 4; chl 2; HOAc 3
IR: SADG 175 3040 2920 2840 1600 1470 1450 1430 1370 1280 1180 1020 840 800 790 750
Raman: SAD 795 3060 3040 3010 2950 2930 2890 1610 1470 1450 1430 1370 1260 1220 1180 1150 1100 1060 1040 1020 1000 980 940 890 870 840 810 770 750 680 650 550 510 470 450 420 260 230 180 130 100
UV: SAD 272 320(1430) 314(856) 306(2400) 300(3470) 288(5540) 278 243 227 MeOH
H NMR: SAD 65 7.2 7.4 7.5 $CDCl_3$
C NMR: JJ 440 FT 30.1 118.9 122.0 127.6 131.5 139.1 145.7 $CDCl_3$
MS: WILEY 928 154(100) 153(86) 152(38) 76(20) 151(14) 155(12) 77(8) 76.5(6)
m/e: 154.08

HODOC No: 00015
CAS No: 4406-87-5
Beilstein No: 5^3, 1784
Name: Acenaphthylene, 1,2-dihydro-5,6-dinitro-
Synonym: Acenaphthene, 5,6-dinitro-
Mol Form: $C_{12}H_8N_2O_4$
M_r: 244.19
mp: 220-4
IR: COB 2563 3050 1620 1600 1580 1520 1440 1330 880 860 800 740
UV: SAD 21675 350(8630) 239(24500) MeOH
m/e: 244.05

HODOC No: 00016
CAS No: 6861-64-9
Beilstein No: 5^1, 276
Name: Acenaphthylene, 1,2-dihydro-5-iodo-
Synonym: Acenaphthene, 5-iodo-
Mol Form: $C_{12}H_9I$
M_r: 280.10
mp: 65
***d*:** $1.6738^{62/4}$
***n*:** 1.6909^{65}
Color: nd (al)
***s*:** Bz 4; EtOH 4
m/e: 279.97

HODOC No: 00017
CAS No: 3807-77-0
Beilstein No: 5^3, 1783
Name: Acenaphthylene, 1,2-dihydro-3-nitro-
Synonym: Acenaphthene, 3-nitro-
Mol Form: $C_{12}H_9NO_2$
M_r: 199.20
mp: 151.5
Color: gr-ye nd (HOAc)
***s*:** H_2O 1; sulf 3; con alk 3; HOAc 3
IR: COB 4935 2900 1600 1510 1490 1400 1350 1300 1200 1150
m/e: 199.06

HODOC No: 00018
CAS No: 602-87-9
Beilstein No: 5^3, 1840
Name: Acenaphthylene, 1,2-dihydro-5-nitro-
Synonym: Acenaphthene, 5-nitro-
Mol Form: $C_{12}H_9NO_2$
M_r: 199.20
mp: 103-4
***s*:** H_2O 3; al 3; eth 3; lig 3
IR: COB 2657 2960 1630 1600 1520 1460 1430 1410 1330 1280 1200 1180 1150 1110 1050 910 860 840 830 820 810 780 730 710
UV: SAD 3364 259 216 MeOH
H NMR: SAD 18055 3.1 6.9 7.0-7.6 8.0 8.1 tfa
m/e: 199.06

HODOC No: 00019
CAS No: 55334-13-9
Name: Acenaphthylene, 1,2-dihydro-5-pentadecyl-
Synonym: Acenaphthene, 5-pentadecyl
Mol Form: $C_{27}H_{40}$
M_r: 364.59
UV: API 519 324 311 305 296 284 231 iso
MS: WILEY 2069 167(100) 364(59) 43(32) 41(24) 29(22) 365(18) 168(16) 165(13)
m/e: 364.31

HODOC No: 00020
CAS No: 55282-69-4
Name: Acenaphthylene, dodecahydro-5-pentadecyl-
Synonym: Acenaphthene, decahydro-5-pentadecyl
Mol Form: $C_{27}H_{50}$
M_r: 374.67
MS: WILEY 2105 121(100) 163(88) 43(66) 81(62) 41(50) 67(49) 122(42) 374(19)
m/e: 374.39

HODOC No: 00021
CAS No: 55334-56-0
Name: Acenaphthylene, 6-dodecyl-1,2,2a,3,4,5-hexahydro-
Mol Form: $C_{24}H_{38}$
M_r: 326.54
MS: WILEY 1944 171(100) 157(44) 326(35) 172(25) 129(14) 128(14) 167(14) 143(14)
m/e: 326.30

HODOC No: 00022
CAS No: 5345-46-0
Beilstein No: 7^3, 2097
Name: 5-Acenaphthylenecarboxaldehyde, 1,2-dihydro-
Synonym: 5-Acenaphthenecarboxaldehyde
Mol Form: $C_{13}H_{10}O$
M_r: 182.21
mp: 107-8
UV: FO 261 332(10000) 253(25119) 245(25119) 232(19953) 227(19953) EtOH
m/e: 182.07

HODOC No: 00023
CAS No: 6833-51-8
Name: 1-Acenaphthylenecarboxylic acid, 1,2-dihydro-
Synonym: 1-Acenaphthenecarboxylic acid
Mol Form: $C_{13}H_{10}O_2$
M_r: 198.21
mp: 161
Color: nd (dil al, w)
m/e: 198.07

HODOC No: 00024
CAS No: 82-86-0
Beilstein No: 7^4, 2498
Name: 1,2-Acenaphthylenedione
Synonym: Acenaphthenequinone
Structure:

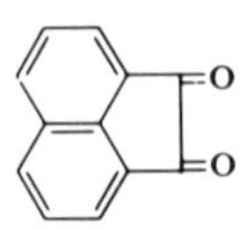

Mol Form: $C_{12}H_6O_2$
M_r: 182.17
bp: sub
mp: 261
Color: ye nd (HOAc)
***s*:** H_2O 1; al 2; bz 2; lig 3; HOAc 2
IR: μSADP 14633 3030 1790 1720 1610 1490 1430 1280 1220 1020 890 830 780
UV: SAD 4140 312 301 225 219 MeOH
H NMR: SAD 18777 7.9 8.1 8.3 tfa
m/e: 182.04

HODOC No: 00025
CAS No: 50817-70-4
Beilstein No: 11^3, 435
Name: 3-Acenaphthylenesulfonic acid, 1,2-dihydro-
Synonym: 3-Acenaphthenesulfonic acid
Structure:

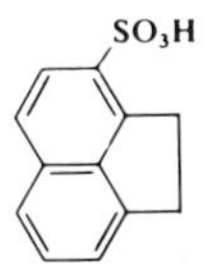

Mol Form: $C_{12}H_{10}O_3S$
M_r: 234.26
mp: 121
Color: nd (al)
***s*:** H_2O 4; Bz 4; EtOH 4; Chl 4
m/e: 234.04

HODOC No: 00026
CAS No: 6306-07-6
Beilstein No: 6^4, 4623
Name: 1-Acenaphthylenol, 1,2-dihydro-
Synonym: 1-Acenaphthenol
Mol Form: $C_{12}H_{10}O$
M_r: 170.20
mp: 147-8
IR: ALD 2:615D
MS: ALD 5:198
m/e: 170.07

HODOC No: 00027
CAS No: 2235-15-6
Beilstein No: 7^4, 1354
Name: 1(2H)-Acenaphthylenone
Synonym: 1-Acenaphthenone
Structure:

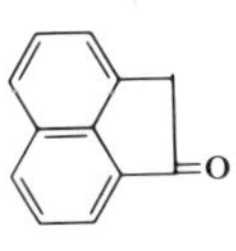

Mol Form: $C_{12}H_8O$
M_r: 168.18
mp: 121
***s*:** H_2O 3,al 3; bz 4; chl 3; sulf 3; NaOH 3
IR: COB 4934 2900 1700 1600 1480 1440 1350 1260 1230 1180 1100
UV: OES 5-295 338(5012) 320(4898) 244(31623) EtOH
m/e: 168.06

HODOC No: 00028
CAS No: 5386-09-4
Beilstein No: 7^4, 1354
Name: 1(2H)-Acenaphthylenone, 6-bromo-
Synonym: 1-Acenaphthenone, 6-bromo-
Mol Form: $C_{12}H_7BrO$
M_r: 246.08
mp: 132-4
IR: COB 4936 2900 1700 1560 1480 1410 1370 1350 1230 1020 810 760
m/e: 245.97

HODOC No: 00029
CAS No: 85925-12-8
Beilstein No: 7^4, 1914
Name: 1(2H)-Acenaphthylenone, 2,2-diphenyl-
Synonym: 1-Acenaphthenone, 2,2-diphenyl-
Structure:

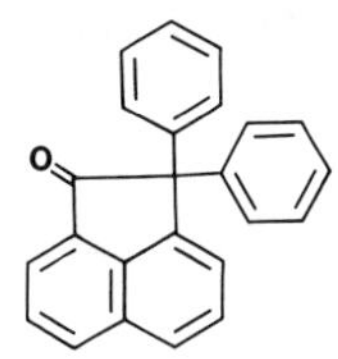

Mol Form: $C_{24}H_{16}O$
M_r: 320.37
mp: 172-3
***s*:** chl 3
IR: SADG 21885 3060 1710 1600 1490 1440 1350 1250 1110 990 830 780 750 700 650
UV: SAD 2713 345(5850) 320(5290) 255(16000) 215(52600) MeOH
H NMR: SAD 11985 7.2 7.3-8.0 8.1 $CDCl_3$
m/e: 320.12

HODOC No: 00030
CAS No: 5386-11-8
Name: 1(2H)-Acenaphthylenone, 5-nitro-
Synonym: 1-Acenaphthcnone, 5-nitro-
Mol Form: $C_{12}H_7NO_3$
M_r: 213.18
mp: 156-7
IR: COB 4938 2930 1720 1610 1510 1490 1400 1350 1250 1200
m/e: 213.04

HODOC No: 00031
CAS No: 75-07-0
Beilstein No: 1^3, 2617
Name: Acetaldehyde
Synonym: Acetic aldehyde
Line Form: CH_3CHO
Mol Form: C_2H_4O
M_r: 44.05
bp: 20.8
mp: -121
***d*:** $0.7834^{18/4}$
***n*:** 1.3316^{20}
***s*:** H_2O 5; al 5; eth 5; bz 5; chl 2
IR: COB 5645 3000 2830 2720 1730 1430 1400 1350 1170 1110 860
UV: OES 5-4 292(10) hp
H NMR: VAR 6 2.2 9.8 $CDCl_3$
C NMR: JJ 5 FT 30.7 199.7 $CDCl_3$
MS: WILEY 12 29(100) 44(77) 43(42) 15(32) 42(12) 14(10) 26(5) 41(5)
m/e: 44.03

HODOC No: 00032
CAS No: 107-20-0
Beilstein No: 1^4, 3134
Name: Acetaldehyde, chloro-
Synonym: Chloroacetaldehyde
Line Form: $ClCH_2CHO$
Mol Form: C_2H_3ClO
M_r: 78.50
bp: 85
***s*:** eth 3
UV: OES 5-3 298(35) cyhex
MS: WILEY 97 50(100) 15(78) 29(45) 78(37) 52(32) 49(30) 14(24) 80(12)
m/e: 77.99

HODOC No: 00033
CAS No: 811-96-1
Name: Acetaldehyde, chlorodifluoro-
Synonym: Chlorodifluoroacetaldehyde
Line Form: ClF_2CCHO
Mol Form: C_2HClF_2O
M_r: 114.48
MS: WILEY 394 29(100) 51(41) 79(39) 31(14) 50(12) 67(11) 28(11) 114(9)
m/e: 113.97

HODOC No: 00034
CAS No: 24547-76-0
Name: Acetaldehyde, [(*p*-chloro-α-methylbenzylidene)amino]-, diethyl acetal
Synonym: Acetaldehyde, 2(4-chloro-α-methylbenzylideneamino), diethyl acetal
Mol Form: $C_{14}H_{20}ClNO_2$
M_r: 269.76
bp: $148\text{-}51^1$
UV: LANG 1945 250(15849) hx
m/e: 269.12

HODOC No: 00035
CAS No: 79-02-7
Beilstein No: 1^4, 3140
Name: Acetaldehyde, dichloro-
Synonym: Chloroaldehyde
Line Form: Cl_2CHCHO
Mol Form: $C_2H_2Cl_2O$
M_r: 112.94
bp: 90-1
***s*:** al 2
IR: µSADP 8040 2940 2700 1750 1350 1220 1060 960 810 730
UV: OES 5-3 302(39) cyhex
m/e: 111.95

HODOC No: 00036
CAS No: 84-83-3
Name: Acetaldehyde, (1,3-dihydro-1,3,3-trimethyl-2H-indol-2-ylidene)-
Synonym: Δ2,α-Indolineacetaldehyde, 1,3,3-trimethyl-
Mol Form: $C_{13}H_{15}NO$
M_r: 201.26
***s*:** chl 3
H NMR: VAR 610 1.7 3.2 5.4 6.7-7.4 10.0 $CDCl_3$
m/e: 201.12

HODOC No: 00037
CAS No: 1019-57-4
Beilstein No: 15^3, 426
Name: Acetaldehyde, (2,4-dinitrophenyl)hydrazone
Synonym: Ethanal-2,4-dinitrophenylhydrazone
Line Form: $2,4\text{-}(NO_2)_2C_6H_3\text{-}NHN{=}CHCH_3$
Mol Form: $C_8H_8N_4O_4$
M_r: 224.17
mp: 168.5
Color: ye sc (al)
***s*:** Bz 4; Et_2O 4; EtOH 4; Chl 4
IR: COB 2094
UV: SAD 1193
H NMR: SAD 18336
MS: ALDN 183
m/e: 224.05

HODOC No: 00038
CAS No: 5314-41-0
Beilstein No: 1^2, 673
Name: Acetaldehyde, di-*sec*-butyl acetal
Synonym: Acetaldehyde, di-*sec*-butyl acetal
Line Form: $CH_3CH\{OCH(CH_3)CH_2CH_3\}_2$
Mol Form: $C_{10}H_{22}O_2$
M_r: 174.27
bp: 171
MS: WILEY 1182 45(100) 101(99) 57(77) 41(24) 29(19) 83(17) 56(11) 174(0)
m/e: 174.16

HODOC No: 00039
CAS No: 22056-82-2
Beilstein No: 1^3, 3181
Name: Acetaldehyde, ethoxy-
Synonym: Ethoxyacetaldehyde
Line Form: $CH_3CH_2OCH_2CHO$
Mol Form: $C_4H_8O_2$
M_r: 88.10
bp: 71-3
***d*:** $0.942^{20/4}$
***n*:** 1.3956^{20}
***s*:** H_2O 4; Ac 4; EtOH 4
m/e: 88.05

HODOC No: 00040
CAS No: 592-56-3
Beilstein No: 1, 609
Name: Acetaldehyde, ethylidenehydrazone
Synonym: Acetaldehyde, azine
Line Form: $CH_3CH{=}NH{=}CHCH_3$
Mol Form: $C_4H_8N_2$
M_r: 84.12
bp: 95-6
***d*:** 0.832^{17}
IR: OES 3-21
UV: OES 3-21 <220 MeOH
MS: WILEY 124 42(100) 15(51) 28(50) 69(41) 27(31) 26(27) 84(25) 41(12)
m/e: 84.07

HODOC No: 00041
CAS No: 141-46-8
Beilstein No: 1^2, 3181
Name: Acetaldehyde, hydroxy-
Synonym: Diose
Line Form: $HOCH_2CHO$
Mol Form: $C_2H_4O_2$
M_r: 60.05
mp: 97
***d*:** 1.366^{100}
***n*:** 1.4772^{19}
Color: pl
***s*:** chl 3
IR: µSADP 17434 3330 2940 1430 1350 1270 1140 1100 1080 1020 900 870 810
MS: WILEY 36 31(100) 32(48) 29(40) 28(13) 60(7) 15(6) 30(6) 42(5)
m/e: 60.02

HODOC No: 00042
CAS No: 10312-83-1
Beilstein No: 1^4, 3955
Name: Acetaldehyde, methoxy-
Synonym: Methoxyacetaldehyde
Line Form: $CH_3O\text{-}CH_2CHO$
Mol Form: $C_3H_6O_2$
M_r: 74.08
bp: $92\text{-}3^{770}$
***d*:** $1.005^{25/4}$
***n*:** 1.3950^{20}
***s*:** H_2O 4; Ac 4; Et_2O 4; EtOH 4
m/e: 74.04

HODOC No: 00043
CAS No: 24547-79-3
Name: Acetaldehyde, [(α-methylbenzylidene)amino]-, diethyl acetal
Synonym: Acetaldehyde, 2(α-methylbenzylideneamino),diethyl acetal
Line Form: $C_6H_5C(CH_3)=NCH_2CH(OCH_2CH_3)_2$
Mol Form: $C_{14}H_{21}NO_2$
M$_r$: 235.31
bp: $124\text{-}5^{1}$
UV: LANG 1943 238(13183) hx
m/e: 235.16

HODOC No: 00044
CAS No: 24547-75-9
Name: Acetaldehyde, [[1-(2-naphthyl)ethylidene]amino]-, diethyl acetal
Synonym: Acetaldehyde, 2(1(2-naphthyl) ethylideneamino),diethyl acetal
Mol Form: $C_{18}H_{23}NO_2$
M$_r$: 285.37
bp: $188\text{-}90^{1}$
UV: LANG 11947 340(1072) 323(1148) 280(9772) 245(43652) hx
m/e: 285.17

HODOC No: 00045
CAS No: 5653-21-4
Beilstein No: 1^{3}, 2678
Name: Acetaldehyde, nitro-, oxime
Synonym: Ethanedial, dioxime *N*-oxide
Line Form: $O_2NCH_2CH=NOH$
Mol Form: $C_2H_4N_2O_3$
M$_r$: 104.06
mp: 79-80
***s*:** H_2O 3; al 3; eth 3; ace 3; bz 3; chl 3
UV: OES 5-4 298(7762) H_2O
m/e: 104.02

HODOC No: 00046
CAS No: 600-26-0
Beilstein No: 2^{4}, 434
Name: Acetaldehyde, 1-nitro-, oxime
Synonym: Acetonitrolic acid
Line Form: $CH_3C(=NOH)NO_2$
Mol Form: $C_2H_4N_2O_3$
M$_r$: 104.06
bp: 87-8d
Color: ye rh (w, al, eth)
***s*:** H_2O 4; Ac 4; Et_2O 4; EtOH 4
m/e: 104.02

HODOC No: 00047
CAS No: 107-29-9
Beilstein No: 1^{4}, 3121
Name: Acetaldehyde, oxime
Synonym: Acetaldoxime
Line Form: $CH_3CH=NOH$
Mol Form: C_2H_5NO
M$_r$: 59.07
bp: 115
mp: 44-6
***d*:** $0.9656^{20/4}$
***n*:** 1.4264^{20}
Color: nd
***s*:** H_2O 3; al 5; eth 5; chl 3
IR: μSADP 15391 3330 2940 1670 1450 1390 1320 1110 990 940 890 830 690
UV: OES 3-4 190(5012)
H NMR: VAR 373 1.9 6.8 7.4 8.9 $CDCl_3$
m/e: 59.04

HODOC No: 00048
CAS No: 24547-78-2
Name: Acetaldehyde, [(*p*,α-dimethylbenzylidene)amino]-, diethyl acetal
Synonym: Toluene, 4(1(2,2-diethoxyethyl)iminoethyl)
Mol Form: $C_{15}H_{23}NO_2$
M$_r$: 249.34
bp: $150\text{-}2^{2}$
UV: LANG 1944 247(14791) hx
m/e: 249.17

HODOC No: 00049
CAS No: 935-07-9
Beilstein No: 15^{3}, 79
Name: Acetaldehyde, phenylhydrazone
Line Form: $CH_3CH=NNHC_6H_5$
Mol Form: $C_8H_{10}N_2$
M$_r$: 134.17
bp: $133\text{-}6^{21}$
mp: 98-101
***s*:** EtOH 4
m/e: 134.08

HODOC No: 00050
CAS No: 66056-06-2
Beilstein No: 19^4, 5643
Name: Acetaldehyde, tetramer
Synonym: Metaldehyde II (tetramer)
Structure:

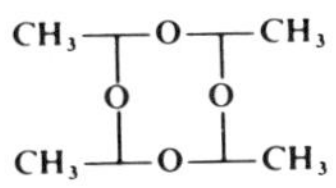

Mol Form: C_2H_4DO
M_r: 46.07
bp: 110, 65^{15}
mp: 47
***s*:** Ac 4; Bz 4; Et_2O 4; EtOH 4
m/e: 46.04

HODOC No: 00051
CAS No: 115-17-3
Beilstein No: 1^4, 3155
Name: Acetaldehyde, tribromo-
Synonym: Bromal
Line Form: Br_3CCHO
Mol Form: C_2HBr_3O
M_r: 277.74
bp: 174, 61^9
***d*:** $2.6650^{25/4}$
***n*:** 1.5939^{20}
***s*:** Ac 4; Et_2O 4; EtOH 4
m/e: 277.76

HODOC No: 00052
CAS No: 75-87-6
Beilstein No: 1^4, 3142
Name: Acetaldehyde, trichloro-
Synonym: Anhydrous chloral
Line Form: Cl_3CCHO
Mol Form: C_2HCl_3O
M_r: 147.39
bp: 97.8
mp: -57.5
***d*:** $1.5121^{20/4}$
***n*:** 1.4580^{20}
***s*:** H_2O 4; al 3; eth 3
IR: μSADP 4626 2860 1750 1350 1020 990 850 740
UV: OES 5-3 290(38) hx
H NMR: SAD 8241 8.9 CCl_4
MS: WILEY 814 82(100) 84(66) 29(55) 47(47) 83(39) 111(35) 85(23) 146(4)
m/e: 145.91

HODOC No: 00053
CAS No: 60-35-5
Beilstein No: 2^4, 399
Name: Acetamide
Synonym: Acetic acid amide
Line Form: CH_3CONH_2
Mol Form: C_2H_5NO
M_r: 59.07
bp: 221.2, 120^2
mp: 80.1
***d*:** $0.9986^{85/4}$
***n*:** 1.4278
Color: trg mcl (al-eth)
***s*:** H_2O 3; al 4; eth 1; bz 2; chl 3; py 3
IR: μCOB 2552 3330 2940 1670 1640 1470 1410 1150 710
Raman: DFB 50 3340 3160 3010 2980 2940 1650 1590 1410 1360 1150 880 580 460 120 90
UV: OES 6-4 185(8710) H_2O
H NMR: SAD 4280 2.0 D_2O
MS: WILEY 33 44(100) 59(93) 43(75) 15(48) 42(38) 28(35) 41(20) 40(12)
m/e: 59.04

HODOC No: 00054
CAS No: 625-77-4
Beilstein No: 2^4, 416
Name: Acetamide, *N*-acetyl-
Synonym: *N*-Acetylacetamide
Line Form: $(CH_3CO)_2NH$
Mol Form: $C_4H_7NO_2$
M_r: 101.10
bp: 223.5, 113^{13}
mp: 79
Color: nd (eth)
***s*:** H_2O 3; al 3; eth 3; chl 3; lig 3
IR: SADG 1854 3250 2990 1740 1710 1700 1500 1380 1300 1230 1030
UV: OES 6-25 206(13490) EtOH
H NMR: SAD 427 2.3 9.5 $CDCl_3$
MS: ALDN 36 42(100) 59(56) 73(22) 42(17) 15(16) 44(15) 101(13) 41(11)
m/e: 101.05

HODOC No: 00055
CAS No: 3027-00-7
Beilstein No: 12, 612
Name: Acetamide, *N*-acetyl-*N*-(4-chlorophenyl)-
Synonym: Diacetanilide, 4″-chloro-
Mol Form: $C_{10}H_{10}ClNO_2$
M$_r$: 211.64
mp: 66-7
***s*:** al 4; eth 4; bz 4
UV: OES 1-317 251(16982) EtOH
m/e: 211.04

HODOC No: 00056
CAS No: 1563-83-3
Beilstein No: 4^4, 352
Name: Acetamide, *N*-acetyl-*N*-ethyl-
Synonym: Diacetamide, *N*-ethyl-
Line Form: $(CH_3CO)_2NC_2H_5$
Mol Form: $C_6H_{11}NO_2$
M$_r$: 129.15
bp: 195-9
***d*:** 1.0092^{20}
***n*:** 1.4513^{20}
***s*:** EtOH 4
m/e: 129.08

HODOC No: 00057
CAS No: 1113-68-4
Beilstein No: 4^4, 183
Name: Acetamide, *N*-acetyl-*N*-methyl-
Synonym: Diacetamide, *N*-methyl-
Line Form: $(CH_3CO)_2NCH_3$
Mol Form: $C_5H_9NO_2$
M$_r$: 115.13
bp: 194.5^{741}, 114.5^{61}
mp: -25
***d*:** $1.0663^{25/4}$
***n*:** 1.4502^{25}
***s*:** H_2O 5; eth 1
UV: SAD 19964
H NMR: SAD 19964 2.4 3.2 $CDCl_3$
MS: ALDN 54 43(100) 73(68) 56(41) 15(34) 30(34) 58(31) 87(27) 115(10)
m/e: 115.06

HODOC No: 00058
CAS No: 2243-69-8
Name: Acetamide, *N*-[4-(acetyloxy)-2-nitrophenyl]-
Synonym: Acetanilide, 4′-hydroxy-2′-nitro-, acetate
Mol Form: $C_{10}H_{10}N_2O_5$
M$_r$: 238.19
mp: 146
IR: µCOB 2881 3330 3130 2940 1750 1700 1590 1520 1470 1370 1350 1270 1220 1140 1020 940 900 850 760
UV: SAD 17921 351(2100) 234(18900) MeOH
m/e: 238.06

HODOC No: 00059
CAS No: 5467-64-1
Name: Acetamide, *N*-[2-(acetyloxy)phenyl]-
Synonym: 2-Acetamidophenol acetate
Mol Form: $C_{10}H_{11}NO_3$
M$_r$: 193.19
***s*:** chl 2
IR: µSADP 12107 3335 2940 1725 1665 1640 1615 1515 1450 1370 1315 1280 1250 1220 1190 1165 1100 1040 935 910 805 745 685
H NMR: SAD 11721 2.1 2.3 7.0-7.4 8.0 $CDCl_3$
m/e: 193.07

HODOC No: 00060
CAS No: 2623-33-8
Beilstein No: 13^3, 1064
Name: Acetamide, *N*-[4-(acetyloxy)phenyl]-
Synonym: Acetanilide, 4′-hydroxy-, acetate (ester)
Line Form: $4\text{-}(CH_3CO_2\text{-}C_6H_4)NHCOCH_3$
Mol Form: $C_{10}H_{11}NO_3$
M$_r$: 193.19
mp: 148-50
***s*:** chl 3
IR: µCOB 3073 3330 2940 1720 1700 1610 1540 1520 1470 1410 1370 1320 1240 1190 1020 920 860 790 690
UV: SAD 1603 245 MeOH
H NMR: SAD 6476 2.1 2.3 7.0 7.5 7.7 $CDCl_3$
m/e: 193.07

HODOC No: 00061
CAS No: 2719-21-3
Name: Acetamide, *N*-(4-acetylphenyl)-
Synonym: Acetanilide, 4′-acetyl-
Mol Form: $C_{10}H_{11}NO_2$
M_r: 177.19
mp: 166-7
***s*:** H_2O 2; al 4; eth 2; bz 2; lig 3; tol 3
IR: µSADP 5097 3230 2860 1670 1590 1520 1450 1390 1350 1320 1250 1180 830 760 730
UV: HBCP 283(33885) al
m/e: 177.08

HODOC No: 00062
CAS No: 1563-87-7
Beilstein No: 12^3, 472
Name: Acetamide, *N*-acetyl-*N*-phenyl-
Synonym: Benzenamine, *N*,*N*-diacetyl-
Line Form: $(CH_3CO)_2NC_6H_5$
Mol Form: $C_{10}H_{11}NO_2$
M_r: 177.19
bp: 200^{100}, 142^{41}
mp: 37-8
Color: ta (lig)
***s*:** H_2O 2; al 3; bz 3; tol 3
UV: HBCP 242(1023) chl
MS: NBS 858
m/e: 177.08

HODOC No: 00063
CAS No: 598-41-4
Beilstein No: 4^4, 2358
Name: Acetamide, 2-amino-
Synonym: Glycinamide
Line Form: $H_2NCH_2CONH_2$
Mol Form: $C_2H_6N_2O$
M_r: 74.08
mp: 67-8
Color: hyg nd (Chl)
***s*:** H_2O 4; al 4; eth 2; ace 3; bz 2; chl 3; MeOH 4; AcOEt 3
UV: OES 2-5 210(126) pH=6 213(158) pH=9
m/e: 74.05

HODOC No: 00064
CAS No: 1668-10-6
Beilstein No: 4^1, 468
Name: Acetamide, 2-amino-, monohydrochloride
Synonym: Glycinamide, monohydrochloride
Line Form: $H_2NCH_2CONH{-}HCl$
Mol Form: $C_2H_7ClN_2O$
M_r: 110.54
mp: dec 204
***s*:** H_2O 4; al 2
IR: SADG 17216 3330 3100 2930 2860 2560 1700 1590 1460 1410 1320 1030 890
H NMR: SAD 9497 3.9 D_2O
m/e: 110.02

HODOC No: 00065
CAS No: 591-07-1
Beilstein No: 3^3, 119
Name: Acetamide, *N*-(aminocarbonyl)-
Synonym: Urea, acetyl-
Line Form: $CH_3CONHCONH_2$
Mol Form: $C_3H_6N_2O_2$
M_r: 102.09
bp: 180-90 sub
mp: 218
***s*:** H_2O 2; al 3; eth 2
IR: µCOB 1806 3450 3330 3230 2860 1670 1470 1430 1370 1250 1100 1030 940 810 760 690
UV: OES 6-10 191(11482) EtOH
MS: NBS 4011
m/e: 102.04

HODOC No: 00066
CAS No: 591-38-8
Beilstein No: 3^4, 142
Name: Acetamide, *N*-[[(aminocarbonyl)amino]carbonyl]-
Synonym: Biuret, 1-acetyl-
Line Form: $CH_3CONHCONHCONH_2$
Mol Form: $C_4H_7N_3O_3$
M_r: 145.11
mp: 193-4
Color: nd (w, al)
***s*:** H_2O 4; EtOH 4
m/e: 145.05

HODOC No: 00067
CAS No: 1448-98-2
Name: Acetamide, *N*-(aminocarbonyl)-2-cyano-
Synonym: Urea, (cyanoacetyl)-
Line Form: $H_2NCONHCOCH_2CN$
Mol Form: $C_4H_5N_3O_2$
M$_r$: 127.10
mp: d 214
***s*:** DMSO 3
IR: COB 6438 3420 3320 3280 2960 2940 2270 1710 1690 1590 1490 1390 1330 1210 1200 1090 970 810 640 580 450
H NMR: SAD 5026 3.8 7.3 10.3 $DMSOd_6$
m/e: 127.04

HODOC No: 00068
CAS No: 760-41-8
Beilstein No: 3^4, 115
Name: Acetamide, *N*-(aminocarbonyl)-2,2,2-trifluoro-
Synonym: Trifluoroacetylurea
Line Form: $(F_3CCO)NHCONH_2$
Mol Form: $C_3H_3F_3N_2O_2$
M$_r$: 156.06
mp: 189
IR: µCOB 3412 3330 1720 1700 1610 1520 1450 1410 1370 1320 1190 1160 1100 950 790 770 740 710
m/e: 156.01

HODOC No: 00069
CAS No: 50609-15-9
Name: Acetamide, *N*-(4-amino-1,6-dihydro-6-oxo-5-pyrimidinyl)-
Synonym: 4-Pyrimidinol, 5-acetamido-6-amino-
Mol Form: $C_6H_8N_4O_2$
M$_r$: 168.15
IR: COB 7439 3360 3150 1655 1605 1545 1460 1380 1305 1245 1005 970 890 790
m/e: 168.06

HODOC No: 00070
CAS No: 38093-62-8
Name: Acetamide, *N*-(2-amino-1,1-diphenyl-2-thioxoethyl)-
Synonym: Acetic acid, amide,thiono,acetamido-diphenyl
Line Form: $(C_6H_5)_2C(NHCOCH_3)C(=S)NH_2$
Mol Form: $C_{16}H_{16}N_2OS$
M$_r$: 284.36
mp: dec 167
UV: LANG 2486 276(7413) 202(37154) EtOH
m/e: 284.10

HODOC No: 00071
CAS No: 1001-53-2
Beilstein No: 4^4, 1193
Name: Acetamide, *N*-(2-aminoethyl)-
Synonym: 2-Acetamido-1-ethanamine
Line Form: $CH_3CONHCH_2CH_2NH_2$
Mol Form: $C_4H_{10}N_2O$
M$_r$: 102.13
bp: 128^3
mp: 51
***s*:** H_2O 3; al 3; eth 1; bz 3
IR: µSADP 19170 3330 2940 1610 1560 1470 1430 1390 1330 1300 1040
m/e: 102.08

HODOC No: 00072
CAS No: 62-47-5
Name: Acetamide, *N*-(2-aminoethyl)-2-mercapto-
Synonym: *N*-(2-Aminoethyl)mercaptoacetamide
Line Form: $HSCH_2CONHCH_2CH_2NH_2$
Mol Form: $C_4H_{10}N_2OS$
M$_r$: 134.19
mp: 120-2
IR: COB 4570 3260 3020 2950 2700 2480 2100 1630 1600 1590 1520 1450 1440 1360 1300 1240 1170 1050 910 870 840 770 700
m/e: 134.05

HODOC No: 00073
CAS No: 6957-50-2
Name: Acetamide, *N*-(7-amino-9H-fluoren-2-yl)-
Synonym: Acetamide, *N*-(7-aminofluoren-2-yl)-
Mol Form: $C_{15}H_{14}N_2O$
M$_r$: 238.27
IR: COB 7006 3360 3250 3050 2930 1670 1610 1590 1540 1470 1370 1310 1270 1130 830 730 710 610
m/e: 238.11

HODOC No: 00074
CAS No: 92245-40-4
Name: Acetamide, 2-amino-2-(*o*-hydroxyphenyl)-2-phenyl-
Synonym: Glycine, amide,2(2-hydroxyphenyl)-2-phenyl
Mol Form: $C_{14}H_{14}N_2O_2$
M$_r$: 242.26
IR: µCOB 5323 3450 3130 1700 1610 1590 1540 1490 1250 1210 1110 1030 970 890 820 760 700
m/e: 242.11

HODOC No: 00075
CAS No: 555-48-6
Beilstein No: 13^3, 41
Name: Acetamide, 2-amino-*N*-phenyl-
Synonym: Acetanilide, 2-amino-
Line Form: $CH_3CONH(C_6H_9NH_2\text{-}2)$
Structure:

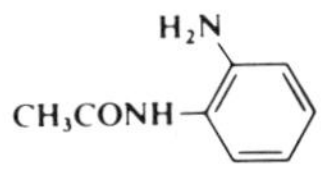

Mol Form: $C_8H_{10}N_2O$
M_r: 150.17
mp: 132
Color: pl (Bz)
***s*:** H_2O 4; Bz 4; EtOH 4
UV: OES 8-95
m/e: 150.08

HODOC No: 00076
CAS No: 34801-09-7
Beilstein No: 13^2, 15
Name: Acetamide, *N*-(2-aminophenyl)-
Synonym: 2-Acetamidoaniline
Mol Form: $C_8H_{10}N_2O$
M_r: 150.17
mp: 132
***s*:** H_2O 3; al 3; eth 2; bz 3
UV: OES 8-95 291(9333) MeOH
m/e: 150.08

HODOC No: 00077
CAS No: 102-28-3
Beilstein No: 13^3, 75
Name: Acetamide, *N*-(3-aminophenyl)-
Synonym: Acetanilide, 3′-amino-
Line Form: $(3\text{-}H_2NC_6H_4)NHCOCH_3$
Mol Form: $C_8H_{10}N_2O$
M_r: 150.17
mp: 87-9
Color: nd or pl (Bz)
***s*:** H_2O 4; al 4; eth 2; ace 4; bz 2
IR: SADG 31875 1670 1610 1550 1490 1450 1370 1330 1260 1160 1020 850 780 690 540 450
UV: OES 8-95 296(2512) MeOH
m/e: 150.08

HODOC No: 00078
CAS No: 59736-00-4
Beilstein No: 13^2, 27
Name: Acetamide, *N*-(3-aminophenyl)-, hydrochloride
Synonym: Acetic acid, amide,*N*(3-aminophenyl), hydrochloride
Mol Form: $C_8H_{11}ClN_2O$
M_r: 186.63
mp: 248-51
***s*:** H_2O 4; al 3; eth 1; bz 1
IR: COB 6285 3400 3250 2850 2630 1650 1590 1560 1540 1490 1450 1440 1380 1330 1310 1300 1270 1170 1030 960 880 790 690 620 530
UV: SAD 1604 225 MeOH
m/e: 186.06

HODOC No: 00079
CAS No: 122-80-5
Beilstein No: 13^3, 166
Name: Acetamide, *N*-(4-aminophenyl)-
Synonym: Acetanilide, 4′-amino-
Line Form: $(4\text{-}H_2NC_6H_4)NHCOCH_3$
Mol Form: $C_8H_{10}N_2O$
M_r: 150.17
bp: 267
mp: 165-8
Color: nd (w)
***s*:** H_2O 3; al 4; eth 4
IR: SADG 281 3370 3300 3120 3060 1670 1600 1550 1520 1430 1370 1320 1260 830 790 700
UV: SAD 510 259(14800) MeOH
m/e: 150.08

HODOC No: 00080
CAS No: 144-80-9
Beilstein No: 14^3, 1949
Name: Acetamide, *N*-[(4-aminophenyl)sulfonyl]-
Synonym: Acetamide, *N*-sulfanilyl-
Line Form: $4\text{-}H_2NC_6H_4SO_2NHCOCH_3$
Mol Form: $C_8H_{10}N_2O_3S$
M_r: 214.23
mp: 182-4
***s*:** H_2O 2; al 3; eth 1; ace 4; alk 4
IR: SADG 20250 3470 3380 3250 3080 2880 1690 1640 1590 1500 1470 1410 1370 1320 1240 1180 11401080 980 850 830 680 630 620 590 530
UV: HBCP 257(22387) al
H NMR: SAD 9808 2.0 6.0 6.7 7.6 DMSOd6
m/e: 214.04

HODOC No: 00081
CAS No: 38093-65-1
Name: Acetamide, *N*-(2-amino-1-phenyl-2-thioxoethyl)-
Synonym: Acetic acid, amide,thiono,acetamido-phenyl
Line Form: $C_6H_5CH(NHCOCH_3)C(=S)NH_2$
Mol Form: $C_{10}H_{12}N_2OS$
M_r: 208.27
mp: 139
UV: LANG 2458 345(63) 273(10000) EtOH
m/e: 208.07

HODOC No: 00082
CAS No: 69986-22-7
Name: Acetamide, 2-[4-(aminosulfonyl)phenoxy]-
Synonym: Acetamide, 2-(*p*-sulfamoylphenoxy)-
Mol Form: $C_8H_{10}N_2O_4S$
M_r: 230.23
mp: 208
***s*:** DMSO 3
IR: SADG 19387 3440 3380 3260 3160 2940 1670 1590 1500 1420 1330 1250 1150 1090 1050 890 840 740 700 650 610 550
UV: SAD 17528 278(640) 268(955) 261(1010) 234(16600) MeOH
H NMR: SAD 9585 4.6 7.1 7.2 7.5 7.8 $DMSOd_6$
m/e: 230.04

HODOC No: 00083
CAS No: 23010-42-6
Beilstein No: 14, 718
Name: Acetamide, *N*-[3-(aminosulfonyl)phenyl]-
Synonym: Acetanilide, 3′-sulfamoyl-
Line Form: $3\text{-}(CH_3CONH)C_6H_4SO_2NH_2$
Mol Form: $C_8H_{10}N_2O_3S$
M_r: 214.23
mp: 216-9
***s*:** ace 3; HOAc 3
UV: HBCP 280(1259) 243(12589) 95% al
m/e: 214.04

HODOC No: 00084
CAS No: 121-61-9
Beilstein No: 14, 702
Name: Acetamide, *N*-[4-(aminosulfonyl)phenyl]-
Synonym: Acetanilide, 4′-sulfamoyl-
Line Form: $4\text{-}(CH_3CONH)C_6H_4SO_2NH_2$
Mol Form: $C_8H_{10}N_2O_3S$
M_r: 214.23
mp: 219-20
Color: nd (HOAc)
***s*:** H_2O 3; al 3; ace 3
UV: HBCP 255(19953) 95% al
m/e: 214.04

HODOC No: 00085
CAS No: 14949-01-0
Beilstein No: 14^1, 722
Name: Acetamide, *N*-[4-(aminosulfonyl)phenyl]-2-chloro-
Synonym: Acetanilide, 2-chloro-4′-sulfamoyl-
Mol Form: $C_8H_9ClN_2O_3S$
M_r: 248.68
mp: 215-7
***s*:** H_2O 2; al 3; ace 3; bz 2; chl 2; DMSO 3; AcOEt 3
IR: SADG 17775 3300 3200 2950 1690 1610 1600 1550 1400 1310 1250 1150 1090 900 830 690 570 540 470 370
UV: SAD 17661 259(21200) 205(20100) MeOH 259(19800) MeOH:KOH
H NMR: SAD 9406 4.3 7.3 7.9 10.6 $DMSOd_6$
m/e: 248.00

HODOC No: 00086
CAS No: 22795-59-1
Name: Acetamide, *N*-[4-(aminosulfonyl)phenyl]-2,2,2-trichloro-
Synonym: Acetanilide, 2,2,2-trichloro-4′-sulfamoyl-
Mol Form: $C_8H_7Cl_3N_2O_3S$
M_r: 317.57
***s*:** DMSO 3
IR: SADG 17783 3300 3100 2980 1730 1590 1530 1400 1330 1150 950 880 810 730 700 650 560 540
UV: SAD 17663 261(17800) MeOH 281(11500) MeOH:KOH
H NMR: SAD 9405 7.4 8.0 11.1 $DMSOd_6$
m/e: 315.92

HODOC No: 00087
CAS No: 59-66-5
Name: Acetamide, *N*-[5-(aminosulfonyl)-1,3,4-thiadiazol-2-yl]-
Synonym: Acetamide, *N*-(5-sulfamoyl-1,3,4-thiadiazol-2-yl)-
Mol Form: $C_4H_6N_4O_3S_2$
M_r: 222.23
mp: 260-1
IR: µCOB 5114 3330 3130 2860 2780 1670 1540 1430 1370 1320 1180 1090 970 910 810 670
UV: LANG 868 264(10965) EtOH
H NMR: SAD 17372 2.2 8.8 DMSOd6
m/e: 221.99

HODOC No: 00088
CAS No: 21674-96-4
Name: Acetamide, *N*-[4-(2-amino-4-thiazolyl)phenyl]-
Synonym: Acetanilide, 4′-(2-amino-4-thiazolyl)-
Mol Form: $C_{11}H_{11}N_3OS$
M_r: 233.28
mp: 239-43
IR: µCOB 920 3330 3130 3030 2940 1640 1540 1520 1470 1410 1350 1190 1040 1020 860 830 800 740 710
m/e: 233.06

HODOC No: 00089
CAS No: 591-08-2
Beilstein No: 3^4, 354
Name: Acetamide, *N*-(aminothioxomethyl)-
Synonym: Urea, 1-acetyl-2-thio-
Line Form: $CH_3CONHCSNH_2$
Mol Form: $C_3H_6N_2OS$
M_r: 118.15
mp: 165
Color: pr (w), rh (al)
***s*:** H_2O 2; al 3; eth 2; DMSOd6 3
IR: SADG 4614 3410 3300 3260 3170 3020 1720 1690 1600 1540 1410 1360 1240 1140 810 710 680
UV: SAD 11381 272(9670) 207(7460) MeOH
H NMR: SAD 9357 2.1 9.2 9.7 DMSOd6
m/e: 118.02

HODOC No: 00090
CAS No: 104-06-3
Name: Acetamide, *N*-[4-[[(aminothioxomethyl)hydrazono]methyl]pheny l]-
Synonym: Acetanilide, 4′-formyl-, 4′-(thiosemicarbazone)
Line Form: $H_2NNHC{=}RNH_2$
Structure:

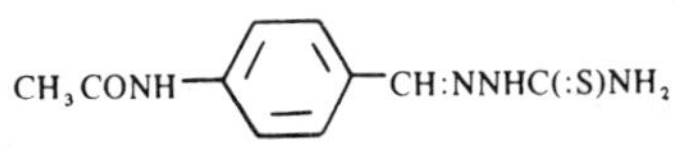

Mol Form: $C_{10}H_{12}N_4OS$
M_r: 236.28
IR: SADG 15153 3320 3260 3160 1670 1610 1580 1530 1410 1360 1310 1260 1170 1130 1100 1030 1010 960 930 880 850 830 810 710 630 590 530 450 430 360
Raman: SAD 929 3300 3200 2920 1670 1610 1580 1540 1500 1410 1370 1310 1260 1230 1180 1110 1040 950 930 870 730 710 670 630 580 530 450 420 410 300 250 120
m/e: 236.07

HODOC No: 00091
CAS No: 6901-13-9
Name: Acetamide, *N*-(5,6,7,7b,8,10a-hexahydro-1,2,3,9-tetramethoxy-8-oxobenzo[a]cyclopenta[3,4]cyclobuta[1,2-c]cyclohep ten-7-yl)-, [7*S*-(7α,7bβ,10aβ)]-
Synonym: Benzo[a]cyclopenta[3,4]cyclobuta[1,2-c]cycloheptene, acetamide deriv.
Structure:

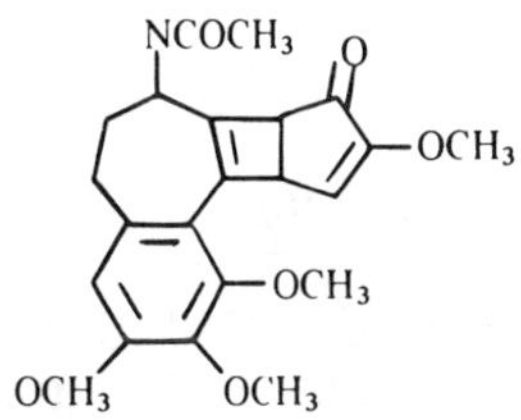

Mol Form: $C_{22}H_{25}NO_6$
M_r: 399.42
***s*:** chl 3
H NMR: VAR 690 1.8-2.3 2.5-2.9 3.6 3.7 3.8 3.9 4.0 4.1 4.8 6.0 6.5 6.7
m/e: 399.17

HODOC No: 00092
CAS No: 29325-49-3
Beilstein No: 13^3, 412
Name: Acetamide, *N,N'*-[1,1'-biphenyl]-2,2'-diylbis-
Synonym: Biphenyl, 2,2'-diacetamido
Line Form: $[2\text{-}(CH_3CONH)C_6H_4\text{-}]_2$
Mol Form: $C_{16}H_{16}N_2O_2$
M_r: 268.30
mp: 164-5
Color: pr (al)
***s*:** Ac 4; Bz 4
IR: SADG 10693
H NMR: SAD 14798
m/e: 268.12

HODOC No: 00093
CAS No: 613-35-4
Beilstein No: 13^3, 437
Name: Acetamide, *N,N'*-[1,1'-biphenyl]-4,4'-diylbis-
Synonym: 4',4'''-Biacetanilide
Line Form: $[4\text{-}CH_3CONHC_6H_4\text{-}]_2$
Mol Form: $C_{16}H_{16}N_2O_2$
M_r: 268.30
mp: 328.3
Color: nd (HOAc)
m/e: 268.12

HODOC No: 00094
CAS No: 2113-47-5
Beilstein No: 12^3, 3125
Name: Acetamide, *N*-[1,1'-biphenyl]-2-yl-
Synonym: Acetanilide, 2'-phenyl-
Line Form: $2(CH_3CONH)C_6H_4\text{-}C_6H_5$
Mol Form: $C_{14}H_{13}NO$
M_r: 211.25
bp: 355
mp: 121
Color: pr or nd (dil al or peth)
***s*:** H_2O 1; al 4; eth 4
IR: SADG 17169 3290 3230 3030 1650 1580 1550 1530 1470 1440 1420 1360 1290 1240 1150 1100 1060 1040 1000 960 930 910 770 750 730 700 650 590 540 510 480
UV: OES 4-507 344(631) 238(19953) MeOH
m/e: 211.10

HODOC No: 00095
CAS No: 2113-54-4
Beilstein No: 12^2, 751
Name: Acetamide, *N*-[1,1'-biphenyl]-3-yl-
Synonym: Acetanilide, 3'-phenyl-
Line Form: $3\text{-}(CH_3CONH)C_6H_4\text{-}C_6H_5$
Mol Form: $C_{14}H_{13}NO$
M_r: 211.25
mp: 149
Color: nd (al)
***s*:** H_2O 1; al 3; bz 4
IR: COB 2673 3370 3010 1660 1590 1530 1440 1370 1310 1020 980 780 750 700
UV: HBCP 242(25119) al
m/e: 211.10

HODOC No: 00096
CAS No: 4075-79-0
Beilstein No: 12^3, 3158
Name: Acetamide, *N*-[1,1'-biphenyl]-4-yl-
Synonym: Acetanilide, 4'-phenyl-
Line Form: $4\text{-}(CH_3CONH)C_6H_4\text{-}C_6H_5$
Mol Form: $C_{14}H_{13}NO$
M_r: 211.25
mp: 172
Color: cr (dil MeOH)
***s*:** H_2O 1; al 4; ace 4; MeOH 4
UV: HBCP 237(19499) al
MS: NBS 1330
m/e: 211.10

HODOC No: 00097
CAS No: 55712-68-0
Name: Acetamide, *N*-[2-[3,4-*bis*(acetyloxy)phenyl]-2-hydroxyethyl]-*N*-methyl-
Synonym: Acetamide, *N*-methyl-*N*-(β,3,4-trihydroxyphenethyl)-, 3,4-diacetate
Mol Form: $C_{15}H_{19}NO_6$
M_r: 309.30
IR: μCOB 5031 3230 2940 1750 1640 1520 1410 1370 1270 1220 1110 1010 1090 970 910 810
m/e: 309.12

HODOC No: 00098
CAS No: 2247-71-4
Name: Acetamide, *N*-[1,1-*bis*(*p*-chlorophenyl)-2,2,2-trifluoroethyl]-
Synonym: Ethane, 1-acetamido-1,1-*bis*(4-chlorophenyl)-2,2,2-trifluoro
Mol Form: $C_{16}H_{12}Cl_2F_3NO$
M_r: 362.16
IR: COB 7608 3290 3040 1670 1540 1500 1290 1210 1170 1100 1020 930 820 740 530
m/e: 361.02

HODOC No: 00099
CAS No: 10479-30-8
Beilstein No: 9^3, 3356
Name: Acetamide, *N,N*-*bis*(phenylmethyl)-
Synonym: Acetamide, *N,N*-dibenzyl-
Line Form: $(C_6H_5CH_2)_2CHCONH_2$
Mol Form: $C_{16}H_{17}NO$
M_r: 239.30
bp: 259^{18}
mp: 129
Color: nd (al, w)
***s*:** Et_2O 4; EtOH 4
m/e: 239.13

HODOC No: 00100
CAS No: 683-57-8
Beilstein No: 2^4, 530
Name: Acetamide, 2-bromo-
Synonym: Bromoacetamide
Line Form: $BrCH_2CONH_2$
Mol Form: C_2H_4BrNO
M_r: 136.96
mp: 91
Color: nd (al, Bz)
***s*:** H_2O 4; Bz 4
m/e: 136.95

HODOC No: 00101
CAS No: 79-15-2
Beilstein No: 2^4, 417
Name: Acetamide, *N*-bromo-
Synonym: Acetobromamide
Line Form: $CH_3CONHBr$
Mol Form: C_2H_4BrNO
M_r: 136.96
mp: 102-5
Color: nd (Chl-hex)
***s*:** Et_2O 4
IR: ALD 3:441F
MS: ALD 3:107B
m/e: 136.95

HODOC No: 00102
CAS No: 5422-53-7
Name: Acetamide, *N*-(2-bromo-4-cyclohexylphenyl)-
Synonym: Acetic acid, amide,*N*(2-bromo-4-cyclohexylphenyl)
Mol Form: $C_{14}H_{18}BrNO$
M_r: 295.19
IR: COB 2677 3300 3000 2900 1650 1600 1530 1450 1420 1370 1320 1270 1000 820
m/e: 295.06

HODOC No: 00103
CAS No: 6335-34-8
Name: Acetamide, 2-bromo-*N,N*-diphenyl-
Synonym: Acetic acid, amide,bromo-*N,N*-diphenyl
Line Form: $BrCH_2CON(C_6H_5)_2$
Mol Form: $C_{14}H_{12}BrNO$
M_r: 289.15
IR: COB 2686 3000 2950 1680 1600 1500 1460 1430 1370 1270 1170 1080 780 750 730 700 690
m/e: 289.01

HODOC No: 00104
CAS No: 430-91-1
Beilstein No: 2, 217
Name: Acetamide, 2-bromo-2-fluoro-
Synonym: Bromofluoroacetamide
Line Form: $(BrCHF)\text{-}CONH_2$
Mol Form: C_2H_3BrFNO
M_r: 154.95
mp: 44
Color: nd (CCl_4)
***s*:** H_2O 4; Et_2O 4; EtOH 4
m/e: 154.94

HODOC No: 00105
CAS No: 5343-65-7
Name: Acetamide, 2-bromo-*N*-(4-methylphenyl)-
Synonym: *p*-Acetotoluidide, 2-bromo-
Mol Form: $C_9H_{10}BrNO$
M_r: 227.08
mp: 118
***s*:** al 4; chl 3; xyl 4
IR: SADG 10489 3280 3020 2920 1660 1600 1580 1520 1380 1360 1290 1050 870 820 800 680 660
H NMR: SAD 2404 2.1 2.2 7.0 7.3 7.7 7.9 $CDCl_3$
m/e: 226.99

HODOC No: 00106
CAS No: 24106-05-6
Beilstein No: 12^2, 456
Name: Acetamide, *N*-(4-bromo-2-methylphenyl)-
Synonym: *o*-Acetotoluidide, 4′-bromo-
Line Form: 4-Br-2-$CH_3C_6H_3NHCOCH_3$
Mol Form: $C_9H_{10}BrNO$
M_r: 227.08
mp: 159-60
Color: nd (dil al, lig)
***s*:** H_2O 3; al 4; chl 2; lig 3
IR: SADG 15588 3260 3020 1660 1600_1520 1480 1390 1370 1290 1270 1180 1120 970 880 870 820 800 710 670
UV: SAD 1605 235 MeOH
H NMR: SAD 6477 2.1 2.2 7.0-7.7 $CDCl_3$
m/e: 226.99

HODOC No: 00107
CAS No: 5326-87-4
Beilstein No: 12^3, 1417
Name: Acetamide, 2-bromo-*N*-phenyl-
Synonym: Acetanilide, 2-bromo-
Line Form: (2-BrC_6H_4)$NHCOCH_3$
Structure:

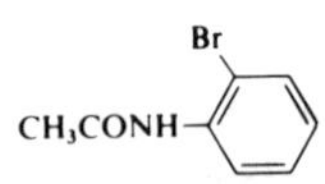

Mol Form: C_8H_8BrNO
M_r: 213.05
mp: 99
Color: nd (al)
***s*:** Et_2O 4; EtOH 4
UV: OES 2-130
m/e: 212.98

HODOC No: 00108
CAS No: 614-76-6
Beilstein No: 12^2, 342
Name: Acetamide, *N*-(2-bromophenyl)-
Synonym: Acetanilide, 2′-bromo-
Mol Form: C_8H_8BrNO
M_r: 213.05
mp: 99
***s*:** H_2O 1; al 3; eth 3
UV: OES 2-130 234(7586) EtOH
m/e: 212.98

HODOC No: 00109
CAS No: 621-38-5
Beilstein No: 12^3, 1424
Name: Acetamide, *N*-(3-bromophenyl)-
Synonym: Acetanilide, 3′-bromo-
Line Form: (3-BrC_6H_4)$NHCOCH_3$
Mol Form: C_8H_8BrNO
M_r: 213.05
mp: 87.5
Color: nd (aq al)
***s*:** al 4; eth 4
UV: OES 2-130 286(851) 278(1096) 246(14125) EtOH
m/e: 212.98

HODOC No: 00110
CAS No: 103-88-8
Beilstein No: 12^3, 1437
Name: Acetamide, *N*-(4-bromophenyl)-
Synonym: Acetanilide, 4′-bromo-
Line Form: (4-BrC_6H_4)$NHCOCH_3$
Mol Form: C_8H_8BrNO
M_r: 213.05
mp: 168
***d*:** 1.717
Color: nd (60% al) or (95% al)
***s*:** H_2O 1; al 3; eth 2; bz 2; chl 3
IR: COB 6065 3290 3180 3050 1670 1600 1590 1540 1490 1400 1310 1260 1070 1010 830 820 740
UV: OES 6-159 257(19953) MeOH
MS: NBS 1352; ALD 1:68C
m/e: 212.98

HODOC No: 00111
CAS No: 1439-12-9
Beilstein No: 12^3, 1078
Name: Acetamide, *N*-bromo-*N*-phenyl-
Synonym: Acetanilide, *N*-bromo-
Line Form: $C_6H_5N(Br)COCH_3$
Mol Form: C_8H_8BrNO
M_r: 213.05
mp: 88
Color: ye pl (peth)
***s*:** Chl 4; lig 4
m/e: 212.98

HODOC No: 00112
CAS No: 7336-54-1
Name: Acetamide, *N*-(5-bromo-2-thiazolyl)-
Synonym: Thiazole, 2-acetamido-5-bromo-
Mol Form: $C_5H_5BrN_2OS$
M_r: 220.07
mp: 213-4
IR: **µCOB 1820** 3450 3130 2940 1720 1560 1430 1390 1280 1250 1240 1160 1000 970 840 780 670
m/e: 219.93

HODOC No: 00113
CAS No: 532-09-2
Name: Acetamide, *N*-(2-butoxyphenyl)-
Synonym: Acetanilide, 2′-butoxy-
Line Form: $2\text{-}(C_4H_9OC_6H_4)NHCOCH_3$
Mol Form: $C_{12}H_{17}NO_2$
M_r: 207.26
mp: 46
***s*:** Et_2O 4; EtOH 4
m/e: 207.13

HODOC No: 00114
CAS No: 1119-49-9
Beilstein No: 4^4, 565
Name: Acetamide, *N*-butyl-
Synonym: Butylacetamide
Line Form: $CH_3CONHC_4H_9$
Mol Form: $C_6H_{13}NO$
M_r: 115.17
bp: 229
***n*:** 1.4388^{25}
IR: **SADG 15174** 3280 2960 2860 1650 1550 1420 1360 1280 730 600
MS: **ALDN 54** 30(100) 43(34) 72(29) 73(21) 44(19) 115(11) 60(10) 86(9)
m/e: 115.10

HODOC No: 00115
CAS No: 53554-45-3
Name: Acetamide, *N*-[4-[[[(butylamino)thioxomethyl]hydrazono]methyl]phenyl]-
Synonym: *p*-Formylacetanilide 4-butyl-3-thiosemicarbazone
Structure:

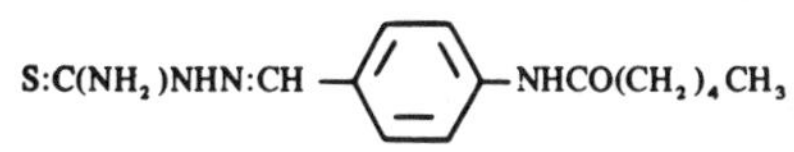

Mol Form: $C_{14}H_{20}N_4OS$
M_r: 292.39
mp: 201
***s*:** DMSO 2
IR: **µSADP 4316** 3225 3125 2940 1670 1590 1515 1450 1410 1370 1335 1250 1165 1090 945 835 805 740
H NMR: **SAD 10598** 0.9 1.0-1.9 2.3 7.7 7.8-8.3 9.9 $DMSOd_6$
m/e: 292.14

HODOC No: 00116
CAS No: 5349-24-6
Beilstein No: 4^4, 567
Name: Acetamide, *N*-butyl-2-chloro-
Synonym: *N*-Butyl-2-chloroacetamide
Line Form: $ClCH_2CONHC_4H_9$
Mol Form: $C_6H_{12}ClNO$
M_r: 149.61
bp: 110^7
***n*:** 1.4665^{25}
m/e: 149.06

HODOC No: 00117
CAS No: 2567-58-0
Beilstein No: 4^4, 567
Name: Acetamide, *N*-butyl-2-chloro-*N*-ethyl-
Synonym: Chloroacetamide, *N*-butyl-*N*-ethyl
Line Form: $ClCH_2CON(C_2H_5)C_4H_9$
Mol Form: $C_8H_{16}ClNO$
M_r: 177.67
bp: $90^{1.5}$
***n*:** 1.4665^{25}
m/e: 177.09

HODOC No: 00118
CAS No: 32322-34-2
Beilstein No: 4^4, 567
Name: Acetamide, *N*-butyl-2-chloro-*N*-(1-methylethyl)-
Synonym: Acetamide, *N*-butyl-2-chloro-*N*-isopropyl-
Line Form: $ClCH_2CON(C_4H_9)CH(CH_3)_2$
Mol Form: $C_9H_{18}ClNO$
M_r: 191.69
bp: $82^{0.5}$
***n*:** 1.4655^{15}
m/e: 191.11

HODOC No: 00119
CAS No: 3663-20-5
Beilstein No: 12^3, 2715
Name: Acetamide, *N*-(4-butylphenyl)-
Synonym: Acetanilide, 4′-butyl-
Line Form: $(4\text{-}C_4H_9\text{-}C_6H_4)NHCOCH_3$
Structure:

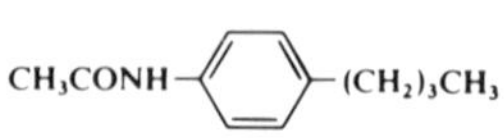

Mol Form: $C_{12}H_{17}NO$
M_r: 191.26
mp: 105
Color: wh pl (al)
***s*:** EtOH 4
m/e: 191.13

HODOC No: 00120
CAS No: 91-49-6
Beilstein No: 12^3, 467
Name: Acetamide, *N*-butyl-*N*-phenyl-
Synonym: Acetanilide, *N*-butyl-
Line Form: $CH_3CON(C_4H_9)C_6H_5$
Mol Form: $C_{12}H_{17}NO$
M_r: 191.26
bp: 281, 141^{10}
mp: 24.5
***d*:** $0.9912^{20/4}$
***n*:** 1.5146^{20}
***s*:** chl 2
IR: SADG 21090 3050 2960 2860 1660 1590 1490 1440 1390 1290 1210 1090 770 700
UV: SAD 1831 238 cyhex
H NMR: VAR 601 0.9 1.4 1.9 3.7 7.0-7.6 $CDCl_3$
m/e: 191.13

HODOC No: 00121
CAS No: 638-20-0
Beilstein No: 3^3, 121
Name: Acetamide, *N,N′*-carbonylbis-
Synonym: Urea, 1,3-diacetyl-
Line Form: $(CH_3CONH)_2CO$
Mol Form: $C_5H_8N_2O_3$
M_r: 144.13
bp: sub 179-80d
mp: 154-5
Color: nd (50% al)
***s*:** H_2O 2; al 2; eth 3; ace 3; bz 3; os 3
UV: OES 6-50 205(21380) EtOH
MS: ALDN 94 43(100) 59(60) 44(40) 70(34) 42(25) 85(16) 28(16) 144(0)
m/e: 144.05

HODOC No: 00122
CAS No: 79-07-2
Beilstein No: 2^4, 490
Name: Acetamide, 2-chloro-
Synonym: Chloracetamide
Line Form: $ClCH_2CONH_2$
Mol Form: C_2H_4ClNO
M_r: 93.51
bp: $224\text{-}5^{747}$
mp: 121
***s*:** H_2O 3; al 4; eth 2
IR: µCOB 287 3330 3130 2940 1640 1470 1410 1370 1270 1180 1100 810 770 650
UV: HBCP 442(13) MeOH
H NMR: SAD 9187 4.2 D_2O
m/e: 93.00

HODOC No: 00123
CAS No: 106738-59-4
Beilstein No: 4^4, 1089
Name: Acetamide, 2-chloro-*N,N*-*bis*(2-chloroallyl)-
Synonym: Chloroacetamide, *N,N*-*bis*(2-chloroallyl)
Line Form: $ClCH_2CON(CH_2CCl{=}CH_2)_2$
Mol Form: $C_8H_{10}Cl_3NO$
M_r: 242.52
bp: $161\text{-}3^{12}$
***n*:** 1.5220^{25}
m/e: 240.98

HODOC No: 00124
CAS No: 106738-34-5
Name: Acetamide, 2-chloro-*N,N-bis*(3-chloroallyl)-
Synonym: Chloroacetamide, *N,N-bis*(3-chloroallyl)
Line Form: $ClCH_2CON(CH_2CH{=}CHCl)_2$
Mol Form: $C_8H_{10}Cl_3NO$
M_r: 242.52
bp: $140\text{-}5^{1}$
***n*:** 1.5220^{25}
m/e: 240.98

HODOC No: 00125
CAS No: 99767-70-1
Beilstein No: 4^4, 499
Name: Acetamide, 2-chloro-*N,N-bis*(2-chloropropyl)-
Synonym: Chloroacetamide, *N,N-bis*(2-chloropropyl)
Line Form: $ClCH_2CON(CH_2CHClCH_3)_2$
Mol Form: $C_8H_{14}Cl_3NO$
M_r: 246.56
bp: $134^{0.7}$
***n*:** 1.5018^{25}
m/e: 245.01

HODOC No: 00126
CAS No: 60850-22-8
Name: Acetamide, 2-chloro-*N,N-bis*(2-ethylhexyl)-
Synonym: Chloroacetamide, *N,N-bis*(2-ethylhexyl)
Line Form: $ClCH_2CON[CH_2CH(C_2H_5)C_4H_9]_2$
Mol Form: $C_{18}H_{36}ClNO$
M_r: 317.92
bp: $154^{0.8}$
***n*:** 1.4622^{25}
MS: WILEY 1257 91(100) 108(55) 90(38) 184(26) 107(21) 51(21) 77(20) 65(17)
m/e: 317.25

HODOC No: 00127
CAS No: 7403-66-9
Beilstein No: 4^4, 516
Name: Acetamide, 2-chloro-*N,N-bis*(1-methylethyl)-
Synonym: Acetamide, 2-chloro-*N,N*-diisopropyl-
Line Form: $ClCH_2CON(i\text{-}C_3H_7)_2$
Mol Form: $C_8H_{16}ClNO$
M_r: 177.67
bp: $86^{2.7}$
mp: 48-9
***n*:** 1.4619^{25}
Color: cr (peth)
m/e: 177.09

HODOC No: 00128
CAS No: 32322-24-0
Beilstein No: 4^4, 7105
Name: Acetamide, 2-chloro-*N,N-bis*(2-methyl-2-propenyl)-
Synonym: Acetamide, 2-chloro-*N,N-bis*(2-methylallyl)-
Line Form: $ClCH_2CON[CH_2C(CH_3){=}CH_2]_2$
Mol Form: $C_{10}H_{16}ClNO$
M_r: 201.69
bp: $133\text{-}5^{20}$
***n*:** 1.4882^{2}
m/e: 201.09

HODOC No: 00129
CAS No: 32322-33-1
Name: Acetamide, 2-chloro-*N,N-bis*(1-methylpropyl)-
Synonym: Acetamide, *N,N*-di-*sec*-butyl-2-chloro-
Line Form: $ClCH_2CON[CH(CH_3)C_2H_5]_2$
Mol Form: $C_{10}H_{20}ClNO$
M_r: 205.72
bp: $92^{0.7}$
***n*:** 1.4681^{25}
m/e: 205.12

HODOC No: 00130
CAS No: 5326-82-9
Beilstein No: 4^4, 633
Name: Acetamide, 2-chloro-*N,N-bis*(2-methylpropyl)-
Synonym: Acetamide, 2-chloro-*N,N*-diisobutyl-
Line Form: $ClCH_2CON[CH_2CH(CH_3)_2]_2$
Mol Form: $C_{10}H_{20}ClNO$
M_r: 205.72
bp: 99^{2}
***n*:** 1.4642^{25}
m/e: 205.12

HODOC No: 00131
CAS No: 2567-51-3
Name: Acetamide, 2-chloro-*N,N-bis*(phenylmethyl)-
Synonym: Acetamide, *N,N*-dibenzyl-2-chloro-
Line Form: $ClCH_2CON(CH_2C_6H_5)_2$
Mol Form: $C_{16}H_{16}ClNO$
M_r: 273.75
bp: $190^{1.8}$
***n*:** 1.5837^{25}
m/e: 273.09

HODOC No: 00132
CAS No: 32322-71-7
Beilstein No: 4^4, 1089
Name: Acetamide, 2-chloro-*N*-(2-chloroallyl)-
Synonym: α-Chloro-*N*-(2-chloroallyl)acetamide
Line Form: $ClCH_2CONH(CH_2\text{-}CCl{=}CH_2)$
Mol Form: $C_5H_7Cl_2NO$
M$_r$: 168.02
bp: $101^{1.4}$
***n*:** 1.5078^{25}
m/e: 166.99

HODOC No: 00133
CAS No: 100130-25-4
Beilstein No: 4^4, 1087
Name: Acetamide, 2-chloro-*N*-3-chloroallyl-
Synonym: Chloroacetamide, *N*-3-chloroallyl
Line Form: $ClCH_2CONH(CH_2CH{=}CHCl)$
Mol Form: $C_5H_7Cl_2NO$
M$_r$: 168.02
bp: $112^{0.5}$
mp: 52-3
Color: cr (peth)
***s*:** peth 4
m/e: 166.99

HODOC No: 00134
CAS No: 34004-41-6
Beilstein No: 13^3, 1664
Name: Acetamide,
2-chloro-*N*-(2-chloro-4-nitrophenyl)-
Synonym: Acetanilide, 2,2′-dichloro-4′-nitro-
Line Form: $(4\text{-}NO_2\text{-}2\text{-}ClC_6H_3)NHCOCH_2Cl$
Mol Form: $C_8H_6Cl_2N_2O_3$
M$_r$: 249.05
mp: 118-9
Color: cr (peth)
m/e: 247.98

HODOC No: 00135
CAS No: 100791-98-8
Beilstein No: 4^4, 499
Name: Acetamide, 2-chloro-*N*-(2-chloropropyl)-
Synonym: Chloroacetamide, *N*-2-chloropropyl
Line Form: $ClCH_2CONH(CH_2CHClCH_3)$
Mol Form: $C_5H_9Cl_2NO$
M$_r$: 170.03
bp: 88^{115}
***n*:** 1.4942^{25}
m/e: 169.01

HODOC No: 00136
CAS No: 100791-92-2
Beilstein No: 4^4, 501
Name: Acetamide, 2-chloro-*N*-(3-chloropropyl)-
Synonym: Chloroacetamide, *N*-3-chloropropyl
Line Form: $ClCH_2CONH\text{-}CH_2CH_2CH_2Cl$
Mol Form: $C_5H_9Cl_2NO$
M$_r$: 170.03
mp: 36-7
m/e: 169.01

HODOC No: 00137
CAS No: 98025-28-6
Name: Acetamide, 2-chloro-*N*-2,3-dichloroallyl-
Synonym: Chloroacetamide, *N*-(2,3-dichlorallyl)
Line Form: $ClCH_2CONH(CH_2\text{-}CCl{=}CHCl)$
Mol Form: $C_5H_6Cl_3NO$
M$_r$: 202.46
bp: $126\text{-}31^{1.8}$
***n*:** 1.5311^{25}
m/e: 200.95

HODOC No: 00138
CAS No: 33560-47-3
Name: Acetamide, 2-chloro-*N*-(2,3-dichlorophenyl)-
Synonym: Acetanilide, 2,2′,3′-trichloro-
Line Form: $ClCH_2CONH(C_6H_3Cl_2\text{-}2,4)$
Mol Form: $C_8H_6Cl_3NO$
M$_r$: 238.49
mp: 101-2
Color: cr (dil al)
***s*:** EtOH 4; peth 4
m/e: 236.95

HODOC No: 00139
CAS No: 22303-26-0
Name: Acetamide, 2-chloro-*N*-(2,5-dichlorophenyl)-
Synonym: Acetanilide, 2,2′,5′-trichloro-
Line Form: $ClCH_2CONH(C_6H_3Cl_2\text{-}2,5)$
Mol Form: $C_8H_6Cl_3NO$
M$_r$: 238.49
mp: 116-7
Color: cr (dil al)
***s*:** EtOH 4
m/e: 236.95

HODOC No: 00140
CAS No: 56978-45-1
Name: Acetamide, 2-chloro-*N*-[(2,4-dichlorophenyl)methyl]-
Synonym: Acetamide, 2-chloro-*N*-(2,4-dichlorobenzyl)-
Line Form: $ClCH_2CONH(CH_2C_6H_3Cl_2\text{-}2,4)$
Mol Form: $C_9H_8Cl_3NO$
M_r: 252.52
mp: 96-7
Color: cr (Bz)
***s*:** Bz 4
m/e: 250.97

HODOC No: 00141
CAS No: 56978-46-2
Name: Acetamide, 2-chloro-*N*-[(3,4-dichlorophenyl)methyl]-
Synonym: Acetamide, 2-chloro-*N*-(3,4-dichlorobenzyl)-
Line Form: $ClCH_2CONH(CH_2C_6H_3Cl_2\text{-}3,4)$
Mol Form: $C_9H_8Cl_3NO$
M_r: 252.52
mp: 105-6
Color: cr (dil al)
***s*:** EtOH 4
m/e: 250.97

HODOC No: 00142
CAS No: 99418-51-6
Beilstein No: 4^4, 502
Name: Acetamide, 2-chloro-*N*-(2,3-dichloropropyl)-
Synonym: Chloroacetamide, *N*-(2,3-dichloropropyl)
Line Form: $ClCH_2CONH(CH_2\text{-}CHCl\text{-}CH_2Cl)$
Mol Form: $C_5H_8Cl_3NO$
M_r: 204.48
mp: 65-6
Color: cr (peth)
***s*:** peth 4
m/e: 202.97

HODOC No: 00143
CAS No: 2315-36-8
Beilstein No: 4^4, 350
Name: Acetamide, 2-chloro-*N,N*-diethyl-
Synonym: *N*-Chloroacetyldiethylamine
Line Form: $ClCH_2CON(C_2H_5)_2$
Mol Form: $C_6H_{12}ClNO$
M_r: 149.61
bp: $190\text{-}5^{25}$
m/e: 149.06

HODOC No: 00144
CAS No: 354-28-9
Beilstein No: 2, 202
Name: Acetamide, 2-chloro-2,2-difluoro-
Synonym: Chlorodifluoroacetamide
Line Form: $ClCF_2\text{-}CONH_2$
Mol Form: $C_2H_2ClF_2NO$
M_r: 129.49
bp: 93^{18}
mp: 78.5
m/e: 128.98

HODOC No: 00145
CAS No: 32322-39-7
Beilstein No: 4^4, 713
Name: Acetamide, 2-chloro-*N,N*-dihexyl-
Synonym: α-Chloro-*N,N*-dihexylacetamide
Line Form: $ClCH_2CON(C_6H_{13})_2$
Mol Form: $C_{14}H_{28}ClNO$
M_r: 261.82
mp: 114-5
Color: cr (peth)
m/e: 261.19

HODOC No: 00146
CAS No: 100395-94-6
Beilstein No: 4^4, 702
Name: Acetamide, 2-chloro-*N,N*-diisopentyl-
Synonym: Chloroacetamide, *N,N-bis*(3-methylbutyl)
Line Form: $ClCH_2CON[CH_2CH_2CH(CH_3)_2]_2$
Mol Form: $C_{12}H_{24}ClNO$
M_r: 233.77
bp: $109^{0.6}$
***n*:** 1.4625^{25}
m/e: 233.15

HODOC No: 00147
CAS No: 10271-73-5
Beilstein No: 4^4, 662
Name: Acetamide, *N*-chloro-*N*-(1,1-dimethylethyl)-
Synonym: Acetamide, *N-tert*-butyl-*N*-chloro-
Line Form: $ClCH_2CONHC(CH_3)_3$
Mol Form: $C_6H_{12}ClNO$
M_r: 149.61
mp: 84
Color: cr (peth)
***s*:** peth 4
m/e: 149.06

HODOC No: 00148
CAS No: 6271-08-5
Name: Acetamide, 2-chloro-*N*-(2,4-dinitrophenyl)-
Synonym: Acetanilide, 2-chloro-2′,4′-dinitro-
Line Form: $ClCH_2CONH[C_6H_3(NO_2)_2\text{-}2,4]_2$
Mol Form: $C_8H_6ClN_3O_5$
M_r: 259.60
mp: 114-5
Color: nd (al)
m/e: 259.00

HODOC No: 00149
CAS No: 2567-60-4
Beilstein No: 4^4, 679
Name: Acetamide, 2-chloro-*N,N*-dipentyl-
Synonym: 2-Chloro-*N,N*-dipentylacetamide
Line Form: $ClCH_2CON(C_5H_{11})_2$
Mol Form: $C_{12}H_{24}ClNO$
M_r: 233.77
bp: 126^1
***n*:** 1.4651^{25}
m/e: 233.15

HODOC No: 00150
CAS No: 93-71-0
Beilstein No: 4^4, 1064
Name: Acetamide, 2-chloro-*N,N*-di-2-propenyl-
Synonym: Acetamide, *N,N*-diallyl-2-chloro-
Line Form: $ClCH_2CON(CH_2CH{=}CH_2)_2$
Mol Form: $C_8H_{12}ClNO$
M_r: 173.63
bp: $92^{0.7}$
***n*:** 1.4932^{25}
m/e: 173.06

HODOC No: 00151
CAS No: 2315-37-9
Beilstein No: 4^4, 476
Name: Acetamide, 2-chloro-*N,N*-dipropyl-
Synonym: α-Chloro-*N,N*-dipropylacetamide
Line Form: $ClCH_2CON(C_3H_7)_2$
Mol Form: $C_8H_{16}ClNO$
M_r: 177.67
bp: 120^8, $90\text{-}2^{0.8}$
***n*:** 1.4670^{20}
m/e: 177.09

HODOC No: 00152
CAS No: 2153-08-4
Name: Acetamide, 2-chloro-*N*-(4-ethoxyphenyl)-
Synonym: *p*-Acetophenetidide, 2-chloro-
Mol Form: $C_{10}H_{12}ClNO_2$
M_r: 213.65
mp: 146
IR: COB 6063 3270 3200 3140 3100 2970 1670 1620 1560 1510 1480 1300 1250 1170 1040 920 820 750 520
UV: SAD 10835 254(12200) MeOH
H NMR: SAD 481 1.5 4.3 4.4 7.1 7.4 8.9 tfa
m/e: 213.06

HODOC No: 00153
CAS No: 53689-25-1
Name: Acetamide, *N*-(5-chloro-2-ethoxyphenyl)-
Synonym: Acetic acid, amide,*N*(5-chloro-2-ethoxyphenyl)
Mol Form: $C_{10}H_{12}ClNO_2$
M_r: 213.65
IR: μCOb 444 3330 3030 2940 1670 1590 1540 1490 1450 1410 1370 1270 1240 1210 1140 1040 800 680
m/e: 213.06

HODOC No: 00154
CAS No: 39086-61-8
Name: Acetamide, 2-chloro-*N*-ethyl-*N*-phenyl-
Synonym: Acetanilide, 2-chloro-*N*-ethyl-
Line Form: $ClCH_2CON(C_6H_5)CH_2CH_3$
Mol Form: $C_{10}H_{12}ClNO$
M_r: 197.65
mp: 30-2
***s*:** ctc 3; CS_2 3
IR: μCOB 3955 3030 2940 2860 1670 1590 1490 1450 1390 1320 1250 1120 790 770 700 670
H NMR: SAD 18978 1.2 3.8 7.1-7.6 CCl_4
m/e: 197.06

HODOC No: 00155
CAS No: 40914-13-4
Beilstein No: 18^4, 7080
Name: Acetamide, 2-chloro-*N*-(2-furanylmethyl)-
Synonym: Acetamide, 2-chloro-*N*-furfuryl-
Line Form: $ClCH_2CONH(CH_2C_4H_3O)$
Mol Form: $C_7H_8ClNO_2$
M_r: 173.59
mp: 58
Color: cr (peth)
m/e: 173.02

HODOC No: 00156
CAS No: 5326-81-8
Beilstein No: 4^4, 713
Name: Acetamide, 2-chloro-*N*-hexyl-
Synonym: α-Chloro-*N*-hexylacetamide
Line Form: $ClCH_2CONH(CH_2)_5\text{-}CH_3$
Mol Form: $C_8H_{16}ClNO$
Mr: 177.67
bp: $95\text{-}105^{0.2}$
mp: 108-9
***s*:** peth 4
m/e: 177.09

HODOC No: 00157
CAS No: 32322-79-5
Beilstein No: 4^2, 647
Name: Acetamide, 2-chloro-*N*-isopentyl-
Synonym: α-Chloro-*N*-isopentylacetamide
Line Form: $ClCH_2CONHCH_2CH_2CH(CH_3)_2$
Mol Form: $C_7H_{14}ClNO$
Mr: 163.64
bp: $134\text{-}5^{13}$
mp: -15
m/e: 163.08

HODOC No: 00158
CAS No: 1709-03-1
Beilstein No: 4^4, 1645
Name: Acetamide, 2-chloro-*N*-(3-methoxypropyl)-
Synonym: α-Chloro-*N*-(3-methoxypropyl)acetamide
Line Form: $CH_3OCH_2CH_2CH_2\text{-}NHCOCH_2Cl$
Mol Form: $C_6H_{12}ClNO_2$
Mr: 165.61
bp: $88^{0.5}$
mp: 30
***n*:** 1.4712^{25}
m/e: 165.06

HODOC No: 00159
CAS No: 2895-21-8
Beilstein No: 4^4, 515
Name: Acetamide, 2-chloro-*N*-(1-methylethyl)-
Synonym: Acetamide, 2-chloro-*N*-isopropyl-
Line Form: $ClCH_2CONHCH(CH_3)_2$
Mol Form: $C_5H_{10}ClNO$
Mr: 135.59
mp: 62
Color: cr (peth)
***s*:** peth 4
m/e: 135.05

HODOC No: 00160
CAS No: 56961-87-6
Beilstein No: 12^3, 1991
Name: Acetamide, *N*-(2-chloro-3-methylphenyl)-
Synonym: Acetanilide, 2-chloro-3-methyl
Line Form: $(2\text{-}Cl\text{-}3\text{-}CH_3C_6H_3)NHCOCH_3$
Structure:

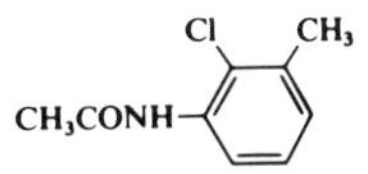

Mol Form: $C_9H_{10}ClNO$
Mr: 183.63
mp: 133-4
Color: nd (al)
***s*:** Bz 4; EtOH 4
m/e: 183.05

HODOC No: 00161
CAS No: 7463-35-6
Beilstein No: 12^2, 1919
Name: Acetamide, *N*-(3-chloro-2-methylphenyl)-
Synonym: *o*-Acetotoluidide, 3′-chloro-
Line Form: $(3\text{-}Cl\text{-}2\text{-}CH_3C_6H_3)NHCOCH_3$
Mol Form: $C_9H_{10}ClNO$
Mr: 183.63
mp: 157-9
Color: nd(dil al)
***s*:** Bz 4; EtOH 4
m/e: 183.05

HODOC No: 00162
CAS No: 7149-79-3
Beilstein No: 12^2, 530
Name: Acetamide, *N*-(3-chloro-4-methylphenyl)-
Synonym: *p*-Acetotoluidide, 3′-chloro-
Line Form: $(3\text{-}Cl\text{-}4\text{-}CH_3)C_6H_3NHCOCH_3$
Mol Form: $C_9H_{10}ClNO$
Mr: 183.63
mp: 105
Color: tcl cr
***s*:** EtOH 4
m/e: 183.05

HODOC No: 00163
CAS No: 56978-43-9
Beilstein No: 12, 871
Name: Acetamide, *N*-(3-chloro-5-methylphenyl)-
Synonym: Acetanilide, 3-chloro-5-methyl
Line Form: $(3\text{-Cl-5-CH}_3)C_6H_3NHCOCH_3$
Mol Form: $C_9H_{10}ClNO$
M_r: 183.63
mp: 151
Color: nd (al)
***s*:** EtOH 4
m/e: 183.05

HODOC No: 00164
CAS No: 56961-88-7
Beilstein No: 12^3, 473
Name: Acetamide, *N*-(4-chloro-3-methylphenyl)-
Synonym: *m*-Acetotoluidide, 4′-chloro-
Line Form: $(4\text{-Cl-3-CH}_3)C_6H_3NHCOCH_3$
Mol Form: $C_9H_{10}ClNO$
M_r: 183.63
mp: 91-2
Color: lf (al)
***s*:** Bz 4; EtOH 4
m/e: 183.05

HODOC No: 00165
CAS No: 32322-70-6
Beilstein No: 4^4, 1105
Name: Acetamide, 2-chloro-*N*-(2-methyl-2-propenyl)-
Synonym: Acetamide, 2-chloro-*N*-(2-methylallyl)-
Line Form: $ClCH_2CONH(CH_2C(CH_3){=}CH_2)$
Mol Form: $C_6H_{10}ClNO$
M_r: 147.60
bp: 96^1
***n*:** 1.4860^{25}
m/e: 147.05

HODOC No: 00166
CAS No: 32322-73-9
Beilstein No: 4^4, 621
Name: Acetamide, 2-chloro-*N*-(1-methylpropyl)-
Synonym: Acetamide, *N*-*sec*-butyl-2-chloro-
Line Form: $ClCH_2CONHCH(CH_3)C_2H_5$
Mol Form: $C_6H_{12}ClNO$
M_r: 149.61
bp: $68^{0.7}$
mp: 45-6
***s*:** peth 4
m/e: 149.06

HODOC No: 00167
CAS No: 881-87-8
Beilstein No: 12^3, 1663
Name: Acetamide, *N*-(2-chloro-4-nitrophenyl)-
Synonym: Acetanilide, 2′-chloro-4′-nitro-
Line Form: $(2\text{-Cl-4-NO}_2C_6H_3)NHCOCH_3$
Structure:

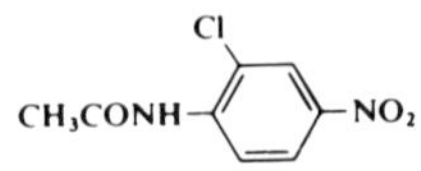

Mol Form: $C_8H_7ClN_2O_3$
M_r: 214.60
mp: 139-40
Color: pr (al)
m/e: 214.01

HODOC No: 00168
CAS No: 712-33-4
Beilstein No: 12^3, 1666
Name: Acetamide, *N*-(3-chloro-4-nitrophenyl)-
Synonym: Acetanilide, 3′-chloro-4′-nitro-
Line Form: $(3\text{-Cl-4-NO}_2)C_6H_3NHCOCH_3$
Structure:

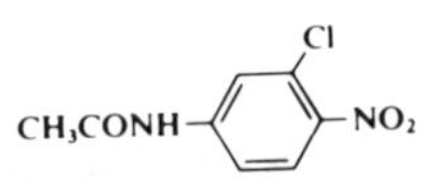

Mol Form: $C_8H_7ClN_2O_3$
M_r: 214.60
mp: 143
Color: pa ye nd (al)
m/e: 214.01

HODOC No: 00169
CAS No: 881-51-6
Beilstein No: 12^3, 1652
Name: Acetamide, *N*-(4-chloro-2-nitrophenyl)-
Synonym: Acetanilide, 4′-chloro-2′-nitro-
Line Form: $(4\text{-}Cl\text{-}2\text{-}NO_2)C_6H_3NHCOCH_3$
Structure:

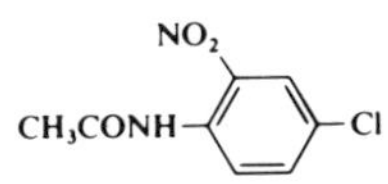

Mol Form: $C_8H_7ClN_2O_3$
M_r: 214.60
mp: 104
Color: ye nd (al)
m/e: 214.01

HODOC No: 00170
CAS No: 5540-60-3
Beilstein No: 12^2, 357
Name: Acetamide, *N*-(4-chloro-3-nitrophenyl)-
Synonym: Acetanilide, 4′-chloro-3′-nitro-
Line Form: $(4\text{-}Cl\text{-}3\text{-}NO_2)C_6H_3NHCOCH_3$
Mol Form: $C_8H_7ClN_2O_3$
M_r: 214.60
mp: 150
Color: ye nd (al)
***s*:** H_2O 1; al 3; DMSO 2
IR: SADG 12741 3300 3100 1660 1590 1530 1470 1380 1360 1340 1310 1250 820 740 710
UV: SAD 15332 325(1390) 243(21200) MeOH
H NMR: SAD 9849 7.1 7.6 7.7 8.0 $DMSOd_6$
m/e: 214.01

HODOC No: 00171
CAS No: 32322-78-4
Beilstein No: 4^4, 678
Name: Acetamide, 2-chloro-*N*-pentyl-
Synonym: α-Chloro-*N*-pentylacetamide
Line Form: $ClCH_2CONHC_5H_{11}$
Mol Form: $C_7H_{14}ClNO$
M_r: 163.64
bp: $82^{0.5}$
***n*:** 1.4665^{25}
m/e: 163.08

HODOC No: 00172
CAS No: 587-65-5
Beilstein No: 12^3, 1288
Name: Acetamide, 2-chloro-*N*-phenyl-
Synonym: Acetanilide, 2-chloro-
Line Form: $(2\text{-}ClC_6H_4)NHCOCH_3$
Mol Form: C_8H_8ClNO
M_r: 169.60
mp: 87-8 (sub)
Color: nd (dil HOAc)
***s*:** Bz 4; Et_2O 4; EtOH 4
IR: SADP 7335
UV: SAD 2070
m/e: 169.03

HODOC No: 00173
CAS No: 533-17-5
Beilstein No: 12^2, 317
Name: Acetamide, *N*-(2-chlorophenyl)-
Synonym: Acetanilide, 2′-chloro-
Mol Form: C_8H_8ClNO
M_r: 169.60
mp: 87-8
***s*:** H_2O 1; al 3; eth 4; bz 3; chl 3
IR: µCOB 1689 3450 3130 1700 1590 1490 1430 1370 1300 1250 1220 1050 1040 1000 750 710
UV: SAD 2070 250 MeOH
H NMR: SAD 6728 2.2 6.8-7.4 7.8 8.2 $CDCl_3$
m/e: 169.03

HODOC No: 00174
CAS No: 588-07-8
Beilstein No: 12^3, 1309
Name: Acetamide, *N*-(3-chlorophenyl)-
Synonym: Acetanilide, 3′-chloro-
Line Form: $(3\text{-}ClC_6H_4)NHCOCH_3$
Mol Form: C_8H_8ClNO
M_r: 169.60
mp: 79
Color: nd
***s*:** H_2O 2; al 4; eth 4; bz 4; chl 3; CS_2 4
IR: SADG 21569 3300 3180 3100 1670 1590 1530 1470 1410 1300 1270 860 780 700 680
UV: SAD 8039 285(814) 277(1640) 244(16200) 210(33700) MeOH
H NMR: SAD 3015 2.2 7.0-7.5 7.7 8.7 $CDCl_3$
m/e: 169.03

HODOC No: 00175
CAS No: 539-03-7
Beilstein No: 12^2, 327
Name: Acetamide, *N*-(4-chlorophenyl)-
Synonym: Acetanilide, 4′-chloro-
Mol Form: C_8H_8ClNO
M_r: 169.60
mp: 179
***d*:** $1.385^{22/4}$
***s*:** H_2O 1; al 3; eth 4; ctc 2
UV: OES 4-157 249(17783) EtOH
m/e: 169.03

HODOC No: 00176
CAS No: 579-11-3
Beilstein No: 12^3, 1077
Name: Acetamide, *N*-chloro-*N*-phenyl-
Synonym: Acetanilide, *N*-chloro-
Line Form: $C_6H_5N(Cl)COCH_3$
Mol Form: C_8H_8ClNO
M_r: 169.60
mp: 91
Color: nd (dil al) pl (peth-Chl)
m/e: 169.03

HODOC No: 00177
CAS No: 19837-92-4
Name: Acetamide, *N*-[4-[[(4-chlorophenyl)amino]sulfonyl]phenyl]-
Synonym: Acetanilide, 4′-[(*p*-chlorophenyl)sulfamoyl]-
Mol Form: $C_{14}H_{13}ClN_2O_3S$
M_r: 324.77
***s*:** DMSO 2
IR: SADG 17795 3360 3220 1695 1595 1530 1490 1450 1400 1325 1310 1250 1150 1090 1010 940 920 830 720 640 615 570 465
UV: SAD 17449 264(30600) 237(17800) MeOH 252(30000) MeOH:KOH
H NMR: SAD 9417 2.1 7.2 7.3 7.8 10.3 $DMSOd_6$
m/e: 324.03

HODOC No: 00178
CAS No: 17743-62-3
Name: Acetamide, *N*-[2-(4-chlorophenyl)-1,1-dimethylethyl]-
Synonym: Acetamide, *N*-(*p*-chloro-α,α-dimethylphenethyl)-
Mol Form: $C_{12}H_{16}ClNO$
M_r: 225.71
mp: 115-7
UV: LANG 2274 276(240) 269(302) 261(245) 220(11482) EtOH
m/e: 225.09

HODOC No: 00179
CAS No: 50610-32-7
Beilstein No: 13, 118
Name: Acetamide, *N*,*N*′-(2-chloro-1,4-phenylene)*bis*-
Synonym: Acetamide, *N*,*N*′-(chloro-*p*-phenylene)*bis*-
Line Form: 2,5-$(CH_3CONH)_2C_6H_3Cl$
Mol Form: $C_{10}H_{11}ClN_2O_2$
M_r: 226.65
mp: 196-7
Color: nd
***s*:** EtOH 4
m/e: 226.05

HODOC No: 00180
CAS No: 60901-60-2
Name: Acetamide, 2-chloro-*N*-phenyl-*N*-2-propenyl-
Synonym: Chloroacetamide, *N*-allyl-*N*-phenyl
Line Form: $ClCH_2CON(C_6H_5)CH_2CH{=}CH_2$
Mol Form: $C_{11}H_{12}ClNO$
M_r: 209.66
bp: $119^{0.5}$
***n*:** 1.5079^{25}
m/e: 209.06

HODOC No: 00181
CAS No: 13916-39-7
Beilstein No: 4^4, 476
Name: Acetamide, 2-chloro-*N*-propyl-
Synonym: α-Chloro-*N*-propylacetamide
Line Form: $ClCH_2CONHC_3H_7$
Mol Form: $C_5H_{10}ClNO$
M_r: 135.59
bp: $105\text{-}6^{10.5}$
mp: 62
***s*:** Ac 4; Et_2O 4; EtOH 4; Chl 4
m/e: 135.05

HODOC No: 00182
CAS No: 4489-33-2
Name: Acetamide, *N*-[4-[[(6-chloro-3-pyridazinyl)amino]sulfonyl]phenyl]-
Synonym: Acetanilide, 4′-[(6-chloro-3-pyridazinyl)sulfamoyl]-
Mol Form: $C_{12}H_{11}ClN_4O_3S$
M_r: 326.75
mp: 218-20
UV: LANG 852 313(1698) 266(24547) EtOH
m/e: 326.02

HODOC No: 00183
CAS No: 32322-48-8
Beilstein No: 4^4, 813
Name: Acetamide, 2-chloro-*N*-tetradecyl-
Synonym: α-Chloro-*N*-tetradecylacetamide
Line Form: $ClCH_2CONHC_{14}H_{29}$
Mol Form: $C_{16}H_{32}ClNO$
M_r: 289.87
mp: 64-5
Color: cr (peth)
m/e: 289.22

HODOC No: 00184
CAS No: 104246-27-7
Name: Acetamide, 2-chloro-*N*-[4-[(2-thiazolylamino)sulfonyl]phenyl]-
Synonym: Benzenesulfonic acid, amide,4(3-chloroacetamido)-*N*(1,3-thiazol-2-yl)
Mol Form: $C_{11}H_{10}ClN_3O_3S_2$
M_r: 331.78
***s*:** DMSO 3
IR: SADG 17800 3320 3280 3090 1705 1610 1595 1560 1540 1435 1400 1320 1270 1130 1080 940 850 830 720 670 650 600 570
UV: SAD 17670 287(23400) 258(20400) MeOH 257(23200) MeOH:KOH
H NMR: SAD 9411 4.3 6.8 7.2 7.8 10.6 DMSOd6
m/e: 330.99

HODOC No: 00185
CAS No: 107-91-5
Beilstein No: 2^4, 1891
Name: Acetamide, 2-cyano-
Synonym: Acetamide, cyano-
Line Form: $NCCH_2CONH_2$
Mol Form: $C_3H_4N_2O$
M_r: 84.07
mp: 121-2
Color: pl (w)
***s*:** H_2O 4
IR: SADP 2237
MS: NBS 3990
m/e: 84.03

HODOC No: 00186
CAS No: 38093-52-6
Name: Acetamide, *N*-(cyanodiphenylmethyl)-
Synonym: *N*-Acetyl-α,α-diphenylglycinonitrile
Line Form: $(C_6H_5)_2C(CN)NHCOCH_3$
Mol Form: $C_{16}H_{14}N_2O$
M_r: 250.28
mp: dec 275
UV: LANG 2483 258(447) EtOH
m/e: 250.11

HODOC No: 00187
CAS No: 38093-55-9
Name: Acetamide, *N*-(cyanodiphenylmethyl)-*N*-methyl-
Synonym: Acetic acid, amide,*N*(cyano-diphenyl-methyl)-*N*-methyl
Line Form: $(C_6H_5)_2C(CN)N(CH_3)COCH_3$
Mol Form: $C_{17}H_{16}N_2O$
M_r: 264.31
mp: 135
UV: LANG 2484 265(398) 260(457) 254(389) EtOH
m/e: 264.13

HODOC No: 00188
CAS No: 4814-80-6
Beilstein No: 4^2, 790
Name: Acetamide, *N*-(cyanomethyl)-
Synonym: Acetamidoacetonitrile
Line Form: $CH_3CONHCH_2CN$
Mol Form: $C_4H_6N_2O$
M_r: 98.10
mp: 79-81
***s*:** H_2O 3
IR: SADG 14379 3250 3050 2940 2240 1660 1560 1410 1370 1300 1100 1030 900 850 700 590 550
H NMR: SAD 8064 2.1 4.2 D_2O
m/e: 98.05

HODOC No: 00189
CAS No: 621-03-4
Beilstein No: 12^3, 558
Name: Acetamide, 2-cyano-*N*-phenyl-
Synonym: Acetanilide, 2-cyano-
Line Form: $C_6H_5NHCOCH_2CN$
Mol Form: $C_9H_8N_2O$
M_r: 160.17
mp: 199-200
Color: nd (al)
m/e: 160.06

HODOC No: 00190
CAS No: 39149-34-3
Name: Acetamide, *N*-(cyanophenylmethyl)-
Synonym: Acetamide, *N*-(α-cyanobenzyl)-
Line Form: $C_6H_5CH(CN)NHCOCH_3$
Mol Form: $C_{10}H_{10}N_2O$
M_r: 174.19
mp: 113
UV: LANG 2457 270(81) 264(158) 259(182) 252(155) EtOH
m/e: 174.08

HODOC No: 00191
CAS No: 42965-13-9
Name: Acetamide, *N*-cyclohexyl-*N*-[(cyclohexylamino)carbonyl]-
Synonym: Urea, 1-acetyl-1,3-dicyclohexyl-
Mol Form: $C_{15}H_{26}N_2O_2$
M_r: 266.37
IR: COB 7740 3280 3050 2940 2860 1700 1640 1540 1440 1380 1340 1240 1080 620
m/e: 266.20

HODOC No: 00192
CAS No: 101057-63-0
Name: Acetamide, *N*-cyclohexyl-2-[(4-oxo-2-phenyl-4H-1-benzopyran-7-yl)oxy]-
Synonym: Flavone, 7(carboxy(amide,*N*-cyclohexyl)methoxy)
Mol Form: $C_{23}H_{23}NO_4$
M_r: 377.42
mp: 203-5
UV: LANG 2415 309(24547) 252(19055) EtOH
m/e: 377.16

HODOC No: 00193
CAS No: 29030-57-7
Beilstein No: 12^3, 2819
Name: Acetamide, *N*-(4-cyclohexylphenyl)-
Synonym: Acetanilide, 4′-cyclohexyl-
Line Form: C_6H_{11}-$C_6H_4NHCOCH_3$
Structure:

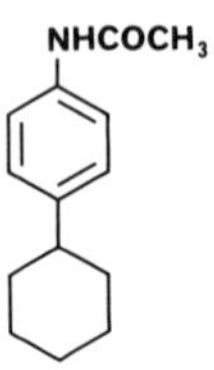

Mol Form: $C_{14}H_{19}NO$
M_r: 217.30
mp: 129
Color: nd (peth)
***s*:** Et_2O 4; EtOH 4; Chl 4
m/e: 217.15

HODOC No: 00194
CAS No: 641-06-5
Beilstein No: 2^4, 416
Name: Acetamide, *N,N*-diacetyl-
Synonym: Triacetamide
Line Form: $(CH_3CO)_3N$
Mol Form: $C_6H_9NO_3$
M_r: 143.14
mp: 79
Color: nd (eth)
***s*:** Et_2O 4
m/e: 143.06

HODOC No: 00195
CAS No: 5834-25-3
Beilstein No: 18^4, 7193
Name: Acetamide, *N*-3-dibenzofuranyl-
Synonym: Dibenzofuran, 3-acetamido-
Line Form: 178
Structure:

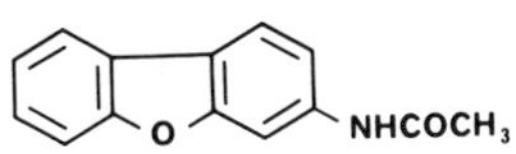

Mol Form: $C_{14}H_{11}NO_2$
M_r: 225.23
mp: 178
***s*:** DMSO 2
IR: µSADP 2454 3225 3030 1670 1615 1540 1495 1450 1430 1370 1300 1280 1250 1190 1125 1020 950 850 820 740 720
UV: SAD 639 313 306 265 217 diox
H NMR: SAD 10854 2.2 7.2-7.7 8.0 8.2 10.2 DMSOd6
MS: NBS 4539
m/e: 225.08

HODOC No: 00196
CAS No: 66178-47-0
Beilstein No: 4, 59
Name: Acetamide, 2,2-dibromo-*N*,*N*-dimethyl-
Synonym: Dibromoacetamide, *N*,*N*-dimethyl
Line Form: $Br_2CHCON(CH_3)_2$
Mol Form: $C_4H_7Br_2NO$
M_r: 242.91
bp: 128^{16}
mp: 79-80
Color: pr(w, eth)
m/e: 242.89

HODOC No: 00197
CAS No: 683-72-7
Beilstein No: 2^4, 505
Name: Acetamide, 2,2-dichloro-
Synonym: Dichloroacetamide
Line Form: $Cl_2CHCONH_2$
Mol Form: $C_2H_3Cl_2NO$
M_r: 127.96
bp: $233\text{-}4^{745}$
mp: 99.4
***s*:** H_2O 3; al 3; eth 3; ace 2
IR: SADG 18587 3320 3150 2810 1670 1400 1230 1210 1130 900 790 760 640
H NMR: SAD 4967 6.3 7.4 ace
m/e: 126.96

HODOC No: 00198
CAS No: 18506-47-3
Name: Acetamide, 2,2-dichloro-*N*-(2,3-dihydro-1,5-dimethyl-3-oxo-2-ph enyl-1H-pyrazol-4-yl)-
Synonym: Acetamide, *N*-antipyrinyl-2,2-dichloro-
Mol Form: $C_{13}H_{13}Cl_2N_3O_2$
M_r: 314.16
mp: 236-8
UV: LANG 615 278(10965) 236(10715) EtOH
m/e: 313.04

HODOC No: 00199
CAS No: 23280-38-8
Name: Acetamide, 2,2-dichloro-*N*-[4-[(dimethylamino)sulfonyl]phenyl]-
Synonym: Acetanilide, 2,2-dichloro-4′-(dimethylsulfamoyl)-
Mol Form: $C_{10}H_{12}Cl_2N_2O_3S$
M_r: 311.17
***s*:** DMSO 3
IR: SADG 17782 3300 3120 2960 1720 1590 1540 1400 1330 1150 960 860 850 810 730 710 660 570 550
H NMR: SAD 9408 2.7 6.7 7.8 8.0 DMSOd6
m/e: 309.99

HODOC No: 00200
CAS No: 56-75-7
Beilstein No: 13^3, 2968
Name: Acetamide, 2,2-dichloro-*N*-[2-hydroxy-1-(hydroxymethyl)-2-(4-n itrophenyl)ethyl]-, [R-(R*,R*)]-
Synonym: Acetamide, 2,2-dichloro-*N*-[β-hydroxy-α-(hydroxymethyl)-*p*-nitrophenethyl]-, *D*-threo-(-)-
Line Form: 4-$NO_2C_6H_4$-CH(OH)-CH(CH_2OH)-$NHCOCH_2Cl$
Structure:

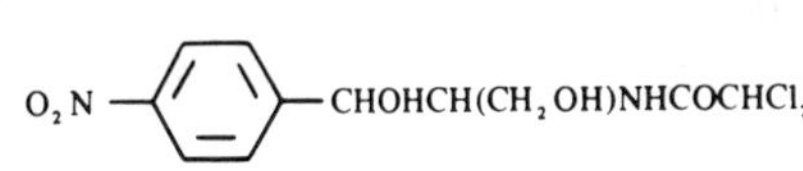

Mol Form: $C_{11}H_{12}Cl_2N_2O_5$
M_r: 323.12
bp: sub (hi vac)
mp: 148-50
α: $[\alpha]^{25/D}$ 19 al,5% $[\alpha]^{25/D}$-25.5 AcOEt

Color: nd or pl (w)
s: H_2O 2; al 4; eth 2; ace 4; bz 1; chl 3
IR: µCOB 5174 3230 3030 1700 1560 1520 1410 1350 1240 1100 1060 970 850 810 760 700
UV: LANG 385 278(10000) H_2O
MS: ALDN 236 153(100) 60(99) 70(85) 155(78) 170(70) 106(46) 77(40) 322(0)
m/e: 322.01

HODOC No: 00201
CAS No: 309-18-2
Name: Acetamide, 2,2-dichloro-*N*-[2-hydroxy-1-(hydroxymethyl)-2-[4-[(trifluoromethyl)sulfonyl]phenyl]ethyl]-
Synonym: Acetamide, 2,2-dichloro-*N*-[β-hydroxy-α-(hydroxymethyl)-*p*-[(trifluoromethyl)sulfonyl]phenethyl]-
Mol Form: $C_{12}H_{12}Cl_2F_3NO_5S$
M$_r$: 410.18
IR: µCOB 7261 3450 3230 3030 1700 1590 1560 1410 1350 1280 1210 1140 1060 1030 900 810 770 710 690 660
m/e: 408.98

HODOC No: 00202
CAS No: 17751-20-1
Name: Acetamide, *N*-(2,6-dichlorophenyl)-2-(diethylamino)-, monohydrochloride
Synonym: Acetanilide, 2′,6′-dichloro-2-(diethylamino)-, monohydrochloride
Mol Form: $C_{12}H_{17}Cl_3N_2O$
M$_r$: 311.63
mp: 174-5
s: H_2O 2
IR: µSADP 7983 3450 3130 2940 2330 1670 1520 1450 1390 1270 1190 1100 960 790
UV: SAD 18341 270(2850) MeOH
H NMR: SAD 11805 1.5 3.4 4.4 7.3-7.7 D_2O
m/e: 310.04

HODOC No: 00203
CAS No: 17751-22-3
Name: Acetamide, *N*-(2,6-dichlorophenyl)-2-(dimethylamino)-, monohydrochloride
Synonym: Acetanilide, 2′,6′-dichloro-2-(dimethylamino)-, monohydrochloride
Mol Form: $C_{10}H_{13}Cl_3N_2O$
M$_r$: 283.58
mp: dec 250
s: H_2O 3
IR: µSADP 7982 2940 2270 1670 1520 1430 1350 1300 1250 1220 1110 1000 960 860 780 690
H NMR: SAD 11804 3.1 4.4 7.2-7.7 D_2O
m/e: 282.01

HODOC No: 00204
CAS No: 77791-64-1
Name: Acetamide, *N*-(2,6-dichlorophenyl)-2-(dipropylamino)-, monohydrochloride
Synonym: Acetic acid, amide,*N*(2,6-dichlorophenyl)-dipropylamino, hydrochloride
Mol Form: $C_{14}H_{21}Cl_3N_2O$
M$_r$: 339.68
mp: 172-3
s: H_2O 2
IR: µSADP 7984 3450 3130 2860 2700 2330 1670 1520 1430 1370 1250 1190 1040 980 890 790 760
H NMR: SAD 11806 1.0 1.9 3.4 4.4 7.2-7.7 D_2O
m/e: 338.07

HODOC No: 00205
CAS No: 685-91-6
Beilstein No: 4[4], 349
Name: Acetamide, *N,N*-diethyl-
Synonym: *N,N*-Diethylacetamide
Line Form: $CH_3CON(C_2H_5)_2$
Mol Form: $C_6H_{13}NO$
M_r: 115.17
bp: 185-6, 91^{30}
***d*:** $0.9130^{17.4/4}$
***n*:** $1.4374^{17.4}$
***s*:** H_2O 3; al 3; eth 5; ace 5; bz 5; ctc 2
IR: μSADP 1874 3130 1670 1450 1390 1370 1280 1220 1180 1100 1090 1040 1000 790
Raman: SAD 922 2980 2940 2880 2740 1640 1490 1460 1370 1320 1280 1230 1170 1090 1040 1000 950 920 710 650 620 590 500 430 340
UV: HBCP <200(9120) H_2O
H NMR: SAD 6701 1.2 2.0 3.3 CCl_4
MS: ALDN 54 58(100) 115(41) 43(31) 44(31) 72(21) 30(14) 29(7) 42(7)
m/e: 115.10

HODOC No: 00206
CAS No: 137-58-6
Beilstein No: 12[3], 2467
Name: Acetamide, 2-(diethylamino)-*N*-(2,6-dimethylphenyl)-
Synonym: 2′,6′-Acetoxylidide, 2-(diethylamino)-
Line Form: $2,6\text{-}(CH_3)_2C_6H_3NHCOCH_2N(C_2H_5)_2$
Mol Form: $C_{14}H_{22}N_2O$
M_r: 234.33
bp: $180\text{-}2^4$
mp: 68-9
Color: nd (Bz, al)
***s*:** Bz 4; Et_2O 4; EtOH 4; Chl 4
MS: NBS 1594
m/e: 234.17

HODOC No: 00207
CAS No: 73-78-9
Name: Acetamide, 2-(diethylamino)-*N*-(2,6-dimethylphenyl)-, monohydrochloride
Synonym: 2′,6′-Acetoxylidide, 2-(diethylamino)-, monohydrochloride
Mol Form: $C_{14}H_{23}ClN_2O$
M_r: 270.79
mp: 127-9
***s*:** H_2O 4
UV: LANG 766 273(263) 264(372) EtOH
m/e: 270.15

HODOC No: 00208
CAS No: 101611-98-7
Name: Acetamide, *N*-[2-(diethylamino)ethyl]-2-[(4-oxo-2-phenyl-4H-1-benzopyran-7-yl)oxy]-
Synonym: Flavone, 7(carboxy(amide,*N*(2-diethylaminoethyl))methoxy)
Mol Form: $C_{23}H_{26}N_2O_4$
M_r: 394.45
mp: 137-8
UV: LANG 2413 307(33113) 252(22909) EtOH 294(23442) 250(19498) cyhex
m/e: 394.19

HODOC No: 00209
CAS No: 5417-52-7
Name: Acetamide, *N*-[4-(diethylamino)-2-methylphenyl]-
Synonym: *o*-Acetotoluidide, 4′-(diethylamino)-
Mol Form: $C_{13}H_{20}N_2O$
M_r: 220.30
mp: 115-6
IR: μCOB 443 3230 3030 2940 1640 1520 1450 1370 1300 1250 1210 1110 1090 1080 1030 960 830 800 710 690
m/e: 220.16

HODOC No: 00210
CAS No: 100030-11-3
Name: Acetamide, *N,N*-diethyl-2-[(4-oxo-2-phenyl-4H-1-benzopyran-7-yl)oxy]-
Synonym: Flavone, 7(carboxy(diethylamide) methoxy)
Mol Form: $C_{21}H_{21}NO_4$
M_r: 351.38
mp: 127-8
UV: LANG 2410 308(25704) 253(20417) EtOH
m/e: 351.15

HODOC No: 00211
CAS No: 4140-70-9
Name: Acetamide, *N,N′*-(4,6-difluoro-2-nitro-*m*-phenylene)*bis*-
Synonym: Benzene, 1,3-diacetamido-4,6-difluoro-2-nitro
Mol Form: $C_{10}H_9F_2N_3O_4$
M_r: 273.19
IR: COB 7811 3380 1670 1625 1510 1420 1370 1270 1145 1020 860 780 685
m/e: 273.06

HODOC No: 00212
CAS No: 83-15-8
Name: Acetamide, *N*-(2,3-dihydro-1,5-dimethyl-3-oxo-2-phenyl-1H-pyrazol-4-yl)-
Synonym: Acetamide, *N*-antipyrinyl-
Mol Form: $C_{13}H_{15}N_3O_2$
Mr: 245.27
s: chl 2
IR: µSADP 10272 3230 3030 1640 1610 1520 1490 1450 1410 1370 1350 1300 1270 1180 1160 1150 1140 1110 1080 1060 100 960 930 900 830 780 760 740 710 690
H NMR: SAD 8332 2.0 2.2 3.1 7.4 9.5 $CDCl_3$
m/e: 245.12

HODOC No: 00213
CAS No: 5702-75-0
Name: Acetamide, *N*-[4-(2,5-dihydro-2,5-dioxo-1H-pyrrol-1-yl)phenyl]-
Synonym: Acetanilide, 4′-maleimido-
Mol Form: $C_{12}H_{10}N_2O_3$
Mr: 230.21
IR: COB 7197 3300 3080 2930 1700 1660 1570 1510 1400 1370 1310 1250 1160 830 680 510
UV: SAD 1431 294(5150) 285(5790) 258(5830) 212(56400) MeOH:KOH 288(7320) 280(7510) 236(4750) 209(41900) MeOH:HCl
m/e: 230.07

HODOC No: 00214
CAS No: 28732-21-0
Name: Acetamide, *N*-[4-[2,5-dihydro-3-(methylthio)-5-oxo-1,2,4-triazin-6-yl]phenyl]-
Synonym: Acetanilide, 4′-[2,5-dihydro-3-(methylthio)-5-oxo-as-triazin-6-yl]-
Mol Form: $C_{12}H_{12}N_4O_2S$
Mr: 276.30
mp: dec 320
UV: LANG 2400 322(19498) 236(20417) EtOH
m/e: 276.07

HODOC No: 00215
CAS No: 16275-44-8
Name: Acetamide, *N*-(2,5-dihydro-5-oxo-2-furanyl)-
Synonym: Acetamide, *N*-(2,5-dihydro-5-oxo-2-furyl)-
Mol Form: $C_6H_7NO_3$
Mr: 141.12
IR: COB 7548 3220 3100 3020 1790 1760 1670 1530 1375 1270 1140 1090 1040 985 920 900 830 790 690
m/e: 141.04

HODOC No: 00216
CAS No: 103260-67-9
Name: Acetamide, *N*-(1,3-dihydro-1-oxo-4-isobenzofuranyl)-
Synonym: Phthalide, 4-acetamido
Mol Form: $C_{10}H_9NO_3$
Mr: 191.18
mp: 183-4
s: DMSO 2
IR: SADG 19329 3340 3060 2950 1750 1700 1620 1530 1490 1430 1390 1320 1270 1250 1240 1090 1000 810 750
UV: SAD 17424 289(2330) 221(2930) MeOH
H NMR: SAD 10079 2.2 5.4 7.4-7.8 8.0 10.1 DMSOd6
m/e: 191.06

HODOC No: 00217
CAS No: 29834-32-0
Name: Acetamide, *N*-(3,4-dihydro-2-phenyl-2H-1-benzopyran-4-yl)-, (2R-*cis*)-
Synonym: Acetamide, *N*-(2-phenyl-4-chromanyl)-, (R,R)-(+)-
Structure:

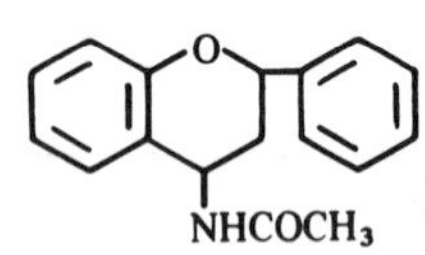

Mol Form: $C_{17}H_{17}NO_2$
Mr: 267.31
mp: 222-3
α: $[\alpha]_D$ 16.7 EtOH
UV: LANG 2587 285(2138) 278(2188) 203(41687) EtOH
m/e: 267.13

HODOC No: 00218
CAS No: 29834-33-1
Name: Acetamide, *N*-(3,4-dihydro-2-phenyl-2H-1-benzopyran-4-yl)-, (2*S*-*cis*)-
Synonym: Acetamide, *N*-(2-phenyl-4-chromanyl)-, (*S*,*S*)-(-)-
Mol Form: $C_{17}H_{17}NO_2$
M_r: 267.31
α: $[\alpha]_D$-14.5 EtOH
UV: LANG 2588 285(2188) 278(2188) 203(40738) EtOH
m/e: 267.13

HODOC No: 00219
CAS No: 127-19-5
Beilstein No: 4^4, 180
Name: Acetamide, *N*,*N*-dimethyl-
Synonym: Acetdimethylamide
Line Form: $CH_3CON(CH_3)_2$
Mol Form: C_4H_9NO
M_r: 87.12
bp: 165, 84^{30}
mp: -20
***d*:** $0.9366^{25/4}$
***n*:** 1.4380^{20}
***s*:** H_20 5; al 5; eth 5; ace 5; bz 5; chl 5
IR: µCOB 1226 2940 1640 1560 1520 1390 1270 1190 1010
Raman: DFB 54 3020 2940 2880 2820 2790 1640 1460 1420 1360 1270 1190 1010 960 740 590 470 420
UV: OES 4-22 218(1000) hp
H NMR: VAR 421 2.1 2.9 3.0 $CDCl_3$
C NMR: JJ 85 cw 21.3 34.5 37.5 169.6 diox
MS: WILEY 151 44(100) 43(52) 87(45) 15(45) 42(29) 45(18) 72(16) 28(12)
m/e: 87.07

HODOC No: 00220
CAS No: 920-54-7
Name: Acetamide, *N*,*N*-dimethyl-, hydrochloride
Synonym: Acetamide, *N*,*N*-dimethyl-, compd. with hydrochloric acid
Line Form: $H_3CCON(CH_3)_2$–HCl
Mol Form: $C_4H_{11}Cl_2NO$
M_r: 160.04
IR: COB 5606 2900 2100 1780 1670 1460 1370 1340 1190 1030 930
m/e: 159.02

HODOC No: 00221
CAS No: 13886-04-9
Name: Acetamide, *N*-[3-[(dimethylamino)methyl]-4-hydroxyphenyl]-
Synonym: *m*-Acetotoluidide, α-(dimethylamino)-4′-hydroxy-
Mol Form: $C_{11}H_{16}N_2O_2$
M_r: 208.25
IR: µCOB 3996 3230 3130 2940 1670 1560 1490 1470 1390 1270 1190 1020 990 840 830 780
m/e: 208.12

HODOC No: 00222
CAS No: 762-84-5
Name: Acetamide, *N*-(1,1-dimethylethyl)-
Synonym: Acetamide, *N*-*tert*-butyl-
Line Form: $H_3CCONC(CH_3)_3$
Mol Form: $C_6H_{13}NO$
M_r: 115.17
IR: SADG 27638 3290 3080 2970 2930 1650 1560 1450 1390 1370 1300 1230 1030 970 780 720 620 470 420 330
Raman: SAD 510 3290 3200 3090 2980 2940 2880 2770 2710 1650 1560 1460 1410 1380 1310 1240 1050 1040 970 930 790 630 480 430 350 200 150
H NMR: SAD 16444 1.4 1.9 5.8 $CDCl_3$
m/e: 115.10

HODOC No: 00223
CAS No: 20330-45-4
Name: Acetamide, *N*-[4-(1,1-dimethylethyl)phenyl]-
Synonym: Acetanilide, 4′-*tert*-butyl-
Mol Form: $C_{12}H_{17}NO$
M_r: 191.26
***s*:** chl 2
IR: SADG 21257 3250 3170 3100 3040 2940 2880 2850 1690 1660 1600 1550 1510 1460 1390 1370 1320 1260 1180 1100 1010 960 930 830 750 600 550 520 450
UV: SAD 2257 245 MeOH
H NMR: SAD 11134 1.3 2.1 7.3 7.4 7.8 $CDCl_3$
m/e: 191.13

HODOC No: 00224
CAS No: 606-38-2
Beilstein No: 12^3, 2491
Name: Acetamide, *N*-(2,4-dimethyl-6-nitrophenyl)-
Synonym: 2′,4′-Acetoxylidide, 6′-nitro-
Line Form: [4,6-$(CH_3)_2$-2-$NO_2C_6H_2$]-NNCOCH3
Structure:

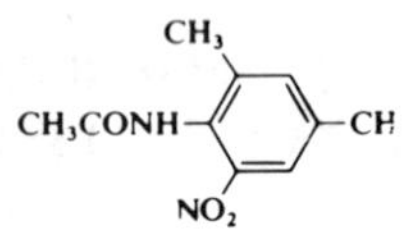

Mol Form: $C_{10}H_{12}N_2O_3$
M_r: 208.21
mp: 172-3
Color: yesh nd (w)
s: H_2O 4; Bz 4; EtOH 4
m/e: 208.08

HODOC No: 00225
CAS No: 5416-12-6
Name: Acetamide, *N*-(2,6-dimethyl-3-nitrophenyl)-
Synonym: 2′,6′-Acetoxylidide, 3′-nitro-
Mol Form: $C_{10}H_{12}N_2O_3$
M_r: 208.21
mp: 170
s: DMSO 3
IR: SADG 19834 3250 3000 2940 1660 1600 1510 1430 1370 1340 1290 1040 820 750 730
UV: SAD 17556 267(4570) MeOH
H NMR: SAD 11636 2.1 2.3 7.3 7.7 8.0 $DMSOd_6$
m/e: 208.08

HODOC No: 00226
CAS No: 2050-43-3
Beilstein No: 12^2, 608
Name: Acetamide, *N*-(2,4-dimethylphenyl)-
Synonym: 2′,4′-Acetoxylidide
Mol Form: $C_{10}H_{13}NO$
M_r: 163.21
bp: 170^{10}
mp: 128-9
s: H_2O 2; al 4; chl 3
IR: μCOB 2886 3230 2940 1670 1560 1470 1370 1300 1040 1020 970 880 810 730 700
UV: SAD 16399 228(6710) MeOH
H NMR: SAD 14171 2.0 2.1 2.2 6.8 6.9 7.2 8.0 $CDCl_3$
m/e: 163.10

HODOC No: 00227
CAS No: 6378-90-1
Name: Acetamide, *N,N′*-(3,3′-dinitro[1,1′-biphenyl]-4,4′-diyl)*bis*-
Synonym: 4′,4‴-Biacetanilide, 2′,2‴-dinitro-
Mol Form: $C_{16}H_{14}N_4O_6$
M_r: 358.29
IR: COB 7535 3350 3080 1710 1620 1580 1500 1450 1370 1340 1270 1230 1160 1080 830 760 700 680 540
m/e: 358.09

HODOC No: 00228
CAS No: 100964-96-3
Name: Acetamide, *N*-(3,7-dinitro-9-oxofluoren-2-yl)-
Synonym: 9-Fluorenone, 2-acetamido-3,7-dinitro
Mol Form: $C_{15}H_9N_3O_6$
M_r: 327.24
IR: COB 7692 3350 3070 2930 1720 1700 1630 1600 1580 1530 1480 1430 1350 1270 1230 1210 1130 1070 930 860 760 730
m/e: 327.05

HODOC No: 00229
CAS No: 60956-27-6
Beilstein No: 12^3, 1680
Name: Acetamide, *N*-(2,3-dinitrophenyl)-
Synonym: Acetanilide, 2′,3′-dinitro-
Line Form: 2,3-$(NO_2)_2C_6H_3$-NHCOCH3
Structure:

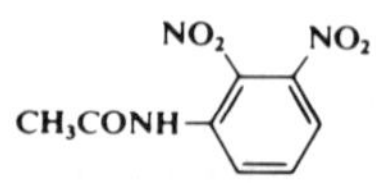

Mol Form: $C_8H_7N_3O_5$
M_r: 225.15
mp: 187
Color: nd (al)
s: al 2; eth 2; bz 2; chl 2
IR: μSADP 11270 3330 3130 1670 1540 1470 1430 1350 1280 1250 1160 850 830 810 760 750 700
UV: SAD 3078 323 230 MeOH
m/e: 225.04

HODOC No: 00230
CAS No: 610-53-7
Beilstein No: 12^3, 1686
Name: Acetamide, *N*-(2,4-dinitrophenyl)-
Synonym: Acetanilide, 2′,4′-dinitro-
Line Form: 2,4-$(NO_2)_2C_6H_3$-$NHCOCH_3$
Mol Form: $C_8H_7N_3O_5$
M_r: 225.15
mp: 125-6
Color: ye nd (al, Bz)
***s*:** H_2O 1; al 4; eth 3; bz 3
IR: µCOB 3194 3330 3130 1720 1610 1520 1470 1350 1270 1220 1140 860 840 750 730
UV: SAD 6707 302(12600) 266(12200) 205 diox
H NMR: SAD 18131 2.4 8.5 8.8 9.0 10.5 $CDCl_3$
m/e: 225.04

HODOC No: 00231
CAS No: 3717-73-5
Beilstein No: 12^3, 1704
Name: Acetamide, *N*-(2,5-dinitrophenyl)-
Synonym: 2′,5′-Dinitroacetanilide
Line Form: 2,5,-$(NO_2)_2$-C_6H_3-$NHCOCH_3$
Mol Form: $C_8H_7N_3O_5$
M_r: 225.15
mp: 121
Color: nd (al)
***s*:** EtOH 4
m/e: 225.04

HODOC No: 00232
CAS No: 73334-01-7
Beilstein No: 12^3, 1705
Name: Acetamide, *N*-(3,4-dinitrophenyl)-
Synonym: Acetanilide, 3′,4′-dinitro-
Line Form: 3,4-$(NO_2)_2C_6H_3$-$NHCOCH_3$
Mol Form: $C_8H_7N_3O_5$
M_r: 225.15
mp: 144-5
Color: ye cr (al), nd (w)
***s*:** H_2O 4; EtOH 4
m/e: 225.04

HODOC No: 00233
CAS No: 38802-18-5
Beilstein No: 12^3, 1706
Name: Acetamide, *N*-(3,5-dinitrophenyl)-
Synonym: Acetanilide, 3′,5′-dinitro-
Line Form: 3,5-$(NO_2)_2C_6H_3$-$NHCOCH_3$
Mol Form: $C_8H_7N_3O_5$
M_r: 225.15
mp: 191
Color: ye nd (dil HOAc,w), (HOAc)
***s*:** H_2O 4; EtOH 4; HOAc 4
m/e: 225.04

HODOC No: 00234
CAS No: 519-87-9
Beilstein No: 12, 247
Name: Acetamide, *N*,*N*-diphenyl-
Synonym: Acetyldiphenylamine
Line Form: $CH_3CON(C_6H_5)_2$
Mol Form: $C_{14}H_{13}NO$
M_r: 211.25
bp: sub
mp: 103
Color: wh cr pwd
***s*:** H_2O 2; al 3; eth 2; chl 2
IR: COB 6362 3070 3040 1680 1600 1500 1460 1370 1300 1070 1030 1000 850 750 700 690 640
UV: SAD 8044 234(10500) MeOH
H NMR: SAD 3016 2.1 7.3 $CDCl_3$
MS: NBS 1330
m/e: 211.10

HODOC No: 00235
CAS No: 1116-24-1
Beilstein No: 4^4, 476
Name: Acetamide, *N*,*N*-dipropyl-
Synonym: Dipropylacetamide
Line Form: $CH_3CON(C_3H_7)_2$
Mol Form: $C_8H_{17}NO$
M_r: 143.22
bp: 209-10, 101^{16}
***d*:** $0.8992^{17.4/4}$
***n*:** $1.4419^{17.4}$
***s*:** EtOH 4
m/e: 143.13

HODOC No: 00236
CAS No: 26249-03-6
Name: Acetamide, *N,N*-ditolyl-
Synonym: *N,N-bis*(methylphenyl)acetamide
Mol Form: $C_{16}H_{17}NO$
M_r: 239.30
IR: **µCOB 442** 3030 2940 1670 1610 1520 1470 1370 1320 1300 1020 830 810 720
m/e: 239.13

HODOC No: 00237
CAS No: 692-93-3
Name: Acetamide, *N,N′*-1,2-ethanediylbis[2-mercapto-
Synonym: Acetamide, *N,N′*-ethylenebis[2-mercapto-
Line Form: $HSCH_2CONHCH_2CH_2NHCOCH_2SH$
Mol Form: $C_6H_{12}N_2O_2S_2$
M_r: 208.29
IR: **COB 4561** 3270 3090 2930 2540 1640 1550 1450 1330 1250 1160 950 720 690
m/e: 208.03

HODOC No: 00238
CAS No: 885-81-4
Beilstein No: 13^3, 1208
Name: Acetamide, *N*-(4-ethoxy-2-nitrophenyl)-
Synonym: *p*-Acetophenetidide, 2′-nitro-
Line Form: $(4\text{-}C_2H_5O\text{-}2\text{-}NO_2)C_6H_3\text{-}NHCOCH_3$
Mol Form: $C_{10}H_{12}N_2O_4$
M_r: 224.21
mp: 104
Color: ye nd (w)
***s*:** al 4; eth 4; ace 4; chl 4
IR: **µCOB 2859** 3330 1700 1590 1520 1470 1370 1350 1320 1280 1250 1220 1150 1110 1080 1040 970 950 860 830 790 760 680
m/e: 224.08

HODOC No: 00239
CAS No: 1777-84-0
Beilstein No: 13^3, 1202
Name: Acetamide, *N*-(4-ethoxy-3-nitrophenyl)-
Synonym: *p*-Acetophenetidide, 3′-nitro-
Line Form: $(4\text{-}CH_3O\text{-}3\text{-}NO_2C_6H_3)NHCOCH_3$
Mol Form: $C_{10}H_{12}N_2O_4$
M_r: 224.21
mp: 123
Color: nd (dil al)
***s*:** Ac 4; Bz 4; EtOH 4
m/e: 224.08

HODOC No: 00240
CAS No: 581-08-8
Beilstein No: 13^3, 779
Name: Acetamide, *N*-(2-ethoxyphenyl)-
Synonym: *o*-Acetophenetidide
Line Form: $2\text{-}C_2H_5OC_6H_4\text{-}NHCOCH_3$
Structure:

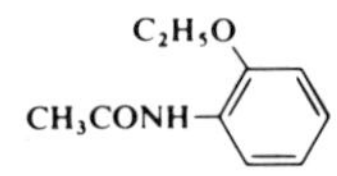

Mol Form: $C_{10}H_{13}NO_2$
M_r: 179.21
bp: 250
mp: 79
Color: lf(dil al)
***s*:** H_2O 1; al 3; eth 3; chl 3; os 3
IR: **COB 5782** 3440 3070 2990 2880 1700 1610 1530 1490 1450 1400 1370 1330 1290 1250 1210 1120 1050 920 750 630
UV: **SAD 16397** 280 243(9700) MeOH
H NMR: **SAD 7939** 1.4 2.1 4.0 6.6-7.1 8.0 8.3 $CDCl_3$
m/e: 179.09

HODOC No: 00241
CAS No: 591-33-3
Beilstein No: 13^3, 950
Name: Acetamide, *N*-(3-ethoxyphenyl)-
Synonym: *m*-Acetophenetidide
Line Form: $3\text{-}C_2H_5OC_6H_4\text{-}NHCOCH_3$
Mol Form: $C_{10}H_{13}NO_2$
M_r: 179.21
mp: 97-9
Color: gy pl (w)
***s*:** H_2O 4; Et_2O 4; EtOH 4
m/e: 179.09

HODOC No: 00242

CAS No: 62-44-2
Beilstein No: 13^3, 1057
Name: Acetamide, *N*-(4-ethoxyphenyl)-
Synonym: *p*-Acetophenetidide
Line Form: 4-C_2H_5O-$C_6H_4NHCOCH_3$
Mol Form: $C_{10}H_{13}NO_2$
M_r: 179.21
mp: 137-8
***n*:** 1.571
Color: mcl pr
***s*:** H_2O 2; al 3; eth 2; ace 3; bz 2; chl 3; py 4
IR: µCOB 5159 3330 3030 1670 1610 1560 1520 1470 1450 1410 1370 1270 1240 1180 1110 1050 930 840 830 790 740
Raman: SAD 519 3280 3080 3040 2980 2930 2880 1680 1660 1620 1600 1540 1520 1480 1460 1440 1420 1400 1370 1330 1310 1270 1240 1190 1160 1120 1110 1050 1020 970 930 900 870 840 810 790 780 720 650 640 610 550 530 460 420 410 390 350 29 0 230 190 120
UV: LANG 548 250(15849) EtOH
MS: ALDN 141 108(100) 109(92) 179(69) 43(48) 137(46) 80(22) 81(18) 53(16)
m/e: 179.09

HODOC No: 00243

CAS No: 625-50-3
Beilstein No: 4^4, 347
Name: Acetamide, *N*-ethyl-
Synonym: Acetamidoethane
Line Form: $CH_3CONHC_2H_5$
Mol Form: C_4H_9NO
M_r: 87.12
bp: 205, $104\text{-}5^{18}$
***d*:** $0.942^{4.5/4}$
***s*:** H_2O 5; al 5; chl 3; HOAc 3
IR: COB 6173 3460 3280 3080 2980 2920 2880 1650 1560 1440 1370 1300 1150 720
H NMR: SAD 146 1.1 2.0 3.3 8.1 $CDCl_3$
MS: ALDN 22 30(100) 87(72) 43(59) 44(48) 72(14) 28(13) 42(11) 29(8)
m/e: 87.07

HODOC No: 00244

CAS No: 1826-56-8
Beilstein No: 12^3, 1596
Name: Acetamide, *N*-ethyl-*N*-(4-nitrophenyl)-
Synonym: Acetanilide, *N*-ethyl-4′-nitro-
Line Form: 4-$NO_2C_6H_4$-N(C_2H_5)$COCH_3$
Mol Form: $C_{10}H_{12}N_2O_3$
M_r: 208.21
mp: 118-9
Color: lf or pr (dil al)
***s*:** H_2O 2; al 4; eth 2; bz 4; lig 1; CS_2 1
IR: SADG 27862 3100 3070 3040 2980 2940 1650 1610 1590 1530 1400 1360 1350 1300 1280 1260 1140 1100 1080 1030 1020 1000 920 880 860 840 780 760 730 700 600 560 500
Raman: SAD 522 3080 1650 1600 1520 1440 1410 1370 1350 1320 1300 1260 1190 1150 920 860 740 640 600 450 420_350 330 190 120
H NMR: SAD 17243 1.2 2.0 3.8 7.4 8.3 $CDCl_3$
m/e: 208.08

HODOC No: 00245

CAS No: 3663-34-1
Name: Acetamide, *N*-(4-ethylphenyl)-
Synonym: Acetanilide, 4′-ethyl-
Mol Form: $C_{10}H_{13}NO$
M_r: 163.21
mp: 90.1-0.7
***s*:** chl 2
IR: SADG 21256 3280 3180 3120 3060 2950 2850 1650 1610 1550 1510 1410 1370 1320 1260 1170 1050 960 820 810 760 610 540 500
UV: SAD 2256 245 MeOH
H NMR: SAD 11133 1.2 2.1 2.6 7.0 7.3 8.2 $CDCl_3$
m/e: 163.10

HODOC No: 00246
CAS No: 529-65-7
Beilstein No: 12^3, 466
Name: Acetamide, *N*-ethyl-*N*-phenyl-
Synonym: Acetanilide, *N*-ethyl-
Line Form: $CH_3CON(C_2H_5)C_6H_5$
Mol Form: $C_{10}H_{13}NO$
M_r: 163.21
bp: 258^{731}
mp: 55
***s*:** H_2O 3; eth 3; ctc 3
IR: COB 5625 3050 2980 2870 1660 1600 1500 1450 1400 1300 1260 1140 1100 1070 1020 770 700 560
UV: OES 4-294 238(5495) hp
H NMR: SAD 1345 1.1 1.7 3.7 7.5 CCl_4
MS: NBS 653
m/e: 163.10

HODOC No: 00247
CAS No: 53-96-3
Beilstein No: 12^3, 3287
Name: Acetamide, *N*-9H-fluoren-2-yl-
Synonym: Acetamide, *N*-fluoren-2-yl-
Structure:

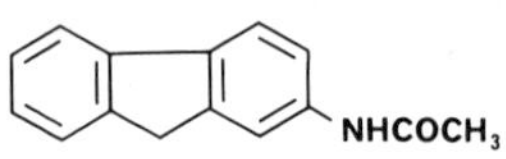

Mol Form: $C_{15}H_{13}NO$
M_r: 223.26
mp: 198
Color: cr (dil al)
***s*:** H_2O 1; al 3; eth 3; HOAc 3
IR: μCOB 6627 3230 3030 2940 1670 1610 1560 1410 1370 1320 870 830 760 730
UV: SAD 4150 288 MeOH
H NMR: SAD 9438 2.1 3.8 7.1-7.8 7.9 9.9 poly D
m/e: 223.10

HODOC No: 00248
CAS No: 5424-77-1
Beilstein No: 12^3, 3299
Name: Acetamide, *N*-9H-fluoren-9-yl-
Synonym: Acetamide, *N*-fluoren-9-yl-
Structure:

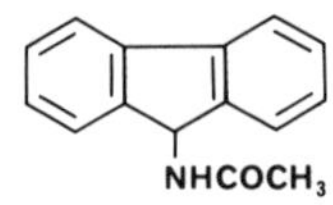

Mol Form: $C_{15}H_{13}NO$
M_r: 223.26
mp: 262
Color: nd (HOAc)
m/e: 223.10

HODOC No: 00249
CAS No: 2805-57-4
Name: Acetamide, *N*-fluoren-1-yl-2,2,2-trifluoro-
Synonym: Acetic acid, amide,*N*(1-fluorenyl)-trifluoro
Mol Form: $C_{15}H_{10}F_3NO$
M_r: 277.23
mp: 154-5
IR: SADG 21864 3280 3060 2890 1720 1620 1580 1540 1450 1430 1340 1260 1170 1160 1060 920 910 890 850 840 820 670 610 530 420
UV: SAD 18399 300(6540) 289(5600) 261(21600) MeOH
H NMR: SAD 11307 3.9 7.2-7.7 7.8-8.1 11.2 $DMSOd_6$
m/e: 277.07

HODOC No: 00250
CAS No: 640-19-7
Beilstein No: 2^4, 454
Name: Acetamide, 2-fluoro-
Synonym: 1081
Line Form: FCH_2CONH_2
Mol Form: C_2H_4FNO
M_r: 77.06
mp: 107-9
IR: ALD 3:439D
MS: ALD 3:103D
m/e: 77.03

HODOC No: 00251
CAS No: 70453-50-8
Name: Acetamide, *N*-(4-fluorophenyl)-2-mercapto-
Synonym: *N-p*-Fluorophenyl-2-mercaptoacetamide
Mol Form: C_8H_8FNOS
M_r: 185.21
***s*:** chl 3
IR: SADG 23990 3280 3160 3080 1640 1610 1550 1500 1410 1400 1330 1200 1130 1090 960 820 780 510 480
H NMR: SAD 11811 2.1 3.3 7.0 7.4 8.9 $CDCl_3$
m/e: 185.03

HODOC No: 00252
CAS No: 13493-47-5
Beilstein No: 14^3, 51
Name: Acetamide, *N*-(2-formylphenyl)-
Synonym: Acetanilide, 2′-formyl-
Line Form: 2-$(CH_3CONH)C_6H_4CHO$
Mol Form: $C_9H_9NO_2$
M_r: 163.17
mp: 70-1
***s*:** H_2O 3; eth 3; ace 3; os 3
UV: HBCP 292(2512) MeOH
m/e: 163.06

HODOC No: 00253
CAS No: 59755-25-8
Beilstein No: 14, 29
Name: Acetamide, *N*-(3-formylphenyl)-
Synonym: *m*-Acetamidobenzaldehyde
Line Form: 3-$(CH_3CONH)C_6H_4CHO$
Mol Form: $C_9H_9NO_2$
M_r: 163.17
mp: 84
Color: pl (Bz)
***s*:** Bz 4; Et_2O 4; EtOH 4
m/e: 163.06

HODOC No: 00254
CAS No: 122-85-0
Beilstein No: 14^2, 25
Name: Acetamide, *N*-(4-formylphenyl)-
Synonym: Acetanilide, 4′-formyl-
Line Form: 4-$(CH_3CONH)C_6H_4CHO$
Mol Form: $C_9H_9NO_2$
M_r: 163.17
mp: 156
Color: pr (w)
***s*:** H_2O 4; Bz 4
m/e: 163.06

HODOC No: 00255
CAS No: 598-42-5
Beilstein No: 3^4, 597
Name: Acetamide, 2-hydroxy-
Synonym: Glycolamide
Line Form: $HOCH_2CONH_2$
Mol Form: $C_2H_5NO_2$
M_r: 75.07
bp: 120
***s*:** H_2O 3
IR: COB 6636 3370 3190 2940 2850 2670 1680 1650 1580 1440 1420 1350 1240 1080 990 690 590 520
H NMR: SAD 10410 4.1 D_2O
m/e: 75.03

HODOC No: 00256
CAS No: 4293-58-7
Name: Acetamide, *N*-(2-hydroxybutyl)-
Synonym: Acetic acid, amide,*N*(2-hydroxybutyl)
Line Form: $CH_3CH_2CH(OH)CH_2NHCOCH_3$
Mol Form: $C_6H_{13}NO_2$
M_r: 131.17
IR: µCOB 4523 3330 3030 2940 1670 1560 1450 1370 1300 1140 1090 960 840
m/e: 131.09

HODOC No: 00257
CAS No: 123-85-3
Name: Acetamide, 2-(2-hydroxyethoxy)-
Synonym: 2-(2-Hydroxyethoxy)acetamide
Line Form: $HOCH_2CH_2OCH_2CONH_2$
Mol Form: $C_4H_9NO_3$
M_r: 119.12
IR: µCOB 4 3450 3330 3130 2940 1670 1470 1370 1330 1140 1100 1060 910 860 710
UV: OES 5-615 270(18) 238(15) EtOH
H NMR: SAD 18509 3.5 3.8 4.6 7.3 $DMSOd_6$
m/e: 119.06

HODOC No: 00258
CAS No: 142-26-7
Beilstein No: 4^4, 1535
Name: Acetamide, *N*-(2-hydroxyethyl)-
Synonym: 2-Acetamidoethanol
Line Form: $CH_3CONHCH_2CH_2OH$
Mol Form: $C_4H_9NO_2$
M$_r$: 103.12
bp: $166\text{-}7^{8}$
mp: 63.5
d: $1.1079^{25/4}$
n: 1.4674^{20}
s: H_2O 5; ace 3; bz 2; lig 2
IR: µCOB 1813 3330 2940 1640 1560 1430 1370 1280 1220 1060
H NMR: SAD 9821 2.0 3.3 3.7 D_2O
m/e: 103.06

HODOC No: 00259
CAS No: 101014-13-5
Name: Acetamide, *N*-(2-hydroxyethyl)-2-[(4-oxo-2-phenyl-4H-1-benzopyran-7-yl)oxy]-
Synonym: Flavone, 7(carboxy(2-hydroxyethylamide)methoxy)
Mol Form: $C_{19}H_{17}NO_5$
M$_r$: 339.33
mp: 210-2
UV: LANG 2411 308(27542) 250(20417) EtOH
m/e: 339.11

HODOC No: 00260
CAS No: 66671-83-8
Name: Acetamide, *N*-[2-hydroxy-1-(hydroxymethyl)-1-methylethyl]-
Synonym: Acetic acid, amide,*N*(*l,l-bis* (hydroxymethyl)ethyl)
Line Form: $HOCH_2C(CH_3)(NHCOCH_3)CH_2OH$
Mol Form: $C_6H_{13}NO_3$
M$_r$: 147.17
mp: 112.5-3.5
IR: COB 4545 3280 3100 2930 2880 2700 2600 1630 1550 1430 1350 1270 1160 1050 940 710
m/e: 147.09

HODOC No: 00261
CAS No: 4857-51-6
Name: Acetamide, 2-(4-hydroxy-*N*-methyl-3,5-dinitroanilino)-
Synonym: Acetic acid, amide,(*N*(3,5-dinitro-4-hydroxyphenyl)methylamino)
Mol Form: $C_9H_{10}N_4O_6$
M$_r$: 270.19
mp: 188-9
s: DMSO 3
IR: SADG 19858 3460 3430 3360 3240 3170 3090 2920 1680 1620 1540 1490 1420 1360 1300 1250 1220 1130 1030 960 920 900 850 780 760 730 650 590 410
H NMR: SAD 11648 3.0 4.0 7.1 7.3 7.4 $DMSOd_6$
m/e: 270.06

HODOC No: 00262
CAS No: 24281-25-2
Name: Acetamide, *N*-hydroxy-2-[[5-methyl-2-(1-methylethyl)cyclohexyl]oxy]-
Synonym: Acetohydroxamic acid, 2-(*p*-menth-3-yloxy)-
Mol Form: $C_{12}H_{23}NO_3$
M$_r$: 229.31
mp: 69-70
IR: SADG 20214 3200 2950 2900 1670 1510 1450 1380 1360 1100 990 840 690
m/e: 229.17

HODOC No: 00263
CAS No: 4146-09-2
Name: Acetamide, *N*-[1-(hydroxymethyl)-2-methylpropyl]-, acetate
Synonym: Acetic acid, amide,*N*(1(hydroxymethyl-2-methylpropyl)),acetate
Line Form: $(CH_3)_2CHCH(CH_2O_2CCH_3)NHCOCH_3$
Mol Form: $C_9H_{17}NO_3$
M$_r$: 187.23
bp: $111\text{-}4^{0.2}$
IR: µCOB 4582 3230 2940 1750 1670 1540 1470 1370 1240 1030
m/e: 187.12

HODOC No: 00264
CAS No: 38423-42-6
Name: Acetamide, *N*-[1-(hydroxymethyl)-2-[4-(methylthio)phenyl]-2-oxoethyl]-
Synonym: Acetamide, *N*-[α-(hydroxymethyl)-*p*-(methylthio)phenacyl]-
Mol Form: $C_{12}H_{15}NO_3S$
M_r: 253.30
mp: 124-5
UV: LANG 2670 312(19055) 232(6026) EtOH
m/e: 253.08

HODOC No: 00265
CAS No: 523-68-2
Beilstein No: 13^3, 1922
Name: Acetamide, *N*-(4-hydroxy-3-methyl-1-naphthalenyl)-
Synonym: Acetamide, *N*-(4-hydroxy-3-methyl-1-naphthyl)-
Line Form: 4-CH_3CONH-2-CH_3-1-$HOC_{10}H_5$
Mol Form: $C_{13}H_{13}NO_2$
M_r: 215.24
mp: 208-10
Color: nd
m/e: 215.09

HODOC No: 00266
CAS No: 25333-55-5
Name: Acetamide, *N*-(*p*-hydroxy-α-methylphenethyl)-, acetate (ester)
Synonym: Acetamide, *N*-(4-hydroxy-α-methylphenethyl)-, acetate
Mol Form: $C_{13}H_{17}NO_3$
M_r: 235.27
IR: μCOB 5207 3330 3130 3030 1750 1640 1560 1520 1450 1370 1220 1190 1160 1020 1010 920 850 760
m/e: 235.12

HODOC No: 00267
CAS No: 117-93-1
Beilstein No: 13^3, 1893
Name: Acetamide, *N*-(2-hydroxy-1-naphthalenyl)-
Synonym: Acetamide, *N*-(2-hydroxy-1-naphthyl)-
Line Form: 1-(CH_3CONH)-2-$HOC_{10}H_6$
Mol Form: $C_{12}H_{11}NO_2$
M_r: 201.21
bp: sub
mp: 235 d
Color: lf (w, dil al)
***s*:** Ac 4; Bz 4; Et_2O 4; EtOH 4
m/e: 201.08

HODOC No: 00268
CAS No: 25351-89-7
Beilstein No: 13^3, 890
Name: Acetamide, *N*-(2-hydroxy-4-nitrophenyl)-
Synonym: Acetanilide, 2′-hydroxy-4′-nitro-
Line Form: (4-NO_2-2-HO)C_6H_3-$NHCOCH_3$
Mol Form: $C_8H_8N_2O_4$
M_r: 196.15
mp: 272
***s*:** DMSO 2
IR: SADG 19287 3380 3100 2750 1660 1640 1590 1540 1500 1420 1320 1260 1070 940 860 800 740 690
UV: SAD 17917 345(9140) 306(7610) 246(7870) MeOH
H NMR: SAD 10009 2.3 7.8 8.4 9.6 $DMSOd_6$
m/e: 196.05

HODOC No: 00269
CAS No: 97-60-9
Beilstein No: 13^3, 881
Name: Acetamide, *N*-(2-hydroxy-5-nitrophenyl)-
Synonym: Acetanilide, 2′-hydroxy-5′-nitro-
Line Form: 5-NO_2-2-HOC_6H_3-$NHCOCH_3$
Mol Form: $C_8H_8N_2O_4$
M_r: 196.15
mp: 279
***s*:** DMSO 2
IR: SADG 19286 3360 3000 2600 1660 1600 1540 1500 1420 1340 1290 1070 890 810 750 690 640 590
UV: SAD 17735 324(6890) 297(7060) 254(14500) 228(11500) MeOH
H NMR: SAD 10008 2.2 6.8 7.1 8.0 9.1 9.5 $DMSOd_6$
m/e: 196.05

HODOC No: 00270
CAS No: 712-34-5
Beilstein No: 13^3, 978
Name: Acetamide, *N*-(3-hydroxy-4-nitrophenyl)-
Synonym: Acetanilide, 3′-hydroxy-4′-nitro-
Line Form: $(4\text{-}NO_2\text{-}3\text{-}HO)C_6H_3\text{-}NHCOCH_4$
Mol Form: $C_8H_8N_2O_4$
M_r: 196.15
mp: 221
***s*:** DMSO 2
IR: SADG 19294 3300 3100 1690 1630 1600 1550 1520 1470 1390 1320 1250 1150 1090 870 810 750 650
UV: SAD 17911 346(13000) MeOH
H NMR: SAD 10018 2.2 7.1 7.7 8.0 10.5 DMSOd6
m/e: 196.05

HODOC No: 00271
CAS No: 7403-75-0
Beilstein No: 13^3, 1208
Name: Acetamide, *N*-(4-hydroxy-2-nitrophenyl)-
Synonym: Acetanilide, 4′-hydroxy-2′-nitro-
Line Form: $(2\text{-}NO_2\text{-}4\text{-}HO)C_6H_3\text{-}NHCOCH_3$
Mol Form: $C_8H_8N_2O_4$
M_r: 196.15
mp: 228
***s*:** DMSO 2
IR: SADG 19297 3340 3200 1660 1590 1520 1440 1360 1330 1280 1240 1130 850 750 700
UV: SAD 17910 359(1920) 233(17900) MeOH
H NMR: SAD 10019 2.1 7.2 7.4 7.5 10.0 10.3 DMSOd6
m/e: 196.05

HODOC No: 00272
CAS No: 67915-26-8
Beilstein No: 13^3, 975
Name: Acetamide, *N*-(5-hydroxy-2-nitrophenyl)-
Synonym: Acetanilide, 5-hydroxy-2-nitro
Line Form: $2\text{-}NO_2\text{-}5\text{-}HOC_6H_3NHCOCH_3$
Mol Form: $C_8H_8N_2O_4$
M_r: 196.15
mp: 270
***s*:** DMSO 2
IR: SADG 19291 3300 3100 2750 1670 1600 1510 1420 1300 1270 1230 1220 1170 1090 880 750 620
UV: SAD 17738 334(8830) 251(12900) 233(14200) MeOH
H NMR: SAD 10014 2.2 6.7 7.7 8.1 10.4 DMSOd6
m/e: 196.05

HODOC No: 00273
CAS No: 4746-61-6
Name: Acetamide, 2-hydroxy-*N*-phenyl-
Synonym: Glycolanilide
Line Form: $C_6H_5NHCOCH_2OH$
Mol Form: $C_8H_9NO_2$
M_r: 151.16
IR: µCOB 4091 3330 3030 1670 1610 1560 1470 1320 1210 1080 990 850 760 690
m/e: 151.06

HODOC No: 00274
CAS No: 614-80-2
Beilstein No: 13^2, 778
Name: Acetamide, *N*-(2-hydroxyphenyl)-
Synonym: Acetanilide, 2′-hydroxy-
Line Form: $(2\text{-}HOC_6H_4)NHCOCH_3$
Structure:

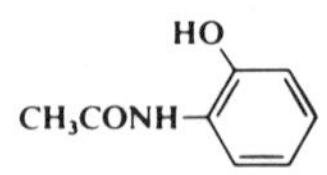

Mol Form: $C_8H_9NO_2$
M_r: 151.16
mp: 209
Color: pl (dil al)
***s*:** H_2O 2; al 4; eth 4; bz 4; DMSOd6 3; alk 4
IR: µCOB 1694 3450 2940 1670 1540 1450 1390 1280 1250 1110 770 670
UV: SAD 8443 283(7410) 242(14300) 207(42300) MeOH
H NMR: SAD 2921 2.1 6.6-7.1 7.7 9.4 9.8 DMSOd6
MS: NBS 468
m/e: 151.06

HODOC No: 00275
CAS No: 621-42-1
Beilstein No: 13^3, 950
Name: Acetamide, *N*-(3-hydroxyphenyl)-
Synonym: Acetanilide, 3′-hydroxy-
Line Form: $(3\text{-}HOC_6H_4)NHCOCH_3$
Mol Form: $C_8H_9NO_2$
M_r: 151.16
mp: 148-9
Color: nd (w)
***s*:** H_2O 4; al 4; eth 2; bz 2; chl 2; DMSO 2
IR: SADG 21411 3340 3060 2920 1660 1610 1570 1500 1380 1350 1280 1170 850 770 690
UV: SAD 6428 290 283 245 212 MeOH
H NMR: SAD 10020 2.1 6.6 7.0-7.2 7.3 9.4 DMSOd6
m/e: 151.06

HODOC No: 00276
CAS No: 103-90-2
Beilstein No: 13^3, 1056
Name: Acetamide, *N*-(4-hydroxyphenyl)-
Synonym: Acetanilide, 4′-hydroxy-
Line Form: $(4\text{-}HOC_6H_4)NHCOCH_3$
Mol Form: $C_8H_9NO_2$
M_r: 151.16
mp: 169-71
***d*:** $1.293^{21/4}$
Color: mcl pr (w)
***s*:** H_2O 1; al 4
IR: **µCOB 5420** 3330 3130 1640 1610 1560 1490 1430 1370 1330 1250 1220 1180 1110 1010 960 860 810 790 710 690
Raman: **SAD 930** 3330 3170 3100 3060 3040 3000 2930 1650 1620 1610 1560 1520 1450 1380 1330 1280 1260 1240 1170 1130 1110 1040 1020 970 950 860 840 800 740 710 690 650 630 600 500 470 410 390 330 210 150
UV: **SAD 2913** 248 MeOH
MS: **AMS 2-630** 109(100) 151(30) 43(22) 80(13) 81(11) 208(10) 86(9) 57(8)
m/e: 151.06

HODOC No: 00277
CAS No: 1202-66-0
Name: Acetamide, *N*-[2-(4-hydroxyphenyl)ethyl]-
Synonym: Acetamide, *N*-(*p*-hydroxyphenethyl)-
Mol Form: $C_{10}H_{13}NO_2$
M_r: 179.21
IR: **µCOB 3624** 3330 3130 1750 1700 1640 1520 1430 1370 1270 1240 1220 1190 1020 1000 980 910 860
m/e: 179.09

HODOC No: 00278
CAS No: 573-27-3
Beilstein No: 13^3, 783
Name: Acetamide, *N*-(2-hydroxyphenyl)-*N*-methyl-
Synonym: 2-*N*-Methylacetamidophenol
Line Form: $(2\text{-}HOC_6H_4)N(CH_3)COCH_3$
Structure:

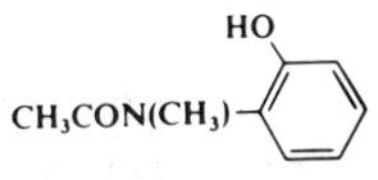

Mol Form: $C_9H_{11}NO_2$
M_r: 165.18
mp: 151
Color: nd (Bz-peth)
***s*:** Bz 4; Et_2O 4; EtOH 4
m/e: 165.08

HODOC No: 00279
CAS No: 579-58-8
Beilstein No: 13^1, 162
Name: Acetamide, *N*-(4-hydroxyphenyl)-*N*-methyl-
Synonym: Acetanilide, 4′-hydroxy-*N*-methyl-
Line Form: $4\text{-}HOC_6H_4\text{-}N(CH_3)COCH_3$
Mol Form: $C_9H_{11}NO_2$
M_r: 165.18
mp: 245
Color: sc (w)
***s*:** Et_2O 4; EtOH 4
IR: **µCOB 6463** 1640 1610 1520 1300 1280 1250 1160 1140 1100 1090 1040 1010 980 830 800
m/e: 165.08

HODOC No: 00280
CAS No: 89169-06-2
Name: Acetamide, *N*-(3-hydroxypropyl)-2-[(4-oxo-2-phenyl-4H-1-benzo pyran-7-yl)oxy]-
Synonym: Flavone, 7(carboxy(3-hydroxypropylamide) methoxy)
Mol Form: $C_{20}H_{19}NO_5$
M_r: 353.35
mp: 174-5
UV: **LANG 2412** 306(25704) 253(19498) EtOH
m/e: 353.13

HODOC No: 00281
CAS No: 26644-97-3
Name: Acetamide, *N*-[4-[(8-hydroxy-5-quinolinyl)azo]phenyl]-, monohydrochloride
Synonym: Quinoline, 5(4-acetamidophenylazo)-8-hydroxy,hydrochloride
Mol Form: $C_{17}H_{15}ClN_4O_2$
Mr: 342.77
IR: µCOB 378 3130 2940 2630 1700 1640 1610 1560 1520 1470 1410 1370 1320 1300 1270 1240 1140 1090 1020 840 810 770 720 660
m/e: 342.09

HODOC No: 00282
CAS No: 51741-72-1
Beilstein No: 25^4, 2031
Name: Acetamide, *N*-1H-imidazol-4-yl-
Synonym: 4-Acetylaminoimidazole
Mol Form: $C_5H_7N_3O$
Mr: 125.13
mp: 226
IR: COB 6906
m/e: 125.06

HODOC No: 00283
CAS No: 144-48-9
Beilstein No: 2^4, 536
Name: Acetamide, 2-iodo-
Synonym: Acetamide, iodo-
Line Form: ICH_2CONH_2
Mol Form: C_2H_4INO
Mr: 184.96
mp: 91-3
s: H_2O 3; tfa 2
IR: µCOB 1827 3330 3230 2940 1670 1470 1180 1160 1090 810 740
UV: OES 4-3 270(398) H_2O
H NMR: SAD 1754 4.0 7.7 tfa
MS: ALD 3:104B
m/e: 184.93

HODOC No: 00284
CAS No: 622-50-4
Beilstein No: 12^3, 1496
Name: Acetamide, *N*-(4-iodophenyl)-
Synonym: Acetanilide, 4′-iodo-
Line Form: $(4\text{-}I\text{-}C_6H_4)NHCOCH_3$
Mol Form: C_8H_8INO
Mr: 261.05
mp: 184-5
Color: ta (w), pr (w, al)
s: H_2O 2; al 4; eth 4; bz 4
IR: SADG 27891 3290 3260 3180 3100 3040 1670 1590 1580 1530 1480 1440 1390 1370 1310 1290 1250 1180 1120 1060 1000 970 840 830 820 740 680 600 550 500 380
Raman: SAD 932 1670 1590 1530 1490 1370 1310 1300 1240 1180 1060 1000 960 850 740 680 510 330 210 150
UV: OES 2-131 254(22909) EtOH
m/e: 260.97

HODOC No: 00285
CAS No: 758-08-7
Name: Acetamide, 2-mercapto-
Synonym: Mercaptoacetamide
Line Form: $HSCH_2CONH_2$
Mol Form: C_2H_5NOS
Mr: 91.13
IR: µCOB 4587 3330 2860 2560 1670 1640 1410 1250 1090 1000 840 810 770
m/e: 91.01

HODOC No: 00286
CAS No: 1190-73-4
Name: Acetamide, *N*-(2-mercaptoethyl)-
Synonym: 2-Acetamidoethanethiol
Line Form: $CH_3CONHCH_2CH_2SH$
Mol Form: C_4H_9NOS
Mr: 119.18
IR: COB 6849 3280 3070 2940 2850 2550 1650 1540 1430 1360 1280 1190 1090 990 700
m/e: 119.04

HODOC No: 00287
CAS No: 93-42-5
Beilstein No: 12^3, 3052
Name: Acetamide, 2-mercapto-*N*-2-naphthalenyl-
Synonym: Acetamide, 2-mercapto-*N*-2-naphthyl-
Line Form: $HSCH_2CONH(2\text{-}C_{10}H_7)$
Mol Form: $C_{12}H_{11}NOS$
M$_r$: 217.27
mp: 111-2
***s*:** H_2O 1; al 4; os 4
IR: SADG 18770 3290 3100 2580 1680 1650 1590 1550 1460 1420 1360 1300 1170 900 870 820 740 680
m/e: 217.06

HODOC No: 00288
CAS No: 4822-44-0
Beilstein No: 12^3, 925
Name: Acetamide, 2-mercapto-*N*-phenyl-
Synonym: Acetanilide, 2-mercapto-
Line Form: $HSCH_2CONHC_6H_5$
Mol Form: C_8H_9NOS
M$_r$: 167.22
mp: 110-1
Color: nd (w,al)
***s*:** H_2O 3; al 4; eth 4; bz 2; chl 2; lig 2
IR: µCOB 3784 3330 2560 1670 1610 1560 1470 1390 1330 1280 1250 1150 1090 1030 960 900 850 800 760 730 690 660
UV: SAD 12029 245(14200) MeOH
H NMR: SAD 1330 2.0 3.3 7.1-7.6 8.6 $CDCl_3$
m/e: 167.04

HODOC No: 00289
CAS No: 31424-45-0
Beilstein No: 12^1, 265
Name: Acetamide, *N*-(mercaptophenyl)-
Synonym: Acetamide, mercapto-*N*-phenyl
Line Form: $HSCH_2CONHC_6H_5$
Mol Form: C_8H_9NOS
M$_r$: 167.22
mp: 110-1
***s*:** Et_2O 4; EtOH 4
m/e: 167.04

HODOC No: 00290
CAS No: 1126-81-4
Beilstein No: 13^2, 1270
Name: Acetamide, *N*-(4-mercaptophenyl)-
Synonym: Acetanilide, 4′-mercapto-
Line Form: $4\text{-}HSC_6H_4NHCOCH_3$
Mol Form: C_8H_9NOS
M$_r$: 167.22
mp: 150-3
IR: ALD 2:935A
MS: NBS 4256; ALD 7:69A
m/e: 167.04

HODOC No: 00291
CAS No: 16332-06-2
Beilstein No: 3^3, 397
Name: Acetamide, 2-methoxy-
Synonym: Acetamide, methoxy
Line Form: $CH_3OCH_2CONH_2$
Mol Form: $C_3H_7NO_2$
M$_r$: 89.09
mp: 96-9
***s*:** chl 3
IR: COB 6062 3390 3190 2980 2940 2820 1640 1450 1430 1360 1200 1110 1090 980 800 720 590 490
H NMR: SAD 845 3.4 3.9 6.6 $CDCl_3$
m/e: 89.05

HODOC No: 00292
CAS No: 66910-71-2
Name: Acetamide, *N*-(2-methoxy-4,5-dinitrophenyl)-
Synonym: *o*-Acetanisidide, 4′,5′-dinitro-
Mol Form: $C_9H_9N_3O_6$
M$_r$: 255.18
mp: 165
IR: SADG 19859 3380 3120 3050 1710 1590 1550 1530 1520 1480 1450 1430 1400 1370 1330 1280 1230 1210 1180 1140 1040 990 940 900 880 820 800 750 740 660 590 580 460 430
UV: SAD 17577 300(7940) 254(15800) 226(12200) MeOH
H NMR: SAD 11649 2.2 4.1 7.8 8.9 9.8 $DMOSd_6$
m/e: 255.05

HODOC No: 00293
CAS No: 73-31-4
Beilstein No: 22^4, 5679
Name: Acetamide, *N*-[2-(5-methoxy-1H-indol-3-yl)ethyl]-
Synonym: Acetamide, *N*-[2-(5-methoxyindol-3-yl)ethyl]-
Structure:

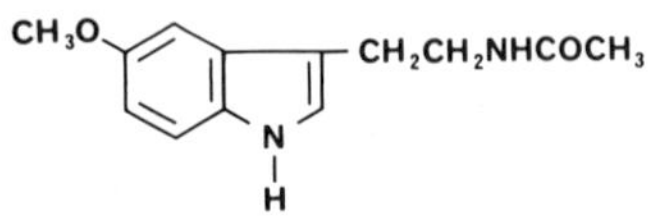

Mol Form: $C_{13}H_{16}N_2O_2$
Mr: 232.27
mp: 116-8
Color: pa ye lf (bz)
IR: ALD 2:1095G
MS: NBS 435; ALD 8:75B
m/e: 232.12

HODOC No: 00294
CAS No: 3900-46-7
Beilstein No: 13^2, 416
Name: Acetamide, *N*-(6-methoxy-2-naphthalenyl)-
Synonym: Acetamide, *N*-(6-methoxy-2-naphthyl)-
Mol Form: $C_{13}H_{13}NO_2$
Mr: 215.24
mp: 159-60
IR: COB 2744 3260 3100 1670 1600 1570 1420 1400 1350 1270 1240 1170 1030 890 860 820
m/e: 215.09

HODOC No: 00295
CAS No: 93-27-6
Beilstein No: 13^3, 890
Name: Acetamide, *N*-(2-methoxy-4-nitrophenyl)-
Synonym: *o*-Acetanisidide, 4′-nitro-
Line Form: $(4\text{-}NO_2\text{-}2\text{-}CH_3O)C_6H_3\text{-}NHCOCH_3$
Structure:

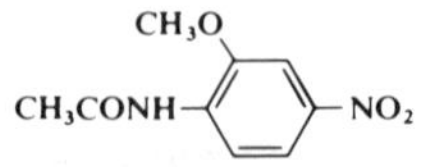

Mol Form: $C_9H_{10}N_2O_4$
Mr: 210.18
mp: 154
Color: pa ye cr (AcOEt)
IR: SADG 20635 3270 3090 1670 1610 1580 1540 1510 1490 1410 1360 1330 1280 1240 1120 1090 1020 870 830 790 740 640 550 450
UV: SAD 18695 339(1030) 222(12600) MeOH
H NMR: SAD 11065 2.3 4.0 7.7 7.9 8.0 8.6 $CDCl_3$
m/e: 210.06

HODOC No: 00296
CAS No: 33721-54-9
Beilstein No: 13^3, 881
Name: Acetamide, *N*-(2-methoxy-5-nitrophenyl)-
Synonym: *o*-Acetanisidide, 5′-nitro-
Line Form: $(5\text{-}NO_2\text{-}2\text{-}CH_3O)C_6H_3\text{-}NHCOCH_3$
Mol Form: $C_9H_{10}N_2O_4$
Mr: 210.18
mp: 178
Color: nd (w)
***s*:** chl 3
IR: SADG 20640 3310 3120 2960 1700 1600 1580 1540 1490 1450 1410 1370 1320 1270 1230 1210 1140 1080 1010 960 860 840 820 750 710 690 670 620 590 520 450
H NMR: SAD 11112 2.3 3.9 6.6 8.1 8.4 10.7 $CDCl_3$
m/e: 210.06

HODOC No: 00297
CAS No: 119-81-3
Beilstein No: 13^3, 1208
Name: Acetamide, *N*-(4-methoxy-2-nitrophenyl)-
Synonym: *p*-Acetanisidide, 2′-nitro-
Line Form: $(2\text{-}NO_2\text{-}4\text{-}CH_3OC_6H_3)NHCOCH_3$
Structure:

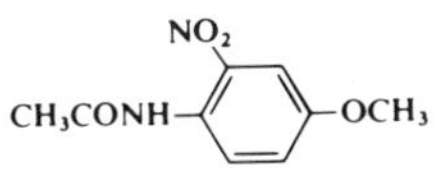

Mol Form: $C_9H_{10}N_2O_4$
Mr: 210.18
mp: 116-7
Color: ye nd (al)
s: H_2O 3; al 3; eth 3; bz 3; HOAc 3
IR: **µSADP 13934** 3330 1700 1590 1520 1470 1320 1250 1030 870 840 830 790 760 660
UV: **SAD 3810** 234 MeOH
H NMR: **SAD 11949** 2.2 3.8 7.2 7.6 8.6 $CDCl_3$
m/e: 210.06

HODOC No: 00298
CAS No: 50651-39-3
Beilstein No: 13^3, 186
Name: Acetamide, *N*-(4-methoxy-3-nitrophenyl)-
Synonym: 4-Acetamido-2-nitroanisole
Line Form: $(3\text{-}NO_2\text{-}4\text{-}CH_3OC_6H_3)NHCOCH_3$
Mol Form: $C_9H_{10}N_2O_4$
Mr: 210.18
mp: 153
Color: og-ye nd (w or dil al)
s: EtOH 4; lig 4
m/e: 210.06

HODOC No: 00299
CAS No: 93-26-5
Beilstein No: 13^3, 778
Name: Acetamide, *N*-(2-methoxyphenyl)-
Synonym: *o*-Acetanisidide
Line Form: $2\text{-}CH_3OC_6H_4\text{-}NHCOCH_3$
Structure:

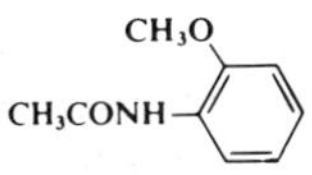

Mol Form: $C_9H_{11}NO_2$
Mr: 165.18
bp: 303-5
mp: 87-8
Color: nd (w)
s: H_2O 4; al 4; eth 3; ace 3; os 3; HOAc 3
IR: **µCOB 1695** 3230 3030 2940 3230 3130 2860 1490 1470 1370 1320 1280 1270 1250 1220 1110 1040 1020 960 930 780 750 710 660
UV: **SAD 2573** 281 244 207 MeOH
H NMR: **SAD 11340** 2.1 3.8 6.7 7.2 7.8 8.3 $CDCl_3$
m/e: 165.08

HODOC No: 00300
CAS No: 588-16-9
Beilstein No: 13^3, 950
Name: Acetamide, *N*-(3-methoxyphenyl)-
Synonym: *m*-Acetanisidide
Line Form: $3\text{-}CH_3OC_6H_4\text{-}NHCOCH_3$
Mol Form: $C_9H_{11}NO_2$
Mr: 165.18
mp: 81
Color: nd or pl (w)
s: H_2O 4; al 4; eth 3; ace 3; os 3
IR: **SADP 9706**
UV: **OES 2-182** 280(3090) 245(11749) EtOH
m/e: 165.08

HODOC No: 00301
CAS No: 51-66-1
Beilstein No: 13^2, 243
Name: Acetamide, *N*-(4-methoxyphenyl)-
Synonym: *p*-Acetanisidide
Mol Form: $C_9H_{11}NO_2$
Mr: 165.18
mp: 130-2
***s*:** H_2O 2; al 3; eth 2; ace 3; chl 3; DMSOd_6 2
IR: SADG 15880 3250 2940 2860 1650 1600 1560 1510 1400 1320 1280 1240 1030 840 810 760
UV: SAD 15958 248(14900) MeOH
H NMR: SAD 6580 2.0 3.7 6.8 7.5 9.7 DMSOd_6
m/e: 165.08

HODOC No: 00302
CAS No: 35813-38-8
Name: Acetamide, *N*-(4-methoxyphenyl)-*N*-methyl-
Synonym: *p*-Acetanisidide, *N*-methyl-
Mol Form: $C_{10}H_{13}NO_2$
Mr: 179.21
mp: 59
***s*:** chl 3
IR: SADG 20634 3030 2930 1660 1580 1510 1440 1420 1370 1350 1290 1240 1160 1140 1100 1080 1020 960 830 790 730 640 590 560 470 360
UV: SAD 18694 281(1300) 275(1470) 228(12400) MeOH
H NMR: SAD 11068 1.9 3.2 3.8 6.9 7.1 $CDCl_3$
m/e: 179.09

HODOC No: 00303
CAS No: 79-16-3
Beilstein No: 4^4, 176
Name: Acetamide, *N*-methyl-
Synonym: Methylacetamide
Line Form: $CH_3CONHCH_3$
Mol Form: C_3H_7NO
Mr: 73.09
bp: 204-6, 95^{14}
mp: 28
***d*:** $0.9371^{25/4}$
***n*:** 1.4301^{20}
***s*:** Ac 4; Bz 4; Et_2O 4; EtOH 4
IR: SADG 21343
UV: OES 1-14
H NMR: SAD 10244
m/e: 73.05

HODOC No: 00304
CAS No: 623-59-6
Beilstein No: 4^4, 207
Name: Acetamide, *N*-[(methylamino)carbonyl]-
Synonym: Urea, 1-acetyl-3-methyl-
Line Form: $CH_3CONHCONHCH_3$
Mol Form: $C_4H_8N_2O_2$
Mr: 116.12
bp: d
mp: 180-1
Color: tcl (w, al), pr (w)
***s*:** H_2O 3; al 2; eth 2; chl 3
IR: µCOB 1809 3330 3230 3130 2860 1670 1540 1520 1450 1410 1370 1270 1250 1160 990 880 860 740 700
H NMR: SAD 7942 2.1 2.9 8.4 10.4 $CDCl_3$
m/e: 116.06

HODOC No: 00305
CAS No: 3073-56-1
Beilstein No: 4^4, 552
Name: Acetamide, *N,N'*-(1-methyl-1,2-ethanediyl)*bis*-
Synonym: Acetamide, *N,N'*-propylenebis-
Line Form: $CH_3CH(NHCOCH_3)$-$CH_2NHCOCH_3$
Mol Form: $C_7H_{14}N_2O_2$
Mr: 158.19
bp: 190^{18}
mp: 138-9
Color: nd (Bz)
***s*:** H_2O 4; EtOH 4; Chl 4
IR: COB 5637 3300 3100 2970 1650 1560 1440 1380 1320 1290 1150 980 740 610
m/e: 158.11

HODOC No: 00306
CAS No: 5461-51-8
Beilstein No: 12^3, 466
Name: Acetamide, *N*-(1-methylethyl)-*N*-phenyl-
Synonym: *N*-Isopropylacetanilide
Line Form: C_6H_5N-$(COCH_3)$-$CH(CH_3)_2$
Mol Form: $C_{11}H_{15}NO$
Mr: 177.24
bp: $262\text{-}3^{712}$
mp: 39
Color: lf (lig)
***s*:** lig 4
m/e: 177.12

HODOC No: 00307
CAS No: 67282-69-3
Name: Acetamide, *N*-(1-methylhexyl)-
Synonym: Heptane, 2-acetamido
Line Form: $CH_3CONHCH(CH_3)(CH_2)_4CH_3$
Mol Form: $C_9H_{19}NO$
M_r: 157.25
IR: **µCOB 695** 3330 3130 2940 2860 1640 1540 1450 1370 1300 1150 1030 970 730
m/e: 157.15

HODOC No: 00308
CAS No: 573-26-2
Beilstein No: 12^3, 1854
Name: Acetamide, *N*-methyl-*N*-(2-methylphenyl)-
Synonym: *o*-Acetotoluidide, *N*-methyl-
Line Form: $(2\text{-}CH_3C_6H_4)N(CH_3)\text{-}COCH_3$
Structure:

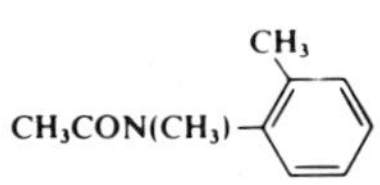

Mol Form: $C_{10}H_{13}NO$
M_r: 163.21
bp: 260
mp: 55-6
***s*:** al 3; chl 3
IR: **COB 5689** 3030 2930 1660 1600 1580 1490 1460 1420 1380 1350 1300 1140 1120 1080 1050 1030 970 770 730
UV: **OES 1-340** 270(316) 263(316) EtOH
H NMR: **VAR 264** 1.8 2.3 3.2 7.1-7.4 $CDCl_3$
m/e: 163.10

HODOC No: 00309
CAS No: 612-03-3
Beilstein No: 12^3, 2052
Name: Acetamide, *N*-methyl-*N*-(4-methylphenyl)-
Synonym: *p*-Acetotoluidide, *N*-methyl-
Line Form: $(4\text{-}CH_3C_6H_4)N(CH_3)COCH_3$
Mol Form: $C_{10}H_{13}NO$
M_r: 163.21
bp: 263
mp: 83
Color: lf (eth-al)
***s*:** al 3; eth 3; lig 3
IR: **COB 5643** 3040 2920 1650 1610 1520 1420 1380 1350 1300 1190 1140 1080 1020 970 830 720
m/e: 163.10

HODOC No: 00310
CAS No: 82898-85-9
Name: Acetamide, *N*-methyl-2-(methylphenylamino)-
Synonym: Acetic acid, amide,(*N*-methyl-phenylamino)-*N*-methyl
Line Form: $C_6H_5N(CH_3)CH_2CONHCH_3$
Mol Form: $C_{10}H_{14}N_2O$
M_r: 178.22
***s*:** chl 3
IR: **µCOB 5236** 3330 2940 1640 1610 1520 1490 1470 1370 1300 1250 1220 1190 1150 1120 1030 980 940 870 750 690
m/e: 178.11

HODOC No: 00311
CAS No: 588-23-8
Beilstein No: 16^3, 387
Name: Acetamide, *N*-[2-methyl-4-[(2-methylphenyl)azo]phenyl]-
Synonym: *o*-Acetotoluidide, 4′-(*o*-tolylazo)-
Mol Form: $C_{16}H_{17}N_3O$
M_r: 267.31
mp: 186-7
Color: red nd (al)
***s*:** Ac 4; Et_2O 4; Chl 4
m/e: 267.14

HODOC No: 00312
CAS No: 88850-40-2
Beilstein No: 16^2, 182
Name: Acetamide, *N*-[4-methyl-2-[(4-methylphenyl)azo]phenyl]-
Synonym: Azobenzene, 2-acetamido-4′-5-dimethyl
Line Form: $2\text{-}(CH_3CONH)\text{-}5\text{-}CH_3C_6H_3N{=}N(C_6H_4CH_3\text{-}4)$
Mol Form: $C_{16}H_{17}N_3O$
M_r: 267.31
mp: 157
Color: ye nd (al-HOAc)
***s*:** Et_2O 4; EtOH 4; Chl 4
m/e: 267.14

HODOC No: 00313
CAS No: 13615-35-5
Name: Acetamide, *N*-(2-methyl-1-naphthalenyl)-
Synonym: Acetamide, *N*-(2-methyl-1-naphthyl)-
Mol Form: $C_{13}H_{13}NO$
M_r: 199.24
mp: 194-5
***s*:** bz 3
IR: COB 2679 3360 3200 1630 1600 1520 1450 1420 1360 1340 1150 1100 820 810 750 730 700
H NMR: SAD 19077 2.2 2.3 7.2-7.6 9.3 poly D
m/e: 199.10

HODOC No: 00314
CAS No: 573-90-0
Beilstein No: 12^3, 2867
Name: Acetamide, *N*-methyl-*N*-1-naphthalenyl-
Synonym: Acetamide, *N*-methyl-*N*-1-naphthyl-
Line Form: $CH_3CON(CH_3)(1\text{-}C_{10}H_7)$
Mol Form: $C_{13}H_{13}NO$
M_r: 199.24
mp: 95-7
Color: nd (lig)
***s*:** H_2O 2; al 3; eth 3; lig 2
IR: µSADP 9768 3030 2940 1670 1590 1520 1410 1300 1250 1140 1010 970 800 780
m/e: 199.10

HODOC No: 00315
CAS No: 612-45-3
Beilstein No: 12^3, 2052
Name: Acetamide, *N*-(4-methyl-2-nitrophenyl)-
Synonym: *p*-Acetotoluidide, 2′-nitro-
Line Form: $(2\text{-}NO_2\text{-}4\text{-}CH_3C_6H_3)NHCOCH_3$
Structure:

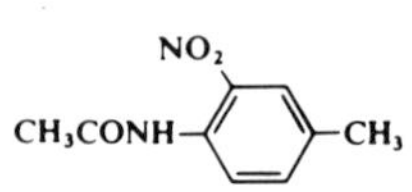

Mol Form: $C_9H_{10}N_2O_3$
M_r: 194.18
bp: 263
mp: 83
Color: lf (eth-al)
***s*:** H_2O 2; al 2; peth 2
IR: SADG 21203 3360 1720 1580 1510 1440 1370 1340 1280 1220 1140 1000 910 820 750
UV: SAD 17792 344(1910) 232(16400) MeOH
H NMR: SAD 17118 2.3 2.4 7.4 8.0 8.6 10.1 $CDCl_3$
m/e: 194.07

HODOC No: 00316
CAS No: 121-95-9
Beilstein No: 12^3, 1595
Name: Acetamide, *N*-methyl-*N*-(4-nitrophenyl)-
Synonym: Acetanilide, *N*-methyl-4′-nitro-
Line Form: $CH_3CON(CH_3)(C_6H_4NO_2\text{-}4)$
Mol Form: $C_9H_{10}N_2O_3$
M_r: 194.18
mp: 153-4
Color: pl (w)
***s*:** H_2O 2; al 3; eth 3
UV: HBCP 288(10965) al
m/e: 194.07

HODOC No: 00317
CAS No: 120-66-1
Beilstein No: 12^3, 1853
Name: Acetamide, *N*-(2-methylphenyl)-
Synonym: *o*-Acetotoluidide
Line Form: $2\text{-}CH_3C_6H_4\text{-}NHCOCH_3$
Structure:

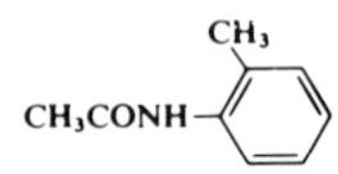

Mol Form: $C_9H_{11}NO$
M_r: 149.18
bp: 296
mp: 110
***d*:** 1.168^{15}
Color: nd (al)
***s*:** H_2O 2; al 3; eth 3; ace 3; bz 2; chl 3; HOAc 3; MeOH 3
IR: SADG 350 3280 3030 2970 1640 1590 1530 1480 1460 1370 1290 1270 1160 1040 1010 970 930 850 750 710 650
UV: SAD 830 230(5930) MeOH
H NMR: SAD 175 2.1 2.2 7.1 7.4 $CDCl_3$
m/e: 149.08

HODOC No: 00318
CAS No: 537-92-8
Beilstein No: 12^3, 1962
Name: Acetamide, *N*-(3-methylphenyl)-
Synonym: *m*-Acetotoluidide
Line Form: (3-$CH_3C_6H_4$)-$NHCOCH_3$
Mol Form: $C_9H_{11}NO$
M_r: 149.18
bp: 303, $182\text{-}3^{14}$
mp: 65.5
***d*:** 1.141^{15}
Color: nd (w)
***s*:** H_2O 2; al 4; eth 4; chl 3
IR: SADG 351 3290 3140 3080 2920 1660 1610 1590 1550 1490 1420 1400 1360 1310 1170 1090 1040 1000 970 890 830 780 750 690
UV: SAD 831 282(628) 244(13800) MeOH
H NMR: SAD 10793 2.1 2.2 6.8 7.1 7.2 7.3 8.7 $CDCl_3$
m/e: 149.08

HODOC No: 00319
CAS No: 103-89-9
Beilstein No: 12^3, 2051
Name: Acetamide, *N*-(4-methylphenyl)-
Synonym: *p*-Acetotoluidide
Line Form: 4-$CH_3C_6H_4$-$NHCOCH_3$
Mol Form: $C_9H_{11}NO$
M_r: 149.18
bp: 307 sub
mp: 148.5
Color: mcl cr or nd (dil al)
***s*:** Bz 4; Et_2O 4; EtOH 4
IR: µSADP 2901 3230 2860 1670 1590 1450 1350 1320 1270 1160 1120 1040 1020 850 810 750
UV: LANG 2664 244(14791) EtOH
m/e: 149.08

HODOC No: 00320
CAS No: 579-10-2
Beilstein No: 12^3, 465
Name: Acetamide, *N*-methyl-*N*-phenyl-
Synonym: Acetanilide, *N*-methyl-
Line Form: $CH_3CON(CH_3)C_6H_5$
Mol Form: $C_9H_{11}NO$
M_r: 149.18
bp: 253^{712}
mp: 102-4
***d*:** $1.0036^{105/4}$
***n*:** 1.576
Color: nd (eth), pr (al)
***s*:** H_2O 3; al 3; eth 3; chl 3; lig 3
IR: µCOB 5937 3030 2940 1670 1610 1490 1430 1390 1350 1300 1140 1090 1020 970 920 790 770 710 690
UV: SAD 8130 224(7870) MeOH
H NMR: SAD 2403 1.9 3.2 7.1-7.6 $CDCl_3$
MS: NBS 444
m/e: 149.08

HODOC No: 00321
CAS No: 21911-76-2
Name: Acetamide, 2-(methylphenylamino)-
Synonym: Acetamide, 2-(*N*-methylanilino)-
Line Form: $C_6H_5N(CH_3)CH_2CONH_2$
Mol Form: $C_9H_{12}N_2O$
M_r: 164.20
IR: µCOB 5343 3330 3130 2940 1640 1610 1490 1470 1390 1370 1330 1320 1250 1210 1120 1090 940 860 750 690
m/e: 164.09

HODOC No: 00322
CAS No: 19837-91-3
Beilstein No: 14, 702
Name: Acetamide, *N*-[4-[[(4-methylphenyl)amino]sulfonyl]phenyl]-
Synonym: Acetanilide, 4′-(*p*-tolylsulfamoyl)-
Mol Form: $C_{15}H_{16}N_2O_3S$
M_r: 304.35
mp: 209
***s*:** H_2O 1; al 3; eth 1; ace 3; DMSO 3
IR: SADG 17790 3360 3200 3020 2920 1690 1600 1530 1400 1380 1330 1260 1150 1090 930 830 720 650 620 590 530 500 410
UV: SAD 17668 263(23200) 231(13600) MeOH 251(13900) MeOH:KOH
H NMR: SAD 9420 2.1 2.2 7.1 7.8 10.3 $DMSOd_6$
m/e: 304.09

HODOC No: 00323
CAS No: 15440-82-1
Name: Acetamide, *N*-[4-[[(4-methylphenyl)sulfonyl]amino]butyl]-
Synonym: Acetamide, *N*-(4-*p*-toluenesulfonamidobutyl)-
Mol Form: $C_{13}H_{20}N_2O_3S$
M_r: 284.36
IR: COB 7826 3370 3150 2940 2860 1650 1550 1440 1360 1310 1150 1060 810 750 650 560
m/e: 284.12

HODOC No: 00324
CAS No: 61996-35-8
Name: Acetamide, *N*-methyl-*N*-2-pyridinyl-
Synonym: Pyridine, 2-*N*-methylacetamido-
Mol Form: $C_8H_{10}N_2O$
M_r: 150.17
UV: LANG 455 262(4467) 225(5754) pH=9.7 293(7943) 233(3996) 5N sulf
m/e: 150.08

HODOC No: 00325
CAS No: 55899-01-9
Name: Acetamide, *N*-methyl-*N*-3-pyridinyl-
Synonym: Pyridine, 3-*N*-methylacetamido-
Mol Form: $C_8H_{10}N_2O$
M_r: 150.17
mp: 62-4
UV: LANG 456 260(3802) pH=9.7 222(8318) 1N sulf
m/e: 150.08

HODOC No: 00326
CAS No: 53209-33-9
Name: Acetamide, *N*-methyl-*N*-4-pyridinyl-
Synonym: Pyridine, 4-*N*-methylacetamido-
Mol Form: $C_8H_{10}N_2O$
M_r: 150.17
mp: 56-7
UV: LANG 457 255(10471) pH=9.7 281(14125) 213(7943) 1N sulf
m/e: 150.08

HODOC No: 00327
CAS No: 22821-80-3
Beilstein No: 13^2, 303
Name: Acetamide, *N*-[4-(methylsulfonyl)phenyl]-
Synonym: Acetanilide, 4′-(methylsulfonyl)-
Mol Form: $C_9H_{11}NO_3S$
M_r: 213.24
mp: 183
IR: COB 7213 3350 3025 3000 2930 1690 1590 1540 1400 1365 1320 1280 1260 1140 1085 970 960 840 825 780 665 590 535 520
UV: OES 2-183 261(22387) EtOH
m/e: 213.05

HODOC No: 00328
CAS No: 39719-87-4
Name: Acetamide, *N*-[4-[[(5-methyl-1,3,4-thiadiazol-2-yl)amino]sulfonyl]phenyl]-
Synonym: Acetanilide, 4′-[(5-methyl-1,3,4-thiadiazol-2-yl)sulfamoyl]-
Mol Form: $C_{11}H_{12}N_4O_3S_2$
M_r: 312.35
***s*:** DMSO 3
IR: SADG 17084 3300 2880 2790 1680 1590 1530 1420 1400 1370 1315 1290 1260 1200 1180 1145 1080 980 915 840 755 710 635 560
UV: SAD 17456 279(24500) 255(18900) MeOH 256(20900) MeOH:KOH
H NMR: SAD 9413 2.2 2.5 6.2 7.8 10.3 $DMSOd_6$
m/e: 312.04

HODOC No: 00329
CAS No: 71119-13-6
Name: Acetamide, *N*-[4-[[(4-methyl-2-thiazolyl)amino]sulfonyl]phenyl]-
Synonym: Acetanilide, 4′-[(4-methyl-2-thiazolyl)sulfamoyl]-
Mol Form: $C_{12}H_{13}N_3O_3S_2$
M_r: 311.36
mp: 253
UV: LANG 56 290(17783) 257(19953) EtOH
m/e: 311.04

HODOC No: 00330
CAS No: 575-36-0
Beilstein No: 12, 1230
Name: Acetamide, *N*-1-naphthalenyl-
Synonym: Acetamide, *N*-1-naphthyl-
Line Form: $CH_3CONH(1-C_{10}H_7)$
Mol Form: $C_{12}H_{11}NO$
M_r: 185.21
mp: 160
***s*:** H_2O 3; al 3; eth 2
IR: SADG 297 3260 3040 1660 1540 1500 1430 1390 1370 1340 1270 1020 960 790 770 760 720
Raman: SAD 926 3060 3010 2960 2930 1630 1580 1510 1460 1440 1420 1380 1350 1310 1260 1220 1190 1160 1140 1080 1050 1020 970 950 930 910 890 880 850 780 730 670 600 550 530 520 500 4770 430 390 370 300 280 220 190 130 110 90
UV: SAD 571 287(6610) 282(6510) 222(64100) MeOH
m/e: 185.08

HODOC No: 00331
CAS No: 581-97-5
Beilstein No: 12, 1284
Name: Acetamide, *N*-2-naphthalenyl-
Synonym: Acetamide, *N*-2-naphthyl-
Line Form: $CH_3CONH(2-C_{10}H_7)$
Mol Form: $C_{12}H_{11}NO$
M_r: 185.21
mp: 134
***s*:** H_2O 3; al 3; chl 2
IR: SADG 298 3280 3080 3050 2820 1670 1590 1560 1500 1470 1430 1400 1360 1280 1210 1180 1130 1020 960 890 860 820 750 640
Raman: SAD 927 3280 3040 3020 2920 1660 1630 1610 1580 1550 1470 1430 1420 1390 1370 1280 1260 1250 1210 1170 1150 1130 1060 1020 980 970 950 920 880 860 810 770 750 650 600 520 500 430 400 320 220 200 180 150 90
UV: SAD 572 330(778) 315(817) 282(7919) 273(6980) 248 242(51100) MeOH
H NMR: SAD 131 2.1 7.3 7.6 8.1 8.2 $CDCl_3$
m/e: 185.08

HODOC No: 00332
CAS No: 17723-04-5
Name: Acetamide, *N*-1-naphthyl-2-(*N*-1-naphthylbenzimidoyl)-
Synonym: Propanoic acid, amide,*N*(1-naphthyl)-3(1-naphthylimino)-3-phenyl
Mol Form: $C_{29}H_{22}N_2O$
M_r: 414.48
mp: 164-5
UV: LANG 1996 353(23442) 216(89125) EtOH
m/e: 414.17

HODOC No: 00333
CAS No: 24626-94-6
Name: Acetamide, *N*-1-naphthyl-2-(*N*-2-naphthylbenzimidoyl)-
Synonym: Propanoic acid, amide,*N*(1-naphthyl)-3(2-naphthylimino)-3-phenyl
Mol Form: $C_{29}H_{22}N_2O$
M_r: 414.48
mp: 163-4
UV: LANG 1997 355(29512) 218(81283) EtOH
m/e: 414.17

HODOC No: 00334
CAS No: 24626-99-1
Name: Acetamide, *N*-2-naphthyl-2-(*N*-1-naphthylbenzimidoyl)-
Synonym: Propanoic acid, amide,*N*(2-naphthyl)-3(1-naphthylimino)-3-phenyl
Mol Form: $C_{29}H_{22}N_2O$
M_r: 414.48
mp: 170-1
UV: LANG 2003 360(33113) 283(16596) 249(30200) 216(74131) EtOH
m/e: 414.17

HODOC No: 00335
CAS No: 17723-05-6
Name: Acetamide, *N*-2-naphthyl-2-(*N*-2-naphthylbenzimidoyl)-
Synonym: Propanoic acid, amide,*N*(2-naphthyl)-3(2-naphthylimino)-3-phenyl
Mol Form: $C_{29}H_{22}N_2O$
M_r: 414.48
mp: 208-9
UV: LANG 2004 362(37154) 253(36308) 217(67608) EtOH
m/e: 414.17

HODOC No: 00336
CAS No: 17772-20-2
Name: Acetamide, *N*-2-naphthyl-2-(*N*-phenylbenzimidoyl)-
Synonym: Propanoic acid, amide,*N*(2-naphthyl)-3-phenyl-3-phenylimino
Mol Form: $C_{25}H_{20}N_2O$
M_r: 364.42
mp: 150-1
UV: LANG 1998 352(35481) 256(28184) 214(46774) EtOH
m/e: 364.16

HODOC No: 00337
CAS No: 75238-15-2
Beilstein No: 2, 175
Name: Acetamide, nitrate
Synonym: Acetamide mononitrate
Line Form: CH_3CONH_2–HNO_3
Mol Form: $C_2H_6N_2O_4$
M_r: 122.08
mp: 98
***s*:** eth 2; chl 3
IR: COB 7373 2580 1880 1620 1560 1460 1360 1320 1200 1100 995 945 810
m/e: 122.03

HODOC No: 00338
CAS No: 5419-82-9
Beilstein No: 12^2, 731
Name: Acetamide, *N*-(1-nitro-2-naphthalenyl)-
Synonym: Acetamide, *N*-(1-nitro-2-naphthyl)-
Line Form: 1-NO_2-2-(CH_3CONH)$C_{10}H_6$
Mol Form: $C_{12}H_{10}N_2O_3$
M_r: 230.21
mp: 126
Color: ye nd (al)
***s*:** H_2O 2; al 3; eth 3; bz 3; HOAc 3; lig 1
IR: µCOB 2786 3130 3030 1670 1590 1540 1470 1370 1330 1280 1220 1140 1000 910 820 750
UV: HBCP 340(4266) cyhex
m/e: 230.07

HODOC No: 00339
CAS No: 2437-30-1
Beilstein No: 12^3, 2969
Name: Acetamide, *N*-(2-nitro-1-naphthalenyl)-
Synonym: Acetamide, *N*-(2-nitro-1-naphthyl)-
Line Form: 2-$NO_2C_{10}H_6$($NHCOCH_3$)-1
Mol Form: $C_{12}H_{10}N_2O_3$
M_r: 230.21
mp: 200
Color: lt ye nd (al, HOAc)
***s*:** EtOH 4; HOAc 4
m/e: 230.07

HODOC No: 00340
CAS No: 24402-72-0
Beilstein No: 12^3, 2972
Name: Acetamide, *N*-(4-nitro-1-naphthalenyl)-
Synonym: Acetamide, *N*-(4-nitro-1-naphthyl)-
Line Form: 4-$NO_2C_{10}H_6$($NHCOCH_3$)-1
Mol Form: $C_{12}H_{10}N_2O_3$
M_r: 230.21
mp: 192-3
Color: pa ye nd (Ace)
***s*:** Ac 4; EtOH 4
m/e: 230.07

HODOC No: 00341
CAS No: 38396-26-8
Beilstein No: 12^3, 2973
Name: Acetamide, *N*-(5-nitro-1-naphthalenyl)-
Synonym: Acetamide, *N*-5-nitro-1-naphthyl-
Line Form: 5-$NO_2C_{10}H_6$($NHCOCH_3$)-1
Mol Form: $C_{12}H_{10}N_2O_3$
M_r: 230.21
mp: 220
Color: br pr (HOAc), ye cr (al)
m/e: 230.07

HODOC No: 00342
CAS No: 43014-68-2
Beilstein No: 12^2, 732
Name: Acetamide, *N*-(5-nitro-2-naphthalenyl)-
Synonym: Acetamide, *N*-5-nitro-2-naphthyl-
Line Form: 5-NO_2-2-(CH_3CONH)$C_{10}H_6$
Mol Form: $C_{12}H_{10}N_2O_3$
M_r: 230.21
mp: 186
Color: ye-br rh (al) ye nd (Bz)
***s*:** EtOH 4; HOAc 4
m/e: 230.07

HODOC No: 00343
CAS No: 6933-57-9
Beilstein No: 12^3, 2976
Name: Acetamide, *N*-(8-nitro-1-naphthalenyl)-
Synonym: Acetamide, *N*-(8-nitro-1-naphthyl)-
Line Form: 8-$NO_2C_{10}H_6(NHCOCH_3)$-1
Mol Form: $C_{12}H_{10}N_2O_3$
M_r: 230.21
mp: 191
Color: nd(w)
m/e: 230.07

HODOC No: 00344
CAS No: 102877-11-2
Beilstein No: 12^2, 733
Name: Acetamide, *N*-6-nitro-2-naphthyl-
Synonym: 2-Naphthylamine, *N*-acetyl-6-nitro
Line Form: 6-NO_2-2-$(CH_3CONH)C_{10}H_6$
Mol Form: $C_{12}H_{10}N_2O_3$
M_r: 230.21
mp: 224
Color: lt ye nd (dil al)
***s*:** Bz 4; EtOH 4; HOAc 4
m/e: 230.07

HODOC No: 00345
CAS No: 102877-13-4
Beilstein No: 12^2, 733
Name: Acetamide, *N*-8-nitro-2-naphthyl-
Synonym: 2-Naphthylamine, *N*-acetyl-8-nitro
Line Form: 8-NO_2-2-$(CH_3CONH)C_{10}H_6$
Mol Form: $C_{12}H_{10}N_2O_3$
M_r: 230.21
mp: 195.5
Color: ye nd (al)
***s*:** HOAc 4
m/e: 230.07

HODOC No: 00346
CAS No: 10151-95-8
Beilstein No: 12^3, 1523
Name: Acetamide, 2-nitro-*N*-phenyl-
Synonym: Acetanilide, 2-nitro-
Line Form: 2-$NO_2C_6H_4$-$NHCOCH_3$
Structure:

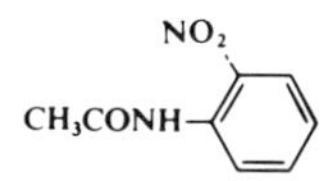

Mol Form: $C_8H_8N_2O_3$
M_r: 180.16
bp: $100^{0.1}$
mp: 94
***d*:** 1.419^{15}
Color: ye pr (lig), lf (dil al)
***s*:** Bz 4; EtOH 4; lig 4
IR: SADG 8419
UV: SAD 1638
H NMR: SAD 3184
MS: NBS 899
m/e: 180.05

HODOC No: 00347
CAS No: 552-32-9
Beilstein No: 12^2, 371
Name: Acetamide, *N*-(2-nitrophenyl)-
Synonym: Acetanilide, 2′-nitro-
Mol Form: $C_8H_8N_2O_3$
M_r: 180.16
bp: $100^{0.1}$
mp: 94
***d*:** 1.419^{15}
***s*:** H_2O 3; al 3; eth 4; bz 3; chl 3; lig 3
IR: SADG 8419 3370 1700 1610 1580 1500 1430 1370 1340 1280 1230 1160 1150 1040 860 790 750 700 650
UV: SAD 1638 338(2400) 231(17500) MeOH
H NMR: SAD 3184 2.3 7.1 7.6 8.2 8.7 10.3 $CDCl_3$
m/e: 180.05

HODOC No: 00348
CAS No: 122-28-1
Beilstein No: 12^3, 1551
Name: Acetamide, *N*-(3-nitrophenyl)-
Synonym: Acetanilide, 3′-nitro-
Line Form: 3-$NO_2C_6H_4$-$NHCOCH_3$
Mol Form: $C_8H_8N_2O_3$
M_r: 180.16
bp: $100^{0.0074}$
mp: 154-6
Color: wh lf (al)
***s*:** H_2O 3; al 3; eth 1; chl 3; tfa 2; alk 3; $phno_2$ 3
IR: SADG 18585 3250 3080 1670 1590 1540 1520 1470 1420 1360 1340 1310 1280 1270 1250 1070 1010 970 880 810 800 740 660
UV: OES 2-132 326(1514) 242(22387) EtOH
H NMR: SAD 5178 2.5 7.7 8.0 8.2 8.6 9.1 tfa
MS: NBS 899
m/e: 180.05

HODOC No: 00349
CAS No: 104-04-1
Beilstein No: 12^3, 1594
Name: Acetamide, *N*-(4-nitrophenyl)-
Synonym: Acetanilide, 4′-nitro-
Line Form: 4-$NO_2C_6H_4$-$NHCOCH_3$
Mol Form: $C_8H_8N_2O_3$
M_r: 180.16
bp: $100^{0.008}$
mp: 216
Color: ye pr (w)
***s*:** H_2O 2; al 3; eth 2; chl 2; tfa 3; HOAc 2; lig 2; alk 3
IR: μSADP 11426 3330 3130 1670 1560 1490 1410 1330 1270 1180 1110 1010 850 750 690
UV: SAD 3138 313 223 MeOH
H NMR: SAD 5129 2.5 7.8 8.4 9.0 tfa
MS: NBS 898
m/e: 180.05

HODOC No: 00350
CAS No: 553-20-8
Beilstein No: 13^3, 881
Name: Acetamide, *N*-(5-nitro-2-propoxyphenyl)-
Synonym: Acetanilide, 5′-nitro-2′-propoxy-
Line Form: (2-C_3H_7O-5-NO_2-C_6H_3)$NHCOCH_3$
Structure:

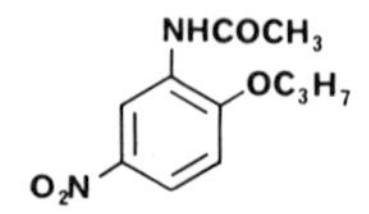

Mol Form: $C_{11}H_{14}N_2O_4$
M_r: 238.23
mp: 102-3
Color: cr (PrOH)
m/e: 238.10

HODOC No: 00351
CAS No: 140-40-9
Name: Acetamide, *N*-(5-nitro-2-thiazolyl)-
Synonym: 2-Acetamido-5-nitrothiazole
Structure:

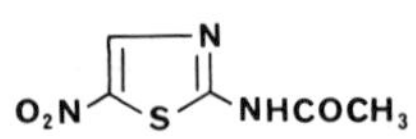

Mol Form: $C_5H_5N_3O_3S$
M_r: 187.17
mp: 264-5
Color: nd (al), pl (HOAc)
IR: μCOB 919 3130 2940 1720 1560 1490 1350 1220 1000 910 820 750 700
UV: SAD 6458 340 236 MeOH
MS: NBS 1003
m/e: 187.01

HODOC No: 00352
CAS No: 37593-11-6
Beilstein No: 12, 1185
Name: Acetamide, *N*-(4-octylphenyl)-
Synonym: *p*-n-Octylacetanilide
Line Form: 4-$(C_8H_{17})C_6H_4NHCOCH_3$
Structure:

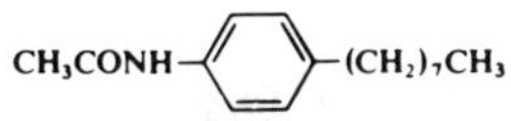

Mol Form: $C_{16}H_{25}NO$
M_r: 247.36
mp: 94
Color: lf or pr (al)
***s*:** Et_2O 4; EtOH 4
m/e: 247.19

HODOC No: 00353
CAS No: 97809-33-1
Name: Acetamide, 2-[(4-oxo-2-phenyl-4H-1-benzopyran-7-yl)oxy]-
Synonym: Flavone, 7(carboxy(amide)methoxy)
Mol Form: $C_{17}H_{13}NO_4$
M_r: 295.28
mp: 263-4.5
UV: LANG 2408 305(23988) 254(18621) EtOH
m/e: 295.08

HODOC No: 00354
CAS No: 90716-89-5
Name: Acetamide, 2,2′-oxybis[*N*-methyl-
Synonym: Ether, dimethyl,1,1′-*bis*(*N*-methylformamido)
Line Form: $HCON(CH_3)CH_2OCH_2N(CH_3)CHO$
Mol Form: $C_6H_{12}N_2O_3$
M_r: 160.17
mp: 64.5-6.0
***s*:** chl 3
IR: SADG 21202 2920 2880 2780 1680 1410 1390 1250 1040 990 700 340
H NMR: SAD 11285 2.9-3.1 4.7-4.8 8.2 $CDCl_3$
m/e: 160.08

HODOC No: 00355
CAS No: 4120-77-8
Beilstein No: 12, 1337
Name: Acetamide, *N*-2-phenanthrenyl-
Synonym: Acetamide, *N*-2-phenanthryl-
Mol Form: $C_{16}H_{13}NO$
M_r: 235.27
mp: 225-6
***s*:** al 3; lig 2; xyl 3; gl HOAc 3
UV: LANG 86 355(214) 338(398) 322(398) 303(9120) 292(18621) 269(56234) 217(15849) 211(17378) EtOH
m/e: 235.10

HODOC No: 00356
CAS No: 4120-78-9
Beilstein No: 12, 1337
Name: Acetamide, *N*-3-phenanthrenyl-
Synonym: Acetamide, *N*-3-phenanthryl-
Mol Form: $C_{16}H_{13}NO$
M_r: 235.27
UV: LANG 88 355(708) 339(794) 310(13490) 298(13490) 279(13804) 260(38019) 237(24547) EtOH
m/e: 235.10

HODOC No: 00357
CAS No: 4235-09-0
Name: Acetamide, *N*-9-phenanthrenyl-
Synonym: Acetamide, *N*-9-phenanthryl-
Mol Form: $C_{16}H_{13}NO$
M_r: 235.27
mp: 195
UV: LANG 84 348(135) 339(155) 331(209) 298(7762) 287(6761) 276(794) 254(39811) 246(29512) 210(29512) EtOH
m/e: 235.10

HODOC No: 00358
CAS No: 621-88-5
Beilstein No: 6^4, 638
Name: Acetamide, 2-phenoxy-
Synonym: Phenoxyacetamide
Line Form: $C_6H_5OCH_2CONH_2$
Mol Form: $C_8H_9NO_2$
M_r: 151.16
mp: 101.5
***s*:** H_2O 1; al 4
IR: µCOB 2889 3450 3130 1700 1610 1590 1490 1430 1350 1300 1250 1180 1160 1120 1090 1060 910 890 820 760 750 690 660
H NMR: SAD 18890 4.5 6.9 7.0 7.3 $CDCl_3$
m/e: 151.06

HODOC No: 00359
CAS No: 103-84-4
Beilstein No: 12^3, 459
Name: Acetamide, *N*-phenyl-
Synonym: Acetamidobenzene
Line Form: $CH_3CONHC_6H_5$
Mol Form: C_8H_9NO
M_r: 135.16
bp: 304
mp: 114.3
***d*:** 1.2190^{15}
***s*:** H_2O 2; al 4; eth 3; ace 4; bz 3; chl 4; MeOH 4; tol 3
IR: COB 5575 3280 3260 3090 3030 1740 1600 1560 1500 1480 1430 1360 1320 1300 1260 1180 1040 1010 1000 960 910 850 840 760 700 660
Raman: SAD 124 3280 3150 3100 3070 2990 2830 1680 1670 1610 1550 1500 1490 1450 1370 1320 1310 1270 1240 1210 1180 1160 1110 1080 1040 1010 970 840 770 750 700 670 620 540 520 420 370 350 280 180 130
UV: SAD 63 241 MeOH
H NMR: SAD 5235 2.1 7.0-7.6 8.0 $CDCl_3$
C NMR: JJ 295 FT 24.1 120.4 124.1 128.7 138.2 169.5 $CDCl_3$
MS: NBS 304
m/e: 135.07

HODOC No: 00360
CAS No: 4128-71-6
Beilstein No: 16^2, 155
Name: Acetamide, *N*-[4-(phenylazo)phenyl]-
Synonym: Acetanilide, 4′-(phenylazo)-
Mol Form: $C_{14}H_{13}N_3O$
M_r: 239.26
mp: 144-6
***s*:** H_2O 2; eth 2; bz 2
UV: OES 6-534 347(23442) EtOH
m/e: 239.11

HODOC No: 00361
CAS No: 2050-85-3
Beilstein No: 13, 20
Name: Acetamide, *N,N′*-1,2-phenylenebis-
Synonym: Acetamide, *N,N′*-*o*-phenylenebis-
Mol Form: $C_{10}H_{12}N_2O_2$
M_r: 192.21
mp: 185-6
***s*:** H_2O 3; al 3; eth 2; ace 3; bz 2; chl 3; HOAc 3
UV: HBCP 278(1000) al
m/e: 192.09

HODOC No: 00362
CAS No: 140-50-1
Beilstein No: 13^1, 30
Name: Acetamide, *N,N′*-1,4-phenylenebis-
Synonym: Acetamide, *N,N′*-*p*-phenylenebis-
Line Form: 1,4-$(CH_3CONH)_2C_6H_4$
Mol Form: $C_{10}H_{12}N_2O_2$
M_r: 192.21
mp: 310-2
IR: µCOB 2845 3330 3130 2940 1670 1560 1520 1470 1410 1370 1320 1250 1000 960 830 750
UV: OES 8-164 264(21878) MeOH
H NMR: SAD 17234 2.5 7.6 9.4 tfa
MS: ALDN 155 108(100) 192(53) 43(34) 107(29) 150(25) 80(11) 109(9) 193(6)
m/e: 192.09

HODOC No: 00363
CAS No: 877-95-2
Name: Acetamide, *N*-(2-phenylethyl)-
Synonym: Acetamide, *N*-phenethyl-
Line Form: $C_6H_5CH_2CH_2NHCOCH_3$
Mol Form: $C_{10}H_{13}NO$
M_r: 163.21
***s*:** chl 3
IR: SADG 3739 3280 3070 2920 2860 1640 1550 1490 1450_1430 1360 1290 1190 1100 1030 980 750 690 600_580 500 450
UV: SAD 11698 268(107) 264(117) 258(150) 254(127) diox
H NMR: VAR 265 1.9 2.8 3.5 6.5 7.3 $CDCl_3$
m/e: 163.10

HODOC No: 00364
CAS No: 588-46-5
Beilstein No: 12^5, 2255
Name: Acetamide, *N*-(phenylmethyl)-
Synonym: Acetamide, *N*-benzyl-
Line Form: $CH_3CONHCH_2C_6H_5$
Mol Form: $C_9H_{11}NO$
M_r: 149.18
bp: 157^2
mp: 61
***s*:** al 4; eth 4
UV: HBCP 258(219) al
MS: NBS 444
m/e: 149.08

HODOC No: 00365
CAS No: 2437-98-1
Beilstein No: 12^3, 466
Name: Acetamide, *N*-phenyl-*N*-propyl-
Synonym: Acetanilide, *N*-propyl-
Line Form: $C_6H_5N(C_3H_7)COCH_3$
Mol Form: $C_{11}H_{15}NO$
M_r: 177.24
bp: 266^{712}
mp: 49
Color: mcl lf (eth, lig)
***s*:** Et_2O 4; EtOH 4; lig 4
m/e: 177.12

HODOC No: 00366
CAS No: 5867-45-8
Beilstein No: 22^4, 4073
Name: Acetamide, *N*-3-pyridinyl-
Synonym: Acetamide, *N*-3-pyridyl-
Line Form: 3-$(CH_3CONH)C_5H_4N$
Mol Form: $C_7H_8N_2O$
M_r: 136.15
bp: 326-7
mp: 133
***s*:** H_2O 3; al 3; eth 2; bz 2; chl 3; lig 1
UV: LANG 453 269(3311) 237(10000) pH=9.7 287(5623) 247(11749) 212(20893) 1N sulf
m/e: 136.06

HODOC No: 00367
CAS No: 5221-42-1
Beilstein No: 22^4, 4108
Name: Acetamide, *N*-4-pyridinyl-
Synonym: Acetamide, *N*-4-pyridyl-
Line Form: 4-$(CH_3CONH)C_5H_4N$
Mol Form: $C_7H_8N_2O$
M_r: 136.15
mp: 148
UV: LANG 454 245(19498) pH=9.7 267(20893) 204(12882) 1N sulf
m/e: 136.06

HODOC No: 00368
CAS No: 5417-50-5
Beilstein No: 22^4, 4607
Name: Acetamide, *N*-3-quinolinyl-
Synonym: Acetamide, *N*-3-quinolyl-
Line Form: 3-$(CH_3CONH)(C_9H_6N)$
Mol Form: $C_{11}H_{10}N_2O$
M_r: 186.20
mp: 166-7
Color: cr (w)
***s*:** H_2O 2; al 3
IR: µSADP 9719 3450 2940 1670 1560 1490 1370 1270 1150 1020 950 900 810 780 750
UV: SAD 21944 334(3840) 322(4100) 249(35900) 213(23900) MeOH 348(5490) 260(22600) 241(32700) 211(17600) MeOH:HCl
H NMR: SAD 18737 2.2 7.3-8.2 8.7 9.0 10.0 poly D
m/e: 186.08

HODOC No: 00369
CAS No: 32433-28-6
Beilstein No: 22^4, 4614
Name: Acetamide, *N*-4-quinolinyl-
Synonym: Acetamide, *N*-4-quinolyl-
Line Form: 4-(CH_3CONH)(C_9H_6N)
Mol Form: $C_{11}H_{10}N_2O$
M_r: 186.20
bp: sub
mp: 176
Color: nd (w+1)
***s*:** EtOH 4
IR: DMS 7769
UV: OES 1-381
m/e: 186.08

HODOC No: 00370
CAS No: 42464-80-2
Beilstein No: 22^4, 4671
Name: Acetamide, *N*-5-quinolinyl-
Synonym: Quinoline, 5-acetamido
Line Form: 5-(CH_3CONH)(C_9H_6N)
Mol Form: $C_{11}H_{10}N_2O$
M_r: 186.20
mp: 178
Color: pl (w), pr (dil al)
***s*:** EtOH 4
m/e: 186.08

HODOC No: 00371
CAS No: 22433-76-7
Beilstein No: 22^2, 355
Name: Acetamide, *N*-6-quinolinyl-
Synonym: Acetamide, *N*-6-quinolyl-
Line Form: 6-(CH_3CONH)(C_9H_6N)
Mol Form: $C_{11}H_{10}N_2O$
M_r: 186.20
mp: 138
Color: nd (w)
***s*:** H_2O 4; al 4; eth 2; bz 3
IR: DMS 7781 3410 3260 2960 1680 1630 1600 1580 1530 1500 1470 1430 1270 1250 1120 1030 1000 880 830
m/e: 186.08

HODOC No: 00372
CAS No: 36164-42-8
Beilstein No: 22^2, 356
Name: Acetamide, *N*-7-quinolinyl-
Synonym: Quinoline, 7-acetamido
Line Form: 7-(CH_3CONH)(C_9H_6N)
Mol Form: $C_{11}H_{10}N_2O$
M_r: 186.20
mp: 167.5
Color: cr (w, al)
m/e: 186.08

HODOC No: 00373
CAS No: 77-46-3
Beilstein No: 13^2, 303
Name: Acetamide, *N*,*N'*-(sulfonyldi-4,1-phenylene)*bis*-
Synonym: Acetanilide, 4',4'''-sulfonylbis-
Mol Form: $C_{16}H_{16}N_2O_4S$
M_r: 332.36
mp: 282-5
***s*:** H_2O 1; al 3
UV: OES 2-494 285(35481) 257(25704) EtOH
m/e: 332.08

HODOC No: 00374
CAS No: 112345-57-0
Name: Acetamide, *N*-(tetrahydro-2,6-dioxo-2H-pyran-3-yl)-, (*S*)-
Synonym: Pentanedioic acid, anhydride,2-acetamido,
Structure:

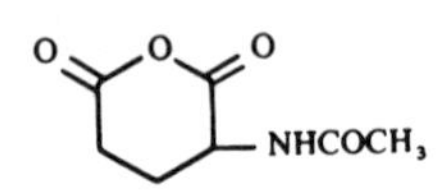

Mol Form: $C_7H_9NO_4$
M_r: 171.15
mp: 128-9
IR: μCOB 5325 3330 3130 2940 1720 1640 1610 1560 1410 1350 1220 1010 950 890 810 710
m/e: 171.05

HODOC No: 00375
CAS No: 477-27-0
Beilstein No: 14^3, 692
Name: Acetamide,
N-(5,6,7,9-tetrahydro-10-hydroxy-1,2,3-trimethoxy-9-oxobenzo[a]heptalen-7-yl)-, (*S*)-
Synonym: Benzo[a]heptalene, acetamide deriv.
Structure:

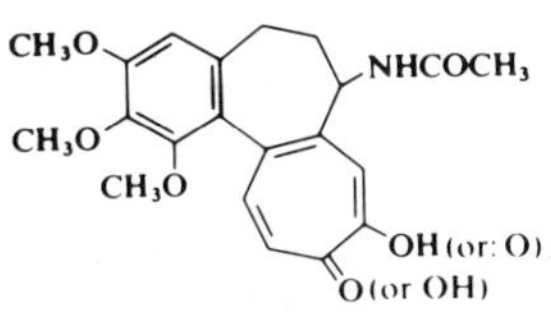

Mol Form: $C_{21}H_{23}NO_6$
M_r: 385.40
mp: 178-9
***d*:** 1.24
α: $[\alpha]^{14/d)-252.5\ CHCl}{}_{3,1.2\%}$
Color: pa ye nd (diox)
***s*:** H_2O 2; al 4; eth 1; bz 1; chl 4
IR: **μCOB 3534** 3230 2940 1670 1610 1540 1490 1450 1350 1320 1270 1220 1140 1090 1040 1000 910 850 700 660
UV: **OES 2-662** 350(16596) 244(29512) EtOH 357(2512) 249(3981) 0.05N NaOH
MS: **NBS 5041**
m/e: 385.15

HODOC No: 00376
CAS No: 28732-19-6
Name: Acetamide,
N-[4-(2,3,4,5-tetrahydro-5-oxo-3-thioxo-1,2,4-triazin-6-yl)phenyl]-
Synonym: Acetanilide, 4′-(2,3,4,5-tetrahydro-5-oxo-3-thioxo-as-triazin-6-yl)-
Mol Form: $C_{11}H_{10}N_4O_2S$
M_r: 262.28
mp: dec 350-2
UV: **LANG 2403** 360(17783) 290(21878) 236(10000) EtOH
m/e: 262.05

HODOC No: 00377
CAS No: 64-86-8
Beilstein No: 14^3, 693
Name: Acetamide,
N-(5,6,7,9-tetrahydro-1,2,3,10-tetramethoxy-9-oxobenzo[a]heptalen-7-yl)-, (*S*)-
Synonym: Benzo[a]heptalene, acetamide deriv.
Structure:

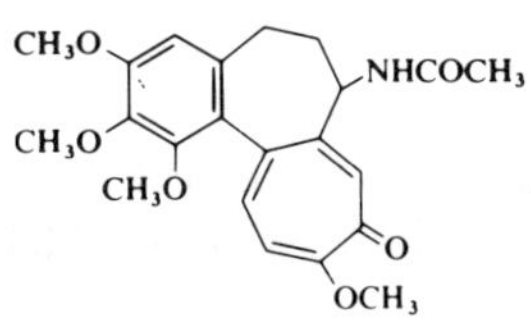

Mol Form: $C_{22}H_{25}NO_6$
M_r: 399.42
mp: 155-7
α: $[\alpha]^{17/D}$ -121 (Chl, c=0.9)
Color: ye pl (w + 1/2) ye cr (Bz)
***s*:** H_2O 4; EtOH 4
IR: **SADP 17152**
UV: **SAD 5410**
MS: **NBS 3098**
m/e: 399.17

HODOC No: 00378
CAS No: 2719-23-5
Name: Acetamide, *N*-2-thiazolyl-
Synonym: 2-Acetamidothiazole
Line Form: 2-$CH_3CONH(C_3H_2NS)$
Mol Form: $C_5H_6N_2OS$
M_r: 142.17
mp: 208
***s*:** H_2O 2
IR: **μCOB 1821** 3450 3130 2940 1700 1560 1370 1300 1240 1180 1060 980 820 780 710 660
UV: **OES 2-32** 269(912) 268(8128) EtOH
m/e: 142.02

HODOC No: 00379
CAS No: 13053-81-1
Beilstein No: 17^{1}, 136
Name: Acetamide, *N*-2-thienyl-
Synonym: 2-Acetamidothiophene
Line Form: 2-$CH_3CONH(C_4H_3S)$
Mol Form: C_6H_7NOS
M_r: 141.18
mp: 161-2
Color: lf (w)
***s*:** H_2O 2; al 4; eth 2; ace 4; bz 2
IR: µCOB 1223 3450 3230 3130 3030 1640 1590 1450 1370 1350 1320 1040 1020 980 860 800 750 710 680
UV: OES 5-64 264(10233) EtOH
m/e: 141.02

HODOC No: 00380
CAS No: 594-65-0
Beilstein No: 2^{4}, 520
Name: Acetamide, 2,2,2-trichloro-
Synonym: Trichloroacetamide
Line Form: Cl_3CCONH_2
Mol Form: $C_2H_2Cl_3NO$
M_r: 162.40
bp: $238\text{-}9^{740}$
mp: 142
***s*:** H_2O 2; al 4; eth 4
IR: µCOB 2333 3330 3230 1700 1610 1390 1350 1110 930 830 750 650
Raman: SAD 523 1700 1680 1610 1400 1120 940 830 750 660 630 450 320 290 210 120
m/e: 160.92

HODOC No: 00381
CAS No: 81012-95-5
Name: Acetamide, *N*-[2,2,2-trichloro-1,1-*bis*(4-chlorophenyl)ethyl]-
Synonym: Acetamide, *N*-[2,2,2-trichloro-1,1-*bis*(*p*-chlorophenyl)ethyl]-
Mol Form: $C_{16}H_{12}ClDNO$
M_r: 271.73
mp: 210-4
IR: COB 7977 3420 3060 1690 1590 1510 1490 1410 1370 1260 1090 1020 890 810 780 760 730 690 550 510
UV: SAD 11539 265(624) 230(19300) MeOH
H NMR: SAD 1570 2.3 7.3 7.6 tfa
m/e: 271.08

HODOC No: 00382
CAS No: 3004-73-7
Name: Acetamide, 2,2,2-trichloro-*N*-(3-chlorophenyl)-
Synonym: Acetanilide, 2,2,2,3′-tetrachloro-
Mol Form: $C_8H_5Cl_4NO$
M_r: 272.94
IR: µCOB 2322 3330 3130 1700 1610 1540 1430 1280 1220 1100 930 880 830 770 680 660
m/e: 270.91

HODOC No: 00383
CAS No: 2877-13-6
Name: Acetamide, 2,2,2-trichloro-*N*-(4-chlorophenyl)-
Synonym: Acetanilide, 2,2,2,4′-tetrachloro-
Mol Form: $C_8H_5Cl_4NO$
M_r: 272.94
IR: µCOB 1692 3330 2940 1700 1540 1490 1470 1410 1250 1090 890 830 760 660
m/e: 270.91

HODOC No: 00384
CAS No: 23144-68-5
Name: Acetamide, 2,2,2-trichloro-*N*-cyclohexyl-
Synonym: *N*-Cyclohexyltrichloroacetamide
Mol Form: $C_8H_{12}Cl_3NO$
M_r: 244.54
IR: µCOB 3154 3450 2940 2860 1720 1520 1450 1320 1250 1220 1090 850 820 740 680
m/e: 243.00

HODOC No: 00385
CAS No: 33560-53-1
Name: Acetamide, 2,2,2-trichloro-*N*-(2,3-dichlorophenyl)-
Synonym: Acetanilide, 2,2,2,2′,3′-pentachloro-
Mol Form: $C_8H_4Cl_5NO$
M_r: 307.38
IR: µCOB 2323 3330 1720 1590 1520 1450 1410 1190 1040 900 810 790 670
m/e: 304.87

HODOC No: 00386
CAS No: 33560-54-2
Name: Acetamide, 2,2,2-trichloro-*N*-(2,4-dichlorophenyl)-
Synonym: Acetanilide, 2,2,2,2′,4′-pentachloro-
Mol Form: $C_8H_4Cl_5NO$
M_r: 307.38
IR: µCOB 2320 3330 3130 1720 1590 1520 1390 1320 1190 1100 1050 880 830 800 710 670
m/e: 304.87

HODOC No: 00387
CAS No: 33715-64-9
Name: Acetamide, 2,2,2-trichloro-*N*-(2,5-dichlorophenyl)-
Synonym: Acetanilide, 2,2,2,2′,5′-pentachloro-
Mol Form: $C_8H_4Cl_5NO$
M_r: 307.38
IR: **µCOB 2324** 3330 3130 1720 1590 1520 1470 1410 1270 1190 1100 1040 930 880 830 810 670
m/e: 304.87

HODOC No: 00388
CAS No: 22303-30-6
Name: Acetamide, 2,2,2-trichloro-*N*-(3,4-dichlorophenyl)-
Synonym: Acetanilide, 2,2,2,3′,4′-pentachloro-
Mol Form: $C_8H_4Cl_5NO$
M_r: 307.38
mp: 123-4
IR: **µCOB 2321** 3230 1700 1590 1520 1470 1370 1250 1210 1040 880 830 830 820 780 670
m/e: 304.87

HODOC No: 00389
CAS No: 2430-00-4
Beilstein No: 4^4, 351
Name: Acetamide, 2,2,2-trichloro-*N*,*N*-diethyl-
Synonym: *N*,*N*-Diethyltrichloroacetamide
Line Form: $Cl_3CCON(C_2H_5)_2$
Mol Form: $C_6H_{10}Cl_3NO$
M_r: 218.50
bp: 109^9
mp: 27
***n*:** 1.4900^{24}
Color: pr
m/e: 216.98

HODOC No: 00390
CAS No: 7291-33-0
Beilstein No: 4^4, 182
Name: Acetamide, 2,2,2-trichloro-*N*,*N*-dimethyl-
Synonym: Dimethyltrichloroacetamide
Line Form: $Cl_3CCON(CH_3)_2$
Mol Form: $C_4H_6Cl_3NO$
M_r: 190.45
bp: 230-3d, 84^4
mp: 12
***d*:** 1.390^{20}
***n*:** 1.5017^{25}
***s*:** Bz 4; Chl 4
m/e: 188.95

HODOC No: 00391
CAS No: 5445-85-2
Name: Acetamide, *N*-(2,2,2-trichloro-1-hydroxyethyl)-
Synonym: *N*-(1-Hydroxy-2,2,2-trichloroethyl)acetamide
Line Form: $CH_3CONHCH(OH)CCl_3$
Mol Form: $C_4H_6Cl_3NO_2$
M_r: 206.45
mp: 157-8
IR: **µCOB 5296** 3330 2940 1670 1540 1390 1120 1090 1020 960 830 810
m/e: 204.95

HODOC No: 00392
CAS No: 14301-34-9
Name: Acetamide, 2,2,2-trichloro-*N*-phenethyl-
Synonym: Acetic acid, amide,*N*-phenethyl-trichloro
Line Form: $C_6H_5CH_2CH_2NHCOCCl_3$
Mol Form: $C_{10}H_{10}Cl_3NO$
M_r: 266.54
IR: **SADG 20779** 3360 2960 2930 2860 1600 1500 1450 1410 1370 1310 1210 1140 1080 1030 1000 960 890 820 750 610 560 530 400
UV: **SAD 1560** 267 264 258 MeOH
H NMR: **SAD 11024** 2.9 3.6 6.8 7.2 $CDCl_3$
m/e: 264.98

HODOC No: 00393
CAS No: 2563-97-5
Beilstein No: 12^3, 464
Name: Acetamide, 2,2,2-trichloro-*N*-phenyl-
Synonym: Acetanilide, 2,2,2-trichloro-
Line Form: $Cl_3CCONHC_6H_5$
Mol Form: $C_8H_6Cl_3NO$
M_r: 238.49
bp: 168-70
mp: 95-7
Color: lf (dil al)
***s*:** EtOH 4
m/e: 236.95

HODOC No: 00394
CAS No: 63051-97-8
Name: Acetamide, *N*-[4-[(tricyclo[3.3.1.13,7]dec-1-ylimino)methyl]phenyl]-
Synonym: Adamantane, 1(4-acetamidobenzylideneamino)
Mol Form: $C_{19}H_{24}N_2O$
Mr: 296.39
mp: 211-2
***s*:** DMSO 3
IR: SADG 20049 3250 3050 2900 2840 1690 1670 1630 1600 1540 1510 1400 1360 1310 1260 1170 1080 830 520
H NMR: SAD 9900 1.7 2.1 2.0-2.3 7.7 8.3 10.1 DMSOd6
m/e: 296.19

HODOC No: 00395
CAS No: 354-38-1
Beilstein No: 2^2, 186
Name: Acetamide, 2,2,2-trifluoro-
Synonym: Trifluoroacetamide
Line Form: F_3CCONH_2
Mol Form: $C_2H_2F_3NO$
Mr: 113.04
bp: 162-3
mp: 70-5
IR: ALD 2:389E
m/e: 113.01

HODOC No: 00396
CAS No: 2247-78-1
Name: Acetamide, *N*-[2,2,2-trifluoro-1,1-*bis*(*p*-fluorophenyl)ethyl]-
Synonym: Ethane, 1-acetamido-1,1-*bis*(4-fluorophenyl)-2,2,2-trifluoro
Mol Form: $C_{16}H_{12}F_5NO$
Mr: 329.25
IR: COB 7609 3260 3050 1670 1550 1510 1300 1240 1170 1090 930 830 600 550 540
m/e: 329.08

HODOC No: 00397
CAS No: 21149-38-2
Name: Acetamide, 2,2,2-trifluoro-*N*,*N*-*bis*(trimethylsilyl)-
Synonym: Bis(trimethylsilyl)trifluoroacetamide
Line Form: $(CH_3)_3SiN{=}C(CF_3)OSi(CH_3)_3$
Mol Form: $C_8H_{18}F_3NOSi_2$
Mr: 257.40
bp: $45\text{-}50^{14}$
***n*:** $1.3839^{20/D}$
***s*:** ctc 3
H NMR: SAD 11616 0.2 0.3 CCl_4
m/e: 257.09

HODOC No: 00398
CAS No: 1547-87-1
Beilstein No: 4^4, 182
Name: Acetamide, 2,2,2-trifluoro-*N*,*N*-dimethyl-
Synonym: Dimethyltrifluoroacetamide
Line Form: $CF_3CON(CH_3)_2$
Mol Form: $C_4H_6F_3NO$
Mr: 141.09
bp: 132-4
***d*:** 1.232^{21}
***n*:** $1.3611^{25.2}$
IR: µCOB 3925 2940 1700 1520 1410 1280 1250 1190 1140 1110_930 760 670
m/e: 141.04

HODOC No: 00399
CAS No: 1995-76-2
Name: Acetamide, *N*-(2,2,2-trifluoro-1,1-diphenylethyl)-
Synonym: Ethane, 1-acetamido-1,1-diphenyl-2,2,2-trifluoro
Line Form: $CF_3C(C_6H_5)_2NHCOCH_3$
Mol Form: $C_{16}H_{14}F_3NO$
Mr: 293.27
IR: COB 7534 3350 3060 1680 1520 1500 1450 1370 1280 1250 1170 1150 1080 940 760 750 700 640 580 540
m/e: 293.10

HODOC No: 00400
CAS No: 344-62-7
Beilstein No: 12^2, 453
Name: Acetamide, *N*-[2-(trifluoromethyl)phenyl]-
Synonym: *o*-Acetotoluidide, α,α,α-trifluoro-
Mol Form: $C_9H_8F_3NO$
M_r: 203.15
mp: 94
IR: COB 7975 3270 1660 1590 1530 1460 1370 1320 1290 1170 1160 1120 1060 1040 950 860 770 695 670 645 605 540
m/e: 203.06

HODOC No: 00401
CAS No: 351-36-0
Beilstein No: 12^3, 1989
Name: Acetamide, *N*-[3-(trifluoromethyl)phenyl]-
Synonym: *m*-Acetotoluidide, α,α,α-trifluoro-
Line Form: $3\text{-}(CH_3CONH)\text{-}C_6H_4CF_3$
Mol Form: $C_9H_8F_3NO$
M_r: 203.15
mp: 103-6
IR: ALD 3:1067F
MS: ALD 7:67A
m/e: 203.06

HODOC No: 00402
CAS No: 404-24-0
Beilstein No: 12^2, 141
Name: Acetamide, 2,2,2-trifluoro-*N*-phenyl-
Synonym: Acetanilide, 2,2,2-trifluoro-
Line Form: $CF_3CONHC_6H_5$
Mol Form: $C_8H_6F_3NO$
M_r: 189.13
mp: 87.6
***s*:** al 4
IR: μCOB 2325 3450 3130 1750 1450 1280 1250 1160 1080 890 760 750 730 690 650
UV: OES 2-122 245(10965) EtOH
MS: NBS 1029
m/e: 189.04

HODOC No: 00403
CAS No: 328-13-2
Name: Acetamide, 2,2,2-trifluoro-*N*-(tetrahydro-2,6-dioxo-2H-pyran-3-yl)-
Synonym: Glutamic anhydride, *N*-(trifluoroacetyl)-
Mol Form: $C_7H_6F_3NO_4$
M_r: 225.12
IR: COB 7405 1820 1765 1725 1560 1420 1220 1190 1085 1035 980 860 725
m/e: 225.02

HODOC No: 00404
CAS No: 109038-67-7
Name: Acetamide, *N*-(β,3,4-trihydroxyphenethyl)-, 3,4-diacetate
Synonym: Toluene, α-acetamidomethyl-α,3,4-trihydroxy,3,4-diacetate
Mol Form: $C_{14}H_{17}NO_6$
M_r: 295.28
IR: μCOB 5218 3450 3330 3130 2940 1790 1640 1560 1520 1430 1390 1210 1180 1100 1010 900 830
m/e: 295.11

HODOC No: 00405
CAS No: 30649-59-3
Name: Acetanilide, 4′-bromo-*N*-β-*D*-glucopyranosyl-, tetraacetate (ester)
Synonym: β-*D*-Glucose, 1(*N*-acetyl-*N*(4-bromophenyl)amino),tetraacetate
Mol Form: $C_{22}H_{26}BrNO_{10}$
M_r: 543.33
mp: 134
UV: LANG 255 226(13183) EtOH
m/e: 543.07

HODOC No: 00406
CAS No: 99854-34-9
Name: Acetanilide, 2-chloro-*N*-2-chloroallyl-
Synonym: Chloroacetamide, *N*-2-chloroallyl-*N*-phenyl
Line Form: $ClCH_2CON(C_6H_5)CH_2CCl{=}CH_2$
Mol Form: $C_{11}H_{11}Cl_2NO$
M_r: 244.11
bp: $138^{0.7}$
***n*:** 1.5602^{25}
m/e: 243.02

HODOC No: 00407
CAS No: 17738-30-6
Name: Acetanilide, 4′-chloro-2-[*N*-(*p*-chlorophenyl)benzimidoyl]-
Synonym: Propanoic acid, amide,*N*(4-chlorophenyl)-3(4-chlorophenylimino)-3-phenyl
Mol Form: $C_{21}H_{16}Cl_2N_2O$
Mr: 383.26
mp: 133-4
UV: LANG 1976 346(31623) 258(18621) EtOH
m/e: 382.06

HODOC No: 00408
CAS No: 22786-85-2
Name: Acetanilide, 2-chloro-4′-[(*p*-chlorophenyl)sulfamoyl]-
Synonym: Benzenesulfonic acid, amide,4(3-chloroacetamido)-*N*(4-chlorophenyl)
Mol Form: $C_{14}H_{12}Cl_2N_2O_3S$
Mr: 359.21
s: DMSO 2
IR: SADG 17796 3250 1685 1680 1590 1515 1485 1425 1395 1370 1325 1250 1215 1155 1085 1010 970 900 825 720 670 575
UV: SAD 17450 266(23100) 237(18400) MeOH 254(32400) MeOH:KOH
H NMR: SAD 9416 4.3 7.1 7.3 7.8 10.7 $DMSOd_6$
m/e: 357.99

HODOC No: 00409
CAS No: 17738-31-7
Name: Acetanilide, 4′-chloro-2-[*N*-(*p*-methoxyphenyl)benzimidoyl]-
Synonym: Propanoic acid, amide,*N*(4-chlorophenyl)-3(4-methoxyphenylimino)-3-phenyl
Mol Form: $C_{22}H_{19}ClN_2O_2$
Mr: 378.84
mp: 153-4
UV: LANG 1977 349(26915) 261(15136) EtOH
m/e: 378.11

HODOC No: 00410
CAS No: 17738-33-9
Name: Acetanilide, 4′-chloro-2-(*N*-1-naphthylbenzimidoyl)-
Synonym: Propanoic acid, amide,*N*(4-chlorophenyl)-3(1-naphthylimino)-3-phenyl
Mol Form: $C_{25}H_{19}ClN_2O$
Mr: 398.87
mp: 196-7
UV: LANG 1979 354(28184) 299(15136) EtOH
m/e: 398.12

HODOC No: 00411
CAS No: 17738-34-0
Name: Acetanilide, 4′-chloro-2-(*N*-2-naphthylbenzimidoyl)-
Synonym: Propanoic acid, amide,*N*(4-chlorophenyl)-3(2-naphthylimino)-3-phenyl
Mol Form: $C_{25}H_{19}ClN_2O$
Mr: 398.87
mp: 186-7
UV: LANG 1980 358(33884) 253(25119) 214(43652) EtOH
m/e: 398.12

HODOC No: 00412
CAS No: 17738-32-8
Name: Acetanilide, 4′-chloro-2-[*N*-(*p*-nitrophenyl)benzimidoyl]-
Synonym: Propanoic acid, amide,*N*(4-chlorophenyl)-3(4-nitrophenylimino)-3- phenyl
Structure:

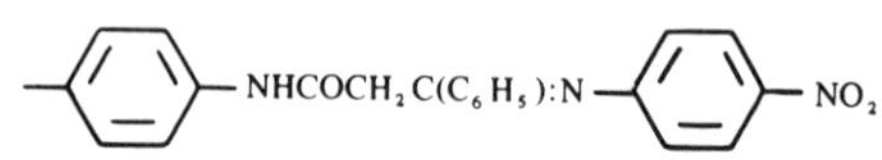

Mol Form: $C_{21}H_{16}ClN_3O_3$
Mr: 393.81
mp: 183-4
UV: LANG 1978 386(33113) 307(14125) 250(15136) EtOH
m/e: 393.09

HODOC No: 00413
CAS No: 17772-19-9
Name: Acetanilide, 4′-chloro-2-(*N*-*p*-tolylbenzimidoyl)-
Synonym: Propanoic acid, amide,*N*(4-chlorophenyl)-3(4-methylphenylimino)-3-phenyl
Structure:

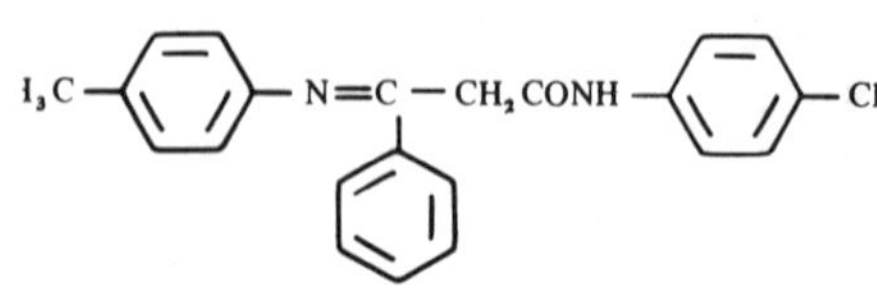

Mol Form: $C_{22}H_{19}ClN_2O$
Mr: 362.84
mp: 122-3
UV: LANG 1975 349(29512) 258(16596) EtOH
m/e: 362.12

HODOC No: 00414
CAS No: 22795-31-9
Name: Acetanilide, 2-chloro-4′-(*p*-tolylsulfamoyl)-
Synonym: Benzenesulfonic acid, amide,4(3-chloroacetamido)-*N*(4-tolyl)
Mol Form: $C_{15}H_{15}ClN_2O_3S$
M_r: 338.79
***s*:** DMSO 3
IR: SADG 17791 3400 3100 2900 1680 1590 1520 1400 1320 1150 1090 820 660 580 530
UV: SAD 17669 265(22400) 231(13200) MeOH 252(26200) MeOH:KOH
H NMR: SAD 9419 2.2 4.3 7.0 7.7 DMSOd6
m/e: 338.05

HODOC No: 00415
CAS No: 102023-81-4
Name: Acetanilide, 2,4′-dichloro-2,2-diphenyl-
Synonym: Acetic acid, amide,chloro-*N*(4-chlorophenyl)-diphenyl
Mol Form: $C_{20}H_{15}Cl_2NO$
M_r: 356.23
mp: 123-4
UV: LANG 2691 252(23442) EtOH
m/e: 355.05

HODOC No: 00416
CAS No: 22961-02-0
Name: Acetanilide, 2,2-dichloro-4′-(phenylsulfamoyl)-
Synonym: Benzenesulfonic acid, amide,4(3,3-dichloroacetamido)-*N*-phenyl
Mol Form: $C_{14}H_{12}Cl_2N_2O_3S$
M_r: 359.21
***s*:** DMSO 3
IR: SADG 17787 3270 3220 1700 1685 1585 1530 1485 1400 1330 1240 1210 1150 1085 920 825 795 745 690 655 545
UV: SAD 17445 267(21300) 226(10900) MeOH 249(21500) MeOH:KOH
H NMR: SAD 9415 6.6 6.9-7.3 7.8 DMSOd6
m/e: 357.99

HODOC No: 00417
CAS No: 32950-57-5
Name: Acetanilide, 2,2-dichloro-4′-(2-thiazolylsulfamoyl)-
Synonym: Benzenesulfonic acid, amide,4(3,3-dichloroacetamido)-*N*(1,3-thiazol-2-yl)
Mol Form: $C_{11}H_9Cl_2N_3O_3S_2$
M_r: 366.23
***s*:** DMSO 3
IR: SADG 17801 3310 1690 1600 1555 1520 1395 1320 1295 1160 1140 1080 930 850 830 790 740 710 680 740 580 560 550
UV: SAD 17453 288(23500) 260(19600) MeOH 288(23400) 261(19600) MeOH:HCl
H NMR: SAD 9409 6.6 6.8 7.3 7.8 7.9 11.0 DMSOd6
m/e: 364.95

HODOC No: 00418
CAS No: 22795-32-0
Name: Acetanilide, 2,2-dichloro-4′-(*p*-tolylsulfamoyl)-
Synonym: Benzenesulfonic acid, amide,4(3,3-dichloroacetamido)-*N*(4-tolyl)
Mol Form: $C_{15}H_{14}Cl_2N_2O_3S$
M_r: 373.24
***s*:** DMSO 3
IR: SADG 17792 3280 3200 2980 1680 1580 1530 1490 1380 1300 1130 1070 890 820 790 660 560 550 510
UV: SAD 17446 268(21400) 230(13100) MeOH 251(21700) MeOH:KOH
H NMR: SAD 9418 2.2 6.7 7.1 7.9 DMSOd6
m/e: 372.01

HODOC No: 00419
CAS No: 441-30-5
Name: Acetanilide, 2′,4′-difluoro-6′-nitro-
Synonym: Acetic acid, amide,*N*(2,4-difluoro-6-nitrophenyl
Mol Form: $C_8H_6F_2N_2O_3$
M_r: 216.14
IR: COB 6934 3220 3170 3000 1670 1610 1550 1500 1420 1350 1260 1150 1080 900 850 750 700
m/e: 216.03

HODOC No: 00420
CAS No: 19577-87-8
Name: Acetanilide, 2′-(1,4-dihydro-1-methyl-4-oxo-2-quinazolinyl)-
Synonym: 4-Quinazolinone, 2(2-acetamidophenyl)-1-methyl
Mol Form: $C_{17}H_{15}N_3O_2$
M_r: 293.31
mp: 279
UV: LANG 2201 307(9772) 232(33884) 206(40738) EtOH
m/e: 293.12

HODOC No: 00421
CAS No: 19578-19-9
Name: Acetanilide, 2′-(1,4-dihydro-1-methyl-4-oxo-2-quinazolinyl)-*N*-methyl-
Synonym: 4-Quinazolinone, 1,4-dihydro-1-methyl-2(2(*N*-methylacetamido)phenyl)
Mol Form: $C_{18}H_{17}N_3O_2$
M_r: 307.33
mp: 201
UV: LANG 1811 307(9333) 280(5754) 232(27542) EtOH
m/e: 307.13

HODOC No: 00422
CAS No: 22721-08-0
Name: Acetanilide, 2′-(3,4-dihydro-4-oxo-2-quinazolinyl)-*N*-methyl-
Synonym: Quinazoline, 4-hydroxy-2(2(*N*-methylacetanido)phenyl)
Mol Form: $C_{17}H_{15}N_3O_2$
M_r: 293.31
mp: 186-7
UV: LANG 2202 283(10965) 230(32359) 204(39811) EtOH
m/e: 293.12

HODOC No: 00423
CAS No: 101588-41-4
Name: Acetanilide, 4′-(4-dimethylamino-*o*-tolylazo)-
Synonym: Acetic acid, amide,*N*(4(4-dimethylamino-2-methylphenylazo)phenyl)
Mol Form: $C_{17}H_{20}N_4O$
M_r: 296.36
mp: 189-90
***s*:** chl 3
IR: SADG 21866 3300 3040 2900 1670 1600 1510 1410 1350 1270 1190 1100 830 530
UV: OES 3-653 545(44668) 355(8511) 1.2N HCl:50% EtOH
H NMR: SAD 11309 2.2 2.7 3.0 6.5-6.7 7.4-8.0 $CDCl_3$
m/e: 296.16

HODOC No: 00424
CAS No: 33560-57-5
Name: Acetanilide, 2,2,2,2′,4′,5′-hexachloro-
Synonym: Acetic acid, amide,trichloro,*N*(2,4,5-trichlorophenyl)
Mol Form: $C_8H_3Cl_6NO$
M_r: 341.83
IR: µCOB 2319 3330 3130 1720 1590 1520 1370 1210 1080 970 940 920 810 720 670
m/e: 338.83

HODOC No: 00425
CAS No: 17738-37-3
Name: Acetanilide, 2-[*N*-(*p*-methoxyphenyl)benzimidoyl]-4′-nitro-
Synonym: Propanoic acid, amide,3(4-methoxyphenylimino)-*N*(4-nitrophenyl)-3-phenyl
Mol Form: $C_{22}H_{19}N_3O_4$
M_r: 389.39
mp: 167-8
UV: LANG 1987 389(36308) 303(14454) EtOH
m/e: 389.14

HODOC No: 00426
CAS No: 4494-47-7
Name: Acetanilide, 2-(*N*-1-naphthylbenzimidoyl)-
Synonym: Propanoic acid, amide,*N*,3-diphenyl-3(1-naphthylimino)
Mol Form: $C_{25}H_{20}N_2O$
M_r: 364.42
mp: 152-3
UV: LANG 1969 356(24547) 298(12589) 216(46774) EtOH
m/e: 364.16

HODOC No: 00427
CAS No: 17694-63-2
Name: Acetanilide, 2-(*N*-2-naphthylbenzimidoyl)-
Synonym: Propanoic acid, amide,*N*,3-diphenyl-3(2-naphthylimino)
Mol Form: $C_{25}H_{20}N_2O$
M_r: 364.42
mp: 188-9
UV: LANG 1970 359(31623) 252(25704) 216(45709) EtOH
m/e: 364.16

HODOC No: 00428
CAS No: 17738-40-8
Name: Acetanilide, 2-(*N*-1-naphthylbenzimidoyl)-4′-nitro-
Synonym: Propanoic acid, amide,3(1-naphthylimino)-*N*(4-nitrophenyl)-3-phenyl
Mol Form: $C_{25}H_{19}N_3O_3$
M_r: 409.42
mp: 213-4
UV: LANG 1989 386(35481) 323(16218) 217(56234) EtOH
m/e: 409.14

HODOC No: 00429
CAS No: 17738-41-9
Name: Acetanilide, 2-(*N*-2-naphthylbenzimidoyl)-4′-nitro-
Synonym: Propanoic acid, amide,3(2-naphthylimino)-*N*(4-nitrophenyl)-3-phenyl
Mol Form: $C_{25}H_{19}N_3O_3$
M_r: 409.42
mp: 211-2
UV: LANG 1990 389(41687) 301(15488) 249(25119) 216(44668) EtOH
m/e: 409.14

HODOC No: 00430
CAS No: 17738-39-5
Name: Acetanilide, 4′-nitro-2-[*N*-(*p*-nitrophenyl)benzimidoyl]-
Synonym: Propanoic acid, amide,*N*(4-nitrophenyl)-3(4-nitrophenylimino)-3-phenyl
Mol Form: $C_{21}H_{16}N_4O_5$
M_r: 404.36
mp: 239-40
UV: LANG 1988 395(47863) EtOH
m/e: 404.11

HODOC No: 00431
CAS No: 17738-35-1
Name: Acetanilide, 4′-nitro-2-(*N*-phenylbenzimidoyl)-
Synonym: Propanoic acid, amide,*N*(4-nitrophenyl)-3-phenyl-3-phenylimino
Mol Form: $C_{21}H_{17}N_3O_3$
M_r: 359.36
mp: 189-90
UV: LANG 1984 385(38019) 295(11482) 238(15136) EtOH
m/e: 359.13

HODOC No: 00432
CAS No: 17738-36-2
Name: Acetanilide, 4′-nitro-2-(*N*-*p*-tolylbenzimidoyl)-
Synonym: Propanoic acid, amide,3(4-methylphenylimino)-*N*(4-nitrophenyl)-3-phenyl
Mol Form: $C_{22}H_{19}N_3O_3$
M_r: 373.39
mp: 194-5
UV: LANG 1985 387(37154) 300(12589) 239(15849) EtOH
m/e: 373.14

HODOC No: 00433
CAS No: 33560-51-9
Name: Acetanilide, 2,2,2′,4′,5′-pentachloro-
Synonym: Acetic acid, amide,dichloro-*N*(2,4,5-trichlorophenyl)
Mol Form: $C_8H_4Cl_5NO$
M_r: 307.38
IR: µCOB 4114 3230 3030 1700 1590 1470 1370 1180 1140 1090 890 830 810 750 710 690 680 650
m/e: 304.87

HODOC No: 00434
CAS No: 3422-71-7
Name: Acetanilide, 2-(*N*-*p*-tolylbenzimidoyl)-
Synonym: Propanoic acid, amide,*N*,3-diphenyl-3(4-methylphenylimino)
Mol Form: $C_{22}H_{20}N_2O$
M_r: 328.39
mp: 173-4
UV: LANG 1965 347(26303) 256(14791) EtOH
m/e: 328.16

HODOC No: 00435
CAS No: 22795-62-6
Name: Acetanilide, 2,2,2-trichloro-4′-(methylsulfamoyl)-
Synonym: *N*1-Methyl-*N*4-(trichloroacetyl)sulfanilamide
Mol Form: $C_9H_9Cl_3N_2O_3S$
M_r: 331.59
***s*:** DMSO 3
IR: SADG 17780 3320 3260 3060 1710 1700 1590 1530 1400 1320 1300 1240 1160 1090 1060 870 820 810 670 630 550 510
UV: SAD 17441 263(18600) MeOH 284(12500) MeOH:KOH
H NMR: SAD 9683 2.5 7.3 7.8 7.9 11.1 DMSOd6
m/e: 329.94

HODOC No: 00436
CAS No: 388-11-4
Name: Acetanilide, 3′,4′,6′-trifluoro-2′-nitro-
Synonym: Benzene, 1-acetamido-2-nitro-3,4,6-trifluoro
Mol Form: $C_8H_5F_3N_2O_3$
M_r: 234.13
IR: COB 7445 3270 3090 1680 1555 1520 1450 1370 1260 1250 1190 1135 1025 965 870 795 770 750 720
m/e: 234.03

HODOC No: 00437
CAS No: 17738-55-5
Name: *p*-Acetanisidide, 2-[*N*-(*p*-chlorophenyl)benzimidoyl]-
Synonym: Propanoic acid, amide,3(4-chlorophenylimino)-*N*(4-methoxyphenyl)-3-phenyl
Mol Form: $C_{22}H_{19}ClN_2O_2$
M_r: 378.84
mp: 144-5
UV: LANG 1340 354(30200) 257(19055) EtOH
m/e: 378.11

HODOC No: 00438
CAS No: 17738-54-4
Name: *p*-Acetanisidide, 2-[*N*-(*p*-methoxyphenyl)benzimidoyl]-
Synonym: Propanoic acid, amide,*N*(4-methoxyphenyl)-3(4-methoxyphenylimino)-3-phenyl
Mol Form: $C_{23}H_{22}N_2O_3$
M_r: 374.42
mp: 123-4
UV: LANG 1342 357(26303) 261(17783) EtOH
m/e: 374.16

HODOC No: 00439
CAS No: 17738-27-1
Name: *p*-Acetanisidide, 2-(*N*-1-naphthylbenzimidoyl)-
Synonym: Propanoic acid, amide,*N*(4-methoxyphenyl)-3(1-naphthylimino)-3-phenyl
Mol Form: $C_{26}H_{22}N_2O_2$
M_r: 394.45
mp: 183-4
UV: LANG 1982 360(25119) EtOH
m/e: 394.17

HODOC No: 00440
CAS No: 17738-28-2
Name: *p*-Acetanisidide, 2-(*N*-2-naphthylbenzimidoyl)-
Synonym: Propanoic acid, amide,*N*(4-methoxyphenyl)-3(2-naphthylimino)-3-phenyl
Mol Form: $C_{26}H_{22}N_2O_2$
M_r: 394.45
mp: 191-2
UV: LANG 1983 360(30200) 251(24547) 215(43652) EtOH
m/e: 394.17

HODOC No: 00441
CAS No: 17738-26-0
Name: *p*-Acetanisidide, 2-[*N*-(*p*-nitrophenyl)benzimidoyl]-
Synonym: Propanoic acid, amide,*N*(4-methoxyphenyl)-3(4-nitrophenylimino)-3-phenyl
Mol Form: $C_{22}H_{19}N_3O_4$
M_r: 389.39
mp: 148-9
UV: LANG 1981 390(33113) 321(14454) 241(17378) EtOH
m/e: 389.14

HODOC No: 00442
CAS No: 17738-52-2
Name: *p*-Acetanisidide, 2-(*N*-phenylbenzimidoyl)-
Synonym: Propanoic acid, amide,*N*(4-methoxyphenyl)-3-phenyl-3-phenylimino
Mol Form: $C_{22}H_{20}N_2O_2$
M_r: 344.39
mp: 136-7
UV: LANG 1336 352(29512) 256(17378) EtOH
m/e: 344.15

HODOC No: 00443
CAS No: 17738-53-3
Name: *p*-Acetanisidide, 2-(*N*-*p*-tolylbenzimidoyl)-
Synonym: Propanoic acid, amide,*N*(4-methoxyphenyl)-3(4-methylphenylimino)-3-phenyl
Mol Form: $C_{23}H_{22}N_2O_2$
Mr: 358.42
mp: 151-2
UV: LANG 1339 355(28184) 258(18197) EtOH
m/e: 358.17

HODOC No: 00444
CAS No: 64-19-7
Beilstein No: 2^4, 94
Name: Acetic acid
Synonym: C2:0
Line Form: CH_3CO_2H
Mol Form: $C_2H_4O_2$
Mr: 60.05
bp: 255.8, 113.5^9
mp: 16.6
***d*:** $1.0492^{20/4}$
***n*:** 1.3720^{20}
***s*:** H_2O 5; al 5; eth 5; ace 5; bz 5; chl 3; os 3; CS_2 3
IR: COB 4819 3000 2930 2650 1710 1410 1350 1290 940 620
Raman: SAD 407 2940 2850 2710 1760 1670 1430 1370 1280 1020 940 900 630 610 4
UV: OES 4-3 208(32) EtOH
H NMR: VAR 8 2.1 11.4 $CDCl_3$
C NMR: JJ 7 FT 20.6 178.1 $CDCl_3$
MS: WILEY 36 45(100) 29(100) 43(99) 60(85) 15(21) 42(12) 28(10) 14(6)
m/e: 60.02

HODOC No: 00445
CAS No: 4977-62-2
Name: Acetic acid, (acetylamino)cyano-, ethyl ester
Synonym: Glycine, *N*-acetyl-2-cyano-, ethyl ester
Line Form: $CH_3CONHCH(CN)CO_2CH_2CH_3$
Mol Form: $C_7H_{10}N_2O_3$
Mr: 170.16
***s*:** chl 3
IR: SADG 21418 3280 3060 3000 2940 1740 1660 1540 1320 1230 1130 1040 870 700 600 540 450
H NMR: VAR 491 1.4 2.1 4.4 5.5 6.7 $CDCl_3$
m/e: 170.07

HODOC No: 00446
CAS No: 100117-83-7
Name: Acetic acid, (5-acetyl-2-ethoxyphenyl)-
Synonym: Phenylacetic acid, 5-acetyl-2-ethoxy
Line Form: $(2\text{-}CH_3CH_2O\text{-}5\text{-}CH_3CO)\text{-}C_6H_3CH_2CO_2H$
Mol Form: $C_{12}H_{14}O_4$
Mr: 222.23
mp: 129
***s*:** chl 2
IR: SADG 21751 3020 2980 2650 1720 1660 1640 1600 1500 1430 1360 1260 1140 1040 930 800 600
UV: SAD 2185 276 225 205 MeOH
H NMR: SAD 11378 1.4 2.5 3.7 4.1 6.9 7.8 7.9 8.9 $CDCl_3$
m/e: 222.09

HODOC No: 00447
CAS No: 5393-75-9
Name: Acetic acid, (5-acetyl-2-methoxyphenyl)
Synonym: Mycotox
Mol Form: $C_{11}H_{12}O_4$
Mr: 208.20
mp: 129.5
***s*:** DMSO 2
IR: SADG 21154 3000 2700 1720 1660 1650 1600 1500 1430 1360 1270 1200 1160 1020 810 800
UV: SAD 2025 272(14700) 222(14400) 202 MeOH
H NMR: SAD 10904 2.5 3.6 3.8 7.0 7.3 7.7 7.8 $DMSOd_6$
m/e: 208.07

HODOC No: 00448
CAS No: 13831-30-6
Beilstein No: 3^4, 576
Name: Acetic acid, (acetyloxy)-
Synonym: Glycolic acid, acetate
Line Form: $CH_3CO_2CH_2CO_2H$
Mol Form: $C_4H_6O_4$
Mr: 118.09
bp: $144\text{-}5^{12}$
mp: 67-8
Color: nd (Bz)
***s*:** H_2O 4; al 4; eth 4; ace 4; bz 1; chl 4
IR: SADG 13914 3100 2600 1740 1420 1370 1210 1070 840 700 640
m/e: 118.03

HODOC No: 00449
CAS No: 623-86-9
Beilstein No: 3^4, 580
Name: Acetic acid, (acetyloxy)-, ethyl ester
Synonym: Glycolic acid, ethyl ester, acetate
Line Form: $CH_3CO_2CH_2CO_2C_2H_5$
Mol Form: $C_6H_{10}O_4$
M$_r$: 146.14
bp: 179
***d*:** $1.0880^{20/4}$
***n*:** 1.4112^{20}
***s*:** Et_2O 4; EtOH 4; HOAc 4
m/e: 146.06

HODOC No: 00450
CAS No: 1190-93-8
Beilstein No: 3^4, 610
Name: Acetic acid, (acetylthio)-
Synonym: Acetic acid, mercapto-, acetate
Line Form: $(CH_3COS)CH_2CO_2H$
Mol Form: $C_4H_6O_3S$
M$_r$: 134.15
bp: $158\text{-}9^{17}$,$115\text{-}8^{2.5}$
Color: ye
***s*:** H_2O 4
m/e: 134.00

HODOC No: 00451
CAS No: 60658-38-0
Name: Acetic acid, amide,*N*(4-amino-2,6-dimethylphenyl)
Synonym: Acetamide, *N*-[*p*-(methylthio)phenacyl]-
Mol Form: $C_{10}H_{14}N_2O$
M$_r$: 178.22
mp: 208
***s*:** DMSO 3
IR: SADG 19841 3390 3340 3270 3010 2960 2910 2840 1650 1600 1520 1480 1430 1360 1340 1280 1220 1150 1020 960 870 850 840 730 700 620 580 560 490
UV: SAD 17561 289(1420) 243(19300) 207 MeOH
H NMR: SAD 11640 2.0 4.7 6.3 8.8 $DMSOd_6$
m/e: 178.11

HODOC No: 00452
CAS No: 795-18-6
Name: Acetic acid, amide,*N*(2-carboxyphenyl)-hydroxy-*N*-methyl
Synonym: β-Alanine, *N*-benzyl-*N*-(carboxymethyl)-, diethyl ester
Mol Form: $C_{10}H_{11}NO_4$
M$_r$: 209.19
IR: µCOB 5372 3130 2940 2630 1700 1670 1590 1450 1370 1270 1100 1080 1040 980 810 780 710
m/e: 209.07

HODOC No: 00453
CAS No: 2322-00-1
Name: Acetic acid, amide,dibutylamino-*N*(2-methoxy-6-methyphenyl),hydrochloride
Synonym: 2-Thenylamine, *N*-methyl-*N*-2-propynyl-
Mol Form: $C_{18}H_{31}ClN_2O_2$
M$_r$: 342.89
mp: 120-2
***s*:** H_2O 2
IR: µSADP 7977 3450 3230 3030 2940 1670 1590 1560 1470 1370 1280 1090 970 910 760
UV: SAD 18340 279(2360) 274(2370) MeOH
H NMR: SAD 11803 1.0 1.1-2.0 2.2 3.3 3.8 4.3 7.0 7.3 D_2O
m/e: 342.21

HODOC No: 00454
CAS No: 7327-60-8
Name: Acetic acid, amide,(diethylamino)-*N*(2-methoxy-6-methylphenyl), hydrochloride
Synonym: Acetonitrile, nitrilotri-
Mol Form: $C_{14}H_{23}ClN_2O_2$
M$_r$: 286.79
***s*:** H_2O 3
IR: µSADP 7975 3450 3130 2940 1700 1610 1540 1490 1390 1280 1090 960 790 710
H NMR: SAD 11801 1.4 2.3 3.4 3.9 4.3 7.0 7.3 D_2O
m/e: 286.14

HODOC No: 00455
CAS No: 50817-44-2
Name: Acetic acid, amide,dimethylamino-*N*(3-methoxy-2-methylphenyl) , hydrochloride
Mol Form: $C_{12}H_{19}ClN_2O_2$
M_r: 258.74
IR: µSADP 7974 3705 3450 3125 3030 1695 1640 1615 1540 1495 1410 1370 1315 1280 1265 1190 1150 1085 1055 980 960 910 870 775 765 705
UV: SAD 18339 278(1920) 274(1940) MeOH
H NMR: SAD 11800 2.3 3.1 3.9 4.3 7.0 7.3 D_2O
m/e: 258.11

HODOC No: 00456
CAS No: 23937-68-0
Name: Acetic acid, amide,dipropylamino-*N*(6-methoxy-2-methylphenyl), hydrochloride
Mol Form: $C_{16}H_{27}ClN_2O_2$
M_r: 314.84
mp: 147-8
***s*:** H_2O 2
IR: µSADP 7976 3450 3125 2940 1450 1250 1085 1010 960 780 750
H NMR: SAD 11802 1.0 1.8 2.2 3.3 3.9 4.3 7.0 7.3 D_2O
m/e: 314.18

HODOC No: 00457
CAS No: 5440-71-1
Name: Acetic acid, amide,*N*(5-hydroxy-2-methoxy-4-nitrophenyl), acetate
Synonym: 3,4,5,6,7,8-Hexahydroanthracen-1(2H)-one
Mol Form: $C_{11}H_{12}N_2O_6$
M_r: 268.22
mp: 187
***s*:** chl 2
IR: SADG 19304 3360 3130 2950 1770 1700 1590 1540 1510 1480 1410 1370 1320 1270 1230 1200 1070 1010 900 800 670
UV: SAD 17926 303(8210) 249(7340) MeOH
H NMR: SAD 10024 2.2 2.4 4.0 7.7 8.1 8.4 $CDCl_3$
m/e: 268.07

HODOC No: 00458
CAS No: 585-05-7
Beilstein No: 3^4, 121
Name: Acetic acid, [(aminocarbonyl)amino]oxo-
Synonym: Oxaluric acid
Line Form: $H_2NCONHCO\text{-}CO_2H$
Mol Form: $C_3H_4N_2O_4$
M_r: 132.07
mp: 208-10d
Color: cr
***s*:** H_2O 4
MS: NBS 279
m/e: 132.02

HODOC No: 00459
CAS No: 928-73-4
Beilstein No: 3^2, 389
Name: Acetic acid, [(aminocarbonyl)hydrazono]-
Synonym: Glyoxylic acid, 2-semicarbazone
Line Form: $H_2NCONHN{=}CHCO_2H$
Mol Form: $C_3H_5N_3O_3$
M_r: 131.09
mp: 128
***s*:** H_2O 2
IR: COB 6645 3450 3280 3180 3070 3080 3000 2920 2600 2530 1710 1660 1580 1510 1420 1350 1180 1060 930 730 660 570 510
UV: OES 4-8 252(12303) pH=6
m/e: 131.03

HODOC No: 00460
CAS No: 25395-22-6
Name: Acetic acid, [2-(aminocarbonyl)phenoxy]-
Synonym: Acetic acid, (*o*-carbamoylphenoxy)-
Mol Form: $C_9H_9NO_4$
M_r: 195.17
IR: SADG 18728 3400 3200 3060 2920 2500 1900 1700 1650 1600 1480 1450 1400 1250 1150 1110 1050 840 750 670 620 490
UV: SAD 7896 287(3100) 230(9360) MeOH
H NMR: SAD 10101 4.9 7.2 7.3-7.6 7.8 8.0 8.3 8.5 DMSO
m/e: 195.05

HODOC No: 00461
CAS No: 7597-91-3
Name: Acetic acid, 2-(5-amino-6-chloro-4-pyrimidinyl)hydrazide
Synonym: Pyrimidine, 4(2-acetylhydrazino)-5-amino-6-chloro
Mol Form: $C_6H_8ClN_5O$
M_r: 201.61
IR: COB 7842 3440 3320 3240 1670 1635 1560 1525 1460 1405 1360 1325 1285 1215 1145 1100 905 840 770
UV: OES 6-80 289(7943) 255(6918) pH=7
m/e: 201.04

HODOC No: 00462
CAS No: 24272-67-1
Name: Acetic acid, [(4-amino-4,5-dihydro-6-methyl-5-oxo-as-triazin-3-yl)thio]-, ethyl ester
Synonym: 1,2,4-Triazine-5-one-3-thioacetic acid,4-amino-4,5-dihydro-6-methyl,ethyl ester
Mol Form: $C_8H_{12}N_4O_3S$
M_r: 244.26
mp: 154-5
UV: LANG 2398 295(7943) 210(11482) EtOH
m/e: 244.06

HODOC No: 00463
CAS No: 471-47-6
Beilstein No: 2^4, 1857
Name: Acetic acid, aminooxo-
Synonym: Aminooxoacetic acid
Line Form: $HO_2C\text{-}CONH_2$
Mol Form: $C_2H_3NO_3$
M_r: 89.05
mp: 210d
Color: cr (w)
***s*:** H_2O 2; al 1; eth 1
IR: µSADP 5854 3330 3130 2780 1640 1430 1370 1220 1050 960 830 820 720 710
MS: ALD 3:111D
m/e: 89.01

HODOC No: 00464
CAS No: 617-36-7
Beilstein No: 2^4, 1857
Name: Acetic acid, aminooxo-, ethyl ester
Synonym: Oxamic acid, ethyl ester
Line Form: $H_2NCOCO_2C_2H_5$
Mol Form: $C_4H_7NO_3$
M_r: 117.10
mp: 114-5
***s*:** H_2O 3; eth 3; bz 1; chl 2
IR: SADG 915 3380 3210 1730 1680 1410 1390 1360 1240 1120 1090 1020 860 810 710 680
H NMR: SAD 2589 1.4 4.4 6.8 $CDCl_3$
m/e: 117.04

HODOC No: 00465
CAS No: 515-96-8
Beilstein No: 2^3, 1594
Name: Acetic acid, aminooxo-, hydrazide
Synonym: Aminooxamide
Line Form: $H_2NNHCOCONH_2$
Mol Form: $C_2H_5N_3O_2$
M_r: 103.08
mp: dec 221
***s*:** H_2O 2; al 1; eth 1; alk 4; ac 4
IR: SADG 4732 3390 3300 1650 1590 1420 1260 1110 990 710 690 650
m/e: 103.04

HODOC No: 00466
CAS No: 2921-14-4
Name: Acetic acid, (aminooxy)-, hydrochloride (2:1)
Synonym: Aminooxyacetic acid hemichloride
Line Form: $(NH_2OCH_2CO_2H)_2\text{–}HCl$
Mol Form: $C_2H_{5.5}ClNO_3$
M_r: 109.30
mp: 152-3
IR: µSADP 2103 2860 1720 1560 1520 1470 1410 1300 1220 1190 1150 1050 890
UV: LANG 2261 <200 H_2O
H NMR: SAD 18681 4.6 D_2O
m/e: 109.02

HODOC No: 00467
CAS No: 20241-76-3
Name: Acetic acid, (4-aminophenoxy),hydrate
Synonym: Anthraquinone, 1-anilino-4,5-dihydroxy-8-nitro-
Mol Form: $C_8H_{11}NO_4$
M_r: 185.17
IR: **µCOB 4026** 3450 2940 2630 2130 1610 1590 1560_1520 1410 1300 1270 1240 1190 1060 920 820 710
m/e: 185.07

HODOC No: 00468
CAS No: 86636-38-6
Name: Acetic acid, [4-(aminosulfonyl)-2,5-dimethylphenoxy]-
Synonym: Acetic acid, [(4-sulfamoyl-2,5-xylyl)oxy]-
Mol Form: $C_{10}H_{13}NO_5S$
M_r: 259.27
mp: 176
***s*:** DMSO 3
IR: **SADG 19406** 3320 3200 3100 2930 2650 1780 1760 1610 1560 1500 1440 1400 1350 1300 1240 1130 1080 900 850 730 670 580 520
UV: **SAD 18632** 282(888) 273(1070) 239(13300) MeOH 282(1010) 276(1250) 242(14100) MeOH:KOH
H NMR: **SAD 9610** 2.2 2.6 4.8 5.2 6.8 7.1 7.7 DMSOd6
m/e: 259.05

HODOC No: 00469
CAS No: 94883-14-4
Name: Acetic acid, [4-(aminosulfonyl)-2-methoxyphenoxy]-
Synonym: Benzenesulfonic acid, amide,4(carboxymethoxy)-3-methoxy
Mol Form: $C_9H_{11}NO_6S$
M_r: 261.24
mp: 205
***s*:** DMSO 3
IR: **SADG 19407** 3300 3250 3150 1750 1590 1510 1430 1410 1310 1260 1200 1140 1100 1010 800 780 620 590
UV: **SAD 17402** 280(3020) 241(11600) MeOH
H NMR: **SAD 9611** 3.9 4.7 6.1 7.1 7.2 7.4 7.5 DMSOd6
m/e: 261.03

HODOC No: 00470
CAS No: 90610-83-6
Name: Acetic acid, [2-(aminosulfonyl)-4-methylphenoxy]-
Synonym: Acetic acid, [(2-sulfamoyl-*p*-tolyl)oxy]-
Mol Form: $C_9H_{11}NO_5S$
M_r: 245.24
mp: 178
***s*:** DMSO 3
IR: **SADG 19402** 3380 3300 3050 2920 2600 1730 1610 1560 1490 1420 1330 1230 1150 1080 900 810 730 680 500
UV: **SAD 17530** 285(3040) 223(9610) MeOH
H NMR: **SAD 9606** 2.3 4.8 6.3 7.0 7.1 7.4 7.6 DMSOd6
m/e: 245.04

HODOC No: 00471
CAS No: 7383-14-4
Name: Acetic acid, [4-(aminosulfonyl)phenoxy]-
Synonym: Acetic acid, (*p*-sulfamoylphenoxy)-
Mol Form: $C_8H_9NO_5S$
M_r: 231.22
mp: 194
***s*:** DMSO 3
IR: **SADG 19385** 3350 3280 3100 2600 1750 1720 1600 1500 1430 1320 1240 1160 1100 1070 910 840 670 580 560
UV: **SAD 17397** 279(668) 269(992) 261(1050) 235(15500) MeOH
H NMR: **SAD 9584** 4.8 5.6 7.1 7.2 7.8 DMSOd6
m/e: 231.02

HODOC No: 00472
CAS No: 2302-88-7
Beilstein No: 3^4, 384
Name: Acetic acid, 2-(aminothioxomethyl)hydrazide
Synonym: Semicarbazide, 1-acetyl-3-thio-
Line Form: $CH_3CONHNHCSNH_2$
Mol Form: $C_3H_7N_3OS$
M_r: 133.17
mp: 165-8
***s*:** DMSO 3
IR: **SADG 18999** 3300 3150 2980 1690 1660 1620 1560 1500 1360 1290 1040 840 670 530
UV: **OES 6-11** 246(12589) EtOH
H NMR: **SAD 9358** 1.9 7.6 9.2 9.6 DMSOd6
m/e: 133.03

HODOC No: 00473
CAS No: 108-24-7
Beilstein No: 2^4, 386
Name: Acetic acid, anhydride
Synonym: Acetic anhydride
Line Form: $(CH_3CO)_2O$
Mol Form: $C_4H_6O_3$
M_r: 102.09
bp: 139.5, 44^{15}
mp: -73.1
***d*:** $1.0820^{20/4}$
***n*:** 1.3901^{20}
***s*:** H_2O 4; al 3; eth 5; bz 3; ctc 2; chl 3
IR: μCOB 5944 3030 2940 1820 1750 1430 1370 1220 1120 1040 1000 890 810 780
Raman: DFB 48 3030 3000 6940 2840 2730 1820 1800 1760 1430 1370 1220 1120 1050 1020 920 810 780 670 650 600 560 530 430 350 200
UV: OES 1-30 225(48) iso
H NMR: SAD 9879 2.2 CCl_4
MS: WILEY 266 43(100) 28(8) 42(6) 44(3) 29(3) 45(2) 41(2) 102(1)
m/e: 102.03

HODOC No: 00474
CAS No: 591-09-3
Beilstein No: 2^4, 435
Name: Acetic acid, anhydride with nitric acid
Synonym: Acetyl nitrate
Line Form: $CH_3CO(ONO_2)$
Mol Form: $C_2H_3NO_4$
M_r: 105.05
bp: 22^{70} exp60
***d*:** $1.24^{15/4}$
m/e: 105.01

HODOC No: 00475
CAS No: 18523-48-3
Beilstein No: 2^4, 547
Name: Acetic acid, azido-
Synonym: Azidoacetic acid
Line Form: $N_3CH_2CO_2H$
Mol Form: $C_2H_3N_3O_2$
M_r: 101.06
bp: 116^{12}, 92^3
mp: 16
***d*:** 1.354^{33}
UV: OES 1-7 285(25) EtOH
m/e: 101.02

HODOC No: 00476
CAS No: 2875-32-3
Name: Acetic acid, (2-benzothiazolyloxy)-
Synonym: BOA
Structure:

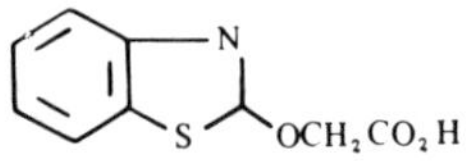

Mol Form: $C_9H_7NO_3S$
M_r: 209.21
IR: μCOB 1658 3130 2630 1720 1610 1470 1350 1250 1210 1000 940 760 730 710 680
m/e: 209.01

HODOC No: 00477
CAS No: 103100-75-0
Name: Acetic acid, (4-benzoyl-3-hydroxyphenoxy)-, dodecyl ester
Mol Form: $C_{27}H_{36}O_5$
M_r: 440.55
IR: μCOB 761 3450 3030 2940 1750 1610 1590 1450 1390 1350_1320 1220 1120 1090 930 850 820 750 730 700
m/e: 440.26

HODOC No: 00478
CAS No: 29193-41-7
Name: Acetic acid, [3-(*N*-benzylbutyramido)-2,4,6-triiodophenyl]-
Synonym: Acetic acid, (3(*N*-benzylbutamido)-2,4,6-triiodophenyl)
Mol Form: $C_{19}H_{18}I_3NO_3$
M_r: 689.05
mp: 102
***s*:** chl 2
IR: SADG 20160 3000 2960 2600 1720 1650 1420 1320 1200 1080 940 870 750 730 700 620
UV: SAD 17599 239(33900) 217(28200) MeOH
H NMR: SAD 10567 0.9 1.2-2.2 4.3 4.7 4.9 7.1 8.3 8.5 $CDCl_3$
m/e: 688.84

HODOC No: 00479
CAS No: 29815-94-9
Beilstein No: 6^4, 858
Name: Acetic acid, *bis*(4-chlorophenoxy)-
Synonym: Acetic acid, *bis*(*p*-chlorophenoxy)-
Line Form: $(4\text{-}ClC_6H_4O)_2CHCO_2H$
Mol Form: $C_{14}H_{10}Cl_2O_4$
M_r: 313.12
mp: 140-2
IR: ALD 3:944B
MS: ALD 6:125D
m/e: 312.00

HODOC No: 00480
CAS No: 5833-19-2
Beilstein No: 9, 675
Name: Acetic acid, *bis*(2,4-dinitrophenyl)-, methyl ester
Synonym: Diphenylacetic acid, 2,2′4,4′-tetranitro, methyl ester
Line Form: $[2,4\text{-}(NO_2)_2C_6H_3]_2CHCOOCH_3$
Mol Form: $C_{15}H_{10}N_4O_{10}$
M_r: 406.25
mp: 159
Color: lf(MeOH-Chl)
***s*:** Chl 4
m/e: 406.04

HODOC No: 00481
CAS No: 109643-29-0
Name: Acetic acid, [*bis*(1-methylethoxy)phosphinyl]-, phenylmethyl ester
Synonym: Acetic acid, phosphono-, benzyl diisopropyl ester
Line Form: $C_6H_5CH_2O_2CCH_2P(=O)(OCH(CH_2)_2)_2$
Mol Form: $C_{15}H_{23}O_5P$
M_r: 314.30
IR: COB 6918 2980 2940 1740 1500 1460 1390 1370 1110 1000 830 750 700
m/e: 314.13

HODOC No: 00482
CAS No: 79-08-3
Beilstein No: 2^4, 526
Name: Acetic acid, bromo-
Synonym: Bromoacetate ion
Line Form: $BrCH_2CO_2H$
Mol Form: $C_2H_3BrO_2$
M_r: 137.95
bp: 208, 127.5^{30}
mp: 50
***d*:** $1.9335^{50/4}$
***n*:** 1.4804^{50}
Color: hex or rh
***s*:** H_2O 5; al 5; eth 5; ace 3; bz 3; chl 2
IR: µCOB 1151 2940 2700 2560 1720 1560 1430 1280 1190 1110 920
H NMR: SAD 8765 3.9 10.6 $CDCl_3$
MS: NBS 327
m/e: 137.93

HODOC No: 00483
CAS No: 18991-98-5
Name: Acetic acid, bromo-, butyl ester
Synonym: Butyl bromoacetate
Line Form: $BrCH_2CO_2(CH_2)_3CH_3$
Mol Form: $C_6H_{11}BrO_2$
M_r: 194.05
MS: WILEY 1347 56(100) 121(35) 123(35) 41(30) 57(30) 29(22) 93(11) 194(0)
m/e: 193.99

HODOC No: 00484
CAS No: 5292-43-3
Beilstein No: 2^3, 482
Name: Acetic acid, bromo-, 1,1-dimethylethyl ester
Synonym: Acetic acid, bromo-, *tert*-butyl ester
Line Form: $BrCH_2CO_2C(CH_3)_3$
Mol Form: $C_6H_{11}BrO_2$
M_r: 194.05
bp: $73\text{-}4^{25}$
***n*:** 1.4430^{20}
***s*:** Et_2O 4; EtOH 4
IR: SADP 6338
m/e: 193.99

HODOC No: 00485
CAS No: 3785-34-0
Name: Acetic acid, bromo-, 1,2-ethanediyl ester
Synonym: Acetic acid, bromo-, ethylene ester
Line Form: $CH_2BrCO_2CH_2CH_2O_2CCH_2Br$
Mol Form: $C_6H_8Br_2O_4$
M_r: 301.93
IR: COB 4322 2960 2890 1750 1460 1430 1360 1280 1140 1110 1040 970 860 670 550
m/e: 301.88

HODOC No: 00486
CAS No: 105-36-2
Beilstein No: 2^4, 527
Name: Acetic acid, bromo-, ethyl ester
Synonym: Antol
Line Form: $BrCH_2CO_2C_2H_5$
Mol Form: $C_4H_7BrO_2$
M_r: 166.00
bp: 168-9, $58\text{-}9^{15}$
***d*:** $1.5059^{20/20}$
***n*:** 1.4489^{20}
***s*:** H_2O 1; al 5; eth 5; ace 3; ctc 2; os 4
IR: µCOB 2227 2940 1720 1470 1450 1430 1370 1280 1210 1160 1110 1030 940 860 710
Raman: SAD 112 3050 2970 2940_2900 2870 2760 2720 1730 1450 1420 1390 1370 1280 1210 1150 1110 1090_1020 940 920 900 850 790 760 700 660 540 440 410 360 330 300 270 200 180 160
H NMR: SAD 6833 1.3 3.8 4.2 CCl_4
MS: NBS 685
m/e: 165.96

HODOC No: 00487
CAS No: 96-32-2
Beilstein No: 2^4, 280
Name: Acetic acid, bromo-, methyl ester
Synonym: Bromoacetic acid methyl ester
Line Form: $CH_3CO_2CH_2Br$
Mol Form: $C_3H_5BrO_2$
M_r: 151.97
bp: $130\text{-}3^{750}$
***d*:** $1.6350^{20/4}$
***n*:** 1.4520^{20}
***s*:** H_2O 1; al 3; eth 3; ace 3; bz 3; os 4
IR: SADG 10988 2960 1740 1430 1280 1190 1160 1100 1000 890 700 670
H NMR: SAD 4205 3.2 3.3 bz
MS: WILEY 900 59(100) 93(84) 95(77) 121(55) 123(54) 72(47) 29(45) 151(1)
m/e: 151.95

HODOC No: 00488
CAS No: 59956-48-8
Beilstein No: 2^1, 96
Name: Acetic acid, bromo-, 2-methylpropyl ester
Synonym: Isobutyl bromoacetate
Line Form: $BrCH_2CO_2CH_2CH(CH_3)_2$
Mol Form: $C_6H_{11}BrO_2$
M_r: 194.05
bp: 188^{752}, 74.5^{10}
***d*:** $1.3269^{20/4}$
***s*:** Ac 4; Et_2O 4; EtOH 4
m/e: 193.99

HODOC No: 00489
CAS No: 620-72-4
Beilstein No: 6^3, 599
Name: Acetic acid, bromo-, phenyl ester
Synonym: Phenyl bromoacetate
Line Form: $BrCH_2CO_2C_6H_5$
Mol Form: $C_8H_7BrO_2$
M_r: 214.04
bp: 140^{20}
mp: 32
Color: pl (al)
***s*:** Et_2O 4; EtOH 4
m/e: 213.96

HODOC No: 00490
CAS No: 35223-80-4
Beilstein No: 2^4, 528
Name: Acetic acid, bromo-, propyl ester
Synonym: Propyl bromoacetate
Line Form: $BrCH_2CO_2C_3H_7$
Mol Form: $C_5H_9BrO_2$
M_r: 180.02
bp: 176^{762}
***d*:** $1.4099^{20/4}$
***n*:** 1.4518^{20}
***s*:** Ac 4; Et_2O 4; EtOH 4
m/e: 179.98

HODOC No: 00491
CAS No: 5589-96-8
Beilstein No: 2^4, 532
Name: Acetic acid, bromochloro-
Synonym: Bromochloroacetic acid
Line Form: $BrCHCl\text{-}CO_2H$
Mol Form: $C_2H_2BrClO_2$
M_r: 172.39
bp: 215, $103\text{-}4^{11}$
mp: 31.5
m/e: 171.89

HODOC No: 00492
CAS No: 22524-32-9
Beilstein No: 2^4, 532
Name: Acetic acid, bromochloro-, ethyl ester
Synonym: Ethyl bromochloroacetate
Line Form: $BrCHCl\text{-}CO_2C_2H_5$
Mol Form: $C_4H_6BrClO_2$
M_r: 200.44
bp: 174d
***d*:** $1.5890^{22/4}$
***n*:** 1.4639^{24}
***s*:** Et_2O 4; EtOH 4
m/e: 199.92

HODOC No: 00493
CAS No: 667-27-6
Name: Acetic acid, bromodifluoro-, ethyl ester
Synonym: Ethyl bromodifluoroacetate
Line Form: $(F_2CBr)\text{-}CO_2C_2H_5$
Mol Form: $C_4H_5BrF_2O_2$
M_r: 201.98
bp: 112^{700}
***d*:** 1.583^{25}
***n*:** 1.3900^{25}
m/e: 201.94

HODOC No: 00494
CAS No: 359-25-1
Beilstein No: 2^4, 531
Name: Acetic acid, bromofluoro-
Synonym: Bromofluoroacetic acid
Line Form: $(BrCHF)CO_2H$
Mol Form: $C_2H_2BrFO_2$
M_r: 155.94
bp: 183, 102^{20}
mp: 48
***s*:** H_2O 4; EtOH 4; Chl 4
m/e: 155.92

HODOC No: 00495
CAS No: 401-55-8
Beilstein No: 2^4, 531
Name: Acetic acid, bromofluoro-, ethyl ester
Synonym: Ethyl bromofluoroacetate
Line Form: $(BrCHF)CO_2C_2H_5$
Mol Form: $C_4H_6BrFO_2$
M_r: 183.99
bp: 154
***d*:** 1.5587^{17}
H NMR: SAD 9113 1.4 4.3 6.5 $CDCl_3$
m/e: 183.95

HODOC No: 00496
CAS No: 6956-82-7
Name: Acetic acid, (4-bromo-2-methylphenoxy)-
Synonym: Acetic acid, (4-bromo-*o*-tolyloxy)-
Mol Form: $C_9H_9BrO_3$
M_r: 244.06
IR: µCOB 1954 3030 2940 2630 1720 1700 1490 1450 1430 1370 1300 1250 1190 1140 1080 910 800 680
m/e: 243.97

HODOC No: 00497
CAS No: 1879-56-7
Beilstein No: 6^4, 1040
Name: Acetic acid, (2-bromophenoxy)-
Synonym: Acetic acid, (*o*-bromophenoxy)-
Line Form: $2\text{-}BrC_6H_4OCH_2CO_2H$
Structure:

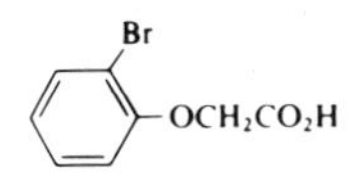

Mol Form: $C_8H_7BrO_3$
M_r: 230.04
mp: 142.5
Color: nd (dil al) cr (w)
***s*:** al 4; eth 4; chl 4; AcOEt 4
UV: OES 4-152 274(2042) 0.1N NaOH
m/e: 229.96

HODOC No: 00498
CAS No: 1878-91-7
Beilstein No: 6^4, 1052
Name: Acetic acid, (4-bromophenoxy)-
Synonym: Acetic acid, (*p*-bromophenoxy)-
Line Form: $4\text{-}BrC_6H_4OCH_2CO_2H$
Mol Form: $C_8H_7BrO_3$
M_r: 230.04
mp: 161-2
Color: pr (al), cr (w)
***s*:** H_2O 2; al 4; eth 4; CS_2 2
IR: µSADP 12439 2860 2560 1720 1700 1590 1490 1450 1430 1370 1280 1240 1180 1090 1060 1000 910 820 800 700
UV: SAD 21977 287(1120) 280(1390) 226(12200) MeOH
H NMR: SAD 18765 4.6 6.9 7.4 11.6 poly D
m/e: 229.96

HODOC No: 00499
CAS No: 3406-76-6
Name: Acetic acid, [(4-bromophenyl)thio]-
Synonym: Acetic acid, [(*p*-bromophenyl)thio]-
Mol Form: $C_8H_7BrO_2S$
Mr: 246.10
mp: 116
IR: μSADP 6154 2940 2330 1700 1470 1430 1300 1190 1090 1000 900 810
UV: SAD 1720 259(10900) MeOH
H NMR: SAD 11013 3.6 7.2 7.4 10.5 $CDCl_3$
m/e: 245.94

HODOC No: 00500
CAS No: 123-86-4
Beilstein No: 2^4, 143
Name: Acetic acid, butyl ester
Synonym: Acetic acid n-butyl ester
Line Form: $CH_3CO_2C_4H_9$
Mol Form: $C_6H_{12}O_2$
Mr: 116.15
bp: 126.5
mp: -77.9
***d*:** $0.8825^{20/4}$
***n*:** 1.3941^{20}
***s*:** H_2O 2; al 5; eth 5; ace 3; chl 3
IR: SADG 10971 2950 2860 1730 1450 1350 1220 1050 1020 940 830 720
Raman: SAD 39 2940 2920 2870 2740 1730 1450 1390 1370 1300 1260 1230 1150 1120 1070 1030 1000 950 920 880 840 810 640 610 510 480 430 330 300_260 230
UV: OES 7-93 214(50) 202(51) iso 208(56) MeOH
H NMR: VAR 140 0.9 1.0-1.8 2.0 4.1 $CDCl_3$
MS: WILEY 432 43(100) 56(34) 41(17) 27(16) 29(15) 72(11) 61(10) 116(1)
m/e: 116.08

HODOC No: 00501
CAS No: 29193-36-0
Name: Acetic acid, [3-(*N*-butylacetamido)-2,4,6-triiodophenyl]-
Mol Form: $C_{14}H_{16}I_3NO_3$
Mr: 626.99
mp: 100
***s*:** chl 2
IR: SADG 20156 3000 2960 2600 1730 1630 1420 1370 1300 1180 930 860 600
UV: SAD 17595 237(38200) MeOH
H NMR: SAD 10563 0.9 1.4 1.8 3.6 4.4 8.4 8.5 $CDCl_3$
m/e: 626.83

HODOC No: 00502
CAS No: 6326-83-6
Beilstein No: 3^4, 612
Name: Acetic acid, 2,2′-[carbonothioylbis(thio)]*bis*-
Synonym: Carbonic acid, trithio-, diester with mercaptoacetic acid
Line Form: $(HO_2CCH_2S)_2CS$
Mol Form: $C_5H_6O_4S_3$
Mr: 226.28
mp: 172-5
IR: ALD 2:325C
MS: ALD 3:39B
m/e: 225.94

HODOC No: 00503
CAS No: 541-88-8
Beilstein No: 2^4, 487
Name: Acetic acid, chloro-, anhydride
Synonym: Chloracetic anhydride
Line Form: $(ClCH_2CO)_2O$
Mol Form: $C_4H_4Cl_2O_3$
Mr: 170.98
bp: 203
mp: 46
***d*:** 1.5497^{20}
Color: pr (Bz)
m/e: 169.95

HODOC No: 00504
CAS No: 590-02-3
Beilstein No: 2^4, 482
Name: Acetic acid, chloro-, butyl ester
Synonym: Butyl chloroacetate
Line Form: $ClCH_2CO_2C_4H_9$
Mol Form: $C_6H_{11}ClO_2$
Mr: 150.60
bp: 183, 94^{38}
***d*:** $1.0704^{20/4}$
***n*:** 1.4297^{20}
***s*:** Et_2O 4; EtOH 4
IR: COB 6464
H NMR: VAR 138
MS: WILEY 874
m/e: 150.04

HODOC No: 00505
CAS No: 89124-06-1
Beilstein No: 2^4, 492
Name: Acetic acid, chloro-, 2-(chloroacetyl)hydrazide
Synonym: Hydrazine, 1,2-*bis*(chloroacetyl)-
Line Form: $(ClCH_2CO)NHNH(COCH_2Cl)$
Mol Form: $C_4H_6Cl_2N_2O_2$
M_r: 185.01
mp: 165-7
IR: COB 4574 3170 3330 1600 1490 1250 1170 950 800
m/e: 183.98

HODOC No: 00506
CAS No: 3848-12-2
Beilstein No: 2^4, 481
Name: Acetic acid, chloro-, 2-chloroethyl ester
Synonym: 2-Chloroethyl chloroacetate
Line Form: $ClCH_2CO_2CH_2CH_2Cl$
Mol Form: $C_4H_6Cl_2O_2$
M_r: 156.99
bp: 101, 89^{10}
***d*:** $1.3600^{25/4}$
***n*:** 1.4619^{25}
***s*:** Et_2O 4
m/e: 155.97

HODOC No: 00507
CAS No: 6974-05-6
Name: Acetic acid, chloro-, decyl ester
Synonym: Decyl chloroacetate
Line Form: $ClCH_2CO_2(CH_2)_9CH_3$
Mol Form: $C_{12}H_{23}ClO_2$
M_r: 234.75
IR: µCOB 5920 2940 1750 1470 1390 1320 1180 980 790 720
m/e: 234.14

HODOC No: 00508
CAS No: 107-59-5
Beilstein No: 2^3, 444
Name: Acetic acid, chloro-, 1,1-dimethylethyl ester
Synonym: Acetic acid, chloro-, *tert*-butyl ester
Line Form: $ClCH_2CO_2C(CH_3)_3$
Mol Form: $C_6H_{11}ClO_2$
M_r: 150.60
bp: $49\text{-}50^{10}$
IR: COB 6623 2990 2940 1760 1730 1480 1460 1410 1400 1370 1310 1210 1150 950 850 770 690 570
m/e: 150.04

HODOC No: 00509
CAS No: 6941-69-1
Beilstein No: 2^3, 448
Name: Acetic acid, chloro-, 1,2-ethanediyl ester
Synonym: Acetic acid, chloro-, ethylene ester
Line Form: $ClCH_2CO_2CH_2CH_2O_2CCH_2Cl$
Mol Form: $C_6H_8Cl_2O_4$
M_r: 215.03
bp: $142\text{-}4^2$
mp: 45-6
Color: pr (eth-peth)
***s*:** H_2O 1; eth 4
IR: µCOB 1185 2940 1750 1450 1390 1320 1160 1050 970 930 890 850 780 690
m/e: 213.98

HODOC No: 00510
CAS No: 105-39-5
Beilstein No: 2^4, 481
Name: Acetic acid, chloro-, ethyl ester
Synonym: Chloroacetic acid ethyl ester
Line Form: $ClCH_2CO_2C_2H_5$
Mol Form: $C_4H_7ClO_2$
M_r: 122.55
bp: 144^{740}, 52^{20}
mp: -76
***d*:** $1.1585^{20/4}$
***n*:** 1.4215^{20}
***s*:** H^2O 1; al 5; eth 5; ace 5; bz 3; chl 3
IR: SADG 234 2980 1760 1470 1450 1410 1370 1310 1260 1190 1100 1030 950 930 860 780 690
Raman: SAD 111 2960 2940 2870 2770 2720 1750 1740 1450 1410 1390 1360 1310 1270 1110 1170 1110 1090 1020 940 860 780 700 590 440 360_340 310 250
H NMR: VAR 73 1.3 4.1 4.3 $CDCl_3$
MS: NBS 196
m/e: 122.01

HODOC No: 00511
CAS No: 30956-28-6
Name: Acetic acid, chloro-, hydrazide
Synonym: Chloroacetic acid hydrazide
Line Form: $ClCH_2CONHNH_2$
Mol Form: $C_2H_5ClN_2O$
M_r: 108.53
mp: 93
IR: µCOB 2335 3230 3030 1610 1490 1390 1240 1160 940 790
m/e: 108.01

HODOC No: 00512
CAS No: 35280-53-6
Beilstein No: 2^3, 447
Name: Acetic acid, chloro-, 2-hydroxyethyl ester
Synonym: Ethylene glycol mono(chloroacetate)
Line Form: $ClCH_2CO_2CH_2CH_2OH$
Mol Form: $C_4H_7ClO_3$
M$_r$: 138.55
bp: 240d, $86^{1.5}$
***d*:** $1.3300^{20/4}$
***n*:** 1.4609^{20}
***s*:** H_2O 4; EtOH 4
m/e: 138.01

HODOC No: 00513
CAS No: 13361-36-9
Beilstein No: 2^3, 447
Name: Acetic acid, chloro-, 2-methoxyethyl ester
Synonym: 2-Methoxyethyl chloracetate
Line Form: $ClCH_2CO_2CH_2CH_2OCH_3$
Mol Form: $C_5H_9ClO_3$
M$_r$: 152.57
bp: $86\text{-}6^9$
***d*:** $1.2015^{20/4}$
***n*:** 1.4382^{20}
***s*:** Et_2O 4
m/e: 152.02

HODOC No: 00514
CAS No: 42831-64-1
Name: Acetic acid, chloro-, 1-methyl-1,2-ethanediyl ester
Synonym: Acetic acid, chloro-, propylene ester
Line Form: $ClCH_2CO_2CH_2CH(CH_3)O_2CCH_2Cl$
Mol Form: $C_7H_{10}Cl_2O_4$
M$_r$: 229.05
IR: **μCOB 1991** 2940 1750 1470 1410 1280 1160 1080 1020 960 790 690
MS: **WILEY 1616** 77(100) 45(49) 120(37) 79(32) 49(29) 42(20) 107(20) 228(1)
m/e: 228.00

HODOC No: 00515
CAS No: 105-48-6
Beilstein No: 2^4, 482
Name: Acetic acid, chloro-, 1-methylethyl ester
Synonym: Acetic acid, chloro-, isopropyl ester
Line Form: $ClCH_2CO_2CH(CH_3)_2$
Mol Form: $C_5H_9ClO_2$
M$_r$: 136.57
bp: 150-1
***d*:** $1.0888^{20/4}$
***n*:** 1.4382^{20}
***s*:** Et_2O 4
m/e: 136.03

HODOC No: 00516
CAS No: 15150-39-7
Beilstein No: 6^2, 378
Name: Acetic acid, chloro-, 4-methylphenyl ester
Synonym: Acetic acid, chloro-, *p*-tolyl ester
Line Form: $ClCH_2CO_2(C_6H_4CH_3\text{-}4)$
Mol Form: $C_9H_9ClO_2$
M$_r$: 184.61
bp: 162^{41}
mp: 32
Color: pl
***s*:** Et_2O 4; EtOH 4
m/e: 184.03

HODOC No: 00517
CAS No: 17696-64-9
Beilstein No: 2^3, 443
Name: Acetic acid, chloro-, 1-methylpropyl ester
Synonym: Acetic acid, chloro-, *sec*-butyl ester
Line Form: $ClCH_2CO_2CH(CH_3)CH_2CH_3$
Mol Form: $C_6H_{11}ClO_2$
M$_r$: 150.60
bp: 163-4
***d*:** $1.062^{20/20}$
***n*:** 1.4251^{19}
***s*:** Et_2O 4; EtOH 4
m/e: 150.04

HODOC No: 00518
CAS No: 13361-35-8
Beilstein No: 2^4, 483
Name: Acetic acid, chloro-, 2-methylpropyl ester
Synonym: Acetic acid, chloro-, isobutyl ester
Line Form: $ClCH_2CO_2CH_2CH(CH_3)_2$
Mol Form: $C_6H_{11}ClO_2$
Mr: 150.60
bp: 170
***d*:** $1.0612^{20/4}$
***n*:** 1.4255^{20}
***s*:** Ac 4; Et_2O 4
m/e: 150.04

HODOC No: 00519
CAS No: 620-73-5
Beilstein No: 6^3, 598
Name: Acetic acid, chloro-, phenyl ester
Synonym: Phenyl chloroacetate
Line Form: $ClCH_2CO_2C_6H_5$
Mol Form: $C_8H_7ClO_2$
Mr: 170.59
bp: 230-5, 114^8
mp: 44-5
***d*:** $1.2202^{44/4}$
***n*:** 1.5146^{44}
Color: nd or pl (al)
***s*:** H_2O 1; al 4; eth 4
IR: **μCOB 2807** 3330 1720 1610 1520 1220 930 890 790 760 690
m/e: 170.01

HODOC No: 00520
CAS No: 140-18-1
Beilstein No: 6^3, 1479
Name: Acetic acid, chloro-, phenylmethyl ester
Synonym: Acetic acid, chloro-, benzyl ester
Line Form: $ClCH_2CO_2CH_2C_6H_5$
Mol Form: $C_9H_9ClO_2$
Mr: 184.61
bp: 147.5^9, $84\text{-}6^{0.4}$
***d*:** $1.1223^{4/4}$
***n*:** 1.5426^{18}
***s*:** Et_2O 4; EtOH 4
m/e: 184.03

HODOC No: 00521
CAS No: 5396-24-7
Beilstein No: 2^4, 482
Name: Acetic acid, chloro-, propyl ester
Synonym: Propyl chloroacetate
Line Form: $ClCH_2CO_2C_3H_7$
Mol Form: $C_5H_9ClO_2$
Mr: 136.57
bp: 161^{764}
***d*:** $1.1022^{20/4}$
***n*:** 1.4261^{20}
***s*:** Et_2O 4
m/e: 136.03

HODOC No: 00522
CAS No: 56952-31-9
Name: Acetic acid, chloro(4-chlorophenoxy)-, ethyl ester
Synonym: Ethyl chloro(*p*-chlorophenoxy)acetate
Mol Form: $C_{10}H_{10}Cl_2O_3$
Mr: 249.08
IR: **μCOB 1977** 2940 1750 1610 1490 1430 1280 1250 1160 1090 1080 1040 960 860 820 790 670
m/e: 248.00

HODOC No: 00523
CAS No: 27143-09-5
Name: Acetic acid, chloro[(4-chlorophenyl)hydrazono]-, ethyl ester
Synonym: Glyoxylic acid, chloro-, ethyl ester, 2-[(*p*-chlorophenyl)hydrazone]
Mol Form: $C_{10}H_{10}Cl_2N_2O_2$
Mr: 261.10
mp: 146
IR: **μSADP 10181** 3330 3130 2940 1720 1560 1520 1490 1370 1320 1300 1280 1240 1160 1090 1010 860 820 790 760 730
UV: **SAD 2757** 317(20200) 288(12300) 216(8290) MeOH
H NMR: **SAD 11728** 1.3 4.3 7.4 10.6 $DMSOd_6$
m/e: 260.01

HODOC No: 00524
CAS No: 76-04-0
Beilstein No: 2^4, 497
Name: Acetic acid, chlorodifluoro-
Synonym: Chlorodifluoroacetic acid
Line Form: $ClF_2C\text{-}CO_2H$
Mol Form: $C_2HClF_2O_2$
M_r: 130.48
bp: 122
mp: 24-6
***n*:** 1.3559^{20}
Color: hyg
***s*:** chl 3
IR: **µSADP 23644** 3130 1790 1430 1180 1140 980 790 700
H NMR: **SAD 14768** 10.7 $CDCl_3$
MS: **WILEY 584** 95(100) 45(82) 51(60) 86(53) 85(48) 50(25) 66(24) 130(0)
m/e: 129.96

HODOC No: 00525
CAS No: 2834-23-3
Name: Acetic acid, chlorodifluoro-, anhydride
Synonym: Chlorodifluoroacetic anhydride
Line Form: $(ClCF_2CO)_2O$
Mol Form: $C_4Cl_2F_4O_3$
M_r: 242.94
bp: 92-3
***n*:** 1.3485^{20}
m/e: 241.92

HODOC No: 00526
CAS No: 91859-33-5
Name: Acetic acid, [2-chloro-4-[(dimethylamino)sulfonyl]phenoxy]-
Synonym: Benzenesulfonic acid, amide,4(carboxymethoxy)-3-chloro-*N*,*N*-dimethyl
Mol Form: $C_{10}H_{12}ClNO_5S$
M_r: 293.71
mp: 142
***s*:** DMSO 3
IR: **SADG 19443** 3050 2900 2600 1750 1720 1590 1490 1430 1390 1340 1240 1160 1100 1080 950 860 810 720 710 650 670 500
H NMR: **SAD 9643** 2.7 5.0 5.6 7.3 7.7 7.8 $DMSOd_6$
m/e: 293.01

HODOC No: 00527
CAS No: 471-44-3
Beilstein No: 2^4, 493
Name: Acetic acid, chlorofluoro-
Synonym: Chlorofluoroacetic acid
Line Form: $ClCHF\text{-}CO_2H$
Mol Form: $C_2H_2ClFO_2$
M_r: 112.49
bp: 162
***d*:** $1.532^{25/4}$
***n*:** 1.4085^{25}
m/e: 111.97

HODOC No: 00528
CAS No: 368-34-3
Beilstein No: 2^3, 453
Name: Acetic acid, chlorofluoro-, butyl ester
Synonym: Chlorofluoroacetic acid, butyl ester
Line Form: $ClCHF\text{-}CO_2C_4H_9$
Mol Form: $C_6H_{10}ClFO_2$
M_r: 168.59
bp: 165-6
***d*:** $1.124^{25/4}$
***n*:** 1.4067^{25}
m/e: 168.04

HODOC No: 00529
CAS No: 401-56-9
Beilstein No: 2^4, 493
Name: Acetic acid, chlorofluoro-, ethyl ester
Synonym: Ethyl chlorofluoroacetate
Line Form: $ClCHF\text{-}CO_2C_2H_5$
Mol Form: $C_4H_6ClFO_2$
M_r: 140.54
bp: 128-30
***d*:** 1.225^{20}
***n*:** 1.3927^{20}
m/e: 140.00

HODOC No: 00530
CAS No: 433-52-3
Beilstein No: 2^4, 493
Name: Acetic acid, chlorofluoro-, methyl ester
Synonym: Methyl chlorofluoroacetate
Line Form: $ClCHF\text{-}CO_2CH_3$
Mol Form: $C_3H_4ClFO_2$
M_r: 126.51
bp: 116
***d*:** $1.323^{25/4}$
***n*:** 1.3903^{25}
m/e: 125.99

HODOC No: 00531
CAS No: 348-70-9
Beilstein No: 2^3, 453
Name: Acetic acid, chlorofluoro-, propyl ester
Synonym: Chlorofluoroacetic acid, propyl ester
Line Form: $ClCHF\text{-}CO_2C_3H_7$
Mol Form: $C_5H_8ClFO_2$
Mr: 154.56
bp: 147
***d*:** $1.170^{25/4}$
***n*:** 1.3994^{25}
m/e: 154.02

HODOC No: 00532
CAS No: 6386-63-6
Name: Acetic acid, [4-chloro-2-(hydroxymethyl)phenoxy]-
Synonym: Acetic acid, [(4-chloro-α-hydroxy-*o*-tolyl)oxy]-
Mol Form: $C_9H_9ClO_4$
Mr: 216.61
mp: 140-3
***s*:** DMSO 3
IR: SADG 19063 3420 2900 2600 1750 1600 1480 1450 1370 1240 1220 1180 1130 1070 1030 910 880 800 690 650 580 550
UV: SAD 17895 285(1520) 280(1560) 227(10700) MeOH
H NMR: SAD 9688 4.6 4.7 5.0 6.9 7.2 7.4 $DMSOd_6$
m/e: 216.02

HODOC No: 00533
CAS No: 13887-55-3
Name: Acetic acid, chloromethoxy-
Line Form: $ClCH_2OCH_2CO_2H$
Mol Form: $C_3H_5ClO_3$
Mr: 124.52
MS: WILEY 515 45(100) 29(36) 36(28) 31(22) 60(21) 15(19) 44(11) 124(0)
m/e: 123.99

HODOC No: 00534
CAS No: 1556-00-9
Beilstein No: 6^4, 2063
Name: Acetic acid, (2-chloro-5-methylphenoxy)-
Synonym: Acetic acid, [(6-chloro-*m*-tolyl)oxy]-
Line Form: $(2\text{-}Cl\text{-}5\text{-}CH_3)C_6H_3OCH_2CO_2H$
Mol Form: $C_9H_9ClO_3$
Mr: 200.61
mp: 137
Color: cr (to)
m/e: 200.02

HODOC No: 00535
CAS No: 19094-75-8
Name: Acetic acid, (2-chloro-6-methylphenoxy)-
Synonym: Acetic acid, [(6-chloro-*o*-tolyl)oxy]-
Mol Form: $C_9H_9ClO_3$
Mr: 200.61
mp: 109.6
IR: µCOB 1164 2940 2560 1720 1610 1470 1370 1300 1270 1220 1180 1030 1050 910 870 830 770 750 690
UV: SAD 4746 274 267 MeOH
H NMR: SAD 12402 2.3 4.6 6.8-7.3 9.9 $CDCl_3$
MS: WILEY 1413 107(100) 28(91) 91(90) 166(80) 141(75) 77(68) 200(51) 121(42)
m/e: 200.02

HODOC No: 00536
CAS No: 94-74-6
Beilstein No: 6^4, 1991
Name: Acetic acid, (4-chloro-2-methylphenoxy)-
Synonym: Acetic acid, [(4-chloro-*o*-tolyl)oxy]-
Line Form: $(4\text{-}Cl\text{-}2\text{-}CH_3\text{-})C_6H_3O\text{-}CH_2CO_2H$
Mol Form: $C_9H_9ClO_3$
Mr: 200.61
mp: 120
Color: pl (Bz, to)
***s*:** H_2O 2; al 4; eth 4; bz 3; ctc 3
IR: µSADP 5667 1700 1590 1490 1300 1240 1190 1140 1080 910 880 800 750
UV: SAD 3794 285 278 227 MeOH
H NMR: SAD 9679 2.2 4.6 6.7 7.1 7.1 8.6 2.5 poly D
m/e: 200.02

HODOC No: 00537
CAS No: 2698-38-6
Beilstein No: 6^4, 1993
Name: Acetic acid, (4-chloro-2-methylphenoxy)-, ethyl ester
Synonym: Acetic acid, [(4-chloro-*o*-tolyl)oxy]-, ethyl ester
Line Form: $(4\text{-}Cl\text{-}2\text{-}CH_3)C_6H_3OCH_2CO_2C_2H_5$
Mol Form: $C_{11}H_{13}ClO_3$
Mr: 228.66
bp: $115\text{-}7^1$
***n*:** 1.5150^{22}
m/e: 228.06

HODOC No: 00538

CAS No: 2698-39-7
Beilstein No: 6^3, 1266
Name: Acetic acid, (4-chloro-2-methylphenoxy)-, propyl ester
Synonym: Acetic acid, [(4-chloro-*o*-tolyl)oxy]-, propyl ester
Line Form: $(4\text{-Cl-2-CH}_3)C_6H_3OCH_2CO_2C_3H_7$
Mol Form: $C_{12}H_{15}ClO_3$
Mr: 242.69
bp: $109\text{-}11^{0.5}$
***n*:** 1.5100^{23}
m/e: 242.07

HODOC No: 00539

CAS No: 4755-77-5
Beilstein No: 2^4, 1853
Name: Acetic acid, chlorooxo-, ethyl ester
Synonym: Glyoxylic acid, chloro-, ethyl ester
Line Form: $ClCO\text{-}CO_2C_2H_5$
Mol Form: $C_4H_5ClO_3$
Mr: 136.53
bp: $137, 30^{10}$
***d*:** $1.2226^{20/4}$
Color: hyg
***s*:** Bz 4; Et_2O 4
IR: SADG 5990 2980 1790 1760 1460 1370 1290 1260 1010 960 760 590
H NMR: SAD 9366 1.5 4.4 CCl_4
m/e: 135.99

HODOC No: 00540

CAS No: 5781-53-3
Name: Acetic acid, chlorooxo-, methyl ester
Synonym: Glyoxylic acid, chloro-, methyl ester
Line Form: $ClCOCO_2CH_3$
Mol Form: $C_3H_3ClO_3$
Mr: 122.51
bp: 118-20
***n*:** $1.4189^{20/D}$
IR: SADG 17846 2960 1760 1430 1260 1160 1010 890 790 490 360
H NMR: SAD 9041 4.0 $CDCl_3$
m/e: 121.98

HODOC No: 00541

CAS No: 614-61-9
Beilstein No: 6^4, 796
Name: Acetic acid, (2-chlorophenoxy)-
Synonym: Acetic acid, (*o*-chlorophenoxy)-
Line Form: $2\text{-ClC}_6H_4OCH_2CO_2H$
Structure:

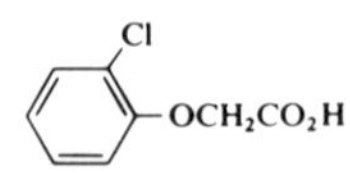

Mol Form: $C_8H_7ClO_3$
Mr: 186.59
mp: 148-9
Color: nd (w, al)
***s*:** H_2O 3; al 3
IR: µCOB 1163 2940 2860 2630 1750 1720 1590 1490 1430 1250 1100 840 760
UV: SAD 3424 283 275 218 MeOH
H NMR: SAD 18672 4.7 6.8-7.5 12.2 poly D
m/e: 186.01

HODOC No: 00542

CAS No: 588-32-9
Beilstein No: 6^4, 816
Name: Acetic acid, (3-chlorophenoxy)-
Synonym: Acetic acid, (*m*-chlorophenoxy)-
Line Form: $3\text{-ClC}_6H_4OCH_2CO_2H$
Mol Form: $C_8H_7ClO_3$
Mr: 186.59
mp: 110
Color: cr (w)
***s*:** H_2O 1
IR: µSADP 13138 3030 2860 2700 2560 1720 1700 1640 1590 1470 1430 1370 1350 1320 1270 1220 1090 1060 960 910 880 850 770 710 670
UV: OES 4-153 279(1380) 272(1514) 0.1N HCl
m/e: 186.01

HODOC No: 00543
CAS No: 122-88-3
Beilstein No: 6^4, 845
Name: Acetic acid, (4-chlorophenoxy)-
Synonym: Acetic acid, (*p*-chlorophenoxy)-
Line Form: 4-$ClC_6H_4OCH_2CO_2H$
Mol Form: $C_8H_7ClO_3$
M_r: 186.59
mp: 156-7
Color: pr or nd (w)
s: H_2O 2; chl 2
IR: μSADP 13139 2860 2630 2560 1700 1590 1490 1430 1370 1300 1280 1240 1180 1110 1080 1010 960 930 840 820 800 700
UV: SAD 457 287(1270) 279(1570) 226(1260) MeOH
H NMR: VAR 500 4.7 6.9 7.3 8.9 $CDCl_3$
MS: NBS 989
m/e: 186.01

HODOC No: 00544
CAS No: 5447-93-8
Beilstein No: 6^3, 694
Name: Acetic acid, (4-chlorophenoxy)-, 2-chloroethyl ester
Synonym: Acetic acid, (*p*-chlorophenoxy)-, 2-chloroethyl ester
Line Form: 4-$ClC_6H_4OCH_2CO_2CH_2CH_2Cl$
Mol Form: $C_{10}H_{10}Cl_2O_3$
M_r: 249.08
bp: $156\text{-}7^{215}$
mp: 41-2
m/e: 248.00

HODOC No: 00545
CAS No: 3685-84-5
Name: Acetic acid, (4-chlorophenoxy)-, 2-(dimethylamino)ethyl ester, hydrochloride
Synonym: Acetic acid, (*p*-chlorophenoxy)-, 2-(dimethylamino)ethyl ester hydrochloride
Mol Form: $C_{12}H_{17}Cl_2NO_3$
M_r: 294.17
mp: 137-9
UV: LANG 1695 279(2291) 226(12023) EtOH 280(1514) chl
m/e: 293.06

HODOC No: 00546
CAS No: 3405-88-7
Name: Acetic acid, [(4-chlorophenyl)thio]-
Synonym: Acetic acid, [(*p*-chlorophenyl)thio]-
Mol Form: $C_8H_7ClO_2S$
M_r: 202.65
IR: SADG 26848 3020 2920 1690 1480 1420 1390 1300 1260 1200 1180 1120 1090 1010 880 810 740 660 540 480 440 370 330
Raman: SAD 833 3070 2940 1650 1580 1400 1300 1270 1170 1100 1080 1010 910 820 750 630 550 430 300 250 120
UV: SAD 20500 258(9040) MeOH 261(11800) MeOH:KOH
H NMR: SAD 15589 3.6 9.4 poly D
m/e: 201.99

HODOC No: 00547
CAS No: 4486-75-3
Name: Acetic acid, (4-chloro-2-sulfamoylphenoxy)-
Synonym: Benzenesulfonic acid, amide,2(carboxymethoxy)-5-chloro
Mol Form: $C_8H_8ClNO_5S$
M_r: 265.66
mp: 200
s: DMSO 3
IR: SADG 19449 3380 3300 3100 2920 2600 1740 1590 1470 1430 1330 1260 1240 1160 1110 1080 900 820 690 650
UV: SAD 17755 290(2910) 229(10200) MeOH
H NMR: SAD 9782 4.9 7.2 7.3 7.6 7.7 DMSO
m/e: 264.98

HODOC No: 00548
CAS No: 22181-43-7
Name: Acetic acid, [(6-chloro-4-sulfamoyl-*m*-tolyl)oxy]-
Synonym: Benzenesulfonic acid, amide,4(carboxymethoxy)-5-chloro-2-methyl
Mol Form: $C_9H_{10}ClNO_5S$
M_r: 279.69
mp: 180
s: DMSO 3
IR: SADG 19453 3500 3400 3300 2950 2600 1740 1590 1480 1440 1380 1300 1240 1150 1060 910 890 720 620 570
UV: SAD 17756 287(1120) 277(1170) 240(12100) 210(45900) MeOH
H NMR: SAD 9786 2.6 4.9 5.5 7.1 7.4 7.9 $DMSOd_6$
m/e: 279.00

HODOC No: 00549
CAS No: 372-09-8
Beilstein No: 2^4, 1888
Name: Acetic acid, cyano-
Synonym: Cyanoacetic acid
Line Form: $NCCH_2CO_2H$
Mol Form: $C_3H_3NO_2$
M$_r$: 85.06
bp: $108^{0.15/d}$
mp: 70-1
***s*:** H_2O 3; al 3; eth 3; chl 2; DMSOd6 3; HOAc 2
IR: μCOB 3796 3130 2270 1720 1280 1180 970 940 880 790
UV: OES 4-7 208(40) EtOH
H NMR: SAD 9355 3.8 11.8 DMSOd6
MS: NBS 38
m/e: 85.02

HODOC No: 00550
CAS No: 5459-58-5
Beilstein No: 2^4, 1891
Name: Acetic acid, cyano-, butyl ester
Synonym: Butyl cyanoacetate
Line Form: $NCCH_2CO_2C_4H_9$
Mol Form: $C_7H_{11}NO_2$
M$_r$: 141.16
bp: 107-8
***n*:** 1.4200^{20}
m/e: 141.08

HODOC No: 00551
CAS No: 105-56-6
Beilstein No: 2^4, 1889
Name: Acetic acid, cyano-, ethyl ester
Synonym: Cyanoacetic acid ethyl ester
Line Form: $NCCH_2CO_2C_2H_5$
Mol Form: $C_5H_7NO_2$
M$_r$: 113.11
bp: 205, 99^{15}
mp: -22.5
***d*:** $1.0654^{20/4}$
***n*:** 1.4175^{20}
***s*:** Et_2O 4; EtOH 4
IR: SADP 2235
MS: NBS 4032
m/e: 113.05

HODOC No: 00552
CAS No: 140-87-4
Beilstein No: 2^4, 1895
Name: Acetic acid, cyano-, hydrazide
Synonym: AB-42
Line Form: $NCCH_2CONHNH_2$
Mol Form: $C_3H_5N_3O$
M$_r$: 99.09
mp: 114-5
Color: pr (al)
***s*:** H_2O 4; EtOH 4
IR: ALD 2:460F
MS: ALD 3:172C
m/e: 99.04

HODOC No: 00553
CAS No: 105-34-0
Beilstein No: 2^4, 1889
Name: Acetic acid, cyano-, methyl ester
Synonym: Cyanoacetic acid methyl ester
Line Form: $NCCH_2CO_2CH_3$
Mol Form: $C_4H_5NO_2$
M$_r$: 99.09
bp: 200-1, 115^{36}
mp: -22.5
***d*:** 1.112^{20}
***n*:** 1.4176^{20}
***s*:** Et_2O 4; EtOH 4
IR: SADP 5247
H NMR: VAR 37
MS: NBS 79
m/e: 99.03

HODOC No: 00554
CAS No: 14447-18-8
Beilstein No: 6^4, 2270
Name: Acetic acid, cyano-, phenylmethyl ester
Synonym: Acetic acid, cyano-, benzyl ester
Line Form: $NCCH_2CO_2CH_2C_6H_5$
Mol Form: $C_{10}H_9NO_2$
M$_r$: 175.18
bp: $141^{0.5}$
***n*:** 1.5206^{19}
m/e: 175.06

HODOC No: 00555
CAS No: 4974-44-1
Name: Acetic acid, cyano-, (phenylmethylene)hydrazide
Synonym: Acetic acid, cyano-, benzylidenehydrazide
Line Form: $NCCH_2CONHNH_2$
Mol Form: $C_{10}H_9N_3O$
M$_r$: 187.19
IR: SADG 27030 3210 3080 2940 2920 2330 2260 2200 1680 1660 1590 1550 1480 1440 1370 1360 1330 1300 1280 1230 1070 1050 990 980 950 920 900 840 740 680 560 490 380
Raman: SAD 129 3080 2960 2940 2260 1670 1620 1610 1580 1570 1500 1450 1340 1320 1300 1270 1220 1180 1160 1090 1070 1030 1000 940 620 580 510 410 350 330 250 170 110
UV: SAD 20596 282(26800) 217(18700) MeOH 296(19500) MeOH:KOH
H NMR: SAD 15728 3.9 7.2-7.5 7.6 8.1 11.4 poly D
m/e: 187.07

HODOC No: 00556
CAS No: 23014-13-3
Name: Acetic acid, [[6-(cyanomethyl)piperonyl]thio]-, methyl ester
Synonym: Acetic acid, (2-cyanomethyl-4,5(methylenedioxy)benzylthio),methyl ester
Structure:

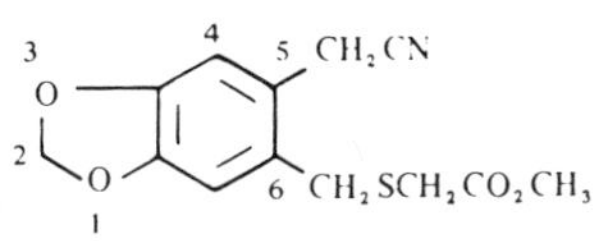

Mol Form: $C_{13}H_{13}NO_4S$
M$_r$: 279.30
mp: 50
s: chl 3
IR: SADG 14849 3000 2950 2250 1730 1610 1500 1480 1430 1370 1260 1160 1030 930 860
H NMR: SAD 8192 3.1 3.6 3.7 3.8 5.9 6.7 6.8 $CDCl_3$
m/e: 279.06

HODOC No: 00557
CAS No: 622-45-7
Beilstein No: 6^4, 36
Name: Acetic acid, cyclohexyl ester
Synonym: Cyclohexanol, acetate
Line Form: $CH_3CO_2C_6H_{11}$
Structure:

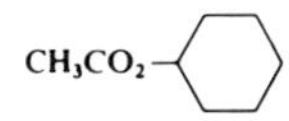

Mol Form: $C_8H_{14}O_2$
M$_r$: 142.19
bp: 173, $62\text{-}3^{12}$
***d*:** $0.9698^{20/4}$
***n*:** 1.4401^{20}
***s*:** Et_2O 4; EtOH 4
IR: SADP 22788
m/e: 142.10

HODOC No: 00558
CAS No: 1552-91-6
Beilstein No: 9^4, 124
Name: Acetic acid, cyclohexylidene-
Synonym: Δ1,α-Cyclohexaneacetic acid
Structure:

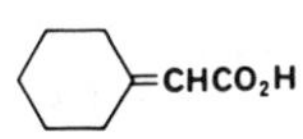

Mol Form: $C_8H_{12}O_2$
M$_r$: 140.17
mp: 92
Color: cr (dil al)
m/e: 140.08

HODOC No: 00559
CAS No: 1552-92-7
Beilstein No: 9^4, 124
Name: Acetic acid, cyclohexylidene-, ethyl ester
Synonym: Δ1,α-Cyclohexaneacetic acid, ethyl ester
Mol Form: $C_{10}H_{16}O_2$
M$_r$: 168.23
m/e: 168.12

HODOC No: 00560
CAS No: 40203-74-5
Beilstein No: 9^4, 124
Name: Acetic acid, cyclohexylidene-, methyl ester
Synonym: Δ1,α-Cyclohexaneacetic acid, methyl ester
Mol Form: $C_9H_{14}O_2$
Mr: 154.20
bp: $98.4\text{-}99.4^{19.5}$
***d*:** $1.0021^{19/4}$
***n*:** 1.4838^{19}
m/e: 154.10

HODOC No: 00561
CAS No: 112-17-4
Beilstein No: 2^4, 168
Name: Acetic acid, decyl ester
Synonym: Acetate C 10
Line Form: $CH_3CO_2(CH_2)_9CH_3$
Mol Form: $C_{12}H_{24}O_2$
Mr: 200.31
bp: 244, 125.8^{155}
mp: -15
***d*:** $0.8671^{20/4}$
***n*:** 1.4273^{20}
***s*:** H_2O 1; al 3; eth 3; bz 3; ctc 3; HOAc 3
IR: µCOB 3718 2940 1750 1470 1370 1240 1040 980 730
H NMR: SAD 6731 0.9 1.3 2.0 4.0 CCl_4
MS: WILEY 1418 70(100) 69(86) 83(70) 84(49) 97(40) 112(35) 68(32) 200(1)
m/e: 200.18

HODOC No: 00562
CAS No: 5435-83-6
Name: Acetic acid, (decylsulfonyl)-
Line Form: $CH_3(CH_2)_9SO_2CH_2CO_2H$
Mol Form: $C_{12}H_{24}O_4S$
Mr: 264.37
mp: 105.7
***s*:** chl 2
IR: SADG 21698 3100 2920 2840 2700 1700 1460 1400 1370 1330 1280 1220 1120 910 760 610 510
H NMR: SAD 10724 0.9 1.3 1.9 3.3 4.0 $CDCl_3$
m/e: 264.14

HODOC No: 00563
CAS No: 623-73-4
Beilstein No: 3^4, 1495
Name: Acetic acid, diazo-, ethyl ester
Synonym: DAAE
Line Form: $N_2CHCO_2C_2H_5$
Mol Form: $C_4H_6N_2O_2$
Mr: 114.10
bp: $140\text{-}1^{750}$ d
mp: -22
***d*:** $1.0852^{18/4}$
***n*:** 1.4605^{20}
Color: ye rh
***s*:** H_2O 2; al 5; eth 5; bz 5; lig 5
IR: µSADP 17518 3130 2940 2080 1700 1450 1390 1350 1250 1190 1040 1020 740
UV: OES 3-18 247(13490) EtOH
m/e: 114.04

HODOC No: 00564
CAS No: 631-64-1
Beilstein No: 2^4, 533
Name: Acetic acid, dibromo-
Synonym: Dibromoacetic acid
Line Form: $Br_2CHCOOH$
Mol Form: $C_2H_2Br_2O_2$
Mr: 215.84
bp: 195^{250}
mp: 48
Color: dlq cr
***s*:** H_2O 4; al 4; eth 3
IR: µCOB 1424 3030 2630 1720 1450 1410 1270 1120 1080 900 870 790 700
H NMR: SAD 14411 5.9 8.5 $CDCl_3$
m/e: 215.84

HODOC No: 00565
CAS No: 617-33-4
Beilstein No: 2^4, 533
Name: Acetic acid, dibromo-, ethyl ester
Synonym: Dibromoacetic acid ethyl ester
Line Form: $Br_2CHCOOC_2H_5$
Mol Form: $C_4H_6Br_2O_2$
Mr: 243.89
bp: 194, 121^{74}
***d*:** $1.9025^{20/20}$
***n*:** 1.5017^{13}
***s*:** H_2O 1; al 5; eth 5
IR: µCOB 953 2940 1750 1720 1470 1450 1390 1370 1270 1210 1140 1030 940 930 860 790 700
H NMR: SAD 17260 1.4 4.3 6.0 CCl_4
m/e: 243.87

HODOC No: 00566
CAS No: 6482-26-4
Beilstein No: 2, 219
Name: Acetic acid, dibromo-, methyl ester
Synonym: Methyl dibromoacetate
Line Form: $Br_2CHCOOCH_3$
Mol Form: $C_3H_4Br_2O_2$
M_r: 229.87
bp: 182-3
***s*:** Et_2O 4; EtOH 4
m/e: 229.86

HODOC No: 00567
CAS No: 565-53-7
Beilstein No: 2^4, 534
Name: Acetic acid, dibromofluoro-, ethyl ester
Synonym: Ethyl dibromofluoroacetate
Line Form: $Br_2CFCOOC_2H_5$
Mol Form: $C_4H_5Br_2FO_2$
M_r: 261.88
bp: 83^{33}
***d*:** 1.894^{25}
***n*:** 1.4633^{25}
m/e: 261.86

HODOC No: 00568
CAS No: 21512-82-3
Name: Acetic acid, (2,4-dibromo-6-formylphenoxy)-
Mol Form: $C_9H_6Br_2O_4$
M_r: 335.94
mp: 176-7
***s*:** DMSO 3
IR: SADG 19064 3060 2920 2860 1750 1690 1570 1550 1445 1420 1380 1350 1270 1250 1220 1140 1090 1045 890 880 740 720 680 570 540 480 340
H NMR: SAD 9689 4.8 7.8 8.2 10.4 $DMSOd_6$
m/e: 335.86

HODOC No: 00569
CAS No: 25168-24-5
Name: Acetic acid, 2,2′-[(dibutylstannylene)*bis*(thio)]*bis*-, diisooctyl ester
Synonym: Acetic acid, [(dibutylstannylene)dithio]di-, diisooctyl ester
Line Form: $\{H_3C(CH_2)_7O_2CCH_2S\}_2Sn\{CH_2(CH_2)_2CH_3\}_2$
Mol Form: $C_{28}H_{56}O_4S_2Sn$
M_r: 639.55
IR: COB 4634 2960 1740 1720 1460 1380 1300 1180 1150 1000 880 700
m/e: 640.26

HODOC No: 00570
CAS No: 79-43-6
Beilstein No: 2^4, 498
Name: Acetic acid, dichloro-
Synonym: DCA
Line Form: $Cl_2CHCOOH$
Mol Form: $C_2H_2Cl_2O_2$
M_r: 128.94
bp: 194
mp: 13.5
***d*:** $1.5634^{20/4}$
***n*:** 1.4658^{20}
***s*:** H_2O 5; al 5; eth 5; ace 3; ctc 2
IR: µSADP 2806 3130 3030 2700 2560 1750 1430 1280 1210 910 820 780 680
H NMR: SAD 166 6.0 11.7 CCl_4
MS: NBS 773
m/e: 127.94

HODOC No: 00571
CAS No: 4124-30-5
Beilstein No: 2^4, 503
Name: Acetic acid, dichloro-, anhydride
Synonym: Dichloroacetic acid anhydride
Line Form: $(Cl_2CHCO)_2O$
Mol Form: $C_4H_2Cl_4O_3$
M_r: 239.87
bp: 214-6d
***d*:** 1.574^{24}
m/e: 237.88

HODOC No: 00572
CAS No: 29003-73-4
Beilstein No: 2^4, 502
Name: Acetic acid, dichloro-, butyl ester
Synonym: Butyl dichloroacetate
Line Form: $Cl_2CHCOOC_4H_9$
Mol Form: $C_6H_{10}Cl_2O_2$
M_r: 185.04
bp: 193-4, 102^{37}
***d*:** $1.1820^{20/4}$
***n*:** 1.4420^{20}
***s*:** Et_2O 4; EtOH 4
m/e: 184.01

HODOC No: 00573
CAS No: 7561-04-8
Beilstein No: 2^4, 502
Name: Acetic acid, dichloro-, ethenyl ester
Synonym: Acetic acid, dichloro-, vinyl ester
Line Form: $Cl_2CHCOOCH{=}{-}CH_2$
Mol Form: $C_4H_4Cl_2O_2$
M$_r$: 154.98
bp: $46\text{-}8^{15}$
IR: **μCOB 1956** 3130 1790 1640 1300 1280 1270 1160 1120 980 940 890 810
m/e: 153.96

HODOC No: 00574
CAS No: 535-15-9
Beilstein No: 2^4, 501
Name: Acetic acid, dichloro-, ethyl ester
Synonym: Dichloroacetic acid ethyl ester
Line Form: $ClCH_2COOC_2H_5$
Mol Form: $C_4H_6Cl_2O_2$
M$_r$: 156.99
bp: 155^{764}, 56^{10}
***d*:** $1.2827^{20/4}$
***n*:** 1.4386^{20}
***s*:** H_2O 2; al 5; eth 5; ace 3; chl 3
IR: **μSADP 5248** 2940 1720 1470 1450 1390 1370 1300 1270 1160 1100 1020 950 860 810 760 710 690
H NMR: **VAR 59** 1.4 4.3 5.9 $CDCl_3$
MS: **WILEY 949** 29(100) 27(30) 83(12) 28(10) 48(9) 85(8) 76(6) 156(0)
m/e: 155.97

HODOC No: 00575
CAS No: 83004-97-1
Name: Acetic acid, dichloro-, heptyl ester
Synonym: Heptyl dichloroacetate
Line Form: $Cl_2CHCO_2(CH_2)_6CH_3$
Mol Form: $C_9H_{16}Cl_2O_2$
M$_r$: 227.12
***s*:** chl 2
IR: **SADG 22191** 2930 2850 1770 1750 1460 1370 1290 1150 980 800 710
H NMR: **SAD 11508** 0.9 1.4 1.7 4.3 6.0 $CDCl_3$
m/e: 226.05

HODOC No: 00576
CAS No: 17704-30-2
Beilstein No: 2^3, 460
Name: Acetic acid, dichloro-, 2-hydroxyethyl ester
Synonym: Ethylene glycol, mono(dichloroacetate)
Line Form: $Cl_2CHCOOCH_2CH_2OH$
Mol Form: $C_4H_6Cl_2O_3$
M$_r$: 172.99
bp: $81\text{-}2^{0.5}$
***d*:** $1.438^{20/4}$
***n*:** 1.4735^{20}
***s*:** EtOH 4
m/e: 171.97

HODOC No: 00577
CAS No: 116-54-1
Beilstein No: 2^4, 501
Name: Acetic acid, dichloro-, methyl ester
Synonym: Dichloroacetic acid methyl ester
Line Form: $Cl_2CHCOOCH_3$
Mol Form: $C_3H_4Cl_2O_2$
M$_r$: 142.97
bp: 142.8, 38^{10}
mp: -51.9
***d*:** $1.3774^{20/4}$
***n*:** 1.4429^{20}
***s*:** H_2O 1; al 3; ctc 3
IR: **μCOB 2229** 2940 1750 1450 1300 1270 1160 1010 1000 810 720
H NMR: **SAD 6590** 3.9 5.9 CCl_4
MS: **WILEY 756** 59(100) 83(57) 85(38) 15(23) 63(11) 87(6) 76(5) 142(1)
m/e: 141.96

HODOC No: 00578
CAS No: 25006-60-4
Beilstein No: 2^4, 502
Name: Acetic acid, dichloro-, 1-methylethyl ester
Synonym: Acetic acid, dichloro-, isopropyl ester
Line Form: $Cl_2CHCOOCH(CH_3)_2$
Mol Form: $C_5H_8Cl_2O_2$
M$_r$: 171.02
bp: 163-4
***d*:** $1.2053^{20/4}$
***n*:** 1.4328^{20}
***s*:** Et_2O 4; EtOH 4
m/e: 169.99

HODOC No: 00579
CAS No: 6957-89-7
Name: Acetic acid, dichloroethoxy-, ethyl ester
Synonym: Dichloroethoxyacetic acid ethyl ester
Line Form: $Cl_2C(OCH_2CH_3)CO_2CH_2CH_3$
Mol Form: $C_6H_{10}Cl_2O_3$
M$_r$: 201.04
MS: WILEY 1412 29(100) 127(21) 63(17) 129(14) 27(13) 28(9) 137(8) 200(0)
m/e: 200.00

HODOC No: 00580
CAS No: 354-19-8
Beilstein No: 2^4, 507
Name: Acetic acid, dichlorofluoro-
Synonym: Dichlorofluoroacetic acid
Line Form: Cl_2CF-COOH
Mol Form: $C_2HCl_2FO_2$
M$_r$: 146.93
bp: 162.5
mp: 162.5
MS: WILEY 814 65(100) 45(81) 110(72) 101(55) 100(47) 66(47) 81(36) 146(0)
m/e: 145.93

HODOC No: 00581
CAS No: 24581-93-9
Name: Acetic acid, (2,4-dichloro-6-formylphenoxy)-, methyl ester
Synonym: 2-Carbomethoxymethoxy-3,5-dichlorobenzaldehyde
Mol Form: $C_{10}H_8Cl_2O_4$
M$_r$: 263.07
mp: 119-20
s: chl 3
IR: SADG 19058 3080 2960 1770 1700 1590 1450 1400 1250 1210 1160 1040 870 660
UV: SAD 17891 313(476) 285(810) 277(737) 249(1630) MeOH
H NMR: SAD 9691 3.8 4.8 7.6 7.7 10.5 $CDCl_3$
m/e: 261.98

HODOC No: 00582
CAS No: 58-54-8
Name: Acetic acid, [2,3-dichloro-4-(2-methylene-1-oxobutyl)phenoxy]-
Synonym: Acetic acid, [2,3-dichloro-4-(2-methylenebutyryl)phenoxy]-
Mol Form: $C_{13}H_{12}Cl_2O_4$
M$_r$: 303.13
mp: 122-3
UV: LANG 1940 334(191) 276(3548) 223(17378) MeOH
m/e: 302.01

HODOC No: 00583
CAS No: 30720-99-1
Name: Acetic acid, [2,3-dichloro-4-(1-oxobutyl)phenoxy]-, methyl ester
Synonym: Acetic acid, (4-butyryl-2,3-dichlorophenoxy)-, methyl ester
Mol Form: $C_{13}H_{14}Cl_2O_4$
M$_r$: 305.14
mp: 40-1
UV: LANG 1530 260(6761) 220(19055) EtOH
m/e: 304.03

HODOC No: 00584
CAS No: 2976-74-1
Beilstein No: 6^4, 883
Name: Acetic acid, (2,3-dichlorophenoxy)-
Synonym: 2,3-Dichlorophenoxyacetic acid
Line Form: (2,3-$Cl_2C_6H_3$-O)-CH_2CO_2H
Mol Form: $C_8H_6Cl_2O_3$
M$_r$: 221.03
mp: 173-5
IR: ALD 3:945F
MS: NBS 4511; ALD 6:127C
m/e: 219.97

HODOC No: 00585
CAS No: 94-75-7
Beilstein No: 6^4, 908
Name: Acetic acid, (2,4-dichlorophenoxy)-
Synonym: B-Selektonon
Line Form: $(2,4\text{-}Cl_2C_6H_3O)CH_2CO_2H$
Structure:

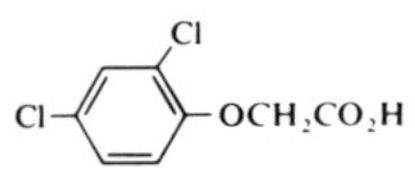

Mol Form: $C_8H_6Cl_2O_3$
Mr: 221.03
bp: $160^{0.4}$
mp: 140-1
Color: cr (Bz)
***s*:** H_2O 1; al 3; bz 2; DMSOd6 2
IR: μCOB 2221 2940 2560 1720 1470 1320 1240 1100 920 870 790 720 690 640
Raman: SAD 828 3080 2980 2950 1670 1590 1580 1480 1450 1390 1370 1310 1260 1230 1160 1120 1110 1050 1030 900 840 810 720 700 690 650 630 460 440 420 400 360 320 300 230 200 170 150 120
UV: SAD 3427 295 286 230 MeOH
H NMR: SAD 10803 4.8 7.0 7.3 7.4 11.9 DMSOd6
MS: NBS 4511
m/e: 219.97

HODOC No: 00586
CAS No: 1928-57-0
Name: Acetic acid, (2,4-dichlorophenoxy)-, 3-(2-butoxyethoxy)propyl ester
Mol Form: $C_{17}H_{24}Cl_2O_5$
Mr: 379.26
IR: μCOB 4089 2940 1750 1490 1390 1320 1210 1110 870 800 720 650
m/e: 378.10

HODOC No: 00587
CAS No: 1929-73-3
Name: Acetic acid, (2,4-dichlorophenoxy)-, 2-butoxyethyl ester
Synonym: Ethanol, 2-butoxy-, (2,4-dichlorophenoxy)acetate
Mol Form: $C_{14}H_{18}Cl_2O_4$
Mr: 321.19
bp: $156\text{-}62^1$
***d*:** 1.232^{20}
IR: μCOB 1150 2940 2860 1790 1750 1560 1490 1390 1350 1300 1190 1120 1110 1090 1040 870 840 800 720 640
m/e: 320.06

HODOC No: 00588
CAS No: 94-80-4
Name: Acetic acid, (2,4-dichlorophenoxy)-, butyl ester
Synonym: Butapon
Mol Form: $C_{12}H_{14}Cl_2O_3$
Mr: 277.14
IR: μCOB 1168 3030 2940 1750 1720 1470 1390 1300 1190 1110 1090 870 840 800 720
m/e: 276.03

HODOC No: 00589
CAS No: 5742-19-8
Name: Acetic acid, (2,4-dichlorophenoxy)-, compd. with 2,2′-iminobis[ethanol] (1:1)
Synonym: Ethanol, 2,2′-iminobis-, (2,4-dichlorophenoxy)acetate (salt)
Mol Form: $C_{12}H_{17}Cl_2NO_5$
Mr: 326.16
IR: μCOB 1189 3330 2940 2700 1610 1490 1430 1330 1280 1270 1100 1060 1030 940 860 790 760
m/e: 325.05

HODOC No: 00590
CAS No: 1928-38-7
Name: Acetic acid, (2,4-dichlorophenoxy)-, methyl ester
Synonym: 2,4-Dichlorophenoxyacetic acid methyl ester
Mol Form: $C_9H_8Cl_2O_3$
Mr: 235.06
mp: 119
IR: μCOB 983 3030 2940 1750 1590 1490 1450 1390 1320 1210 1110 1090 870 840 800 760 720
m/e: 233.99

HODOC No: 00591
CAS No: 94-11-1
Beilstein No: 6^4, 910
Name: Acetic acid, (2,4-dichlorophenoxy)-, 1-methylethyl ester
Synonym: Acetic acid, (2,4-dichlorophenoxy)-, isopropyl ester
Line Form: $(2,4\text{-}Cl_2C_6H_3O)CH_2CO_2CH(CH_3)_2$
Mol Form: $C_{11}H_{12}Cl_2O_3$
Mr: 263.11
bp: $139\text{-}40^1$
IR: μCOB 1148 3030 2940 1750 1720 1470 1390 1370 1300 1210 1150 1100 1090 950 870 840 830 800 720
m/e: 262.02

HODOC No: 00592
CAS No: 3035-67-4
Name: Acetic acid, (2,4-dichlorophenoxy)-, 6-methylheptyl ester
Mol Form: $C_{16}H_{22}Cl_2O_3$
M_r: 333.24
IR: **µCOB 1974** 2940 1750 1720 1590 1470 1390 1370 1300 1190 1110 1090 870 840 810 720
m/e: 332.09

HODOC No: 00593
CAS No: 1928-44-5
Beilstein No: 6^3, 706
Name: Acetic acid, (2,4-dichlorophenoxy)-, octyl ester
Synonym: 2,4-*D* octyl ester
Line Form: $(2,4\text{-}Cl_2C_6H_3O)CH_2CO_2C_8H_{17}$
Mol Form: $C_{16}H_{22}Cl_2O_3$
M_r: 333.24
bp: $173\text{-}4^1$
IR: **µCOB 1975** 2940 1750 1720 1590 1470 1390 1300 1190 1100 1090 870 840 800 720
m/e: 332.09

HODOC No: 00594
CAS No: 588-22-7
Beilstein No: 6^4, 954
Name: Acetic acid, (3,4-dichlorophenoxy)-
Synonym: 3,4-*D*
Line Form: $(3,4\text{-}Cl_2C_6H_3O)CH_2CO_2H$
Mol Form: $C_8H_6Cl_2O_3$
M_r: 221.03
mp: 141-2
***s*:** ace 3
IR: **µSADP 13144** 2940 2560 1700 1590 1560 1470 1430 1390 1370 1320 1270 1220 1150 1120 1090 1020 910 890 860 790 730 680
UV: **SAD 450** 291 283 230 223 MeOH
H NMR: **SAD 4276** 4.8 7.1 7.3 7.4 ace
MS: **WILEY 1556** 32(100) 118(74) 123(70) 117(57) 103(48) 91(48) 125(46) 220(36)
m/e: 219.97

HODOC No: 00595
CAS No: 6065-82-3
Beilstein No: 3^4, 1494
Name: Acetic acid, diethoxy-, ethyl ester
Synonym: Glyoxylic acid, ethyl ester, 2-(diethyl acetal)
Line Form: $(C_2H_5O)_2CHCO_2C_2H_5$
Mol Form: $C_8H_{16}O_4$
M_r: 176.20
bp: 199
***d*:** 0.985
***n*:** 1.4100^{20}
IR: **ALD 2:346C**
MS: **ALD 3:52A**
m/e: 176.10

HODOC No: 00596
CAS No: 1067-74-9
Beilstein No: 4^2, 976
Name: Acetic acid, (diethoxyphosphinyl)-, methyl ester
Synonym: Acetic acid, phosphono-, P,P-diethyl methyl ester
Line Form: $(CH_3CH_2O)_2P(=O)CH_2CO_2CH_3$
Mol Form: $C_7H_{15}O_5P$
M_r: 210.16
bp: $131.5\text{-}132^9$
***s*:** H_2O 3; os 3
IR: **COB 7407** 2990 1745 1445 1400 1280 1165 1125 1055 1030 970 895 845
m/e: 210.07

HODOC No: 00597
CAS No: 50283-86-8
Name: Acetic acid, [4-[(diethylamino)sulfonyl]phenoxy]-
Synonym: Acetic acid, [*p*-(diethylsulfamoyl)phenoxy]-
Mol Form: $C_{12}H_{17}NO_5S$
M_r: 287.32
mp: 149
***s*:** DMSO 3
IR: **SADG 19390** 2980 1740 1715 1595 1580 1490 1465 1430 1325 1300 1295 1245 1195 1150 1110 1085 1070 1010 925 840 725 685 650 565
UV: **SAD 17286** 279(707) 240(15700) MeOH
H NMR: **SAD 9588** 1.1 3.2 4.8 6.2 7.1 7.7 $DMSOd_6$
m/e: 287.08

HODOC No: 00598
CAS No: 93568-54-8
Name: Acetic acid, [[4-(diethylsulfamoyl)-*o*-tolyl]oxy]-
Synonym: Benzenesulfonic acid, amide,4(carboxymethoxy)-*N*,*N*-diethyl-3-methyl
Mol Form: $C_{13}H_{19}NO_5S$
Mr: 301.34
mp: 99
***s*:** chl 3
IR: SADG 20613 3180 2980 2960 1770 1750 1715 1595 1495 1435 1315 1270 1230 1195 1180 1130 1095 1070 1015 940 810 720 705 600 575 520
H NMR: SAD 9603 1.1 2.3 3.2 4.7 6.8 7.6 $CDCl_3$
m/e: 301.10

HODOC No: 00599
CAS No: 381-73-7
Beilstein No: 2^4, 455
Name: Acetic acid, difluoro-
Synonym: Difluoroacetic acid
Line Form: $F_2CH\text{-}CO_2H$
Mol Form: $C_2H_2F_2O_2$
Mr: 96.03
bp: 132-4, $67\text{-}70^{20}$
mp: -1
***d*:** 1.526
***n*:** 1.3470^{20}
IR: ALD 2:267H
MS: ALD 2:164C
m/e: 96.00

HODOC No: 00600
CAS No: 454-31-9
Beilstein No: 2^4, 455
Name: Acetic acid, difluoro-, ethyl ester
Synonym: Ethyl difluoroacetate
Line Form: $F_2CHCO_2C_2H_5$
Mol Form: $C_4H_6F_2O_2$
Mr: 124.08
bp: 99.2^{750}
***d*:** $1.1893^{9.8}$
***s*:** H_2O 1
IR: µCOB 1959 3030 1790 1450 1390 1300 1220 1140 1090 1020 940
H NMR: SAD 14699 1.4 4.3 5.9 CCl_4
MS: WILEY 517 29(100) 51(44) 27(22) 81(21) 45(17) 28(14) 31(5) 124(0)
m/e: 124.03

HODOC No: 00601
CAS No: 433-53-4
Name: Acetic acid, difluoro-, methyl ester
Synonym: Methyl difluoroacetate
Line Form: $F_2CHCO_2CH_3$
Mol Form: $C_3H_4F_2O_2$
Mr: 110.06
***s*:** chl 2
IR: µSADP 56 2940 1785 1450 1410 1315 1250 1135 1085 1010 980 805
H NMR: SAD 11478 3.9 5.9 $CDCl_3$
m/e: 110.02

HODOC No: 00602
CAS No: 23939-20-0
Name: Acetic acid, [(1,6-dihydro-6-oxo-4-tricyclo[3.3.1.13,7]dec-1-yl-2-pyrimidinyl)thio]-
Synonym: Acetic acid, [[4-(1-adamantyl)-6-hydroxy-2-pyrimidinyl]thio]-
Structure:

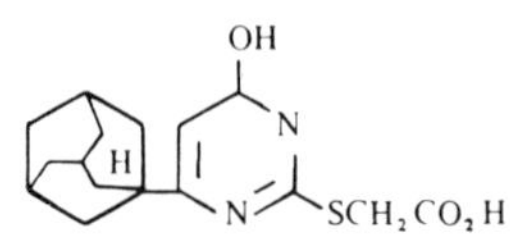

Mol Form: $C_{16}H_{20}N_2O_3S$
Mr: 320.39
mp: 222-4
***s*:** DMSO 2
IR: SADG 19751 3100 2900 2600 1740 1640 1580 1450 1390 1250 1190 1000 860 560
UV: SAD 17962 283(7530) 235(7050) MeOH
H NMR: SAD 10985 1.8 2.0 3.9 5.8 $DMSOd_6$
m/e: 320.12

HODOC No: 00603
CAS No: 598-89-0
Beilstein No: 2^4, 537
Name: Acetic acid, diiodo-
Synonym: Diiodoacetic acid
Line Form: I_2CHCO_2H
Mol Form: $C_2H_2I_2O_2$
Mr: 311.84
mp: 110
Color: lf,ye cr or wh nd(bz)
***s*:** Bz 4; Et_2O 4; EtOH 4
m/e: 311.81

HODOC No: 00604
CAS No: 1421-89-2
Beilstein No: 4^4, 1431
Name: Acetic acid, 2-(dimethylamino)ethyl ester
Synonym: Dimethylaminoethanol acetate
Line Form: CH_3CO_2-$CH_2CH_2N(CH_3)_2$
Mol Form: $C_6H_{13}NO_2$
M_r: 131.17
bp: 153-5
m/e: 131.09

HODOC No: 00605
CAS No: 39910-60-6
Name: Acetic acid, [[[(1,1-dimethylethoxy)carbonyl]amino]oxy]-, pentachlorophenyl ester
Synonym: Carbamic acid, (carboxy(pentachlorophenyl ester)methoxy),*tert*-butyl ester*
Line Form: $C_6Cl_5O_2CCH_2ONHCO_2C(CH_3)_3$
Mol Form: $C_{13}H_{12}Cl_5NO_5$
M_r: 439.49
mp: 172-3
UV: LANG 2312 326(537) 298(537) 212(66069) EtOH
m/e: 436.92

HODOC No: 00606
CAS No: 540-88-5
Beilstein No: 2^4, 151
Name: Acetic acid, 1,1-dimethylethyl ester
Synonym: Acetic acid, *tert*-butyl ester
Line Form: CH_3CO_2-$C(CH_3)_3$
Mol Form: $C_6H_{12}O_2$
M_r: 116.15
bp: 97-8
***d*:** $0.8665^{20/4}$
***n*:** 1.3855^{20}
***s*:** al 3; eth 3; chl 3; HOAc 3
IR: µSADP 18668 2940 1720 1450 1370 1250 1160 1020 940 920 840 760
UV: OES 7-94 213(55) MeOH
H NMR: VAR 141 1.5 2.0 $CDCl_3$
C NMR: JJ 194 FT 22.3 28.1 79.9 170.2 $CDCl_3$
MS: WILEY 432 43(100) 57(50) 41(35) 59(30) 56(26) 29(22) 27(12) 116(0)
m/e: 116.08

HODOC No: 00607
CAS No: 1798-04-5
Name: Acetic acid, [4-(1,1-dimethylethyl)phenoxy]-
Synonym: Acetic acid, (*p-tert*-butylphenoxy)-
Mol Form: $C_{12}H_{16}O_3$
M_r: 208.25
IR: µCOB 1075 3030 2940 2560 1720 1610 1520 1450 1370 1300 1220 1190 1090 930 830 800
UV: SAD 1817 280 273 222 MeOH
H NMR: SAD 6844 1.3 4.6 6.9 7.3 9.6 $CDCl_3$
m/e: 208.11

HODOC No: 00608
CAS No: 3344-19-2
Name: Acetic acid, [4-(1,1-dimethylethyl)phenoxy]-, ethyl ester
Synonym: Acetic acid, (*p-tert*-butylphenoxy)-, ethyl ester
Mol Form: $C_{14}H_{20}O_3$
M_r: 236.30
bp: $110\text{-}35^{1.5}$
IR: µCOB 1074 3030 2940 1750 1610 1520 1470 1390 1370 1240 1190 1060 960 830
MS: WILEY 1664 87(100) 43(27) 101(18) 135(9) 88(4) 236(3) 41(3) 91(3)
m/e: 236.14

HODOC No: 00609
CAS No: 25141-25-7
Beilstein No: 6^2, 244
Name: Acetic acid, (2,4-dinitrophenoxy)-
Synonym: 2,4-Dinitrophenoxyacetic acid
Line Form: $[2,4\text{-}(NO_2)_2C_6H_3O]CH_2CO_2H$
Structure:

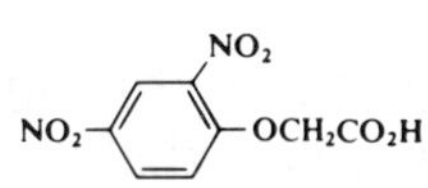

Mol Form: $C_8H_6N_2O_7$
M_r: 242.14
mp: 147-8
Color: pa ye pr (w)
***s*:** EtOH 4
m/e: 242.02

HODOC No: 00610
CAS No: 85196-33-4
Beilstein No: 6^4, 1381
Name: Acetic acid, (2,4-dinitrophenoxy)-, ethyl ester
Synonym: Phenoxyacetic acid, 2,4-dinitro, ethyl ester
Line Form: $[2,4\text{-}(NO_2)_2C_6H_3O]CH_2CO_2C_2H_5$
Mol Form: $C_{10}H_{10}N_2O_7$
M_r: 270.19
mp: 77-8
IR: µ **SADP 3694** 2940 1750 1610 1560 1520 1450 1330 1220 1080 910 830 750 680
UV: **SAD 1128** 287(11100) 252(8280) 214(14900) diox
H NMR: **SAD 10653** 1.3 4.3 4.9 7.2 8.4 8.7 $CDCl_3$
m/e: 270.05

HODOC No: 00611
CAS No: 42288-38-0
Beilstein No: 6, 259
Name: Acetic acid, (3,5-dinitrophenoxy)-
Synonym: Phenoxyacetic acid, 3,5-dinitro
Line Form: $[3,5\text{-}(NO_2)_2C_6H_3O]CH_2CO_2H$
Mol Form: $C_8H_6N_2O_7$
M_r: 242.14
mp: 207
Color: pa br cr pw
***s*:** EtOH 4
m/e: 242.02

HODOC No: 00612
CAS No: 2719-07-5
Name: Acetic acid, 2-(2,4-dinitrophenyl)hydrazide
Synonym: Acetic acid 2,4-dinitrophenylhydrazide
Mol Form: $C_8H_8N_4O_5$
M_r: 240.17
IR: **µCOB 3178** 3330 3130 2940 1670 1610 1590 1520 1490 1470 1430 1370 1330 1300 1150 1010 920 830 750 730
m/e: 240.05

HODOC No: 00613
CAS No: 6854-84-8
Name: Acetic acid, 2,2′-dithiobis-, dihydrazide
Synonym: Acetic acid, dithiodi-, dihydrazide
Line Form: $H_2NNHCOCH_2SSCH_2CONHNH_2$
Mol Form: $C_4H_{10}N_4O_2S_2$
M_r: 210.27
mp: 104-6
IR: **µCOB 4521** 3330 3130 2940 1640 1540 1470 1390 1330 1180 1020 1010 960 730 690
m/e: 210.02

HODOC No: 00614
CAS No: 112-66-3
Beilstein No: 2^4, 170
Name: Acetic acid, dodecyl ester
Synonym: Dodecanol acetate
Line Form: $CH_3CO_2C_{12}H_{25}$
Mol Form: $C_{14}H_{28}O_2$
M_r: 228.36
bp: 260-70
***d*:** $1.4329\text{-}1.4339^{20/D}$
***n*:** 1.4439^{20}
IR: **SADG 15134** 2920 2850 1740 1460 1355 1230 1030 715 595
Raman: **SAD 868** 2940 2900 2880 2860 2720 1740 1440 1370 1300 1130 1080 1070 1030 960 940 890 850 640 610 410 350 220
H NMR: **SAD 6394** 0.9 1.3 1.9 4.0 CCl_4
MS: **WILEY 1628** 43(100) 55(50) 69(44) 83(42) 56(38) 70(38) 61(36) 228(1)
m/e: 228.21

HODOC No: 00615
CAS No: 627-03-2
Beilstein No: 2^4, 574
Name: Acetic acid, ethoxy-
Synonym: Ethoxyacetic acid
Line Form: $C_2H_5OCH_2CO_2H$
Mol Form: $C_4H_8O_3$
M_r: 104.10
bp: 206-7, 111^{25}
***d*:** $1.1021^{20/4}$
***n*:** 1.4194^{20}
***s*:** H_2O 4; al 4; eth 4; chl 3
IR: **µSADP 7392** 3450 2940 2560 1720 1430 1220 1110 1030 1010 880 850 670
H NMR: **VAR 417** 1.3 3.7 4.1 11.0 $CDCl_3$
MS: **WILEY 296** 31(100) 29(62) 18(61) 59(52) 60(21) 45(21) 27(21) 104(0)
m/e: 104.05

HODOC No: 00616
CAS No: 817-95-8
Beilstein No: 3^4, 581
Name: Acetic acid, ethoxy-, ethyl ester
Synonym: Ethoxyacetic acid ethyl ester
Line Form: $C_2H_5OCH_2CO_2C_2H_5$
Mol Form: $C_6H_{12}O_3$
Mr: 132.15
bp: 158, 52^{12}
***d*:** $0.9702^{20/4}$
***n*:** 1.4039^{20}
***s*:** al 3; eth 3; ace 3
IR: SADG 10593 3480 3370 1650 1530 1430 1410 1360 1340 1270 1220 1180 1050 900 880 790 770 730 580 550
Raman: SAD 92 2940 2870 2820 1740 1480 1460 1450 1360 1280 1240 1200 1130 1070 1050 1000 960 930 910 850 810 630 610 510 440 420 300
UV: OES 1-112 214(83) EtOH
H NMR: SAD 8149 1.2 1.3 3.6 4.1 4.3 CCl_4
MS: WILEY 625 31(100) 59(84) 29(76) 88(41) 43(32) 45(27) 132(1)
m/e: 132.08

HODOC No: 00617
CAS No: 17640-26-5
Beilstein No: 3^4, 578
Name: Acetic acid, ethoxy-, methyl ester
Synonym: Ethoxyacetic acid methyl ester
Line Form: $C_2H_5OCH_2CO_2CH_3$
Mol Form: $C_5H_{10}O_3$
Mr: 118.13
bp: $147\text{-}8^{784}$
***d*:** 1.0112^{15}
***s*:** Ac 4; Et_2O 4; EtOH 4
m/e: 118.06

HODOC No: 00618
CAS No: 579-94-2
Beilstein No: 6^2, 47
Name: Acetic acid, ethoxy-, 5-methyl-2-(1-methylethyl)cyclohexyl ester, (1α,2β,5α)-
Synonym: Acetic acid, ethoxy-, *p*-menth-3-yl ester
Structure:

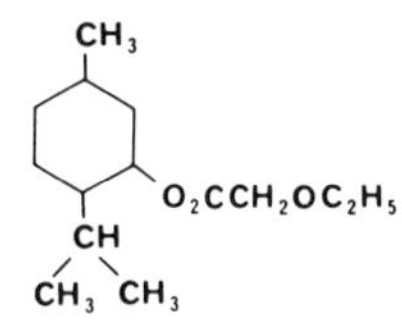

Mol Form: $C_{14}H_{26}O_3$
Mr: 242.34
bp: 155^{20}
***d*:** $0.9545^{20/4}$
α: $[\alpha]^{20/D}$ -66.3
***s*:** Et_2O 4; EtOH 4; Chl 4
m/e: 242.19

HODOC No: 00619
CAS No: 59376-54-4
Name: Acetic acid, (ethoxysulfonyl)-, ethyl ester
Synonym: Acetic acid, sulfo, diethyl ester
Line Form: $(C_2H_5O_3S)CH_2CO_2C_2H_5$
Mol Form: $C_6H_{12}O_5S$
Mr: 196.21
bp: $116^{0.5}$
***d*:** 1.232^{16}
***n*:** 1.436^{21}
IR: COB 2206
m/e: 196.04

HODOC No: 00620
CAS No: 10031-87-5
Beilstein No: 2^4, 161
Name: Acetic acid, 2-ethylbutyl ester
Synonym: 2-Ethylbutyl acetate
Line Form: CH_3CO_2-$CH_2CH(C_2H_5)_2$
Mol Form: $C_8H_{16}O_2$
M_r: 144.21
bp: 162-3, 63^{20}
mp: 100
***d*:** $0.8790^{20/4}$
***n*:** 1.4109^{20}
***s*:** H_2O 1; al 3; eth 3; ctc 3
IR: SADG 15002 2960 2880 1750 1460 1380 1360 1230 1030 770
Raman: SAD 458 2940 2910 2880 2740 1740 1450 1390 1360 1310 1280 1240 1160 1130 1060 1040 960 950 920 890 860 830 780 750 640 630 610 550_550 450 400 370 330 290 200
H NMR: SAD 7305 0.9 1.1-1.7 2.0 3.9 CCl_4
m/e: 144.12

HODOC No: 00621
CAS No: 103-09-3
Beilstein No: 2^4, 166
Name: Acetic acid, 2-ethylhexyl ester
Synonym: 2-Ethyl-1-hexanol acetate
Line Form: CH_3CO_2-$CH_2CH(C_2H_5)C_4H_9$
Mol Form: $C_{10}H_{20}O_2$
M_r: 172.26
bp: 199, 95^{20}
mp: -93
***d*:** $0.8734^{20/20}$
***n*:** 1.4204^{20}
***s*:** H_2O 1; al 3; eth 3
MS: WILEY 1151 43(100) 70(60) 57(38) 55(33) 41(29) 56(20) 29(16) 172(0)
m/e: 172.15

HODOC No: 00622
CAS No: 627-04-3
Beilstein No: 3^4, 603
Name: Acetic acid, (ethylthio)-
Synonym: *S*-(Carboxymethyl)ethyl mercaptan
Line Form: $C_2H_5SCH_2CO_2H$
Mol Form: $C_4H_8O_2S$
M_r: 120.16
bp: 164^{83}, 109^5
mp: -8.5
***d*:** $1.1497^{20/4}$
***s*:** H_2O 4; al 4; eth 4
IR: DMS 8047 3500 2960 2620 2580 1700 1430 1410 1380 1300 1280 1230 1200 1140 1040 930 800 700 680 660
H NMR: SAD 17160 1.3 2.7 3.2 12.0 CCl_4
MS: NBS 189
m/e: 120.02

HODOC No: 00623
CAS No: 144-49-0
Beilstein No: 2^4, 446
Name: Acetic acid, fluoro-
Synonym: Cymonic acid
Line Form: FCH_2CO_2H
Mol Form: $C_2H_3FO_2$
M_r: 78.04
bp: 165
mp: 35.2
***d*:** 1.3693^{36}
Color: nd
***s*:** H_2O 3; al 3
IR: µSADP 17442 3330 2940 2500 1720 1610 1430 1330 1300 1250 1060 1050 1030 940 890 690
MS: WILEY 98 18(100) 45(70) 33(51) 31(28) 61(27) 17(24) 29(19) 78(3)
m/e: 78.01

HODOC No: 00624
CAS No: 406-74-6
Beilstein No: 2^4, 450
Name: Acetic acid, fluoro-, 1,1-dimethylethyl ester
Synonym: Acetic acid, fluoro-, *tert*-butyl ester
Line Form: $FCH_2CO_2C(CH_3)_3$
Mol Form: $C_6H_{11}FO_2$
M_r: 134.14
bp: 130-1
***d*:** 0.9904^{20}
***n*:** 1.386^{20}
m/e: 134.07

HODOC No: 00625
CAS No: 459-72-3
Beilstein No: 2^4, 447
Name: Acetic acid, fluoro-, ethyl ester
Synonym: Ethyl fluoroacetate
Line Form: $FCH_2CO_2C_2H_5$
Mol Form: $C_4H_7FO_2$
M_r: 106.09
bp: 119.3^{753}
***d*:** 1.098
***n*:** 1.3755^{20}
***s*:** H_2O 4
IR: ALD 2:340H
MS: ALD 3:43D
m/e: 106.04

HODOC No: 00626
CAS No: 453-18-9
Beilstein No: 2^4, 447
Name: Acetic acid, fluoro-, methyl ester
Synonym: Methyl fluoroacetate
Line Form: $FCH_2CO_2CH_3$
Mol Form: $C_3H_5FO_2$
M_r: 92.07
bp: 104-5
***d*:** 1.1613^{15}
m/e: 92.03

HODOC No: 00627
CAS No: 405-79-8
Beilstein No: 6^4, 776
Name: Acetic acid, (4-fluorophenoxy)-
Synonym: Acetic acid, (*p*-fluorophenoxy)-
Line Form: $4\text{-F-}C_6H_4\text{-}OCH_2CO_2H$
Mol Form: $C_8H_7FO_3$
M_r: 170.13
mp: 104-5
IR: ALD 3:945C
MS: ALD 6:125B
m/e: 170.04

HODOC No: 00628
CAS No: 24612-99-5
Name: Acetic acid, (2-formyl-4-methoxyphenoxy)-, methyl ester
Mol Form: $C_{11}H_{12}O_5$
M_r: 224.20
mp: 73-4
***s*:** chl 3
IR: SADG 19073 2980 2910 1775 1685 1620 1595 1505 1445 1400 1300 1220 1165 1080 1035 980 935 870 805 745 700 635 590 440
UV: SAD 17329 291(1140) 283(1030) MeOH
H NMR: SAD 9692 3.8 4.7 6.9 7.1 7.3 10.6 $CDCl_3$
m/e: 224.07

HODOC No: 00629
CAS No: 88521-64-6
Name: Acetic acid, (2-formyl-4-methylphenoxy)-
Mol Form: $C_{10}H_{10}O_4$
M_r: 194.18
mp: 150-1
***s*:** DMSO 3
IR: SADG 19071 3060 2920 2750 2600 1740 1700 1690 1580 1470 1420 1390 1250 1200 1050 910 810 770 670
H NMR: SAD 9767 2.3 4.9 5.4 7.1 7.5 7.6 10.5 $DMSOd_6$
m/e: 194.06

HODOC No: 00630
CAS No: 6965-69-1
Name: Acetic acid, (2-formyl-4-nitrophenoxy)-
Mol Form: $C_9H_7NO_6$
M_r: 225.15
mp: 190-2
***s*:** DMSO 3
IR: SADG 21419 3100 2880 2720 2600 1750 1690 1610 1590 1520 1480 1430 1330 1280 1250 1240 1170 1080 940 840 750 700 570
UV: SAD 7805 300(10100) 229(9410) MeOH 308(10800) 231(9370) MeOH:KOH 300(9910) 228(8570) MeOH:Cl
H NMR: SAD 9766 5.1 5.2 7.4 8.4 8.5 10.4 $DMSOd_6$
m/e: 225.03

HODOC No: 00631
CAS No: 40359-34-0
Beilstein No: 8, 45
Name: Acetic acid, (2-formylphenoxy)-, methyl ester
Synonym: 2-Carbomethoxymethoxybenzaldehyde
Mol Form: $C_{10}H_{10}O_4$
Mr: 194.18
mp: 55-6
***s*:** chl 3
IR: SADG 19075 3080 2950 2860 2750 1760 1690 1600 1480 1450 1390 1280 1200 1150 1100 1060 820 750
H NMR: SAD 9690 3.8 4.8 6.9 7.2 7.5 7.8 13.8 $CDCl_3$
m/e: 194.06

HODOC No: 00632
CAS No: 22042-71-3
Beilstein No: 8^3, 227
Name: Acetic acid, (4-formylphenoxy)-
Synonym: Acetic acid, (*p*-formylphenoxy)-
Line Form: 4-HCO-C_6H_4-OCH_2CO_2H
Mol Form: $C_9H_8O_4$
Mr: 180.15
mp: 197-9
m/e: 180.04

HODOC No: 00633
CAS No: 51264-69-8
Name: Acetic acid, (4-formylphenoxy)-, ethyl ester
Synonym: Acetic acid, (*p*-formylphenoxy)-, ethyl ester
Mol Form: $C_{11}H_{12}O_4$
Mr: 208.20
MS: WILEY 1479 208(100) 135(73) 105(40) 77(28) 29(27) 207(21) 28(20) 79(16)
m/e: 208.07

HODOC No: 00634
CAS No: 112-06-1
Beilstein No: 2^4, 162
Name: Acetic acid, heptyl ester
Synonym: Heptyl acetate
Line Form: $CH_3CO_2(CH_2)_6CH_3$
Mol Form: $C_9H_{18}O_2$
Mr: 158.23
bp: 192-4, 96^{28}
mp: -50.2
***d*:** $0.8750^{15/4}$
***n*:** 1.4150^{20}
***s*:** H_2O 1; al 3; eth 3; ctc 3
IR: SADG 15518 2920 2850 1740 1460 1380 1360 1230 1030 720
H NMR: SAD 6786 0.9 1.4 1.6 2.0 4.0 CCl_4
MS: WILEY 990 43(100) 56(38) 70(29) 41(25) 55(24) 61(20) 42(17) 158(0)
m/e: 158.13

HODOC No: 00635
CAS No: 142-92-7
Beilstein No: 2^4, 159
Name: Acetic acid, hexyl ester
Synonym: Hexyl acetate
Line Form: $CH_3CO_2(CH_2)_5CH_3$
Mol Form: $C_8H_{16}O_2$
Mr: 144.21
bp: 171.5, 61.5^{12}
mp: -80.9
***d*:** $0.8779^{15/4}$
***n*:** 1.4092^{20}
***s*:** Et_2O 4; EtOH 4
IR: COB 4799
H NMR: SAD 217
MS: WILEY 800
m/e: 144.12

HODOC No: 00636
CAS No: 1068-57-1
Beilstein No: 2^4, 435
Name: Acetic acid, hydrazide
Synonym: Acethydrazide
Line Form: $CH_3CONHNH_2$
Mol Form: $C_2H_6N_2O$
Mr: 74.08
bp: 129^{18}
mp: 67
***s*:** H_2O 3; al 3; eth 2
IR: µSADP 11891 3450 3330 2940 1670 1610 1470 1370 1270 990 880
H NMR: SAD 9497 3.9 D_2O
m/e: 74.05

HODOC No: 00637
CAS No: 79-14-1
Beilstein No: 3^4, 571
Name: Acetic acid, hydroxy-
Synonym: Glycolic acid
Line Form: $HOCH_2CO_2H$
Mol Form: $C_2H_4O_3$
Mr: 76.05
bp: d
mp: 80
Color: rh nd (w) lf (eth)
***s*:** H_2O 3; al 3; eth 3
IR: COB 6254 3270 2930 2850 2640 2560 1720 1430 1260 1090 930 890 680
H NMR: SAD 6411 4.3 D_2O
MS: NBS 24
m/e: 76.02

HODOC No: 00638
CAS No: 626-25-5
Beilstein No: 3^1, 92
Name: Acetic acid, hydroxy-, anhydride
Synonym: Glycolic anhydride
Line Form: $(HOCH_2CO)_2O$
Mol Form: $C_4H_6O_5$
Mr: 134.08
bp: d
mp: 128-30
Color: pw
m/e: 134.02

HODOC No: 00639
CAS No: 623-50-7
Beilstein No: 3^4, 580
Name: Acetic acid, hydroxy-, ethyl ester
Synonym: Glycolic acid, ethyl ester
Line Form: $HOCH_2CO_2C_2H_5$
Mol Form: $C_4H_8O_3$
Mr: 104.10
bp: 160, 69^{25}
***d*:** $1.0826^{23/4}$
***n*:** 1.4180^{20}
***s*:** Et_2O 4; EtOH 4
IR: SADG 27960 3410 2980 2920 1740 1440 1380 1270 1220 1090 1030 990 930 860 810 700 570 360
Raman: SAD 85 3500 2980 2950 2880 1740 1450 1400 1360 1300 1280 1230 1170 1110 1030 920 870 800 710 580 480 470 370
m/e: 104.05

HODOC No: 00640
CAS No: 96-35-5
Beilstein No: 3^4, 578
Name: Acetic acid, hydroxy-, methyl ester
Synonym: Glycolic acid, methyl ester
Line Form: $HOCH_2CO_2CH_3$
Mol Form: $C_3H_6O_3$
Mr: 90.08
bp: 52^{17}
***d*:** $1.677^{18/4}$
***s*:** H_2O 3; al 5; eth 5
IR: µSADP 695 3330 2940 1720 1450 1240 1100 980 890 850 700
H NMR: SAD 19832 3.7 3.8 4.1 CCl_4
MS: NBS 53
m/e: 90.03

HODOC No: 00641
CAS No: 72928-38-2
Name: Acetic acid, hydroxy-, 2-phenylethyl ester
Synonym: Acetic acid, hydroxy,phenethyl ester
Line Form: $HOCH_2CO_2CH_2CH_2C_6H_5$
Mol Form: $C_{10}H_{12}O_3$
Mr: 180.19
IR: µCOB 4095 3450 3030 2940 1750 1520 1470 1280 1240 1090 1050 860 750 700
m/e: 180.08

HODOC No: 00642
CAS No: 90357-58-7
Beilstein No: 3^4, 588
Name: Acetic acid, hydroxy-, propyl ester
Synonym: Hydroxyacetic acid, propyl ester
Line Form: $HOCH_2CO_2C_3H_7$
Mol Form: $C_5H_{10}O_3$
M_r: 118.13
bp: 170-1
***d*:** $1.0631^{18/4}$
***n*:** 1.4231^{18}
IR: μCOB 4096 3450 2940 1750 1470 1390 1220 1100 970
m/e: 118.06

HODOC No: 00643
CAS No: 36691-76-6
Name: Acetic acid, [4-[[(2-hydroxyethyl)amino]sulfonyl]phenoxy]-
Synonym: Benzenesulfonic acid, amide,4(carboxymethoxy)-*N*(2-hydroxyethyl)
Mol Form: $C_{10}H_{13}NO_6S$
M_r: 275.27
mp: 134
***s*:** DMSO 3
IR: SADG 19392 3450 3280 2900 2650 1700 1600 1500 1400 1310 1250 1150 1080 1060 960 860 840 650 590 570
UV: SAD 19715 279(699) 239(17000) MeOH
H NMR: SAD 9590 2.9 3.5 4.8 7.1 7.2 7.4 7.8 $DMSOd_6$
m/e: 275.05

HODOC No: 00644
CAS No: 97388-49-3
Beilstein No: 6, 893
Name: Acetic acid, [2-(hydroxymethyl)phenoxy]-
Synonym: Phenoxyacetic acid, 2(hydroxymethyl)
Line Form: $2\text{-}HOCH_2\text{-}C_6H_4OCH_2CO_2H$
Mol Form: $C_9H_{10}O_4$
M_r: 182.17
mp: 119-21
***s*:** DMSO 3
IR: SADG 19076 3420 2950 2600 1790 1740 1600 1500 1460 1390 1370 1240 1220 1120 1080 1030 920 750 690 620
UV: SAD 17337 303(10300) 227(9090) MeOH
H NMR: SAD 9863 4.6 4.7 5.9 6.9 7.0 7.1 7.4 $DMSOd_6$
m/e: 182.06

HODOC No: 00645
CAS No: 24589-96-6
Name: Acetic acid, [(α-hydroxy-4-nitro-*o*-tolyl)oxy]-
Synonym: Acetic acid, (2(hydroxymethyl)-4-nitrophenoxy)
Mol Form: $C_9H_9NO_6$
M_r: 227.16
mp: 141-2
***s*:** DMSO 3
IR: SADG 19068 3350 2900 2600 1750 1730 1600 1540 1490 1430 1350 1260 1090 1060 1030 900 820 750 690
UV: SAD 17330 302(811) 251(3590) MeOH
H NMR: SAD 9768 4.6 4.9 7.1 8.2 8.3 $DMSOd_6$
m/e: 227.04

HODOC No: 00646
CAS No: 1878-83-7
Beilstein No: 6, 817
Name: Acetic acid, (3-hydroxyphenoxy)-
Synonym: Acetic acid, (*m*-hydroxyphenoxy)-
Mol Form: $C_8H_8O_4$
M_r: 168.14
mp: 158-9
***s*:** H_2O 3; al 3; ace 2; tol 3
IR: SADG 2468 3470 3110 3020 2900 2560 1720 1610 1590 1510 1450 1430 1330 1270 1250 1180 1150 1090 970 900 810 790 680
UV: SAD 10026 376(484) 360(652) 280(2020) 273(2270) MeOH
H NMR: SAD 1603 4.6 6.2-6.7 7.1 $aceD_6$
m/e: 168.04

HODOC No: 00647
CAS No: 1878-84-8
Beilstein No: 6^4, 5744
Name: Acetic acid, (4-hydroxyphenoxy)-
Synonym: Acetic acid, (*p*-hydroxyphenoxy)-
Line Form: $4\text{-}HOC_6H_4OCH_2CO_2H$
Mol Form: $C_8H_8O_4$
M_r: 168.14
mp: 154(hyd)
Color: nd (to), Pr (+1/2w)
m/e: 168.04

HODOC No: 00648
CAS No: 5941-05-9
Name: Acetic acid, [(2-hydroxyphenyl)methylene]hydrazide
Synonym: Acetic acid, salicylidenehydrazide
Mol Form: $C_9H_{10}N_2O_2$
M_r: 178.18
mp: 203.5-4.5
UV: LANG 2322 322(10000) 289(18197) 280(19055) 225(14454) MeOH 362(9333) 278(9333) 226(16596) 0.1N KOH
m/e: 178.07

HODOC No: 00649
CAS No: 32253-74-0
Name: Acetic acid, iminodi-, monoammonium salt
Synonym: Acetic acid, iminodi,ammonium salt
Line Form: $HN(CH_2CO_2H)(CH_2CO_2NH_4)$
Mol Form: $C_4H_{10}N_2O_4$
M_r: 150.13
IR: µCOB 2304 2940 2630 1610 1470 1390 1280 1060 860 710
m/e: 150.06

HODOC No: 00650
CAS No: 102-15-8
Beilstein No: 12^1, 248
Name: Acetic acid, [[imino(phenylamino)methyl]thio]-
Synonym: Acetic acid, [(phenylamidino)thio]-
Line Form: $C_6H_5N{=}C(NH_2)SCH_2CO_2H$
Mol Form: $C_9H_{10}N_2O_2S$
M_r: 210.24
bp: 175^{760}
mp: 159-61
***s*:** H_2O 3; al 2; eth 2; bz 2
IR: µSADP 9519 3450 3330 3030 1670 1610 1540 1450 1330 1190 1160 760 710 690
m/e: 210.05

HODOC No: 00651
CAS No: 64-69-7
Beilstein No: 2^4, 534
Name: Acetic acid, iodo-
Synonym: Iodoacetic acid
Line Form: ICH_2CO_2H
Mol Form: $C_2H_3IO_2$
M_r: 185.95
bp: DEC
mp: 82-3
***s*:** H'i2O 3; al 3; eth 2; chl 2; peth 3
IR: SADG 21053 3050 2630 1710 1540 1430 1250 1150 1080 890 840
UV: OES 1-6 261(457) H_2O 268(309) pH=1 260(380) 0.05N NaOH 273(457) 224(513) EtOH 271(331) hx
H NMR: SAD 214 3.7 10.8 $CDCl_3$
MS: WILEY 1277 186(100) 59(26) 141(22) 127(19) 142(11) 15(9) 42(6) 43(5)
m/e: 185.92

HODOC No: 00652
CAS No: 623-48-3
Beilstein No: 2^4, 535
Name: Acetic acid, iodo-, ethyl ester
Synonym: Ethyl iodoacetate
Line Form: $ICH_2CO_2C_2H_5$
Mol Form: $C_4H_7IO_2$
M_r: 214.00
bp: 178-80, 73^{16}
***d*:** $1.8173^{13/4}$
***n*:** 1.5079^{13}
Color: oil
***s*:** al 3; eth 3
IR: µSADP 22423 2940 1720 1430 1370 1280 1160 1100 1030
m/e: 213.95

HODOC No: 00653
CAS No: 89775-61-1
Name: Acetic acid, isocyanato-, isobutyl ester
Synonym: Acetic acid, isocyano,2-isobutyl ester
Line Form: $(CH_3)_2CHCH_2O_2CCH_2N{=}C{=}O$
Mol Form: $C_7H_{11}NO_3$
M_r: 157.16
bp: 76^6
IR: µCOB 2173 2940 2270 1750 1470 1390 1220 1000 940 890 720
m/e: 157.07

HODOC No: 00654
CAS No: 68-11-1
Beilstein No: 3^4, 600
Name: Acetic acid, mercapto-
Synonym: Glycolic acid, 2-thio-
Line Form: $HSCH_2CO_2H$
Mol Form: $C_2H_4O_2S$
M_r: 92.11
bp: 120^{20}
mp: -16.5
***d*:** 1.3253^{20}
***n*:** 1.5080^{20}
***s*:** H_2O 5; al 5; eth 5; chl 2
IR: SADG 21703 3050 2650 2560 1720 1360 1280 1140 980 890 750 680
Raman: SAD 8 2930 2570 1680 1400 1290 1240 1170 990 900 810 760 670 580 490 460 300 220
UV: OES 3-4 240(1820) pH=10 244(4677) pH=10.6 245(7413) pH=12.5
H NMR: SAD 10969 2.1 3.3 9.4 $CDCl_3$
MS: WILEY 184 47(100) 45(52) 92(43) 46(40) 74(12) 28(10) 42(9) 29(9)
m/e: 91.99

HODOC No: 00655
CAS No: 689-85-0
Name: Acetic acid, mercapto-, 2-acetylhydrazide
Synonym: Hydrazine, 1-acetyl-2-(mercaptoacetyl)-
Line Form: $HSCH_2CO$-NHNH-$COCH_3$
Mol Form: $C_4H_8N_2O_2S$
M_r: 148.18
mp: 121-2
IR: COB 4529 3220 3050 2930 2570 1590 1490 1360 1210 1120 1000 910 710
m/e: 148.03

HODOC No: 00656
CAS No: 10193-99-4
Name: Acetic acid, mercapto-, 2,2-*bis*[[(mercaptoacetyl)oxy]methyl]-1,3-propanediyl ester
Synonym: Acetic acid, mercapto-, neopentanetetrayl ester
Line Form: $\{HSCH_2CO_2CH_2\}_4C$
Mol Form: $C_{13}H_{20}O_8S_4$
M_r: 432.52
IR: COB 7202 2960 2570 1740 1470 1410 1280 1150 1020
m/e: 432.00

HODOC No: 00657
CAS No: 10047-28-6
Beilstein No: 3^4, 620
Name: Acetic acid, mercapto-, butyl ester
Synonym: Butyl mercaptoacetate
Line Form: $HSCH_2CO_2C_4H_9$
Mol Form: $C_6H_{12}O_2S$
M_r: 148.21
bp: $85\text{-}8^{16}$
***d*:** $1.03^{20/4}$
IR: SADG 23973 2980 2580 1740 1460 1420 1380 1290 1150 1060 1020 950
H NMR: SAD 9734 1.0 1.1-1.9 2.1 3.3 4.2 $CDCl_3$
m/e: 148.06

HODOC No: 00658
CAS No: 91362-36-6
Name: Acetic acid, mercapto-, compd. with 2,2′-iminobis[ethanol] (1:1)
Synonym: Ethanol, 2,2′-iminobis-, mercaptoacetate (salt)
Line Form: $HSCH_2CO_2H$–$(HOCH_2CH_2)_2NH$
Mol Form: $C_6H_{15}NO_4S$
M_r: 197.24
IR: μCOB 3697 3230 2860 2780 2560 1560 1370 1240 1050 950 750 700
m/e: 197.07

HODOC No: 00659
CAS No: 13220-80-9
Name: Acetic acid, mercapto-, compd. with 2,2′,2″-nitrilotris[ethanol] (1:1)
Synonym: Ethanol, 2,2′,2″-nitrilotris-, mercaptoacetate (salt)
Line Form: $HSCH_2CO_2H$–$(HOCH_2CH_2)_3N$
Mol Form: $C_8H_{19}NO_5S$
M_r: 241.29
IR: μCOB 3698 3230 2860 2560 1590 1370 1240 1100 1030 1000 920
m/e: 241.10

HODOC No: 00660
CAS No: 45163-99-3
Name: Acetic acid, mercapto-, decyl ester
Synonym: Decyl mercaptoacetate
Line Form: $HSCH_2CO_2(CH_2)_9CH_3$
Mol Form: $C_{12}H_{24}O_2S$
M$_r$: 232.37
***s*:** chl 3
IR: SADG 23974 2940 2860 1740 1470 1410 1380 1290 1150 1010 980 720
H NMR: SAD 9735 0.9 1.3 2.0 3.2 4.2 $CDCl_3$
m/e: 232.15

HODOC No: 00661
CAS No: 3746-39-2
Beilstein No: 3^3, 432
Name: Acetic acid, mercapto-, dodecyl ester
Synonym: Dodecyl mercaptoacetate
Line Form: $HSCH_2CO_2C_{12}H_{25}$
Mol Form: $C_{14}H_{28}O_2S$
M$_r$: 260.42
mp: 3
***d*:** $0.43^{20/4}$
IR: SADG 23975 2940 2860 1740 1470 1415 1380 1290 1150 1015 720
H NMR: SAD 9736 0.9 1.3 2.0 3.2 4.2 $CDCl_3$
m/e: 260.18

HODOC No: 00662
CAS No: 123-81-9
Name: Acetic acid, mercapto-, 1,2-ethanediyl ester
Synonym: Acetic acid, mercapto-, ethylene ester
Line Form: $HSCH_2CO_2CH_2CH_2O_2CCH_2SH$
Mol Form: $C_6H_{10}O_4S_2$
M$_r$: 210.26
bp: $137\text{-}9^{1\text{-}2}$
IR: SADG 15772 2960 2480 1740 1450 1405 1370 1285 1140 1055 1030 965 875 695 575 490 415
H NMR: SAD 6443 2.1 3.3 4.4 $CDCl_3$
MS: WILEY 1492 45(100) 44(83) 60(75) 18(67) 43(61) 14(52) 28(50) 210(1)
m/e: 210.00

HODOC No: 00663
CAS No: 623-51-8
Beilstein No: 3^4, 617
Name: Acetic acid, mercapto-, ethyl ester
Synonym: Ethyl mercaptoacetate
Line Form: $HSCH_2CO_2C_2H_5$
Mol Form: $C_4H_8O_2S$
M$_r$: 120.16
bp: 156-8, 55^{17}
***d*:** 1.0964^{15}
***n*:** 1.4582^{20}
***s*:** al 3; eth 3; ctc 2
IR: μSADP 22483 2940 2560 1720 1470 1430 1370 1300 1150 1030 860
H NMR: SAD 9804 1.3 2.0 3.3 4.2 $CDCl_3$
MS: ALDN 61 47(100) 29(97) 120(25) 74(24) 46(21) 27(18) 45(13) 75(11)
m/e: 120.02

HODOC No: 00664
CAS No: 7659-86-1
Name: Acetic acid, mercapto-, 2-ethylhexyl ester
Synonym: 2-Ethylhexyl mercaptoacetate
Line Form: $HSCH_2CO_2[CH_2CH(C_2H_5)CH_2CH_2CH_2CH_3]$
Mol Form: $C_{10}H_{20}O_2S$
M$_r$: 204.32
bp: 133.5
***d*:** $0.97^{20/4}$
H NMR: SAD 9806 0.9 1.1-1.7 2.0 3.2 4.1 $CDCl_3$
m/e: 204.12

HODOC No: 00665
CAS No: 22811-02-5
Beilstein No: 3^4, 622
Name: Acetic acid, mercapto-, hexadecyl ester
Synonym: Cetyl thioglycolate
Line Form: $HSCH_2CO_2(CH_2)_{15}CH_3$
Mol Form: $C_{18}H_{36}O_2S$
M$_r$: 316.52
mp: 20-3
***s*:** chl 3
IR: SADG 23976 2930 2860 1740 1470 1380 1290 1150 1010 720
H NMR: SAD 9737 0.9 1.3 2.0 3.2 4.1 $CDCl_3$
m/e: 316.24

HODOC No: 00666
CAS No: 760-30-5
Name: Acetic acid, mercapto-, hydrazide
Synonym: Acetic acid, hydrazide,mercapto
Line Form: $HSCH_2CONHNH_2$
Mol Form: $C_2H_6N_2OS$
M_r: 106.14
IR: COB 7372 3250 1655 1520 1425 1330 1235 1185 1105 1005 910
m/e: 106.02

HODOC No: 00667
CAS No: 62-48-6
Name: Acetic acid, mercapto-, 2-(mercaptoacetyl)hydrazide
Synonym: Hydrazine, 1,2-*bis*(mercaptoacetyl)-
Line Form: $HSCH_2CONHNHCOCH_2SH$
Mol Form: $C_4H_8N_2O_2S_2$
M_r: 180.24
mp: 155-7
IR: µCOB 4588 3130 3030 2560 1590 1490 1410 1220 1150 940 800 750 700
m/e: 180.00

HODOC No: 00668
CAS No: 10193-98-3
Name: Acetic acid, mercapto-, 2-[[(mercaptoacetyl)oxy]methyl]-2-methyl-1,3-propan ediyl ester
Synonym: Acetic acid, mercapto-, triester with 2-(hydroxymethyl)-2-methyl-1,3-propanediol
Line Form: $CH_3C(CH_2O_2CCH_2SH)_3$
Mol Form: $C_{11}H_{18}O_6S_3$
M_r: 342.43
IR: COB 7092 2930 2580 1740 1470 1410 1380 1280 1150 1020
m/e: 342.03

HODOC No: 00669
CAS No: 19788-48-8
Name: Acetic acid, mercapto-, 2-methoxyethyl ester
Line Form: $HSCH_2CO_2CH_2CH_2OCH_3$
Mol Form: $C_5H_{10}O_3S$
M_r: 150.19
IR: µSADP 9052 2940 2560 1720 1450 1410 1280 1120 1030 860 830
H NMR: SAD 9801 2.1 3.3 3.4 3.6 4.3 $CDCl_3$
C NMR: JJ 127 FT 26.3 58.7 64.4 70.2 170.7 $CDCl_3$
m/e: 150.04

HODOC No: 00670
CAS No: 2365-48-2
Beilstein No: 3^4, 614
Name: Acetic acid, mercapto-, methyl ester
Synonym: Mercaptoacetic acid methyl ester
Line Form: $HSCH_2CO_2CH_3$
Mol Form: $C_3H_6O_2S$
M_r: 106.14
bp: $42\text{-}3^{10}$
***n*:** 1.4657^{20}
***s*:** Et_2O 4; EtOH 4
IR: µSADP_22018 2940 2560 1720 1430 1320 1190 1160 1020 870
H NMR: SAD 9160 1.9 3.2 3.7 CCl_4
m/e: 106.01

HODOC No: 00671
CAS No: 7383-61-1
Beilstein No: 3^3, 430
Name: Acetic acid, mercapto-, 1-methylethyl ester
Synonym: Acetic acid, mercapto-, isopropyl ester
Line Form: $HSCH_2CO_2CH(CH_3)_2$
Mol Form: $C_5H_{10}O_2S$
M_r: 134.19
bp: $80\text{-}8^{45}$
***d*:** $1.05^{20/4}$
H NMR: SAD 9733 1.3 2.0 3.2 5.1 $CDCl_3$
MS: ALDN 80 43(100) 47(59) 27(39) 41(38) 75(36) 92(28) 45(19) 134(16)
m/e: 134.04

HODOC No: 00672
CAS No: 7664-80-4
Beilstein No: 3^3, 431
Name: Acetic acid, mercapto-, octyl ester
Synonym: Octyl mercaptoacetate
Line Form: $HSCH_2CO_2C_8H_{17}$
Mol Form: $C_{10}H_{20}O_2S$
M_r: 204.32
bp: 125^{17}
***n*:** 1.4606^{21}
IR: SADG 18437 2950 1730 1450 1400 1370 1270 1130 1000
H NMR: SAD 9391 0.9 1.1-1.8 2.0 3.2 4.1 CCl_4
m/e: 204.12

HODOC No: 00673
CAS No: 625-45-6
Beilstein No: 3^4, 574
Name: Acetic acid, methoxy-
Synonym: Methoxyacetic acid
Line Form: $CH_3OCH_2CO_2H$
Mol Form: $C_3H_6O_3$
M_r: 90.08
bp: 203-4, 96^{13}
***d*:** $1.1768^{20/4}$
***n*:** 1.4168^{20}
Color: hyg
***s*:** H_2O 3; al 3; eth 3
IR: SADP 20584
MS: WILEY 175 45(100) 60(27) 29(24) 15(16) 31(11) 42(5) 46(3) 90(0)
m/e: 90.03

HODOC No: 00674
CAS No: 3938-96-3
Beilstein No: 3^3, 381
Name: Acetic acid, methoxy-, ethyl ester
Synonym: Ethyl methoxyacetate
Line Form: $CH_3OCH_2CO_2C_2H_5$
Mol Form: $C_5H_{10}O_3$
M_r: 118.13
bp: 142, $44\text{-}5^9$
***d*:** 1.0118^{15}
***n*:** 1.4050^{20}
***s*:** H_2O 2; al 4; eth 4; ctc 3; os 3
IR: SADG 8490 3000 2930 2820 1750 1450 1380 1200 1130 1030 950 840
H NMR: SAD 2863 1.3 3.4 3.9 4.1 CCl_4
m/e: 118.06

HODOC No: 00675
CAS No: 6290-49-9
Beilstein No: 3^4, 578
Name: Acetic acid, methoxy-, methyl ester
Synonym: Methoxyacetic acid methyl ester
Line Form: $CH_3OCH_2CO_2CH_3$
Mol Form: $C_4H_8O_3$
M_r: 104.10
bp: 131, 57^{50}
***d*:** $1.0511^{20/4}$
***n*:** 1.3962^{20}
***s*:** H_2O 2; al 4; eth 4; ace 3; os 3
IR: µSADP 6265 2940 1720 1450 1370 1280 1190 1120 1020 990 930 890 840 700
H NMR: SAD 14705 3.4 3.7 3.9 CCl_4
MS: ALDN 41 45(100) 74(42) 29(26) 15(26) 60(10) 75(8) 42(8) 104(0)
m/e: 104.05

HODOC No: 00676
CAS No: 1878-85-9
Beilstein No: 6^3, 4234
Name: Acetic acid, (2-methoxyphenoxy)-
Synonym: Acetic acid, (*o*-methoxyphenoxy)-
Line Form: $2\text{-}CH_3O\text{-}C_6H_4OCH_2CO_2H$
Structure:

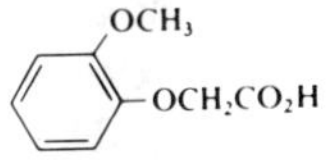

Mol Form: $C_9H_{10}O_4$
M_r: 182.17
mp: 123-5
Color: nd (w)
***s*:** Bz 4; Et_2O 4; EtOH 4
m/e: 182.06

HODOC No: 00677
CAS No: 2088-24-6
Beilstein No: 6^4, 5674
Name: Acetic acid, (3-methoxyphenoxy)-
Synonym: Acetic acid, (*m*-methoxyphenoxy)-
Line Form: $3\text{-}CH_3O\text{-}C_6H_4OCH_2CO_2H$
Mol Form: $C_9H_{10}O_4$
M_r: 182.17
mp: 118
Color: nd (w)
m/e: 182.06

HODOC No: 00678
CAS No: 79-20-9
Beilstein No: 2^4, 122
Name: Acetic acid, methyl ester
Synonym: Devoton
Line Form: $CH_3CO_2CH_3$
Mol Form: $C_3H_6O_2$
M_r: 74.08
bp: 57
mp: -98.1
***d*:** $0.9330^{20/4}$
***n*:** 1.3595^{20}
***s*:** H_2O 4; al 5; eth 5; ace 4; bz 3; ctc 3; chl 4
IR: µSADP 2228 2940 1750 1450 1370 1250 1050 980 840
Raman: DFB 46 3020 2940 2850 2720 1740 1450 1370 1190 1160 1050 980 840 640 610 440 300
UV: OES 7-11 209(58) iso 203(62) H_2O
H NMR: SAD 10261 2.0 3.7 CCl_4
MS: WILEY 81 43(100) 74(15) 29(10) 42(10) 59(6) 31(3) 44(3) 30(1)
m/e: 74.04

HODOC No: 00679
CAS No: 33445-07-7
Beilstein No: 3^4, 574
Name: Acetic acid, (1-methylethoxy)-
Synonym: Acetic acid, isopropoxy-
Line Form: $(CH_3)_2CHO-CH_2CO_2H$
Mol Form: $C_5H_{10}O_3$
M_r: 118.13
bp: 113^{21}
IR: µCOB 3867 3130 3030 2630 2560 1720 1430 1390 1140 910 830 800
m/e: 118.06

HODOC No: 00680
CAS No: 108-21-4
Beilstein No: 2^4, 141
Name: Acetic acid, 1-methylethyl ester
Synonym: Acetic acid, isopropyl ester
Line Form: $CH_3CO_2-CH(CH_3)_2$
Mol Form: $C_5H_{10}O_2$
M_r: 102.13
bp: 90
mp: -73.4
***d*:** $0.8718^{20/4}$
***n*:** 1.3773^{20}
***s*:** H_2O 3; al 3; eth 5; ace 3; chl 3
IR: SADG 10939 2950 2900 1700 1430 1330 1200 1130 1090 1060 960 900 830 760
Raman: AC 22-1074 * 1780 1739 1469 1359 1168 1134 981 840 658 458 336
UV: OES 7-51 204(62) H_2O 209(59) MeOH
H NMR: SAD 14814 0.9 1.1-1.8 1.6 2.4 $CDCl_3$
C NMR: JJ 126 cw 20.8 21.8 67.4 169.6 diox
MS: WILEY 277 43(100) 45(32) 27(14) 61(12) 41(11) 42(9) 39(7) 102(1)
m/e: 102.07

HODOC No: 00681
CAS No: 40248-63-3
Beilstein No: 6^1, 25
Name: Acetic acid, [[5-methyl-2-(1-methylethyl)cyclohexyl]oxy]-, [1R-(1α,2β,5α)]-
Synonym: *l*-Methoxyacetic acid
Structure:

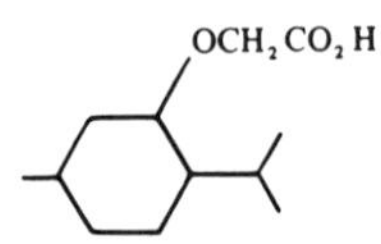

Mol Form: $C_{12}H_{22}O_3$
M_r: 214.29
bp: 171^{11}
mp: 53-5
α: $[\alpha]^{20/D}$-92.93 MeOH
***s*:** H_2O 2; al 4; eth 3; ace 4; os4
IR: µCOB 1754 2940 2630 1720 1450 1430 1350 1280 1250 1210 1140 940 690
H NMR: SAD 10902 3.4 4.1 D_2O
m/e: 214.16

HODOC No: 00682
CAS No: 5333-40-4
Beilstein No: 6^3, 1902
Name: Acetic acid, [5-methyl-2-(1-methylethyl)phenoxy]-
Synonym: Acetic acid, (thymyloxy)-
Line Form: $3\text{-}(CH_3)_2CH\text{-}5\text{-}CH_3C_6H_3OCH_2CO_2H$
Mol Form: $C_{12}H_{16}O_3$
M_r: 208.25
mp: 149-50
Color: nd (dil al), cr (Bz)
***s*:** H_2O 2; al 4; eth 4; bz 3
IR: µSADP 20584 2940 2560 1720 1610 1590 1520 1450 1430 1370 1300 1250 1180 1110 1090 950 930 840 810 760 680
UV: SAD 7915 280(1770) 274(1810) 217(8390) MeOH
m/e: 208.11

HODOC No: 00683
CAS No: 36685-69-5
Name: Acetic acid, [2-methyl-4-(4-morpholinylsulfonyl)phenoxy]-
Synonym: Acetic acid, (4(morpholinosulfonyl)-2-methylphenoxy)
Mol Form: $C_{13}H_{17}NO_6S$
M_r: 315.33
mp: 176
***s*:** DMSO 2
IR: SADG 17821 3310 3150 2920 2840 2780 1540 1505 1450 1375 1305 1280 1225 1150 1005 980 870 810 600
UV: SAD 17401 279(1160) 269(1701) 241(15500) MeOH
H NMR: SAD 9604 2.3 2.9 3.7 4.9 6.5 7.1 7.6 $DMSOd_6$
m/e: 315.08

HODOC No: 00684
CAS No: 36685-72-0
Name: Acetic acid, [3-methyl-4-(4-morpholinylsulfonyl)phenoxy]-
Synonym: Acetic acid, 4(morpholinosulfonyl)-3-methylphenoxy
Mol Form: $C_{13}H_{17}NO_6S$
M_r: 315.33
mp: 186
***s*:** DMSO 3
IR: SADG 19401 2960 2850 1740 1715 1600 1565 1485 1460 1430 1335 1325 1290 1255 1240 1160 1140 1105 1090 1070 940 810 730 650 615 575 515
UV: SAD 19718 241(16300) MeOH 245(16700) MeOH:KOH
H NMR: SAD 9605 2.6 3.0 3.6 4.8 6.8 7.0 7.1 7.8 $DMSOd_6$
m/e: 315.08

HODOC No: 00685
CAS No: 92565-72-5
Name: Acetic acid, [(1-methylnonyl)oxy]-
Line Form: $CH_3(CH_2)_7CH(CH_3)OCH_2CO_2H$
Mol Form: $C_{12}H_{24}O_3$
M_r: 216.31
***s*:** chl 3
H NMR: SAD 9571 0.9 1.2 1.3 3.5 4.1 9.6 $CDCl_3$
m/e: 216.17

HODOC No: 00686
CAS No: 1878-49-5
Beilstein No: 6^4, 1964
Name: Acetic acid, (2-methylphenoxy)-
Synonym: Acetic acid, (*o*-tolyloxy)-
Line Form: $2\text{-}CH_3\text{-}C_6H_4O\text{-}CH_2CO_2H$
Mol Form: $C_9H_{10}O_3$
M_r: 166.17
mp: 157
Color: lf (w)
***s*:** H_2O 3; al 3; ace 2; bz 2; os 2; con sulf 3
IR: µCOB 2252 2560 1720 1700 1590 1490 1430 1300 1270 1250 1190 1160 1140 1090 1050 970 940 920 850 780 760 710 690
UV: SAD 15872 276(1280) 270(1400) 214(6600) MeOH
H NMR: SAD 7149 2.3 4.7 5.6 6.7-7.3 $aceD_6$
m/e: 166.06

HODOC No: 00687
CAS No: 1643-15-8
Beilstein No: 6^4, 2051
Name: Acetic acid, (3-methylphenoxy)-
Synonym: Acetic acid, (*m*-tolyloxy)-
Line Form: $3\text{-}CH_3\text{-}C_6H_4\text{-}OCH_2CO_2H$
Mol Form: $C_9H_{10}O_3$
M_r: 166.17
mp: 103-4
Color: nd (w)
***s*:** H_2O 3; al 3; bz 3
IR: μSADP 17479 2940 2560 1750 1610 1590 1490 1430 1300 1270 1160 1100 930 860 780 760 690
UV: SAD 5550 278 271 216 MeOH
m/e: 166.06

HODOC No: 00688
CAS No: 940-64-7
Beilstein No: 6^4, 2119
Name: Acetic acid, (4-methylphenoxy)-
Synonym: Acetic acid, (*p*-tolyloxy)-
Line Form: $4\text{-}CH_3C_6H_4OCH_2CO_2H$
Mol Form: $C_9H_{10}O_3$
M_r: 166.17
mp: 136
Color: nd (w)
***s*:** Bz 4; EtOH 4
m/e: 166.06

HODOC No: 00689
CAS No: 533-18-6
Beilstein No: 6^4, 1960
Name: Acetic acid, 2-methylphenyl ester
Synonym: Acetic acid, *o*-tolyl ester
Line Form: $CH_3CO_2\text{-}(C_6H_4CH_3\text{-}2)$
Structure:

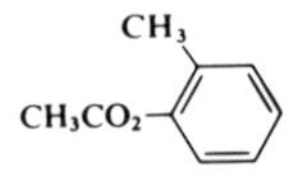

Mol Form: $C_9H_{10}O_2$
M_r: 150.17
bp: 208, 89^{10}
***d*:** 1.0533^{15}
***n*:** 1.5002^{20}
***s*:** Et_2O 4; EtOH 4
IR: μCOB 7943 3030 2940 1750 1590 1490 1370 1210 1180 1110 1040 1010 940 910 830 780 740 700 650
Raman: SAD 3060 2940 2860 2740 2590 1750 1610 1580 1490 1440 1380 1280 1220 1170 1150 1110 1040 1010 980 940 910 830 770 750 710 650 560 520 440 340 290 190
H NMR: SAD 19062 2.1 poly D
m/e: 150.07

HODOC No: 00690
CAS No: 122-46-3
Beilstein No: 6^4, 2047
Name: Acetic acid, 3-methylphenyl ester
Synonym: Acetic acid, *m*-tolyl ester
Line Form: $CH_3CO_2\text{-}(C_6H_4CH_3\text{-}3)$
Mol Form: $C_9H_{10}O_2$
M_r: 150.17
bp: 212
***d*:** $1.043^{20/4}$
***n*:** 1.4978^{20}
***s*:** Bz 4; Et_2O 4; EtOH 4
IR: SADP 8043
UV: SAD 2175
H NMR: SAD 18724
m/e: 150.07

HODOC No: 00691
CAS No: 140-39-6
Beilstein No: 6^4, 2112
Name: Acetic acid, 4-methylphenyl ester
Synonym: Acetic acid, *p*-tolyl ester
Line Form: CH_3CO_2-($C_6H_4CH_3$-4)
Mol Form: $C_9H_{10}O_2$
M$_r$: 150.17
bp: 212.5, $107\text{-}9^{5}$
d: $1.0512^{17/4}$
n: 1.5163^{22}
s: H_2O 2; al 3; eth 3; ctc 2; chl 3
IR: SADG 267 3040 2930 1760 1620 1600 1510 1430 1370 1220 1200 1170 1100 1050 1020 940 910 850 810 760 700 640
Raman: SAD 68 3070 3040 3000 2940 2860 2730 1750 1610 1590 1500 1430 1400 1380 1300 1210 1160 1100 1000 940 900 830 800 750 700 640 620 590 500 490 460 410 340 320 190
UV: SAD 1458 271(406) 265(446) MeOH
H NMR: SAD 10397 2.1 2.3 6.8 7.0 CCl_4
MS: WILEY 880 108(100) 107(43) 15(20) 77(16) 51(12) 28(10) 150(9) 40(9)
m/e: 150.07

HODOC No: 00692
CAS No: 105-46-4
Beilstein No: 2^2, 141
Name: Acetic acid, 1-methylpropyl ester
Synonym: Acetic acid, *sec*-butyl ester
Line Form: $CH_3CO_2CH(CH_3)CH_2CH_3$
Mol Form: $C_6H_{12}O_2$
M$_r$: 116.15
bp: 112.2^{760}
d: $0.8716^{20/4}$
n: 1.3888^{20}
s: H_2O 1; al 3; eth 3; ctc 2
IR: µSADP 5250 2940 1720 1470 1370 1250 1120 1110 1100 1030 1020 940 860
Raman: AC 22-1074 *1780 1763 1469 1395 1368 972 894 865 658 366 207
H NMR: SAD 6787 0.9 1.2 1.5 2.0 4.8 CCl_4
MS: WILEY 428 43(100) 56(15) 87(15) 15(13) 29(12) 41(12) 27(10) 116(1)
m/e: 116.08

HODOC No: 00693
CAS No: 54657-08-8
Beilstein No: 2^3, 241
Name: Acetic acid, 1-methylpropyl ester, (R)-
Synonym: (2R)-2-Acetoxybutane
Line Form: CH_3CO_2-$CH(CH_3)CH_2CH_3$
Mol Form: $C_6H_{12}O_2$
M$_r$: 116.15
bp: 116-7
d: $0.8730^{19/4}$
n: 1.3899^{18}
α: $[\alpha]^{D}$ -20.2_{546}
s: Ac 4; Et_2O 4; EtOH 4
m/e: 116.08

HODOC No: 00694
CAS No: 66610-38-6
Beilstein No: 2^3, 341
Name: Acetic acid, 1-methylpropyl ester, (*S*)-
Synonym: Acetic acid, *sec*-butyl ester (*d*)
Line Form: $CH_3CO_2CH(CH_3)CH_2CH_3$
Mol Form: $C_6H_{12}O_2$
M$_r$: 116.15
bp: 112
d: $0.8758^{16/4}$
n: 1.3877^{20}
α: $[\alpha]^{20/D}$ +25.43
s: Ac 4; Et_2O 4; EtOH 4
m/e: 116.08

HODOC No: 00695
CAS No: 110-19-0
Beilstein No: 2^4, 149
Name: Acetic acid, 2-methylpropyl ester
Synonym: Acetic acid, isobutyl ester
Line Form: CH_3CO_2-$CH_2CH(CH_3)_2$
Mol Form: $C_6H_{12}O_2$
Mr: 116.15
bp: 117.2
mp: -98.6
***d*:** $0.8712^{20/4}$
***n*:** 1.3902^{20}
***s*:** H_2O 2; al 5; eth 5; ace 3; ctc 2
IR: SADG 50 2950 2860 1740 1470 1390 1380 1360 1230 1180 1030 990 960 940 920 900 820 630
Raman: SAD 451 2940 2920 2880 2790 2760 2730 1730 1470 1460 1400 1370 1350 1300 1260 1180 1140 1040 990 970 950 930 910 830 810 640 610 580 510 410 390 330 280
UV: OES 7-94 214(56) 208(50) iso 208(58) MeOH
H NMR: SAD 8590 0.9 1.9 2.0 3.8 CCl_4
MS: WILEY 430 43(100) 56(23) 15(15) 73(15) 41(13) 27(11) 29(9) 116(1)
m/e: 116.08

HODOC No: 00696
CAS No: 80638-53-5
Beilstein No: 4, 162
Name: Acetic acid, [(1-methylpropyl)amino]oxo-
Synonym: Oxamic acid, *N-sec*-butyl
Line Form: HO_2C-$CONHCH(CH_3)CH_2CH_3$
Mol Form: $C_6H_{11}NO_3$
Mr: 145.15
mp: 88-9
Color: cr (eth)
***s*:** Et_2O 4
m/e: 145.07

HODOC No: 00697
CAS No: 2444-37-3
Beilstein No: 3^4, 603
Name: Acetic acid, (methylthio)-
Synonym: *S*-Methylmercaptoacetate
Line Form: $CH_3SCH_2CO_2H$
Mol Form: $C_3H_6O_2S$
Mr: 106.14
bp: $130\text{-}1^{27}$
***d*:** $1.223^{20/20}$
***n*:** 1.495^{20}
m/e: 106.01

HODOC No: 00698
CAS No: 36691-78-8
Name: Acetic acid, [4-(4-morpholinylsulfonyl)phenoxy]-
Synonym: Acetic acid, (4(morpholinosulfonyl)phenoxy)
Mol Form: $C_{12}H_{15}NO_6S$
Mr: 301.30
mp: 214
***s*:** DMSO 2
IR: SADG 19394 2850 1745 1710 1590 1580 1500 1455 1430 1340 1320 1310 1290 1245 1160 1105 1065 935 840 735 700 655 565 535
UV: SAD 19716 227(768) 240(18400) MeOH 242(18700) MeOH:KOH
H NMR: SAD 9592 2.9 3.7 4.8 5.4 7.2 7.7 $DMSOd_6$
m/e: 301.06

HODOC No: 00699
CAS No: 2976-75-2
Beilstein No: 6^4, 4220
Name: Acetic acid, (1-naphthalenyloxy)-
Synonym: Acetic acid, (1-naphthyloxy)-
Line Form: 1-$C_{10}H_7OCH_2CO_2H$
Mol Form: $C_{12}H_{10}O_3$
Mr: 202.20
mp: 190
Color: pr
***s*:** H_2O 2; al 4; eth 4; $DMSOd_6$ 2
IR: SADG 8422 3030 2941 2564 1754 1724 1587 1515 1471 1449 1408 1266 1250 1124 1031 926 775 671
UV: SAD 1646 319(1820) 288(5490) 282(5410) 255(3440) 215(50000) MeOH
H NMR: SAD 10486 4.9 6.9 7.3-7.7 7.8 8.3 $DMSOd_6$
m/e: 202.06

HODOC No: 00700
CAS No: 120-23-0
Beilstein No: 6^4, 4274
Name: Acetic acid, (2-naphthalenyloxy)-
Synonym: Acetic acid, (2-naphthyloxy)-
Line Form: 2-$C_{10}H_7OCH_2CO_2H$
Mol Form: $C_{12}H_{10}O_3$
M_r: 202.20
mp: 156
Color: pr(w)
***s*:** H_2O 3; al 3; eth 3; DMSOd6 2; HOAc 3
IR: COB 6255 3060 2920 2780 2720 2580 1740 1630 1600 1510 1440 1390 1250 1180 1080 910 850 750 670 550 480
UV: SAD 1647 325 317 311 227 MeOH
H NMR: SAD 10485 4.8 7.1-7.6 7.8 8.7 DMSOd6
m/e: 202.06

HODOC No: 00701
CAS No: 625-75-2
Beilstein No: 2^4, 537
Name: Acetic acid, nitro-
Synonym: Nitroacetic acid
Line Form: $O_2NCH_2CO_2H$
Mol Form: $C_2H_3NO_4$
M_r: 105.05
mp: 92-3d
Color: nd (Chl)
***s*:** Bz 4; Et_2O 4; EtOH 4; Chl 4
m/e: 105.01

HODOC No: 00702
CAS No: 626-35-7
Beilstein No: 2^4, 537
Name: Acetic acid, nitro-, ethyl ester
Synonym: Ethyl nitroacetate
Line Form: $O_2NCH_2CO_2C_2H_5$
Mol Form: $C_4H_7NO_4$
M_r: 133.10
bp: $105\text{-}7^{25}$, 83^6
***d*:** $1.1953^{20/4}$
***n*:** 1.4250^{20}
***s*:** H_2O 2; al 5; eth 4; dil alk 3
UV: HBCP 270(158) MeOH
m/e: 133.04

HODOC No: 00703
CAS No: 2483-57-0
Beilstein No: 2^4, 537
Name: Acetic acid, nitro-, methyl ester
Synonym: Methyl nitroacetate
Line Form: $NO_2CH_2CO_2CH_3$
Mol Form: $C_3H_5NO_4$
M_r: 119.07
bp: 107^{28}
***d*:** $1.320^{0/4}$
m/e: 119.02

HODOC No: 00704
CAS No: 1878-87-1
Beilstein No: 6^4, 1261
Name: Acetic acid, (2-nitrophenoxy)-
Synonym: Acetic acid, (*o*-nitrophenoxy)-
Line Form: 2-NO_2-$C_6H_4OCH_2CO_2H$
Mol Form: $C_8H_7NO_5$
M_r: 197.14
mp: 158
Color: pr (w)
***s*:** Et_2O 4; EtOH 4
m/e: 197.03

HODOC No: 00705
CAS No: 610-69-5
Beilstein No: 6^4, 1256
Name: Acetic acid, 2-nitrophenyl ester
Synonym: Acetic acid, *o*-nitrophenyl ester
Line Form: CH_3CO_2-($C_6H_4NO_2$-2)
Structure:

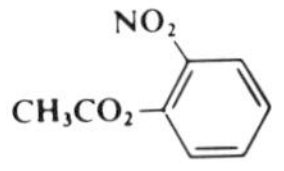

Mol Form: $C_8H_7NO_4$
M_r: 181.14
bp: 253d
mp: 40-1
Color: nd or pr (lig)
***s*:** H_2O 3; al 4; eth 4; ace 4; bz 4; lig 2
UV: OES 2-128 259(5623) EtOH
m/e: 181.04

HODOC No: 00706
CAS No: 1523-06-4
Beilstein No: 6^4, 1273
Name: Acetic acid, 3-nitrophenyl ester
Synonym: Acetic acid, *m*-nitrophenyl ester
Line Form: $CH_3CO_2(C_6H_4NO_2\text{-}3)$
Mol Form: $C_8H_7NO_4$
M_r: 181.14
mp: 55-6
***s*:** H_2O 3; al 3; lig 3
UV: HBCP 257(7943) 95% al
m/e: 181.04

HODOC No: 00707
CAS No: 830-03-5
Beilstein No: 6^4, 1298
Name: Acetic acid, 4-nitrophenyl ester
Synonym: Acetic acid, *p*-nitrophenyl ester
Line Form: $CH_3CO_2(C_6H_4NO_2\text{-}4)$
Mol Form: $C_8H_7NO_4$
M_r: 181.14
mp: 81-2
Color: lf (dil al)
***s*:** H_2O 4; al 3; bz 4; chl 3; os 3; lig 3
IR: SADG 18359 3110 3080 2850 1770 1610 1590 1520 1480 1360 1240 1190 1150 1100 1010 910 860 850 740 700 670
UV: SAD 7321 271(7230) MeOH
H NMR: SAD 8790 2.4 7.3 8.3 $CDCl_3$
MS: ALD 7:40A
m/e: 181.04

HODOC No: 00708
CAS No: 143-13-5
Beilstein No: 2^4, 167
Name: Acetic acid, nonyl ester
Synonym: Acetic acid n-nonyl ester
Line Form: $CH_3CO_2\text{-}(CH_2)_8CH_3$
Mol Form: $C_{11}H_{22}O_2$
M_r: 186.28
bp: 208-12
***d*:** $0.8785^{15/4}$
***n*:** 1.426^{20}
m/e: 186.16

HODOC No: 00709
CAS No: 822-23-1
Beilstein No: 2^4, 171
Name: Acetic acid, octadecyl ester
Synonym: 1-Octadecanol acetate
Line Form: $CH_3CO_2(CH_2)_{17}CH_3$
Mol Form: $C_{20}H_{40}O_2$
M_r: 312.52
bp: $222\text{-}3^{15}$
mp: 34-5
***s*:** EtOH 4
m/e: 312.30

HODOC No: 00710
CAS No: 112-14-1
Beilstein No: 2^4, 165
Name: Acetic acid, octyl ester
Synonym: Caprylyl acetate
Line Form: $CH_3CO_2\text{-}(CH_2)_7CH_3$
Mol Form: $C_{10}H_{20}O_2$
M_r: 172.26
bp: 210, $112\text{-}3^{30}$
mp: -38.5
***d*:** $0.8705^{20/4}$
***n*:** 1.4150^{20}
***s*:** H_2O 1; al 3; eth 3; ctc 2; os 3
IR: µSADP 3875 2940 2860 1720 1450 1350 1220 1030 720
Raman: SAD 57 2940 2900 2870 2730 1740 1450 1390 1360 1300 1270 1240 1210 1150 1120 1080 1060 1040 1000 950 930 890 870 850 820 760 730 640 600 490 400 350 300 260
H NMR: SAD 6542 0.9 1.3 1.6 2.0 4.0 CCl_4
MS: WILEY 1155 43(100) 57(40) 70(40) 41(36) 55(32) 29(30) 56(28) 172(1)
m/e: 172.15

HODOC No: 00711
CAS No: 298-12-4
Beilstein No: 3^4, 1489
Name: Acetic acid, oxo-
Synonym: Formic acid, formyl-
Line Form: $HCO\text{-}CO_2H$
Mol Form: $C_2H_2O_3$
M_r: 74.03
mp: 98
Color: rh pr (w+1/2)
***s*:** H_2O 4; al 2; eth 2; bz 2
UV: HBCP 584(8) 580(10) 576(8) 565(1) H_2O pH=7.1
MS: ALDN 13 29(100) 28(76) 30(46) 44(45) 45(44) 46(16) 56(7) 74(2)
m/e: 74.00

HODOC No: 00712
CAS No: 52709-42-9
Beilstein No: 10^2, 455
Name: Acetic acid, oxo-, phenylmethyl ester
Synonym: Benzyl glyoxylate
Line Form: $C_6H_5COCO_2CH_3$
Mol Form: $C_9H_8O_3$
M_r: 164.15
bp: 246-8 137^{14}
***n*:** 1.5268^{20}
***s*:** ctc 2
IR: µSADP 23016 2940 1750 1700 1590 1450 1330 1210 1180 1110 1030 1010 940 920 830 750 690
UV: SAD 17655 254(9270) MeOH
H NMR: SAD 7039 3.9 7.5 7.9 CCl_4
m/e: 164.05

HODOC No: 00713
CAS No: 500-72-1
Beilstein No: 12^2, 550
Name: Acetic acid, oxo(phenylamino)-
Synonym: Oxanilic acid
Line Form: $HO_2CCONHC_6H_5$
Mol Form: $C_8H_7NO_3$
M_r: 165.14
mp: 150
Color: nd (Bz)
***s*:** Et_2O 4; EtOH 4; Chl 4
IR: COB 2842
m/e: 165.04

HODOC No: 00714
CAS No: 1457-85-8
Beilstein No: 12^3, 550
Name: Acetic acid, oxo(phenylamino)-, ethyl ester
Synonym: Oxanilic acid, ethyl ester
Line Form: $C_6H_5NHCO\text{-}CO_2C_2H_5$
Mol Form: $C_{10}H_{11}NO_3$
M_r: 193.19
bp: 260-300
mp: 65-6
Color: pl or pr (al) nd (w)
***s*:** Ac 4; Bz 4; Et_2O 4; EtOH 4
m/e: 193.07

HODOC No: 00715
CAS No: 97980-65-9
Name: Acetic acid, [(4-oxo-2-phenyl-4H-1-benzopyran-7-yl)oxy]-
Synonym: Flavone, 7(carboxymethoxy)
Mol Form: $C_{17}H_{12}O_5$
M_r: 296.26
mp: 274-6
UV: LANG 2405 307(26915) 250(21380) EtOH
m/e: 296.07

HODOC No: 00716
CAS No: 119-41-5
Name: Acetic acid, [(4-oxo-2-phenyl-4H-1-benzopyran-7-yl)oxy]-, ethyl ester
Synonym: Angorlisin
Mol Form: $C_{19}H_{16}O_5$
M_r: 324.31
mp: 123.5-124
***s*:** chl 3
IR: SADG 13886 3060 2980 1750 1640 1610 1580 1450 1370 1290 1250 1190 1100 1040 920 860 790 710 690
UV: SAD 16183 304(18200) 251(13500) MeOH
H NMR: SAD 8520 1.3 4.3 4.7 6.7 6.9 7.0 7.3-7.6 7.8 8.1 $CDCl_3$
m/e: 324.10

HODOC No: 00717
CAS No: 27654-54-2
Name: Acetic acid, [(4-oxo-2-phenyl-4H-1-benzopyran-7-yl)oxy]-, methyl ester
Synonym: Flavone, 7(carboxy(methyl ester) methoxy)
Mol Form: $C_{18}H_{14}O_5$
M_r: 310.29
mp: 134-5
UV: LANG 2406 307(27542) 248(21878) EtOH 293(25704) 240(22387) cyhex
m/e: 310.08

HODOC No: 00718
CAS No: 53117-17-2
Name: Acetic acid, oxo(3-pyridinylamino)-, ethyl ester
Synonym: Oxalic acid, monoamide,*N*-pyrid-3-yl, ethyl ester
Mol Form: $C_9H_{10}N_2O_3$
M_r: 194.18
IR: COB 7347 3340 1730 1705 1690 1595 1555 1480 1415 1370 1305 1170 1020 945 850 805 710 680 550 485
m/e: 194.07

HODOC No: 00719
CAS No: 110-99-6
Beilstein No: 3^4, 577
Name: Acetic acid, 2,2′-oxybis-
Synonym: Acetic acid, oxydi-
Line Form: $O(CH_2CO_2H)_2$
Mol Form: $C_4H_6O_5$
M_r: 134.08
bp: d
mp: 148
Color: mcl pr (w + 1)
***s*:** H_2O 4; Et_2O 4; EtOH 4
IR: ALD 2:308G
MS: ALD 2:175B
m/e: 134.02

HODOC No: 00720
CAS No: 13909-29-0
Name: Acetic acid, [*p*-[3-(2-chloroethyl)-3-nitrosoureido]phenyl]-
Synonym: Urea, 1(4(carboxymethyl)phenyl)-3 (2-chloroethyl)-3-nitroso
Mol Form: $C_{11}H_{12}ClN_3O_4$
M_r: 285.68
mp: dec122
IR: COB 7786 3320 1740 1690 1595 1540 1480 1400 1335 1225 1090 970 795
m/e: 285.05

HODOC No: 00721
CAS No: 99076-71-8
Name: Acetic acid, [*p*-(dimethylsulfamoyl)phenoxy]-
Synonym: Benzenesulfonic acid, amide,4(carboxymethoxy)-*N*,*N*-dimethyl
Mol Form: $C_{10}H_{13}NO_5S$
M_r: 259.27
mp: 177
***s*:** DMSO 3
IR: SADG 19389 3400 3100 2920 2600 1750 1720 1600 1500 1430 1330 1250 1150 1090 950 910 830 730 690 650 570 540
H NMR: SAD 9587 2.6 4.8 6.3 7.2 7.7 $DMSOd_6$
m/e: 259.05

HODOC No: 00722
CAS No: 91013-44-4
Name: Acetic acid, [*p*-(dimethylsulfamoyl)phenyl]-
Synonym: Benzenesulfonic acid, amide,4(carboxymethyl)-*N*,*N*-dimethyl
Mol Form: $C_{10}H_{13}NO_4S$
M_r: 243.27
mp: 87
***s*:** chl 3
IR: SADG 19442 3050 2960 2650 1700 1600 1420 1340 1230 1160 1090 950 760 710 680 590
UV: SAD 17750 274(538) 229(10200) MeOH
H NMR: SAD 9642 2.7 3.7 7.5 7.6 9.0 $CDCl_3$
m/e: 243.06

HODOC No: 00723
CAS No: 628-63-7
Beilstein No: 2^1, 60
Name: Acetic acid, pentyl ester
Synonym: Amyl acetate
Line Form: $CH_3CO_2(CH_2)_4CH_3$
Mol Form: $C_7H_{14}O_2$
M_r: 130.18
bp: 149.2
mp: -70.8
***d*:** $0.8756^{20/4}$
***n*:** 1.4023^{20}
***s*:** H_2O 2; al 5; eth 5; ctc 3
IR: COB 4801 2960 2940 2880 1740 1470 1370 1240 1120 1040 970 950 920 820 770 730
H NMR: SAD 6710 0.9 1.4 2.0 4.0 CCl_4
C NMR: JJ 273 cw 14.1 20.5 22.8 28.6 28.9 64.3 170.1 diox
MS: NBS 268
m/e: 130.10

HODOC No: 00724
CAS No: 122-59-8
Beilstein No: 6^4, 634
Name: Acetic acid, phenoxy-
Synonym: Glycolic acid phenyl ether
Line Form: $C_6H_5OCH_2CO_2H$
Mol Form: $C_8H_8O_3$
M_r: 152.14
bp: 285d
mp: 98-9
Color: nd or pl (w)
***s*:** H_2O 3; al 4; eth 4; bz 4; CS_2 4; HOAc 4
IR: μCOB 2393 2940 2560 1700 1590 1490 1450 1370 1320 1280 1240 1190 1160 1100 1080 960 840 750 690
UV: SAD 1637 276(1190) 269(1430) 264(1050) 218(7060) MeOH
H NMR: SAD 264 4.7 6.9 7.3 8.8 ace
MS: WILEY 904 107(100) 77(93) 152(81) 94(24) 79(21) 51(21) 65(15) 39(12)
m/e: 152.05

HODOC No: 00725
CAS No: 14316-61-1
Beilstein No: 6, 162
Name: Acetic acid, phenoxy-, anhydride
Synonym: Phenoxyacetic anhydride
Line Form: $(C_6H_5OCH_2CO)_2O$
Mol Form: $C_{16}H_{14}O_5$
M_r: 286.27
mp: 67-9
Color: lf (eth)
***s*:** Bz 4
m/e: 286.08

HODOC No: 00726
CAS No: 2555-49-9
Beilstein No: 6^4, 575
Name: Acetic acid, phenoxy-, ethyl ester
Synonym: Ethyl phenoxyacetate
Line Form: CH_3CO_2-$CH_2CH_2OC_6H_5$
Mol Form: $C_{10}H_{12}O_3$
M_r: 180.19
bp: $109\text{-}12^3$
***d*:** 1.1082^{20}
***n*:** 1.5080^{20}
IR: μCOB 1073 3030 2940 1750 1610 1490 1470 1370 1240 1180 1060 960 880 750 690
m/e: 180.08

HODOC No: 00727
CAS No: 2065-23-8
Beilstein No: 6^4, 635
Name: Acetic acid, phenoxy-, methyl ester
Synonym: Methyl phenoxyacetate
Line Form: $C_6H_5OCH_2CO_2CH_3$
Mol Form: $C_9H_{10}O_3$
M_r: 166.17
bp: 245, 130^{14}
***d*:** $1.1493^{20/4}$
***n*:** 1.5155^{20}
***s*:** Et_2O 4; EtOH 4
MS: NBS 696
m/e: 166.06

HODOC No: 00728
CAS No: 122-79-2
Beilstein No: 6^4, 611
Name: Acetic acid, phenyl ester
Synonym: Phenol acetate
Line Form: $CH_3CO_2C_6H_5$
Mol Form: $C_8H_8O_2$
M_r: 136.14
bp: 195-7, $75\text{-}6^8$
***d*:** $1.0780^{20/4}$
***n*:** 1.5035^{20}
***s*:** H_2O 2; al 5; eth 5; ctc 3; chl 5
IR: COB 5703 3070 3050 2940 1760 1740 1710 1600 1490 1480 1460 1430 1370 1210 1200 1160 1070 1030 1000 920 890 820 750 680 670
Raman: SAD 64 3190 3090 3030 2950 2870 1770 1740 1600 1460 1430 1370 1200 1160 1030 1010 930 900 820 750 700 670 620 600 530 500 370 340 270
UV: OES 8-90 260(302) EtOH
H NMR: SAD 6777 2.1 6.9-7.4 CCl_4
MS: NBS 312
m/e: 136.05

HODOC No: 00729
CAS No: 102-39-6
Beilstein No: 6^3, 4324
Name: Acetic acid, 2,2′-[1,3-phenylenebis(oxy)]*bis*-
Synonym: Acetic acid, (*m*-phenylenedioxy)di-
Line Form: 1,3-$(HO_2CCH_2O)_2C_6H_4$
Mol Form: $C_{10}H_{10}O_6$
M_r: 226.18
mp: 195
Color: nd (HOAc, w)
***s*:** H_2O 4; HOAc 4
m/e: 226.05

HODOC No: 00730
CAS No: 103-45-7
Beilstein No: 6^4, 3073
Name: Acetic acid, 2-phenylethyl ester
Synonym: Acetic acid, phenethyl ester
Line Form: $CH_3CO_2\text{-}CH_2CH_2C_6H_5$
Mol Form: $C_{10}H_{12}O_2$
M$_r$: 164.19
bp: 232.6, 153^{76}
mp: -31.1
***d*:** $1.0883^{20/4}$
***n*:** 1.5171^{20}
***s*:** Et_2O 4; EtOH 4
IR: µCOB 1703 3030 2940 1750 1610 1490 1450 1390 1370 1250 1050 1030 980 860 750 700
Raman: SAD 72 3090 3060 3000 2970 2940 2860 1730 1600 1580 1440 1390 1360 1340 1280 1200 1170 1150 1080 1060 1030 1000 930 900 850 810 760 630 610 580 570 490 400 320 240 170
UV: SAD 1172 280(2470) 225(5590) 219(5490) MeOH
H NMR: VAR 261 2.0 2.9 4.3 7.3 $CDCl_3$
MS: WILEY 1061 104(100) 43(82) 91(21) 105(13) 65(9) 39(9) 51(8) 164(0)
m/e: 164.08

HODOC No: 00731
CAS No: 66970-44-3
Name: Acetic acid, 1-phenylhydrazide, hydrazone with 1-C-(phenylazo)-*D*-galactose 2,3,4,5,6-pentaacetate
Synonym: *D*-Galactose, 1(*N*-acetyl-diphenylformazanyl), pentaacetate
Line Form: $CH_3CON(C_6H_5)N{=}C(N{=}NC_6H_5)\{CH(O_2CCH_3)\}_4CH_2O_2CCH_3$
Mol Form: $C_{30}H_{34}N_4O_{11}$
M$_r$: 626.59
mp: 170
UV: LANG 2320 357(13804) 231(16982) EtOH
m/e: 626.22

HODOC No: 00732
CAS No: 114-83-0
Beilstein No: 15^2, 92
Name: Acetic acid, 2-phenylhydrazide
Synonym: Acetic acid phenylhydrazone
Line Form: $CH_3CONHNHC_6H_5$
Mol Form: $C_8H_{10}N_2O$
M$_r$: 150.17
mp: 130-2
Color: hex pr (eth)
***s*:** H_2O 4; al 4; eth 2; bz 3; chl 3; tfa 2
IR: SADG 8474 3280 3230 3020 2940 1640 1600 1550 1500 1380 1300 1250 970 760 720 700
UV: SAD 1801 281(1570) 234(11800) MeOH
H NMR: SAD 3285 2.3 7.4 tfa
m/e: 150.08

HODOC No: 00733
CAS No: 42806-50-8
Name: Acetic acid, [[[(phenylmethoxy)carbonyl]amino]oxy]-, pentachlorophenyl ester
Synonym: Carbamic acid, *N*(carboxy(pentachlorophenyl ester)methoxy),benzyl ester
Line Form: $C_6H_5CH_2O_2CNHOCH_2CO_2C_6Cl_5$
Mol Form: $C_6H_{10}Cl_5NO_5$
M$_r$: 353.41
mp: 170-1
UV: LANG 2313 325(741) 305(2042) 260(2630) 214(63096) EtOH
m/e: 350.90

HODOC No: 00734
CAS No: 140-11-4
Beilstein No: 6^4, 2262
Name: Acetic acid, phenylmethyl ester
Synonym: Acetic acid, benzyl ester
Line Form: $CH_3CO_2CH_2C_6H_5$
Mol Form: $C_9H_{10}O_2$
M_r: 150.17
bp: 215.5, $93\text{-}4^{10}$
mp: -51.3
***d*:** $1.0550^{20/4}$
***n*:** 1.5232^{20}
***s*:** H_2O 2; al 5; eth 3; ace 3; chl 3
IR: µCOB 3715 3030 1720 1450 1370 1350 1160 1020 960 900 830 750 700
Raman: SAD 470 3210 3170 3070 3010 2940 2900 2850 2710 1740 1610 1590 1500 1460 1380 1360 1220 1180 1160 1080 1040 1010 940 910 840 830 760 700 650 630 580 500 440 310 280 230 190
UV: SAD 260 267(114) 263(174) 261 257(214) 252(174) MeOH
H NMR: VAR 530 2.1 5.1 7.3 $CDCl_3$
C NMR: JJ 345 FT 20.7 66.1 128.1 128.4 136.2 170.5 $CDCl_3$
MS: WILEY 879 108(100) 91(52) 90(42) 43(39) 150(30) 79(23) 107(17) 77(13)
m/e: 150.07

HODOC No: 00735
CAS No: 103-46-8
Beilstein No: 6^4, 2702
Name: Acetic acid, [(phenylmethyl)thio]-
Synonym: Acetic acid, (benzylthio)-
Line Form: $C_6H_5CH_2S\text{-}CH_2CO_2H$
Mol Form: $C_9H_{10}O_2S$
M_r: 182.23
mp: 62-3
IR: COB 5293
UV: SAD 5473 267 MeOH
H NMR: VAR 532 3.1 3.9 7.3 11.0 $CDCl_3$
MS: NBS 935
m/e: 182.04

HODOC No: 00736
CAS No: 103-04-8
Beilstein No: 6^4, 1538
Name: Acetic acid, (phenylthio)-
Synonym: Carboxymethyl phenyl sulfide
Line Form: $C_6H_5SCH_2CO_2H$
Mol Form: $C_8H_8O_2S$
M_r: 168.20
mp: 64-6
IR: ALD 2:818F
MS: ALD 6:111B
m/e: 168.02

HODOC No: 00737
CAS No: 4408-78-0
Beilstein No: 4^4, 3612
Name: Acetic acid, phosphono-
Synonym: Carboxymethanephosphonic acid
Line Form: $(HO_2CCH_2)PO(OH)_2$
Mol Form: $C_2H_5O_5P$
M_r: 140.03
mp: 142-3
***s*:** al 2; eth 1; ace 2; bz 1; chl 1
IR: COB 6855 2900 2500 1710 1400 1250 1120 1000 940 860 700
m/e: 139.99

HODOC No: 00738
CAS No: 591-87-7
Beilstein No: 2^4, 180
Name: Acetic acid, 2-propenyl ester
Synonym: Acetic acid, allyl ester
Line Form: $CH_3CO_2CH_2CH{=}CH_2$
Mol Form: $C_5H_8O_2$
M_r: 100.11
bp: 103.5
***d*:** $0.9275^{20/4}$
***n*:** 1.4049^{20}
***s*:** H_2O 2; al 5; eth 5; ace 3
IR: COB 5686 3080 2950 1750 1650 1450 1430 1370 1240 1030 980 930 600
Raman: SAD 101 3090 3020 2980 2940 2890 2840 2710 1740 1650 1450 1420 1350 1290 1240 1160 1100 1030 1000 980 960 920 890 830 620 570 470 410 370 340 290
H NMR: SAD 18713 2.0 4.5 5.2 5.3 5.9 CCl_4
MS: WILEY 246 43(100) 58(11) 41(11) 39(9) 57(8) 15(5) 29(3) 100(1)
m/e: 100.05

HODOC No: 00739
CAS No: 119-45-9
Name: Acetic acid, [2-[(2-propenylamino)carbonyl]phenoxy]-
Synonym: Acetic acid, [*o*-(allylcarbamoyl)phenoxy]-
Mol Form: $C_{12}H_{13}NO_4$
M_r: 235.23
IR: **µCOB 1231** 3330 3030 2940 2630 1750 1610 1560 1490 1430 1320 1250 1220 1160 1120 1080 1000 930 860 760 730 690
m/e: 235.08

HODOC No: 00740
CAS No: 109-60-4
Beilstein No: 2^4, 138
Name: Acetic acid, propyl ester
Synonym: 1-Acetoxypropane
Line Form: CH_3CO_2-$CH_2CH_2CH_3$
Mol Form: $C_5H_{10}O_2$
M_r: 102.13
bp: 101.6
mp: -95
***d*:** $0.8878^{20/4}$
***n*:** 1.3842^{20}
***s*:** H_2O 2; al 5; eth 5; ctc 3
IR: **COB 6176** 2960 2880 1740 1460 1390 1360 1240 1060 1040 1020 960 910 890
Raman: **SAD 33** 2940 2880 2740 1730 1450 1390 1370 1350 1280 1250 1150 1130 1120 1100 1060 1050 1020 980 970 910 890 880 840 760 630 490 410 350 310 260
UV: **OES 7-51** 203(51) H_2O 209(55) MeOH
H NMR: **SAD 115** 0.9 1.6 2.0 4.0 CCl_4
MS: **WILEY 277** 43(100) 61(32) 73(15) 42(9) 41(6) 59(5) 27(5) 102(1)
m/e: 102.07

HODOC No: 00741
CAS No: 54515-77-4
Name: Acetic acid, *sec*-octyl ester
Synonym: *sec*-Octyl acetate
Line Form: $CH_3CO_2CH(CH_3)(CH_2)_5CH_3$
Mol Form: $C_{10}H_{20}O_2$
M_r: 172.26
IR: **µCOB 201** 2940 1720 1470 1370 1250 1120 1030 940 730
MS: **WILEY 1152** 43(100) 87(37) 112(17) 70(16) 55(14) 41(12) 58(12) 172(1)
m/e: 172.15

HODOC No: 00742
CAS No: 91013-73-9
Name: Acetic acid, (*p*-sulfamoylphenoxy)-, ethyl ester
Synonym: Benzenesulfonic acid, amide,4(carboxy(ethyl ester)methoxy)
Mol Form: $C_{10}H_{13}NO_5S$
M_r: 259.27
mp: 146
***s*:** DMSO 3
IR: **SADG 19386** 3360 3280 3100 3000 1750 1600 1500 1430 1320 1300 1220 1150 1090 1070 900 830 680 580 550
UV: **SAD 17285** 279(708) 268(1000) 262(1050) 235(15800) MeOH
H NMR: **SAD 9583** 1.2 4.2 4.9 7.1 7.2 7.8 $DMSOd_6$
m/e: 259.05

HODOC No: 00743
CAS No: 90091-32-0
Name: Acetic acid, (*p*-sulfamoylphenoxy)-, hydrazide
Synonym: Benzenesulfonic acid, amide,4(carboxymethoxy),hydrazide
Mol Form: $C_8H_{11}N_3O_4S$
M_r: 245.25
mp: 209
***s*:** DMSO 3
IR: **SADG 19388** 3300 3200 3100 2900 1650 1600 1540 1500 1430 1330 1240 1140 1060 910 810 670 580 540
UV: **SAD 18628** 278(600) 268(933) 262(1000) 235(18600) MeOH 235(17200) MeOH:KOH
H NMR: **SAD 9586** 3.4 4.6 7.1 7.2 7.8 9.5 $DMSOd_6$
m/e: 245.05

HODOC No: 00744
CAS No: 96117-14-5
Name: Acetic acid, [(6-sulfamoylthymyl)oxy]-
Synonym: Benzenesulfonic acid, amide,4(carboxymethoxy)-5-isopropyl-2-methyl
Mol Form: $C_{12}H_{17}NO_5S$
M_r: 287.32
IR: **SADG 19409** 3420 3370 3300 3280 1720 1605 1560 1520 1500 1435 1390 1295 1250 1225 1175 1145 1120 1095 980 895 830 730 705 655 645 570 550 520
UV: **SAD 17287** 281(908) 271(1140) 239(12100) 207(46700) MeOH
H NMR: **SAD 9613** 1.2 2.6 3.3 4.8 6.8 7.1 7.8 9.3 $DMSOd_6$
m/e: 287.08

HODOC No: 00745
CAS No: 91004-91-0
Name: Acetic acid, [(4-sulfamoyl-*o*-tolyl)oxy]-
Synonym: Benzenesulfonic acid, amide,4(carboxymethoxy)-3-methyl
Mol Form: $C_9H_{11}NO_5S$
M_r: 245.24
mp: 183
***s*:** DMSO 3
IR: SADG 19395 3360 3250 3100 2900 2600 1750 1710 1600 1490 1430 1330 1240 1140 1130 1100 930 810 650 600 580
UV: SAD 18629 280(996) 271(1270) 238(14500) MeOH 281(1170) 272(1500) 240(14900) MeOH:KOH
H NMR: SAD 9576 2.3 4.8 5.8-6.8 7.0 7.1 7.6 7.7 $DMSOd_6$
m/e: 245.04

HODOC No: 00746
CAS No: 91248-67-8
Name: Acetic acid, [(4-sulfamoyl-*o*-tolyl)oxy]-, ethyl ester
Synonym: Benzenesulfonic acid, amide,4(carboxy(ethyl ester)methoxy)-3-methyl
Mol Form: $C_{11}H_{15}NO_5S$
M_r: 273.29
mp: 127
***s*:** DMSO 2
IR: SADG 19396 3360 3260 1750 1595 1545 1495 1445 1330 1310 1220 1155 1140 1100 1065 965 820 610 590
UV: SAD 19717 280(1160) 270(1500) 264(1420) 238(15000) MeOH
H NMR: SAD 9582 1.2 2.3 4.2 4.9 6.2 7.0 7.6 7.7 $DMSOd_6$
m/e: 273.07

HODOC No: 00747
CAS No: 90648-83-2
Name: Acetic acid, [(4-sulfamoyl-*o*-tolyl)oxy]-, hydrazide
Synonym: Benzenesulfonic acid, amide,4(carboxy(hydrazide)methoxy)-3-methyl
Mol Form: $C_9H_{13}N_3O_4S$
M_r: 259.27
mp: 192
***s*:** DMSO 3
IR: SADG 1937 3300 3260 3100 1650 1640 1530 1490 1430 1320 1250 1140 1090 910 770 590
UV: SAD 17529 279(1010) 271(1260) 236(15300) 203(47300) MeOH
H NMR: SAD 9581 2.3 3.8 4.6 7.0 7.2 7.6 $DMSOd_6$
m/e: 259.06

HODOC No: 00748
CAS No: 91013-74-0
Name: Acetic acid, [(5-sulfamoyl-2,4-xylyl)oxy]-
Synonym: Benzenesulfonic acid, amide,5(carboxymethoxy)-2,4-dimethyl
Mol Form: $C_{10}H_{13}NO_5S$
M_r: 259.27
mp: 174
***s*:** DMSO 3
IR: SADG 19404 3300 3250 2920 2600 1750 1610 1500 1490 1440 1370 1300 1250 1200 1170 1150 1080 900 850 690 650 550
UV: SAD 18631 285(2490) 232(8970) MeOH 285(2520) 233(8870) MeOH:KOH
H NMR: SAD 9608 2.2 2.5 4.7 6.1 7.1 7.2 7.3 $DMSOd_6$
m/e: 259.05

HODOC No: 00749
CAS No: 123-43-3
Beilstein No: 4^4, 102
Name: Acetic acid, sulfo-
Synonym: Sulfoacetic acid
Line Form: $HO_3SCH_2CO_2H$
Mol Form: $C_2H_4O_5S$
M_r: 140.11
bp: 245d
mp: 84-6
Color: hyg ta (w+1)
***s*:** H_2O 4; Ac 4; EtOH 4
m/e: 139.98

HODOC No: 00750
CAS No: 116401-69-5
Name: Acetic acid, sulfo-, compd. with 1-butanamine (1:1)
Synonym: 1-Butanamine, sulfoacetate
Line Form: $HO_2CCH_2SO_2(ONH_3CH_2CH_2CH_2CH_3)$
Mol Form: $C_6H_{15}NO_5S$
M$_r$: 213.24
IR: **µCOB 1257** 2940 2630 1750 1590 1490 1370 1270 1190 1050 1040 820 800
m/e: 213.07

HODOC No: 00751
CAS No: 123-45-5
Beilstein No: 3^3, 424
Name: Acetic acid, 2,2′-sulfonylbis-
Synonym: Acetic acid, sulfonyldi-
Line Form: $HO_2CCH_2S(=O)_2CH_2CO_2H$
Mol Form: $C_4H_6O_6S$
M$_r$: 182.14
mp: 187
***s*:** H_2O 4; al 4; eth 3; sulf 3
IR: **COB 7211** 3000 2950 2515 1700 1405 1380 1340 1270 1225 1140 900 760 715 680 575 550 510
m/e: 181.99

HODOC No: 00752
CAS No: 562-90-3
Beilstein No: 2^4, 445
Name: Acetic acid, tetraanhydride with silicic acid (H4SiO4)
Synonym: Acetic acid, tetraanhydride with H4SiO4
Line Form: $(CH_3CO_2)_4Si$
Mol Form: $C_8H_{12}O_8Si$
M$_r$: 264.26
bp: 148^{5-6}
mp: 110
IR: **COB 5528**
m/e: 264.03

HODOC No: 00753
CAS No: 14711-84-3
Name: Acetic acid, (tetradecyloxy)-
Synonym: Tetradecyloxyacetic acid
Line Form: $CH_3(CH_2)_{13}OCH_2CO_2H$
Mol Form: $C_{16}H_{32}O_3$
M$_r$: 272.41
mp: 61.5
***s*:** chl 3
H NMR: **SAD 9572** 0.9 1.3 3.6 4.1 $CDCl_3$
m/e: 272.24

HODOC No: 00754
CAS No: 10405-54-6
Name: Acetic acid, thio-, *S,S′*-diester with *N*-[1,1-*bis*(mercaptomethyl)ethyl]-*p*-toluenesulfonamide
Synonym: *p*-Toluenesulfonamide, *N*-[1,1-*bis*(mercaptomethyl)ethyl]-, diacetate
Mol Form: $C_{15}H_{21}NO_4S_3$
M$_r$: 375.50
mp: 80-1
IR: **COB 4576** 3260 2930 1680 1590 1400 1340 1300 1260 1100 980 950 900 850 800
m/e: 375.06

HODOC No: 00755
CAS No: 90611-37-3
Name: Acetic acid, thio-, *S*-2-norbornyl ester
Synonym: 2-Norbornanethiol, acetate
Structure:

Mol Form: $C_9H_{14}OS$
M$_r$: 170.26
bp: 113^{11}
***n*:** 1.5130^{20}
***s*:** ctc 3
IR: **COB 6584** 2960 2870 1700 1470 1450 1350 1300 1270 1110 1050 950 840 760 630 580
H NMR: **SAD 1582** 1.1-2.1 2.2 3.3 CCl_4
m/e: 170.08

HODOC No: 00756
CAS No: 123-93-3
Beilstein No: 3^4, 612
Name: Acetic acid, 2,2′-thiobis-
Synonym: Acetic acid, thiodi-
Line Form: $(HO_2CCH_2)_2S$
Mol Form: $C_4H_6O_4S$
M$_r$: 150.15
mp: 129
Color: cr (w)
***s*:** H_2O 2; al 4; bz 3
IR: **SADG 514** 3060 2910 2720 2600 1700 1430 1390 1300 1230 1200 920 870 840 790 660
H NMR: **SAD 6440** 3.5 D_2O
MS: **NBS 4143**
m/e: 150.00

HODOC No: 00757
CAS No: 75-96-7
Beilstein No: 2^4, 534
Name: Acetic acid, tribromo-
Synonym: Tribromoacetic acid
Line Form: Br_3CCO_2H
Mol Form: $C_2HBr_3O_2$
M$_r$: 293.74
bp: 245
mp: 135
Color: mcl
***s*:** H_2O 3; al 3; eth 3
IR: SADG 21648 3100 2600 1720 1400 860 770 680
m/e: 293.75

HODOC No: 00758
CAS No: 599-99-5
Beilstein No: 2^3, 485
Name: Acetic acid, tribromo-, ethyl ester
Synonym: Ethyl tribromoacetate
Line Form: $Br_3CCO_2C_2H_5$
Mol Form: $C_4H_5Br_3O_2$
M$_r$: 321.79
bp: 225, 148^{73}
***d*:** $2.2300^{20/20}$
***n*:** 1.5438^{13}
***s*:** al 3; eth 3
IR: µSADP 7387 3030 1750 1470 1450 1390 1370 1300 1210 1100 1020 950 860 790 760 710
m/e: 321.78

HODOC No: 00759
CAS No: 84998-84-5
Beilstein No: 6^4, 1068
Name: Acetic acid, (2,4,6-tribromophenoxy)-
Synonym: (2,4,6-Tribromophenoxy)acetic acid
Line Form: $(2,4,6\text{-}Br_3C_6H_2O)CH_2CO_2H$
Structure:

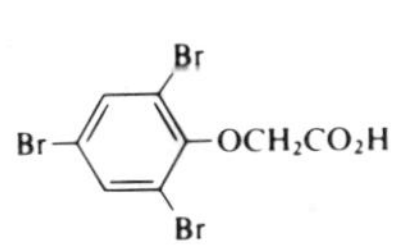

Mol Form: $C_8H_5Br_3O_3$
M$_r$: 385.83
mp: 200
Color: nd (dil al)
***s*:** Et_2O 4; EtOH 4
m/e: 385.78

HODOC No: 00760
CAS No: 76-03-9
Beilstein No: 2^4, 508
Name: Acetic acid, trichloro-
Synonym: Aceto-Caustin
Line Form: Cl_3CCO_2H
Mol Form: $C_2HCl_3O_2$
M$_r$: 163.39
bp: 197.5, $141\text{-}2^{25}$
mp: 58
***d*:** $1.6218^{64/1}$
***n*:** 1.4603^{61}
Color: dlq cr
***s*:** H_2O 4; al 3; eth 3; ctc 2
IR: µCOB 2376 3030 2630 2500 1750 1410 1250 1150 950 870 830 700 670
UV: OES 1-6 <260 H_2O
H NMR: SAD 6 11.4 CCl_4
MS: WILEY 1026 44(100) 83(72) 85(47) 36(32) 28(14) 47(14) 38(11) 162(0)
m/e: 161.90

HODOC No: 00761
CAS No: 4124-31-6
Beilstein No: 2^4, 518
Name: Acetic acid, trichloro-, anhydride
Synonym: Trichloracetic anhydride
Line Form: $(Cl_3CCO)_2O$
Mol Form: $C_4Cl_6O_3$
M$_r$: 308.76
bp: 222-4d, $98\text{-}100^{12}$
***d*:** 1.6908^{20}
***s*:** Et_2O 4; HOAc 4
m/e: 305.80

HODOC No: 00762
CAS No: 3657-07-6
Beilstein No: 2^4, 515
Name: Acetic acid, trichloro-, butyl ester
Synonym: Butyl trichloroacetate
Line Form: $Cl_3CCO_2C_4H_9$
Mol Form: $C_6H_9Cl_3O_2$
M$_r$: 219.49
bp: 203-5, $97\text{-}9^{19}$
***d*:** $1.2778^{20/4}$
***n*:** 1.4525^{25}
***s*:** ctc 3
IR: SADG 22115 2960 1770 1460 1370 1230 1010 970 820 670
H NMR: SAD 11413 1.0 1.2-2.1 4.4 CCl_4
m/e: 217.97

HODOC No: 00763
CAS No: 78987-65-2
Name: Acetic acid, trichloro-, 2-butyloctyl ester
Line Form: $CCl_3CO_2CH_2CH(CH_2CH_2CH_2CH_3)(CH_2)_5CH_3$
Mol Form: $C_{14}H_{25}Cl_3O_2$
M$_r$: 331.70
***s*:** ctc 3
IR: SADG 22146 2930 2850 1770 1460 1370 1230 980 820 740 670
H NMR: SAD 11472 0.9 1.3 1.8 4.2 CCl_4
m/e: 330.09

HODOC No: 00764
CAS No: 4974-21-4
Beilstein No: 2^4, 515
Name: Acetic acid, trichloro-, 2-chloroethyl ester
Synonym: β-Chloroethyl trichloroacetate
Line Form: $Cl_3CCO_2CH_2CH_2Cl$
Mol Form: $C_4H_4Cl_4O_2$
M$_r$: 225.88
bp: 217, 100^{14}
***d*:** $1.5357^{20/4}$
***n*:** 1.4813^{20}
***s*:** Et_2O 4
m/e: 223.90

HODOC No: 00765
CAS No: 29228-78-2
Name: Acetic acid, trichloro-, compd. with 2,2′,2″-nitrilotris[ethanol] (1:1)
Synonym: Ethanol, 2,2′,2″-nitrilotris-, trichloroacetate (salt)
Line Form: Cl_3CCO_2H–$(HOCH_2CH_2)_3N$
Mol Form: $C_8H_{16}Cl_3NO_5$
M$_r$: 312.57
IR: μCOB 3166 3330 3130 2940 1670 1470 1430 1390 1330 1220 1090 1020 910 840 820 750 680
m/e: 311.01

HODOC No: 00766
CAS No: 23399-81-7
Name: Acetic acid, trichloro-, 2,4-dichlorophenyl ester
Synonym: 2,4-Dichlorophenyl trichloroacetate
Mol Form: $C_8H_3Cl_5O_2$
M$_r$: 308.37
bp: 104^{12}
IR: μCOB 2000 3130 1790 1590 1470 1390 1250 1220 1140 1100 1060 950 860 830 820 780 680
m/e: 305.86

HODOC No: 00767
CAS No: 515-84-4
Beilstein No: 2^4, 514
Name: Acetic acid, trichloro-, ethyl ester
Synonym: Ethyl trichloroacetate
Line Form: $Cl_3CCO_2C_2H_5$
Mol Form: $C_4H_5Cl_3O_2$
M$_r$: 191.44
bp: 167-8, 62^{12}
***d*:** $1.3836^{20/4}$
***n*:** 1.4505^{20}
***s*:** H_2O 1; al 3; eth 3; bz 3; chl 2
IR: μCOB 5676 2940 1790 1470 1450 1370 1300 1240 1100 1020 960 890 840 750 680
Raman: SAD 110 2980 2950 2890 2730 1770 1460 1310 1280 1250 1120 1110 1030 970 890 840 750 700 440 390 380 340 290 270 250 220 210
H NMR: SAD 8684 1.4 4.4 $CDCl_3$
C NMR: JJ 51 FT 13.7 65.4 89.9 161.1 $CDCl_3$
MS: WILEY 1317 29(100) 117(17) 119(16) 27(14) 82(6) 121(5) 47(4) 190(0)
m/e: 189.94

HODOC No: 00768
CAS No: 65611-31-6
Name: Acetic acid, trichloro-, heptyl ester
Synonym: Heptyl trichloroacetate
Line Form: $Cl_3CCO_2(CH_2)_6CH_3$
Mol Form: $C_9H_{15}Cl_3O_2$
M$_r$: 261.57
***s*:** ctc 3
IR: SADG 22116 2930 2850 1770 1460 1370 1230 980 810 740 670
H NMR: SAD 11414 0.9 1.4 1.8 4.3 CCl_4
m/e: 260.01

HODOC No: 00769
CAS No: 33560-17-7
Beilstein No: 2^3, 474
Name: Acetic acid, trichloro-, 2-hydroxyethyl ester
Synonym: β-Hydroxyethyl trichloroacetate
Line Form: $Cl_3CCO_2CH_2CH_2OH$
Mol Form: $C_4H_5Cl_3O_3$
M$_r$: 207.44
bp: $130\text{-}4^{12}$
***d*:** $1.532^{20/4}$
***n*:** 1.4775^{20}
***s*:** EtOH 4
m/e: 205.93

HODOC No: 00770
CAS No: 35449-34-4
Beilstein No: 2^3, 474
Name: Acetic acid, trichloro-, 2-methoxyethyl ester
Synonym: 2-Methoxyethyl trichloroacetate
Line Form: $Cl_3CCO_2CH_2CH_2OCH_3$
Mol Form: $C_5H_7Cl_3O_3$
M$_r$: 221.46
bp: $98\text{-}9^{17}$
mp: 14.5
***d*:** $1.3826^{20/4}$
***n*:** 1.4563^{20}
***s*:** Bz 4; Et_2O 4; EtOH 4
m/e: 219.95

HODOC No: 00771
CAS No: 598-99-2
Beilstein No: 2^4, 513
Name: Acetic acid, trichloro-, methyl ester
Synonym: Methyl trichloroacetate
Line Form: $Cl_3CCO_2CH_3$
Mol Form: $C_3H_3Cl_3O_2$
M$_r$: 177.41
bp: 153.8, 44.5^{12}
mp: -17.5
***d*:** $1.4874^{20/4}$
***n*:** 1.4572^{20}
***s*:** H_2O 1; al 4; eth 4; ctc 3
IR: SADG 16809 3000 2950 2840 1760 1430 1330 1230 1170 990 900 820 740 670
H NMR: SAD 8233 4.0 CCl_4
MS: WILEY 1186 59(100) 117(32) 119(30) 15(20) 82(14) 121(10) 84(9) 176(0)
m/e: 175.92

HODOC No: 00772
CAS No: 3974-99-0
Beilstein No: 2^3, 471
Name: Acetic acid, trichloro-, 1-methylethyl ester
Synonym: Acetic acid, trichloro-, isopropyl ester
Line Form: $Cl_3CCO_2CH(CH_3)_2$
Mol Form: $C_5H_7Cl_3O_2$
M$_r$: 205.46
bp: 173.5^{747} $65.5\text{-}67^{15}$
***d*:** $1.3034^{20/4}$
***n*:** 1.4428^{20}
***s*:** H_2O 1; al 3; eth 3; bz 3
IR: µSADP 927 3030 1750 1470 1390 1330 1250 1180 1140 1100 980 900 860 830 810 740 680
H NMR: SAD 12445 1.4 5.1 CCl_4
m/e: 203.95

HODOC No: 00773
CAS No: 33560-15-5
Beilstein No: 2^4, 515
Name: Acetic acid, trichloro-, 2-methylpropyl ester
Synonym: Acetic acid, trichloro-, isobutyl ester
Line Form: $Cl_3CCO_2CH_2CH(CH_3)_2$
Mol Form: $C_6H_9Cl_3O_2$
M$_r$: 219.49
bp: 187-9, $93\text{-}4^{20}$
***d*:** $1.2636^{20/4}$
***n*:** 1.4483^{20}
***s*:** Bz 4; Et_2O 4; EtOH 4
m/e: 217.97

HODOC No: 00774
CAS No: 65611-32-7
Name: Acetic acid, trichloro-, nonyl ester
Synonym: Nonyl trichloroacetate
Line Form: $CCl_3CO_2(CH_2)_8CH_3$
Mol Form: $C_{11}H_{19}Cl_3O_2$
M$_r$: 289.62
***s*:** ctc 3
IR: SADG 22402 2920 2850 1770 1460 1370 1220 1010 970 890 820 740 670 420
H NMR: SAD 11491 0.9 1.3 1.8 4.3 CCl_4
m/e: 288.05

HODOC No: 00775
CAS No: 16958-78-4
Name: Acetic acid, trichloro-, octyl ester
Synonym: Octyl trichloroacetate
Line Form: $CCl_3CO_2(CH_2)_7CH_3$
Mol Form: $C_{10}H_{17}Cl_3O_2$
M$_r$: 275.59
***s*:** ctc 3
IR: SADG 22144 2920 2850 1770 1460 1370 1240 980 820 740 670
H NMR: SAD 11470 0.9 1.3 1.8 4.3 CCl_4
m/e: 274.03

HODOC No: 00776
CAS No: 78987-68-5
Name: Acetic acid, trichloro-, 3-phenylpropyl ester
Line Form: $C_6H_5(CH_2)_3O_2CCCl_3$
Mol Form: $C_{11}H_{11}Cl_3O_2$
M$_r$: 281.56
***s*:** ctc 2
IR: SADG 22017 3040 2960 1770 1600 1500 1450 1240 1000 820 740 690 670
H NMR: SAD 11427 2.0 2.7 4.3 7.2 CCl_4
m/e: 279.98

HODOC No: 00777
CAS No: 13313-91-2
Beilstein No: 2^3, 471
Name: Acetic acid, trichloro-, propyl ester
Synonym: Propyl trichloroacetate
Line Form: $Cl_3CCO_2CH_2CH_2CH_3$
Mol Form: $C_5H_7Cl_3O_2$
M_r: 205.46
bp: 187 69^{10}
***d*:** $1.3221^{20/4}$
***n*:** 1.4501^{20}
***s*:** H_2O 1; al 3; eth 3; ctc 3
H NMR: SAD 11412 1.0 1.8 4.3 CCl_4
m/e: 203.95

HODOC No: 00778
CAS No: 51698-77-2
Name: Acetic acid, trichloro-, 2-propynyl ester
Synonym: Propargyl trichloroacetate
Line Form: $CCl_3CO_2CH_2C{\equiv}CH$
Mol Form: $C_5H_3Cl_3O_2$
M_r: 201.43
IR: μCOB 1184 3450 2940 2130 1790 1430 1370 1220 1020 990 950 830 750 680
m/e: 199.92

HODOC No: 00779
CAS No: 6135-29-1
Beilstein No: 3^3, 35
Name: Acetic acid, trichloro-, trichloromethyl ester
Synonym: Methanol, trichloro-, trichloroacetate
Line Form: $Cl_3CCO_2CCl_3$
Mol Form: $C_3Cl_6O_2$
M_r: 280.75
bp: 191-2, $73\text{-}4^{10}$
mp: 34
***s*:** Bz 4; Et_2O 4; Chl 4; lig 4
m/e: 277.80

HODOC No: 00780
CAS No: 25056-70-6
Name: Acetic acid, trichloro-, 2-(2,4,5-trichlorophenoxy)ethyl ester
Synonym: Hexanate
Mol Form: $C_{10}H_6Cl_6O_3$
M_r: 386.86
IR: μCOB 1976 2940 1790 1590 1490 1450 1350 1280 1240 1140 1090 1040 980 880 830 680
m/e: 383.84

HODOC No: 00781
CAS No: 25438-23-7
Name: Acetic acid, trichloro-, 2,4,5-trichlorophenyl ester
Synonym: Phenol, 2,4,5-trichloro-, trichloroacetate
Mol Form: $C_8H_2Cl_6O_2$
M_r: 342.81
IR: μCOB 1979 3130 1790 1470 1350 1220 1190 1120 1090 960 880 850 820 790 710 680
m/e: 339.82

HODOC No: 00782
CAS No: 4007-00-5
Beilstein No: 6^4, 962
Name: Acetic acid, (2,3,6-trichlorophenoxy)-
Synonym: 2,3,6-Trichlorophenoxyacetic acid
Line Form: $(2,3,6\text{-}Cl_3C_6H_2)OCH_2CO_2H$
Mol Form: $C_8H_5Cl_3O_3$
M_r: 255.48
mp: 148-9
IR: μCOB 2238 3030 2700 2560 1720 1560 1450 1430 1410 1270 1180 1060 910 800 750 690
UV: SAD 3431 284 276 MeOH
m/e: 253.93

HODOC No: 00783
CAS No: 93-76-5
Beilstein No: 6^4, 973
Name: Acetic acid, (2,4,5-trichlorophenoxy)-
Synonym: BCF-Bushkiller
Line Form: $(2,4,5\text{-}Cl_3\text{-}C_6H_2\text{-})OCH_2CO_2H$
Structure:

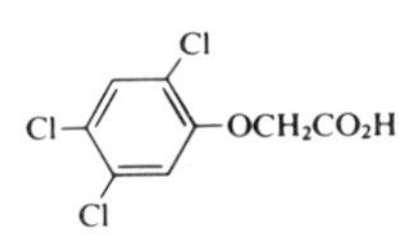

Mol Form: $C_8H_5Cl_3O_3$
M_r: 255.48
mp: 157-8
Color: cr (Bz)
***s*:** H_2O 1; al 3; bz 4
IR: SADG 8175 3030 2857 2564 1724 1587 1471 1429 1351 1282 1220 1136 1087 1075 943 870 781 676
UV: SAD 3432 297(2330) 289(2530) 206(47900) MeOH
H NMR: SAD 18671 4.8 7.2 7.5 11.8 poly D
m/e: 253.93

HODOC No: 00784
CAS No: 2545-59-7
Name: Acetic acid, (2,4,5-trichlorophenoxy)-, 2-butoxyethyl ester
Synonym: Ethanol, 2-butoxy-, (2,4,5-trichlorophenoxy)acetate
Mol Form: $C_{14}H_{17}Cl_3O_4$
M_r: 355.63
bp: $163\text{-}6^{1}$
***d*:** 1.280^{20}
***s*:** ctc 3
IR: μCOB 3164 2940 2860 1750 1590 1470 1450 1350 1280 1190 1140 1090 870 730 680
m/e: 354.02

HODOC No: 00785
CAS No: 93-79-8
Name: Acetic acid, (2,4,5-trichlorophenoxy)-, butyl ester
Synonym: Arboricid
Mol Form: $C_{12}H_{13}Cl_3O_3$
M_r: 311.58
IR: μCOB 1149 2940 1750 1720 1590 1490 1370 1350 1300 1240 1190 1140 1100 880 730 680
m/e: 309.99

HODOC No: 00786
CAS No: 40583-17-3
Name: Acetic acid, (2,4,5-trichlorophenoxy)-, 4-hydroxybutyl ester
Synonym: 2,4,5-T Butylglycol ester
Mol Form: $C_{12}H_{13}Cl_3O_4$
M_r: 327.58
IR: μCOB 1986 3450 2940 1750 1590 1470 1350 1300 1210 1140 1090 880 730 680
m/e: 325.99

HODOC No: 00787
CAS No: 93-78-7
Name: Acetic acid, (2,4,5-trichlorophenoxy)-, 1-methylethyl ester
Synonym: Acetic acid, (2,4,5-trichlorophenoxy)-, isopropyl ester
Mol Form: $C_{11}H_{11}Cl_3O_3$
M_r: 297.55
mp: 45
***s*:** chl 3
IR: μSADP 1325 3125 2940 1755 1585 1490 1390 1350 1300 1220 1135 1110 1085 945 875 835 735 680
UV: SAD 395 296 287 cyhex
H NMR: SAD 10406 1.3 4.6 5.1 6.9 7.4 $CDCl_3$
m/e: 295.98

HODOC No: 00788
CAS No: 61792-07-2
Name: Acetic acid, (2,4,5-trichlorophenoxy)-, 1-methylpropyl ester
Synonym: Acetic acid, 2,4,5-trichlorophenoxy,*sec*-butyl ester
Mol Form: $C_{12}H_{13}Cl_3O_3$
M_r: 311.58
IR: μCOB 1966 3130 2940 1750 1590 1470 1390 1350 1300 1210 1140 1090 960 880 830 730 680
m/e: 309.99

HODOC No: 00789
CAS No: 1928-40-1
Name: Acetic acid, (2,4,5-trichlorophenoxy)-, propyl ester
Synonym: Propyl 2,4,5-trichlorophenoxyacetate
Mol Form: $C_{11}H_{11}Cl_3O_3$
M_r: 297.55
IR: μCOB 2395 2940 1750 1590 1470 1370 1350 1300 1250 1190 1140 1090 970 880 830 730 680
m/e: 295.98

HODOC No: 00790
CAS No: 575-89-3
Beilstein No: 6^4, 1011
Name: Acetic acid, (2,4,6-trichlorophenoxy)-
Synonym: 2,4,6-T
Line Form: (2,4,6-Cl_3-C_6H_2)OCH_2CO_2H
Structure:

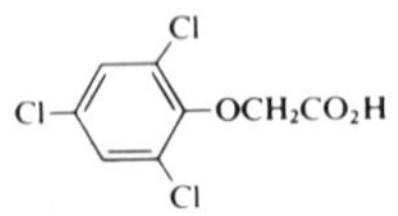

Mol Form: $C_8H_5Cl_3O_3$
Mr: 255.48
mp: 177
Color: cr (al)
s: H_2O 2; al 3; bz 2; lig 2
IR: µ **SADP** 3030 2560 1700 1560 1470 1430 1370 1280 1250 1220 1140 1090 1040 890 860 820 800 700
UV: **SAD 3433** 289 MeOH
m/e: 253.93

HODOC No: 00791
CAS No: 6064-76-2
Name: Acetic acid, (tridecyloxy)-
Synonym: 3-Oxahexadecanoic acid
Line Form: $CH_3(CH_2)_{12}OCH_2CO_2H$
Mol Form: $C_{15}H_{30}O_3$
Mr: 258.39
mp: 56.5
s: chl 3
H NMR: **SAD 9569** 0.9 1.3 1.6 3.6 4.1 $CDCl_3$
m/e: 258.22

HODOC No: 00792
CAS No: 76-05-1
Beilstein No: 2^4, 458
Name: Acetic acid, trifluoro-
Synonym: Perfluoroacetic acid
Line Form: F_3CCO_2H
Mol Form: $C_2HF_3O_2$
Mr: 114.02
bp: 72.4
mp: -15.2
d: 1.5351^0
s: H_2O 3; al 3; eth 3; ace 3
IR: **DMS 3697** 3587 3134 2992 2597 1830 1792 1466 1415 1240 1199 1180 1122 903 781 706 665
H NMR: **SAD 19890** 11.3 $CDCl_3$
C NMR: **JJ 1** FT 115.0 163.0 $CDCl_3$
MS: **NBS 147**
m/e: 113.99

HODOC No: 00793
CAS No: 407-25-0
Beilstein No: 2^4, 469
Name: Acetic acid, trifluoro-, anhydride
Synonym: Bis(trifluoroacetic) anhydride
Line Form: $(F_3CCO)_2O$
Mol Form: $C_4F_6O_3$
Mr: 210.03
bp: 39-40
mp: -65
d: $1.490^{25/4}$
n: 1.269^{25}
IR: **SADP 3096**
m/e: 209.98

HODOC No: 00794
CAS No: 31528-89-9
Name: Acetic acid, trifluoro-, 1,4-butanediyl ester
Synonym: 1,4-Butanediol, *bis*(trifluoroacetate)
Line Form: $CF_3CO_2(CH_2)_4O_2CCF_3$
Mol Form: $C_8H_8F_6O_4$
Mr: 282.13
s: chl 3
IR: **SADG 20693** 2960 1790 1550 1470 1400 1350 1220 1150 1020 920 770 730 700 520 330
H NMR: **SAD 10714** 1.9 4.4 $CDCl_3$
m/e: 282.03

HODOC No: 00795
CAS No: 367-64-6
Beilstein No: 2^4, 464
Name: Acetic acid, trifluoro-, butyl ester
Synonym: Butyl trifluoroacetate
Line Form: CF_3CO_2-C_4H_9
Mol Form: $C_6H_9F_3O_2$
M_r: 170.13
bp: 100.2
***d*:** $1.0268^{22/4}$
***n*:** 1.353^{22}
***s*:** chl 3
IR: μSADP 21369 2940 1790 1470 1410 1350 1240 1160 940 780 730
H NMR: SAD 11996 1.0 1.2-2.0 4.4 $CDCl_3$
m/e: 170.06

HODOC No: 00796
CAS No: 334-41-8
Name: Acetic acid, trifluoro-, compd. with dibutylamine (1:1)
Synonym: Amine, dibutyl,trifluoroacetate
Line Form: $CH_3(CH_2)_3NH(CH_2)_3CH_3$–$CF_3CO_2H$
Mol Form: $C_{10}H_{20}F_3NO_2$
M_r: 243.26
mp: 138-9
***s*:** H_2O 2
IR: SADG 21200 3040 2960 2940 2860 1670 1590 1460 1420 1380 1190 1160 1140 820 790 720
H NMR: SAD 11286 0.9 1.1-2.0 3.1 D_2O
m/e: 243.14

HODOC No: 00797
CAS No: 1549-45-7
Name: Acetic acid, trifluoro-, cyclohexyl ester
Synonym: Cyclohexanol trifluoroacetate
Mol Form: $C_8H_{11}F_3O_2$
M_r: 196.16
***s*:** chl 3
H NMR: SAD 10716 1.0-2.2 5.0 $CDCl^3$
m/e: 196.07

HODOC No: 00798
CAS No: 333-88-0
Name: Acetic acid, trifluoro-, decyl ester
Synonym: 1-Decanol trifluoroacetate
Line Form: $CF_3CO_2(CH_2)_9CH_3$
Mol Form: $C_{12}H_{21}F_3O_2$
M_r: 254.28
***s*:** chl 3
IR: SADG 22174 2920 2850 1790 1460 1400 1340 1210 1150 770 720
H NMR: SAD 11497 0.9 1.3 1.7 4.3 $CDCl^3$
m/e: 254.15

HODOC No: 00799
CAS No: 83088-72-6
Name: Acetic acid, trifluoro-, (2,6-dichlorophenyl)methyl ester
Synonym: 2,6-Dichlorobenzyl trifluoroacetate
Mol Form: $C_9H_5Cl_2F_3O_2$
M_r: 273.03
***s*:** chl 2
IR: SADG 20945 2960 1790 1580 1560 1430 1380 1330 1210 1130 1080 980 920 890 810 760 720 610 510 390 370
H NMR: SAD 11515 5.7 7.4 $CDCl^3$
m/e: 271.96

HODOC No: 00800
CAS No: 101567-63-9
Name: Acetic acid, trifluoro-, 1,3-dimethylbutyl ester
Line Form: $CF_3CO_2CH(CH_3)CH_2CH(CH_3)_2$
Mol Form: $C_8H_{13}F_3O_2$
M_r: 198.18
***s*:** chl 3
H NMR: SAD 11011 0.9 1.3 1.5-1.9 5.2 $CDCl_3$
m/e: 198.09

HODOC No: 00801
CAS No: 400-52-2
Beilstein No: 2^4, 465
Name: Acetic acid, trifluoro-, 1,1-dimethylethyl ester
Synonym: Acetic acid, trifluoro-, *tert*-butyl ester
Line Form: $F_3CCO_2C(CH_3)_3$
Mol Form: $C_6H_9F_3O_2$
M_r: 170.13
bp: 83
***n*:** 1.3300^{25}
***s*:** chl 3
IR: μSADP 7140 2940 1790 1470 1390 1370 1240 1160 890
H NMR: SAD 10721 1.6 $CDCl^3$
MS: WILEY 1126 41(100) 155(73) 69(59) 57(52) 29(19) 39(9) 61(9) 170(0)
m/e: 170.06

HODOC No: 00802
CAS No: 28745-07-5
Name: Acetic acid, trifluoro-, 3,7-dimethyloctyl ester
Synonym: 1-Octanol, 3,7-dimethyl-, trifluoroacetate
Line Form: $CF_3CO_2CH_2CH_2CH(CH_3)CH_2CH_2CH_2CH(CH_3)_2$
Mol Form: $C_{12}H_{21}F_3O_2$
M_r: 254.28
***s*:** chl 3
IR: SADG 22149 2950 2860 1790 1460 1390 1350 1210 1150 930 770 720
H NMR: SAD 11473 0.9 1.0 1.1-1.5 1.7 4.4 $CDCl_3$
m/e: 254.15

HODOC No: 00803
CAS No: 40336-13-8
Name: Acetic acid, trifluoro-, 1,1-dimethyl-2-phenylethyl ester
Line Form: $CF_3CO_2C(CH_3)_2CH_2C_6H_5$
Mol Form: $C_{12}H_{13}F_3O_2$
M_r: 246.22
***s*:** chl 3
IR: SADG 22190 3020 2980 1780 1600 1490 1450 1360 1210 1150 1100 830 770 730 710 690
H NMR: SAD 11506 1.5 3.1 7.2 $CDCl_3$
m/e: 246.09

HODOC No: 00804
CAS No: 7556-79-8
Name: Acetic acid, trifluoro-, 2,2-dimethylpropyl ester
Synonym: Acetic acid, trifluoro-, neopentyl ester
Line Form: $(CH_3)_3CCH_2O_2CCF_3$
Mol Form: $C_7H_{11}F_3O_2$
M_r: 184.15
***s*:** chl 3
H NMR: SAD 11995 1.0 4.1 $CDCl_3$
m/e: 184.07

HODOC No: 00805
CAS No: 52195-70-7
Name: Acetic acid, trifluoro-, 2,2-dimethylpropylidene ester
Synonym: 1,3-Propanediol,2,2-dimethyl,*bis*(trifluoroacetate)
Line Form: $(CH_3)_2C(CH_2O_2CCF_3)_2$
Mol Form: $C_9H_{10}F_6O_4$
M_r: 296.16
***s*:** chl 3
H NMR: SAD 11514 1.1 4.2 $CDCl_3$
m/e: 296.05

HODOC No: 00806
CAS No: 6222-00-0
Name: Acetic acid, trifluoro-, dodecyl ester
Synonym: 1-Dodecanol trifluoroacetate
Line Form: $CF_3CO_2(CH_2)_{11}CH_3$
Mol Form: $C_{14}H_{25}F_3O_2$
M_r: 282.33
***s*:** chl 3
IR: SADG 20894 2920 2850 1790 1460 1400 1350 1210 1150 940 770 720
H NMR: SAD 11509 0.9 1.3 1.8 4.4 $CDCl_3$
m/e: 282.18

HODOC No: 00807
CAS No: 2613-44-7
Name: Acetic acid, trifluoro-, 1,2-ethanediyl ester
Synonym: Acetic acid, trifluoro-, ethylene ester
Line Form: $CF_3CO_2CH_2CH_2O_2CCF_3$
Mol Form: $C_6H_4F_6O_4$
M_r: 254.08
***s*:** chl 3
IR: SADG 20688 2970 1800 1450 1390 1360 1330 1220 1150 860 770 720 520
H NMR: SAD 10707 4.6 $CDCl_3$
m/e: 254.00

HODOC No: 00808
CAS No: 31528-86-6
Name: Acetic acid, trifluoro-, 1,2-ethanediylbis(oxy-2,1-ethanediyl) ester
Synonym: Triethylene glycol, ditrifluoroacetate
Line Form: $CF_3CO_2CH_2CH_2OCH_2CH_2OCH_2CH_2O_2CCF_3$
Mol Form: $C_{10}H_{12}F_6O_6$
M$_r$: 342.18
***s*:** chl 3
H NMR: SAD 10709 3.9 4.0 4.6 $CDCl_3$
m/e: 342.05

HODOC No: 00809
CAS No: 30668-92-9
Name: Acetic acid, trifluoro-, 2-ethoxyethyl ester
Line Form: $CH_3CH_2OCH_2CH_2O_2CCF_3$
Mol Form: $C_6H_9F_3O_3$
M$_r$: 186.12
***s*:** chl 3
IR: SADG 20691 2980 2880 1790 1440 1390 1340 1210 1150 850 770 720 690
H NMR: SAD 10712 1.2 3.5 4.5 $CDCl_3$
m/e: 186.05

HODOC No: 00810
CAS No: 383-63-1
Beilstein No: 2^4, 463
Name: Acetic acid, trifluoro-, ethyl ester
Synonym: Ethyl perfluoroacetate
Line Form: $F_3CCO_2C_2H_5$
Mol Form: $C_4H_5F_3O_2$
M$_r$: 142.07
bp: 60-2
***d*:** $1.194^{20/4}$
***n*:** 1.308^{20}
IR: μSADP 3082 2940 1790 1490 1470 1410 1390 1350 1300 1240 1150 1020 860 780 730
H NMR: SAD 8663 1.4 4.4 $CDCl_3$
C NMR: JJ 52 FT 13.8 64.7 115.3 158.1 $CDCl_3$
MS: WILEY 761 29(100) 69(52) 27(21) 28(16) 99(16) 128(5) 51(3) 142(2)
m/e: 142.02

HODOC No: 00811
CAS No: 1893-99-8
Name: Acetic acid, trifluoro-, 1-ethylhexyl ester
Line Form: $CF_3CO_2CH(CH_2CH_3)(CH_2)_4CH_3$
Mol Form: $C_{10}H_{17}F_3O_2$
M$_r$: 226.23
***s*:** chl 3
H NMR: SAD 11416 0.9 1.0 1.3 1.7 5.0 $CDCl_3$
m/e: 226.12

HODOC No: 00812
CAS No: 6222-03-3
Name: Acetic acid, trifluoro-, hexadecyl ester
Synonym: 1-Hexadecyl trifluoroacetate
Line Form: $CF_3CO_2(CH_2)_{15}CH_3$
Mol Form: $C_{18}H_{33}F_3O_2$
M$_r$: 338.44
***s*:** chl 3
IR: SADG 22413 2920 2850 1790 1460 1400 1340 1210 1150 770 720
H NMR: SAD 11512 0.9 1.3 1.7 4.3 $CDCl_3$
m/e: 338.24

HODOC No: 00813
CAS No: 816-51-3
Name: Acetic acid, trifluoro-, hexamethylene ester
Synonym: 1,6-Hexanediol, *bis*(trifluoroacetate)
Line Form: $CF_3CO_2(CH_2)_6O_2CCF_3$
Mol Form: $C_{10}H_{12}F_6O_4$
M$_r$: 310.18
***s*:** chl 3
IR: SADG 20695 2940 1790 1460 1400 1350 1210 1150 770 720
H NMR: SAD 10718 1.2 2.1 4.3 $CDCl_3$
m/e: 310.06

HODOC No: 00814
CAS No: 687-10-5
Name: Acetic acid, trifluoro-, 5-hexenyl ester
Synonym: 5-Hexen-1-ol, trifluoroacetate
Line Form: $CF_3CO_2(CH_2)_4CH{=}CH_2$
Mol Form: $C_8H_{11}F_3O_2$
M$_r$: 196.16
***s*:** chl 3
H NMR: SAD 11458 1.3-2.0 2.1 4.4 5.0 5.1 5.8 $CDCl_3$
m/e: 196.07

HODOC No: 00815
CAS No: 400-61-3
Name: Acetic acid, trifluoro-, hexyl ester
Synonym: Hexyl trifluoroacetate
Line Form: $CH_3(CH_2)_5O_2CCF_3$
Mol Form: $C_8H_{13}F_3O_2$
M_r: 198.18
***s*:** chl 3
H NMR: SAD 10719 0.9 1.1-1.6 1.8 4.3 $CDCl_3$
m/e: 198.09

HODOC No: 00816
CAS No: 41017-81-6
Name: Acetic acid, trifluoro-, 2-methoxyethyl ester
Synonym: β-Methoxyethyl trifluoroacetate
Line Form: $CF_3CO_2CH_2CH_2OCH_3$
Mol Form: $C_5H_7F_3O_3$
M_r: 172.10
***s*:** chl 3
IR: SADG 22397 2930 2880 1790 1450 1400 1380 1350 1210 1160 1020 860 770 730
H NMR: SAD 11487 3.4 3.7 4.5 $CDCl_3$
m/e: 172.03

HODOC No: 00817
CAS No: 431-47-0
Beilstein No: 2^4, 463
Name: Acetic acid, trifluoro-, methyl ester
Synonym: Methyl trifluoroacetate
Line Form: $F_3CCO_2CH_3$
Mol Form: $C_3H_3F_3O_2$
M_r: 128.05
***d*:** $1.28^{20/4}$
IR: SADP 2502
H NMR: SAD 8662
m/e: 128.01

HODOC No: 00818
CAS No: 327-69-5
Name: Acetic acid, trifluoro-, 3-methylbutyl ester
Synonym: Acetic acid, trifluoro-, isopentyl ester
Line Form: $(CH_3)_2CHCH_2CH_2O_2CCF_3$
Mol Form: $C_7H_{11}F_3O_2$
M_r: 184.15
***s*:** chl 3
H NMR: SAD 10720 1.0 1.4-2.0 4.3 $CDCl_3$
m/e: 184.07

HODOC No: 00819
CAS No: 400-39-5
Name: Acetic acid, trifluoro-, 1-methylethenyl ester
Synonym: Acetic acid, trifluoro-, isopropenyl ester
Line Form: $CF_3CO_2C(CH_3){=}CH_2$
Mol Form: $C_5H_5F_3O_2$
M_r: 154.08
bp: 63-5
MS: WILEY 923 69(100) 43(47) 97(31) 58(18) 40(12) 41(9) 28(7) 154(5)
m/e: 154.02

HODOC No: 00820
CAS No: 90991-68-7
Name: Acetic acid, trifluoro-, 1-methylhexyl ester
Synonym: 2-Heptyl trifluoroacetate
Line Form: $CF_3CO_2CH(CH_3)(CH_2)_4CH_3$
Mol Form: $C_9H_{15}F_3O_2$
M_r: 212.20
***s*:** chl 2
IR: SADG 22027 2930 2870 1790 1460 1380 1330 1210 1160 1100 850 770 720
H NMR: SAD 11463 0.9 1.1-1.8 5.1 $CDCl_3$
m/e: 212.10

HODOC No: 00821
CAS No: 763-50-8
Name: Acetic acid, trifluoro-, 1-methylpentyl ester
Line Form: $CF_3CO_2CH(CH_3)(CH_2)_3CH_3$
Mol Form: $C_8H_{13}F_3O_2$
M_r: 198.18
***s*:** chl 3
H NMR: SAD 11451 0.9 1.0 1.1-2.0 5.1 $CDCl_3$
m/e: 198.09

HODOC No: 00822
CAS No: 10042-29-2
Name: Acetic acid, trifluoro-, 2-methylpentyl ester
Line Form: $CF_3CO_2CH_2CH(CH_3)CH_2CH_2CH_3$
Mol Form: $C_8H_{13}F_3O_2$
M_r: 198.18
bp: 138-42
***s*:** chl 2
H NMR: SAD 11439 0.9 1.0 1.1-1.6 2.0 4.2 $CDCl_3$
m/e: 198.09

HODOC No: 00823
CAS No: 1536-78-3
Name: Acetic acid, trifluoro-, 1-methylpropyl ester
Synonym: Acetic acid, trifluoro-, *sec*-butyl ester
Line Form: $F_3CCO_2CH(CH_3)CH_2CH_3$
Mol Form: $C_6H_9F_3O_2$
M_r: 170.13
***s*:** chl 3
H NMR: SAD 10715 1.0 1.3 1.6 5.0 $CDCl_3$
m/e: 170.06

HODOC No: 00824
CAS No: 71151-84-3
Name: Acetic acid, trifluoro-, 1-methyl-1-propylhexyl ester
Synonym: 4-Methyl-4-nonanyl trifluoroacetate
Line Form: $CF_3CO_2C(CH_3)(CH_2CH_2CH_3)(CH_2)_4CH_3$
Mol Form: $C_{12}H_{21}F_3O_2$
M_r: 254.28
***s*:** chl 3
IR: SADG 22173 2960 2870 1780 1460 1370 1210 1160 1120 770
H NMR: SAD 11496 0.9 1.1-1.9 $CDCl_3$
m/e: 254.15

HODOC No: 00825
CAS No: 26891-95-2
Name: Acetic acid, trifluoro-, 1-methyltrimethylene ester
Synonym: 1,3-Butanediol, *bis*(trifluoroacetate)
Line Form: $CF_3CO_2CH(CH_3)CH_2CH_2O_2CCF_3$
Mol Form: $C_8H_8F_6O_4$
M_r: 282.13
***s*:** chl 3
IR: SADG 20694 2970 1790 1550 1450 1380 1340 1220 1150 1090 940 910 850 770 730 720 620 330
H NMR: SAD 10717 1.4 2.2 4.4 5.2 $CDCl_3$
m/e: 282.03

HODOC No: 00826
CAS No: 658-78-6
Name: Acetic acid, trifluoro-, 4-nitrophenyl ester
Synonym: Acetic acid, trifluoro-, *p*-nitrophenyl ester
Line Form: $CF_3CO_2(C_6H_4NO_2\text{-}4)$
Mol Form: $C_8H_4F_3NO_4$
M_r: 235.11
bp: 120^{12}
mp: 35-9
IR: ALD 2.908G
MS: ALD 7:40C
m/e: 235.01

HODOC No: 00827
CAS No: 83088-71-5
Name: Acetic acid, trifluoro-, (3-nitrophenyl)methyl ester
Synonym: *m*-Nitrobenzyl trifluoroacetate
Mol Form: $C_9H_6F_3NO_4$
M_r: 249.14
***s*:** chl 3
H NMR: SAD 11421 5.5 7.5-7.9 8.3 $CDCl_3$
m/e: 249.02

HODOC No: 00828
CAS No: 405-84-5
Name: Acetic acid, trifluoro-, (4-nitrophenyl)methyl ester
Synonym: Acetic acid, trifluoro-, *p*-nitrobenzyl ester
Mol Form: $C_9H_6F_3NO_4$
M_r: 249.14
***s*:** chl 3
H NMR: SAD 11420 5.5 7.6 8.3 $CDCl_3$
m/e: 249.02

HODOC No: 00829
CAS No: 30767-14-7
Name: Acetic acid, trifluoro-, nonyl ester
Synonym: n-Nonyl trifluoroacetate
Line Form: $CF_3CO_2(CH_2)_8CH_3$
Mol Form: $C_{11}H_{19}F_3O_2$
M_r: 240.25
***s*:** chl 3
IR: SADG 22164 2930 2860 1790 1460 1400 1340 1210 1150 770 720
H NMR: SAD 11490 0.9 1.3 1.8 4.4 $CDCl_3$
m/e: 240.13

HODOC No: 00830
CAS No: 2561-21-9
Name: Acetic acid, trifluoro-, octyl ester
Synonym: 1-Octanol trifluoroacetate
Line Form: $CF_3CO_2(CH_2)_7CH_3$
Mol Form: $C_{10}H_{17}F_3O_2$
M_r: 226.23
***s*:** chl 3
IR: SADG 22143 2940 2860 1780 1460 1390 1340 1210 1150 1040 850 770 720 520
H NMR: SAD 11469 0.9 1.3 1.8 4.4 $CDCl_3$
m/e: 226.12

HODOC No: 00831
CAS No: 327-70-8
Name: Acetic acid, trifluoro-, pentyl ester
Synonym: Amyl trifluoroacetate
Line Form: $CH_3CH_2(CH_2)_2CH_2O_2CCF_3$
Mol Form: $C_7H_{11}F_3O_2$
M$_r$: 184.15
s: chl 3
IR: SADG 20893 2960 2940 2850 1780 1590 1550 1460 1390 1340 1210 1150 920 850 770 720 510
H NMR: SAD 11997 0.9 1.4 1.8 4.3 $CDCl_3$
m/e: 184.07

HODOC No: 00832
CAS No: 27912-97-6
Name: Acetic acid, trifluoro-, 2-phenoxyethyl ester
Synonym: Ethanol, 2-phenoxy-, trifluoroacetate
Line Form: $CF_3CO_2CH_2CH_2OC_6H_5$
Mol Form: $C_{10}H_9F_3O_3$
M$_r$: 234.16
s: chl 3
IR: SADG 22158 3040 2930 1790 1590 1490 1450 1340 1210 1150 1070 910 740 720 680
H NMR: SAD 11483 4.1 4.5 6.8 6.9 7.2 $CDCl_3$
m/e: 234.05

HODOC No: 00833
CAS No: 55419-66-4
Name: Acetic acid, trifluoro-, 2-phenylethyl ester
Synonym: 2-Phenylethyl trifluoroacetate
Line Form: $C_6H_5CH_2CH_2O_2CCF_3$
Mol Form: $C_{10}H_9F_3O_2$
M$_r$: 218.17
s: chl 2
IR: SADG 20897 3300 2970 1790 1500 1450 1390 1340 1210 1150 770 740 720 690 490
H NMR: SAD 11425 3.0 4.5 7.2 $CDCl_3$
m/e: 218.06

HODOC No: 00834
CAS No: 19692-35-4
Name: Acetic acid, trifluoro-, 1-phenylpropyl ester
Synonym: Acetic acid, trifluoro-, α-ethylbenzyl ester
Line Form: $C_6H_5(CH_2)_3O_2CCF_3$
Mol Form: $C_{11}H_{11}F_3O_2$
M$_r$: 232.19
s: chl 3
IR: SADG 20899 3060 3020 2940 2860 1780 1600 1550 1490 1450 1390 1340 1210 1140 1020 990 940 870 760 730 720 690 510 480 350 310
H NMR: SAD 11999 1.9 2.6 4.2 7.2 $CDCl_3$
m/e: 232.07

HODOC No: 00835
CAS No: 104237-19-6
Name: Acetic acid, trifluoro-, 2-phenylpropyl ester
Line Form: $C_6H_5CH(CH_3)CH_2O_2CCF_3$
Mol Form: $C_{11}H_{11}F_3O_2$
M$_r$: 232.19
s: chl 3
H NMR: SAD 11426 1.3 3.1 4.4 7.2 $CDCl_3$
m/e: 232.07

HODOC No: 00836
CAS No: 7647-95-2
Name: Acetic acid, trifluoro-, 1,3-propanediyl ester
Synonym: Acetic acid, trifluoro-, trimethylene ester
Line Form: $CF_3CO_2CH_2CH_2CH_2O_2CCF_3$
Mol Form: $C_7H_6F_6O_4$
M$_r$: 268.11
s: chl 3
IR: SADG 22414 2970 1780 1540 1460 1390 1340 1330 1210 1140 1020 920 770 720 510 320
H NMR: SAD 11513 2.2 4.5 $CDCl_3$
m/e: 268.02

HODOC No: 00837
CAS No: 383-67-5
Name: Acetic acid, trifluoro-, 2-propenyl ester
Synonym: Acetic acid, trifluoro-, allyl ester
Line Form: $CH_2{=}CHCH_2O_2CCF_3$
Mol Form: $C_5H_5F_3O_2$
M$_r$: 154.08
s: chl 2
IR: SADG 21321 3070 2930 1780 1640 1550 1440 1380 1320 1210 1160 980 930 900 820 760 720 640 510
H NMR: SAD 10722 4.8 5.3 5.4 5.8 $CDCl_3$
m/e: 154.02

HODOC No: 00838
CAS No: 383-66-4
Beilstein No: 2^4, 464
Name: Acetic acid, trifluoro-, propyl ester
Synonym: Propyl trifluoroacetate
Line Form: $F_3CCO_2C_3H_7$
Mol Form: $C_5H_7F_3O_2$
M_r: 156.10
bp: 82.5
***d*:** $1.1285^{25/4}$
***n*:** $1.3233^{22.5}$
***s*:** chl 3
IR: SADP 13148
UV: SAD 3432
H NMR: SAD 11411 1.0 1.8 4.3 $CDCl_3$
m/e: 156.04

HODOC No: 00839
CAS No: 7556-83-4
Name: Acetic acid, trifluoro-, propylene ester
Synonym: 1,2-Propanediol, *bis*(trifluoroacetate)
Line Form: $CF_3CO_2CH(CH_3)CH_2O_2CCF_3$
Mol Form: $C_7H_6F_6O_4$
M_r: 268.11
***s*:** chl 3
IR: SADG 20692 2990 1800 1450 1380 1330 1220 1160 1070 860 770 730 520
H NMR: SAD 10713 1.4 4.4 4.5 5.4 $CDCl_3$
m/e: 268.02

HODOC No: 00840
CAS No: 6222-02-2
Name: Acetic acid, trifluoro-, tetradecyl ester
Synonym: 1-Tetradecanol trifluoroacetate
Line Form: $CF_3CO_2(CH_2)_{13}CH_3$
Mol Form: $C_{16}H_{29}F_3O_2$
M_r: 310.38
***s*:** chl 3
IR: SADG 22412 2920 2850 1790 1460 1390 1340 1210 1150 760 720
H NMR: SAD 11511 0.9 1.3 1.7 4.4 $CDCl_3$
m/e: 310.21

HODOC No: 00841
CAS No: 407-38-5
Name: Acetic acid, trifluoro-, 2,2,2-trifluoroethyl ester
Synonym: Ethanol, 2,2,2-trifluoro-, trifluoroacetate
Line Form: $CF_3CO_2CH_2CF_3$
Mol Form: $C_4H_2F_6O_2$
M_r: 196.05
bp: 55
***s*:** chl 2
IR: µSADP 7150 2941 1786 1408 1351 1282 1176 1136 1087 980 893 826 769 735
H NMR: SAD11484 4.7 $CDCl_3$
m/e: 196.00

HODOC No: 00842
CAS No: 63012-94-2
Name: Acetic acid, trifluoro-, 2,2,4-trimethylpentyl ester
Line Form: $CF_3CO_2CH_2C(CH_3)_2CH_2CH(CH_3)_2$
Mol Form: $C_{10}H_{17}F_3O_2$
M_r: 226.23
***s*:** chl 3
IR: SADG 22124 2960 2880 1790 1470 1380 1340 1210 1150 940 770 720
H NMR: SAD 11442 0.9 1.0 1.3 1.7 4.1 $CDCl_3$
m/e: 226.12

HODOC No: 00843
CAS No: 379-30-6
Name: Acetic acid, trifluoro-, triphenylmethyl ester
Synonym: Acetic acid, trifluoro-, trityl ester
Line Form: $F_3CCO_2C(C_6H_5)_3$
Mol Form: $C_{21}H_{15}F_3O_2$
M_r: 356.32
***s*:** chl 3
H NMR: SAD 11422 7.3 $CDCl_3$
m/e: 356.10

HODOC No: 00844
CAS No: 53800-01-4
Name: Acetic acid, trifluoro-, undecyl ester
Synonym: 1-Undecanol, trifluoroacetate
Line Form: $CF_3CO_2(CH_2)_{10}CH_3$
Mol Form: $C_{13}H_{23}F_3O_2$
M_r: 268.31
***s*:** chl 3
IR: SADG 22185 2930 2850 1790 1460 1400 1340 1210 1150 770 720
H NMR: SAD 11502 0.9 1.3 1.7 4.3 $CDCl_3$
m/e: 268.17

HODOC No: 00845
CAS No: 13171-00-1
Name: Acetic acid, trifluoro,1-ethyl-1-methylbutyl ester
Synonym: Ketone, 6-*tert*-butyl-1,1-dimethyl-4-indanyl methyl
Line Form: $CF_3CO_2C(CH_3)(CH_2CH_3)CH_2CH_2CH_3$
Mol Form: $C_9H_{15}F_3O_2$
M$_r$: 212.20
s: chl 2
IR: SADG 22024 2960 2880 1780 1460 1360 1210 1160 1130 820 770
H NMR: SAD 11456 0.9 1.4 1.5 1.9 $CDCl_3$
m/e: 212.10

HODOC No: 00846
CAS No: 3246-66-0
Name: Acetic acid, trifluoro,1-methyldecyl ester
Synonym: Phenol, *p*-[*N*-(*p*-nitrophenyl)formimidoyl]-
Line Form: $CF_3CO_2CH(CH_3)(CH_2)_8CH_3$
Mol Form: $C_{13}H_{23}F_3O_2$
M$_r$: 268.31
s: chl 3
IR: SADG 22187 2930 2850 1790 1460 1380 1330 1210 1150 1110 850 770 720
H NMR: SAD 11503 0.9 1.3 1.4 1.6 5.1 $CDCl_3$
m/e: 268.17

HODOC No: 00847
CAS No: 594-68-3
Beilstein No: 2^4, 537
Name: Acetic acid, triiodo-
Synonym: Triiodoacetic acid
Line Form: CI_3CO_2H
Mol Form: $C_2HI_3O_2$
M$_r$: 437.74
mp: 150d
Color: ye lf
s: H_2O 4; Et_2O 4; EtOH 4
m/e: 437.71

HODOC No: 00848
CAS No: 29193-37-1
Name: Acetic acid, [2,4,6-triiodo-3-(*N*-methylbutyramido)phenyl]-
Synonym: Acetic acid, (3(*N*-methylbutyramido)-2,4,6-triiodophenyl)
Mol Form: $C_{13}H_{14}I_3NO_3$
M$_r$: 612.96
mp: 167
s: chl 2
IR: SADG 20157 3000 2960 2600 1710 1650 1420 1380 1320 1240 1200 1140 1050 930 870 610
UV: SAD 17596 236(38700) MeOH
H NMR: SAD 10564 0.9 1.3-2.1 3.1 4.3 8.4 11.2 $CDCl_3$
m/e: 612.81

HODOC No: 00849
CAS No: 29193-34-8
Name: Acetic acid, [2,4,6-triiodo-3-(*N*-propylacetamido)phenyl]-
Synonym: Acetic acid, (3(*N*-propylacetamido)2,4,6-triiodophenyl
Mol Form: $C_{13}H_{14}I_3NO_3$
M$_r$: 612.96
mp: 110
s: chl 2
IR: SADG 20155 3000 2960 2600 1730 1620 1420 1370 1300 1180 930 870 600
UV: SAD 17594 237(37800) MeOH
H NMR: SAD 10562 0.9 1.2-2.0 3.6 4.4 8.4 8.9 $CDCl_3$
m/e: 612.81

HODOC No: 00850
CAS No: 108-05-4
Beilstein No: 2^4, 176
Name: Acetic acid ethenyl ester
Synonym: Acetic acid vinyl ester
Line Form: $CH_3CO_2CH{=}CH_2$
Mol Form: $C_4H_6O_2$
M$_r$: 86.09
bp: 72.3
mp: -93.2
***d*:** $0.9317^{20/4}$
***n*:** 1.3959^{20}
***s*:** H_2O 1; al 5; eth 3; ace 3; bz 3; ctc 3; chl 3; os 3
IR: SADG 8112 3100 2980 1760 1650 1470 1290 1220 1130 1020 970 950 880 850 790
Raman: SAD 498 3130 3100 3050 3000 2940 2850 2730 1760 1690 1650 1620 1430 1390 1380 1300 1220 1140 980 960 880 850 710 640 470 410 230
UV: OES 1-29 258(1) hx
H NMR: VAR 65 2.1 4.6 4.9 7.3 $CDCl_3$
C NMR: JJ 61 cw 20.2 96.8 141.8 167.6 diox
MS: WILEY 136 43(100) 15(15) 28(10) 27(8) 42(8) 86(5) 44(5) 14(4)
m/e: 86.04

HODOC No: 00851
CAS No: 141-78-6
Beilstein No: 2^4, 127
Name: Acetic acid ethyl ester
Synonym: Acetic acid, ethyl ester
Line Form: $CH_3CO_2C_2H_5$
Mol Form: $C_4H_8O_2$
M$_r$: 88.10
bp: 77
mp: -83.6
***d*:** $0.9003^{20/4}$
***n*:** 1.3723^{20}
***s*:** H_2O 3; al 5; eth 5; ace 4; bz 4; chl 5; os 4
IR: COB 4802 2980 1740 1480 1460 1440 1370 1300 1230 1100 1050 940 850 780
Raman: SAD 28 2970 2940 2880 2780 2720 1730 1450 1400 1350 1300 1270 1240 1170 1110 1090 1040 1000 940 920 840 780 630 460 440 380 310 200
UV: OES 4-21 209(72) MeOH
H NMR: VAR 79 1.3 2.0 4.1 $CDCl_3$
MS: WILEY 158 43(100) 29(20) 45(14) 61(12) 15(18) 27(11) 70(9) 88(2)
m/e: 88.05

HODOC No: 00852
CAS No: 758-12-3
Name: Acetic acid-*d*
Synonym: Acetic acid, *O*-deutero
Line Form: H_3CCO_2D
Mol Form: $C_2H_3DO_2$
M$_r$: 61.06
***n*:** $1.3715^{20/D}$
***s*:** ctc 3
H NMR: SAD 10236 2.1 CCl_4
m/e: 61.03

HODOC No: 00853
CAS No: 53897-38-4
Name: Acetic-*d*3 acid, butyl ester
Synonym: Acetic acid, trideutero,butyl ester
Line Form: $CD_3CO_2(CH_2)_3CH_3$
Mol Form: $C_6H_9D_3O_2$
M$_r$: 119.18
MS: WILEY 471 46(100) 56(56) 41(19) 64(16) 28(15) 79(13) 29(9) 119(1)
m/e: 119.11

HODOC No: 00854
CAS No: 53778-70-4
Name: Acetic-*d*3 acid, 1-methylethyl ester
Synonym: Acetic-*d*3 acid, isopropyl ester
Line Form: $CD_3CO_2CH(CH_3)_2$
Mol Form: $C_5H_7D_3O_2$
M$_r$: 105.15
MS: WILEY 313 46(100) 43(18) 41(18) 64(17) 27(10) 42(10) 39(10) 105(1)
m/e: 105.09

HODOC No: 00855
CAS No: 53897-35-1
Name: Acetic-*d*3 acid, 1-methylpropyl ester
Synonym: Acetic acid, trideutero,*sec*-butyl ester
Line Form: $CD_3CO_2CH(CH_3)CH_2CH_3$
Mol Form: $C_6H_9D_3O_2$
M$_r$: 119.18
MS: WILEY 471 46(100) 56(80) 90(24) 41(12) 29(10) 45(6) 57(6) 119(1)
m/e: 119.11

HODOC No: 00856
CAS No: 30564-13-7
Name: Acetoacetamide, 2,2′-methylenebis[*N*-cyclohexyl-
Synonym: Glutaramide, 2,4-diacetyl-*N*,*N*′-dicyclohexyl-
Mol Form: $C_{21}H_{34}N_2O_4$
Mr: 378.49
mp: dec 207
IR: **µCOB 4412** 3330 2940 1720 1640 1540 1450 1350 1150 970 890 700
m/e: 378.25

HODOC No: 00857
CAS No: 7145-38-2
Name: Acetoacetamide, *N*-(4-methyl-2-thiazolyl)-
Synonym: 2-Acetoacetamido-4-methylthiazole
Mol Form: $C_8H_{10}N_2O_2S$
Mr: 198.23
s: DMSO 3
IR: **SADG 13915** 3180 3050 2980 1730 1660 1560 1420 1330 1310 1170 1160 950 860 790 740 600 540
H NMR: **SAD 8335** 2.2 2.3 3.7 6.7 DMSOd6
m/e: 198.05

HODOC No: 00858
CAS No: 13236-39-0
Name: Acetoacetic acid, 2,4-*bis*(2,4-dichlorophenoxy)-, ethyl ester
Synonym: Butanoic acid, 3-oxo,2,4-*bis*(2,4-dichlorophenoxy),ethyl ester
Mol Form: $C_{18}H_{14}Cl_4O_5$
Mr: 452.10
mp: 96-8
s: chl 2
IR: **µSADP 3690** 3125 2940 1725 1615 1495 1315 1300 1250 1085 1020 870 800 745 715
UV: **SAD 1124** 282(4870) 229(20300) diox
H NMR: **SAD 10651** 1.3 4.3 5.1 5.3 6.7 6.8 7.1 7.3 $CDCl_3$
m/e: 449.96

HODOC No: 00859
CAS No: 1869-63-2
Name: Acetoacetic acid, 4,4-difluoro-, ethyl ester, (2,4-dinitrophenyl)hydrazone
Synonym: Butanoic acid, 4,4-difluoro-3-oxo,2,4-dinitrophenylhydrazone,ethyl ester
Structure:

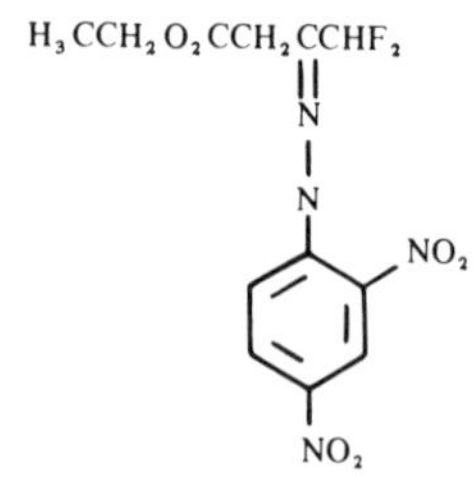

Mol Form: $C_{12}H_{12}F_2N_4O_6$
Mr: 346.24
s: chl 2
H NMR: **VAR 293** 1.3 3.7 4.3 6.3 8.0 8.4 9.1 11.8 $CDCl_3$
m/e: 346.07

HODOC No: 00860
CAS No: 15780-57-1
Name: Acetoacetic acid, 1-isopropylcyclohexyl ester
Synonym: Cyclohexanol, 1-isopropyl-, acetoacetate
Mol Form: $C_{13}H_{22}O_3$
Mr: 226.30
MS: **WILEY 1608** 43(100) 57(65) 82(38) 41(24) 44(23) 39(19) 67(19) 226(0)
m/e: 226.16

HODOC No: 00861
CAS No: 15780-65-1
Name: Acetoacetic acid, 1-thio-, *S*-allyl ester
Synonym: Butanoic acid, thiolo,3-oxo,allyl ester
Line Form: $CH_3COCH_2C(=O)SCH_2CH=CH_2$
Mol Form: $C_7H_{10}O_2S$
Mr: 158.21
MS: **WILEY 979** 43(100) 39(80) 41(64) 45(56) 27(39) 38(30) 26(28) 158(2)
m/e: 158.04

HODOC No: 00862
CAS No: 15780-64-0
Name: Acetoacetic acid, 1-thio-, *S-sec*-butyl ester
Synonym: Butanoic acid, thiolo,3-oxo,*sec*-butyl ester
Line Form: $CH_3COCH_2C(=O)SCH(CH_3)CH_2CH_3$
Mol Form: $C_8H_{14}O_2S$
M_r: 174.25
MS: WILEY 1172 85(100) 43(66) 57(20) 174(18) 41(13) 29(11) 61(10) 90(10)
m/e: 174.07

HODOC No: 00863
CAS No: 764-29-4
Name: Acetohydroxamic acid, 2,2′-dithiodi-
Synonym: Disulfide, 1,2-*bis*(2-hydroxyamino-2-oxoethyl)
Line Form: $HONHCOCH_2SSCH_2CONHOH$
Mol Form: $C_4H_8N_2O_4S_2$
M_r: 212.23
mp: 141
IR: COB 4527 3200 1640 1500 1380 1200 1140 1050 980 880 710
m/e: 211.99

HODOC No: 00864
CAS No: 772-68-9
Name: Acetohydroxamic acid, 2-[(tetrahydro-2H-pyran-2-yl)thio]-
Synonym: Acetohydroxamic acid, 2-[(tetrahydropyran-2-yl)thio]-
Structure:

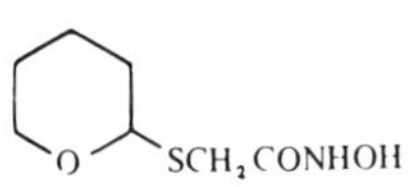

Mol Form: $C_7H_{13}NO_3S$
M_r: 191.24
mp: 100-1
IR: COB 4535 3150 2930 1620 1400 1320 1290 1140 1010 980 890 720
m/e: 191.06

HODOC No: 00865
CAS No: 108473-88-7
Beilstein No: 8^3, 1137
Name: 2′-Acetonaphthone, 4′-hydroxy-
Synonym: 1-Naphthol, 3-acetyl
Line Form: $3\text{-}CH_3CO\text{-}1\text{-}HOC_{10}H_6$
Mol Form: $C_{12}H_{10}O_2$
M_r: 186.20
mp: 173-4
Color: nd (Bz)
***s*:** EtOH 4; HOAc 4
m/e: 186.07

HODOC No: 00866
CAS No: 24764-47-4
Beilstein No: 8^3, 1135
Name: 2′-Acetonaphthone, 1′-hydroxy-4′-nitro-
Synonym: 1-Naphthol, 2-acetyl-4-nitro
Line Form: $2\text{-}CH_3CO\text{-}4\text{-}NO_2\text{-}1\text{-}HOC_{10}H_5$
Mol Form: $C_{12}H_9NO_4$
M_r: 231.20
mp: 159
Color: ye nd (al)
***s*:** Bz 4; Et_2O 4
m/e: 231.05

HODOC No: 00867
CAS No: 28613-56-1
Name: Acetone, 4-(1-adamantyl)thiosemicarbazone
Synonym: Semicarbazide, 4-(1-adamantyl)-1-isopropylidene-3-thio-
Mol Form: $C_{14}H_{23}N_3S$
M_r: 265.40
mp: dec 200
***s*:** chl 2
IR: SADG 19917 3280 2980 2900 2840 1600 1530 1480 1360 1300 1250 1190 1130 1100 990 960 800 710 680 650 590 550
H NMR: SAD 10505 1.7 1.9 2.0 2.1 2.3 7.4 8.0 $CDCl_3$
m/e: 265.16

HODOC No: 00868
CAS No: 25978-95-4
Name: Acetone, [5-(*o*-hydroxyphenyl)-*s*-triazol-3-yl]hydrazone
Synonym: 1,2,4-Triazole, 1-hydro-3(2-hydroxyphenyl)-5(isopropylidenehydazino)
Mol Form: $C_{11}H_{13}N_5O$
M_r: 231.25
mp: 280-2
IR: SADG 22073 3100 1630 1610 1450 1380 1240 1100 1070 1020 820 750
UV: SAD 18900 300(7590) 243(27800) 209(40800) MeOH
H NMR: SAD 11789 1.9 2.0 6.9 7.0 7.3 8.0 10.1 $DMSOd_6$
m/e: 231.11

HODOC No: 00869
CAS No: 75-05-8
Beilstein No: 2^4, 419
Name: Acetonitrile
Synonym: Cyanomethane
Line Form: CH_3CN
Mol Form: C_2H_3N
M_r: 41.05
bp: 81.6
mp: -45.7
***d*:** 0.7857^{20}
***n*:** 1.3442^{30}
***s*:** H_2O 5; al 5; eth 5; ace 5; bz 5; ctc 5
IR: μSADP 269 2940 2270 2220 1430 1410 1370 1030 920 790 750
Raman: DFB 63 3010 2950 2890 2850 2730 2290 2260 1450 1380 920 750 380
UV: OES 4-2 338(126) C_2Cl_4
H NMR: SAD 9154 1.9 CCl_4
C NMR: JJ 4 cw 1.3 117.7 diox
MS: WILEY 9 41(100) 40(52) 39(19) 38(12) 14(11) 12(6) 26(3) 13(3)
m/e: 41.03

HODOC No: 00870
CAS No: 540-61-4
Beilstein No: 4^4, 2363
Name: Acetonitrile, amino-
Synonym: AAN
Line Form: H_2NCH_2CN
Mol Form: $C_2H_4N_2$
M_r: 56.07
bp: 58^{15}
***s*:** EtOH 4
m/e: 56.04

HODOC No: 00871
CAS No: 24272-68-2
Name: Acetonitrile, [(4-amino-4,5-dihydro-6-methyl-5-oxo-1,2,4-triazin-3-yl)thio]-
Synonym: Acetonitrile, [(4-amino-4,5-dihydro-6-methyl-5-oxo-as-triazin-3-yl)thio]-
Mol Form: $C_6H_7N_5OS$
M_r: 197.21
mp: 193
UV: LANG 2397 288(7413) 207(12023) EtOH
m/e: 197.04

HODOC No: 00872
CAS No: 590-17-0
Beilstein No: 2^4, 531
Name: Acetonitrile, bromo-
Synonym: Bromoacetonitrile
Line Form: $BrCH_2CN$
Mol Form: C_2H_2BrN
M_r: 118.95
bp: $60\text{-}2^{24}$
m/e: 118.94

HODOC No: 00873
CAS No: 3010-04-6
Beilstein No: 4^4, 1139
Name: Acetonitrile, (butylamino)-
Synonym: Butylaminoacetonitrile
Line Form: $CH_3(CH_2)_3NH\text{-}CH_2CN$
Mol Form: $C_6H_{12}N_2$
M_r: 112.17
bp: 85^9
***d*:** 0.891
***n*:** 1.4337^{20}
IR: ALD 2:454G
MS: ALD 3:167B
m/e: 112.10

HODOC No: 00874
CAS No: 107-14-2
Beilstein No: 2^4, 492
Name: Acetonitrile, chloro-
Synonym: Chloracetonitrile
Line Form: $ClCH_2CN$
Mol Form: C_2H_2ClN
M$_r$: 75.50
bp: 126-7, $30\text{-}2^{15}$
***d*:** 1.1930^{20}
***n*:** 1.4202^{25}
***s*:** Et_2O 4; EtOH 4
IR: COB 7886 3015 2970 2260 1690 1420 1270 930 740 490
H NMR: SAD 6783 4.1 CCl_4
MS: NBS 23
m/e: 74.99

HODOC No: 00875
CAS No: 421-05-6
Name: Acetonitrile, chlorodifluoro-
Synonym: Chlorodifluoroacetonitrile
Line Form: F_2CClCN
Mol Form: C_2ClF_2N
M$_r$: 111.48
MS: WILEY 365 76(100) 85(11) 50(7) 31(5) 92(5) 87(4) 77(2) 111(1)
m/e: 110.97

HODOC No: 00876
CAS No: 1074-58-4
Beilstein No: 12, 7
Name: Acetonitrile, (cyclohexylamino)-
Synonym: Cyclohexylaminoacetonitrile
Line Form: $C_6H_{11}NHCH_2CN$
Structure:

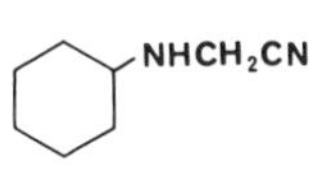

Mol Form: $C_8H_{14}N_2$
M$_r$: 138.20
bp: $74\text{-}6^1$
mp: 18
IR: µCOB 3391 3330 2940 2860 1450 1410 1370 1250 1120 1030 920 890 860 750
m/e: 138.12

HODOC No: 00877
CAS No: 4435-18-1
Beilstein No: 9^4, 124
Name: Acetonitrile, cyclohexylidene-
Synonym: Δ1,α-Cyclohexaneacetonitrile
Line Form: C_6H_{10}=CHCN
Structure:

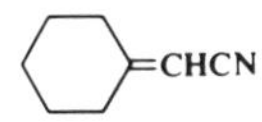

Mol Form: $C_8H_{11}N$
M$_r$: 121.17
bp: $107\text{-}8^{22}$
***d*:** $0.9483^{15/4}$
***n*:** 1.4382^{25}
***s*:** Et_2O 4; EtOH 4
m/e: 121.09

HODOC No: 00878
CAS No: 3252-43-5
Beilstein No: 2, 219
Name: Acetonitrile, dibromo-
Synonym: Dibromoacetonitrile
Line Form: Br_2CHCN
Mol Form: C_2HBr_2N
M$_r$: 196.84
bp: $67\text{-}9^{24}$
***d*:** 2.296
***n*:** 1.5393^{20}
IR: COB 6434 3000 1580 1220 1120 970 690 620 450
MS: ALD 3:162C
m/e: 196.85

HODOC No: 00879
CAS No: 6698-74-4
Name: Acetonitrile, dibromofluoro-
Synonym: Dibromofluoroacetonitrile
Line Form: Br_2CF-CN
Mol Form: C_2Br_2FN
M$_r$: 214.83
bp: $90\text{-}1^{758}$
***n*:** 1.4620^{20}
IR: µCOB 3236 1050 810 790 740 700
m/e: 214.84

HODOC No: 00880
CAS No: 3018-12-0
Beilstein No: 2^4, 506
Name: Acetonitrile, dichloro-
Synonym: Dichloroacetonitrile
Line Form: Cl_2CHCN
Mol Form: C_2HCl_2N
M_r: 109.94
bp: 112-3
***d*:** 1.369^{20}
***n*:** 1.4391^{25}
***s*:** MeOH 3
IR: COB 4639 2990 2250 1260 1200 1000 960 770 710
m/e: 108.95

HODOC No: 00881
CAS No: 353-82-2
Beilstein No: 2^4, 508
Name: Acetonitrile, dichlorofluoro-
Synonym: Dichlorofluoroacetonitrile
Line Form: Cl_2CF-CN
Mol Form: C_2Cl_2FN
M_r: 127.93
bp: 32.3^{738}
mp: -110
***d*:** 1.3909^{25}
***n*:** 1.3682^{20}
IR: μCOB 3258 2270 1110 1090 1010 970 860 830
MS: WILEY 547 92(100) 66(9) 31(6) 47(4) 73(4) 57(4) 94(3) 127(0)
m/e: 126.94

HODOC No: 00882
CAS No: 3010-02-4
Beilstein No: 4^4, 2378
Name: Acetonitrile, (diethylamino)-
Synonym: Diethylaminoacetonitrile
Line Form: $(C_2H_5)_2NCH_2CN$
Mol Form: $C_6H_{12}N_2$
M_r: 112.17
bp: 170, 70^{24}
***n*:** 1.4260^{20}
***s*:** H_2O 3
IR: μCOB 384 2940 2780 2170 1670 1470 1430 1390 1350 1320 1300 1270 1210 1160 1120 1100 1080 980 890 860 820 810 770 760
m/e: 112.10

HODOC No: 00883
CAS No: 51054-41-2
Name: Acetonitrile, (3,4-dihydro-6,7-dimethoxy-1(2H)-isoquinolinylidene)-
Synonym: Δ1(2H),α-Isoquinolineacetonitrile, 3,4-dihydro-6,7-dimethoxy-
Mol Form: $C_{13}H_{14}N_2O_2$
M_r: 230.25
mp: 173
UV: LANG 2170 325(15849) 274(9772) 226(22909) EtOH 326(17783) 277(14454) chl
m/e: 230.11

HODOC No: 00884
CAS No: 926-64-7
Beilstein No: 4^4, 2369
Name: Acetonitrile, (dimethylamino)-
Synonym: *N*-(Cyanomethyl)dimethylamine
Line Form: $(CH_3)_2NCH_2CN$
Mol Form: $C_4H_8N_2$
M_r: 84.12
bp: 137-8, 42^{22}
***d*:** $0.8650^{20/0}$
***n*:** 1.4095^{20}
***s*:** H_2O 4; EtOH 4
m/e: 84.07

HODOC No: 00885
CAS No: 24426-40-2
Beilstein No: 4^4, 2374
Name: Acetonitrile, (ethylamino)-
Synonym: (Ethylamino)acetonitrile
Line Form: $C_2H_5NHCH_2CN$
Mol Form: $C_4H_8N_2$
M_r: 84.12
bp: 166-7, $81\text{-}3^{29}$
***s*:** Et_2O 4; EtOH 4
m/e: 84.07

HODOC No: 00886
CAS No: 107-16-4
Beilstein No: 3^4, 598
Name: Acetonitrile, hydroxy-
Synonym: Cyanomethanol
Line Form: $HOCH_2CN$
Mol Form: C_2H_3NO
M$_r$: 57.05
bp: 183d, 119^{24}
mp: 72
***n*:** 1.4117^{19}
***s*:** H_2O 4; al 4; eth 4; bz 1; chl 1
IR: µSADP 5141 3450 2940 2220 1640 1430 1350 1240 1050 980 890 710
MS: WILEY 26 18(100) 29(83) 28(81) 27(78) 30(65) 17(23) 26(12) 57(1)
m/e: 57.02

HODOC No: 00887
CAS No: 628-87-5
Beilstein No: 4^4, 2428
Name: Acetonitrile, 2,2′-iminobis-
Synonym: Acetonitrile, iminodi-
Line Form: $CNCH_2$-NH-CH_2CN
Mol Form: $C_4H_5N_3$
M$_r$: 95.10
mp: 77-9
***s*:** H_2O 3; al 3; eth 2; bz 2; chl 2
IR: µCOB 5822 3330 2860 2220 1450 1390 1330 1300 1120 1080 930 900 860 750
UV: OES 2-15 271(9) EtOH 267(10) aa
H NMR: SAD 4312 1.8 3.7 $CDCl_3$
m/e: 95.05

HODOC No: 00888
CAS No: 624-75-9
Beilstein No: 2^2, 206
Name: Acetonitrile, iodo-
Synonym: Cyanoiodomethane
Line Form: ICH_2CN
Mol Form: C_2H_2IN
M$_r$: 166.95
bp: $182\text{-}4^{720}$
***d*:** 2.307
***n*:** 1.5744^{20}
IR: ALD 2:450A
MS: ALD 3:162B
m/e: 166.92

HODOC No: 00889
CAS No: 1738-36-9
Beilstein No: 3, 242
Name: Acetonitrile, methoxy-
Synonym: Cyanomethyl methyl ether
Line Form: CH_3OCH_2CN
Mol Form: C_3H_5NO
M$_r$: 71.08
bp: 118.1^{731}
***d*:** $0.9492^{20/4}$
***n*:** 1.3831^{20}
***s*:** H_2O 2; al 3; eth 3; ace 3; chl 3; ac 3; alk 3; os 3
IR: µSADP 19936 2940 2220 1470 1350 1280 1190 1110 1010 960 920 890
H NMR: VAR 28 3.5 4.2 $CDCl_3$
m/e: 71.04

HODOC No: 00890
CAS No: 109-82-0
Name: Acetonitrile, (methyleneamino)-
Synonym: Glycinonitrile, *N*-methylene-
Line Form: CH_2=NCH_2CN
Mol Form: $C_3H_4N_2$
M$_r$: 68.08
IR: µCOB 820 2940 2860 2630 2220 1470 1430 1350 1320 1300 1250 1180 1160 1120 1090 1020 970 940 900 860 690
m/e: 68.04

HODOC No: 00891
CAS No: 1116-43-4
Name: Acetonitrile, 2,2′,2″,2‴-(methylenedinitrilo)tetrakis-
Synonym: Acetonitrile, (methylenedinitrilo)tetra-
Line Form: $(NCCH_2)_2NCH_2N(CH_2CN)_2$
Mol Form: $C_9H_{10}N_6$
M$_r$: 202.21
IR: COB 5823 2940 2240 1420 1340 1280 1180 1150 1130 1020 950 850
m/e: 202.10

HODOC No: 00892
CAS No: 20661-61-4
Name: Acetonitrile, (methylnitroamino)-
Synonym: Acetonitrile, (methylnitramino)-

HODOC No: 00893
CAS No: 14321-25-6
Name: Acetonitrile, [methyl(phenylmethyl)amino]-
Synonym: Acetonitrile, (benzylmethylamino)-
Line Form: $C_6H_5CH_2N(CH_3)CH_2CN$
Mol Form: $C_{10}H_{12}N_2$
M_r: 160.21
IR: **μCOB 3353** 3030 2940 2780 2220 1610 1490 1450 1410 1370 1320 1120 1080 1030 1020 980 860 840 740 690
m/e: 160.10

HODOC No: 00894
CAS No: 56095-81-9
Name: Acetonitrile, [(1-methylpropyl)amino]-
Synonym: *N-sec*-Butylaminoacetonitrile
Line Form: $CH_3CH_2CH(CH_3)NHCH_2CN$
Mol Form: $C_6H_{12}N_2$
M_r: 112.17
IR: **μCOB 3399** 3450 3030 2270 1470 1430 1390 1180 1150 1090 960 870 780 740
m/e: 112.10

HODOC No: 00895
CAS No: 35120-10-6
Beilstein No: 3^4, 630
Name: Acetonitrile, (methylthio)-
Synonym: Cyanomethyl methyl sulfide
Line Form: CH_3SCH_2CN
Mol Form: C_3H_5NS
M_r: 87.14
bp: $61\text{-}3^{15}$
***d*:** 1.039
***n*:** 1.4826^{20}
IR: ALD 2:451G
Raman: SAD 979 2970 2930 2830 2800 2250 1430 1400 1320 1240 1190 990 920 850 750 700 500 370 290 190
H NMR: SAD 13379 2.3 3.3 CCl_4
MS: ALD 3:153B
m/e: 87.01

HODOC No: 00896
CAS No: 3598-14-9
Beilstein No: 6^4, 640
Name: Acetonitrile, phenoxy-
Synonym: Phenoxyacetonitrile
Line Form: $C_6H_5OCH_2CN$
Mol Form: C_8H_7NO
M_r: 133.14
bp: 239-40, 128^{17}
***d*:** $1.0991^{20/4}$
***n*:** 1.5246^{20}
***s*:** Et_2O 4; EtOH 4
m/e: 133.05

HODOC No: 00897
CAS No: 3009-97-0
Beilstein No: 12^3, 196
Name: Acetonitrile, (phenylamino)-
Synonym: Acetonitrile, anilino-
Line Form: $C_6H_5NHCH_2CN$
Mol Form: $C_8H_8N_2$
M_r: 132.16
mp: 48
***s*:** H_2O 2; al 3; eth 2; bz 3; chl 2; lig 2
IR: SADG 18497 3400 3060 2950 2240 1600 1500 1440 1310 1250 1180 1100 880 750 690
UV: SAD 2096 289 240 MeOH
H NMR: SAD 6201 3.8 3.9 6.6 6.8 7.2 $CDCl_3$
m/e: 132.07

HODOC No: 00898
CAS No: 545-06-2
Beilstein No: 2^4, 524
Name: Acetonitrile, trichloro-
Synonym: Cyanotrichloromethane
Line Form: Cl_3CCN
Mol Form: C_2Cl_3N
M_r: 144.39
bp: 85.7
mp: -42
***d*:** $1.4403^{25/4}$
***n*:** 1.4409^{20}
***s*:** H_2O 1
IR: **μCOB 3256** 2270 1030 970 780
MS: NBS 383
m/e: 142.91

HODOC No: 00899
CAS No: 353-85-5
Beilstein No: 2^4, 472
Name: Acetonitrile, trifluoro-
Synonym: Cyanotrifluoromethane
Line Form: CF_3CN
Mol Form: C_2F_3N
M_r: 95.02
bp: -64
IR: µSADP 2503 2380 2220 2000 1430 1330 1250 1090 920 810 790
MS: NBS 4000
m/e: 95.00

HODOC No: 00900
CAS No: 630-72-8
Beilstein No: 2, 229
Name: Acetonitrile, trinitro-
Synonym: Trinitroacetonitrile
Line Form: $(NO_2)_3CCN$
Mol Form: $C_2N_4O_6$
M_r: 176.04
bp: 220 exp
mp: 41.5
Color: wax
***s*:** Et_2O 4
m/e: 175.98

HODOC No: 00901
CAS No: 23699-41-4
Name: Acetophenone, 4′-(*o*-aminophenoxy)-
Synonym: Acetophenone, 4(2-aminophenoxy)
Mol Form: $C_{14}H_{13}NO_2$
M_r: 227.25
mp: 96-7
IR: µSADP 4499 3450 3330 3130 2940 1640 1610 1590 1560 1490 1410 1350 1300 1270 1220 1190 1160 950 880 840 820 760
UV: SAD 1252 259(17700) MeOH
H NMR: SAD 10592 2.5 3.7 6.6-7.1 7.8 $CDCl_3$
m/e: 227.09

HODOC No: 00902
CAS No: 21290-60-8
Name: Acetophenone, 2-(benzyl-*tert*-butylamino)-3′,4′-*bis*(benzyloxy)-
Synonym: Acetophenone, α(benzyl-*tert*-butylamino)-3,4-dibenzyloxy
Structure:

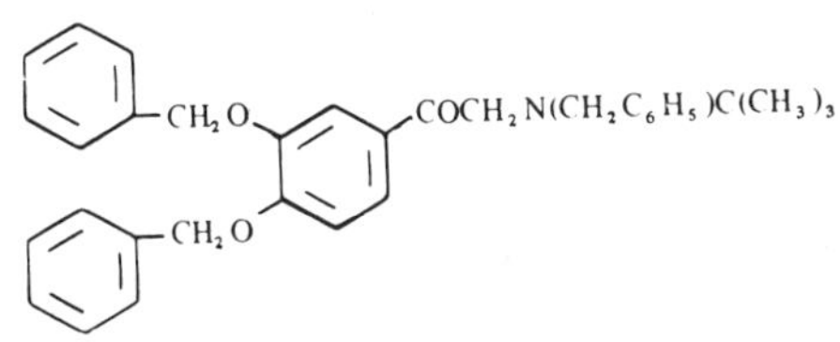

Mol Form: $C_{33}H_{35}NO_3$
M_r: 493.61
mp: 83.5-84
***s*:** chl 3
H NMR: SAD 9225 1.2 3.8 5.1 5.2 6.8 7.0 7.5 $CDCl_3$
m/e: 493.26

HODOC No: 00903
CAS No: 21185-61-5
Name: Acetophenone, 3′,4′-*bis*(benzyloxy)-2-(dibenzylamino)-
Synonym: Acetophenone, α(dibenzylamino)-3,4-dibenzoxy
Mol Form: $C_{36}H_{33}NO_3$
M_r: 527.63
bp: 96-7.5
IR: SADG 17499 3040 2900 2840 1680 1590 1580 1510 1450 1430 1380 1250 1140 1000 850 770 750 700
H NMR: SAD 9222 3.7 5.1 5.2 6.8 7.1-7.6 $CDCl_3$
m/e: 527.25

HODOC No: 00904
CAS No: 25555-22-0
Name: Acctophenone, 4′-chloro-2-(1-pyrrolidinylimino)-
Synonym: Acetophenone, 4-chloro-α(pyrrolidinylimino)
Mol Form: $C_{12}H_{13}ClN_2O$
M_r: 236.69
mp: 88-9
IR: SADG 20529 2970 2960 2850 1620 1580 1560 1510 1470 1440 1380 1330 1260 1210 1160 1150 1090 1000 850 840 800 780 740 670 510 470 380
UV: SAD 18232 335(20700) 253(9590) MeOH
H NMR: SAD 11158 2.0 3.4 7.0 7.4 7.9 $CDCl_3$
m/e: 236.07

HODOC No: 00905
CAS No: 109311-05-9
Name: Acetophenone, 2′,4′-dihydroxy-, benzoate
Synonym: Acetophenone, 2,4-dihyroxy,4-benzoate
Mol Form: $C_{15}H_{12}O_4$
M_r: 256.24
mp: 106
UV: LANG 2144 324(4266) 261(13804) 227(15849) 216(14791) cyhex
m/e: 256.07

HODOC No: 00906
CAS No: 108972-73-2
Name: Acetophenone, 4′-(4-dimethylamino-*o*-tolylazo)-
Synonym: Azobenzene, 4-acetyl-4′(dimethylamino)-2′-methyl
Mol Form: $C_{17}H_{19}N_3O$
M_r: 281.34
mp: 152.5-4.0
IR: SADG 21165 2900 1670 1590 1520 1420 1360 1330 1290 1280 1210 1190 1090 1010 950 870 850 800 680 590 460
UV: SAD 17868 450(35200) 286(14000) 230(10800) MeOH
H NMR: SAD 11193 2.6 3.0 6.5 7.8 8.0 $CDCl_3$
m/e: 281.15

HODOC No: 00907
CAS No: 23975-31-7
Name: Acetophenone, 2′-hydroxy-2-(phenylsulfinyl)-
Synonym: Acetophenone, 2-hydroxy-α(phenylsulfinyl)
Mol Form: $C_{14}H_{12}O_3S$
M_r: 260.29
mp: 116
s: chl 2
IR: SADG 19682 3400 3040 2940 1630 1480 1440 1330 1300 1250 1210 1150 1040 830 750 700 680 580
UV: SAD 17308 331(4900) 258(14900) 211(30200) MeOH
H NMR: SAD 10271 4.3 4.6 6.9 7.0 7.3-7.8 $CDCl_3$
m/e: 260.05

HODOC No: 00908
CAS No: 25561-48-2
Name: Acetophenone, 4′-methoxy-2-[(4-methyl-1-piperazinyl)imino]-
Synonym: Acetophenone, 4-methoxy-α(4-methyl-1-piperazinylimino)
Structure:

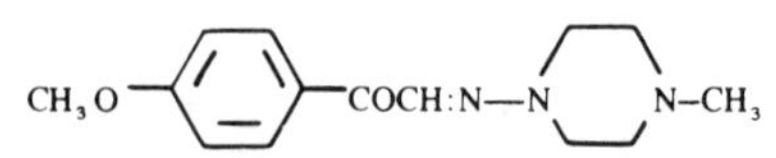

Mol Form: $C_{14}H_{19}N_3O_2$
M_r: 261.31
mp: 73-5
IR: SADG 20583 2930 2830 2790 1620 1590 1560 1540 1510 1440 1410 1360 1330 1280 1240 1160 1130 1120 1070 1020 990 830 810 740 610 510
UV: SAD 18240 329(22500) 227(9250) MeOH
H NMR: SAD 11166 2.3 2.6 3.4 3.8 6.9 7.3 8.1 $CDCl_3$
m/e: 261.15

HODOC No: 00909
CAS No: 25555-21-9
Name: Acetophenone, 4′-methoxy-2-(1-pyrrolidinylimino)-
Synonym: Acetophenone, 4-methoxy-α(pyrrolidinylimino)
Structure:

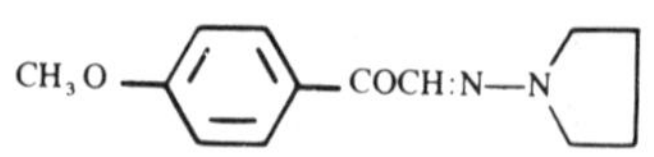

Mol Form: $C_{13}H_{16}N_2O_2$
M_r: 232.27
mp: 82-3
IR: SADG 20528 2980 2940 2860 2840 1600 1570 1520 1480 1450 1410 1390 1340 1280 1240 1170 1090 1020 970 860 830 810 750 720 670 600 560 510 410
UV: SAD 18231 341(26000) 224(9710) MeOH
H NMR: SAD 11157 2.0 3.5 3.8 6.9 7.0 8.0 $CDCl_3$
m/e: 232.12

HODOC No: 00910
CAS No: 25555-20-8
Name: Acetophenone, 4′-methyl-2-(1-pyrrolidinylimino)-
Synonym: Acetophenone, 4-methyl-α-pyrrolidinylimino
Mol Form: $C_{13}H_{16}N_2O$
M_r: 216.27
mp: 84-5
***s*:** chl 2
IR: SADG 20527 3020 2960 1600 1510 1450 1390 1270 1090 860 830 740 720
UV: SAD 18230 332(20800) 265(7570) MeOH
H NMR: SAD 11156 2.0 2.4 3.4 7.1 7.3 7.9 $CDCl_3$
m/e: 216.13

HODOC No: 00911
CAS No: 25561-43-7
Name: Acetophenone, 2-(morpholinoimino)-4′-phenyl-
Synonym: Acetophenone, 4-phenyl-α-morpholinoimino
Mol Form: $C_{18}H_{18}N_2O_2$
M_r: 294.34
mp: 157
***s*:** chl 2
IR: SADG 20537 3040 2960 2860 1640 1600 1500 1440 1330 1240 1160 1100 1000 900 830 770 740 700
UV: SAD 18239 326(21800) 226(11500) MeOH
H NMR: SAD 11165 3.3 3.8 7.2-7.8 8.1 $CDCl_3$
m/e: 294.14

HODOC No: 00912
CAS No: 107920-11-6
Name: Acetophenone, 3′-[*p*-(ethylmethylamino)phenylazo]-
Synonym: Azobenzene, 3-acetyl-4′(*N*-ethylmethylamino)
Mol Form: $C_{17}H_{19}N_3O$
M_r: 281.34
mp: 79-80
IR: μSADP 7269 2940 1700 1610 1520 1430 1390 1350 1250 1150 990 820 790 690
UV: SAD 18294 424(32100) 273(11600) 240(19700) MeOH
H NMR: SAD 11197 1.1 2.6 2.9 3.4 6.7 7.4 7.7-8.1 8.3 $CDCl_3$
m/e: 281.15

HODOC No: 00913
CAS No: 107920-10-5
Name: Acetophenone, 4′-[*p*-(ethylmethylamino)phenylazo]-
Synonym: Azobenzene, 4-acetyl-4′(*N*-ethylmethylamino)
Mol Form: $C_{17}H_{19}N_3O$
M_r: 281.34
mp: 150.0-0.5
IR: SADG 21146 2970 2880 1670 1590 1510 1410 1370 1340 1300 1250 1210 1140 1130 1070 990 950 840 820 790 680 630 590 530 460 390
UV: SAD 2013 282(12600) 230(8720) MeOH
H NMR: SAD 11009 1.5 2.8 3.6 4.0 7.2 7.7 8.0 8.2 tfa
m/e: 281.15

HODOC No: 00914
CAS No: 25555-25-3
Name: Acetophenone, 4′-phenoxy-2-(1-pyrrolidinylimino)-
Synonym: Acetophenone, 4-phenoxy-α-pyrrolidinylimino
Mol Form: $C_{18}H_{18}N_2O_2$
M_r: 294.34
mp: 104-5
***s*:** chl 2
IR: SADG 20532 3050 2970 2860 1620 1580 1490 1450 1330 1210 1160 1020 900 860 800 750 700
UV: SAD 18235 339(23800) 222(14700) MeOH
H NMR: SAD 11160 2.0 3.4 6.8-7.5 8.0 $CDCl_3$
m/e: 294.14

HODOC No: 00915
CAS No: 25555-26-4
Name: Acetophenone, 4′-(phenylthio)-2-(1-pyrrolidinylimino)-
Synonym: Acetophenone, 4-phenylthio-α-pyrrolidinylimino
Mol Form: $C_{18}H_{18}N_2OS$
M_r: 310.40
mp: 110-1
***s*:** chl 3
IR: SADG 20533 3040 2970 2860 1620 1590 1500 1450 1340 1220 1080 1010 900 810 760
UV: SAD 18236 347(25400) 240(12500) MeOH
H NMR: SAD 11161 2.0 3.4 7.0 7.1-7.6 7.9 $CDCl_3$
m/e: 310.11

HODOC No: 00916
CAS No: 28567-96-6
Name: Acetophenone, 2-(1,2,3,4-tetrahydro-3-phenyl-2-thioxo-4-quinazolinyl)-
Synonym: 2(1H)-Quinazolinethione, 3,4-dihydro-4-phenacyl-3-phenyl-
Mol Form: $C_{22}H_{18}N_2OS$
M_r: 358.44
mp: dec 253
UV: LANG 2602 283(20417) 249(19055) EtOH
m/e: 358.11

HODOC No: 00917
CAS No: 17738-48-6
Name: *p*-Acetotoluidide, 2-[*N*-(*p*-chlorophenyl)benzimidoyl]-
Synonym: Propanoic acid, amide,3(4-chlorophenylimino)-3-phenyl-*N*(4-toly
Mol Form: $C_{22}H_{19}ClN_2O$
M_r: 362.84
mp: 155-6
UV: LANG 1338 355(30903) 258(18197) EtOH
m/e: 362.12

HODOC No: 00918
CAS No: 17738-56-6
Name: *p*-Acetotoluidide, 2-[*N*-(*p*-methoxyphenyl)benzimidoyl]-
Synonym: Propanoic acid, amide,3(4-methoxyphenylimino)-*N*(4-methylphenyl)-3-phenyl
Mol Form: $C_{23}H_{22}N_2O_2$
M_r: 358.42
mp: 176-7
UV: LANG 1341 355(27542) 261(19055) EtOH
m/e: 358.17

HODOC No: 00919
CAS No: 17738-50-0
Name: *p*-Acetotoluidide, 2-(*N*-1-naphthylbenzimidoyl)-
Synonym: Propanoic acid, amide,*N*(4-methylphenyl)-3(1-naphthylamino)-3-phenyl
Mol Form: $C_{26}H_{22}N_2O$
M_r: 378.45
mp: 195-6
UV: LANG 1972 355(25119) EtOH
m/e: 378.17

HODOC No: 00920
CAS No: 17738-51-1
Name: *p*-Acetotoluidide, 2-(*N*-2-naphthylbenzimidoyl)-
Synonym: Propanoic acid, amide,*N*(4-methylphenyl)-3(2-naphthylimino)-3-phenyl
Mol Form: $C_{26}H_{22}N_2O$
M_r: 378.45
mp: 205-6
UV: LANG 1973 357(30200) 252(24547) 215(42658) EtOH
m/e: 378.17

HODOC No: 00921
CAS No: 17738-49-7
Name: *p*-Acetotoluidide, 2-[*N*-(*p*-nitrophenyl)benzimidoyl]-
Synonym: Propanoic acid, amide,*N*(4-methylphenyl)-3(4-nitrophenylimino)-3-phenyl
Mol Form: $C_{22}H_{19}N_3O_3$
M_r: 373.39
mp: 169-70
UV: LANG 1971 389(33113) 318(14454) 243(16218) EtOH
m/e: 373.14

HODOC No: 00922
CAS No: 17738-24-8
Name: *p*-Acetotoluidide, 2-(*N*-phenylbenzimidoyl)-
Synonym: Propanoic acid, amide,*N*(4-methylphenyl)-3-phenyl-3-phenylimino
Mol Form: $C_{22}H_{20}N_2O$
M_r: 328.39
mp: 130-1
UV: LANG 1335 351(31623) 254(19498) EtOH
m/e: 328.16

HODOC No: 00923
CAS No: 17738-25-9
Name: *p*-Acetotoluidide, 2-(*N*-*p*-tolylbenzimidoyl)-
Synonym: Propanoic acid, amide,*N*(4-methylphenyl)-3(4-methylphenylimino)-3-phenyl
Mol Form: $C_{23}H_{22}N_2O$
M_r: 342.42
mp: 127-9
UV: LANG 1337 354(29512) 258(18621) EtOH
m/e: 342.17

HODOC No: 00924

CAS No: 506-96-7
Beilstein No: 2^4, 398
Name: Acetyl bromide
Synonym: Acetic acid, bromide
Line Form: CH_3COBr
Mol Form: C_2H_3BrO
M_r: 121.95
bp: 76
mp: -96
***d*:** $1.6625^{16/4}$
***n*:** 1.4486^{20}
α: 1.45376^{16}
***s*:** eth 5; ace 3; bz 5; chl 5
IR: COB 1032 3020 2930 1880 1810 1710 1420 1350 1290 1080 1010 940
UV: OES 7-3 253(95) 213(331) cy 235(89) 206(135) sulf
H NMR: SAD 4980 2.8 $CDCl_3$
MS: WILEY 493 43(100) 15(84) 14(19) 79(13) 42(12) 81(12) 13(9) 122(1)
m/e: 121.94

HODOC No: 00925

CAS No: 598-21-0
Beilstein No: 2^4, 536
Name: Acetyl bromide, bromo-
Synonym: Bromoacetyl bromide
Line Form: $BrCH_2COBr$
Mol Form: $C_2H_2Br_2O$
M_r: 199.84
bp: 147-50
***d*:** $2.317^{22/22}$
***n*:** 1.5449^{20}
***s*:** ace 3; ctc 3
IR: μSADP 9717 3030 1890 1790 1390 1240 1150 1000 900 850 710
H NMR: SAD 9167 4.4 CCl_4
m/e: 199.85

HODOC No: 00926

CAS No: 10588-31-5
Beilstein No: 2^3, 485
Name: Acetyl bromide, tribromo-
Synonym: Tribromoacetyl bromide
Line Form: Br_3CCOBr
Mol Form: C_2Br_4O
M_r: 355.64
bp: 210-5
***s*:** Bz 4; Et_2O 4; Chl 4
m/e: 355.67

HODOC No: 00927

CAS No: 34069-94-8
Beilstein No: 2^3, 476
Name: Acetyl bromide, trichloro-
Synonym: Trichloroacetyl bromide
Line Form: Cl_3CCOBr
Mol Form: C_2BrCl_3O
M_r: 225.28
bp: 143
***d*:** $1.900^{15/15}$
***s*:** Bz 4; Et_2O 4
m/e: 223.82

HODOC No: 00928

CAS No: 75-36-5
Beilstein No: 2^4, 395
Name: Acetyl chloride
Synonym: Acetic chloride
Line Form: CH_3COCl
Mol Form: C_2H_3ClO
M_r: 78.50
bp: 52
mp: -112
***d*:** $1.1051^{20/4}$
***n*:** 1.3886^{20}
***s*:** eth 5; ace 5; bz 5; ctc 3; chl 5
IR: COB 6165 3600 3020 2940 2760 1900 1810 1720 1420 1360 1090 1020 950
UV: OES 7-3 242(41) 198(24) cy 213(39) sulf
H NMR: SAD 8780 2.7 CCl_4
MS: WILEY 97 43(100) 15(56) 14(14) 42(11) 35(11) 63(10) 13(8) 78(1)
m/e: 77.99

HODOC No: 00929

CAS No: 22118-09-8
Beilstein No: 2^1, 97
Name: Acetyl chloride, bromo-
Synonym: Bromoacetyl chloride
Line Form: BCH_2COCl
Mol Form: C_2H_2BrClO
M_r: 156.39
bp: 127
***d*:** 1.908^0
IR: COB 6635 3010 2960 1810 1790 1400 1210 1160 1010 970 860 730 660 520 440
m/e: 155.90

HODOC No: 00930
CAS No: 79-04-9
Beilstein No: 2^4, 488
Name: Acetyl chloride, chloro-
Synonym: Chloracetyl chloride
Line Form: $ClCH_2COCl$
Mol Form: $C_2H_2Cl_2O$
M_r: 112.94
bp: 107
mp: -22
***d*:** $1.4202^{20/4}$
***n*:** 1.4530^{20}
***s*:** eth 5; ace 3; ctc 3
IR: µCOB 2239 2940 1820 1790 1540 1410 1030 970 780 710
H NMR: SAD 9147 4.5 CCl_4
MS: NBS 130
m/e: 111.95

HODOC No: 00931
CAS No: 359-32-0
Beilstein No: 2^3, 453
Name: Acetyl chloride, chlorofluoro-
Synonym: Chlorofluoroacetyl chloride
Line Form: ClCHF-COCl
Mol Form: C_2HCl_2FO
M_r: 130.93
bp: 69.5
***d*:** $1.468^{25/4}$
***n*:** 1.3992^{25}
m/e: 129.94

HODOC No: 00932
CAS No: 4025-77-8
Beilstein No: 4^4, 103
Name: Acetyl chloride, (chlorosulfonyl)-
Synonym: Chlorosulfonylacetyl chloride
Line Form: $ClSO_2CH_2COCl$
Mol Form: $C_2H_2Cl_2O_3S$
M_r: 177.00
bp: 98^{10}
m/e: 175.91

HODOC No: 00933
CAS No: 79-36-7
Beilstein No: 3^4, 504
Name: Acetyl chloride, dichloro-
Synonym: Dichloracetyl chloride
Line Form: $Cl_2CHCOCl$
Mol Form: C_2HCl_3O
M_r: 147.39
bp: 108-10
***d*:** $1.5315^{16/4}$
***n*:** 1.4591^{20}
***s*:** H_2O 6; al 6; eth 5
IR: µCOB 2378 3030 1820 1790 1210 1060 980 800 740
MS: WILEY 813 83(100) 85(64) 63(39) 48(39) 35(37) 28(33) 47(28) 146(2)
m/e: 145.91

HODOC No: 00934
CAS No: 774-74-3
Name: Acetyl chloride, (2,4-dichlorophenoxy)-
Synonym: (2,4-Dichlorophenoxy)acetyl chloride
Mol Form: $C_8H_5Cl_3O_2$
M_r: 239.48
bp: 122
IR: COB 1031 3070 2920 1810 1780 1550 1480 1300 1270 1230 1110 1050 940 870 800 770 700 650 570
MS: WILEY 1667 28(100) 175(85) 162(83) 36(53) 164(52) 177(52) 161(31) 238(28)
m/e: 237.94

HODOC No: 00935
CAS No: 34164-06-2
Name: Acetyl chloride, (dichlorophosphinyl)-
Synonym: Acetic acid, chloride,dichlorophosphinyl
Line Form: $Cl_2P(=O)CH_2COCl$
Mol Form: $C_2H_2Cl_3O_2P$
M_r: 195.37
IR: COB 6854 2920 1790 1240 1050 1000 840 790
m/e: 193.89

HODOC No: 00936
CAS No: 14077-58-8
Beilstein No: 3^3, 396
Name: Acetyl chloride, ethoxy-
Synonym: Ethoxyacetyl chloride
Line Form: $C_2H_5OCH_2COCl$
Mol Form: $C_4H_7ClO_2$
M_r: 122.55
bp: 123-4, $49\text{-}50^{37}$
***d*:** 1.1170
***n*:** 1.4204^{20}
***s*:** Ac 4; Et_2O 4
m/e: 122.01

HODOC No: 00937
CAS No: 38870-89-2
Beilstein No: 3^3, 396
Name: Acetyl chloride, methoxy-
Synonym: Methoxyacetyl chloride
Line Form: CH_3OCH_2COCl
Mol Form: $C_3H_5ClO_2$
M_r: 108.52
bp: 112-3, $46\text{-}9^{62}$
***d*:** $1.1871^{20/4}$
***n*:** 1.4199^{20}
***s*:** eth 3; ace 3; ctc 3; chl 4
IR: µCOB 2228 2940 1750 1450 1430 1430 1320 1270 1210 1180 1010 930 890 790
Raman: SAD 916 3000 2960 2900 2850 1750 1450 1420 1320 1180 1000 930 890 800 710 690 600 430 400 320 250 190
H NMR: SAD 19658 3.5 4.3 CCL_4
m/e: 108.00

HODOC No: 00938
CAS No: 15356-62-4
Beilstein No: 6^1, 25
Name: Acetyl chloride, [[5-methyl-2-(1-methylethyl)cyclohexyl]oxy]-, [1R-(1α,2β,5α)]-
Synonym: Acetyl chloride, (*p*-menth-3-yloxy)-, (-)-
Structure:

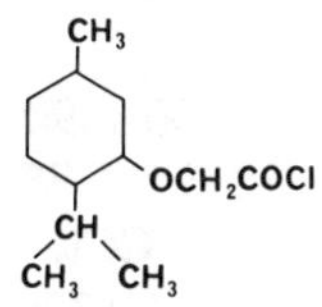

Mol Form: $C_{12}H_{21}ClO_2$
M_r: 232.74
bp: $128\text{-}31^{11}$
***d*:** $^{\alpha 13/D}$ -84.8 chl
α: $[\alpha]^{13/D}$ -84.8
***s*:** eth 3
IR: µSADP 25206 2940 1790 1450 1410 1210 1140 1000 950 910 750
m/e: 232.12

HODOC No: 00939
CAS No: 21062-20-4
Beilstein No: 3, 240
Name: Acetyl chloride, 2,2′-oxybis-
Synonym: Acetyl chloride, oxydi-
Line Form: $O(CH_2COCl)_2$
Mol Form: $C_4H_4Cl_2O_3$
M_r: 170.98
bp: 116^{15}
***s*:** Chl 4
m/e: 169.95

HODOC No: 00940
CAS No: 701-99-5
Beilstein No: 6^3, 613
Name: Acetyl chloride, phenoxy-
Synonym: Phenoxyacetyl chloride
Line Form: $C_6H_5O-CH_2COCl$
Mol Form: $C_8H_7ClO_2$
Mr: 170.59
bp: 225-6, 111^{13}
***s*:** eth 3
IR: **μSADP 4723** 2940 1820 1590 1490 1430 1320 1270 1220 1180 1090 940 860 790 760 690
H NMR: **SAD 18695** 4.8 6.7-7.4 CCl_4
MS: **WILEY 1121** 107(100) 77(75) 170(34) 51(18) 28(18) 79(15) 65(12) 172(11)
m/e: 170.01

HODOC No: 00941
CAS No: 76-02-8
Beilstein No: 2^4, 519
Name: Acetyl chloride, trichloro-
Synonym: Trichloroacetic acid chloride
Line Form: Cl_3CCOCl
Mol Form: C_2Cl_4O
Mr: 181.83
bp: 118
***d*:** $1.6202^{20/4}$
***n*:** 1.4695^{20}
***s*:** eth 5
IR: **μCOB 391** 1790 1750 1560 1020 850 800 740
m/e: 179.87

HODOC No: 00942
CAS No: 354-32-5
Beilstein No: 2^4, 470
Name: Acetyl chloride, trifluoro-
Synonym: Perfluoroacetyl chloride
Line Form: CF_3COCl
Mol Form: C_2ClF_3O
Mr: 132.47
bp: -18
m/e: 131.96

HODOC No: 00943
CAS No: 557-99-3
Beilstein No: 2^4, 393
Name: Acetyl fluoride
Synonym: Methylcarbonyl fluoride
Line Form: CH_3COF
Mol Form: C_2H_3FO
Mr: 62.04
bp: 20.8
mp: 60
***d*:** 1.032
***n*:** $1.002^{15/4}$
***s*:** al 5; eth 5; bz 3; chl 3; CS_2 2
IR: **μSADP 11912** 3030 1850 1390 1190 1080 1050 1000 820
UV: **OES 7-3** 212(34) cyhex
m/e: 62.02

HODOC No: 00944
CAS No: 507-02-8
Beilstein No: 2^4, 399
Name: Acetyl iodide
Line Form: CH_3COI
Mol Form: C_2H_3IO
Mr: 169.95
bp: 108, 36^{50}
***d*:** $2.0674^{20/4}$
***n*:** 1.5491^{20}
***s*:** Et_2O 4
m/e: 169.92

HODOC No: 00945
CAS No: 13250-46-9
Beilstein No: 3^3, 278
Name: Acetyl isothiocyanate
Synonym: Acetic acid, anhydride with isothiocyanic acid
Line Form: CH_3CONCS
Mol Form: C_3H_3NOS
Mr: 101.12
bp: 132-3, 30^9
***d*:** $1.1523^{13/4}$
***n*:** 1.5231^{18}
***s*:** eth 3; CS_2 3
UV: **OES 6-7** 333(21) EtOH
m/e: 100.99

HODOC No: 00946
CAS No: 25867-19-0
Beilstein No: 21^4, 4564
Name: Aconitane-7,8-diol,
4-[[[2-(2,5-dioxo-1-pyrrolidinyl)benzoyl]oxy]methyl]
-20-ethyl-1,6,14,16-tetramethoxy-, (1α,6β,14α,16β)-
Synonym: 2H-12,3,6a-Ethanylylidene-7,9-methanonaphth[2,3-b]azocine, aconitane-7,8-diol deriv.
Mol Form: $C_{36}H_{48}N_2O_{10}$
M$_r$: 668.75
mp: 111-4
α: $[\alpha]^{D}$ +31.5
Color: amor
s: Bz 4; EtOH 4; Chl 4; peth 4
m/e: 668.33

HODOC No: 00947
CAS No: 509-20-6
Beilstein No: 21^4, 2899
Name: Aconitane-3,8,13,14,15-pentol,
20-ethyl-1,6,16-trimethoxy-4-(methoxymethyl)-,
(1α,3α,6α,14α,15α,16β)-
Synonym: 11aH-12,3,6a-Ethanylylidene-7,9-methanonaphth[2,3-b]azocine, aconitane-3,8,13,14,15-pentol deriv.
Structure:

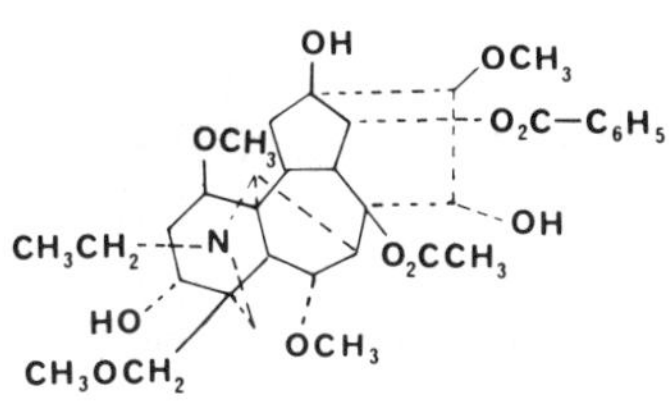

Mol Form: $C_{25}H_{41}NO_9$
M$_r$: 499.58
mp: 132
d: $^{\alpha}$ 23
α: $[\alpha]^{20/D}$ +23
Color: amor
s: H_2O 3; al 3; eth 2; chl 3; lig 2
UV: HBCP <220 IN HCl
m/e: 499.28

HODOC No: 00948
CAS No: 302-27-2
Beilstein No: 21^4, 2901
Name: Aconitane-3,8,13,14,15-pentol,
20-ethyl-1,6,16-trimethoxy-4-(methoxymethyl)-,
8-acetate 14-benzoate, (1α,3α,6α,14α,15α,16β)-
Synonym: 2H-12,3,6a-Ethanylylidene-7,9-methanonaphth[2,3-b]azocine, aconitane-3,8,13,14,15-pentol deriv.
Structure:

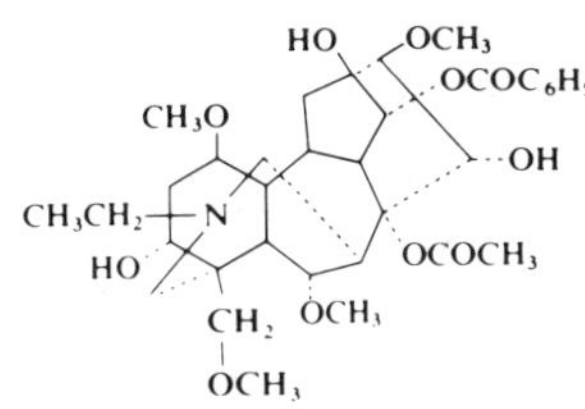

Mol Form: $C_{34}H_{47}NO_{11}$
M$_r$: 645.71
mp: 204
α: $[\alpha]^{20/D}$ +19 (Chl)
Color: rh lf
s: Bz 4; EtOH 4; Chl 4
IR: SADP 7536
m/e: 645.31

HODOC No: 00949
CAS No: 4491-19-4
Beilstein No: 21^4, 2889
Name: Aconitane-3,8,13,14-tetrol,
20-ethyl-1,6,16-trimethoxy-4-(methoxymethyl)-,
8-acetate 14-benzoate, (1α,3α,6α,14α,16β)-
Synonym: 2H-12,3,6a-Ethanylylidene-7,9-methanonaphth[2,3-b]azocine, aconitane-3,8,13,14-tetrol deriv.
Mol Form: $C_{34}H_{47}NO_{10}$
M$_r$: 629.71
mp: 202-3d
Color: cr
s: Et_2O 4; EtOH 4; Chl 4
m/e: 629.32

HODOC No: 00950
CAS No: 127-29-7
Beilstein No: 21[4], 2890
Name: Aconitane-3,8,13,14-tetrol, 20-ethyl-1,6,16-trimethoxy-4-(methoxymethyl)-, 8-acetate 14-(3,4-dimethoxybenzoate), (1α,3α,6α,14α,16β)-
Synonym: 2H-12,3,6a-Ethanylylidene-7,9-methanonaphth[2,3-b]azocine, aconitane-3,8,13,14-tetrol deriv.
Structure:

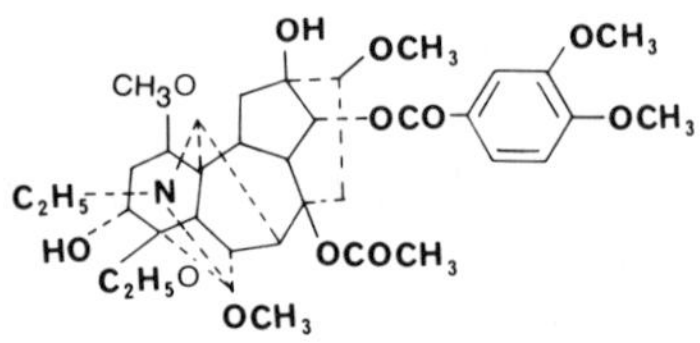

Mol Form: $C_{36}H_{51}NO_{12}$
Mr: 689.76
mp: 214
α: $[\alpha]^{15/D}$ +13.4 (al)
Color: tcl (MeOH)
s: Et_2O 4; EtOH 4
m/e: 689.34

HODOC No: 00951
CAS No: 32854-75-4
Beilstein No: 21[4], 2850
Name: Aconitane-4,8,9-triol, 20-ethyl-1,14,16-trimethoxy-, 4-[2-(acetylamino)benzoate], (1α,14α,16β)-
Synonym: 2H-12,3,6a-Ethanylylidene-7,9-methanonaphth[2,3-b]azocine, aconitane-4,8,9-triol deriv.
Structure:

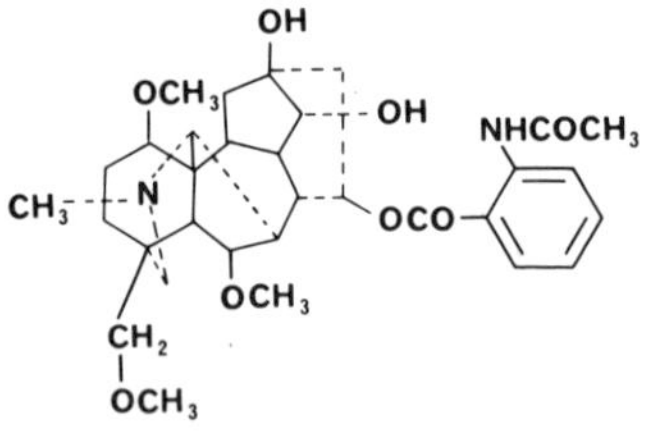

Mol Form: $C_{32}H_{44}N_2O_8$
Mr: 584.68
mp: 217-8
α: $[\alpha]^{18/D}$ +27 (Chl)
Color: hex pl (al)
m/e: 584.31

HODOC No: 00952
CAS No: 6078-26-8
Beilstein No: 21[4], 2869
Name: Aconitane-8,13,14-triol, 20-ethyl-1,6,16-trimethoxy-4-(methoxymethyl)-, 8-acetate 14-(3,4-dimethoxybenzoate), (1α,6α,14α,16β)-
Synonym: 2H-12,3,6a-Ethanylylidene-7,9-methanonaphth[2,3-b]azocine, aconitane-8,13,14-triol deriv.
Structure:

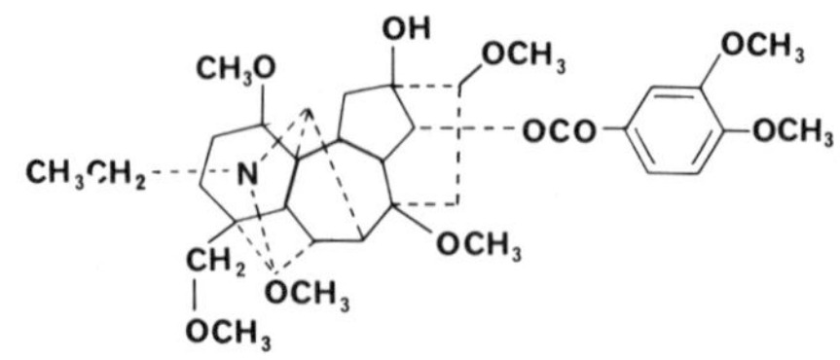

Mol Form: $C_{36}H_{51}NO_{11}$
Mr: 673.77
mp: 118-23
α: $[\alpha]^{D}$ +12(al)
s: Et_2O 4; EtOH 4; Chl 4
m/e: 673.35

HODOC No: 00953
CAS No: 561-07-9
Name: Aconitane-8,13,14-triol, 1,6,16-trimethoxy-4-(methoxymethyl)-20-methyl-, 8-acetate 14-benzoate, (1α,6α,14α,16β)-
Synonym: 2H-12,3,6a-Ethanylylidene-7,9-methanonaphth[2,3-b]azocine, aconitane-8,13,14-triol deriv.
Structure:

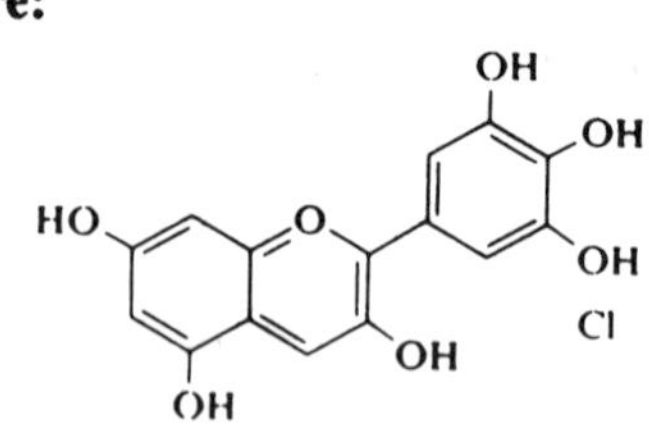

Mol Form: $C_{33}H_{45}NO_9$
Mr: 599.69
mp: dec 198- 200
α: $[\alpha]^{20/D}$ 25 al
s: H_2O 1; al 3; eth 3; ace 3; chl 3
UV: OES 1-1110 230(12589) EtOH
m/e: 599.31

HODOC No: 00954
CAS No: 561-05-7
Beilstein No: 21[4], 6794
Name: Aconitan-15-one, 14-(benzoyloxy)-20-ethyl-3,13-dihydroxy-1,6,16-trimethoxy-4-(methoxymethyl)-, (1α,3α,6α,14α,16β)-
Synonym: 2H-12,3,6a-Ethanylylidene-7,9-methanonaphth[2,3-b]azocine, aconitan-15-one deriv.
Mol Form: $C_{32}H_{43}NO_9$
M_r: 585.66
mp: 171
Color: nd
m/e: 585.29

HODOC No: 00955
CAS No: 578-06-3
Beilstein No: 22[4], 4982
Name: 1-Acridinamine
Synonym: Acridine, 1-amino-
Mol Form: $C_{13}H_{10}N_2$
M_r: 194.22
mp: 183
Color: ye nd (dil al)
m/e: 194.08

HODOC No: 00956
CAS No: 581-28-2
Beilstein No: 22[4], 4984
Name: 2-Acridinamine
Synonym: Acridine, 2-amino-
Mol Form: $C_{13}H_{10}N_2$
M_r: 194.22
mp: 213-4
Color: ye nd (w or al)
s: H_2O 3; al 4; eth 3; ctc 3; dil ac 3
UV: OES 7-446 462(3548) 371(8318) 356(6457) 273(48978) 256(31623) pH=2.5 355(5754) 339(3631) 260(77625) pH=11.0
m/e: 194.08

HODOC No: 00957
CAS No: 581-29-3
Beilstein No: 22[4], 4987
Name: 3-Acridinamine
Synonym: Acridine, 3-amino-
Mol Form: $C_{13}H_{10}N_2$
M_r: 194.22
mp: 224
Color: ye nd (w or al)
s: H_2O 3; eth 4; ace 3; bz 3; dil ac 4
UV: OES 7-446 410(6166) 353(8318) 337(4467) 321(2239) 262(67608) 237(28840) pH=11.0 454(12589) 365(14125) 349(10715) 274(44668) 233(41687) pH=2.5
m/e: 194.08

HODOC No: 00958
CAS No: 578-07-4
Beilstein No: 22[4], 4989
Name: 4-Acridinamine
Synonym: Acridine, 4-amino-
Mol Form: $C_{13}H_{10}N_2$
M_r: 194.22
bp: 183-4
mp: 108
Color: red-br pr (peth) og nd (MeOH)
s: Ac 4; Bz 4; Et_2O 4; EtOH 4
m/e: 194.08

HODOC No: 00959
CAS No: 90-45-9
Beilstein No: 21[4], 4174
Name: 9-Acridinamine
Synonym: Acridine, 9-amino-
Mol Form: $C_{13}H_{10}N_2$
M_r: 194.22
mp: 241
Color: ye nd (Ace or al)
s: al 3; ace 3; $DMSOd_6$ 2; dil HCl 4
IR: µSADP 7591 3330 3230 1670 1610 1560 1490 1450 1390 1370 1250 1160 1120 1020 950 910 790 760
UV: SAD 2125 332(1050) 315(641) 260(59800) 218(19400) MeOH
H NMR: SAD 11293 7.2-7.8 8.0 8.5 $DMSOd_6$
m/e: 194.08

HODOC No: 00960
CAS No: 260-94-6
Beilstein No: 20^4, 3987
Name: Acridine
Synonym: 9-Azaanthracene
Structure:

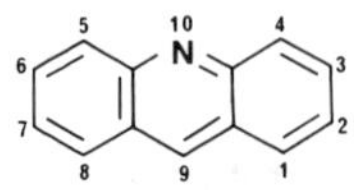

Mol Form: $C_{13}H_9N$
M_r: 179.21
bp: 345-6
mp: 111
***d*:** $1.005^{20/4}$
Color: rh nd or pr (al)
***s*:** H_2O 2; al 4; eth 4; bz 4; ctc 2; CS_2 4
IR: **µSADP 1153** 3030 1610 1560 1520 1470 1410 1150 1000 960 930 910 860 810 790 760 740
UV: **SAD 342** 348 339 250 212 MeOH
H NMR: **SAD 81** 6.5-7.5 CCl_4
C NMR: **STOT 262** 125.7 126.7 128.5 129.7 130.5 136.1 149.3
MS: **WILEY 1216** 179(100) 178(14) 180(14) 89(12) 89.5(11) 76(8) 177(7) 63(7)
m/e: 179.07

HODOC No: 00961
CAS No: 20378-06-7
Beilstein No: 20^4, 3293
Name: Acridine, 1,2,3,4,4a,9,9a,10-octahydro-, *cis*-
Synonym: Acridan, 1,2,3,4,4a,9a-hexahydro-, *cis*-
Structure:

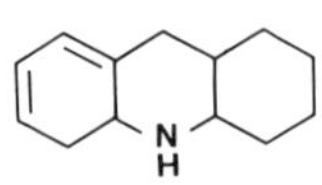

Mol Form: $C_{13}H_{17}N$
M_r: 187.27
mp: 71-2
***s*:** chl 2
IR: **µSADP 14493** 3330 3030 2940 1610 1470 1430 1370 1320 1280 1250 1060 1030 740 690
UV: **SAD 4121** 305 254 MeOH
H NMR: **SAD 11855** 1.2-2.2 2.5 2.9 3.4 3.5 6.3-7.2 $CDCl_3$
m/e: 187.14

HODOC No: 00962
CAS No: 1207-69-8
Beilstein No: 20^4, 3995
Name: Acridine, 9-chloro-
Synonym: 9-Chloroacridine
Mol Form: $C_{13}H_8ClN$
M_r: 213.65
bp: sub
mp: 122
Color: nd (al)
***s*:** H_2O 4; EtOH 4
m/e: 213.03

HODOC No: 00963
CAS No: 86-38-4
Beilstein No: 21^4, 1553
Name: Acridine, 6,9-dichloro-2-methoxy-
Synonym: 3,9-Dichloro-7-methoxyacridine
Mol Form: $C_{14}H_9Cl_2NO$
M_r: 278.12
mp: 160-1
Color: nd (Bz)
***s*:** tfa 3
IR: **µCOB 596** 2940 1640 1560 1470 1370 1300 1270 1220 1060 1020 940 890 820 800 780 760 650
UV: **SAD 6362** 336(4800) 320(2590) 264(150000) 230(16400) MeOH
H NMR: **SAD 13655** 4.2 7.9-8.5 8.8 tfa
m/e: 277.01

HODOC No: 00964
CAS No: 92-81-9
Beilstein No: 20^4, 3885
Name: Acridine, 9,10-dihydro-
Synonym: Acridan
Mol Form: $C_{13}H_{11}N$
M_r: 181.22
bp: 300d (sub)
mp: 169-71
Color: pr or pl (al)
***s*:** H_2O 1; al 4; eth 3; ace 3
IR: **µDMS 7224** 3330 3030 1610 1590 1470 1450 1410 1300 1160 1120 1030 890 760 750 710
UV: **OES 6-461** 287(15849) EtOH
m/e: 181.09

HODOC No: 00965
CAS No: 6267-02-3
Name: Acridine, 9,10-dihydro-9,9-dimethyl-
Synonym: Acridan, 9,9-dimethyl-
Mol Form: $C_{15}H_{15}N$
M_r: 209.28
MS: WILEY 1489 194(100) 195(16) 193(14) 209(9) 96.5(7) 192(7) 83.5(6) 191(4)
m/e: 209.12

HODOC No: 00966
CAS No: 5836-06-6
Name: Acridine, 9-[4-(hexyloxy)phenyl]-9,10-dihydro-9-methoxy-10-methyl-
Synonym: Acridan, 9-[*p*-(hexyloxy)phenyl]-9-methoxy-10-methyl-
Mol Form: $C_{27}H_{31}NO_2$
M_r: 401.52
mp: 77-8
***s*:** chl 3
IR: µCOB 3530 2940 1590 1490 1470 1350 1300 1240 1160 1140 1090 1060 940 860 830
m/e: 401.24

HODOC No: 00967
CAS No: 7613-08-3
Name: Acridine, 2-isothiocyanato-
Synonym: Isothiocyanic acid, 2-acridinyl ester
Mol Form: $C_{14}H_8N_2S$
M_r: 236.28
mp: 167
UV: LANG 1066 373(8913) 276(42658) 250(61660) MeOH
m/e: 236.04

HODOC No: 00968
CAS No: 613-15-0
Beilstein No: 20^4, 4037
Name: Acridine, 2-methyl-
Synonym: 2-Methylacridine
Mol Form: $C_{14}H_{11}N$
M_r: 193.23
mp: 134
Color: ye nd (dil al)
***s*:** H_2O 2; al 3; eth 3; bz 3; con sulf 3; lig 2
IR: µAPI 2326 3030 1640 1560 1520 1470 1390 1150 1030 1000 980 960 930 860 820 760 750 740
m/e: 193.09

HODOC No: 00969
CAS No: 602-56-2
Beilstein No: 20^4, 4355
Name: Acridine, 9-phenyl-
Synonym: 9-Phenylacridine
Mol Form: $C_{19}H_{13}N$
M_r: 255.30
bp: 401 sub
mp: 184
Color: ye nd, lf (al)
***s*:** H_2O 1; al 2; eth 3; bz 4; con sulf 4
UV: OES 4-751 390(3981) 355(12589) 251(199526) 33% eth:EtOH
m/e: 255.10

HODOC No: 00970
CAS No: 92-62-6
Beilstein No: 22^4, 5487
Name: 3,6-Acridinediamine
Synonym: Acridine, 3,6-diamino-
Mol Form: $C_{13}H_{11}N_3$
M_r: 209.24
mp: 284-6
Color: ye nd (al or w)
***s*:** H_2O 3; al 4; eth 2; bz 2; con sulf 3
UV: OES 7-457 444(38905) 261(53703) 211(20893) pH=7.0
m/e: 209.10

HODOC No: 00971
CAS No: 65-61-2
Name: 3,6-Acridinediamine, *N,N,N',N'*-tetramethyl-, monohydrochloride
Synonym: Acridine, 3,6-*bis*(dimethylamino)-, monohydrochloride
Mol Form: $C_{17}H_{20}ClN_3$
M_r: 301.80
IR: µCOB 3791 2940 2630 1640 1610 1520 1470 1390 1350 1180 1120 1060 930 860 830 810 780 690
m/e: 301.13

HODOC No: 00972
CAS No: 442-16-0
Beilstein No: 22^4, 6679
Name: 3,9-Acridinediamine, 7-ethoxy-
Synonym: Acridine, 6,9-diamino-2-ethoxy-
Mol Form: $C_{15}H_{15}N_3O$
M_r: 253.29
mp: 226
Color: og-ye cr or sc (50% al)
m/e: 253.12

HODOC No: 00973
CAS No: 22817-17-0
Name: 2-Acridinol
Synonym: 2-Hydroxyacridine
Mol Form: $C_{13}H_9NO$
M_r: 195.21
mp: >250
UV: FO 409 398(5012) 357(7943) 296(1995) 255(100000) EtOH
m/e: 195.07

HODOC No: 00974
CAS No: 7132-70-9
Name: 3-Acridinol
Synonym: Acridine, 3-hydroxy
Mol Form: $C_{13}H_9NO$
M_r: 195.21
mp: sealed tube 285
UV: FO 410 485(3981) 397(3981) 350(6310) 285(19953) 259(79433) EtOH
m/e: 195.07

HODOC No: 00975
CAS No: 18123-20-1
Beilstein No: 21^4, 1562
Name: 4-Acridinol
Synonym: 4-Hydroxyacridine
Mol Form: $C_{13}H_9NO$
M_r: 195.21
mp: 117
***s*:** al 3; eth 3
UV: FO 411 394(3981) 358(3162) 258(63096) EtOH
m/e: 195.07

HODOC No: 00976
CAS No: 643-62-9
Name: 9-Acridinol
Synonym: 9-Hydroxyacridine
Mol Form: $C_{13}H_9NO$
M_r: 195.21
UV: FO 412 400(7943) 380(6310) 307(1259) 294(1585) 252(50119) EtOH
m/e: 195.07

HODOC No: 00977
CAS No: 578-95-0
Beilstein No: 21^4, 4171
Name: 9(10H)-Acridinone
Synonym: Acridanone
Structure:

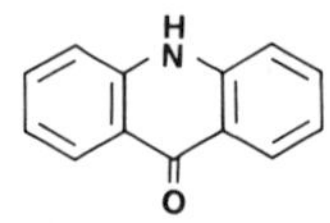

Mol Form: $C_{13}H_9NO$
M_r: 195.21
bp: 354
mp: 154
Color: ye lf (al)
***s*:** H_2O 1; al 2; eth 1; bz 1; chl 1; alk 3; HOAc 3; con sulf 3
IR: μSADP 7560 3130 2940 1640 1610 1540 1470 1350 1270 1180 1160 1020 940 810 750 700 670
UV: SAD 2117 294(1860) 253(59700) 214(27000) MeOH
MS: ALDN 158 195(100) 167(53) 166(24) 196(19) 139(19) 140(18) 83.5(9) 113(8)
m/e: 195.07

HODOC No: 00978
CAS No: 66073-40-3
Beilstein No: 22^4, 6500
Name: 9(10H)-Acridinone, 1-amino-
Synonym: 1-Aminoacridan-9-one
Mol Form: $C_{13}H_{10}N_2O$
M_r: 210.22
mp: 285
Color: ye nd (al)
***s*:** EtOH 4; HOAc 4
m/e: 210.08

HODOC No: 00979
CAS No: 27918-14-5
Beilstein No: 22^4, 6503
Name: 9(10H)-Acridinone, 2-amino-
Synonym: 9-Acridanone, 2-amino-
Mol Form: $C_{13}H_{10}N_2O$
M_r: 210.22
mp: 298
Color: ye nd (al)
***s*:** Ac 4; EtOH 4
m/e: 210.08

HODOC No: 00980
CAS No: 50433-64-2
Beilstein No: 22^4, 6508
Name: 9(10H)-Acridinone, 3-amino-
Synonym: 9-Acridanone, 3-amino-
Mol Form: $C_{13}H_{10}N_2O$
M_r: 210.22
mp: 301-3
Color: pa ye nd (al)
m/e: 210.08

HODOC No: 00981
CAS No: 719-54-0
Beilstein No: 21^4, 4187
Name: 9(10H)-Acridinone, 10-methyl-
Synonym: 9-Acridanone, 10-methyl-
Mol Form: $C_{14}H_{11}NO$
M_r: 209.23
mp: 205-7
IR: ALD 3:904C
m/e: 209.08

HODOC No: 00982
CAS No: 949-53-1
Name: Acrylanilide, 2,3,3,3′-tetrachloro-
Synonym: Propenoic acid, amide,*N*(3-chlorophenyl)-2,3,3-trichloro
Mol Form: $C_9H_5Cl_4NO$
M_r: 284.95
IR: µCOB 3987 3330 3130 1670 1610 1540 1430 1320 1300 1270 940 920 860 790 780 680
m/e: 282.91

HODOC No: 00983
CAS No: 3881-57-0
Name: Acrylic acid, trichloro-, pentachlorophenyl ester
Synonym: Propenoic acid, trichloro,pentachlorophenyl ester
Mol Form: $C_9Cl_8O_2$
M_r: 423.71
IR: :gmCOB 4109 1790 1610 1390 1370 1210 1140 920 860 810 710
m/e: 419.74

HODOC No: 00984
CAS No: 2224-98-8
Name: Acrylic acid, trichloro-, 2,4,5-trichlorophenyl ester
Synonym: Propenoic acid, trichloro,2,4,5-tricholorophenyl ester
Mol Form: $C_9H_2Cl_6O_2$
M_r: 354.82
IR: µCOB 4110 3130 1750 1560 1470 1350 1280 1210 1090 1040 940 910 890 790 680
m/e: 351.82

HODOC No: 00985
CAS No: 58-61-7
Beilstein No: 26^4, 3597
Name: Adenosine
Synonym: A
Structure:

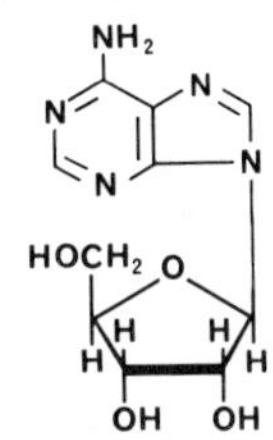

Mol Form: $C_{10}H_{13}N_5O_4$
M_r: 267.23
mp: 235-6 anh
α: $[\alpha]^{20/D}$-60.0 H_2O,1%$[\alpha]^{20/D}$-43.5 10%-HCl,2%
Color: n(w+3/2)
***s*:** H_2O 2; al 1
IR: COB 6798 3330 3160 2880 1610 1580 1510 1470 1430 1300 1200 1080 1040 960 790
UV: SAD 1497 259(14100) 207(18500) MeOH
C NMR: JJ 376 FT 61.8 70.8 73.6 86.0 88.1 119.4 140.1 149.1 152.4 156.1 DMSOd6
MS: ALDN 210 135(100) 136(77) 164(75) 178(35) 108(23) 43(20) 33(17) 267(7)
m/e: 267.10

HODOC No: 00986
CAS No: 146-77-0
Name: Adenosine, 2-chloro-
Synonym: Antibiotic AT 265B
Structure:

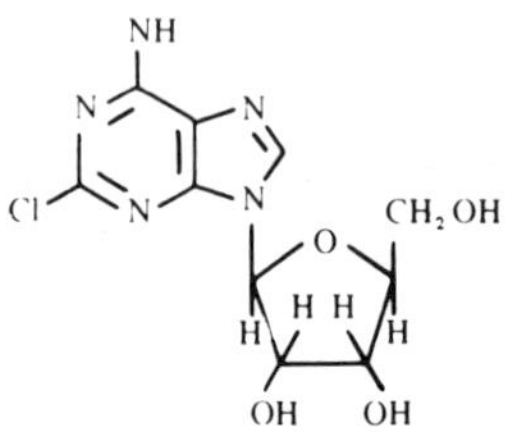

Mol Form: $C_{10}H_{12}ClN_5O_4$
M$_r$: 301.68
IR: COB 6938 3300 3150 2900 1630 1600 1430 1350 1310 1250 1200 990 940 690
UV: OES 7-283 265(15136) pH=13 264(15136) pH=7 265(14454) pH=1
m/e: 301.06

HODOC No: 00987
CAS No: 958-09-8
Name: Adenosine, 2′-deoxy-
Synonym: Adenine deoxyribonucleoside
Mol Form: $C_{10}H_{13}N_5O_3$
M$_r$: 251.23
s: H_2O 2
UV: OES 5-233 259(14791) H_2O
H NMR: VAR 563 2.6 2.7 3.8 3.9 4.2 4.6 6.4 8.1 8.3 D_2O
m/e: 251.10

HODOC No: 00988
CAS No: 29886-19-9
Beilstein No: 26^4, 3603
Name: Adenosine, 2′,3′-diacetate
Synonym: 2′,3′-Diacetyladenosine
Mol Form: $C_{14}H_{17}N_5O_6$
M$_r$: 351.30
mp: 184-6
IR: ALD 2:1130F
MS: ALD 8:122B
m/e: 351.12

HODOC No: 00989
CAS No: 24723-77-1
Name: Adenosine, 2-methoxy-
Synonym: 2-Methoxyadenosine
Mol Form: $C_{11}H_{15}N_5O_5$
M$_r$: 297.26
IR: COB 7395 3360 3200 2940 1640 1610 1520 1490 1380 1290 1210 1140 1050 1030 970 860 790 740 670
m/e: 297.11

HODOC No: 00990
CAS No: 13364-95-9
Name: Adenosine, 2-(methylamino)-
Synonym: 9H-Purine, 6-amino-2-(methylamino)-9-β-*D*-ribofuranosyl-
Structure:

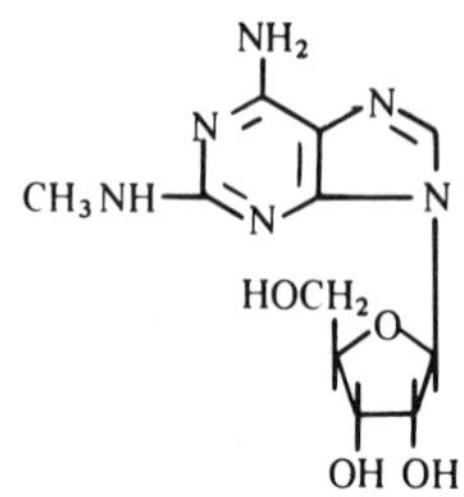

Mol Form: $C_{11}H_{16}N_6O_4$
M$_r$: 296.27
IR: COB 7403 3350 3150 2920 1660 1600 1540 1390 1300 1120 1090 900 790 700
UV: OES 4-352 299(9333) 254(13804) pH=1 287(8710) 258(10965) pH=7 288(8913) 257(10715) pH=13
m/e: 296.12

HODOC No: 00991
CAS No: 61-19-8
Beilstein No: 31, 27
Name: 5′-Adenylic acid
Synonym: Adenosine 5′-(dihydrogen phosphate)
Structure:

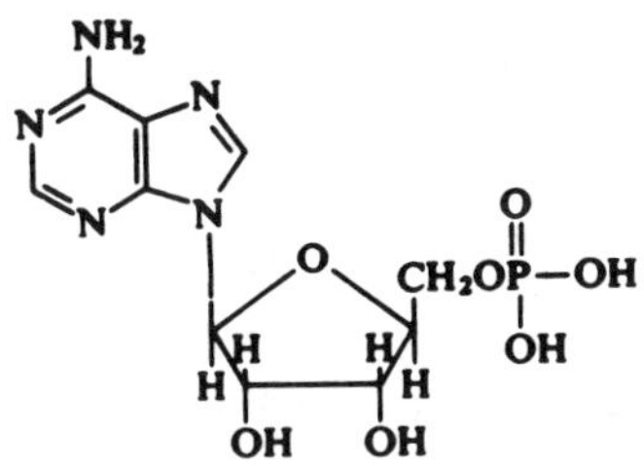

Mol Form: $C_{10}H_{14}N_5O_7P$
M_r: 347.21
mp: dec 195
α: $[\alpha]^{50/D}$ -41.78 H_2O
***s*:** H_2O 4; al 3; 10% HCl 3
IR: µSADP 14657 3450 3330 3130 1700 1590 1490 1410 1350 1280 1220 1110 1080 1010 940 830 730 710 690
UV: OES 8-173 258(11220) 2N HCl 259(14454) pH=7.0
C NMR: STOT 472 64.4 71.4 75.4 85.1 87.9 118.9 140.8 149.3 153.3 155.9
m/e: 347.06

HODOC No: 00992
CAS No: 4360-12-7
Beilstein No: 23^4, 3212
Name: Ajmalan-17,21-diol, (17R,21α)-
Synonym: 5H-6,10:11,12a-Dimethanoindolo[3,2-b]quinolizine, ajmalan-17,21-diol deriv.
Structure:

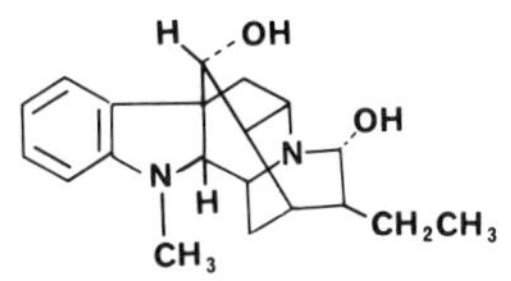

Mol Form: $C_{20}H_{26}N_2O_2$
M_r: 326.42
bp: 250-7(anh)
mp: 158-60(hyd)
α: $[\alpha]^{20/D}$ 144 chl
Color: pl (+3.5w) (aq AcOEt)
***s*:** H_2O 1; al 3; eth 2; bz 2; chl 3; AcOEt 3
IR: µSADP 23488 3450 2940 1610 1490 1370 1300 1240 1210 1110 1050 1020 1000 960 910 860 790 760 750 690
UV: OES 8-532 291(3162) 248(9120) 50% EtOH
MS: NBS 2512
m/e: 326.20

HODOC No: 00993
CAS No: 107-95-9
Beilstein No: 4^4, 2526
Name: β-Alanine
Synonym: Abufene
Line Form: $H_2NCH_2CH_2CO_2H$
Mol Form: $C_3H_7NO_2$
M_r: 89.09
mp: 207d
***d*:** 1.437^{19}
Color: nd, rh pr (al)
***s*:** H_2O 3; al 2; eth 1; ace 1
IR: SADG 15288 3448 3333 2941 2632 2222 1613 1563 1493 1449 1408 1316 1282 1250 1149 1053 990 952 877 840
H NMR: SAD 10221 2.5 3.2 D_2O
m/e: 89.05

HODOC No: 00994
CAS No: 338-69-2
Beilstein No: 4^4, 2480
Name: *D*-Alanine
Synonym: Alanine, *D*-
Line Form: $CH_3CH(NH_2)CO_2H$
Mol Form: $C_3H_7NO_2$
M_r: 89.09
bp: sub
mp: 314d
α: $[\alpha]^{25/D}$-13.6 6N HCl,1%
Color: nd (w, al)
***s*:** H_2O 3; al 2; eth 1
H NMR: SAD 16069 1.8 4.4 7.5 tfa
C NMR: JJ 36 ini 17.2 51.5 176.5 H_2O
m/e: 89.05

HODOC No: 00995
CAS No: 302-72-7
Beilstein No: 4^4, 2481
Name: *DL*-Alanine
Synonym: Alanine, *DL*-
Line Form: $CH_3CH(NH_2)CO_2H$
Mol Form: $C_3H_7NO_2$
M_r: 89.09
bp: sub 258
mp: 295-6d
***d*:** 1.424
Color: orh pr or nd (w)
***s*:** H_2O. EtOH 4
IR: SADG 164
UV: OES 2-10
H NMR: SAD 7552
m/e: 89.05

HODOC No: 00996
CAS No: 56-41-7
Beilstein No: 4^4, 2480
Name: *L*-Alanine
Synonym: Alanine, *L*-
Line Form: $CH_3CH(NH_2)CO_2H$
Mol Form: $C_3H_7NO_2$
M_r: 89.09
bp: sub 160-3
mp: 314d
***d*:** 1.432^{22}
α: $[\alpha]^{25/D}$ +2.8 (w,c=6)
Color: rh (w)
***s*:** H_2O 3; al 2; eth 1; ace 1; py 2
IR: SADG 164 3080 3000 2930 2810 2730 2600 1620 1590 1520 1450 1410 1350 1300 1120 1020 920 850 770 650 550 420
Raman: SAD 425 3080 3000 2980 2960 2940 2890 2750 1640 1620 1600 1550 1520 1480 1470 1420 1390 1370 1310 1250 1170 1150 1120 1030 1020 930 860 780 650 550 490 420 330 290 270 240 190 160 120
UV: OES 2-10 205(5) pH=1.0
H NMR: SAD 7553 1.5 3.8 4.8 D_2O
C NMR: JJ 36 ini 17.2 51.5 176.5 H_2O
MS: NBS 51
m/e: 89.05

HODOC No: 00997
CAS No: 817-11-8
Beilstein No: 4^4, 2568
Name: β-Alanine, *N,N-bis*(2-carboxyethyl)-
Synonym: Propionic acid, 3,3′,3″-nitrilotri-
Line Form: $N(CH_2CH_2CO_2H)_3$
Mol Form: $C_9H_{15}NO_6$
M_r: 233.21
mp: 186-8d
IR: ALD 2:303D
m/e: 233.09

HODOC No: 00998
CAS No: 505-47-5
Name: β-Alanine, *N*-(2-carboxyethyl)-
Synonym: Propionic acid, 3,3′-iminodi-
Line Form: $HN(CH_2CH_2CO_2H)_2$
Mol Form: $C_6H_{11}NO_4$
M_r: 161.15
mp: 151-1.5
IR: μCOB 3993 2940 2630 2500 1700 1610 1560 1470 1430 1390 1320 1280 1220 1020 970 860 810 680
m/e: 161.07

HODOC No: 00999
CAS No: 38470-24-5
Name: β-Alanine, *N*-(2-carboxyethyl)-*N*-(4-chlorophenyl)-
Synonym: 3,3′-(4-Chlorophenylimino)dipropionic acid
Structure:

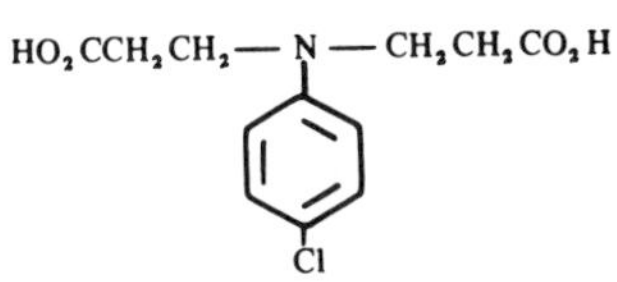

Mol Form: $C_{12}H_{14}ClNO_4$
M$_r$: 271.69
mp: 133.0-3.5
***s*:** DMSO 2
IR: μSADP 4252 2940 2630 1725 1695 1615 1515 1430 1370 1220 980 910 805
H NMR: SAD 10582 2.5 3.6 6.7 7.2 DMSOd6
m/e: 271.06

HODOC No: 01000
CAS No: 59039-83-7
Name: β-Alanine, *N*-(cyanomethyl)-*N*-methyl-, methyl ester
Line Form: $NCCH_2N(CH_3)CH_2CH_2CO_2CH_3$
Mol Form: $C_7H_{12}N_2O_2$
M$_r$: 156.18
IR: μCOB 3463 2940 2780 1720 1430 1270 1160 1040 840 780
m/e: 156.09

HODOC No: 01001
CAS No: 79-83-4
Beilstein No: 4^4, 2569
Name: β-Alanine, *N*-(2,4-dihydroxy-3,3-dimethyl-1-oxobutyl)-, (R)-
Synonym: Pantothenic acid, *D*-
Line Form: $HOCH_2C(CH_3)_2CH(OH)CONHCH_2CH_2CO_2H$
Mol Form: $C_9H_{17}NO_5$
M$_r$: 219.23
α: $[\alpha]^{25/D}$ +37.5 (w)
Color: ye visc oil
***s*:** H_2O 4; Bz 4; Et_2O 4
m/e: 219.11

HODOC No: 01002
CAS No: 4244-84-2
Beilstein No: 4^1, 499
Name: β-Alanine, ethyl ester, hydrochloride
Synonym: 3-Aminopropionic acid ethyl ester hydrochloride
Line Form: $NH_2CH_2CH_2CO_2CH_2CH_2$–HCl
Mol Form: $C_5H_{12}ClNO_2$
M$_r$: 153.60
mp: 65.5
***s*:** H^2O 3; eth 1
IR: SADG 16169 3000 2700 2020 1720 1490 1400 1360 1330 1200 1010 960 930 770 390
H NMR: SAD 8551 1.3 2.9 3.4 4.2 D_2O
m/e: 153.06

HODOC No: 01003
CAS No: 2679-14-3
Beilstein No: 4^3, 1264
Name: β-Alanine, *N*-methyl-
Synonym: *N*-Methyl-β-alanine
Line Form: $CH_3NHCH_2CH_2CO_2H$
Mol Form: $C_4H_9NO_2$
M$_r$: 103.12
mp: 146 (anh)
Color: pl (aq al, +1w)
***s*:** H_2O 4; EtOH 4
m/e: 103.06

HODOC No: 01004
CAS No: 2213-08-3
Beilstein No: 4^3, 1264
Name: β-Alanine, *N*-methyl-, ethyl ester
Synonym: Ethyl 3-(methylamino)propanoate
Line Form: $CH_3NHCH_2CH_2CO_2C_2H_5$
Mol Form: $C_6H_{13}NO_2$
M$_r$: 131.17
bp: 80^{21}
***d*:** $1.0082^{20/20}$
***n*:** 1.4443^{20}
***s*:** Et_2O 4; EtOH 4
m/e: 131.09

HODOC No: 01005
CAS No: 5652-38-0
Beilstein No: 12^3, 933
Name: β-Alanine, *N*-phenyl-
Synonym: 3-Anilinopropanoic acid
Line Form: $C_6H_5NHCH_2CH_2CO_2H$
Mol Form: $C_9H_{11}NO_2$
M_r: 165.18
mp: 92
Color: nd (Bz or dil al), lf (dil al, Bz, eth)
***s*:** Bz 4; Et_2O 4; EtOH 4
m/e: 165.08

HODOC No: 01006
CAS No: 2304-94-1
Name: β-Alanine, *N*-[(phenylmethoxy)carbonyl]-
Synonym: β-Alanine, *N*-carboxy-, *N*-benzyl ester
Line Form: $C_6H_5CH_2O_2CNHCH_2CH_2CO_2H$
Mol Form: $C_{11}H_{13}NO_4$
M_r: 223.22
***s*:** chl 2
IR: SADG 16165 3350 3050 2960 2600 1690 1530 1410 1310 1250 1210 1130 1010 950 770 740 720 690 600 560
H NMR: SAD 8547 2.6 3.4 5.1 7.3 8.2 $CDCl_3$
m/e: 223.08

HODOC No: 01007
CAS No: 23574-01-8
Name: β-Alanine, *N*-(phenylmethyl)-, methyl ester
Synonym: β-Alanine, *N*-benzyl-, methyl ester
Line Form: $C_6H_5CH_2NHCH_2CH_2CO_2CH_3$
Mol Form: $C_{11}H_{15}NO_2$
M_r: 193.23
IR: μCOB 3479 3330 3030 2940 2780 1720 1590 1490 1450 1430 1350 1250 1160 1120 1030 1010 740 700
m/e: 193.11

HODOC No: 01008
CAS No: 515-94-6
Beilstein No: 4^4, 2581
Name: Alanine, 3-amino-
Synonym: Propionic acid, 2,3-diamino-
Line Form: $H_2NCH_2CH(NH_2)$-CO_2H
Mol Form: $C_3H_8N_2O_2$
M_r: 104.11
mp: ca 110
Color: hyg rosettes
***s*:** H_2O 4
m/e: 104.06

HODOC No: 01009
CAS No: 24164-44-1
Name: Alanine, 3-(benzylthio)-*N*-carboxy-, *N*-benzyl pentachlorophenyl ester, *DL*-
Synonym: Phenol, pentachloro-, ester with 3-(benzylthio)-*N*-carboxy-*DL*-alanine *N*-benzyl ester
Mol Form: $C_{24}H_{18}Cl_5NO_4S$
M_r: 593.71
mp: 171-3
UV: LANG 2142 325(55) 298(56) 212(9333) EtOH
m/e: 590.94

HODOC No: 01010
CAS No: 25129-13-9
Name: Alanine, 3-[(diphenylmethyl)thio]-*N*-(methylsulfonyl)-, methyl ester, *L*-
Synonym: *L*-*S*-Diphenylmethyl-*N*-methanesulfonylcysteine methyl ester
Line Form: $(C_6H_5)_2CHSCH_2CH_2CH(CO_2CH_3)NHSO_2CH_3$
Mol Form: $C_{18}H_{21}NO_4S_2$
M_r: 379.47
mp: 76.5-7.5
***s*:** chl 3
IR: SADG 19155 3280 3020 2960 1750 1600 1490 1450 1330 1210 1140 970 740 700 520
H NMR: SAD 9929 2.8 3.0 3.8 4.4 5.3 5.4 7.2-7.6 $CDCl_3$
m/e: 379.09

HODOC No: 01011
CAS No: 62-57-7
Beilstein No: 4^4, 2616
Name: Alanine, 2-methyl-
Synonym: AIB
Line Form: $(CH_3)_2C(NH_2)CO_2H$
Mol Form: $C_4H_9NO_2$
M_r: 103.12
bp: sub, 280
mp: 335 (sealed tube)
Color: mcl pr
***s*:** H_2O 4; al 2; eth 1
IR: μCOB 5374 2940 2780 2080 1640 1560 1540 1450 1410 1350 1280 1270 1190 1090 990 890 800 780
H NMR: SAD 6500 1.5 D_2O
MS: NBS 101; ALD 3:3D
m/e: 103.06

HODOC No: 01012
CAS No: 17966-60-8
Beilstein No: 9^4, 794
Name: *D*-Alanine, *N*-benzoyl-
Synonym: Alanine, *N*-benzoyl-, *D*-
Line Form: $CH_3CH(NHCOC_6H_5)CO_2H$
Mol Form: $C_{10}H_{11}NO_3$
M_r: 193.19
mp: 151
α: $[\alpha]^{20/D}$ +24 (w, c=1)
IR: COB 2614
UV: SAD 13329
m/e: 193.07

HODOC No: 01013
CAS No: 35554-98-4
Beilstein No: 4^4, 3296
Name: *D*-Alanine, 3-sulfo-
Synonym: *D*-Cysteic acid
Line Form: $HO_2CCH(NH_2)CH_2SO_3H$
Mol Form: $C_3H_7NO_5S$
M_r: 169.15
mp: 260d
α: $[\alpha]^{20/D}$ +8.66 (w)
Color: oct cr or nd (dil al) pr or nd (w+1)
***s*:** H_2O 4
m/e: 169.00

HODOC No: 01014
CAS No: 1115-69-1
Beilstein No: 4^4, 2497
Name: *DL*-Alanine, *N*-acetyl-
Synonym: Alanine, *N*-acetyl-, *DL*-
Line Form: $CH_3CH(NHCOCH_3)CO_2H$
Mol Form: $C_5H_9NO_3$
M_r: 131.13
mp: 137-8
***s*:** H_2O 3; al 3; eth 1
IR: μSADP 10918 3330 2860 2440 2270 1890 1700 1610 1540 1450 1370 1330 1300 1270 1220 1150 1090 1020 980 910 850 750 700
UV: OES 2-38 215(398) pH=1.0
H NMR: SAD 11583 1.4 2.0 4.4 D_2O
MS: ALDN 76 ini 44(100) 43(54) 86(33) 42(19) 28(17) 45(9) 27(7) 131(2)
m/e: 131.06

HODOC No: 01015
CAS No: 1205-02-3
Beilstein No: 9^4, 794
Name: *DL*-Alanine, *N*-benzoyl-
Synonym: Alanine, *N*-benzoyl-, *DL*-
Line Form: $CH_3CH(NHCOC_6H_5)CO_2H$
Mol Form: $C_{10}H_{11}NO_3$
M_r: 193.19
bp: d
mp: 165-6
Color: pl, pr or lf (eth)
***s*:** H_2O 3; al 3; eth 2; DMSO 2
IR: :gmCOB 2772 3333 2632 2532 1724 1613 1587 1563 1493 1471 1389 1299 1220 1176 1111 926 893 877 820 714 690
UV: OES 1-326 229(10965)
H NMR: SAD 8549 1.5 4.5 7.3-7.6 7.9 7.0-8.2 8.6 $DMSOd_6$
m/e: 193.07

HODOC No: 01016
CAS No: 600-21-5
Beilstein No: 4^4, 2492
Name: *DL*-Alanine, *N*-methyl-
Synonym: Alanine, *N*-methyl-, *DL*-
Line Form: $CH_3CH(NHCH_3)CO_2H$
Mol Form: $C_4H_9NO_2$
M_r: 103.12
bp: sub, 292
mp: 280d
Color: rh pr (alk al)
m/e: 103.06

HODOC No: 01017
CAS No: 3024-83-7
Beilstein No: 4^3, 1714
Name: *DL*-Alanine, 3-sulfo-
Synonym: Alanine, 3-sulfo-, *DL*-
Line Form: $HO_2CCH(NH_2)CH_2SO_3H$
Mol Form: $C_3H_7NO_5S$
M_r: 169.15
mp: 272-4d
Color: pr (w)
***s*:** H_2O 3; al 1; bz 1
IR: SADG 16801 3150 3000 2980 1750 1600 1500 1240 1150 1090 1040 890 740
m/e: 169.00

HODOC No: 01018
CAS No: 1948-31-8
Beilstein No: 4^4, 2519
Name: *L*-Alanine, *N*-*L*-alanyl-
Synonym: Alanine, *N*-*L*-alanyl-, *L*-
Line Form: $H_2NCH(CH_3)CONHCH(CH_3)CO_2H$
Mol Form: $C_6H_{12}N_2O_3$
M_r: 160.17
mp: 298
α: $[\alpha]^{20/D}$ -21.6 (H_2O)
s: H_2O 4
UV: HBCP
MS: NBS 596
m/e: 160.08

HODOC No: 01019
CAS No: 5874-90-8
Beilstein No: 4^4, 2522
Name: *L*-Alanine, *N*-(*N*-*L*-alanyl-*L*-alanyl)-
Synonym: Alanine, *N*-(*N*-*L*-alanyl-*L*-alanyl)-, *L*-
Line Form: $CH_3CH(NH_2)[CONHCH(CH_3)]_2CO_2H$
Mol Form: $C_9H_{17}N_3O_4$
M_r: 231.24
mp: 224d
α: $[\alpha]^{24/D}$ -78.4 (2N HCl, c=4)
IR: ALD 2:405F
m/e: 231.12

HODOC No: 01020
CAS No: 1482-97-9
Beilstein No: 4^2, 830
Name: *L*-Alanine, 3-amino-, monohydrochloride
Synonym: Propionic acid, 2,3-diamino-, monohydrochloride, *L*-
Line Form: $NH_2CH_2CH(NH_2)CO_2H–HCl$
Mol Form: $C_3H_9ClN_2O_2$
M_r: 140.57
α: $[\alpha]^{20/D}$ 25.0 1N HCl
s: H_2O 2
H NMR: SAD 8222 3.5 4.1 D_2O
m/e: 140.04

HODOC No: 01021
CAS No: 2198-64-3
Beilstein No: 9^4, 793
Name: *L*-Alanine, *N*-benzoyl-
Synonym: Alanine, *N*-benzoyl-, *L*-
Line Form: $C_6H_5CONH\text{-}CH(CH_3)CO_2H$
Mol Form: $C_{10}H_{11}NO_3$
M_r: 193.19
mp: 152-4
α: $[\alpha]^{20/D}$ -2.4 (w, c=1)
Color: pl (w)
IR: μCOB 2614 3333 2632 2532 1724 1613 1587 1563 1493 1471 1389 1299 1220 1176 1111 926 893 877 794 714 690
UV: SAD 13329 ini 226(11100) MeOH
m/e: 193.07

HODOC No: 01022
CAS No: 7244-67-9
Name: *L*-Alanine, *N*-benzoyl-, methyl ester
Synonym: Alanine, *N*-benzoyl-, methyl ester, *L*-
Line Form: $C_6H_5CONH\text{-}CH(CH_3)CO_2CH_3$
Mol Form: $C_{11}H_{13}NO_3$
M_r: 207.22
mp: 60-2
IR: ALD 2:895F
MS: NBS 1273
m/e: 207.09

HODOC No: 01023
CAS No: 78627-20-0
Name: *L*-Alanine, *N*-[[(4-chlorophenyl)methoxy]carbonyl]-
Synonym: Alanine, *N*-carboxy(4-chlorobenzyl ester)
Mol Form: $C_{11}H_{12}ClNO_4$
M_r: 257.66
mp: 101-2
UV: LANG 1702 274(182) 266(257) 219(11220) EtOH
m/e: 257.05

HODOC No: 01024
CAS No: 3082-75-5
Beilstein No: 4^4, 2485
Name: *L*-Alanine, ethyl ester
Synonym: Alanine, ethyl ester, *L*-
Line Form: $CH_3CH(NH_2)CO_2C_2H_5$
Mol Form: $C_5H_{11}NO_2$
M_r: 117.14
bp: 96.5-8.5
mp: 96-8
UV: SAD 1398 >210 MeOH
MS: WILEY 444 44(100) 29(9) 42(8) 45(4) 43(3) 74(2) 41(2) 117(1)
m/e: 117.08

HODOC No: 01025
CAS No: 1142-20-7
Beilstein No: 6^4, 2321
Name: *L*-Alanine, *N*-[(phenylmethoxy)carbonyl]-
Synonym: Alanine, *N*-carboxy-, *N*-benzyl ester, *L*-
Line Form: $C_6H_5CH_2O_2CNHCH(CH_3)CO_2H$
Mol Form: $C_{11}H_{13}NO_4$
M_r: 223.22
mp: 82-4
α: $[\alpha]^{23/D}$ -14.2 (aa, c=2)
IR: SADG 4683
UV: SAD 11888
H NMR: SAD 8814
m/e: 223.08

HODOC No: 01026
CAS No: 556-27-4
Beilstein No: 4^4, 3147
Name: *L*-Alanine, 3-(2-propenylsulfinyl)-, (*S*)-
Synonym: Alanine, 3-(allylsulfinyl)-, (*S*)-*L*-
Line Form: $CH_2{=}CHCH_2SOCH_2CH(NH_2)CO_2H$
Mol Form: $C_6H_{11}NO_3S$
M_r: 177.21
mp: 164-6
α: $[\alpha]^{20/D}$ +63.5 (c=2)
Color: nd (dil ac)
***s*:** H_2O 4
m/e: 177.05

HODOC No: 01027
CAS No: 23537-25-9
Name: *L*-Alanine, 3-sulfo-, monohydrate
Synonym: Alanine, 3-sulfo-, monohydrate, *L*-
Line Form: $HO_3SCH_2CH(NH_2)CO_2H{-}H_2O$
Mol Form: $C_3H_9NO_6S$
M_r: 187.16
IR: SADG 2445 3480 3400 3150 2950 2470 1760 1700 1610 1580 1530 1490 1410 1380 1340 1270 1230 1160 1130 1080 1060 1050 1020 950 840 780 720 570 540 510 450 370
Raman: SAD 815 3160 3060 3000 2960 2940 1760 1720 1610 1510 1410 1370 1230 1200 1140 1080 1060 1040 1030 960 850 780 730 710 630 550 460 320 220 130
H NMR: SAD 2802 4.5 D_2O
m/e: 187.02

HODOC No: 01028
CAS No: 527-00-4
Beilstein No: 3^3, 1116
Name: Allaric acid
Synonym: Allomucic acid
Line Form: $HO_2C(CHOH)_4CO_2H$
Mol Form: $C_6H_{10}O_8$
M_r: 210.13
mp: 198-200d
Color: nd or pr (w)
***s*:** H_2O 4
IR: SADP 21948
m/e: 210.04

HODOC No: 01029
CAS No: 488-44-8
Beilstein No: 1^4, 2839
Name: Allitol
Synonym: Allit
Line Form: $HOCH_2(CHOH)_4CH_2OH$
Structure:

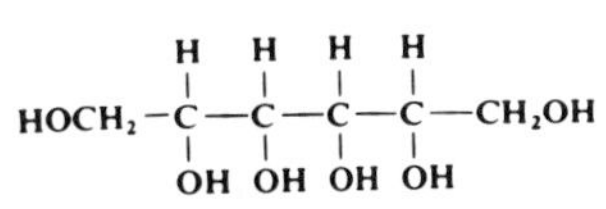

Mol Form: $C_6H_{14}O_6$
M_r: 182.17
mp: 150-1
***s*:** H_2O 4
m/e: 182.08

HODOC No: 01030
CAS No: 3107-04-8
Name: *DL*-Alloisoleucine
Synonym: Alloisoleucine, *DL*-
Line Form: $CH_3CH_2CH(CH_3)CH(NH_2)CO_2H$
Mol Form: $C_6H_{13}NO_2$
M_r: 131.17
IR: SADG 18606 2970 2610 1610 1580 1510 1460 1420 1390 1340 1320 1300 1180 1120 1080 1030 980 960 920 860 790 760 710 670 550 530 440 390 330
Raman: SAD 819 2960 2910 2840 2720 1640 1620 1590 1510 1450 1440 1400 1350 1320 1280 1250 1190 1130 1060 1050 990 930 900 850 800 750 670 540 420 350 230 110
UV: SAD 371 338(1680) 290(8480) 282(9530) 246(50300) 208(21600) MeOH
m/e: 131.09

HODOC No: 01031
CAS No: 1509-34-8
Beilstein No: 4^4, 2777
Name: *L*-Alloisoleucine
Synonym: Alloisoleucine, *L*-
Line Form: $CH_3CH_2CH(CH_3)CH(NH_2)CO_2H$
Mol Form: $C_6H_{13}NO_2$
M_r: 131.17
mp: 280-1d
***d*:** $^{\alpha 20/D}$-14.2 H_2O,2%
α: $[\alpha]^{20/D}$ -14.2 (w, 2%)
***s*:** H_2O 3; al 2; tfa 3
H NMR: SAD 520 1.1 1.2 1.6 2.3 4.4 7.3 tfa
m/e: 131.09

HODOC No: 01032
CAS No: 2595-97-3
Beilstein No: 1^4, 4298
Name: *D*-Allose
Synonym: Allose, *D*-
Line Form: $HOCH_2(CHOH)_4CHO$
Structure:

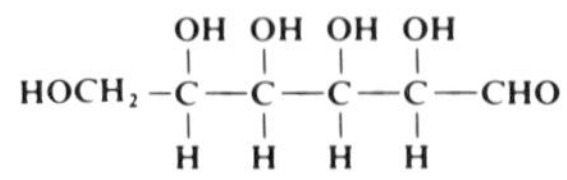

Mol Form: $C_6H_{12}O_6$
M_r: 180.15
mp: 128
α: $[\alpha]^D$ 0.58 → 14.4 (w, c=5)
Color: cr (w)
***s*:** H_2O 4
m/e: 180.06

HODOC No: 01033
CAS No: 7635-11-2
Beilstein No: 1^4, 4300
Name: *L*-Allose
Synonym: Allose, *L*-
Line Form: $HOCH_2(CHOH)_4CHO$
Mol Form: $C_6H_{12}O_6$
M_r: 180.15
mp: 128-9
α: $[\alpha]^{20/D}$ -0.58 (al)
Color: pr (dil al)
***s*:** H_2O 4
m/e: 180.06

HODOC No: 01034
CAS No: 643-03-8
Beilstein No: 1^4, 2839
Name: *D*-Altritol
Synonym: Altritol, *D*-
Structure:

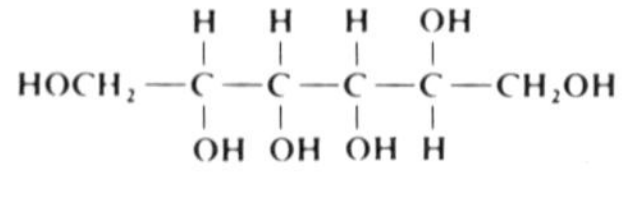

Mol Form: $C_6H_{14}O_6$
M_r: 182.17
mp: 87-8
Color: pr (al)
***s*:** H_2O 4; EtOH 4
m/e: 182.08

HODOC No: 01035
CAS No: 1990-29-0
Beilstein No: 1[4], 4300
Name: *D*-Altrose
Synonym: Altrose, *D*-
Structure:

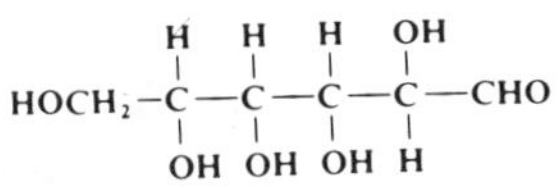

Mol Form: $C_6H_{12}O_6$
M_r: 180.15
mp: 103.5
α: $[\alpha]^{20/D}$ 11.7 → 33.1 (w, c=7)
Color: pr (MeOH,al)
s: H_2O 4
m/e: 180.06

HODOC No: 01036
CAS No: 18917-91-4
Name: Aluminum, *tris*(2-hydroxypropanoato-*O*1,*O*2)-
Synonym: Propanoic acid, 2-hydroxy-, aluminum complex
Line Form: $\{CH_3CH(OH)CO_2\}_3Al$
Mol Form: $C_9H_{15}AlO_9$
M_r: 294.19
s: H_2O 3
IR: SADG 16794 3500 3000 2500 1620 1490 1400 1300 1260 1110 1050 940 870 630 590 450 400
H NMR: SAD 8228 1.5 4.4 D_2O
m/e: 294.05

HODOC No: 01037
CAS No: 3873-54-9
Name: Ammonium, [3-(dodecylamino)-2-hydroxypropyl]trimethyl-, *p*-toluenesulfonate (salt)
Synonym: [3-(Dodecylamino)-2-hydroxypropyl]trimethylammonium *p*-toluenesulfonate
Mol Form: $C_{25}H_{48}N_2O_4S$
M_r: 472.70
IR: μCOB 4420 3230 2940 1490 1330 1220 1180 1120 1030 1010 970 940 910 810 690
m/e: 472.33

HODOC No: 01038
CAS No: 3971-58-2
Name: Ammonium, [3-(hexadecylthio)-2-hydroxypropyl]trimethyl-, *p*-toluenesulfonate (salt)
Synonym: [3-(Hexadecylthio)-2-hydroxypropyl]trimethylammonium *p*-toluenesulfonate
Mol Form: $C_{29}H_{55}NO_4S_2$
M_r: 545.85
IR: μCOB 4418 3330 2940 1470 1220 1160 1120 1040 1010 940 930 820 690
m/e: 545.36

HODOC No: 01039
CAS No: 16704-90-8
Name: Ammonium, [3-(hexylamino)-2-hydroxypropyl]trimethyl-, *p*-toluenesulfonate (salt)
Synonym: [3-(Hexylamino)-2-hydroxypropyl]trimethylammonium *p*-toluenesulfonate
Mol Form: $C_{19}H_{36}N_2O_4S$
M_r: 388.55
mp: 117.5-20.0
IR: μCOB 4422 3230 2940 1470 1330 1220 1180 1160 1030 1010 970 940 810 690
m/e: 388.24

HODOC No: 01040
CAS No: 16704-91-9
Name: Ammonium, [2-hydroxy-3-(octylamino)propyl]trimethyl-, *p*-toluenesulfonate (salt)
Synonym: [2-Hydroxy-3-(octylamino)propyl]trimethylammonium *p*-toluenesulfonate
Mol Form: $C_{21}H_{40}N_2O_4S$
M_r: 416.60
mp: 114.5-5.5
IR: μCOB 4419 3330 2940 1490 1330 1220 1180 1120 1030 1010 970 940 910 810 710 690
m/e: 416.27

HODOC No: 01041
CAS No: 5178-20-1
Beilstein No: 4^4, 1671
Name: Ammonium, (2-hydroxypropyl)trimethyl-, chloride, acetate, (R)-(-)-
Synonym: 2-Acetoxypropyl trimethyl ammonium chloride (*L*)
Line Form: $CH_3CH(O_2CCH_3)CH_2N^+(CH_3)_3Cl^-$
Mol Form: $C_8H_{18}ClNO_2$
Mr: 195.68
bp: d
mp: 172-3
α: $[\alpha]^D$ -41.9
m/e: 195.10

HODOC No: 01042
CAS No: 3763-33-5
Name: Ammonium, [2-hydroxy-3-(tetradecylamino)propyl]trimethyl-, *p*-toluenesulfonate
Synonym: [2-Hydroxy-3-(tetradecylamino)propyl]trimethylammonium *p*-toluenesulfonate
Mol Form: $C_{27}H_{52}N_2O_4S$
Mr: 500.75
IR: **μCOB 4421** 3230 2940 1490 1470 1330 1220 1190 1120 1100 1030 1010 970 940 910 810 690
m/e: 500.36

HODOC No: 01043
CAS No: 897-06-3
Beilstein No: 7^4, 2449
Name: Androsta-1,4-diene-3,17-dione
Synonym: Androstadienedione
Structure:

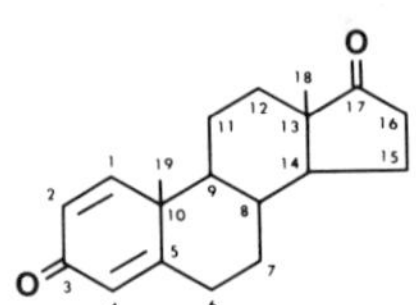

Mol Form: $C_{19}H_{24}O_2$
Mr: 284.38
mp: 138-40
s: chl 3
IR: **ALD 2:1281F**
UV: **LANG 1179** 242(15136) EtOH
H NMR: **SAD 11885** 1.0 1.3 1.3-2.7 6.0 6.2 7.1 $CDCl_3$
MS: **NBS 2091**
m/e: 284.18

HODOC No: 01044
CAS No: 7738-93-4
Beilstein No: 7^4, 2813
Name: Androsta-1,4-diene-3,11,17-trione
Synonym: 1,4-Androstadiene-3,11,17-trione
Mol Form: $C_{19}H_{22}O_3$
Mr: 298.36
mp: 198-201
IR: **COB 5451** 2960 2940 2840 1740 1700 1660 1620 1600 1450 1380 1290 1220 1150 1080 1040 1000 920 880 820 700 620
UV: **OES 4-777** 243(15136) H_2O 238(13804) EtOH
H NMR: **SAD 19140** 0.9 1.5 1.7-2.3 2.2-2.7 2.4 6.1 6.1 7.6 $CDCl_3$
m/e: 298.16

HODOC No: 01045
CAS No: 19041-66-8
Beilstein No: 9^3, 751
Name: Androsta-1,4-dien-3-one, 17-(benzoyloxy)-, (17β)-
Synonym: Androsta-1,4-dien-3-one, 17β-hydroxy-, benzoate
Mol Form: $C_{26}H_{30}O_3$
Mr: 390.50
mp: 218-9
s: chl 3
H NMR: **SAD 11881** 1.0 1.2 1.2-2.7 4.8 6.1 6.2 7.0 7.2-7.6 8.0 $CDCl_3$
m/e: 390.22

HODOC No: 01046
CAS No: 846-48-0
Beilstein No: 8^4, 1111
Name: Androsta-1,4-dien-3-one, 17-hydroxy-, (17β)-
Synonym: Androsta-1,4-dien-3-one, 17β-hydroxy-
Structure:

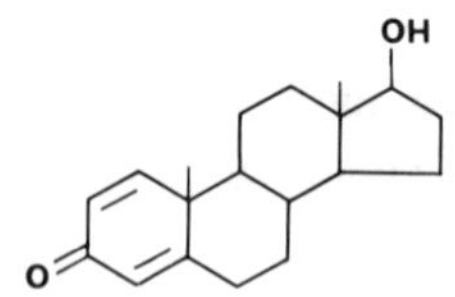

Mol Form: $C_{19}H_{26}O_2$
Mr: 286.40
mp: 164-6
α: $[\alpha]^{25/D}$ +25 (Chl)
m/e: 286.19

HODOC No: 01047
CAS No: 72-63-9
Beilstein No: 8^4, 1119
Name: Androsta-1,4-dien-3-one, 17-hydroxy-17-methyl-, (17β)-
Synonym: Androsta-1,4-dien-3-one, 17β-hydroxy-17-methyl-
Structure:

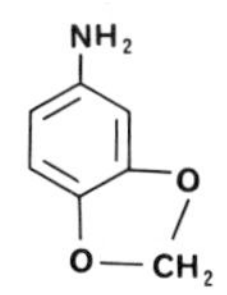

Mol Form: $C_{20}H_{28}O_2$
Mr: 300.42
mp: 165-7
UV: LANG 1896 243(15849) MeOH 248(15136) chl
MS: NBS 2243
m/e: 300.21

HODOC No: 01048
CAS No: 3834-18-2
Name: 5α-Androstane, 3,3,17,17-tetrafluoro-
Structure:

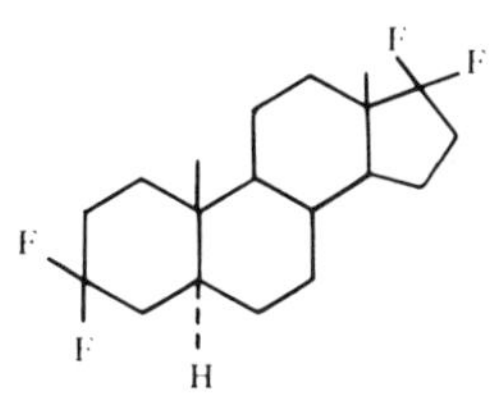

Mol Form: $C_{19}H_{28}F_4$
Mr: 332.41
s: chl 3
IR: μCOB 5256 2940 1450 1370 1320 1280 1160 1080 1060 1000 960 940
m/e: 332.21

HODOC No: 01049
CAS No: 65137-78-2
Name: Androstane-16-acetic acid, 3,17-dihydroxy-, γ-lactone, (3β,5α,16β,17β)-
Synonym: 8H-Naphth[2′,1′:4,5]indeno[1,2-b]furan, androstane-16-acetic acid deriv.
Mol Form: $C_{21}H_{32}O_3$
Mr: 332.46
mp: 240-2
s: chl 3
IR: μCOB 5248 3570 2940 1750 1450 1390 1350 1160 1040 1020 950 850
m/e: 332.24

HODOC No: 01050
CAS No: 96191-75-2
Name: 5β-Androstane-16β-acetic acid, 3β,17β-dihydroxy-, γ-lactone
Synonym: 5β-Androstane-3gb,17β-diol-16β-ylacetic acid,γ-lactone
Mol Form: $C_{21}H_{32}O_3$
Mr: 332.46
mp: 196-7
s: chl 3
IR: μCOB 5274 3570 2940 1750 1450 1390 1330 1300 1180 1050 1030 960 880 830
m/e: 332.24

HODOC No: 01051
CAS No: 438-08-4
Beilstein No: 9^3, 2644
Name: Androstane-17-carboxylic acid, (5β,17β)-
Synonym: 5β-Androstane-17β-carboxylic acid
Structure:

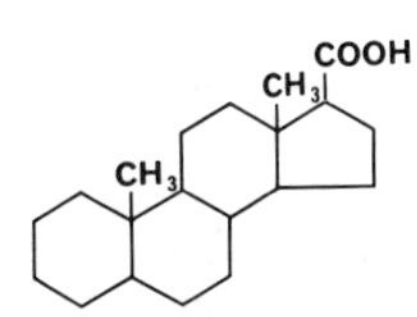

Mol Form: $C_{20}H_{32}O_2$
Mr: 304.45
bp: sub $160^{0.002}$
mp: 228-9
Color: nd (gl HOAc)
m/e: 304.24

HODOC No: 01052
CAS No: 15173-56-5
Name: Androstane-17-carboxylic acid, 3-(acetyloxy)-, (3β,5α,17β)-
Synonym: 5α-Androstane-17β-carboxylic acid, 3β-hydroxy-, acetate
Structure:

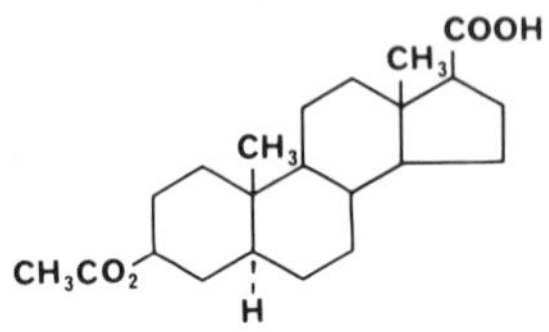

Mol Form: $C_{22}H_{34}O_4$
Mr: 362.49
mp: 249-52
s: Chl 4
H NMR: SAD 9985
m/e: 362.25

HODOC No: 01053
CAS No: 3330-50-5
Beilstein No: 10^3, 669
Name: Androstane-17-carboxylic acid, 3-(acetyloxy)-, methyl ester, (3β,5α,17β)-
Synonym: 5α-Androstane-17β-carboxylic acid, 3β-hydroxy-, methyl ester, acetate
Mol Form: $C_{23}H_{36}O_4$
Mr: 376.51
mp: 147-50
s: chl 2
H NMR: SAD 9986 0.7 0.9 0.8-2.5 2.0 3.7 4.7 $CDCl_3$
m/e: 376.26

HODOC No: 01054
CAS No: 739-27-5
Name: Androstane-11,17-dione, 3-hydroxy-, (3α,5β)-
Synonym: 5β-Androstane-11,17-dione, 3α-hydroxy-
Mol Form: $C_{19}H_{28}O_3$
Mr: 304.41
IR: COB 5444 3470 2970 2920 2860 1720 1700 1450 1380 1300 1170 1050 930 850 610
m/e: 304.20

HODOC No: 01055
CAS No: 1236-49-3
Name: Androstan-17-ol, acetate, (5α,17β)-
Synonym: 5α-Androstan-17β-ol, acetate
Mol Form: $C_{12}H_{10}NNaO_3S$
Mr: 277.25
IR: μCOB 1460 3330 3030 1610 1520 1490 1320 1240 1190 1140 1050 840 750 700 690 650
UV: SAD 18414 294(25200) MeOH
H NMR: SAD 11659 6.7-7.3 7.8 D_2O
m/e: 271.03

HODOC No: 01056
CAS No: 36216-01-0
Name: Androstan-3-ol, (3β)-
Synonym: Androstan-3β-ol
Mol Form: $C_{19}H_{32}O$
Mr: 276.44
IR: μSADP 10840 3230 2940 1450 1370 1160 1140 1090 1040 990 950 890 730
UV: OES 2-616 312(5495) 97% sulf
H NMR: SAD 9994 0.7 0.8 0.9-2.0 1.6 3.6 $CDCl_3$
m/e: 276.25

HODOC No: 01057
CAS No: 36378-49-1
Beilstein No: 7^4, 853
Name: Androstan-17-one
Synonym: Androstane-17-one
Mol Form: $C_{19}H_{30}O$
Mr: 274.43
mp: 122-4
s: Chl 4
UV: OES 7-853
H NMR: SAD 9954
MS: ALDN 215
m/e: 274.23

HODOC No: 01058
CAS No: 5953-69-5
Name: Androstan-17-one, 3-(benzoyloxy)-, (3α,5α)-
Synonym: Androsterone, benzoate
Mol Form: $C_{26}H_{34}O_3$
Mr: 394.53
H NMR: SAD 9964 1.0 1.1 1.1-2.5 2.1-2.5 4.9 7.3-7.6 4.0 $CDCl_3$
m/e: 394.25

HODOC No: 01059
CAS No: 7152-51-4
Name: Androstan-17-one, 11-hydroxy-, (5α,11β)-
Synonym: 5α-Androstan-17-one, 11β-hydroxy-
Mol Form: $C_{19}H_{30}O_2$
M_r: 290.43
IR: COB 5433 3540 2900 2840 1730 1450 1400 1370 1290 1230 1160 1020 960 860 750 680
m/e: 290.22

HODOC No: 01060
CAS No: 481-29-8
Beilstein No: 8^4, 641
Name: Androstan-17-one, 3-hydroxy-, (3β,5α)-
Synonym: 5α-Androstan-17-one, 3β-hydroxy-
Structure:

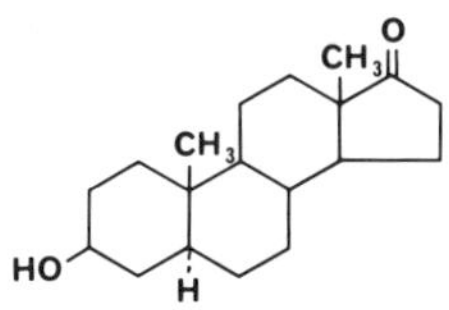

Mol Form: $C_{19}H_{30}O_2$
M_r: 290.43
mp: 177-9
α: $[\alpha]^{20/D}$ +108 (MeOH)
Color: cr (Bz-peth, Ace)
IR: COB 5457 3460 2920 2840 1730 1470 1440 1410 1370 1300 1140 1080 1050 980 900 840 610
UV: OES 3-761
MS: ALDN 222
m/e: 290.22

HODOC No: 01061
CAS No: 53-41-8
Beilstein No: 7^4, 642
Name: Androstan-17-one, 3-hydroxy-, (3α,5α)-
Synonym: Androkinin
Structure:

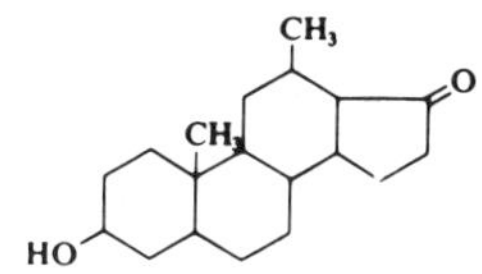

Mol Form: $C_{19}H_{30}O_2$
M_r: 290.43
mp: 185
α: $[\alpha]^{20/D}{}_D$94.6 abs al,0.7%
Color: lf or nd (al, Ace)
s: H_2o 2; al 3; eth 3; ace 3; bz 3; chl 2; os 3
IR: μSADP 718 3450 2780 1720 1450 1370 1280 1250 1160 1120 1090 1060 1030 1000 970 910 850 830
UV: OES 2-616 480(407) 402(1259) 310(7762) 97% sulf
H NMR: SAD 11896 0.8 0.9 2.0 2.0-2.6 4.0 $CDCl_3$
MS: WILEY 1890 67(100) 79(92) 91(88) 107(85) 93(82) 81(81) 109(66) 290(44)
m/e: 290.22

HODOC No: 01062
CAS No: 53-42-9
Name: Androstan-17-one, 3-hydroxy-, (3α,5β)-
Synonym: 5β-Androstan-17-one, 3α-hydroxy-
Structure:

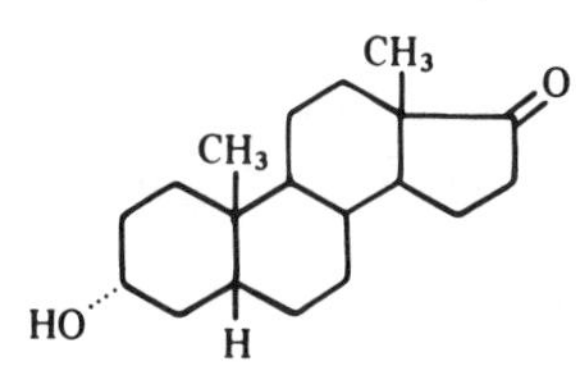

Mol Form: $C_{19}H_{30}O_2$
M_r: 290.43
MS: WILEY 1888 278(100) 218(56) 109(55) 161(53) 257(52) 121(49) 147(49) 290(19)
m/e: 290.22

HODOC No: 01063
CAS No: 571-31-3
Name: Androstan-17-one, 3-hydroxy-, (3β,5β)-
Synonym: 5β-Androstan-17-one, 3β-hydroxy-
Structure:

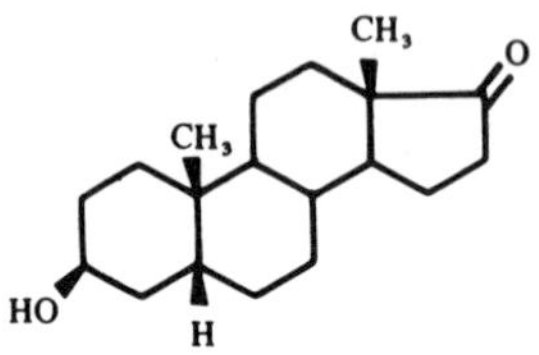

Mol Form: $C_{19}H_{30}O_2$
M_r: 290.43
MS: WILEY 1889 107(100) 108(99) 79(87) 93(87) 81(83) 290(77) 91(76) 95(55)
m/e: 290.22

HODOC No: 01064
CAS No: 4589-93-9
Name: Androstan-17-one, 3-hydroxy-, (2,4-dinitrophenyl)hydrazone, (3α,5β)-
Synonym: 5β-Androstan-17-one, 3α-hydroxy-, (2,4-dinitrophenyl)hydrazone
Mol Form: $C_{25}H_{34}N_4O_5$
M_r: 470.54
IR: COB 5487 3550 3320 3090 2920 2850 1610 1590 1510 1420 1330 1310 1220 1130 1040 910 830 740 710 630
m/e: 470.25

HODOC No: 01065
CAS No: 1057-07-4
Beilstein No: 9^3, 736
Name: Androstan-3-one, 17-(benzoyloxy)-, (5α,17β)-
Synonym: 5α-Androstan-3-one, 17β-hydroxy-, benzoate
Mol Form: $C_{26}H_{34}O_3$
M_r: 394.53
mp: 194-8
s: chl 3
IR: ALD 2:1285G
H NMR: SAD 11884 0.9 1.0 1.1-2.1 2.1-2.5 4.8 7.1-7.6 8.0 $CDCl_3$
m/e: 394.25

HODOC No: 01066
CAS No: 1649-46-3
Name: Androstan-3-one, 2-fluoro-17-hydroxy-, (2α,5α,17β)-
Synonym: 5α-Androstan-3-one, 2α-fluoro-17β-hydroxy-
Mol Form: $C_{19}H_{29}FO_2$
M_r: 308.42
IR: COB 5456 3510 2930 2840 1720 1470 1440 1390 1380 1190 1140 1080 1020 990 650 610
m/e: 308.22

HODOC No: 01067
CAS No: 521-18-6
Beilstein No: 8^4, 634
Name: Androstan-3-one, 17-hydroxy-, (5α,17β)-
Synonym: 5α-Androstan-3-one, 17β-hydroxy-
Structure:

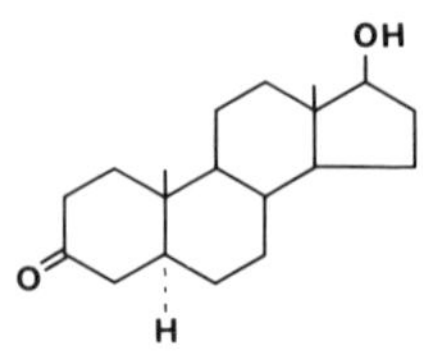

Mol Form: $C_{19}H_{30}O_2$
M_r: 290.43
bp: sub, $135^{0.01}$
mp: 179-83
α: $[\alpha]^{20/D}$ +33 (diox, c=1)
IR: COB 5455 3440 2900 2840 1700 1470 1440 1330 1280 1140 1050 950 610
H NMR: SAD 9944 0.8 1.0 1.1-2.1 1.9 2.1-2.6 3.6 $CDCl_3$
MS: ALD 10:103A
m/e: 290.22

HODOC No: 01068
CAS No: 521-11-9
Name: Androstan-3-one, 17-hydroxy-17-methyl-, (5α,17β)-
Synonym: 5α-Androstan-3-one, 17β-hydroxy-17-methyl-
Mol Form: $C_{20}H_{32}O_2$
M_r: 304.45
mp: 192-3
s: AcOEt 2
UV: HBCP 283(14125)
m/e: 304.24

HODOC No: 01069
CAS No: 521-12-0
Name: Androstan-3-one, 2-methyl-17-(1-oxopropoxy)-, (2α,5α,17β)-
Synonym: 5α-Androstan-3-one, 17β-hydroxy-2α-methyl-, propionate
Mol Form: $C_{23}H_{36}O_3$
M_r: 360.51
mp: 126-30
α: $[\alpha]_D$ 24
H NMR: SAD 9956 0.8 0.9 1.0 1.1 0.5-2.7 2.3 4.6 $CDCl_3$
m/e: 360.27

HODOC No: 01070
CAS No: 633-35-2
Name: Androsta-1,4,6-triene-3,17-dione
Synonym: 1,4,6-Androstatriene-3,17-dione
Structure:

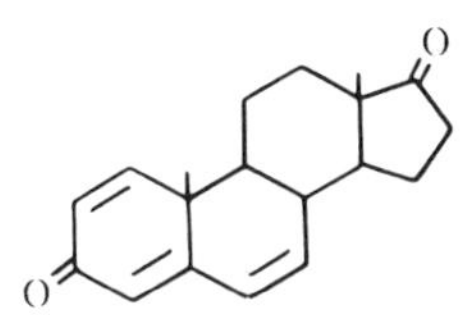

Mol Form: $C_{19}H_{22}O_2$
M_r: 282.36
***s*:** chl 2
UV: OES 1-822 298(18621) 256(13490) 222(15488) EtOH
H NMR: VAR 335 1.0 1.2 1.3-2.7 6.0-6.5 7.1 $CDCl_3$
m/e: 282.16

HODOC No: 01071
CAS No: 1852-61-5
Name: Androst-4-ene-3,17-diol, (3α,17β)-
Synonym: Androst-4-ene-3α,17β-diol
Mol Form: $C_{19}H_{30}O_2$
M_r: 290.43
IR: COB 5429 3330 2930 1660 1440 1360 1270 1130 1050 1020 990 900 870 860 710
m/e: 290.22

HODOC No: 01072
CAS No: 3642-89-5
Beilstein No: 6^4, 6394
Name: Androst-5-ene-3,16-diol, (3β,16α)-
Synonym: Androst-5-ene-3β,16α-diol
Mol Form: $C_{19}H_{30}O_2$
M_r: 290.43
mp: 216.7
α: $[\alpha]^{22/D}$ -110 (95%al,c=1.09)
Color: cr (aq MeOH)
m/e: 290.22

HODOC No: 01073
CAS No: 1259-22-9
Name: Androst-5-ene-3,17-diol, 3-acetate 17-(4-methylbenzenesulfonate), (3β,17β)-
Synonym: Androst-5-ene-3β,17β-diol, 3-acetate 17-*p*-toluenesulfonate
Mol Form: $C_{28}H_{38}O_5S$
M_r: 486.64
IR: µCOB 5356 2940 1720 1590 1450 1370 1330 1250 1160 1090 1020 940 910 890 860 810 690
m/e: 486.24

HODOC No: 01074
CAS No: 2099-26-5
Name: Androst-5-ene-3,17-diol, diacetate, (3β,17β)-
Synonym: Androst-5-ene-3β,17β-diol, diacetate
Mol Form: $C_{23}H_{34}O_4$
M_r: 374.50
IR: COB 6153 2880 2840 2820 1730 1470 1420 1370 1240 1130 1040 960 900 840 800 620
UV: OES 2-726 408(13490) 307(7244) 97% sulf
m/e: 374.25

HODOC No: 01075
CAS No: 2297-30-5
Name: Androst-5-ene-3,17-diol, dipropanoate, (3β,17β)-
Synonym: Androst-5-ene-3β,17β-diol, dipropionate
Mol Form: $C_{25}H_{38}O_4$
M_r: 402.55
mp: 112-4
***s*:** chl 3
H NMR: SAD 11882 0.8 1.0 1.1 1.2-2.1 2.3 4.6 5.4 $CDCl_3$
m/e: 402.28

HODOC No: 01076
CAS No: 3593-85-9
Name: Androst-5-ene-3,17-diol, 17-methyl-, dipropanoate, (3β,17β)-
Synonym: Androst-5-ene-3β,17β-diol, 17-methyl-, dipropionate
Mol Form: $C_{26}H_{40}O_4$
M$_r$: 416.58
mp: 82-6
α: $[\alpha]^{20/D}$-54 to-48 EtOH,c=1
s: chl 3
H NMR: SAD 9969 0.9 1.1 1.2 1.4 1.4-2.2 2.2-2.5 4.6 5.4 $CDCl_3$
m/e: 416.29

HODOC No: 01077
CAS No: 21507-41-5
Name: Androst-1-ene-3,17-dione
Synonym: 1-Androstene-3,17-dione
Mol Form: $C_{19}H_{26}O_2$
M$_r$: 286.40
IR: μCOB 5448 3040 2960 2920 2840 1740 1680 1470 1450 1380 1250 1160 1050 1010 820 780
UV: OES 7-847 230(10965) EtOH
H NMR: SAD 9955 0.9 1.0 1.1-2.7 5.9 7.2 $CDCl_3$
m/e: 286.19

HODOC No: 01078
CAS No: 63-05-8
Name: Androst-4-ene-3,17-dione
Synonym: Androstenedione
Mol Form: $C_{19}H_{26}O_2$
M$_r$: 286.40
IR: COB 5443 2960 2920 2850 1730 1660 1610 1450 1430 1420 1380 1270 1220 1060 1010 870 780 680
UV: OES 7-847 240(16218) EtOH
H NMR: SAD 9959 0.9 1.2 1.3-2.6 2.2-2.6 5.8 $CDCl_3$
C NMR: JJ 482 FT 13.6 17.3 20.2 21.6 30.7 31.2 32.4 33.8 35.0 35.6 38.5 47.3 50.7 53.7 123.9 170.0 198.6 219.7 $CDCl_3$
m/e: 286.19

HODOC No: 01079
CAS No: 382-44-5
Name: Androst-4-ene-3,17-dione, 11-hydroxy-, (11β)-
Synonym: Androst-4-ene-3,17-dione, 11β-hydroxy-
Mol Form: $C_{19}H_{26}O_3$
M$_r$: 302.39
IR: COB 5467 3400 2920 2840 1740 1730 1650 1610 1450 1420 1410 1380 1230 1160 1010 940 870 710
UV: OES 8-504 241(14125) EtOH
m/e: 302.19

HODOC No: 01080
CAS No: 63-00-3
Name: Androst-4-ene-3,17-dione, 6-hydroxy-, (6β)-
Synonym: Androst-4-ene-3,17-dione, 6β-hydroxy-
Mol Form: $C_{19}H_{26}O_3$
M$_r$: 302.39
IR: COB 6155 3260 2930 2850 1740 1690 1610 1470 1450 1420 1370 1330 1260 1180 1040 1020 920 870 720 690
UV: OES 6-850 238(10715) H_2O
m/e: 302.19

HODOC No: 01081
CAS No: 382-45-6
Beilstein No: 7^4, 2796
Name: Androst-4-ene-3,11,17-trione
Synonym: Adrenosterone
Structure:

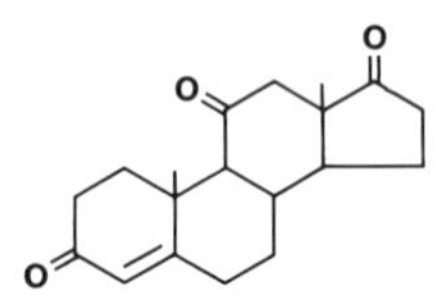

Mol Form: $C_{19}H_{24}O_3$
M$_r$: 300.38
bp: sub in high voe
mp: 220-4
α: $[\alpha]^{20/D}$ 262 abs-al,1% $[\alpha]^{25/D}$ 281 ace
Color: nd (al)
s: H_2O 2; al 3; eth 3; ace 3; chl 3
IR: μDMS 2476 2860 1750 1700 1670 1610 1450 1370 1330 1220 1150 1090 1050 1010 940 880 780 690
UV: OES 3-754 285 sulf
MS: NBS 2238
m/e: 300.17

HODOC No: 01082
CAS No: 1045-69-8
Beilstein No: 8^4, 976
Name: Androst-4-en-3-one, 17-(acetyloxy)-, (17β)-
Synonym: Testosterone, acetate
Mol Form: $C_{21}H_{30}O_3$
M_r: 330.45
mp: 138-44
***s*:** chl 3
UV: OES 2-676 296(18621) 97% sulf
H NMR: SAD 9966 0.9 1.2 1.0-2.6 2.1 2.2-2.6 4.6 5.7 $CDCl_3$
MS: ALDN 239 ini 43(100) 124(82) 147(71) 288(31) 228(28) 55(26) 91(25) 330(15)
m/e: 330.22

HODOC No: 01083
CAS No: 2088-71-3
Beilstein No: 9^4, 470
Name: Androst-4-en-3-one, 17-(benzoyloxy)-, (17β)-
Synonym: Testosterone, benzoate
Structure:

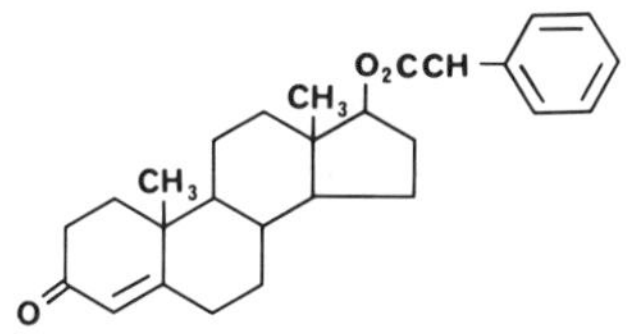

Mol Form: $C_{26}H_{32}O_3$
M_r: 392.51
mp: 191-4
***s*:** chl 3
IR: SADG 17886 3020 2940 1700 1670 1610 1450 1380 1300 1260 1100 1060 990 860 710
H NMR: SAD 11880 1.0 1.2 1.2-2.2 2.2-2.6 4.8 5.7 7.2-7.6 8.0 $CDCl_3$
MS: ALDN 260 105(100) 77(17) 147(12) 124(11) 106(9) 228(5) 41(5) 392(3)
m/e: 392.24

HODOC No: 01084
CAS No: 1093-58-9
Beilstein No: 8^4, 989
Name: Androst-4-en-3-one, 4-chloro-17-hydroxy-, (17β)-
Synonym: Androst-4-en-3-one, 4-chloro-17β-hydroxy-
Mol Form: $C_{19}H_{27}ClO_2$
M_r: 322.86
mp: 188-90
α: $[\alpha]^{20/D}$148 chl
IR: COB 5453 3400 2930 1690 1680 1580 1440 1410 1370 1290 1180 1070 1050 980 850 810
m/e: 322.17

HODOC No: 01085
CAS No: 14191-92-5
Name: Androst-4-en-3-one, 17-[(cyclohexylcarbonyl)oxy]-, (17β)-
Synonym: Cyclohexanecarboxylic acid, ester with testosterone
Mol Form: $C_{26}H_{38}O_3$
M_r: 398.56
mp: 128.5-129
***s*:** chl 3
UV: OES 1-1020 242(13183) EtOH
H NMR: SAD 9977 0.9 1.2 0.9-2.2 2.3 4.7 5.8 $CDCl_3$
m/e: 398.28

HODOC No: 01086
CAS No: 1807-02-9
Beilstein No: 8^4, 2056
Name: Androst-4-en-3-one, 11,17-dihydroxy-17-methyl-, (11α,17β)-
Synonym: Androst-4-en-3-one, 11α,17β-dihydroxy-17-methyl-
Structure:

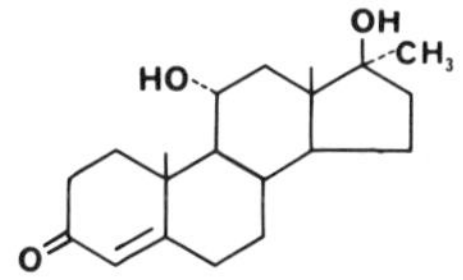

Mol Form: $C_{20}H_{30}O_3$
M_r: 318.44
mp: 153-5
MS: ALD 10:105C
m/e: 318.22

HODOC No: 01087
CAS No: 76-43-7
Name: Androst-4-en-3-one, 9-fluoro-11,17-dihydroxy-17-methyl-, (11β,17β)-
Synonym: Androst-4-en-3-one, 9-fluoro-11β,17β-dihydroxy-17-methyl-
Mol Form: $C_{20}H_{29}FO_3$
Mr: 336.43
mp: 270
IR: **µCOB 5398** 3330 1670 1430 1350 1280 1240 1180 1140 1090 1030 950 940 930 900 860
m/e: 336.21

HODOC No: 01088
CAS No: 1649-25-8
Name: Androst-4-en-3-one, 4-fluoro-17-hydroxy-, (17β)-
Synonym: Androst-4-en-3-one, 4-fluoro-17β-hydroxy-
Mol Form: $C_{19}H_{27}FO_2$
Mr: 306.40
IR: **COB 5454** 3370 2940 2870 1690 1670 1630 1440 1360 1320 1200 1150 1080 1060 990 870 610
m/e: 306.20

HODOC No: 01089
CAS No: 1852-58-0
Name: Androst-4-en-3-one, 6-fluoro-17-hydroxy-, (6β,17β)-
Synonym: Androst-4-en-3-one, 6β-fluoro-17β-hydroxy-
Mol Form: $C_{19}H_{27}FO_2$
Mr: 306.40
IR: **COB 5462** 3470 2960 2890 2860 1670 1470 1450 1430 1410 1370 1340 1240 1190 1070 1000 960 880 720
m/e: 306.20

HODOC No: 01090
CAS No: 58-22-0
Beilstein No: 8^4, 974
Name: Androst-4-en-3-one, 17-hydroxy-, (17β)-
Synonym: Androlin
Structure:

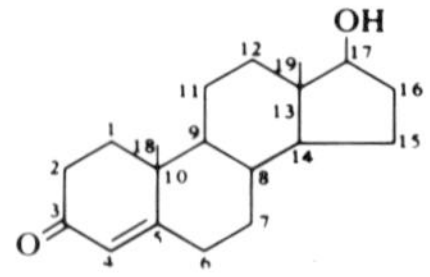

Mol Form: $C_{19}H_{28}O_2$
Mr: 288.41
mp: 155
α: $[\alpha]^{24/D}$ 109 al,4%
Color: nd (dil Ace)
s: H_2O 1; al 3; eth 3; ace 3; os 3
IR: **COB 5450** 3400 2930 1740 1660 1610 1430 1410 1380 1360 1350 1330 1280 1230 1200 1130 1110 1060 1050 1020 960 940 870 830 690
UV: **OES 7-851** 241(15136) MeOH
H NMR: **SAD 11900** 0.8 1.2 0.8-2.1 2.1-2.6 3.0 3.6 5.7 $CDCl_3$
C NMR: **JJ 483** FT 11.0 17.3 20.6 23.2 30.1 31.5 32.7 33.8 35.6 36.4 38.6 42.7 50.4 53.9 81.2 123.6 171.4 199.4 $CDCl_3$
MS: **ALDN 221** 246(100) 288(82) 248(66) 290(50) 247(40) 249(30) 289(28) 291(24)
m/e: 288.21

HODOC No: 01091
CAS No: 58-18-4
Beilstein No: 8[4], 1010
Name: Androst-4-en-3-one, 17-hydroxy-17-methyl-, (17β)-
Synonym: Androst-4-en-3-one, 17β-hydroxy-17-methyl-
Structure:

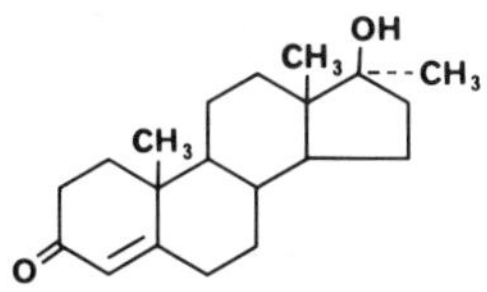

Mol Form: $C_{20}H_{30}O_2$
Mr: 302.44
mp: 161-6
α: $[\alpha]^{20/D}$ +82 (al)
s: Et_2O 4; EtOH 4
IR: µCOB 5051 3570 2860 1670 1610 1450 1370 1330 1270 1220 1180 1090 940 930 860
UV: OES 4-837 241(16218) EtOH
MS: ALD 10:105D
m/e: 302.22

HODOC No: 01092
CAS No: 57-85-2
Beilstein No: 8[4], 977
Name: Androst-4-en-3-one, 17-(1-oxopropoxy)-, (17β)-
Synonym: Testosterone, propionate
Structure:

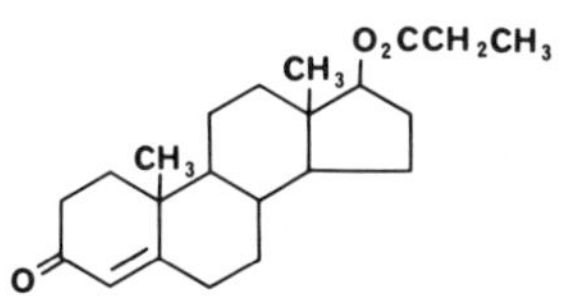

Mol Form: $C_{22}H_{32}O_3$
Mr: 344.47
mp: 118-22
α: $[\alpha]^{25/D}$ 83-90 (diox,_c=1)
s: Et_2O 4; Py 4; EtOH 4
UV: LANG 1889
m/e: 344.24

HODOC No: 01093
CAS No: 99099-00-0
Name: Androst-5-en-17-one, 3β-hydroxy-, nitrite
Synonym: 5-Androstene-3β-ol-17-one, nitrite
Mol Form: $C_{19}H_{27}NO_3$
Mr: 317.41
mp: 144-6
s: chl 3
IR: µCOB 6143 2940 1750 1640 1470 1370 1140 1020 860 820
UV: OES 6-852 372(69) 357(69) 346(54) 336(38) 229(2399) cy 373(68) 360(68) py
m/e: 317.20

HODOC No: 01094
CAS No: 2283-82-1
Beilstein No: 8[4], 994
Name: Androst-5-en-17-one, 3-hydroxy-, (3α)-
Synonym: Androst-5-en-17-one, 3α-hydroxy-
Structure:

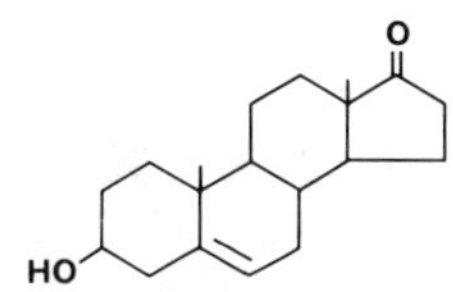

Mol Form: $C_{19}H_{28}O_2$
Mr: 288.41
mp: 149-51
IR: ALD 2:1279E
MS: ALD 10:103B
m/e: 288.21

HODOC No: 01095
CAS No: 65628-35-5
Name: Androst-5-en-17-one, 3-hydroxy-, (aminothioxomethyl)hydrazone, (3β)-
Synonym: 5-Androstene-3β-ol-17-one, thiosemicarbazone
Mol Form: $C_{20}H_{31}N_3OS$
Mr: 361.53
IR: COB 5491 3600 3410 3350 3000 1660 1570 1520 1430 1400 1350 1280 1220 1130 1040 960 820 730
m/e: 361.22

HODOC No: 01096
CAS No: 7745-40-6
Name: Androst-5-en-17-one, 3-hydroxy-, cyclic 1,2-ethanediyl acetal
Synonym: Spiro[17H-cyclopenta[a]phenanthrene-17,2′-[1,3]dioxolane], androst-5-en-17-one deriv.
Mol Form: $C_{21}H_{32}O_3$
M$_r$: 332.46
***s*:** chl 3
IR: μCOB 3380 3570 3450 2940 1470 1430 1390 1300 1160 1100 1040 950 900 880 830
m/e: 332.24

HODOC No: 01097
CAS No: 956-34-3
Name: Aniline, *o*-chloro-*N*-[[(*p*-chlorophenyl)thio]methyl]-
Synonym: Methane, (2-chlorophenylamino)-(4-chlorophenylthio)
Mol Form: $C_{13}H_{11}Cl_2NS$
M$_r$: 284.19
mp: 64-5
IR: SADG 21976 3400 3060 2920 1590 1500 1470 1440 1380 1320 1290 1270 1230 1210 1150 1130 1080 1030 1010 920 820 730 680 640 580 540 490 430
UV: SAD 2899 245(18400) MeOH
H NMR: SAD 11292 4.7 6.7 6.8 7.0-7.3 $CDCl_3$
m/e: 283.00

HODOC No: 01098
CAS No: 16330-03-3
Name: Aniline, 4-chloro-*N*-methyl-3-nitro-
Mol Form: $C_7H_7ClN_2O_2$
M$_r$: 186.59
mp: 53
***s*:** chl 3
IR: SADG 19613 3420 3060 2900 2800 1610 1530 1420 1350 1330 1270 1160 1130 1060 900 840 800 750 650 400
UV: SAD 17374 383(1130) 251(20600) MeOH
H NMR: SAD 10033 2.8 4.1 6.7 7.0 7.2 $CDCl_3$
m/e: 186.02

HODOC No: 01099
CAS No: 25287-41-6
Name: Aniline, *p*-chloro-*N*-[1-(2-naphthyl)ethylidene]-
Synonym: Naphthalene, 2-acetyl,imine, *N*(4-chlorophenyl)
Mol Form: $C_{18}H_{14}ClN$
M$_r$: 279.75
mp: 135-6
UV: LANG 2521 294(13490) 285(13490) 247(50119) EtOH
m/e: 279.08

HODOC No: 01100
CAS No: 25287-23-4
Name: Aniline, *p*-chloro-*N*-(*p*,α-dimethylbenzylidene)-
Synonym: Acetophenone, imine,*N*(4-chlorophenyl)-4-methyl
Mol Form: $C_{15}H_{14}ClN$
M$_r$: 243.72
mp: 95-6
UV: LANG 2348 315(3162) 255(19953) EtOH
m/e: 243.08

HODOC No: 01101
CAS No: 615-86-1
Beilstein No: 12, 662
Name: Aniline, 2,4,5-dibromo-
Synonym: Aniline, 2,4,5-tribromo-
Line Form: 2,4,5-$Br_3C_6H_2NH_2$
Mol Form: $C_6H_4Br_3N$
M$_r$: 326.81
mp: 80-1
Color: nd (al)
***s*:** al 4; eth 4; bz 4; dil ac 3
UV: HBCP 284(112) 274(200) 268(159) al
m/e: 326.79

HODOC No: 01102
CAS No: 95766-63-5
Beilstein No: 15^3, 876
Name: Aniline, 3,3′-hydrazodi-
Synonym: Hydrazobenzene, 3,3′-diamino-
Line Form: (3-$H_2NC_6H_4$)-NHNH-($C_6H_4NH_2$-3)
Mol Form: $C_{12}H_{14}N_4$
M$_r$: 214.26
mp: 151
Color: sym (al)
m/e: 214.12

HODOC No: 01103
CAS No: 104177-72-2
Beilstein No: 12^3, 272
Name: Aniline, *p*-isopentyl-
Synonym: Aniline, 4-isopentyl
Line Form: $C_6H_5NHCH_2CH_2CH(CH_3)_2$
Mol Form: $C_{11}H_{17}N$
M_r: 163.25
bp: 254.5, $126\text{-}7^{14}$
***d*:** $0.928^{15/4}$
***n*:** 1.5305^{20}
m/e: 163.14

HODOC No: 01104
CAS No: 25349-12-6
Name: Aniline, *N*-(*p*-methoxy-α-methylbenzylidene)-*m*-nitro-
Synonym: Acetophenone, imine,4-methoxy-*N*(3-nitrophenyl)
Mol Form: $C_{15}H_{14}N_2O_3$
M_r: 270.27
mp: 112
UV: LANG 2360 345(2399) 275(23988) EtOH
m/e: 270.10

HODOC No: 01105
CAS No: 25287-21-2
Name: Aniline, *N*-(α-methylbenzylidene)-*m*-nitro-
Synonym: Acetophenone, imine,*N*(3-nitrophenyl)
Mol Form: $C_{14}H_{12}N_2O_2$
M_r: 240.25
mp: 87
UV: LANG 2342 313(3311) 250(23988) EtOH
m/e: 240.09

HODOC No: 01106
CAS No: 91252-74-3
Name: Aniline, 5-(methylsulfonyl)-2-propoxy-
Mol Form: $C_{10}H_{15}NO_3S$
M_r: 229.28
IR: μCOB 7317 3330 1610 1560 1520 1280 1240 1150 1120 1060 970 950 860 810 760
m/e: 229.08

HODOC No: 01107
CAS No: 25287-39-2
Name: Aniline, *N*-[1-(2-naphthyl)ethylidene]-
Synonym: Naphthalene, 2-acetyl,imine,*N*,-phenyl
Mol Form: $C_{18}H_{15}N$
M_r: 245.31
mp: 136-7
UV: LANG 2519 326(3631) 294(12882) 283(15488) 251(51286) cyhex
m/e: 245.12

HODOC No: 01108
CAS No: 5461-23-4
Name: Aniline, *p*-[(*p*-nitrophenyl)sulfinyl]-
Synonym: 4-Amino-4′-nitrodiphenylsulfoxide
Mol Form: $C_{12}H_{10}N_2O_3S$
M_r: 262.27
mp: 132
***s*:** H_2O 2; al 4
UV: HBCP 277(21379) al
m/e: 262.04

HODOC No: 01109
CAS No: 25287-25-6
Name: Aniline, *N*-(*p*,α-dimethylbenzylidene)-*m*-nitro-
Synonym: Acetophenone, imine,4-methyl-*N*(3-nitrophenyl)
Mol Form: $C_{15}H_{14}N_2O_2$
M_r: 254.27
mp: 95
UV: LANG 2359 325(3548) 259(25119) EtOH
m/e: 254.11

HODOC No: 01110
CAS No: 4880-33-5
Name: Aniline, *N*-[1-(5,6,7,8-tetrahydro-2-naphthyl)ethylidene]-
Synonym: Tetralin, 6(1(phenylimno)ethyl)
Mol Form: $C_{18}H_{19}N$
M_r: 249.34
mp: 31
UV: LANG 2515 259(17378) EtOH
m/e: 249.15

HODOC No: 01111
CAS No: 21598-82-3
Name: Aniline, 2,2′-trimethylenedi-
Synonym: Propane, 1,3-*bis*(2-aminophenyl)
Mol Form: $C_{15}H_{18}N_2$
Mr: 226.31
mp: 72
***s*:** chl 3
IR: SADG 17841 3400 3320 3220 3020 2960 2880 1630 1600 1580 1490 1450 1250 870 860 750 700 500 440
UV: SAD 17033 232(7220)MeOH
H NMR: SAD 9403 1.9 2.5 3.4 6.5-7.2 $CDCl_3$
m/e: 226.15

HODOC No: 01112
CAS No: 6325-62-8
Name: *p*-Anisidine, *N*-(*p*-methoxy-α-methylbenzylidene)-
Synonym: Acetophenone, imine,4-methoxy-*N*(**4-methoxyphenyl)**
Mol Form: $C_{16}H_{17}NO_2$
Mr: 255.30
mp: 128-9
UV: LANG 2357 316(4467) 271(17783) EtOH
m/e: 255.13

HODOC No: 01113
CAS No: 25287-42-7
Name: *p*-Anisidine, *N*-[1-(2-naphthyl)ethylidene]-
Synonym: Naphthalene, 2-acetyl,imine,*N*(4-methoxyphenyl)
Mol Form: $C_{19}H_{17}NO$
Mr: 275.33
mp: 153
UV: LANG 2522 329(5012) 292(13183) 284(15136) 249(45709) EtOH
m/e: 275.13

HODOC No: 01114
CAS No: 108123-07-5
Name: Anisole, 2,4-dimethyl-6-(phenylsulfonyl)-
Synonym: Sulfone, diphenyl,3,5-dimethyl-2-methoxy
Mol Form: $C_{15}H_{16}O_3S$
Mr: 276.34
mp: 103
IR: µSADP 13983 2940 1560 1470 1450 1390 1300 1210 1150 1120 1090 990 860 770 760 740 710 690
UV: SAD 3829 286(2480) 274(2110) 267(1680) 260(1260) 237(13500) 203 MeOH
H NMR: SAD 11952 2.2 2.3 3.8 7.2 7.5 7.8 8.0 $CDCl_3$
m/e: 276.08

HODOC No: 01115
CAS No: 14168-83-3
Beilstein No: 6^3, 3497
Name: Anisole, *m*-styryl-, (Z)-
Synonym: Stilbene, 3-methoxy(*cis*)
Mol Form: $C_{15}H_{14}O$
Mr: 210.26
bp: $122\text{-}3^{0.8}$ 96^1
***s*:** H_2O 2; al 3; eth 3; ace 3
UV: JACSA 90-4633-68 274(10000) 95% al
MS: ALDN 172 ini 210(100) 165(37) 179(30) 178(27) 209(25) 167(20) 152(18)
m/e: 210.10

HODOC No: 01116
CAS No: 107775-84-8
Beilstein No: 6, 794
Name: Anisole, 2,2′-sulfonyldi-
Synonym: Sulfone, diphenyl,2,2′-dimethoxy
Mol Form: $C_{14}H_{14}O_4S$
Mr: 278.31
mp: 157-8
***s*:** al 4; bz 2; HOAc 3; con sulf 3; xyl 2; peth 1
IR: µSADP 17826 2940 1590 1470 1430 1300 1280 1150 1080 1020 810 760 730 710
UV: SAD 5654 288 207 MeOH
m/e: 278.06

HODOC No: 01117
CAS No: 7539-25-5
Beilstein No: 6^3, 873
Name: Anisole, 2,3,5-trinitro-
Synonym: 2,3,5-Trinitroanisole
Line Form: 2,3,5-$(NO_2)_3$ - $C_6H_2OCH_3$
Mol Form: $C_7H_5N_3O_7$
Mr: 243.13
mp: 106.8
***d*:** 1.618^{15}
Color: ye nd (al)
***s*:** Ac 4; Bz 4; EtOH 4
IR: SADP 20343
UV: SAD 6837
m/e: 243.01

HODOC No: 01118
CAS No: 116-71-2
Beilstein No: 7^4, 2748
Name:
Anthra[9,1,2-cde]benzo[rst]pentaphene-5,10-dione
Synonym: Ahcovat Dark Blue BO
Structure:

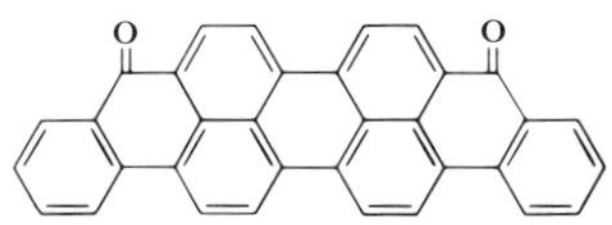

Mol Form: $C_{34}H_{16}O_2$
M_r: 456.47
mp: 490-5d
Color: Vt-bl or bk nd (Ph NO_2)
***s*:** al 1; bz 1; xy 3; py 3; sulf 3; HOAc 1
UV: OES 8-723 680(13804) 620(14791)
m/e: 456.12

HODOC No: 01119
CAS No: 610-49-1
Beilstein No: 12^3, 3335
Name: 1-Anthracenamine
Synonym: 1-Anthramine
Line Form: 1-$H_2NC_{14}H_9$
Mol Form: $C_{14}H_{11}N$
M_r: 193.23
mp: 130
Color: gold-ye nd (al)
***s*:** H_2O 1; al 4; ace 2; con HCl 1
IR: SADG 10379 3448 3333 3030 1613 1563 1538 1449 1389 1351 1316 1282 1136 1000 952 901 870 787 735
UV: SAD 6851 257(49300) 240(64000) MeOH
H NMR: SAD 2331 6.2-8.3 ace
m/e: 193.09

HODOC No: 01120
CAS No: 613-13-8
Beilstein No: 12^3, 3355
Name: 2-Anthracenamine
Synonym: 2-Anthramine
Line Form: 2-$H_2NC_{14}H_9$
Mol Form: $C_{14}H_{11}N$
M_r: 193.23
bp: sub
mp: 238-41
Color: ye lf (al)
***s*:** H_2O 1; al 3; os 2; con sulf 1
IR: µSADP 18155 3450 3330 3030 1640 1590 1470 1220 1180 950 890 810 740
UV: SAD 5762 336 260 MeOH
m/e: 193.09

HODOC No: 01121
CAS No: 779-03-3
Beilstein No: 7^4, 2361
Name: 9-Anthracenamine
Synonym: 9-Anthramine
Line Form: 9-$H_2NC_{14}H_9$
Mol Form: $C_{14}H_{11}N$
M_r: 193.23
mp: 145-50
Color: ye lf (dil al) br Bz)
***s*:** al 3; eth 3; bz 3; chl 3
UV: OES 5-400 410(6310) 380(6310) 270(31623) $CHCl_3$
m/e: 193.09

HODOC No: 01122
CAS No: 32666-78-7
Name: 2-Anthracenamine, hydrochloride
Synonym: 2-Anthramine, hydrochloride
Mol Form: $C_{14}H_{12}ClN$
M_r: 229.69
UV: FO 399 377(5012) 357(6310) 339(3981) 324(3162) 308(1585) 253(251189) 2N HCl,50% EtOH
m/e: 229.07

HODOC No: 01123
CAS No: 120-12-7
Beilstein No: 5^4, 2281
Name: Anthracene
Synonym: Anthracin
Structure:

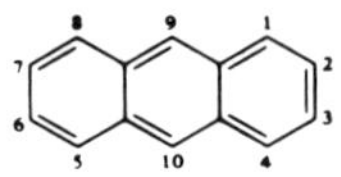

Mol Form: $C_{14}H_{10}$
Mr: 178.22
bp: 340, 226.5^{53}
mp: 214-6
***d*:** $1.283^{25/4}$
Color: ta or mcl pr (al)
***s*:** H_2O 1; al 2; eth 2; ace 2; bz 2; ctc 2; chl 2; tol 2; CS_2 2
IR: **µCOB 6027** 3030 1610 1540 1450 1320 1270 1160 1140 950 890 740
Raman: **SAD 1181** 3070 3050 3020 1670 1630 1580 1560 1500 1480 1410 1380 1350 1260 1190 1170 1010 980 960 920 780 760 630 520 480 400 290 250 130
UV: **SAD 394** 376(7590) 355(7770) 338(5290) 322(2750) 309(1230) 296(531) 250(20000) 220(11800) 218(11700) MeOH
C NMR: **STOT 90** 125.7 130.3 132.4 132.8
MS: **ALDN 141** 178(100) 176(17) 179(16) 89(14) 76(11) 88(9) 151(6) 152(6)
m/e: 178.08

HODOC No: 01124
CAS No: 3613-42-1
Beilstein No: 5^4, 2819
Name: Anthracene, 9,10-*bis*(phenylmethyl)-
Synonym: Anthracene, 9,10-dibenzyl-
Line Form: 9,10-$(C_6H_5CH_2)_2C_{14}H_8$
Structure:

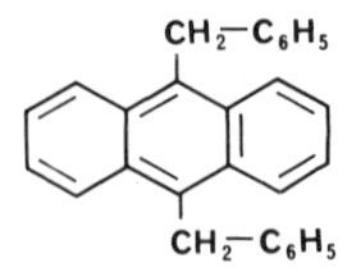

Mol Form: $C_{28}H_{22}$
Mr: 358.45
mp: 245
***d*:** 0.1787^{16}
***s*:** H_2O 1; eth 2; bz 2; lig 1; CS_2 2
IR: **µCOB 3557** 3030 1610 1490 1470 1450 1370 1080 1030 960 800 750 720 690
UV: **OES 7-1150** 400(7079) 380(7079) 360(4467) 343(2951) 265(66069) $CHCl_3$
m/e: 358.17

HODOC No: 01125
CAS No: 1498-69-7
Beilstein No: 5^2, 598
Name: Anthracene, 9-butyl-
Synonym: 9-Butylanthracene
Mol Form: $C_{18}H_{18}$
Mr: 234.32
mp: 49-50
UV: **API 192** 388 368 349 332 317 256 243 222 218 iso
MS: **WILEY 1657** 191(100) 234(37) 192(19) 189(16) 235(7) 190(6) 165(6) 202(5)
m/e: 234.14

HODOC No: 01126
CAS No: 55133-89-6
Name: Anthracene, 9-butyltetradecahydro-
Synonym: Anthracene, 9-butyl-perhydro
Mol Form: $C_{18}H_{32}$
Mr: 248.43
MS: **WILEY 1743** 95(100) 191(96) 41(69) 109(59) 67(56) 81(52) 55(51) 248(18)
m/e: 248.25

HODOC No: 01127
CAS No: 4985-70-0
Beilstein No: 5^4, 2292
Name: Anthracene, 1-chloro-
Synonym: 1-Chloroanthracene
Line Form: $1\text{-}ClC_{14}H_9$
Mol Form: $C_{14}H_9Cl$
M_r: 212.66
mp: 83.5
***d*:** $1.1707^{100/4}$
***n*:** 1.6959^{100}
Color: lf (HOAc)
***s*:** H_2O 1; al 3; eth 3; bz 3; ctc 3
UV: OES 5-393 385(10000) 365(10965) 340(10000) 320(10000) hx
m/e: 212.04

HODOC No: 01128
CAS No: 17135-78-3
Beilstein No: 5^4, 2292
Name: Anthracene, 2-chloro-
Synonym: 2-Chloroanthracene
Line Form: $2\text{-}ClC_{14}H_9$
Mol Form: $C_{14}H_9Cl$
M_r: 212.66
mp: 215
Color: nd or lf
***s*:** H_2O 1; ctc 3
IR: OES 5-393
UV: OES 5-393 385(7943) 365(8913) 355(7943) 325(7943) hx
m/e: 212.04

HODOC No: 01129
CAS No: 716-53-0
Beilstein No: 5^4, 2292
Name: Anthracene, 9-chloro-
Synonym: 9-Chloroanthracene
Line Form: $9\text{-}ClC_{14}H_9$
Mol Form: $C_{14}H_9Cl$
M_r: 212.66
mp: 106
Color: gold-ye nd (al)
***s*:** H_2O 1; al 3; eth 3; bz 3; ctc 3; chl 2; CS_2 3; con sulf 3
IR: SADG 17696 3040 1620 1520 1440 1390 1310 1300 1260 1140 1000 940 880 830 770 730
UV: HBCP 392(7943) 340(2512) 255(316230) al
H NMR: SAD 8648 7.2-7.7 7.8 8.2 8.4 $CDCl_3$
m/e: 212.04

HODOC No: 01130
CAS No: 24463-19-2
Beilstein No: 5^4, 2313
Name: Anthracene, 9-(chloromethyl)-
Synonym: 9-Anthracylmethyl chloride
Line Form: $9\text{-}ClCH_2\text{-}C_{14}H_9$
Mol Form: $C_{15}H_{11}Cl$
M_r: 226.69
mp: 138-40
IR: ALD 3:620H
m/e: 226.05

HODOC No: 01131
CAS No: 55255-70-4
Name: Anthracene, 9-cyclohexyltetradecahydro-
Synonym: Anthracene, 9-cylohexyl-perhydro
Mol Form: $C_{20}H_{34}$
M_r: 274.47
MS: WILEY 1846 41(100) 191(88) 55(81) 109(74) 67(68) 81(53) 190(41) 274(9)
m/e: 274.27

HODOC No: 01132
CAS No: 523-27-3
Beilstein No: 5^4, 2295
Name: Anthracene, 9,10-dibromo-
Synonym: 9,10-Dibromoanthracene
Line Form: $9,10\text{-}Br_2C_{14}H_8$
Mol Form: $C_{14}H_8Br_2$
M_r: 334.01
bp: sub
mp: 226
Color: ye nd (to or xyl)
***s*:** H_2O 1; al 2; eth 2; bz 2; chl 3
IR: µCOB 446 2860 1610 1470 1390 1300 1250 1030 930 840 750 670 660
UV: SAD 1084 401(12900) 380(12900) 360(7490) 342(3230) 258(15300) 221(16300) 217(16800) MeOH
H NMR: VAR 618 7.6 8.6 $CDCl_3$
m/e: 333.90

HODOC No: 01133
CAS No: 66259-12-9
Beilstein No: 5^2, 575
Name: Anthracene, 1,4-dichloro-
Synonym: 1,4-Dichloroanthracene
Line Form: 1,4-Cl_2-$C_{14}H_8$
Mol Form: $C_{14}H_8Cl_2$
M_r: 247.11
mp: 180
Color: ye cr (Ace)
m/e: 246.00

HODOC No: 01134
CAS No: 6406-96-8
Beilstein No: 5^4, 2293
Name: Anthracene, 1,5-dichloro-
Synonym: 1,5-Dichloroanthracene
Line Form: 1,5-Cl_2-$C_{14}H_8$
Mol Form: $C_{14}H_8Cl_2$
M_r: 247.11
mp: 187
Color: ye nd (BuOH)
m/e: 246.00

HODOC No: 01135
CAS No: 14381-66-9
Beilstein No: 5^3, 2133
Name: Anthracene, 1,8-dichloro-
Synonym: 1,8-Dichloroanthracene
Line Form: 1,8-Cl_2-$C_{14}H_8$
Mol Form: $C_{14}H_8Cl_2$
M_r: 247.11
mp: 185
Color: ye nd (to)
m/e: 246.00

HODOC No: 01136
CAS No: 613-07-0
Beilstein No: 5^3, 2133
Name: Anthracene, 2,3-dichloro-
Synonym: 2,3-Dichloroanthracene
Line Form: 2,3-Cl_2-$C_{14}H_8$
Mol Form: $C_{14}H_8Cl_2$
M_r: 247.11
mp: 261
Color: pa ye lf
***s*:** AcOEt 4; HOAc 4
m/e: 246.00

HODOC No: 01137
CAS No: 605-48-1
Beilstein No: 5^4, 2293
Name: Anthracene, 9,10-dichloro-
Synonym: 9,10-Dichloroanthracene
Line Form: 9,10-$Cl_2C_{14}H_8$
Mol Form: $C_{14}H_8Cl_2$
M_r: 247.11
mp: 212
Color: ye nd (MeCOEt or CCl_4)
***s*:** al 2; eth 2; bz 3; chl 2
IR: µCOB 6478 3030 1610 1450 1320 1270 1030 950 750 700
UV: OES 5-391 400(1259) 380(1259) 360(7943) 340(3162) peth
H NMR: VAR 619 7.6 8.5 $CDCl_3$
m/e: 246.00

HODOC No: 01138
CAS No: 613-31-0
Beilstein No: 5^4, 2182
Name: Anthracene, 9,10-dihydro-
Synonym: Dihydroanthracene
Structure:

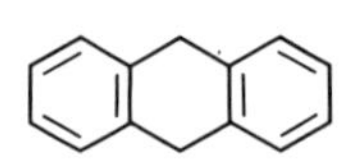

Mol Form: $C_{14}H_{12}$
M_r: 180.24
bp: 305, 165-70^{13}
mp: 111
***d*:** $0.8976^{11/4}$
Color: ta or pr
***s*:** H_2O 1; al 3; eth 3; bz 3; chl 3
IR: COB 7021 3025 2940 2850 2815 1800 1580 1485 1475 1450 1425 1280 1205 1180 1120 1040 955 895 755 745 730 605
UV: DMS D10-43 299(4350) 265(17200) 209(45500) lpet
H NMR: VAR 307 3.9 7.2 $CDCl_3$
C NMR: STOT 99 37.7 128.0 137.3
MS: ALDN 143 179(100) 180(96) 178(67) 89(26) 165(24) 176(17) 76(16) 181(15)
m/e: 180.09

HODOC No: 01139
CAS No: 610-46-8
Beilstein No: 5^4, 2327
Name: Anthracene, 1,3-dimethyl-
Synonym: 1,3-Dimethylanthracene
Line Form: 1,3-$(CH_3)_2$-$C_{14}H_8$
Mol Form: $C_{16}H_{14}$
Mr: 206.27
bp: $140\text{-}5^2$
mp: 83
Color: pa br lf (eth) cr (al)
s: Et_2O 4; EtOH 4
IR: SADP 19880
UV: SAD 6579
m/e: 206.11

HODOC No: 01140
CAS No: 613-06-9
Beilstein No: 5^4, 2328
Name: Anthracene, 2,3-dimethyl-
Synonym: 2,3-Dimethylanthracene
Line Form: 2,3-$(CH_3)_2$-$C_{14}H_8$
Mol Form: $C_{16}H_{14}$
Mr: 206.27
mp: 252
Color: bl or gr lf (Bz)
s: H_2O 1; al 3; bz 4
IR: µSADP 19875 2940 1790 1640 1540 1470 1450 1390 1350 1280 1160 1120 1020 1010 950 900 860 740
UV: SAD 6574 341 326 311 256 226 MeOH
m/e: 206.11

HODOC No: 01141
CAS No: 781-43-1
Beilstein No: 5^4, 2329
Name: Anthracene, 9,10-dimethyl-
Synonym: 9,10-Dimethylanthracene
Line Form: 9,10-$(CH_3)_2$-$C_{14}H_8$
Mol Form: $C_{16}H_{14}$
Mr: 206.27
mp: 182-4
IR: µCOB 3565 3030 2860 1610 1450 1390 1370 1020 980 840 810 750 740
UV: SAD 9144 340(2620) 259(171000) 252(88700) 217(13500) MeOH
MS: NBS 4457; ALD 4:30C
m/e: 206.11

HODOC No: 01142
CAS No: 26979-27-1
Name: Anthracene, 9,10-di-1-naphthalenyl-
Synonym: Anthracene, 9,10-di-1-naphthyl-
Mol Form: $C_{34}H_{22}$
Mr: 430.51
IR: SADG 10587 3060 1500 1440 1360 930 800 780 760 660 560
UV: FO 405 394(7943) 373(10000) 355(5012) 305(2512) 255(63096) 228(125893) EtOH
m/e: 430.17

HODOC No: 01143
CAS No: 33685-60-8
Beilstein No: 5^4, 2296
Name: Anthracene, 9,10-dinitro-
Synonym: 9,10-Dinitroanthracene
Line Form: 9,10-$(NO_2)_2$-$C_{14}H_8$
Mol Form: $C_{14}H_8N_2O_4$
Mr: 268.21
mp: 294
s: H_2O 1; al 1; eth 1; chl 1; HOAc 1; $PhNO_2$ 2; $PhNH_2$ 2
UV: OES 4-483 387(3802) 367(4169) 352(3162) 335(2188) EtOH
m/e: 268.05

HODOC No: 01144
CAS No: 1499-10-1
Beilstein No: 5^4, 2807
Name: Anthracene, 9,10-diphenyl-
Synonym: 9,10-Diphenylanthracene
Line Form: 9.10-$(C_6H_5)_2$-$C_{14}H_8$
Mol Form: $C_{26}H_{18}$
Mr: 330.40
mp: 245-8
IR: SADG 1359 3060 1600 1450 1390 1160 1030 940 760 750 700 660 610
UV: FO 403 399(7943) 379(7943) 358(5012) 338(3162) 262(100000) EtOH
H NMR: SAD 1611 7.3 7.6 $CDCl_3$
MS: NBS 2563
m/e: 330.14

HODOC No: 01145
CAS No: 2883-70-7
Name: Anthracene, 9-dodecyl-
Synonym: 9-Dodecylanthracene
Mol Form: $C_{26}H_{34}$
M_r: 346.53
UV: API 1008 383 364 345 330 255 215 EtOH
MS: WILEY 1990 191(100) 346(46) 192(18) 29(14) 347(13) 43(13) 41(12) 189(7)
m/e: 346.27

HODOC No: 01146
CAS No: 2026-16-6
Beilstein No: 5^4, 2414
Name: Anthracene, 2-ethenyl-
Synonym: Anthracene, 2-vinyl-
Line Form: 2-CH_2=CH-$C_{14}H_9$
Mol Form: $C_{16}H_{12}$
M_r: 204.26
mp: 125
UV: LANG 2023 386(6310) 369(6310) 350(4898) 255(134896) 217(10471) EtOH
m/e: 204.09

HODOC No: 01147
CAS No: 2444-68-0
Name: Anthracene, 9-ethenyl-
Synonym: Anthracene, 9-vinyl-
Mol Form: $C_{16}H_{12}$
M_r: 204.26
mp: 64-5
UV: LANG 1175 388(6026) 368(7244) 350(5370) 256(32359) 220(16982) EtOH
m/e: 204.09

HODOC No: 01148
CAS No: 66291-35-8
Name: Anthracene, 1-ethyl-
Synonym: 1-Ethylanthracene
Mol Form: $C_{16}H_{14}$
M_r: 206.27
UV: OES 6-655 376(7762) 358(8511) 341(5888) 326(3020) 311(1318) 254(169824) iso
MS: WILEY 1463 206(100) 191(84) 189(23) 207(18) 192(15) 165(13) 190(10) 89(10)
m/e: 206.11

HODOC No: 01149
CAS No: 52251-71-5
Beilstein No: 5^3, 2164
Name: Anthracene, 2-ethyl-
Synonym: 2-Ethylanthracene
Line Form: 2-CH_3CH_2-$C_{14}H_9$
Mol Form: $C_{16}H_{14}$
M_r: 206.27
mp: 152-3
IR: ALD 3:5796G
MS: NBS 4457
m/e: 206.11

HODOC No: 01150
CAS No: 605-83-4
Beilstein No: 5^4, 2326
Name: Anthracene, 9-ethyl-
Synonym: 9-Ethylanthracene
Line Form: 9-C_2H_5-$C_{14}H_9$
Mol Form: $C_{16}H_{14}$
M_r: 206.27
mp: 59
***d*:** $1.0413^{99/4}$
***n*:** 1.6767^{99}
Color: bl flr lf (al)
***s*:** H_2O 1; al 3; eth 3
IR: µDMS 5873 2940 1610 1490 1450 1370 1350 1160 1060 1010 940 890 850 790 730
UV: OES 6-655 383(10715) 378(5495) 364(10715) 347(6457) 331(3020) 316(1175) 256(213796) 248(95499) iso
m/e: 206.11

HODOC No: 01151
CAS No: 605-82-3
Beilstein No: 5^2, 560
Name: Anthracene, 9-ethyl-9,10-dihydro-
Synonym: 9-Ethyl-9,10-dihydroanthracene
Line Form: 9-C_2H_5-$C_{14}H_{11}$
Mol Form: $C_{16}H_{16}$
M_r: 208.29
bp: 320-3
***d*:** $1.049^{18/18}$
***s*:** Bz 4; Et_2O 4; EtOH 4
m/e: 208.13

HODOC No: 01152
CAS No: 6109-22-4
Beilstein No: 5^2, 472
Name: Anthracene, 1,2,3,4,5,6-hexahydro-
Structure:

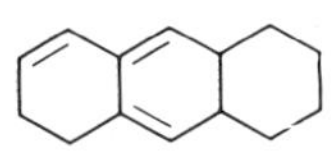

Mol Form: $C_{14}H_{16}$
M_r: 184.27
bp: 160^{15}
mp: 67
Color: lf (MeOH) cr (al)
***s*:** Bz 4
m/e: 184.13

HODOC No: 01153
CAS No: 7613-10-7
Name: Anthracene, 2-isothiocyanato-
Synonym: Isothiocyanic acid, 2-anthryl ester
Mol Form: $C_{15}H_9NS$
M_r: 235.29
mp: 199-200
UV: LANG 958 386(6310) 366(8511) 346(7244) 290(75858) 278(54954) 250(25119) MeOH
m/e: 235.05

HODOC No: 01154
CAS No: 610-48-0
Beilstein No: 5^4, 2310
Name: Anthracene, 1-methyl-
Synonym: 1-Methylanthracene
Line Form: $1\text{-}CH_3\text{-}C_{14}H_9$
Mol Form: $C_{15}H_{12}$
M_r: 192.25
bp: 199-200
mp: 85-6
***d*:** $1.0471^{99/4}$
***n*:** 1.6802^{99}
Color: bl nd (MeOH) lf (al)
***s*:** H_2O 1; al 3; eth 3; bz 3; chl 3; sulf 3
UV: OES 2-435 377(8318) 359(8511) 354(5129) 343(5495) 326(2818) 311(1202) 299(692) 253(165959) 212(36308) iso
m/e: 192.09

HODOC No: 01155
CAS No: 613-12-7
Beilstein No: 5^4, 2311
Name: Anthracene, 2-methyl-
Synonym: 2-Methylanthracene
Line Form: $2\text{-}CH_3\text{-}C_{14}H_9$
Mol Form: $C_{15}H_{12}$
M_r: 192.25
bp: sub
mp: 209
***d*:** $1.80^{0/4}$
Color: gr bl flr lf (sub)
***s*:** H_2O 1; al 2; eth 2; ace 1; bz 3; chl 3; CS_2 3; MeOH 2; HOAc 2
IR: µCOB 2608 2940 1640 1540 1470 1450 1390 1320 1280 1180 1040 950 890 870 800 740
UV: SAD 9681 376(4920) 356(5590) 338(4170) 296(1380) 252(21700) MeOH
MS: WILEY 1336 192(100) 191(34) 189(22) 193(17) 178(15) 165(12) 82.5(10) 96(10)
m/e: 192.09

HODOC No: 01156
CAS No: 779-02-2
Beilstein No: 5^4, 2312
Name: Anthracene, 9-methyl-
Synonym: 9-Methylanthracene
Line Form: $9\text{-}CH_3\text{-}C_{14}H_9$
Mol Form: $C_{15}H_{12}$
M_r: 192.25
bp: $196\text{-}7^{12}$
mp: 81.5
***d*:** $1.065^{99/4}$
***n*:** 1.6959^{99}
Color: yesh nd (dil al) pr (Bz, al)
***s*:** al 3; eth 3; ace 3; bz 3; chl 3; os 4
IR: COB 7023 3055 2935 1680 1625 1445 1410 1390 1350 1315 1280 1160 1040 995 950 880 835 775 730 600 530
UV: SAD 6940 347(5550) 331(2820) 316(1310) 255(280000) 249(155000) MeOH
H NMR: VAR 317 3.0 $CDCl_3$
MS: NBS 4388
m/e: 192.09

HODOC No: 01157
CAS No: 23674-14-8
Name: Anthracene, 9-(4-methylphenyl)-
Synonym: Anthracene, 9-*p*-tolyl-
Mol Form: $C_{21}H_{16}$
M_r: 268.34
***s*:** chl 3
IR: µSADP 4033 1640 1515 1450 1370 1220 1175 1110 1020 935 895 845 805 795 735
H NMR: SAD 11033 2.4 7.0-7.4 7.5-8.0 8.3 $CDCl_3$
m/e: 268.13

HODOC No: 01158
CAS No: 602-60-8
Beilstein No: 5^4, 2296
Name: Anthracene, 9-nitro-
Synonym: 5-Nitroanthracene
Line Form: 9-NO_2-$C_{14}H_9$
Mol Form: $C_{14}H_9NO_2$
M_r: 223.22
bp: 275^{17}
mp: 146
Color: ye nd (al) pr (HOAc or xyl)
***s*:** H_2O 1; al 2; ace 4; chl 2; CS_2 4; HOAc 2
IR: SADG 18720 3050 1630 1510 1440 1360 1310 1260 890 770 730
UV: SAD 6850 347(3800) 332(2590) 250(150000) 219(15100) 219(15100) 216(15600) 213(15500) MeOH
H NMR: SAD 6101 7.5 7.9 8.5 $CDCl_3$
MS: ALDN 182 223(100) 176(42) 177(38) 165(29) 88(25) 208(22) 151(18) 224(16)
m/e: 223.06

HODOC No: 01159
CAS No: 1079-71-6
Beilstein No: 5^4, 1584
Name: Anthracene, 1,2,3,4,5,6,7,8-octahydro-
Synonym: Octahydroanthracene
Structure:

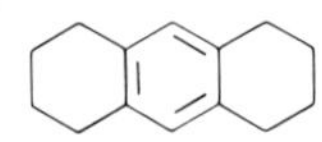

Mol Form: $C_{14}H_{18}$
M_r: 186.28
bp: 293-5, 167^{12}
mp: 78
***d*:** $0.9703^{80/4}$
***n*:** 1.5372^{80}
Color: pl (al)
***s*:** H_2O 1; al 3; bz 4; ctc 2; HOAc 3
IR: SADG 628 2941 2587 1613 1493 1429 1351 1299 1250 1136 980 926 847
UV: DMS D3-6 285(2000) 280(1400) 276(1600) 271(990) 267(750) 226(8000) 222(9300) 206(43500) 1pet
H NMR: SAD 550 1.7 2.6 6.6 7.1-7.7 CCl_4
C NMR: STOT 99 25.3 30.7 130.8 134.9
m/e: 186.14

HODOC No: 01160
CAS No: 1981-38-0
Beilstein No: 5^4, 2628
Name: Anthracene, 2-phenyl-
Synonym: 2-Phenylanthracene
Line Form: 2-C_6H_5-$C_{14}H_9$
Mol Form: $C_{20}H_{14}$
M_r: 254.31
mp: 207
IR: µCOB 3569 1610 1540 1520 1490 1450 1430 1160 960 910 890 870 760 740 690
UV: OES 7-861 389(4677) 368(6607) 350(5012) 334(2818) 319(1413) 281(70795) 261(52481) CH_2Cl_2
m/e: 254.11

HODOC No: 01161
CAS No: 602-55-1
Beilstein No: 5^4, 2628
Name: Anthracene, 9-phenyl-
Synonym: 9-Phenylanthracene
Line Form: 9-C_6H_5-$C_{14}H_9$
Mol Form: $C_{20}H_{14}$
M_r: 254.31
bp: 417
mp: 156
Color: bl lf (al)(HOAc)
***s*:** H_2O 1; al 3; eth 3; bz 3; chl 3; CS_2 3
IR: µCOB 3558 3030 1610 1490 1450 1370 1180 1080 1020 950 940 880 850 790 750 740 700
UV: OES 7-861 387(8511) 368(9333) 350(6026) 332(2884) 318(1230) 259(134896) CH_2Cl2
m/e: 254.11

HODOC No: 01162
CAS No: 6596-35-6
Beilstein No: 5^4, 492
Name: Anthracene, tetradecahydro-
Synonym: Perhydroanthracene
Structure:

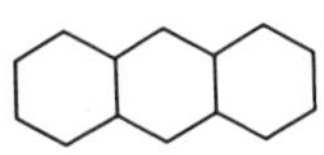

Mol Form: $C_{14}H_{24}$
M_r: 192.33
bp: $128^{11/}$
mp: 93
MS: WILEY 1342 41(100) 135(99) 192(95) 67(85) 81(65) 55(59) 96(47) 95(47)
m/e: 192.19

HODOC No: 01163
CAS No: 2141-42-6
Beilstein No: 5^4, 1909
Name: Anthracene, 1,2,3,4-tetrahydro-
Synonym: 1,2,3,4-Tetrahydroanthracene
Structure:

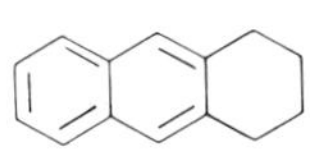

Mol Form: $C_{14}H_{14}$
M_r: 182.25
mp: 106-7
UV: FO 216 324(794) 309(794) 296(3162) 286(5012) 276(5012) 230(79433) cyhex
m/e: 182.11

HODOC No: 01164
CAS No: 1210-12-4
Beilstein No: 9^1, 304
Name: 9-Anthracenecarbonitrile
Synonym: 9-Anthronitrile
Line Form: 9-$C_{14}H_9CN$
Mol Form: $C_{15}H_9N$
M_r: 203.23
mp: 175-7
IR: ALD 3:1140A
UV: FO 401 403(6310) 382(10000) 365(10000) 348(5012) 256(158489) EtOH
MS: NBS 4435; ALD 7:138C
m/e: 203.07

HODOC No: 01165
CAS No: 642-31-9
Beilstein No: 7^4, 1738
Name: 9-Anthracenecarboxaldehyde
Synonym: 9-Anthraldehyde
Line Form: 9-$C_{14}H_9CHO$
Structure:

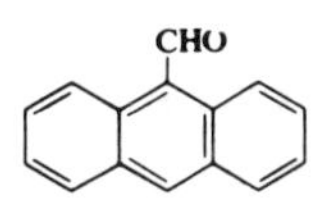

Mol Form: $C_{15}H_{10}O$
M_r: 206.23
mp: 104-5
Color: og nd (dil HOAc)
***s*:** H_2O 1; bz 3; HOAc 3
UV: OES 8-348 328(759) 320(759) 295(10965) 280(11482) 251(53703) EtOH
MS: ALDN 168 206(100) 178(66) 205(48) 177(40) 176(40) 88(26) 89(20) 76(17)
m/e: 206.07

HODOC No: 01166
CAS No: 10527-16-9
Beilstein No: 7^3, 2529
Name: 9-Anthracenecarboxaldehyde, 10-chloro-
Synonym: 9-Anthraldehyde, 10-chloro-
Line Form: 10-Cl-9-$C_{14}H_8CHO$
Mol Form: $C_{15}H_9ClO$
M_r: 240.67
mp: 215-8
IR: ALD 2:812D
m/e: 240.03

HODOC No: 01167
CAS No: 607-42-1
Beilstein No: 9^3, 3492
Name: 1-Anthracenecarboxylic acid
Synonym: 1-Anthroic acid
Structure:

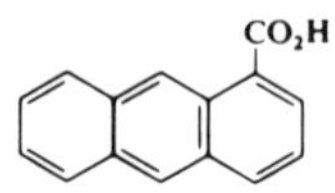

Mol Form: $C_{15}H_{10}O_2$
Mr: 222.23
bp: sub
mp: 251-2
Color: ye nd (HOAc) ye pr (al, AcOEt))
s: H_2O 1; al 3; eth 3; bz 2; chl 2; alk 3; HOAc 3
UV: HBCP 375(5012) 355(5012) 250(100000) al
m/e: 222.07

HODOC No: 01168
CAS No: 613-08-1
Beilstein No: 9^3, 3495
Name: 2-Anthracenecarboxylic acid
Synonym: 2-Anthroic acid
Mol Form: $C_{15}H_{10}O_2$
Mr: 222.23
bp: sub
mp: 281
Color: ye lf (al) nd, lf (sub)
s: HOAc 4
m/e: 222.07

HODOC No: 01169
CAS No: 723-62-6
Beilstein No: 9^3, 3494
Name: 9-Anthracenecarboxylic acid
Synonym: 9-Anthroic acid
Line Form: 9-$C_{14}H_9CO_2H$
Mol Form: $C_{15}H_{10}O_2$
Mr: 222.23
bp: sub
mp: 214d
s: H_2O 1; al 3
IR: COB 7592 1680 1560 1525 1490 1450 1425 1345 1290 1265 1255 1230 1150 1020 915 890 855 790 720 640 560 515
UV: SAD 8465 344(5000) 329(2560) 253(16100) 221(12300) 218(12200) MeOH
MS: ALD 6:189D
m/e: 222.07

HODOC No: 01170
CAS No: 602-69-7
Beilstein No: 10^3, 3633
Name: 1-Anthracenecarboxylic acid, 9,10-dihydro-9,10-dioxo-
Synonym: 1-Anthroic acid, 9,10-dihydro-9,10-dioxo-
Mol Form: $C_{15}H_8O_4$
Mr: 252.21
mp: 293-4
Color: pa ye nd (HOAc)
m/e: 252.04

HODOC No: 01171
CAS No: 6416-55-3
Beilstein No: 10^3, 4787
Name: 2-Anthracenecarboxylic acid, 9,10-dihydro-1,4-dihydroxy-9,10-dioxo-
Synonym: 2-Anthroic acid, 9,10-dihydro-1,4-dihydroxy-9,10-dioxo-
Mol Form: $C_{15}H_8O_6$
Mr: 284.21
mp: 249-50
Color: ye br or red nd
s: Ac 4; EtOH 4
m/e: 284.03

HODOC No: 01172
CAS No: 478-43-3
Beilstein No: 10^3, 4789
Name: 2-Anthracenecarboxylic acid, 9,10-dihydro-4,5-dihydroxy-9,10-dioxo-
Synonym: 2-Anthroic acid, 9,10-dihydro-4,5-dihydroxy-9,10-dioxo-
Mol Form: $C_{15}H_8O_6$
Mr: 284.21
bp: sub
mp: 321
Color: ye or og nd (MeOH, Py)
s: H_2O 2; al 2; eth 2; ace 2; bz 2; chl 2; py 4; sulf 3; alk 3
UV: OES 4-542 430(1585) 260(1995) 230(5012) MeOH
m/e: 284.03

HODOC No: 01173
CAS No: 17636-18-9
Beilstein No: 10^3, 4829
Name: 2-Anthracenecarboxylic acid, 9,10-dihydro-4,5-dihydroxy-7-methoxy-9,10-dioxo-
Synonym: 2-Anthroic acid, 9,10-dihydro-4,5-dihydroxy-7-methoxy-9,10-dioxo-
Mol Form: $C_{16}H_{10}O_7$
Mr: 314.24
mp: 300
Color: red-br or ye nd (sub)
m/e: 314.04

HODOC No: 01174
CAS No: 117-78-2
Beilstein No: 10^3, 3640
Name: 2-Anthracenecarboxylic acid, 9,10-dihydro-9,10-dioxo-
Synonym: 2-Anthroic acid, 9,10-dihydro-9,10-dioxo-
Structure:

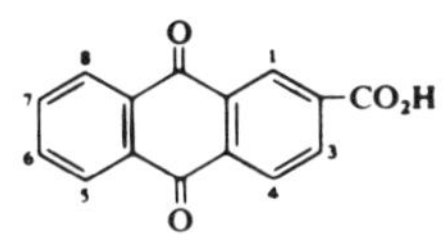

Mol Form: $C_{15}H_8O_4$
Mr: 252.21
bp: sub
mp: 290-2
Color: ye nd (HOAc)
s: al 2; eth 1; ace 3; bz 1; HOAc 2
UV: BCSJA 35-1690-62 355(3311) sulf
MS: ALDN 210 ini 267(100) 223(58) 249(50) 44(29) 69(24) 55(21) 57(21) 43(21)
m/e: 252.04

HODOC No: 01175
CAS No: 128-67-6
Beilstein No: 13^3, 3650
Name: 2-Anthracenecarboxylic acid, 9,10-dihydro-1-nitro-9,10-dioxo-
Synonym: 2-Anthroic acid, 9,10-dihydro-1-nitro-9,10-dioxo-
Mol Form: $C_{15}H_7NO_6$
Mr: 297.21
mp: 288d
Color: nd (al or HOAc)
s: al 3; eth 1; ace 3; HOAc 3; lig 1
IR: µDMS 6155 2860 1700 1590 1540 1470 1350 1300 1270 980 880 800 740 710
m/e: 297.03

HODOC No: 01176
CAS No: 61157-85-5
Beilstein No: 10^3, 3653
Name: 2-Anthracenecarboxylic acid, 9,10-dihydro-5-nitro-9,10-dioxo-
Synonym: Anthraquinone-2-carboxylic acid, 5-nitro
Mol Form: $C_{15}H_7NO_6$
Mr: 297.21
Color: yesh nd (HOAc)
s: HOAc 4
m/e: 297.03

HODOC No: 01177
CAS No: 1260-17-9
Beilstein No: 10^3, 4874
Name: 2-Anthracenecarboxylic acid, 7-β-*D*-glucopyranosyl-9,10-dihydro-3,5,6,8-tetrahydroxy-1-methyl-9,10-dioxo-
Synonym: 2-Anthroic acid, 7-*D*-glucopyranosyl-9,10-dihydro-3,5,6,8-tetrahydroxy-1-methyl-9,10-dioxo-
Structure:

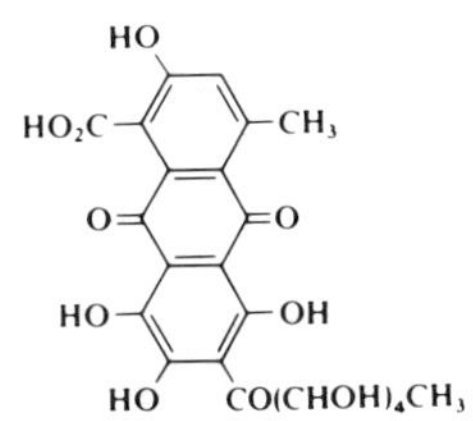

Mol Form: $C_{22}H_{20}O_{13}$
Mr: 492.37
mp: 136d
α: $[\alpha]^{15/6450}$ 51.6 H_20,1%
Color: red mcl pr (aq, MeOH)
s: H_2O 3; al 3; eth 2; bz 1; chl 1
IR: µSADP 23370 3450 3030 2940 2630 1640 1450 1410 1300 1250 1090 1050 1000
UV: SAD 22011 499(5610) 312(7630) 277(24100) 222(13900) MeOH 560(5790) 336(13800) 300(15400) 28(10500) MeOH:KOH
m/e: 492.09

HODOC No: 01178
CAS No: 53760-37-5
Beilstein No: 14^3, 291
Name: 9,10-Anthracenediamine
Synonym: 9,10-Diaminoanthracene
Line Form: 9,10-$(NH_2)_2$-$C_{14}H_8$
Mol Form: $C_{14}H_{12}N_2$
M_r: 208.25
mp: 196
Color: red cr
m/e: 208.10

HODOC No: 01179
CAS No: 1217-45-4
Beilstein No: 9^3, 4650
Name: 9,10-Anthracenedicarbonitrile
Synonym: DCA
Line Form: 9,10-$C_{14}H_8(CN)_2$
Mol Form: $C_{16}H_8N_2$
M_r: 228.24
mp: 337d
m/e: 228.07

HODOC No: 01180
CAS No: 7044-91-9
Beilstein No: 7^4, 2615
Name: 9,10-Anthracenedicarboxaldehyde
Synonym: 9,10-Anthracenedialdehyde
Line Form: 9,10-$C_{14}H_8(CHO)_2$
Mol Form: $C_{16}H_{10}O_2$
M_r: 234.24
mp: 245-6
m/e: 234.07

HODOC No: 01181
CAS No: 7475-29-8
Beilstein No: 10^3, 4082
Name: 1,2-Anthracenedicarboxylic acid, 9,10-dihydro-9,10-dioxo-
Synonym: 1,2-Anthraquinonedicarboxylic acid
Mol Form: $C_{16}H_8O_6$
M_r: 296.22
mp: 270
***s*:** EtOH 4; HOAc 4
m/e: 296.03

HODOC No: 01182
CAS No: 10253-28-8
Beilstein No: 10^3, 4085
Name: 1,6-Anthracenedicarboxylic acid, 9,10-dihydro-9,10-dioxo-
Synonym: 1,6-Anthraquinonedicarboxylic acid
Mol Form: $C_{16}H_8O_6$
M_r: 296.22
mp: 336-8
Color: lt ye nd (al)
m/e: 296.03

HODOC No: 01183
CAS No: 27485-15-0
Beilstein No: 10^2, 643
Name: 2,3-Anthracenedicarboxylic acid, 9,10-dihydro-9,10-dioxo-
Synonym: 2,3-Anthraquinonedicarboxylic acid
Mol Form: $C_{16}H_8O_6$
M_r: 296.22
mp: 340-2
Color: ye nd (HOAc)
m/e: 296.03

HODOC No: 01184
CAS No: 577-95-7
Beilstein No: 6^4, 6889
Name: 1,2-Anthracenediol
Synonym: 1,2-Anthradiol
Line Form: 1,2-$(HO)_2C_{14}H_8$
Mol Form: $C_{14}H_{10}O_2$
M_r: 210.22
mp: 160-2
Color: pa gr lf
***s*:** Et_2O 4; EtOH 4; HOAc 4
m/e: 210.07

HODOC No: 01185
CAS No: 607-62-5
Beilstein No: 6^2, 998
Name: 1,5-Anthracenediol
Synonym: Anthracene, 1,5-dihydroxy
Line Form: 1,5-$(HO)_2$-$C_{14}H_8$
Mol Form: $C_{14}H_{10}O_2$
M_r: 210.22
mp: 265d
Color: ye nd
***s*:** Bz 4; Et_2O 4; EtOH 4; HOAc 4
m/e: 210.07

HODOC No: 01186
CAS No: 607-61-4
Beilstein No: 6, 1033
Name: 1,8-Anthracenediol
Synonym: Chrysazol
Line Form: 1,8-$(HO)_2C_{14}H_8$
Mol Form: $C_{14}H_{10}O_2$
M_r: 210.22
mp: 225d
Color: ye nd (dil al) lf (al-HOAc)
***s*:** Bz 4; Et_2O 4; EtOH 4
m/e: 210.07

HODOC No: 01187
CAS No: 101488-73-7
Beilstein No: 6^4, 6892
Name: 2,6-Anthracenediol
Synonym: 2,6-Anthradiol
Line Form: 2,6-$(HO)_2$-$C_{14}H_8$
Mol Form: $C_{14}H_{10}O_2$
M_r: 210.22
mp: 295-300d
Color: pa ye lf (al)
***s*:** Et_2O 4; EtOH 4; HOAc 4
m/e: 210.07

HODOC No: 01188
CAS No: 4981-66-2
Beilstein No: 6^4, 6892
Name: 9,10-Anthracenediol
Synonym: 9,10-Anthradiol
Line Form: 9,10-$(HO)_2$-$C_{14}H_8$
Mol Form: $C_{14}H_{10}O_2$
M_r: 210.22
mp: 180
Color: br or ye nd
***s*:** Et_2O 4; EtOH 4
m/e: 210.07

HODOC No: 01189
CAS No: 655-04-9
Beilstein No: 7^4, 2563
Name: 1,2-Anthracenedione
Synonym: 1,2-Anthraquinone
Mol Form: $C_{14}H_8O_2$
M_r: 208.20
mp: 185-90d
Color: og-br cr
***s*:** EtOH 4; Chl 4
m/e: 208.05

HODOC No: 01190
CAS No: 635-12-1
Beilstein No: 7^4, 2564
Name: 1,4-Anthracenedione
Synonym: 1,4-Anthracenequinone
Mol Form: $C_{14}H_8O_2$
M_r: 208.20
mp: 218d
Color: ye nd
MS: NBS 4465
m/e: 208.05

HODOC No: 01191
CAS No: 84-65-1
Beilstein No: 7^4, 2556
Name: 9,10-Anthracenedione
Synonym: Anthracene-9,10-quinone
Structure:

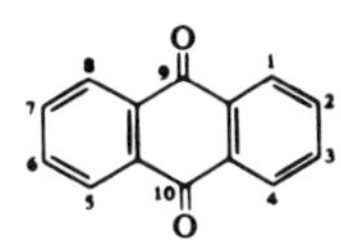

Mol Form: $C_{14}H_8O_2$
M_r: 208.20
bp: 377
mp: 286sub
***d*:** $1.438^{20/4}$
Color: ye rh nd (al, Bz)
***s*:** H_2O 1; al 2; eth 2; bz 2; chl 2
IR: SADG 279 1675 1590 1580 1330 1285 1170 940 810 690 630 390
Raman: SAD 997 3080 1670 1650 1630 1600 1580 1560 1480 1320 1220 1180 1150 1080 1030 980 930 820 790 770 680 480 450 370 300 250 150
UV: LANG 815 420(60) 399(89) 380(110) 319(5370) 272(16982) 251(56234) 206(19498) cy 325(4266) 273(14125) 252(42658) 205(24547) EtOH
MS: WILEY 1478 208(100) 180(77) 152(51) 76(42) 151(25) 209(15) 207(14) 150(12)
m/e: 208.05

HODOC No: 01192
CAS No: 16872-65-4
Beilstein No: 7[4], 2561
Name: 1,4-Anthracenedione, 2,3-dibromo-
Synonym: 2,3-Dibromo-1,4-anthraquinone
Mol Form: $C_{14}H_6Br_2O_2$
M_r: 363.99
bp: sub
mp: 283
Color: ye nd (to)
***s*:** Bz 4; Chl 4
m/e: 363.87

HODOC No: 01193
CAS No: 82-45-1
Beilstein No: 14[3], 407
Name: 9,10-Anthracenedione, 1-amino-
Synonym: Anthraquinone, 1-amino-
Line Form: 1-H_2N-$C_{14}H_7O_2$
Mol Form: $C_{14}H_9NO_2$
M_r: 223.22
bp: sub
mp: 253-4
Color: red nd (al)
***s*:** Ac 4; Bz 4; EtOH 4; Chl 4
IR: COB 280
UV: SAD 3787
MS: ALDN 182
m/e: 223.06

HODOC No: 01194
CAS No: 117-79-3
Beilstein No: 14[3], 429
Name: 9,10-Anthracenedione, 2-amino-
Synonym: Anthraquinone, 2-amino-
Line Form: 2-H_2N-$C_{14}H_7O_2$
Mol Form: $C_{14}H_9NO_2$
M_r: 223.22
bp: sub
mp: 303-6
Color: red nd (al, HOAc)
***s*:** H_2O 1; al 2; eth 1; ace 3; bz 3; chl 3
IR: SADG 8630 3448 3333 3030 1667 1613 1587 1493 1333 1299 1282 926 847 714
UV: SAD 3788 445(3420) 297(17400) 240(23800) MeOH
MS: ALDN 182 223(100) 237(36) 139(25) 220(22) 167(16) 195(15) 224(15) 236(12)
m/e: 223.06

HODOC No: 01195
CAS No: 6337-19-5
Beilstein No: 14, 482
Name: 9,10-Anthracenedione, 1-amino-2-benzoyl-
Synonym: Anthraquinone, 1-amino-2-benzoyl-
Line Form: 1-NH_2-2-(C_6H_5CO)-$C_{14}H_6O_2$
Mol Form: $C_{21}H_{13}NO_3$
M_r: 327.32
mp: 190
Color: red nd (HOAc)
***s*:** HOAc 4
m/e: 327.09

HODOC No: 01196
CAS No: 3300-23-0
Beilstein No: 14[3], 423
Name: 9,10-Anthracenedione, 1-amino-2-bromo-
Synonym: Anthraquinone, 1-amino-2-bromo-
Line Form: (2-Br-1-H_2N)-$C_{14}H_6O_2$
Mol Form: $C_{14}H_8BrNO_2$
M_r: 301.11
mp: 182
Color: ye red nd (HOAc) nd (xyl)
***s*:** H_2O 1; al 2; eth 2; ace 2; bz 2; py 4; HOAc 3; tol 3; lig 2
UV: OES 4-482 469(7586) 308(4365) EtOH
m/e: 300.97

HODOC No: 01197
CAS No: 6375-37-7
Beilstein No: 14[3], 423
Name: 9,10-Anthracenedione, 1-amino-3-bromo-
Synonym: Anthraquinone, 1-amino-3-bromo-
Line Form: (3-Br-1-H_2N)-$C_{14}H_6O_2$
Mol Form: $C_{14}H_8BrNO_2$
M_r: 301.11
mp: 243
Color: red nd (to)
***s*:** H_2O 1; al 1; eth 1; ace 1; bz 2; py 3; $PhNO_2$ 3; tol 2
IR: OES 4-482 471(6918) EtOH
UV: OES 4-482
m/e: 300.97

HODOC No: 01198
CAS No: 81-62-9
Beilstein No: 14^3, 423
Name: 9,10-Anthracenedione, 1-amino-4-bromo-
Synonym: Anthraquinone, 1-amino-4-bromo-
Line Form: (4-Br-1-H_2N)-$C_{14}H_6O_2$
Mol Form: $C_{14}H_8BrNO_2$
M_r: 301.11
mp: 170
***s*:** H_2O 1
UV: LANG 1374 472(8318) 307(5495) 244(36308) MeOH 484(6457) 308(5754) dmf
m/e: 300.97

HODOC No: 01199
CAS No: 117-01-1
Name: 9,10-Anthracenedione, 2-amino-1-bromo-3-chloro-
Synonym: Anthraquinone, 2-amino-1-bromo-3-chloro-
Mol Form: $C_{14}H_7BrClNO_2$
M_r: 335.56
mp: 228-9
IR: COB 6077 3490 3390 3070 1670 1660 1600 1580 1560 1480 1350 1260 1140 1030 920 810 710 630
UV: SAD 10858 418(4580) 300(27500) 246(32800) 229(41600) MeOH
m/e: 334.93

HODOC No: 01200
CAS No: 81-50-5
Beilstein No: 14^3, 3444
Name: 9,10-Anthracenedione, 1-amino-4-bromo-2-methyl-
Synonym: Anthraquinone, 1-amino-4-bromo-2-methyl-
Line Form: (4-Br-2-CH_3-1-H_2N)-$C_{14}H_5O_2$
Mol Form: $C_{15}H_{10}BrNO_2$
M_r: 315.14
mp: 232
IR: COB 4613
m/e: 314.99

HODOC No: 01201
CAS No: 128-83-6
Name: 9,10-Anthracenedione, 1-amino-2-bromo-4-[(4-methylphenyl)amino]-
Synonym: Anthraquinone, 1-amino-2-bromo-4-*p*-toluidino-
Mol Form: $C_{21}H_{15}BrN_2O_2$
M_r: 406.25
IR: µCOB 4621 3330 3030 1560 1520 1410 1330 1280 1030 860 820 740 730
m/e: 406.03

HODOC No: 01202
CAS No: 117-11-3
Name: 9,10-Anthracenedione, 1-amino-5-chloro-
Synonym: Anthraquinone, 1-amino-5-chloro-
Mol Form: $C_{14}H_8ClNO_2$
M_r: 257.66
IR: µCOB 4624 3450 3230 1670 1640 1610 1590 1540 1470 1300 1270 1160 810 760 740 700
UV: SAD 5758 304 244 diox
m/e: 257.02

HODOC No: 01203
CAS No: 82-27-9
Name: 9,10-Anthracenedione, 2-amino-1-chloro-
Synonym: Anthraquinone, 2-amino-1-chloro-
Mol Form: $C_{14}H_8ClNO_2$
M_r: 257.66
IR: µCOB 4625 3330 3130 1670 1610 1560 1490 1330 1270 1000 840 760 710
UV: OES 1-543 440(4786) py
m/e: 257.02

HODOC No: 01204
CAS No: 84-46-8
Beilstein No: 14^3, 435
Name: 9,10-Anthracenedione, 2-amino-3-chloro-
Synonym: Anthraquinone, 2-amino-3-chloro-
Line Form: (3-Cl-2-H_2N)-$C_{14}H_6O_2$
Mol Form: $C_{14}H_8ClNO_2$
M_r: 257.66
mp: 280-3
IR: µCOB 281 3450 3330 3030 1670 1610 1560 1470 1350 1270 1020 710
m/e: 257.02

HODOC No: 01205
CAS No: 81-49-2
Name: 9,10-Anthracenedione, 1-amino-2,4-dibromo-
Synonym: Anthraquinone, 1-amino-2,4-dibromo-
Mol Form: $C_{14}H_7Br_2NO_2$
M_r: 379.01
IR: **µCOB 4611** 3330 3230 1670 1590 1520 1370 1330 1270 1140 1010 890 800 740 720
m/e: 378.88

HODOC No: 01206
CAS No: 81-51-6
Beilstein No: 14^3, 681
Name: 9,10-Anthracenedione, 1-amino-2,4-dihydroxy-
Synonym: Anthraquinone, 1-amino-2,4-dihydroxy-
Line Form: $2,4\text{-}(HO)_2\text{-}1\text{-}H_2N\text{-}C_{14}H_5O_2$
Mol Form: $C_{14}H_9NO_4$
M_r: 255.22
mp: 316d
IR: **COB 7641** 3400 3250 1610 1580 1530 1460 1300 1220 950 850 730
m/e: 255.05

HODOC No: 01207
CAS No: 3963-78-8
Beilstein No: 14^2, 185
Name: 9,10-Anthracenedione, 3-amino-1,2-dihydroxy-
Synonym: Anthraquinone, 3-amino-1,2-dihydroxy-
Line Form: $[1,2\text{-}(HO)_2\text{-}3\text{-}H_2N]\text{-}C_{10}H_5O_2$
Mol Form: $C_{14}H_9NO_4$
M_r: 255.22
bp: sub d
mp: 300
Color: dk red pr (HOAc)
***s*:** aq NH_3 4
m/e: 255.05

HODOC No: 01208
CAS No: 2835-53-2
Beilstein No: 14^2, 185
Name: 9,10-Anthracenedione, 4-amino-1,2-dihydroxy-
Synonym: Anthraquinone, 4-amino-1,2-dihydroxy-
Mol Form: $C_{14}H_9NO_4$
M_r: 255.22
mp: d
Color: gr bl nd (al)
***s*:** EtOH 4
m/e: 255.05

HODOC No: 01209
CAS No: 568-98-9
Beilstein No: 14^3, 658
Name: 9,10-Anthracenedione, 1-amino-2-hydroxy-
Synonym: 1-Amino-2-hydroxyanthraquinone
Line Form: $(2\text{-}HO\text{-}1\text{-}H_2N)\text{-}C_{14}H_6O_2$
Mol Form: $C_{14}H_9NO_3$
M_r: 239.22
mp: 250
Color: pr or nd (al)
m/e: 239.06

HODOC No: 01210
CAS No: 116-85-8
Beilstein No: 14^3, 652
Name: 9,10-Anthracenedione, 1-amino-4-hydroxy-
Synonym: Anthraquinone, 1-amino-4-hydroxy-
Line Form: $(4\text{-}HO\text{-}1\text{-}H_2N)\text{-}C_{14}H_6O_2$
Mol Form: $C_{14}H_9NO_3$
M_r: 239.22
mp: 215
***s*:** al 3; ace 3
IR: **SADG 21068** 3400 3290 1620 1580 1530 1460 1435 1410 1360 1325 1295 1255 1165 1020 960 830 790 720 555 440
UV: **LANG 1375** 559(2291) 527(2754) 287(6026) 227(45709) MeOH
m/e: 239.06

HODOC No: 01211
CAS No: 63572-76-9
Beilstein No: 14, 273
Name: 9,10-Anthracenedione, 1-amino-8-hydroxy-
Synonym: 1-Amino-8-hydroxyanthraquinone
Line Form: $(8\text{-}HO\text{-}1\text{-}H_2N)\text{-}C_{14}H_6O_2$
Mol Form: $C_{14}H_9NO_3$
M_r: 239.22
mp: 230
Color: nd (Bz)
***s*:** Bz 4; EtOH 4
m/e: 239.06

HODOC No: 01212
CAS No: 568-99-0
Beilstein No: 14^3, 651
Name: 9,10-Anthracenedione, 2-amino-1-hydroxy-
Synonym: Anthraquinone, 2-amino-1-hydroxy
Line Form: (2-H_2N-1-HO)-$C_{14}H_6O_2$
Mol Form: $C_{14}H_9NO_3$
M_r: 239.22
bp: sub
mp: 226-7
Color: red br nd (al)
***s*:** Bz 4; Et_2O 4; EtOH 4
m/e: 239.06

HODOC No: 01213
CAS No: 2379-90-0
Name: 9,10-Anthracenedione, 1-amino-4-hydroxy-2-methoxy-
Synonym: Anthraquinone, 1-amino-4-hydroxy-2-methoxy-
Mol Form: $C_{15}H_{11}NO_4$
M_r: 269.24
***s*:** chl 2
H NMR: VAR 635 4.0 6.5 7.3 7.7 8.3 14.2 $CDCl_3$
m/e: 269.07

HODOC No: 01214
CAS No: 82-28-0
Beilstein No: 14^3, 492
Name: 9,10-Anthracenedione, 1-amino-2-methyl-
Synonym: Anthraquinone, 1-amino-2-methyl-
Line Form: (2-CH_3-1-H_2N)-$C_{10}H_6O_2$
Mol Form: $C_{15}H_{11}NO_2$
M_r: 237.24
mp: 205-6
***s*:** H_2O 1; al 3; eth 2; bz 3; chl 3; HOAc 3; $PhNO_2$ 3
IR: µSADP 19527 3450 3330 3030 2940 1670 1610 1560 1470 1430 1390 1330 1280 1190 1160 1000 960 900 850 820 750 710
UV: SAD 7792 305(6140) 247(36900) 202 MeOH
MS: ALDN 193 237(100) 236(20) 180(18) 238(17) 222(12) 208(10) 165(8) 76(7)
m/e: 237.08

HODOC No: 01215
CAS No: 4947-16-4
Beilstein No: 14^2, 123
Name: 9,10-Anthracenedione, 1-amino-4-methyl-
Synonym: Anthraquinone, 1-amino-4-methyl-
Line Form: (4-CH_3-I-H_2N)-$C_{14}H_6O_2$
Mol Form: $C_{15}H_{11}NO_2$
M_r: 237.24
mp: 181
UV: LANG 1371
m/e: 237.08

HODOC No: 01216
CAS No: 116-77-8
Name: 9,10-Anthracenedione, 1-amino-2-methyl-4-[(4-methylphenyl)amino]-
Synonym: Anthraquinone, 1-amino-2-methyl-4-*p*-toluidino-
Mol Form: $C_{22}H_{18}N_2O_2$
M_r: 342.38
IR: µCOB 4622 3330 3030 1560 1520 1410 1280 1110 820 730
m/e: 342.14

HODOC No: 01217
CAS No: 6937-74-2
Beilstein No: 14^3, 427
Name: 9,10-Anthracenedione, 1-amino-4-nitro-
Synonym: Anthraquinone, 1-amino-4-nitro-
Line Form: (4-NO_2-1-H_2N)-$C_{14}H_6O_2$
Mol Form: $C_{14}H_8N_2O_4$
M_r: 268.21
mp: 298-300
***s*:** H_2O 1; al 2; $PhNO_2$ 3
UV: OES 1-544 474(200) EtOH
m/e: 268.05

HODOC No: 01218
CAS No: 2475-44-7
Name: 9,10-Anthracenedione, 1,4-*bis*(methylamino)-
Synonym: Anthraquinone, 1,4-*bis*(methylamino)-
Mol Form: $C_{16}H_{14}N_2O_2$
M_r: 266.28
***s*:** chl 2
H NMR: VAR 652 3.0 7.1 7.7 8.3 10.5 $CDCl_3$
m/e: 266.11

HODOC No: 01219
CAS No: 632-83-7
Beilstein No: 7^3, 4076
Name: 9,10-Anthracenedione, 1-bromo-
Synonym: Anthraquinone, 1-bromo-
Mol Form: $C_{14}H_7BrO_2$
M_r: 286.10
bp: sub
mp: 188
Color: ye nd (Bz)
***s*:** Bz 4; EtOH 4
m/e: 285.96

HODOC No: 01220
CAS No: 572-83-8
Beilstein No: 7^3, 4076
Name: 9,10-Anthracenedione, 2-bromo-
Synonym: Anthraquinone, 2-bromo-
Mol Form: $C_{14}H_7BrO_2$
M_r: 286.10
bp: sub
mp: 204-5
Color: ye nd (to)
***s*:** Bz 4
m/e: 285.96

HODOC No: 01221
CAS No: 18084-26-9
Beilstein No: 8^3, 3772
Name: 9,10-Anthracenedione, 3-bromo-1,2-dihydroxy-
Synonym: Anthraquinone, 3-bromo-1,2-dihydroxy-
Mol Form: $C_{14}H_7BrO_4$
M_r: 318.10
bp: sub
mp: 260-1
Color: br-red nd (to)
m/e: 317.95

HODOC No: 01222
CAS No: 128-93-8
Beilstein No: 14^3, 424
Name: 9,10-Anthracenedione, 1-bromo-4-(methylamino)-
Synonym: Anthraquinone, 1-bromo-4-(methylamino)-
Line Form: 4-CH_3NH-1-Br-$C_{14}H_6O_2$
Mol Form: $C_{15}H_{10}BrNO_2$
M_r: 315.14
mp: 194
Color: br-red nd (py)
***s*:** Py 4
IR: **μCOB 282** 3330 3030 1670 1640 1610 1370 1320 1270 1190 1160 1120 1000 830 790 730 720 680
UV: **OES 4-544** 502(8128) 274(9772)
m/e: 314.99

HODOC No: 01223
CAS No: 6374-84-1
Beilstein No: 14^1, 446
Name: 9,10-Anthracenedione, 2-bromo-1-(methylamino)-
Synonym: Anthraquinone, 2-bromo-1-(methylamino)-
Line Form: (2-Br-1-CH_3NH)-$C_{14}H_6O_2$
Mol Form: $C_{15}H_{10}BrNO_2$
M_r: 315.14
mp: 170-2
Color: br nd (HOAc)
***s*:** HOAc 4
m/e: 314.99

HODOC No: 01224
CAS No: 82-44-0
Beilstein No: 7^4, 2559
Name: 9,10-Anthracenedione, 1-chloro-
Synonym: Anthraquinone, 1-chloro-
Mol Form: $C_{14}H_7ClO_2$
M_r: 242.65
bp: sub
mp: 162
Color: ye nd (to or al)
***s*:** H_2O 1; al 2; eth 5; bz 3; ctc 2; $PhNO_2$ 3; HOAc 3; AmOH 3
IR: **SADG 21039** 3050 1680 1600 1570 1440 1420 1310 1260 1230 1170 1130 960 800 720 700
UV: **SAD 859** 334 253 209 MeOH
H NMR: **SAD 182** 7.5-8.0 8.1-8.4 CCl_4
MS: **ALDN 196** 74.5(100) 242(76) 214(50) 186(32) 151(32) 75(30) 224(24) 150(22)
m/e: 242.01

HODOC No: 01225
CAS No: 131-09-9
Beilstein No: 7^4, 2559
Name: 9,10-Anthracenedione, 2-chloro-
Synonym: Anthraquinone, 2-chloro-
Mol Form: $C_{14}H_7ClO_2$
M_r: 242.65
bp: sub
mp: 211
Color: pa ye nd (al, HOAc)
s: H_2O 1; al 2; eth 1; bz 2; tol 4; $PhNO_2$ 3; con sulf 3
IR: SADG 15166 3030 1667 1587 1471 1316 1282 1149 926 847 709
UV: SAD 15813 324(5000) 274(17000) 255(50800) MeOH
MS: ALD 6:68D
m/e: 242.01

HODOC No: 01226
CAS No: 129-35-1
Beilstein No: 7^4, 2575
Name: 9,10-Anthracenedione, 1-chloro-2-methyl-
Synonym: Anthraquinone, 1-chloro-2-methyl-
Mol Form: $C_{15}H_9ClO_2$
M_r: 256.67
mp: 170-1
s: al 1; eth 1; py 2
IR: µCOB 4609 3330 1670 1590 1560 1370 1320 1270 1120 1030 980 850 760 710
MS: ALDN 204 256(100) 165(43) 258(34) 228(32) 193(30) 163(23) 221(20) 257(20)
m/e: 256.03

HODOC No: 01227
CAS No: 129-40-8
Beilstein No: 7^2, 720
Name: 9,10-Anthracenedione, 1-chloro-5-nitro-
Synonym: Anthraquinone, 1-chloro-5-nitro-
Mol Form: $C_{14}H_6ClNO_4$
M_r: 287.64
mp: 314
s: H_2O 1; al 1; eth 1; bz 2; lig 1; py 3
IR: µCOB 299 3030 1670 1560 1540 1370 1320 1270 1220 1140 990 880 820 760 700 690
m/e: 287.00

HODOC No: 01228
CAS No: 521-62-0
Beilstein No: 17^4, 2535
Name: 9,10-Anthracenedione, 3-[(6-deoxy-α-*L*-mannopyranosyl)oxy]-1,8-dihydroxy-6-methyl-
Synonym: Frangulin A
Structure:

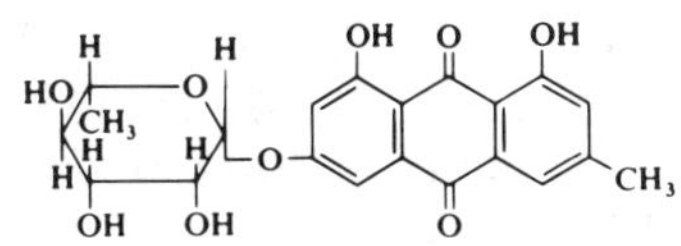

Mol Form: $C_{21}H_{20}O_9$
M_r: 416.36
mp: 228
Color: ye or red (al)
s: H_2O 1; al 3; bz 3; chl 3; HOAc 3
UV: HBCP 430(11221) 300(9333) 282(14125) 264(19055)
MS: NBS 2525
m/e: 416.11

HODOC No: 01229
CAS No: 1758-68-5
Beilstein No: 14^3, 3438
Name: 9,10-Anthracenedione, 1,2-diamino-
Synonym: Anthraquinone, 1,2-diamino-
Line Form: 1,2-$(H_2N)_2$-$C_{14}H_6O_2$
Mol Form: $C_{14}H_{10}N_2O_2$
M_r: 238.23
mp: 303-4
Color: vt nd
s: al 2; eth 2; chl 2; py 3; con sulf 3; con HCl 1; dil HCl 2; xyl 2
IR: ALD 3:901G
UV: OES 4-491 360(6607) 262(52481) sulf
MS: ALD 6:71B
m/e: 238.07

HODOC No: 01230
CAS No: 10146-44-8
Beilstein No: 14^3, 439
Name: 9,10-Anthracenedione, 1,3-diamino-
Synonym: Anthraquinone, 1,3-diamino-
Mol Form: $C_{14}H_{10}N_2O_2$
M_r: 238.23
mp: 290
Color: red
s: Py 4
m/e: 238.07

HODOC No: 01231
CAS No: 128-95-0
Beilstein No: 14^3, 3439
Name: 9,10-Anthracenedione, 1,4-diamino-
Synonym: Anthraquinone, 1,4-diamino-
Mol Form: $C_{14}H_{10}N_2O_2$
M_r: 238.23
mp: 268
Color: dk vt nd (py)
***s*:** H_2O 2; al 3; bz 3; $PhNO_2$ 3; py 4; HOAc 3
IR: µSADP 14895 3450 3330 1670 1590 1540 1450 1410 1320 1270 1180 1120 1020 830 790 730
UV: SAD 4314 248 MeOH
MS: ALDN 193 238(100) 239(17) 237(12) 44(10) 18(8) 119(8) 240(7) 77(6)
m/e: 238.07

HODOC No: 01232
CAS No: 129-44-2
Beilstein No: 14^3, 466
Name: 9,10-Anthracenedione, 1,5-diamino-
Synonym: Anthraquinone, 1,5-diamino-
Mol Form: $C_{14}H_{10}N_2O_2$
M_r: 238.23
bp: sub
mp: 319
Color: dk red nd (al, HOAc)
***s*:** H_2O 1; al 2; eth 2; ace 2; bz 2; chl 2; con sulf 3; $PhNO_2$ 3
IR: SADG 8283 3448 3333 3030 1613 1538 1471 1389 1282 1163 1053 862 800 769 709
UV: SAD 858 275 229 MeOH
MS: ALDN 193 238(100) 239(16) 44(12) 210(12) 119(10) 28(7) 182(6) 237(6)
m/e: 238.07

HODOC No: 01233
CAS No: 1758-64-1
Beilstein No: 14^2, 119
Name: 9,10-Anthracenedione, 1,6-diamino-
Synonym: Anthraquinone, 1,6-diamino-
Mol Form: $C_{14}H_{10}N_2O_2$
M_r: 238.23
mp: 297
Color: red nd (HOAc)
m/e: 238.07

HODOC No: 01234
CAS No: 3927-72-8
Beilstein No: 14^1, 470
Name: 9,10-Anthracenedione, 1,7-diamino-
Synonym: Anthraquinone, 1,7-diamino-
Mol Form: $C_{14}H_{10}N_2O_2$
M_r: 238.23
mp: 290
Color: red nd
m/e: 238.07

HODOC No: 01235
CAS No: 129-42-0
Beilstein No: 14^3, 478
Name: 9,10-Anthracenedione, 1,8-diamino-
Synonym: Anthraquinone, 1,8-diamino-
Mol Form: $C_{14}H_{10}N_2O_2$
M_r: 238.23
mp: 265
Color: red nd (al, HOAc)
***s*:** H_2O 1; al 3; eth 2; py 3; $PhNO_2$ 3; HOAc 2
UV: OES 1-551 580(100) 490(10000) 458(12589) 435(10000) 250(25119) diox
MS: ALDN 193 238(100) 239(16) 210(10) 119(9) 44(6) 154(5) 77(5) 91(5)
m/e: 238.07

HODOC No: 01236
CAS No: 605-22-1
Beilstein No: 14^3, 480
Name: 9,10-Anthracenedione, 2,3-diamino-
Synonym: Anthraquinone, 2,3-diamino-
Mol Form: $C_{14}H_{10}N_2O_2$
M_r: 238.23
mp: 353
Color: red
***s*:** Py 4
m/e: 238.07

HODOC No: 01237
CAS No: 131-14-6
Beilstein No: 14^3, 480
Name: 9,10-Anthracenedione, 2,6-diamino-
Synonym: Anthraquinone, 2,6-diamino-
Mol Form: $C_{14}H_{10}N_2O_2$
M_r: 238.23
mp: 320d
Color: red-br pr (aq-py)
***s*:** H_2O 2; al 3; chl 3; con sulf 3; xyl 3; py 3
IR: SADG 18647 3440 3340 3210 1670 1640 1580 1450 1330 1290 1240 1160 1080 890 850 750 670
UV: OES 1-551 384(5012) 333(12589) 281(10000) diox
MS: ALDN 193 238(100) 239(16) 210(14) 182(12) 91(12) 154(9) 181(9) 223(7)
m/e: 238.07

HODOC No: 01238
CAS No: 605-44-7
Beilstein No: 14^3, 483
Name: 9,10-Anthracenedione, 2,7-diamino-
Synonym: Anthraquinone, 2,7-diamino-
Mol Form: $C_{14}H_{10}N_2O_2$
M_r: 238.23
bp: sub
mp: 320
Color: og-ye nd (al) dk red nd (sub)
m/e: 238.07

HODOC No: 01239
CAS No: 128-94-9
Name: 9,10-Anthracenedione, 1,8-diamino-4,5-dihydroxy-
Synonym: Anthraquinone, 1,8-diamino-4,5-dihydroxy-
Mol Form: $C_{14}H_{10}N_2O_4$
M_r: 270.23
mp: sub
UV: LANG 1585 293(8913) 239(38905) MeOH
m/e: 270.06

HODOC No: 01240
CAS No: 2872-48-2
Beilstein No: 14^3, 660
Name: 9,10-Anthracenedione, 1,4-diamino-2-methoxy-
Synonym: Anthraquinone, 1,4-diamino-2-methoxy-
Mol Form: $C_{15}H_{12}N_2O_3$
M_r: 268.26
mp: 234-6
IR: COB 7666 3400 3250 3070 2970 1600 1570 1530 1450 1290 1220 1160 990 850 740 590
m/e: 268.08

HODOC No: 01241
CAS No: 82-33-7
Beilstein No: 14^1, 4675
Name: 9,10-Anthracenedione, 1,4-diamino-5-nitro-
Synonym: Anthraquinone, 1,4-diamino-5-nitro-
Mol Form: $C_{14}H_9N_3O_4$
M_r: 283.23
mp: 278
IR: COB 7672 3450 3320 3080 1650 1600 1570 1530 1440 1390 1280 1180 830 800 730 710 570
m/e: 283.06

HODOC No: 01242
CAS No: 605-42-5
Beilstein No: 7^2, 718
Name: 9,10-Anthracenedione, 2,7-dibromo-
Synonym: Anthraquinone, 2,7-dibromo-
Mol Form: $C_{14}H_6Br_2O_2$
M_r: 363.99
bp: sub
mp: 248
Color: lt ye lf
***s*:** Bz 4; HOAc 4
m/e: 363.87

HODOC No: 01243
CAS No: 602-73-3
Beilstein No: 7^3, 4068
Name: 9,10-Anthracenedione, 1,3-dichloro-
Synonym: Anthraquinone, 1,3-dichloro-
Mol Form: $C_{14}H_6Cl_2O_2$
M_r: 277.09
mp: 209-10
Color: ye nd (HOAc)
***s*:** HOAc 4
m/e: 275.97

HODOC No: 01244
CAS No: 602-25-5
Beilstein No: 7^4, 2559
Name: 9,10-Anthracenedione, 1,4-dichloro-
Synonym: Anthraquinone, 1,4-dichloro-
Mol Form: $C_{14}H_6Cl_2O_2$
M_r: 277.09
mp: 187-8
Color: og ye nd (HOAc)
***s*:** H_2O 1; al 2; eth 2; bz 2; $PhNO_2$ 4; py 4; HOAc 3
UV: HBCP 357(39811) 256(39811) 213(28184) cyhex
m/e: 275.97

HODOC No: 01245
CAS No: 82-46-2
Beilstein No: 7^4, 2559
Name: 9,10-Anthracenedione, 1,5-dichloro-
Synonym: Anthraquinone, 1,5-dichloro-
Mol Form: $C_{14}H_6Cl_2O_2$
M_r: 277.09
mp: 252
Color: ye nd (to)
***s*:** H_2O 1; al 2; ace 2; bz 2 $PhNO_2$ 3; con sulf 3; HOAc 3
IR: μCOB 4626 3130 1700 1590 1450 1430 1320 1270 1210 1160 1140 980 810 750 710 700
UV: SAD 6238 342 256 diox
m/e: 275.97

HODOC No: 01246
CAS No: 727-51-5
Beilstein No: 7^2, 715
Name: 9,10-Anthracenedione, 1,6-dichloro-
Synonym: Anthraquinone, 1,6-dichloro-
Mol Form: $C_{14}H_6Cl_2O_2$
M_r: 277.09
mp: 202.5-3.5
***s*:** al 2; eth 2; ace 2; bz 3; HOAc 3; tol 3; $PhNO_2$ 3
UV: LANG 820 422(135) 402(135) 331(5248) 274(15849) 257(41687) 212(30200) cy 336(4786) 258(39811) 210(28184) EtOH
m/e: 275.97

HODOC No: 01247
CAS No: 1594-69-0
Beilstein No: 7^3, 4070
Name: 9,10-Anthracenedione, 1,7-dichloro-
Synonym: Anthraquinone, 1,7-dichloro-
Mol Form: $C_{14}H_6Cl_2O_2$
M_r: 277.09
mp: 213-4
Color: ye nd (HOAc)
***s*:** H_2O 1; al 2; $PhNO_2$ 3; HOAc 3
UV: HBCP 282(19953) 270(19953) 263(50119) 213(26303) cyhex
m/e: 275.97

HODOC No: 01248
CAS No: 82-43-9
Beilstein No: 7^4, 2560
Name: 9,10-Anthracenedione, 1,8-dichloro-
Synonym: Anthraquinone, 1,8-dichloro-
Mol Form: $C_{14}H_6Cl_2O_2$
M_r: 277.09
mp: 202-3
Color: ye nd (HOAc)
***s*:** H_2O 1; al 2; bz 3; tol 3; $PhNO_2$ 3
IR: μCOB 298 3030 1670 1560 1450 1430 1320 1250 1180 1140 980 860 840 790 760 740 660
UV: SAD 6244 339 253 diox
m/e: 275.97

HODOC No: 01249
CAS No: 84-45-7
Beilstein No: 7^4, 2560
Name: 9,10-Anthracenedione, 2,3-dichloro-
Synonym: Anthraquinone, 2,3-dichloro-
Mol Form: $C_{14}H_6Cl_2O_2$
M_r: 277.09
mp: 271
Color: ye nd (HOAc)
***s*:** al 2; bz 2; HOAc 2
UV: LANG 822 426(76) 403(115) 383(138) 326(4786) 280(18197) 261(52481) 211(28840) cy 330(4571) 263(50119) 210(30903) EtOH
m/e: 275.97

HODOC No: 01250
CAS No: 605-40-3
Beilstein No: 7^4, 2560
Name: 9,10-Anthracenedione, 2,6-dichloro-
Synonym: Anthraquinone, 2,6-dichloro-
Mol Form: $C_{14}H_6Cl_2O_2$
M_r: 277.09
mp: 291
Color: ye nd (al or HOAc)
s: Bz 4; EtOH 4; HOAc 4
UV: LANG 823 327(5370) 280(24547) 270(24547) 260(54954) 252(34674) 213(28840) cy 325(5623) 280(19055) 262(46774) 210(29512) EtOH
m/e: 275.97

HODOC No: 01251
CAS No: 605-43-6
Beilstein No: 7^4, 2560
Name: 9,10-Anthracenedione, 2,7-dichloro-
Synonym: Anthraquinone, 2,7-dichloro-
Mol Form: $C_{14}H_6Cl_2O_2$
M_r: 277.09
mp: 212
Color: yesh nd
s: Et_2O 4
UV: LANG 824 423(74) 403(129) 381(162) 328(4677) 280(20417) 263(72444) 212(33884) cy 330(4169) 278(15849) 261(52481) 209(33113) EtOH
m/e: 275.97

HODOC No: 01252
CAS No: 72-48-0
Beilstein No: 8^4, 3256
Name: 9,10-Anthracenedione, 1,2-dihydroxy-
Synonym: Anthraquinone, 1,2-dihydroxy-
Structure:

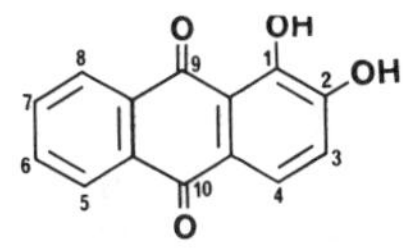

Mol Form: $C_{14}H_8O_4$
M_r: 240.20
bp: 430 sub
mp: 289-90
Color: og or red tcl nd or pr (al, sub)
s: H_2O 2; al 3; eth 3; ace 3; bz 3; chl 1; py 5; CS_2 3; MeOH 3
UV: HBCP 435(6310) 248(25119) al
MS: ALDN 195 240(100) 241(15) 239(12) 212(8) 92(6) 77(6) 18(5) 138(5)
m/e: 240.04

HODOC No: 01253
CAS No: 518-83-2
Beilstein No: 8^4, 3259
Name: 9,10-Anthracenedione, 1,3-dihydroxy-
Synonym: Anthraquinone, 1,3-dihydroxy-
Mol Form: $C_{14}H_8O_4$
M_r: 240.20
mp: 268-70
Color: ye nd, lf
s: H_2O 1; al 3; ace 4; bz 3; $PhNO_2$ 3; alk 3; HOAc 3
UV: HBCP 415(6166) 284(22908) 246(26916) al
m/e: 240.04

HODOC No: 01254
CAS No: 81-64-1
Beilstein No: 8^4, 3260
Name: 9,10-Anthracenedione, 1,4-dihydroxy-
Synonym: Anthraquinone, 1,4-dihydroxy-
Mol Form: $C_{14}H_8O_4$
M_r: 240.20
mp: 200
Color: ye red lf (eth) dk red nd
***s*:** H_2O 3; al 3; eth 3; bz 3; KOH 3; sulf 3
IR: µSADP 8810 3450 1610 1590 1450 1390 1350 1270 1240 1160 1140 1060 970 770 720 690
UV: OES 5-392 450(8511) 325(2455) 279(10233) 249(30903) 225(21878) EtOH
MS: ALDN 195 240(100) 239(20) 241(11) 128(7) 102(6) 44(6) 51(6) 77(5)
m/e: 240.04

HODOC No: 01255
CAS No: 117-12-4
Beilstein No: 8^4, 3268
Name: 9,10-Anthracenedione, 1,5-dihydroxy-
Synonym: Anthraquinone, 1,5-dihydroxy-
Mol Form: $C_{14}H_8O_4$
M_r: 240.20
bp: sub
mp: 280
Color: pa ye pl (gl HOAc)
***s*:** H_2O 1; al 2; eth 2; ace 2; bz 3; con sulf 3; CS_2 2
IR: SADG 16448 3420 3050 1630 1600 1570 1470 1440 1360 1290 1230 1170 1080 1040 860 780 710
UV: OES 1-545 420(20893) 250(39811) MeOH 422(7943) 255(15849) diox
MS: ALDN 195 240(100) 244(18) 92(14) 241(14) 184(10) 223(9) 91(8) 212(8)
m/e: 240.04

HODOC No: 01256
CAS No: 569-10-8
Beilstein No: 8^4, 3270
Name: 9,10-Anthracenedione, 1,6-dihydroxy-
Synonym: Anthraquinone, 1,6-dihydroxy-
Mol Form: $C_{14}H_8O_4$
M_r: 240.20
mp: 276
Color: og-ye nd (gl HOAc)
***s*:** Ac 4
m/e: 240.04

HODOC No: 01257
CAS No: 569-08-4
Beilstein No: 8^4, 3270
Name: 9,10-Anthracenedione, 1,7-dihydroxy-
Synonym: Anthraquinone, 1,7-dihydroxy-
Mol Form: $C_{14}H_8O_4$
M_r: 240.20
bp: sub
mp: 292-3
Color: ye nd (sub)
***s*:** H_2O 1; al 3; eth 3; bz 3; chl 3; CS_2 3; HOAc 3
MS: ALDN 195 240(100) 212(25) 184(22) 241(17) 155(12) 92(9) 239(9) 106(9)
m/e: 240.04

HODOC No: 01258
CAS No: 117-10-2
Beilstein No: 8^4, 3270
Name: 9,10-Anthracenedione, 1,8-dihydroxy-
Synonym: Anthraquinone, 1,8-dihydroxy-
Mol Form: $C_{14}H_8O_4$
M_r: 240.20
bp: sub
mp: 193
Color: red or red-ye nd or lf (al)
***s*:** H; iO 1; al 3; eth 3; ace 3; chl 3; alk 3; HOAc 3 $PhNO_2$ 2
IR: µCOB 5147 3330 1670 1610 1590 1490 1450 1370 1350 1270 1210 1160 1080 1050 1030 970 850 840 810 780 750
UV: SAD 4318 428(11100) 282(10800) 251(22100) 224(45000) MeOH
MS: ALDN 195 240(100) 241(15) 212(11) 184(10) 138(7) 92(6) 63(5) 223(5)
m/e: 240.04

HODOC No: 01259
CAS No: 483-35-2
Beilstein No: 8^4, 3272
Name: 9,10-Anthracenedione, 2,3-dihydroxy-
Synonym: Anthraquinone, 2,3-dihydroxy-
Mol Form: $C_{14}H_8O_4$
M_r: 240.20
bp: sub
mp: 330
Color: ye-br nd (HOAc) ye nd (sub)
m/e: 240.04

HODOC No: 01260
CAS No: 84-60-6
Beilstein No: 8^4, 3272
Name: 9,10-Anthracenedione, 2,6-dihydroxy-
Synonym: Anthraquinone, 2,6-dihydroxy-
Mol Form: $C_{14}H_8O_4$
M_r: 240.20
mp: 360d
Color: ye nd (al)
***s*:** H_2O 2; al 2; eth 1; bz 1; chl 1; con sulf 3; alk 3
IR: COB 7239 3285 1660 1560 1445 1295 1240 1145 1080 880 835 745 730 570 555
UV: OES 1-545 341(7943) 296(15849) 269(39811) diox
m/e: 240.04

HODOC No: 01261
CAS No: 572-93-0
Beilstein No: 8^4, 3273
Name: 9,10-Anthracenedione, 2,7-dihydroxy-
Synonym: Anthraquinone, 2,7-dihydroxy-
Mol Form: $C_{14}H_8O_4$
M_r: 240.20
bp: sub
mp: 350-5
Color: ye nd (+1w, dil al) nd (sub)
***s*:** H_2O 1; al 3; eth 2; bz 2; chl 2; HOAc 3
IR: µSADP 3588 3330 2860 1670 1610 1560 1470 1370 1320 1250 940 890 860 800 750
m/e: 240.04

HODOC No: 01262
CAS No: 57990-92-8
Name: 9,10-Anthracenedione, 1,5-dihydroxy-2,4-dinitro-
Synonym: 9,10-Anthraquinone, 1,5-dihydroxy-2,4-dinitro
Mol Form: $C_{14}H_6N_2O_8$
M_r: 330.20
IR: µCOB 992 3230 3030 1640 1610 1540 1450 1370 1300 1220 1160 1150 1100 860 780 740 660
m/e: 330.01

HODOC No: 01263
CAS No: 81-55-0
Beilstein No: 8^4, 3272
Name: 9,10-Anthracenedione, 1,8-dihydroxy-4,5-dinitro-
Synonym: Anthraquinone, 1,8-dihydroxy-4,5-dinitro-
Mol Form: $C_{14}H_6N_2O_8$
M_r: 330.20
mp: 296
IR: COB 7627 3150 3090 1690 1630 1600 1550 1460 1370 1260 1190 1100 850 800 770 720 670 450
m/e: 330.01

HODOC No: 01264
CAS No: 481-72-1
Beilstein No: 8^4, 3578
Name: 9,10-Anthracenedione, 1,8-dihydroxy-3-(hydroxymethyl)-
Synonym: Anthraquinone, 1,8-dihydroxy-3-(hydroxymethyl)-
Line Form: $[1,8\text{-}(HO)_2\text{-}3\text{-}HOCH_2]\text{-}C_{14}H_5O_2$
Mol Form: $C_{15}H_{10}O_5$
M_r: 270.23
bp: sub
mp: 223-4
Color: og ye nd (to, al)
***s*:** Bz 4; Et_2O 4; EtOH 4
UV: OES 4-518
m/e: 270.05

HODOC No: 01265
CAS No: 602-63-1
Beilstein No: 8^4, 3276
Name: 9,10-Anthracenedione, 1,2-dihydroxy-3-methyl-
Synonym: Anthraquinone, 1,2-dihydroxy-3-methyl-
Mol Form: $C_{15}H_{10}O_4$
M_r: 254.23
bp: sub
mp: 245
Color: og nd
m/e: 254.06

HODOC No: 01266
CAS No: 481-74-3
Beilstein No: 8^4, 3277
Name: 9,10-Anthracenedione, 1,8-dihydroxy-3-methyl-
Synonym: Anthraquinone, 1,8-dihydroxy-3-methyl-
Mol Form: $C_{15}H_{10}O_4$
M_r: 254.23
bp: sub
mp: 196
***d*:** 0.92
Color: ye hex or mcl nd (sub)
***s*:** Ac 4; Bz 4; HOAc 4
IR: SADP 3579
m/e: 254.06

HODOC No: 01267
CAS No: 568-93-4
Beilstein No: 8^4, 3259
Name: 9,10-Anthracenedione, 1,2-dihydroxy-3-nitro-
Synonym: Anthraquinone, 1,2-dihydroxy-3-nitro-
Mol Form: $C_{14}H_7NO_6$
M_r: 285.20
bp: sub
mp: 244d
Color: og nd or pl (HOAc)
***s*:** H_2O 2; al 3; bz 3; chl 3; sulf 3; HOAc 3
IR: μCOB 341 3230 2940 1670 1640 1590 1540 1470 1430 1350 1250 1160 1060 1030 950 930 880 830 800 760 740 710 680 660
m/e: 285.03

HODOC No: 01268
CAS No: 2243-71-2
Beilstein No: 8^4, 3259
Name: 9,10-Anthracenedione, 1,2-dihydroxy-4-nitro-
Synonym: Anthraquinone, 1,2-dihydroxy-4-nitro-
Mol Form: $C_{14}H_7NO_6$
M_r: 285.20
bp: subd
mp: 289d
Color: gold-ye nd (HOAc or al)
***s*:** Bz 4; EtOH 4; Chl 4; HOAc 4
m/e: 285.03

HODOC No: 01269
CAS No: 517-92-0
Beilstein No: 8^4, 3270
Name: 9,10-Anthracenedione, 1,8-dihydroxy-2,4,5,7-tetranitro-
Synonym: Anthraquinone, 1,8-dihydroxy-2,4,5,7-tetranitro-
Mol Form: $C_{14}H_4N_4O_{12}$
M_r: 420.19
bp: d
mp: exp
Color: ye pl or lf
***s*:** Et_2O 4; EtOH 4
m/e: 419.98

HODOC No: 01270
CAS No: 6448-90-4
Beilstein No: 8^1, 720
Name: 9,10-Anthracenedione, 1,5-dimethoxy-
Synonym: Anthraquinone, 1,5-dimethoxy-
Mol Form: $C_{16}H_{12}O_4$
M_r: 268.25
UV: LANG 962 387(8710) 254(24547) 220(42658) MeOH
m/e: 268.07

HODOC No: 01271
CAS No: 73453-71-1
Beilstein No: 8^4, 3270
Name: 9,10-Anthracenedione, 1,6-dimethoxy-
Synonym: Anthraquinone, 1,6-dimethoxy
Mol Form: $C_{16}H_{12}O_4$
M_r: 268.25
mp: 185
***s*:** DMSO 2
IR: SADG 19292 3360 3150 2700 1660 1610 1560 1530 1430 1350 1300 1270 1210 1160 980 860 790 740 670 560
UV: SAD 17918 293(3850) 239(22000) 207(25500) MeOH
H NMR: SAD 10016 2.2 7.3 7.6 8.1 10.3-10.6 $DMSOd_6$
m/e: 268.07

HODOC No: 01272
CAS No: 75312-36-6
Beilstein No: 8^4, 3270
Name: 9,10-Anthracenedione, 1,7-dimethoxy-
Synonym: Anthraquinone, 1,7-dimethoxy-
Mol Form: $C_{16}H_{12}O_4$
M_r: 268.25
mp: 191
UV: LANG 964 374(7244) 266(29512) 218(28184) MeOH
m/e: 268.07

HODOC No: 01273
CAS No: 6407-55-2
Beilstein No: 8^4, 3271
Name: 9,10-Anthracenedione, 1,8-dimethoxy-
Synonym: Anthraquinone, 1,8-dimethoxy-
Mol Form: $C_{16}H_{12}O_4$
M_r: 268.25
mp: 221
s: al 3; eth 1; bz 3; chl 3; peth 1; gl HOAc 3
IR: µSADP 3566 2940 1670 1590 1470 1430 1320 1280 1240 1190 1060 980 900 840 790 750
UV: LANG 965 387(7586) 255(23988) 219(39811) MeOH
MS: NBS 4713
m/e: 268.07

HODOC No: 01274
CAS No: 963-96-2
Beilstein No: 8^4, 3273
Name: 9,10-Anthracenedione, 2,6-dimethoxy-
Synonym: Anthraquinone, 2,6-dimethoxy-
Mol Form: $C_{16}H_{12}O_4$
M_r: 268.25
mp: 256
UV: LANG 966
m/e: 268.07

HODOC No: 01275
CAS No: 73453-68-6
Beilstein No: 8^4, 3273
Name: 9,10-Anthracenedione, 2,7-dimethoxy-
Synonym: Anthraquinone, 2,7-dimethoxy-
Mol Form: $C_{16}H_{12}O_4$
M_r: 268.25
mp: 215
UV: LANG 967 274(47863) 211(26915) MeOH
m/e: 268.07

HODOC No: 01276
CAS No: 3285-98-1
Beilstein No: 7^4, 2583
Name: 9,10-Anthracenedione, 1,2-dimethyl-
Synonym: Anthraquinone, 1,2-dimethyl-
Mol Form: $C_{16}H_{12}O_2$
M_r: 236.25
mp: 156
Color: nd (Ace or HOAc)
s: Ac 4; Et_2O 4; EtOH 4; HOAc 4
MS: NBS 1616
m/e: 236.08

HODOC No: 01277
CAS No: 3285-97-0
Beilstein No: 7^4, 2583
Name: 9,10-Anthracenedione, 1,3-dimethyl-
Synonym: Anthraquinone, 1,3-dimethyl-
Mol Form: $C_{16}H_{12}O_2$
M_r: 236.25
mp: 162
Color: nd (HOAc)
s: H_2O 1; al 2; bz 2; HOAc 3
IR: µDMS 6137 2860 1670 1590 1430 1390 1370 1320 1280 1250 1120 1050 960 890 790 730 710
H NMR: VAR 650 2.5 2.8 7.3 7.7 8.0 8.2 $CDCl_3$
m/e: 236.08

HODOC No: 01278
CAS No: 1519-36-4
Beilstein No: 7^4, 2583
Name: 9,10-Anthracenedione, 1,4-dimethyl-
Synonym: Anthraquinone, 1,4-dimethyl-
Mol Form: $C_{16}H_{12}O_2$
M_r: 236.25
bp: sub
mp: 140-1
Color: ye nd (al, sub)
s: H_2O 1; al 2; bz 3; xyl 3; HOAc 3
IR: µDMS 6138 2860 1640 1590 1540 1450 1370 1320 1250 1150 1060 1030 970 850 820 720
m/e: 236.08

HODOC No: 01279
CAS No: 6531-35-7
Beilstein No: 7^3, 4131
Name: 9,10-Anthracenedione, 2,3-dimethyl-
Synonym: Anthraquinone, 2,3-dimethyl-
Mol Form: $C_{16}H_{12}O_2$
M_r: 236.25
bp: sub
mp: 210
Color: ye nd (al, to, or xyl), cr (HOAc)
***s*:** al 3; bz 3; chl 2; xyl 3; HOAc 3
IR: SADG 10560 3030 2941 1667 1587 1471 1449 1389 1316 1299 1220 952 719
UV: LANG 1369 329(4898) 258(44668) MeOH
H NMR: SAD 2738 2.4 7.8 8.0 8.3 $CDCl_3$
m/e: 236.08

HODOC No: 01280
CAS No: 3837-38-5
Beilstein No: 7^3, 4132
Name: 9,10-Anthracenedione, 2,6-dimethyl-
Synonym: Anthraquinone, 2,6-dimethyl-
Mol Form: $C_{16}H_{12}O_2$
M_r: 236.25
bp: sub
mp: 242
Color: ye nd (HOAc or al)
***s*:** EtOH 4; HOAc 4; to 4
m/e: 236.08

HODOC No: 01281
CAS No: 3286-01-9
Beilstein No: 7^3, 4133
Name: 9,10-Anthracenedione, 2,7-dimethyl-
Synonym: Anthraquinone, 2,7-dimethyl-
Mol Form: $C_{16}H_{12}O_2$
M_r: 236.25
mp: 170
Color: yesh nd (al)
***s*:** EtOH 4
m/e: 236.08

HODOC No: 01282
CAS No: 84-47-9
Beilstein No: 7^4, 2594
Name: 9,10-Anthracenedione, 2-(1,1-dimethylethyl)-
Synonym: Anthraquinone, 2-*tert*-butyl-
Mol Form: $C_{18}H_{16}O_2$
M_r: 264.31
mp: 98-100
***s*:** ctc 3; CS_2 3
IR: COB 6274 3070 2960 2900 2870 1670 1660 1590 1480 1460 1450 1390 1360 1330 1300 1270 1260 1170 980 930 860 850 800 730 710 670 640 570
MS: ALD 6:69B
m/e: 264.12

HODOC No: 01283
CAS No: 68833-53-4
Name: 9,10-Anthracenedione, 1,3-dinitro-
Synonym: Anthraquinone, 1,3-dinitro
Mol Form: $C_{14}H_6N_2O_6$
M_r: 298.20
mp: 246-50
Color: ye nd (HNO_3)
m/e: 298.02

HODOC No: 01284
CAS No: 82-35-9
Beilstein No: 7^4, 2562
Name: 9,10-Anthracenedione, 1,5-dinitro-
Synonym: Anthraquinone, 1,5-dinitro-
Mol Form: $C_{14}H_6N_2O_6$
M_r: 298.20
bp: sub
mp: 422
Color: pa ye nd (xyl) ye-cr (sub)
***s*:** H_2O 1; al 2; eth 2; bz 2; $PhNO_2$ 4; con sulf 2; HOAc 2
IR: µDMS 6147 3330 3230 1640 1590 1520 1450 1370 1250 1160 1050 830 790 740 690
UV: OES 1-543 322(5012) 258(39811) diox
m/e: 298.02

HODOC No: 01285
CAS No: 1604-42-8
Beilstein No: 7^3, 4082
Name: 9,10-Anthracenedione, 1,6-dinitro-
Synonym: Anthraquinone, 1,6-dinitro-
Mol Form: $C_{14}H_6N_2O_6$
M_r: 298.20
mp: 257-9
***s*:** $PhNO_2$ 3
UV: LANG 1382 315(4786) 259(38019) 219(26915) MeOH 325(4677) thf
m/e: 298.02

HODOC No: 01286
CAS No: 1604-43-9
Beilstein No: 7^3, 4082
Name: 9,10-Anthracenedione, 1,7-dinitro-
Synonym: Anthraquinone, 1,7-dinitro-
Mol Form: $C_{14}H_6N_2O_6$
M_r: 298.20
mp: 295
UV: LANG 1383 322(4169) 259(37154) MeOH
m/e: 298.02

HODOC No: 01287
CAS No: 129-39-5
Beilstein No: 7^4, 2562
Name: 9,10-Anthracenedione, 1,8-dinitro-
Synonym: Anthraquinone, 1,8-dinitro-
Mol Form: $C_{14}H_6N_2O_6$
M_r: 298.20
mp: 312
UV: LANG 1384 320(3802) 254(33884) MeOH
m/e: 298.02

HODOC No: 01288
CAS No: 84-51-5
Beilstein No: 7^4, 2583
Name: 9,10-Anthracenedione, 2-ethyl-
Synonym: Anthraquinone, 2-ethyl-
Mol Form: $C_{16}H_{12}O_2$
M_r: 236.25
mp: 108-11
IR: ALD 2:786E
MS: NBS 4586; ALD 6:69A
m/e: 236.08

HODOC No: 01289
CAS No: 56600-58-9
Beilstein No: 7^2, 743
Name: 9,10-Anthracenedione, 2-ethyl-1-nitro-
Synonym: 2-Ethyl-1-nitroanthraquinone
Mol Form: $C_{16}H_{11}NO_4$
M_r: 281.25
mp: 226
Color: yesh br (HOAc)
m/e: 281.07

HODOC No: 01290
CAS No: 569-06-2
Beilstein No: 7^4, 2558
Name: 9,10-Anthracenedione, 1-fluoro-
Synonym: Anthraquinone, 1-fluoro-
Mol Form: $C_{14}H_7FO_2$
M_r: 226.19
mp: 228-31
IR: SADG 23126 1675 1585 1445 1325 1275 1255 1165 1005 870 830 800 725 705
UV: LANG 2373 329(5754) 250(36308) 208(28184) MeOH
m/e: 226.04

HODOC No: 01291
CAS No: 31297-82-2
Name: 9,10-Anthracenedione, 2-(β-*D*-glucopyranosyloxy)-1-hydroxy-
Synonym: Alizarin-2-β-*D*-glucopyranoside
Structure:

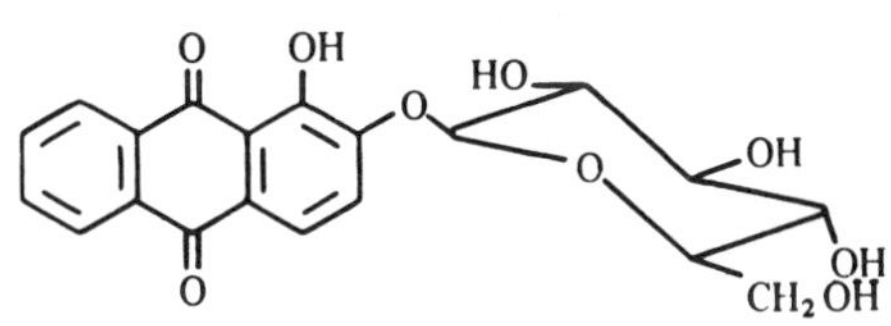

Mol Form: $C_{20}H_{18}O_9$
M_r: 402.34
mp: 236-7
UV: LANG 416 416(6310) 329(3090) 250(28184) 230(18197) EtOH
m/e: 402.10

HODOC No: 01292
CAS No: 82-12-2
Beilstein No: 8^4, 3750
Name: 9,10-Anthracenedione, 1,2,3,5,6,7-hexahydroxy-
Synonym: Anthraquinone, 1,2,3,5,6,7-hexahydroxy-
Mol Form: $C_{14}H_8O_8$
Mr: 304.20
bp: sub (d)
Color: red rh, red-ye nd (sub)
***s*:** H_2O 1; al 2; eth 2; ace 3; alk 3
UV: HBCP 438(8318) 349(8318) 295(43652) 258(7080) 222(15136) al
m/e: 304.02

HODOC No: 01293
CAS No: 129-43-1
Beilstein No: 8^4, 2586
Name: 9,10-Anthracenedione, 1-hydroxy-
Synonym: Anthraquinone, 1-hydroxy-
Mol Form: $C_{14}H_8O_3$
Mr: 224.20
bp: sub
mp: 194-5
Color: red-og nd (al)
***s*:** H_2O 1; al 3; eth 3; bz 3; liq NH_3 2
IR: μSADP 11414 3450 3030 1670 1640 1590 1450 1350 1280 1270 1220 1150 1060 1010 880 830 780 730 700
UV: OES 4-484 402(5495) dmf
MS: ALDN 183 224(100) 139(19) 223(18) 168(16) 225(14) 196(10) 44(7) 140(5)
m/e: 224.05

HODOC No: 01294
CAS No: 605-32-3
Beilstein No: 8^4, 2598
Name: 9,10-Anthracenedione, 2-hydroxy-
Synonym: Anthraquinone, 2-hydroxy-
Mol Form: $C_{14}H_8O_3$
Mr: 224.20
bp: sub
mp: 306
Color: ye pl or nd (al or HOAc)
***s*:** H_2O 1; al 3; eth 3; aq NH_3 3; KOH 3
IR: μSADP 11256 3450 1670 1590 1490 1450 1330 1300 930 850 710 700
UV: SAD 3068 270 244 239 MeOH
MS: ALDN 183 224(100) 139(30) 168(29) 196(27) 225(13) 223(11) 84(9) 195(9)
m/e: 224.05

HODOC No: 01295
CAS No: 6003-11-8
Beilstein No: 8^4, 3257
Name: 9,10-Anthracenedione, 1-hydroxy-2-methoxy-
Synonym: Anthraquinone, 1-hydroxy-2-methoxy-
Mol Form: $C_{15}H_{10}O_4$
Mr: 254.23
mp: 230-1
UV: LANG 415
MS: ALDN 203
m/e: 254.06

HODOC No: 01296
CAS No: 82-39-3
Beilstein No: 8^4, 2586
Name: 9,10-Anthracenedione, 1-methoxy-
Synonym: Anthraquinone, 1-methoxy-
Mol Form: $C_{15}H_{10}O_3$
Mr: 238.23
mp: 169.5
***s*:** al 2; bz 4; chl 4
IR: SADG 1115 1675 1580 1470 1450 1310 1275 1075 970 800 700 395
UV: LANG 960 382(5012) 332(2692) 252(36308) MeOH
MS: ALDN 194 238(100) 209(41) 152(27) 224(22) 180(19) 76(18) 181(17) 239(16)
m/e: 238.06

HODOC No: 01297
CAS No: 3274-20-2
Beilstein No: 8^4, 2598
Name: 9,10-Anthracenedione, 2-methoxy-
Synonym: Anthraquinone, 2-methoxy-
Mol Form: $C_{15}H_{10}O_3$
Mr: 238.23
mp: 196
***s*:** H_2O 1; al 2; bz 3; gl HOAc 3
UV: LANG 961 330(3981) 262(33884) 210(24547) MeOH
MS: ALDN 194 238(100) 237(22) 139(21) 239(19) 209(15) 208(14) 195(9) 167(9)
m/e: 238.06

HODOC No: 01298
CAS No: 84-54-8
Beilstein No: 7^4, 2574
Name: 9,10-Anthracenedione, 2-methyl-
Synonym: Anthraquinone, 2-methyl-
Mol Form: $C_{15}H_{10}O_2$
M_r: 222.23
bp: sub
mp: 182-3, 177 sub
Color: yesh nd (al, HOAc)
***s*:** Bz 4; EtOH 4; HOAc 4
IR: SADG 8459
UV: SAD 1746
H NMR: SAD 3200
MS: NBS 1459
m/e: 222.07

HODOC No: 01299
CAS No: 82-38-2
Beilstein No: 14^3, 408
Name: 9,10-Anthracenedione, 1-(methylamino)-
Synonym: Anthraquinone, 1-(methylamino)-
Mol Form: $C_{15}H_{11}NO_2$
M_r: 237.24
mp: 170
Color: ye-red nd
***s*:** al 3; bz 3; chl 3; HOAc 3
IR: µCOB 300 3330 2940 1670 1610 1590 1520 1390 1350 1320 1270 1180 1160 1080 990 860 830 810 780 740 710 660
UV: SAD 5830 272 244 MeOH
H NMR: SAD 4221 3.0 7.0 7.3-7.9 8.2 9.6 $CDCl_3$
MS: ALDN 193 237(100) 220(66) 223(44) 236(31) 152(21) 180(17) 30(14) 151(12)
m/e: 237.08

HODOC No: 01300
CAS No: 20327-68-8
Beilstein No: 14^3, 430
Name: 9,10-Anthracenedione, 2-(methylamino)-
Synonym: Anthraquinone, 2-(methylamino)-
Mol Form: $C_{15}H_{11}NO_2$
M_r: 237.24
mp: 226-7
Color: red nd (HOAc)
***s*:** Et_2O 4; EtOH 4; HOAc 4; to 4
m/e: 237.08

HODOC No: 01301
CAS No: 129-15-7
Beilstein No: 7^4, 2576
Name: 9,10-Anthracenedione, 2-methyl-1-nitro-
Synonym: Anthraquinone, 2-methyl-1-nitro-
Mol Form: $C_{15}H_9NO_4$
M_r: 267.23
mp: 270-1
Color: pa ye nd (HOAc)
***s*:** H_2O 1; al 1; eth 2; bz 2; chl 2; $PhNO_2$ 3; HOAc 2
IR: µCOB 4628 3030 1700 1610 1540 1470 1410 1390 1330 1300 1160 1150 990 860 820 800 780 750 710
MS: NBS 4709
m/e: 267.05

HODOC No: 01302
CAS No: 82-34-8
Beilstein No: 7^4, 2561
Name: 9,10-Anthracenedione, 1-nitro-
Synonym: Anthraquinone, 1-nitro-
Mol Form: $C_{14}H_7NO_4$
M_r: 253.20
bp: 270^7
mp: 230
Color: nd (HOAc) ye pr (Ace)
***s*:** H_2O 1; al 2; eth 2; ace 3; bz 3; chl 3; HOAc 3
IR: µDMS 6146 3030 1670 1560 1540 1450 1370 1320 1280 1150 970 870 830 810 740 700
UV: OES 1-543 322(6310) 254(50119) diox
m/e: 253.04

HODOC No: 01303
CAS No: 605-27-6
Beilstein No: 7^4, 2561
Name: 9,10-Anthracenedione, 2-nitro-
Synonym: Anthraquinone, 2-nitro-
Mol Form: $C_{14}H_7NO_4$
M_r: 253.20
bp: sub
mp: 184-5
Color: ye nd (HOAc or al)
***s*:** H_2O 1; al 2; eth 2; ace 3; chl 3; sulf 3; HOAc 3; lig 2
UV: OES 2-379 420(20) 323(5248) 258(38911) MeOH
m/e: 253.04

HODOC No: 01304
CAS No: 5413-65-0
Beilstein No: 8^4, 3732
Name: 9,10-Anthracenedione, 1,2,4,5,8-pentahydroxy-
Synonym: Anthraquinone, 1,2,4,5,8-pentahydroxy-
Mol Form: $C_{14}H_8O_7$
M_r: 288.20
bp: sub
mp: d
m/e: 288.03

HODOC No: 01305
CAS No: 80468-92-4
Beilstein No: 7^3, 4078
Name: 9,10-Anthracenedione, 1,3,6,8-tetrabromo-
Synonym: Anthraquinone, 2,4,5,7-tetrabromo
Mol Form: $C_{14}H_4Br_4O_2$
M_r: 519.79
mp: 312
Color: og-ye nd (Bz)
m/e: 519.69

HODOC No: 01306
CAS No: 34234-10-1
Beilstein No: 8^3, 3783
Name: 9,10-Anthracenedione, 1,2,3,4-tetrachloro-5,8-dihydroxy-
Synonym: Anthraquinone, 1,2,3,4-tetrachloro-5,8-dihydroxy-
Mol Form: $C_{14}H_4Cl_4O_4$
M_r: 377.98
mp: 270
Color: red pl (HOAc)
***s*:** Bz 4; HOAc 4
m/e: 375.89

HODOC No: 01307
CAS No: 75313-07-4
Beilstein No: 8^3, 4288
Name: 9,10-Anthracenedione, 1,2,4,6-tetrahydroxy-
Synonym: Anthraquinone, 1,2,4,6-tetrahydroxy
Mol Form: $C_{14}H_8O_6$
M_r: 272.20
Color: dk red nd (sub)
m/e: 272.03

HODOC No: 01308
CAS No: 632-77-9
Beilstein No: 8^3, 4288
Name: 9,10-Anthracenedione, 1,2,5,6-tetrahydroxy-
Synonym: Anthraquinone, 1,2,5,6-tetrahydroxy-
Mol Form: $C_{14}H_8O_6$
M_r: 272.20
bp: sub
mp: 340
Color: og-red nd (Py)
***s*:** Py 4; EtOH 4; HOAc 4
m/e: 272.03

HODOC No: 01309
CAS No: 81-61-8
Beilstein No: 8^4, 3683
Name: 9,10-Anthracenedione, 1,2,5,8-tetrahydroxy-
Synonym: Anthraquinone, 1,2,5,8-tetrahydroxy-
Mol Form: $C_{14}H_8O_6$
M_r: 272.20
mp: 275
Color: og nd
***s*:** H; iO2 1; al 2; eth 2; ace 2; bz 2; alk 2; os 2; HOAc 2
IR: **μCOB 4614** 3450 1610 1590 1470 1280 1160 1080 850 810 770 730
UV: **OES 2-381** 575(19498) sulf
m/e: 272.03

HODOC No: 01310
CAS No: 68844-87-1
Beilstein No: 8^2, 584
Name: 9,10-Anthracenedione, 1,2,6,7-tetrahydroxy-
Synonym: Anthraquinone, 1,2,6,7-tetrahydroxy
Mol Form: $C_{14}H_8O_6$
M_r: 272.20
mp: 330
m/e: 272.03

HODOC No: 01311
CAS No: 68844-88-2
Beilstein No: 8^2, 585
Name: 9,10-Anthracenedione, 1,2,7,8-tetrahydroxy-
Synonym: Anthraquinone, 1,2,7,8-tetrahydroxy
Mol Form: $C_{14}H_8O_6$
M_r: 272.20
mp: 318d
Color: red nd or pr (HOAc)
m/e: 272.03

HODOC No: 01312
CAS No: 632-82-6
Beilstein No: 8^4, 3683
Name: 9,10-Anthracenedione, 1,3,5,7-tetrahydroxy-
Synonym: Anthraquinone, 1,3,5,7-tetrahydroxy-
Mol Form: $C_{14}H_8O_6$
M_r: 272.20
bp: sub
mp: 150-60d,360anh
Color: yesh nd (al + 2w)
s: Ac 4; Bz 4; EtOH 4; HOAc 4
m/e: 272.03

HODOC No: 01313
CAS No: 81-60-7
Beilstein No: 8^4, 3684
Name: 9,10-Anthracenedione, 1,4,5,8-tetrahydroxy-
Synonym: Anthraquinone, 1,4,5,8-tetrahydroxy-
Mol Form: $C_{14}H_8O_6$
M_r: 272.20
bp: sub
mp: 300
Color: gr nd (HOAc) br nd (Bz-lig)
s: H_2O 1; al 3; alk 3; Ac_2O 2
UV: OES 4-486 562(31623) 548(25119) 533(12589) 523(25119) 510(19953) 488(12589) 478(10000) peth
MS: ALDN 213 272(100) 273(15) 136(10) 274(3) 108(2) 142(2) 243(2) 271(2)
m/e: 272.03

HODOC No: 01314
CAS No: 602-64-2
Beilstein No: 8^4, 3567
Name: 9,10-Anthracenedione, 1,2,3-trihydroxy-
Synonym: Anthraquinone, 1,2,3-trihydroxy-
Mol Form: $C_{14}H_8O_5$
M_r: 256.20
bp: sub (290)
mp: 313
Color: ye nd (dil al)
s: H_2O 2; al 3; eth 3; CS_2 3; HOAc 3; sulf 2; alk 2
IR: µSADP 21103 3450 1640 1590 1470 1370 1280 1140 710
UV: OES 2-381 414(6457) 287(30903) 245(19953) 241(19055) 213(17783) EtOH
m/e: 256.04

HODOC No: 01315
CAS No: 81-54-9
Beilstein No: 8^4, 3568
Name: 9,10-Anthracenedione, 1,2,4-trihydroxy-
Synonym: Anthraquinone, 1,2,4-trihydroxy-
Mol Form: $C_{14}H_8O_5$
M_r: 256.20
bp: sub
mp: 259
Color: og red, dk red or og-ye nd (al)
s: H_2O 2; al 4; eth 3; bz 4; HOAc 4
IR: SADG 16449 3399 3050 1620 1580 1430 1280 1170 1060 960 770 740 700
m/e: 256.04

HODOC No: 01316
CAS No: 6486-93-7
Beilstein No: 8^4, 3568
Name: 9,10-Anthracenedione, 1,2,5-trihydroxy-
Synonym: Anthraquinone, 1,2,5-trihydroxy-
Mol Form: $C_{14}H_8O_5$
M_r: 256.20
mp: 278
Color: red nd (gl HOAc)
s: Et_2O 4
m/e: 256.04

HODOC No: 01317
CAS No: 82-29-1
Beilstein No: 8^4, 3568
Name: 9,10-Anthracenedione, 1,2,6-trihydroxy-
Synonym: Anthraquinone, 1,2,6-trihydroxy-
Mol Form: $C_{14}H_8O_5$
M_r: 256.20
bp: 459d
mp: 330 (sub)
Color: ye nd (al)
s: H_2O 3; al 3; eth 2; bz 3
IR: µSADP 3598 3330 2860 1670 1590 1450 1330 1280 1190 1040 880 840 750 710
MS: NBS 1815
m/e: 256.04

HODOC No: 01318
CAS No: 602-65-3
Beilstein No: 8^4, 3569
Name: 9,10-Anthracenedione, 1,2,7-trihydroxy-
Synonym: Anthraquinone, 1,2,7-trihydroxy-
Mol Form: $C_{14}H_8O_5$
M_r: 256.20
bp: 462
mp: 374
Color: og nd (al)
s: Bz 4; EtOH 4; HOAc 4
m/e: 256.04

HODOC No: 01319
CAS No: 51030-24-1
Beilstein No: 8^4, 3569
Name: 9,10-Anthracenedione, 1,2,8-trihydroxy-
Synonym: Anthraquinone, 1,2,8-trihydroxy-
Mol Form: $C_{14}H_8O_5$
M_r: 256.20
bp: sub
mp: 239-40
Color: red nd (HOAc, sub)
m/e: 256.04

HODOC No: 01320
CAS No: 52431-74-0
Beilstein No: 8^4, 3570
Name: 9,10-Anthracenedione, 1,3,8-trihydroxy-
Synonym: Anthraquinone, 1,3,8-trihydroxy-
Mol Form: $C_{14}H_8O_5$
M_r: 256.20
mp: 287-8
Color: brt red nd (bz)
s: H_2O 2; con sulf 3; con alk 3
UV: OES 4-485 450(7943) 292(12589) 256(15849) EtOH
m/e: 256.04

HODOC No: 01321
CAS No: 2961-04-8
Beilstein No: 8^4, 3570
Name: 9,10-Anthracenedione, 1,4,5-trihydroxy-
Synonym: Anthraquinone, 1,4,5-trihydroxy-
Mol Form: $C_{14}H_8O_5$
M_r: 256.20
mp: 271
Color: red br nd or lf, dk red nd (Py)
s: H_2O 2; al 2; HOAc 3
IR: HBCP 529(10965) 516(10965) 495(13489) 480(12023) cy
UV: HBCP
m/e: 256.04

HODOC No: 01322
CAS No: 7475-11-8
Beilstein No: 8^2, 558
Name: 9,10-Anthracenedione, 1,4,6-trihydroxy-
Synonym: Anthraquinone, 1,4,6-trihydroxy
Mol Form: $C_{14}H_8O_5$
M_r: 256.20
mp: 300
Color: bt red nd (al) br (py)
s: Py 4; EtOH 4
m/e: 256.04

HODOC No: 01323
CAS No: 478-29-5
Beilstein No: 8^4, 3572
Name: 9,10-Anthracenedione, 1,2,5-trihydroxy-6-methyl-
Synonym: Anthraquinone, 1,2,5-trihydroxy-6-methyl-
Mol Form: $C_{15}H_{10}O_5$
M_r: 270.23
mp: 282
Color: og-red nd (to)
s: Bz 4; Et_2O 4; Py 4; EtOH 4
m/e: 270.05

HODOC No: 01324
CAS No: 518-82-1
Beilstein No: 8^4, 3575
Name: 9,10-Anthracenedione, 1,3,8-trihydroxy-6-methyl-
Synonym: Anthraquinone, 1,3,8-trihydroxy-6-methyl-
Mol Form: $C_{15}H_{10}O_5$
M_r: 270.23
bp: sub
mp: 256-7
Color: og-red mcl nd (HOAc)
***s*:** Et_2O 4; EtOH 4
m/e: 270.05

HODOC No: 01325
CAS No: 82-48-4
Beilstein No: 11^3, 635
Name: 1,8-Anthracenedisulfonic acid, 9,10-dihydro-9,10-dioxo-
Synonym: 1,8-Anthraquinonedisulfonic acid
Mol Form: $C_{14}H_8O_8S_2$
M_r: 368.32
mp: 293-4d
Color: ye nd (+5w)
***s*:** H_2O 4; EtOH 4
m/e: 367.97

HODOC No: 01326
CAS No: 903-46-8
Name: 1,8-Anthracenedisulfonic acid, 9,10-dihydro-9,10-dioxo-, disodium salt
Synonym: Anthraquinone-1,8-disulfonic acid disodium salt
Mol Form: $C_{14}H_6Na_2O_8S_2$
M_r: 424.28
IR: μCOB 4617 3450 3030 1670 1640 1590 1320 1210 1040 1370 790 760 740
m/e: 411.93

HODOC No: 01327
CAS No: 84-50-4
Beilstein No: 11^2, 198
Name: 2,6-Anthracenedisulfonic acid, 9,10-dihydro-9,10-dioxo-
Synonym: Anthraquinone-2,6-disulfonic acid
Mol Form: $C_{14}H_8O_8S_2$
M_r: 368.32
***s*:** H_2O 4; al 3; eth 1; bz 1
UV: CCCCA 23-130-58 520(10) 333(6310) 262(63096) al
m/e: 367.97

HODOC No: 01328
CAS No: 84-49-1
Beilstein No: 11^2, 198
Name: 2,7-Anthracenedisulfonic acid, 9,10-dihydro-9,10-dioxo-
Synonym: 2,7-Anthraquinonedisulfonic acid
Mol Form: $C_{14}H_8O_8S_2$
M_r: 368.32
***s*:** H_2O 4; al 3; eth 1; bz 1
m/e: 367.97

HODOC No: 01329
CAS No: 114191-96-7
Name: 2-Anthraceneethanol
Synonym: Anthracene, 2(2-hydroxyethyl)
Mol Form: $C_{16}H_{14}O$
M_r: 222.27
mp: 121-2
UV: LANG 2022 385(6761) 364(8913) 345(5754) 330(2754) 256(144544) 218(10233) EtOH
m/e: 222.10

HODOC No: 01330
CAS No: 72948-52-8
Beilstein No: 6^3, 3786
Name: 9-Anthracenemethanol, α-phenyl-
Synonym: 9-(α-Hydroxybenzyl)anthracene
Line Form: 9-[$C_6H_5CH(OH)$]-$C_{14}H_9$
Structure:

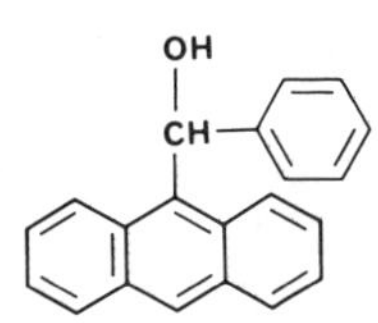

Mol Form: $C_{21}H_{16}O$
M_r: 284.34
mp: 114
UV: LANG 1356 385(7244) 364(7943) 347(5370) 332(3162) 255(109648) MeOH
m/e: 284.12

HODOC No: 01331
CAS No: 34817-14-6
Beilstein No: 11^3, 627
Name: 1-Anthracenesulfonic acid, 5-chloro-9,10-dihydro-9,10-dioxo-
Synonym: 1-Chloroanthraquinone-5-sulfonic acid
Mol Form: $C_{14}H_7ClO_5S$
M_r: 322.71
mp: 236-7
Color: ye rh pr (HCl or HOAc, +5w)
***s*:** H_2O 4
m/e: 321.97

HODOC No: 01332
CAS No: 82-49-5
Beilstein No: 11^3, 626
Name: 1-Anthracenesulfonic acid, 9,10-dihydro-9,10-dioxo-
Synonym: α-Anthraquinonesulfonic acid
Structure:

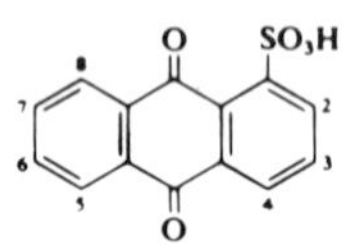

Mol Form: $C_{14}H_8O_5S$
M_r: 288.26
mp: 214
Color: lf (HOAc) ye lf (con HCl, +3w)
***s*:** H_2O 4; al 3; HOAc 4
IR: **µSADP 18664** 3130 1700 1590 1490 1320 1280 1220 1180 960 830 810 760 730 700
UV: **OES 3-427** 328(4266) 257(40738) H_2O
m/e: 288.01

HODOC No: 01333
CAS No: 128-56-3
Name: 1-Anthracenesulfonic acid, 9,10-dihydro-9,10-dioxo-, sodium salt
Synonym: 9,10-Anthraquinone-1-sulfonate sodium salt
Mol Form: $C_{14}H_7NaO_5S$
M_r: 316.24
IR: **µCOB 4618** 3450 3030 1700 1590 1320 1270 1210 1050 950 830 800 750 700 660
m/e: 309.99

HODOC No: 01334
CAS No: 82-50-8
Name: 1-Anthracenesulfonic acid, 9,10-dihydro-5-nitro-9,10-dioxo-
Synonym: 1-Anthraquinonesulfonic acid, 5-nitro-
Mol Form: $C_{14}H_7NO_7S$
M_r: 333.26
IR: **µCOB 3762** 3030 2630 2170 1700 1610 1590 1540 1330 1280 1220 1150 1040 990 880 820 780 700
m/e: 332.99

HODOC No: 01335
CAS No: 6258-06-6
Name: 2-Anthracenesulfonic acid, 1-amino-4-bromo-9,10-dihydro-9,10-dioxo-, monosodium salt
Synonym: 1-Amino-4-bromoanthraquinone-2-sodium sulfonate
Mol Form: $C_{14}H_7BrNNaO_5S$
M_r: 409.15
IR: **µCOB 4616** 3450 3030 1670 1590 1540 1390 1330 1210 1140 1050 1020 950 890 810 750 720 670
m/e: 402.91

HODOC No: 01336
CAS No: 84-48-0
Beilstein No: 11^3, 628
Name: 2-Anthracenesulfonic acid, 9,10-dihydro-9,10-dioxo-
Synonym: Anthraquinone-2-sulfonate
Structure:

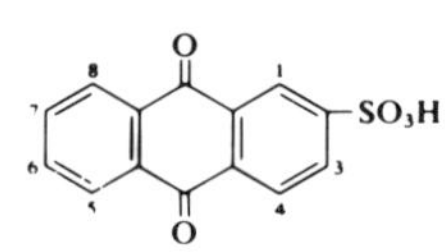

Mol Form: $C_{14}H_8O_5S$
M_r: 288.26
Color: ye lf (+3w)
***s*:** H_2O 4; al 3; eth 1
IR: **µSADP 7620** 3570 3030 1670 1590 1470 1330 1280 1210 1160 1120 1090 1040 960 930 860 710
UV: **OES 3-427** 330(5248) 256(46774) H_2O
m/e: 288.01

HODOC No: 01337
CAS No: 6483-86-9
Beilstein No: 11^3, 629
Name: 2-Anthracenesulfonic acid, 9,10-dihydro-5-nitro-9,10-dioxo-
Synonym: 2-Anthraquinonesulfonic acid, 5-nitro-
Mol Form: $C_{14}H_7NO_7S$
M_r: 333.26
mp: 255d
Color: yesh pl (dil HNO_3)
***s*:** H_2O 4
m/e: 332.99

HODOC No: 01338
CAS No: 476-60-8
Beilstein No: 8^3, 3706
Name: 1,4,9,10-Anthracenetetrol
Synonym: 1,4,9,10-Anthracenetetraol
Mol Form: $C_{14}H_{10}O_4$
M_r: 242.22
mp: 147-9
IR: ALD 2:598B
MS: ALD 4:149A
m/e: 242.06

HODOC No: 01339
CAS No: 17534-14-4
Beilstein No: 6^4, 4933
Name: 9-Anthracenethiol
Synonym: 9-Mercaptoanthracene
Line Form: 9-HS-$C_{14}H_9$
Mol Form: $C_{14}H_{10}S$
M_r: 210.28
mp: 90-3
UV: LANG 1754 380(6310) 258(91201) cyhex
m/e: 210.05

HODOC No: 01340
CAS No: 577-33-3
Beilstein No: 6^4, 7601
Name: 1,2,10-Anthracenetriol
Synonym: Anthrarobin
Line Form: 1,2,10-$(HO)_3$-$C_{14}H_7$
Mol Form: $C_{14}H_{10}O_3$
M_r: 226.22
mp: 208
Color: ye lf, nd (al-w)
***s*:** H_2O 2; al 4; eth 4; ace 4; bz 3; HOAc 4
IR: μ SADP 18451 3330 1590 1470 1320 1100 1040 880 830 770 690
UV: SAD 5899 303 274 244 207 MeOH
m/e: 226.06

HODOC No: 01341
CAS No: 115374-35-1
Beilstein No: 8^3, 2799
Name: 1,4,9-Anthracenetriol
Synonym: Anthracene, 1,4,9-trihydroxy
Line Form: 1,4,9-$(HO)_3$-$C_{14}H_7$
Mol Form: $C_{14}H_{10}O_3$
M_r: 226.22
mp: 156
Color: og-red nd (al)
***s*:** EtOH 4
m/e: 226.06

HODOC No: 01342
CAS No: 1143-38-0
Beilstein No: 6^4, 7602
Name: 1,8,9-Anthracenetriol
Synonym: 9(10H)-Anthracenone, 1,8-dihydroxy-
Line Form: 1,8,9-$(HO)_3$-$C_{14}H_7$
Mol Form: $C_{14}H_{10}O_3$
M_r: 226.22
mp: 178-80
Color: ye pl or nd (lig)
***s*:** H_2O 1; al 3; eth 2; ace 3; bz 3; chl 4; py 4; dil NaOH 3; lig 3
IR: μCOB 5152 3450 3030 1610 1590 1490 1470 1450 1390 1370 1330 1270 1220 1160 1080 1020 920 780 730 720 680
UV: SAD 7706 288(9290) 256(11400) MeOH
H NMR: SAD 6053 4.2 6.8 7.4 12.2 $CDCl_3$
m/e: 226.06

HODOC No: 01343
CAS No: 27354-06-9
Beilstein No: 8^2, 372
Name: 1,9,10-Anthracenetriol
Synonym: Anthracene, 1,9,10-trihydroxy (enol form)
Mol Form: $C_{14}H_{10}O_3$
M_r: 226.22
mp: 204-6
Color: gr nd (eth)
***s*:** Et_2O 4; EtOH 4
m/e: 226.06

HODOC No: 01344
CAS No: 610-50-4
Beilstein No: 6^4, 4928
Name: 1-Anthracenol
Synonym: 1-Anthrol
Line Form: 1-HO-$C_{14}H_9$
Mol Form: $C_{14}H_{10}O$
Mr: 194.22
bp: 234^{13}
mp: 158
Color: cr (Bz), br nd or lf (al or HOAc)
s: H_2O 1; al 4; eth 4; NaOH 3; os 2
IR: µSADP 11383 3330 3030 1610 1540 1450 1410 1350 1280 1190 1140 1050 1020 940 880 860 830 790 750 730
UV: OES 4-492 395(3981) dmf
m/e: 194.07

HODOC No: 01345
CAS No: 613-14-9
Beilstein No: 6^4, 4929
Name: 2-Anthracenol
Synonym: 2-Anthrol
Line Form: 2-HO-$C_{14}H_9$
Mol Form: $C_{14}H_{10}O$
Mr: 194.22
mp: 255
Color: ye (Bz),br lf or nd (dil al)
s: Ac 4; Bz 4; Et_2O 4; EtOH 4
m/e: 194.07

HODOC No: 01346
CAS No: 529-86-2
Beilstein No: 6^4, 4930
Name: 9-Anthracenol
Synonym: Anthranol
Line Form: 9-HO-$C_{14}H_9$
Mol Form: $C_{14}H_{10}O$
Mr: 194.22
mp: 160-4
Color: ye red lf (dil al)
m/e: 194.07

HODOC No: 01347
CAS No: 90-44-8
Beilstein No: 6^4, 4930
Name: 9(10H)-Anthracenone
Synonym: Anthracene, 9,10-dihydro-9-oxo-
Structure:

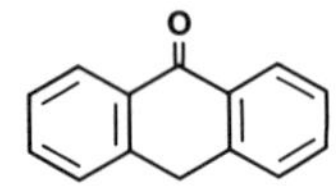

Mol Form: $C_{14}H_{10}O$
Mr: 194.22
mp: 155
Color: nd (Bz-lig, HOAc)
s: ace 3; bz 3; con sulf 3; dil alk 3
IR: SADG 475 3030 1667 1613 1471 1408 1316 1176 1087 1042 962 943 719 671
UV: SAD 1491 307(5750) 263(29800) diox
H NMR: VAR 621 4.3 8.4 $CDCl_3$
MS: NBS 4395
m/e: 194.07

HODOC No: 01348
CAS No: 1415-73-2
Name: 9(10H)-Anthracenone, 10-β-*D*-glucopyranosyl-1,8-dihydroxy-3-(hydroxymethyl)-, (R)-
Synonym: Aloin
Structure:

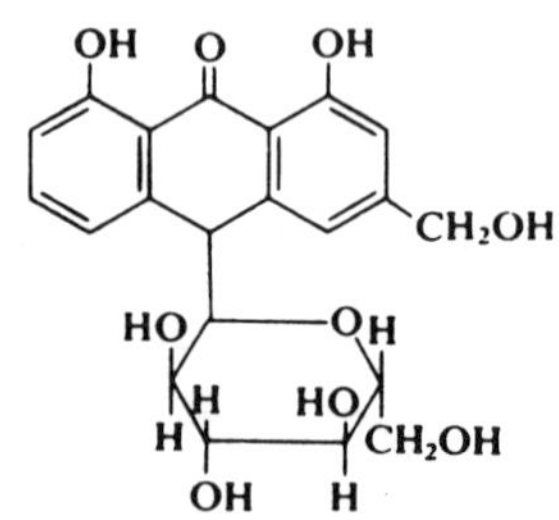

Mol Form: $C_{21}H_{22}O_9$
Mr: 418.38
mp: 148.5
α: $[\alpha]_D$-8.3 dil-al $[\alpha]_D$-10.4 AcOE
s: H_2O 3; al 3; eth 2; ace 3; bz 2; chl 1; KOH 3
IR: µSADP 3555 3230 2860 1610 1470 1390 1300 1240 1190 1020 910 860 830 780
UV: OES 6-930 354(10715) 297(9333) 269(7943)
m/e: 418.13

HODOC No: 01349
CAS No: 6313-44-6
Beilstein No: 7^3, 2371
Name: 9(10H)-Anthracenone, 10-nitro-
Synonym: Anthrone, 10-nitro-
Mol Form: $C_{14}H_9NO_3$
M_r: 239.22
mp: 140
Color: cr (Bz-lig) nd(CS_2)
***s*:** Bz 4; EtOH 4
m/e: 239.06

HODOC No: 01350
CAS No: 6437-03-2
Name: Anthranilic acid, *N*-(3-butyl-3,4-dihydro-4-oxo-2-quinazolinyl)-
Synonym: 4-Quinazolinone, 3-butyl-2(2-carboxyanilino)-3,4-dihydro
Mol Form: $C_{19}H_{19}N_3O_3$
M_r: 337.36
mp: 215-7
UV: LANG 1410 327(17378) 292(20893) 219(43652) EtOH
m/e: 337.14

HODOC No: 01351
CAS No: 100614-29-7
Name: Anthranilic acid, *N*-chloroacetyl-, butyl ester
Synonym: Benzoic acid, 2(2-chloroacetamido),butyl ester
Mol Form: $C_{13}H_{16}ClNO_3$
M_r: 269.71
mp: 46-8
IR: µSADP 13822 3230 2940 1700 1590 1520 1450 1390 1300 1270 1140 1090 760 690
UV: SAD 3726 304 252 222 MeOH
H NMR: SAD 11917 1.0 1.1-2.1 4.2 4.3 7.1 7.5 8.0 8.7 11.9 $CDCl_3$
m/e: 269.08

HODOC No: 01352
CAS No: 19589-46-9
Name: Anthranilic acid, *N*-(2-chloro-4-quinazolinyl)-
Synonym: Quinazoline, 4(2-carboxyphenylamino)-2-chloro
Structure:

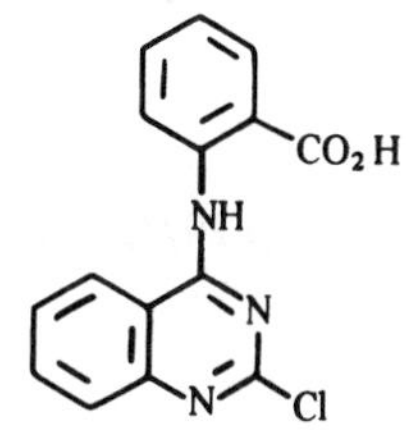

Mol Form: $C_{15}H_{10}ClN_3O_2$
M_r: 299.70
mp: 308-10
UV: LANG 2063 360(20893) 348(22909) 299(7943) 256(11220) 223(36308) EtOH
m/e: 299.05

HODOC No: 01353
CAS No: 33717-41-8
Name: Anthranilic acid, *N*-(*N*,*N*-diethyl-β-alanyl)-, ethyl ester, monohydrochloride
Synonym: Benzoic acid, 2(3(diethylamino)propamido),ethyl ester,hydrochloride
Mol Form: $C_{16}H_{25}ClN_2O_3$
M_r: 328.82
mp: 163-5
***s*:** chl 2
IR: µSADP 13838 2940 2630 1700 1610 1540 1450 1350 1280 1250 1090 760 690
UV: SAD 3740 305(4870) 252(13600) 222(26900) MeOH
H NMR: SAD 11932 1.4 3.0-3.6 4.4 7.0-7.7 8.1 8.8 11.1 $CDCl_3$
m/e: 328.16

HODOC No: 01354
CAS No: 6511-74-6
Name: Anthranilic acid, *N*-[3,4-dihydro-3-(2-hydroxyethyl)-4-oxo-2-quinazolinyl]-
Synonym: 4-Quinazolinone, 2(2-carboxyphenylamino)-3(2-hydroxyethyl)
Mol Form: $C_{17}H_{15}N_3O_4$
M$_r$: 325.31
mp: 183-4
UV: LANG 1260 327(16982) 292(20417) 219(44668) EtOH
m/e: 325.11

HODOC No: 01355
CAS No: 6437-04-3
Name: Anthranilic acid, *N*-(3,4-dihydro-3-methyl-4-oxo-2-quinazolinyl)-, methyl ester
Synonym: 4-Quinazolinone, 2(2-carboxy(methyl ester) phenylamino)-3-methyl
Mol Form: $C_{17}H_{15}N_3O_3$
M$_r$: 309.31
mp: 236-7
UV: LANG 1259 331(16596) 290(18197) 219(41687) EtOH
m/e: 309.11

HODOC No: 01356
CAS No: 19589-48-1
Name: Anthranilic acid, *N*-(2,3-dihydro-3-methyl-2-oxo-4(1H)-quinazolinylidene)-, methyl ester
Synonym: Quinazoline, 4(2-carboxy(methyl ester) phenylimino)-3,4-dihydro-2-hydroxy-3-met*
Mol Form: $C_{17}H_{15}N_3O_3$
M$_r$: 309.31
mp: 226-7
UV: LANG 2042 318(4786) 213(26303) EtOH
m/e: 309.11

HODOC No: 01357
CAS No: 33709-18-1
Name: Anthranilic acid, *N*-(*N*-isobutylglycyl)-, ethyl ester, monohydrochloride
Synonym: Glycine, amide,*N*(2-carboxy(ethyl ester)phenyl-*N*2-isobutyl,hydrochloride
Mol Form: $C_{15}H_{23}ClN_2O_3$
M$_r$: 314.80
mp: 180-3
***s*:** chl 2
IR: µSADP 13834 2940 2630 1720 1700 1610 1590 1540 1450 1370 1300 1270 1160 1090 930 760 690
UV: SAD 3737 304(23200) 251(14100) 222(27100) MeOH
H NMR: SAD 11919 1.1 1.4 2.3 3.0 4.1 4.4 7.1 7.5 8.0 8.5 11.3 $CDCl_3$
m/e: 314.14

HODOC No: 01358
CAS No: 33709-23-8
Name: Anthranilic acid, *N*-(morpholinoacetyl)-, ethyl ester, monohydrochloride
Synonym: Anthranilic acid, *N*-morpholinoacetyl-, ethyl ester, hydrochloride
Mol Form: $C_{15}H_{21}ClN_2O_4$
M$_r$: 328.78
mp: 129-33
***s*:** chl 2
IR: µSADP 13840 3130 2940 2630 1700 1610 1540 1450 1390 1300 1250 1120 1090 1010 960 870 770 700
UV: SAD 3742 304(4540) 252(13200) 222(25500) MeOH
H NMR: SAD 11934 1.4 3.7 4.1 4.2 4.4 6.9-7.6 7.9 8.3 11.3 $CDCl_3$
m/e: 328.12

HODOC No: 01359
CAS No: 24833-14-5
Name: Anthraquinone, 1-(diphenylphosphinyl)-
Synonym: Phosphine, anthraquinon-1-yl-diphenyl,oxide
Structure:

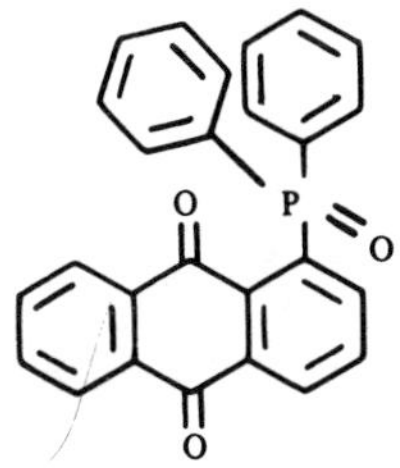

Mol Form: $C_{26}H_{17}O_3P$
M_r: 408.37
mp: 239-40
UV: LANG 1349 322(4169) 256(40738) 223(24547) EtOH
m/e: 408.09

HODOC No: 01360
CAS No: 81-77-6
Beilstein No: 24^4, 2193
Name: 5,9,14,18-Anthrazinetetrone, 6,15-dihydro-
Synonym: Anthraquinone blue
Structure:

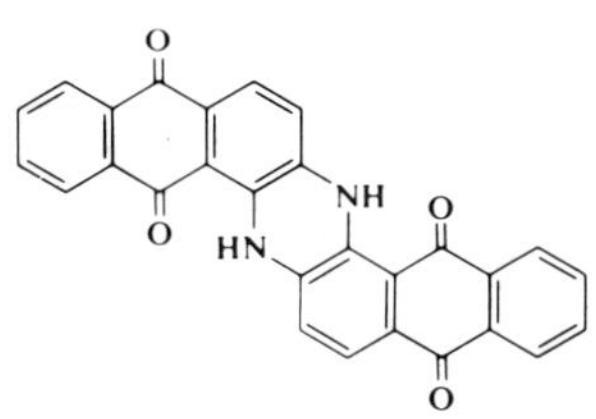

Mol Form: $C_{28}H_{14}N_2O_4$
M_r: 442.40
mp: 470-500d
Color: bl nd
***s*:** H_2O 1; al 1; eth 1; ace 1; bz 1; dil alk 3; $PhNO_2$ 3
UV: OES 3-1017 371(16218) 293(147911) 255(83176) 248(81283) 216(95499) EtOH
m/e: 442.10

HODOC No: 01361
CAS No: 650-69-1
Beilstein No: 8^3, 1599
Name: 8′-Apo-β,ψ-carotenal, 3-hydroxy-, (3R)-
Synonym: 8′-Apo-β-carotenal, 3-hydroxy-, (R)-all-*trans*-
Structure:

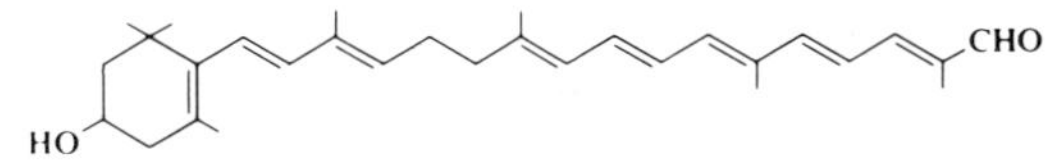

Mol Form: $C_{30}H_{40}O_2$
M_r: 432.62
mp: 147
Color: pl (Bz-peth),cr (al)
***s*:** H_2O 1; al 4; eth 4; ace 4; bz 4; CS_2 4; lig 2
UV: HBCP 497
m/e: 432.30

HODOC No: 01362
CAS No: 608-54-8
Beilstein No: 3^4, 1265
Name: *L*-Arabinaric acid
Synonym: Pentaric acid, meso-(+)-
Line Form: $HO_2C(CHOH)_3CO_2H$
Mol Form: $C_5H_8O_7$
M_r: 180.11
mp: 128
α: $[\alpha]^{20/D}$ +22.2
Color: cr (Ace), pl (w)
***s*:** H_2O 4; EtOH 4
m/e: 180.03

HODOC No: 01363
CAS No: 488-82-4
Beilstein No: 1^4, 2832
Name: *D*-Arabinitol
Synonym: Arabinitol, *D*-
Structure:

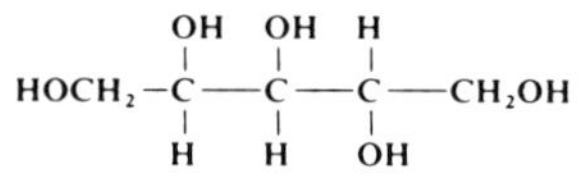

Mol Form: $C_5H_{12}O_5$
M_r: 152.14
mp: 103-4
α: $[\alpha]^{20/D}$ 11.8 borax sol, 9.5%
Color: Pr (dil al)
***s*:** H_2O 4; al 2; eth 1
IR: SADG 1563 3340 2940 1620 1430 1310 1210 1100 1080 1030 940 900 860 730 690 640
H NMR: SAD 14950 3.1-3.9 4.4 DSMOd6
m/e: 152.07

HODOC No: 01364
CAS No: 6018-27-5
Beilstein No: 1^4, 2832
Name: *DL*-Arabinitol
Synonym: Arabinitol, *DL*-
Mol Form: $C_5H_{12}O_5$
M_r: 152.14
mp: 106
Color: Pr (90% al) cr (al-HOAc)
***s*:** H_2O 4
m/e: 152.07

HODOC No: 01365
CAS No: 7643-75-6
Beilstein No: 1^4, 2832
Name: *L*-Arabinitol
Synonym: Arabinitol, *L*-
Mol Form: $C_5H_{12}O_5$
M_r: 152.14
mp: 102-3
α: $[\alpha]_D$-5.4 borax sol
***s*:** H_2O 4; al 2; eth 1
IR: SADG 18654 3340 2940 1620 1430 1310 1210 1100 1080 1020 940 900 860 730 690 640
H NMR: SAD 6502 3.4-4.1 D_2O
m/e: 152.07

HODOC No: 01366
CAS No: 13752-83-5
Beilstein No: 3^4, 1205
Name: Arabinonic acid
Synonym: Arabonic acid
Structure:

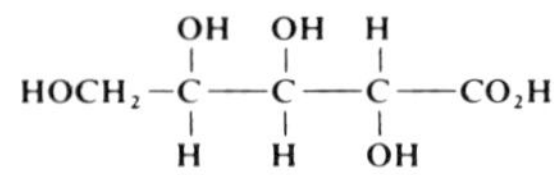

Mol Form: $C_5H_{10}O_6$
M_r: 166.13
mp: 114-6
α: $[\alpha]^{25/D}$ +10.5 (c=6)
Color: (dil HOAc)
***s*:** H_2O 4; EtOH 4
m/e: 166.05

HODOC No: 01367
CAS No: 608-53-7
Beilstein No: 3^2, 979
Name: *L*-Arabinonic acid
Synonym: Arabinonic acid, *L*-
Mol Form: $C_5H_{10}O_6$
M_r: 166.13
α: $[\alpha]^{20/D}$ -9.6 (w, c=2.5)
Color: cr (al)
m/e: 166.05

HODOC No: 01368
CAS No: 608-45-7
Beilstein No: 1^4, 4215
Name: α-*D*-Arabinopyranose
Synonym: Arabinopyranose, α-*D*-
Structure:

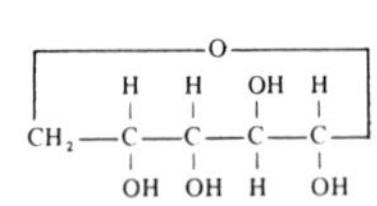

Mol Form: $C_5H_{10}O_5$
M_r: 150.13
mp: 155-6
***d*:** 1.585
Color: cr (MeOH)
IR: SADG 17087
m/e: 150.05

HODOC No: 01369
CAS No: 7296-55-1
Beilstein No: 1^4, 4218
Name: α-*L*-Arabinopyranose
Synonym: Arabinopyranose, α-*L*-
Structure:

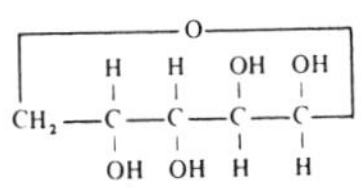

Mol Form: $C_5H_{10}O_5$
M_r: 150.13
mp: 159-60
***d*:** $1.585^{20/4}$
Color: orth cr (Py) cr (dil al)
m/e: 150.05

HODOC No: 01370
CAS No: 6748-95-4
Beilstein No: 1^4, 4215
Name: β-*D*-Arabinopyranose
Synonym: Arabinopyranose, β-*D*-
Mol Form: $C_5H_{10}O_5$
M_r: 150.13
mp: 155-6
***d*:** 1.625
α: $[\alpha]^D$ -175→_-108^{20}(mut)
Color: cr (MeOH)
***s*:** H_2O 4
IR: SADG 17087
m/e: 150.05

HODOC No: 01371
CAS No: 7296-56-2
Beilstein No: 1^4, 4218
Name: β-*L*-Arabinopyranose
Synonym: Arabinopyranose, β-*L*-
Mol Form: $C_5H_{10}O_5$
M_r: 150.13
mp: 159-60
***d*:** $1.625^{20/4}$
α: $[\alpha]^D$ 190.5 → +104.5^{20}
m/e: 150.05

HODOC No: 01372
CAS No: 5329-50-0
Name: β-*D*-Arabinopyranoside, phenylmethyl
Synonym: Arabinopyranoside, benzyl, β-*D*-
Structure:

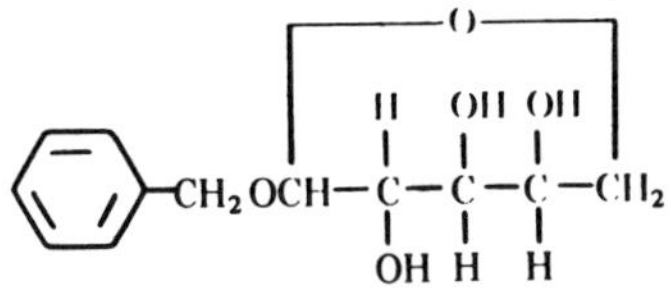

Mol Form: $C_{12}H_{16}O_5$
M_r: 240.24
mp: 169-70
UV: LANG 543 270(89) 265(148) 258(191) 253(145) 211(6457) EtOH
m/e: 240.10

HODOC No: 01373
CAS No: 31178-68-4
Beilstein No: 31, 32
Name: α-*D*-Arabinose
Synonym: Arabinose, α-*D*-
Structure:

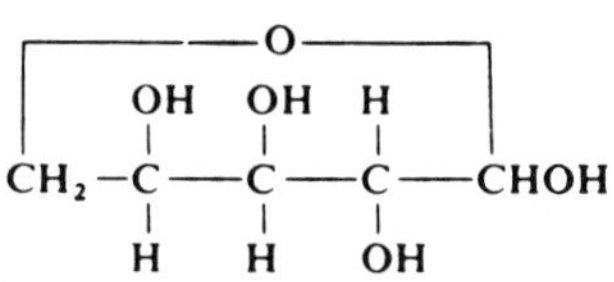

Mol Form: $C_5H_{10}O_5$
M_r: 150.13
mp: 155.5-6.5
***d*:** 1.585
α: $[\alpha]^{20/D}$-175 to -105
***s*:** H_2O 4; al 2; eth 1; ace 1; MeOH 1
IR: SADG 17087 ini 3340 3225 2950 1460 1350 1310 1250 1130 1110 1100 1080 1050 1000 890 840 780
C NMR: STOT 461 69.6 72.9 73.4 97.7
m/e: 150.05

HODOC No: 01374
CAS No: 31178-69-5
Beilstein No: 31, 32
Name: β-*D*-Arabinose
Synonym: Arabinose, β-*D*-
Mol Form: $C_5H_{10}O_5$
M_r: 150.13
mp: 155.5-6.5
***d*:** 1.625
α: $[\alpha]^{20/D}$-175 to -108
***s*:** H_2O 4; al 2; eth 1; ace 1; MeOH 1
IR: SADG 17087 ini 3340 3225 2950 1460 1350 1310 1250 1130 1110 1100 1080 1050 1000 890 840 780 680
C NMR: STOT 461 63.3 67.1 69.7 93.5
m/e: 150.05

HODOC No: 01375
CAS No: 20235-19-2
Beilstein No: 1^4, 4223
Name: *DL*-Arabinose
Synonym: Arabinose, *DL*-
Mol Form: $C_5H_{10}O_5$
M_r: 150.13
mp: 164-5
***d*:** $1.585^{20/4}$
Color: pr, nd (al)
***s*:** H_2O 4; al 2; eth 1; bz 1
IR: SADG 1562 3340 3225 2950 1460 1350 1310 1250 1130 1110 1100 1080 1050 1000 890 840 780 680
m/e: 150.05

HODOC No: 01376
CAS No: 5149-15-5
Beilstein No: 3^1, 44
Name: Arabinose, diphenylhydrazone, *L*-
Synonym: Arabinose, diphenylhydrazone
Mol Form: $C_{17}H_{20}N_2O_4$
M_r: 316.34
mp: 204.5
α: $[\alpha]^{20/D}$ +18.5 py
Color: orth pr, nd (dil al)
m/e: 316.14

HODOC No: 01377
CAS No: 94916-06-0
Name: Arabinose, (*p*-nitrophenyl)hydrazone, 2,3,4,5-tetraacetate
Synonym: Arabinose, 4-nitrophenylhydrazone, tetraacetate
Mol Form: $C_{19}H_{23}N_3O_{10}$
M_r: 453.39
mp: 133-4
IR: SADG 7035 3340 1750 1615 1600 1535 1510 1480 1370 1315 1275 1225 1140 1120 1055 975 845 750
UV: LANG 390 373(19055) 245(8710) EtOH
H NMR: SAD 2808 2.1 4.3 5.0-5.9 7.0 8.2 8.6 $CDCl_3$
m/e: 453.14

HODOC No: 01378
CAS No: 7200-25-1
Beilstein No: 4^3, 1359
Name: *DL*-Arginine
Synonym: Arginine, *DL*-
Line Form: $H_2NC(=NH)NH(CH_2)_3CH(NH_2)CO_2H$
Mol Form: $C_6H_{14}N_4O_2$
M_r: 174.20
mp: 217-8
***s*:** H:i2O 1; al 1; eth 1; bz 1
IR: SADG 1565 3340 3120 2950 1680 1625 1450 1470 1410 1330 1190 1140 985 660
UV: OES 6-93 208(331) pH=8
C NMR: STOT 479 24.6 28.2 41.2 54.8 157.2 174.9 H_2O
m/e: 174.11

HODOC No: 01379
CAS No: 74-79-3
Beilstein No: 4^4, 2648
Name: *L*-Arginine
Synonym: Arginine, *L*-
Line Form: $H_2NC(=NH)NH(CH_2)_3CHNH_2CO_2H$
Mol Form: $C_6H_{14}N_4O_2$
M_r: 174.20
mp: 244d
α: $[\alpha]^{20/D}$ 12.5 H_2O,3.5%
***s*:** H_2O 3; al 2; eth 1
UV: HBCP 205(1905) H_2O
m/e: 174.11

HODOC No: 01380
CAS No: 97790-61-9
Beilstein No: 11^3, 547
Name: Arginine, 8-hydroxy-5,7-dinitro-2-naphthalenesulfonate
Synonym: Arginine, flavinate
Line Form: $C_6H_{14}N_4O_2{\cdot}C_{10}H_6N_2O_8S$
Mol Form: $C_{16}H_{20}N_6O_{10}S$
M$_r$: 488.41
mp: 258-60d
Color: ye-og lf
IR: SADG 18844
UV: SAD 16015
m/e: 488.10

HODOC No: 01381
CAS No: 116295-81-9
Beilstein No: 6^4, 1450
Name: Arginine, picrate (*d*)
Synonym: Benzenesulfonic acid, 3,4-dichloro-, compd. with *D*-histidine (2:1)
Line Form: $C_6H_{14}N_4O_2{\cdot}C_6H_3O_7{\cdot}2H_2O$
Mol Form: $C_{12}H_{19}F_2N_7O_{10}$
M$_r$: 459.31
mp: 217-8d
Color: nd (w)
m/e: 459.12

HODOC No: 01382
CAS No: 116295-93-3
Beilstein No: 6^4, 1450
Name: *DL*-Arginine, compd. with 2,4,6-trinitrophenol (1:1)
Synonym: Arginine, picrate (*dl*)
Line Form: $C_6H_{14}N_4O_2{\cdot}C_6H_3N_3O_7$
Mol Form: $C_{12}H_{17}N_7O_9$
M$_r$: 403.30
m/e: 403.11

HODOC No: 01383
CAS No: 116295-79-5
Beilstein No: 6^4, 1450
Name: *DL*-Arginine, compd. with 2,4,6-trinitrophenol (1:2)
Synonym: Arginine, dipicrate (*dl*)
Line Form: $C_6H_{14}N_4O_2(C_6H_3N_3O_7)_2$
Mol Form: $C_{18}H_{20}N_{10}O_{16}$
M$_r$: 632.40
mp: 196d
m/e: 632.11

HODOC No: 01384
CAS No: 154-92-7
Beilstein No: 9^4, 871
Name: *L*-Arginine, *N*2-benzoyl-
Synonym: Arginine, *N*2-benzoyl-, *L*-
Line Form: $H_2NC(=NH)NH(CH_2)_3CH(NHCOC_6H_5)CO_2H$
Mol Form: $C_{13}H_{18}N_4O_3$
M$_r$: 278.30
mp: 285d
IR: ALD 2:876F
MS: ALD 7:19B
m/e: 278.14

HODOC No: 01385
CAS No: 2645-08-1
Name: *L*-Arginine, *N*2-benzoyl-, ethyl ester, monohydrochloride
Synonym: Arginine, *N*2-benzoyl-, ethyl ester, monohydrochloride, *L*-
Line Form: $H_2NC(=NH)NH(CH_2)_3CH(CO_2CH_2CH_3)NHCOC_6H_5HCl$
Mol Form: $C_{15}H_{23}ClN_4O_3$
M$_r$: 342.81
mp: 129
IR: COB 7978 3100 1730 1640 1580 1530 1490 1440 1340 1310 1220 1160 1020 710 620 510
m/e: 342.15

HODOC No: 01386
CAS No: 116295-80-8
Beilstein No: 6^4, 1450
Name: *L*-Arginine, compd. with 2,4,6-trinitrophenol (1:2)
Synonym: Arginine, dipicrate (*l*)
Line Form: $C_6H_{14}N_4O_2(C_6H_3N_3O_7)_2$
Mol Form: $C_{18}H_{20}N_{10}O_{16}$
M$_r$: 632.40
mp: 200d
m/e: 632.11

HODOC No: 01387
CAS No: 73791-38-5
Name: *L*-Arginine, *N*2-[(4-methylphenyl)sulfonyl]-, trihydrate
Synonym: Arginine, *N*2(4-tolysulfonyl),trihydrate(*l*)
Mol Form: $C_{13}H_{26}N_4O_7S$
Mr: 382.42
mp: dec 252
IR: COB 7529 3350 3150 2870 1660 1600 1560 1450 1400 1330 1160 1090 820 670 570
m/e: 382.15

HODOC No: 01388
CAS No: 82364-21-4
Beilstein No: 4^2, 847
Name: *L*-Arginine, mono(8-hydroxy-5,7-dinitro-2-naphthalenesulfonate)
Synonym: 2-Naphthalenesulfonic acid, 8-hydroxy-5,7-dinitro-, compd. with *L*-arginine (1:1)
Mol Form: $C_{16}H_{20}N_6O_{10}S$
Mr: 488.41
mp: dec 258-60
***s*:** H_2O 1; al 1; eth 1
IR: SADG 18844 3440 3350 3200 1750 1670 1620 1580 1510 1330 1300 1230 1200 1160 1020 940 830 660
UV: SAD 16015 342(32700) 260(19000) 211(27000) MeOH
m/e: 488.10

HODOC No: 01389
CAS No: 1234-35-1
Beilstein No: 6^4, 2442
Name: *L*-Arginine, *N*2-[(phenylmethoxy)carbonyl]-
Synonym: Arginine, *N*2-carboxy-, *N*2-benzyl ester, *L*-
Line Form: $HN{=}C(NH_2)\text{-}NH(CH_2)_3CH(CO_2H)\text{-}NHCO_2CH_2C_6H_5$
Mol Form: $C_{14}H_{20}N_4O_4$
Mr: 308.32
mp: 173-5
α: $[\alpha]^{23/D}$ -9.3 (0.2N HCl, c=5.3)
IR: ALD 2:878F
MS: NBS 4852; ALD 7:20B
m/e: 308.15

HODOC No: 01390
CAS No: 15606-95-8
Beilstein No: 1^4, 1358
Name: Arsenic acid (H3AsO4), triethyl ester
Synonym: Ethyl arsenate ((EtO)3AsO)
Line Form: $(C_2H_5O)_3AsO$
Mol Form: $C_6H_{15}AsO_4$
Mr: 226.10
bp: 235-8, $118\text{-}20^{15}$
***d*:** $1.3023^{20/0}$
***n*:** 1.4343^{20}
MS: NBS 4541
m/e: 226.02

HODOC No: 01391
CAS No: 3141-12-6
Beilstein No: 1^4, 1358
Name: Arsenous acid, triethyl ester
Synonym: Arsenious acid (H3AsO3), triethyl ester
Line Form: $(C_2H_5O)_3As$
Mol Form: $C_6H_{15}AsO_3$
Mr: 210.10
bp: 165-6, $66\text{-}7^{12}$
***d*:** $1.2239^{20/4}$
***n*:** 1.4369^{13}
MS: NBS 1306
m/e: 210.02

HODOC No: 01392
CAS No: 692-42-2
Beilstein No: 4^3, 1797
Name: Arsine, diethyl-
Line Form: $(C_2H_5)_2AsH$
Mol Form: $C_4H_{11}As$
Mr: 134.05
bp: 105
***d*:** $1.1338^{24/4}$
***n*:** 1.4709
***s*:** Ac 4; Bz 4; Et_2O 4; EtOH 4
m/e: 134.01

HODOC No: 01393
CAS No: 368-97-8
Beilstein No: 16^3, 958
Name: Arsine, difluorophenyl-
Line Form: $C_6H_5AsF_2$
Mol Form: $C_6H_5AsF_2$
Mr: 190.02
bp: 110^{48}
mp: 42
Color: wax
m/e: 189.96

HODOC No: 01394
CAS No: 593-57-7
Beilstein No: 4^4, 3665
Name: Arsine, dimethyl-
Synonym: Dimethylarsine
Line Form: $(CH_3)_2AsH$
Mol Form: C_2H_7As
M_r: 106.00
bp: 36
***d*:** $1.213^{29/29}$
Color: ign in air
***s*:** Ac 4; Bz 4; Et_2O 4; EtOH 4
m/e: 105.98

HODOC No: 01395
CAS No: 829-83-4
Beilstein No: 16^3, 918
Name: Arsine, diphenyl-
Synonym: Diphenylarsine
Line Form: $(C_6H_5)_2AsH$
Mol Form: $C_{12}H_{11}As$
M_r: 230.13
bp: 174^{25}
***d*:** $1.30^{25/25}$
Color: oil
***s*:** Et_2O 4; EtOH 4
m/e: 230.01

HODOC No: 01396
CAS No: 593-59-9
Beilstein No: 4^3, 1797
Name: Arsine, ethyl-
Synonym: Ethylarsine
Line Form: $C_2H_5AsH_2$
Mol Form: C_2H_7As
M_r: 106.00
bp: 36
***d*:** $1.217^{22/22}$
***s*:** Et_2O 4; EtOH 4
m/e: 105.98

HODOC No: 01397
CAS No: 430-40-0
Beilstein No: 4^3, 1799
Name: Arsine, ethyldifluoro-
Synonym: Arsine, difluoro-ethyl
Line Form: $C_2H_5AsF_2$
Mol Form: $C_2H_5AsF_2$
M_r: 141.98
bp: 94.3, 74^{100}
mp: -38.7
***d*:** 1.708^{17}
Color: fume (air)
m/e: 141.96

HODOC No: 01398
CAS No: 5410-89-9
Beilstein No: 16^2, 448
Name: Arsine, (4-methoxyphenyl)oxo-
Synonym: Arsine, 4-methoxyphenyl, oxide
Line Form: $(4-CH_3OC_6H_4)AsO$
Structure:

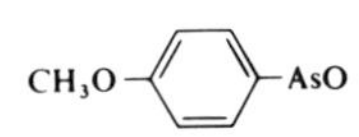

Mol Form: $C_7H_7AsO_2$
M_r: 198.05
mp: 114-6(anh)
Color: cr (Chl-eth) or (Bz-peth)
***s*:** Bz 4; Chl 4
m/e: 197.97

HODOC No: 01399
CAS No: 593-52-2
Beilstein No: 4^3, 1795
Name: Arsine, methyl-
Synonym: Methylarsine
Line Form: CH_3AsH_2
Mol Form: CH_5As
M_r: 91.97
bp: 2
mp: -143
***s*:** Ac 4; Et_2O 4; EtOH 4
m/e: 91.96

HODOC No: 01400
CAS No: 431-76-5
Name: Arsine, methylbis(trifluoromethyl)-
Synonym: Arsine, *bis*(trifluoromethyl)methyl
Line Form: $(CF_3)_2AsCH_3$
Mol Form: $C_3H_3AsF_6$
M_r: 227.97
bp: 52^{760}
IR: **µDMS 56** 2940 2860 2220 1890 1430 1370 1270 1180 1160 1110 1050 860 730
m/e: 227.94

HODOC No: 01401
CAS No: 593-58-8
Beilstein No: 4^3, 1817
Name: Arsine, methyloxo-
Synonym: Arsenosomethane
Line Form: CH_3AsO
Mol Form: CH_3AsO
M_r: 105.95
bp: 275d
mp: 95
Color: pr (CS_2)(al)
***s*:** Bz 4; EtOH 4; Chl 4
m/e: 105.94

HODOC No: 01402
CAS No: 637-03-6
Beilstein No: 16^3, 956
Name: Arsine, oxophenyl-
Synonym: Arsenosobenzene
Line Form: C_6H_5AsO
Mol Form: C_6H_5AsO
M_r: 168.02
mp: 144-6
Color: cr (Bz-eth) or (Chl-eth)
***s*:** H_2O 1; al 2; eth 1; bz 4; chl 4
UV: **OES 2-51** 234(5248) 216(9333) EtOH
m/e: 167.96

HODOC No: 01403
CAS No: 15925-05-0
Name: Arsine, *p*-phenylenebis[diphenyl-
Synonym: Arsine, 1,4-phenylenebis-diphenyl
Structure:

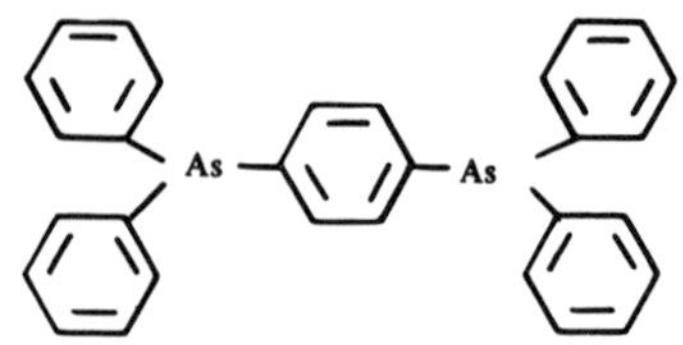

Mol Form: $C_{30}H_{24}As_2$
M_r: 534.33
mp: 145-7
UV: **LANG 1151** 253(21878) 205(89125) EtOH
m/e: 534.03

HODOC No: 01404
CAS No: 617-75-4
Beilstein No: 4^4, 3666
Name: Arsine, triethyl-
Synonym: Triethylarsenic
Line Form: $(C_2H_5)_3As$
Mol Form: $C_6H_{15}As$
M_r: 162.10
bp: 138-9
***d*:** $1.150^{20/4}$
***n*:** 1.467^{20}
***s*:** Ac 4; Et_2O 4; EtOH 4
m/e: 162.04

HODOC No: 01405
CAS No: 420-42-8
Beilstein No: 3^4, 266
Name: Arsine, (trifluoromethyl)-
Synonym: (Trifluoromethyl)arsine
Line Form: CF_3AsH_2
Mol Form: CH_2AsF_3
M_r: 145.94
bp: -11.6^{781}
MS: **ALDN 97** 146(100) 76(60) 51(59) 69(57) 75(37) 94(36) 144(31) 33(23)
m/e: 145.93

HODOC No: 01406
CAS No: 593-88-4
Beilstein No: 4^4, 3665
Name: Arsine, trimethyl-
Synonym: Trimethylarsenic
Line Form: $(CH_3)_3As$
Mol Form: C_3H_9As
Mr: 120.02
bp: 52
mp: -87.3
***d*:** 1.144^{15}
***s*:** Bz 4; Et_2O 4; EtOH 4
MS: NBS 188
m/e: 119.99

HODOC No: 01407
CAS No: 603-32-7
Beilstein No: 16^3, 921
Name: Arsine, triphenyl-
Synonym: Triphenylarsane
Line Form: $(C_6H_5)_3As$
Mol Form: $C_{18}H_{15}As$
Mr: 306.22
mp: 61
***d*:** $1.2634^{18/4}$
***n*:** 1.6888^{21}
***s*:** H_2O 1; al 2; eth 4; bz 4; chl 3
IR: µSADP 23128 3030 1590 1470 1430 1300 1080 1020 1000 740 690
UV: SAD 22010 248(13600) MeOH
H NMR: SAD 8930 7.1-7.5 $CDCl_3$
MS: ALDN 229 152(100) 306(72) 227(47) 151(22) 154(18) 51(16) 307(15) 229(13)
m/e: 306.04

HODOC No: 01408
CAS No: 359-69-3
Name: Arsine, *tris*(pentafluoroethyl)-
Synonym: Tris(pentafluoroethyl)arsine
Line Form: $(C_2F_5)_3As$
Mol Form: C_6AsF_{15}
Mr: 431.96
bp: 96
m/e: 431.90

HODOC No: 01409
CAS No: 432-02-0
Beilstein No: 3^4, 266
Name: Arsine, *tris*(trifluoromethyl)-
Synonym: Tris(trifluoromethyl)arsine
Line Form: $(CF_3)_3As$
Mol Form: C_3AsF_9
Mr: 281.94
bp: 33.3
***s*:** H_2O 1; eth 3
IR: µDMS 52 2220 2170 1320 1270 1220 1180 1150 1110 740
MS: NBS 2060
m/e: 281.91

HODOC No: 01410
CAS No: 1153-05-5
Beilstein No: 16^3, 1021
Name: Arsine oxide, triphenyl-
Synonym: Triphenylarsenic oxide
Line Form: $(C_6H_5)_3AsO$
Mol Form: $C_{18}H_{15}AsO$
Mr: 322.22
mp: 192
IR: SADG 31361 3040 1470 1440 1340 1310 1180 1150 1070 1020 980 930 870 730 680 670 470 340
UV: LANG 792 270(1445) 263(1698) 258(1413) 253(1122) 220(23442) EtOH
H NMR: SAD 8697 7.5 7.8 $CDCl_3$
m/e: 322.03

HODOC No: 01411
CAS No: 3937-40-4
Beilstein No: 16^3, 1026
Name: Arsine sulfide, triphenyl-
Synonym: Triphenylarsine sulfide
Line Form: $(C_6H_5)_3AsS$
Mol Form: $C_{18}H_{15}AsS$
Mr: 338.28
mp: 162
***s*:** H_2O 1; al 2; eth 1
IR: µSADP 23130 3030 1490 1450 1320 1190 1160 1090 1060 1020 1000 750 740 690
UV: LANG 793 <200 EtOH
H NMR: SAD 14100 7.5-7.9 DMSO
m/e: 338.01

HODOC No: 01412
CAS No: 75-60-5
Beilstein No: 4^2, 993
Name: Arsinic acid, dimethyl-
Synonym: Arsine oxide, hydroxydimethyl-
Line Form: $(CH_3)_2AsO_2H$
Mol Form: $C_2H_7AsO_2$
M$_r$: 138.00
mp: 200
***s*:** H_2O 4; al 3; eth 1
IR: SADG 18107 3010 2740 2380 1730 1400 1300 1250 910 880 860 750 650 600 350 320 280
H NMR: SAD 6487 2.0 D_2O
m/e: 137.97

HODOC No: 01413
CAS No: 4656-80-8
Beilstein No: 16^2, 443
Name: Arsinic acid, diphenyl-
Synonym: Arsine oxide, hydroxydiphenyl-
Line Form: $(C_6H_5)_2AsO_2H$
Mol Form: $C_{12}H_{11}AsO_2$
M$_r$: 262.13
mp: 178
Color: nd, pr
***s*:** H_2O 2; al 3; eth 2; bz 2; chl 3
UV: OES 2-302 263(1380) 220(18197) 0.1N HCl
m/e: 262.00

HODOC No: 01414
CAS No: 88373-40-4
Name: Arsinic acid, (4-methylphenyl)-
Synonym: 4-Methylphenylarsinic acid
Structure:

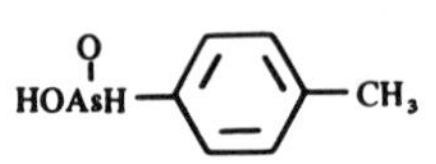

Mol Form: $C_7H_9AsO_2$
M$_r$: 200.06
mp: 300
IR: µSADP 8487 2780 2270 1610 1390 1190 1100 870 810 790 770
UV: SAD 18378 271(470) 265(578) 260(690) 254(618) 223(13000) MeOH
H NMR: SAD 11277 2.4 7.5 7.7 7.8 DMSOd6
m/e: 199.98

HODOC No: 01415
CAS No: 503-80-0
Beilstein No: 4^3, 1814
Name: Arsinous acid, dimethyl-, anhydride
Synonym: Arsine, oxybis[dimethyl-
Line Form: $(CH_3)_2As\text{-}O\text{-}As(CH_3)_2$
Mol Form: $C_4H_{12}As_2O$
M$_r$: 225.98
bp: 150
mp: -25
***d*:** 1.4816^{15}
***n*:** 1.5225^{9}
***s*:** Et_2O 4; EtOH 4
m/e: 225.93

HODOC No: 01416
CAS No: 557-89-1
Beilstein No: 4^4, 3670
Name: Arsinous chloride, dimethyl-
Synonym: Arsine, chlorodimethyl-
Line Form: $(CH_3)_2AsCl$
Mol Form: C_2H_6AsCl
M$_r$: 140.44
bp: 109
mp: 45
***d*:** $1.5046^{12/4}$
***n*:** 1.5203^{12}
***s*:** EtOH 4
m/e: 139.94

HODOC No: 01417
CAS No: 712-48-1
Beilstein No: 16^3, 944
Name: Arsinous chloride, diphenyl-
Synonym: Arsine, chlorodiphenyl-
Line Form: $(C_6H_5)_2AsCl$
Mol Form: $C_{12}H_{10}AsCl$
M$_r$: 264.57
bp: 337, 193^{20}
mp: 44
***d*:** $1.4820^{16/4}$
***n*:** 1.6332^{56}
Color: rh pl (peth)
***s*:** Ac 4; Bz 4; Et_2O 4; EtOH 4
m/e: 263.97

HODOC No: 01418
CAS No: 359-54-6
Name: Arsinous fluoride, *bis*(trifluoromethyl)-
Synonym: Arsine, fluorobis(trifluoromethyl)-
Line Form: $(CF_3)_2AsF$
Mol Form: C_2AsF_7
Mr: 231.93
bp: 25^{760}
IR: **µDMS 53** 2220 1370 1250 1220 1160 1120 1030 830 800 740 690
m/e: 231.91

HODOC No: 01419
CAS No: 359-76-2
Name: Arsinous iodide, *bis*(pentafluoroethyl)-
Synonym: Arsine, iodobis(pentafluoroethyl)-
Line Form: $(CF_3CF_2)_2AsI$
Mol Form: $C_4AsF_{10}I$
Mr: 439.85
bp: 120^{760}
IR: **µDMS 54** 2130 1190 1180 1110 740
m/e: 439.81

HODOC No: 01420
CAS No: 359-55-7
Beilstein No: 3^4, 267
Name: Arsinous iodide, *bis*(trifluoromethyl)-
Synonym: Arsine, iodobis(trifluoromethyl)-
Line Form: $(CF_3)_2AsI$
Mol Form: C_2AsF_6I
Mr: 339.84
bp: 92
***n*:** 1.425^{23}
Color: ye oil
***s*:** H_2O 1; eth 3
UV: **HBCP** 287(1205) eth
m/e: 339.82

HODOC No: 01421
CAS No: 121-59-5
Beilstein No: 16^2, 497
Name: Arsonic acid, [4-[(aminocarbonyl)amino]phenyl]-
Synonym: Arsanilic acid, *N*-carbamoyl-
Line Form: $4\text{-}(H_2NCONH)C_6H_4AsO(OH)_2$
Mol Form: $C_7H_9AsN_2O_4$
Mr: 260.07
mp: 174
Color: nd (w)
***s*:** H_2O 2; al 2; eth 1; chl 1 $DMSOd_6$ 2; alk 3
IR: **SADG 10057** 3448 3226 2778 2381 1667 1587 1493 1408 1351 1220 1163 1087 901 806 769
UV: **SAD 5482** 257 MeOH
H NMR: **SAD 4441** 6.2 7.7 8.1 9.1 $DMSOd_6$
m/e: 259.98

HODOC No: 01422
CAS No: 2163-77-1
Beilstein No: 16^2, 521
Name: Arsonic acid, (3-amino-4-hydroxyphenyl)-
Synonym: *m*-Arsanilic acid, 4-hydroxy-
Mol Form: $C_6H_8AsNO_4$
Mr: 233.05
mp: dec 290
***s*:** H_2O 1; al 2; os 2
IR: **µSADP 18414** 2940 2630 1640 1590 1490 1430 1370 1320 1250 1140 1080 910 860 830 760
UV: **SAD 7600** 299(4160) 254(5970) 219(32800) MeOH
m/e: 232.97

HODOC No: 01423
CAS No: 2045-00-3
Beilstein No: 16^3, 1099
Name: Arsonic acid, (2-aminophenyl)-
Synonym: *o*-Aminobenzenearsonic acid
Line Form: $2\text{-}H_2NC_6H_4AsO(OH)_2$
Mol Form: $C_6H_8AsNO_3$
Mr: 217.05
mp: 153-4
Color: nd (al-eth)
***s*:** H_2O 4; al 4; eth 2; alk 3; HOAc 3
IR: **µCOB 1465** 3330 2630 2270 1640 1610 1560 1490 1320 1160 1090 860 790 760
UV: **SAD 1622** 315 246 MeOH
m/e: 216.97

HODOC No: 01424
CAS No: 2038-72-4
Beilstein No: 16^3, 1104
Name: Arsonic acid, (3-aminophenyl)-
Synonym: (*m*-Aminophenyl)arsonic acid
Line Form: 3-$H_2NC_6H_4AsO(OH)_2$
Mol Form: $C_6H_8AsNO_3$
M_r: 217.05
mp: 215
Color: pr (w)
***s*:** H_2O 4
m/e: 216.97

HODOC No: 01425
CAS No: 98-50-0
Beilstein No: 16^2, 491
Name: Arsonic acid, (4-aminophenyl)-
Synonym: *p*-Aminobenzenearsonic acid
Line Form: 4-$H_2NC_6H_4AsO(OH)_2$
Mol Form: $C_6H_8AsNO_3$
M_r: 217.05
mp: 232
***d*:** 1.9571^{10}
Color: mcl nd (w, al)
***s*:** H_2O 3; al 2; eth 3; ace 1; bz 1; chl 1; DMSOd_6 2; HOAc 2
IR: COB 6399 2560 2400 2160 1620 1590 1500 1410 1230 1180 1100 850 750 640
UV: SAD 1505 268(21700) MeOH
H NMR: SAD 10551 6.5 6.7 7.4 DMSO
m/e: 216.97

HODOC No: 01426
CAS No: 590-72-7
Beilstein No: 4^4, 3682
Name: Arsonic acid, butyl-
Synonym: 1-Butanearsonic acid
Line Form: $CH_3CH_2CH_2CH_2$-$AsO(OH)_2$
Mol Form: $C_4H_{11}AsO_3$
M_r: 182.05
mp: 160
***s*:** H_2O 4; al 3; eth 1
IR: µCOB 5945 3130 2860 2630 2330 2220 1450 1370 1220 1190 940 870 770 710
H NMR: SAD 6486 1.0 1.1-2.1 2.6 D_2O
m/e: 181.99

HODOC No: 01427
CAS No: 33781-24-7
Beilstein No: 16^2, 457
Name: Arsonic acid, (2-chlorophenyl)-
Synonym: Benzenearsonic acid, *o*-chloro-
Line Form: 2-$ClC_6H_4AsO(OH)_2$
Mol Form: $C_6H_6AsClO_3$
M_r: 236.48
mp: 186-7
Color: nd (w or dil al)
***s*:** H_2O 4; EtOH 4
m/e: 235.92

HODOC No: 01428
CAS No: 5410-26-4
Beilstein No: 16^2, 457
Name: Arsonic acid, (3-chlorophenyl)-
Synonym: Benzenearsonic acid, *m*-chloro-
Line Form: 3-$ClC_6H_4AsO(OH)_2$
Mol Form: $C_6H_6AsClO_3$
M_r: 236.48
mp: 175
Color: cr (w)
***s*:** H_2O 3; al 3; MeOH 3
IR: µSADP 15166 2630 2270 1560 1470 1390 1210 1120 990 890 780 760 680
UV: SAD 4477 280 272 266 MeOH
m/e: 235.92

HODOC No: 01429
CAS No: 5440-04-0
Beilstein No: 16^3, 1058
Name: Arsonic acid, (4-chlorophenyl)-
Synonym: Benzenearsonic acid, *p*-chloro-
Line Form: 4-$ClC_6H_4AsO(OH)_2$
Mol Form: $C_6H_6AsClO_3$
M_r: 236.48
mp: 283-5
Color: wh nd (al)
***s*:** al 3; eth 3; alk 4
IR: µSADP 19961 2860 2380 1640 1560 1470 1390 1250 1090 1020 910 820 760
UV: SAD 7842 273(501) 263(721) 256(680) 228(17700) MeOH
m/e: 235.92

HODOC No: 01430
CAS No: 622-68-4
Name: Arsonic acid, [4-[[4-(dimethylamino)phenyl]azo]phenyl]-
Synonym: Benzenearsonic acid, *p*-[[*p*-(dimethylamino)phenyl]azo]-
Mol Form: $C_{14}H_{16}AsN_3O_3$
M$_r$: 349.21
***s*:** DMSO 2
IR: **µCOB 249** 3030 2940 1640 1610 1540 1470 1390 1250 1160 1090 930 890 820 760 750 730
Raman: **SAD 61** 2930 2900 2880 2850 2730 1740 1440 1370 1300 1160 1120 1080 1060 1010 920 880 850 640 610 430 350 300 220 160
H NMR: **SAD 12796** 3.1 6.2 6.9 7.9 8.0 DMSOd$_6$
m/e: 349.04

HODOC No: 01431
CAS No: 507-32-4
Beilstein No: 4^3, 1823
Name: Arsonic acid, ethyl-
Synonym: Ethanearsonic acid
Line Form: $CH_3CH_2AsO(OH)_2$
Mol Form: $C_2H_7AsO_3$
M$_r$: 154.00
bp: $209\text{-}11^{12}$
mp: 99.5
Color: nd (al), rh nd (w)
***s*:** H_2O 4; EtOH 4
m/e: 153.96

HODOC No: 01432
CAS No: 6299-08-7
Beilstein No: 16^2, 464
Name: Arsonic acid, (2-hydroxyphenyl)-
Synonym: Benzenearsonic acid, *o*-hydroxy-
Line Form: $2\text{-}HOC_6H_4AsO(OH)_2$
Mol Form: $C_6H_7AsO_4$
M$_r$: 218.03
mp: 190-1
Color: nd (w)
***s*:** H_2O 4; Bz 4; EtOH 4
IR: **SADP 18164**
UV: **SAD 5760**
m/e: 217.96

HODOC No: 01433
CAS No: 98-14-6
Beilstein No: 16^3, 1076
Name: Arsonic acid, (4-hydroxyphenyl)-
Synonym: Benzenearsonic acid, *p*-hydroxy-
Line Form: $4\text{-}HOC_6H_4AsO(OH)_2$
Mol Form: $C_6H_7AsO_4$
M$_r$: 218.03
mp: 170-4
Color: nd (Ace)
***s*:** H_2O 4; al 4; eth 2; HOAc 2
IR: **µSADP 478** 3130 1590 1490 1450 1390 1280 1210 1160 1100 890 830 770
UV: **SAD 169** 279 238 MeOH
H NMR: **SAD 6643** 4.9 7.1 7.7 D_2O
m/e: 217.96

HODOC No: 01434
CAS No: 3969-57-1
Beilstein No: 16^3, 1060
Name: Arsonic acid, (2-methylphenyl)-
Synonym: *o*-Toluenearsonic acid
Line Form: $(2\text{-}CH_3C_6H_4)AsO(OH)_2$
Structure:

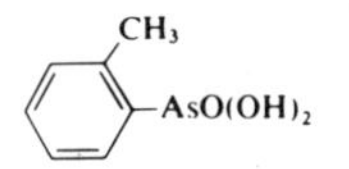

Mol Form: $C_7H_9AsO_3$
M$_r$: 216.06
mp: 163-4
Color: nd (w)
***s*:** EtOH 4
m/e: 215.98

HODOC No: 01435
CAS No: 5410-37-7
Beilstein No: 16^3, 1061
Name: Arsonic acid, (3-methylphenyl)-
Synonym: *m*-Toluenearsonic acid
Line Form: $3\text{-}CH_3C_6H_4AsO(OH)_2$
Mol Form: $C_7H_9AsO_3$
M$_r$: 216.06
mp: 150
Color: nd (w)
***s*:** EtOH 4
m/e: 215.98

HODOC No: 01436
CAS No: 3969-54-8
Beilstein No: 16^3, 1061
Name: Arsonic acid, (4-methylphenyl)-
Synonym: *p*-Toluenearsonic acid
Line Form: $4\text{-}CH_3C_6H_4AsO(OH)_2$
Mol Form: $C_7H_9AsO_3$
M_r: 216.06
bp: 105-10d, (-w)
Color: nd (w)
***s*:** EtOH 4
m/e: 215.98

HODOC No: 01437
CAS No: 5410-29-7
Beilstein No: 16^3, 1059
Name: Arsonic acid, (2-nitrophenyl)-
Synonym: Benzenearsonic acid, *o*-nitro-
Line Form: $2\text{-}NO_2C_6H_4AsO(OH)_2$
Mol Form: $C_6H_6AsNO_5$
M_r: 247.03
mp: 233d
***s*:** H_2O 2; al 2; ace 2; HOAc 3
IR: µSADP 20253 2940 1610 1540 1450 1350 1180 1120 910 890 860 790 740 710
UV: SAD 6784 261(5500) MeOH
m/e: 246.95

HODOC No: 01438
CAS No: 618-07-5
Beilstein No: 16^3, 1059
Name: Arsonic acid, (3-nitrophenyl)-
Synonym: Benzenearsonic acid, *m*-nitro-
Line Form: $3\text{-}NO_2C_6H_4AsO(OH)_2$
Mol Form: $C_6H_6AsNO_5$
M_r: 247.03
mp: 200
Color: lf ye lf (w)
***s*:** H_2O 2; eth 1; bz 2; chl 2
IR: µCOB 3805 3130 1610 1540 1470 1370 1350 1280 1220 1140 1090 950 880 780
UV: OES 4-67 252(7762) EtOH
m/e: 246.95

HODOC No: 01439
CAS No: 98-72-6
Beilstein No: 16^3, 1059
Name: Arsonic acid, (4-nitrophenyl)-
Synonym: Benzenearsonic acid, *p*-nitro-
Line Form: $4\text{-}NO_2C_6H_4AsO(OH)_2$
Mol Form: $C_6H_6AsNO_5$
M_r: 247.03
mp: 310
Color: lf or nd (w)
***s*:** H_2O 2; al 2 DMSOd6 2
IR: SADG 18382 3125 2857 2273 1613 1515 1351 1075 847 781 758
H NMR: SAD 8798 8.2 8.5 9.3 DMSO
m/e: 246.95

HODOC No: 01440
CAS No: 98-05-5
Beilstein No: 16^2, 457
Name: Arsonic acid, phenyl-
Synonym: Benzenearsonic acid
Structure:

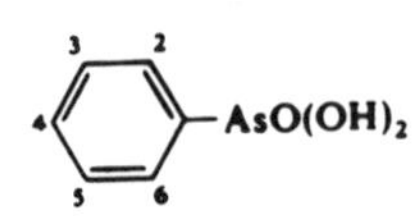

Mol Form: $C_6H_7AsO_3$
M_r: 202.04
mp: dec 158-62
***s*:** H_2O 3; al 4; tfa 2
IR: µCOB 6367 2860 2270 1640 1490 1450 1350 1190 1090 870 790 750
UV: SAD 1628 269(830) 262(1010) 256(816) 216(10100) MeOH
H NMR: SAD 672 7.8-7.9 tfa
m/e: 201.96

HODOC No: 01441
CAS No: 620-27-9
Beilstein No: 16^3, 1057
Name: Arsonic acid, (phenylmethyl)-
Synonym: α-Toluenearsonic acid
Line Form: $C_6H_5CH_2AsO(OH)_2$
Mol Form: $C_7H_9AsO_3$
M_r: 216.06
mp: 167-8
Color: nd (al)
***s*:** H_2O 2; al 2; ctc 2; chl 4; AcOEt 4; CS_2 2
IR: µCOB 3759 2941 2273 1515 1471 1190 1176 1124 1075 926 909 893 775 699 690 660
m/e: 215.98

HODOC No: 01442
CAS No: 590-34-1
Beilstein No: 4^3, 1826
Name: Arsonic acid, 2-propenyl-
Synonym: 2-Propene-1-arsonic acid
Line Form: $(CH_2{=}CHCH_2)AsO(OH)_2$
Mol Form: $C_3H_7AsO_3$
M_r: 166.01
mp: 130-1
***s*:** H_2O 4; al 4
IR: **µCOB 3807** 2700 2330 1640 1470 1430 1390 1240 1060 1000 940 920 820 770 710 660
m/e: 165.96

HODOC No: 01443
CAS No: 107-34-6
Beilstein No: 4^4, 3682
Name: Arsonic acid, propyl-
Synonym: 1-Propanearsonic acid
Line Form: $CH_3CH_2CH_2AsO_3H_2$
Mol Form: $C_3H_9AsO_3$
M_r: 168.02
mp: 134-5
Color: nd (al), pl (w)
***s*:** H_2O 4; al 4; eth 1
IR: **µSADP 9707** 2860 2270 1590 1470 1370 1220 920 830 810 770
m/e: 167.98

HODOC No: 01444
CAS No: 421-30-7
Beilstein No: 3^4, 267
Name: Arsonous dibromide, (trifluoromethyl)-
Synonym: Arsine, dibromo(trifluoromethyl)-
Line Form: CF_3AsBr_2
Mol Form: $CAsBr_2F_3$
M_r: 301.73
bp: 118^{715}
***d*:** 1.528^{20}
m/e: 301.75

HODOC No: 01445
CAS No: 50361-05-2
Beilstein No: 4^3, 1810
Name: Arsonous dichloride, (2-chloroethenyl)-, (E)-
Synonym: Dichloro(*trans*-2-chlorovinyl)arsine
Line Form: $ClCH{=}CHAsCl_2$
Mol Form: $C_2H_2AsCl_3$
M_r: 207.32
bp: 196d, 93^{20}
mp: 0.1
***d*:** $1.888^{20/4}$
***s*:** Et_2O 4; EtOH 4
m/e: 205.84

HODOC No: 01446
CAS No: 696-28-6
Beilstein No: 16^3, 958
Name: Arsonous dichloride, phenyl-
Synonym: Arsine, dichlorophenyl-
Line Form: $C_6H_5AsCl_2$
Mol Form: $C_6H_5AsCl_2$
M_r: 222.93
bp: 254-7, 131^{14}
***d*:** $1.6516^{20/4}$
***n*:** 1.6386^{15}
***s*:** Bz 4; Et_2O 4; EtOH 4
m/e: 221.90

HODOC No: 01447
CAS No: 421-32-9
Beilstein No: 3^4, 268
Name: Arsonous dichloride, (trifluoromethyl)-
Synonym: Arsine, dichloro(trifluoromethyl)-
Line Form: CF_3AsCl_2
Mol Form: $CAsCl_2F_3$
M_r: 214.83
bp: 71
***d*:** 1.431^{20}
***s*:** Ac 4; EtOH 4
MS: **Al 174**
m/e: 213.85

HODOC No: 01448
CAS No: 420-24-6
Beilstein No: 4^{3}, 1796
Name: Arsonous difluoride, methyl-
Synonym: Arsine, difluoromethyl-
Line Form: CH_3AsF_2
Mol Form: CH_3AsF_2
Mr: 127.95
bp: 76.5
mp: -29.7
***d*:** 1.924^{18}
Color: fume (air)
m/e: 127.94

HODOC No: 01449
CAS No: 6380-34-3
Beilstein No: 16^{2}, 412
Name: Arsonous diiodide, phenyl-
Synonym: Arsine, diiodophenyl-
Line Form: $C_6H_5AsI_2$
Mol Form: $C_6H_5AsI_2$
Mr: 405.83
bp: 205.5^{14} 185^{10}
mp: 15
IR: **COB 7181** 3055 1470 1435 1330 1300 1180 1070 1000 910 840 730 690
m/e: 405.77

HODOC No: 01450
CAS No: 353-91-3
Beilstein No: 3^{4}, 268
Name: Arsonous diiodide, (trifluoromethyl)-
Synonym: Arsine, diiodo(trifluoromethyl)-
Line Form: CF_3AsI_2
Mol Form: $CAsF_3I_2$
Mr: 397.74
bp: 183d, 100^{48}
m/e: 397.73

HODOC No: 01451
CAS No: 10504-35-5
Beilstein No: 18^{4}, 3046
Name: *D*-Ascorbic acid
Synonym: *D*-threo-Hex-2-enonic acid, γ-lactone
Structure:

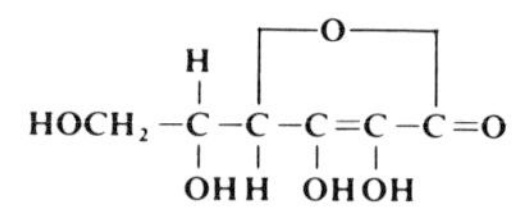

Mol Form: $C_6H_8O_6$
Mr: 176.12
mp: 192d
α: $[\alpha]^{18/D}$-48 MeOH,1%$[\alpha]^{20/D}$-23.8 H_20,3%
Color: Pl or mcl nd
***s*:** H_2O 4; al 3; eth 1; bz 1; chl 1
UV: **HBCP** 254(7943) al
C NMR: **JJ 171** ini 63.2 69.9 77.1 118.8 156.3 174.0 H_2O
m/e: 176.03

HODOC No: 01452
CAS No: 62624-30-0
Beilstein No: 18^{4}, 3047
Name: *DL*-Ascorbic acid
Synonym: Ascorbic acid (*DL*)
Mol Form: $C_6H_8O_6$
Mr: 176.12
mp: 168-9
***s*:** H_2O 4; al 3; eth 1; bz 1; chl 1
UV: **OES 6-83** 280(398) H_2O
H NMR: **VAR 464** 3.8 4.1 5.0 D_2O
C NMR: **JJ 171** ini 63.2 69.9 77.1 118.8 156.3 174.0 H_2O
m/e: 176.03

HODOC No: 01453
CAS No: 50-81-7
Beilstein No: 18^4, 3038
Name: *L*-Ascorbic acid
Synonym: Adenex
Mol Form: $C_6H_8O_6$
M_r: 176.12
mp: 190-2d
***d*:** 1.65
α: $[\alpha]^{20/D}$ 24 H_2O,1%
***s*:** H_2O 4; al 3; eth 1; bz 1; chl 1; peth 1
IR: SADG 470 3520 3410 3320 3220 3030 2740 1705 1675 1500 1335 1270 1140 1120 1020 990 870 820 755
UV: SAD 3455 247 H_2O
H NMR: SAD 3126 3.7 3.8 4.1 5.0 D_2O
C NMR: JJ 171 ini 63.2 69.9 77.1 118.8 156.3 174.0 H_2O
MS: NBS 838
m/e: 176.03

HODOC No: 01454
CAS No: 528-81-4
Beilstein No: 18^4, 2301
Name: *L*-Ascorbic acid, 6-deoxy-
Synonym: Ascorbic acid, 6-deoxy-, *L*-
Structure:

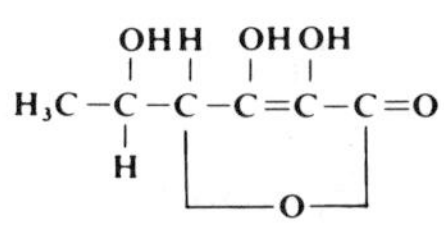

Mol Form: $C_6H_8O_5$
M_r: 160.12
bp: sub, $160^{0.001}$
mp: 168
α: $[\alpha]^{22/D}$ +36.7 (0.1N HCl,c=1)
Color: Pr (AcOEt)
***s*:** H_2O 4; Ac 4; EtOH 4
m/e: 160.04

HODOC No: 01455
CAS No: 70-47-3
Beilstein No: 4^4, 3005
Name: *L*-Asparagine
Synonym: Asparagine, *L*-
Line Form: $H_2NCOCH_2CH(NH_2)CO_2H$
Mol Form: $C_4H_8N_2O_3$
M_r: 132.12
mp: 236(anh)
***d*:** $1.543^{15/4}$
α: $[\alpha]^{20/D}$-5.42 H_2O,1.3%
Color: rh (w+1)
***s*:** H_2O 3; al 1; eth 1; MeOH 1
IR: μSADP 17219 3330 2940 1700 1640 1590 1520 1430 1300 1250 1150 1110 1040 990 900 880 760
UV: HBCP 217(316) H_2O pH=10.6
m/e: 132.05

HODOC No: 01456
CAS No: 5794-24-1
Beilstein No: 4^1, 531
Name: *D*-Asparagine, monohydrate
Synonym: Asparagine, monohydrate, *D*-
Line Form: $H_2NCOCH_2CH(NH_2)CO_2H–H_2O$
Mol Form: $C_4H_{10}N_2O_4$
M_r: 150.13
mp: anh 234.5
***d*:** 1.523(15/4)
α: $[\alpha]^{20/D}$5.41 H_2O,1.3%
***s*:** H_2O 2; al 1; eth 1; bz 1; MeOH 1
IR: μSADP 15266 3333 3030 2941 2500 1667 1626 1563 1515 1429 1389 1351 1316 1235 1149 1099 1075 833 800
m/e: 150.06

HODOC No: 01457
CAS No: 69833-18-7
Beilstein No: 4^2, 900
Name: *DL*-Asparagine, monohydrate
Synonym: β-Asparagine, hydrate(*DL*)
Line Form: $H_2NCOCH_2CH(NH_2)CO_2H–H_2O$
Mol Form: $C_4H_{10}N_2O_4$
Mr: 150.13
bp: dec 213-5
mp: 182-3
***d*:** 1.4540(15/4)
***s*:** H_2O 2; al 1; eth 1; bz 1; MeOH 1
IR: μSADP 6629 3333 3030 2941 2500 1667 1626 1563 1515 1429 1389 1351 1316 1235 1149 1099 1075 833 800
UV: HBCP 220(100) H_2O
C NMR: STOT 479 35.9 52.6 174.5 175.7 D_2O
m/e: 150.06

HODOC No: 01458
CAS No: 5794-13-8
Beilstein No: 4^4, 2998
Name: *L*-Asparagine, monohydrate
Synonym: Asparagine, monohydrate, *L*-
Line Form: $H_2NCOCH_2CH(NH_2)CO_2H{\cdot}H_2O$
Mol Form: $C_4H_{10}F_2N_2O_4$
Mr: 188.13
mp: 233-5
***d*:** $1.543^{15/4}$
α: $[\alpha]^{20/D}$ +31 (3N HCl, c=10)
MS: ALD 3:10A
m/e: 188.06

HODOC No: 01459
CAS No: 1783-96-6
Beilstein No: 4^4, 2998
Name: *D*-Aspartic acid
Synonym: Aspartic acid, *D*-
Line Form: $HO_2CCH_2CH(NH_2)CO_2H$
Mol Form: $C_4H_7NO_4$
Mr: 133.10
mp: 269-71
***d*:** $1.6613^{13/13}$
α: [α](20/D)-25.5 HCl
***s*:** H_2O 3; al 1; eth 1; bz 1; dil HCl 3
IR: COB 5818 3000 2660 2510 1690 1640 1600 1520 1420 1360 1340 1300 1250 1150 1110 1050 990 900 750 650 600 550 450 420 360
m/e: 133.04

HODOC No: 01460
CAS No: 617-45-8
Beilstein No: 4^4, 3000
Name: *DL*-Aspartic acid
Synonym: Aspartic acid, *DL*-
Line Form: $HO_2CCH_2CH(NH_2)CO_2H$
Mol Form: $C_4H_7NO_4$
Mr: 133.10
mp: 275 (sealed tube)
***d*:** $1.6632^{13/13}$
Color: mcl pr (w)
***s*:** H_2O 2; al 1; eth 1; bz 1; py 1
UV: OES 1-30 <200 pH=7.4
C NMR: STOT 479 37.3 52.9 175.2 178.5 D_2O
m/e: 133.04

HODOC No: 01461
CAS No: 56-84-8
Beilstein No: 4^4, 2998
Name: *L*-Aspartic acid
Synonym: Aspartic acid, *L*-
Line Form: $HO_2CCH_2CH(NH_2)CO_2H$
Mol Form: $C_4H_7NO_4$
Mr: 133.10
mp: 324d, 270 (sealed tube)
***d*:** $1.6613^{13/13}$
α: $[\alpha]^{20/D}$4.36 H_2O $[\alpha]^{24/D}$24.6 6N HCl, 2%
Color: rh lf (w)
***s*:** H_2O 2; al 1; eth 1; bz 1; dil HCl 3; py 1
IR: SADG 18085 3000 2660 2510 1690 1640 1600 1520 1420 1360 1340 1300 1250 1150 1110 1050 990 900 750 650 600 550 450 420 360
MS: NBS 290
m/e: 133.04

HODOC No: 01462
CAS No: 97021-39-1
Name: Aspartic acid, *N*-(benzylcarbamoyl)-, diethyl ester
Mol Form: $C_{16}H_{22}N_2O_5$
Mr: 322.35
mp: 82
IR: μCOB 2159 3330 2940 1750 1610 1560 1470 1410 1370 1350 1330 1280 1190 1100 1050 1030 960 870 760 700
m/e: 322.15

HODOC No: 01463
CAS No: 13253-91-3
Name: Aspartic acid, dibutyl ester, hydrochloride, *L*-
Synonym: Aspartic acid, dibutyl ester, hydrochloride
Line Form: $CH_3(CH_2)_3O_2CCH_2CH(NH_2)CO_2(CH_2)_3CH_3{-}HCl$
Mol Form: $C_{12}H_{24}ClNO_4$
Mr: 281.77
mp: 113.5-4.0
IR: COB 4573 2900 2600 1740 1600 1490 1450 1250 1190 1090 930 840 760
m/e: 281.14

HODOC No: 01464
CAS No: 687-30-9
Name: Aspartic acid, *N*-(trifluoroacetyl)-, 1-ethyl ester
Line Form: $CH_3CH_2O_2CCH(NHCOCF_3)CH_2CO_2H$
Mol Form: $C_8H_{10}F_3NO_5$
Mr: 257.16
IR: COB 7416 1735 1700 1555 1370 1180 1025 905 855 730
m/e: 257.05

HODOC No: 01465
CAS No: 43101-48-0
Beilstein No: 4^3, 1524
Name: *DL*-Aspartic acid, diethyl ester
Synonym: Diethyl *DL*-aspartate
Line Form: $H_5C_2O_2C{-}CH(NH_2)CH_2CO_2C_2H_5$
Mol Form: $C_8H_{15}NO_4$
Mr: 189.20
bp: 110^5
m/e: 189.10

HODOC No: 01466
CAS No: 4515-21-3
Name: *DL*-Aspartic acid, *N*-[(phenylmethoxy)carbonyl]-
Synonym: Aspartic acid, *N*-carboxy-, *N*-benzyl ester, *DL*-
Line Form: $C_6H_5CH_2O_2CNHCH(CO_2H)CH_2CO_2H$
Mol Form: $C_{12}H_{13}NO_6$
Mr: 267.23
mp: 115.5-6.5
s: ace 3
H NMR: SAD 8965 2.9 4.7 5.1 6.6 7.3 9.6 $aceD_6$
m/e: 267.07

HODOC No: 01467
CAS No: 7611-43-0
Beilstein No: 4^3, 1521
Name: *L*-Aspartic acid, *N*-[2-[(2-amino-2-oxoethyl)amino]-2-carboxyethyl]-
Synonym: Lycomarasmine
Line Form: $HO_2CCH_2CH(CO_2H)NHCH_2CH(CO_2H)NHCH_2CONH_2$
Structure:

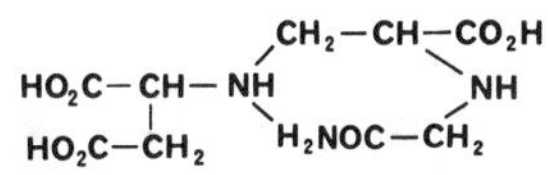

Mol Form: $C_9H_{15}N_3O_7$
Mr: 277.23
mp: 227-9d
α: $[\alpha]^{20/D}$ -48 (w)
m/e: 277.09

HODOC No: 01468
CAS No: 4631-12-3
Beilstein No: 9^4, 839
Name: *L*-Aspartic acid, *N*-benzoyl-
Synonym: Aspartic acid, *N*-benzoyl-, *L*-
Line Form: $HO_2CCH_2CH(NHCOC_6H_5)CO_2H$
Mol Form: $C_{11}H_{11}NO_5$
Mr: 237.20
mp: 171-3
α: $[\alpha]^{20/D}$ 37 0.76N NaOH
Color: nd or lf
s: H_2O 2; al 2; eth 1; chl 1
IR: μSADP 6793 3230 2940 2560 1700 1640 1590 1520 1470 1410 1320 1250 1180 1080 900 690
m/e: 237.06

HODOC No: 01469
CAS No: 6327-59-9
Name: *L*-Aspartic acid, *bis*(phenylmethyl) ester, hydrochloride
Synonym: Aspartic acid, dibenzyl ester, hydrochloride
Line Form: $C_6H_5CH_2O_2CCH_2CH(NH_2)CO_2CH_2C_6H_5{-}HCl$
Mol Form: $C_{18}H_{20}ClNO_4$
Mr: 349.80
IR: COB 7355 2900 2000 1750 1730 1500 1450 1390 1360 1250 1190 1090 730 700
m/e: 349.11

HODOC No: 01470
CAS No: 1152-61-0
Beilstein No: 6[4], 2395
Name: *L*-Aspartic acid, *N*-[(phenylmethoxy)carbonyl]-
Synonym: Aspartic acid, *N*-carboxy-, *N*-benzyl ester, *L*-
Line Form: $HO_2CCH_2CH(CO_2H)\text{-}NHCO_2CH_2C_6H_5$
Mol Form: $C_{12}H_{13}NO_6$
M_r: 267.23
mp: 117-9
α: $[\alpha]^{23/D}$ +8.6 (aa, c=7)
IR: SADG 4690 3350 3030 2940 1700 1590 1530 1465 1450 1415 1320 1300 1270 1225 1185 1085 1060 915 775 745 695 635 595 460
UV: LANG 1699 264(155) 258(191) 252(148) 206(8511) EtOH
H NMR: SAD 17027 2.8 4.5 5.1 7.2 7.4 12.0 poly D
MS: NBS 4709; ALD 7:18B
m/e: 267.07

HODOC No: 01471
CAS No: 466-49-9
Beilstein No: 23[4], 2688
Name: Aspidospermidine, 1-acetyl-17-methoxy-
Synonym: 1H-Indolizino[8,1-cd]carbazole, aspidospermidine deriv.
Structure:

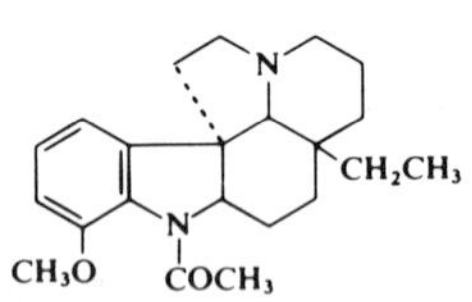

Mol Form: $C_{22}H_{30}N_2O_2$
M_r: 354.47
bp: 220^2
mp: 212
α: $[\alpha]^{15/D}$-100.2 al
Color: nd or pr (al) nd (peth)
***s*:** H_2O 2; al 3; eth 2; bz 3; chl 3
UV: OES 6-1002 290(2818) 255(11749) 218(33884) EtOH
MS: NBS 2774
m/e: 354.23

HODOC No: 01472
CAS No: 466-43-3
Name: Atisine
Synonym: 9H,12cH-8a,11-Ethano-6,12b-propano-5H-benz[h]oxazolo[2,3-a]isoquinoline, atisine deriv.
Structure:

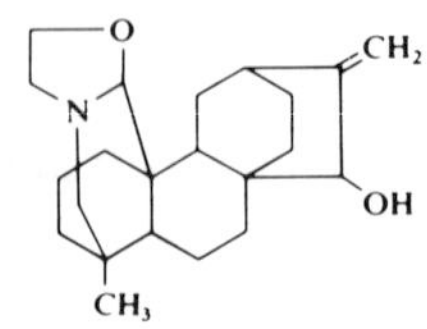

Mol Form: $C_{22}H_{33}NO_2$
M_r: 343.49
mp: 57-60
Color: rh bipym
***s*:** Et_2O 4; EtOH 4; Chl 4
m/e: 343.25

HODOC No: 01473
CAS No: 286-18-0
Beilstein No: 20[4], 1937
Name: 7-Azabicyclo[4.1.0]heptane
Synonym: 2,3-Tetramethyleneaziridine
Structure:

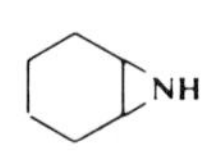

Mol Form: $C_6H_{11}N$
M_r: 97.15
bp: $48\text{-}51^{22}$
mp: 20-2
***s*:** Bz 4; Et_2O 4; EtOH 4
m/e: 97.09

HODOC No: 01474
CAS No: 552-70-5
Beilstein No: 21^4, 3315
Name: 9-Azabicyclo[3.3.1]nonan-3-one, 9-methyl-
Synonym: 3-Granataninone, 9-methyl-
Structure:

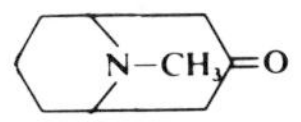

Mol Form: $C_9H_{15}NO$
M_r: 153.22
bp: 246
mp: 54
***d*:** 1.001^{100}
***n*:** 1.4760^{100}
Color: orh pr (peth)
***s*:** H_2O 4; Et_2O 4; EtOH 4
MS: NBS 501
m/e: 153.12

HODOC No: 01475
CAS No: 6530-11-6
Name: 1-Azabicyclo[2.2.2]octan-3-amine, *N*-(phenylmethyl)-
Synonym: Quinuclidine, 3-(benzylamino)-
Mol Form: $C_{14}H_{20}N_2$
M_r: 216.31
bp: $130\text{-}3^{0.4}$
***n*:** $1.5562^{20/D}$
IR: COB 7505 3250 3030 2940 2860 1640 1600 1490 1450 1320 1130 1060 980 740 700 610
m/e: 216.16

HODOC No: 01476
CAS No: 100-76-5
Beilstein No: 20^4, 1966
Name: 1-Azabicyclo[2.2.2]octane
Synonym: 1,4-Ethanopiperidine
Structure:

Mol Form: $C_7H_{13}N$
M_r: 111.18
mp: 158 (sealed tube)
Color: (eth)
***s*:** H_2O 4; Ac 4; Et_2O 4; EtOH 4
MS: NBS 129
m/e: 111.10

HODOC No: 01477
CAS No: 529-17-9
Beilstein No: 20^4, 1963
Name: 8-Azabicyclo[3.2.1]octane, 8-methyl-
Synonym: Tropane
Structure:

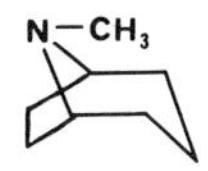

Mol Form: $C_8H_{15}N$
M_r: 125.21
bp: 163-9
***d*:** $0.9259^{15/15}$
MS: NBS 217
m/e: 125.12

HODOC No: 01478
CAS No: 10344-62-4
Beilstein No: 22^4, 200
Name: 1-Azabicyclo[2.2.2]octane-2-carboxylic acid, (*S*)-
Synonym: 2-Quinuclidinecarboxylic acid
Structure:

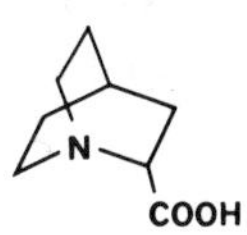

Mol Form: $C_8H_{13}NO_2$
M_r: 155.19
mp: 280d
Color: col hyg cr (al-Ace)
m/e: 155.09

HODOC No: 01479
CAS No: 529-38-4
Beilstein No: 22, 202
Name: 8-Azabicyclo[3.2.1]octane-2-carboxylic acid, 3-(benzoyloxy)-8-methyl-, ethyl ester, [1R-(*exo*,*exo*)]-
Synonym: Cocaethylene
Structure:

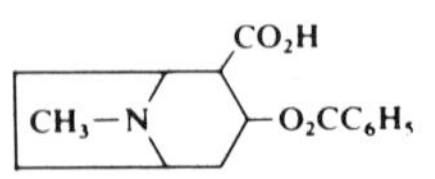

Mol Form: $C_{18}H_{23}NO_4$
M_r: 317.37
mp: 109
Color: pr (Eth)
***s*:** Et_2O 4; EtOH 4
m/e: 317.16

HODOC No: 01480
CAS No: 519-09-5
Beilstein No: 22^4, 2098
Name: 8-Azabicyclo[3.2.1]octane-2-carboxylic acid, 3-(benzoyloxy)-8-methyl-, [1R-(*exo*,*exo*)]-
Synonym: 1αH,5αH-Tropane-2β-carboxylic acid, 3β-hydroxy-, benzoate (ester)
Mol Form: $C_{16}H_{19}NO_4$
M_r: 289.32
mp: 195
α: $[\alpha]^{14/D}$ -63.5 (w, p =1.7)
Color: nd (w)
***s*:** Bz 4; EtOH 4
IR: DMS 398
MS: NBS 2138
m/e: 289.13

HODOC No: 01481
CAS No: 21030-43-3
Beilstein No: 22^2, 156
Name: 8-Azabicyclo[3.2.1]octane-2-carboxylic acid, 3-(benzoyloxy)-8-methyl-, methyl ester, (endo,endo)-(±)-
Synonym: 1αH,5αH-Tropane-2α-carboxylic acid, 3α-hydroxy-, methyl ester, benzoate (ester), (±)-
Mol Form: $C_{17}H_{21}NO_4$
M_r: 303.34
mp: 79-80
***s*:** H_2O 1; al 4; eth 4; ace 3; bz 4; ctc 4; chl 4; lig 4
IR: μCOB 5005 3030 1750 1720 1470 1280 1190 1110 1040 880 780 730 710 680
UV: OES 3-657 281(794) 274(1000) 229(15849) EtOH
MS: AMS 2-654 ini 182(100) 82(95) 83(41) 105(35) 94(33) 77(31) 96(27) 303(25)
m/e: 303.15

HODOC No: 01482
CAS No: 21206-60-0
Beilstein No: 22^4, 2103
Name: 8-Azabicyclo[3.2.1]octane-2-carboxylic acid, 3-(benzoyloxy)-8-methyl-, methyl ester, (*exo*,*exo*)-(±)-
Synonym: 1αH,5αH-Tropane-2β-carboxylic acid, 3β-hydroxy-, methyl ester, benzoate (ester), (±)-
Mol Form: $C_{17}H_{21}NO_4$
M_r: 303.34
mp: 79-80
Color: rh bipym pr (peth)
***s*:** Ac 4; Bz 4; Et_2O 4; EtOH 4
IR: COB 5005
UV: OES 3-657
MS: AMS 2-654
m/e: 303.15

HODOC No: 01483
CAS No: 50-36-2
Beilstein No: 22^4, 2101
Name: 8-Azabicyclo[3.2.1]octane-2-carboxylic acid, 3-(benzoyloxy)-8-methyl-, methyl ester, [1R-(*exo*,*exo*)]-
Synonym: 1αH,5αH-Tropane-2β-carboxylic acid, 3β-hydroxy-, methyl ester, benzoate (ester)
Mol Form: $C_{17}H_{21}NO_4$
M$_r$: 303.34
bp: 187-8 (sub,vac)
mp: 98
***n*:** 1.5022^{98}
α: $[\alpha]^{20/4}$-16.3 chl,4%$[\alpha]^{18/D}$-35 50%-al,1%
Color: mcl pr (al)
***s*:** H_2O 2; al 4; eth 4; ace 3; bz 4; chl 4; py 4; CS_2 3
UV: HBCP 281(794) 274(1000) 229(15849) al
MS: AMS 2-654 ini 182(100) 82(95) 83(41) 105(35) 94(33) 77(31) 96(27) 303(25)
m/e: 303.15

HODOC No: 01484
CAS No: 47195-07-3
Beilstein No: 22^2, 156
Name: 8-Azabicyclo[3.2.1]octane-2-carboxylic acid, 3-(benzoyloxy)-8-methyl-, methyl ester, [1*S*-(*exo*,*exo*)]-
Synonym: (+)-Cocaine
Structure:

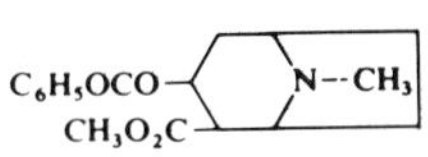

Mol Form: $C_{17}H_{21}NO_4$
M$_r$: 303.34
bp: $187\text{-}8^{0.1}$
mp: 98
α: $[\alpha]^{20/D}$ 15.8 chl,10%*
Color: mcl pr (eth)
***s*:** H_2O 1; al 4; eth 4; ace 3; bz 3
UV: HBCP 281(794) 274(1000) 229(15849) al
MS: AMS 2-654 ini 182(100) 82(95) 83(41) 105(35) 94(33) 77(31) 96(27) 303(25)
m/e: 303.15

HODOC No: 01485
CAS No: 79735-02-7
Beilstein No: 22^2, 156
Name: 8-Azabicyclo[3.2.1]octane-2-carboxylic acid, 3-hydroxy-8-methyl-, (2-*exo*,3-endo)-(±)-
Synonym: (±)-Alloecgonine
Structure:

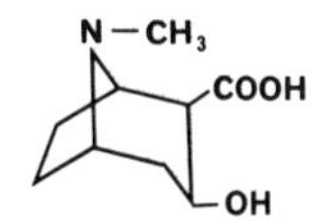

Mol Form: $C_9H_{15}NO_3$
M$_r$: 185.21
mp: 203
Color: pl (w + 3)
***s*:** H_2O 4; EtOH 4
UV: OES 3-217
MS: NBS 981
m/e: 185.11

HODOC No: 01486
CAS No: 481-37-8
Beilstein No: 22^4, 2097
Name: 8-Azabicyclo[3.2.1]octane-2-carboxylic acid, 3-hydroxy-8-methyl-, [1R-(*exo*,*exo*)]-
Synonym: 1αH,5αH-Tropane-2β-carboxylic acid, 3β-hydroxy-
Mol Form: $C_9H_{15}NO_3$
M$_r$: 185.21
mp: 205
α: $[\alpha]^{15/D}$ -45.5 (w, c=5)
Color: mcl pr
MS: NBS 981
m/e: 185.11

HODOC No: 01487
CAS No: 5796-31-6
Beilstein No: 22^2, 150
Name: 8-Azabicyclo[3.2.1]octane-2-carboxylic acid, 3-hydroxy-8-methyl-, hydrochloride, [1R-(*exo*,*exo*)]-
Synonym: 1αH,5αH-Tropane-2β-carboxylic acid, 3β-hydroxy-, hydrochloride
Mol Form: $C_9H_{16}ClNO_3$
M$_r$: 221.67
mp: 246
α: $[\alpha]^{15/D}$-59 H_2O,10%
***s*:** H_2O 3; al 3; MeOH 4; i-AmOH 2
UV: HBCP 550(3162) 480(6310) 465(5012) 400(5012) H_2O
m/e: 221.08

HODOC No: 01488
CAS No: 102558-41-8
Name: 8-Azabicyclo[3.2.1]octane-2-methanol, 3-hydroxy-8-methyl-α,α-diphenyl-
Synonym: 2-Tropanemethanol, 3-hydroxy-α,α-diphenyl-
Structure:

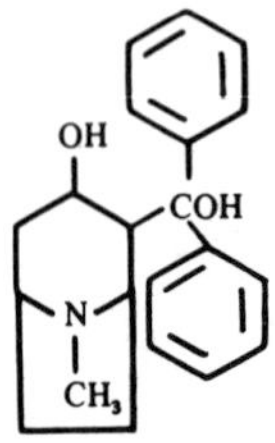

Mol Form: $C_{21}H_{25}NO_2$
M_r: 323.41
IR: **μCOB 882** 3570 3030 2940 1610 1470 1430 1390 1220 1060 1010 960 890 700
m/e: 323.19

HODOC No: 01489
CAS No: 1619-34-7
Beilstein No: 21^4, 237
Name: 1-Azabicyclo[2.2.2]octan-3-ol
Synonym: 3-Quinuclidinol
Structure:

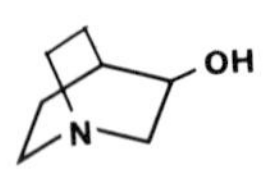

Mol Form: $C_7H_{13}NO$
M_r: 127.18
mp: 221
Color: cr (Bz)
***s*:** ace 3
IR: **SADG 7672** 3100 2940 2860 1460 1350 1310 1130 1120 1050 990 950 820 800 780 640
H NMR: **SAD 17500** 0.9-2.3 2.3-3.4 3.8 5.3 $CDCl_3$
MS: **NBS 236**
m/e: 127.10

HODOC No: 01490
CAS No: 7432-10-2
Beilstein No: 21^4, 168
Name: 8-Azabicyclo[3.2.1]octan-3-ol, 8-methyl-
Synonym: 3-Tropanol
Structure:

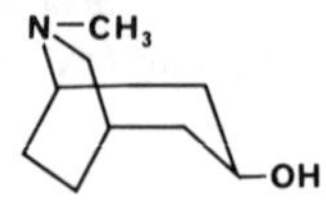

Mol Form: $C_8H_{15}NO$
M_r: 141.21
bp: 229
mp: 64
***s*:** H_2O 4; al 4; eth 3; chl 3
IR: **μSADP 21887** 3130 2940 1470 1450 1430 1320 1270 1240 1110 1060 1040 980 950 910 860 810 770
H NMR: **SAD 10555** 1.5-2.2 2.3 3.0-3.4 4.0 $CDCl_3$
MS: **NBS 4111**
m/e: 141.12

HODOC No: 01491
CAS No: 19145-60-9
Beilstein No: 21^4, 173
Name: 8-Azabicyclo[3.2.1]octan-3-ol, 8-methyl-, benzoate (ester), endo-
Synonym: 1αH,5αH-Tropan-3α-ol, benzoate (ester)
Structure:

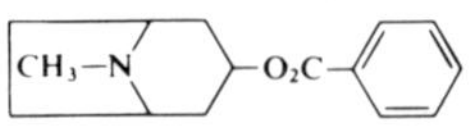

Mol Form: $C_{15}H_{19}NO_2$
M_r: 245.31
bp: 175-80
mp: 41-2
Color: cr (eth)
***s*:** H_2O 4; Et_2O 4; EtOH 4
m/e: 245.14

HODOC No: 01492
CAS No: 537-26-8
Beilstein No: 21^4, 174
Name: 8-Azabicyclo[3.2.1]octan-3-ol, 8-methyl-, benzoate (ester), *exo-*
Synonym: 1αH,5αH-Tropan-3β-ol, benzoate (ester)
Structure:

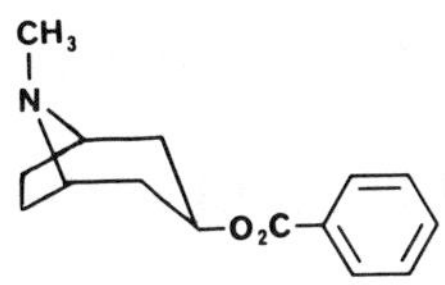

Mol Form: $C_{15}H_{19}NO_2$
Mr: 245.31
bp: d
mp: 49
***d*:** $1.0426^{100/4}$
***n*:** 1.5080^{100}
Color: pl or tab
***s*:** Bz 4; Et_2O 4; EtOH 4; peth 4
m/e: 245.14

HODOC No: 01493
CAS No: 120-29-6
Beilstein No: 21^2, 17
Name: 8-Azabicyclo[3.2.1]octan-3-ol, 8-methyl-, endo-
Synonym: 1αH,5αH-Tropan-3α-ol
Mol Form: $C_8H_{15}NO$
Mr: 141.21
bp: 229
mp: 64
Color: hyg pl (eth)
***s*:** H_2O 4; Et_2O 4; EtOH 4
IR: SADP 21887
H NMR: SAD 10555
MS: NBS 4111
m/e: 141.12

HODOC No: 01494
CAS No: 135-97-7
Beilstein No: 2^4, 169
Name: 8-Azabicyclo[3.2.1]octan-3-ol, 8-methyl-, *exo-*
Synonym: 1αH,5αH-Tropan-3β-ol
Structure:

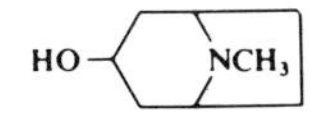

Mol Form: $C_8H_{15}NO$
Mr: 141.21
bp: 24^{760}
mp: 109
Color: rh ta or cr (eth), rh bipym (peth-Bz)
***s*:** H_2O 4; al 4; eth 2; bz 3; chl 3
IR: µSADP 24059 3130 2940 2780 1450 1350 1300 1240 1140 1060 1040 1020 970 920 860 790 780 740
H NMR: SAD 10411 1.4-2.2 2.3 3.0-3.7 3.9 $CDCl_3$
m/e: 141.12

HODOC No: 01495
CAS No: 24123-89-5
Beilstein No: 21^4, 4101
Name: 1-Azabicyclo[2.2.2]octan-3-one, 2-(phenylmethylene)-
Synonym: 3-Quinuclidinone, 2-benzylidene-
Structure:

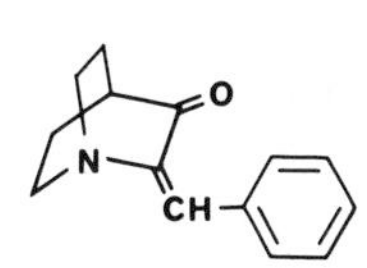

Mol Form: $C_{14}H_{15}NO$
Mr: 213.27
mp: 130-3
***s*:** chl 2
IR: COB 6487 2950 2870 1700 1630 1490 1470 1440 1330 1250 1170 1100 940 860 730 690 500
UV: SAD 11857 394(277) 296(20600) 222(8510) MeOH
H NMR: SAD 323 2.0 2.6 3.1 7.0 7.4 8.0 $CDCl_3$
m/e: 213.12

HODOC No: 01496
CAS No: 532-24-1
Beilstein No: 21[4], 3299
Name: 8-Azabicyclo[3.2.1]octan-3-one, 8-methyl-
Synonym: 1αH,5αH-Tropan-3-one
Structure:

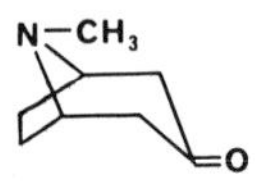

Mol Form: $C_8H_{13}NO$
Mr: 139.19
bp: $224\text{-}5^{714}$, 113^{25}
mp: 42-4
***d*:** $1.9872^{100/4}$
***n*:** 1.4598^{100}
***s*:** al 3; eth 3; ace 3; bz 3; chl 2; os 4; peth 3
IR: **μSADP 21888** 2940 1720 1470 1410 1350 1320 1280 1240 1210 1150 1110 1050 1010 990 950 910 840 770 730 710
H NMR: **SAD 10365** 1.4-3.0 3.5 $CDCl_3$
MS: **NBS 342**
m/e: 139.10

HODOC No: 01497
CAS No: 484-93-5
Beilstein No: 22[4], 284
Name: 8-Azabicyclo[3.2.1]oct-2-ene-2-carboxylic acid, 8-methyl-, (1R)-
Synonym: Anhydroecgonine
Structure:

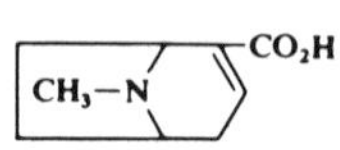

Mol Form: $C_9H_{13}NO_2$
Mr: 167.20
mp: 226-30d
Color: cr (MeOH) (MeOH-eth)
***s*:** H_2O 4
m/e: 167.09

HODOC No: 01498
CAS No: 1202-71-7
Name: Azacyclododecan-2-one
Synonym: 1-Aza-2-cyclododecanone
Mol Form: $C_{11}H_{21}NO$
Mr: 183.28
mp: 156-7
***s*:** chl 2
IR: **SADG 16507** 3300 2920 1640 1540 1460 1440 1270 730 680
H NMR: **SAD 8421** 1.2-1.9 2.2 3.3 6.4 $CDCl_3$
m/e: 183.16

HODOC No: 01499
CAS No: 32566-77-1
Name: Azacyclododecan-2-one, 3-bromo-
Synonym: Undecanoic acid, 11-amino-2-bromo,α-lactam
Structure:

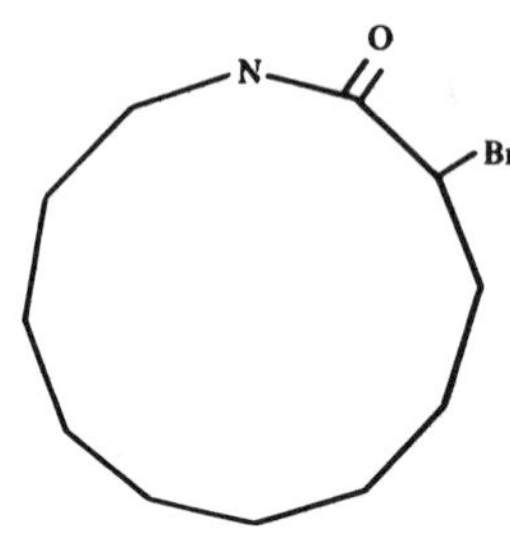

Mol Form: $C_{11}H_{20}BrNO$
Mr: 261.18
mp: 159-60
***s*:** chl 2
H NMR: **SAD 8423** 1.2-1.9 2.2 3.2 3.6 4.4 6.5 $CDCl_3$
m/e: 261.07

HODOC No: 01500
CAS No: 32566-75-9
Name: Azacyclododecan-2-one, 3,3-dibromo-
Synonym: Undecanoic acid, 11-amino-2,2-dibromo,ϖ-lactam
Mol Form: $C_{11}H_{19}Br_2NO$
Mr: 339.07
mp: 180.0-0.5
***s*:** chl 3
IR: **SADG 16533** 3300 3050 2940 1660 1530 1460 1350 1260 790 740 670 650
H NMR: **SAD 8424** 1.2-1.8 2.7 3.4 6.8 $CDCl_3$
m/e: 338.98

HODOC No: 01501
CAS No: 947-04-6
Beilstein No: 21^4, 3284
Name: Azacyclotridecan-2-one
Synonym: 12-Aminododecanoic acid lactam
Structure:

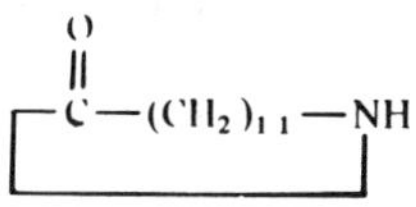

Mol Form: $C_{12}H_{23}NO$
M_r: 197.31
mp: 151-2
IR: **µCOB 1812** 3330 3030 2940 1640 1560 1450 1350 1280 1270 1240 1190 750 690
m/e: 197.18

HODOC No: 01502
CAS No: 14026-69-8
Name: Azacycloundecane-2-carboxylic acid
Structure:

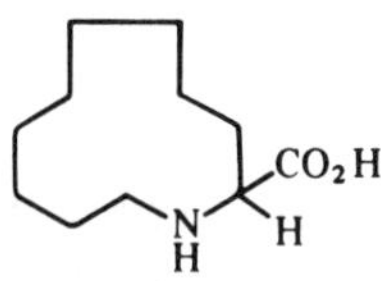

Mol Form: $C_{11}H_{21}NO_2$
M_r: 199.28
mp: 143-4d
***s*:** H_2O 2
IR: **SADG 16503** 3000 2940 2840 2500 1600 1470 1390 1310 790 710 480
H NMR: **SAD 8417** 1.5 1.9 2.8-3.6 3.7 D_2O
m/e: 199.16

HODOC No: 01503
CAS No: 32566-73-7
Name: Azacycloundecan-2-one, 3-bromo-
Synonym: Decanoic acid, 10-amino-2-bromo, ω-(lactam)
Structure:

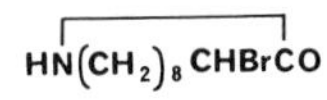

Mol Form: $C_{10}H_{18}BrNO$
M_r: 247.15
mp: 172-3
***s*:** chl 2
IR: **SADG 16526** 3250 3100 2930 1650 1560 1460 1420 1370 1250 1200 720 660
H NMR: **SAD 8418** 1.4 2.2 3.2 3.6 4.4 6.6 $CDCl_3$
m/e: 247.06

HODOC No: 01504
CAS No: 32566-72-6
Name: Azacycloundecan-2-one, 3-chloro-
Synonym: Decanoic acid, *l*0-amino-2-chloro, ω-lactam
Mol Form: $C_{10}H_{18}ClNO$
M_r: 203.70
mp: 149-50
***s*:** chl 2
IR: **SADG 16512** 3300 3100 2920 1650 1550 1460 1440 1360 1220 680
H NMR: **SAD 8420** 1.1-2.0 2.2 3.1 3.7 4.5 6.9 $CDCl_3$
m/e: 203.11

HODOC No: 01505
CAS No: 32566-71-5
Name: Azacycloundecan-2-one, 3,3-dibromo-
Synonym: Decanoic acid, 10-amino, 2,2-dibromo, ω-lactam
Mol Form: $C_{10}H_{17}Br_2NO$
M_r: 325.05
mp: 156
IR: **SADG 16532**
H NMR: **SAD 8419**
m/e: 324.97

HODOC No: 01506
CAS No: 32566-70-4
Name: Azacycloundecan-2-one, 3,3-dichloro-
Synonym: Decanoic acid, 10-amino-2,2-dichloro, ω-lactam
Mol Form: $C_{10}H_{17}Cl_2NO$
M$_r$: 238.15
mp: 165-6
***s*:** chl 2
IR: SADG 16519 3350 3050 2950 1670 1540 1460 1440 1260 1040 820 700 660
H NMR: SAD 8422 1.1-2.0 2.6 3.5 6.9 $CDCl_3$
m/e: 237.07

HODOC No: 01507
CAS No: 32566-66-8
Name: 2(1H)-Azecinone, 3,3-dichlorooctahydro-
Synonym: Nonanoic acid, 9-amino-2,2-dichloro,ω-lactam
Mol Form: $C_9H_{15}Cl_2NO$
M$_r$: 224.12
mp: 172-4
***s*:** chl 3
IR: SADG 16518 3360 2930 2860 1670 1510 1460 1430 1360 970 810 780 750 700 660
H NMR: SAD 8416 1.2-2.0 2.6 3.4 7.0 $CDCl_3$
m/e: 223.05

HODOC No: 01508
CAS No: 6142-53-6
Name: 2(1H)-Azecinone, octahydro-
Synonym: Nonanoic acid, 9-amino-, lactam
Structure:

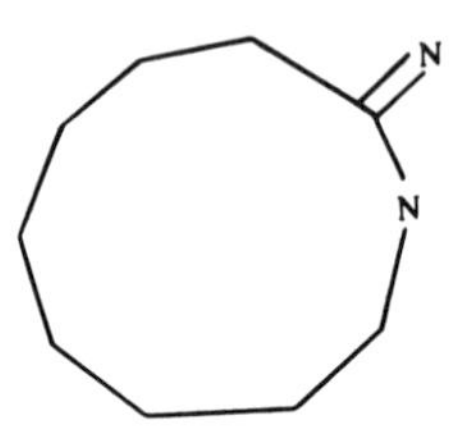

Mol Form: $C_9H_{17}NO$
M$_r$: 155.23
mp: 137.5-139
***s*:** chl 3
IR: SADG 16505 3300 2930 2860 1640 1550 1460 1430 1360 1290 750 700
H NMR: SAD 8415 1.2-2.0 2.3 3.3 6.1 $CDCl_3$
m/e: 155.13

HODOC No: 01509
CAS No: 1931-83-5
Name: Azelaic acid, *bis*(2,2,4-trimethylpentyl) ester
Synonym: Bis(2,2,4-trimethylpentyl) azelaate
Mol Form: $C_{25}H_{48}O_4$
M$_r$: 412.63
IR: μCOB 963 2940 1750 1470 1410 1370 1250 1180 1100 1020 730
m/e: 412.36

HODOC No: 01510
CAS No: 100996-43-8
Name: Azelaic acid, diester with 8,8-dimethyl-2-oxabicyclo[4.2.0]octan-7-ol
Synonym: Nonanedioic acid, *bis*(8,8-dimethyl-2-oxabicyclo(4.2.0)octan-7-yl) ester
Structure:

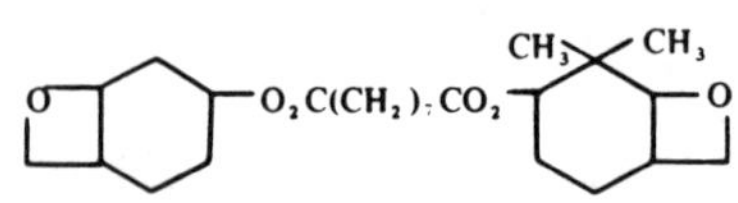

Mol Form: $C_{27}H_{44}O_6$
M$_r$: 464.62
***n*:** 1.4830^{20}
IR: μCOB 1735 2940 1720 1470 1370 1250 1180 1120 1080 1050 1020 990 950 900 730
m/e: 464.31

HODOC No: 01511
CAS No: 111-49-9
Beilstein No: 20^4, 1406
Name: 1H-Azepine, hexahydro-
Synonym: Azacycloheptane
Structure:

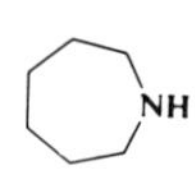

Mol Form: $C_6H_{13}N$
M$_r$: 99.17
bp: 138^{749}
***d*:** $0.8643^{22/4}$
***n*:** 1.4631^{20}
***s*:** H_2O 3; al 4; eth 4
IR: SADP 23773
MS: ALDN 34 43(100) 70(94) 30(65) 56(62) 28(47) 57(44) 99(41) 44(33)
m/e: 99.10

HODOC No: 01512
CAS No: 55045-05-1
Name: 1H-Azepine, hexahydro-1-(2-naphthalenyl)-
Synonym: Azepine, *N*(2-naphthyl)-perhydro
Mol Form: $C_{16}H_{19}N$
Mr: 225.32
MS: WILEY 1601 225(100) 127(55) 196(47) 141(45) 224(28) 41(27) 154(24) 128(23)
m/e: 225.15

HODOC No: 01513
CAS No: 55256-26-3
Name: 1H-Azepine, hexahydro-1-[1,2,3,4-tetrahydro-4-methyl-6-(1-methy lethyl)-2-naphthalenyl]-
Synonym: Azepine, *N*(6-isopropyl-4-methyl-1,2,3,4-tetrahydronaphth-2-yl)-perhydro
Structure:

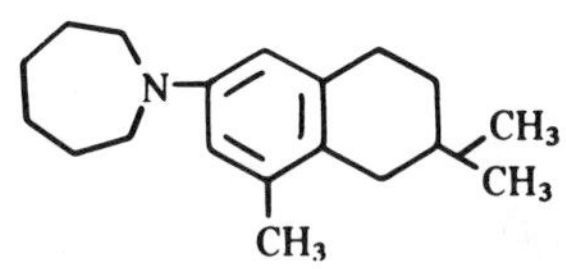

Mol Form: $C_{20}H_{31}N$
Mr: 285.45
MS: WILEY 1873 281(100) 270(89) 41(85) 112(70) 285(63) 43(55) 283(54) 266(49)
m/e: 285.25

HODOC No: 01514
CAS No: 23853-67-0
Name: 1H-Azepine, hexahydro-1-(1,2,3,4-tetrahydro-4,5,8-trimethyl-2-nap hthyl)-
Synonym: Azepine, perhydro-*N*(1,2,3,4-tetrahydro-4,5,8-trimethylnapth-2-yl)
Mol Form: $C_{19}H_{29}N$
Mr: 271.43
MS: WILEY 1837 269(100) 267(49) 254(46) 41(28) 270(25) 157(22) 141(19) 271(8)
m/e: 271.23

HODOC No: 01515
CAS No: 2525-16-8
Name: 2H-Azepine, 3,4,5,6-tetrahydro-7-methoxy-
Synonym: 1-Aza-2-methoxy-1-cycloheptene
Mol Form: $C_7H_{13}NO$
Mr: 127.18
IR: µCOB 3437 2860 1670 1430 1350 1330 1240 1180 1140 1090 1030 1000 840 800 740 680
Raman: SAD 1086 2940 2930 2850 2710 2660 1680 1470 1450 1360 1340 1290 1270 1250 1220 1200 1150 1100 1080 1070 1040 1000 980 950 890 870 840 800 750 680 600 510 480 400 350 260 190
m/e: 127.10

HODOC No: 01516
CAS No: 5227-53-2
Name: 1H-Azepine-2-carboxylic acid, hexahydro-
Synonym: Azepine-2-carboxylic acid, hexahydro-
Mol Form: $C_7H_{13}NO_2$
Mr: 143.18
mp: dec 205-8
***s*:** H_2O 3
IR: SADG 16499 3100 2950 2860 1620 1390 1310 860 740
H NMR: SAD 8403 1.5-2.3 3.3 3.8 D_2O
m/e: 143.09

HODOC No: 01517
CAS No: 3457-66-7
Name: 2H-Azepin-2-one, 3-bromohexahydro-
Synonym: α-Bromocaprolactam
Structure:

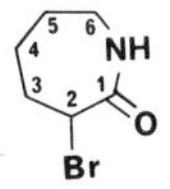

Mol Form: $C_6H_{10}BrNO$
Mr: 191.05
mp: 113-5
***s*:** chl 3
IR: SADG 16521 3200 2940 1670 1630 1480 1420 1350 1340 1250 1220 1170 970 870 800 750 600 500 420 360
H NMR: SAD 8404 1.4-2.3 2.9-3.9 4.7 7.5 $CDCl_3$
m/e: 190.99

HODOC No: 01518
CAS No: 1468-55-9
Name: 2H-Azepin-2-one, 3-chlorohexahydro-
Synonym: α-Chlorocaprolactam
Mol Form: $C_6H_{10}ClNO$
M_r: 147.60
mp: 93-4
IR: SADG 16508 3200 3080 2940 1680 1640 1480 1430 1350 1330 970 810 770 620 460 370
H NMR: SAD 8406 1.1-2.3 2.8-3.9 4.6 7.6 $CDCl_3$
m/e: 147.05

HODOC No: 01519
CAS No: 1709-14-4
Name: 2H-Azepin-2-one, 3,3-dichlorohexahydro-
Synonym: 2,2-Dichlorocaprolactam
Mol Form: $C_6H_9Cl_2NO$
M_r: 182.04
mp: 124-6
***s*:** chl 3
IR: SADG 16515 3200 2950 1670 1460 1490 1420 1410 1360 1320 1260 1120 1100 1080 980 890 800 620 520 510 400
H NMR: SAD 8407 1.4-2.3 2.6 3.4 7.7 $CDCl_3$
m/e: 181.01

HODOC No: 01520
CAS No: 105-60-2
Beilstein No: 21^4, 3196
Name: 2H-Azepin-2-one, hexahydro-
Synonym: Aminocaproic lactam
Structure:

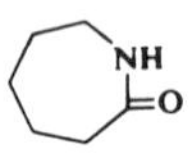

Mol Form: $C_6H_{11}NO$
M_r: 113.15
bp: 139^{12}
mp: 69-71
Color: lf (lig)
***s*:** H_2O 4; Bz 4; EtOH 4; Chl 4
IR: COB 1228
UV: OES 6-91
H NMR: SAD 6492
MS: NBS 145
m/e: 113.08

HODOC No: 01521
CAS No: 2073-32-7
Beilstein No: 21, 243
Name: 2H-Azepin-2-one, hexahydro-3-methyl-
Synonym: α-Methylcaprolactam
Mol Form: $C_7H_{13}NO$
M_r: 127.18
mp: 105-6
α: $[\alpha]^{20/D}$ -36.1
Color: cr (Bz-peth)
***s*:** H_2O 4; Bz 4
m/e: 127.10

HODOC No: 01522
CAS No: 2210-07-3
Beilstein No: 21, 242
Name: 2H-Azepin-2-one, hexahydro-5-methyl-
Synonym: γ-Methylcaprolactam
Mol Form: $C_7H_{13}NO$
M_r: 127.18
mp: 68-9
α: $[\alpha]^{20/D}$ -22.2
Color: cr (peth, Bz-lig)
***s*:** H_2O 4; Et_2O 4
m/e: 127.10

HODOC No: 01523
CAS No: 29184-60-9
Name: 2H-Azepin-2-one, hexahydro-7-(2-quinolyl)-
Synonym: Hexanoic acid, 6(quinol-2-ylamino),ε-lactam
Structure:

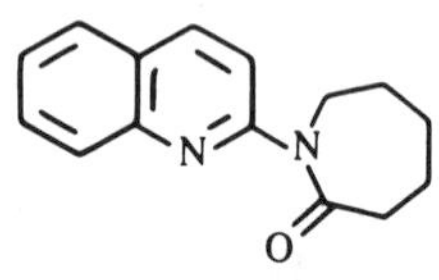

Mol Form: $C_{15}H_{16}N_2O$
M_r: 240.29
IR: COB 7568 3060 2930 1670 1620 1600 1500 1430 1400 1320 1270 1210 1200 1130 970 820 780
m/e: 240.13

HODOC No: 01524
CAS No: 503-29-7
Beilstein No: 20^4, 53
Name: Azetidine
Synonym: Azacyclobutane
Structure:

Mol Form: C_3H_7N
M$_r$: 57.09
bp: 63^{748}
***d*:** $0.8436^{20/4}$
***n*:** 1.4287^{25}
***s*:** Ac 4; Bz 4; Et_2O 4; EtOH 4
MS: NBS 7
m/e: 57.06

HODOC No: 01525
CAS No: 2517-04-6
Beilstein No: 22^4, 6
Name: 2-Azetidinecarboxylic acid
Mol Form: $C_4H_7NO_2$
M$_r$: 101.10
mp: 217d
α: $[\alpha]^{20/D}$ -108 (c=3.6)
Color: cr (95% MeOH)
IR: ALD 3:347G
MS: ALD 3:16B
m/e: 101.05

HODOC No: 01526
CAS No: 930-21-2
Beilstein No: 21^4, 3141
Name: 2-Azetidinone
Synonym: β-Alanine, lactam
Mol Form: C_3H_5NO
M$_r$: 71.08
bp: 106^{15}
mp: 73-4
***s*:** Et_2O 4; EtOH 4; Chl 4
m/e: 71.04

HODOC No: 01527
CAS No: 18775-73-0
Name: 2-Azetidinone, 1-amino-3,3-diphenyl-
Structure:

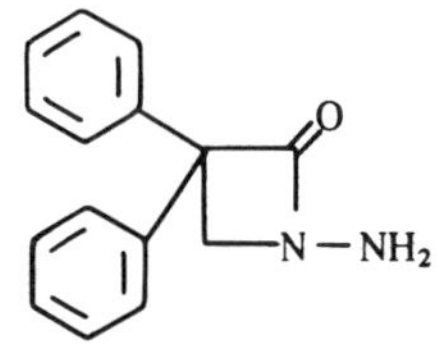

Mol Form: $C_{15}H_{14}N_2O$
M$_r$: 238.27
mp: 155-7
***s*:** chl 3
IR: SADG 14938 3320 3250 3060 2900 1740 1660 1600 1490 1440 1060 970 830 750 700
H NMR: SAD 8427 4.0 4.1 7.1-7.5 $CDCl_3$
m/e: 238.11

HODOC No: 01528
CAS No: 22085-98-9
Name: 2-Azetidinone, 3-chloro-4-(*p*-chlorophenyl)-1-phenyl-
Synonym: Propanoic acid, 3-amino-2-chloro-3(4-chlorophenyl)-*N*-phenyl,B-lactam*(cis)*
Structure:

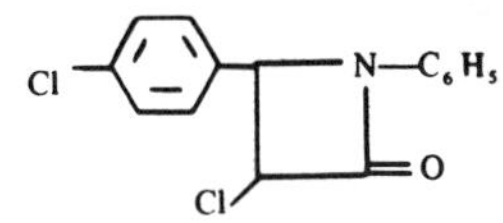

Mol Form: $C_{15}H_{11}Cl_2NO$
M$_r$: 292.15
IR: SADG 27366 1760 1600 1490 1460 1420 1390 1350 1210 1170 1160 1130 1090 1020 930 900 880 830 750 720 690 570 510 460
Raman: SAD 527 3070 3000 2960 1760 1600 1500 1460 1420 1390 1250 1230 1210 1180 1160 1090 1030 1020 1000 930 880 830 760 670 650 630 620 570 470 310 270 180
UV: SAD 20759 255(14600) 224(16100) MeOH 248(12800) 223(15800) MeOH:KOH
H NMR: SAD 15956 5.2 5.4 7.0-7.5 $CDCl_3$
m/e: 291.02

HODOC No: 01529
CAS No: 151-56-4
Beilstein No: 20^4, 3
Name: Aziridine
Synonym: Azacyclopropane
Structure:

Mol Form: C_2H_5N
M$_r$: 43.07
bp: 56^{756}
***d*:** $0.8321^{20/4}$
***s*:** H_2O 5; al 3; eth 4; chl 2; os 5
IR: μCOB 2839 3230 3030 2940 1640 1470 1270 1220 1090 1020 930 860 780
H NMR: VAR 372 0.0 1.6 $CDCl_3$
MS: WILEY 11 42(100) 28(79) 43(55) 15(36) 41(20) 40(15) 14(14) 27(10)
m/e: 43.04

HODOC No: 01530
CAS No: 2549-67-9
Name: Aziridine, 2-ethyl-
Synonym: 1,2-Butylenimine
Mol Form: C_4H_9N
M$_r$: 71.12
MS: WILEY 66 28(100) 42(79) 56(56) 30(51) 43(39) 70(32) 41(32) 71(12)
m/e: 71.07

HODOC No: 01531
CAS No: 1072-44-2
Beilstein No: 20^4, 5
Name: Aziridine, 1-methyl-
Synonym: *N*-Methylaziridine
Mol Form: C_3H_7N
M$_r$: 57.09
bp: 27.5
***d*:** $0.7572^{19.5}$
***n*:** 1.3885^{19}
***s*:** H_2O 5
MS: WILEY 27 42(100) 15(49) 57(33) 28(27) 27(18) 41(11) 26(10) 14(9)
m/e: 57.06

HODOC No: 01532
CAS No: 75-55-8
Beilstein No: 20^4, 56
Name: Aziridine, 2-methyl-
Synonym: 2-Methylaziridine
Structure:

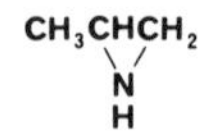

Mol Form: C_3H_7N
M$_r$: 57.09
bp: $66\text{-}7^{751}$
***d*:** 0.812^{16}
IR: μSADP 14580 3230 2940 1450 1410 1370 1240 1090 1060 1020 940 820
H NMR: SAD 16993 0.6 1.2 1.3 1.7 2.0 $CDCl_3$
MS: WILEY 27 28(100) 56(46) 30(32) 57(30) 29(25) 27(22) 15(15) 42(14)
m/e: 57.06

HODOC No: 01533
CAS No: 63021-45-4
Name: Aziridine, 1-(2-naphthalenylcarbonyl)-
Synonym: Aziridine, 1-(2-naphthoyl)-
Mol Form: $C_{13}H_{11}NO$
M$_r$: 197.22
IR: COB 7674 3050 2930 1670 1660 1630 1360 1330 1180 1150 1120 1050 950 910 820 780 770 480
m/e: 197.08

HODOC No: 01534
CAS No: 3164-46-3
Name: Aziridine, 1-(2-phenylethyl)-
Synonym: Aziridine, 1-phenethyl-
Mol Form: $C_{10}H_{13}N$
M$_r$: 147.21
MS: WILEY 846 56(100) 28(15) 104(14) 146(14) 91(11) 42(11) 147(8) 15(8)
m/e: 147.10

HODOC No: 01535
CAS No: 23853-53-4
Name: Aziridine, 1-(1,2,3,4-tetrahydro-2-naphthyl)-
Synonym: Aziridine, 1(1,2,3,4-tetrahydronaphth-2-yl)
Mol Form: $C_{12}H_{15}N$
M$_r$: 173.25
MS: WILEY 1165 130(100) 129(60) 115(46) 104(44) 28(40) 128(39) 131(33) 173(31)
m/e: 173.12

HODOC No: 01536
CAS No: 598-60-7
Beilstein No: 21, 368
Name: Aziridinedione
Synonym: Oximide
Structure:

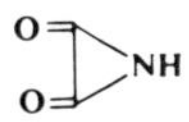

Mol Form: C_2HNO_2
M_r: 71.03
Color: pr (al)
m/e: 71.00

HODOC No: 01537
CAS No: 1072-52-2
Beilstein No: 20^4, 9
Name: 1-Aziridineethanol
Synonym: *N*-(β-Hydroxyethyl)aziridine
Mol Form: C_4H_9NO
M_r: 87.12
bp: 168
***d*:** 1.088
***n*:** 1.4560^{20}
IR: ALD 2:186H
m/e: 87.07

HODOC No: 01538
CAS No: 1121-92-2
Beilstein No: 20^4, 1550
Name: Azocine, octahydro-
Synonym: Azacyclooctane
Mol Form: $C_7H_{15}N$
M_r: 113.20
bp: $51\text{-}3^{15}$
mp: 29
***d*:** 0.896
***n*:** 1.4720^{20}
IR: ALD 2:77B
MS: NBS 146; ALD 2:200C
m/e: 113.12

HODOC No: 01539
CAS No: 14026-63-2
Name: 2-Azocinecarboxylic acid, octahydro-
Synonym: 2-Azocinecarboxylic acid, perhydro
Structure:

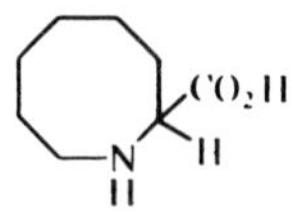

Mol Form: $C_8H_{15}NO_2$
M_r: 157.20
mp: dec 168-71
***s*:** H_2O 2
IR: SADG 16500 3000 2940 2860 2700 1610 1430 1380 1070 810 520
H NMR: SAD 8408 1.5-2.6 3.3 3.8 D_2O
m/e: 157.11

HODOC No: 01540
CAS No: 32566-59-9
Name: 2(1H)-Azocinone, 3,3-dibromohexahydro-
Synonym: Heptanoic acid, 7-amino-2,2-dibromo, ω-lactam
Structure:

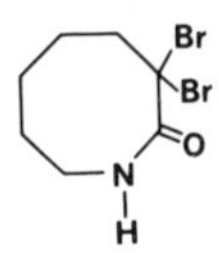

Mol Form: $C_7H_{11}Br_2NO$
M_r: 282.97
mp: 147-8
***s*:** chl 3
IR: SADG 16529 3200 2940 1660 1490 1440 1390 1360 1330 1310 1190 960 830 800 760 660 600 540 460 410 350
H NMR: SAD 8410 1.4-2.1 3.1 3.6 6.6 $CDCl_3$
m/e: 282.92

HODOC No: 01541
CAS No: 26228-96-6
Name: 2(1H)-Azocinone, 3,3-dichlorohexahydro-
Synonym: Heptanoic acid, 7-amino-2,2-dichloro, ω-lactam
Structure:

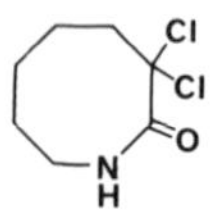

Mol Form: $C_7H_{11}Cl_2NO$
Mr: 196.07
mp: 95-7
***s*:** chl 3
IR: SADG 16516 3220 2940 2880 1650 1490 1450 1400 1360 1350 1320 1260 1200 1040 970 870 790 670 640 540 490
H NMR: SAD 8409 1.4-2.1 2.8 3.6 6.7 $CDCl_3$
m/e: 195.02

HODOC No: 01542
CAS No: 673-66-5
Beilstein No: 21^4, 3227
Name: 2(1H)-Azocinone, hexahydro-
Synonym: 7-Aminoheptanoic acid lactam
Structure:

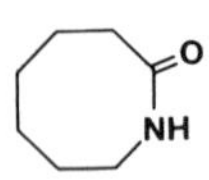

Mol Form: $C_7H_{13}NO$
Mr: 127.18
bp: $148\text{-}50^{10}$
mp: 35-8
IR: ALD 2:417F
MS: ALD 3:129D
m/e: 127.10

HODOC No: 01543
CAS No: 80-49-9
Beilstein No: 21^4, 199
Name: 8-Azoniabicyclo[3.2.1]octane, 3-[(hydroxyphenylacetyl)oxy]-8,8-dimethyl-, bromide, endo-(±)-
Synonym: 1αH,5αH-Tropanium, 3α-hydroxy-8-methyl-, bromide, mandelate
Structure:

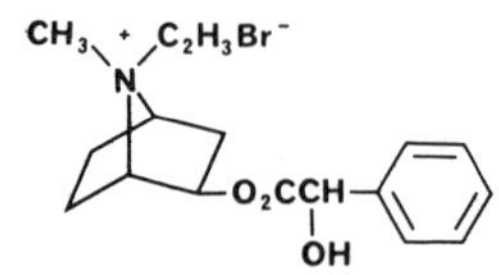

Mol Form: $C_{17}H_{24}BrNO_3$
Mr: 369.27
mp: 192
***s*:** DMSO 2
IR: SADG 21004 3450 3020 2950 1740 1450 1400 1350 1340 1250 1180 1110 1060 1050 940 740 700 600 570
H NMR: SAD 10238 1.2-3.0 3.1 3.2 3.9 5.0 5.2 6.1 7.4 DMSOd6
m/e: 369.09

HODOC No: 01544
CAS No: 5661-71-2
Beilstein No: 20^4, 1606
Name: 1H-Azonine, octahydro-
Synonym: Azacyclononane
Structure:

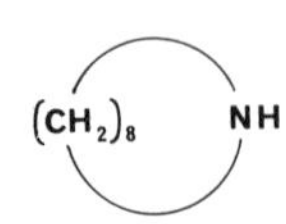

Mol Form: $C_8H_{17}N$
Mr: 127.22
bp: 123
***d*:** 0.833
***n*:** 1.4760^{20}
IR: ALD 2:200F
MS: NBS 239; ALD 2:77C
m/e: 127.14

HODOC No: 01545
CAS No: 32566-65-7
Name: 2H-Azonin-2-one, 3-bromooctahydro-
Synonym: Caprylic acid, 8-amino-2-bromo, ω-lactam
Mol Form: $C_8H_{14}BrNO$
M_r: 219.10
mp: 146-7
***s*:** chl 3
IR: SADG 16523 3260 2940 1660 1560 1450 1360 1210 1020 880 730 690
H NMR: SAD 8412 1.2-2.4 2.8-4.1 4.5 6.4 $CDCl_3$
m/e: 219.03

HODOC No: 01546
CAS No: 32566-64-6
Name: 2H-Azonin-2-one, 3-chlorooctahydro-
Synonym: Caprylic acid, 8-amino-2-chloro, ω-lactam
Structure:

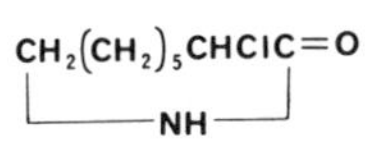

Mol Form: $C_8H_{14}ClNO$
M_r: 175.65
mp: 132-3
***s*:** chl 3
IR: SADG 16510 3260 2950 1660 1560 1450 1360 1230 790 710 690
H NMR: SAD 8414 1.5 2.1 3.3 4.5 6.7 $CDCl_3$
m/e: 175.08

HODOC No: 01547
CAS No: 32566-63-5
Name: 2H-Azonin-2-one, 3,3-dibromooctahydro-
Synonym: Caprylic acid, 8-amino-2,2-dibromo, ω-lactam
Mol Form: $C_8H_{13}Br_2NO$
M_r: 297.00
mp: 156
***s*:** chl 3
IR: SADG 16530 3260 2940 1660 1560 1450 1360 1210 730 690
H NMR: SAD 8413 1.6 2.7 3.4 6.8 $CDCl_3$
m/e: 296.94

HODOC No: 01548
CAS No: 935-30-8
Beilstein No: 21^4, 3260
Name: 2H-Azonin-2-one, octahydro-
Synonym: 8-Aminooctanoic acid lactam
Structure:

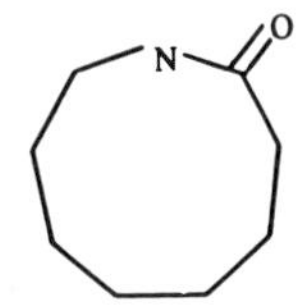

Mol Form: $C_8H_{15}NO$
M_r: 141.21
mp: 76-7
***s*:** chl 3
IR: SADG 16504 3300 3060 2970 2940 2860 1650 1530 1440 1330 1280 1150 1030 860 790 730
UV: OES 3-171 <219 EtOH
H NMR: SAD 8411 1.6 2.4 3.4 7.0 $CDCl_3$
m/e: 141.12

HODOC No: 01549
CAS No: 275-51-4
Beilstein No: 5^4, 1636
Name: Azulene
Synonym: Bicyclo[5.3.0]decapentaene
Structure:

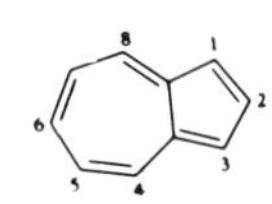

Mol Form: $C_{10}H_8$
M_r: 128.16
bp: 270d, $115\text{-}35^{10}$
mp: 99-100
Color: bl or grsh bk lf (al)
***s*:** H_2O 1; al 3; eth 3; ace 3; chl 2; ac 3; os 3
IR: µSADP 10783 2940 1590 1540 1450 1430 1390 1280 1210 1050 960 950 890 790 760 730
UV: OES 6-262 700(95) 633(257) 580(316) 352(501) 341(3981) 326(2818) 295(3548) 279(43652) 273(47863) 238(14791) MeOH
H NMR: VAR 551 7.1 7.4 7.6 7.9 8.3 $CDCl_3$
C NMR: JJ 367 FT 117.9 122.6 136.4 136.9 137.0 140.1 $CDCl_3$
MS: WILEY 553 128(100) 51(16) 102(14) 127(12) 129(11) 64(10) 63(10) 50(9)
m/e: 128.06

HODOC No: 01550
CAS No: 489-84-9
Beilstein No: 5^4, 1751
Name: Azulene, 1,4-dimethyl-7-(1-methylethyl)-
Synonym: Azulene, 7-isopropyl-1,4-dimethyl-
Mol Form: $C_{15}H_{18}$
M_r: 198.29
bp: $167\text{-}8^{12}$
mp: 31.5
***d*:** $0.973^{20/0}$
Color: bl-vt pl (al)
***s*:** al 3; eth 3; AcOEt 3
IR: μSADP 17428 3330 2940 1700 1610 1560 1470 1390 1210 1160 1050 1020 960 920 810 770 710
UV: SAD 7335 288(17500) 245(14700) MeOH
MS: NBS 1166
m/e: 198.14

HODOC No: 01551
CAS No: 529-08-8
Beilstein No: 5^4, 1754
Name: Azulene, 4,8-dimethyl-2-(1-methylethyl)-
Synonym: Azulene, 2-isopropyl-4,8-dimethyl-
Mol Form: $C_{15}H_{18}$
M_r: 198.29
bp: $140\text{-}60^{2}$
mp: 31-2
Color: red-vt nd (al)
***s*:** Et_2O 4; EtOH 4
UV: HBCP
m/e: 198.14

HODOC No: 01552
CAS No: 941-81-1
Beilstein No: 5^4, 1721
Name: Azulene, 4,6,8-trimethyl-
Synonym: 4,6,8-Trimethylazulene
Mol Form: $C_{13}H_{14}$
M_r: 170.24
mp: 79-81
***s*:** chl 3
IR: μCOB 3553 2940 1590 1540 1490 1430 1370 1330 1220 1080 850 750 710 670
UV: OES 5-369 350(5012) 344(3981) 335(3981) 291(50119) 286(50119) 245(31623) cy 649(182) 588(407) 565(447) 547(479) hx
H NMR: SAD 8699 2.5 2.8 6.9 7.3 7.6 $CDCl_3$
m/e: 170.11

HODOC No: 01553
CAS No: 1055-26-1
Name: Azulene, 1,2,3-triphenyl-
Synonym: 1,2,3-Triphenylazulene
Mol Form: $C_{28}H_{20}$
M_r: 356.44
IR: SADG 17744 3200 3000 2940 1580 1530 1480 1400 1330 1260 1160 1090 940 840 760 710 610 580
UV: OES 5-888 303(12882) 246(48978)
MS: WILEY 2025 356(100) 357(31) 278(9) 276(9) 139(8) 277(6) 358(5) 163(5)
m/e: 356.16

HODOC No: 01554
CAS No: 489-86-1
Beilstein No: 6^3, 412
Name: 5-Azulenemethanol, 1,2,3,4,5,6,7,8-octahydro-α,α,3,8-tetramethyl-, [3*S*-(3α,5α,8α)]-
Synonym: Guai-1(5)-en-11-ol
Structure:

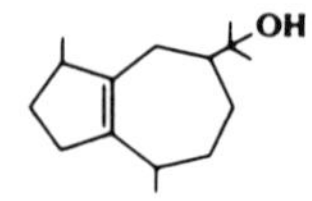

Mol Form: $C_{15}H_{26}O$
M_r: 222.36
bp: 288d, 165^{17}
mp: 91
***d*:** $0.9074^{100/4}$
***n*:** 1.4716^{100}
α: $[\alpha]^{20/D}$-30 al,4%
Color: trg pr (al)
***s*:** H_2O 1; al 3; eth 3
MS: WILEY 1582 59(100) 161(89) 107(62) 91(58) 43(54) 105(53) 41(53) 222(16)
m/e: 222.20

HODOC No: 01555
CAS No: 6606-18-4
Name: Barbituric acid, 5-allyl-5-(1-cyclohexen-1-yl)-2-thio-, monosodium salt
Synonym: Barbituric acid, 5-allyl-5(1-cyclohexenyl)-2-thio,sodium salt
Mol Form: $C_{13}H_{11}N_2NaO_2S$
M_r: 288.28
UV: LANG 1424 296(18197) 239(6026) EtOH
m/e: 282.04

HODOC No: 01556
CAS No: 102747-81-9
Beilstein No: 24^4, 2035
Name: Barbituric acid, 5,5-diallyl-1,3-dibenzyl-
Mol Form: $C_{24}H_{24}N_2O_3$
M_r: 388.44
mp: 89
***s*:** chl 3
IR: µSADP 4243 3030 2940 1670 1490 1450 1410 1350 1300 1190 1080 1010 930 760 750 690
UV: SAD 20810 267(253) 263(385) 257(495) MeOH
H NMR: SAD 11002 2.7 4.9 5.0 5.3 7.1-7.5 $CDCl_3$
m/e: 388.18

HODOC No: 01557
CAS No: 103279-05-6
Name: Barbituric acid, 5,5-diallyl-1-octadecyl-
Synonym: Barbituric acid, 5,5-dially-1-octadecyl
Mol Form: $C_{28}H_{48}N_2O_3$
M_r: 460.67
mp: 162-3
***s*:** chl 3
IR: µSADP 4394 3225 3125 2940 1725 1665 1450 1410 1390 1315 1205 1085 1000 925 815 720
H NMR: SAD 10839 0.9 1.3 2.7 3.8 5.1 5.5 $CDCl_3$
m/e: 460.37

HODOC No: 01558
CAS No: 202-94-8
Name: 11H-Benz[bc]aceanthrylene
Synonym: Benz[a]anthracene, 1,12-methylene-
Mol Form: $C_{19}H_{12}$
M_r: 240.29
UV: FO 516 390(3162) 369(6310) 360(5012) 349(10000) 339(6310) 333(6310) 321(5012) 306(6310) 293(39811) 288(50119) 281(50119) 272(31623) 256(39811) EtOH
m/e: 240.09

HODOC No: 01559
CAS No: 479-23-2
Beilstein No: 5^3, 2469
Name: Benz[j]aceanthrylene, 1,2-dihydro-
Synonym: Cholanthrene
Structure:

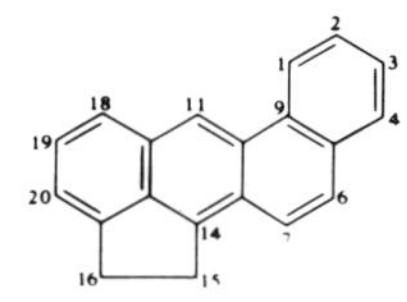

Mol Form: $C_{20}H_{14}$
M_r: 254.31
bp: sub $210^{0.2}$
mp: 174-5d
Color: pa ye lf (Bz-al)
***s*:** H_2O 1; al 3; bz 3; HOAc 3; lig 3; tol 3
UV: OES 1-842 295(79433) 285(63096) 274(39811) 262(39811) EtOH
m/e: 254.11

HODOC No: 01560
CAS No: 56-49-5
Beilstein No: 5^3, 2484
Name: Benz[j]aceanthrylene, 1,2-dihydro-3-methyl-
Synonym: Cholanthrene, 3-methyl-
Mol Form: $C_{21}H_{16}$
M_r: 268.34
bp: 280^{80}
mp: 180
***d*:** 1.28^{20}
Color: yesh nd (Bz)
***s*:** H_2O 1
UV: OES 2-653 188(25119) hp
m/e: 268.13

HODOC No: 01561
CAS No: 205-99-2
Beilstein No: 5^3, 2516
Name: Benz[e]acephenanthrylene
Synonym: 3,4-Benz[e]acephenanthrylene
Structure:

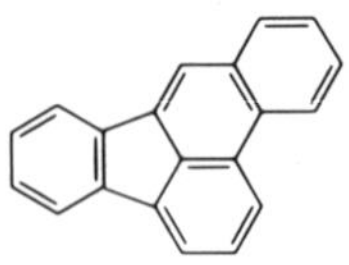

Mol Form: $C_{20}H_{12}$
M_r: 252.30
mp: 168
Color: nd (Bz)
s: H_2O 1; bz 5
IR: μCOB 3545 1590 1490 1410 1280 1210 1150 950 890 820 780 750
UV: HBCP 369(6918) 338(9120) 301(34674) 256(34674) al
MS: NBS 1780
m/e: 252.09

HODOC No: 01562
CAS No: 225-11-6
Beilstein No: 20^4, 4292
Name: Benz[a]acridine
Synonym: 7-Azabenz[a]anthracene
Structure:

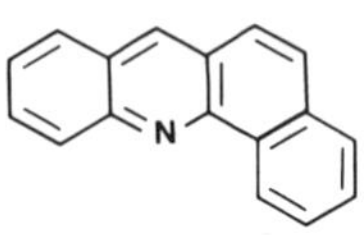

Mol Form: $C_{17}H_{11}N$
M_r: 229.26
mp: 108
Color: nd (dil al)
s: Bz 4; Et_2O 4; EtOH 4
m/e: 229.09

HODOC No: 01563
CAS No: 257-89-6
Beilstein No: 20^4, 4288
Name: Benz[b]acridine
Synonym: 5-Azanaphthacene
Structure:

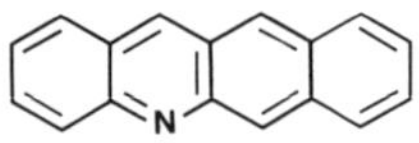

Mol Form: $C_{17}H_{11}N$
M_r: 229.26
mp: 225-6
Color: pr (Chl)
s: Bz 4; Et_2O 4; EtOH 4
m/e: 229.09

HODOC No: 01564
CAS No: 225-51-4
Beilstein No: 20^4, 4291
Name: Benz[c]acridine
Synonym: 12-Azabenz[a]anthracene
Structure:

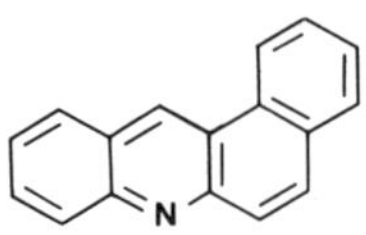

Mol Form: $C_{17}H_{11}N$
M_r: 229.26
mp: 132
Color: nd (dil al)
s: Bz 4; Et_2O 4; EtOH 4
m/e: 229.09

HODOC No: 01565
CAS No: 55044-74-1
Name: Benz[b]acridine, 1,2,3,4,7,8,9,10-octahydro-12-methyl-
Synonym: 2,3:6,7-Dibenzoquinoline 12-methyl-1,2,3,4,7,8,9,10-octahydro
Structure:

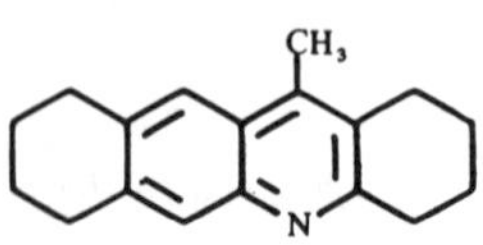

Mol Form: $C_{18}H_{21}N$
M_r: 251.35
MS: WILEY 1754 251(100) 250(93) 236(37) 252(19) 39(8) 248(8) 41(8) 222(8)
m/e: 251.17

HODOC No: 01566
CAS No: 55044-73-0
Name: Benz[c]acridine, 1,2,3,4,8,9,10,11-octahydro-7-methyl-
Synonym: 2,3:7,8-Dibenzoquinoline, 1,2-methyl- 1,2,3,4,6,7,8,9-octahydro
Structure:

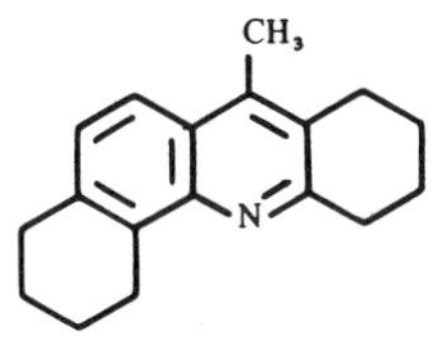

Mol Form: $C_{18}H_{21}N$
Mr: 251.35
MS: **WILEY 1753** 251(100) 250(92) 236(36) 252(19) 222(8) 39(7) 248(7) 237(7)
m/e: 251.17

HODOC No: 01567
CAS No: 100-52-7
Beilstein No: 7^4, 505
Name: Benzaldehyde
Synonym: Artificial Almond Oil
Line Form: C_6H_5CHO
Structure:

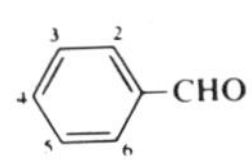

Mol Form: C_7H_6O
Mr: 106.12
bp: 178, 62^{10}
mp: -26
***d*:** $1.0415^{10/4}$
***n*:** 1.5463^{20}
***s*:** H_2O 2; al 5; eth 5; ace 4; bz 4; chl 3; liq NH_3 3; lig 4
IR: **SADG 364** 3030 2778 2703 1695 1667 1587 1449 1389 1299 1205 1163 1075 1020 833 746 690
UV: **OES 6-106** 290(1000) 283(1259) 241(15849) hx
H NMR: **VAR 151** 7.4-8.0 10.0 $CDCl_3$
C NMR: **JJ 229** FT 128.9 129.5 134.2 136.4 192.0 $CDCl_3$
MS: **WILEY 315** 77(100) 105(92) 106(90) 51(51) 50(32) 78(25) 52(12) 74(12)
m/e: 106.04

HODOC No: 01568
CAS No: 84333-02-8
Name: Benzaldehyde, 4-(acetyloxy)-2-chloro-5-methoxy-
Synonym: Benzaldehyde, 2-chloro-4-hydroxy-5-methoxy,acetate
Mol Form: $C_{10}H_9ClO_4$
Mr: 228.62
mp: 97.0-7.5
***s*:** chl 2
IR: **µSADP 4448** 3030 2940 1725 1665 1585 1495 1390 1315 1280 1205 1150 1055 990 915 875 860 835 750
H NMR: **SAD 10586** 2.3 3.9 7.1 7.4 10.3 $CDCl_3$
m/e: 228.02

HODOC No: 01569
CAS No: 881-68-5
Beilstein No: 8^3, 2030
Name: Benzaldehyde, 4-(acetyloxy)-3-methoxy-
Synonym: Vanillin, acetate
Line Form: $[4\text{-}(CH_3CO_2)\text{-}3\text{-}CH_3O]\text{-}C_6H_3CHO$
Mol Form: $C_{10}H_{10}O_4$
Mr: 194.18
mp: 102-3
***s*:** H_2O 2; al 4; eth 4
UV: **OES 2-215** 305(4266) 257(11220) EtOH
MS: **WILEY 1347** 151(100) 43(90) 51(19) 153(18) 109(16) 152(15) 79(13) 194(13)
m/e: 194.06

HODOC No: 01570
CAS No: 51072-66-3
Name: Benzaldehyde, 4(1-adamanty)-3-thiosemicarbazone
Synonym: 2-Bromo-4,5-dimethoxynitrobenzene
Mol Form: $C_{18}H_{23}N_3S$
Mr: 313.44
mp: dec 200
***s*:** chl 2
IR: **SADG 19916** 3300 3150 2900 1540 1510 1410 1360 1300 1220 1200 1140 940 750 690 610
UV: **SAD 17611** 317(33900) 212(17400) MeOH
H NMR: **SAD 10504** 1.7 2.1 2.3 7.2-7.7 7.9 10.1 $CDCl_3$
m/e: 313.16

HODOC No: 01571
CAS No: 529-23-7
Beilstein No: 14^3, 47
Name: Benzaldehyde, 2-amino-
Synonym: *o*-Aminobenzaldehyde
Line Form: 2-$H_2NC_6H_4CHO$
Mol Form: C_7H_7NO
M_r: 121.13
mp: 39-42
Color: silv lf
***s*:** H_2O 2; al 4; eth 4; bz 3; chl 3; lig 1
IR: µSADP 15304 3450 3330 2860 1670 1610 1490 1450 1210 1120 1060 1020 970 890 790 750 710 670
UV: DMS D9-37 371(5000) 260(6000) 232(21000) 227(20000) EtOH
m/e: 121.05

HODOC No: 01572
CAS No: 3398-07-0
Beilstein No: 14^3, 50
Name: Benzaldehyde, 2-amino-, oxime
Synonym: Anthranilaldehyde, oxime
Line Form: 2-H_2N-C_6H_4CH=NOH
Structure:

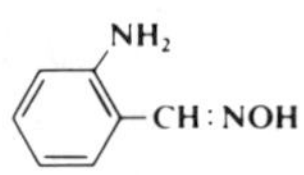

Mol Form: $C_7H_8N_2O$
M_r: 136.15
mp: 135-6
Color: nd (Bz)
***s*:** Bz 4; Et_2O 4; EtOH 4; HOAc 4
m/e: 136.06

HODOC No: 01573
CAS No: 1709-44-0
Beilstein No: 14^3, 353
Name: Benzaldehyde, 3-amino-
Synonym: Benzaldehyde, *m*-amino-
Line Form: 3-$H_2NC_6H_4CHO$
Mol Form: C_7H_7NO
M_r: 121.13
mp: 28-30
Color: nd (AcOEt)
***s*:** eth 3; ac 3
IR: µSADP 15643 3450 3330 2860 1700 1610 1560 1470 1430 1350 1220 1140 970 890 790 690
UV: OES 4-113 256(7586) 245(8913) hx
m/e: 121.05

HODOC No: 01574
CAS No: 2835-66-7
Beilstein No: 14, 28
Name: Benzaldehyde, 3-amino-, oxime
Synonym: Benzaldehyde, *m*-amino-, oxime
Line Form: 3-$H_2NC_6H_4CH$=NOH
Structure:

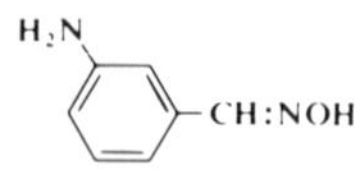

Mol Form: $C_7H_8N_2O$
M_r: 136.15
mp: 88
Color: nd (Bz)
***s*:** Et_2O 4; EtOH 4
MS: NBS 310
m/e: 136.06

HODOC No: 01575
CAS No: 556-18-3
Beilstein No: 14^3, 57
Name: Benzaldehyde, 4-amino-
Synonym: Benzaldehyde, *p*-amino-
Line Form: 4-$H_2NC_6H_4CHO$
Mol Form: C_7H_7NO
M_r: 121.13
mp: 71-2
Color: pl (w)
***s*:** H_2O 3; al 3; eth 3; ac 3
UV: OES 6-111 332(26303) EtOH
m/e: 121.05

HODOC No: 01576
CAS No: 3419-18-9
Beilstein No: 14^3, 58
Name: Benzaldehyde, 4-amino-, oxime
Synonym: Benzaldehyde, *p*-amino-, oxime
Line Form: 4-H_2-NC_6H_4CH=NOH
Mol Form: $C_7H_8N_2O$
M_r: 136.15
mp: 124
Color: ye cr (w)
MS: NBS 310
m/e: 136.06

HODOC No: 01577
CAS No: 109466-84-4
Beilstein No: 14^3, 52
Name: Benzaldehyde, 2-amino-4-nitro-
Line Form: $(4\text{-}NO_2\text{-}2\text{-}H_2N)C_6H_3CHO$
Mol Form: $C_7H_6N_2O_3$
M$_r$: 166.13
mp: 124
Color: nd (w)
m/e: 166.04

HODOC No: 01578
CAS No: 56008-61-8
Beilstein No: 14^3, 53
Name: Benzaldehyde, 2-amino-5-nitro-
Synonym: 2-Amino-5-nitrobenzaldehyde
Line Form: $(5\text{-}NO_2\text{-}2\text{-}H_2N)C_6H_3CHO$
Mol Form: $C_7H_6N_2O_3$
M$_r$: 166.13
mp: 200-5
Color: ye pr (al)
m/e: 166.04

HODOC No: 01579
CAS No: 51818-99-6
Beilstein No: 14^2, 27
Name: Benzaldehyde, 4-amino-3-nitro-
Synonym: 4-Amino-3-nitrobenzaldehyde
Line Form: $(3\text{-}NO_2\text{-}4\text{-}H_2N)C_6H_3CHO$
Mol Form: $C_7H_6N_2O_3$
M$_r$: 166.13
mp: 190-1
Color: ye nd (w)
***s*:** EtOH 4
m/e: 166.04

HODOC No: 01580
CAS No: 1208-03-3
Name: Benzaldehyde, 4-[*bis*(2-chloroethyl)amino]-
Synonym: Benzaldehyde, *p*-[*bis*(2-chloroethyl)amino]-
Mol Form: $C_{11}H_{13}Cl_2NO$
M$_r$: 246.13
mp: 86-7
***s*:** chl 3
IR: SADG 1021 3080 2960 2750 1670 1590 1560 1520 1400 1370 1290 1240 1170 960 820 750 710 610
H NMR: SAD 9841 3.7 3.8 6.8 7.8 9.9 $CDCl_3$
m/e: 245.04

HODOC No: 01581
CAS No: 26459-95-0
Name: Benzaldehyde, 4-[*bis*(2-chloroethyl)amino]-2-methyl-
Synonym: *o*-Tolualdehyde, 4-[*bis*(2-chloroethyl)amino]-
Mol Form: $C_{12}H_{15}Cl_2NO$
M$_r$: 260.15
***s*:** chl 2
IR: SADG 1022 3020 2760 2740 1660 1600 1540 1450 1390 1350 1200 1130 970 850 810 730 710 630
UV: SAD 8897 327(29400) 247(7810) 206(19000) MeOH 328(22800) 250(8220) 205(20200) MeOH:HCl
H NMR: SAD 9842 2.6 3.7 3.8 6.5 6.6 7.7 10.1 CDC_3
m/e: 259.05

HODOC No: 01582
CAS No: 1620-98-0
Beilstein No: 8^4, 601
Name: Benzaldehyde, 3,5-*bis*(1,1-dimethylethyl)-4-hydroxy-
Synonym: Benzaldehyde, 3,5-di-*tert*-butyl-4-hydroxy-
Line Form: $3,5\text{-}[(CH_3)_3C]_2\text{-}4\text{-}HOC_6H_2CHO$
Structure:

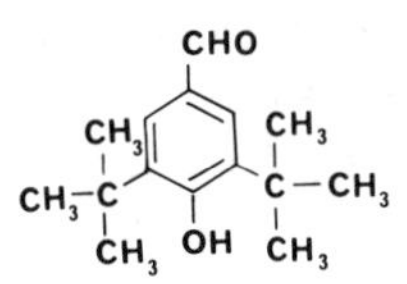

Mol Form: $C_{15}H_{22}O_2$
M$_r$: 234.32
mp: 186
***s*:** chl 3
IR: µCOB 1577 3450 2940 1670 1590 1560 1430 1390 1350 1320 1300 1270 1240 1210 1100 1030 970 930 890 820 790 760 690
UV: SAD 9219 288(14200) 227(16500) 205(18500) MeOH
H NMR: SAD 6132 1.5 5.9 7.8 9.9 $CDCl_3$
m/e: 234.16

HODOC No: 01583
CAS No: 5447-02-9
Beilstein No: 8^3, 2029
Name: Benzaldehyde, 3,4-*bis*(phenylmethoxy)-
Synonym: Benzaldehyde, 3,4-*bis*(benzyloxy)-
Line Form: 3,4-$(C_6H_5CH_2O)_2$-C_6H_3CHO
Mol Form: $C_{21}H_{18}O_3$
M_r: 318.35
mp: 88-90
IR: ALD 3:927B
MS: ALD 6:102A
m/e: 318.13

HODOC No: 01584
CAS No: 6630-33-7
Beilstein No: 7^4, 578
Name: Benzaldehyde, 2-bromo-
Synonym: Benzaldehyde, *o*-bromo-
Line Form: 2-BrC_6H_4CHO
Mol Form: C_7H_5BrO
M_r: 184.01
bp: 230, 118^{12}
mp: 21-2
***n*:** $1.5925^{20/5}$
***s*:** H_2O 1; al 4; bz 4; ctc 2
IR: μSADP 9988 3130 2860 1700 1590 1470 1450 1390 1270 1190 1110 1040 1030 820 750 680
UV: SAD 18427 282(687) MeOH
H NMR: SAD 8591 7.3-7.7 7.8 10.3 CCl_4
m/e: 183.95

HODOC No: 01585
CAS No: 3132-99-8
Beilstein No: 7^4, 579
Name: Benzaldehyde, 3-bromo-
Synonym: Benzaldehyde, *m*-bromo-
Line Form: 3-BrC_6H_4CHO
Mol Form: C_7H_5BrO
M_r: 184.01
bp: 233-6
***n*:** 1.5935^{20}
***s*:** H_2O 1; al 4; eth 4; ctc 2
IR: μSADP 19509 2860 1700 1560 1470 1430 1390 1280 1190 1060 1030 880 860 790 750 700 680
UV: SAD 17179 291(790) 245(6500) MeOH
H NMR: SAD 8592 7.4 7.6-7.9 8.0 10.0 CCl_4
m/e: 183.95

HODOC No: 01586
CAS No: 1122-91-4
Beilstein No: 7^4, 579
Name: Benzaldehyde, 4-bromo-
Synonym: Benzaldehyde, *p*-bromo-
Line Form: 4-BrC_6H_4CHO
Mol Form: C_7H_5BrO
M_r: 184.01
bp: $66\text{-}8^2$
mp: 67
Color: lf (dil al)
***s*:** H_2O 1; al 4; bz 4; chl 2
IR: SADG 10081 3030 2857 1695 1587 1471 1408 1389 1299 1282 1190 1163 1064 1000 826 806 680
UV: SAD 5724 246 MeOH
H NMR: SAD 4047 7.7 10.0 $CDCl_3$
m/e: 183.95

HODOC No: 01587
CAS No: 5392-10-9
Beilstein No: 8^3, 2054
Name: Benzaldehyde, 2-bromo-4,5-dimethoxy-
Synonym: Veratraldehyde, 6-bromo-
Line Form: [6-Br-3,4-$(CH_3O)_2$]-C_6H_2CHO
Mol Form: $C_9H_9BrO_3$
M_r: 244.06
mp: 150-1
***s*:** chl 3
IR: SADG 22380 3060 3000 2960 2860 1670 1580 1500 1460 1440 1380 1330 1260 1210 1170 1140 1030 1010 970 860 800 730 650 580 440 400
H NMR: SAD 11775 3.9 4.0 7.1 7.4 10.2 $CDCl_3$
m/e: 243.97

HODOC No: 01588
CAS No: 6948-30-7
Beilstein No: 8^3, 2051
Name: Benzaldehyde, 3-bromo-4,5-dimethoxy-
Synonym: Veratraldehyde, 5-bromo-
Line Form: 5-Br-3,4-$(CH_3O)_2C_6H_2CHO$
Mol Form: $C_9H_9BrO_3$
M_r: 244.06
mp: 150-1
IR: ALD 2:809D
MS: ALD 6:99B
m/e: 243.97

HODOC No: 01589
CAS No: 22532-60-1
Beilstein No: 8^2, 74
Name: Benzaldehyde, 2-bromo-4-hydroxy-
Synonym: 2-Bromo-4-hydroxybenzaldehyde
Line Form: (2-Br-4-HO)C_6H_3CHO
Mol Form: $C_7H_5BrO_2$
M_r: 200.01
mp: 159.5
Color: pa ye nd (w)
m/e: 199.95

HODOC No: 01590
CAS No: 2973-80-0
Beilstein No: 8^4, 248
Name: Benzaldehyde, 2-bromo-5-hydroxy-
Synonym: 2-Bromo-5-hydroxybenzaldehyde
Line Form: (2-Br-5-HO)C_6H_3CHO
Mol Form: $C_7H_5BrO_2$
M_r: 200.01
mp: 135
Color: nd (w)
***s*:** Ac 4; Bz 4; Et_2O 4; EtOH 4
m/e: 199.95

HODOC No: 01591
CAS No: 2973-78-6
Beilstein No: 8^3, 290
Name: Benzaldehyde, 3-bromo-4-hydroxy-
Synonym: 3-Bromo-4-hydroxybenzaldehyde
Line Form: (3-Br-4-HO)C_6H_3CHO
Mol Form: $C_7H_5BrO_2$
M_r: 200.01
mp: 124
Color: lf (w)
***s*:** H_2O 2; al 4; eth 4; ace 4; bz 3; chl 3; HOAc 3; lig 2
MS: ALDN 163 201(100) 199(96) 200(65) 202(63) 74(23) 171(17) 173(17) 63(16)
m/e: 199.95

HODOC No: 01592
CAS No: 22532-62-3
Beilstein No: 8^4, 231
Name: Benzaldehyde, 4-bromo-2-hydroxy-
Synonym: Salicylaldehyde, 4-bromo-
Line Form: (4-Br-2-HO(C_6H_3CHO
Mol Form: $C_7H_5BrO_2$
M_r: 200.01
mp: 52
Color: nd (dil al)
***s*:** Ac 4; Bz 4; Et_2O 4; EtOH 4
m/e: 199.95

HODOC No: 01593
CAS No: 20035-32-9
Beilstein No: 8^2, 56
Name: Benzaldehyde, 4-bromo-3-hydroxy-
Line Form: (4-Br-3-HO)C_6H_3CHO
Mol Form: $C_7H_5BrO_2$
M_r: 200.01
mp: 131.5
***s*:** Ac 4; Bz 4; Chl 4
m/e: 199.95

HODOC No: 01594
CAS No: 1761-61-1
Beilstein No: 8^4, 228
Name: Benzaldehyde, 5-bromo-2-hydroxy-
Synonym: Salicylaldehyde, 5-bromo-
Line Form: (5-Br-2-HO)C_6H_3CHO
Mol Form: $C_7H_5BrO_2$
M_r: 200.01
mp: 105-6
Color: nd (al), lf (eth)
***s*:** H_2O 1; al 3; eth 3; chl 2
IR: µCOB 4058 3230 2860 2780 1670 1610 1590 1470 1370 1320 1280 1220 1180 890 880 830 770 730 710
UV: SAD 1609 336 251 221 MeOH
H NMR: SAD 6363 6.9 7.6 7.7 9.9 $CDCl_3$
m/e: 199.95

HODOC No: 01595
CAS No: 2973-76-4
Beilstein No: 8^4, 1781
Name: Benzaldehyde, 3-bromo-4-hydroxy-5-methoxy-
Synonym: Vanillin, 5-bromo-
Line Form: (5-Br-4-HO-3-CH_3O)-C_6H_2CHO
Mol Form: $C_8H_7BrO_3$
M_r: 230.04
mp: 164-6
Color: pl (HOAc), nd, pl (al)
***s*:** H_2O 1; al 3; eth 2; bz 2 $DMSOd_6$ 3
IR: SADG 1632 3333 3030 2857 1667 1587 1493 1471 1429 1351 1282 1149 1042 847 833 787 680
UV: SAD 9321 286(10100) 232(15900) 214(17600) MeOH
H NMR: SAD 9200 4.0 7.4 7.7 9.8 $DMSOd_6$
m/e: 229.96

HODOC No: 01596
CAS No: 25016-01-7
Beilstein No: 8^3, 186
Name: Benzaldehyde, 5-bromo-2-methoxy-
Synonym: *o*-Anisaldehyde, 5-bromo-
Line Form: (5-Br-2-CH_3O)C_6H_3CHO
Mol Form: $C_8H_7BrO_2$
M_r: 214.04
mp: 116-9
IR: SADG 20813 3100 2880 2760 1680 1590 1470 1450 1380 1250 1230 1170 1110 1010 870 810 640 610 510
H NMR: SAD 11359 3.9 6.9 7.6 7.9 10.4 $CDCl_3$
MS: ALD 6:88D
m/e: 213.96

HODOC No: 01597
CAS No: 96458-82-1
Name: Benzaldehyde, (bromotriphenylphosphoranyl)hydrazone
Synonym: Phosphonium, (benzylidenehydrazino)-triphenyl,bromide
Line Form: $(C_6H_5)_3PBrNHN{=}CHC_6H_5$
Mol Form: $C_{25}H_{22}BrN_2P$
M_r: 460.32
mp: 208
***s*:** chl 3
IR: SADG 16435 3060 2900 2820 2680 1590 1480 1440 1210 1110 1030 990 970 950 840 750 730 690 550 520
UV: SAD 17046 274(21000) MeOH
H NMR: SAD 8913 7.2-8.1 9.2 $CDCl_3$
m/e: 460.07

HODOC No: 01598
CAS No: 5736-88-9
Beilstein No: 8^4, 255
Name: Benzaldehyde, 4-butoxy-
Synonym: Benzaldehyde, *p*-butoxy-
Line Form: 4-(C_4H_9O)-C_6H_4CHO
Mol Form: $C_{11}H_{14}O_2$
M_r: 178.22
bp: $148\text{-}9^{10}$
m/e: 178.10

HODOC No: 01599
CAS No: 1200-14-2
Beilstein No: 7^4, 743
Name: Benzaldehyde, 4-butyl-
Synonym: Benzaldehyde, *p*-butyl-
Line Form: 4-C_4H_9-C_6H_4CHO
Mol Form: $C_{11}H_{14}O$
M_r: 162.22
bp: 123^7
***n*:** 1.5265
m/e: 162.10

HODOC No: 01600
CAS No: 89-98-5
Beilstein No: 7^4, 561
Name: Benzaldehyde, 2-chloro-
Synonym: Benzaldehyde, *o*-chloro-
Line Form: 2-ClC_6H_4CHO
Mol Form: C_7H_5ClO
M_r: 140.56
bp: 211.9, 84.3^{10}
mp: 12.4
***d*:** $1.2483^{20/4}$
***n*:** 1.5662^{20}
Color: nd
***s*:** H_2O 2; al 3; eth 3; ace 3; bz 3; ctc 3
IR: µ COB 1344 3030 2860 1700 1610 1470 1450 1390 1270 1210 1160 1110 1050 1040 850 820 760 710
Raman: SAD 567 3070 2870 2760 1700 1650 1600 1470 1450 1400 1300 1280 1210 1170 1120 1070 1040 1010 960 880 830 720 690 640 490 420 300 250 210 170
UV: SAD 208 294 246 cyhex
H NMR: SAD 50 7.4 7.8 10.5 CCl_4
MS: WILEY 737 138(100) 139(79) 111(44) 140(38) 50(32) 76(20) 77(20) 113(15)
m/e: 140.00

HODOC No: 01601
CAS No: 34158-76-4
Beilstein No: 15^3, 86
Name: Benzaldehyde, 2-chloro-, phenylhydrazone
Synonym: Benzaldehyde, *o*-chloro-, phenylhydrazone
Line Form: $2\text{-ClC}_6H_4CH{=}NNHC_6H_5$
Mol Form: $C_{13}H_{11}ClN_2$
M_r: 230.68
mp: 84
IR: SADG 20292 3300 3040 1600 1560 1520 1500 1460 1360 1310 1250 1140 1120 1030 910 880 750 710 690 630 510 450
UV: SAD 17890 352(17200) 299(7310) 249(12900) MeOH
H NMR: SAD 10461 6.6-7.4 7.6 7.9 8.0 $CDCl_3$
m/e: 230.06

HODOC No: 01602
CAS No: 587-04-2
Beilstein No: 7^4, 566
Name: Benzaldehyde, 3-chloro-
Synonym: Benzaldehyde, *m*-chloro-
Line Form: 3-ClC_6H_4CHO
Mol Form: C_7H_5ClO
M_r: 140.56
bp: 213-4, 55^1
mp: 17-8
***d*:** $1.2410^{20/4}$
***n*:** 1.5650^{20}
Color: pr
***s*:** H_2O 2; al 3; eth 3; ace 3; bz 3; chl 2
IR: SADG 18339 3060 2840 2720 1700 1600 1570 1470 1430 1380 1270 1190 1070 900 870 780 720 680
UV: SAD 15920 290(854) 243(7160) MeOH
H NMR: SAD 8786 7.6 7.8 7.9 10.0 $CDCl_3$
MS: WILEY 737 139(100) 140(81) 111(57) 141(38) 142(27) 75(24) 77(22) 50(19)
m/e: 140.00

HODOC No: 01603
CAS No: 104-88-1
Beilstein No: 7^4, 568
Name: Benzaldehyde, 4-chloro-
Synonym: Benzaldehyde, *p*-chloro-
Line Form: 4-ClC_6H_4CHO
Mol Form: C_7H_5ClO
M_r: 140.56
bp: 213-4, $72\text{-}5^3$
mp: 47.5
***d*:** $1.196^{61/4}$
***n*:** $1.555^{61/2}$
Color: pl
***s*:** H_2O 3; al 4; eth 4; ace 3; bz 4; chl 3; CS_2 3; HOAc 3
IR: μCOB 3958 2860 2780 1700 1590 1490 1390 1300 1210 1160 1100 1080 1010 830 820 790 700
Raman: SAD 568 3090 3070 2860 2760 1740 1700 1690 1680 1630 1590 1480 1420 1390 1290 1250 1210 1160 1130 1100 1090 1010 880 840 820 800 740 700 670 630 550 490 360 320 180
UV: SAD 207 254 MeOH
H NMR: VAR 146 7.5 7.8 10.0 $CDCl_3$
MS: WILEY 736 139(100) 140(73) 111(58) 141(39) 75(24) 142(24) 50(20) 113(19)
m/e: 140.00

HODOC No: 01604
CAS No: 3101-04-0
Name: Benzaldehyde, 4-chloro-, (4-chlorophenyl)hydrazone
Synonym: Benzaldehyde, *p*-chloro-, (*p*-chlorophenyl)hydrazone
Mol Form: $C_{13}H_{10}Cl_2N_2$
M_r: 265.13
IR: μCOB 1473 3330 3030 1610 1520 1490 1410 1250 1090 1010 920 830
m/e: 264.02

HODOC No: 01605
CAS No: 1424-66-4
Beilstein No: 14^2, 26
Name: Benzaldehyde, 2-chloro-4-(dimethylamino)-
Synonym: *o*-Chloro-*p*-dimethylaminobenzaldehyde
Line Form: $[4\text{-}(CH_3)_2N\text{-}2\text{-Cl}]C_6H_3CHO$
Mol Form: $C_9H_{10}ClNO$
M_r: 183.63
mp: 80-2
IR: ALD 3:920D
MS: ALD 6:92D
m/e: 183.05

HODOC No: 01606
CAS No: 56962-10-8
Beilstein No: 8^4, 246
Name: Benzaldehyde, 2-chloro-3-hydroxy-
Synonym: 2-Chloro-3-hydroxybenzaldehyde
Line Form: (2-Cl-3-HO)-C_6H_3CHO
Mol Form: $C_7H_5ClO_2$
M_r: 156.56
mp: 139.5
Color: cr (aq, HOAc)
***s*:** EtOH 4; HOAc 4
m/e: 156.00

HODOC No: 01607
CAS No: 56962-11-9
Beilstein No: 8^4, 296
Name: Benzaldehyde, 2-chloro-4-hydroxy-
Synonym: *o*-Chloro-*p*-hydroxybenzaldehyde
Line Form: (2-Cl-4-HO)C_6H_3CHO
Mol Form: $C_7H_5ClO_2$
M_r: 156.56
mp: 147-8
Color: nd (w or HOAc)
***s*:** Et_2O 4; EtOH 4; HOAc 4
m/e: 156.00

HODOC No: 01608
CAS No: 7310-94-3
Beilstein No: 8^1, 526
Name: Benzaldehyde, 2-chloro-5-hydroxy-
Synonym: 2-Chloro-5-hydroxybenzaldehyde
Line Form: (2-Cl-5-HO)C_6H_3CHO
Mol Form: $C_7H_5ClO_2$
M_r: 156.56
mp: 110-1
Color: nd (w or HOAc)
***s*:** EtOH 4; HOAc 4
m/e: 156.00

HODOC No: 01609
CAS No: 1927-94-2
Beilstein No: 8^2, 44
Name: Benzaldehyde, 3-chloro-2-hydroxy-
Synonym: Salicylaldehyde, 3-chloro-
Mol Form: $C_7H_5ClO_2$
M_r: 156.56
mp: 55
***s*:** os 3
UV: LANG 2121 336(2570) 263(6310) 258(6310) 218(12882) cy 390(1122) 338(1047) 262(2291) 216(4898) EtOH
m/e: 156.00

HODOC No: 01610
CAS No: 2420-16-8
Beilstein No: 8^4, 296
Name: Benzaldehyde, 3-chloro-4-hydroxy-
Synonym: 3-Chloro-4-hydroxybenzaldehyde
Line Form: (3-Cl-4-HO)C_6H_3CHO
Mol Form: $C_7H_5ClO_2$
M_r: 156.56
bp: $149\text{-}50^{14}$
mp: 139
Color: nd (w)
***s*:** Et_2O 4; EtOH 4
m/e: 156.00

HODOC No: 01611
CAS No: 2420-26-0
Beilstein No: 8^4, 223
Name: Benzaldehyde, 4-chloro-2-hydroxy-
Synonym: Salicylaldehyde, 4-chloro-
Line Form: (4-Cl-2-HO)C_6H_3CHO
Mol Form: $C_7H_5ClO_2$
M_r: 156.56
mp: 52.5
Color: nd (al or aq HOAc)
***s*:** Ac 4; Bz 4; Et_2O 4
m/e: 156.00

HODOC No: 01612
CAS No: 56962-12-0
Beilstein No: 8^4, 247
Name: Benzaldehyde, 4-chloro-3-hydroxy-
Synonym: 4-Chloro-3-hydroxybenzaldehyde
Line Form: (4-Cl-3-HO)C_6H_3CHO
Mol Form: $C_7H_5ClO_2$
M_r: 156.56
mp: 121
Color: nd (al)
***s*:** Ac 4; Bz 4; Et_2O 4; EtOH 4
m/e: 156.00

HODOC No: 01613
CAS No: 635-93-8
Beilstein No: 8^4, 224
Name: Benzaldehyde, 5-chloro-2-hydroxy-
Synonym: Salicylaldehyde, 5-chloro-
Line Form: (5-Cl-2-HO)C_6H_3CHO
Mol Form: $C_7H_5ClO_2$
M_r: 156.56
bp: 105^{12}
mp: 99-100
Color: pl (al)
s: H_2O 1; al 4; eth 3; alk 3
IR: µCOB 330 3230 2940 2860 1700 1670 1610 1560 1520 1470 1390 1320 1280 1220 1160 1110 1090 1020 950 910 890 840 830 770 730 700 660 640
UV: OES 2-84 390(6026) 234(19055) 0.1N NaOH 338(3090) 255(8913) 222(23442) pH=3
H NMR: SAD 9850 7.0 7.4 7.5 9.8 10.9 $CDCl_3$
MS: WILEY 953 156(100) 155(87) 157(36) 158(32) 110(18) 127(16) 99(14) 138(13)
m/e: 156.00

HODOC No: 01614
CAS No: 1595-14-8
Beilstein No: 8^3, 183
Name: Benzaldehyde, 5-chloro-2-hydroxy-, oxime
Synonym: Salicylaldehyde, 5-chloro-, oxime
Line Form: (5-Cl-2-HO)C_6H_3CH=NOH
Mol Form: $C_7H_6ClNO_2$
M_r: 171.58
mp: 128
Color: nd (w)
s: EtOH 4
m/e: 171.01

HODOC No: 01615
CAS No: 19463-48-0
Beilstein No: 8^4, 1780
Name: Benzaldehyde, 3-chloro-4-hydroxy-5-methoxy-
Synonym: Vanillin, 5-chloro-
Line Form: (5-Cl-4-HO-3-CH_3O)C_6H_2CHO
Mol Form: $C_8H_7ClO_3$
M_r: 186.59
mp: 165
Color: tetr
s: H_2O 1; al 3; HOAc 3
IR: SADG 977 3333 3030 2857 1667 1587 1493 1471 1429 1351 1282 1149 1053 971 847 794 676
UV: SAD 8579 286(10500) 233(17200) 212(17500) MeOH
H NMR: SAD 14781 3.5 3.9 7.4 7.5 8.0 poly D
m/e: 186.01

HODOC No: 01616
CAS No: 23731-06-8
Beilstein No: 8^4, 433
Name: Benzaldehyde, 5-(chloromethyl)-2-hydroxy-
Synonym: 2,5-Cresotaldehyde, α-chloro-
Line Form: (3-ClCH_2-6-HO)-C_6H_3CHO
Mol Form: $C_8H_7ClO_2$
M_r: 170.59
mp: 88
IR: µCOB 3968 3130 2860 1670 1590 1490 1370 1280 1270 1210 1150 940 830 770 740 690
m/e: 170.01

HODOC No: 01617
CAS No: 5568-33-2
Beilstein No: 7^4, 610
Name: Benzaldehyde, 2-chloro-4-nitro-
Synonym: 2-Chloro-4-nitrobenzaldehyde
Line Form: (2-Cl-4-NO_2)C_6H_3CHO
Mol Form: $C_7H_4ClNO_3$
M_r: 185.56
mp: 74
Color: nd (lig)
m/e: 184.99

HODOC No: 01618
CAS No: 6361-21-3
Beilstein No: 7^4, 609
Name: Benzaldehyde, 2-chloro-5-nitro-
Synonym: 2-Chloro-5-nitrobenzaldehyde
Line Form: (2-Cl-5-NO_2)C_6H_3CHO
Mol Form: $C_7H_4ClNO_3$
M_r: 185.56
mp: 80
Color: cr (al)
s: EtOH 4; Chl 4
IR: ALD 3:920H
MS: ALD 6:93C
m/e: 184.99

HODOC No: 01619
CAS No: 6361-22-4
Beilstein No: 7^4, 609
Name: Benzaldehyde, 2-chloro-6-nitro-
Synonym: 2-Chloro-6-nitrobenzaldehyde
Line Form: (2-Cl-6-NO_2)C_6H_3CHO
Mol Form: $C_7H_4ClNO_3$
M_r: 185.56
mp: 69-71
IR: ALD 3:917A
MS: ALD 6:87B
m/e: 184.99

HODOC No: 01620
CAS No: 5551-11-1
Beilstein No: 7^4, 609
Name: Benzaldehyde, 4-chloro-2-nitro-
Synonym: 4-Chloro-2-nitrobenzaldehyde
Line Form: (4-Cl-2-NO_2)-C_6H_3CHO
Mol Form: $C_7H_4ClNO_3$
M_r: 185.56
mp: 67-8
Color: ye nd
***s*:** Ac 4; Bz 4; Et_2O 4; EtOH 4
m/e: 184.99

HODOC No: 01621
CAS No: 16588-34-4
Beilstein No: 7^4, 609
Name: Benzaldehyde, 4-chloro-3-nitro-
Synonym: 4-Chloro-3-nitrobenzaldehyde
Line Form: (4-Cl-3-NO_2)-C_6H_3CHO
Mol Form: $C_7H_4ClNO_3$
M_r: 185.56
mp: 64-5
***s*:** H_2O 2; chl 3
IR: SADG 10694 3080 2850 2720 1700 1600 1530 1350 1200 1050 920 830 820
UV: SAD 8817 215(15500) MeOH
H NMR: SAD 9754 7.8 8.1 8.4 10.1 $CDCl_3$
m/e: 184.99

HODOC No: 01622
CAS No: 6628-86-0
Beilstein No: 7^3, 919
Name: Benzaldehyde, 5-chloro-2-nitro-
Synonym: 3-Chloro-6-nitrobenzaldehyde
Line Form: (5-Cl-2-NO_2)-C_6H_3CHO
Mol Form: $C_7H_4ClNO_3$
M_r: 185.56
mp: 65-9
IR: ALD 2:806A
m/e: 184.99

HODOC No: 01623
CAS No: 27241-93-6
Name: Benzaldehyde, *o*-chloro-, (*p*-chlorophenyl)hydrazone
Synonym: 2-Chlorobenzaldehyde-*p*-chlorphenyl-hydrazone
Mol Form: $C_{13}H_{10}Cl_2N_2$
M_r: 265.13
IR: µCOB 1472 3330 3030 1590 1560 1520 1280 1250 1150 1090 1050 1030 900 820 750 720
m/e: 264.02

HODOC No: 01624
CAS No: 24083-16-7
Name: Benzaldehyde, 4-(decyloxy)-
Synonym: Benzaldehyde, *p*-(decyloxy)-
Mol Form: $C_{17}H_{26}O_2$
M_r: 262.38
IR: SADG 26705 2920 2850 2720 1690 1600 1580 1510 1460 1420 1390 1310 1250 1210 1150 1100 1010940 850 830 720 640 620 510
Raman: SAD 562 3080 2930 2860 1700 1610 1580 1430 1400 1320 1260 1220 1170 860 840 720 660 650620 340 220
UV: SAD 20379 284(17800) 220(12400) MeOH
H NMR: SAD 15497 0.9 1.1-1.6 1.8 4.0 6.9 7.7 9.8 CCl_4
m/e: 262.19

HODOC No: 01625
CAS No: 90-59-5
Beilstein No: 8^3, 188
Name: Benzaldehyde, 3,5-dibromo-2-hydroxy-
Synonym: Salicylaldehyde, 3,5-dibromo-
Line Form: (3,5-Br_2-2-HO)C_6H_2CHO
Mol Form: $C_7H_4Br_2O_2$
M_r: 277.91
mp: 82-3
***s*:** chl 3
IR: µCOB 4135 3130 2860 1670 1610 1450 1430 1280 1210 1150 960 920 870 740 710 680
UV: SAD 15240 343(2704) 297(840) 260(5514) 223(20400) MeOH
H NMR: SAD 5748 7.6 7.8 9.8 11.4 $CDCl_3$
m/e: 277.86

HODOC No: 01626
CAS No: 6334-18-5
Beilstein No: 7^4, 575
Name: Benzaldehyde, 2,3-dichloro-
Synonym: 2,3-Dichlorobenzaldehyde
Line Form: 2,3-$Cl_2C_6H_3CHO$
Mol Form: $C_7H_4Cl_2O$
M_r: 175.01
mp: 65-7
Color: cr (dil al)
***s*:** Et_2O 4; EtOH 4
m/e: 173.96

HODOC No: 01627
CAS No: 874-42-0
Beilstein No: 7[4], 575
Name: Benzaldehyde, 2,4-dichloro-
Synonym: *o,p*-Dichlorobenzaldehyde
Line Form: 2,4-$Cl_2C_6H_3CHO$
Mol Form: $C_7H_4Cl_2O$
M_r: 175.01
mp: 72
Color: pr
***s*:** H_2O 1; al 3; eth 3; bz 3; chl 3; HOAc 3
IR: **μCOB 3189** 2860 1700 1590 1470 1370 1280 1250 1210 1140 1100 1050 870 850 820 760 690 660
Raman: **SAD 565** 3080 2890 2790 1680 1650 1630 1590 1530 1470 1410 1380 1290 1250 1210 1160 1140 1110 1050 1020 890 850 830 760 710 700 680 580 560 470 420 400 340 310 220 190_160
UV: **SAD 205** 292 256 216 cyhex
H NMR: **SAD 3151** 7.3 7.4 7.8 10.4 CDC_3
MS: **WILEY 1166** 173(100) 175(70) 174(61) 176(39) 145(26) 75(18) 147(17) 74(15)
m/e: 173.96

HODOC No: 01628
CAS No: 6361-23-5
Beilstein No: 7[4], 576
Name: Benzaldehyde, 2,5-dichloro-
Synonym: 2,5-Dichlorobenzaldehyde
Line Form: 2,5-$Cl_2C_6H_3CHO$
Mol Form: $C_7H_4Cl_2O$
M_r: 175.01
bp: 231-3
mp: 58
Color: nd (al)
***s*:** Bz 4; Et_2O 4; EtOH 4; Chl 4
m/e: 173.96

HODOC No: 01629
CAS No: 83-38-5
Beilstein No: 7[4], 576
Name: Benzaldehyde, 2,6-dichloro-
Synonym: *o,o'*-Dichlorobenzaldehyde
Line Form: 2,6-$Cl_2C_6H_3CHO$
Mol Form: $C_7H_4Cl_2O$
M_r: 175.01
mp: 71
Color: nd (lig)
***s*:** Et_2O 4; EtOH 4; lig 4
IR: **ALD 3:916C**
m/e: 173.96

HODOC No: 01630
CAS No: 25185-95-9
Name: Benzaldehyde, 2,6-dichloro-, oxime
Synonym: 2,6-Dichlorobenzaldehyde oxime
Mol Form: $C_7H_5Cl_2NO$
M_r: 190.02
IR: **SADG 26836** 3290 1580 1560 1460 1440 1410 1290 1270 1210 1180 1090 1070 980 910 870 770 710 640 520 460 350 330
Raman: **SAD 572** 3120 3080 3000 1630 1580 1560 1500 1440 1300 1270 1220 1190 1160 1090 1080 920 880 720 670 530 470 420 320 210 200 180
UV: **SAD 20495** 214(2210) MeOH 272(584) MeOH:KOH
H NMR: **SAD 16603** 7.0-7.5 8.3 11.5 poly D
m/e: 188.97

HODOC No: 01631
CAS No: 6287-38-3
Beilstein No: 7[4], 576
Name: Benzaldehyde, 3,4-dichloro-
Synonym: 3,4-Dichlorobenzaldehyde
Line Form: 3,4-$Cl_2C_6H_3CHO$
Mol Form: $C_7H_4Cl_2O$
M_r: 175.01
bp: 247-8
mp: 44
***s*:** H_2O 1; al 3; eth 3; ctc 2
IR: **μCOB 3188** 2860 2700 1700 1590 1560 1470 1410 1370 1280 1270 1190 1140 1030 920 890 880 830 810 740 700 680
Raman: **SAD 566** 3090 3060 2850 2800 2760 2730 1740 1700 1670 1620 1590 1400 1290 1260 1230 1200 1160 1130 1030 1010 960 920 870 830 730 690 560 550 490 450 420 350 320 300 220*
UV: **SAD 2066** 280 MeOH
H NMR: **SAD 8761** 7.6 7.7 7.9 10.0 CCl_4
MS: **WILEY 1166** 173(100) 175(70) 174(58) 145(40) 176(39) 147(25) 75(20) 74(17)
m/e: 173.96

HODOC No: 01632
CAS No: 10203-08-4
Beilstein No: 7^4, 577
Name: Benzaldehyde, 3,5-dichloro-
Synonym: 3,5-Dichlorobenzaldehyde
Line Form: $3,5\text{-}Cl_2C_6H_3CHO$
Mol Form: $C_7H_4Cl_2O$
M_r: 175.01
bp: $235\text{-}40^{748}$
mp: 65
Color: nd or lf (dil HOAc)
s: Ac 4; Bz 4; Et_2O 4; EtOH 4
m/e: 173.96

HODOC No: 01633
CAS No: 56962-13-1
Beilstein No: 8^2, 55
Name: Benzaldehyde, 2,4-dichloro-3-hydroxy-
Line Form: $(2,4\text{-}Cl_2\text{-}3\text{-}HO)C_6H_2CHO$
Mol Form: $C_7H_4Cl_2O_2$
M_r: 191.01
mp: 141
Color: cr (Ace)
s: lig 4
m/e: 189.96

HODOC No: 01634
CAS No: 56962-15-3
Beilstein No: 8^2, 56
Name: Benzaldehyde, 2,4-dichloro-5-hydroxy-
Synonym: Benzaldehyde, 4,6-dichloro-3-hydroxy
Line Form: $(4,6\text{-}Cl_2\text{-}3\text{-}HO)C_6H_2CHO$
Mol Form: $C_7H_4Cl_2O_2$
M_r: 191.01
mp: 130
Color: nd
s: Bz 4; Et_2O 4; EtOH 4
m/e: 189.96

HODOC No: 01635
CAS No: 56962-14-2
Beilstein No: 8^3, 204
Name: Benzaldehyde, 2,6-dichloro-3-hydroxy-
Line Form: $(2,6\text{-}Cl_2\text{-}3\text{-}HO)C_6H_2CHO$
Mol Form: $C_7H_4Cl_2O_2$
M_r: 191.01
mp: 142
Color: cr (w)
s: HOAc 4
m/e: 189.96

HODOC No: 01636
CAS No: 90-60-8
Beilstein No: 8^4, 227
Name: Benzaldehyde, 3,5-dichloro-2-hydroxy-
Synonym: Salicylaldehyde, 3,5-dichloro-
Line Form: $(3,5\text{-}Cl_2\text{-}2\text{-}HO)C_6H_2CHO$
Mol Form: $C_7H_4Cl_2O_2$
M_r: 191.01
mp: 95
Color: ye rh (HOAc)
s: H_2O 1
MS: ALDN 153 190(100) 189(93) 191(67) 192(64) 63(26) 133(21) 172(19) 144(17)
m/e: 189.96

HODOC No: 01637
CAS No: 5331-93-1
Beilstein No: 8, 54
Name: Benzaldehyde, 3,5-dichloro-2-hydroxy-, oxime
Synonym: Salicylaldehyde, 3,5-dichloro-, oxime
Line Form: $(3,5\text{-}Cl_2\text{-}2\text{-}HO)C_6H_2CH{=}NOH$
Mol Form: $C_7H_5Cl_2NO_2$
M_r: 206.02
mp: 195-6
Color: nd (dil al)
s: Bz 4; Et_2O 4; EtOH 4
m/e: 204.97

HODOC No: 01638
CAS No: 2314-36-5
Beilstein No: 8, 81
Name: Benzaldehyde, 3,5-dichloro-4-hydroxy-
Synonym: 3,5-Dichloro-4-hydroxybenzaldehyde
Line Form: $(3,5\text{-}Cl_2\text{-}4\text{-}HO)C_6H_2CHO$
Mol Form: $C_7H_4Cl_2O_2$
M_r: 191.01
mp: 158-9
Color: nd (Chl, dil al)
s: Et_2O 4; EtOH 4
m/e: 189.96

HODOC No: 01639
CAS No: 24454-82-8
Beilstein No: 8^4, 1748
Name: Benzaldehyde, 2,3-diethoxy-
Synonym: 2,3-Diethoxybenzaldehyde
Line Form: $2,3\text{-}(C_2H_5O)_2\text{-}C_6H_3CHO$
Mol Form: $C_{11}H_{14}O_3$
M_r: 194.22
bp: 169^{37}
s: Et_2O 4; EtOH 4
m/e: 194.09

HODOC No: 01640
CAS No: 2029-94-9
Beilstein No: 8^4, 1766
Name: Benzaldehyde, 3,4-diethoxy-
Synonym: 3,4-Diethoxybenzaldehyde
Line Form: 3,4-$(C_2H_5O)_2$-C_6H_3CHO
Mol Form: $C_{11}H_{14}O_3$
M_r: 194.22
bp: 278-30
***s*:** EtOH 4
m/e: 194.09

HODOC No: 01641
CAS No: 120-21-8
Beilstein No: 14^3, 71
Name: Benzaldehyde, 4-(diethylamino)-
Synonym: Benzaldehyde, *p*-(diethylamino)-
Line Form: 4-$(C_2H_5)_2N$-C_6H_4CHO
Mol Form: $C_{11}H_{15}NO$
M_r: 177.24
bp: 174^7
mp: 41
Color: ye nd (w)
***s*:** H_2O 4; al 3; eth 3; bz 3; ctc 3
IR: SADG 542 3030 2941 2703 1667 1587 1563 1538 1408 1351 1316 1282 1250 1190 1176 1075 1020 826 714
UV: SAD 2044 243(6340) MeOH
H NMR: SAD 534 1.2 3.4 6.6 7.6 9.6 CCl_4
MS: WILEY 1198 162(100) 177(36) 134(33) 77(13) 163(12) 106(10) 51(6) 132(6)
m/e: 177.12

HODOC No: 01642
CAS No: 17754-90-4
Name: Benzaldehyde, 4-(diethylamino)-2-hydroxy-
Synonym: Salicylaldehyde, 4-(diethylamino)-
Line Form: 4-$(C_2H_5)_2N$-2-HOC_6H_3CHO
Mol Form: $C_{11}H_{15}NO_2$
M_r: 193.23
mp: 62-4
m/e: 193.11

HODOC No: 01643
CAS No: 24677-78-9
Beilstein No: 8^3, 1979
Name: Benzaldehyde, 2,3-dihydroxy-
Synonym: *o*-Pyrocatechualdehyde
Line Form: 2,3-$(HO)_2C_6H_3CHO$
Mol Form: $C_7H_6O_3$
M_r: 138.12
bp: $119\text{-}20^{16}$
mp: 108
Color: ye nd
***s*:** Ac 4; EtOH 4; HOAc 4
m/e: 138.03

HODOC No: 01644
CAS No: 95-01-2
Beilstein No: 8^4, 1753
Name: Benzaldehyde, 2,4-dihydroxy-
Synonym: β-Resorcylaldehyde
Line Form: 2,4-$(HO)_2C_6H_3CHO$
Mol Form: $C_7H_6O_3$
M_r: 138.12
bp: $220\text{-}8^{22}$
mp: 201-2
Color: nd (eth-lig)
***s*:** H_2O 3; al 4; eth 4; bz 2; chl 4; HOAc 3
IR: μSADP 14539 3450 3030 1610 1590 1490 1430 1390 1320 1240 1160 1120 970 850 800 690
UV: OES 3-100 326(16982) 280(12023) pH=6.98
H NMR: SAD 17186 6.4 6.5 7.4 9.7 9.9-11.8 poly D
m/e: 138.03

HODOC No: 01645
CAS No: 5399-68-8
Beilstein No: 8^4, 1756
Name: Benzaldehyde, 2,4-dihydroxy-, oxime
Synonym: β-Resorcylaldehyde, oxime
Line Form: (2,4-$(HO)_2$-C_6H_3)-CH=NOH
Mol Form: $C_7H_7NO_3$
M_r: 153.13
mp: 192
***s*:** H_2O 3; al 4; eth 4
IR: μNRC 1879 3230 2940 1640 1610 1540 1490 1450 1370 1350 1320 1270 1220 1180 1120 1000 980 950 930 860 850 830 800 750 740 710 660
H NMR: SAD 18933 7.1 8.2 10.3 poly D
m/e: 153.04

HODOC No: 01646
CAS No: 1194-98-5
Beilstein No: 8^4, 1759
Name: Benzaldehyde, 2,5-dihydroxy-
Synonym: Gentisaldehyde
Line Form: 2,5-$(HO)_2C_6H_3CHO$
Mol Form: $C_7H_6O_3$
M_r: 138.12
mp: 99
Color: ye nd (Bz)
***s*:** H_2O 4; EtOH 4; Chl 4
IR: SADP 19175
UV: SAD 6271
m/e: 138.03

HODOC No: 01647
CAS No: 139-85-5
Beilstein No: 8^4, 1762
Name: Benzaldehyde, 3,4-dihydroxy-
Synonym: Protocatechualdehyde
Line Form: 3,4-$(HO)_2C_6H_3CHO$
Mol Form: $C_7H_6O_3$
M_r: 138.12
mp: 153-4
Color: lf (w, to)
***s*:** H_2O 3; al 4; eth 4
IR: SADG 10196 3333 3030 2857 1639 1587 1538 1471 1429 1389 1282 1176 1163 1111 870 800 741 671
UV: SAD 6272 312 280 231 MeOH
H NMR: SAD 4283 7.0 7.3 7.4 9.8 ace
m/e: 138.03

HODOC No: 01648
CAS No: 3343-59-7
Beilstein No: 8^4, 1774
Name: Benzaldehyde, 3,4-dihydroxy-, oxime
Synonym: Protocatechualdehyde, oxime
Line Form: 3,4-$(HO)_2C_6H_3CH{=}NOH$
Mol Form: $C_7H_7NO_3$
M_r: 153.13
mp: 157d
Color: nd (w)
***s*:** H_2O 4; EtOH 4
m/e: 153.04

HODOC No: 01649
CAS No: 6248-20-0
Beilstein No: 8^3, 2132
Name: Benzaldehyde, 2,4-dihydroxy-3-methyl-
Synonym: β-Resorcylaldehyde, 3-methyl-
Line Form: [2,4-$(HO)_2$-3-CH_3]-C_6H_2CHO
Mol Form: $C_8H_8O_3$
M_r: 152.14
mp: 156
UV: LANG 2122 258(18197) 232(7762) 217(16596) EtOH 330(5370) 286(14125) 213(16218) H_2O
m/e: 152.05

HODOC No: 01650
CAS No: 526-37-4
Beilstein No: 8^3, 2138
Name: Benzaldehyde, 2,6-dihydroxy-4-methyl-
Synonym: γ-Resorcylaldehyde, 4-methyl-
Line Form: [2,6-$(HO)_2$-4-CH_3]C_6H_2CHO
Mol Form: $C_8H_8O_3$
M_r: 152.14
mp: 124
Color: ye nd (w)
***s*:** H_2O 4; EtOH 4; Chl 4
m/e: 152.05

HODOC No: 01651
CAS No: 73635-75-3
Beilstein No: 8^3, 2064
Name: Benzaldehyde, 4,5-dihydroxy-2-nitro-
Synonym: 4,5-Dihydroxy-2-nitrobenzaldehyde
Line Form: [3,4-$(HO)_2$-6-NO_2]-C_6H_2CHO
Mol Form: $C_7H_5NO_5$
M_r: 183.11
mp: 209
***s*:** ace 2
UV: SAD 17389 319(5680) 264(13400) MeOH
H NMR: SAD 10044 7.3 7.6 8.9 10.3 ace
m/e: 183.02

HODOC No: 01652
CAS No: 86-51-1
Name: Benzaldehyde, 2,3-dimethoxy-
Synonym: *o*-Veratraldehyde
Mol Form: $C_9H_{10}O_3$
M$_r$: 166.17
IR: **µSADP 1294** 3030 2860 2780 2700 1670 1590 1470 1390 1250 1160 1060 1000 910
Raman: **SAD 969** 3070 3010 2940 2860 2840 2750 1690 1600 1470 1440 1390 1320 1260 1250 1220 1190 1170 1150 1090 1000 970 920 810 800 770 750 660 580 540 500 390 350 290 250 230 180 110
UV: **SAD 389** 317 258 MeOH
H NMR: **SAD 10405** 3.8 3.9 6.9 7.3 9.8 $CDCl_3$
MS: **AMS 1-92** 166(100) 77(36) 76(31) 51(30) 151(28) 65(24) 52(23) 108(22)
m/e: 166.06

HODOC No: 01653
CAS No: 613-45-6
Beilstein No: 8^4, 1754
Name: Benzaldehyde, 2,4-dimethoxy-
Synonym: 2,4-Dimethoxybenzaldehyde
Line Form: 2,4-$(CH_3O)_2$-C_6H_3CHO
Mol Form: $C_9H_{10}O_3$
M$_r$: 166.17
bp: 165^{10}
mp: 72
Color: nd (al or lig)
***s*:** H_2O 1; al 3; eth 3; bz 3; chl 2; lig 3
IR: **SADG 18362** 2860 1670 1600 1500 1440 1330 1250 1210 1170 1090 1010 820 670
UV: **SAD 7344** 314(9490) 273(13200) 231(14100) 205(13600) MeOH
H NMR: **SAD 8791** 3.8 3.9 6.5 7.8 10.3 $CDCl_3$
MS: **AMS 2-90** 166(100) 165(83) 149(50) 135(29) 63(23) 120(19) 106(19) 122(18)
m/e: 166.06

HODOC No: 01654
CAS No: 93-02-7
Beilstein No: 8^4, 1759
Name: Benzaldehyde, 2,5-dimethoxy-
Synonym: 2,5-Dimethoxybenzaldehyde
Line Form: 2,5-$(CH_3O)_2C_6H_3CHO$
Mol Form: $C_9H_{10}O_3$
M$_r$: 166.17
bp: 146^{10}
mp: 52
***s*:** H_2O 2; al 3; eth 3
UV: **LANG 1520** 355(5129) 257(8913) 225(16596) EtOH 349(5012) 252(7943) 224(19055) cyhex
MS: **NBS 693**
m/e: 166.06

HODOC No: 01655
CAS No: 120-14-9
Beilstein No: 8^4, 1765
Name: Benzaldehyde, 3,4-dimethoxy-
Synonym: Veratraldehyde
Line Form: 3,4-$(CH_3O)_2C_6H_3CHO$
Mol Form: $C_9H_{10}O_3$
M$_r$: 166.17
bp: 258, $172\text{-}5^{18}$
mp: 44
Color: nd (eth, lig, to)
***s*:** H_2O 2; al 4; eth 4; chl 2
IR: **µCOB 6480** 2940 2860 1700 1590 1520 1470 1430 1390 1350 1270 1250 1160 1150 1030 870 810 740
UV: **SAD 1179** 307(9170) 273(11400) 229(17300) MeOH
H NMR: **VAR 236** 3.9 4.0 7.0 7.4 7.5 10.0 $CDCl_3$
MS: **WILEY 1082** 166(100) 165(49) 15(28) 95(28) 51(26) 77(23) 79(18) 167(14)
m/e: 166.06

HODOC No: 01656
CAS No: 38257-04-4
Name: Benzaldehyde, 3,4-dimethoxy-, (4-nitrophenyl)hydrazone
Mol Form: $C_{15}H_{15}N_3O_4$
M$_r$: 301.29
mp: 209-10
***s*:** DMSO 3
IR: **µSADP 13385** 3230 2940 1590 1490 1300 1270 1160 1110 830 750
UV: **SAD 3504** 331(8090) 301(9610) 203 MeOH
H NMR: **SAD 11982** 3.8 3.9 6.9-7.4 8.0 8.1 11.1 $DMSOd_6$
m/e: 301.11

HODOC No: 01657
CAS No: 7311-34-4
Beilstein No: 8^4, 1786
Name: Benzaldehyde, 3,5-dimethoxy-
Synonym: 3,5-Dimethoxybenzaldehyde
Line Form: 3,5-$(CH_3O)_2C_6H_3CHO$
Mol Form: $C_9H_{10}O_3$
M_r: 166.17
bp: 151^{16}
mp: 45.5
***s*:** H_2O 2; al 3; bz 3; peth 2
IR: COB 7706 3000 2960 2840 1705 1605 1550 1465 1430 1385 1350 1325 1300 1205 1155 1065 925 845 710 675
UV: LANG 1516 325(2512) 270(6457) EtOH 320(2291) 271(6607) 263(5129) cyhex
C NMR: JJ 346 FT 55.4 107.0 138.4 161.2 191.6 $CDCl_3$
m/e: 166.06

HODOC No: 01658
CAS No: 20357-25-9
Beilstein No: 8^4, 1786
Name: Benzaldehyde, 4,5-dimethoxy-2-nitro-
Synonym: Veratraldehyde, 6-nitro-
Line Form: 3,4-$(CH_3O)_2$-6-$NO_2C_6H_2CHO$
Mol Form: $C_9H_9NO_5$
M_r: 211.17
mp: 131-3
IR: ALD 3:924E
MS: ALD 6:100A
m/e: 211.05

HODOC No: 01659
CAS No: 15764-16-6
Beilstein No: 7^4, 706
Name: Benzaldehyde, 2,4-dimethyl-
Synonym: 2,4-Dimethylbenzaldehyde
Line Form: 2,4$(CH_3)_2C_6H_3CHO$
Mol Form: $C_9H_{10}O$
M_r: 134.17
bp: 218, 99^{10}
mp: -9
***s*:** al 3; eth 3; ace 3; bz 3; chl 2
IR: SADG 17860 3000 2920 2840 2720 1690 1610 1570 1450 1370 1300 1230 1190 1130 810 780 710
UV: SAD 17514 258(13400) 209(19800) MeOH
H NMR: SAD 9504 2.4 2.6 7.0 7.1 7.6 10.2 $CDCl_3$
MS: NBS 299
m/e: 134.07

HODOC No: 01660
CAS No: 5779-94-2
Beilstein No: 7^4, 707
Name: Benzaldehyde, 2,5-dimethyl-
Synonym: 2,5-Dimethylbenzaldehyde
Line Form: 2,5-$(CH_3)_2C_6H_3CHO$
Mol Form: $C_9H_{10}O$
M_r: 134.17
bp: 220, 100^{10}
***s*:** al 4; eth 3; ace 3; bz 3; ctc 3; os 3
IR: SADG 17861 3000 2940 2860 2720 1690 1610 1570 1500 1400 1380 1280 1240 1150 930 810 770 670
UV: SAD 17091 303(1810) 252(11200) MeOH
H NMR: SAD 9505 2.3 2.6 7.0 7.2 7.4 10.2 CCl_4
MS: NBS 298
m/e: 134.07

HODOC No: 01661
CAS No: 5973-71-7
Beilstein No: 7^4, 707
Name: Benzaldehyde, 3,4-dimethyl-
Synonym: 3,4-Dimethylbenzaldehyde
Line Form: 3,4-$(CH_3)_2C_6H_3CHO$
Mol Form: $C_9H_{10}O$
M_r: 134.17
bp: 223-5
***s*:** Ac 4; Et_2O 4; EtOH 4
MS: NBS 299
m/e: 134.07

HODOC No: 01662
CAS No: 5779-95-3
Beilstein No: 7^4, 707
Name: Benzaldehyde, 3,5-dimethyl-
Synonym: 3,5-Dimethylbenzaldehyde
Line Form: 3,5-$(CH_3)_2C_6H_3CHO$
Mol Form: $C_9H_{10}O$
M_r: 134.17
bp: 220-2
mp: 9
***s*:** Ac 4; Bz 4; Et_2O 4; EtOH 4
m/e: 134.07

HODOC No: 01663
CAS No: 579-72-6
Beilstein No: 14, 25
Name: Benzaldehyde, 2-(dimethylamino)-
Synonym: Anthranilaldehyde, *N,N*-dimethyl-
Line Form: 2-$(CH_3)_2N$-C_6H_4CHO
Mol Form: $C_9H_{11}NO$
Mr: 149.18
bp: 244, 120^{11}
***s*:** ace 3; bz 3; chl 3; os 3
UV: OES 2-181 330(31623) bz 380(2512) 240(12589) EtOH
m/e: 149.08

HODOC No: 01664
CAS No: 619-22-7
Beilstein No: 14^3, 53
Name: Benzaldehyde, 3-(dimethylamino)-
Synonym: Benzaldehyde, *m*-(dimethylamino)-
Line Form: 3-$(CH_3)_2N$-C_6H_4CHO
Mol Form: $C_9H_{11}NO$
Mr: 149.18
bp: 138^9
Color: pa ye oil
***s*:** Bz 4; Et_2O 4; EtOH 4
m/e: 149.08

HODOC No: 01665
CAS No: 100-10-7
Beilstein No: 14^3, 58
Name: Benzaldehyde, 4-(dimethylamino)-
Synonym: Benzaldehyde, *p*-(dimethylamino)-
Line Form: 4-$(CH_3)_2N$-C_6H_4CHO
Mol Form: $C_9H_{11}NO$
Mr: 149.18
bp: 176-7^{17}
mp: 74-5
Color: lf (w)
***s*:** H_2O 2; al 3; eth 3; ace 3; bz 3; chl 2; os 3
IR: µCOB 6440 2860 2780 1700 1610 1540 1490 1450 1370 1320 1250 1180 1080 1010 940 810 730
Raman: SAD 151 1670 1660 1600 1560 1440 1420 1400 1380 1250 1180 1170 1130 950 840 740 640 600 530 500 360 340 180
UV: SAD 1838 340 242 MeOH
H NMR: VAR 238 3.1 6.7 7.7 9.7 $CDCl_3$
C NMR: JJ 349 FT 39.7 110.8 124.9 131.6 154.1 189.7 $CDCl_3$
MS: Wiley 869 148(100) 149(77) 77(13) 132(9) 42(9) 150(8) 105(6) 104(5)
m/e: 149.08

HODOC No: 01666
CAS No: 6300-80-7
Name: Benzaldehyde, *p*-(dimethylamino)-, purin-6-ylhydrazone
Synonym: Purine, 6(2(4(dimethylamino) phenylmethylene)hydrazino)-9-hydro
Mol Form: $C_{14}H_{15}N_7$
Mr: 281.31
IR: COB 7099 3450 2900 2850 1600 1520 1370 1180 1130 810
m/e: 281.14

HODOC No: 01667
CAS No: 2725-53-3
Beilstein No: 8^4, 547
Name: Benzaldehyde, 5-(1,1-dimethylethyl)-2-hydroxy-
Synonym: 5-*tert*-Butyl-2-hydroxybenzaldehyde
Line Form: 5-$(CH_3)_3C$-2-HO-C_6H_3CHO
Structure:

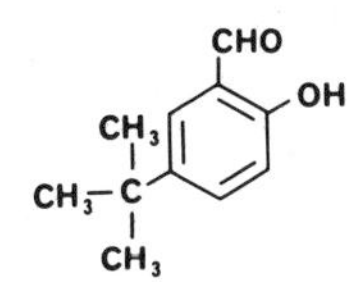

Mol Form: $C_{11}H_{14}O_2$
Mr: 178.22
bp: 251-2
***d*:** 1.039^{20}
m/e: 178.10

HODOC No: 01668
CAS No: 528-75-6
Beilstein No: 7^3, 923
Name: Benzaldehyde, 2,4-dinitro-
Synonym: 2,4-Dinitrobenzaldehyde
Line Form: 2,4-$(NO_2)_2C_6H_3CHO$
Mol Form: $C_7H_4N_2O_5$
Mr: 196.11
bp: 190-210^{10-20}
mp: 72
Color: pa ye pr (al), pl (Bz)
***s*:** H_2O 2; al 3; eth 3; bz 3; chl 2; lig 2
IR: SADG 16691 3100 3050 2880 1700 1600 1530 1470 1400 1350 1250 1170 1050 930 860 810 740 710 670
UV: SAD 16924 238(11200) MeOH
H NMR: SAD 8845 8.2 8.7 9.0 10.5 $CDCl_3$
m/e: 196.01

HODOC No: 01669
CAS No: 606-31-5
Beilstein No: 7^4, 611
Name: Benzaldehyde, 2,6-dinitro-
Synonym: 2,6-Dinitrobenzaldehyde
Line Form: 2,6-$(NO_2)_2C_6H_3CHO$
Mol Form: $C_7H_4N_2O_5$
Mr: 196.11
mp: 123
Color: lf (dil HOAc)
***s*:** H_2O 3; al 3; eth bz; 3 chl; 3 HOAc; 3 CS_2 2
IR: SADG 16711 3100 3050 2900 1720 1580 1530 1450 1340 1200 1180 980 900 840 820 750 730 700
UV: SAD 16927 226(12600) MeOH
H NMR: SAD 18549 8.0 8.5 10.5 poly D
m/e: 196.01

HODOC No: 01670
CAS No: 613-69-4
Beilstein No: 8^4, 180
Name: Benzaldehyde, 2-ethoxy-
Synonym: Benzaldehyde, *o*-ethoxy-
Line Form: 2-C_2H_5O-C_6H_4CHO
Mol Form: $C_9H_{10}O_2$
Mr: 150.17
bp: 247-9, 143-7^{25}
mp: 20-2
***s*:** al 5; eth 5; chl 2
IR: SADG 27974 3060 2970 2930 2850 2750 1670 1590 1480 1470 1450 1380 1280 1240 1190 1160 1110 1090 1040 920 830 750 650 590 490 430
H NMR: SAD 11782 1.4 4.1 6.9 7.0 7.5 7.8 10.5 $CDCl_3$
MS: NBS 4147
m/e: 150.07

HODOC No: 01671
CAS No: 22924-15-8
Beilstein No: 8, 60
Name: Benzaldehyde, 3-ethoxy-
Synonym: Benzaldehyde, *m*-ethoxy-
Line Form: 3-C_2H_5O-C_6H_4CHO
Mol Form: $C_9H_{10}O_2$
Mr: 150.17
bp: 245.5
***d*:** $1.0768^{20/4}$
***n*:** 1.5408^{20}
***s*:** al 3; eth 3; bz 3
MS: AMS 1-341 121(100) 150(49) 65(29) 39(27) 29(24) 27(21) 93(20) 51(13)
m/e: 150.07

HODOC No: 01672
CAS No: 10031-82-0
Beilstein No: 8^4, 254
Name: Benzaldehyde, 4-ethoxy-
Synonym: Benzaldehyde, *p*-ethoxy-
Line Form: 4-C_2H_5O-C_6H_4CHO
Mol Form: $C_9H_{10}O_2$
Mr: 150.17
bp: 249, 140^{20}
mp: 13-4
***d*:** $1.08^{21/21}$
***s*:** al 4; eth 4; bz 4
IR: SADG 26089 2980 2940 2820 2760 1680 1590 1570 1500 1470 1390 1310 1250 1210 1150 1110 1040 920 830 770 650 610 500 420 320
UV: SAD 20034 282(16900) 277(16700) 220(12700) MeOH
H NMR: SAD 15055 1.4 4.0 6.9 7.7 7.8 CCl_4
MS: AMS 1-343 121(100) 150(38) 172(27) 65(25) 39(23) 29(18) 27(15) 93(14)
m/e: 150.07

HODOC No: 01673
CAS No: 492-88-6
Beilstein No: 8^4, 1748
Name: Benzaldehyde, 3-ethoxy-2-hydroxy-
Synonym: Salicylaldehyde, 3-ethoxy-
Line Form: (3-C_2H_5O-2-HO)C_6H_3CHO
Mol Form: $C_9H_{10}O_3$
Mr: 166.17
bp: 263-4^{740}
mp: 64-5
IR: SADG 33446 3070 2990 2950 2900 1650 1590 1480 1460 1400 1380 1330 1270 1220 1170 1120 1090 1070 1010 890 830 770 740 650 530 360 320
Raman: SAD 967 3070 2980 2930 2880 1650 1630 1580 1470 1400 1370 1330 1280 1220 1170 1140 1120 1090 1020 1000 900 840 750 660 550 440 340 210 150 90
UV: SAD 3281 342(2930) 264(10100) 220(18100) MeOH 281(5890) 237(18900) MeOH:KOH 342(1010) 268(4360) 219(11300) MeOH:HCl
H NMR: SAD 19493 1.5 4.1 6.8-7.4 9.9 11.0 $CDCl_3$
m/e: 166.06

HODOC No: 01674
CAS No: 121-32-4
Beilstein No: 8^4, 1765
Name: Benzaldehyde, 3-ethoxy-4-hydroxy-
Synonym: Bourbonal
Line Form: $(3\text{-}C_2H_5O\text{-}4\text{-}HO)C_6H_3CHO$
Mol Form: $C_9H_{10}O_3$
M_r: 166.17
bp: 77-8
mp: 77-8
***s*:** H_2O 2; al 3; eth 3; bz 3; chl 3
IR: µCOB 3725 3230 2940 1670 1590 1490 1430 1390 1300 1270 1160 1120 1040 990 900 870 850 810 750
Raman: SAD 560 3370 3090 3070 2940 1690 1670 1600 1580 1520 1380 1280 1240 1200 1180 1170 1120 1040 980 890 800 780 740 630 590 530 460 420 350 320 220 190 140 100
UV: SAD 17152 308(9310) 277(10200) 231(15600) MeOH
H NMR: SAD 1320 1.4 4.2 6.6 7.0 7.4 9.8 $CDCl_3$
MS: WILEY 1083 137(100) 28(47) 138(38) 27(26) 29(24) 81(13) 109(12) 166(5)
m/e: 166.06

HODOC No: 01675
CAS No: 43057-77-8
Beilstein No: 8^3, 1992
Name: Benzaldehyde, 4-ethoxy-2-hydroxy-
Synonym: 4-Ethoxy-2-hydroxybenzaldehyde
Line Form: $(4\text{-}C_2H_5O\text{-}2\text{-}HO)C_6H_3CHO$
Mol Form: $C_9H_{10}O_3$
M_r: 166.17
mp: 35
Color: nd (dil al)
***s*:** EtOH 4
m/e: 166.06

HODOC No: 01676
CAS No: 2539-53-9
Beilstein No: 8^4, 1765
Name: Benzaldehyde, 4-ethoxy-3-hydroxy-
Synonym: 4-Ethoxy-3-hydroxybenzaldehyde
Line Form: $(4\text{-}C_2H_5O\text{-}3\text{-}HO)C_6H_3CHO$
Mol Form: $C_9H_{10}O_3$
M_r: 166.17
mp: 126-8
m/e: 166.06

HODOC No: 01677
CAS No: 80832-54-8
Beilstein No: 8^3, 2004
Name: Benzaldehyde, 5-ethoxy-2-hydroxy-
Line Form: $(5\text{-}C_2H_5O\text{-}2\text{-}HO)C_6H_3CHO$
Mol Form: $C_9H_{10}O_3$
M_r: 166.17
bp: 230
mp: 51.5
Color: ye pr
***s*:** Et_2O 4; EtOH 4; Chl 4
m/e: 166.06

HODOC No: 01678
CAS No: 120-25-2
Beilstein No: 8^4, 1766
Name: Benzaldehyde, 4-ethoxy-3-methoxy-
Synonym: 4-Ethoxy-*m*-anisaldehyde
Line Form: $4\text{-}C_2H_5O\text{-}3\text{-}CH_3O\text{-}C_6H_3CHO$
Mol Form: $C_{10}H_{12}O_3$
M_r: 180.19
bp: sub
mp: 64-5
Color: mcl pr
***s*:** H_2O 2; al 3; eth 3; bz 3; chl 3; HOAc 3
IR: SADG 10048 3030 2941 1667 1587 1493 1449 1408 1389 1333 1250 1136 1020 901 862 794 763 725
UV: SAD 5155 307 275 229 MeOH
H NMR: SAD 4218 1.5 3.9 4.2 7.0 7.4 9.8 $CDCl_3$
m/e: 180.08

HODOC No: 01679
CAS No: 56917-14-7
Beilstein No: 8, 98
Name: Benzaldehyde, 4-ethoxy-3-methyl-
Synonym: *m*-Tolualdehyde, 4-ethoxy
Line Form: $4\text{-}C_2H_5O\text{-}3\text{-}CH_3C_6H_3CHO$
Mol Form: $C_{10}H_{12}O_2$
M_r: 164.19
bp: 258-60
mp: 33-4
m/e: 164.08

HODOC No: 01680
CAS No: 22927-13-5
Beilstein No: 7^4, 696
Name: Benzaldehyde, 2-ethyl-
Synonym: Benzaldehyde, *o*-ethyl-
Line Form: 2-(CH_3CH_2)-C_6H_4CHO
Mol Form: $C_9H_{10}O$
Mr: 134.17
bp: 209, 86-7^{10}
***d*:** 1.0216$^{17/4}$
***n*:** 1.5425^{17}
Color: oil
MS: NBS 299
m/e: 134.07

HODOC No: 01681
CAS No: 34246-54-3
Name: Benzaldehyde, 3-ethyl-
Synonym: Benzaldehyde, *m*-ethyl-
Line Form: 3-(CH_3CH_2)-C_6H_4CHO
Mol Form: $C_9H_{10}O$
Mr: 134.17
bp: 212^{762}
Color: oil
MS: NBS 299
m/e: 134.07

HODOC No: 01682
CAS No: 4748-78-1
Beilstein No: 7^4, 701
Name: Benzaldehyde, 4-ethyl-
Synonym: Benzaldehyde, *p*-ethyl-
Line Form: 4-(CH_3CH_2)-C_6H_4CHO
Mol Form: $C_9H_{10}O$
Mr: 134.17
bp: 221, 109-10^{10}
MS: NBS 299
m/e: 134.07

HODOC No: 01683
CAS No: 446-52-6
Beilstein No: 7^4, 559
Name: Benzaldehyde, 2-fluoro-
Synonym: Benzaldehyde, *o*-fluoro-
Line Form: 2-FC_6H_4CHO
Mol Form: C_7H_5FO
Mr: 124.11
bp: 175, 90-1^{46}
mp: -44.5
***d*:** 1.178
***n*:** 1.5234^{20}
IR: ALD 2:797H
MS: ALD 6:79C
m/e: 124.03

HODOC No: 01684
CAS No: 456-48-4
Beilstein No: 7^4, 559
Name: Benzaldehyde, 3-fluoro-
Synonym: Benzaldehyde, *m*-fluoro-
Line Form: 3-FC_6H_4CHO
Mol Form: C_7H_5FO
Mr: 124.11
bp: 173, 66-8^{20}
***d*:** 1.17
***n*:** 1.5206^{20}
IR: ALD 2:799B
m/e: 124.03

HODOC No: 01685
CAS No: 459-57-4
Beilstein No: 7^4, 559
Name: Benzaldehyde, 4-fluoro-
Synonym: Benzaldehyde, *p*-fluoro-
Line Form: 4-FC_6H_4CHO
Mol Form: C_7H_5FO
Mr: 124.11
bp: 181.5, 104-5^{74}
mp: -10
m/e: 124.03

HODOC No: 01686
CAS No: 351-54-2
Beilstein No: 8^4, 296
Name: Benzaldehyde, 3-fluoro-4-methoxy-
Synonym: *p*-Anisaldehyde, 3-fluoro-
Line Form: (3-F-4-CH_3O)-C_6H_3CHO
Mol Form: $C_8H_7FO_2$
Mr: 154.13
mp: 34-5
IR: ALD 3:918A
m/e: 154.04

HODOC No: 01687
CAS No: 618-65-5
Beilstein No: 17^4, 3010
Name: Benzaldehyde, 2-(β-*D*-glucopyranosyloxy)-
Synonym: Helicin
Structure:

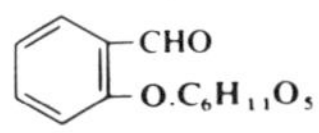

Mol Form: $C_{13}H_{16}O_7$
Mr: 284.25
mp: 175
α: $[\alpha]^{20/D}$ -60.4^{20} (w)
Color: nd (w)
s: H_2O 4; EtOH 4
IR: ALD 2:798F
m/e: 284.09

HODOC No: 01688
CAS No: 27893-41-0
Beilstein No: 8^3, 224
Name: Benzaldehyde, 4-(heptyloxy)-
Synonym: Benzaldehyde, *p*-(heptyloxy)-
Line Form: 4-[$CH_3(CH_2)_6O$]-C_6H_4CHO
Mol Form: $C_{14}H_{20}O_2$
Mr: 220.30
bp: $143^{0.9}$
m/e: 220.15

HODOC No: 01689
CAS No: 49763-69-1
Name: Benzaldehyde, 4-hexyl-
Synonym: *p*-Hexylbenzaldehyde
Line Form: 4-C_6H_{13}-C_6H_4CHO
Mol Form: $C_{13}H_{18}O$
Mr: 190.27
bp: 136^7
m/e: 190.14

HODOC No: 01690
CAS No: 7162-59-6
Name: Benzaldehyde, 2-(hexyloxy)-
Synonym: Benzaldehyde, *o*-(hexyloxy)-
Mol Form: $C_{13}H_{18}O_2$
Mr: 206.27
IR: μCOB 5291 2940 2860 1700 1610 1490 1450 1390 1280 1240 1190 1160 1100 1040 1020 940 830 760
Raman: SAD 561 3070 2930 2870 1690 1600 1580 1510 1470 1430 1390 1310 1250 1220 1160 1010 900 860 830 720 650 640 620 340 260 220
m/e: 206.13

HODOC No: 01691
CAS No: 5736-94-7
Beilstein No: 8^3, 224
Name: Benzaldehyde, 4-(hexyloxy)-
Synonym: Benzaldehyde, *p*-(hexyloxy)-
Line Form: 4-$C_6H_{13}O$-C_6H_4CHO
Mol Form: $C_{13}H_{18}O_2$
Mr: 206.27
bp: 166^{12}
m/e: 206.13

HODOC No: 01692
CAS No: 5281-18-5
Beilstein No: 7^4, 530
Name: Benzaldehyde, hydrazone
Synonym: Benzalhydrazone
Line Form: $C_6H_5CH{=}NNH_2$
Mol Form: $C_7H_8N_2$
Mr: 120.15
bp: 140^{14}
mp: 16
Color: lf
s: al 3
UV: OES 7-118 275(19498) $CHCl_3$
m/e: 120.07

HODOC No: 01693
CAS No: 90-02-8
Beilstein No: 8^4, 176
Name: Benzaldehyde, 2-hydroxy-
Synonym: Salicylaldehyde
Line Form: 2-HOC_6H_4CHO
Mol Form: $C_7H_6O_2$
M_r: 122.12
bp: 197, 93^{25}
mp: -7
***d*:** $1.1674^{20/4}$
***n*:** 1.5740^{20}
***s*:** H_2O 2; al 5; eth 5; ace 4; bz 4; chl 2
IR: SADG 8567 3226 3030 2857 2778 2500 1667 1639 1587 1493 1471 1389 1333 1282 1235 1205 1149 1124 1031 952 901 769 725 671
Raman: SAD 150 3070 2850 1660 1640 1620 1590 1580 1450 1390 1320 1280 1230 1150 1110 1070 1030 880 860 770 660 560 450 410 290 270 210 150
UV: SAD 2646 328 259 cyhex
H NMR: SAD 3232 7.0 7.4 9.8 11.0 $CDCl_3$
C NMR: JJ 231 cw 117.4 119.6 121.0 133.6 136.6 161.4 196.7 diox
MS: WILEY 502 122(100) 121(96) 39(40) 65(32) 76(20) 93(18) 104(15) 38(14)
m/e: 122.04

HODOC No: 01694
CAS No: 959-36-4
Beilstein No: 8^2, 43
Name: Benzaldehyde, 2-hydroxy-, [(2-hydroxyphenyl)methylene]hydrazone
Synonym: Salicylaldehyde, azine
Line Form: 2-HOC_6H_4CH=NN=CH(C_6H_4OH-2)
Mol Form: $C_{14}H_{12}N_2O_2$
M_r: 240.25
mp: 214
***s*:** H_2O 1; al 3; bz 4; chl 3; alk 4
IR: µCOB 426 3030 1640 1590 1490 1390 1330 1320 1280 1220 1160 1150 1120 1040 990 980 890 790 750 730 680
UV: SAD 1586 292(26500) 219(23300) MeOH
m/e: 240.09

HODOC No: 01695
CAS No: 94-67-7
Beilstein No: 8^4, 203
Name: Benzaldehyde, 2-hydroxy-, oxime
Synonym: Salicylaldehyde, oxime
Line Form: 2-HOC_6H_4CH=NOH
Mol Form: $C_7H_7NO_2$
M_r: 137.13
mp: 63
***s*:** H_2O 2; al 4; eth 4; bz 4,chl 3; lig 1
IR: µCOB 1679 3450 2940 1610 1590 1490 1470 1430 1300 1270 1210 1160 990 960 900 860 790 770 740 730 700 650
UV: HBCP 305(3981) 265(6310) al
H NMR: VAR 156 8.2 10.1 $CDCl_3$
m/e: 137.05

HODOC No: 01696
CAS No: 100-83-4
Beilstein No: 8^4, 240
Name: Benzaldehyde, 3-hydroxy-
Synonym: Benzaldehyde, *m*-hydroxy-
Line Form: 3-HOC_6H_4CHO
Mol Form: $C_7H_6O_2$
M_r: 122.12
bp: 240, 161^{20}
mp: 108
Color: nd (w)
***s*:** H_2O 2; al 3; eth 3; ace 3; bz 3; lig 1
IR: µCOB 1576 3230 2700 1670 1590 1490 1350 1320 1280 1250 1180 1160 1000 940 880 790 710 680 660
UV: SAD 562 315 253 MeOH
H NMR: SAD 7840 7.1-7.5 7.9 10.0 13.2 ace
MS: NBS 198
m/e: 122.04

HODOC No: 01697
CAS No: 18428-76-7
Beilstein No: 8^3, 203
Name: Benzaldehyde, 3-hydroxy-, [(3-hydroxyphenyl)methylene]hydrazone
Synonym: Benzaldehyde, *m*-hydroxy-, azine
Line Form: 3-HOC_6H_4CH=NN=CH(C_6H_4OH-3)
Mol Form: $C_{14}H_{12}N_2O_2$
M_r: 240.25
mp: 162
Color: ye pl (al)
***s*:** H_2O 1; al 2; bz 2; chl 2
UV: OES 1-561 330(21380) 306(28184) 298(28840) EtOH
MS: NBS 4603
m/e: 240.09

HODOC No: 01698
CAS No: 22241-18-5
Beilstein No: 8, 61
Name: Benzaldehyde, 3-hydroxy-, oxime
Synonym: Benzaldehyde, *m*-hydroxy-, oxime
Line Form: 3-HOC_6H_4CH=NOH
Mol Form: $C_7H_7NO_2$
M_r: 137.13
mp: 90
Color: (Bz)
***s*:** H_2O 4; Et_2O 4; EtOH 4
m/e: 137.05

HODOC No: 01699
CAS No: 123-08-0
Beilstein No: 8^4, 251
Name: Benzaldehyde, 4-hydroxy-
Synonym: Benzaldehyde, *p*-hydroxy-
Line Form: 4-HOC_6H_4CHO
Mol Form: $C_7H_6O_2$
M_r: 122.12
mp: 117
***d*:** $1.129^{130/4}$
***n*:** 1.5105^{130}
Color: nd (w)
***s*:** H_2O 2; al 4; eth 4; ace 2; bz 3
IR: μCOB 4315 3130 2860 1920 1670 1590 1540 1470 1430 1320 1280 1220 1160 860 830 790 730 640
UV: SAD 1266 315(2640) 253(8550) 219(17100) MeOH
H NMR: SAD 7349 7.0 7.8 8.7 9.9 ace
MS: AMS 1-303 121(100) 122(84) 93(36) 65(35) 39(27) 28(11) 63(10) 38(8)
m/e: 122.04

HODOC No: 01700
CAS No: 5466-23-9
Beilstein No: 8^3, 239
Name: Benzaldehyde, 4-hydroxy-, [(4-hydroxyphenyl)methylene]hydrazone
Synonym: Benzaldehyde, *p*-hydroxy-, azine
Line Form: 4-HOC_6H_4CH=NH=CH(C_6H_4OH-4)
Mol Form: $C_{14}H_{12}N_2O_2$
M_r: 240.25
mp: 239-40
Color: ye (al)
***s*:** H_2O 2; al 4; bz 4; ctc 2; lig 2
UV: OES 1-561 335(43652) EtOH
m/e: 240.09

HODOC No: 01701
CAS No: 699-06-9
Name: Benzaldehyde, 4-hydroxy-, oxime
Synonym: Benzaldehyde, *p*-hydroxy-, oxime
Mol Form: $C_7H_7NO_2$
M_r: 137.13
IR: μCOB 1673 3330 3030 1610 1520 1430 1320 1280 1220 1160 1110 950 880 830 810
m/e: 137.05

HODOC No: 01702
CAS No: 1948-40-9
Beilstein No: 8^4, 301
Name: Benzaldehyde, 4-hydroxy-3,5-diiodo-
Synonym: 3,5-Diiodo-4-hydroxybenzaldehyde
Line Form: (3,5-I_2-4-HO)C_6H_2CHO
Mol Form: $C_7H_4I_2O_2$
M_r: 373.91
mp: 206.5
Color: nd (al)
***s*:** Ac 4; EtOH 4; HOAc 4
m/e: 373.83

HODOC No: 01703
CAS No: 708-76-9
Beilstein No: 8^4, 2717
Name: Benzaldehyde, 2-hydroxy-4,6-dimethoxy-
Synonym: Salicylaldehyde, 4,6-dimethoxy-
Line Form: 4,6-$(CH_3O)_2$-2-HOC_6H_2CHO
Mol Form: $C_9H_{10}O_4$
M_r: 182.17
mp: 68-70
IR: ALD 2:808C
MS: ALD 6:97C
m/e: 182.06

HODOC No: 01704

CAS No: 32246-34-7
Name: Benzaldehyde, 3-hydroxy-2,4-dimethoxy-
Synonym: 3-Hydroxy-2,4-dimethoxybenzaldehyde
Mol Form: $C_9H_{10}O_4$
M_r: 182.17
mp: 101-4
***s*:** chl 3
IR: SADG 16752 3200 3000 2980 2960 2840 2680 1660 1600 1570 1500 1470 1430 1370 1340 1280 1230 1160 1080 980 890 810 780 520
UV: SAD 16698 285(12200) 235(16800) 210(12400) MeOH 367(3350) 299(8740) 254(22500) MeOH:KOH
H NMR: SAD 8841 4.0 4.1 6.5 6.8 7.4 10.3 $CDCl_3$
m/e: 182.06

HODOC No: 01705

CAS No: 29865-90-5
Beilstein No: 8^4, 2718
Name: Benzaldehyde, 3-hydroxy-4,5-dimethoxy-
Synonym: Veratraldehyde, 5-hydroxy-
Line Form: $[3,4\text{-}(CH_3O)_2\text{-}5\text{-}HO]C_6H_2CHO$
Mol Form: $C_9H_{10}O_4$
M_r: 182.17
bp: $177\text{-}80^{12}$
mp: 62-3
***s*:** Bz 4; EtOH 4; lig 4
m/e: 182.06

HODOC No: 01706

CAS No: 22080-96-2
Beilstein No: 8^3, 3365
Name: Benzaldehyde, 4-hydroxy-2,6-dimethoxy-
Synonym: 4-Hydroxy-2,6-dimethoxybenzaldehyde
Line Form: $[2,6\text{-}(CH_3O)_2\text{-}4\text{-}HO]\text{-}C_6H_2CHO$
Mol Form: $C_9H_{10}O_4$
M_r: 182.17
mp: 70-1
Color: nd or pl (MeOH), pr (Bz)
m/e: 182.06

HODOC No: 01707

CAS No: 134-96-3
Beilstein No: 8^4, 2718
Name: Benzaldehyde, 4-hydroxy-3,5-dimethoxy-
Synonym: Cedar aldehyde
Line Form: $[3,5\text{-}(CH_3O)_2\text{-}4\text{-}HO]C_6H_2CHO$
Mol Form: $C_9H_{10}O_4$
M_r: 182.17
bp: $192\text{-}3^{14}$
mp: 113
Color: br nd (lig)
***s*:** H_2O 2; al 4; eth 4; bz 4; chl 4; HOAc 4; lig 2
UV: OES 2-180 309(14454) 231(17783) 215(20417) EtOH
m/e: 182.06

HODOC No: 01708

CAS No: 5438-36-8
Beilstein No: 8^3, 2059
Name: Benzaldehyde, 4-hydroxy-3-iodo-5-methoxy-
Synonym: 4-Hydroxy-5-iodo-3-methoxybenzaldehyde
Line Form: $(5\text{-}I\text{-}3\text{-}CH_3O\text{-}4\text{-}HO)C_6H_2CHO$
Mol Form: $C_8H_7IO_3$
M_r: 278.04
mp: 180
Color: pa ye
***s*:** H_2O 1; al 2; eth 2
IR: SADG 10320 3226 3030 2857 1667 1587 1493 1449 1408 1351 1299 1250 1163 1042 855 781 671
UV: SAD 6703 295 226 MeOH
m/e: 277.94

HODOC No: 01709
CAS No: 148-53-8
Beilstein No: 8^4, 1747
Name: Benzaldehyde, 2-hydroxy-3-methoxy-
Synonym: *m*-Anisaldehyde, 2-hydroxy-
Line Form: $(3\text{-}CH_3O\text{-}2\text{-}HO)C_6H_3CHO$
Mol Form: $C_8H_8O_3$
M_r: 152.14
bp: 265-6, 128^{10}
mp: 44-5
Color: lt ye lf, gr nd (w, lig)
***s*:** H_2O 2; al 4; eth 4; ctc 4; lig 2
IR: µCOB 2408 2780 2700 1670 1590 1470 1390 1320 1280 1250 1220 1190 1080 950 830 780 740 720
Raman: SAD 966 3070 3010 2940 2880 2840 1640 1590 1510 1470 1440 1370 1330 1280 1220 1180 1100 1070 1030 950 890 840 800 780 770 730 720 650 570 540 510 430 370 330 290 250 220 190 100
UV: SAD 383 340(2820) 263(9810) 219(18000) MeOH
H NMR: SAD 528 3.8 6.7-7.2 9.9 10.8 CCl_4
MS: WILEY 906 152(100) 106(39) 151(36) 109(31) 81(25) 53(20) 29(19) 27(14)
m/e: 152.05

HODOC No: 01710
CAS No: 673-22-3
Beilstein No: 8^4, 1754
Name: Benzaldehyde, 2-hydroxy-4-methoxy-
Synonym: *p*-Anisaldehyde, 2-hydroxy-
Line Form: $(4\text{-}CH_3O\text{-}2\text{-}HO)C_6H_3CHO$
Mol Form: $C_8H_8O_3$
M_r: 152.14
mp: 40-2
Color: nd (w), cr (al)
***s*:** al 3; eth 3; bz 3; lig 3
UV: LANG 2119 314(6918) 278(19953) 230(12882) 211(23988) cy 316(6026) 280(16218) 232(10715) 213(14125) EtOH
m/e: 152.05

HODOC No: 01711
CAS No: 672-13-9
Beilstein No: 8^4, 1759
Name: Benzaldehyde, 2-hydroxy-5-methoxy-
Synonym: *m*-Anisaldehyde, 6-hydroxy-
Line Form: $(5\text{-}CH_3O\text{-}2\text{-}HO)C_6H_3CHO$
Mol Form: $C_8H_8O_3$
M_r: 152.14
bp: 247-8
mp: 4
Color: ye (w)
***s*:** Et_2O 4; EtOH 4
m/e: 152.05

HODOC No: 01712
CAS No: 621-59-0
Beilstein No: 8^4, 1764
Name: Benzaldehyde, 3-hydroxy-4-methoxy-
Synonym: *p*-Anisaldehyde, 3-hydroxy-
Line Form: $(4\text{-}CH_3O\text{-}3\text{-}HO)C_6H_3CHO$
Mol Form: $C_8H_8O_3$
M_r: 152.14
bp: 179^{15}
mp: 113-5
***d*:** 1.196
***s*:** H_2O 2; al 3; eth 3; bz 3; chl 4; HOAc 3
IR: SADG 18370 3200 2900 1680 1610 1580 1510 1450 1340 1280 1120 1020 870 830 790 770
UV: SAD 6183 274 230 MeOH
MS: WILEY 903 151(100) 152(63) 18(37) 123(18) 18(16) 51(14) 109(12) 53(11)
m/e: 152.05

HODOC No: 01713
CAS No: 57179-35-8
Beilstein No: 8^3, 2073
Name: Benzaldehyde, 3-hydroxy-5-methoxy-
Synonym: 3-Hydroxy-5-methoxybenzaldehyde
Line Form: $(5\text{-}CH_3O\text{-}3\text{-}HO)C_6H_3CHO$
Mol Form: $C_8H_8O_3$
M_r: 152.14
bp: 110-20 (sub)
mp: 130-1
***s*:** Bz 4; EtOH 4; HOAc 4
m/e: 152.05

HODOC No: 01714
CAS No: 18278-34-7
Beilstein No: 8^4, 1754
Name: Benzaldehyde, 4-hydroxy-2-methoxy-
Synonym: *o*-Anisaldehyde, 4-hydroxy-
Line Form: $(2\text{-}CH_3O\text{-}4\text{-}HO)C_6H_3CHO$
Mol Form: $C_8H_8O_3$
M_r: 152.14
mp: 153
Color: lf (Bz), nd (w)
***s*:** Et_2O 4; EtOH 4; Chl 4
m/e: 152.05

HODOC No: 01715
CAS No: 121-33-5
Beilstein No: 8^4, 1763
Name: Benzaldehyde, 4-hydroxy-3-methoxy-
Synonym: 4-Formyl-2-methoxyphenol
Line Form: $(3\text{-}CH_3O\text{-}4\text{-}HO)C_6H_3CHO$
Mol Form: $C_8H_8O_3$
M_r: 152.14
bp: 285, 170^{15}
mp: 81-2
***d*:** 1.056
Color: tetr (w, lig)
***s*:** H_2O 2; al 4; eth 4; ace 4; bz 3; chl 4; CS_2 4; lig 3
IR: μCOB 2530 3330 2940 1670 1590 1520 1450 1430 1370 1300 1270 1210 1150 1120 1030 820 780 730
UV: OES 3-153 309(10471) 279(10233) 232(14454) 208(12023) EtOH 301(9333) 273(12303) 228(16596) 206(13804) eth
H NMR: VAR 197 3.9 6.5 7.0 7.4 9.8 $CDCl_3$
C NMR: JJ 292 FT 56.0 109.4 114.8 127.4 129.5 147.5 152.3 191.3 $CDCl_3$
MS: WILEY 905 151(100) 152(96) 18(95) 15(50) 81(42) 29(39) 51(32) 52(30)
m/e: 152.05

HODOC No: 01716
CAS No: 54700-36-6
Beilstein No: 8^3, 2133
Name: Benzaldehyde, 2-hydroxy-4-methoxy-3-methyl-
Synonym: 2-Hydroxy-4-methoxy-3-methylbenzaldehyde
Line Form: $(3\text{-}CH_3\text{-}4\text{-}CH_3O\text{-}2\text{-}HO)C_6H_2CHO$
Structure:

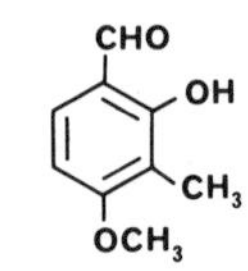

Mol Form: $C_9H_{10}O_3$
M_r: 166.17
mp: 64
UV: LANG 2123 315(3467) 285(21878) 216(19055) cy 287(22387) 216(20417) EtOH
m/e: 166.06

HODOC No: 01717
CAS No: 2450-26-2
Beilstein No: 8^2, 289
Name: Benzaldehyde, 4-hydroxy-3-methoxy-2-nitro-
Synonym: Vanillin, 2-nitro-
Line Form: $(2\text{-}NO_2\text{-}3\text{-}CH_3O\text{-}4\text{-}HO)C_6H_2CHO$
Mol Form: $C_8H_7NO_5$
M_r: 197.14
mp: 137
***s*:** H_2O 2; al 3; eth 3; ace 3; bz 2; chl 2; HOAc 3
IR: SADG 1510 3240 2950 1670 1590 1570 1530 1520 1510 1430 1370 1310 1270 1200 1040 970 900 830 810 780
UV: SAD 9262 273(8400) 224(14300) MeOH
m/e: 197.03

HODOC No: 01718
CAS No: 6635-20-7
Name: Benzaldehyde, 4-hydroxy-3-methoxy-5-nitro-
Synonym: Vanillin, 5-nitro-
Mol Form: $C_8H_7NO_5$
M_r: 197.14
IR: SADG 33564 3220 3090 2990 2960 2890 1680 1620 1580 1550 1470 1410 1370 1340 1270 1230 1190 1170 1110 1060 920 880 820 770 630 590 540 370
Raman: SAD 968 3200 1750 1680 1610 1580 1550 1470 1450 1430 1400 1370 1330 1270 1220 1100 1050 970 910 820 760 450 350 240 190
UV: SAD 6802 276(7950) 246(8300) 218(31500) MeOH 329(9110) 236(10700) MeOH:KOH 286(5050) 226(13100) MeOH:HCl
H NMR: SAD 2309 4.1 7.9 8.5 9.9 tfa
m/e: 197.03

HODOC No: 01719
CAS No: 698-27-1
Beilstein No: 8^4, 434
Name: Benzaldehyde, 2-hydroxy-4-methyl-
Synonym: 2,4-Cresotaldehyde
Line Form: 2-HO-4-$CH_3C_6H_3CHO$
Mol Form: $C_8H_8O_2$
M_r: 136.14
bp: $219\text{-}21^{726}$
mp: 60-1
Color: nd (al, w)
***s*:** Et_2O 4; EtOH 4; Chl 4
m/e: 136.05

HODOC No: 01720
CAS No: 613-84-3
Beilstein No: 8^4, 432
Name: Benzaldehyde, 2-hydroxy-5-methyl-
Synonym: 2,5-Cresotaldehyde
Line Form: 6-HO-3-$CH_3C_6H_3CHO$
Mol Form: $C_8H_8O_2$
M_r: 136.14
bp: 217-8
mp: 56
***d*:** $1.0913^{59.2/4}$
***n*:** $1.547^{59.2}$
Color: pl (aq,_al)
***s*:** Et_2O 4; EtOH 4; Chl 4
m/e: 136.05

HODOC No: 01721
CAS No: 18362-36-2
Beilstein No: 8^4, 430
Name: Benzaldehyde, 2-hydroxy-6-methyl-
Synonym: 2,6-Cresotaldehyde
Line Form: 6-HO-2-$CH_3C_6H_3CHO$
Mol Form: $C_8H_8O_2$
M_r: 136.14
bp: $228\text{-}9^{728}$
mp: 31.6
Color: nd (w)
***s*:** Bz 4
m/e: 136.05

HODOC No: 01722
CAS No: 41438-18-0
Beilstein No: 8^4, 429
Name: Benzaldehyde, 4-hydroxy-2-methyl-
Synonym: 4,2-Cresotaldehyde
Line Form: 4-HO-2-$CH_3C_6H_3CHO$
Mol Form: $C_8H_8O_2$
M_r: 136.14
mp: 110
Color: pl (w)
***s*:** Et_2O 4; EtOH 4; Chl 4
m/e: 136.05

HODOC No: 01723
CAS No: 15174-69-3
Beilstein No: 8^4, 431
Name: Benzaldehyde, 4-hydroxy-3-methyl-
Synonym: 4,3-Cresotaldehyde
Line Form: 4-HO-3-$CH_3C_6H_3CHO$
Mol Form: $C_8H_8O_2$
M_r: 136.14
bp: 250-1, 140^{20}
mp: 118
Color: pr (w)
m/e: 136.05

HODOC No: 01724
CAS No: 536-32-3
Name: Benzaldehyde, 2-hydroxy-4-(1-methylethyl)-
Synonym: Salicylaldehyde, 4-isopropyl-
Mol Form: $C_{10}H_{12}O_2$
M_r: 164.19
***s*:** chl 3
H NMR: VAR 259 1.2 2.9 6.8 6.9 7.4 9.8 11.0 $CDCl_3$
m/e: 164.08

HODOC No: 01725
CAS No: 59893-28-6
Name: Benzaldehyde, 2-hydroxy-5-(1-methylpropyl)-
Synonym: 5-*sec*-Butylsalicylaldehyde
Mol Form: $C_{11}H_{14}O_2$
M_r: 178.22
IR: **µCOB 950** 3230 2940 1670 1590 1470 1370 1280 1240 1210 1160 1140 930 890 830 770 740 710
m/e: 178.10

HODOC No: 01726
CAS No: 5274-70-4
Beilstein No: 8^4, 232
Name: Benzaldehyde, 2-hydroxy-3-nitro-
Synonym: Salicylaldehyde, 3-nitro-
Line Form: $(3\text{-}NO_2\text{-}2\text{-}HO)C_6H_3CHO$
Mol Form: $C_7H_5NO_4$
M_r: 167.11
mp: 109-10
Color: nd (HOAc)
***s*:** Bz 4; EtOH 4
m/e: 167.02

HODOC No: 01727
CAS No: 97-51-8
Beilstein No: 8^4, 234
Name: Benzaldehyde, 2-hydroxy-5-nitro-
Synonym: Salicylaldehyde, 5-nitro-
Line Form: $(5\text{-}NO_2\text{-}2\text{-}HO)C_6H_3CHO$
Mol Form: $C_7H_5NO_4$
M_r: 167.11
mp: 128
Color: (dil HOAc)
***s*:** ace 3
IR: **µCOB 345** 3130 1670 1640 1610 1470 1370 1350 1300 1240 1190 1100 930 920 830 780 750 720 630
UV: **OES 7-103** 310(10000) 258(2089) 230(8318) EtOH
H NMR: **SAD 6813** 7.2 8.4 8.8 9.7 10.2 ace
MS: **ALDN 27** 167(100) 65(32) 39(27) 166(24) 53(14) 120(14) 63(13) 75(11)
m/e: 167.02

HODOC No: 01728
CAS No: 1595-15-9
Beilstein No: 8^3, 193
Name: Benzaldehyde, 2-hydroxy-5-nitro-, oxime
Synonym: Salicylaldehyde, 5-nitro-, oxime
Line Form: $(5\text{-}NO_2\text{-}2\text{-}HO)C_6H_3CH{=}NOH$
Mol Form: $C_7H_6N_2O_4$
M_r: 182.13
mp: 225
m/e: 182.03

HODOC No: 01729
CAS No: 42123-33-1
Beilstein No: 8^4, 249
Name: Benzaldehyde, 3-hydroxy-2-nitro-
Synonym: 3-Hydroxy-2-nitrobenzaldehyde
Line Form: $(2\text{-}NO_2\text{-}3\text{-}HO)C_6H_3CHO$
Mol Form: $C_7H_5NO_4$
M_r: 167.11
mp: 152
Color: nd or pl (Bz, lig)
***s*:** al 3; bz 3; lig 2
IR: **µDMS 6174** 3130 1670 1590 1520 1470 1330 1300 1210 1160 1080 1030 860 790 770 740
m/e: 167.02

HODOC No: 01730
CAS No: 704-13-2
Beilstein No: 8^3, 211
Name: Benzaldehyde, 3-hydroxy-4-nitro-
Synonym: 3-Hydroxy-4-nitrobenzaldehyde
Line Form: $(4\text{-}NO_2\text{-}3\text{-}HO)C_6H_3CHO$
Mol Form: $C_7H_5NO_4$
M_r: 167.11
mp: 128
Color: ye lf
***s*:** Bz 4; Et_2O 4; EtOH 4
IR: **ALD 3:921D**
MS: **ALD 6:94B**
m/e: 167.02

HODOC No: 01731
CAS No: 90151-04-5
Beilstein No: 8^3, 253
Name: Benzaldehyde, 4-hydroxy-2-nitro-
Line Form: $(2\text{-}NO_2\text{-}4\text{-}HO)C_6H_3CHO$
Mol Form: $C_7H_5NO_4$
M_r: 167.11
mp: 67
Color: ye nd
***s*:** Bz 4; Et_2O 4; EtOH 4
m/e: 167.02

HODOC No: 01732
CAS No: 3011-34-5
Beilstein No: 8^4, 303
Name: Benzaldehyde, 4-hydroxy-3-nitro-
Synonym: 4-Formyl-2-nitrophenol
Line Form: $(3\text{-}NO_2\text{-}4\text{-}HO)C_6H_3CHO$
Mol Form: $C_7H_5NO_4$
M_r: 167.11
mp: 144.3
Color: dk ye nd (al, w)
***s*:** H_2O 3; al 3; eth 2; ace 2; chl 2
IR: SADG 13552 3226 3030 2941 1695 1613 1563 1538 1493 1429 1370 1333 1266 1149 1124 855 833 714
UV: SAD 16111 324(2820) 254(15400) 210(10800) MeOH
H NMR: SAD 7122 7.4 8.2 8.7 10.1 ace
m/e: 167.02

HODOC No: 01733
CAS No: 26879-83-4
Beilstein No: 8^3, 254
Name: Benzaldehyde, 4-hydroxy-3-nitro-, oxime
Synonym: 4-Hydroxy-3-nitrobenzaldoxime
Line Form: $(3\text{-}NO_2\text{-}4\text{-}HO)C_6H_3CH{=}NOH$
Mol Form: $C_7H_6N_2O_4$
M_r: 182.13
mp: 189
Color: pr or nd (al-Chl)
***s*:** Et_2O 4; EtOH 4; HOAc 4
m/e: 182.03

HODOC No: 01734
CAS No: 42454-06-8
Beilstein No: 8^4, 250
Name: Benzaldehyde, 5-hydroxy-2-nitro-
Synonym: 5-Hydroxy-2-nitrobenzaldehyde
Line Form: $(5\text{-}HO\text{-}2\text{-}NO_2)\text{-}C_6H_3CHO$
Mol Form: $C_7H_5NO_4$
M_r: 167.11
mp: 166-8
IR: ALD 3:975H
m/e: 167.02

HODOC No: 01735
CAS No: 24019-66-7
Beilstein No: 8^1, 559
Name: Benzaldehyde, 2-hydroxy-3-(2-propenyl)-
Synonym: Salicylaldehyde, 3-allyl-
Mol Form: $C_{10}H_{10}O_2$
M_r: 162.18
bp: $245.5\text{-}246^{755}$
***d*:** 1.098^{15}
IR: µSADP 25030 3030 2860 1670 1610 1490 1450 1390 1280 1240 1220 990 920 850 750
UV: SAD 8895 330(2980) 259(10900) 216(20600) MeOH 272(5620) 233(15800) MeOH:HCl
H NMR: SAD 14824 3.4 5.0 5.9 6.8 6.9 7.3 9.7 11.2 CCl_4
MS: WILEY 1035 162(100) 133(45) 115(40) 134(36) 116(30) 77(29) 91(26) 105(25)
m/e: 162.07

HODOC No: 01736
CAS No: 1043-91-0
Name: Benzaldehyde, imine,3,4-dimethoxy-*N*(2-pyrid-2-ylethyl)
Mol Form: $C_{16}H_{18}N_2O_2$
M_r: 270.32
***s*:** ctc 3
IR: SADG 14832 3040 2920 2820 1630 1580 1500 1460 1430 1370 1320 1250 1130 1020 800 750
H NMR: SAD 8320 3.1 3.7 3.8 3.9 6.7 7.0 7.3 7.5 8.0 8.5 CCl_4
m/e: 270.14

HODOC No: 01737
CAS No: 26260-02-6
Beilstein No: 7^4, 582
Name: Benzaldehyde, 2-iodo-
Synonym: Benzaldehyde, *o*-iodo-
Line Form: $2\text{-}IC_6H_4CHO$
Mol Form: C_7H_5IO
M_r: 232.01
bp: 129^{14}
mp: 37
***s*:** H_2O 2; ace 3
IR: DMS 3407 2840 2730 1690 1640 1580 1460 1390 1260 1200 1020 750 670
UV: OES 4-100 301(1738) 292(2089) 245(12589) hx
m/e: 231.94

HODOC No: 01738
CAS No: 696-41-3
Beilstein No: 7^4, 583
Name: Benzaldehyde, 3-iodo-
Synonym: Benzaldehyde, *m*-iodo-
Line Form: 3-IC_6H_4CHO
Mol Form: C_7H_5IO
M$_r$: 232.01
bp: $124\text{-}5^{13}$
mp: 57
***s*:** al 3; ace 3; bz 3
IR: DMS 3406 3050 2860 2740 1700 1580 1570 1420 1380 1270 1200 1160 880 850 780 700 680
UV: OES 5-84 305(1259) 298(1288) 228(25119) cyhex
m/e: 231.94

HODOC No: 01739
CAS No: 15164-44-0
Beilstein No: 7^4, 584
Name: Benzaldehyde, 4-iodo-
Synonym: Benzaldehyde, *p*-iodo-
Line Form: 4-IC_6H_4CHO
Mol Form: C_7H_5IO
M$_r$: 232.01
bp: $264\text{-}5^{725}$
mp: 77-8
***s*:** H_2O 2; al 3; bz 3
IR: DMS 3405 3050 2820 2720 1690 1660 1580 1560 1380 1280 1200 1160 1050 1010 830 810
UV: OES 5-84 290(3981) 283(14454) 276(15849) 268(19498) cyhex
m/e: 231.94

HODOC No: 01740
CAS No: 135-02-4
Beilstein No: 8^4, 180
Name: Benzaldehyde, 2-methoxy-
Synonym: *o*-Anisaldehyde
Line Form: 2-CH_3O-C_6H_4CHO
Mol Form: $C_8H_8O_2$
M$_r$: 136.14
bp: 243-4, $124\text{-}5^{18}$
mp: 37-8
***d*:** $1.1326^{20/4}$
***n*:** 1.5600^{20}
Color: pr
***s*:** H_2O 1; al 3; eth 4; ace 4; bz 3; ctc 3; chl 4
IR: µCOB 329 2860 2780 1700 1610 1490 1450 1390 1280 1250 1190 1160 1110 1040 1020 950 830 760 650
UV: SAD 3456 308 247 cyhex
H NMR: SAD 1895 3.9 6.8-7.9 10.4 CCl_4
MS: ALDN 84 136(100) 77(81) 135(54) 76(46) 39(43) 65(35) 51(35) 119(31)
m/e: 136.05

HODOC No: 01741
CAS No: 29577-53-5
Name: Benzaldehyde, 2-methoxy-, oxime
Synonym: *o*-Anisaldehyde, oxime
Mol Form: $C_8H_9NO_2$
M$_r$: 151.16
IR: SADG 27005 3160 3050 2990 2860 1630 1590 1570 1490 1460 1440 1320 1300 1250 1190 1170 1110 1050 1020 970 870 780 740 640 570 530 470 410
Raman: SAD 569 3080 3040 3000 2940 2840 1630 1600 1500 1470 1440 1320 1310 1290 1260 1220 1200 1170 1110 1060 1030 970 950 930 880 790 750 740 640 580 480 370 200
UV: SAD 20576 277(1390) 244(16000) MeOH
H NMR: SAD 15716 3.8 6.9 6.9 7.3 7.6 8.5 9.3 $CDCl_3$
m/e: 151.06

HODOC No: 01742
CAS No: 591-31-1
Beilstein No: 8^4, 241
Name: Benzaldehyde, 3-methoxy-
Synonym: *m*-Anisaldehyde
Line Form: 3-CH_3O-C_6H_4CHO
Mol Form: $C_8H_8O_2$
M$_r$: 136.14
bp: 230, 62^1
***d*:** $1.1187^{20/4}$
***n*:** 1.5530^{20}
***s*:** H_2O l; al 3; eth 4; ace 4; bz 3; ctc 2; chl 4
IR: SADG 17084 3080 3020 2960 2850 2740 1710 1580 1480 1460 1430 1380 1320 1250 1190 1140 1030 920 900 860 780 730 680
UV: SAD 16856 310(2570) 252(5250) 219(21600) MeOH
H NMR: SAD 8849 3.8 7.0-7.5 9.9 CCl_4
MS: ALDN 83 136(100) 135(94) 107(35) 77(34) 65(18) 92(16) 39(12) 63(12)
m/e: 136.05

HODOC No: 01743
CAS No: 123-11-5
Beilstein No: 8^4, 252
Name: Benzaldehyde, 4-methoxy-
Synonym: *p*-Anisaldehyde
Line Form: 4-CH_3O-C_6H_4CHO
Mol Form: $C_8H_8O_2$
M$_r$: 136.14
bp: 249.5, 83^2
mp: 0
***d*:** $1.191^{15/4}$
***n*:** 1.5730^{20}
***s*:** H; 2O 1; al 5; eth 5; ace 4; bz 3; chl 4
IR: µCOB 1081 3030 2940 2860 2700 1700 1610 1590 1520 1470 1450 1430 1350 1250 1220 1180 1160 1110 1030 860 830 810 760 760
Raman: SAD 152 3070 3010 2930 2840 1690 1680 1600 1570 1450 1420 1390 1320 1250 1210 1180 1160 1100 1020 1000 850 830 770 760 710 640 630 610 600 510 480 390 370 330 270 180
UV: SAD 536 273 MeOH
H NMR: VAR 194 3.9 7.0 7.8 9.9 $CDCl_3$
MS: WILEY 694 135(100) 136(69) 76(33) 77(33) 92(20) 39(20) 29(16) 63(16)
m/e: 136.05

HODOC No: 01744
CAS No: 3235-04-9
Beilstein No: 8^4, 268
Name: Benzaldehyde, 4-methoxy-, oxime
Synonym: *p*-Anisaldehyde, oxime
Line Form: 4-CH_3O-C_6H_4CH=NOH
Mol Form: $C_8H_9NO_2$
M$_r$: 151.16
mp: 64-5(anti), 133(syn)
Color: nd (Bz)
***s*:** bz 2
UV: HBCP 265(17378) al
MS: NBS 469
m/e: 151.06

HODOC No: 01745
CAS No: 38998-17-3
Beilstein No: 8^4, 482
Name: Benzaldehyde, 4-methoxy-2,3-dimethyl-
Synonym: *p*-Anisaldehyde, 2,3-dimethyl-
Line Form: 2,3-$(CH_3)_2$-4-(CH_3O)-C_6H_2CHO
Mol Form: $C_{10}H_{12}O_2$
M$_r$: 164.19
mp: 49-52
***s*:** chl 3
IR: ALD 3:924G
H NMR: SAD 9046 2.2 2.6 3.8 6.8 7.6 10.1 $CDCl_3$
MS: ALD 6:100C
m/e: 164.08

HODOC No: 01746
CAS No: 6745-75-1
Beilstein No: 8^4, 484
Name: Benzaldehyde, 4-methoxy-2,5-dimethyl-
Synonym: *p*-Anisaldehyde, 2,5-dimethyl-
Line Form: 2,5-$(CH_3)_2$-4-CH_3O-C_6H_2CHO
Mol Form: $C_{10}H_{12}O_2$
M$_r$: 164.19
bp: $147\text{-}9^{12}$
mp: 31-5
IR: ALD 3:924A
MS: ALD 6:99A
m/e: 164.08

HODOC No: 01747
CAS No: 57415-35-7
Beilstein No: 8, 102
Name: Benzaldehyde, 2-methoxy-4-methyl-
Synonym: 2-Methoxy-4-methylbenzaldehyde
Line Form: $2\text{-}CH_3O\text{-}4\text{-}CH_3C_6H_3CHO$
Mol Form: $C_9H_{10}O_2$
M_r: 150.17
bp: $263\text{-}4^{720}$
mp: 42-3
m/e: 150.07

HODOC No: 01748
CAS No: 32723-67-4
Beilstein No: 8^4, 432
Name: Benzaldehyde, 4-methoxy-3-methyl-
Synonym: *p*-Anisaldehyde, 3-methyl-
Line Form: $(4CH_3O\text{-}3\text{-}CH_3)C_6H_3CHO$
Mol Form: $C_9H_{10}O_2$
M_r: 150.17
bp: $80\text{-}5^1$
***d*:** 1.025
***n*:** 1.5670^{20}
IR: SADG 17676 3000 2950 2840 2720 1690 1600 1500 1450 1390 1370 1320 1250 1120 1020 980 810 780 740 640 620
UV: SAD 18605 208(3620) 274(3580) 227(10600) MeOH
H NMR: SAD 9134 2.2 3.9 6.9 7.5 7.6 9.8 CCl_4
MS: ALD 6:89A
m/e: 150.07

HODOC No: 01749
CAS No: 22065-49-2
Beilstein No: 8^3, 190
Name: Benzaldehyde, 2-methoxy-3-nitro-
Synonym: *o*-Anisaldehyde, 3-nitro-
Line Form: $3\text{-}NO_2\text{-}2\text{-}CH_3O\text{-}C_6H_3CHO$
Mol Form: $C_8H_7NO_4$
M_r: 181.14
mp: 89-90
Color: ye pr (dil al), nd (Bz)
***s*:** Et_2O 4; EtOH 4
m/e: 181.04

HODOC No: 01750
CAS No: 53055-05-3
Beilstein No: 8^3, 210
Name: Benzaldehyde, 3-methoxy-2-nitro-
Synonym: *m*-Anisaldehyde, 2-nitro-
Line Form: $(3\text{-}CH_3O\text{-}2\text{-}NO_2)\text{-}C_6H_3CHO$
Mol Form: $C_8H_7NO_4$
M_r: 181.14
mp: 102-3
IR: ALD 3:917B
MS: ALD 6:93A
m/e: 181.04

HODOC No: 01751
CAS No: 2426-87-1
Beilstein No: 8^2, 283
Name: Benzaldehyde, 3-methoxy-4-(phenylmethoxy)-
Synonym: Benzaldehyde, 4-(benzyloxy)-3-methoxy-
Line Form: $(4\text{-}C_6H_5O\text{-}3\text{-}CH_3O)C_6H_3CHO$
Mol Form: $C_{15}H_{14}O_3$
M_r: 242.26
bp: $213\text{-}4^5$
mp: 64-5
IR: SADG 27075 3060 3030 2940 2830 2720 1680 1580 1510 1460 1420 1390 1340 1260 1230 1190 1160 1130 1080 1030 920 860 810 780 730 690 650 590 560 530 460 380
Raman: SAD 153 3080 2950 1700 1670 1600 1590 1510 1500 1470 1450 1430 1400 1390 1350 1330 1270 1200 1130 1030 1000 990 970 860 800 750 730 660 620 600 530 430 340 230 160 110
H NMR: SAD 15746 3.8 5.1 6.9 7.2-7.5 9.8 $CDCl_3$
m/e: 242.09

HODOC No: 01752
CAS No: 529-20-4
Beilstein No: 7^4, 667
Name: Benzaldehyde, 2-methyl-
Synonym: *o*-Tolualdehyde
Line Form: $2\text{-}CH_3C_6H_4CHO$
Mol Form: C_8H_8O
M_r: 120.14
bp: 200, 94^{10}
***d*:** $1.0386^{19/4}$
***n*:** 1.5481^{20}
***s*:** H_2O 2; al 3; eth 3; ace 4; bz 3; ctc 2; chl 3
IR: SADG 9147 3030 2857 1695 1587 1563 1493 1449 1370 1266 1205 862 833 752
UV: SAD 13722 280(872) 249(4550) MeOH
H NMR: SAD 3936 2.7 7.3 7.7 10.2 CCl_4
MS: WILEY 478 91(100) 120(78) 119(77) 65(28) 63(21) 51(15) 92(12) 50(12)
m/e: 120.06

HODOC No: 01753
CAS No: 1773-44-0
Name: Benzaldehyde, 2-methyl-, (2,4-dinitrophenyl)hydrazone
Synonym: *o*-Tolualdehyde, (2,4-dinitrophenyl)hydrazone
Mol Form: $C_{14}H_{12}N_4O_4$
M_r: 300.26
IR: μCOB 2062 3330 3130 2940 1610 1590 1520 1470 1430 1390 1330 1300 1220 1140 1090 970 910 830 760 740 720
MS: ALDN 226 300(100) 117(52) 91(38) 118(31) 121(20) 301(18) 90(15) 65(15)
m/e: 300.09

HODOC No: 01754
CAS No: 620-23-5
Beilstein No: 7^4, 669
Name: Benzaldehyde, 3-methyl-
Synonym: *m*-Tolualdehyde
Line Form: $3\text{-}CH_3C_6H_4CHO$
Mol Form: C_8H_8O
M_r: 120.14
bp: 199, $93\text{-}4^{17}$
***d*:** $1.0189^{21/4}$
***n*:** 1.5413^{21}
***s*:** H_2O 2; al 5; eth 5; ace 4; bz 3; chl 3
IR: SADG 16692 3050 2940 2840 2740 1710 1600 1480 1380 1280 1240 1140 1030 900 780 740 690
UV: OES 5-136 288(1514) 244(13490) iso
H NMR: SAD 9244 2.4 7.4 7.6 9.9 CCl_4
MS: WILEY 475 91(100) 119(85) 120(78) 65(29) 63(20) 64(16) 51(15) 50(12)
m/e: 120.06

HODOC No: 01755
CAS No: 2880-05-9
Name: Benzaldehyde, 3-methyl-, (2,4-dinitrophenyl)hydrazone
Synonym: *m*-Tolualdehyde, (2,4-dinitrophenyl)hydrazone
Line Form: $3\text{-}CH_3C_6H_4CH{=}NNH[C_6H_3(NO_2)_2\text{-}2,4]$
Structure:

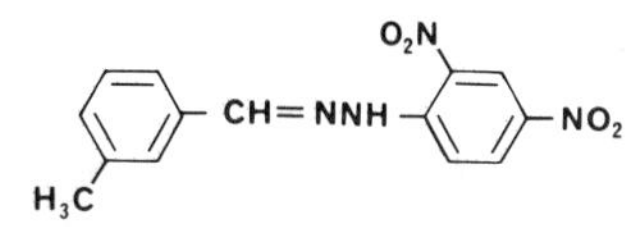

Mol Form: $C_{14}H_{12}N_4O_4$
M_r: 300.26
mp: 194-5
IR: μCOB 2061 3330 3130 2940 1610 1590 1520 1470 1430 1330 1300 1270 1220 1140 1090 950 850 790 740 720 690
UV: SAD 2341 227(22000) 207 diox
MS: ALDN 226 300(100) 121(39) 91(38) 93(21) 79(20) 65(19) 78(18) 178(18)
m/e: 300.09

HODOC No: 01756
CAS No: 104-87-0
Beilstein No: 7^4, 672
Name: Benzaldehyde, 4-methyl-
Synonym: *p*-Tolualdehyde
Line Form: $4\text{-}CH_3C_6H_4CHO$
Mol Form: C_8H_8O
M_r: 120.14
bp: 204-5, 106^{10}
***d*:** $1.0194^{17/4}$
***n*:** 1.5454^{20}
***s*:** H_2O 2; al 5; eth 5; ace 5; chl 4
IR: COB 6182 3040 2920 2810 2720 1700 1600 1580 1380 1300 1210 1170 840 800 760
UV: SAD 15952 255(21300) MeOH
H NMR: SAD 6900 2.4 7.3 7.8 10.0 $CDCl_3$
MS: WILEY 477 120(100) 119(12) 91(11) 65(3) 39(2) 63(1) 133(1) 92(1)
m/e: 120.06

HODOC No: 01757
CAS No: 2571-00-8
Beilstein No: 15^3, 472
Name: Benzaldehyde, 4-methyl-, (2,4-dinitrophenyl)hydrazone
Synonym: *p*-Tolualdehyde, (2,4-dinitrophenyl)hydrazone
Line Form: $4\text{-}CH_3C_6H_4CH{=}NNH[C_6H_3(NO_2)_2\text{-}2,4]$
Mol Form: $C_{14}H_{12}N_4O_4$
M_r: 300.26
mp: 233-4
IR: µCOB 2093 3330 3130 2940 1610 1590 1520 1470 1430 1330 1320 1270 1220 1140 1090 960 930 910 850 930 780 740 720 690
MS: NBS 4815
m/e: 300.09

HODOC No: 01758
CAS No: 3046-82-0
Beilstein No: 8, 436
Name: Benzaldehyde, 3,3′-methylenebis[6-hydroxy-
Synonym: Salicylaldehyde, 5,5′-methylenedi-
Mol Form: $C_{15}H_{12}O_4$
M_r: 256.24
mp: 142-3
***s*:** HOAc 3
IR: µCOB 3980 3130 1670 1470 1390 1280 1150 920 850 830 770 750 670
m/e: 256.07

HODOC No: 01759
CAS No: 122-03-2
Beilstein No: 7^4, 723
Name: Benzaldehyde, 4-(1-methylethyl)-
Synonym: Benzaldehyde, *p*-isopropyl-
Line Form: $4\text{-}(CH_3)_2CH\text{-}C_6H_4CHO$
Mol Form: $C_{10}H_{12}O$
M_r: 148.19
bp: 235-6, $103\text{-}4^{10}$
***d*:** $0.9755^{20/4}$
***n*:** 1.5301^{20}
***s*:** H_2O l; al 3; eth 3; ctc 2
IR: µCOB 328 2940 2700 1700 1610 1590 1470 1430 1390 1320 1280 1220 1160 1110 1050 1020 830 730 700
UV: SAD 1680 255 MeOH
H NMR: SAD 6825 1.3 3.0 7.3 7.8 9.9 CCl_4
MS: WILEY 854 133(100) 105(74) 148(71) 119(40) 79(31) 91(31) 77(21) 103(17)
m/e: 148.09

HODOC No: 01760
CAS No: 552-89-6
Beilstein No: 7^4, 584
Name: Benzaldehyde, 2-nitro-
Synonym: Benzaldehyde, *o*-nitro-
Line Form: $2\text{-}NO_2\text{-}C_6H_4CHO$
Mol Form: $C_7H_5NO_3$
M_r: 151.11
bp: 153^{23}
mp: 43-4
***d*:** $1.2844^{20/4}$
Color: ye nd (w)
***s*:** H_2O 2; al 4; eth 4; ace 4; bz 4; chl 2; os 4
IR: µCOB 1790 2940 1700 1590 1540 1450 1390 1350 1320 1280 1190 860 820 790 740 690
UV: SAD 1091 252(4400) MeOH
H NMR: VAR 148 7.9 10.4 $CDCl_3$
MS: NBS 466
m/e: 151.03

HODOC No: 01761
CAS No: 99-61-6
Beilstein No: 7^4, 591
Name: Benzaldehyde, 3-nitro-
Synonym: Benzaldehyde, *m*-nitro-
Line Form: $3\text{-}NO_2C_6H_4CHO$
Mol Form: $C_7H_5NO_3$
M_r: 151.11
bp: 164^{23}
mp: 58-9
***d*:** $1.2792^{20/4}$
Color: lt ye nd (w)
***s*:** H_2O 2; al 3; eth 3; ace 4; bz 4; chl 3; os 3
IR: SADG 15356 3030 2857 1695 1613 1587 1538 1471 1449 1351 1299 1205 1087 935 909 820 730 685
UV: SAD 626 233(20900) H_2O
H NMR: SAD 6718 7.9 8.3 8.5 8.7 10.2 $CDCl_3$
MS: WILEY 898 151(100) 150(85) 77(74) 51(66) 105(50) 50(28) 76(22) 104(16)
m/e: 151.03

HODOC No: 01762
CAS No: 555-16-8
Beilstein No: 7^4, 598
Name: Benzaldehyde, 4-nitro-
Synonym: Benzaldehyde, *p*-nitro-
Line Form: $4\text{-}NO_2C_6H_4CHO$
Mol Form: $C_7H_5NO_3$
M_r: 151.11
bp: sub
mp: 106-8
***d*:** 1.496
Color: lf, pr (w)
***s*:** H_2O 2; al 4; eth 2; bz 3; chl 3; HOAc 3; lig 2
IR: SADG 384 3030 2857 1587 1538 1449 1389 1351 1282 1250 1149 1042 893 833 769 741 671
UV: SAD 1092 265(10600) MeOH
H NMR: SAD 6027 8.1 8.4 10.2 $CDCl_3$
MS: WILEY 898 151(100) 150(94) 77(62) 51(58) 50(27) 105(24) 104(19) 76(17)
m/e: 151.03

HODOC No: 01763
CAS No: 74663-99-3
Beilstein No: 7^4, 584
Name: Benzaldehyde, 4-nitroso-
Synonym: Benzaldehyde, *p*-nitroso-
Line Form: $4\text{-}ON\text{-}C_6H_4CHO$
Mol Form: $C_7H_5NO_2$
M_r: 135.12
mp: 137-8
***s*:** H_2O 2; al 3; HOAc 3
UV: ZEELA 60-1142-56 800(50) 330(8913) 275(10000) H_2O
m/e: 135.03

HODOC No: 01764
CAS No: 49763-66-8
Name: Benzaldehyde, 4-octyl-
Synonym: *p*-Octylbenzaldehyde
Line Form: $4\text{-}CH_3(CH_2)_7\text{-}C_6H_4CHO$
Mol Form: $C_{15}H_{22}O$
M_r: 218.32
bp: 181^{14}
***n*:** 1.5113^{20}
m/e: 218.17

HODOC No: 01765
CAS No: 24083-13-4
Beilstein No: 8^4, 256
Name: Benzaldehyde, 4-(octyloxy)-
Synonym: Benzaldehyde, *p*-(octyloxy)-
Line Form: $4\text{-}CH_3(CH_2)_7O\text{-}C_6H_4CHO$
Mol Form: $C_{15}H_{22}O_2$
M_r: 234.32
bp: $131^{0.5}$
m/e: 234.16

HODOC No: 01766
CAS No: 622-31-1
Beilstein No: 7^4, 527
Name: Benzaldehyde, oxime, (E)-
Synonym: (E)-Benzaldehyde oxime
Line Form: $C_6H_5CH{=}NOH$
Mol Form: C_7H_7NO
M_r: 121.13
mp: 130
***d*:** $1.145^{20/4}$
Color: nd (eth)
***s*:** H_2O 3; al 4; eth 4
IR: SADG 1574 3320 3060 2980 2880 1630 1580 1490 1450 1300 1280 1210 1170 1070 950 870 750 690 640 510 460 380 310
UV: OES 4-113 247(15849) EtOH
H NMR: SAD 14953 ini 7.5 8.2 9.9 $CDCl_3$
MS: WILEY 488 18(100) 103(46) 17(24) 76(20) 77(18) 28(16) 50(12) 121(12)
m/e: 121.05

HODOC No: 01767
CAS No: 622-32-2
Beilstein No: 7^4, 527
Name: Benzaldehyde, oxime, (Z)-
Synonym: *cis*-Benzaldehyde oxime
Line Form: $C_6H_5CH{=}NOH$
Mol Form: C_7H_7NO
M_r: 121.13
bp: 200, $118\text{-}9^{10}$
mp: 36-7
***d*:** $1.1111^{20/4}$
***n*:** 1.5908^{20}
Color: pr
***s*:** Bz 4; Et_2O 4; EtOH 4
IR: COB 1674
UV: SAD 9290
H NMR: SAD 14953
MS: WILEY 488
m/e: 121.05

HODOC No: 01768
CAS No: 19635-52-0
Beilstein No: 7^4, 578
Name: Benzaldehyde, pentachloro-
Synonym: Pentachlorobenzaldehyde
Line Form: C_6Cl_5CHO
Mol Form: C_7HCl_5O
M_r: 278.34
mp: 202.5
Color: nd (Bz, al)
***s*:** al 2; eth 4; ace 4; bz 4; CS_2 4
UV: OES 1-115 290(794) 262(6310) 225(31623) iso
m/e: 275.85

HODOC No: 01769
CAS No: 653-37-2
Beilstein No: 7^4, 561
Name: Benzaldehyde, pentafluoro-
Synonym: Formylpentafluorobenzene
Line Form: C_6F_5CHO
Mol Form: C_7HF_5O
M_r: 196.07
bp: 164-6
mp: 20
***n*:** 1.4506^{20}
IR: ALD 2:810D
MS: ALD 6:101A
m/e: 195.99

HODOC No: 01770
CAS No: 39515-51-0
Beilstein No: 8^4, 242
Name: Benzaldehyde, 3-phenoxy-
Synonym: *m*-Phenoxybenzaldehyde
Line Form: $3\text{-}C_6H_5O\text{-}C_6H_4CHO$
Mol Form: $C_{13}H_{10}O_2$
M_r: 198.21
bp: $140^{0.1}$
***d*:** 1.147
***n*:** 1.5954^{20}
m/e: 198.07

HODOC No: 01771
CAS No: 588-64-7
Beilstein No: 15^2, 57
Name: Benzaldehyde, phenylhydrazone
Synonym: Benzalphenylhydrazine
Line Form: $C_6H_5CH{=}NNHC_6H_5$
Mol Form: $C_{13}H_{12}N_2$
M_r: 196.24
mp: 156
Color: nd (lig), pr
***s*:** al 2; eth 2; ace 3; bz 3; liq NH_3 3
IR: µSADP 20678 3330 3030 1700 1590 1490 1450 1370 1320 1270 1140 1060 940 890 750 690
UV: SAD 7948 343(17200) 300(8990) 236(12900) MeOH
H NMR: SAD 3606 6.5-7.7 9.2 ace
m/e: 196.10

HODOC No: 01772
CAS No: 1700-37-4
Beilstein No: 8^3, 145
Name: Benzaldehyde, 3-(phenylmethoxy)-
Synonym: Benzaldehyde, *m*-(benzyloxy)-
Line Form: 3-$C_6H_5CH_2O$-C_6H_4CHO
Mol Form: $C_{14}H_{12}O_2$
M_r: 212.23
mp: 56-8
IR: ALD 2:810G
MS: ALD 6:101C
m/e: 212.08

HODOC No: 01773
CAS No: 4397-53-9
Beilstein No: 8^4, 258
Name: Benzaldehyde, 4-(phenylmethoxy)-
Synonym: Benzaldehyde, *p*-(benzyloxy)-
Line Form: 4-$C_6H_5CH_2O$-C_6H_4CHO
Mol Form: $C_{14}H_{12}O_2$
M_r: 212.23
mp: 72-4
IR: ALD 2:810H
MS: ALD 6:101D
m/e: 212.08

HODOC No: 01774
CAS No: 588-68-1
Beilstein No: 7^4, 530
Name: Benzaldehyde, (phenylmethylene)hydrazone
Synonym: Benzaldehyde, azine
Line Form: $C_6H_5CH{=}NN{=}CHC_6H_5$
Mol Form: $C_{14}H_{12}N_2$
M_r: 208.25
mp: 93
Color: ye pr (al)
***s*:** H_2O 1; al 3; eth 3; ace 3; bz 3; ctc 2; chl 3
IR: SADG 485 3030 1613 1587 1493 1449 1299 1075 962 862 758 694 671
UV: SAD 1584 299(21300) 252(12100) MeOH
H NMR: SAD 702 7.3 7.8 8.5 CCl_4
MS: ALDN 170 103(100) 77(91) 131(89) 51(62) 208(62) 104(54) 207(46) 76(42)
m/e: 208.10

HODOC No: 01775
CAS No: 28752-82-1
Beilstein No: 8^1, 519
Name: Benzaldehyde, 2-(2-propenyloxy)-
Synonym: Benzaldehyde, *o*-(allyloxy)-
Line Form: 2-($CH_2{=}CHCH_2O$)-C_6H_4CHO
Mol Form: $C_{10}H_{10}O_2$
M_r: 162.18
bp: 130^{10}
***d*:** 1.094^{15}
IR: µSADP 25029 2860 1700 1590 1490 1470 1430 1390 1280 1240 1220 1190 1160 1100 1040 1010 1000 940 840 760
UV: SAD 8894 317(3890) 253(8840) 215(18200) MeOH
H NMR: SAD 9840 4.6 5.3 5.4 6.0 7.0 7.4 7.7 9.9 CCl_4
MS: WILEY 1036 41(100) 39(50) 121(29) 120(28) 18(16) 65(14) 76(14) 162(1)
m/e: 162.07

HODOC No: 01776
CAS No: 40663-68-1
Beilstein No: 8^4, 256
Name: Benzaldehyde, 4-(2-propenyloxy)-
Synonym: *p*-(Allyloxy)benzaldehyde
Line Form: 4-(CH_2=$CHCH_2O$)-C_6H_4CHO
Mol Form: $C_{10}H_{10}O_2$
Mr: 162.18
bp: 142^{10}
IR: µ**SADP 23094** 2780 2700 1700 1610 1560 1520 1430 1320 1270 1240 1220 1160 1110 1010 1000 940 860 830
UV: SAD 20975 282(17600) 276(17800) 220(12700) MeOH
H NMR: SAD 9828 4.6 5.3 5.4 5.9 6.9 7.7 9.8 CCl_4
MS: WILEY 1036 41(100) 162(50) 39(23) 121(18) 65(10) 161(8) 18(8) 133(7)
m/e: 162.07

HODOC No: 01777
CAS No: 28785-06-0
Name: Benzaldehyde, 4-propyl-
Synonym: Benzaldehyde, *p*-propyl-
Line Form: 4-$CH_3CH_2CH_2$-C_6H_4CHO
Mol Form: $C_{10}H_{12}O$
Mr: 148.19
bp: 108^9
***n*:** 1.5320^{20}
m/e: 148.09

HODOC No: 01778
CAS No: 56962-16-4
Beilstein No: 7^2, 181
Name: Benzaldehyde, 2,3,4,5-tetrachloro-
Synonym: 2,3,4,5-Tetrachlorobenzaldehyde
Line Form: 2,3,4,5Cl_4C_6HCHO
Mol Form: $C_7H_2Cl_4O$
Mr: 243.90
mp: 106
m/e: 241.89

HODOC No: 01779
CAS No: 29344-95-4
Beilstein No: 7^3, 1147
Name: Benzaldehyde, 2,3,4,5-tetramethyl-
Synonym: 2,3,4,5-Tetramethylbenzaldehyde
Line Form: 2,3,4,5-$(CH_3)_4C_6HCHO$
Mol Form: $C_{11}H_{14}O$
Mr: 162.22
bp: 142-4^{15}
mp: 29-30
m/e: 162.10

HODOC No: 01780
CAS No: 5780-08-5
Beilstein No: 7^4, 571
Name: Benzaldehyde, 2,3,4,6-tetramethyl-
Synonym: Isoduraldehyde
Line Form: 2,3,4,6-$(CH_3)_4$-C_6H-CHO
Mol Form: $C_{11}H_{14}O$
Mr: 162.22
bp: 136^{10}
mp: 15(fp)
m/e: 162.10

HODOC No: 01781
CAS No: 17432-37-0
Beilstein No: 7^3, 1148
Name: Benzaldehyde, 2,3,5,6-tetramethyl-
Synonym: Durene aldehyde
Line Form: 2,3,5,6-$(CH_3)_4C_6H$-CHO
Mol Form: $C_{11}H_{14}O$
Mr: 162.22
bp: 135^{11}
mp: 20(fp)
***n*:** 1.5560^{30}
m/e: 162.10

HODOC No: 01782
CAS No: 19361-59-2
Beilstein No: 7^4, 577
Name: Benzaldehyde, 2,3,4-trichloro-
Synonym: 2,3,4-Trichlorobenzaldehyde
Line Form: 2,3,4-Cl_3-C_6H_2CHO
Mol Form: $C_7H_3Cl_3O$
Mr: 209.45
mp: 91
Color: nd (al)
***s*:** EtOH 4
m/e: 207.92

HODOC No: 01783
CAS No: 56961-75-2
Beilstein No: 7^2, 180
Name: Benzaldehyde, 2,3,5-trichloro-
Synonym: 2,3,5-Trichlorobenzaldehyde
Line Form: 2,3,5Cl_3-C_6H_2CHO
Mol Form: $C_7H_3Cl_3O$
Mr: 209.45
mp: 56
Color: nd (al)
***s*:** Ac 4; Bz 4; Et_2O 4; EtOH 4
m/e: 207.92

HODOC No: 01784
CAS No: 4659-47-6
Beilstein No: 7^4, 577
Name: Benzaldehyde, 2,3,6-trichloro-
Synonym: 2,3,6-Trichlorobenzaldehyde
Line Form: 2,3,6-$Cl_3C_6H_2CHO$
Mol Form: $C_7H_3Cl_3O$
M$_r$: 209.45
mp: 86-7
Color: nd (lig)
***s*:** Ac 4; Bz 4; Et_2O 4
m/e: 207.92

HODOC No: 01785
CAS No: 35696-87-8
Beilstein No: 7^4, 577
Name: Benzaldehyde, 2,4,5-trichloro-
Synonym: 2,4,5-Trichlorobenzaldehyde
Line Form: 2,4,5-$Cl_3C_6H_2CHO$
Mol Form: $C_7H_3Cl_3O$
M$_r$: 209.45
mp: 112-3
Color: nd (al)
***s*:** H_2O 2; al 4; eth 4; ace 4; bz 4; chl 4; CS_2 4
MS: WILEY 1473 207(100) 208(56) 211(35) 179(22) 181(22) 74(19) 109(18) 212(17)
m/e: 207.92

HODOC No: 01786
CAS No: 24473-00-5
Beilstein No: 7^4, 578
Name: Benzaldehyde, 2,4,6-trichloro-
Synonym: 2,4,6-Trichlorobenzaldehyde
Line Form: 2,4,6-$Cl_3C_6H_2CHO$
Mol Form: $C_7H_3Cl_3O$
M$_r$: 209.45
mp: 58-9
***s*:** eth 4; lig 2
UV: HBCP 346(32) 304(631) 260(5012) al
m/e: 207.92

HODOC No: 01787
CAS No: 56961-76-3
Beilstein No: 7^2, 180
Name: Benzaldehyde, 3,4,5-trichloro-
Line Form: 3,4,5-$Cl_3C_6H_2CHO$
Mol Form: $C_7H_3Cl_3O$
M$_r$: 209.45
mp: 90-1
Color: nd (al)
***s*:** Ac 4; Bz 4; Et_2O 4; EtOH 4
m/e: 207.92

HODOC No: 01788
CAS No: 455-19-6
Beilstein No: 7^4, 678
Name: Benzaldehyde, 4-(trifluoromethyl)-
Synonym: *p*-Tolualdehyde, α,α,α-trifluoro-
Line Form: 4-F_3C-C_6H_4CHO
Mol Form: $C_8H_5F_3O$
M$_r$: 174.11
bp: $80\text{-}1^{25}$
***n*:** 1.4630^{20}
m/e: 174.03

HODOC No: 01789
CAS No: 2144-08-3
Beilstein No: 8^3, 3357
Name: Benzaldehyde, 2,3,4-trihydroxy-
Synonym: 2,3,4-Trihydroxybenzaldehyde
Line Form: 2,3,4-$(HO)_3C_6H_2CHO$
Mol Form: $C_7H_6O_4$
M$_r$: 154.12
mp: 159-60
m/e: 154.03

HODOC No: 01790
CAS No: 487-70-7
Beilstein No: 8, 390
Name: Benzaldehyde, 2,4,6-trihydroxy-
Synonym: Formylphloroglucinol
Mol Form: $C_7H_6O_4$
M$_r$: 154.12
***s*:** bz 1; chl 1
IR: COB 7547 3390 3280 1650 1605 1580 1525 1480 1400 1380 1320 1275 1225 1160 1080 1010 830 815 795
UV: SAD 20203 456(2070) 293(10000) 226(7880) 209(10400) MeOH 497(2600) 323(14300) MeOH:KOH
H NMR: SAD 15233 5.9 7.7 10.0 DMSO
m/e: 154.03

HODOC No: 01791
CAS No: 2103-57-3
Beilstein No: 8^4, 2714
Name: Benzaldehyde, 2,3,4-trimethoxy-
Synonym: 2,3,4-Trimethoxybenzaldehyde
Line Form: 2,3,4-$(CH_3O)_3$-C_6H_2CHO
Mol Form: $C_{10}H_{12}O_4$
M$_r$: 196.19
bp: $120\text{-}5^{0.5}$
***n*:** 1.5547^{20}
H NMR: SAD 9120 3.8 3.9 4.0 6.8 7.6 10.3 $CDCl_3$
m/e: 196.07

HODOC No: 01792
CAS No: 4460-86-0
Beilstein No: 8^4, 2715
Name: Benzaldehyde, 2,4,5-trimethoxy-
Synonym: Asaraldehyde
Line Form: 2,4,5-$(CH_3O)_3$-C_6H_2CHO
Mol Form: $C_{10}H_{12}O_4$
M_r: 196.19
bp: 140^4, sub
mp: 114
***s*:** H_2O 3; eth 3; chl 3; lig 3
IR: SADG 11349 2960 2920 2840 2820 1650 1600 1510 1460 1400 1340 1280 1260 1250 1200 1170 1110 1030 1010 850 810 750
H NMR: SAD 5407 3.8 3.9 4.0 6.5 7.3 10.3 $CDCl_3$
m/e: 196.07

HODOC No: 01793
CAS No: 830-79-5
Beilstein No: 8^4, 2717
Name: Benzaldehyde, 2,4,6-trimethoxy-
Synonym: 2,4,6-Trimethoxybenzaldehyde
Line Form: 2,4,6-$(CH_3O)_3$-C_6H_2CHO
Mol Form: $C_{10}H_{12}O_4$
M_r: 196.19
mp: 118-20
IR: ALD 2:807D
MS: ALD 6:95D
m/e: 196.07

HODOC No: 01794
CAS No: 86-81-7
Beilstein No: 8^4, 2719
Name: Benzaldehyde, 3,4,5-trimethoxy-
Synonym: 3,4,5-Trimethoxybenzaldehyde
Line Form: 3,4,5-$(CH_3O)_3$-C_6H_2CHO
Mol Form: $C_{10}H_{12}O_4$
M_r: 196.19
mp: 71-4
***s*:** chl 3
IR: SADG 18729 3020 2970 2850 2750 1680 1580 1500 1450 1420 1380 1320 1230 1110 990 840 750 620 590
UV: SAD 6955 286 217 MeOH 285(9420) MeOH:KOH 272(1510) MeOH:HCl
H NMR: SAD 8680 3.9 7.1 9.8 $CDCl_3$
m/e: 196.07

HODOC No: 01795
CAS No: 34341-28-1
Beilstein No: 7^4, 729
Name: Benzaldehyde, 2,3,4-trimethyl-
Synonym: 2,3,4-Trimethylbenzaldehyde
Line Form: 2,3,4-$(CH_3)_3$-C_6H_2CHO
Mol Form: $C_{10}H_{12}O$
M_r: 148.19
bp: 121.5^{11}
mp: 7-8
m/e: 148.09

HODOC No: 01796
CAS No: 34341-29-2
Beilstein No: 7^4, 730
Name: Benzaldehyde, 2,3,6-trimethyl-
Synonym: 2,3,6-Trimethylbenzaldehyde
Line Form: 2,3,6-$(CH_3)_3$-C_6H_2CHO
Mol Form: $C_{10}H_{12}O$
M_r: 148.19
bp: $113\text{-}4^{10}$
m/e: 148.09

HODOC No: 01797
CAS No: 5779-72-6
Beilstein No: 7^4, 731
Name: Benzaldehyde, 2,4,5-trimethyl-
Synonym: Duraldehyde
Line Form: 2,4,5-$(CH_3)_3$-C_6H_2CHO
Mol Form: $C_{10}H_{12}O$
M_r: 148.19
bp: 243, 121^{10}
mp: 43.5
Color: pl (al)
m/e: 148.09

HODOC No: 01798
CAS No: 487-68-3
Beilstein No: 7^4, 730
Name: Benzaldehyde, 2,4,6-trimethyl-
Synonym: Mesitaldehyde
Line Form: 2,4,6-$(CH_3)_3$-C_6H_2CHO
Mol Form: $C_{10}H_{12}O$
M_r: 148.19
bp: 237-40, 192^{50}
mp: 14
***s*:** H_2O 1; al 3; eth 3; ace 3; bz 3
IR: SADG 16693 2950 2860 2780 1680 1610 1560 1440 1380 1280 1210 1150 1030 940 850 780 720
Raman: API 79 2920 1687 1608 1477 1382 1295 1208 1148 573 526
UV: SAD 16666 264(10600) 213(17600) MeOH
MS: WILEY 855 147(100) 148(73) 119(50) 91(19) 77(18) 105(18) 63(9) 65(9)
m/e: 148.09

HODOC No: 01799
CAS No: 606-34-8
Beilstein No: 7^3, 926
Name: Benzaldehyde, 2,4,6-trinitro-
Synonym: TNBAL
Line Form: 2,4,6-$(NO_2)_3$-C_6H_2CHO
Mol Form: $C_7H_3N_3O_7$
M_r: 241.11
mp: 119
Color: dk gr pl (Bz)
***s*:** H_2O 1; al 3; eth 3; ace 3; bz 3; chl 3; HOAc 3
IR: SADG 16694 3100 2880 1720 1600 1560 1400 1340 1190 1150 1070 910 840 770 730 700
UV: SAD 16667 222(21800) MeOH
m/e: 241.00

HODOC No: 01800
CAS No: 55-21-0
Beilstein No: 9^3, 1064
Name: Benzamide
Synonym: Benzoic acid amide
Line Form: $C_6H_5CONH_2$
Structure:

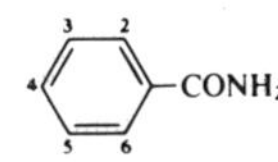

Mol Form: C_7H_7NO
M_r: 121.13
bp: 290
mp: 132-3
***d*:** $1.0792^{130/4}$
Color: mcl pr or pl (w)
***s*:** H_2O 2; al 4; eth 2; bz 2; ctc 4; CS_2 4
IR: SADG 340 3448 3226 1667 1639 1587 1449 1370 1124 781 694
Raman: SAD 123 3270 3160 3070 3000 1690 1610 1570 1500 1450 1420 1180 1150 1120 1080 1030 1010 850 800 780 640 620 400 250 180 140 110
UV: SAD 744 268(691) 225(11000) MeOH
H NMR: SAD 689 7.5-8.1 tfa
MS: WILEY 489 103(100) 76(40) 50(27) 51(16) 39(11) 75(10) 52(9) 121(0)
m/e: 121.05

HODOC No: 01801
CAS No: 19958-58-8
Name: Benzamide, *N*-(acetyloxy)-*N*-phenyl-
Synonym: Hydroxylamine, *O*-acetyl-*N*-benzoyl-*N*-phenyl-
Line Form: $C_6H_5CON(C_6H_5)O_2CCH_3$
Mol Form: $C_{15}H_{13}NO_3$
M_r: 255.26
IR: COB 7008 3060 2930 1780 1670 1580 1490 1450 1380 1340 1270 1180 1000 920 850 790 700 690 650 560
m/e: 255.09

HODOC No: 01802
CAS No: 88-68-6
Beilstein No: 14^3, 889
Name: Benzamide, 2-amino-
Synonym: Benzamide, *o*-amino-
Line Form: 2-$H_2NC_6H_4CONH_2$
Mol Form: $C_7H_8N_2O$
M_r: 136.15
bp: 300
mp: 110-1
***s*:** H_2O 3; al 3; eth 2; bz 2; AcOEt 4
IR: SADG 8588 3448 3333 3226 1695 1639 1515 1389 1299 1266 1235 800
UV: SAD 2987 248 213 MeOH
MS: ALDN 83 119(100) 136(82) 92(58) 65(28) 39(12) 64(10) 63(10) 52(8)
m/e: 136.06

HODOC No: 01803
CAS No: 3544-24-9
Beilstein No: 4^3, 998
Name: Benzamide, 3-amino-
Synonym: Benzamide, *m*-amino-
Line Form: 3-$H_2NC_6H_4CONH_2$
Mol Form: $C_7H_8N_2O$
M_r: 136.15
mp: 79-80(hyd), 113-4(anh)
Color: ye mcl nd (+1w) nd (Bz)
***s*:** H_2O 4; Et_2O 4; EtOH 4
UV: HBCP
MS: ALDN 83
m/e: 136.06

HODOC No: 01804
CAS No: 2835-68-9
Beilstein No: 14^3, 1061
Name: Benzamide, 4-amino-
Synonym: Benzamide, *p*-amino-
Line Form: 4-$H_2NC_6H_4CONH_2$
Mol Form: $C_7H_8N_2O$
M_r: 136.15
mp: 183
Color: ye cr (+1/4w)
***s*:** H_2O 2; al 3; eth 3
UV: HBCP 285(19953) al
MS: ALDN 83 120(100) 136(80) 92(44) 65(38) 39(16) 137(8) 121(8) 91(8)
m/e: 136.06

HODOC No: 01805
CAS No: 614-22-2
Beilstein No: 9^3, 1115
Name: Benzamide, *N*-(aminocarbonyl)-
Synonym: Urea, benzoyl-
Line Form: $C_6H_5CONHCONH_2$
Mol Form: $C_8H_8N_2O_2$
M_r: 164.16
mp: 214-5
Color: fl or lf (al)
***s*:** H_2O 3; al 3; eth 1
IR: SADG 10160 3380 3350 3230 1700 1660 1610 1470 1390 1260 1100 950 780 710 700 640
UV: SAD 5968 232 MeOH
MS: NBS 4240
m/e: 164.06

HODOC No: 01806
CAS No: 51884-03-8
Beilstein No: 9^1, 163
Name: Benzamide, *N*-(aminocarbonyl)-4-nitro-
Synonym: *N*-*p*-Nitrobenzoylurea
Mol Form: $C_8H_7N_3O_4$
M_r: 209.15
mp: 243-5
IR: μCOB 4132 3330 3230 3030 1670 1540 1470 1430 1350 1280 1090 960 880 850 780 710
m/e: 209.04

HODOC No: 01807
CAS No: 840-10-8
Name: Benzamide, 4-amino-2-chloro-*N*-[2-(diethylamino)ethyl]-
Synonym: Chloroprocainamid
Mol Form: $C_{13}H_{20}ClN_2O$
M_r: 255.75
IR: μCOB 696 3330 2940 2860 1640 1520 1490 1370 1280 1060 910 830 690 660
m/e: 255.13

HODOC No: 01808
CAS No: 51-06-9
Beilstein No: 14^3, 1077
Name: Benzamide,
4-amino-*N*-[2-(diethylamino)ethyl]-
Synonym: Benzamide, *p*-amino-*N*-[2-(diethylamino)ethyl]-
Line Form: $4\text{-}H_2NC_6H_4CONHCH_2CH_2N(C_2H_5)_2$
Mol Form: $C_{13}H_{21}N_3O$
M_r: 235.32
bp: $210\text{-}5^2$
mp: 46-8
MS: NBS 1601
m/e: 235.17

HODOC No: 01809
CAS No: 614-39-1
Name: Benzamide,
4-amino-*N*-[2-(diethylamino)ethyl]-,
monohydrochloride
Synonym: Benzamide, *p*-amino-*N*-[2-(diethylamino)ethyl]-, monohydrochloride
Mol Form: $C_{13}H_{22}ClN_3O$
M_r: 271.78
mp: 165-9
***s*:** H_2O 4; al 3; eth 1; bz 1; chl 2
IR: µCOB 5406 3330 2940 2630 1610 1540 1490 1280 1190 1020 850 770
UV: HBCP 278 H_2O
m/e: 271.15

HODOC No: 01810
CAS No: 81-46-9
Name: Benzamide,
N-(4-amino-9,10-dihydro-9,10-dioxo-1-anthracenyl)-
Synonym: Benzamide, *N*-(4-amino-1-anthraquinonyl)-
Mol Form: $C_{21}H_{14}N_2O_3$
M_r: 342.33
mp: 283
IR: µCOB 4619 3450 3230 3030 1670 1590 1490 1410 1390 1270 1250 1180 1020 830 790 730 700
UV: LANG 1587 538(813) dmf
m/e: 342.10

HODOC No: 01811
CAS No: 117-06-6
Name: Benzamide,
N-(5-amino-9,10-dihydro-9,10-dioxo-1-anthracenyl)-
Synonym: Benzamide, *N*-(5-amino-1-anthraquinonyl)-
Mol Form: $C_{21}H_{14}N_2O_3$
M_r: 342.33
IR: µCOB 4620 3450 3230 3030 1670 1640 1590 1520 1410 1330 1270 1210 1020 860 810 770 700
m/e: 342.10

HODOC No: 01812
CAS No: 38093-63-9
Name: Benzamide,
N-(2-amino-1,1-diphenyl-2-thioxoethyl)-
Synonym: Acetic acid, amide,thiono,benzamido-diphenyl
Line Form: $(C_6H_5)_2C(NHCOC_6H_5)C(=S)NH_2$
Mol Form: $C_{21}H_{18}N_2OS$
M_r: 346.43
mp: dec 193
UV: LANG 2509 336(91) 329(87) 270(9772) EtOH
m/e: 346.11

HODOC No: 01813
CAS No: 10146-67-5
Name: Benzamide,
N-(4-amino-3-methyl-1-anthraquinonyl)-
Synonym: Anthraquinone, 1-amino-4-benzamido-2-methyl-
Mol Form: $C_{22}H_{16}N_2O_3$
M_r: 356.36
mp: 267
UV: LANG 1588 539(8913) dmf
m/e: 356.12

HODOC No: 01814
CAS No: 4424-17-3
Name: Benzamide, 2-amino-*N*-phenyl-
Synonym: Benzanilide, 2-amino-
Mol Form: $C_{13}H_{12}N_2O$
M_r: 212.24
***s*:** DMSO 2
IR: µSADP 13326 3450 3230 3030 1640 1590 1560 1520 1430 1320 1250 1150 910 890 750 690
UV: SAD 3479 328 253 MeOH
H NMR: SAD 11681 6.3 6.6 6.8 7.0-7.5 7.6-7.9 9.9 DMSO
m/e: 212.09

HODOC No: 01815
CAS No: 782-45-6
Beilstein No: 14^3, 1062
Name: Benzamide, 4-amino-*N*-phenyl-
Synonym: Benzanilide, 4-amino-
Line Form: 4-$H_2NC_6H_4CONHC_6H_5$
Mol Form: $C_{13}H_{12}N_2O$
M_r: 212.24
mp: 135-6
***s*:** EtOH 4
m/e: 212.09

HODOC No: 01816
CAS No: 127-71-9
Beilstein No: 14^3, 1958
Name: Benzamide, *N*-[(4-aminophenyl)sulfonyl]-
Synonym: Benzamide, *N*-sulfanilyl-
Line Form: 4-H_2N-$C_6H_4SO_2NHCO$-C_6H_5
Mol Form: $C_{13}H_{12}N_2O_3S$
M_r: 276.30
mp: 181-2
Color: hex pr (60% al)
m/e: 276.06

HODOC No: 01817
CAS No: 614-23-3
Beilstein No: 9^3, 1120
Name: Benzamide, *N*-(aminothioxomethyl)-
Synonym: Urea, 1-benzoyl-2-thio-
Line Form: $C_6H_5CONHCSNH_2$
Mol Form: $C_8H_8N_2OS$
M_r: 180.22
mp: 171
Color: pr (dil al)
***s*:** EtOH 4
m/e: 180.04

HODOC No: 01818
CAS No: 7500-78-9
Beilstein No: 10^3, 3295
Name: Benzamide, 2-benzoyl-
Synonym: Benzamide, *o*-benzoyl-
Line Form: 2-$(C_6H_5CO)C_6H_4CONH_2$
Mol Form: $C_{14}H_{11}NO_2$
M_r: 225.23
mp: 165
Color: nd (to)
***s*:** Bz 4; EtOH 4; To 4
m/e: 225.08

HODOC No: 01819
CAS No: 614-28-8
Beilstein No: 9^4, 768
Name: Benzamide, *N*-benzoyl-
Synonym: *N*-Benzoylbenzamide
Line Form: C_6H_5CONH-COC_6H_5
Mol Form: $C_{14}H_{11}NO_2$
M_r: 225.23
mp: 152-3
***s*:** H_2O 1; al 3; bz 3; chl 3
UV: OES 1-556 278(5248) 230(15849) 0.01N NaOH:EtOH
MS: ALDN 184 105(100) 77(65) 225(38) 51(30) 197(10) 50(10) 106(9) 226(8)
m/e: 225.08

HODOC No: 01820
CAS No: 3027-01-8
Beilstein No: 12^3, 520
Name: Benzamide, *N*-benzoyl-*N*-phenyl-
Synonym: Dibenzamide, *N*-phenyl-
Line Form: $(C_6H_5CO)_2NC_6H_5$
Mol Form: $C_{20}H_{15}NO_2$
M_r: 301.32
mp: 163-4
***s*:** H_2O 4; al 3; eth 3
IR: µSADP 12160 2940 1700 1640 1590 1490 1450 1390 1320 1270 1180 1140 1090 1060 1030 1000 940 920 850 790 760 730 710 690
MS: ALDN 227 105(100) 77(63) 197(48) 51(23) 198(10) 39(10) 65(9) 301(1)
m/e: 301.11

HODOC No: 01821
CAS No: 100725-06-2
Name: Benzamide, *N,N-bis*(2-cyanoethyl)-*p*-nitro-
Synonym: Benzoic acid, amide,*N,N-bis*(2-cyanoethyl)-4-nitro
Mol Form: $C_{13}H_{12}N_4O_3$
M_r: 272.25
IR: µCOB 3977 3030 2940 2220 1640 1610 1540 1470 1450 1350 1330 1270 1160 1150 1040 870 860 760 730 700
m/e: 272.09

HODOC No: 01822
CAS No: 4001-73-4
Beilstein No: 9^3, 1387
Name: Benzamide, 2-bromo-
Synonym: Benzamide, *o*-bromo-
Line Form: 2-$BrC_6H_4CONH_2$
Mol Form: C_7H_6BrNO
M_r: 199.03
bp: sub
mp: 160-1
Color: nd (w)
***s*:** H_2O 2; al 3; eth 2; tfa 2
IR: SADG 2241 3448 3226 1639 1587 1493 1389 1136 1042 758 671
UV: SAD 9845 267(406) MeOH
H NMR: SAD 457 7.6 tfa
m/e: 198.96

HODOC No: 01823
CAS No: 22726-00-7
Beilstein No: 9^3, 1399
Name: Benzamide, 3-bromo-
Synonym: Benzamide, *m*-bromo-
Line Form: 3-$BrC_6H_4CONH_2$
Mol Form: C_7H_6BrNO
M_r: 199.03
bp: sub
mp: 155.3
Color: lf (w or al)
***s*:** H_2O 2; al 3
IR: COB 7510 3350 3175 1660 1625 1565 1430 1395 1275 1150 1130 1070 1030 1000 930 905 795 745 675 660 560 460
UV: OES 5-85 247(10000) 59% sulf
m/e: 198.96

HODOC No: 01824
CAS No: 698-67-9
Beilstein No: 9^3, 1418
Name: Benzamide, 4-bromo-
Synonym: Benzamide, *p*-bromo-
Line Form: 4-$BrC_6H_4CONH_2$
Mol Form: C_7H_6BrNO
M_r: 199.03
mp: 192
Color: nd or pl (w)
***s*:** H_2O 3; al 4; eth 2; bz 1; chl 3; CS_2 1; lig 1; HOAc 3
IR: µSADP 12372 3330 3130 2940 1670 1590 1470 1410 1370 1150 1120 1010 850 780 750
UV: OES 5-85 242(13490) H_2O
m/e: 198.96

HODOC No: 01825
CAS No: 5798-94-7
Name: Benzamide, 5-bromo-*N*,2-dihydroxy-
Synonym: Salicylohydroxamic acid, 5-bromo-
Line Form: 5-Br-2-HOC_6H_3-C(OH)=NOH
Mol Form: $C_7H_6BrNO_3$
M_r: 231.03
mp: 232d
Color: cr (al)
m/e: 230.95

HODOC No: 01826
CAS No: 70301-29-0
Name: Benzamide, 4-bromo-*N*-3-pyridinyl-
Synonym: Benzoic acid, amide,4-bromo-*N*(pyrid-3-yl)
Mol Form: $C_{12}H_9BrN_2O$
M_r: 276.11
mp: 169-70
UV: LANG 1010 270(15136) EtOH
m/e: 275.99

HODOC No: 01827
CAS No: 14547-77-4
Name: Benzamide, 4-bromo-*N*-4-pyridinyl-
Synonym: Benzamide, *p*-bromo-*N*-4-pyridyl-
Mol Form: $C_{12}H_9BrN_2O$
M_r: 276.11
mp: 219-20
UV: LANG 1017 270(23988) EtOH
m/e: 275.99

HODOC No: 01828
CAS No: 575-74-6
Name: Benzamide, *N*-butyl-4-chloro-2-hydroxy-
Synonym: Salicylamide, *N*-butyl-4-chloro-
Line Form: 4-Cl-2-$HOC_6H_3CONHC_4H_9$
Mol Form: $C_{11}H_{14}ClNO_2$
M_r: 227.68
mp: 91-2
UV: LANG 1690 300(5495) 245(12589) EtOH 306(6607) chl
m/e: 227.07

HODOC No: 01829
CAS No: 609-66-5
Beilstein No: 9^3, 1338
Name: Benzamide, 2-chloro-
Synonym: Benzamide, *o*-chloro-
Line Form: 2-$ClC_6H_4CONH_2$
Mol Form: C_7H_6ClNO
M_r: 155.58
mp: 142-4
Color: rh nd (w)
***s*:** H_2O 3; al 3; eth 3
IR: µSADG 8398 3330 3230 1640 1590 1470 1120 1050 950 820 780 740 710
H NMR: SAD 17362 7.2-7.6 poly D
m/e: 155.01

HODOC No: 01830
CAS No: 618-48-4
Beilstein No: 9^3, 1349
Name: Benzamide, 3-chloro-
Synonym: Benzamide, *m*-chloro-
Line Form: 3-$ClC_6H_4CONH_2$
Mol Form: C_7H_6ClNO
M_r: 155.58
mp: 135-7
Color: nd
***s*:** H_2O 2; al 3; eth 3
UV: HBCP 248(9550) 60% sulf
m/e: 155.01

HODOC No: 01831
CAS No: 619-56-7
Beilstein No: 9^3, 1363
Name: Benzamide, 4-chloro-
Synonym: Benzamide, *p*-chloro-
Line Form: 4-$ClC_6H_4CONH_2$
Mol Form: C_7H_6ClNO
M_r: 155.58
mp: 179-80
***s*:** H_2O 3; al 4; eth 4
IR: SADG 15750 3333 3226 1667 1613 1563 1493 1389 1099 1010 847 800
UV: OES 5-85 237(12589) H_2O
H NMR: SAD 7947 7.6 8.0 7.1-8.3 DMSO
MS: NBS 531
m/e: 155.01

HODOC No: 01832
CAS No: 21721-78-8
Name: Benzamide, 2-[(chloroacetyl)amino]-
Synonym: Acetanilide, 2′-carbamoyl-2-chloro-
Mol Form: $C_9H_9ClN_2O_2$
M_r: 212.63
IR: µCOB 5373 3330 3130 2940 1700 1640 1610 1590 1520 1390 1300 1110 920 760 750 730
m/e: 212.04

HODOC No: 01833
CAS No: 57338-87-1
Name: Benzamide, 2-chloro-*N*,*N*-*bis*(2-cyanoethyl)-
Synonym: Benzamide, *o*-chloro-*N*,*N*-*bis*(2-cyanoethyl)-
Mol Form: $C_{13}H_{12}ClN_3O$
M_r: 261.70
mp: 83
IR: µCOB 4017 2940 2270 1610 1590 1560 1540 1470 1430 1370 1350 1330 1280 1270 1250 1220 1160 1150 1060 1030 1000 910 780 750 740 700
m/e: 261.07

HODOC No: 01834
CAS No: 106472-11-1
Name: Benzamide, *p*-chloro-*N*,*N*-*bis*(2-cyanoethyl)-
Synonym: Benzoic acid, amide,*N*,*N*-*bis*(2-cyanoethyl)-4-chloro
Mol Form: $C_{13}H_{12}ClN_3O$
M_r: 261.70
IR: µCOB 4054 2940 2270 1640 1470 1390 1330 1270 1150 1100 1040 1020 840 830 760 750
m/e: 261.07

HODOC No: 01835
CAS No: 22192-81-0
Name: Benzamide, *o*-chloro-*N*-(2-cyanoethyl)-
Synonym: Benzoic acid, amide,2-chloro-*N*(2-cyanoethyl)
Mol Form: $C_{10}H_9ClN_2O$
M_r: 208.64
IR: µCOB 4083 3230 3030 2940 2270 1670 1540 1470 1370 1320 1090 1040 870 750 730 710 690
m/e: 208.04

HODOC No: 01836
CAS No: 22192-82-1
Name: Benzamide, *p*-chloro-*N*-(2-cyanoethyl)-
Synonym: Benzoic acid, amide,4-chloro-*N*(2-cyanoethyl)
Mol Form: $C_{10}H_9ClN_2O$
M_r: 208.64
IR: **µCOB 4086** 3230 3130 2270 1640 1610 1560 1450 1370 1330 1250 1100 1010 860 850 730 690
m/e: 208.04

HODOC No: 01837
CAS No: 642-84-2
Name: Benzamide, 5-chloro-*N*-(3,4-dichlorophenyl)-2-hydroxy-
Synonym: Salicylanilide, 3′,4′,5-trichloro-
Line Form: $C_{13}H_8Cl_3NO_2$
Structure:

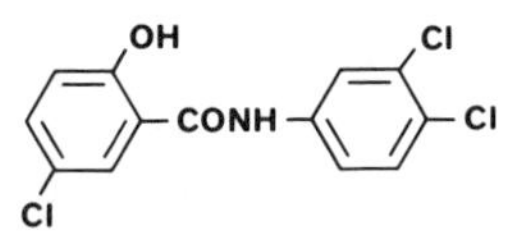

Mol Form: $C_{13}H_8Cl_3NO_2$
M_r: 316.56
mp: 246-8
m/e: 314.96

HODOC No: 01838
CAS No: 81-45-8
Name: Benzamide, *N*-(4-chloro-9,10-dihydro-9,10-dioxo-1-anthracenyl)-
Synonym: Benzamide, *N*-(4-chloro-1-anthraquinonyl)-
Mol Form: $C_{21}H_{12}ClNO_3$
M_r: 361.76
IR: **COB 2603** 3200 1680 1640 1590 1520 1500 1380 1340 1260 1170 1010 860 820 810 750 700
UV: **LANG 1586** 420(6607) 250(15488) dmf
m/e: 361.05

HODOC No: 01839
CAS No: 26385-07-9
Beilstein No: 9^4, 728
Name: Benzamide, *N*-(2-chloroethyl)-
Synonym: *N*-(2-Chloroethyl)benzamide
Line Form: $C_6H_5CONH\text{-}CH_2CH_2Cl$
Mol Form: $C_9H_{10}ClNO$
M_r: 183.63
mp: 103-6
m/e: 183.05

HODOC No: 01840
CAS No: 13909-28-9
Name: Benzamide, 4-[[[(2-chloroethyl)nitrosoamino]carbonyl]amino]-*N*,*N*-dimethyl-
Synonym: Urea, 1-(2-chloroethyl)-3-[*p*-(dimethylcarbamoyl)phenyl]-1-nitroso-
Mol Form: $C_{12}H_{15}ClN_4O_3$
M_r: 298.72
IR: **COB 7730** 3220 3100 2940 1720 1610 1520 1490 1400 1340 1230 1080 950 850 760 710 650 570
m/e: 298.08

HODOC No: 01841
CAS No: 7461-41-8
Name: Benzamide, 4-chloro-*N*-(1-methylethyl)-
Synonym: Benzamide, *p*-chloro-*N*-isopropyl-
Line Form: $4\text{-}ClC_6H_4CONH\text{-}CH(CH_3)_2$
Mol Form: $C_{10}H_{12}ClNO$
M_r: 197.65
mp: 142-5
***s*:** chl 3
IR: **SADG 16577** 3300 3060 2980 1630 1530 1480 1450 1380 1360 1090 1010 840 760 670
UV: **SAD 16643** 234(13700) MeOH
H NMR: **SAD 8890** 1.2 4.2 6.7 7.2 7.6 $CDCl_3$
m/e: 197.06

HODOC No: 01842
CAS No: 64160-38-9
Beilstein No: 12, 733
Name: Benzamide, *N*-(2-chloro-4-nitrophenyl)-
Synonym: Benzanilide, 2′-chloro-4′-nitro-
Mol Form: $C_{13}H_9ClN_2O_3$
M_r: 276.67
mp: 161
IR: **COB 2678** 3400 3100 1700 1610 1600 1550 1520 1470 1410 1350 1320 1280 1250 1150 1100 1080 900 840 830 740 720 700
m/e: 276.03

HODOC No: 01843
CAS No: 32853-24-0
Name: Benzamide, *N*-(2-chloro-4-nitrophenyl)-2-hydroxy-
Synonym: Salicylanilide, 2′-chloro-4′-nitro-
Mol Form: $C_{13}H_9ClN_2O_4$
M_r: 292.67
mp: 185-6
UV: **LANG 2012** 331(16218) 240(8128) EtOH
m/e: 292.03

HODOC No: 01844
CAS No: 1020-39-9
Beilstein No: 12^3, 1289
Name: Benzamide, *N*-(2-chlorophenyl)-
Synonym: Benzanilide, 2′-chloro-
Line Form: $C_6H_5CONH(C_6H_5Cl\text{-}2)$
Structure:

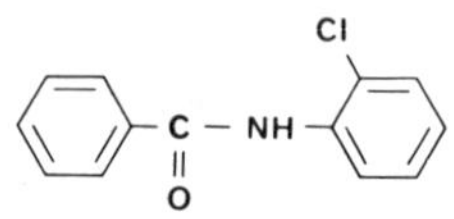

Mol Form: $C_{13}H_{10}ClNO$
M_r: 231.67
mp: 99
***s*:** DMSO 3
IR: SADG 21184 3200 3050 1650 1510 1470 1420 1300 1050 900 740 710 680
UV: SAD 2080 <212 MeOH
H NMR: SAD 11291 7.2-7.8 8.1 10.0 DMSO
m/e: 231.05

HODOC No: 01845
CAS No: 2866-82-2
Beilstein No: 12^3, 1338
Name: Benzamide, *N*-(4-chlorophenyl)-
Synonym: Benzanilide, 4′-chloro-
Line Form: $C_6H_5CONH(C_6H_4\text{-}Cl\text{-}4)$
Mol Form: $C_{13}H_{10}ClNO$
M_r: 231.67
mp: 192-3
UV: LANG 2150 270(16218) 230(12033) EtOH
m/e: 231.05

HODOC No: 01846
CAS No: 5014-47-1
Beilstein No: 12^2, 296
Name: Benzamide, *N*-chloro-*N*-phenyl-
Synonym: Benzanilide, *N*-chloro-
Line Form: $C_6H_5CON(Cl)C_6H_5$
Mol Form: $C_{13}H_{10}ClNO$
M_r: 231.67
mp: 82
Color: nd (lig)
***s*:** Et_2O 4; EtOH 4
m/e: 231.05

HODOC No: 01847
CAS No: 18453-17-3
Name: Benzamide, *N*-(4-chlorophenyl)-4-[(1-methylethyl)sulfonyl]-
Synonym: Benzanilide, 4′-chloro-4-(isopropylsulfonyl)-
Mol Form: $C_{16}H_{16}ClNO_3S$
M_r: 337.80
mp: 205-6
***s*:** DMSO 3
IR: SADG 16573 3380 3100 2980 1680 1590 1530 1490 1400 1310 1290 1250 1130 1070 830 740 690 600 540
UV: SAD 16641 280(12400) 230(20600) MeOH
H NMR: SAD 9025 1.2 3.4 7.4 7.8 8.0 8.2 10.6 DMSOd6
m/e: 337.05

HODOC No: 01848
CAS No: 14547-79-6
Name: Benzamide, 4-chloro-*N*-3-pyridinyl-
Synonym: Benzamide, *p*-chloro-*N*-3-pyridyl-
Mol Form: $C_{12}H_9ClN_2O$
M_r: 232.66
mp: 155-6
UV: LANG 1009 271(14454) 245(12589) EtOH
m/e: 232.04

HODOC No: 01849
CAS No: 14547-76-3
Name: Benzamide, 4-chloro-*N*-4-pyridinyl-
Synonym: Benzamide, *p*-chloro-*N*-4-pyridyl-
Mol Form: $C_{12}H_9ClN_2O$
M_r: 232.66
mp: 212-3
UV: LANG 1016 269(19498) EtOH
m/e: 232.04

HODOC No: 01850
CAS No: 17174-98-0
Beilstein No: 9^3, 4199
Name: Benzamide, 2-cyano-
Synonym: Benzamide, *o*-cyano-
Line Form: $2\text{-}NCC_6H_4CONH_2$
Mol Form: $C_8H_6N_2O$
M_r: 146.14
mp: 175
Color: nd (MeOH)
***s*:** Ac 4; EtOH 4
m/e: 146.05

HODOC No: 01851
CAS No: 94963-68-5
Name: Benzamide, *N*-(cyanodiphenylmethyl)-
Synonym: Benzoic acid, amide,*N*(cyanodiphenyl-methyl)
Line Form: $(C_6H_5)_2C(CN)NHCOC_6H_5$
Mol Form: $C_{21}H_{16}N_2O$
M_r: 312.35
mp: 203
UV: LANG 2507 2200(19953) EtOH
m/e: 312.13

HODOC No: 01852
CAS No: 38093-58-2
Name: Benzamide, *N*-(cyanodiphenylmethyl)-*N*-methyl-
Synonym: Glycinonitrile, *N*-benzoyl-2,2-diphenyl-*N*-methyl
Line Form: $(C_6H_5)_2C(CN)N(CH_3)COC_6H_5$
Mol Form: $C_{22}H_{18}N_2O$
M_r: 326.38
mp: 136
UV: LANG 2508 <200 EtOH
m/e: 326.14

HODOC No: 01853
CAS No: 1141-61-3
Name: Benzamide, *N*-(2-cyanoethyl)-4-nitro-
Synonym: Benzamide, *N*-(2-cyanoethyl)-*p*-nitro-
Mol Form: $C_{10}H_9N_3O_3$
M_r: 219.19
IR: μCOB 4059 3330 3130 2270 1670 1610 1560 1520 1470 1350 1320 1090 880 850 710 690
MS: WILEY 1546 125(100) 127(32) 220(23) 139(22) 218(17) 103(17) 77(11) 51(10)
m/e: 219.06

HODOC No: 01854
CAS No: 5692-27-3
Beilstein No: 9^4, 791
Name: Benzamide, *N*-(cyanomethyl)-
Synonym: Benzamidoacetonitrile
Line Form: $C_6H_5CONHCH_2CN$
Mol Form: $C_9H_8N_2O$
M_r: 160.17
mp: 142
***s*:** DMSO 2
IR: SADG 18114 3250 3060 2940 1630 1590 1580 1530 1480 1410 1350 1320 1310 1290 1260 1150 1070 1020 980 930 890 830 790 720 690 650 580 420 320 300
UV: SAD 15829 225(10300) MeOH
H NMR: SAD 11019 4.4 7.5 7.9 9.1 $DMSOd_6$
m/e: 160.06

HODOC No: 01855
CAS No: 5692-26-2
Name: Benzamide, *N*-(cyanophenylmethyl)-
Synonym: Benzamide, *N*-(α-cyanobenzyl)-
Line Form: $C_6H_5CONHCH(CN)C_6H_5$
Mol Form: $C_{15}H_{12}N_2O$
M_r: 236.26
mp: 159
UV: LANG 2498 227(14791) EtOH
m/e: 236.09

HODOC No: 01856
CAS No: 3080-42-0
Name: Benzamide, *N*-cyclohexyl-*N*-[(cyclohexylamino)carbonyl]-
Synonym: Urea, 1-benzoyl-1,3-dicyclohexyl-
Mol Form: $C_{20}H_{28}N_2O_2$
M_r: 328.43
IR: COB 7703 3300 3060 2940 2860 1700 1640 1540 1440 1340 1240 1120 1080 890 780 700 630
m/e: 328.22

HODOC No: 01857
CAS No: 92042-04-1
Name: Benzamide, *N*-cyclopropyl-3,4,5-trimethoxy-
Synonym: Benzoic acid, amide,*N*-cyclopropyl-3,4,5-trimethoxy
Mol Form: $C_{13}H_{17}NO_4$
M_r: 251.27
***s*:** chl 3
IR: μCOB 3359 3450 3330 3030 2860 1670 1590 1490 1470 1410 1330 1220 1120 1000 860
m/e: 251.12

HODOC No: 01858
CAS No: 87-10-5
Name: Benzamide, 3,5-dibromo-*N*-(4-bromophenyl)-2-hydroxy-
Synonym: Salicylanilide, 3,4′,5-tribromo-
Mol Form: $C_{13}H_8Br_3NO_2$
M_r: 446.91
mp: 226-8
UV: LANG 1694 330(9772) 283(13804) 220(33884) EtOH
m/e: 446.81

HODOC No: 01859
CAS No: 1154-59-2
Name: Benzamide, 3,5-dichloro-*N*-(3,4-dichlorophenyl)-2-hydroxy-
Synonym: Salicylanilide, 3,3′,4′,5-tetrachloro-
Line Form: $C_{13}H_7Cl_4NO_2$
Structure:

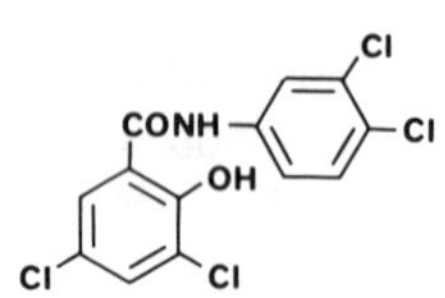

Mol Form: $C_{13}H_7Cl_4NO_2$
M_r: 351.00
mp: 161
IR: μCOB 2876 3230 3030 1640 1590 1540 1470 1390 1320 1280 1250 1180 1030 920 870 820 780 780 740 690 680
UV: SAD 5079 278 217 MeOH
m/e: 348.92

HODOC No: 01860
CAS No: 17705-40-7
Name: Benzamide, 3,5-dichloro-2-hydroxy-*N*-(4-nitrophenyl)-
Synonym: Salicylanilide, 3,5-dichloro-4′-nitro-
Mol Form: $C_{13}H_8Cl_2N_2O_4$
M_r: 327.11
mp: 212
UV: LANG 2014 319(7943) EtOH
m/e: 325.99

HODOC No: 01861
CAS No: 21122-75-8
Name: Benzamide, *N*-(2,4-dichlorophenyl)-2-[[3-(trifluoromethyl)phenyl]amino]-
Synonym: Benzanilide, 2′,4′-dichloro-2-(α,α,α-trifluoro-*m*-toluidino)-
Mol Form: $C_{20}H_{13}Cl_2F_3N_2O$
M_r: 425.22
mp: 105-7
***s*:** chl 3
UV: SAD 16877 339(6380) 286(18600) MeOH
H NMR: SAD 9306 6.9 7.1-7.5 7.7 8.4 9.4 $CDCl_3$
m/e: 424.04

HODOC No: 01862
CAS No: 304-84-7
Name: Benzamide, *N*,*N*-diethyl-4-hydroxy-3-methoxy-
Synonym: Vanillamide, *N*,*N*-diethyl-
Line Form: $(4\text{-}HO\text{-}3\text{-}CH_3O\text{-}C_6H_3\text{-})CON(C_2H_5)_2$
Mol Form: $C_{12}H_{17}NO_3$
M_r: 223.26
mp: 96-9
***s*:** chl 3
IR: μCOB 5209 2940 1610 1520 1490 1430 1390 1370 1270 1240 1030 890 820 730
UV: SAD 9444 282(4160) 205(36400) MeOH
H NMR: SAD 406 1.2 3.4 3.8 6.6 6.9 $CDCl_3$
m/e: 223.12

HODOC No: 01863
CAS No: 20884-72-4
Name: Benzamide, *N*,*N*-diethyl-*o*-(isopropylsulfonyl)-
Synonym: Benzoic acid, amide,*N*,*N*-diethyl-2(isopropylsulfonyl)
Mol Form: $C_{14}H_{21}NO_3S$
M_r: 283.37
mp: 88-90
***s*:** chl 3
IR: SADG 17883 2980 2930 1620 1590 1565 1490 1450 1430 1380 1350 1300 1280 1250 1210 1140 1115 1095 1050 930 870 780 760 730 690 650 580 460
m/e: 283.12

HODOC No: 01864
CAS No: 20884-71-3
Name: Benzamide, *N,N*-diethyl-*p*-(isopropylsulfonyl)-
Synonym: Benzoic acid, amide,*N,N*-diethyl-4(isopropylsulfonyl)
Mol Form: $C_{14}H_{21}NO_3S$
M_r: 283.37
mp: 129-31
***s*:** chl 3
IR: SADG 16563 3080 2980 1620 1430 1380 1360 1300 1270 1140 1030 750 720 670 590
H NMR: SAD 8960 1.2 1.3 3.0-3.7 7.5 7.9 $CDCl_3$
m/e: 283.12

HODOC No: 01865
CAS No: 134-62-3
Beilstein No: 9^4, 1716
Name: Benzamide, *N,N*-diethyl-3-methyl-
Synonym: *m*-Toluamide, *N,N*-diethyl-
Line Form: $3\text{-}CH_3C_6H_4CON(C_2H_5)_2$
Mol Form: $C_{12}H_{17}NO$
M_r: 191.26
bp: 160^{19},111^1
***d*:** $0.996^{20/4}$
***n*:** 1.5212^{20}
***s*:** H_2O 4; Bz 4; Et_2O 4; EtOH 4
IR: ALD 2:942F
MS: ALD 7:74C
m/e: 191.13

HODOC No: 01866
CAS No: 20884-78-0
Name: Benzamide, *N,N*-diethyl-*p*-(methylsulfonyl)-
Synonym: Benzoic acid, amide,*N,N*-diethyl-4(methysulfonyl)
Mol Form: $C_{12}H_{17}NO_3S$
M_r: 255.32
mp: 67-9
***s*:** chl 3
IR: SADG 16570 3080 2990 1630 1430 1390 1310 1210 1150 1090 950 840 770 580 550 510
UV: SAD 16638 218(11200) MeOH
H NMR: SAD 8885 1.2 3.1 3.4 7.5 7.9 $CDCl_3$
m/e: 255.09

HODOC No: 01867
CAS No: 3553-11-5
Name: Benzamide, *N*-(2,3-dihydro-1,5-dimethyl-3-oxo-2-phenyl-1H-pyrazol-4-yl)-3,4,5-trimethoxy-
Synonym: Antipyrine, 4-(3,4,5-trimethoxybenzamido)-
Structure:

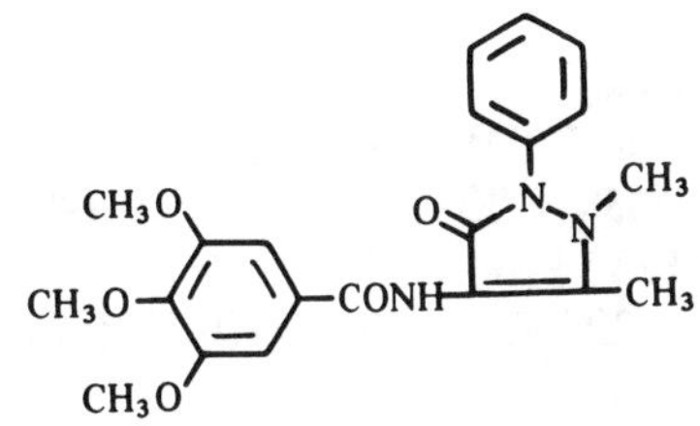

Mol Form: $C_{21}H_{23}N_3O_5$
M_r: 397.41
mp: 218-9
UV: LANG 616 277(18197) 211(43652) EtOH
m/e: 397.16

HODOC No: 01868
CAS No: 2987-68-0
Beilstein No: 14^3, 454
Name: Benzamide, *N,N'*-(9,10-dihydro-9,10-dioxo-1,4-anthracenediyl)*bis*
Synonym: Benzamide, *N,N'*-1,4-anthraquinonylenebis-
Line Form: $1,4(C_6H_5CONH)_2\text{-}C_{14}H_6O_2$
Mol Form: $C_{28}H_{18}N_2O_4$
M_r: 446.43
mp: 271
UV: LANG 1589 479(7079) 390(19953) dmf
m/e: 446.13

HODOC No: 01869
CAS No: 52405-73-9
Name: Benzamide, 2,5-dihydroxy-
Synonym: 2,5-Dihydroxybenzamide
Mol Form: $C_7H_7NO_3$
M_r: 153.13
IR: µCOB 3177 3330 3030 1670 1590 1490 1370 1280 1240 1060 940 790 670
m/e: 153.04

HODOC No: 01870
CAS No: 89-73-6
Beilstein No: 10^3, 160
Name: Benzamide, *N*,2-dihydroxy-
Synonym: Salicylohydroxamic acid
Line Form: 2-$HOC_6H_4CONHOH$
Mol Form: $C_7H_7NO_3$
M_r: 153.13
mp: 168
Color: nd (HOAc)
***s*:** H_2O 2; al 4; eth 4; $DMSOd_6$ 2; HOAc 3
IR: μSADP 24858 3130 2940 1610 1520 1490 1450 1350 1320 1250 1160 1100 1050 1030 910 780 750
UV: SAD 8811 299(4280) 236(6710) 204(42100) MeOH
H NMR: SAD 8810 7.0 7.5 7.8 DMSO
m/e: 153.04

HODOC No: 01871
CAS No: 42020-21-3
Beilstein No: 10^2, 258
Name: Benzamide, 2,5-dimethoxy-
Synonym: 2,5-Dimethoxybenzamide
Line Form: 2,5-$(CH_3O)_2C_6H_3CONH_2$
Mol Form: $C_9H_{11}NO_3$
M_r: 181.18
mp: 141-2
Color: lf (Bz-peth), nd (w)
***s*:** Bz 4; Et_2O 4; EtOH 4; Chl 4
m/e: 181.07

HODOC No: 01872
CAS No: 1521-41-1
Beilstein No: 10^2, 261
Name: Benzamide, 3,4-dimethoxy-
Synonym: Veratrimidic acid
Line Form: 3,4-$(CH_3O)_2C_6H_3CONH_2$
Mol Form: $C_9H_{11}NO_3$
M_r: 181.18
mp: 164
Color: cr (w)
***s*:** H_2O 4; eth 3; bz 3; tfa 2
IR: SADG 10241 3448 3333 3125 2941 2857 1667 1587 1515 1471 1429 1389 1282 1250 1149 1136 1020 862
UV: SAD 6456 289 257 212 MeOH
H NMR: SAD 2218 4.1 7.2 7.7 7.8 tfa
m/e: 181.07

HODOC No: 01873
CAS No: 17213-58-0
Beilstein No: 10^3, 1449
Name: Benzamide, 3,5-dimethoxy-
Synonym: 3,5-Dimethoxybenzamide
Line Form: 3,5-$(CH_3O)_2C_6H_3CONH_2$
Mol Form: $C_9H_{11}NO_3$
M_r: 181.18
mp: 148-9
Color: nd (Bz)
***s*:** H_2O 3; al 4; eth 4; bz 4; lig 4
IR: SADG 16710 3390 3180 2940 2840 1650 1630 1590 1460 1440 1420 1400 1340 1280 1250 1200 1150 1120 1100 1070 1050 990 930 910 870 845 810 755 680 620 540 420
UV: HBCP 229(2344) 247(4786) MeOH
H NMR: SAD 8834 3.9 6.8 7.2 7.5 8.1 $DMSOd_6$
m/e: 181.07

HODOC No: 01874
CAS No: 611-74-5
Beilstein No: 9^4, 727
Name: Benzamide, *N*,*N*-dimethyl-
Synonym: Dimethylbenzamide
Line Form: $C_6H_5CON(CH_3)_2$
Mol Form: $C_9H_{11}NO$
M_r: 149.18
mp: 43-4
MS: NBS 444
m/e: 149.08

HODOC No: 01875
CAS No: 3765-62-6
Name: Benzamide, 4-(dimethylamino)-*N*-phenyl-
Synonym: Benzanilide, 4-(dimethylamino)-
Mol Form: $C_{15}H_{16}N_2O$
M_r: 240.29
mp: dec 183-4
UV: LANG 1324 315(29512) 225(10715) EtOH
m/e: 240.13

HODOC No: 01876
CAS No: 21122-70-3
Name: Benzamide, 2-[(2,3-dimethylphenyl)amino]-*N*-(4-ethoxyphenyl)-
Synonym: *p*-Benzophenetidide, 2-(2,3-xylidino)-
Mol Form: $C_{23}H_{24}N_2O_2$
M_r: 360.43
mp: 148-9
***s*:** chl 3
IR: SADG 17583 3360 3260 3040 2980 1630 1580 1500 1460 1400 1380 1310 1230 1160 1040 810 780 740 580
UV: SAD 16873 347(8870) 282(19900) MeOH
H NMR: SAD 9309 1.4 2.2 2.3 4.0 6.5-7.3 7.4-7.7 7.9 9.1 $CDCl_3$
m/e: 360.18

HODOC No: 01877
CAS No: 121-81-3
Beilstein No: 9^4, 1351
Name: Benzamide, 3,5-dinitro-
Synonym: 3,5-Dinitrobenzamide
Line Form: 3,5-$(NO_2)_2$-$C_6H_3CONH_2$
Mol Form: $C_7H_5N_3O_5$
M_r: 211.13
mp: 183-5
Color: lf (w)
***s*:** H_2O 4
IR: ALD 2:943G
MS: ALD 7:76A
m/e: 211.02

HODOC No: 01878
CAS No: 4051-56-3
Beilstein No: 12^3, 511
Name: Benzamide, *N,N*-diphenyl-
Synonym: *N*-Benzoyldiphenylamine
Line Form: $C_6H_5CON(C_6H_5)_2$
Mol Form: $C_{19}H_{15}NO$
M_r: 273.32
mp: 180
Color: rh pr (al), nd
***s*:** H_2O 2; al 2; eth 2
IR: µSADP 20850 3030 1670 1590 1490 1450 1350 1320 790 760 730 700 690
UV: SAD 7290 268(26200) 205(49200) MeOH
m/e: 273.12

HODOC No: 01879
CAS No: 644-33-7
Beilstein No: 9^3, 1210
Name: Benzamide, *N,N'*-1,2-ethanediylbis-
Synonym: Benzamide, *N,N'*-ethylenebis-
Line Form: $C_6H_5CONHCH_2CH_2NHCOC_6H_5$
Mol Form: $C_{16}H_{16}N_2O_2$
M_r: 268.30
mp: 248
Color: pr, nd (al)
IR: µCOB 237 3330 3030 1640 1560 1490 1450 1390 1300 1240 1090 930 890 810 700 670
UV: SAD 10644 226(19400) MeOH
m/e: 268.12

HODOC No: 01880
CAS No: 938-73-8
Beilstein No: 10^2, 58
Name: Benzamide, 2-ethoxy-
Synonym: Benzamide, *o*-ethoxy-
Line Form: 2-C_2H_5O-$C_6H_4CONH_2$
Mol Form: $C_9H_{11}NO_2$
M_r: 165.18
mp: 132-4
Color: nd (w, al)
***s*:** H_2O 2; al 4; eth 4; chl 2
IR: µSADG 19676 3330 3130 1670 1590 1490 1450 1410 1300 1250 1180 1120 1040 930 820 780 760 720
UV: SAD 6477 293 233 MeOH
H NMR: SAD 6054 1.5 4.2 7.0 7.5 7.9 8.2 $CDCl_3$
m/e: 165.08

HODOC No: 01881
CAS No: 55836-69-6
Beilstein No: 10^1, 64
Name: Benzamide, 3-ethoxy-
Synonym: Benzamide, *m*-ethoxy-
Line Form: 3-C_2H_5O-$C_6H_4CONH_2$
Mol Form: $C_9H_{11}NO_2$
M_r: 165.18
mp: 139
Color: nd (w)
m/e: 165.08

HODOC No: 01882
CAS No: 55836-71-0
Beilstein No: 10^3, 342
Name: Benzamide, 4-ethoxy-
Synonym: Benzamide, *p*-ethoxy-
Line Form: 4-C_2H_5O-$C_6H_4CONH_2$
Mol Form: $C_9H_{11}NO_2$
M_r: 165.18
mp: 206
Color: pr (dil al)
***s*:** EtOH 4
m/e: 165.08

HODOC No: 01883
CAS No: 6760-09-4
Name: Benzamide, 2-ethoxy-*N*-2-pyridinyl-
Synonym: Benzamide, *o*-ethoxy-*N*-2-pyridyl-
Mol Form: $C_{14}H_{14}N_2O_2$
M_r: 242.26
mp: 94-5
UV: LANG 1629 292(16218) 251(12023) EtOH 305(18621) 249(11220) 7.1% HCl
m/e: 242.11

HODOC No: 01884
CAS No: 6760-08-3
Name: Benzamide, 2-ethoxy-*N*-3-pyridinyl-
Synonym: Benzamide, *o*-ethoxy-*N*-3-pyridyl-
Mol Form: $C_{14}H_{14}N_2O_2$
M_r: 242.26
mp: 67-8
UV: LANG 1632 265(13490) EtOH 295(10965) 267(15849) 7.1% HCl
m/e: 242.11

HODOC No: 01885
CAS No: 6760-07-2
Name: Benzamide, 2-ethoxy-*N*-4-pyridinyl-
Synonym: Benzamide, *o*-ethoxy-*N*-4-pyridyl-
Mol Form: $C_{14}H_{14}N_2O_2$
M_r: 242.26
mp: 82-3
UV: LANG 1635 301(6918) 269(18621) EtOH 285(23988) 7.1% HCl
m/e: 242.11

HODOC No: 01886
CAS No: 6760-06-1
Name: Benzamide, 3-ethoxy-*N*-2-pyridinyl-
Synonym: Benzamide, *m*-ethoxy-*N*-2-pyridyl-
Mol Form: $C_{14}H_{14}N_2O_2$
M_r: 242.26
mp: 112-3
UV: LANG 1630 289(15136) 256(11749) 214(25704) EtOH 305(20417) 244(12303) 7.1% HCl
m/e: 242.11

HODOC No: 01887
CAS No: 6760-05-0
Name: Benzamide, 3-ethoxy-*N*-3-pyridinyl-
Synonym: Benzamide, *m*-ethoxy-*N*-3-pyridyl-
Mol Form: $C_{14}H_{14}N_2O_2$
M_r: 242.26
mp: 94-5
UV: LANG 1633 269(14454) EtOH 264(16982) 7.1% HCl
m/e: 242.11

HODOC No: 01888
CAS No: 6760-04-9
Name: Benzamide, 3-ethoxy-*N*-4-pyridinyl-
Synonym: Benzamide, *m*-ethoxy-*N*-4-pyridyl-
Mol Form: $C_{14}H_{14}N_2O_2$
M_r: 242.26
mp: 119-20
UV: LANG 1636 268(19953) EtOH 284(26915) 7.1% HCl
m/e: 242.11

HODOC No: 01889
CAS No: 6760-03-8
Name: Benzamide, 4-ethoxy-*N*-2-pyridinyl-
Synonym: Benzamide, *p*-ethoxy-*N*-2-pyridyl-
Mol Form: $C_{14}H_{14}N_2O_2$
M_r: 242.26
mp: 113-4
UV: LANG 1631 290(23988) EtOH 314(30903) 247(10000) 230(10965) 7.1% HCl
m/e: 242.11

HODOC No: 01890
CAS No: 6854-06-4
Name: Benzamide, 4-ethoxy-*N*-3-pyridinyl-
Synonym: Benzamide, *p*-ethoxy-*N*-3-pyridyl-
Mol Form: $C_{14}H_{14}N_2O_2$
Mr: 242.26
mp: 159
UV: LANG 1634 280(23988) EtOH 289(21878) 7.1% HCl
m/e: 242.11

HODOC No: 01891
CAS No: 6760-02-7
Name: Benzamide, 4-ethoxy-*N*-4-pyridinyl-
Synonym: Pyridine, 4-(*p*-ethoxybenzamido)-
Mol Form: $C_{14}H_{14}N_2O_2$
Mr: 242.26
mp: 153-5
UV: LANG 1637 284(27542) EtOH 302(25119) 7.1% HCl
m/e: 242.11

HODOC No: 01892
CAS No: 28029-41-6
Name: Benzamide, *m*-ethoxy-*N*-4-pyridylthio-
Synonym: Benzoic acid, amide,thiono,3-ethoxy-*N*(4-pyridyl)
Structure:

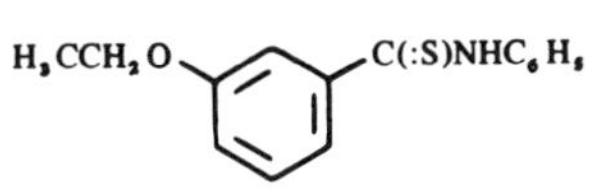

Mol Form: $C_{14}H_{14}NOS$
Mr: 244.32
mp: 163-4
UV: LANG 2383 454(316) 329(15488) 289(11482) 230(13183) EtOH
m/e: 244.08

HODOC No: 01893
CAS No: 605-90-3
Beilstein No: 22, 503
Name: Benzamide, *N*-(8-ethoxy-5-quinolinyl)-
Synonym: Benzamide, *N*-(8-ethoxy-5-quinolyl)-
Structure:

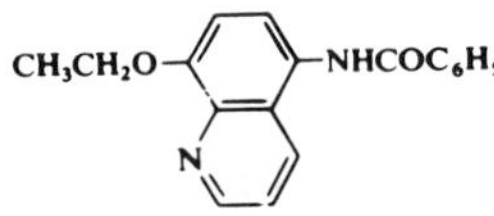

Mol Form: $C_{18}H_{16}N_2O_2$
Mr: 292.32
mp: 206
Color: ye nd (al)
m/e: 292.12

HODOC No: 01894
CAS No: 614-17-5
Beilstein No: 9^4, 727
Name: Benzamide, *N*-ethyl-
Synonym: *N*-Ethylbenzamide
Line Form: $C_6H_5CONHCH_2CH_3$
Mol Form: $C_9H_{11}NO$
Mr: 149.18
bp: 285^{745}
mp: 70-1
Color: nd (w)
MS: NBS 444
m/e: 149.08

HODOC No: 01895
CAS No: 445-28-3
Beilstein No: 9^1, 136
Name: Benzamide, 2-fluoro-
Synonym: Benzamide, *o*-fluoro-
Line Form: $2\text{-}FC_6H_4CONH_2$
Mol Form: C_7H_6FNO
Mr: 139.12
mp: 112-5
IR: ALD 3:1077H
MS: ALD 7:74A
m/e: 139.04

HODOC No: 01896
CAS No: 455-37-8
Beilstein No: 9^1, 137
Name: Benzamide, 3-fluoro-
Synonym: Benzamide, *m*-fluoro-
Line Form: 3-$FC_6H_4CONH_2$
Mol Form: C_7H_6FNO
M_r: 139.12
mp: 129-32
IR: COB 7429 3500 3410 1670 1630 1575 1530 1460 1415 1300 1260 1215 1150 930 890 830 765
m/e: 139.04

HODOC No: 01897
CAS No: 824-75-9
Beilstein No: 9^4, 954
Name: Benzamide, 4-fluoro-
Synonym: Benzamide, *p*-fluoro-
Line Form: 4-$FC_6H_4CONH_2$
Mol Form: C_7H_6FNO
M_r: 139.12
mp: 155-7
IR: ALD 3:1079C
m/e: 139.04

HODOC No: 01898
CAS No: 456-07-5
Name: Benzamide, 4-fluoro-*N*-hydroxy-
Synonym: Benzohydroxamic acid, *p*-fluoro-
Mol Form: $C_7H_6FNO_2$
M_r: 155.12
mp: 165
***s*:** DMSO 2
IR: SADG 21756 3280 3030 2770 1640 1610 1580 1500 1430 1390 1320 1280 1230 1150 1090 1030 1000 890 840 740 610 520 400 340
UV: OES 4-104 230(8511) 0.1N HCl 265(5623) 219(8318) 0.005N NaOH
H NMR: SAD 11560 7.3 7.8 DMSO
m/e: 155.04

HODOC No: 01899
CAS No: 101772-33-2
Name: Benzamide, 4-(hexyloxy)-
Synonym: Benzamide, *p*-(hexyloxy)-
Mol Form: $C_{13}H_{19}NO_2$
M_r: 221.29
mp: 149-51
IR: COB 5140 3400 3000 2930 1670 1600 1500 1460 1360 1250 1170 840
m/e: 221.14

HODOC No: 01900
CAS No: 65-45-2
Beilstein No: 10^3, 152
Name: Benzamide, 2-hydroxy-
Synonym: Salicylamide
Line Form: 2-$HOC_6H_4CONH_2$
Mol Form: $C_7H_7NO_2$
M_r: 137.13
bp: 181.5^{14}
mp: 142
***d*:** $1.175^{140/4}$
***s*:** H_2O 2; al 3; eth 2; $DMSOd_6$ 2
IR: μCOB 1696 3450 3230 3030 1670 1640 1610 1490 1450 1430 1350 1250 1120 1090 880 850 770 750 650
UV: SAD 973 302(3810) 235(7120) MeOH
H NMR: SAD 6740 7.0 7.5 8.0 8.2 13.1 $DMSOd_6$
MS: ALDN 86 120(100) 97(85) 137(78) 65(22) 39(21) 121(16) 64(15) 63(15)
m/e: 137.05

HODOC No: 01901
CAS No: 618-49-5
Beilstein No: 10^3, 254
Name: Benzamide, 3-hydroxy-
Synonym: Benzamide, *m*-hydroxy-
Line Form: 3-$HOC_6H_4CONH_2$
Mol Form: $C_7H_7NO_2$
M_r: 137.13
mp: 170.5
Color: pl (w)
***s*:** H_2O 2; al 3; eth 3; bz 1; chl 1; CS_2 1
IR: μSADP 17370 3450 3330 3130 1720 1670 1590 1470 1270 1110 930 890 760 730 680
UV: SAD 5510 295 233 MeOH
m/e: 137.05

HODOC No: 01902
CAS No: 619-57-8
Beilstein No: 10^3, 340
Name: Benzamide, 4-hydroxy-
Synonym: Benzamide, *p*-hydroxy-
Line Form: 4-$HOC_6H_4CONH_2$
Mol Form: $C_7H_7NO_2$
M_r: 137.13
mp: 162
Color: nd (w +1)
***s*:** H_2O 2; al 3; eth 3
UV: OES 6-111 252(14454) EtOH:HCl
m/e: 137.05

HODOC No: 01903
CAS No: 495-18-1
Beilstein No: 9[4], 912
Name: Benzamide, *N*-hydroxy-
Synonym: Benzenehydroxamic acid
Line Form: $C_6H_5CONHOH$
Structure:

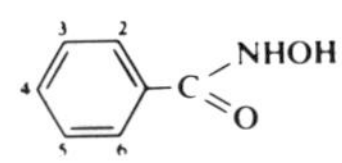

Mol Form: $C_7H_7NO_2$
M_r: 137.13
mp: 131 exp
Color: rh ta, lf (eth)
***s*:** H_2O 3; al 3; eth 2; bz 2
IR: SADG 10011 3333 3030 2703 1667 1563 1493 1449 1316 1149 1020 901 781 680
Raman: SAD 125 3400 3170 3070 1630 1590 1500 1450 1440 1370 1300 1250 1170 1150 1130 1080 1060 1040 950 850 790 750 570 520 450 430 400 370 350 270 180 150 120 100
UV: SAD 2190 224 MeOH
H NMR: SAD 18725 7.2-7.6 7.8 10.2 poly D
MS: NBS 4096
m/e: 137.05

HODOC No: 01904
CAS No: 1778-08-1
Name: Benzamide, 2-hydroxy-*N*,*N*-dimethyl-
Synonym: Salicylamide, *N*,*N*-dimethyl-
Line Form: $2\text{-}HOC_6H_4CON(CH_3)_2$
Mol Form: $C_9H_{11}NO_2$
M_r: 165.18
mp: 164-6
UV: LANG 2287 282(2818) EtOH
m/e: 165.08

HODOC No: 01905
CAS No: 15789-03-4
Name: Benzamide, 3-hydroxy-*N*,*N*-dimethyl-
Synonym: Benzamide, *m*-hydroxy-*N*,*N*-dimethyl-
Line Form: $3\text{-}HOC_6H_4CON(CH_3)_2$
Mol Form: $C_9H_{11}NO_2$
M_r: 165.18
mp: 128-30
UV: LANG 2292 284(2455) 215(12589) EtOH
m/e: 165.08

HODOC No: 01906
CAS No: 20876-99-7
Name: Benzamide, 4-hydroxy-*N*,*N*-dimethyl-
Synonym: Benzamide, *p*-hydroxy-*N*,*N*-dimethyl-
Line Form: $4\text{-}HOC_6H_4CON(CH_3)_2$
Mol Form: $C_9H_{11}NO_2$
M_r: 165.18
mp: 204-5
UV: LANG 2279 265(12303) EtOH
m/e: 165.08

HODOC No: 01907
CAS No: 20884-74-6
Name: Benzamide, *N*-(2-hydroxyethyl)-*p*-(isopropylsulfonyl)-
Synonym: Benzoic acid, amide,*N*(2-hydroxyethyl)-4(isopropylsulfonyl)
Mol Form: $C_{12}H_{17}NO_4S$
M_r: 271.32
mp: 106-7
***s*:** chl 3
IR: SADG 16565 3400 3300 3080 2980 1640 1550 1460 1390 1290 1130 1050 850 740 680 600
H NMR: SAD 8882 1.3 3.2 3.4-3.9 7.6 7.8 7.9 $CDCl_3$
m/e: 271.09

HODOC No: 01908
CAS No: 97153-34-9
Name: Benzamide, *N*-[3-hydroxy-1-(hydroxymethyl)propyl]-, dibenzoate
Synonym: Benzoic acid, amide,*N*(3-hydroxy-1(hydroxymethyl)propyl),dibenzoate
Line Form: $C_6H_5CO_2CH_2CH_2CH(NHCOCi6H_5)CH_2O_2CC_6H_5$
Mol Form: $C_{25}H_{23}NO_5$
M_r: 417.44
mp: 153-4
IR: COB 4546 3310 3050 1720 1710 1640 1540 1320 1280 1120 720
m/e: 417.16

HODOC No: 01909
CAS No: 34661-33-1
Name: Benzamide, *N*-hydroxy-4-methoxy-*N*-(3-methylphenyl)-
Synonym: *N*-*m*-Tolyl-*p*-methoxybenzohydroxamic acid
Mol Form: $C_{15}H_{15}NO_3$
Mr: 257.27
mp: 112.5
***s*:** chl 3
IR: SADG 16974 3200 3020 2920 1620 1610 1580 1490 1380 1250 1180 1030 830 780 760 750 700 680
H NMR: SAD 8542 2.3 3.7 6.6 6.7-7.2 7.3 $CDCl_3$
m/e: 257.11

HODOC No: 01910
CAS No: 4146-06-9
Name: Benzamide, *N*-[1-(hydroxymethyl)butyl]-, benzoate
Synonym: Benzoic acid, amide,*N*(1-hydroxymethyl-butyl),benzoate
Line Form: $C_6H_5CONH(CH_2CH_2CH_3)CH_2O_2CC_6H_5$
Mol Form: $C_{19}H_{21}NO_3$
Mr: 311.36
IR: µCOB 4522 3330 2940 1720 1640 1540 1470 1320 1280 1140 710
m/e: 311.15

HODOC No: 01911
CAS No: 551-35-9
Beilstein No: 10^2, 58
Name: Benzamide, 2-hydroxy-*N*-(1-methylethyl)-
Synonym: Salicylamide, *N*-isopropyl-
Line Form: $2\text{-}HOC_6H_4CONHCH(CH_3)_2$
Mol Form: $C_{10}H_{13}NO_2$
Mr: 179.21
mp: 66-7
Color: cr_(dil al)
m/e: 179.09

HODOC No: 01912
CAS No: 5395-85-7
Beilstein No: 12, 1300
Name: Benzamide, 2-hydroxy-*N*-2-naphthalenyl-
Synonym: Salicylamide, *N*-2-naphthyl-
Mol Form: $C_{17}H_{13}NO_2$
Mr: 263.28
mp: 188-9
***s*:** al 3
IR: COB 7624 3400 3050 1630 1600 1490 1420 1310 1230 960 840 810 740 480
m/e: 263.09

HODOC No: 01913
CAS No: 1152-51-8
Name: Benzamide, 2-hydroxy-*N*-(4-nitrophenyl)-
Synonym: Salicylanilide, 4′-nitro-
Mol Form: $C_{13}H_{10}N_2O_4$
Mr: 258.22
mp: 231
UV: LANG 1539 331(19498) EtOH 326(21380) diox
m/e: 258.06

HODOC No: 01914
CAS No: 87-17-2
Beilstein No: 12^3, 944
Name: Benzamide, 2-hydroxy-*N*-phenyl-
Synonym: Salicylanilide
Line Form: $2\text{-}HOC_6H_4CONHC_6H_5$
Mol Form: $C_{13}H_{11}NO_2$
Mr: 213.22
mp: 136-7
Color: pr (w,_al)
***s*:** H_2O 3; al 2; eth 2; bz 2; chl 2
IR: ALD 2:944D
UV: HBCP 295(145) 265(13489) chl
MS: ALD 7:76C
m/e: 213.08

HODOC No: 01915
CAS No: 27559-45-1
Beilstein No: 12^1, 269
Name: Benzamide, 3-hydroxy-*N*-phenyl-
Synonym: Benzanilide, 3-hydroxy-
Line Form: $3\text{-}HOC_6H_4CONHC_6H_5$
Mol Form: $C_{13}H_{11}NO_2$
M_r: 213.22
mp: 156
Color: nd or pl (w or dil al)
***s*:** H_2O 2; al 3; eth 2; bz 2; chl 1; tfa 2
IR: COB 6265 3300 3030 2930 2830 2690 2560 1620 1600 1560 1500 1460 1380 1340 1240 1200 1160 890 750 710 690
UV: SAD 2903 303 266 240 MeOH
H NMR: SAD 3235 7.1 7.2 7.4 7.6 7.9 9.9 tfa
m/e: 213.08

HODOC No: 01916
CAS No: 14121-97-2
Beilstein No: 12^3, 947
Name: Benzamide, 4-hydroxy-*N*-phenyl-
Synonym: Benzanilide, 4-hydroxy-
Line Form: $4\text{-}HOC_6H_4CONHC_6H_5$
Mol Form: $C_{13}H_{11}NO_2$
M_r: 213.22
mp: 201-2
Color: pl or nd (w)
***s*:** EtOH 4
m/e: 213.08

HODOC No: 01917
CAS No: 3743-28-0
Beilstein No: 13, 416
Name: Benzamide, *N*-(3-hydroxyphenyl)-
Synonym: Benzanilide, 3′-hydroxy-
Line Form: $3\text{-}HOC_6H_4NH\text{-}COC_6H_5$
Mol Form: $C_{13}H_{11}NO_2$
M_r: 213.22
mp: 174
Color: nd (to)
***s*:** Et_2O 4; EtOH 4
m/e: 213.08

HODOC No: 01918
CAS No: 15457-50-8
Beilstein No: 13^2, 249
Name: Benzamide, *N*-(4-hydroxyphenyl)-
Synonym: Benzanilide, 4′-hydroxy-
Line Form: $4\text{-}HOC_6H_4NH\text{-}COC_6H_5$
Mol Form: $C_{13}H_{11}NO_2$
M_r: 213.22
mp: 216-7
Color: nd
***s*:** HOAc 4
m/e: 213.08

HODOC No: 01919
CAS No: 20884-70-2
Name: Benzamide, *N*-isopropyl-*o*-(isopropylsulfonyl)-
Synonym: Benzoic acid, amide,*N*-isopropyl-2(isopropylsulfonyl)
Mol Form: $C_{13}H_{19}NO_3S$
M_r: 269.35
mp: 115-7
***s*:** chl 3
IR: SADG 16562 3300 3080 2980 1650 1630 1550 1460 1380 1360 1300 1250 1140 1050 780 730 670 580
H NMR: SAD 8958 1.2 4.0 6.5 7.3-7.6 7.8 $CDCl_3$
m/e: 269.11

HODOC No: 01920
CAS No: 20884-77-9
Name: Benzamide, *N*-isopropyl-*p*-(methylsulfonyl)-
Synonym: Benzoic acid, amide,*N*-isopropyl-4(methylsulfonyl)
Mol Form: $C_{11}H_{15}NO_3S$
M_r: 241.29
mp: 185-7
***s*:** DMSO 3
IR: SADG 16569 3340 3020 2980 1640 1530 1450 1390 1320 1280 1140 1080 960 840 780 750 650 560 510
H NMR: SAD 8880 1.2 3.2 4.1 8.0 8.4 $DMSOd_6$
m/e: 241.08

HODOC No: 01921
CAS No: 2439-77-2
Beilstein No: 12^3, 155
Name: Benzamide, 2-methoxy-
Synonym: *o*-Anisamide
Line Form: 2-CH_3O-$C_6H_4CONH_2$
Mol Form: $C_8H_9NO_2$
M_r: 151.16
mp: 129
Color: nd (Bz), pr (eth), pl (w)
***s*:** H_2O 2; eth 4; bz 4; tfa 3; sulf 3
IR: SADG 1706 3448 3226 2857 1639 1563 1471 1408 1282 1250 1111 752
UV: SAD 9342 289(3240) 232(8300) MeOH
H NMR: SAD 825 4.0 7.1 7.6 8.0 tfa
m/e: 151.06

HODOC No: 01922
CAS No: 5813-86-5
Beilstein No: 10^3, 254
Name: Benzamide, 3-methoxy-
Synonym: *m*-Anisamide
Line Form: 3-CH_3O-$C_6H_4CONH_2$
Mol Form: $C_8H_9NO_2$
M_r: 151.16
mp: 133-6
IR: ALD 3:1079A
MS: ALD 7:74D
m/e: 151.06

HODOC No: 01923
CAS No: 3424-93-9
Beilstein No: 12^3, 341
Name: Benzamide, 4-methoxy-
Synonym: *p*-Anisamide
Line Form: 4-CH_3O-$C_6H_4CONH_2$
Mol Form: $C_8H_9NO_2$
M_r: 151.16
bp: 295, 183^{40}
mp: 166-7
Color: nd or ta (w)
***s*:** H_2O 4; EtOH 4
m/e: 151.06

HODOC No: 01924
CAS No: 14547-75-2
Name: Benzamide, 4-methoxy-*N*-4-pyridinyl-
Synonym: *p*-Anisamide, *N*-4-pyridyl-
Mol Form: $C_{13}H_{12}N_2O_2$
M_r: 228.24
mp: 145-6
UV: LANG 1018 280(23988) EtOH
m/e: 228.09

HODOC No: 01925
CAS No: 527-85-5
Beilstein No: 9^4, 1701
Name: Benzamide, 2-methyl-
Synonym: *o*-Toluamide
Line Form: 2-$CH_3C_6H_4CONH_2$
Mol Form: C_8H_9NO
M_r: 135.16
mp: 147
***s*:** H_2O 2; al 4; eth 2; bz 2; tfa 2
IR: SADG 8236 3448 3226 3030 2941 1667 1639 1493 1389 1149 752
Raman: SAD 925 3350 3160 3080 3050 2960 2920 1680 1610 1590 1580 1490 1460 1440 1400 1380 1300 1200 1170 1160 1140 1060 990 830 740 640 530 470 380 230 160
UV: SAD 566 269 MeOH
H NMR: SAD 3163 2.5 7.5 tfa
MS: NBS 307
m/e: 135.07

HODOC No: 01926
CAS No: 618-47-3
Beilstein No: 9^4, 1716
Name: Benzamide, 3-methyl-
Synonym: *m*-Toluamide
Line Form: 3-$CH_3C_6H_4CONH_2$
Mol Form: C_8H_9NO
M_r: 135.16
mp: 97
***s*:** H_2O 2; al 3; eth 2; bz 2; chl 2
IR: µSADP 2154 3330 3130 1640 1590 1470 1430 1370 1110 880 770 740
UV: SAD 567 276 229 MeOH
H NMR: SAD 692 2.3 6.5 7.3 7.6 $CDCl_3$
m/e: 135.07

HODOC No: 01927
CAS No: 619-55-6
Beilstein No: 9^4, 1733
Name: Benzamide, 4-methyl-
Synonym: *p*-Toluamide
Line Form: $4-CH_3C_6H_4CONH_2$
Mol Form: C_8H_9NO
M_r: 135.16
mp: 160
***s*:** H_2O 2; al 4; eth 4; bz 2; chl 2; tfa 3
IR: **μCOB 1810** 3330 3130 2940 1670 1610 1560 1410 1390 1190 1140 1120 1020 830 790 740 730 670
UV: **SAD 568** 235 MeOH
H NMR: **SAD 693** 2.5 7.4 7.9 11.7 tfa
m/e: 135.07

HODOC No: 01928
CAS No: 613-93-4
Beilstein No: 9^3, 1068
Name: Benzamide, *N*-methyl-
Synonym: *N*-Methylbenzamide
Line Form: $C_6H_5CONHCH_3$
Mol Form: C_8H_9NO
M_r: 135.16
bp: 291^{765}, 167^{12}
mp: 82
***s*:** al 3; ace 3
IR: **COB 6028** 3430 3290 3060 3020 2980 1690 1540 1460 1420 1390 1290 1200 1020 990 900 710 610 580
UV: **OES 5-140** 242(12023) 59% sulf
MS: **NBS 304**
m/e: 135.07

HODOC No: 01929
CAS No: 56042-78-5
Name: Benzamide, 2-(methylamino)-*N*-(phenylmethyl)-
Synonym: Benzamide, *N*-benzyl-*o*-methylamino-
Mol Form: $C_{15}H_{16}N_2O$
M_r: 240.29
***s*:** ctc 3; CS_2 3
IR: **μCOB 5292** 3330 3030 2860 2780 1640 1590 1520 1470 1410 1320 1270 1240 1160 1030 840 740 690
m/e: 240.13

HODOC No: 01930
CAS No: 18453-16-2
Name: Benzamide, *N*-(1-methylethyl)-3-[(1-methylethyl)sulfonyl]-
Synonym: Benzamide, *N*-isopropyl-*m*-(isopropylsulfonyl)-
Mol Form: $C_{13}H_{19}NO_3S$
M_r: 269.35
mp: 122-4
***s*:** chl 3
IR: **SADG 16561** 3320 3080 2980 1640 1550 1460 1380 1360 1300 1280 1120 690 550
UV: **SAD 16636** 289(4220) 281(4890) 220(28600) MeOH
H NMR: **SAD 8956** 1.3 3.2 4.2 7.2 7.5 7.9 8.1 8.3 $CDCl_3$
m/e: 269.11

HODOC No: 01931
CAS No: 18507-23-8
Name: Benzamide, *N*-(1-methylethyl)-4-[(1-methylethyl)sulfonyl]-
Synonym: Benzamide, *N*-isopropyl-*p*-(isopropylsulfonyl)-
Mol Form: $C_{13}H_{19}NO_3S$
M_r: 269.35
mp: 136-8
***s*:** chl 3
IR: **SADG 16560** 3340 3030 2980 1640 1530 1460 1380 1360 1300 1280 1140 1130 1090 860 740 670 600
H NMR: **SAD 8883** 1.3 3.2 4.2 7.1 7.7 7.9 $CDCl_3$
m/e: 269.11

HODOC No: 01932
CAS No: 6876-66-0
Beilstein No: 12, 929
Name: Benzamide, 4-methyl-*N*-(4-methylphenyl)-
Synonym: *p*-Tolu-*p*-toluidide
Mol Form: $C_{15}H_{15}NO$
M_r: 225.28
mp: 158-9
IR: **μCOB 235** 3330 3030 2940 1640 1590 1520 1470 1410 1370 1320 1250 810 750 700 660
m/e: 225.12

HODOC No: 01933
CAS No: 584-70-3
Beilstein No: 12^3, 1858
Name: Benzamide, *N*-(2-methylphenyl)-
Synonym: *o*-Benzotoluidide
Line Form: $C_6H_5CONH(C_6H_4CH_3\text{-}2)$
Mol Form: $C_{14}H_{13}NO$
M$_r$: 211.25
mp: 145-6
***d*:** 1.205^{15}
Color: rh nd (AcOEt-Ace)
***s*:** H_2O 2; al 3; eth 3
IR: **µDMS 6548** 3230 3030 1670 1610 1590 1540 1490 1450 1370 1320 1280 1250 1120 1080 1030 920 850 790 750 720 690
UV: **SAD 1100** 224(13900) MeOH
H NMR: **SAD 12560** 2.3 7.0-7.5 7.3-7.6 $CDCl_3$
m/e: 211.10

HODOC No: 01934
CAS No: 582-77-4
Beilstein No: 12^3, 1964
Name: Benzamide, *N*-(3-methylphenyl)-
Synonym: *m*-Benzotoluidide
Line Form: $C_6H_5CONH(C_6H_4CH_3\text{-}3)$
Mol Form: $C_{14}H_{13}NO$
M$_r$: 211.25
mp: 125
***d*:** 1.170^{15}
Color: mcl pr (dil al)
***s*:** EtOH 4
m/e: 211.10

HODOC No: 01935
CAS No: 582-78-5
Beilstein No: 12^3, 2062
Name: Benzamide, *N*-(4-methylphenyl)-
Synonym: *p*-Benzotoluidide
Line Form: $C_6H_5CONH(C_6H_4CH_3\text{-}4)$
Mol Form: $C_{14}H_{13}NO$
M$_r$: 211.25
bp: 232
mp: 158
***d*:** 1.202^{15}
Color: rh nd (al)
***s*:** Et_2O 4; EtOH 4
H NMR: **SAD 10571**
m/e: 211.10

HODOC No: 01936
CAS No: 14547-74-1
Name: Benzamide, 4-methyl-*N*-4-pyridinyl-
Synonym: *p*-Toluamide, *N*-4-pyridyl-
Mol Form: $C_{13}H_{12}N_2O$
M$_r$: 212.24
mp: 184-5
UV: **LANG 1015** 269(19498) EtOH
m/e: 212.09

HODOC No: 01937
CAS No: 65442-07-1
Name: Benzamide, *N*-methyl-*N*-3-pyridinyl-
Synonym: Pyridine, 3-*N*-methylbenzamido-
Mol Form: $C_{13}H_{12}N_2O$
M$_r$: 212.24
mp: 92-3
UV: **LANG 461** 263(15488) pH=9.7 268(7943) 259(7762) 223(11482) 1N sulf
m/e: 212.09

HODOC No: 01938
CAS No: 37529-67-2
Name: Benzamide, *N*-(4-methyl-2-thiazolyl)-
Synonym: Thiazole, 2-benzamido-4-methyl-
Structure:

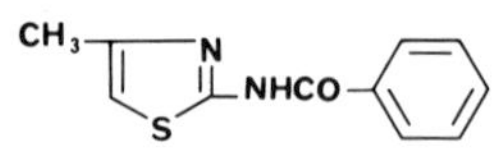

Mol Form: $C_{11}H_{10}N_2OS$
M$_r$: 218.26
mp: 151
***s*:** chl 2
IR: **µSADP 3636** 3330 3130 2940 1670 1560 1450 1320 1150 1080 1010 980 890 860 790 750 710 700
H NMR: **SAD 10511** 1.9 6.5 7.2-7.6 7.9 $CDCl_3$
m/e: 218.05

HODOC No: 01939
CAS No: 634-42-4
Beilstein No: 12^3, 2870
Name: Benzamide, *N*-1-naphthalenyl-
Synonym: Benzamide, *N*-1-naphthyl-
Line Form: $C_6H_5CONH(1\text{-}C_{10}H_7)$
Structure:

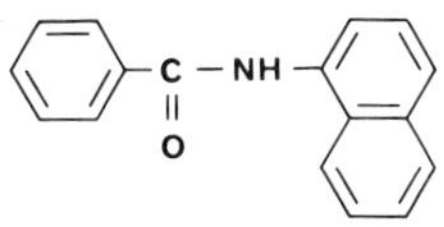

Mol Form: $C_{17}H_{13}NO$
M_r: 247.28
mp: 161-2
Color: nd (al or HOAc)
***s*:** al 2; bz 3; HOAc 3
IR: µSADP 3630 3230 3030 1640 1540 1490 1390 1350 1280 790 770 710
UV: SAD 1102 290 221 MeOH
m/e: 247.10

HODOC No: 01940
CAS No: 18271-22-2
Name: Benzamide, *N*-2-naphthalenyl-
Synonym: Benzamide, *N*-2-naphthyl-
Mol Form: $C_{17}H_{13}NO$
M_r: 247.28
UV: FO 285 332(2512) 298(12589) 262(25119) 226(50119) cyhex
m/e: 247.10

HODOC No: 01941
CAS No: 610-15-1
Beilstein No: 9^3, 1477
Name: Benzamide, 2-nitro-
Synonym: Benzamide, *o*-nitro-
Line Form: $2\text{-}NOC_6H_4CONH_2$
Mol Form: $C_7H_6N_2O_3$
M_r: 166.13
bp: 317
mp: 176.6
Color: nd (aq al)
***s*:** H_2O 3; al 3; eth 3
UV: OES 2-89 250(5012) EtOH
MS: NBS 689
m/e: 166.04

HODOC No: 01942
CAS No: 645-09-0
Beilstein No: 9^3, 1515
Name: Benzamide, 3-nitro-
Synonym: Benzamide, *m*-nitro-
Line Form: $3\text{-}NO_2C_6H_4CONH_2$
Mol Form: $C_7H_6N_2O_3$
M_r: 166.13
bp: 310-5
mp: 142.7
***s*:** H_2O 3; al 3; eth 3
IR: ALD 3:1079B
UV: OES 5-87 215(20417) H_2O
m/e: 166.04

HODOC No: 01943
CAS No: 619-80-7
Beilstein No: 9^3, 1710
Name: Benzamide, 4-nitro-
Synonym: Benzamide, *p*-nitro-
Line Form: $4\text{-}NO_2C_6H_4CONH_2$
Mol Form: $C_7H_6N_2O_3$
M_r: 166.13
mp: 201.4
Color: nd (w)
***s*:** H_2O 1; al 3; eth 3
IR: SADG 18119 3480 3430 3310 3200 3110 1670 1620 1590 1530 1510 1400 1380 1340 1320 1110 870 760 690 640 610 500
UV: SAD 1725 260 MeOH
m/e: 166.04

HODOC No: 01944
CAS No: 6333-15-9
Beilstein No: 12^3, 1600
Name: Benzamide, 4-nitro-*N*-(4-nitrophenyl)-
Synonym: Benzanilide, 4,4′-dinitro-
Line Form: $4\text{-}NO_2C_6H_4CONH(C_6H_4NO_2\text{-}4)$
Mol Form: $C_{13}H_9N_3O_5$
M_r: 287.22
mp: 266-7d
UV: LANG 2492 320(18621) 257(9550) EtOH 457(234) 333(17378) dmf
m/e: 287.05

HODOC No: 01945
CAS No: 2385-27-5
Beilstein No: 12^3, 506
Name: Benzamide, 2-nitro-*N*-phenyl-
Synonym: Benzanilide, 2-nitro-
Line Form: $2\text{-}NO_2C_6H_4CONHC_6H_5$
Mol Form: $C_{13}H_{10}N_2O_3$
M_r: 242.22
mp: 155
Color: nd (al or Bz)
***s*:** H_2O 1; al 4; eth 2; bz 3; chl 3; lig 2
IR: SADG 20314 3360 3060 1730 1690 1610 1590 1550 1520 1460 1340 1270 1250 1150 1100 1070 950 850 780 740 700 610 520
UV: LANG 2491 245(17783) EtOH 245(15849) chl
H NMR: SAD 10376 7.1 7.3-7.7 8.0 8.1 8.9 $CDCl_3$
m/e: 242.07

HODOC No: 01946
CAS No: 2243-73-4
Beilstein No: 12^3, 506
Name: Benzamide, 3-nitro-*N*-phenyl-
Synonym: Benzanilide, 3-nitro-
Line Form: $3\text{-}NO_2C_6H_4CONHC_6H_5$
Mol Form: $C_{13}H_{10}N_2O_3$
M_r: 242.22
mp: 154
***s*:** H_2O 2; al 3; eth 3; bz 3
IR: µCOB 4124 3330 3030 1670 1540 1470 1430 1350 1270 1160 1080 930 880 820 800 740 710 690 670
UV: LANG 2490 261(15849) EtOH
H NMR: SAD 12556 7.0-8.0 8.4 8.8 DMSO
m/e: 242.07

HODOC No: 01947
CAS No: 3460-11-5
Beilstein No: 12^2, 153
Name: Benzamide, 4-nitro-*N*-phenyl-
Synonym: Benzanilide, 4-nitro-
Line Form: $4\text{-}NO_2C_6H_4CONHC_6H_5$
Mol Form: $C_{13}H_{10}N_2O_3$
M_r: 242.22
mp: 211
***s*:** H_2O 1; al 3; eth 3; abs al 3
UV: LANG 2489 290(9333) 243(15849) EtOH 325(1585) 263(23442) chl
m/e: 242.07

HODOC No: 01948
CAS No: 728-90-5
Beilstein No: 12^4, 1525
Name: Benzamide, *N*-(2-nitrophenyl)-
Synonym: Benzanilide, 2′-nitro-
Line Form: $C_6H_5CONH(C_6H_4NO_2\text{-}2)$
Mol Form: $C_{13}H_{10}N_2O_3$
M_r: 242.22
mp: 98
Color: gold-ye nd (al)
***s*:** H_2O 2; al 3; eth 4
UV: LANG 2336 352(3388) 247(18197) EtOH 365(5623) 257(23442) chl
m/e: 242.07

HODOC No: 01949
CAS No: 4771-08-8
Beilstein No: 12^3, 1552
Name: Benzamide, *N*-(3-nitrophenyl)-
Synonym: Benzanilide, 3′-nitro-
Line Form: $C_6H_5CONH(C_6H_4NO_2\text{-}3)$
Mol Form: $C_{13}H_{10}N_2O_3$
M_r: 242.22
mp: 157
Color: pl
***s*:** H_2O 1; al 2; eth 4; chl 4; DMSO 2
IR: µCOB 4124 3330 3030 1670 1540 1470 1430 1350 1270 1160 1080 930 880 820 800 740 710 690 670
UV: SAD 3421 259 MeOH
H NMR: SAD 9817 7.5-7.9 7.9-8.4 8.9 10.9 DMSO
m/e: 242.07

HODOC No: 01950
CAS No: 3393-96-2
Beilstein No: 12^3, 1597
Name: Benzamide, *N*-(4-nitrophenyl)-
Synonym: Benzanilide, 4′-nitro-
Line Form: $C_6H_5CONH(C_6H_4NO_2\text{-}4)$
Mol Form: $C_{13}H_{10}N_2O_3$
M_r: 242.22
mp: 199
Color: ye nd (AcOEt)
***s*:** H_2O 1; al 1; chl 1
UV: LANG 2335 320(18621) 230(13490) EtOH 320(18621) chl
m/e: 242.07

HODOC No: 01951
CAS No: 64398-84-1
Name: Benzamide, *N*-(5-nitro-2-thiazolyl)-
Synonym: 2-Benzamido-5-nitrothiazole
Mol Form: $C_{10}H_7N_3O_3S$
M_r: 249.23
mp: 257-60
IR: COB 6580 3020 2990 2930 1670 1600 1550 1480 1350 1240 1170 1100 1070 940 820 720 680 500
UV: SAD 9456 341(16800) 240(12700) MeOH
H NMR: SAD 16133 7.2-7.8 8.1 8.5 11.6 DMSO
m/e: 249.02

HODOC No: 01952
CAS No: 93-98-1
Beilstein No: 12^3, 502
Name: Benzamide, *N*-phenyl-
Synonym: Benzanilide
Line Form: $C_6H_5CONHC_6H_5$
Mol Form: $C_{13}H_{11}NO$
M_r: 197.22
bp: $117\text{-}9^{10}$, sub
mp: 163
***d*:** 1.315
Color: lf (al)
***s*:** H_2O 1; al 2; eth 2; HOAc 2
IR: SADG 385 3340 3040 1650 1600 1530 1930 1320 1250 740 710 680 640 500
Raman: SAD 514 3070 1660 1600 1530 1450 1440 1320 1260 1250 1180 1170 1150 1080 1000 930 920 890 850 830 800 760 720 700 680 620 580 510 490 420 410 370 310 260 210 190 130
UV: OES 7-455 266(12882) EtOH
MS: ALDN 160 105(100) 197(32) 77(19) 91(18) 106(9) 198(7) 120(5) 181(5)
m/e: 197.08

HODOC No: 01953
CAS No: 744-38-7
Beilstein No: 13^1, 8
Name: Benzamide, *N,N'*-1,2-phenylenebis-
Synonym: Benzamide, *N,N'-o*-phenylenebis-
Line Form: $1,2\text{-}(C_6H_5CONH)_2C_6H_4$
Mol Form: $C_{20}H_{16}N_2O_2$
M_r: 316.34
mp: 306
IR: COB 2755
m/e: 316.12

HODOC No: 01954
CAS No: 5467-04-9
Beilstein No: 13, 98
Name: Benzamide, *N,N'*-1,4-phenylenebis-
Synonym: Benzamide, *N,N'-p*-phenylenebis-
Mol Form: $C_{20}H_{16}N_2O_2$
M_r: 316.34
mp: >300
***s*:** al 2; eth 2
IR: COB 2756 3350 3080 1650 1620 1580 1550 1520 1480 1420 1320 1280 1030 900 830 720 690
m/e: 316.12

HODOC No: 01955
CAS No: 1485-70-7
Beilstein No: 12^3, 2259
Name: Benzamide, *N*-(phenylmethyl)-
Synonym: Benzamide, *N*-benzyl-
Line Form: $C_6H_5CONHCH_2C_6H_5$
Mol Form: $C_{14}H_{13}NO$
M_r: 211.25
mp: 104-6
***s*:** H_2O 2; al 3; eth 2; HOAc 3
UV: LANG 1329 225(12589) 50% 0.1N HCl:50% MeOH 219(12023) 50% 0.1N NaOH:50% MeOH
m/e: 211.10

HODOC No: 01956
CAS No: 15563-53-8
Name: Benzamide, *N*-(phenyl-1-piperidinylmethyl)-
Synonym: Benzamide, *N*-(α-piperidinobenzyl)-
Structure:

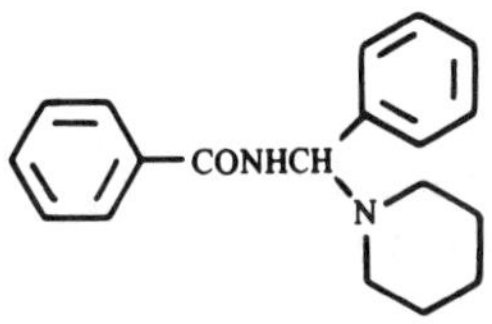

Mol Form: $C_{19}H_{22}N_2O$
M_r: 294.38
IR: μCOB 875 3450 3030 2940 2780 1670 1590 1520 1470 1320 1250 1100 1030 990 710 700 660
m/e: 294.17

HODOC No: 01957
CAS No: 21122-71-4
Name: Benzamide, *N*-phenyl-2-[[3-(trifluoromethyl)phenyl]amino]-
Synonym: Benzanilide, 2-(α,α,α-trifluoro-*m*-toluidino)-
Mol Form: $C_{20}H_{15}F_3N_2O$
M$_r$: 356.33
mp: 124-5
***s*:** chl 3
IR: SADG 17584 3340 3260 3050 1640 1590 1520 1460 1420 1320 1280 1160 1110 1060 990 890 790 740 690
H NMR: SAD 9310 6.7-7.7 8.0 9.3 $CDCl_3$
m/e: 356.11

HODOC No: 01958
CAS No: 4589-12-2
Beilstein No: 22^2, 329
Name: Benzamide, *N*-2-pyridinyl-
Synonym: Benzamide, *N*-2-pyridyl-
Mol Form: $C_{12}H_{10}N_2O$
M$_r$: 198.21
mp: 87
***s*:** H_2O 2; al 2; eth 2; ace 2; bz 2
IR: SADG 18428 3160 3050 1675 1595 1575 1520 1430 1295 1270 1235 1170 1150 1120 1090 1065 1050 1015 990 920 890 780 770 715 690 670 600 520 500
UV: LANG 458 278(12589) 239(11749) pH=9.7
H NMR: SAD 6555 7.0 7.3-7.6 7.6-8.2 8.4 $CDCl_3$
m/e: 198.08

HODOC No: 01959
CAS No: 5221-40-9
Name: Benzamide, *N*-3-pyridinyl-
Synonym: Pyridine, 3-benzamido-
Mol Form: $C_{12}H_{10}N_2O$
M$_r$: 198.21
mp: 110-2
UV: LANG 459 260(15136) pH=9.7
m/e: 198.08

HODOC No: 01960
CAS No: 5221-44-3
Beilstein No: 22^2, 341
Name: Benzamide, *N*-4-pyridinyl-
Synonym: Pyridine, 4-benzamido-
Mol Form: $C_{12}H_{10}N_2O$
M$_r$: 198.21
mp: 202
UV: LANG 460 264(16596) pH=9.7 281(30903) 1N sulf
m/e: 198.08

HODOC No: 01961
CAS No: 20884-75-7
Name: Benzamide, *N*-*tert*-butyl-*p*-(isopropylsulfonyl)-
Synonym: Benzoic acid, amide,*N*-*tert*-butyl-4 (isopropylsulfonyl)
Mol Form: $C_{14}H_{21}NO_3S$
M$_r$: 283.37
mp: 149-51
***s*:** chl 3
IR: SADG 16566 3300 3060 2980 1640 1540 1460 1390 1360 1300 1260 1140 1130 1050 850 730 670 600 550
UV: SAD 16637 337(14100) MeOH
H NMR: SAD 8889 1.3 1.5 3.2 6.4 7.8 $CDCl_3$
m/e: 283.12

HODOC No: 01962
CAS No: 618-73-5
Beilstein No: 10^2, 346
Name: Benzamide, 3,4,5-trihydroxy-
Synonym: Gallamide
Line Form: $3,4,5-(HO)_3C_6H_2CONH_2$
Mol Form: $C_7H_7NO_4$
M$_r$: 169.13
mp: 244-5d
Color: lf (w +1)
***s*:** H_2O 4; EtOH 4; Chl 4
m/e: 169.04

HODOC No: 01963
CAS No: 6262-27-7
Beilstein No: 12^2, 263
Name: Benzamide, 3,4,5-trihydroxy-*N*-phenyl-
Synonym: Gallanilide
Line Form: 3,4,5-$(HO)_3C_6H_2CONHC_6H_5$
Mol Form: $C_{13}H_{11}NO_4$
M_r: 245.22
mp: 207
Color: lf (+ 2w, dil al)
***s*:** EtOH 4; HOAc 4
m/e: 245.07

HODOC No: 01964
CAS No: 6639-19-6
Beilstein No: 9, 465
Name: Benzamide, *N*,*N*,2-trimethyl-
Synonym: *N*,*N*-Dimethyl-*o*-toluamide
Line Form: 2-$CH_3C_6H_4CON(CH_3)_2$
Mol Form: $C_{10}H_{13}NO$
M_r: 163.21
bp: 147^{15}
***d*:** 1.033^{25}
UV: LANG 2286 265(447) EtOH 265(427) H_2O
m/e: 163.10

HODOC No: 01965
CAS No: 6935-65-5
Beilstein No: 9, 477
Name: Benzamide, *N*,*N*,3-trimethyl-
Synonym: *m*-Toluamide, *N*,*N*-dimethyl-
Line Form: 3-$CH_3C_6H_4CON(CH_3)_2$
Mol Form: $C_{10}H_{13}NO$
M_r: 163.21
bp: 148^{12}
***d*:** 1.043^{15}
UV: LANG 2291 275(5248) 235(4571) EtOH 265(589) 230(5623) H_2O
m/e: 163.10

HODOC No: 01966
CAS No: 14062-78-3
Beilstein No: 9, 486
Name: Benzamide, *N*,*N*,4-trimethyl-
Synonym: *p*-Toluamide, *N*,*N*-dimethyl-
Line Form: 4-$CH_3C_6H_4CON(CH_3)_2$
Mol Form: $C_{10}H_{13}NO$
M_r: 163.21
bp: 156^{10}
mp: 41
***s*:** H_2O 2; al 2
UV: LANG 2278 235(6310) EtOH 234(8128) H_2O
m/e: 163.10

HODOC No: 01967
CAS No: 1230-64-4
Name: Benzanilide, 4,4′-*bis*(dimethylamino)-
Synonym: Benzoic acid, amide,4-dimethylamino-*N*(4-dimethylaminophenyl)
Mol Form: $C_{17}H_{21}N_3O$
M_r: 283.36
mp: dec 232-4
UV: LANG 1325 318(33884) 247(7413) EtOH
m/e: 283.17

HODOC No: 01968
CAS No: 21122-64-5
Name: Benzanilide, 4′-chloro-2′-nitro-2-(2,3-xylidino)-
Synonym: Benzoic acid, amide,*N*(4-chloro-2-nitrophenyl)-2(2,3-dimethylphenylamino)
Mol Form: $C_{21}H_{18}ClN_3O_3$
M_r: 395.82
mp: 148-9
***s*:** chl 3
IR: SADG 17576 3350 3060 2940 1660 1630 1600 1570 1500 1450 1340 1260 1170 760 750
UV: SAD 17494 394(4290) 378(4540) 255(24500) 217(17900) MeOH
H NMR: SAD 9308 2.2 2.3 6.6-7.4 7.6 8.2 8.9 9.3 11.1 $CDCl_3$
m/e: 395.10

HODOC No: 01969
CAS No: 21122-72-5
Name: Benzanilide,
2′-chloro-2-(α,α,α-trifluoro-*m*-toluidino)-
Synonym: Benzoic acid, amide,*N*(2-chlorophenyl)-2(3(trifluoromethyl)phenylamino)
Mol Form: $C_{20}H_{14}ClF_3N_2O$
M_r: 390.77
mp: 90-2
***s*:** chl 3
IR: SADG 17585 3440 3280 1650 1570 1510 1440 1330 1300 1210 1110 1080 850 800 730 700 530
H NMR: SAD 9307 6.9 7.1 7.2-7.5 7.7 8.4 8.5 9.5 $CDCl_3$
m/e: 390.07

HODOC No: 01970
CAS No: 21122-74-7
Name: Benzanilide,
2′,3′-dichloro-2-(α,α,α-trifluoro-*m*-toluidino)-
Synonym: Benzoic acid, amide,*N*(2,3-dichlorophenyl)-2(3(trifluoromethyl)phenylamino*
Mol Form: $C_{20}H_{13}Cl_2F_3N_2O$
M_r: 425.22
mp: 138-40
***s*:** chl 2
IR: SADG 17588 3420 3340 3060 1670 1610 1580 1500 1450 1400 1330 1290 1250 1220 1170 1110 1070 930 880 790 780 750 700 600 570
UV: SAD 16876 341(6170) 286(16800) MeOH
H NMR: SAD 9303 7.0 7.1-7.6 7.7 8.4 8.6 9.4 $CDCl_3$
m/e: 424.04

HODOC No: 01971
CAS No: 21239-99-6
Name: Benzanilide,
2′,5′-dichloro-2-(α,α,α-trifluoro-*m*-toluidino)-
Synonym: Benzoic acid, amide,*N*(2,5-dichlorophenyl)-2(3(trifluoromethyl)phenylamino*
Mol Form: $C_{20}H_{13}Cl_2F_3N_2O$
M_r: 425.22
mp: 119-21
***s*:** chl 3
IR: SADG 17590 3420 3340 3120 1660 1570 1500 1440 1400 1320 1260 1210 1160 1110 1060 920 880 800 790 750 690 590 550
UV: SAD 16878 242(7950) 286(16700) 265(15600) 221(33400) 206(47200) MeOH
H NMR: SAD 9305 6.9 7.0 7.2-7.5 7.7 8.5 8.6 9.4 $CDCl_3$
m/e: 424.04

HODOC No: 01972
CAS No: 21122-76-9
Name: Benzanilide,
2′-nitro-2-(α,α,α-trifluoro-*m*-toluidino)-
Synonym: Benzoic acid, amide,*N*(2-nitrophenyl)-2(3(trifluoromethyl)phenylamino)
Mol Form: $C_{20}H_{14}F_3N_3O_3$
M_r: 401.32
mp: 108-10
***s*:** chl 3
IR: SADG 17593 3360 3306 3060 1650 1580 1510 1450 1330 1270 1160 1110 740 700 600 520
UV: SAD 17499 352(7100) 286(19400) 251(21900) MeOH
H NMR: SAD 9302 6.8-7.1 7.1-7.6 7.7 8.3 8.9 9.6 11.2 $CDCl_3$
m/e: 401.10

HODOC No: 01973
CAS No: 21122-69-0
Name: *p*-Benzanisidide, 2-(2,3-xylidino)-
Synonym: Benzoic acid, amide,2(2,3-dimethyl-phenylamino)-*N*(4-methoxyphenyl)
Mol Form: $C_{22}H_{22}N_2O_2$
M_r: 346.41
mp: 153-4
***s*:** chl 3
IR: SADG 17582 3320 3250 3060 2950 2830 1620 1580 1500 1460 1400 1300 1240 1030 820 740
UV: SAD 17496 346(9420) 281(20600) MeOH
H NMR: SAD 9304 2.2 2.3 3.7 6.8 6.5-7.6 7.4 7.8 9.0 $CDCl_3$
m/e: 346.17

HODOC No: 01974
CAS No: 59765-98-9
Name: Benz[a]anthracen-11-amine
Synonym: 1,2-Benzanthracene, 11-amino
Mol Form: $C_{18}H_{13}N$
M_r: 243.29
UV: FO 527 395(7943) 379(7943) 305(50119) 286(50119) 267(79433) EtOH
m/e: 243.10

HODOC No: 01975
CAS No: 56-55-3
Beilstein No: 5[4], 2549
Name: Benz[a]anthracene
Synonym: Benzanthracene
Structure:

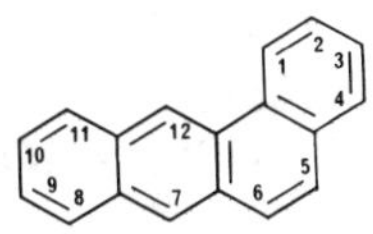

Mol Form: $C_{18}H_{12}$
Mr: 228.28
bp: 435 sub
mp: 162
Color: ye br pl or lf (al-HOAc)
s: H_2O 1; al 2; eth 3; ace 3; bz 4; DMSOd6 2; os 3; HOAc 2
IR: SADG 15114 3030 1613 1493 943 877 794 741 680
UV: SAD 335 341 300 288 277 MeOH
H NMR: SAD 6018 7.5-8.4 9.1 9.4 DMSOd6
MS: WILEY 1620 228(100) 229(19) 114(18) 113(13) 101(7) 100(6) 122(5)
m/e: 228.09

HODOC No: 01976
CAS No: 7225-65-2
Name: 1H-Benz[de]anthracene, hexadecahydro-6-octyl-
Synonym: 4,4*a*,10-Benzanthracene, 6-octyl
Mol Form: $C_{25}H_{44}$
Mr: 344.60
MS: WILEY 1987 135(100) 344(63) 231(60) 95(54) 67(48) 81(42) 43(40) 149(30)
m/e: 344.34

HODOC No: 01977
CAS No: 4389-09-7
Name: 4H-Benz[de]anthracene, 5,6-dihydro-
Synonym: 5,6-Benz-peri-naphthane
Mol Form: $C_{17}H_{14}$
Mr: 218.28
UV: FO 364 354(1259) 346(398) 337(1259) 323(794) 303(15849) 291(12589) 281(12589) 257(79433) EtOH
m/e: 218.11

HODOC No: 01978
CAS No: 6366-24-1
Name: Benz[a]anthracene, 10-chloro-7-methyl-
Synonym: Benzanthracene, 10-chloro-7-methyl
Mol Form: $C_{19}H_{13}Cl$
Mr: 276.75
UV: FO 526 392(1585) 366(6310) 356(7943) 349(7943) 339(6310) 326(5012) 305(7943) 294(100000) 282(79433) 273(39811) 260(31623) 233(31623) 225(31623) EtOH,95%
m/e: 276.07

HODOC No: 01979
CAS No: 7496-10-8
Name: Benz[a]anthracene, 9-chloro-7-methyl-
Synonym: 1,2-Benzanthracene, 9-chloro-7-methyl
Mol Form: $C_{19}H_{13}Cl$
Mr: 276.75
UV: FO 525 390(501) 374(5012) 356(7943) 340(6310) 325(5012) 295(100000) 284(79433) 274(39811) 255(31623) 237(31623) 224(31623) EtOH,95%
m/e: 276.07

HODOC No: 01980
CAS No: 16434-59-6
Beilstein No: 5[3], 2305
Name: Benz[a]anthracene, 7,12-dihydro-
Synonym: 7,12-Dihydrobenz[a]anthracene
Structure:

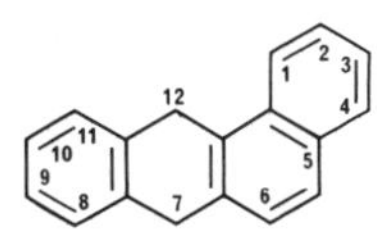

Mol Form: $C_{18}H_{14}$
Mr: 230.29
mp: 112
Color: lf (Bz, al, HOAc)
MS: NBS 4565
m/e: 230.11

HODOC No: 01981
CAS No: 313-74-6
Name: Benz[a]anthracene, 1,12-dimethyl-
Synonym: 1,12-Dimethylbenz[a]anthracene
Mol Form: $C_{20}H_{16}$
Mr: 256.33
MS: WILEY 1776 256(100) 241(75) 239(59) 119.5(38) 240(36) 120(28) 257(22) 242(15)
m/e: 256.13

HODOC No: 01982
CAS No: 35187-28-1
Name: Benz[a]anthracene, 7,11-dimethyl-
Synonym: 7,11-DMBA
Mol Form: $C_{20}H_{16}$
M_r: 256.33
UV: **FO 512** 390(1259) 372(6310) 356(10000) 338(7943) 323(5012) 303(10000) 294(100000) 284(63096) 274(39811) 261(39811) EtOH
m/e: 256.13

HODOC No: 01983
CAS No: 57-97-6
Beilstein No: 5^4, 2587
Name: Benz[a]anthracene, 7,12-dimethyl-
Synonym: 7,12-Dimethylbenz[a]anthracene
Line Form: 9.10-$(CH_3)_2C_{18}H_{10}$
Structure:

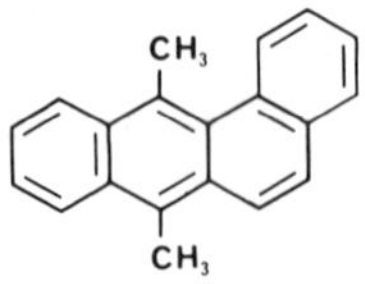

Mol Form: $C_{20}H_{16}$
M_r: 256.33
mp: 122-3
Color: pa ye pl (al, HOAc)
s: Ac 4; Bz 4
IR: SADG 23860
H NMR: SAD 15010
MS: NBS 1823
m/e: 256.13

HODOC No: 01984
CAS No: 3697-30-1
Name: Benz[a]anthracene, 7-ethyl-
Synonym: 7-Ethylbenz[a]anthracene
Mol Form: $C_{20}H_{16}$
M_r: 256.33
MS: WILEY 1774 241(100) 256(76) 239(26) 242(22) 257(16) 120.5(12) 114(10) 253(9)
m/e: 256.13

HODOC No: 01985
CAS No: 6366-20-7
Name: Benz[a]anthracene, 7-methoxy-
Synonym: 1,2-Benzanthracene, 7-methoxy
Mol Form: $C_{19}H_{14}O$
M_r: 258.30
UV: **FO 520** 388(1259) 370(6310) 353(10000) 337(6310) 321(5012) 302(6310) 291(100000) 280(79433) 270(39811) 258(39811) EtOH
m/e: 258.10

HODOC No: 01986
CAS No: 63020-61-1
Name: Benz[a]anthracene, 8-methoxy-7-methyl-
Synonym: 1,2-Benzanthracene, 8-methoxy-7-methyl
Mol Form: $C_{20}H_{16}O$
M_r: 272.33
UV: **FO 519** 367(7943) 299(50119) 288(50119) 273(39811) 238(31623) 225(31623) EtOH,95%
m/e: 272.12

HODOC No: 01987
CAS No: 2381-15-9
Beilstein No: 5^4, 2564
Name: Benz[a]anthracene, 10-methyl-
Synonym: 7-Methyl-1,2-benzanthracene
Mol Form: $C_{19}H_{14}$
M_r: 242.30
mp: 141
s: H_2O 1; al 3; HOAc 3
IR: SADP 9263
UV: **OES 6-814** 330(14125) 290(109648) 280(89125) iso
MS: WILEY 1696 28(100) 242(90) 241(30) 239(20) 243(18) 119.5(12) 14(11) 240(9)
m/e: 242.11

HODOC No: 01988
CAS No: 6111-78-0
Beilstein No: 5^4, 2566
Name: Benz[a]anthracene, 11-methyl-
Synonym: 8-Methyl-1,2-benzanthracene
Mol Form: $C_{19}H_{14}$
M_r: 242.30
bp: 360d
mp: 160
Color: ye mcl pl (Bz-al)
***s*:** H_2O 1; al 3; HOAc 3
IR: µSADP 9261 3030 1610 1490 1470 1240 890 810 790 750 690
UV: SAD 2405 317(4340) 301(6740) 289(82600) 278(68800) 268(42200) 230(36300) MeOH
MS: WILEY 1692 242(100) 28(88) 243(21) 239(18) 121(13) 107.5(10) 14(10) 119.5(8)
m/e: 242.11

HODOC No: 01989
CAS No: 2422-79-9
Beilstein No: 5^4, 2567
Name: Benz[a]anthracene, 12-methyl-
Synonym: 9-Methyl-1,2-benzanthracene
Mol Form: $C_{19}H_{14}$
M_r: 242.30
mp: 150-1
Color: pl (al)
***s*:** H_2O 1; al 3; CS_2 3; HOAc 3
IR: µSADP 9262 3030 1610 1490 890 890 790 750 680
UV: SAD 2406 315(4930) 301(11500) 287(79600) 279(65400) 269(37400) 256(33900) MeOH
MS: WILEY 1709 242(100) 243(20) 241(18) 239(15) 121(12) 107.5(8) 226(5) 106.5(5)
m/e: 242.11

HODOC No: 01990
CAS No: 2498-76-2
Name: Benz[a]anthracene, 2-methyl-
Synonym: 2-Methylbenz[a]anthracene
Mol Form: $C_{19}H_{14}$
M_r: 242.30
mp: 150.7-1.7
IR: µSADP 9267 3030 1610 1520 1470 1450 1390 1330 1150 1010 950 880 820 750 680
UV: FO 500 386(1259) 376(631) 365(3162) 359(5012) 348(6310) 342(7943) 332(6310) 327(6310) 302(7943) 291(79433) 280(63096) 270(39811) 259(31623) 226(39811) EtOH,95%
MS: WILEY 1704 242(100) 239(27) 241(26) 243(20) 240(14) 119.5(13) 120(9) 212(8)
m/e: 242.11

HODOC No: 01991
CAS No: 2498-75-1
Name: Benz[a]anthracene, 3-methyl-
Synonym: 3-Methylbenz[a]anthracene
Mol Form: $C_{19}H_{14}$
M_r: 242.30
bp: $160^{0.1}$
mp: 156-7
***s*:** H_2O 1; al 3; ace 3; bz 3; xyl 3; HOAc 3; py 3; peth 3
IR: µSADP 9268 3030 1610 1490 1370 940 890 880 820 810 740 680
UV: SAD 2412 327(6510) 314(4880) 302(9850) 289(74400) 278(61700) 268(34000) MeOH
MS: WILEY 1700 242(100) 243(21) 239(18) 241(15) 121(12) 107.5(7) 240(6) 120.5(5)
m/e: 242.11

HODOC No: 01992
CAS No: 316-49-4
Name: Benz[a]anthracene, 4-methyl-
Synonym: 4-Methylbenz[a]anthracene
Mol Form: $C_{19}H_{14}$
M_r: 242.30
mp: 126.2-7.2
***s*:** H_2O 1; al 3; eth 3; xyl 3
IR: µSADP 9269 3030 1610 1470 1330 1280 950 890 800 760 740 690
UV: OES 6-813 295(100000) 280(79433) iso
MS: WILEY 1706 242(100) 243(20) 239(18) 241(15) 121(11) 240(7) 107.5(6) 120.5(5)
m/e: 242.11

HODOC No: 01993
CAS No: 2541-69-7
Beilstein No: 5^4, 2567
Name: Benz[a]anthracene, 7-methyl-
Synonym: 3,4-Benz-9-methylanthracene
Mol Form: $C_{19}H_{14}$
M_r: 242.30
mp: 183-4
Color: ye pl (al)
***s*:** H_2O 1; al 3; eth 3; ace 3; ctc 3; CS_2 3; HOAc 3
IR: μSADP 9263 3030 1610 1490 1030 890 790 750 690
UV: SAD 2407 302(9670) 288(95400) 277(75100) 267(44000) 257(41700) MeOH
MS: WILEY 1705 242(100) 243(20) 241(18) 239(15) 121(11) 107.5(8) 119.5(6) 226(5)
m/e: 242.11

HODOC No: 01994
CAS No: 2381-31-9
Beilstein No: 5^4, 2563
Name: Benz[a]anthracene, 8-methyl-
Synonym: 5-Methyl-1,2-benzanthracene
Mol Form: $C_{19}H_{14}$
M_r: 242.30
bp: $160^{0.1}$, sub
mp: 156-7
Color: pl (Bz-al), nd (Bz-lig)
***s*:** H_2O 1; al 3; eth 3; bz 3; xyl 3
IR: μSADP 9259 3030 1610 1520 1450 1390 1250 1220 950 900 890 780 760 760 700
UV: OES 6-813 385(1413) 340(8913) 290(95499) 280(85114) iso
MS: WILEY 1697 242(100) 243(20) 239(18) 241(18) 121(11) 107.5(8) 215(8) 240(6)
m/e: 242.11

HODOC No: 01995
CAS No: 2381-16-0
Beilstein No: 5^4, 2564
Name: Benz[a]anthracene, 9-methyl-
Synonym: 6-Methyl-1,2-benzanthracene
Mol Form: $C_{19}H_{14}$
M_r: 242.30
mp: 126-7
Color: nd (al)
***s*:** H_2O 1; al 3; eth 3; ctc 3; chl 3; xyl 3; CS_2 3
IR: μSADP 9260 3030 1610 1490 1430 1330 1190 960 890 780 750
UV: SAD 2404 331(6000) 302(6140) 288(79600) 277(68500) 267(41000) 257(38900) MeOH
MS: WILEY 1690 242(100) 243(20) 239(19) 226(18) 241(16) 119.5(11) 227(11) 113(9)
m/e: 242.11

HODOC No: 01996
CAS No: 63020-49-5
Name: Benz[a]anthracene, 10-(1-methylethyl)-
Synonym: 1,2-Benzanthracene, 10-isopropyl
Mol Form: $C_{21}H_{18}$
M_r: 270.35
UV: FO 511 389(1000) 369(7943) 350(10000) 336(2512) 323(5012) 303(10000) 292(125893) 280(100000) 271(50119) 259(39811) 232(39811) 221(39811) EtOH
m/e: 270.14

HODOC No: 01997
CAS No: 54889-82-6
Name: Benz[a]anthracene, 8-propyl-
Synonym: 1,2-Benzanthracene, 8-propyl
Mol Form: $C_{21}H_{18}$
M_r: 270.35
MS: WILEY 1831 141(100) 270(82) 239(26) 242(23) 271(18) 120.5(10) 107.5(9) 121(8)
m/e: 270.14

HODOC No: 01998
CAS No: 6366-18-3
Name: Benz[a]anthracene, 1,2,3,4-tetrahydro-7-methyl-
Synonym: 1,2-Benzanthracene, 7-methyl-1,2,3,4-tetrahydro
Mol Form: $C_{19}H_{18}$
M_r: 246.33
UV: FO 391 389(10000) 368(7943) 351(6310) 335(3162) 261(199526) abs al
m/e: 246.14

HODOC No: 01999
CAS No: 57652-76-3
Beilstein No: 9^3, 3625
Name: Benz[a]anthracene-1-acetic acid
Synonym: 1,2-Benzanthracene-1-acetic acid
Structure:

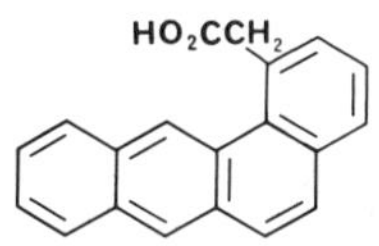

Mol Form: $C_{20}H_{14}O_2$
Mr: 286.31
mp: 204-5
Color: lf (Bz-peth)
m/e: 286.10

HODOC No: 02000
CAS No: 20316-12-5
Beilstein No: 9^3, 3625
Name: Benz[a]anthracene-7-acetic acid
Synonym: 1,2-Benzanthracene-7-acetic acid
Mol Form: $C_{20}H_{14}O_2$
Mr: 286.31
mp: 270-4
Color: nd (aq HOAc)
m/e: 286.10

HODOC No: 02001
CAS No: 13728-58-0
Beilstein No: 9^3, 3626
Name: Benz[a]anthracene-8-acetic acid
Synonym: 1,2-Benzanthracene-8-acetic acid
Mol Form: $C_{20}H_{14}O_2$
Mr: 286.31
mp: 232-3
Color: nd (Ace-Bz)
m/e: 286.10

HODOC No: 02002
CAS No: 6366-23-0
Name: Benz[a]anthracene-10-carbonitrile, 7-methyl-
Synonym: 1,2-Benzanthracene-10-carbonitrile,7-methyl
Mol Form: $C_{20}H_{13}N$
Mr: 267.31
UV: FO 529 383(6310) 366(7943) 351(6310) 299(100000) 288(79433) 278(39811) 228(31623) EtOH,95%
m/e: 267.10

HODOC No: 02003
CAS No: 63018-70-2
Name: Benz[a]anthracene-8-carbonitrile, 7-methyl-
Synonym: 1,2-Benzanthracene-8-carbonitrile,7-methyl
Mol Form: $C_{20}H_{13}N$
Mr: 267.31
UV: FO 528 399(6310) 380(7943) 300(79433) 288(63096) 278(31623) 240(31623) 225(31623) EtOH,95%
m/e: 267.10

HODOC No: 02004
CAS No: 7505-62-6
Beilstein No: 7^3, 2806
Name: Benz[a]anthracene-7-carboxaldehyde
Synonym: Benz[a]anthracene-7-carbaldehyde
Structure:

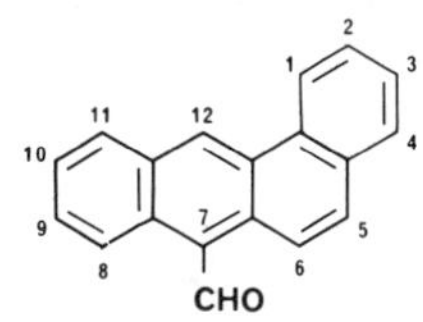

Mol Form: $C_{19}H_{12}O$
Mr: 256.28
mp: 148
Color: ye pr or nd
***s*:** Ac 4; Bz 4; Et_2O 4; EtOH 4
UV: HBCP
m/e: 256.09

HODOC No: 02005
CAS No: 18508-00-4
Beilstein No: 7^4, 2645
Name: Benz[a]anthracene-5,6-dione
Synonym: Benz[a]anthracene-5,6-quinone
Structure:

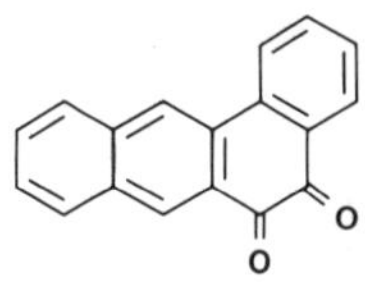

Mol Form: $C_{18}H_{10}O_2$
Mr: 258.26
mp: 262-3d
Color: red nd (to)
***s*:** al 2; bz 2; sulf 3; tol 2; HOAc 2
UV: HBCP 400(6026) 256(36308) 234(25119) al
m/e: 258.07

HODOC No: 02006
CAS No: 2498-66-0
Beilstein No: 7^4, 2644
Name: Benz[a]anthracene-7,12-dione
Synonym: Benzanthracene-7,12-dione
Structure:

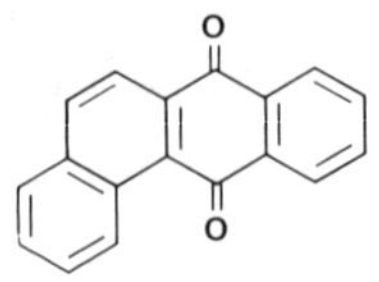

Mol Form: $C_{18}H_{10}O_2$
M$_r$: 258.26
mp: 170-1
s: al 2; eth 2; ace 3; bz 4; chl 4; HOAc 3; sulf 3; lig 2
UV: HBCP 384(631) 372(501) 340(5012) 325(5012) 300(10000) 288(79433) 277(50119) 266(50119) 255(31623) 224(5119) al
MS: NBS 1844
m/e: 258.07

HODOC No: 02007
CAS No: 5133-12-0
Name: Benz[a]anthracen-4-ol
Synonym: 4-Hydroxybenz[a]anthracene
Mol Form: $C_{18}H_{12}O$
M$_r$: 244.27
mp: 225
s: bz 3; py 3; tol 3; HOAc 4
UV: HBCP 315(19953) 288(50119) 278(50119) al
m/e: 244.09

HODOC No: 02008
CAS No: 56183-17-6
Beilstein No: 6^4, 5083
Name: Benz[a]anthracen-6-ol
Synonym: 6-Hydroxybenz[a]anthracene
Mol Form: $C_{18}H_{12}O$
M$_r$: 244.27
mp: 225
Color: pa ye or og nd (Bz)
s: Bz 4; HOAc 4
m/e: 244.09

HODOC No: 02009
CAS No: 110334-95-7
Beilstein No: 7^2, 942
Name: 4H-Benz[de]anthracen-7-ol, 5,6-dihydro-
Synonym: 4,4*a*,10-Benzanthracene, 7-hydroxy-4,5,6-trihydro
Structure:

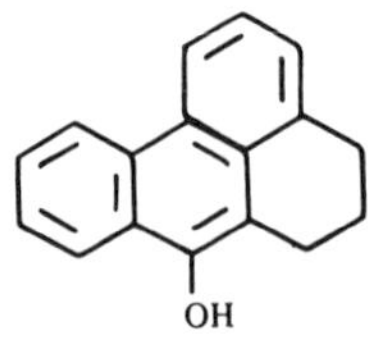

Mol Form: $C_{17}H_{14}O$
M$_r$: 234.28
mp: 151-2
s: al 3; eth 2; ace 3; bz 1; peth 1
UV: FO 375 365(1259) 348(1259) 304(12589) 255(50119) EtOH
m/e: 234.10

HODOC No: 02010
CAS No: 17526-35-1
Name: Benz[a]anthracen-7-ol, 12-methyl-, acetate
Synonym: 1,2-Benzanthracene, 7-hydroxy-12-methyl,acetate
Mol Form: $C_{21}H_{16}O_2$
M$_r$: 300.34
UV: FO 523 375(6310) 358(7943) 342(6310) 293(79433) 283(63096) 273(39811) 263(31623) 235(25119) 223(31623) EtOH,95%
m/e: 300.12

HODOC No: 02011
CAS No: 30468-02-1
Beilstein No: 7^4, 1820
Name: 7H-Benz[de]anthracen-7-one, 11-chloro-
Synonym: 1,9-Benzanthr-7-one, 11-chloro
Mol Form: $C_{17}H_9ClO$
M$_r$: 264.69
mp: 182-3
Color: ye nd (HOAc)
m/e: 264.03

HODOC No: 02012
CAS No: 61994-44-3
Beilstein No: 8^4, 1476
Name: 7H-Benz[de]anthracen-7-one, 2-hydroxy-
Synonym: 1,9-Benzanthr-7-one, 2-hydroxy
Mol Form: $C_{17}H_{10}O_2$
M_r: 246.25
mp: 294
Color: ye nd
m/e: 246.07

HODOC No: 02013
CAS No: 246-02-6
Beilstein No: 5^4, 2280
Name: Benz[a]azulene
Synonym: 1,2-Benzazulene
Structure:

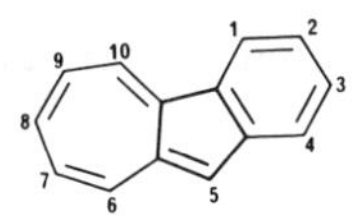

Mol Form: $C_{14}H_{10}$
M_r: 178.22
mp: 176
UV: FO 406 402(2512) 382(3981) 363(3981) 346(2512) 322(25119) 300(63096) 256(10000) 245(10000) cyhex
m/e: 178.08

HODOC No: 02014
CAS No: 62-53-3
Beilstein No: 12^3, 217
Name: Benzenamine
Synonym: Aminobenzene
Line Form: $C_6H_5NH_2$
Structure:

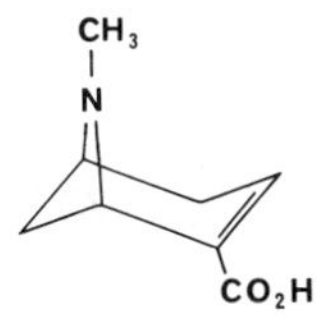

Mol Form: C_6H_7N
M_r: 93.12
bp: 184, 68.3^{10}
mp: -6.3
***d*:** $1.0217^{20/4}$
***n*:** 1.5863^{20}
***s*:** H_2O; 3:al; 5:eth 5; ace 5; bz 5; ctc 3; lig 3
IR: μCOB 4834 3450 3330 1610 1490 1470 1280 1180 1120 1060 1030 1000 880 750 690
UV: SAD 1129 287 235 cyhex
H NMR: SAD 191 3.3 6.4 6.6 7.0 CCl_4
C NMR: STOT 197 116.3 119.2 130.0 147.9
MS: WILEY 192 93(100) 66(33) 65(18) 39(18) 92(11) 46.5(9) 28(8) 38(8)
m/e: 93.06

HODOC No: 02015
CAS No: 632-99-5
Name: Benzenamine,
4-[(4-aminophenyl)(4-imino-2,5-cyclohexadien-1-ylid ene)methyl]-2-methyl-, monohydrochloride
Synonym: C.I. Basic Violet 14, monohydrochloride
Structure:

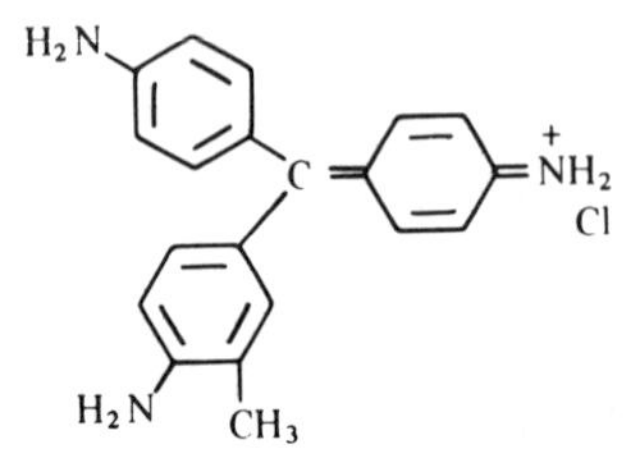

Mol Form: $C_{20}H_{20}ClN_3$
M$_r$: 337.83
mp: dec 200
***s*:** H_2O 2; al 2; eth 1
IR: **μCOB 3770** 3230 3130 2630 1640 1590 1540 1470 1370 1330 1280 1160 910 840 820
m/e: 337.13

HODOC No: 02016
CAS No: 641-29-2
Beilstein No: 13^3, 565
Name: Benzenamine,
4,4′-[(2-aminophenyl)methylene]*bis*-
Synonym: Aniline, 2,4′,4″-methylidynetri-
Line Form: $(4\text{-}H_2NC_6H_4)_2CH(C_6H_4NH_2\text{-}2)$
Structure:

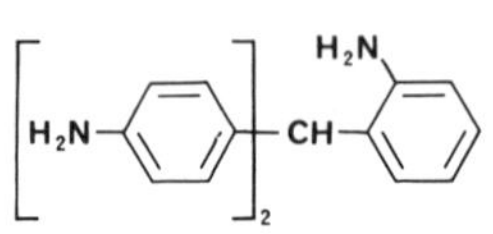

Mol Form: $C_{19}H_{19}N_3$
M$_r$: 289.36
mp: 165
Color: cr (al)
***s*:** EtOH 4
m/e: 289.16

HODOC No: 02017
CAS No: 554-55-2
Beilstein No: 16^2, 148
Name: Benzenamine, 2,2′-azobis-
Synonym: Aniline, 2,2′-azodi-
Line Form: $2\text{-}H_2NC_6H_4N{=}N(C_6H_4NH_2\text{-}2)$
Mol Form: $C_{12}H_{12}N_4$
M$_r$: 212.24
mp: 134
Color: red pl (al or Bz)
***s*:** Et_2O 4
m/e: 212.11

HODOC No: 02018
CAS No: 538-41-0
Beilstein No: 16^3, 375
Name: Benzenamine, 4,4′-azobis-
Synonym: Aniline, 4,4′-azodi-
Line Form: $4\text{-}H_2NC_6H_4N{=}N(C_6H_4NH_2\text{-}4)$
Mol Form: $C_{12}H_{12}N_4$
M$_r$: 212.24
mp: 250-1
Color: gold-ye nd (al), og-ye pr (al)
***s*:** H_2O 2; al 3; bz 4; chl 4; lig 2
IR: **SADG 31432** 3460 3370 3340 3200 1610 1580 1570 1500 1430 1290 1230 1150 1140 1110 1050 940 830 720 640 540 520 450
UV: **OES 6-407** 399(30903) EtOH
m/e: 212.11

HODOC No: 02019
CAS No: 6257-64-3
Beilstein No: 16^2, 174
Name: Benzenamine, 4,4′-azobis[*N*,*N*-dimethyl-
Synonym: Aniline, 4,4′-azobis[*N*,*N*-dimethyl-
Mol Form: $C_{16}H_{20}N_4$
M$_r$: 268.35
mp: 273
***s*:** al 2; eth 3; bz 4; chl 4
UV: **OES 3-604** 505(50119) 309(6918) 1.2N HCl:50% EtOH 485(6310) 405(33113) 319(10965) 95% sulf:EtOH
m/e: 268.17

HODOC No: 02020
CAS No: 603-28-1
Beilstein No: 13^3, 565
Name: Benzenamine, 3-[*bis*(4-aminophenyl)methyl]-
Synonym: Triphenylmethane, 3,4′,4″-triamino
Line Form: $(4\text{-}H_2NC_6H_4)_2CH(C_6H_4NH_2\text{-}3)$
Mol Form: $C_{19}H_{19}N_3$
M_r: 289.36
***s*:** EtOH 4
m/e: 289.16

HODOC No: 02021
CAS No: 24544-04-5
Beilstein No: 12, 168
Name: Benzenamine, 2,6-*bis*(1-methylethyl)-
Synonym: Aniline, 2,6-diisopropyl-
Line Form: $2,6((CH_3)_2CH\text{-})_2C_6H_3NH_2$
Mol Form: $C_{12}H_{19}N$
M_r: 177.28
bp: 257
mp: -45
***d*:** 0.94
***n*:** 1.5332^{20}
IR: ALD 2:641C
MS: ALD 5:61C
m/e: 177.15

HODOC No: 02022
CAS No: 328-74-5
Beilstein No: 12^3, 2498
Name: Benzenamine, 3,5-*bis*(trifluoromethyl)-
Synonym: 3,5-Xylidine, α,α,α,α′,α′,α′-hexafluoro-
Line Form: $3,5\text{-}(CF_3)_2C_6H_3NH_2$
Mol Form: $C_8H_5F_6N$
M_r: 229.12
bp: 85^{15}
***d*:** 1.487
***n*:** 1.4335^{20}
m/e: 229.03

HODOC No: 02023
CAS No: 615-36-1
Beilstein No: 12^3, 1415
Name: Benzenamine, 2-bromo-
Synonym: Aniline, *o*-bromo-
Line Form: $2\text{-}BrC_6H_4NH_2$
Mol Form: C_6H_6BrN
M_r: 171.02
bp: 229, 110.5^{19}
mp: 32
***d*:** $1.578^{20/4}$
***n*:** 1.6113^{20}
***s*:** H_2O 1; al 3:eth 3
IR: μSADP 9779 3450 3330 1610 1490 1450 1300 1250 1160 1140 1060 1040 1010 930 840 740
UV: OES 6-68 300(1585) 237(10000) MeOH
H NMR: SAD 17165 4.0 6.5 6.6 7.0 7.3 $CDCl_3$
m/e: 170.97

HODOC No: 02024
CAS No: 591-19-5
Beilstein No: 12^3, 1674
Name: Benzenamine, 3-bromo-
Synonym: Aniline, *m*-bromo-
Line Form: $3\text{-}BrC_6H_4NH_2$
Mol Form: C_6H_6BrN
M_r: 171.02
bp: 251, 130^{12}
mp: 18.5
***d*:** $1.5793^{20.4/4}$
***n*:** $1.6260^{20.4}$
***s*:** H_2O 2; al 3; eth 3
IR: μCOB 5758 3450 3330 1610 1590 1490 1450 1320 1300 1270 1160 1060 990 870 850 770 680
UV: OES 6-68 300(1585) 247(10000) MeOH
H NMR: SAD 17190 3.6 6.4 6.7 6.8-7.1 $CDCl_3$
m/e: 170.97

HODOC No: 02025
CAS No: 106-40-1
Beilstein No: 12^3, 1429
Name: Benzenamine, 4-bromo-
Synonym: Aniline, *p*-bromo-
Line Form: $4\text{-BrC}_6H_4NH_2$
Mol Form: C_6H_6BrN
M$_r$: 171.02
bp: d
mp: 66.4
***d*:** $1.4970^{100/4}$
Color: rh bipym nd (60% al)
***s*:** H_2O 1; al 3; eth 3; chl 2
IR: **µCOB 6112** 3450 3330 3030 1610 1590 1490 1430 1280 1180 1120 1080 1010 820
UV: **SAD 1829** 297 245 MeOH
H NMR: **SAD 278** 3.5 6.6 7.2 $CDCl_3$
m/e: 170.97

HODOC No: 02026
CAS No: 38762-41-3
Beilstein No: 12, 652
Name: Benzenamine, 4-bromo-2-chloro-
Synonym: Aniline, 4-bromo-2-chloro-
Line Form: $(4\text{-Br-2-Cl})C_6H_3NH_2$
Mol Form: C_6H_5BrClN
M$_r$: 205.46
mp: 70-2
IR: **ALD 2:645C**
MS: **ALD 5:71A**
m/e: 204.93

HODOC No: 02027
CAS No: 697-86-9
Beilstein No: 12^2, 335
Name: Benzenamine, 2-bromo-4,6-dichloro-
Synonym: Aniline, 2-bromo-4,6-dichloro-
Line Form: $(2\text{-Br-4,6-Cl}_2)C_6H_2NH_2$
Mol Form: $C_6H_4BrCl_2N$
M$_r$: 239.91
bp: 273
mp: 83.5
Color: nd (al)
***s*:** Bz 4; EtOH 4; Chl 4
m/e: 238.89

HODOC No: 02028
CAS No: 56978-48-4
Beilstein No: 12, 653
Name: Benzenamine, 4-bromo-2,3-dichloro-
Synonym: 4-Bromo-2,3-dichlorobenzenamine
Line Form: $(4\text{-Br-2,3-Cl}_2)C_6H_2NH_2$
Mol Form: $C_6H_4BrCl_2N$
M$_r$: 239.91
mp: 77.5
Color: nd
***s*:** Bz 4; Et_2O 4; EtOH 4
m/e: 238.89

HODOC No: 02029
CAS No: 1940-27-8
Beilstein No: 12^3, 1471
Name: Benzenamine, 4-bromo-2,5-dichloro-
Synonym: Aniline, 4-bromo-2,5-dichloro
Line Form: $(4\text{-Br-2,5-Cl}_2)C_6H_2NH_2$
Mol Form: $C_6H_4BrCl_2N$
M$_r$: 239.91
mp: 91
Color: nd
***s*:** Et_2O 4; EtOH 4
m/e: 238.89

HODOC No: 02030
CAS No: 53142-19-1
Beilstein No: 12^1, 315
Name: Benzenamine, 3-bromo-*N*,*N*-diethyl-
Synonym: Aniline, *N*,*N*-diethyl-3-bromo
Line Form: $3\text{-BrC}_6H_4N(C_2H_5)_2$
Mol Form: $C_{10}H_{14}BrN$
M$_r$: 227.12
bp: 142^{10}
***s*:** Ac 4; EtOH 4
m/e: 227.03

HODOC No: 02031
CAS No: 2052-06-4
Beilstein No: 12^2, 347
Name: Benzenamine, 4-bromo-*N*,*N*-diethyl-
Synonym: Aniline, *p*-bromo-*N*,*N*-diethyl-
Line Form: 4-Br-C_6H_4-$N(CH_2CH_3)_2$
Mol Form: $C_{10}H_{14}BrN$
Mr: 227.12
bp: 270
mp: 38
Color: nd or pr
***s*:** H_2O 1; al 4; eth 4
IR: µSADP 12473 2940 1590 1490 1350 1270 1210 1160 1080 1010 810 740
UV: SAD 21979 315(2130) 267(19700) MeOH
H NMR: SAD 17180 1.1 3.2 6.5 7.2 $CDCl_3$
m/e: 227.03

HODOC No: 02032
CAS No: 698-00-0
Beilstein No: 12^3, 1416
Name: Benzenamine, 2-bromo-*N*,*N*-dimethyl-
Synonym: Aniline, *o*-bromo-*N*,*N*-dimethyl-
Line Form: 2-BrC_6H_4-$N(CH_3)_2$
Mol Form: $C_8H_{10}BrN$
Mr: 199.07
bp: $107\text{-}8^{14}$
***d*:** $1.3880^{25/25}$
***n*:** 1.5748^{25}
***s*:** al 4
IR: µSADP 12410 2940 2860 1590 1470 1450 1320 1190 1110 1050 1020 940 760 730
UV: OES 1-204 255(5012) 216(15849) 188(39811) hp
m/e: 199.00

HODOC No: 02033
CAS No: 16518-62-0
Beilstein No: 12^3, 1422
Name: Benzenamine, 3-bromo-*N*,*N*-dimethyl-
Synonym: Aniline, *m*-bromo-*N*,*N*-dimethyl-
Line Form: 3-BrC_6H_4-$N(CH_3)_2$
Mol Form: $C_8H_{10}BrN$
Mr: 199.07
bp: 259, $100\text{-}4^2$
mp: 11
***s*:** al 4; HOAc 4; con ac 3
UV: OES 4-172 303(2692) 254(12882) cyhex
m/e: 199.00

HODOC No: 02034
CAS No: 24596-19-8
Beilstein No: 12^2, 605
Name: Benzenamine, 4-bromo-2,6-dimethyl-
Synonym: 2,6-Xylidine, 4-bromo-
Line Form: 4-Br-2,6-$(CH_3)_2$-$C_6H_2NH_2$
Mol Form: $C_8H_{10}BrN$
Mr: 199.07
mp: 50-2
IR: ALD 3:736E
m/e: 199.00

HODOC No: 02035
CAS No: 586-77-6
Beilstein No: 12^2, 345
Name: Benzenamine, 4-bromo-*N*,*N*-dimethyl-
Synonym: Aniline, *p*-bromo-*N*,*N*-dimethyl-
Line Form: 4-BrC_6H_4-$N(CH_3)_2$
Mol Form: $C_8H_{10}BrN$
Mr: 199.07
bp: 264
mp: 52
***s*:** H_2O_1; al 3; eth 4; os 3
IR: µSADP 12442 2860 1610 1490 1450 1350 1240 1190 1160 1090 1060 1000 940 810 760 690
UV: SAD 21978 310(1990) 261(18000) MeOH 262(920) 257(933) 222(9880) MeOH:HCl
H NMR: SAD 17177 2.9 6.5 7.2 $CDCl_3$
m/e: 199.00

HODOC No: 02036
CAS No: 1817-73-8
Beilstein No: 12^3, 1724
Name: Benzenamine, 2-bromo-4,6-dinitro-
Synonym: Aniline, 2-bromo-4,6-dinitro-
Line Form: 2-Br-4,6-$(NO_2)_2$-$C_6H_2NH_2$
Mol Form: $C_6H_4BrN_3O_4$
Mr: 261.01
bp: sub
mp: 153-4
Color: ye nd (al or HOAc)
***s*:** al 4; ace 4; HOAc 3
IR: SADG 9014 3448 3333 3030 1613 1587 1515 1471 1408 1333 1316 1266 1124 1064 935 901 730 699
UV: SAD 14764 336(12000) 268(8700) 229(10100) MeOH
H NMR: SAD 4317 7.7 8.5 8.9 ace
m/e: 260.94

HODOC No: 02037
CAS No: 19056-41-8
Name: Benzenamine, 3-bromo-4-methoxy-
Synonym: *p*-Anisidine, 3-bromo-
Line Form: $(3\text{-}Br\text{-}4\text{-}CH_3O)C_6H_3NH_2$
Mol Form: C_7H_8BrNO
M_r: 201.04
mp: 64
Color: lf
***s*:** Et_2O 4; EtOH 4
m/e: 200.98

HODOC No: 02038
CAS No: 6358-77-6
Beilstein No: 13^3, 858
Name: Benzenamine, 5-bromo-2-methoxy-
Synonym: *o*-Anisidine, 5-bromo-
Line Form: $(5\text{-}Br\text{-}2\text{-}CH_3O)\text{-}C_6H_2NH_2$
Mol Form: C_7H_8BrNO
M_r: 201.04
mp: 97-8
Color: pr
***s*:** Bz 4; Et_2O 4
m/e: 200.98

HODOC No: 02039
CAS No: 583-68-6
Beilstein No: 12^3, 2158
Name: Benzenamine, 2-bromo-4-methyl-
Synonym: *p*-Toluidine, 2-bromo-
Line Form: $2\text{-}Br\text{-}4\text{-}CH_3C_6H_3NH_2$
Mol Form: C_7H_8BrN
M_r: 185.04
bp: 240, $120\text{-}2^{30}$
mp: 26
***d*:** 1.510^{20}
***n*:** 1.5999^{20}
Color: lf
***s*:** H_2O 1; al 3; eth 3
IR: **μCOB 5718** 3570 3450 3030 2940 1610 1590 1540 1520 1300 1270 1210 1150 1080 1030 990 860 810 760 670
UV: **HBCP** 298(2138) H_2O
H NMR: **SAD 17211** 2.2 3.8 6.6 6.8 7.2 $CDCl_3$
m/e: 184.98

HODOC No: 02040
CAS No: 53078-85-6
Beilstein No: 12^2, 474
Name: Benzenamine, 2-bromo-5-methyl-
Synonym: 2-Bromo-5-methylaniline
Line Form: $6\text{-}Br\text{-}3\text{-}CH_3\text{-}C_6H_3NH_2$
Mol Form: C_7H_8BrN
M_r: 185.04
bp: $129\text{-}30^{15}$
mp: 46
***d*:** $1.474^{25/25}$
Color: pr
***s*:** Et_2O 4; EtOH 4
m/e: 184.98

HODOC No: 02041
CAS No: 53848-17-2
Beilstein No: 12^3, 1922
Name: Benzenamine, 2-bromo-6-methyl-
Synonym: 2-Bromo-6-methylaniline
Line Form: $6\text{-}Br\text{-}2\text{-}CH_3C_6H_3NH_2$
Mol Form: C_7H_8BrN
M_r: 185.04
bp: 130^{16}
***s*:** Et_2O 4
IR: **SADP 4720**
UV: **SAD 1301**
m/e: 184.98

HODOC No: 02042
CAS No: 55289-36-6
Beilstein No: 12^1, 390
Name: Benzenamine, 3-bromo-2-methyl-
Synonym: 2-Amino-6-bromotoluene
Line Form: $3\text{-}Br\text{-}2\text{-}CH_3C_6H_3NH_2$
Mol Form: C_7H_8BrN
M_r: 185.04
bp: 253-5
m/e: 184.98

HODOC No: 02043
CAS No: 7745-91-7
Beilstein No: 12^3, 2137
Name: Benzenamine, 3-bromo-4-methyl-
Synonym: *p*-Toluidine, 3-bromo-
Line Form: 3-Br-4-$CH_3C_6H_3NH_2$
Mol Form: C_7H_8BrN
M_r: 185.04
bp: 254-7
mp: 26
***s*:** Et_2O 4
IR: SADP 21023
m/e: 184.98

HODOC No: 02044
CAS No: 583-75-5
Beilstein No: 12^2, 456
Name: Benzenamine, 4-bromo-2-methyl-
Synonym: *o*-Toluidine, 4-bromo-
Line Form: 4-Br-2-$CH_3C_6H_3NH_2$
Mol Form: C_7H_8BrN
M_r: 185.04
bp: 240
mp: 59.5
Color: cr (al)
***s*:** H_2O 2; al 3; eth 4; chl 2; HOAc 4
IR: µCOB 402 3330 2940 1640 1590 1410 1280 1190 1150 1060 1040 870 810 690
UV: SAD 8169 296(1810) 243(10400) MeOH
H NMR: SAD 11050 2.1 3.4 6.5 7.0 7.1 $CDCl_3$
m/e: 184.98

HODOC No: 02045
CAS No: 6933-10-4
Beilstein No: 12^3, 1992
Name: Benzenamine, 4-bromo-3-methyl-
Synonym: *m*-Toluidine, 4-bromo-
Line Form: 4-Br-3-$CH_3C_6H_3NH_2$
Mol Form: C_7H_8BrN
M_r: 185.04
bp: 240
mp: 81
Color: pl_(50% al) cr (al)
***s*:** H_2O 1:al 4; chl 2
IR: SADG 20811 3410 3320 2950 1630 1600 1570 1470 1440 1370 1300 1250 1010 860 810 730 670
UV: SAD 18783 296(1570) 244(12400) MeOH
H NMR: SAD 11357 2.2 3.5 6.3 6.5 7.2 $CDCl_3$
m/e: 184.98

HODOC No: 02046
CAS No: 39478-78-9
Beilstein No: 12^3, 1923
Name: Benzenamine, 5-bromo-2-methyl-
Synonym: 2-Amino-4-bromotoluene
Line Form: 5-Br-2-$CH_3C_6H_3NH_2$
Mol Form: C_7H_8BrN
M_r: 185.04
bp: 253-7d, 139^{17}
mp: 33
Color: lf
***s*:** al 3; eth 4; ac 3
IR: HBCP
UV: HBCP 293(1738) H_2O
m/e: 184.98

HODOC No: 02047
CAS No: 13296-94-1
Beilstein No: 12^2, 403
Name: Benzenamine, 2-bromo-4-nitro-
Synonym: Aniline, 2-bromo-4-nitro-
Line Form: (2-Br-4-NO_2)$C_6H_3NH_2$
Mol Form: $C_6H_5BrN_2O_2$
M_r: 216.02
mp: 104-5
Color: ye nd (al)
***s*:** al 3; DMSOd_6 2; HOAc 4
IR: SADG 1449 3448 3333 3030 1613 1587 1471 1316 1250 1111 885 813 735 694 633
UV: SAD 11540 360(14200) 205(15800) MeOH
H NMR: SAD 10028 6.7 6.9 8.0 8.2 DMSO
m/e: 215.95

HODOC No: 02048
CAS No: 10403-47-1
Beilstein No: 12^3, 1674
Name: Benzenamine, 2-bromo-5-nitro-
Synonym: Aniline, 2-bromo-5-nitro-
Line Form: (2-Br-5-NO_2)$C_6H_3NH_2$
Mol Form: $C_6H_5BrN_2O_2$
M_r: 216.02
mp: 141
Color: pa ye nd (al)
***s*:** al 3; eth 4; bz 4; chl 4; lig 2
IR: µSADP 1517 3450 3330 1640 1590 1540 1470 1430 1350 1030 940 870 820 800 740
UV: OES 6-62 370(2291) 285(5495) 254(13183) 229(13804) EtOH
m/e: 215.95

HODOC No: 02049
CAS No: 59255-95-7
Beilstein No: 12^2, 402
Name: Benzenamine, 2-bromo-6-nitro-
Synonym: Aniline, 2-bromo-6-nitro-
Line Form: (2-Br-6-NO_2)$C_6H_3NH_2$
Mol Form: $C_6H_5BrN_2O_2$
M_r: 216.02
mp: 74.5
***d*:** 1.988
Color: og or ye nd (dil al)
***s*:** EtOH 4
m/e: 215.95

HODOC No: 02050
CAS No: 40787-96-0
Beilstein No: 12^3, 1676
Name: Benzenamine, 3-bromo-4-nitro-
Synonym: Aniline, 3-bromo-4-nitro-
Line Form: 3-Br-4-(NO_2)$C_6H_3NH_2$
Mol Form: $C_6H_5BrN_2O_2$
M_r: 216.02
mp: 175-6
Color: ye nd (al)
***s*:** EtOH 4
m/e: 215.95

HODOC No: 02051
CAS No: 875-51-4
Beilstein No: 12^3, 1670
Name: Benzenamine, 4-bromo-2-nitro-
Synonym: Aniline, 4-bromo-2-nitro-
Line Form: (4-Br-2-NO_2)-$C_6H_3NH_2$
Mol Form: $C_6H_5BrN_2O_2$
M_r: 216.02
bp: sub
mp: 111.5
Color: og-ye nd (w)
***s*:** EtOH 4
m/e: 215.95

HODOC No: 02052
CAS No: 53324-38-2
Beilstein No: 12^3, 1674
Name: Benzenamine, 4-bromo-3-nitro-
Synonym: Aniline, 4-bromo-3-nitro-
Line Form: (4-Br-3-NO_2)$C_6H_3NH_2$
Mol Form: $C_6H_5BrN_2O_2$
M_r: 216.02
mp: 132
Color: nd (al)
***s*:** Et_2O 4; EtOH 4; Chl 4; HOAc 4
m/e: 215.95

HODOC No: 02053
CAS No: 5228-61-5
Beilstein No: 12^3, 1671
Name: Benzenamine, 5-bromo-2-nitro-
Synonym: Aniline, 5-bromo-2-nitro-
Line Form: (3-Br-6-NO_2)$C_6H_3NH_2$
Mol Form: $C_6H_5BrN_2O_2$
M_r: 216.02
mp: 151-2
Color: red-ye nd (dil al)
***s*:** EtOH 4
m/e: 215.95

HODOC No: 02054
CAS No: 780-20-1
Beilstein No: 12^2, 347
Name: Benzenamine, 4-bromo-*N*-(phenylmethylene)-
Synonym: Aniline, *N*-benzylidene-*p*-bromo-
Line Form: C_6H_5CH=N(C_6H_4Br-4)
Mol Form: $C_{13}H_{10}BrN$
M_r: 259.12
mp: 67
IR: SADG 760 1630 1580 1480 1190 1165 1070 1005 1075 880 820 755 690 670 545
UV: LANG 36 309(12023) 263(19953) 220(15849) EtOH
m/e: 259.00

HODOC No: 02055
CAS No: 5877-51-0
Name: Benzenamine, *N*-[(4-bromophenyl)methylene]-
Synonym: Aniline, *N*-(*p*-bromobenzylidene)-
Line Form: 4-BrC_6H_4CH=NC_6H_5
Mol Form: $C_{13}H_{10}BrN$
M_r: 259.12
mp: 74
UV: LANG 586 315(10715) 273(20893) EtOH
m/e: 259.00

HODOC No: 02056
CAS No: 30293-62-0
Name: Benzenamine, 4-bromo-*N*-(3-phenyl-2-propenyl)-
Synonym: Cinnamylamine, *N*-(*p*-bromophenyl)-
Mol Form: $C_{15}H_{14}BrN$
M_r: 287.17
***s*:** chl 3
IR: SADG 16611 3400 3020 2820 1580 1480 1430 1380 1290 1220 1160 1100 1050 950 800 730 680 490
H NMR: SAD 8488 3.6 3.8 6.2-7.8 6.5 7.2 7.3 $CDCl_3$
m/e: 287.03

HODOC No: 02057
CAS No: 17708-39-3
Beilstein No: 12^1, 318
Name: Benzenamine, 4-bromo-*N*-(3-phenyl-2-propenylidene)-
Synonym: Aniline, *p*-bromo-*N*-cinnamylidene-
Mol Form: $C_{15}H_{12}BrN$
M_r: 285.16
mp: 120
***s*:** chl 3
IR: SADG 16600 3020 1610 1590 1560 1470 1440 1140 1060 970 940 810 740 680 490
UV: SAD 16915 306(13500) 228(6030) MeOH 282(956) 251(12000) 224(8500) MeOH:HCl
H NMR: SAD 8483 6.9-7.6 8.1 $CDCl_3$
m/e: 285.02

HODOC No: 02058
CAS No: 4344-55-2
Beilstein No: 13^3, 998
Name: Benzenamine, 4-butoxy-
Synonym: Aniline, *p*-butoxy-
Line Form: $4\text{-}C_4H_9O\text{-}C_6H_4NH_2$
Mol Form: $C_{10}H_{15}NO$
M_r: 165.23
bp: 132^4
m/e: 165.12

HODOC No: 02059
CAS No: 2696-85-7
Beilstein No: 12^3, 2714
Name: Benzenamine, 2-butyl-
Synonym: Aniline, *o*-butyl-
Line Form: $2\text{-}CH_3(CH_2)_3\text{-}C_6H_4NH_2$
Mol Form: $C_{10}H_{15}N$
M_r: 149.23
bp: $122\text{-}5^{12}$
***d*:** $0.953^{20/4}$
Color: ye oil
m/e: 149.12

HODOC No: 02060
CAS No: 5369-17-5
Beilstein No: 12^2, 633
Name: Benzenamine, 3-butyl-
Synonym: Aniline, *m*-butyl-
Line Form: $3\text{-}H_2NC_6H_4(CH_2)_3CH_3$
Mol Form: $C_{10}H_{15}N$
M_r: 149.23
bp: $125\text{-}8^9$
Color: pa ye oil
m/e: 149.12

HODOC No: 02061
CAS No: 104-13-2
Beilstein No: 12^3, 2715
Name: Benzenamine, 4-butyl-
Synonym: Aniline, *p*-butyl-
Line Form: $4\text{-}H_2N\text{-}C_6H_4\text{-}CH_2CH_2CH_2CH_3$
Mol Form: $C_{10}H_{15}N$
M_r: 149.23
bp: 261, $133\text{-}4^{14}$
***d*:** $0.945^{20/4}$
Color: pa ye
***s*:** ctc 2
IR: SADG 16695 3450 3360 3200 3000 2920 2840 1620 1510 1460 1370 1270 1170 1120 820 680
H NMR: SAD 8604 0.9 1.1-1.8 2.5 3.3 6.4 6.9 CCl_4
m/e: 149.12

HODOC No: 02062
CAS No: 1126-78-9
Beilstein No: 12^3, 247
Name: Benzenamine, *N*-butyl-
Synonym: Aniline, *N*-butyl-
Line Form: $C_6H_5NHC_4H_9$
Mol Form: $C_{10}H_{15}N$
M_r: 149.23
bp: 241-6, $118\text{-}20^{15}$
mp: -14.4
***d*:** $0.9323^{20/4}$
***n*:** 1.5341^{20}
***s*:** Et_2O 4; EtOH 4
IR: SADG 532
UV: SAD 1836
H NMR: SAD 280
MS: Al 103
m/e: 149.12

HODOC No: 02063
CAS No: 939-61-7
Name: Benzenamine, *N*-butyl-2-chloro-
Synonym: Aniline, *N*-butyl-*o*-chloro-
Mol Form: $C_{10}H_{14}ClN$
M_r: 183.67
***s*:** ctc 3; CS_2 3
IR: µCOB 4050 3450 3030 2940 2860 1610 1520 1470 1430 1330 1300 1160 1150 1120 1040 740
m/e: 183.08

HODOC No: 02064
CAS No: 492-80-8
Beilstein No: 14^3, 227
Name: Benzenamine, 4,4′-carbonimidoylbis[*N*,*N*-dimethyl-
Synonym: Auramine base
Line Form: $[(CH_3)_2N\text{-}C_6H_4]_2C{=}NH$
Structure:

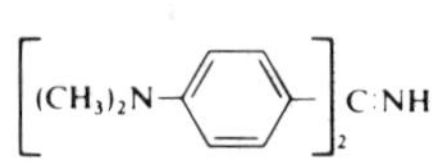

Mol Form: $C_{17}H_{21}N_3$
M_r: 267.36
mp: 136
Color: ye or col pl (al)
***s*:** H_2O 1; al 3; eth 2
UV: HBCP 441(5495) 372(3467) 270(3020) aa
m/e: 267.17

HODOC No: 02065
CAS No: 95-51-2
Beilstein No: 12^2, 314
Name: Benzenamine, 2-chloro-
Synonym: Aniline, *o*-chloro-
Mol Form: C_6H_6ClN
M_r: 127.57
bp: 208.84
mp: -14
***d*:** $1.21253^{20/4}$
***n*:** 1.5895^{20}
***s*:** H_2O 1; al 5; eth 3; ace 3; os 3
IR: µCOB 6960 3450 3330 1610 1490 1450 1320 1270 1160 1090 1050 1020 940 830 750 680
Raman: SAD 681 3380 3200 3150 3060 3030 1610 1590 1490 1450 1420 1310 1260 1160 1150 1090 1050 1030 880 840 750 680 560 480 380 260 170
UV: SAD 254 290 236 cyhex
H NMR: SAD 59 3.8 6.4-7.3 CCl_4
MS: ALDN 69 ini 127(100) 65(37) 129(34) 92(26) 28(24) 64(20) 63(17) 39(17)
m/e: 127.02

HODOC No: 02066
CAS No: 108-42-9
Beilstein No: 12^3, 1303
Name: Benzenamine, 3-chloro-
Synonym: Aniline, *m*-chloro-
Line Form: $3\text{-}ClC_6H_4NH_2$
Mol Form: C_6H_6ClN
M_r: 127.57
bp: 230, 101.2^{10}
mp: -10.3
***d*:** $1.2161^{20/4}$
***n*:** 1.5941^{20}
***s*:** H_2O 1; al 5; eth 5; ace 5; bz 5; ctc 5; chl 3; os 3
IR: µCOB 5761 3450 3330 3030 1610 1590 1490 1450 1320 1300 1270 1160 1080 990 890 850 760 680
Raman: SAD 682 3370 3060 1620 1590 1480 1440 1370 1300 1270 1200 1160 1080 990 880 770 690 530 430 400 240 190
UV: SAD 376 294 242 MeOH
H NMR: SAD 88 3.6 6.4-7.1 $CDCl_3$
MS: WILEY 547 127(100) 129(32) 65(25) 51(22) 92(19) 28(16) 100(11) 39(9)
m/e: 127.02

HODOC No: 02067
CAS No: 141-85-5
Beilstein No: 12^2, 320
Name: Benzenamine, 3-chloro-, hydrochloride
Synonym: Aniline, *m*-chloro-, hydrochloride
Mol Form: $C_6H_7Cl_2N$
Mr: 164.03
mp: 221.5-2.5
s: H_2O 4; al 4
IR: µSADP 16545 3450 3330 2500 1560 1470 1430 1300 1280 1080 1000 980 860 780 670
UV: SAD 5193 293 272 265 241 MeOH
m/e: 163.00

HODOC No: 02068
CAS No: 106-47-8
Beilstein No: 12^3, 1325
Name: Benzenamine, 4-chloro-
Synonym: Aniline, *p*-chloro-
Line Form: $4\text{-}ClC_6H_4NH_2$
Mol Form: C_6H_6ClN
Mr: 127.57
bp: 232
mp: 72.5
d: $1.429^{19/4}$
n: 1.5546^{87}
Color: rh pr
s: H_2O 3; al 3; eth 3; chl 3; os 3
IR: µCOB 5546 3570 3450 3030 1610 1490 1430 1280 1180 1120 1090 1000 810 650
Raman: SAD 239 3390 3060 1620 1600 1490 1320 1290 1190 1100 1010 840 820 700 640 400 380 330 130
UV: SAD 393 296 243 MeOH
H NMR: VAR 123 3.6 6.6 7.1 $CDCl_3$
MS: NBS 234
m/e: 127.02

HODOC No: 02069
CAS No: 33663-37-5
Name: Benzenamine, 4-chloro-, hydrobromide
Synonym: Aniline, *p*-chloro-, hydrobromide
Mol Form: C_6H_7BrClN
Mr: 207.48
IR: COB 4350 2900 2570 2530 1600 1550 1490 1450 1360 1090 1000 810
m/e: 206.95

HODOC No: 02070
CAS No: 779-76-0
Beilstein No: 12^3, 1284
Name: Benzenamine, 2-chloro-*N*-(2-chlorophenyl)-
Synonym: Diphenylamine, 2,2′-dichloro-
Line Form: $(2\text{-}ClC_6H_4)_2NH$
Mol Form: $C_{12}H_9Cl_2N$
Mr: 238.10
bp: $200\text{-}2^{27}$
mp: 30-1
m/e: 237.01

HODOC No: 02071
CAS No: 15979-79-0
Beilstein No: 12^3, 1330
Name: Benzenamine, 3-chloro-*N*-(4-chlorophenyl)-
Synonym: Diphenylamine, 3,4′-dichloro-
Line Form: $3\text{-}ClC_6H_4NH(C_6H_4Cl\text{-}4)$
Mol Form: $C_{12}H_9Cl_2N$
Mr: 238.10
bp: 235^{25}
mp: 63-4
Color: pr(al)
m/e: 237.01

HODOC No: 02072
CAS No: 2873-89-4
Beilstein No: 12^3, 1329
Name: Benzenamine, 4-chloro-*N*,*N*-diethyl-
Synonym: Aniline, *p*-chloro-*N*,*N*-diethyl-
Line Form: $4\text{-}Cl\text{-}C_6H_4\text{-}N(C_2H_5)_2$
Mol Form: $C_{10}H_{14}ClN$
Mr: 183.67
bp: 251-3, $95\text{-}6^{1.5}$
mp: 45-6
Color: nd (al)
s: al 3
IR: µSADP 9752 2940 1850 1670 1610 1490 1350 1270 1190 1150 1100 1080 1010 920 800 740 690
m/e: 183.08

HODOC No: 02073
CAS No: 6358-64-1
Name: Benzenamine, 4-chloro-2,5-dimethoxy-
Synonym: Aniline, 4-chloro-2,5-dimethoxy-
Mol Form: $C_8H_{10}ClNO_2$
Mr: 187.62
IR: COB 7156 3440 3300 2940 2840 1620 1590 1515 1410 1310 1270 1210 1065 1030 840 815 730 660
UV: SAD 6146 241 MeOH
H NMR: SAD 2129 4.0 7.2 7.3 tfa
m/e: 187.04

HODOC No: 02074
CAS No: 97-50-7
Beilstein No: 13, 786
Name: Benzenamine, 5-chloro-2,4-dimethoxy-
Synonym: Aniline, 5-chloro-2,4-dimethoxy-
Line Form: $[4,6\text{-}(CH_3O)_2\text{-}3\text{-}Cl]C_6H_2NH_2$
Mol Form: $C_8H_{10}ClNO_2$
M$_r$: 187.62
mp: 90-2
IR: ALD 2:649E
MS: ALD 5:78D
m/e: 187.04

HODOC No: 02075
CAS No: 698-01-1
Beilstein No: 12^2, 316
Name: Benzenamine, 2-chloro-*N,N*-dimethyl-
Synonym: Aniline, *o*-chloro-*N,N*-dimethyl-
Line Form: $2\text{-}ClC_6H_4\text{-}N(CH_3)_2$
Mol Form: $C_8H_{10}ClN$
M$_r$: 155.62
bp: 207-8
***d*:** $1.1067^{20/4}$
***n*:** 1.5528^{20}
***s*:** ace 3; bz 4
UV: OES 1-205 250(5012) EtOH 256(7943) 211(19953) 186(39811) hp
m/e: 155.05

HODOC No: 02076
CAS No: 6848-13-1
Beilstein No: 12^3, 1304
Name: Benzenamine, 3-chloro-*N,N*-dimethyl-
Synonym: Aniline, *m*-chloro-*N,N*-dimethyl-
Line Form: $3\text{-}ClC_6H_4\text{-}N(CH_3)_2$
Mol Form: $C_8H_{10}ClN$
M$_r$: 155.62
bp: 232, 90^2
***s*:** H_2O 2; al 3; ace 3; bz 3
IR: SADG 947 3080 2900 2810 1600 1560 1500 1450 1350 1230 1170 1100 990 960 830 810 760 680
UV: SAD 8552 305(2500) 258(16200) 210(25200) MeOH
H NMR: SAD 2600 2.9 6.6 7.1 CCl_4
m/e: 155.05

HODOC No: 02077
CAS No: 698-69-1
Beilstein No: 12^3, 1329
Name: Benzenamine, 4-chloro-*N,N*-dimethyl-
Synonym: Aniline, *p*-chloro-*N,N*-dimethyl-
Line Form: $4\text{-}Cl\text{-}C_6H_4N(CH_3)_2$
Mol Form: $C_8H_{10}ClN$
M$_r$: 155.62
bp: 231
mp: 35.5
Color: nd (al)
***s*:** al 3
UV: OES 4-172 308(1995) 257(14791) cyhex
m/e: 155.05

HODOC No: 02078
CAS No: 5388-62-5
Beilstein No: 12^3, 1720
Name: Benzenamine, 4-chloro-2,6-dinitro-
Synonym: Aniline, 4-chloro-2,6-dinitro-
Line Form: $4\text{-}Cl\text{-}2,6\text{-}(NO_2)_2\text{-}C_6H_2NH_2$
Mol Form: $C_6H_4ClN_3O_4$
M$_r$: 217.56
mp: 147
Color: og-ye nd (al)
***s*:** al 3
IR: µSADP 21003 3450 3330 3030 1640 1590 1540 1450 1410 1350 1270 1240 900 770 740 730
UV: SAD 8154 438(8190) 318(744) 229(26000) MeOH
m/e: 216.99

HODOC No: 02079
CAS No: 4316-56-7
Name: Benzenamine, 4-chloro-*N,N*-diphenyl-
Synonym: Triphenylamine, 4-chloro-
Mol Form: $C_{18}H_{14}ClN$
M$_r$: 279.75
mp: 108-9
UV: LANG 1147 298(23988) 207(40738) EtOH
m/e: 279.08

HODOC No: 02080
CAS No: 15793-48-3
Name: Benzenamine, 5-chloro-2-ethoxy-
Synonym: *o*-Phenetidine, 5-chloro-
Mol Form: $C_8H_{10}ClNO$
M$_r$: 171.62
IR: µCOB 32 3450 3330 2940 1610 1520 1430 1390 1270 1250 1140 1110 1040 930 890 850 790
m/e: 171.05

HODOC No: 02081
CAS No: 13519-74-9
Beilstein No: 12^3, 1284
Name: Benzenamine, 2-chloro-*N*-ethyl-
Synonym: Aniline, *o*-chloro-*N*-ethyl-
Line Form: 2-ClC_6H_4NH-CH_2CH_3
Mol Form: $C_8H_{10}ClN$
M_r: 155.62
bp: $219.^{726}$
***d*:** 1.104^{25}
MS: NBS 532
m/e: 155.05

HODOC No: 02082
CAS No: 15258-44-3
Beilstein No: 12, 604
Name: Benzenamine, 3-chloro-*N*-ethyl-
Synonym: Aniline, *m*-chloro-*N*-ethyl-
Line Form: 3-ClC_6H_4-$NHCH_2CH_3$
Mol Form: $C_8H_{10}ClN$
M_r: 155.62
bp: 243-4
***d*:** 1.12^{19}
m/e: 155.05

HODOC No: 02083
CAS No: 18505-77-6
Name: Benzenamine, *N*-(2-chloroethyl)-2-nitro-
Synonym: Aniline, *N*-(2-chloroethyl)-*o*-nitro-
Mol Form: $C_8H_9ClN_2O_2$
M_r: 200.62
mp: 59
UV: LANG 2473 415(5888) 282(4467) 232(19953) EtOH
m/e: 200.04

HODOC No: 02084
CAS No: 367-21-5
Beilstein No: 12^3, 1389
Name: Benzenamine, 3-chloro-4-fluoro-
Synonym: Aniline, 3-chloro-4-fluoro-
Line Form: (4-F-3-Cl)$C_6H_3NH_2$
Mol Form: C_6H_5ClFN
M_r: 145.56
mp: 45-7
m/e: 145.01

HODOC No: 02085
CAS No: 10468-17-4
Beilstein No: 15^2, 8
Name: Benzenamine, 3-chloro-*N*-hydroxy-
Synonym: Hydroxylamine, *N*-(*m*-chlorophenyl)-
Line Form: 3-(HONH)-C_6H_4Cl
Mol Form: C_6H_6ClNO
M_r: 143.57
mp: 49
Color: pl (Bz)
***s*:** Bz 4; peth 4
m/e: 143.01

HODOC No: 02086
CAS No: 823-86-9
Beilstein No: 15^2, 9
Name: Benzenamine, 4-chloro-*N*-hydroxy-
Synonym: Hydroxylamine, *N*-(*p*-chlorophenyl)-
Line Form: 4-(HONH)C_6H_4Cl
Mol Form: C_6H_6ClNO
M_r: 143.57
mp: 87-8
Color: lf (dil al)
***s*:** Bz 4; EtOH 4; Chl 4
m/e: 143.01

HODOC No: 02087
CAS No: 5606-39-3
Name: Benzenamine, 4-chloro-2-(iminophenylmethyl)-
Synonym: Aniline, 2-benzimidoyl-4-chloro-
Mol Form: $C_{13}H_{11}ClN_2$
M_r: 230.68
mp: 74
***s*:** chl 3
IR: SADG 17766 3440 3260 1610 1580 1570 1530 1470 1440 1400 1370 1320 1240 1200 1150 1025 910 895 815 780 755 700 660 640 565 530
UV: SAD 17023 361(3430) 220(25900) MeOH 440(2690) 274(15200) 248(10400) MeOH:HCl
H NMR: SAD 9485 6.3 6.5 7.0 7.3 9.4 $CDCl_3$
m/e: 230.06

HODOC No: 02088
CAS No: 2401-24-3
Name: Benzenamine, 2-chloro-5-methoxy-
Synonym: *m*-Anisidine, 6-chloro-
Line Form: (6-Cl-3-CH_3O)-$C_6H_3NH_2$
Mol Form: C_7H_8ClNO
M_r: 157.59
bp: 140-5^{18}, 136-7^{10}
mp: 27
***n*:** 1.5848$^{25.5}$
m/e: 157.03

HODOC No: 02089
CAS No: 51114-68-2
Beilstein No: 13^3, 854
Name: Benzenamine, 3-chloro-2-methoxy-
Synonym: 1-Amino-3-chloro-2-methoxybenzene
Line Form: (3-Cl-2-CH_3O)$C_6H_3NH_2$
Mol Form: C_7H_8ClNO
M_r: 157.59
bp: 246
***n*:** 1.5742^{20}
IR: SADG 16747 3460 3380 3060 3000 2940 2840 1610 1570 1480 1450 1310 1290 1220 1040 990 900 760 730 630
UV: SAD 16697 290(2150) 241(7360) MeOH 274(658) 268(672) 212(7970) MeOH:HCl
H NMR: SAD 8615 3.8 3.9 6.4-6.8 CCl_4
m/e: 157.03

HODOC No: 02090
CAS No: 93-50-5
Beilstein No: 13^2, 184
Name: Benzenamine, 4-chloro-2-methoxy-
Synonym: *o*-Anisidine, 4-chloro-
Line Form: (4-Cl-2-CH_3O)$C_6H_3NH_2$
Mol Form: C_7H_8ClNO
M_r: 157.59
bp: 260
mp: 52
Color: nd or pr (dil al)
***s*:** al 3; eth 3; bz 3; chl 3
IR: μSADP 8926 3450 3330 2940 1610 1590 1490 1450 1270 1220 1180 1140 1100 1020 890 860 850 790 790 760
UV: SAD 2291 294 243 MeOH
H NMR: SAD 11220 3.7 3.8 6.6 $CDCl_3$
m/e: 157.03

HODOC No: 02091
CAS No: 95-03-4
Beilstein No: 13^2, 183
Name: Benzenamine, 5-chloro-2-methoxy-
Synonym: *o*-Anisidine, 5-chloro-
Line Form: (3-Cl-6-CH_3O)$C_6H_3NH_2$
Mol Form: C_7H_8ClNO
M_r: 157.59
mp: 84
Color: nd (dil al)
***s*:** al 3; lig 2
IR: SADG 31039 3460 3370 3020 2970 2930 2840 1690 1610 1580 1500 1460 1440 1420 1280 1240 1220 1180 1130 1100 1080 1020 900 850 840 790 640 590 440
UV: SAD 2291 294 243 MeOH
m/e: 157.03

HODOC No: 02092
CAS No: 615-65-6
Beilstein No: 12^3, 2152
Name: Benzenamine, 2-chloro-4-methyl-
Synonym: *p*-Toluidine, 2-chloro-
Line Form: 2-Cl-4-$CH_3C_6H_3NH_2$
Mol Form: C_7H_8ClN
M_r: 141.59
bp: 219^{732}
mp: 7
***d*:** 1.151^{20}
***n*:** 1.5748^{22}
***s*:** al 2; bz 2
IR: μSADP 18467 3450 3330 3030 2940 1640 1520 1470 1300 1270 1220 1160 1050 1000 890 880 810 710 690
UV: SAD 5910 296 237 MeOH
MS: NBS 361
m/e: 141.03

HODOC No: 02093
CAS No: 95-81-8
Beilstein No: 12^3, 1991
Name: Benzenamine, 2-chloro-5-methyl-
Synonym: *m*-Toluidine, 6-chloro-
Line Form: 6-Cl-3-$CH_3C_6H_3NH_2$
Mol Form: C_7H_8ClN
M_r: 141.59
bp: 228-30
mp: 29-30
Color: pl
***s*:** EtOH 4
m/e: 141.03

HODOC No: 02094
CAS No: 932-32-1
Beilstein No: 12^1, 298
Name: Benzenamine, 2-chloro-*N*-methyl-
Synonym: Aniline, *o*-chloro-*N*-methyl-
Line Form: $2\text{-ClC}_6H_4\text{-NHCH}_3$
Mol Form: C_7H_8ClN
M_r: 141.59
bp: 218, $95\text{-}6^{12}$
***d*:** $1.1735^{11.5}$
***n*:** 1.5780^{25}
***s*:** al 3; ace 3; bz 3
UV: OES 1-146 300(2512) 246(10000) EtOH
m/e: 141.03

HODOC No: 02095
CAS No: 87-60-5
Beilstein No: 12^3, 1919
Name: Benzenamine, 3-chloro-2-methyl-
Synonym: *o*-Toluidine, 3-chloro-
Line Form: $3\text{-Cl-2-CH}_3C_6H_3NH_2$
Mol Form: C_7H_8ClN
M_r: 141.59
bp: 245, $96\text{-}9^{10}$
mp: 0-2
***n*:** 1.5880^{20}
***s*:** H_2O 3; al 3; eth 1; bz 1
IR: μSADP 4720 3450 3230 3030 2940 1610 1560 1450 1370 1300 1210 1150 1100 1010 860 760 700
UV: SAD 1301 290 241 207 MeOH
H NMR: SAD 6400 2.2 3.6 6.5 6.8 6.9 $CDCl_3$
m/e: 141.03

HODOC No: 02096
CAS No: 95-74-9
Beilstein No: 12^3, 2151
Name: Benzenamine, 3-chloro-4-methyl-
Synonym: *p*-Toluidine, 3-chloro-
Line Form: $3\text{-Cl-4-CH}_3\text{-C}_6H_3NH_2$
Mol Form: C_7H_8ClN
M_r: 141.59
bp: 242-4, $112\text{-}3^3$
mp: 26
***s*:** al 3; ctc 2
IR: SADG 8699 3440 3360 3220 3020 2920 1630 1570 1500 1450 1420 1290 1250 1200 1040 990 900 850 800 690
UV: SAD 4325 296 237 cyhex
H NMR: SAD 3266 2.2 3.4 6.3 6.5 6.9 CCl_4
m/e: 141.03

HODOC No: 02097
CAS No: 7006-52-2
Beilstein No: 12^2, 301
Name: Benzenamine, 3-chloro-*N*-methyl-
Synonym: Aniline, *m*-chloro-*N*-methyl-
Line Form: $3\text{-ClC}_6H_4\text{-NHCH}_3$
Mol Form: C_7H_8ClN
M_r: 141.59
bp: 235
***d*:** $1.174^{11.5}$
m/e: 141.03

HODOC No: 02098
CAS No: 95-69-2
Beilstein No: 12^3, 1914
Name: Benzenamine, 4-chloro-2-methyl-
Synonym: *o*-Toluidine, 4-chloro-
Line Form: $4\text{-Cl-2-CH}_3C_6H_3NH_2$
Mol Form: C_7H_8ClN
M_r: 141.59
bp: 241^{730}
mp: 29-30
Color: lf (al)
***s*:** al 3; ctc 2
IR: SADG 669 3460 3380 3220 2970 2930 1620 1490 1450 1400 1280 1200 1150 1090 1000 890 810 650
UV: SAD 5411 296(2550) 242(13300) MeOH
H NMR: SAD 558 2.1 3.5 6.5 7.0 CCl_4
m/e: 141.03

HODOC No: 02099
CAS No: 7149-75-9
Beilstein No: 12^3, 1998
Name: Benzenamine, 4-chloro-3-methyl-
Synonym: *m*-Toluidine, 4-chloro-
Line Form: $4\text{-Cl-3-CH}_3C_6H_3NH_2$
Mol Form: C_7H_8ClN
M_r: 141.59
bp: 241
mp: 83-4
Color: nd (peth)
***s*:** Ac 4; Bz 4; EtOH 4
m/e: 141.03

HODOC No: 02100
CAS No: 932-96-7
Beilstein No: 12^3, 1329
Name: Benzenamine, 4-chloro-*N*-methyl-
Synonym: Aniline, *p*-chloro-*N*-methyl-
Line Form: $4\text{-}ClC_6H_4\text{-}NHCH_3$
Mol Form: C_7H_8ClN
M_r: 141.59
bp: 240
***d*:** $1.169^{11.5}$
***n*:** 1.5835^{20}
***s*:** al 3; ace 3; bz 3
UV: OES 1-146 295(1514) 245(10715) H_2O
m/e: 141.03

HODOC No: 02101
CAS No: 95-79-4
Beilstein No: 12^2, 453
Name: Benzenamine, 5-chloro-2-methyl-
Synonym: *o*-Toluidine, 5-chloro-
Mol Form: C_7H_8ClN
M_r: 141.59
bp: 237^{722} 140^{38}
mp: 26
***s*:** al 3
IR: µCOB 2315 3570 3450 3030 2940 1610 1590 1490 1450 1430 1300 1270 1210 1140 1100 1060 1030 990 900 840 790 650
UV: SAD 1302 292(2580) 240(7820) 206 MeOH
MS: ALDN 90 141(100) 106(91) 140(35) 143(33) 77(26) 28(16) 52(12) 79(11)
m/e: 141.03

HODOC No: 02102
CAS No: 6259-42-3
Name: Benzenamine, 5-chloro-2-methyl-, hydrochloride
Synonym: *o*-Toluidine, 5-chloro-, hydrochloride
Mol Form: $C_7H_9Cl_2N$
M_r: 178.05
mp: 265-7
IR: µCOB 1536 2940 2560 1640 1520 1490 1370 1180 890 870 810
m/e: 177.01

HODOC No: 02103
CAS No: 3970-41-0
Beilstein No: 12^3, 1659
Name: Benzenamine, 2-chloro-3-nitro-
Synonym: Aniline, 2-chloro-3-nitro-
Line Form: $(2\text{-}Cl\text{-}3\text{-}NO_2)C_6H_3NH_2$
Mol Form: $C_6H_5ClN_2O_2$
M_r: 172.56
mp: 95-6
m/e: 172.00

HODOC No: 02104
CAS No: 121-87-9
Beilstein No: 12^3, 1667
Name: Benzenamine, 2-chloro-4-nitro-
Synonym: Aniline, 2-chloro-4-nitro-
Line Form: $(2\text{-}Cl\text{-}4\text{-}NO_2)C_6H_3NH_2$
Mol Form: $C_6H_5ClN_2O_2$
M_r: 172.56
mp: 108
Color: ye nd (w)
***s*:** Et_2O 4; EtOH 4; HOAc 4
IR: SADG 8634
UV: SAD 3799
H NMR: SAD 3257
m/e: 172.00

HODOC No: 02105
CAS No: 6283-25-6
Beilstein No: 12^3, 1661
Name: Benzenamine, 2-chloro-5-nitro-
Synonym: Aniline, 2-chloro-5-nitro-
Line Form: $(2\text{-}Cl\text{-}5\text{-}NO_2)C_6H_3NH_2$
Mol Form: $C_6H_5ClN_2O_2$
M_r: 172.56
mp: 121
Color: ye nd (lig)
***s*:** Ac 4; Et_2O 4; EtOH 4; HOAc 4
IR: ALD 2:723F
MS: ALD 5:172D
m/e: 172.00

HODOC No: 02106
CAS No: 769-11-9
Beilstein No: 12^3, 1658
Name: Benzenamine, 2-chloro-6-nitro-
Synonym: Aniline, 2-chloro-6-nitro-
Line Form: $(2\text{-Cl-6-}NO_2)C_6H_3NH_2$
Mol Form: $C_6H_5ClN_2O_2$
M_r: 172.56
mp: 76
Color: ye nd (dil al)
***s*:** al 3; 50% al 3
UV: OES 4-59 382(5754) 275(5495) 231(16596) cyhex
m/e: 172.00

HODOC No: 02107
CAS No: 825-41-2
Beilstein No: 12^3, 1665
Name: Benzenamine, 3-chloro-4-nitro-
Synonym: Aniline, 3-chloro-4-nitro-
Line Form: $(3\text{-Cl-4-}NO_2)C_6H_3NH_2$
Mol Form: $C_6H_5ClN_2O_2$
M_r: 172.56
mp: 156-7
Color: ye lf (Bz)
***s*:** al 3; eth 3; HOAc 3; lig 1; CS_2 1
IR: µSADP 21922 3450 3330 1640 1590 1560 1490 1470 1350 1320 1250 1140 1040 920 850 830 750 690
UV: OES 4-59 374(13490) 233(6026) EtOH
H NMR: SAD 17371 6.4 6.6 6.7 8.0 poly D
m/e: 172.00

HODOC No: 02108
CAS No: 5344-44-5
Beilstein No: 12, 732
Name: Benzenamine, 3-chloro-5-nitro-
Synonym: Aniline, 3-chloro-5-nitro-
Line Form: $(3\text{-Cl-5-}NO_2)C_6H_3NH_2$
Mol Form: $C_6H_5ClN_2O_2$
M_r: 172.56
mp: 133-4
Color: og-ye nd (al)
***s*:** Et_2O 4; EtOH 4
m/e: 172.00

HODOC No: 02109
CAS No: 89-63-4
Beilstein No: 12^3, 1649
Name: Benzenamine, 4-chloro-2-nitro-
Synonym: Aniline, 4-chloro-2-nitro-
Line Form: $(4\text{-Cl-2-}NO_2)\text{-}C_6H_3NH_2$
Mol Form: $C_6H_5ClN_2O_2$
M_r: 172.56
mp: 116-7
Color: dk og-ye pr (dil al)
***s*:** al 4; eth 4; ace 2; HOAc 4; lig 2
IR: µSADP 10564 3450 3330 1640 1610 1560 1520 1470 1410 1370 1330 1250 1160 1120 900 820 770
UV: SAD 2909 415(5190) 234(23200) MeOH
H NMR: SAD 3237 7.0 7.1 7.3 8.0 ace
MS: ALDN 133 172(100) 126(91) 99(62) 63(55) 90(47) 28(39) 174(32) 128(28)
m/e: 172.00

HODOC No: 02110
CAS No: 635-22-3
Beilstein No: 12^3, 1660
Name: Benzenamine, 4-chloro-3-nitro-
Synonym: Aniline, 4-chloro-3-nitro-
Line Form: $(4\text{-Cl-3-}NO_2)\text{-}C_6H_3NH_2$
Mol Form: $C_6H_5ClN_2O_2$
M_r: 172.56
mp: 103
Color: ye nd or pr (w) nd (peth)
***s*:** H_2O 3; al 4; eth 3; chl 3; lig 2
IR: µSADP 21913 3330 3230 1610 1520 1490 1350 1300 870 820 750 680
UV: OES 5-51 250(4074) 0.1N sulf
H NMR: SAD 17370 5.6 6.8 7.2 7.2 poly D
m/e: 172.00

HODOC No: 02111
CAS No: 1635-61-6
Beilstein No: 12^3, 1655
Name: Benzenamine, 5-chloro-2-nitro-
Synonym: Aniline, 5-chloro-2-nitro-
Line Form: $(3\text{-Cl-6-}NO_2)\text{-}C_6H_3NH_2$
Mol Form: $C_6H_5ClN_2O_2$
M_r: 172.56
bp: sub
mp: 126.5
Color: gold ye nd (CS_2)ye lf (al, Bz)
***s*:** Et_2O 4; EtOH 4
m/e: 172.00

HODOC No: 02112
CAS No: 16611-15-7
Name: Benzenamine, 4-chloro-2-nitro-*N*-phenyl-
Synonym: Diphenylamine, 4-chloro-2-nitro-
Mol Form: $C_{12}H_9ClN_2O_2$
Mr: 248.66
mp: 61.5-2.5
***s*:** ctc 2; CS_2 3
IR: COB 6743 3360 3100 3050 1620 1600 1570 1510 1420 1350 1260 1150 1070 900 810 760 690 500
UV: SAD 10864 433(6840) 256(18400) 235(14700) MeOH
H NMR: SAD 1381 7.3 8.2 9.4 CCl_4
m/e: 248.04

HODOC No: 02113
CAS No: 25781-92-4
Name: Benzenamine, 5-chloro-2-nitro-*N*-phenyl-
Synonym: 5-Chloro-2-nitrodiphenylamine
Mol Form: $C_{12}H_9ClN_2O_2$
Mr: 248.66
mp: 108-11
IR: SADG 20825 3330 3100 3060 1610 1590 1570 1490 1480 1410 1330 1310 1280 1240 1220 1150 1060 930 840 830 820 740 700 680 520 480 430
UV: SAD 18792 415(7540) 263(16200) 224(15200) MeOH
H NMR: SAD 11363 6.7 7.1 7.2-7.7 8.1 9.5 $CDCl_3$
m/e: 248.04

HODOC No: 02114
CAS No: 22760-87-8
Name: Benzenamine, 4-chloro-*N*-[1-(4-nitrophenyl)ethylidene]-
Synonym: Aniline, *p*-chloro-*N*-(α-methyl-*p*-nitrobenzylidene)-
Mol Form: $C_{14}H_{11}ClN_2O_2$
Mr: 274.69
mp: 141
UV: LANG 2356 337(3311) 276(19055) 229(14791) EtOH
m/e: 274.05

HODOC No: 02115
CAS No: 93-67-4
Name: Benzenamine, 5-chloro-2-phenoxy-
Synonym: Aniline, 5-chloro-2-phenoxy-
Mol Form: $C_{12}H_{10}ClNO$
Mr: 219.66
IR: SADG 8825 3470 3420 3360 3080 1610 1580 1490 1420 1280 1210 1180 1150 1130 1080 1070 1020 920 850 810 740 690 660 610 540 490 450
Raman: SAD 1065 3440 3330 3060 1620 1590 1500 1430 1310 1280 1220 1190 1170 1070 1020 1010 920 820 780 750 730 670 620 580 480 350 230 220 120
UV: SAD 5769 298 MeOH
H NMR: SAD 2014 3.7 6.4-7.5 CCl_4
m/e: 219.05

HODOC No: 02116
CAS No: 101-17-7
Beilstein No: 12^2, 320
Name: Benzenamine, 3-chloro-*N*-phenyl-
Synonym: Diphenylamine, 3-chloro-
Line Form: $3\text{-}ClC_6H_4NHC_6H_5$
Mol Form: $C_{12}H_{10}ClN$
Mr: 203.66
bp: $335\text{-}6^{724}$
***d*:** 1.2
***n*:** 1.6513^{20}
***s*:** H_2O 1; al 3; eth 3; bz 3; gl HOAc 3
IR: μCOB 7334 3330 3030 1590 1490 1470 1410 1320 1080 990 920 770 750 700 680
MS: ALD 5:87B
m/e: 203.05

HODOC No: 02117
CAS No: 1205-71-6
Beilstein No: 12^3, 1330
Name: Benzenamine, 4-chloro-*N*-phenyl-
Synonym: Diphenylamine, 4-chloro-
Line Form: $4\text{-}ClC_6H_4NHC_6H_5$
Mol Form: $C_{12}H_{10}ClN$
Mr: 203.66
bp: $334\text{-}5^{726}$
mp: 74
Color: cr(MeOH)
***s*:** Bz 4; Et_2O 4; EtOH 4; lig 4
m/e: 203.05

HODOC No: 02118
CAS No: 3789-77-3
Beilstein No: 16^3, 341
Name: Benzenamine, 4-[(3-chlorophenyl)azo]-*N*,*N*-dimethyl-
Synonym: Aniline, *p*-[(*m*-chlorophenyl)azo]-*N*,*N*-dimethyl-
Line Form: $4\text{-}[(CH_3)_2N]C_6H_4N{=}N(C_6H_4Cl\text{-}3)$
Mol Form: $C_{14}H_{14}ClN_3$
M$_r$: 259.72
mp: 103-4
IR: SADG 21164 3060 2900 1600 1510 1440 1390 1360 1220 1140 810 790 680
UV: SAD 2058 325 312 257 cyhex
H NMR: SAD 11965 3.0 6.7 7.3 7.7 7.8 $CDCl_3$
m/e: 259.09

HODOC No: 02119
CAS No: 25287-19-8
Name: Benzenamine, 4-chloro-*N*-(1-phenylethylidene)-
Synonym: Aniline, *p*-chloro-*N*-(α-methylbenzylidene)-
Mol Form: $C_{14}H_{12}ClN$
M$_r$: 229.69
mp: 77-8
UV: LANG 2340 314(2455) 246(19953) EtOH
m/e: 229.07

HODOC No: 02120
CAS No: 17508-50-8
Name: Benzenamine, *N*-[1-(4-chlorophenyl)ethylidene]-
Synonym: Aniline, *N*-(*p*-chloro-α-methylbenzylidene)-
Mol Form: $C_{14}H_{12}ClN$
M$_r$: 229.69
mp: 94-5
UV: LANG 2344 328(2042) 255(21878) cy 312(2089) 253(20893) EtOH
m/e: 229.07

HODOC No: 02121
CAS No: 25287-48-3
Name: Benzenamine, *N*-[1-(4-chlorophenyl)ethylidene]-4-methoxy-
Synonym: *p*-Anisidine, *N*-(*p*-chloro-α-methylbenzylidene)-
Mol Form: $C_{15}H_{14}ClNO$
M$_r$: 259.72
mp: 94
UV: LANG 2354 332(3020) 254(17783) EtOH
m/e: 259.08

HODOC No: 02122
CAS No: 76553-86-1
Name: Benzenamine, 2-chloro-*N*-(3-phenyl-2-propenylidene)-
Synonym: Cinnamaldehyde, imine*N*(2-chlorophenyl)
Mol Form: $C_{15}H_{12}ClN$
M$_r$: 241.70
***s*:** chl 2
H NMR: SAD 10296 7.0 7.8 8.3 $CDCl_3$
m/e: 241.07

HODOC No: 02123
CAS No: 15286-50-7
Beilstein No: 12^1, 305
Name: Benzenamine, 4-chloro-*N*-(3-phenyl-2-propenylidene)-
Synonym: Aniline, *p*-chloro-*N*-cinnamylidene-
Mol Form: $C_{15}H_{12}ClN$
M$_r$: 241.70
mp: 107
***s*:** chl 3
IR: SADG 16599 3080 3020 1610 1590 1560 1470 1430 1150 1090 970 930 820 740 680 480
H NMR: SAD 8482 6.9-7.6 8.1 $CDCl_3$
m/e: 241.07

HODOC No: 02124
CAS No: 121-50-6
Name: Benzenamine, 2-chloro-5-(trifluoromethyl)-
Synonym: *m*-Toluidine, 6-chloro-α,α,α-trifluoro-
Line Form: $4\text{-}Cl\text{-}3H_2NC_6H_3CF_3$
Mol Form: $C_7H_5ClF_3N$
M$_r$: 195.56
***d*:** 1.428
***n*:** 1.4975^{20}
IR: ALD 2:645H
MS: NBS 1108; ALD 5:71D
m/e: 195.01

HODOC No: 02125
CAS No: 320-51-4
Name: Benzenamine, 4-chloro-3-(trifluoromethyl)-
Synonym: *m*-Toluidine, 4-chloro-α,α,α-trifluoro-
Line Form: $5\text{-}NH_2\text{-}2\text{-}Cl\text{-}C_6H_3CF_3$
Mol Form: $C_7H_5ClF_3N$
M_r: 195.56
mp: 34-6
IR: ALD 2:645G
MS: ALD 4:71C
m/e: 195.01

HODOC No: 02126
CAS No: 6373-50-8
Beilstein No: 12^3, 2819
Name: Benzenamine, 4-cyclohexyl-
Synonym: Aniline, *p*-cyclohexyl-
Line Form: $4\text{-}C_6H_{11}\text{-}C_6H_4NH_2$
Structure:

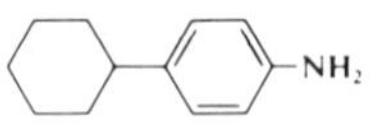

Mol Form: $C_{12}H_{17}N$
M_r: 175.26
mp: 55
Color: pl (lig)
***s*:** Bz 4; EtOH 4
MS: NBS 836
m/e: 175.14

HODOC No: 02127
CAS No: 1821-36-9
Beilstein No: 12^2, 98
Name: Benzenamine, *N*-cyclohexyl-
Synonym: Cyclohexylamine, *N*-phenyl-
Line Form: $C_6H_5NHC_6H_{11}$
Structure:

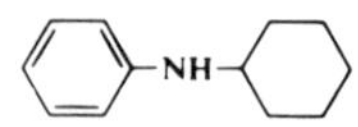

Mol Form: $C_{12}H_{17}N$
M_r: 175.26
bp: 279^{764}, $191\text{-}2^{73}$
mp: 16
***d*:** $1.0155^{20/4}$
***n*:** 1.5610^{20}
Color: mcl pr
***s*:** H_2O 1; al 3; eth 3; bz 3
UV: SAD 20900 294(1380) 246(8530) MeOH 260(201) 254(251) 248(236) 238(206) MeOH:HCl
MS: ALDN 138 132(100) 32(38) 175(28) 93(17) 110(17) 118(13) 77(12) 133(11)
m/e: 175.14

HODOC No: 02128
CAS No: 608-22-0
Beilstein No: 12, 655
Name: Benzenamine, 2,3-dibromo-
Synonym: 2,3-Dibromoaniline
Line Form: $2,3\text{-}Br_2\text{-}C_6H_3NH_2$
Mol Form: $C_6H_5Br_2N$
M_r: 248.91
mp: 43
Color: pl (dil al)
***s*:** Ac 4; Et_2O 4; EtOH 4
m/e: 248.88

HODOC No: 02129
CAS No: 615-57-6
Beilstein No: 12^3, 1471
Name: Benzenamine, 2,4-dibromo-
Synonym: Aniline, 2,4-dibromo-
Line Form: 2,4-Br_2-$C_6H_3NH_2$
Mol Form: $C_6H_5Br_2N$
Mr: 248.91
bp: 156^{74}
mp: 79-80
***d*:** 2.260^{20}
Color: rh bipym (Chl) nd or lf (dil al)
***s*:** al 3; eth 3; chl 3; HOAc 3
IR: SADG 18853 3470 3390 3050 1610 1560 1470 1390 1290 1240 1140 1090 1020 860 830 800 680
H NMR: SAD 2378 4.0 6.6 7.2 7.5 $CDCl_3$
m/e: 248.88

HODOC No: 02130
CAS No: 3638-73-1
Beilstein No: 12^3, 1474
Name: Benzenamine, 2,5-dibromo-
Synonym: Aniline, 2,5-dibromo-
Line Form: 2,5-$Br_2C_6H_3NH_2$
Mol Form: $C_6H_5Br_2N$
Mr: 248.91
mp: 53-5
Color: pr (al)
***s*:** al 4; eth 3; chl 2
IR: SADG 884 3448 3333 3030 1613 1587 1471 1408 1282 1250 1136 1053 1010 877 826 763 645
UV: SAD 8517 302(3520) 211(37400) MeOH
H NMR: SAD 2568 4.1 6.7 6.9 7.3 $CDCl_3$
m/e: 248.88

HODOC No: 02131
CAS No: 608-30-0
Beilstein No: 12^3, 1474
Name: Benzenamine, 2,6-dibromo-
Synonym: Aniline, 2,6-dibromo-
Line Form: 2,6-$Br_2C_6H_3NH_2$
Mol Form: $C_6H_5Br_2N$
Mr: 248.91
bp: 262-4
mp: 87-8
Color: nd (al)
***s*:** al 4; eth 4; bz 4; chl 4
IR: µSADP 19145 3450 3330 1610 1560 1470 1300 1050 860 760 760 710
UV: SAD 6258 247 208 MeOH
H NMR: SAD 18119 4.5 6.4 7.3 $CDCl_3$
m/e: 248.88

HODOC No: 02132
CAS No: 615-55-4
Beilstein No: 12^1, 329
Name: Benzenamine, 3,4-dibromo-
Synonym: Aniline, 3,4-dibromo-
Line Form: 3,4-$Br_2C_6H_3NH_2$
Mol Form: $C_6H_5Br_2N$
Mr: 248.91
bp: sub
mp: 80-1
Color: cr (dil al)
***s*:** EtOH 4
m/e: 248.88

HODOC No: 02133
CAS No: 626-40-4
Beilstein No: 12^3, 1475
Name: Benzenamine, 3,5-dibromo-
Synonym: Aniline, 3,5-dibromo-
Line Form: 3,5-$Br_2C_6H_3NH_2$
Mol Form: $C_6H_5Br_2N$
Mr: 248.91
mp: 57
Color: nd (dil al)
***s*:** al 4; eth 4; bz 4
UV: OES 3-57 305(3162) 250(10000) 212(31623) EtOH
m/e: 248.88

HODOC No: 02134
CAS No: 38573-62-5
Beilstein No: 12^2, 356
Name: Benzenamine, 2,4-dibromo-*N*-(2,4-dibromophenyl)-
Synonym: Bis(2,4-dibromophenyl)amine
Mol Form: $C_{12}H_7Br_4N$
Mr: 480.80
mp: 187.5
***s*:** H_2O 1; al 2; bz 3; chl 3
IR: SADG 20312 3370 3060 1580 1510 1470 1440 1360 1320 1080 1030 860 800 740
UV: SAD 18218 203(18900) 242(7780) MeOH
m/e: 480.73

HODOC No: 02135
CAS No: 41727-70-2
Beilstein No: 13^3, 1195
Name: Benzenamine, 3,5-dibromo-4-methoxy-
Synonym: *p*-Anisidine, 3,5-dibromo-
Line Form: $(3,5\text{-}Br_2\text{-}4\text{-}CH_3O)C_6H_2NH_2$
Mol Form: $C_7H_7Br_2NO$
M_r: 278.94
mp: 66
Color: pl (lig)
***s*:** al 3; eth 3; ace 3; bz 3; chl 3; HOAc 3
UV: HBCP 303(4898) H_2O
m/e: 278.89

HODOC No: 02136
CAS No: 827-23-6
Beilstein No: 12^2, 403
Name: Benzenamine, 2,4-dibromo-6-nitro-
Synonym: Aniline, 2,4-dibromo-6-nitro-
Line Form: $(2,4\text{-}Br_2\text{-}6\text{-}NO_2)C_6H_2NH_2$
Mol Form: $C_6H_4Br_2N_2O_2$
M_r: 293.91
mp: 128
Color: ye cr
m/e: 293.86

HODOC No: 02137
CAS No: 827-94-1
Beilstein No: 12^3, 1678
Name: Benzenamine, 2,6-dibromo-4-nitro-
Synonym: Aniline, 2,6-dibromo-4-nitro-
Line Form: $(2,6\text{-}Br_2\text{-}4\text{-}NO_2)C_6H_2NH_2$
Mol Form: $C_6H_4Br_2N_2O_2$
M_r: 293.91
mp: 207
Color: ye nd (al, HOAc)
***s*:** H_2O 2; HOAc 3
IR: μSADP 21004 3450 3330 1610 1520 1470 1330 1300 1280 1140 900 750 700
UV: SAD 8155 351(12400) 248(5450) 211(15300) MeOH
H NMR: SAD 18172 6.3 8.2 poly D
m/e: 293.86

HODOC No: 02138
CAS No: 613-29-6
Beilstein No: 12^3, 769
Name: Benzenamine, *N,N*-dibutyl-
Synonym: Aniline, *N,N*-dibutyl-
Line Form: $C_6H_5N(C_4H_9)_2$
Mol Form: $C_{14}H_{23}N$
M_r: 205.33
bp: 274.8, 138.8^{10}
mp: -32.2
***d*:** $0.9037^{20/4}$
***n*:** 1.5186^{20}
***s*:** H_2O 1; al 5; eth 5; ace 4; bz 4; ctc 3; ac 3
IR: SADG 321 3333 2941 2857 1587 1493 1449 1370 1282 1220 1042 752 694
UV: SAD 612 305(1910) 259(13300) MeOH
H NMR: SAD 147 1.0 1.1-1.9 3.2 6.3-7.3 CCl_4
m/e: 205.18

HODOC No: 02139
CAS No: 608-27-5
Beilstein No: 12, 621
Name: Benzenamine, 2,3-dichloro-
Synonym: Aniline, 2,3-dichloro-
Line Form: $2,3\text{-}Cl_2C_6H_3NH_2$
Mol Form: $C_6H_5Cl_2N$
M_r: 162.01
bp: 252
mp: 24
Color: nd (lig)
***s*:** al 3; eth 4; ace 3; bz 2; ctc 2; lig 2
IR: μSADP 19095 3450 3330 1610 1590 1470 1450 1300 1210 1100 1030 900 770 740 700
UV: SAD 6233 293 240 cyhex
H NMR: SAD 5005 4.0 6.5 6.7 6.9 CCl_4
MS: WILEY 1025 161(100) 163(62) 90(16) 99(11) 63(11) 165(10) 126(10) 127(8)
m/e: 160.98

HODOC No: 02140
CAS No: 554-00-7
Beilstein No: 12^3, 1389
Name: Benzenamine, 2,4-dichloro-
Synonym: Aniline, 2,4-dichloro-
Line Form: 2,4-$Cl_2C_6H_3NH_2$
Mol Form: $C_6H_5Cl_2N$
M_r: 162.01
bp: 245
mp: 63-4
***d*:** 1.567^{20}
Color: pr (Ace) nd (dil al) (lig)
***s*:** H_2O 2; al 3; eth 3; chl 2
IR: µCOB 5787 3450 3330 1610 1590 1470 1410 1300 1250 1150 1110 1060 1050 860 860 810 730
UV: SAD 16957 305(1540) 245(10900) MeOH
H NMR: SAD 8851 4.0 6.6 7.0 7.2 $CDCl_3$
MS: NBS 619
m/e: 160.98

HODOC No: 02141
CAS No: 95-82-9
Beilstein No: 12^3, 1397
Name: Benzenamine, 2,5-dichloro-
Synonym: Aniline, 2,5-dichloro-
Line Form: 2,5-$Cl_2C_6H_3NH_2$
Mol Form: $C_6H_5Cl_2N$
M_r: 162.01
bp: 251
mp: 50
Color: nd (lig)
***s*:** H_2O 2; al 3; eth 3; bz 3; chl 3; CS_2 3
IR: SADG 8093 3448 3333 3030 1613 1493 1429 1299 1282 1136 1087 1042 909 855 800 671
Raman: SAD 238 3340 3090 3070 1620 1590 1570 1550 1490 1310 1270 1150 1100 1050 710 590 550 450 390 340 320 280 250 220 110
UV: SAD 253 296 241 cyhex
H NMR: SAD 58 4.0 6.7 7.1 $CDCl_3$
MS: NBS 619
m/e: 160.98

HODOC No: 02142
CAS No: 608-31-1
Beilstein No: 12^2, 337
Name: Benzenamine, 2,6-dichloro-
Synonym: Aniline, 2,6-dichloro-
Line Form: 2,6-$Cl_2C_6H_3NH_2$
Mol Form: $C_6H_5Cl_2N$
M_r: 162.01
mp: 39
***s*:** H_2O 2; al 3; eth 3
IR: SADG 8093 3500 3400 3100 1620 1490 1420 1280 1090 910 850 800 650 580
UV: SAD 253 296 241 cyhex
H NMR: SAD 58 4.0 6.7 7.1 $CDCl_3$
m/e: 160.98

HODOC No: 02143
CAS No: 95-76-1
Beilstein No: 12^3, 1402
Name: Benzenamine, 3,4-dichloro-
Synonym: Aniline, 3,4-dichloro-
Line Form: 3,4-$Cl_2C_6H_3NH_2$
Mol Form: $C_6H_5Cl_2N$
M_r: 162.01
bp: 272, 145^{15}
mp: 72
Color: nd (lig)
***s*:** al 3; eth 3; bz 2; chl 2
IR: µSADP 6459 3450 3330 1640 1610 1560 1470 1430 1280 1240 1140 1020 890 860 820 690
UV: SAD 1794 247 MeOH
H NMR: SAD 271 3.6 6.4 6.7 7.1 $CDCl_3$
MS: WILEY 1025 161(100) 163(64) 99(14) 90(12) 165(10) 126(10) 63(8) 162(8)
m/e: 160.98

HODOC No: 02144
CAS No: 626-43-7
Beilstein No: 12^3, 337
Name: Benzenamine, 3,5-dichloro-
Synonym: Aniline, 3,5-dichloro-
Line Form: $3,5\text{-}Cl_2C_6H_3NH_2$
Mol Form: $C_6H_5Cl_2N$
M_r: 162.01
bp: 260^{741}
mp: 51-3
Color: nd (lig, dil al)
s: H_2O 1; al 3; eth 3; ctc 3; lig 3
IR: SADG 8973 3571 3448 3030 1639 1613 1587 1449 1299 1111 990 935 820 794 667
UV: SAD 6259 300 250 213 MeOH
H NMR: SAD 2164 3.7 6.4 6.7 CCl_4
m/e: 160.98

HODOC No: 02145
CAS No: 35113-90-7
Beilstein No: 12^2, 334
Name: Benzenamine, 2,4-dichloro-*N,N*-dimethyl-
Synonym: Aniline, 2,4-dichloro-*N,N*-dimethyl-
Line Form: $2,4\text{-}Cl_2\text{-}C_6H_3\text{-}N(CH_3)_2$
Mol Form: $C_8H_9Cl_2N$
M_r: 190.06
bp: 234, $92\text{-}5^2$
mp: 169-70
m/e: 189.01

HODOC No: 02146
CAS No: 58566-66-8
Beilstein No: 12, 626
Name: Benzenamine, 3,4-dichloro-*N,N*-dimethyl-
Synonym: Aniline, 3,4-dichloro-*N,N*-dimethyl-
Line Form: $3,4\text{-}Cl_2C_6H_3N(CH_3)_2$
Mol Form: $C_8H_9Cl_2N$
M_r: 190.06
mp: 38-9
m/e: 189.01

HODOC No: 02147
CAS No: 64063-37-2
Beilstein No: 12, 872
Name: Benzenamine, 2,6-dichloro-3-methyl-
Synonym: 2,6-Dichloro-3-methylaniline
Line Form: $(2,6\text{-}Cl_2\text{-}3\text{-}CH_3)\text{-}C_6H_2NH_2$
Mol Form: $C_7H_7Cl_2N$
M_r: 176.04
mp: 37-9
IR: ALD 3:738E
m/e: 175.00

HODOC No: 02148
CAS No: 2683-43-4
Beilstein No: 12^3, 1667
Name: Benzenamine, 2,4-dichloro-6-nitro-
Synonym: Aniline, 2,4-dichloro-6-nitro-
Line Form: $(2,4\text{-}Cl_2\text{-}6\text{-}NO_2)C_6H_2NH_2$
Mol Form: $C_6H_4Cl_2N_2O_2$
M_r: 207.01
mp: 101-3
s: ctc 3; chl 2; CS_2 3
IR: ALD 3:833C
UV: SAD 10866 411(5590) 233(19700) MeOH
H NMR: SAD 855 6.6 7.5 8.1 $CDCl_3$
MS: ALD 5:180D
m/e: 205.96

HODOC No: 02149
CAS No: 99-30-9
Beilstein No: 12^3, 1669
Name: Benzenamine, 2,6-dichloro-4-nitro-
Synonym: Aniline, 2,6-dichloro-4-nitro-
Line Form: $(2,6\text{-}Cl_2\text{-}4\text{-}NO_2)C_6H_2NH_2$
Mol Form: $C_6H_4Cl_2N_2O_2$
M_r: 207.01
mp: 191
Color: ye nd (al, HOAc)
s: al 3; $DMSOd_6$ 2; ac 3
IR: SADG 21083 3490 3350 3090 1610 1570 1480 1350 1320 1270 1140 1250 890 790 730 700
UV: SAD 1814 350 245 MeOH
H NMR: SAD 9851 6.9 8.1 $DMSOd_6$
MS: NBS 1251
m/e: 205.96

HODOC No: 02150
CAS No: 58373-59-4
Beilstein No: 12, 621
Name: Benzenamine, 2,4-dichloro-*N*-phenyl-
Synonym: 2,4-Dichlorodiphenylamine
Line Form: $2,4\text{-}Cl_2C_6H_3NHC_6H_5$
Mol Form: $C_{12}H_9Cl_2N$
M_r: 238.10
mp: 64
Color: nd(dil MeOH)
m/e: 237.01

HODOC No: 02151
CAS No: 39052-12-5
Beilstein No: 13^1, 308
Name: Benzenamine, 3,4-diethoxy-
Synonym: Aniline, 3,4-diethoxy-
Line Form: 3,4-$(C_2H_5O)_2C_6H_3NH_2$
Mol Form: $C_{10}H_{15}NO_2$
Mr: 181.22
bp: $132\text{-}5^{0.45}$
mp: 47-9
***s*:** Bz 4; Et_2O 4; EtOH 4
IR: COB 7664 3455 3370 2975 1620 1595 1500 1480 1445 1390 1320 1295 1270 1230 1175 1140 1090 1045 985 900 820 790
UV: SAD 13825 294(2790) 239(7340) MeOH 278(2700) 230(6820) MeOH:HCl
H NMR: SAD 425 1.3 1.4 3.4 4.0 6.2 6.7 $CDCl_3$
m/e: 181.11

HODOC No: 02152
CAS No: 579-66-8
Name: Benzenamine, 2,6-diethyl-
Synonym: Aniline, 2,6-diethyl-
Line Form: 2,6-$(C_2H_5)_2$-$C_6H_3NH_2$
Mol Form: $C_{10}H_{15}N$
Mr: 149.23
bp: 243
mp: 3
***d*:** 0.906
***n*:** 1.5452^{20}
IR: ALD 2:641B
MS: ALD 5:61B
m/e: 149.12

HODOC No: 02153
CAS No: 91-66-7
Beilstein No: 12^3, 260
Name: Benzenamine, *N,N*-diethyl-
Synonym: Aniline, *N,N*-diethyl-
Line Form: $C_6H_5N(C_2H_5)_2$
Mol Form: $C_{10}H_{15}N$
Mr: 149.23
bp: 216.3, 92^{10}
mp: -38.8
***d*:** $0.9307^{20/4}$
***n*:** 1.5409^{20}
Color: ye oil
***s*:** H_2O 2; al 3; eth 4; ace 3; ctc 3; chl 4
IR: μCOB 6952 2940 1610 1520 1470 1410 1350 1270 1210 1160 1100 1080 1040 1010 990 860 790 750 690
Raman: SAD 247 3190 3150 3070 2960 2930 2870 1600 1480 1450 1400 1370 1350 1330 1310 1260 1190 1150 1070 1030 1010 980 910 890 800 740 710 610 520 410 390 310 210
UV: SAD 83 304(195) 260(1390) MeOH
H NMR: SAD 716 1.1 3.3 6.6 7.1 CCl_4
C NMR: JJ 390 FT 12.5 44.2 112.0 115.5 129.1 147.8 $CDCl_3$
MS: ALDN 103 139(100) 106(50) 149(35) 77(30) 104(12) 79(11) 51(11) 28(10)
m/e: 149.12

HODOC No: 02154
CAS No: 5882-45-1
Name: Benzenamine, *N,N*-diethyl-, hydrochloride
Synonym: Aniline, *N,N*-diethyl-, hydrochloride
Line Form: $C_6H_5N\{CH_2CH_3\}_2$–HCl
Mol Form: $C_{10}H_{16}ClN$
Mr: 185.69
***s*:** chl 3
H NMR: SAD 11418 1.3 3.7 7.5-7.8 8.0 $CDCl_3$
m/e: 185.10

HODOC No: 02155
CAS No: 6287-43-0
Name: Benzenamine, *N,N*-diethyl-2,4-dimethyl-
Synonym: 2,4-Xylidine, *N,N*-diethyl-
Mol Form: $C_{12}H_{19}N$
Mr: 177.28
IR: μCOB 1399 2940 2780 1610 1520 1450 1390 1300 1250 1240 1180 1110 1090 1030 870 820 770 740
m/e: 177.15

HODOC No: 02156
CAS No: 3995-37-7
Name: Benzenamine, *N,N*-diethyl-2,5-dimethyl-
Synonym: 2,5-Xylidine, *N,N*-diethyl-
Line Form: 2,5-$(CH_3)_2$-$C_6H_3N(C_2H_5)_2$
Mol Form: $C_{12}H_{19}N$
M$_r$: 177.28
bp: $95\text{-}6^{11}$
IR: **µCOB 6969** 2940 2780 1670 1610 1590 1520 1450 1390 1250 1180 1110 990 870 800 740 700
m/e: 177.15

HODOC No: 02157
CAS No: 3995-38-8
Name: Benzenamine, *N,N*-diethyl-2,6-dimethyl-
Synonym: 2,6-Xylidine, *N,N*-diethyl-
Mol Form: $C_{12}H_{19}N$
M$_r$: 177.28
IR: **µCOB 4178** 3030 2940 1470 1370 1250 1240 1190 1100 1030 770
m/e: 177.15

HODOC No: 02158
CAS No: 4928-41-0
Name: Benzenamine, *N,N*-diethyl-4-[(4-ethylphenyl)azo]-
Synonym: Aniline, *N,N*-diethyl-*p*-[(*p*-ethylphenyl)azo]-
Line Form: 4-[$(C_2H_5)_2N$]-C_6H_4-N=N-(C_6H_4-C_2H_5-4)
Mol Form: $C_{18}H_{23}N_3$
M$_r$: 281.38
mp: 131-2
IR: SADG 21721
UV: SAD 17870
H NMR: SAD 11299
m/e: 281.19

HODOC No: 02159
CAS No: 606-46-2
Beilstein No: 12^3, 1844
Name: Benzenamine, *N,N*-diethyl-2-methyl-
Synonym: *o*-Toluidine, *N,N*-diethyl-
Line Form: 2-$CH_3C_6H_4N(C_2H_5)_2$
Mol Form: $C_{11}H_{17}N$
M$_r$: 163.25
bp: $208\text{-}9^{755}$
***s*:** H_2O 2; al 3; eth 3
IR: **µNRC 649** 3130 2940 2780 1610 1590 1470 1450 1390 1330 1300 1250 1180 1110 890 810 790 760 730 700
Raman: **SAD 248** 3060 2980 2940 2880 2830 2740 1600 1580 1450 1380 1360 1300 1270 1250 1200 1160 1150 1110 1070 1060 1020 980 930 900 860 800 760 730 710 600 570 530 500 460 440 340 300 200
m/e: 163.14

HODOC No: 02160
CAS No: 613-48-9
Beilstein No: 12^2, 492
Name: Benzenamine, *N,N*-diethyl-4-methyl-
Synonym: *p*-Toluidine, *N,N*-diethyl-
Line Form: 4-$CH_3C_6H_4N(C_2H_5)_2$
Mol Form: $C_{11}H_{17}N$
M$_r$: 163.25
bp: 229^{770}
***d*:** 0.9242^{16}
***s*:** H_2O 2; al 5; eth 5
IR: **µNRC 650** 3130 2860 1850 1640 1590 1520 1450 1370 1270 1190 1160 1100 1080 1010 920 900 800 780 700 670
m/e: 163.14

HODOC No: 02161
CAS No: 3010-53-5
Name: Benzenamine, *N,N*-diethyl-4-[(3-methylphenyl)azo]-
Synonym: Aniline, *N,N*-diethyl-*p*-(*m*-tolylazo)-
Line Form: 4-$(C_2H_5)_2$-C_6H_4N=$N(C_6H_4CH_3$-3)
Mol Form: $C_{17}H_{21}N_3$
M_r: 267.36
mp: 72-3
***s*:** chl 3
IR: µSADP 7486 2940 1615 1565 1515 1450 1410 1370 1315 1265 1190 1150 1125 1075 1010 910 875 830 785 725 690
UV: OES 3-657 520(11220) 323(18621) 1.2N HCl:50% EtOH
H NMR: SAD 11298 1.2 2.4 3.4 6.7 7.2 7.4 7.6 7.9 $CDCl_3$
m/e: 267.17

HODOC No: 02162
CAS No: 2216-16-2
Beilstein No: 12^3, 1545
Name: Benzenamine, *N,N*-diethyl-3-nitro-
Synonym: Aniline, *N,N*-diethyl-*m*-nitro-
Line Form: 3-NO_2-$C_6H_4N(C_2H_5)_2$
Mol Form: $C_{10}H_{14}N_2O_2$
M_r: 194.22
bp: 288-90
Color: ye
m/e: 194.11

HODOC No: 02163
CAS No: 2216-15-1
Beilstein No: 12^3, 1585
Name: Benzenamine, *N,N*-diethyl-4-nitro-
Synonym: Aniline, *N,N*-diethyl-*p*-nitro-
Line Form: 4-NO_2-$C_6H_4N(C_2H_5)_2$
Mol Form: $C_{10}H_{14}N_2O_2$
M_r: 194.22
mp: 77-8
***d*:** 1.225
Color: ye nd (lig) pl (al)
***s*:** al 3; lig 2
IR: µDMS 5696 3030 1610 1540 1490 1450 1430 1320 1280 1190 1120 1080 1010 830 750 730 690
UV: OES 7-298 415(23988)
H NMR: SAD 19915 1.2 3.5 6.6 8.1 $CDCl_3$
m/e: 194.11

HODOC No: 02164
CAS No: 64462-51-7
Name: Benzenamine, *N,N*-diethyl-4-(2-nitroethenyl)-
Synonym: Aniline, *N,N*-diethyl-*p*-(2-nitrovinyl)-
Mol Form: $C_{12}H_{16}N_2O_2$
M_r: 220.26
mp: 98
***s*:** chl 2
IR: µSADP 3325 3125 3030 2940 1640 1615 1540 1495 1390 1335 1265 1205 1100 980 815 745
UV: SAD 1014 325 312 266 MeOH
H NMR: SAD 10845 1.2 3.4 6.6 7.3 7.4 7.9 $CDCl_3$
m/e: 220.12

HODOC No: 02165
CAS No: 367-25-9
Beilstein No: 12^1, 2978
Name: Benzenamine, 2,4-difluoro-
Synonym: Aniline, 2,4-difluoro-
Line Form: 2,4-$F_2C_6H_3NH_2$
Mol Form: $C_6H_5F_2N$
M_r: 129.10
bp: 170^{753}
mp: -7.5
***d*:** 1.268
***n*:** 1.5063^{20}
IR: ALD 2:643F
MS: ALD 5:66A
m/e: 129.04

HODOC No: 02166
CAS No: 5509-65-9
Name: Benzenamine, 2,6-difluoro-
Synonym: Aniline, 2,6-difluoro-
Line Form: 2,6-F_2-$C_6H_3NH_2$
Mol Form: $C_6H_5F_2N$
M_r: 129.10
bp: $51\text{-}2\text{-}^{15}$
***d*:** 1.199
***n*:** 1.5084^{20}
m/e: 129.04

HODOC No: 02167
CAS No: 61033-71-4
Name: Benzenamine, 4-(4,5-dihydro-1H-imidazol-2-yl)-
Synonym: 2-Imidazoline, 2-(*p*-aminophenyl)-
Mol Form: $C_9H_{11}N_3$
M_r: 161.20
IR: COB 7143 1605 1560 1525 1490 1465 1445 1310 1285 1180 1170 1090 985 950 830 720 670 515
m/e: 161.10

HODOC No: 02168
CAS No: 533-70-0
Beilstein No: 12^3, 1508
Name: Benzenamine, 2,4-diiodo-
Synonym: Aniline, 2,4-diiodo-
Line Form: $2,4\text{-}I_2\text{-}C_6H_3NH_2$
Mol Form: $C_6H_5I_2N$
M_r: 344.92
mp: 95-6
***d*:** 2.748
Color: br nd or rh cr (al)
***s*:** Ac 4; Bz 4; Et_2O 4; EtOH 4
m/e: 344.85

HODOC No: 02169
CAS No: 64085-53-6
Beilstein No: 12, 675
Name: Benzenamine, 2,5-diiodo-
Synonym: Aniline, 2,5-diiodo
Line Form: $2,5\text{-}I_2\text{-}C_6H_3NH_2$
Mol Form: $C_6H_5I_2N$
M_r: 344.92
mp: 88-9
Color: nd (al)
***s*:** Ac 4; Et_2O 4; EtOH 4
m/e: 344.85

HODOC No: 02170
CAS No: 49865-78-3
Beilstein No: 12, 675
Name: Benzenamine, 2,6-diiodo-
Synonym: Aniline, 2,6-diiodo
Line Form: $2,6\text{-}I_2C_6H_3NH_2$
Mol Form: $C_6H_5I_2N$
M_r: 344.92
mp: 122
Color: nd (al)
***s*:** Ac 4; Bz 4; Et_2O 4; EtOH 4
m/e: 344.85

HODOC No: 02171
CAS No: 42016-95-5
Beilstein No: 12, 675
Name: Benzenamine, 3,4-diiodo-
Synonym: 3,4-Diiodoaniline
Line Form: $3,4\text{-}I_2\text{-}C_6H_3NH_2$
Mol Form: $C_6H_5I_2N$
M_r: 344.92
mp: 94.5
Color: pa ye lf or pr (Bz-peth)
***s*:** Bz 4; Et_2O 4; EtOH 4
m/e: 344.85

HODOC No: 02172
CAS No: 35122-96-4
Beilstein No: 12^1, 377
Name: Benzenamine, 3,5-diiodo-
Synonym: Aniline, 3,5-diiodo-
Line Form: $3,5\text{-}I_2\text{-}C_6H_3NH_2$
Mol Form: $C_6H_5I_2N$
M_r: 344.92
mp: 110
Color: nd (al)
***s*:** Et_2O 4; EtOH 4; Chl 4
m/e: 344.85

HODOC No: 02173
CAS No: 5398-27-6
Beilstein No: 12^3, 1680
Name: Benzenamine, 2,6-diiodo-4-nitro-
Synonym: Aniline, 2,6-diiodo-4-nitro-
Line Form: $(2,6\text{-}I_2\text{-}4\text{-}NO_2)\text{-}C_6H_2NH_2$
Mol Form: $C_6H_4I_2N_2O_2$
M_r: 389.91
mp: 245
Color: pa ye lf or nd (Bz)
***s*:** al 2; bz 3
IR: µSADP 24026 3450 3330 1610 1540 1490 1300 1270 1110 900 740 680
UV: SAD 19651 358(14600) 270(8580) 234(13900) 215(16700) MeOH
MS: ALD 5:180C
m/e: 389.84

HODOC No: 02174
CAS No: 6299-67-8
Beilstein No: 13^3, 2104
Name: Benzenamine, 2,3-dimethoxy-
Synonym: Aniline, 2,3-dimethoxy-
Line Form: $2,3\text{-}(CH_3O)_2\text{-}C_6H_3NH_2$
Mol Form: $C_8H_{11}NO_2$
M_r: 153.17
bp: 137^{15}
***s*:** H_2O 4
m/e: 153.08

HODOC No: 02175
CAS No: 2735-04-8
Beilstein No: 13^3, 2131
Name: Benzenamine, 2,4-dimethoxy-
Synonym: Aniline, 2,4-dimethoxy-
Line Form: $2,4-(CH_3O)_2-C_6H_3NH_2$
Mol Form: $C_8H_{11}NO_2$
M_r: 153.17
mp: 33.5
Color: pl (lig)
***s*:** H_2O 2; al 3; eth 3; bz 3; chl 2; lig 3
IR: SADG 21408 3440 3350 2940 2830 1600 1520 1470 1440 1280 1230 1200 1150 1030 920 830 790 690
UV: SAD 7705 296(3770) 236(8190) MeOH
H NMR: SAD 5012 3.5 3.7 3.8 6.3 6.5 6.6 $CDCl_3$
m/e: 153.08

HODOC No: 02176
CAS No: 102-56-7
Beilstein No: 13^3, 475
Name: Benzenamine, 2,5-dimethoxy-
Synonym: Aniline, 2,5-dimethoxy-
Line Form: $2,5-(CH_3O)_2-C_6H_3NH_2$
Mol Form: $C_8H_{11}NO_2$
M_r: 153.17
bp: 270
mp: 84
***s*:** H_2O 3; al 3; chl 3; lig 3
IR: SADG 21010 3460 3370 3000 2960 1620 1520 1460 1310 1230 1040 840 790
UV: SAD 2087 295 235 208 MeOH
H NMR: VAR 508 3.7 3.8 6.2 6.3 6.7 $CDCl_3$
MS: NBS 499
m/e: 153.08

HODOC No: 02177
CAS No: 2734-70-5
Beilstein No: 13^3, 2128
Name: Benzenamine, 2,6-dimethoxy-
Synonym: Aniline, 2,6-dimethoxy-
Line Form: $2,6-(CH_3O)_2C_6H_3NH_2$
Mol Form: $C_8H_{11}NO_2$
M_r: 153.17
bp: 146^{23}
mp: 75
Color: pl (al), lf (peth)
***s*:** Bz 4; Et_2O 4; EtOH 4; lig 4
m/e: 153.08

HODOC No: 02178
CAS No: 6315-89-5
Beilstein No: 12^3, 2105
Name: Benzenamine, 3,4-dimethoxy-
Synonym: Aniline, 3,4-dimethoxy-
Line Form: $3,4-(CH_3O)_2C_6H_3NH_2$
Mol Form: $C_8H_{11}NO_2$
M_r: 153.17
mp: 87-8
Color: lf (eth)
***s*:** eth 3; chl 3
IR: SADG 8944 3333 3226 3030 2941 1587 1493 1449 1316 1266 1220 1149 1111 1010 935 833 775 741 685
UV: SAD 6158 295 239 MeOH
H NMR: SAD 2138 3.5 3.8 6.2 6.3 6.7 $CDCl_3$
m/e: 153.08

HODOC No: 02179
CAS No: 6313-37-7
Beilstein No: 13, 789
Name: Benzenamine, 2,5-dimethoxy-4-nitro-
Synonym: Aniline, 2,5-dimethoxy-4-nitro-
Line Form: $[3,5-(CH_3O)_2-4-NO_2]-C_6H_2NH_2$
Mol Form: $C_8H_{10}N_2O_4$
M_r: 198.17
mp: 152.5
IR: ALD 3:835A
m/e: 198.06

HODOC No: 02180
CAS No: 27895-67-6
Name: Benzenamine, *N*-[(3,4-dimethoxyphenyl)methylene]-
Synonym: Aniline, *N*-veratrylidene-
Mol Form: $C_{15}H_{15}NO_2$
M_r: 241.27
***s*:** ctc 3; CS_2 3
IR: μCOB 3965 3030 2940 2860 1610 1590 1520 1470 1430 1390 1330 1280 1250 1220 1160 1140 1030 970 900 870 850 810 760 690
m/e: 241.11

HODOC No: 02181
CAS No: 87-59-2
Beilstein No: 12^3, 2438
Name: Benzenamine, 2,3-dimethyl-
Synonym: 2,3-Xylidine
Line Form: 2,3-$(CH_3)_2C_6H_3NH_2$
Mol Form: $C_8H_{11}N$
M_r: 121.17
bp: 221-2, 106^{15}
mp: 15
***d*:** 0.9931^{20}
***n*:** 1.5684^{20}
***s*:** H_2O 2; al 4; eth 4; ctc 3
IR: SADG 663 3448 3333 3030 2857 1613 1587 1471 1429 1370 1282 1099 990 775 714 654
UV: SAD 5403 286 236 cyhex
H NMR: SAD 893 2.0 2.2 3.4 6.5 6.8 CCl_4
C NMR: JJ 303 FT 12.4 20.3 113.1 120.4 120.6 125.9 136.8 144.5 $CDCl_3$
m/e: 121.09

HODOC No: 02182
CAS No: 95-68-1
Beilstein No: 12^3, 2469
Name: Benzenamine, 2,4-dimethyl-
Synonym: 2,4-Xylidine
Line Form: 2,4-$(CH_3)_2C_6H_3NH_2$
Mol Form: $C_8H_{11}N$
M_r: 121.17
bp: 214, 91^{10}
mp: -14.3
***d*:** $0.9723^{20/4}$
***n*:** 1.5569^{20}
***s*:** H_2O 2; al 3; eth 3; bz 3; ctc 2
IR: SADG 10778 3448 3333 3030 2941 1613 1515 1493 1471 1370 1266 1235 1149 1031 1000 870 806 725 680
UV: SAD 1288 287 235 MeOH
H NMR: SAD 237 2.0 2.2 3.4 6.4 6.7 CCl_4
MS: WILEY 490 ini 120(100) 121(70) 77(25) 51(16) 105(13) 104(11) 42(10) 50(7)
m/e: 121.09

HODOC No: 02183
CAS No: 21436-96-4
Name: Benzenamine, 2,4-dimethyl-, hydrochloride
Synonym: 2,4-Xylidine, hydrochloride
Mol Form: $C_8H_{12}ClN$
M_r: 157.63
IR: μCOB 172 2860 2560 1920 1590 1490 1390 1210 1040 810
m/e: 157.07

HODOC No: 02184
CAS No: 95-78-3
Beilstein No: 12^3, 2503
Name: Benzenamine, 2,5-dimethyl-
Synonym: 2,5-Xylidine
Line Form: 2,5-$(CH_3)_2C_6H_3NH_2$
Mol Form: $C_8H_{11}N$
M_r: 121.17
bp: 214, $97\text{-}100^{10}$
mp: 15.5
***d*:** $0.9790^{21/4}$
***n*:** 1.5591^{21}
Color: ye lf (lig)
***s*:** H_2O 2; eth 3; ctc 3
IR: μCOB 6973 3330 2940 1610 1560 1490 1450 1390 1300 1000 860 790
UV: SAD 5399 288 237 cyhex
H NMR: SAD 3276 2.0 2.2 3.2 6.3 6.8 CCl_4
m/e: 121.09

HODOC No: 02185
CAS No: 87-62-7
Beilstein No: 12^3, 2462
Name: Benzenamine, 2,6-dimethyl-
Synonym: 2,6-Xylidine
Line Form: 2,6-$(CH_3)_2C_6H_3NH_2$
Mol Form: $C_8H_{11}N$
M_r: 121.17
bp: 214^{739}
mp: 11.2
***d*:** 0.9842^{20}
***n*:** 1.5610^{20}
***s*:** H_2O 1; al 3; eth 3; ctc 3
IR: μCOB 4180 3450 3330 3030 2940 1610 1490 1450 1390 1320 1280 1240 1090 760 740
UV: SAD 15776 283(2020) 233(7490) MeOH
H NMR: SAD 6006 2.0 3.2 6.3-6.9 CCl_4
m/e: 121.09

HODOC No: 02186
CAS No: 95-64-7
Beilstein No: 12^3, 2443
Name: Benzenamine, 3,4-dimethyl-
Synonym: 3,4-Xylidine
Line Form: $3,4\text{-}(CH_3)_2C_6H_3NH_2$
Mol Form: $C_8H_{11}N$
M_r: 121.17
bp: 228
mp: 51
***d*:** 1.076^{18}
Color: pl or pr (lig)
***s*:** H_2O 2; eth 3; chl 2; lig 4
IR: SADG 679 3448 3333 3226 3030 2941 1613 1587 1515 1449 1370 1299 1266 1163 1020 877 820 709
UV: SAD 5525 293 238 cyhex
H NMR: SAD 5577 2.2 3.4 6.5 6.9 $CDCl_3$
m/e: 121.09

HODOC No: 02187
CAS No: 108-69-0
Beilstein No: 12^3, 2495
Name: Benzenamine, 3,5-dimethyl-
Synonym: 3,5-Xylidine
Line Form: $3,5\text{-}(CH_3)_2C_6H_3NH_2$
Mol Form: $C_8H_{11}N$
M_r: 121.17
bp: 220-1, $99\text{-}100^{20}$
mp: 9.8
***d*:** $0.9706^{20/4}$
***n*:** 1.5581^{20}
***s*:** H_2O 2; eth 3; ctc 3
IR: SADG 12106 3448 3333 3226 3030 2941 1613 1471 1370 1316 1176 1031 943 826 685
UV: SAD 15178 288(2420) 239(11000) 207(35900) MeOH
H NMR: SAD 5580 2.2 3.3 6.1 6.3 CCl_4
m/e: 121.09

HODOC No: 02188
CAS No: 611-21-2
Beilstein No: 12^3, 1842
Name: Benzenamine, *N*,2-dimethyl-
Synonym: *o*-Toluidine, *N*-methyl-
Line Form: $2\text{-}CH_3C_6H_4NHCH_3$
Mol Form: $C_8H_{11}N$
M_r: 121.17
bp: 207-8, 99^{17}
***d*:** $0.9709^{20/4}$
***n*:** 1.5649^{20}
***s*:** H_2O 1; al 5; eth 5; ace 3
IR: µSADP 21401 3450 2940 2780 1610 1590 1520 1470 1450 1320 1300 1270 1160 1060 1040 990 750 710
UV: OES 1-216 294(2512) 244(12589) EtOH
H NMR: SAD 17212 2.1 2.9 3.4 6.6 6.7 7.1 7.1 $CDCl_3$
MS: WILEY 490 121(100) 106(82) 91(33) 39(20) 65(19) 77(18) 51(15) 104(12)
m/e: 121.09

HODOC No: 02189
CAS No: 696-44-6
Beilstein No: 12^3, 1953
Name: Benzenamine, *N*,3-dimethyl-
Synonym: *m*-Toluidine, *N*-methyl-
Line Form: $3\text{-}CH_3C_6H_4NHCH_3$
Mol Form: $C_8H_{11}N$
M_r: 121.17
bp: 206-7, $120\text{-}1^{40}$
***n*:** 1.5557^{25}
***s*:** H_2O 1; al 5; eth 5; ace 3
IR: µNRC 559 3450 3030 2860 2780 1610 1490 1320 1270 1190 1180 1090 990 920 890 840 770 690
H NMR: SAD 18885 2.2 2.7 3.4 6.5 7.1 $CDCl_3$
m/e: 121.09

HODOC No: 02190
CAS No: 623-08-5
Beilstein No: 12^3, 2025
Name: Benzenamine, *N*,4-dimethyl-
Synonym: *p*-Toluidine, *N*-methyl-
Line Form: $4\text{-}CH_3C_6H_4NHCH_3$
Mol Form: $C_8H_{11}N$
M_r: 121.17
bp: 209-11, 102^{20}
***d*:** $0.9348^{55/4}$
***n*:** 1.5568^{20}
***s*:** H_2O 1; al 5; eth 5; ace 3
IR: µSADP 8409 3450 2940 2780 1610 1520 1490 1450 1320 1250 1180 1150 1110 1050 810 700
UV: OES 3-164 302(2239) 246(13183) iso
H NMR: SAD 17131 2.2 2.6 3.3 6.4 6.9 $CDCl_3$
m/e: 121.09

HODOC No: 02191
CAS No: 121-69-7
Beilstein No: 12^3, 245
Name: Benzenamine, *N*,*N*-dimethyl-
Synonym: Aniline, *N*,*N*-dimethyl-
Line Form: $C_6H_5N(CH_3)_2$
Mol Form: $C_8H_{11}N$
M_r: 121.17
bp: 194,77^{13}
mp: 2.5
***d*:** $0.9557^{20/4}$
***n*:** 1.5582^{20}
Color: pa ye
***s*:** H_2O 2; al 3; eth 3; ace 3; bz 3; ctc 3; chl 4; os 3
IR: µCOB 6951 2860 1610 1520 1450 1350 1240 1190 1160 1140 1060 1030 990 940 860 750 690
Raman: SAD 246 3160 3070 2920 2840 2790 1600 1510 1480 1450 1410 1340 1190 1030 990 950 860 800 840 610 510 470 400 300 70
UV: SAD 3 298(2290) 251(11600) MeOH
H NMR: SAD 1 2.9 6.4-6.7 7.1 CCl_4
MS: WILEY 490 120(100) 121(70) 77(25) 51(16) 105(13) 104(11) 42(10) 50(7)
m/e: 121.09

HODOC No: 02192
CAS No: 5882-44-0
Beilstein No: 12^3, 251
Name: Benzenamine, *N*,*N*-dimethyl-, hydrochloride
Synonym: Aniline, *N*,*N*-dimethyl-, hydrochloride
Line Form: $C_6H_5N(CH_3)_2{\cdot}HCl$
Mol Form: $C_8H_{12}ClFN_{28}$
M_r: 554.81
mp: 85-95
***d*:** $1.1156^{19/4}$
Color: hyg pl (w, Bz)
***s*:** H_2O 4; EtOH 4; Chl 4
m/e: 554.15

HODOC No: 02193
CAS No: 874-52-2
Beilstein No: 12^2, 88
Name: Benzenamine, *N*,*N*-dimethyl-, *N*-oxide
Synonym: Aniline, *N*,*N*-dimethyl-, *N*-oxide
Line Form: $C_6H_5N(O)(CH_3)_2$
Mol Form: $C_8H_{11}NO$
M_r: 137.17
mp: 152-3
***s*:** H_2O 4; al 4; eth 2; chl 4
UV: OES 3-164 290(3) 255(251) EtOH
m/e: 137.08

HODOC No: 02194
CAS No: 61149-61-9
Beilstein No: 12^3, 1704
Name: Benzenamine, *N*,*N*-dimethyl-2,5-dinitro-
Synonym: Aniline, *N*,*N*-dimethyl-2,5-dinitro
Line Form: $2,5\text{-}(NO_2)_2C_6H_3N(CH_3)_2$
Mol Form: $C_8H_9N_3O_4$
M_r: 211.17
mp: 113
IR: SADG 19848 3100 2950 2870 2800 1620 1530 1450 1430 1410 1340 1300 1270 1200 1170 1130 1100 1060 980 870 830 810 770 730 700 620 550 430 390
UV: SAD 17567 442(2800) 251(19700) MeOH
H NMR: SAD 11642 3.0 7.6 7.8 $CDCl_3$
m/e: 211.06

HODOC No: 02195
CAS No: 35998-95-9
Beilstein No: 12^3, 1682
Name: Benzenamine, *N,N*-dimethyl-3,4-dinitro-
Synonym: Aniline, *N,N*-dimethyl-3,4-dinitro-
Line Form: 3,4-$(NO_2)_2$-$C_6H_3N(CH_3)_2$
Mol Form: $C_8H_9N_3O_4$
M_r: 211.17
mp: 175-6
IR: SADG 21262 3120 2900 2680 1600 1560 1530 1490 1430 1410 1380 1360 1310 1260 1220 1190 1140 1050 870 840 820 780 740 680 560 340
UV: SAD 17568 390(14500) 228(12300) MeOH
H NMR: SAD 11542 3.1 6.9 7.1 8.1 $DMSOd_6$
m/e: 211.06

HODOC No: 02196
CAS No: 5393-49-7
Beilstein No: 12, 203
Name: Benzenamine, *N,N'*-(1,2-dimethyl-1,2-ethanediylidene)*bis*-
Synonym: Aniline, *N,N'*-(dimethylethanediylidene)di-
Line Form: C_6H_5N=$C(CH_3)C(CH_3)$=NC_6H_5
Mol Form: $C_{16}H_{16}N_2$
M_r: 236.30
mp: 139
***s*:** al 2; eth 2; lig 2
UV: LANG 1332 332(2884) 226(32359) EtOH
m/e: 236.13

HODOC No: 02197
CAS No: 6310-21-0
Beilstein No: 12^3, 2726
Name: Benzenamine, 2-(1,1-dimethylethyl)-
Synonym: Aniline, *o-tert*-butyl-
Line Form: 2-$H_2NC_6H_4$-$C(CH_3)_3$
Mol Form: $C_{10}H_{15}N$
M_r: 149.23
bp: 233-5
***d*:** 0.977^{15}
***n*:** 1.5453^{20}
***s*:** al 4; eth 4; bz 4; ac 3
UV: HBCP 286(2239) 237(7763) iso
m/e: 149.12

HODOC No: 02198
CAS No: 5369-19-7
Beilstein No: 12^2, 637
Name: Benzenamine, 3-(1,1-dimethylethyl)-
Synonym: Aniline, *m-tert*-butyl-
Line Form: 3-$H_2NC_6H_4$-$C(CH_3)_3$
Mol Form: $C_{10}H_{15}N$
M_r: 149.23
bp: 229^{708}
***s*:** Bz 4; Et_2O 4; EtOH 4
m/e: 149.12

HODOC No: 02199
CAS No: 769-92-6
Beilstein No: 12^3, 2728
Name: Benzenamine, 4-(1,1-dimethylethyl)-
Synonym: Aniline, *p-tert*-butyl-
Line Form: 4-$H_2NC_6H_4$-$C(CH_3)_3$
Mol Form: $C_{10}H_{15}N$
M_r: 149.23
bp: 240^{740}
mp: 17
***d*:** $0.9525^{15/4}$
***n*:** 1.5380^{20}
α: 1.5380^{20}
Color: ye rd (peth)
***s*:** H_2O 2; al 5; eth 5; bz 4; ctc 3
IR: SADG 15799 3448 3333 3030 2941 1613 1515 1471 1389 1370 1266 1190 1124 833 741
UV: SAD 4680 291 237 cyhex
H NMR: SAD 6449 1.3 3.2 6.4 7.0 CCl_4
m/e: 149.12

HODOC No: 02200
CAS No: 937-33-7
Beilstein No: 12^3, 270
Name: Benzenamine, *N*-(1,1-dimethylethyl)-
Synonym: Aniline, *N-tert*-butyl-
Line Form: $C_6H_5NHC(CH_3)_3$
Mol Form: $C_{10}H_{15}N$
M_r: 149.23
bp: $214\text{-}6^{753}$,$93\text{-}8^{19}$
***n*:** 1.5270^{20}
***s*:** al 3; ace 4; bz 4; chl 4; os 5
UV: OES 3-265 260(162) 254(195) 248(155) 0.04N HCl:48% EtOH 235(7586) 0.04N NaOH:48% EtOH
m/e: 149.12

HODOC No: 02201
CAS No: 15408-59-0
Name: Benzenamine, *N*-(1,1-dimethylethyl)-, hydrochloride
Synonym: Aniline, *N*-*tert*-butyl-, hydrochloride
Line Form: $C_6H_5NHC(CH_3)_3$–HCl
Mol Form: $C_{10}H_{16}ClN$
Mr: 185.69
IR: **µCOB 5066** 2940 2630 2080 1610 1490 1390 1370 1280 1180 780 730 700
m/e: 185.10

HODOC No: 02202
CAS No: 607-59-0
Beilstein No: 16^3, 361
Name: Benzenamine, *N,N*-dimethyl-4-(1-naphthalenylazo)-
Synonym: Aniline, *N,N*-dimethyl-*p*-(1-naphthylazo)-
Line Form: 4-$(CH_3)_2N$-C_6H_4-N=N-(1-$C_{10}H_7$)
Mol Form: $C_{18}H_{17}N_3$
Mr: 275.34
mp: 192-4
IR: COB 3883
UV: OES 3-689
m/e: 275.14

HODOC No: 02203
CAS No: 613-65-0
Beilstein No: 16^3, 362
Name: Benzenamine, *N,N*-dimethyl-4-(2-naphthalenylazo)-
Synonym: Aniline, *N,N*-dimethyl-*p*-(2-naphthylazo)-
Line Form: 4-$(CH_3)_2NC_6H_4N$=N(2-$C_{10}H_7$)
Mol Form: $C_{18}H_{17}N_3$
Mr: 275.34
mp: 174
***s*:** Bz 4
m/e: 275.14

HODOC No: 02204
CAS No: 80879-86-3
Beilstein No: 12^3, 2442
Name: Benzenamine, 2,3-dimethyl-4-nitro-
Synonym: 2,3-Xylidine, 4-nitro-
Line Form: (2,3-$(CH_3)_2$-4-NO_2)$C_6H_2NH_2$
Mol Form: $C_8H_{10}N_2O_2$
Mr: 166.17
mp: 114
Color: ye pr (al)
***s*:** al 4
UV: OES 4-175 382(9772) EtOH
m/e: 166.07

HODOC No: 02205
CAS No: 31167-04-1
Beilstein No: 12^2, 612
Name: Benzenamine, 2,4-dimethyl-3-nitro-
Synonym: 2,4-Xylidine, 3-nitro-
Line Form: 2,4-$(CH_3)_2$-3-NO_2-$C_6H_2NH_2$
Mol Form: $C_8H_{10}N_2O_2$
Mr: 166.17
mp: 81-2
Color: ye nd
***s*:** EtOH 4; lig 4
m/e: 166.07

HODOC No: 02206
CAS No: 2124-47-2
Beilstein No: 12^2, 612
Name: Benzenamine, 2,4-dimethyl-5-nitro-
Synonym: 2,4-Xylidine, 5-nitro-
Line Form: 2,4-$(CH_3)_2$-5-NO_2-$C_6H_2NH_2$
Mol Form: $C_8H_{10}N_2O_2$
Mr: 166.17
mp: 123
Color: og ye nd (al)
***s*:** EtOH 4
m/e: 166.07

HODOC No: 02207
CAS No: 1635-84-3
Beilstein No: 12^2, 612
Name: Benzenamine, 2,4-dimethyl-6-nitro-
Synonym: 2,4-Xylidine, 6-nitro-
Line Form: 2,4-$(CH_3)_2$-6-NO_2-$C_6H_2NH_2$
Mol Form: $C_8H_{10}N_2O_2$
Mr: 166.17
mp: 76
Color: og red nd or pl (lig)
***s*:** EtOH 4
m/e: 166.07

HODOC No: 02208
CAS No: 67083-28-7
Beilstein No: 12^2, 605
Name: Benzenamine, 2,6-dimethyl-3-nitro-
Synonym: 2,6-Dimethyl-3-nitroaniline
Line Form: 2,6-$(CH_3)_2$-3-$NO_2C_6H_2NH_2$
Mol Form: $C_8H_{10}N_2O_2$
Mr: 166.17
mp: 81-2
Color: ye nd (dil al)
***s*:** al 3
UV: HBCP 380(14454) al
m/e: 166.07

HODOC No: 02209
CAS No: 16947-63-0
Beilstein No: 12^3, 2469
Name: Benzenamine, 2,6-dimethyl-4-nitro-
Synonym: 2,6-Xylidine, 4-nitro-
Line Form: $4\text{-}NO_2\text{-}2,6\text{-}(CH_3)_2C_6H_2NH_2$
Mol Form: $C_8H_{10}N_2O_2$
M_r: 166.17
mp: 76
Color: og-red nd or pl(lig)
***s*:** chl 2
IR: SADG 19835 3490 3400 2960 2800 1630 1590 1480 1460 1420 1320 1300 1280 1230 1090 880 740 580 460 380
UV: SAD 17557 376(15100) MeOH
H NMR: SAD 11637 2.2 4.3 7.9 $CDCl_3$
m/e: 166.07

HODOC No: 02210
CAS No: 64823-23-0
Beilstein No: 12^3, 2454
Name: Benzenamine, 3,4-dimethyl-2-nitro-
Synonym: 3,4-Dimethyl-2-nitroaniline
Line Form: $3,4\text{-}(CH_3)_2\text{-}2\text{-}NO_2\text{-}C_6H_2NH_2$
Mol Form: $C_8H_{10}N_2O_2$
M_r: 166.17
mp: 65-6
Color: red pr (al)
***s*:** EtOH 4
m/e: 166.07

HODOC No: 02211
CAS No: 64823-22-9
Beilstein No: 12, 1106
Name: Benzenamine, 3,4-dimethyl-5-nitro-
Synonym: 3,4-Xylidine, 5-nitro-
Line Form: $3,4(CH_3)_2\text{-}5\text{-}NO_2\text{-}C_6H_2NH_2$
Mol Form: $C_8H_{10}N_2O_2$
M_r: 166.17
mp: 74-5
Color: og lf (al)
***s*:** EtOH 4
m/e: 166.07

HODOC No: 02212
CAS No: 34761-82-5
Beilstein No: 12^3, 2500
Name: Benzenamine, 3,5-dimethyl-4-nitro-
Synonym: 3,5-Dimethyl-*p*-nitroaniline
Line Form: $(3,5\text{-}(CH_3)_2\text{-}4\text{-}NO_2)C_6H_2NH_2$
Mol Form: $C_8H_{10}N_2O_2$
M_r: 166.17
mp: 133
Color: og pr (Bz) lf (lig)
***s*:** Bz 4; EtOH 4
m/e: 166.07

HODOC No: 02213
CAS No: 6972-71-0
Beilstein No: 12^3, 2455
Name: Benzenamine, 4,5-dimethyl-2-nitro-
Synonym: 3,4-Xylidine, 6-nitro-
Line Form: $(4,5\text{-}(CH_3)_2\text{-}2\text{-}NO_2)C_6H_2NH_2$
Mol Form: $C_8H_{10}N_2O_2$
M_r: 166.17
mp: 140
Color: red-br pr (al)
***s*:** eth 2; ace 3; bz 3; os 4; lig 1
IR: SADG 16696 3500 3400 2990 1640 1580 1500 1410 1360 1310 1240 1150 1060 870 840 760 730
m/e: 166.07

HODOC No: 02214
CAS No: 610-17-3
Beilstein No: 12^3, 1516
Name: Benzenamine, *N,N*-dimethyl-2-nitro-
Synonym: Aniline, *N,N*-dimethyl-*o*-nitro-
Line Form: $2\text{-}NO_2\text{-}C_6H_4N(CH_3)_2$
Mol Form: $C_8H_{10}N_2O_2$
M_r: 166.17
bp: 146^{20}
mp: -20
***d*:** $1.1794^{20/4}$
***n*:** 1.6102^{20}
Color: ye-og
***s*:** H_2O 3; al 4; eth 3; chl 4
UV: OES 1-207 429(1995) 246(12589) EtOH
m/e: 166.07

HODOC No: 02215
CAS No: 619-31-8
Beilstein No: 12^3, 1544
Name: Benzenamine, *N,N*-dimethyl-3-nitro-
Synonym: Aniline, *N,N*-dimethyl-*m*-nitro-
Line Form: $3\text{-}NO_2\text{-}C_6H_4N(CH_3)_2$
Mol Form: $C_8H_{10}N_2O_2$
M$_r$: 166.17
bp: 280-5
mp: 60-1
***d*:** 1.313^{17}
Color: red mcl pr (eth)
***s*:** H_2O 1; al 3; eth 3
IR: **µCOB 1418** 2940 1640 1540 1470 1350 1240 1110 1060 1000 970 880 840 780 780 740 670
UV: **SAD 8994** 250(2020) MeOH
H NMR: **SAD 14846** 3.0 6.9 7.3 $CDCl_3$
MS: **ALD 5:151B**
m/e: 166.07

HODOC No: 02216
CAS No: 100-23-2
Beilstein No: 12^3, 1583
Name: Benzenamine, *N,N*-dimethyl-4-nitro-
Synonym: Aniline, *N,N*-dimethyl-*p*-nitro-
Line Form: $4\text{-}NO_2\text{-}C_6H_4N(CH_3)_2$
Mol Form: $C_8H_{10}N_2O_2$
M$_r$: 166.17
mp: 164-5
Color: ye nd (al)
***s*:** H_2O 1; al 3; eth 3; HOAc 3
IR: **SADG 18646** 2940 1600 1530 1490 1450 1380 1320 1230 1190 1120 1060 940 810 750 690
UV: **SAD 4313** 231 MeOH
m/e: 166.07

HODOC No: 02217
CAS No: 4584-57-0
Beilstein No: 12^3, 3311
Name: Benzenamine, *N,N*-dimethyl-4-[2-(4-nitrophenyl)ethenyl]-
Synonym: 4-Stilbenamine, *N,N*-dimethyl-4′-nitro-
Line Form: $4\text{-}(CH_3)_2N\text{-}C_6H_4CH{=}CH\text{-}(C_6H_4NO_2\text{-}4)$
Mol Form: $C_{16}H_{16}N_2O_2$
M$_r$: 268.30
mp: 256-9
m/e: 268.12

HODOC No: 02218
CAS No: 2844-15-7
Beilstein No: 12, 553
Name: Benzenamine, *N,N*-dimethyl-4-[2-(4-nitrophenyl)ethenyl]-, (E)-
Synonym: 4-Stilbenamine, *N,N*-dimethyl-4′-nitro-, (E)-
Mol Form: $C_{16}H_{16}N_2O_2$
M$_r$: 268.30
mp: 250
***s*:** H_2O 1; al 3; bz 3; chl 3; HOAc 3
UV: **OES 6-670** 432(30200) bz
H NMR: **SAD 19307** ini 3.6 7.3 7.7 7.7 8.2 tfa
m/e: 268.12

HODOC No: 02219
CAS No: 14064-72-3
Name: Benzenamine, *N,N*-dimethyl-4-[2-(4-nitrophenyl)ethenyl]-, (Z)-
Synonym: 4-Stilbenamine, *N,N*-dimethyl-4′-nitro-, (Z)-
Mol Form: $C_{16}H_{16}N_2O_2$
M$_r$: 268.30
mp: 104
***s*:** H_2O 1; al 3; bz 3; chl 3; HOAc 3
IR: **SADG 31630** ini 2900 2810 1600 1580 1520 1500 1450 1360 1330 1220 1190 1110 1060 960 940 870 840 800 750 680 520
UV: **OES 6-670** 418(8710) bz
H NMR: **SAD 19307** ini 3.6 7.3 7.7 7.7 8.2 tfa
m/e: 268.12

HODOC No: 02220
CAS No: 138-89-6
Beilstein No: 12^3, 1509
Name: Benzenamine, *N,N*-dimethyl-4-nitroso-
Synonym: Aniline, *N,N*-dimethyl-*p*-nitroso-
Line Form: $4\text{-}ON\text{-}C_6H_4N(CH_3)_2$
Mol Form: $C_8H_{10}N_2O$
M$_r$: 150.17
mp: 92-3
***d*:** 1.145^{20}
Color: gr pl (eth)
***s*:** H_2O 2; al 3; eth 3; chl 3; $HCONH_2$ 3
IR: **µSADP 5891** 2860 1590 1520 1450 1370 1330 1270 1220 1190 1090 820 700
UV: **SAD 2908** 273 MeOH
H NMR: **SAD 6829** 3.1 6.6 7.7 $CDCl_3$
MS: **ALDN 104** 150(100) 42(95) 77(69) 120(50) 119(42) 79(39) 104(39) 78(33)
m/e: 150.08

HODOC No: 02221
CAS No: 41165-12-2
Beilstein No: 16^1, 323
Name: Benzenamine, 2,6-dimethyl-4-(phenylazo)-
Synonym: 3,5-Dimethyl-4-aminoazobenzene
Mol Form: $C_{14}H_{15}N_3$
Mr: 225.28
mp: 101
***s*:** H_2O 2; al 3; eth 3; ace 3; bz 3; chl 3; os 3; lig 3
UV: CHBEA 89-2756-56 385(7943) 270(50119) al
m/e: 225.13

HODOC No: 02222
CAS No: 60-11-7
Beilstein No: 16^3, 340
Name: Benzenamine, *N,N*-dimethyl-4-(phenylazo)-
Synonym: Brilliant Fast Oil Yellow
Line Form: $4\text{-}(CH_3)_2N\text{-}C_6H_4N{=}NC_6H_5$
Mol Form: $C_{14}H_{15}N_3$
Mr: 225.28
bp: d
mp: 117
Color: ye lf (al)
***s*:** H_2O 1; al 4; eth 3; chl 2; py 4; ac 3; lig 2
IR: µSADP 347 2940 1610 1520 1470 1450 1410 1370 1320 1240 1150 1140 1060 940 920 820 760 730 680
UV: SAD 133 313 253 cyhex
H NMR: SAD 10351 3.0 6.7 7.2-7.6 7.8 $CDCl_3$
m/e: 225.13

HODOC No: 02223
CAS No: 838-95-9
Beilstein No: 12^3, 3306
Name: Benzenamine, *N,N*-dimethyl-4-(2-phenylethenyl)-, (E)-
Synonym: 4-Stilbenamine, *N,N*-dimethyl-, (E)-
Mol Form: $C_{16}H_{17}N$
Mr: 223.30
bp: $145^{0.1}$
mp: 150
***s*:** H_2O 1; al 3; chl 3
UV: TETRA 23-173-67 345(30903) MeOH
m/e: 223.14

HODOC No: 02224
CAS No: 14301-11-2
Name: Benzenamine, *N,N*-dimethyl-4-(2-phenylethenyl)-, (Z)-
Synonym: 4-Stilbenamine, *N,N*-dimethyl-, (Z)-
Line Form: $4\text{-}(CH_3)_2N\text{-}C_6H_4CH{=}CH\text{-}C_6H_5$
Mol Form: $C_{16}H_{17}N$
Mr: 223.30
mp: 28-9
***s*:** H_2O 1; al 3
UV: HCACA 48-517-65 320(13489) 80% al
m/e: 223.14

HODOC No: 02225
CAS No: 42881-80-1
Name: Benzenamine, *N,N*-dimethyl-4-(phenylthio)-
Synonym: Aniline, *N,N*-dimethyl-*p*-(phenylthio)-
Mol Form: $C_{14}H_{15}NS$
Mr: 229.33
mp: 66.7-7.3
IR: µSADP 9233 3130 2940 1590 1520 1490 1450 1370 1240 1190 1110 1060 1030 940 820 750 690
UV: SAD 2388 277(25600) MeOH
H NMR: SAD 11815 2.9 6.6 7.1 7.3 $CDCl_3$
m/e: 229.09

HODOC No: 02226
CAS No: 2049-92-5
Beilstein No: 12, 1179
Name: Benzenamine, 4-(1,1-dimethylpropyl)-
Synonym: *p-tert*-Amylaniline
Mol Form: $C_{11}H_{17}N$
Mr: 163.25
bp: 259-62
IR: µCOB 175 3330 3230 3030 2940 1610 1520 1470 1390 1370 1280 1190 1010 830 780 750 690
m/e: 163.14

HODOC No: 02227
CAS No: 52610-17-0
Name: Benzenamine, *N,N*-dimethyl-4-(1H-1,2,4-triazol-3-ylazo)-
Synonym: 1,2,4-Triazole, 3(4-dimethylamino-phenylazo)-1-hydro
Mol Form: $C_{10}H_{12}N_6$
Mr: 216.24
IR: SADG 10859 3100 2890 1600 1520 1460 1360 1250 1130 930 800 620 530
UV: SAD 13287 444(28100) 272(9170) MeOH
H NMR: SAD 11716 3.1 3.4 6.9 7.8 8.4 $DMSOd_6$
m/e: 216.11

HODOC No: 02228
CAS No: 328-96-1
Name: Benzenamine, *N,N*-dimethyl-4-[[3-(trifluoromethyl)phenyl]azo]-
Synonym: Aniline, *N,N*-dimethyl-*p*-[(α,α,α-trifluoro-*m*-tolyl)azo]-
Line Form: 4-$(CH_3)_2N$-C_6H_4N=$N(C_6H_4CF_3$-3)
Mol Form: $C_{15}H_{14}F_3N_3$
M_r: 293.28
bp: 76^{8}
mp: 77-9
***s*:** chl 3
IR: µSADP 343 2940 1610 1520 1470 1370 1330 1140 1060 940 820 690
UV: LANG 2427 285(11482) 231(23988) MeOH 297(12882) 215(28840) 0.01N KOH
H NMR: SAD 10418 2.9 6.6 7.5 7.8 8.1 $CDCl_3$
m/e: 293.11

HODOC No: 02229
CAS No: 602-03-9
Beilstein No: 12^{3}, 1680
Name: Benzenamine, 2,3-dinitro-
Synonym: Aniline, 2,3-dinitro-
Line Form: 2,3-$(NO_2)_2C_6H_3NH_2$
Mol Form: $C_6H_5N_3O_4$
M_r: 183.12
mp: 178
***d*:** 1.646^{50}
***s*:** H_2O 1; al 3; eth 2
IR: µSADP 11388 3570 3450 3130 1610 1590 1560 1520 1470 1450 1350 1280 1160 1120 1040 860 810 800 730 700
UV: SAD 3120 227 MeOH
m/e: 183.03

HODOC No: 02230
CAS No: 97-02-9
Beilstein No: 12^{3}, 1681
Name: Benzenamine, 2,4-dinitro-
Synonym: Aniline, 2,4-dinitro-
Line Form: 2,4-$(NO_2)_2C_6H_3NH_2$
Mol Form: $C_6H_5N_3O_4$
M_r: 183.12
mp: 180
***d*:** 1.615^{14}
Color: ye nd (dil Ace) gr ye ta (al)
***s*:** H_2O 1; al 2; ace 2; HCl 2
IR: µCOB 6260 3450 3330 1640 1590 1520 1490 1430 1390 1330 1270 1120 1060 940 840 750 720 700 640
UV: SAD 387 336(1480) 257(969) 225(1110) 210(1120) MeOH
H NMR: SAD 3290 3.1 7.2 8.2 8.9 $aceD_6$
MS: NBS 953
m/e: 183.03

HODOC No: 02231
CAS No: 619-18-1
Beilstein No: 12^{3}, 1704
Name: Benzenamine, 2,5-dinitro-
Synonym: Aniline, 2,5-dinitro-
Line Form: 2,5-$(NO_2)_2$-$C_6H_3NH_2$
Mol Form: $C_6H_5N_3O_4$
M_r: 183.12
mp: 137
Color: og nd (al)
***s*:** EtOH 4
m/e: 183.03

HODOC No: 02232
CAS No: 606-22-4
Beilstein No: 12^{3}, 1704
Name: Benzenamine, 2,6-dinitro-
Synonym: Aniline, 2,6-dinitro-
Line Form: 2,6-$(NO_2)_2$-$C_6H_3NH_2$
Mol Form: $C_6H_5N_3O_4$
M_r: 183.12
mp: 141-2
Color: gold lf (50% HOAc) ye nd (al)
***s*:** H_2O 1; al 2; eth 3; bz 3; lig 1
IR: µSADP 17483 3450 3330 1640 1560 1520 1430 1350 1270 1100 890 740
UV: SAD 5551 317 250 221 MeOH
MS: NBS 953
m/e: 183.03

HODOC No: 02233
CAS No: 618-87-1
Beilstein No: 12^3, 1705
Name: Benzenamine, 3,5-dinitro-
Synonym: Aniline, 3,5-dinitro-
Line Form: $3,5\text{-}(NO_2)_2C_6H_3NH_2$
Mol Form: $C_6H_5N_3O_4$
Mr: 183.12
mp: 163
***d*:** 1.601^{50}
Color: ye nd (dil al)
***s*:** H_2O 1; al 3; eth 3; ace 2; bz 2
IR: SADG 778 3500 3400 3100 1640 1620 1530 1450 1350 1310 920 880 810 730
UV: SAD 9712 395(2180) 255(11900) 227(26900) MeOH
H NMR: SAD 1836 6.0 7.8 8.1 $aceD_6$
m/e: 183.03

HODOC No: 02234
CAS No: 961-68-2
Beilstein No: 12^3, 1683
Name: Benzenamine, 2,4-dinitro-*N*-phenyl-
Synonym: Diphenylamine, 2,4-dinitro-
Line Form: $2,4\text{-}(NO_2)_2C_6H_3NHC_6H_5$
Structure:

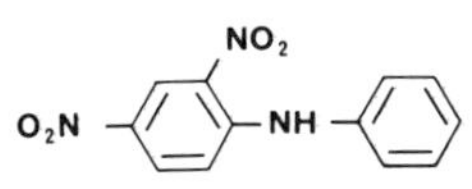

Mol Form: $C_{12}H_9N_3O_4$
Mr: 259.21
mp: 157
Color: ye red nd (al)
***s*:** H_2O 1; al 3; eth 2; ace 3; bz 2; chl 3 $DMSOd_6$ 2; py 3
IR: SADG 15685 3333 3125 3030 1695 1613 1587 1515 1493 1429 1316 1266 1220 1136 1053 917 840 746 690
UV: SAD 15779 352(17600) 230(13200) MeOH
H NMR: SAD 6045 7.2 7.5 8.3 8.9 10.2 $DMSOd_6$
MS: NBS 1847
m/e: 259.06

HODOC No: 02235
CAS No: 13744-81-5
Beilstein No: 12^3, 1705
Name: Benzenamine, 2,6-dinitro-*N*-phenyl-
Synonym: Diphenylamine, 2,6-dinitro-
Line Form: $2,6\text{-}(NO_2)_2C_6H_3NHC_6H_5$
Mol Form: $C_{12}H_9N_3O_4$
Mr: 259.21
mp: 107-8
Color: og lf(al, HOAc)
***s*:** EtOH 4; HOAc 4
m/e: 259.06

HODOC No: 02236
CAS No: 2908-76-1
Beilstein No: 12^3, 1684
Name: Benzenamine, *N*-(2,4-dinitrophenyl)-2,4-dinitro-
Synonym: Diphenylamine, 2,2′,4,4′-tetranitro-
Line Form: $[2,4\text{-}(NO_2)_2C_6H_3]_2NH$
Mol Form: $C_{12}H_7N_5O_8$
Mr: 349.20
mp: 201
Color: ye nd or cr (HOAc)
***s*:** H_2O 1; al 2; eth 2; ace 2; bz 2; chl 2; AcOEt 2; HOAc 3; py 3
UV: OES 7-373 413(21380) H_2O
m/e: 349.03

HODOC No: 02237
CAS No: 603-34-9
Beilstein No: 12^3, 292
Name: Benzenamine, *N,N*-diphenyl-
Synonym: *N,N*-Diphenylaniline
Line Form: $(C_6H_5)_3N$
Mol Form: $C_{18}H_{15}N$
Mr: 245.31
bp: 365
mp: 127
***d*:** $0.774^{0/0}$
***n*:** 1.353^{16}
Color: mcl (MeOH, Bz)
***s*:** H_2O 1; al 2; eth 3; bz 3; MeOH 3
IR: COB 7593 3065 3035 1580 1505 1460 1445 1330 1310 1280 1170 1150 1075 1030 995 890 750 700 625 615 520 510
UV: SAD 20992 296(24800) MeOH
H NMR: SAD 14449 6.8-7.4 $CDCl_3$
m/e: 245.12

HODOC No: 02238
CAS No: 7510-33-0
Beilstein No: 12^2, 125
Name: Benzenamine, *N,N'*-(1,2-diphenyl-1,2-ethanediylidene)*bis*-
Synonym: Aniline, *N,N'*-(diphenylethanediylidene)di-
Line Form: $C_6H_5N=C(C_6H_5)C(C_6H_5)=NC_6H_5$
Mol Form: $C_{26}H_{20}N_2$
Mr: 360.43
mp: 144
***s*:** Bz 4; Chl 4
UV: LANG 1333
m/e: 360.16

HODOC No: 02239
CAS No: 652-45-9
Beilstein No: 12^2, 790
Name: Benzenamine, 3-(diphenylmethyl)-
Synonym: Triphenylmethane, 3-amino
Line Form: $(3\text{-}H_2NC_6H_4)CH\text{-}(C_6H_5)_2$
Mol Form: $C_{19}H_{17}N$
Mr: 259.33
mp: 120
Color: nd (eth)
***s*:** Et_2O 4; EtOH 4
m/e: 259.14

HODOC No: 02240
CAS No: 603-38-3
Beilstein No: 12^2, 790
Name: Benzenamine, 4-(diphenylmethyl)-
Synonym: *p*-(Diphenylmethyl)aniline
Line Form: $(4\text{-}H_2NC_6H_4)CH(C_6H_5)_2$
Mol Form: $C_{19}H_{17}N$
Mr: 259.33
bp: ca 248^{12}
mp: 84-5
Color: pr or lf (eth, lig) ta (peth)
***s*:** Bz 4; Et_2O 4; lig 4
m/e: 259.14

HODOC No: 02241
CAS No: 574-45-8
Name: Benzenamine, *N*-(diphenylmethylene)-
Synonym: Aniline, *N*-(diphenylmethylene)-
Line Form: $(C_6H_5)_2C=NC_6H_5$
Mol Form: $C_{19}H_{15}N$
Mr: 257.32
mp: 113
IR: SADG 8561 3050 1610 1590 1570 1490 1480 1445 1315 1295 1275 1220 1140 1070 1025 960 925 775 760 715 705 695 680 650 590 550
UV: LANG 208 344(2399) 244(20417) os
m/e: 257.12

HODOC No: 02242
CAS No: 86926-11-6
Name: Benzenamine, 2-(diphenylphosphino)-*N,N*-diethyl-
Synonym: Phosphine, 2-diethylaminophenyl-diphenyl
Line Form: $[2\text{-}(C_2H_5)_2NC_6H_4]\text{-}P(C_6H_5)_2$
Mol Form: $C_{22}H_{24}NP$
Mr: 333.39
mp: 121-2
UV: LANG 1344
m/e: 333.16

HODOC No: 02243
CAS No: 5931-54-4
Name: Benzenamine, 3-(diphenylphosphino)-*N,N*-dimethyl-
Synonym: Aniline, *m*-(diphenylphosphino)-*N,N*-dimethyl-
Mol Form: $C_{20}H_{20}NP$
Mr: 305.34
mp: 77-8
UV: LANG 1343 314(2754) 262(20417) EtOH
m/e: 305.13

HODOC No: 02244
CAS No: 2217-07-4
Beilstein No: 12^3, 268
Name: Benzenamine, *N,N*-dipropyl-
Synonym: Aniline, *N,N*-dipropyl-
Line Form: $C_6H_5N(C_3H_7)_2$
Mol Form: $C_{12}H_{19}N$
M_r: 177.28
bp: 245, 127^{10}
***d*:** 0.9104^{20}
***n*:** 1.5271^{20}
Color: ye lf
***s*:** H_2O 1; al 3; eth 3; ace 3; bz 3; ctc 2
IR: SADG 2180 2941 2857 1587 1493 1471 1351 1282 1220 1190 1149 1099 1031 980 877 741 680
UV: SAD 9792 305(2330) 259(16900) MeOH
H NMR: SAD 431 0.9 1.6 3.2 6.4-7.2 CCl_4
m/e: 177.15

HODOC No: 02245
CAS No: 63870-44-0
Beilstein No: 13^1, 128
Name: Benzenamine, 2,2′-diselenobis-
Synonym: Diselenide, *bis*(2-aminophenyl)
Structure:

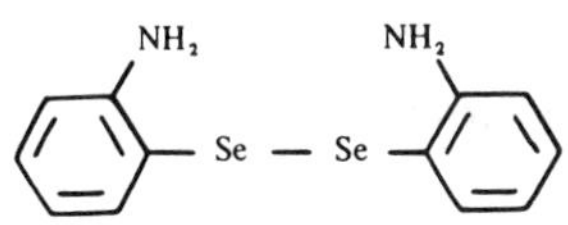

Mol Form: $C_{12}H_{12}N_2Se_2$
M_r: 342.15
mp: 81
***s*:** H_2O 2; al 3; eth 3; bz 3; chl 3; lig 2
UV: LANG 285 362(3981) 235(19055) EtOH
m/e: 343.93

HODOC No: 02246
CAS No: 1141-88-4
Beilstein No: 13^3, 907
Name: Benzenamine, 2,2′-dithiobis-
Synonym: Aniline, 2,2′-dithiodi-
Line Form: $(2\text{-}H_2NC_6H_4)_2S_2$
Mol Form: $C_{12}H_{12}N_2S_2$
M_r: 248.35
mp: 93
***s*:** H_2O 1; al 4; eth 3
IR: µSADP 8895 3450 3330 3030 2940 1610 1470 1450 1300 1250 1150 1030 860 850 750 700
UV: SAD 2289 335 MeOH
H NMR: SAD 11085 4.2 6.5 6.6 7.1 7.2 $CDCl_3$
m/e: 248.04

HODOC No: 02247
CAS No: 722-27-0
Beilstein No: 13^3, 1248
Name: Benzenamine, 4,4′-dithiobis-
Synonym: Aniline, 4,4′-dithiodi-
Line Form: $(4\text{-}H_2NC_6H_4)_2S_2$
Mol Form: $C_{12}H_{12}N_2S_2$
M_r: 248.35
mp: 85
***s*:** H_2O 3; al 4; eth 4; bz 2; chl 4; lig 2
IR: COB 4353 3420 3350 3020 1630 1600 1500 1430 1180 1090 830
UV: HBCP 294(12882) 256(16982) al
m/e: 248.04

HODOC No: 02248
CAS No: 104-42-7
Beilstein No: 12^3, 2776
Name: Benzenamine, 4-dodecyl-
Synonym: Aniline, *p*-dodecyl-
Line Form: $4\text{-}H_2NC_6H_4C_{12}H_{25}$
Mol Form: $C_{18}H_{31}N$
M_r: 261.43
mp: 41-2
IR: COB 3093
m/e: 261.25

HODOC No: 02249
CAS No: 621-95-4
Beilstein No: 13^3, 470
Name: Benzenamine, 4,4′-(1,2-ethanediyl)*bis*-
Synonym: 1,2-Bis(*p*-aminophenyl)ethane
Line Form: $4\text{-}H_2NC_6H_4CH_2CH_2(C_6H_4NH_2\text{-}4)$
Structure:

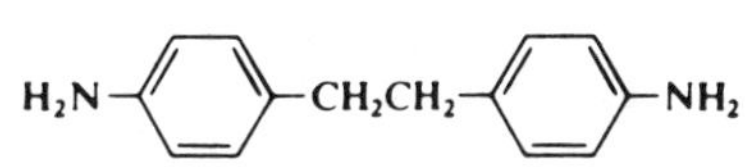

Mol Form: $C_{14}H_{16}N_2$
M_r: 212.28
bp: sub
mp: 136-8
Color: pl (w)
***s*:** H_2O 1; al 4
IR: µSADP 20065 3450 3330 3030 2940 1640 1520 1450 1280 1270 1180 1150 1110 830 760 730 700
MS: AMS 2-382 106(100) 107(8) 212(7) 77(6) 79(3) 78(2) 210(2) 140(2)
m/e: 212.13

HODOC No: 02250
CAS No: 30834-75-4
Name: Benzenamine, *N,N'*-1,2-ethanediylidenebis-
Synonym: Aniline, *N,N'*-ethanediylidenedi-
Line Form: $C_6H_5N{=}CHCH{=}NC_6H_5$
Mol Form: $C_{14}H_{12}N_2$
M$_r$: 208.25
mp: 42-3
UV: LANG 1331 342(2089) 285(5370) 244(18197) EtOH 340(1862) 285(5248) 246(12890) chl
m/e: 208.10

HODOC No: 02251
CAS No: 617-20-9
Beilstein No: 13^1, 85
Name: Benzenamine, 2,2'-(1,2-ethenediyl)*bis*-
Synonym: 2,2'-Stilbenediamine
Mol Form: $C_{14}H_{14}N_2$
M$_r$: 210.26
mp: 176
***s*:** al 3; ace 3
UV: HBCP 307(28184) 295(28840) 228(16218) al
m/e: 210.12

HODOC No: 02252
CAS No: 28096-87-9
Beilstein No: 13^3, 510
Name: Benzenamine, 2,2'-(1,2-ethenediyl)*bis*-, (E)-
Synonym: 2,2'-Stilbenediamine, (E)-
Line Form: $2\text{-}H_2NC_6H_4CH{=}CH(C_6H_4NH_2\text{-}2)$
Mol Form: $C_{14}H_{14}N_2$
M$_r$: 210.26
mp: 176
Color: gold-ye pr (al)
***s*:** Ac 4
UV: HBCP
m/e: 210.12

HODOC No: 02253
CAS No: 7314-06-9
Beilstein No: 13^3, 513
Name: Benzenamine, 4,4'-(1,2-ethenediyl)*bis*-, (E)-
Synonym: 4,4'-Stilbenediamine, (E)-
Line Form: $4\text{-}H_2NC_6H_4CH{=}CH(C_6H_4NH_2\text{-}4)$
Mol Form: $C_{14}H_{14}N_2$
M$_r$: 210.26
bp: sub
mp: 231
Color: ye nd or lf (al)
***s*:** H_2O 2; bz 2; MeOH 3; CS_2 2
UV: OES 1-573 328(39811) H_2O 328(38019) aa
m/e: 210.12

HODOC No: 02254
CAS No: 3867-18-3
Beilstein No: 12^3, 2785
Name: Benzenamine, 2-ethenyl-
Synonym: Aniline, *o*-vinyl-
Line Form: $2\text{-}H_2NC_6H_4CH{=}CH_2)$
Mol Form: C_8H_9N
M$_r$: 119.16
bp: 112^{20}, $97\text{-}8^{8}$
***d*:** $1.0181^{20/4}$
***n*:** 1.6124^{20}
***s*:** ace 3; bz 3
UV: OES 2-139 316(2512) 230(12589) EtOH
m/e: 119.07

HODOC No: 02255
CAS No: 15411-43-5
Beilstein No: 12^3, 2786
Name: Benzenamine, 3-ethenyl-
Synonym: Aniline, *m*-vinyl-
Line Form: $3\text{-}H_2NC_6H_4CH{=}CH_2$
Mol Form: C_8H_9N
M$_r$: 119.16
bp: $112\text{-}5^{12}$
***d*:** $1.0216^{20/20}$
***n*:** 1.6069^{26}
***s*:** ace 3; bz 3
UV: OES 8-92 225(21380) H_2O
m/e: 119.07

HODOC No: 02256
CAS No: 1520-21-4
Beilstein No: 12^3, 2786
Name: Benzenamine, 4-ethenyl-
Synonym: Aniline, *p*-vinyl-
Line Form: $4\text{-}H_2NC_6H_4CH{=}CH_2$
Mol Form: C_8H_9N
M$_r$: 119.16
bp: 116^{9}
mp: 23.5
***d*:** $1.012^{20/20}$
***n*:** 1.6250^{22}
***s*:** ace 3; bz 3
UV: OES 6-168 277(15849) EtOH
m/e: 119.07

HODOC No: 02257
CAS No: 94-70-2
Beilstein No: 13^3, 756
Name: Benzenamine, 2-ethoxy-
Synonym: *o*-Phenetidine
Line Form: 2-C_2H_5O-$C_6H_4NH_2$
Mol Form: $C_8H_{11}NO$
M_r: 137.17
bp: 232.5, $127\text{-}8^{14}$
mp: 21
***n*:** 1.5560^{20}
***s*:** H_2O 2; al 3; eth 3; ctc 2
IR: µCOB 1464 3570 3330 3230 3130 3030 2940 1640 1520 1490 1470 1390 1350 1280 1220 1150 1120 1050 930 780 740
UV: SAD 1285 288 238 cyhex
H NMR: SAD 236 1.4 3.6 3.9 6.6 CCl_4
m/e: 137.08

HODOC No: 02258
CAS No: 621-33-0
Beilstein No: 13^3, 932
Name: Benzenamine, 3-ethoxy-
Synonym: *m*-Phenetidine
Line Form: 3-C_2H_5O-$C_6H_4NH_2$
Mol Form: $C_8H_{11}NO$
M_r: 137.17
bp: 248, $127\text{-}8^{11}$
***s*:** Et_2O 4; EtOH 4
m/e: 137.08

HODOC No: 02259
CAS No: 156-43-4
Beilstein No: 13^3, 996
Name: Benzenamine, 4-ethoxy-
Synonym: *p*-Phenetidine
Line Form: 4-C_2H_5O-$C_6H_4NH_2$
Mol Form: $C_8H_{11}NO$
M_r: 137.17
bp: 254, 125^{11}
mp: 2.4
***d*:** $1.0652^{16/4}$
***n*:** 1.5528^{20}
***s*:** H_2O 2; al 3; eth 3; chl 3
IR: µCOB 2231 3450 3330 3230 2940 2860 1610 1520 1470 1390 1240 1110 1050 920 820
UV: SAD 1286 304 238 cyhex
H NMR: VAR 208 1.4 3.3 3.9 6.6 6.7 $CDCl_3$
MS: AMS 1-331 108(100) 109(64) 137(59) 80(48) 41(46) 53(24) 27(19) 82(16)
m/e: 137.08

HODOC No: 02260
CAS No: 637-56-9
Beilstein No: 13^2, 224
Name: Benzenamine, 4-ethoxy-, hydrochloride
Synonym: *p*-Phenetidine, hydrochloride
Mol Form: $C_8H_{12}ClNO$
M_r: 173.63
mp: 236-7
***s*:** H_2O 2
IR: µCOB 3190 3330 2860 2630 2560 1610 1520 1470 1370 1270 1180 1040 930 830 810 750 680
H NMR: SAD 17295 1.4 4.1 7.1 7.5 D_2O
m/e: 173.06

HODOC No: 02261
CAS No: 16790-23-1
Beilstein No: 13, 365
Name: Benzenamine, 2-ethoxy-*N,N*-diethyl-
Synonym: *o*-Phenetidine, *N,N*-diethyl-
Line Form: 2-C_2H_5O-$C_6H_4N(C_2H_5)_2$
Mol Form: $C_{12}H_{19}NO$
M_r: 193.28
bp: 231-3
***s*:** Bz 4; Et_2O 4; EtOH 4
m/e: 193.15

HODOC No: 02262
CAS No: 1864-92-2
Beilstein No: 13^3, 942
Name: Benzenamine, 3-ethoxy-*N,N*-diethyl-
Synonym: *m*-Phenetidine, *N,N*-diethyl-
Line Form: 3-C_2H_5O-$C_6H_4N(C_2H_5)_2$
Mol Form: $C_{12}H_{19}NO$
M_r: 193.28
bp: 268-70, 145^{14}
***n*:** 1.5325^{25}
***s*:** al 3; bz 3; HOAc 3
UV: HBCP 400(21379) 236(9333) al
m/e: 193.15

HODOC No: 02263
CAS No: 16383-89-4
Beilstein No: 13^3, 888
Name: Benzenamine, 2-ethoxy-4-nitro-
Synonym: *o*-Phenetidine, 4-nitro-
Line Form: (4-NO_2-2-C_2H_5O)$C_6H_3NH_2$
Mol Form: $C_8H_{10}N_2O_3$
M_r: 182.17
mp: 91
Color: ye nd (dil al)
***s*:** Ac 4; Et_2O 4; EtOH 4
m/e: 182.07

HODOC No: 02264
CAS No: 616-86-4
Beilstein No: 13^2, 286
Name: Benzenamine, 4-ethoxy-2-nitro-
Synonym: *p*-Phenetidine, 2-nitro-
Line Form: $(2\text{-}NO_2\text{-}4\text{-}C_2H_5O)\text{-}C_6H_3NH_2$
Mol Form: $C_8H_{10}N_2O_3$
M$_r$: 182.17
mp: 113
***s*:** al 2; eth 4; chl 4
IR: μCOB 3835 3450 3330 2940 1670 1590 1520 1470 1430 1390 1350 1280 1220 1160 1100 1050 960 840 830 790 760 680
m/e: 182.07

HODOC No: 02265
CAS No: 1777-87-3
Beilstein No: 13^2, 284
Name: Benzenamine, 4-ethoxy-3-nitro-
Synonym: *p*-Phenetidine, 3-nitro-
Line Form: $(3\text{-}NO_2\text{-}4\text{-}C_2H_5O)C_6H_3NH_2$
Mol Form: $C_8H_{10}N_2O_3$
M$_r$: 182.17
mp: 41
Color: og-ye nd (Bz, dil al)
***s*:** Ac 4; Bz 4
m/e: 182.07

HODOC No: 02266
CAS No: 578-54-1
Beilstein No: 12^3, 2374
Name: Benzenamine, 2-ethyl-
Synonym: Aniline, *o*-ethyl-
Line Form: $2\text{-}C_2H_5\text{-}C_6H_4NH_2$
Mol Form: $C_8H_{11}N$
M$_r$: 121.17
bp: 209-10
mp: -43
***d*:** $0.983^{22/4}$
***n*:** 1.5584^{22}
***s*:** H_2O 2; al 4; eth 4; chl 2
IR: μSADP 1601 3450 3330 2940 2860 1640 1590 1490 1450 1390 1280 1150 1060 1030 960 930 840 750
UV: SAD 3239 285 235 cyhex
H NMR: SAD 111 1.2 2.4 3.3 6.3-7.1 $CDCl_3$
C NMR: JJ 305 FT 12.9 23.9 115.2 118.5 126.6 127.8 128.2 144.1 $CDCl_3$
m/e: 121.09

HODOC No: 02267
CAS No: 587-02-0
Beilstein No: 12^3, 2379
Name: Benzenamine, 3-ethyl-
Synonym: Aniline, *m*-ethyl-
Line Form: $3\text{-}C_2H_5\text{-}C_6H_4NH_2$
Mol Form: $C_8H_{11}N$
M$_r$: 121.17
bp: $214\text{-}5^{764}$, $93\text{-}5^{6}$
mp: -64
***d*:** 0.9896
***s*:** Et_2O 4; EtOH 4
m/e: 121.09

HODOC No: 02268
CAS No: 589-16-2
Beilstein No: 12^3, 2380
Name: Benzenamine, 4-ethyl-
Synonym: Aniline, *p*-ethyl-
Line Form: $4\text{-}C_2H_5\text{-}C_6H_4NH_2$
Mol Form: $C_8H_{11}N$
M$_r$: 121.17
bp: 217-8, 92.3^{10}
mp: -4.8
***d*:** $0.9679^{20/4}$
***n*:** 1.5554^{20}
***s*:** H_2O 2; al 4; eth 4; ctc 2
IR: μCOB 308 3570 3450 2940 1610 1520 1280 1190 1120 820
UV: SAD 11849 289(1540) 236(10300) MeOH
H NMR: SAD 493 1.1 2.5 3.2 6.4 6.8 CCl_4
m/e: 121.09

HODOC No: 02269
CAS No: 103-69-5
Beilstein No: 12^3, 255
Name: Benzenamine, *N*-ethyl-
Synonym: Aniline, *N*-ethyl-
Line Form: $C_6H_5NHC_2H_5$
Mol Form: $C_8H_{11}N$
M_r: 121.17
bp: 204.7, 97.5^{18}
mp: -63.5
***d*:** $0.9625^{20/4}$
***n*:** 1.5559^{20}
***s*:** H_2O 1; al 5; eth 5; ace 4; bz 4; ctc 3; os 3
IR: µCOB 6271 3450 3030 2940 2860 1610 1520 1450 1430 1390 1320 1270 1180 1150 1100 1080 1030 990 750 690
Raman: SAD 683 3400 3060 2970 2920 2870 1600 1490 1450 1420 1320 1270 1250 1180 1150 1120 1100 1060 1030 990 930 900 830 760 650 620 470 410 340 220
UV: SAD 2378 295 246 MeOH
H NMR: SAD 1890 1.1 3.0 3.2 6.3-7.2 CCl_4
MS: WILEY 489 106(100) 121(36) 77(23) 51(15) 39(12) 28(11) 27(11) 53(11)
m/e: 121.09

HODOC No: 02270
CAS No: 4348-19-0
Beilstein No: 12^3, 257
Name: Benzenamine, *N*-ethyl-, hydrochloride
Synonym: Aniline, *N*-ethyl-, hydrochloride
Line Form: $C_6H_5NHC_2H_5 \cdot HCl$
Mol Form: $C_8H_{12}ClFN_{28}$
M_r: 554.81
mp: 178.5
***d*:** $1.0085^{182/4}$
Color: nd
***s*:** H_2O 3; al 3; eth 1; chl 3
IR: µSADP 10644 3330 2780 2440 1590 1490 1430 1110 1030 1010 880 800 750 690
UV: SAD 21966 293(142) 248(962) MeOH 294(1810) 246(2210) MeOH:KOH
H NMR: SAD 18747 1.4 3.5 D_2O
m/e: 554.15

HODOC No: 02271
CAS No: 3846-50-2
Beilstein No: 12^3, 1682
Name: Benzenamine, *N*-ethyl-2,4-dinitro-
Synonym: Aniline, *N*-ethyl-2,4-dinitro-
Line Form: 2,4-$(NO_2)_2$-C_6H_3-NHC_2H_5
Mol Form: $C_8H_9N_3O_4$
M_r: 211.17
mp: 110-3
IR: SADG 19853 3370 3100 2980 1620 1580 1520 1500 1470 1420 1330 1290 1230 1150 1130 1110 1090 1040 930 830 740 650 590 510
UV: SAD 17572 348(17700) 258(9120) 212(13700) MeOH
H NMR: SAD 11646 1.4 3.5 7.0 8.3 8.5 9.1 $CDCl_3$
m/e: 211.06

HODOC No: 02272
CAS No: 64653-59-4
Beilstein No: 12^2, 593
Name: Benzenamine, 2-ethyl-*N*-(2-ethylphenyl)-
Synonym: 2,2′-Diethyldiphenylamine
Line Form: $NH(CH_2CH_2C_6H_5)_2$
Mol Form: $C_{16}H_{19}N$
M_r: 225.32
bp: $335\text{-}7^{603}$ 190^{15}
mp: 28-30
***n*:** 1.5550^{25}
***s*:** H_2O 1; al 4; eth 4; chl 2; ac 3
IR: SADG 15028 3333 3030 2941 1613 1493 1449 1351 1124 1031 752 694
UV: SAD 110 293 258 MeOH
H NMR: SAD 6917 1.5 2.8 2.9 7.2 $CDCl_3$
m/e: 225.15

HODOC No: 02273
CAS No: 24549-06-2
Name: Benzenamine, 2-ethyl-6-methyl-
Synonym: *o*-Toluidine, 6-ethyl-
Mol Form: $C_9H_{13}N$
M_r: 135.20
bp: 231
mp: -33
***d*:** 0.968
***n*:** 1.5525^{20}
IR: ALD 2:641A
MS: ALD 5:61A
m/e: 135.10

HODOC No: 02274
CAS No: 94-68-8
Beilstein No: 12^3, 1843
Name: Benzenamine, *N*-ethyl-2-methyl-
Synonym: *o*-Toluidine, *N*-ethyl-
Line Form: $2\text{-}CH_3C_6H_4NHC_2H_5$
Mol Form: $C_9H_{13}N$
M$_r$: 135.20
bp: 218, 95.5^{10}
mp: 15
***d*:** $0.948^{25/4}$
***n*:** 1.5456^{20}
***s*:** al 3; eth 3
IR: µSADP 8090 3450 3030 2940 2860 1610 1590 1520 1470 1450 1370 1320 1270 1190 1160 1050 990 920 830 750 710
UV: HBCP 291(2138) 244(7244) al
H NMR: SAD 17125 1.2 2.0 3.0 3.2 6.5 6.6 7.0 7.1 $CDCl_3$
MS: WILEY 679 120(100) 135(33) 91(23) 106(16) 65(12) 39(11) 77(11) 27(9)
m/e: 135.10

HODOC No: 02275
CAS No: 102-27-2
Beilstein No: 12^3, 1954
Name: Benzenamine, *N*-ethyl-3-methyl-
Synonym: *m*-Toluidine, *N*-ethyl-
Line Form: $3\text{-}CH_3C_6H_4NHC_2H_5$
Mol Form: $C_9H_{13}N$
M$_r$: 135.20
bp: 221, $111\text{-}2^{20}$
***n*:** 1.5451^{20}
***s*:** al 3; eth 3
IR: µCOB 981 3450 3030 2940 2860 1610 1590 1490 1450 1430 1370 1330 1300 1280 1270 1180 1160 1150 1100 1060 990 950 860 840 770 690
UV: SAD 4577 295 247 cyhex
H NMR: SAD 18786 1.1 2.2 3.0 3.2 6.1 6.5 7.0 CCl_4
m/e: 135.10

HODOC No: 02276
CAS No: 622-57-1
Beilstein No: 12^3, 2027
Name: Benzenamine, *N*-ethyl-4-methyl-
Synonym: *p*-Toluidine, *N*-ethyl-
Line Form: $4\text{-}CH_3C_6H_4NHC_2H_5$
Mol Form: $C_9H_{13}N$
M$_r$: 135.20
bp: 217
***d*:** 0.9391^{16}
***s*:** al 3; eth 3
IR: µSADP 8313 3450 3030 2940 2860 1640 1590 1520 1490 1450 1410 1370 1320 1300 1250 1180 1150 1060 940 810
UV: SAD 21922 299(1560) 245(9870) MeOH 328(134) 269(183) 260(268) 253(250) 244(251) MeOH:HCl
H NMR: SAD 17127 1.1 2.2 3.0 3.2 6.4 7.0 $CDCl_3$
m/e: 135.10

HODOC No: 02277
CAS No: 613-97-8
Beilstein No: 12^3, 258
Name: Benzenamine, *N*-ethyl-*N*-methyl-
Synonym: Aniline, *N*-ethyl-*N*-methyl-
Line Form: $C_6H_5N(CH_3)C_2H_5$
Mol Form: $C_9H_{13}N$
M$_r$: 135.20
bp: 203-5, $93\text{-}5^{13}$
***d*:** $0.9193^{55/4}$
***s*:** H_2O 1; al 5; eth 5; ctc 3
IR: µCOB 5925 3030 1670 1610 1520 1450 1370 1270 1220 1190 1160 1090 1030 1000 890 860 790 750 690
H NMR: SAD 1920 1.0 2.8 3.3 6.4-7.3 CCl_4
MS: ALDN 82 120(100) 18(63) 135(34) 77(30) 32(25) 17(17) 106(16) 51(14)
m/e: 135.10

HODOC No: 02278
CAS No: 612-64-6
Beilstein No: 12[3], 1119
Name: Benzenamine, *N*-ethyl-*N*-nitroso-
Synonym: Aniline, *N*-ethyl-*N*-nitroso-
Line Form: $C_6H_5N(NO)C_2H_5$
Mol Form: $C_8H_{10}N_2O$
Mr: 150.17
bp: $119\text{-}20^{15}$
***d*:** $1.0874^{20/4}$
Color: yesh
***s*:** H_2O 1; HOAc 3
UV: **HBCP** 272(6761) al
m/e: 150.08

HODOC No: 02279
CAS No: 606-99-5
Beilstein No: 12[3], 291
Name: Benzenamine, *N*-ethyl-*N*-phenyl-
Synonym: Diphenylamine, *N*-ethyl-
Line Form: $(C_6H_5)_2NC_2H_5$
Mol Form: $C_{14}H_{15}N$
Mr: 197.27
bp: 295-7, 148^{13}
***d*:** $1.0396^{20/20}$
***n*:** 1.6095^{20}
m/e: 197.12

HODOC No: 02280
CAS No: 348-54-9
Beilstein No: 12[3], 1273
Name: Benzenamine, 2-fluoro-
Synonym: Aniline, *o*-fluoro-
Line Form: $2\text{-}FC_6H_4NH_2$
Mol Form: C_6H_6FN
Mr: 111.11
bp: $174\text{-}6^{757}$, $55\text{-}6^{12}$
mp: -34.6
***d*:** 1.1513^{21}
***n*:** 1.5421^{20}
Color: pa ye
***s*:** H_2O 1; al 3; eth 3; ctc 2
IR: **µCOB 6308** 3450 3330 1640 1590 1490 1470 1330 1320 1280 1210 1150 1140 1020 920 860 760 740
UV: **OES 8-41** 277(1660) pH=9.2 259(794) 2N HCl
H NMR: **SAD 5031** 3.5 6.3-7.1 CCl_4
m/e: 111.05

HODOC No: 02281
CAS No: 372-19-0
Beilstein No: 12[3], 1274
Name: Benzenamine, 3-fluoro-
Synonym: Aniline, *m*-fluoro-
Line Form: $3\text{-}FC_6H_4NH_2$
Mol Form: C_6H_6FN
Mr: 111.11
bp: 187-9, $82\text{-}3^{18}$
***d*:** 1.1561^{19}
***n*:** 1.5436^{20}
***s*:** H_2O 2; al 3; eth 3; chl 2
IR: **SADG 715** 3448 3333 3030 1613 1493 1316 1299 1163 1136 1000 943 840 763 676
UV: **SAD 9650** 284(1800) 237(10100) MeOH
H NMR: **SAD 5032** 3.6 6.2-6.6 7.0 $CDCl_3$
C NMR: **JJ 158** FT 102.0 104.8 110.8 130.5 148.6 163.9 $CDCl_3$
m/e: 111.05

HODOC No: 02282
CAS No: 371-40-4
Beilstein No: 12[3], 1276
Name: Benzenamine, 4-fluoro-
Synonym: Aniline, *p*-fluoro-
Line Form: $4\text{-}FC_6H_4NH_2$
Mol Form: C_6H_6FN
Mr: 111.11
bp: $181\text{-}2^{757}$, 85^{19}
mp: -0.8
***d*:** $1.1725^{20/4}$
***n*:** 1.5195^{20}
Color: pa ye
***s*:** H_2O 2; al 3; eth 3; ctc 2
IR: **µSADP 19617** 3450 3330 1640 1520 1350 1280 1240 1160 1120 830 750
UV: **SAD 6459** 290(1730) 230(6100) MeOH
H NMR: **SAD 5033** 3.3 6.4 6.8 CCl_4
MS: **WILEY 366** 111(100) 84(30) 83(22) 145(17) 110(8) 112(7) 57(7) 64(5)
m/e: 111.05

HODOC No: 02283
CAS No: 393-56-6
Name: Benzenamine, 2-fluoro-*N*,*N*-dimethyl-
Synonym: Aniline, *o*-fluoro-*N*,*N*-dimethyl-
Mol Form: $C_8H_{10}FN$
Mr: 139.16
UV: **FO 92** 284(1995) 250(12589) hp
m/e: 139.08

HODOC No: 02284
CAS No: 452-84-6
Name: Benzenamine, 2-fluoro-5-methyl-
Synonym: *m*-Toluidine, 6-fluoro-
Mol Form: C_7H_8FN
M_r: 125.14
IR: COB 7683 3485 3400 3030 2925 2870 1630 1605 1520 1435 1305 1275 1210 1200 1160 1075 945 920 850 785 770 725
m/e: 125.06

HODOC No: 02285
CAS No: 452-77-7
Name: Benzenamine, 3-fluoro-4-methyl-
Synonym: *p*-Toluidine, 3-fluoro-
Mol Form: C_7H_8FN
M_r: 125.14
IR: SADG 33536 3440 3380 3230 3050 2940 2870 1640 1590 1520 1460 1390 1310 1280 1200 1160 1120 990 940 840 810 740 620 570 460 320
Raman: SAD 1066 3430 3310 3210 3060 3040 2920 2860 2740 1630 1590 1510 1460 1430 1380 1310 1280 1200 1110 1000 950 810 760 740 700 620 560 510 450 330 250 120
UV: SAD 6464 290 235 MeOH
H NMR: SAD 13664 2.1 3.4 6.2-6.5 6.8 $CDCl_3$
m/e: 125.06

HODOC No: 02286
CAS No: 452-71-1
Beilstein No: 12^2, 453
Name: Benzenamine, 4-fluoro-2-methyl-
Synonym: *o*-Toluidine, 4-fluoro-
Line Form: $(4\text{-F-2-CH}_3)C_6H_3NH_2$
Mol Form: C_7H_8FN
M_r: 125.14
bp: 94^{16}
mp: 14.2
***d*:** 1.1263^{18}
***n*:** 1.5363^{18}
***s*:** eth 3; ace 3; bz 3; ctc 3
IR: SADG 18931 3450 3380 3030 2930 1630 1500 1460 1420 1270 1230 1140 990 950 860 800 710 680 560
UV: SAD 6465 290 231 MeOH
H NMR: SAD 5038 2.0 3.3 6.2-6.8 CCl_4
m/e: 125.06

HODOC No: 02287
CAS No: 367-29-3
Beilstein No: 12^3, 1909
Name: Benzenamine, 5-fluoro-2-methyl-
Synonym: *o*-Toluidine, 5-fluoro-
Line Form: $5\text{-F-2-CH}_3C_6H_3NH_2$
Mol Form: C_7H_8FN
M_r: 125.14
mp: 38-40
IR: SADG 33537 3490 3400 3230 3050 2980 2940 2910 2870 1630 1510 1470 1430 1380 1310 1290 1280 1210 1160 1130 1070 1000 960 840 790 730 620 490 460 450 340
Raman: SAD 1067 3420 3320 3210 3080 3050 2880 2820 2760 1650 1610 1500 1460 1430 1390 1320 1300 1280 1210 1160 1130 1080 1040 1000 960 930 850 790 760 730 600 570 530 500 460 360 220 140
UV: SAD 6466 285 234 MeOH
H NMR: SAD 2226 2.0 3.5 6.2 6.3 6.8 CCl_4
MS: ALD 5:65D
m/e: 125.06

HODOC No: 02288
CAS No: 364-76-1
Beilstein No: 12^3, 1647
Name: Benzenamine, 4-fluoro-3-nitro-
Synonym: Aniline, 4-fluoro-3-nitro-
Line Form: $(4\text{-F-3-NO}_2)C_6H_3NH_2$
Mol Form: $C_6H_5FN_2O_2$
M_r: 156.11
mp: 97-8
***s*:** chl 3
IR: COB 6616 3460 3340 3220 3070 1630 1590 1530 1440 1350 1320 1220 1140 1070 940 820 760 690 550 480
Raman: OES 8-37 250(6761) 5N HCl 357(1698) PH=9.2
UV: OES 8-37
H NMR: SAD 13924 4.0 6.8-7.4 $CDCl_3$
m/e: 156.03

HODOC No: 02289
CAS No: 542-11-0
Beilstein No: 12^3, 232
Name: Benzenamine, hydrobromide
Synonym: Aniline, hydrobromide
Line Form: $C_6H_5NH_2{\cdot}HBr$
Mol Form: $C_6H_8BrFN_{28}$
Mr: 570.21
mp: 286
IR: COB 4349 2900 2600 1600 1560 1500 1450 1200 1100 1030 750 680
UV: SAD 17283 7.5 D_2O
H NMR: SAD 17283
m/e: 570.07

HODOC No: 02290
CAS No: 142-04-1
Beilstein No: 12^3, 232
Name: Benzenamine, hydrochloride
Synonym: Aniline, hydrochloride
Line Form: $C_6H_5NH_2{\cdot}HCl$
Mol Form: $C_6H_8ClFN_{28}$
Mr: 526.76
bp: 245
mp: 198
***d*:** 1.2215^4
Color: lf or nd
***s*:** H_2O 4; al 4; eth 1; chl 1; DMSOd6 2
IR: μCOB 1247 2940 2630 2000 1590 1520 1490 1210 1110 1030 740 690
UV: OES 7-76 250(151)
H NMR: SAD 9379 7.5 9.8 DMSOd6
MS: NBS 257
m/e: 526.12

HODOC No: 02291
CAS No: 100-65-2
Beilstein No: 15^3, 5
Name: Benzenamine, *N*-hydroxy-
Synonym: *N*-Hydroxyaniline
Line Form: C_6H_5NHOH
Mol Form: C_6H_7NO
Mr: 109.12
mp: 83-4
Color: nd (w, Bz, peth)
***s*:** Bz 4; Et_2O 4; EtOH 4; Chl 4
UV: OES 1-94
MS: NBS 119
m/e: 109.05

HODOC No: 02292
CAS No: 611-22-3
Beilstein No: 15^3, 15
Name: Benzenamine, *N*-hydroxy-2-methyl-
Synonym: Hydroxylamine, *N*-*o*-tolyl-
Line Form: $2\text{-}CH_3C_6H_4NHOH$
Mol Form: C_7H_9NO
Mr: 123.15
mp: 44
Color: nd_(eth-Bz)
***s*:** Bz 4; Et_2O 4; EtOH 4
m/e: 123.07

HODOC No: 02293
CAS No: 620-25-7
Beilstein No: 15^3, 17
Name: Benzenamine, *N*-hydroxy-3-methyl-
Synonym: Hydroxylamine, *N*-*m*-tolyl-
Line Form: $3\text{-}CH_3C_6H_4NHOH$
Mol Form: C_7H_9NO
Mr: 123.15
mp: 68.5
Color: lf (Bz-peth)
***s*:** Et_2O 4; EtOH 4; Chl 4
m/e: 123.07

HODOC No: 02294
CAS No: 623-10-9
Beilstein No: 15^3, 18
Name: Benzenamine, *N*-hydroxy-4-methyl-
Synonym: Hydroxylamine, *N*-*p*-tolyl-
Line Form: $4\text{-}CH_3C_6H_4NHOH$
Mol Form: C_7H_9NO
Mr: 123.15
bp: 115-20d
mp: 98
Color: lf (Bz)
***s*:** Et_2O 4; EtOH 4; Chl 4
IR: SADP 8122
UV: SAD 2189
m/e: 123.07

HODOC No: 02295
CAS No: 5805-95-8
Beilstein No: 15^2, 14
Name: Benzenamine, *N*-hydroxy-2-methyl-3-nitro-
Synonym: Hydroxylamine, *N*-(3-nitro-*o*-tolyl)-
Line Form: 2-HONH-6-$NO_2C_6H_3CH_3$
Mol Form: $C_7H_8N_2O_3$
Mr: 168.15
mp: 120-1
Color: col or ye (Bz)
***s*:** Et_2O 4; EtOH 4
m/e: 168.05

HODOC No: 02296
CAS No: 43192-03-6
Beilstein No: 15^2, 16
Name: Benzenamine, *N*-hydroxy-4-methyl-3-nitro-
Synonym: 4-Hydroxylamino-2-nitrotoluene
Line Form: 4-(HONH)-2-$NO_2C_6H_3CH_3$
Mol Form: $C_7H_8N_2O_3$
Mr: 168.15
mp: 108-9
Color: ye (Bz)
***s*:** Bz 4
m/e: 168.05

HODOC No: 02297
CAS No: 135-20-6
Name: Benzenamine, *N*-hydroxy-*N*-nitroso-, ammonium salt
Synonym: Hydroxylamine, *N*-nitroso-*N*-phenyl-, ammonium salt
Line Form: $C_6H_5N(NO)ONH_4$
Mol Form: $C_6H_9N_3O_2$
Mr: 155.15
mp: 163-4
***s*:** DMSO 2
IR: **µSADP 7196** 3030 1450 1390 1330 1270 1220 1050 910 750 690
UV: **SAD 2040** 292 221 MeOH
H NMR: **SAD 11144** 7.2-7.5 7.7 7.9 DMSO
m/e: 155.07

HODOC No: 02298
CAS No: 615-43-0
Beilstein No: 12^3, 1487
Name: Benzenamine, 2-iodo-
Synonym: Aniline, *o*-iodo-
Line Form: 2-$IC_6H_4NH_2$
Mol Form: C_6H_6IN
Mr: 219.02
mp: 60-1
Color: nd (dil al)
***s*:** H_2O 2; al 4; eth 4; ace 3; os 3
IR: **µSADP 17451** 3450 3330 1640 1590 1470 1450 1320 1270 1160 1010 940 750
UV: **OES 5-57** 293(3236) 238(8318) cyhex
H NMR: **SAD 18795** 4.0 6.5 6.7 7.1 7.6 $CDCl_3$
m/e: 218.95

HODOC No: 02299
CAS No: 626-01-7
Beilstein No: 12^3, 1489
Name: Benzenamine, 3-iodo-
Synonym: Aniline, *m*-iodo-
Line Form: 3-$IC_6H_4NH_2$
Mol Form: C_6H_6IN
Mr: 219.02
bp: $145\text{-}6^{15}$
mp: 33
***n*:** 1.6811^{20}
Color: lf
***s*:** H_2O 1; al 3; chl 3
IR: **µSADP 17800** 3450 3330 3030 1610 1590 1470 1430 1320 1270 1160 1060 990 850 770 680
UV: **SAD 5637** 295(2210) 243(8150) 215(33700) MeOH
m/e: 218.95

HODOC No: 02300
CAS No: 540-37-4
Beilstein No: 12^3, 1493
Name: Benzenamine, 4-iodo-
Synonym: Aniline, *p*-iodo-
Line Form: 4-$IC_6H_4NH_2$
Mol Form: C_6H_6IN
Mr: 219.02
mp: 67-8
Color: nd (w)
***s*:** H_2O 2; al 3; eth 3; os 3; peth 2
IR: **µCOB 5935** 3570 3450 3030 1610 1590 1490 1410 1280 1180 1120 1060 1000 750
UV: **SAD 6181** 296 249 MeOH
H NMR: **SAD 17203** 3.6 6.4 7.4 $CDCl_3$
m/e: 218.95

HODOC No: 02301
CAS No: 29289-13-2
Beilstein No: 12², 533
Name: Benzenamine, 2-iodo-4-methyl-
Synonym: *p*-Toluidine, 2-iodo-
Line Form: 2-I-4-$CH_3C_6H_3NH_2$
Mol Form: C_7H_8IN
M_r: 233.04
bp: d
mp: 40
Color: pr
s: Ac 4; Bz 4; Et_2O 4; EtOH 4
m/e: 232.97

HODOC No: 02302
CAS No: 13194-69-9
Beilstein No: 12², 474
Name: Benzenamine, 2-iodo-5-methyl-
Synonym: *m*-Toluidine, 6-iodo-
Line Form: 6-I-3-$CH_3C_6H_3NH_2$
Mol Form: C_7H_8IN
M_r: 233.04
mp: 48
Color: nd_(dil al) (br in air)
s: Et_2O 4; EtOH 4; Chl 4
m/e: 232.97

HODOC No: 02303
CAS No: 35944-64-0
Beilstein No: 12², 533
Name: Benzenamine, 3-iodo-4-methyl-
Synonym: 3-Iodo-*p*-toluidine
Line Form: 3-I-4-CH_3-$C_6H_3NH_2$
Mol Form: C_7H_8IN
M_r: 233.04
mp: 39-40
Color: nd (dil al, peth)
s: Ac 4; Et_2O 4; EtOH 4
m/e: 232.97

HODOC No: 02304
CAS No: 74586-54-2
Beilstein No: 12¹, 406
Name: Benzenamine, 3-iodo-5-methyl-
Synonym: 3-Iodo-5-methylaniline
Line Form: 5-I-3-$CH_3C_6H_3NH_2$
Mol Form: C_7H_8IN
M_r: 233.04
mp: 78.5
Color: nd (peth)
s: Ac 4; Et_2O 4; EtOH 4
m/e: 232.97

HODOC No: 02305
CAS No: 13194-68-8
Beilstein No: 12³, 1927
Name: Benzenamine, 4-iodo-2-methyl-
Synonym: *o*-Toluidine, 4-iodo-
Line Form: 4-I-2-$CH_3C_6H_3NH_2$
Mol Form: C_7H_8IN
M_r: 233.04
mp: 91-2
Color: nd (dil al) pr (lig)
s: H_2O 3; al 4; eth 4; bz 4; HOAc 3; lig 3
IR: μSADP 11627 3450 2940 1610 1560 1470 1390 1300 1270 1190 1150 1050 1030 990 870 860 810 760 690
UV: SAD 3228 269(2110) 247(16200) MeOH
H NMR: SAD 18478 2.0 4.6 6.4 7.1 7.2 poly D
m/e: 232.97

HODOC No: 02306
CAS No: 83863-33-6
Beilstein No: 12¹, 391
Name: Benzenamine, 5-iodo-2-methyl-
Synonym: *o*-Toluidine, 5-iodo
Line Form: 5-I-2-$CH_3C_6H_3NH_2$
Mol Form: C_7H_8IN
M_r: 233.04
bp: 273d
mp: 48-9
Color: nd (aq al)
s: EtOH 4
IR: SADP 11627
UV: SAD 3228
m/e: 232.97

HODOC No: 02307
CAS No: 10394-64-6
Name: Benzenamine, 3-iodo-5-nitro-
Synonym: Aniline, 3-iodo-5-nitro-
Line Form: 3-I-5-$NO_2C_6H_3NH_2$
Mol Form: $C_6H_5IN_2O_2$
M_r: 264.02
bp: 135.5-9.0
mp: 136-9
IR: COB 6570 3430 3350 3090 1630 1600 1570 1520 1440 1340 1280 1100 990 870 750 660
m/e: 263.94

HODOC No: 02308
CAS No: 2131-64-8
Beilstein No: 13^3, 53
Name: Benzenamine, 4-isothiocyanato-*N*,*N*-dimethyl-
Synonym: Isothiocyanic acid, *p*-(dimethylamino)phenyl ester
Mol Form: $C_9H_{10}N_2S$
M_r: 178.24
mp: 67
IR: **µCOB 3211** 3030 2940 2080 1610 1520 1450 1370 1240 1180 1060 930 810 760
UV: **OES 4-224** 310(25704) 297(28840) 256(5623) cyhex
m/e: 178.06

HODOC No: 02309
CAS No: 622-16-2
Beilstein No: 12^3, 906
Name: Benzenamine, *N*,*N*′-methanetetraylbis-
Synonym: Carbodiimide, diphenyl-
Line Form: $C_6H_5N{=}C{=}NC_6H_5$
Mol Form: $C_{13}H_{10}N_2$
M_r: 194.22
bp: 331, 218^{71}
mp: 168-70
***s*:** H_2O 2; al 2; eth 2; bz 3
IR: **DMS 10082** 3080 2900 2150 1590 1490 1460 1290 1240 1200 1170 1100 1070 1030 900 800 760 690 620 600 520
m/e: 194.08

HODOC No: 02310
CAS No: 20220-77-3
Beilstein No: 13, 378
Name: Benzenamine, *N*,*N*′-methanetetraylbis[2-methoxy-
Synonym: Carbodiimide, *bis*(*o*-methoxyphenyl)-
Mol Form: $C_{15}H_{14}N_2O_2$
M_r: 254.27
mp: 56
IR: **COB 7512** 3010 2980 2950 2150 2110 2040 1590 1500 1460 1310 1260 1210 1110 1040 1030 840 750 610 580
m/e: 254.11

HODOC No: 02311
CAS No: 726-42-1
Beilstein No: 12^3, 2111
Name: Benzenamine, *N*,*N*′-methanetetraylbis[4-methyl-
Synonym: Carbodiimide, di-*p*-tolyl-
Line Form: $(4\text{-}CH_3C_6H_4)N{=}C{=}N(C_6H_4CH_3\text{-}4)$
Mol Form: $C_{15}H_{14}N_2$
M_r: 222.27
bp: $221\text{-}3^{20}$
mp: 56-8
IR: **ALD 2:997B**
MS: **ALD 7:144A**
m/e: 222.12

HODOC No: 02312
CAS No: 90-04-0
Beilstein No: 13^3, 754
Name: Benzenamine, 2-methoxy-
Synonym: *o*-Anisidine
Line Form: $2\text{-}CH_3O\text{-}C_6H_4NH_2$
Mol Form: C_7H_9NO
M_r: 123.15
bp: 224, 90^4
mp: 6.2
***d*:** $1.0923^{20/4}$
***n*:** 1.5715^{10}
***s*:** H_2O 2; al 3; eth 3; ace 3; bz 3
IR: **µNRC 2153** 3450 3330 3230 2940 2860 2700 1610 1490 1470 1350 1270 1220 1190 1140 1080 1050 1020 910 840 760 730 660
UV: **OES 5-107** 237(8511) cyhex
MS: **ALDN 65** 108(100) 80(71) 123(18) 53(13) 52(7) 109(6) 92(4) 81(4)
m/e: 123.07

HODOC No: 02313
CAS No: 536-90-3
Beilstein No: 13^3, 932
Name: Benzenamine, 3-methoxy-
Synonym: *m*-Anisidine
Line Form: $3\text{-}CH_3O\text{-}C_6H_4NH_2$
Mol Form: C_7H_9NO
M$_r$: 123.15
bp: 251
mp: -1
***d*:** $1.096^{20/4}$
***n*:** 1.5794^{20}
***s*:** H_2O 2; al 3; eth 3; ace 3; bz 3; ctc 2
IR: µSADP 15296 3450 3330 2940 1610 1490 1470 1330 1280 1210 1160 1040 940 830 770 690
UV: SAD 4566 285 cyhex
H NMR: SAD 9380 3.5 3.6 6.0-6.3 7.0 CCl_4
MS: NBS 204
m/e: 123.07

HODOC No: 02314
CAS No: 104-94-9
Beilstein No: 13^3, 994
Name: Benzenamine, 4-methoxy-
Synonym: *p*-Anisidine
Line Form: $4\text{-}CH_3O\text{-}C_6H_4NH_2$
Mol Form: C_7H_9NO
M$_r$: 123.15
bp: 243, 115^{13}
mp: 57.2
***d*:** $1.071^{57/4}$
***n*:** 1.5559^{60}
Color: ta (w) rh pl
***s*:** H_2O 3; al 4; eth 4; ace 3; bz 3
IR: µCOB 5672 3450 3330 2860 1640 1520 1470 1450 1330 1300 1280 1250 1180 1030 820 740
Raman: SAD 236 3430 3360 3070 3020 2980 2950 2850 1620 1470 1460 1410 1330 1300 1280 1240 1180 1150 1040 1010 850 810 710 650 540 440 380 250
UV: SAD 392 300 235 MeOH
MS: ALDN 65 108(100) 123(99) 80(25) 53(11) 109(10) 124(9) 65(7) 52(7)
m/e: 123.07

HODOC No: 02315
CAS No: 20265-97-8
Beilstein No: 13^3, 994
Name: Benzenamine, 4-methoxy-, hydrochloride
Synonym: *p*-Anisidine, hydrochloride
Line Form: $4\text{-}CH_3O\text{-}C_6H_4NH_2{\cdot}HCl$
Mol Form: $C_7H_{10}ClFNO_{28}$
M$_r$: 610.59
mp: 216-8
IR: SADG 20750 2900 2600 1910 1620 1510 1450 1300 1260 1170 1020 830 510
UV: SAD 18758 281(1390) 274(1630) 222(9640) MeOH
H NMR: SAD 11658 3.8 7.1 7.4 D_2O
m/e: 609.91

HODOC No: 02316
CAS No: 700-75-4
Name: Benzenamine, 2-methoxy-*N*,*N*-dimethyl-
Synonym: *o*-Anisidine, *N*,*N*-dimethyl-
Mol Form: $C_9H_{13}NO$
M$_r$: 151.20
bp: 209
***s*:** chl 3
IR: SADG 20631 3060 2940 2820 2780 1590 1500 1450 1320 1290 1230 1180 1150 1110 1050 1020 940 790 730 590 540 480 370
UV: SAD 18691 281(2650) 242(7140) MeOH
H NMR: SAD 11571 2.7 3.8 6.5 6.8 CCl_4
m/e: 151.10

HODOC No: 02317
CAS No: 701-56-4
Name: Benzenamine, 4-methoxy-*N*,*N*-dimethyl-
Synonym: *p*-Anisidine, *N*,*N*-dimethyl-
Mol Form: $C_9H_{13}NO$
M$_r$: 151.20
mp: 47
***s*:** chl 3
IR: SADG 20632 2940 2820 2790 1610 1510 1450 1440 1330 1290 1230 1170 1150 1130 1030 940 810 680 520
UV: SAD 18692 307(1750) 246(11700) MeOH
H NMR: SAD 11064 2.9 3.8 6.8 $CDCl_3$
m/e: 151.10

HODOC No: 02318
CAS No: 90437-76-6
Beilstein No: 13[1], 121
Name: Benzenamine, 2-methoxy-*N*,*N*-dimethyl-4-nitro-
Synonym: *o*-Anisidine, *N*,*N*-dimethyl-4-nitro-
Line Form: $(4\text{-}NO_2\text{-}2\text{-}CH_3O)C_6H_3N(CH_3)_2$
Mol Form: $C_9H_{12}N_2O_3$
M_r: 196.20
mp: 99
IR: SADG 20642
UV: SAD 18700
H NMR: SAD 11113
m/e: 196.08

HODOC No: 02319
CAS No: 101-70-2
Name: Benzenamine, 4-methoxy-*N*-(4-methoxyphenyl)-
Synonym: Diphenylamine, 4,4′-dimethoxy-
Mol Form: $C_{14}H_{15}NO_2$
M_r: 229.26
IR: µCOB 2521 3030 2940 1520 1470 1370 1300 1250 1180 1110 1080 1030 760 720
UV: LANG 410 289(10715) 251(7413) 226(7413) EtOH
m/e: 229.11

HODOC No: 02320
CAS No: 2615-05-6
Name: Benzenamine, 2-methoxy-4-[(2-methoxyphenyl)azo]-
Synonym: *o*-Anisidine, 4-[(*o*-methoxyphenyl)azo]-
Mol Form: $C_{14}H_{15}N_3O_2$
M_r: 257.28
mp: 115-6
***s*:** chl 3
IR: µCOB 1466 3330 2940 1640 1610 1520 1470 1390 1320 1280 1240 1140 1110 1030 860 810 800 750
H NMR: SAD 13436 3.8 3.9 4.2 6.7 7.0 7.2 7.5 7.6 $CDCl_3$
m/e: 257.12

HODOC No: 02321
CAS No: 120-71-8
Beilstein No: 13[2], 338
Name: Benzenamine, 2-methoxy-5-methyl-
Synonym: *o*-Anisidine, 5-methyl-
Line Form: $(2\text{-}CH_3O\text{-}5\text{-}CH_3)\text{-}C_6H_3NH_2$
Mol Form: $C_8H_{11}NO$
M_r: 137.17
bp: 235
mp: 52-4
***s*:** H_2O 2; al 3; eth 3; bz 3; chl 2; peth 3
IR: SADG 15482 3410 3310 3200 2930 2830 1620 1520 1460 1300 1230 1160 1040 880 870 800
UV: SAD 15769 290(3620) 236(7120) MeOH
H NMR: SAD 6399 2.2 3.6 3.8 6.5 6.6 6.7 $CDCl_3$
MS: ALD 5:72B
m/e: 137.08

HODOC No: 02322
CAS No: 16452-01-0
Beilstein No: 13[1], 213
Name: Benzenamine, 3-methoxy-4-methyl-
Synonym: 4-Amino-2-methoxytoluene
Line Form: $3\text{-}CH_3O\text{-}4\text{-}CH_3C_6H_3NH_2$
Mol Form: $C_8H_{11}NO$
M_r: 137.17
bp: 250-2
mp: 58
***s*:** Bz 4; Et_2O 4; EtOH 4; lig 4
m/e: 137.08

HODOC No: 02323
CAS No: 102-50-1
Beilstein No: 13[2], 330
Name: Benzenamine, 4-methoxy-2-methyl-
Synonym: *p*-Anisidine, 2-methyl-
Line Form: $(4\text{-}CH_3O\text{-}2\text{-}CH_3)\text{-}C_6H_3NH_2$
Mol Form: $C_8H_{11}NO$
M_r: 137.17
bp: 248-9
mp: 13-14
***d*:** 1.065
***n*:** 1.5647^{20}
***s*:** al 3; ctc 3; lig 3
IR: SADG 18724 3440 3380 3230 3030 2950 2840 1610 1500 1460 1410 1310 1280 1230 1150 1040 860 800 710
UV: SAD 6909 297(2740) 233(8740) MeOH
H NMR: SAD 6105 2.0 3.2 3.6 6.5 CCl_4
MS: ALD 5:72A
m/e: 137.08

HODOC No: 02324
CAS No: 5961-59-1
Beilstein No: 13^3, 1009
Name: Benzenamine, 4-methoxy-*N*-methyl-
Synonym: *p*-Anisidine, *N*-methyl-
Line Form: $4\text{-}CH_3O\text{-}C_6H_4NHCH_3$
Mol Form: $C_8H_{11}NO$
M_r: 137.17
mp: 36
UV: LANG 2674 308(2089) 240(10471) EtOH 272(1259) 220(9120) 0.1N HCl:EtOH
MS: NBS 325
m/e: 137.08

HODOC No: 02325
CAS No: 50868-72-9
Beilstein No: 13^3, 1573
Name: Benzenamine, 5-methoxy-2-methyl-
Synonym: 3-Amino-4-methylanisole
Line Form: $5\text{-}CH_3O\text{-}2\text{-}CH_3C_6H_3NH_2$
Mol Form: $C_8H_{11}NO$
M_r: 137.17
bp: 253, 140^{20}
mp: 47
Color: nd (w)
***s*:** Et_2O 4
IR: SADP 20460
UV: SAD 6909
m/e: 137.08

HODOC No: 02326
CAS No: 25287-24-5
Name: Benzenamine, 4-methoxy-*N*-[1-(4-methylphenyl)ethylidene]-
Synonym: *p*-Anisidine, *N*-(*p*,α-dimethylbenzylidene)-
Mol Form: $C_{16}H_{17}NO$
M_r: 239.30
mp: 79-80
UV: LANG 2350 326(3467) 254(19055) EtOH
m/e: 239.13

HODOC No: 02327
CAS No: 85-45-0
Beilstein No: 13^2, 195
Name: Benzenamine, 2-methoxy-3-nitro-
Synonym: *o*-Anisidine, 3-nitro-
Line Form: $(3\text{-}NO_2\text{-}2\text{-}CH_3O)C_6H_3NH_2$
Mol Form: $C_7H_8N_2O_3$
M_r: 168.15
mp: 67
Color: pa ye nd (lig)
***s*:** EtOH 4; lig 4
m/e: 168.05

HODOC No: 02328
CAS No: 97-52-9
Beilstein No: 13^3, 888
Name: Benzenamine, 2-methoxy-4-nitro-
Synonym: *o*-Anisidine, 4-nitro-
Line Form: $(4\text{-}NO_2\text{-}2\text{-}CH_3O)C_6H_3NH_2$
Mol Form: $C_7H_8N_2O_3$
M_r: 168.15
mp: 139
***s*:** DMSO 3
IR: SADG 20636 3470 3350 3190 3010 2840 1620 1580 1520 1470 1440 1380 1300 1280 1270 1230 1150 1090 1020 930 860 810 780 740 720 640 580 460
UV: SAD 18696 388(14700) 258(5580) MeOH
H NMR: SAD 11066 3.9 6.3 6.7 7.6 7.8 $DMSOd_6$
m/e: 168.05

HODOC No: 02329
CAS No: 99-59-2
Beilstein No: 13^3, 878
Name: Benzenamine, 2-methoxy-5-nitro-
Synonym: *o*-Anisidine, 5-nitro-
Line Form: $(5\text{-}NO_2\text{-}2\text{-}CH_3O)C_6H_3NH_2$
Mol Form: $C_7H_8N_2O_3$
M_r: 168.15
mp: 118
***d*:** $1.2068^{15.6}$
***s*:** H_2O 3; al 4; eth 3; ace 4; bz 4; AcOEt 4; HOAc 4; lig 2
IR: ALD 3:826G
UV: SAD 17651 374(3900) 302(5300) 257(14200) 226(11300) MeOH
H NMR: SAD 11061 4.0 5.4 7.0 7.5 7.6 $DMSOd_6$
MS: ALDN 127 168(100) 153(21) 122(18) 52(14) 79(12) 95(11) 65(10) 169(10)
m/e: 168.05

HODOC No: 02330
CAS No: 16554-45-3
Beilstein No: 13^2, 191
Name: Benzenamine, 2-methoxy-6-nitro-
Synonym: *o*-Anisidine, 6-nitro-
Line Form: $(6\text{-}NO_2\text{-}2\text{-}CH_3O)C_6H_3NH_2$
Mol Form: $C_7H_8N_2O_3$
M$_r$: 168.15
mp: 76
***s*:** al 3; ace 3
UV: HBCP 393(4571) 284(5012) 236(14125) cyhex
H NMR: SAD 11658 3.8 7.1 7.4 D_2O
m/e: 168.05

HODOC No: 02331
CAS No: 16554-47-5
Beilstein No: 13^3, 974
Name: Benzenamine, 3-methoxy-2-nitro-
Synonym: *m*-Anisidine, 2-nitro-
Line Form: $(2\text{-}NO_2\text{-}3\text{-}CH_3O)C_6H_3NH_2$
Mol Form: $C_7H_8N_2O_3$
M$_r$: 168.15
mp: 124-5
***s*:** al 3; ace 3
UV: SAD 18697 375(1890) 300(2150) 232(10500) MeOH
H NMR: SAD 10552 3.8 6.0 6.3 6.5 7.2 DMSOd6
m/e: 168.05

HODOC No: 02332
CAS No: 16292-88-9
Beilstein No: 13^3, 975
Name: Benzenamine, 3-methoxy-4-nitro-
Synonym: *m*-Anisidine, 4-nitro-
Line Form: $(4\text{-}NO_2\text{-}3\text{-}CH_3O)\text{-}C_6H_3NH_2$
Mol Form: $C_7H_8N_2O_3$
M$_r$: 168.15
bp: sub
mp: 169
Color: ye nd (al)
***s*:** Ac 4; EtOH 4
m/e: 168.05

HODOC No: 02333
CAS No: 586-10-7
Beilstein No: 13^3, 977
Name: Benzenamine, 3-methoxy-5-nitro-
Synonym: *m*-Anisidine, 5-nitro-
Line Form: $(5\text{-}NO_2\text{-}3\text{-}CH_3O)C_6H_3NH_2$
Mol Form: $C_7H_8N_2O_3$
M$_r$: 168.15
mp: 120
***s*:** DMSO 2
IR: SADG 20639 3440 3350 3080 2970 1640 1580 1520 1450 1340 1310 1200 1160 1080 1040 960 840 780 740 660 580 550
UV: SAD 18698 305(3180) 235(16700) MeOH
H NMR: SAD 11111 3.8 5.3 6.6 7.1 DMSOd6
m/e: 168.05

HODOC No: 02334
CAS No: 96-96-8
Beilstein No: 13^3, 1203
Name: Benzenamine, 4-methoxy-2-nitro-
Synonym: *p*-Anisidine, 2-nitro-
Line Form: $(2\text{-}NO_2\text{-}4\text{-}CH_3O)C_6H_4NH_2$
Mol Form: $C_7H_8N_2O_3$
M$_r$: 168.15
mp: 129
Color: dk red pr (w or al)
***s*:** H_2O 4; Ac 4; Et_2O 4; EtOH 4
IR: SADG 15483
UV: SAD 1299
H NMR: SAD 10932
MS: i
m/e: 168.05

HODOC No: 02335
CAS No: 577-72-0
Beilstein No: 13^2, 284
Name: Benzenamine, 4-methoxy-3-nitro-
Synonym: *p*-Anisidine, 3-nitro-
Line Form: $(4\text{-}CH_3\text{-}3\text{-}NO_2)C_6H_3NH_2$
Mol Form: $C_7H_8N_2O_3$
M$_r$: 168.15
mp: 57
***s*:** Ac 4; Et_2O 4; EtOH 4
UV: OES 5-102
m/e: 168.05

HODOC No: 02336
CAS No: 16133-49-6
Beilstein No: 13^3, 975
Name: Benzenamine, 5-methoxy-2-nitro-
Synonym: *m*-Anisidine, 6-nitro-
Line Form: $(2\text{-}NO_2\text{-}5\text{-}CH_3O)C_6H_3NH_2$
Mol Form: $C_7H_8N_2O_3$
M_r: 168.15
bp: sub
mp: 131
Color: br nd
***s*:** EtOH 4
m/e: 168.05

HODOC No: 02337
CAS No: 22760-85-6
Name: Benzenamine, 4-methoxy-*N*-[1-(4-nitrophenyl)ethylidene]-
Synonym: *p*-Anisidine, *N*-(α-methyl-*p*-nitrobenzylidene)-
Mol Form: $C_{15}H_{14}N_2O_3$
M_r: 270.27
mp: 98
UV: LANG 2358 370(3548) 273(18621) EtOH
m/e: 270.10

HODOC No: 02338
CAS No: 2501-03-3
Beilstein No: 13^1, 155
Name: Benzenamine, 4-methoxy-*N*-[(2-nitrophenyl)methylene]-
Synonym: *p*-Anisidine, *N*-(*o*-nitrobenzylidene)-
Mol Form: $C_{14}H_{12}N_2O_3$
M_r: 256.25
mp: 80-1
IR: μCOB 993 3030 2940 1610 1590 1560 1540 1520 1370 1350 1300 1270 1250 1190 1160 1110 1030 860 840 830 790 780 770 750 730 700 680
m/e: 256.08

HODOC No: 02339
CAS No: 1207-92-7
Beilstein No: 13, 366
Name: Benzenamine, 2-methoxy-*N*-phenyl-
Synonym: *o*-Anisidine, *N*-phenyl-
Line Form: $2\text{-}CH_3OC_6H_4NHC_6H_5$
Mol Form: $C_{13}H_{13}NO$
M_r: 199.24
bp: $325\text{-}6^{732}$
***s*:** al 5; eth 5
UV: LANG 565 238(22387) 98% sulf
m/e: 199.10

HODOC No: 02340
CAS No: 1208-86-2
Beilstein No: 13^3, 1021
Name: Benzenamine, 4-methoxy-*N*-phenyl-
Synonym: *p*-Anisidine, *N*-phenyl-
Line Form: $4\text{-}CH_3OC_6H_4NH\text{-}C_6H_5$
Mol Form: $C_{13}H_{13}NO$
M_r: 199.24
bp: 195^{12}
mp: 105
UV: LANG 407 285(16982) 240(6457) 219(10715) EtOH 320(257) 297(468) 265(1096) 220(7586) sulf
m/e: 199.10

HODOC No: 02341
CAS No: 76655-66-8
Name: Benzenamine, 4-[(4-methoxyphenyl)azo]-3-methyl-
Synonym: Azobenzene, 4-amino-4′-methoxy-2-methyl
Line Form: $(2\text{-}CH_3\text{-}4\text{-}H_2N)\text{-}C_6H_3N{=}N\text{-}(C_6H_4\text{-}OCH_3\text{-}4)$
Mol Form: $C_{14}H_{15}N_3O$
M_r: 241.28
UV: LANG 1729
m/e: 241.12

HODOC No: 02342
CAS No: 2743-00-2
Name: Benzenamine, 4-methoxy-*N*-(1-phenylethylidene)-
Synonym: *p*-Anisidine, *N*-(α-methylbenzylidene)-
Mol Form: $C_{15}H_{15}NO$
M_r: 225.28
mp: 86
UV: LANG 2341 325(2630) 245(20417) EtOH
m/e: 225.12

HODOC No: 02343
CAS No: 25287-28-9
Name: Benzenamine, *N*-[1-(4-methoxyphenyl)ethylidene]-
Synonym: Aniline, *N*-(*p*-methoxy-α-methylbenzylidene)-
Mol Form: $C_{15}H_{15}NO$
M_r: 225.28
mp: 100-1
UV: LANG 2345 315(3981) 273(20417) EtOH
m/e: 225.12

HODOC No: 02344
CAS No: 3369-37-7
Beilstein No: 12^3, 370
Name: Benzenamine, *N*-[(2-methoxyphenyl)methylene]-
Synonym: Aniline, *N*-(*o*-methoxybenzylidene)-
Line Form: 2-CH_3O-C_6H_4CH=NC_6H_5
Mol Form: $C_{14}H_{13}NO$
M_r: 211.25
bp: $204\text{-}5^{25}$
mp: 33-4
UV: LANG 399 330(12882) 266(14454) EtOH
m/e: 211.10

HODOC No: 02345
CAS No: 836-41-9
Name: Benzenamine, *N*-[(4-methoxyphenyl)methylene]-
Synonym: Aniline, *N*-(*p*-methoxybenzylidene)-
Mol Form: $C_{14}H_{13}NO$
M_r: 211.25
mp: 62
UV: LANG 199 310(14454) 285(15488) EtOH
m/e: 211.10

HODOC No: 02346
CAS No: 3246-78-4
Name: Benzenamine, *N*-[(4-methoxyphenyl)methylene]-4-methyl-
Synonym: *p*-Toluidine, *N*-(*p*-methoxybenzylidene)-
Mol Form: $C_{15}H_{15}NO$
M_r: 225.28
mp: 92
UV: LANG 588 322(16596) 285(17378) 224(15136) EtOH
m/e: 225.12

HODOC No: 02347
CAS No: 95-53-4
Beilstein No: 12^3, 1837
Name: Benzenamine, 2-methyl-
Synonym: *o*-Toluidine
Line Form: 2-$CH_3C_6H_4NH_2$
Mol Form: C_7H_9N
M_r: 107.15
bp: 200.2, 80.1^{10}
mp: -14.7
***d*:** $0.9984^{20/4}$
***n*:** 1.5725^{20}
***s*:** H_2O 2; al 5; eth 5; ctc 5
IR: SADG 8160 3470 3380 3220 3020 2920 1620 1580 1500 1470 1440 1380 1300 1270 1140 1060 1030 980 930 850 750 710
Raman: SAD 677 3370 3210 3040 2980 2940 2910 2880 2850 2740 2620 1620 1600 1580 1490 1460 1440 1380 1310 1300 1270 1200 1150 1140 1080 1030 980 850 740 580 530 510 440 420 31 260 180
UV: SAD 442 285 233 cyhex
H NMR: SAD 107 2.0 3.2 6.3-7.1 CCl_4
MS: WILEY 330 106(100) 107(83) 77(17) 79(12) 39(12) 52(10) 53(10) 28(9)
m/e: 107.07

HODOC No: 02348
CAS No: 636-21-5
Beilstein No: 12^2, 432
Name: Benzenamine, 2-methyl-, hydrochloride
Synonym: *o*-Toluidine, hydrochloride
Mol Form: $C_7H_{10}ClN$
M_r: 143.61
bp: 242.2^{760}
mp: 215
***s*:** H_2O 4; al 3; eth 1; bz 1
IR: µSADP 6330 2860 2630 2000 1610 1490 1470 1320 1180 1120 1060 1040 750 700
UV: SAD 1740 284 266 258 MeOH
H NMR: SAD 10946 2.5 7.4 D_2O
MS: WILEY 783 106(100) 107(81) 36(33) 77(14) 79(12) 38(12) 108(6) 143(0)
m/e: 143.05

HODOC No: 02349
CAS No: 108-44-1
Beilstein No: 12^3, 1949
Name: Benzenamine, 3-methyl-
Synonym: *m*-Toluidine
Line Form: $3\text{-}CH_3C_6H_4NH_2$
Mol Form: C_7H_9N
M_r: 107.15
bp: 203.3, 82.3^{10}
mp: -30.4
***d*:** $0.9889^{20/4}$
***n*:** 1.5681^{20}
***s*:** H_2O 2; al 5; eth 5; ace 5; bz 5; ctc 5; hp 5
IR: SADG 8161 3450 3360 3220 3040 2940 1620 1590 1500 1470 1310 1290 1170 1000 930 870 850 770 690
Raman: SAD 234 3430 3350 3220 3040 3010 2910 2860 2730 1620 1580 1490 1460 1370 1310 1290 1160 1110 1080 990 920 840 770 730 530 510 430 290 230 210
UV: SAD 443 286(1460) 236(8900) MeOH
H NMR: SAD 108 2.2 3.3 6.2-6.6 6.9 CCl_4
m/e: 107.07

HODOC No: 02350
CAS No: 638-03-9
Beilstein No: 12, 856
Name: Benzenamine, 3-methyl-, hydrochloride
Synonym: *m*-Toluidine, hydrochloride
Mol Form: $C_7H_{10}ClN$
M_r: 143.61
bp: 250^{760}
mp: 228
***s*:** H_2O 4; al 4
IR: µSADP 9746 2940 1750 1610 1520 1470 1370 1190 1040 1010 940 910 860 790 700
UV: SAD 21950 286(80) 268(267) 261(309) 238(470) MeOH 287(1500) 236(7180) MeOH:KOH
H NMR: SAD 18740 2.3 7.0-7.4 10.3 poly D
MS: WILEY 784 107(100) 106(96) 36(40) 79(15) 77(15) 38(15) 108(8) 143(0)
m/e: 143.05

HODOC No: 02351
CAS No: 106-49-0
Beilstein No: 12^3, 2017
Name: Benzenamine, 4-methyl-
Synonym: *p*-Toluidine
Line Form: $4\text{-}CH_3C_6H_4NH_2$
Mol Form: C_7H_9N
M_r: 107.15
bp: 200.5, 79.6^{10}
mp: 44.5
***d*:** $0.9619^{20/4}$
***n*:** 1.5534^{45}
Color: lf (w+1)
***s*:** H_2O 2; al 4; eth 3; ace 3; ctc 3; py 4
IR: COB 4318 3460 3380 3020 2920 1620 1510 1450 1260 1180 1120 1040 810
Raman: SAD 233 3440 3350 3230 3060 3040 3020 2930 2860 2740 1620 1380 1320 1280 1220 1180 850 720 650 470 420 340 260 130
UV: SAD 1161 294 291 288 237 cyhex
H NMR: SAD 3172 2.2 3.3 6.3 6.8 CCl_4
MS: NBS 113
m/e: 107.07

HODOC No: 02352
CAS No: 540-23-8
Beilstein No: 12^2, 487
Name: Benzenamine, 4-methyl-, hydrochloride
Synonym: *p*-Toluidine, hydrochloride
Mol Form: $C_7H_{10}ClN$
M_r: 143.61
bp: 257.5^{760}
mp: 243
***s*:** H_2O 4; al 4; eth 1; bz 1; tfa 3; CS_2 1
IR: SADG 8458 3040 2920 2590 1560 1510 1210 1120 1140 820
UV: SAD 1745 269(306) 260(391) 239(832) MeOH
H NMR: SAD 3288 2.4 7.3 8.8 tfa
MS: WILEY 784 106(100) 107(68) 77(14) 79(9) 28(9) 36(9) 53(7) 143(0)
m/e: 143.05

HODOC No: 02353
CAS No: 100-61-8
Beilstein No: 12^3, 240
Name: Benzenamine, *N*-methyl-
Synonym: Aniline, *N*-methyl-
Line Form: $C_6H_5NHCH_3$
Mol Form: C_7H_9N
M_r: 107.15
bp: 196.2, 86^{15}
mp: -57
***d*:** $0.9891^{20/4}$
***n*:** 1.5684^{20}
***s*:** H_2O 1; al 3; eth 3; ctc 3; chl 3
IR: µCOB 6959 3450 2940 1610 1520 1430 1320 1270 1180 1150 1080 990 870 750 690
UV: SAD 1754 293 245 MeOH
H NMR: SAD 3203 2.7 3.3 6.4 6.6 7.1 CCl_4
C NMR: JJ 250 cw 30.2 112.3 116.7 129.2 150.2 diox
MS: WILEY 329 28(100) 106(85) 107(66) 77(24) 32(18) 51(16) 39(15) 79(11)
m/e: 107.07

HODOC No: 02354
CAS No: 2739-12-0
Beilstein No: 12^3, 244
Name: Benzenamine, *N*-methyl-, hydrochloride
Synonym: Aniline, *N*-methyl-, hydrochloride
Line Form: $C_6H_5NHCH_3{\cdot}HCl$
Mol Form: $C_7H_{10}ClFN_{28}$
M_r: 540.79
bp: 122-3
mp: 122.5-123
***d*:** $1.0660^{131/4}$
Color: nd (Chl-eth)
***s*:** H_2O 4; al 3; eth 1; bz 1; chl 4
IR: µSADP 10647 3330 2630 2380 1610 1560 1490 1410 1280 1120 1080 1030 980 850 790 760 690
m/e: 540.13

HODOC No: 02355
CAS No: 92-36-4
Beilstein No: 27^2, 436
Name: Benzenamine, 4-(6-methyl-2-benzothiazolyl)-
Synonym: Benzothiazole, 2-(*p*-aminophenyl)-6-methyl-
Mol Form: $C_{14}H_{12}N_2S$
M_r: 240.31
bp: 434
mp: 194.8
***s*:** al 2; eth 2; bz 2; HOAc 2
IR: COB 7147 3460 3320 3200 3050 2950 1630 1610 1480 1460 1310 1230 1180 980 820 570
UV: SAD 20837 342(32600) 223(27100) MeOH 412(17100) 314(14300) 257(8030) 252(8170) 227(20600)MeOH:HCl
H NMR: SAD 16534 2.4 5.2 6.7 7.2 7.6 7.7 poly D
MS: ALDN 195 240(100) 118(32) 121(29) 239(18) 122(14) 77(11) 78(11) 69(7)
m/e: 240.07

HODOC No: 02356
CAS No: 2716-62-3
Name: Benzenamine, *N*-(1-methylbutyl)-
Synonym: Aniline, *N*-(1-methylbutyl)-
Line Form: $C_6H_5NHC(CH_3)CH_2CH_2CH_3$
Mol Form: $C_{11}H_{17}N$
M_r: 163.25
IR: µCOB 4181 3030 2940 1610 1520 1430 1370 1330 1270 1180 1150 1090 870 750 690
m/e: 163.14

HODOC No: 02357
CAS No: 2051-84-5
Beilstein No: 12^3, 272
Name: Benzenamine, *N*-(3-methylbutyl)-
Synonym: Aniline, *N*-isopentyl-
Line Form: $C_6H_5NH\text{-}CH_2CH_2CH(CH_3)_2$
Mol Form: $C_{11}H_{17}N$
M_r: 163.25
bp: 254-5, $126\text{-}7^{14}$
***d*:** $0.8912^{55/4}$
***n*:** 1.5305^{20}
***s*:** Et_2O 4; EtOH 4
IR: µCOB 5874 3450 3030 2940 1610 1520 1370 1320 1250 1180 1150 1090 860 750 690
H NMR: SAD 17282 0.9 1.2-2.1 3.0 3.4 6.5 6.6 7.1 $CDCl_3$
m/e: 163.14

HODOC No: 02358
CAS No: 19406-51-0
Beilstein No: 12^3, 2184
Name: Benzenamine, 4-methyl-3,5-dinitro-
Synonym: *p*-Toluidine, 3,5-dinitro-
Line Form: $3,5\text{-}(NO_2)_2\text{-}4\text{-}CH_3C_6H_2NH_2$
Mol Form: $C_7H_7N_3O_4$
M_r: 197.14
mp: 171
Color: ye nd (w)
***s*:** Ac 4; Bz 4; Et_2O 4; EtOH 4
m/e: 197.04

HODOC No: 02359
CAS No: 101-77-9
Beilstein No: 13^3, 454
Name: Benzenamine, 4,4′-methylenebis-
Synonym: Aniline, 4,4′-methylenedi-
Line Form: $(4\text{-}H_2NC_6H_4)_2CH_2$
Mol Form: $C_{13}H_{14}N_2$
M_r: 198.25
bp: $398\text{-}9^{768}$, 257^{18}
mp: 92-3
Color: pl or nd (w) pl (bz)
***s*:** H_2O 2; al 4; eth 4; bz 4
IR: COB 6249 3450 3430 3140 1630 1510 1290 1280 1180 820
UV: SAD 18317 289(3120) 243(22000) MeOH
H NMR: SAD 11592 3.6 4.6 6.5 6.8 $DMSOd_6$
m/e: 198.12

HODOC No: 02360
CAS No: 101-14-4
Name: Benzenamine, 4,4′-methylenebis[2-chloro-
Synonym: Aniline, 4,4′-methylenebis[2-chloro-
Mol Form: $C_{13}H_{12}Cl_2N_2$
M_r: 267.14
***s*:** ctc 3
IR: COB 6719 3440 3360 3040 3020 2930 1620 1500 1410 1300 1260 1200 1040 930 810 720 620 560 500
UV: SAD 2914 298 247 MeOH
H NMR: SAD 6404 3.7 3.8 6.6 6.8 7.0 CCl_4
m/e: 266.04

HODOC No: 02361
CAS No: 101-61-1
Beilstein No: 13^3, 454
Name: Benzenamine, 4,4′-methylenebis[*N*,*N*-dimethyl-
Synonym: Aniline, 4,4′-methylenebis[*N*,*N*-dimethyl-
Line Form: $[4\text{-}(CH_3)_2N\text{-}C_6H_4]_2CH_2$
Mol Form: $C_{17}H_{22}N_2$
M_r: 254.36
bp: 390d, $182\text{-}5^3$
mp: 91-2
Color: pl or ta (al, lig)
***s*:** H_2O 1; al 2; eth 4; bz 4; CS_2 4; ac 3
IR: COB 5649 3060 2880 2800 1610 1520 1440 1340 1230 1190 1170 1070 940 830 790
H NMR: SAD 9462 2.8 3.8 6.6 7.0 $CDCl_3$
MS: ALDN 203 254(100) 253(62) 134(29) 210(26) 255(20) 118(15) 126(14) 137(14)
m/e: 254.18

HODOC No: 02362
CAS No: 1223-20-7
Name: Benzenamine, 4,4′-methylenebis[2-methoxy-
Synonym: *o*-Anisidine, 4,4′-methylenedi-
Mol Form: $C_{15}H_{18}N_2O_2$
M_r: 258.31
IR: µCOB 3994 3450 3330 2940 1610 1590 1520 1470 1450 1430 1300 1270 1240 1160 1030 960 860 830 810 750
m/e: 258.14

HODOC No: 02363
CAS No: 17474-44-1
Beilstein No: 13^2, 113
Name: Benzenamine, 4,4′-methylenebis[2-nitro-
Synonym: Aniline, 4,4′-methylenebis[2-nitro-
Line Form: $(3\text{-}NO_2\text{-}4\text{-}H_2NC_6H_3)_2CH_2$
Mol Form: $C_{13}H_{12}N_4O_4$
M_r: 288.25
mp: 232-3
Color: red nd
m/e: 288.09

HODOC No: 02364
CAS No: 26946-33-8
Beilstein No: 13^2, 118
Name: Benzenamine, 4,4′-methylenebis[3-nitro-
Synonym: Aniline, 4,4′-methylenebis[3-nitro-
Line Form: $(2\text{-}NO_2\text{-}4\text{-}H_2NC_6H_3)_2CH_2$
Mol Form: $C_{13}H_{12}N_4O_4$
M_r: 288.25
mp: 205
Color: og pl (al)
***s*:** EtOH 4; HOAc 4
m/e: 288.09

HODOC No: 02365
CAS No: 52562-19-3
Beilstein No: 12^3, 2793
Name: Benzenamine, 2-(1-methylethenyl)-
Synonym: *o*-Isopropenylaniline
Line Form: $2\text{-}H_2NC_6H_4\text{-}C(CH_3){=}CH_2$
Mol Form: $C_9H_{11}N$
M_r: 133.18
bp: 95^{13}
***d*:** 0.976
***n*:** 1.5722^{20}
m/e: 133.09

HODOC No: 02366
CAS No: 643-28-7
Beilstein No: 12^3, 2683
Name: Benzenamine, 2-(1-methylethyl)-
Synonym: Aniline, *o*-isopropyl-
Line Form: $2\text{-}H_2N\text{-}C_6H_4\text{-}CH(CH_3)_2$
Mol Form: $C_9H_{13}N$
M_r: 135.20
bp: $220\text{-}1^{745}$, 95^{13}
***d*:** $0.9760^{12/4}$
***s*:** H_2O 1; eth 3; bz 3; ctc 3
IR: µSADP 15599 3450 3330 2940 1610 1590 1490 1450 1350 1300 1250 1150 1030 930 750
UV: SAD 4676 285 235 cyhex
H NMR: SAD 4676 1.1 2.6 3.3 6.3-7.1 CCl_4
m/e: 135.10

HODOC No: 02367
CAS No: 99-88-7
Name: Benzenamine, 4-(1-methylethyl)-
Synonym: 4-Aminocumene
Line Form: $4\text{-}H_2NC_6H_4\text{-}CH(CH_3)_2$
Mol Form: $C_9H_{13}N$
M_r: 135.20
bp: 225
***d*:** $0.953^{20/4}$
m/e: 135.10

HODOC No: 02368
CAS No: 768-52-5
Beilstein No: 12^3, 266
Name: Benzenamine, *N*-(1-methylethyl)-
Synonym: Aniline, *N*-isopropyl-
Line Form: $C_6H_5NHCH(CH_3)_2$
Mol Form: $C_9H_{13}N$
M_r: 135.20
bp: 203, $72\text{-}5^{14}$, $226\text{-}7^{745}$
***d*:** 0.9526
***n*:** 1.5380^{20}
***s*:** al 3; eth 3; ace 3; bz 3
IR: SADG 6293 3333 2941 1613 1515 1471 1429 1370 1316 1250 1000 862 746 690
UV: SAD 12762 295(1640) 248(11000) MeOH
H NMR: SAD 1335 1.1 3.2 3.5 6.3-7.2 CCl_4
m/e: 135.10

HODOC No: 02369
CAS No: 2479-47-2
Beilstein No: 13^3, 495
Name: Benzenamine, 4,4′-(1-methylethylidene)*bis*-
Synonym: Aniline, 4,4′-isopropylidenedi-
Line Form: $(4\text{-}H_2NC_6H_4)_2C(CH_3)_2$
Mol Form: $C_{15}H_{18}N_2$
M_r: 226.31
bp: $200\text{-}30^{55}$
mp: 128-9
IR: µCOB 4051 3450 3330 3230 2940 1610 1520 1470 1370 1280 1180 830 810 740
H NMR: SAD 14567 1.6 3.5 6.6 7.1 $CDCl_3$
m/e: 226.15

HODOC No: 02370
CAS No: 5650-10-2
Name: Benzenamine, 4-(1-methylethyl)-*N*-phenyl-
Synonym: Cumidine, *N*-phenyl-
Mol Form: $C_{15}H_{17}N$
M_r: 211.29
MS: WILEY 1505 196(100) 211(42) 193(16) 197(16) 167(8) 168(8) 51(8) 39(8)
m/e: 211.14

HODOC No: 02371
CAS No: 548-61-8
Beilstein No: 13^3, 566
Name: Benzenamine, 4,4′,4″-methylidynetris-
Synonym: Aniline, 4,4′,4″-methylidynetri-
Line Form: $(4\text{-}H_2NC_6H_4)_3CH$
Mol Form: $C_{19}H_{19}N_3$
M_r: 289.36
mp: 208
Color: lf (w, al, bz)
***s*:** Et_2O 4; EtOH 4
m/e: 289.16

HODOC No: 02372
CAS No: 603-48-5
Beilstein No: 13^3, 566
Name: Benzenamine, 4,4′,4″-methylidynetris[*N*,*N*-dimethyl-
Synonym: Aniline, 4,4′,4″-methylidynetris[*N*,*N*-dimethyl-
Line Form: $[4\text{-}(CH_3)_2NC_6H_4]_3CH$
Mol Form: $C_{25}H_{31}N_3$
M_r: 373.52
mp: 175
Color: lf (al), nd (bz, lig)
***s*:** Bz 4; Et_2O 4; Chl 4
IR: μCOB 3855 3030 2940 1610 1560 1520 1470 1450 1350 1240 1210 1160 1120 1060 940 870 830 810 790 750 740
H NMR: VAR 360 0.9 1.0 2.2 2.3 5.4 $CDCl_3$
m/e: 373.25

HODOC No: 02373
CAS No: 2051-53-8
Beilstein No: 12^3, 2733
Name: Benzenamine, 2-methyl-5-(1-methylethyl)-
Synonym: 2-Amino-*p*-cymene
Line Form: $2\text{-}CH_3\text{-}5\text{-}(CH_3)_2CH\text{-}C_6H_3NH_2$
Mol Form: $C_{10}H_{15}N$
M_r: 149.23
bp: 241, 118^{12}
mp: -16
***d*:** $0.9942^{20/4}$
***n*:** 1.5387^{20}
***s*:** ctc 3; CS_2 3
IR: COB 6760 3470 3390 3000 2950 2920 2860 1620 1580 1510 1460 1420 1300 1290 1270 1260 1050 990 850 800 710 630 450
m/e: 149.12

HODOC No: 02374
CAS No: 10436-75-6
Beilstein No: 12, 905
Name: Benzenamine, 4-methyl-*N*-(1-methylethyl)-
Synonym: *p*-Toluidine, *N*-isopropyl-
Line Form: $4\text{-}CH_3C_6H_4NHCH(CH_3)_2$
Mol Form: $C_{10}H_{15}N$
M_r: 149.23
bp: 219-21
***d*:** 0.9226^{20}
***n*:** 1.5332
m/e: 149.12

HODOC No: 02375
CAS No: 2437-39-0
Beilstein No: 12^3, 2741
Name: Benzenamine, 5-methyl-2-(1-methylethyl)-
Synonym: Aniline, 2-isopropyl-5-methyl-
Line Form: $2\text{-}(CH_3)_2CH\text{-}5\text{-}CH_3\text{-}C_6H_3NH_2$
Mol Form: $C_{10}H_{15}N$
M_r: 149.23
bp: 238-42
Color: oil
***s*:** Et_2O 4; EtOH 4
UV: HBCP
m/e: 149.12

HODOC No: 02376
CAS No: 6950-79-4
Beilstein No: 12^3, 2741
Name: Benzenamine, *N*-methyl-4-(1-methylethyl)-
Synonym: *p*-Isopropyl-*N*-methylaniline
Line Form: 4-$(CH_3)_2CH$-$C_6H_4NHCH_3$
Mol Form: $C_{10}H_{15}N$
Mr: 149.23
bp: $225\text{-}7^{734}, 111\text{-}2^{11}$
Color: oil
***s*:** Et_2O 4; EtOH 4
m/e: 149.12

HODOC No: 02377
CAS No: 617-00-5
Beilstein No: 12^2, 437
Name: Benzenamine, 2-methyl-*N*-(2-methylphenyl)-
Synonym: Di-*o*-tolylamine
Line Form: $(2\text{-}CH_3C_6H_4)_2NH$
Structure:

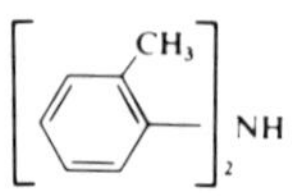

Mol Form: $C_{14}H_{15}N$
Mr: 197.27
bp: 312^{727}, 192^{23}
mp: 52-3
Color: cr
***s*:** H_2O 2
UV: HBCP 240(6918) 95% al
m/e: 197.12

HODOC No: 02378
CAS No: 626-13-1
Beilstein No: 12^2, 467
Name: Benzenamine, 3-methyl-*N*-(3-methylphenyl)-
Synonym: Di-*m*-tolylamine
Line Form: $(3\text{-}CH_3C_6H_4)_2NH$
Mol Form: $C_{14}H_{15}N$
Mr: 197.27
bp: 319-20
mp: 53
Color: pa ye (peth)
***s*:** H_2O 2; al 3; eth 3; peth 3
UV: HBCP 259(7943) 95% al
m/e: 197.12

HODOC No: 02379
CAS No: 620-93-9
Beilstein No: 12^3, 2033
Name: Benzenamine, 4-methyl-*N*-(4-methylphenyl)-
Synonym: Di-*p*-tolylamine
Line Form: $(4\text{-}CH_3C_6H_4)_2NH$
Mol Form: $C_{14}H_{15}N$
Mr: 197.27
bp: 330.5
mp: 79
Color: nd (peth)
***s*:** Et_2O 4; peth 4
m/e: 197.12

HODOC No: 02380
CAS No: 97-56-3
Beilstein No: 16^2, 386
Name: Benzenamine,
2-methyl-4-[(2-methylphenyl)azo]-
Synonym: *o*-Aminoazotoluene
Line Form: $(4\text{-}H_2N\text{-}3\text{-}CH_3)C_6H_3N{=}N(C_6H_4CH_3\text{-}2)$
Mol Form: $C_{14}H_{15}N_3$
Mr: 225.28
mp: 101-3
Color: ye lf (al)
***s*:** Et_2O 4; EtOH 4
m/e: 225.13

HODOC No: 02381
CAS No: 2834-78-8
Beilstein No: 16^3, 389
Name: Benzenamine,
2-methyl-4-[(4-methylphenyl)azo]-
Synonym: *o*-Toluidine, 4-*p*-tolylazo-
Line Form: $(4\text{-}H_2N\text{-}3\text{-}CH_3)C_6H_3N{=}N(C_6H_4CH_3\text{-}4)$
Mol Form: $C_{14}H_{15}N_3$
Mr: 225.28
mp: 128
Color: og-ye nd (lig) ye pr (al)
m/e: 225.13

HODOC No: 02382
CAS No: 5014-86-8
Beilstein No: 16^2, 180
Name: Benzenamine,
3-methyl-4-[(2-methylphenyl)azo]-
Synonym: *m*-Toluidine, 4-(*o*-tolylazo)-
Line Form: (4-H_2N-2-CH_3)C_6H_3N=N($C_6H_4CH_3$-2)
Mol Form: $C_{14}H_{15}N_3$
M_r: 225.28
mp: 116-7
Color: ye nd (lig)
***s*:** EtOH 4; lig 4
m/e: 225.13

HODOC No: 02383
CAS No: 3398-09-2
Beilstein No: 16^3, 392
Name: Benzenamine,
3-methyl-4-[(3-methylphenyl)azo]-
Synonym: *m*-Toluidine, 4-(*m*-tolylazo)-
Line Form: (4-H_2N-2-CH_3)C_6H_3N=N($C_6H_4CH_3$-3)
Mol Form: $C_{14}H_{15}N_3$
M_r: 225.28
bp: 80
Color: ye-gold nd (al) ye-br nd (lig)
***s*:** EtOH 4
MS: NBS 1495
m/e: 225.13

HODOC No: 02384
CAS No: 3963-79-9
Beilstein No: 16, 348
Name: Benzenamine,
3-methyl-4-[(4-methylphenyl)azo]-
Synonym: 4-Amino-2,4′-dimethylazobenzene
Line Form: (4-H_2N-2-CH_3)C_6H_3N=N($C_6H_4CH_3$-4)
Mol Form: $C_{14}H_{15}N_3$
M_r: 225.28
mp: 127
Color: ye pr (al), gold-ye (lig)
***s*:** EtOH 4; lig 4
m/e: 225.13

HODOC No: 02385
CAS No: 58010-91-6
Beilstein No: 16^3, 396
Name: Benzenamine,
4-methyl-2-[(4-methylphenyl)azo]-
Synonym: *p*-Toluidine, 2-(*p*-tolylazo)-
Line Form: (5-CH_3-2-H_2N)C_6H_3N=N($C_6H_4CH_3$-4)
Mol Form: $C_{14}H_{15}N_3$
M_r: 225.28
mp: 118.5
***s*:** al 3; bz 4; AcOEt 3
UV: JOCEA 23-621-57 350(23442) 1.2N HCl
MS: NBS 1495
m/e: 225.13

HODOC No: 02386
CAS No: 2058-62-0
Beilstein No: 16^3, 360
Name: Benzenamine,
N-methyl-4-[(3-methylphenyl)azo]-
Synonym: Aniline, *N*-methyl-*p*-(*m*-tolylazo)-
Line Form: 4-CH_3NH-C_6H_4N=N($C_6H_4CH_3$-3)
Mol Form: $C_{14}H_{15}N_3$
M_r: 225.28
mp: 89-90
***s*:** chl 2
IR: SADG 15268 3350 3060 2900 1600 1580 1530 1400 1350 1300 1270 1220 1140 900 810 770 740 700
UV: OES 3-469 511(42658) 321(7413) 1.2N HCl:50% EtOH
H NMR: SAD 10742 2.0 2.7 3.8 6.5 7.2-7.5 7.6-8.0 $CDCl_3$
m/e: 225.13

HODOC No: 02387
CAS No: 603-83-8
Beilstein No: 12^3, 1944
Name: Benzenamine, 2-methyl-3-nitro-
Synonym: *o*-Toluidine, 3-nitro-
Line Form: $3\text{-}NO_2\text{-}2\text{-}CH_3C_6H_3NH_2$
Mol Form: $C_7H_8N_2O_2$
$\mathbf{M_r}$: 152.15
bp: 305d
mp: 92
Color: ye rh nd (w), ye lf (al)
***s*:** H_2O 2; al 3; eth 3; bz 3; chl 3
IR: SADG 4606 3400 3330 3230 3000 2960 2920 1650 1620 1580 1520 1480 1470 1390 1360 1320 1300 1200 1130 1050 920 900 800 780 730 690
UV: SAD 11379 348(1400) 235(14000) MeOH
H NMR: SAD 17386 2.1 5.2 poly D
m/e: 152.06

HODOC No: 02388
CAS No: 99-52-5
Beilstein No: 12^3, 1938
Name: Benzenamine, 2-methyl-4-nitro-
Synonym: *o*-Toluidine, 4-nitro-
Line Form: $(4\text{-}NO_2\text{-}2\text{-}CH_3)C_6H_3NH_2$
Mol Form: $C_7H_8N_2O_2$
$\mathbf{M_r}$: 152.15
mp: 131-3
***d*:** $1.1586^{140/4}$
***s*:** H_2O 2; al 3; bz 3; $DMSOd_6$ 2; HOAc 3
IR: SADG 8625 3470 3360 3230 2920 1640 1600 1590 1510 1480 1310 1270 1100 900 830 750 650
UV: SAD 3619 225 MeOH
H NMR: SAD 8033 2.2 6.4 6.7 7.8 7.9 $DMSOd_6$
MS: NBS 477; ALD 5:171A
m/e: 152.06

HODOC No: 02389
CAS No: 99-55-8
Beilstein No: 12^3, 1932
Name: Benzenamine, 2-methyl-5-nitro-
Synonym: *o*-Toluidine, 5-nitro-
Line Form: $(5\text{-}NO_2\text{-}2\text{-}CH_3)C_6H_3NH_2$
Mol Form: $C_7H_8N_2O_2$
$\mathbf{M_r}$: 152.15
mp: 104-7
***s*:** H_2O 2; al 3; eth 3; ace 3; bz 3; chl 3
IR: SADG 8626 3490 3400 3090 2980 1630 1500 1380 1350 1300 1280 1100 950 870 810 740
UV: SAD 3620 286 249 226 MeOH
H NMR: SAD 3252 2.2 3.9 7.1 7.4 7.5 $CDCl_3$
MS: ALDN 107 77(100) 79(91) 78(22) 94(18) 51(18) 122(17) 104(15) 152(13)
m/e: 152.06

HODOC No: 02390
CAS No: 570-24-1
Beilstein No: 12^3, 1929
Name: Benzenamine, 2-methyl-6-nitro-
Synonym: *o*-Toluidine, 6-nitro-
Line Form: $(6\text{-}NO_2\text{-}2\text{-}CH_3)C_6H_3NH_2$
Mol Form: $C_7H_8N_2O_2$
$\mathbf{M_r}$: 152.15
mp: 94-6
***d*:** $1.1900^{100/4}$
***s*:** H_2O 2; al 3; eth 3; bz 3; chl 3
IR: SADG 4703 3490 3380 2980 2920 1620 1590 1570 1520 1460 1440 1420 1380 1360 1320 1250 1190 1080 890 730
UV: SAD 8592 284(4880) 225(14900) MeOH
H NMR: SAD 18233 2.2 6.5 6.9 7.2 7.8 poly D
MS: ALD 5:158B
m/e: 152.06

HODOC No: 02391
CAS No: 601-87-6
Beilstein No: 12^2, 476
Name: Benzenamine, 3-methyl-2-nitro-
Synonym: 3-Methyl-2-nitroaniline
Mol Form: $C_7H_8N_2O_2$
$\mathbf{M_r}$: 152.15
mp: 108
***s*:** H_2O 2; al 4; eth 3
UV: HBCP 404(3162) 283(3386) al
m/e: 152.06

HODOC No: 02392
CAS No: 611-05-2
Beilstein No: 12^3, 2008
Name: Benzenamine, 3-methyl-4-nitro-
Synonym: *m*-Toluidine, 4-nitro-
Line Form: $4\text{-}NO_2\text{-}3\text{-}CH_3C_6H_3NH_2$
Mol Form: $C_7H_8N_2O_2$
M_r: 152.15
mp: 138
Color: lt ye nd (w, dil al)
***s*:** H_2O 2; al 3; eth 3; ac 4
IR: µSADP 21815 3450 3330 3230 2940 1640 1560 1470 1350 1300 1270 1180 1080 860 820 760 690
UV: OES 2-103 380(6026) H_2O
H NMR: SAD 18212 2.5 6.1 6.4 6.5 7.9 poly D
m/e: 152.06

HODOC No: 02393
CAS No: 618-61-1
Beilstein No: 12^3, 2007
Name: Benzenamine, 3-methyl-5-nitro-
Synonym: *m*-Toluidine, 5-nitro-
Line Form: $5\text{-}NO_2\text{-}3\text{-}CH_3C_6H_3NH_2$
Mol Form: $C_7H_8N_2O_2$
M_r: 152.15
mp: 98
Color: ye-red or red-br nd (al)
***s*:** H_2O 2; al 3; eth 4; bz 3
UV: HBCP 375(1175) 287(3891) 235(15136) al
m/e: 152.06

HODOC No: 02394
CAS No: 89-62-3
Beilstein No: 12^3, 2174
Name: Benzenamine, 4-methyl-2-nitro-
Synonym: *p*-Toluidine, 2-nitro-
Line Form: $(2\text{-}NO_2\text{-}4\text{-}CH_3)C_6H_3NH_2$
Mol Form: $C_7H_8N_2O_2$
M_r: 152.15
mp: 115-6
***d*:** $1.164^{121/4}$
***s*:** H_2O 2; al 3; chl 3
IR: SADG 611 3460 3340 3170 2920 1640 1600 1560 1510 1450 1400 1370 1340 1280 1240 1200 1170 1100 920 870 820 780 770 680
UV: SAD 3621 418(5010) 279(4970) 229(19200) MeOH
H NMR: SAD 543 2.2 5.9 6.7 7.2 7.9 $CDCl_3$
MS: ALD 5:170B
m/e: 152.06

HODOC No: 02395
CAS No: 119-32-4
Beilstein No: 12^3, 2165
Name: Benzenamine, 4-methyl-3-nitro-
Synonym: *p*-Toluidine, 3-nitro-
Line Form: $(3\text{-}NO_2\text{-}4\text{-}CH_3)C_6H_3NH_2$
Mol Form: $C_7H_8N_2O_2$
M_r: 152.15
mp: 77-9
***s*:** H_2O 2; al 3; eth 3; bz 3; CS_2 2
IR: µSADP 15634 3450 3330 2940 1640 1490 1430 1390 1350 1300 1100 940 870 820 740
UV: SAD 4691 289 251 228 MeOH
MS: ALDN 107 77(100) 135(75) 107(73) 152(71) 79(62) 106(49) 105(45) 104(36)
m/e: 152.06

HODOC No: 02396
CAS No: 578-46-1
Beilstein No: 12^3, 2004
Name: Benzenamine, 5-methyl-2-nitro-
Synonym: *m*-Toluidine, 6-nitro-
Line Form: $6\text{-}NO_2\text{-}3\text{-}CH_3C_6H_3NH_2$
Mol Form: $C_7H_8N_2O_2$
M_r: 152.15
mp: 112
Color: ye lf (w), pr (dil al)
***s*:** H_2O 3; al 3; eth 3; bz 3; chl 3
UV: HBCP 374(7413) 233(12882) al
MS: ALDN 107 152(100) 77(67) 79(54) 106(40) 135(15) 78(15) 104(13) 122(12)
m/e: 152.06

HODOC No: 02397
CAS No: 612-28-2
Beilstein No: 12^3, 1516
Name: Benzenamine, *N*-methyl-2-nitro-
Synonym: Aniline, *N*-methyl-*o*-nitro-
Line Form: $2\text{-}NO_2C_6H_4NHCH_3$
Mol Form: $C_7H_8N_2O_2$
M_r: 152.15
bp: d
mp: 38
Color: red or og nd (peth)
***s*:** H_2O 2; al 3; eth 3; ace 3; bz 3; os 3; con ac 3; lig 2
IR: SADG 31359 3390 3070 2900 2810 1610 1560 1500 1430 1420 1410 1340 1320 1250 1220 1170 1120 1070 1040 1020 850 760 730 680 650 540 510
UV: OES 7-119 429(5012) EtOH
H NMR: SAD 19120 3.0 6.6 6.8 7.4 8.0 8.1 $CDCl_3$
m/e: 152.06

HODOC No: 02398
CAS No: 619-26-1
Beilstein No: 12^3, 1544
Name: Benzenamine, *N*-methyl-3-nitro-
Synonym: Aniline, *N*-methyl-*m*-nitro-
Line Form: 3-NO_2-$C_6H_4NHCH_3$
Mol Form: $C_7H_8N_2O_2$
M_r: 152.15
mp: 68
Color: red ye nd or pr (al) cr (lig)
***s*:** Bz 4; Et_2O 4; EtOH 4
m/e: 152.06

HODOC No: 02399
CAS No: 100-15-2
Beilstein No: 12^3, 1584
Name: Benzenamine, *N*-methyl-4-nitro-
Synonym: Aniline, *N*-methyl-*p*-nitro-
Line Form: 4-NO_2-$C_6H_4NHCH_3$
Mol Form: $C_7H_8N_2O_2$
M_r: 152.15
bp: d
mp: 152
***d*:** $1.201^{155/4}$
Color: br-ye pr (al) cr (eth)
***s*:** H_2O 1; al 3; eth 2; bz 3; chl 3; lig 2
IR: μSADP 18329 3330 2940 1610 1560 1470 1320 1190 1160 1110 1060 1000 830 750 700
UV: SAD 5832 229 MeOH
H NMR: VAR 489 3.0 4.6 6.6 8.1 $CDCl_3$
m/e: 152.06

HODOC No: 02400
CAS No: 22760-88-9
Name: Benzenamine, 4-methyl-*N*-[1-(4-nitrophenyl)ethylidene]-
Synonym: *p*-Toluidine, *N*-(α-methyl-*p*-nitrobenzylidene)-
Mol Form: $C_{15}H_{14}N_2O_2$
M_r: 254.27
mp: 131-2
UV: LANG 2352 347(2884) 276(16218) 224(15136) EtOH
m/e: 254.11

HODOC No: 02401
CAS No: 10595-51-4
Beilstein No: 7^3, 3370
Name: Benzenamine, *N*-methyl-4-nitroso-
Synonym: Aniline, *N*-methyl-*p*-nitroso-
Line Form: 4-ON-$C_6H_4NHCH_3$
Mol Form: $C_7H_8N_2O$
M_r: 136.15
mp: 118
Color: bl pl (Bz)
***s*:** H_2O 2; al 3; eth 3; bz 2; chl 3
IR: μSADP 20537 3230 2940 1610 1540 1450 1350 1280 1250 1140 830 780
H NMR: VAR 488 3.5 4.1 $CDCl_3$
MS: ALDN 83 136(100) 77(64) 106(57) 79(49) 65(38) 78(22) 105(18) 50(18)
m/e: 136.06

HODOC No: 02402
CAS No: 614-00-6
Beilstein No: 12^3, 1129
Name: Benzenamine, *N*-methyl-*N*-nitroso-
Synonym: Aniline, *N*-methyl-*N*-nitroso-
Line Form: $C_6H_5N(NO)CH_3$
Mol Form: $C_7H_8N_2O$
M_r: 136.15
bp: 225d, 121^{13}
mp: 14.7
***d*:** $1.1240^{20/4}$
***n*:** 1.5769^{20}
Color: ye
***s*:** H_2O 1; al 3; eth 3
IR: SADG 27904 3020 2940 1590 1490 1470 1440 1390 1310 1290 1200 1090 1030 1000 950 820 760 690 520
UV: OES 8-67 269(7413) H_2O
H NMR: SAD 17273 3.4 7.3-7.6 $CDCl_3$
C NMR: JJ 244 FT 31.1 119.0 127.1 129.3 142.2 $CDCl_3$
MS: ALDN 83 106(100) 77(44) 107(30) 136(21) 79(16) 78(11) 51(9) 104(9)
m/e: 136.06

HODOC No: 02403
CAS No: 1205-39-6
Beilstein No: 12^2, 436
Name: Benzenamine, 2-methyl-*N*-phenyl-
Synonym: *o*-Toluidine, *N*-phenyl-
Line Form: 2-$CH_3C_6H_4$-NH-C_6H_5
Mol Form: $C_{13}H_{13}N$
Mr: 183.24
bp: 305^{727}
mp: 41
m/e: 183.10

HODOC No: 02404
CAS No: 552-82-9
Beilstein No: 12^3, 290
Name: Benzenamine, *N*-methyl-*N*-phenyl-
Synonym: Diphenylamine, *N*-methyl-
Line Form: $(C_6H_5)_2NCH_3$
Mol Form: $C_{13}H_{13}N$
Mr: 183.24
bp: 293-4, 145^{10}
mp: -7.5
***d*:** $1.0476^{20/4}$
***n*:** 1.6193^{20}
***s*:** H_2O 1; al 2; ctc 3; MeOH 2
IR: µCOB 3797 3030 2860 1590 1490 1350 1250 1190 1140 1100 1080 1030 990 860 750 690
UV: OES 6-474 292(11220) 245(9120) 206(29512) EtOH
H NMR: SAD 356 3.2 7.0 CCl_4
MS: ALDN 145 183(100) 182(62) 167(26) 104(18) 77(17) 184(14) 51(12) 91(6)
m/e: 183.10

HODOC No: 02405
CAS No: 4209-15-8
Name: Benzenamine, 4-methyl-*N*-(1-phenylethylidene)-
Synonym: *p*-Toluidine, *N*-(α-methylbenzylidene)-
Mol Form: $C_{15}H_{15}N$
Mr: 209.28
mp: 31
UV: LANG 2339 315(2455) 246(18621) EtOH
m/e: 209.12

HODOC No: 02406
CAS No: 4209-16-9
Name: Benzenamine, *N*-[1-(4-methylphenyl)ethylidene]-
Synonym: Aniline, *N*-(*p*,α-dimethylbenzylidene)-
Mol Form: $C_{15}H_{15}N$
Mr: 209.28
mp: 48-50
UV: LANG 2343 310(2239) 256(17378) EtOH
m/e: 209.12

HODOC No: 02407
CAS No: 5877-55-4
Beilstein No: 12^3, 1848
Name: Benzenamine, 2-methyl-*N*-(phenylmethylene)-
Synonym: *o*-Toluidine, *N*-benzylidene-
Line Form: C_6H_5CH=N($C_6H_4CH_3$-2)
Mol Form: $C_{14}H_{13}N$
Mr: 195.25
bp: 314, 176^{15}
***d*:** $1.041^{20/4}$
***n*:** 1.6310^{25}
***s*:** H_2O 1; ace 3
UV: TETRA 19-445-63 325(4266) 261(12303) 244(9772) 235(8318) 226(7763) cyhex
m/e: 195.10

HODOC No: 02408
CAS No: 5877-58-7
Beilstein No: 12^3, 1958
Name: Benzenamine, 3-methyl-*N*-(phenylmethylene)-
Synonym: *m*-Toluidine, *N*-benzylidene-
Line Form: C_6H_5CH=N($C_6H_4CH_3$-3)
Mol Form: $C_{14}H_{13}N$
Mr: 195.25
bp: 315^{775}
mp: 30-2
***n*:** 1.6353^{25}
***s*:** ace 3; os 3
UV: TETRA 19-445-63 319(6166) 262(15136) 251(13804) 242(14125) 233(12882) 225(12023) hx
m/e: 195.10

HODOC No: 02409
CAS No: 2272-45-9
Beilstein No: 12^3, 2096
Name: Benzenamine, 4-methyl-*N*-(phenylmethylene)-
Synonym: *p*-Toluidine, *N*-benzylidene-
Line Form: $C_6H_5CH{=}N(C_6H_4CH_3\text{-}4)$
Mol Form: $C_{14}H_{13}N$
Mr: 195.25
bp: 318^{755}, 178^{11}
mp: 35
Color: ye cr
***s*:** Ac 4
UV: LANG 587 315(9550) 263(14125) 224(12023) EtOH
MS: ALDN 158 194(100) 77(95) 195(90) 51(61) 178(28) 39(27) 91(25) 118(16)
m/e: 195.10

HODOC No: 02410
CAS No: 15286-48-3
Beilstein No: 12^1, 416
Name: Benzenamine, 4-methyl-*N*-(3-phenyl-2-propenylidene)-
Synonym: *p*-Toluidine, *N*-cinnamylidene-
Mol Form: $C_{16}H_{15}N$
Mr: 221.29
mp: 83
***s*:** al 3; eth 2; bz 3; chl 3; peth 2
IR: SADG 16596 3020 2900 1620 1570 1490 1440 1370 1140 970 820 800 730 680 520 490
UV: SAD 16916 327(23000) 297(24400) 229(11500) MeOH 372(586) 282(9290) 250(12800) 224(7920) MeOH:HCl
H NMR: sad 8490 2.3 7.1 7.0-7.6 8.1 $CDCl_3$
m/e: 221.12

HODOC No: 02411
CAS No: 55751-54-7
Beilstein No: 12^3, 2721
Name: Benzenamine, 2-(1-methylpropyl)-
Synonym: Aniline, *o*-*sec*-butyl-
Line Form: $2\text{-}H_2NC_6H_4CH(CH_3)CH_2CH_3$
Mol Form: $C_{10}H_{15}N$
Mr: 149.23
bp: $120\text{-}2^{16}$
***d*:** 0.9574^{20}
***s*:** al 3; ace 3; bz 3; ctc 2
IR: µSADP 15601 3330 3030 2940 1610 1590 1490 1450 1370 1300 1270 1160 750
UV: SAD 4678 286 235 cyhex
H NMR: SAD 6447 0.9 1.2 1.5 2.5 3.3 6.5 6.6-7.1 CCl_4
m/e: 149.12

HODOC No: 02412
CAS No: 30273-11-1
Beilstein No: 12^2, 635
Name: Benzenamine, 4-(1-methylpropyl)-
Synonym: Aniline, *p*-*sec*-butyl-
Line Form: $4\text{-}H_2NC_6H_4\text{-}CH(CH_3)CH_2CH_3$
Mol Form: $C_{10}H_{15}N$
Mr: 149.23
bp: 238^{762}, 118^{15}
***d*:** $0.949^{15/4}$
***n*:** 1.5360^{29}
***s*:** Bz 4; Et_2O 4
m/e: 149.12

HODOC No: 02413
CAS No: 30090-17-6
Beilstein No: 12^3, 2715
Name: Benzenamine, 4-(2-methylpropyl)-
Synonym: Aniline, *p*-isobutyl-
Line Form: $(4\text{-}H_2NC_6H_4)\text{-}CH_2CH(CH_3)_2$
Mol Form: $C_{10}H_{15}N$
Mr: 149.23
bp: 238^{762}, 112^{11}
***d*:** $0.949^{15/4}$
Color: pa ye
***s*:** Ac 4; Bz 4; EtOH 4
m/e: 149.12

HODOC No: 02414
CAS No: 54837-90-0
Beilstein No: 12^3, 2028
Name: Benzenamine, 4-methyl-*N*-propyl-
Synonym: *N*-Propyl-*p*-toluidine
Line Form: $4\text{-}CH_3C_6H_4NHCH_2CH_2CH_3$
Mol Form: $C_{10}H_{15}N$
Mr: 149.23
bp: 235
***d*:** 0.9243^{20}
***n*:** 1.5367
m/e: 149.12

HODOC No: 02415
CAS No: 6068-69-5
Beilstein No: 12^3, 769
Name: Benzenamine, *N*-(1-methylpropyl)-
Synonym: Aniline, *N-sec*-butyl-
Line Form: $C_6H_5NHCH(CH_3)C_2H_5$
Mol Form: $C_{10}H_{15}N$
Mr: 149.23
bp: 225, $112\text{-}4^{22}$
***n*:** 1.5333^{20}
***s*:** chl 2; os 3
IR: SADG 15874 3448 3030 2941 1613 1515 1429 1370 1316 1250 1149 870 752 694
UV: SAD 15955 296(1760) 249(12000) MeOH
H NMR: VAR 568 1.0 1.2 1.5 3.4 $CDCl_3$
C NMR: JJ 389 FT 10.2 20.1 29.6 49.7 113.1 116.6 129.1 147.7 $CDCl_3$
m/e: 149.12

HODOC No: 02416
CAS No: 588-47-6
Beilstein No: 12^3, 269
Name: Benzenamine, *N*-(2-methylpropyl)-
Synonym: Aniline, *N*-isobutyl-
Line Form: $C_6H_5NHCH_2CH(CH_3)_2$
Mol Form: $C_{10}H_{15}N$
Mr: 149.23
bp: 231-3, $109\text{-}10^{13}$
***d*:** $0.940^{18/4}$
***n*:** 1.5328^{20}
***s*:** H_2O 1; eth 4; bz 4
UV: OES 3-265 260(158) 254(219) 248(195) 0.04N HCl:48% EtOH 292(1585) 245(10471) 0.004N NaOH:48% EtOH
m/e: 149.12

HODOC No: 02417
CAS No: 479-45-8
Beilstein No: 12^3, 1738
Name: Benzenamine, *N*-methyl-*N*,2,4,6-tetranitro-
Synonym: Aniline, *N*-methyl-*N*,2,4,6-tetranitro-
Line Form: $2,4,6\text{-}(NO_2)_3\text{-}C_6H_2\text{-}N(NO_2)CH_3$
Structure:

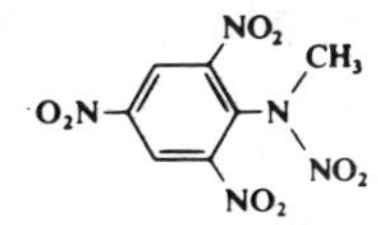

Mol Form: $C_7H_5N_5O_8$
Mr: 287.14
bp: exp 187
mp: 131-2
***d*:** 1.57^{10}
Color: ye pr (al)
***s*:** H_2O 1; al 2; eth 2; ace 3; bz 3; chl 2; py 3; CS_2 1
IR: SADG 750 3125 3030 1613 1563 1449 1429 1333 1282 1075 971 926 806 794 763 725
UV: SAD 9663 224(22800) MeOH
H NMR: SAD 18185 3.6 9.2 poly D
MS: ALDN 221 241(100) 181(33) 91(31) 66(31) 30(29) 224(28) 90(28) 287(4)
m/e: 287.01

HODOC No: 02418
CAS No: 104-96-1
Beilstein No: 13^3, 1221
Name: Benzenamine, 4-(methylthio)-
Synonym: Aniline, *p*-(methylthio)-
Line Form: $4\text{-}CH_3S\text{-}C_6H_4NH_2$
Mol Form: C_7H_9NS
Mr: 139.21
bp: 272-3, 140^{15}
***d*:** $1.1379^{20/4}$
***n*:** 1.6395^{20}
***s*:** al 3; eth 3; ace 3; bz 3
IR: µSADP 19425 3450 3330 2940 1640 1610 1490 1430 1280 1190 1140 1100 970 820 730
UV: OES 7-125 320(631) 300(1514) 264(10715) 238(7586) hx 321(794) 300(1820) 263(15136) 241(7943) EtOH
m/e: 139.05

HODOC No: 02419
CAS No: 23153-09-5
Name: Benzenamine, 4-(methylthio)-2-nitro-
Synonym: Aniline, 4-(methylthio)-2-nitro-
Mol Form: $C_7H_8N_2O_2S$
M_r: 184.21
mp: 75
***s*:** chl 3
IR: µSADP 7493 3570 3350 2940 1640 1585 1565 1515 1470 1410 1370 1335 1280 1250 1175 1125 960 900 860 820 765 730 720
H NMR: SAD 11301 2.5 6.0 6.8 7.3 8.1 $CDCl_3$
m/e: 184.03

HODOC No: 02420
CAS No: 2524-67-6
Beilstein No: 27^4, 450
Name: Benzenamine, 4-(4-morpholinyl)-
Synonym: Morpholine, 4-(*p*-aminophenyl)-
Structure:

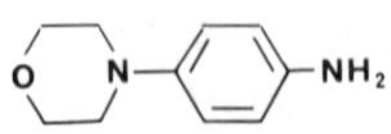

Mol Form: $C_{10}H_{14}N_2O$
M_r: 178.22
mp: 132.5
m/e: 178.11

HODOC No: 02421
CAS No: 542-15-4
Beilstein No: 12^3, 232
Name: Benzenamine, nitrate
Synonym: Aniline, nitrate
Line Form: $C_6H_5NH_2{\cdot}HNO_3$
Mol Form: $C_6H_8FN_{29}O_3$
M_r: 553.31
mp: 190d
***d*:** 1.356^4
Color: rh
***s*:** H_2O 4; Et_2O 4; EtOH 4
m/e: 553.13

HODOC No: 02422
CAS No: 88-74-4
Beilstein No: 12^3, 1513
Name: Benzenamine, 2-nitro-
Synonym: Aniline, *o*-nitro-
Line Form: $2\text{-}NO_2C_6H_4NH_2$
Mol Form: $C_6H_6N_2O_2$
M_r: 138.12
bp: 284, $165\text{-}6^{28}$
mp: 71.5
***d*:** 1.442^{15}
***s*:** H_2O 2; al 3; eth 4; ace 4; bz 4; chl 4
IR: SADG 229 3448 3333 3030 1613 1563 1493 1429 1333 1250 1163 1111 781 746 671
Raman: SAD 237 3380 1630 1590 1510 1450 1400 1340 1320 1290 1180 1110 960 870 820 640 540 400 150
UV: SAD 385 277 232 MeOH
H NMR: SAD 6495 6.3 6.7 6.9 7.4 8.1 $CDCl_3$
MS: ALDN 87 138(100) 65(80) 39(23) 92(17) 52(12) 63(11) 64(11) 66(9)
m/e: 138.04

HODOC No: 02423
CAS No: 99-09-2
Beilstein No: 12^3, 154
Name: Benzenamine, 3-nitro-
Synonym: Aniline, *m*-nitro-
Line Form: $3\text{-}NO_2\text{-}C_6H_4NH_2$
Mol Form: $C_6H_6N_2O_2$
M_r: 138.12
bp: 305-7d, $100^{0.6}$
mp: 114
***d*:** $1.1747^{160/4}$, $0.991^{25/4}$
***s*:** H_2O 2; al 3; eth 3; ace 3; bz 2; chl 3; MeOH 4
IR: SADG 70 3448 3333 3030 1613 1563 1515 1493 1351 1266 1087 1075 1000 926 870 820 794 741 667
UV: SAD 80 371(1090) 233(12000) MeOH
MS: ALDN 87 138(100) 39(35) 65(23) 92(23) 66(14) 63(12) 28(11) 80(11)
m/e: 138.04

HODOC No: 02424
CAS No: 100-01-6
Beilstein No: 12^3, 1580
Name: Benzenamine, 4-nitro-
Synonym: Aniline, *p*-nitro-
Line Form: $4\text{-}NO_2\text{-}C_6H_4NH_2$
Mol Form: $C_6H_6N_2O_2$
M$_r$: 138.12
bp: 331.7, $106^{0.03}$
mp: 148-9
***d*:** $1.424^{20/4}$
Color: pa ye mcl nd (w)
***s*:** H_2O 1; al 3; eth 3; ace 3; bz 2; chl 3; $DMSOd_6$ 2; MeOH 4; tol 3
IR: μCOB 2257 3570 3450 1640 1610 1520 1470 1450 1320 1190 1110 840 750 700
UV: SAD 2243 228 MeOH
H NMR: SAD 9378 6.7 6.8 8.1 $DMSOd_6$
MS: ALDN 87 65(100) 108(52) 138(47) 39(42) 92(39) 63(18) 52(16) 80(16)
m/e: 138.04

HODOC No: 02425
CAS No: 645-55-6
Beilstein No: 16^2, 343
Name: Benzenamine, *N*-nitro-
Synonym: Aniline, *N*-nitro-
Line Form: $C_6H_5NHNO_2$
Mol Form: $C_6H_6N_2O_2$
M$_r$: 138.12
bp: exp
mp: 48
Color: lf (peth)
***s*:** H_2O 4; Bz 4; Et_2O 4; EtOH 4
m/e: 138.04

HODOC No: 02426
CAS No: 18264-71-6
Beilstein No: 12^3, 1518
Name: Benzenamine, 2-nitro-*N*-(2-nitrophenyl)-
Synonym: Diphenylamine, 2,2′-dinitro-
Line Form: $(2\text{-}NO_2C_6H_4)_2NH$
Structure:

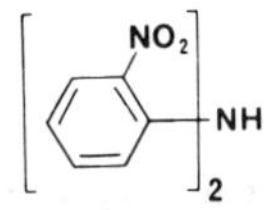

Mol Form: $C_{12}H_9N_3O_4$
M$_r$: 259.21
mp: 169
Color: ye lf(al, ace), cr(al,HOAc)
***s*:** al 4; ace 3; MeOH 3; HOAc 4
UV: OES 8-232 420(9333) 380(7079) 290(12023) 264(16596) 225(15136) EtOH
H NMR: NBS 1847
m/e: 259.06

HODOC No: 02427
CAS No: 612-36-2
Beilstein No: 12^3, 1587
Name: Benzenamine, 2-nitro-*N*-(4-nitrophenyl)-
Synonym: Diphenylamine, 2,4′-dinitro-
Line Form: $2\text{-}O_2NC_6H_4NH(C_6H_4NO_2\text{-}4)$
Mol Form: $C_{12}H_9N_3O_4$
M$_r$: 259.21
mp: 222-3
Color: red nd(HOAc)
***s*:** to 4; Chl 4
MS: NBS 1847
m/e: 259.06

HODOC No: 02428
CAS No: 15979-87-0
Beilstein No: 12^3, 1587
Name: Benzenamine, 3-nitro-*N*-(4-nitrophenyl)-
Synonym: Diphenylamine, 3,4′-dinitro-
Line Form: $3\text{-}NO_2C_6H_4NH(C_6H_4NO_2\text{-}4)$
Mol Form: $C_{12}H_9N_3O_4$
M$_r$: 259.21
mp: 217
Color: pa ye(Chl, aq py)
***s*:** Ac 4; Chl 4
m/e: 259.06

HODOC No: 02429
CAS No: 1821-27-8
Beilstein No: 12^3, 1587
Name: Benzenamine, 4-nitro-*N*-(4-nitrophenyl)-
Synonym: Diphenylamine, 4,4′-dinitro-
Line Form: $(4\text{-}NO_2C_6H_4)_2NH$
Mol Form: $C_{12}H_9N_3O_4$
M_r: 259.21
mp: 216
Color: ye nd(al)
***s*:** H_2O 1; al 2; ace 3; bz 2; tol 1; HOAc 3
UV: OES 5-301 560(31623) EtOH
MS: NBS 1847
m/e: 259.06

HODOC No: 02430
CAS No: 119-75-5
Beilstein No: 12^3, 1517
Name: Benzenamine, 2-nitro-*N*-phenyl-
Synonym: Diphenylamine, 2-nitro-
Line Form: $2\text{-}NO_2C_6H_4NHC_6H_5$
Mol Form: $C_{12}H_{10}N_2O_2$
M_r: 214.21
mp: 75.5
***s*:** H_2O 1; al 3; ctc 2
IR: SADG 232 3333 3030 1613 1587 1563 1493 1449 1408 1333 1266 1149 1075 735 699
UV: SAD 388 424(6770) 258(14400) 218(13700) MeOH
H NMR: SAD 681 6.5-7.4 8.1 9.4 CCl_4
MS: NBS 1365
m/e: 214.07

HODOC No: 02431
CAS No: 4531-79-7
Beilstein No: 12^3, 1545
Name: Benzenamine, 3-nitro-*N*-phenyl-
Synonym: Diphenylamine, 3-nitro-
Line Form: $3\text{-}NO_2C_6H_4NHC_6H_5$
Mol Form: $C_{12}H_{10}N_2O_2$
M_r: 214.21
mp: 114
***s*:** al 4; eth 4; bz 4; lig 2
UV: OES 4-385 367(1479) 282(16218) 265(18197) EtOH
MS: NBS 1365
m/e: 214.07

HODOC No: 02432
CAS No: 836-30-6
Beilstein No: 12^3, 1586
Name: Benzenamine, 4-nitro-*N*-phenyl-
Synonym: Diphenylamine, 4-nitro-
Line Form: $4\text{-}NO_2C_6H_4NHC_6H_5$
Mol Form: $C_{12}H_{10}N_2O_2$
M_r: 214.21
bp: 211^{30}
mp: 133-4
***s*:** H_2O 1; al 4; ace 2; HOAc 4; con sulf 3
IR: SADG 12506 3333 3030 1587 1538 1471 1282 1176 1111 1000 833 735 685
UV: SAD 15263 390(16900) 258(8330) MeOH
H NMR: SAD 5762 7.1 7.0-7.5 8.1 8.2 ace
MS: NBS 1365
m/e: 214.07

HODOC No: 02433
CAS No: 7297-52-1
Beilstein No: 12, 1332
Name: Benzenamine, 4-[2-(4-nitrophenyl)ethenyl]-, (E)-
Synonym: 4-Stilbenamine, 4′-nitro-, (E)-
Mol Form: $C_{14}H_{12}N_2O_2$
M_r: 240.25
***s*:** H_2O 1; al 2; eth 2; bz 2; py 3
UV: OES 1-561 405(25119) 285(10000) EtOH 400(25119) bz
m/e: 240.09

HODOC No: 02434
CAS No: 22760-89-0
Name: Benzenamine, *N*-[1-(4-nitrophenyl)ethylidene]-
Synonym: Aniline, *N*-(α-methyl-*p*-nitrobenzylidene)-
Mol Form: $C_{14}H_{12}N_2O_2$
M_r: 240.25
mp: 119
UV: LANG 2353 314(3020) 256(22387) EtOH
m/e: 240.09

HODOC No: 02435
CAS No: 17064-77-6
Beilstein No: 12^3, 321
Name: Benzenamine, *N*-[(2-nitrophenyl)methylene]-
Synonym: Aniline, *N*-(*o*-nitrobenzylidene)-
Line Form: 2-$NO_2C_6H_4CH{=}NC_6H_5$
Mol Form: $C_{13}H_{10}N_2O_2$
M_r: 226.22
mp: 64.5
UV: LANG 202 339(5888) 264(13804) 223(19055) os
MS: ALDN 185 77(100) 51(94) 209(49) 179(42) 152(33) 50(32) 39(26) 226(21)
m/e: 226.07

HODOC No: 02436
CAS No: 5676-82-4
Name: Benzenamine, *N*-[(3-nitrophenyl)methylene]-
Synonym: Aniline, *N*-(*m*-nitrobenzylidene)-
Mol Form: $C_{13}H_{10}N_2O_2$
M_r: 226.22
mp: 65
IR: µCOB 1795 3130 2940 1640 1590 1540 1490 1350 1190 1090 1080 970 900 810 760 740 690 680
UV: SAD 3156 257 238 MeOH
m/e: 226.07

HODOC No: 02437
CAS No: 785-80-8
Beilstein No: 12^3, 322
Name: Benzenamine, *N*-[(4-nitrophenyl)methylene]-
Synonym: Aniline, *N*-(*p*-nitrobenzylidene)-
Line Form: 4-$NO_2C_6H_4CH{=}NC_6H_5$
Mol Form: $C_{13}H_{10}N_2O_2$
M_r: 226.22
mp: 93
UV: LANG 200 341(11482) 292(13490) EtOH
MS: ALDN 185 226(100) 77(54) 179(48) 225(38) 51(34) 104(29) 227(17) 152(12)
m/e: 226.07

HODOC No: 02438
CAS No: 22856-77-5
Beilstein No: 12^1, 346
Name: Benzenamine, 3-nitro-*N*-(3-phenyl-2-propenylidene)-
Synonym: Aniline, *N*-cinnamylidene-*m*-nitro-
Mol Form: $C_{15}H_{12}N_2O_2$
M_r: 252.26
mp: 92-3
***s*:** chl 3
IR: SADG 16601 3080 1620 1580 1510 1440 1340 1150 980 950 870 800 740 730 680 660
H NMR: SAD 8484 7.0-7.6 7.9 8.2 $CDCl_3$
m/e: 252.09

HODOC No: 02439
CAS No: 659-49-4
Beilstein No: 7^3, 3370
Name: Benzenamine, 4-nitroso-
Synonym: Aniline, *p*-nitroso-
Line Form: 4-$ONC_6H_4NH_2$
Mol Form: $C_6H_6N_2O$
M_r: 122.12
bp: dec
mp: 173-4
Color: bl nd (Bz)
***s*:** H_2O 3; al 3; bz 2
UV: OES 2-58 420(31623) 270(3981) H_2O 415(31623) 270(3981) MeOH 330(15849) 230(1585) pH=1.4 355(15849) 270(3162) pH=12
m/e: 122.05

HODOC No: 02440
CAS No: 156-10-5
Beilstein No: 12^3, 347
Name: Benzenamine, 4-nitroso-*N*-phenyl-
Synonym: Diphenylamine, 4-nitroso-
Line Form: 4-$ONC_6H_4NHC_6H_5$
Mol Form: $C_{12}H_{10}N_2O$
M_r: 198.21
mp: 143
***s*:** H_2O 2; al 4; eth 4; bz 4; chl 4; lig 2
IR: µSADP 9744 2940 1610 1540 1470 1350 1280 1220 1160 1110 830 740 690
UV: SAD 17154 274(4950) 228(8870) MeOH
MS: NBS 1156
m/e: 198.08

HODOC No: 02441
CAS No: 86-30-6
Beilstein No: 12^3, 1120
Name: Benzenamine, *N*-nitroso-*N*-phenyl-
Synonym: Diphenylamine, *N*-nitroso-
Line Form: $(C_6H_5)_2N$-NO
Mol Form: $C_{12}H_{10}N_2O$
M_r: 198.21
mp: 66.5
Color: ye pl(lig)
***s*:** al 2; bz 3; chl 2
IR: SADG 21687 3050 1590 1490 1480 1440 1300 1180 1150 1080 1040 990 900 750 740 690
UV: SAD 153 262 MeOH
H NMR: SAD 10374 7.0 7.2-7.6 $CDCl_3$
MS: ALDN 161 169(100) 30(74) 168(54) 167(30) 70(13) 51(13) 31(12) 198(0)
m/e: 198.08

HODOC No: 02442
CAS No: 16245-79-7
Beilstein No: 12^3, 2770
Name: Benzenamine, 4-octyl-
Synonym: Aniline, *p*-octyl-
Line Form: 4-C_8H_{17}-$C_6H_4NH_2$
Mol Form: $C_{14}H_{23}N$
M_r: 205.33
bp: 310-1,170-2^{17}
mp: 20
***s*:** Et_2O 4
m/e: 205.18

HODOC No: 02443
CAS No: 101-67-7
Name: Benzenamine, 4-octyl-*N*-(4-octylphenyl)-
Synonym: Diphenylamine, 4,4′-dioctyl-
Mol Form: $C_{28}H_{43}N$
M_r: 393.63
IR: COB 5504 3420 2950 2900 1610 1520 1470 1390 1360 1330 1240 1190 820 760
m/e: 393.34

HODOC No: 02444
CAS No: 101-80-4
Beilstein No: 13^3, 1006
Name: Benzenamine, 4,4′-oxybis-
Synonym: Aniline, 4,4′-oxydi-
Line Form: $(4\text{-}H_2NC_6H_4)_2O$
Mol Form: $C_{12}H_{12}N_2O$
M_r: 200.23
mp: dec 188-90
IR: µCOB 1463 3450 3330 3230 3030 1610 1490 1280 1220 870 830 730 690
H NMR: SAD 19057 4.6 DMSO
m/e: 200.09

HODOC No: 02445
CAS No: 13665-98-0
Beilstein No: 12^3, 1487
Name: Benzenamine, 2,3,4,5,6-pentabromo-
Synonym: Aniline, 2,3,4,5,6-pentabromo-
Line Form: $C_6Br_5NH_2$
Mol Form: $C_6H_2Br_5N$
M_r: 482.60
mp: 265-6
Color: nd (al-tol)
***s*:** EtOH 4
m/e: 482.61

HODOC No: 02446
CAS No: 527-20-8
Beilstein No: 12^3, 1415
Name: Benzenamine, 2,3,4,5,6-pentachloro-
Synonym: Aniline, 2,3,4,5,6-pentachloro-
Line Form: $C_6Cl_5NH_2$
Mol Form: $C_6H_2Cl_5N$
M_r: 265.35
mp: 232
Color: nd (al)
***s*:** Et_2O 4; EtOH 4; lig 4
m/e: 262.86

HODOC No: 02447
CAS No: 771-60-8
Name: Benzenamine, 2,3,4,5,6-pentafluoro-
Synonym: Aniline, 2,3,4,5,6-pentafluoro-
Line Form: $C_6F_5NH_2$
Mol Form: $C_6H_2F_5N$
Mr: 183.07
bp: 153-4
mp: 34
IR: COB 6598 3500 3410 3200 2670 2650 2470 2430 2090 1670 1600 1530 1350 1310 1170 1130 1000 950 850 560
MS: NBS 953
m/e: 183.01

HODOC No: 02448
CAS No: 2243-30-3
Beilstein No: 12^3, 2758
Name: Benzenamine, 2,3,4,5,6-pentamethyl-
Synonym: Pentamethylaniline
Line Form: $C_6(CH_3)_5NH_2$
Mol Form: $C_{11}H_{17}N$
Mr: 163.25
bp: 277-8
mp: 152-3
Color: mcl pr (al)
***s*:** Et_2O 4; EtOH 4
MS: NBS 655
m/e: 163.14

HODOC No: 02449
CAS No: 33228-44-3
Beilstein No: 12^3, 2747
Name: Benzenamine, 4-pentyl-
Synonym: Aniline, *p*-pentyl-
Line Form: $4\text{-}H_2N\text{-}C_6H_4\text{-}(CH_2)_4CH_3$
Mol Form: $C_{11}H_{17}N$
Mr: 163.25
bp: 133^7
m/e: 163.14

HODOC No: 02450
CAS No: 2655-27-8
Beilstein No: 12^3, 271
Name: Benzenamine, *N*-pentyl-
Synonym: Aniline, *N*-pentyl-
Line Form: $C_6H_5NH(CH_2)_4CH_3$
Mol Form: $C_{11}H_{17}N$
Mr: 163.25
bp: 260
IR: μCOB 176 3450 3130 2940 1610 1520 1470 1430 1390 1320 1270 1180 1150 1000 870 750 690
H NMR: SAD 17036 1.3 2.7 $CDCl_3$
m/e: 163.14

HODOC No: 02451
CAS No: 39905-50-5
Beilstein No: 13^3, 998
Name: Benzenamine, 4-(pentyloxy)-
Synonym: Aniline, *p*-pentyloxy-
Line Form: $4\text{-}H_2N\text{-}C_6H_4\text{-}O\text{-}(CH_2)_4CH_3$
Mol Form: $C_{11}H_{17}NO$
Mr: 179.25
bp: 144^5
m/e: 179.13

HODOC No: 02452
CAS No: 2688-84-8
Beilstein No: 13^3, 758
Name: Benzenamine, 2-phenoxy-
Synonym: Aniline, *o*-phenoxy-
Line Form: $2\text{-}C_6H_5O\text{-}C_6H_4NH_2$
Mol Form: $C_{12}H_{11}NO$
Mr: 185.21
bp: $307\text{-}8^{728}$, $172\text{-}3^{14}$
mp: 44-5
Color: cr (lig)
***s*:** al 3; eth 3; ace 3; bz 3; os 4
IR: μSADP 4493 3450 2940 1610 1590 1470 1370 1300 1270 1210 1150 1080 1020 930 890 880 810 790 750 690
UV: SAD 1246 288(2580) 279(2500) 271(2240) MeOH
m/e: 185.08

HODOC No: 02453
CAS No: 3586-12-7
Beilstein No: 13^3, 933
Name: Benzenamine, 3-phenoxy-
Synonym: Aniline, *m*-phenoxy-
Line Form: 3-C_6H_5O-$C_6H_4NH_2$
Mol Form: $C_{12}H_{11}NO$
M_r: 185.21
bp: 315, 190^{14}
mp: 37
Color: pr (lig)
***s*:** al 3; eth 3; ace 3; bz 3; os 3; lig 2
IR: µSADP 4494 3450 3330 2940 1610 1590 1470 1370 1280 1220 1160 1150 1080 1020 1000 960 900 850 820 790 760 730 690
UV: SAD 1247 288(2730) 277(3000) 270(2790) MeOH
H NMR: SAD 18692 3.5 6.3 6.3 6.8-7.5 $CDCl_3$
m/e: 185.08

HODOC No: 02454
CAS No: 139-59-3
Beilstein No: 13^3, 999
Name: Benzenamine, 4-phenoxy-
Synonym: Aniline, *p*-phenoxy-
Line Form: 4-C_6H_5O-$C_6H_4NH_2$
Mol Form: $C_{12}H_{11}NO$
M_r: 185.21
bp: 187-9^{14}
mp: 85-6
Color: nd (w), cr (dil al)
***s*:** H_2O 3; al 4; eth 4; lig 2
IR: µSADP 4495 3450 3330 2940 1640 1610 1490 1390 1220 1160 1090 1060 910 870 830 790 710 690
UV: SAD 1248 283 MeOH
m/e: 185.08

HODOC No: 02455
CAS No: 122-39-4
Beilstein No: 12^3, 284
Name: Benzenamine, *N*-phenyl-
Synonym: Anilinobenzene
Line Form: $(C_6H_5)_2NH$
Structure:

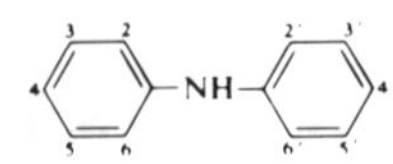

Mol Form: $C_{12}H_{11}N$
M_r: 169.21
bp: 302, 179^{22}
mp: 54-5
***d*:** $1.160^{22/20}$
Color: mcl lf(dil al)
***s*:** H_2O 1; al 4; eth 3; ace 4; bz 4; ctc 4; chl 2; py 4; AcOEt 4; HOAc 3ilig3
IR: µCOB 4833 3450 3030 1590 1490 1450 1410 1300 1240 1180 1080 1020 1000 870 750 690
Raman: SAD 241 3400 3080 1610 1520 1510 1470 1420 1310 1230 1180 1160 1040 1000 890 850 760 710 650 620 570 520 420 350 320 250 220 140
UV: SAD 30 282 cyhex
H NMR: SAD 11 5.4 6.9-7.4 $CDCl_3$
MS: ALDN 130 169(100) 168(48) 167(29) 83(20) 44(16) 170(13) 77(11) 51(10)
m/e: 169.09

HODOC No: 02456
CAS No: 537-67-7
Name: Benzenamine, *N*-phenyl-, hydrochloride
Synonym: Diphenylamine, hydrochloride
Line Form: $(C_6H_5)_2NH$–HCl
Mol Form: $C_{12}H_{12}ClN$
M_r: 205.67
IR: SADG 18214 3040 2600 2400 1960 1590 1550 1490 1470 1410 1160 1020 750 730 700 680 610 480
UV: SAD 15859 284(21100) MeOH
H NMR: SAD 11725 6.9 7.1-7.4 10.3 DMSO
m/e: 205.07

HODOC No: 02457
CAS No: 2835-58-7
Beilstein No: 16^3, 334
Name: Benzenamine, 2-(phenylazo)-
Synonym: Aniline, *o*-(phenylazo)-
Line Form: $2\text{-}H_2NC_6H_4N{=}NC_6H_5$
Mol Form: $C_{12}H_{11}N_3$
M_r: 197.23
mp: 59
Color: red nd or pr (al)
***s*:** H_2O 1; al 2; eth 4; ace 4; bz 4; os 4
UV: OES 3-342 328(20417) 1N HCl:50% EtOH 427(10000) 312(25119) EtOH 328(20417) 1N HCl
m/e: 197.10

HODOC No: 02458
CAS No: 2835-59-8
Beilstein No: 16^3, 335
Name: Benzenamine, 3-(phenylazo)-
Synonym: Aniline, *m*-(phenylazo)-
Line Form: $3\text{-}H_2NC_6H_4N{=}NC_6H_5$
Mol Form: $C_{12}H_{11}N_3$
M_r: 197.23
mp: (i)69-70, (ii)90-1
Color: (i) og-ye nd (peth) (ii) br-red cr
***s*:** H_2O 1; al 3; eth 3; ace 3; bz 3; chl 3
IR: µSADP 7698 3450 3330 1610 1490 1450 1330 1320 1280 1150 1080 1000 920 870 790 690
UV: OES 3-342 301(15849) EtOH 430(759) 322(19953) 1.2N HCl:50% EtOH 430(759) 322(19953) 1N HCl
m/e: 197.10

HODOC No: 02459
CAS No: 60-09-3
Beilstein No: 16^2, 149
Name: Benzenamine, 4-(phenylazo)-
Synonym: *p*-Aminoazobenzene
Line Form: $4\text{-}H_2NC_6H_4N{=}NC_6H_5$
Mol Form: $C_{12}H_{11}N_3$
M_r: 197.23
bp: 360
mp: 127
Color: og mcl nd (al)
***s*:** H_2O 2; al 3; eth 3; bz 3; chl 3; lig 2
IR: SADG 8032 3448 3333 3030 1639 1587 1493 1471 1429 1316 1205 1136 840 769 690
UV: SAD 142 384(25200) 245(9270) MeOH
H NMR: SAD 17244 6.5-8.0 $CDCl_3$
MS: ALDN 160 92(100) 197(40) 65(36) 18(32) 77(30) 120(21) 51(15) 39(13)
m/e: 197.10

HODOC No: 02460
CAS No: 27652-35-3
Beilstein No: 12^2, 781
Name: Benzenamine, 2-(2-phenylethenyl)-, (E)-
Synonym: 2-Stilbenamine, (E)-
Mol Form: $C_{14}H_{13}N$
M_r: 195.25
mp: 106
***s*:** H_2O 2; al 3; MeOH 3
UV: OES 3-454 286(16596) 240(14791) EtOH
m/e: 195.10

HODOC No: 02461
CAS No: 62058-64-4
Beilstein No: 12^2, 781
Name: Benzenamine, 2-(2-phenylethenyl)-, (Z)-
Synonym: 2-Amino-*cis*-stilbene
Mol Form: $C_{14}H_{13}N$
M_r: 195.25
bp: $180\text{-}1^{11}$ $104.4^{0.2}$
***s*:** H_2O 1; ac 3
UV: OES 3-454 275(7943) 237(15849) EtOH
m/e: 195.10

HODOC No: 02462
CAS No: 14064-82-5
Name: Benzenamine, 3-(2-phenylethenyl)-, (E)-
Synonym: 3-Stilbenamine, (E)-
Mol Form: $C_{14}H_{13}N$
M_r: 195.25
mp: 120
***s*:** H_2O 2; al 3; bz 2
UV: TETRA 23-173-67 297(25119) cyhex
m/e: 195.10

HODOC No: 02463
CAS No: 4309-66-4
Beilstein No: 12^2, 782
Name: Benzenamine, 4-(2-phenylethenyl)-, (E)-
Synonym: 4-Stilbenamine, (E)-
Mol Form: $C_{14}H_{13}N$
M_r: 195.25
mp: 151
***s*:** H_2O 1; al 2; bz 2; HOAc 2
UV: OES 4-506 324(31623) 30% MeOH
m/e: 195.10

HODOC No: 02464
CAS No: 5697-85-8
Beilstein No: 12^3, 3242
Name: Benzenamine, 2-(2-phenylethyl)-
Synonym: Aniline, *o*-phenethyl-
Line Form: 2-$H_2NC_6H_4CH_2CH_2C_6H_5$
Mol Form: $C_{14}H_{15}N$
M_r: 197.27
bp: $173\text{-}83^{11}$
mp: 33
m/e: 197.12

HODOC No: 02465
CAS No: 5369-22-2
Beilstein No: 12^3, 3242
Name: Benzenamine, 3-(2-phenylethyl)-
Synonym: Aniline, *m*-phenethyl-
Line Form: 3-$H_2NC_6H_4CH_2CH_2C_6H_5$
Mol Form: $C_{14}H_{15}N$
M_r: 197.27
bp: 180^6
mp: 51
Color: pl (peth)
m/e: 197.12

HODOC No: 02466
CAS No: 13024-49-2
Beilstein No: 12^3, 3243
Name: Benzenamine, 4-(2-phenylethyl)-
Synonym: Aniline, *p*-phenethyl-
Line Form: 4-$H_2NC_6H_4CH_2CH_2C_6H_5$
Mol Form: $C_{14}H_{15}N$
M_r: 197.27
mp: 48
Color: pl (eth-peth)
m/e: 197.12

HODOC No: 02467
CAS No: 1749-19-5
Name: Benzenamine, *N*-(1-phenylethylidene)-
Synonym: Aniline, *N*-(α-methylbenzylidene)-
Line Form: $C_6H_5C(CH_3){=}NC_6H_5$
Mol Form: $C_{14}H_{13}N$
M_r: 195.25
mp: 41
UV: LANG 207 309(2138) 245(22387) EtOH
m/e: 195.10

HODOC No: 02468
CAS No: 1484-26-0
Beilstein No: 13^3, 933
Name: Benzenamine, 3-(phenylmethoxy)-
Synonym: Aniline, *m*-(benzyloxy)-
Line Form: 3-$H_2NC_6H_4OCH_2C_6H_5$
Mol Form: $C_{13}H_{13}NO$
M_r: 199.24
mp: 63-67
IR: ALD 3:746F
MS: ALD 5:86B
m/e: 199.10

HODOC No: 02469
CAS No: 28059-64-5
Beilstein No: 12^3, 3214
Name: Benzenamine, 2-(phenylmethyl)-
Synonym: *o*-Toluidine, α-phenyl-
Line Form: 2-NH_2-C_6H_4-CH_2-C_6H_5
Mol Form: $C_{13}H_{13}N$
M_r: 183.24
bp: $172\text{-}3^{12}$
mp: 55-7
m/e: 183.10

HODOC No: 02470
CAS No: 61424-26-8
Beilstein No: 12^3, 3215
Name: Benzenamine, 3-(phenylmethyl)-
Synonym: 3-Benzylaniline
Line Form: 3-$H_2NC_6H_4CH_2C_6H_5$
Mol Form: $C_{13}H_{13}N$
M_r: 183.24
mp: 46
Color: cr (lig)
***s*:** lig 4
m/e: 183.10

HODOC No: 02471
CAS No: 1135-12-2
Beilstein No: 12^3, 3215
Name: Benzenamine, 4-(phenylmethyl)-
Synonym: *p*-Toluidine, α-phenyl-
Line Form: 4-$H_2NC_6H_4CH_2C_6H_5$
Mol Form: $C_{13}H_{13}N$
M_r: 183.24
bp: 300
mp: 34-5
***d*:** 1.038^{25}
Color: mcl (lig)
***s*:** Et_2O 4; EtOH 4; lig 4
m/e: 183.10

HODOC No: 02472
CAS No: 538-51-2
Beilstein No: 12^3, 319
Name: Benzenamine, *N*-(phenylmethylene)-
Synonym: Aniline, *N*-benzylidene-
Line Form: $C_6H_5N{=}CHC_6H_5$
Mol Form: $C_{13}H_{11}N$
M_r: 181.22
bp: 310
mp: 54
***d*:** $1.038^{55/4}$
***n*:** 1.600^{100}
Color: pa ye nd (CS_2) pl (dil al)
***s*:** H_2O 1; al 3; eth 3; chl 2; liq SO_2 3; NH_3 3
IR: **μCOB 6257** 3030 2860 1640 1610 1590 1490 1450 1370 1320 1190 1180 1080 1030 980 910 880 770 690
UV: SAD 743 261(15200) 236(9230) MeOH
H NMR: SAD 1208 7.3 7.9 8.4 $CDCl_3$
MS: ALDN 144 180(100) 181(96) 77(47) 51(15) 182(13) 104(12) 78(11) 76(5)
m/e: 181.09

HODOC No: 02473
CAS No: 603-40-7
Beilstein No: 13^3, 529
Name: Benzenamine, 4,4′-(phenylmethylene)*bis*-
Synonym: Aniline, 4,4′-benzylidenedi-
Line Form: $(4\text{-}H_2NC_6H_4)_2CHC_6H_5$
Mol Form: $C_{19}H_{18}N_2$
M_r: 274.35
mp: 139-40
Color: pr_(Bz,eth)
***s*:** Et_2O 4; EtOH 4; lig 4
m/e: 274.15

HODOC No: 02474
CAS No: 129-73-7
Beilstein No: 13^3, 529
Name: Benzenamine, 4,4′-(phenylmethylene)*bis*[*N*,*N*-dimethyl-
Synonym: Aniline, 4,4′-benzylidenebis[*N*,*N*-dimethyl-
Line Form: $[4\text{-}(CH_3)_2NC_6H_4]_2CHC_6H_5$
Mol Form: $C_{23}H_{26}N_2$
M_r: 330.45
mp: 102
Color: nd or lf (al, Bz)
***s*:** Bz 4; Et_2O 4
m/e: 330.21

HODOC No: 02475
CAS No: 101-75-7
Beilstein No: 16^3, 343
Name: Benzenamine, *N*-phenyl-4-(phenylazo)-
Synonym: Diphenylamine, 4-(phenylazo)-
Line Form: $4\text{-}(C_6H_5NH)\text{-}C_6H_4N{=}NC_6H_5$
Mol Form: $C_{18}H_{15}N_3$
M_r: 273.32
mp: 82
Color: ye pl or pr
***s*:** H_2O 1; al 4; eth 4; lig 4
UV: HBCP 450(63096) 380(15488) 260(12589) al
m/e: 273.13

HODOC No: 02476
CAS No: 953-21-9
Beilstein No: 12^1, 173
Name: Benzenamine, *N*-(3-phenyl-2-propenylidene)-
Synonym: Aniline, *N*-cinnamylidene-
Line Form: $C_6H_5CH{=}CHCH{=}NC_6H_5$
Mol Form: $C_{15}H_{13}N$
M_r: 207.26
mp: 109
***s*:** H_2O 2; al 2; eth 3; chl 3
IR: SADG 16595 3040 1610 1560 1470 1430 1130 980 730 670
H NMR: SAD 8481 7.0-7.6 8.2 $CDCl_3$
m/e: 207.10

HODOC No: 02477
CAS No: 21229-95-8
Beilstein No: 13, 534
Name: Benzenamine, 4-(phenylsulfinyl)-
Synonym: Aniline, *p*-(phenylsulfinyl)-
Mol Form: $C_{12}H_{11}NOS$
M_r: 217.27
mp: 152
***s*:** H_2O 3; al 4; eth 4
UV: HBCP 278(10233) al
m/e: 217.06

HODOC No: 02478
CAS No: 4273-98-7
Beilstein No: 13, 339
Name: Benzenamine, 2-(phenylsulfonyl)-
Synonym: Aniline, *o*-(phenylsulfonyl)-
Mol Form: $C_{12}H_{11}NO_2S$
Mr: 233.27
mp: 122-4
***s*:** H_2O 1; al 4; bz 4; HOAc 4
UV: HBCP 319(4169) 280(251) 274(871) 228(21379) cyhex
m/e: 233.05

HODOC No: 02479
CAS No: 7019-01-4
Beilstein No: 13^2, 298
Name: Benzenamine, 4-(phenylsulfonyl)-
Synonym: Aniline, *p*-(phenylsulfonyl)-
Mol Form: $C_{12}H_{11}NO_2S$
Mr: 233.27
mp: 176
***s*:** H_2O 1; al 3; bz 4; HOAc 4
UV: HBCP 270(18621) cyhex
m/e: 233.05

HODOC No: 02480
CAS No: 1134-94-7
Beilstein No: 13^3, 904
Name: Benzenamine, 2-(phenylthio)-
Synonym: Aniline, *o*-(phenylthio)-
Line Form: $2\text{-}H_2NC_6H_4SC_6H_5$
Mol Form: $C_{12}H_{11}NS$
Mr: 201.27
bp: 258^{100}
mp: 35-6
Color: pl (al)
m/e: 201.06

HODOC No: 02481
CAS No: 1135-14-4
Beilstein No: 13^3, 1224
Name: Benzenamine, 4-(phenylthio)-
Synonym: Aniline, *p*-(phenylthio)-
Line Form: $4\text{-}H_2NC_6H_4SC_6H_5$
Mol Form: $C_{12}H_{11}NS$
Mr: 201.27
bp: 282^{100}
mp: 96
Color: cr(lig)
m/e: 201.06

HODOC No: 02482
CAS No: 37832-42-1
Beilstein No: 12, 117
Name: Benzenamine, phosphate (1:1)
Synonym: Monobasic aniline orthophosphate
Line Form: $C_6H_5NH_2\text{–}H_3PO_4$
Mol Form: $C_6H_{10}NO_4P$
Mr: 191.12
mp: 180
***s*:** H_2O 4; al 4; eth 1
IR: µSADP 6291 2860 2560 2330 2080 1610 1560 1490 1190 1060 960 900 740 690
UV: SAD 1731 287(2720) 234(15000) MeOH
m/e: 191.03

HODOC No: 02483
CAS No: 36357-49-0
Beilstein No: 16, 789
Name: Benzenamine, 3,3′,3″-phosphinylidynetris-
Synonym: Tris(3-aminophenyl)phosphine oxide
Mol Form: $C_{18}H_{18}N_3OP$
Mr: 323.32
mp: 258
***s*:** H_2O 2; ace 3
UV: LANG 1083 307(6026) 249(16982) 217(52481) EtOH
m/e: 323.12

HODOC No: 02484
CAS No: 589-09-3
Beilstein No: 12^2, 277
Name: Benzenamine, *N*-2-propenyl-
Synonym: Aniline, *N*-allyl-
Line Form: $CH_2\text{=}CHCH_2NH\text{-}C_6H_5$
Mol Form: $C_9H_{11}N$
Mr: 133.18
bp: $217\text{-}9^{736}$, $105\text{-}8^{12}$
***d*:** $0.982^{25/4}$
***n*:** 1.563^{20}
***s*:** H_2O 2; al 3; eth 5; ace 3
IR: µSADP 13702 3330 3030 1590 1490 1320 1270 1160 990 920 870 750 690
H NMR: SAD 19344 3.6 3.6 5.1 5.1 5.8 6.5 6.6 7.1 $CDCl_3$
m/e: 133.09

HODOC No: 02485
CAS No: 1821-39-2
Beilstein No: 12^3, 2657
Name: Benzenamine, 2-propyl-
Synonym: Aniline, *o*-propyl-
Line Form: $2\text{-}C_3H_7\text{-}C_6H_4NH_2$
Mol Form: $C_9H_{13}N$
M$_r$: 135.20
bp: 226^{758}, 116^{15}
***d*:** $0.9602^{20/4}$
***n*:** 1.5427^{20}
***s*:** H_2O 1; al 3; eth 3
UV: OES 4-240 288(2188) cyhex
m/e: 135.10

HODOC No: 02486
CAS No: 2696-84-6
Beilstein No: 12^3, 2658
Name: Benzenamine, 4-propyl-
Synonym: Aniline, *p*-propyl-
Line Form: $4\text{-}C_3H_7\text{-}C_6H_4NH_2$
Mol Form: $C_9H_{13}N$
M$_r$: 135.20
bp: 224-6
m/e: 135.10

HODOC No: 02487
CAS No: 622-80-0
Beilstein No: 12^3, 264
Name: Benzenamine, *N*-propyl-
Synonym: Aniline, *N*-propyl-
Line Form: $C_6H_5NHC_3H_7$
Mol Form: $C_9H_{13}N$
M$_r$: 135.20
bp: 222, 100^{11}
***d*:** $0.9443^{20/4}$
***n*:** 1.5428^{20}
***s*:** Et_2O 4; EtOH 4
IR: COB 7463
H NMR: SAD 1919
m/e: 135.10

HODOC No: 02488
CAS No: 542-16-5
Beilstein No: 12^2, 66
Name: Benzenamine, sulfate (2:1)
Synonym: Aniline, sulfate (2:1)
Line Form: $C_6H_5NH_2\text{–}1/2H_2SO_4$
Mol Form: $C_6H_8NO_2S$
M$_r$: 142.16
***d*:** 1.377^4
***s*:** H_2O 3; al 2; eth 1; tfa 2
IR: SADG 8397 3030 2941 2564 1587 1563 1515 1493 1333 1190 1111 1031 741 685
UV: SAD 1498 285(1570) 234(8570) MeOH
H NMR: SAD 1217 7.5 tfa
m/e: 142.04

HODOC No: 02489
CAS No: 1122-83-4
Beilstein No: 19^3, 1116
Name: Benzenamine, *N*-sulfinyl-
Synonym: Aniline, *N*-sulfinyl-
Line Form: $C_6H_5N{=}SO$
Mol Form: C_6H_5NOS
M$_r$: 139.17
bp: 200
***d*:** 1.236
***n*:** 1.6270^{20}
IR: COB 5757 3060 1480 1450 1300 1290 1170 1160 1070 920 760 680
UV: SAD 8823 310(2870) 235(8290) MeOH
MS: ALDN 88 139(100) 39(66) 64(62) 111(55) 84(49) 63(48) 67(48) 38(44)
m/e: 139.01

HODOC No: 02490
CAS No: 119-59-5
Beilstein No: 13^3, 1246
Name: Benzenamine, 4,4′-sulfinylbis-
Synonym: Aniline, 4,4′-sulfinyldi-
Line Form: $(4\text{-}H_2NC_6H_4)_2SO$
Mol Form: $C_{12}H_{12}N_2OS$
M$_r$: 232.29
mp: 175d
Color: pr (w, al)
***s*:** H_2O 3; al 3
UV: HBCP 300(17783) 264(14791) al
m/e: 232.07

HODOC No: 02491
CAS No: 599-61-1
Beilstein No: 13^3, 984
Name: Benzenamine, 3,3′-sulfonylbis-
Synonym: Aniline, 3,3′-sulfonyldi-
Line Form: $(3H_2NC_6H_4-)_2SO_2$
Mol Form: $C_{12}H_{12}N_2O_2S$
M_r: 248.29
mp: 167-70
***s*:** H_2O 4; al 4
IR: SADG 18338 3470 3390 1610 1590 1480 1450 1310 1280 1140 1080 980 780 710 680 670
UV: SAD 3522 313 227 MeOH
MS: ALD 10:25A
m/e: 248.06

HODOC No: 02492
CAS No: 80-08-0
Beilstein No: 13^3, 1246
Name: Benzenamine, 4,4′-sulfonylbis-
Synonym: Aniline, 4,4′-sulfonyldi-
Line Form: $(4-H_2NC_6H_4-)_2SO_2$
Mol Form: $C_{12}H_{12}N_2O_2S$
M_r: 248.29
mp: 175-6
Color: cr (95% al)
***s*:** al 3; $DMSOd_6$ 2
IR: COB 6054 3460 3400 3370 3350 3250 1630 1590 1500 1440 1280 1150 1110 1070 830 730 700 650 630
UV: SAD 864 296 260 MeOH
H NMR: SAD 6921 5.9 6.6 7.4 $DMSOd_6$
MS: ALD 10:24D
m/e: 248.06

HODOC No: 02493
CAS No: 634-83-3
Beilstein No: 12^2, 340
Name: Benzenamine, 2,3,4,5-tetrachloro-
Synonym: Aniline, 2,3,4,5-tetrachloro-
Line Form: $2,3,4,5-Cl_4C_6HNH_2$
Mol Form: $C_6H_3Cl_4N$
M_r: 230.90
mp: 118-20
Color: nd (al)
***s*:** Bz 4; Et_2O 4; EtOH 4
MS: NBS 1538
m/e: 228.90

HODOC No: 02494
CAS No: 3481-20-7
Beilstein No: 12^3, 1414
Name: Benzenamine, 2,3,5,6-tetrachloro-
Synonym: Aniline, 2,3,5,6-tetrachloro-
Line Form: $2,3,5,6-Cl_4C_6HNH_2$
Mol Form: $C_6H_3Cl_4N$
M_r: 230.90
mp: 108
Color: nd (lig, al)
***s*:** H_2O 1; al 3; eth 4; CS_2 4; lig 4
IR: µDMS 5777 3570 3450 2940 1610 1560 1470 1430 1300 1270 1120 1090 970 820 680
m/e: 228.90

HODOC No: 02495
CAS No: 363-73-5
Name: Benzenaminc, 2,3,4,6-tetrafluoro-
Synonym: Aniline, 2,3,4,6-tetrafluoro-
Line Form: $2,3,4,6-F_4-C_6HNH_3$
Mol Form: $C_6H_3F_4N$
M_r: 165.08
bp: 65^{20}
***n*:** 1.4622^{20}
m/e: 165.02

HODOC No: 02496
CAS No: 700-17-4
Name: Benzenamine, 2,3,5,6-tetrafluoro-
Synonym: Aniline, 2,3,5,6-tetrafluoro-
Line Form: $2,3,5,6-F_4C_6HNH_2$
Mol Form: $C_6H_3F_4N$
M_r: 165.08
bp: 158
mp: 31-2
***s*:** ctc 3
IR: COB 6569 3500 3400 3100 1660 1530 1500 1400 1270 1180 1090 950 880 800 710 670 560
UV: SAD 12720 230(12200) 214(7130) MeOH
H NMR: SAD 3701 4.0 6.3 CCl_4
MS: ALDN 124 165(100) 145(35) 118(23) 137(17) 99(13) 88(10) 138(10) 75(9)
m/e: 165.02

HODOC No: 02497
CAS No: 2217-45-0
Beilstein No: 12^3, 2743
Name: Benzenamine, 2,3,4,5-tetramethyl-
Synonym: Aniline, 2,3,4,5-tetramethyl
Line Form: 2,3,4,5-$(CH_3)_4$-C_6HNH_2
Mol Form: $C_{10}H_{15}N$
M_r: 149.23
bp: 259-60
mp: 70
Color: lf (w)
***s*:** Et_2O 4; EtOH 4; lig 4
m/e: 149.12

HODOC No: 02498
CAS No: 488-71-1
Beilstein No: 12^3, 2744
Name: Benzenamine, 2,3,4,6-tetramethyl-
Synonym: Aniline, 2,3,4,6-tetramethyl-
Line Form: 2,3,4,6-$(CH_3)_4$-C_6HNH_2
Mol Form: $C_{10}H_{15}N$
M_r: 149.23
bp: 255
mp: 23-4
***d*:** 0.978^{24}
***s*:** EtOH 4
m/e: 149.12

HODOC No: 02499
CAS No: 2217-46-1
Beilstein No: 12^3, 2745
Name: Benzenamine, 2,3,5,6-tetramethyl-
Synonym: Aniline, 2,3,5,6-tetramethyl-
Line Form: 2,3,5,6-$(CH_3)_4$-C_6HNH_2
Mol Form: $C_{10}H_{15}N$
M_r: 149.23
bp: 261-2
mp: 75
Color: pr (w, aq MeOH)
m/e: 149.12

HODOC No: 02500
CAS No: 769-06-2
Name: Benzenamine, *N*,*N*,2,6-tetramethyl-
Synonym: 2,6-Xylidine, *N*,*N*-dimethyl-
Mol Form: $C_{10}H_{15}N$
M_r: 149.23
bp: 82-3
IR: SADG 31545 3020 2970 2920 2860 2790 1590 1470 1420 1370 1320 1220 1150 1100 1050 950 770 570 490 370
UV: FO 89 225(1995) hp
H NMR: SAD 13993 2.3 2.8 6.9 $CDCl_3$
m/e: 149.12

HODOC No: 02501
CAS No: 4913-13-7
Beilstein No: 12^3, 2496
Name: Benzenamine, *N*,*N*,3,5-tetramethyl-
Synonym: 3,5-Xylidine, *N*,*N*-dimethyl-
Line Form: 3,5-$(CH_3)_2C_6H_3$-$N(CH_3)_2$
Mol Form: $C_{10}H_{15}N$
M_r: 149.23
bp: 88^4
m/e: 149.12

HODOC No: 02502
CAS No: 3698-54-2
Beilstein No: 12^2, 427
Name: Benzenamine, 2,3,4,6-tetranitro-
Synonym: Aniline, 2,3,4,6-tetranitro-
Line Form: 2,3,4,6-$(NO_2)_4$-C_6HNH_2
Mol Form: $C_6H_3N_5O_8$
M_r: 273.11
mp: 220
***s*:** H_2O 1; ace 2; bz 2; chl 2; HOAc 3; $PhNO_2$ 3; lig 2
UV: OES 7-56 402(6026) 320(10471) diox
m/e: 273.00

HODOC No: 02503
CAS No: 61977-46-6
Beilstein No: 27^2, 194
Name: Benzenamine, *N*-2-thiazolidinylidene-
Synonym: 2-(Phenylimino)thiazolidine
Mol Form: $C_9H_{10}N_2S$
M_r: 178.24
mp: 160
***s*:** H_2O 2; eth 3; bz 2; ac 3
MS: WILEY 1201 177(100) 118(84) 178(82) 77(34) 104(22) 150(18) 91(18) 51(17)
m/e: 178.06

HODOC No: 02504
CAS No: 139-65-1
Beilstein No: 13^3, 1246
Name: Benzenamine, 4,4′-thiobis-
Synonym: Aniline, 4,4′-thiodi-
Line Form: $(4\text{-}H_2NC_6H_4)_2S$
Mol Form: $C_{12}H_{12}N_2S$
Mr: 216.29
mp: 108-9
Color: nd (w)
***s*:** H_2O 2; al 4; eth 4; bz 4; tfa 3
IR: SADG 8866 3490 3440 3370 3230 3060 1620 1590 1490 1290 1260 1170 820 720 680 650
UV: SAD 6007 264 MeOH
H NMR: SAD 2074 7.5 tfa
m/e: 216.07

HODOC No: 02505
CAS No: 147-82 0
Beilstein No: 12^3, 1477
Name: Benzenamine, 2,4,6-tribromo-
Synonym: Aniline, 2,4,6-tribromo-
Line Form: $2,4,6\text{-}Br_3C_6H_2NH_2$
Mol Form: $C_6H_4Br_3N$
Mr: 326.81
bp: 300
mp: 122
***d*:** $2.35^{20/20}$
Color: nd (al, Bz)
***s*:** H_2O 1; al 2; eth 3; chl 3
IR: μCOB 2313 3450 3330 1610 1560 1540 1450 1390 1280 1240 1090 1050 860 850 730 700 680
UV: SAD 2576 313 251 diox
H NMR: SAD 7689 4.6 7.5 $CDCl_3$
MS: NBS 2519
m/e: 326.79

HODOC No: 02506
CAS No: 609-16-5
Beilstein No: 12^3, 1487
Name: Benzenamine, 3,4,5-tribromo-
Synonym: Aniline, 3,4,5-tribromo-
Line Form: $3,4,5\text{-}Br_3C_6H_2NH_2$
Mol Form: $C_6H_4Br_3N$
Mr: 326.81
mp: 123
Color: nd (al)
***s*:** Et_2O 4; EtOH 4
m/e: 326.79

HODOC No: 02507
CAS No: 634-67-3
Beilstein No: 12, 626
Name: Benzenamine, 2,3,4-trichloro-
Synonym: Aniline, 2,3,4-trichloro-
Line Form: $2,3,4\text{-}Cl_3C_6H_2NH_2$
Mol Form: $C_6H_4Cl_3N$
Mr: 196.46
bp: 292^{774}
mp: 73
Color: nd (lig)
***s*:** EtOH 4
m/e: 194.94

HODOC No: 02508
CAS No: 636-30-6
Beilstein No: 12^3, 1409
Name: Benzenamine, 2,4,5-trichloro-
Synonym: Aniline, 2,4,5-trichloro-
Line Form: $2,4,5\text{-}Cl_3C_6H_2NH_2$
Mol Form: $C_6H_4Cl_3N$
Mr: 196.46
bp: 270
mp: 96.5
Color: nd (lig)
***s*:** al 3; eth 3; CS_2 4; lig 2
IR: μCOB 2314 3570 3450 1610 1470 1390 1300 1280 1250 1140 1090 1060 940 920 870 840 740 680
UV: SAD 7882 312(3240) 248(10500) 213(34600) MeOH
MS: NBS 1107
m/e: 194.94

HODOC No: 02509
CAS No: 634-93-5
Beilstein No: 12^3, 1409
Name: Benzenamine, 2,4,6-trichloro-
Synonym: Aniline, 2,4,6-trichloro-
Line Form: $2,4,6\text{-}Cl_3C_6H_2NH_2$
Mol Form: $C_6H_4Cl_3N$
Mr: 196.46
bp: 262^{746}
mp: 78.5
Color: cr (al), nd (lig or peth)
***s*:** H_2O 1; al 3; eth 3; chl 3; CS_2 4; lig 3
IR: μCOB 5660 3450 3330 1610 1560 1470 1390 1300 1080 850 790 740 700
UV: SAD 16958 310(3160) 244(8900) 209(31700) MeOH
H NMR: SAD 8850 4.4 7.2 $CDCl_3$
m/e: 194.94

HODOC No: 02510
CAS No: 634-91-3
Beilstein No: 12^3, 1414
Name: Benzenamine, 3,4,5-trichloro-
Synonym: Aniline, 3,4,5-trichloro-
Line Form: 3,4,5-$Cl_3C_6H_2NH_2$
Mol Form: $C_6H_4Cl_3N$
M_r: 196.46
mp: 89-91
m/e: 194.94

HODOC No: 02511
CAS No: 363-81-5
Name: Benzenamine, 2,4,6-trifluoro-
Synonym: Aniline, 2,4,6-trifluoro-
Line Form: 2,4,6-F_3-$C_6H_2NH_2$
Mol Form: $C_6H_4F_3N$
M_r: 147.09
bp: 57^{22}
mp: 33-7
m/e: 147.03

HODOC No: 02512
CAS No: 88-17-5
Beilstein No: 12^2, 453
Name: Benzenamine, 2-(trifluoromethyl)-
Synonym: *o*-Toluidine, α,α,α-trifluoro-
Line Form: 2-$H_2NC_6H_4CF_3$
Mol Form: $C_7H_6F_3N$
M_r: 161.12
bp: 68^{15}
mp: 34-7
***d*:** 1.282
***n*:** 1.4810^{20}
IR: ALD 2:632H
MS: ALD 5:43B
m/e: 161.05

HODOC No: 02513
CAS No: 98-16-8
Beilstein No: 12^3, 1988
Name: Benzenamine, 3-(trifluoromethyl)-
Synonym: *m*-Toluidine, α,α,α-trifluoro-
Line Form: 3-$H_2NC_6H_4CF_3$
Mol Form: $C_7H_6F_3N$
M_r: 161.12
bp: 187.5^{764}, $74\text{-}5^{10}$
mp: 5-6
***d*:** $1.3047^{12.5}$
***n*:** 1.4787^{20}
***s*:** H_2O 2; al 3; eth 3
IR: μCOB 1416 3330 3230 3030 1640 1520 1470 1350 1250 1160 1140 1060 1000 890 870 790 750 690
Raman: SAD 680 3390 3230 3060 3040 1620 1600 1500 1470 1350 1260 1180 1120 1070 1040 1000 900 840 790 760 700 670 590 540 450 360 330 230 180 150
UV: SAD 263 295 238 cyhex
MS: NBS 620
m/e: 161.05

HODOC No: 02514
CAS No: 455-14-1
Name: Benzenamine, 4-(trifluoromethyl)-
Synonym: *p*-Toluidine, α,α,α-trifluoro-
Line Form: 4-NH_2-$C_6H_4CF_3$
Mol Form: $C_7H_6F_3N$
M_r: 161.12
bp: 117.5^{60}
mp: 38
***d*:** 1.283^{27}
***n*:** 1.4815^{25}
m/e: 161.05

HODOC No: 02515
CAS No: 84483-25-0
Beilstein No: 12, 676
Name: Benzenamine, 2,3,5-triiodo-
Synonym: 2,3,5-Triiodobenzenamine
Line Form: 2,3,5-I_3-$C_6H_2NH_2$
Mol Form: $C_6H_4I_3N$
M_r: 470.81
mp: 116
Color: nd
***s*:** Bz 4; EtOH 4
m/e: 470.75

HODOC No: 02516
CAS No: 24154-37-8
Beilstein No: 12^3, 1508
Name: Benzenamine, 2,4,6-triiodo-
Synonym: Aniline, 2,4,6-triiodo-
Line Form: 2,4,6-I_3-$C_6H_2NH_2$
Mol Form: $C_6H_4I_3N$
Mr: 470.81
mp: 185.5
Color: ye nd or pl (al) pr (HOAc)
***s*:** al 2; AcOEt 4; CS_2 3; HOAc 3
UV: OES 2-48 315(4266) 229(31623) MeOH
m/e: 470.75

HODOC No: 02517
CAS No: 108673-30-9
Beilstein No: 12, 676
Name: Benzenamine, 3,4,5-triiodo-
Synonym: 3,4,5-Triiodoaniline
Line Form: 3,4,5-I_3-$C_6H_2NH_2$
Mol Form: $C_6H_4I_3N$
Mr: 470.81
mp: 174d
Color: nd (al-Ace)
***s*:** Ac 4; Bz 4; Et_2O 4; EtOH 4
m/e: 470.75

HODOC No: 02518
CAS No: 24313-88-0
Beilstein No: 13^3, 2368
Name: Benzenamine, 3,4,5-trimethoxy-
Synonym: Aniline, 3,4,5-trimethoxy-
Line Form: 3,4,5-$(CH_3O)_3$-$C_6H_2NH_2$
Mol Form: $C_9H_{13}NO_3$
Mr: 183.20
mp: 110-3
IR: ALD 2:648G
MS: ALD 5:77C
m/e: 183.09

HODOC No: 02519
CAS No: 137-17-7
Beilstein No: 12^3, 2700
Name: Benzenamine, 2,4,5-trimethyl-
Synonym: Aniline, 2,4,5-trimethyl-
Line Form: 2,4,5-$(CH_3)_3$-$C_6H_2NH_2$
Mol Form: $C_9H_{13}N$
Mr: 135.20
bp: 234-5
mp: 68
***d*:** 0.957
Color: nd (w)
***s*:** EtOH 4
m/e: 135.10

HODOC No: 02520
CAS No: 88-05-1
Beilstein No: 12^2, 631
Name: Benzenamine, 2,4,6-trimethyl-
Synonym: Aniline, 2,4,6-trimethyl-
Mol Form: $C_9H_{13}N$
Mr: 135.20
bp: 232-3
mp: -5
***d*:** 0.9633
***n*:** 1.5495^{20}
***s*:** ctc 2
IR: µSADP 5555 3450 3330 2940 1640 1490 1430 1370 1300 1250 1150 1010 860 730
UV: SAD 1549 288(2050) 234(7770) MeOH
H NMR: SAD 8779 2.1 2.2 3.2 6.6 CCl_4
m/e: 135.10

HODOC No: 02521
CAS No: 609-72-3
Beilstein No: 12^3, 1843
Name: Benzenamine, *N*,*N*,2-trimethyl-
Synonym: *o*-Toluidine, *N*,*N*-dimethyl-
Line Form: 2-$CH_3C_6H_4N(CH_3)_2$
Mol Form: $C_9H_{13}N$
Mr: 135.20
bp: 185.3, $70\text{-}2^{15}$
mp: -60
***d*:** $0.9286^{20/4}$
***n*:** 1.5152^{20}
***s*:** Et_2O 4; EtOH 4
IR: COB 1564
UV: SAD 1118
H NMR: SAD 6578
m/e: 135.10

HODOC No: 02522
CAS No: 121-72-2
Beilstein No: 12^3, 1953
Name: Benzenamine, *N*,*N*,3-trimethyl-
Synonym: *m*-Toluidine, *N*,*N*-dimethyl-
Line Form: $3\text{-}CH_3C_6H_4N(CH_3)_2$
Mol Form: $C_9H_{13}N$
M_r: 135.20
bp: 212
***d*:** $0.9410^{20/4}$
***n*:** 1.5492^{20}
***s*:** al 5; eth 5
IR: μCOB 3875 3030 2940 1610 1520 1450 1350 1330 1240 1180 1150 1140 1100 1060 1010 1000 960 830 770 700 690
UV: SAD 22019 299(2000) 252(1220) MeOH 268(385) 261(418) MeOH:HCl
H NMR: SAD 14636 2.2 2.8 7.0 CCl_4
m/e: 135.10

HODOC No: 02523
CAS No: 99-97-8
Beilstein No: 12^3, 2026
Name: Benzenamine, *N*,*N*,4-trimethyl-
Synonym: *p*-Toluidine, *N*,*N*-dimethyl-
Line Form: $4\text{-}CH_3C_6H_4N(CH_3)_2$
Mol Form: $C_9H_{13}N$
M_r: 135.20
bp: 211, 89^{12}
***d*:** $0.9366^{20/4}$
***n*:** 1.5366^{20}
***s*:** H_2O 1; al 5; eth 5; ctc 3
IR: SADG 8303 3060 2900 2850 2790 1620 1560 1520 1480 1450 1340 1220 1190 1160 1130 1060 950 800 690
UV: SAD 1119 304 253 cyhex
H NMR: SAD 3171 2.2 2.8 6.5 6.9 CCl_4
m/e: 135.10

HODOC No: 02524
CAS No: 54-88-6
Beilstein No: 16^3, 391
Name: Benzenamine, *N*,*N*,3-trimethyl-4-(phenylazo)-
Synonym: *m*-Toluidine, *N*,*N*-dimethyl-4-(phenylazo)-
Line Form: $[4\text{-}(CH_3)_2N\text{-}2\text{-}CH_3]C_6H_3N{=}NC_6H_5$
Mol Form: $C_{15}H_{17}N_3$
M_r: 239.31
mp: 67-9
IR: ALD 2:1263F
m/e: 239.14

HODOC No: 02525
CAS No: 489-98-5
Beilstein No: 12^3, 1737
Name: Benzenamine, 2,4,6-trinitro-
Synonym: Aniline, 2,4,6-trinitro-
Line Form: $2,4,6\text{-}(NO_2)_3\text{-}C_6H_2NH_2$
Mol Form: $C_6H_4N_4O_6$
M_r: 228.11
bp: exp
mp: 192-5
***d*:** 1.762^{10}
Color: dk ye pr (HOAc)
***s*:** H_2O 1; al 2; eth 2; ace 3; bz 3; chl 2; AcOEt 3; HOAc 3
UV: OES 8-36 410(8128) 317(12023) 240(16218) EtOH:15% aa
H NMR: SAD 14209 9.0 9.1 DMSO
m/e: 228.01

HODOC No: 02526
CAS No: 131-73-7
Beilstein No: 12^3, 1734
Name: Benzenamine, 2,4,6-trinitro-*N*-(2,4,6-trinitrophenyl)-
Synonym: Diphenylamine, 2,2′,4,4′,6,6′-hexanitro-
Line Form: $[2,4,6\text{-}(NO_2)_3C_6H_2]_2NH$
Structure:

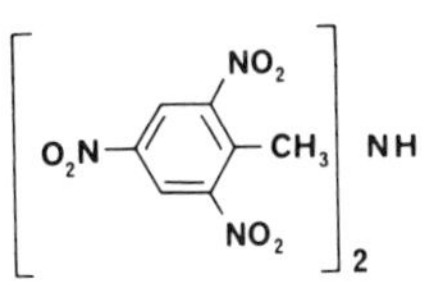

Mol Form: $C_{12}H_5N_7O_{12}$
M_r: 439.20
mp: 244d
Color: pa ye pr(HOAc)
***s*:** H_2O 1; al 1; eth 2; ace 2; bz 1; ctc 1; aq alk 3; py 4; tol 1
IR: SADG 15728 3448 3030 1587 1538 1493 1333 1299 1075 926 714
UV: SAD 15786 384(18900) 205(34200) MeOH
H NMR: SAD 6046 8.8 $DMSOd_6$
m/e: 439.00

HODOC No: 02527
CAS No: 114-80-7
Name: Benzenaminium, 3-[[(dimethylamino)carbonyl]oxy]-*N*,*N*,*N*-trimethyl-, bromide
Synonym: Ammonium, (*m*-hydroxyphenyl)trimethyl-, bromide, dimethylcarbamate
Mol Form: $C_{12}H_{19}BrN_2O_2$
Mr: 302.19
mp: dec 167
***s*:** H_2O 3; al 3
UV: LANG 1154 266(457) 259(501) H_2O
m/e: 302.06

HODOC No: 02528
CAS No: 16056-11-4
Beilstein No: 12^2, 88
Name: Benzenaminium, *N*,*N*,*N*-trimethyl-, bromide
Synonym: Ammonium, trimethylphenyl-, bromide
Line Form: $C_6H_5N(CH_3)_3Br$
Mol Form: $C_9H_{14}BrN$
Mr: 215.11
mp: 213-4
***s*:** H_2O 4; al 3; eth 1; HOAc 3; MeOH 4
IR: SADG 10615 3030 2941 2564 1587 1493 1471 1408 1111 1020 990 943 847 769 690
UV: SAD 13339 264(168) 258(246) 254(276) 248(220) MeOH
m/e: 215.03

HODOC No: 02529
CAS No: 71-43-2
Beilstein No: 5^4, 583
Name: Benzene
Synonym: [6]Annulene
Structure:

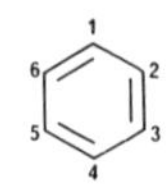

Mol Form: C_6H_6
Mr: 78.11
bp: 80.1
mp: 5.5
***d*:** $0.8765^{20/4}$
***n*:** 1.5011^{20}
Color: rh pr
***s*:** H_2O 2; al 5; eth 5; ace 5; ctc 3; chl 5; HOAc 5
IR: SADG 136 3080 3030 1965 1815 1480 1035 670
Raman: DFB 66 3060 3040 2950 1610 1580 1180 990 610
UV: SAD 198 268 260 254 248 243 cyhex
H NMR: SAD 3429 7.2 CCl_4
C NMR: STOT 49 128.7
MS: WILEY 102 78(100) 52(19) 51(18) 50(16) 77(14) 39(13) 79(6) 76(6)
m/e: 78.05

HODOC No: 02530
CAS No: 55030-46-1
Name: Benzene, (1 hexyl-1-heptenyl)-
Synonym: 6-Tridecene, 7-phenyl
Line Form: $CH_3(CH_2)_4CH{=}C(C_6H_5)(CH_2)_5CH_3$
Mol Form: $C_{19}H_{30}$
Mr: 258.43
UV: OES 1-835 242(9550) iso
MS: WILEY 1790 118(100) 117(72) 91(55) 41(33) 131(23) 173(22) 43(21) 258(8)
m/e: 258.23

HODOC No: 02531
CAS No: 6279-48-7
Name: Benzene, 1-amino-4-dimethylamino-2,5-dinitro
Mol Form: $C_8H_{10}N_4O_4$
M_r: 226.18
mp: 84
IR: SADG 19849 3480 3360 2950 1640 1610 1530 1510 1420 1350 1330 1240 1160 1090 1020 930 880 840 810 750 420
UV: SAD 17569 430(3030) 246(19400) MeOH
H NMR: SAD 11643 2.7 5.9 7.1 7.9 $CDCl_3$
m/e: 226.07

HODOC No: 02532
CAS No: 622-37-7
Beilstein No: 5^4, 795
Name: Benzene, azido-
Synonym: Azidobenzene
Line Form: $C_6H_5N_3$
Mol Form: $C_6H_5N_3$
M_r: 119.12
bp: 70^{11}
mp: -27.5
***d*:** $1.0880^{20/20}$
***n*:** 1.5589^{25}
Color: pa ye oil
***s*:** H_2O 1; al 2; eth 2
IR: SADG 17119 2130 2100 1590 1480 1460 1280 1270 1120 1070 1020 980 800 740 680 660 520 480
UV: OES 4-65 275(5129) 259(5623) EtOH
MS: ALDN 60 91(100) 64(84) 119(51) 63(39) 38(31) 39(28) 65(20) 51(17)
m/e: 119.05

HODOC No: 02533
CAS No: 2101-88-4
Beilstein No: 5^4, 761
Name: Benzene, 1-azido-4-bromo-
Synonym: *p*-Azidobromobenzene
Line Form: $4\text{-}N_3C_6H_4Br$
Mol Form: $C_6H_4BrN_3$
M_r: 197.02
bp: 105^{10}
mp: 20
Color: pl
***s*:** H_2O 1; al 2; eth 4; bz 4
IR: DMS 7579 3100 2065 2035 1560 1463 1278 1119 1064 1002 817 690
UV: OES 6-59 293(1514) 283(1995) 251(12023) iso
m/e: 196.96

HODOC No: 02534
CAS No: 3296-05-7
Beilstein No: 5^4, 760
Name: Benzene, 1-azido-4-chloro-
Synonym: *p*-Azidochlorobenzene
Line Form: $4\text{-}N_3C_6H_4Cl$
Mol Form: $C_6H_4ClN_3$
M_r: 153.56
bp: 96^{20}
mp: 20
***s*:** H_2O 1; eth 3
UV: OES 4-53 279(7079) 267(6607) EtOH
m/e: 153.01

HODOC No: 02535
CAS No: 1965-25-9
Beilstein No: 5^2, 208
Name: Benzene, 1-azido-2,4-dichloro-
Synonym: 2,4-Dichlorophenyl azide
Line Form: $2,4\text{-}Cl_2C_6H_3N_3$
Mol Form: $C_6H_3Cl_2N_3$
M_r: 188.01
mp: 54
Color: ye nd (al) pr (Ace, Bz)
***s*:** Bz 4; Et_2O 4; EtOH 4; peth 4
m/e: 186.97

HODOC No: 02536
CAS No: 54467-95-7
Beilstein No: 5, 278
Name: Benzene, 1-azido-2-iodo-
Synonym: 1-Azido-2-iodobenzene
Line Form: $2\text{-}N_3C_6H_4I$
Mol Form: $C_6H_4IN_3$
M_r: 245.02
bp: $90\text{-}100^{0.9}$
***d*:** $1.8893^{25/4}$
***n*:** 1.6631^{25}
Color: ye
***s*:** Ac 4; EtOH 4
m/e: 244.94

HODOC No: 02537
CAS No: 622-79-7
Beilstein No: 5^4, 759
Name: Benzene, (azidomethyl)-
Synonym: Toluene, α-azido-
Line Form: $C_6H_5CH_2$-N=N≡N
Mol Form: $C_7H_7N_3$
M_r: 133.15
bp: 108^{23}, 74^{11}
***d*:** 1.0655^{25}
***n*:** 1.5341^{25}
***s*:** H_2O 1; al 5; eth 5
IR: COB 7422 3035 2920 2865 1705 1610 1505 1465 1360 1260 1205 1080 1030 875 745 700
m/e: 133.06

HODOC No: 02538
CAS No: 31656-92-5
Beilstein No: 5^4, 876
Name: Benzene, 1-azido-2-methyl-
Synonym: Toluene, *o*-azido-
Line Form: 2-$CH_3C_6H_4N_3$
Mol Form: $C_7H_7N_3$
M_r: 133.15
bp: 90.5^{30}
mp: 10
Color: pa ye liq
***s*:** Et_2O 4
MS: NBS 291
m/e: 133.06

HODOC No: 02539
CAS No: 4113-72-8
Beilstein No: 5^4, 876
Name: Benzene, 1-azido-3-methyl-
Synonym: Toluene, *m*-azido-
Line Form: 3-$CH_3C_6H_4N_3$
Mol Form: $C_7H_7N_3$
M_r: 133.15
bp: $92\text{-}5^{31}$
***s*:** Et_2O 4
MS: NBS 291
m/e: 133.06

HODOC No: 02540
CAS No: 2101-86-2
Beilstein No: 5^4, 876
Name: Benzene, 1-azido-4-methyl-
Synonym: Toluene, *p*-azido-
Line Form: 4-$CH_3C_6H_4N_3$
Mol Form: $C_7H_7N_3$
M_r: 133.15
bp: d 180, 80^{10}
***d*:** $1.0527^{23/4}$
***s*:** Et_2O 4; EtOH 4
MS: NBS 4080
m/e: 133.06

HODOC No: 02541
CAS No: 20615-75-2
Beilstein No: 5^3, 773
Name: Benzene, 1-azido-4-methyl-2-nitro-
Synonym: Toluene, 4-azido-3-nitro-
Line Form: 4-N_3-3-$NO_2C_6H_3CH_3$
Mol Form: $C_7H_6N_4O_2$
M_r: 178.14
mp: 38
Color: ye nd or pl (lig, dil al)
***s*:** EtOH 4
m/e: 178.05

HODOC No: 02542
CAS No: 1516-58-1
Beilstein No: 5, 278
Name: Benzene, 1-azido-2-nitro-
Synonym: 1-Azido-2-nitrobenzene
Mol Form: $C_6H_4N_4O_2$
M_r: 164.12
mp: 53-5
***s*:** al 4; bz 4; chl 3; HOAc 4
IR: μSADP 1879 3030 1610 1590 1540 1490 1350 1210 1150 1140 1020 890 830 750 740
UV: SAD 525 210 MeOH
m/e: 164.03

HODOC No: 02543
CAS No: 1516-59-2
Beilstein No: 5, 278
Name: Benzene, 1-azido-3-nitro-
Synonym: 1-Azido-3-nitrobenzene
Mol Form: $C_6H_4N_4O_2$
M_r: 164.12
mp: 56
***s*:** H_2O 1; al 4; eth 4; bz 4; CS_2 4
IR: μSADP 1884 3130 2080 1610 1540 1450 1350 1300 1150 1080 900 880 810 790 740 710
UV: OES 6-61 315(1585) 242(20417) iso
m/e: 164.03

HODOC No: 02544
CAS No: 1516-60-5
Beilstein No: 5, 278
Name: Benzene, 1-azido-4-nitro-
Synonym: *p*-Azidonitrobenzene
Mol Form: $C_6H_4N_4O_2$
M_r: 164.12
mp: 75
***s*:** H_2O 1; al 4; eth 4; bz 4; chl 2; CS_2 4
IR: μSADP 1885 3030 2130 1590 1520 1470 1350 1280 1190 1110 860 850 810 750
UV: OES 6-61 299(16982) 220(14125) iso
H NMR: SAD 7301 7.2 8.3 $CDCl_3$
m/e: 164.03

HODOC No: 02545
CAS No: 22001-17-8
Name: Benzene, 1-azido-3-(trifluoromethyl)-
Synonym: Toluene, *m*-azido-α,α,α-trifluoro-
Mol Form: $C_7H_4F_3N_3$
M_r: 187.12
bp: $51^{8.5}$
IR: SADG 17142 3080 2160 2100 1610 1590 1490 1450 1320 1280 1170 1120 1060 880 790 690
UV: SAD 18597 249(18200) MeOH
H NMR: SAD 9293 7.0-7.7 CCl_4
m/e: 187.04

HODOC No: 02546
CAS No: 108974-19-2
Name: Benzene, 2-(benzyloxy)-1-methoxy-4-propenyl-
Synonym: Ether, benzyl (2-methoxy-5-propenylphenyl)
Mol Form: $C_{17}H_{18}O_2$
M_r: 254.31
IR: SADG 15355 3020 2930 1600 1580 1510 1450 1410 1370 1290 1260 1220 1150 1130 1020 960 910 850 780 730 690 610 470
Raman: SAD 1101 3070 3050 3000 2930 2910 2880 2850 2730 1650 1600 1520 1450 1420 1390 1340 1320 1260 1230 1220 1200 1160 1140 1040 1010 990 970 920 850 820 790 750 640 620 580 410 340 270 200 160
UV: SAD 438 258(18300) 211(31500) MeOH
H NMR: SAD 6695 1.8 3.8 5.0 6.3 7.3 $CDCl_3$
m/e: 254.13

HODOC No: 02547
CAS No: 20273-27-2
Name: Benzene, [1,1′-bicyclohexyl]-4-yl-
Synonym: Bicyclohexyl, 4-phenyl-
Mol Form: $C_{18}H_{26}$
M_r: 242.39
mp: 84
UV: LANG 1900 268(186) 260(245) 252(195) 210(9772) EtOH
MS: AMS 2-230 117(100) 242(96) 159(74) 91(68) 83(42) 104(40) 55(34) 160(33)
m/e: 242.20

HODOC No: 02548
CAS No: 91-13-4
Beilstein No: 5^4, 929
Name: Benzene, 1,2-*bis*(bromomethyl)-
Synonym: *o*-Xylene, α,α′-dibromo-
Line Form: 1,2-$(BrCH_2)_2$-C_6H_4
Mol Form: $C_8H_8Br_2$
M_r: 261.95
bp: $128\text{-}30^{4.5}$
mp: 95
***d*:** 1.988^0
Color: rh (Chl)
***s*:** H_2O 1; al 3; eth 3; ctc 3; chl 3; lig 3; peth 3
IR: μCOB 1905 3030 1590 1490 1450 1430 1240 1210 1180 760 740
H NMR: SAD 1882 4.6 7.3 CCl_4
m/e: 261.90

HODOC No: 02549
CAS No: 626-15-3
Beilstein No: 5^4, 946
Name: Benzene, 1,3-*bis*(bromomethyl)-
Synonym: *m*-Xylene, α,α′-dibromo-
Line Form: 1,3-$(BrCH_2)_2$-C_6H_4
Mol Form: $C_8H_8Br_2$
M_r: 261.95
bp: $135\text{-}40^{20}$
mp: 77
***d*:** 1.959^0
Color: nd (Chl), pr (Ace)
***s*:** H_2O 1; al 3; eth 3; chl 3; lig 3
IR: **µSADP 19167** 3030 1590 1490 1450 1220 1160 1090 900 800 750 700
UV: **SAD 6269** <210 MeOH
H NMR: **SAD 13652** 4.6 $CDCl_3$
m/e: 261.90

HODOC No: 02550
CAS No: 623-24-5
Beilstein No: 5^4, 970
Name: Benzene, 1,4-*bis*(bromomethyl)-
Synonym: *p*-Xylene, α,α′-dibromo-
Line Form: 1,4-$(BrCH_2)_2$-C_6H_4
Mol Form: $C_8H_8Br_2$
M_r: 261.95
bp: 245, $155\text{-}8^{14}$
mp: 145-7
***d*:** 2.012^0
Color: mcl pr (al), cr (Chl, Bz)
***s*:** H_2O 1; al 4; eth 2; bz 3; chl 4
IR: **µSADP 7615** 2940 1610 1520 1450 1430 1250 1220 1140 1090 1020 860 850 750
UV: **SAD 1364** 237(8300) 222(8040) MeOH
m/e: 261.90

HODOC No: 02551
CAS No: 19294-04-3
Beilstein No: 5^4, 1027
Name: Benzene, 1,3-*bis*(bromomethyl)-5-methyl-
Synonym: Mesitylene, α,α′-dibromo-
Line Form: 3,5-$(BrCH_2)_2C_6H_3CH_3$
Mol Form: $C_9H_{10}Br_2$
M_r: 275.98
mp: 66
Color: pr
***s*:** Et_2O 4; EtOH 4; lig 4
m/e: 275.91

HODOC No: 02552
CAS No: 612-12-4
Beilstein No: 5^4, 927
Name: Benzene, 1,2-*bis*(chloromethyl)-
Synonym: *o*-Xylene, α,α′-dichloro-
Line Form: 1,2-$(ClCH_2)$-C_6H_4
Mol Form: $C_8H_8Cl_2$
M_r: 175.05
bp: 239-40, $130\text{-}5^{10}$
mp: 55
***d*:** 1.393^0
Color: mcl (lig)
***s*:** H_2O 1:al 4; eth 4; ctc 3; chl 4; lig 4
IR: **SADG 4620** 3030 2941 2857 1639 1613 1471 1449 1299 1250 1220 1190 840 769 676
H NMR: **SAD 497** 4.7 7.3 CCl_4
m/e: 174.00

HODOC No: 02553
CAS No: 626-16-4
Beilstein No: 5^4, 944
Name: Benzene, 1,3-*bis*(chloromethyl)-
Synonym: *m*-Xylene, α,α′-dichloro-
Line Form: 1,3-$(ClCH_2)_2C_6H_4$
Mol Form: $C_8H_8Cl_2$
M_r: 175.05
bp: 250-5
mp: 34.2
***d*:** 1.302^{20}
Color: cr
MS: **NBS 809**
m/e: 174.00

HODOC No: 02554
CAS No: 623-25-6
Beilstein No: 5^4, 967
Name: Benzene, 1,4-*bis*(chloromethyl)-
Synonym: *p*-Xylene, α,α′-dichloro-
Line Form: 1,4-$(ClCH_2)_2C_6H_4$
Mol Form: $C_8H_8Cl_2$
M_r: 175.05
bp: 240-50d, 135^{16}
mp: 100
***d*:** 1.417^0
Color: pl (al)
***s*:** H_2O 1; al 4; eth 4; ace 4; chl 4; HOAc 2
IR: µCOB 3346 2940 1590 1540 1470 1410 1270 1110 1030 830 760 690
Raman: SAD 1164 3050 3010 2960 1620 1440 1320 1280 1260 1210 1180 1150 960 920 840 820 750 690 670 640 520 370 350 140 110
UV: SAD 3624 364 227 MeOH
H NMR: SAD 3254 4.5 7.3 $CDCl_3$
m/e: 174.00

HODOC No: 02555
CAS No: 25445-34-5
Beilstein No: 6^4, 3136
Name: Benzene, 2,4-*bis*(chloromethyl)-1-methoxy-
Synonym: Anisole, 2,4-*bis*(chloromethyl)-
Line Form: 2,4-$(ClCH_2)_2C_6H_3OCH_3$
Mol Form: $C_9H_{10}Cl_2O$
M_r: 205.07
mp: 61
IR: COB 7657 1615 1510 1465 1455 1320 1270 1165 1130 1030 960 910 820 760 660 520
m/e: 204.01

HODOC No: 02556
CAS No: 3022-16-0
Name: Benzene, 1,4-*bis*(chloromethyl)-2,3,5,6-tetramethyl-
Synonym: Bis(chloromethyl)durene
Mol Form: $C_{12}H_{16}Cl_2$
M_r: 231.15
IR: SADG 7674 3000 2910 1440 1380 1370 1270 1230 1190 1010 820 730 700 630 600 560 530 500
Raman: SAD 1166 3010 2910 1570 1450 1390 1370 1300 1250 1160 1080 770 660 590 500 450 380
UV: SAD 19701 295(1020) 244(4370) 212(15500) MeOH
H NMR: SAD 2146 2.3 4.6 CCl_4
m/e: 230.06

HODOC No: 02557
CAS No: 13209-15-9
Beilstein No: 5^4, 929
Name: Benzene, 1,2-*bis*(dibromomethyl)-
Synonym: *o*-Xylene, α,α,α′,α′-tetrabromo-
Line Form: 1,2-$(Br_2CH)_2C_6H_4$
Mol Form: $C_8H_6Br_4$
M_r: 417.74
mp: 116-7
Color: mcl
***s*:** H_2O 2; chl 4; lig 1
IR: SADG 10412 3030 2941 1587 1493 1449 1316 1220 1136 862 781 763 654
UV: SAD 7890 289(1270) 280(1560) 208(2850) MeOH
H NMR: SAD 2362 7.1 7.3 7.6 $CDCl_3$
m/e: 417.72

HODOC No: 02558
CAS No: 36323-28-1
Beilstein No: 5^3, 839
Name: Benzene, 1,3-*bis*(dibromomethyl)-
Synonym: 1,3-Bis(dibromomethyl)benzene
Line Form: 1,3-$(Br_2CH)_2C_6H_4$
Mol Form: $C_8H_6Br_4$
M_r: 417.74
mp: 107
Color: nd (al), pr (Chl)
***s*:** H_2O 1; al 3; bz 4; chl 4; lig 3; HOAc 3
IR: SADG 17892 3000 1580 1480 1430 1310 1200 1140 1070 930 880 790 760 680 660
H NMR: SAD 9596 6.7 7.4-7.6 7.7 $CDCl_2$
m/e: 417.72

HODOC No: 02559
CAS No: 1592-31-0
Beilstein No: 5^3, 860
Name: Benzene, 1,4-*bis*(dibromomethyl)-
Synonym: *p*-Xylene, α,α,α′,α′-tetrabromo-
Line Form: 1,4-$(Br_2CH)_2C_6H_4$
Mol Form: $C_8H_6Br_4$
M_r: 417.74
mp: 172
Color: mcl pr (Chl)
***s*:** Bz 4
m/e: 417.72

HODOC No: 02560
CAS No: 7398-82-5
Beilstein No: 5^3, 857
Name: Benzene, 1,4-*bis*(dichloromethyl)-
Synonym: *p*-Xylene, α,α,α′,α′-tetrachloro-
Line Form: 1,4-$(Cl_2CH)_2C_6H_4$
Mol Form: $C_8H_6Cl_4$
M$_r$: 243.94
bp: 1.606^0
mp: 9.5
***d*:** 1.606^0
***s*:** H_2O 1; al 3; eth 3; bz 3; chl 3
IR: **μCOB 1913** 3030 1590 1520 1430 1250 1210 1190 1110 1020 850 810 760 710
H NMR: **SAD 7394** 6.7 7.6 $CDCl_3$
m/e: 241.92

HODOC No: 02561
CAS No: 1014-60-4
Name: Benzene, 1,3-*bis*(1,1-dimethylethyl)-
Synonym: Benzene, *m*-di-*tert*-butyl-
Mol Form: $C_{14}H_{22}$
M$_r$: 190.31
UV: **API 236** 267 261 iso
MS: **WILEY 1331** 175(100) 57(60) 41(31) 29(25) 190(15) 176(14) 65(11) 39(10)
m/e: 190.17

HODOC No: 02562
CAS No: 17177-98-9
Name: Benzene, 2,4-*bis*(1,1-dimethylethyl)-1-methoxy-
Synonym: Anisole, 2,4-di-*tert*-butyl-
Mol Form: $C_{15}H_{24}O$
M$_r$: 220.34
MS: **WILEY 1565** 191(100) 205(23) 57(21) 206(20) 192(15) 74(7) 41(7) 220(0)
m/e: 220.18

HODOC No: 02563
CAS No: 15181-11-0
Name: Benzene, 1,3-*bis*(1,1-dimethylethyl)-5-methyl-
Synonym: Toluene, 3,5-di-*tert*-butyl-
Mol Form: $C_{15}H_{24}$
M$_r$: 204.34
MS: **WILEY 1451** 189(100) 57(57) 41(30) 29(21) 204(16) 190(14) 91(11) 39(10)
m/e: 204.19

HODOC No: 02564
CAS No: 4023-79-4
Name: Benzene, 1,4-*bis*(*N*-ethyl-naphth-2-ylamino
Synonym: 1,3-Butanedione, 1-*p*-tolyl-
Mol Form: $C_{30}H_{28}N_2$
M$_r$: 416.54
IR: **μCOB** 3030 2940 1640 1610 1520 1470 1390 1370 1280 1250 1190 1150 1120 1090 830 810 760 750
m/e: 416.23

HODOC No: 02565
CAS No: 31173-90-7
Name: Benzene, 1,4-*bis*(1-ethylpropyl)-
Synonym: Benzene, *p*-*bis*(1-ethylpropyl)-
Mol Form: $C_{16}H_{26}$
M$_r$: 218.37
MS: **WILEY 1552** 189(100) 43(68) 29(47) 44(36) 41(32) 91(30) 27(29) 218(13)
m/e: 218.20

HODOC No: 02566
CAS No: 13073-35-3
Name: Benzene, 1,4-*bis*(2-hydroxybenzylideneamino)-2-nitro
Synonym: Butyric acid, 2-amino-4-(ethylthio)-, *L*-
Mol Form: $C_{20}H_{15}N_3O_4$
M$_r$: 361.34
mp: 194-6
UV: **LANG 1561** 431(7079) 358(29512) 272(21878) 232(28840) EtOH
m/e: 361.11

HODOC No: 02567
CAS No: 1605-18-1
Beilstein No: 5^4, 1557
Name: Benzene, 1,4-*bis*(1-methylethenyl)-
Synonym: Benzene, *p*-diisopropenyl-
Line Form: 1,4-$[H_2C{=}C(CH_3)_2]C_6H_4$
Structure:

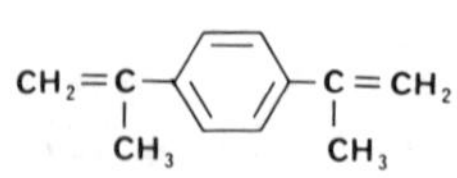

Mol Form: $C_{12}H_{14}$
M$_r$: 158.23
mp: 63-4
***d*:** 1.0649^{22}
MS: **WILEY 985** 158(100) 143(45) 128(22) 115(16) 117(14) 159(14) 28(8) 129(8)
m/e: 158.11

HODOC No: 02568
CAS No: 577-55-9
Beilstein No: 5^4, 1125
Name: Benzene, 1,2-*bis*(1-methylethyl)-
Synonym: Benzene, *o*-diisopropyl-
Line Form: 1,2-$[(CH_3)_2CH]_2C_6H_4$
Mol Form: $C_{12}H_{18}$
M_r: 162.26
bp: 204, 115.4^{50}
mp: -57
***d*:** $0.8701^{20/4}$
***n*:** 1.4960^{20}
***s*:** H_2O 1; al 5; eth 5; ace 5; bz 5; ctc 5
IR: μCOB 1261 2940 1610 1490 1390 1370 1080 1030 940 750
UV: HBCP 271(195) 268(178) 263(246) 257(191) iso
MS: WILEY 1047 147(100) 115(9) 39(8) 43(8) 77(8) 79(7) 117(7) 162(4)
m/e: 162.14

HODOC No: 02569
CAS No: 99-62-7
Beilstein No: 5^4, 1125
Name: Benzene, 1,3-*bis*(1-methylethyl)-
Synonym: Benzene, *m*-diisopropyl-
Line Form: 1,3-$[(CH_3)_2CH-]_2C_6H_4$
Mol Form: $C_{12}H_{18}$
M_r: 162.26
bp: 203.2, 75^9
mp: -61, -63.1 (fp)
***d*:** $0.8559^{20/4}$
***n*:** 1.4883^{20}
***s*:** H_2O 1; al 5; eth 5; ace 5; bz 5; ctc 5
IR: SADG 356 3030 2941 1587 1471 1370 1163 1053 926 885 787 714
Raman: SAD 394 3060 2970 2940 2910 2870 2760 2710 1600 1590 1470 1450 1380 1350 1310 1290 1250 1240 1210 1180 1150 1110 1090 1050 1020 1010 960 930 890 850 800 710 640 580 550 480 440 320 280 150 130
UV: SAD 15755 270(236) 263(282) 257(232) 211(8970) MeOH
H NMR: SAD 7836 1.2 2.8 6.7-7.2 CCl_4
MS: WILEY 1052 147(100) 119(39) 162(35) 43(30) 105(23) 91(23) 41(17) 27(16)
m/e: 162.14

HODOC No: 02570
CAS No: 100-18-5
Beilstein No: 5^4, 1126
Name: Benzene, 1,4-*bis*(1-methylethyl)-
Synonym: Benzene, *p*-diisopropyl-
Line Form: 1,4-$[(CH_3)_2CH]_2C_6H_4$
Mol Form: $C_{12}H_{18}$
M_r: 162.26
bp: 210.3, 120^{50}
mp: -17
***d*:** $0.8568^{20/4}$
***n*:** 1.4898^{20}
***s*:** H_2O 1; al 5; eth 5; ace 5; bz 5; ctc 5
IR: μCOB 3609 2940 1670 1520 1450 1430 1390 1100 1060 1050 1010 830
Raman: SAD 393 3060 3010 2970 2940 2910 2880 2760 2720 1620 1460 1450 1310 1300 1280 1220 1200 1150 1100 1060 1050 1000 960 920 900 840 790 750 450 270 170
UV: SAD 15754 271(1440) 263(1460) 217(26500) 212(27700) MeOH
H NMR: SAD 7835 1.2 2.8 7.0 CCl_4
MS: WILEY 1052 147(100) 119(28) 162(25) 91(17) 43(16) 105(14) 41(14) 148(12)
m/e: 162.14

HODOC No: 02571
CAS No: 85417-71-6
Name: Benzene, 1,2-*bis*(1-methylpropoxy)-
Synonym: Benzene, 1,2-di-*sec*-butoxy
Line Form: 1,2-$(CH_3CH_2CH(CH_3)-O-)_2-C_6H_4$
Mol Form: $C_{14}H_{22}O_2$
M_r: 222.31
mp: 28-9
UV: LANG 2472 278(2344) 225(7413) 208(11749) EtOH
m/e: 222.16

HODOC No: 02572
CAS No: 1079-96-5
Name: Benzene, 1,3-*bis*(1-methylpropyl)-
Synonym: Benzene, *m*-di-*sec*-butyl-
Mol Form: $C_{14}H_{22}$
M_r: 190.31
MS: WILEY 1331 161(100) 29(46) 57(30) 41(24) 27(19) 43(19) 91(15) 190(14)
m/e: 190.17

HODOC No: 02573
CAS No: 3316-23-2
Name: Benzene, *p-bis*(2-nitropropenyl)-
Synonym: Benzene, 1,4-*bis*(2-nitroprpenyl)
Mol Form: $C_{12}H_{12}N_2O_4$
M_r: 248.23
mp: 120.2-1.5
IR: **µSADP 3353** 2940 1665 1515 1470 1430 1390 1335 1300 1220 990 925 860 840 785 720
UV: **SAD 1039** 332 227 MeOH
H NMR: **SAD 11010** 2.5 7.5 8.0 $CDCl_3$
m/e: 248.08

HODOC No: 02574
CAS No: 3179-16-6
Name: Benzene, *m-bis*(2-nitrovinyl)-
Synonym: Benzene, 1,3-*bis*(2-nitrovinyl)
Mol Form: $C_{10}H_8N_2O_4$
M_r: 220.17
mp: 210
***s*:** DMSO 2
IR: **SADG 18037** 3110 3060 1640 1510 1330 1270 1160 960 840 790
UV: **SAD 1018** 301 MeOH
H NMR: **SAD 10699** 7.4-8.4 DMSO
m/e: 220.05

HODOC No: 02575
CAS No: 2455-43-8
Name: Benzene, 1,4-*bis*(4-phenoxyphenoxy)-
Synonym: Benzene, *p-bis*(*p*-phenoxyphenoxy)-
Mol Form: $C_{30}H_{22}O_4$
M_r: 446.47
IR: **COB 7677** 3040 1430 1200 1160 1090 1010 780 750 690 500
m/e: 446.15

HODOC No: 02576
CAS No: 5368-75-2
Name: Benzene, *p-bis*(4-phenylbutyl)-
Synonym: Benzene, 1,4-*bis*(4-phenylbutyl)
Mol Form: $C_{26}H_{30}$
M_r: 342.50
UV: **API 908** 274 268 265 262 260 218 cyhex
MS: **WILEY 1976** 91(100) 223(44) 342(40) 131(32) 104(20) 117(18) 105(16) 92(14)
m/e: 342.23

HODOC No: 02577
CAS No: 10403-73-3
Beilstein No: 6^4, 5572
Name: Benzene, 1,2-*bis*(phenylmethoxy)-
Synonym: Benzene, *o-bis*(benzyloxy)-
Line Form: 1,2-$(C_6H_5CH_2O)_2C_6H_4$
Mol Form: $C_{20}H_{18}O_2$
M_r: 290.34
mp: 63-4
Color: yesh nd or pr (al)
***s*:** Et_2O 4; EtOH 4; peth 4
m/e: 290.13

HODOC No: 02578
CAS No: 621-91-0
Beilstein No: 6^4, 5729
Name: Benzene, 1,4-*bis*(phenylmethoxy)-
Synonym: Benzene, *p-bis*(benzyloxy)-
Line Form: 1,4-$(C_6H_5CH_2O)_2C_6H_4$
Mol Form: $C_{20}H_{18}O_2$
M_r: 290.34
mp: 130
Color: pl (al)
***s*:** Et_2O 4; EtOH 4; HOAc 4
m/e: 290.13

HODOC No: 02579
CAS No: 793-23-7
Beilstein No: 5^4, 2512
Name: Benzene, 1,4-*bis*(phenylmethyl)-
Synonym: *p*-Xylene, α,α′-diphenyl-
Line Form: 1,4-$(C_6H_5CH_2)_2C_6H_4$
Mol Form: $C_{20}H_{18}$
M_r: 258.34
bp: 225^{18}
mp: 87-8
***s*:** al 3; eth 2; bz 3; chl 3; CS_2 3; HOAc 3
IR: **SADG 10490** 3080 3040 2910 2850 1590 1500 1480 1440 1420 1100 1070 1030 850 760 720 700
UV: **SAD 8131** 274(562) 269(906) 265(898) 262(969) 260(943) MeOH
H NMR: **SAD 2406** 3.9 7.0 7.1 $CDCl_3$
m/e: 258.14

HODOC No: 02580
CAS No: 402-31-3
Beilstein No: 5^4, 965
Name: Benzene, 1,3-*bis*(trifluoromethyl)-
Synonym: *m*-Xylene, α,α,α,α′,α′,α′-hexafluoro-
Line Form: 1,3-$(F_3C)_2C_6H_4$
Mol Form: $C_8H_4F_6$
M_r: 214.10
bp: 116
***d*:** 1.3790^{25}
***n*:** 1.3916^{25}
IR: SADG 8021 3070 1630 1610 1460 1350 1320 1270 1170 1140 1070 910 830 810 750 700 680 650 570 430 390 340
Raman: SAD 749 3090 3040 2980 2900 2800 1630 1610 1500 1460 1360 1180 1130 1100 1070 1040 1010980 810 760 650 570 460 390 290 120
UV: SAD 9754 267(211) 260(137) 256(110) 254(110) 203(2030) MeOH
H NMR: SAD 1476 7.2 7.4-8.0 CCl_4
MS: NBS 4487
m/e: 214.02

HODOC No: 02581
CAS No: 433-19-2
Name: Benzene, 1,4-*bis*(trifluoromethyl)-
Synonym: *p*-Xylene, α,α,α,α′,α′,α′-hexafluoro-
Mol Form: $C_8H_4F_6$
M_r: 214.10
IR: μSADP 864 3130 1410 1320 1160 1140 1100 1080 1010 850 750
MS: ALDN 175 214(100) 145(88) 195(83) 69(28) 164(27) 50(26) 75(25) 74(11)
m/e: 214.02

HODOC No: 02582
CAS No: 108-86-1
Beilstein No: 5^4, 670
Name: Benzene, bromo-
Synonym: Bromobenzene
Line Form: C_6H_5Br
Mol Form: C_6H_5Br
M_r: 156.00
bp: 156, 43^{18}
mp: -30.8
***d*:** $1.4950^{20/4}$
***n*:** 1.5597^{20}
***s*:** H_2O 1; al 4; eth 4; bz 4; ctc 3
IR: μCOB 6476 3030 1590 1470 1450 1080 1020 1000 910 740 690
Raman: DFB 73 3160 3070 1570 1170 1160 1070 1020 1000 680 610 310 250 180
UV: SAD 954 271(1190) 264(1710) 261(1730) 257(1590) 250(1390) MeOH
H NMR: SAD 1212 7.1 7.4 CCl_4
C NMR: JJ 152 FT 122.4 126.7 129.8 131.4 $CDCl_3$
MS: WILEY 947 77(100) 156(94) 158(92) 50(17) 78(11) 38(9) 51(7) 157(7)
m/e: 155.96

HODOC No: 02583
CAS No: 3433-80-5
Beilstein No: 5^4, 836
Name: Benzene, 1-bromo-2-(bromomethyl)-
Synonym: Toluene, *o*,α-dibromo-
Line Form: 2-$BrC_6H_4CH_2Br$
Mol Form: $C_7H_6Br_2$
M_r: 247.93
bp: 129^{19}
mp: 31
Color: cr (al, lig)
***s*:** Et_2O 4; EtOH 4; HOAc 4
m/e: 247.88

HODOC No: 02584
CAS No: 823-78-9
Beilstein No: 5^4, 836
Name: Benzene, 1-bromo-3-(bromomethyl)-
Synonym: Toluene, *m*,α-dibromo-
Line Form: $3\text{-}BrC_6H_4CH_2Br$
Mol Form: $C_7H_6Br_2$
M_r: 247.93
mp: 41-3
Color: nd or lf
***s*:** chl 3
IR: SADG 20561 3060 2960 1560 1470 1420 1210 1060 990 880 840 770 680 610 550
H NMR: SAD 11588 4.3 7.0 7.1-7.4 7.5 $CDCl_3$
m/e: 247.88

HODOC No: 02585
CAS No: 589-15-1
Beilstein No: 5^4, 836
Name: Benzene, 1-bromo-4-(bromomethyl)-
Synonym: Toluene, *p*,α-dibromo-
Line Form: $4\text{-}BrC_6H_4CH_2Br$
Mol Form: $C_7H_6Br_2$
M_r: 247.93
mp: 63
Color: nd (al)
***s*:** H_2O 2; al 3; eth 4; bz 3; chl 3; CS_2 4; HOAc 4
IR: SADG 8631 3030 2970 1600 1490 1440 1400 1220 1200 1090 1060 1010 820 800 710
UV: SAD 3793 238 cyhex
H NMR: SAD 3256 4.4 7.2 7.4 $CDCl_3$
m/e: 247.88

HODOC No: 02586
CAS No: 18066-89-2
Name: Benzene, 1-bromo-4-(bromophenylmethyl)-
Synonym: Methane, bromo(*p*-bromophenyl)phenyl-
Mol Form: $C_{13}H_{10}Br_2$
M_r: 324.02
IR: μCOB 1953 3030 1590 1490 1470 1410 1160 1080 1010 860 840 790 750 720 700 650
m/e: 323.91

HODOC No: 02587
CAS No: 62692-40-4
Beilstein No: 5^3, 1209
Name: Benzene, (3-bromo-3-butenyl)-
Synonym: 1-Butene, 2-bromo-4-phenyl
Line Form: $C_6H_5CH_2CH_2CBr{=}CH_2$
Mol Form: $C_{10}H_{11}Br$
M_r: 210.09
bp: $117\text{-}8^{21}$, $90\text{-}1^{5}$
***d*:** $1.2907^{20/4}$
***n*:** 1.5450^{20}
***s*:** Ac 4
m/e: 210.00

HODOC No: 02588
CAS No: 40734-75-6
Beilstein No: 5^4, 1377
Name: Benzene, (4-bromo-2-butenyl)-
Synonym: 2-Butene, 1-bromo-4-phenyl-
Line Form: $C_6H_5CH_2CH{=}CHCH_2Br$
Mol Form: $C_{10}H_{11}Br$
M_r: 210.09
bp: $126\text{-}30^{10.5}$
***d*:** $1.2660^{20/4}$
***n*:** 1.5678^{20}
***s*:** Et_2O 4
m/e: 210.00

HODOC No: 02589
CAS No: 1200-03-9
Beilstein No: 6^4, 558
Name: Benzene, (4-bromobutoxy)-
Synonym: Ether, 4-bromobutyl phenyl
Line Form: $C_6H_5\text{-}O\text{-}(CH_2)_4Br$
Mol Form: $C_{10}H_{13}BrO$
M_r: 228.11
bp: $153\text{-}6^{18}$
mp: 41
Color: cr (al)
***s*:** al 2; ctc 2
IR: SADG 9852 3030 2941 1587 1493 1449 1429 1370 1282 1235 1163 1075 1031 926 870 741 680
H NMR: SAD 4585 1.9 3.3 3.8 6.8 7.2 CCl_4
m/e: 228.01

HODOC No: 02590
CAS No: 694-80-4
Beilstein No: 5^4, 680
Name: Benzene, 1-bromo-2-chloro-
Synonym: *o*-Bromochlorobenzene
Line Form: 2-ClC_6H_4Br
Mol Form: C_6H_4BrCl
M_r: 190.45
bp: 204^{765}
mp: -12.3
***d*:** $1.6387^{25/4}$
***n*:** 1.5809^{20}
***s*:** H_2O 1; bz 4; ctc 2
IR: SADG 10311 3030 1563 1449 1429 1235 1124 1020 943 741 719 649
UV: SAD 6684 278 270 263 256 MeOH
H NMR: SAD 2268 6.8-7.6 CCl_4
MS: WILEY 1316 192(100) 190(76) 111(50) 75(25) 194(24) 113(16) 50(10) 74(8)
m/e: 189.92

HODOC No: 02591
CAS No: 108-37-2
Beilstein No: 5^4, 680
Name: Benzene, 1-bromo-3-chloro-
Synonym: *m*-Bromochlorobenzene
Line Form: 3-ClC_6H_4Br
Mol Form: C_6H_4BrCl
M_r: 190.45
bp: 196
mp: -21.5
***d*:** $1.6302^{20/4}$
***n*:** 1.5771^{20}
***s*:** H_2O 1; al 4; eth 4
IR: µCOB 1179 3030 1560 1470 1410 1390 1160 1110 1100 1080 1060 1000 870 770 760 670
UV: OES 5-45 278(324) 271(398) 264(302) 257(166) 251(120) cyhex
H NMR: SAD 17191 7.0 7.5 CCl_4
MS: WILEY 1316 192(100) 190(78) 111(61) 75(26) 194(24) 113(19) 28(16) 50(13)
m/e: 189.92

HODOC No: 02592
CAS No: 106-39-8
Beilstein No: 5^4, 681
Name: Benzene, 1-bromo-4-chloro-
Synonym: *p*-Bromochlorobenzene
Line Form: 4-ClC_6H_4Br
Mol Form: C_6H_4BrCl
M_r: 190.45
bp: 196^{756}
mp: 68
***d*:** $1.576^{71/4}$
***n*:** 1.5531^{70}
Color: nd or pl (al, eth)
***s*:** H_2O 1; al 2; eth 3; bz 3; ctc 3; chl 3
IR: µCOB 4782 3130 1610 1470 1390 1090 1080 1010 810
Raman: SAD 333 3150 3070 1630 1570 1560 1370 1290 1170 1090 1070 990 810 730 690 620 330 290 260 90
UV: SAD 3519 273 cyhex
H NMR: SAD 3251 7.1 7.4 CCl_4
MS: WILEY 1317 192(100) 190(76) 111(57) 75(28) 194(24) 113(18) 50(13) 74(11)
m/e: 189.92

HODOC No: 02593
CAS No: 55162-34-0
Beilstein No: 6^3, 742
Name: Benzene, 1-bromo-4-(2-chloroethoxy)-
Synonym: 4-Bromo-β-chlorophenetole
Line Form: (4-BrC_6H_4)OCH_2CH_2Cl
Mol Form: C_8H_8BrClO
M_r: 234.50
mp: 55-7
MS: WILEY 1652 172(100) 174(98) 236(59) 63(59) 234(46) 65(29) 27(19) 238(14)
m/e: 233.94

HODOC No: 02594
CAS No: 578-51-8
Beilstein No: 5^4, 832
Name: Benzene, 1-bromo-2-(chloromethyl)-
Synonym: Toluene, *o*-bromo-α-chloro-
Line Form: 2-$BrC_6H_4CH_2Cl$
Mol Form: C_7H_6BrCl
M_r: 204.47
bp: $124\text{-}6^{20}$
***s*:** Et_2O 4; EtOH 4
m/e: 203.93

HODOC No: 02595
CAS No: 97329-43-6
Beilstein No: 5^1, 155
Name: Benzene, 1-bromo-2-chloro-3-methyl-
Synonym: 3-Bromo-2-chloromethylbenzene
Line Form: $2\text{-Cl-}3\text{-BrC}_6H_3CH_3$
Mol Form: C_7H_6BrCl
M_r: 204.47
bp: $125\text{-}35^{50}$
mp: 18
m/e: 203.93

HODOC No: 02596
CAS No: 932-77-4
Beilstein No: 5^4, 832
Name: Benzene, 1-bromo-3-(chloromethyl)-
Synonym: Toluene, *m*-bromo-α-chloro-
Line Form: $3\text{-BrC}_6H_4CH_2Cl$
Mol Form: C_7H_6BrCl
M_r: 204.47
bp: 119^{18}
mp: 32-3
***s*:** EtOH 4
m/e: 203.93

HODOC No: 02597
CAS No: 62356-27-8
Beilstein No: 5^4, 832
Name: Benzene, 1-bromo-3-chloro-2-methyl-
Synonym: 1-Bromo-3-chloro-2-methylbenzene
Line Form: $6\text{-Cl-}2\text{-BrC}_6H_3CH_3$
Mol Form: C_7H_6BrCl
M_r: 204.47
bp: $118\text{-}20^{40}$
m/e: 203.93

HODOC No: 02598
CAS No: 589-17-3
Beilstein No: 5^4, 832
Name: Benzene, 1-bromo-4-(chloromethyl)-
Synonym: Toluene, *p*-bromo-α-chloro-
Line Form: $4\text{-BrC}_6H_4CH_2Cl$
Mol Form: C_7H_6BrCl
M_r: 204.47
bp: 236, $110\text{-}1^9$
mp: 50
Color: nd (al, peth)
***s*:** H_2O 1; al 4; eth 4; peth 3
IR: SADG 10851 3030 2960 1590 1480 1440 1400 1250 1210 1190 1090 1060 1000 820 720 670
UV: SAD 1261 230(11100) MeOH
m/e: 203.93

HODOC No: 02599
CAS No: 14495-51-3
Beilstein No: 5^3, 713
Name: Benzene, 1-bromo-4-chloro-2-methyl-
Synonym: Toluene, 2-bromo-5-chloro-
Line Form: $5\text{-Cl-}2\text{-BrC}_6H_3CH_3$
Mol Form: C_7H_6BrCl
M_r: 204.47
bp: $98\text{-}100^{25}$
m/e: 203.93

HODOC No: 02600
CAS No: 69190-56-3
Beilstein No: 5^1, 155
Name: Benzene, 2-bromo-1-chloro-3-methyl-
Synonym: Toluene, 2-bromo-3-chloro
Line Form: $3\text{-Cl-}2\text{-BrC}_6H_3CH_3$
Mol Form: C_7H_6BrCl
M_r: 204.47
bp: $103\text{-}5^{25}$
mp: 4-5
m/e: 203.93

HODOC No: 02601
CAS No: 27139-97-5
Beilstein No: 5^4, 831
Name: Benzene, 2-bromo-4-chloro-1-methyl-
Synonym: Toluene, 2-bromo-4-chloro-
Line Form: $2\text{-Br-}4\text{-Cl-C}_6H_3CH_3$
Mol Form: C_7H_6BrCl
M_r: 204.47
bp: $112\text{-}4^{12}$
MS: WILEY 1438 206(100) 204(77) 125(72) 171(42) 169(41) 89(40) 208(24) 127(24)
m/e: 203.93

HODOC No: 02602
CAS No: 54932-72-8
Beilstein No: 5^4, 832
Name: Benzene, 4-bromo-1-chloro-2-methyl-
Synonym: Toluene, 5-bromo-2-chloro-
Line Form: $5\text{-Br-}2\text{-ClC}_6H_3CH_3$
Mol Form: C_7H_6BrCl
M_r: 204.47
bp: $127\text{-}9^{45}$
MS: WILEY 1483 206(100) 204(77) 125(49) 169(29) 171(29) 89(27) 208(16) 63(16)
m/e: 203.93

HODOC No: 02603
CAS No: 29682-39-1
Beilstein No: 5^{1}, 132
Name: Benzene, 1-bromo-2-chloro-4-nitro-
Synonym: 3-Chloro-4-bromonitrobenzene
Line Form: (4-NO_2-2-Cl)-C_6H_3Br
Mol Form: $C_6H_3BrClNO_2$
M_r: 235.45
mp: 62
IR: **SADG 19301** 3100 1600 1570 1530 1450 1370 1340 1280 1140 1020 890 740 660 500 430
UV: **SAD 17924** 273(10700) 210 MeOH
H NMR: **SAD 10022** 7.9 8.1 8.4 $CDCl_3$
m/e: 234.90

HODOC No: 02604
CAS No: 41513-04-6
Beilstein No: 5^{4}, 731
Name: Benzene, 1-bromo-4-chloro-2-nitro-
Synonym: 1-Bromo-4-chloro-2-nitrobenzene
Line Form: (2-NO_2-4-Cl)-C_6H_3Br
Mol Form: $C_6H_3BrClNO_2$
M_r: 235.45
mp: 67-9
***n*:** 1.5832^{70}
IR: **SADG 20795** 3080 1560 1530 1450 1350 1240 1150 1090 1030 870 820 750 740 600 500 470
UV: **SAD 18774** 306(1230) 220(17700) MeOH
H NMR: **SAD 11352** 7.4 7.7 7.8 $CDCl_3$
m/e: 234.90

HODOC No: 02605
CAS No: 16588-24-2
Name: Benzene, 4-bromo-1-chloro-2-nitro-
Synonym: 1-Bromo-4-chloro-3-nitrobenzene
Line Form: (4-Cl-3-NO_2)-C_6H_3Br
Mol Form: $C_6H_3BrClNO_2$
M_r: 235.45
mp: 72
***n*:** 1.5839^{70}
m/e: 234.90

HODOC No: 02606
CAS No: 30203-84-0
Name: Benzene, 1-bromo-4-[(4-chlorophenyl)methyl]-
Synonym: Methane, (*p*-bromophenyl)(*p*-chlorophenyl)-
Mol Form: $C_{13}H_{10}BrCl$
M_r: 280.57
mp: 51-2
IR: **COB 7893** 3030 2910 1490 1440 1410 1090 1070 1010 840 800 790 500 480
H NMR: **SAD 13373** 3.9 6.9-7.3 7.4 $CDCl_3$
m/e: 279.97

HODOC No: 02607
CAS No: 445-01-2
Name: Benzene, 4-bromo-1-chloro-2-(trifluoromethyl)-
Synonym: Toluene, 5-bromo-2-chloro-α,α,α-trifluoro-
Mol Form: $C_7H_3BrClF_3$
M_r: 258.45
***n*:** $1.5062^{20/4}$
H NMR: **SAD 9132** 7.3 7.5 7.8 CCl_4
m/e: 257.91

HODOC No: 02608
CAS No: 56961-77-4
Beilstein No: 5, 209
Name: Benzene, 1-bromo-2,3-dichloro-
Synonym: 1-Bromo-2,3-dichlorobenzene
Line Form: 2,3-$Cl_2C_6H_3Br$
Mol Form: $C_6H_3BrCl_2$
M_r: 224.89
bp: 243^{705}
mp: 60
***s*:** Bz 4; Et_2O 4; Chl 4
m/e: 223.88

HODOC No: 02609
CAS No: 1193-72-2
Beilstein No: 5^{3}, 564
Name: Benzene, 1-bromo-2,4-dichloro-
Synonym: 1-Bromo-2,4-dichlorobenzene
Line Form: 2,4-$Cl_2C_6H_3Br$
Mol Form: $C_6H_3BrCl_2$
M_r: 224.89
bp: 235, 111^{21}
mp: 25
***s*:** Bz 4; Et_2O 4; Chl 4
m/e: 223.88

HODOC No: 02610
CAS No: 19752-55-7
Beilstein No: 5^2, 162
Name: Benzene, 1-bromo-3,5-dichloro-
Synonym: Bromo-3,5-dichlorobenzene
Line Form: 3,5-$Cl_2C_6H_3Br$
Mol Form: $C_6H_3BrCl_2$
M_r: 224.89
bp: 232^{757}
mp: 82-4
Color: pr (al)
***s*:** H_2O 1; al 3; eth 3; bz 4; chl 3
MS: WILEY 1592 226(100) 224(62) 145(47) 228(45) 74(43) 147(30) 75(28) 109(24)
m/e: 223.88

HODOC No: 02611
CAS No: 19393-92-1
Beilstein No: 5, 210
Name: Benzene, 2-bromo-1,3-dichloro-
Synonym: 1-Bromo-2,6-dichlorobenzene
Line Form: 2,6-$Cl_2C_6H_3Br$
Mol Form: $C_6H_3BrCl_2$
M_r: 224.89
bp: 242^{765}
mp: 65
Color: pr (al)
***s*:** Bz 4; Et_2O 4; Chl 4
m/e: 223.88

HODOC No: 02612
CAS No: 1435-50-3
Beilstein No: 5^3, 564
Name: Benzene, 2-bromo-1,4-dichloro-
Synonym: 1-Bromo-2,5-dichlorobenzene
Line Form: 2,5-$Cl_2C_6H_3Br$
Mol Form: $C_6H_3BrCl_2$
M_r: 224.89
bp: 225^{731}, 119^{20}
mp: 35
Color: pr or nd (al)
***s*:** Bz 4; Et_2O 4; EtOH 4; lig 4
IR: COB 986
UV: SAD 8491
MS: ALDN 183
m/e: 223.88

HODOC No: 02613
CAS No: 18282-59-2
Beilstein No: 5^3, 564
Name: Benzene, 4-bromo-1,2-dichloro-
Synonym: 1-Bromo-3,4-dichlorobenzene
Line Form: 3,4-$Cl_2C_6H_3Br$
Mol Form: $C_6H_3BrCl_2$
M_r: 224.89
bp: 237, 124^{33}
mp: 25
Color: pr
***s*:** H_2O 1; al 2; eth 4; bz 4; chl 4
IR: μCOB 987 3130 1890 1820 1720 1560 1450 1370 1250 1140 1090 1080 1030 870 810 790 670
MS: ALDN 183 226(100) 224(62) 228(45) 145(33) 147(22) 109(15) 74(14) 75(9)
m/e: 223.88

HODOC No: 02614
CAS No: 2859-78-1
Beilstein No: 6^4, 5621
Name: Benzene, 4-bromo-1,2-dimethoxy-
Synonym: 1-Bromo-3,4-dimethoxybenzene
Line Form: 3,4-$(CH_3O)_2C_6H_3Br$
Mol Form: $C_8H_9BrO_2$
M_r: 216.05
mp: 255-6
***d*:** 1.702
***n*:** 1.5743^{20}
IR: ALD 2:562B
MS: ALD 4:92D
m/e: 215.98

HODOC No: 02615
CAS No: 576-23-8
Beilstein No: 5^4, 928
Name: Benzene, 1-bromo-2,3-dimethyl-
Synonym: *o*-Xylene, 3-bromo-
Line Form: 2,3-$(CH_3)_2C_6H_3Br$
Mol Form: C_8H_9Br
M_r: 184.06
bp: 214, 83^{11}
***d*:** $1.365^{20/4}$
***s*:** Ac 4; Bz 4; Et_2O 4
m/e: 183.99

HODOC No: 02616
CAS No: 583-70-0
Beilstein No: 5^4, 945
Name: Benzene, 1-bromo-2,4-dimethyl-
Synonym: *m*-Xylene, 4-bromo-
Line Form: 2,4-$(CH_3)_2C_6H_3Br$
Mol Form: C_8H_9Br
Mr: 184.06
bp: 205, 84^{13}
mp: 0
***n*:** 1.5501^{20}
***s*:** H_2O 1; al 4; eth 4; ace 4
IR: SADG 31320 3000 2940 2920 2850 2730 1590 1460 1370 1220 1150 1130 1020 910 860 790 710 680580 530 430
MS: WILEY 1253 105(100) 184(91) 186(89) 51(22) 77(22) 103(19) 79(18) 185(16)
m/e: 183.99

HODOC No: 02617
CAS No: 556-96-7
Beilstein No: 5^4, 945
Name: Benzene, 1-bromo-3,5-dimethyl-
Synonym: *m*-Xylene, 5-bromo-
Line Form: 3,5-$(CH_3)_2C_6H_3Br$
Mol Form: C_8H_9Br
Mr: 184.06
bp: 204, 83-9^{12}
***d*:** 1.362^{20}
***n*:** 1.5462^{22}
***s*:** eth 4; ace 3; bz 3
UV: ZPCHA 25B-223-34 274(204) 268(240) 262(182) hx
MS: WILEY 1254 ini 105(100) 184(82) 186(79) 51(21) 77(20) 103(19) 79(17) 185(14)
m/e: 183.99

HODOC No: 02618
CAS No: 576-22-7
Beilstein No: 5^4, 945
Name: Benzene, 2-bromo-1,3-dimethyl-
Synonym: *m*-Xylene, 2-bromo-
Line Form: 2,6-$(CH_3)_2C_6H_3Br$
Mol Form: C_8H_9Br
Mr: 184.06
bp: 203-4, 98-9^{20}
***n*:** 1.5552^{20}
***s*:** eth 4; ace 3; bz 3
IR: SADG 846 3060 2960 1580 1460 1440 1380 1030 770 700 570
UV: SAD 8492 270 264 MeOH
MS: WILEY 1254 ini 105(100) 184(82) 186(79) 51(21) 77(20) 103(19) 79(17) 185(14)
m/e: 183.99

HODOC No: 02619
CAS No: 553-94-6
Beilstein No: 5^4, 969
Name: Benzene, 2-bromo-1,4-dimethyl-
Synonym: *p*-Xylene, 2-bromo-
Line Form: 2,5-$(CH_3)_2C_6H_3Br$
Mol Form: C_8H_9Br
Mr: 184.06
bp: 199-200, 88-9^{13}
mp: 9
***d*:** 1.3582^{18}
***n*:** 1.5514^{18}
Color: lf or pl
***s*:** H_2O 1; al 4; bz 3
IR: µSADP 9736 2940 1610 1560 1490 1430 1370 1280 1210 1150 1040 980 870 860 800 730 690 670
UV: SAD 20875 278(658) 271(545) 264(424) MeOH
H NMR: SAD 18739 2.2 2.3 6.9 7.0 7.3 CCl_4
MS: WILEY 1255 105(100) 184(87) 186(86) 51(24) 77(20) 103(18) 79(16) 185(16)
m/e: 183.99

HODOC No: 02620
CAS No: 583-71-1
Beilstein No: 5^4, 928
Name: Benzene, 4-bromo-1,2-dimethyl-
Synonym: *o*-Xylene, 4-bromo-
Line Form: $3,4\text{-}(CH_3)_2C_6H_3Br$
Mol Form: C_8H_9Br
Mr: 184.06
bp: 214-5
mp: -0.2
***d*:** $1.3708^{20/4}$
***n*:** 1.5530^{20}
***s*:** H_2O 1; al 4; eth 4
IR: **µSADP 20131** 2940 1590 1490 1450 1390 1280 1180 1140 1120 1090 1010 990 860 830 810 770 750 690
UV: **SAD 6710** 279 271 MeOH
MS: **WILEY 1254** 105(100) 184(94) 186(92) 51(22) 77(20) 79(16) 185(16) 103(16)
m/e: 183.99

HODOC No: 02621
CAS No: 3972-65-4
Beilstein No: 5^4, 1050
Name: Benzene, 1-bromo-4-(1,1-dimethylethyl)-
Synonym: Benzene, 1-bromo-4-*tert*-butyl-
Line Form: $4\text{-}Br\text{-}C_6H_4\text{-}C(CH_3)_3$
Mol Form: $C_{10}H_{13}Br$
Mr: 212.11
bp: 231-2, 103^{10}
mp: 19
***d*:** $1.2286^{20/4}$
***n*:** 1.5436^{20}
***s*:** H_2O 1; eth 3; bz 3; chl 3
IR: **µSADP 2107** 2940 1640 1610 1490 1390 1370 1270 1210 1180 1110 1080 1010 920 820 730
m/e: 212.02

HODOC No: 02622
CAS No: 19613-76-4
Beilstein No: 5^4, 749
Name: Benzene, 1-bromo-2,3-dinitro-
Synonym: 1-Bromo-2,3-dinitrobenzene
Line Form: $2,3\text{-}(NO_2)_2C_6H_3Br$
Mol Form: $C_6H_3BrN_2O_4$
Mr: 246.00
bp: 320
mp: 101-2
Color: pa ye pl (al)
***s*:** EtOH 4
m/e: 245.93

HODOC No: 02623
CAS No: 584-48-5
Beilstein No: 5^4, 749
Name: Benzene, 1-bromo-2,4-dinitro-
Synonym: 1-Bromo-2,4-dinitrobenzene
Line Form: $2,4\text{-}(NO_2)_2C_6H_3Br$
Mol Form: $C_6H_3BrN_2O_4$
Mr: 246.00
mp: 75
Color: ye nd (al)
***s*:** al 4
IR: **µCOB 2899** 3130 1610 1560 1470 1350 1250 1150 1030 920 900 850 830 760 740 730 680 650
UV: **OES 2-43** 265(10233) pH=7.4
H NMR: **SAD 11593** 8.1 8.4 8.7 $CDCl_3$
m/e: 245.93

HODOC No: 02624
CAS No: 4185-79-9
Beilstein No: 5^4, 749
Name: Benzene, 2-bromo-1,3-dinitro-
Synonym: 1-Bromo-2,6-dinitrobenzene
Line Form: $2,6\text{-}(NO_2)_2C_6H_3Br$
Mol Form: $C_6H_3BrN_2O_4$
Mr: 246.00
mp: 107
Color: ye pr (al)
***s*:** EtOH 4
m/e: 245.93

HODOC No: 02625
CAS No: 13296-65-6
Beilstein No: 5^4, 750
Name: Benzene, 2-bromo-1,4-dinitro-
Synonym: Benzene, 1-bromo-2,5-dinitro
Line Form: $2,5\text{-}(NO_2)_2C_6H_3Br$
Mol Form: $C_6H_3BrN_2O_4$
Mr: 246.00
mp: 70
Color: nd (al), pr (al-eth)
***s*:** Et_2O 4; EtOH 4
m/e: 245.93

HODOC No: 02626
CAS No: 610-38-8
Beilstein No: 5^3, 640
Name: Benzene, 4-bromo-1,2-dinitro-
Synonym: 3,4-Dinitrobromobenzene
Line Form: $3,4\text{-}(NO_2)_2C_6H_3Br$
Mol Form: $C_6H_3BrN_2O_4$
M_r: 246.00
mp: 34-5
Color: nd (al), pl (al-eth)
***s*:** Et_2O 4; EtOH 4
m/e: 245.93

HODOC No: 02627
CAS No: 98-81-7
Beilstein No: 5^4, 1349
Name: Benzene, (1-bromoethenyl)-
Synonym: Styrene, α-bromo-
Line Form: $C_6H_5CBr{=}CH_2$
Mol Form: C_8H_7Br
M_r: 182.04
bp: $86\text{-}7^{14}$, 71^3
mp: -44
***d*:** 1.4025^{23}
***n*:** 1.5881^{20}
m/e: 181.97

HODOC No: 02628
CAS No: 2039-88-5
Beilstein No: 5^4, 1349
Name: Benzene, 1-bromo-2-ethenyl-
Synonym: Styrene, *o*-bromo-
Line Form: $2\text{-}BrC_6H_4CH{=}CH_2$
Mol Form: C_8H_7Br
M_r: 182.04
bp: 209.2, 98^{20}
mp: -52.8 (fp)
***d*:** $1.4160^{20/4}$
***n*:** 1.5927^{20}
IR: COB 7026 3070 1835 1630 1590 1470 1435 1415 1275 1200 1120 1025 985 915 760 730 660
UV: OES 6-153 254(7244) EtOH
m/e: 181.97

HODOC No: 02629
CAS No: 2039-86-3
Beilstein No: 5^3, 1176
Name: Benzene, 1-bromo-3-ethenyl-
Synonym: Styrene, *m*-bromo-
Line Form: $3\text{-}BrC_6H_4CH{=}CH_2$
Mol Form: C_8H_7Br
M_r: 182.04
bp: $90\text{-}4^{20}$
***d*:** $1.4059^{20/4}$
***n*:** 1.5933^{20}
UV: OES 8-85 248(13804) MeOH
m/e: 181.97

HODOC No: 02630
CAS No: 2039-82-9
Beilstein No: 5^4, 1349
Name: Benzene, 1-bromo-4-ethenyl-
Synonym: Styrene, *p*-bromo-
Line Form: $4\text{-}BrC_6H_4CH{=}CH_2$
Mol Form: C_8H_7Br
M_r: 182.04
bp: 212, 109^{30}
mp: 7.7 (fp)
***d*:** 1.3984^{20}
***n*:** 1.5947^{20}
***s*:** H_2O 1; chl 4; HOAc 3
IR: μCOB 157 3030 1610 1590 1470 1390 1100 1060 1000 980 910 830 780 730
UV: SAD 6702 296 287 254 216 205 MeOH
m/e: 181.97

HODOC No: 02631
CAS No: 588-72-7
Beilstein No: 5^4, 1346
Name: Benzene, (2-bromoethenyl)-, (E)-
Synonym: Styrene, β-bromo-, (E)-
Line Form: $C_6H_5CH{=}CHBr$
Structure:

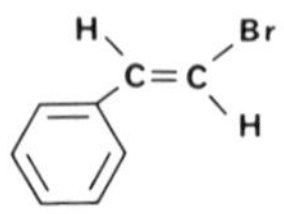

Mol Form: C_8H_7Br
Mr: 182.04
bp: 219d, 108^{20}
mp: 7
***d*:** 1.4269^{16}
***n*:** 1.6093^{20}
***s*:** H_2O 1; al 5; eth 5; chl 3
IR: SADG 8158 3080 3020 1610 1580 1450 1330 1280 1220 1180 1070 1030 940 840 770 730 690
UV: SAD 435 295 285 258 208 MeOH
H NMR: VAR 497 6.8 7.1 7.3 $CDCl_3$
MS: WILEY 1236 182(100) 184(98) 104(94) 77(46) 51(34) 103(16) 50(13) 183(9)
m/e: 181.97

HODOC No: 02632
CAS No: 588-73-8
Beilstein No: 5^4, 1349
Name: Benzene, (2-bromoethenyl)-, (Z)-
Synonym: Styrene, β-bromo-, (Z)-
Line Form: $C_6H_5CH{=}CHBr$
Structure:

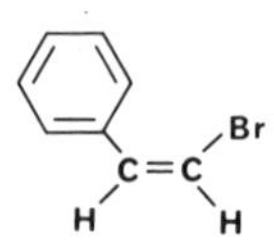

Mol Form: C_8H_7Br
Mr: 182.04
bp: $55\text{-}6^2$
mp: -7
***d*:** 1.4322^{10}
***n*:** 1.5990^{22}
IR: SADG 8158
UV: SAD 435
H NMR: VAR 497
MS: NBS 934
m/e: 181.97

HODOC No: 02633
CAS No: 13249-58-6
Beilstein No: 5^4, 2173
Name: Benzene, 1,1′-(bromoethenylidene)*bis*-
Synonym: Ethylene, 2-bromo-1,1-diphenyl-
Line Form: $(C_6H_5)_2C{=}CHBr$
Mol Form: $C_{14}H_{11}Br$
Mr: 258.13
bp: $175\text{-}6^{11.5}$
mp: 41-2
***s*:** al 2; eth 3; ace 3; CS_2 3
IR: COB 7614 3060 3040 1590 1500 1450 1320 1220 1080 930 810 770 740 700 610
UV: OES 5-399 264(16982) cy 259(15488) MeOH 259(12882) EtOH
m/e: 258.00

HODOC No: 02634
CAS No: 1607-57-4
Beilstein No: 5^4, 2577
Name: Benzene, 1,1′,1″-(1-bromo-1-ethenyl-2-ylidene)*tris*-
Synonym: Ethylene, bromotriphenyl-
Line Form: $C_6H_5CBr{=}C(C_6H_5)_2$
Mol Form: $C_{20}H_{15}Br$
Mr: 334.22
mp: 116-7
Color: nd (HOAc)
***s*:** H_2O 1; chl 3; HOAc 3
IR: SADG 12456 3080 3060 1600 1490 1440 1150 1070 950 840 760 740 690 590
UV: SAD 15235 219(7980) 229(19400) MeOH
H NMR: SAD 5738 6.9-7.5 $CDCl_3$
m/e: 334.04

HODOC No: 02635
CAS No: 583-19-7
Beilstein No: 6^4, 1035
Name: Benzene, 1-bromo-2-ethoxy-
Synonym: Phenetole, *o*-bromo-
Line Form: $2\text{-}(C_2H_5O)C_6H_4Br$
Mol Form: C_8H_9BrO
Mr: 200.05
bp: 222-6
***s*:** Et_2O 4; EtOH 4
m/e: 199.98

HODOC No: 02636
CAS No: 2655-84-7
Beilstein No: 6^3, 739
Name: Benzene, 1-bromo-3-ethoxy-
Synonym: Phenetole, *m*-bromo-
Line Form: 3-(C_2H_5O)C_6H_4Br
Mol Form: C_8H_9BrO
Mr: 200.05
bp: 222
***s*:** H_2O 1; al 4; eth 4
MS: AMS 1-313 172(100) 174(98) 65(57) 93(46) 63(44) 200(42) 29(42) 202(41)
m/e: 199.98

HODOC No: 02637
CAS No: 588-96-5
Beilstein No: 6^4, 1045
Name: Benzene, 1-bromo-4-ethoxy-
Synonym: Phenetole, *p*-bromo-
Line Form: 4-(C_2H_5O)C_6H_4Br
Mol Form: C_8H_9BrO
Mr: 200.05
bp: 230-2, 109^{17}
mp: 4
***d*:** $1.4071^{25/4}$
***n*:** 1.5517^{20}
***s*:** H_2O 1; al 4; eth 4; chl 3
IR: SADG 317 3030 2941 1587 1471 1370 1266 1235 1163 1099 1053 1000 952 826 645
UV: SAD 607 288(1080) 281(1330) 228(11200) MeOH
H NMR: VAR 198 1.4 3.9 6.7 7.3 $CDCl_3$
MS: WILEY 1411 172(100) 174(98) 200(42) 202(41) 65(17) 93(16) 173(8) 63(8)
m/e: 199.98

HODOC No: 02638
CAS No: 589-10-6
Beilstein No: 6^4, 556
Name: Benzene, (2-bromoethoxy)-
Synonym: Phenetole, β-bromo-
Line Form: $C_6H_5OCH_2CH_2Br$
Mol Form: C_8H_9BrO
Mr: 200.05
bp: 240-50d, $125\text{-}30^{20}$
mp: 39
***s*:** H_2O 1; al 4; eth 4
IR: μCOB 5738 3030 2940 1610 1590 1490 1430 1300 1250 1180 1090 1040 1020 880 750 690
H NMR: VAR 199 3.8 4.2 $CDCl_3$
MS: WILEY 1411 107(100) 109(88) 94(77) 200(51) 202(50) 77(25) 27(24) 65(23)
m/e: 199.98

HODOC No: 02639
CAS No: 38661-81-3
Beilstein No: 5^4, 906
Name: Benzene, (1-bromoethyl)-, (±)-
Synonym: (±)-(1-Bromoethyl)benzene
Line Form: $C_6H_5CHBrCH_3$
Mol Form: C_8H_9Br
Mr: 184.06
bp: 202-3, 85^{13}
***d*:** $1.3605^{20/4}$
***n*:** 1.5612^{20}
***s*:** al 3; eth 3; bz 3; ctc 2
IR: μCOB 2362 3030 1590 1490 1450 1370 1220 1180 1040 1020 960 910 760 690
UV: SAD 1279 225(5360) MeOH
H NMR: SAD 223 2.0 5.1 7.3 CCl_4
MS: WILEY 1253 ini 105(100) 104(16) 103(14) 77(13) 79(12) 51(12) 106(9) 184(2)
m/e: 183.99

HODOC No: 02640
CAS No: 585-71-7
Beilstein No: 5^4, 906
Name: Benzene, (1-bromoethyl)-
Synonym: (α-Bromoethyl)benzene
Line Form: $C_6H_5CHBr\text{-}CH_3$
Mol Form: C_8H_9Br
Mr: 184.06
bp: $217\text{-}8^{734}$, 92^{11}
***d*:** $1.3535^{25/4}$
***n*:** 1.5543^{25}
MS: NBS 962
m/e: 183.99

HODOC No: 02641
CAS No: 1459-14-9
Beilstein No: 5^4, 906
Name: Benzene, (1-bromoethyl)-, (R)-
Synonym: (R)-(1-Bromoethyl)benzene
Line Form: C_6H_5-$CHBrCH_3$
Mol Form: C_8H_9Br
M_r: 184.06
bp: 203, 86-8^{15}
***d*:** 1.3108^{23}
***n*:** 1.5612^{20}
α: $[\alpha]^{14/D}$ +1.5
MS: WILEY 1253
m/e: 183.99

HODOC No: 02642
CAS No: 1973-22-4
Beilstein No: 5^4, 906
Name: Benzene, 1-bromo-2-ethyl-
Synonym: *o*-Bromoethylbenzene
Line Form: 2-Br-$C_6H_4CH_2CH_3$
Mol Form: C_8H_9Br
M_r: 184.06
bp: 199.3, 97.6^{30}
mp: -67.9, (fp)
***d*:** 1.3548^{20}
***n*:** 1.5472^{20}
***s*:** Ac 4; Bz 4; Et_2O 4; EtOH 4
IR: μCOB 158 3030 2940 1470 1450 1370 1040 1020 860 790 750 650
H NMR: SAD 19453 1.2 2.2 6.7-7.3 7.4 CCl_4
MS: WILEY 1254 169(100) 171(98) 105(78) 184(65) 186(64) 51(26) 77(24) 90(18)
m/e: 183.99

HODOC No: 02643
CAS No: 2725-82-8
Beilstein No: 5^4, 906
Name: Benzene, 1-bromo-3-ethyl-
Synonym: *m*-Bromoethylbenzene
Line Form: 3-$BrC_6H_4CH_2CH_3$
Mol Form: C_8H_9Br
M_r: 184.06
bp: 200-8, 85-6^{20}
***d*:** $1.3493^{20/4}$
***n*:** 1.5465^{20}
m/e: 183.99

HODOC No: 02644
CAS No: 1585-07-5
Beilstein No: 5^4, 906
Name: Benzene, 1-bromo-4-ethyl-
Synonym: *p*-Bromoethylbenzene
Line Form: 4-$BrC_6H_4C_2H_5$
Mol Form: C_8H_9Br
M_r: 184.06
bp: 204, 103.6^{30}
mp: -43.5 (fp)
***d*:** 1.3423^{20}
***n*:** 1.5445^{20}
***s*:** Ac 4; Bz 4; Et_2O 4; EtOH 4
IR: μSADP 20130 3030 2940 2860 1590 1560 1490 1450 1390 1370 1320 1210 1180 1110 1060 1010 960 820 780 750 710 690
UV: SAD 6709 276 268 261 255 220 MeOH
MS: WILEY 1252 169(100) 171(98) 184(58) 186(57) 105(56) 90(18) 77(18) 185(14)
m/e: 183.99

HODOC No: 02645
CAS No: 103-63-9
Beilstein No: 5^4, 907
Name: Benzene, (2-bromoethyl)-
Synonym: β-Bromoethylbenzene
Line Form: $C_6H_5CH_2CH_2Br$
Mol Form: C_8H_9Br
M_r: 184.06
bp: 217-8^{734}
***d*:** $1.3587^{20/4}$
***n*:** 1.5372^{20}
***s*:** H_2O 1; eth 3; bz 3; ctc 2
IR: μCOB 1004 2940 1890 1610 1490 1080 1010 800
UV: SAD 1289 264(281) 258(378) 252(385) 247(366) MeOH
H NMR: SAD 6030 1.2 2.7 6.8-7.2 7.5 CCl_4
MS: WILEY 1253 91(100) 105(48) 184(28) 186(27) 51(13) 77(13) 79(8) 103(8)
m/e: 183.99

HODOC No: 02646
CAS No: 6529-53-9
Name: Benzene, 1-(2-bromoethyl)-4-chloro-
Synonym: 1-Bromo-2-(4-chlorophenyl)ethane
Mol Form: C_8H_8BrCl
M_r: 218.50
IR: μCOB 4060 3030 1610 1520 1470 1450 1430 1270 1220 1100 890 810 730 670
m/e: 217.95

HODOC No: 02647
CAS No: 932-87-6
Beilstein No: 5^4, 1529
Name: Benzene, (bromoethynyl)-
Synonym: (Bromoethynyl)benzene
Line Form: $BrC \equiv CC_6H_5$
Mol Form: C_8H_5Br
M_r: 180.02
bp: $88\text{-}90^{13}$, $68\text{-}9^{3}$
***d*:** $1.456^{24.2}$
Raman: **API 141** 3062 2201 1597 1229 1178 999 837 622 356 267
m/e: 179.96

HODOC No: 02648
CAS No: 766-46-1
Beilstein No: 5^3, 1351
Name: Benzene, 1-bromo-2-ethynyl-
Synonym: *o*-Bromophenylacetylene
Line Form: $2\text{-}BrC_6H_4\text{-}C \equiv CH$
Mol Form: C_8H_5Br
M_r: 180.02
bp: $92\text{-}3^{20}$
***d*:** 1.4434^{25}
***n*:** 1.5962^{25}
m/e: 179.96

HODOC No: 02649
CAS No: 766-81-4
Beilstein No: 5^3, 1351
Name: Benzene, 1-bromo-3-ethynyl-
Synonym: *m*-Bromophenylacetylene
Line Form: $3\text{-}BrC_6H_4\text{-}C \equiv CH$
Mol Form: C_8H_5Br
M_r: 180.02
bp: $85\text{-}6^{16}$
***d*:** 1.4466^{25}
***n*:** 1.5896^{25}
m/e: 179.96

HODOC No: 02650
CAS No: 766-96-1
Beilstein No: 5^4, 1529
Name: Benzene, 1-bromo-4-ethynyl-
Synonym: 1-Bromo-4-ethynylbenzene
Line Form: $4\text{-}BrC_6H_4\text{-}C \equiv CH$
Mol Form: C_8H_5Br
M_r: 180.02
bp: $88\text{-}90^{16}$
mp: 64-5
***s*:** chl 3
IR: **µSADP 7604** 2940 2325 1640 1585 1495 1470 1410 1390 1265 1235 1175 1100 1065 1055 1020 1010 820 805 770 695
H NMR: **SAD 11266** 3.1 7.2-7.6 $CDCl_3$
m/e: 179.96

HODOC No: 02651
CAS No: 1072-85-1
Beilstein No: 5^4, 678
Name: Benzene, 1-bromo-2-fluoro-
Synonym: *o*-Bromofluorobenzene
Line Form: $2\text{-}F\text{-}C_6H_4Br$
Mol Form: C_6H_4BrF
M_r: 173.99
bp: 151-2, 57^{22}
***n*:** 1.5337^{20}
IR: **SADG 33157** 3060 1580 1490 1470 1440 1370 1260 1230 1150 1110 1040 1020 930 820 750 650 540460 430
Raman: **SAD 1154** 3170 3070 1590 1480 1450 1390 1280 1260 1230 1150 1110 1060 1030 870 820 750 700 650 550 470 300 260 160
UV: **SAD 6685** 271 264 258 210 MeOH
H NMR: **SAD 2269** 6.7-7.6 CCl_4
MS: **NBS 807**
m/e: 173.95

HODOC No: 02652
CAS No: 1073-06-9
Beilstein No: 5^4, 678
Name: Benzene, 1-bromo-3-fluoro-
Synonym: *m*-Bromofluorobenzene
Line Form: $3\text{-}FC_6H_4Br$
Mol Form: C_6H_4BrF
Mr: 173.99
bp: 149-51
***n*:** $1.5257^{20/D}$
***s*:** ctc 3
IR: μCOB 4092 3030 1590 1470 1430 1270 1220 1050 1000 860 780 670
Raman: SAD 1155 3180 3080 1590 1480 1430 1350 1290 1270 1220 1150 1110 1080 1060 1000 860 780 670 620 560 530 440 350 310 250 210 180
UV: SAD 6165 273 266 259 213 cyhex
H NMR: SAD 5001 7.0 7.1-7.4 CCl_4
MS: NBS 807
m/e: 173.95

HODOC No: 02653
CAS No: 460-00-4
Beilstein No: 5^4, 678
Name: Benzene, 1-bromo-4-fluoro-
Synonym: *p*-Bromofluorobenzene
Line Form: $4\text{-}FC_6H_4Br$
Mol Form: C_6H_4BrF
Mr: 173.99
bp: 151-2
mp: -17.4
***d*:** 1.593^{15}
***n*:** 1.5310^{15}
***s*:** H_2O 1; al 3; eth 3; chl 3
IR: μCOB 5656 3030 1890 1590 1490 1410 1390 1220 1150 1080 1010 820
Raman: SAD 1156 3160 3080 1660 1600 1590 1490 1380 1290 1230 1160 1070 1030 1010 830 810 690 660 630 600 580 550 500 480 420 330 300 230
UV: OES 5-46 278(891) 271(1047) 265(794) 21(7079) cyhex
H NMR: VAR 121 6.9 7.4 $CDCl_3$
MS: NBS 807
m/e: 173.95

HODOC No: 02654
CAS No: 30752-19-3
Beilstein No: 6^4, 1047
Name: Benzene, 1-bromo-4-(hexyloxy)-
Synonym: Ether, *p*-bromophenyl hexyl
Line Form: $(4\text{-}BrC_6H_4)\text{-}O\text{-}(CH_2)_5CH_3$
Mol Form: $C_{12}H_{17}BrO$
Mr: 256.16
bp: $156\text{-}7^{13}$
***d*:** 1.2306^{20}
***n*:** 1.5262^{20}
IR: μCOB 3485 2860 2780 1590 1490 1470 1280 1240 1160 1060 1020 1000 820
m/e: 256.05

HODOC No: 02655
CAS No: 583-55-1
Beilstein No: 5^4, 648
Name: Benzene, 1-bromo-2-iodo-
Synonym: *o*-Bromoiodobenzene
Line Form: $2\text{-}IC_6H_4Br$
Mol Form: C_6H_4BrI
Mr: 281.90
bp: 257^{754}, $120\text{-}1^{15}$
mp: 9-10
***d*:** $2.2571^{25/4}$
***n*:** 1.6618^{25}
***s*:** H_2O 1; al 2; ace 3; HOAc 2
IR: μCOB 5655 3030 1430 1250 1100 1000 750 690
UV: SAD 1590 263 239 cyhex
H NMR: SAD 6466 6.9 7.1 7.7 7.8 CCl_4
MS: NBS 2061
m/e: 281.85

HODOC No: 02656
CAS No: 591-18-4
Beilstein No: 5^4, 698
Name: Benzene, 1-bromo-3-iodo-
Synonym: *m*-Bromoiodobenzene
Line Form: $3\text{-}IC_6H_4Br$
Mol Form: C_6H_4BrI
Mr: 281.90
bp: 252^{754}, 120^{18}
mp: -9.3
***s*:** H_2O 1; al 2; HOAc 2
UV: OES 5-46 269(575) 263(603) 253(646) 235(10000) 230(8913) 210(25119) cyhex
H NMR: SAD 19894 6.9 7.4 7.6 7.8 $CDCl_3$
C NMR: .
MS: NBS 2061
m/e: 281.85

HODOC No: 02657
CAS No: 589-87-7
Beilstein No: 5^4, 698
Name: Benzene, 1-bromo-4-iodo-
Synonym: *p*-Bromoiodobenzene
Line Form: $4\text{-}IC_6H_4Br$
Mol Form: C_6H_4BrI
M_r: 281.90
bp: 252^{754}
mp: 92
Color: pr or pl (eth-al)
***s*:** H_2O 1; al 2; eth 3; chl 2
IR: µCOB 5910 3130 1890 1560 1470 1410 1390 1250 1100 1060 1000 800
UV: SAD 1764 239(18700) MeOH
H NMR: SAD 6488 7.2 7.5 $CDCl_3$
MS: NBS 2061
m/e: 281.85

HODOC No: 02658
CAS No: 97456-81-0
Beilstein No: 5, 315
Name: Benzene, 1-bromo-3-iodo-2-methyl-
Synonym: Toluene, 2-bromo-6-iodo
Line Form: $2\text{-}Br\text{-}6\text{-}I\text{-}C_6H_3CH_3$
Mol Form: C_7H_6BrI
M_r: 295.93
bp: $135\text{-}40^{15}$
***d*:** $2.044^{20/20}$
m/e: 295.87

HODOC No: 02659
CAS No: 71838-16-9
Beilstein No: 5^3, 727
Name: Benzene, 2-bromo-1-iodo-4-methyl-
Synonym: Toluene, 3-bromo-4-iodo-
Line Form: $3\text{-}Br\text{-}4\text{-}IC_6H_3CH_3$
Mol Form: C_7H_6BrI
M_r: 295.93
bp: 265-7
***d*:** 2.044^{20}
m/e: 295.87

HODOC No: 02660
CAS No: 32337-96-5
Name: Benzene, 1-bromo-2-iodo-3-nitro-
Synonym: 3-Bromo-2-iodonitrobenzene
Mol Form: $C_6H_3BrINO_2$
M_r: 326.90
***s*:** chl 2
IR: SADG 20629 3060 1530 1410 1370 1260 1190 860 790 740 700
H NMR: SAD 10642 7.3 7.5 7.8 $CDCl_3$
m/e: 326.84

HODOC No: 02661
CAS No: 2493-02-9
Beilstein No: 12^1, 321
Name: Benzene, 1-bromo-4-isocyanato-
Synonym: Isocyanic acid, *p*-bromophenyl ester
Line Form: $4\text{-}BrC_6H_4NCO$
Mol Form: C_7H_4BrNO
M_r: 197.01
bp: 226, 158^{14}
Color: nd
***s*:** Et_2O 4
m/e: 196.95

HODOC No: 02662
CAS No: 1985-12-2
Beilstein No: 12^3, 1463
Name: Benzene, 1-bromo-4-isothiocyanato-
Synonym: Isothiocyanic acid, *p*-bromophenyl ester
Line Form: $4\text{-}BrC_6H_4NCS$
Mol Form: C_7H_4BrNS
M_r: 213.07
mp: 60-1
Color: nd
***s*:** EtOH 4
m/e: 212.92

HODOC No: 02663
CAS No: 578-57-4
Beilstein No: 6^4, 1037
Name: Benzene, 1-bromo-2-methoxy-
Synonym: Anisole, *o*-bromo-
Line Form: $2\text{-}BrC_6H_4OCH_3$
Mol Form: C_7H_7BrO
M_r: 186.03
bp: 216, 94^{10}
mp: 2.5
***d*:** $1.5018^{20/4}$
***n*:** 1.5727^{20}
***s*:** H_2O 1; al 4; eth 4
IR: COB 7958 3070 3010 2940 2840 1590 1485 1470 1450 1435 1300 1270 1240 1160 1120 1055 1030 920 795 750 660 570
UV: OES 5-92 289(3020) 284(2818) 277(2754) cyhex
H NMR: SAD 12532 3.8 6.8 6.9 7.2 7.5 $CDCl_3$
MS: WILEY 1277 186(100) 188(98) 143(42) 145(41) 171(34) 173(34) 77(15) 63(15)
m/e: 185.97

HODOC No: 02664
CAS No: 2398-37-0
Beilstein No: 6^4, 1043
Name: Benzene, 1-bromo-3-methoxy-
Synonym: Anisole, *m*-bromo-
Line Form: $3\text{-}BrC_6H_4OCH_3$
Mol Form: C_7H_7BrO
M_r: 186.03
bp: $210\text{-}1^{752}$, 105^{16}
***n*:** 1.5635^{20}
***s*:** H_2O 1; al 3; eth 3; bz 3; CS_2 3
IR: SADG 23441 3060 3000 2950 2820 1580 1570 1470 1440 1420 1280 1240 1220 1180 1150 1080 1060 1040 990 830 820 760 680 550 430
UV: OES 5-92 283(2291) 276(2344) cyhex
H NMR: SAD 13769 3.7 6.8 6.9-7.2 $CDCl_3$
m/e: 185.97

HODOC No: 02665
CAS No: 104-92-7
Beilstein No: 6^4, 1044
Name: Benzene, 1-bromo-4-methoxy-
Synonym: Anisole, *p*-bromo-
Line Form: $4\text{-}BrC_6H_4OCH_3$
Mol Form: C_7H_7BrO
M_r: 186.03
bp: 215, 100^{16}
mp: 13-4
***d*:** $1.4564^{20/4}$
***n*:** 1.5642^{20}
***s*:** H_2O 2; al 4; eth 4; ctc 3; chl 4
IR: µCOB 3237 2940 1590 1490 1280 1250 1180 1080 1040 1000 820
UV: SAD 605 257 MeOH
H NMR: SAD 141 3.7 6.7 7.3 CCl_4
MS: NBS 4356
m/e: 185.97

HODOC No: 02666
CAS No: 100-39-0
Beilstein No: 5^4, 829
Name: Benzene, (bromomethyl)-
Synonym: Toluene, α-bromo-
Line Form: $C_6H_5CH_2Br$
Mol Form: C_7H_7Br
M_r: 170.03
bp: 201, 114^{15}
mp: -3
***d*:** 1.4380^{25}
***n*:** 1.5752^{20}
Color: pr
***s*:** H_2O 1; al 5; eth 5,ctc 3
IR: SADG 327 3020 2920 1700 1500 1450 1220 1200 1070 1030 810 750 690
UV: SAD 660 225(7060) MeOH
H NMR: SAD 154 4.3 7.2 CCl_4
MS: WILEY 1117 91(100) 92(36) 39(25) 65(17) 51(16) 63(13) 50(12) 170(2)
m/e: 169.97

HODOC No: 02667
CAS No: 95-46-5
Beilstein No: 5^4, 825
Name: Benzene, 1-bromo-2-methyl-
Synonym: Toluene, *o*-bromo-
Line Form: 2-$BrC_6H_4CH_3$
Mol Form: C_7H_7Br
M_r: 170.03
bp: 181.7, 59^{10}
mp: -27.8
***d*:** $1.4232^{20/4}$
***n*:** 1.5565^{20}
***s*:** H_2O 1; al 4; eth 4,bz 4; ctc 5
IR: COB 4314 3070 2920 1670 1480 1470 1460 1380 1270 1030 1000 800 740 660
UV: SAD 738 287(39) 273(257) 266(305) 259(250) 252 MeOH
H NMR: SAD 161 2.4 7.1 7.5 CCl_4
MS: WILEY 1119 91(100) 170(90) 172(88) 90(19) 89(18) 65(15) 171(14) 63(10)
m/e: 169.97

HODOC No: 02668
CAS No: 591-17-3
Beilstein No: 5^4, 827
Name: Benzene, 1-bromo-3-methyl-
Synonym: Toluene, *m*-bromo-
Line Form: 3-$BrC_6H_4CH_3$
Mol Form: C_7H_7Br
M_r: 170.03
bp: 183.7
mp: -39.8
***d*:** $1.4099^{20/4}$
***n*:** 1.5510^{20}
***s*:** H_2O 1; al 3; eth 5; ace 3; ctc 2; chl 3
IR: SADG 8259 3050 2920 1600 1570 1470 1210 1070 1000 870 830 770 680 670
Raman: DFB 82 3140 3060 2920 2860 2730 1600 1570 1380 1270 1210 1160 1090 1070 1030 1000 830 670 520 380 310 230 200 170
UV: SAD 739 274(248) 267(303) 260(237) MeOH
H NMR: SAD 162 2.3 7.0-7.3 CCl_4
MS: WILEY 1118 91(100) 170(100) 172(98) 65(22) 171(15) 63(13) 89(12) 39(11)
m/e: 169.97

HODOC No: 02669
CAS No: 106-38-7
Beilstein No: 5^4, 827
Name: Benzene, 1-bromo-4-methyl-
Synonym: Toluene, *p*-bromo-
Line Form: 4-$BrC_6H_4CH_3$
Mol Form: C_7H_7Br
M_r: 170.03
bp: 184.3, 61.9^{10}
mp: 28.5
***d*:** $1.3995^{20/4}$
***n*:** 1.5477^{20}
Color: cr (al)
***s*:** H_2O 1; al 3; eth 3; ace 3; bz 3; ctc 2; chl 3
IR: COB 1005 3030 2920 1490 1110 1070 1010 800
Raman: SAD 349 3060 3020 2970 2920 2870 2730 2540 1590 1480 1450 1380 1300 1270 1210 1170 1070 1040 1010 960 820 790 690 660 630 590 580 540 480 410 370 290 210
UV: SAD 740 277(411) 269(494) 261(489) 220(10600) MeOH
H NMR: SAD 163 2.3 7.0 7.3 CCl_4
MS: WILEY 1118 91(100) 170(79) 172(77) 65(19) 171(17) 63(12) 89(11) 169(11)
m/e: 169.97

HODOC No: 02670
CAS No: 611-17-6
Beilstein No: 5^4, 833
Name: Benzene, 1-(bromomethyl)-2-chloro-
Synonym: Toluene, α-bromo-*o*-chloro-
Line Form: 2-$ClC_6H_4CH_2Br$
Mol Form: C_7H_6BrCl
M_r: 204.47
bp: 109^{10}
m/e: 203.93

HODOC No: 02671
CAS No: 766-80-3
Beilstein No: 5^4, 833
Name: Benzene, 1-(bromomethyl)-3-chloro-
Synonym: Toluene, α-bromo-*m*-chloro-
Line Form: 3-$ClC_6H_4CH_2Br$
Mol Form: C_7H_6BrCl
M_r: 204.47
bp: 109^{10}
***d*:** $1.5652^{25/25}$
m/e: 203.93

HODOC No: 02672
CAS No: 622-95-7
Beilstein No: 5^4, 833
Name: Benzene, 1-(bromomethyl)-4-chloro-
Synonym: Toluene, α-bromo-*p*-chloro-
Line Form: $4\text{-}ClC_6H_4CH_2Br$
Mol Form: C_7H_6BrCl
M$_r$: 204.47
mp: 51
Color: cr (al)
***s*:** Bz 4; Et_2O 4; EtOH 4; HOAc 4
m/e: 203.93

HODOC No: 02673
CAS No: 42533-63-1
Beilstein No: 5, 335
Name: Benzene, 1-(bromomethyl)-2-chloro-4-nitro-
Synonym: 2-Chloro-4-nitrobenzyl bromide
Line Form: $(2\text{-}Cl\text{-}4\text{-}NO_2)C_6H_3CH_2Br$
Mol Form: $C_7H_5BrClNO_2$
M$_r$: 249.47
mp: 50
IR: ALD 2:717B
m/e: 248.92

HODOC No: 02674
CAS No: 27129-86-8
Beilstein No: 5^4, 1027
Name: Benzene, 1-(bromomethyl)-3,5-dimethyl-
Synonym: Mesitylene, α-bromo-
Line Form: $3,5\text{-}(CH_3)_2C_6H_3CH_2Br$
Mol Form: $C_9H_{11}Br$
M$_r$: 198.08
bp: 118^{22}, $229\text{-}31^{740}$
mp: 40
Color: nd (eth)
***s*:** Bz 4; Et_2O 4; EtOH 4; Chl 4
m/e: 198.00

HODOC No: 02675
CAS No: 5411-53-0
Name: Benzene, 1-bromo-5-methyl-2,4-dinitro-
Synonym: 2,4-Dinitro-5-bromotoluene
Mol Form: $C_7H_5BrN_2O_4$
M$_r$: 260.02
mp: 114
***s*:** bz 2
IR: SADG 20311 2940 2850 1600 1580 1520 1460 1370 1340 1250 1200 1030 910 830 740 650 540 450 400
UV: SAD 18130 251(13200) 207 MeOH
H NMR: SAD 10357 2.7 7.8 8.5 $CDCl_3$
m/e: 259.94

HODOC No: 02676
CAS No: 776-74-9
Beilstein No: 5^4, 1850
Name: Benzene, 1,1′-(bromomethylene)*bis*-
Synonym: Methane, bromodiphenyl-
Line Form: $(C_6H_5)_2CHBr$
Mol Form: $C_{13}H_{11}Br$
M$_r$: 246.12
bp: 184^{20}, $150\text{-}5^2$
mp: 45
***s*:** al 3; bz 4; chl 3
IR: SADG 15344 3060 3020 1600 1490 1450 1180 1150 1080 1030 750 700
UV: SAD 528 264 258 253 MeOH
H NMR: VAR 606 6.3 7.2-7.6 $CDCl_3$
m/e: 246.00

HODOC No: 02677
CAS No: 7073-94-1
Beilstein No: 5, 395
Name: Benzene, 1-bromo-2-(1-methylethyl)-
Synonym: Cumene, *o*-bromo-
Line Form: $2\text{-}BrC_6H_4\text{-}CH(CH_3)_2$
Mol Form: $C_9H_{11}Br$
M$_r$: 198.08
bp: $205\text{-}7^{740}$
MS: WILEY 1387 183(100) 185(96) 104(71) 198(38) 200(37) 103(31) 77(22) 51(18)
m/e: 198.00

HODOC No: 02678
CAS No: 5433-01-2
Beilstein No: 5^4, 994
Name: Benzene, 1-bromo-3-(1-methylethyl)-
Synonym: Cumene, *m*-bromo-
Line Form: $3\text{-}BrC_6H_4CH(CH_3)_2$
Mol Form: $C_9H_{11}Br$
M_r: 198.08
bp: 208-10, 94-6^{20}
m/e: 198.00

HODOC No: 02679
CAS No: 586-61-8
Beilstein No: 5^2, 307
Name: Benzene, 1-bromo-4-(1-methylethyl)-
Synonym: Cumene, *p*-bromo-
Line Form: $4\text{-}BrC_6H_4\text{-}CH(CH_3)_2$
Mol Form: $C_9H_{11}Br$
M_r: 198.08
bp: 218.7, 97-8^5
mp: -22.5
***d*:** $1.3145^{20/4}$
***n*:** 1.5569^{20}
***s*:** H_2O 1; eth 3; bz 3; ctc 2; chl 3
IR: µSADP 21358 2940 1590 1490 1410 1080 1010 830 760 710
H NMR: SAD 9390 1.2 2.8 7.1 7.4 CCl_4
MS: WILEY 1387 183(100) 185(97) 104(54) 198(41) 200(36) 199(15) 77(15) 103(14)
m/e: 198.00

HODOC No: 02680
CAS No: 1459-00-3
Beilstein No: 5^4, 995
Name: Benzene, (2-bromo-1-methylethyl)-
Synonym: Cumene, β-bromo-
Line Form: $C_6H_5CH(CH_3)CH_2Br$
Mol Form: $C_9H_{11}Br$
M_r: 198.08
bp: 188-9, 117-8^{20}
m/e: 198.00

HODOC No: 02681
CAS No: 456-41-7
Beilstein No: 5^2, 238
Name: Benzene, 1-(bromomethyl)-3-fluoro-
Synonym: Toluene, α-bromo-*m*-fluoro-
Line Form: $3\text{-}FC_6H_4CH_2Br$
Mol Form: C_7H_6BrF
M_r: 188.02
bp: 88^{20}
***n*:** 1.5474^{20}
IR: ALD 2:530C
MS: ALD 4:49B
m/e: 187.96

HODOC No: 02682
CAS No: 596-43-0
Beilstein No: 5^4, 2500
Name: Benzene, 1,1′,1″-(bromomethylidyne)*tris*-
Synonym: Methane, bromotriphenyl-
Line Form: $(C_6H_5)_3CBr$
Mol Form: $C_{19}H_{15}Br$
M_r: 322.21
bp: 230^{15}
mp: 152-4
IR: ALD 3:615F
MS: NBS 2479; ALD 4:75D
m/e: 322.04

HODOC No: 02683
CAS No: 3913-23-3
Name: Benzene, 2-(bromomethyl)-1-methoxy-4-nitro-
Synonym: Anisole, 2-(bromomethyl)-4-nitro-
Line Form: $(5\text{-}NO_2\text{-}2\text{-}CH_3O)C_6H_3CH_2Br$
Mol Form: $C_8H_8BrNO_3$
M_r: 245.05
mp: 78-80
IR: ALD 2:721A
MS: ALD 5:166B
m/e: 244.97

HODOC No: 02684
CAS No: 89-92-9
Beilstein No: 5^4, 928
Name: Benzene, 1-(bromomethyl)-2-methyl-
Synonym: *o*-Xylene, α-bromo-
Line Form: $2\text{-}CH_3C_6H_4CH_2Br$
Mol Form: C_8H_9Br
M_r: 184.06
bp: $216\text{-}7^{742}$, 108^{16}
mp: 21
***d*:** 1.3811^{23}
***n*:** 1.5730^{20}
Color: pr
***s*:** H_2O 1; al 3; eth 3; ace 3; bz 3; os 3
IR: SADG 441 3020 2970 1600 1490 1460 1380 1230 1210 1190 1050 830 760 750 720 610 590 520 490
MS: WILEY 1253 105(100) 79(11) 77(11) 51(10) 106(10) 103(9) 184(8) 168(8)
m/e: 183.99

HODOC No: 02685
CAS No: 620-13-3
Beilstein No: 5^2, 293
Name: Benzene, 1-(bromomethyl)-3-methyl-
Synonym: *m*-Xylene, α-bromo-
Line Form: $3\text{-}CH_3C_6H_4CH_2Br$
Mol Form: C_8H_9Br
M_r: 184.06
bp: 212-3, 105^{13}
***d*:** 1.3711^{23}
***n*:** 1.5660^{20}
***s*:** H_2O 1; al 4; eth 4; os 3
IR: μSADP 18666 3030 2940 1610 1490 1450 1390 1270 1220 1090 890 790 740 700
UV: SAD 5997 232 cyhex
H NMR: SAD 17202 2.3 4.4 $CDCl_3$
MS: WILEY 1255 105(100) 184(12) 186(12) 77(11) 79(10) 106(9) 51(9) 103(8)
m/e: 183.99

HODOC No: 02686
CAS No: 104-81-4
Beilstein No: 5^3, 858
Name: Benzene, 1-(bromomethyl)-4-methyl-
Synonym: *p*-Xylene, α-bromo-
Line Form: $4\text{-}CH_3C_6H_4CH_2Br$
Mol Form: C_8H_9Br
M_r: 184.06
bp: $218\text{-}20^{740}$
mp: 35
***d*:** 1.324
Color: nd (al)
***s*:** H_2O 1; al 3; eth 4; chl 4
IR: SADG 8460 3020 2960 2920 1630 1530 1450 1390 1240 1210 1190 1110 850 820 750 730 610 500 470
UV: HBCP 236(5012) hx
MS: WILEY 1255 105(100) 77(10) 79(10) 106(9) 51(9) 103(8) 184(7) 186(7)
m/e: 183.99

HODOC No: 02687
CAS No: 2437-76-5
Beilstein No: 5^4, 1063
Name: Benzene, 2-bromo-1-methyl-4-(1-methylethyl)-
Synonym: *p*-Cymene, 2-bromo-
Line Form: $4\text{-}(CH_3)_2CH\text{-}2\text{-}BrC_6H_3CH_3$
Mol Form: $C_{10}H_{13}Br$
M_r: 212.11
bp: 234.3, 99^9
***d*:** $1.2689^{18/4}$
***n*:** 1.5360^{20}
***s*:** Et_2O 4; EtOH 4; Chl 4
m/e: 212.02

HODOC No: 02688
CAS No: 4478-10-8
Beilstein No: 5^3, 958
Name: Benzene, 2-bromo-4-methyl-1-(1-methylethyl)-
Synonym: *p*-Cymene, 3-bromo-
Line Form: $3\text{-}Br\text{-}4\text{-}(CH_3)_2CH\text{-}C_6H_3CH_3$
Mol Form: $C_{10}H_{13}Br$
M_r: 212.11
bp: $232\text{-}3^{741}$
MS: NBS 1338
m/e: 212.02

HODOC No: 02689
CAS No: 3958-60-9
Beilstein No: 5^3, 752
Name: Benzene, 1-(bromomethyl)-2-nitro-
Synonym: Toluene, α-bromo-*o*-nitro-
Line Form: $2\text{-}O_2NC_6H_4CH_2Br$
Mol Form: $C_7H_6BrNO_2$
Mr: 215.03
mp: 46-7
Color: pl (dil al)
***s*:** H_2O 1; al 3; eth 3; bz 3; chl 2; lig 3
IR: SADG 9758 3070 2860 1600 1570 1520 1470 1420 1330 1290 1210 1190 1150 1140 1100 1060 950 840 790 770 730 690 650
UV: SAD 14910 255(4860) 214(13900) MeOH
H NMR: SAD 4628 4.8 7.6 8.0 $CDCl_3$
m/e: 214.96

HODOC No: 02690
CAS No: 7149-70-4
Beilstein No: 5^4, 860
Name: Benzene, 1-bromo-2-methyl-4-nitro-
Synonym: 4-Bromo-3-methylnitrobenzene
Line Form: $2\text{-Br-}5\text{-}NO_2C_6H_3CH_3$
Mol Form: $C_7H_6BrNO_2$
Mr: 215.03
bp: $140\text{-}3^{17}$
mp: 78
Color: cr (al)
***s*:** Et_2O 4
m/e: 214.96

HODOC No: 02691
CAS No: 3958-57-4
Beilstein No: 5^4, 860
Name: Benzene, 1-(bromomethyl)-3-nitro-
Synonym: Toluene, α-bromo-*m*-nitro-
Line Form: $3\text{-}O_2NC_6H_4CH_2Br$
Mol Form: $C_7H_6BrNO_2$
Mr: 215.03
bp: $153\text{-}4^{8}$
mp: 58-9
Color: nd or pl (al)
***s*:** H_2O 1; al 3
IR: µSADP 20270 3030 2940 1540 1470 1450 1350 1320 1240 1210 1160 1100 1080 930 900 870 820 810 740 690
UV: SAD 6793 260(7070) 214(15500) MeOH
H NMR: SAD 18820 4.6 7.5 7.8 8.2 8.2 $CDCl_3$
m/e: 214.96

HODOC No: 02692
CAS No: 52414-97-8
Beilstein No: 5^4, 860
Name: Benzene, 1-bromo-3-methyl-2-nitro-
Synonym: 3-Bromo-2-nitrotoluene
Line Form: $3\text{-Br-}2\text{-}NO_2C_6H_3CH_3$
Mol Form: $C_7H_6BrNO_2$
Mr: 215.03
bp: $152\text{-}4^{30}$,$129\text{-}30^{10}$
mp: 27.9
Color: pa ye nd
***s*:** Et_2O 4; EtOH 4
m/e: 214.96

HODOC No: 02693
CAS No: 52488-28-5
Beilstein No: 5^3, 752
Name: Benzene, 1-bromo-3-methyl-5-nitro-
Synonym: Toluene, 3-bromo-5-nitro-
Line Form: $3\text{-Br-}5\text{-}NO_2C_6H_3CH_3$
Mol Form: $C_7H_6BrNO_2$
Mr: 215.03
bp: 269-70
mp: 84
Color: pa ye nd or pr (MeOH)
***s*:** Et_2O 4; EtOH 4
m/e: 214.96

HODOC No: 02694
CAS No: 100-11-8
Beilstein No: 5^4, 861
Name: Benzene, 1-(bromomethyl)-4-nitro-
Synonym: Toluene, α-bromo-*p*-nitro-
Line Form: $4\text{-}O_2NC_6H_4CH_2Br$
Mol Form: $C_7H_6BrNO_2$
Mr: 215.03
mp: 99-100
Color: nd (al)
***s*:** H_2O 2; al 4; eth 4; chl 2; HOAc 3
IR: SADG 8530 3050 2920 2860 1620 1600 1550 1500 1420 1350 1330 1300 1230 1200 1110 1100 1020 860 810 760 700
UV: SAD 2238 271(11600) MeOH
H NMR: SAD 3222 4.5 7.6 8.2 $CDCl_3$
m/e: 214.96

HODOC No: 02695
CAS No: 5326-34-1
Beilstein No: 5^4, 860
Name: Benzene, 1-bromo-4-methyl-2-nitro-
Synonym: Toluene, 4-bromo-3-nitro-
Line Form: $4\text{-Br-3-NO}_2C_6H_3CH_3$
Mol Form: $C_7H_6BrNO_2$
Mr: 215.03
mp: 35
***n*:** 1.5682^{20}
Color: pa ye nd (MeOH)
m/e: 214.96

HODOC No: 02696
CAS No: 41085-43-2
Beilstein No: 5^4, 860
Name: Benzene, 2-bromo-1-methyl-3-nitro-
Synonym: 1-Bromo-2-methyl-6-nitrobenzene
Line Form: $2\text{-Br-3-NO}_2\text{-}C_6H_3\text{-}CH_3$
Mol Form: $C_7H_6BrNO_2$
Mr: 215.03
bp: 157^{22}
mp: 41-2
Color: ye pr (al)
***s*:** EtOH 4
m/e: 214.96

HODOC No: 02697
CAS No: 7745-93-9
Beilstein No: 5^4, 861
Name: Benzene, 2-bromo-1-methyl-4-nitro-
Synonym: Toluene, 2-bromo-4-nitro-
Line Form: $2\text{-Br-4-NO}_2C_6H_3CH_3$
Mol Form: $C_7H_6BrNO_2$
Mr: 215.03
bp: $150\text{-}1^{20}$
mp: 78
Color: nd (al)
***s*:** Et_2O 4
IR: SADP 20880
UV: SAD 8060
H NMR: SAD 9753
m/e: 214.96

HODOC No: 02698
CAS No: 40385-54-4
Beilstein No: 5^4, 861
Name: Benzene, 2-bromo-4-methyl-1-nitro-
Synonym: 2-Bromo-4-methyl-1-nitrobenzene
Line Form: $4\text{-NO}_2\text{-3-Br-}C_6H_3CH_3$
Mol Form: $C_7H_6BrNO_2$
Mr: 215.03
bp: $154\text{-}5^{20}$
mp: 37
Color: pa ye pr or nd (MeOH)
***s*:** Et_2O 4; EtOH 4
m/e: 214.96

HODOC No: 02699
CAS No: 60956-26-5
Beilstein No: 5^4, 860
Name: Benzene, 4-bromo-1-methyl-2-nitro-
Synonym: 4-Bromo-2-nitrotoluene
Line Form: $4\text{-Br-2-NO}_2C_6H_3CH_3$
Mol Form: $C_7H_6BrNO_2$
Mr: 215.03
bp: 256-7, 130^{12}
mp: 47
Color: pa ye nd (dil al)
***s*:** Et_2O 4; EtOH 4
m/e: 214.96

HODOC No: 02700
CAS No: 52414-98-9
Beilstein No: 5^4, 860
Name: Benzene, 4-bromo-2-methyl-1-nitro-
Synonym: 4-Bromo-6-methylnitrobenzene
Line Form: $5\text{-Br-2-NO}_2C_6H_3CH_3$
Mol Form: $C_7H_6BrNO_2$
Mr: 215.03
bp: 267, 143^{10}
mp: 56
Color: cr (al)
m/e: 214.96

HODOC No: 02701
CAS No: 1765-40-8
Name: Benzene, (bromomethyl)pentafluoro-
Synonym: Toluene, α-bromo-2,3,4,5,6-pentafluoro-
Line Form: $C_6F_5CH_2Br$
Mol Form: $C_7H_2BrF_5$
Mr: 259.98
bp: 174-5
***n*:** 1.4710^{20}
MS: NBS 4683
m/e: 259.93

HODOC No: 02702
CAS No: 17293-49-1
Name: Benzene, 1,1′-[1-(bromomethyl)-1,3-propanediyl]*bis*-, (*S*)-
Synonym: Butane, 1-bromo-2,4-diphenyl-, (+)-
Line Form: $C_6H_5CH_2CH_2CH(C_6H_5)CH_2Br$
Mol Form: $C_{16}H_{17}Br$
Mr: 288.20
bp: $122\text{-}30^{0.02}$
***n*:** 1.5812^{25}
α: $[\alpha]^{20/D}16.8$ chl,c=10
***s*:** ctc 3
IR: SADG 14728 3070 3030 2940 2860 1600 1490 1450 1210 1070 1020 750 690 650 530
H NMR: SAD 8500 1.5-2.6 2.9 3.4 6.8-7.4 CCl_4
m/e: 288.05

HODOC No: 02703
CAS No: 19614-16-5
Beilstein No: 6^4, 1647
Name: Benzene, 1-bromo-2-(methylthio)-
Synonym: Sulfide, *o*-bromophenyl methyl
Line Form: $2\text{-}BrC_6H_4SCH_3$
Mol Form: C_7H_7BrS
Mr: 202.09
mp: 38-40
m/e: 201.95

HODOC No: 02704
CAS No: 577-19-5
Beilstein No: 5^4, 728
Name: Benzene, 1-bromo-2-nitro-
Synonym: *o*-Bromonitrobenzene
Line Form: $2\text{-}NO_2C_6H_4Br$
Mol Form: $C_6H_4BrNO_2$
Mr: 201.00
bp: 258^{756}
mp: 43
***d*:** $1.6245^{80/4}$
Color: pa ye (al)
***s*:** H_2O 1; al 4; eth 3; ace 3; bz 3; chl 2
IR: µCOB 346 3030 1590 1540 1490 1350 1250 1150 1040 950 860 780
UV: SAD 8128 291(1210) 252(2850) 208(13100) MeOH
H NMR: SAD 2426 7.2-8.0 $CDCl_3$
m/e: 200.94

HODOC No: 02705
CAS No: 585-79-5
Beilstein No: 5^4, 729
Name: Benzene, 1-bromo-3-nitro-
Synonym: *m*-Bromonitrobenzene
Line Form: $3\text{-}NO_2C_6H_4Br$
Mol Form: $C_6H_4BrNO_2$
Mr: 201.00
bp: 265
mp: 56
***d*:** $1.7036^{20/4}$
***n*:** 1.5979^{20}
Color: rh
***s*:** H_2O 2; al 3; eth 3; bz 3
IR: µCOB 1219 3030 1590 1540 1470 1350 1270 1110 1060 1000 890 870 800 730
UV: HBCP 255(7079) 293(1288) 302(1318) hx
m/e: 200.94

HODOC No: 02706
CAS No: 586-78-7
Beilstein No: 5^4, 729
Name: Benzene, 1-bromo-4-nitro-
Synonym: *p*-Bromonitrobenzene
Line Form: $4\text{-}NO_2C_6H_4Br$
Mol Form: $C_6H_4BrNO_2$
Mr: 201.00
bp: 256
mp: 127
***d*:** 1.948
Color: rh or mcl pr (al)
***s*:** H_2O 1; al 3; eth 3; bz 3; chl 2
IR: µSADP 7897 3130 1610 1520 1330 1280 1180 1110 1060 1020 960 860 840 740 680
UV: SAD 505 276 214 MeOH
H NMR: SAD 10377 7.6 8.0 $CDCl_3$
C NMR: JJ 147 FT 124.9 129.8 132.5 147.0 $CDCl_3$
m/e: 200.94

HODOC No: 02707
CAS No: 19519-75-6
Beilstein No: 5^4, 706
Name: Benzene, 1-bromo-2-nitroso-
Synonym: *o*-Bromonitrosobenzene
Line Form: $2\text{-}ONC_6H_4Br$
Mol Form: C_6H_4BrNO
Mr: 185.00
mp: 98
Color: nd
m/e: 184.95

HODOC No: 02708
CAS No: 13125-68-3
Beilstein No: 5^4, 706
Name: Benzene, 1-bromo-3-nitroso-
Synonym: 1-Bromo-3-nitrosobenzene
Line Form: $3\text{-BrC}_6\text{H}_4\text{NO}$
Mol Form: C_6H_4BrNO
M$_r$: 185.00
mp: 78
m/e: 184.95

HODOC No: 02709
CAS No: 3623-23-2
Beilstein No: 5^4, 706
Name: Benzene, 1-bromo-4-nitroso-
Synonym: *p*-Bromonitrosobenzene
Line Form: $4\text{-ON-C}_6\text{H}_4\text{Br}$
Mol Form: C_6H_4BrNO
M$_r$: 185.00
mp: 95
Color: nd (al)
***s*:** al 3; bz 3; chl 4; lig 3
IR: **µDMS 5369** 1590 1490 1450 1390 1300 1180 1110 1060 1010 830 690
UV: **OES 3-51** 750(42) 316(12023) 293(10000) 227(6310) EtOH
m/e: 184.95

HODOC No: 02710
CAS No: 349-03-1
Name: Benzene, 1-bromo-2-nitro-4-(trifluoromethyl)-
Synonym: Toluene, 4-bromo-α,α,α-trifluoro-3-nitro-
Mol Form: $C_7H_3BrF_3NO_2$
M$_r$: 269.00
IR: **COB 7649** 1615 1545 1480 1360 1325 1255 1180 1145 1085 1040 905 835 790 710 665 460
m/e: 268.93

HODOC No: 02711
CAS No: 344-04-7
Beilstein No: 5^4, 680
Name: Benzene, bromopentafluoro-
Synonym: Bromopentafluorobenzene
Line Form: $\text{Br-C}_6\text{F}_5$
Mol Form: C_6BrF_5
M$_r$: 245.96
bp: 137
mp: -31
***d*:** 1.981
***n*:** 1.4490^{20}
IR: ALD 2:542B
MS: NBS 1703
m/e: 245.91

HODOC No: 02712
CAS No: 101-55-3
Beilstein No: 6^4, 1047
Name: Benzene, 1-bromo-4-phenoxy-
Synonym: Ether, *p*-bromophenyl phenyl
Line Form: $(4\text{-BrC}_6\text{H}_4)\text{OC}_6\text{H}_5$
Mol Form: $C_{12}H_9BrO$
M$_r$: 248.09
bp: $124\text{-}8^{3\text{-}4}$
mp: 18.72
***d*:** 1.6088^{20}
***n*:** 1.6084^{20}
***s*:** H_2O 1; eth 3; ctc 3
IR: **COB 4317** 3080 3040 1580 1490 1240 1160 1070 1010 870 830 750 690
UV: **SAD 608** 288 278 270 235 cyhex
H NMR: **SAD 6715** 6.8 7.3 CCl_4
MS: **WILEY 1739** 248(99) 250(98) 141(44) 77(31) 169(24) 51(19) 249(15) 168()14)
m/e: 247.98

HODOC No: 02713
CAS No: 54737-45-0
Beilstein No: 5^4, 2164
Name: Benzene, 1-bromo-2-(2-phenylethenyl)-, (E)-
Synonym: *trans*-2-Bromostilbene
Line Form: $2\text{-BrC}_6\text{H}_4\text{CH=CHC}_6\text{H}_5$
Mol Form: $C_{14}H_{11}Br$
M$_r$: 258.13
bp: $145^{0.5}$
mp: 34
***n*:** 1.6822^{25}
***s*:** H_2O 1; al 3; peth 3
UV: **JACSA 78-475-56** 296(25119) 95% al
m/e: 258.00

HODOC No: 02714
CAS No: 4877-77-4
Beilstein No: 5^{4}, 2164
Name: Benzene, 1-bromo-2-(2-phenylethenyl)-, (Z)-
Synonym: Stilbene, 2-bromo-, (Z)-
Line Form: $2\text{-BrC}_6H_4CH{=}CHC_6H_5$
Mol Form: $C_{14}H_{11}Br$
M_r: 258.13
bp: $121^{0.5}$
***n*:** 1.6404^{25}
***s*:** H_2O 1; al 3; peth 3
UV: JACSA 78-475-56 268(10233) 95% al
m/e: 258.00

HODOC No: 02715
CAS No: 14064-45-0
Beilstein No: 5^{3}, 1954
Name: Benzene, 1-bromo-3-(2-phenylethenyl)-, (E)-
Synonym: Stilbene, 3-bromo-, (E)-
Mol Form: $C_{14}H_{11}Br$
M_r: 258.13
mp: 89-90
UV: TETRA 23-173-67 296(28184) cyhex
m/e: 258.00

HODOC No: 02716
CAS No: 13041-70-8
Name: Benzene, 1-bromo-4-(2-phenylethenyl)-, (E)-
Synonym: Stilbene, 4-bromo-, (E)-
Mol Form: $C_{14}H_{11}Br$
M_r: 258.13
mp: 139
UV: OES 8-313 322(21380) 315(35481) 301(39811) 227(19055) hx
m/e: 258.00

HODOC No: 02717
CAS No: 2116-36-1
Name: Benzene, 1-bromo-4-(phenylmethyl)-
Synonym: Methane, (*p*-bromophenyl)phenyl-
Mol Form: $C_{13}H_{11}Br$
M_r: 246.12
IR: COB 7901 3060 3030 2910 1600 1490 1460 1400 1070 1010 840 790 750 700 480
m/e: 246.00

HODOC No: 02718
CAS No: 65662-88-6
Beilstein No: 6^{3}, 1049
Name: Benzene, 1-bromo-4-(phenylthio)-
Synonym: *p*-Bromophenyl phenyl sulfide
Line Form: $(4\text{-BrC}_6H_4)SC_6H_5$
Mol Form: $C_{12}H_9BrS$
M_r: 264.16
bp: 196^{18}
mp: 25-7
Color: cr (peth)
m/e: 263.96

HODOC No: 02719
CAS No: 4489-23-0
Beilstein No: 8^{3}, 1188
Name: Benzene, 1-bromo-4-(1-propenyl)-
Synonym: Benzene, 1-bromo-4-propenyl-
Line Form: $4\text{-BrC}_6H_4CH{=}CHCH_3$
Mol Form: C_9H_9Br
M_r: 196.07
bp: 238-9, 110^{12}
mp: 35
***d*:** $1.332^{36/4}$
***n*:** 1.590^{36}
m/e: 195.99

HODOC No: 02720
CAS No: 2294-43-1
Beilstein No: 5^{4}, 1363
Name: Benzene, 1-bromo-4-(2-propenyl)-
Synonym: Benzene, 1-allyl-4-bromo-
Line Form: $4\text{-BrC}_6H_4CH_2CH{=}CH_2$
Mol Fotm: C_9H_9Br
M_r: 196.07
bp: $222\text{-}3^{730}$,96^{12}
***d*:** $1.324^{15/4}$
***n*:** 1.559^{15}
m/e: 195.99

HODOC No: 02721
CAS No: 4392-24-9
Beilstein No: 5^3, 1188
Name: Benzene, (3-bromo-1-propenyl)-
Synonym: Benzene, (3-bromopropenyl)-
Line Form: $C_6H_5(CH{=}CHCH_2Br$
Mol Form: C_9H_9Br
Mr: 196.07
bp: 130^{10}
mp: 34
***d*:** 1.3428^{30}
***n*:** 1.613^{20}
Color: nd (al, eth)
***s*:** EtOH 4
m/e: 195.99

HODOC No: 02722
CAS No: 25244-30-8
Beilstein No: 6^1, 105
Name: Benzene, 1-bromo-4-(2-propenyloxy)-
Synonym: Ether, allyl *p*-bromophenyl
Mol Form: C_9H_9BrO
Mr: 212.06
bp: 126^{14}
MS: WILEY 1506 41(100) 212(45) 214(44) 133(33) 39(18) 172(17) 174(16) 105(16)
m/e: 211.98

HODOC No: 02723
CAS No: 588-63-6
Beilstein No: 6^3, 549
Name: Benzene, (3-bromopropoxy)-
Synonym: Ether, 3-bromopropyl phenyl
Line Form: $C_6H_5OCH_2CH_2CH_2Br$
Mol Form: $C_9H_{11}BrO$
Mr: 214.08
bp: 127^{18}
mp: 10.0-1.5
***d*:** $1.365^{16/16}$
***s*:** Et_2O 4
IR: µCOB 445 3030 2940 1610 1490 1470 1450 1300 1250 1180 1090 1040 930 890 810 750 690
Raman: SAD 282 3070 2970 2910 2880 2760 1600 1590 1490 1470 1420 1380 1340 1300 1280 1260 1240 1200 1170 1150 1080 1030 1000 930 890 810 770 750 650 610 590 560 510 490 460 420 360 330 300 250 220 160
H NMR: SAD 17275 2.2 3.5 3.9 6.8 6.8 7.2 CCl_4
m/e: 214.00

HODOC No: 02724
CAS No: 63790-14-7
Beilstein No: 5^4, 981
Name: Benzene, (1-bromopropyl)-, (±)-
Synonym: Benzene, (1-bromopropyl) (*dl*)
Line Form: $C_6H_5(CHBrCH_2CH_3)$
Mol Form: $C_9H_{11}Br$
Mr: 198.08
bp: 105^{17}
***d*:** $1.3098^{19/4}$
***n*:** 1.5517^{19}
m/e: 198.00

HODOC No: 02725
CAS No: 637-59-2
Beilstein No: 5^4, 982
Name: Benzene, (3-bromopropyl)-
Synonym: 1-Bromo-3-phenylpropane
Line Form: $C_6H_5CH_2CH_2CH_2Br$
Mol Form: $C_9H_{11}Br$
Mr: 198.08
bp: 219-20, 110^{12}
***d*:** $1.3106^{25/4}$
***n*:** 1.5440^{25}
***s*:** H_2O 1; eth 4
IR: µSADP 9057 3030 2940 2860 1610 1490 1450 1330 1270 1240 1190 1180 1080 1030 960 900 860 800 740 690
UV: OES 1-265 268(174) 265(186) 261(229) 259(240) 254(214) 248(170) 243(135) 237(117) EtOH
H NMR: VAR 237 2.2 2.8 3.4 7.2 $CDCl_3$
C NMR: JJ 347 cw 32.9 34.0 34.3 126.0 128.4 140.6 diox
MS: WILEY 1388 91(100) 198(20) 200(20) 92(16) 65(7) 39(4) 51(3) 117(3)
m/e: 198.00

HODOC No: 02726
CAS No: 1794-48-5
Beilstein No: 5^2, 408
Name: Benzene, (3-bromo-1-propynyl)-
Synonym: 1-Bromo-3-phenyl-2-propyne
Line Form: $C_6H_5C{\equiv}CCH_2Br$
Mol Form: C_9H_7Br
Mr: 194.05
bp: 135^{15}
Raman: API 144 3064 2950 2265 2226 1599 1489 1476 1206 1186 987 609
m/e: 193.97

HODOC No: 02727
CAS No: 1125-52-6
Name: Benzene, 1-bromo-2,3,4,5-tetrachloro-
Line Form: 2,3,4,5-Cl_4C_6HBr
Mol Form: C_6HBrCl_4
M_r: 293.78
mp: 98
IR: µCOB 2359 3130 1390 1330 1180 1080 860 810 660
m/e: 291.80

HODOC No: 02728
CAS No: 392-83-6
Beilstein No: 5^4, 831
Name: Benzene, 1-bromo-2-(trifluoromethyl)-
Synonym: Toluene, *o*-bromo-α,α,α-trifluoro-
Line Form: 2-$BrC_6H_4CF_3$
Mol Form: $C_7H_4BrF_3$
M_r: 224.00
bp: 167-8
***d*:** 1.652
***n*:** 1.4817^{20}
IR: SADG 33170 3060 1630 1590 1570 1470 1440 1420 1310 1260 1170 1130 1110 1080 1030 1020 950 870 760 720 680 630 590 440 420 360
Raman: SAD 1157 3150 3080 1600 1580 1480 1450 1320 1260 1170 1120 1100 1050 950 790 750 690 650 590 400 270 190
UV: SAD 7986 276(1030) 268(1130) 262(784) MeOH
H NMR: SAD 19518 7.1-7.4 7.5-7.8 $CDCl_3$
MS: ALD 4:45C
m/e: 223.94

HODOC No: 02729
CAS No: 401-78-5
Beilstein No: 5^4, 831
Name: Benzene, 1-bromo-3-(trifluoromethyl)-
Synonym: Toluene, *m*-bromo-α,α,α-trifluoro-
Line Form: 3-$BrC_6H_4CF_3$
Mol Form: $C_7H_4BrF_3$
M_r: 224.00
bp: 151-2
mp: 1
***d*:** 1.613
***n*:** 1.4716^{20}
IR: ALD 2:534F
MS: ALD 4:50C
m/e: 223.94

HODOC No: 02730
CAS No: 402-43-7
Beilstein No: 5^4, 831
Name: Benzene, 1-bromo-4-(trifluoromethyl)-
Synonym: Toluene, *p*-bromo-α,α,α-trifluoro-
Line Form: 4-$BrC_6H_4CF_3$
Mol Form: $C_7H_4BrF_3$
M_r: 224.00
bp: 154-5
***d*:** 1.607^{25}
***n*:** 1.4705^{25}
IR: µSADP 20764 2940 1590 1490 1390 1330 1160 1140 1100 1060 1010 830 780 720
Raman: SAD 1158 3080 1600 1490 1440 1330 1290 1180 1130 1080 1070 1000 810 780 690 630 590 500 440 410 380 280 230 210 160
H NMR: SAD 9131 7.5 7.6 CCl_4
m/e: 223.94

HODOC No: 02731
CAS No: 31053-99-3
Beilstein No: 5^4, 1014
Name: Benzene, 1-bromo-2,3,5-trimethyl-
Line Form: 2,3,5-$(CH_3)_3C_6H_2Br$
Mol Form: $C_9H_{11}Br$
M_r: 198.08
bp: 238, 117^{17}
mp: 15
***n*:** 1.5516^{20}
***s*:** Bz 4
m/e: 198.00

HODOC No: 02732
CAS No: 5469-19-2
Beilstein No: 5^4, 1014
Name: Benzene, 1-bromo-2,4,5-trimethyl-
Synonym: 5-Bromopseudocumene
Line Form: 2,4,5-$(CH_3)_3C_6H_2Br$
Mol Form: $C_9H_{11}Br$
M_r: 198.08
bp: 233-5
mp: 73
Color: nd (al)
***s*:** H_2O 1; al 4
IR: µSADP 18806 3130 1610 1490 1450 1390 1270 1190 1150 1020 960 880 770 720
UV: SAD 6078 281 MeOH
m/e: 198.00

HODOC No: 02733
CAS No: 576-83-0
Beilstein No: 5^4, 1027
Name: Benzene, 2-bromo-1,3,5-trimethyl-
Synonym: Mesitylene, 2-bromo-
Line Form: 2,4,6-$(CH_3)_3C_6H_2Br$
Mol Form: $C_9H_{11}Br$
M_r: 198.08
bp: 225, 105-7^{16}
mp: -1
***d*:** 1.3191^{10}
***n*:** 1.5510^{20}
***s*:** H_2O 1; eth 4; bz 3; ctc 2
IR: µCOB 7502 2940 1470 1370 1180 1030 1010 950 690
UV: SAD 1591 268 cyhex
H NMR: SAD 670 2.2 2.3 6.8 CCl_4
MS: WILEY 1387 183(100) 185(97) 104(54) 198(41) 200(36) 199(15) 77(15) 103(14)
m/e: 198.00

HODOC No: 02734
CAS No: 4185-53-9
Beilstein No: 5^3, 646
Name: Benzene, 2-bromo-1,3,5-trinitro-
Synonym: 2-Bromo-1,3,5-trinitrobenzene
Line Form: 2,4,6-$(NO_2)_3C_6H_2Br$
Mol Form: $C_6H_2BrN_3O_6$
M_r: 291.00
mp: 123
***s*:** H_2O 1; al 3; bz 3; chl 3; HOAc 3
UV: TETRA 18-161-62 641 552 502 453 liq NH_3
m/e: 290.91

HODOC No: 02735
CAS No: 538-81-8
Beilstein No: 5^2, 589
Name: Benzene, 1,1′-(1,3-butadiene-1,4-diyl)*bis*-, (E,E)-
Synonym: 1,3-Butadiene, 1,4-diphenyl-, (E,E)-
Line Form: $C_6H_5CH{=}CHCH{=}CHC_6H_5$
Mol Form: $C_{16}H_{14}$
M_r: 206.27
bp: 350^{720}
mp: 152.5
***s*:** H_2O 1; al 3; eth 2; bz 3; chl 3; peth 3
IR: SADG 8298 3050 3010 1600 1490 1440 1080 1000 980 920 830 750 690 500
UV: SAD 1088 319 cyhex
H NMR: SAD 5140 6.6 6.9 6.6 6.9 7.1-7.6 $CDCl_3$
m/e: 206.11

HODOC No: 02736
CAS No: 5807-76-1
Beilstein No: 5^4, 2318
Name: Benzene, 1,1′-(1,3-butadiene-1,4-diyl)*bis*-, (Z,Z)-
Synonym: 1,3-Butadiene, 1,4-diphenyl-, (Z,Z)-
Line Form: $C_6H_5CH{=}CHCH{=}CHC_6H_5$
Mol Form: $C_{16}H_{14}$
M_r: 206.27
mp: 70.5
***d*:** 0.9697^{100}
***n*:** 1.6183^{100}
Color: lf or nd (al, MeOH)
***s*:** H_2O 1; al 2; eth 4; bz 4; chl 4; peth 3; HOAc 3
IR: DMS 882 3020 1610 1500 1450 1370 1180 1080 1020 920 800 780 750 700
UV: OES 1-669 306(26915) bz 299(29512) hx 300(31623) peth
m/e: 206.11

HODOC No: 02737
CAS No: 1450-63-1
Beilstein No: 5^4, 2816
Name: Benzene, 1,1′,1″,1‴-(1,3-butadiene-1,4-diylidene)tetrakis-
Synonym: 1,3-Butadiene, 1,1,4,4-tetraphenyl-
Line Form: $(C_6H_5)_2C{=}CHCH{=}C(C_6H_5)_2$
Mol Form: $C_{28}H_{22}$
M_r: 358.45
mp: 203-4
***s*:** al 3; bz 3; chl 3; HOAc 3
IR: SADG 21330 3030 1600 1580 1490 1440 1350 1070 1030 1000 900 770 700 620 580
Raman: SAD 1131 3180 3050 2860 1610 1590 1570 1540 1490 1450 1370 1330 1300 1280 1270 1240 1180 1150 1130 1110 1030 1000 960 940 900 850 830 780 760 740 700 640 620 610 570 500470 400 390 300 260 220 200 110 90
UV: OES 7-1150 341(37154) thf
H NMR: VAR 699 6.8 7.2 7.4 $CDCl_3$
MS: NBS 2812
m/e: 358.17

HODOC No: 02738
CAS No: 806-71-3
Beilstein No: 5^4, 2817
Name: Benzene, 1,1′,1″,1‴-(1,3-butadiene-1,2,3,4-tetrayl)tetrakis-
Synonym: 1,3-Butadiene, 1,2,3,4-tetraphenyl-
Line Form: $C_6H_5CH{=}C(C_6H_5)C(C_6H_5){=}CHC_6H_5$
Mol Form: $C_{28}H_{22}$
M$_r$: 358.45
mp: 183
IR: COB 4774 3060 1600 1490 1440 1350 1070 1030 930 870 790 760 710 700
UV: OES 1-1044 356(41687) diox
MS: WILEY 2030 267(100) 358(72) 167(50) 281(30) 268(23) 280(22) 265(20) 279(18)
m/e: 358.17

HODOC No: 02739
CAS No: 16939-57-4
Beilstein No: 5^4, 1536
Name: Benzene, 1,3-butadienyl-, (E)-
Synonym: 1,3-Butadiene, 1-phenyl-, (E)-
Line Form: $C_6H_5CH{=}CHCH{=}CH_2$
Mol Form: $C_{10}H_{10}$
M$_r$: 130.18
bp: 76^{11}
mp: 4.5 (fp)
***d*:** $0.9286^{20/4}$
***n*:** 1.6089^{25}
***s*:** H_2O l; al 3; eth 3; ace 3; bz 3; os 3
IR: DMS 8571 3090 3000 1810 1640 1610 1500 1450 1420 1310 1300 1160 1070 1030 1000 940 900 860 760 690 640
UV: OES 3-237 306(4467) 280(22387) 233(6457) iso
m/e: 130.08

HODOC No: 02740
CAS No: 31915-94-3
Beilstein No: 5^3, 1363
Name: Benzene, 1,3-butadienyl-, (Z)-
Synonym: 1,3-Butadiene, 1-phenyl-, (Z)-
Line Form: $CH_2{=}CHCH{=}CHC_6H_5$
Mol Form: $C_{10}H_{10}$
M$_r$: 130.18
bp: 86^{11}
mp: -57 (fp)
***d*:** $0.9334^{20/4}$
***n*:** 1.6095^{20}
m/e: 130.08

HODOC No: 02741
CAS No: 886-66-8
Beilstein No: 5^4, 2462
Name: Benzene, 1,1′-(1,3-butadiyne-1,4-diyl)*bis*-
Synonym: Butadiyne, diphenyl-
Line Form: $C_6H_5C{\equiv}CC{\equiv}CC_6H_5$
Mol Form: $C_{16}H_{10}$
M$_r$: 202.24
mp: 86-7
IR: ALD 2:1257D
MS: ALD 10:131D
m/e: 202.08

HODOC No: 02742
CAS No: 1083-56-3
Beilstein No: 5^4, 1937
Name: Benzene, 1,1′-(1,4-butanediyl)*bis*-
Synonym: Butane, 1,4-diphenyl-
Line Form: $C_6H_5(CH_2)_4C_6H_5$
Mol Form: $C_{16}H_{18}$
M$_r$: 210.30
bp: 317, $108\text{-}9^{0.1}$
mp: 52.5
***s*:** H_2O 1; al 3; eth 3; chl 3
IR: SADG 17431 3080 3060 3030 2950 2860 1600 1490 1450 1020 740 690
UV: OES 4-635 268(398) 265(398) 261(501) 259(501) 253(398) 248(251) iso
H NMR: SAD 9297 1.6 2.6 7.2 $CDCl_3$
MS: ALDN 172 91(100) 40(99) 92(66) 44(63) 210(51) 65(34) 104(23) 117(20)
m/e: 210.14

HODOC No: 02743
CAS No: 6282-65-1
Name: Benzene, 1,1′-[1,4-butanediylbis(oxymethylene)]*bis*-
Synonym: Butane, 1,4-*bis*(benzyloxy)
Line Form: $C_6H_5CH_2O(CH_2)_4OCH_2C_6H_5$
Mol Form: $C_{18}H_{22}O_2$
M$_r$: 270.35
MS: WILEY 1832 91(100) 28(87) 107(51) 77(46) 105(45) 106(42) 92(22) 270(0)
m/e: 270.16

HODOC No: 02744
CAS No: 1005-64-7
Beilstein No: 5^4, 1374
Name: Benzene, 1-butenyl-, (E)-
Synonym: 1-Butene, 1-phenyl-, (E)-
Line Form: $CH_3CH_2CH{=}CHC_6H_5$
Mol Form: $C_{10}H_{12}$
M_r: 132.20
bp: 198.7, $91\text{-}2^{23}$
mp: -43.1
***d*:** $0.9019^{20/4}$
***n*:** 1.5420^{20}
***s*:** H_2O 1; al 3; eth 3; bz 3; ctc 3; chl 3
IR: µSADP 7852 3030 2940 1610 1490 1450 1390 1080 1050 1030 960 940 750 690
UV: JACSA 79-4795-57 251(16218) eth
m/e: 132.09

HODOC No: 02745
CAS No: 1560-09-4
Beilstein No: 5^1, 1205
Name: Benzene, 1-butenyl-, (Z)-
Synonym: 1-Butene, 1-phenyl-, (Z)-
Line Form: $CH_3CH_2CH{=}CHC_6H_5$
Mol Form: $C_{10}H_{12}$
M_r: 132.20
bp: 189, $84\text{-}5^{23}$
***d*:** 0.9065^{25}
***n*:** 1.5390^{20}
***s*:** H_2O 1; al 3; eth 3; bz 3; ctc 3; chl 3
UV: OES 6-288 290(204) 242(11482) EtOH
m/e: 132.09

HODOC No: 02746
CAS No: 1560-06-1
Beilstein No: 5^4, 1375
Name: Benzene, 2-butenyl-
Synonym: 2-Butene, 1-phenyl-
Line Form: $CH_3CH{=}CHCH_2C_6H_5$
Mol Form: $C_{10}H_{12}$
M_r: 132.20
bp: 176, $81\text{-}2^{21}$
***d*:** $0.8831^{20/4}$
***n*:** 1.5101^{20}
IR: SADG 33398 3090 3080 3040 2980 2930 2870 1610 1490 1450 1380 1310 1190 1140 1070 1030 970 740 690 580 540 450
Raman: SAD 1129 3200 3160 3050 3000 2940 2910 2860 2730 1670 1600 1580 1450 1440 1380 1300 1200 1150 1030 1010 910 880 820 750 620 540 450 330 220
UV: SAD 2155 264 cyhex
H NMR: SAD 9429 1.7 3.3 7.1 CCl_4
m/e: 132.09

HODOC No: 02747
CAS No: 768-56-9
Beilstein No: 5^4, 1378
Name: Benzene, 3-butenyl-
Synonym: 1-Butene, 4-phenyl-
Line Form: $C_6H_5CH_2CH_2CH{=}CH_2$
Mol Form: $C_{10}H_{12}$
M_r: 132.20
bp: 177, 64^{10}
mp: -70
***d*:** $0.8831^{20/4}$
***n*:** 1.5059^{20}
***s*:** H_2O 1; eth 3; bz 3
IR: SADG 23803 3090 3080 3040 2990 2940 2860 1640 1610 1500 1460 1030 1000 920 750 700 630 570 490
Raman: SAD 724 3220 3180 3060 3010 2980 2930 2900 2860 1650 1610 1590 1500 1450 1420 1340 1300 1280 1260 1210 1180 1160 1120 1070 1040 1010 940 900 850 830 810 770 750 700 650 630 580 560 520 500 440 350 290 250 230
UV: HCACA 19-1270-36 262(200)
H NMR: SAD 14337 2.4 2.7 4.97 5.01 5.8 7.2 $CDCl_3$
MS: WILEY 630 91(100) 132(18) 39(14) 65(13) 92(10) 51(8) 27(7) 104(5)
m/e: 132.09

HODOC No: 02748
CAS No: 1703-88-4
Name: Benzene,
2-(1-butenyl)-5-(1,1-dimethylethyl)-1,3-dimethyl-
Synonym: 1-Butene, 1-(4-*tert*-butyl-2,6-xylyl)-
Mol Form: $C_{16}H_{24}$
Mr: 216.35
MS: WILEY 1544 201(100) 216(28) 41(11) 29(10) 159(8) 57(8) 128(7) 27(7)
m/e: 216.19

HODOC No: 02749
CAS No: 1726-14-3
Beilstein No: 5^4, 2209
Name: Benzene, 1,1′-(1-butenylidene)*bis*-
Synonym: 1-Butene, 1,1-diphenyl-
Line Form: $CH_3CH_2CH{=}C(C_6H_5)_2$
Mol Form: $C_{16}H_{16}$
Mr: 208.29
bp: 295-7, $157\text{-}8^{14}$
mp: -1.2
***d*:** 0.9934^{20}
***n*:** 1.5604^{20}
m/e: 208.13

HODOC No: 02750
CAS No: 1126-79-0
Beilstein No: 5^4, 558
Name: Benzene, butoxy-
Synonym: Ether, butyl phenyl
Line Form: $C_6H_5OC_4H_9$
Mol Form: $C_{10}H_{14}O$
Mr: 150.21
bp: 210, 95^{17}
mp: -19.4
***d*:** $0.9351^{20/4}$
***n*:** 1.4969^{20}
***s*:** eth 3; ace 3
IR: SADG 334 3040 2950 2860 1600 1490 1470 1390 1240 1170 1080 1030 880 750 690 510
Raman: API 74 3068 2937 2911 2870 1600 1246 1173 1029 999 615
MS: WILEY 895 94(100) 29(17) 150(16) 39(15) 27(11) 41(10) 65(10) 77(9)
m/e: 150.10

HODOC No: 02751
CAS No: 4157-77-1
Name: Benzene, (1-butoxyethyl)-
Synonym: Ether, butyl α-methylbenzyl
Line Form: $CH_3(CH_2)_3OCH(CH_3)C_6H_5$
Mol Form: $C_{12}H_{18}O$
Mr: 178.26
bp: $90\text{-}7^{8}$
***s*:** ctc 3; CS_2 3
IR: μCOB 3161 2940 2860 1590 1490 1450 1370 1350 1210 1100 1030 910 760 700
m/e: 178.14

HODOC No: 02752
CAS No: 588-67-0
Beilstein No: 6^2, 410
Name: Benzene, (butoxymethyl)-
Synonym: Ether, benzyl butyl
Line Form: $C_6H_5CH_2O(CH_2)_3CH_3$
Mol Form: $C_{11}H_{16}O$
Mr: 164.24
bp: 220.5^{744}
***d*:** $0.9310^{10/4}$
IR: COB 7128 3040 2970 2880 1500 1460 1370 1210 1100 1030 730 700
H NMR: SAD 19879 0.9 1.1-1.8 3.4 4.4 CCl_4
m/e: 164.12

HODOC No: 02753
CAS No: 2052-13-3
Beilstein No: 6^3, 1247
Name: Benzene, 1-butoxy-2-methyl-
Synonym: Ether, butyl *o*-tolyl
Line Form: $(2\text{-}CH_3C_6H_4)\text{-}O\text{-}C_4H_9$
Mol Form: $C_{11}H_{16}O$
Mr: 164.24
bp: 223
***d*:** $0.943^{0/0}$
m/e: 164.12

HODOC No: 02754
CAS No: 23079-65-4
Beilstein No: 6^4, 2040
Name: Benzene, 1-butoxy-3-methyl-
Synonym: Ether, butyl *m*-tolyl
Line Form: $(3\text{-}CH_3C_6H_4)\text{-}O\text{-}C_4H_9$
Mol Form: $C_{11}H_{16}O$
Mr: 164.24
bp: 229.2
***d*:** $0.9407^{70/0}$
***n*:** 1.4970^{20}
***s*:** Et_2O 4
m/e: 164.12

HODOC No: 02755
CAS No: 10519-06-9
Beilstein No: 6^3, 1354
Name: Benzene, 1-butoxy-4-methyl-
Synonym: Ether, butyl *p*-tolyl
Line Form: $(4\text{-}CH_3C_6H_4)\text{-}O\text{-}C_4H_9$
Mol Form: $C_{11}H_{16}O$
Mr: 164.24
bp: 229.5, 88^3
***d*:** $0.9232^{25/25}$
***n*:** 1.4970^{20}
***s*:** eth 3
UV: HBCP 266(219) 260(245) 254(204) 218(6761) al
m/e: 164.12

HODOC No: 02756
CAS No: 54815-20-2
Name: Benzene, (1-butoxy-1-methylethyl)-
Synonym: Ether, butyl α,α-dimethylbenzyl
Line Form: $C_6H_5C(CH_3)_2O(CH_2)_3CH_3$
Mol Form: $C_{13}H_{20}O$
Mr: 192.29
MS: WILEY 1341 121(100) 119(78) 177(55) 43(36) 91(33) 41(19) 120(15) 192(0)
m/e: 192.15

HODOC No: 02757
CAS No: 104-51-8
Beilstein No: 5^4, 1033
Name: Benzene, butyl-
Synonym: Butylbenzene
Line Form: $C_6H_5C_4H_9$
Mol Form: $C_{10}H_{14}$
Mr: 134.21
bp: 183.3, 62.2^{10}
mp: -88 (fp)
***d*:** $0.8601^{20/4}$
***n*:** 1.4898^{20}
***s*:** H_2O 1; al 5; eth 5; ace 5; bz 5; ctc 5; peth 5
IR: μCOB 3597 2940 1590 1490 1450 1390 1110 1030 740 700
Raman: DFB 68 3050 3000 2960 2940 2920 2870 2850 2740 1600 1580 1450 1440 1330 1300 1200 1180 1150 1100 1050 1030 1000 990 970 920 780 620 500 280 220
UV: SAD 1739 261 cyhex
H NMR: SAD 3425 0.9 1.1-1.8 2.6 7.1 CCl_4
C NMR: JJ 380 cw 13.9 22.5 33.9 35.8 125.7 128.3 142.7 diox
MS: WILEY 663 91(100) 92(55) 134(24) 27(12) 39(10) 65(10) 105(8) 51(7)
m/e: 134.11

HODOC No: 02758
CAS No: 42548-38-9
Name: Benzene, 1-butyl-2-ethyl-
Synonym: 1-Ethyl-2-butylbenzene
Line Form: $C_6H_5CH_2CH(CH_2CH_3)_2$
Mol Form: $C_{12}H_{18}$
Mr: 162.26
MS: WILEY 1051 92(100) 43(77) 91(64) 29(25) 71(20) 27(20) 162(18) 41(15)
m/e: 162.14

HODOC No: 02759
CAS No: 2400-04-6
Beilstein No: 5^{4}, 1223
Name: Benzene, (1-butylhexadecyl)-
Synonym: Eicosane, 5-phenyl-
Line Form: $CH_3(CH_2)_3CH(C_6H_5)(CH_2)_{14}CH_3$
Mol Form: $C_{26}H_{46}$
Mr: 358.63
bp: 197^{1}
mp: 30.2
***d*:** $0.8549^{20/4}$
***n*:** 1.4796^{20}
UV: API 1021 267 262 257 252 205 190 cyhex
MS: WILEY 2045 91(100) 147(32) 43(23) 41(16) 105(12) 57(12) 29(12) 358(2)
m/e: 358.36

HODOC No: 02760
CAS No: 100144-29-4
Name: Benzene, 2-butyl-1-hydroxy-3-methyl
Mol Form: $C_{11}H_{16}O$
Mr: 164.24
MS: WILEY 1068 149(100) 121(34) 164(31) 150(11) 91(10) 115(6) 116(5) 109(5)
m/e: 164.12

HODOC No: 02761
CAS No: 719-79-9
Beilstein No: 5^{4}, 1944
Name: Benzene, 1,1′-butylidenebis-
Synonym: Butane, 1,1-diphenyl-
Line Form: $CH_3CH_2CH_2CH(C_6H_5)_2$
Mol Form: $C_{16}H_{18}$
Mr: 210.30
bp: 286-8, $161\text{-}3^{20}$
mp: 27
***d*:** $0.9928^{20/4}$
***n*:** 1.5664^{20}
***s*:** H_2O 1; al 3; eth 3; bz 3; chl 3
IR: µSADP 9980 3030 2940 1610 1490 1450 1370 1330 1110 1100 1060 1030 910 840 790 760 750 730 690
UV: OES 7-671 269(479) 262(617) 260(603) 255(447) cyhex
m/e: 210.14

HODOC No: 02762
CAS No: 1595-11-5
Beilstein No: 5^{3}, 999
Name: Benzene, 1-butyl-2-methyl-
Synonym: Toluene, *o*-butyl-
Line Form: $2\text{-}CH_3C_6H_4C_4H_9$
Mol Form: $C_{11}H_{16}$
Mr: 148.24
bp: 208, 81^{10}
***d*:** $0.8710^{20/4}$
***n*:** 1.4960^{20}
***s*:** Ac 4; Bz 4; Et_2O 4
m/e: 148.13

HODOC No: 02763
CAS No: 1595-04-6
Beilstein No: 5^{3}, 999
Name: Benzene, 1-butyl-3-methyl-
Synonym: Toluene, *m*-butyl-
Line Form: $2\text{-}CH_3C_6H_4H_9$
Mol Form: $C_{11}H_{16}$
Mr: 148.24
bp: 205, 79^{10}
***d*:** $0.8590^{20/4}$
***n*:** 1.4910^{20}
***s*:** Ac 4; Bz 4; Et_2O 4
m/e: 148.13

HODOC No: 02764
CAS No: 1595-05-7
Beilstein No: 5^{4}, 1042
Name: Benzene, 1-butyl-4-methyl-
Synonym: Toluene, *p*-butyl-
Line Form: $4\text{-}CH_3C_6H_4C_4H_9$
Mol Form: $C_{11}H_{16}$
Mr: 148.24
bp: 207, 80.6^{10}
mp: -85
***d*:** $0.8586^{20/4}$
***n*:** 1.4916^{20}
***s*:** Ac 4; Bz 4; Et_2O 4
m/e: 148.13

HODOC No: 02765
CAS No: 7137-55-5
Beilstein No: 5^3, 931
Name: Benzene, 1-butyl-2-nitro-
Synonym: *o*-Nitrobutylbenzene
Line Form: $2\text{-}NO_2C_6H_4C_4H_9$
Mol Form: $C_{10}H_{13}NO_2$
M$_r$: 179.21
bp: 260, $131\text{-}3^{15}$
***d*:** $1..071^{20/4}$
m/e: 179.09

HODOC No: 02766
CAS No: 20651-76-7
Beilstein No: 5^2, 318
Name: Benzene, 1-butyl-3-nitro-
Synonym: 1-Butyl-3-nitrobenzene
Line Form: $3\text{-}NO_2C_6H_4C_4H_9$
Mol Form: $C_{10}H_{13}NO_2$
M$_r$: 179.21
bp: 275^{752}
***s*:** Py 4; peth 4; Chl 4
m/e: 179.09

HODOC No: 02767
CAS No: 20651-75-6
Beilstein No: 5^3, 931
Name: Benzene, 1-butyl-4-nitro-
Synonym: *p*-Butylnitrobenzene
Line Form: $4\text{-}NO_2C_6H_4C_4H_9$
Mol Form: $C_{10}H_{13}NO_2$
M$_r$: 179.21
bp: $143\text{-}5^{15}$
***d*:** $1.065^{20/4}$
m/e: 179.09

HODOC No: 02768
CAS No: 2719-63-3
Name: Benzene, (1-butyloctyl)-
Synonym: Dodecane, 5-phenyl-
Line Form: $CH_3(CH_2)_6CH(C_6H_5)CH_2CH_2CH_2CH_3$
Mol Form: $C_{18}H_{30}$
M$_r$: 246.42
MS: WILEY 1731 91(100) 147(20) 41(13) 105(12) 189(9) 43(9) 92(9) 246(5)
m/e: 246.23

HODOC No: 02769
CAS No: 62155-44-6
Name: Benzene, 1-butyl-4-(phenylmethyl)-
Synonym: Methane, (*p*-butylphenyl)phenyl-
Mol Form: $C_{17}H_{20}$
M$_r$: 224.33
bp: 326.41^{760}
***d*:** 0.95696^{20}
IR: μSADP 8242 3030 2940 1610 1490 1450 1370 1210 1180 1110 1080 1030 930 830 790 730 690
UV: API 805 274 269 265 262 260 221 iso
m/e: 224.16

HODOC No: 02770
CAS No: 1126-80-3
Name: Benzene, (butylthio)-
Synonym: Sulfide, butyl phenyl
Line Form: $C_6H_5S(CH_2)_3CH_3$
Mol Form: $C_{10}H_{14}S$
M$_r$: 166.27
bp: 112^{15}
UV: LANG 1109 257(7762) hp
m/e: 166.08

HODOC No: 02771
CAS No: 622-76-4
Beilstein No: 5^2, 413
Name: Benzene, 1-butynyl-
Synonym: 1-Butyne, 1-phenyl-
Line Form: $C_6H_5C{\equiv}CCH_2CH_3$
Mol Form: $C_{10}H_{10}$
M$_r$: 130.18
bp: 201-3
***d*:** 0.923^{21}
Raman: API 57 2914 2238 2206 1599 1258 1184 1161 999 893 539 383
m/e: 130.08

HODOC No: 02772
CAS No: 16520-62-0
Beilstein No: 5^4, 1535
Name: Benzene, 3-butynyl-
Synonym: 1-Butyne, 4-phenyl-
Line Form: $C_6H_5CH_2CH_2C{\equiv}CH$
Mol Form: $C_{10}H_{10}$
M$_r$: 130.18
bp: 190, 83^{15}
***d*:** $0.9258^{20/4}$
***n*:** 1.5208^{20}
m/e: 130.08

HODOC No: 02773
CAS No: 108-90-7
Beilstein No: 5^4, 640
Name: Benzene, chloro-
Synonym: Chlorobenzene
Line Form: C_6H_5Cl
Mol Form: C_6H_5Cl
M_r: 112.55
bp: 132, 22^{10}
mp: -45.6
***d*:** $1.1058^{20/4}$
***n*:** 1.5241^{20}
***s*:** H_2O 1; al 5; eth 5; bz 4; ctc 4; chl 4; CS_2 4
IR: µCOB 5726 3030 1590 1490 1450 1140 1090 1030 890 750 690
Raman: DFB 72 3160 3060 1580 1170 1160 1080 1020 1000 700 610 420 290 190
UV: OES 6-63 263(191) H_2O
H NMR: SAD 714 7.2 CCl_4
C NMR: JJ 153 FT 126.5 128.6 128.8 134.3 diox
MS: WILEY 370 112(100) 77(45) 114(33) 51(12) 50(10) 113(7) 36(6) 75(5)
m/e: 112.01

HODOC No: 02774
CAS No: 80438-67-1
Name: Benzene, 1-chloro-3,5-*bis*(1,1-dimethylethyl)-
Synonym: 1-Chloro-3,5-di-*tert*-butylbenzene
Mol Form: $C_{14}H_{21}Cl$
M_r: 224.76
IR: COB 4327 3080 2960 1600 1570 1480 1420 1400 1360 1280 1250 870 850 780 700
MS: WILEY 1596 ini 209(100) 211(33) 57(32) 224(16) 210(15) 41(10) 181(6) 226(5)
m/e: 224.13

HODOC No: 02775
CAS No: 54845-36-2
Name: Benzene, 2-chloro-1,3-*bis*(1-methylethyl)-
Synonym: 1-Chloro-2,6-diisopropylbenzene
Mol Form: $C_{12}H_{17}Cl$
M_r: 196.71
MS: WILEY 1369 181(100) 183(32) 43(27) 196(25) 182(13) 153(10) 41(9) 139(9)
m/e: 196.10

HODOC No: 02776
CAS No: 27260-64-6
Name: Benzene, 1-chloro-2,4-*bis*(trichloromethyl)-
Synonym: *m*-Xylene, α,α,α,α′,α′,α′,4-heptachloro-
Mol Form: $C_8H_3Cl_7$
M_r: 347.27
bp: $170\text{-}80^8$
IR: µCOB 1910 3130 1470 1390 1280 1220 1160 1050 970 830 810 790 780 760 750
m/e: 343.81

HODOC No: 02777
CAS No: 10388-10-0
Name: Benzene, 2-chloro-1,4-*bis*(trichloromethyl)-
Synonym: *p*-Xylene, α,α,α,α′,α′,α′,2-heptachloro-
Mol Form: $C_8H_3Cl_7$
M_r: 347.27
mp: 78-80
IR: µCOB 1908 3130 1470 1390 1280 1210 1060 850 790 750 710
m/e: 343.81

HODOC No: 02778
CAS No: 7501-11-3
Name: Benzene, 1-chloro-2-[2-(2-chloroethoxy)ethoxy]-
Synonym: Ethane, 1(2-chloroethoxy)-2(2-chlorophenoxy)
Mol Form: $C_{10}H_{12}Cl_2O_2$
M_r: 235.10
MS: WIlEY 1653 63(100) 45(44) 128(40) 107(38) 65(34) 234(24) 27(19) 236(15)
m/e: 234.02

HODOC No: 02779
CAS No: 20001-64-3
Name: Benzene, 1-chloro-2-(1-chloroethyl)-
Synonym: 1-Chloro-2-(α-chloroethyl)benzene
Mol Form: $C_8H_8Cl_2$
M_r: 175.05
MS: WILEY 1169 139(100) 103(44) 141(33) 174(20) 125(18) 77(17) 51(15) 176(13)
m/e: 174.00

HODOC No: 02780
CAS No: 611-19-8
Beilstein No: 5^4, 816
Name: Benzene, 1-chloro-2-(chloromethyl)-
Synonym: Toluene, *o*,α-dichloro-
Line Form: $2\text{-}ClC_6H_4CH_2Cl$
Mol Form: $C_7H_6Cl_2$
M_r: 161.02
bp: 217, $94\text{-}5^{10}$
mp: -17
***d*:** $1.2699^{0/4}$
***n*:** 1.5530^{20}
***s*:** H_2O 1; al 2; eth 4; bz 4; ctc 2; HOAc 4; CS_2 3
IR: μCOB 144 3030 2940 1610 1590 1470 1450 1280 1270 1210 1120 1050 1040 820 760 740 680 670
Raman: SAD 325 3060 3020 2960 2860 1590 1570 1470 1440 1280 1260 1210 1150 1120 1040 1010 900 860 820 760 730 670 570 500 450 440 380 360 290 250 230 170 110
UV: SAD 206 278(263) 271(339) MeOH
H NMR: SAD 1182 4.6 7.1-7.5 CCl_4
MS: WILEY 1003 125(100) 127(32) 160(22) 89(15) 162(15) 63(10) 62(9) 139(9)
m/e: 159.98

HODOC No: 02781
CAS No: 620-20-2
Beilstein No: 5^4, 816
Name: Benzene, 1-chloro-3-(chloromethyl)-
Synonym: Toluene, *m*,α-dichloro-
Line Form: $3\text{-}ClC_6H_4CH_2Cl$
Mol Form: $C_7H_6Cl_2$
M_r: 161.02
bp: $215\text{-}6^{753}$, $110\text{-}1^{25}$
***d*:** $1.2695^{15/4}$
***n*:** 1.5554^{20}
***s*:** EtOH 4
MS: NBS 597
m/e: 159.98

HODOC No: 02782
CAS No: 104-83-6
Beilstein No: 5^4, 816
Name: Benzene, 1-chloro-4-(chloromethyl)-
Synonym: Toluene, *p*,α-dichloro-
Line Form: $4\text{-}ClC_6H_4CH_2Cl$
Mol Form: $C_7H_6Cl_2$
M_r: 161.02
bp: 223, 117^{20}
mp: 31
Color: nd (dil al)
***s*:** ctc 2
IR: SADG 142 3030 2960 1700 1600 1490 1440 1400 1090 1010 830 810 720 680 640
Raman: SAD 326 3190 3090 3060 3020 2960 2870 1600 1490 1440 1360 1310 1260 1210 1170 1150 1090 1050 1010 1000 940 910 830 800 720 680 620 500 410 350 320 250 170
UV: SAD 209 273 277 268 261 256 226 cyhex
H NMR: SAD 1183 4.5 7.3 CCl_4
m/e: 159.98

HODOC No: 02783
CAS No: 57403-35-7
Beilstein No: 5^4, 858
Name: Benzene, 1-chloro-4-(chloromethyl)-2-nitro-
Synonym: 4-Chloro-3-nitrobenzyl chloride
Line Form: $(4\text{-}Cl\text{-}3\text{-}NO_2)C_6H_3CH_2Cl$
Mol Form: $C_7H_5Cl_2NO_2$
M_r: 206.02
bp: 122^{10}
***d*:** 1.32
***n*:** 1.5855^{20}
m/e: 204.97

HODOC No: 02784
CAS No: 50274-95-8
Name: Benzene, 2-chloro-1-(chloromethyl)-4-nitro-
Synonym: 2-Chloro-4-nitrobenzyl chloride
Line Form: $(2\text{-}Cl\text{-}4\text{-}NO_2)C_6H_3CH_2Cl$
Mol Form: $C_7H_5Cl_2NO_2$
M_r: 206.02
mp: 48-9
IR: ALD 3:817E
m/e: 204.97

HODOC No: 02785
CAS No: 7205-90-5
Beilstein No: 6^4, 1593
Name: Benzene, 1-chloro-4-[(chloromethyl)thio]-
Synonym: Sulfide, chloromethyl *p*-chlorophenyl
Line Form: 4-ClC_6H_4-S-CH_2Cl
Mol Form: $C_7H_6Cl_2S$
M_r: 193.08
bp: $128\text{-}9^{12}$
***d*:** 1.346
***n*:** 1.6055^{20}
IR: ALD 2:626F
MS: ALD 5:31D
m/e: 191.96

HODOC No: 02786
CAS No: 77008-62-9
Name: Benzene, 1-chloro-2-[1-(4-chlorophenyl)ethyl]-
Synonym: Ethane, 1(2-chlorophenyl)-1(4-chlorophenyl)
Mol Form: $C_{14}H_{12}Cl_2$
M_r: 251.14
IR: COB 7065 3080 2980 2880 1590 1490 1440 1380 1090 1040 1010 830 750 690 550
m/e: 250.03

HODOC No: 02787
CAS No: 77008-64-1
Name: Benzene, 1-chloro-3-[1-(4-chlorophenyl)ethyl]-
Synonym: Ethane, 1(3-chlorophenyl)-1(4-chlorophenyl)
Mol Form: $C_{14}H_{12}Cl_2$
M_r: 251.14
IR: μCOB 1177 3030 2940 1590 1490 1430 1410 1370 1100 1020 880 850 830 780 730 690
m/e: 250.03

HODOC No: 02788
CAS No: 134-83-8
Name: Benzene, 1-chloro-4-(chlorophenylmethyl)-
Synonym: Methane, chloro(*p*-chlorophenyl)phenyl-
Mol Form: $C_{13}H_{10}Cl_2$
M_r: 237.12
bp: 145^3
IR: μSADP 6293 3030 1640 1590 1490 1450 1390 1210 1090 1010 840 790 760 740 690
UV: SAD 1733 229 cyhex
H NMR: SAD 11016 5.9 7.2 CCl_4
m/e: 236.02

HODOC No: 02789
CAS No: 70288-97-0
Name: Benzene, 1-chloro-3-(dibromomethyl)-
Synonym: Toluene, α,α-dibromo-*m*-chloro-
Mol Form: $C_7H_5Br_2Cl$
M_r: 282.37
IR: COB 7631 1595 1575 1480 1435 1425 1230 1180 1140 1080 1000 890 880 785 690 655 585 570 530
m/e: 281.84

HODOC No: 02790
CAS No: 88-66-4
Beilstein No: 5^3, 699
Name: Benzene, 1-chloro-2-(dichloromethyl)-
Synonym: Toluene, *o*,α,α-trichloro-
Line Form: 2-$ClC_6H_4CHCl_2$
Mol Form: $C_7H_5Cl_3$
M_r: 195.47
bp: 228.5
***d*:** 1.399^{15}
m/e: 193.95

HODOC No: 02791
CAS No: 54965-40-1
Name: Benzene, 1-chloro-4-(2,2-dichloro-1-methylethyl)-
Synonym: Benzene, 1-chloro-4(1,1-dischloroisopropyl)
Mol Form: $C_9H_9Cl_3$
M_r: 223.52
MS: WILEY 1569 126(100) 128(33) 127(17) 222(16) 224(16) 150(14) 91(10) 89(8)
m/e: 221.98

HODOC No: 02792
CAS No: 608-23-1
Beilstein No: 5^3, 834
Name: Benzene, 1-chloro-2,3-dimethyl-
Synonym: *o*-Xylene, 3-chloro-
Line Form: 2,3-$(CH_3)_2C_6H_3Cl$
Mol Form: C_8H_9Cl
M_r: 140.60
bp: 185-7, $62\text{-}3^{12}$
***d*:** 1.053^{20}
***n*:** 1.526^{20}
***s*:** Bz 4; Chl 4
MS: WILEY 741
m/e: 140.04

HODOC No: 02793
CAS No: 95-66-9
Beilstein No: 5^4, 943
Name: Benzene, 1-chloro-2,4-dimethyl-
Synonym: *m*-Xylene, 4-chloro-
Line Form: 2,4-$(CH_3)_2C_6H_3Cl$
Mol Form: C_8H_9Cl
M_r: 140.60
bp: 187-8, 89^{24}
***d*:** $1.0598^{20/20}$
***n*:** 1.5230^{25}
***s*:** H_2O 1; ace 3; bz 4
UV: HBCP 375(741) 370(832) 358(871) hx
MS: ALDN 89 105(100) 140(50) 51(16) 77(15) 103(13) 125(10) 106(9) 139(9)
m/e: 140.04

HODOC No: 02794
CAS No: 556-97-8
Beilstein No: 5^3, 834
Name: Benzene, 1-chloro-3,5-dimethyl-
Synonym: *m*-Xylene, 5-chloro-
Line Form: 3,5-$(CH_3)_2C_6H_3Cl$
Mol Form: C_8H_9Cl
M_r: 140.60
bp: 187-8, 66^{12}
***s*:** H_2O 1; ace 3; bz 3
UV: HBCP 386(251) 375(302) 365(257) hx
m/e: 140.04

HODOC No: 02795
CAS No: 6781-98-2
Beilstein No: 5^4, 943
Name: Benzene, 2-chloro-1,3-dimethyl-
Synonym: *m*-Xylene, 2-chloro-
Line Form: 2,6-$(CH_3)_2C_6H_3Cl$
Mol Form: C_8H_9Cl
M_r: 140.60
bp: 185-7, 62-3^{12}
***d*:** 1.053^{20}
***n*:** 1.526^{20}
***s*:** H_2O 1; ace 3; bz 4; chl 3
UV: ZPCBA 25-217-34 277(155) 272(263) 266(275) hx
MS: WILEY 741 105(100) 140(44) 51(15) 103(14) 77(14) 106(10) 79(8) 78(7)
m/e: 140.04

HODOC No: 02796
CAS No: 95-72-7
Beilstein No: 5^4, 965
Name: Benzene, 2-chloro-1,4-dimethyl-
Synonym: *p*-Xylene, 2-chloro-
Line Form: 2,5-$(CH_3)_2C_6H_3Cl$
Mol Form: C_8H_9Cl
M_r: 140.60
bp: 187
mp: 1.6
***d*:** $1.0589^{15/4}$
***s*:** H_2O 1; ace 3; bz 4; ctc 3
IR: SADG 15756 2941 1613 1471 1370 1190 1042 870 794 680
UV: SAD 3625 278 269 264 215 MeOH
H NMR: SAD 6402 2.3 6.9 7.0 7.1 CCl_4
MS: WILEY 738 91(100) 140(18) 51(9) 92(8) 77(7) 65(6) 142(6) 39(4)
m/e: 140.04

HODOC No: 02797
CAS No: 615-60-1
Beilstein No: 5^3, 815
Name: Benzene, 4-chloro-1,2-dimethyl-
Synonym: *o*-Xylene, 4-chloro-
Line Form: 3,4-$(CH_3)_2C_6H_3Cl$
Mol Form: C_8H_9Cl
M_r: 140.60
bp: 194^{755}
mp: -6
***d*:** $1.0692^{15/15}$
***s*:** H_2O 1; ace 3; bz 4; ctc 3
IR: µSADP 23015 2940 1720 1590 1490 1450 1280 1210 1180 1120 1100 1020 880 830 810 770 700
UV: SAD 17654 278(226) 269(270) 265(231) 215(8150) MeOH
H NMR: SAD 7038 2.2 2.3 6.8-7.2 CCl_4
MS: ALDN 89 105(100) 140(46) 51(29) 125(25) 77(21) 39(15) 142(14) 103(13)
m/e: 140.04

HODOC No: 02798
CAS No: 3972-56-3
Beilstein No: 5^4, 1048
Name: Benzene, 1-chloro-4-(1,1-dimethylethyl)-
Synonym: Benzene, 1-*tert*-butyl-4-chloro-
Line Form: $4\text{-}ClC_6H_4C(CH_3)_3$
Mol Form: $C_{10}H_{13}Cl$
M_r: 168.66
bp: 211
***d*:** $1.0075^{18/4}$
***n*:** 1.5123^{20}
m/e: 168.07

HODOC No: 02799
CAS No: 515-40-2
Beilstein No: 5^4, 1048
Name: Benzene, (2-chloro-1,1-dimethylethyl)-
Synonym: (β-Chloro-*tert*-butyl)benzene
Line Form: $C_6H_5\text{-}C(CH_3)_2CH_2Cl$
Mol Form: $C_{10}H_{13}Cl$
M_r: 168.66
bp: 222^{741}, $104\text{-}5^{18}$
***d*:** $1.047^{20/4}$
***n*:** 1.5247^{20}
***s*:** Ac 4; Bz 4; Et_2O 4; EtOH 4
m/e: 168.07

HODOC No: 02800
CAS No: 602-02-8
Beilstein No: 5^4, 744
Name: Benzene, 1-chloro-2,3-dinitro-
Line Form: $2,3\text{-}(NO_2)_2C_6H_3Cl$
Mol Form: $C_6H_3ClN_2O_4$
M_r: 202.55
mp: 78
Color: pr (al) cr (MeOH)
***s*:** Et_2O 4
m/e: 201.98

HODOC No: 02801
CAS No: 97-00-7
Beilstein No: 5^4, 744
Name: Benzene, 1-chloro-2,4-dinitro-
Synonym: 1-Chloro-2,4-dinitrobenzene
Line Form: $2,4\text{-}(NO_2)C_6H_3Cl$
Mol Form: $C_6H_3ClN_2O_4$
M_r: 202.55
bp: 315, $158\text{-}60^2$
mp: 53
***d*:** $1.4982^{75/4}$
***n*:** 1.5857^{60}
Color: ye rh (eth) nd (al) ye cr
***s*:** H_2O 1; al 2; eth 3; bz 3 CS_2 3
IR: μCOB 347 2940 1610 1540 1470 1350 1250 1040 920 900 850 830 730
Raman: SAD 746 3090 1590 1550 1470 1390 1360 1300 1250 1190 1160 1140 1100 1090 1050 920 910 880 840 760 730 690 660 540 520 480 370 330 320 280 220 180 160
UV: SAD 5204 238 206 MeOH
MS: ALDN 164 30(100) 202(61) 75(55) 110(39) 74(36) 109(27) 63(22) 204(20)
m/e: 201.98

HODOC No: 02802
CAS No: 618-86-0
Beilstein No: 5^3, 637
Name: Benzene, 1-chloro-3,5-dinitro-
Synonym: 1-Chloro-3,5-dinitrobenzene
Line Form: $3,5\text{-}(NO_2)_2C_6H_3Cl$
Mol Form: $C_6H_3ClN_2O_4$
M_r: 202.55
mp: 59
Color: nd (al, peth)
***s*:** Et_2O 4; EtOH 4
m/e: 201.98

HODOC No: 02803
CAS No: 606-21-3
Beilstein No: 5^4, 744
Name: Benzene, 2-chloro-1,3-dinitro-
Synonym: 1-Chloro-2,6-dinitrobenzene
Line Form: 2,6-$(NO_2)_2C_6H_3Cl$
Mol Form: $C_6H_3ClN_2O_4$
M_r: 202.55
bp: 315
mp: 88
***d*:** 1.6867^{16}
Color: ye nd (al, HOAc)
***s*:** H_2O 1; al 3; eth 3; chl 2; tol 3
IR: SADG 1087 3030 1587 1538 1471 1370 1333 1064 909 826 704
UV: SAD 8976 289(1350) 208(17100) MeOH
H NMR: SAD 8593 7.8 8.1 $CDCl_3$
MS: ALDN 164 30(100) 75(71) 202(43) 63(38) 74(36) 110(31) 84(17) 91(17)
m/e: 201.98

HODOC No: 02804
CAS No: 610-40-2
Beilstein No: 5^4, 744
Name: Benzene, 4-chloro-1,2-dinitro-
Synonym: 1-Chloro-3,4-dinitrobenzene
Line Form: 3,4-$(NO_2)_2C_6H_3Cl$
Mol Form: $C_6H_3ClN_2O_4$
M_r: 202.55
bp: 315d, 160^4
mp: 40-1
Color: mcl or rh nd (eth or lig)
***s*:** Ac 4; Bz 4; Et_2O 4
m/e: 201.98

HODOC No: 02805
CAS No: 393-75-9
Name: Benzene, 2-chloro-1,3-dinitro-5-(trifluoromethyl)-
Synonym: Toluene, 4-chloro-α,α,α-trifluoro-3,5-dinitro-
Line Form: 4-Cl-3,5-$(NO_2)_2$-$C_6H_2CF_3$
Mol Form: $C_7H_2ClF_3N_2O_4$
M_r: 270.55
mp: 56-8
m/e: 269.97

HODOC No: 02806
CAS No: 948-98-1
Beilstein No: 5^4, 2162
Name: Benzene, 1,1′-(1-chloro-1,2-ethenediyl)*bis*-, (E)-
Synonym: Stilbene, α-chloro-, (E)-
Line Form: $C_6H_5CCl{=}CHC_6H_5$
Mol Form: $C_{14}H_{11}Cl$
M_r: 214.68
bp: 320-4
mp: 53-4
***s*:** H_2O 1; al 3; eth 3; bz 3; chl 3; peth 4; HOAc 3
UV: JACSA 87-863-65 284(21878) 220(12882) al
MS: NBS 1368
m/e: 214.05

HODOC No: 02807
CAS No: 948-99-2
Beilstein No: 5^4, 2161
Name: Benzene, 1,1′-(1-chloro-1,2-ethenediyl)*bis*-, (Z)-
Synonym: Stilbene, α-chloro-, (Z)-
Line Form: $C_6H_5CCl{=}CHC_6H_5$
Mol Form: $C_{14}H_{11}Cl$
M_r: 214.68
bp: $160\text{-}2^{12}$
***n*:** 1.6281^{19}
***s*:** H_2O 1; al 3; eth 3; bz 3; chl 3
UV: JACSA 87-863-65 271(10965) 223(19953) al
MS: NBS 1368
m/e: 214.05

HODOC No: 02808
CAS No: 618-34-8
Beilstein No: 5^4, 1345
Name: Benzene, (1-chloroethenyl)-
Synonym: Styrene, α-chloro-
Line Form: $C_6H_5CCl{=}CH_2$
Mol Form: C_8H_7Cl
M_r: 138.59
bp: 199, 73^{16}
mp: -23
***d*:** $1.1016^{18/4}$
***n*:** 1.5612^{20}
***s*:** Et_2O 4; EtOH 4
m/e: 138.02

HODOC No: 02809

CAS No: 2039-87-4
Beilstein No: 5^4, 1345
Name: Benzene, 1-chloro-2-ethenyl-
Synonym: Styrene, *o*-chloro-
Line Form: 2-$ClC_6H_4CH{=}CH_2$
Mol Form: C_8H_7Cl
M_r: 138.59
bp: 188.7, 64.6^{10}
mp: -63.1 (fp)
***d*:** 1.1000^{20}
***n*:** 1.5649^{20}
***s*:** al 3; eth 3; ace 3; ctc 3; peth 5; HOAc 3
IR: µCOB 184 3130 1640 1590 1560 1470 1450 1410 1280 1210 1120 1040 990 920 860 790 760 740 680
UV: SAD 3862 313 300 290 281 244 229 cyhex
H NMR: SAD 18061 5.3 5.6 7.1 7.0 7.3 7.4 CCl_4
m/e: 138.02

HODOC No: 02810

CAS No: 2039-85-2
Beilstein No: 5^4, 1345
Name: Benzene, 1-chloro-3-ethenyl-
Synonym: Styrene, *m*-chloro-
Line Form: 3-$ClC_6H_4CH{=}CH_2$
Mol Form: C_8H_7Cl
M_r: 138.59
bp: $62\text{-}3^6$
***d*:** $1.1168^{20/4}$
***n*:** 1.5625^{20}
***s*:** H_2O 1; al 3; eth 3
IR: COB 7094 3090 1830 1630 1595 1565 1475 1415 1200 1165 1080 985 915 880 850 785 710 685 650
UV: SAD 3863 297(716) 287(1090) 252(44000) 216(25600) 210(41000) 205 MeOH
m/e: 138.02

HODOC No: 02811

CAS No: 1073-67-2
Beilstein No: 5^4, 1345
Name: Benzene, 1-chloro-4-ethenyl-
Synonym: Styrene, *p*-chloro-
Mol Form: C_8H_7Cl
M_r: 138.59
bp: 192, 66.3^{10}
mp: 15.9
***d*:** $1.0868^{20/4}$
***n*:** 1.5660^{20}
***s*:** H_2O 1; al 3; eth 3; ace 5; bz 5; ctc 5; peth 5
IR: µCOB 136 3030 1640 1590 1540 1490 1390 1110 1090 1010 980 910 830 790 730
UV: SAD 21981 310(64) 304(66) 274(362) 269(358) MeOH
H NMR: VAR 498 5.3 5.7 6.7 7.3 $CDCl_3$
MS: WILEY 718 138(100) 103(55) 140(32) 51(22) 77(18) 102(14) 139(11) 50(10)
m/e: 138.02

HODOC No: 02812

CAS No: 4110-77-4
Beilstein No: 5^4, 1346
Name: Benzene, (2-chloroethenyl)-, (E)-
Synonym: Styrene, β-chloro-, (E)-
Line Form: $C_6H_5CH{=}CHCl$
Mol Form: C_8H_7Cl
M_r: 138.59
bp: 199, 90^{18}
***d*:** $1.1095^{18/4}$
***n*:** 1.5648^{20}
***s*:** H_2O 1; al 3; eth 3; ace 3
UV: OES 1-178 254(19055) cyhex
H NMR: SAD 15613 ini 6.1 6.4 6.5 6.7 7.1-7.4 7.6 CCl_4
MS: WILEY 717 103(100) 138(74) 51(50) 77(43) 140(24) 50(22) 102(16) 75(12)
m/e: 138.02

HODOC No: 02813
CAS No: 4604-28-8
Beilstein No: 5^4, 1346
Name: Benzene, (2-chloroethenyl)-, (Z)-
Synonym: Styrene, β-chloro-, (Z)-
Line Form: $C_6H_5CH{=}CHCl$
Mol Form: C_8H_7Cl
M_r: 138.59
bp: $59^{3.5}$
***d*:** $1.1046^{25/4}$
***n*:** 1.5762^{25}
UV: OES 1-178
H NMR: SAD 15613
MS: NBS 330
m/e: 138.02

HODOC No: 02814
CAS No: 18084-97-4
Beilstein No: 5^4, 2577
Name: Benzene, 1,1′,1″-(1-chloro-1-ethenyl-2-ylidene)*tris*-
Synonym: Ethylene, chlorotriphenyl-
Line Form: $(C_6H_5)_2C{=}CClC_6H_5$
Mol Form: $C_{20}H_{15}Cl$
M_r: 290.77
mp: 117
***s*:** H_2O 1; al 3; eth 3; ace 3; bz 3; chl 4; peth 3
UV: OES 3-768 288(10000) 228(19953) EtOH
m/e: 290.09

HODOC No: 02815
CAS No: 569-57-3
Name: Benzene, 1,1′,1″-(1-chloro-1-ethenyl-2-ylidene)*tris*[4-methoxy-
Synonym: Ethylene, chlorotris(*p*-methoxyphenyl)-
Mol Form: $C_{23}H_{21}ClO_3$
M_r: 380.85
mp: 114-6
IR: μCOB 5144 3077 2899 2857 1613 1522 1464 1443 1282 1250 1176 1149 1111 1026 971 866 847 837 826 813 772
m/e: 380.12

HODOC No: 02816
CAS No: 10477-40-4
Name: Benzene, (1-chloroethoxy)-
Synonym: (1-Chloroethoxy)benzene
Line Form: $C_6H_5OCH(Cl)CH_3$
Mol Form: C_8H_9ClO
M_r: 156.60
MS: WILEY 955 94(100) 156(41) 107(29) 77(22) 63(19) 65(15) 158(13) 27(13)
m/e: 156.03

HODOC No: 02817
CAS No: 614-72-2
Beilstein No: 6^4, 785
Name: Benzene, 1-chloro-2-ethoxy-
Synonym: Phenetole, *o*-chloro-
Line Form: $2\text{-}C_2H_5O\text{-}C_6H_4Cl$
Mol Form: C_8H_9ClO
M_r: 156.60
bp: 210, $97\text{-}8^{15}$
***d*:** $1.1288^{15/4}$
***n*:** 1.5284^{25}
***s*:** Bz 4; Et_2O 4; EtOH 4
MS: WILEY 956
m/e: 156.03

HODOC No: 02818
CAS No: 2655-83-6
Beilstein No: 6^4, 811
Name: Benzene, 1-chloro-3-ethoxy-
Synonym: Phenetole, *m*-chloro-
Line Form: $3\text{-}C_2H_5O\text{-}C_6H_4Cl$
Mol Form: C_8H_9ClO
M_r: 156.60
bp: $204\text{-}5^{717}$
***d*:** $1.1712^{20/4}$
***s*:** H_2O 1; al 4; eth 4; bz 4; HOAc 4
IR: μSADP 13922 2940 1610 1490 1390 1300 1240 1160 1090 1040 890 830 750 690
UV: SAD 3802 270 219 cyhex
MS: AMS 1-311 128(100) 131(32) 156(28) 65(21) 29(19) 27(18) 13(16) 64(12)
m/e: 156.03

HODOC No: 02819
CAS No: 622-61-7
Beilstein No: 6^4, 823
Name: Benzene, 1-chloro-4-ethoxy-
Synonym: Phenetole, *p*-chloro-
Line Form: 4-C_2H_5O-C_6H_4Cl
Mol Form: C_8H_9ClO
M_r: 156.60
bp: 212-4, 98^{17}
mp: 21
***d*:** $1.1254^{20/4}$
***n*:** $1.5252^{20/2}$
***s*:** al 3; eth 3; bz 4; ctc 2; HOAc 3
IR: µCOB 5691 2940 1610 1490 1410 1300 1240 1160 1120 1120 1040 1000 930 820
Raman: SAD 281 3070 2980 2940 2880 2770 2730 1600 1580 1490 1470 1450 1440 1390 1370 1280 1240 1170 1110 1090 1040 1000 920 900 830 800 780 690 660 630 510 480 410 380 360 330 300
UV: SAD 1284 288(2770) 281(3430) 299(25200) MeOH
H NMR: SAD 235 1.4 3.9 6.7 7.1 CCl_4
MS: WILEY 955 128(100) 130(33) 156(33) 65(11) 158(10) 75(6) 29(6) 63(6)
m/e: 156.03

HODOC No: 02820
CAS No: 622-86-6
Beilstein No: 6^4, 556
Name: Benzene, (2-chloroethoxy)-
Synonym: Phenetole, β-chloro-
Line Form: $C_6H_5OCH_2CH_2Cl$
Mol Form: C_8H_9ClO
M_r: 156.60
bp: 217-20, $100\text{-}2^{12}$
mp: 28
***s*:** H_2O 1; al 4; eth 4; ace 4; bz 4; ctc 2; lig 4
IR: µCOB 6783 3030 1590 1490 1450 1390 1270 1240 1150 1120 1040 1030 930 730
UV: SAD 1283 282(1850) 275(2090) 219(7250) MeOH
H NMR: SAD 234 1.4 4.0 6.6-7.4 CCl_4
m/e: 156.03

HODOC No: 02821
CAS No: 2243-44-9
Beilstein No: 6^3, 568
Name: Benzene, [2-(2-chloroethoxy)ethoxy]-
Synonym: Ethane, 1-(2-chloroethoxy)-2-phenoxy-
Line Form: $ClCH_2CH_2O$-CH_2CH_2-OC_6H_5
Mol Form: $C_{10}H_{13}ClO_2$
M_r: 200.66
bp: 149^{10}
***d*:** $1.149^{15/15}$
MS: NBS 1180
m/e: 200.06

HODOC No: 02822
CAS No: 21120-79-6
Beilstein No: 6^2, 329
Name: Benzene, 1-(2-chloroethoxy)-2-methyl-
Synonym: Phenetole, β-chloro-*o*-methyl-
Mol Form: $C_9H_{11}ClO$
M_r: 170.63
bp: $227\text{-}9^{762}$
MS: WILEY 1125 108(100) 107(39) 91(24) 63(23) 121(21) 172(18) 77(17) 170(6)
m/e: 170.05

HODOC No: 02823
CAS No: 32669-06-0
Name: Benzene, 1,1′-[(2-chloroethoxy)methylene]*bis*-
Synonym: Ether, 2-chloroethyl diphenylmethyl
Line Form: $ClCH_2CH_2OCH(C_6H_5)_2$
Mol Form: $C_{15}H_{15}ClO$
M_r: 246.72
IR: SADG 26935 3060 3020 2960 2860 1600 1580 1490 1450 1430 1390 1340 1290 1250 1180 1100 1070 1050 1030 970 920 870 840 830 750 740 690 670 650 610 560 480 460
Raman: SAD 1116 3080 3020 2970 2880 2780 1610 1590 1500 1460 1430 1400 1300 1260 1190 1160 1120 1080 1040 1010 930 880 840 760 670 660 620 570 270 90
H NMR: SAD 15667 5.2 7.0-7.4 CCl_4
m/e: 246.08

HODOC No: 02824
CAS No: 38661-82-4
Beilstein No: 5^4, 898
Name: Benzene, (1-chloroethyl)-, (±)-
Synonym: (±)-1-Chloro-1-phenylethane
Line Form: $C_6H_5CHClCH_3$
Mol Form: C_8H_9Cl
M$_r$: 140.60
bp: $81\text{-}2^{17}$
***d*:** $1.0620^{20/4}$
***n*:** 1.5276^{20}
***s*:** al 3; eth 4; bz 3
IR: SADG 31370 3060 3030 2980 2920 2860 1590 1480 1440 1370 1220 1190 1040 1020 960 900 750 680 630 610 570 510
UV: OES 1-197 260(200) 218(5370) cyhex
H NMR: SAD 19128 1.8 5.0 7.2-7.6 $CDCl_3$
MS: WILEY 740 ini 105(100) 140(19) 77(13) 103(12) 79(11) 106(9) 125(8) 104(8)
m/e: 140.04

HODOC No: 02825
CAS No: 1459-15-0
Beilstein No: 5^4, 898
Name: Benzene, (1-chloroethyl)-, (R)-
Synonym: Benzene, (1-chloroethyl)-, (R)-(+)-
Line Form: $C_6H_5CHClCH_3$
Mol Form: C_8H_9Cl
M$_r$: 140.60
bp: 85^{20}
***d*:** $1.0631^{20/4}$
***n*:** 1.5250^{25}
α: $[\alpha]^{20/D}$ +54.6 (undil)
***s*:** Bz 4; Et_2O 4; EtOH 4
MS: WILEY 740
m/e: 140.04

HODOC No: 02826
CAS No: 3756-41-0
Beilstein No: 5^4, 898
Name: Benzene, (1-chloroethyl)-, (*S*)-
Synonym: Benzene, (1-chloroethyl)-, (-)-
Line Form: $C_6H_5CHClCH_3$
Mol Form: C_8H_9Cl
M$_r$: 140.60
bp: 85^{20}
***d*:** $1.0632^{20/4}$
α: $[\alpha]^{20/D}$ -30.1 (undil)
***s*:** Bz 4; Et_2O 4; EtOH 4
MS: WILEY 740
m/e: 140.04

HODOC No: 02827
CAS No: 89-96-3
Beilstein No: 5^4, 897
Name: Benzene, 1-chloro-2-ethyl-
Synonym: *o*-Chloroethylbenzene
Line Form: $2\text{-}ClC_6H_4CH_2CH_3$
Mol Form: C_8H_9Cl
M$_r$: 140.60
bp: 178.4, 81.7^{30}
mp: -82.7
***d*:** $1.0569^{20/4}$
***n*:** 1.5218^{20}
***s*:** H_2O 1; ace 3; bz 3; ctc 3; chl 3
IR: SADG 26 3060 2960 2930 2870 1570 1470 1440 1130 1060 1020 790 750 680
Raman: SAD 771 3190 3150 3060 3020 2970 2940 2910 2880 2850 2730 1600 1570 1450 1380 1330 1310 1280 1240 1210 1160 1140 1100 1080 1060 1040 990 970 950 920 860 850 790 760 730 690 620 580 530 490 460 450 410 380 330 270 200 140
UV: SAD 32 273(222) 266(267) 259(203) 212(940) MeOH
H NMR: SAD 1111 1.2 2.7 7.2 CCl_4
MS: WILEY 743 125(100) 105(56) 140(41) 127(32) 51(15) 77(13) 142(13) 103(10)
m/e: 140.04

HODOC No: 02828
CAS No: 620-16-6
Beilstein No: 5^4, 897
Name: Benzene, 1-chloro-3-ethyl-
Synonym: *m*-Chloroethylbenzene
Line Form: $3\text{-}ClC_6H_4CH_2CH_3$
Mol Form: C_8H_9Cl
M$_r$: 140.60
bp: 183.8, 86.1^{30}
mp: -55 (fp)
***d*:** 1.0529^{20}
***n*:** 1.5195^{20}
***s*:** Ac 4; Bz 4; Et_2O 4; EtOH 4
MS: WILEY 740 125(100) 105(88) 140(52) 127(32) 142(17) 51(16) 77(15) 103(13)
m/e: 140.04

HODOC No: 02829
CAS No: 622-98-0
Beilstein No: 5^4, 897
Name: Benzene, 1-chloro-4-ethyl-
Synonym: *p*-Chloroethylbenzene
Line Form: $4\text{-}ClC_6H_4CH_2CH_3$
Mol Form: C_8H_9Cl
M_r: 140.60
bp: 184.4, 63.5^{10}
mp: -62.6
***d*:** $1.0455^{20/4}$
***n*:** 1.5175^{20}
***s*:** H^2O 1; al 5; eth 5; ace 5; peth 5; HOAc 3
MS: WILEY 741 125(100) 105(48) 140(39) 127(32) 51(13) 142(12) 77(11) 103(9)
m/e: 140.04

HODOC No: 02830
CAS No: 622-24-2
Beilstein No: 5^4, 899
Name: Benzene, (2-chloroethyl)-
Synonym: (β-Chloroethyl)benzene
Line Form: $C_6H_5CH_2CH_2Cl$
Mol Form: C_8H_9Cl
M_r: 140.60
bp: 197-8, 92^{20}
***d*:** $1.069^{25/4}$
***n*:** 1.5276^{20}
***s*:** H_2O 1; al 3; eth 3; ace 3; bz 3; ctc 2; CS_2 3; lig 3
IR: µSADP 8637 3030 1920 1610 1490 1450 1320 1250 1080 1030 920 890 820 750 710
UV: OES 2-139 209(7413) EtOH
H NMR: SAD 8594 3.0 3.6 7.2 CCl_4
MS: WILEY 738 105(100) 140(48) 142(15) 51(15) 77(14) 103(13) 139(10) 106(9)
m/e: 140.04

HODOC No: 02831
CAS No: 54789-30-9
Name: Benzene, 1-(1-chloroethyl)-4-(chloromethyl)-
Synonym: 4-(1-Chloroethyl)benzyl chloride
Mol Form: $C_9H_{10}Cl_2$
M_r: 189.08
IR: µCOB 1911 2940 1520 1450 1430 1390 1270 1240 1040 970 840 830 740 690
m/e: 188.02

HODOC No: 02832
CAS No: 332-43-4
Name: Benzene, 1-(2-chloroethyl)-4-fluoro-
Synonym: *p*-Fluorophenethyl chloride
Mol Form: C_8H_8ClF
M_r: 158.59
***n*:** $1.5077^{20/D}$
***s*:** ctc 3
IR: SADG 20782 3040 2960 2860 1600 1500 1430 1310 1220 1150 1080 1010 930 890 820 780 710 690 650 540 520 480 410
H NMR: SAD 11124 3.0 3.6 6.7-7.3 CCl_4
m/e: 158.03

HODOC No: 02833
CAS No: 873-73-4
Name: Benzene, 1-chloro-4-ethynyl-
Synonym: 1-Chloro-4-ethynylbenzene
Mol Form: C_8H_5Cl
M_r: 136.57
mp: 45.4-6.1
MS: WILEY 684 136(100) 138(32) 101(20) 75(11) 50(10) 137(9) 74(9) 100(6)
m/e: 136.01

HODOC No: 02834
CAS No: 348-51-6
Beilstein No: 5^4, 652
Name: Benzene, 1-chloro-2-fluoro-
Synonym: *o*-Chlorofluorobenzene
Line Form: $2\text{-}FC_6H_4Cl$
Mol Form: C_6H_4ClF
M_r: 130.54
bp: 137.6
mp: -43
***d*:** $1.2233^{30/4}$
***n*:** 1.4918^{30}
***s*:** H; iO2 1; ace 3; bz 3
IR: COB 6472 3070 3060 1900 1780 1590 1510 1480 1450 1380 1260 1240 1120 1070 1030 930 820 750 680 550 490
UV: OES 4-53 272(1514) 266(1413) 260(1000) 214(10000) 210(10000) cyhex
H NMR: SAD 16093 6.8-7.2 7.4 CCl_4
MS: WILEY 586 130(100) 95(43) 132(32) 75(18) 50(9) 94(8) 13(17) 74(6)
m/e: 130.00

HODOC No: 02835
CAS No: 625-98-9
Beilstein No: 5^4, 652
Name: Benzene, 1-chloro-3-fluoro-
Synonym: *m*-Chlorofluorobenzene
Line Form: $3\text{-}FC_6H_4Cl$
Mol Form: C_6H_4ClF
M$_r$: 130.54
bp: 127.6
***d*:** 1.221^{25}
***n*:** 1.4911
IR: COB 6385 3070 1590 1580 1530 1520 1480 1470 1430 1420 1330 1290 1260 1240 1220 1150 1130 1080 1060 1000 960 880 860 770 680 670 520
UV: OES 4-52 272(1000) 265(1000) 258(603) 215(8511) 211(8511) 211(8511) cyhex
H NMR: SAD 16094 6.7-7.4 CCl_4
MS: WILEY 586 ini 112(100) 77(54) 114(32) 78(29) 96(22) 51(20) 50(17) 130(1)
m/e: 130.00

HODOC No: 02836
CAS No: 352-33-0
Beilstein No: 5^4, 652
Name: Benzene, 1-chloro-4-fluoro-
Synonym: *p*-Chlorofluorobenzene
Line Form: $4\text{-}FC_6H_4Cl$
Mol Form: C_6H_4ClF
M$_r$: 130.54
bp: 130^{757}
mp: -26.8
***d*:** 1.4990^{15}
***n*:** 1.4990^{15}
***s*:** H_2O 1; al 3; eth 3; bz 3
IR: µCOB 6279 3030 1890 1590 1490 1430 1220 1150 1100 830
UV: OES 4-53 277(1148) 271(1096) 263(776) 215(5495) 209(5495) cyhex
H NMR: SAD 16095 6.9 7.2 $CDCl_3$
MS: NBS 264
m/e: 130.00

HODOC No: 02837
CAS No: 452-09-5
Name: Benzene, 4-chloro-2-fluoro-1-methoxy-
Synonym: Anisole, 4-chloro-2-fluoro-
Mol Form: C_7H_6ClFO
M$_r$: 160.57
MS: WILEY 1004 160(100) 145(76) 117(39) 162(32) 147(24) 119(12) 161(8) 81(7)
m/e: 160.01

HODOC No: 02838
CAS No: 443-83-4
Beilstein No: 5^3, 691
Name: Benzene, 1-chloro-3-fluoro-2-methyl-
Synonym: Toluene, 2-chloro-6-fluoro-
Line Form: $2\text{-}Cl\text{-}6\text{-}FC_6H_3CH_3$
Mol Form: C_7H_6ClF
M$_r$: 144.57
bp: 154-6
***d*:** 1.191
***n*:** 1.5026^{20}
IR: ALD 2:535G
MS: ALD 4:57B
m/e: 144.01

HODOC No: 02839
CAS No: 452-73-3
Beilstein No: 5^3, 691
Name: Benzene, 2-chloro-4-fluoro-1-methyl-
Synonym: Toluene, 2-chloro-4-fluoro-
Line Form: $2\text{-}Cl\text{-}4\text{-}FC_6H_3CH_3$
Mol Form: C_7H_6ClF
M$_r$: 144.57
bp: 152-3
***d*:** $1.1972^{20/4}$
***n*:** 1.4985^{25}
m/e: 144.01

HODOC No: 02840
CAS No: 350-31-2
Beilstein No: 5^2, 185
Name: Benzene, 1-chloro-2-fluoro-4-nitro-
Synonym: 1-Chloro-2-fluoro-4-nitrobenzene
Line Form: $2\text{-}F\text{-}4\text{-}NO_2C_6H_3Cl$
Mol Form: $C_6H_3ClFNO_2$
M$_r$: 175.54
bp: $114\text{-}6^{24}$
mp: 63-4
Color: pa ye lf (lig)
m/e: 174.98

HODOC No: 02841
CAS No: 345-17-5
Beilstein No: 5^4, 724
Name: Benzene, 1-chloro-4-fluoro-2-nitro-
Synonym: 1-Chloro-4-fluoro-2-nitrobenzene
Line Form: 4-F-2-NO_2-C_6H_3Cl
Mol Form: $C_6H_3ClFNO_2$
M_r: 175.54
bp: 238.5
mp: 37.3
Color: pr
m/e: 174.98

HODOC No: 02842
CAS No: 350-30-1
Beilstein No: 5^4, 725
Name: Benzene, 2-chloro-1-fluoro-4-nitro-
Synonym: 3-Chloro-4-fluoronitrobenzene
Line Form: 2-Cl-4-$NO_2C_6H_3F$
Mol Form: $C_6H_3ClFNO_2$
M_r: 175.54
bp: 227-32
mp: 41.5
m/e: 174.98

HODOC No: 02843
CAS No: 345-18-6
Beilstein No: 5^4, 725
Name: Benzene, 4-chloro-1-fluoro-2-nitro-
Synonym: 5-Chloro-2-fluoronitrobenzene
Line Form: 4-Cl-2-$NO_2C_6H_3F$
Mol Form: $C_6H_3ClFNO_2$
M_r: 175.54
bp: 138.5^{29}
mp: 10.2
Color: pr (lig)
m/e: 174.98

HODOC No: 02844
CAS No: 615-41-8
Beilstein No: 5^4, 695
Name: Benzene, 1-chloro-2-iodo-
Synonym: *o*-Chloroiodobenzene
Line Form: 2-IC_6H_4Cl
Mol Form: C_6H_4ClI
M_r: 238.45
bp: 234-5, 110^{18}
mp: 0.7
***d*:** 1.9515^{25}
***n*:** 1.6331^{25}
***s*:** H_2O 1; ace 3; ctc 2
IR: µCOB 1175 3130 1560 1450 1430 1250 1110 1100 1030 1010 750 720
UV: SAD 825 276(2970) 224(8840) MeOH
H NMR: SAD 690 6.9 7.4 7.8 CCl_4
MS: NBS 1624
m/e: 237.90

HODOC No: 02845
CAS No: 625-99-0
Beilstein No: 5^4, 695
Name: Benzene, 1-chloro-3-iodo-
Synonym: *m*-Chloroiodobenzene
Line Form: 3-IC_6H_4Cl
Mol Form: C_6H_4ClI
M_r: 238.45
bp: 230
***s*:** H_2O 1; ace 3
UV: OES 5-47 281(479) 274(646) 266(724) 258(724) 252(646) 236(10965) 231(12023) 224(8913) cyhex
MS: NBS 1624
m/e: 237.90

HODOC No: 02846
CAS No: 637-87-6
Beilstein No: 5^4, 696
Name: Benzene, 1-chloro-4-iodo-
Synonym: *p*-Chloroiodobenzene
Line Form: $4\text{-}IC_6H_4Cl$
Mol Form: C_6H_4ClI
M_r: 238.45
bp: 227, 108^{13}
mp: 57
***d*:** $1.886^{27/4}$
Color: lf (Ace, al)
***s*:** H_2O 1; al 3; chl 2 $phno_2$ 3
IR: COB 7225 1890 1470 1410 1385 1260 1090 1055 1005 810 720 475
UV: SAD 10886 236(17400) MeOH
H NMR: SAD 8783 7.1 7.6 $CDCl_3$
MS: NBS 1624
m/e: 237.90

HODOC No: 02847
CAS No: 2401-22-1
Beilstein No: 5^3, 726
Name: Benzene, 1-chloro-2-iodo-4-methyl-
Synonym: Toluene, 4-chloro-3-iodo-
Line Form: $3\text{-}I\text{-}4\text{-}ClC_6H_3CH_3$
Mol Form: C_7H_6ClI
M_r: 252.48
bp: 249, $120\text{-}5^{14.5}$
mp: -10
***d*:** 1.7940^{15}
***n*:** 1.614^{23}
m/e: 251.92

HODOC No: 02848
CAS No: 42048-11-3
Beilstein No: 5, 315
Name: Benzene, 1-chloro-3-iodo-2-methyl-
Synonym: Toluene, 2-chloro-6-iodo
Line Form: $6\text{-}I\text{-}2\text{-}ClC_6H_3CH_3$
Mol Form: C_7H_6ClI
M_r: 252.48
bp: $132\text{-}3^{25}$
***d*:** $1.844^{20/20}$
m/e: 251.92

HODOC No: 02849
CAS No: 83846-48-4
Beilstein No: 5^4, 842
Name: Benzene, 2-chloro-4-iodo-1-methyl-
Synonym: Toluene, 2-chloro-4-iodo-
Line Form: $4\text{-}I\text{-}2\text{-}ClC_6H_3CH_3$
Mol Form: C_7H_6ClI
M_r: 252.48
bp: $136\text{-}47^{16}$
mp: 8
***n*:** 1.624^{23}
m/e: 251.92

HODOC No: 02850
CAS No: 33184-48-4
Beilstein No: 5^3, 726
Name: Benzene, 4-chloro-2-iodo-1-methyl-
Synonym: Toluene, 4-chloro-2-iodo-
Line Form: $2\text{-}I\text{-}4\text{-}ClC_6H_3CH_3$
Mol Form: C_7H_6ClI
M_r: 252.48
bp: 242-3, $133\text{-}7^{7.2}$
mp: -25
***d*:** 1.8358^{15}
***n*:** 1.620^{23}
m/e: 251.92

HODOC No: 02851
CAS No: 3320-83-0
Beilstein No: 12^3, 1296
Name: Benzene, 1-chloro-2-isocyanato-
Synonym: Isocyanic acid, *o*-chlorophenyl ester
Line Form: $2\text{-}ClC_6H_4NCO$
Mol Form: C_7H_4ClNO
M_r: 153.56
bp: $115\text{-}7^{43}$
mp: 30-1
***s*:** ctc 2
IR: SADG 15887 3060 2240 1590 1510 1440 1160 1050 750
UV: SAD 17641 284(554) 276(715) 268(672) 235(12200) MeOH
H NMR: SAD 8150 7.1-7.5 7.6 CCl_4
MS: WILEY 920 ini 153(100) 125(41) 63(35) 90(34) 155(33) 37(20) 38(19) 62(19)
m/e: 153.00

HODOC No: 02852
CAS No: 2909-38-8
Beilstein No: 12[3], 1316
Name: Benzene, 1-chloro-3-isocyanato-
Synonym: Isocyanic acid, *m*-chlorophenyl ester
Line Form: 3-ClC_6H_4NCO
Mol Form: C_7H_4ClNO
M_r: 153.56
bp: $113\text{-}4^{43}$
***s*:** chl 2
IR: μSADP 22270 3030 2270 1590 1520 1450 1160 1090 1080 880 830 780 680
UV: SAD 17648 286(846) 278(1060) 238(16200) MeOH
H NMR: SAD 2527 7.2-7.8 $CDCl_3$
m/e: 153.00

HODOC No: 02853
CAS No: 104-12-1
Beilstein No: 12[3], 1376
Name: Benzene, 1-chloro-4-isocyanato-
Synonym: Isocyanic acid, *p*-chlorophenyl ester
Line Form: 4-ClC_6H_4NCO
Mol Form: C_7H_4ClNO
M_r: 153.56
bp: $115\text{-}7^{45}$
mp: 30-1
IR: μSADP 22271 3030 2270 1590 1520 1140 1110 1090 1010 830
UV: OES 4-96 287(692) 279(813) 273(661) 271(661) 235(16596) hx
MS: NBS 4169
m/e: 153.00

HODOC No: 02854
CAS No: 2131-55-7
Beilstein No: 12[2], 330
Name: Benzene, 1-chloro-4-isothiocyanato-
Synonym: Isothiocyanic acid, *p*-chlorophenyl ester
Mol Form: C_7H_4ClNS
M_r: 169.62
bp: 249-50
mp: 45
***s*:** H_2O 1; al 3
IR: SADG 20699 3080 2080 1590 1480 1090 1010 930 820
UV: OES 6-95 283(8318) MeOH
H NMR: SAD 11174 7.1 7.3 $CDCl_3$
MS: WILEY 1115 ini 169(100) 171(37) 69(33) 108(26) 63(16) 82(14) 134(13) 39(11)
m/e: 168.98

HODOC No: 02855
CAS No: 19241-35-1
Name: Benzene, 1-chloro-3-isothiocyanato-2-methyl-
Synonym: Isothiocyanic acid, 3-chloro-*o*-tolyl ester
Mol Form: C_8H_6ClNS
M_r: 183.65
***s*:** ctc 3
IR: SADG 20696 3060 2920 2070 1560 1440 1370 1270 1240 1200 1180 1140 1060 1010 940 770 730 690 610 540 440 390
H NMR: SAD 11171 2.3 6.9-7.3 CCl_4
m/e: 182.99

HODOC No: 02856
CAS No: 19241-37-3
Name: Benzene, 2-chloro-4-isothiocyanato-1-methyl-
Synonym: Isothiocyanic acid, 3-chloro-*p*-tolyl ester
Mol Form: C_8H_6ClNS
M_r: 183.65
***s*:** ctc 3
IR: SADG 20697 3060 2920 2070 1590 1550 1480 1430 1390 1370 1270 1230 1130 1040 990 950 860 800 690 620 560 510 430 410 330
H NMR: SAD 11172 2.3 7.0 7.1 7.2 CCl_4
m/e: 182.99

HODOC No: 02857
CAS No: 766-51-8
Beilstein No: 6[4], 785
Name: Benzene, 1-chloro-2-methoxy-
Synonym: Anisole, *o*-chloro-
Line Form: 2-$ClC_6H_4OCH_3$
Mol Form: C_7H_7ClO
M_r: 142.58
bp: 198.5, $90\text{-}1^{16}$
mp: -26.8
***d*:** $1.1911^{20/4}$
***n*:** 1.5480^{20}
***s*:** H_2O 1; al 3; eth 3; chl 2
IR: SADG 21856 3030 2941 2857 1587 1471 1449 1299 1282 1266 1136 1053 1020 794 746 690
UV: SAD 2281 283 218 cyhex
H NMR: SAD 11222 3.9 6.8 6.9 7.1-7.5 $CDCl_3$
MS: NBS 4113
m/e: 142.02

HODOC No: 02858
CAS No: 2845-89-8
Beilstein No: 6^4, 811
Name: Benzene, 1-chloro-3-methoxy-
Synonym: Anisole, *m*-chloro-
Line Form: $3\text{-}ClC_6H_4OCH_3$
Mol Form: C_7H_7ClO
M_r: 142.58
bp: 193-4, 70^{9}
***d*:** $1.1759^{12/4}$
***n*:** 1.5365^{20}
***s*:** H_2O 1:al 3; eth 3
IR: OES 5-93 282(2138) 275(2138) 219(7244) cy
UV: OES 5-93
m/e: 142.02

HODOC No: 02859
CAS No: 623-12-1
Beilstein No: 6^4, 822
Name: Benzene, 1-chloro-4-methoxy-
Synonym: Anisole, *p*-chloro-
Line Form: $4\text{-}ClC_6H_4OCH_3$
Mol Form: C_7H_7ClO
M_r: 142.58
bp: 197.5, 75^{10}
mp: 18
***d*:** $1.201^{20/4}$
***n*:** 1.5390^{20}
***s*:** H_2O 1; al 4; eth 4; ctc 3; chl 4
IR: SADG 338 3030 2857 1587 1471 1282 1176 1087 1031 826
UV: SAD 741 288(1360) 281(1660) 227(11200) MeOH
H NMR: SAD 164 3.7 6.7 7.2 CCl_4
C NMR: JJ 235 FT 55.3 115.2 125.4 129.2 158.2 $CDCl_3$
MS: NBS 4113
m/e: 142.02

HODOC No: 02860
CAS No: 57356-27-1
Beilstein No: 6^4, 1385
Name: Benzene, 1-chloro-5-methoxy-2,4-dinitro-
Synonym: 1-Chloro-5-methoxy-2,4-dinitrobenzene
Line Form: $5\text{-}CH_3O\text{-}2,4\text{-}(NO_2)_2\text{-}C_6H_2Cl$
Mol Form: $C_7H_5ClN_2O_5$
M_r: 232.57
mp: 105
IR: SADG 19608
UV: SAD 17370
H NMR: SAD 10030
m/e: 231.99

HODOC No: 02861
CAS No: 3587-60-8
Beilstein No: 6^4, 2253
Name: Benzene, [(chloromethoxy)methyl]-
Synonym: Ether, benzyl chloromethyl
Line Form: $C_6H_5CH_2OCH_2Cl$
Mol Form: C_8H_9ClO
M_r: 156.60
bp: 103^{13}
***d*:** $1.1350^{20/4}$
***n*:** 1.5192^{20}
m/e: 156.03

HODOC No: 02862
CAS No: 5472-99-1
Beilstein No: 6^4, 1347
Name: Benzene, 1-chloro-3-methoxy-2-nitro-
Synonym: Anisole, 3-chloro-2-nitro-
Line Form: $(3\text{-}Cl\text{-}2\text{-}NO_2)C_6H_3OCH_3$
Mol Form: $C_7H_6ClNO_3$
M_r: 187.58
mp: 56
***s*:** Bz 4; Et_2O 4; Chl 4
m/e: 187.00

HODOC No: 02863
CAS No: 10298-80-3
Beilstein No: 6^4, 1352
Name: Benzene, 1-chloro-4-methoxy-2-nitro-
Synonym: Anisole, 4-chloro-3-nitro-
Line Form: $(4\text{-}Cl\text{-}3\text{-}NO_2)C_6H_3OCH_3$
Mol Form: $C_7H_6ClNO_3$
M_r: 187.58
mp: 43-5
IR: ALD 3:821A
MS: ALD 5:166C
m/e: 187.00

HODOC No: 02864
CAS No: 89-21-4
Beilstein No: 6^2, 226
Name: Benzene, 4-chloro-1-methoxy-2-nitro-
Synonym: Anisole, 4-chloro-2-nitro-
Mol Form: $C_7H_6ClNO_3$
M_r: 187.58
mp: 98
***s*:** al 3; eth 2; chl 2; MeOH 3; lig 2
IR: µSADP 4597 2940 1610 1520 1470 1450 1350 1270 1180 1120 1010 890 810 710
UV: OES 5-86 342(2344) EtOH
H NMR: SAD 7082 4.0 7.1 7.6 7.8 $CDCl_3$
m/e: 187.00

HODOC No: 02865
CAS No: 100-44-7
Beilstein No: 5^4, 809
Name: Benzene, (chloromethyl)-
Synonym: Toluene, α-chloro-
Line Form: $C_6H_5CH_2Cl$
Mol Form: C_7H_7Cl
M_r: 126.58
bp: 179.3, 66^{11}
mp: -39
***d*:** $1.1002^{20/20}$
***n*:** 1.5391^{20}
***s*:** H_2O 1; al 5; eth 5; ctc 2; chl 5
IR: SADG 6 3060 3030 1500 1460 1270 1210 1070 1030 810 760 700 670
Raman: SAD 769 3060 3000 2960 1600 1580 1440 1260 1210 1180 1150 1030 1000 980 810 800 760 700 670 610 550 460 330 270 120
UV: SAD 5 265(211) 259(231) 254(190) 217(6970) MeOH
H NMR: SAD 3 4.5 7.3 CCl_4
MS: WILEY 531 91(100) 126(23) 2(15) 39(13) 65(12) 63(10) 92(9) 128(7)
m/e: 126.02

HODOC No: 02866
CAS No: 95-49-8
Beilstein No: 5^4, 805
Name: Benzene, 1-chloro-2-methyl-
Synonym: Toluene, *o*-chloro-
Line Form: $2\text{-}ClC_6H_4CH_3$
Mol Form: C_7H_7Cl
M_r: 126.58
bp: 159.1, 42.6^{10}
mp: -35.1
***d*:** $1.0825^{20/4}$
***n*:** 1.5268^{20}
***s*:** H_2O 1; al 3; eth 5; ace 5; bz 3; ctc 5; chl 5; hp 5
IR: µCOB 3794 3030 2940 1590 1490 1470 1450 1390 1280 1140 1060 1040 1020 990 940 860 800 750 700 680
Raman: DFB 79 3150 3060 3020 2990 2920 2910 2860 2740 2570 1590 1570 1440 1400 1380 1280 1210 1160 1130 1090 1050 1040 1020 990 800 740 680 550 510 440 360 240 160
UV: SAD 37 273(250) 265(288) 259 212(9390) MeOH
H NMR: SAD 13 2.4 7.2 $CDCl_3$
C NMR: JJ 232 FT 19.9 126.4 127.0 129.0 130.9 134.4 135.9 $CDCl_3$
MS: WILEY 530 91(100) 126(28) 128(9) 65(9) 92(8) 63(6) 39(5) 89(5)
m/e: 126.02

HODOC No: 02867
CAS No: 108-41-8
Beilstein No: 5^4, 806
Name: Benzene, 1-chloro-3-methyl-
Synonym: Toluene, *m*-chloro-
Line Form: $3\text{-}ClC_6H_4CH_3$
Mol Form: C_7H_7Cl
M_r: 126.58
bp: 162
mp: -47.8
***d*:** $1.0722^{20/4}$
***n*:** 1.5214^{19}
***s*:** H_2O 1; al 3; eth 5; bz 3; ctc 3; chl 3
IR: COB 5671 3070 2930 1610 1600 1580 1480 1220 1160 1100 1080 1000 890 870 860 770 680
UV: SAD 1116 274(172) 265(200) 260(150) 213(5560) MeOH
H NMR: SAD 3168 2.3 7.1 CCl_4
C NMR: JJ 233 FT 21.0 125.5 127.1 129.1 129.3 134.0 139.7 $CDCl_3$
MS: WILEY 533 91(100) 126(48) 128(16) 125(14) 63(10) 89(9) 65(8) 92(8)
m/e: 126.02

HODOC No: 02868
CAS No: 106-43-4
Beilstein No: 5^4, 806
Name: Benzene, 1-chloro-4-methyl-
Synonym: Toluene, *p*-chloro-
Line Form: $4\text{-}ClC_6H_4CH_3$
Mol Form: C_7H_7Cl
M_r: 126.58
bp: 162, 44^{10}
mp: 7.5
***d*:** $1.0697^{20/4}$
***n*:** 1.5150^{20}
***s*:** H_2O 1; al 3; eth 5; ctc 3; chl 3; HOAc 3
IR: COB 4280 3040 2920 2870 1880 1630 1600 1490 1450 1400 1380 1100 1020 800 640 490 380
Raman: DFB 85 3190 3100 3060 3040 2970 2920 2870 2730 1600 1580 1380 1300 1210 1170 1090 1040 820 800 690 640 380 300 250
UV: SAD 212 277(507) 269(527) 263(388) 256 244 220(11000) MeOH
H NMR: SAD 51 2.3 7.0 7.1 CCl_4
C NMR: JJ 234 FT 20.7 128.3 130.4 131.2 136.2 $CDCl_3$
MS: WILEY 532 91(100) 126(42) 125(17) 128(13) 63(9) 89(8) 92(8) 65(7)
m/e: 126.02

HODOC No: 02869
CAS No: 91244-98-3
Name: Benzene, 3-(chloromethyl)-1,4-dimethoxy-2,5-dimethyl-
Synonym: Benzene, 3(chlormethyl)-1-4-dimethoxy-2,5-dimethyl
Mol Form: $C_{11}H_{15}ClO_2$
Mr: 214.68
IR: SADG 14131 2960 2930 2840 1510 1460 1430 1400 1310 1250 1220 1170 1130 1040 910 870 820 730 680 600 480
Raman: SAD 1115 3050 3010 3000 2930 2850 1610 1590 1460 1450 1410 1380 1330 1280 1260 1230 1190 1160 1070 1020 1000 950 910 860 780 700 680 650 630 540 500 470 340 210 160
UV: SAD 16230 293(3150) MeOH
H NMR: SAD 7722 2.3 3.8 4.7 6.6 $CDCl_3$
m/e: 214.08

HODOC No: 02870
CAS No: 824-55-5
Beilstein No: 5^4, 1012
Name: Benzene, 1-(chloromethyl)-2,4-dimethyl-
Synonym: 1-(Chloromethyl)-2,4-dimethylbenzene
Line Form: 2,4-$(CH_3)_2C_6H_3CH_2Cl$
Mol Form: $C_9H_{11}Cl$
Mr: 154.63
bp: 215-6, $86\text{-}7^{12}$
***s*:** Bz 4; Et_2O 4; EtOH 4
m/e: 154.05

HODOC No: 02871
CAS No: 90-99-3
Beilstein No: 12, 604
Name: Benzene, 1,1′-(chloromethylene)*bis*-
Synonym: Methane, chlorodiphenyl-
Line Form: $(C_6H_5)_2CHCl$
Mol Form: $C_{13}H_{11}Cl$
Mr: 202.67
bp: 140^3
mp: 15-7
***d*:** 1.140
***n*:** 1.5951^{20}
***s*:** chl 3
IR: SADG 8548 3040 1600 1490 1460 1220 1080 1030 820 750 690
Raman: SAD 772 3070 2950 1610 1590 1500 1460 1270 1220 1180 1160 1070 1040 1010 920 850 830 780 760 720 700 640 620 590 510 410 380 350 310 270 240 180 100
UV: SAD 2403 218 cyhex
H NMR: VAR 176 6.1 7.3 $CDCl_3$
MS: NBS 1210; ALD 5:74D
m/e: 202.05

HODOC No: 02872
CAS No: 1712-70-5
Name: Benzene, 1-chloro-4-(1-methylethenyl)-
Synonym: Styrene, *p*-chloro-α-methyl-
Mol Form: C_9H_9Cl
Mr: 152.61
IR: μCOB 7012 3130 2940 1640 1590 1490 1450 1390 1370 1320 1110 1100 1010 890 830 760 740
m/e: 152.04

HODOC No: 02873
CAS No: 3360-55-2
Beilstein No: 5, 485
Name: Benzene, (2-chloro-1-methylethenyl)-
Synonym: Styrene, β-chloro-α-methyl-
Line Form: $C_6H_5C(CH_3)=CHCl$
Mol Form: C_9H_9Cl
Mr: 152.61
bp: 213-5, $102\text{-}6^{14}$
m/e: 152.04

HODOC No: 02874
CAS No: 934-53-2
Beilstein No: 5^4, 992
Name: Benzene, (1-chloro-1-methylethyl)-
Synonym: Cumene, α-chloro-
Line Form: $C_6H_5CCl(CH_3)_2$
Mol Form: $C_9H_{11}Cl$
M_r: 154.63
bp: 98^{1}
***d*:** 1.192^{25}
***n*:** 1.5290^{25}
m/e: 154.05

HODOC No: 02875
CAS No: 2077-13-6
Beilstein No: 5^4, 992
Name: Benzene, 1-chloro-2-(1-methylethyl)-
Synonym: Cumene, *o*-chloro-
Line Form: $2\text{-}ClC_6H_4CH(CH_3)_2$
Mol Form: $C_9H_{11}Cl$
M_r: 154.63
bp: 191.1, 92^{30}
mp: -74.4 (fp)
***d*:** 1.0341^{20}
***n*:** 1.5168^{20}
***s*:** Ac 4; Bz 4; Et_2O 4; EtOH 4
MS: WILEY 926 139(100) 103(34) 141(32) 154(30) 77(17) 51(11) 156(10) 140(9)
m/e: 154.05

HODOC No: 02876
CAS No: 1467-05-6
Beilstein No: 5^4, 1004
Name: Benzene, 1-(chloromethyl)-4-ethyl-
Synonym: Toluene, α-chloro-*p*-ethyl-
Line Form: $4\text{-}C_2H_5\text{-}C_6H_4CH_2Cl$
Structure:

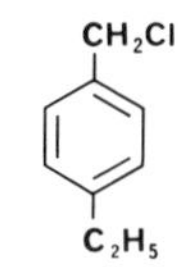

Mol Form: $C_9H_{11}Cl$
M_r: 154.63
bp: $95\text{-}6^{15}$
***n*:** 1.5290^{25}
***s*:** Bz 4; EtOH 4; Chl 4
IR: COB 4930 3030 2960 1600 1500 1450 1260 1030 900 740 700
MS: WILEY 923 ini 119(100) 154(31) 91(16) 103(13) 104(12) 139(12) 105(12) 156(10)
m/e: 154.05

HODOC No: 02877
CAS No: 2621-46-7
Beilstein No: 5^4, 992
Name: Benzene, 1-chloro-4-(1-methylethyl)-
Synonym: Cumene, *p*-chloro-
Line Form: $4\text{-}ClC_6H_4CH(CH_3)_2$
Mol Form: $C_9H_{11}Cl$
M_r: 154.63
bp: 198.3, 74^{10}
mp: -12.3
***d*:** $1.0208^{20/4}$
***n*:** 1.5117^{20}
***s*:** H_2O 1; al 5; eth 5; ace 5; bz 4; ctc 5; peth
IR: μCOB 189 2940 1590 1540 1490 1410 1090 1050 1010 830 760
MS: WILEY 924 103(100) 141(77) 154(72) 77(33) 51(25) 139(24) 156(24) 140(22)
m/e: 154.05

HODOC No: 02878
CAS No: 824-47-5
Beilstein No: 5^4, 992
Name: Benzene, (2-chloro-1-methylethyl)-
Synonym: Cumene, β-chloro-
Line Form: $C_6H_5CH(CH_3)CH_2Cl$
Mol Form: $C_9H_{11}Cl$
M_r: 154.63
bp: 85^{13}
***d*:** 1.047^{19}
***n*:** 1.5245^{19}
MS: NBS 510
m/e: 154.05

HODOC No: 02879
CAS No: 345-35-7
Beilstein No: 5^4, 813
Name: Benzene, 1-(chloromethyl)-2-fluoro-
Synonym: Toluene, α-chloro-*o*-fluoro-
Line Form: $2\text{-}FC_6H_4CH_2Cl$
Mol Form: C_7H_6ClF
M_r: 144.57
bp: 86^{40}
***d*:** 1.216
***n*:** 1.5150^{20}
IR: ALD 2:527H
MS: ALD 4:44A
m/e: 144.01

HODOC No: 02880
CAS No: 352-11-4
Beilstein No: 5^4, 813
Name: Benzene, 1-(chloromethyl)-4-fluoro-
Synonym: Toluene, α-chloro-*p*-fluoro-
Line Form: $4\text{-}FC_6H_4CH_2Cl$
Mol Form: C_7H_6ClF
M_r: 144.57
bp: 82^{26}
***d*:** 1.207
***n*:** 1.5130
IR: ALD 2:532C
MS: ALD 4:54A
m/e: 144.01

HODOC No: 02881
CAS No: 76-83-5
Beilstein No: 5^4, 2497
Name: Benzene, 1,1′,1″-(chloromethylidyne)*tris*-
Synonym: Methane, chlorotriphenyl-
Line Form: $(C_6H_5)_3CCl$
Mol Form: $C_{19}H_{15}Cl$
M_r: 278.76
bp: 310, $230\text{-}5^{20}$
mp: 113-4
Color: nd or pr (bz-peth)
***s*:** H_2O 1; al 2; eth 4; ace 3; bz 4; ctc 4; chl 4; CS_2 4
IR: SADG 8336 3070 1490 1450 1320 1160 1040 1000 900 760 700 660
UV: SAD 1259 259(822) 253(781) MeOH
H NMR: SAD 213 7.2 CCl_4
m/e: 278.09

HODOC No: 02882
CAS No: 824-94-2
Beilstein No: 6^4, 2137
Name: Benzene, 1-(chloromethyl)-4-methoxy-
Synonym: Anisole, *p*-(chloromethyl)-
Line Form: $4\text{-}CH_3OC_6H_4CH_2Cl$
Mol Form: C_8H_9ClO
M_r: 156.60
bp: $116\text{-}20^{15}$, $83\text{-}4^2$
***d*:** 1.159^{20}
***n*:** 1.553
IR: μSADP 19934 3030 2940 2860 1610 1590 1520 1470 1450 1430 1330 1320 1250 1180 1110 1030 910 830 730
UV: SAD 6608 282 274 229 MeOH
m/e: 156.03

HODOC No: 02883
CAS No: 93-06-1
Name: Benzene, 2-(chloromethyl)-1-methoxy-4-nitro-
Synonym: Anisole, 2-(chloromethyl)-4-nitro-
Mol Form: $C_8H_8ClNO_3$
M_r: 201.60
mp: 76-80
***s*:** chl 3
IR: SADG 14376 3060 3020 2960 2850 1610 1590 1500 1490 1450 1330 1270 1090 1010 920 820 750 670 630
UV: SAD 16331 304(9030) 229(8210) MeOH
H NMR: SAD 8062 4.0 4.6 7.0 8.1 8.2 $CDCl_3$
m/e: 201.02

HODOC No: 02884
CAS No: 552-45-4
Beilstein No: 5^4, 926
Name: Benzene, 1-(chloromethyl)-2-methyl-
Synonym: *o*-Xylene, α-chloro-
Line Form: $2\text{-}CH_3C_6H_4CH_2Cl$
Mol Form: C_8H_9Cl
Mr: 140.60
bp: 197-9, $95\text{-}6^{25}$
***d*:** 1.063
***n*:** 1.5408^{20}
***s*:** H_2O 1; al 5; eth 5
IR: SADG 16698 3020 2960 1600 1490 1460 1375 1285 1255 1210 1190 1150 1100 1040 1030 830 760 740 720 670 590 530 490 440
H NMR: SAD 8605 2.4 4.5 7.2 $CDCl_3$
MS: WILEY 739 105(100) 140(27) 51(13) 77(11) 103(9) 79(9) 106(9) 142(9)
m/e: 140.04

HODOC No: 02885
CAS No: 620-19-9
Beilstein No: 5^4, 943
Name: Benzene, 1-(chloromethyl)-3-methyl-
Synonym: *m*-Xylene, α-chloro-
Line Form: $3\text{-}CH_3C_6H_4CH_2Cl$
Mol Form: C_8H_9Cl
Mr: 140.60
bp: 195-6, 101-2
***d*:** 1.064^{20}
***n*:** 1.5345^{20}
***s*:** H_2O 1; al 3; eth 3
MS: WILEY 743 105(100) 140(25) 77(10) 51(10) 106(9) 142(8) 79(8) 103(8)
m/e: 140.04

HODOC No: 02886
CAS No: 104-82-5
Beilstein No: 5^4, 966
Name: Benzene, 1-(chloromethyl)-4-methyl-
Synonym: *p*-Xylene, α-chloro-
Line Form: $4\text{-}CH_3C_6H_4CH_2Cl$
Mol Form: C_8H_9Cl
Mr: 140.60
bp: 200-2, 81^{15}
***d*:** $1.0512^{20/4}$
***n*:** 1.5380
***s*:** H_2O 1; al 3; eth 5
IR: SADG 178 3020 2960 1620 1520 1450 1380 1270 1110 840 820 750 730 670 520 470
MS: WILEY 740 102(100) 140(25) 51(13) 77(12) 79(10) 103(10) 106(9) 142(8)
m/e: 140.04

HODOC No: 02887
CAS No: 4395-79-3
Beilstein No: 5^4, 1062
Name: Benzene, 2-chloro-1-methyl-4-(1-methylethyl)-
Synonym: *p*-Cymene, 2-chloro-
Line Form: $2\text{-}Cl\text{-}4\text{-}(CH_3)_2CH\text{-}C_6H_3CH_3$
Mol Form: $C_{10}H_{13}Cl$
Mr: 168.66
bp: 216-7, 104^{20}
***d*:** $1.0104^{25/4}$
***n*:** 1.5078^{20}
***s*:** H_2O 1; ace 3; bz 3
MS: WILEY 1100 153(100) 155(32) 168(30) 117(25) 115(17) 91(11) 170(11) 154(10)
m/e: 168.07

HODOC No: 02888
CAS No: 4395-80-6
Beilstein No: 5^3, 956
Name: Benzene, 2-chloro-4-methyl-1-(1-methylethyl)-
Synonym: *p*-Cymene, 3-chloro-
Line Form: $2\text{-}Cl\text{-}4\text{-}CH_3C_6H_3CH(CH_3)_2$
Mol Form: $C_{10}H_{13}Cl$
Mr: 168.66
bp: $213\text{-}4^{736}$, 113^{35}
***d*:** 1.018^{15}
***n*:** 1.5180^{20}
m/e: 168.07

HODOC No: 02889
CAS No: 60468-54-4
Name: Benzene, 1-(chloromethyl)-2-methyl-3-nitro-
Synonym: α1-Chloro-3-nitro-*o*-xylene
Line Form: $(3\text{-}NO_2\text{-}2\text{-}CH_3)C_6H_3CH_2Cl$
Mol Form: $C_8H_8ClNO_2$
M_r: 185.60
mp: 43-6
m/e: 185.02

HODOC No: 02890
CAS No: 84540-59-0
Beilstein No: 5^2, 302
Name: Benzene, 4-(chloromethyl)-1-methyl-2-nitro-
Synonym: *p*-Xylene, α4-chloro-2-nitro-
Line Form: $(3\text{-}NO_2\text{-}4\text{-}CH_3)C_6H_3CH_2Cl$
Mol Form: $C_8H_8ClNO_2$
M_r: 185.60
bp: $87\text{-}8^{0.05}$
mp: 48-50
m/e: 185.02

HODOC No: 02891
CAS No: 612-23-7
Beilstein No: 5^4, 854
Name: Benzene, 1-(chloromethyl)-2-nitro-
Synonym: Toluene, α-chloro-*o*-nitro-
Line Form: $2\text{-}NO_2C_6H_4CH_2Cl$
Mol Form: $C_7H_6ClNO_2$
M_r: 171.58
mp: 50-2
***n*:** 1.5557^{62}
Color: cr (lig)
***s*:** H_2O 1; al 3; eth 3; ace 4; bz 4; AcOEt 4; HOAc 3
IR: SADG 11091 3030 2840 1600 1560 1490 1440 1420 1360 1320 1290 1250 1180 1150 1120 850 790 770 750 690 660
UV: SAD 6794 265(10700) MeOH
H NMR: SAD 18138 4.9 7.3-7.7 8.0 $CDCl_3$
m/e: 171.01

HODOC No: 02892
CAS No: 83-42-1
Beilstein No: 5^4, 854
Name: Benzene, 1-chloro-2-methyl-3-nitro-
Synonym: Toluene, 2-chloro-6-nitro-
Line Form: $2\text{-}Cl\text{-}6\text{-}NO_2\text{-}C_6H_3CH_3$
Mol Form: $C_7H_6ClNO_2$
M_r: 171.58
bp: 238
mp: 37-40
***n*:** 1.5377^{69}
Color: nd_(dil al)
***s*:** H_2O 1; al 3
IR: SADG 436 3090 2870 1600 1530 1450 1350 1280 1200 1110 1020 870 800 750 730 700
UV: SAD 1291 253(6060) 210(2580) MeOH
m/e: 171.01

HODOC No: 02893
CAS No: 13290-74-9
Beilstein No: 5^4, 855
Name: Benzene, 1-chloro-2-methyl-4-nitro-
Synonym: Toluene, 2-chloro-5-nitro-
Line Form: $2\text{-}Cl\text{-}5\text{-}NO_2C_6H_3CH_3$
Mol Form: $C_7H_6ClNO_2$
M_r: 171.58
bp: 248^{711}
mp: 41
Color: ye cr
***s*:** Et_2O 4
m/e: 171.01

HODOC No: 02894
CAS No: 619-23-8
Beilstein No: 5^4, 855
Name: Benzene, 1-(chloromethyl)-3-nitro-
Synonym: Toluene, α-chloro-*m*-nitro-
Line Form: $3\text{-}NO_2C_6H_4CH_2Cl$
Mol Form: $C_7H_6ClNO_2$
M_r: 171.58
bp: $175\text{-}83^{30\text{-}5}$
mp: 45-7
***n*:** 1.5577^{62}
Color: pa ye nd (lig)
***s*:** Ac 4; Bz 4; Et_2O 4; EtOH 4
m/e: 171.01

HODOC No: 02895
CAS No: 5367-26-0
Beilstein No: 5^4, 853
Name: Benzene, 1-chloro-3-methyl-2-nitro-
Synonym: Toluene, 3-chloro-2-nitro-
Line Form: $3\text{-}Cl\text{-}2\text{-}NO_2C_6H_3CH_3$
Mol Form: $C_7H_6ClNO_2$
M_r: 171.58
mp: 23
Color: ye cr
m/e: 171.01

HODOC No: 02896
CAS No: 16582-38-0
Beilstein No: 5^4, 855
Name: Benzene, 1-chloro-3-methyl-5-nitro-
Synonym: Toluene, 3-chloro-5-nitro-
Line Form: $3\text{-}Cl\text{-}5\text{-}NO_2C_6H_3CH_3$
Mol Form: $C_7H_6ClNO_2$
M_r: 171.58
mp: 61
***n*:** 1.5404^{69}
Color: ye nd (al)
***s*:** EtOH 4
m/e: 171.01

HODOC No: 02897
CAS No: 100-14-1
Beilstein No: 5^4, 856
Name: Benzene, 1-(chloromethyl)-4-nitro-
Synonym: Toluene, α-chloro-*p*-nitro-
Line Form: $4\text{-}NO_2C_6H_4CH_2Cl$
Mol Form: $C_7H_6ClNO_2$
M_r: 171.58
mp: 71
***n*:** 1.5647^{62}
Color: pl or nd (al)
***s*:** H_2O 1; al 3; eth 3; ace 4; bz 4; chl 3; AcOEt 4; MeOH 3
IR: COB 5827 3100 3080 2850 1610 1530 1480 1450 1420 1350 1310 1260 1210 1170 1100 1010 870 860 700 670
UV: SAD 3800 260 cyhex
H NMR: SAD 3258 4.6 7.6 8.2 $CDCl_3$
m/e: 171.01

HODOC No: 02898
CAS No: 89-60-1
Beilstein No: 5^4, 855
Name: Benzene, 1-chloro-4-methyl-2-nitro-
Synonym: Toluene, 4-chloro-3-nitro-
Line Form: $4\text{-}Cl\text{-}3\text{-}NO_2C_6H_3CH_3$
Mol Form: $C_7H_6ClNO_2$
M_r: 171.58
bp: 260^{745}, 118^{11}
mp: 7
***n*:** 1.5572^{20}
***s*:** H_2O 1; ctc 3
IR: µSADP 18798 3030 2940 2860 1540 1470 1350 1330 1280 1270 1150 1140 1050 920 890 820 800 750 690
UV: SAD 6072 298 250 cyhex
H NMR: SAD 2107 2.4 7.3 7.5 CCl_4
m/e: 171.01

HODOC No: 02899
CAS No: 3970-40-9
Beilstein No: 5^4, 854
Name: Benzene, 2-chloro-1-methyl-3-nitro-
Synonym: Toluene, 2-chloro-3-nitro-
Line Form: $2\text{-}Cl\text{-}3\text{-}NO_2C_6H_3CH_3$
Mol Form: $C_7H_6ClNO_2$
M_r: 171.58
bp: 263
mp: 21-2
m/e: 171.01

HODOC No: 02900
CAS No: 121-86-8
Beilstein No: 5^4, 855
Name: Benzene, 2-chloro-1-methyl-4-nitro-
Synonym: Toluene, 2-chloro-4-nitro-
Line Form: $2\text{-}Cl\text{-}4\text{-}NO_2C_6H_3CH_3$
Mol Form: $C_7H_6ClNO_2$
M_r: 171.58
bp: 260
mp: 68
***n*:** 1.5470^{69}
Color: nd (al)
***s*:** H_2O 2; al 3; eth 3; chl 2; HOAc 3
IR: µSADP 18905 3130 2940 2860 1540 1490 1450 1390 1350 1280 1210 1160 1120 1040 890 820 800 760 710
UV: SAD 6140 304 254 MeOH
H NMR: SAD 5000 2.5 7.4 8.0 8.2 $CDCl_3$
m/e: 171.01

HODOC No: 02901
CAS No: 38939-88-7
Beilstein No: 5^4, 856
Name: Benzene, 2-chloro-4-methyl-1-nitro-
Synonym: 1-Chloro-5-methyl-2-nitrobenzene
Line Form: 3-Cl-4-$NO_2C_6H_3CH_3$
Mol Form: $C_7H_6ClNO_2$
Mr: 171.58
bp: 146^{19}
mp: 24
Color: pa ye nd
s: H_2O 1; ctc 3
IR: SADG 15876 3100 3040 2930 2860 1600 1530 1480 1350 1260 1220 1160 1050 1000 830 750 690
H NMR: SAD 6119 2.4 7.2 7.3 7.7 CCl_4
m/e: 171.01

HODOC No: 02902
CAS No: 89-59-8
Beilstein No: 5^4, 853
Name: Benzene, 4-chloro-1-methyl-2-nitro-
Synonym: Toluene, 4-chloro-2-nitro-
Line Form: 4-Cl-2-$NO_2C_6H_3CH_3$
Mol Form: $C_7H_6ClNO_2$
Mr: 171.58
bp: 240^{720}, 115.5^{11}
mp: 38
d: 1.2559^{80}
Color: mcl nd
s: H_2O 1; al 3; eth 3; chl 2
IR: SADG 21056 3070 2930 1520 1480 1340 1110 880 810 790 750 700
UV: SAD 3293 298 250 cyhex
H NMR: SAD 6772 2.6 7.3 7.4 7.9 $CDCl_3$
m/e: 171.01

HODOC No: 02903
CAS No: 5367-28-2
Beilstein No: 5^4, 854
Name: Benzene, 4-chloro-2-methyl-1-nitro-
Synonym: Toluene, 5-chloro-2-nitro-
Line Form: 5-Cl-2-$NO_2C_6H_3CH_3$
Mol Form: $C_7H_6ClNO_2$
Mr: 171.58
mp: 25-6
n: 1.5496^{65}
Color: ye
s: H_2O 1; ctc 3
IR: SADG 13701 3110 2990 2940 2850 1600 1570 1520 1470 1440 1380 1340 1290 1200 1100 890 820 750 680
UV: SAD 15700 265(7230) MeOH
H NMR: SAD 7397 2.6 7.3 7.9 CCl_4
m/e: 171.01

HODOC No: 02904
CAS No: 484-65-1
Beilstein No: 5^4, 1139
Name: Benzene, (chloromethyl)pentamethyl-
Synonym: (Chloromethyl)pentamethylbenzene
Line Form: $(CH_3)_5C_6CH_2Cl$
Mol Form: $C_{12}H_{17}Cl$
Mr: 196.71
mp: 82-4
m/e: 196.10

HODOC No: 02905
CAS No: 4039-92-3
Name: Benzene, 1-(chloromethyl)-4-phenoxy-
Synonym: Ether, α-chloro-*p*-tolyl phenyl
Mol Form: $C_{13}H_{11}ClO$
Mr: 218.67
bp: $125\text{-}30^{0.2}$
IR: µCOB 1984 3030 1590 1490 1240 1160 1100 870 830 750 690
m/e: 218.05

HODOC No: 02906
CAS No: 24033-03-2
Name: Benzene, 1-(chloromethyl)-3-(phenylmethoxy)-
Synonym: Ether, benzyl α-chloro-*m*-tolyl
Line Form: 3-$(C_6H_5CH_2O)C_6H_4CH_2Cl$
Mol Form: $C_{14}H_{13}ClO$
Mr: 232.70
mp: 77-9
IR: ALD 2:560D
MS: ALD 4:89D
m/e: 232.07

HODOC No: 02907
CAS No: 1754-74-1
Name: Benzene, (2-chloro-2-methylpropyl)-
Synonym: 1,1-Dimethyl-2-phenylethyl chloride
Line Form: $C_6H_5CH_2CCl(CH_3)_2$
Mol Form: $C_{10}H_{13}Cl$
M_r: 168.66
***s*:** chl 3
H NMR: VAR 263 1.6 3.1 7.3 $CDCl_3$
m/e: 168.07

HODOC No: 02908
CAS No: 1084-17-9
Beilstein No: 5^3, 772
Name: Benzene, 2-chloro-4-methyl-1,3,5-trinitro-
Synonym: Toluene, 3-chloro-2,4,6-trinitro-
Mol Form: $C_7H_4ClN_3O_6$
M_r: 261.57
mp: 150-1
***s*:** H_2O 1; al 2; eth 2; ace 3; bz 3; chl 3; HOAc 3; lig 2
IR: µSADP 22179 3030 1587 1538 1429 1389 1351 1220 1176 1064 980 917 901 769 746 735 725
UV: SAD 19577 340(2200) 210(23300) MeOH
m/e: 260.98

HODOC No: 02909
CAS No: 88-73-3
Beilstein No: 5^4, 721
Name: Benzene, 1-chloro-2-nitro-
Synonym: *o*-Chloronitrobenzene
Line Form: $2\text{-}NO_2C_6H_4Cl$
Mol Form: $C_6H_4ClNO_2$
M_r: 157.55
bp: 246, 119^{8}
mp: 34-5
***d*:** 1.368^{242}
Color: mcl nd
***s*:** H_2O 1; al 3; eth 3; ace 4; bz 3; ctc 3; tol 4; py 4; MeOH 3
IR: µSADP 3231 3030 1590 1540 1470 1350 1300 1270 1180 1140 1080 1010 890 880 810 750 740
UV: OES 5-47 286(1288) 244(3981) 216(14125) 210(15849) cyhex
H NMR: SAD 10548 7.2-7.6 7.8 $CDCl_3$
m/e: 156.99

HODOC No: 02910
CAS No: 121-73-3
Beilstein No: 5^4, 722
Name: Benzene, 1-chloro-3-nitro-
Synonym: *m*-Chloronitrobenzene
Line Form: $3\text{-}NO_2C_6H_4Cl$
Mol Form: $C_6H_4ClNO_2$
M_r: 157.55
bp: 235-6
mp: 46
***d*:** $1.343^{50/4}$
***n*:** $1.5374^{80/\alpha}$
Color: pa ye rh pr (al)
***s*:** H_2O 1; al 3; eth 3; bz 3; chl 3; CS_2 3; HOAc 3
IR: µCOB 6359 3030 1590 1540 1470 1350 1300 1140 1060 880 790 740 730
Raman: SAD 304 3100 1580 1550 1530 1470 1400 1360 1300 1270 1240 1170 1130 1010 1000 880 820 750 660 540 500 420 390 340 220 190 120 100
UV: OES 2-46 257(6310) EtOH
H NMR: SAD 6923 7.5 7.7 8.1 $CDCl_3$
MS: WILEY 974 111(100) 157(73) 75(52) 113(32) 159(23) 74(10) 29(10) 76(8)
m/e: 156.99

HODOC No: 02911
CAS No: 100-00-5
Beilstein No: 5^4, 723
Name: Benzene, 1-chloro-4-nitro-
Synonym: *p*-Chloronitrobenzene
Line Form: $4\text{-}NO_2C_6H_4Cl$
Mol Form: $C_6H_4ClNO_2$
M_r: 157.55
bp: 242, 113^{8}
mp: 83.6
***d*:** $1.2979^{90.5}$
***n*:** $1.5376^{100/\alpha}$
Color: mcl pr
***s*:** H_2O l; al 2; eth 3; chl 3; CS_2 3
IR: µCOB 1409 3130 1610 1590 1540 1470 1430 1350 1320 1280 1180 1110 1100 1010 850 790 740 670
Raman: SAD 305 3100 1610 1580 1520 1480 1430 1360 1350 1180 1110 1100 860 750 740 630 530 320 290 210
UV: SAD 1290 270(10300) 215(8140) MeOH
H NMR: VAR 122 7.5 8.2 $CDCl_3$
MS: ALDN 114 157(100) 111(94) 75(91) 127(40) 99(30) 50(28) 113(27) 159(25)
m/e: 156.99

HODOC No: 02912
CAS No: 21969-11-9
Name: Benzene, 1-chloro-4-[(4-nitrophenyl)thio]-
Synonym: Sulfide, *p*-chlorophenyl *p*-nitrophenyl
Mol Form: $C_{12}H_8ClNO_2S$
M_r: 265.70
IR: SADG 27135 3090 3060 3020 2910 2830 1590 1570 1510 1470 1420 1390 1360 1340 1290 1260 1210 1190 1170 1110 1080 1010 950 940 850 820 740 720 700 680 630 560 540 490 400
Raman: SAD 293 3080 1600 1580 1510 1480 1430 1400 1370 1360 1340 1300 1190 1180 1110 1100 1080 1070 1020 860 820 750 730 630 560 550 410 350 270 190
H NMR: SAD 15780 7.2 8.0 $CDCl_3$
m/e: 265.00

HODOC No: 02913
CAS No: 932-33-2
Beilstein No: 5^4, 705
Name: Benzene, 1-chloro-2-nitroso-
Synonym: *o*-Chloronitrosobenzene
Line Form: 2-ONC_6H_4Cl
Mol Form: C_6H_4ClNO
M_r: 141.55
mp: 65-6
Color: nd (al)
***s*:** Bz 4; Et_2O 4; EtOH 4; peth 4
m/e: 141.00

HODOC No: 02914
CAS No: 932-78-5
Beilstein No: 5^4, 705
Name: Benzene, 1-chloro-3-nitroso-
Synonym: *m*-Chloronitrosobenzene
Line Form: 3-ONC_6H_4Cl
Mol Form: C_6H_4ClNO
M_r: 141.55
mp: 72
Color: nd (Bz)
***s*:** Ac 4; Bz 4; Et_2O 4; EtOH 4
m/e: 141.00

HODOC No: 02915
CAS No: 932-98-9
Beilstein No: 5^4, 705
Name: Benzene, 1-chloro-4-nitroso-
Synonym: *p*-Chloronitrosobenzene
Line Form: 4-ONC_6H_4Cl
Mol Form: C_6H_4ClNO
M_r: 141.55
mp: 92-3
Color: (HOAc or al)
***s*:** al 3; chl 4; HOAc 3
IR: DMS 5367 1587 1504 1462 1402 1316 1174 1085 1010 822
UV: OES 3-52 750(46) 313(12023) 288(10715) 226(7413) EtOH
MS: NBS 4111
m/e: 141.00

HODOC No: 02916
CAS No: 121-17-5
Beilstein No: 5^4, 857
Name: Benzene, 1-chloro-2-nitro-4-(trifluoromethyl)-
Synonym: Toluene, 4-chloro-α,α,α-trifluoro-3-nitro-
Line Form: 4-Cl-3-$NO_2C_6H_3CF_3$
Mol Form: $C_7H_3ClF_3NO_2$
M_r: 225.55
bp: 222
***d*:** 1.511
***n*:** 1.4893^{20}
IR: SADG 80 3080 1620 1580 1560 1530 1480 1400 1370 1320 1280 1240 1170 1150 1120 1080 1050910 830 780 750 720 670 520
Raman: SAD 742 3090 1620 1580 1550 1500 1490 1400 1360 1300 1260 1180 1140 1080 1060 900 800 750 670 520 490 360 280 140
UV: SAD 93 213(16100) MeOH 212(17600) MeOH:KOH
H NMR: SAD 466 8.1 CCl^4
MS: ALD 5:165C
m/e: 224.98

HODOC No: 02917
CAS No: 777-37-7
Beilstein No: 5^4, 857
Name: Benzene, 1-chloro-4-nitro-2-(trifluoromethyl)-
Synonym: Toluene, 2-chloro-α,α,α-trifluoro-5-nitro-
Line Form: $5\text{-}NO_2\text{-}2\text{-}ClC_6H_3CF_3$
Mol Form: $C_7H_3ClF_3NO_2$
M_r: 225.55
bp: 232, 129^{30}
mp: 22 (fp)
***d*:** 1.527
***n*:** 1.5083^{26}
IR: SADG 33049 3100 1610 1580 1530 1470 1410 1350 1300 1280 1250 1170 1140 1120 1040 920 890 840 790 740 730 680 660 610 530 450 430
Raman: SAD 1146 3090 1620 1590 1530 1480 1420 1360 1250 1180 1130 1040 890 800 730 660 540 440 320 260 140
UV: SAD 20850 263(9460) MeOH
H NMR: SAD 19481 7.8 8.4 8.6 $CDCl_3$
MS: ALD 5:165B
m/e: 224.98

HODOC No: 02918
CAS No: 118-83-2
Beilstein No: 5^3, 748
Name: Benzene, 4-chloro-1-nitro-2-(trifluoromethyl)-
Synonym: Toluene, 5-chloro-α,α,α-trifluoro-2-nitro-
Line Form: $5\text{-}Cl\text{-}2\text{-}NO_2C_6H_3CF_3$
Mol Form: $C_7H_3ClF_3NO_2$
M_r: 225.55
***d*:** 1.526
***n*:** 1.4980^{20}
IR: ALD 2:718G
MS: ALD 5: 165D
m/e: 224.98

HODOC No: 02919
CAS No: 344-07-0
Name: Benzene, chloropentafluoro-
Synonym: Chloropentafluorobenzene
Mol Form: C_6ClF_5
M_r: 202.51
bp: $122.0\text{-}3.0^{750}$
***d*:** 1.568^{25}
***n*:** 1.4256^{20}
IR: COB 4309 1740 1640 1620 1510 1460 1440 1360 1340 1240 1150 1100 1060 1010 980 880 830 710 310
UV: SAD 11720 258(447) MeOH
MS: ALDN 164 202(100) 204(33) 117(21) 167(20) 133(8) 203(7) 93(5) 101(5)
m/e: 201.96

HODOC No: 02920
CAS No: 2689-07-8
Name: Benzene, 1-chloro-2-phenoxy-
Synonym: Ether, *o*-chlorophenyl phenyl
Mol Form: $C_{12}H_9ClO$
M_r: 204.64
UV: OES 4-375 282(794) 276(1698) 269(1585) EtOH
MS: WILEY 1442 169(100) 204(66) 141(30) 76(30) 206(22) 51(20) 170(14) 70(11)
m/e: 204.03

HODOC No: 02921
CAS No: 6452-49-9
Name: Benzene, 1-chloro-3-phenoxy-
Synonym: Ether, *m*-chlorophenyl phenyl
Mol Form: $C_{12}H_9ClO$
M_r: 204.64
MS: WILEY 1442 204(100) 141(44) 206(33) 77(28) 169(20) 51(18) 205(16) 70(14)
m/e: 204.03

HODOC No: 02922
CAS No: 7005-72-3
Beilstein No: 6^4, 826
Name: Benzene, 1-chloro-4-phenoxy-
Synonym: Ether, *p*-chlorophenyl phenyl
Line Form: $(4\text{-}ClC_6H_4)OC_6H_5$
Mol Form: $C_{12}H_9ClO$
M_r: 204.64
bp: 284-5
***d*:** 1.2026^{15}
***n*:** 1.599
UV: OES 4-375 289(794) 279(1514) 272(1413) EtOH
MS: WILEY 1441 204(100) 206(34) 141(29) 77(23) 205(14) 51(12) 169(7) 75(6)
m/e: 204.03

HODOC No: 02923
CAS No: 39145-47-6
Name: Benzene, 1-(4-chlorophenoxy)-2-nitro-
Synonym: Ether, *p*-chlorophenyl *o*-nitrophenyl
Mol Form: $C_{12}H_8ClNO_3$
M_r: 249.64
mp: 42-5
IR: SADG 20797 3080 1610 1580 1530 1480 1450 1400 1340 1300 1230 1190 1150 1080 1000 870 840 770 740 660 590 490
UV: SAD 18776 308(2100) 223(19200) MeOH
H NMR: SAD 11353 6.9-7.7 7.9 $CDCl_3$
m/e: 249.02

HODOC No: 02924
CAS No: 1657-52-9
Beilstein No: 5^4, 2161
Name: Benzene, 1-chloro-2-(2-phenylethenyl)-, (E)-
Synonym: Stilbene, 2-chloro-, (E)-
Line Form: $2\text{-}ClC_6H_4CH{=}CHC_6H_5$
Mol Form: $C_{14}H_{11}Cl$
M_r: 214.68
bp: $208\text{-}10^{30}$,$138\text{-}40^{2}$
mp: 39-40
***s*:** H_2O 1; al 3; chl 3
UV: OES 7-512 295(19498) EtOH
MS: NBS 1368
m/e: 214.05

HODOC No: 02925
CAS No: 1657-51-8
Beilstein No: 5^4, 2160
Name: Benzene, 1-chloro-2-(2-phenylethenyl)-, (Z)-
Synonym: Stilbene, 2-chloro-, (Z)-
Line Form: $2\text{-}ClC_6H_4CH{=}CHC_6H_5$
Mol Form: $C_{14}H_{11}Cl$
M_r: 214.68
bp: $118\text{-}20^{2}$
mp: $118\text{-}20^{1.5}$
Color: ye oil
***s*:** H_2O 1; al 3; chl 3
UV: OES 7-512 276(11220) EtOH
MS: NBS 1368
m/e: 214.05

HODOC No: 02926
CAS No: 14064-43-8
Beilstein No: 5^4, 2161
Name: Benzene, 1-chloro-3-(2-phenylethenyl)-, (E)-
Synonym: Stilbene, 3-chloro-, (E)-
Line Form: $3\text{-}ClC_6H_4CH{=}CHC_6H_5$
Mol Form: $C_{14}H_{11}Cl$
M_r: 214.68
bp: $175\text{-}80^{0.2}$
mp: 73-4
***s*:** H_2O 1; al 3; chl 3
UV: TETRA 23-173-67 300(26925) cyhex
MS: NBS 1368
m/e: 214.05

HODOC No: 02927
CAS No: 14064-67-6
Name: Benzene, 1-chloro-3-(2-phenylethenyl)-, (Z)-
Synonym: Stilbene, 3-chloro-, (Z)-
Line Form: $3\text{-}ClC_6H_4CH{=}CHC_6H_5$
Mol Form: $C_{14}H_{11}Cl$
M_r: 214.68
bp: $175\text{-}80^{0.2}$
mp: 73-4
MS: NBS 1368
m/e: 214.05

HODOC No: 02928
CAS No: 4714-23-2
Beilstein No: 5^{4}, 2161
Name: Benzene, 1-chloro-4-(2-phenylethenyl)-
Synonym: Stilbene, 4-chloro-
Line Form: 4-$ClC_6H_4CH=CHC_6H_5$
Mol Form: $C_{14}H_{11}Cl$
M_r: 214.68
mp: 129
Color: lf (peth)
MS: NBS 1368
m/e: 214.05

HODOC No: 02929
CAS No: 1657-50-7
Beilstein No: 5^{3}, 1961
Name: Benzene, 1-chloro-4-(2-phenylethenyl)-, (E)-
Synonym: Stilbene, 4-chloro-, (E)-
Mol Form: $C_{14}H_{11}Cl$
M_r: 214.68
mp: 129
***s*:** H_2O 1; al 2; chl 3; peth 2; HOAc 2
UV: OES 4-494 283(10965) EtOH
m/e: 214.05

HODOC No: 02930
CAS No: 1657-49-4
Name: Benzene, 1-chloro-4-(2-phenylethenyl)-, (Z)-
Synonym: Stilbene, 4-chloro-, (Z)-
Mol Form: $C_{14}H_{11}Cl$
M_r: 214.68
bp: $118\text{-}20^{1.5}$ 108^{1}
***s*:** H_2O 1; al 3; chl 3
UV: TETRA 23-187-61 276(10715) MeOH
m/e: 214.05

HODOC No: 02931
CAS No: 831-81-2
Beilstein No: 5^{4}, 1847
Name: Benzene, 1-chloro-4-(phenylmethyl)-
Synonym: Methane, (*p*-chlorophenyl)phenyl-
Line Form: (4-ClC_6H_4)$CH_2C_6H_5$
Mol Form: $C_{13}H_{11}Cl$
M_r: 202.67
bp: 298^{742}, $147\text{-}8^{8}$
mp: 7.5
***d*:** $1.1247^{20/4}$
***s*:** Ac 4
m/e: 202.05

HODOC No: 02932
CAS No: 80-00-2
Beilstein No: 6^{4}, 1587
Name: Benzene, 1-chloro-4-(phenylsulfonyl)-
Synonym: Sulfone, *p*-chlorophenyl phenyl
Line Form: 4-ClC_6H_4-SO_2-C_6H_5
Mol Form: $C_{12}H_9ClO_2S$
M_r: 252.70
mp: 94
***s*:** H_2O 1; al 2; eth 3; ace 4; bz 4; os 3
IR: µSADP 17818 3030 1560 1490 1450 1390 1320 1280 1240 1150 1100 1060 1010 840 830 820 760 730 700 690
UV: SAD 5649 274 267 242 MeOH
MS: NBS 4653
m/e: 252.00

HODOC No: 02933
CAS No: 35673-03-1
Beilstein No: 5^{2}, 372
Name: Benzene, (1-chloro-1-propenyl)-
Synonym: Benzene, (1-chloropropenyl)-
Line Form: $C_6H_5CCl=CHCH_3$
Mol Form: C_9H_9Cl
M_r: 152.61
bp: $90.5^{0.9}$
***d*:** $1.085^{20/4}$
***n*:** 1.5635^{15}
***s*:** H_2O 1; ace 3; bz 3; tol 3
UV: HBCP 253(1906) cyhex
m/e: 152.04

HODOC No: 02934
CAS No: 23204-80-0
Beilstein No: 5^3, 1186
Name: Benzene, 1-chloro-3-(1-propenyl)-, (E)-
Synonym: Benzene, 1-chloro-3-propenyl-, (E)-
Line Form: $ClCH_2CH{=}CHC_6H_5$
Mol Form: C_9H_9Cl
M$_r$: 152.61
bp: 106.7^{13}
mp: 8-9
***d*:** $1.0926^{20/4}$
***n*:** 1.5851^{20}
***s*:** H_2O 1; al 5; eth 5; ace 4; bz 3
IR: SADG 27940 3050 3010 2950 1700 1650 1590 1570 1490 1450 1440 1300 1250 1200 1150 1080 1030 960 840 750 690 600 500
UV: HBCP 253(20417) cyhex
H NMR: SAD 17294 ini 4.1 6.2 6.6 7.3 CCl_4
MS: WILEY 902 ini 177(100) 115(36) 152(21) 116(11) 118(11) 57.5(9) 58(8) 154(7)
m/e: 152.04

HODOC No: 02935
CAS No: 13099-50-8
Beilstein No: 5^3, 1186
Name: Benzene, (2-chloro-1-propenyl)-
Synonym: Benzene, (2-chloropropenyl)-
Line Form: $C_6H_5CH{=}CClCH_3$
Mol Form: C_9H_9Cl
M$_r$: 152.61
bp: $118\text{-}23^{28}$
IR: SADP 17183
m/e: 152.04

HODOC No: 02936
CAS No: 2687-12-9
Beilstein No: 5^4, 1359
Name: Benzene, (3-chloro-1-propenyl)-
Synonym: Benzene, (3-chloropropenyl)-
Line Form: $C_6H_5CH{=}CHCH_2Cl$
Mol Form: C_9H_9Cl
M$_r$: 152.61
bp: 140^{37}, 115^{13}
mp: 8-9
Color: nd
m/e: 152.04

HODOC No: 02937
CAS No: 6268-37-7
Beilstein No: 5^4, 1363
Name: Benzene, (3-chloro-2-propenyl)-
Synonym: Benzene, (3-chloroallyl)-
Line Form: $C_6H_5CH_2CH{=}\text{-}CHCl$
Mol Form: C_9H_9Cl
M$_r$: 152.61
bp: 212-4, 97^{18}
***d*:** $1.073^{13/4}$
***n*:** 1.545^{14}
***s*:** H_2O 1; eth 3; ace 3; bz 3; tol 3
Raman: API 147 3054 2967 1660 1600 1303 1251 1212 1157 1000 678 670 605
UV: HBCP 253(3020) hx
MS: WILEY 901 117(100) 152(73) 154(28) 119(23) 91(21) 118(16) 115(12) 58(10)
m/e: 152.04

HODOC No: 02938
CAS No: 54644-23-4
Beilstein No: 6^2, 529
Name: Benzene, 1-(3-chloro-2-propenyl)-4-methoxy-
Synonym: Benzene, 1(3-chloroallyl)-4-methoxy
Mol Form: $C_{10}H_{11}ClO$
M$_r$: 182.64
bp: 120^{11}
***d*:** $1.155^{8/4}$
***n*:** $1.553^{8/D}$
MS: WILEY 1238 147(100) 182(47) 91(25) 115(23) 184(16) 131(13) 77(13) 148(11)
m/e: 182.05

HODOC No: 02939
CAS No: 13997-70-1
Beilstein No: 6^4, 825
Name: Benzene, 1-chloro-4-(2-propenyloxy)-
Synonym: Ether, allyl *p*-chlorophenyl
Line Form: $4\text{-}ClC_6H_4\text{-}O\text{-}CH_2CH{=}CH_2$
Mol Form: C_9H_9ClO
M$_r$: 168.61
bp: $106\text{-}7^{12}$
***d*:** 1.131^{15}
***n*:** 1.5348^{25}
***s*:** Bz 4; Et_2O 4; EtOH 4
m/e: 168.03

HODOC No: 02940
CAS No: 53299-53-9
Name: Benzene, [(2-chloro-2-propenyl)oxy]-
Synonym: Ether, 2-chloroallyl phenyl
Line Form: $CH_2=C(Cl)CH_2OC_6H_5$
Mol Form: C_9H_9ClO
M_r: 168.61
MS: WILEY 1096 133(100) 131(72) 94(50) 105(40) 39(38) 77(36) 132(35) 168(29)
m/e: 168.03

HODOC No: 02941
CAS No: 54410-95-6
Name: Benzene, [(3-chloro-2-propenyl)oxy]-
Synonym: 3-Chloroallyl phenyl ether
Line Form: $ClCH=CHCH_2OC_6H_5$
Mol Form: C_9H_9ClO
M_r: 168.61
MS: WILEY 1096 75(100) 77(46) 94(38) 133(36) 39(28) 131(22) 168(19) 51(14)
m/e: 168.03

HODOC No: 02942
CAS No: 54644-21-2
Name: Benzene, 1-[(3-chloro-2-propenyl)oxy]-2-methyl-
Synonym: Toluene, 2(3-chloroallyloxy)
Mol Form: $C_{10}H_{11}ClO$
M_r: 182.64
MS: WILEY 1237 75(100) 108(86) 107(76) 77(62) 147(33) 39(32) 79(32) 182(18)
m/e: 182.05

HODOC No: 02943
CAS No: 54644-22-3
Name: Benzene, 1-[(3-chloro-2-propenyl)oxy]-3-methyl-
Synonym: Toluene, 3(3-chloroallyloxy)
Mol Form: $C_{10}H_{11}ClO$
M_r: 182.64
MS: WILEY 1238 75(100) 108(55) 147(54) 77(52) 107(50) 39(28) 91(23) 182(22)
m/e: 182.05

HODOC No: 02944
CAS No: 51496-26-5
Name: Benzene, 4-chloro-2-(2-propenyl)-1-(2-propenyloxy)-
Synonym: 2-Allyloxy-3-allylchlorobenzene
Mol Form: $C_{12}H_{13}ClO$
M_r: 208.68
IR: µCOB 1987 3030 2940 1640 1590 1490 1430 1280 1250 1180 1140 1020 990 920 870 800 690
m/e: 208.07

HODOC No: 02945
CAS No: 1730-86-5
Name: Benzene, 1-chloro-2-propyl-
Synonym: *o*-Chloropropylbenzene
Line Form: $2\text{-}ClC_6H_4CH_2CH_2CH_3$
Mol Form: $C_9H_{11}Cl$
M_r: 154.63
bp: 79^{10}
***d*:** $1.0367^{17/4}$
***n*:** 1.5213^{16}
MS: WILEY 924 125(100) 127(32) 154(24) 119(10) 126(9) 156(8) 91(7) 89(7)
m/e: 154.05

HODOC No: 02946
CAS No: 52944-34-0
Beilstein No: 5^4, 979
Name: Benzene, 1-chloro-4-propyl-
Synonym: *p*-Chloropropylbenzene
Line Form: $4\text{-}ClC_6H_4CH_2CH_2CH_3$
Mol Form: $C_9H_{11}Cl$
M_r: 154.63
bp: 203
***n*:** $1.5151^{17.9}$
m/e: 154.05

HODOC No: 02947
CAS No: 55449-46-2
Beilstein No: 5^4, 980
Name: Benzene, (2-chloropropyl)-, (R)-
Synonym: R-(-)-2-Chloro-1-phenylpropane
Line Form: $CH_3CHCl\text{-}CH_2C_6H_5$
Mol Form: $C_9H_{11}Cl$
Mr: 154.63
bp: 94^{17}
***d*:** $1.038^{19/4}$
***n*:** 1.5198^{22}
α: $[\alpha]_{5461}$ -24.9[α]-16.2 al,5%
***s*:** H_2O 1; ace 3; bz 3; chl 3
MS: WILEY 925 ini 91(100) 154(15) 92(10) 119(8) 107(8) 117(6) 65(6) 77(5)
m/e: 154.05

HODOC No: 02948
CAS No: 16583-73-6
Beilstein No: 5^4, 980
Name: Benzene, (2-chloropropyl)-, (*S*)-
Synonym: Benzene, (2-chloropropyl)-, (+)-
Line Form: $CH_3CHCl\text{-}CH_2C_6H_5$
Mol Form: $C_9H_{11}Cl$
Mr: 154.63
bp: 94^{17}
***d*:** $1.038^{19/4}$
***n*:** 1.5198^{20}
α: $[\alpha]_{43}{}^{59}$ 21.2
***s*:** H_2O 1; ace 3; bz 3; chl 3
MS: WILEY 925 ini 91(100) 154(15) 92(10) 119(8) 107(8) 117(6) 65(6) 77(5)
m/e: 154.05

HODOC No: 02949
CAS No: 104-52-9
Beilstein No: 5^4, 980
Name: Benzene, (3-chloropropyl)-
Synonym: 1-Chloro-3-phenylpropane
Line Form: $C_6H_5CH_2CH_2CH_2Cl$
Mol Form: $C_9H_{11}Cl$
Mr: 154.63
bp: 219-20, 110^{21}
***d*:** $1.056^{21/4}$
***n*:** 1.5160^{25}
***s*:** ctc 2
IR: µCOB 7936 3030 2940 1490 1450 1280 1270 1080 1030 970 910 820 740 700
UV: SAD 2054 267 263 260 258 252 247 242 iso
H NMR: SAD 6860 2.0 2.7 3.4 7.1 CCl_4
MS: WILEY 925 91(100) 92(27) 154(22) 65(7) 156(7) 105(5) 39(4) 51(4)
m/e: 154.05

HODOC No: 02950
CAS No: 4911-65-3
Beilstein No: 6^3, 981
Name: Benzene, [(3-chloropropyl)thio]-
Synonym: Sulfide, 3-chloropropyl phenyl
Line Form: $C_6H_5SCH_2CH_2CH_2Cl$
Mol Form: $C_9H_{11}ClS$
Mr: 186.69
bp: $116\text{-}7^4$
***d*:** $1.1536^{20/4}$
***n*:** $1.5752^{20/2}$
***s*:** Ac 4; Py 4
m/e: 186.03

HODOC No: 02951
CAS No: 17061-92-6
Name: Benzene, 1-chloro-2-(2-propynyloxy)-
Synonym: Ether, *o*-chlorophenyl 2-propynyl
Mol Form: C_9H_7ClO
Mr: 166.60
IR: µCOB 4328 3330 3130 2940 2860 2130 1590 1490 1450 1370 1320 1280 1220 1180 1120 1080 1050 1010 920 750 670 650 630
UV: OES 6-203 281(2138) 274(2399) 225(8128) EtOH
m/e: 166.02

HODOC No: 02952
CAS No: 789-02-6
Beilstein No: 5^4, 1885
Name: Benzene, 1-chloro-2-[2,2,2-trichloro-1-(4-chlorophenyl)ethyl]-
Synonym: Ethane, 1,1,1-trichloro-2-(*o*-chlorophenyl)-2-(*p*-chlorophenyl)-
Line Form: $(2\text{-}ClC_6H_5)CH(4\text{-}ClC_6H_5)CCl_3$
Structure:

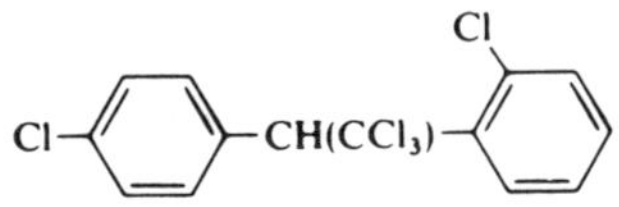

Mol Form: $C_{14}H_9Cl_5$
M_r: 354.48
bp: 74-4.5
mp: 73-5
IR: COB 6714 3070 2950 1590 1570 1490 1470 1450 1410 1090 1040 1020 840 830 780 750 720 650 610 500 460
UV: OES 1-547 280(501) 275(631) 235(15849) 80% EtOH
MS: ALD 4:76C
m/e: 351.91

HODOC No: 02953
CAS No: 2136-89-2
Beilstein No: 5^4, 823
Name: Benzene, 1-chloro-2-(trichloromethyl)-
Synonym: Toluene, *o*,α,α,α-tetrachloro-
Line Form: $2\text{-}ClC_6H_4CCl_3$
Mol Form: $C_7H_4Cl_4$
M_r: 229.91
bp: 264.3, 129.5^{13}
mp: 30, 29.4(fp)
***d*:** $1.5187^{20/4}$
***n*:** 1.5836^{20}
***s*:** H_2O 1; eth 3; ace 3; ctc 2; os 3
IR: SADG 271 3060 1590 1570 1460 1440 1270 1180 1130 1050 1040 870 790 730 700
UV: SAD 493 281(801) 274(920) 224(7080) MeOH
H NMR: SAD 1190 7.3 8.1 CCl_4
MS: WILEY 1615 193(100) 195(95) 197(30) 123(19) 61(12) 159(11) 96.5(10) 228(7)
m/e: 227.91

HODOC No: 02954
CAS No: 2136-81-4
Beilstein No: 5^4, 823
Name: Benzene, 1-chloro-3-(trichloromethyl)-
Synonym: Toluene, *m*,α,α,α-tetrachloro-
Line Form: $3\text{-}ClC_6H_4CCl_3$
Mol Form: $C_7H_4Cl_4$
M_r: 229.91
bp: 255
***d*:** 1.495^{14}
***n*:** 1.4461^{20}
***s*:** H_2O 1; eth 3; ace 3; ctc 2; os 3
IR: SADG 8852 3090 1590 1570 1470 1410 1180 1100 1080 1000 910 810 750 700 680 630
UV: SAD 5918 283 276 269 cyhex
H NMR: SAD 2051 7.1-8.0 CCl_4
m/e: 227.91

HODOC No: 02955
CAS No: 5216-25-1
Beilstein No: 5^4, 802
Name: Benzene, 1-chloro-4-(trichloromethyl)-
Synonym: Toluene, *p*,α,α,α-tetrachloro-
Line Form: $4\text{-}ClC_6H_4CCl_3$
Mol Form: $C_7H_4Cl_4$
M_r: 229.91
bp: 245, $108\text{-}12^8$
***d*:** 1.4463^{20}
***s*:** Ac 4; Et_2O 4
IR: SADG 140
UV: SAD 204
H NMR: SAD 1181
m/e: 227.91

HODOC No: 02956
CAS No: 88-16-4
Name: Benzene, 1-chloro-2-(trifluoromethyl)-
Synonym: Toluene, *o*-chloro-α,α,α-trifluoro-
Mol Form: $C_7H_4ClF_3$
M_r: 180.55
bp: 152.2
***s*:** chl 3
Raman: SAD 750 3160 3090 1600 1580 1480 1460 1360 1320 1290 1270 1170 1120 1100 1090 1010 790 770 650 590 470 430 350 320 290 200 180 130
UV: SAD 85 275(737) 267(227) 261(560) 217(6040) 212(5920) MeOH
H NMR: SAD 10137 7.1-7.5 7.6 CCl_4
m/e: 180.00

HODOC No: 02957
CAS No: 98-56-6
Beilstein No: 5^4, 815
Name: Benzene, 1-chloro-4-(trifluoromethyl)-
Synonym: Toluene, *p*-chloro-α,α,α-trifluoro-
Line Form: 4-$ClC_6H_4CF_3$
Mol Form: $C_7H_4ClF_3$
M_r: 180.55
bp: 138-9
mp: -33
***n*:** 1.4431^{30}
IR: SADG 75 1610 1580 1490 1400 1330 1280 1160 1130 1090 1060 1010 950 830 780 720 590 510 490 410 380
Raman: SAD 751 3080 2990 1610 1580 1500 1450 1330 1280 1180 1100 1070 1000 820 780 710 630 500 410 380 280 170
UV: SAD 86 272(117) 266(160) 261(207) 255(184) 219(6680) MeOH
H NMR: SAD 1177 7.4 7.5 CCl_5
MS: ALD 4:56C
m/e: 180.00

HODOC No: 02958
CAS No: 54625-53-5
Beilstein No: 6^2, 1067
Name: Benzene, 1-chloro-2,3,4-trimethoxy-
Synonym: 2,3,4-Trimethoxychlorobenzene
Mol Form: $C_9H_{11}ClO_3$
M_r: 202.63
bp: $252\text{-}6^{748.5}$
***s*:** al 2; eth 2; ctc 2
MS: WILEY 1424 202(100) 187(47) 204(34) 168(33) 144(31) 153(21) 189(16) 127(14)
m/e: 202.04

HODOC No: 02959
CAS No: 1667-04-5
Beilstein No: 5^4, 1026
Name: Benzene, 2-chloro-1,3,5-trimethyl-
Synonym: Mesitylene, 2-chloro-
Line Form: 2,4,6-$(CH_3)_3C_6H_2Cl$
Mol Form: $C_9H_{11}Cl$
M_r: 154.63
bp: 204-6, 104^{25}
mp: 20
***d*:** 1.0337^{30}
***n*:** 1.5212^{30}
***s*:** Et_2O 4; EtOH 4
IR: μCOB 1914 3030 2940 1590 1470 1370 1180 1050 950 850 700
UV: OES 3-204 268(224) EtOH
MS: WILEY 925
m/e: 154.05

HODOC No: 02960
CAS No: 56961-16-1
Beilstein No: 5^4, 757
Name: Benzene, 1-chloro-2,4,5-trinitro-
Synonym: 1-Chloro-2,4,5-trinitrobenzene
Line Form: 2,4,5-$(NO_2)_3C_6H_2Cl$
Mol Form: $C_6H_2ClN_3O_6$
M_r: 247.55
mp: 116
Color: ye lf (al)
***s*:** Bz 4; EtOH 4
m/e: 246.96

HODOC No: 02961
CAS No: 88-88-0
Beilstein No: 5^4, 757
Name: Benzene, 2-chloro-1,3,5-trinitro-
Synonym: 1-Chloro-2,4,6-trinitrobenzene
Line Form: 2,4,6-$(NO_2)_3C_6H_2Cl$
Mol Form: $C_6H_2ClN_3O_6$
M_r: 247.55
mp: 83
***d*:** 1.797^{20}
Color: wh nd or pl (Chl, al-lig)
***s*:** H_2O 1; al 3; eth 2; ace 4; bz 3; chl 4; tol 4; py 4; peth 2
IR: SADG 10107 3080 1600 1560 1540 1530 1340 1060 910 720
UV: SAD 7579 344(1040) 226(8260) MeOH
H NMR: SAD 4220 8.9 $CDCl_3$
MS: ALDN 199 109(100) 30(95) 74(92) 247(62) 111(35) 73(34) 97(28) 62(25)
m/e: 246.96

HODOC No: 02962
CAS No: 55538-62-0
Name: Benzene, 2-chloro-1,3,5-*tris*(1-methylethyl)-
Synonym: 1-Chloro-2,4,6-triisopropylbenzene
Mol Form: $C_{15}H_{23}Cl$
Mr: 238.79
MS: WILEY 1675 223(100) 225(33) 238(31) 43(30) 224(16) 28(15) 181(14) 195(11)
m/e: 238.15

HODOC No: 02963
CAS No: 3306-02-3
Name: Benzene, 1,1′-(1-cyclobutene-1,2-diyl)*bis*-
Synonym: Cyclobutene, 1,2-diphenyl-
Mol Form: $C_{16}H_{14}$
Mr: 206.27
mp: 54.5
***s*:** H_2O 1
UV: OES 7-650 297(18197) 227(23988) 223(22387) hx
m/e: 206.11

HODOC No: 02964
CAS No: 4392-30-7
Beilstein No: 5^4, 1387
Name: Benzene, cyclobutyl-
Synonym: Cyclobutane, phenyl-
Structure:

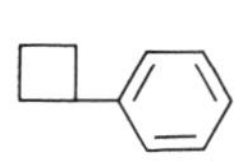

Mol Form: $C_{10}H_{12}$
Mr: 132.20
bp: 190-1^{155},101-2^{41}
***d*:** 0.9378$^{20/4}$
***n*:** 1.5277^{20}
MS: NBS 289
m/e: 132.09

HODOC No: 02965
CAS No: 21072-37-7
Beilstein No: 5^4, 2242
Name: Benzene, 1,1′-(1,2-cyclohexanediyl)*bis*-, *cis*-
Synonym: Cyclohexane, 1,2-diphenyl-, *cis*-
Line Form: 1,2-$(C_6H_5)_2$-C_6H_{10}
Mol Form: $C_{18}H_{20}$
Mr: 236.34
mp: 46-7
Color: nd
***s*:** Et_2O 4
m/e: 236.16

HODOC No: 02966
CAS No: 21072-38-8
Beilstein No: 5^4, 2242
Name: Benzene, 1,1′-(1,2-cyclohexanediyl)*bis*-, *trans*-
Synonym: Cyclohexane, 1,2-diphenyl-, *trans*-
Line Form: 1,2-$(C_6H_5)_2$-C_6H_{10}
Mol Form: $C_{18}H_{20}$
Mr: 236.34
bp: 155-70^{7}
mp: 55-6
m/e: 236.16

HODOC No: 02967
CAS No: 10470-04-9
Beilstein No: 5^4, 2242
Name: Benzene, 1,1′-(1,4-cyclohexanediyl)*bis*-
Synonym: Cyclohexane, 1,4-diphenyl-
Line Form: 1,4-$(C_6H_5)_2C_6H_{10}$
Structure:

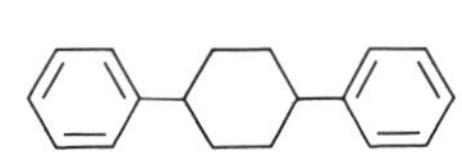

Mol Form: $C_{18}H_{20}$
Mr: 236.34
bp: 190^{29}
mp: 170
Color: cr (al)
m/e: 236.16

HODOC No: 02968
CAS No: 771-98-2
Beilstein No: 5^4, 1557
Name: Benzene, 1-cyclohexen-1-yl-
Synonym: 1-Phenylcyclohexene
Line Form: 1-C_6H_5-C_6H_9
Mol Form: $C_{12}H_{14}$
Mr: 158.23
bp: 251-3, 125-6^{14}
mp: -11
***d*:** 0.9939$^{20/4}$
***n*:** 1.5718^{20}
***s*:** MeOH 4
UV: OES 4-405 250(12023) iso
MS: WILEY 984 104(100) 158(27) 105(9) 91(7) 78(5) 103(5) 77(5) 115(5)
m/e: 158.11

HODOC No: 02969
CAS No: 15232-96-9
Beilstein No: 5^4, 1558
Name: Benzene, 2-cyclohexen-1-yl-
Synonym: 3-Phenylcyclohexene
Line Form: 3-C_6H_5-C_6H_9
Mol Form: $C_{12}H_{14}$
M$_r$: 158.23
bp: $115\text{-}7^{16}$
***d*:** $0.9800^{20/4}$
***n*:** 1.5530^{20}
UV: OES 3-353 268(398) 264(457) 261(603) 258(617) 253(661) 248(631) EtOH
MS: WILEY 984 158(100) 103(62) 129(56) 130(50) 143(37) 115(34) 91(33) 67(25)
m/e: 158.11

HODOC No: 02970
CAS No: 4994-16-5
Name: Benzene, 3-cyclohexen-1-yl-
Synonym: Benzene, (3-cyclohexen-1-yl)-
Mol Form: $C_{12}H_{14}$
M$_r$: 158.23
***s*:** ctc 3; CS_2 3
IR: μCOB 3156 3030 2940 1610 1490 1450 1140 1080 1030 880 750 730 700 680 650
UV: OES 3-353 248(776) EtOH
MS: ALDN 116 158(100) 103(62) 129(56) 130(50) 143(37) 115(34) 91(33) 67(25)
m/e: 158.11

HODOC No: 02971
CAS No: 54774-82-2
Name: Benzene, [1-(1-cyclohexen-1-yl)ethyl]-
Synonym: [1-(1-Cyclohexen-1-yl)ethyl]benzene
Mol Form: $C_{14}H_{18}$
M$_r$: 186.28
MS: WILEY 1282 105(100) 106(30) 80(26) 81(21) 79(12) 91(11) 77(8) 186(4)
m/e: 186.14

HODOC No: 02972
CAS No: 827-52-1
Beilstein No: 5^4, 1424
Name: Benzene, cyclohexyl-
Synonym: 1,1′-Biphenyl, 1,2,3,4,5,6-hexahydro-
Line Form: $C_6H_5C_6H_{11}$
Mol Form: $C_{12}H_{16}$
M$_r$: 160.25
bp: 235-6, $127\text{-}8^{30}$
mp: 7-8
***d*:** $0.9502^{20/4}$
***n*:** 1.5329^{20}
Color: pl
***s*:** H_2O 1; al 4; eth 3; ctc 2
IR: SADG 8520 3030 2941 1587 1493 1449 1075 1031 1000 735 694
Raman: SAD 399 3070 3010 2940 2930 2900 2860 1600 1580 1440 1400 1370 1330 1300 1260 1230 1200 1180 1150 1130 1050 1030 1000 860 830 7770 620 460 430 270 220
UV: SAD 2172 259 cyhex
H NMR: SAD 291 1.0-2.1 2.4 7.1 CCl_4
MS: WILEY 1016 104(100) 117(80) 160(80) 91(52) 56(32) 84(28) 41(24) 81(18)
m/e: 160.13

HODOC No: 02973
CAS No: 1504-96-7
Name: Benzene, 1-cyclohexyl-4-ethoxy-
Synonym: Phenetole, *p*-cyclohexyl-
Mol Form: $C_{14}H_{20}O$
M$_r$: 204.30
IR: μCOB 27 3130 2940 1610 1520 1450 1390 1280 1250 1180 1110 1050 930 820 800
m/e: 204.15

HODOC No: 02974
CAS No: 4413-16-5
Beilstein No: 5^4, 1454
Name: Benzene, (1-cyclohexylethyl)-
Synonym: Ethane, 1-cyclohexyl-1-phenyl-
Line Form: $C_6H_5CH(CH_3)C_6H_{11}$
Mol Form: $C_{14}H_{20}$
M$_r$: 188.30
bp: 264-6
***d*:** 0.9773^{17}
***n*:** 1.549^{17}
UV: API 868 268 265 262 259 255 253 248 cyhex
MS: WILEY 1309 105(100) 106(86) 41(44) 39(39) 27(36) 104(32) 91(30) 188(20)
m/e: 188.16

HODOC No: 02975
CAS No: 1603-61-8
Name: Benzene, (2-cyclohexylethyl)-
Synonym: Ethane, 1-cyclohexyl-2-phenyl-
Mol Form: $C_{14}H_{20}$
M_r: 188.30
UV: **API 869** 269 265 262 260 255 254 249 cyhex
MS: **WILEY 1312** 92(100) 91(36) 55(34) 188(16) 41(16) 97(10) 65(8) 105(6)
m/e: 188.16

HODOC No: 02976
CAS No: 55334-30-0
Name: Benzene, [3-(2-cyclohexylethyl)-6-cyclopentylhexyl]-
Synonym: Hexane, 3(2-cyclohexylethyl)-6-cyclopentyl-1-phenyl
Mol Form: $C_{25}H_{40}$
M_r: 340.57
UV: **API 870** 269 265 262 260 255 254 249 cyhex
MS: **WILEY 1971** 92(100) 91(48) 55(38) 41(25) 83(23) 69(18) 97(12) 340(4)
m/e: 340.31

HODOC No: 02977
CAS No: 2206-48-6
Beilstein No: 6^4, 3899
Name: Benzene, 1-cyclohexyl-2-methoxy-
Synonym: Anisole, *o*-cyclohexyl-
Line Form: $2\text{-}C_6H_{11}\text{-}C_6H_4OCH_3$
Mol Form: $C_{13}H_{18}O$
M_r: 190.27
bp: $267\text{-}8^{749}$
***d*:** 1.007^{18}
***n*:** 1.5365^{18}
MS: **WILEY 1326** 147(100) 190(100) 121(56) 91(47) 134(27) 108(15) 119(15) 107(14)
m/e: 190.14

HODOC No: 02978
CAS No: 613-36-5
Beilstein No: 6^4, 3904
Name: Benzene, 1-cyclohexyl-4-methoxy-
Synonym: Anisole, *p*-cyclohexyl-
Line Form: $4\text{-}C_6H_{11}\text{-}C_6H_4OCH_3$
Mol Form: $C_{13}H_{18}O$
M_r: 190.27
bp: $275\text{-}6^{748}$
mp: 57-8
UV: **FO 56** 282(1995) 277(2512) cyhex
m/e: 190.14

HODOC No: 02979
CAS No: 7137-56-6
Beilstein No: 5^3, 1259
Name: Benzene, 1-cyclohexyl-2-nitro-
Synonym: Cyclohexane, 2-nitrophenyl
Line Form: $2\text{-}NO_2C_6H_4\text{-}C_6H_{11}$
Mol Form: $C_{12}H_{15}NO_2$
M_r: 205.24
bp: 174^{16}, $113^{0.5}$
mp: 45
***d*:** $1.111^{23/4}$
***n*:** 1.5472^{25}
m/e: 205.11

HODOC No: 02980
CAS No: 5458-48-0
Beilstein No: 5^4, 1426
Name: Benzene, 1-cyclohexyl-4-nitro-
Synonym: *p*-Cyclohexylnitrobenzene
Line Form: $4\text{-}NO_2C_6H_4\text{-}C_6H_{11}$
Mol Form: $C_{12}H_{15}NO_2$
M_r: 205.24
bp: 210^{25}
mp: 58.5
Color: ye pl (al)
***s*:** Bz 4; Et_2O 4
m/e: 205.11

HODOC No: 02981
CAS No: 2206-38-4
Beilstein No: 6^4, 565
Name: Benzene, (cyclohexyloxy)-
Synonym: Ether, cyclohexyl phenyl
Line Form: $C_6H_{11}\text{-}O\text{-}C_6H_5$
Mol Form: $C_{12}H_{16}O$
M_r: 176.25
bp: 260-2, 128^{15}
***d*:** $1.0077^{20/4}$
***n*:** 1.520^{22}
***s*:** H_2O 1; ace 3; bz 3
MS: **WILEY 1194** 94(100) 176(13) 55(8) 95(8) 41(7) 67(6) 39(4) 77(4)
m/e: 176.12

HODOC No: 02982
CAS No: 15570-45-3
Name: Benzene,
1,1′,1″,1‴-(1,3-cyclopentadiene-1,2,3,4-tetrayl)tetrak is-
Synonym: 1,3-Cyclopentadiene, 1,2,3,4-tetraphenyl-
Structure:

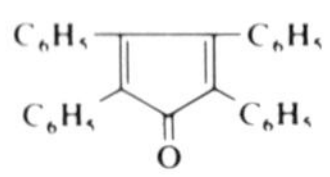

Mol Form: $C_{29}H_{22}$
Mr: 370.46
mp: 177-80
IR: ALD 2:512H
MS: ALD 4:34A
m/e: 370.17

HODOC No: 02983
CAS No: 2175-90-8
Beilstein No: 5^3, 2300
Name: Benzene,
(2,4-cyclopentadien-1-ylidenephenylmethyl)-
Synonym: Methane, 2,4-cyclopentadien-1-ylidenediphenyl-
Structure:

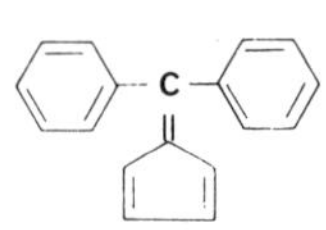

Mol Form: $C_{18}H_{14}$
Mr: 230.29
mp: 82-3
IR: µCOB 1096 3030 1590 1560 1490 1450 1370 1330 1080 1030 1000 930 900 790 760 700 680
MS: NBS 1558; ALD 4:18C
m/e: 230.11

HODOC No: 02984
CAS No: 700-88-9
Beilstein No: 5^4, 1409
Name: Benzene, cyclopentyl-
Synonym: Cyclopentane, phenyl-
Line Form: $C_6H_5C_5H_9$
Structure:

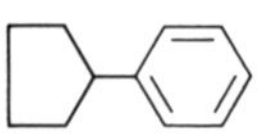

Mol Form: $C_{11}H_{14}$
Mr: 146.22
bp: 219, 102^{18}
***d*:** $0.9462^{20/4}$
***n*:** 1.5280^{20}
***s*:** Et_2O 4
UV: JACSA 72-5261-50 269(191) 260(10965) 259(275) 248(240) cyhex
MS: NBS 4133
m/e: 146.11

HODOC No: 02985
CAS No: 55334-31-1
Name: Benzene,
[6-cyclopentyl-3-(3-cyclopentylpropyl)hexyl]-
Synonym: Heptane, 1,7-dicyclopentyl-4-phenethyl
Mol Form: $C_{25}H_{40}$
Mr: 340.57
UV: API 871 269 265 262 260 255 254 249 cyhex
MS: WILEY 1970 92(100) 91(37) 41(24) 55(20) 69(16) 83(15) 97(10) 340(1)
m/e: 340.31

HODOC No: 02986
CAS No: 2883-12-7
Beilstein No: 5^4, 1455
Name: Benzene, (3-cyclopentylpropyl)-
Synonym: Propane, 1-cyclopentyl-3-phenyl-
Line Form: $C_5H_9(CH_2)_3C_6H_5$
Structure:

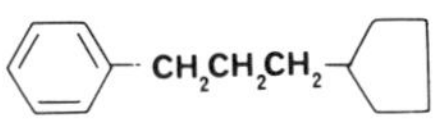

Mol Form: $C_{14}H_{20}$
Mr: 188.30
bp: 271^{743}
***d*:** $0.9233^{19/4}$
***n*:** 1.5130^{19}
IR: API 867
UV: API 867 269 265 262 260 255 254 249 cyhex
MS: WILEY 1307 92(100) 91(82) 41(35) 55(33) 188(27) 39(19) 104(13) 69(12)
m/e: 188.16

HODOC No: 02987
CAS No: 55191-62-3
Name: Benzene, 1,1′-[3-(3-cyclopentylpropyl)-1,5-pentanediyl]*bis*-
Synonym: Pentane, 3(3-cyclopentylpropyl)-1,5-diphenyl
Mol Form: $C_{25}H_{34}$
Mr: 334.52
UV: API 508 268 265 262 260 255 254 248 iso
MS: WILEY 1962 92(100) 91(100) 41(17) 105(15) 55(13) 104(11) 69(10) 334(10)
m/e: 334.27

HODOC No: 02988
CAS No: 29881-14-9
Name: Benzene, 1,1′-(1,2-cyclopropanediyl)*bis*-
Synonym: Cyclopropane, 1,2-diphenyl-
Mol Form: $C_{15}H_{14}$
Mr: 194.26
IR: SADG 26922 3060 3020 1930 1860 1790 1730 1590 1570 1490 1450 1440 1370 1350 1270 1200 1190 1170 1150 1110 1060 1020 990 980 930 890 820 760 740 720 680 610 560 510
Raman: SAD 382 3070 3010 1610 1510 1470 1440 1400 1330 1300 1230 1190 1160 1070 1030 1010 990 950 910 840 780 760 740 620 590 410 360 240
UV: SAD 20550 275(812) 268(1230) 229(15420) MeOH
H NMR: SAD 15644 2.1 2.4 CCl_4
m/e: 194.11

HODOC No: 02989
CAS No: 873-49-4
Beilstein No: 5^3, 1200
Name: Benzene, cyclopropyl-
Synonym: Cyclopropane, phenyl-
Line Form: C_3H_5-C_6H_5
Mol Form: C_9H_{10}
Mr: 118.17
bp: 173.6, 80^{37}
mp: -31
***d*:** $0.9317^{20/4}$
***n*:** 1.5285^{20}
***s*:** H_2O 1; eth 3; ace 3; chl 3
IR: µSADP 12108 3030 2940 1590 1490 1450 1430 1370 1300 1220 1180 1160 1100 1080 1040 1020 1000 960 890 810 750 690
UV: OES 7-217 275(631) 265(631) 220(10000) 190(31623) EtOH
H NMR: VAR 528 0.7 0.8 1.8 7.1 $CDCl_3$
MS: ALDN 60 117(100) 118(70) 91(42) 115(27) 38(22) 51(19) 63(13) 103(12)
m/e: 118.08

HODOC No: 02990
CAS No: 1667-00-1
Name: Benzene, (cyclopropylmethyl)-
Synonym: Benzylcyclopropane
Mol Form: $C_{10}H_{12}$
Mr: 132.20
IR: SADG 33155 3080 3030 3000 2910 2850 1600 1580 1490 1450 1430 1370 1330 1290 1170 1070 1040 1030 1020 970 930 830 800 770 730 700 580 480
Raman: SAD 1188 3060 3000 2910 2840 1600 1580 1460 1430 1380 1290 1230 1200 1150 1120 1070 1030 1010 970 930 880 830 810 770 740 620 580 490 410 320 250
UV: SAD 6522 270 266 264 261 MeOH
H NMR: SAD 4144 0.1-0.7 2.5 7.2 $CDCl_3$
m/e: 132.09

HODOC No: 02991
CAS No: 104-72-3
Beilstein No: 5^4, 1193
Name: Benzene, decyl-
Synonym: Decane, 1-phenyl-
Line Form: $C_6H_5(CH_2)_9CH_3$
Mol Form: $C_{16}H_{26}$
M_r: 218.37
bp: 300, 156^{10}
mp: -14.4 (fp)
***d*:** 0.8555^{20}
***n*:** 1.4832^{20}
***s*:** Ac 4; Bz 4; Et_2O 4; EtOH 4
H NMR: SAD 5830 0.9 1.3 2.6 7.1 CCl_4
MS: NBS 4507
m/e: 218.20

HODOC No: 02992
CAS No: 55334-72-0
Name: Benzene, (2-decyldodecyl)-
Synonym: Heneicosane, 11-benzyl
Line Form: $CH_3(CH_2)_9CH(CH_2C_6H_5)(CH_2)_9CH_3$
Mol Form: $C_{28}H_{50}$
M_r: 386.68
UV: OES 5-900 269(316) 265(398) 262(501) 260(501) 254(501) 249(501) 244(398) cyhex
MS: WILEY 2132 43(100) 91(88) 57(75) 41(64) 55(45) 71(40) 85(27) 386(1)
m/e: 386.39

HODOC No: 02993
CAS No: 55373-90-5
Name: Benzene, 2-(1-decyl-1-undecenyl)-1,4-dimethyl-
Synonym: 10-Heneicosene, 11(2,5-dimethylphenyl)
Mol Form: $C_{29}H_{50}$
M_r: 398.69
UV: API 396 276 269 iso
MS: WILEY 2157 119(100) 133(46) 145(26) 146(24) 105(18) 69(17) 157(14) 398(3)
m/e: 398.39

HODOC No: 02994
CAS No: 55373-91-6
Name: Benzene, 2-(1-decylundecyl)-1,4-dimethyl-
Synonym: Heneicosane, 11(2,5-dimethylphenyl)
Mol Form: $C_{29}H_{52}$
M_r: 400.70
UV: API 1029 276 268 262 215 200 cyhex
MS: WILEY 2161 119(100) 43(18) 259(17) 41(12) 120(11) 55(8) 57(5) 400(0)
m/e: 400.41

HODOC No: 02995
CAS No: 883-40-9
Beilstein No: 7^4, 1372
Name: Benzene, 1,1′-(diazomethylene)*bis*-
Synonym: Methane, diazodiphenyl-
Line Form: $(C_6H_5)_2CN_2$
Mol Form: $C_{13}H_{10}N_2$
M_r: 194.22
bp: exp
mp: 30-2
Color: bl-red nd (peth)
***s*:** Ac 4; Et_2O 4; EtOH 4
m/e: 194.08

HODOC No: 02996
CAS No: 583-53-9
Beilstein No: 5^4, 682
Name: Benzene, 1,2-dibromo-
Synonym: Benzene, *o*-dibromo-
Line Form: $2\text{-}BrC_6H_4Br$
Mol Form: $C_6H_4Br_2$
M_r: 233.90
bp: 225, 92^{10}
mp: 7.1
***d*:** $1.9843^{20/4}$
***n*:** 1.6155^{20}
***s*:** H_2O 1; al 3; eth 5; ace 5; bz 5; ctc 5; peth 5
IR: μCOB 4276 3030 1590 1450 1430 1250 1100 1010 740 700
UV: OES 5-46 280(229) 272(295) 265(263) 220(12589) cyhex
H NMR: SAD 11724 7.1 7.5 CCl_4
MS: WILEY 1651 236(100) 234(51) 238(49) 155(30) 157(29) 75(18) 50(13) 76(12)
m/e: 233.87

HODOC No: 02997
CAS No: 108-36-1
Beilstein No: 5^4, 682
Name: Benzene, 1,3-dibromo-
Synonym: Benzene, *m*-dibromo-
Line Form: 1,3-$Br_2C_6H_4$
Mol Form: $C_6H_4Br_2$
M_r: 233.90
bp: 218, 66^5
mp: -7
***d*:** $1.9523^{20/4}$
***n*:** 1.6083^{17}
***s*:** H_2O 1; al 3; eth 5
IR: **µCOB 6879** 3030 1560 1450 1410 1080 1060 990 880 790 740 680
UV: **HBCP** 278(427) 270(589) 264(417) 257(214) hx
H NMR: **SAD 18683** 7.0 7.3 7.6 CCl_4
MS: **NBS 264**
m/e: 233.87

HODOC No: 02998
CAS No: 106-37-6
Beilstein No: 5^4, 683
Name: Benzene, 1,4-dibromo-
Synonym: Benzene, *p*-dibromo-
Line Form: 1,4-$Br_2C_6H_4$
Mol Form: $C_6H_4Br_2$
M_r: 233.90
bp: 218-9
mp: 87.3
***d*:** 2.261^{17}
***n*:** 1.5742
Color: pl
***s*:** H_2O 1; al 3; eth 4; ace 4; bz 3; CS_2 4; tol 3
IR: **µCOB 4278** 3130 1610 1490 1470 1390 1080 1000 820
Raman: **SAD 348** 3140 3070 1630 1570 1550 1370 1300 1180 1070 820 710 630 310 280 220 100
UV: **OES 5-46** 283(263) 275(389) 267(363) 260(316) 258(309) 229(15136) cyhex
MS: **WILEY 1651** 236(100) 234(51) 238(49) 155(30) 157(29) 156(25) 158(19) 75(18)
m/e: 233.87

HODOC No: 02999
CAS No: 56908-88-4
Beilstein No: 5^3, 719
Name: Benzene, 1,3-dibromo-5-(bromomethyl)-
Synonym: 3,5-Dibromobenzyl bromide
Line Form: 3,5-$Br_2C_6H_3CH_2Br$
Mol Form: $C_7H_5Br_3$
M_r: 325.82
bp: 173^{19}
mp: 96
Color: pl or nd (al)
***s*:** EtOH 4
m/e: 325.79

HODOC No: 03000
CAS No: 104514-49-0
Beilstein No: 5^1, 117
Name: Benzene, 1,2-dibromo-3-chloro-
Synonym: 1,2-Dibromo-3-chlorobenzene
Line Form: 2,3-$Br_2C_6H_3Cl$
Mol Form: $C_6H_3Br_2Cl$
M_r: 268.34
bp: 264, 142^{23}
mp: 73.5
Color: lf
***s*:** Bz 4; Et_2O 4; Chl 4
m/e: 267.83

HODOC No: 03001
CAS No: 60956-24-3
Beilstein No: 5, 212
Name: Benzene, 1,2-dibromo-4-chloro-
Synonym: 1,2-Dibromo-4-chlorobenzene
Line Form: 3,4-$Br_2C_6H_3Cl$
Mol Form: $C_6H_3Br_2Cl$
M_r: 268.34
bp: 256, 121^{19}
mp: 35.5
Color: pr (al)
***s*:** Bz 4; Et_2O 4; Chl 4
m/e: 267.83

HODOC No: 03002
CAS No: 19230-27-4
Beilstein No: 5^3, 563
Name: Benzene, 1,3-dibromo-2-chloro-
Synonym: 1,3-Dibromo-2-chlorobenzene
Line Form: 2,6-Br_2-C_6H_3Cl
Mol Form: $C_6H_3Br_2Cl$
Mr: 268.34
bp: 265
mp: 71
Color: pl
s: Bz 4; Et_2O 4; Chl 4
m/e: 267.83

HODOC No: 03003
CAS No: 3460-24-0
Beilstein No: 5, 212
Name: Benzene, 1,4-dibromo-2-chloro-
Synonym: 1,4-Dibromo-2-chlorobenzene
Line Form: 2,5-$Br_2C_6H_3Cl$
Mol Form: $C_6H_3Br_2Cl$
Mr: 268.34
bp: 259, 121^{24}
mp: 40-1
Color: pr
s: Bz 4; Et_2O 4; Chl 4
m/e: 267.83

HODOC No: 03004
CAS No: 24932-49-8
Name: Benzene, 1,2-dibromo-3,4-dimethyl-
Synonym: *o*-Xylene, 3,4-dibromo-
Line Form: 1,2-$(CH_3)_2$-3,4-$Br_2C_6H_2$
Mol Form: $C_8H_8Br_2$
Mr: 261.95
bp: 277
mp: 7
d: $1.784^{15/15}$
s: EtOH 4
m/e: 261.90

HODOC No: 03005
CAS No: 24932-48-7
Name: Benzene, 1,2-dibromo-4,5-dimethyl-
Synonym: 1,2-Dibromo-4,5-dimethylbenzene
Line Form: 1,2-$(CH_3)_2$-4,5-$Br_2C_6H_2$
Mol Form: $C_8H_8Br_2$
Mr: 261.95
bp: 278
mp: 88
Color: lf (al)
m/e: 261.90

HODOC No: 03006
CAS No: 1074-24-4
Beilstein No: 5^3, 859
Name: Benzene, 1,4-dibromo-2,5-dimethyl-
Synonym: *p*-Xylene, 2,5-dibromo-
Line Form: 1,4-$(CH_3)_2$-2,5-Br_2-C_6H_2
Mol Form: $C_8H_8Br_2$
Mr: 261.95
bp: 261, 150^{21}
mp: 75
Color: mcl cr
m/e: 261.90

HODOC No: 03007
CAS No: 615-87-2
Beilstein No: 5^3, 839
Name: Benzene, 1,5-dibromo-2,4-dimethyl-
Synonym: *m*-Xylene, 4,6-dibromo-
Line Form: 1,5-Br_2-2,4-$(CH_3)_2C_6H_2$
Mol Form: $C_8H_8Br_2$
Mr: 261.95
bp: 255-6
mp: 68
Color: pl (al)
s: EtOH 4
m/e: 261.90

HODOC No: 03008
CAS No: 5789-30-0
Beilstein No: 5^3, 1817
Name: Benzene, 1,1′-(1,2-dibromo-1,2-ethanediyl)*bis*-
Synonym: Bibenzyl, α,α′-dibromo-
Line Form: $C_6H_5CHBrCHBrC_6H_5$
Structure:

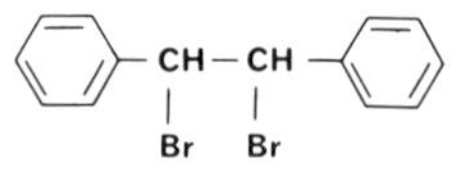

Mol Form: $C_{14}H_{12}Br_2$
Mr: 338.04
mp: 243-4
s: Et_2O 4; xyl 4
IR: ALD 2:547G
UV: OL
m/e: 337.93

HODOC No: 03009
CAS No: 13440-24-9
Beilstein No: 5^3, 1817
Name: Benzene, 1,1′-(1,2-dibromo-1,2-ethanediyl)*bis*-, (R*,*S**)-
Synonym: Bibenzyl, α,α′-dibromo-, meso-
Line Form: $C_6H_5CH(Br)CH(Br)C_6H_5$
Mol Form: $C_{14}H_{12}Br_2$
Mr: 338.04
mp: 243-4
***s*:** H_2O 1; al 2; eth 3; CS2 3; xyl 3
UV: BSCFA 1348-55 250(6310) al
m/e: 337.93

HODOC No: 03010
CAS No: 24162-64-9
Name: Benzene, 1,2-dibromo-4-ethenyl-
Synonym: Styrene, 3,4-dibromo-
Line Form: $3,4\text{-}Br_2C_6H_3CH{=}CH_2$
Mol Form: $C_8H_6Br_2$
Mr: 259.94
bp: $101\text{-}3^{22}$
***n*:** 1.6251^{22}
m/e: 259.88

HODOC No: 03011
CAS No: 32917-57-0
Name: Benzene, 1,4-dibromo-2-ethenyl-
Synonym: Styrene, 2,5-dibromo-
Line Form: $2,5\text{-}Br_2C_6H_3CH{=}CH_2$
Mol Form: $C_8H_6Br_2$
Mr: 259.94
bp: $101\text{-}2^{20}$
***d*:** 1.7742^{20}
***n*:** 1.6225^{20}
m/e: 259.88

HODOC No: 03012
CAS No: 7436-90-0
Beilstein No: 5^3, 1179
Name: Benzene, (2,2-dibromoethenyl)-
Synonym: Styrene, β,β-dibromo-
Line Form: $C_6H_5CH{=}CBr_2$
Mol Form: $C_8H_6Br_2$
Mr: 259.94
bp: 144^{24}
***d*:** 1.819^{22}
m/e: 259.88

HODOC No: 03013
CAS No: 93-52-7
Beilstein No: 5^4, 909
Name: Benzene, (1,2-dibromoethyl)-
Synonym: α,β-Dibromoethylbenzene
Line Form: $C_6H_5(CHBrCH_2Br)$
Mol Form: $C_8H_8Br_2$
Mr: 261.95
bp: $139\text{-}51^{15}$
mp: 74
***s*:** al 3; eth 3; bz 3; chl 3; HOAc 3; lig 3; MeOH 3
IR: SADG 10850 3020 2970 1490 1450 1420 1350 1220 1180 1140 1120 1060 890 750 680 650
UV: SAD 3295 223 cyhex
H NMR: VAR 503 4.0 4.1 5.2 7.4 $CDCl_3$
C NMR: JJ 286 FT 34.9 50.8 127.5 128.6 128.9 138.4 $CDCl_3$
m/e: 261.90

HODOC No: 03014
CAS No: 1435-53-6
Beilstein No: 5^3, 569
Name: Benzene, 2,4-dibromo-1-fluoro-
Synonym: 1,3-Dibromo-4-fluorobenzene
Line Form: $2,4\text{-}Br_2C_6H_3F$
Mol Form: $C_6H_3Br_2F$
Mr: 251.89
bp: $103\text{-}5^{32}$
mp: 69
***d*:** 2.047^{20}
***n*:** 1.5840^{20}
IR: COB 6573 3100 1580 1480 1420 1380 1260 1230 1070 870 810 680 620 550 490
m/e: 251.86

HODOC No: 03015
CAS No: 5456-94-0
Beilstein No: 6^2, 358
Name: Benzene, 1,5-dibromo-2-methoxy-4-methyl-
Synonym: Anisole, 2,4-dibromo-5-methyl-
Mol Form: $C_8H_8Br_2O$
Mr: 277.95
mp: 75-6
IR: COB 7627 2960 2940 2850 1580 1470 1460 1440 1380 1360 1285 1250 1190 1170 1060 1005 940 870 835 720 660 570
m/e: 277.89

HODOC No: 03016
CAS No: 618-31-5
Beilstein No: 5^4, 836
Name: Benzene, (dibromomethyl)-
Synonym: Toluene, α,α-dibromo-
Line Form: $C_6H_5CHBr_2$
Mol Form: $C_7H_6Br_2$
M_r: 247.93
bp: 156^{23}
***d*:** 1.51
***n*:** 1.6147^{20}
***s*:** H_2O 1; al 5; eth 5
IR: µCOB 6597 3030 1590 1490 1450 1220 1150 830 770 690
UV: OES 1-125 276(708) iso
MS: ALD 4:33D
m/e: 247.88

HODOC No: 03017
CAS No: 61563-25-5
Beilstein No: 5^4, 835
Name: Benzene, 1,2-dibromo-3-methyl-
Synonym: 2,3-Dibromotoluene
Line Form: $2,3\text{-}Br_2C_6H_3CH_3$
Mol Form: $C_7H_6Br_2$
M_r: 247.93
bp: $126\text{-}8^{19}$
mp: 30-1
***d*:** $1.8234^{15/4}$
***n*:** 1.5984^{25}
m/e: 247.88

HODOC No: 03018
CAS No: 60956-23-2
Beilstein No: 5^4, 836
Name: Benzene, 1,2-dibromo-4-methyl-
Synonym: 3,4-Dibromotoluene
Line Form: $3,4\text{-}Br_2C_6H_3CH_3$
Mol Form: $C_7H_6Br_2$
M_r: 247.93
bp: 290-1, $123\text{-}4^{19}$
mp: -10 (fp)
***d*:** $1.8197^{25/4}$
***n*:** 1.5979^{25}
m/e: 247.88

HODOC No: 03019
CAS No: 69321-60-4
Beilstein No: 5^4, 836
Name: Benzene, 1,3-dibromo-2-methyl-
Synonym: 2,6-Dibromotoluene
Line Form: $2,6\text{-}Br_2C_6H_3CH_3$
Mol Form: $C_7H_6Br_2$
M_r: 247.93
bp: 246, 122^{23}
mp: 5-6
***d*:** 1.812^{22}
m/e: 247.88

HODOC No: 03020
CAS No: 615-59-8
Beilstein No: 5^4, 835
Name: Benzene, 1,4-dibromo-2-methyl-
Synonym: Toluene, 2,5-dibromo-
Line Form: $2,5\text{-}Br_2C_6H_3CH_3$
Mol Form: $C_7H_6Br_2$
M_r: 247.93
bp: 236, $135\text{-}6^{35}$
mp: 5.6 (fp)
***d*:** 1.8127^{17}
***n*:** 1.5982^{18}
***s*:** H_2O 1
IR: µNRC 410 3030 2940 1590 1470 1430 1370 1270 1250 1190 1140 1110 1090 1060 1020 940 860 850 840 800 770 680
H NMR: SAD 18915 2.3 7.1 7.2 7.3 $CDCl_3$
m/e: 247.88

HODOC No: 03021
CAS No: 31543-75-6
Beilstein No: 5^4, 835
Name: Benzene, 2,4-dibromo-1-methyl-
Synonym: Toluene, 2,4-dibromo-
Line Form: $2,4\text{-}Br_2C_6H_3CH_3$
Mol Form: $C_7H_6Br_2$
M_r: 247.93
bp: $103\text{-}4^{11}$
mp: -9.7 (fp)
***d*:** $1.8176^{25/4}$
***n*:** 1.5964^{25}
m/e: 247.88

HODOC No: 03022
CAS No: 619-75-0
Beilstein No: 5^4, 862
Name: Benzene, 1-(dibromomethyl)-4-nitro-
Synonym: Toluene, α,α-dibromo-*p*-nitro-
Line Form: $4\text{-}NO_2C_6H_4CHBr_2$
Mol Form: $C_7H_5Br_2NO_2$
Mr: 292.92
mp: 84
Color: nd (al)
***s*:** H_2O 1; al 4; eth 4
IR: µSADP 4558 3030 1610 1520 1410 1330 1220 1140 1100 1010 960 860 810 750 690 670
UV: SAD 1260 270(13000) MeOH
H NMR: SAD 17069 6.8 7.8 8.2 $CDCl_3$
m/e: 292.87

HODOC No: 03023
CAS No: 26429-41-4
Beilstein No: 5^4, 731
Name: Benzene, 1,2-dibromo-3-nitro-
Synonym: 2,3-Dibromonitrobenzene
Line Form: $2,3\text{-}Br_2C_6H_3NO_2$
Mol Form: $C_6H_3Br_2NO_2$
Mr: 278.90
mp: 85
Color: mcl pr
***s*:** Ac 4; Et_2O 4; Chl 4
m/e: 278.85

HODOC No: 03024
CAS No: 5411-50-7
Beilstein No: 5^4, 732
Name: Benzene, 1,2-dibromo-4-nitro-
Synonym: 1,2-Dibromo-4-nitrobenzene
Line Form: $3,4\text{-}Br_2\text{-}C_6H_3NO_2$
Mol Form: $C_6H_3Br_2NO_2$
Mr: 278.90
bp: 296, 180^{20}
mp: 58-9
***d*:** 2.354^{8}
***n*:** 1.9835^{111}
Color: nd (al,HOAc) mcl pr
***s*:** Bz 4; EtOH 4; HOAc 4
m/e: 278.85

HODOC No: 03025
CAS No: 13402-32-9
Beilstein No: 5^3, 621
Name: Benzene, 1,3-dibromo-2-nitro-
Synonym: 1,3-Dibromo-2-nitrobenzene
Line Form: $2,6\text{-}Br_2\text{-}C_6H_3NO_2$
Mol Form: $C_6H_3Br_2NO_2$
Mr: 278.90
bp: sub
mp: 84
***d*:** 2.211^{8},1.9211^{111}
Color: nd (al), mcl pr (al)
***s*:** Ac 4; Bz 4; EtOH 4
m/e: 278.85

HODOC No: 03026
CAS No: 6311-60-0
Beilstein No: 5^3, 621
Name: Benzene, 1,3-dibromo-5-nitro-
Synonym: 1,3-Dibromo-5-nitrobenzene
Line Form: $1,3\text{-}Br_2\text{-}5\text{-}NO_2C_6H_3$
Mol Form: $C_6H_3Br_2NO_2$
Mr: 278.90
mp: 106
***d*:** 1.9341^{111},2.363^{8}
Color: pl or pr (eth) lf or nd (al)
***s*:** Bz 4; Et_2O 4; EtOH 4
m/e: 278.85

HODOC No: 03027
CAS No: 3460-18-2
Beilstein No: 5^4, 739
Name: Benzene, 1,4-dibromo-2-nitro-
Synonym: 1,4-Dibromo-2-nitrobenzene
Line Form: $1,4\text{-}Br_2\text{-}2\text{-}NO_2C_6H_3$
Mol Form: $C_6H_3Br_2NO_2$
Mr: 278.90
mp: 85-6
***d*:** 1.9416^{111},2.368^{8}
Color: yesh pl (Ace)
***s*:** H_2O 1; ace 3; bz 3; chl 2
IR: µSADP 24259 3030 1590 1540 1470 1350 1280 1160 1090 1040 880 830 750 740
UV: SAD 8520 301(3710) 261(12500) 218(37400) MeOH 303(3390) 256(12500) 217(38900) MeOH:KOH 302(3540) 261(12100) 219(36200) MeOH:HCl
H NMR: SAD 2571 7.6 8.0 $CDCl_3$
m/e: 278.85

HODOC No: 03028
CAS No: 51686-78-3
Beilstein No: 5^3, 620
Name: Benzene, 2,4-dibromo-1-nitro-
Synonym: 1,3-Dibromo-4-nitrobenzene
Line Form: 2,4-$Br_2C_6H_3NO_2$
Mol Form: $C_6H_3Br_2NO_2$
M_r: 278.90
mp: 62
***d*:** 1.9581^{111},2.356^8
Color: ye pl or pr (al)
***s*:** Ac 4; Bz 4
m/e: 278.85

HODOC No: 03029
CAS No: 63076-62-0
Name: Benzene, 2,4-dibromo-1-(pentyloxy)-
Synonym: Ether, 2,4-dibromophenyl pentyl
Mol Form: $C_{11}H_{14}Br_2O$
M_r: 320.03
IR: **µCOB 4097** 2940 1560 1470 1390 1280 1270 1250 1040 870 800 690
m/e: 319.94

HODOC No: 03030
CAS No: 827-08-7
Name: Benzene, 1,2-dibromo-3,4,5,6-tetrafluoro-
Synonym: Benzene, 1,2-dibromotetrafluoro-
Line Form: 1,2-$Br_2C_6F_4$
Mol Form: $C_6Br_2F_4$
M_r: 305.86
bp: 198
mp: 12
***d*:** 2.260^{20}
***n*:** 1.5151^{25}
MS: NBS 2296
m/e: 305.83

HODOC No: 03031
CAS No: 1559-87-1
Name: Benzene, 1,3-dibromo-2,4,5,6-tetrafluoro-
Synonym: Benzene, 1,3-dibromotetrafluoro-
Line Form: 1,3-$Br_2C_6F_4$
Mol Form: $C_6Br_2F_4$
M_r: 305.86
bp: $86\text{-}8^{18}$
***n*:** 1.5178^{23}
MS: NBS 2296
m/e: 305.83

HODOC No: 03032
CAS No: 344-03-6
Beilstein No: 5^4, 685
Name: Benzene, 1,4-dibromo-2,3,5,6-tetrafluoro-
Synonym: Benzene, 1,4-dibromotetrafluoro-
Line Form: 1,4-$Br_2C_6F_4$
Mol Form: $C_6Br_2F_4$
M_r: 305.86
mp: 78-9
MS: NBS 2296
m/e: 305.83

HODOC No: 03033
CAS No: 7657-09-2
Name: Benzene, 1,4-dibromo-2-(trifluoromethyl)-
Synonym: Toluene, 2,5-dibromo-α,α,α-trifluoro-
Mol Form: $C_7H_3Br_2F_3$
M_r: 301.90
mp: 49-50
IR: COB 7974 1585 1465 1390 1305 1270 1255 1170 1155 1130 1100 1030 890 820 720 655 520 480
H NMR: SAD 13804 7.5 7.8 $CDCl_3$
m/e: 301.86

HODOC No: 03034
CAS No: 6942-99-0
Beilstein No: 5^3, 921
Name: Benzene, 2,4-dibromo-1,3,5-trimethyl-
Synonym: Mesitylene, 2,4-dibromo-
Line Form: 2,4-Br_2-1,3,5-$(CH_3)_3C_6H$
Mol Form: $C_9H_{10}Br_2$
M_r: 275.98
bp: 285
mp: 65.5
Color: nd (al)
***s*:** ctc 2
IR: SADG 10518 3030 2941 1587 1493 1449 1316 1282 1149 1087 730 699
H NMR: SAD 2548 2.3 2.6 7.0 CCl_4
MS: NBS 2007
m/e: 275.91

HODOC No: 03035
CAS No: 20367-35-5
Beilstein No: 6^4, 5566
Name: Benzene, 1,2-dibutoxy-
Synonym: Benzene, *o*-dibutoxy-
Line Form: 1,2-$(C_4H_9O)_2C_6H_4$
Mol Form: $C_{14}H_{22}O_2$
M_r: 222.31
bp: 241^{765}, $135\text{-}8^{12}$
Color: ye
m/e: 222.16

HODOC No: 03036
CAS No: 104-36-9
Beilstein No: 6^4, 5721
Name: Benzene, 1,4-dibutoxy-
Synonym: Benzene, *p*-dibutoxy-
Line Form: 1,4-$(C_4H_9O)_2C_6H_4$
Mol Form: $C_{14}H_{22}O_2$
M_r: 222.31
bp: 158^{15}
mp: 45-6
***s*:** ctc 3
IR: COB 4842 3050 2950 2930 2870 1510 1470 1430 1390 1230 1100 1070 970 820 770 650 520
UV: SAD 246 290(2650) 227(10500) MeOH
H NMR: SAD 708 1.0 1.6 3.8 6.7 CCl_4
MS: NBS 4523
m/e: 222.16

HODOC No: 03037
CAS No: 95-50-1
Beilstein No: 5^4, 654
Name: Benzene, 1,2-dichloro-
Synonym: Benzene, *o*-dichloro-
Line Form: 1,2-$Cl_2C_6H_4$
Mol Form: $C_6H_4Cl_2$
M_r: 147.00
bp: 180.5, 86^{18}
mp: -17
***d*:** $1.3048^{20/4}$
***n*:** 1.5515^{20}
***s*:** H_2O 1; al 3; eth 3; ace 5; bz 5; ctc 5; lig 5
IR: COB 4275 3070 1570 1480 1460 1440 1130 1040 750 660
Raman: DFB 80 3150 3070 3030 3000 1570 1390 1280 1160 1130 1040 1020 660 490 470 430 330 240 200 160
UV: SAD 303 269 220 iso
H NMR: SAD 746 6.9-7.6 CCl_4
C NMR: JJ 148 FT 127.7 130.5 132.6 $CDCl_3$
MS: WILEY 817 146(100) 148(64) 111(39) 75(23) 50(17) 74(15) 113(12) 73(10)
m/e: 145.97

HODOC No: 03038
CAS No: 541-73-1
Beilstein No: 5^4, 657
Name: Benzene, 1,3-dichloro-
Synonym: Benzene, *m*-dichloro-
Line Form: 1,3-$Cl_2C_6H_4$
Mol Form: $C_6H_4Cl_2$
M_r: 147.00
bp: 173, 53^{10}
mp: -24.7
***d*:** $1.2884^{20/4}$
***n*:** 1.5459^{20}
***s*:** H_2O 1; al 3; eth 3; ace 5; bz 3; ctc 5; lig 5
IR: µCOB 5654 3130 1560 1470 1410 1120 1090 1000 860 780
Raman: DFB 83 3160 3080 1580 1130 1110 1070 1000 660 430 400 370 220 200 180
UV: SAD 1671 278(2800) 270(3340) 263(2330) 256(1420) 250(830) 216(12800) MeOH
H NMR: SAD 8596 7.2 7.4 CCl_4
C NMR: JJ 149 cw 127.0 128.9 130.6 135.1 diox
MS: WILEY 816 146(100) 148(68) 111(42) 75(34) 50(30) 74(22) 73(15) 37(15)
m/e: 145.97

HODOC No: 03039
CAS No: 106-46-7
Beilstein No: 5^4, 685
Name: Benzene, 1,4-dichloro-
Synonym: Benzene, *p*-dichloro-
Line Form: 1,4-$Cl_2C_6H_4$
Mol Form: $C_6H_4Cl_2$
M_r: 147.00
bp: 174, 55^{10}
mp: 53.1
***d*:** $1.2475^{20/4}$
***n*:** 1.5285^{20}
Color: mcl pr, lf (Ace)
***s*:** H_2O 1; al 5; eth 3; ace 5; bz 3; ctc 3; chl 3; CS_2 3
IR: **μCOB 4279** 3030 1890 1610 1470 1410 1390 1100 1020 810
Raman: **DFB 86** 3150 3080 1580 1380 1290 1170 1110 1080 1070 750 630 350 330 300
UV: **SAD 55** 280(301) 272(377) 264(287) 224(12800) MeOH
H NMR: **SAD 715** 7.2 CCl_4
MS: **WILEY 818** 146(100) 148(65) 111(32) 75(19) 74(12) 150(10) 113(10) 73(8)
m/e: 145.97

HODOC No: 03040
CAS No: 2142-29-2
Name: Benzene, 1,4-dichloro-2,5-*bis*(trichloromethyl)-
Synonym: *p*-Xylene, α,α,α,α′,α′,α′,2,5-octachloro-
Mol Form: $C_8H_2Cl_8$
M_r: 381.72
mp: 193
IR: **μCOB 1907** 3130 1470 1330 1190 1090 860 790 780
UV: **OES 5-120** 302(1950) 294(1995) 242(10965) 219(30903)
m/e: 377.77

HODOC No: 03041
CAS No: 54935-00-1
Name: Benzene, 1,4-dichloro-2-(2-chloroethenyl)-
Synonym: Styrene, β,2,5-trichloro-
Mol Form: $C_8H_5Cl_3$
M_r: 207.48
MS: **WILEY 1459** 206(100) 208(98) 171(78) 173(51) 136(42) 210(32) 135(27) 85(25)
m/e: 205.95

HODOC No: 03042
CAS No: 45892-47-5
Name: Benzene, 2,4-dichloro-1-(2-chloroethenyl)-
Synonym: Styrene, β,2,4-trichloro
Mol Form: $C_8H_5Cl_3$
M_r: 207.48
MS: **WILEY 1460** 206(100) 171(97) 208(97) 173(66) 136(42) 135(34) 210(31) 85(30)
m/e: 205.95

HODOC No: 03043
CAS No: 54965-01-4
Name: Benzene, 1,2-dichloro-4-(1-chloroethyl)-
Synonym: 3,4-Dichloro-α-methylbenzylchloride
Mol Form: $C_8H_7Cl_3$
M_r: 209.49
MS: **WILEY 1475** 173(100) 175(66) 137(25) 208(21) 102(20) 210(20) 174(15) 101(14)
m/e: 207.96

HODOC No: 03044
CAS No: 54965-00-3
Name: Benzene, 1,4-dichloro-2-(1-chloroethyl)-
Synonym: Benzene, 2(1-chloroethyl)-1,4-dichloro
Mol Form: $C_8H_7Cl_3$
M_r: 209.49
MS: **WILEY 1476** 173(100) 175(64) 208(25) 210(24) 102(22) 137(22) 101(18) 159(11)
m/e: 207.96

HODOC No: 03045
CAS No: 54965-02-5
Name: Benzene, 1,4-dichloro-2-(2-chloroethyl)-
Synonym: Benzene, 2(2-chloroethyl)-1,4-dichloro
Mol Form: $C_8H_7Cl_3$
M_r: 209.49
MS: **WILEY 1474** 159(40) 161(26) 208(9) 210(9) 163(4) 160(3) 212(3) 101(2)
m/e: 207.96

HODOC No: 03046
CAS No: 102-47-6
Beilstein No: 5^4, 820
Name: Benzene, 1,2-dichloro-4-(chloromethyl)-
Synonym: Toluene, α,3,4-trichloro-
Line Form: 3,4-$Cl_2C_6H_3CH_2Cl$
Mol Form: $C_7H_5Cl_3$
M_r: 195.47
bp: 241
mp: 37.5
***s*:** H_2O 1; al 3; ctc 3; os 3
IR: SADG 15060 3070 2970 1600 1570 1480 1450 1400 1280 1210 1140 1040 930 880 820 730 700 680 660
UV: SAD 21877 284(307) 275(370) 268(287) 229(96300) MeOH
H NMR: SAD 6971 4.4 7.1 7.3 7.4 CCl_4
MS: WILEY 1346 159(100) 161(64) 194(21) 196(20) 61(12) 163(10) 123(10) 89(8)
m/e: 193.95

HODOC No: 03047
CAS No: 2014-83-7
Beilstein No: 5^4, 820
Name: Benzene, 1,3-dichloro-2-(chloromethyl)-
Synonym: Toluene, α,2,6-trichloro-
Line Form: 2,6-$Cl_2C_6H_3CH_2Cl$
Mol Form: $C_7H_5Cl_3$
M_r: 195.47
bp: $117\text{-}9^{14}$
mp: 39-40
Color: cr (lig, eth, al-eth)
***s*:** al 4; eth 4; lig 3
IR: μSADP 25215 3030 1590 1450 1270 1210 1160 1140 1100 970 920 830 780 760 750 680
UV: SAD 21003 286(495) 278(553) MeOH
H NMR: SAD 14853 4.8 7.0-7.5 $CDCl_3$
m/e: 193.95

HODOC No: 03048
CAS No: 3290-06-0
Beilstein No: 5^3, 699
Name: Benzene, 1,3-dichloro-5-(chloromethyl)-
Synonym: Toluene, α,3,5-trichloro-
Line Form: 3,5-$Cl_2C_6H_3CH_2Cl$
Mol Form: $C_7H_5Cl_3$
M_r: 195.47
bp: $60^{0.35}$
mp: 36
Color: cr (MeOH)
***s*:** EtOH 4
m/e: 193.95

HODOC No: 03049
CAS No: 94-99-5
Beilstein No: 5^4, 819
Name: Benzene, 2,4-dichloro-1-(chloromethyl)-
Synonym: Toluene, α,2,4-trichloro-
Line Form: 2,4-Cl_2-$C_6H_3CH_2Cl$
Mol Form: $C_7H_5Cl_3$
M_r: 195.47
bp: 120^{13}
IR: SADG 139 3080 2960 1585 1555 1468 1440 1385 1265 1095 1050 905 865 845 820 740 720 670 650 555 460 410 395
Raman: SAD 766 3070 3020 2970 2950 2870 1590 1560 1510 1470 1450 1430 1390 1300 1270 1210 1200 1150 1100 1050 1030 850 820 770 740 730 680 650 560 520 460 420 400 340 310 270 210 180
UV: SAD 203 283(234) 274(317) 229(10100) MeOH
H NMR: SAD 308 4.6 7.4 $CDCl_3$
MS: WILEY 1345 159(100) 161(64) 194(20) 196(19) 61(12) 123(11) 163(10) 62(9)
m/e: 193.95

HODOC No: 03050
CAS No: 56961-84-3
Beilstein No: 5^3, 702
Name: Benzene, 1,2-dichloro-4-(dichloromethyl)-
Synonym: Benzylidene chloride, 3,4-dichloro
Line Form: 3,4-$Cl_2C_6H_3CHCl_2$
Mol Form: $C_7H_4Cl_4$
M_r: 229.91
bp: 257
***d*:** $1.518^{22/22}$
***s*:** Bz 4; Et_2O 4; EtOH 4
m/e: 227.91

HODOC No: 03051
CAS No: 56961-85-4
Beilstein No: 5^3, 703
Name: Benzene, 1,3-dichloro-5-(dichloromethyl)-
Synonym: Benzylidene chloride, 3,5-dichloro
Line Form: 3,5-$Cl_2C_6H_3CHCl_2$
Mol Form: $C_7H_4Cl_4$
M_r: 229.91
mp: 36.5
Color: cr (MeOH or dil HOAc)
***s*:** Ac 4; Et_2O 4
m/e: 227.91

HODOC No: 03052
CAS No: 56961-83-2
Beilstein No: 5^3, 702
Name: Benzene, 1,4-dichloro-2-(dichloromethyl)-
Synonym: 1,4-Dichloro-2-(dichloromethyl)benzene
Line Form: 2,5-$Cl_2C_6H_3CHCl_2$
Mol Form: $C_7H_4Cl_4$
M_r: 229.91
mp: 42
Color: cubic cr (Chl)
***s*:** Bz 4; Et_2O 4; EtOH 4
MS: NBS 1522
m/e: 227.91

HODOC No: 03053
CAS No: 54932-84-2
Name: Benzene, 1,4-dichloro-2,5-diethyl-
Mol Form: $C_{10}H_{12}Cl_2$
M_r: 203.10
MS: WILEY 1424 187(100) 189(54) 202(53) 167(38) 204(34) 115(20) 169(12) 116(12)
m/e: 202.03

HODOC No: 03054
CAS No: 38204-89-6
Beilstein No: 5^3, 855
Name: Benzene, 1,3-dichloro-2,5-dimethyl-
Synonym: 2,6-Dichloro-*p*-xylene
Line Form: 1,4-$(CH_3)_2$-2,6-Cl_2-C_6H_2
Mol Form: $C_8H_8Cl_2$
M_r: 175.05
bp: 222^{770}
mp: 15
m/e: 174.00

HODOC No: 03055
CAS No: 1124-05-6
Beilstein No: 5^3, 855
Name: Benzene, 1,4-dichloro-2,5-dimethyl-
Synonym: *p*-Xylene, 2,5-dichloro-
Line Form: 2,5-Cl_2-1,4-$(CH_3)_2$-C_6H_5
Mol Form: $C_8H_8Cl_2$
M_r: 175.05
bp: 222
mp: 71
***s*:** chl 3
IR: µSADP 13673 2940 1470 1450 1370 1350 1180 1060 1030 980 880 760
Raman: SAD 1163 3050 2990 2960 2920 2850 2740 1600 1540 1440 1380 1270 1230 1030 940 710 510 470 350 330 250 100
UV: SAD 3623 283 MeOH
H NMR: SAD 3253 2.2 7.1 $CDCl_3$
MS: WILEY 1170 139(100) 174(63) 176(40) 141(32) 103(23) 51(21) 77(15) 140(11)
m/e: 174.00

HODOC No: 03056
CAS No: 2084-45-9
Beilstein No: 5^3, 835
Name: Benzene, 1,5-dichloro-2,4-dimethyl-
Synonym: *m*-Xylene, 4,6-dichloro-
Line Form: 1,3-$(CH_3)_2$-4,6-Cl_2-C_6H_2
Mol Form: $C_8H_8Cl_2$
M_r: 175.05
bp: 223
mp: 68-9
Color: lf
***s*:** Bz 4; Et_2O 4; Chl 4
m/e: 174.00

HODOC No: 03057
CAS No: 34840-79-4
Beilstein No: 5^3, 855
Name: Benzene, 2,3-dichloro-1,4-dimethyl-
Synonym: 2,3-Dichloro-1,4-dimethylbenzene
Line Form: 1,4-$(CH_3)_2$-2,3-Cl_2-C_6H_2
Mol Form: $C_8H_8Cl_2$
M_r: 175.05
bp: 230
mp: -2
***d*:** $1.2332^{20/20}$
m/e: 174.00

HODOC No: 03058
CAS No: 55191-16-7
Name: Benzene,
1,3-dichloro-5-(1,1-dimethylethyl)-2-methoxy-
Synonym: Benzene, 1-*tert*-butyl-3,5-dichloro-4-methoxy
Mol Form: $C_{11}H_{14}Cl_2O$
M_r: 233.13
MS: WILEY 1646 217(100) 219(64) 232(22) 234(14) 189(14) 41(12) 218(12) 22(11)
m/e: 232.04

HODOC No: 03059
CAS No: 951-86-0
Beilstein No: 5^3, 1963
Name: Benzene,
1,1′-(1,2-dichloro-1,2-ethenediyl)*bis*-, (E)-
Synonym: Stilbene, α,α′-dichloro-, (E)-
Mol Form: $C_{14}H_{10}Cl_2$
M_r: 249.13
bp: 180^{18} $106\text{-}7^{0.005}$
mp: 143-4
***s*:** H_2O 1; al 4; eth 4
UV: OES 3-433 270(10000) cyhex
m/e: 248.02

HODOC No: 03060
CAS No: 5216-32-0
Beilstein No: 5^3, 1963
Name: Benzene,
1,1′-(1,2-dichloro-1,2-ethenediyl)*bis*-, (Z)-
Synonym: Stilbene, α,α′-dichloro-, (Z)-
Line Form: $C_6H_5CCl{=}CClC_6H_5$
Mol Form: $C_{14}H_{10}Cl_2$
M_r: 249.13
bp: 180^{18}
mp: 67.5-8.0
***s*:** H_2O 1; al 3; eth 3
UV: MOCHA 1146-53 287(6310) hx
m/e: 248.02

HODOC No: 03061
CAS No: 6607-45-0
Beilstein No: 5^4, 1348
Name: Benzene, (1,2-dichloroethenyl)-
Synonym: Styrene, α,β-dichloro-
Line Form: $C_6H_5CCl{=}CHCl$
Mol Form: $C_8H_6Cl_2$
M_r: 173.03
bp: 221
m/e: 171.98

HODOC No: 03062
CAS No: 2123-28-6
Beilstein No: 5^3, 1174
Name: Benzene, 1,2-dichloro-3-ethenyl-
Synonym: Styrene, 2,3-dichloro-
Line Form: $2,3\text{-}Cl_2C_6H_3CH{=}CH_2$
Mol Form: $C_8H_6Cl_2$
M_r: 173.03
bp: $92\text{-}4^{4.5}$
***d*:** $1.2849^{20/20}$
***n*:** 1.5848^{20}
m/e: 171.98

HODOC No: 03063
CAS No: 2039-83-0
Beilstein No: 5^4, 1347
Name: Benzene, 1,2-dichloro-4-ethenyl-
Synonym: Styrene, 3,4-dichloro-
Line Form: $3,4\text{-}Cl_2C_6H_3CH{=}CH_2$
Mol Form: $C_8H_6Cl_2$
M_r: 173.03
bp: $69\text{-}70^4$, 95^5
***d*:** $1.256^{20/4}$
***n*:** 1.5857^{20}
IR: μCOB 185 3030 1610 1560 1470 1390 1120 1030 980 910 880 820 690
H NMR: SAD 6345 2.4 CCl_4
m/e: 171.98

HODOC No: 03064
CAS No: 26140-94-3
Name: Benzene, 1,2-dichloro-4-ethenyl-, homopolymer
Synonym: Styrene, 3,4-dichloro-, polymers
Mol Form: $C_8H_6Cl_2$
M_r: 173.03
IR: μCOB 2341 3030 2940 1560 1470 1410 1140 1030 890 820 750 710 690
m/e: 171.98

HODOC No: 03065
CAS No: 28469-92-3
Beilstein No: 5^3, 1174
Name: Benzene, 1,3-dichloro-2-ethenyl-
Synonym: Styrene, 2,6-dichloro-
Line Form: 2,6-Cl_2-C_6H_3-CH=CH_2
Mol Form: $C_8H_6Cl_2$
M_r: 173.03
bp: $64\text{-}5^3$
mp: 88-90
***d*:** $1.2631^{20/4}$
***n*:** 1.5752^{20}
IR: µCOB 4856 3130 1640 1560 1450 1410 1190 1090 980 940 780 740
UV: SAD 11543 239(6810) 213(30700) MeOH
H NMR: SAD 1340 5.6 5.7 6.7 7.2 CCl_4
MS: ALD 4:70B
m/e: 171.98

HODOC No: 03066
CAS No: 2155-42-2
Beilstein No: 5^3, 1175
Name: Benzene, 1,3-dichloro-5-ethenyl-
Synonym: Styrene, 3,5-dichloro-
Line Form: 3,5-$Cl_2C_6H_3CH$=CH_2
Mol Form: $C_8H_6Cl_2$
M_r: 173.03
bp: 59^1
***d*:** $1.225^{25/4}$
***n*:** 1.5745^{25}
IR: µCOB 186 3030 2940 1610 1590 1540 1470 1410 1370 1030 990 900 850 720
UV: OES 1-5 369(5) 351(16) 347(16) 340(17) 335(23) 330(22) 320(25) 316(23) 306(23) hp
m/e: 171.98

HODOC No: 03067
CAS No: 1123-84-8
Beilstein No: 5^4, 1347
Name: Benzene, 1,4-dichloro-2-ethenyl-
Synonym: Styrene, 2,5-dichloro-
Line Form: 2,5-$Cl_2C_6H_3CH$=CH_2
Mol Form: $C_8H_6Cl_2$
M_r: 173.03
bp: $72\text{-}3^2$
***d*:** $1.4045^{20/4}$
***n*:** 1.5798^{20}
IR: µSADP 11404 3130 1610 1590 1560 1470 1410 1390 1250 1190 1140 1090 1040 960 920 880 860 810 700
MS: WILEY 1145 172(100) 174(64) 137(56) 102(23) 101(21) 139(18) 75(12) 68(12)
m/e: 171.98

HODOC No: 03068
CAS No: 25084-93-9
Name: Benzene, 1,4-dichloro-2-ethenyl-, homopolymer
Synonym: Styrene, 2,5-dichloro-, polymers
Mol Form: $C_8H_6Cl_2$
M_r: 173.03
IR: µCOB 982 3030 2940 1610 1560 1490 1450 1410 1190 1120 1090 1040 1010 940 830 750 690 660
m/e: 171.98

HODOC No: 03069
CAS No: 698-88-4
Beilstein No: 5^4, 1348
Name: Benzene, (2,2-dichloroethenyl)-
Synonym: Styrene, β,β-dichloro-
Line Form: C_6H_5CH=CCl_2
Mol Form: $C_8H_6Cl_2$
M_r: 173.03
bp: 225, $85\text{-}6^5$
***d*:** $1.2531^{20/4}$
***n*:** 1.5852^{20}
***s*:** ace 3; chl 3
UV: OES 3-133 260(15849) hp
m/e: 171.98

HODOC No: 03070
CAS No: 2123-27-5
Beilstein No: 5^4, 1347
Name: Benzene, 2,4-dichloro-1-ethenyl-
Synonym: Styrene, 2,4-dichloro-
Line Form: 2,4-$Cl_2C_6H_3CH=CH_2$
Mol Form: $C_8H_6Cl_2$
M_r: 173.03
bp: 81^6
***d*:** $1.243^{25/4}$
***n*:** 1.5828^{20}
IR: μCOB 1952 3130 1610 1590 1560 1470 1410 1370 1100 1050 980 920 870 830 820 700
m/e: 171.98

HODOC No: 03071
CAS No: 72-55-9
Beilstein No: 5^4, 2176
Name: Benzene, 1,1′-(dichloroethenylidene)*bis*[4-chloro-
Synonym: Ethylene, 1,1-dichloro-2,2-*bis*(*p*-chlorophenyl)-
Line Form: $(4\text{-}ClC_6H_4)_2C=CCl_2$
Mol Form: $C_{14}H_8Cl_4$
M_r: 318.02
mp: 88-90
IR: ALD 2:550G
MS: ALD 4:72C
m/e: 315.94

HODOC No: 03072
CAS No: 5392-86-9
Beilstein No: 6^4, 885
Name: Benzene, 2,4-dichloro-1-ethoxy-
Synonym: Phenetole, 2,4-dichloro-
Line Form: 2,4-$Cl_2C_6H_3OC_2H_5$
Mol Form: $C_8H_8Cl_2O$
M_r: 191.05
bp: 237
***s*:** H_2O 1; al 3; eth 3; bz 3
IR: μCOB 1983 2940 1590 1470 1410 1280 1250 1110 1060 1040 940 870 830 800 740
MS: WILEY 1322 159(100) 161(64) 139(46) 174(39) 176(24) 103(14) 141(14) 190(1)
m/e: 190.00

HODOC No: 03073
CAS No: 1074-11-9
Beilstein No: 5^4, 901
Name: Benzene, (1,2-dichloroethyl)-
Synonym: α,β-Dichloroethylbenzene
Line Form: $C_6H_5CHClCH_2Cl$
Mol Form: $C_8H_8Cl_2$
M_r: 175.05
bp: 233-4, 115^{15}
***d*:** $1.240^{15/4}$
***n*:** 1.5544^{15}
***s*:** H_2O 1; ace 3; bz 3
MS: WILEY 1167 125(100) 103(39) 127(32) 51(27) 104(22) 139(20) 77(20) 174(15)
m/e: 174.00

HODOC No: 03074
CAS No: 54484-61-6
Name: Benzene, 1,2-dichloro-3-ethyl-
Synonym: 2,3-Dichloroethylbenzene
Mol Form: $C_8H_8Cl_2$
M_r: 175.05
IR: μCOB 188 3030 2940 1590 1470 1430 1180 1080 1040 840 780 750 740 710
Raman: SAD 38 2970 2940 2870 2770 2730 1760 1450 1400 1370 1310 1270 1160 1110 1090 1020 900 870 810 780 760 380 330
MS: WILEY 1171 159(100) 161(63) 139(62) 174(42) 176(27) 141(20) 103(19) 51(12)
m/e: 174.00

HODOC No: 03075
CAS No: 6623-59-2
Name: Benzene, 1,2-dichloro-4-ethyl-
Synonym: 3,4-Dichloroethylbenzene
Mol Form: $C_8H_8Cl_2$
M_r: 175.05
IR: μCOB 187 3030 2940 1470 1410 1140 1030 880 850 820 690
MS: WILEY 1168 159(100) 161(63) 139(48) 174(40) 176(26) 141(16) 163(10) 160(8)
m/e: 174.00

HODOC No: 03076
CAS No: 33407-02-2
Name: Benzene, 1,3-dichloro-2-ethyl-
Synonym: 1,3-Dichloro-2-ethylbenzene
Mol Form: $C_8H_8Cl_2$
M_r: 175.05
MS: WILEY 1169 159(100) 161(64) 174(29) 176(18) 139(14) 163(11) 103(10) 160(8)
m/e: 174.00

HODOC No: 03077
CAS No: 54484-63-8
Beilstein No: 5, 354
Name: Benzene, 1,4-dichloro-2-ethyl-
Synonym: 2,5-Dichloroethylbenzene
Mol Form: $C_8H_8Cl_2$
M_r: 175.05
bp: 213.5
***d*:** 1.239^{0}
***s*:** eth 2; bz 3; chl 2; peth 2
IR: µCOB 138 2940 1470 1390 1100 1060 880 850 810
UV: OES 1-84 283(251) 274(316) $CHCl_3$
MS: WILEY 1168 159(100) 161(64) 139(52) 174(45) 176(28) 141(16) 103(16) 51(11)
m/e: 174.00

HODOC No: 03078
CAS No: 72-54-8
Beilstein No: 5^4, 1884
Name: Benzene, 1,1′-(2,2-dichloroethylidene)*bis*[4-chloro-
Synonym: Ethane, 1,1-dichloro-2,2-*bis*(*p*-chlorophenyl)-
Line Form: $(4\text{-}ClC_6)H_4)_2CHCHCl_2$
Structure:

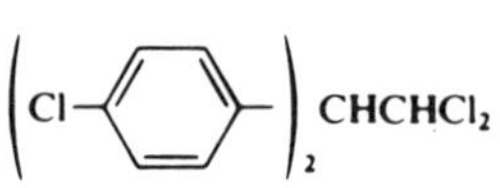

Mol Form: $C_{14}H_{10}Cl_4$
M_r: 320.03
mp: 109-10
***s*:** chl 2
IR: COB 1001 3100 3020 2990 1590 1490 1410 1090 1020 850 825 800 760 700
UV: SAD 5898 268 231 MeOH
H NMR: SAD 2040 4.5 6.3 7.3 $CDCl_3$
m/e: 317.95

HODOC No: 03079
CAS No: 498-67-9
Beilstein No: 5^4, 818
Name: Benzene, (dichlorofluoromethyl)-
Synonym: Toluene, α,α-dichloro-α-fluoro-
Line Form: $C_6H_5CCl_2F$
Mol Form: $C_7H_5Cl_2F$
M_r: 179.01
bp: 178-80
***d*:** 1.3138^{11}
***n*:** 1.5180^{11}
***s*:** EtOH 4
m/e: 177.98

HODOC No: 03080
CAS No: 3032-81-3
Beilstein No: 5^3, 581
Name: Benzene, 1,3-dichloro-5-iodo-
Synonym: 1,3-Dichloro-5-iodobenzene
Line Form: $3,5\text{-}Cl_2C_6H_3I$
Mol Form: $C_6H_3Cl_2I$
M_r: 272.89
mp: 56-8
m/e: 271.87

HODOC No: 03081
CAS No: 29682-41-5
Beilstein No: 5^4, 580
Name: Benzene, 1,4-dichloro-2-iodo-
Synonym: 1,4-Dichloro-2-iodobenzene
Line Form: $2,5\text{-}Cl_2C_6H_3I$
Mol Form: $C_6H_3Cl_2I$
M_r: 272.89
bp: 255-6, 134^{12}
mp: 23
Color: pl (al)
***s*:** Bz 4; Et_2O 4; EtOH 4; Chl 4
m/e: 271.87

HODOC No: 03082
CAS No: 1984-59-4
Beilstein No: 6^2, 178
Name: Benzene, 1,2-dichloro-3-methoxy-
Synonym: Anisole, 2,3-dichloro-
Line Form: $2,3\text{-}Cl_2\text{-}C_6H_3OCH_3$
Mol Form: $C_7H_6Cl_2O$
M_r: 177.02
mp: 31-3
IR: ALD 3:629A
MS: ALD 4:90D
m/e: 175.98

HODOC No: 03083
CAS No: 1984-65-2
Name: Benzene, 1,3-dichloro-2-methoxy-
Synonym: Anisole, 2,6-dichloro-
Mol Form: $C_7H_6Cl_2O$
M_r: 177.02
IR: SADG 22935 3000 2930 2840 2810 1560 1460 1430 1410 1240 1190 1060 990 780 760 730 600 550 530
Raman: SAD 716 3140 3080 3010 2950 2860 2830 1570 1470 1260 1190 1160 1110 1080 1000 810 740 610 560 400 360 210
H NMR: SAD 13055 3.8 6.9 7.2 CCl_4
m/e: 175.98

HODOC No: 03084
CAS No: 33719-74-3
Beilstein No: 6^3, 715
Name: Benzene, 1,3-dichloro-5-methoxy-
Synonym: Anisole, 3,5-dichloro-
Line Form: 3,5-Cl_2-$C_6H_3OCH_3$
Mol Form: $C_7H_6Cl_2O$
M_r: 177.02
mp: 40-2
IR: ALD 3:631C
MS: ALD 4:94B
m/e: 175.98

HODOC No: 03085
CAS No: 1984-58-3
Beilstein No: 6^4, 942
Name: Benzene, 1,4-dichloro-2-methoxy-
Synonym: Anisole, 2,5-dichloro-
Line Form: 2,5-Cl_2-$C_6H_3OCH_3$
Mol Form: $C_7H_6Cl_2O$
M_r: 177.02
bp: 140^{40}
mp: 24
IR: µCOB 3261 3030 2940 2860 1590 1490 1470 1450 1410 1270 1140 1060 880 800 720 710
MS: WILEY 1187 ini 176(100) 178(64) 133(58) 161(35) 135(29) 163(23) 63(13) 180(12)
m/e: 175.98

HODOC No: 03086
CAS No: 553-82-2
Beilstein No: 6^4, 855
Name: Benzene, 2,4-dichloro-1-methoxy-
Synonym: Anisole, 2,4-dichloro-
Line Form: 2,4-Cl_2-$C_6H_3OCH_3$
Mol Form: $C_7H_6Cl_2O$
M_r: 177.02
bp: 235^{740}, 125^{10}
mp: 28-9
Color: pr
***s*:** chl 2
IR: SADG 20988 3000 2940 2840 1580 1490 1380 1290 1260 1100 1050 1010 860 830 800 700 640
UV: SAD 18864 293(2360) 285(2610) 229(9320) MeOH
H NMR: SAD 11587 3.8 6.8 7.2 7.3 $CDCl_3$
MS: WILEY 1187 ini 176(100) 178(64) 133(58) 161(35) 135(29) 163(23) 63(13) 180(12)
m/e: 175.98

HODOC No: 03087
CAS No: 98-87-3
Beilstein No: 5^4, 817
Name: Benzene, (dichloromethyl)-
Synonym: Toluene, α,α-dichloro-
Line Form: $C_6H_5CHCl_2$
Mol Form: $C_7H_6Cl_2$
M_r: 161.02
bp: 205
mp: -16.4
***d*:** 1.2557^{14}
***n*:** 1.5502^{20}
***s*:** H_2O 1; al 3; eth 3; ctc 2
IR: µSADP 2489 3130 3030 1490 1470 1330 1250 1220 1190 1080 1030 1010 880 840 810 700
UV: SAD 657 288 272 265 259 247 240 220 MeOH
H NMR: SAD 685 6.6 7.4 CCl_4
MS: WILEY 1002 125(100) 127(32) 160(14) 89(13) 63(9) 162(9) 126(8) 62(7)
m/e: 159.98

HODOC No: 03088
CAS No: 32768-54-0
Beilstein No: 5[4], 815
Name: Benzene, 1,2-dichloro-3-methyl-
Synonym: Toluene, 2,3-dichloro-
Line Form: 2,3-$Cl_2C_6H_3CH_3$
Mol Form: $C_7H_6Cl_2$
M_r: 161.02
bp: 207-8, 61-2[3]
***n*:** 1.5511[20]
***s*:** Bz 4
MS: WILEY 1002
m/e: 159.98

HODOC No: 03089
CAS No: 95-75-0
Beilstein No: 5[4], 815
Name: Benzene, 1,2-dichloro-4-methyl-
Synonym: Toluene, 3,4-dichloro-
Line Form: 3,4-$Cl_2C_6H_3CH_3$
Mol Form: $C_7H_6Cl_2$
M_r: 161.02
bp: 208.9, 81.8[10]
mp: -15.2
***d*:** 1.2564[20/4]
***n*:** 1.5471[20]
***s*:** H_2O 1; al 5; eth 5; ace 5; bz 5; ctc 5; lig 5
IR: SADG 143 3020 2920 1600 1560 1470 1380 1260 1210 1130 1030 870 800 680 640
Raman: SAD 768 3060 3040 2920 2870 2740 2540 1600 1570 1470 1450 1380 1280 1260 1220 1150 1140 1100 1040 1010 880 840 820 730 720 690 650 570 550 500 470 440 410 380 370 320 260 210 130
UV: SAD 210 281(519) 273(513) 266(328) 218(8060) MeOH
H NMR: SAD 574 2.3 6.7-7.3 CCl_4
MS: WILEY 1002 ini 125(100) 160(68) 162(43) 127(32) 161(11) 126(10) 99(6) 124(6)
m/e: 159.98

HODOC No: 03090
CAS No: 118-69-4
Beilstein No: 5[4], 815
Name: Benzene, 1,3-dichloro-2-methyl-
Synonym: Toluene, 2,6-dichloro-
Line Form: 2,6-$Cl_2C_6H_3CH_3$
Mol Form: $C_7H_6Cl_2$
M_r: 161.02
bp: 198
***n*:** 1.5507[20]
***s*:** H_2O 1; chl 3
IR: SADG 664 3050 2920 1580 1550 1430 1370 1260 1200 1080 1000 770 690
UV: SAD 5404 274 266 262 259 cyhex
H NMR: SAD 894 2.4 6.8-7.3 CCl_4
MS: WILEY 1003 125(100) 160(46) 127(31) 162(30) 89(27) 63(13) 62(13) 126(11)
m/e: 159.98

HODOC No: 03091
CAS No: 25186-47-4
Beilstein No: 5[3]694
Name: Benzene, 1,3-dichloro-5-methyl-
Synonym: Toluene, 3,5-dichloro-
Mol Form: $C_7H_6Cl_2$
M_r: 161.02
bp: 201-2[760]
mp: 26
***s*:** H_2O 1
Raman: DFB 92 3060 3010 2910 2860 1590 1570 1380 1230 1210 1120 1060 1040 1000 520 430 400 240 230 220 200 160
MS: WILEY 1002 ini 125(100) 160(68) 162(43) 127(32) 161(11) 126(10) 99(6) 124(6)
m/e: 159.98

HODOC No: 03092
CAS No: 19398-61-9
Beilstein No: 5[4], 694
Name: Benzene, 1,4-dichloro-2-methyl-
Synonym: Toluene, 2,5-dichloro-
Line Form: 2,5-$Cl_2C_6H_3CH_3$
Mol Form: $C_7H_6Cl_2$
M_r: 161.02
bp: 200[770]
mp: 5
***d*:** 1.2535[20]
***n*:** 1.5449[20]
***s*:** H_2O 1; bz 3
MS: WILEY 1002 ini 125(100) 160(68) 162(43) 127(32) 161(11) 126(10) 99(6) 124(6)
m/e: 159.98

HODOC No: 03093
CAS No: 95-73-8
Beilstein No: 5^4, 815
Name: Benzene, 2,4-dichloro-1-methyl-
Synonym: Toluene, 2,4-dichloro-
Line Form: 2,4-$Cl_2C_6H_3CH_3$
Mol Form: $C_7H_6Cl_2$
M_r: 161.02
bp: 196-7
mp: -13.5
***d*:** $1.2498^{20/20}$
***n*:** 1.5511^{20}
***s*:** H_2O 1; ctc 3
IR: SADG 144 3060 2920 1590 1560 1470 1440 1380 1260 1110 1050 990 860 840 800 700 690 650
Raman: SAD 767 3070 3020 2960 2930 2860 2740 2550 1600 1570 1480 1450 1390 1300 1280 1260 1210 1150 1130 1110 1060 880 840 810 750 720 690 650 470 440 410 390 320 280 210 190 130
UV: SAD 211 281(588) 273(588) 266(378) 218(3740) MeOH
H NMR: SAD 1185 2.3 7.0 7.3 CCl_4
MS: WILEY 1003 125(100) 160(46) 127(32) 162(30) 89(20) 159(12) 63(12) 62(12)
m/e: 159.98

HODOC No: 03094
CAS No: 2051-90-3
Beilstein No: 5^4, 1848
Name: Benzene, 1,1′-(dichloromethylene)*bis*-
Synonym: Methane, dichlorodiphenyl-
Line Form: $(C_6H_5)_2CCl_2$
Mol Form: $C_{13}H_{10}Cl_2$
M_r: 237.12
bp: 305d, 190^{21}
***d*:** 1.235^{18}
***s*:** eth 3; bz 3; ctc 3
IR: µSADP 24272 3030 1490 1450 1220 1160 860 790 750 730 690
UV: SAD 8525 250(5120) MeOH
H NMR: SAD 2385 7.2-7.8 CCl_4
m/e: 236.02

HODOC No: 03095
CAS No: 41575-21-7
Name: Benzene, 1,2-dichloro-4-(1-methylethenyl)-
Synonym: 3,4-Dichloro-α-methylstyrene
Mol Form: $C_9H_8Cl_2$
M_r: 187.06
MS: WILEY 1278 186(100) 188(68) 151(39) 115(39) 116(35) 173(29) 171(25) 136(20)
m/e: 186.00

HODOC No: 03096
CAS No: 54789-29-6
Name: Benzene, 1-(dichloromethyl)-4-ethyl-
Synonym: 4-Ethylbenzal chloride
Mol Form: $C_9H_{10}Cl_2$
M_r: 189.08
MS: WILEY 1296 153(100) 117(38) 115(38) 155(31) 151(22) 188(16) 154(12) 116(9)
m/e: 188.02

HODOC No: 03097
CAS No: 4132-70-1
Name: Benzene, 1,4-dichloro-2-(1-methylethyl)-
Synonym: Cumene, 2,5-dichloro-
Mol Form: $C_9H_{10}Cl_2$
M_r: 189.08
MS: WILEY 1294 173(100) 175(63) 188(28) 137(23) 190(18) 102(18) 101(13) 177(10)
m/e: 188.02

HODOC No: 03098
CAS No: 23063-36-7
Beilstein No: 5^3, 855
Name: Benzene, 1-(dichloromethyl)-4-methyl-
Synonym: *p*-Xylene, α,α-dichloro-
Line Form: 4-$CH_3C_6H_4CHCl_2$
Mol Form: $C_8H_8Cl_2$
M_r: 175.05
bp: $98\text{-}9^{10}$
mp: 48-9
***s*:** al 2; bz 2; chl 2
IR: µCOB 1951 3030 2940 1610 1520 1450 1410 1250 1210 1180 1110 850 790 770 730 690
MS: WILEY 1170 139(100) 141(33) 119(26) 91(24) 120(22) 103(19) 174(16) 51(15)
m/e: 174.00

HODOC No: 03099
CAS No: 81686-41-1
Beilstein No: 5^3, 957
Name: Benzene, 1,4-dichloro-2-methyl-5-(1-methylethyl)-
Synonym: *p*-Cymene, 2,5-dichloro
Line Form: 2,5-Cl_2-4-$(CH_3)_2CH$-$C_6H_2CH_3$
Mol Form: $C_{10}H_{12}Cl_2$
M_r: 203.10
bp: 240-2
***d*:** $1.165^{20/4}$
m/e: 202.03

HODOC No: 03100
CAS No: 619-28-3
Beilstein No: 5^3, 750
Name: Benzene, 1-(dichloromethyl)-3-nitro-
Synonym: Toluene, α,α-dichloro-*m*-nitro-
Line Form: $3\text{-}NO_2C_6H_4CHCl_2$
Mol Form: $C_7H_5Cl_2NO_2$
M$_r$: 206.02
mp: 65
Color: mcl (al)
***s*:** Et_2O 4; EtOH 4
m/e: 204.97

HODOC No: 03101
CAS No: 29682-46-0
Beilstein No: 5, 332
Name: Benzene, 1,3-dichloro-2-methyl-4-nitro-
Synonym: Toluene, 2,6-dichloro-3-nitro-
Line Form: $2,6\text{-}Cl_2\text{-}3\text{-}NO_2C_6H_2CH_3$
Mol Form: $C_7H_5Cl_2NO_2$
M$_r$: 206.02
mp: 53-5
IR: ALD 2:728A
MS: ALD 5:179A
m/e: 204.97

HODOC No: 03102
CAS No: 619-78-3
Beilstein No: 5^4, 859
Name: Benzene, 1-(dichloromethyl)-4-nitro-
Synonym: Toluene, α,α-dichloro-*p*-nitro-
Line Form: $4\text{-}NO_2C_6H_4CH_2Cl_2$
Mol Form: $C_7H_5Cl_2NO_2$
M$_r$: 206.02
mp: 46
Color: pr (al)
***s*:** H_2O 1; al 4; eth 4
UV: OES 2-85 256(14454) hx
m/e: 204.97

HODOC No: 03103
CAS No: 3209-22-1
Beilstein No: 5^4, 725
Name: Benzene, 1,2-dichloro-3-nitro-
Synonym: 1,2-Dichloro-3-nitrobenzene
Line Form: $2,3\text{-}Cl_2C_6H_3NO_2$
Mol Form: $C_6H_3Cl_2NO_2$
M$_r$: 192.00
bp: 257-8
mp: 61-2
***d*:** 1.721^{14}
Color: mcl nd (peth, HOAc)
***s*:** H_2O 1; al 3; eth 3; ace 3; bz 3; chl 2; peth 3; HOAc 3; os 4
IR: µCOB 1217 3130 1590 1540 1430 1350 1280 1160 1050 880 790 760 720
UV: SAD 6270 215 MeOH
H NMR: SAD 4282 7.2-7.8 $CDCl_3$
MS: NBS 4379
m/e: 190.95

HODOC No: 03104
CAS No: 99-54-7
Beilstein No: 5^4, 726
Name: Benzene, 1,2-dichloro-4-nitro-
Synonym: 1,2-Dichloro-4-nitrobenzene
Line Form: $3,4\text{-}Cl_2C_6H_3NO_2$
Mol Form: $C_6H_3Cl_2NO_2$
M$_r$: 192.00
bp: 255-6, 189^{100}
mp: 43
***d*:** $1.4558^{75/4}$
Color: nd (al)
***s*:** H_2O 1; al 3; eth 3; ctc 2
IR: µCOB 1218 3130 1590 1540 1450 1390 1350 1150 1030 890 890 830 770 740 680
Raman: SAD 1149 3140 3080 2910 2680 2570 2480 1660 1640 1600 1580 1540 1520 1480 1460 1430 1380 1350 1280 1260 1240 1150 1100 1030 890 770 740 680 570 550 460 350 320 300 210 190 110
UV: SAD 1114 287(11400) 225(6130) MeOH
H NMR: SAD 687 7.6 8.0 8.3 CCl_4
m/e: 190.95

HODOC No: 03105
CAS No: 601-88-7
Beilstein No: 5^4, 726
Name: Benzene, 1,3-dichloro-2-nitro-
Synonym: 1,3-Dichloro-2-nitrobenzene
Line Form: 2,6-$Cl_2C_6H_3NO_2$
Mol Form: $C_6H_3Cl_2NO_2$
Mr: 192.00
bp: 130^8
mp: 72.5
***d*:** 1.603^{17},1.4094^{80}
Color: nd or pr (al, CS_2)
***s*:** Et_2O 4; EtOH 4
m/e: 190.95

HODOC No: 03106
CAS No: 618-62-2
Beilstein No: 5^4, 727
Name: Benzene, 1,3-dichloro-5-nitro-
Synonym: 1,3-Dichloro-5-nitrobenzene
Line Form: 3,5-$Cl_2C_6H_3NO_2$
Mol Form: $C_6H_3Cl_2NO_2$
Mr: 192.00
mp: 65.4
***d*:** 1.4000^{100}
***n*:** 1.4000^{100}
Color: mcl pr or lf (HOAc, al)
***s*:** H_2O 1; al 3; eth 3
IR: µSADP 19143 3030 1560 1430 1350 1300 1180 1160 1100 900 890 820 750 750
UV: OES 4-50 307(1202) 257(6166) hp
H NMR: SAD 2163 7.7 8.2 $CDCl_3$
m/e: 190.95

HODOC No: 03107
CAS No: 89-61-2
Beilstein No: 5^4, 726
Name: Benzene, 1,4-dichloro-2-nitro-
Synonym: 1,4-Dichloro-2-nitrobenzene
Line Form: 2,5-$Cl_2C_6H_3NO_2$
Mol Form: $C_6H_3Cl_2NO_2$
Mr: 192.00
bp: 267
mp: 56
***d*:** 1.439^{75}
***n*:** $1.4390^{75/4}$
Color: pl or pr (al) pl (AcOEt)
***s*:** H_2O 1; al 3; eth 3; bz 3; ctc 2; chl 3; CS_2 3
IR: µCOB 5766 3030 1590 1540 1450 1350 1280 1270 1150 1100 1060 890 820 760
UV: SAD 1115 304(1170) 220(18200) MeOH
H NMR: SAD 688 7.5 7.8 CCl_4
m/e: 190.95

HODOC No: 03108
CAS No: 611-06-3
Beilstein No: 5^4, 726
Name: Benzene, 2,4-dichloro-1-nitro-
Synonym: 1,3-Dichloro-4-nitrobenzene
Line Form: 2,4-$Cl_2C_6H_3NO_2$
Mol Form: $C_6H_3Cl_2NO_2$
Mr: 192.00
bp: 258.5, $100\text{-}1^4$
mp: 34
***d*:** 1.4790^{80}
***n*:** $1.5512^{70/\alpha}$
Color: nd (al)
***s*:** H_2O 1; al 3; eth 3; chl 2
IR: SADG 18360 3030 1563 1493 1449 1316 1124 1087 1042 862 820 781 741
UV: SAD 5521 260(4800) MeOH
H NMR: SAD 8677 1.6-2.4 3.4-4.4 D_2O
MS: NBS 4379
m/e: 190.95

HODOC No: 03109
CAS No: 18984-16-2
Beilstein No: 5^4, 1353
Name: Benzene, 1,2-dichloro-4-(2-nitroethenyl)-
Synonym: Styrene, 3,4-dichloro-β-nitro-
Line Form: 3,4-$Cl_2C_6H_3CH=CHNO_2$
Mol Form: $C_8H_5Cl_2NO_2$
M_r: 218.03
mp: 96
***s*:** chl 2
UV: SAD 1007 305 233 MeOH
H NMR: SAD 10844 7.2-7.6 7.9 $CDCl_3$
m/e: 216.97

HODOC No: 03110
CAS No: 6974-46-5
Name: Benzene, 1,2-dichloro-4-(2-nitropropenyl)-
Mol Form: $C_9H_7Cl_2NO_2$
M_r: 232.06
mp: 81
***s*:** chl 2
IR: μSADP 3346 3030 1670 1610 1520 1470 1410 1330 1280 1220 1140 1040 990 920 870 840 820 760 730 690
UV: SAD 1033 300 228 MeOH
H NMR: SAD 10848 2.4 7.2 7.4 7.5 7.9 $CDCl_3$
m/e: 230.99

HODOC No: 03111
CAS No: 18982-44-0
Name: Benzene, 2,4-dichloro-1-(2-nitro-1-propenyl)-
Synonym: Benzene, 2,4-dichloro-1-(2-nitropropenyl)-
Mol Form: $C_9H_7Cl_2NO_2$
M_r: 232.06
mp: 84-5
***s*:** chl 2
IR: μSADP 3345 3030 1670 1610 1540 1470 1450 1390 1330 1220 1150 1110 1060 990 900 870 840 820 780 730
UV: SAD 1032 284 cyhex
H NMR: SAD 10847 2.3 7.2 7.4 7.5 7.9 $CDCl_3$
m/e: 230.99

HODOC No: 03112
CAS No: 55538-69-7
Name: Benzene, 1,2-dichloro-4-phenoxy-
Synonym: 3,4-Dichlorodiphenyl ether
Mol Form: $C_{12}H_8Cl_2O$
M_r: 239.09
MS: WILEY 1670 238(100) 240(66) 77(53) 175(31) 168(27) 51(26) 239(17) 242(11)
m/e: 238.00

HODOC No: 03113
CAS No: 24910-69-8
Name: Benzene, 1,4-dichloro-2-phenoxy-
Synonym: Ether, 2,5-dichlorophenyl phenyl
Mol Form: $C_{12}H_8Cl_2O$
M_r: 239.09
MS: WILEY 1669 168(100) 238(65) 203(63) 77(56) 205(55) 240(43) 51(32) 204(18)
m/e: 238.00

HODOC No: 03114
CAS No: 51892-26-3
Name: Benzene, 2,4-dichloro-1-phenoxy-
Synonym: 2,4-Dichlorodiphenyl ether
Mol Form: $C_{12}H_8Cl_2O$
M_r: 239.09
MS: WILEY 1669 168(100) 238(100) 77(84) 240(65) 203(53) 51(43) 175(22) 205(18)
m/e: 238.00

HODOC No: 03115
CAS No: 1657-47-2
Beilstein No: 5^3, 1962
Name: Benzene, 2,4-dichloro-1-(2-phenylethenyl)-, (E)-
Synonym: Stilbene, 2,4-dichloro-, (E)-
Mol Form: $C_{14}H_{10}Cl_2$
M_r: 249.13
mp: 79-80
***s*:** H_2O 1; al 3; bz 3
UV: JOCEA 30-1473-65 330(18197) 95% al
m/e: 248.02

HODOC No: 03116
CAS No: 1657-69-8
Name: Benzene, 2,4-dichloro-1-(2-phenylethenyl)-, (Z)-
Synonym: 2,4-Dichloro-*cis*-stilbene
Mol Form: $C_{14}H_{10}Cl_2$
M_r: 249.13
bp: $140\text{-}50^3$
***s*:** H_2O 1; al 3; bz 3
UV: OES 7-507 290(7586) EtOH
m/e: 248.02

HODOC No: 03117
CAS No: 38862-78-1
Beilstein No: 5^3, 1188
Name: Benzene, (3,3-dichloro-2-propenyl)-
Synonym: Benzene, (3,3-dichloroallyl)-
Line Form: $C_6H_5CHCH{=}CHCl_2$
Mol Form: $C_9H_8Cl_2$
M$_r$: 187.06
bp: $142\text{-}3^{30}$ $121\text{-}2^{14}$
mp: 59
***s*:** eth 3; bz 3; chl 3; peth 3
UV: HBCP 295(1995) 258(19055) al
m/e: 186.00

HODOC No: 03118
CAS No: 5441-16-7
Name: Benzene, 2,4-dichloro-1-(2-propenyloxy)-
Synonym: Ether, allyl 2,4-dichlorophenyl
Mol Form: $C_9H_8Cl_2O$
M$_r$: 203.06
IR: µCOB 4263 3080 2920 2860 1650 1580 1480 1420 1390 1280 1260 1100 1060 1010 990 920 870 800 750 640 560
m/e: 202.00

HODOC No: 03119
CAS No: 17061-90-4
Name: Benzene, 2,4-dichloro-1-(2-propynyloxy)-
Synonym: Ether, 2,4-dichlorophenyl 2-propynyl
Mol Form: $C_9H_6Cl_2O$
M$_r$: 201.04
IR: µCOB 4326 3230 2940 2860 2130 1490 1390 1350 1320 1280 1250 1110 1080 1030 940 890 830 790 760 720 690 650
m/e: 199.98

HODOC No: 03120
CAS No: 13014-24-9
Beilstein No: 5^4, 824
Name: Benzene, 1,2-dichloro-4-(trichloromethyl)-
Synonym: Toluene, α,α,α,3,4-pentachloro-
Line Form: $3,4\text{-}Cl_2C_6H_3CCl_3$
Mol Form: $C_7H_3Cl_5$
M$_r$: 264.36
bp: 283.1, 167.3^{30}
mp: 25.8 (fp)
***d*:** 1.5913^{20}
***n*:** 1.5886^{20}
IR: SADG 146 3090 1580 1560 1470 1450 1370 1270 1190 1150 1130 1030 920 880 810 760 730 690 670 530 430 360
Raman: SAD 764 3180 3080 1600 1390 1280 1200 1160 1140 1040 930 820 760 740 680 530 420 390 240 200 160
UV: SAD 213 286(550) 278(640) 270(490) 236(10200) MeOH
H NMR: SAD 1186 7.4 7.7 8.0 CCl_4
m/e: 261.87

HODOC No: 03121
CAS No: 13014-18-1
Name: Benzene, 2,4-dichloro-1-(trichloromethyl)-
Synonym: Toluene, α,α,α,2,4-pentachloro-
Mol Form: $C_7H_3Cl_5$
M$_r$: 264.36
IR: SADG 18006 3090 1570 1550 1460 1430 1370 1250 1170 1140 1100 1040 870 790 750 680 560 480 460 410 370
Raman: SAD 763 3090 3070 1590 1560 1460 1380 1270 1220 1190 1160 1130 1110 1060 800 770 670 500 470 390 200 170
UV: SAD 2157 277 234 MeOH
H NMR: SAD 706 7.2 7.5 8.1 CCl_4
m/e: 261.87

HODOC No: 03122
CAS No: 55191-17-8
Beilstein No: 5, 448
Name: Benzene, 1,4-dichloro-2,3,5-triethyl-
Mol Form: $C_{12}H_{16}Cl_2$
M$_r$: 231.15
***d*:** 1.131^0
***s*:** eth 3; bz 2; chl 3; CS_2 3
MS: WILEY 1632 215(100) 230(66) 217(63) 232(42) 195(27) 128(15) 115(15) 216(12)
m/e: 230.06

HODOC No: 03123
CAS No: 320-60-5
Name: Benzene, 2,4-dichloro-1-(trifluoromethyl)-
Synonym: Toluene, 2,4-dichloro-α,α,α-trifluoro-
Mol Form: $C_7H_3Cl_2F_3$
M$_r$: 215.00
***n*:** $1.4802^{20/D}$
IR: SADG 17856 3100 2920 1590 1560 1480 1380 1300 1240 1170 1130 1100 1090 1020 870 820 740 600 520 460
H NMR: SAD 9125 7.3 7.5 7.6 CCl_4
m/e: 213.96

HODOC No: 03124
CAS No: 57386-83-1
Beilstein No: 5^3, 919
Name: Benzene, 2,4-dichloro-1,3,5-trimethyl-
Synonym: 2,4-Dichloromesitylene
Line Form: $2,4\text{-}Cl_2\text{-}1,3,5(CH_3)_3C_6H$
Mol Form: $C_9H_{10}Cl_2$
M$_r$: 189.08
bp: 243-4
mp: 59
***s*:** al 3; eth 2; bz 2
IR: µCOB 4163 2940 1470 1410 1390 1240 1080 980 860 700
m/e: 188.02

HODOC No: 03125
CAS No: 1630-09-7
Beilstein No: 5^4, 758
Name: Benzene, 2,4-dichloro-1,3,5-trinitro-
Synonym: Dichloro-1,3,5-trinitrobenzene
Line Form: $1,3\text{-}Cl_2\text{-}2,4\text{-}6\text{-}(NO_2)_3C_6H$
Mol Form: $C_6HCl_2N_3O_6$
M$_r$: 281.99
mp: 128-9
IR: µSADP 2476 2940 1590 1540 1390 1330 1190 960 900 780 720
UV: SAD 656 <215 MeOH
H NMR: SAD 10677 9.2 DMSO
m/e: 280.92

HODOC No: 03126
CAS No: 1087-02-1
Beilstein No: 5^4, 1607
Name: Benzene, 1,4-dicyclohexyl-
Synonym: Benzene, *p*-dicyclohexyl-
Line Form: $4\text{-}C_6H_{11}C_6H_4C_6H_{11}$
Structure:

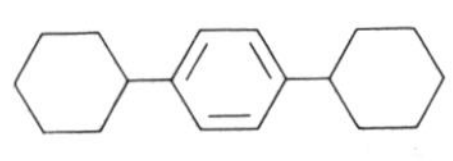

Mol Form: $C_{18}H_{26}$
M$_r$: 242.39
bp: 120-3
mp: 103-5
IR: µSADP 8195 2940 2860 1520 1450 1430 1350 1270 1180 1140 1100 1000 920 890 880 820 790 760
UV: SAD 2197 271 263 257 218 213 MeOH
MS: ALDN 197 117(100) 55(98) 91(94) 83(71) 159(64) 41(59) 242(41) 129(38)
m/e: 242.20

HODOC No: 03127
CAS No: 55191-38-3
Name: Benzene, 1,3-didecyl-
Mol Form: $C_{26}H_{46}$
M$_r$: 358.63
UV: API 862 272 265 214 cyhex
MS: WILEY 2041 105(100) 43(48) 106(34) 358(31) 41(26) 91(24) 232(23) 57(18)
m/e: 358.36

HODOC No: 03128
CAS No: 2655-95-0
Name: Benzene, 1,4-didecyl-
Synonym: Benzene, *p*-didecyl-
Mol Form: $C_{26}H_{46}$
M$_r$: 358.63
UV: API 1024 272 263 258 218 213 EtOH
MS: WILEY 2049 105(100) 358(45) 43(33) 104(26) 91(22) 41(19) 29(19) 57(17)
m/e: 358.36

HODOC No: 03129
CAS No: 91-14-5
Beilstein No: 5^4, 1540
Name: Benzene, 1,2-diethenyl-
Synonym: Benzene, *o*-divinyl-
Line Form: 1,2-$(CH_2{=}CH)_2C_6H_4$
Mol Form: $C_{10}H_{10}$
M$_r$: 130.18
bp: 82^{14}
***d*:** $0.9325^{22/4}$
***n*:** $1,5767^{20}$
***s*:** ace 3; bz 3
UV: OES 4-273 257(15849) 230(19953) cyhex
MS: ALDN 76 ini 130(100) 128(28) 129(27) 115(23) 127(13) 131(11) 51(10) 77(9)
m/e: 130.08

HODOC No: 03130
CAS No: 108-57-6
Beilstein No: 5^3, 1367
Name: Benzene, 1,3-diethenyl-
Synonym: Benzene, *m*-divinyl-
Line Form: 1,3-$(CH_2{=}CH)_2C_6H_4$
Mol Form: $C_{10}H_{10}$
M$_r$: 130.18
bp: 121^{76}, 52^3
mp: -52.3
***d*:** $0.9294^{20/4}$
***n*:** 1.5760^{20}
***s*:** ace 3; bz 3
MS: WILEY 592 130(100) 129(35) 115(33) 128(33) 127(15) 51(13) 13(11) 77(10)
m/e: 130.08

HODOC No: 03131
CAS No: 105-06-6
Beilstein No: 5^4, 1540
Name: Benzene, 1,4-diethenyl-
Synonym: Benzene, *p*-divinyl-
Line Form: 1,4-$(CH_2{=}CH)_2C_6H_4$
Mol Form: $C_{10}H_{10}$
M$_r$: 130.18
bp: $95\text{-}6^{18}$, $34^{0.2}$
mp: 31
***d*:** 0.913^{40}
***n*:** 1.5835^{25}
***s*:** ace 3; bz 3
IR: µCOB 207 3030 1640 1520 1410 1300 1120 1020 980 910 850
MS: WILEY 591 130(100) 128(28) 129(27) 115(23) 127(13) 131(11) 51(10) 77(9)
m/e: 130.08

HODOC No: 03132
CAS No: 2050-46-6
Beilstein No: 6^4, 5565
Name: Benzene, 1,2-diethoxy-
Synonym: Benzene, *o*-diethoxy-
Line Form: 1,2-$(C_2H_5O)_2C_6H_4$
Mol Form: $C_{10}H_{14}O_2$
M$_r$: 166.21
bp: 219
mp: 43-5
***d*:** $1.0075^{20/4}$
***n*:** 1.5083^{25}
Color: pr (peth, dil al)
***s*:** al 3; eth 4; ctc 3
IR: SADG 347 3030 2941 1587 1471 1389 1235 1124 1042 935 730
Raman: SAD 267 3090 3070 2990 2940 2890 2790 2730 1590 1510 1480 1460 1440 1400 1370 1320 1300 1280 1250 1160 1130 1110 1050 930 850 820 780 750 730 590 520 460 380 330 200
UV: SAD 823 282(701) 243(15300) MeOH
H NMR: SAD 171 1.4 4.0 6.7 CCl_4
m/e: 166.10

HODOC No: 03133
CAS No: 2049-73-2
Beilstein No: 6^4, 5664
Name: Benzene, 1,3-diethoxy-
Synonym: Benzene, *m*-diethoxy-
Line Form: 1,3-$(C_2H_5O)_2C_6H_4$
Mol Form: $C_{10}H_{14}O_2$
M$_r$: 166.21
bp: 235^{756}
mp: 12.4
Color: pr
***s*:** H_2O 1; al 3; eth 3
IR: SADG 27992 2970 2930 2870 1590 1490 1470 1390 1360 1330 1290 1260 1180 1150 1110 1090 1040 990 980 910 840 820 760 680 600 580 450
UV: OES 3-263 275(1905) 225(6607) EtOH
m/e: 166.10

HODOC No: 03134
CAS No: 122-95-2
Beilstein No: 6^4, 5720
Name: Benzene, 1,4-diethoxy-
Synonym: Benzene, *p*-diethoxy-
Line Form: 1,4-$(C_2H_5O)_2C_6H_4$
Mol Form: $C_{10}H_{14}O_2$
M$_r$: 166.21
bp: 246
mp: 72
Color: pl (dil al)
s: al 4; eth 3; bz 3; ctc 3; chl 3
IR: SADG 157 3030 2941 1887 1613 1493 1449 1408 1266 1220 1124 1064 935 826
Raman: SAD 699 3080 3060 2980 2930 2880 2780 2730 1650 1620 1590 1480 1460 1450 1400 1380 1310 1280 1260 1170 1120 1050 930 850 830 800 710 640 590 460 390 290 200
UV: SAD 235 290(2840) 226(10200) MeOH
H NMR: SAD 309 1.4 3.9 6.7 CCl_4
MS: ALDN 126 110(100) 166(62) 109(62) 138(16) 81(13) 29(10) 137(9) 27(8)
m/e: 166.10

HODOC No: 03135
CAS No: 6314-97-2
Name: Benzene, (2,2-diethoxyethyl)-
Synonym: Acetaldehyde, phenyl-, diethyl acetal
Line Form: $C_6H_5CH_2CH(OCH_2CH_3)_2$
Mol Form: $C_{12}H_{18}O_2$
M$_r$: 194.26
MS: WILEY 1353 75(100) 47(18) 91(17) 135(13) 103(7) 65(6) 31(5) 194(0)
m/e: 194.13

HODOC No: 03136
CAS No: 14444-77-0
Beilstein No: 6^4, 610
Name: Benzene, (diethoxymethoxy)-
Synonym: Orthoformic acid, diethyl phenyl ester
Line Form: $C_6H_5OCH(OC_2H_5)_2$
Mol Form: $C_{11}H_{16}O_3$
M$_r$: 196.24
bp: $103\text{-}4^{10}$
d: 1.014
n: 1.4822^{20}
IR: ALD 2:571G
MS: ALD 4:105D
m/e: 196.11

HODOC No: 03137
CAS No: 41901-72-8
Beilstein No: 6^3, 4449
Name: Benzene, 1,4-diethoxy-2-methyl-
Synonym: 2,5-Diethoxytoluene
Line Form: 2,5-$(C_2H_5O)_2C_6H_3CH_3$
Mol Form: $C_{11}H_{16}O_2$
M$_r$: 180.24
bp: 247-9
mp: 24-5
d: 1.0134^{15}
Color: nd (lig)
s: Bz 4; Et_2O 4; EtOH 4
m/e: 180.12

HODOC No: 03138
CAS No: 35364-90-0
Name: Benzene, 1-(diethoxymethyl)-4-(1-methylethyl)-
Synonym: *p*-Isopropylbenzaldehyde diethyl acetal
Line Form: 4-$(CH_3)_2CH\text{-}C_6H_4CH(OC_2H_5)_2$
Mol Form: $C_{14}H_{22}O_2$
M$_r$: 222.31
bp: 263^{724}, $142\text{-}3^{18.5}$
d: $0.944^{19/4}$
n: 1.484^{19}
m/e: 222.16

HODOC No: 03139
CAS No: 87406-33-5
Name: Benzene, (1,1-diethoxy-2-propynyl)-
Synonym: Propiolophenone, diethyl acetal
Line Form: $C_6H_5C{\equiv}CCH(OCH_2CH_3)_2$
Mol Form: $C_{13}H_{16}O_2$
M$_r$: 204.26
bp: $99\text{-}100^2$
s: chl 3
MS: WILEY 1443 131(100) 159(92) 103(36) 77(27) 29(25) 102(20) 28(17) 204(5)
m/e: 204.12

HODOC No: 03140
CAS No: 135-01-3
Beilstein No: 5^4, 1065
Name: Benzene, 1,2-diethyl-
Synonym: Benzene, *o*-diethyl-
Line Form: 1,2-$(C_2H_5)_2C_6H_4$
Mol Form: $C_{10}H_{14}$
M_r: 134.21
bp: 183.5, 63^{10}
mp: -31.2
***d*:** 0.8800^{20}
***n*:** 1.5035^{20}
***s*:** H_2O 1; al 5; eth 5; ace 5; bz 5; ctc 5; lig 5
IR: µCOB 6365 2940 1590 1490 1450 1370 1060 960 790 750
Raman: API 224 3042 2967 2934 2892 2869 1605 1210 1058 1039 710
UV: SAD 8528 271(255) 263(296) 256(254) 211(8680) MeOH
H NMR: SAD 19653 1.2 2.6 7.0 CCl_4
MS: WILEY 671 105(100) 119(98) 134(47) 91(25) 77(17) 39(16) 27(15) 51(12)
m/e: 134.11

HODOC No: 03141
CAS No: 141-93-5
Beilstein No: 5^4, 1066
Name: Benzene, 1,3-diethyl-
Synonym: Benzene, *m*-diethyl-
Line Form: 1,3-$(C_2H_5)_2C_6H_4$
Mol Form: $C_{10}H_{14}$
M_r: 134.21
bp: 181.1, 60^{10}
mp: -83.9
***d*:** $0.8602^{20/4}$
***n*:** 1.4955^{20}
***s*:** H; iO 1; al 5; eth 5; ace 5; bz 5; ctc 5; lig 5
IR: µSADP 18713 2940 1610 1490 1450 1370 1320 1190 890 860 790 700
Raman: API 167 3055 2966 2935 2908 2875 1610 1443 1235 1004 716
UV: SAD 4993 1.2 2.6 6.9 CCl_4
H NMR: SAD 4993 1.2 2.6 6.9 CCl_4
MS: WILEY 663 119(100) 105(92) 134(43) 91(19) 27(16) 39(15) 77(15) 51(12)
m/e: 134.11

HODOC No: 03142
CAS No: 105-05-5
Beilstein No: 5^4, 1067
Name: Benzene, 1,4-diethyl-
Synonym: Benzene, *p*-diethyl-
Line Form: 1,4-$(C_2H_5)_2C_6H_4$
Mol Form: $C_{10}H_{14}$
M_r: 134.21
bp: 183.8, 63^{10}
mp: -42.8 (fp)
***d*:** $0.8620^{20/4}$
***n*:** 1.4967^{20}
***s*:** H'i2O 1; al 5; eth 5; ace 5; bz 5; ctc 5; lig 5
IR: µCOB 6326 2940 1890 1670 1520 1450 1390 1320 1120 1060 970 810
Raman: API 168 3050 3028 3007 2965 2933 2906 2891 2869 1198 800
UV: SAD 8529 273(513) 267(481) 264(503) 259(418) 218(7980) MeOH
MS: WILEY 669 119(100) 105(75) 134(40) 91(21) 27(14) 39(14) 73(13) 51(10)
m/e: 134.11

HODOC No: 03143
CAS No: 54410-75-2
Name: Benzene, 1,2-diethyl-3,4-dimethyl-
Mol Form: $C_{12}H_{18}$
M_r: 162.26
MS: WILEY 1043 147(100) 162(37) 133(24) 148(12) 91(8) 119(8) 105(6) 117(6)
m/e: 162.14

HODOC No: 03144
CAS No: 21411-35-8
Name: Benzene, 1,1′-(1,4-diethyl-1,4-dimethyl-1,4-butanediyl)*bis*-
Synonym: Octane, 3,6-dimethyl-3,6-diphenyl-
Line Form: $CH_3CH_2C(CH_3)(C_6H_5)CH_2CH_2C(C_6H_5)(CH_3)CH_2CH_3$
Mol Form: $C_{22}H_{30}$
M_r: 294.46
bp: $126^{0.1}$
***s*:** chl 3
IR: SADG 17434 3060 3030 2960 2890 1600 1500 1440 1370 1030 760 700 570
H NMR: SAD 9299 0.6 1.2 1.3-1.8 7.1 CCl_4
m/e: 294.23

HODOC No: 03145
CAS No: 13632-93-4
Beilstein No: 5^4, 1105
Name: Benzene, 1,2-diethyl-3-methyl-
Synonym: Toluene, 2,3-diethyl-
Line Form: 2,3-$(C_2H_5)_2C_6H_3CH_3$
Mol Form: $C_{11}H_{16}$
M_r: 148.24
bp: 206.6, 104.4^{30}
***d*:** 0.8910^{20}
***n*:** 1.5105^{20}
***s*:** Ac 4; Bz 4; Et_2O 4; EtOH 4
MS: NBS 441
m/e: 148.13

HODOC No: 03146
CAS No: 13732-80-4
Beilstein No: 5^4, 1106
Name: Benzene, 1,2-diethyl-4-methyl-
Synonym: Toluene, 3,4-diethyl-
Line Form: 3,4-$(C_2H_5)_2C_6H_3CH_3$
Mol Form: $C_{11}H_{16}$
M_r: 148.24
bp: 203.6, 101.9^{30}
***d*:** 0.8762^{20}
***n*:** 1.5039^{20}
***s*:** Ac 4; Bz 4; Et_2O 4; EtOH 4
MS: NBS 441
m/e: 148.13

HODOC No: 03147
CAS No: 13632-95-6
Beilstein No: 5^4, 1105
Name: Benzene, 1,3-diethyl-2-methyl-
Synonym: Toluene, 2,6-diethyl-
Line Form: 2,5-$(C_2H_5)_2C_6H_3CH_3$
Mol Form: $C_{11}H_{16}$
M_r: 148.24
bp: 208.8, 106.2^{30}
***d*:** 0.8907^{20}
***n*:** 1.5106^{20}
***s*:** Ac 4; Bz 4; Et_2O 4; EtOH 4
MS: NBS 441
m/e: 148.13

HODOC No: 03148
CAS No: 2050-24-0
Beilstein No: 5^4, 1106
Name: Benzene, 1,3-diethyl-5-methyl-
Synonym: Toluene, 3,5-diethyl-
Line Form: 3,5-$(C_2H_5)_2C_6H_3CH_3$
Mol Form: $C_{11}H_{16}$
M_r: 148.24
bp: 205, 79^{10}
mp: -74.1
***d*:** $0.8748^{20/4}$
***n*:** 1.5027^{20}
***s*:** H_2O 1; al 5; eth 5; ace 5; bz 5; ctc 5; lig 5
IR: μSADP 19517 2940 1610 1470 1390 1180 1060 920 890 850 700
m/e: 148.13

HODOC No: 03149
CAS No: 13632-94-5
Beilstein No: 5^4, 1106
Name: Benzene, 1,4-diethyl-2-methyl-
Synonym: Toluene, 2,5-diethyl-
Line Form: 2,5-$(C_2H_5)_2C_6H_3CH_3$
Mol Form: $C_{11}H_{16}$
M_r: 148.24
bp: 207.1, 104.8^{30}
***d*:** 0.8758^{20}
***n*:** 1.5034^{20}
***s*:** Ac 4; Bz 4; Et_2O 4; EtOH 4
MS: NBS 441
m/e: 148.13

HODOC No: 03150
CAS No: 1758-85-6
Beilstein No: 5^4, 1105
Name: Benzene, 2,4-diethyl-1-methyl-
Synonym: Toluene, 2,4-diethyl-
Line Form: 2,4-$(C_2H_5)_2C_6H_3CH_3$
Mol Form: $C_{11}H_{16}$
M_r: 148.24
bp: 205, 103.1^{30}
***d*:** 0.8748^{20}
***n*:** 1.5027^{20}
***s*:** Ac 4; Bz 4; Et_2O 4; EtOH 4
MS: NBS 441
m/e: 148.13

HODOC No: 03151
CAS No: 367-11-3
Beilstein No: 5^3, 523
Name: Benzene, 1,2-difluoro-
Synonym: Benzene, *o*-difluoro-
Line Form: 1,2-$F_2C_6H_4$
Mol Form: $C_6H_4F_2$
M_r: 114.09
bp: $91\text{-}2^{751}$
mp: -34
***d*:** $1.1599^{18/4}$
***n*:** 1.4451^{18}
***s*:** H_2O 1; ace 3; bz 3; chl 3
IR: COB 6473 3180 1620 1510 1470 1460 1400 1290 1270 1210 1200 1020 930 860 840 760 750
H NMR: SAD 14733 6.6-7.1 7.3 $CDCl_3$
m/e: 114.03

HODOC No: 03152
CAS No: 372-18-9
Beilstein No: 5^3, 523
Name: Benzene, 1,3-difluoro-
Synonym: Benzene, *m*-difluoro-
Line Form: 1,3-$F_2C_6H_4$
Mol Form: $C_6H_4F_2$
M_r: 114.09
bp: 83
mp: -59
***d*:** $1.1572^{20/4}$
***n*:** 1.4374^{20}
***s*:** H_2O 1; ace 3; bz 3
IR: µSADP 18756 3130 1610 1490 1450 1280 1270 1160 1120 950 860 770 740 680
UV: OES 4-54 265(1000) 260(1000) 254(708) 203(7943) cyhex
H NMR: SAD 14733 6.6-7.1 7.3 $CDCl_3$
m/e: 114.03

HODOC No: 03153
CAS No: 540-36-3
Beilstein No: 5^4, 637
Name: Benzene, 1,4-difluoro-
Synonym: Benzene, *p*-difluoro-
Line Form: 1,4-$F_2C_6H_4$
Mol Form: $C_6H_4F_2$
M_r: 114.09
bp: 89
mp: -13
***d*:** $1.1688^{20/4}$
***n*:** 1.4422^{20}
***s*:** H_2O 1; ace 3; bz 3; ctc 2
IR: µSADP 18772 3030 1520 1220 1190 1090 830 740
Raman: SAD 1152 3080 3000 1620 1600 1390 1290 1250 1150 1080 1020 890 860 850 800 770 700 640 460 380 170
UV: SAD 6058 272 266 263 260 258 250 MeOH
H NMR: SAD 4994 7.0 CCl_4
C NMR: JJ 150 FT 116.5 159.1 $CDCl_3$
MS: NBS 150
m/e: 114.03

HODOC No: 03154
CAS No: 327-92-4
Beilstein No: 5^4, 743
Name: Benzene, 1,5-difluoro-2,4-dinitro-
Synonym: 1,3-Difluoro-4,6-dinitrobenzene
Line Form: 1,3-F_2-4,6-$(NO_2)_2$-C_6H_2
Mol Form: $C_6H_2F_2N_2O_4$
M_r: 204.08
mp: 75-6
***s*:** al 2
IR: µ COB 7469 3130 1640 1560 1350 1330 1300 1220 1180 1050 920 880 840 780 750 740 700 660 640
Raman: SAD 1144 3100 3070 2660 1640 1620 1550 1530 1490 1430 1410 1390 1350 1320 1300 1210 1170 1150 1050 910 870 840 770 740 710 660 640 620 600 490 450 420 370 340 320 250 160
H NMR: SAd 14738 7.4 8.9 $CDCl_3$
m/e: 204.00

HODOC No: 03155
CAS No: 405-42-5
Name: Benzene, (2,2-difluoroethenyl)-
Synonym: Styrene, β,β-difluoro-
Line Form: $C_6H_5CH=CF_2$
Mol Form: $C_8H_6F_2$
M_r: 140.12
MS: WILEY 744 140(100) 89(73) 139(19) 141(19) 114(11) 120(7) 119(7) 63(6)
m/e: 140.04

HODOC No: 03156
CAS No: 455-31-2
Beilstein No: 5^4, 801
Name: Benzene, (difluoromethyl)-
Synonym: Toluene, α,α-difluoro-
Line Form: $C_6H_5CHF_2$
Mol Form: $C_7H_6F_2$
M_r: 128.11
bp: 140
***d*:** 1.1370^{19}
***n*:** 1.4577^{20}
***s*:** H_2O 1; al 3
UV: OES 1-127 284(6) 267(224) 261(295) 256(257) 251(186) 239(105) iso
m/e: 128.04

HODOC No: 03157
CAS No: 446-35-5
Beilstein No: 5^4, 720
Name: Benzene, 2,4-difluoro-1-nitro-
Synonym: 1,3-Difluoro-4-nitrobenzene
Line Form: $2,4\text{-}F_2C_6H_3NO_2$
Mol Form: $C_6H_3F_2NO_2$
M_r: 159.09
bp: 207
mp: 9.8
***d*:** 1.4571^{14}
***n*:** 1.5149^{14}
***s*:** chl 2
IR: SADG 8010 3100 1610 1540 1490 1350 1280 1150 1090 980 860 750 620
H NMR: SAD 8239 6.7-7.2 8.0 $CDCl_3$
m/e: 159.01

HODOC No: 03158
CAS No: 393-38-4
Name: Benzene, 1,4-difluoro-2-(trifluoromethyl)-
Synonym: Toluene, α,α,α,2,5-pentafluoro-
Mol Form: $C_7H_3F_5$
M_r: 182.09
UV: OES 1-116 272(6310) 204(7762) 183(48978) hp
m/e: 182.02

HODOC No: 03159
CAS No: 55517-88-9
Name: Benzene, 1,4-dihexadecyl-
Synonym: Benzene, *p*-dihexadecyl-
Mol Form: $C_{38}H_{70}$
M_r: 526.94
MS: WILEY 2250 105(100) 43(70) 57(41) 526(39) 41(27) 55(19) 104(19) 527(17)
m/e: 526.55

HODOC No: 03160
CAS No: 615-42-9
Beilstein No: 5^4, 700
Name: Benzene, 1,2-diiodo-
Synonym: Benzene, *o*-diiodo-
Line Form: $1,2\text{-}I_2C_6H_4$
Mol Form: $C_6H_4I_2$
M_r: 329.90
bp: 286^{750}, 100^3
mp: 27
***d*:** 2.54^{20}
***n*:** 1.7179^{20}
Color: pl or pr (lig)
***s*:** H_2O 2; al 2
UV: LANG 1906 234(9550) 219(18197) EtOH
MS: NBS 2541
m/e: 329.84

HODOC No: 03161
CAS No: 626-00-6
Beilstein No: 5^4, 700
Name: Benzene, 1,3-diiodo-
Synonym: Benzene, *m*-diiodo-
Line Form: $1.3\text{-}I_2C_6H_4$
Mol Form: $C_6H_4I_2$
M_r: 329.90
bp: 285
mp: 40.4
***d*:** 2.47^{25}
Color: rh pl or pr (eth-al)
***s*:** Et_2O 4; EtOH 4; Chl 4
MS: NBS 2542
m/e: 329.84

HODOC No: 03162
CAS No: 624-38-4
Beilstein No: 5^4, 700
Name: Benzene, 1,4-diiodo-
Synonym: Benzene, *p*-diiodo-
Line Form: 1,4-$I_2C_6H_4$
Mol Form: $C_6H_4I_2$
M_r: 329.90
bp: 285 (sub)
mp: 131-2
Color: rh lf (al)
s: H_2O 1; al 3; eth 4; chl 2
IR: µCOB 6474 3130 1610 1470 1370 1060 980 800
UV: SAD 1258 225 MeOH
H NMR: SAD 207 7.4 CDC_3
MS: NBS 2542
m/e: 329.84

HODOC No: 03163
CAS No: 104-49-4
Name: Benzene, 1,4-diisocyanato-
Synonym: Isocyanic acid, *p*-phenylene ester
Structure:

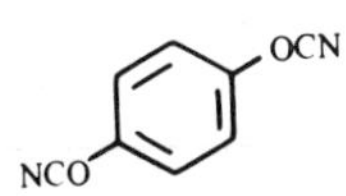

Mol Form: $C_8H_4N_2O_2$
M_r: 160.12
MS: WILEY 1005 160(100) 104(38) 77(33) 52(30) 54(30) 132(26) 51(24) 26(23)
m/e: 160.03

HODOC No: 03164
CAS No: 584-84-9
Beilstein No: 13^3, 289
Name: Benzene, 2,4-diisocyanato-1-methyl-
Synonym: Isocyanic acid, 4-methyl-*m*-phenylene ester
Line Form: 2,4-(OCN)$C_6H_3CH_3$
Mol Form: $C_9H_6N_2O_2$
M_r: 174.15
bp: 251, 126^{11}
mp: 20-1
***d*:** $1.2244^{20/4}$
s: Ac 4; Bz 4; Et_2O 4
IR: SADG 29893 2940 2250 1620 1570 1530 1440 1380 1270 1210 1170 1130 1070 990 890 870 810 740 700 630 600 560 450 330
Raman: SAD 557 3070 2920 2860 2740 2560 2270 1610 1580 1500 1430 1380 1210 1160 1110 1070 890 740 700 560 410 340 240 150
H NMR: SAD 17271 2.3 6.7 7.0 7.1 CCl_4
MS: NBS 815
m/e: 174.04

HODOC No: 03165
CAS No: 4044-65-9
Beilstein No: 13^3, 231
Name: Benzene, 1,4-diisothiocyanato-
Synonym: Isothiocyanic acid, *p*-phenylene ester
Line Form: 1,4-$(SCN-)_2C_6H_4$
Mol Form: $C_8H_4N_2S_2$
M_r: 192.25
mp: 132
Color: nd (Ace, HOAc)
m/e: 191.98

HODOC No: 03166
CAS No: 91-16-7
Beilstein No: 6^4, 5564
Name: Benzene, 1,2-dimethoxy-
Synonym: Benzene, *o*-dimethoxy-
Line Form: 1,2-$(CH_3O)_2C_6H_4$
Mol Form: $C_8H_{10}O_2$
M_r: 138.16
bp: 206, 90^{10}
mp: 22.5
***d*:** $1.0842^{25/25}$
***n*:** 1.5827^{21}
***s*:** H_2O 2; al 3; eth 3; ctc 3
IR: SADG 8189 2941 2857 1587 1493 1449 1333 1235 1163 1124 1031 901 806 741
Raman: SAD 1094 3180 3080 3000 2930 2840 1590 1500 1450 1330 1300 1260 1180 1160 1050 1030 810 760 580 540 480 380 330 210 170
UV: SAD 476 275(3920) 225(11700) MeOH
H NMR: SAD 1112 3.7 6.8 CCl_4
MS: WILEY 722 138(100) 95(61) 123(41) 77(36) 52(35) 51(25) 65(23) 41(22)
m/e: 138.07

HODOC No: 03167
CAS No: 151-10-0
Beilstein No: 6^4, 5663
Name: Benzene, 1,3-dimethoxy-
Synonym: Benzene, *m*-dimethoxy-
Line Form: 1,3-$(CH_3O)_2C_6H_4$
Mol Form: $C_8H_{10}O_2$
M_r: 138.16
bp: 217-8
mp: -52
***d*:** $1.0552^{25/25}$
***n*:** 1.5231^{20}
***s*:** H_2O 2; al 3; eth 3; bz 3; ctc 3; sulf 3
IR: µCOB 7456 2940 1610 1490 1470 1430 1330 1280 1270 1210 1150 1080 1050 930 910 830 760 690
UV: SAD 2999 273 MeOH
H NMR: SAD 1891 3.6 5.9-7.4 CCl_4
MS: NBS 331
m/e: 138.07

HODOC No: 03168
CAS No: 150-78-7
Beilstein No: 6^4, 5718
Name: Benzene, 1,4-dimethoxy-
Synonym: Benzene, *p*-dimethoxy-
Line Form: 1,4-$(CH_3O)_2C_6H_4$
Mol Form: $C_8H_{10}O_2$
M_r: 138.16
bp: 212.6, 109^{20}
mp: 58-60
***d*:** $1.0526^{55/55}$
Color: lf (w)
***s*:** H_2O 2; al 3; eth 4; bz 4; chl 3
IR: µSADP 1508 2940 1890 1640 1520 1430 1280 1220 1180 1110 1030 940 830 710
Raman: SAD 697 3120 3080 3060 3020 2960 2950 2910 2840 2600 1660 1620 1590 1500 1470 1460 1420 1310 1270 1220 1180 1170 1150 1060 1030 850 830 750 710 680 640 600 560 450 410 390 280 220
UV: SAD 17874 289(3080) 225(9670) MeOH
H NMR: SAD 10801 3.7 6.8 $CDCl_3$
MS: NBS 331
m/e: 138.07

HODOC No: 03169
CAS No: 101-48-4
Beilstein No: 7^4, 664
Name: Benzene, (2,2-dimethoxyethyl)-
Synonym: Acetaldehyde, phenyl-, dimethyl acetal
Line Form: $C_6H_5CH_2CH(OCH_3)_2$
Mol Form: $C_{10}H_{14}O_2$
M_r: 166.21
bp: 193-4
IR: µCOB 3708 2940 2860 1610 1490 1450 1350 1190 1120 1060 970 840 820 750 700
UV: SAD 1408 267 263 258 252 246 241 cyhex
MS: NBS 699
m/e: 166.10

HODOC No: 03170
CAS No: 4463-33-6
Beilstein No: 6^4, 5860
Name: Benzene, 1,2-dimethoxy-3-methyl-
Synonym: Toluene, 2,3-dimethoxy-
Line Form: 2,3-$(CH_3O)_2C_6H_3CH_3$
Mol Form: $C_9H_{12}O_2$
Mr: 152.18
bp: 202-3, $92\text{-}3^{18}$
***d*:** $1.0335^{20/4}$
***n*:** 1.5121^{25}
***s*:** Et_2O 4; EtOH 4
m/e: 152.08

HODOC No: 03171
CAS No: 494-99-5
Beilstein No: 6^4, 5873
Name: Benzene, 1,2-dimethoxy-4-methyl-
Synonym: Toluene, 3,4-dimethoxy-
Line Form: 3,4-$(CH_3O)_2C_6H_3CH_3$
Mol Form: $C_9H_{12}O_2$
Mr: 152.18
bp: 219-21, $122\text{-}4^{27}$
mp: 24
***d*:** $1.0509^{25/4}$
***n*:** 1.5257^{25}
Color: pr (eth)
***s*:** H_2O 1; ctc 2; os 4
IR: SADG 4718 2940 2840 1620 1590 1510 1460 1410 1330 1260 1240 1200 1150 1130 1030 920 850 800 760
UV: SAD 8604 280(2730) 227(7410) MeOH
H NMR: SAD 2626 2.2 3.8 6.6 CCl_4
m/e: 152.08

HODOC No: 03172
CAS No: 5673-07-4
Beilstein No: 6^4, 5877
Name: Benzene, 1,3-dimethoxy-2-methyl-
Synonym: Toluene, 2,6-dimethoxy-
Line Form: 2,6-$(CH_3O)_2C_6H_3CH_3$
Mol Form: $C_9H_{12}O_2$
Mr: 152.18
mp: 39-41
IR: SADG 33532 3010 2940 2850 1590 1470 1430 1370 1320 1270 1250 1180 1170 1130 1090 1040 860 760 700 690 590 490
Raman: SAD 1098 3090 3060 3010 2920 2860 2830 2720 1600 1450 1440 1370 1286 1190 1150 1100 1040 870 850 800 770 740 700 620 520 500 370 300 210
UV: SAD 6278 278 272 269 MeOH
H NMR: SAD 5015 2.1 3.8 6.5 7.1 $CDCl_3$
MS: ALD 4:91B
m/e: 152.08

HODOC No: 03173
CAS No: 4179-19-5
Beilstein No: 6^3, 4533
Name: Benzene, 1,3-dimethoxy-5-methyl-
Synonym: Toluene, 3,5-dimethoxy-
Line Form: 3,5-$(CH_3O)_2C_6H_3CH_3$
Mol Form: $C_9H_{12}O_2$
Mr: 152.18
bp: 244, 102^8
***d*:** 1.0478^{15}
***n*:** 1.5234^{20}
***s*:** Bz 4; Et_2O 4; EtOH 4
m/e: 152.08

HODOC No: 03174
CAS No: 24599-58-4
Beilstein No: 6^4, 5867
Name: Benzene, 1,4-dimethoxy-2-methyl-
Synonym: Toluene, 2,5-dimethoxy-
Line Form: 2,5-$(CH_3O)_2C_6H_3CH_3$
Mol Form: $C_9H_{12}O_2$
Mr: 152.18
mp: 20-2
m/e: 152.08

HODOC No: 03175
CAS No: 38064-90-3
Beilstein No: 6^4, 5864
Name: Benzene, 2,4-dimethoxy-1-methyl-
Synonym: 1,3-Dimethoxy-4-methylbenzene
Line Form: $2,4\text{-}(CH_3O)_2C_6H_3CH_3$
Mol Form: $C_9H_{12}O_2$
Mr: 152.18
bp: 211, $110\text{-}20^{30}$
s: H_2O 2; al 3; eth 3
IR: µSADP 11031 2940 2860 1590 1470 1280 1210 1150 1120 1040 990 930 830 760 710 690
UV: SAD 3000 327 280 275 cyhex
m/e: 152.08

HODOC No: 03176
CAS No: 4132-76-7
Name: Benzene, 1,2-dimethoxy-4-(1-methylethyl)-
Synonym: Cumene, 3,4-dimethoxy-
Line Form: $(3,4\text{-}(CH_3O)_2C_6H_3)CH(CH_3)_2$
Mol Form: $C_{11}H_{16}O_2$
Mr: 180.24
bp: 234.6
d: 1.0203^0
m/e: 180.12

HODOC No: 03177
CAS No: 4132-71-2
Beilstein No: 6^4, 5990
Name: Benzene, 1,4-dimethoxy-2-(1-methylethyl)-
Synonym: Cumene, 2,5-dimethoxy-
Line Form: $2,5\text{-}(CH_3O)_2C_6H_3CH(CH_3)_2$
Mol Form: $C_{11}H_{16}O_2$
Mr: 180.24
bp: $114\text{-}6^{15}$
d: $1.0129^{17/4}$
n: 1.5103^{17}
s: Bz 4; Et_2O 4
m/e: 180.12

HODOC No: 03178
CAS No: 90-87-9
Name: Benzene, (2,2-dimethoxy-1-methylethyl)-
Synonym: Hydratropaldehyde, dimethyl acetal
Line Form: $CH(OCH_3)_2CH(C_6H_5)CH_3$
Mol Form: $C_{11}H_{16}O_2$
Mr: 180.24
IR: µCOB 3732 2940 1590 1490 1270 1370 1180 1140 1060 960 940 760 700
H NMR: SAD 19582 1.7 1.8 1.9 11.8 DMSOd6
m/e: 180.12

HODOC No: 03179
CAS No: 67785-72-2
Name: Benzene, 1-(dimethoxymethyl)-4-(1-methylethyl)-
Synonym: Benzaldehyde, 4-isopropyl, dimethyl acetal
Line Form: $4\text{-}(CH_3)_2CH\text{-}C_6H_4CH(OCH_3)_2$
Mol Form: $C_{12}H_{18}O_2$
Mr: 194.26
bp: 244-5
d: 1.0633^{15}
m/e: 194.13

HODOC No: 03180
CAS No: 20627-73-0
Beilstein No: 7^4, 585
Name: Benzene, 1-(dimethoxymethyl)-2-nitro-
Synonym: Benzaldehyde, *o*-nitro-, dimethyl acetal
Line Form: $2\text{-}NO_2C_6H_4CH(OCH_3)_2$
Structure:

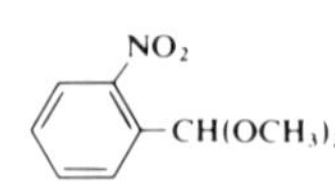

Mol Form: $C_9H_{11}NO_4$
Mr: 197.18
bp: 274-6, $138\text{-}9^{11}$
Color: gr-ye
s: Bz 4
m/e: 197.07

HODOC No: 03181
CAS No: 3395-79-7
Beilstein No: 7^4, 592
Name: Benzene, 1-(dimethoxymethyl)-3-nitro-
Synonym: Benzaldehyde, *m*-nitro-, dimethyl acetal
Line Form: $3\text{-}NO_2C_6H_4CH(OCH_3)_2$
Structure:

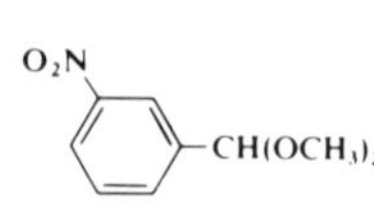

Mol Form: $C_9H_{11}NO_4$
Mr: 197.18
bp: $162\text{-}4^{19}$
d: 1.209^{15}
s: Bz 4
m/e: 197.07

HODOC No: 03182
CAS No: 881-67-4
Beilstein No: 7^4, 600
Name: Benzene, 1-(dimethoxymethyl)-4-nitro-
Synonym: Benzaldehyde, *p*-nitro-, dimethyl acetal
Line Form: $4\text{-}NO_2C_6H_4CH(OCH_3)_2$
Mol Form: $C_9H_{11}NO_4$
M$_r$: 197.18
bp: $294\text{-}6^{774}$
mp: 23-5
***s*:** Bz 4
m/e: 197.07

HODOC No: 03183
CAS No: 709-09-1
Beilstein No: 6^3, 4264
Name: Benzene, 1,2-dimethoxy-4-nitro-
Synonym: 1,2-Dimethoxy-4-nitrobenzene
Line Form: $3,4\text{-}(CH_3O)_2C_6H_3NO_2$
Mol Form: $C_8H_9NO_4$
M$_r$: 183.16
bp: 230^{15}
mp: 98
***d*:** $1.1888^{133/4}$
Color: ye nd (al-w)
***s*:** H_2O 1; al 4; eth 4; chl 3; lig 2
IR: SADG 18930 2941 2857 1587 1493 1471 1316 1266 1235 1176 1087 1020 847 730
UV: HBCP 323(7943) al
H NMR: SAD 5028 3.9 6.9 7.7 7.9 $CDCl_3$
m/e: 183.05

HODOC No: 03184
CAS No: 6665-97-0
Beilstein No: 6^3, 4344
Name: Benzene, 1,3-dimethoxy-2-nitro-
Synonym: 2-Nitroresorcinol dimethyl ether
Line Form: $2,6\text{-}(CH_3O)_2\text{-}C_6H_3NO_2$
Mol Form: $C_8H_9NO_4$
M$_r$: 183.16
mp: 131
Color: ye nd (al or HOAc)
***s*:** Bz 4; EtOH 4; Chl 4
m/e: 183.05

HODOC No: 03185
CAS No: 89-39-4
Beilstein No: 6^4, 5786
Name: Benzene, 1,4-dimethoxy-2-nitro-
Synonym: 1,4-Dimethoxy-2-nitrobenzene
Line Form: $2,5\text{-}(CH_3O)_2C_6H_3NO_2$
Mol Form: $C_8H_9NO_4$
M$_r$: 183.16
bp: 169^{13} sub
mp: 72-3
***d*:** $1.1666^{132/4}$
Color: gold-ye nd (dil al)
***s*:** H_2O 1; al 3; bz 3; chl 3; sulf 3
IR: SADG 1074 3030 2941 2857 1587 1538 1471 1351 1282 1235 1205 1031 1020 909 833
UV: OES 8-93 357(2455) MeOH
H NMR: SAD 8156 3.9 4.0 7.3 7.6 $CDCl_3$
m/e: 183.05

HODOC No: 03186
CAS No: 4920-84-7
Beilstein No: 6^4, 5691
Name: Benzene, 2,4-dimethoxy-1-nitro-
Synonym: 1,3-Dimethoxy-4-nitrobenzene
Line Form: $2,4\text{-}(CH_3O)_2C_6H_3NO_2$
Mol Form: $C_8H_9NO_4$
M$_r$: 183.16
mp: 76-7
***d*:** $1.1876^{132/4}$
Color: nd (al)
***s*:** H_2O 1; al 3; os 3; lig 1
IR: µSADP 19579 3030 2860 1610 1590 1520 1350 1280 1220 1100 1040 830 750 710
UV: OES 5-141 321(6026) 282(5248) EtOH
m/e: 183.05

HODOC No: 03187
CAS No: 2815-67-0
Name: Benzene, 1,2-dimethoxy-3-(2-nitroethenyl)-
Synonym: Styrene, 2,3-dimethoxy-β-nitro-
Mol Form: $C_{10}H_{11}NO_4$
M$_r$: 209.19
***s*:** chl 2
H NMR: VAR 257 3.8 3.9 7.1 7.7 8.2 $CDCl_3$
m/e: 209.07

HODOC No: 03188
CAS No: 5529-38-4
Beilstein No: 6^2, 987
Name: Benzene, 1,4-dimethoxy-2-[2-(4-nitrophenyl)ethenyl]-
Synonym: Stilbene, 2,5-dimethoxy-4′-nitro-
Line Form: 2,5-$(CH_3O)_2C_6H_3CH$=$CH(C_6H_4NO_2$-4)
Mol Form: $C_{16}H_{15}NO_4$
M_r: 285.28
mp: 117-9
IR: ALD 2:734D
MS: ALD 5:186C
m/e: 285.10

HODOC No: 03189
CAS No: 122-47-4
Name: Benzene, 1,2-dimethoxy-4-(2-nitro-1-propenyl)-
Synonym: Benzene, 1,2-dimethoxy-4-(2-nitropropenyl)-
Mol Form: $C_{11}H_{13}NO_4$
M_r: 223.22
mp: 73
***s*:** chl 2
IR: μSADP 3342 3030 2940 1665 1615 1540 1470 1370 1315 1280 1175 1150 1020 935 880 855 805 775
UV: SAD 1030 244 MeOH
H NMR: SAD 10846 2.5 3.9 6.9 7.0 7.1 8.0 $CDCl_3$
m/e: 223.08

HODOC No: 03190
CAS No: 27983-18-2
Name: Benzene, 1,2-dimethoxy-3-phenoxy-
Synonym: 3-Phenoxyveratrole
Line Form: 2,3-$(CH_3O)_2C_6H_3OC_6H_5$
Mol Form: $C_{14}H_{14}O_3$
M_r: 230.25
bp: 152^2
mp: 48.5
m/e: 230.09

HODOC No: 03191
CAS No: 93-16-3
Beilstein No: 6^4, 6325
Name: Benzene, 1,2-dimethoxy-4-(1-propenyl)-
Synonym: Benzene, 1,2-dimethoxy-4-propenyl-
Line Form: 3,4-$(CH_3O)_2C_6H_3$-CH=$CHCH_3$
Mol Form: $C_{11}H_{14}O_2$
M_r: 178.22
bp: 270.5, $138\text{-}40^{12}$
***d*:** $1.0521^{20/4}$
***n*:** 1.5616^{20}
IR: SADG 23807
UV: OES 7-354
H NMR: SAD 19570
m/e: 178.10

HODOC No: 03192
CAS No: 6380-24-1
Beilstein No: 6^4, 6325
Name: Benzene, 1,2-dimethoxy-4-(1-propenyl)-, (Z)-
Synonym: Benzene, 1,2-dimethoxy-4-propenyl-, (Z)-
Line Form: 3,4-$(CH_3O)_2$-$C_6H_3(CH$=$CHCH_3)$
Mol Form: $C_{11}H_{14}O_2$
M_r: 178.22
bp: 270.5, $138\text{-}40^{12}$
***d*:** $1.0521^{20/4}$
***n*:** 1.5616^{20}
***s*:** H_2O 1; ace 3; bz 3
IR: SADG 23807 3000 2950 2850 1610 1590 1520 1470 1450 1430 1380 1340 1310 1270 1240 1200 1170 1150 1040 970 930 870 820 790 770 620
Raman: SAD 708 ini 3080 3010 2940 2920 2840 2740 1660 1610 1520 1450 1420 1380 1340 1320 1300 1270 1240 1200 1160 1150 1039 970 920 860 830 770 740 620 580 510 470 390 320 210 130
UV: OES 7-354 283(3802) 260(15136) EtOH
H NMR: SAD 19570 ini 1.9 3.8 3.9 5.9 6.4 $CDCl_3$
MS: NBS 874
m/e: 178.10

HODOC No: 03193
CAS No: 93-15-2
Beilstein No: 6^4, 6337
Name: Benzene, 1,2-dimethoxy-4-(2-propenyl)-
Synonym: Benzene, 4-allyl-1,2-dimethoxy-
Line Form: 3,4-$(CH_3O)_2C_6H_3$-CH_2CH=CH_2
Structure:

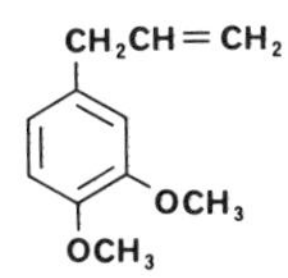

Mol Form: $C_{11}H_{14}O_2$
M$_r$: 178.22
bp: 254.7, 104^2
mp: -4
***d*:** $1.0396^{20/4}$
***n*:** 1.5340^{20}
Color: cr (hx)
***s*:** H_2O 1; al 3; eth 3
IR: **µCOB 3729** 2940 1670 1640 1590 1520 1450 1410 1250 1240 1150 1140 1030 910 860 810 770 750
UV: **SAD 5557** 280 230 MeOH
H NMR: **SAD 4257** 3.3 3.7 5.0 5.9 6.6 CCl_4
MS: **WILEY 1270** 178(100) 147(28) 163(28) 91(16) 103(16) 151(14) 179(12) 107(11)
m/e: 178.10

HODOC No: 03194
CAS No: 4364-06-1
Name: Benzene, (3,3-dimethoxy-1-propenyl)-
Synonym: Cinnamaldehyde, dimethyl acetal
Line Form: $(H_3CO)_2CHCH$=CHC_6H_5
Mol Form: $C_{11}H_{14}O_2$
M$_r$: 178.22
IR: **µCOB 3733** 3030 2940 2780 2700 1670 1610 1590 1560 1490 1450 1280 1190 1120 1060 1050 970 910 840 750 690
m/e: 178.10

HODOC No: 03195
CAS No: 5438-54-0
Beilstein No: 6^1, 540
Name: Benzene, 1,3-dimethoxy-2-(2-propenyloxy)-
Synonym: Allyl 2,6-dimethoxyphenyl ether
Mol Form: $C_{11}H_{14}O_3$
M$_r$: 194.22
bp: $140\text{-}1^{14}$
UV: **OES 1-404** 270(794) diox
MS: **WILEY 1352** 153(100) 194(38) 125(24) 110(24) 93(19) 95(15) 39(13) 154(12)
m/e: 194.09

HODOC No: 03196
CAS No: 95-47-6
Beilstein No: 5^4, 917
Name: Benzene, 1,2-dimethyl-
Synonym: *o*-Xylene
Line Form: 1,2-$(CH_3)_2C_6H_4$
Mol Form: C_8H_{10}
M$_r$: 106.16
bp: 144.4, 32^{10}
mp: -25.2
***d*:** $0.8802^{10/4}$
***n*:** 1.5055^{20}
***s*:** H_2O 1; al 5; eth 5; ace 5; bz 5; ctc 5; peth 5
IR: **µCOB 3576** 3030 2940 1590 1470 1390 1210 1120 1060 1030 750
Raman: **DFB 78** 3080 3050 2980 2920 2880 2860 2730 1610 1580 1450 1380 1220 1160 1050 980 740 580 510 260 180
UV: **SAD 7** 270(211) 262(254) MeOH
H NMR: **VAR 201** 2.2 7.1 $CDCl_3$
MS: **WILEY 322** 91(100) 106(60) 105(24) 39(16) 51(15) 77(13) 27(10) 65(8)
m/e: 106.08

HODOC No: 03197
CAS No: 108-38-3
Beilstein No: 5^4, 932
Name: Benzene, 1,3-dimethyl-
Synonym: *m*-Xylene
Line Form: 1,3-$(CH_3)_2C_6H_4$
Mol Form: C_8H_{10}
M$_r$: 106.16
bp: 139.1, 28.1^{10}
mp: -48
***d*:** $0.8642^{20/4}$
***n*:** 1.4972^{10}
***s*:** H_2O 1; al 5; eth 5; ace 5; bz 5; chl 3; os 5; peth 5
IR: µCOB 3578 3030 2940 1590 1470 1370 1160 1100 1050 890
Raman: DFB 81 3060 3010 2920 2860 2730 1620 1590 1380 1260 1250 1170 1090 1040 1000 830 640 620 280 230 210
UV: SAD 317 265 cyhex
H NMR: VAR 202 2.3 7.0 $CDCl_3$
C NMR: JJ 297 cw 21.2 126.2 128.3 130.0 137.6 diox
MS: WILEY 323 91(100) 106(65) 105(29) 39(18) 51(15) 77(13) 27(10) 78(8)
m/e: 106.08

HODOC No: 03198
CAS No: 106-42-3
Beilstein No: 5^4, 951
Name: Benzene, 1,4-dimethyl-
Synonym: *p*-Xylene
Line Form: 1,4-$(CH_3)_2C_6H_4$
Mol Form: C_8H_{10}
M$_r$: 106.16
bp: 138.3, 27^{10}
mp: 13.3
***d*:** $0.8611^{20/4}$
***n*:** 1.4958^{20}
Color: mcl pr (al)
***s*:** H_2O 1; al 5; eth 5; ace 5; bz 5; chl 3; os 5; peth 5
IR: µCOB 3575 2940 2860 1890 1790 1640 1490 1430 1390 1210 1110 1020 780
Raman: DFB 84 3060 3030 3010 2910 2860 2720 1620 1580 1450 1380 1310 1210 1180 830 810 650 460 410
UV: SAD 609 275 212 cyhex
H NMR: VAR 203 2.3 7.1 $CDCl_3$
MS: WILEY 326 91(100) 106(61) 105(30) 77(12) 51(10) 92(8) 79(7) 103(6)
m/e: 106.08

HODOC No: 03199
CAS No: 10375-96-9
Beilstein No: 5^4, 1168
Name: Benzene, 1,4-dimethyl-2,5-*bis*(1-methylethyl)-
Synonym: *p*-Xylene, 2,5-diisopropyl-
Line Form: 1.4-$[(CH_3)_2CH]_2$-2,3-$(CH_3)_2$-C_6H_2
Structure:

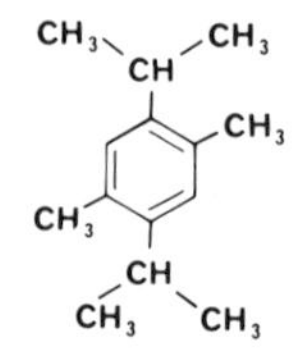

Mol Form: $C_{14}H_{22}$
M$_r$: 190.31
mp: 37-8
***s*:** chl 3
IR: SADG 1692 3020 2960 2920 2860 1610 1460 1445 1400 1385 1365 1275 1130 1105 1080 1040 995 895 885 790 670 465
UV: SAD 9338 278(731) 269(677) 217(9670) MeOH
H NMR: SAD 424 1.2 2.2 3.0 6.8 CCl_4
MS: WILEY 1330 175(100) 190(24) 176(14) 133(14) 147(6) 119(6) 191(4) 91(4)
m/e: 190.17

HODOC No: 03200
CAS No: 3548-85-4
Beilstein No: 5^1, 295
Name: Benzene, 1,1′-(1,4-dimethyl-1,4-butanediyl)*bis*-
Synonym: Hexane, 2,5-diphenyl-
Line Form: $C_6H_5CH(CH_3)$-CH_2CH_2-$CH(CH_3)C_6H_5$
Mol Form: $C_{18}H_{22}$
M$_r$: 238.35
bp: $140\text{-}2^{1.5}$
IR: SADG 17432 3020 2960 1600 1490 1440 1370 750 690 530
H NMR: SAD 9295 1.2 1.5 2.6 7.1 CCl_4
m/e: 238.17

HODOC No: 03201
CAS No: 52161-54-3
Name: Benzene, 1,1′-(1,4-dimethyl-1-butene-1,4-diyl)*bis*-
Synonym: 2,5-Diphenyl-2-hexene
Line Form: $CH_3CH(C_6H_5)CH_2CH{=}C(CH_3)C_6H_5$
Mol Form: $C_{18}H_{20}$
M_r: 236.34
MS: WILEY 1665 119(100) 91(53) 221(42) 41(30) 143(28) 236(25) 39(18) 77(16)
m/e: 236.16

HODOC No: 03202
CAS No: 27854-40-6
Name: Benzene, (1,1-dimethyldecyl)-
Synonym: Undecane, 2-methyl-2-phenyl-
Line Form: $CH_3(CH_2)_8C(CH_3)_2C_6H_5$
Mol Form: $C_{18}H_{30}$
M_r: 246.42
MS: WILEY 1735 119(100) 91(19) 120(10) 44(9) 41(8) 105(6) 43(5) 246(2)
m/e: 246.23

HODOC No: 03203
CAS No: 55134-08-2
Name: Benzene, (2,3-dimethyldecyl)-
Synonym: Decane, 2,3-dimethyl-1-phenyl
Line Form: $CH_3(CH_2)_6CH(CH_3)CH(CH_3)CH_2C_6H_5$
Mol Form: $C_{18}H_{30}$
M_r: 246.42
MS: WILEY 1735 57(100) 43(56) 71(56) 91(52) 92(47) 85(40) 41(25) 246(7)
m/e: 246.23

HODOC No: 03204
CAS No: 55134-09-3
Name: Benzene, (3,3-dimethyldecyl)-
Synonym: Decane, 3,3-dimethyl-1-phenyl
Line Form: $CH_3(CH_2)_6C(CH_3)_2CH_2CH_2C_6H_5$
Mol Form: $C_{18}H_{30}$
M_r: 246.42
MS: WILEY 1728 91(100) 57(35) 43(27) 71(24) 105(24) 85(22) 147(20) 246(12)
m/e: 246.23

HODOC No: 03205
CAS No: 54986-45-7
Name: Benzene, (9,9-dimethyldecyl)-
Synonym: Decane, 9,9-dimethyl-1-phenyl
Line Form: $(CH_3)_3C(CH_2)_8C_6H_5$
Mol Form: $C_{18}H_{30}$
M_r: 246.42
MS: WILEY 1731 57(100) 91(49) 92(38) 246(16) 41(16) 190(13) 43(13) 56(10)
m/e: 246.23

HODOC No: 03206
CAS No: 603-06-5
Beilstein No: 5^3, 822
Name: Benzene, 1,2-dimethyl-3,4-dinitro-
Synonym: 3,4-Dinitro-*o*-xylene
Line Form: $1,2\text{-}(CH_3)_2\text{-}3,4\text{-}(NO_2)_2\text{-}C_6H_2$
Mol Form: $C_8H_8N_2O_4$
M_r: 196.15
mp: 82
Color: nd (al)
***s*:** Bz 4; Et_2O 4
m/e: 196.05

HODOC No: 03207
CAS No: 616-69-3
Beilstein No: 5^3, 823
Name: Benzene, 1,2-dimethyl-3,5-dinitro-
Synonym: *o*-Xylene, 3,5-dinitro-
Line Form: $1,2\text{-}(CH_3)_2\text{-}3,5\text{-}(NO_2)_2C_6H_2$
Mol Form: $C_8H_8N_2O_4$
M_r: 196.15
mp: 77
Color: ye nd (al, peth)
***s*:** Ac 4; Bz 4; EtOH 4
m/e: 196.05

HODOC No: 03208
CAS No: 610-23-1
Beilstein No: 5^3, 823
Name: Benzene, 1,2-dimethyl-4,5-dinitro-
Synonym: 1,2-Dimethyl-4,5-dinitrobenzene
Line Form: $1,2\text{-}(CH_3)_2\text{-}4,5\text{-}(NO_2)_2C_6H_2$
Mol Form: $C_8H_8N_2O_4$
M_r: 196.15
mp: 118
Color: nd (al, Bz)
***s*:** Ac 4; Bz 4; Et_2O 4
m/e: 196.05

HODOC No: 03209
CAS No: 603-02-1
Beilstein No: 5^3, 844
Name: Benzene, 1,3-dimethyl-2,4-dinitro-
Synonym: *m*-Xylene, 2,4-dinitro-
Line Form: 1,3-$(CH_3)_2$-2,4-$(NO_2)_2$-C_6H_2
Mol Form: $C_8H_8N_2O_4$
M_r: 196.15
mp: 83-4
Color: lf (al)
***s*:** Bz 4; Et_2O 4
m/e: 196.05

HODOC No: 03210
CAS No: 74709-95-8
Beilstein No: 5^3, 295
Name: Benzene, 1,3-dimethyl-2,5-dinitro-
Synonym: *m*-Xylene, 2,5-dinitro-
Line Form: 1,3-$(CH_3)_2$-2,5-$(NO_2)_2C_6H_2$
Mol Form: $C_8H_8N_2O_4$
M_r: 196.15
mp: 101
Color: ye cr (al)
***s*:** Ac 4; Bz 4; Et_2O 4; EtOH 4
m/e: 196.05

HODOC No: 03211
CAS No: 711-41-1
Beilstein No: 5^4, 972
Name: Benzene, 1,4-dimethyl-2,3-dinitro-
Synonym: *p*-Xylene, 2,3-dinitro-
Line Form: 1,4-$(CH_3)_2$-2,3-$(NO_2)_2C_6H_2$
Mol Form: $C_8H_8N_2O_4$
M_r: 196.15
mp: 93
Color: mcl pr (al)
***s*:** H_2O 1; al 4; eth 3; ace 3; bz 3; chl 3
UV: HBCP 295(1445) hx
m/e: 196.05

HODOC No: 03212
CAS No: 712-32-3
Beilstein No: 5^4, 973
Name: Benzene, 1,4-dimethyl-2,5-dinitro-
Synonym: *p*-Xylene, 2,5-dinitro-
Line Form: 1,4-$(CH_3)_2$-2,5-$(NO_2)_2$-C_6H_2
Mol Form: $C_8H_8N_2O_4$
M_r: 196.15
mp: 147-8
Color: ye nd (al)
***s*:** Ac 4; Bz 4; Chl 4
m/e: 196.05

HODOC No: 03213
CAS No: 616-72-8
Beilstein No: 5^4, 950
Name: Benzene, 1,5-dimethyl-2,4-dinitro-
Synonym: *m*-Xylene, 4,6-dinitro-
Line Form: 1,3-$(CH_3)_2$-4,6-$(NO_2)_2$-C_6H_2
Mol Form: $C_8H_8N_2O_4$
M_r: 196.15
mp: 94
Color: cr (al)
***s*:** chl 3
IR: SADG 14347 3100 2990 1610 1590 1520 1450 1370 1340 1260 910 850 840 740 690
Raman: SAD 739 3080 3050 2980 2940 1620 1590 1530 1490 1440 1420 1380 1360 1350 1310 1250 1170 1130 1040 1010 920 840 730 600 510 440 370 250 160
UV: SAD 16316 250(14400) 206(12500) MeOH
H NMR: SAD 8091 2.8 7.6 8.8 $CDCl_3$
m/e: 196.05

HODOC No: 03214
CAS No: 610-03-7
Beilstein No: 5^1, 181
Name: Benzene, 2,3-dimethyl-1,4-dinitro-
Synonym: 3,6-Dinitro-*o*-xylene
Line Form: 1,2-$(CH_3)_2$-3,6-$(NO_2)_2C_6H_2$
Mol Form: $C_8H_8N_2O_4$
M_r: 196.15
mp: 89-90
Color: nd (al)
***s*:** Ac 4; Bz 4; Et_2O 4; EtOH 4
m/e: 196.05

HODOC No: 03215
CAS No: 609-92-7
Beilstein No: 5^4, 973
Name: Benzene, 2,5-dimethyl-1,3-dinitro-
Synonym: *p*-Xylene, 2,6-dinitro-
Line Form: 1.4-$(CH_3)_2$-2,6-$(NO_2)_2$-C_6H_2
Mol Form: $C_8H_8N_2O_4$
M_r: 196.15
mp: 123-4
Color: nd (al)
***s*:** H_2O 1; al 3; eth 3; ace 3; bz 3; chl 3
UV: HBCP 295(1514) hx
m/e: 196.05

HODOC No: 03216
CAS No: 4613-11-0
Beilstein No: 5^4, 1940
Name: Benzene,
1,1′-(1,2-dimethyl-1,2-ethanediyl)*bis*-, (R*,*S**)-
Synonym: Bibenzyl, α,α′-dimethyl-, erythro-
Line Form: $CH_3CH(C_6H_5)CH(C_6H_5)CH_3$
Mol Form: $C_{16}H_{18}$
M$_r$: 210.30
mp: 128
***s*:** al 3; eth 3; chl 3
UV: OES 5-518 268(316) 264(398) 261(501) 258(501) 253(501) 248(316) iso
m/e: 210.14

HODOC No: 03217
CAS No: 2726-21-8
Beilstein No: 5^4, 1941
Name: Benzene,
1,1′-(1,2-dimethyl-1,2-ethanediyl)*bis*-, (R*,R*)-(±)-
Synonym: Bibenzyl, α,α′-dimethyl-, (±)-
Line Form: $CH_3CH(C_6H_5)$-$CH(C_6H_5)$-CH_3
Mol Form: $C_{16}H_{18}$
M$_r$: 210.30
bp: 283-4^{752}, 140^{10}
mp: 8
***d*:** $0.9757^{20/4}$
***n*:** 1.5552^{20}
m/e: 210.14

HODOC No: 03218
CAS No: 54889-78-0
Name: Benzene,
1,1′-[(1,2-dimethyl-1,2-ethanediyl)*bis*(oxymethylene)]*bis*-
Synonym: Butane, 2,3-*bis*(benzyloxy)
Line Form: $C_6H_5CH_2OCH(CH_3)CH(CH_3)OCH_2C_6H_5$
Mol Form: $C_{18}H_{22}O_2$
M$_r$: 270.35
MS: WILEY 1831 91(100) 105(42) 77(34) 106(28) 92(28) 28(23) 107(21) 270(0)
m/e: 270.16

HODOC No: 03219
CAS No: 782-06-9
Beilstein No: 5^3, 2010
Name: Benzene,
1,1′-(1,2-dimethyl-1,2-ethenediyl)*bis*-, (E)-
Synonym: Stilbene, α,α′-dimethyl-, (E)-
Mol Form: $C_{16}H_{16}$
M$_r$: 208.29
mp: 107
***d*:** 0.987^{20}
***n*:** $1.6173^{20/HE}$
***s*:** H_2O 1; al 2; eth 3; MeOH 2
UV: MOCHA 1146-53 242(10000) hx
m/e: 208.13

HODOC No: 03220
CAS No: 782-05-8
Beilstein No: 5^3, 2009
Name: Benzene,
1,1′-(1,2-dimethyl-1,2-ethenediyl)*bis*-, (Z)-
Synonym: Stilbene, α,α′-dimethyl-, (Z)-
Line Form: $C_6H_5C(CH_3)=C(CH_3)C_6H_5$
Mol Form: $C_{16}H_{16}$
M$_r$: 208.29
mp: 67-8
***d*:** 1.004^{20} $0.9537^{78/4}$
***n*:** 1.5612^{78}
***s*:** H_2O 1; al 3; HOAc 3; MeOH 3
UV: MOCHA 1146-53 270(7943) 218(10000) hx
m/e: 208.13

HODOC No: 03221
CAS No: 6669-13-2
Beilstein No: 6^4, 559
Name: Benzene, (1,1-dimethylethoxy)-
Synonym: Ether, *tert*-butyl phenyl
Line Form: $(CH_3)_3C$-O-C_6H_5
Mol Form: $C_{10}H_{14}O$
M$_r$: 150.21
bp: 185-6
***d*:** 0.9214^{20}
IR: μCOB 202 3030 2940 1590 1540 1490 1390 1370 1240 1160 1060 1020 930 910 890 810 780 710 690
UV: LANG 2648 280(708) 271(933) EtOH
MS: WILEY 895 94(100) 41(11) 39(11) 57(10) 77(4) 65(4) 51(4) 150(1)
m/e: 150.10

HODOC No: 03222
CAS No: 98-06-6
Beilstein No: 5^4, 1045
Name: Benzene, (1,1-dimethylethyl)-
Synonym: Benzene, *tert*-butyl-
Line Form: $C_6H_5\text{-}C(CH_3)_3$
Mol Form: $C_{10}H_{14}$
M_r: 134.21
bp: 169.1, 50.7^{10}
mp: -57.8 (fp)
***d*:** $0.8665^{20/4}$
***n*:** 1.4927^{20}
***s*:** H_2O 1; al 4; eth 4; ace 5; bz 5; ctc 5; peth 5
IR: µCOB 3599 2940 1590 1470 1370 1270 1210 1100 1030 750
Raman: DFB 70 3060 2960 2920 2900 2860 2770 2710 1600 1580 1470 1450 1270 1210 1190 1160 1120 1030 1000 930 910 840 710 620 540 450 420 140
UV: SAD 228 267 257 cyhex
H NMR: SAD 53 1.3 7.0-7.4 CCl_4
MS: WILEY 665 119(100) 91(48) 41(24) 134(24) 39(14) 51(12) 77(11) 79(10)
m/e: 134.11

HODOC No: 03223
CAS No: 98-19-1
Beilstein No: 5^4, 1130
Name: Benzene, 1-(1,1-dimethylethyl)-3,5-dimethyl-
Synonym: *m*-Xylene, 5-*tert*-butyl-
Line Form: $3,5\text{-}(CH_3)_2C_6H_3\text{-}C(CH_3)_3$
Mol Form: $C_{12}H_{18}$
M_r: 162.26
bp: 207
mp: 17
***s*:** ctc 3
IR: SADG 15549 2950 2910 2860 1605 1480 1460 1395 1365 1300 1230 1200 1040 910 850 715
UV: SAD 1461 265 iso
H NMR: SAD 6546 1.3 2.3 6.7 6.9 CCl_4
MS: WILEY 1046 147(100) 162(25) 41(22) 119(21) 107(20) 91(13) 39(13) 148(12)
m/e: 162.14

HODOC No: 03224
CAS No: 81-15-2
Beilstein No: 5^4, 1132
Name: Benzene, 1-(1,1-dimethylethyl)-3,5-dimethyl-2,4,6-trinitro-
Synonym: *m*-Xylene, 5-*tert*-butyl-2,4,6-trinitro-
Line Form: $3,5\text{-}(CH_3)_2\text{-}2,4,6\text{-}(NO_2)_3(C_6)C(CH_3)_3$
Structure:

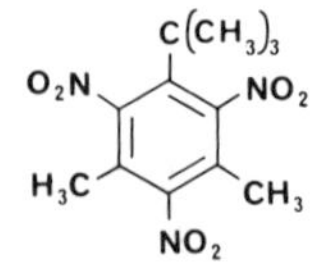

Mol Form: $C_{12}H_{15}N_3O_6$
M_r: 297.26
mp: 110
Color: pl, nd (al)
***s*:** H_2O 1; al 2; eth 3; chl 3
IR: µCOB 2871 2860 1560 1470 1390 1350 1280 1210 1160 1030 950 890 860 780
Raman: SAD 1143 3030 2980 2950 2930 2880 1580 1540 1510 1480 1450 1440 1390 1360 1280 1210 1200 1170 1040 960 930 890 870 830 790 740 700 650 630 590 560 530 480 450 420 370 320 260 240 230 190 160 140 120
UV: SAD 422 331(4230) MeOH
H NMR: SAD 6497 1.5 2.2 $CDCl_2$
m/e: 297.10

HODOC No: 03225
CAS No: 1746-23-2
Name: Benzene, 1-(1,1-dimethylethyl)-4-ethenyl-
Synonym: Styrene, *p*-*tert*-butyl-
Mol Form: $C_{12}H_{16}$
M_r: 160.25
MS: WILEY 1014 145(100) 160(26) 117(22) 146(12) 105(11) 91(8) 41(8) 128(6)
m/e: 160.13

HODOC No: 03226
CAS No: 17269-94-2
Name: Benzene, 1-(1,1-dimethylethyl)-4-ethoxy-
Synonym: Phenetole, *p*-*tert*-butyl-
Mol Form: $C_{12}H_{18}O$
M_r: 178.26
MS: WILEY 1211 163(100) 135(38) 178(26) 107(15) 164(12) 41(7) 91(7) 77(6)
m/e: 178.14

HODOC No: 03227
CAS No: 54889-97-3
Name: Benzene, 1-(1,1-dimethylethyl)-4-(2-ethoxyethoxy)-
Synonym: Ethane, 1(4-*tert*-butylphenoxy)-2-ethoxy
Mol Form: $C_{14}H_{22}O_2$
M_r: 222.31
MS: WILEY 1574 135(100) 207(92) 45(52) 222(34) 73(26) 209(15) 107(14) 136(10)
m/e: 222.16

HODOC No: 03228
CAS No: 14411-56-4
Name: Benzene, 1-(1,1-dimethylethyl)-3-ethyl-
Synonym: Benzene, 1-*tert*-butyl-3-ethyl-
Mol Form: $C_{12}H_{18}$
M_r: 162.26
UV: API 525 271 267 263 262 257 iso
MS: WILEY 1043 ini 147(100) 162(20) 119(18) 91(17) 148(12) 41(6) 79(6) 117(4)
m/e: 162.14

HODOC No: 03229
CAS No: 7364-19-4
Beilstein No: 5, 446
Name: Benzene, 1-(1,1-dimethylethyl)-4-ethyl-
Synonym: Benzene, 1-*tert*-butyl-4-ethyl-
Mol Form: $C_{12}H_{18}$
M_r: 162.26
bp: 209-13
UV: API 526 272 263 258 251 iso
MS: WILEY 1043 ini 147(100) 162(20) 119(18) 91(17) 148(12) 41(6) 79(6) 117(4)
m/e: 162.14

HODOC No: 03230
CAS No: 5396-38-3
Beilstein No: 6^4, 3297
Name: Benzene, 1-(1,1-dimethylethyl)-4-methoxy-
Synonym: Anisole, *p*-*tert*-butyl-
Line Form: $4\text{-}CH_3O\text{-}C_6H_4\text{-}C(CH_3)_3$
Mol Form: $C_{11}H_{16}O$
M_r: 164.24
bp: $235\text{-}7^{730}$
***d*:** 0.9436^{20}
***n*:** 1.5039^{20}
IR: COB 3095
UV: SAD 10848 282(1500) 276(1730) 223(8830) MeOH
H NMR: SAD 713 1.3 3.7 6.7 7.2 CCl_4
MS: WILEY 1067 149(100) 164(23) 121(16) 150(14) 109(11) 135(9) 91(8) 41(6)
m/e: 164.12

HODOC No: 03231
CAS No: 60772-80-7
Name: Benzene, 1-(1,1-dimethylethyl)-2-methoxy-3-methyl-
Synonym: Toluene, 2-methoxy-3-*tert*-butyl
Mol Form: $C_{12}H_{18}O$
M_r: 178.26
IR: µCOB 5384 2940 1590 1450 1410 1350 1250 1220 1100 1010 780 760
m/e: 178.14

HODOC No: 03232
CAS No: 83-66-9
Beilstein No: 6^4, 3402
Name: Benzene,
1-(1,1-dimethylethyl)-2-methoxy-4-methyl-3,5-dinitro

Synonym: Anisole, 6-*tert*-butyl-3-methyl-2,4-dinitro-
Structure:

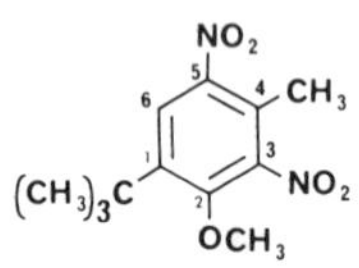

Mol Form: $C_{12}H_{16}N_2O_5$
Mr: 268.26
bp: 135^{16}
mp: 84-6
Color: pa ye lf (al)
s: H_2O 1; al 2; eth 3; chl 3
IR: SADG 8877 3030 2941 1587 1538 1471 1351 1265 1053 943 870 826
UV: SAD 6029 265 MeOH
H NMR: SAD 2085 1.4 2.4 3.9 8.0 $CDCl_3$
m/e: 268.11

HODOC No: 03233
CAS No: 1074-92-6
Beilstein No: 5^4, 1096
Name: Benzene, 1-(1,1-dimethylethyl)-2-methyl-
Synonym: Toluene, *o-tert*-butyl-
Line Form: $2\text{-}CH_3C_6H_4C(CH_3)_3$
Mol Form: $C_{11}H_{16}$
Mr: 148.24
bp: 200.4, 99.4^{20}
mp: -50.3 (fp)
d: 0.8897^{20}
n: 1.5076^{20}
s: Ac 4; Bz 4; Et_2O 4; EtOH 4
Raman: API 181 3064 3018 2963 2933 2915 2875 1602 1196 1058 680
UV: API 275 268 262 255 iso
MS: WILEY 868 ini 133(100) 105(34) 148(29) 93(13) 134(11) 91(10) 115(5) 77(4)
m/e: 148.13

HODOC No: 03234
CAS No: 1075-38-3
Beilstein No: 5^4, 1096
Name: Benzene, 1-(1,1-dimethylethyl)-3-methyl-
Synonym: Toluene, *m-tert*-butyl-
Line Form: $3\text{-}CH_3C_6H_4C(CH_3)_3$
Mol Form: $C_{11}H_{16}$
Mr: 148.24
bp: 189.3, 90.4^{30}
mp: -41.4 (fp)
d: 0.8657^{20}
n: 1.4944^{20}
s: Ac 4; Bz 4; Et_2O 4; EtOH 4
IR: COB 3067 3040 2960 2910 2870 1610 1500 1480 1460 1360 1270 1200 1100 900 870 790 700
UV: OES 2-273 271(200) 267(158) 264(251) 262(200) 257(158) iso
MS: WILEY 864 133(100) 105(33) 148(24) 41(20) 93(14) 39(14) 91(13) 134(11)
m/e: 148.13

HODOC No: 03235
CAS No: 98-51-1
Beilstein No: 5^4, 1097
Name: Benzene, 1-(1,1-dimethylethyl)-4-methyl-
Synonym: Toluene, *p-tert*-butyl-
Line Form: $4\text{-}CH_3C_6H_4C(CH_3)_3$
Mol Form: $C_{11}H_{16}$
Mr: 148.24
bp: 193, 70^{10}
mp: -52
d: $0.8612^{20/4}$
n: 1.4918^{20}
s: H_2O 1; al 2; eth 4; ace 3; bz 3; chl 4
IR: COB 3605 3100 3060 3030 2970 2870 2740 1900 1650 1580 1510 1460 1360 1270 1199 1110 1020 810 720
Raman: API 183 3065 3034 3015 2983 2959 2924 2903 2877 1606 796
UV: OES 1-409 272(347) 266(355) 264(407) 259(282) 212(8313) iso
C NMR: JJ 422 FT 20.7 31.4 34.2 125.0 128.7 134.5 148.0 $CDCl_3$
MS: WILEY 867 133(100) 105(30) 148(22) 41(19) 39(13) 91(12) 93(12) 134(11)
m/e: 148.13

HODOC No: 03236
CAS No: 4132-49-4
Name: Benzene, 1-(1,1-dimethylethyl)-4-(1-methylethyl)-
Synonym: Cumene, *p-tert*-butyl-
Mol Form: $C_{13}H_{20}$
M_r: 176.29
IR: **μCOB 1097** 3030 2940 1520 1470 1410 1370 1270 1110 1050 1020 830
m/e: 176.16

HODOC No: 03237
CAS No: 54932-87-5
Name: Benzene, 1-(1,1-dimethylethyl)-4-[(2-methyl-2-propenyl)oxy]-
Synonym: Benzene, 1-*tert*-butyl-4-methallyloxy
Mol Form: $C_{14}H_{20}O$
M_r: 204.30
MS: **WILEY 1446** 135(100) 189(84) 55(38) 173(37) 107(29) 204(22) 150(20) 41(18)
m/e: 204.15

HODOC No: 03238
CAS No: 547-94-4
Beilstein No: 5^3, 1003
Name: Benzene, 2-(1,1-dimethylethyl)-4-methyl-1,3,5-trinitro-
Synonym: Toluene, 3-*tert*-butyl-2,4,6-trinitro-
Line Form: $3\text{-}CH_3\text{-}2,4,6\text{-}(NO_2)_3C_6HC(CH_3)_3$
Structure:

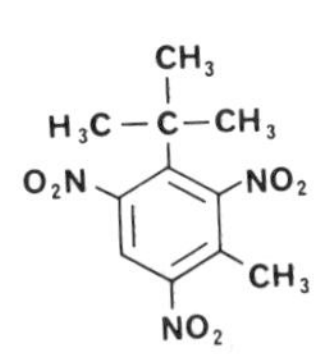

Mol Form: $C_{11}H_{13}N_3O_6$
M_r: 283.23
mp: 96-7
Color: ye nd (al)
***s*:** Bz 4; Et_2O 4; EtOH 4; Chl 4
m/e: 283.08

HODOC No: 03239
CAS No: 1886-57-3
Beilstein No: 5^4, 1051
Name: Benzene, 1-(1,1-dimethylethyl)-2-nitro-
Synonym: Benzene, 1-*tert*-butyl-2-nitro-
Line Form: $2\text{-}NO_2C_6H_4\text{-}C(CH_3)_3$
Mol Form: $C_{10}H_{13}NO_2$
M_r: 179.21
bp: 250.5^{765},114.5^{10}
***d*:** 1.074^{15}
Color: ye liq
m/e: 179.09

HODOC No: 03240
CAS No: 23132-52-7
Beilstein No: 5^4, 1052
Name: Benzene, 1-(1,1-dimethylethyl)-3-nitro-
Synonym: Benzene, 1-*tert*-butyl-3-nitro-
Line Form: $3\text{-}NO_2C_6H_4C(CH_3)_3$
Mol Form: $C_{10}H_{13}NO_2$
M_r: 179.21
bp: $250\text{-}2^{704}$
mp:
Color: ye-red liq
m/e: 179.09

HODOC No: 03241
CAS No: 26459-52-9
Name: Benzene, 1-(1,1-dimethylethyl)-3-phenoxy-
Synonym: Ether, *m-tert*-butylphenyl phenyl
Mol Form: $C_{16}H_{18}O$
M_r: 226.30
MS: **WILEY 1608** 211(100) 226(36) 183(28) 28(20) 212(17) 32(14) 178(14) 91.5(13)
m/e: 226.14

HODOC No: 03242
CAS No: 5331-28-2
Name: Benzene, 1-(1,1-dimethylethyl)-4-phenoxy-
Synonym: 4-*tert*-Butyldiphenyl ether
Mol Form: $C_{16}H_{18}O$
M_r: 226.30
MS: **WILEY 1607** 211(100) 225(31) 212(17) 92(7) 77(6) 226(6) 117(5) 118(5)
m/e: 226.14

HODOC No: 03243
CAS No: 48144-15-6
Name: Benzene, 1-(1,1-dimethylethyl)-4-(2-propynyloxy)-
Synonym: *p-tert*-Butylphenyl 2-propynyl ether
Mol Form: $C_{13}H_{16}O$
M$_r$: 188.26
IR: COB 4330 3300 2960 1600 1520 1450 1400 1360 1300 1270 1210 1180 1110 1050 930 840 800 750 660 620
m/e: 188.12

HODOC No: 03244
CAS No: 145-39-1
Beilstein No: 5^3, 1055
Name: Benzene, 1-(1,1-dimethylethyl)-3,4,5-trimethyl-2,6-dinitro-
Synonym: Benzene, 1-*tert*-butyl-3,4,5-trimethyl-2,6-dinitro-
Structure:

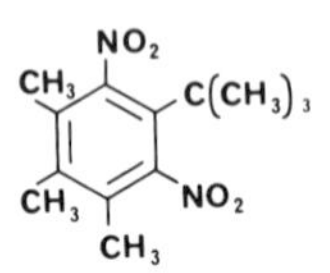

Mol Form: $C_{13}H_{18}N_2O_4$
M$_r$: 266.28
mp: 135-6
Color: pa ye pr (al)
***s*:** H_2O; al 2; eth 3
IR: µSADP 581 2940 1590 1470 1390 1350 1190 1100 1020 890
m/e: 266.13

HODOC No: 03245
CAS No: 22539-65-7
Name: Benzene, 1,2-dimethyl-3-(1-methylethyl)-
Synonym: Cumene, 2,3-dimethyl-
Mol Form: $C_{11}H_{16}$
M$_r$: 148.24
***s*:** ctc 3; CS_2 3
IR: µCOB 3981 3030 2940 1590 1470 1450 1390 1370 1110 1030 780 730
m/e: 148.13

HODOC No: 03246
CAS No: 4132-77-8
Beilstein No: 5^4, 1105
Name: Benzene, 1,2-dimethyl-4-(1-methylethyl)-
Synonym: Cumene, 3,4-dimethyl-
Line Form: $3,4\text{-}(CH_3)_2C_6H_3CH(CH_3)_2$
Mol Form: $C_{11}H_{16}$
M$_r$: 148.24
bp: 199, $86\text{-}7^{16}$
***d*:** $0.8710^{20/4}$
***n*:** 1.4951^{20}
IR: µCOB 7118 2940 1520 1470 1370 1020 1000 880 820 720
UV: SAD 9448 274(457) 265(499) 212(9100) MeOH
H NMR: SAD 420 1.2 2.2 2.8 6.9 CCl_4
m/e: 148.13

HODOC No: 03247
CAS No: 14411-75-7
Beilstein No: 5^4, 1103
Name: Benzene, 1,3-dimethyl-2-(1-methylethyl)-
Synonym: Cumene, 2,6-dimethyl-
Line Form: $2,6\text{-}(CH_3)_2C_6H_3\text{-}CH(CH_3)_2$
Mol Form: $C_{11}H_{16}$
M$_r$: 148.24
bp: 199, 98.2^{30}
***d*:** 0.890^{20}
***n*:** 1.509^{20}
***s*:** Ac 4; Bz 4; Et_2O 4; EtOH 4
m/e: 148.13

HODOC No: 03248
CAS No: 4706-90-5
Beilstein No: 5^4, 1103
Name: Benzene, 1,3-dimethyl-5-(1-methylethyl)-
Synonym: Cumene, 3,5-dimethyl-
Line Form: $3,5\text{-}(CH_3)_2C_6H_3CH(CH_3)_2$
Mol Form: $C_{11}H_{16}$
M$_r$: 148.24
bp: $83\text{-}5^{17}$
***n*:** 1.4935^{25}
MS: NBS 441
m/e: 148.13

HODOC No: 03249
CAS No: 4132-72-3
Beilstein No: 5^4, 1104
Name: Benzene, 1,4-dimethyl-2-(1-methylethyl)-
Synonym: Cumene, 2,5-dimethyl-
Line Form: 2,5-$(CH_3)_2C_6H_3CH(CH_3)_2$
Mol Form: $C_{11}H_{16}$
M_r: 148.24
bp: 195
IR: SADG 2262 3040 3010 2950 2910 2860 1610 1495 1455 1380 1355 1275 1165 1075 980 875 810 800 725 455
MS: WILEY 865 133(100) 148(25) 134(11) 105(10) 91(8) 115(6) 117(6) 131(5)
m/e: 148.13

HODOC No: 03250
CAS No: 4706-89-2
Beilstein No: 5^4, 1104
Name: Benzene, 2,4-dimethyl-1-(1-methylethyl)-
Synonym: Cumene, 2,4-dimethyl-
Line Form: 2,4-$(CH_3)_2C_6H_3CH(CH_3)_2$
Mol Form: $C_{11}H_{16}$
M_r: 148.24
bp: 194-5, 77^{13}
***n*:** 1.4998^{25}
m/e: 148.13

HODOC No: 03251
CAS No: 1483-60-9
Beilstein No: 5^4, 1129
Name: Benzene, 2,4-dimethyl-1-(1-methylpropyl)-
Synonym: 2,4-Dimethyl-1-(1-methylpropyl)benzene
Line Form: 2,4-$(CH_3)_2C_6H_3CH(CH_3)C_2H_5$
Mol Form: $C_{12}H_{18}$
M_r: 162.26
bp: 84^8
***d*:** $0.8680^{20/20}$
***n*:** 1.4939^{25}
MS: WILEY 1045 133(100) 162(17) 134(12) 27(11) 91(10) 119(8) 29(8) 105(8)
m/e: 162.14

HODOC No: 03252
CAS No: 83-41-0
Beilstein No: 5^4, 930
Name: Benzene, 1,2-dimethyl-3-nitro-
Synonym: 1,2-Dimethyl-3-nitrobenzene
Line Form: 2,3-$(CH_3)_2$-C_6H_3-NO_2
Mol Form: $C_8H_9NO_2$
M_r: 151.16
bp: 240, 131^{20}
mp: 15
***d*:** 1.1402^{20}
***n*:** 1.5441^{20}
Color: nd (al)
***s*:** H_2O 1; al 3; ctc 3
IR: SADG 10357 3030 2941 1613 1515 1471 1351 1299 1205 1099 885 794 741
UV: SAD 6805 258 207 MeOH
H NMR: SAD 2311 2.4 7.0-7.3 7.5 CCl_4
m/e: 151.06

HODOC No: 03253
CAS No: 99-51-4
Beilstein No: 5^4, 930
Name: Benzene, 1,2-dimethyl-4-nitro-
Synonym: *o*-Xylene, 4-nitro-
Line Form: 3,4-$(CH_3)_2$-C_6H_3-NO_2
Mol Form: $C_8H_9NO_2$
M_r: 151.16
bp: 250^{750}, 143^{21}
mp: 30-1
***d*:** 1.112^{15}
***n*:** 1.5202^{20}
Color: ye pr (al)
***s*:** H_2O 1; al 5
IR: SADG 18399 3030 2941 1613 1515 1449 1333 1299 1075 1010 909 885 820 730
UV: LANG 555 211(15136) hx 215(13183) EtOH
MS: NBS 4154
m/e: 151.06

HODOC No: 03254
CAS No: 81-20-9
Beilstein No: 5^4, 348
Name: Benzene, 1,3-dimethyl-2-nitro-
Synonym: *m*-Xylene, 2-nitro-
Line Form: $2,6\text{-}(CH_3)_2C_6H_3NO_2$
Mol Form: $C_8H_9NO_2$
M_r: 151.16
bp: 225^{744}
mp: 14-6
***d*:** 1.112^{15}
***n*:** 1.5202^{20}
***s*:** H_2O 1; al 4; ctc 3
IR: SADG 775 3030 2941 1587 1538 1471 1389 1176 1053 847 800
UV: SAD 9710 251(1310) 207(12200) MeOH
H NMR: SAD 320 2.3 7.1 CCl_4
MS: ALDN 106 77(100) 79(73) 105(63) 91(59) 120(48) 119(45) 134(44) 151(22)
m/e: 151.06

HODOC No: 03255
CAS No: 99-12-7
Beilstein No: 5^4, 948
Name: Benzene, 1,3-dimethyl-5-nitro-
Synonym: *m*-Xylene, 5-nitro-
Line Form: $3,5\text{-}(CH_3)_2C_6H_3NO_2$
Mol Form: $C_8H_9NO_2$
M_r: 151.16
bp: 273^{739}
mp: 75
Color: nd (al)
***s*:** H_2O 1; al 4; eth 4
UV: HBCP 271(7413) al
m/e: 151.06

HODOC No: 03256
CAS No: 89-58-7
Beilstein No: 5^4, 971
Name: Benzene, 1,4-dimethyl-2-nitro-
Synonym: *p*-Xylene, 2-nitro-
Line Form: $2,5\text{-}(CH_3)_2C_6H_3NO_2$
Mol Form: $C_8H_9NO_2$
M_r: 151.16
bp: 240-1, $64\text{-}5^{0.35}$
***d*:** 1.132^{15}
***n*:** 1.5413^{20}
Color: pa ye
***s*:** H_2O 1; al 3
IR: μSADP 17346 2940 1610 1540 1470 1450 1350 1300 1160 1040 920 810
UV: SAD 5498 296 255 cyhex
H NMR: SAD 9747 2.4 2.5 7.1 7.3 7.7 CCl_4
MS: ALDN 106 134(100) 79(85) 77(83) 106(44) 91(34) 103(33) 39(27) 151(24)
m/e: 151.06

HODOC No: 03257
CAS No: 89-87-2
Beilstein No: 5^4, 948
Name: Benzene, 2,4-dimethyl-1-nitro-
Synonym: *m*-Xylene, 4-nitro-
Line Form: $2,4\text{-}(CH_3)_2C_6H_3NO_2$
Mol Form: $C_8H_9NO_2$
M_r: 151.16
bp: 246^{744}, 122^{18}
mp: 9
***d*:** 1.135^{15}
***n*:** 1.5473^{25}
***s*:** H_2O 1; eth 3; ace 3; bz 3; chl 3
IR: μSADP 2397 3106 3030 2924 1618 1600 1529 1515 1449 1370 1351 1316 1250 1176 1163 1085 1042 855 840 832 781 752 746 685
MS: ALDN 106 134(100) 79(82) 77(73) 106(33) 103(25) 78(22) 51(19) 151(14)
m/e: 151.06

HODOC No: 03258
CAS No: 55191-25-8
Name: Benzene, (1,1-dimethylnonyl)-
Synonym: Decane, 2-methyl-2-phenyl-
Line Form: $CH_3(CH_2)_7C(CH_3)_2C_6H_5$
Mol Form: $C_{17}H_{28}$
M_r: 232.39
MS: WILEY 1649 119(100) 91(20) 120(10) 41(9) 105(6) 44(4) 118(4) 232(3)
m/e: 232.22

HODOC No: 03259
CAS No: 55191-39-4
Name: Benzene, 1,4-dimethyl-2-octadecyl-
Synonym: Octadecane, 1-(2,5-xylyl)-
Mol Form: $C_{26}H_{46}$
$\mathbf{M_r}$: 358.63
MS: WILEY 2051 119(100) 358(47) 120(44) 43(34) 41(23) 359(14) 57(12) 55(11)
m/e: 358.36

HODOC No: 03260
CAS No: 54518-00-2
Name: Benzene, (2,4-dimethylpentyl)-
Synonym: Pentane, 2,4-dimethyl-1-phenyl
Line Form: $C_6H_5CH_2CH(CH_3)CH_2CH(CH_3)_2$
Mol Form: $C_{13}H_{20}$
$\mathbf{M_r}$: 176.29
MS: WILEY 1198 43(100) 91(39) 85(38) 92(31) 41(25) 27(21) 28(19) 176(10)
m/e: 176.16

HODOC No: 03261
CAS No: 55521-12-5
Name: Benzene, 1-[2-(3,4-dimethylphenyl)ethyl]-2,3-dimethyl-
Synonym: Ethane, 1(2,3-dimethylphenyl)-2(3,4-dimethylphenyl)
Mol Form: $C_{18}H_{22}$
$\mathbf{M_r}$: 238.35
MS: WILEY 1677 223(100) 238(36) 224(19) 193(14) 91(11) 208(10) 132(10) 77(8)
m/e: 238.17

HODOC No: 03262
CAS No: 2049-95-8
Beilstein No: 5^4, 1090
Name: Benzene, (1,1-dimethylpropyl)-
Synonym: Benzene, *tert*-pentyl-
Line Form: $CH_3CH_2C(C_6H_5)(CH_3)_2$
Mol Form: $C_{11}H_{16}$
$\mathbf{M_r}$: 148.24
bp: 192.4, 92.9^{30}
***d*:** 0.8748^{20}
***n*:** 1.4958^{20}
IR: COB 3604 3100 3060 3040 2970 2930 2890 1600 1500 1460 1450 1370 1360 1300 1090 1040 1000 790 760 700
UV: OES 1-409 268(110) 264(151) 261(158) 258(195) 253(155) 248(123) iso
MS: WILEY 867 105(100) 148(13) 106(12) 91(11) 77(9) 27(9) 79(7) 39(6)
m/e: 148.13

HODOC No: 03263
CAS No: 4481-30-5
Beilstein No: 5^4, 1092
Name: Benzene, (1,2-dimethylpropyl)-
Synonym: Butane, 2-methyl-3-phenyl-
Line Form: $(CH_3)_2CHCH(CH_3)C_6H_5$
Mol Form: $C_{11}H_{16}$
$\mathbf{M_r}$: 148.24
bp: 188, 89.3^{30}
***d*:** 0.870^{20}
***n*:** 1.486^{20}
MS: NBS 441
m/e: 148.13

HODOC No: 03264
CAS No: 17059-45-9
Beilstein No: 5^4, 1102
Name: Benzene, 1,3-dimethyl-2-propyl-
Synonym: *m*-Xylene, 2-propyl-
Line Form: $2,6\text{-}(CH_3)_2C_6H_3C_3H_7$
Mol Form: $C_{11}H_{16}$
$\mathbf{M_r}$: 148.24
bp: 207.6, 105.2^{30}
***d*:** 0.8856^{20}
***n*:** 1.5063^{20}
***s*:** Ac 4; Bz 4; Et_2O 4; EtOH 4
m/e: 148.13

HODOC No: 03265
CAS No: 3982-64-7
Beilstein No: 5^4, 1102
Name: Benzene, 1,3-dimethyl-5-propyl-
Synonym: *m*-Xylene, 5-propyl-
Line Form: $3,5\text{-}(CH_3)_2C_6H_3C_3H_7$
Mol Form: $C_{11}H_{16}$
$\mathbf{M_r}$: 148.24
bp: 202.2, 100.8^{30}
mp: -59.1
***d*:** 0.8607^{20}
***n*:** 1.4952^{20}
***s*:** Ac 4; Bz 4; Et_2O 4; EtOH 4
m/e: 148.13

HODOC No: 03266
CAS No: 3042-50-0
Name: Benzene, 1,4-dimethyl-2-propyl-
Synonym: *p*-Xylene, 2-propyl-
Line Form: 2,5-$(CH_3)_2$-C_6H_3-C_3H_7
Mol Form: $C_{11}H_{16}$
M_r: 148.24
bp: 204.3, 102.5^{30}
***d*:** 0.8717^{20}
***n*:** 1.4999^{20}
***s*:** Ac 4; Bz 4; Et_2O 4; EtOH 4
m/e: 148.13

HODOC No: 03267
CAS No: 1007-26-7
Beilstein No: 5^3, 936
Name: Benzene, (2,2-dimethylpropyl)-
Synonym: Benzene, neopentyl-
Line Form: C_6H_5-$CH_2C(CH_3)_3$
Mol Form: $C_{11}H_{16}$
M_r: 148.24
bp: 186^{758}
***d*:** $0.8581^{18/4}$
***n*:** 1.4884^{18}
MS: WILEY 862 57(100) 92(73) 91(40) 41(28) 29(23) 39(14) 65(10) 148(10)
m/e: 148.13

HODOC No: 03268
CAS No: 61827-85-8
Beilstein No: 5^4, 1102
Name: Benzene, 2,4-dimethyl-1-propyl-
Synonym: *m*-Xylene, 4-propyl-
Line Form: 2,4-$(CH_3)_2C_6H_3C_3H_7$
Mol Form: $C_{11}H_{16}$
M_r: 148.24
bp: 206.6, 104.4^{30}
***d*:** 0.8723^{20}
***n*:** 1.4998^{20}
***s*:** Ac 4; Bz 4; Et_2O 4; EtOH 4
m/e: 148.13

HODOC No: 03269
CAS No: 2050-03-5
Name: Benzene, 1-(1,1-dimethylpropyl)-4-methoxy-
Synonym: Anisole, *p-tert*-pentyl-
Mol Form: $C_{12}H_{18}O$
M_r: 178.26
bp: $235\text{-}7^{730}$
***d*:** 0.9436^{20}
***n*:** 1.5039^{20}
IR: µCOB 3095 3030 2940 1610 1590 1520 1470 1390 1370 1300 1250 1190 1110 1040 830 810 780 730 660 650
Raman: SAD 701 3080 3010 2970 2940 2920 2880 2840 2770 2720 1660 1620 1580 1520 1460 1410 1390 1370 1330 1310 1250 1240 1190 1120 1060 1040 1020 940 920 840 800 780 730 670 650 640 560 530 490 470 440 380 370 320 280 180
H NMR: SAD 17327 0.7 1.2 1.6 3.7 6.7 6.7 7.1 CCl_4
m/e: 178.14

HODOC No: 03270
CAS No: 33125-57-4
Name: Benzene, 1-(1,1-dimethylpropyl)-3-methyl-
Synonym: Toluene, *m-tert*-pentyl-
Mol Form: $C_{12}H_{18}$
M_r: 162.26
bp: $208\text{-}9^{760}$
***d*:** $0.8671^{20/4}$
UV: API 523 272 264 262 258 iso
m/e: 162.14

HODOC No: 03271
CAS No: 4237-70-1
Name: Benzene, 1-(1,1-dimethylpropyl)-4-methyl-
Synonym: Toluene, *p-tert*-pentyl-
Mol Form: $C_{12}H_{18}$
M_r: 162.26
UV: API 524 273 266 264 258 257 iso
m/e: 162.14

HODOC No: 03272
CAS No: 602-27-7
Beilstein No: 5^3, 864
Name: Benzene, 1,4-dimethyl-2,3,5-trinitro-
Synonym: *p*-Xylene, 2,3,5-trinitro-
Line Form: 1,4-$(CH_3)_2$-2,3,5-$(NO_2)_3$-C_6H
Structure:

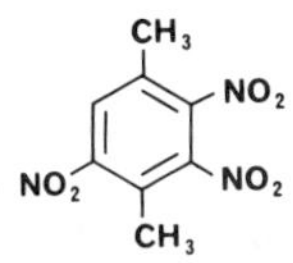

Mol Form: $C_8H_7N_3O_6$
M_r: 241.15
bp: 410 exp
mp: 139-40
***d*:** 1.59^{19}
Color: mcl nd (al) lf (al-Bz)
***s*:** EtOH 4
m/e: 241.03

HODOC No: 03273
CAS No: 632-92-8
Beilstein No: 5^4, 950
Name: Benzene, 2,4-dimethyl-1,3,5-trinitro-
Synonym: *m*-Xylene, 2,4,6-trinitro-
Line Form: 1,3-$(CH_3)_2$-2,4,6-$(NO_2)_3C_6H$
Structure:

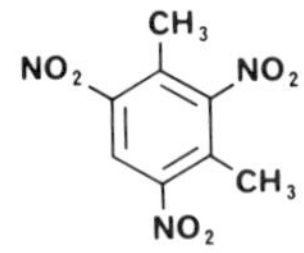

Mol Form: $C_8H_7N_3O_6$
M_r: 241.15
mp: 184
***d*:** 1.604^{19}
Color: pa ye pr or lf (al-Bz)
***s*:** H_2O 1; al 2; eth 2; ace 3; bz 4; chl 4; py 4; AcOEt 3; HNO_3 2
UV: **HBCP** 600(10000) 413(16596) py
m/e: 241.03

HODOC No: 03274
CAS No: 528-29-0
Beilstein No: 5^4, 738
Name: Benzene, 1,2-dinitro-
Synonym: Benzene, *o*-dinitro-
Line Form: 1,2-$(NO_2)_2C_6H_4$
Mol Form: $C_6H_4N_2O_4$
M_r: 168.10
bp: 319^{775}, 194^{30}
mp: 118.5
***d*:** $1.3119^{120/4}$
***n*:** 1.565^{17}
Color: nd (Bz), pl (al)
***s*:** H_2O 1; al 3; bz 3; chl 3; $DMSOd_6$ 2; MeOH 3; AcOEt 3; tol 3
IR: **μCOB 6608** 1590 1540 1450 1350 1150 1000 860 840 790 760 730
H NMR: **SAD 7871** 8.1 8.3 $DMSOd_6$
m/e: 168.02

HODOC No: 03275
CAS No: 99-65-0
Beilstein No: 5^4, 379
Name: Benzene, 1,3-dinitro-
Synonym: Benzene, *m*-dinitro-
Line Form: 1,3-$(NO_2)_2C_6H_4$
Mol Form: $C_6H_4N_2O_4$
M_r: 168.10
bp: 291^{756}, 167^{14}
mp: 90
***d*:** $1.5751^{18/4}$
Color: rh pl (al)
***s*:** H_2O 2; al 4; eth 3; ace 4; bz 4; chl 3; py 4; tol 3; MeOH 3
IR: **μCOB 4836** 3030 1610 1520 1450 1350 1080 930 730
UV: **SAD 33** 233(18500) MeOH
H NMR: **SAD 10134** 7.9 8.6 9.0 $CDCl_3$
MS: **ALDN 127** 168(100) 75(60) 76(55) 30(50) 18(41) 50(39) 122(30) 92(21)
m/e: 168.02

HODOC No: 03276
CAS No: 100-25-4
Beilstein No: 5^4, 741
Name: Benzene, 1,4-dinitro-
Synonym: Benzene, *p*-dinitro-
Line Form: 1,4-$(NO_2)_2C_6H_4$
Mol Form: $C_6H_4N_2O_4$
M_r: 168.10
bp: 298^{777}, 183^{34}
mp: 174
***d*:** $1.625^{18/4}$
Color: nd (al)
***s*:** H_2O 1; al 2; ace 3; bz 3; chl 2; HOAc 3; tol 3; AcOEt 3
IR: SADG 15462 3125 1587 1538 1471 1351 1111 1000 862 840 704
UV: OES 6-60 260(12882) EtOH
H NMR: SAD 6746 8.5 ace
MS: ALDN 127 30(100) 75(75) 168(69) 76(52) 50(47) 122(28) 92(28) 18(26)
m/e: 168.02

HODOC No: 03277
CAS No: 13286-95-8
Beilstein No: 5^4, 1878
Name: Benzene, 1,1′-(1,2-dinitro-1,2-ethanediyl)*bis*-, (R*,*S**)-
Synonym: Bibenzyl, α,α′-dinitro-, meso-
Line Form: $C_6H_5CH(NO_2)CH(NO_2)C_6H_5$
Mol Form: $C_{14}H_{12}N_2O_4$
M_r: 272.25
mp: 235-6
Color: nd (HOAc)
***s*:** Ac 4
m/e: 272.08

HODOC No: 03278
CAS No: 13440-25-0
Beilstein No: 5^4, 1878
Name: Benzene, 1,1′-(1,2-dinitro-1,2-ethanediyl)*bis*-, (R*,R*)-(±)-
Synonym: Bibenzyl, α,α′-dinitro-, (±)-
Line Form: $C_6H_5CH(NO_2)CH(NO_2)C_6H_5$
Mol Form: $C_{14}H_{12}N_2O_4$
M_r: 272.25
mp: 154-5
Color: pr (al,HOAc)
***s*:** Ac 4; Bz 4; Et_2O 4; EtOH 4
m/e: 272.08

HODOC No: 03279
CAS No: 16906-54-0
Beilstein No: 5^4, 2172
Name: Benzene, 1,1′-(1,2-dinitro-1,2-ethenediyl)*bis*-, (E)-
Synonym: Stilbene, α,α′-dinitro-, (E)-
Line Form: $C_6H_5C(NO_2)=C(NO_2)C_6H_5$
Mol Form: $C_{14}H_{10}N_2O_4$
M_r: 270.23
mp: 187-8
Color: pa ye nd or pr (al)
***s*:** H_2O 1; al 2; eth 2; ace 3; bz 3; chl 3; HOAc 3; MeOH 2
UV: HBCP 262(25119) al
m/e: 270.06

HODOC No: 03280
CAS No: 1796-05-0
Beilstein No: 5^4, 2171
Name: Benzene, 1,1′-(1,2-dinitro-1,2-ethenediyl)*bis*-, (Z)-
Synonym: Stilbene, α,α′-dinitro-, (Z)-
Line Form: $C_6H_5C(NO_2)=C(NO_2)C_6H_5$
Mol Form: $C_{14}H_{10}N_2O_4$
M_r: 270.23
bp: d
mp: 108-9
Color: ye pym pr (al)
***s*:** H_2O 1; al 3; eth 3; ace 4; bz 4; chl 4; MeOH 3
UV: HBCP 265(31623) al
m/e: 270.06

HODOC No: 03281
CAS No: 21160-03-2
Beilstein No: 5^3, 1797
Name: Benzene, 1,1′-(dinitromethylene)*bis*-
Synonym: Methane, dinitrodiphenyl-
Line Form: $(C_6H_5)_2C(NO_2)_2$
Mol Form: $C_{13}H_{10}N_2O_4$
M_r: 258.22
mp: 79-80
Color: pl (dil al)
***s*:** Et_2O 4; EtOH 4; Bz 4; Chl 4
m/e: 258.06

HODOC No: 03282
CAS No: 2486-07-9
Beilstein No: 6^4, 1375
Name: Benzene, 2,4-dinitro-1-phenoxy-
Synonym: Ether, 2,4-dinitrophenyl phenyl
Line Form: $[2,4\text{-}(NO_2)_2C_6H_3]OC_6H_5$
Mol Form: $C_{12}H_8N_2O_5$
Mr: 260.19
bp: $230\text{-}50^{27}$
mp: 71
Color: pl (al), nd (al-Ace)
***s*:** Et_2O 4; EtOH 4
m/e: 260.04

HODOC No: 03283
CAS No: 7508-74-9
Name: Benzene, 1-(2,4-dinitrophenoxy)-2-methoxy-4-(2-propenyl)-
Synonym: Benzene, 4-allyl-1-(2,4-dinitrophenoxy)-2-methoxy-
Mol Form: $C_{16}H_{13}N_2O_6$
Mr: 329.27
mp: 113-5
***s*:** chl 2
IR: SADG 20291 3100 2980 1620 1530 1500 1460 1340 1310 1290 1190 1150 1120 1030 920 830 740
UV: SAD 17743 282(11900) MeOH
H NMR: SAD 10460 3.4 3.7 5.1 5.9 6.7-7.3 8.2 8.8 $CDCl_3$
m/e: 329.08

HODOC No: 03284
CAS No: 98751-32-7
Beilstein No: 5^4, 2171
Name: Benzene, 1,3-dinitro-2-(2-phenylethenyl)-, (E)-
Synonym: Stilbene, 2,6-dinitro (*trans*)
Line Form: $2,6\text{-}(NO_2)_2C_6H_3CH{=}CHC_6H_5$
Mol Form: $C_{14}H_{10}N_2O_4$
Mr: 270.23
mp: 114
Color: ye nd (Bz, HOAc)
***s*:** H_2O 1; bz 3; chl 3
UV: HBCP 330(5012) bz
m/e: 270.06

HODOC No: 03285
CAS No: 56456-42-9
Beilstein No: 5^4, 2170
Name: Benzene, 2,4-dinitro-1-(2-phenylethenyl)-, (E)-
Synonym: Stilbene, 2,4-dinitro (*trans*)
Line Form: $2,4\text{-}(NO_2)_2C_6H_3CH{=}CHC_6H_5$
Mol Form: $C_{14}H_{10}N_2O_4$
Mr: 270.23
bp: 412 exp
mp: 143-5
Color: pa ye (HOAc)
***s*:** al 1; bz 2; chl 3; CS_2 3; xyl 3
IR: SADG 10295 3120 1600 1530 1510 1340 1150 970 950 900 830 750 720 690
UV: SAD 6665 240(15500) 205(29100) diox
H NMR: SAD 2256 7.1 7.4 7.6 7.9 8.4 8.7 $CDCl_3$
MS: ALDN 212 ini 91(100) 90(68) 105(65) 104(61) 77(52) 63(46) 165(45) 270(8)
m/e: 270.06

HODOC No: 03286
CAS No: 75634-10-5
Beilstein No: 5^2, 541
Name: Benzene, 2,4-dinitro-1-(2-phenylethenyl)-, (Z)-
Synonym: Stilbene, 2,4-dinitro (*cis*)
Line Form: $2,4\text{-}(NO_2)_2C_6H_3CH{=}CHC_6H_5$
Mol Form: $C_{14}H_{10}N_2O_4$
Mr: 270.23
mp: 127
Color: ye pl (HOAc)
***s*:** bz 4; chl 3; HOAc 3
MS: ALDN 212 ini 91(100) 90(68) 105(65) 104(61) 77(52) 63(46) 165(45) 270(8)
m/e: 270.06

HODOC No: 03287
CAS No: 7259-71-4
Beilstein No: 5, 232
Name: Benzene, 1,3-dinitroso-
Synonym: Benzene, *m*-dinitroso-
Mol Form: $C_6H_4N_2O_2$
Mr: 136.10
***s*:** H_2O 1; al 3; eth 2; bz 3; HOAc 3
UV: OES 3-54 300(4467) 268(1000) EtOH
m/e: 136.03

HODOC No: 03288
CAS No: 3379-37-1
Name: Benzene, 1,2-diphenoxy-
Synonym: Benzene, *o*-diphenoxy-
Mol Form: $C_{18}H_{14}O_2$
M_r: 262.29
MS: WILEY 1799 262(100) 169(32) 142(27) 77(22) 263(20) 168(16) 28(15) 51(12)
m/e: 262.10

HODOC No: 03289
CAS No: 3379-38-2
Beilstein No: 6, 815
Name: Benzene, 1,3-diphenoxy-
Synonym: Benzene, *m*-diphenoxy-
Line Form: 1,3-$(C_6H_5O)_2$-C_6H_4
Mol Form: $C_{18}H_{14}O_2$
M_r: 262.29
mp: 60-2
***s*:** chl 3
IR: μCOB 7486 3030 1590 1470 1330 1270 1220 1160 1120 1080 1020 970 860 810 760 730 690
UV: SAD 18918 277(3270) 270(3280) MeOH
H NMR: SAD 11582 6.7 6.9-7.4 $CDCl_3$
MS: ALDN 208 262(100) 46(65) 29(63) 27(54) 141(49) 43(42) 77(25) 263(21)
m/e: 262.10

HODOC No: 03290
CAS No: 3061-36-7
Name: Benzene, 1,4-diphenoxy-
Synonym: Benzene, *p*-diphenoxy-
Mol Form: $C_{18}H_{14}O_2$
M_r: 262.29
MS: WILEY 1798 262(100) 77(28) 263(19) 192(18) 111(15) 190(14) 141(12) 185(12)
m/e: 262.10

HODOC No: 03291
CAS No: 6280-98-4
Beilstein No: 6^4, 5566
Name: Benzene, 1,2-dipropoxy-
Synonym: Benzene, *o*-dipropoxy-
Line Form: 1,2-$(C_3H_7O)_2C_6H_4$
Mol Form: $C_{12}H_{18}O_2$
M_r: 194.26
bp: 234-7, $117\text{-}20^{2}$
***d*:** $0.9554^{33/4}$
***n*:** 1.4950^{27}
***s*:** chl 2
IR: SADG 346 2941 2857 1587 1471 1389 1266 1136 1053 990 758
UV: SAD 821 277(2500) 222(8110) MeOH
H NMR: SAD 10792 1.0 1.8 3.9 6.8 $CDCl_3$
m/e: 194.13

HODOC No: 03292
CAS No: 56106-37-7
Beilstein No: 6^3, 4308
Name: Benzene, 1,3-dipropoxy-
Synonym: 1,3-Dipropoxybenzene
Line Form: 1,3-$(C_3H_7O)_2C_6H_4$
Mol Form: $C_{12}H_{18}O_2$
M_r: 194.26
bp: 251, $127\text{-}8^{12}$
***d*:** $1.035^{20/21}$
***n*:** 1.5138^{83}
m/e: 194.13

HODOC No: 03293
CAS No: 3898-41-7
Name: Benzene, 1,4-dipropoxy-
Synonym: Benzene, *p*-dipropoxy-
Mol Form: $C_{12}H_{18}O_2$
M_r: 194.26
IR: μCOB 4852 2940 1590 1560 1470 1390 1220 1110 1080 1050 1020 980 820 740
m/e: 194.13